Comprehensive Organometallic Chemistry II

A Review of the Literature 1982–1994

Comprehensive Organometallic Chemistry II

A Review of the Literature 1982–1994

Editors-in-Chief

Edward W. Abel
University of Exeter, UK

F. Gordon A. Stone
Baylor University, Waco, TX, USA

Geoffrey Wilkinson
Imperial College of Science, Technology and Medicine, London, UK

Volume 1

LITHIUM, BERYLLIUM, AND BORON GROUPS

Volume Editor

Catherine E. Housecroft
Institut für Anorganische Chemie der Universität Basel, Switzerland

PERGAMON

UK Elsevier Science Ltd., The Boulevard, Langford Lane, Kidlington, Oxford OX5 1GB, UK

USA Elsevier Science Inc., 660 White Plains Road, Tarrytown, New York 10591-5153, USA

JAPAN Elsevier Science Japan, Tsunashima Building Annex, 3-20-12 Yushima, Bunkyo-ku, Tokyo 113, Japan

First edition 1995

Library of Congress Cataloging in Publication Data
Comprehensive organometallic chemistry II : a review of the literature 1982–1994 / editors-in-chief, Edward W. Abel, F. Gordon A. Stone, Geoffrey Wilkinson
p. cm.
Includes indexes.
1. Organometallic chemistry. I. Abel, Edward W. II. Stone, F. Gordon A. III. Wilkinson, Geoffrey.
QD411.C652 1995
547'.05—dc20 95–7030

British Library Cataloguing in Publication Data
A catalogue record for this book is available from the British Library.

ISBN 0–08–040608–4 (set : alk. paper)
ISBN 0–08–042308–6 (Volume 1)

Important note
For safety reasons, readers should always consult the list of abbreviations on p. xi before making use of the experimental details provided.

∞™ The paper used in this publication meets the minimum requirements of the American National Standard for Information Sciences—Permanence of Paper for Printed Library Materials, ANSI Z39.48–1984.

Chemical structures drawn by Synopsys Scientific Systems Ltd., Leeds, UK.

Printed and bound in Great Britain by BPC Wheatons Ltd., Exeter, UK.

Contents

Preface		vii
Preface to 'Comprehensive Organometallic Chemistry'		vii
Contributors to Volume 1		ix
Abbreviations		xi
Contents of All Volumes		xv
1	Alkali Metals D. S. WRIGHT and M. A. BESWICK, *University of Cambridge, UK*	1
2	Beryllium N. A. BELL, *Sheffield Hallam University, UK*	35
3	Magnesium, Calcium, Strontium and Barium W. E. LINDSELL, *Heriot-Watt University, Edinburgh, UK*	57
4	Compounds with Three- or Four-coordinate Boron, Emphasizing Cyclic Systems C. E. HOUSECROFT, *Institut für Anorganische Chemie der Universität Basel, Switzerland*	129
5	Boron Rings Ligated to Metals G. E. HERBERICH, *Technische Hochschule Aachen, Germany*	197
6	Polyhedral Carbaboranes T. ONAK, *California State University, Los Angeles, CA, USA*	217
7	Main-group Heteroboranes L. J. TODD, *Indiana University, Bloomington, IN, USA*	257
8	Metallaboranes L. BARTON and D. K. SRIVASTAVA, *University of Missouri–St. Louis, MO, USA*	275
9	Transition Metal Metallacarbaboranes R. N. GRIMES, *University of Virginia, Charlottesville, VA, USA*	373
10	Aluminum J. J. EISCH, *State University of New York at Binghamton, NY, USA*	431
11	Gallium, Indium and Thallium, Excluding Transition Metal Derivatives M. A. PAVER, C. A. RUSSELL and D. S. WRIGHT, *University of Cambridge, UK*	503
12	Transition Metal Complexes of Aluminum, Gallium, Indium and Thallium C. J. CARMALT and N. C. NORMAN, *University of Newcastle upon Tyne, UK*, and L. M. CLARKSON, *E. I. DuPont de Nemours & Co. Inc., Deepwater, NJ, USA*	545
Author Index		573
Subject Index		613

Preface

'Comprehensive Organometallic Chemistry', published in 1982, was well received and remains very highly cited in the primary journal literature. Since its publication, studies on the chemistry of molecules with carbon–metal bonds have continued to expand rapidly. This is due to many factors, ranging from the sheer intellectual challenge and excitement provided by the continuing production of novel results, which demand new ideas, through to the successful application of organometallic species in organic syntheses, the generation of living catalysts for polymerization, and the synthesis of precursors for materials employed in the electronic and ceramic industries. For many reasons, therefore, we judged it timely to update 'Comprehensive Organometallic Chemistry' with a new work.

Due to the scope and depth of this area of chemistry, to have merely revised each of the original nine volumes did not seem the most user-friendly or cost-effective procedure to follow. As a consequence of the sheer bulk of the literature of the subject, a revised edition would necessarily require either the elimination of much chemistry of archival value but which is still important, or the production of a set of volumes significantly larger in number than the original nine. Accordingly, we decided it would be best to use the original work as a basis for new volumes focusing on organometallic chemistry reported since 1982, with reference back to the original work when necessary. For ease of use the new volumes maintain the same general structure as employed previously but reflect the changes in substance and direction the field has undergone in the last ten years. Thus it is not surprising that the largest volume in the new work concerns the role of the transition elements in metal-mediated organic syntheses.

The expansion of organometallic chemistry since the early 1980s also led us to decide that an updating of 'Comprehensive Organometallic Chemistry' would be more effectively accomplished if each volume had one or two editors who would be responsible both for recruiting experts for the Herculean task of writing the many chapters of each volume and for overseeing the content. We are deeply indebted to the volume editors and their authors for the time and effort they have given to the project.

As with the original 'Comprehensive Organometallic Chemistry', published some thirteen years ago, we hope this new version will serve as a pivotal reference point for new work and will function to generate new ideas and perceptions for the continued advance of what will surely continue as a vibrant area of chemistry.

Edward W. Abel
Exeter, UK

F. Gordon A. Stone
Waco, Texas, USA

Geoffrey Wilkinson
London, UK

Preface to 'Comprehensive Organometallic Chemistry'

Although the discovery of the platinum complex that we now know to be the first π-alkene complex, $K[PtCl_3(C_2H_4)]$, by Zeise in 1827 preceded Frankland's discovery (1849) of diethylzinc, it was the latter that initiated the rapidly developing interest during the latter half of the nineteenth century in compounds with organic groups bound to the elements. This era may be considered to have reached its apex in the discovery by Grignard of the magnesium reagents which occupy a special place because of their ease of synthesis and reactivity. With the exception of trimethylplatinum chloride discovered by Pope, Peachy and Gibson in 1907 by use of the Grignard reagent, attempts to make stable transition metal alkyls and aryls corresponding to those of main group elements met with little success, although it is worth recalling that even in 1919 Hein and his co-workers were describing the 'polyphenyl-chromium' compounds now known to be arene complexes.

The other major area of organometallic compounds, namely metal compounds of carbon monoxide, originated in the work starting in 1868 of Schützenberger and later of Mond and his co-workers and was subsequently developed especially by Hieber and his students. During the first half of this century, aided by the use of magnesium and, later, lithium reagents the development of main group organo chemistry was quite rapid, while from about 1920 metal carbonyl chemistry and catalytic reactions of carbon monoxide began to assume importance.

In 1937 Krause and von Grosse published their classic book 'Die Chemie der Metallorganischen Verbindungen'. Almost 1000 pages in length, it listed scores of compounds, mostly involving metals of the main groups of the periodic table. Compounds of the transition elements could be dismissed in 40 pages. Indeed, even in 1956 the stimulating 197-page monograph 'Organometallic Compounds' by Coates adequately reviewed organo transition metal complexes within 27 pages.

Although exceedingly important industrial processes in which transition metals were used for catalysis of organic reactions were developed in the 1930s, mainly in Germany by Reppe, Koch, Roelen, Fischer and Tropsch and others, the most dramatic growth in our knowledge of organometallic chemistry, particularly of transition metals, has stemmed from discoveries made in the middle years of this century. The introduction in the same period of physical methods of structure determination (infrared, nuclear magnetic resonance, and especially single-crystal X-ray diffraction) as routine techniques to be used by preparative chemists allowed increasingly sophisticated exploitation of discoveries. Following the recognition of the structure of ferrocene, other major advances quickly followed, including the isolation of a host of related π-complexes, the synthesis of a plethora of organometallic compounds containing metal–metal bonds, the characterization of low-valent metal species in which hydrocarbons are the only ligands, and the recognition from dynamic NMR spectra that ligand site exchange and tautomerism were common features in organometallic and metal carbonyl chemistry. The discovery of alkene polymerization using aluminium alkyl–titanium chloride systems by Ziegler and Natta and of the Wacker palladium–copper catalysed ethylene oxidation led to enormous developments in these areas.

In the last two decades, organometallic chemistry has grown more rapidly in scope than have the classical divisions of chemistry, leading to publications in journals of all national chemical societies, the appearance of primary journals specifically concerned with the topic, and the growth of annual review volumes designed to assist researchers to keep abreast of accelerating developments.

Organometallic chemistry has become a mature area of science which will obviously continue to grow. We believe that this is an appropriate time to produce a comprehensive review of the subject, treating organo derivatives in the widest sense of both main group and transition elements. Although advances in transition metal chemistry have appeared to dominate progress in recent years, spectacular progress has, nevertheless, also been made in our knowledge of organo compounds of main group elements such as aluminium, boron, lithium and silicon.

In these Volumes we have assembled a compendium of knowledge covering contemporary organometallic and carbon monoxide chemistry. In addition to reviewing the chemistry of the elements individually, two Volumes survey the use of organometallic species in organic synthesis and in catalysis, especially of industrial utility. Within the other Volumes are sections devoted to such diverse topics as the nature of carbon–metal bonds, the dynamic behaviour of organometallic compounds in solution, heteronuclear metal–metal bonded compounds, and the impact of organometallic compounds on the environment. The Volumes provide a unique record, especially of the intensive studies conducted during the past 25 years. The last Volume of indexes of various kinds will assist readers seeking information on the properties and synthesis of compounds and on earlier reviews.

As Editors, we are deeply indebted to all those who have given their time and effort to this project. Our Contributors are among the most active research workers in those areas of the subject that they have reviewed and they have well justified international reputations for their scholarship. We thank them sincerely for their cooperation.

Finally, we believe that 'Comprehensive Organometallic Chemistry', as well as providing a lasting source of information, will provide the stimulus for many new discoveries since we do not believe it possible to read any of the articles without generating ideas for further research.

E. W. ABEL
Exeter

F. G. A. STONE
Bristol

G. WILKINSON
London

Contributors to Volume 1

Professor L. Barton
Department of Chemistry, University of Missouri–St. Louis, 8001 Natural Bridge Road, St. Louis, MO 63121, USA

Dr N. A. Bell
Department of Chemistry, Sheffield Hallam University, Pond Street, Sheffield, S1 1WB, UK

Dr M. A. Beswick
Department of Chemistry, University of Cambridge, Lensfield Road, Cambridge, CB2 1EW, UK

Dr C. J. Carmalt
Department of Chemistry, University of Newcastle upon Tyne, Newcastle upon Tyne, NE1 7RU, UK

Dr L. M. Clarkson
E I DuPont de Nemours & Co. Inc., Jackson Laboratory, Chambers Works, Deepwater, NJ 08023, USA

Professor J. J. Eisch
Department of Chemistry, State University of New York at Binghamton, PO Box 6000, Binghamton, NY 13902-6000, USA

Professor R. N. Grimes
Department of Chemistry, University of Virginia, McCormick Road, Charlottesville, VA 22901, USA

Professor G. E. Herberich
Institut für Anorganische Chemie, Technische Hochschule Aachen, Templergraben 55, D-52056 Aachen, Germany

Dr C. E. Housecroft
Institut für Anorganische Chemie der Universität Basel, Spitalstrasse 51, CH-4056 Basel, Switzerland

Dr W. E. Lindsell
Department of Chemistry, Heriot-Watt University, Riccarton, Edinburgh, EH14 4AS, UK

Professor T. Onak
Department of Chemistry and Biochemistry, California State University, Los Angeles, 5151 State University Drive, Los Angeles, CA 90032-8208, USA

Dr N. C. Norman
Department of Chemistry, Bedson Building, University of Newcastle upon Tyne, Newcastle upon Tyne, NE1 7RU, UK

Dr M. A. Paver
Department of Chemistry, University of Cambridge, Lensfield Road, Cambridge, CB2 1EW, UK

Dr C. A. Russell
Department of Chemistry, University of Cambridge, Lensfield Road, Cambridge, CB2 1EW, UK

Dr D. K. Srivastava
Department of Chemistry, University of Missouri-St. Louis, 8001 Natural Bridge Road, St. Louis, MO 63121, USA

Professor L. J. Todd
Department of Chemistry and Biochemistry, Indiana University, Bloomington, IN 47405, USA

Dr D. S. Wright
Department of Chemistry, University of Cambridge, Lensfield Road, Cambridge, CB2 1EW, UK

Abbreviations

The abbreviations used throughout 'Comprehensive Organometallic Chemistry II' are consistent with those used in 'Comprehensive Organometallic Chemistry' and with other standard texts in this area. The abbreviations in some instances may differ from those commonly used in other branches of chemistry.

Ac	acetyl
acac	acetylacetonate
AIBN	2,2'-azobisisobutyronitrile
Ar	aryl
arphos	1-(diphenylphosphino)-2-(diphenylarsino)ethane
Azb	azobenzene
9-BBN	9-borabicyclo[3.3.1]nonyl
9-BBN-H	9-borabicyclo[3.3.1]nonane
BHT	2,6-di-*t*-butyl-4-methylphenol (butylated hydroxytoluene)
bipy	2,2'-bipyridyl
t-BOC	*t*-butoxycarbonyl
bsa	*N,O*-bis(trimethylsilyl)acetamide
bstfa	*N,O*-bis(trimethylsilyl)trifluoroacetamide
btaf	benzyltrimethylammonium fluoride
Bz	benzyl
can	ceric ammonium nitrate
cbd	cyclobutadiene
1,5,9-cdt	cyclododeca-1,5,9-triene
chd	cyclohexadiene
chpt	cycloheptatriene
[Co]	cobalamin
(Co)	cobaloxime $[Co(DMG)_2]$ derivative
cod	1,5-cyclooctadiene
cot	cyclooctatetraene
Cp	η^5-cyclopentadienyl
Cp*	pentamethylcyclopentadienyl
18-crown-6	1,4,7,10,13,16-hexaoxacyclooctadecane
CSA	camphorsulfonic acid
csi	chlorosulfonyl isocyanate
Cy	cyclohexyl
dabco	1,4-diazabicyclo[2.2.2]octane
dba	dibenzylideneacetone
dbn	1,5-diazabicyclo[4.3.0]non-5-ene
dbu	1,8-diazabicyclo[5.4.0]undec-7-ene
dcc	dicyclohexylcarbodiimide
dcpe	1,2-bis(dicyclohexylphosphino)ethane
ddq	2,3-dichloro-5,6-dicyano-1,4-benzoquinone
deac	diethylaluminum chloride
dead	diethyl azodicarboxylate
depe	1,2-bis(diethylphosphino)ethane
depm	1,2-bis(diethylphosphino)methane
det	diethyl tartrate (+ or –)

DHP	dihydropyran
diars	1,2-bis(dimethylarsino)benzene
dibal-H	diisobutylaluminum hydride
dien	diethylenetriamine
DIGLYME	bis(2-methoxyethyl)ether
diop	2,3-*O*-isopropylidene-2,3-dihydroxy-1,4-bis(diphenylphosphino)butane
dipt	diisopropyl tartrate (+ or –)
dma	dimethylacetamide
dmac	dimethylaluminum chloride
DMAD	dimethyl acetylenedicarboxylate
dmap	4-dimethylaminopyridine
DME	dimethoxyethane
DMF	*N,N'*-dimethylformamide
DMG	dimethylglyoximate
DMI	*N,N'*-dimethylimidazalone
dmpe	1,2-bis(dimethylphosphino)ethane
dmpm	bis(dimethylphosphino)methane
DMSO	dimethyl sulfoxide
dmtsf	dimethyl(methylthio)sulfonium fluoroborate
dpam	bis(diphenylarsino)methane
dppb	1,4-bis(diphenylphosphino)butane
dppe	1,2-bis(diphenylphosphino)ethane
dppf	1,1'-bis(diphenylphosphino)ferrocene
dpph	1,6-bis(diphenylphosphino)hexane
dppm	bis(diphenylphosphino)methane
dppp	1,3-bis(diphenylphosphino)propane
eadc	ethylaluminum dichloride
edta	ethylenediaminetetraacetate
eedq	*N*-ethoxycarbonyl-2-ethoxy-1,2-dihydroquinoline
en	ethylene-1,2-diamine
Et_2O	diethyl ether
F_6 acac	hexafluoroacetylacetonate
Fc	ferrocenyl
Fp	$Fe(CO)_2Cp$
HFA	hexafluoroacetone
hfacac	hexafluoroacetylacetonate
hfb	hexafluorobut-2-yne
HMPA	hexamethylphosphoramide
hobt	hydroxybenzotriazole
$IpcBH_2$	isopinocampheylborane
Ipc_2BH	diisopinocampheylborane
kapa	potassium 3-aminopropylamide
K-selectride	potassium tri-*s*-butylborohydride
LAH	lithium aluminum hydride
LDA	lithium diisopropylamide
LICA	lithium isopropylcyclohexylamide
LITMP	lithium tetramethylpiperidide
L-selectride	lithium tri-*s*-butylborohydride
LTA	lead tetraacetate
mcpba	*m*-chloroperbenzoic acid
MeCN	acetonitrile
MEM	methoxyethoxymethyl
MEM-Cl	β-methoxyethoxymethyl chloride

Mes	mesityl
mma	methyl methacrylate
mmc	methylmagnesium carbonate
MOM	methoxymethyl
Ms	methanesulfonyl
MSA	methanesulfonic acid
MsCl	methanesulfonyl chloride
nap	1-naphthyl
nbd	norbornadiene
NBS	*N*-bromosuccinimide
NCS	*N*-chlorosuccinimide
nmo	*N*-methylmorpholine *N*-oxide
NMP	*N*-methyl-2-pyrrolidone
Nu^-	nucleophile
ox	oxalate
pcc	pyridinium chlorochromate
pdc	pyridinium dichromate
phen	1,10-phenanthroline
phth	phthaloyl
ppa	polyphosphoric acid
ppe	polyphosphate ester
$[PPN]^+$	$[(Ph_3P)_2N]^+$
ppts	pyridinium *p*-toluenesulfonate
py	pyridine
pz	pyrazolyl
Red-Al	sodium bis(2-methoxyethoxy)aluminum dihydride
sal	salicylaldehyde
salen	*N,N'*-bis(salicylaldehydo)ethylenediamine
SEM	β-trimethylsilylethoxymethyl
tas	tris(diethylamino)sulfonium
tasf	tris(diethylamino)sulfonium difluorotrimethylsilicate
tbaf	tetra-*n*-butylammonium fluoride
TBDMS	*t*-butyldimethylsilyl
TBDMS-Cl	*t*-butyldimethylsilyl chloride
TBDPS	*t*-butyldiphenylsilyl
tbhp	*t*-butyl hydroperoxide
TCE	2,2,2-trichloroethanol
TCNE	tetracyanoethene
TCNQ	7,7,8,8-tetracyanoquinodimethane
terpy	2,2':6',2"-terpyridyl
tes	triethylsilyl
Tf	triflyl (trifluoromethanesulfonyl)
TFA	trifluoracetic acid
TFAA	trifluoroacetic anhydride
tfacac	trifluoroacetylacetonate
THF	tetrahydrofuran
THP	tetrahydropyranyl
tipbs-Cl	2,4,6-triisopropylbenzenesulfonyl chloride
tips-Cl	1,3-dichloro-1,1,3,3-tetraisopropyldisiloxane
TMEDA	tetramethylethylenediamine [1,2-bis(dimethylamino)ethane]
TMS	trimethylsilyl
TMS-Cl	trimethylsilyl chloride
TMS-CN	trimethylsilyl cyanide
Tol	tolyl
tpp	*meso*-tetraphenylporphyrin

Tr	trityl (triphenylmethyl)
tren	2,2',2''-triaminotriethylamine
trien	triethylenetetraamine
triphos	1,1,1-tris(diphenylphosphinomethyl)ethane
Ts	tosyl
TsMIC	tosylmethyl isocyanide
ttfa	thallium trifluoroacetate

Contents of All Volumes

Volume 1 Lithium, Beryllium, and Boron Groups
1 Alkali Metals
2 Beryllium
3 Magnesium, Calcium, Strontium and Barium
4 Compounds with Three- or Four-coordinate Boron, Emphasizing Cyclic Systems
5 Boron Rings Ligated to Metals
6 Polyhedral Carbaboranes
7 Main-group Heteroboranes
8 Metallaboranes
9 Transition Metal Metallacarbaboranes
10 Aluminum
11 Gallium, Indium and Thallium, Excluding Transition Metal Derivatives
12 Transition Metal Complexes of Aluminum, Gallium, Indium and Thallium
Author Index
Subject Index

Volume 2 Silicon Group, Arsenic, Antimony, and Bismuth
1 Organosilanes
2 Carbacyclic Silanes
3 Organopolysilanes
4 Silicones
5 Germanium
6 Tin
7 Lead
8 Arsenic, Antimony and Bismuth
Author Index
Subject Index

Volume 3 Copper and Zinc Groups
1 Gold
2 Copper and Silver
3 Mercury
4 Cadmium and Zinc
Author Index
Subject Index

Volume 4 Scandium, Yttrium, Lanthanides and Actinides, and Titanium Group
1 Zero Oxidation State Complexes of Scandium, Yttrium and the Lanthanide Elements
2 Scandium, Yttrium, and the Lanthanide and Actinide Elements, Excluding their Zero Oxidation State Complexes
3 Titanium Complexes in Oxidation States Zero and Below
4 Titanium Complexes in Oxidation States +2 and +3
5 Titanium Complexes in Oxidation State +4
6 Zirconium and Hafnium Complexes in Oxidation States Zero and Below
7 Metallocene(II) Complexes of Zirconium and Hafnium
8 Zirconium and Hafnium Compounds in Oxidation State +3
9 Bis(cyclopentadienyl)zirconium and -hafnium Halide Complexes in Oxidation State +4
10 Bis(cyclopentadienyl) Metal(IV) Compounds with Si, Ge, Sn, N, P, As, Sb, O, S, Se, Te or Transition Metal-centred Ligands
11 Zirconium and Hafnium Complexes in Oxidation State +4

12 Cationic Organozirconium and Organohafnium Complexes
13 Cyclooctatetraene Complexes of Zirconium and Hafnium
Author Index
Subject Index

Volume 5 Vanadium and Chromium Groups
1 Vanadium
2 Niobium and Tantalum
3 Hexacarbonyls and Carbonyl Complexes of Carbon σ-Bonded Ligands of Chromium, Molybdenum and Tungsten
4 Carbonyl Complexes of Noncarbon σ-Bonded Ligands of Chromium, Molybdenum and Tungsten
5 Organometallic Complexes of Chromium, Molybdenum and Tungsten without Carbonyl Ligands
6 π-Complexes of Chromium, Molybdenum and Tungsten, Excluding those of Cyclopentadienyls and Arenes
7 Cyclopentadienyl Complexes of Chromium, Molybdenum and Tungsten
8 Arene and Heteroarene Complexes of Chromium, Molybdenum and Tungsten
Author Index
Subject Index

Volume 6 Manganese Group
1 Manganese Carbonyls and Manganese Carbonyl Halides
2 Manganese Alkyls and Hydrides
3 Manganese Complexes Containing Nonmetallic Elements
4 Manganese Hydrocarbon Complexes Excluding Cyclopentadienyl
5 Cyclopentadienyl Manganese Complexes
6 Manganese Nitrosyl and Isonitrile Complexes
7 High-valent Organomanganese Compounds
8 Technetium
9 Low-valent Organorhenium Compounds
10 High-valent Organorhenium Compounds
Author Index
Subject Index

Volume 7 Iron, Ruthenium, and Osmium
1 Iron Compounds without Hydrocarbon Ligands
2 Mononuclear Iron Compounds with η^1–η^6 Hydrocarbon Ligands
3 Dinuclear Iron Compounds with Hydrocarbon Ligands
4 Polynuclear Iron Compounds with Hydrocarbon Ligands
5 Introduction to Organoruthenium and Organoosmium Chemistry
6 Mononuclear Complexes of Ruthenium and Osmium Containing η^1 Carbon Ligands
7 Complexes of Ruthenium and Osmium Containing η^2–η^6 Hydrocarbon Ligands: (i) Complexes not Containing Cyclobutadiene, Cyclopentadienyl or η-Arene Coligands
8 Complexes of Ruthenium and Osmium Containing η^2–η^6 Hydrocarbon Ligands: (ii) Complexes Containing Four- and Five-membered Rings (Including MCp(arene) Complexes)
9 Complexes of Ruthenium and Osmium Containing η^2–η^6 Hydrocarbon Ligands: (iii) Complexes Containing Six-, Seven- and Eight-membered Rings
10 Ruthenocenes and Osmocenes
11 Binuclear Complexes of Ruthenium and Osmium Containing Metal–Metal Bonds
12 Trinuclear Clusters of Ruthenium and Osmium: (i) Introduction and Simple Neutral, Anionic and Hydrido Clusters
13 Trinuclear Clusters of Ruthenium and Osmium: (ii) Hydrocarbon Ligands on Metal Clusters
14 Trinuclear Clusters of Ruthenium and Osmium: (iii) Clusters with Metal–Carbon Bonds to Heteroatom Ligands
15 Tetranuclear Clusters of Ruthenium and Osmium
16 Medium- and High-nuclearity Clusters of Ruthenium and Osmium
Author Index
Subject Index

Volume 8 Cobalt, Rhodium, and Iridium
1 Cobalt
2 Rhodium
3 Iridium
4 Cluster Complexes of Cobalt, Rhodium, and Iridium
Author Index
Subject Index

Volume 9 Nickel, Palladium, and Platinum
1 Nickel Complexes with Carbonyl, Isocyanide, and Carbene Ligands
2 Nickel–Carbon σ-Bonded Complexes
3 Nickel–Carbon π-Bonded Complexes
4 Palladium Complexes with Carbonyl, Isocyanide and Carbene Ligands
5 Palladium–Carbon σ-Bonded Complexes
6 Palladium–Carbon π-Bonded Complexes
7 Platinum Complexes with Carbonyl, Isocyanide and Carbene Ligands
8 Platinum–Carbon σ-Bonded Complexes
9 Platinum–Carbon π-Bonded Complexes
Author Index
Subject Index

Volume 10 Heteronuclear Metal–Metal Bonds
1 Synthesis of Compounds Containing Heteronuclear Metal–Metal Bonds
2 Heterodinuclear Compounds
3 Heteronuclear Clusters Containing C_1, C_2, C_3, ..., C_n Acyclic Hydrocarbyl Ligands
4 Binary Carbonyls, Carbonyls plus Hydrides, Carbonyls plus Phosphines, Cyclic Hydrocarbyls and Main-group Ligands without Acyclic Hydrocarbyls
5 Cluster Complexes with Bonds Between Transition Elements and Copper, Silver and Gold
6 Cluster Complexes with Bonds Between Transition Elements and Zinc, Cadmium, and Mercury
7 Catalysis and Related Reactions with Compounds Containing Heteronuclear Metal–Metal Bonds
Author Index
Subject Index

Volume 11 Main-group Metal Organometallics in Organic Synthesis
1 Lithium
2 Sodium and Potassium
3 Magnesium
4 Zinc and Cadmium
5 Boron
6 Aluminum
7 Silicon
8 Tin
9 Mercury
10 Thallium
11 Lead
12 Antimony and Bismuth
13 Selenium
14 Tellurium
Author Index
Subject Index

Volume 12 Transition Metal Organometallics in Organic Synthesis
1 Introduction and Fundamentals
2 Transition Metal Hydrides: Hydrocarboxylation, Hydroformylation, and Asymmetric Hydrogenation
3.1 Transition Metal Alkyl Complexes from Hydrometallation
3.2 Transition Metal Alkyl Complexes from RLi and CuX
3.3 Transition Metal Alkyl Complexes: Main-group Transmetallation and Insertion Chemistry
3.4 Transition Metal Alkyl Complexes: Oxidative Addition and Transmetallation

3.5 Transition Metal Alkyl Complexes: Oxidative Addition and Insertion
3.6 Transition Metal Alkyl Complexes: Multiple Insertion Cascades
3.7 Transition Metal Alkyl Complexes: Reductive Dimerization of Alkenes and Alkynes
4 Transition Metal Carbonyl Complexes
5.1 Transition Metal Carbene Complexes: Cyclopropanation
5.2 Transition Metal Carbene Complexes: Diazodecomposition, Ylide, and Insertion
5.3 Transition Metal Carbene Complexes: Alkyne and Vinyl Ketene Chemistry
5.4 Transition Metal Carbene Complexes: Photochemical Reactions of Carbene Complexes
5.5 Transition Metal Carbene Complexes: Tebbe's Reagent and Related Nucleophilic Alkylidenes
6.1 Transition Metal Alkene, Diene, and Dienyl Complexes: Nucleophilic Attack on Alkene Complexes
6.2 Transition Metal Alkene, Diene, and Dienyl Complexes: Complexation of Dienes for Protection
6.3 Transition Metal Alkene, Diene, and Dienyl Complexes: Nucleophilic Attack on Diene and Dienyl Complexes
7.1 Transition Metal Alkyne Complexes: Transition Metal-stabilized Propargyl Systems
7.2 Transition Metal Alkyne Complexes: Pauson–Khand Reaction
7.3 Transition Metal Alkyne Complexes: Transition Metal-catalyzed Cyclotrimerization
7.4 Transition Metal Alkyne Complexes: Zirconium–Benzyne Complexes
8.1 Transition Metal Allyl Complexes: Telomerization of Dienes
8.2 Transition Metal Allyl Complexes: Pd, W, Mo-assisted Nucleophilic Attack
8.3 Transition Metal Allyl Complexes: Intramolecular Alkene and Alkyne Insertions
8.4 Transition Metal Allyl Complexes: Trimethylene Methane Complexes
8.5 Transition Metal Allyl Complexes: π-Allylnickel Halides and Other π-Allyl Complexes Excluding Palladium
9.1 Transition Metal Arene Complexes: Nucleophilic Addition
9.2 Transition Metal Arene Complexes: Ring Lithiation
9.3 Transition Metal Arene Complexes: Side-chain Activation and Control of Stereochemistry
10 Synthetically Useful Coupling Reactions Promoted by Ti, V, Nb, W, Mo Reagents
11.1 Transition Metal-catalyzed Oxidations: Asymmetric Epoxidation
11.2 Transition Metal-catalyzed Oxidations: Asymmetric Hydroxylation
11.3 Transition Metal-catalyzed Oxidations: Other Oxidations
12.1 Transition Metals in Polymer Synthesis: Ziegler–Natta Reaction
12.2 Transition Metals in Polymer Synthesis: Ring-opening Metathesis Polymerization and Other Transition Metal Polymerization Techniques
Author Index
Subject Index

Volume 13 Structure Index
Structures of Organometallic Compounds Determined by Diffraction Methods

Volume 14 Cumulative Indexes
Cumulative Formula Index
Cumulative Subject Index

1
Alkali Metals

MICHAEL A. BESWICK and DOMINIC S. WRIGHT
University of Cambridge, UK

1.1 INTRODUCTION 1

1.2 METALLOORGANIC COMPLEXES (CONTAINING E–M BONDS; E = HETEROATOM) 2

1.2.1 Group 15–Alkali Metal Bonded Complexes 2

1.2.1.1 Amido- and imino-metal complexes 2

1.2.1.2 P–M, As–M and Sb–M bonded complexes 10

1.2.2 Group 16–Alkali Metal Bonded Complexes 11

1.2.2.1 O–M bonded complexes 11

1.2.2.2 S–M, Se–M and Te–M bonded complexes 13

1.2.3 Si–M, Sn–M and Pb–M Bonded Complexes 14

1.2.4 Multimetallated Complexes 16

1.3 ORGANOMETALLIC COMPLEXES (CONTAINING C–M BONDS) 18

1.3.1 σ- and π-Bonded Complexes 18

1.3.2 Mixed-metal Compounds 25

1.3.2.1 Mixed-metal transition metal complexes 25

1.3.2.2 Mixed-metal main group complexes 29

1.4 THEORETICAL CALCULATIONS 29

1.5 REFERENCES 30

1.1 INTRODUCTION

The 10 years from 1982 to 1992 have been an eventful and exciting period in the history of alkali metal organometallic compounds. This has been a period in which new structural principles and theoretical interpretations have given fresh insights into the factors affecting both the modes of aggregation and solvation of such alkali metal complexes and of the nature of bonding (e.g., ionic vs. covalent) occurring within them. These advances have constituted a genuine maturing of an area which owes its roots to the early pioneering work of Schlenk in 1914.[1] The First International Conference on the Chemistry of the Alkali and Alkaline Earth Metals in 1994 (Cambridge, UK) illustrated the sheer diversity and enormous quantity of work being carried out in the area of alkali metal organometallics throughout the world.[2]

The major factors affecting the new understanding in the area of alkali metal organometallics since 1982 are advances both in the handling of these air- and moisture-sensitive compounds and particularly the increased application and availability of x-ray crystallography, which now makes the complete characterization of these complexes standard. Indeed, the greatest single difference between the survey of this area in *COMC-I* (by Wardell)[3] and that in this chapter is the stress on structurally characterized compounds. A further significant factor in the continued development in this area is the development of ever more powerful calculational methods (used both as a predictive and interpretational tool).

Although this chapter principally concerns organometallic complexes (strictly speaking containing C–M bonds, M being an alkali metal), any review in this area would not be complete without a survey of the chemistry of the metalloorganic derivatives (containing, e.g., N–M, P–M, O–M, S–M, etc. bonds).

Indeed, it is now commonplace for *s*-block chemists to use the term 'organometallic' in a generic sense to describe the whole gambit of both ('genuinely') organometallic and metalloorganic compounds. Underlying this expanded definition ('...not only compounds with C–Li bonds...but also (alkali metal) derivatives of organic molecules in general'[4]) has been the realization since 1982 of the close structural link between all these species. This structural link (so-called 'ring-stacking' and 'ring-laddering'[4–5]) forms one of the major themes of this chapter.

Up to the mid-1990s, the vast majority of studies have concerned organolithium complexes (whether unsolvated $[RLi]_n$ or Lewis base donor complexes $[RLi{\cdot}xL]_n$; R = organic group bonded to Li; x = number of donor ligands (L); n = aggregation number, the number of monomer units per molecule). Undoubtedly, this situation is due mainly to the difficulty in handling the generally far more air-sensitive heavier alkali metal derivatives. Additionally, the far lower solubility of these species in organic solvents makes them more difficult to crystallize or examine in solution. A 1993 review by Weiss gives an excellent overview of the current state of development of the structures of organo alkali metal and related complexes.[6] In addition, organolithium complexes (mainly C–Li bonded) have been comprehensively reviewed by Setzer and Schleyer,[7] and the heavier alkali metals (mainly C–M bonded) have been reviewed by Schade and Schleyer.[8] Recent developments in the area of metalloorganic derivatives have been reviewed by Gregory *et al.* (N–Li bonded complexes),[4] Mulvey (N–Li bonded),[5] Seebach (O–Li bonded),[9] and by Boche (S–Li bonded).[10] Among the many workers who have greatly improved the understanding of the structures and properties of alkali metal organometallics since 1982 have been Collum (USA), Boche (Marburg, Germany), Jackman (USA), Lappert (Sussex, UK), Mulvey (Strathclyde, UK), Power (USA), Schleyer (Erlangen, Germany), Seebach (ETH, Zürich, Switzerland), Snaith (Cambridge, UK), Streitweiser (USA), Wade (Durham, UK), and Weiss (Hamburg, Germany).

1.2 METALLOORGANIC COMPLEXES (CONTAINING E–M BONDS; E = HETEROATOM)

Although metalloorganic complexes which contain a variety of E–M bonds (E = N, P, O, S, etc.) are not strictly speaking organometallic, it is now recognized that the structural trends within these compounds are similar to those occurring in C–M bonded complexes (see Section 1.3). For this reason a discussion of the state of development in this area is appropriate here. However, this discussion will be mainly restricted to systems containing 'simple' organic anions with no additional functionalities, for example, amido-metal derivatives, containing R_2N^- anions, where R does not contain any donor (O, S, etc.) centres. This section attempts to give a broad coverage of the structurally characterized compounds containing E–M bonds (E = main group element (excluding halogens) other than carbon).

1.2.1 Group 15–Alkali Metal Bonded Complexes

1.2.1.1 Amido- and imino-metal complexes

Since the early 1980s the structures of several uncomplexed iminolithium complexes, $[R^1R^2C{=}NLi]_n$ ((**1**)–(**4**)), have been investigated in the solid state (Table 1).

Table 1 Structures of several uncomplexed iminolithium complexes.

$[R^1R^2C{=}NLi]_n$	*Ref.*
$[Bu^t_2C{=}NLi]_6$ (**1**)	11–14
$[(Me_2N)_2C{=}NLi]_6$ (**2**)	12–14
$[Bu^t(Ph)C{=}NLi]_6$ (**3**)	13,14
$[Me_2N(Ph)C{=}NLi]_6$ (**4**)	13,14

All of the complexes are hexamers in the solid state. This structural motif is frequently observed in the organometallic complexes of lithium (see Section 1.3).[6] Detailed examination of the core N–Li distances within these hexamers suggested to Snaith, Mulvey and Wade that, rather than being regarded as a Li_6 chair complexed by six imino groups as such hexamers had previously been regarded, the hexameric structures are best regarded as a 'stack' of two trimeric Li_3N_3 rings (e.g., (**3**)).[4–5] This view provides the basis for the so-called 'ring-stacking' and 'ring-laddering' models which furnish a wholistic view of the structures of a host of organometallic (C–Li bonded, see Section 1.3) and metalloorganic complexes (O–Li, S–Li, P–Li, etc.; see later in this section).

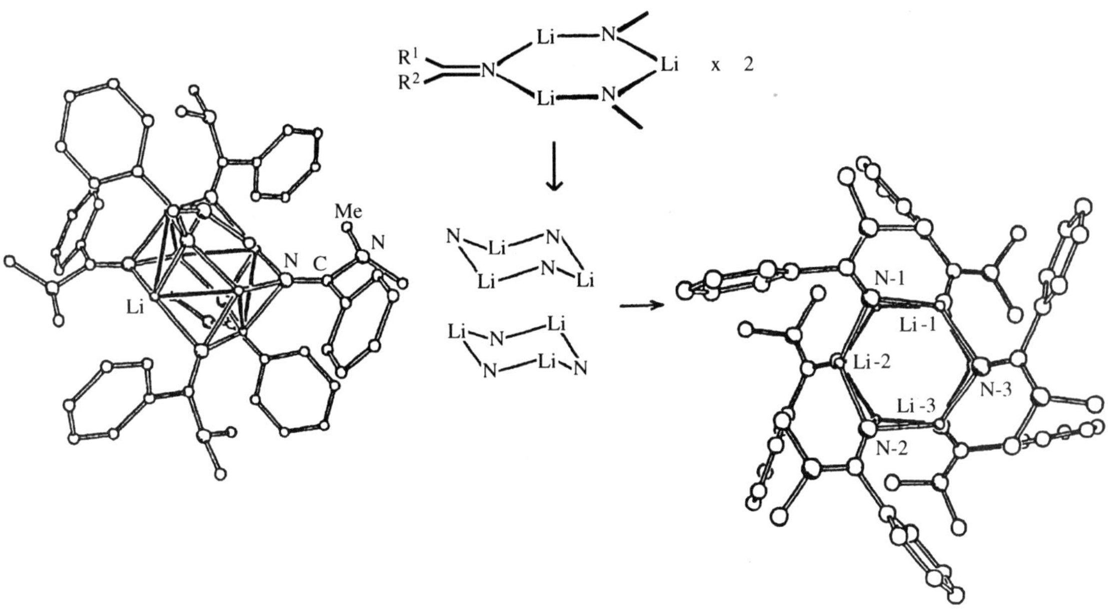

(**3**) viewed as a Li_6 chair

(**3**) viewed as a stack of two Li_3N_3 rings

The diarylimidolithiums ($[R_2C{=}NLi]_n$; R = Ph (**5**)[15–17] or *p*-MeC_6H_4 (**6**))[16] are amorphous powders. This results from the fact that the aryl groups are flat and sterically undemanding and trimeric Li_3N_3 units can therefore stack infinitely ($n = \infty$) (Scheme 1).

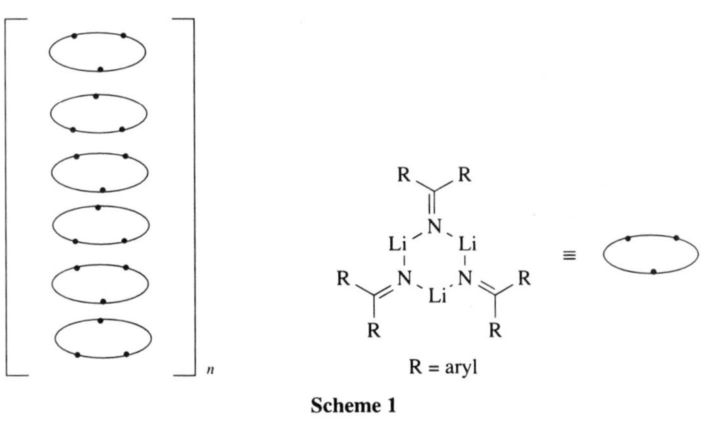

Scheme 1

Addition of Lewis base donor molecules to hexameric and polymeric imidolithiums can produce adducts $[R_2C{=}NLi{\cdot}xL]_n$, although the hexamers (**1**)–(**4**) are particularly resistent to complexation. Only in the case of addition of the powerful Lewis base donor HMPA (hexamethylphosphoramide) to (**1**) is such an adduct formed from a hexamer, in this case $[Bu^t_2C{=}NLi{\cdot}HMPA]_2$ (**7**).[18–19] In contrast, the diaryliminolithiums are more receptive to donation and addition of pyridine to (**5**) produces the tetrameric cubane $[Ph_2C{=}NLi{\cdot}py]_4$ (**8**) (a 'stack' of two dimers).[18,20] A key point of interest is that the Lewis base adducts (**7**) and (**8**) are based on dimeric building blocks rather than trimeric ones. The reasons for this can be seen from a coordination arc argument (Scheme 2). Solvation of dimeric Li_2N_2 rings leads to a less sterically crowded situation than if solvation of intact trimeric rings was to occur. For steric reasons the dimer units of (**7**) cannot stack into a cubane structure akin to (**8**). Solvation of (**5**) by HMPA produces an unusual pentametallic 'ate' complex $[(Ph_2C{=}NLi)_6Li_5{\cdot}HMPA]^-[Li(HMPA)_4]^+$ (**9**) in which a trigonal bipyramidal arrangement of metal centres is present in the anion.[21]

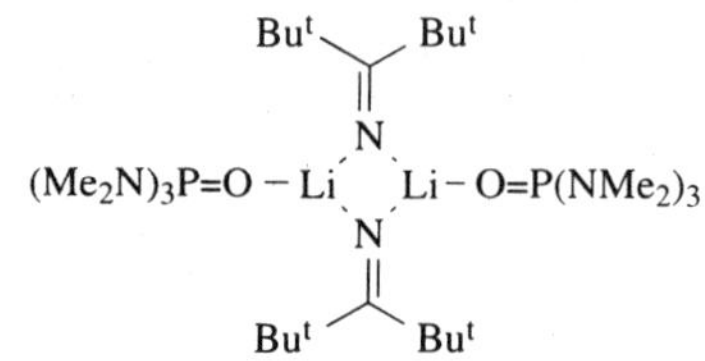

(7)

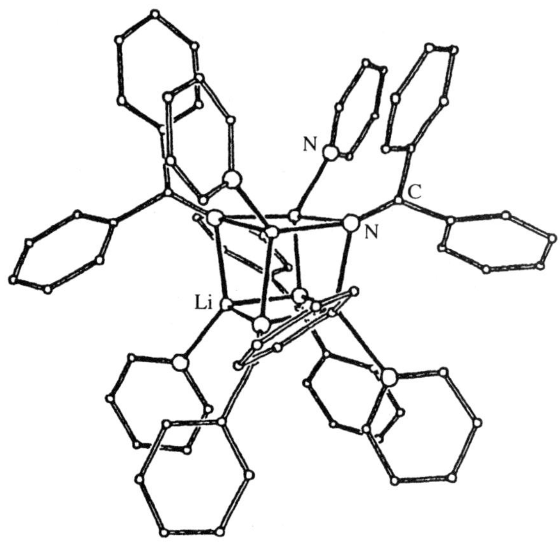

(8)

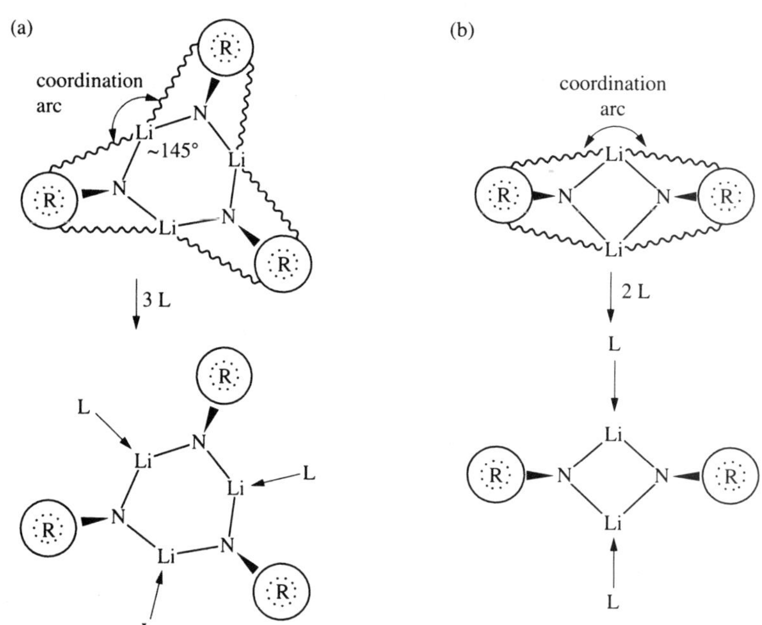

Scheme 2

Since the mid-1980s, mixed-metal imino complexes have been investigated. The x-ray structures of the cubane $[\{(Me_2N)_2C{=}N\}_4LiNa_3{\cdot}3HMPA]$ **(10)**[22] and of the hexametallic 'triple-stack' $[\{Bu^t(Ph)C{=}N\}_6Li_4Na_2]$ **(11)**[23] have been determined. Distortions within the Na_3LiN_4 core of (**10**) leave the lithium centre very sterically crowded and as a consequence it is not complexed by HMPA. Somewhat surprisingly, (**11**) is a stack of three dimeric rings (cf. complexes (**1**)–(**4**)). The latter probably arises due to the disparity between N–Li and N–Na bond lengths, which makes association of dimeric rings more favoured.

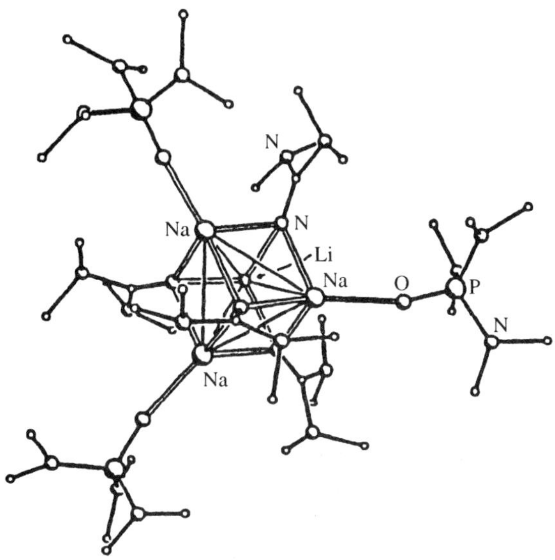

(10)

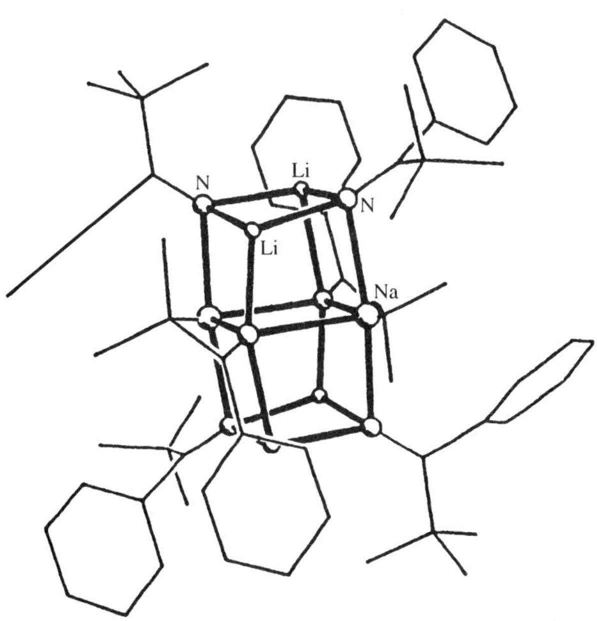

(11)

Organonitrogen–lithium compounds, particularly lithium amides $[R_2NLi]_n$, are commonly used as proton abstractors in organic synthesis.[24–8] The power of these species as reagents, especially where the organic groups (R) are bulky, relies principally on their low nucleophilicity compared with C–Li bonded complexes (e.g., MeLi). Thus proton abstraction is favoured over nucleophilic addition to an organic substrate (e.g., ketones can be enolized without attack at the C=O functionality). Such deprotonation reactions can also be regio- and/or enantiospecific.[27] A number of R_2NLi bases are now commercially available (e.g., lithium diisopropylamide, Pr^i_2NLi (LDA), and *N*-lithiohexamethyldisilazide $[(TMS)_2NLi]$).[28]

Amidolithium compounds $[R_2NLi]_n$, in which there is no complexation of the metal centres by Lewis base donor molecules, form ring or ladder structures. Further stacking association of rings, which is a common feature in organometallics $[RLi]_n$ and iminolithium compounds $[R_2C{=}NLi]_n$, is precluded since the R groups of the R_2N units project above and below the ring plane. Table 2 lists the structurally characterized uncomplexed simple lithium amides (containing no additional functionalities within their R groups) ((**12**)–(**17**)). Structurally elucidated uncomplexed lithium amides are rare since bulky substituents (R) are required to prevent polymerization into insoluble powders which probably have infinite ladder structures.

Table 2 Some structurally characterized uncomplexed simple lithium amides.

$[R_2NLi]_n$	*Method*	*Ref.*
$[Me_2C(CH_2)_3CMe_2NLi]_4$ (**12**)	x-ray	29
$[Bz_2NLi]_3$ (**13**)	x-ray	30
$[(TMS)_2NLi]_3$ (**14**)	x-ray	31
$[(TMS)_2NLi]_2$ (**15**)	electron diffraction	32
$[(Me_3Ge)_2NLi]_3$ (**16**)	x-ray	33
$[H_2C(CH_2)_5NLi]_6$ (**17**)	x-ray	34

The association numbers (n) in the above complexes appear to be largely dictated by the size of the R groups. Complex (**12**), which contains large R groups, is noteworthy in being the only ring tetramer known in amidolithium chemistry. The trimeric ring complex (**13**) is of interest in that the formally two-coordinate Li^+ cations within the ring are additionally stabilized by agostic C–H···Li interactions (shortest C···Li, 0.260 nm). Rather than being a 'stack' structure as may be suggested by its N_6Li_6 core, complex (**17**) is best regarded as a 'cyclized ladder' as can be seen by detailed examination of the N–Li distances within the cage.

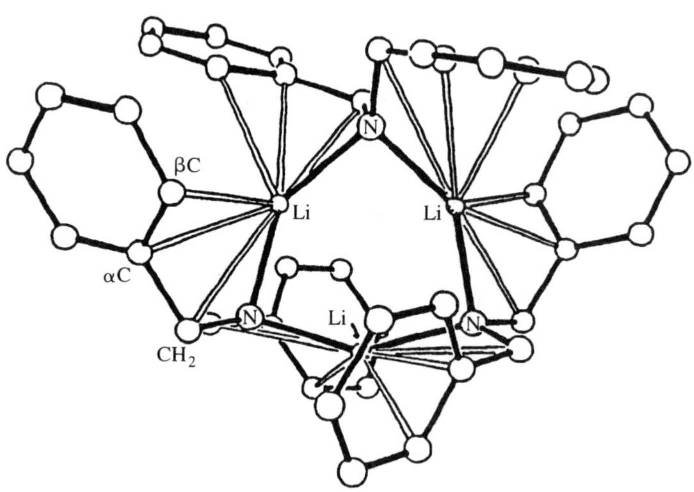

(**13**)

Heavier alkali metal analogues of (**14**) are also known. The structures of $[(TMS)_2NNa]_\infty$ (**18a**), an infinite polymer with an Na–N–Na–N–Na zig-zag backbone,[35] and $[(TMS)_2NK{\cdot}toluene]_2$ (**18b**), in which the K^+ ions form ion–dipole interactions with the aromatic solvent molecules,[36] have been determined.

Complexation of amidolithiums with added Lewis base donor molecules, as expected, leads to lower association numbers. The extent of association of these species (monomers, dimers, or molecular ladders) depends largely on the donicity and ratio of donor (L) to lithium employed.

When a ratio of L:Li of less than 1:1 is used, ladder structures can result. Lithiation of pyrrolidine, $[H_2C(CH_2)_3NH]$, in hexane produces a polymeric powder which is assumed to have an infinite ladder structure (cf. the cyclized ladder arrangement of (**17**)). If the lithiation is carried out in the presence of the ligands TMEDA or PMDETA ($(Me_2NCH_2CH_2)_2NMe$) then crystalline molecular ladders, $[\{H_2C(CH_2)_3NLi\}_2{\cdot}TMEDA]_2$ (**19**) and $[\{H_2C(CH_2)_3NLi\}_3{\cdot}PMDETA]_2$ (**20**), are isolated.[37,38a] A similar complex, though with an alternative folded four rung ladder segment, is that of $[H_2C(CH_2)_4NLi{\cdot}H_2C(CH_2)_4NH]_4$ (**21**) in which unmetallated amine molecules function as Lewis base ligands and prevent infinite lateral association.[38b] These complexes can be seen as segments of the uncomplexed infinite ladder and provide indirect evidence that, except where bulky substituents are present (e.g., complexes (**12**)–(**17**)), infinite ladder polymers are formed for the majority of uncomplexed amidolithiums.

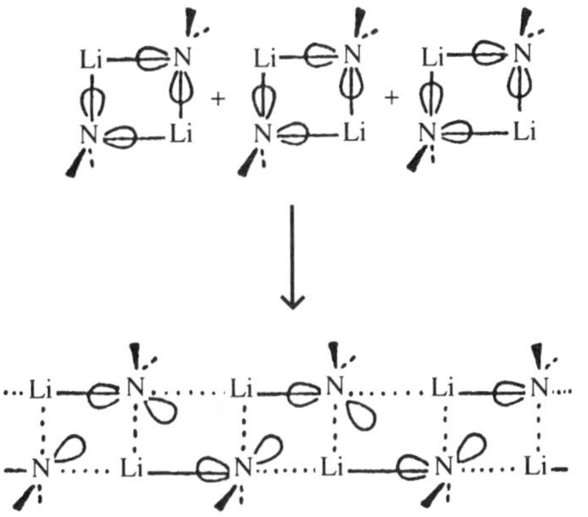

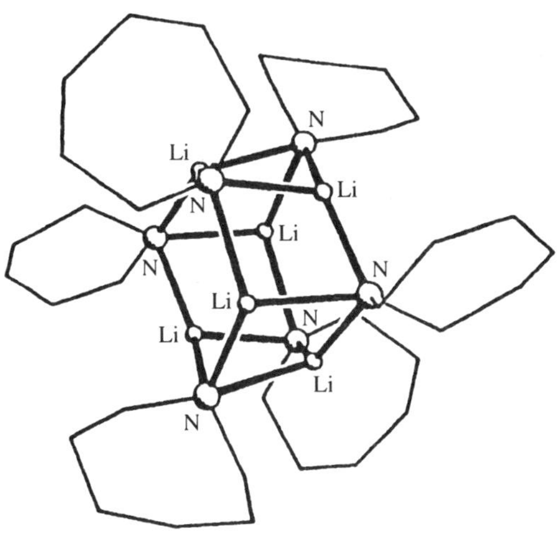

(17)

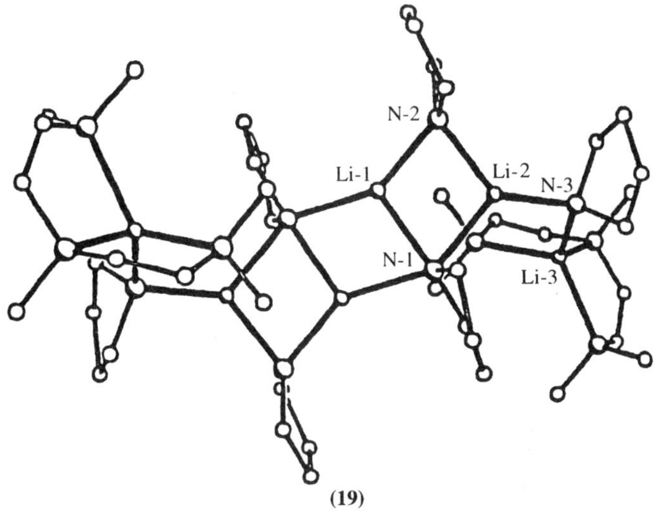

(19)

Increasing the donor ratio to 1:1 and employing a monodentate or bidentate Lewis base donor molecule gives rise to dimers. Table 3 lists the examples of dimeric amidolithium complexes which have been structurally characterized ((**22**)–(**29**)). The formation of dimers can be seen as occurring by progressive disintegration of the parent infinite ladder of the uncomplexed amidolithium. Lateral

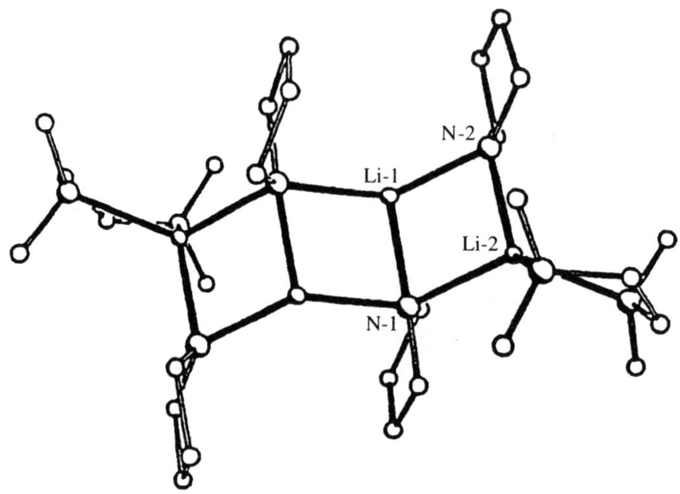

(20)

association into such a ladder structure or vertical association into a stack structure is prevented by coordination of the Li^+ cations by the Lewis base donor ligands, as seen, for example, in the structure of $[Bz_2NLi{\cdot}OEt_2]_2$ (**22**).

Table 3 Some dimeric amidolithium complexes which have been structurally characterized.

$[R_2NLi{\cdot}xL]_2$	*Method*	*Ref.*
$[Bz_2NLi{\cdot}OEt_2]_2$ (**22**)	x-ray	30a
$[Bz_2NLi{\cdot}HMP]_2$ (**23**)	x-ray	30b
$[(TMS)_2NLi{\cdot}OEt_2]_2$ (**24**)	x-ray	29,39
$[(TMS)_2NLi{\cdot}THF]_2$ (**25**)	x-ray	40
$[2,4,6\text{-}Bu^t{}_3C_6H_2N(H)Li{\cdot}OEt_2]_2$ (**26**)	x-ray	41
$[Mes_2BN(H)Li{\cdot}OEt_2]_2$ (**27**)	x-ray	42
$[1,8\text{-}C_{10}H_6NHSiMe_2NLi{\cdot}OEt_2]_2$ (**28**)	x-ray	43
$[PhN(Me)Li{\cdot}TMEDA]_2$ (**29**)	x-ray	44

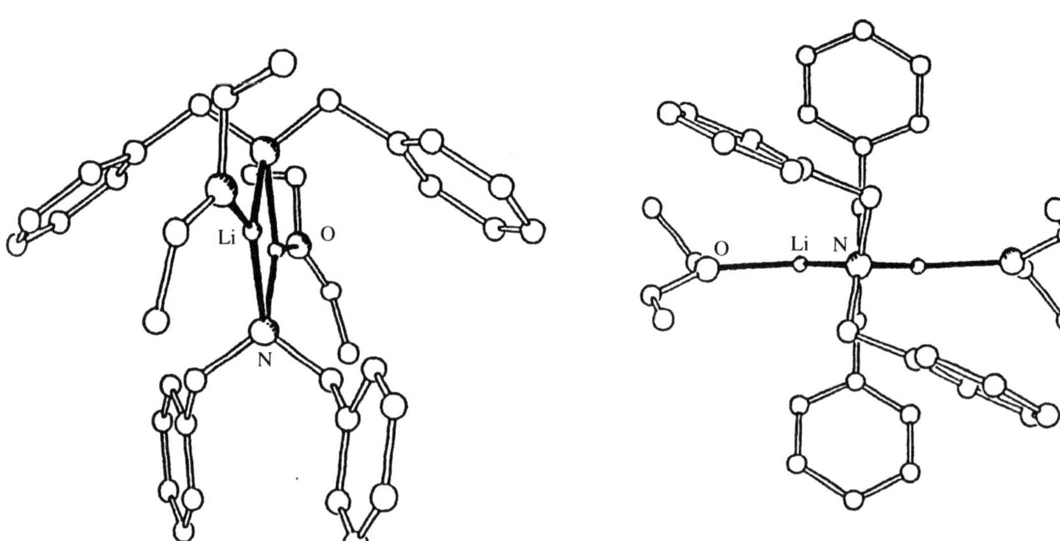

(22)

In all of the above dimers the relatively short Li···Li contacts (0.24–0.25 nm) have no real significance and are purely a consequence of the acute μ-N angles (74–8°). The presence of a bidentate Lewis base attached to the lithium centres of dimer (**29**) is unusual. Bidentate donor molecules usually lead, for steric reasons, to monomers containing three-coordinate lithium. The reasons for the formation of a dimer as opposed to a monomer for (**29**) stem from the sterically uncrowded arrangement in which the methyl and phenyl groups are aligned perpendicular to the Li_2N_2 ring plane.

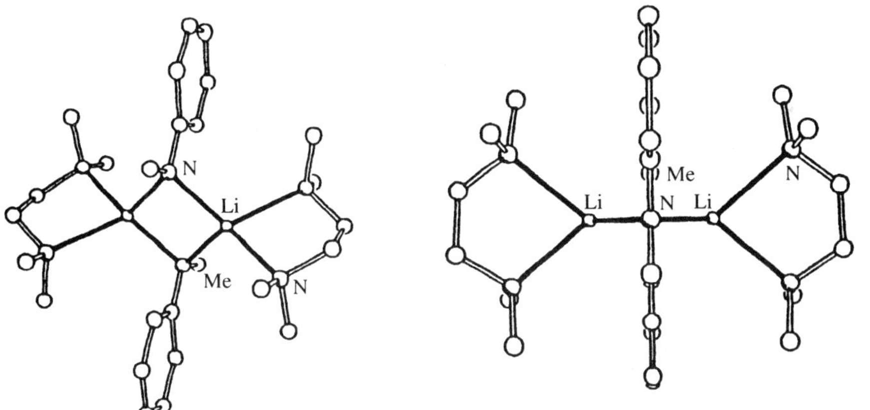

(**29**)

Table 4 gives details of monomeric amidolithium complexes which have been structurally characterized ((**30**)–(**48**)). Almost all are contact ion pairs (except for (**40**), (**41**) and (**48**) in which the Li^+ cations are encased by ligand coordination).

Table 4 Some monomeric amidolithium complexes which have been structurally characterized.

$R_2NLi{\cdot}xL$	*Ref.*
Ph(naphthyl)NLi·TMEDA (**30**)	44
Ph(naphthyl)NLi·PMDETA (**31**)	44
2,4,6-$Bu^t_3C_6H_2N(H)Li$·TMEDA (**32**)	32
2,4,6-$Bu^t_3C_6H_2[Si(CHMe_2)_2Cl]NLi{\cdot}2THF$ (**33**)	45
$C_{12}H_8SNLi{\cdot}3THF$ (**34**)	46
$C_{12}H_8ONLi{\cdot}3THF$ (**35**)	46
$Mes_2B(Ph)NLi{\cdot}2OEt_2$ (**36**)	47
$Mes_2B(Mes)NLi{\cdot}2OEt_2$ (**37**)	48
$Bu^t_2BN(Bu^t)Li$·TMEDA (**38**)	49
$MeBN(Bu^t)BMe_2N(Bu^t)Li$·TMEDA (**39**)	49
$[Mes_2BNBMes_2]^-[Li{\cdot}3OEt_2]^+$ (**40**)	42
$[Mes_2BNSiPh_3]^-[Li(12\text{-crown-}4)_2]^+$ (**41**)	50
$(TMS)_2NLi$·TMEDA (**42**)	51
$(TMS)_2NLi{\cdot}2THF$ (**43**)	52
$(Ph_2MeSi)_2NLi{\cdot}2THF$ (**44**)	52
$(TMS)_2NLi$·(12-crown-4) (**45**)	50,53
Ph_2NLi·(12-crown-4) (**46**)	50,54
$(Ph_2MeSi)_2NLi$·(12-crown-4) (**47**)	52
$[(Ph_3Si)_2N]^-[Li(12\text{-crown-}4)_2]^+$ (**48**)	50,52,55

The extent of aggregation in the above monomers is limited by steric crowding about the lithium centres which prevents further association of the monomeric units by bridging of the amido nitrogen centres. Even where the Li^+ cations are three-coordinate (as in (**30**), (**32**), (**36**)–(**39**) and (**42**)–(**44**)), monomeric structures are sustained. However, an interesting compromise is reached in the structure of (**30**). Although monomeric units occur in the solid state, these units form weakly linked dimers by agostic (*ortho* and *meta*) aryl C–H···Li interactions (~ 0.31 nm). Shorter intramolecular C–H···Li interactions also occur in (**32**) and (**44**) (*ortho*-C–H <0.29 nm). Thus the lithium centres make up for their coordinatively unsaturated environment by robbing electron density from a neighbouring (aromatic) group.

For the heavier alkali metal complexes the structural features are similar to those outlined for amidolithium compounds although these have been less extensively studied. However, the larger sizes of the alkali metal cations commonly leads to higher coordination numbers for the metals. As can be seen from Table 5, which lists some complexes x-ray characterized since the late 1980s, the structures of heavier alkali metal amides are commonly dimers ((**49**)–(**58**)).

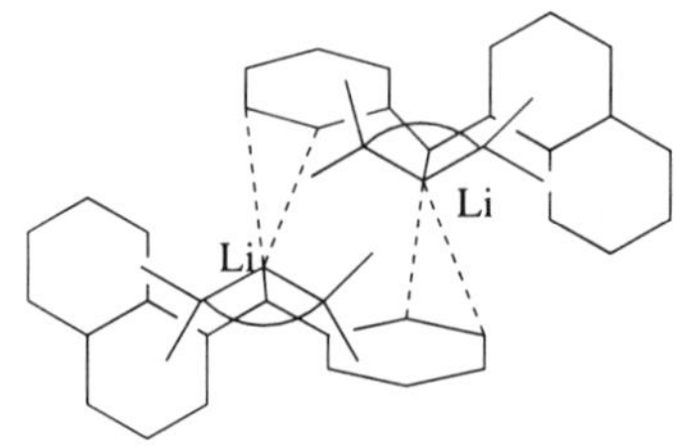

(30)

Table 5 Some heavier alkali metal amides which have been structurally characterized.

$[R_2NM{\cdot}xL]_n$	X-ray structure	Ref.
$[Na\{NPh(py)\}{\cdot}PMDETA]_2$ (**49**)	dimer	56
$[NaN{=}CBu^t_2]_4[HN{=}CBu^t_2]_2$ (**50**)	cubane	57
$[Na_4(NMe_2)_4(HNMe_2){\cdot}TMEDA]_2$ (**51**)	stack	58
$[Na_{10}(NMe_2)_{10}{\cdot}4TMEDA]$ (**52**)	stack	58
$[Na_{10}(NMe_2)_{12}{\cdot}4TMEDA]$ (**53**)	stack	58
$[Na_{10}(NMe_2)_{10}{\cdot}(p\text{-xylyl})_2{\cdot}4TMEDA]$ (**54**)	stack	58
$[Na(indolyl){\cdot}TMEDA]_2$ (**55**)	dimer	59
$[Na(indolyl){\cdot}PMDETA]_2$ (**56**)	dimer	59
$[M(carbazolyl){\cdot}PMDETA]_2$ (M = K (**57**), M = Cs (**58**))	dimers	60

Finally, mixed amidolithium/inorganic salt complexes which have superior activity and stereoselectivity over the parent amide bases have been studied since the late 1980s.[61–2] Such interest has been spurred by the characterizations of two LDA complexes in the solid state[63–4] and by the promise that greater reagent selectivities may yet be possible with an increased understanding of the structures and dynamics of these species.[9] Structurally characterized examples of these complexes include $[(C_6H_4)_2(NH)NLi{\cdot}LiCl{\cdot}4THF]$ (**59**) (a Li_2NCl ring in the solid state)[65] and the LDA cocomplex $[\{(Me_2CH)_2NLi\}_2{\cdot}LiCl]$ (**60**) (which forms a ladder-type structure in the solid state).[66]

1.2.1.2 P–M, As–M and Sb–M bonded complexes

Many of the structural features outlined earlier (Section 1.2.1.1) for amido-metal complexes have been observed in the solid-state structures of the heavier group 15 analogues. The majority of the species so far characterized are uncomplexed or Lewis base complexed phosphides or arsenides of lithium ($[R_2ELi]_n$ and $[R_2ELi{\cdot}xL]_n$). Table 6 lists structurally characterized P–Li bonded complexes ((**61**)–(**72**)).

Table 6 Some structurally characterized P–Li bonded complexes.

Compound	Structure (x-ray)	Ref.
$[LiPPh_2{\cdot}PMDETA]$ (**61**)	monomer	67
$[Li(12\text{-crown-}4)_2]^+ [PPh_2]^-$ (**62**)	ion-separated	68
$[LiP(CH\text{-}TMS)_2]_2$ (**63**)	dimer	69
$[LiP(TMS)_2{\cdot}2THF]_2$ (**64**)	dimer	70
$[LiPPh(TMS){\cdot}TMEDA]_2$ (**65**)	dimer	71
$[LiPBu^t_2{\cdot}THF]_4$ (**66**)	four-rung ladder	72
$[LiP(TMS)_2]_4{\cdot}2THF$ (**67**)	four-rung ladder	70
$[Li(P\text{-}TMS)_2]_6$ (**68**)	six-rung ladder	73
$[LiPH(2,4,6\text{-}Me_3C_6H_2){\cdot}2THF]_\infty$ (**69**)	helical chain polymer (both enantiomers)	74
$[LiPPh_2{\cdot}OEt_2]_\infty$ (**70**)	helical chain polymer	75
$[LiP(C_6H_{11})_2{\cdot}THF]_\infty$ (**71**)	helical chain polymer	75
$[LiPPh_2{\cdot}2THF]_\infty$ (**72**)	helical chain polymer	75

The uncomplexed complexes (**63**) and (**68**) make an interesting comparison. Reduction in the steric bulk of the group attached to the anionic phosphorus centre allows lateral association from a dimer for (**63**) to a remarkable six-rung ladder structure for (**68**). Evidently there is still sufficient steric crowding at the termini of the ladder to prevent cyclization into a structure similar to the amidolithium complex (**17**) (Section 1.2.1.1). The dimeric structure of (**63**) can also be compared to the trimeric structure observed for the uncomplexed arsenic analogue (**78**). Presumably, the increase in the size of the ring in the latter stems from the accommodation of a larger heteroatomic centre. The structures of the four-rung ladders (**66**) and (**67**) illustrate that the ratio of the Lewis base ligand (L) to lithium depends on the steric bulk of the organic groups attached to the phosphorus. In (**66**), where less sterically demanding PBu^t_2 groups are present, the ladder terminates with doubly solvated lithium centres and in (**67**), where the more sterically demanding $P(TMS)_2$ groups are present, only one THF is required to terminate the ladder. Direct ^{31}P–^{7}Li coupling constants fall in the range 36–122 Hz in such P–Li bonded complexes.

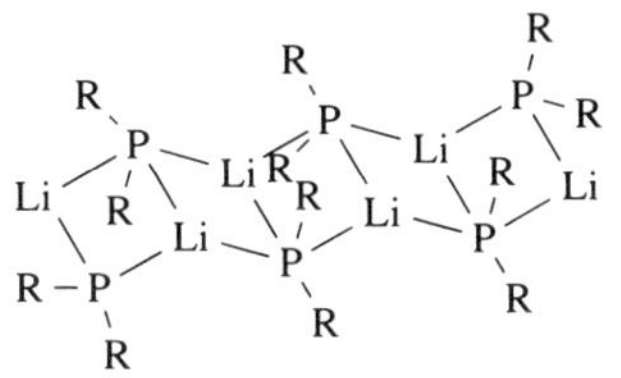

(68)

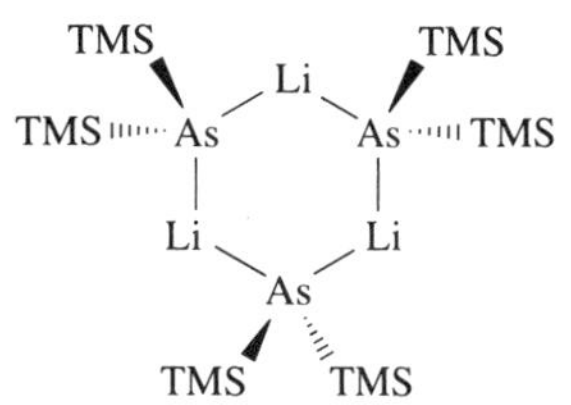

(78)

The complexes [$LiAsPh_2$·3(1,4-dioxane)] (**73**),[54] [$LiAs(1,3,5\text{-}Pr^i_3C_6H_2Si)_2(F)Si(Pr^i_3Si)$·2THF] (**74**)[76] are As–Li bonded monomers. The only molecular Sb–Li or Bi–Li bonded monomer is that of [$LiSb(C_6H_{11})_2$·PMDETA] (**75**).[77] The complexes [$LiAsPh_2$·$2OEt_2$]$_2$ (**76**)[54] and [$LiAs(AsBu^t_2)Bu^t$·THF]$_2$ (**77**)[78] are dimers and the largest oligomer is that of uncomplexed complex [$LiAs(CH\text{-}TMS)_2$]$_3$ (**78**).[79] The structure of the latter can be compared to those of unsolvated trimeric amidolithium complexes (complexes (**13**), (**14**) and (**16**)) which have symmetrical and planar Li_3N_3 cores. The Li_3As_3 core of (**78**) is nonplanar and exhibits large angular distortions. Apart from (**75**), the only other Sb–Li or Bi–Li bonded complexes which have been characterized in the solid state are the polymers [$LiE\{C(TMS)_3\}_2$·$MeOCH_2CH_2OMe$]$_\infty$ (**79**) (E = Sb,[80] Bi[81]).

1.2.2 Group 16–Alkali Metal Bonded Complexes

1.2.2.1 O–M bonded complexes

The discovery of high-temperature superconductivity has given rise to increased interest in the structures and properties of alkoxides of the alkali metals.[82] Up to the mid-1990s, comparatively few complexes had been structurally characterized.[83] In contrast, the structural chemistry of alkali metal enolates is more extensive and the structures and reactivity of lithium enolates have been excellently reviewed by Seebach who has made a major contribution in this area.[9] Boche has reviewed the structures of lithium compounds containing sulfones, sulfoxides (which includes information on the structures of complexes of thioethers and 1,3-dithianes, nitriles, nitro compounds and hydrazones)[10] and further structural information can be obtained from the reviews by Setzer and Schleyer[7] and Schade and Schleyer.[8]

By 1982 the structures of a number of simple unsolvated alkoxides of the alkali metals had been determined. These mainly have layer structures except where more bulky substituents are present, for example, MOMe (M = Li–Cs)[84] are all layer structures, whereas $MOBu^t$ (M = Na–Cs) are molecular.[85] The most remarkable of these is that of [$NaOBu^t$]$_n$ (**80**), which crystallizes with two independent molecules in the asymmetric unit, a hexamer and a nonamer.

As with imidolithium compounds, stacking of trimeric or Lewis base solvated dimeric rings is the norm, and ring dimers, tetrameric cubanes and hexameric stacks are the most prevalent structural motifs for molecular alkali metal alkoxides and enolates. Table 7 lists some of the recently structurally characterized alkoxides and enolates ((**81**)–(**104**)).

As with the amido- and imino-complexes discussed earlier (Section 1.2.1.1), the association numbers (n) of the alkoxides and enolates reflect the extent of Lewis base solvation of the metals and the steric bulk of the organic substituents present. M_2O_2 ring dimers result for complexes (**81**)–(**85**) and (**89**)–(**92**) where the metal centres are coordinatively saturated (giving four-coordinate lithium centres and a five-coordinate sodium in (**90**)) and for (**86**)–(**88**) where the steric bulk of the organic substituents is the major factor preventing stacking into tetrameric structures (even though the metal centres are only

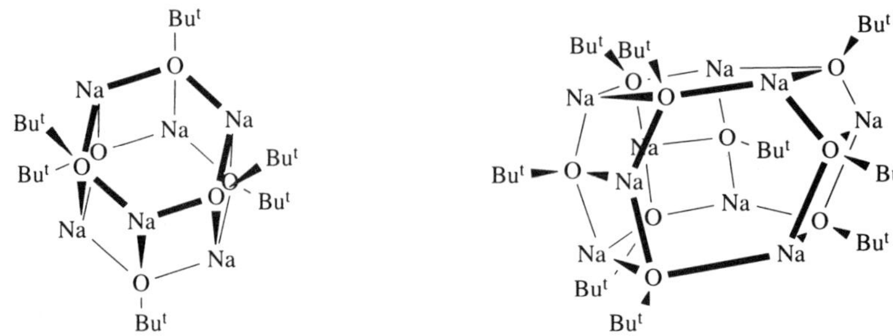

(80)

Table 7 Some structurally characterized alkoxides and enolates.

Compound	*X-ray structure*	*Ref.*
$[Bu^tC(=CH_2)OLi{\cdot}TMEDA]_2$ (**81**)	dimer	86
$[Bu^tO{\cdot}C(=CMe_2)OLi{\cdot}TMEDA]_2$ (**82**)	dimer	87
$[Bu^tO{\cdot}C(=CHMe)OLi{\cdot}TMEDA]_2$ (**83**)	dimer	87
$[MeCH=C(NMe_2)OLi{\cdot}TMEDA]_2$ (**84**)	dimer	86
$[HC=CH{\cdot}(CH=CH)_2C=C(NMe_2)OLi{\cdot}2THF]_2$ (**85**)	dimer	88
$[Bu^t{}_3COLi{\cdot}THF]_2$ (**86**)	dimer	89
$[2,6\text{-}Bu^t{}_2\text{-}4\text{-}MeC_6H_2OLi{\cdot}OEt_2]_2$ (**87**)	dimer	90
$[Bu^tC(=CH_2)CH=C(Bu^t)OLi{\cdot}DMPU]_2$[a] (**88**)	dimer	91
$[p\text{-}BrC_6H_4OLi{\cdot}TMEDA]_2$ (**89**)	dimer	92
$[PhONa{\cdot}PMDETA]_2$ (**90**)	dimer	92
$[2,6\text{-}Bu^t{}_2C_6H_3OLi{\cdot}OEt_2]_2$ (**91**)	dimer	90
$[2,4,6\text{-}(CF_3)_3C_6H_2ONa{\cdot}2THF]_2$ (**92**)	dimer	93
$[2,6\text{-}(Me_2NCH_2)_2C_6H_3OLi]_3$ (**93**)	trimer	94
$[TMS\text{-}OK{\cdot}\eta\text{-}C_6H_6]_4$ (**94**)	tetramer	95
$[Bu^tC(=CH_2)OLi{\cdot}THF]_4$ (**95**)	tetramer	96
$[H_2C(CH_2)_2CH=COLi{\cdot}THF]_4$ (**96**)	tetramer	96
$[2\text{-}Me_2NCH_2C_6H_4C(=CH_2)OLi]_4$ (**97**)	tetramer	97
$[Bu^tC(=O)CH_2CH(Bu^t)OLi]_4$ (**98**)	tetramer	98
$[Bu^t{}_2Si(NH_2)OLi]_4$ (**99**)	tetramer	99
$[MeOC(=CHBu^t)OLi{\cdot}THF]_4$ (**100**)	tetramer	87
$[Bu^tC(=CH_2)ONa{\cdot}O=(Bu^t)Me]_4$ (**101**)	tetramer	100
$[Bu^tC(=CH_2)OLi]_6$ (**102**)	hexamer	100,101
$[Et_2NCH=C(OEt)OLi]_6$ (**103**)	hexamer	102
$[Bu^tC(=CH_2)OK{\cdot}THF]_6$ (**104**)	hexamer	101

[a] DMPU = $[(CH_2)_4(NMe)_2]C=O$.

monosolvated and three-coordinate). In the sodium complex (**92**) the presence of Na···F interactions with the *ortho*-CF_3 groups also plays a role in satisfying the coordination sphere of the metal and in preventing further association of the dimeric units. The association of dimeric units (in which monosolvation of the metal centres is present) into tetrameric complexes occurs where less sterically demanding organic groups are found. In complexes (**97**)–(**99**) intramolecular solvation by donor arms present on the organic substituents give the metal centres four-coordinate pseudotetrahedral geometries. Similar four-coordination of the metal centres is found in complexes (**94**)–(**96**), (**100**) and (**101**), where solvation of the metal centres by added Lewis base donor molecules occurs. Complex (**94**) is worthy of note in that the K^+ cations of the K_4O_4 cubane are loosely coordinated by η^6-benzene ligands. The weakness of this coordination is indicated by the fact that the tetrameric cubane structure of unsolvated $[KO\text{-}TMS]_4$ is not fragmented (e.g., into the K_2O_2 dimeric constituents) by such arene solvation.[103]

Less sterically demanding organic groups give hexameric structures (complexes (**102**)–(**104**)). In (**102**) the presence of a π-interaction between the enol CH_2=group and lithium presumably prevents further stacking of the Li_3O_3 rings. In (**103**) and (**104**) further stacking is precluded by intra- and intermolecular solvation of the metal centres, respectively. The hexameric structure of the potassium complex (**104**) can be compared to that of the tetrameric sodium complex (**101**) in which the same enolate anion is present. The preference for the formation, and subsequent association, of trimeric K_3O_3 rings in (**104**) appears to be dictated by the larger size of K^+ compared to Na^+ and possibly also by the steric requirements of the Lewis base donor molecules involved. The observation of a trimeric Na_3O_3 ring structure in (**93**) is of particular interest since similar M_3O_3 rings are the immediate precursors to the formation of hexameric structures. Vertical association into such a hexameric stacked structure is

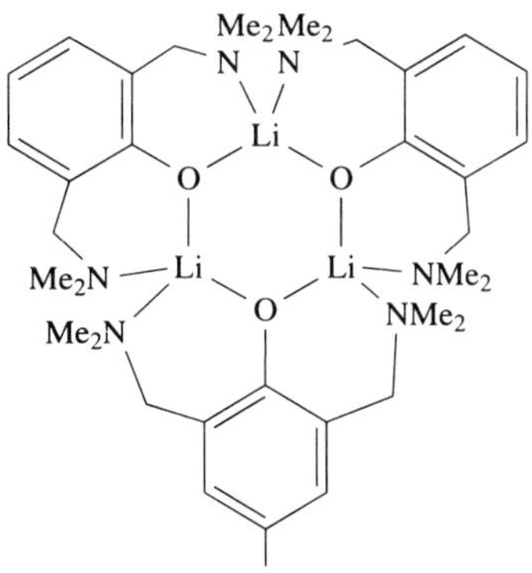

(93)

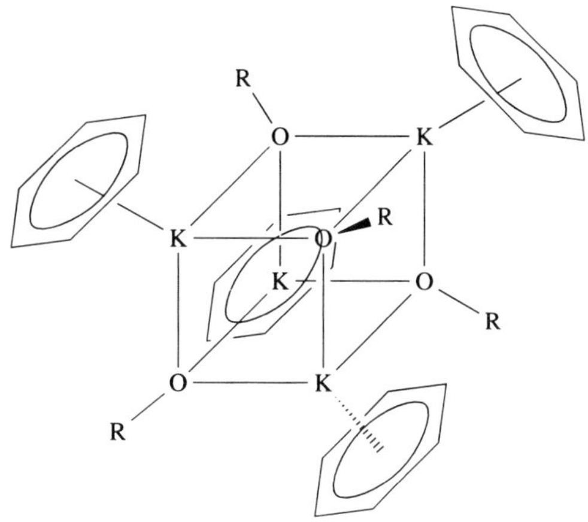

R = TMS

(94)

prevented in (**93**) by the intramolecular coordination of the metal centres by both *ortho*-CH_2NMe_2 donor arms.

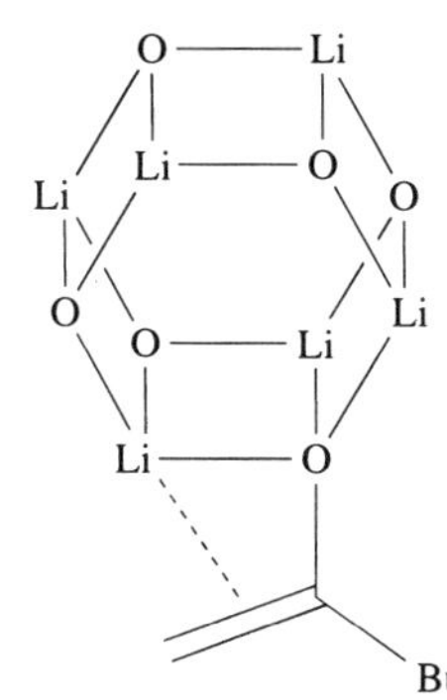

(102)

Much of the current interest in enolates and alkoxides of the alkali metals centres on their synthetic applications and as product-forming intermediates in organic synthesis. The addition of Bu^tO^- anions (or amines as Lewis bases) to organolithiums produces 'superbases' with enhanced reactivity (e.g., towards reaction with C–H bonds in organic substrates).[104] Since the late 1980s the tetrameric superbase complexes $[Bu^nLi{\cdot}Bu^tOLi]_4$ (**105**)[105] and $[(4,6\text{-}Me_2\text{-}2\text{-}NaCH_2C_6H_2)OLi]_4$ (**106**)[106] (containing a Li_4O_4 cubane core) have been structurally characterized. There has also been considerable interest in alkoxide/alkali metal salt complexes, which may also have applications as synthetic organic reagents. The cubanes $[(PhOLi)_3{\cdot}LiNCS{\cdot}4HMPA]$ (**107a**)[107] and $[\{2,6\text{-}(Me_2NCH_2)_2{\cdot}C_6H_3OLi\}_2(LiI)_2]$ (**107b**)[94] have been prepared and structurally elucidated. Complex (**107a**) is synthesized by the *in situ* reaction of Bu^nLi, PhOH and NH_4NCS (4:3:1) and HMPA, whereas cocomplexation of the trimeric complex (**93**) with anhydrous $[LiI]_\infty$ occurs directly to give (**107b**). Interestingly, the NCS^- anion of (**107**) does not form part of the cubane core (by μ_3-bridging of three lithium centres), but instead the NCS^- anion is terminally attached and a HMPA ligand completes the Li_4O_4 core. The latter illustrates the ylidic nature of the phosphine oxide functionality of HMPA which functions like an alkoxide in (**107a**) ($[Me_2N]_3P^-\text{–}O^+$ rather than $[Me_2N]_3P{=}O$).

1.2.2.2 S–M, Se–M and Te–M bonded complexes

There are few structurally characterized examples of simple complexes containing chalconide–alkali metal bonds (unsolvated $[REM]_n$ or solvated $[REM{\cdot}xL]_n$; E = S, Se, Te). Those that have been structurally elucidated have many similarities with the alkoxides and enolates discussed above (Section 1.2.2.1). Table 8 lists a representative cross-section of these compounds ((**108**)–(**122**)).

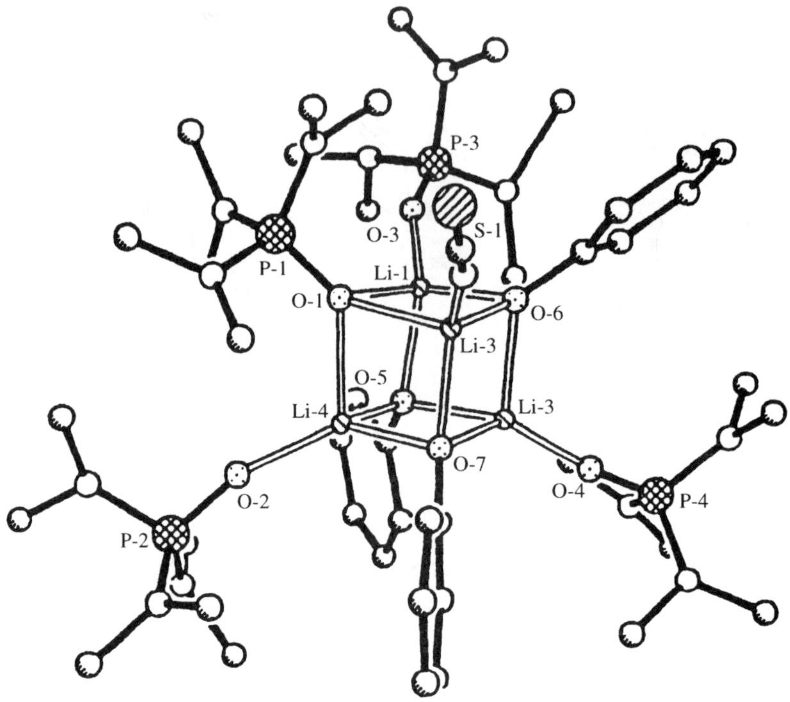

(107a)

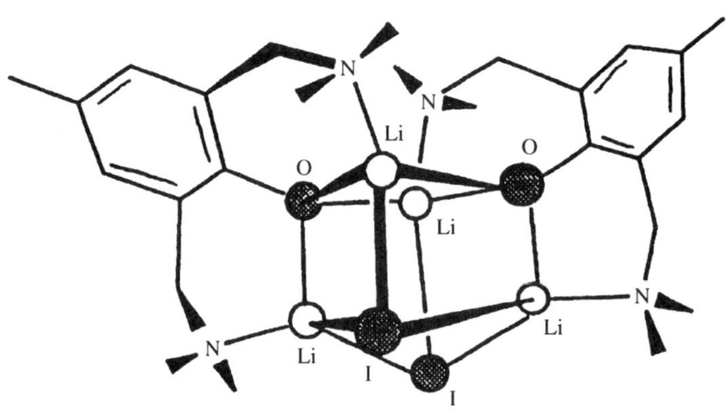

(107b)

Although complexes (**115**)–(**117**) do not fall strictly into the class of simple thiolate complexes, they have been included in Table 8 as examples of more highly functionalized complexes or organometallics containing S–Li bonds. Further information on such complexes can be found in the review by Boche.[10]

Those complexes which are particularly worthy of note are (**120**), which contains a Te_6Li_6 core and is the only structurally characterized heavier chalconide hexamer, and (**122**), which has a folded ladder arrangement in the solid state.

1.2.3 Si–M, Sn–M and Pb–M Bonded Complexes

In recent years there has been an interest in the occurrence of bonds between the heaviest (most metallic) elements in group 14 and the alkali metals. Organic chemists have frequently used such species as sources of 'soft' nucleophiles in organic reactions, for example, triorganostannates, commonly formulated as $R_3Sn^-M^+$ or $R_3Sn–M$.[119] However, only since the mid-1980s has such metal–metal bonding been broadly demonstrated in the solid state.

Table 8 Some structurally characterized complexes containing E–M bonds (E = S, Se, Te).

Compound[a]	*X-ray structure*	*Ref.*
[2,4,6-$Bu^t_3C_6H_2SLi\cdot 3THF$] (**108**)	monomer	108
[2,4,6-$Bu^t_3C_6H_2SeLi\cdot 3THF$] (**109**)	monomer	109
[2,4,6-$Bu^t_3C_6H_2TeLi\cdot 3THF$] (**110**)	monomer	110
[2-$Me_2NCH_2C_6H_3TeLi\cdot DME$] (**111**)	monomer	110
[2-$MeC_6H_4SLi\cdot 3py$] (**112**)	monomer	111
$[(TMS)_2CHSLi\cdot 2THF]_2$ (**113**)	dimer	112
$[(TMS)_3CSLi\cdot 2THF]_2$ (**114**)	dimer	112
$[PhC(=O)SLi\cdot TMEDA]_2$ (**115**)	dimer	113
[2-Me-(1,3-dithiane)Li·TMEDA]$_2$ (**116**)	dimer	114
[$PhSCH_2Li\cdot TMEDA$] (**117**)	dimer	115
$[(TMS)_3SiTeLi\cdot 2THF]_2$ (**118**)	dimer	116
$[(TMS)_3SiTeLi\cdot DME]_2$ (**119**)	dimer	117
$[(TMS)_3SiTeLi]_6$ (**120**)	hexamer (stack)	118
$[PhSLi\cdot Py]_\infty$ (**121**)	chain polymer	111
$[BzSLi\cdot Py]_\infty$ (**122**)	folded ladder polymer	111

[a] Although having a S_2Li_2 ring dimer structure, due to fractional occupancy there are only 3.5 THF ligands in the crystal.

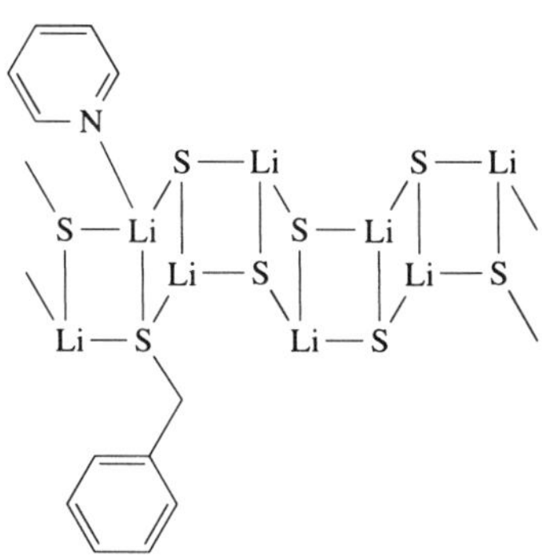

(122)

Currently only a few silicon–alkali metal bonded complexes have been structurally characterized.[120] These complexes form similar aggregates to their C–M bonded relatives; for example, the largest aggregate is $[TMS\text{-}Li]_6$ (**123**), which has a hexameric stack structure reminiscent of many organometallic complexes $[RLi]_6$.[120b] There so far appear to be no examples of simple germanium–alkali metal bonded complexes which have been characterized in the solid state. The only structurally characterized tin or lead complexes to contain metal–metal bonding are [Ph_3E–Li·PMDETA] (E = Sn (**124**),[121a] Pb (**125**)[122]) and [$(Me_2CH)_3Sn$–K·3toluene] (**126**).[123] The latter is of interest in that the K^+ cation is solvated not by Lewis base donor molecules but by three η-toluene ligands. Interestingly, the Ph_3E^- anions present in (**124**) and (**125**) also remain pyramidal upon ion separation and appear to use essentially *p*-orbitals in their bonding to carbon (rather than being sp^3 hybridized).[124] This situation contrasts with that occurring in Ph_3C^- complexes where ion separation is accompanied by increased planarity and essentially sp^2 character at the carbon centre.[125] The difference is no doubt a consequence of the increased effective energetic separation between the *s*- and *p*-orbitals on descending group 14. Values of ^{119}Sn–^7Li coupling constants of up to 466 Hz have been observed in solution for Sn–Li bonded complexes.[121] The magnitude of these coupling constants may well reflect the fact that tin is using essentially an *s*-orbital in its lone pair, which is directed towards lithium in the Sn–Li bonds.

The cyclopentadienyl derivatives of groups 13 and 14 have also proved useful in the synthesis of anionic organometallic π-complexes. Nucleophilic addition of Cp^- to CpTl and to Cp_2E (E = Sn, Pb) produces a range of ion-separated and ion-paired species. These include ion-paired [Cp_2E(μ-Cp)Na·PMDETA] (E = Sn (**127**),[126] Pb (**128**)[127]), ion-separated $2[Cp_3E]^-[Mg(THF)_6]^{2+}$ (E = Sn (**129**),[128] Pb (**130**)[127]), and ion-paired [CpTl(μ-Cp)Li·PMDETA] (**131**).[129] There are no heavy *p*-block metal–alkali metal bonds within these species, but instead E(μ-Cp)M bridges link the cationic and anionic halves together.

A review by Veith, who has contributed a great deal to this area, on heavy *p*-block metal cage complexes is a valuable source for synthetic and structural work carried out on *p*-block metal anion complexes.[130]

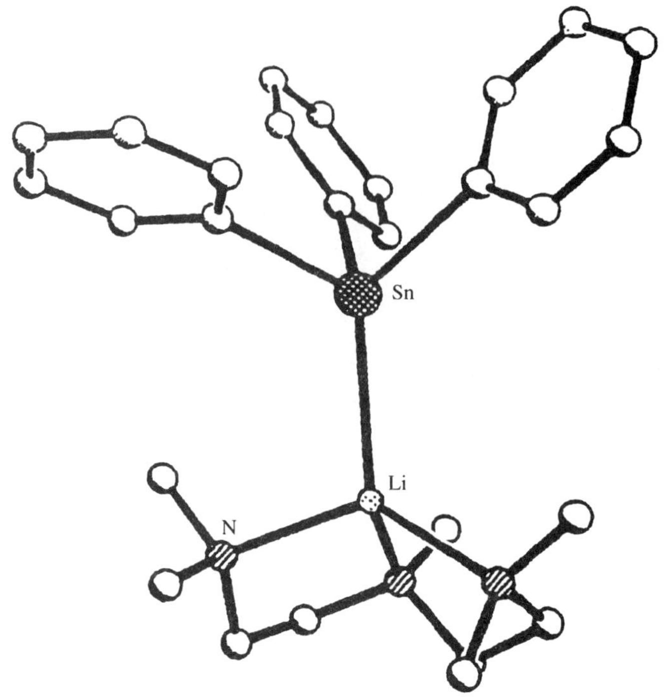

(124)

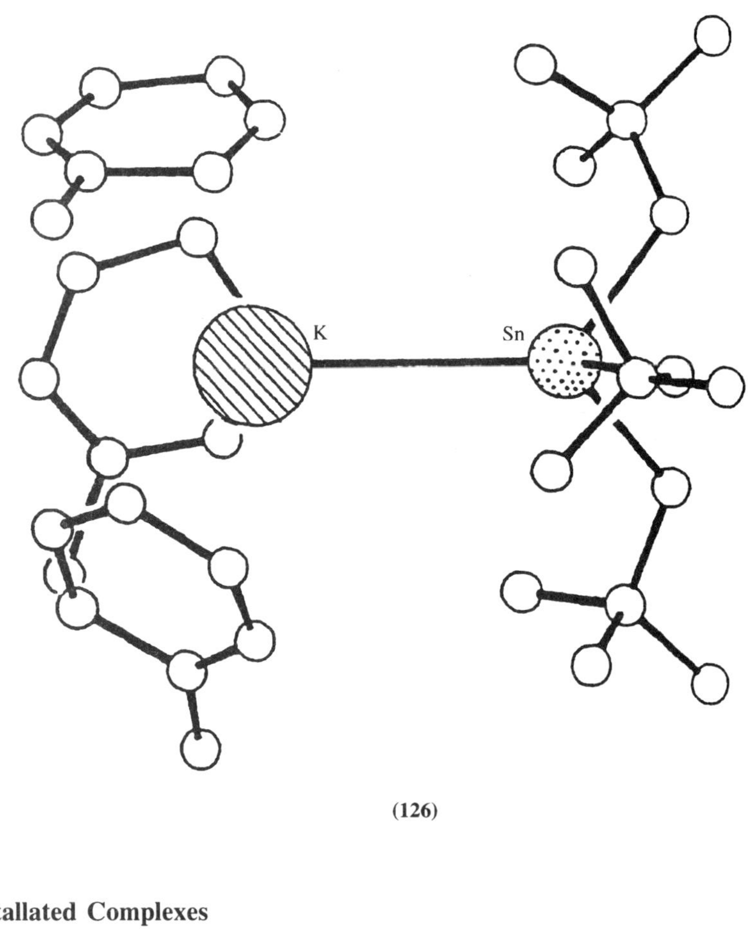

(126)

1.2.4 Multimetallated Complexes

Complexes containing simple functionalities in which more than one deprotonation of the acidic centres has occurred are rare. A few examples have been structurally characterized since the mid-1980s.

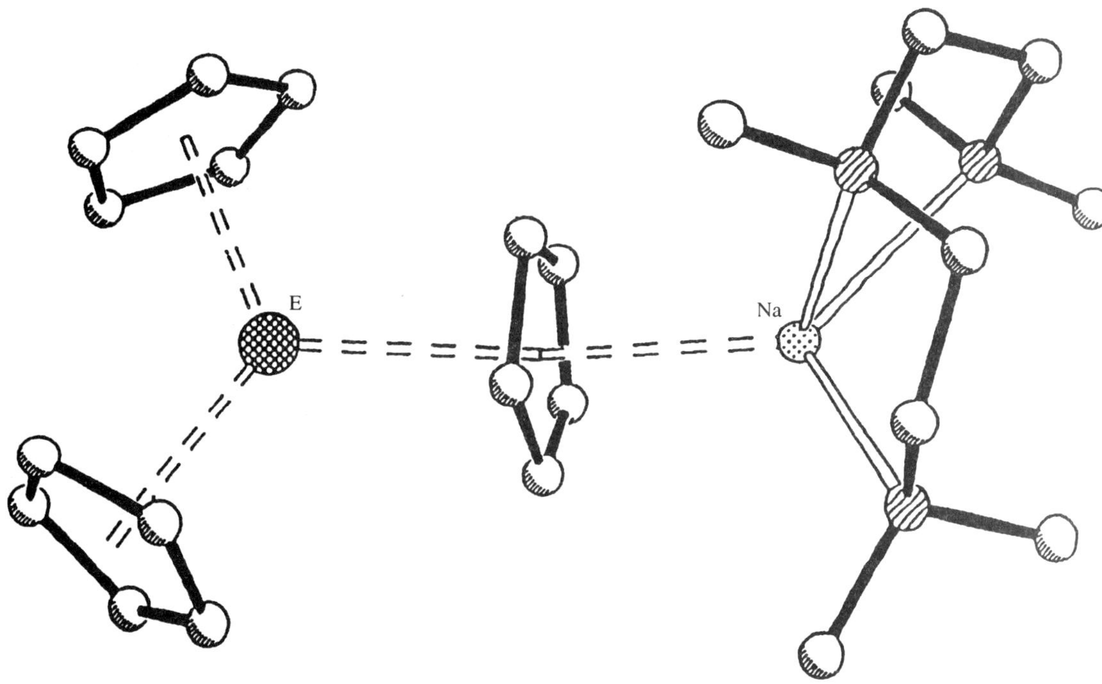

(**127**) E = Sn
(**128**) E = Pb

These tend to be highly aggregated cages whose structures are diverse and rather unpredictable. A few examples will serve to demonstrate the nature of these complexes.

Dilithiation of 1-aminonaphthalene with 2 equiv. Bu^nLi in Et_2O produces a Li_{20} cage complex $[\{Li_2N(naphthyl)\}_{10}{\cdot}6Et_2O]{\cdot}OEt_2$ (**132**) whose core is that of two interlocked Li_8 cubes sharing one edge.[131] The latter is paramagnetic in the solid state and in solution.

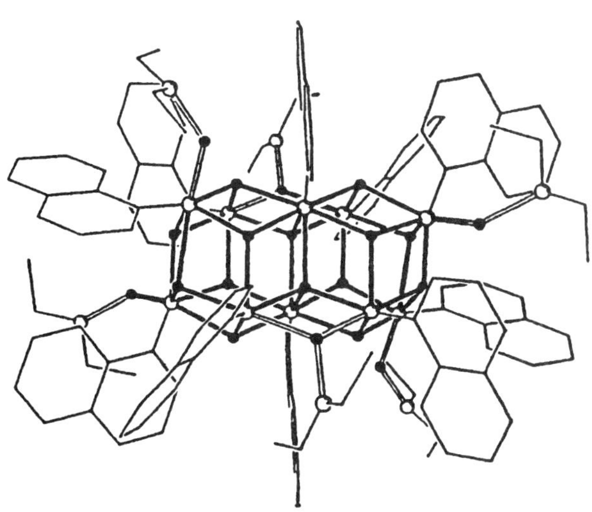

(**132**)

Dilithiation of $PhCH_2N(H)CH_2CH_2NMe_2$ occurs smoothly at the acidic N–H proton and at one *ortho*-C–H bond of the phenyl ring.[132] The product is that of the Li_8 cage complex $[o\text{-}LiPh_2CH_2N(Li)CH_2CH_2NMe_2]_4$ (**133**), which is schematically constructed from three interlocked Li_4 tetrahedra sharing opposite edges.

Dilithiated sulfones and other more heavily functionalized species are also known.[10] Among these $[o\text{-}LiPhSC(TMS)_2O_2Li{\cdot}2OEt_2]_2$ (**134**),[133] $[\{(TMS\text{-}CCN)Li_2\}_{12}{\cdot}6OEt_2]$ (**135**),[134] $[\{(TMS\text{-}CSO_2Ph)Li_2\}_6\text{-}Li_2O{\cdot}10THF]$ (**136**)[135] and $[\{(TMS)_2C(Li_2)CN\}_{12}{\cdot}6OEt_2{\cdot}C_6H_{14}]$ (**137**)[10] have been structurally characterized.[136]

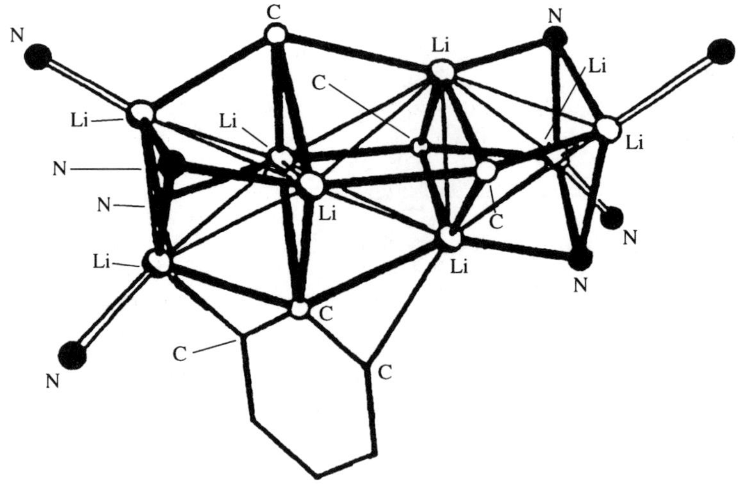

(133)

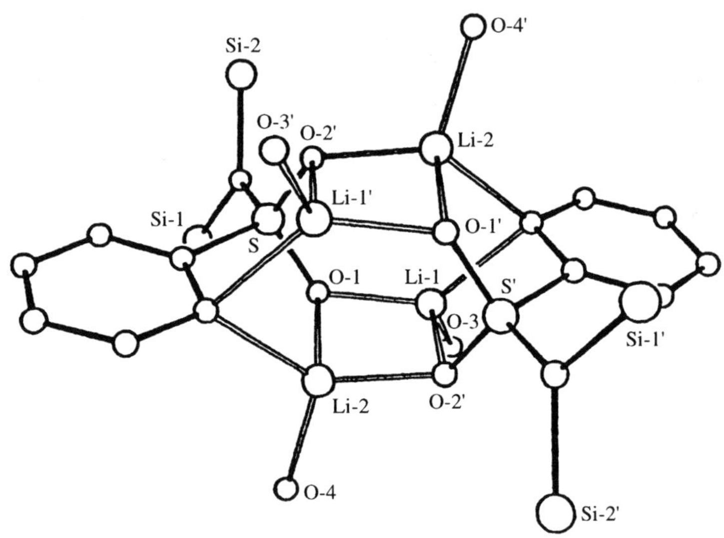

(134)

1.3 ORGANOMETALLIC COMPLEXES (CONTAINING C–M BONDS)

1.3.1 σ- and π-Bonded Complexes

In an attempt to review work since 1982 in the area of alkali metal organometallic chemistry, the historical work carried out at the beginning of the century by Schlenk[1c] should still be mentioned, as his initial work on alkyl- and aryllithium chemistry must be perceived as a foundation stone of modern chemistry in this area. With the ever-increasing availability of low-temperature crystallography to the synthetic chemist the study of alkali metal organics has led to the structural diversity known in the mid-1990s. Although some investigations of lithium organometallic compounds have been made in solution by NMR spectroscopy and can be used to determine the degree of aggregation and coupling, this information is often used solely to complement the x-ray single crystal data. The importance of x-ray crystallography as a tool in the discovery of the plethora of structural types now known for the alkali metal organometallics has been portrayed and reviewed by Setzer and Schleyer,[7] and Schade and Schleyer.[8] X-ray structural investigations have shown, if nothing else, that the frequent textbook treatment of organo alkali metal compounds simply as carbanions with monomeric (RM) or ion-separated (R^-M^+) structures is a gross oversimplification of the true nature of these highly aggregated and solvated compounds. More recently, neutron and synchrotron radiation have been used successfully in the probing of the structures of organometallic alkali metal compounds.

For the purposes of this section organometallic alkali metal compounds are defined as those containing σ or π alkali metal–carbon bonds. Metalloorganic complexes (containing O–M, S–M, N–M, P–M, etc., bonds; M = alkali metal) have been described in Section 1.2. When considering the heavier alkali metals, the vast majority entered in the Cambridge Crystallographic Database are metalloorganic derivatives and comparatively few show metal–carbon bonding. This is due to the enhanced reactivity of these compounds over those of lithium, which leads to thermal decomposition and handling problems.

The vast literature now available on organolithium compounds illustrates that solvation of the metal ions by Lewis bases has the greatest impact on their physical and chemical behaviour.[26,137] It is also important to note that solvation often lowers the degree of aggregation, although organolithium compounds may remain associated even in polar solvents (e.g., THF). Table 9 illustrates the latter point and lists the structures of a number of organolithium complexes ((**138**)–(**184**)).

Table 9 The structures of a number of organolithium compounds.

Compound formula	*Association number (n)*	*Solvent, crystal, base adduct*	*Ref.*
[LiBz·TMEDA·THF] (**138**)	monomer	base adduct	138
[Li{CH(TMS)Ph}·TMEDA] (**139**)	monomer	base adduct	139
[Li(CPh_3)·TMEDA] (**140**)	monomer	base adduct	140
[Li(CPh_3)·2OEt_2] (**141**)	monomer	base adduct	141
[Li{$CH(TMS)_2$}·PMDETA] (**142**)	monomer	base adduct	142
[LiPh·PMDETA] (**143**)	monomer	base adduct	143
[Li(2,4,6-$Bu^t_3C_6H_2$)·TMPDA] (**144**)	monomer	base adduct	144
[Li(η^2-allyl)·PMDETA] (**145**)	monomer	base adduct	145
[Li{η^3-1,3-bis(trimethylsilyl)allyl}·TMEDA] (**146**)	monomer	base adduct	146
[Li(η^5-indenyl)·TMEDA] (**147**)	monomer	base adduct	147
LiBz (**148**)	monomer	Et_2O, THF	148
LiPr (**149**)	monomer	Et_2O, THF	148
$LiBu^t$ (**150**)	monomer/dimer	Et_2O, THF	148
LiPh (**151**)	dimer	Et_2O, THF	148
$[LiC(TMS)_3]_2$ (**152**)	dimer	crystal	149
$[LiPh·TMEDA]_2$ (**153**)	dimer	base adduct	150
[LiC≡CPh·TMPDA]$_2$ (**154**)[b]	dimer	base adduct	151
$[LiBu^t·2OEt_2]_2$ (**155**)	dimer	base adduct	152
$[LiBu^n]_6$ (**156**)	dimer/tetramer	THF	148
[Li{2,6-$(NMe_2)_2C_6H_3$}]$_3$ (**157**)	trimer, ring	crystal	153
LiMe (**158**)	tetramer	Et_2O, THF	148
LiEt (**159**)	tetramer	Et_2O, THF	148
$LiBu^t$ (**160**)	tetramer	hydrocarbon	148
$[LiMe]_4$ (**161**)	tetramer	crystal	154
$[LiEt]_4$ (**162**)	tetramer	crystal	155a
$[LiBu^t]_4$ (**163**)	tetramer	crystal	152
$[LiCH_2CH_2CH_2OMe]_4$ (**164**)	tetramer	base adduct	156
$[LiCH_2CH_2CH_2NMe_2]_4$ (**165**)	tetramer	base adduct	157
$[LiPh·Et_2O]_4$ (**166**)	tetramer	base adduct	158
$[LiPh(Me_2S)]_4$ (**167**)	tetramer	base adduct	159
[(LiC≡CBu^t)$_4$(THF)$_4$] (**168**)	tetramer	base adduct	160
[(LiC≡CPh)$_4$(TMHDA)$_2$] (**169**)[c]	chain of tetramers (double helix)	base adduct	161
$[(LiMe)_4(TMEDA)_2]_n$ (**170**)	three-dimentional net of tetramers	base adduct	162
LiEt (**171**)	hexamer	hydrocarbon	148
$[LiBu^n]_6$ (**172**)	hexamer	hydrocarbon	148
$[LiBu^n]_6$ (**173**)	hexamer, cluster	crystal	154
[Li(cyclohexyl)]$_6$ (**174**)	hexamer, cluster	crystal	163
[Li(tetramethyl cyclopropylmethyl)]$_6$ (**175**)	hexamer, cluster	crystal	164
[Li(CH_2-TMS)]$_6$ (**176**)	hexamer, ring	crystal	165
[(LiC≡CBu^t)$_{12}$(THF)$_4$] (**177**)	stack	base adduct	160
[Li(allyl)·TMEDA]$_n$ (**178**)	chain	base adduct	166
[Li(η^3-1,3-diphenylallyl)·Et_2O]$_n$ (**179**)	chain	base adduct	167
[LiBz·dabco]$_n$[a] (**180**)	chain	base adduct	168
$[LiBz·2Et_2O]_n$ (**181**)	chain	base adduct	169
[Li([12]crown-4)$_2$][$CHPh_2$] (**182**)	ion pair	base adduct	170
[Li([12]crown-4)$_2$][CPh_3] (**183**)	ion pair	base adduct	170
[Li(THF)$_3$][Li{$C(TMS)_3$}$_2$] (**184**)	ion pair	base adduct	171

[a] dabco = $N[(CH_2)_2]_3N$. [b] TMPDA = $(Me_2N)CH_2(NMe_2)$. [c] TMHDA = $(Me_2N)(CH_2)_6(Me_2N)$.

The initial investigations, by x-ray powder diffraction[155a] and single crystal methods,[155b] of the methyl and ethyl alkali metal compounds were carried out in the early 1960s. Until the late 1980s it was thought that no structural anomalies arose in the solid state for the methyl derivatives of the alkali metals (MeLi through to MeCs) and that all possessed the same tetrameric structure, $[MeM]_4$. In the case of $[MeLi]_4$ (**161**)[154,155a] the tetramers have idealized T_d symmetry and form a body-centred-cubic lattice. The low volatility and nonmelting characteristics of methyllithium have been directly related to the C–Li interactions occurring between each tetramer unit and eight neighbouring units. The structure of methyllithium has recently been further refined by the use of neutron diffraction studies[154] and there are no phase transitions down to 1.5 K.

Initial studies of the structure of methylsodium showed it to adopt a tetrameric structure reminiscent of that of methyllithium. Due to the larger size of sodium there is a greater possibility of interactions with neighbouring molecules and as a result a marked decrease in the solubility in hydrocarbons is observed. However, this initial structural solution was a result of the incorporation of varying amounts of MeLi into the interstitial sites within an MeNa host lattice. The synthesis of MeNa containing a range of Na:Li (from 3:1 to 36:1) ratios has been achieved from the metal exchange reactions of $NaOBu^t$ added to MeLi in differing ratios in hexane/ether solution. Each of these mixed MeLi/MeNa compositions has the same basic structure. The structure is that of a zeolite-type host–guest compound in which 24 tetrameric $[MeNa]_4$ units aggregate in a giant cubic cell producing a cavity which can be filled with up to eight $[MeLi]_4$ units. The latter is thought to be the first example of a supramolecular organometallic compound. Work since the late 1980s has shown pure methylsodium to have a more compact structure.[172] The order of addition of MeLi to $NaOBu^t$ (or vice versa) is critical in its preparation. Only if MeLi solution is added to $NaOBu^t$ is pure MeNa formed (rather than MeLi solution to the $NaOBu^t$ solution as in the synthesis of the mixed MeLi/MeNa phase). The dependence on the order of addition of the reagents possibly stems from the envelopment of the $[MeNa]_4$ tetramers that are formed by the ether-solvated $[MeLi]_4$ tetramers, thus the formation of the alternative host–guest compound is avoided.[173] With a combination of neutron diffraction and x-ray synchrotron diffraction on $NaCD_3$ powder samples[174] it was possible to solve the complicated structure of methylsodium. The structure shows that half the ions forming $[MeNa]_4$ tetramers have slightly distorted T_d symmetry but are very similar to those of methyllithium, the remaining eight sodium and eight methyl ions connect the tetramers through Na–C contacts. This structure, with tetrameric aggregates and a single ion within the same ionic lattice and in which optimal ionic packing is therefore achieved, can be regarded as intermediate between the structure of methyllithium and the methyl derivatives of the heaviest alkali metals (K, Rb, Cs).

Methylpotassium, methylrubidium and methylcaesium were prepared by reactions of $MOBu^t$ (M = K, Rb, Cs) with methyllithium.[175–6] The structure of each has the same NiAs-type lattice. Initial powder diffraction studies on MeK showed the trigonal prismatic coordination of Me^- ions to six K^+ ions, although the exact position of the hydrogen atom sites could not be located. More recent neutron diffraction studies on KCD_3 powder samples[177] have verified the presence of pyramidal methyl anions. These recent measurements show the structure to have an orthorhombic structure where each methyl group is surrounded by six K^+ ions in a distorted trigonal prism (rather than the smaller hexagonal cell initially thought). The observed distortion of the geometry around the Me^- ions is a result of the pyramidal shape of the Me^- ions; the K^+ ions on the side of the Me^- lone pair more closely contact the carbon centre than the three K^+ ions on the opposite side. As was the case with MeK, reexamination of older powder data[176] on methylrubidium and methylcaesium showed that these also crystallize in the same orthorhombic structure.[177]

The structures of the classical reagents $[Bu^nLi]_6$ (**172**), $[Bu^tLi]_4$ (**160**) and $[Bu^tLi{\cdot}2OEt_2]$ (**155**) have been obtained by Stalke using low-temperature single crystal x-ray diffraction.[152] Since all three complexes are highly air-sensitive and/or melt or decompose at room temperature, great care had to be taken to ensure no phase changes, decomposition or melting of the sample occurred; the optimized temperature was found to be −80 °C for the preparation of crystalline samples. The three compounds were shown to be discrete aggregates, Bu^nLi consisting of six lithium atoms in a trigonal antiprismatic conformation approximating to D_{3d} symmetry, Bu^tLi consisting of tetrameric units of almost T_d symmetry, and $[Bu^tLi{\cdot}2OEt_2]$ (**155**) a dimer with a central four-membered ring consisting of two lithium atoms and the two bridging α-carbon atoms of the *t*-butyl groups. The hexameric structure of $[Bu^nLi]_6$ (**172**) can be regarded as a stack of two Li_3C_3 ring trimers and that of $[Bu^tLi]_4$ (**160**) as a stack of two dimers. The reason for the different association states observed in the solid-state structures stems from the higher steric bulk of the *t*-butyl groups in (**160**) than the *n*-butyl groups in (**172**). In both complexes short agostic C–H···Li interactions further stabilize the lithium centres which are formally only three-coordinate.

Alkali metal ethynides MC≡CH (M = Na, K, Rb)[178] and propynides[179] have related structures and have been known for some time. Their low volatility and insolubility in nonpolar solvents is a result of their layered structures. The alkali metal acetylides MC≡CM (M = Li, Na, K) have also been investigated,[180–1] and it was shown that the length of added Lewis base diamine ligands has a large influence on the structure and composition of these compounds. A similar effect is observed in the structures of phenylethynyllithium compounds containing additional diamine ligands; in $[(LiC{\equiv}CPh)_4(TMHDA)_2]$ (**169**)[161] a tetrameric cubane structure is formed in which the separate nitrogen centres intermolecularly link the Li^+ ions of separate cubanes into a polymer. Stacking of the dimer units of $[LiC{\equiv}CPh{\cdot}TMPDA]_2$ (**154**)[151] is precluded by solvation of the lithium centres by chelation of both the nitrogen centres of the TMPDA ligand, which has a shorter carbon spacer than TMHDA. Evidently this ligand-dictated behaviour is entropically based.

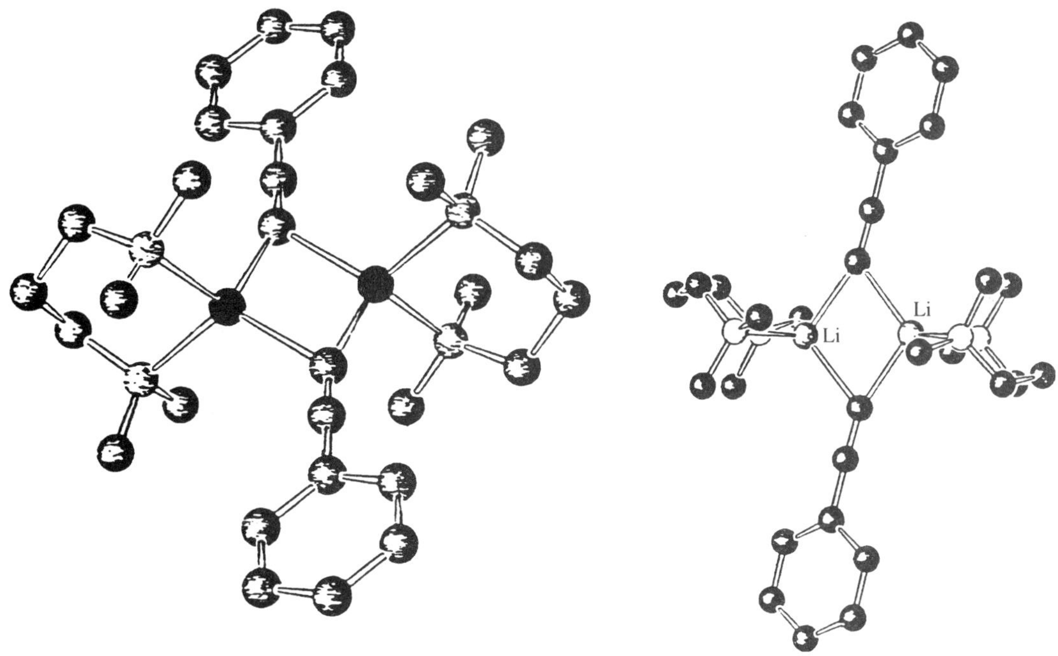

(154)

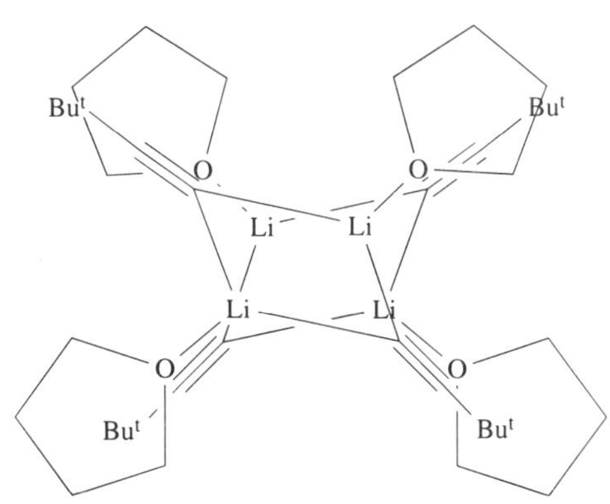

(168)

The structures of Lewis base adduct complexes of $LiC{\equiv}CBu^t$ appear to be dependent on the method of crystallization. Two complexes, the tetramer $[LiC{\equiv}CBu^t{\cdot}THF]_4$ (**168**) and the dodecamer $[LiC{\equiv}CBu^t)_{12}{\cdot}THF]$ (**177**), can be isolated from THF solutions.[160] The structures of both complexes echo the stacking principle outlined earlier (Section 1.2) with (**168**) being a stack of two dimeric rings and (**177**) being a stack of six dimeric rings.

Polylithiated hydrocarbons have been structurally investigated, and theoretical calculations have shown that the gas-phase *cis*-planar structure of dilithiomethane (Li_2CH_2) is more stable than the tetrahedral structure.[182] Up to the mid-1990s, only solid-state structural determinations have taken place and the neutron diffraction study showing bent CH_2 anions coordinated in a complex way to six lithium ions in a highly distorted antifluorite structure.[183]

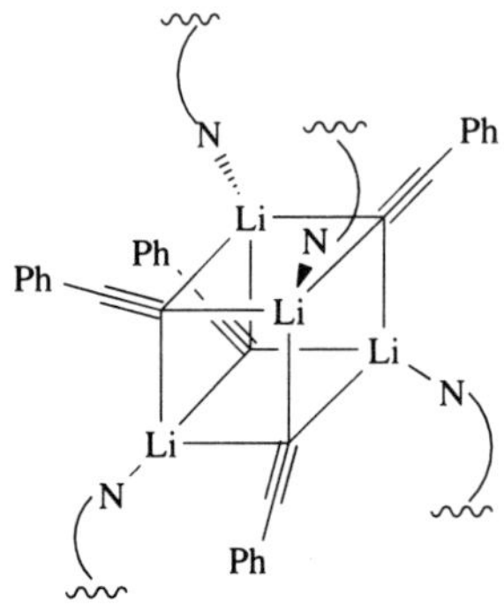

(169)

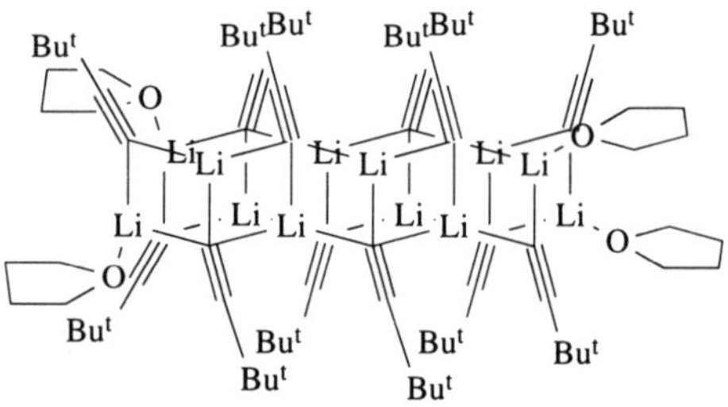

(177)

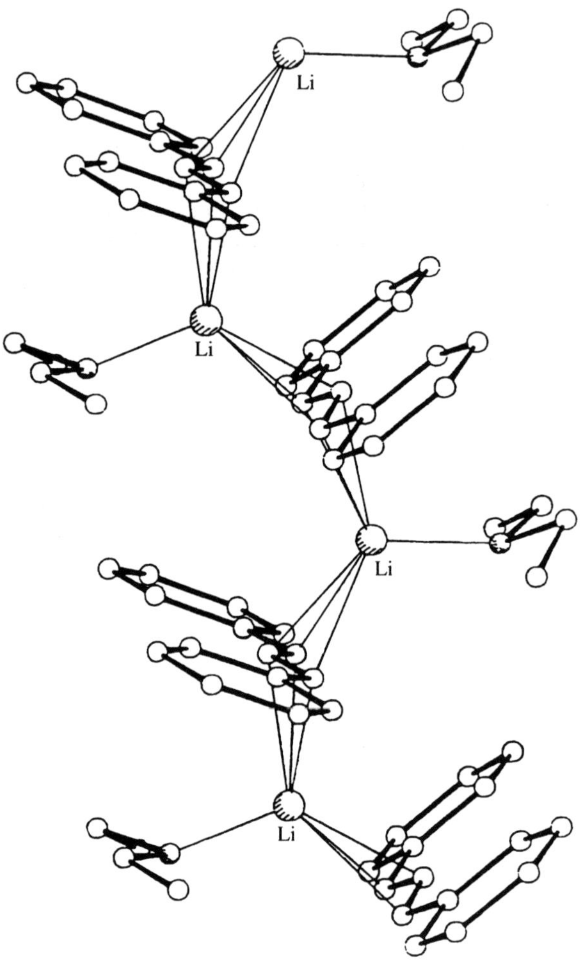

(179)

Table 10 lists some structurally characterized organo alkali metal complexes in which π-bonding of the carbanions to the heavier metals (Na–Cs) is observed ((**185**)–(**202**)). Some related organolithium complexes are listed in Table 9.

Recent additions to the cyclopentadienyl alkali metal derivatives include $[K\{\eta^5\text{-}C_5Bz_5\}\cdot(THF)_3]$ (**189**)[188] and the first unsolvated example $[K(\eta^5\text{-}C_5H_4TMS)]_n$ (**203**).[195]

Of the new discoveries made since 1982 the simplest metallocenes were those of the lithocene anion $[LiCp_2]^-$ (**204**)[196] and the related anion $[Li(isodiCp)_2]^-$ (**205**).[197] The Li–$Cp_{centroid}$ distances are 0.200 nm and 0.201 nm, respectively, in these complexes. The preparation and crystal structures of sodium and potassium pentamethylcyclopentadienyl pyridine solvates are known. The crystal structure of [NaCp*·3py] (**187**) has a piano-stool monomer structure and $[KCp^*\cdot 2py]_\infty$ (**200**) has a linear zig-zag chain structure.[186]

Table 10 Some structurally characterized complexes containing π-bonding to the alkali metal.

Compound	*Structure*	*Ref.*
[Na(1-phenyl allyl)·PMDETA] (**185**)	monomer	184
[Na(fluorenyl)·PMDETA] (**186**)	monomer	185
[Na(Cp*)·3py] (**187**)	monomer	186
[Na($CHPh_2$)·PMDETA] (**188**)	monomer	187
[K{C_5Bz_5}·3THF] (**189**)	monomer	188
$[NaPh·PMDETA]_2$ (**190**)	dimer	187
$[K(fluorenyl)·TMEDA]_2$ (**191**)	bridging	189
$[Na(Bz)·TMEDA]_4$ (**192**)	tetramer	190
[Na(*o*-xylyl)·TMEDA]$_4$ (**193**)	tetramer	191
$[Na(CHPh_2)·TMEDA]_4$ (**194**)	tetramer	191
$[Na(indenyl)·TMEDA]_n$ (**195**)	chain	192
$[Na(fluorenyl)·TMEDA]_n$ (**196**)	chain	185
[Na(η^1-Bz)·PMDETA]$_n$ (**197**)	chain	191
$[Na(CPh_3)·TMEDA]_n$ (**198**)	chain	193
[Na(fluorenyl)·TMPDA] (**199**)	ring	185
$[K(Cp^*)·2py]_\infty$ (**200**)	chain	186
[K(η-Bz)·PMDETA]$_n$ (**201**)	chain	194
$[Rb(fluorenyl)·PMDETA]_n$ (**202**)	chain	185

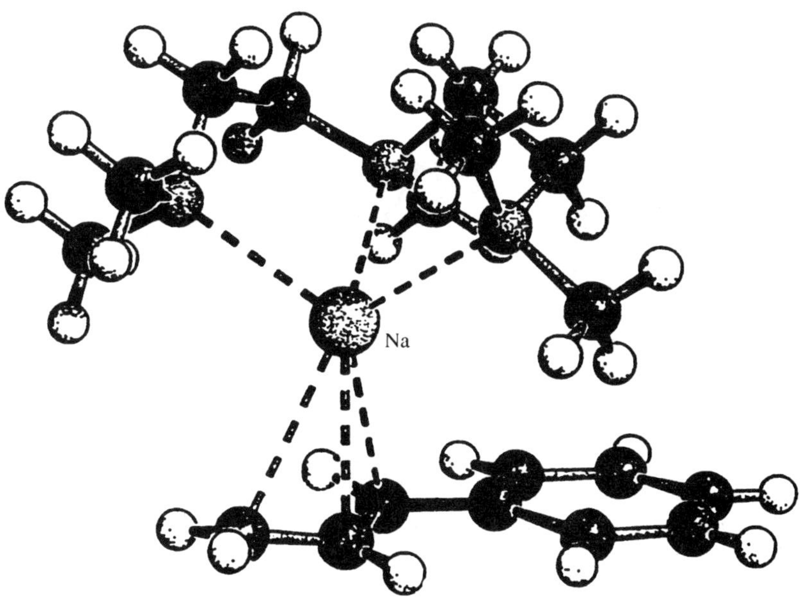

(185)

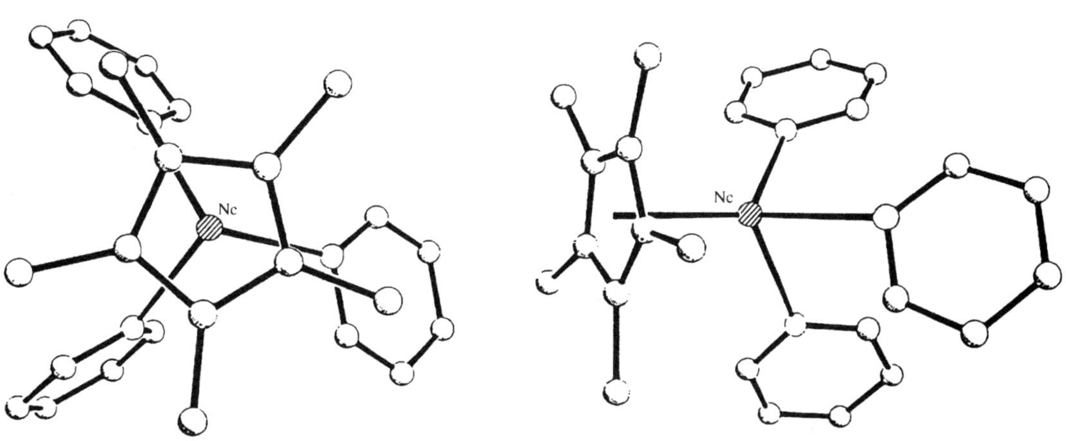

(187)

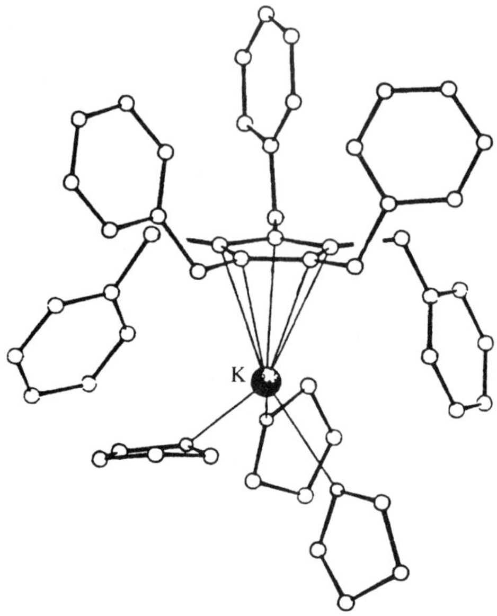

(189)

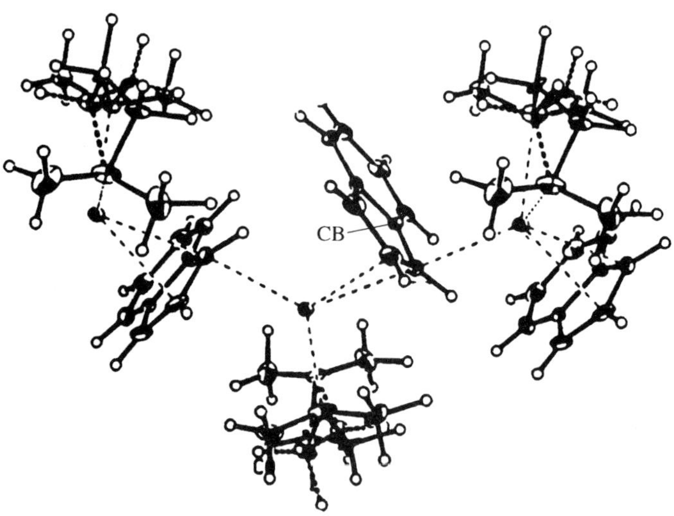

(195)

The study of the metallation of allyl compounds has yielded some interesting compounds and occurred as a natural progression from the immense study of transition metal complexes containing this η^3 moiety. In the crystal structure of [Li(allyl)·TMEDA]$_\infty$ (**178**) polymer chains are observed with lithium atoms rendered formally four-coordinate by bonding to TMEDA and by σ-bonding to the terminal allylic carbon atoms of the two neighbouring anions (i.e., ···CH_2⩦CH⩦CH_2–Li-(TMEDA)–CH_2⩦CH⩦CH_2···).[166] In contrast, in [(1,3-diphenylallyl)Li·Et_2O] (**179**) η^3-allyllithium moieties are observed in the solid state. Monomer units aggregate to form polymeric chains with the lithium atoms lying almost symmetrically above and below the allyl groups.[167]

Few examples of allyllithium compounds have been investigated so far by x-ray methods and fewer still for the heavier alkali metals. A study in 1991 showed that the reaction of allylbenzene with Bu^{n}Na gave 1-phenylallylsodium which, on addition of PMDETA, yielded the monomer [Na(1-PhC_3H_4)·PMDETA] (**185**).[184] The crystal structure showed it to be a contact ion-pair with η^3-coordination of the allyl system.

[Na(indenyl)·TMEDA]$_n$ (**195**) has been shown to crystallize in an infinite polymeric zig-zag chain structure with the sodium atoms sandwiched between two indenyl units which exhibit η^1- and η^2-bonding between the cation and anion sites.[192] In the analogous lithium compound [Li(indenyl)·TMEDA] (**206**),[198] the indenyl ligand was shown to favour η^5-coordination in this monomeric structure.

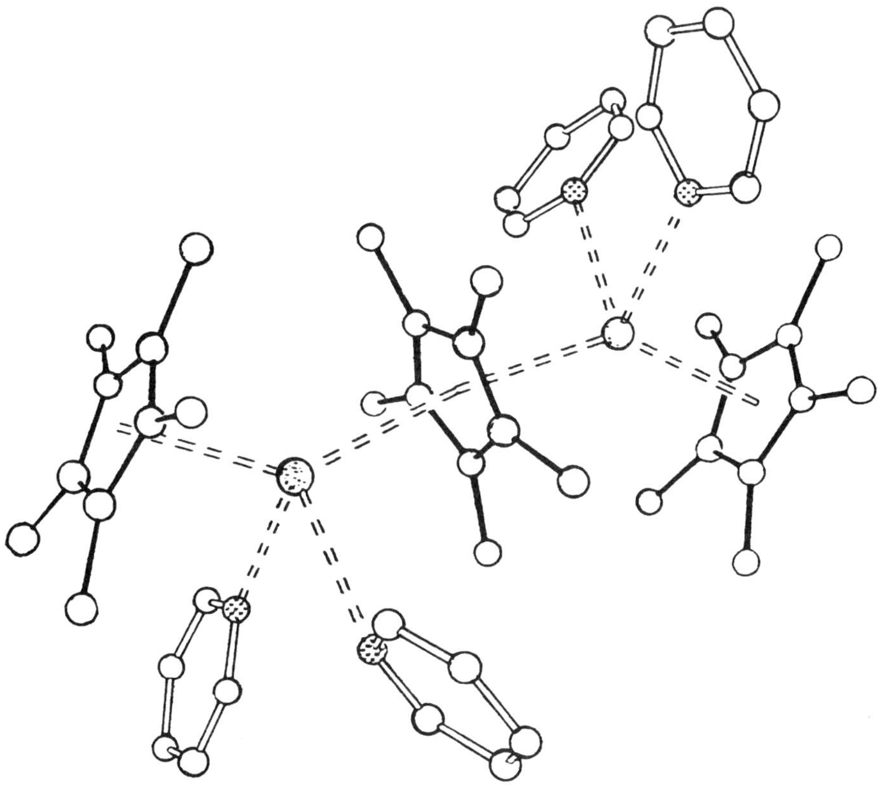

(200)

In donor-solvated fluorenyl alkali metal compounds it is clear that as the alkali metal group is descended there is a marked difference in structural type. The association states are dictated by the nature of the donor ligands and the metal sizes. Hence, a monomer is observed for [Na(fluorenyl)·PMDETA] (**186**),[185] a four-membered ring for $[Na(fluorenyl)\cdot TMPDA]_4$ (**199**),[185] and $[Na(fluorenyl)\cdot TMEDA]_\infty$ (**196**)[185] is a polymer. $[K(fluorenyl)\cdot TMEDA]_2$ (**191**)[189] exists as a complicated dimer structure in which the large potassium atom cannot be coordinated in a chelating mode by the donor molecule and as a result the TMEDA ligands bridge between two metal cations (cf. the polymeric chains found in its sodium analogue (**196**) where the smaller metal cation can be chelated by TMEDA). In the case of $[Rb(fluorenyl)\cdot PMDETA]_\infty$ (**202**), prepared by the transmetallation of [Li(fluorenyl)] with $RbOBu^t$ in THF, the structure consists of 'square-wave-like' polymeric chains in which the Rb^+ cations are coordinated by PMDETA in a chelating mode and where the fluorenyl ligands link the monomer units together.[185]

1.3.2 Mixed-metal Compounds

1.3.2.1 Mixed-metal transition metal complexes

Organometallic compounds of alkali metals containing mixed-metal compositions have been known for over a century but it was not until recently, with the availability of x-ray crystallography, that they have been looked at in detail. Complexes of this type have been examined mainly for lithium and there are fewer examples for the heavier alkali metals. In general, these compounds can be divided into two catagories according to the organo alkali metal–transition metal coordination mode: (i) those exhibiting π-interactions often observed in ion-paired systems; and (ii) those with σ-interactions often resulting in aggregate formation.

Table 11 illustrates some examples of the different alkali metal–carbon interactions observed ((**207**)–(**218**)). In $[K(DME)]_2[Yb(cot)_2]$ (**207**),[147] which is prepared by the stoichiometric reaction of cyclooctatetraene (cotH), potassium and ytterbium in liquid ammonia, the crystal structure shows the ytterbium atom to be sandwiched between two [8]annulene rings. Each potassium is coordinated by DME and in a π-manner to each of the [8]annulene ring faces thereby sandwiching the $[Yb(cot)_2]^{2-}$ moiety.

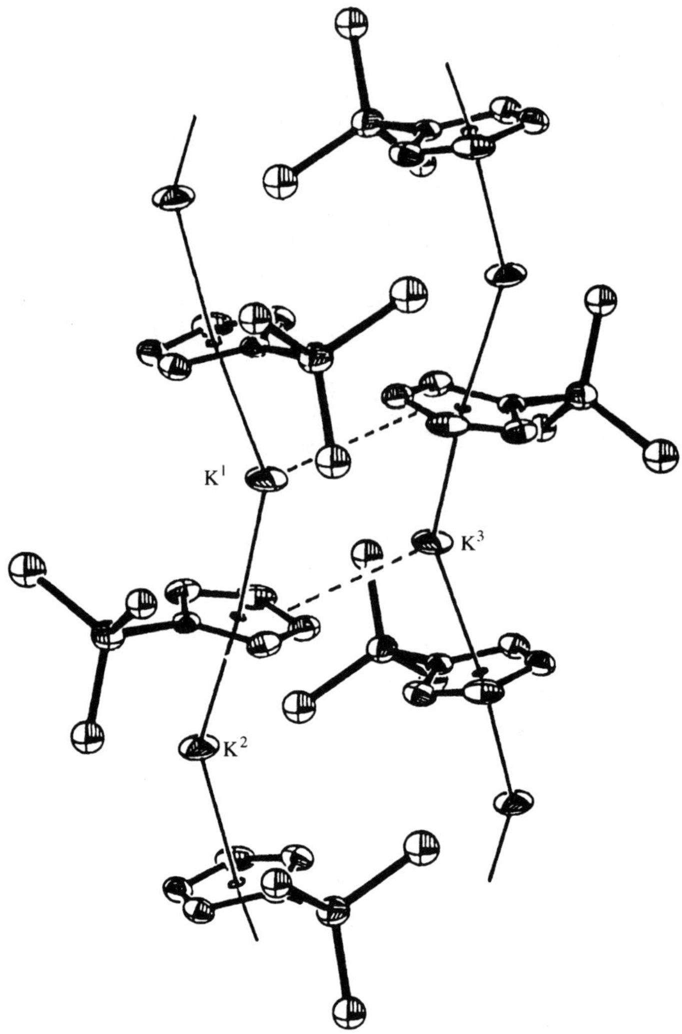

(203)

Table 11 Examples of two different types of alkali metal–carbon interaction.

Compound	*Ref.*	*Compound*	*Ref.*
π-interactions		*σ-interactions*	
$[K(DME)]_2[Yb(cot)_2]$ (**207**)	147	$[Na_4(THF)_5][Ph_2Ni(C_2H_4)]_2$ (**213**)	204
$[K]_2[(Ph_3P)_2Ph_2PC_6H_4RuH_2]_2$ (**208**)	199	$NaEu[N(TMS)_2]_3$ (**214**)	205
$[Li][Co(C_2H_4)(PMe_3)_3]$ (**209**)	200	$NaY[N(TMS)_2]_3$ (**215**)	205
$[Li(PMDETA)][Ni_2(C_2H_4)_2(\mu\text{-}H)]$ (**210**)	201	$[\{Li(OEt_2)\}(CuPh_2)]_2$ (**216**)	206
$[K\{OsH_3(PMe_2Ph)_3\}]_2$ (**211**)	202	$[Li_2Cu_2Me_4]$ (**217**)	207
$[K\{Co(C_2H_4)(PMe_3)_2\}]_2$ (**212**)	203	$[Li_6(OEt_2)_3Cu(C_2Ph)_{10}]$ (**218**)	208

This type of bonding is similar to that found in $[K(diglyme)][Ce(cot)_2]$ (**219**),[209] but in this case only one annulene ring is capped by one potassium cation. In $K_2[(Ph_3P)_2(Ph_2PC_6H_4)RuH_2]_2$ (**208**)[199] the two potassium atoms link two $[(Ph_3P)_2(Ph_2PC_6H_4)RuH_2]^-$ units, only one of which is sandwiched by two orthometallated triphenylphosphino groups. This alkali metal encapsulation by two phenyl rings is also observed in $[K\{OsH_3(PMe_2Ph)_3\}]_2$ (**211**).[202] A study of the space-filling model of the latter shows that the potassium ions are surrounded by a lipophilic sheath, aiding its solubility in aromatic hydrocarbons.

The compounds $Li[Co(C_2H_4)(PMe_3)_3]$ (**209**),[200] $[K\{Co(C_2H_4)(PMe_3)_2\}]_2$ (**212**)[203] and $[Li(PMDETA)][Ni_2(C_2H_4)_2(\mu\text{-}H)]$ (**210**)[201] all have alkali metal–alkene interactions.

In the dimer $[\{Li(OEt_2)\}(CuPh_2)]_2$ (**216**),[206] ether-solvated Li^+ ions hold together two linear diphenylcuprate units. Each of the phenyl groups bridges a copper and a lithium centre unsymmetrically. This asymmetry is consistent with the interaction energy (and covalent character) being stronger towards

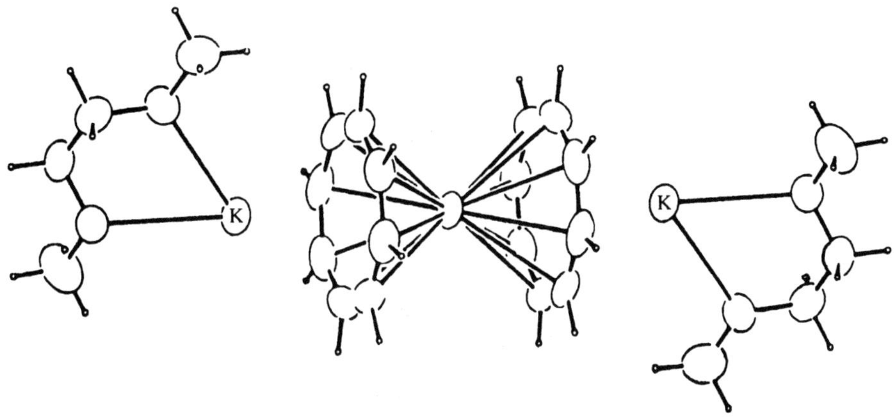

(207)

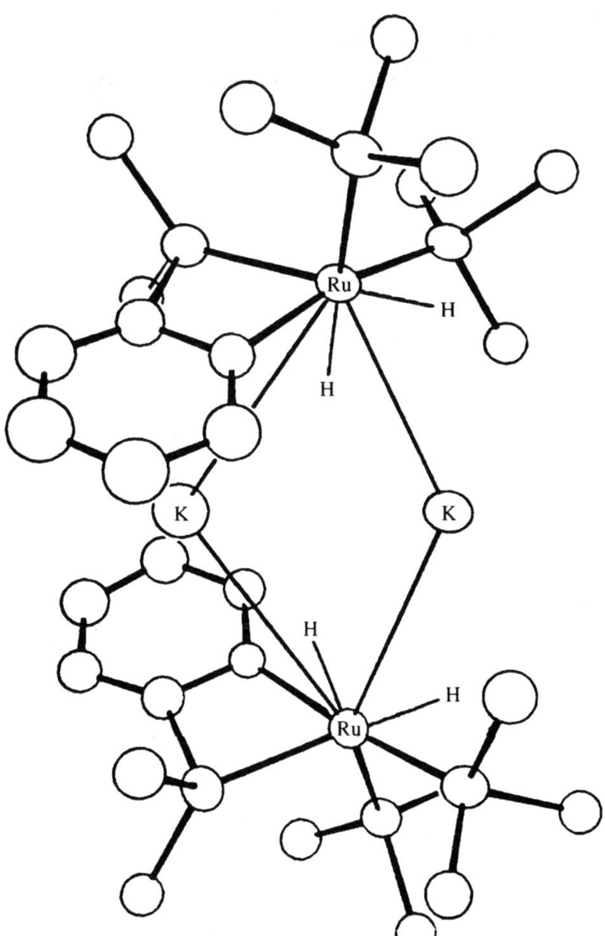

(208)

the transition metal side than to the alkali metal. A similar situation is observed in $[Na_4(THF)_5][Ph_2Ni(C_2H_4)]_2$ (**213**),[204] where the unsymmetrically bridging phenyl groups are bent towards the nickel centre.

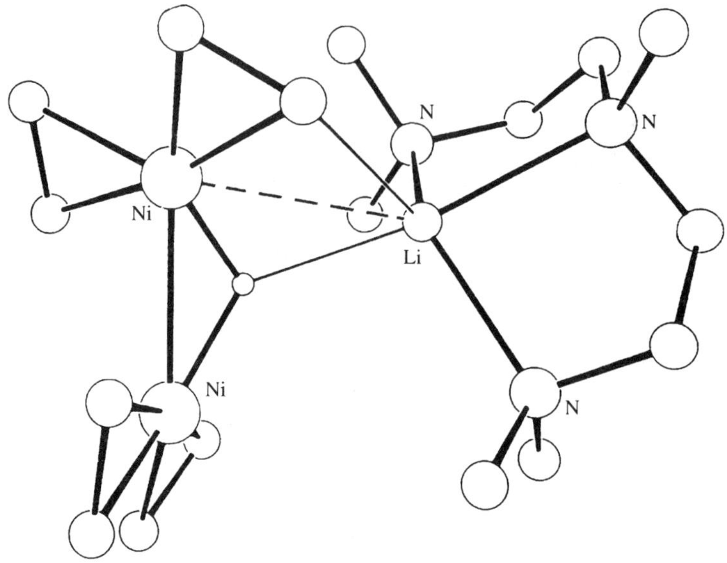

(210)

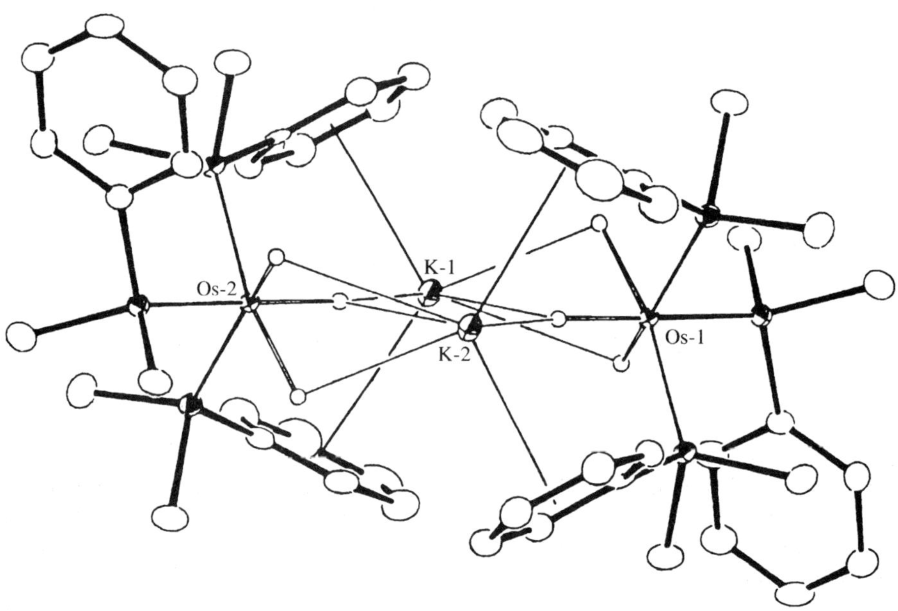

(211)

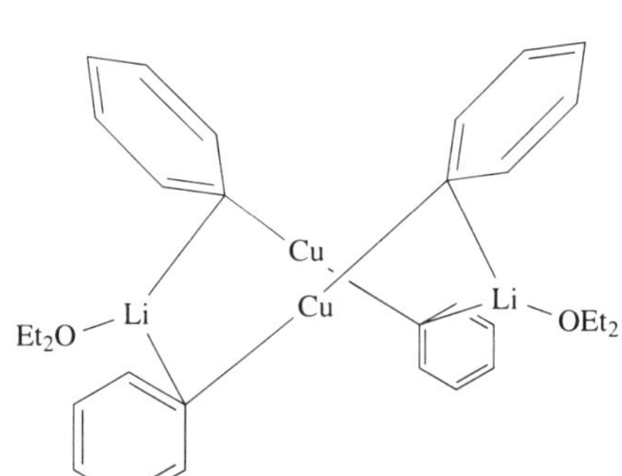

(216)

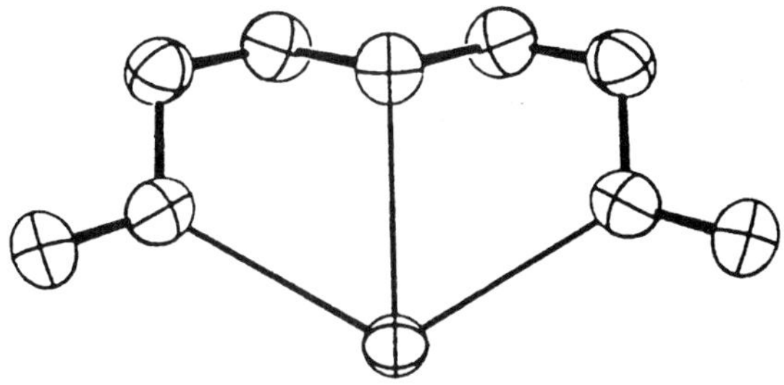

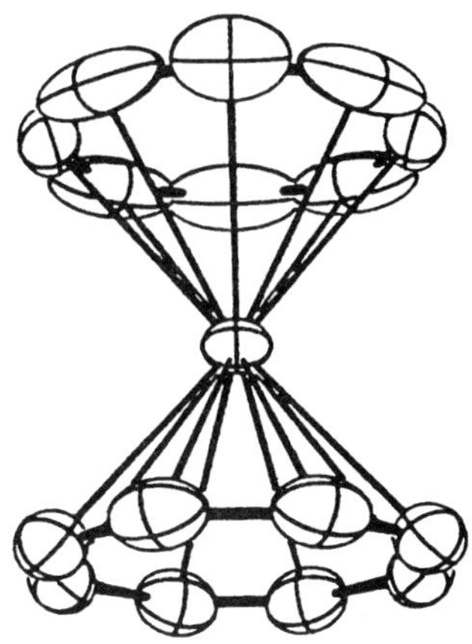

(219)

1.3.2.2 Mixed-metal main group complexes

This catagory represents one of the largest and longest known group of compounds in this area. These ionic systems exist as solvent-separated ion-pairs or ion-pairs in THF solutions. Even poorly chelating ligands are able to separate the sodium ions from $Na[BPh_4]$ in the solid state or in solution. A typical feature in all of these complexes is that carbanion methyl and phenyl groups can act as bridging ligands. This area will not be reviewed fully but a brief survey will be provided as a reference point for the reader.

The 'ate' compounds found in this group are magnesates,[210] zincates,[211] cadmates,[211b] borates,[212] aluminates,[213] gallates[213a] and indates.[214] Most of the structures contain isolated four-coordinate metallate anions and alkali metal cations, often exhibiting relatively weak alkali metal–carbon interactions.

A few mixed-salt complexes should also be mentioned in this context. The complexes $[(PhLi{\cdot}OEt_2)_3{\cdot}LiBr]$ (**220**)[158] and $[\{(cyclopropyl)Li{\cdot}OEt_2\}_2{\cdot}(LiBr{\cdot}OEt_2)_2]$ (**221**)[215] have cubane structures in the solid state. In (**220**) only three of the Li^+ cations are coordinated by Et_2O as a consequence of the steric crowding around the Li_4C_4 core. The same cubane structural motif (a stack of two solvated dimers) is observed in the parent structure of $[PhLi{\cdot}OEt_2]_4$ (**166**)[158] so that it can be seen that this unit provides a template for the incorporation of one LiBr monomer in (**220**).

1.4 THEORETICAL CALCULATIONS

Theoretical calculations of alkali metal organometallic and metalloorganic compounds have now become an invaluable tool to experimental chemists. *Ab initio* MO calculations have most commonly been applied to small aggregates (e.g., monomers), where the highest calculational levels can be used. Since lithium organometallics generally contain light atoms, these species are particularly accessible to *ab initio* calculations and, for this reason, these have been the subject of most theoretical studies.[216]

The most general and without doubt the most fundamental conclusion which can be drawn from the host of theoretical studies on organometallic compounds (containing C–Li bonds) and metalloorganic complexes (N–Li, O–Li, etc.) is that the alkali metal bonding is essentially ionic. Accordingly organolithium compounds are best termed 'ionic molecules' or 'supramolecules'. A further conclusion

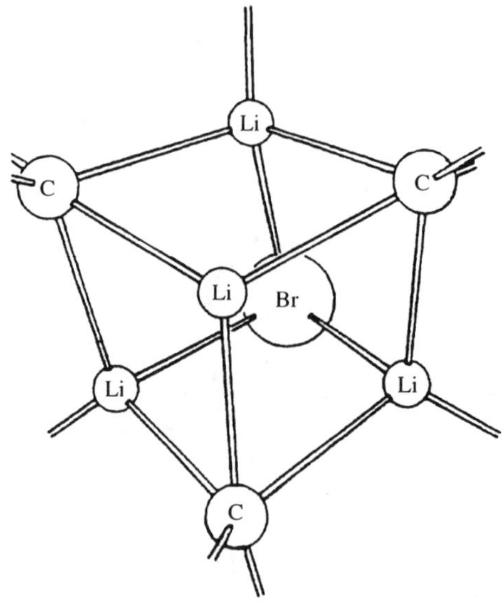

(220)

that can be drawn is that the bonding within the plethora of $[EM]_n$ (E = C, N, O, etc.; M = alkali metal) rings, stacks and ladders described in Sections 1.2 and 1.3 is noncluster bonding, in that it occurs at the periphery of the structures and with very little or no metal–metal interactions. This view is classically illustrated by calculational studies on the simplest organolithium compound $[MeLi]_4$. When first published, the structure of $[MeLi]_4$ was rationalized in terms of a qualitative MO approach[137] and refinement of this model has led to the common textbook description of the bonding within it as being 'electron deficient'; the interaction of the Li_3 faces of the Li_4 tetrahedron with a methyl group being described as a two-electron three-centre bond.[217] This largely covalent view of the bonding in $[MeLi]_4$ was overestimated as a result of the observation of ^{13}C–^{7}Li NMR coupling[218] and has been further strengthened by the generally low melting points of organolithium compounds and by their frequently high solubility in organic solvents. However, Streitweiser developed an alternative model in which purely electrostatic point charges can explain the structure of $[MeLi]_4$.[219] The extension of this model, by describing the atoms as hard spheres, soon followed (Hard Sphere Electrostatic Model).[220] The essential correctness of this model has become increasingly validated by MO calculations which illustrate that C–Li bonds are about 80–8% ionic and that there is an even greater ionic contribution for the heavier alkali metal–carbon bonds.[8,221–2] Schleyer has undoubtedly made the greatest contribution in this area in publicizing and developing this (now accepted) ionic model.

Calculational studies on amido- and iminolithium complexes can be obtained from the review by Gregory *et al.* in which the thermodynamics of ring, ladder and stack formation is examined in detail.[10] Calculations on these species also reflect the high ionicity of the N–Li bonding involved, a fact which is particularly apparent from the high positive charges developed on lithium (typically +0.6–0.7 e) and high negative charges on the nitrogen centres (typically −0.8–0.9 e).[4–5]

1.5 REFERENCES

1. (a) W. Schlenk, J. Appenrodt, A. Michael and A. Thal, *Ber. Dtsch. Chem. Ges.*, 1914, **47**, 473; (b) W. Shlcenk and R. Ochs, *ibid.*, 1916, **49**, 608; (c) W. Schlenk and J. Holtz, *ibid.*, 1917, **50**, 262.
2. 'The First International Conference on the Alkali and Alkaline Earth Metals', Cambridge, 1994.
3. J. Wardell, in 'COMC-I', vol. 1, p. 43.
4. K. Gregory, P. von R. Schleyer and R. Snaith, *Adv. Inorg. Chem.*, 1994, **37**, 47.
5. R. E. Mulvey, *Chem. Rev.*, 1991, **20**, 167.
6. E. Weiss, *Angew. Chem.*, 1993, **105**, 1588; *Angew. Chem., Int. Ed. Engl.*, 1993, **32**, 1501.
7. W. N. Setzer and P. v. R. Schleyer, *Adv. Organomet. Chem.*, 1985, **24**, 353.
8. C. Schade and P. v. R. Schleyer, *Adv. Organomet. Chem.*, 1987, **27**, 169.
9. D. Seebach, *Angew. Chem.*, 1988, **100**, 1685; *Angew. Chem., Int. Ed. Engl.*, 1988, **27**, 1624.
10. G. Boche, *Angew. Chem.*, 1989, **101**, 286; *Angew. Chem., Int. Ed. Engl.*, 1989, **28**, 277.
11. H. M. M. Shearer, K. Wade and G. J. Whitehead, *J. Chem. Soc., Chem. Commun.*, 1979, 943.
12. W. Clegg, R. Snaith, H. M. M. Shearer, K. Wade and G. J. Whitehead, *J. Chem. Soc., Dalton Trans.*, 1983, 1309.
13. D. Barr, W. Clegg, R. E. Mulvey, R. Snaith, K. Wade and D. Reed, *J. Chem. Soc., Chem. Commun.*, 1986, 295.
14. D. R. Armstrong *et al.*, *J. Chem. Soc., Dalton Trans.*, 1987, 1071.

15. L.-H. Chan and E. G. Rochow, *J. Organomet. Chem.*, 1967, **9**, 231.
16. P. L. Pickard and T. L. Tolbot, *J. Org. Chem.*, 1961, **26**, 4886.
17. I. Pattison, K. Wade and B. K. Wyatt, *J. Chem. Soc. (A)*, 1968, 837.
18. D. Barr, R. Snaith, W. Clegg, R. E. Mulvey and K. Wade, *J. Chem. Soc., Dalton Trans.*, 1987, 2141.
19. D. Barr, R. Snaith, W. Clegg, R. E. Mulvey and K. Wade, *Angew. Chem.*, 1985, **97**, 322; *Angew. Chem., Int. Ed. Engl.*, 1985, **24**, 328.
20. D. Barr, R. Snaith, W. Clegg and R. E. Mulvey, *J. Chem. Soc., Chem. Commun.*, 1984, 79.
21. D. Barr, R. Snaith, W. Clegg, R. E. Mulvey and K. Wade, *J. Chem. Soc., Chem. Commun.*, 1984, 226.
22. W. Clegg, R. E. Mulvey, R. Snaith, G. E. Toogood and K. Wade, *J. Chem. Soc., Chem. Commun.*, 1986, 1740.
23. D. Barr, W. Clegg, R. E. Mulvey and R. Snaith, *J. Chem. Soc., Chem. Commun.*, 1989, 57.
24. M. Fieser, 'Reagents for Organic Synthesis', Wiley, New York, 1990, vol 15 and proceeding volumes give specific details of the applications of specific lithium amides.
25. J. C. Stowell, 'Carbanions in Organic Synthesis', Wiley, New York, 1979.
26. B. J. Wakefield, 'The Chemistry of Organolithium Compounds', Pergamon, Oxford, 1974.
27. C. H. Heathcock, in 'Asymmetric Synthesis' ed. J. D. Morrison, Academic Press, New York, 1984, vol. 3B.
28. G. B. Gill and D. A. Whiting, *Aldrichim. Acta*, 1986, **19**, 31 for further preparative details, see J. L. Wardell, in 'The Chemistry of the Metal–Carbon Bond', ed. F. R. Hartley, Wiley, Chichester, 1987, vol. 4; L. Brandsma and H. D. Verkruijsse, 'Preparative Polar Organometallic Chemistry', Elsevier, Amsterdam, 1986, vol. 3, pp. 352–456.
29. M. F. Lappert, M. J. Slade, A. Singh, J. L. Atwood, R. D. Rogers and R. J. Shakir, *J. Am. Chem. Soc.*, 1983, **302**, 105.
30. (a) D. R. Armstrong, W. Clegg, R. E. Mulvey, D. Barr and R. Snaith, *J. Chem. Soc., Chem. Commun.*, 1984, 285; (b) D. R. Armstrong, R. E. Mulvey, G. T. Walker, D. Barr and R. Snaith, *J. Chem. Soc., Dalton Trans.*, 1988, 617.
31. R. D. Rogers, J. L. Atwood and R. J. Grüning, *J. Organomet. Chem.*, 1978, **157**, 229; D. Mootz, A. Zinnius and B. Böttcher, *Angew. Chem.*, 1969, **81**, 398; *Angew. Chem., Int. Ed. Engl.*, 1969, **8**, 378.
32. T. Fjeldberg, P. B. Hitchcock, M. F. Lappert and A. J. Thorne, *J. Chem. Soc., Chem. Commun.*, 1984, 822.
33. M. Rannenberg, H.-D. Hausen and J. Weidlein, *J. Organomet. Chem.*, 1989, **376**, C27.
34. R. Snaith *et al.*, *Angew. Chem.*, 1989, **101**, 1279; *Angew. Chem., Int. Ed. Engl.*, 1989, **28**, 1241.
35. R. Grüning and J. L. Atwood, *J. Organomet. Chem.*, 1977, **137**, 101.
36. P. G. Williard, *Acta Crystallogr., Part C*, 1988, **44**, 277.
37. D. R. Armstrong, D. Barr, W. Clegg, S. M. Hodgson, R. E. Mulvey, D. Reed, R. Snaith and D. S. Wright, *J. Am. Chem. Soc.*, 1989, **111**, 4719.
38. (a) D. R. Armstrong *et al.*, *J. Chem. Soc., Chem. Commun.*, 1986, 869; (b) G. Boche, I. Langlotz, M. Marsch, K. Harm and N. E. S. Nudelman, *Angew. Chem.*, 1992, **104**, 1239; *Angew. Chem., Int. Ed. Engl.*, 1992, **31**, 1205.
39. L. M. Englehardt, A. S. May, C. L. Raston and A. H. White, *J. Chem. Soc., Dalton Trans.*, 1983, 1671.
40. L. M. Englehardt, B. S. Jolly, P. C. Junk, C. L. Raston, B. W. Skelton and A. H. White, *Aust. J. Chem. Soc.*, 1986, **39**, 1337.
41. B. Cetinkaya, P. B. Hitchcock, M. F. Lappert, M. C. Misra and A. J. Thorne, *J. Chem. Soc., Chem. Commun.*, 1984, 148.
42. R. A. Bartlett, H. Chen, H. V. R. Dias, M. M. Olmstead, P. P. Power and K. J. Weese, *J. Am. Chem. Soc.*, 1988, **110**, 446.
43. J. L. Atwood, S. G. Bott, S. M. Hawkins and M. F. Lappert, unpublished results in Ref. 4.
44. D. Barr, W. Clegg, R. E. Mulvey, R. Snaith and D. S. Wright, *J. Chem. Soc., Chem. Commun.*, 1987, 716.
45. R. Boese and U. Klingebiel, *J. Organomet. Chem.*, 1986, **315**, C17.
46. M. Bremer, R. Flock and P. v. R. Schleyer, unpublished results in Ref. 4.
47. R. A. Bartlett, X. Feng, M. M. Olmstead, P. P. Power and K. J. Weese, *J. Am. Chem. Soc.*, 1987, **109**, 4851.
48. H. Chen, R. A. Bartlett, M. M. Olmstead, P. P. Power and S. C. Shoner, *J. Am. Chem. Soc.*, 1990, **112**, 1048.
49. P. Paetzold, C. Pelzer and R. Boese, *Chem. Ber.*, 1988, **121**, 51.
50. P. P. Power, *Acc. Chem. Res.*, 1988, **21**, 147.
51. J. L. Atwood, M. F. Lappert, W.-P. Leung and H. Zhang, unpublished results in Ref. 4.
52. H. Chen, R. A. Bartlett, H. V. R. Dias, M. M. Olmstead and P. P. Power, *J. Am. Chem. Soc.*, 1989, **111**, 4338.
53. P. P. Power and X. Xiaojie, *J. Chem. Soc., Chem. Commun.*, 1984, 358.
54. R. A. Bartlett, H. V. R. Dias, H. Hope, B. D. Murray, M. M. Olmstead and P. P. Power, *J. Am. Chem. Soc.*, 1986, **108**, 6921.
55. R. A. Bartlett and P. P. Power, *J. Am. Chem. Soc.*, 1987, **109**, 6509.
56. P. C. Andrews, W. Clegg and R. E. Mulvey, *Angew. Chem.*, 1990, **102**, 1480; *Angew. Chem., Int. Ed. Engl.*, 1990, **29**, 1440.
57. W. Clegg, M. MacGregor, R. E. Mulvey and P. A. O'Neil, *Angew. Chem.*, 1992, **104**, 74; 252, correction; *Angew. Chem., Int. Ed. Engl.*, 1992, **31**, 93.
58. N. P. Lorenzen, J. Kopf, F. Olrich, U. Schümann and E. Weiss, *Angew. Chem.*, 1990, **102**, 1481; *Angew. Chem., Int. Ed. Engl.*, 1990, **29**, 1441.
59. K. Gregory, M. Bremer, P. v. R. Schleyer, N. P. Lorenzen, J. Kopf and E. Weiss, *Organometallics*, 1990, **9**, 1485.
60. K. Gregory, M. Bremer, P. v. R. Schleyer, P. A. A. Klusener and L. Brandsma, *Angew. Chem.*, 1989, **101**, 1261; *Angew. Chem., Int. Ed. Engl.*, 1989, **28**, 1224.
61. P. L. Hall, J. H. Gilchrist and D. B. Collum, *J. Am. Chem. Soc.*, 1991, **113**, 9571.
62. A. S. Galiano-Roth, Y. J. Kim, J. H. Gilchrist, A. T. Harrison, D. J. Fuller and D. B. Collum, *J. Am. Chem. Soc.*, 1991, **113**, 5053.
63. N. D. R. Barnett, R. E. Mulvey, W. Clegg and P. A. O'Neil, *J. Am. Chem. Soc.*, 1991, **113**, 8187.
64. M. P. Bernstein *et al.*, *J. Am. Chem. Soc.*, 1992, **114**, 5100.
65. L. M. Englehardt, G. E. Jacobsen, A. H. White and C. L. Raston, *Inorg. Chem.*, 1991, **30**, 3978.
66. F. S. Mair, W. Clegg and P. A. O'Neil, *J. Am. Chem. Soc.*, 1993, **115**, 3388.
67. R. E. Mulvey, K. Wade, D. R. Armstrong, G. T. Walker, R. Snaith, W. Clegg and D. Reed, *Polyhedron*, 1986, **5**, 987.
68. H. Hope, M. M. Olmstead, P. P. Power, X. Xu, *J. Am. Chem. Soc.*, 1984, **106**, 819.
69. P. B. Hitchcock, M. F. Lappert, P. P. Power and S. J. Smith, *J. Chem. Soc., Chem. Commun.*, 1984, 1669.
70. E. Hey, P. B. Hitchcock, M. F. Lappert and A. K. Rai, *J. Organomet. Chem.*, 1987, **325**, 1.
71. E. Hey, C. L. Raston, B. W. Skelton and A. H. White, *J. Organomet. Chem.*, 1989, **362**, 1.
72. R. A. Jones, A. L. Stuart and T. C. Wright, *J. Am. Chem. Soc.*, 1983, **105**, 7459.
73. E. Hey-Hawkins and E. Sattler, *J. Chem., Soc., Chem. Commun.*, 1992, 775.
74. E. Hey and F. Weller, *J. Chem. Soc., Chem. Commun.*, 1992, 220.
75. R. A. Bartlett, M. M. Olmstead and P. P. Power, *Inorg. Chem.*, 1986, **25**, 1243.

76. M. Driess and H. Pritzkow, *Angew. Chem.*, 1992, **104**, 350; *Angew. Chem. Int. Ed. Engl.*, 1992, **31**, 316; see also G. Becker and C. Whitthauer, *Z. Anorg. Allg. Chem.*, 1982, **492**, 28.
77. A. J. Edwards, D. J. Kane, M. A. Paver, P. R. Raithby, C. A. Russell and D. S. Wright, unpublished results in Ref. 4.
78. A. M. Arif, R. A. Jones and K. B. Kidd, *J. Chem. Soc. Chem. Commun.*, 1986, 1440.
79. P. B. Hitchcock, M. F. Lappert and S. J. Smith, *J. Organomet. Chem.*, 1987, **320**, C27.
80. G. Becker, A. Münch and C. Whitthauer, *Z. Anorg. Allg. Chem.*, 1982, **492**, 15.
81. O. Mundt, G. Becker, M. Rössler and C. Whitthauer, *Z. Anorg. Allg. Chem.*, 1983, **506**, 42.
82. (a) D. Bradley, *Chem. Rev.*, 1989, **89**, 1317; (b) K. C. Caulton, L. G. Hubert-Pfalzgraf, *Chem. Rev.*, 1990, **90**, 969.
83. For a review of phenoxides, see K. C. Malhotra and R. L. Martin, *J. Organomet. Chem.*, 1982, **239**, 159.
84. (a) P. J. Wheatley, *J. Chem. Soc.*, 1961, 4270; (b) H. Dunken, J. Krause, *Z. Chem.*, 1961, **1**, 27; (c) E. Weiss, *Z. Anorg. Allg. Chem.*, 1964, **332**, 197; (d) E. Weiss and W. Büchner, *Angew. Chem.*, 1963, **75**, 1116; (e) E. Weiss, *Helv. Chim. Acta*, 1963, **46**, 2051; (f) E. Weiss and H. Alsdorf, *Z. Anorg. Allg. Chem.*, 1970, **372**, 206.
85. (a) T. Greiser and E. Weiss, *Chem. Ber.*, 1977, **110**, 3388; (b) J. E. Davies, J. Kopf and E. Weiss, *Acta Crystallogr., Part B*, 1982, **38**, 2251; (c) E. Weiss, H. Alsdorf, H. Kühr, *Angew. Chem.*, 1967, **79**, 816; *Angew. Chem., Int. Ed. Engl.*, 1967, **6**, 801; (d) E. Weiss, H. Alsdorf, H. Kühr and H.-F. Grützmacher, *Chem. Ber.*, 1968, **101**, 3777.
86. T. Laube, J. D. Dunitz and D. Seebach, *Helv. Chim. Acta*, 1985, **68**, 1373.
87. D. Seebach, R. Amstutz, T. Laube, W. B. Schweizer and J. D. Dunitz, *J. Am. Chem. Soc.*, 1985, **107**, 5403.
88. W. Bauer, T. Laube and D. Seebach, *Chem. Ber.*, 1985, **118**, 764.
89. J. Hvoslef, H. Hope, B. D. Murray and P. P. Power, *J. Chem. Soc., Chem. Commun.*, 1983, 1438.
90. B. Cetinkaya, I. Gümrükcü, M. F. Lappert, J. L. Atwood and R. Shakir, *J. Am. Chem. Soc.*, 1980, **102**, 2086.
91. R. Amstutz, J. D. Dunitz, T. Laube, W. B. Schweizer and D. Seebach, *Chem. Ber.*, 1986, **119**, 434.
92. M. Geissler and E. Weiss, in press.
93. S. Brooker, F. T. Edelmann, T. Kottke, H. W. Roesky, G. M. Sheldrick and K. H. Whitmire, *J. Chem. Soc., Chem. Commun.*, 1991, 144.
94. P. A. van der Schaaf, M. P. Hogerheide, D. M. Grove, A. L. Spek and G. van Koten, *J. Chem. Soc., Chem. Commun.*, 1992, 1703.
95. G. R. Fuentes, P. S. Coan, W. E. Streih and K. G. Caulton, *Polyhedron*, 1991, **10**, 2371.
96. R. Amstutz, W. B. Schweizer, D. Seebach and J. D. Dunitz, *Inorg. Chim. Acta*, 1981, **64**, 2617.
97. J. T. B. H. Jastrzebski, G. van Koten, M. N. Christophersen and C. H. Stam, *J. Organomet. Chem.*, 1985, **292**, 319.
98. P. G. Williard and J. M. Salvino, *Tetrahedron Lett.*, 1985, **26**, 3931.
99. O. Graalmann, U. Klingebiel, W. Glegg, M. Haase and G. M. Sheldrick, *Angew. Chem.*, 1984, **96**, 904; *Angew. Chem., Int. Ed. Engl.*, 1984, **23**, 891.
100. P. G. Williard and G. B. Carpenter, *J. Am. Chem. Soc.*, 1986, **108**, 462.
101. P. G. Williard and G. B. Carpenter, *J. Am. Chem. Soc.*, 1985, **107**, 3345.
102. J. T. B. H. Jastrzebski, G. van Koten, W. F. van der Mieroop, *Helv. Chim. Acta*, 1988, **142**, 169.
103. E. Weiss, K. Hoffman, H.-F. Grützmacher, *Chem. Ber.*, 1970, **103**, 1190.
104. For the activation of organolithiums by the addition of Lewis bases or alkoxides, see (a) L. Lochmann, J. Posisil and D. Lim, *Tetrahedron Lett.*, 1966, 257; (b) L. Lochmann and J. Trekoval, *Collect. Czech. Chem. Commun.*, 1988, **53**, 76; (c) M. Schlosser, *J. Organomet. Chem.*, 1967, **8**, 9; (d) M. Schlosser and S. Strunk, *Tetrahedron Lett.*, 1984, **25**, 741; (e) M. Schlosser, *Pure Appl. Chem.*, 1988, **60**, 1627; (f) C. Agami, *Bull. Soc. Chim. Fr.*, 1970, **37**, 1620; (g) T. Durst, *Stud. Org. Soc. Chem. (Amsterdam)*, 1984, **5**, 239; (h) R. R. Fraser and T. Mansour, *Tetrahedron Lett.*, 1986, **27**, 331; (i) W. Bauer and E. Lochmann, *J. Am. Chem. Soc.*, 1992, **114**, 7482.
105. M. Marsch, K. Harms, L. Lochmann and G. Boche, *Angew. Chem.*, 1990, **102**, 334; *Angew. Chem., Int. Ed. Engl.*, 1990, **29**, 308.
106. S. Harder and A. Streitweiser, *Angew. Chem.*, 1993, **105**, 1108; *Angew. Chem., Int. Ed. Engl.*, 1993, **32**, 1066.
107. P. R. Raithby, D. Reed, R. Snaith and D. S. Wright, *Angew. Chem.*, 1991, **103**, 1040; *Angew. Chem., Int. Ed. Engl.*, 1991, **30**, 1011.
108. K. Ruhlandte-Senge and P. P. Power, *Inorg. Chem.*, 1991, **30**, 2633.
109. W.-W. du Mont, S. Kubiniok, L. Lange, S. Pohl, W. Saak and I. Wagner, *Chem. Ber.*, 1991, **124**, 1315; K. Ruhlandte-Senge and P. P. Power, *Inorg. Chem.*, 1991, **30**, 3683.
110. J. Bonasia and J. Arnold, *J. Organomet. Chem.*, 1993, **449**, 147.
111. A. J. Banister, W. Clegg and W. R. Gill, *J. Chem. Soc., Chem. Commun.*, 1987, 850.
112. M. Aslam, R. A. Bartlett, E. Block, M. M. Olmstead, P. P. Power and G. E. Sigel, *J. Chem. Soc., Chem. Commun.*, 1985, 1674.
113. D. R. Armstrong, A. J. Banister, W. Clegg and W. R. Gill, *J. Chem. Soc., Chem. Commun.*, 1986, 1672.
114. R. Amstutz, D. Seebach, P. Seiler, W. B. Schweizer and J. D. Dunitz, *Angew. Chem.*, 1980, **92**, 59; *Angew. Chem., Int. Ed. Engl.*, 1980, **19**, 53.
115. R. Amstutz, T. Laube, W. B. Schweizer, D. Seebach and J. D. Dunitz, *Helv. Chim. Acta*, 1984, **67**, 224.
116. J. Bonasia, D. E. Gindelberger, B. O. Dobbousi and J. Arnold, *J. Am. Chem. Soc.*, 1992, **114**, 5209.
117. G. Becker, K. W. Klinkhammer, S. Lartiges, P. Böttcher and W. Poll, *Z. Anorg. Allg. Chem.*, 1992, **613**, 7.
118. G. Becker, K. W. Klinkhammer and W. Massa, *Z. Anorg. Allg. Chem.*, 1993, **619**, 628.
119. (a) M. Pereyre, J.-P. Quintard and R. Rahm, 'Tin in Organic Synthesis', Butterworths, London, 1987; (b) Y. Yamamoto (ed.), *Tetrahedron*, 1989, **45**, 909, and references therein; (c) P. G. Harrison, 'Chemistry of Tin', Chapman and Hall, New York, 1989; (d) I. Omae, 'Organotin Chemistry', Elsevier, Amsterdam, 1989.
120. (a) T. F. Schaaf, W. Butler, M. D. Glick and J. P. Oliver, *J. Am. Chem. Soc.*, 1974, **96**, 7593; (b) W. H. Isley, T. F. Schaaf, M. D. Glick and J. P. Oliver, *J. Am. Chem. Soc.*, 1980, **102**, 3769; (c) B. Tecle, W. H. Isley and J. P. Oliver, *Organometallics*, 1982, **1**, 875, (d) G. Becker, H.-M. Hartmann, H. Münch Hartmann, E. Hengge and F. Schrank, *Z. Anorg. Allg. Chem.*, 1989, **572**, 63; (e) H. V. R. Dais, M. M. Olmstead, K. Ruhlandte-Senge and P. P. Power, *J. Organomet. Chem.*, 1993, **462**, 1.
121. (a) D. Reed, D. Stalke and D. S. Wright, *Angew. Chem.*, 1991, **103**, 1539; *Angew. Chem., Int. Ed. Engl.*, 1991, **30**, 1459; (b) H. J. Reich, J. B. Borst and R. R. Dykstra, *Organometallics*, 1993, **13**, 1.
122. D. R. Armstrong, M. G. Davidson, D. Moncrieff, D. Stalke and D. S. Wright, *J. Chem. Soc., Chem. Commun.*, 1992, 1413.
123. P. B. Hitchcock, M. F. Lappert, G. A. Lawless and B. Royo, *J. Chem. Soc., Chem. Commun.*, 1993, 554.

124. For example, see: T. Birchall and J. Vetrone, *J. Chem. Soc., Chem. Commun.*, 1988, 877.
125. J. J. Brooks and G. D. Stucky, *J. Am. Chem. Soc.*, 1972, **94**, 273.
126. M. G. Davidson, D. Stalke and D. S. Wright, *Angew. Chem.*, 1992, **104**, 1265; *Angew. Chem., Int. Ed. Engl.*, 1992, **31**, 1226.
127. D. R. Armstrong *et al.*, in press.
128. A. J. Edwards *et al.*, *J. Chem. Soc., Dalton Trans*, 1993, 1465.
129. D. R. Armstrong *et al.*, *Angew. Chem.*, 1993, **105**, 1807; *Angew. Chem., Int. Ed. Engl.*, 1993, **32**, 1774.
130. M. Veith, *Chem. Rev.*, 1990, **90**, 3.
131. D. R. Armstrong *et al.*, *Angew. Chem.*, 1991, **103**, 1702; *Angew. Chem., Int. Ed. Engl.*, 1991, **30**, 1707.
132. D. Barr *et al. J. Chem. Soc., Chem. Commun.*, 1988, 367.
133. W. Hollstein, K. Harms, M. Marsch and G. Boche, *Angew. Chem.*, 1987, **99**, 1279; *Angew. Chem., Int. Ed. Engl.*, 1987, **26**, 1287.
134. W. Zarges, M. Marsch, K. Harms and G. Boche, *Chem. Ber.*, 1989, **122**, 1307.
135. H.-J. Gais, J. Vollhardte, H. Günther, D. Moskau, H. J. Lindner and S. Braun, *J. Am. Chem. Soc.*, 1988, **110**, 978.
136. For a review on dilithiated compounds, see: A. Maercker and M. Theis, *Top. Curr. Chem.*, 1987, **138**, 1.
137. (a) M. Schlosser, 'Struktur und Reactivitat polarer Organomtalle', Springer, Berlin, 1973; (b) M. Schlosser *Angew. Chem.*, 1964, **76**, 124; 258; *Angew. Chem. Int. Ed. Engl.*, 1964, **3**, 287; 362; (c) H. F. Ebel, *Fortschr. Chem. Forsch.*, 1969, **12**, 387; (d) P. West, *MTP Int. Rev. Sci. Inorg. Chem. Ser. Onee.*, 1972, **4**, 1.
138. W. Zarges, M. Marsch, K. Harms and G. Boche, *Chem. Ber.*, 1989, **122**, 2303.
139. W. Zarges, M. Marsch, K. Harms, W. Koch, G. Frenking and G. Boche, *Chem. Ber.*, 1991, **124**, 543.
140. J. J. Brooks and G. D. Stucky, *J. Am. Chem. Soc.*, 1970, **92**, 1150.
141. R. A. Bartlett, H. V. R. Dias and P. P. Power, *J. Organomet. Chem.*, 1988, **341**, 1.
142. M. F. Lappert, L. M. Englehardt, C. L. Raston and A. H. White, *J. Chem. Soc., Chem. Commun.*, 1982, 1323.
143. U. Schümann, J. Kopf and E. Weiss, *Angew. Chem.*, 1985, **97**, 222.
144. T. Maetzke and D. Seemann, *Helv. Chim. Acta.*, 1989, **72**, 624.
145. U. Schümann, E. Weiss, H. Dietrich and W. Mahdi, *J. Organomet. Chem.*, 1987, **322**, 299.
146. G. Boche *et al.*, *J. Am. Chem. Soc.*, 1992, **114**, 1562.
147. S. A. Kinsley, A. Streitweiser and Z. Zalkin, *Organometallics*, 1985, **4**, 52.
148. Yearly report: (a) *Organomet. Chem.* 1972 ff; (b) *Annu. Ref. Prog. Chem.* 1950, **47** to 1966, **63**; *Annu. Rep. Prog. Chem., Sect. A.* in subsequent years.
149. W. Hiller, M. Layh and W. Uhl, *Angew. Chem.*, 1991, **103**, 339; *Angew. Chem. Int. Ed. Engl.*, 1991, **30**, 324.
150. D. Theonnes and E. Weiss, *Chem. Ber.*, 1978, **111**, 3157.
151. B. Schubert and E. Weiss, *Chem. Ber.*, 1983, **116**, 3212.
152. T. Kottke and D. Stalke, *Angew. Chem.* 1993, **105**, 619; *Angew. Chem. Int. Ed. Engl.*, 1993, **32**, 580; see also T. Kottke, D. Stalke, *J. Appl. Crystallogr.*, 1993, **26**, 615.
153. S. Harder, J. Boersma, L. Brandsma, J. A. Kanters, W. Bauer and P. von R. Schleyer, *Organometallics*, 1989, **8**, 1696.
154. (a) E. Weiss and G. Hencken, *J. Organomet. Chem.*, 1969, **21**, 265; (b) E. Weiss, T. Lambertsen, B. Schubert, J. K. Cockcroft and A. Weidenmann, *Chem. Ber.*, 1990, **123**, 79.
155. (a) E. Weiss and E. A. C. Lucken, *J. Organomet. Chem.*, 1964, **2**, 197; (b) H. Dietrich, *Acta Crystallogr.*, 1963, **16**, 681; *J. Organomet. Chem.*, 1981, **205**, 291.
156. G. W. Klump, P. J. A. Geurink, N. J. R. van Eikema Hommes, F. J. J. de Kanter, M. Vos and A. L. Spek, *Recl. Trav. Chim. Pays-Bas.*, 1986, **10**, 398.
157. K. S. Lee, P. G. Willard and J. W. Suggs, *J. Organomet. Chem.*, 1986, **299**, 311.
158. H. Hope and P. P. Power, *J. Am. Chem. Soc.*, 1983, **105**, 5320.
159. M. M. Olmstead and P. P. Power, *J. Am. Chem. Soc.*, 1990, **112**, 8008.
160. M. Geissler, J. Kopf, B. Schubert, E. Weiss, W. Neugebauer and P. von R. Schleyer, *Angew. Chem.*, 1987, **99**, 569; *Angew. Chem. Int. Ed. Engl.*, 1987, **26**, 587.
161. B. Schubert and E. Weiss, *Angew. Chem.*, 1983, **95**, 499; *Angew. Chem. Int. Ed. Engl.*, 1983, **22**, 496; *Angew. Chem. Suppl.*, 1983, 703.
162. H. Köster, D. Theonnes and E. Weiss, *J. Organomet. Chem.*, 1978, **160**, 1.
163. R. Zerger, W. Rhine and G. Stucky, *J. Am. Chem. Soc.*, 1974, **96**, 6048.
164. A. Maercker, M. Bsata, W. Buchmeiser and B. Engelen, *Chem. Ber.*, 1984, **117**, 2547.
165. B. Tecle, A. F. M. M. Rahman and J. P. Oliver, *J. Organomet. Chem.*, 1986, **317**, 267.
166. H. Köster and E. Weiss, *Chem. Ber.*, 1982, **115**, 3422.
167. G. Boche *et al.*, *Angew. Chem.*, 1986, **98**, 84; *Angew. Chem. Int. Ed. Engl.*, 1986, **25**, 104.
168. S. P. Patterman, I. L. Karle and G. D. Stucky, *J. Am. Chem. Soc.*, 1970, **92**, 1150.
169. M. A. Beno, H. Hope, M. M. Olmstead and P. P. Power, *Organometallics*, 1985, **4**, 2117.
170. M. M. Olmstead and P. P. Power, *J. Am. Chem. Soc.*, 1985, **107**, 2174.
171. C. Eaborn, P. B. Hitchcock, J. D. Smith and A. C. Sullivan, *J. Chem. Soc., Chem. Commun.*, 1983, 827.
172. (a) E. Weiss, S. Corbelin, J. K. Cockcroft and A. N. Fitch, *Angew. Chem.* 1990, **102**, 728; *Angew. Chem. Int. Ed. Engl.*, 1990, **29**, 650; (b) E. Weiss, S. Corbelin, J. K. Cockcroft and A. N. Fitch, *Chem. Ber.*, 1990, **123**, 1629.
173. P. West and R. Waack, *J. Am, Chem. Soc.*, 1967, **89**, 4395.
174. A. K. Cheetham and A. P. Wilkinson, *Angew. Chem.*, 1992, **104**, 1594; *Angew. Chem. Int. Ed. Engl.*, 1992, **31**, 1557.
175. (a) E. Weiss and G. Sauermann, *Angew. Chem.*, 1968, **80**, 123; *Angew. Chem. Int. Ed. Engl.*, 1968, **7**, 133; (b) E. Weiss and G. Sauermann, *Chem. Ber.*, 1970, **103**, 265.
176. E. Weiss and H. Köster, *Chem. Ber.*, 1977, **110**, 717.
177. E. Weiss, T. Lambertsen, B. Schubert and J. K. Cockcroft, *J. Organomet. Chem.*, 1988, **358**, 1.
178. E. Weiss and H. Plass, *Chem. Ber.*, 1968, **101**, 2947.
179. M. Atoji, *J. Chem. Phys.*, 1972, **56**, 4947.
180. (a) R. Juza, V. Werle, H. U. Schuster, *Z. Anorg. Allg. Chem.*, 1967, **352**, 252; (b) M. Atoji, *J. Chem. Phys.*, 1974, **60**, 3324.
181. H. Foppl, *Angew. Chem.*, 1958, **70**, 401.
182. J. B. Collins *et al.*, *J. Am. Chem. Soc.*, 1976, **89**, 5419.
183. G. D. Stucky, M. M. Eddy, W. H. Harrison, R. Lagow, H. Kawa and D. E. Cox, *J. Am. Chem. Soc.*, 1990, **112**, 2425.

184. S. Corbelin, J. Kopf, N. P. Lorenzen and E. Weiss, *Angew. Chem.*, 1991, **103**, 875; *Angew. Chem., Int. Ed. Engl.*, 1991, **30**, 825.
185. S. Corbelin, J. Kopf and E. Weiss, *Chem. Ber.*, 1991, **124**, 2417.
186. G. Rabe, H. W. Roesky, D. Stalke, F. Pauer and G. M. Sheldrick, *J. Organomet. Chem.*, 1991, **403**, 11.
187. U. Schümann, U. Behrens and E. Weiss, *Angew. Chem.*, 1989, **101**, 481; *Angew. Chem. Int. Ed. Engl.*, 1989, **28**, 476.
188. J. Lorberth, S. H. Shin, S. Wocadlo and W. Massa, *Angew. Chem.*, 1989, **101**, 793; *Angew. Chem., Int. Ed. Engl.*, 1989, **28**, 735.
189. R. Zerger, W. Rhine and G. D. Stucky, *J. Am. Chem. Soc.*, 1974, **96**, 5441.
190. C. Schade, P. von R. Schleyer, H. Dietrich and W. Mahdi, *J. Am. Chem. Soc.*, 1986, **108**, 2484.
191. S. Corbelin, N. P. Lorenzen, J. Kopf and E. Weiss, *J. Organomet. Chem.*, 1991, **415**, 293.
192. C. Schade, P. von R. Schleyer, P. Gregory, H. Dietrich and W. Mahdi, *J. Organomet. Chem.*, 1988, **341**, 19.
193. K. Köster and E. Weiss, *J. Organomet. Chem.*, 1979, **168**, 273.
194. D. Hoffman *et al.*, *J. Am. Chem. Soc.*, 1994, **116**, 528.
195. P. Jutzi, W. Leffers, B. Hampel, S. Pohl and W. Saak, *Angew. Chem.*, 1987, **99**, 563; *Angew. Chem., Int. Ed. Engl.*, 1987, **26**, 583.
196. S. Harder and M. H. Prosenc, *Angew. Chem.*, 1994, **106**, 1830; *Angew. Chem., Int. Ed. Engl.*, 1994, **33**, 1744.
197. F. Zaegel, J. C. Gallucci, P. Meunier, B. Gautheron, M. R. Sivik and L. A. Paquette, *J. Am. Chem. Soc.*, 1994, **116**, 6466.
198. W. E. Rhine and G. D. Stucky, *J. Am. Chem. Soc.*, 1975, **97**, 737.
199. G. P. Pez, R. A. Grey and J. Corsi, *J. Am. Chem. Soc.*, 1981, **103**, 7528.
200. H. F. Klein, H. Witty and U. Schubert, *J. Chem. Soc., Chem. Commun.*, 1983, 231.
201. R. Goddard, C. Kruger, K. R. Porrschke and G. Wilke, *J. Organomet. Chem.*, 1986, **308**, 85.
202. J. C. Huffman, M. A. Green, S. L. Kaiser and K. G. Caulton, *J. Am. Chem. Soc.*, 1985, **107**, 5111.
203. H. F. Klein, J. Gross, J.-M. Basset and U. Schubert, *Z. Naturforsch., Teil B*, 1980, **35**, 614.
204. D. J. Brauer, C. Kruger, P. J. Roberts and Y.-H. Tsay, *Angew. Chem.*, 1976, **88**, 52; *Angew. Chem. Int. Ed. Engl.*, 1976, **15**, 48.
205. T. D. Tilley, R. A. Anderson and A. Zalkin, *Inorg. Chem.*, 1984, **23**, 2271.
206. N. P. Lorenzen and E. Weiss, *Angew. Chem.* 1990, **102**, 322; *Angew. Chem., Int. Ed. Engl.*, 1990, **29**, 300.
207. R. G. Pearson and C. D. Gregory, *J. Am. Chem. Soc.*, 1976, **98**, 4098.
208. F. Olbrich, J. Kopf and E. Weiss, *Angew. Chem.*, 1993, **105**, 1136; *Angew. Chem., Int. Ed. Engl.*, 1993, **32**, 1077.
209. K. O. Hodgson and K. N. Raymond, *Inorg. Chem.*, 1972, **11**, 3030.
210. E. P. Squiller, R. R. Whittle and H. G. Richey, Jr., *J. Am. Chem. Soc.*, 1985, **107**, 432; E. P. Squiller, R. R. Whittle and H. G. Richey, *Organometallics*, 1985, **4**, 1154; A. D. Pajerski, M. Parvez and H. G. Richey, Jr., *J. Am. Chem. Soc.*, 1988, **110**, 2660; T. Greiser, J. Kopf, D. Theonnes and E. Weiss, *Chem. Ber.*, 1981, **114**, 209; D. Theonnes and E. Weiss, *Chem. Ber.*, 1978, **111**, 3726; B. Schubert and E. Weiss, *Chem. Ber.*, 1984, **117**, 366.
211. (a) E. Weiss and R. Wolfrum, *Chem. Ber.*, 1984, **117**, 35; (b) E. Weiss and H. Plass, *J. Organomet. Chem.*, 1968, **14**, 21.
212. W. E. Rhine, G. D. Stucky and S. W. Peterson, *J. Am. Chem. Soc.*, 1975, **97**, 6401; Ya. Ozol, S. Vimba and A. Ievins, *Sov. Phys. Crystallogr.*, 1962, **7**, 289; Ya. Ozol, S. Vimba and A. Ievins, *Latv. PSR Zinat. Akad. Vestis, Kim. Ser.*, 1975, 517; K. Hoffman and E. Weiss, *J. Organomet. Chem.*, 1974, **67**, 221.
213. (a) R. Wolfrum, G. Sauermann and E. Weiss, *J. Organomet. Chem.*, 1969, **18**, 27; (b) R. L. Gerteis, R. E. Dickerson and T. L. Brown, *Inorg. Chem.*, 1964, **3**, 872; (c) G. Hencken and E. Weiss, *J. Organomet. Chem.*, 1974, **73**, 35.
214. K. Hoffman and E. Weiss, *J. Organomet. Chem.*, 1972, **37**, 1; K. Hoffman and E. Weiss, *J. Organomet. Chem.*, 1973, **50**, 17; K. Hoffman and E. Weiss, *J. Organomet. Chem.*, 1973, **50**, 25.
215. H. Smidbaur, A. Schier and U. Schubert, *Chem. Ber.*, 1983, **116**, 1938.
216. P. v. R. Schleyer, *Pure Appl. Chem.*, 1984, **56**, 151; P. v. R. Schleyer, *ibid.*, 1983, **55**, 355.
217. For example, see: F. A. Cotton and G. Wilkinson, 'Advanced Inorganic Chemistry', 5th edn., Wiley, New York, 1988, p. 202.
218. (a) L. D. McKeever, R. Waack, M. A. Doran and E. B. Baker, *J. Am. Chem. Soc.*, 1968, **90**, 3244; (b) *ibid.*, 1969, **91**, 1057; (c) L. M. Seitz and T. L. Brown, *ibid.*, 1966, **88**, 2174; (d) T. L. Brown, *ibid.*, 1966, **88**, 4140.
219. (a) A. Streitweiser, Jr., *J. Organomet. Chem.*, 1978, **156**, 1; (b) A. Streitweiser, J. E. Williams, S. Alexandratos and J. M. McKelvey, *J. Am. Chem. Soc.*, 1976, **98**, 4778.
220. (a) R. Bushby and H. L. Steel, *J. Organomet. Chem.*, 1987, **336**, C25; (b) *J. Chem. Soc., Perkin Trans. 2*, 1990, 1143; 1169, and references therein.
221. E. Kaufmann, K. Raghavachari, A. E. Reed and P. v. R. Schleyer, *Organometallics*, 1988, 1597, and references therein.
222. C. Lambert and P. v. R. Schleyer, *Angew. Chem.*, in press.

2
Beryllium

NORMAN A. BELL
Sheffield Hallam University, UK

2.1 INTRODUCTION 35
2.2 ALKYL, ARYL, ALKENYL AND ALKYNYL COMPOUNDS 36
2.2.1 Alkyl and Aryl Derivatives 36
2.2.2 Alkenyl Derivatives 38
2.2.3 Alkynyl Compounds 39
2.3 COORDINATION COMPLEXES 39
2.4 ANIONIC COMPLEXES 42
2.5 ALKOXY, AMINO, THIO AND PHOSPHINO COMPOUNDS 42
2.6 CYCLOPENTADIENYLBERYLLIUM COMPOUNDS 44
2.6.1 Dicyclopentadienylberyllium and Related Compounds 44
2.6.2 Miscellaneous Beryllium Cyclopentadienyls 48
2.7 HYDRIDES 50
2.8 MISCELLANEOUS REACTIONS 52
2.9 REFERENCES 54

2.1 INTRODUCTION

Since the last chapter was published in *COMC-I* in 1982, the number of publications on synthetic and structural aspects of organoberyllium compounds has plunged dramatically. Much of what was then known about the coordination chemistry of σ-bonded alkyl, aryl and hydride systems, in terms of structural and association studies, stemmed from the laboratories of Coates at Durham and Wyoming in the 1960s and 1970s, and it is no small coincidence that since his retirement this diminution has occurred. A commemorative issue[1] to mark Coates' contributions to organometallic chemistry was published by *The Journal of Organometallic Chemistry* in 1987 on the occasion of his 70th birthday. Most of the activity since 1980 has taken place in the area of cyclopentadienylberyllium chemistry where the controversy over the nature of dicyclopentadienylberyllium in the solid, solution and vapour states appears to have been resolved.

Beryllium has the potential to provide a rich lode of new chemistry. In general, beryllium chemistry is relatively unexplored compared with that of its less toxic counterparts. The notoriety of beryllium compounds as poisons (it is now cited in the Guinness Book of Records[2] as the most toxic nonradioactive element) deters the experimentalist but it does not inhibit the computer chemist. Indeed, the number of publications on theoretical studies of real and imaginary systems containing beryllium has increased dramatically; however, this area is considered to be outside the scope of this review, except where it relates directly to known compounds. No review articles on the organometallic chemistry of beryllium have appeared since 1980.

2.2 ALKYL, ARYL, ALKENYL AND ALKYNYL COMPOUNDS

Cyclopentadienylberyllium compounds are somewhat unique and with one exception are considered separately in Section 2.6.

Since the previous survey[3] no new general routes for the preparation of soluble organoberyllium compounds R_2Be and RBeX (X = halogen) have been developed. In the few cases where new organoberyllium compounds have been synthesized, the Grignard reagent or organoalkali metal reaction with a beryllium dihalide is the usual method chosen. Some chiral organoberyllium compounds (Section 2.8) have been prepared, but not isolated, and have been used *in situ* in ethereal solutions to study reduction reactions of ketones.

2.2.1 Alkyl and Aryl Derivatives

Two further spirobicyclic organoberyllium compounds have been prepared, this time derived from boronatophosphonium ylides, thus extending the range from the two simple carbon–oxygen and carbon–sulfur spiran type compounds prepared in the 1950s.[3] Reaction of the bifunctional phosphine–borane species (**1**) with beryllium chloride in THF (Scheme 1) produces the sublimable and moisture-sensitive complex (**2**), containing the robust boron–phosphorus backbone, in which internal coordination via Be ← O=P bonding is reflected in the reduction of $\nu_{P=O}$ in the infrared spectrum; the ^{13}C NMR spectrum illustrates the nonequivalence of the two methyl groups on phosphorus. As aged samples at room temperature become nonvolatile, the complex appears to undergo oligomerization on standing. A study of the sulfur analogue appears to have been confined to zinc and cadmium.[4–5] A similar reaction (Scheme 2) results in the formation of the first nonionic tetraalkylberyllate (**3**) but this has not been studied in detail.[6]

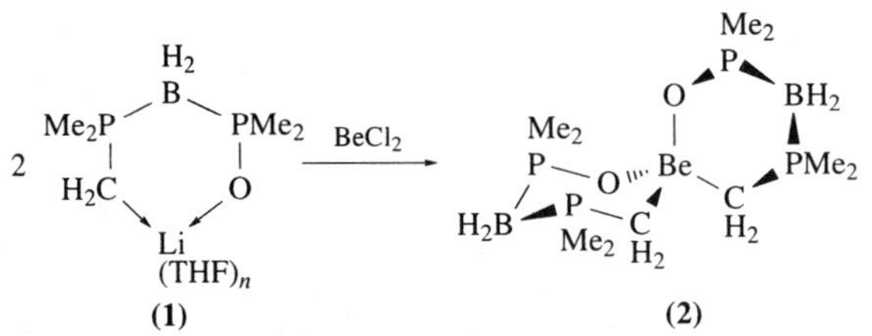

Scheme 1

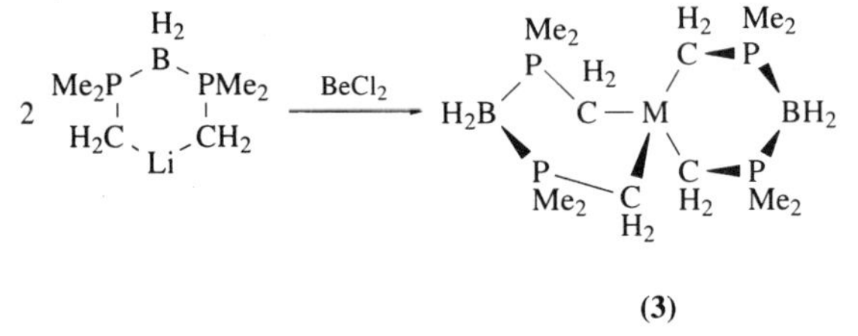

M = Be

Scheme 2

Reaction between dimethylsulfoxonium methylide and beryllium chloride produces a 4:1 complex for which the monomeric tetrahedral arrangement (**4**) has been proposed on the basis of conductimetric and ^{1}H NMR spectral evidence.[7] However, in view of the affinity of beryllium for hard oxygen donors, this formulation is somewhat doubtful and electron donation from sulfoxide to beryllium is more rational.

A novel crystalline polymeric organoberyllium compound results from the reaction of the anthracene radical anion with beryllium chloride which did not form an extended array of C–Be π bonds, as was expected. Lithium anthracene (radical anion) in the presence of excess anthracene reacts with beryllium chloride in THF to precipitate a yellow nonvolatile, insoluble air-sensitive compound which strong chelating agents could not solubilize. Although elemental analysis showed the presence of beryllium, the overall stoichiometry was unsatisfactory and so indirect characterization was achieved by hydrolysis, yielding 9,9',10,10'-tetrahydro-9,9'-bianthryl (**5**). NMR spectroscopic assignment of the pseudoequatorial deuterium atoms, combined with the assumed hydrolysis stereochemistry of the deuterium-

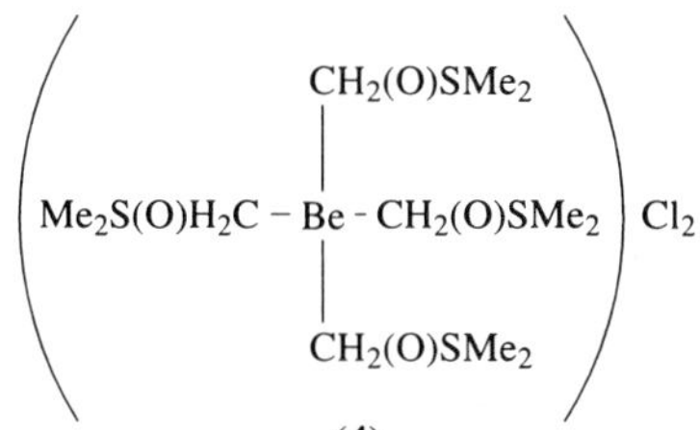

(4)

labelled analogue (**6**) provide a consistent structural hypothesis for the beryllium–anthracene polymer being formulated as (**7**). Although the polymer appears to contain two-coordinate beryllium, as found with bulky groups bonded to beryllium, additional coordination may occur with loosely bound THF molecules to explain the nonstoichiometric elemental analysis.[8]

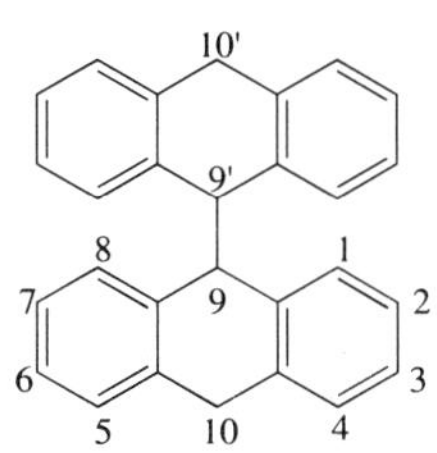

(5)

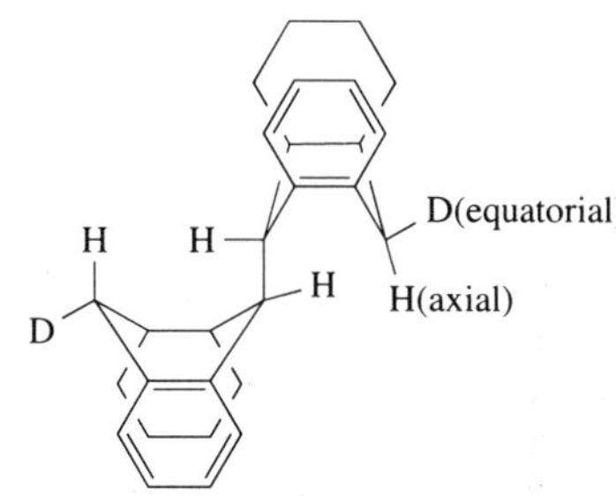

(6)

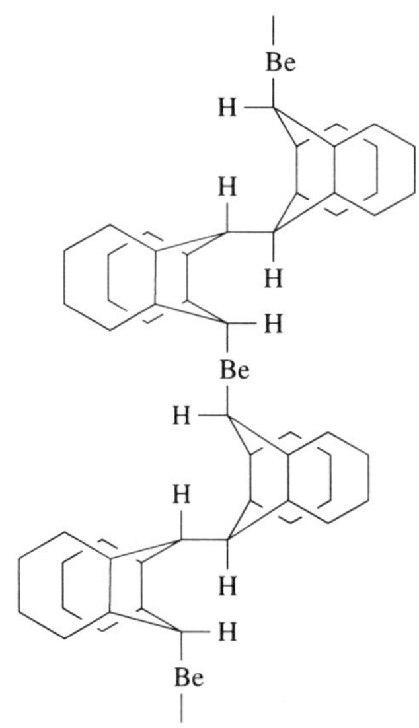

(7)

When lithium anthracenide reacts with beryllium chloride in the absence of excess anthracene, an insoluble product similar in appearance to the bianthryl complex is formed. However, hydrolysis yields the dialkylated species 9,10-bis(4-hydroxybutyl)-9,10-dihydroanthracene (**8**) resulting from cleavage of THF, facilitated by the strong Lewis acidic character of beryllium chloride, by the lithium anthracene dianion before reaction with beryllium chloride to form a beryllium alkoxide. Thus, in order to produce the bianthryl polymeric beryllium material, the cleavage reaction must be thwarted by the use of excess anthracene to act as an 'electron diluent' and shift the equilibrium towards the monoanion radical (Equation (1)).[8]

Marynick has carried out MO calculations on a large number of monomeric and dimeric compounds and found that molecules of type CpBeX (X = H, Me, Ph or F) should exist as monomers in the gas phase. Both Me and fluorine in the terminal position tend to destabilize dimers regardless of the bridging group (H, Me or Ph). While Ph is a highly favoured bridging group, BH_4 provides only small dimer stabilization.[9] Beryllium-9 NMR spectroscopic studies provide no support for the existence of the dimeric species $(MeBH_4)_2$, suggested[3] from IR studies to be present in equilibrium with monomer.[10]

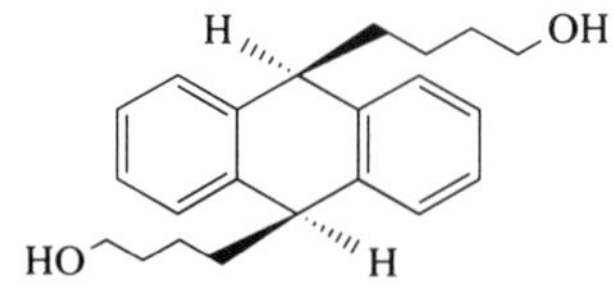

(8)

$$2A^- \rightleftharpoons A + A^{2-} \qquad (1)$$

2.2.2 Alkenyl Derivatives

A series of bis(dienyl) metal compounds of beryllium and zinc has been prepared[11] by metal exchange reactions involving the potassium pentadienide and metal chloride in THF in a 2:1 ratio for five different dienyl anions of open-chain structure (**9**) and two cyclic ones. The formation of the beryllium compounds requires temperatures above 40 °C, whereas the zinc analogues are much more readily formed at 0 °C. The products were not fully analysed but the NMR spectra and GLC of the hydrolysate showed the presence of two THF molecules per metal atom. Crystalline 1:1 coordination compounds, which are monomeric in benzene solution, are obtained on addition of TMEDA. When a 1:1 molar ratio of reactants is used, although RZnCl·2THF can be isolated, only $R_2Be \cdot 2THF$ is obtained, presumably due to the tendency of RBeCl to disproportionate to R_2Be and $BeCl_2$. The beryllium compounds are much more stable than those of magnesium[12] and zinc, and the greater stability of the beryllium and magnesium compounds, relative to zinc, is also found for the octadienyl as well as the pentadienyl series. The ^{1}H NMR spectra of all the pentadienyl metal THF complexes show simple averaged AB_2X patterns due to rapid 1,3 rearrangement and the chemical shifts of the protons at C-1 and C-3 increase with increasing covalent character of the M–C bond (Zn–C > Be–C > Mg–C). This fluxionality can be frozen out at room temperature by the addition of chelating diamines, and bis(dienyl)beryllium compounds of open-chain structure from the pentadienyl, 2-methyl and 3-methylpentadienyl anions assume the terminally σ-bonded (*E*)-2,4-pentadienyl structure (**10**) in solution as evidenced by ^{1}H and ^{13}C NMR spectroscopy. However, the zinc compounds exist as *E–Z* mixtures of terminally bonded structures (**10**) and (**11**). The 2,4-hexadienide and 2,4-dimethylpentadienyl compounds exist as mixtures of σ-terminally bonded isomers. Surprisingly, hydrolysis studies of the TMEDA complexes yielded predominantly 1,4 dienes, rather than the 1,3 isomers.

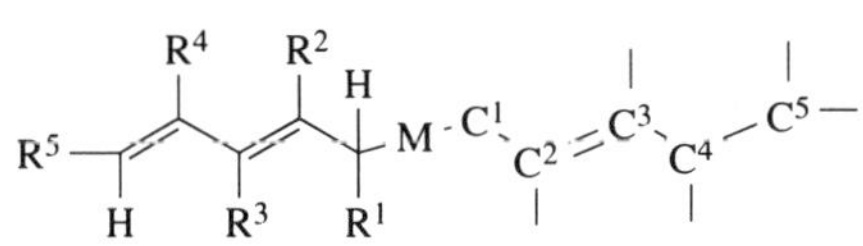

(9)

$R^1 = R^2 = R^3 = R^4 = R^5 = H$

$R^3 = Me, R^1 = R^2 = R^4 = R^5 = H$

$R^4 = Me, R^1 = R^2 = R^3 = R^5 = H$

$R^5 = Me, R^1 = R^2 = R^3 = R^4 = H$

$R^2 = R^4 = Me, R^1 = R^3 = R^5 = H$

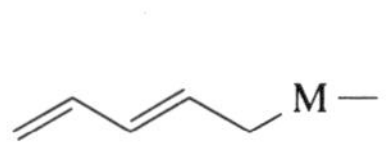

(10)

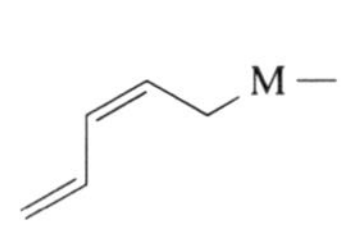

(11)

Bis(cycloheptadienyl)beryllium and the cycloctadienyl analogue are fluxional at room temperature. Both assume the (*Z*)-2,4-dienyl-type structure (**12**), contrary to electron density distribution considerations, and show unexpectedly simple ^{1}H NMR spectra. This can be explained by the rapid transannular 1,5 metallotropic rearrangement induced by σ–π conjugation, arising from the close proximity of C-1 and C-5 in the twisted structure.[11]

(12)

2.2.3 Alkynyl Compounds

Both coordination and anionic complexes of beryllium alkynyls have been studied and theoretical studies predict berylliryne (**13**) to be a stable molecule.[13]

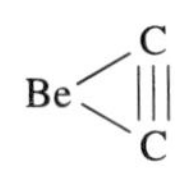

(13)

Alkyl complexes of general formula R_8Al_2M (M = Mg, Ca, Sr or Ba) are known, the former existing in solution in equilibrium with a complex of formula $(R_5AlMg)_2$.[14] An even weaker complex is formed with Me_2Be and Me_3Al which cannot be isolated from solution in which it exists in equilibrium with the reactants.[3] However, in view of the better bridging nature of alkynyl groups compared with alkyl groups, it is not surprising that the reaction between $Me_2(MeC{\equiv}C)Al$ and $(MeC{\equiv}C)_2Be$ in 1,4-dioxane yields tetra-μ-propyn-1-yl(tetramethyl)dialuminoberyllium (Equation (2)) and the corresponding magnesium complex may be similarly prepared. The beryllium compound is monomeric in benzene and the low conductance indicates its nonionic nature. Although it is involatile at room temperature, it decomposes at raised temperature *in vacuo* and dioxane does not coordinate to it, in contrast with the precursors, indicating coordinative saturation in the complex. Only the 1:2 complex can be isolated and when the reactants are mixed in a 1:1 ratio the product contains free $(MeC{\equiv}C)_2Be$ together with the 1:2 complex. Of three possible structural models proposed, only the one (**14**) which minimizes steric hindrance between the methyl groups on aluminum and that of the propynyl group, satisfies the IR and ^1H NMR spectral data. Thus, it appears that alkynyl groups bridge the two metal centres with three-centre, two-electron bonds, together with some interaction between Al(sp^3) hybrid atomic orbitals and the π(C≡C) groups.[15] In view of other examples, particularly in group 13 chemistry where there is clear structural evidence for the bridging ligand acting as a three-electron donor so that it is strongly σ-bonded to one metal atom and weakly π-bonded to the other, resulting in the bridging carbon atoms being asymmetrically placed between two metal atoms,[14] crystallographic studies of this compound would have been particularly interesting.

$$[Me_2AlC{\equiv}CMe]_2 + (MeC{\equiv}C)_2Be \longrightarrow Me_4(MeC{\equiv}C)_4Al_2Be \qquad (2)$$

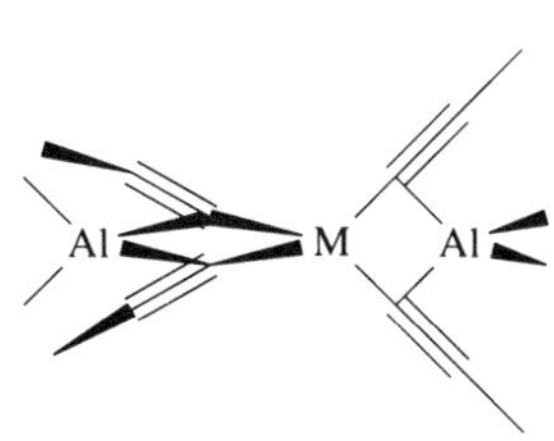

(14)

2.3 COORDINATION COMPLEXES

Evidence for π-type interactions similar to those mentioned above is less clear for group 2 systems. In $[MeBe(C{\equiv}CMe)NMe_3]_2$, although the bridging propynyl groups are tilted slightly towards one of the beryllium atoms, the structure is best considered as being essentially of type (**15**) with the propynyl group acting as a one-electron donor.[3] Although there is no experimental evidence for π-bonding between a R_2Be unit and a simple alkene or alkyne ligand, x-ray diffraction studies of the dimeric 1:1 coordination complex formed between $(MeC{\equiv}C)_2Be$ and Me_3N provide very strong evidence for π-type interactions involving the bridging alkynyl groups (**16**). Not only is this compound one of a limited number containing the Be_2C_2 ring system, but the molecule also contains two independent centrosymmetric dimers with the structural parameters of both (**15**) and (**16**), thus affording a unique opportunity for comparison of the two different types of bonding within the same molecule. In both structures (dimers A and B Figures 1 and 2),[16] the Be_2C_2 rings are crystallographically constrained to be planar, leading to a highly distorted tetrahedral coordination about beryllium (angles ranging from 103.4° to 115.0° in dimer A, and from 96.2° to 121.1° in dimer B). Within the two dimers the μ-alkynyl groups are found to adopt quite different orientations with respect to the beryllium atoms (Table 1). Thus, in dimer B, not only does the bridging carbon C-1 adopt a marked asymmetric position with respect to the metal atoms, but the Be–μ(C)–CMe angles are highly asymmetric, affording a closer side-on approach of the alkynyl groups to the second beryllium atom. One result is that the bonding in dimer B has a significantly higher contribution from (**16**) than for dimer A. This is supported by comparing the $Be–C_{bridge}$ and $Be–C_{terminal}$ bond lengths. Whereas in dimer A the Be–C-1 distance is significantly longer (electron-deficient) than the terminal bond length Be–C-4, in dimer B the values are very similar, but no significant differences in alkyne bridging bond lengths are found. The greater the contribution from (**16**) to the bonding, the more electron-precise is the Be_2C_2 ring, thereby reducing the need for Be···Be interactions. In dimer B this distance is the longest known for Be_2X_2 bridged systems (apart from $BeCl_2$) and is significantly longer than in dimer A. This provides further evidence that the Be_2C_2 bridge in dimer B is the more electron-precise, and results in an opening up of the C-1–Be–C-1' angle from 96.2° for dimer B to 103.4° in dimer A. The crystallographic data therefore show that dimer A has a predominantly electron-deficient constitution of type (**15**) in which significant cross-ring metal–metal bonding is present, as shown in band structure calculations[17] for $(Me_2Be)_n$. Dimer B, however, tends towards an electron-precise type-(**16**) structure with the bridging alkyl groups acting as three-electron donors, being σ-bonded to one beryllium atom and π-bonded to the other. Packing effects do not explain the presence of these two structural types and MO calculations show that the Be'–C-2 bond index in dimer B is greater than in dimer A.[18] However, other MO calculations indicate that the alkyne MOs are not strongly involved in π-bonding to the metal.[19]

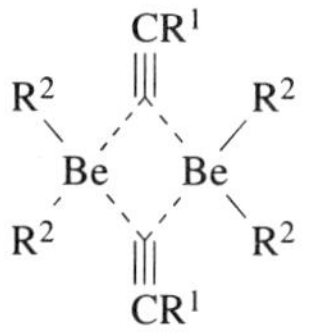

(15)

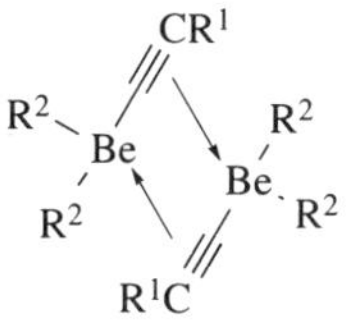

(16)

Table 1 Bond lengths (nm) for $[(MeC{\equiv}C)_2BeNMe_3]_2$.

Parameter	*Figure 1 (dimer A)*	*Figure 2 (dimer B)*
Be–C-1	0.1836(6)	0.1763(7)
Be–C-1'	0.1904(6)	0.2042(6)
Be–C-2	0.2812(7)	0.2538(7)
Be–C-4	0.1719(6)	0.1735(6)
Be···Be'	0.2319(6)	0.2549(6)
Be–C-1–C-2	0.1549(4)	0.1686(4)
Be'–C-1–C-2	0.1285(3)	0.1002(3)

Coordination compounds of reduced *N*-heterocycles have been identified for beryllium and the organometallic radical complexes of reduced pyrazine[20] and the 2,2'- and 4,4'-bipyridyls[21–2] have been characterized by ESR spectroscopy. A solution of 4,4'-bipyridyl and diphenylberyllium in THF forms a weakly coloured diamagnetic solution. On contact with a potassium mirror, reduction occurs to yield a very persistent and intensely blue 1:2 radical complex (the metallaviolene) (**17**) which was characterized by its metal isotope hyperfine splitting in which the twofold coordination of the organoberyllium cation leads to a septet.[22] Pyrazine and Ph_2Be also undergo electron transfer on reduction with potassium or sodium to form the 1:2 anion radical complex (**18**) with the phenylmetal cation (Scheme 3). In contrast, Ph_2Mg forms a 1:1 radical complex while the higher homologues for zinc, cadmium and mercury lead to reduction to metals.[20] A 1:1 complex (**19**) is also formed for beryllium with 2,2'-bipyridyl where the stereochemistry of the ligand allows chelation of the metal to take place. Although in all cases structures have been depicted showing two-coordinate beryllium, no doubt coordinative saturation of the metal by THF also occurs.

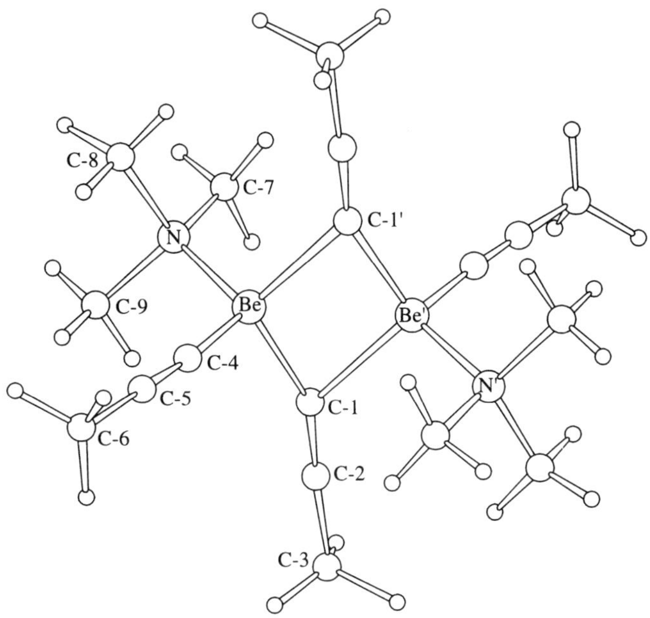

Figure 1 Structure of [MeC≡C)$_2$BeNMe$_3$]$_2$, type (**15**), dimer A (reproduced from *J. Organomet. Chem.*, 1984, **273**, 183).

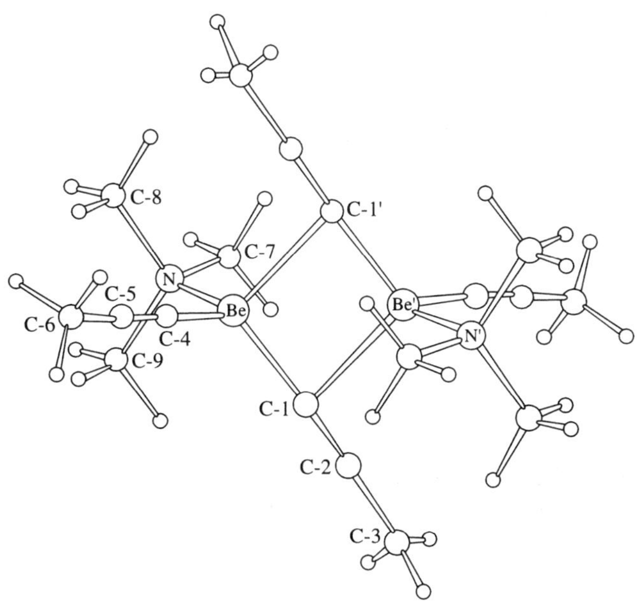

Figure 2 Structure of [MeC≡C)$_2$BeNMe$_3$]$_2$, type (**16**), dimer B (reproduced from *J. Organomet. Chem.*, 1984, **273**, 183).

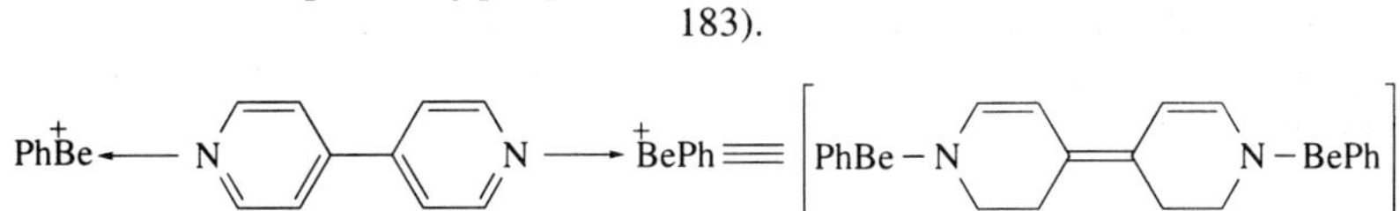

(17)

A range of similar complexes has been prepared with a variety of other main group organometallic cations, and ESR coupling constants correlate excellently with results from HMO calculations to exemplify the continuous variation from genuine paramagnetic ion pairs to mostly covalently bonded radical complexes. The metal coupling constants of the metallaviolenes and the corresponding pyrazine complexes exhibit a ratio of approximately 1:2; however, such a general proportionality does not exist for the parameters of the corresponding 2,2'-bipyridyl radical complexes where chelate effects play an important role.[20–2]

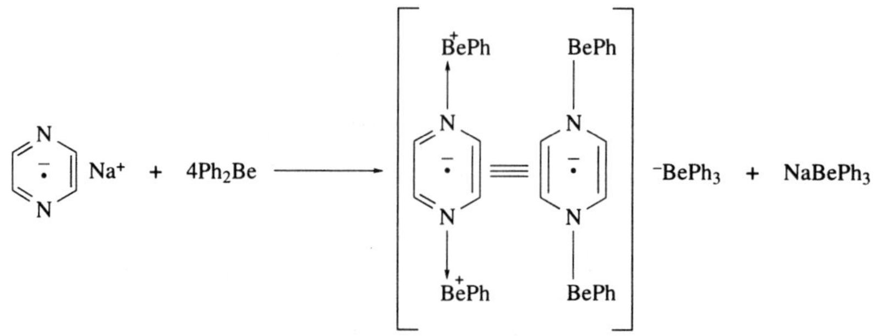

(18)

Scheme 3

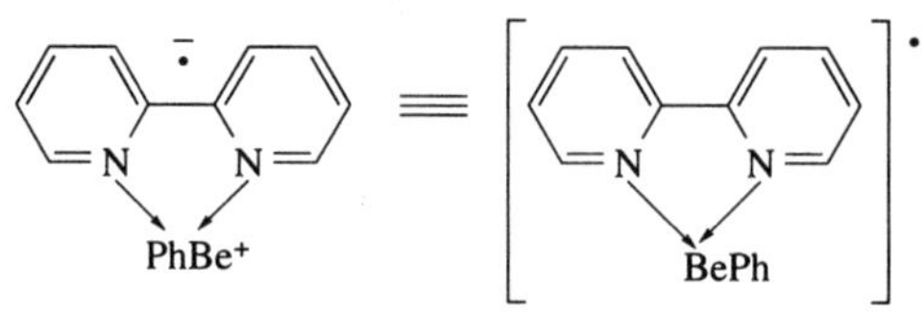

(19)

2.4 ANIONIC COMPLEXES

Only one anionic complex has been prepared since the last survey, namely lithium tri-*t*-butylberyllate; its structure has also been determined.[23] The only previous structural characterization of an ionic organoberyllium complex was for Li_2BeMe_4, deduced from powder diffraction data[3] as distorted tetrahedral anionic around beryllium. Di-*t*-butylberyllium and *t*-butyllithium react rapidly in equimolar quantities in pentane at room temperature to form needles of $LiBeBu^t_3$. When the reaction is performed in ethereal solvents the ^{9}BeNMR spectral resonance of this solution is ~2 ppm upfield from that of $LiBeBu^t_3$ in ether-free solvents, but the nature of the product in ethereal solvents has not been established.

The structure of $LiBeBu^t_3$, determined by x-ray diffraction methods at −149 °C, shows the presence of a dimeric unit (Figure 3) in which the geometry around beryllium is trigonal planar, with beryllium and three tertiary carbon atoms of the *t*-butyl groups lying in one plane; the planarity is clearly indicated by the sum of the three C–Be–C angles being 359.9°. The lithium atom resides above this plane in a cavity, having close contacts with Be, C-2, C-3, C-8, C-10, H-8B, H-8C and H-10B, and with C-6A, H-6A and H-6B of the second molecule. The Be–C distances are not all equal: Be–C-1 (0.1812(4) nm) is similar to those found in $(NaBeEt_2H{\cdot}OEt_2)_2$ and Li_2BeMe_4[3] but is shorter than the Be–C-2 and Be–C-3 distances (0.1854(4) and 0.1864(3) nm, respectively) since the latter are both involved in interactions with the lithium atom which also results in a distortion of the trigonal planar angle to 117.1(2)°. The Be–Li distance is 0.2227(6) nm and the four Li–C distances within the molecular unit (average 0.2228 nm) are similar to values in MeLi and EtLi, and are shorter than the Li–C-6A distance of (0.2407(4) nm) to the second molecular unit of the dimer. The exact nature of the interaction is not clearly understood because, while the trigonal planar arrangement of alkyl groups around beryllium indicates that lithium has an electrostatic interaction with the organoberyllium anion, the fact that the compound sublimes and dissolves in hydrocarbon solvents suggests a more covalent interaction.[23]

2.5 ALKOXY, AMINO, THIO AND PHOSPHINO COMPOUNDS

In the 1960s and 1970s elegant association studies by Coates dominated both this area and the coordination chemistry of beryllium. Since then, little progress has been made in this area and only one compound of this type, a diorganophosphino beryllium, has appeared in the literature as a preliminary communication.[24] It is surprising that diorganophosphino derivatives of beryllium have remained elusive, whereas so much is now known about transition metal analogues, and alkali metal derivatives are widely used as PR_2-transfer reagents.

The reaction between Cp*BeCl and $LiPBu^t_2$ at −78 °C in ether results in the formation of crystalline $Cp{*}BePBu^t_2$ (Equation (3)) and x-ray diffraction studies show the compound to be monomeric in the

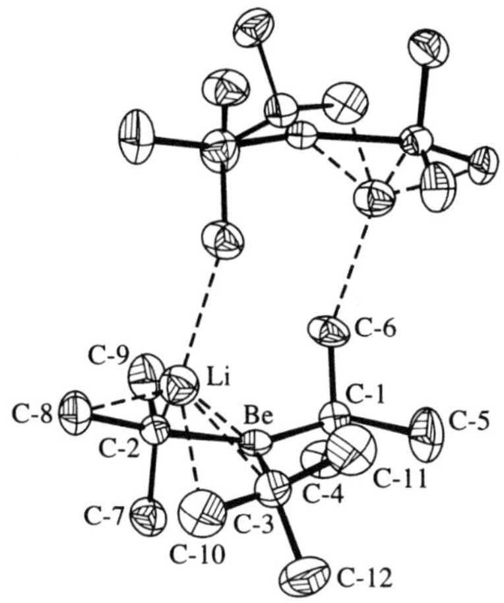

Figure 3 Structure of $LiBeBu^t_3$ (reproduced by permission of the American Chemical Society from *Organometallics*, 1988, **7**, 2421).

solid state (Figure 4). This is unique for $R^1BeXR^2_n$ systems (X = O, S; n = 1 or X = N; n = 2), doubtless due to the steric bulk of both the Cp* and Bu^t_2P moieties and the coordination characteristics of the former. The Cp* unit is bound in the familiar η^5 fashion, while the phosphorus atom has a trigonal pyramidal geometry with an average Be–P–C angle of 105.1° and thus behaves as a one-electron donor giving beryllium an electron count of eight. The Be–P distance of 0.2083(6) nm has no precedent and the P–Be–$C_{centroid}$ angle of 168.3(7)° reflects some steric strain between the Bu^t groups and the methyl substituents on the Cp* ring.

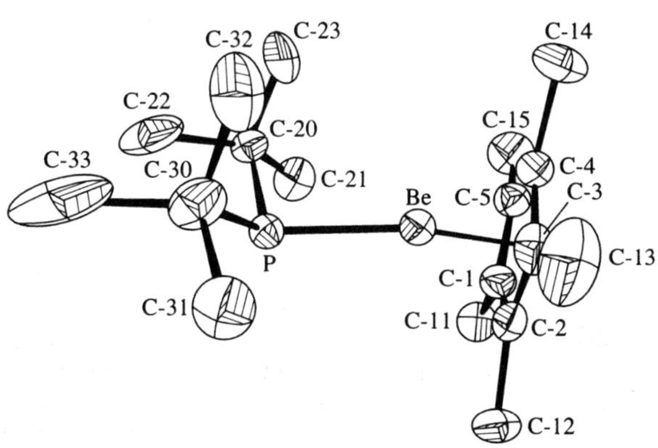

Figure 4 Structure of $Cp^*BePBu^t_2$ (reproduced by permission of the Royal Society of Chemistry from *J. Chem. Soc., Chem. Commun.*, 1990, 693).

$$Cp^*BeCl \ + \ LiPBu^t_2 \longrightarrow Cp^*BePBu^t_2 \ + \ LiCl \qquad (3)$$

The $^{31}P\{^1H\}$ NMR spectrum in C_6D_6 at ambient temperature consists of a 1:1:1:1 quartet at $\delta = 0.07$ (relative to H_3PO_4) with a large Be–P coupling constant of 50.0 Hz in contrast with the much smaller values, typically 4–6.2 Hz, for $^1J_{Be-P}$ coupling in beryllium–phosphine adducts.

The phosphorus atom acts as a Lewis base and reacts with $(Me_3Al)_2$ to produce a compound formulated as $Cp^*BePBu^t_2{\cdot}AlMe_3$ for which even larger Be–P coupling exists ($^1J_{Be-P} = 67$ Hz). On heating the complex in a sealed tube for 2 h, disproportionation occurs into Cp*BeMe and $[Me_2Al(\mu\text{-}PBu^t_2)]_2$, while reaction with ethanol yields Cp*BeOEt and Bu^t_2PH.[24]

2.6 CYCLOPENTADIENYLBERYLLIUM COMPOUNDS

2.6.1 Dicyclopentadienylberyllium and Related Compounds

The compound Cp_2Be, commonly called beryllocene, is the lightest member of the ferrocene family of compounds Cp_2M^{II}. Its structure has a long and controversial history since its initial preparation in 1959. Spectral and diffraction studies have often been in conflict and, until recently, there has been no consistent description of its structure and bonding in solid, solution and gaseous phases.[3]

A redetermination of its solid-state structure at −145 °C using a four-circle diffractometer[25] confirms the earlier structure[3] obtained from photographic data at −120 °C and these more precise bonding parameters limit the possible descriptions for the molecular bonding present. The structure (Figure 5) shows that the two rings are parallel and crystallographically indistinguishable, even at this low temperature, due to disorder of the beryllium atom between the two sites (Be and Be') which are separated by 0.1303 nm. The beryllium atom lies above the centre of one ring and almost directly below carbon atom C-5 of the other ring at a distance of 0.1891(6) nm. In the peripherally bonded ring, C-5 and H-5 deviate significantly from the plane formed by C-1–C-4 with C-5 moved closer to Be, while H-5 is bent away; both features are indicative of a C-5–Be bonding interaction. Additionally, the C-5–C-1 and C-5–C-4 bond lengths are significantly longer than others in the ring. Calculation of the bond lengths in the peripherally bonded ring in the absence of disorder show the alternation (Figure 6), clearly consistent with the ring having partial diene character, thus ruling out η^3 bonding with allyl character. Although the inequality in bond lengths shows the same pattern as in cyclopentadiene itself and in other σ-bonded metal cyclopentadienyls, it is not as marked. Thus, it appears that the delocalized structure of the peripheral ring is only slightly perturbed towards a σ-bonded structure, consistent with the small angles between beryllium and the ring (Be–C-5–C-1, 94.4° and Be–C-5–C-4, 91.4°) so that the ring has essentially an sp^2 hybridized carbon atom with the bond to beryllium involving a p_π orbital. The structural evidence therefore indicates that the peripherally bonded ring retains a largely π-delocalized character with the negative charge only partly localized on the carbon directly bonded to beryllium. This accounts for the absence of characteristic diene features in the Raman spectrum.[25]

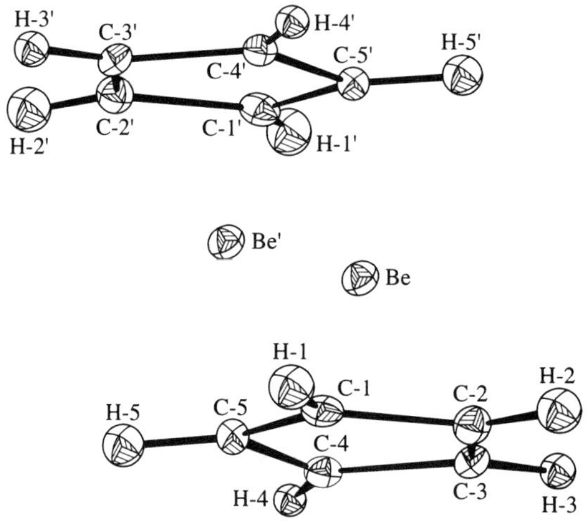

Figure 5 Structure of beryllocene at −145 °C (reproduced by permission of the Commonwealth Science and Industry Research Organisation from *Aust. J. Chem.*, 1984, **37**, 1604).

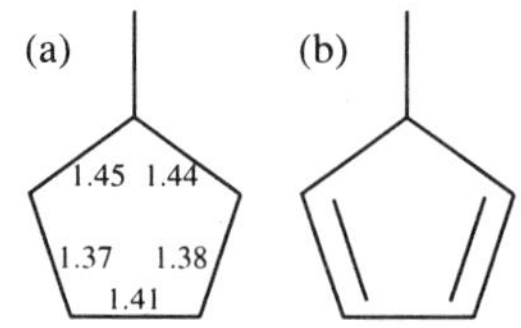

Figure 6 Ordered structure for the peripheral ring of beryllocene showing (a) estimated C–C bond lengths, (b) bonding character (reproduced by permission of the Commonwealth Science and Industry Research Organisation from *Aust. J. Chem.*, 1984, **37**, 1605).

Although early electron diffraction data indicated that Cp_2Be had quite a different structure in the vapour phase, a redetermination showed good agreement with the solid-state model. However, vapour-phase infrared data differed from the solid state and solution spectra and were interpreted[3] on the basis of the ionic model $Cp^-Be^{2+}Cp^-$. A reinvestigation of the infrared spectrum of Cp_2Be in the vapour phase differs from that reported previously.[3] The spectrum is much weaker than that of CpBeCl and shows seven bands in the 3200–3000 cm^{-1} region, thus excluding the original C_{5v} unsymmetrical sandwich structure and a σ–π-type structure. This complexity of bands in the C–H stretching region indicates that the solid state slipped sandwich structure persists in the vapour phase with Be–C stretching modes occurring at 598 cm^{-1} and 313 cm^{-1} for the η^1 and η^5 rings, respectively. Lack of bands in the 1500–1600 cm^{-1} region indicates the absence of full double bonds in the rings. Bands at ~740 cm^{-1} and 970 cm^{-1} are considered to be diagnostic of the presence of a peripherally bonded ring and have necessitated reassignment of similar bands observed in the solution spectrum.[26]

The dipolar nature of Cp_2Be indicates that it does not have the typical ferrocene type structure in solution. A more recent study of its infrared spectrum shows agreement with earlier results, with no change on varying the solvent and is similar to that of the gas phase. By comparison with other pentahapto and monohapto metal cyclopentadienyls, it can be concluded that beryllium is pentahapto bonded to one ring and monohapto to the other in solution.[27] Thus, the similarity in intensity and positions assigned to the peripherally bonded ring in the infrared spectrum of Cp_2Be in solid, solution and vapour phases indicate that the (η^1-Cp)(η^5-Cp)Be structure does not differ significantly between the different states.

In view of the flexible nature of Cp_2Be in solution due to the dynamic behaviour of the beryllium atom which oscillates between two equivalent positions relative to the rings and of the rings themselves which interchange and are fluxional,[3] a reinvestigation of the dipolar nature in solution was deemed necessary by a microwave dielectric method.[28] Previous assumptions made to determine the dipole moment were felt to be invalid. Results prove that Cp_2Be has a significantly smaller dipole moment than that obtained earlier, although it clearly does not have the nondipolar ferrocene-type structure. A model with nonparallel rings is believed to account better for the measured dielectric absorption in solution, and is also favoured from theoretical studies. No evidence could be found for the existence of the 2:1 benzene complex $[Cp_2Be]_2{\cdot}C_6H_6$, proposed on the basis of NMR spectroscopic studies, nor for the formation of a coordination complex with dioxane.[3] These results further attest to the coordination saturation of beryllium in Cp_2Be, although solvating benzene or dioxane molecules may be partly organized through statistically favoured stereospecific solute–solvent interactions. The stereospecific solvation by benzene is well illustrated with CpBe halides in which the cyclopentadienyl protons experience a large benzene-induced upfield NMR spectral shift.[28]

Earlier NMR spectroscopic studies have shown that Cp_2Be is fluxional in solution with the low-temperature ^1H NMR spectrum comprising a single resonance, indicating that the protons of both rings are magnetically equivalent and implying that movement of the beryllium atom between the two sites is rapid, even at −135 °C.[3] The ^{13}C NMR spectrum in both cyclohexane and ether also displays a single resonance with fine structure, however, arising from incomplete decoupling of the quadrupolar ^9Be nucleus ($I = 3/2$). The ^{13}C proton-decoupled NMR spectrum at different temperatures indicates that inversion of the molecular dipole caused by exchange of beryllium between the two rings (i.e., the monohapto becomes pentahapto and vice versa) occurs with a correlation time of the order of 10^{-10} s and an apparent activation energy of 5.2 kJ mol^{-1} at room temperature, and this provides for the first time an order of magnitude of the beryllium inversion frequency in solution. Inversion occurs by the interchange of the centrally bonded η^5 and peripheral η^1 bonding roles of the two rings and causes hydrogen and the carbon atoms each to be dynamically averaged in their respective NMR spectra, resulting in the equivalence of all carbon and hydrogen atoms on the NMR spectroscopic timescale at all accessible temperatures. Migration of the beryllium atom around the peripherally bonded ring can occur by the extremely rapid mechanism of ring interchange combined with rotation of the centrally bonded ring. This leads to scrambling of the positions in the peripherally bonded ring with a timescale similar to that for the interchange of the ring roles. The rapidity of the movement of the beryllium atom and the low activation energy indicates that the fluxional process of this compound in solution cannot be reduced by lowering the temperature in the accessible range in order to obtain the static NMR spectrum.[29–30]

Beattie and Nugent have reviewed and rationalized the variety of structures found for main group metal cyclopentadienyls, from centrally bonded ferrocene-type compounds to peripherally bonded compounds, with a continuous range of structures in between (Figure 7). In an η^5 structure the metal is positioned at point C where the metal–ring angle is approximately 60°, whereas in the σ-bonded structure the metal is located at point A with a corresponding angle of ~125°. In the case of the beryllium compound, the metal makes an angle of 97° with the peripheral ring, considerably less than if the geometry about C-5 were tetrahedral (i.e., σ-bonded) and, therefore, the beryllium atom is located at position B, intermediate between the two extremes of η^5 and η^1 bonding. As the bonding changes from pentahapto (point C) to monohapto (point A) the metal–carbon bond becomes longer and weaker and the ionic character of the interaction increases while the increased delocalization of the ring causes the carbon–carbon bonds to approach equal lengths and the geometry around C-5 to change from tetrahedral to trigonal so that the C-5–H-5 bond moves into the plane of the ring. The overall energy of the system is composed of two opposing forces: one part is the contribution of the metal–carbon bond which favours peripheral bonding and the other is the delocalization energy of the ring π-system which favours central bonding. The unusual slipped-sandwich structure of beryllocene may result when neither of these two forces is dominant,[31] although solid-angle calculations indicate that steric factors play a role in explaining why the η^5–η^1 structure exists rather than the Cp_2Be structure.[32]

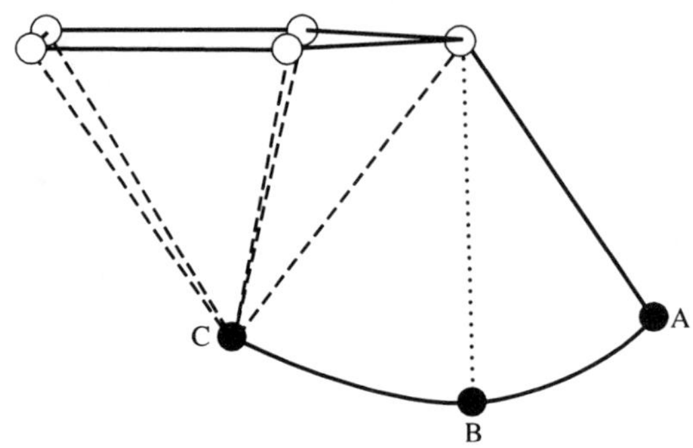

Figure 7 The range of structures from monohapto (A) to pentahapto (C) (reproduced from *Inorg. Chim. Acta*, 1992, **198–200**, 312).

To provide further insight into the beryllocene story, attempts to prepare the more voluminous decamethyl derivative by reaction of MCp^*_n (n = 1, M = Li[33] and n = 2, M = Mg[34]) with $BeCl_2$ produced only the one ring compound Cp*BeCl as a sublimable crystalline material. When Cp*BeCl and beryllocene were mixed in C_6D_{12} or $CDCl_3$, the peaks due to reactants diminished with time and three new peaks appeared which may be explained by the equilibrium of Equation (4) which was reached faster in $CDCl_3$ than in C_6D_{12} and lies to the right in the polar solvent. Attempts to prepare a pure sample of the mixed-ring compound by reaction of CpBeCl with LiCp* in a solvent were hampered by this equilibrium between CpBeCl and the mixed-ring product. To overcome this, the reaction was carried out in the absence of solvent at 60 °C (Equation (5)). In view of the differing bonding natures of the separate rings in beryllocene, CpBeCp* also proved to be an interesting structural problem. The infrared spectrum in C_6D_6 and CS_2 shows virtually all the spectral characteristics in solution of a pentahapto Me_5C_5 group. It may also be reasonably inferred that due to the absence of the two very distinct peaks in CpBeX (X = Cp, Cl or Br) at ~1015 cm^{-1} and ~1125 cm^{-1} and the combinations and overtones between 1800 cm^{-1} and 2700 cm^{-1} (characteristic of a pentahapto ring), that the C_5H_5 ring is not pentahapto. In support of the formulation Cp*Be(η^1-Cp) is the presence of the strongest peak, at ~660 cm^{-1}, assigned to a vibration of the η^1 ring in beryllocene, together with other peaks which closely match those of the η^1 ring in beryllocene. The ^1H NMR spectrum shows the expected 3:1 ratio for the two sharp peaks observed, thus indicating fluxional behaviour of the monohapto ring. Although low-temperature studies were unable to freeze out the dynamic behaviour, peak broadening indicated a slowing down of the ring-whizzing at low temperatures (–138 °C).[33]

$$\text{CpBe} + \text{Cp}^*\text{BeCl} \rightleftharpoons \text{Cp}^*\text{BeCp} + \text{CpBeCl} \quad (4)$$

$$\text{Cp}^*\text{Li} + \text{CpBeCl} \longrightarrow \text{Cp}^*\text{BeCp} + \text{LiCl} \quad (5)$$

Dihydro-1H-azaborolyl rings are isostructural and isoelectronic with the cyclopentadienyl system (**20**). As a further contribution to the beryllocene story Ab_2Be (Ab = 1-t-butyldihydro-2-methyl-1H-azaborolyl) and the mixed-ring compound AbBeCp have been prepared. The former is obtained as a crystalline material by reaction of AbLi and $BeCl_2$ in a 2:1 molar ratio in ether and the latter can only be isolated as a yellow oil by reaction of beryllocene and Ab_2Be in a 1:1 ratio in ether. For Ab_2Be the prochiral nature of the Ab ligand leads to diastereoisomerism. If the two rings in one molecule use the same site for coordination, the resulting isomer has the atomic sequence BN/NB and can be crystallized from ether at –30 °C, while the isomer having different sites, the BN/BN isomer, remains in solution.

The x-ray crystal structure at −150 °C of the BN/NB isomer (Figure 8) shows one η^1 and one η^5 coordinated Ab ring[35], similar to that found for the low-temperature structure of beryllocene[25] but without disorder of the beryllium atoms. Comparison of bond distances of both compounds[35] shows that the η^1-coordinated ring behaves as a diene, as found in the calculated ordered structure of beryllocene,[25] and that the C-3–B and C-3–C-4 distances are longer than in the η^5 ring while C-4–C-5 is significantly shorter. In contrast with beryllocene but in agreement with theoretical calculations on beryllocene,[3] the Ab rings are not coplanar, the interplanar angle being 14.5°, with disorder being absent due to the noncoplanarity of the Ab rings. The Be–$C_{centroid}$ distance of 0.147 nm is somewhat shorter than in beryllocene (0.1505 nm) and the beryllium is bonded in the η^1 ring to the carbon adjacent to boron with a Be–C-3/1 distance of 0.1755 nm, significantly shorter than in beryllocene (0.1826 nm). The crystal structure shows two enantiomeric forms (*R,R* and *S,S*) alternating in the unit cell for the BN/NB diastereoisomer.

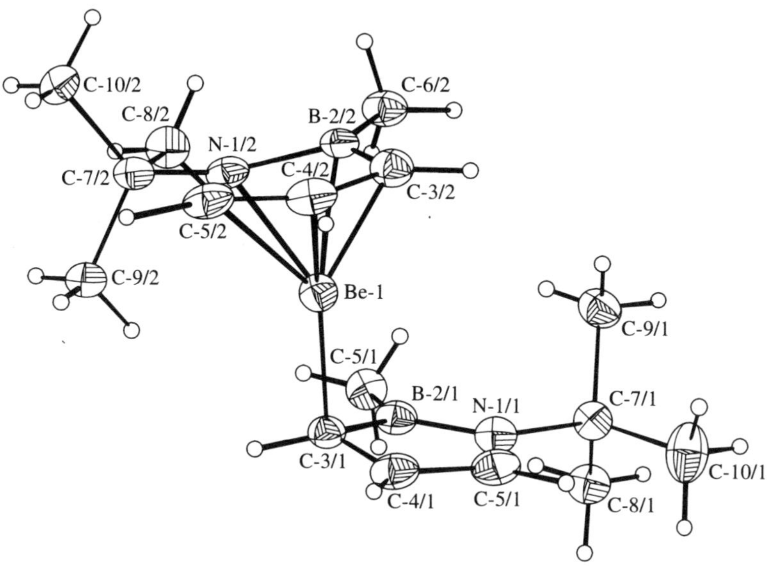

Figure 8 Structure of Ab_2Be (reproduced by permission of the American Chemical Society from *Organometallics*, 1987, **6**, 437).

(20)

The ^{1}H NMR spectrum of Ab_2Be as a mixture of BN/BN and BN/NB isomers in $CDCl_3$ or d_8 toluene at room temperature shows the signals of both compounds simultaneously and shows no indication of the terminally bonded C-3 atom. Solutions of the structurally characterized isomer in $CDCl_3$ or d_8 toluene show only three signals for the ring protons but, after 5 days, the original 1:1 mixture of the two isomers is reformed, an effect only possible if some dissociation of one ring takes place (Equation (6)) to allow ring-flipping; the low conductivity of solutions may explain the slow interconversion.

$$Ab_2Be \rightleftharpoons AbBe^+ + Ab^- \qquad (6)$$

A flip of Ab and/or Cp in AbBeCp solutions is unobservable by ^{1}H NMR spectroscopy as only enantiomers are present. AbBeCp is more stable in THF than Ab_2Be and shows remarkable conductivity, a 0.004 M solution having a conductivity greater than that for LiBr. This is in accordance with the ring-flip observed in Ab_2Be solutions, with dissociation of the η^5-coordinated cyclopentadienyl ring occurring in the mixed ring compound, while the η^1-bonded Ab ring remains bonded to beryllium. With increasing temperature, the Ab ring becomes more η^5-coordinated so that the fluxionality becomes similar to that of Ab_2Be at room temperature. The ^{11}B NMR spectroscopic results of both compounds are in accordance with the ^{1}H NMR spectroscopic results, the chemical shift for Ab_2Be indicating a state between σ-bonding and π-bonding. As a result of these studies it is suggested that beryllocene in solution should exist in equilibrium with $CpBe^+$ and Cp^-, the rate of dissociation depending on the solvent. The fact that in AbBeCp the η^5-bonded cyclopentadienyl ring dissociates, whereas the Ab ring remains η^1-bonded, indicates that η^1 coordination is stronger than η^5, in agreement with the ^1H NMR spectroscopic results that at about 90 °C half of the molecules show η^5-coordinated Ab rings.[35]

2.6.2 Miscellaneous Beryllium Cyclopentadienyls

Although the vibrational spectra of both solid and solution phases of CpBeCl indicated η^5-C_{5v} symmetry, as found from electron diffraction studies on the gas phase, mass-spectral evidence appeared to show the presence of dimeric and trimeric species.[3] However, x-ray crystallographic studies[36] clearly show that there are no intermolecular interactions and that CpBeCl consists of discrete monomeric units, separated by distances greater than the van der Waals radii. The C_{5v} axis of symmetry passes through the centroid of the C_5 ring and the beryllium and chlorine atoms, with Cl–Be–$C_{centroid}$ being essentially linear (178.4(3)°). The Be–$C_{centroid}$ distance (0.1451(5) nm) is shorter than that found (0.1484(7) nm) from electron diffraction studies on the vapour, as are the Be–C distances (0.1867–0.1875 nm cf. 0.1916(6) nm). The absence of association and the failure to form coordination complexes, as is the case for other CpBeX compounds, may be attributed to the ability of the η^5-cyclopentadienyl ligand to satisfy the electronic requirement of the metal. The Be–Cl distance of (0.1869(3) nm) is shorter than that found in the four coordinate $(MeCN)_2BeCl_2$ (0.198 nm) and comparable with the average Be–Cl distance of 0.188(1) nm in $Cl_2Be_3(OBu^t)_4$,[37] and is thus indicative of some π interaction. The hydrogen atoms, although not previously located in the electron diffraction study, do show a slight tendency to bend towards the beryllium atom, in agreement with theoretical studies.[38]

For both CpBeCl and the corresponding hydride the HOMO derived from SCF-X_α scattered wave calculations (X_α-SW) have been computed to be a doubly degenerate orbital that is located on the Cp ring and is therefore principally a Cp–ring π-MO with minor contribution from the beryllium moiety. The lone pair and σ_{Be-X}MOs participate only to a minor extent in interstitial bonding and thus it is suggested that the electronic structure of these simple molecules is more consistent with a six- rather than eight-electron counting procedure. Revision of the UV photoelectron spectroscopic (PES) assignments for CpBeCl has been suggested on the basis of these X_α-SW calculations,[39] and the MO sequence for derivatives of the type CpBeR have also been investigated by the partial retention of diatomic differential overlap (PRDDO) method.[9]

Reaction of Cp*BeCl with MeLi results in the formation of Cp*BeMe for which the dipole moment lies in the direction of the methyl group so that the methyl anion can then act as a σ-donor to a Lewis acid. Thus, $Cp^*_2Yb(\mu\text{-}Me)BeCp^*$ may be prepared as dark orange prisms from Cp*BeMe and Cp^*_2Yb in pentane. The x-ray crystallographic structure (Figure 9) at −80 °C shows that the complex may be viewed as being formed by bringing up the Cp*BeMe fragment to Cp^*_2Yb in a linear fashion, the Yb–C(Me)–Be angle being 177.2(3)°. The geometry of the fragments in the solid state is only very weakly perturbed relative to the individual parts, the principal perturbations being that the Yb–CCp* distance lengthens by 6 pm and the angles between the ring centroids around ytterbium close by ~13°. The most interesting feature is the geometry of the methyl group which bridges the two metal centres. The hydrogen atoms on C-1 are close to the ytterbium atom with an average Yb···H distance of 0.259(8) nm, with angles around the bridging atom of Yb–C–H (96.7(5.7)°), Be–C–H (110.6(4.7)°) and H–C–H (108.3(8.2)°). Thus, the structural data are consistent with the HOMO of Cp*BeMe acting as a σ-donor to the Lewis acid Cp^*_2Yb (Figure 10). The ^1H NMR spectrum in solution shows the bridging methyl group is deshielded only by ~0.1 ppm relative to the beryllium reactant, indicative of the weak nature of the interaction, and the complex undergoes rapid intermolecular chemical exchange at −70 °C.[40]

CpBeCl has been used as a beryllium insertion reagent into the bridging position of pentaborane(9) to form $[\mu\text{-}(Cp)Be]B_5H_8$ as a low-melting colourless solid (Equation (7)). The ambient ^{11}B FT NMR spectrum is appropriate for a μ-substituted B_5H_9 derivative consisting of three doublets of intensity 2:2:1. The most notable feature of the spectrum is the large upfield shift of the resonances associated with two of the basal boron atoms, suggesting a higher concentration of electron density while the other chemical shifts are not significantly different from those of B_5H_9, although the average of the chemical shifts for the two sets of basal boron resonances is the same as that observed for $B_5H_8^-$ salts. The static structure, solubility in hydrocarbons and volatility, however, confirm the covalency of the B–Be–B bonding. The x-ray structure (Figure 11) provides more information than is discernible from the NMR spectral data. The shortening of the B-2–B-3 distance suggests a high concentration of electron density in this region, in agreement with the assignment of the high-field resonance of the boron atoms adjacent to beryllium in the ^{11}B NMR spectrum. The tilt of H(4–5) under the base of the square pyramid and of

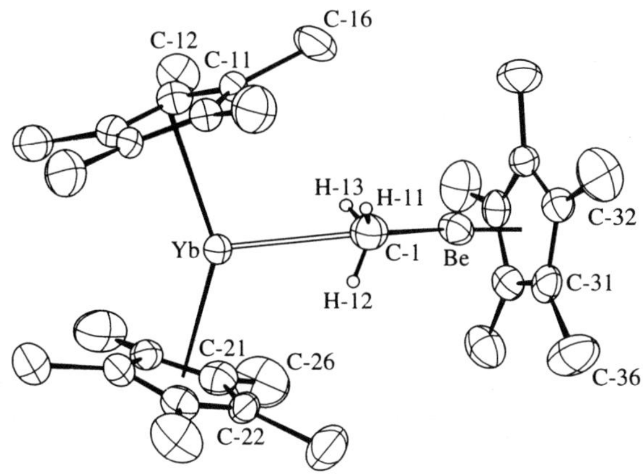

Figure 9 Structure of $Cp^*_2Yb(\mu\text{-}Me)BeCp^*$ (reproduced by permission of the American Chemical Society from *J. Am. Chem. Soc.*, 1987, **109**, 5854).

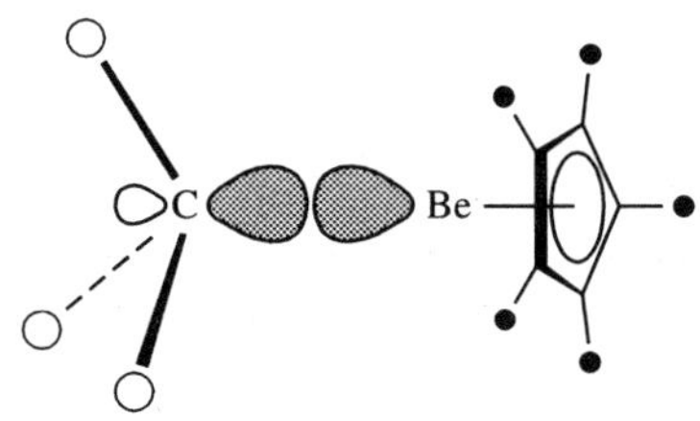

Figure 10 Interaction of HOMO of Cp*BeMe with Cp*Yb (reproduced by permission of the American Chemical Society from *J. Am. Chem. Soc.*, 1987, **109**, 5855).

H-1 and Cp–Be towards one another is most likely due to unknown electronic effects.[41] Again, the 4° tilt of ring hydrogen towards beryllium is consistent with the prediction of a calculation on CpBeH that the compactness of the beryllium orbitals causes such a distortion in order to maximize overlap of ring orbitals with those on beryllium.[38] The Be–C distances (0.1877(3)–0.1894(3) nm) are in the same range as found in other beryllium cyclopentadienyls.[3]

$$KB_5H_8 + CpBeCl \longrightarrow (\mu\text{-}CpBe)B_5H_8 + KCl \qquad (7)$$

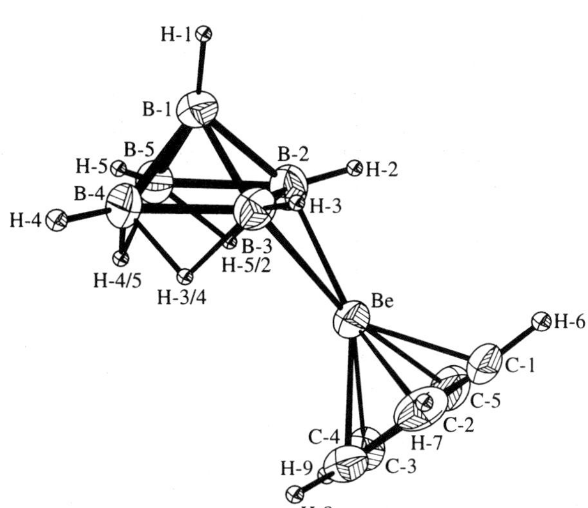

Figure 11 Structure of $[\mu\text{-}(Cp)Be]B_5H_8$ (reproduced by permission of the American Chemical Society from *Inorg. Chem.*, 1981, **20**, 2186).

The compound has surprisingly high thermal stability and reacts rapidly with trimethylamine to form Me_3NBH_3 as the only observable boron-containing product. With dimethylzinc in ether, deprotonation occurs to form methane and what is tentatively suggested as $[\mu_{2,3}\text{-}CpBe]$ $(\mu_{3,4}\text{-}CH_3Zn)B_5H_7$. This latter reaction is much slower in C_6D_6, requiring more than a month to complete in the dark; although the reaction much faster in the light, it is complicated by side reactions to form CpBeMe and possibly (μ-MeZn)B_5H_8 (Equation (8)). Brönsted acids react rapidly and cleanly with (μ-CpBe)B_5H_8 in nonbasic solvents to produce B_5H_9 (Equation (9)).[41]

$$Me_2Zn + (\mu\text{-CpBe})B_5H_8 \longrightarrow CpBeMe + (\mu\text{-MeZn})B_5H_8 \quad (8)$$

$$(\mu\text{-CpBe})B_5H_8 + HX \longrightarrow CpBeX + B_5H_9 \ (X = Cl, Br, OH) \quad (9)$$

Beryllium-9 resonances for a wide range of compounds have been determined.[10] η^5-Cyclopentadienylberyllium compounds (CpBeX) exhibit very high upfield shifts, ascribed largely to paramagnetic ring-current effects with only small differences as X is varied, the differences being ascribed to neighbour anisotropy effects as electronegativity does not explain the trend. $CpBeBH_4$ is unique in exhibiting a very well-resolved ^{9}Be NMR multiplet spectrum and is the only known example of ^{9}Be–^{11}B coupling (3.6 Hz) as a 1:1:1:1 quartet and an unresolved ^{9}Be–^{10}B septet of equal intensity of 1.2 Hz is also observed.[10]

Thermodynamic functions of gaseous CpBeX (X = Cl, Br or BH_4) have been calculated from structural and spectroscopic data. The significantly high rotational barrier in $CpBeBH_4$ has been explained by a consistent overlap between the $2p$ orbital of beryllium and the π orbital of the ring, and is supported by the shortness of the Be–ring distance.[42] The bonding in CpBeX may be rationalized by considering the linear combination of Cp and BeX fragments. The degenerate Cp π-orbitals (e_1) interact with the p_x and p_y orbitals of an sp-hybridized BeX fragment, while the lowest energy Cp π-orbital (a_1) interacts with the BeX sp hybrid. The resulting three bonding molecular orbitals are occupied by six electrons giving rise to a 'three-dimensional aromaticity'.[38]

2.7 HYDRIDES

Although details of routes aimed at the preparation of organoberyllium hydrides (RBeH) and anionic organoberyllium hydrides have been reported, only in a few cases have solvent-free compounds been isolated.[3] Exchange reactions involving Sn–H and Al–H proved unsatisfactory. In order to provide a more comprehensive study of exchange reactions involving beryllium compounds, reactions of sodium trialkylhydridoboronates ($NaBR_3H$, R = Me or Et) with organo- and haloberyllium compounds in a variety of proportions have been studied.[43] Exchange between Et_2Be and $NaBEt_3H$ in diethylether does not take place to any appreciable extent beyond the EtBeH stage, even when the Be:B ratio is 1:4. Separation of EtBeH from ether proved impossible; this was also the case resulting from the reaction of EtBeCl and $NaBEt_3H$ in ether following separation of NaCl and evaporation of Et_3B (Equation (10)). In the latter case the organoberyllium hydride was isolated as its trimethylamine complex $(EtBeH{\cdot}NMe_3)_2$, for which the enthalpy of coordination of the amine has been shown to be ~40 kJ mol^{-1}. Similar routes involving $BeCl_2$ and $NaBR_3H$ in 1:1, 1:2 and 1:4 ratios were unsuccessful in yielding pure beryllium hydrides, although reaction of $NaBMe_3H$ with $BeCl_2$ in ether in a 2:1 ratio yielded a product of approximate composition $HBe(Me_3BH)(OEt_2)_{1.16}$.

$$NaBEt_3H + EtBeCl \longrightarrow EtBeH + NaCl + Et_3B \quad (10)$$

It was once believed that, in the solid state, beryllium hydrides were associated and that hydrogen played a bridging role between two metal atoms.[3] Structural characterization of the two beryllium hydrides tris(3-*t*-butylpyrazolyl)hydroboratoberyllium hydride[44] and 2-dimethylamino-*N*-methylethylamidoberyllium hydride[45] has dispelled both these myths. The former is prepared by metathesis of the corresponding bromide or chloride (prepared from BeX_2 (X = Cl, Br) and Tl[η^3-HB(3-Bu^{t}pz)$_3$]) with $LiAlH_4$ in ether, the reaction of the bromide proceeding faster than the chloride. The reactivity scheme leading to its formation and some of its chemical transformations is shown in Scheme 4. The corresponding derivative of the less sterically demanding tris(3,5-dimethylpyrazolyl)hydroborato ligand could not be prepared by this method. The compound is monomeric, resembling its four-coordinate magnesium and zinc analogues, doubtless due to the sterically demanding nature of the tris(3-*t*-butylpyrazolyl)hydroborato ligand (HB(3-Bu^{t}pz)$_3$) and, in terms of electron count the ligand is formally analogous to η^5-C_5H_5. Thus, chelation of the three nitrogen atom donors provides a well-defined coordination environment for a series of half-sandwich-bonded compounds of beryllium. The solid-state structure of [η^3-HB(3-Bu^{t}pz)$_3$]BeH (Figure 12) and its bromide analogue (Figure 13) confirm both the monomeric nature of the complexes and also the η^3 coordination mode of the pyrazolyl ligand. Both complexes show almost identical coordination environments, despite the significant difference in size between hydrogen and bromine, exemplified by the similarity in the average Be–N distances of 0.1778(3) and 0.180(1) nm, average N–Be–N angles of 100.2(2)° and 102(1)°, and H–Be–N (117.6(2)°) and Br–Be–N (116(1)°) angles in the hydride and bromide, respectively.

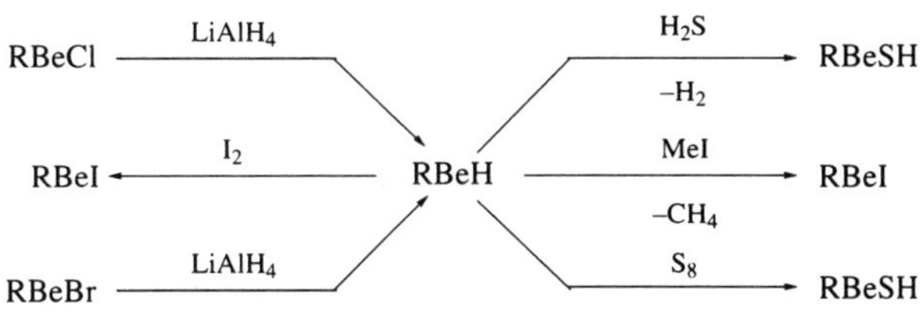

(R = η^3-HB (3-Bu^{t}pz)$_3$)

Scheme 4

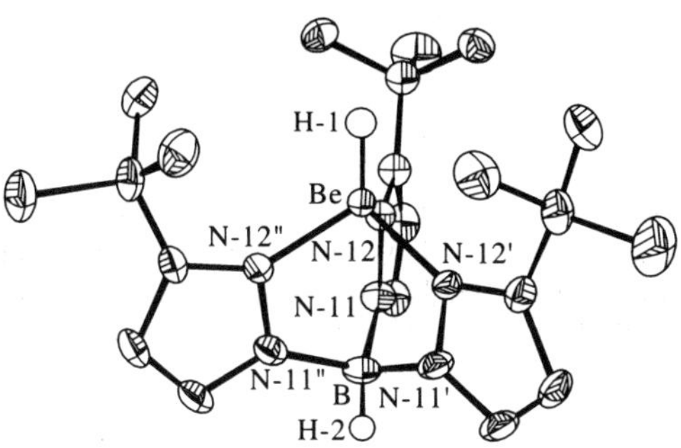

Figure 12 Structure of [η^3-HB(3-Bu^{t}pz)$_3$]BeH (reproduced by permission of the American Chemical Society from *Inorg. Chem.*, 1992, **31**, 985).

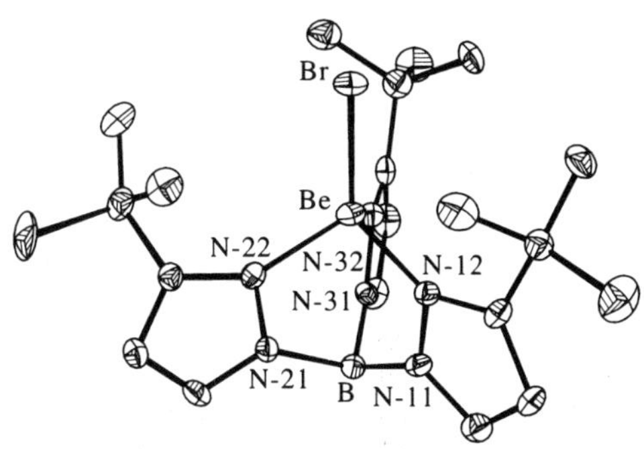

Figure 13 Structure of [η^3-HB(3-Bu^{t}pz)$_3$]BeBr (reproduced by permission of the American Chemical Society from *Inorg. Chem.*, 1992, **31**, 984).

Although claimed to be the first x-ray structural characterization of a compound containing a terminal Be–H bond,[44] some eight years earlier the crystal structure of [HBeN(Me)$C_2H_4NMe_2$]$_2$, prepared from BeH_2 and *N,N,N'*-trimethylethylenediamine (Equation (11)), clearly showed the presence of a terminal Be–H bond, but on this occasion in a dimeric species (Figure 14).[45] The two halves of the dimer are linked by a centrosymmetric four-membered Be_2N_2 ring with one nitrogen atom of the bidentate ligand located in the bridge and the other forming a coordinate bond to the metal. This results in the formation of two, puckered five-membered rings, which enable both beryllium and nitrogen to achieve four-coordination. The enhancement of the donor strength of three-coordinate nitrogen when attached to an electropositive metal, relative to that of a tertiary amine, and the increase in acceptor strength of beryllium when bonded to an electronegative nitrogen atom, relative to the metal in R_2Be or RBeH, accounts for nitrogen rather than hydrogen acting as the bridging group. The Be–H distance (0.139(2) nm) is shorter than the average Be–H bridge distance[3] (0.148(3) nm) in [$NaBeEt_2H{\cdot}OEt_2$]$_2$ but is slightly longer than that in [η^3-HB(3-Bu^{t}pz)$_3$]BeH (0.123(7) nm)[44] and in CpBeH (0.132 nm) determined by microwave spectroscopy on the gas-phase monomer.[3] The Be···Be' cross-ring nonbonding contact of 0.2316(3) nm[45] is similar to the short Be···Be distances found with other three-electron-bonding ligands.[44]

$$2BeH_2 + 2Me_2NC_2H_4NHMe \longrightarrow [HBeN(Me)C_2H_4NMe_2]_2 + 2H_2 \quad (11)$$

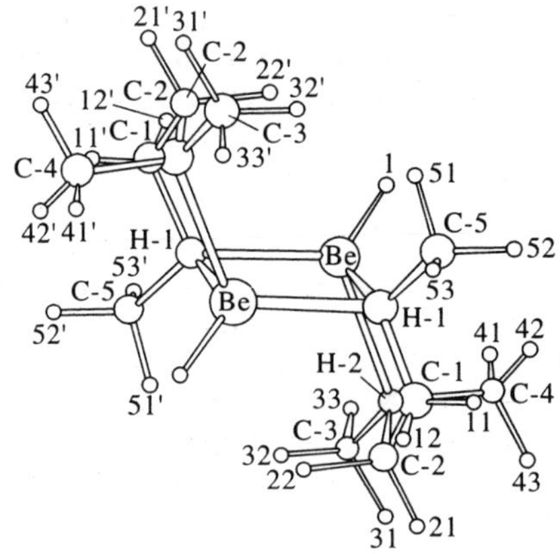

Figure 14 Structure of $[HBeN(Me)C_2H_4NMe_2]_2$ (reproduced by permission of the International Union of Crystallography from *Acta Crystallogr., Sect. C*, 1984, **40**, 609).

In the bromide (Figure 13) the Be–Br bond length of 0.2122(7) nm[44] is similar to that of 0.219(3) nm found in $[BrBeOBu^t{\cdot}OEt_2]_2$.[46] In order to clearly assign the terminal Be–H stretching frequency, which could not be detected for $[HBeN(Me)C_2H_4NMe_2]_2$,[3] the corresponding deuteride was prepared by reaction of $LiAlD_4$ using benzene, rather than ether, as solvent since in ether the hydride results as the major derivative even though RBeD ($R = \eta^3$-HB(3-Bu^{t}pz)$_3$) does not exchange with ether. The terminal Be–H stretching frequency at 1865 cm^{-1} shifts to 1410 cm^{-1} on deuteration ($\nu_H/\nu_D = 1.32$) compared with 1720 cm^{-1} and 1275 cm^{-1} in CpBeH and CpBeD, respectively,[47] whereas the corresponding absorption in Cp*BeH is much higher at 1948 cm^{-1}.[48] In the ^{1}H NMR spectrum the Be–H resonance is observed as two broad peaks ($\delta = 5.0$ ppm) at room temperature, close to the B–H resonance ($\delta = 4.5$ ppm). The relative assignments have been confirmed by ^{2}H NMR spectroscopy on the isotopomer, RBeD, and the one-bond $^1J_{^9Be-H}$ and much larger $^1H_{^{11}B-H}$ coupling constants of 28 Hz and 113 Hz, respectively, have been determined. This is the first coupling constant to be measured for a terminal Be–H bond since the corresponding resonance could not be observed in CpBeH[44] or $[HBeN(Me)C_2H_4NMe_2]_2$.[3]

Reactions of RBeH ($R = \eta^3$-HB(3-Bu^{t}pz)$_3$) (Scheme 4) with MeI produces RBeI, also formed from RBeH and I_2, contrasting with CpBeH which is unreactive to MeI. It also reacts with H_2S to form the thiol RBeSH, alternatively prepared by direct insertion of elemental sulfur into the Be–H bond. This series of RBeX compounds (X = H, Cl, Br, I or SH) allows a comparison of substituent effects on the ^{9}Be resonances which lie in the range typically found for four-coordinate beryllium complexes and confirms that increasing the electronegativity of the substituent on beryllium shifts the ^{9}Be resonance to lower field.[44]

Reaction of Cp_2ReH with MCl_2(M = Be or Mg) (Equation (12)) results in the first type of compound for which it is proposed that M is directly bonded to another metal and also contains a hydrogen bridge between M and the other metal (**21**) on the basis of the very large downfield shift in the ^{1}H NMR spectra relative to Cp_2ReH and from infrared spectral data. As this type of bonding is unique for beryllium, crystallographic studies would have been particularly interesting. In contrast, the zinc and cadmium analogues do not appear to contain the hydride bridge which is in keeping with the fact that hydrogen bridging in zinc and cadmium compounds is unknown.[49]

$$Cp_2ReH + BeCl_2 \longrightarrow Cp_2Re\overset{H}{\longrightarrow}BeCl_2 \quad (12)$$

(21)

2.8 MISCELLANEOUS REACTIONS

As part of an extensive programme investigating alkylmetal asymmetric reduction reactions, Italian workers have followed the stereochemical course of the reduction of prochiral ketones (Equation (13)) by the optically active alkylberyllium compounds R_2Be and RBeCl (R = (*R*)2-methylbutyl, *cis*-myrtanyl, [{(1*S*,2*S*)-6,6-dimethylbicyclo[3.1.1]heptan-2-yl}methyl] and *endo*-(1*R*)-camphanyl) in a variety of solvents.[50–2] The main product is always the reduction carbinol and the addition product is only formed in low or negligible yields. For R = (*R*)2-methylbutyl, alkylphenyl and α-alkynyl, ketones are reduced to the corresponding optically active carbinols within short reaction times and in high yield. The extent of asymmetric reduction increases with steric hindrance of the alkyl group on the carbonyl moiety and, while (*S*)-alkylphenylcarbinols result from alkylphenylketones, the reduction of α-alkynylketones affords (*R*)-alcohols. Reactions with R_2Be (R = *cis*-myrtanyl and *endo*-(1*R*)-camphanyl) are much faster than those of the corresponding RBeCl with both chemical and optical yields being moderately high. The capability of enantioface discrimination of R_2Be and RBeCl was roughly the same, the halogen only determining the extent of the asymmetric induction. The absolute configuration of products was *R* in all cases except one for R = *cis*-myrtanyl and *S* in all cases for R = *endo*-(1*R*)-camphanyl, in contrast with reactions involving the corresponding organomagnesium and organoaluminum derivatives. It is clear from the comparative study of these reactions with those involving other organometallic compounds that a simple mechanistic model cannot be presented to reflect a unifying view of the mechanism of this type of reaction with the metals studied.

$$R^2R^1CO \xrightarrow{[H]} R^2R^1\dot{C}HOH \qquad (13)$$

Although acylberyllium bromides, prepared from an acylbromide and beryllium metal in ether, have been known since 1966, nothing is known of their nature. They have not been isolated and are used *in situ* to effect synthetic transformations.[3] Their synthetic use has been extended to convert ArCH=CHCHO to 1,6-diaryl 1,3,5-hexatrienes (**22**) and terphenyls (**23**) via the mechanism shown in Scheme 5,[53] and to convert monosubstituted benzaldehydes in the presence of $HgCl_2$ to the diacetyl derivatives of the corresponding dihydrobenzoin.[54]

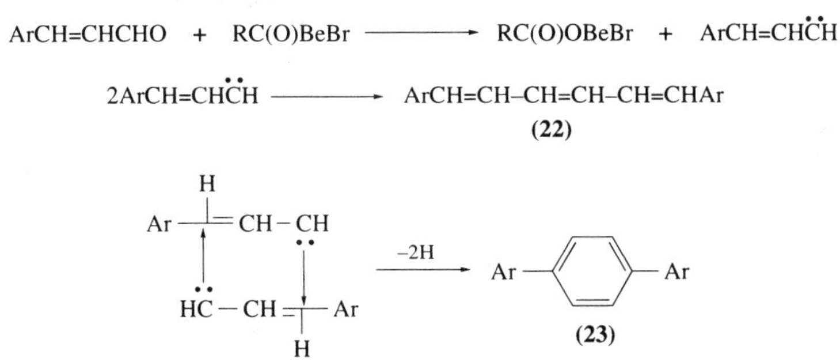

Scheme 5

The nature of the dimeric product formed from reaction of acetone with MeBeBr in diethylether has been shown to have a similar structure[46] to that formed with the corresponding Grignard reagent.[55] The structure (Figure 15) contains a centrosymmetric Be_2O_2 four-membered ring and each beryllium atom becomes four-coordinate by also bonding to a terminal bromine and forming a coordinate bond to an ether oxygen atom. This provides further confirmation that oxygen is a better bridging group for group 2 elements, due to its enhanced donor capability when attached to an electropositive metal, than halogen, alkyl or aryl. Thus, compounds of type R^2BeOR^1 are associated, the degree of association depending on the nature of R^2 and R^1, and the presence of donor solvents.[3] As for the magnesium analogue, the *t*-butoxy and ether oxygen atoms adopt a trigonal arrangement. The three Be–O distances are the same within experimental error, in contrast with the magnesium analogue where the Mg–O ring distance is shorter than the Mg–O ether bond length, and the Be_2O_2 ring is essentially square (with angles of 89(2)° and 91(2)°), resulting in the other angles at beryllium being considerably larger than tetrahedral. The replacement of magnesium by the much smaller beryllium atom in the structure results in a larger ring angle at beryllium of (89(2)°) (83.3(6)° in the magnesium case) and in consequence the distortion of the other angles away from tetrahedral is much less pronounced at beryllium. The short Be···Be cross-ring distance of 0.230(4) nm is similar to the value found with other three-electron bridging ligands.[14]

The inability of (metal organic vapour phase epitaxy) (MOVPE) to grow abrupt high–low uncompensated *p–n* junctions in group 13–15 materials has been a key obstacle to its use in epitaxial growth of certain device structures and has been primarily due to the lack of suitable sources of low vapour pressure *p*-type dopants. Magnesium, zinc and cadmium are the most commonly used *p*-dopants in MOVPE but suffer from possessing high vapour pressures. The low vapour pressure of beryllium and its low diffusivity below 800 °C make it the ideal *p*-type dopant for MOVPE growth of group 13–15

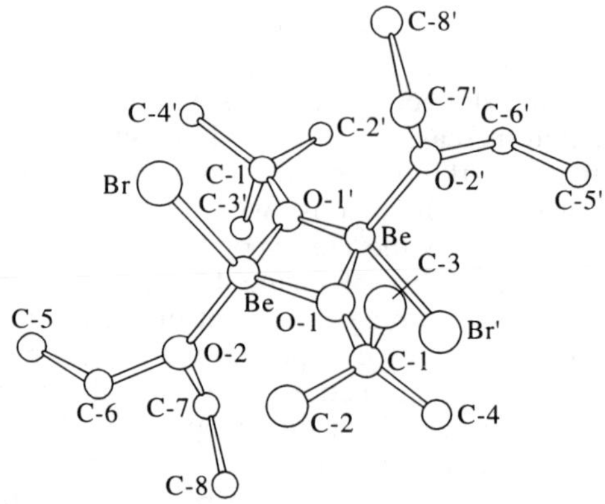

Figure 15 Structure of $[BrBeOBu^t \cdot OEt_2]_2$ (reproduced by permission of the International Union of Crystallography from *Acta Crystallogr., Sect. C*, 1984, **40**, 609).

materials. However, the barrier to its use in these applications, due to the absence of a high-purity beryllium source which could be transported in the reaction chamber in a controllable fashion, has now come down and growth of beryllium-doped gallium–arsenic and In–Ga–As epilayers at atmospheric pressure has been achieved. Diethylberyllium has been found to be a clean and controllable convenient source of beryllium, having the following advantages over other sources: (i) it is liquid below room temperature, (ii) it has a high enough vapour pressure to allow vapourization below room temperature, and (iii) it undergoes no reaction which can adversely affect doping efficiency in the gas phase or at the gas–epilayer interface. This has led to the commercial availability of high-purity diethylberyllium in which impurity concentration profiles measured by secondary-ion mass spectrometry (SIMS) in epilayers showed the most significant impurity is silicon.[56–60]

Surprisingly, and despite the fact that the toxicity of beryllium will certainly limit the commercial importance of products, there are still many claims for the effectiveness of organoberyllium compounds as dimerization–polymerization catalysts.

2.9 REFERENCES

1. *Journal of Organometallic Chemistry*, 1987, **325** (tribute edition to G.E. Coates).
2. P. Matthews (ed.), 'Guinness Book of Records', Guinness Publishing, 1994, 75.
3. N. A. Bell, in 'COMC-I', vol. 1, p. 121.
4. H. Schmidbaur and E. Weiss, *Angew. Chem. Int. Ed. Engl.*, 1981, **20**, 283.
5. H. Schmidbaur, E. Weiss and W. Graf, *Organometallics*, 1985, **4**, 1233.
6. H. Schmidbaur and G. Müller, *Monatsh. Chem.*, 1980, **111**, 1233.
7. Y. Yamamoto, *Bull. Chem. Soc. Jpn.*, 1987, **60**, 1189.
8. C. M. Berke and A. Streitwieser, Jr, *J. Organomet. Chem.*, 1980, **197**, 123.
9. D. S. Marynick, *J. Am. Chem. Soc.*, 1981, **103**, 1328.
10. D. F. Gaines, K. M. Coleson and D. F. Hillenbrand, *J. Magn. Reson.*, 1981, **44**, 84.
11. H. Yasuda, Y. Ohnuma, A. Nakamura, Y. Kai, N. Yasuoka and N. Kasai, *Bull. Chem. Soc. Jpn*, 1980, **53**, 1101.
12. H. Yasuda, M. Yamauchi, A. Nakamura, T. Sei, Y. Kai, N. Yasuoka and N. Kasai, *Bull. Chem. Soc. Jpn.*, 1980, **53**, 1089.
13. G. Frenking, *Chem. Phys. Lett.*, 1984, **111**, 529.
14. M. E. O'Neill and K. Wade, in 'COMC-I', vol. 1, p. 1.
15. K. B. Starowieyski and A. Chwojnowski, *J. Organomet. Chem.*, 1981, **215**, 151.
16. N. A. Bell, I. W. Nowell, G. E. Coates and H. M. M. Shearer, *J. Organomet. Chem.*, 1984, **273**, 179.
17. E. Canadell and O. Eisenstein, *Inorg. Chem.*, 1983, **22**, 3856.
18. R. Snaith, unpublished observations, cited in Ref. 16.
19. P. N. V. P. Kumar and E. D. Jemmis, *J. Am. Chem. Soc.*, 1988, **110**, 125.
20. W. Kaim, *Z. Naturforsch., Teil B*, 1981, **36**, 1110.
21. W. Kaim, *Chem. Ber.*, 1981, **114**, 3789.
22. W. Kaim, *J. Organomet. Chem.*, 1983, **241**, 157.
23. J. R. Wermer, D. F. Gaines and H. A. Harris, *Organometallics*, 1988, **7**, 2421.
24. J. L. Atwood, S. G. Bott, R. A. Jones and S. U. Koschmieder, *J. Chem. Soc., Chem. Commun.*, 1990, 692.
25. K. W. Nugent, J. K. Beattie, T. W. Hambley and M. R. Snow, *Aust. J. Chem.*, 1984, **37**, 1601.
26. K. W. Nugent and J. K. Beattie, *Inorg. Chem.*, 1988, **27**, 4269.
27. S. J. Pratten, M. K. Cooper and M. J. Aroney, *Polyhedron*, 1984, **3**, 1347.
28. S. J. Pratten, M. K. Cooper and M. J. Aroney, *J. Chem. Soc., Dalton Trans.*, 1985, 1761.
29. K. W. Nugent and J. K. Beattie, *J. Chem. Soc., Chem. Commun.*, 1986, 186.
30. K. W. Nugent, J. K. Beattie and L. D. Field, *J. Phys. Chem.*, 1989, **93**, 5371.
31. J. K. Beattie and K. W. Nugent, *Inorg. Chim. Acta*, 1992, **198–200**, 309.
32. E. B. Lobkovsky, *J. Organomet. Chem.*, 1984, **277**, 53.

33. S. J. Pratten, M. K. Cooper and M. J. Aroney, *J. Organomet. Chem.*, 1990, **381**, 147.
34. C. J. Burns and R. A. Andersen, *J. Organomet. Chem.*, 1987, **325**, 31.
35. G. Schmid, O. Boltsch and R. Boese, *Organometallics*, 1987, **6**, 435.
36. R. Goddard, J. Akhtar and K. B. Starowieyski, *J. Organomet. Chem.*, 1985, **282**, 149.
37. N. A. Bell, H. M. M. Shearer and J. Twiss, *Acta Crystallogr., Sect, C*, 1984, **40**, 610.
38. E. D. Jemmis and P. v. R. Schleyer, *J. Am. Chem. Soc.*, 1982, **104**, 4781.
39. M. Lattman and A. H. Cowley, *Inorg. Chem.*, 1984, **23**, 241.
40. C. J. Burns and R. A. Andersen, *J. Am. Chem. Soc.*, 1987, **109**, 5853.
41. D. F. Gaines, K. M. Coleson and J. C. Calabrese, *Inorg. Chem.*, 1981, **20**, 2185.
42. L. Bencivenni, R. Gigli and M. Pelino, *Thermochim. Acta*, 1983, **63**, 317.
43. N. A. Bell, G. E. Coates and J. A. Heslop, *J. Organomet. Chem.*, 1987, **329**, 287.
44. R. Han and G. Parkin, *Inorg. Chem.*, 1992, **31**, 983.
45. N. A. Bell, G. E. Coates, M. L. Schneider and H. M. M. Shearer, *Acta Crystallogr., Sect. C*, 1984, **40**, 608.
46. N. A. Bell, H. M. M. Shearer and J. Twiss, *Acta Crystallogr., Sect. C*, 1984, **40**, 605.
47. H. Schmidbaur, *Gmelin, Be-Org.*, 1987, **1**, 155.
48. R. A. Andersen, unpublished observations, cited in Ref. 44.
49. V. M. Ishchenko, T. M. Arkhireeva, B. M. Bulychev, G. L. Soloveichik and S. N. Nikolaeva, *Koord. Khim., (Eng)* 1986, **12**, 57.
50. G. Giacomelli, M. Falorni and L. Lardicci, *Gazz. Chim. Ital.*, 1985, **115**, 289.
51. M. Falorni, L. Lardicci, C. Rosini and G. Giacomelli, *J. Org. Chem.*, 1986, **51**, 2030.
52. M. Falorni, L. Lardicci, and G. Giacomelli, *J. Org. Chem.*, 1989, **54**, 2383.
53. I. I. Lapkin, N. E. Evstafeeva, S. V. Sinani, *Zh. Obshch. Khim.*, 1979, **49**, 167 (*Chem. Abstr.*, 1979, **90**, 168 173).
54. I. I. Lapkin and S. V. Sinani, *Zh. Obshch. Khim*, 1987, **57**, 372 (*Chem. Abstr.*, 1988, **108**, 94 166).
55. N. A. Bell, P. T. Moseley and H. M. M. Shearer, *Acta Crystallogr., Sect. C*, 1984, **40**, 602.
56. J. D. Parsons, L. S. Lichtmann, F. G. Krajenbrink and D. W. Brown, *J. Cryst. Growth*, 1986, **77**, 32.
57. J. D. Parsons and F. G. Krajenbrink, *J. Electrochem. Soc.*, 1983, **130**, 1782.
58. N. Ohtsuka, K. Kodama, M. Ozeki and Y. Sakuma, *J. Cryst. Growth*, 1991, **115**, 460.
59. R. Mellet, R. Azoulay, L. Dugrand, E. V. K. Rao and A. Mircea, *Conf. Ser. Inst. Phys.*, 1982, **63**, 583 (*Chem. Abstr.*, 1982, **97**, 154 652).
60. R. S. Sillmon, S. M. Hues, D. K. Gaskill, N. Bottka and P. E. Thompson, *J. Electron. Mat.*, 1989, **18**, 501.

3

Magnesium, Calcium, Strontium and Barium

W. EDWARD LINDSELL

Heriot-Watt University, Edinburgh, UK

3.1 INTRODUCTION 58

3.2 MAGNESIUM 58

3.2.1 Organomagnesium Halides 58

3.2.1.1 Preparation 58

3.2.1.2 Structures and physical properties 72

3.2.1.3 Reactions 78

3.2.2 Diorganylmagnesium Compounds 85

3.2.2.1 Preparation 85

3.2.2.2 Structures and physical properties 87

3.2.2.3 Reactions 94

3.2.3 Addition Compounds of Magnesium with Unsaturated Organic Molecules 96

3.2.3.1 Alkenes, alkynes and dienes 97

3.2.3.2 Cyclooctatetraene 98

3.2.3.3 Anthracene and related aromatic compounds 99

3.2.4 Organomagnesium Hydrides 101

3.2.5 Organomagnesium Alkoxides, Amides and Related Compounds 101

3.2.5.1 Organomagnesium alkoxides and related compounds with oxygen-bonded ligands 101

3.2.5.2 Organomagnesium amides and related compounds with nitrogen-bonded ligands 103

3.2.6 Heterobimetallic Compounds 104

3.2.6.1 Alkali metal compounds 105

3.2.6.2 Zinc compounds 106

3.2.6.3 Aluminum compounds 106

3.3 CALCIUM, STRONTIUM AND BARIUM 107

3.3.1 Organometal Halides 107

3.3.1.1 Preparation 107

3.3.1.2 Structures, properties and reactivity 109

3.3.2 Diorganylmetal Compounds 110

3.3.2.1 Dialkyl and diaryl metal compounds 110

3.3.2.2 Dicyclopentadienylmetal and related compounds 112

3.3.3 Addition Compounds with Unsaturated Organic Molecules 117

3.3.4 Organometal Hydrides, Oxides and Amides 118

3.3.4.1 Organocalcium hydrides 118

3.3.4.2 Organometal oxides 118

3.3.4.3 Organometal amides 118

3.3.5 Heterobimetallic Compounds 119

3.3.5.1 Alkali metal compounds 119

3.3.5.2 Zinc compounds 119

3.3.5.3 Aluminum compounds 119

3.3.5.4 Transition metal compounds 120

3.4 REFERENCES 120

3.1 INTRODUCTION

The organometallic chemistry of magnesium and the heavier group 2 metals has developed considerably since the review in *COMC-I*.[1] Much of the extensive literature concerning magnesium compounds has involved applications in organic synthesis, but in this area new reagents and reaction pathways for C–C bond formation have emerged, including highly regio- and stereoselective reactions and new transition metal catalysed procedures. Full structural characterizations of a range of organomagnesium derivatives in the solid state,[2] together with investigations of solutions by newer physical techniques, have had significant impact on the understanding of the structure and reactivity of these reagents. Advances in the organic chemistry of heavier alkaline earth metals have also been founded on full structural characterizations which were not available in earlier work.[3–4] An impetus to the increased interest in organometallics of these heavier metals is the search for new volatile precursors, via chemical vapour deposition (CVD), to oxide or halide thin films containing alkaline earth metals such as superconducting cuprates, with potential applications in electronic and optical devices.

Characterized, isolable organometallic compounds all contain the metals in formal 2+ oxidation states, the simple derivatives having the basic formulae MR_2 or RMX, with R and X being uninegative carbon and heteroatom ligands, respectively. Coordinative unsaturation of these basic units is relieved by the addition of Lewis bases or by the formation of associated structures, and recent studies have indicated more clearly the nature and importance of this coordination, including intramolecular interactions, and of steric effects of ligands on the structure and reactivity of these organometallics.[2,4–5] Noteworthy, since the last review, is the full characterization of a new class of adducts between magnesium and anthracene derivatives that have versatile applications in organometallic chemistry.[6]

Oxidation state 1+ organometallic compounds have not been definitively characterized under normal conditions, although they have been proposed as intermediates in the preparation and reactions of compounds in solution. However, gas-phase radicals MR (R = Me, C≡CH, η-Cp, η-pyrrolate; M = Mg, Ca and/or Sr) have been prepared by reactions of vapourized metal atoms and studied by laser spectroscopy, either under thermal conditions[7] or after cooling by supersonic expansion:[8] these metal(I) derivatives contain an unpaired electron primarily in a metal-centred orbital. Mg^+–alkene complexes have also been studied in the gas phase by ion cyclotron resonance.[9] Moreover, under matrix isolation conditions at 9 K, cluster organomagnesium and calcium species have been observed and assigned the formulae MeM_xX (M = Mg, Ca; X = halogen; $x \geq 2$).[10]

This chapter will concentrate on the more orthodox organometallic compounds, and magnesium is covered separately from calcium, strontium and barium since, although all metals form stoichiometrically related compounds, the heavier metals differ in their chemistry from magnesium.

3.2 MAGNESIUM

3.2.1 Organomagnesium Halides

Grignard reagents are the most commonly handled organomagnesium compounds and will normally be represented in this chapter by the simple formula RMgX (R = organic group, X = halogen), although solvation, disproportionation, aggregation and deviations from this stoichiometry do occur, as discussed in appropriate sections. These derivatives are mainly employed as intermediates in syntheses of organic and organometallic products, and new developments of preparative methods for a range of reagents are important.

3.2.1.1 Preparation

(i) Direct reaction

The direct oxidative addition reaction of organic halide to magnesium metal (Equation (1)) is the standard route to Grignard reagents, with many variations in reaction conditions as discussed previously.[1] Recent modifications in procedures and studies of this reaction will be discussed here. Grignard reagents have many applications in large-scale, commercial processes, and a short account of the direct preparation, handling, transfer and disposal of large quantities of the reagents has been published.[11] Also, a continuous, high-yield process for the preparation of Grignard reagents, in particular pure allylmagnesium chloride, has been presented.[12]

$$RX + Mg \longrightarrow RMgX \qquad (1)$$

(a) Initiation and activation of magnesium. The well-established process of activation of magnesium by prior reaction with iodine has been outlined,[13] and it should be noted that heating the iodine to sublimation, if necessary, is effective.[14] Also, many specific examples have been listed of the use of 1,2-dibromoethane as an 'entrainer' for sluggish Grignard reactions,[13] and the catalytic effect of an added, preformed Grignard reagent has been studied.[15] Activation of the magnesium by amalgamation has been found to be useful for the formation of the dimagnesium compound $CH_2(MgBr)_2$.[16]

A review of newer procedures for the activation of metals, including magnesium, has recently been published.[17] Simple stirring of magnesium turnings under an inert atmosphere activates the metal for the clean synthesis of reactive allylic and benzylic organomagnesium chlorides.[18] This 'dry stirring' can be carried out under an argon as well as a nitrogen atmosphere, ruling out the formation of surface nitrides as a factor in the activation process, and optical and electron microscopic examinations of the resulting grey–black metal reveal the formation of small (<1 μm) microcrystalline particles of high, oxide-free surface area.[18] Ultrasonic irradiation of the magnesium is another simple, useful method of reducing the induction period of some Grignard reactions, and is especially effective in the presence of reaction-inhibiting amounts of water or alcohols.[19–20] The main effects of sonication are the removal of adsorbed water/alcohol, the mechanical destruction of the oxide layer and the prevention of by-product deposition on the surface of the magnesium. Slurries of pure, finely divided magnesium in THF or other ether, formed by thermal evaporation and condensation of the metal into the solvent at −110 °C *in vacuo*, are active for low-temperature preparations; this method affords cyclopropylmethyl, benzocyclobutenylmethyl and various 2-alkenyl organomagnesium reagents in good yields at ~−70 °C.[21] One advantage of the methods of activation discussed here is that the products are not contaminated with other activating chemicals.

Rieke's method of forming activated magnesium by reduction of magnesium halides with potassium or sodium metal in refluxing THF or diglyme, especially in the presence of KI,[1,22] has found application in many syntheses (e.g., see Lai,[13] Fürstner[17] and Rieke *et al.*[22]). A safer, room temperature procedure to form metal of comparable reactivity is the reduction of $MgCl_2$ in THF or glyme by lithium in the presence 10% naphthalene or biphenyl as an electron carrier.[22–3] The reduction takes 6–24 h, and the aromatic compound can be removed, if necessary, from the final grey–black powder by washing with fresh solvent.

Magnesium–anthracene compounds, for example, $Mg(anthracene)(THF)_3$[6,24] (see Section 3.2.3.3), have provided a useful source of reactive magnesium for Grignard syntheses. The parent complex, and silylanthracene analogues, can be used directly in reaction with some organic halides to afford Grignard reagents.[25,26a] Benzyl, allyl and propargyl halides, and $(TMS)_3CCl$, give good yields of organometallic product in THF, toluene or even hexane solvent at −10 °C to +20 °C, but simple alkyl and aryl halides give mainly organic products derived from the anthracene group reacting as a nucleophile or radical. Magnesium adducts of methyl- and phenyl-substituted anthracenes react similarly to the parent complex but give lower yields of Grignard reagents.[26b] A polystyrene-supported magnesium–anthracene adduct also yields benzylic magnesium halides (>90%) on reaction with the corresponding organic halide.[27] Formation of Grignards from allyl or propargyl chloride and commercial magnesium powder is promoted by small amounts of magnesium–anthracene complex acting under phase transfer conditions.[26a] $Mg(anthracene)(THF)_3$ decomposes on heating, either in an inert solvent or neat *in vacuo*, or on ultrasonication in a hydrocarbon solvent to give highly reactive, pyrophoric magnesium powders with high surface areas.[28] Similar highly active magnesium is formed by the thermal decomposition of catalytically prepared MgH_2,[28] and both forms of metal are useful in the preparation of a wide range of Grignard reagents under mild conditions in THF, diethyl ether, toluene or *n*-heptane as the solvent.[21c,28] Active, 'organically solvated' magnesium can also be formed from commercial magnesium powder in THF by the catalytic action of anthracene.[29]

Co-condensation of magnesium with a 50–300-fold excess of fluorobenzene at 77 K and subsequent heating of the matrix quantitatively produces solvent-free phenylmagnesium fluoride in a one-stage reaction.[30] Matrix isolation studies of the reaction between magnesium, deposited as magnesium atoms, and Mg_2, Mg_3 and Mg_x aggregates, in argon/MeX (X = I, Br, Cl, F) mixtures at 9 K have been studied spectroscopically, and the formation of cluster organomagnesium reagents, such as $MeMg_xBr$ ($x \geq 2$), occurs;[10] it is noteworthy that the larger metallic clusters are most reactive, $Mg_x \approx Mg_4 > Mg_3 \approx Mg_2 > Mg$, with halide reactivity following the trend MeI > MeF ≈ MeBr > MeCl, and magnesium atoms or MeCl are inert in all cases. Studies of the chemisorption and subsequent decomposition of MeBr on an Mg(0001) single-crystal surface showed the formation of surface bromide and gas phase hydrocarbons, including ethane, but even at 123 K no evidence for stable Mg–C bond-formation could be deduced.[31]

(b) Solvents and added ligands. Various solvents have been employed in the direct reaction[1] and, as noted above, activated magnesium has been employed to form Grignards in ethereal and hydrocarbon solvents.[25,28] By-products derived from radical reactions with diethyl ether have been noted in the direct preparations of phenylmagnesium bromide[32] and adamantylmagnesium bromides.[33–4]

Hexamethylphosphoric triamide (HMPT) has found application as a polar solvent in Grignard synthesis,[1] but care should be taken of its toxicity. Sulfamides and sulfonamides, including tetraethylsulfamide ($SO_2(NEt_2)_2$) and *N,N*-diethyltrimethylmethanesulfonamide ($Me_3CSO_2NEt_2$), have also been proposed as suitable inert, polar solvents for the preparation and reaction of Grignards.[35] Acetonitrile is normally reactive to Grignard reagents, but electrochemical oxidation of magnesium in this solvent containing alkyl or aryl halide (RX) and 2,2'-bipyridyl yields the adducts RMgX(bipy), or related dimagnesium reagents with α,ω-dihalides.[36] With solutions containing R^1X and R^2_4NX the oxidation products are salts ($R^2_4N[R^1MgX_2(MeCN)]$), but both these ionic derivatives and the neutral bipyridyl adducts lack typical Grignard reactivity.[36] All types of Grignard reagents are precipitated from solutions in diethyl ether by the complexing agent $N(CH_2CH_2OCH_2CH_2OMe)_3$ to form thermally stable, easily handled powders that show chemical behaviour typical of conventional Grignard reagents although with lower reaction rates.[37]

(c) Kinetics and mechanism. It has been previously established that the rate of reaction between an alkyl halide and magnesium in ethereal solutions is normally pseudo first order with respect to the alkyl halide and proportional to the area of the active metal surface,[1] but new kinetic studies have been made on some reaction systems. Further investigations of the reaction of bromocyclopentane with a rotating disc of magnesium in diethyl ether have been reported.[38–9] One study indicates that a mass transfer limited reaction occurs at 0 °C with a reaction rate proportional to $\omega^{1/2}$ (ω = angular velocity), with more limited data showing $k \propto (\eta)^{-5/6}(\rho)^{1/6}$ (η is the shear viscosity of the solution and ρ is the density of solution) and $k \propto (D)^{2/3}$ (*D* is the diffusion coefficient of alkyl bromide).[38] Other studies on the same system support a reaction for which the overall rate constant is influenced but not limited by mass transport, and at low temperatures (<–5 °C) with high rotating speeds of the disc (>2500 rpm) the reaction rate is limited by the chemical reaction with an activation energy of 51 kJ mol^{-1}.[39]

1- and 2-Bromoadamantane are reported to give reasonable yields of the corresponding Grignard reagents only in a 'static', unstirred reaction with magnesium,[33] although other workers have obtained good yields of 1-adamantylmagnesium bromide under normal, stirred conditions.[34] Under 'static' conditions the reaction rate is greater for the secondary derivative but both 1- and 2-bromoadamantane show zero-order kinetics for the reaction and a black deposit is formed on the metal surface that x-ray photoelectron spectroscopy showed to comprise mainly magnesium(0) and biadamantyl in its lower and intermediate regions.[33] Under stirred conditions, however, the reaction is initially first order in bromoadamantane but becomes zero order as a deposit of biadamantyl and finely dispersed magnesium builds up on the surface of the metal; at this stage, formation of biadamantyl ceases, and it is proposed that the limited access of organic bromide to the surface causes the rate of Grignard production to fall and show the zero-order dependence on the substrate.[34]

Tuulmets and co-workers[15,20,40–1] have reported various kinetic studies on the formation of Grignard reagents from alkyl halides, particularly Bu^nBr, in different media. After an induction period, reactions in toluene containing small amounts of ether or other bases (<1 mol per mole of alkyl halide) show a rapid formation of the monosolvated Grignard reagent followed by a slow reaction of zero-order kinetics, catalysed by the solvated organomagnesium compound. In the rapid stage the catalytic activity of bases follows the sequence Et_2O > diglyme > Et_3N > THF > Bu^n_2O.[40] The introduction of a different Grignard reagent as a catalyst for the slow stage of the reaction may produce an oscillating rate of reaction.[15] The effects of inhibitors and of sonication on the induction period have been monitored: inhibition is attributed to reaction with any formed Grignard that would clean the magnesium surface during initiation rather than to radical capture; sonication reduces the effects of inhibitors and also increases the rate of the slow stage of the reaction.[20,41]

The participation of radicals in the normal preparative Grignard reaction has been well established and is supported by more recent research. The reaction of bromocycloheptane (RBr) with magnesium in the presence of the radical scavenger 2,2,6,6-tetramethylpiperidine-*N*-oxyl (TMPO·), in diethyl ether containing *t*-amyl alcohol and LiBr, forms *N*-cycloheptoxy-2,2,6,6-tetramethylpiperidine (TMPOR) in a 93% or better yield, based on RBr.[42] The amyl alcohol was introduced to protonate any Grignard reagent formed and thus prevent its reaction with TMPO, while the LiBr was added to solubilize $TMPO^-$, formed on reduction of TMPO· by magnesium(0) in a side reaction. In the absence of TMPO·, cycloheptane is formed in a 95% or better yield, and it was concluded that both this product and TMPOR arise from the common intermediate cycloheptyl radical that is involved in the major pathway to cycloheptylmagnesium bromide.[42] Other evidence for radical intermediates includes further observations of CIDNP effects,[43] ring opening of the 2-phenylcyclobutylmethyl radical,[44] cyclization of radicals from 5-hexenyl halides,[45–6] and radical trapping by dicyclohexylphosphine (DCPH)[45,47] during Grignard formation under various conditions. Evidence for radical anion intermediates in reactions of Grignard formation from polychloroaromatic compounds has been adduced from cyclizations of ω-bromoalkoxopolychloroarenes and -heteroarenes.[48]

Although a radical mechanism is generally accepted for the Grignard reaction, an issue of current controversy is whether the intermediate radicals diffuse freely in solution or remain absorbed at the magnesium surface. Garst and co-workers[46,49–50] propose a 'diffusion' (D) model for typical alkyl halides (Scheme 1), including step 4 by which all radicals R• leave the surface before forming the Grignard reagent and may react with solvent, or undergo normal coupling, isomerization, and related radical reactions: a mathematical analysis of this reaction scheme using kinetic data from homogeneous solution measurements quantitatively predicts the extent of isomerization of 5-hexenyl bromide, and other primary alkyl bromides, under differing reaction conditions.[49] The only arbitrary parameter in this mathematical analysis is δ, the trapping efficiency of the magnesium surface for diffusing radicals. 'Atypical' organic halides such as strained small-ring or unsaturated compounds may behave differently and form the anion radicals $RX^{\bullet -}$.[46,49] The action of *n*-hexyl bromide as an 'entrainer' for the formation of cyclopropylmagnesium bromide can be explained using this D model via an S_H2 reaction of the free cyclopropyl radical on the more readily formed hexyl Grignard reagent (Equation (2)).[50a] The formation of 20–30% cyclopropane during the normal preparation of cyclopropylmagnesium bromide has also been attributed to solvent attack by freely diffusing cyclopropyl radicals,[50b] although another study of this reaction in $(C_2D_5)_2O$ or in the presence of deuterated dicyclohexylphosphine (DCPD) found little deuterium incorporation in the cyclopropane, and this is inconsistent with the D model.[47b]

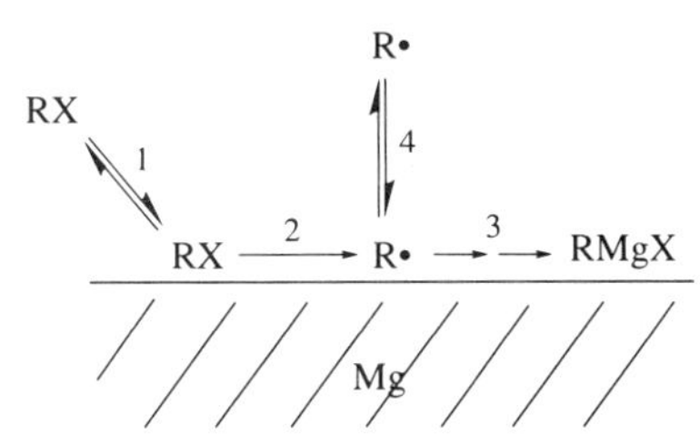

Scheme 1

$$C_3H_5^{\bullet} + C_6H_{13}MgBr \longrightarrow C_3H_5MgBr + C_6H_{13}^{\bullet} \qquad (2)$$

Walborsky and co-workers[47,51–3] have presented older[1] and also more recent evidence, based on product analyses, stereochemistry and radical trapping experiments, for a reaction mainly confined to the magnesium surface. This 'absorption' (A) model basically comprises Scheme 1 omitting step 4, but the variations shown in Scheme 2 have been proposed to explain experimental observations: electron transfer occurs from magnesium to the organic halide, via rate-determining steps 1 and/or 2 of Scheme 2, in which the radical anion formed by step 1 may be a transition state rather than a true intermediate, and is followed by the various steps 3–7.[51] A reaction sequence with steps 1 and 4 would lead to retention of configuration at the reacting carbon centre. Formation of a loose, adsorbed radical pair via steps 2 or 3 would generate random stereochemistry at this carbon in the resulting products. Reactions of the organic bromides (**1**)–(**5**) with magnesium give Grignard reagents with, at least, partial retention of stereochemistry at carbon, as determined by hydrolysis/deuterolysis or carbonation;[47,52] the chiral bromides form Grignard reagents with decreasing optical purity in the order (**2**) > (**1**) > (**4**) ≈ (**3**), whereas freely diffusing radicals would give racemic products, and, although these systems could be considered 'atypical', it should be noted that (**4**) is totally aliphatic.[47c] High degrees of stereochemical retention have also been reported for reactions of magnesium with other cyclopropyl bromides.[54] By using Rieke magnesium, reactions can be carried out at –65/–70 °C with an increase in configurational retention,[51] and under these conditions *exo*-(**5**) gives exclusively the *exo* product but *endo*-(**5**) gives a 50:50 mixture of *endo* and *exo* products.[47a] Further evidence for the lack of significant numbers of freely diffusing radicals during the Grignard preparations of these specific systems has been adduced from the low incorporation of deuterium in the hydrocarbon by-product RH/D when reactions were conducted in deuterated solvents or in the presence of DCPD as a radical trap.[47,52] Also, the distributions of side products formed during reactions are not those of radicals in homogeneous solutions.[51] Moreover, it has been demonstrated that the chemistry of the 1,1-dimethyl-5-hexenyl radical, generated by photolysis in diethyl ether solution, is not affected by the presence of magnesium and that this radical does not react with various activated forms of magnesium to give a Grignard reagent.[53]

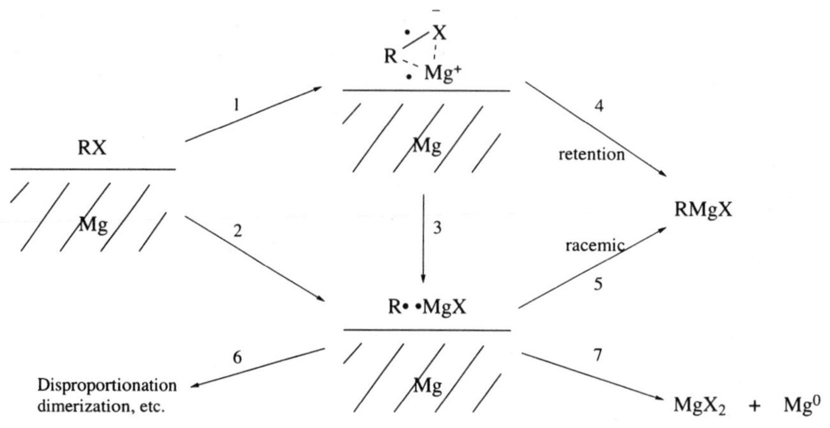

Scheme 2

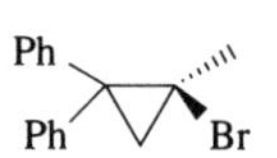

(1)

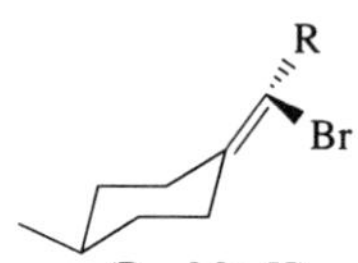

(R = Me, H)
(2)

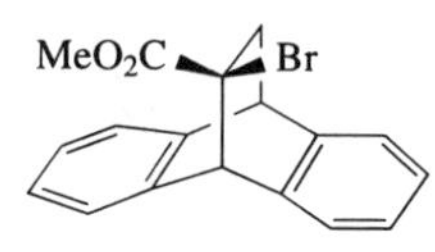

(3)

(4)

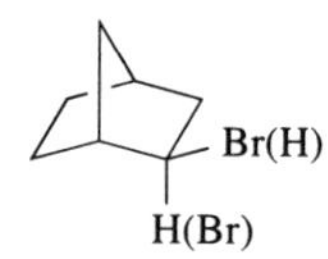

(5)

Whether the 'D' or 'A' mechanism is the better model for the formation of Grignard reagents with normal organic functions is still in dispute. The D model does enable quantitative predictions to be made and so should be subject to testing;[46,55] it does require all radicals to diffuse freely into the solution, but being produced at the metal surface these radicals will have higher concentrations in this region of the solution (~10^{-3} M) and hence be readily trapped, and thus differ in behaviour from normal radicals formed homogeneously.[46] Moreover, there is evidence for a proportion,[46–7] if small, of radicals reacting normally in side reactions. However, the systems cited and studied by Walborsky do appear to behave in a different manner, and formation of 'surface-adsorbed' radicals is an explanation, but such species would be difficult to characterize.

Under ultrahigh vacuum (UHV) conditions, reactions of methyl bromide with a clean magnesium surface produce a surface bromide and gas phase hydrocarbons but give no evidence for the retention of surface methyl moieties; for this system, no stable organomagnesium compound is formed even in the presence of co-adsorbed dimethyl ether, and this implies that a simple, direct, surface-mediated C–Br cleavage is not a viable route to the Grignard reagent under these experimental conditions.[31] The hydrocarbon products could arise from the loss of weakly adsorbed methyl radicals into the gas phase. Co-condensation of organic halides with magnesium is a pathway to unsolvated Grignards,[1,30] and two types of ESR spectra obtained from the condensate with phenyl or *n*-alkyl halides at 77 K have been assigned to phenyl or alkyl radicals R•, respectively, and/or ion radical pairs ($RX^{\bullet -}$, $Mg^{\bullet +}$), the latter occurring with fluoro and chloro derivatives and giving singlet ESR resonances.[56] The two types of radical correspond to those proposed in the initial steps of Scheme 2, and this ESR study, backed by chromatographic product analyses, suggests that stronger carbon–halogen bonds and greater aggregation of the magnesium in the condensate promote ion radical formation.[56] Radical, CCl_3•, and dichlorocarbene intermediates have been implicated in related reactions of co-condensed magnesium atoms with carbon tetrachloride.[57]

Matrix isolation studies at 9 K have also shown that clusters of magnesium (Mg_x ($x \geq 4$)) react more readily than smaller metal fragments or magnesium atoms with alkyl halides and that organomagnesium cluster compounds may form.[10] The stability of the clusters $R(Mg)_xX$ (x = 2, 3) is also supported by theoretical calculations,[58] and such species may initially be formed at the metal surface during normal Grignard preparations. *Ab initio* theoretical studies of the gas phase reactions of a magnesium atom with MeX and C_2H_3X (X = F, Cl) to form RMgX show that activation energies are lower for X = F than for X = Cl;[59] however, these are concerted insertions into the C–X bonds, not involving radical intermediates, and are not directly relevant to normal preparative reactions.

Carbanions have not often been considered as intermediates in the formation of Grignard reagents, but these could be formed by reduction of initial radicals (Equation (3)); such anions have been postulated in the reactions of certain aryl bromides (R = Ar).[60–1] The Grignard reaction of the *o*-bromobenzyltrimethylstannane (**6**) gives the four products of Scheme 3, identified by derivatization, and these are believed to arise via rearrangements of the intermediate aryl carbanion.[60] Ring cleavage side reactions that occur during the preparations of the crown ether Grignard reagents (**7**) (A = MgBr) from the corresponding bromides (**7**) (A = Br) have also been attributed to the intermediacy of the aryl carbanions, and related evidence is presented for the formation of phenyl anions in the reaction between bromobenzene and magnesium in the presence of (**7**) (A = MgBr, *n* = 0).[61]

$$\underset{}{R^\bullet} + \underset{(MgX^\bullet)}{\text{'Mg}^{I}\text{'}} \longrightarrow R^- + \underset{(MgX^+)}{\text{'Mg}^{II}\text{'}} \qquad (3)$$

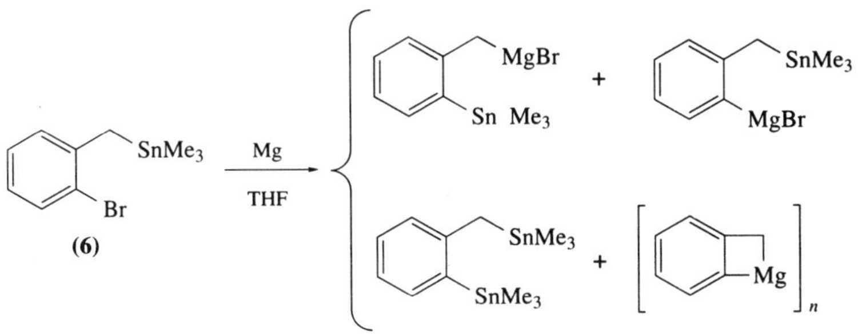

Scheme 3

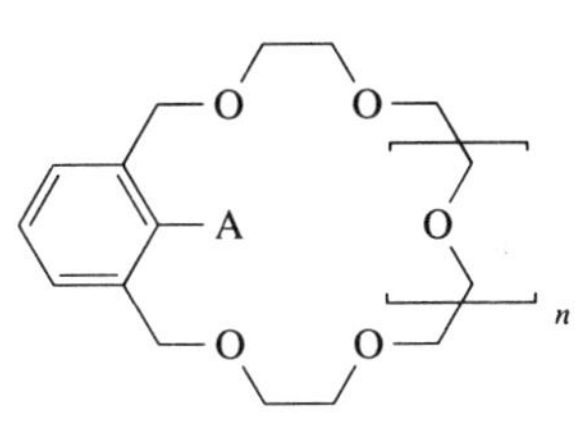

(**7**) *n* = 0 or 1

Magnesium–anthracene compounds are a source of reactive magnesium metal but also react directly with certain organic halides to form Grignard reagents (see above). The mechanism of the latter reactions probably involves electron transfer from [Mg(anthracene)$(THF)_3$] or $[\text{anthracene}]^{\bullet-}$ to the organic halide in solution, and the catalytic effect of small amounts of anthracene on the direct preparation of Grignards is to promote such homogeneous rather than surface electron transfer reactions.[25a]

(ii) Exchange reactions

(a) Metal–hydrogen exchange. Few new applications of the metallation of acidic hydrocarbons by Grignard reagents (Equation (4)) have been reported. Details have been given for the monometallation of ethyne by Bu^nMgCl to form ethynylmagnesium chloride in THF.[62] η-Cp*MgCl(THF) can be prepared from pentamethylcyclopentadiene by exchange with Pr^iMgCl in toluene at 110 °C, followed by crystallization from THF.[63] The enthalpies of metallation of various hydrocarbons by pentylmagnesium chloride in diethyl ether have been determined and show a linear correlation with pK_a values of the carbon acids, with ΔH varying from -148 kJ mol^{-1} to $+9.2$ kJ mol^{-1} from most acidic cyclopentadiene to cyclohexane, respectively.[64] The secondary α-deuterium isotope effect for the reaction of phenylacetylene with the deuteriated Grignard reagent 1-bromomagnesiopentane-1,1-d_2 is very small (0.99 ± 0.03) and is mechanistically nondiscriminating.[65]

$$R^1H + R^2MgX \longrightarrow R^1MgX + R^2H \qquad (4)$$

The aromatic ring of the 1,3-xylyl crown ether (**7**) (A = H, *n* = 0) is metallated at the 2-position by PhMgBr, or diphenylmagnesium;[66] initially, a 1:1 complex is formed between the organomagnesium reagent and the crown ether at room temperature, but on heating this undergoes intramolecular metallation to form (**7**) (A = MgBr, *n* = 0). A heterolytic mechanism is shown in Scheme 4 and is supported by observations of related reactions of metal–halogen exchange, (**7**) (A = Br), and cleavage of ethers, (**7**) (A = OR).[5] Heterolytic dissociation of the initial 'side-on' complex (**7a**) forms (**7b**), and this is followed by a combined intramolecular electrophilic attack at the aromatic carbon by coordinated MgY^+ and nucloceophilic attack at the hydrogen by the magnesiate anion $[MgPh_2Y]^-$ to cause metallation. Complexation of the polydentate oxygen ligand to magnesium enhances its reactivity in this case; however, the homologous crown-5 (**7**) (A = H, *n* = 1) with a larger cavity is not metallated and only forms 1:1 complexes, which in the case of the diphenylmagnesium reactant can adopt a rotaxane structure.[66]

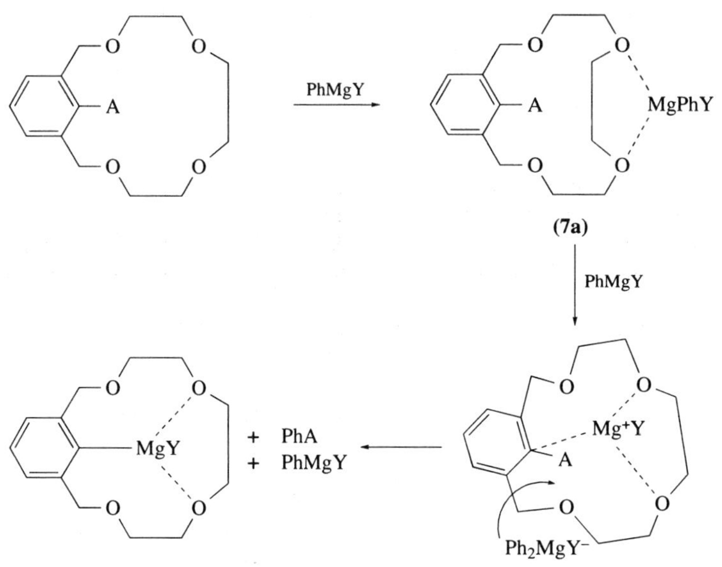

(**7a**)

(**7b**)

Scheme 4

Hauser bases R_2NMgBr (R_2NH = 2,2,6,6,-tetramethylpiperidine (TMPH), or diisopropylamine (DAH)) are stable even in boiling THF, unlike the lithium analogues, and have found use in metallation of activated, amidocarboxy-substituted aromatic or cubane systems to form Grignard reagents as synthetic intermediates.[67] *Ortho* metallation of diethylbenzamide occurs in refluxing THF. '*Ortho*magnesiations' of cubanes include reactions of (**8**) (A = B = H) with the amidomagnesium bases in refluxing THF to give the dimetallated product (**8**) (A = B = MgBr) but only the monometallated product (A = MgBr, B = H) at room temperature; other activated cubanes, even cubane monoamides, undergo up to three *ortho* metallation steps.[67] Magnesium diamides such as $Mg(TMP)_2$ are also useful metallating agents but produce organomagnesium amides.

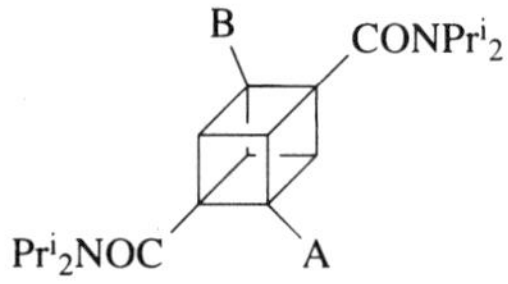

(**8**)

(b) Metal–halogen exchange. Metal–halogen exchange (Equation (5)) is often useful for preparation of Grignard reagents from halides of more electronegative organic groups,[1] such as perfluoroalkyl derivatives—which can be prepared from perfluoroalkyl iodides and phenyl- or ethylmagnesium bromide at low temperatures, as outlined in a recent review.[68] Polyhalobenzenes react at room temperature with excess aryl- or alkenylmagnesium halides to give organic products containing new aryl–aryl or aryl–alkenyl bonds via a sequence of reactions, including primary metal–halogen exchange to form the halogenated Grignard followed by elimination of magnesium halide to give an aryne

intermediate; the less basic, *sp*-hybridized, alkynylmagnesium halides are inactive unless alkyl Grignard reagents are added to initiate the organometallic exchange.[69]

$$R^1X^1 + R^2MgX^2 \longrightarrow R^1MgX^2 + R^2X^1 \quad (5)$$

Various bromoanisoles react with an excess of an alkylmagnesium halide in THF to form aromatic Grignards: 2- and 4-bromoanisole require a large excess of the more basic reactant Pr^iMgX to give relatively poor conversions (64% and 28%, respectively), whereas 2,4- and 2,6-dibromoanisole and 2,4,6-tribromoanisole are easily converted into the mono-Grignard reagent by metal exchange with the bromine atom *ortho* to the methoxy group.[70] Methylmagnesium bromide reacts relatively slowly in these reactions, probably as it is a weaker base, and addition of other solvents to THF also retards the reaction.[70] 2-, 3- or 4-Pyridylmagnesium bromide can be prepared in a high yield by exchange between the corresponding iodopyridine and EtMgBr in THF at room temperature.[71] Partial exchange between (*Z*)- or (*E*)-β-bromostyrene and Bu^nMgBr occurs at ambient temperature in the mixed solvent DME/THF, but not in THF alone, and the resulting alkenyl Grignard reagents are formed with a highly stereospecific retention of the alkene configuration, as shown by subsequent carbonation.[72]

Bromine–magnesium exchange occurs between the crown ethers (**7**) (A = Br, *n* = 0, 1) and PhMgBr (or faster with $MgPh_2$) at room temperature in apolar solvents to afford the corresponding organomagnesium compound in a more facile reaction than the analogous metallation of (**7**) (A = H), and a related heterolytic mechanism involving combined intramolecular electrophilic attack and external nucleophilic attack is proposed (Scheme 4 with A = Br).[66] This reaction does not proceed in THF since coordination of magnesium by the excess THF competes with the initial complexation by the crown ether.

The effects of solvent variation, complexation and basicity of the Grignard reagent, and electron-withdrawing substituents on the organic halides support a heterolytic mechanism for these metal–halogen exchanges in most cases. However, in iron-catalysed 'Kharasch reactions' between alkyl halides and Grignard reagents there is CIDNP evidence for radical intermediates and, among other steps, an S_H2 reaction of alkyl radicals with the Grignard reagent, related to Equation (2), effects metal–halogen exchange.[73]

(c) Miscellaneous exchange reactions. Transmetallation reactions are sometimes useful for the preparation of Grignard reagents. The exchange between an organolithium derivative and one equivalent of anhydrous magnesium halide (Equation (6)) is often a convenient method since the lithium precursor may be more easily obtained by a reaction such as metallation or lithium–halogen exchange. The exchange reaction can be carried out at low temperature if required and occurs with retention of stereochemistry. Some recent synthetic applications of this technique, employing $MgBr_2$ or $MgBr_2{\bullet}Et_2O$ in THF, are given elsewhere.[74–8] A related exchange involving both lithium and potassium centres is shown in Equation (7).[79]

$$LiR + MgX_2 \longrightarrow RMgX + LiX \quad (6)$$

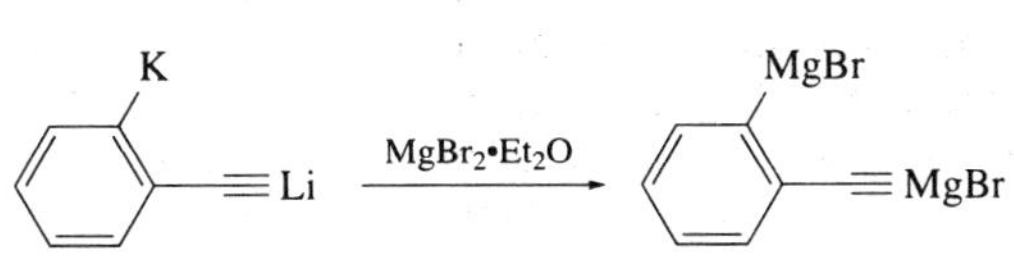

(7)

Stable organomercury compounds can usually be obtained in high purity by crystallization so that subsequent quantitative exchange with magnesium (Equation (8)), can provide pure Grignard reagents.[80] The formation of the corresponding mercury compound from an impure Grignard and regeneration of the organomagnesium by mercury–magnesium exchange is a useful purification procedure. However, it is to be noted that transmetallation of magnesium with the organomercury species (**7**) (A = HgBr) does not occur.[78] An alternative mercury–magnesium exchange reaction between organomercury(II) chloride and MeMgBr has been employed to form the Grignard reagent of cubane (**8**) (A = B = MgBr) and of an amide-substituted cyclopropane.[81] The mechanism of these reactions probably involves the initial formation of a diorganomercury compound followed by reaction with metallic magnesium or an alkyl Grignard reagent.

$$RHgBr + Mg \longrightarrow RMgBr + Mg/Hg \quad (8)$$

Alkyl and aryl pyridyl sulfoxides react with Grignard reagents such as PhMgBr to afford the corresponding 3- and 4-pyridyl and 4-quinolyl Grignard reagents via sulfur–magnesium exchange in

THF at room temperature.[82] Diorganotellurium compounds react with Bu^nMgCl or PhMgCl at 20 °C in THF to form a range of alkenyl-, alkynyl-, aryl-, allyl- and benzylmagnesium chlorides by the exchange of Equation (9) ($R^2 = Bu^n$ or Ph), and, if Bu^tMgCl is used as the reactant, Bu^nMgCl is produced from $TeBu^n_2$.[83]

$$R^1TeR^2 + R^2MgCl \longrightarrow R^1MgCl + TeR^2_2 \quad (9)$$

Equimolar amounts of diorganomagnesium compounds and magnesium halide redistribute to form solutions in normal solvents of the corresponding Grignard reagent via the dynamic Schlenk equilibrium (Equation (10)). Thus, α,ω-di-Grignard reagents are formed by ring opening of magnesacyclopentane, or its higher homologues, on reaction with $MgBr_2$ in THF.[84]

$$MgR_2 + MgX_2 \rightleftharpoons 2RMgCl \quad (10)$$

(iii) Reactions with unsaturated bonds

The addition of Grignard reagent to an unsaturated bond is a common organic synthetic procedure, and this process may form a new organometallic compound, although not always identified. Multiple bonds involving heteroatoms or conjugated with heteroatom functions, as in α,β-unsaturated carbonyl compounds, give addition products containing a magnesium atom primarily associated with the heteroatom and will not be considered in this section. Only intermolecular reactions between Grignards and C–C multiple bonds will be discussed here; intramolecular transformations are covered in Section 3.2.1.3(ii). Also, polymerizations of alkenes initiated by Grignard reagents will be excluded, although the products may be polymeric Grignard reagents.

(a) Carbomagnesiation. Earlier work on the addition of preformed Grignard reagents to alkenes[1] (Equation (11)) has been reviewed by Lehmkuhl.[85] Uncatalysed carbomagnesiation reactions of isolated double bonds occur most readily with allylic or *t*-butyl Grignard reagents and with terminal or strained cyclic alkenes in diethyl ether or a hydrocarbon solvent; coordinating solvents such as THF reduce the reaction rate; conjugated alkenes react more readily. Such observations, together with stereochemical and kinetic data,[1,85] support a nonionic, concerted mechanism with a highly ordered transition state, and with allylic Grignard reagents a six-centred cyclic pathway (**9**) has been proposed to account for the inversion of the organic group during addition.

$$R^2R^3C{=}CR^4R^5 + R^1MgX \longrightarrow R^1R^2R^3C{-}CR^4R^5MgX \quad (11)$$

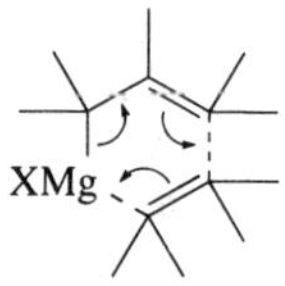

(9)

Syn addition of simple alkyl, phenyl and vinyl Grignard reagents to methylcyclopropene, 3,3-dimethylcyclopropene and 1,3,3-trimethylcyclopropane occurs in a high yield, with magnesium being attached to C-2; sterically hindered 1,2,3-triphenylcyclopropane is inert to addition and alkynyl Grignard reagents are unreactive.[86] Further addition of cyclopropylmagnesium halide to cyclopropene can occur,[86] as in the reaction of 1-isobutenylmagnesium bromide with 3,3-dimethylcyclopropane, which gives a mixture of organometallic products (**10**).[87] A kinetic study of additions to a cyclopropene, spiro[2,4]hept-1-ene, showed that EtMgBr or $MgEt_2$ react at approximately the same rate in a reaction that is first order in each reactant, probably involving a four-centre transition state; relative reaction rates for organometallics MgR_2 (R = Me, Et, Pr^i, Bu^t) are 1, 10.3, 32 and 8, respectively, probably reflecting opposing steric and electronic effects.[88]

Additions to alkenes functionally substituted with hydroxy, amino, alkoxy, fluoro or silyl groups have been reviewed.[89] Electron-withdrawing groups activate alkenes to carbomagnesiation, but reactions of alkyl and aryl Grignard reagents RMgX with perfluoro-2-methyl-2-pentene (C_6F_{12}) cause α- and γ-fluorine elimination, as MgFX, giving two isomeric, substituted fluoroalkenes (RC_6F_{11}) and no organometallic products.[90] Activation of alkenes by silyl groups is demonstrated by the regioselective formation of TMS-CH(MgX)CH_2R from vinyltrimethylsilylane and alkyl or 2-enyl Grignard reagents

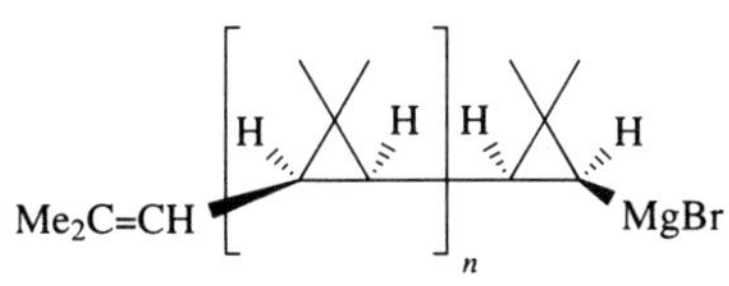

(10) $n = 0,1,2$

RMgX.[91] Addition of Pr^iMgBr also occurs to vinyl(alkoxy)silanes ($(RO)_nMe_{3-n}SiCH{=}CH_2$ (n = 1 or 2)) for a range of alkoxy groups, and a frontier orbital analysis suggests that replacement of alkyl by alkoxy groups on silicon should promote the reaction, but with the primary reagent Bu^nMgBr a competing substitution of the alkoxy groups may predominate and reactions are more subject to steric hindrance.[92]

Activation of alkenes by neighbouring nitrogen functions is displayed in the regioselective addition of 2-butenylmagnesium chloride to the isolated double bond of 2-(1-methyl-2-propen-1-yl)pyridine and related 2-(α-methylallylic) aza-aromatic compounds,[93] and also in the additions of allylic Grignard reagents to *N*-allylaniline, cinnamylamine and *N,N*-dimethylcinnamylamine.[94] Activation by interaction of the nitrogen with magnesium is proposed, possibly facilitating the formation of a six-centre transition state;[93] activation by metallation of the primary and secondary amine groups is greater than by coordination with a tertiary amine although less effective than comparable metallated hydroxyl groups, and the products may be amido derivatives rather than Grignard reagents.[94] Hydroxyl groups assist additions of allylic Grignard reagents to *o*-allylphenol[95] and various homoallylic alcohols,[96] but the organometallic intermediates are probably cyclic oxomagnesium compounds. The addition of phenylmagnesium bromide to unhindered vinylic sulfones may also be noted.[97]

Reviews covering earlier work on the carbomagnesiation of 1,3-dienes,[85,98] alkynes[99] and conjugated enynes[100] have been published. Grignard reagents do not add readily to simple alkynes in the absence of transition metal catalysts unless proximal nitrogen or oxygen functional groups are present.[89,99] Allylmagnesium chloride adds to 3-(dimethylamino)-1-phenyl-1-propyne in refluxing THF, with magnesium being attached to C-1.[94] *Anti* additions of Grignard reagents to propargylic and homopropargylic alcohols,[99] and to 2-butyne-1,4-diol,[101] also occur but intermediate organometallic cyclic oxomagnesium compounds are formed.[99] Simple, internal enynes undergo 1,4-addition with allylic Grignard reagents, forming allenyl products in low yields, and functionalization at the position α to the unsaturation facilitates addition but may promote addition to the triple bond, forming conjugated 1,3-dienyl products.[100]

Cp_2ZrCl_2 or related zirconocene derivatives are effective catalysts for carbomagnesiation of 1-alkenes, and some bi- and tricyclic alkenes, under mild reaction conditions (<25 °C).[102–6] The reaction of aliphatic 1-alkenes with EtMgX is highly regioselective (Equation (12)), with internal double bonds being unreactive;[102] 1-alkenes with functional groups (e.g., OR, OH, SR, NR_2, TMS) on the chain react similarly[102] and, contrary to earlier reports, allylic alcohols and ethers are active substrates.[104] Addition to 1-alkenes with aromatic substituents, such as styrene, mainly occurs with opposite regioselectivity to that of Equation (12) to form a benzylic Grignard, and the disubstituted β-methylstyrene is also ethylmagnesiated.[103,105] $MgEt_2$ is more active than EtMgX (with varying activity: X = Cl > Br > I) in these zirconium-catalysed reactions,[102] although Grignard reagents do give good yields of the products in diethyl ether or THF.[103–5] Hydrocarbon solvents are suitable if the stoichiometric complexes $R^1MgX{\cdot}NR^2_3$ are employed.[104] Mechanistic studies have implicated alkene–zirconocene complexes (zirconacyclopropanes) and zirconacyclopentanes (**11**) as intermediates in these reactions, and one possible pathway for the catalytic reaction is illustrated in Scheme 5 (cf. Takahashi and co-workers[103]). However, variations on Scheme 5 must occur: the formation of 1,4-di-Grignard reagents under some conditions[105a] (e.g., Equation (13))[106] may involve Zr/Mg transmetallation and cleavage of both Zr–C bonds of zirconacyclopentane (**11**); the formation of unsaturated Grignards in reactions of higher alkylmagnesium reagents (Equation (14)) must involve a β-hydrogen elimination step after head-to-tail coupling of alkene units to form the intermediate zirconacyclopentane.[103b] Various alkenylmagnesium products have been noted in zirconium-catalysed reactions of 1-octene and Pr^nMgBr.[107]

The oxygen functions of allylic and homoallylic alcohols, or the corresponding ethers, can afford regio- and/or diastereocontrol in zirconium-catalysed carbomagnesiation reactions (e.g., Equation (15)) in which coordination of oxygen to magnesium directs the mode of metallacyclopentane cleavage.[104] Related zirconium-catalysed reactions are cyclomagnesiations of 1,7-octadiene and of various 1,6-heptadienes using *n*-butylmagnesium chloride (e.g., Equation (16)) occurring via intermediate (**12**), for which the stereoselectivity varies with the substrate and solvent.[105]

Nickel compounds, in the presence of $AlMe_3$ as a reductant, catalyse the predominantly *syn* additions of methyl Grignard reagents to alkynylsilanes, although the catalyst also promotes isomerization of the

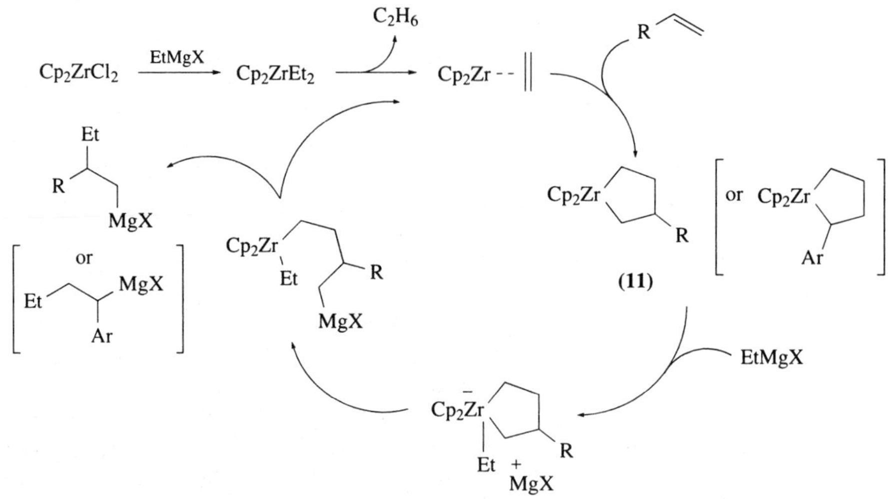

Scheme 5

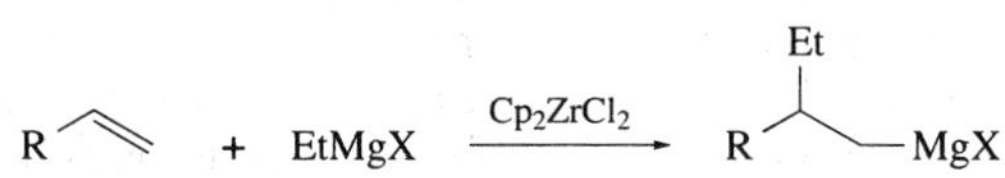

(12)

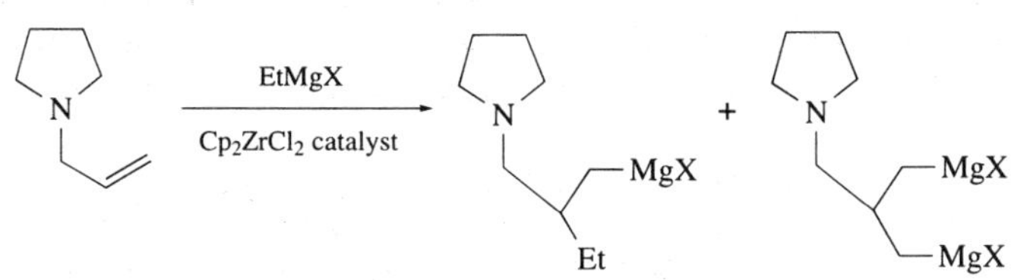

(13)

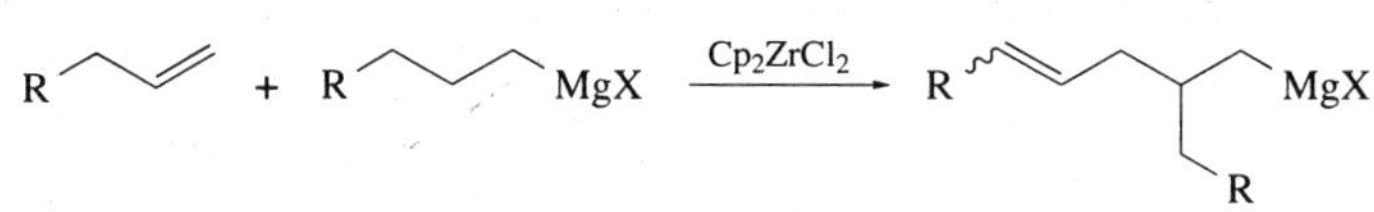

(14)

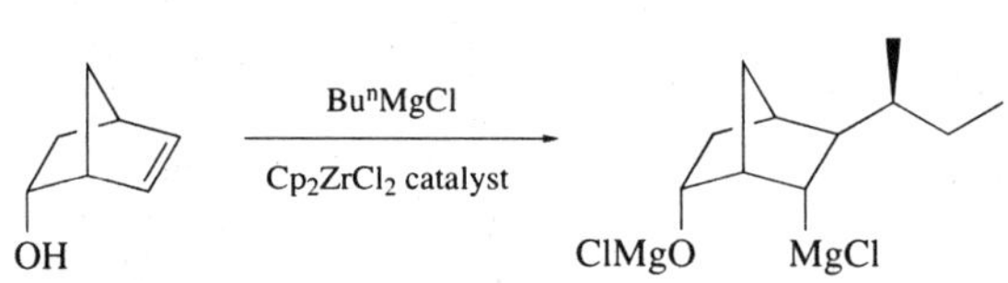

(15)

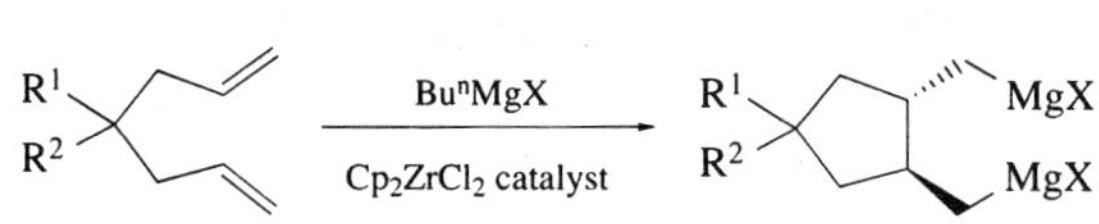

(16)

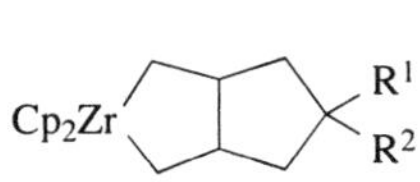

(12)

product (Equation (17)), particularly if proximal oxygen or nitrogen functions X are present;[108] higher Grignards with β-hydrogen atoms afford mainly hydrometallation products under these conditions.[99] Copper(I)-catalysed Grignard additions to alkynes have been reviewed:[99,109] ethyne undergoes *syn* addition (35–40%) (Equation (18)) rather than metallation, but propargylic alcohols in diethyl ether normally undergo heteroatom-assisted *anti* additions, forming oxomagnesium compounds. Grignard reagents add to disubstituted alkynes in the presence of $PdCl_2$ with HMPT as the cosolvent, but the intermediates are probably not organomagnesium compounds.[110]

Additions of arylmagnesium bromides to the triple bond of conjugated enynes are catalysed by various transition metal compounds, particularly nickel(II) derivatives such as $Ni(acac)_2$.[111] *Syn* adducts are formed primarily, but these isomerize under the reaction conditions. Related reactions of conjugated dienes have been described but mixtures of addition and coupling products are formed.[112]

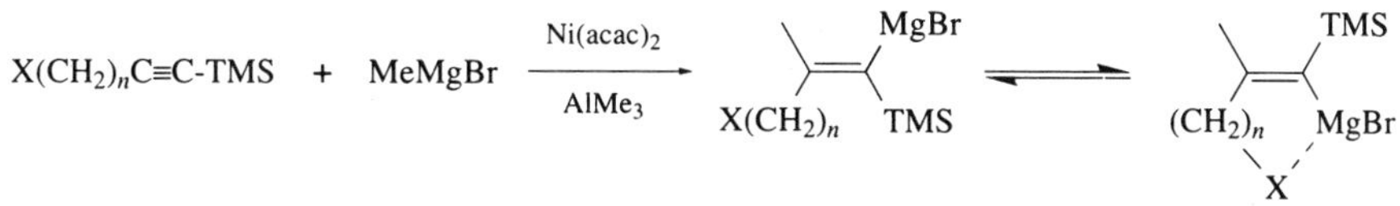

(17)

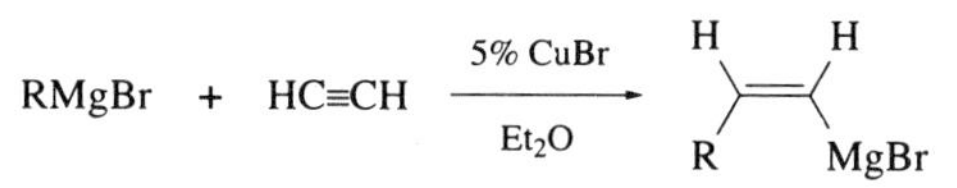

(18)

(b) Hydromagnesiation. Titanium catalysts ($TiCl_4$, $Ti(OPr^i)_4$ or Cp_2TiCl_2) cause Grignard reagents with β-hydrogens to react with the vinyl group of 1-alkenes, but not internal alkenes or vinylidene groups, as in Equation (19).[113–14] Styrene and related vinyl aromatic compounds[115] yield α-arylethylmagnesium halides predominantly. The reversible reaction is promoted by loss of the volatile alkene product. A range of Grignard reagents, other than MeMgX, can be used and other transition metal catalysts including $NiCl_2$ are also effective.[114–15] Moreover, 4-hydroxy functions of homoallylic alcohols are tolerated in this reaction.[116] In the hydromagnesiation of 1-octene, with 5% Cp_2TiCl_2 as the catalyst, Bu^iMgCl was found to be the best Grignard reagent (53% yield), although halide-free $MgBu^i_2$ and MeMgH, both giving non-Grignard products, are better.[113] Isomerization of the 1-alkene to the 2-alkene (≥30%) by the titanium catalyst is a competing side reaction. Mechanisms of these titanium-catalysed reactions probably involve the formation of a Ti–H bond by β-hydrogen elimination from an intermediate alkyltitanium compound, insertion of the 1-alkene to form a new Ti–C bond and Ti–Mg exchange to liberate the Grignard reagent. Titanium(III) species (Cp_2TiR and Cp_2TiH) are believed to be active in systems with Cp_2TiCl_2.[113–14]

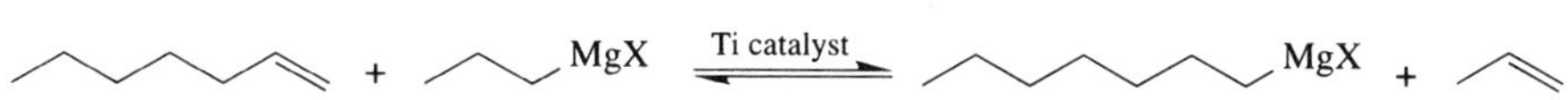

(19)

Cp_2TiCl_2, but not $TiCl_4$, is a catalyst for the room temperature hydromagnesiation of 1,3-butadiene and 2-substituted 1,3-butadienes, regioselectively as in Equation (20), where R = alkyl or TMS however, 4-substituted 1,3-butadienes do not react.[114] Allylic Grignard reagents formed by this method have found various applications in organic synthesis.[114] Disubstituted alkynes are *syn* hydrometallated by Bu^iMgBr in diethyl ether at room temperature in the presence of catalytic amounts of Cp_2TiCl_2 to form alkenyl Grignard reagents in high yields; low regioselectivity is observed for reaction of dialkylalkynes, but when R^1 = Ph and R^2 = alkyl or TMS, or R^1 = TMS and R^2 = alkyl, the product of Equation (21) is selectively formed. Propargylic alcohols also react with two molar equivalents of Bu^iMgBr under Cp_2TiCl_2 catalysis to give the *syn* addition product (**13**), and this contrasts with the normal *anti* pathway of copper(I)-catalysed carbomagnesiation reactions. However, isomerization of (*Z*) to (*E*) forms of alkenyl Grignards derived from 3-trimethylsilylpropargylic alcohols occurs.[114]

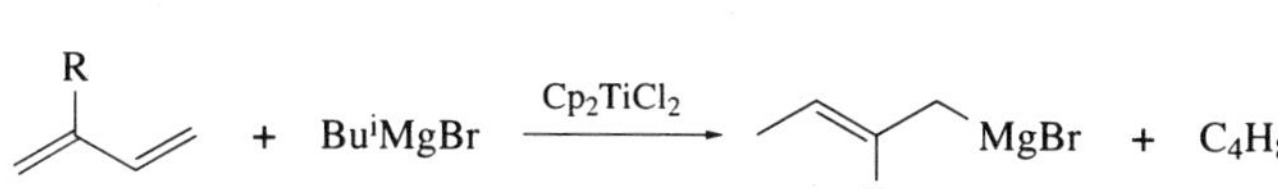

(20)

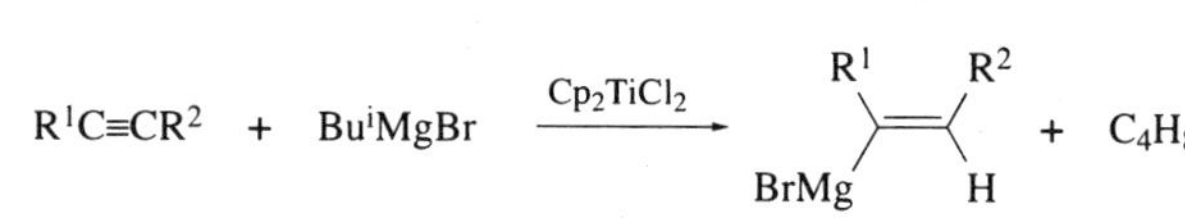

(21)

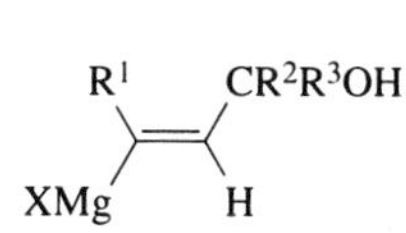

(13)

(c) Other addition reactions. Phosphorus and arsenic ylides react with $MgCl_2$ to form dimeric 1:1 adducts, assigned ionic structures containing carbon– and chlorine–magnesium bonds (Equation (22)).[117]

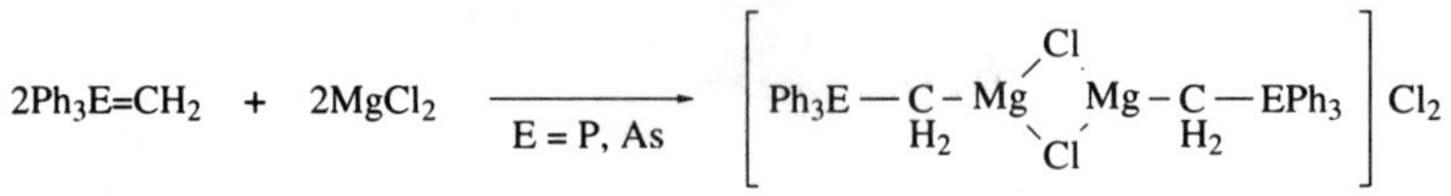

(22)

(iv) Preparation of functionalized Grignard reagents

Various examples of Grignard reagents bearing silicon-, nitrogen-, oxygen-, sulfur- or halogen-containing functions or unsaturated C–C bonds, formed by direct reaction, exchange reactions or addition to unsaturated organic compounds, have been mentioned in the preceding sections. In this section, further examples of interest will be presented.

Allylmagnesium halides can be produced by several methods, but (*Z*) and (*E*) forms of γ-substituted reagents isomerize rapidly even at low temperatures. The direct reaction between 'Rieke' magnesium and the γ-disubstituted (*E*) (geranyl) or (*Z*) (neryl) allylic chloride forms (**14**) with extremely high retention of stereochemistry only at temperatures below –95 °C, no selectivity is observed above –65 °C and Grignard reagents from γ-monosubstituted 2-decenyl chlorides still undergo rapid stereoisomerization at –100 °C.[118] The direct synthesis of allenylmagnesium chloride from propargyl chloride needs special procedures or the use of magnesium–anthracene,[26a] but it may be noted that $Me_2C{=}C{=}CHMgBr$ is easily prepared from $Me_2CBrC{\equiv}CH$ and magnesium activated by the presence of a little mercuric chloride.[119]

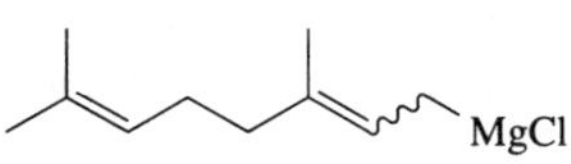

(**14**) (*E*) or (*Z*)

Halogenated Grignard reagents of the type $CF_3CXYMgX$ can be made smoothly by metal–halogen exchange between CF_3CXYZ (X = F, Cl, Br; Y, Z = Cl, Br) and alkyl- or arylmagnesium halide at low temperatures.[120] However, direct reaction of $CF_3CHClBr$ with magnesium at –53 °C gave $CF_3CClBrMgBr$ as the intermediate organometallic formed by secondary metallation of $CF_3CHClBr$ by $CF_3CHClMgBr$.[121] Magnesium–iodine exchange has been used to prepare the functionalized perfluoroalkyl Grignard reagents $RO_2C(CF_2)_4OCF_2CF_2MgBr$[122] and $FSO_2(CF_2)_nMgBr$.[123] Direct reactions of benzotrihalides ($PhCX_3$; X = Cl, Br) with magnesium give diphenylacetylene and other organics as products; when X = F no reaction was reported, but trifluoromethylphenyl halides can explode during normal Grignard preparations, probably through promoted reactions of the CF_3 group.[124]

High-yield, direct preparations have been described for the functionalized methyl Grignard reagents $RSCH_2MgCl$ (R = Me, *p*-Tol)[125] and $(TMS)_3CMgBr$.[25,126] The Grignard reagent $PhO(CH_2)_3MgBr$ is formed, without accompanying cyclopropane, by reaction of the corresponding γ-bromoether with Rieke magnesium at –78 °C.[127]

Various *ortho*-functionalized aromatic Grignard reagents have been synthesized. The phosphine species $(o\text{-}BrMgC_6H_4CH_2)_2PMe$ can be prepared from magnesium and $(o\text{-}BrC_6H_4CH_2)_2PMe$;[128] other syntheses involve *ortho* metallation,[66–7] *ortho* halogen–metal exchange[66,70] or transmetallation of *ortho*-lithiated derivatives with $MgBr_2$,[75] including formation of the homochiral acetal (**15**).[129]

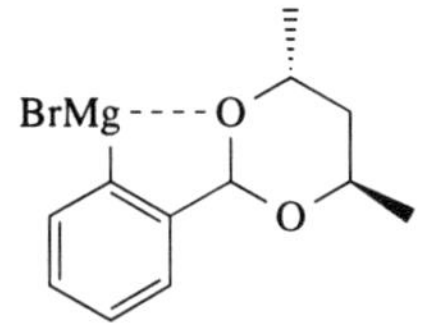

(**15**)

The trimethylsilyl group is useful for protection of terminal alkyne[130] or phenol[131] functions during Grignard reactions and also stabilizes the furfuryl Grignard reagents (**16**) (R = H, Me, Bu) against ring opening.[132] Silylated methylmagnesium chloride ($(Pr^iO)Me_2SiCH_2MgCl$) is a hydroxymethyl anion equivalent, via oxidative cleavage of the C–Si bond in subsequent organic products.[133] Primary amine group protection is provided by the tetramethyldisilylazacyclopentane ring in the Grignard reagent (**17**), prepared by direct reaction of the protected aminopropyl bromide.[134]

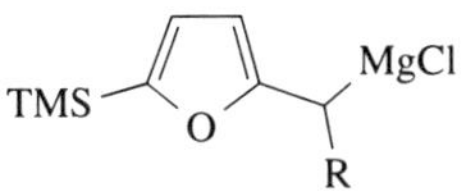

(16)

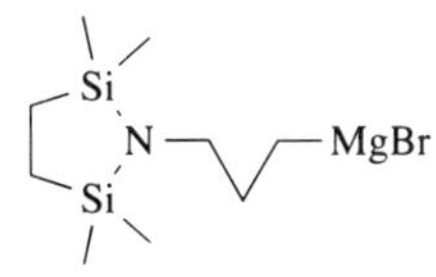

(17)

(v) Di- and poly-Grignard reagents

Although aliphatic di-Grignard reagents with metal centres separated by at least four carbon atoms, as in $XMg(CH_2)_nMgX$ ($n \geq 4$), are prepared normally, the synthesis of 1,1-, 1,2- and 1,3-reagents have presented problems.[1] Bickelhaupt and co-workers have reported many significant advances in this area, outlined in recent reviews.[135–6] The 1,1-di-Grignard $CH_2(MgBr)_2$ can best be prepared as an insoluble product in a 60–80% yield by reaction of dibromomethane with magnesium amalgam in diisopropyl ether and can be purified by dissolution in diethyl ether–benzene.[16,137] The silylated reagents $(TMS)RC(MgBr)_2$ (R = H or TMS)[138] can be prepared similarly. However, $(TMS)_2C(MgBr)_2$ is soluble in diisopropyl ether and difficult to separate from the mono-Grignard coproduct, and simple direct reaction of $(TMS)_2CBr_2$ with triply sublimed magnesium in diethyl ether yields pure crystals; *in situ* transmetallation using lithium-4,4'-di-*t*-butylbiphenyl (LiDBB) and $MgBr_2$ can also be used.[138b]

vic-Di-Grignard reagents have not been prepared from open-chain 1,2-haloalkanes, since elimination of the metal halide to form an alkene predominates.[135] However, trapping evidence shows that 1,2-bis(halomagnesio)ethane is formed in admixture with other dimetal species during the exchange and coupling reaction of LiDBB/$MgBr_2$ and CH_2I_2 at −100 °C (Equation (23)).[139] Pure *cis*-1,2-bis(bromomagnesio)cyclopropane (**18**) is prepared in ~10% yield by reaction of either *cis*- or *trans*-1,2-dibromocyclopropane with magnesium;[140] β elimination to form cyclopropene is less favourable although this is still the main product.[135]

(18)

$$CH_2I_2 \xrightarrow[MgBr_2]{LiDBB} CH_2(MgX)_2 + XMg(CH_2)_2MgX + XMg(CH_2)_3MgX + XMg(CH_2)_4MgX \quad (23)$$

The 1,3-di-Grignard reagent (**19**) (R = H), previously obtained by an indirect route to avoid the competing formation of cyclopropane,[1] can be prepared directly in a 30% yield by slow addition of 1,3-dibromopropane to excess magnesium in diethyl ether.[135,141a] Precipitation as oligomeric magnesacyclobutane and reconstitution by addition of one equivalent of $MgBr_2$ affords a stable product, but decomposition by elimination of HMgBr is promoted by the presence of excess $MgBr_2$. The disubstituted reagent (**19**) (R = Me) can be prepared similarly[141] and, lacking β-hydrogens, is more stable.[135] Cyclic 1,3-dibromides (**20**) (<9% yield) and (**21**) (<45% yield) can also be formed by direct reaction, but (**21**) decomposes by elimination to form 2-cyclopentenylmagnesium bromide.[135,142] Higher α,ω-bis(bromomagesio)alkanes ($(CH_2)_n(MgBr)_2$ (n = 4, 5;[84] n = 6, 9[143a])), *o*-$(BrMgCH_2CH_2)_2C_6H_4$[143a] and oxygen-containing derivatives, including $O[(CH_2)_nMgBr]_2$ (n = 2, 4),[143b] have been prepared and studied, but the latter species disproportionate completely in THF to oxamagnesacycloalkanes and $MgBr_2$. 1,4-Di-Grignard reagents are also formed in some reactions catalysed by Cp_2ZrCl_2 (see Equations (13)[106] and (16)[105]). The transition metal dimagnesium reagents (**22**) (M = Ti, Zr, Hf)[137a] are formed from Cp_2MCl_2 and $CH_2(MgBr)_2$.

2,2'-Bis(chloromagnesiomethyl)biphenyl,[144] and a range of other benzylic di-Grignard reagents, as well as the trimetallic derivative (**23**), have been prepared from the corresponding benzyl halide and a magnesium–anthracene adduct.[25a] The aliphatic trimetallic (**24**) can be prepared in a good yield by direct reaction with normal magnesium turnings.[145] The 1,3-di-Grignard reagent *o*-$(BrMgCH_2)C_6H_4MgBr$ is unstable to formation of 1-magnesacyclobutabenzene[146] (cf. Scheme 3), but *o*-$BrMg(CH_2)_3C_6H_4MgBr$ is preparable.[143a] The synthesis of a phosphine di-Grignard reagent ($(o\text{-}BrMgC_6H_4CH_2)_2PMe$),[128] and of an alkynyl–aryl derivative by transmetallation (see Equation (7)),[79] may be noted. Also, dialkynyl Grignard reagents have been formed by metallation of diethynylsilanes.[147]

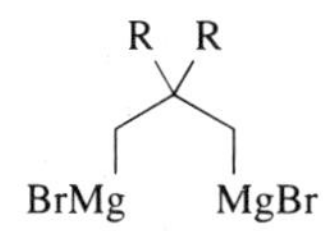

(19)

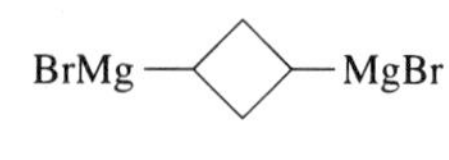

(20)

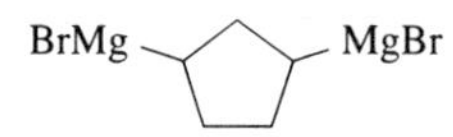

(21)

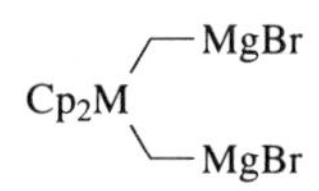

(22)

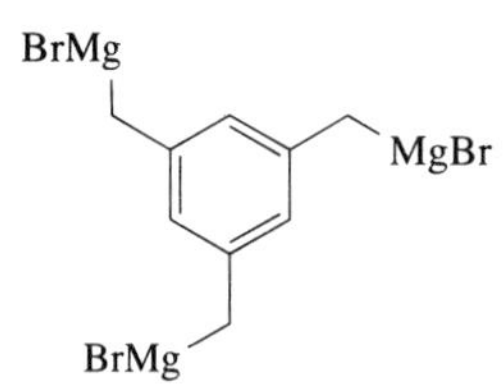

(23)

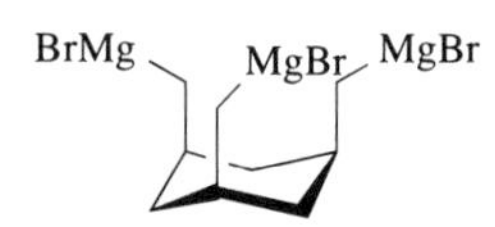

(24)

o- and *p*-Dichlorobenzene react with anthracene-activated magnesium to give the corresponding di-Grignard reagents, with a 70% yield for the 1,2 derivative.[148] Cubane di-Grignards, including (**8**) (A = B = MgBr), are formed by metallation or transmetallation reactions, and trimagnesium cubane derivatives can also be formed.[67,81]

Insoluble, cross-linked poly(chloromethyl)styrene is converted (90%) into a poly-Grignard reagent by reaction with [magnesium(anthracene)$(THF)_3$] in THF, and other brominated polymers give similar products.[27,149] Also, homopolymerization of *m*-vinylbenzylmagnesium chloride with benzylmagnesium chloride produces a poly-Grignard reagent.[150]

3.2.1.2 Structures and physical properties

(i) Solid state

Although Grignard reagents are mainly prepared and utilized in solution, the structures of crystalline compounds are of interest.[2] Definitive characterizations of the crystal and molecular structures using x-ray diffraction techniques give an insight into the possible coordination geometries and associated structures in solution, and provide structural parameters for theoretical calculations. Information from solid-state structures, in combination with results from studies using less rigorous physical and spectroscopic techniques for structural determinations in solution, should also give a clearer understanding of the stereochemical pathways of Grignard reactions.

The diorganylmagnesium compound $Mg[C(TMS)_3]_2$ is a linear molecule in the solid state, and any tendency to increase the coordination number of 2 by association or solvation is prevented by the sterically demanding ligands.[151] However, simple isolated two-coordinate species RMgX have not been observed; the small, electronegative halide ligand presents little steric hindrance to higher coordination and also enhances the Lewis acidity of the metal relative to that in diorganylmagnesium compounds. Solid Grignard reagents are commonly four-coordinate (i.e., (**25**)),[1–2] with two additional coordination sites occupied by solvent, other Lewis base or bridging ligand species. Although tricoordinate magnesium has been claimed in $\{MgCl[CH(TMS)_2](OEt_2)\}_n$ ($n = 1$),[2] this species appears to be a chloride-bridged dimer of type (**26**), that is, $n = 2$,[152] so that no Grignard reagent with a coordination number less than four has, as yet, been structurally characterized.

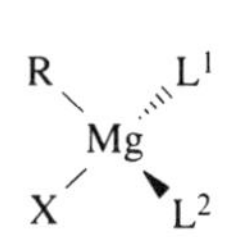

(25)

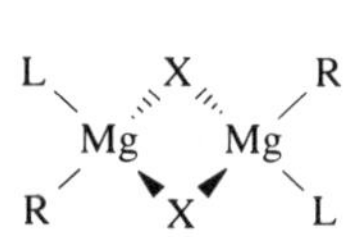

(26)

Some geometric parameters of four-coordinate, pseudotetrahedral Grignard reagents, studied since the previous review, are given in Table 1 and it may be noted that angles C–Mg–X are somewhat larger than the ideal tetrahedral angle, whereas angles L–Mg–L are smaller. Four derivatives with chiral, bidentate sparteine ligands of general structure (**27**) are included in Table 1: EtMgBr[(−)-sparteine] (**27**) (Z = Et, Y = Br, R = H), Bu^{t}MgCl[(−)-sparteine] (**27**) (Y = Bu^t, Z = Cl, R = H), EtMgBr[(−)-6-benzylsparteine] (**27**); (Y = Et, Z = Br, R = Bz) and EtMgBr[(−)-α-isosparteine].[153–4] In these chiral complexes, sparteine ligands adopt an all chair conformation with a Λ-shaped cavity and (−)-α-isosparteine is similar but less open. Ligand Y is in a more open position than Z so that the less crowded EtMgBr((−)-sparteine) with the halogen ligand at site Y is the most active in asymmetric selective polymerizations of (*R*,*S*)-α-methylbenzyl methacrylate.[153]

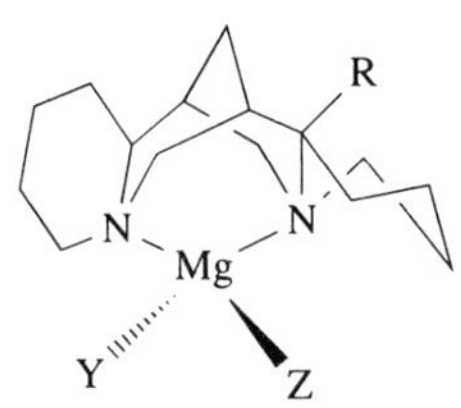

(27)

Table 1 Selected structural parameters of crystalline, four-coordinate complexes of organomagnesium halides, [RMgX(L^1)L^2].

	Bond lengths (nm)		*Bond angles* (°)		
Complex	*Mg–C*	*Mg–X*	*C–Mg–X*	*L^1–Mg–L^2*	*Ref.*
EtMgBr[(−)-sparteine]	0.224(3)	0.2482(9)	115.1(8)	84.0(8)	153
Bu^{t}MgCl[(−)-sparteine]	0.219(2)	0.2332(7)	114.7(5)	83.9(5)	153
EtMgBr[(−)-6-benzylsparteine]	0.234(5)	0.2506(16)	111.4(13)	83.1(14)	153
EtMgBr[(−)-α-isosparteine]	0.2224(15)	0.2506(6)	112.1(5)	83.9(5)	154
$Ph_3CMgBr(OEt_2)_2$	0.225(1)	0.2465(5)	116.4(3)	101.5 (4)	155
$(TMS)_2C[MgBr(THF)_2]_2$[a]	0.210(4)		121(1)	93(1) mean	156
	0.214(4)		122(1)		
$[(TMS)_2HCMg(\mu\text{-}Cl)(OEt_2)]_2$	0.2131(8)				152

[a] Angles: Mg–C–Mg 112(2)°, Si–C–Si 113(2)°.

The colourless triphenylmethyl Grignard reagent $Ph_3CMgBr(OEt_2)_2$ obtained from Et_2O–benzene has a normal four-coordinate structure of type (**25**) with a Ph_3C–Mg σ bond and two ether ligands.[155] This contrasts with the highly coloured triphenylmethyl alkali metal derivatives such as [Ph_3CM(TMEDA)] (M = Li, Na) with *polyhapto*-carbanion bonding to the metal. The 1,1-di-Grignard reagent $(TMS)_2C[MgBr(THF)_2]_2$ (**28**) containing four bulky groups around the central carbon is also, surprisingly, monomeric with two normal pseudotetrahedral magnesium centres.[156] The Mg–C bonds in (**28**) are slightly shorter than in other four-coordinate systems, possibly due to an electrostatic strengthening arising from the attachment of four electropositive groups to the central carbon which, combined with the steric crowding, may account for the low reactivity of this reagent.[156]

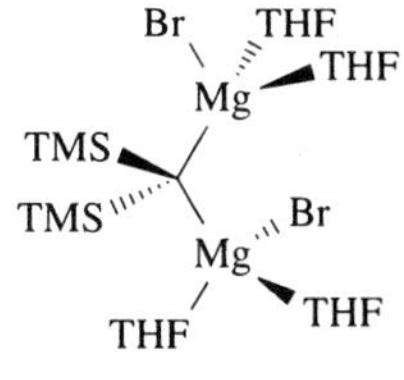

(28)

Nonsolvated Grignard reagents prepared in hydrocarbon solvents are normally deficient in halogen, and isolated solids often have a composition close to R_3Mg_2X.[1] During studies on dineopentylmagnesium, crystallization from toluene–heptane, in which some magnesium bromide remained, gave a polymeric unsolvated product of the same composition (Np_3Mg_2Br).[157] This species was shown by x-ray analysis to possess the chain structure (**29**) involving alternating groups of two $MgNp_2$ and two NpMgBr fragments connected by bridging alkyl and bromide ligands in a *spiro* fashion, producing four-coordinate magnesium atoms. Unlike the highly symmetrical chains of $(MgMe_2)_n$ or $(MgEt_2)_n$,[1] structure (**29**) contains unsymmetrical Mg–Np–Mg bridges (bonds lengths 0.220–0.229 nm and 0.233–0.242 nm), long Mg–Br bonds (0.2803(4) nm and 0.2818(6) nm) and pairs of magnesium atoms alternately positioned along two parallel straight lines.[157]

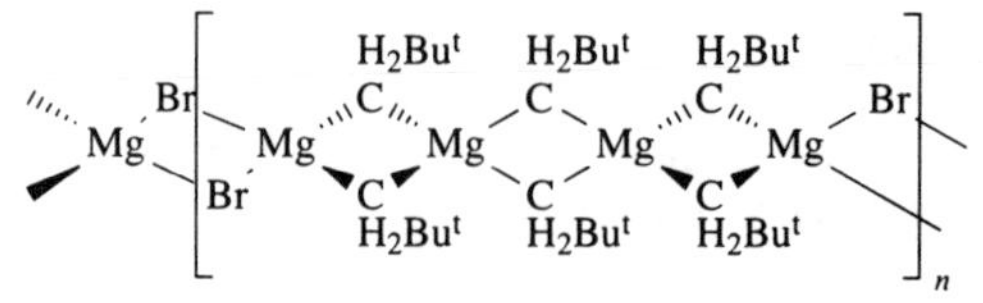

(**29**)

Although coordination numbers of more than four for alkylmagnesium halides have previously been found, for example, in five-coordinate $MeMgBr(THF)_3$ which contains a small methyl ligand, they were uncommon.[1] The results of more recent studies indicate that a major obstacle to higher coordination numbers is steric hindrance and that suitably constructed ligands, with added chelate stability, will increase coordination numbers to five or six. Thus, the substituted phenyl Grignard reagents *o*-$[MeO(CH_2CH_2O)_nCH_2]C_6H_4MgBr$ (n = 0, 1, 2 or 3) crystallize from THF with the structures indicated diagramatically in (**30**)–(**33**).[80] With only one oxygen atom in the substituent (n = 0), a dimeric solid structure is formed containing five-coordinate magnesiums in a distorted, trigonal-bipyramidal geometry, (**30**). When n = 1, 2 or 3, the monomeric six-coordinate, pseudooctahedral complexes (**31**), (**32**) or (**33**), respectively, are formed, with solvation by external THF decreasing as the number of intramolecularly coordinated oxygen donors increases. Compound (**33**) is unsolvated, showing only intramolecular 'solvation',[5] and its solubility in THF is very low.[80] The Mg–C bond of (**30**) is in the normal range (0.2110(8) nm), but in the pseudooctahedral complexes the Mg–C bonds are somewhat longer ((**31**) 0.216(1) nm; (**32**) 0.2147(7) nm; (**33**) 0.2151(9) nm) as a result of the higher coordination number; other structural parameters also reflect the geometric constraints of the polydentate ligands.[80] Another pseudooctahedral species is the tris-THF solvate of 1-bromomagnesio-2-pivaloyl-tetrahydroisoquinoline (**34**); again, six coordination is promoted by formation of a five-membered chelate ring, and in this α-metallated amide the Mg–C bond (0.2245(11) nm) may be more polar.[74] Higher coordination numbers and unusual geometries are produced by intramolecular complexation of macrocyclic ethers as in the unsolvated aryl Grignard reagents of 1,3-xylyl-crown-4 (**35**) and 1,3-xylyl-crown-5 (**36**) derivatives.[2] Both species show six coordination: (**35**) possesses two shorter and two longer Mg–O interactions in a distorted pentagonal pyramid with an Mg–C distance of 0.210(1) nm;[61] (**36**) has an unsymmetrical, distorted octahedral structure with only four of the five oxygens coordinated and has slightly longer Mg–C (0.2176(5) nm) and Mg–Br bonds than (**35**).[78]

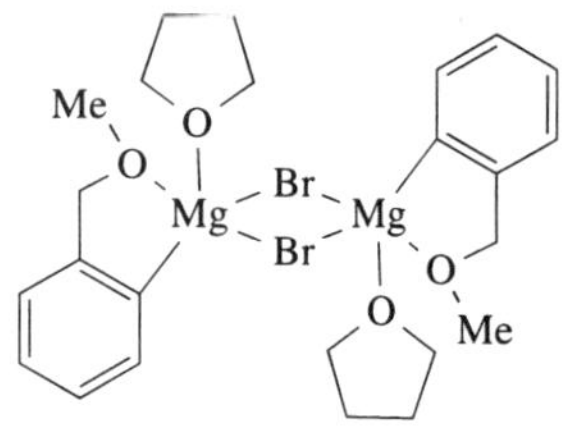

(**30**)

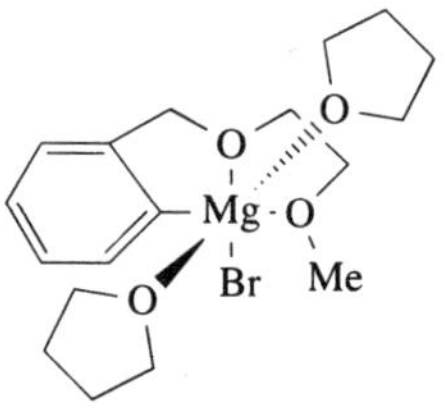

(**31**)

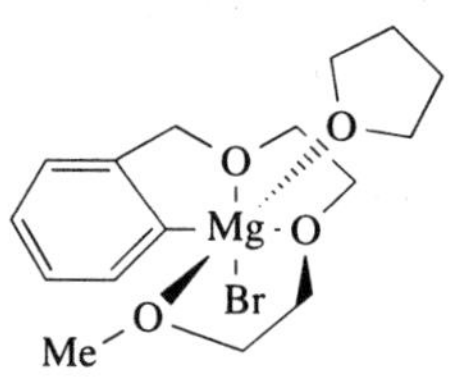

(**32**)

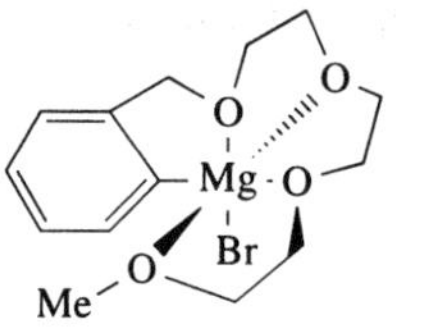

(**33**)

Allylmagnesium chloride forms a crystalline complex with TMEDA that has the dimeric structure (**37**) with η^1-bonded allyl ligands (Mg–C 0.2179(3) nm),[158] and this is consistent with the proposed structure in solution.[1] Each magnesium atom in (**37**) is five-coordinate with two markedly different bridging Mg–Cl distances.[158]

All characterized cyclopentadienylmagnesium compounds have η^5-bonded rings.[1] The 1,2,4-tris-trimethylsilylcyclopentadienyl derivative $[\{\eta\text{-}C_5H_2(TMS)_3\}MgBr(TMEDA)]$ has the structure (**38**),[159] analogous to that of $[CpMgBr(Et_2NCH_2CH_2NEt_2)]$,[1] which can be considered to be tetrahedral if the cyclopentadienyl ring is assumed to occupy one ligand site. Three dimeric, solvated cyclopentadienyl Grignard reagents, $[CpMg(\mu\text{-}Cl)(OEt_2)]_2$,[160] and $[Cp^*Mg(\mu\text{-}Cl)L]_2$ ($Cp^* = \eta\text{-}C_5Me_5$; $L = OEt_2$[160] or THF[161]) have structures of type (**26**), in which R = η^5-ring, X = Cl and L = OEt_2 or THF, and have symmetrically bridging chloride ligands, unlike (**37**). In all these cyclopentadienylmagnesium halides the η^5-ring–magnesium distances are somewhat longer than in the corresponding magnesocenes.[159–61]

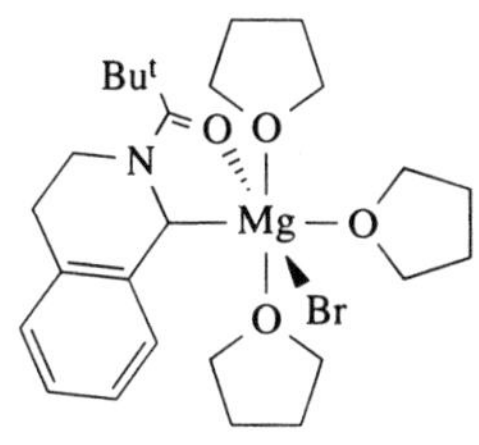

(34)

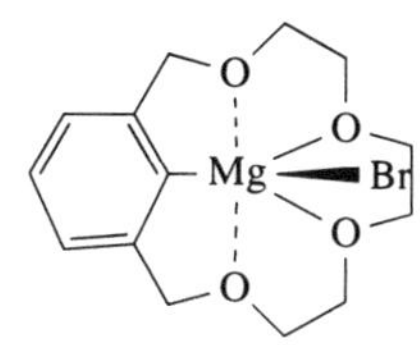

(35)

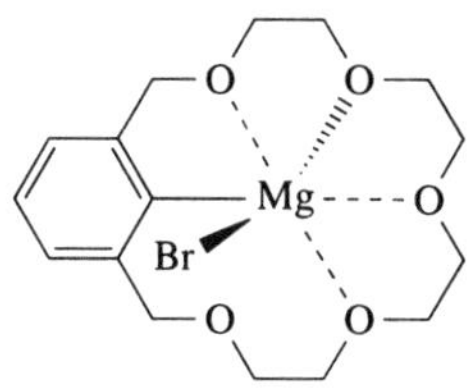

(36)

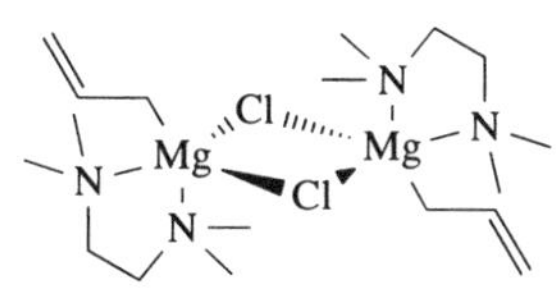

(37)

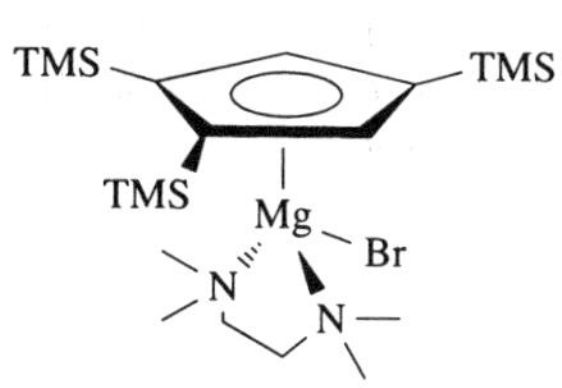

(38)

(ii) Solution

(a) Composition and structures. In ethers or other Lewis base solvents, organomagnesium halides are normally solvated and are believed to have coordination numbers and geometries related to those found in the solid state. Extensive earlier studies on solutions of Grignard reagents by a variety of techniques have established that the solvated derivatives RMgX are in equilibrium with other species and participate in various group exchange, association and ionization processes, including disproportionation according to the Schlenk equilibrium (Equation (24)).[1] More recent investigations have, in general, supported these conclusions, although new observations have been made.

$$2\text{RMgX(solvent)}_x \rightleftharpoons \text{MgR}_2\text{(solvent)}_y + \text{MgX}_2\text{(solvent)}_z \qquad (24)$$

From a chemical study of alkylmagnesium chlorides in THF, including the formation of $[NBu_4]_2[MgCl_4]$ on addition of $[NBu_4][BF_4]$, it has been proposed that the composition of the solution is best described as an equilibrium between $[RMgCl(THF)_x]$, $[(THF)_4Mg(\mu\text{-}Cl)_2MgR_2]$ and $[RMg_2(\mu\text{-}Cl)_3(THF)_5]$, that is, solvated $MgCl_2$ associates with both organometallic compounds of Equation (24).[162]

Structural studies by large-angle x-ray scattering (LAXS) and extended x-ray absorption fine structure (EXAFS) spectroscopy of the bromo and iodo Grignard reagents RMgX (R = Me, Et, Ph; X = Br, I) in diethyl ether have suggested that the species are octahedrally coordinated.[163] Monomeric and dimeric forms have been identified with dimerization constants $K_{di}(= [(RMgX)_2][RMgX]^{-2})$ in the range 0.2–0.8 M^{-1}. Data analyses, including derived bond lengths for Mg–I (0.274–0.281 nm) or Mg–Br (~0.254 nm) and other interatomic distances, are best for six coordination, both for monomers and dimers.[163] Definitive confirmation of these conclusions would be desirable since it has often been assumed that, as in the crystalline state, pseudotetrahedral coordination is most likely in diethyl ether, whereas a higher inner-sphere coordination number might be expected in better donor solvents or with less sterically hindered systems. Diffraction studies have also indicated that MgI_2 in THF is mainly dissociated into solvated MgI^+ and I^-, with possible implications for Schlenk equilibria involving this halide.[163]

Magnesium-25 NMR spectra of Grignard reagents support the Schlenk equilibrium in THF, and for RMgBr (R = Et, Cp) separate, temperature-dependent resonances assignable to solvated $MgBr_2$ and to overlapping RMgBr/MgR_2 species give approximate equilibrium constants in line with earlier studies.[164] Variable-temperature ^1H NMR spectra of isobutyl- and *t*-butylmagnesium bromides in THF, supplemented by ^{13}C NMR, also support equilibria of unassociated species (Equation (24)), although association equilibria do occur at higher concentrations, lower temperatures and in the presence of excess bromide.[165] Thermodynamic parameters for Schlenk equilibria of some butyl[165] and aryl[166] Grignard reagents are given in Table 2. Negative values for enthalpy and entropy changes are consistent with a net increase in solvation on disproportionation of RMgBr (i.e., in Equation (24)), $(y + z) > 2x$) but the data for 2,6-$(MeOCH_2)_2C_6H_3MgBr$ are exceptional since intramolecular coordination of four ether side chains in (2,6-$(MeOCH_2)_2C_6H_3)_2Mg$ gives a nonsolvated, six-coordinate diarylmagnesium; similar intramolecular coordination also influences Schlenk equilibria of arylmagnesium bromides with *o*-amine substituents.[166] Activation parameters ($\Delta S^* = +13$ J K^{-1} mol^{-1} and $\Delta H^* = +61$ kJ mol^{-1}) for disproportionation of Bu^iMgBr can be rationalized by the release of THF during formation of a more ordered bimolecular transition state (**39**).[165]

Table 2 Selected thermodynamic parameters for the Schlenk equilibrium (Equation (24)) in THF, determined by NMR spectroscopy.

Grignard reagent	*Temperature range* (K)	ΔH (kJ mol^{-1})	ΔS (J K^{-1} mol^{-1})	*K*	*Ref.*
Bu^iMgBr	250–290	−31.3	−121.1	0.40	165
Bu^tMgBr	285–330	−63.6	−221	0.14	165
PhMgBr	209–236	−18.7	−79.3	0.134	166
4-$MeOCH_2C_6H_4MgBr$	203–241	−16.5	−64.0	0.352	166
2-$MeOCH_2C_6H_4MgBr$	234–292	−13.7	−83.9	0.010	166
2,6-$(MeOCH_2)_2C_6H_3MgBr$	208–238	+6.8	+23.0	1.061	166

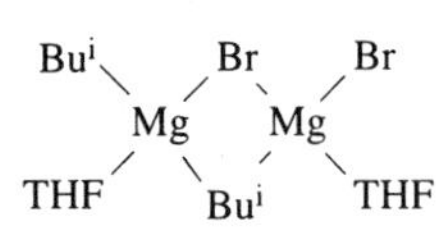

(39)

The dimagnesium derivative $BrMg(CH_2)_nMgBr$ ($n = 4$) in THF is in a Schlenk-type equilibrium with dimerized 1,6-dimagnesacyclodecane (Equation (25)), with thermodynamic parameters $\Delta H = -63$ kJ mol^{-1} and $\Delta S = -247$ J K^{-1} mol^{-1}, similar to those for the higher homologue with $n = 5$ for which monomeric and dimeric cyclic forms are also in equilibrium.[84] In diethyl ether no disproportionation of compounds $BrMg(CH_2)_nMgBr$ occurs.[135] The Schlenk equilibrium of the compound (*o*-BrMg-$C_6H_4CH_2)_2PMe$ with the corresponding magnesacycle can be monitored by ^{31}P NMR spectroscopy.[128] However, for some di-Grignard reagents, cyclization is particularly favourable and the Schlenk equilibrium is displaced completely on the side of the diorganylmagnesium product.[135] Thus, 1,5-bis(2-bromomagnesiophenyl)-3-oxapentane and related aliphatic organomagnesium compounds have intramolecular oxygen interactions which stabilize the monomeric ring (Equation (26)),[143b,166] and 2-(bromomagnesiomethyl)phenylmagnesium bromide in THF cyclizes completely to sparingly soluble, associated 1-magnesacyclobutabenzene.[146] Also, in THF solutions *o*-bis(chloromagnesiomethyl)benzene affords only chloride-free cyclic products, including trimeric $[\{Mg(1,2\text{-}(CH_2)C_6H_4(CH_2))(THF)_2\}_3]$,[167] suggesting extensive disproportionation of the di-Grignard reagent, and it is of interest that *p*-bis(chloromagnesiomethyl)benzene gives a chloride-depleted product of composition $[Mg_{1.5}Cl(C_8H_8)(THF)_2]$.[168]

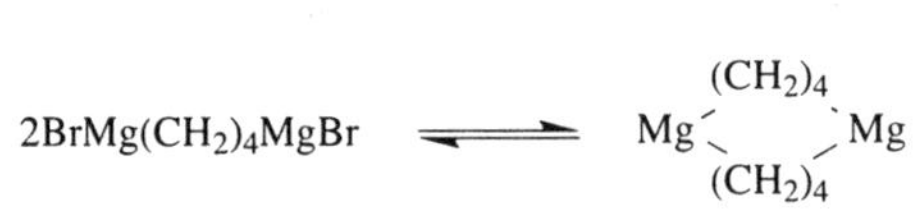

(25)

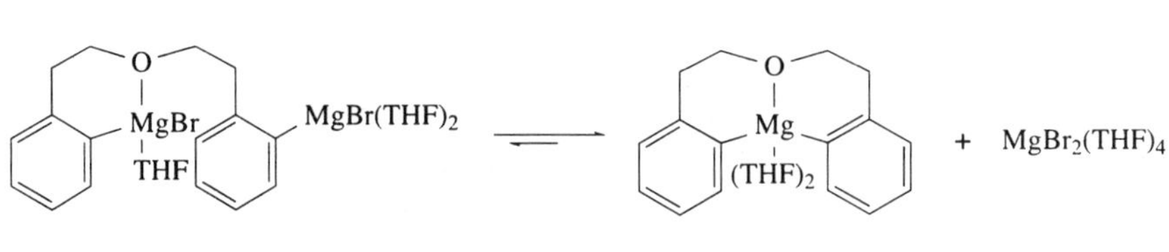

(26)

Proton NMR spectra of Grignard reagents with oxygen atoms in ether side chains[80] or in macrocyclic crowns[61,78] support the intramolecular coordination of oxygen atoms to the magnesium, and structures related to those observed the solid state, (**30**)–(**33**), (**35**) and (**36**), are probably maintained in solution.

(b) Stereochemical changes. As previously discussed,[1] inversion of stereochemistry occurs readily at α-carbon atoms of primary alkyl Grignard reagents, whereas secondary α chiral centres, especially in cyclic systems, are configurationally more stable. Carbon-13 NMR spectroscopic studies of a noncyclic, secondary organometallic ($PhPr^iCHMgCl$), provided an upper limit on the first-order rate constant (0.02 s^{-1}) for the relatively slow interconversion of (*R*) and (*S*) enantiomers in diethyl ether at 25 °C.[18] The *cis* and *trans* isomers of the cyclic secondary Grignard reagent (**40**) are reported to be configurationally stable to inversion at the α-carbon in refluxing diethyl ether, in spite of the presence of a metallated hydroxymethyl group.[54a] Other cyclic Grignard reagents with α-carbon configurational stability include derivatives formed from the organic halides (**1**)–(**5**).[47,51]

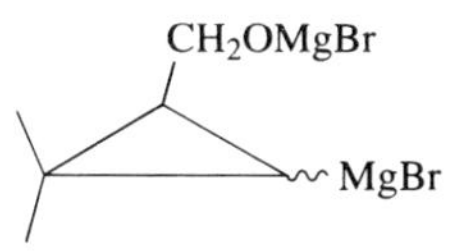

(40)

There is normally a rapidly established equilibrium between (*E*) and (*Z*) forms of 3-substituted 2-enyl Grignard reagents even at −80 °C.[1] The 3-disubstituted reagents (*E*) (geranyl) and (*Z*) (neryl) ($Me_2C{=}CH(CH_2)_2CMe{=}CHCH_2)MgCl$ do interconvert readily above − 60 °C to give an (*E*):(*Z*) ratio of ~3:1, but below −95 °C both isomers retain their stereochemistry.[118] Rapid (*E*)/(*Z*) stereoisomerization is reported for a 3-monosubstituted 2-enyl Grignard reagent even at −100 °C,[118] although there is evidence for a high degree of stereoretention by (*E*)-$NC(CH_2)_3CH{=}CHCH_2MgBr$ under similar conditions.[76]

Proton NMR spectra in THF of the crown ether Grignard reagents with solid state structures (**35**) and (**36**) indicate that, at ambient temperatures, fast exchange processes interconvert the coordinated and noncoordinated benzylic oxygen atoms of (**36**) and also create a mirror plane containing the benzene ring in both molecules.[61,78] A plausible explanation for the latter observation is that fast ionic dissociation and recombination of Br^- occurs with formation of a symmetrical intermediate organomagnesium cation.

(c) Spectroscopic properties. Few new, purely spectroscopic studies of Grignard reagents in solution have been reported, but more routine use of spectroscopy, especially NMR, is commonplace. Of interest is the application of ^{25}Mg NMR spectroscopy, and the chemical shifts for some organomagnesium compounds are given in Table 3; shifts range over ~200 ppm with the broadened resonances of σ-bonded organometallics occurring to high frequency of $Mg(aq)^{2+}$.[164] Carbon-13 NMR spectra of series of saturated and unsaturated organomagnesium compounds have been analysed[169] (e.g., see Table 3), and other reports include ^{13}C NMR data for benzyl and allyl Grignard reagents[18] and diarylmagnesium compounds.[166]

Table 3 Some NMR spectroscopic parameters of organomagnesium compounds in solution.

Compound	*Solvent*	$\delta(^{25}Mg)^a$ (ppm)	$\Delta w_{1/2}{}^a$ (Hz)	$\delta(^{13}C)$, (*a-C*)^b (ppm)	$^1J(^1H{-}^{13}C)$, (*a-C–H*)^b (Hz)
MeMgI	Et_2O			−13.4	108.4
MeMgBr	THF	67.8	1900		
EtMgBr	Et_2O			0.9	107.1
	THF	56.2	1100		
Bu^tMgBr	Et_2O			−1.8	
PhMgBr	THF			167.0	
$CH_2{=}CHMgBr$	THF			170.1	112.0
$CH_2{=}CHCH_2MgBr$	THF	29.7	1000	60.4	134.0
CpMgBr	THF	−26.8	60	105.9	163.0
$MgEt_2$	Et_2O			−1.1	104.5
	THF	99.2	3200		
$MgCp_2$	THF	−33.8	90	106.0	165.0
$MgBr_2$	THF	13.9	36		

[a] Source: Benn *et al.*[164] δ relative to external $MgSO_4{\cdot}7H_2O$ in D_2O. [b] Source: Khalilov *et al.*, Muslukhov *et al.*[169]

(iii) Thermochemical and theoretical studies

The Mg–C bond dissociation energies of 29 organomagnesium bromides have been derived from thermochemical measurements combined with one independent bond energy determined from the thermal decomposition of Bu^tMgBr.[170] The Mg–C bond strength of saturated alkyl derivatives (RMgBr) decreases with increasing number of β-hydrogens in group R, as follows (mean energies in kJ mol^{-1} are given in parentheses): Me (253); Me_3CCH_2 (226); Bu^i (213); Bu^n and higher *n*-alkyl (209); Et (205); Et_2CH (187); Bu^s (186); Pr^i (184); Bu^t (174). For benzyl and allyl derivatives, Mg–C dissociation energies are ~200 kJ mol^{-1}, whereas bonds to sp^2 carbons of phenyl or vinyl groups are stronger (289 kJ mol^{-1}). For cycloalkyl Grignards, bond energies fall from R = cyclopropyl (234 kJ mol^{-1}) to larger ring derivatives (e.g., cyclohexyl, 190 kJ mol^{-1}).[170] Heats of formation of Grignard reagents are also reported.[64,170]

Several *ab initio* calculations have been reported on isolated, unsolvated Grignard reagents, including MeMgX, giving structural parameters,[58,171] vibrational frequencies,[171] exothermicities of formation from organic halides and magnesium atoms[58–9,171] and activation energies and transition state properties for concerted insertion reactions.[59] Comparison with observations on matrix-isolated species[10] are of interest. Reactions between MeMgX (X = Cl, F) and magnesium to form linear dinuclear species MeMgMgX (Mg–Mg 0.289 nm, Mg–C 0.2099 nm) are calculated to be exothermic by 23 kJ mol^{-1}, attributable mainly to Mg–Mg bonding.[58] Formation of RMg_3X by insertion into RMg_2X is favourable (~8 kJ mol^{-1}), but insertion of further magnesium atoms is predicted to yield negligible stabilization.[58] *Ab initio* calculations on a model for dimagnesium reagents of carboxylic acids (Ivanov reagents), such as $CH_2CO_2(MgX)_2$, support a structure of type (**41**).[172]

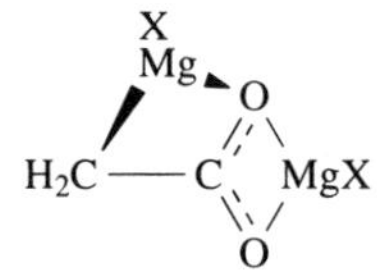

(41)

3.2.1.3 Reactions

The many important reactions of Grignard reagents applicable to organic synthesis are presented in Chapter 3, Volume 11. Also, some reactions have already been discussed in Section 3.2.1.1 as indirect preparations of organomagnesium halides, so that only more fundamental reactions and some new mechanistic aspects of Grignard reactivity will be covered in this section. Applications in organometallic synthesis will also be mentioned.

(i) Simple reactions and analytical procedures

Direct fluorination of some Grignard reagents RMgX in ethereal solvents at – 60 °C by reaction with helium-diluted fluorine gas affords moderate yields (25–37%) of alkyl or aryl fluorides RF, but not when R = Bu^t.[173] Reactions of RMgBr in Et_2O with XeF_2 give a complex mixture of products, but predominantly RH, RCHMe(OEt) and R_2.[174] The rate constants for carbonation by CO_2 of a series of Grignard reagents RMgX have been measured in an optimization of reaction conditions for radioactive carbon-11 labelling procedures; rates fall in the sequence R = Me > Et > Pr^n ≈ Bu^n ≈ Pr^i > Bu^i > Bu^s, and rate constants show a good correlation with Taft E_s values.[175]

The generation and trapping of radicals R• from reactions of organomagnesium halides (RMgX) has been studied.[176] Radicals are formed by electron transfer reactions from RMgX to an indole bisnitrone, and high electron transfer rate constants for these reactions have been derived from the Marcus theory.[176a] Radicals R· have also been produced by S_H2 displacement using triphenylimidazolyl or diphenylaminyl radicals.[176b]

Solution concentrations of Grignard reagents can be analysed by direct titration using 1-pyreneacetic acid, with the end point being indicated by the intense red colour of the dianion formed by metallation of the benzylic carbon atom; this method has been compared with earlier procedures involving double deprotonations to coloured dianions.[177] Direct titration of diphenyl ditelluride with a Grignard reagent according to Equation (27) gives a distinct red to pale-yellow colour change.[178] *N*-Phenyl-1-naphthylamine, which forms a yellow–orange amide on deprotonation, is an alternative indicator for titration of solutions of Grignard reagents with *s*-butanol[179] (cf. earlier procedures).[1]

$$\text{PhTeTePh} + \text{RMgX} \longrightarrow \text{RTePh} + \text{PhTeMgX} \tag{27}$$

(ii) Skeletal rearrangements

Cycloalkylmethyl and open-chain alkenyl Grignard reagents can interconvert according to Equation (28),[1] with the reverse reaction being an intramolecular carbomagnesiation (see Section 3.2.1.1(iii)(a)). For simple cyclopropyl[21a] and cyclobutyl systems the equilibrium of Equation (28) lies in favour of the open, unsaturated organometallic unless the ring is especially stabilized, whereas cyclopentylmethyl and cyclohexylmethyl Grignard reagents are more stable than the corresponding acyclic alkenyl derivatives.[1,85] For the cyclobutylmethyl/4-pentenyl system, *gem*-dimethyl substitution somewhat stabilizes ((3,3-methylcyclobutyl)methyl)magnesium bromide, giving an equilibrium constant $K = 500$ at 100 °C in THF for Equation (28) ($n = 3$), and the rate of ring cleavage is retarded by factor of five relative to the unsubstituted compound.[180] A 2-phenyl substituent in the cyclobutylmethyl Grignard reagent accelerates the rate of cleavage[44] but 1-aryl or 1-methyl substitution reduces the rate, probably by steric destabilization of the transition state;[181] kinetic studies with differing 1-aryl substituents support a concerted mechanism with a cyclic transition state having significant polar character (**42**).[181a]

$$\text{(CH}_2)_n\ \text{CHCH}_2\text{MgX} \rightleftharpoons \text{H}_2\text{C=CH(CH}_2)_n\text{MgX} \qquad n = 2,3,4 \text{ or } 5 \tag{28}$$

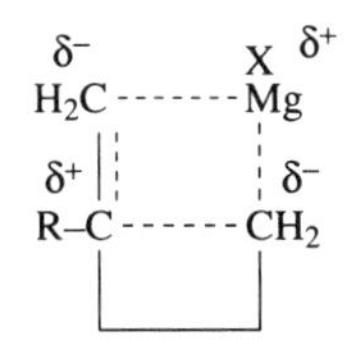

(42)

The cyclization of Equation (29) occurs diastereoselectively at 67 °C by intramolecular, suprafacial addition to the trimethylsilylethene moiety from the less hindered side.[182] A related intramolecular carbomagnesiation of a silylethyne group (Equation (30), $n = 4$) yields a cyclopentylidene, and high stereoselectively is observed in formation of substituted rings.[183] The reaction of Equation (30) also occurs in a high yield for the higher homologue with $n = 5$, but only to a small extent for cyclobutylidene species ($n = 3$).[183] Organometallic products formed by thermal rearrangement and cleavage of *endo*- and *exo*-5-(chloromagnesiomethyl)norbornene in THF are depicted in Scheme 6, and include cyclopentadienyl Grignard reagents formed by loss of dihydrogen or propene units from the initially formed allylcyclopentenylmagnesium chloride.[184]

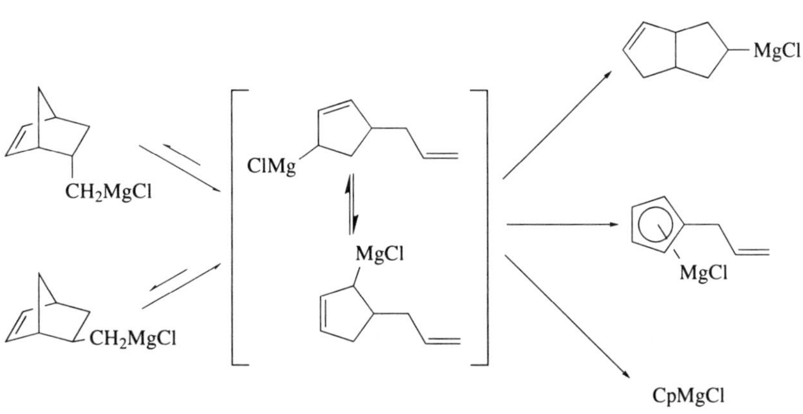

Scheme 6

Oppolzer has developed two types of synthetically useful, intramolecular cyclizations of 2-enyl Grignard reagents, termed 'magnesium–ene' reactions and illustrated by the general reactions of

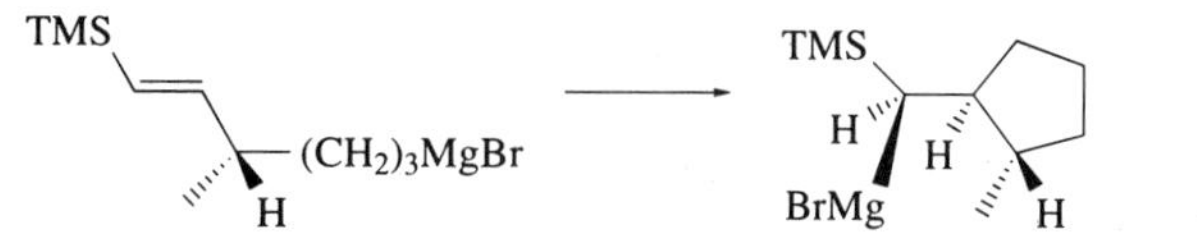

(29)

(30)

Equations (31) and (32).[185] The allylic Grignard reagent is prepared normally or at low temperature using activated magnesium,[21b] and the cyclizations occur on heating. Type I reactions (Equation (31)), occur most favourably for formation of five-membered rings, although examples of six-membered rings are known (e.g., Equation (33)); under kinetic control the magnesium donor and acceptor sites become *cis* in the cyclic product, but this stereoselectivity can be reduced by steric hindrance, and at higher temperatures (>100 °C) reversibility of the cyclization can lead to thermodynamic control and the production of *trans* configuration. The ease of cyclization by type II reactions (Equation (32)) decreases with ring size in the order $6 > 5 \approx 7 \gg 8$, and the C–C bond is preferentially formed at the more substituted ene terminal; in the absence of severe steric effects, dialkyl-substituted allylmagnesium units react in the (Z) form to give a *cis* orientation of the ring substituent with respect to the magnesium acceptor site in the product, and transition state (**43**) is proposed for the formation of a six-membered ring.[185]

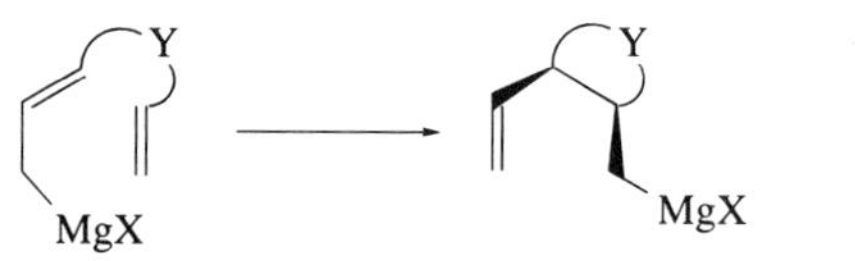

(31)

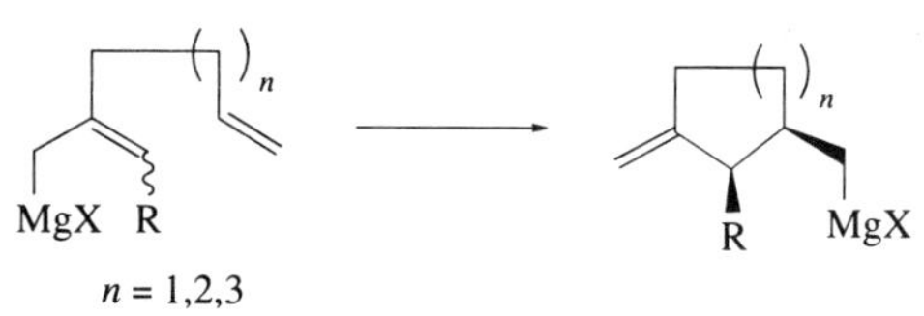

(32)

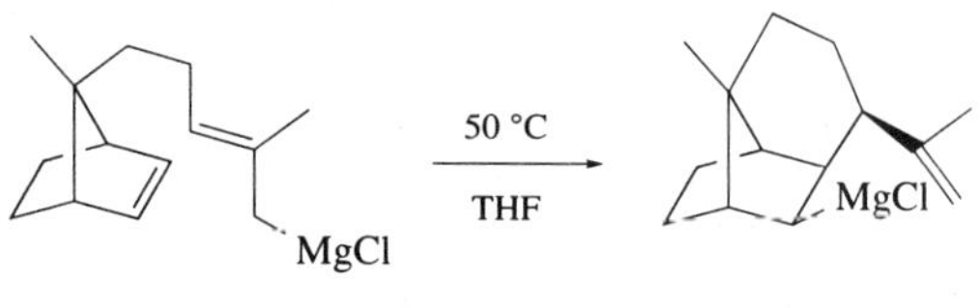

(33)

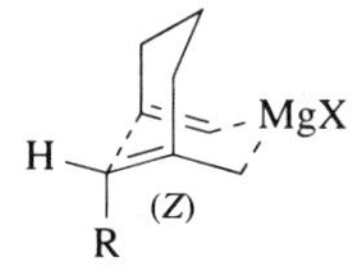

(**43**)

(iii) Reactions with organic compounds

Reactions between Grignard reagents and organic compounds containing a carbonyl function can follow heterolytic or homolytic, single-electron transfer (SET) pathways, and evidence for these mechanisms from earlier studies has been reviewed.[186–8] *Ab initio* SCF MO calculations of model Grignard reactions with carbonyl groups have identified a polar pathway and support an SET process as the HOMO energy level of the Grignard reagent increases and the LUMO of the ketone is reduced.[189] The heterolytic, concerted mechanism for 1,2-addition, which normally operates with alkyl ketones such as acetone, proceeds through a four-centre transition state (**44**) or a six-centre transition state (**45**) for normal γ addition of allylic Grignard reagents, and a similar cyclic transition state (**46**) is possible for concerted polar reductions.[186–8] Further studies of SET mechanisms have been reported,[186–94] and some of the intermediates that have been proposed are depicted simply in Scheme 7; other species of the

Schlenk equilibrium, associated organomagnesium reagents, and coordinated solvent molecules are omitted from this scheme. The mechanism operating in any particular system is dependent on the substrate, the solvent and the Grignard reagent. Studies indicate that SET is the dominating overall process in reactions of aromatic compounds such as benzophenone, although there are significant differences in the interpretation of the detailed mechanistic steps by various research groups.[186–94]

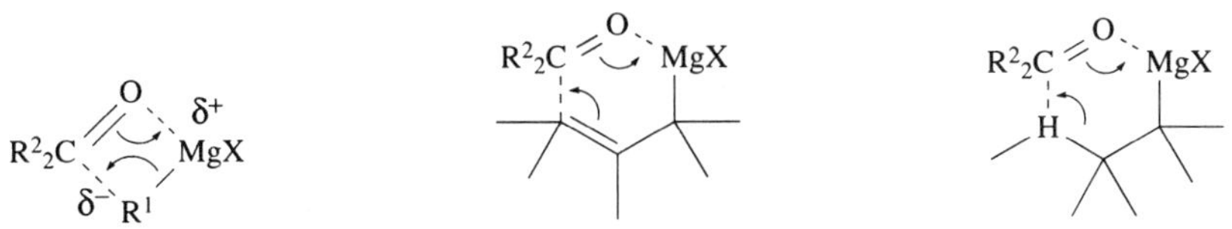

(44) **(45)** **(46)**

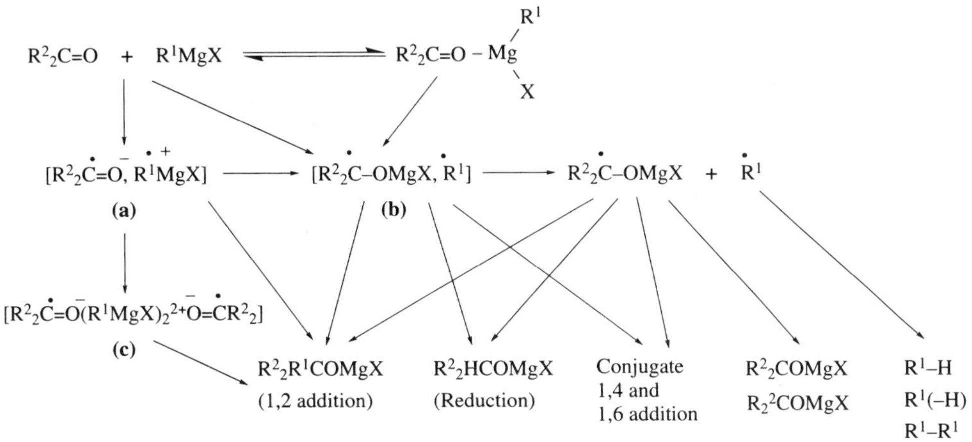

Scheme 7

Tertiary Grignard reagents are more prone to SET reactions, but formation of cycloalkylated products from substituted 5-hexenyl reagents also provides direct evidence for radical intermediates in reactions of primary alkylmagnesium halides with aromatic ketones.[186b] Reactions of benzophenone with primary, secondary and tertiary alkylmagnesium halides have been assigned inner-sphere SET mechanisms by Holm, involving concerted electron transfer and magnesium transfer from carbon to oxygen, possibly via a cyclic transition state, to form a caged radical pair of type (**b**) in Scheme 7 without formation of a radical ion pair precursor of type (**a**).[187,190] Evidence supporting this mechanism has been deduced from carbonyl group ^{13}C kinetic isotope effects for reactions of RMgX (R = Me, Bu^t), evaluation of activation energies, and secondary β-deuterium kinetic isotope effects of Grignard reagents.[190] Alternatively, Japanese workers have reported conflicting ^{14}C kinetic isotope effects for related reactions and, in conjunction with comparative studies of aromatic substituent effects on reactivity and product distribution, have proposed alternative SET steps:[191] a rapid reversible outer-sphere electron transfer reaction followed either by a rate-determining alkyl transfer from an intermediate of type (**a**) for RMgX (R = Me, Bu^n, Ph, Bz) or an isomerization of the radical ion pair (**a**) for Bu^{t}MgX.[191] Moreover, allylic Grignard reagents have been deduced to react with benzophenone by an SET mechanism with electron transfer as the rate-determining step,[191] whereas Holm's results support a concerted, heterolytic pathway, probably with a six-centre transition state (**45**).[190]

Ketyl radical intermediates in Grignard reactions of aromatic ketones can be detected by ESR and electronic spectroscopy, but similar spectroscopic evidence is lacking for radical counter cations of ion pairs (**a**) in Scheme 7.[192–3] Maruyama *et al.*[193] have explained their spectroscopic results by the formation of dimeric species of type (**c**), which give triplet ESR spectra at 77 K, and they propose that such dimeric ion pairs are implicated in alkyl transfer to the ketone. Generally, it has been assumed that normal addition products from SET Grignard reactions with carbonyl groups arise chiefly from combination of radicals within a solvent cage and that freely diffusing radicals generate mainly side products. However, a kinetic analysis by Walling[194] suggests that free ketyl and alkyl radicals can give the normal product distributions and that long-lived caged species may not be important.

The mechanism for enolization of hindered alkyl mesityl ketones by Grignard reagents involves initial coordination of magnesium to the carbonyl oxygen atom in a rapid preequilibrium, followed by a rate-determining abstraction of an α-hydrogen atom via a six-membered cyclic transition state (**47**) incorporating a moderate amount of C–H stretching.[195] Benzylic Grignard reagents with aldehydes and ketones give *ortho* and normal products as a result of a reversible rearrangement step (Equation (34)) during the reaction pathway, and diol products can result from less sterically hindered aldehydes by a subsequent Prins-type reaction.[196] Additions of Grignard reagents to α,β-unsaturated carbonyl compounds can also take concerted polar or homolytic mechanisms;[197] only substrates with a *cis* conformation of C=C and C=O bonds can give 1,4-addition products by a polar mechanism, as occurs with PhMgX; Bu^tMgX undergoes 1,4-addition by homolysis, whereas primary and secondary Grignard reagents normally react by polar, concerted mechanisms but may also produce some 1,4-addition product by a homolytic pathway even from *trans*-1,3-enones.[197] Similarly, polar and SET mechanisms have been established for additions to α-diketones to give carbon-attacked (1,2) and oxygen-attacked (1,4) products.[198]

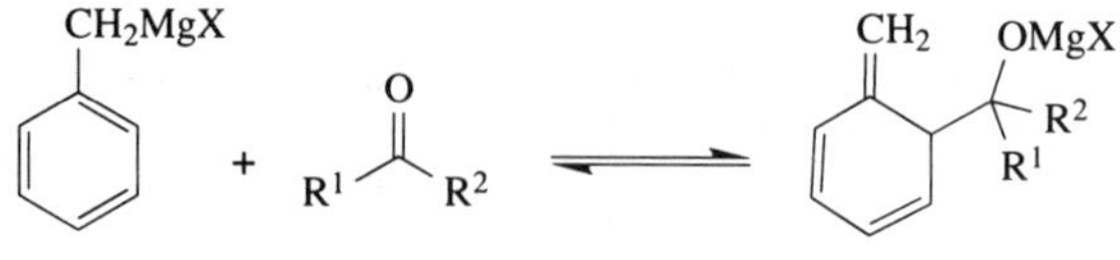

(34)

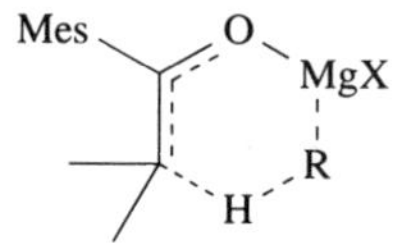

(47)

Addition of Grignard reagents to a prochiral carbonyl function generates a new asymmetric carbon centre, and in a chiral reaction environment, enantioselective or diastereoselective additions can take place. Reactions between simple aldehydes and Grignard reagents in the presence of simple chiral ethereal solvents[199] give products with relatively low *ee* values. Addition of certain chiral aprotic bidentate nitrogen or oxygen ligands to reaction systems, for example, (**48**) in reactions between aromatic aldehydes and arylmagnesium bromides, can give chiral alcohols in up to 75% *ee*, and the addition of a protic chiral auxiliary such as an α-D-glucofuranose derivative, which reacts with excess Grignard reagent and probably forms a polynuclear chiral organomagnesium species, also causes enantioselective alkylation.[200] Addition of chiral 2,2'-dihydroxy-1,1'-binaphthyl (binaphthol) to Grignard reagents and $[TiCl(OPr^i)_3]$ can give alkylation products with very high *ee* values, but these are essentially reactions of organotitanium compounds.[200] More recently, it has been found that the chiral diols α,α,α',α'-tetraaryl-2,2-dimethyl-1,3-dioxolane-4,5-dimethanol (**49**) (TADDOL) with three equivalents of primary Grignard reagent form diolate–alkyl magnesium systems that alkylate ketones with very high enantioselectivities (up to >99:1) to form chiral tertiary alcohols.[201] There are many examples of Grignard reagents incorporating chiral organic groups reacting with achiral carbonyl substrates,[74,129,199,202–3] or of achiral reagents reacting with chiral carbonyl compounds,[204–6] to form products diastereoselectively.

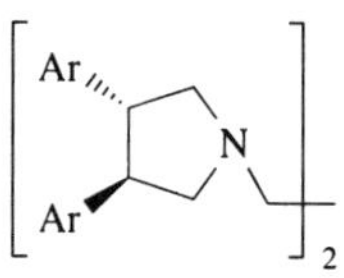

(48)

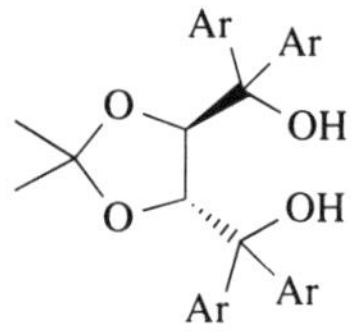

(49) Ar = Ph, 2-naphthyl

The asymmetry of the above reactions has been explained in terms of a concerted mechanism and a favoured steric pathway for substrate coordination to the magnesium, with chelation of the magnesium by a chiral substrate,[205] chiral auxiliary[200] or achiral polyether[204] giving higher enantio- or diastereoselectivity. Thus, based on the x-ray determined structure of (**34**), it is shown that displacement of the more exposed '*exo*' THF ligand by acetophenone and the formation of a four-centre transition state in the least hindered orientation generates the observed diastereomeric alcohol.[74]

Asymmetric conjugate addition to α,β-unsaturated carbonyl substrates by Grignard reagents also occurs, with high stereoselectivities resulting when chelating chiral complexes can be formed as intermediates, and reactions catalysed by copper or zinc reagents are particularly effective.[206] As noted in Section 3.2.1.2(i), structurally characterized chiral sparteine complexes of the general structure (**27**) are initiators for asymmetric selective polymerization of racemic acrylates, and the high activity of EtMgBr[(−)-sparteine] can be rationalized by steric arguments.[153]

Addition of Grignard reagent to a hindered cyano group is facilitated by high pressures (5 kbar).[207] Additions to pyridinium cations appear to be polar in nature, with the exception of reactions of Bu^tMgX, and can be regio- and stereoselective,[208a] whereas reactions of Grignard reagents with pyridazine[208b] or with nitroarenes[209] occur via homolytic SET processes. Noncatalysed, cross-coupling reactions of Grignard reagents with allylic halides, and with alkenyl or alkynyl halides or arylsulfones, can exhibit homolytic or nucleophilic characteristics.[210] An interesting self-coupling of aryl Grignard reagents, used to form bi- and polyaryl products, occurs by electron transfer to 1,4-dihalo-2-butene, giving 1,3-butadiene as the coproduct.[211] Coupling of polyhalobenzenes with Grignard reagents occurs via carbomagnesiation of an aryne intermediate[69] (cf. Section 3.2.1.1(ii)(b)).

Grignard reactions with organic molecules are often carried out in the presence of catalytic or stoichiometric amounts of transition metal compounds, as the latter can modify or improve the utility of the magnesium derivatives, including applications in asymmetric syntheses. The mechanisms of such reactions are not always clear, but normally depend on the formation and reactions of intermediate organotransition metal species and so will not be discussed in this chapter. Some representative references to the use of compounds of the following metals in catalytic systems may be noted: cerium,[212] titanium,[213] manganese,[214] iron,[215] nickel,[97,216–17] palladium,[216,218] copper[109,206,219] and zinc.[206]

(iv) Synthesis of organometallic compounds

Grignard reagents are widely used to form metal–carbon bonds by transfer of alkyl groups to metal centres.[1] Some recent examples of this technique and of related reactions with organometallic compounds are presented here.

Borate esters ($B(OR^1)_3$) are converted in good yields by two equivalents of R^2MgBr into dialkylborinic esters ($R^2{}_2B(OR^1)$), and 1-alkenylboronic esters, such as $BuCH{=}CHB(OPr^i)_2$, give excellent yields of alkyl-1-alkenylborinic esters on reaction with two to three equivalents of a Grignard reagent.[220] Six-coordinate silicon and germanium complexes such as (**50**) are readily obtained from silica gel or GeO_2, respectively, and are converted on reaction with the reagents RMgX into the organometallic products MR_4 or $R_3M(1,2\text{-}OC_6H_4OH)$ (M = Si, Ge), depending on the alkyl group R and the counter-cation; in the presence of Cp_2TiCl_2, reduction to R_3MH occurs.[221] Reactions of PhMgBr with $GeBr_4$ give the linear polygermanes Ge_3Ph_8 (up to 11%) and Ge_4Ph_{10} (up to 18%) as by-products separable from the main products $GePh_4$ and Ge_2Ph_6.[222] Aryl Grignard reagents react readily with MCl_4 (M = Si, Sn) or ECl_3 (E = P, As), with Li_2CuCl_4 or CuX as a catalyst, to afford the fully arylated products MAr_4 or EAr_3 in good yields.[223] Alkylations of aminophosphazenes by Grignard reagents have also been studied.[224]

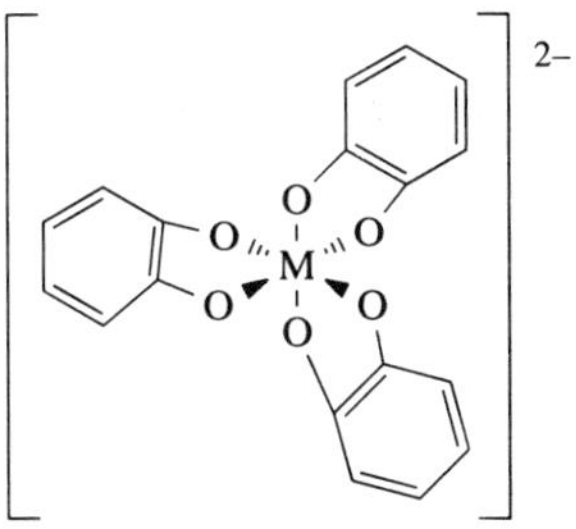

(**50**) M = Si, Ge

Di-Grignard reagents, or the corresponding cyclic organomagnesium compounds formed by elimination of magnesium halides as a result of the Schlenk equilibrium in solution, have been employed to produce new dimetallic or metallacyclic organometallics.[135–7,144,146,225] The methylene species $CH_2(MgBr)_2$ reacts with Me_3MCl (M = Si, Sn) to form $CH_2(MMe_3)_2$; reaction with Me_2MCl_2 gives a mixture of the cyclic products $(Me_2MCH_2)_n$ (M = Ge, n = 2, 3, 4; M = Sn, n = 3, 4), but use of a reagent with the approximate composition $CH_2Mg{\cdot}CH_2(MgBr)_2$ affords higher yields of the 1,3-digermacyclobutane (**51**).[135,137b] The 1,3-di-Grignard reagents $BrMgCH_2CR_2CH_2MgBr$ (R = H, Me), similarly, will form acyclic 1,3-distannanes, the metallacyclobutanes (**52**) (M = $GeMe_2$, $SnMe_2$) or the *spiro*-bicyclic compounds (**53**) (M = Si, Ge) by reaction with the group 14 compounds Me_3SnCl, Me_2MCl_2 or MCl_4, respectively.[135–6,141] The cyclic and *spiro*-bicyclic organometallics (**54**) and (**55**)

(M = Si, Ge, Sn) are similarly formed from the titanium di-Grignard (**22**).[135,137a] Dimagnesium compounds are also precursors to silicon and tin metallacycles of types (**56**)[144] and (**57**)[225] (M = $SiMe_2$, $SnPh_2$; R = H, TMS), and *spiro* products.[225]

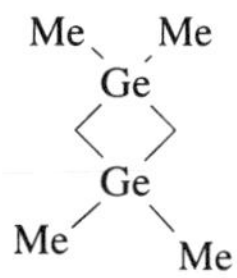

(**51**)

(**52**)

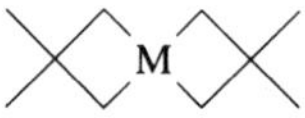

(**53**)

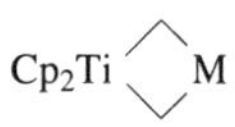

(**54**)

(**55**)

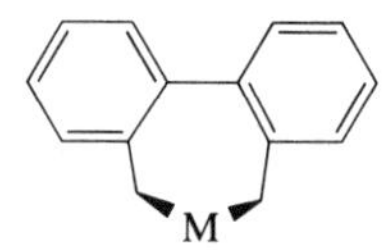

(**56**)

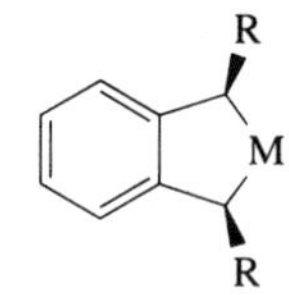

(**57**)

In alkylations of titanium or zirconium tetrachloride by Grignard reagents RMgX, the by-product $MgCl_2$ plays an important role; transient magnesium–chlorotitanium intermediates are observed during formation of TiR_4, and in the synthesis of ZrR_4 the formation of magnesium salts of $[ZrCl_6]^{2-}$ and $[ZrCl_5(THF)]^-$, which are insoluble in THF, causes a reduction in the yield of the organometallic product.[226] Alkyl Grignard reagents promote cyclization reactions of Cp_2ZrCl_2 with unsaturated hydrocarbons, as in Equation (35),[227] and cause stoichiometric transformations of zirconacyclopentanes related to catalytic processes outlined in Scheme 5.[103a] The transformations of Equation (36) produce a magnesium–titanium *spiro* compound that is a precursor for titanium–carbene chemistry, as illustrated by the reaction with trimethylphosphine.[136] Di-Grignard reagents will form various group 4 metallacycles of types (**52**) and (**54**) (M = $TiCp_2$, $ZrCp_2$, $HfCp_2$) and 1,3-dizirconium- and dihafniumcyclobutanes; metallacycles (**52**) (M = VCp_2, $MoCp_2$, WCp_2, ReO(η-Cp*)) and related rhodium or platinum species can also be prepared by similar reactions.[135–6] The use of pentamethylcyclopentadienylmagnesium chloride to introduce the η-Cp* ligand is exemplified by formation of $(\eta\text{-Cp*})_2ThCl_2$ from $ZrCl_4$.[63]

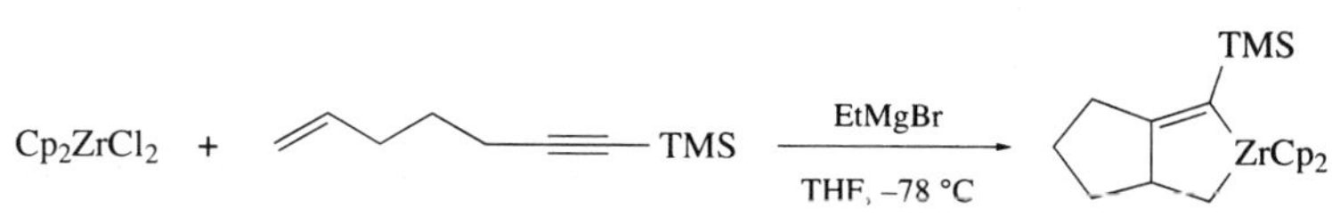

(35)

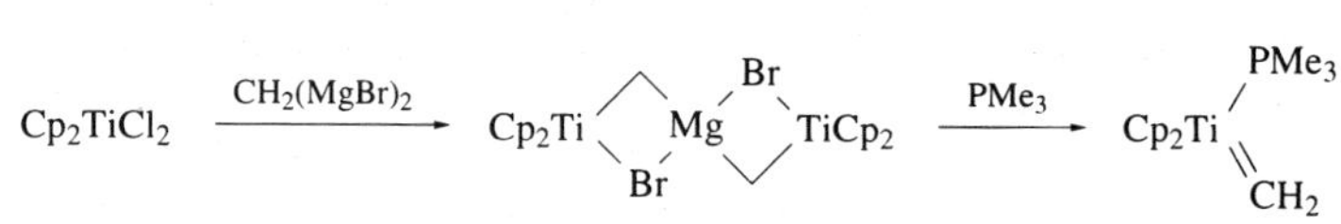

(36)

Allylmagnesium bromide and $W(OPh)_6$ in THF at 60 °C afford $[W(\eta\text{-}C_3H_5)_4]$ in a 61% yield, and reactions of $[WCl(\eta\text{-}C_3H_5)_3]$ with various reagents RMgX in ether at −78 °C produce the derivatives $[WR(\eta\text{-}C_3H_5)_3]$ containing alkyl ligands with agostic β-hydrogen atoms.[228] In the presence of $MgCl_2$, benzyl and allylic Grignard reagents alkylate the cyclopentadienyl ring of Cp_2WCl_2 (e.g., Equation (37)), whereas MeMgCl yields Cp_2WMe_2 under these conditions.[229] Attack by RMgBr at a carbonyl group of pentacarbonyliron forms the acyl anion $[Fe(COR)(CO)_4]^-$, which on protonation gives the aldehyde RCHO in a greater than 90% yield,[230] and EtMgBr with the methacrylate complex $[Fe(CH_2{=}CHCOOMe)(CO)_4]$ at −22 °C produces $EtCOCH_2CH_2COOMe$ after quenching, but only in a modest yield.[231]

The mesitylene complex $[OsCl_2(CO)(\eta\text{-Mes})]$ is transformed by the Grignard reagents RMgX into $[OsXR(CO)(\eta\text{-Mes})]$ (R = alkyl, Ph) or $[OsR_2(CO)(\eta\text{-Mes})]$ (R = alkyl), but cyclohexylmagnesium bromide forms mainly the cyclohexene complex $[Os(CO)(\eta\text{-}C_6H_{10})(\eta\text{-Mes})]$.[232] A route to η^6-arene–bis(fluoroaryl)metal complexes of cobalt and nickel is illustrated by Equation (38).[233] Reaction of the *o*-xylidene di-Grignard reagent with $[PtI_2(cod)]$ affords the metallacycle (**57**) (M = Pt(cod); R = H),[167] and this reagent is also useful for the syntheses of metallaindenes of other transition metals.[234]

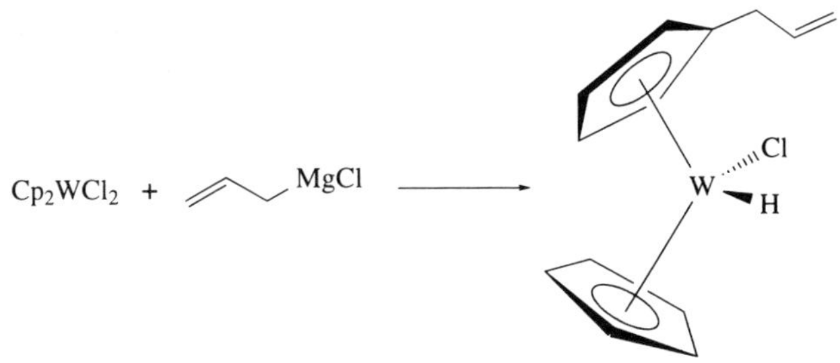

(37)

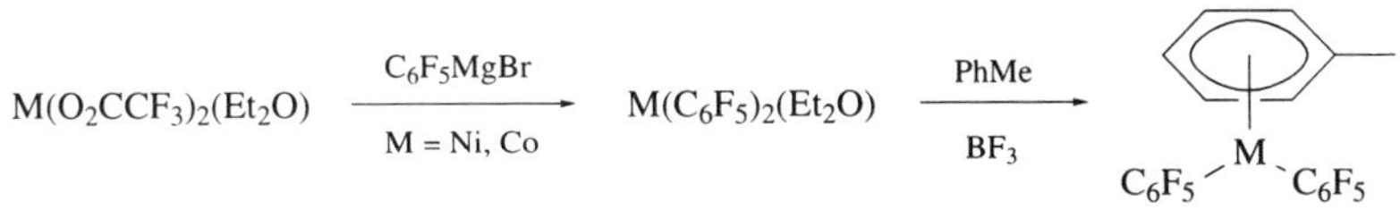

(38)

3.2.2 Diorganylmagnesium Compounds

3.2.2.1 Preparation

Preformed Grignard reagents are useful precursors for diorganylmagnesium compounds. Displacement of the Schlenk equilibrium (Equation (24)) to the right and separation of the magnesium halide affords the organometallic product; addition of dioxane followed by removal of precipitated $MgX_2(dioxane)_2$ is a common procedure.[1] Acyclic polyethers are reported to be better than dioxane or pyridine for inducing this disproportionation, and dropwise addition of $Me(OCH_2CH_2)_nOMe$ (e.g., $n = 2$, 5) to ethereal solutions of RMgBr at 0–10 °C precipitates polyether complexes of $MgBr_2$, leaving solutions of MgR_2 in 95% or better yields.[235] Also, it is claimed that simple addition of THF to solutions of alkyl- or arylmagnesium iodides in diethyl ether selectively precipitates $[Mg(THF)_6]I_2$ and, after cooling and separation of this solid by filtration, leaves MgR_2 in solution (e.g., 80% $MgBu^n{}_2$).[236] The Schlenk equilibrium of several di-Grignard reagents lies in favour of the cyclized diorganylmagnesium compound, and separations of components based on differences in solubility yield the pure halogen-free organometallic: thus, precipitation or crystallization from THF solutions of the di-Grignard reagent (**18**),[140] 1,8-di(halomagnesio)naphthalene,[237] *o*-(bromomagnesio)benzylmagnesium bromide (cf. Scheme 3),[146,238] 1,2-bis(chloromagnesiomethyl)benzene[167,234] or 1,5-bis(2-bromomagnesiophenyl)-3-oxapentane (Equation (26))[143b,166] affords the corresponding cyclic diorganylmagnesium. The volatile, unsolvated compound $Mg[C(TMS)_3]_2$ is formed as a sublimate on thermolysis of the mixed-metal organomagnesium–lithium bromide $[Li(THF)_2(\mu\text{-}Br)_2Mg\{C(TMS)_3\}(THF)]$ under vacuum (170 °C, 0.08 torr).[151]

Reactions of diorganylmercury compounds with magnesium metal (Equation (39)) produce samples of diorganylmagnesium free of residual halide.[1] Diorganylmercury precursors can be formed from Grignard reagents and $HgBr_2$, either in a single step or via the organylmercury bromide and subsequent reaction with aqueous $SnCl_2$/NaOH.[166,239] Several pure *ortho*-substituted diarylmagnesium compounds have been synthesized quantitatively from the corresponding diarylmercury compounds by stirring with a large excess of magnesium in THF for 1–2 weeks.[166] Similar reactions in THF of excess magnesium with cyclic mercury compounds have produced high yields of magnesacycloalkanes, such as $[Mg(CH_2)_5]_n$ ($n = 1$–2),[143,239] and tetrameric *o*-phenylenemagnesium ($[MgC_6H_4]_4$) (Equation (40)).[237]

$$HgR_2 + Mg \longrightarrow MgR_2 + Hg \qquad (39)$$

(40)

A two-stage process for the synthesis of alkane-soluble *n*-butyl(*s*-butyl)magnesium has been described.[240] Initial direct reaction in a C_8–C_9 alkane of activated magnesium powder with *n*-butyl chloride containing 10% *n*-octyl chloride to give the unsolvated Grignard reagent, followed by reaction with one molar equivalent of *s*-butyllithium, produces, after filtration, a solution of dialkylmagnesiums (85%), mainly Bu^nMgBu^s. The presence of some octylmagnesium entities in the hydrocarbon solution prevents precipitation of butylmagnesium species over extended periods of time.[240] Solvated diarylmagnesium compounds of the types $[MgAr_2(ether)_2]$ or $[MgAr^1Ar^2(ether)_2]$ have been formed by arylation of magnesium halide etherates or arylmagnesium halides with aryllithium in diethyl ether or THF, but only the complexes with Et_2O and not THF could be completely desolvated.[241] A lithium reagent is also used in the preparation of the hydrocarbon-soluble $\{Mg[CH(TMS)_2]_2\}_n$ (Equation (41)).[152]

$$Mg(OC_6H_2Bu^t\text{-}2,6\text{-}Me\text{-}4)_2 \xrightarrow{Li[CH(TMS)_2]} \{Mg[CH(TMS)_2]_2\}_n \quad (41)$$

Active magnesium, prepared by the reduction of $MgCl_2$ with potassium metal,[22] reacts directly with cyclopentadiene, indene or fluorene in boiling THF to give bis(cyclopentadienyl)-, bis-(indenyl)- or bis(fluorenyl)magnesium, respectively.[242] Metallation of acidic hydrocarbons with dialkylmagnesium is another useful synthetic route to the former and related organometallics. The unsolvated compounds MgR_2 (R = cyclopentadienyl, mono- or disubstituted cyclopentadienyl, indenyl, fluorenyl or 1-alkynyl) are produced by reaction of *n*-butyl(*s*-butyl)magnesium with the corresponding hydrocarbon acids in alkane solvents at temperatures between 0 °C and 110 °C;[243–5] similarly, reaction with the pyridyl compound $2\text{-}(TMS)_2CHC_5H_4N$ forms (**58**),[246] and diphenyl sulfone is *ortho* metallated.[243a] However, sterically hindered 1,2,4-tris(trimethylsilyl)cyclopentadiene is metallated by dimethylmagnesium in diethyl ether only in the presence of TMEDA, giving, at room temperature, $[MgMe\{\eta\text{-}C_5H_2(TMS)_3\}(TMEDA)]$ or, on prolonged reaction at 35 °C, $[Mg\{\eta\text{-}C_5H_2\text{-}(TMS)_3\}_2(TMEDA)]$.[159] Complexation by a crown ether can activate $MgPh_2$ to metallate an adjacent aromatic C–H bond, as illustrated in Scheme 4 (A = H; Y = Ph), in which a mixed diarylmagnesium compound is formed, and the related halogen–magnesium exchange reaction (Scheme 4: A = Br; Y = Ph) also occurs more readily to yield the same product.[66] However, the larger 1,3-xylylene-18-crown-5 (**7**) (n = 1) is inactive to metallation by $MgPh_2$ (A = H), and much less active to bromine–magnesium exchange (A = Br).[66]

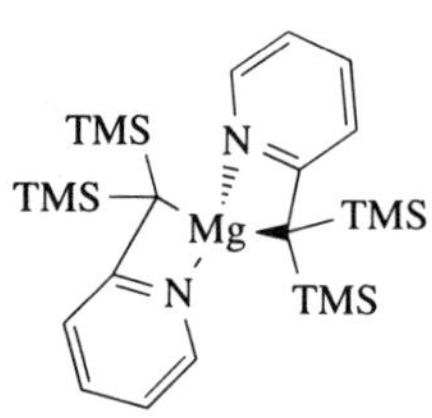

(58)

Redistribution of organic groups between two diorganylmagnesium compounds is a useful method for the preparation of organometallics containing two different organic ligands. Equimolar mixtures of $MgCp_2$ and MgR_2 afford the derivatives CpMgR (R = neopentyl[247] or 2-methylallyl[248]), whereas dimeric products are obtained by the reactions shown in Equations (42) and (43).[249]

$$MgEt_2 + Mg(C_nH_{2n}NR^1R^2)_2 \longrightarrow [MgEt(C_nH_{2n}NR^1R^2)]_2 \quad (42)$$

$$MgEt_2 + Mg(C_4H_8OR)_2 \longrightarrow [MgEt(C_4H_8OR)]_2 \quad (43)$$

The catalysed addition of active MgH_2 to 1-alkenes (Equation (44)) is an alternative route to dialkylmagnesiums.[113,250] Active MgH_2 can be produced from $MgEt_2$ and $LiAlH_4$[113] or, more directly in THF via reaction of H_2 (60 bar) and magnesium powder catalysed by magnesium–anthracene/$TiCl_4$ or magnesium–anthracene/$CrCl_3$,[250a] and promoted by the addition of $MgCl_2$ as a cocatalyst.[250b] The $TiCl_4$ catalyst can be employed *in situ* for subsequent hydromagnesiation (Equation (44)), but $ZrCl_4$ is more active for this reaction and may be combined with MgH_2 prepared using the $CrCl_3$ catalyst: the reaction shown by Equation (44) occurs regiospecifically (up to 99.7%) with the addition of magnesium to C-1 and gives high yields of isolable organometallics (e.g., $MgEt_2$ 97%, $Mg(Bu^n)_2$ 86%, $Mg(C_8H_{17\text{-}n})_2$ 87%).[250] Also, compounds (**59**) or (**60**) are preparable via Equation (44) (alkene = $R^1R^2N(CH_2)_nCH{=}CH_2$ (n = 1, 2) or $RO(CH_2)_2CH{=}CH_2$, respectively).[250] Cp_2TiCl_2 is an alternative catalyst for Equation (44), and for related hydromagnesiations of 1-alkene by R^1MgH (R^1 = Me, Et, Bu^t, Ph, Cp) or by MgR^2_2 (R^2 = alkyl group with a β-hydrogen atom).[113] The mechanisms of these reactions probably involve an intermediate transition metal hydride, as proposed for the corresponding reactions of Grignard reagents (see Section 3.2.1.1(iii)(b)).

$$MgH_2 + 2H_2C{=}CHR \xrightarrow{\text{catalyst}} Mg(CH_2CH_2R)_2 \quad (44)$$

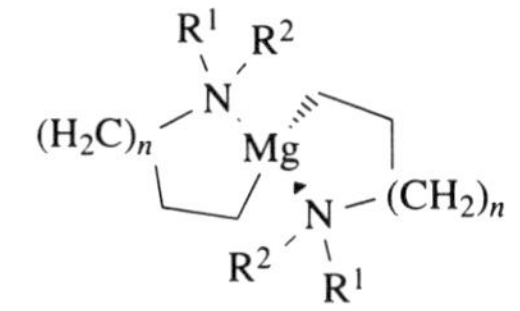

(59)

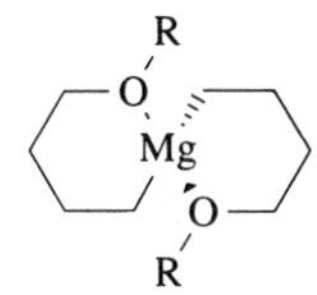

(60)

Carbomagnesiations of alkenes by dialkylmagnesiums to form new organometallics (Equation (11); $X = R^1$) show similar features to related reactions of Grignard reagents (see Section 3.2.1.1(iii)(a)). Uncatalysed additions of MgR_2 to the cyclopropene spiro[2,4]hept-1-ene vary in reaction rates with alkyl group R in the order $Pr^i > Et > Bu^t > Me$, and probably occur via four-centred transition states.[88] Dzhemilev has studied the zirconium-catalysed carbomagnesiation of 1-alkenes, and some norbornenes, by ethylmagnesium derivatives (Equation (12); X = various groups);[102,251] diethylmagnesium is the most active organometallic species but adds only one ethyl group in ether solvents and does not react with acyclic internal double bonds (e.g., Equation (45)). The activity of ethylmagnesium reagents in Cp_2ZrCl_2-catalysed carbomagnesiations of 1-alkenes decreases in the sequence: $MgEt_2 \approx (TMS)MgEt \gg CpMgEt > Ph(Et)NMgEt > EtMgCl > EtMgBr \approx Et_2NMgEt > EtMgI$.[102] Zirconacyclopentanes (**11**) (see Scheme 5) are intermediates in these reactions, and under some conditions 1-alkenes can afford 1,4-dimagnesium species (or the corresponding cyclized magnesacyclopentanes). Thus, with $MgEt_2$ as the reactant, instead of a Grignard reagent, the dimagnesium species (X = Et) is the main product of Equation (13) in diethyl ether solvent at 0 °C, whereas the monomagnesium product predominates in THF at 15–20 °C.[106] Cp_2ZrCl_2-catalysed reactions of higher dialkylmagnesiums (MgR_2; R = Pr^n, Bu^n, *n*-hexyl) with styrenes or 1-alkenes can also form magnesacyclopentanes of type (**61**) in high yields,[252] although alkenylorganomagnesiums may be coproducts.[107]

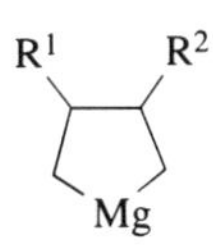

(61)

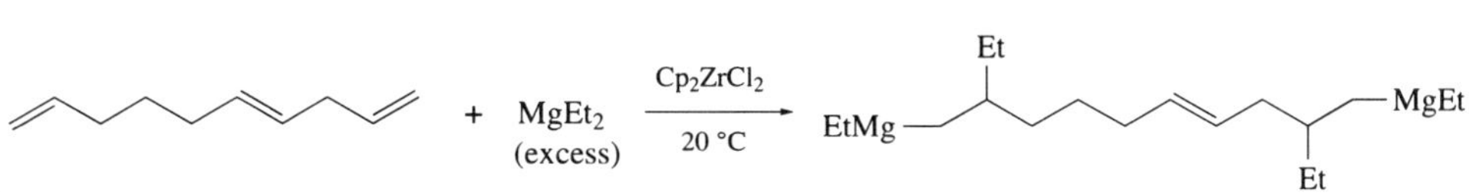

(45)

3.2.2.2 Structures and physical properties

(i) Solid state

The structures of a range of crystalline diorganylmagnesium compounds have been determined by diffraction techniques, and this topic has been reviewed.[2] Structural parameters of some species are listed in Table 4. Dineopentylmagnesium ($Mg(Np)_2$) was shown to exist as a two-coordinate monomer in the gas phase, but the unsolvated compound is associated in condensed phases.[1] More recently, the first linear two-coordinate structure for a crystalline organomagnesium has been established for $Mg[C(TMS)_3]_2$ (**62**), containing bulky tris(trimethylsilyl)methyl groups: the magnesium atom is formally *sp* hybridized and Mg–C bonds are only slightly shorter (0.2116 nm) than bonds in derivatives with higher coordination numbers.[151] Low-temperature neutron and x-ray diffraction studies have revealed a novel polymeric structure for $\{Mg[CH(TMS)_2]_2\}_n$, a species with slightly less bulky ligands.[152] Each $Mg[CH(TMS)_2]_2$ unit, containing two normal Mg–C bonds (mean Mg–C bond length 0.2111 nm, angle C–Mg–C 140.0°), also interacts intermolecularly with a methyl group of a neighbouring bis(trimethylsilyl)methyl ligand as illustrated for one centre in (**63**).[152] The weaker, intermolecular agostic interaction (distances: Mg–C^2 0.2535(4) nm; Mg–H 0.2333(4) nm, 0.2414(5) nm and 0.2516(4) nm) generates a three-coordinate magnesium ($MgC^1_2C^2$) and creates the pseudo-trigonal-bipyramidal arrangement $Si{-}C^2H_3{\cdots}Mg$ around the methyl carbon. A linear three-centre, two-electron bond involving the Si–C···Mg atoms has been suggested, but interaction with C–H bonds may also be significant.[152]

Table 4 Structural parameters of some crystalline diorganylmagnesiums containing two terminal Mg–C bonds at each magnesium centre.

Complex	*Bond length(s), Mg–C* (nm)	*Bond angles* (°) C–Mg–C	O–Mg–O	*Ref.*
$Mg[C(TMS)_3]_2$	0.2116(2)	180		151
$\{Mg[CH(TMS)_2]_2\}_n$	0.2117(4), 0.2105(4)	140.0(2)		152
$Mg[CH(TMS)_2]_2(OEt_2)$	0.217(1)			152
$\{Mg[CH(TMS)_2]_2\}_2$(μ-*p*-dioxane)	0.2118(5)			152
Mg(2-Me-1,2-$C_2B_{10}H_{10})_2$(*p*-dioxane)$_2$	0.2156(5)	123.5(4)	99.5(5)	253
$MgPh_2(THF)_2$	0.2127(4)	122.4(1)	94.2(1)	254
Mg(*p*-Tol)$_2$(THF)$_2$	0.2132(8), 0.2126(7)	124.4(3)	96.7(2)	254
[Mg($CH_2Bu^t)_2$(μ-*p*-dioxane)]$_n$	0.2133(2)	139.95(9)	89.47(6)	255
[Mg(μ-*o*-$CH_2C_6H_4CH_2$)(THF)$_2$]$_3$	0.2154(8)–0.2170(8)	126.8(3), 130.5(3)	94.1(2), 94.0(2)	167
$MgEt_2$(18-crown-6)[a]	0.2104(2)	180		256
$MgPh_2$(1,3-xylyl-18-crown-5)[a]	0.2189(5), 0.2190(5)	163.8(2)		257

[a] Rotaxane structure.

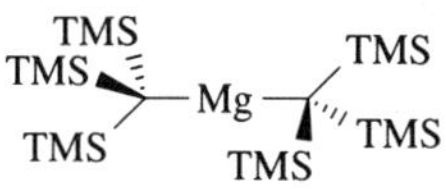

(62)

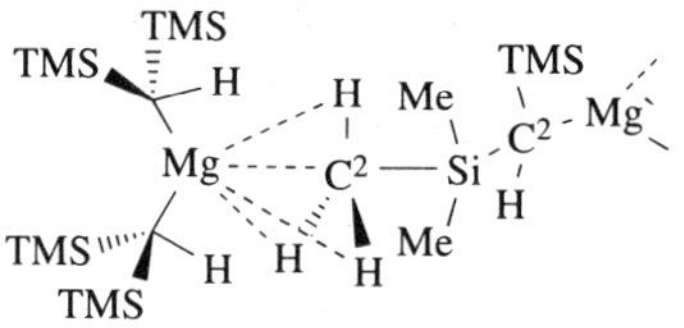

(63)

An attempt to characterize dineopentylmagnesium in the solid state resulted in the crystallization of the polymeric unsolvated mixed adduct $\{(MgNp_2)_2[Mg(Np)Br]_2\}_n$ (**29**) with tetrahedrally coordinated spiromagnesium atoms, including fragments connected only by four bridging neopentyl ligands.[157] Unlike the symmetrical alkyl bridges in unsolvated chains of $(MgR_2)_n$ (R = Me, Et),[1] the electron-deficient neopentyl bridges are unsymmetrical (see Section 3.2.1.2(i)).[157] Unsolvated $(MgPh_2)_n$ forms linear chains (**64**), related to the structures of $(MgR_2)_n$ (R = Me, Et), containing symmetrical μ,η^1-bridging phenyl groups with an Mg–C bond length of 0.2261(2) nm and an Mg–C–Mg angle of 77.73(9)°.[254]

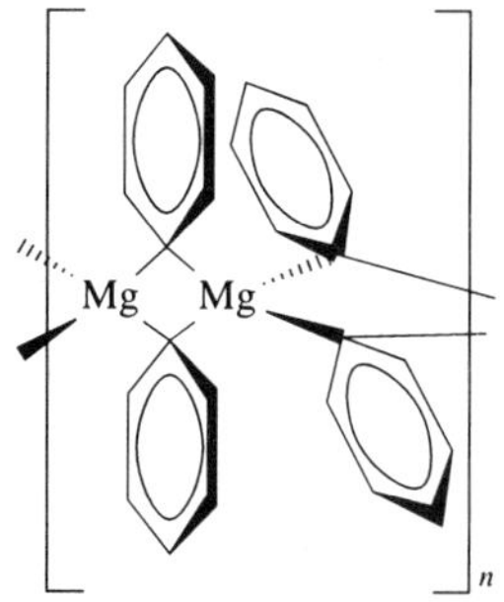

(64)

Solvation of (**63**) by ethers gives three-coordinate, trigonal-planar complexes $Mg[CH(TMS)_2]_2(OEt_2)$ or $\{Mg[CH(TMS)_2]_2\}_2$(μ-*p*-dioxane), with a geometry of type (**65**) at each magnesium centre.[152] The more common, distorted tetrahedral geometry (**66**), previously established for some TMEDA, bis(quinuclidine) or bis(THF) complexes of diorganylmagnesium compounds,[1–2] is also found in the *closo*-dicarbadecaboranyl derivative (Mg(2-Me-1,2-$C_2B_{10}H_{10})_2$(*p*-dioxane)$_2$),[253] in [Mg($CH_2Bu^t)_2$(μ-*p*-dioxane)]$_n$,[255] $MgPh_2(THF)_2$,[254] and the mononuclear form of Mg(*p*-Tol)$_2$(THF)$_2$;[254] *cyclo*-tris[(μ-*o*-xylidene)bis(tetrahydrofuran)magnesium] (**67**) contains similar tetrahedral magnesium atoms in a 15-membered metallacycle.[167] In these four-coordinate complexes, C–Mg–C angles are greater than O–Mg–O (see Table 4), and this can be rationalized by Bent's rule for increasing *s*-character of magnesium orbitals directed towards the less-electronegative carbon ligands, possibly combined with steric factors.[2] Crystals of solvated di-*p*-tolylmagnesium also contain a four-coordinate, dinuclear unit (**68**) in which two aryl ligands are μ,η^1-bridging, as in polymeric $(MgPh_2)_n$, and structural parameters include the following: Mg–C_{term} distance of 0.2130(7) nm, Mg–C_μ distances of 0.2245(7) and 0.2313(7) nm; Mg–C_μ–Mg angle of 77.5(2)°.[254]

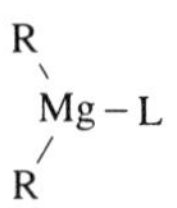

(65)

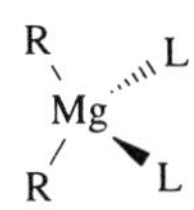

(66)

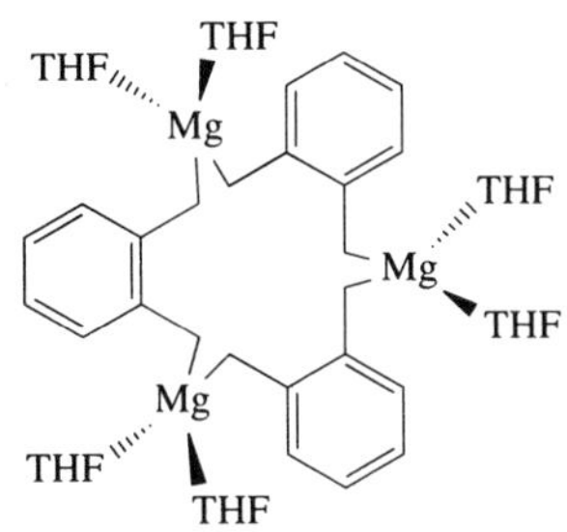

(67)

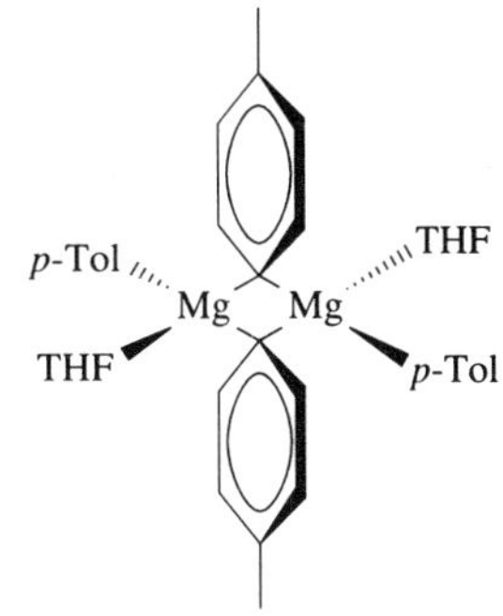

(68)

Complexes of polyethers with diorganylmagnesium compounds exhibit coordination numbers greater than four. The three oxygen atoms of diglyme are bound to magnesium in $Mg(C_6H_4Bu^t\text{-}p)_2[MeO(CH_2CH_2O)_2Me]$ to give a five-coordinate structure (**69**) (R = Me), with slightly longer Mg–O bonds than in the distorted tetrahedral ether complexes, and with an Mg–C bond length of 0.2157(2) nm and a C–Mg–C angle of 112.6(1)°.[2] Related five-coordinate structures are found for the tetraglyme adduct $Mg(C_6H_4Bu^t\text{-}p)_2[MeO(CH_2CH_2O)_4Me]$ (i.e., (**69**), R = $(CH_2CH_2O)_2Me$)), and for the two $Mg(C_6H_4Bu^t\text{-}p)_2$ units coordinated peripherally to separate, equivalent sets of three adjacent oxygen atoms in the large crown ether complex $[Mg(C_6H_4Bu^t\text{-}p)_2]_2$(1,3,16,18-dixylyl-30-crown-8).[2] Of particular interest are the structural characterizations of 'threaded' or rotaxane diorganylmagnesium complexes with crown ethers. The centrosymmetric structure of $MgEt_2$(18-crown-6) (**70**) includes magnesium at the inversion centre within a plane of the six oxygen atoms and with short Mg–C bonds (0.2104(2) nm) lying perpendicularly above and below the plane; the six oxygen atoms are almost equidistant from the magnesium and create a hexagonal bipyramidal geometry but the rather long Mg–O distances (0.2767–0.2792 nm) suggest rather weak interactions between $MgEt_2$ and the crown ether.[256] The rotaxane structure of $MgPh_2$(1,3-xylylene-18-crown-5) (**71**) (Ar = Ph),[257] and of the very closely similar derivative (**71**) (Ar = $p\text{-}Bu^tC_6H_4$),[2] is less symmetrical, with only four oxygen atoms bound to the magnesium, two rather tightly (0.220–0.222 nm) and two less strongly (0.252 nm), and the two Mg–C bonds deviating slightly from linearity (see Table 4). This gives a six-coordinate complex.

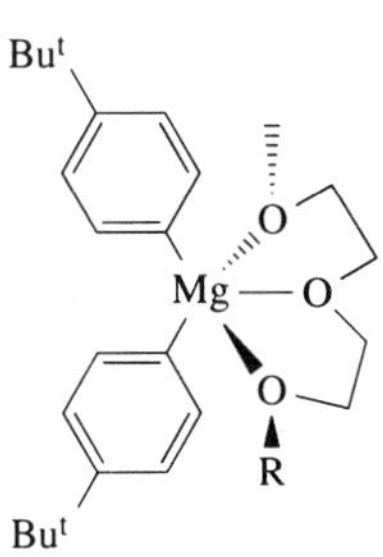

(69)

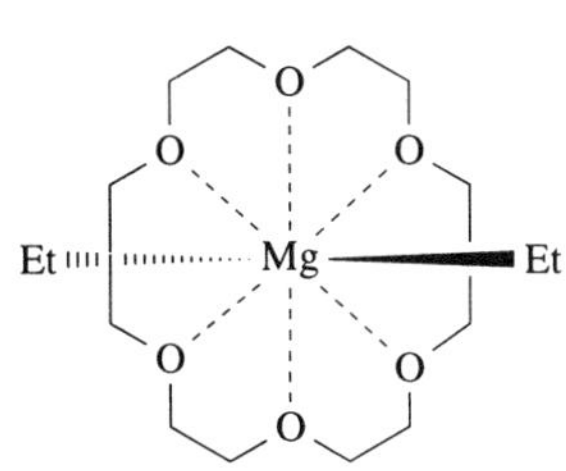

(70)

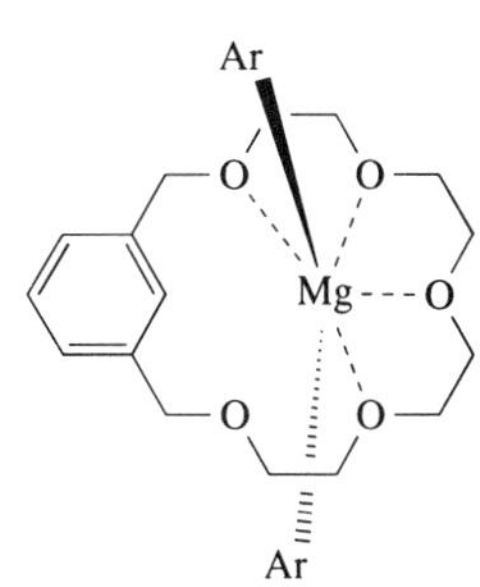

(71)

Crown ethers or cryptands may cause diorganylmagnesiums to disproportionate into ion pairs. Thus, 15-crown-5 and dimethylmagnesium form an ion-separated system $[MeMg(15\text{-crown-}5)^+Mg_2Me_5^-]_n$ with structure (**72**) in the solid state.[258] In (**72**) the cation $[MeMg(15\text{-crown-}5)]^+$ contains magnesium coordinated to the five oxygen atoms in the equatorial plane and to an apical methyl group (Mg–C 0.2140(7) nm); the anion is a chain of $[Mg_2Me_5]^-$ units linked by single methyl μ_2 bridges with the two magnesium atoms in the dinuclear units being bridged by two methyl groups to form a four-membered Mg_2C_2 ring; the ions are not completely separated since a weak interaction occurs between the cation, through the other apical site of the pentagonal pyramid, and a terminal methyl group of the polymeric anion (Mg–C 0.328 nm).[258] Cryptands dissociate $MgEt_2$ into ions, and structure (**73**) containing a 2:1

ratio of separated cations [EtMg(2,2,1-cryptand)]$^+$ and anions $[Mg_2Et_4(\mu\text{-}Et)_2]^{2-}$ has been determined in the crystalline state.[259] In (**73**) the magnesium in the cation is six-coordinate, bonding to one ethyl group and to two nitrogen atoms but only three oxygen donor atoms of the cryptand; the centrosymmetric dinuclear dianion contains four-coordinate magnesiums with relatively long Mg–C bonds (Mg–C_{term} 0.2209(9) nm and 0.2238(9) nm, Mg–C_μ 0.2359(9) nm and 0.2364(9) nm).[259] The solid 1:2 complex (**74**) between 2,1,1-cryptand and dineopentylmagnesium ($MgNp_2$) is also ionic. The structure comprises cations [NpMg(2,1,1-cryptand)]$^+$, with all six heteroatoms of the cryptand coordinated to magnesium, giving a pentagonal bipyramidal geometry, and mononuclear anions $[MgNp_3]^-$ of slightly distorted trigonal planar geometry (C–Mg–C angles 116.9(5)°, 117.6(5)° and 124.8(5)°) and varying Mg–C bond lengths (0.2125(12) nm, 0.2240(12) nm and 0.2296(16) nm).[259]

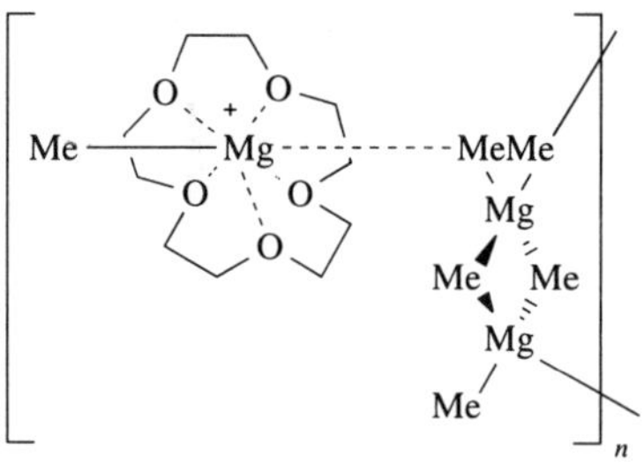

(**72**)

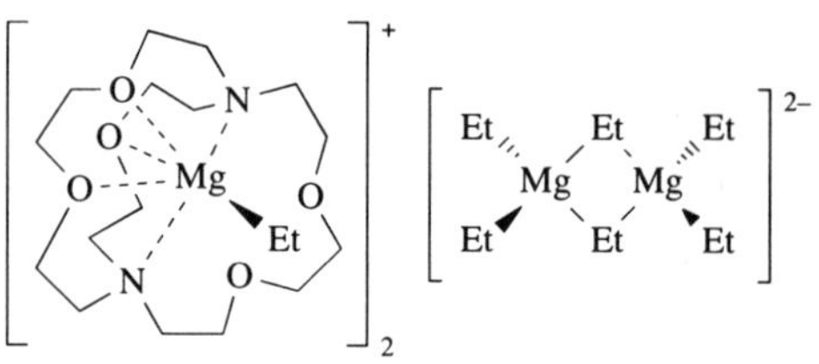

(**73**)

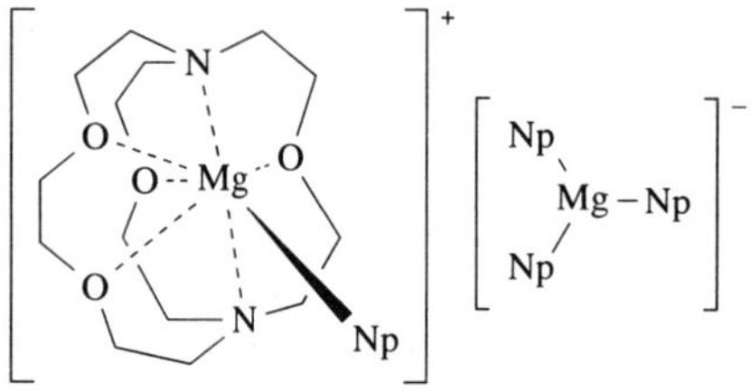

(**74**)

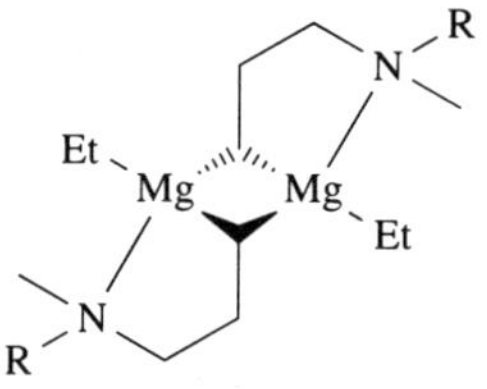

(**75**)

Intramolecular coordination of suitably positioned electron-donating heteroatoms in the organic group of diorganylmagnesiums can form stabilized, solvent-free structures. A highly distorted tetrahedral geometry is present in solid $[Mg\{(TMS)_2(o\text{-pyridyl})\}_2]$ (**58**), which contains two C,N chelates as four-membered rings and a wide C–Mg–C angle (157.0°).[246] In bis(4-methoxybutyl)magnesium (**60**) (R = Me), the two C,O chelating ligands form six-membered rings, and the distortion from tetrahedral geometry around the magnesium atom is less, with a C–Mg–C angle of 140.2(2)°.[260] The mixed-alkyl dimers $[MgEt(CH_2CH_2CH_2NR^1R^2)]_2$, prepared via Equation (42), are inner complexes containing five-membered C,N chelate rings, with the carbon atoms also participating in three-centre, two-electron bridges between the metal centres to form a centrosymmetric structure with a central Mg_2C_2 ring (**75**).[249] Each magnesium atom in (**75**) has a coordination number of four, and in crystals containing R^1 = Me, R^2 = Me or Cy, bond lengths include Mg–C_{term} distances of 0.2142(3) nm or 0.2129(4) nm and Mg–C_μ distances are 0.2294(2) nm, 0.2287(3) nm and 0.2267(3) nm, respectively.[249]

Crystals of $[Mg(o\text{-anisyl})_2(THF)]_2$ contain a very unsymmetrical dimer, (**76**), in which one anisyl group acts as a simple C,O bridge, whereas another bridges through the carbon atom and also chelates with the oxygen atom to one metal centre.[261] The coordination spheres around the magnesium atoms are each completed by a carbon-bonded anisyl ligand and a THF molecule of solvation to create one four-coordinate magnesium with a formal positive charge, and one formally negative with five coordination.[261] Pseudo-trigonal-bipyramidal geometry is found for (**77**), in which intramolecular oxygen coordination from the 10-membered oxamagnesacycle creates two six-membered chelate rings; the THF molecules occupy an axial and an equatorial site and the equatorial angle C–Mg–C is 130.0(2)°.[166] Intramolecular coordination in $Mg[2,6\text{-}C_6H_3(OMe)_2]_2$ gives the nonsolvated, pseudo-octahedral structure (**78**) in the solid state, in which the C–Mg–C angle is almost linear (173.4(2)°).[166] The structures of the solvent-free, mixed-diaryl magnesium compounds (**79**)[262] and (**80**),[66] containing intramolecularly coordinated crown ether side chains, have been determined and can be compared with structures of the related Grignard reagents (**35**)[61] and (**36**),[78] respectively. Structure (**79**) is practically identical to (**35**), with two stronger and two weaker Mg–O interactions giving a six-coordinate geometry. However, structure (**80**) is markedly different from (**36**), and is irregular with an approximate trigonal-bipyramidal geometry. Oxygen O-4 is outside the coordination sphere, O-1 is essentially nonbonded (Mg–O-1 0.2713(4) nm), the two Mg–C and Mg–O-3 bonds are in equatorial sites and O-2 and O-5 are axial.[66] In both (**79**) and (**80**) the xylyl–magnesium bond (0.2127 nm or 0.2128 nm) is shorter than the other Mg–C(aryl) bond (0.2154 nm).

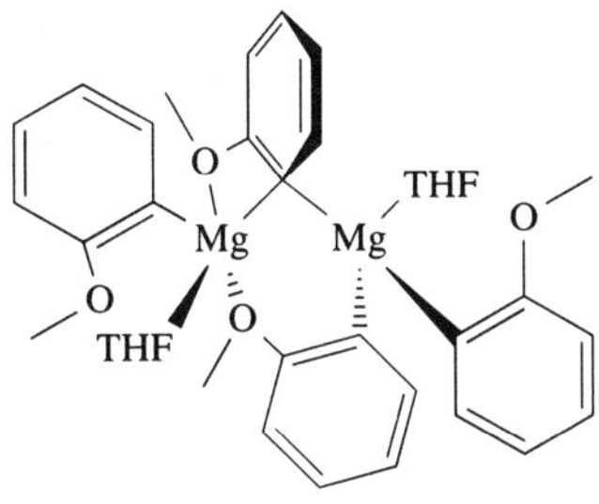

(76)

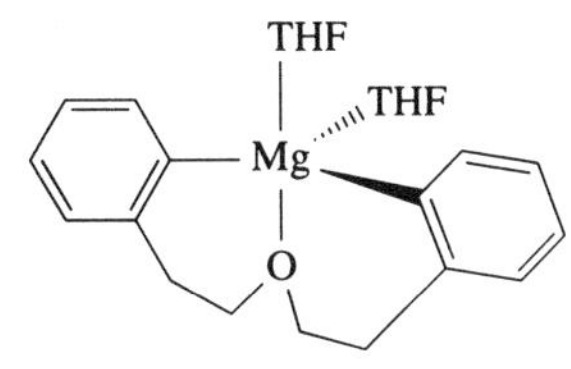

(77)

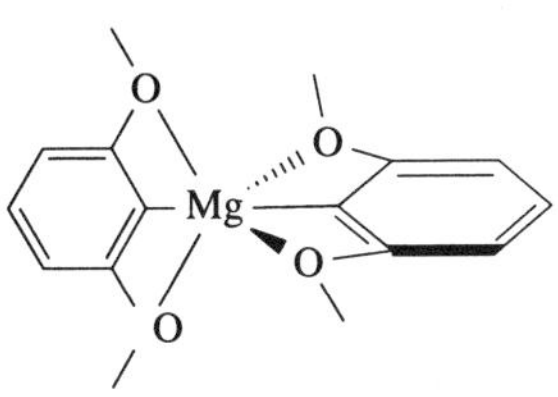

(78)

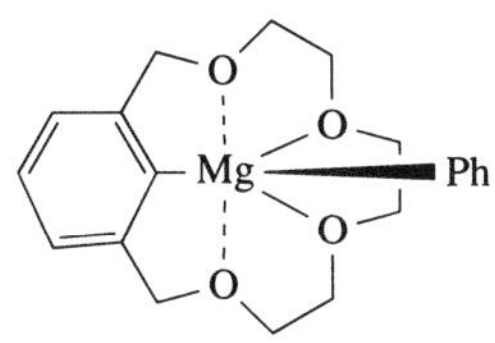

(79)

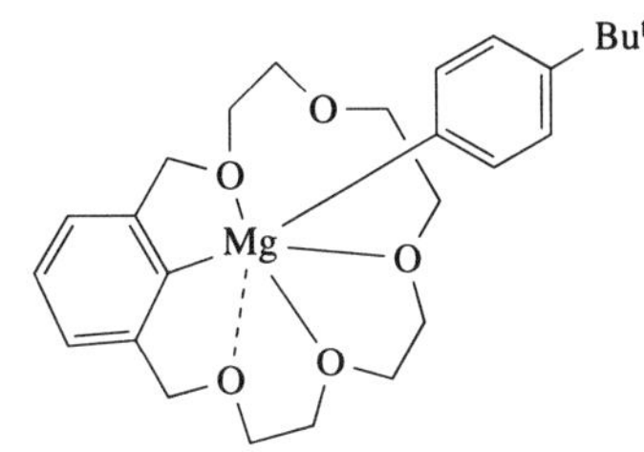

(80)

Three related, novel structures of tetrameric organomagnesium clusters have been determined.[237] Tetrakis[(μ_3-*o*-phenylene)tetrahydrofuran–magnesium], tetrakis[(μ_3-1,8-naphthalenediyl)tetrahydrofuran–magnesium] and tetrakis[(μ_3-*cis*-diphenylvinylene)tetrahydrofuran–magnesium] contain distorted tetrahedral cores of four magnesium atoms. Each face of this tetrahedron is bridged by an unsaturated 1,2-dicarbanionic organic ligand forming one single C–Mg bond and one three-centre, two-electron bridge between the other carbon and two magnesium atoms. An arrangement of the four ligands giving a cluster with C_2 symmetry, (**81**), is found for the *o*-phenylene species, but an arrangement with S_4 symmetry, (**82**), occurs for the other derivatives. Each magnesium has a pseudotetrahedral coordination sphere comprising one THF molecule, one single bonded carbon atom and two bridging carbon atoms. Single Mg–C bonds (0.214–0.218 nm) and bridging Mg–Cμ bonds (0.224–0.231 nm) are within expected ranges for diarylmagnesium compounds.[237,349]

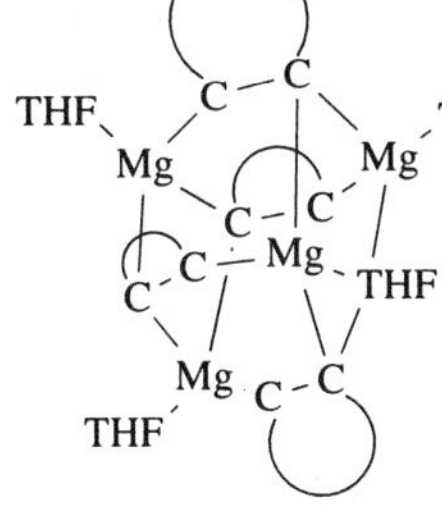

(81)

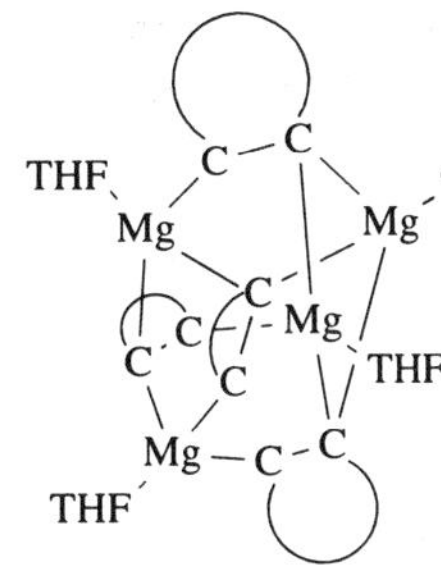

(82)

Dialkynylmagnesium adducts with TMEDA ([Mg(C≡CR)$_2$(TMEDA)$_2$]; R = Bu^t, Ph) crystallize as six-coordinate, pseudo-octahedral complexes with *trans*-alkynyl groups (**83**).[263–4] The high coordination number in such simple species probably results from the low steric demand of the alkynyl ligands and the greater electronegativity of the *sp* carbon atoms. Mg–C bond lengths in these derivatives (0.2176 nm and 0.2200 nm) are normal.

The crystal structures of two substituted derivatives of dicyclopentadienylmagnesium have been reported. Molecules of Mg(η-Bu^{t}C$_5$H$_4$)$_2$ are monomers with parallel cyclopentadienyl rings in a staggered conformation, as in the parent compound,[1] but the magnesium–ring(centroid) distance is 0.200

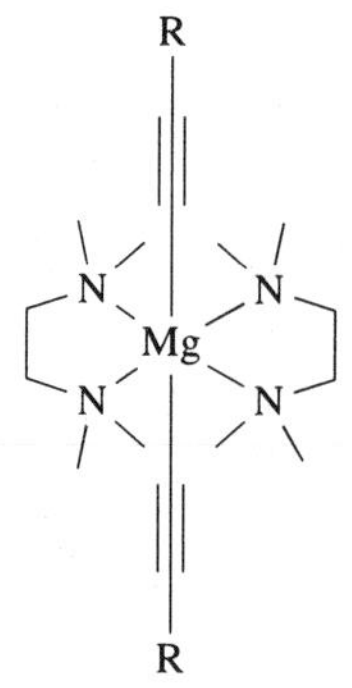

(83)

nm compared with 0.196 nm in solid $MgCp_2$; the two bulky Bu^t groups adopt a *trans* conformation.[245] $Mg[\eta\text{-}1,2,4\text{-}(TMS)_3C_5H_2]_2$ has eclipsed rings, although only one pair of TMS groups is eclipsed, and the slight bending of the two rings from parallel (7.8°) may result from steric interaction between this pair of substituents; all silicon atoms are bent away by a mean of 8° from the ring plane, and the average magnesium–centroid distance (0.203 nm) is longer than in less substituted derivatives.[159]

(ii) Solution

Unlike Grignard reagents, the halogen-free diorganylmagnesium compounds do not exhibit the complications of Schlenk equilibria in solution. In coordinating solvents such as THF, mononuclear forms are observed but association via alkyl bridges may occur, and in nonsolvating media the species MgR_2 are commonly aggregated. Even dineopentylmagnesium, which can be monomeric in the gas phase, is trimeric in benzene, and di-*s*-butylmagnesium is dimeric in pentane.[1] Bulky $Mg[C(TMS)_3]_2$ is monomeric in solution, with the solid state structure (**62**) probably being retained.[151] Also, $Mg[CH(TMS)_2]_2$ is monomeric in hydrocarbon solvents with no evidence for the intermolecular association observed in the solid state (see (**63**)).[152]

The clusters $[Mg(o\text{-}C_6H_4)]_4$ and $[Mg(1,8\text{-naphthalenediyl})]_4$ with crystal structures of types (**81**) and (**82**), respectively, remain tetrameric in THF; the solution structures are presumably related to those of the solid state, but ^{13}C NMR spectra indicate the presence of symmetrical organic bridges even at low temperatures, and a fast intramolecular exchange involving rotation of the ligands about the faces of the Mg_4 tetrahedron has been proposed.[237] Magnesacyclohexane has a dimeric structure in the solid state[1] and shows a monomer–dimer equilibrium in THF (Equation (46), $Z = (CH_2)_5$), favouring 1,7-dimagnesacyclododecane with $\Delta H = -48 \pm 3$ kJ mol^{-1} and $\Delta S = -106 \pm -10$ J K^{-1} mol^{-1}, and the related 2,3-benzo derivative exhibits similar behaviour.[84,135,143] Other simple magnesacycles have a higher tendency to dimerization in THF, and species $[Mg(CH_2)_n]_2$ ($n = 4$, 6, 9) are completely dimerized.[84,135,143] One factor promoting dimerization is the relief of strain in the smaller ring, caused by the large C–Mg–C angle (e.g., 141.5° in crystalline $[Mg(CH_2)_5(THF)_2]_2$) of the solvated, tetracoordinate $Mg(THF)_2R_2$ centres that are more readily accommodated in the larger ring. However, other conformational effects and entropic changes also play a part, and it may be noted that unlike dimeric $[Mg(CH_2)_6]_2$ the corresponding 4,5-benzo derivative $[Mg(CH_2CH_2)_2\text{-}1,2\text{-}C_6H_4]$ is monomeric in solution. The inclusion of an oxygen atom in magnesacycles $[Mg\{(CH_2)_n\}O]$ ($n = 3$, 4) and related derivatives stabilizes the monomeric ring by an intramolecular coordination of oxygen to the magnesium that is stronger than intramolecular solvation by THF (cf. structure (**77**)).[143b]

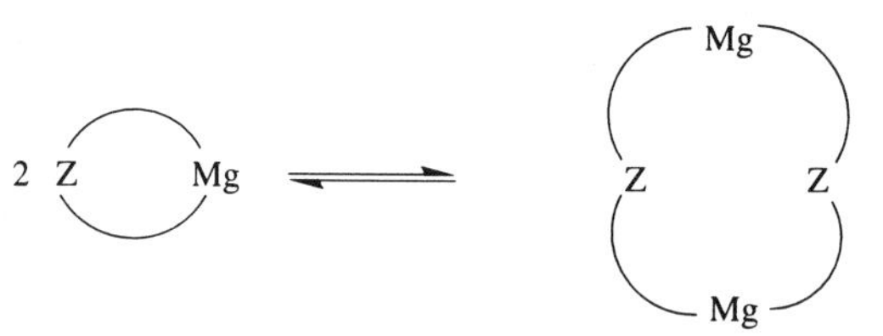

(46)

Various NMR studies on solutions of diorganylmagnesium compounds have been reported. Carbon-13 NMR spectra of alkyl, aryl and unsaturated species have been tabulated.[169] Magnesium-25 NMR chemical shifts of σ-bonded compounds occur to low field of $MgSO_4 \cdot 7H_2O$, for example, $\delta = 99.2$ ppm for $MgEt_2$ in THF; bis(2-methylallyl)magnesium with $\delta(^{25}Mg) = 70.4$ ppm must also be η^1 bonded.[164] The ^{25}Mg NMR chemical shift of bis(cyclopentadienyl)magnesium,[265] or its ring-substituted derivatives,[159,245,265] occur at high fields ($\delta \sim -80$ ppm). For $MgCp_2$ in noncoordinating toluene the ^{25}Mg resonance occurs at $\delta = -85.4$ ppm, but addition of Lewis base donor solvents causes downfield shifts in this resonance, correlating with related $^{13}C(Cp)$ shifts, and from these observations an increasing strength of donor complexation has been assigned: $Pr^i_2O \approx NEt_3 < Et_2O < PMe_3 <$ 1,4-dioxane < DME < THF < TMEDA (e.g., $\delta(^{25}Mg)$ for the TMEDA adduct is −25.9 ppm).[265] It is of interest that phosphine donor complexes of $MgCp_2$ are indicated by these studies, although they have not been isolated. $MgCp^*_2$ shows negligible changes of $\delta(^{25}Mg)$ and $\delta(^{13}C(Cp^*))$ shifts in toluene or THF solvents, suggesting weak complexation with THF.[265] In a related synthetic study, the only reported isolable adduct of $MgCp^*_2$ was $Cp^*_2Mg(bipy)$.[266]

Optical spectral investigations of the interactions of polyamines with *n*-butyl(9-*n*-propylfluorenyl-)magnesium in cyclohexane have suggested that only loose ion pair complexes are formed, such as (Bu^nMg^+, TMEDA, $[PFl]^-$) (PFl = 9(*n*-propyl)fluorenyl), with $\lambda_{max} = 383$ nm.[267]

Fluxionality of diorganylmagnesium compounds in solution has been probed by ^1H and ^{13}C NMR spectroscopy. η-Cyclopentadienyl(2-methylallyl)magnesium contains a σ-bonded allyl group that exchanges terminal allylic carbon atoms, probably via a bimolecular transition state (**84**), with activation parameters $E_a = 54.3$ kJ mol^{-1} and $\Delta S^* = 27$ J K^{-1} mol^{-1}, determined from lineshape analysis of variable-temperature $^{13}C(^1H)$ NMR spectra in THF.[248] NMR spectra indicate that the mixed-diaryl compounds $MgAr^1Ar^2(THF)_2$ exchange aryl groups, probably via aryl-bridged binuclear intermediates.[241] Inversion of alkyl groups bonded to nitrogen in inner complexes of types (**59**)[260] and (**75**)[249] has also been studied by variable-temperature NMR.

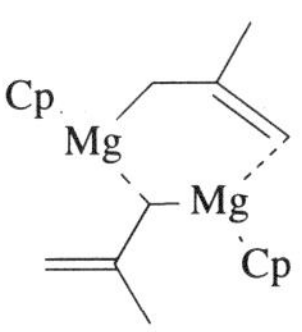

(84)

Addition of crown ethers or cryptands to diorganylmagnesium compounds in solution produces various new species observable by NMR spectroscopy.[66,256–9,268] Disproportionation into ionic forms can occur (Equation (47)), and the anions may be in equilibrium with associated forms, as in Equations (48) and (49). These ions are related to the solid-state species (**72**)–(**74**). Cryptands are more effective in such ionizations, causing disproportionation of diarylmagnesium[268] and dialkylmagnesium[259] compounds. Crown ethers also promote ionization of dialkylmagnesiums in benzene but not in the more solvating ethereal solvents, and the species $[NpMg(15\text{-crown-}5)]^+$, with corresponding magnesiate anions, has been identified by NMR.[258] Significant ion pair formation has not been observed from interactions between crown ethers and diarylmagnesiums in solution, but threaded complexes ([Mg(*p*-Tol)$_2$(crown)] (crown = 18-crown-6, 15-crown-5)) with structures analogous to (**70**) appear to be formed;[268] similar complexes may also be present in solutions of crown ethers and dineopentylmagnesium.[258] Threaded complexes of diarylmagnesiums and 1,3-xylylene-18-crown-5 (**71**) also form in toluene solution, and these 'rotaxanes' are in equilibrium with externally coordinated complexes ($MgAr_2$(crown)).[5,257] The rotaxanes cannot be produced by direct penetration by the bulky $MgAr_2$ species (Ar = Ph, $C_6H_4Bu^t$-*p*) of the crown ring, and the proposed mechanism of formation is Mg–C bond ionization to form intermediate ions ($[ArMg(crown)]^+$ and $[MgAr_3]^-$), which then reform the threaded complex from the opposite side of the cavity.[5,257] NMR spectroscopy in solution has also established that species formed by metallation reactions of 1,3-xylylene crown ethers have structures related to the solid-state species (**79**) and (**80**), although the apparent higher symmetry of polydentate coordination suggests ligand flexibility or dynamic processes.[66]

$$2MgR_2 + C \rightleftharpoons [RMg(C)]^+ + [MgR_3]^- \qquad (47)$$

C = cryptand, crown ether

$$MgR_2 + [MgR_3]^- \rightleftharpoons [Mg_2R_5]^- \qquad (48)$$

$$2[MgR_3]^- \rightleftharpoons [Mg_2R_6]^{2-} \qquad (49)$$

(iii) Gas phase

Dineopentylmagnesium ($MgNp_2$) is volatile and forms a monomer in the gas phase that has previously been shown to have a linear structure by electron diffraction.[1] The mixed-ligand species CpMgNp is also volatile, and gas-phase electron diffraction indicates the presence of a η^5-cyclopentadienyl and a single-bonded neopentyl ligand in a monomeric molecule (**85**).[247] The Mg–C(Np) bond of 0.212(2) nm and the magnesium–Cp parameters, with Mg–C(ring) 0.2328(7) nm, are indistinguishable from the corresponding structural parameters in gaseous $MgNp_2$ and $MgCp_2$, respectively.[247]

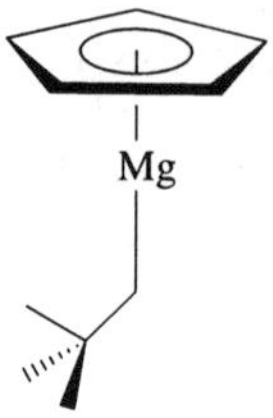

(85)

The thermal average molecular structure of $MgCp^*_2$ has been determined by gas-phase electron diffraction.[269] $MgCp^*_2$ has a parallel sandwich structure in which the rings are predominately eclipsed, although a dynamic equilibrium with some staggered form cannot be excluded, and the distances Mg–Cp(centroid) and Mg–C are 0.2011(8) nm and 0.2341(6) nm, respectively, with methyl groups bent from the ring plane towards the metal (5(2)°).

(iv) Thermochemical and theoretical studies

Mean bond dissociation energies of monomeric and dimeric $MgNp_2$ have been derived from their respective sublimation enthalpies and from the enthalpy of formation of the solid compound ($\Delta H_f^0 = -235.8 \pm 5.5$ kJ mol^{-1}). For the monomer, the mean Mg–C bond energy is 133.7 ± 5.0 kJ mol^{-1}, which is lower than reported for other alkylmagnesium compounds.[270] The enthalpy difference between $[MgNp_2]_2$ and $MgNp_2$ is 142.0 ± 5.0 kJ mol^{-1}, and the mean bond dissociation energy of the dimeric bridging system Mg–Np–Mg is relatively high (205 ± 3 kJ mol^{-1}).[270]

Geometrical parameters, IR and Raman frequencies and orbital energies of $MgCp_2$ have been derived from by *ab initio* LCAO–MO calculations at the Hartree–Fock level, which indicate that bonding is mainly between magnesium 3*s* and 3*p* orbitals and the ring π system, leaving a high local effective charge on the magnesium atom.[271]

3.2.2.3 Reactions

Thermal decompositions of simple diorganylmagnesium compounds at elevated temperatures occur primarily by elimination of an alkene from species with β-hydrogen atoms, as discussed previously.[1] Of interest is the thermolysis of the heteroligand, inner complexes $\{EtMg[(CH_2)_nX]\}_2$, with $X = NR_2$, $n = 3$, 4 (cf. **75**), at 210–220 °C, or X = OR, $n = 4$ at 190–200 °C, in which the elimination of an unsaturated amine or ether, respectively, followed by ethane (route (i), Scheme 8) is preferred to the alternative elimination of ethene with a saturated amine or ether (route (ii), Scheme 8).[249]

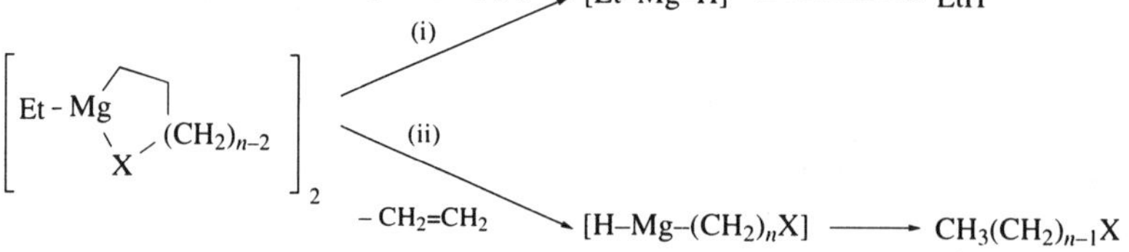

Scheme 8

Reactions of carbon monoxide at atmospheric pressure with $MgEt_2$ give, after hydrolysis, the various products of Equation (50), with 4-ethylhexan-3-one predominating; the relative amounts of products are dependent on the concentration of $MgEt_2$ and on the solvent polarity. Studies indicated that, whereas Et_2CO arises from unionized $MgEt_2$, $EtCOCHEt_2$ is formed from the ionic disproportionation species $[MgEt_3]^-$ and, by using ^{13}CO, the intermediate $Et^{13}CO^{13}C(Et)_2MgEt$ was identified.[272] A carbene and radical mechanism was proposed for this C–C coupling of CO since reaction of $MgEt_2$ with nitrosodurene produces paramagnetic ArN–NArOMgEt (R = 2,3,5,6-Me_4C_6H), probably via coupling of the nitrene ArN and the radical ArNOMgEt.[272b] Electron transfer from MgR_2 compounds to bifunctional nitrogen heterocycles with homolytic loss of one organic ligand generates organometallic radicals, formally containing magnesium(II) and radical anion ligands, as in Equations (51) and (52).[273] Complexation by the nitrogen heterocycle precedes electron transfer, and reduction of the primary complex by electron transfer from potassium metal gives a symmetrical anion radical (Equation (53)) rather than the unsymmetrical, neutral radical of Equation (52).[273]

$$MgEt_2 + CO \longrightarrow EtCOCHEt_2 + EtCOC(OH)Et_2 + EtCOCH(OH)Et + Et_2CO + Et_2CH(OH) \quad (50)$$

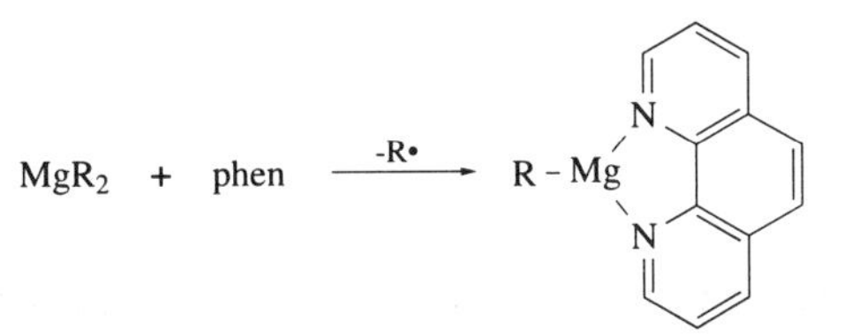

(51)

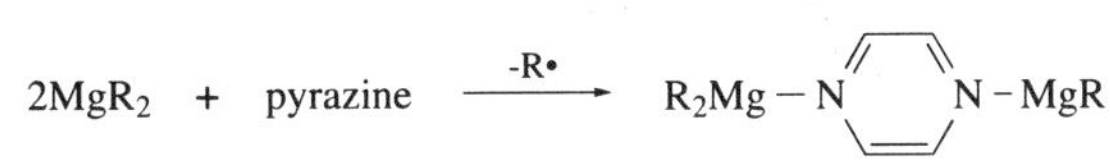

(52)

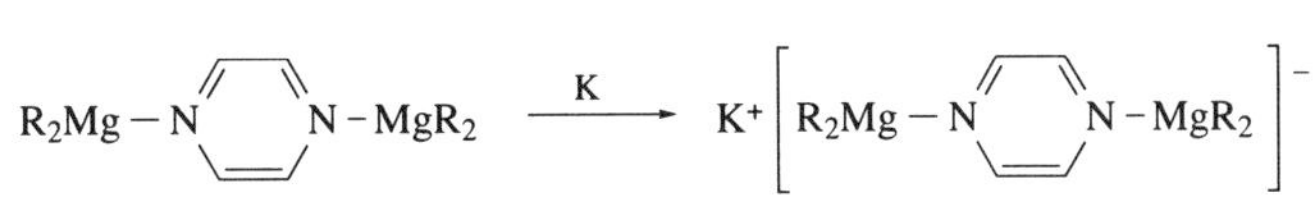

(53)

Reactions of organic functional groups with diorganylmagnesium compounds are similar to those of Grignard reagents, and the latter species in solution normally comprise a proportion of MgR_2 species as a consequence of the Schlenk equilibrium. However, the absence of halogen and the presence of two organic ligands per magnesium can influence the course of a reaction, and the reactivity of MgR_2 compounds is usually greater than that of RMgX. Moreover, compounds MgR_2 are often soluble in hydrocarbon solvents, so that reactions can be carried out more easily in less polar media. Metallation of compounds with active hydrogen atoms,[243–5] and uncatalysed[88] and catalysed[102,106–7,251] carbomagnesiations of unsaturated compounds have been mentioned in Section 3.2.2.1. Reductive cyclization of 1,7- or 1,6-dienes by $MgBu_2$, catalysed by Cp_2ZrCl_2, should also be noted[105a] (cf. Equation (16), X = Bu^n).

Differences in reactivity of the simple ethyl ligand and the chelating heteroorganic ligand of compounds $[EtMg(CH_2)_nX]_2$ are evident in the magnesiation of ethene at 140 °C and 80 bar, when X = OR and n = 4, since up to six C_2H_4 units are preferentially inserted into the magnesium–Et bond, but insertion of C_2H_4 occurs approximately equally into both Mg–C bonds when X = NR_2 and n = 3, 4.[249] No difference has been observed in the reactivity of the Mg–C bonds in these mixed-ligand complexes to carbonation with CO_2, but addition of the ethyl group to benzaldehyde occurs more rapidly.[249]

Chelation has often been invoked, without direct evidence, to explain the stereochemistry of reactions of alkoxyketones with organomagnesium reagents. NMR studies have provided kinetic evidence that reactions of nonsterically hindered α-alkoxyketones with $MgMe_2$ to form methylmagnesium alkoxides proceed through chelation in the transition state, which increases the rate of reaction, but that chelation is of minor importance in related reactions of β-alkoxyketones.[274] The amount of reduction accompanying alkyl addition in reactions of dialkylmagnesium compounds with ketones is decreased by the addition of certain alkali metal alkoxides (including LiOMe and NaOMe) or tetraalkylammonium halide salts to the reaction mixtures; a salt:MgR_2 ratio of at least 1:1 is best, and it is probable that magnesiate anions of the type $[R_2Mg(\mu\text{-}OMe)_2MgR_2]^{2-}$ promote the addition process.[275]

Crown ethers and cryptands cause dissociation of MgR_2 compounds, as in Equation (47) and the related Equations (48) and (49), forming ions which may exhibit greater reactivity than the neutral compounds. Although acyclic polyethers and other crown ethers cause a small increase in the rate of metallation of fluorene by $MgEt_2$, it is found that 15-crown-5 significantly increases the rate of this reaction, either in THF or more considerably in diethyl ether solutions, and the reactive species is probably a magnesiate anion.[276] Specific activation of $MgEt_2$ by 15-crown-5 is also noted in its addition to pyridine, giving the 1,4-addition product in greater amounts than the 1,2-addition product,[276] but only the latter is obtained in the absence of the crown ether; similar activation of the addition of $MgEt_2$ to pyridine is caused by the presence of small amounts of 2,1,1-cryptand, with the larger 2,2,1- and 2,2,2-cryptands showing related reactivity.[259] Both 1,4 and 1,2 addition products are also produced from pyridine and mixtures of $MgEt_2$ and LiEt, and this supports the involvement of magnesiate anions in these reactions.[277]

Intramolecular metallation of 1,3-xylylene-15-crown-4 (**7**) (A = H, $n = 0$) by $MgPh_2$ (Scheme 4; A = H, Y = Ph) is promoted by a favourable interaction of magnesium with the crown ether and the formation of nucleophilic anions $[MgPh_3]^-$ and an electrophilic cationic centre; this reaction occurs in diethyl ether and not in the more coordinating solvent THF, and, since the larger-ring crown ether (**7**) (A = H, $n = 1$) is not metallated, activation by a specific cavity size is evident.[66,262] Related halogen–metal exchange between crown ethers (**7**) (A = Br, $n = 0$, 1) and $MgPh_2$ also occurs, more readily for the smaller ring and faster than the corresponding metallation reactions. The influence of the crown ether on these reactions is clearly illustrated in the course of the reaction shown by Equation (54), in which no reaction occurs at the remote bromine substituent although this bromine exchanges readily with *n*-butyllithium.[66] The facile ether cleavage reaction of (2-methoxy-1,3-xylylene)-15-crown-4 (**7**) (A = OMe, $n = 0$) by $MgPh_2$ to form (**86**) is another example of activation by a crown ether: a reaction mechanism related to that of Scheme 4 is possible, in which combined S_N2 attack of $MgPh_3^-$ on the methyl group to form toluene and electrophilic attack on the oxygen by $Mg(crown)Ph^+$ (i.e., (**7b**; A = OMe, Y = Ph) gives (**86**).[5,278] Ether cleavage to form (**86**), which only proceeds in diethyl ether solution and not THF, occurs at a considerably slower rate for the 2-ethoxy derivative, while the 2-isopropoxy or 2-methylthio derivatives are unreactive.[278] Cleavage of an ether link of 2,1,1-cryptand by dineopentylmagnesium, accompanied by proton abstraction, occurs readily in heated diethyl ether–benzene solutions and affords a neopentylmagnesium alkoxide with the dimeric structure (**87**) in the solid state.[279]

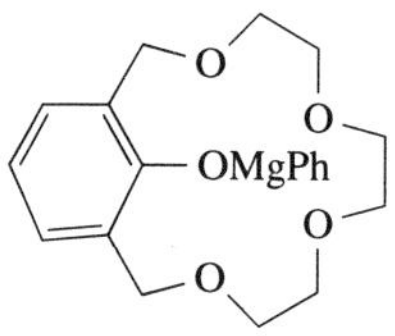

(86)

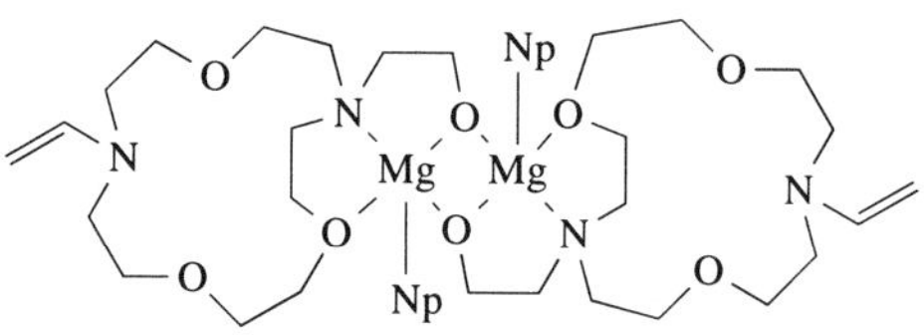

(87)

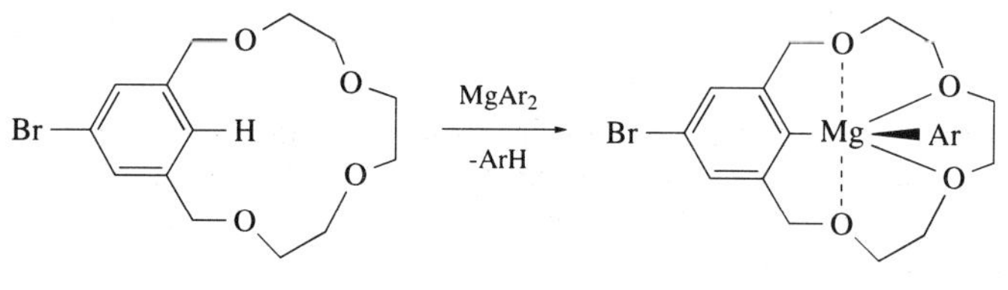

(54)

Diorganylmagnesium compounds have found many applications, in similar reactions to their Grignard analogues, for the synthesis of organometallic compounds.[1] Magnesacycles[135] have been used for the preparation of organometallics of main group and transition metals; for example, 1,2-dihydro-1-magnesabenzocyclobutene (see Scheme 3) as a precursor for cyclic silicon, germanium and tin compounds.[146] Commercially available hydrocarbon solutions of Bu^nMgBu^s have proved to be useful in organometallic syntheses and in the formation of supported transition metal catalysts for polymerizations.[240] Silyl-substituted dicyclopentadienylmagnesium compounds, formed by metallation using Bu^nMgBu^s, are also useful ligand transfer reagents, as in the synthesis of $[Zr\{\eta\text{-}C_5H_3(TMS)_2\}Cl_3]$.[244]

3.2.3 Addition Compounds of Magnesium with Unsaturated Organic Molecules

Magnesium metal reacts by formal two-electron transfer with various unsaturated organic compounds to give products which had previously been incompletely characterized.[1] More recently, structural characterizations of several of these products have been reported, and some species may be considered to be cyclic diorganylmagnesiums, related to compounds discussed in Section 3.2.2. However, since these addition products have a tendency to revert to unsaturated hydrocarbon and magnesium by a formal homolytic cleavage of the two Mg–C bonds, they will be discussed in this section.

3.2.3.1 Alkenes, alkynes and dienes

Magnesium metal adds to the activated monoenes, styrene and 1,1-diphenylethene, in liquid ammonia or HMPT but generally does not form addition products with simple alkenes.[1] Condensation of magnesium vapour into a solution of diphenylethyne in THF at −100 °C, followed by warming to room temperature, affords a mixture of addition products, derivatives of *cis*- and *trans*-diphenylvinylenemagnesium (**88**), which on reaction with D_2O, give *cis*- and *trans*-PhCD=CPhD in a 3:2 ratio. The more soluble *cis* product (**88a**), which produces only *cis*-PhCD=CPhD on deuterolysis, can be isolated from the THF solution as pure, yellow crystals containing solvated, tetrameric clusters $[Mg(\mu_3\text{-}C_2Ph_2)(THF)]_4$ of type (**82**), structurally related to tetrakis[(μ_3-*o*-phenylene)tetrahydrofuran–magnesium] and tetrakis[(μ_3-1,8-naphthalenediyl)tetrahydrofuran–magnesium], as discussed in Section 3.2.2.2(i).[237] Alkyne ligands are also η^2 bonded to a magnesium atom in the structurally characterized complexes (**89**)[280a] and (**90**),[280b] formed from reactions of bis(cyclopentadienyl)titanium derivatives with metallic magnesium and 1,4-bis(trimethylsilyl)-1,3-butadiyne or 1,2-bis(trimethylsilyl)ethyne.

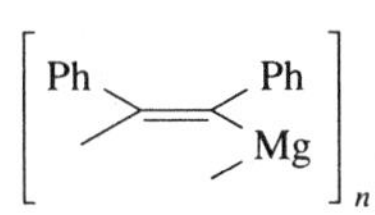

(a)

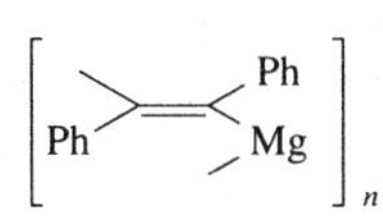

(b)

(88)

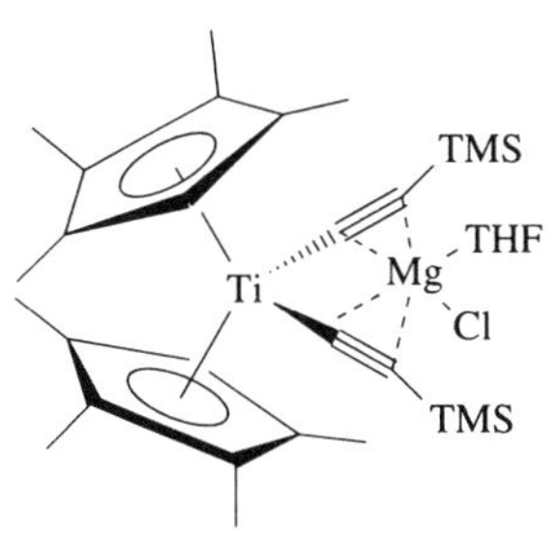

(89)

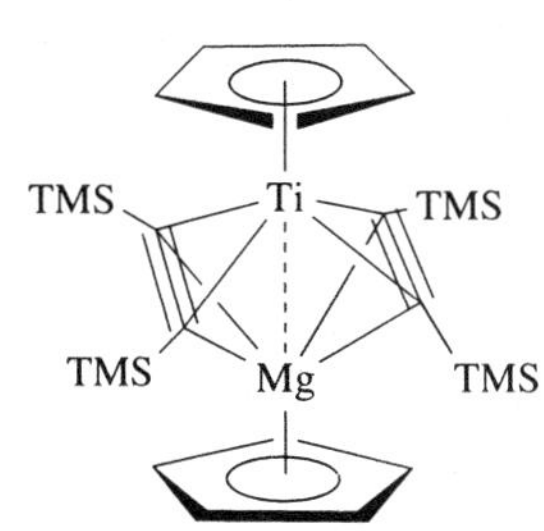

(90)

Activated magnesium reacts with 1,3-dienes, including 1,3-butadiene, isoprene, myrcene and (*E*),(*E*)-1,4-diphenyl-1,3-butadiene, to form addition products.[1,98,281] Highly activated magnesium, prepared using Rieke's method of reducing anhydrous $MgCl_2$ in THF with preformed lithium naphthalenide or with lithium using naphthalene as an electron carrier, is very effective in the syntheses of magnesium compounds of various acyclic, substituted 1,3-butadienes,[282] and of 1,2-bis(methylene)cycloalkanes, forming adducts of type (**91**) (n = 5, 6, 7).[283–5] Magnesium–diene compounds of 1:1 stoichiometry have been reported for all these species. Butadiene and isoprene give the products $[Mg(diene)_x]_n$, with the variation of x = 1–4 arising from coupling of diene under varying reaction conditions, and with degrees of association n >1, as discussed previously.[1,98]

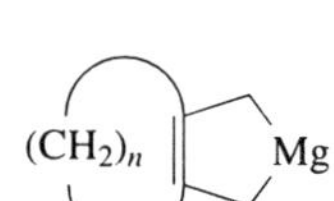

(91)

In the solid state, (*E*),(*E*)-1,4-diphenyl-1,3-butadienemagnesium forms a mononuclear solvated structure (**92**) containing magnesium 0.171 nm from the plane of the C_4 ligand and coordinated to three THF molecules and to the two outer carbon atoms in a 2-butene-1,4-diylmagnesium ring, with Mg–C distances of 0.226 nm and 0.232 nm.[286] However, weak interactions with the central carbon atoms

(Mg–C = 0.252 nm and 0.256 nm) suggest a tendency towards η^4 coordination. Other diene–magnesium compounds have not been structurally characterized. The ^{13}C NMR spectrum of the 1:1 complex 1,2-bis(methylene)cyclohexane–magnesium in THF solution indicates a symmetrical organic ligand[283] and is consistent with the metallocyclic structure (**91**) (n = 4), but oligomeric structures are also possible. Structure (**67**) of a related aromatic species may be noted.[167]

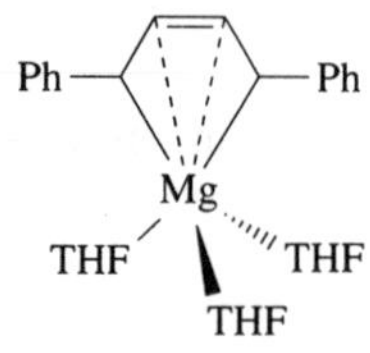

(92)

In toluene at 50 °C, $[Mg(butadiene)(THF)_2]_n$ is partially desolvated to form $[Mg(butadiene)(THF)_{0.5}]_n$, and in boiling toluene it decomposes to give active magnesium and butadiene.[287] Oxidation of diene–magnesium compounds with dry oxygen also releases the coordinated diene;[283,287] hydrolysis forms dihydro products, for example, 1,2-dimethylcyclohex-1-ene and 1-methyl-2-methylenecyclohexane in the ratio 91:9 from (**91**) (n = 4).[283]

Applications of diene–magnesium complexes in organic syntheses include coupling with α,ω-dibromoalkanes to form carbocycles,[282] as illustrated by the spiroannulation of the 1,2-bis(methylene)cycloalkane–magnesium reagent (Equation (55)).[283] Other organic reactions of diene–magnesium compounds with electrophiles have afforded functionalized ketones, alcohols and γ-lactones, including spirocyclic derivatives.[98,282–5] Diene–magnesium reagents have also been employed in transition metal-catalysed coupling reactions,[98] and in the formation of cyclic organoboron,[288] organophosphorus[289] and organosilicon[282,290] compounds, including spiro compounds such as (**93**) from $SiCl_4$.[282] The reagents have proved to be useful for the introduction of diene ligands into complexes of early transition metals, such as $Cp_2Zr(\eta$-diene).[281,291]

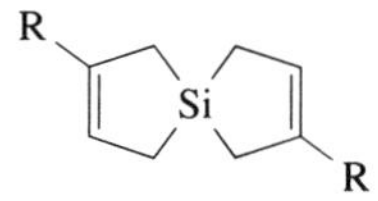

(93)

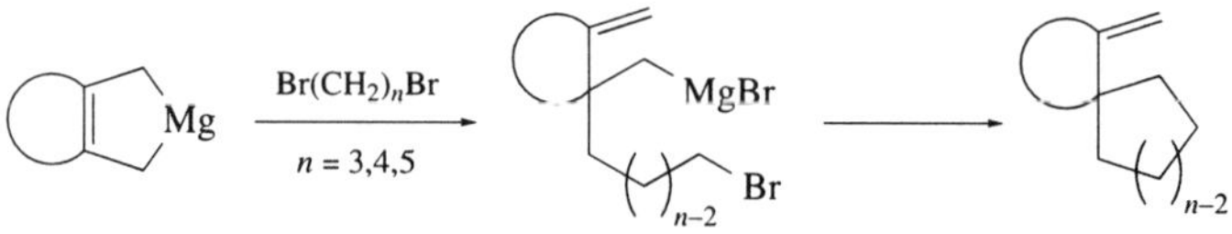

(55)

In crystalline Mg_2C_3 each magnesium atom is bonded to three terminal carbon atoms of three linear C_3 units and also in an η^2 mode to two carbon atoms of a fourth C_3 unit, creating an approximately tetrahedral coordination geometry around the metal.[292] Also, it may be noted that SCF MO calculations on the η^2 interaction of MgH_2 with ethene give a low bond dissociation energy of 22–25 kJ mol^{-1} and Mg–C distances of 0.290 nm.[293]

3.2.3.2 Cyclooctatetraene

Highly activated magnesium formed by potassium reduction of $MgCl_2$,[242] magnesium activated by sonication with catalytic amounts of anthracene,[294] or preformed anthracene–magnesium compounds[294] react with cyclooctatetraene in THF to provide convenient syntheses of [Mg(cot)] in high yields. Cyclooctatetraaene–magnesium gives a single ^{13}C NMR resonance in solution (δ = 90.54 ppm)[242] and also in the solid state as $[Mg(cot)(THF)_{2.5}]$ (δ = 91.5 ppm);[295] this is consistent with an aromatic dianion and an ionic or contact ion pair structure in the solid form. Copper-catalysed couplings of [Mg(cot)] with allyl or propargyl halides give 1,5-diallyl- or 1,5-dipropargyl-1,3,6-cyclooctatriene, respectively,[242] and conversions of [Mg(cot)] into cyclic organosilicon and organophosphorus compounds have been reported.[98,296]

3.2.3.3 Anthracene and related aromatic compounds

The orange-coloured compound [Mg(anthracene)$(THF)_3$] is prepared in high yield by reaction of magnesium powder, activated by a small amount of alkyl bromide, with anthracene in THF at 20–60 °C over 24 h (Scheme 9);[24,294,297] the rate of formation depends on the concentration of anthracene and on the surface area of magnesium.[297] This direct reaction is reversible, and raising the temperature or reducing the anthracene and/or THF concentrations shifts the equilibrium in favour of the reactants.[6] [Mg(anthracene)$(THF)_3$] is also formed by reaction of sodium–anthracene with $MgBr_2$.[298] The substituted anthracene derivatives [Mg(An)$(THF)_3$], including An = 2-methyl, 1,4-dimethyl, 9-methyl, 9-ethyl, 9,10-dimethyl and 9-phenyl anthracene,[26b,299] are formed, in 50–90% yields, by reaction of magnesium with the appropriate aromatic derivative in THF at ambient temperature. Under similar reaction conditions, 9-trimethylsilyl- and 9,10-bis(trimethylsilyl)anthracene give [Mg(An)$(THF)_2$],[294,300] and anthracenes substituted with ether or amino groups at the 9 or 9,10 positions form organometallic products in which inner complexation may replace coordinated THF (e.g., (**94**)–(**97**)).[301] Protolyses of all the above 1:1 magnesium–anthracene derivatives yield the corresponding 9,10-dihydroanthracenes.

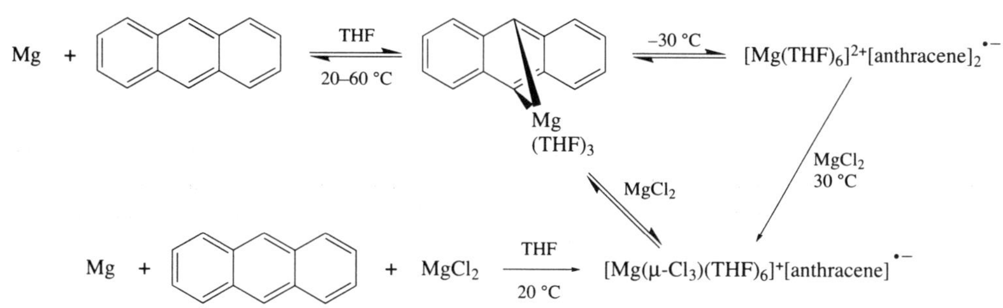

Scheme 9

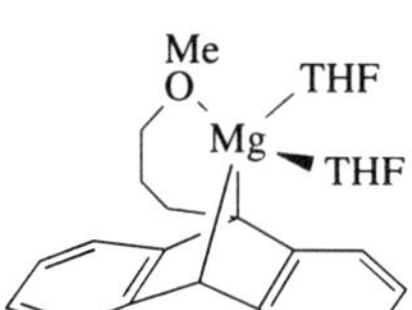

(94)

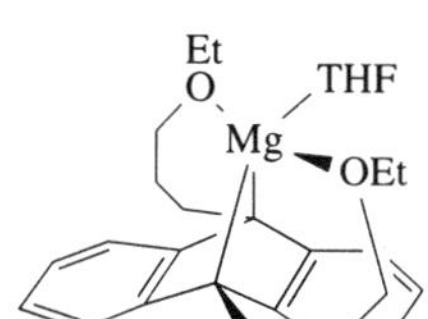

(95)

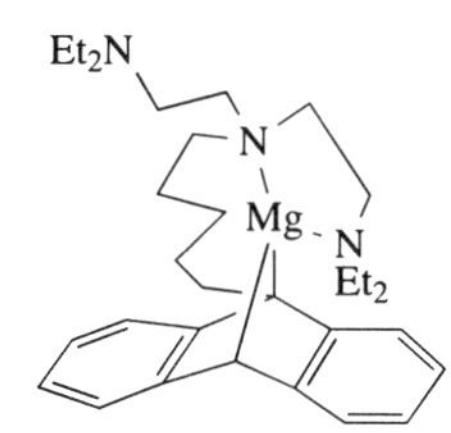

(96)

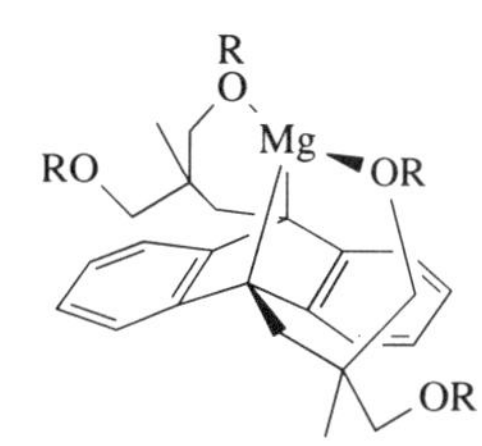

(97)

9,10-Diphenylanthracene reacts with magnesium powder in THF at 20 °C to give the deep-blue compound $Mg(THF)_6[C_{14}H_8(Ph_2\text{-}9,10)]_2$, a salt of the 9,10-diphenylanthracene radical anion; only on heating at 60 °C is this reversibly converted into $\{Mg[C_{14}H_8(Ph_2\text{-}9,10)](THF)_3\}$, but prolonged heating causes decomposition into active magnesium metal and free 9,10-diphenylanthracene.[26b] The radical anion compound of unsubstituted anthracene is only stable in equilibria in THF solutions at –25 °C or below, but the monoanion may be isolated as $[Mg_2(\mu\text{-}Cl_3)(THF)_6]^+[C_{14}H_{10}]^{\bullet-}$ after addition of $MgCl_2$ (Scheme 9).[6,299] Radical anions are identifiable by their ESR and UV–visible spectra, and afford 1:1 mixtures of 9,10-dihydroanthracene and the parent anthracene on protolysis. $[Mg_2\text{-}(\mu\text{-}Cl_3)(THF)_6]^+[C_{14}H_{10}]^{\bullet-}$ forms a crystal lattice of discrete ions, and the bond lengths and electron deformation density of the anion indicate that this radical is formed by the addition of an electron to the LUMO of neutral anthracene.[299] A polymer-supported magnesium–anthracene reagent, containing a mixture of dianion and radical anion derivatives, is formed by exchange between anthracene-functionalized polystryrene and [Mg(anthracene)$(THF)_3$].[27] Also, fluoranthrene (Fla) reacts with magnesium in THF to form insoluble $[Mg(THF)_6(Fla)_2]$, containing the radical anions $[Fla]^{\bullet-}$.[294]

Crystal structures of 1:1 anthracene–magnesium adducts show that magnesium is bonded to C-9 and C-10 of an anthracene ligand folded along the C-9–C-10 axis, as in (**98**). In compounds with $R^1 = R^2 = H$,[155] or $R^1 = Me$, $R^2 = H$,[299] the mean Mg–C distances are 0.230 or 0.232 nm, C–Mg–C angles 71.4° or 72.6°, folding angles 28.6° or 40.95°, respectively, and $n = 3$. In the silylated derivative with $R^1 = H$, $R^2 = TMS$, magnesium is bonded to only two THF molecules ($n = 2$) and mean geometric parameters are an Mg–C distance of 0.223 nm, a C–Mg–C angle of 78.2° and a folding angle of 45°;[300]

crystals containing a 1:1 mixture of $[Mg\{C_{14}H_8(TMS)_2\text{-}9,10\}L_2]$ ($L_2 = (THF)_2$ or TMEDA) show similar parameters for the two discrete molecules.[294] Four coordination of magnesium in the yellow, 9,10-disilylated derivatives may arise partly from steric hindrance, but the shorter Mg–C distances and other parameters suggest that electronic effects of the TMS groups also cause stronger, localized Mg–C interactions. Alternatively, more delocalization and ionic character would increase the electrophilicity of magnesium and its tendency to bind more THF molecules, as in the five-coordinate, orange-coloured derivatives. A high-frequency shift of solid-state ^{13}C NMR resonances of C-9 and C-10 nuclei in solid magnesium–anthracene derivatives, associated with a colour change from orange to yellow, has been attributed to increasing sp^3 carbon character.[295]

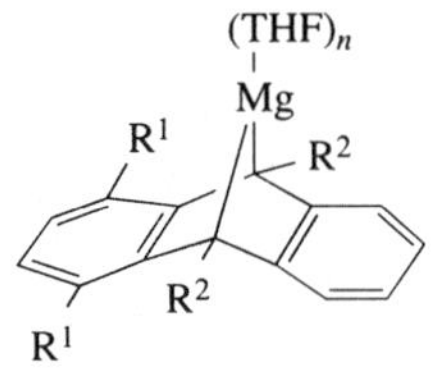

(98)

$[Mg(anthracene)(THF)_3]$ slowly decomposes in diethyl ether or hydrocarbon solvents to form magnesium and anthracene,[287,294] and graphite accelerates this decomposition,[295] which occurs via the yellow intermediate $[Mg(anthracene)(THF)_2]$. Silylation at the 9/10-positions increases the stability, and (**98**) (R^1 = H, R^2 = TMS) is stable indefinitely in hydrocarbon solvents.[294] THF can be exchanged for other oxygen- or nitrogen-bonded ligands, and reported derivatives, $[Mg(anthracene)L_n]$, include L = 1,4-dioxane (n = 3), DME (n = 2), $EtN(CH_2CH_2OMe)_2$ (n = 1), $MeN(CH_2CH_2NMe_2)_2$ (n = 1) or (−)-sparteine (n = 1).[287,294] Inner complexes, including (**94**)–(**97**), formed by partial or total replacement of solvent ligands with ether or amine functionalities, have been characterized by solid-state ^{13}C NMR and IR spectra.[301] Exchange of anthracene with substituted anthracene ligands can also occur.[26b,294]

1:1 Magnesium–anthracene adducts are formally cyclic diorganylmagnesium compounds but participate in the dynamic equilibria of Scheme 9. They can act as simple organic nucleophiles, as single electron donors, or as a source of active magnesium metal, and these properties have been used in various stoichiometric and catalytic reactions.[6] Most studies have involved $[Mg(anthracene)(THF)_3]$ but substituted anthracene reagents do show related but modified reactivities.[26b] Nucleophilic behaviour is shown by cleavage of THF at elevated temperatures, giving 9,10-bis(4-hydroxybutyl)-9,10-dihydroanthracene as the main product at 120 °C after protolysis,[297] and additions to acetone, ethyl acetate and CO_2 have been reported.[6] Insertion of ethene forms the crystalline products (**99**) (R = H; Me from the 9-Me organometallic), which have been structurally characterized as typical diorganylmagnesium compounds with a primary C–Mg bond (0.211 nm) but a longer, more reactive Mg–C-9 bond.[6,302] Magnesium–anthracene compounds react with organic halides, variously, to yield mono- and disubstituted dihydroanthracenes, reduction products or Grignard reagents (see Section 3.2.1.1(a));[6,25–7] dehalogenation reactions of *gem*-dihalocyclopropanes, α,α-dichlorocyclobutanones and haloketones have also been studied.[303]

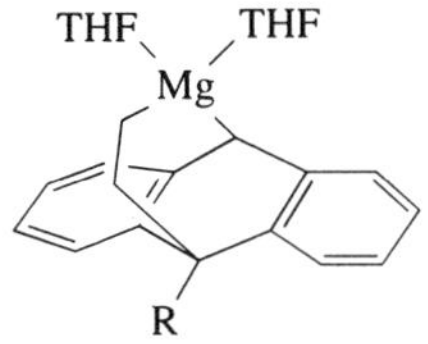

(99)

$[Mg(anthracene)(THF)_3]$ has found various stoichiometric and catalytic applications in organometallic synthesis,[6] including the introduction of anthracene ligands, as in the synthesis of $[Ni(PR_3)_2\text{-}(\eta^4\text{-anthracene})]$,[304] and as a reducing agent for reactions of other donor ligands with transition metal halides.[6] Highly active magnesium (Mg*), formed by decomposition of $[Mg(anthracene)(THF)_3]$[28] or from commercial magnesium with catalytic amounts of anthracene,[29] and reactive powders of groups 8–12 metals, formed by reaction of Mg* with transition metal salts,[29] are useful reagents.[6] The catalytic synthesis of active MgH_2* should also be noted (cf. Section 3.2.2.1).[6,250]

3.2.4 Organomagnesium Hydrides

Mixing of equimolar quantities of MgR_2 and MgH_2 is a general synthetic method for organylmagnesium hydrides of empirical formula RMgH.[1,113] Cyclopentadienyl magnesium hydride can be formed by this conproportionation reaction,[113] but is also conveniently prepared by reaction of MgH_2 with cyclopentadiene in THF.[305] CpMgH is isolated as a crystalline solvate ($[CpMgH(THF)]_n$), and in THF solutions the dimeric structure (**100**) has been proposed on the basis of molecular association measurements and ^1H NMR and IR spectra.[305] *Ab initio* SCF MO calculations on monomeric CpMgH give bond distances Mg–H of 0.173 nm and Mg–Cp(centroid) of 0.201 nm, and, as for $MgCp_2$,[271] Mg–Cp bonding is attributed to interactions of Mg $3p_{x,y}$ and $3s$ orbitals with the e_1 and a_1 π orbitals of the ring.[247] Calculations on allylmagnesium hydride favour η^1 bonding of the organic ligand.[306]

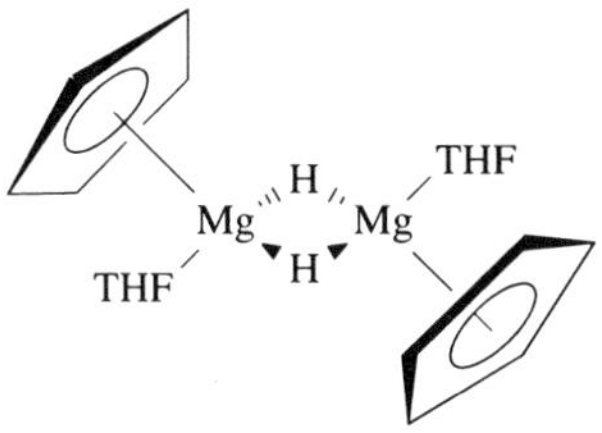

(100)

CpMgH reacts with cyclopentadiene, HCl, ROH or R_2NH to give CpMX (X = Cp, Cl, OR or NR_2, respectively). Aromatic ketones, polynuclear hydrocarbons and trityl halides are reduced by CpMgH, and a single-electron transfer mechanism is indicated by the observation of paramagnetic intermediates in these reactions.[305] Hydrometallations of 1-octene by compounds RMgH in THF are catalysed by 5% Cp_2TiCl_2, and the yields of product fall with variation of the organic group in the order R = Me > Cp > Ph > Bu^t > Et:[113] 1-deuterooctane (68%) is obtained from MeMgH after quenching with D_2O (Equation (56)), and the isomerization products *cis*- and *trans*-2-octene account for most of the remaining material. The hydrometallation mechanism probably involves $[Cp_2TiH]$ as an intermediate (see Section 3.2.1.1(iii)(b)).[113]

$$n\text{-}C_6H_{13}CH{=}CH_2 + MeMgH \xrightarrow{Cp_2TiCl_2} n\text{-}C_6H_{13}CH_2CH_2MgMe \xrightarrow{D_2O} n\text{-}C_7H_{15}CH_2D \quad (56)$$

$$4MgMe_2 + LAH \xrightarrow{THF} 4MeMgH + LiAlMe_4 \quad (57)$$

$$2MgMe_2 + LAH \xrightarrow{THF} MeMg_2H_3 + LiAlMe_3 \quad (58)$$

Reactions between $MgMe_2$ and LAH in diethyl ether produce insoluble MgH_2 as the only magnesium product, but in THF, at a reactant ratio of 4:1, soluble MeMgH is formed (Equation (57)), and at a reactant ratio of 2:1, $MeMg_2H_3$ is produced (Equation (58)).[307]

3.2.5 Organomagnesium Alkoxides, Amides and Related Compounds

3.2.5.1 Organomagnesium alkoxides and related compounds with oxygen-bonded ligands

The 1:1 reaction of a diorganylmagnesium MgR^1_2 with an alcohol or phenol is a simple preparative route to organometallics of type $R^1Mg(OR^2)$.[1] Monomeric (**101**) or polymeric $[CpMg(OR^2)]_n$ ($R^2 = OCH_2CH_2X$ (X = OEt, NMe_2), $O(CH_2CH_2O)_2Et$, $O(CH_2CH_2NMe)_2Me$) are isolable from reactions of $MgCp_2$ with 2,4-dimethyl-6-*t*-butylphenol or the appropriate polyfunctional alcohol, respectively, in toluene.[308] $[CpMg(OEt)]_4$ (**102**) can be formed from $MgCp_2$ and ethanol but is also preparable by conproportionation of an equimolar mixture of $MgCp_2$ and $Mg(OEt)_2$ in toluene.[265] Cleavage of allyl aryl ethers by normal magnesium turnings in THF gives high yields of allylmagnesium aryloxides, as exemplified by the formation of $PhCH{=}CHCH_2MgOPh$ (90%) from cinnamyl phenyl ether.[309] Cleavage of a range of allylic or benzylic ethers in refluxing THF with activated Rieke magnesium, formed by reduction of $MgCl_2$ with potassium[310a] or with lithium/naphthalene,[22] gives allyl or benzylmagnesium alkoxides in reasonable to high yields. $[Mg(anthracene)(THF)_3]$ also reacts with benzyl ethers, as in the formation of the dimagnesium species *o*- and *p*-$(MeOMgCH_2)_2C_6H_4$ in yields of 70% and 40%, respectively, from *o*- and *p*-bis(methoxymethyl)benzene in THF (20 °C, 16 h).[25c] Active magnesium, obtained by dehydrogention of catalytically prepared MgH_2 or by thermolysis of $[Mg(anthracene)(THF)_3]$,[28] or magnesium activated by addition of catalytic amounts of reduced 3*d* transition metal chlorides, especially $TiCl_4$,[310b] inserts into a C–O bond of boiling THF to form (**103**).

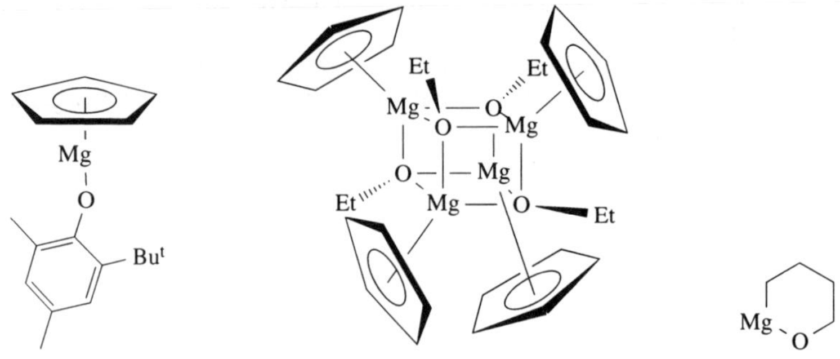

(**101**) (**102**) (**103**)

Complexation with crown ethers or cryptands accompanied by ionic dissociation (Equations (47)–(49)) can activate diorganylmagnesium compounds to cleave suitably positioned ether bonds to produce organomagnesium alkoxides or aryloxides: (**86**) is formed from $MgPh_2$ and (2-methoxy-1,3-xylylene)-15-crown ether,[278] and (**87**) from $MgNp_2$ and 2,1,1-cryptand[279] (see Section 3.2.2.3). Additions of diorganylmagnesium or Grignard reagents, especially allylic reagents, to carbon–carbon multiple bonds are anchimerically assisted by neighbouring hydroxyl groups,[89,95,99] and the products or intermediates may be cyclic organomagnesium alkoxides such as (**104**), formed by addition to propargylic alcohols.[99] Compounds of type (**104**) are useful in organic synthesis.[311]

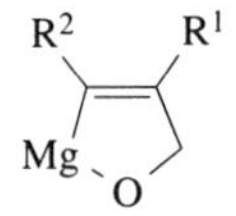

(**104**)

Crystalline $[CpMg(OEt)]_4$ has a cubane-type structure (**102**) with μ_3 ethoxy bridges and Mg–O bond lengths in the range 0.2059–0.2075 nm. Each magnesium atom is bonded to a cyclopentadienyl ligand with an average magnesium–ring(centroid) distance of 0.210 45 nm, which is somewhat longer than found in $MgCp_2$.[265] In THF solution, the tetramer dissociates to the mononuclear, solvated species $[CpMg(OEt)(THF)_2]$.[265] Steric hindrance appears to inhibit association of mononuclear (**101**), but compounds $[CpMg(OR^2)]_n$ ($R^2 = OCH_2CH_2X$ (X = OEt, NMe_2), $O(CH_2CH_2O)_2Et$, $O(CH_2CH_2NMe)_2Me$) do not form monomeric inner complexes and are polymeric, probably with alkoxy bridges. However, addition of pyridine to $[CpMg(OR^2)]_n$ ($R^2 = OCH_2CH_2X$ (X = OEt, NMe_2)) displaces the bridging groups to form $[CpMg(OR^2)(py)_2]$.[308]

Dimeric (**87**), formed by cleavage of 2,1,1-cryptand, has a centrosymmetric crystal structure with two μ_2 alkoxy groups (Mg–O 0.2010(4) nm and 0.1979(4) nm). The five-coordinate magnesium atoms are each bonded also to the neopentyl ligand (Mg–C 0.2181(6) nm), and to one oxygen and one nitrogen donor site of the heterocyclic rings.[279] The remaining heteroatoms of the potentially polydentate macrocycles do not displace the alkoxy bridges. A mixed organometallic ($[MgPh(OAr)\cdot MgPh_2]_2$) with a central structure related to that of (**87**) has been briefly mentioned; it is formed by Me–OAr bond cleavage of 2-$MeOC_6H_3$-1,3-$[CH_2O(CH_2CH_2O)_2Me]_2$ with $MgPh_2$.[5]

Phenylmagnesium carboxylates of the formula PhMg(OOCR)·THF (R = alkyl or chloroalkyl) have been prepared by reaction of PhMgBr and RCOONa in THF, and analysis of IR spectra suggest that the carboxylate groups are bridging.[312] Both carboxylate and sulfonate derivatives (RMgL ($L = OSO_2CF_3$, OSO_2Ar, OCOR)) have been formed from LiR and the respective salts MgL_2; although not structurally characterized, these derivatives give high stereoselectivities in addition reactions to some prochiral organic carbonyl compounds.[313]

3.2.5.2 Organomagnesium amides and related compounds with nitrogen-bonded ligands

A simple synthetic route to compounds of type $R^1Mg(NR^2R^3)$ is the reaction of equimolar amounts of MgR^1_2 and a secondary amine, HNR^2R^3. By this method, sterically hindered 1,3,6,8-tetra-*t*-butylcarbazole and diethylmagnesium in THF afford (**105**) as the bistetrahydrofuran adduct,[314] and hexamethyldisilazane ($HN(TMS)_2$) and $MgBu^nBu^s$ in heptane form nonsolvated (**106**).[315] Metallation of activated C–H groups by the amide bases $Mg(TMP)_2$ (TMPH = 2,2,6,6-tetramethylpiperidine) or $Mg(NPr^i_2)_2$ can also produce organomagnesium amides: *ortho*magnesiations of aromatics (e.g., Equation (59)) and related α metallations of amidocarboxy-substituted cyclopropanes and cubanes, such as (**8**) (A = B = H), occur under mild reaction conditions.[67] Reactions of metal amide (MNR^2R^3 (M = Li or Na)) with the appropriate Grignard reagent (Equation (60)) have produced (**107**)[316] and (**108**)[317] but, unlike $M[N(CH_2CH_2NMe_2)Bz]$, which forms dimeric (**107**), M[N(2-pyridyl)Ph] and Bu^nMgCl in THF only yield the monomeric bis(amide) ($Mg[N(2\text{-pyridyl})Ph]_2(THF)_2$), probably via THF-induced disproportionation of $Bu^nMg[N(2\text{-pyridyl})Ph]$.[316]

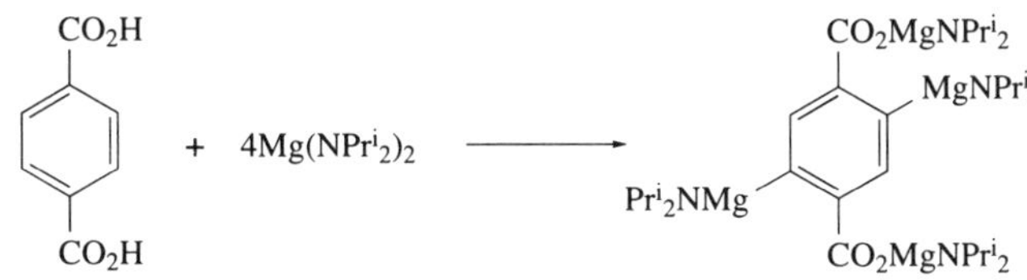

(59)

$$MNR^2R^3 + R^1MgCl \longrightarrow R^1Mg(NR^2R^3) + MCl \qquad (60)$$

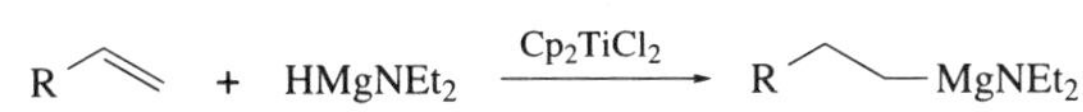

(61)

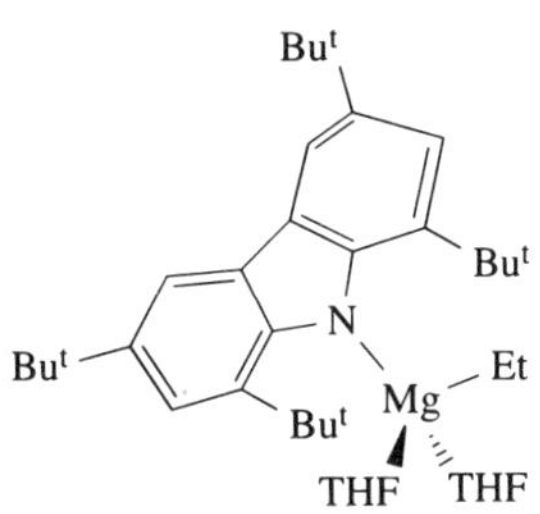

(**105**)

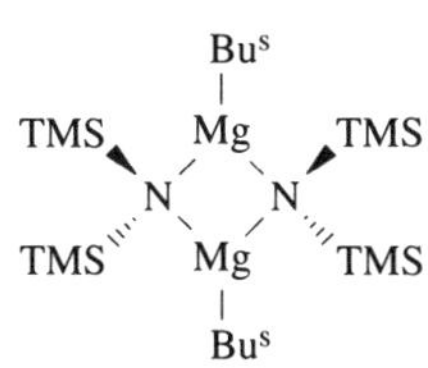

(**106**)

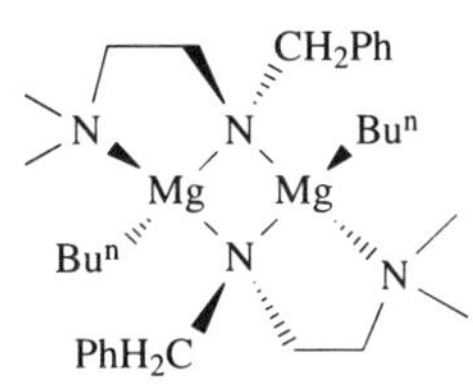

(**107**)

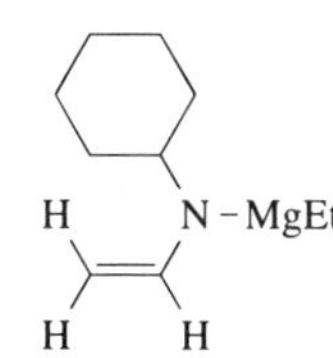

(**108**)

Hydromagnesiations of various 1-alkenes by diethylamidomagnesium hydride catalysed by Cp_2TiCl_2 (Equation (61)) give organomagnesium diethylamides in up to 95% yields.[318] Carbomagnesiations of 1-alkenes by compounds $EtMgNR_2$ catalysed by Cp_2ZrCl_2 also give organomagnesium amides (Equation (12)) ($X = NR_2$)) but the yields are relatively low.[102b]

Crystal structures have been determined for compounds (**105**),[314] (**106**)[315] and (**107**).[316] The latter two compounds are dimers with bridging amido ligands, whereas such bridging is prevented in the monomer (**105**) by the steric effects of the 1- and 8-Bu^t substituents of the carbazolyl ligand. Magnesium is four-coordinate in (**105**), with Mg–C and Mg–N bond distances of 0.2141(4) nm and 0.2087(3) nm, respectively, and coordination to the carbazolyl nitrogen atom is distorted.[314] (**106**) contains an Mg_2N_2 ring (Mg–N 0.2118(4) nm, Mg–N–Mg 87.1(2)°), and bulky trimethylsilyl substituents and the Bu^s ligands (Mg–C 0.208(1) nm) may stabilize the three coordination of the magnesium atoms.[315] The magnesium atoms in (**107**) achieve the more common four coordination by internal complexation of tertiary amine groups,[316] and the structure is analogous to that of $[MeMgNMe(CH_2CH_2NMe_2)]_2$, previously reported.[1] Proton NMR spectra indicate that the rigid, dimeric structure of (**107**) is retained in solution.[316] Variable-temperature ^1H NMR studies of the aza-allyl compound (**108**) in hexane solution show that rotation about the C=C bond occurs with a free energy barrier of 60.6 ± 1.3 kJ mol^{-1}.[317]

Paramagnetic organomagnesium compounds containing ligands with Mg–N bonds are formed from diorganylmagnesium compounds and nitrogen heterocycles by electron transfer and homolytic loss of an alkyl or aryl radical, as illustrated by Equations (51) and (52); ionic radical complexes can also be generated, as in Equation (53), and related species are also formed from 2,2'- or 4,4'-dipyridines.[273,319] These polar species formally contain magnesium(II) and π* radical anion ligands, and their electronic stuctures have been studied by ESR and electron nuclear double resonance (ENDOR) spectroscopy.

Organomagnesium thiocyanate and selenocyanate compounds of the formula RMgNCY (R = Et, Ph; Y = S, Se) have been prepared in THF by metathesis from RMgBr and KYCN with precipitation of KBr, and have been isolated as the solvates $RMg(NCY)(THF)_2$.[320] RMgNCS species show evidence for a Schlenk equilibrium in THF solution (Equation (24) (X = NCS)) with $[Mg(NCS)_2(THF)_4]$ crystallizing from solution below −10 °C. IR spectra support nonbridging, nitrogen-bonded pseudohalogen ligands in these compounds, which show typical Grignard reactivity towards alcohols, aldehydes and ketones.[320] Cobalt-containing organomagnesium diazenides are formed by reaction of $K[CoN_2(PMe_3)_3]$ with RMgX in diethyl ether and have been structurally characterized by x-ray diffraction ($R = Bu^t$) as dimeric species (**109**).[321] The Mg_2N_2 ring of (**109**) involves bridging sp^2 nitrogen atoms (Mg–N 0.2097(3) nm, Mg–N–Mg 95.8(1)°). The magnesium atom is typically four coordinate, being remote from the cobalt atom. In THF solutions the compounds display Schlenk-type equilibria by forming $Mg(THF)_4$-$[N_2Co(PMe_3)_3]_2$.

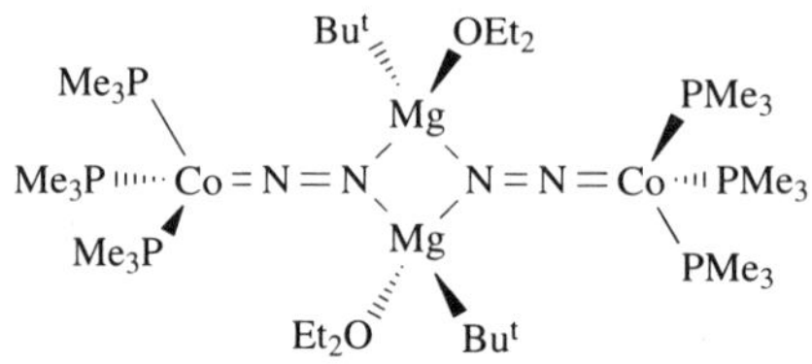

(109)

A series of organomagnesium tris(pyrazolyl)hydroborates have been prepared and fully characterized.[322] Reactions of MgR^1_2 with $M[HB(3\text{-}R^2\text{-}5\text{-}R^3\text{-pz})_3]$ (M = K or Tl; $R^2 = R^3 = Me$ or $R^2 = Bu^t$, $R^3 = H$), in Et_2O–THF form the unsolvated, monomeric complexes (**110**) (R^1 = alkyl, Ph, vinyl) in good yields; reactions of Grignard reagents with $M[HB(3\text{-}R^2\text{-}5\text{-}R^3\text{-pz})_3]$ are more complex but, if potassium salts (M = K) are used, the principal products from primary and secondary Grignard reagents are also (**110**). Crystal structure determinations of the three derivatives (**110**) (R^1 = Me or Pr^i, $R^2 = Bu^t$, $R^3 = H$; $R^1 = CH_2TMS$, $R^2 = R^3 = Me$) show the presence of a four-coordinate magnesium atom with a tridentate tris(pyrazolylborate) and a monodentate alkyl ligand. Proton and carbon-13 NMR spectra suggest that structure (**110**) is retained in hydrocarbon solution. On heating at 80–120 °C in benzene, complexes $[HB(3,5\text{-}Me_2pz)_3]MgR$ undergo disproportionation, via a Schlenk-type equilibrium, to form $Mg[HB(3,5\text{-}Me_2\text{-pz})_3]_2$, but complexes $[HB(3\text{-}Bu^t\text{-pz})_3]MgR$, with the sterically demanding Bu^t substituents, are thermally stable. Reactions of the more stable derivatives $[HB(3\text{-}Bu^t\text{-pz})_3]MgR$ have produced new four-coordinate complexes ($[HB(3\text{-}Bu^t\text{-pz})_3]MgX$) by metallation of H–X bonds (including X = C≡CR from alkynes, and $X = OCR{=}CH_2$ from enolization of ketones) and by insertion into the Mg–R bond of CO_2 ($X = O_2CMe$) or oxygen (X = OOR; or X = O-TMS, after Si–C cleavage of $R = CH_2TMS$).[322]

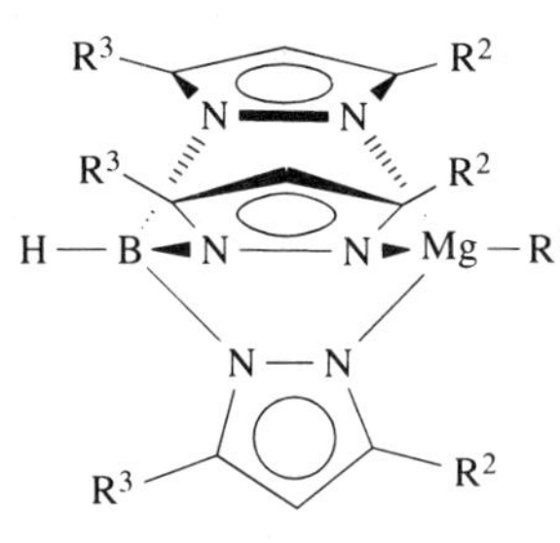

(110)

3.2.6 Heterobimetallic Compounds

Magnesium forms organometallic compounds containing other metal centres that show varying polarity, depending on the relative electronegativities of the two metals, with magnesium associated with either the more positively or negatively polarized units. In ionic species, magnesium cations or magnesiate anions are present. Compounds with direct transition metal–magnesium bonds are presented in Volume 10, and other transition metal derivatives, such as cuprates, are discussed in volumes covering the specific metals.

3.2.6.1 Alkali metal compounds

Complexes of formulae Li_nMgR_{2+n} (n = 1, 2, 3) are formed by mixing the appropriate stoichiometric quantities of MgR_2 and LiR in diethyl ether, and species Li_nMgMe_{2+n} (n = 2, 3) have been identified in solution by ^1H and ^7Li NMR spectroscopy.[1] Mixing 2:1 or 4:1 ratios of $MgMe_2$ and $(LiMe)_4$ in the presence of complexed TMEDA affords the crystalline complexes $[Li(TMEDA)]_2MgMe_4$ (**111**) or $[Li(TMEDA)MgMe_3]_2$ (**112**). In the crystal structure of (**111**) a central, tetrahedral $MgMe_4$ unit (mean Mg–C distance 0.226 nm), formally a magnesiate anion, is connected by pairs of bridging methyl groups to two $Li(TMEDA)^+$ cations (mean Li–C distance 0.228 nm, Li–C–Mg 70.3°) with short Mg–Li distances (mean 0.2616 nm).[323] Structure (**112**), proposed[323] by the analogy with that determined previously for crystalline $[\{Li(TMEDA)\}_2Mg_2Ph_6]$,[1] also contains four-coordinate lithium and magnesium atoms with dimethyl bridges and a formal $Mg_2Me_6^{2-}$ magnesiate centre. Reaction of phenylsodium with $MgPh_2$ in the presence of pentamethyldiethylenetriamine (PMDTA) gives crystals of structurally characterized (**113**), in which tetrahedral $MgPh_4^{2-}$, containing typical σ Mg–C bonds (mean distance 0.229 nm), is bonded by asymmetrically bridging phenyl groups to two $Na(PMDTA)^+$ groups producing five-coordination at the larger sodium centre; the Na–C(*ipso*) interactions are orthogonal to the aromatic plane and must involve the π electrons.[263]

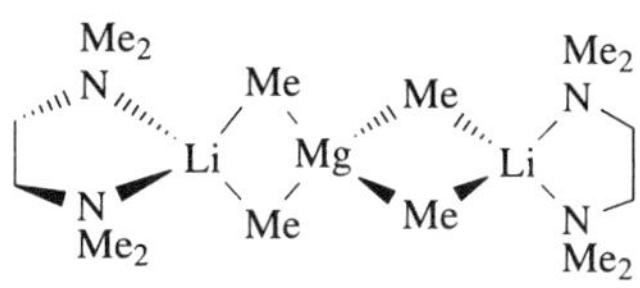

(**111**)

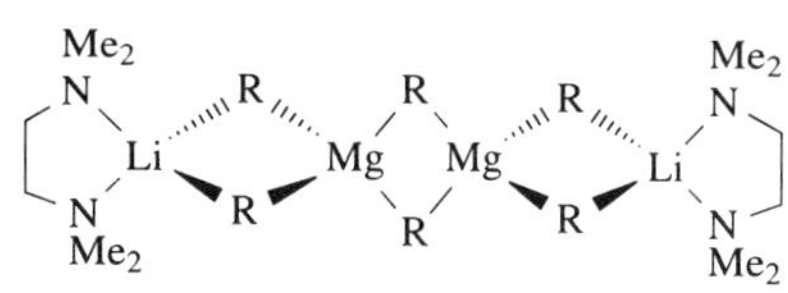

(**112**)

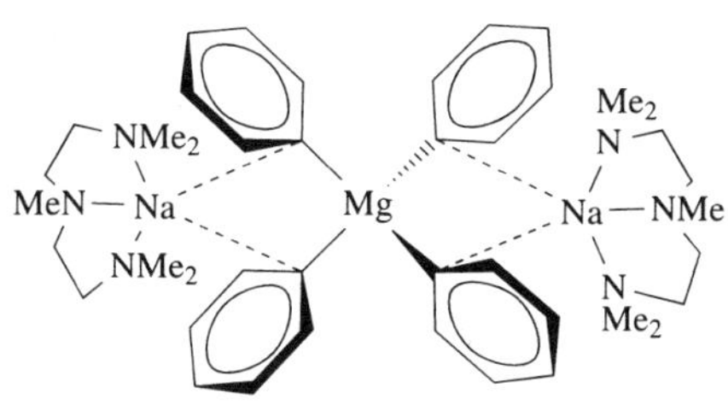

(**113**)

Reaction in hexane of ethyllithium and diethylmagnesium in a 2:1 ratio with excess toluene forms the isolable complexes Li_2MgBz_4, but in the presence of increasing amounts of TMEDA, $\{[Li(TMEDA)]_2MgBz_4\}$ and $[Li_2(TMEDA)_3MgBz_4]$ are also formed.[324] Crystals of $[Li_2(TMEDA)_3MgBz_4]$ contain the discrete four-coordinate cations $[Li(TMEDA)_2]^+$ and the heterometallic anions (**114**), which contain a distorted tetrahedral magnesium centre with two terminal Mg–C bonds (mean distance 0.224 nm) and two bridging benzyl groups (mean Mg–C distance 0.2318 nm) to Li(TMEDA).[324] Alkynyl complexes, such as $[\{Li(TMEDA)\}_2Mg(C{\equiv}CPh)_4]$ containing diaminolithium entities, can be prepared,[324] but also derivatives $M_2[Mg(diamine)(C{\equiv}CR)_3]_2$ (**115**) (M = Li, R = Ph, diamine = TMEDA;[324] M = Na, R = Bu^t, diamine = TMEDA or η^2-PMDTA[263]) in which a bidentate diamine is coordinated to magnesium. The molecular structures of the lithium and sodium species (**115**) are similar in the crystalline state: five-coordinate magnesium centres have pseudo-trigonal-bipyramidal geometry with normal σ bonds to one axial (Mg–C 0.231 nm or 0.226 nm) and two equatorial (Mg–C 0.219 or 0.216 nm) alkynyl ligands; each lithium[324] or sodium[263] ion interacts with the π electrons of α-carbon atoms of four alkynyl groups, gaining a coordination number of four, and forming four-membered M_2C_2 rings.

$MgEt_2$–LiEt mixtures in Et_2O solution produce both 4- and 2-ethylpyridine in reaction with pyridine, and considerable amounts of the 1,4-addition product in reaction with 2-cyclohexene-1-one. This reactivity is different from that of either parent organometallic and has been attributed to the presence of magnesiate anions in solution, as formed from MgR_2 compounds with crown ethers or cryptands (cf. Equations (47)–(49)).[277]

Reaction between $(TMS)_3CLi$ and $MgBr_2$ in THF gives the organometallic halide (**116**), containing typical four-coordinate magnesium and lithium centres and two bridging bromine atoms in the solid state.[325] NMR spectra of 1:1 species formed in solution by mixing compounds MgR^1_2 and alkali metal alkoxides MOR^2 suggest an ion pair structure of magnesiate anions (**117**) with associated cations ($M = Na^+$ or K^+).[326] Products of mixtures MgR^1_2/MOR^2 or $MR^1/Mg(OR^2)_2$ (M = Li, Na, K;

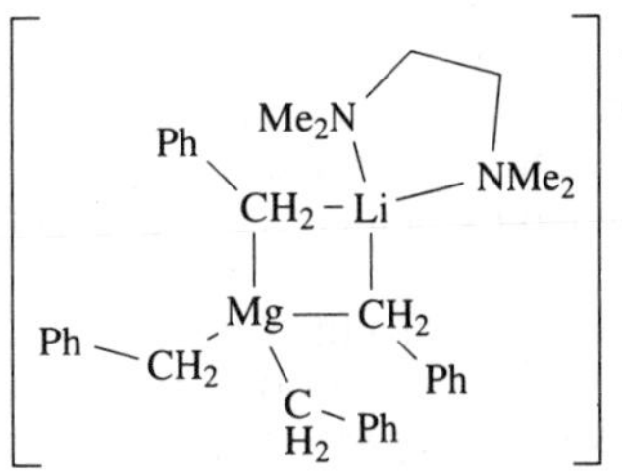

(114)

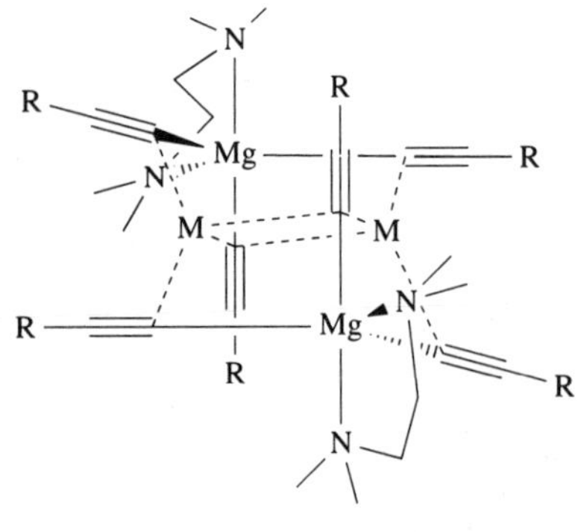

(115)

$R^2 = OCH_2CH_2OEt$) in hydrocarbon solvents have also been studied by ^1H and ^{13}C NMR spectroscopy.[327] Useful modifications in chemical reactivity of the organometallic reagents can be achieved by employing mixed magnesium–alkali metal organometallic systems containing some alkoxide[275,327] or dialkylamide[328] groups, and the use of 1:1 species (MgR_2/(*S*)-di-*O*-lithiobinaphthol) in the highly enantioselective alkylation of carbonyl compounds is of interest.[200]

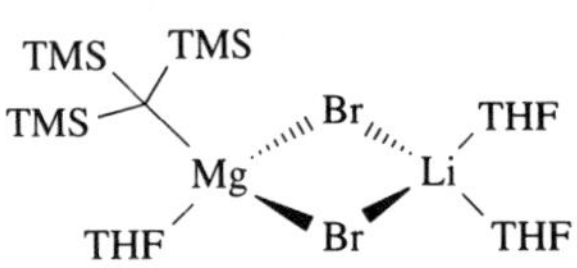

(116)

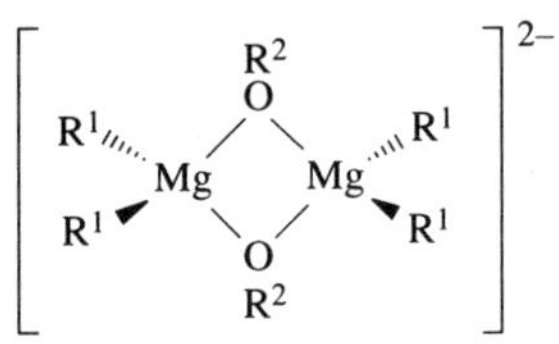

(117)

3.2.6.2 *Zinc compounds*

Unlike MgR_2 (Equation (47)), the compounds ZnR_2 are not disproportionated into ions by cryptands or crown ethers, but mixtures of the two organometallics ionize (Equation (62)), especially if C = 1,4,8,11-tetramethyl-1,4,8,11-tetraazacyclotetradecane (14N4).[329] A related reaction of the crown ether complex (**79**) with $ZnPh_2$ forms zincate anions ($[Zn_nPh_{2n+1}]^-$ (n = 1, 2)) and the cation [Mg(1,3-xylylene-15-crown-4)]$^-$.[66]

$$MgR_2 + ZnR_2 + C \rightleftharpoons [RMg(C)]^+ + [ZnR_3]^- \qquad (62)$$

C = cryptand, crown ether and other macrocycles

Reactions of mixed species of composition $R^2R^1{}_2ZnMgX$, obtained from $ZnR^1{}_2$ and the Grignard reagents R^2MgX, have been studied.[330] Also, formally 1,1-dimetallic compounds of magnesium and zinc, formed by a carbometallation reaction (Equation (63)) have been used in organic syntheses.[331]

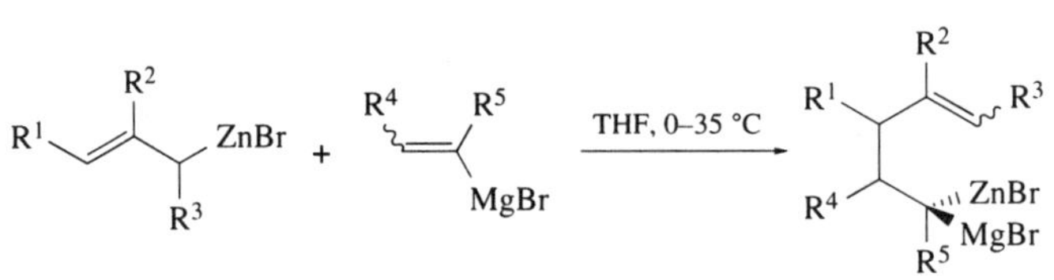

(63)

3.2.6.3 *Aluminum compounds*

Hydrocarbon solutions of species $(MgBu^n{}_2)_n(AlR_3)$ (R = Bu^i, $n \geq 1$;[332] R = Et, n = 8[333]) are obtained from reactions between AlR_3 and the unsolvated Grignard reagent Bu^nMgCl, prepared by direct reaction of magnesium and Bu^nCl in heptane. These magnesium–aluminum complexes can be used in metallation reactions,[333] and partial hydrolyses of such species form oligomeric or polymeric alumoxanes containing structural units of the type $[-Al(OMgR)O-]_n$.[334]

$[Me_2Al(C{\equiv}CMe)]_2$ and $Mg(C{\equiv}CMe)_2$ combine in 1,4-dioxane to give the isolable, unsolvated complex $[(Me_2Al)_2Mg(C{\equiv}CMe)_4]$, for which spectroscopic studies suggest a structure with a central magnesium atom, tetrahedrally coordinated by a four σ-bonded alkynyl ligands, and two $AlMe_2$ groups, each bonded by two alkynyl bridges, probably via interactions with the π electrons.[335]

Magnesium–anthracene derivatives react with organoaluminum compounds to yield aluminum or heterobimetallic products.[6,336–7] Insertion of R_2AlX (X = H, R = H, Me, Et, Bu^i; X = OEt, R = Et) into one Mg–C bond of [Mg(anthracene)$(THF)_3$] in THF forms bimetallic products with analogous crystal structures (**118**) (X = H or OEt). The magnesium in (**118**) is five-coordinate and, for X = H, the unsymmetrical Mg–H–Al bridge (Mg–H 0.196 nm, Al–H 0.162 nm) is consistent with a cationic Mg^+ centre interacting with an aluminate anion.[336] The nonsolvated compounds [Mg(μ-R)AlR_2(anthracene)], formed from magnesium–anthracene and AlR_3 (R = Me, Et) in toluene, may have a structure related to (**118**) with a bridging alkyl group (X = R), but these are cleaved by ethereal solvents to liberate $[AlR_3(ether)]$.[336] 9,10-Bis(trimethylsilyl)anthracene–magnesium reacts with $AlMe_3$ in cyclohexane to give a 1:2 derivative, assigned structure (**119**), which is also cleaved by THF.[337]

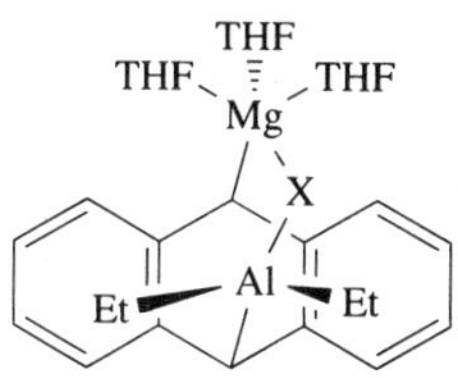

(118)

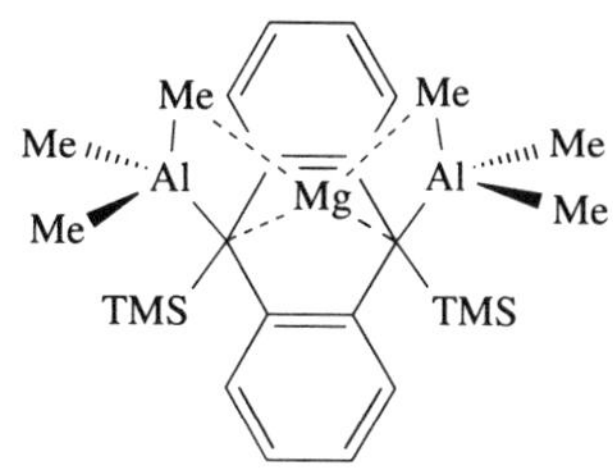

(119)

3.3 CALCIUM, STRONTIUM AND BARIUM

The larger size:charge ratios and ionic radii of calcium, strontium and barium ions, relative to those of magnesium, impart a greater ionicity on compounds of the heavier metals and promote higher coordination numbers. The increased polarity enhances ligand lability and the nucleophilicity of simple organic groups, causing the organometallic derivatives to be more reactive and unstable. Coordinative unsaturation, a consequence of the larger metal size, may lead to oligomerization via intermolecular coordination or to associative decomposition reactions. Many of the significant advances in this area of organometallic chemistry since 1982 have involved fully characterized derivatives of more stable delocalized organic ligands, such as cyclopentadienyl species, and have included sterically bulky groups to reduce the tendency to molecular aggregation and to increase stability towards decomposition.[3–4]

3.3.1 Organometal Halides

3.3.1.1 Preparation

Several new methods have been reported for activation of heavier group 2 metals for the synthesis of the Grignard analogues RMX by direct reaction of the metal with an organic halide RX. 'Ultrafine particles' of calcium, produced by evaporation of the metal under vacuum, react when mixed with alkyl or aryl iodides or bromides in THF to form RCaX in reasonable yields, often after short reaction times and at low temperatures.[338–9] In an alternative procedure, co-condensation of calcium vapour with THF at 77 K, followed by warming to room temperature, produces an active metal slurry that also affords organocalcium halides from various alkyl and aryl halides under mild conditions.[339–40] However, both these forms of vaporized calcium only give low yields of organometallic products in aromatic solvents or in reactions with organic chlorides, are inert to organic fluorides, and cause Wurtz-type coupling of benzyl halides.

Reduction of the iodides MI_2 (M = Ca, Sr or Ba) with metallic potassium in THF precipitates finely divided alkaline earth metals that react rapidly with triphenylmethyl chloride to form the compounds $Ph_3CMCl(THF)_n$ in high yields, although the reactivity of these powders with simple alkyl halides was not reported.[341] An active form of calcium is precipitated from THF by reduction of anhydrous CaX_2 (X = Br or I) with lithium naphthalenide, whilst a similar reaction of CaX_2 with lithium biphenylide produces a soluble form of highly reactive metal that is very effective in the direct synthesis of organocalcium reagents.[342] These sources of 'activated calcium',[342–3] and a related soluble anthracene reagent,[344] are probably calcium–aromatic complexes (see Section 3.3.3). The soluble calcium–biphenyl complex reacts in THF with organic bromides and chlorides, including tertiary systems such as 1-bromoadamantane, at temperatures as low as −78 °C to give the products RCaX in high yields.[342] Even aryl fluorides oxidatively add to form species ArCaF at room temperature,[342] and polymeric organocalcium reagents ((**120**) (X = Br, Cl, F) or (**121**)) are formed quantitatively in reactions of calcium–biphenyl or calcium–anthracene reagents with halogenated, cross-linked polystyrene resins.[344] A related reactive form of barium, produced as a brown powder from BaI_2 and two equivalents of lithium biphenylide in THF, generates allylbarium chlorides in high yields at −78 °C (Equation (64)) with greater than 98% retention of double bond stereochemistry.[345]

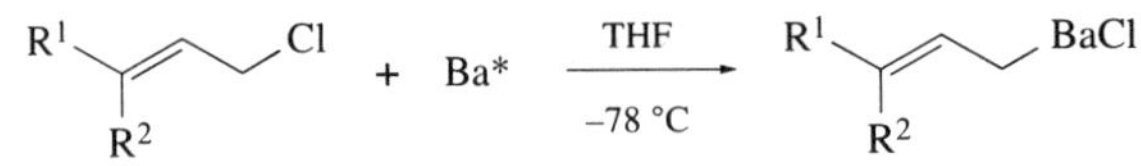

(64)

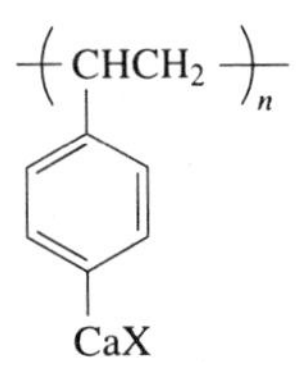

(120)

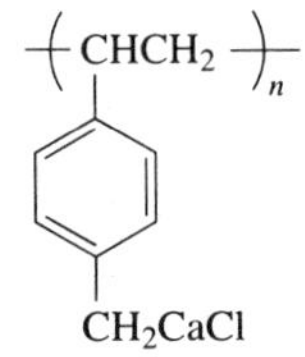

(121)

The syntheses of organocalcium halides from calcium–biphenyl or calcium–naphthalene complexes and alkyl halides probably follow a single-electron transfer (SET) mechanism (Scheme 10) since alkyl radical intermediates are implicated in systems involving chiral cyclopropyl halides or 6-chloro-6-methyl-1-heptene as reactants. The complete racemization of (*S*)-(+)-(**1**), and its chloro analogue, in the formation of configurationally stable cyclopropylcalcium halides is assigned to the rapid inversion of the intermediate cyclopropyl radical R• in Scheme 10; also, the large amounts of 1,1,2-trimethylcyclopentyl product formed from 6-chloro-6-methyl-1-heptene is consistent with the radical intermediates and cyclization of Equation (65).[343] Calcium in liquid ammonia reduces and cleaves (**1**) via a radical mechanism, but the solid metallic bronze $Ca(NH_3)_6$ and chiral (**1**) in THF afford the quenched cyclopropylcalcium product with partial retention of chirality, and this has been attributed to reaction of surface-adsorbed radicals in a mechanism related to that proposed controversially for magnesium (see Scheme 2).[343]

$$\text{R–X} + \text{Ca/arene} \xrightarrow{\text{SET}} [\text{R–X}^{\bullet -}\ \text{Ca}^{\bullet +}\text{/arene}] \xrightarrow{-\text{XCa}^{\bullet}\text{/arene}} \text{R}\bullet \xrightarrow{\text{SET}} \text{RCaX}$$

Scheme 10

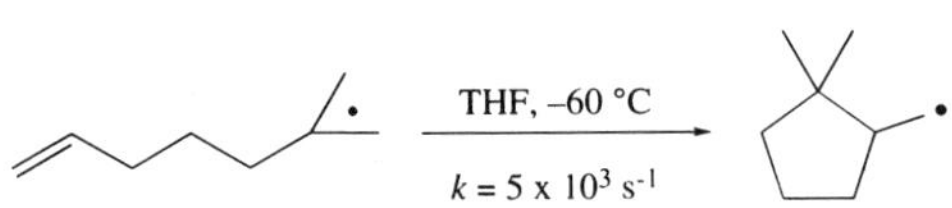

(65)

Solvent-free arylcalcium halides, formulated as ArCaX (X = F, Cl or Br), are formed as reactive brown solids in ~40–50% yield by vaporization and co-condensation at 77 K of calcium metal with excess aryl halide.[339,346] Reactions of calcium vapour co-condensed with CCl_4 at 77 K occur via Cl• abstraction to form CCl_3•.[57] The deposition of calcium vapour with argon or with argon–methyl halide under matrix isolation conditions at 9 K generates calcium atoms and the clusters Ca_2 and Ca_x ($x \geq 3$), which are more reactive towards the organic halides than corresponding magnesium species.[10] Calcium atoms react only with MeI and MeF, whereas the calcium clusters are more reactive and form species formulated as $MeCa_xX$ (X = F, Cl, Br or I; $x \geq 2$) from all methyl halides. The reactivity trend for oxidative addition of MeX to matrix-isolated calcium species is MeI > MeF > MeBr > MeCl, with the anomalous position of MeF being consistent with a Ca–F bond strength considerably higher than that of C–F.[10]

Transmetallation of a group 1 organometallic compound with a calcium halide has been employed for the synthesis of some organocalcium halides. The optically pure organolithium compound, formed from (*S*)-(+)-(**1**) and *t*-butyllithium, reacts with $CaBr_2$ in Et_2O at −15 °C to form chiral (1-methyl-2,2-

diphenylcyclopropyl)calcium bromide, which gave carbonation products with complete retention of stereochemistry; this establishes that the metal exchange occurs with retention and that the cyclopropylcalcium bond maintains its configuration under the reaction conditions.[343] Metathetical exchange of KCp* with CaI_2 in a 3:2 molar ratio in THF affords a mixture of the complexes $Cp^*CaI(THF)_2$ and $Cp^*_2Ca(THF)_2$ from which the latter can be removed by dissolution in hexane, leaving the organocalcium halide in a 57% yield.[347] Reactions of $K(C_5H_nPr^i_{5-n})$ ($n = 2, 1$) with CaI_2 in THF, even in 2:1 molar ratios, also yield similar mixtures of solvated mono- and dicyclopentadienyl products that are not easily separated, whereas in Et_2O as the solvent, nonsolvated $Ca(C_5H_nPr^i_{5-n})_2$ is the main product.[4,348] Ligand exchange reactions between RCaI (R = Me, Bu^n or Ph) and R^2TeR^1 at −70 °C in THF give the more stable organocalcium compounds R^2CaI (R^2 = vinyl, allyl or benzyl), and these products have been trapped by addition to aldehydes in reactions of up to 80% yield.[83]

Equimolar mixtures of $Cp^*_2Ca(THF)_2$ or $(\eta\text{-}C_5Me_4Et)_2Ca(THF)$ and CaI_2 in THF undergo a Schlenk-type equilibrium, and the mixed organometallic iodide, $Cp^*CaI(THF)_2$ or $(\eta\text{-}C_5Me_4Et)CaI(THF)_2$, respectively, can be isolated from the reactants.[347,349] However, in some direct reactions of an organic halide with calcium metal, the existence of a Schlenk-type equilibrium may lead to disproportionation of the initially formed organocalcium halide to produce a diorganylcalcium, as in the syntheses of $CaPh_2$[66] and $Ca[CH(TMS)_2]_2$[350] in THF. Differences in solubility between the calcium species in equilibrium may influence the composition of the product in solution.

The addition of alkaline earth metal compounds RMI to conjugated enynes[1,100] forms new organometallic iodides, but only the products of subsequent hydrolysis have been reported. The applications of the compounds $Ph_3CMX(THF)_n$ (M = Ca, Sr, Ba; X = Cl, Br, I)[351] and other organocalcium halides[352] in the initiation of anionic polymerizations of unsaturated monomers, including 1,3-butadiene, styrene and vinylpyridines, also produces intermediate polymeric organometallic species.

3.3.1.2 Structures, properties and reactivity

The only Grignard-type compounds of the heavier group 2 metals to be structurally characterized by x-ray diffraction in the solid state are the cyclopentadienyl species $[Cp^*Ca(\mu\text{-}I)(THF)_2]_2$[347] (**122**) and, briefly noted, $[(\eta\text{-}C_5HPr^i_4)Ca(\mu\text{-}I)(THF)]_2$.[4] Dimeric (**122**) has two, slightly asymmetrically bridging iodide ligands, and each calcium atom is coordinated to a η^5-pentamethylcyclopentadienyl group (mean Ca–C(ring) distance 0.267 nm) and two THF ligands. The coordination number of calcium in (**122**) is higher than in $[Cp^*Mg(\mu\text{-}Cl)(THF)]_2$[161] or in related, structurally characterized magnesium compounds.[160]

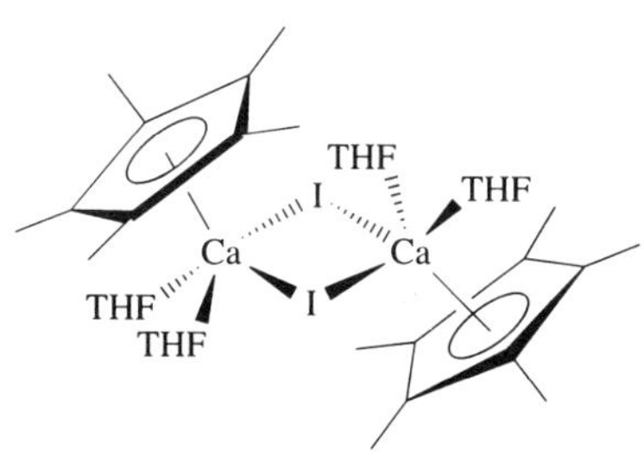

(122)

Few new studies of simple alkyl or aryl metal halides in solution have been reported. Symmetrizations of the initially formed species PhCaI[66] and $[(TMS)_2CH]CaBr$[350] into the diorganylcalcium derivatives and calcium halides suggest that a Schlenk-type equilibrium exists in THF. Also, the *t*-C–Mg bond of chiral 1-methyl-2,2-diphenylcyclopropylcalcium bromide is stable to inversion in THF at −15 °C.[343] Organocalcium halides, formed by reactions of calcium–aromatic complexes[342,344] or by exchange with organotellurium derivatives,[83] give good yields of addition products on reaction with organic carbonyl compounds, and various other high-yielding functionalization reactions of polymeric organocalcium reagents have been described.[344] Solvent-free arylcalcium reagents formed by vapour synthesis give moderate yields of Grignard-type products in direct reactions with organic electrophiles.[346] The very high α selectivity in additions of allylbarium chlorides to carbonyl compounds in THF at −78 °C is noteworthy and, when combined with retention of double bond configuration in preparations from allyl chlorides (Equation (64)), suggest that these barium reagents may have useful organic synthetic applications since magnesium or calcium reagents give primarily γ products and lithium reagents are less selective.[345]

Alkylated cyclopentadienylcalcium iodides show evidence of a Schlenk equilibrium in solution.[3,347,349] The position of the equilibrium is dependent on the organic group, the solvent and temperature: for the pentamethylcyclopentadienyl species in THF (Equation (66)) the equilibrium constant ($K \approx 4$) is consistent with ligands undergoing complete scrambling,[347] whereas the related equilibrium for ethyltetramethylcyclopentadienyl species lies more in favour of (η-C_5Me_4Et)-$CaI(THF)_2$.[349] In dioxane, Cp*CaI is symmetrized to $CaCp^*_2$ and insoluble $CaI_2(dioxane)_n$. In benzene, NMR studies show that dilute hydrocarbon solutions of $Cp^*CaI(THF)_2$ slowly transform into $Cp^*_2Ca(THF)_2$, probably via dissociation of THF, but that ligand exchange is slow on the NMR timescale at room temperature.[347]

$$CaCp^*_2(THF)_2 + CaI_2(THF)_2 \overset{THF}{\rightleftharpoons} 2Cp^*CaI(THF)_2 \quad (66)$$

3.3.2 Diorganylmetal Compounds

3.3.2.1 Dialkyl and diaryl metal compounds

Co-condensation of calcium vapour with $(TMS)_2CHBr$ in THF at 77 K affords $Ca[CH(TMS)_2]_2(THF)_3$, probably via Schlenk symmetrization of $[(TMS)_2CH]CaBr$, and subsequent reaction with 1,4-dioxane gives crystalline (**123**).[350] Compound (**123**) is the only simple dialkylcalcium that has been structurally characterized and contains a four-coordinate calcium atom (Ca–C 0.2373(4) nm, C–Ca–C 133.7(2)°), with the bulky bis(trimethylsilyl)methyl groups preventing a more normal, higher coordination number; in solution, the high-field ^1H NMR shift of the methyne hydrogen ($\delta = -1.74$) suggests considerable anionic character for the alkyl ligand.[350] *Ab initio* pseudopotential studies on various monomeric molecules MX_2 (M = Ca, Sr, Ba), including the dimethyl derivatives, have indicated a bent geometry for barium and strontium species, as found experimentally for the dihalides in the gas phase. $BaMe_2$ is calculated to have a small C–M–C angle of 115.9° with a linearization energy of ~33 kJ mol^{-1}, and angles increase with different metals in the order Ba < Sr < Ca; $SrMe_2$ has a calculated angle of 130.6°, whereas $CaMe_2$ is predicted to be 'quasilinear'.[353] The bending of these simple divalent molecules results from minor covalent contributions of metal *d* orbitals to the overall bonding and from polarization of the metal cation by the anions.

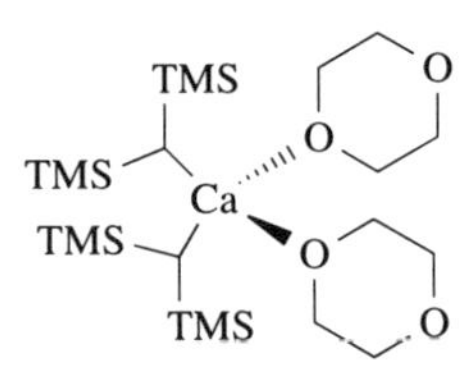

(123)

Reaction of iodobenzene with calcium metal in THF at −20 °C, followed by evaporation to dryness and extraction of the solid with Et_2O, yields a solution of $CaPh_2$ that slowly decomposes at room temperature over a period of days.[66] Halogen–calcium exchange using $CaPh_2$ (Equation (67)) quantitatively produces an internal crown ether complex characterized by the formation of phenyltrimethyltin and 2-(trimethylstannyl)-1,3-xylylene-18-crown-5 on reaction with Me_3SnCl.[66]

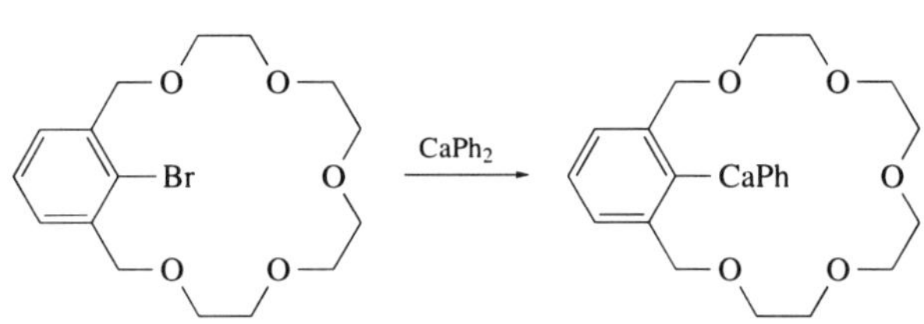

(67)

α-Arylalkylmetal compounds of the group 2 metals are more stable and have been prepared by various reactions. Solutions of dicumylbarium ($Ba(CMe_2Ph)_2$) are obtained from the direct reaction of finely divided barium in THF with $PhCMe_2(OMe)$ after separation from the insoluble by-product $Ba(OMe)_2$,[1,354] and $Ba(CMe_2Ph)_2$ can be transformed into $Ba(CPh_3)_2$ by metallation of triphenylmethane.[354] Addition of $Ba(CMe_2Ph)_2$ to (**124**), gives the monoadduct

$Ba[CPh(CH_2CMe_2Ph)C_6H_4C_6H_4CPh{=}CH_2)]_2$, or, more slowly with excess reagent, the diadduct $Ba[CPh(CH_2CMe_2Ph)C_6H_4C_6H_4CPh(CH_2CMe_2Ph)]$.[355] Dicumylbarium also acts as a monofunctional initiator for anionic polymerization of styrene or vinylpyridines, giving polymers with α-arylalkylbarium 'living' end groups,[356] and the bifunctional initiator $Ba(CPh_2CH_2CH_2CPh_2)$, prepared from 1,1-diphenylethene and barium metal, gives similar living polymer systems with α,ω-dicarbanionic ends.[356–7] Dibenzylmercury reacts with 1,1-diphenylethene and a barium mirror in benzene at 90 °C to form bis(1,1,3-triphenylpropyl)barium, probably via a radical mechanism, and calcium or strontium react similarly (Equation (68), M = Ba, Sr, Ca).[358] Bis(triphenylmethyl)metal compounds are formed by various reactions,[1] including oxidative addition of hexaphenylethane to strontium or barium metals in aromatic hydrocarbon solvents.[357a,359] Also, triphenylmethane reacts with calcium, strontium or barium in ammonia gas-saturated ethereal solvents at room temperature or, more rapidly, with barium in mixed ammonia–ethereal solvents at – 40 °C to give the corresponding products $M(CPh_3)_2$ in good yields; solvated electrons are implicated in the mechanism of these reactions and metal reactivity falls in the order Ba > Sr > Ca.[360]

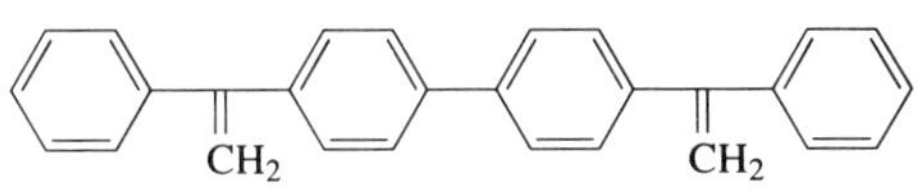

(124)

$$Hg(CH_2Ph)_2 + 2Ph_2C{=}CH_2 + M \longrightarrow M(CPh_2CH_2CH_2Ph)_2 + Hg \qquad (68)$$

No definitive structural characterizations of α-arylalkylmetal compounds have been reported, but the physical properties and colours of the species suggest ionic constitutions. Triphenylmethyl derivatives are likely to possess polyhapto ligands, as found in related alkali metal compounds, rather than the simple σ-bonded ligand of the magnesium derivative $(Ph_3CMgBr(OEt_2)_2)$.[155] The complexes $[Ba(CPh_3)_2L_2]$ (L = THF, HMPT) have been isolated and characterized by IR spectra and by ^1H and ^{13}C NMR spectra, although species solvated by more THF molecules can also be obtained. Fast atom bombardment (FAB) positive-ion mass spectra of the complexes $[Ba(CPh_3)_2L_2]$ show parent molecular ions for the monomeric, protonated ions $[HBa(CPh_3)_2L_n]^+$ (n = 1 or 2) in the gas phase.[360]

Electronic spectra, with λ_{max}(THF) = 338 nm, and conductance studies on $Ba(CMe_2Ph)_2$ in THF or tetrahydropyran solutions support the presence of ion pairs, free ions and triple ions in the equilibria of Equations (69)–(71) (R = CMe_2Ph).[354] The dissociation constant for $Ba(CMe_2Ph)_2$ of Equation (69) in THF varies between 1×10^{-9} M and 2.3×10^{-9} M in the temperature range +15 °C to –70 °C, and the metallation of triphenylmethane by $Ba(CMe_2Ph)_2$ shows a rate consistent with a mechanism involving exclusively the free ions.[354] Equilibria of the type shown in Equations (69)–(71) also occur with organobarium end groups of 'living' poly-2- or poly-4-vinylpyridine, and polymerization of the monomer is propagated by ion pair sites.[356] The shift of the absorbance maximum from 540 nm at 25 °C to 660 nm at –105 °C for the monoanionic adduct of (**124**) ($Ba[CPh(CH_2CMe_2Ph)$-$C_6H_4C_6H_4CPh{=}CH_2)]_2$) in THF is attributed to equilibria between contact (tight) and solvent-separated (loose) ion pairs (Equation (72)), with the latter predominating at low temperatures.[355] Similar solvent- and temperature-dependent equilibria of contact and solvent-separated ion pairs are proposed for $Ba(CPh_3)_2$ in solution.[360]

$$Ba^{2+}[R^-]_2 \rightleftharpoons Ba^{2+}R^- + R^- \qquad (69)$$

$$2Ba^{2+}[R^-]_2 \rightleftharpoons Ba^{2+}R^- + Ba^{2+}[R^-]_3 \qquad (70)$$

$$Ba^{2+}[R^-]_3 \rightleftharpoons Ba^{2+}[R^-]_2 + R^- \qquad (71)$$

$$R^-Ba^{2+}R^- \rightleftharpoons R^- \parallel Ba^{2+}R^- \rightleftharpoons R^- \parallel Ba^{2+} \parallel R^- \qquad (72)$$

Previously studied dialkynylmetal compounds[1] ($[M(C{\equiv}CPh)_2(THF)_4]$ (M = Ba, Sr, Ca)) can be prepared in high yields by reactions of the appropriate metal with phenylethyne in ammonia gas-saturated THF.[360]

3.3.2.2 Dicyclopentadienylmetal and related compounds

(i) Dicyclopentadienylmetal compounds

The thermally stable, unsubstituted metallocenes MCp_2 (M = Ca, Sr, Ba) were first characterized in 1956–61[1] and, except for modified syntheses (see below),[361] no significant developments in their chemistry have been reported since the early 1980s. The low solubility and volatility of the MCp_2 compounds, especially for M = Sr and Ba, preclude many chemical studies and presumably stem from intermolecular interactions between molecules, as found in the polymeric structure of crystalline $CaCp_2$.[1] Recent investigations of dicyclopentadienylmetal compounds have concentrated on derivatives with alkyl or trimethylsilyl ring substituents that sterically reduce the tendency to molecular association, limit coordination of additional Lewis bases and, in the case of extremely bulky substituents, stabilize the organometallics to oxidative decomposition.[3–4]

Direct reaction of a metal with a substituted cyclopentadiene is a useful route to new metallocenes. Granular barium, strontium or calcium metal in ammonia gas-saturated THF reacts with Cp* to give good yields (>75%) of $MCp^*_2(THF)_2$ (M = Ba, Sr, Ca).[360] $BaCp^*_2(THF)_2$ is also formed rapidly in THF from C_5Me_5H and an activated barium powder, made by reduction of BaI_2 with potassium metal.[341] Reaction of calcium dissolved in liquid ammonia with methylcyclopentadiene affords the corresponding metallocene, which can be isolated as $Ca(\eta\text{-}C_5H_4Me)_2(DME)$ after crystallization from dimethoxyethane.[362] Co-condensation of the metal vapour with hexane and $Bu^tC_5H_5$[245] or 1,3-$(TMS)_2C_5H_4$[363] at −196 °C and subsequent warming produces the base-free products $M(\eta\text{-}C_5H_4Bu^t)_2$ or $M[\eta\text{-}C_5H_3\text{-}1,3\text{-}(TMS)_2]_2$ (M = Ca, Sr, Ba), respectively, which can be converted into adducts with THF.

Hydrogen–metal exchange between alkyl-substituted cyclopentadiene and calcium amide ($Ca(NH_2)_2$) in liquid ammonia, followed by addition of THF, affords $CaCp^*_2(THF)_2$ or $Ca[\eta\text{-}C_5H_3\text{-}1,3\text{-}(Bu^t)_2]_2(THF)$.[364] However, the related reaction using the trimethylsilyl-substituted cyclopentadienes $(TMS)_nC_5H_{6-n}$ (n = 1, 2) in liquid ammonia causes cleavage of the C–Si bonds and results in the isolation of $CaCp_2(THF)_2$ (Equation (73)); only by reacting $Ca(NH_2)_2$ and $(TMS)_nC_5H_{6-n}$ (n = 1, 2) directly in THF can the trimethylsilylated organometallics be obtained (Equation (74)). Diindenylcalcium in THF also reacts with trimethylsilylcyclopentadiene to give $Ca[C_5H_4(TMS)]_2(THF)$.[364]

$$Ca(NH)_2 \ + \ 2(TMS)_nC_5H_{6-n} \xrightarrow{\text{liquid/NH}_3\text{/THF}} CaCp_2(THF)_2 \tag{73}$$

$$Ca(NH)_2 \ + \ 2(TMS)_nC_5H_{6-n} \xrightarrow{\text{THF}} Ca[\eta\text{-}C_5H_{5-n}(TMS)_n]_2(THF) \tag{74}$$

Metathesis between two molar equivalents of an alkali metal cyclopentadienide and an alkaline earth metal iodide (Equation (75)) can give bis(cyclopentadienyl)metal compounds in high yields.[3] If the alkaline earth metallocene is soluble in an ethereal or hydrocarbon solvent, potassium or sodium cyclopentadienide is a good reagent since insoluble KI or NaI can be easily removed by filtration: using this method, $CaCp_2(THF)_2$ and $Ca(\eta\text{-}C_5H_4Me)_2(THF)_2$,[361] $MCp^*_2(THF)_2$ (M = Ca, Sr, Ba),[266,361] $MCp^*_2(OEt_2)$ (M = Ca, Sr),[266,269] $Ca(\eta\text{-}C_5Me_4Et)_2(THF)$,[349] $M(\eta\text{-}C_5H_2\text{-}1,2,4\text{-}Pr^i_3)_2(THF)_n$ (M = Ca, Sr, Ba, n = 0–2),[348] $M(\eta\text{-}C_5HPr^i_4)_2$ (M = Ca, Ba)[365] and $Ca[\eta\text{-}C_5Me_4(CH_2CH_2NMe_2)]_2$[366] have been prepared. For preparations of $MCp_2(THF)_n$ (M = Sr, $n \approx 1$; M = Ba, $n \approx 0.25$) and related methylcyclopentadienyl species, which are virtually insoluble in ethereal solvents, lithium cyclopentadienide is the best reagent since the by-product LiI can be removed by dissolving in THF.[361] Monocyclopentadienylmetal iodides are intermediates in these exchanges and, as noted earlier, 2:1 reactions of $K(C_5H_nPr^i_{5-n})$ (n = 2, 1) and CaI_2 in THF yield mixtures of solvated mono- and dicyclopentadienyl products that are not easily separated, whereas in Et_2O as the solvent the base-free products $Ca(C_5H_nPr^i_{5-n})_2$ are obtained.[4,348]

$$2M^2(C_5H_{5-n}R_n) \ + \ M^1I_2 \longrightarrow M^1(\eta\text{-}C_5H_{5-n}R_n)_2 \ + \ 2M^2I \tag{75}$$

The unsolvated derivatives MCp^*_2 (M = Ca, Sr) can be obtained by refluxing a solution of the adducts $MCp^*_2(OEt_2)$ in toluene either under vacuum[266,269] or at atmospheric pressure,[367] and $BaCp^*_2(THF)_2$ can also be desolvated by the same method[266] or by slow sublimation under a dynamic vacuum.[367] Only one THF molecule is removed from $CaCp^*_2(THF)_2$ by such methods,[364,367] reflecting the general ease of desolvation of the metals: Ba > Sr > Ca. Other base-free species ($M(\eta\text{-}C_5H_4Bu^t)_2$,[245] $M[\eta\text{-}C_5H_3\text{-}1,3\text{-}(TMS)_2]_2$[363] and $M(\eta\text{-}C_5H_2\text{-}1,2,4\text{-}Pr^i_3)_2$[348] (M = Ca, Sr, Ba), and $M(\eta\text{-}C_5HPr^i_4)_2$ (M = Ca, Ba)[365]) can be synthesized directly or by thermal desolvation of THF adducts. The volatility of Ba-$(\eta\text{-}C_5HPr^i_4)_2$, which sublimes at 90 °C, 0.01 torr, is notable.[365]

The base-free metallocenes MCp^*_2 have been structurally characterized in the gas phase by electron diffraction (M = Ca,[269,368] Sr,[369] Ba[369]) and in crystals by x-ray diffraction (M = Ca, Ba)[367] (Table 5). In both phases there are discrete MCp^*_2 units, unlike the extended polymeric network of crystalline $CaCp_2$ in which each calcium interacts with four Cp rings; also, the molecules MCp^*_2 have bent sandwich structures (**125**) ($\alpha < 180°$), in contrast with the regular structures of all characterized bis(cyclopentadienyl)magnesium compounds with $\alpha = 180°$ (see Section 3.2.2.2). Electron diffraction studies of MCp^*_2 give $\alpha \approx 150°$, decreasing slightly in the order Ca > Sr > Ba, but these are thermal average angles and, although they may reflect the equilibrium structure, the apparent bending could possibly be the result of large-amplitude deformation modes from an equilibrium structure with $\alpha = 180°$.[269,369] In the solid state, the bulky pentamethylcyclopentadienyl rings of $CaCp^*_2$ prevent the polymerization observed for $CaCp_2$, and isolated molecules have $\alpha = 147°$, with no obvious crystal packing reasons for this bending; bending is greater in $BaCp^*_2$ ($\alpha = 131°$) but weak interactions with methyl groups on neighbouring $BaCp^*_2$ units (**126**) (shortest Ba–C(Me) distance 0.3349(5) nm) do create 'quasipolymeric' chains that could influence this angle.[367] It may be noted that solid-state ^{13}C NMR spectra of crystalline $M(\eta\text{-}C_5H_4Bu^t)_2$ (M = Ca, Sr) show resonances for at least two types of cyclopentadienyl ring and suggest aggregated structures.[245]

Table 5 Selected structural parameters of ring-substituted bis(cyclopentadienyl) derivatives of the heavier alkaline earth metals.

Complex	*Phase (technique)*[a]	*Average M–C distance* (nm)	*Ring(centroid)–M–ring (centroid) angle, α* (°)	*Ref.*
$CaCp^*_2$	Gas (E)	0.2609(6)	154(3)	269,368
$CaCp^*_2$	Solid (X)	0.264(2)	147.0	367
$Ca(\eta\text{-}C_5H_2\text{-}1,2,4\text{-}Pr^i_3)_2$	Solid (X)	0.262(2)	169.7	4
$Ca(\eta\text{-}C_5HPr^i_4)_2$	Solid (X)	0.264(1)	162.3	365
$SrCp^*_2$	Gas (E)	0.2750(8)	149(3)	369
$BaCp^*_2$	Gas (E)	0.2898(10)	148(6)	369
$BaCp^*_2$	Solid (X)	0.299(2)	131.0	367
$Ba(\eta\text{-}C_5HPr^i_4)_2$	Solid (X)	0.294(2)	154.2	365
$Ca(\eta\text{-}C_5H_4Me)_2(DME)$	Solid (X)	0.2676(9)	134.8	362
$Ca(\eta\text{-}C_5H_4Bu^t)_2(THF)_2$	Solid (X)	0.2733	133.3	245
$Ca[\eta\text{-}C_5H_3\text{-}1,3\text{-}(TMS)_2]_2(THF)$	Solid (X)	0.268	135.3	363,364
$CaCp^*_2((TMS)C{\equiv}C\text{–}C{\equiv}C(TMS))$	Solid (X)	0.265(3)	144.4	370
$[CaCp^*_2(\mu\text{-}Cl)(HfClCp^*_2)]$	Solid (X)	0.265(3)	144.2	371
$Sr(\eta\text{-}C_5H_4Bu^t)_2(THF)_2$	Solid (X)	0.2864	133.1	245
$Sr[\eta\text{-}C_5H_3\text{-}1,3\text{-}(TMS)_2]_2(THF)$	Solid (X)	0.281	134	363
$Ba(\eta\text{-}C_5H_2\text{-}1,2,4\text{-}Pr^i_3)_2(THF)_2$	Solid (X)	0.303(2)	131.7	348
$[(BaCp^*_2)_2(\mu\text{-}pyrazine)]$	Solid (X)	0.296(2)	138.0	372

[a] X, x-ray diffraction; E, electron diffraction.

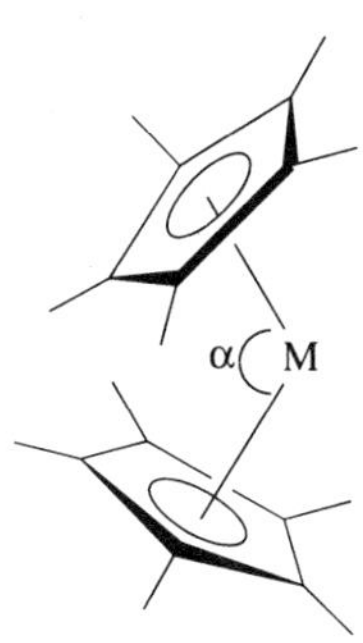

(125)

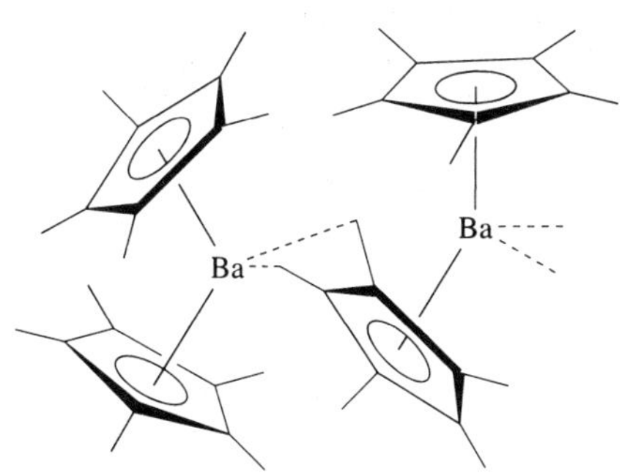

(126)

The compounds $M(\eta\text{-}C_5HPr^i_4)_2$ (M = Ca, Ba)[365] and $Ca(\eta\text{-}C_5H_2\text{-}1,2,4\text{-}Pr^i_3)_2$[4] form monomeric units in the solid state in which the bulky isopropyl substituents of the two rings intermesh to create a cage around the metal preventing oligomerization and suppressing reactivity. These sterically hindered species are also bent, with bending increasing from calcium to barium derivatives, although the angles α are wider than in the corresponding MCp^*_2 compounds (see Table 5).[365] The compounds $M(\eta\text{-}C_5HPr^i_4)_2$ are tight ion pairs in toluene solution, and dynamic NMR studies show that hindered rotation occurs, either about the C–Pr^i or metal–ring bonds with $\Delta G^{\ddagger} \leq 46$ kJ mol^{-1}.[365]

Various theoretical studies on dicyclopentadienyl compounds of alkaline earth or lanthanide metals have failed to predict a bent equilibrium structure.[369] Recent *ab initio* pseudopotential calculations suggest that $\alpha \approx 147°$ for $BaCp_2$ but that the linearization energy is less than 1.5 kJ mol^{-1} and that $CaCp_2$ is genuinely linear.[373] Low bending potentials are consistent with 'floppy' MCp_2 compounds possessing large-amplitude bending motions, and the mainly π-type covalent contributions to the strongly ionic metal–ring bonding cause less bending than metal *d* orbital involvement in the σ-bonded compounds MX_2 (X = F, Me).[373] The bonding in the radical CaCp has also been subjected to calculations using electron propagator theory.[374]

Dicyclopentadienylmetal compounds form adducts with Lewis bases and, unless highly sterically hindered, such as $M(\eta\text{-}C_5HPr^i_4)_2$ (M = Ca, Ba),[365] they are obtained as solvates from ethereal solvents, including THF, DME and Et_2O. Table 5 lists some structural parameters for complexes of type (**127**) with two oxygen donor groups (i.e., $Ca(\eta\text{-}C_5H_4Me)_2(DME)$,[362] $M(\eta\text{-}C_5H_4Bu^t)_2(THF)_2$ (M = Ca, Sr)[245] and $Ba(\eta\text{-}C_5H_2\text{-}1,2,4\text{-}Pr^i_3)_2(THF)_2$[348]) and of type (**128**) (i.e., $M[\eta\text{-}C_5H_3\text{-}1,3\text{-}(TMS)_2]_2(THF)$, M = Ca, Sr),[363–4] in which the bent metallocene is coordinated by one additional oxygen donor ligand. Steric effects of the alkylated cyclopentadienyl rings largely determine the degree of solvation and, for related systems, higher coordination numbers are found for larger metals; for example, $M(\eta\text{-}C_5H_2\text{-}1,2,4\text{-}Pr^i_3)_2(THF)_n$ (M = Ca, $n \leq 1$; M = Sr, Ba, $n \leq 2$).[348]

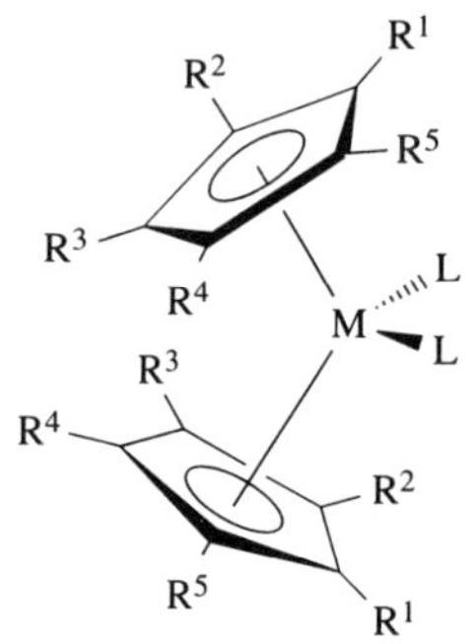

(**127**)

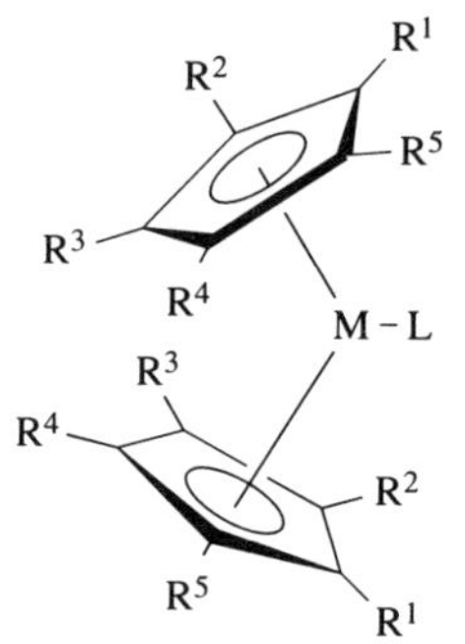

(**128**)

Adducts of bis(pentamethylcyclopentadienyl)metal compounds with various Lewis base ligands have been reported.[3] Highly coloured complexes with aromatic amine ligands include the characterized species $Cp^*_2M(bipy)$ (M = Ca, Sr, Ba)[266] and the binuclear pyrazine-bridged complexes $[(Cp^*_2M)_2(\mu\text{-}N_2C_4R_4)]$ (M = Ca, Ba; R = H, Me),[372] with structure (**129**) determined by x-ray diffraction in the solid state for the derivative with M = Ba, R = H (Table 5). Intramolecular coordination of both amine groups in $Ca[\eta\text{-}C_5Me_4(CH_2CH_2NMe_2)]_2$ produces an unsolvated inner complex with a structure of type (**127**).[366] Solid complexes with two isocyanide ligands (Cp^*_2M(2,6-xylylisocyanide)$_2$) or one phosphine ligand ($Cp^*_2M(PEt_3)$ (M = Ca, Sr, Ba)) can be isolated, although NMR spectra suggest that some dissociation of these ligands occurs in solution.[266] Moreover, $CaCp^*_2$ and 1,4-bis(trimethylsilyl)-1,3-butadiene interact in toluene forming a deep-orange solution from which crystals of $Cp^*_2Ca((TMS)C{\equiv}C\text{–}C{\equiv}C(TMS))$ (**130**) have been structurally characterized (Table 5): two crystallographically independent molecules containing the diyne coordinated in differing conformations to the bent metallocene are observed and closest Ca–C(diyne) distances range from 0.2878 nm to 0.3102 nm.[370] A 1:1 adduct of $CaCp^*_2$ with 2-butyne has also been briefly described.[3]

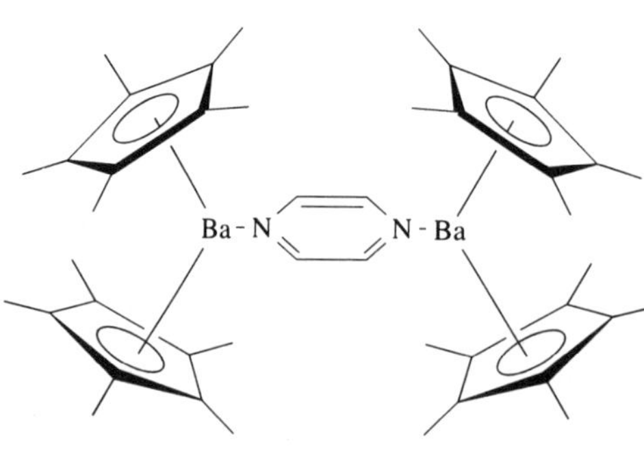

(**129**)

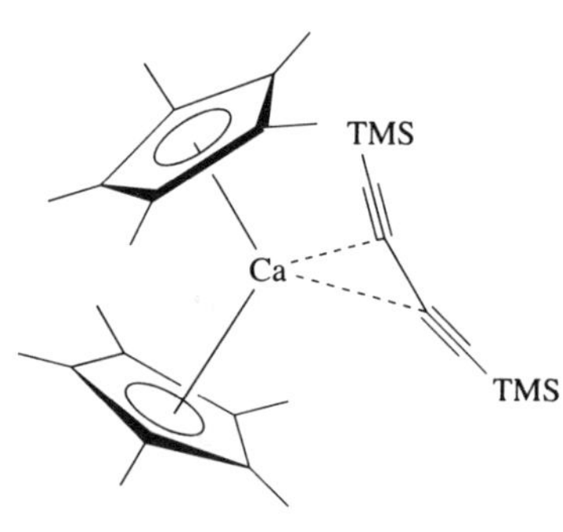

(**130**)

Ligand exchange between the transition metal halides [$NiCl_2$(dppe)] or YbI_2 and $CaCp^*_2$ produces $NiCp^*_2$ or $YbCp^*_2$(THF), respectively.[371] However, the complexes $Cp^*_2MCl_2$ (M = Zr, Hf, Th) will not undergo further Cp*–Cl exchange but form bimetallic adducts (Equation (76)) linked by a single chloride bridge, with the geometry of structure (**128**) (L = Cl) around the calcium, as shown by a structural determination of crystalline [$Cp^*_2Ca(\mu$-Cl)(HfClCp$^*_2$)] (Table 5).[371] A related complex ($Cp^*_2Ca(\mu$-Cl)YCp$^*_2$) has been mentioned.[4] The carbonyl ligands of some transition metal complexes will also function as Lewis bases to $CaCp^*_2$ by forming isocarbonyl bridging interactions (Ca–O≡C–M^1), and isolated adducts include [$Cp^*_2Ca(OC)_2ZrCp^*_2$] and tetrametallic [Cp^*_2Ca-(μ-OC)$_2$Cr(CO)(η-mesitylene)]$_2$ (**131**), containing a 12-membered ring of two calcium atoms, two chromium atoms and four carbonyl groups.[371]

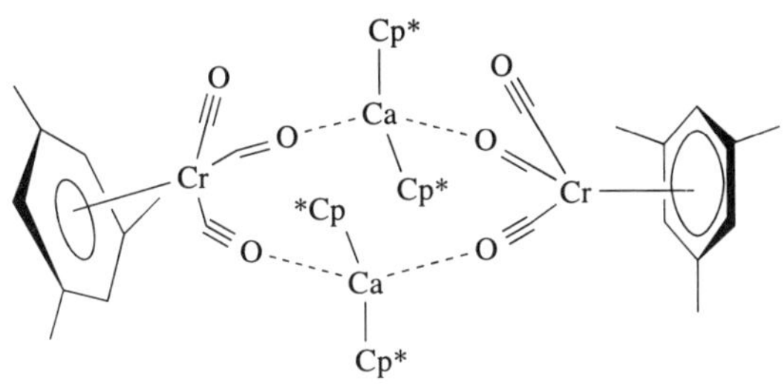

(131)

$$Cp^*_2Ca + Cp^*_2MCl_2 \xrightarrow[\text{M = Zr, Hf, Th}]{\text{toluene}} Cp^*_2Ca(\mu\text{-Cl})MClCp^*_2 \quad (76)$$

A soluble mono(pentamethylcyclopentadienyl)alkylcalcium derivative with the bulky $(TMS)_2CH$ ligand is formed from $CaCp^*_2$ (Equation (77)), and is easily separated from the precipitated LiCp*.[349] However, $BaCp^*_2$ under similar conditions forms a bimetallic species (Li[$Cp^*_2BaCH(TMS)_2$](THF)$_2$). The mixed-ligand species [Cp*Ca{CH(TMS)$_2$}(THF)$_3$] is fairly stable in benzene solution, but removal of THF ligands can induce disproportionation.[349] $CaCp^*_2$ has been transformed also into various other mono(pentamethylcyclopentadienyl)calcium species with heteroligands (see Section 3.3.4).

$$Cp^*_2Ca(THF)_2 + LiCH(TMS)_2 \xrightarrow{\text{THF}} Cp^*Ca[CH(TMS)_2](THF)_3 + LiCp^* \quad (77)$$

(ii) Diindenyl and difluorenyl metal compounds

Indene reacts with barium granules in mixed ammonia–THF, or ammonia–THF containing HMPT, at − 40 °C to form the isolable bis(indenyl)barium complexes [$Ba(C_9H_7)_2L_2$] (L = THF or HMPT, respectively), and NMR studies suggest the presence of contact ion pairs in solutions of these organometallics.[360]

Reaction of fluorene with barium in liquid ammonia at −80 °C has yielded orange–yellow [$Ba(Fl)_2(NH_3)_4$] (Fl = fluorenyl ($C_{13}H_9$)) after crystallization from THF,[375] and similar reactions in mixed ammonia–ethereal solvents at − 40 °C afford the complexes [$Ba(Fl)_2L_2$] (L = Et_2O, THF, HMPT or L_2 = DME).[360] An x-ray study of [$Ba(Fl)_2(NH_3)_4$] in the solid state reveals two conformers of bent sandwich structure with the barium-atom coordinated to four additional ammonia ligands, (**132**).[375] Ring(Fl centroid)–barium–ring(Fl centroid) angles are 112.3° and 116.4°, and average barium–ring(centroid) distances are 0.294 nm to the five-membered rings, which are essentially η^5 bonded, although one conformer includes a fluorenyl ligand tending towards η^3 coordination. The bending is attributed to the stereochemical influence of the NH_3 ligands as well as *d* orbital participation and polarization effects on the largely ionic bonding.[375]

Tilted ion pair structures have previously been proposed for bis(fluorenyl)barium complexes in solutions.[1] More recent UV–visible and NMR investigations of the complexes [$Ba(Fl)_2L_2$] in benzene, pyridine or THF solutions support the existence of such contact ion pairs at room temperature, but indicate the formation of solvent-separated ion pairs (Equation (72)) (R = Fl) at lower temperatures or on addition of more polar, coordinating solvents such as diglyme or HMPA. Interestingly, the monomeric molecular ions [$HBa(C_{13}H_9)_2L]^+$ are formed in the gas phase of FAB mass spectra of the species [$Ba(Fl)_2L_2$].[360] Compounds of the linked dicarbanionic α,ω-bis(9-fluorenyl)alkane ligands [$M^{2+}\{Fl'(CH_2)_nFl'\}^{2-}$] (M = Sr, Ba; Fl' = $C_{13}H_8$; n = 2, 4, 6) are contact (tight) ion pairs in THF at room

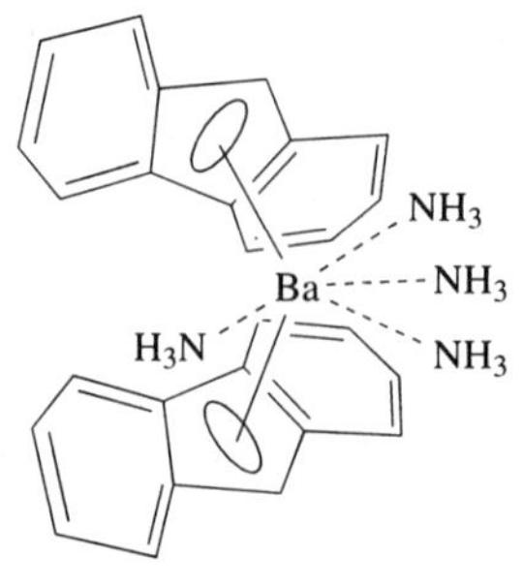

(132)

temperature, but the species behave differently on cooling the solution: when M = Sr and $n = 4$ or 6, fully solvent-separated ion pairs form below −30 °C; when M = Ba and $n = 4$ or 6, mixed tight–loose ion pairs form at −80 °C; when M = Ba and $n = 2$, tight ion pairs remain at −110 °C, and tight ion pairs of the corresponding strontium compound undergo slow solvation at −100 °C.[376] Note that a barium derivative of a 9-fluorenylide ($Ba[Fl'PPh_2CHPPh_2]_2$ ($Fl'PPh_2CHPPh_2$ = (**133**)) has been prepared.[377]

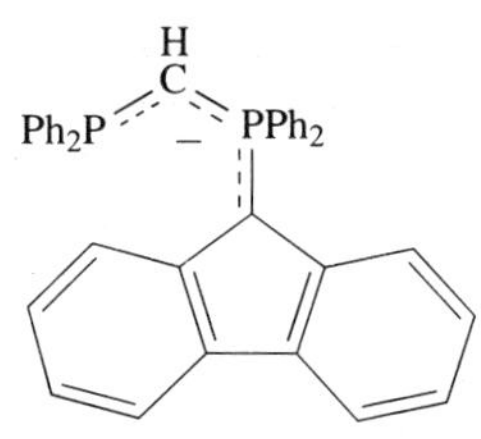

(133)

(iii) Metallacarbaboranes

closo-Metallacarbaboranes are obtained by reaction of MI_2 (M = Ca, Sr) with $Na_2[$*nido*$\text{-}7,9\text{-}C_2B_{10}H_{12}]$ in THF, and recrystallization of the products from MeCN–Et_2O affords adducts with MeCN that have been structurally characterized.[378] In the calcium compound (**134**), the calcium atom is coordinated to four MeCN ligands and asymmetrically caps the open hexagonal face of [*nido*-7,9-$C_2B_{10}H_{12}$] with Ca–C distances of 0.2701 nm and 0.2895 nm. In the strontium species [*closo*-1,1,1-$(MeCN)_3$-1,2,4-$SrC_2B_{10}H_{12}]_n$) each strontium atom also caps the hexagonal face of the *nido*-carbaborane, with Sr–C distances of 0.2997 nm and 0.3176 nm, but, unlike the calcium complex (**134**), each strontium atom is coordinated to only three MeCN ligands and additional coordination to adjacent carbaborane units via a pair of Sr–H–C-2 and Sr–H–B-12 interactions creates spiral polymeric chains.[378]

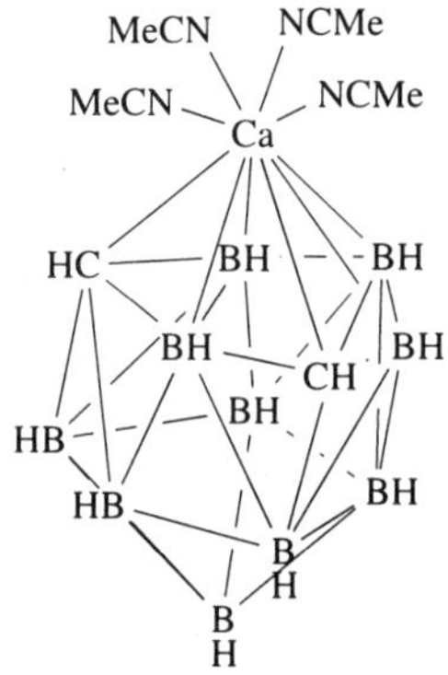

(134)

3.3.3 Addition Compounds with Unsaturated Organic Molecules

Highly reactive calcium metal, formed by reduction of CaI_2 with biphenyllithium, reacts with 1,3-dienes in THF to form addition compounds, analogous to magnesium derivatives (see Section 3.2.3.1), which can formally be represented as the metallacycles (**135**) although no definitive structural information is available.[342] Diene–calcium species of type (**135**) with $R^1 = H$, $R^2 = Ph$ or $R^1 = Me$, $R^2 = H$ have been studied and undergo dialkylation reactions *in situ* with α,ω-dihaloalkanes or dichlorosilanes to form three-, five- or six-membered ring products; the reactivity and product yields of these diene–calcium compounds are generally higher than those of magnesium analogues, and the reactions are stereospecific and highly regioselective.[342]

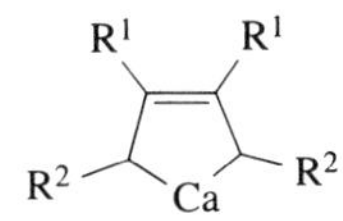

(135)

A preparative route to the addition compound Ca(cot) is by reaction of calcium with cyclooctatetraene in liquid ammonia.[362] The derivatives M(cot) (M = Ca, Sr, Ba) can also be produced by desolvation at 60 °C of the adducts $M(cot)(THF)_n$, formed by co-condensation of cyclooctatetraene with the metal vapour and THF at −196 °C and warming of the condensate to room temperature.[379] The thermally stable compounds M(cot) are sparingly soluble in ethereal solvents, form the adducts $M(cot)(py)_n$ with pyridine, act as a source of metal for synthesis of $Ph_3CMCl(THF)_n$ from triphenylmethyl chloride, and react with $RPCl_2$ ($R = Ph_2C(C_5H_4N\text{-}2)$) to give the phosphacyclopropane $RP(C_8H_8)$.[379] The combined reaction of calcium and potassium metals with cyclooctatetraene in liquid ammonia affords $K_2[Ca(cot)_2]$, which can be converted into the stable solvate $[K(DME)]_2[Ca(cot)_2]$, isostructural with $[K(DME)]_2[Y(cot)_2]$ and, hence, probably containing the eclipsed sandwich anion (**136**) with parallel-planar C_8H_8 rings.[380]

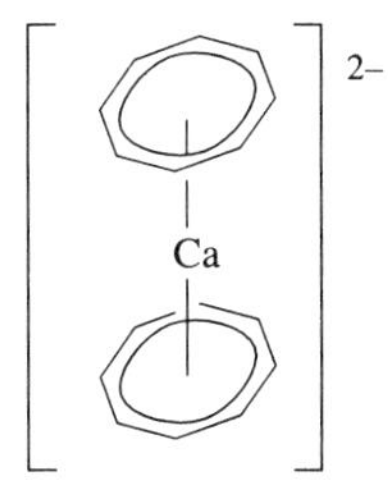

(136)

Reactions of anthracene with excess heavy alkaline earth metals in THF solution give the highly coloured, sparingly soluble derivatives $[M(THF)_n(anthracene)]$ (M = Ca, n = 4; M = Sr, n = 2; M = Ba, n = 1).[381] $[Ca(THF)_4(anthracene)]$ can also be formed by metal exchange between $[Mg(anthracene)(THF)_3]$ and calcium metal. UV–visible, IR and solid state ^{13}C NMR spectral studies indicate that these are essentially ionic compounds, containing $[anthracene]^{2-}$ anions with increasing ionicity in the order Ca < Sr < Ba. Addition of electrophiles to the 9,10-positions occurs on protolysis, carbonation or cleavage of THF at 60 °C. Formation of $[Ca(THF)_4(anthracene)]$ involves two SET steps with $Ca(anthracene)_2$ as an intermediate, and the violet radical anion species $[Ca(THF)_6(anthracene)_2]$ can be prepared from calcium and concentrated anthracene solutions in THF but disproportionates in dilute solutions (Equation (78)).[381] The radical anion complexes $[M(NH_3)_2{}^{2+}(arene)_2{}^{\bullet -}]$ (M = Ca, Sr, Ba; arene = naphthalene, anthracene, tetracene, perylene, etc.) can also be formed from the arene and metal in liquid ammonia with heats of formation decreasing with the metal in the order Ca > Sr > Ba, but lattice energies do not vary extensively with size of the hydrocarbon radical anion.[382] On pyrolysis, the complexes $[M(NH_3)_2{}^{2+}(arene)_2{}^{-}]$ containing smaller arenes degrade to metal, arene and ammonia but complexes with larger arenes liberate hydrogen and the hydrocarbon is polymerized.[382] The use of naphthalene–, biphenyl–, and anthracene–calcium derivatives in the preparation of organocalcium halides should be noted.[342–4]

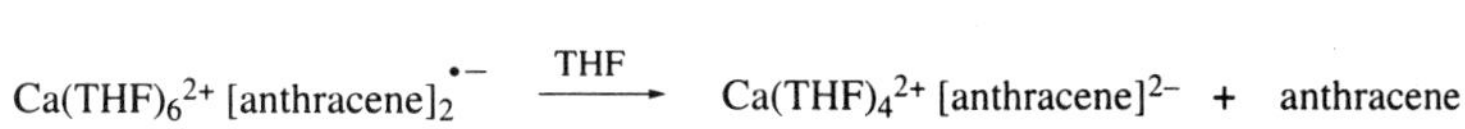

(78)

3.3.4 Organometal Hydrides, Oxides and Amides

3.3.4.1 Organocalcium hydrides

Calcium atoms insert into aromatic C–H bonds on co-condensation from the vapour phase with aromatic hydrocarbons at 77 K and subsequent warming to room temperature. By this method, organometallics of the type ArCaH can be formed from benzene, mono- and dialkyl benzenes, and trimethylsilylbenzene.[339,383] Aromatic radical anions are intermediates in the formation of ArCaH from the arene–calcium matrix and have been directly observed by ESR spectroscopy at cryogenic temperatures.[383–4] In THF solution, PhCaH displays a ^{1}H NMR resonance assigned to the hydridic hydrogen atom at $\delta = 0.2$ ppm.[383] Organocalcium hydrides react with trimethylchlorosilane (Equation (79)) to form the reduction product TMS-H and aryltrimethylsilanes, the latter being mainly *m*-(TMS)C_6H_4R from the alkyl-substituted species RC_6H_4CaH but *p*-$(TMS)_2C_6H_4$ from (TMS)C_6H_4CaH. Reactions of PhCaH with various organic substrates also give phenylated and reduced products.[383]

$$\text{ArCaH} \xrightarrow[\text{ii, H}_2\text{O}]{\text{i, TMS-Cl}} \text{TMS-Ar} + \text{TMS-H} \qquad (79)$$

3.3.4.2 Organometal oxides

(η-C_5Me_4Et)CaI$(THF)_2$ reacts with potassium aryl oxides containing bulky *ortho* substituents (KOAr (Ar = OC_6H_2-2,6-Bu^t_2-4-Me or OC_6H_2-2,4,6-$(CH_2NMe_2)_3$)) to give unsolvated cyclopentadienylcalcium aryl oxides in quantitative yields (Equation (80)).[349] It is probable that the *o*-amine substituents of [(η-C_5Me_4Et)Ca{OC_6H_2-2,4,6-$(CH_2NMe_2)_3$}] coordinate intramolecularly to the metal centre. Oxygen reacts with Ca[η-C_5H_3-1,3-$(TMS)_2]_2$ to form the stable, binuclear biradical (**137**), structurally characterized by x-ray diffraction.[385] In (**137**) each calcium atom is bonded to a normal η^5-cyclopentadienyl ligand (mean Ca–C distance = 0.2657 nm) and connected by a pair of oxygen-bridging enolate ligands with a slightly asymmetric $Ca_2(\mu\text{-}O)_2$ core. Interestingly, an unpaired electron is situated on each oxygenated cyclopentadienyl ring and a superexchange interaction between the two electrons is indicated by resonances at $g = 9.01$, 4.21 and 3.87 of variable-temperature ESR spectra in toluene solutions. Formation of (**137**) may involve initial complexation of O_2 to calcium centres in two molecules of Ca[η-C_5H_3-1,3-$(TMS)_2]_2$.[385]

$$(\eta\text{-C}_5\text{Me}_4\text{Et})\text{CaI(THF)}_2 + \text{KOAr} \xrightarrow{\text{toluene}} (\eta\text{-C}_5\text{Me}_4\text{Et})\text{Ca(OAr)} + \text{KI} \qquad (80)$$

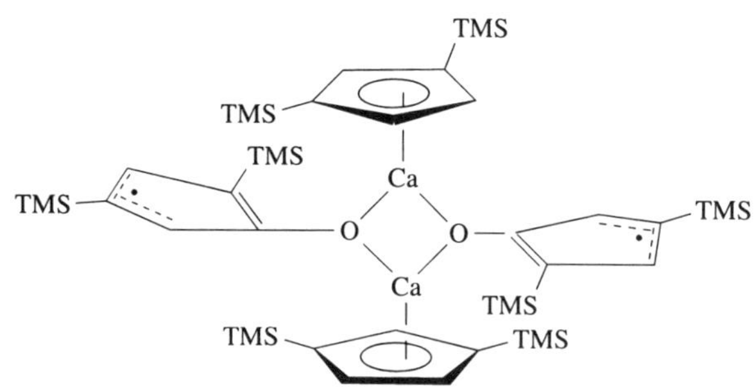

(137)

3.3.4.3 Organometal amides

Careful addition of methylcyclopentadiene to calcium in liquid ammonia has produced [(η-C_5H_4Me)$Ca(NH_2)]_n$, which may be an intermediate in the preparation of Ca(η-$C_5H_4Me)_2$ under these reaction conditions.[362] The mixed cyclopentadienylcalcium amide Cp*Ca[N$(TMS)_2$]$(THF)_3$ is formed according to Equation (81), and is the only product even if two molar equivalents of LiN$(TMS)_2$ are used.[349] However, Cp*Ca[N$(TMS)_2$]$(THF)_3$ partially disproportionates in solution and the Schlenk-type equilibrium (Equation (82)) has $K = 9.8 \times 10^{-3}$ in a saturated benzene-d_6 solution.

$$Cp^*_2Ca(THF)_2 + LiN(TMS)_2 \xrightarrow{THF} Cp^*_2Ca[N(TMS)_2(THF)_3 + LiCp^* \quad (81)$$

$$2Cp^*Ca[N(TMS)_2](THF)_3 \rightleftharpoons Cp^*_2Ca(THF)_2 + Ca[N(TMS)_2]_2(THF)_2 \quad (82)$$

Reaction of (η-C_5Me_4Et)CaI$(THF)_2$ in THF with the potassium salt of the deprotonated cyclic amine 2,2,5,5-tetramethyl-2,5-disila-1-azacyclopentane forms the binuclear amide (**138**).[349] An x-ray investigation of crystalline (**138**) established the approximately centrosymmetric dimeric structure containing two bridging amide ligands with a slightly asymmetric, planar $Ca_2(\mu\text{-}N)_2$ core. Evidence for steric crowding in (**138**) is adduced from the relatively long Ca–C distances (mean 0.270 nm) and some short intramolecular atom contacts. The Ca···Ca separation is 0.3654 nm, and it is of interest that (**138**) fluoresces under 254 nm UV radiation. Neither (**138**) nor the amide product of Equation (81) are hydrogenolysed by H_2 (1 atm, 22 °C) in toluene.[349]

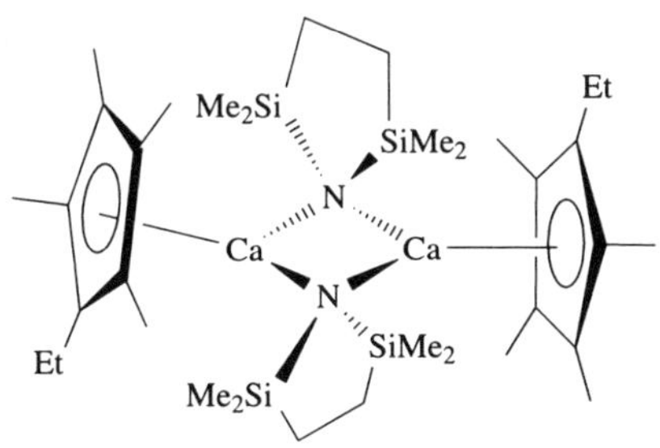

(138)

3.3.5 Heterobimetallic Compounds

3.3.5.1 Alkali metal compounds

Whereas the lithium reagents Li[CH$(TMS)_2$] or LiN$(TMS)_2$ displace a Cp* group from $CaCp^*_2$ to form mixed-ligand complexes (Equations (77) and (81)), these lithium reagents combine with $BaCp^*_2$ in THF to give isolable, hydrocarbon-insoluble complexes formulated as barates (Li[$BaCp^*_2$(E)](THF)$_2$ (E = CH$(TMS)_2$ or N$(TMS)_2$)).[349] These lithium–barium aggregates tend to disproportionate in THF,[3] and their structures are unknown. The potassium bis(octatetraene)calcate $K_2[Ca(cot)_2]$ and the solvated species $[K(L_2)]_2[Ca(cot)_2]$ (L = THF or L_2 = DME) should be noted; the latter species are ionic and contain $[Ca(cot)_2]^{2-}$ with the regular sandwich structure (**136**) (see Section 3.3.3).[380]

3.3.5.2 Zinc compounds

Alkaline earth metal zincates,[1] formed by reaction of calcium, strontium or barium metal with $ZnEt_2$ in benzene at 70–105 °C, add to trimethylsilyl(vinyl)ethyne ((TMS)C≡C–CH=CH_2) in 1:1 proportions to form organometallic species that on hydrolysis produce equal amounts of (TMS)C≡CCH_2CH_2Et and (TMS)CH=C=$CHCH_2Et$. Similar reactions with alkyl(vinyl)ethynes form only allenes, and the difference is attributed to electronic effects of the silicon atom.[386] The applications in polymerization reactions of zincates ($MZnR_4$ (M = Ca, Sr, Ba)) and other bimetallic species of the alkaline earth metals have been reviewed.[352,357a]

3.3.5.3 Aluminum compounds

Trimethylaluminum abstracts the THF ligands from $Cp_2Ca(THF)_2$ in toluene solution to form $AlMe_3$(THF) and the base-free polymer $(CaCp_2)_n$. The same reaction with $Cp^*_2Ca(THF)_2$ removes the THF but, since the sterically bulky permethylated rings prevent oligomerization, two $AlMe_3$(THF) groups remain coordinated to two $CaCp^*_2$ units via bridging methyl interactions to form the tetrametallic product (**139**) (see Equation (83)).[387] The centrosymmetric structure of (**139**) in the solid state contains aluminum atoms in a slightly distorted tetrahedral coordination geometry and calcium atoms in a typical geometry, (**127**) (L = Me), with Ca–C(μ-Me) distances of 0.2948(7) nm and 0.2999(7) nm and mean Ca–C(ring) distances of 0.270(1) nm.[387]

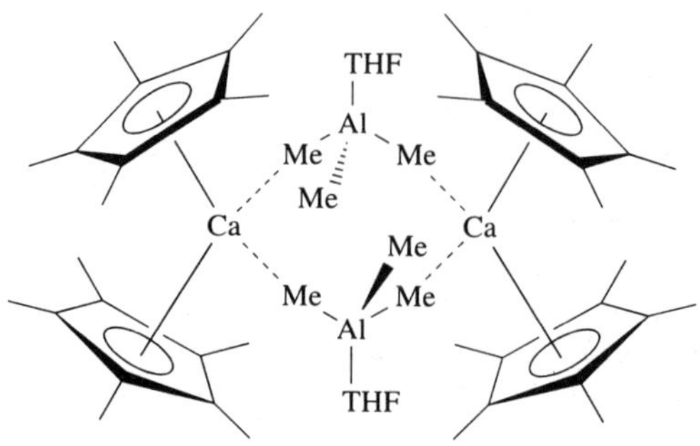

(139)

$$2Cp^*Ca(THF)_2 + 2Al_2Me_6 \xrightarrow{\text{toluene}} [Cp^*_2Ca(\mu\text{-}Me)_2AlMe(THF)]_2 + 2AlMe_3(THF) \quad (83)$$

3.3.5.4 Transition metal compounds

Additions of soluble copper(I) salts, including CuCN·2LiBr, to organocalcium halides RCaX form uncharacterized organometallics, presumed to be calcium cuprates, which show distinctive and useful reactivity as reagents.[342] Cross-coupling of these cuprates with benzoyl chloride forms ketones in high yields and conjugate-addition reactions occur with 1,3-enones. Similar calcium cuprate reagents are formed from polymeric organocalcium halides, and these react with various electrophiles to form highly derivatized polymers.[344]

Bimetallic adducts of $CaCp^*_2$ with early transition metal halides, for example, $[Cp^*_2Ca(\mu\text{-}Cl)MClCp^*_2]$ (M = Zr, Hf, Th) (Equation (76)), and with some transition metal carbonyl complexes, for example, $[Cp^*_2Ca(OC)_2ZrCp^*_2]$ and the chromium derivative (**131**), may also be noted here (see Section 3.3.2.2).[371]

3.4 REFERENCES

1. W. E. Lindsell in 'COMC-I', vol. 1, p. 155 and references therein.
2. P. R. Markies, O. S. Akkerman, F. Bickelhaupt, W. J. J. Smeets and A. L. Spek, *Adv. Organomet. Chem.*, 1991, **32**, 147.
3. T. P. Hanusa, *Polyhedron*, 1990, **9**, 1345.
4. T. P. Hanusa, *Chem. Rev.* 1993, **93**, 1023.
5. F. Bickelhaupt, *Acta Chem. Scand.*, 1992, **46**, 409.
6. B. Bogdanovic, *Acc. Chem. Res.*, 1988, **21**, 261.
7. L. C. O'Brien and P. F. Bernath, *J. Am. Chem. Soc.*, 1986, **108**, 5017; C. R. Brazier and P. F. Bernath, *J. Chem. Phys.*, 1989, **91**, 4548; A. M. R. P. Bopegedera, C. R. Brazier and P. F. Bernath, *J. Mol. Spectrosc.*, 1988, **129**, 268.
8. E. S. J. Robles, A. M. Ellis and T. A. Miller, *J. Phys. Chem.*, 1992, **96**, 8791; *J. Am. Chem. Soc.*, 1992, **114**, 7171.
9. J. S. Uppal and R. H. Staley, *J. Am. Chem. Soc.*, 1982, **104**, 1229.
10. K. J. Klabunde and A. Whetten, *J. Am. Chem. Soc.*, 1986, **108**, 6529; Y. Imizu and K. J. Klabunde, *Inorg. Chem.*, 1984, **23**, 3602.
11. P. E. Rakita, J. F. Aultman and L. Stapleton, *Chem. Eng.*, 1990, **97**, 110.
12. J. R. Jennings, *J. Organomet. Chem.*, 1987, **325**, 25.
13. Y.-H. Lai, *Synthesis*, 1981, 585.
14. T. S. Eckert, *J. Chem. Educ.*, 1987, **64**, 179; K. L. Williamson, *ibid.*, 1988, **65**, 376.
15. A. Tuulmets, M. Horak, A. Ong and M. Limberg, *Org. React. (Tartu)*, 1988, **25**, 116.
16. B. J. J. van de Heisteeg, G. Schat, O. S. Akkerman and F. Bickelhaupt, in 'Organometallic Syntheses', eds. R. B. King and J. J. Eisch, Elsevier, Amsterdam, 1988, vol. 4. p. 389.
17. A. Fürstner, *Angew. Chem., Int. Ed. Engl.*, 1993, **32**, 164.
18. K. V. Baker, J. M. Brown, N. Hughes, A. J. Skarnulis and A. Sexton, *J. Org. Chem.*, 1991, **56**, 698.
19. J. D. Sprich and G. S. Lewandos, *Inorg. Chim. Acta*, 1983, **76**, L241; S. Clough, E. Goldman, S. Williams and B. George, *J. Chem. Educ.*, 1986, **63**, 176.
20. (a) A. Tuulmets, M. Horak and K. Kaubi, *Org. React. (Tartu)*, 1986, **23**, 397; (b) A. Tuulmets and K. Heinoja, *ibid.*, 1990, **27**, 63.
21. (a) E. P. Kündig and C. Perret, *Helv. Chim. Acta*, 1981, **64**, 2606; (b) W. Oppolzer, E. P. Kündig, P. M. Bishop and C. Perret, *Tetrahedron Lett.*, 1982, **23**, 3901; (c) W. Oppolzer and P. Schneider, *ibid.*, 1984, **25**, 3305.
22. R. D. Rieke, T. P. Burns, R. M. Wehmeyer and B. E. Kahn, *ACS Symp. Ser.*, 1987, **333**, 223.
23. R. D. Rieke, P. T.-J. Li, T. P. Burns and S. T. Uhm, *J. Org. Chem.*, 1981, **46**, 4323.
24. B. Bogdanovic, S.-T. Liao, K. Schlichte and U. Westeppe, in 'Organometallic Syntheses', eds. R. B. King and J. J. Eisch, Elsevier, Amsterdam, 1988, vol. 4. p. 410.

25. (a) S. Harvey, P. C. Junk, C. L. Raston and G. Salem, *J. Org. Chem.*, 1988, **53**, 3134; (b) L. M. Engelhardt, R. I. Papasergio, C. L. Raston, G. Salem and A. H. White, *J. Chem. Soc., Dalton Trans.*, 1986, 789; (c) M. J. Gallagher, S. Harvey, C. L. Raston and R. E. Sue, *J. Chem. Soc., Chem. Commun.*, 1988, 289.
26. (a) B. Bogdanovic, N. Janke and H.-G. Kinzelmann, *Chem. Ber.*, 1990, **123**, 1507; (b) B. Bogdanovic, N. Janke, H.-G. Kinzelmann, K. Seevogel and J. Treber, *ibid.*, 1990, **123**, 1529.
27. S. Harvey and C. L. Raston, *J. Chem. Soc., Chem. Commun.*, 1988, 652.
28. E. Bartmann *et al.*, *Chem. Ber.*, 1990, **123**, 1517.
29. H. Bönnemann, B. Bogdanovic, R. Brinkmann and B. Spliethoff, *J. Organomet. Chem.*, 1993, **451**, 23.
30. G. B. Sergeev, V. V. Smirnov and F. Z. Badaev, *J. Organomet. Chem.*, 1982, **224**, C29.
31. R. G. Nuzzo and L. H. Dubois, *J. Am. Chem. Soc.*, 1986, **108**, 2881.
32. L. A. Jones, S. L. Kirby, D. M. Kean and G. L. Campbell, *J. Organomet. Chem.*, 1985, **284**, 159.
33. G. Molle, P. Bauer and J. E. Dubois, *J. Org. Chem.*, 1982, **47**, 4120.
34. A. G. Yurchenko, T. V. Fedorenko and V. N. Rodionov, *J. Org. Chem. USSR (Engl. Transl.)*, 1985, **21**, 1529.
35. H. G. Richey, Jr. and J. Farkas, Jr., *J. Org. Chem.*, 1987, **52**, 479.
36. P. C. Hayes, A. Osman, N. Seudeal and D. G. Tuck, *J. Organomet. Chem.*, 1985, **291**, 1.
37. A. Boudin, G. Cerveau, C. Chuit, R. J. P. Corriu and C. Reye, *Tetrahedron*, 1989, **45**, 171.
38. K. S. Root, J. Deutch and G. M. Whitesides, *J. Am. Chem. Soc.*, 1981, **103**, 5475.
39. P. Hasler and W. Richarz, *Ind. Eng. Chem. Res.*, 1989, **28**, 38; W. W. Hammerschmidt and W. Richarz, *ibid.*, 1991, **30**, 82.
40. A. Tuulmets, M. Horak, E. Pill and A. Riikoja, *Org. React. (Tartu)*, 1985, **22**, 93; A. Tuulmets, M. Horak, K. Sarv and E. Aaresild, *ibid.*, 1985, *22*, 332; A. Tuulmets, M. Horak, E. Aaresild and K. Sarv, *ibid.*, 1985, **22**, 460.
41. A. Tuulmets and K. Heinoja, *Org. React. (Tartu)*, 1990, **27**, 42; A. Tuulmets, M. Horak, E. Hansen and V. Palm, *ibid.*, 1987, **24**, 412.
42. K. S. Root, C. L. Hill, L. M. Lawrence and G. M. Whitesides, *J. Am. Chem. Soc.*, 1989, **111**, 5405.
43. Y. Liu, H. Dang, S. Kong, P. Liu and X. Sun, *Kexue Tongbao*, 1987, **32**, 167 (*Chem. Abstr.*, 1988, **108**, 150 524).
44. E. A. Hill, C. L. Harder, R. Wagner, D. Meh and R. P. Bowman, *J. Organomet. Chem.*, 1986, **302**, 5.
45. E. C. Ashby and J. Oswald, *J. Org. Chem.*, 1988, **53**, 6068.
46. J. F. Garst, *Acc. Chem. Res.*, 1991, **24**, 95.
47. (a) H. M. Walborsky and M. Topolski, *J. Am. Chem. Soc.*, 1992, **114**, 3455; (b) H. M. Walborsky and C. Zimmermann, *ibid.*, 1992, **114**, 4996; (c) C. Hamdouchi, M. Topolski, V. Goedken and H. M. Walborsky, *J. Org. Chem.*, 1993, **58**, 3148.
48. B. J. Wakefield, J. P. Whitten and P. S. Farley, *J. Chem. Soc., Perkin Trans. 1*, 1982, 93.
49. J. F. Garst, B. L. Swift, *J. Am. Chem. Soc.*, 1989, **111**, 241; J. F. Garst, B. L. Swift and D. W. Smith, *ibid.*, 1989, **111**, 234; J. F. Garst, J. E. Deutch and G. M. Whitesides, *ibid.*, 1986, **108**, 2490.
50. (a) J. F. Garst, F. Ungvary, R. Batlaw and K. E. Lawrence, *J. Am. Chem. Soc.*, 1991, **113**, 6697; (b) J. F. Garst, F. Ungvary, R. Batlaw and K. E. Lawrence, *ibid.*, 1991, **113**, 5392.
51. H. M. Walborsky, *Acc. Chem. Res.*, 1990, **23**, 286; *Chem. Unser Zeit*, 1991, **25**, 108.
52. J. Rachon and H. M. Walborsky, *Tetrahedron Lett.*, 1989, **30**, 7345; H. M. Walborsky and J. Rachon, *J. Am. Chem. Soc.*, 1989, **111**, 1896.
53. H. M. Walborsky, M. Topolski, C. Hamdouchi and J. Pankowski, *J. Org. Chem.*, 1992, **57**, 6188.
54. (a) H. G. Richey, Jr. and L. M. Moses, *J. Org. Chem.*, 1983, **48**, 4013; (b) W. Kirmse, J. Rode and K. Rode, *Chem. Ber.*, 1986, **119**, 3672.
55. C. Walling, *Acc. Chem. Res.*, 1991, **24**, 255.
56. G. B. Sergeev, V. V. Zagorsky and F. Z. Badaev, *J. Organomet. Chem.*, 1983, **243**, 123.
57. G. B. Sergeev, V. V. Smirnov, V. V. Zagorskii and A. M. Kosolapov, *Dokl. Akad. Nauk SSSR*, 1981, **256**, 1169.
58. P. G. Jasien and C. E. Dykstra, *J. Am. Chem. Soc.*, 1983, **105**, 2089; *ibid.*, 1985, **107**, 1891.
59. S. R. Davis, *J. Am. Chem. Soc.*, 1991, **113**, 4145; L. Liu and S. R. Davis, *J. Phys. Chem.*, 1991, **95**, 8619.
60. H. J. R. de Boer, O. S. Akkerman and F. Bickelhaupt, *Angew. Chem., Int. Ed. Engl.*, 1988, **27**, 687.
61. P. R. Markies, O. S. Akkerman, F. Bickelhaupt, W. J. J. Smeets and A. L. Spek, *J. Am. Chem., Soc.*, 1988, **110**, 4284.
62. A. B. Holmes and C. N. Sporikou, *Org. Synth.*, 1987, **65**, 61.
63. P. J. Fagan, J. M. Manriquez, E. A. Maatta, A. M. Seyam and T. J. Marks, *J. Am. Chem. Soc.*, 1981, **103**, 6650.
64. T. Holm, *Acta Chem. Scand., Ser. B*, 1983, **37**, 797.
65. E. A. Hill and D. C. Link, *Organometallics*, 1982, **1**, 1501.
66. P. R. Markies *et al.*, *Organometallics*, 1991, **10**, 3826.
67. P. E. Eaton, *Angew. Chem., Int. Ed. Engl.*, 1992, **31**, 1421; P. E. Eaton, Y. Xiong and J. P. Zhou, *J. Org. Chem.*, 1992, **57**, 4277; P. E. Eaton, C.-H. Lee and Y. Xiong, *J. Am. Chem. Soc.*, 1989, **111**, 8016.
68. D. J. Burton and Z.-Y. Yang, *Tetrahedron*, 1992, **48**, 189.
69. C.-J. F. Du and H. Hart, *J. Org. Chem.*, 1987, **52**, 4311.
70. H. Nishiyama *et al.*, *J. Org. Chem.*, 1992, **57**, 407.
71. N. Furukawa, T. Shibutani and H. Fujihara, *Tetrahedron Lett.*, 1987, **28**, 5845.
72. T. Sugita, Y. Sakabe, T. Sasahara, M. Tsukada and K. Ichikawa, *Bull. Chem. Soc. Jpn.*, 1984, **57**, 2319.
73. G. F. Lehr and R. G. Lawler, *J. Am. Chem. Soc.*, 1984, **106**, 4048.
74. D. Seebach, J. Hansen, P. Seiler and J. M. Gromek, *J. Organomet. Chem.*, 1985, **285**, 1.
75. (a) M. P. Sibi, M. A. J. Miah and V. Snieckus, *J. Org. Chem.*, 1984, **49**, 737; M. P. Sibi, S. Chattopadhyay, J. W. Dankwardt and V. Snieckus, *J. Am. Chem. Soc.*, 1985, **107**, 6312; (b) B. M. Trost and W. H. Pearson, *Tetrahedron Lett.*, 1983, **24**, 269.
76. C. E. Tucker, T. N. Majid and P. Knochel, *J. Am. Chem. Soc.*, 1992, **114**, 3983.
77. H. Yasuda, T. Nishi, S. Miyanaga and A. Nakamura, *Organometallics*, 1985, **4**, 359.
78. P. R. Markies, A. Villena, O. S. Akkerman, F. Bickelhaupt, W. J. J. Smeets and A. L. Spek, *J. Organomet. Chem.*, 1993, **463**, 7.
79. H. Hommes, H. D. Verkruijsse and L. Brandsma, *Tetrahedron Lett.*, 1981, **22**, 2495.
80. P. R. Markies *et al.*, *Organometallics*, 1991, **10**, 1531.
81. P. E. Eaton, G. T. Cunkle, G. Marchioro and R. M. Martin, *J. Am. Chem. Soc.*, 1987, **109**, 948.
82. N. Furukawa, T. Shibutani, K. Matsumura, H. Fujihara and S. Oae, *Tetrahedron Lett.*, 1986, **27**, 3899.
83. T. Kanda, T. Sugino, N. Kambe and N. Sonoda, *Phosphorus Sulfur Silicon*, 1992, **67**, 103.
84. H. C. Holtkamp, G. Schat, C. Blomberg and F. Bickelhaupt, *J. Organomet. Chem.*, 1982, **240**, 1.

85. H. Lehmkuhl, *Bull. Soc. Chim. Fr., Ser. II*, 1981, 87.
86. T. Yu. Rudashevskaya and O. A. Nesmeyanova, *Bull. Acad. Sci. USSR, Div. Chem. Sci. (Engl. Transl.)*, 1983, **32**, 1647.
87. H. Lehmkuhl and K. Mehler, *Liebigs Ann. Chem.*, 1982, 2244.
88. E. K. Watkins and H. G. Richey, Jr., *Organometallics*, 1992, **11**, 3785.
89. J. V. N. Vara Prasad and C. N. Pillai, *J. Organomet. Chem.*, 1983, **259**, 1.
90. N. Ishikawa, S. Butler and M. Maruta, *Bull. Chem. Soc. Jpn.*, 1981, **54**, 3084.
91. H. Lehmkuhl, K. Hauschild and M. Bellenbaum, *Chem. Ber.*, 1984, **117**, 383.
92. K. Tamao, R. Kanatani and M. Kumada, *Tetrahedron Lett.*, 1984, **25**, 1905; 1913.
93. R. Lazzaroni, D. Pini, S. Bertozzi and G. Fatti, *J. Org. Chem.*, 1986, **51**, 505.
94. H. G. Richey, Jr., L. M. Moses, M. S. Domalski, W. F. Erickson and A. S. Heyn, *J. Org. Chem.*, 1981, **46**, 3773.
95. H. G. Richey, Jr. and M. S. Domalski, *J. Org. Chem.*, 1981, **46**, 3780.
96. J. Kang, *Organometallics*, 1984, **3**, 525.
97. J.-L. Fabre, M. Julia and J.-N. Verpeaux, *Bull. Soc. Chim. Fr.*, 1985, 762; 772.
98. U. M. Dzhemilev, A. G. Ibragimov and G. A. Tolstikov, *J. Organomet. Chem.*, 1991, **406**, 1.
99. J. F. Normant and A. Alexakis, *Synthesis*, 1981, 841.
100. L. Miginiac, *J. Organomet. Chem.*, 1982, **238**, 235.
101. Y. Ishino, K. Wakamoto and T. Hirashima, *Chem. Lett.*, 1984, 765.
102. (a) U. M. Dzhemilev, O. S. Vostrikova and G. A. Tolstikov, *J. Organomet. Chem.*, 1986, **304**, 17; (b) U. M. Dzhemilev and O. S. Vostrikova, *ibid.*, 1985, **285**, 43.
103. (a) T. Takahashi, T. Seki, Y. Nitto, M. Saburi, C. J. Rousset and E. Negishi, *J. Am. Chem. Soc.*, 1991, **113**, 6266; (b) C. J. Rousset, E. Negishi, N. Suzuki and T. Takahashi, *Tetrahedron Lett.*, 1992, **33**, 1965.
104. A. H. Hoveyda and Z. Xu, *J. Am. Chem. Soc.*, 1991, **113**, 5079; A. H. Hoveyda, Z. Xu, J. P. Morken and A. F. Houri, *ibid.*, 1991, **113**, 8950; A. H. Hoveyda, J. P. Morken, A. F. Houri and Z. Xu, *ibid.*, 1992, **114**, 6692.
105. (a) K. S. Knight and R. M. Waymouth, *J. Am. Chem. Soc.*, 1991, **113**, 6268; (b) U. Wischmeyer, K. S. Knight and R. M. Waymouth, *Tetrahedron Lett.*, 1992, **33**, 7735.
106. D. P. Lewis, P. M. Muller, R. J. Whitby and R. V. H. Jones, *Tetrahedron Lett.*, 1991, **32**, 6797.
107. U. M. Dzhemilev, O. S. Vostrikova, R. M. Sultanov and I. N. Batalina, *Bull. Russ. Acad. Sci., Div. Chem. Sci. (Engl. Transl.)* 1992, **41**, 936.
108. R. S. E. Conn, M. Karras and B. B. Snider, *Isr. J. Chem.*, 1984, **24**, 108.
109. E. Erdik, *Tetrahedron*, 1984, **40**, 641.
110. C. Broquet and H. Riviere, *J. Organomet. Chem.*, 1982, **226**, 1.
111. L. M. Zubritskii, T. N. Fomina and Kh. V. Bal'yan, *J. Org. Chem. USSR (Engl. Transl.)*, 1981, **17**, 63; *ibid.*, 1982, **18**, 1209.
112. L. M. Zubritskii, N. D. Romashchenkova and A. A. Petrov, *Zh. Org. Khim.*, 1983, **19**, 2465 (*Chem. Abstr.*, 1984, **101**, 6700).
113. E. C. Ashby and R. D. Ainslie, *J. Organomet. Chem.*, 1983, **250**, 1.
114. F. Sato, *J. Organomet. Chem.*, 1985, **285**, 53 and references therein.
115. T. Amano *et al.*, *Bull. Chem. Soc. Jpn.*, 1986, **59**, 1656.
116. Y. Kobayashi *et al.*, *Synth. Commun.*, 1986, **16**, 597.
117. Y. Yamamoto, *Bull. Soc. Chem. Jpn.*, 1983, **56**, 1772; *ibid.*, 1984, **57**, 2835.
118. A. Yanagisawa, S. Habaue and H. Yamamoto, *J. Am. Chem. Soc.*, 1991, **113**, 5893.
119. F. Barbot and Ph. Miginiac, *J. Organomet. Chem.*, 1992, **440**, 249.
120. I. Hemer, A. Posta and V. Dedek, *J. Fluorine Chem.*, 1984, **26**, 467.
121. T. Takagi *et al.*, *J. Org. Chem.*, 1992, **57**, 3921.
122. L. S. Chen, G. J. Chen, M. T. Ryan and C. Tamborski, *J. Fluorine Chem.*, 1987, **34**, 299.
123. Q. Chen and J. Chen, *Youji Huaxue*, 1990, **10**, 179 (*Chem. Abstr.*, 1990, **113**, 23 968)
124. E. C. Ashby and D. M. Al-Fekri, *J. Organomet. Chem.*, 1990, **390**, 275.
125. K. Ogura, M. Fujita, K. Takahashi and H. Iida, *Chem. Lett.*, 1982, 1697.
126. C. L. Smith, L. M. James and K. L. Sibley, *Organometallics*, 1992, **11**, 2938.
127. T. P. Burns and R. D. Rieke, *J. Org. Chem.*, 1983, **48**, 4141.
128. K. Jurkschat and H.-P. Abicht, *Z. Chem.*, 1985, **25**, 338.
129. M. Kaino, K. Ishihara and H. Yamamoto, *Bull. Chem. Soc. Jpn.*, 1989, **62**, 3736.
130. N. Plavac and M. Bersohn, *Synth. Commun.*, 1983, **13**, 867.
131. R. D. Bindal, S. Durani, R. S. Kapil and N. Anand, *Synthesis*, 1982, 405.
132. K. Takanishi, H. Urabe and I. Kuwajima, *Tetrahedron Lett.*, 1987, **28**, 2281.
133. K. Tamao and N. Ishida, *Tetrahedron Lett.*, 1984, **25**, 4245.
134. F. Z. Basha and J. F. DeBernardis, *Tetrahedron Lett.*, 1984, **25**, 5271.
135. F. Bickelhaupt, *Angew. Chem., Int. Ed. Engl.*, 1987, **26**, 990; *Pure Appl. Chem.*, 1986, **58**, 537.
136. F. Bickelhaupt, *Pure Appl. Chem.*, 1990, **62**, 699.
137. (a) B. J. J. van de Heisteeg, G. Schat, O. S. Akkerman and F. Bickelhaupt, *J. Organomet. Chem.*, 1986, **308**, 1; (b) J. W. Bruin, G. Schat, O. S. Akkerman and F. Bickelhaupt, *ibid.*, 1985, **288**, 13.
138. (a) B. J. J. van de Heisteeg, G. Schat, M. A. G. M. Tinga, O. S. Akkerman and F. Bickelhaupt, *Tetrahedron Lett.*, 1986, **27**, 6123; (b) M. Hogenbirk, N. J. R. van Eikema Hommes, G. Schat, O. S. Akkerman, F. Bickelhaupt and G. W. Klumpp, *ibid.*, 1989, **30**, 6195.
139. N. J. R. van Eikema Hommes, F. Bickelhaupt and G. W. Klumpp, *Recl. Trav. Chim. Pays-Bas*, 1988, **107**, 393.
140. J. W. F. L. Seetz, O. S. Akkerman and F. Bickelhaupt, *Tetrahedron Lett.*, 1981, **22**, 4857.
141. (a) J. W. F. L. Seetz, B. J. J. van de Heisteeg, G. Schat, O. S. Akkerman and F. Bickelhaupt, *J. Organomet. Chem.*, 1984, **277**, 319; (b) J. W. F. L. Seetz, B. J. J. van de Heisteeg, H. J. R. de Boer, G. Schat, O. S. Akkerman and F. Bickelhaupt, in 'Organometallic Syntheses', eds. R. B. King and J. J. Eisch, Elsevier, Amsterdam, 1988, vol. 4, p. 392.
142. J. W. F. L. Seetz, H. Ent, R. B. Rookhuizen, O. S. Akkerman and F. Bickelhaupt, *Recl. Trav. Chim. Pays-Bas*, 1988, **107**, 160.
143. (a) F. J. M. Freijee, G. Schat, O. S. Akkerman and F. Bickelhaupt, *J. Organomet. Chem.*, 1982, **240**, 217; (b) F. J. M. Freijee, G. van der Wal, G. Schat, O. S. Akkerman and F. Bickelhaupt, *ibid.*, 1982, **240**, 229.

144. L. M. Engelhardt, W.-P. Leung, C. L. Raston, P. Twiss and A. H. White, *J. Chem. Soc., Dalton Trans.*, 1984, 331.
145. P. Boudjouk, R. Sooriyakumaran and C. A. Kapfer, *J. Organomet. Chem.*, 1985, **281**, C21.
146. H. J. R. de Boer, O. S. Akkerman and F. Bickelhaupt, *J. Organomet. Chem.*, 1987, **321**, 291.
147. O. G. Yarosh, Z. G. Ivanova, T. D. Burnashova and M. G. Voronkov, *Bull. Acad. Sci. USSR, Div. Chem. Sci. (Engl.Transl.)*, 1984, **33**, 2394.
148. S. Zhang, S. Liao, S. Yu and D. Yu, *Huaxue Tongbao*, 1988, 34 (*Chem. Abstr.*, 1989, **110**, 8258); M. Wang, S. Liao, Y. Xu and D. Yu, *Youji Huaxue*, 1991, **11**, 99 (*Chem. Abstr.*, 1991, **114**, 207 310).
149. S. Itsuno, G. D. Darling, H. D. H. Stöver and J. M. J. Fréchet, *J. Org. Chem.*, 1987, **52**, 4644.
150. K. Hatada, H. Nakanishi, K. Ute and T. Kitayama, *Polymer J. (Tokyo)*, 1986, **18**, 581.
151. S. S. Al-Juaid, C. Eaborn, P. B. Hitchcock, C. A. McGeary and J. D. Smith, *J. Chem. Soc., Chem. Commun.*, 1989, 273.
152. P. B. Hitchcock, J. A. K. Howard, M. F. Lappert, W.-P. Leung and S. A. Mason, *J. Chem. Soc., Chem. Commun.*, 1990, 847.
153. H. Kageyama, K. Miki, N. Tanaka, N. Kasai, Y. Okamoto and H. Yuki, *Bull. Chem. Soc. Jpn.*, 1983, **56**, 1319; H. Kageyama, K. Miki, Y. Kai, N. Kasai, Y. Okamoto and H. Yuki, *ibid.*, 1983, **56**, 2411; *ibid.*, 1984, **57**, 1189.
154. H. Kageyama, K. Miki, Y. Kai, N. Kasai, Y. Okamoto and H. Yuki, *Acta Crystallogr., Sect. B*, 1982, **38**, 2264.
155. L. M. Engelhardt, S. Harvey, C. L. Raston and A. H. White, *J. Organomet. Chem.*, 1988, **341**, 39.
156. M. Hogenbirk, G. Schat, O. S. Akkerman, F. Bickelhaupt, W. J. J. Smeets and A. L. Spek, *J. Am. Chem. Soc.*, 1992, **114**, 7302.
157. P. R. Markies *et al.*, *J. Organomet. Chem.*, 1989, **375**, 11.
158. M. Marsch, K. Harms, W. Massa and G. Boche, *Angew. Chem., Int. Ed. Engl.*, 1987, **26**, 696.
159. C. P. Morley, P. Jutzi, C. Krüger and J. M. Wallis, *Organometallics*, 1987, **6**, 1084.
160. C. Dohmeier, D. Loos, C. Robl, H. Schnöckel and D. Fenske, *J. Organomet. Chem.*, 1993, **448**, 5.
161. R. E. Cramer, P. N. Richmann and J. W. Gilje, *J. Organomet. Chem.*, 1991, **408**, 131.
162. P. Sobota and B. Duda, *J. Organomet. Chem.*, 1987, **332**, 239.
163. (a) A. Wellmar and I. Persson, *J. Organomet. Chem.*, 1991, **415**, 155; A. Wellmar, A. Hallberg and I. Persson, *ibid.*, 1991, **415**, 167; (b) A. Ericson and I. Persson, *ibid.*, 1987, **326**, 151; A. Wellmar and I. Persson, *ibid.*, 1991, **415**, 143.
164. R. Benn, H. Lehmkuhl, K. Mehler and A. Rufinska, *Angew. Chem., Int. Ed. Engl.*, 1984, **23**, 534.
165. P. E. M. Allen, S. Hagias, S. F. Lincoln, C. Mair and E. H. Williams, *Ber. Bunsenges. Phys. Chem.*, 1982, **86**, 515; P. E. M. Allen, S. Hagias, C. Mair and E. H. Williams, *ibid.*, 1984, **88**, 623.
166. P. R. Markies *et al.*, *J. Organomet. Chem.*, 1991, **402**, 289.
167. M. F. Lappert, T. R. Martin, C. L. Raston, B. W. Skelton and A. H. White, *J. Chem. Soc., Dalton Trans.*, 1982, 1959.
168. W.-P. Leung, C. L. Raston, B. W. Skelton and A. H. White, *J. Chem. Soc., Dalton Trans.*, 1984, 1801.
169. L. M. Khalilov, A. A. Panasenko, R. R. Muslukhov, A. G. Ibragimov, G. A. Tolstikov and U. M. Dzhemilev, *Bull. Acad. Sci. USSR, Div. Chem. Sci. (Engl. Trans.)*, 1988, **37**, 458; R. R. Muslukhov, L. M. Khalilov, A. A. Ibragimov, U. M. Dzhemilev and A. A. Panasenko, *Organomet. Chem. USSR (Engl. Transl.)*, 1988, **1**, 378.
170. T. Holm, *J. Chem. Soc., Perkin Trans. 2*, 1981, 464.
171. S. Sakai and K. D. Jordan, *J. Am. Chem. Soc.*, 1982, **104**, 4019.
172. J. Kaneti, P. von R. Schleyer and A. J. Kos, *J. Chem. Soc., Chem. Commun.*, 1985, 1014
173. J. DeYoung, H. Kawa and R. J. Lagow, *J. Chem. Soc., Chem. Commun.*, 1992, 811.
174. Z. Bregar, S. Stavber and M. Zupan, *J. Fluorine Chem.*, 1989, **44**, 187.
175. H. Yamazaki and N. Hayashi, *Chem. Lett.*, 1993, 525.
176. (a) P. Carloni, L. Greci, P. Stipa and L. Eberson, *J. Org. Chem.*, 1991, **56**, 4733; (b) B. S. Tanaseichuk, A. I. Belozerov, E. P. Sanaeva and K. P. Butin, *J. Org. Chem. USSR (Engl. Transl.)*, 1983, **19**, 494.
177. H. Kiljunen and T. A. Hase, *J. Org. Chem.*, 1991, **56**, 6950.
178. Y. Aso, H. Yamashita, T. Otsubo and F. Ogura, *J. Org. Chem.*, 1989, **54**, 5627.
179. D. E. Bergbreiter and E. Pendergrass, *J. Org. Chem.*, 1981, **46**, 219.
180. E. A. Hill, D. C. Link and P. Donndelinger, *J. Org. Chem.*, 1981, **46**, 1177.
181. (a) E. A. Hill and B. Li, *J. Organomet. Chem.*, 1993, **448**, 9; (b) E. A. Hill and M. W. Hallade, *ibid.*, 1988, **352**, 263.
182. K. Utimoto, K. Imi, H. Shiragami, S. Fujikura and H. Nozaki, *Tetrahedron Lett.*, 1985, **26**, 2101.
183. S. Fujikura, M. Inoue, K. Utimoto and H. Nozaki, *Tetrahedron Lett.*, 1984, **25**, 1999.
184. E. A. Hill, K. Hsieh, K. Condroski, H. Sonnentag, D. Skalitzky and D. Gagas, *J. Org. Chem.*, 1989, **54**, 5286.
185. W. Oppolzer, *Angew. Chem., Int. Ed. Engl.*, 1989, **28**, 38 and references therein.
186. (a) E. C. Ashby, *Pure Appl. Chem.*, 1980, **52**, 545; (b) E. C. Ashby and J. R. Bowers, Jr., *J. Am. Chem. Soc.*, 1981, **103**, 2242.
187. T. Holm, *Acta Chem. Scand., Ser. B*, 1983, **37**, 567.
188. M. Dagonneau, *Bull. Soc. Chim. Fr., Ser. II*, 1982, 269.
189. S. Nagase and Y. Uchibori, *Tetrahedron Lett.*, 1982, **23**, 2585.
190. T. Holm, *J. Am. Chem. Soc.*, 1993, **115**, 916; *Acta Chem. Scand., Ser B*, 1988, **42**, 685; T. Holm and J. Ø. Madsen, *ibid.*, 1992, **46**, 985.
191. H. Yamataka, T. Matsuyama and T. Hanafusa, *J. Am. Chem. Soc.*, 1989, **111**, 4912; H. Yamataka, N. Miyano and T. Hanafusa, *J. Org. Chem.*, 1991, **56**, 2573.
192. (a) E. C. Ashby and A. B. Goel, *J. Am. Chem. Soc.*, 1981, **103**, 4983; (b) T. Holm, *Acta Chem. Scand., Ser. B*, 1982, **36**, 266
193. K. Maruyama and T. Katagiri, *J. Phys. Org. Chem.*, 1991, **4**, 381; *ibid.*, 1989, **2**, 205; *ibid.*, 1988, **1**, 21.
194. C. Walling, *J. Am. Chem. Soc.*, 1988, **110**, 6846.
195. A. G. Pinkus and A. Sabesan, *J. Chem. Soc., Perkin Trans. 2*, 1981, 473.
196. C. Bernardon, *J. Organomet. Chem.*, 1989, **367**, 11; R. A. Benkeser and D. C. Snyder, *J. Org. Chem.*, 1982, **47**, 1243.
197. T. Holm, *Acta Chem. Scand.*, 1991, **45**, 925.
198. (a) T. Holm, *Acta Chem. Scand., Ser. B*, 1987, **41**, 278; (b) K. Maruyama and T. Katagiri, *J. Phys. Org. Chem.*, 1991, **4**, 178.
199. L. Jalander and R. Strandberg, *Acta Chem. Scand., Ser B*, 1983, **37**, 15.
200. R. Noyori and M. Kitamura, *Angew. Chem., Int. Ed. Engl.*, 1991, **30**, 49.
201. B. Weber and D. Seebach, *Angew. Chem., Int. Ed. Engl.*, 1992, **31**, 84.

202. M. Falorni, L. Lardicci, G. Uccello Barretta and G. Giacomelli, *Gazz. Chim. Ital.*, 1988, **118**, 495.
203. W. H. Miles, S. L. Rivera and J. D. Delrosario, *Tetrahedron Lett.*, 1992, **33**, 305.
204. B. Reitstoen, L. Kilaas and T. Anthonsen, *Acta Chem. Scand., Ser. B*, 1986, **40**, 440; Y. Yamamoto and K. Maruyama, *J. Am. Chem. Soc.*, 1985, **107**, 6411.
205. (a) B. F. Bonini, S. Masiero, G. Mazzanti and P. Zani, *Tetrahedron Lett.*, 1991, **32**, 6801; (b) E. Buschmann, B. Zeeh, N. Götz, L. Sproesser and N. Walker, *Liebigs Ann. Chem.*, 1987, 349.
206. B. E. Rossiter and N. M. Swingle, *Chem. Rev.*, 1992, **92**, 771.
207. N. E. Agafonov, G. A. Stashina and V. M. Zhulin, *Bull. Acad Sci. USSR, Div. Chem. Sci. (Engl. Transl.)*, 1991, **40**, 1724.
208. (a) A. G. Schultz and L. Flood, *J. Org. Chem.*, 1986, **51**, 838; T. Holm, *Acta Chem. Scand.*, 1991, **45**, 276; (b) T. Holt, *ibid.*, 1990, **44**, 279.
209. G. Bartoli, *Acc. Chem. Res.*, 1984, **17**, 109; G. Bartoli, R. Dalpozzo and L. Grossi, *J. Chem. Soc., Perkin Trans. 2*, 1989, 573.
210. K. Muraoka, M. Nojima, S. Kusabayashi and S. Nagase, *J. Chem. Soc., Perkin Trans. 2*, 1986, 761; G. A. Russell and P. Ngoviwatchai, *Tetrahedron Lett.*, 1987, **28**, 6113; J. J. Eisch, M. Behrooz and J. E. Galle, *ibid.*, 1984, **25**, 4851.
211. (a) S. K. Taylor, S. G. Bennett, K. J. Heinz and L. K. Lashley, *J. Org. Chem.*, 1981, **46**, 2194; (b) U. Anton and K. Müllen, *Makromol. Chem. Rapid Commun.*, 1993, **14**, 223.
212. T. Imamoto, T. Hatajima and K. Ogata, *Tetrahedron Lett.*, 1991, **32**, 2787; M. T. Crimmins and D. Dedopoulou, *Synth. Commun.*, 1992, **22**, 1953.
213. Y. M. Zhang and T. C. Liu, *Synth. Commun.*, 1988, **18**, 2173.
214. G. Cahiez and B. Laboue, *Tetrahedron Lett.*, 1992, **33**, 4439.
215. A. Yanagisawa, N. Nomura and H. Yamamoto, *Synlett.*, 1991, 513; F. Babudri, A. Dettole, V. Fiandanese, G. Marchese and F. Naso, *J. Organomet. Chem.*, 1991, **405**, 53.
216. T. Hayashi and M. Kumada, *Acc. Chem. Res.*, 1982, **15**, 395; M. Sawamura and Y. Ito, *Chem. Rev.* 1992, **92**, 857.
217. G. Consiglio and A. Indolese, *Organometallics*, 1991, **10**, 3425; P. A. Ashworth, N. J. Dixon, P. J. Kocienski and S. N. Wadman, *J. Chem. Soc., Perkin Trans. 1*, 1992, 3431; W. L. Cheng and T. Y. Luh, *J. Chem. Soc., Chem. Commun.*, 1992, 1392.
218. P. L. Castle and D. A. Widdowson, *Tetrahedron Lett.*, 1986, **27**, 6013; K. Yuan and W. J. Scott, *J. Org. Chem.*, 1990, **55**, 6188.
219. H. Kotsuki, I. Kadota and M. Ochi, *J. Org. Chem.*, 1990, **55**, 4417; G. Cahiez and M. Alami, *Tetrahedron Lett.*, 1990, **31**, 7425; H. Sakata, Y. Aoki and I. Kuwajima, *ibid.*, 1990, **31**, 1161.
220. T. E. Cole and B. D. Haly, *Organometallics*, 1992, **11**, 652; H. C. Brown, N. Vasumathi and N. N. Joshi, *ibid.*, 1993, **12**, 1058.
221. A. Boudin, G. Cerveau, C. Chuit, R. J. P. Corriu and C. Reye, *Angew. Chem., Int. Ed. Engl.*, 1986, **25**, 473; G. Cerveau, C. Chuit, R. J. P. Corriu and C. Reye, *Organometallics*, 1991, **10**, 1510.
222. S. Roller, D. Simon and M. Dräger, *J. Organomet. Chem.*, 1986, **301**, 27.
223. L. I. Zakharkin and E. V. Anikina, *Bull. Acad. Sci. USSR, Div. Chem. Sci. (Engl. Transl.)*, 1990, **39**, 635.
224. H. R. Allcock, J. L. Desorcie and L. J. Wagner, *Inorg. Chem.*, 1985, **24**, 333.
225. M. F. Lappert, W.-P. Leung, C. L. Raston, B. W. Skelton and A. H. White, *J. Chem. Soc., Dalton Trans.*, 1992, 775.
226. P. Sobota, J. Utko and Z. Janas, *J. Organomet. Chem.*, 1986, **316**, 19.
227. E. Negishi, F. E. Cederbaum and I. Takahashi, *Tetrahedron Lett.*, 1986, **27**, 2829.
228. R. Benn *et al.*, *Polyhedron*, 1990, **9**, 11; R. Benn, T. H. Brock, P. W. Jolly, A. Rufinska and G. Schroth, *ibid.*, 1990, **9**, 23.
229. T. C. Forschner and N. J. Cooper, *J. Am. Chem. Soc.*, 1989, **111**, 7420.
230. M. Yamashita, K. Miyoshi, Y. Nakazono and R. Suemitsu, *Bull. Chem. Soc. Jpn.*, 1982, **55**, 1663.
231. D. Rakshit and S. E. Thomas, *J. Organomet. Chem.*, 1987, **333**, C3.
232. U. Wecker and H. Werner, *J. Organomet. Chem.*, 1992, **424**, 199; H. Werner, S. Stahl and M. Schulz, *Chem. Ber.*, 1991, **124**, 707.
233. S.-T. Lin, R. N. Narske and K. J. Klabunde, *Organometallics*, 1985, **4**, 571.
234. M. F. Lappert, T. R. Martin and C. L. Ralston, *Inorg. Synth.*, 1989, **26**, 144.
235. Y. Saheki, K. Sasada, N. Satoh, N. Kawaichi and K. Negoro, *Chem. Lett.*, 1987, 2299.
236. K. Lühder, D. Nehls and K. Madeja, *J. Prakt. Chem.*, 1983, **325**, 1027.
237. M. A. G. M. Tinga *et al.*, *J. Am. Chem. Soc.*, 1993, **115**, 2808.
238. H. J. R. de Boer, O. S. Akkerman and F. Bickelhaupt, in 'Organometallic Syntheses', eds. R. B. King and J. J. Eisch, Elsevier, Amsterdam, 1988, vol. 4, p. 396.
239. F. Bickelhaupt and O. S. Akkerman, in 'Organometallic Syntheses', eds. R. B. King and J. J. Eisch, Elsevier, Amsterdam, 1986, vol. 3, p. 403.
240. C. W. Kamienski and B. J. McElroy, in 'Organometallic Syntheses', eds. R. B. King and J. J. Eisch, Elsevier, Amsterdam, 1986, vol. 3, p. 395.
241. C. G. Screttas and M. Micha-Screttas, *J. Organomet. Chem.*, 1985, **292**, 325.
242. U. M. Dzhemilev, A. G. Ibragimov, E. V. Gribanova and L. M. Khalilov, *Bull. Acad. Sci. USSR, Div. Chem. Sci. (Engl. Transl.)*, 1988, **37**, 347.
243. (a) J. J. Eisch and R. Sanchez, *J. Organomet. Chem.*, 1985, **296**, C27; (b) R. Sanchez and J. J. Eisch, in 'Organometallic Syntheses', eds. R. B. King and J. J. Eisch, Elsevier, Amsterdam, 1986, vol. 3, pp. 391; 393; 400.
244. A. W. Duff, P. B. Hitchcock, M. F. Lappert, R. G. Taylor and J. A. Segal, *J. Organomet. Chem.*, 1985, **293**, 271.
245. M. G. Gardiner, C. L. Raston and C. H. L. Kennard, *Organometallics*, 1991, **10**, 3680.
246. M. J. Henderson, R. I. Papasergio, C. L. Raston, A. H. White and M. F. Lappert, *J. Chem. Soc., Chem. Commun.*, 1986, 672.
247. R. A. Andersen, R. Blom, A. Haaland, B. E. R. Schilling and H. V. Volden, *Acta Chem. Scand., Ser. A*, 1985, **39**, 563.
248. R. Benn, H. Lehmkuhl, K. Mehler and A. Rufinska, *J. Organomet. Chem.*, 1985, **293**, 1.
249. B. Bogdanovic, G. Koppetsch, C. Krüger and R. Mynott, *Z. Naturforsch., Teil B*, 1986, **41**, 617.
250. (a) B. Bogdanovic, *Angew. Chem., Int. Ed. Engl.*, 1985, **24**, 262; B. Bogdanovic, M. Schwickardi and P. Sikorski, *ibid*, 1982, **21**, 199; (b) B. Bogdanovic, M. Schwickardi and U. Westeppe, in 'Organometallic Syntheses', eds. R. B. King and J. J. Eisch, Elsevier, Amsterdam, 1988, vol. 4, p. 399; B. Bogdanovic, G. Koppetsch and M. Schwickardi, *ibid.*, vol. 4, p. 404.

251. U. M. Dzhemilev, O. S. Vostrikova, R. M. Sultanov, A. G. Kukovinets and L. M. Khalilov, *Bull. Acad. Sci. USSR, Div. Chem. Sci. (Engl. Transl.)*, 1984, **33**, 1873; U. M. Dzhemilev, O. S. Vostrikova and R. M. Sultanov, *ibid.*, 1983, **32**, 193.
252. U. M. Dzhemilev, R. M. Sultanov, R. G. Gaimaldinov, R. R. Muslukhov, S. I. Lomakina and G. A. Tolstikov, *Bull. Russ. Acad. Sci., Div. Chem. Sci. (Eng. Transl.)*, 1992, **41**, 770; U. M. Dzhemilev, R. M. Sultanov, R. G. Gaimaldinov and G. A. Tolstikov, *Bull. Acad. Sci. USSR, Div. Chem. Sci. (Engl. Transl.)*, 1991, **40**, 1229;
253. W. Clegg, D. A. Brown, S. J. Bryan and K. Wade, *J. Organomet. Chem.*, 1987, **325**, 39.
254. P. R. Markies *et al.*, *J. Organomet. Chem.*, 1990, **393**, 315.
255. M. Parvez, A. D. Pajerski and H. G. Richey, Jr., *Acta Crystallogr., Sect. C*, 1988, **44**, 1212.
256. A. D. Pajerski, G. L. BergStresser, M. Parvez and H. G. Richey, Jr., *J. Am. Chem. Soc.*, 1988, **110**, 4844.
257. P. R. Markies, T. Nomoto, O. S. Akkerman, F. Bickelhaupt, W. J. J. Smeets and A. L. Spek, *J. Am. Chem. Soc.*, 1988, **110**, 4845.
258. A. D. Pajerski, M. Parvez and H. G. Richey, Jr., *J. Am. Chem. Soc.*, 1988, **110**, 2660.
259. E. P. Squiller, R. R. Whittle and H. G. Richey, Jr., *J. Am. Chem. Soc.*, 1985, **107**, 432.
260. K. Angermund *et al.*, *Z. Naturforsch., Teil B*, 1986, **41**, 455.
261. P. R. Markies *et al.*, *J. Organomet. Chem.*, 1991, **411**, 291.
262. P. R. Markies, T. Nomoto, O. S. Akkerman, F. Bickelhaupt, W. J. J. Smeets and A. L. Spek, *Angew. Chem., Int. Ed. Engl.*, 1988, **27**, 1084.
263. M. Geissler, J. Kopf and E. Weiss, *Chem. Ber.*, 1989, **122**, 1395.
264. B. Schubert, U. Behrens and E. Weiss, *Chem. Ber.*, 1981, **114**, 2640.
265. H. Lehmkuhl, K. Mehler, R. Benn, A. Rufinska and C. Krüger, *Chem. Ber.*, 1986, **119**, 1054.
266. C. J. Burns and R. A. Andersen, *J. Organomet. Chem.*, 1987, **325**, 31.
267. G. Helary, L. Lefevre-Jenot, M. Fontanille and J. Smid, *J. Organomet. Chem.*, 1981, **205**, 139.
268. H. G. Richey, Jr. and D. M. Kushlan, *J. Am. Chem. Soc.*, 1987, **109**, 2510.
269. R. A. Andersen, R. Blom, J. M. Boncella, C. J. Burns and H. V. Volden, *Acta Chem. Scand., Ser. A*, 1987, **41**, 24.
270. O. S. Akkerman, G. Schat, E. A. I. M. Evers and F. Bickelhaupt, *Recl. Trav. Chim. Pays-Bas*, 1983, **102**, 109.
271. K. Faegri, Jr., J. Almlöf and H. P. Lüthi, *J. Organomet. Chem.*, 1983, **249**, 303.
272. (a) P. Sobota and M. Nowak, *J. Organomet. Chem.*, 1988, **340**, 1; (b) P. Sobota, M. Nowak and W. Kramarz, *ibid.*, 1984, **275**, 161.
273. W. Kaim, *Acc. Chem. Res.*, 1985, **18**, 160.
274. S. V. Frye, E. L. Eliel and R. Cloux, *J. Am. Chem. Soc.*, 1987, **109**, 1862.
275. H. G. Richey, Jr. and J. P. DeStephano, *J. Org. Chem.*, 1990, **55**, 3281.
276. H. G. Richey, Jr. and B. A. King, *J. Am. Chem. Soc.*, 1982, **104**, 4672.
277. H. G. Richey, Jr. and J. Farkas, Jr., *Organometallics*, 1990, **9**, 1778
278. G.-J. M. Gruter, G. P. M. van Klink, O. S. Akkerman and F. Bickelhaupt, *Organometallics*, 1993, **12**, 1180.
279. E. P. Squiller, R. R. Whittle and H. G. Richey, Jr., *Organometallics*, 1985, **4**, 1154.
280. (a) S. I. Troyanov, V. Varga and K. Mach, *Organometallics*, 1993, **12**, 2820; (b) V. Varga, K. Mach, G. Schmid and U. Thewalt, *J. Organomet. Chem.*, 1993, **454**, C1.
281. H. Yasuda, Y. Kajihara, K. Mashima, K. Nagasuna, K. Lee and A. Nakamura, *Organometallics*, 1982, **1**, 388.
282. R. D. Rieke and H. Xiong, *J. Org. Chem.*, 1991, **56**, 3109; H. Xiong and R. D. Rieke, *ibid.*, 1989, **54**, 3247.
283. R. D. Rieke and H. Xiong, *J. Org. Chem.*, 1992, **57**, 6560; H. Xiong and R. D. Rieke, *Tetrahedron Lett.*, 1991, **32**, 5269.
284. H. Xiong and R. D. Rieke, *J. Am. Chem. Soc.*, 1992, **114**, 4415.
285. H. Xiong and R. D. Rieke, *J. Org. Chem.*, 1992, **57**, 7007.
286. Y. Kai *et al.*, *Chem. Lett.*, 1982, 1277.
287. B. Bogdanovic, N. Janke, H.-G. Kinzelmann and U. Westeppe, *Chem. Ber.*, 1988, **121**, 33.
288. G. E. Herberich *et al.*, *Chem. Ber.* 1986, **119**, 420.
289. W. J. Richter, *Chem. Ber.*, 1983, **116**, 3293.
290. W. J. Richter, *J. Organomet. Chem.*, 1985, **289**, 45.
291. H. Yasuda, K. Tatsumi and A. Nakamura, *Acc. Chem. Res.*, 1985, **18**, 120; G. Erker, C. Krüger and G. Müller, *Adv. Organomet. Chem.* , 1985, **24**, 1.
292. H. Fjellvåg and P. Karen, *Inorg. Chem.*, 1992, **31**, 3260.
293. O. Gropen, A. Haaland and D. Defrees, *Acta Chem. Scand., Ser. A*, 1985, **39**, 367.
294. T. Alonso, S. Harvey, P. C. Junk, C. L. Raston, B. W. Skelton and A. H. White, *Organometallics*, 1987, **6**, 2110.
295. W. M. Brooks, C. L. Raston, R. E. Sue, F. J. Lincoln and J. J. McGinnity, *Organometallics*, 1991, **10**, 2098.
296. W. J. Richter, *Chem. Ber.*, 1985, **118**, 97.
297. B. Bogdanovic, S.-T. Liao, R. Mynott, K. Schlichte and U. Westeppe, *Chem. Ber.*, 1984, **117**, 1378.
298. P. K. Freeman and L. L. Hutchinson, *J. Org. Chem.*, 1983, **48**, 879.
299. B. Bogdanovic, N. Janke, C. Krüger, R. Mynott, K. Schlichte and U. Westeppe, *Angew. Chem., Int. Ed. Engl.*, 1985, **24**, 960.
300. H. Lehmkuhl, A. Shakoor, K. Mehler, C. Krüger, K. Angermund and Y.-H. Tsay, *Chem. Ber.*, 1985, **118**, 4239.
301. R. Benn *et al.*, *Chem. Ber.*, 1993, **126**, 225.
302. B. Bogdanovic, N. Janke, C. Krüger, K. Schlichte and J. Treber, *Angew. Chem., Int. Ed. Engl.*, 1987, **26**, 1025.
303. B. Bogdanovic, K. Schlichte and U. Westeppe, *Chem. Ber.*, 1988, **121**, 27.
304. A. Stanger and K. P. C. Vollhardt, *Organometallics*, 1992, **11**, 317.
305. A. B. Goel and E. C. Ashby, *J. Organomet. Chem.*, 1981, **214**, C1.
306. T. Clark, C. Rohde and P. von R. Schleyer, *Organometallics*, 1983, **2**, 1344.
307. A. B. Goel and E. C. Ashby, *Inorg. Chim. Acta*, 1984, **87**, 61.
308. O. N. D. Mackey and C. P. Morley, *J. Organomet. Chem.*, 1992, **426**, 279.
309. A. Maerker, in 'Organometallic Syntheses', eds. R. B. King and J. J. Eisch, Elsevier, Amsterdam, 1986, vol. 3, p. 398.
310. (a) E. Bartmann, *J. Organomet. Chem.*, 1987, **332**, 19; (b) E. Bartmann, *ibid.*, 1985, **284**, 149.
311. J.-G. Duboudin, B. Jousseaume and E. Thoumazeau, *Bull. Soc. Chim. Fr., Ser. II*, 1983, 105.
312. P. N. Kapoor, A. K. Bhagi, H. K. Sharma and R. N. Kapoor, *J. Organomet. Chem.*, 1989, **369**, 281.
313. M. T. Reetz, N. Harmat and R. Mahrwald, *Angew. Chem., Int. Ed. Engl.*, 1992, **31**, 342.
314. N. Kuhn, M. Schulten, R. Boese and D. Bläser, *J. Organomet. Chem.*, 1991, **421**, 1.

315. L. M. Engelhardt, B. S. Jolly, P. C. Junk, C. L. Raston, B. W. Skelton and A. H. White, *Aust. J. Chem.*, 1986, **39**, 1337.
316. K. W. Henderson, R. E. Mulvey, W. Clegg and P. A. O'Neil, *J. Organomet. Chem.*, 1992, **439**, 237.
317. J. J. Lalonde, D. E. Bergbreiter and M. Newcomb, *J. Org. Chem.*, 1986, **51**, 1369.
318. O. S. Vostrikova, R. M. Sultanov and U. M. Dzhemilev, *Bull. Acad. Sci. USSR, Div. Chem. Sci. (Engl. Transl.)*, 1983, **32**, 1724.
319. W. Kaim, *Angew. Chem., Int. Ed. Engl.*, 1982, **21**, 140; *J. Am. Chem. Soc.*, 1982, **104**, 3833; *J. Organomet. Chem.*, 1983, **241**, 157.
320. H. Böhland, F. R. Hofmann, W. Hanay and H. J. Berner, *Z. Anorg. Allg. Chem.*, 1989, **577**, 53.
321. H.-F. Klein, H. König, S. Koppert, K. Ellrich and J. Riede, *Organometallics*, 1987, **6**, 1341.
322. R. Han and G. Parkin, *Organometallics*, 1991, **10**, 1010; *J. Am. Chem. Soc.*, 1992, **114**, 748; *ibid.*, 1990, **112**, 3662.
323. T. Greiser, J. Kopf, D. Thoennes and E. Weiss, *Chem. Ber.*, 1981, **114**, 209.
324. B. Schubert and E. Weiss, *Chem. Ber.*, 1984, **117**, 366.
325. N. H. Buttrus *et al.*, *J. Chem. Soc., Dalton Trans.*, 1988, 381.
326. E. M. Hanawalt and H. G. Richey, Jr., *J. Am. Chem. Soc.*, 1990, **112**, 4983.
327. C. G. Screttas and M. Micha-Screttas, *J. Organomet. Chem.*, 1985, **290**, 1; *ibid.*, 1986, **316**, 1; C. G. Screttas and B. R. Steele, *J. Org. Chem.*, 1988, **53**, 5151.
328. C. Fehr and J. Galindo, *Helv. Chim. Acta*, 1986, **69**, 228.
329. R. M. Fabicon, A. D. Pajerski and H. G. Richey, Jr., *J. Am. Chem. Soc.*, 1991, **113**, 6680.
330. R. A. Kjonaas and R. K. Hoffer, *J. Org. Chem.*, 1988, **53**, 4133; W. Tückmantel, K. Oshima and H. Nozaki, *Chem. Ber.*, 1986, **119**, 1581.
331. P. Knochel and J. F. Normant, *Tetrahedron Lett.*, 1986, **27**, 4427; 5727; P. Knochel, C. Xiao and M. C. P. Yeh, *ibid.*, 1988, **29**, 6697.
332. N. N. Korneev, S. L. Gershkokhen and G. I. Shcherbakova, *J. Gen. Chem. USSR (Engl. Transl.)*, 1982, **52**, 535.
333. V. V. Gavrilenko, L. A. Chekulaeva and L. I. Zakharkin, *Bull. Acad. Sci. USSR, Div. Chem. Sci. (Engl. Transl.)*, 1987, **36**, 605.
334. S. L. Gershkokhen, N. N. Korneev and G. I. Shcherbakova, *J. Gen. Chem. USSR (Engl. Transl.)*, 1982, **52**, 2246; N. N. Korneev, S. L. Gershkokhen and G. I. Shcherbakova, *ibid.*, 1982, **52**, 2351.
335. K. B. Starowieyski and A. Chwojnowski, *J. Organomet. Chem.*, 1981, **215**, 151.
336. H. Lehmkuhl, K. Mehler, R. Benn, A. Rufinska, G. Schroth and C. Krüger, *Chem. Ber.*, 1984, **117**, 389.
337. H. Lehmkuhl *et al.*, *Chem. Ber.*, 1985, **118**, 4248.
338. K. Mochida, S. Ogura and T. Yamanishi, *Bull. Chem. Soc. Jpn.*, 1986, **59**, 2633.
339. K. Mochida, *J. Synth. Org. Chem. Jpn.*, 1988, **46**, 704.
340. K. Mochida and T. Yamanishi, *J. Organomet. Chem.*, 1987, **332**, 247.
341. M. J. McCormick, K. B. Moon, S. R. Jones and T. P. Hanusa, *J. Chem. Soc., Chem. Commun.*, 1990, 778.
342. T.-C. Wu, H. Xiong and R. D. Rieke, *J. Org. Chem.*, 1990, **55**, 5045.
343. H. M. Walborsky and C. Hamdouchi, *J. Org. Chem.*, 1993, **58**, 1187.
344. R. A. O'Brien, T. Chen and R. D. Rieke, *J. Org. Chem.*, 1992, **57**, 2667.
345. A. Yanagisawa, S. Habaue and H. Yamamoto, *J. Am. Chem. Soc.*, 1991, **113**, 8955.
346. K. Mochida and H. Ogawa, *J. Organomet. Chem.*, 1983, **243**, 131.
347. M. J. McCormick, S. C. Sockwell, C. E. H. Davies, T. P. Hanusa and J. C. Huffman, *Organometallics*, 1989, **8**, 2044.
348. D. J. Burkey, R. A. Williams and T. P. Hanusa, *Organometallics*, 1993, **12**, 1331.
349. S. C. Sockwell, T. P. Hanusa and J. C. Huffman, *J. Am. Chem. Soc.*, 1992, **114**, 3393.
350. F. G. N. Cloke, P. B. Hitchcock, M. F. Lappert, G. A. Lawless and B. Royo, *J. Chem. Soc., Chem. Commun.*, 1991, 724.
351. W. E. Lindsell, F. C. Robertson and I. Soutar, *Eur. Polym. J.*, 1983, **19**, 115.
352. A. A. Arest-Yakubovich, *Russ. Chem. Rev. (Engl. Transl.)*, 1981, **50**, 601.
353. M. Kaupp and P. von R. Schleyer, *J. Am. Chem. Soc.*, 1992, **114**, 491.
354. L. C. Tang, C. Mathis and B. François, *J. Organomet. Chem.*, 1983, **243**, 359.
355. R. Nuffer and B. François, *J. Solution Chem.*, 1982, **11**, 91.
356. L. C. Tang and B. François, *Eur. Polym. J.*, 1983, **19**, 707; 715.
357. (a) A. A. Arest-Yakubovich, R. V. Basova, B. I. Nakhmanovich and E. V. Kristal'nyi, *Acta Polymerica*, 1984, **35**, 1; (b) B. I. Nakhmanovich and A. A. Arest-Yakubovich, *Vysokomol. Soedin., Ser. B*, 1984, **26**, 476.
358. R. V. Basova, Z. M. Baidakova, E. V. Kristal'nyi, B. I. Nakhmanovich and A. A. Arest-Yakubovich, *Organomet. Chem. USSR (Engl. Transl.)*, 1988, **1**, 432.
359. B. I. Nakhmanovich, E. V. Kristal'nyi, R. V. Basova, Z. M. Baidakova, and A. A. Arest-Yakubovich, *Dokl. Akad. Nauk SSSR*, 1980, **255**, 364.
360. S. R. Drake and D. J. Otway, *J. Chem. Soc., Chem. Commun.*, 1991, 517; *Polyhedron*, 1992, **11**, 745.
361. M. J. McCormick, R. A. Williams, L. J. Levine and T. P. Hanusa, *Polyhedron*, 1988, **7**, 725.
362. A. Hammel, W. Schwarz and J. Weidlein, *J. Organomet. Chem.*, 1989, **378**, 347.
363. L. M. Engelhardt, P. C. Junk, C. L. Raston and A. H. White, *J. Chem. Soc., Chem. Commun.*, 1988, 1500.
364. P. Jutzi, W. Leffers, G. Müller and B. Huber, *Chem. Ber.*, 1989, **122**, 879.
365. R. A. Williams, K. F. Tesh and T. P. Hanusa, *J. Am. Chem. Soc.*, 1991, **113**, 4843.
366. P. Jutzi, J. Dahlhaus and M. O. Kristen, *J. Organomet. Chem.*, 1993, **450**, C1.
367. R. A. Williams, T. P. Hanusa and J. C. Huffman, *Organometallics*, 1990, **9**, 1128; *J. Chem. Soc., Chem. Commun.*, 1988, 1045.
368. R. A. Andersen, J. M. Boncella, C. J. Burns, R. Blom, A. Haaland and H. V. Volden, *J. Organomet. Chem.*, 1986, **312**, C49.
369. R. Blom, K. Faegri, Jr. and H. V. Volden, *Organometallics*, 1990, **9**, 372; R. A. Andersen, R. Blom, C. J. Burns and H. V. Volden, *J. Chem. Soc., Chem. Commun.*, 1987, 768.
370. R. A. Williams, T. P. Hanusa and J. C. Huffman, *J. Am. Chem. Soc.*, 1990, **112**, 2454.
371. S. C. Sockwell, P. S. Tanner and T. P. Hanusa, *Organometallics*, 1992, **11**, 2634.
372. R. A. Williams, T. P. Hanusa and J. C. Huffman, *J. Organomet. Chem.*, 1992, **429**, 143.
373. M. Kaupp, P. von R. Schleyer, M. Dolg and H. Stoll, *J. Am. Chem. Soc.*, 1992, **114**, 8202.
374. J. V. Ortiz, *J. Am. Chem. Soc.*, 1991, **113**, 3593.
375. C. Mösges, F. Hampel and P. von R. Schleyer, *Organometallics*, 1992, **11**, 1769.

376. C. Mathis, B. François and J. Smid, *J. Organomet. Chem.*, 1981, **217**, 291.
377. H. Schmidbaur, U. Deschler and D. Seyferth, *Z. Naturforsch., Teil B*, 1982, **37**, 950.
378. (a) R. Khattar, C. B. Knobler and M. F. Hawthorne, *J. Am. Chem. Soc.*, 1990, **112**, 4962; (b) R. Khattar, C. B. Knobler and M. F. Hawthorne, *Inorg. Chem.*, 1990, **29**, 2191.
379. D. S. Hutchings, P. C. Junk, W. C. Patalinghug, C. L. Raston and A. H. White, *J. Chem. Soc., Chem. Commun.*, 1989, 973.
380. S. A. Kinsley, A. Streitwieser, Jr. and A. Zalkin, *Organometallics*, 1985, **4**, 52.
381. H. Bönnemann *et al.*, *Main Group Met. Chem.*, 1990, **13**, 341.
382. G. R. Stevenson, L. E. Schock, R. Concepcion and R. A. Peterson, *J. Phys. Chem.*, 1985, **89**, 175; G. R. Stevenson and L. E. Schock, *J. Am. Chem. Soc.*, 1983, **105**, 3742; G. R. Stevenson, M. P. Espe and T. Emanuelson, *J. Org. Chem.*, 1985, **50**, 4289.
383. K. Mochida, Y. Hiraga, H. Takeuchi and H. Ogawa, *Organometallics*, 1987, **6**, 2293.
384. K. Mochida and Y. Mizuno, *Chem. Lett.*, 1986, 1125; *Bull. Chem. Soc. Jpn.*, 1987, **60**, 273.
385. M. G. Gardiner, G. R. Hanson, P. C. Junk, C. L. Raston, B. W. Skelton and A. H. White, *J. Chem. Soc., Chem. Commun.*, 1992, 1154.
386. V. N. Cherkasov and L. N. Cherkasov, *Organomet. Chem. USSR (Engl. Transl.)*, 1989, **2**, 147.
387. P. S. Tanner, R. A. Williams and T. P. Hanusa, *Inorg. Chem.*, 1993, **32**, 2234.

4

Compounds with Three- or Four-coordinate Boron, Emphasizing Cyclic Systems

CATHERINE E. HOUSECROFT
Institut für Anorganische Chemie der Universität Basel, Switzerland

4.1 INTRODUCTION 130

4.2 THEORETICAL ASPECTS 131

4.3 DERIVATIVES OF CYCLOALKANES 133

4.3.1 Three-membered Rings including Bicyclic Systems 133
4.3.2 Four-membered Rings 135
4.3.3 Five-membered Rings 135
4.3.4 Six-membered Rings 136
4.3.5 Seven-membered and Larger Rings 137
4.3.6 Bicyclic Systems (excluding Three-membered Rings) 138
4.3.6.1 9-Borabicyclo[3.3.1]nonane (9-BBN) and its derivatives 138
4.3.6.2 Borabicyclononanes (excluding 9-BBN) and borabicyclodecanes 143
4.3.6.3 Miscellaneous simple borabicycloalkanes 143
4.3.7 Boratricycloalkanes and Larger Systems 144
4.3.8 Boraadamantanes 145

4.4 DERIVATIVES OF CYCLOALKENES AND BORA AROMATICS 148

4.4.1 Three-membered Rings 148
4.4.2 Four-membered Rings 149
4.4.3 Five-membered Rings 150
4.4.4 Six-membered Rings 152
4.4.5 Seven- and Eight-membered Rings 153
4.4.6 Bicyclic, Tricyclic and Larger Systems including Boron adjacent to an Aromatic Ring 154

4.5 RINGS CONTAINING BORON AND SILICON ATOMS 157

4.6 RINGS CONTAINING BORON AND TIN, GERMANIUM AND LEAD ATOMS 158

4.7 RINGS CONTAINING BORON AND NITROGEN ATOMS 158

4.7.1 Three-membered Rings 158
4.7.2 Four-membered Rings 159
4.7.3 Five-membered Rings 160
4.7.4 Six-membered Rings (excluding Borazines) 163
4.7.5 Rings in which the Boron Atom is Four Coordinate 165

4.8 RINGS CONTAINING BORON AND PHOSPHORUS ATOMS 169

4.8.1 Three-membered Rings 169
4.8.2 Four-membered Rings 169
4.8.3 Five-membered Rings 169
4.8.4 Six-membered Rings 170
4.8.5 Bicyclic Systems 170
4.8.6 Rings in which the Boron Atom is Four Coordinate 170

4.9 RINGS CONTAINING BORON AND OXYGEN ATOMS 171

4.9.1 Three-membered Rings 171
4.9.2 Four-membered Rings 171
4.9.3 Five-membered Rings 172
4.9.3.1 Monocyclic five-membered ring systems 172
4.9.3.2 Five-membered heterocycles adjacent to aromatic rings 174
4.9.4 Six-membered Rings 174
4.9.5 Eight-membered Rings 177
4.9.6 Bicyclic Systems (excluding Five-membered Heterocycles adjacent to Aromatic Rings) 177
4.9.7 Rings in which the Boron Atom is Four Coordinate 178

4.10 RINGS CONTAINING BORON AND SULFUR OR SELENIUM ATOMS 179

4.10.1 Four-membered Rings 179
4.10.2 Five-membered Rings 179
4.10.3 Six-membered Rings 181

4.11 RINGS CONTAINING BORON AND MORE THAN ONE OTHER HETEROATOM 181

4.11.1 Boron with Silicon or Tin and Nitrogen 181
4.11.2 Boron with Silicon and Oxygen 181
4.11.3 Boron with Nitrogen and Phosphorus, Arsenic or Antimony 182
4.11.4 Boron with Nitrogen and Oxygen 182
4.11.5 Boron with Phosphorus and Oxygen 184
4.11.6 Boron with Nitrogen, Phosphorus and Oxygen 184
4.11.7 Boron with Nitrogen and Sulfur or Selenium 184
4.11.8 Boron with Oxygen and Sulfur 185

4.12 REFERENCES 186

4.1 INTRODUCTION

This chapter concentrates on cyclic boron-containing compounds; it follows the review of Morris[1] and covers the years 1982–1992. No attempt has been made to cover the applications of boranes in organic synthesis; readers are guided to relevant reviews and texts.[2–8] The large number of complexes exhibiting four-coordinate boron has meant that some such compounds have had to be omitted from this review. However, it is hoped that sufficient coverage has been given to enable readers at least to find their way into the relevant literature. Neither has it been possible to give full coverage of noncyclic systems of boron. References to many such compounds are found amongst surveys of organoboron compounds,[2–8] as well as in a survey by Niedenzu[9] and in timely articles by Meller,[10] Berndt[11] and Paetzold[12] (Table 1). Reviews[9–28] on, or including, individual topics within the field of cyclic boron-containing compounds are listed in Table 1. Generally, however, in the present work, suitable secondary sources were not readily available for the period 1982–1992 and the primary literature has been cited.

Table 1 Review articles.

Topic of, or included in, the review article	*Ref.*
Annual survey for boron 1982	9
Borylation of aromatics: use of R_2NBX_2 compounds	10
Classical and nonclassical methyleneboranes	11
Iminoboranes	12
Three-membered rings containing boron	13
Photorearrangements in the synthesis of borirenes and boracarbenoids	14
Benzoannulated cycloboranes	15
Boron with oxygen and sulfur	16
Rings with boron–phosphorus and boron–arsenic bonds (1969–1979 literature)	17
Rings with boron–oxygen bonds (1969–1979 literature)	18
Rings with boron–sulfur and boron–selenium bonds (1969–1979 literature)	19
Chemistry of borinium and borenium ions	20
Chemistry of boraadamantane	21
Chemistry of 9-borabicyclo[3.3.1]nonane dimer, $\{9\text{-BBN}\}_2$	22
Electrochemistry of boron compounds	23
Phosphorus–boron containing compounds	24
Structural types of borates and polyborates	25
Pyrazole derivatives of boron	26
Boron–nitrogen chemistry	27
Boron nitride and its precursors	28

Structural figures in this review are schematic and, for the sake of consistency, π-interactions between boron and adjacent atoms (e.g., nitrogen) have not been shown if the boron atom is otherwise three-coordinate.

4.2 THEORETICAL ASPECTS

Theoretical studies are important in many ways, but in particular in that they may direct the practical chemist towards new goals. Between 1982 and 1992, this has been the case for small ring boron chemistry.[11,13] In this section, research of a primarily theoretical nature is considered, although theoretical studies which overlap with experimental work are incorporated into relevant sections throughout the review.

As part of a study of delocalization in small carbon rings, the relative stabilities of compounds (**1**)–(**3**) have been calculated; of the three isomers, (**1**) is the most stable, in line with expectations from Hückel theory.[29] The results of SINDO1 calculations on a range of heteroatomic $(4n + 2)$ π-ring systems show that (**4**) and (**5**) are highly aromatic, (**6**) possesses slight aromatic character, and (**7**) is antiaromatic. An aromatic index, AI, is determined for each ring, for example, for (**4**), AI = 1.614 and for (**5**), AI = 1.606, while for (**6**) and (**7**), values of AI are 1.347 and 0.899, respectively.[30] Budzelaar and Schleyer have shown that (**6**) possesses a resonance energy of 230 kJ mol^{-1} although the boron–boron π-interaction is very weak.[31] The destabilizing antiaromatic character of the 4π electron ring $(BH)(NH)_2$ can be relieved by the presence of an amino substituent on the boron atom.[31] In the four-membered rings (**8**) and (**9**), remote boron atoms lead to ring folding (calculated to be 56.1°), whereas adjacent boron atoms result in the prediction of a planar ring.[30] Studies of the possible isomers for $C_2B_2H_4$ include systems with four-membered rings and the heterocycle (**10**).[32–4] Modelled by (**10**), the structure of 2,4-diboramethylenecyclopropane (a derivative of which has been synthetically realized)[11,35] is predicted to have the unusual geometry (**11**); the global minimum exhibits a strong B=C interaction.[33,34]

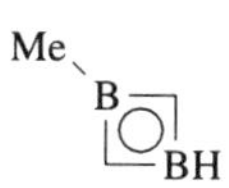
(**1**)

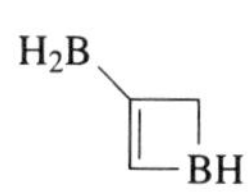
(**2**)

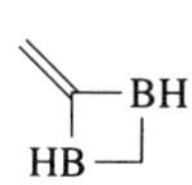
(**3**)

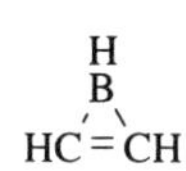
(**4**)

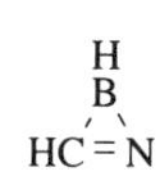
(**5**)

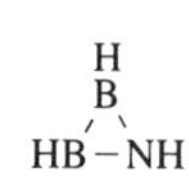
(**6**)

(**7**)

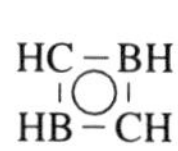
(**8**)

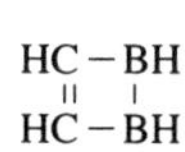
(**9**)

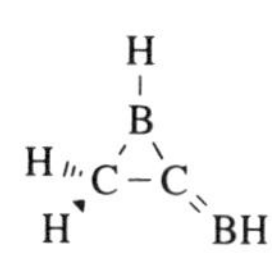
(**10**)

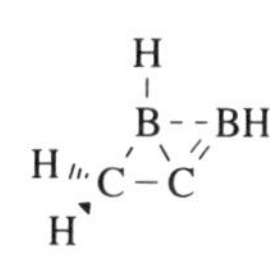
(**11**)

The dimerization of (**4**) may follow one of two possible pathways, both of which involve nonplanar transition states. One path leads to a classical structure for the dimer $C_4B_2H_6$ while the other gives a *nido*-carbaborane. The C_2 path has an activation energy of 46 kJ mol^{-1}.[36] At the 3–21 G level, Frenking[37] has found that structure (**4**) does not represent a potential energy minimum. Independent work by P. von R. Schleyer and co-workers[38] has looked at strain energies and substituent effects in borirenes; for (**4**), strain and resonance energies of 288 and 199 kJ mol^{-1}, respectively, have been calculated at the 6–31G* level. The saturated borirane ring has a higher strain energy (by 75 kJ mol^{-1}) than cyclopropane.[38]

Studies of four-membered rings have included diboretene (**8**), and diaminodiboretane (RHF and GVB levels with split-valence 3–21G basis set). The calculated C···C distance in (**8**) is 0.1964 nm (nonbonding) and this shortens on going to the amino derivative; nonplanar geometries are preferable. Comparisons with related experimental structures are favourable.[39,40] Optimization of the structure of 1,2-dihydroborete (**12**), (MNDO/3) indicates a nonplanar ring; at the 6–31G* level, an internal fold angle (folded along B···C) of 152° is determined. These studies have been extended to include the spirocyclic molecule (**13**),[41] derivatives of which have been synthesized. Related to (**12**) is the boretane C_3H_6BH which also favours a folded-ring structure; cyclic C_3H_6BH has been studied as part of an *ab initio* investigation of allylborane and its isomers.[42]

Ab initio calculations on (**14**) (STO–3G basis set)[43] and on the five-membered heterocycle C_3H_4BF have been carried out.[44] The structure of borabenzene (**15**) with no L, has been optimized (MNDO) by Raabe *et al.*; the B–C bond distance is 0.141 nm. These results show that, due to the presence of a low lying σ* orbital, borabenzene is a strong Lewis acid.[45] The work was extended using *ab initio* (STO–3G and 4–31G basis sets) methods; the stabilization energy is ~126 kJ mol^{-1} less than that of benzene.[46]

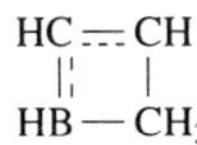

(12)

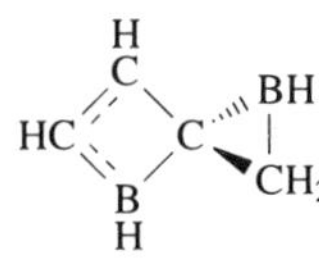

(13)

Vibrational spectral properties of borabenzene have been determined.[46] Cioslowski *et al.* have probed the electronic structures of (**15**) with no L, or L = CO or N_2, at the HF/6–31G* and MP2/6–31G* levels and show that the three compounds should all be stable with planar geometries. Their data support σ-donation and π-back donation in the adducts.[47] The optimized geometry of borepin (**16**) (STO–3G and 4–31G basis sets) is that of a planar ring[48,49] with a weakly conjugated system; optimized bond parameters are: C–C = 0.1345 nm, 0.1451 nm and 0.1344 nm, B–C = 0.1523 nm and C–B–C angle = 124.7°.[48] The relative stability[49,50] of (**16**) with respect to its valence isomers (**17**) and (**18**) is indicated in Table 2.[49] Aspects of the thermochemistry of borabenzene and borepin have been studied (6–31G* level).[51]

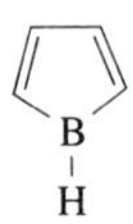

(14)

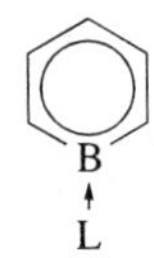

(15)

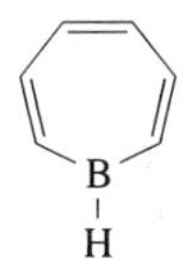

(16)

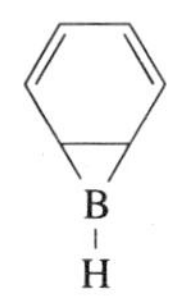

(17)

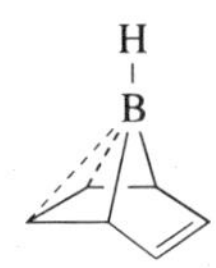

(18)

Table 2 Relative energies (kJ mol^{-1}) for borepin (**16**) and its valence isomers.[48]

Isomer	*Level of calculation*			
	STO-3G(SCF)	4-31G(SCF)	4-31G(RMP2)	6-31G*(SCF)
(**16**)	0.0	0.0	0.0	0.0
(**17**)	107.6	182.9	187.5	157.8
(**18**)	6.7	156.6	112.6	98.8

The calculated (MNDO) heats of formation of the isomers (**19**), (**20**) and (**21**) are −34.1, +48.6, and +69.7 kJ mol^{-1}, respectively. There is evidence that each of (**19**) and (**20**) possesses localized bond character although (**21**) exhibits delocalization; the latter is not uniform, however, and isomer (**21**) is described in terms of two fused allylic-type systems. The cyclotrimer $\{C_4H_4BN\}_3$ has also been studied.[52] The results of SCF-MO calculations on 11 unsaturated boron–nitrogen heterocycles, for example, (**22**), provide an insight into ring current contributions in these systems. Going from a CC to a BN unit causes significant reduction in the aromaticity of a given molecule.[53]

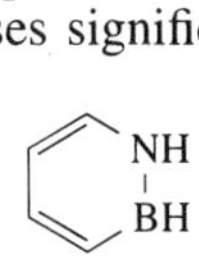

(19)

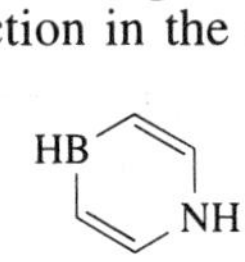

(20)

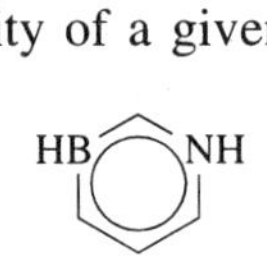

(21)

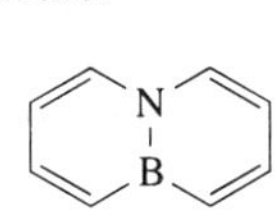

(22)

Structures for pyrazolylborane (H_2Bpz) with the BH_2 group coplanar with, perpendicular to, and bridging the N–N edge of the pyrazole ring have been optimized. For the latter, the B–N distances are 0.1590 nm (6–31G* basis set) and a classical bonding picture for the BNN-ring is suggested.[54] Compounds of type (**23**) are discussed in Section 4.7.5. In 1985, CNDO calculations were used to assess the feasibility of the reaction between R_2BH and pyrazole (Hpz) and the subsequent dimerization of R_2Bpz ($R_2 = H_2$ or $MeNCH_2CH_2NMe$). The results indicate that the reactions are indeed energetically favourable for R = H, but for $R_2 = MeNCH_2CH_2NMe$, although the reaction of R_2BH with Hpz is exothermic, dimerization is dependent on the geometry of the B_2N_4 ring of the dimer.[55]

A discussion of five- and six-membered rings would not be complete without a mention of the possibilities of boron analogues of C_{60}.[56] Theoretical studies on the hypothetical $C_{12}B_{24}N_{24}$ have been carried out; this may be a realistic synthetic target.[57]

Members of a group of molecules (**24**), termed 'trefoil aromatics', exhibit a three-centre–two-electron interaction at the centre of an annulene system. Despite the fact that other valence tautomers might, or will, be preferred to the trefoil structures, results of MNDO/3 and *ab initio* calculations suggest that some of the structures may be reactive intermediates, perhaps isolable.[58]

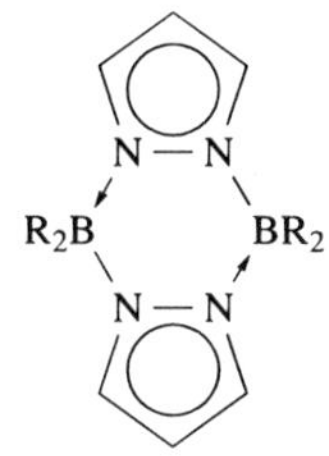

(23)

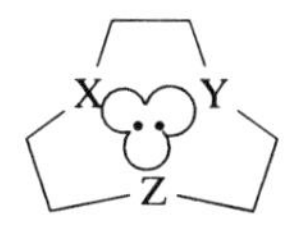

XYZ = CCB
CBB
BBB
CBN
BBN
BNN

(24)

Synthetic studies of bicyclic compounds related to models (**25**) are numerous. Heats of formation for the four compounds represented by (**25**) have been determined and B–N bond energies lie in the range 53.9–59.0 kJ mol^{-1}.[59]

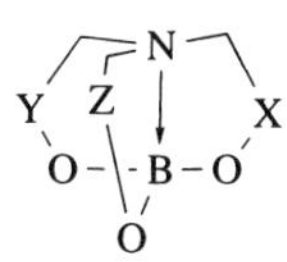

$X = (CH_2)_x$ $\quad x = y = z = 1, 2$
$Y = (CH_2)_y$ $\quad x = z = 1, y = 2$
$Z = (CH_2)_z$ $\quad x = z = 2, y = 1$

(25)

4.3 DERIVATIVES OF CYCLOALKANES

4.3.1 Three-membered Rings including Bicyclic Systems

Reactions of aromatic compounds with sodium and $Cl_2BNPr^i_2$[10] in $MeOCH_2CH_2OMe$ (e.g., Equation (1)) are routes to products such as (**26**). Carbenoid intermediates have been proposed.[60,61] The first example of a three-membered heterocycle containing a four-coordinate boron atom, (**27**), was prepared by the photolysis of $[Me_4N][Ph_3BCH{=}CHPh]$ in acetonitrile.[62] Anion (**27**) has also been prepared by reducing the corresponding borirene (see Section 4.4.1).[63] The $[Me_4N]^+$ salt of (**27**) is a stable, colourless solid, but is air and moisture sensitive. Anion (**27**) reacts with O_2 according to Equation (2). The crystal structures of $[Ph_4As]$[(**27**)] (in the anion, C–B = 0.161(3) nm, C–C = 0.149(3) nm[62,63] and $[Bu_4N]$[(**27**)] have been determined.[63] Photolysis of $[Ph_3BC_6H_4\text{-}4\text{-}Ph]^-$ leads to anion (**28**), isolated and structurally characterized as the blood-red $[Me_4N]^+$ salt. In solution, (**28**) is fluxional.[64]

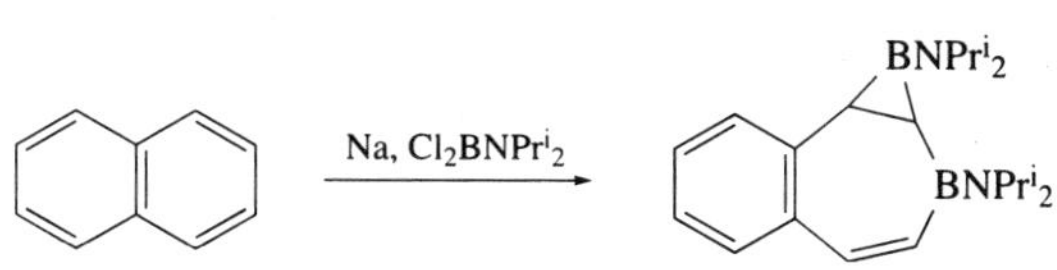

(26) (1)

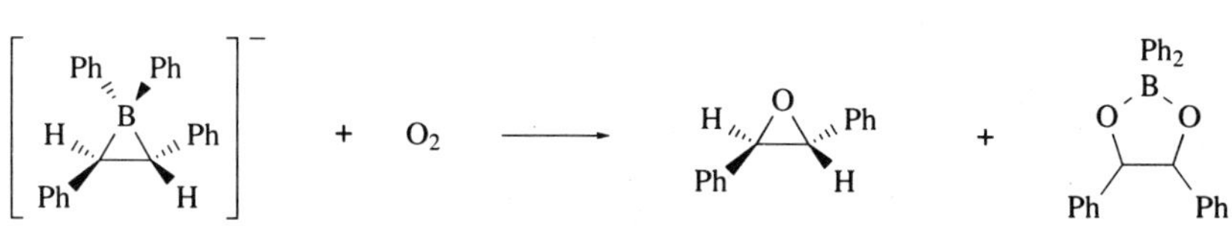

(27) (2)

The reaction of $(MesClB)_2C{=}C(TMS)Ph$ with activated magnesium leads to a complex, X, which exhibits a solution equilibrium between classical and nonclassical structures. With THF, X forms an adduct (**29**) in which the classical structure is confirmed crystallographically; the angle between the planes of the three- and five-membered rings is 114.4°.[65]

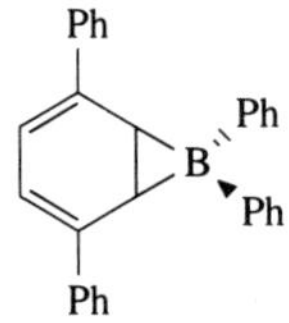

(28)

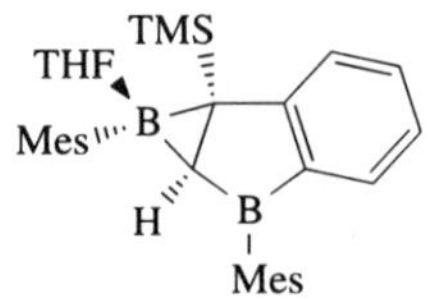

(29)

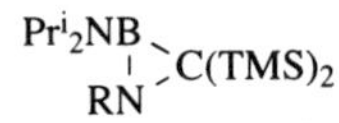

(30)
R = Ph, CH_2Ph

The reactions of PhN_3 and $PhCH_2N_3$ with $Pr^i{}_2N{=}B{=}C(TMS)_2$ lead to the cyclic molecules (**30**); four-membered rings are formed when $Pr^i{}_2N{=}B{=}C(TMS)_2$ reacts with carbonyl compounds or isonitriles (see Section 4.3.2).[66]

Classical and nonclassical methyleneboranes have been reviewed by Berndt.[11] The synthesis of a 2,4-diboramethylenecyclopropane by the reaction of $(TMS)_2C{=}C(BClBu^t)_2$ with Na/K alloy in boiling pentane was first reported by Berndt and Klusik.[35] A three-membered ring structure with an exocyclic C–B group was proposed on the basis of NMR spectroscopic data, but a bridged structure akin to (**11**) was suggested after calculations on model compounds (see Section 4.2).[32,34] Crystallographic confirmation comes from the trimethylgermyl derivative (**31**). This and three related compounds have been prepared from $R^1{}_2C{=}C(BR^2Cl)_2$ (R^1 = $GeMe_3$, R^2 = Dur; R^1 = TMS, R^2 = Bu^t; R^1 = TMS, R^2 = Dur (Dur = 2,3,5,6-Me_4C_6H)) and activated (Bogdanović) magnesium. Bond distances within the framework are B-1–B-2 = 0.1839(9) nm, B-2–C-2 = 0.1351(9) nm, B-1–C-1 = 0.1617(8) nm, B-1–C-2 = 0.1545(9) nm and C-1–C-2 = 0.1475(8) nm.[67] These data agree well with theoretical predictions. The reduction of (**31**) with lithium yields Li_2[(**32**)]. Sodium, potassium, caesium and magnesium salts of (**32**) can be obtained from $(TMS)_2C{=}C(BClBu^t)_2$. Anion (**32**) has been characterized by ^{13}C and ^{11}B NMR spectroscopy. Crystal structural data for K_2[(**32**)] indicate that the two π-electrons are delocalized; distances in the ring are C–B = 0.150 nm and B–B = 0.158 nm. In the lattice, the K^+ ions interact with the exocyclic C=C bond and one of the metal ions also interacts with the boron atoms of the next anion; a consequence of this is the formation of helical chains running along the *c*-direction.[68] Addition reactions of $HN(TMS)_2$ to (**31**) and of HCl to the *B*-duryl analogue of (**31**) give borirane products; crystallographic characterization confirms the presence of the saturated three-membered ring.[69]

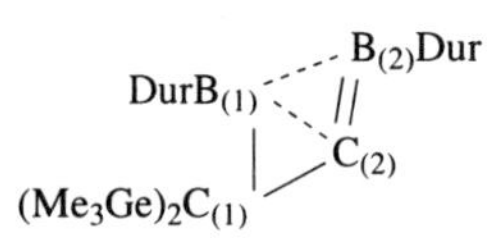

(31)

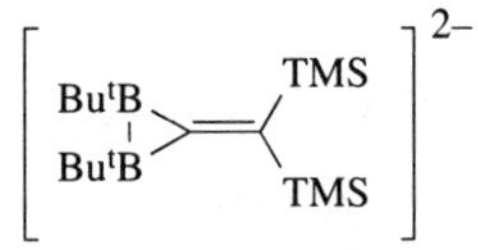

(32)

The unusual compound (**33**) with a 'trapezoidal' geometry results indirectly from the reaction of $(Dur)_2B_2F_2$ with $Li_2[(Dur)B{\equiv}CB(Dur){=}C(TMS)_2]$. The initial four-membered ring product rearranges to (**33**) on standing. Of the five atoms involved in the three three-membered rings, four lie in a plane. The B···B distances are 0.1933(6) nm, 0.1662(6) nm and 0.1704(6) nm.[70] The reaction of Et_2NBCl_2 with potassium in boiling cyclohexane yields, as one of the products, cyclic $\{Et_2NB\}_3$. Tetrameric and hexameric compounds are also formed.[71] Although it is straying into the realm of clusters, the tetrahedral cluster $Bu^t{}_4B_4$ (Equation (3)) is appropriately included here. In $CDCl_3$ solution, $Bu^t{}_4B_4$ is characterized by a signal at δ +135.4 in its ^{11}B NMR spectrum.[72]

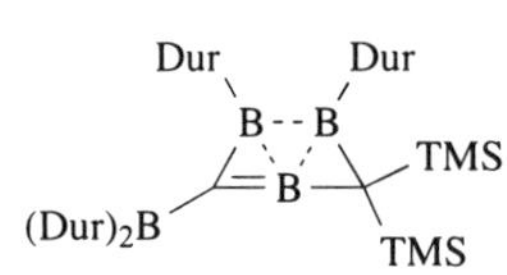

(33)
Dur = 2,3,5,6-Me_4C_6H

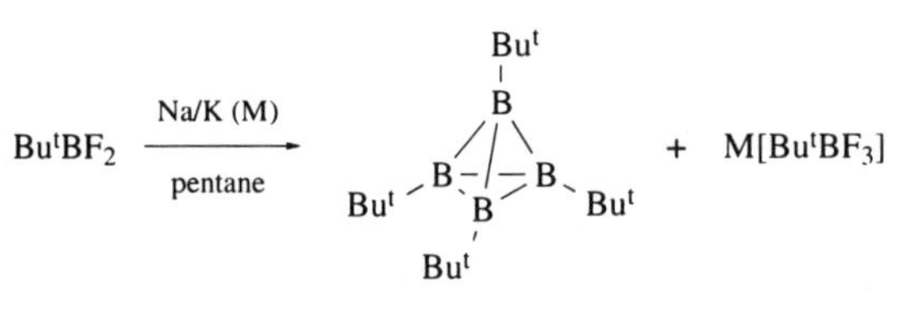

(3)

4.3.2 Four-membered Rings

Derivatives of 1,3-diboretane are no longer molecules of theory. Compound (**34**) forms when (**31**) is heated with TMS-C≡C-TMS in pentane at reflux. There is evidence that a carbene intermediate is involved.[73] Related stanna- and germa-derivatives (**35**) and (**36**) have also been prepared. Compound (**35**) has been characterized spectroscopically and by x-ray diffraction.[74] The crystal structure of the air-sensitive complex (**36**) has been determined; HCl adds to the C=Ge double bond of (**36**).[75] Heterocycles (**37**) are formed by treating $Li_2[RB{\equiv}CBR{=}C(TMS)_2]$ with RBF_2 in pentane. For R = Dur, the crystal structure of (**37**) has been determined and the B_2C_2 ring is planar. The Lewis acidity of the external boron atom is apparent in its ability to form adducts, for example, with THF and Et_2O.[76] The reaction of $Li_2[RB{\equiv}CBR{=}C(TMS)_2]$ with Me_3SnCl or MeI leads to 1,3-diboretanes $\{(TMS)_2C\}_2(R^1B)_2(CR^2{}_2)$ (R^1 = Dur or Mes; R^2 = $SnMe_3$ or Me). Ring opening occurs upon heating.[77]

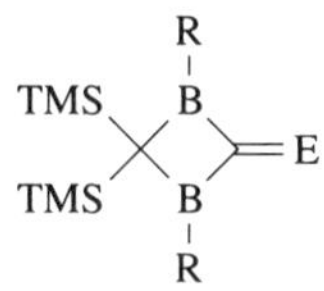

(**34**) R = Bu^t; E = C=C(TMS)$_2$
(**35**) R = Bu^t; E = Sn{CH(TMS)$_2$}$_2$
(**36**) R = Bu^t; E = Ge{N(TMS)$_2$}$_2$
(**37**) R = Mes or Dur; E = B(Mes) or B(Dur)

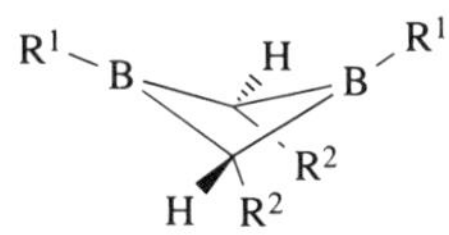

(**38**)

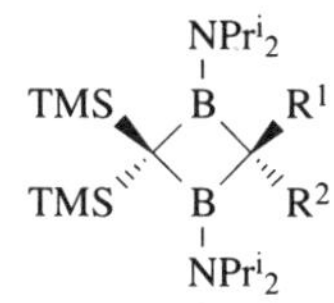

(**39**)
$R^1 = R^2$ = H or $AuPPh_3$
R^1 = H, R^2 = $AuPPh_3$

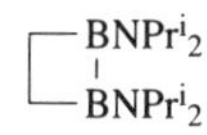

(**40**)

Starting from 1,3-dihydro-1,3-boretes, the 1,3-diboretanes (**38**) ($R^1 = NMe_2$ and R^2 = Bu^t, or R^1 = NPr$^i{}_2$ and R^2 = TMS) have been prepared along with gold(I) phosphine derivatives. A folded ring structure has been confirmed crystallographically for R^1 = NPr$^i{}_2$ and R^2 = TMS; the exocyclic B–C bond distances lie in the range 0.158–0.159 nm. On the other hand, derivatives (**39**) exhibit planar B_2C_2-rings.[78,79]

Propene is eliminated and cyclization occurs when Pr$^i{}_2$N=B=C(TMS)$_2$ is heated (500–600 °C). Four-membered rings are formed when Pr$^i{}_2$N=B=C(TMS)$_2$ reacts with R^1R^2CO (R^1 = H, R^2 = Me, Bu^t, or CMe=CH$_2$Me; R^1 = Me or Ph, R^2 = Ph; R^1 = Me, R^2 = OEt) or Bu^{t}N≡C.[66] The noncyclic product Cl(Pr$^i{}_2$N)BCH$_2$CH$_2$B(NPr$^i{}_2$)Cl is obtained when $Cl_2BCH_2CH_2BCl_2$ reacts with Pr$^i{}_2$NH in pentane. Cyclization to compound (**40**) (91% yield) occurs after treatment with Na/K alloy;(**40**) has been characterized structurally.[80] Dehydrohalogenation of aminofluoren-9-yl boranes yields monomeric complexes which dimerize to compounds of the type exemplified by (**41**). In (**41**), the central B_2C_2-ring has a folded conformation.[81]

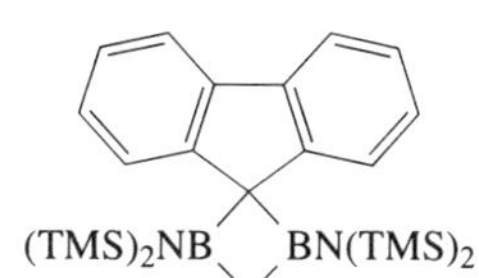

(**41**)

4.3.3 Five-membered Rings

A range of η^1-C_5Me_5 derivatives of boron has been synthesized and these include $Me_5C_5B(CH_2)_4$.[82] The use of borane adducts as precursors for various cyclic boron-containing systems is a successful approach. The method is exemplified in Equation (4), choice of an appropriate diene allows the ring size to be varied. The benzyl iodide is needed to remove the PPh_3 in the form of a sparingly soluble phosphonium salt.[83] Another efficient synthetic method is given in Equation (5); the products are highly air sensitive.[84]

The thermal isomerizations of (**42**) (n = 0–2) at 150 °C in diglyme have been studied. The rate of isomerization decreases as a function of the steric requirements of the cyclic structure.[85]

Five-membered rings containing two adjacent boron atoms are newly accessed compounds. Herberich *et al.* have achieved this goal by reacting $CH_2{=}C\{CH_2B(NR_2)_2\}_2$ (R = Et or Pr^i) with HCl in

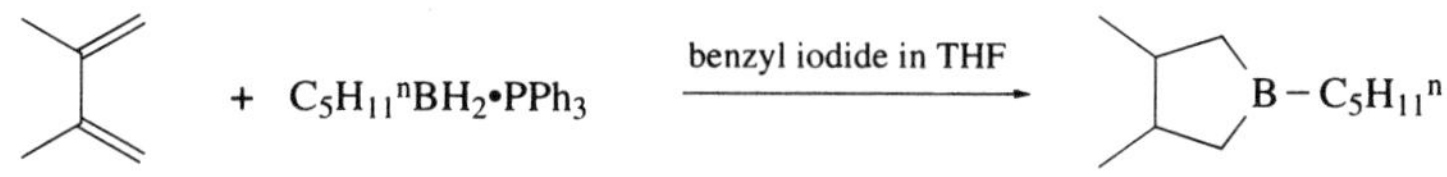

(4)

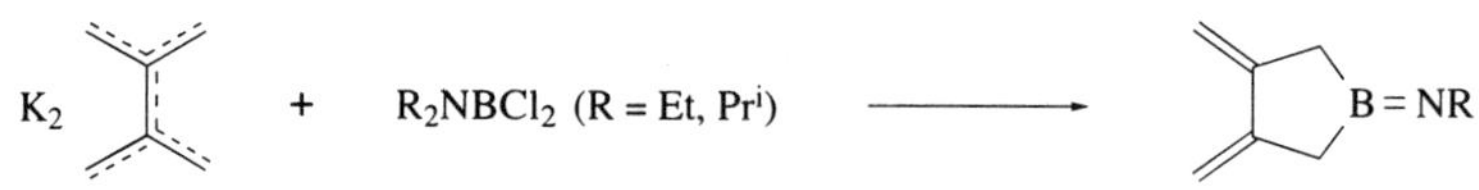

(5)

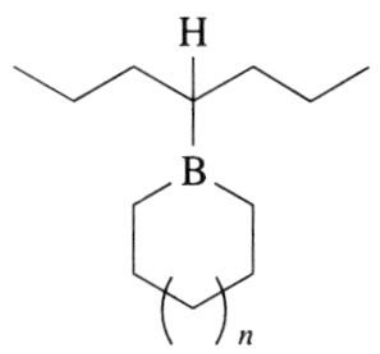
(42)

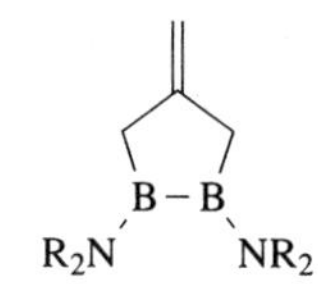
(43) R = Et or Pr^i

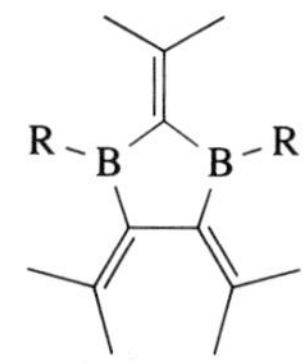
(44) R = Cl
(45) R = Me
(46) R = NMe$_2$

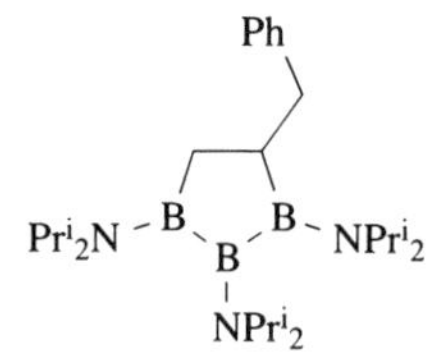
(47)

ether, followed by treatment with Na/K alloy. For R = Pr^i, the crystal structure of the product (**43**) has been determined and the ring exhibits a twist conformation; the B–B distance is 0.1723(8) nm.[86] A second ring-closure route is the reaction of $Cl_2B(CH_2)_3BCl_2$ with TMS–NEt_2 followed by treatment of the product with Na/K alloy. The five-membered ring so formed undergoes insertion when it reacts with a range of alkynes in the presence of a catalytic amount of $BF_3{\cdot}OEt_2$ (Equation (6)).[87]

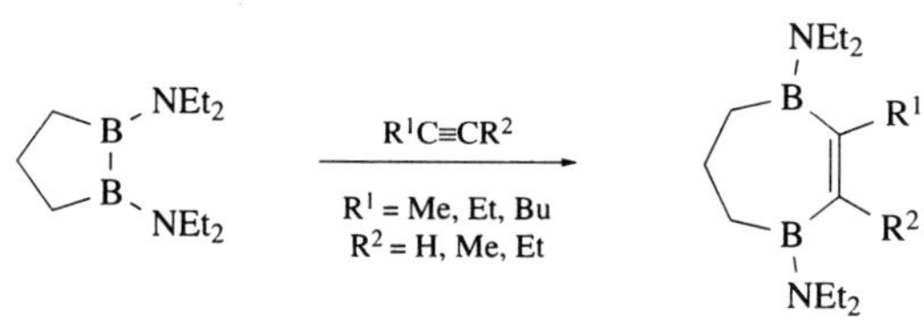

(6)

Five-membered C_3B_2-rings with two remote boron atoms are represented by compounds (**44**)–(**46**). Compound (**44**) is formed by reacting $\{Me_2C{=}C(BCl_2)\}_2$ with $Me_2C{=}C(SnMe_3)_2$. The conversion of (**44**) into (**45**) or (**46**) is accomplished by reaction with MeLi or Me_2N-TMS, respectively. The crystal structure of (**46**) has been determined.[88] The C_2B_3 system is exemplified by compound (**47**), the preparation and characterization (by NMR and mass spectroscopies and an x-ray diffraction study) of which have been reported.[89]

4.3.4 Six-membered Rings

A simple and convenient method (Equation (7)) of preparing borinane (**48**) has been described by Brown and co-workers.[90,91] Separation of the final products by fractional crystallization is not successful but triethylamine preferentially complexes with (**48**). Adduct formation can be reversed by treatment with $BF_3{\cdot}OEt_2$. Once pure borinane is available, the supply can be increased easily (Equation (8)).[91] The IR spectrum of (**48**) in solution indicates that a dimeric species is present; an absorption at 1560 cm^{-1} is characteristic of dimer formation.[91] The reaction of borinane with Bu^{n}C≡CI followed by sodium hydroxide leads to ring expansion.[92] The thermal isomerization of (**42**) (n = 0–2) at 150 °C has been studied (see Section 4.3.3).[85]

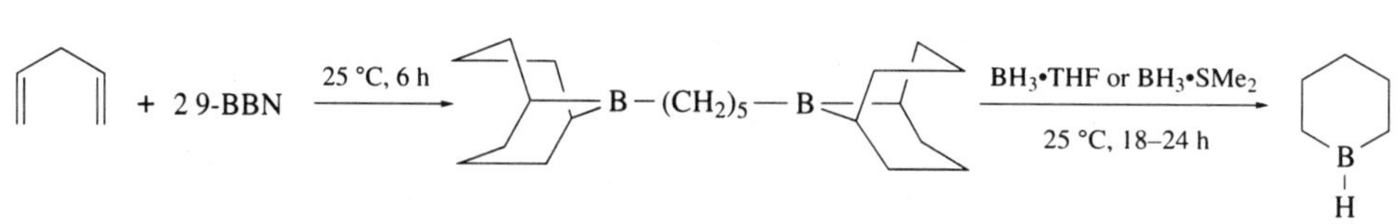

(7)

(48)

+

2 9-BBN

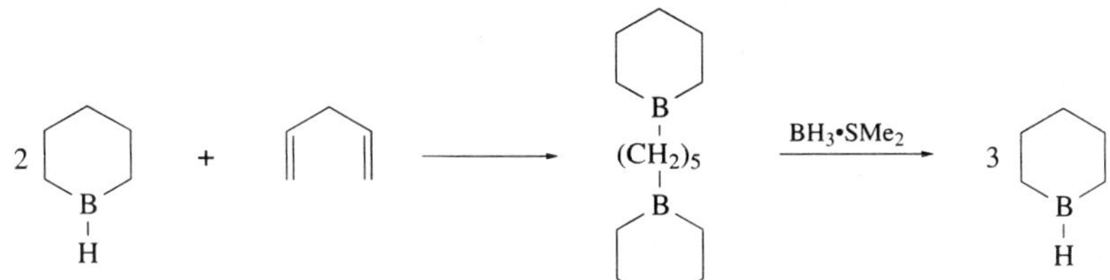

(8)

Compounds with the C_4B_2 framework are now known for 1,2- 1,3- and 1,4-arrangements. Ring closure occurs when $Cl_2B(CH_2)_4BCl_2$ is treated with TMS-NEt_2, followed by Na/K alloy. The product is (**49**) and, as in the five-membered analogue in Equation (6), the six-membered ring undergoes expansion upon reaction with alkynes.[87] Related to the formation of (**43**) is that of the six-membered ring (**50**). The first stage in the synthesis of (**50**) follows that of (**43**), but treatment of the intermediate compound with $Cp_2Ti(CH_2)(AlMe_2)Cl$ in pyridine replaces reaction with Na/K alloy (see Section 4.3.3). The crystal structure of (**50**) (R = Pr^i) confirms the positions of the two boron atoms; the ring contains a mirror plane and has a boat confirmation.[93] The elimination of lithium chloride (Equation (9)) drives the formation of (**51**). This C_4B_2-ring exhibits a chair conformation in the solid state (R = Bu^t).[94]

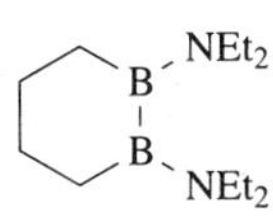

(**49**)

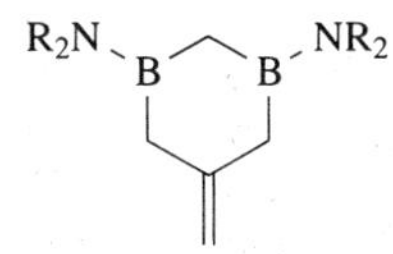

(**50**) R = Et or Pr^i

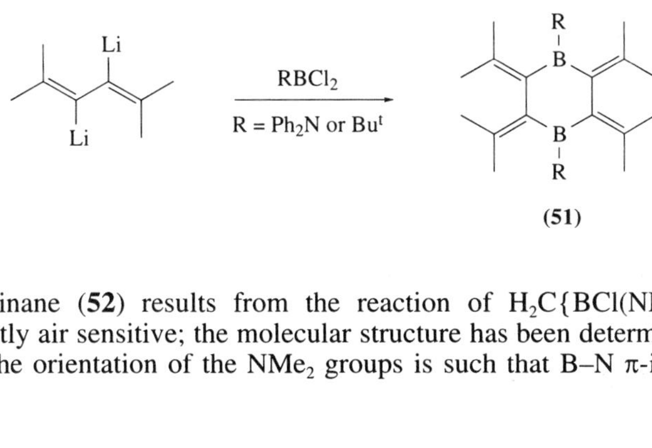

(**51**)

(9)

The 1,2,4,5-tetraborinane (**52**) results from the reaction of $H_2C\{BCl(NMe_2)\}_2$ with potassium. Compound (**52**) is slightly air sensitive; the molecular structure has been determined. The C_2B_4-ring has a chair conformation; the orientation of the NMe_2 groups is such that B–N π-interactions are efficient (B–N = 0.1405 nm).[95]

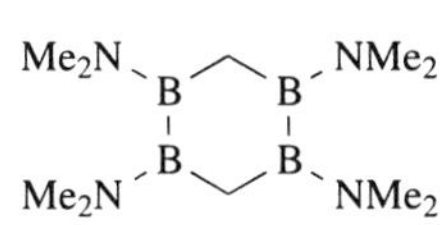

(**52**)

4.3.5 Seven-membered and Larger Rings

Studies of pheromone synthesis via organoboranes have included the stereospecific synthesis of (*Z*)-7-alken-1-ols. The method reported[96,97] makes use of borepane, (**53**), and the *in situ* preparation given in Equation (10). Methods for preparing boracyclanes with nine to twelve ring atoms have not been easy, but Brown *et al.* have reported the successful sequential one-carbon homologation of *B*-methoxyboracyclanes. This allows the ring in $(MeO)B(CH_2)_n$ to be expanded from $n = 5$ to 12 by using *in situ* $LiCH_2Cl$; yields are >75%.[98,99] A second strategy for constructing large boron-containing rings is to use $Li_2[C(CH_2)_3]$; with R_2NBCl_2 (R = Me, Et, Pr^i), for example, (**54**) can be accessed.[84,93] The crystal structure of (**54**) for R = Pr^i has been determined; the centrosymmetric molecule exhibits a puckered C_6B_2-ring and the environment at each boron atom is trigonal planar (B–N = 0.1408(1) nm, B–C = 0.1588(2) and 0.1614(1) nm).[93] The thermal isomerization of (**42**) with $n = 2$ (150 °C) has been studied (see Section 4.3.3).[85]

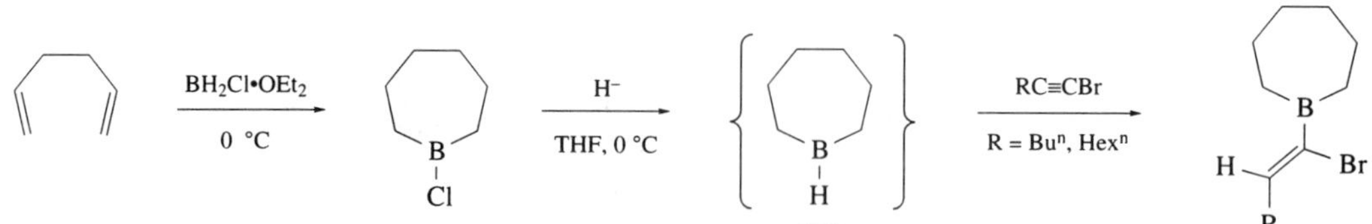

(53)

(10)

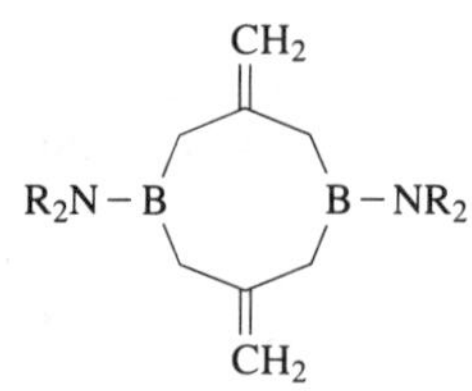

(54) R = Me, Et, Pr^i

4.3.6 Bicyclic Systems (excluding Three-membered Rings)

4.3.6.1 9-Borabicyclo[3.3.1]nonane (9-BBN) and its derivatives

The gas phase thermolysis of 9-BBN dimer (400–620 °C) has been investigated. At 490 °C, (**55**) and (**56**) are formed in almost equal quantities; it is assumed that (**56**) is dimeric or oligomeric. Heating (**56**) in the condensed phase leads to compound (**57**) which is stable to air and water; the crystal structure of (**57**) has been determined.[100]

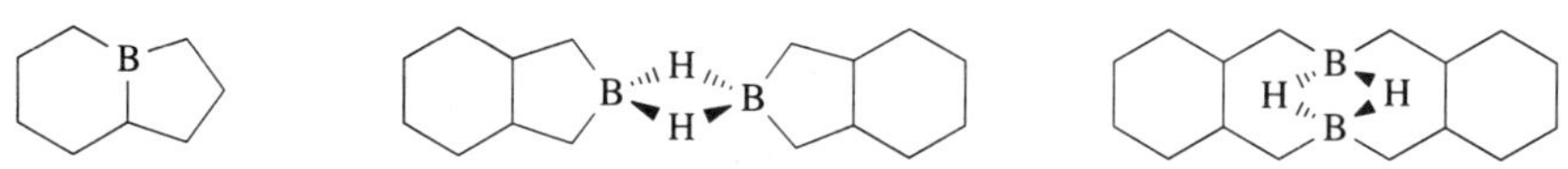

(55) (56) (57)

A comprehensive survey of derivatives of 9-BBN is too great to be included here, and this survey attempts to highlight some species. Review articles[2–4,7,22] include uses of derivatives of 9-BBN in organic syntheses. Anionic derivatives of 9-BBN have been prepared by several routes. The reaction of 9-BBN dimer with Li-TMS gives Li[(**58**)] ($R^1 = R^2 = H$, and $R^1 = R^2 =$ TMS). No mixed species ($R^1 = H$, $R^2 =$ TMS) is produced.[101] Li[(**58**)] ($R^1 = R^2 =$ TMS) is also formed in the reaction of *B*-OMe-9-BBN with Li-TMS.[101] Alkali metal salts of (**58**) ($R^1 = R^2 = H$) result after treating $(9\text{-BBN})_2$ with MH (M = Li, Na, K) in THF and heating ($T \leq 150$ °C). Selected spectral properties of the salts are listed in Table 3.[102] The salts K[(**58**)] in which $R^1 = H$ and $R^2 = Bu^n$, Bu^t, Ph, Cy or thexyl ($CHMe_2CMe_2$) can be obtained from *B*-R^2-9-BBN and potassium hydride. The stereoselectivities of the reagents tend to increase as a function of the steric requirements of R^2.[103] The reaction of *B*-Et-9-BBN with lithium, sodium or potassium (M) hydroxide yields M[(**58**)] ($R^1 = OH$ and $R^2 = Et$). Each salt M[(**58**)] undergoes ethane elimination (at 100 °C and with a Na[$BHEt_3$] catalyst) to give *B*-MO-9-BBN. Association occurs above 160 °C.[104]

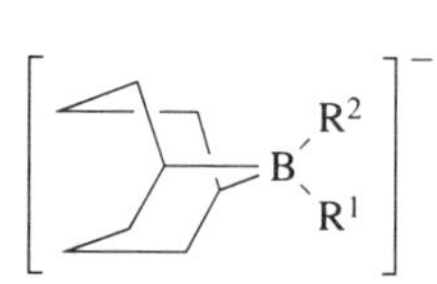

(**58**)
$R^1 = R^2 =$ H or TMS
$R^1 = H$; $R^2 = Bu^n$, Bu^t, Ph, Cy, thexyl
$R^1 = OH$; $R^2 = Et$

Two methods of producing the alkylated derivatives *B*-R-9-BBN in high yields are the reactions of 9-BBN with R_2CuLi and $R_2CuMgBr$ (R = Me, Pr^n, Pr^i, Bu^n, Bu^i, Bu^t, Ph).[105] The ethyl derivative *B*-Et-9-BBN has been synthesized by a method similar to that shown in Equation (4); cyclooctadiene is treated with $EtBH_2{\cdot}PPh_3$ in the presence of methyl or benzyl iodide.[83] The reaction of Me_5C_5Li with

Table 3 Selected spectroscopic properties of M[(**58**)] for $R^1 = R^2 = H$.

	M=*Li*	*M*=*Na*	*M*=*K*
IR ν(BH) in THF (cm^{-1})	2110	2080	2085
^{11}B NMR δ	−17.4	−18.3	−16.3
J(BH) (Hz)	74	74	74

B-X-9-BBN (X = Cl or Br) provides a route to *B*-Me_5C_5-9-BBN.[82] The preparation of *B*-$(TMS)_3Si$-9-BBN has been reported.[106] A high-yield synthesis of *B*-F-9-BBN has been achieved in the reaction of $(9\text{-BBN})_2$ with SbF_3 or SbF_5; IR and ^1H, ^{13}C, ^{11}B and ^{19}F NMR spectroscopic results have been reported. Crystallographic data for *B*-F-9-BBN illustrate that the unit cell contains three different molecules, two of which are associated together to form an asymmetric dimer supported by a B–F–B bridge.[107]

Derivative (**59**) has been synthesized and its electrochemistry studied. No reversible reduction process is observed in keeping with the general result for simple trialkylboranes. However, this is not the case for $Mes_2BC_6H_4$-4-$BMes_2$ in which each unoccupied boron 2*p* orbital interacts significantly with the π-system of the C_6H_4-unit.[108]

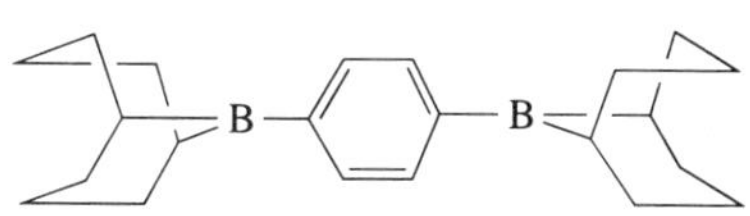

(59)

The reaction of Bu^tCHO with $(9\text{-BBN})_2$ proceeds with H_2 elimination to give *B*-$Bu^tC(O)$-9-BBN. Further treatment with $(9\text{-BBN})_2$ yields complex (**60**), a crystal structure determination of which shows that the CO_2B_2H-ring is essentially planar. Upon heating, (**60**) undergoes a phase change.[109] The boron-chelate complexes (**61**) are made by reacting $(9\text{-BBN})_2$ with $R^1C(O)N(OH)R^2$. A range of related compounds have also been prepared and spectroscopically characterized.[110]

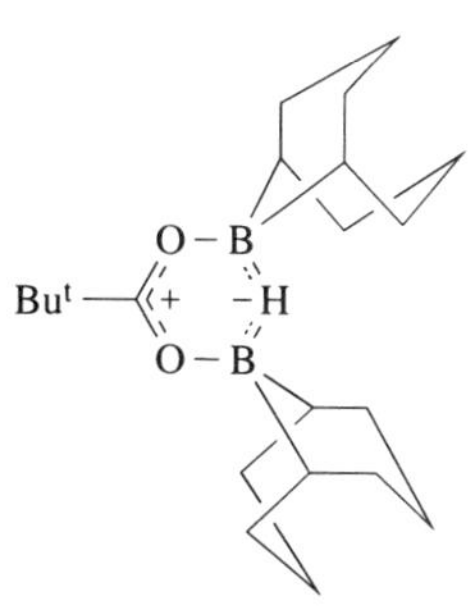

(60)

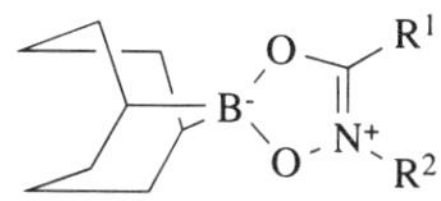

(61)
$R^1 = R^2 = Me$
$R^1 = R^2 = Ph$
$R^1 = Me$; $R^2 = Ph$

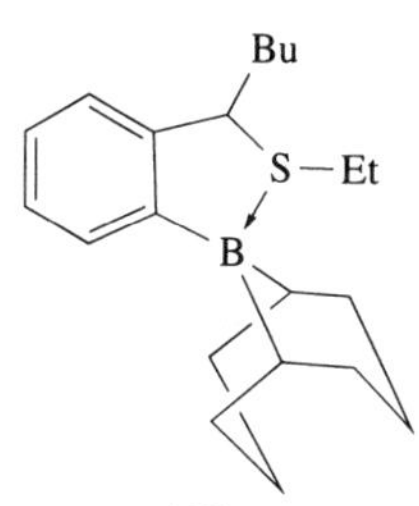

(62)

Several novel metal cluster compounds involving 9-BBN substituents have been synthesized. These include $Co_2(CO)_6(CBC_8H_{14})(CBu^t)$ which possesses a tetrahedral Co_2C_2-core,[111] and a series of related triosmium clusters, for example, $(\mu\text{-H})_3Os_3(CO)_9\{\mu_3,\eta^2\text{-}C(OBC_8H_{14})$.[112]

The oxidation of 9-BBN by dioxygen in THF is a chemiluminescent radical process, the product of which is a stable hydroperoxide.[113] With Pr^iCSSH, 9-BBN reacts in toluene at room temperature to give the corresponding *B*-thioacylthio-9-BBN which is a useful thioacylating agent.[114] The crystal structure of the 9-BBN derivative (**62**) has been elucidated; the intramolecular S–B distance is 0.2071(2) nm and the butyl and *S*-ethyl groups are in an *anti*-configuration. In solution, diastereoisomerization is fast on the NMR spectroscopic timescale.[115]

Exchange reactions between *B*-R-9-BBN (R = Me or Et) and $BH_3{\cdot}L$ (L = THF or Me_2S) have been studied by ^{11}B NMR spectroscopy. For L = THF in THF solution, the production of R_3B is immediate.[116] The mechanism of bromination of *B*-Pr^i-9-BBN has been studied; changes in the reaction conditions lead to a change in pathway from radical substitution at the α-hydrogen atom of the Pr^i group to electrophilic cleavage of the B–C(Pr^i) bond.[117]

Reactions of 9-BBN with alkynyl stannanes and derivative chemistry have been studied by Wrackmeyer and co-workers. With $Me_3SnC{\equiv}CBu^t$, 9-BBN reacts to give (**63**) (R^1 = H, R^2 = Bu^t), but if $Me_3SnC{\equiv}CEMe_3$ (E = Si or Sn) is used, the products are (**64**). These products have been characterized by mass and ^{119}Sn, ^{29}Si, ^{1}H, ^{11}B and ^{13}C NMR spectroscopies.[118,119] Further derivatives in these series have been made.[119] Reactions between *B*-Et-9-BBN and $Me_3SnC{\equiv}CR^2$ (R^2 = H or Me) (at −78 °C and warming to RT) give the compounds (**65**) (R^1 = Et; R^2 = H or Me) which, when heated at 70 °C, isomerize to (**63**) (R^1 = Et; R^2 = H or Me).[120] When *B*-Me-9-BBN reacts with $Me_3SnC{\equiv}CH$, the (*E*)-isomer, (**65**) (R^1 = Me; R^2 = H), is selectively formed. On irradiation, this compound transforms into the (*Z*)-isomer giving the first example of *trans*-boryl and stannyl substituents.[121] Tricyclic derivatives are formed in the reactions of *B*-R^1-9-BBN (R^1 = Me, Et, Pr^i) and $Me_2Sn(C{\equiv}CH)_2$; products (**66**) (R^1 = Me, Et, Pr^i; R^2 = H) have been spectroscopically characterized.[122] An extension of this chemistry[123] includes reactions between *B*-R^1-9-BBN (R^1 = Me, Et, Pr^i) and $Me_2Sn(C{\equiv}CR^2)_2$ (R^2 = Bu^t, TMS) in addition to analogous systems reported earlier.[122] Isomeric products exemplified by the pairs (**66**) (R^1 = Me or Et; R^2 = TMS) and (**67**) (R = Me or Et) have been characterized.[123]

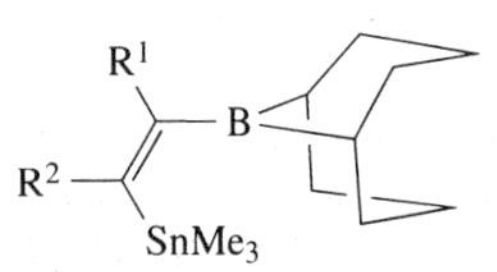

(63)
R^1 = H; e.g. R^2 = Bu^t
R^1 = Et; R^2 = H, Me

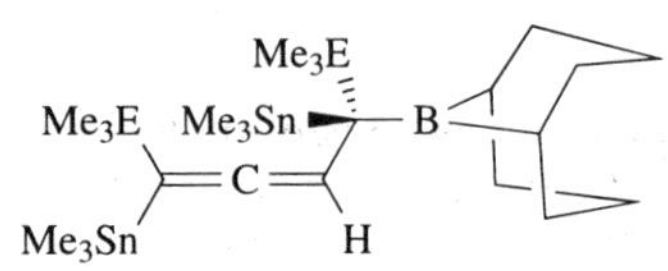

(**64**) E = Si or Sn

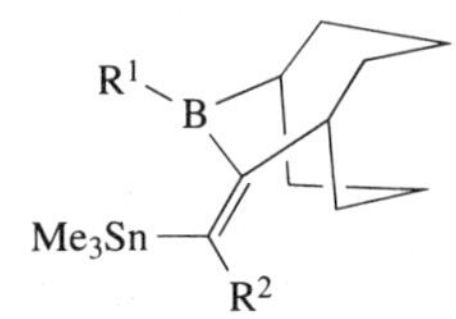

(65)
R^1 = Et; R^2 = H, Me
R^1 = Me; R^2 = H

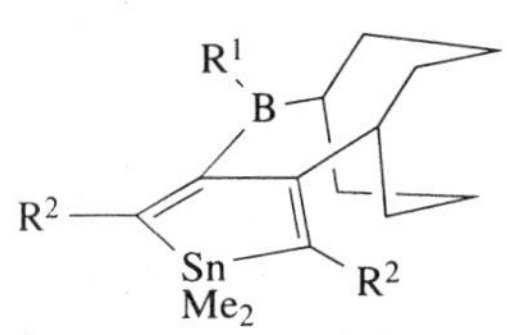

(66)
R^1 = Me, Et, Pr^i; R^2 = H
R^1 = Me, Et; R^2 = TMS

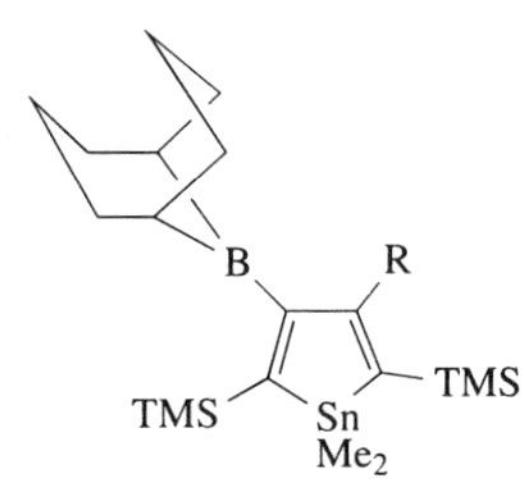

R = Me, Et

(67)

Derivatives of 9-BBN involving nitrogen-donor ligands have received detailed attention. The results of a study of the complex formation between $(9\text{-BBN})_2$ and a range of amines, L, indicate that the dimer is cleaved in a symmetrical fashion to give the adducts 9-BBN·L. For L = py, the adduct formation (25 °C) obeys second-order kinetics and the rate of adduct formation exceeds that of the dissociation of the dimer. The scheme proposed is given in Equations (11) and (12). For very bulky amines (e.g., Bu^tNH_2, Bu^nNH_2, 2-Me-py), the mechanism of adduct formation is a dissociative one, exhibiting first-order kinetics.[124] The interaction between *B*-OSO_2CF_3-9-BBN and 2,6-Me_2py (L) yields the borenium salt $[B\text{-L-9-BBN}]^+[OSO_2CF_3]^-$, although with py or 2,4-Me_2py, adducts are formed. Adducts are produced when *B*-Cl-9-BBN is treated with all three of these amines.[125] The salt [(**68**)][PF_6] forms when *B*-Cl-9-BBN reacts with 2,2'-bipyridine (bipy) in the presence of $[NH_4][PF_6]$ (81% yield). There is a dramatic shift in the ^{11}B NMR spectral resonance from δ +8.9 (d_6-DMSO) for *B*-Cl-9-BBN to δ +82.0 ($CDCl_3$) for cation (**68**). Complex cation (**68**) is a member of a series of species which has been prepared with the aim of providing a constrained and planar bipy ligand and, therefore, a planar π-system. In keeping with that of other members of this series, the electrochemistry of (**68**) reveals a two-step redox system with a persistent and thermally stable radical ion.[126] Related studies focus on complex (**69**) which also exhibits a two-step redox property.[127]

The first examples of monomeric diorgano(pyrazolyl)boranes have been prepared and fully characterized. Sterically hindered 3,5-di- and 3,4,5-trisubstituted pyrazoles react with $(9\text{-BBN})_2$ to yields adducts (**70**), isolated as low-melting, crystalline solids. The crystal structure of (**70**) for R = Et has been determined; significant steric repulsions exist in the molecule and the B–N π-interaction cannot be optimized (B–N = 0.1466 nm). In solution, NMR spectral data suggest that the boryl group exchanges between the two nitrogen donor atoms, although the process is frozen out in the ^{13}C NMR spectrum at

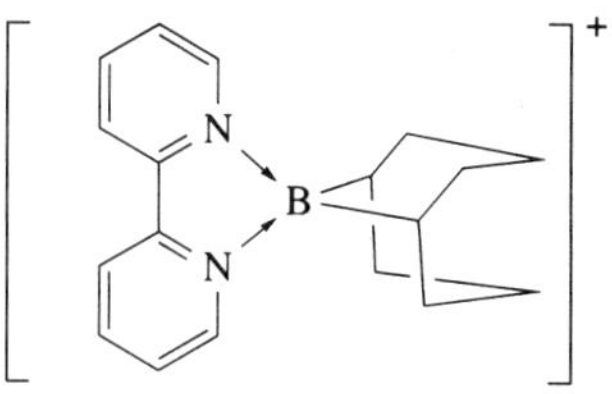

(68)

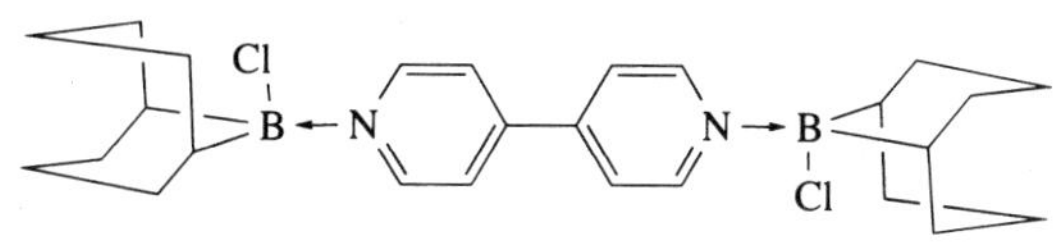

(69)

$$(9\text{-BBN})_2 + \text{amine (L)} \underset{k'_{-2}}{\overset{k'_2}{\rightleftharpoons}} 9\text{-BBN}\bullet\text{L} + 9\text{-BBN} \qquad (11)$$

$$9\text{-BBN} + \text{L} \xrightarrow{\text{fast}} 9\text{-BBN}\bullet\text{L} \qquad (12)$$

−100 °C.[128] The reaction of *B*-Me_2N-9-BBN with pyrazole (Hpz) produces *B*-pz-9-BBN·NMe_2H; further treatment with Hpz results in Hpz for NMe_2H displacement in the adduct. *B*-pz-9-BBN·Hpz can also be prepared from Hpz and $(9\text{-BBN})_2$. Deprotonation of *B*-pz-9-BBN·Hpz yields $[B,B\text{-pz}_2\text{-9-BBN}]^-$ and reactions with palladium(II) compounds have been studied. When *B*-pz-9-BBN·Hpz is treated with 9-BBN, complex (**71**) is formed.[129] Quinuclidine reacts with 9-BBN at 150 °C to give two isomeric adducts (**72**) and (**73**). The isomers can be separated and both have been crystallographically characterized. Complex (**73**) reacts with $BF_3{\cdot}OEt_2$ to yield equal amounts of *syn*- and *anti*-bis-(9-borabicyclo[4.2.1]nonane).[130]

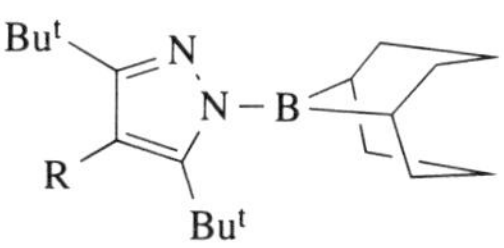

(70) R = H, Me, Et

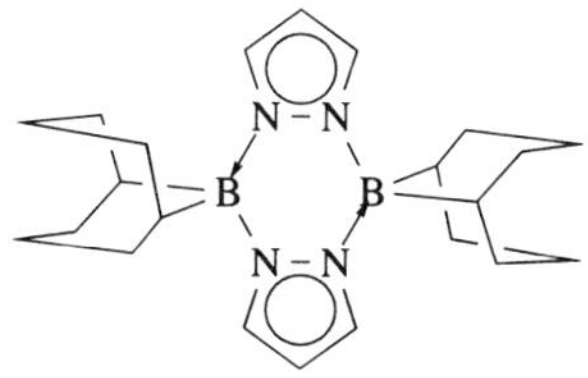

(71)

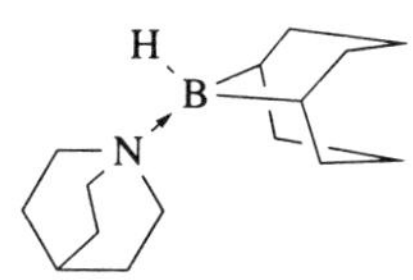

(72)

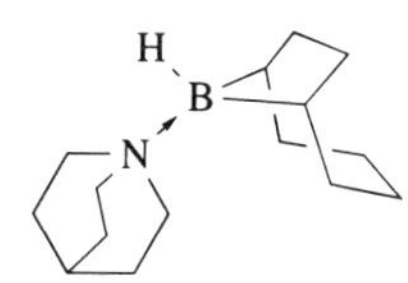

(73)

An interesting transannular hydroboration reaction (Equation (13)) occurs in the reaction of 9-BBN and (**74**). The *B*-hydrogen atom of 9-BBN is transferred regio- and stereospecifically to the C=C double bond which lies parallel to the N=N bridge. The structure of isomer (**75**) has been determined by x-ray diffraction methods; the B–N distance is 0.1395(3) nm. A second reaction involving a related azo-bridged system has also been studied.[131]

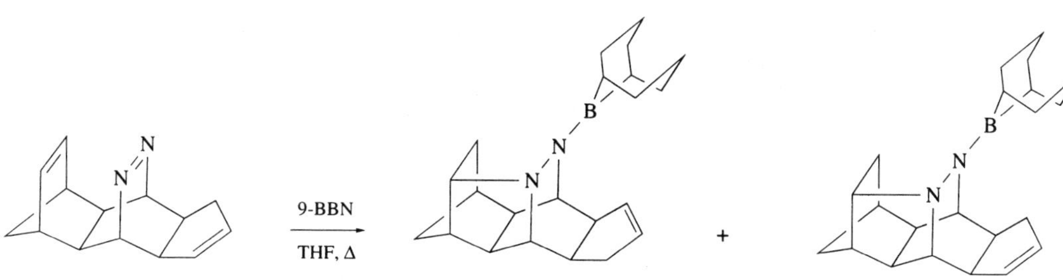

(13)

The reaction of $(EtO)_2PCl$ with *B*-(NBu^{t}Li)-9-BBN leads to complex (**76**).[132] Continuous irradiation of a cyclopropane solution of (**76**) containing a dialkyl peroxide (Et_2O_2 or $Bu^t_2O_2$) leads to the formation of the borylaminyl radical (**77**) (R = Bu^t) which is detected by ESR spectroscopy.[132,133] Analogous species for R = Pr^i and Bu^tCH_2 have also been studied in both cyclopropane (179 K) and ethene (198 K) solvents. Well-resolved ESR spectra have been recorded and parameters are given in Table 4.[133]

A series of compounds involving two 9-BBN units[22] bridged by chalcogen atoms has been the subject of study. An organoborane model for the Fischer–Tropsch synthesis is seen in the quantitative reductive C_1-polymerization shown in Equation (14). Products for $n \leq 10$ have been observed.[134] Compound (**78**) reacts with aluminum trichloride or tribromide to yield *B*-X-9-BBN (X = Cl or Br) and a bridged dimer $\{B\text{-(OAlX}_2\text{)-9-BBN}\}_2$. The crystal structure of the dimer for X = Br has been determined; the B–O bond

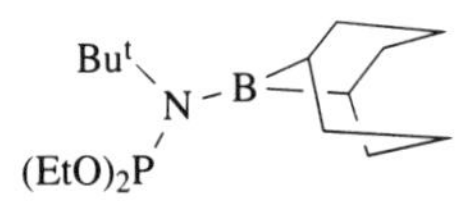

(76)

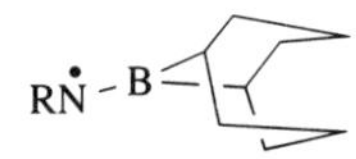

(77) R = Bu^t, Pr^i, Bu^{t}CH$_2$

Table 4 ESR spectral parameters for radicals **(77)**.[133]

R	*Solvent*	*Temp.* (K)	*g-factor*	*Hyperfine splittings* (G)
Bu^t	ethene	198	2.0037	16.2(^{14}N), 16.2(^{11}B), 5.0(4H$_\delta$), 2.6(2H$_\gamma$)
Pr^i	cyclopropane	179	2.0039	16.2(^{14}N), 16.2(^{11}B), 23.3^a (1H$_\beta$), 5.0 (4H$_\delta$), 2.7 (2H$_\gamma$)
Bu^{t}CH$_2$	cyclopropane	179	2.0039	16.2(^{14}N), 16.2(^{11}B), 37.3 (2H$_\beta$), 5.0 (4H$_\delta$), 2.7 (2H$_\gamma$)

[a] Difficulties in locating the wing-lines in the spectrum may mean that the value is larger.[133]

distance is 0.1392(8) nm.[135] The sulfur analogue of **(78)** can be prepared by reacting (9-BBN)$_2$ with sulfur. This product is an efficient sulfidation reagent for generating thio compounds by sulfur-for-oxygen exchange. Further reaction of the sulfide with sulfur gives the disulfide **(79)** (X = S).[136] The reaction of (9-BBN)$_2$ with selenium (150 °C) yields the selenido analogue of **(78)**, and continued treatment with selenium (120 °C) leads to **(79)** (X = Se). The compound *B*-SeH-9-BBN can be synthesized by reacting the selenido bridged species with phenylamine.[137] The interaction of **(78)** with various bases (e.g., py, 4-Mepy, 4-Etpy, 4-Bu^{t}py, 3,5-Me$_2$py, 4-Me$_2$Npy and quinuclidine) yields stable 1:1 adducts. In solution, the amine ligand, L, exchanges between the two equivalent boron sites; the rate of this process is controlled by the basicity and steric requirements of L.[138] Crystallographic studies have been carried out on the adducts **(78)**·L for L = py and quinuclidine, and molecules in both adducts are packed in the crystal lattice in head-to-tail pairs.[139] Each of **(78)**·py and **(78)**·Q (Q = quinuclidine) exhibits an unexpectedly large premelting endothermic phase transition; at >90 °C and >120 °C, respectively, **(78)**·Q and **(78)**·py transform to opaque, waxy materials. Complex **(78)**·Q has been studied by ^{11}B NMR spectroscopy both in the solid state and in the mesophase. Comparisons of the lattice packing in **(78)**·Q and **(78)**·py with those in **(78)**·L (L = 4-Mepy, 3-Brpy and 4-Me$_2$Npy) indicate that the unique packing in the former two adducts facilitates intermolecular exchange of L between pairs of molecules of **(78)**. This phenomenon brings order to the so-called 'fluxional solid phase'.[139,140]

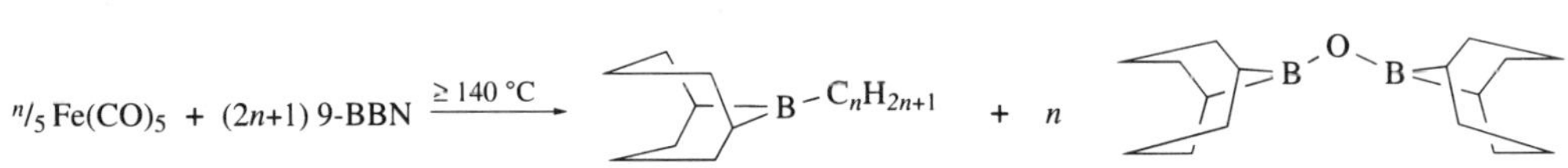

(14)

(78)

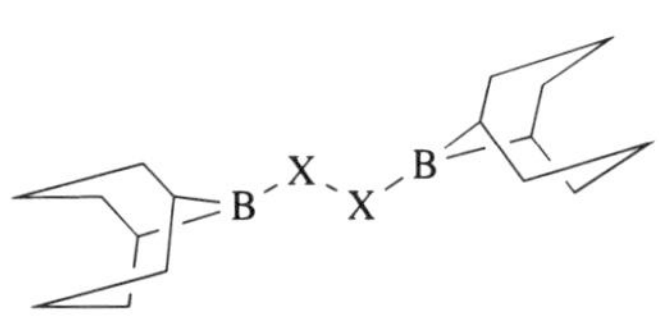

(79) X = S, Se

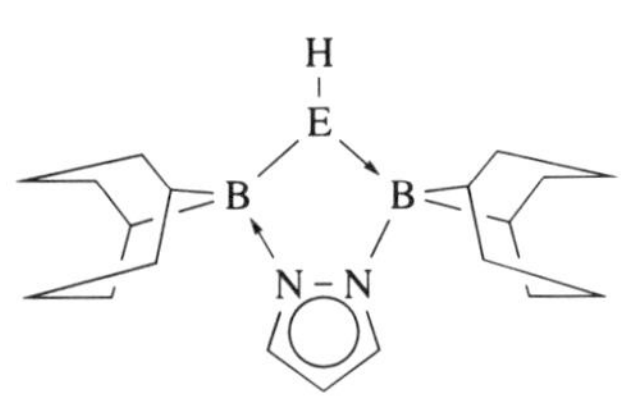

(80) E = O, S, Se

With pyrazole (Hpz), the oxide **(78)** and corresponding sulfide and selenide complexes react to give bridged species **(80)**. Complexes **(80)** (E = O and S) have been structurally characterized.[141] This work has been extended for **(78)** and its sulfide analogue to include reactions with 3-Mepz,[141,142] indazole,[141] 3,5-Me$_2$pz, 3,5-Ph$_2$pz, 3,5-Bu$^t_2$pz, and 3-Me-5-Phpz.[142] For E = S and base = 3-Mepz, the product of type **(80)** rearranges via hydrogen migration to give **(81)**, the structure of which has been confirmed by x-ray crystallography (B–S = 0.1955(1) and 0.1736(1) nm).[142]

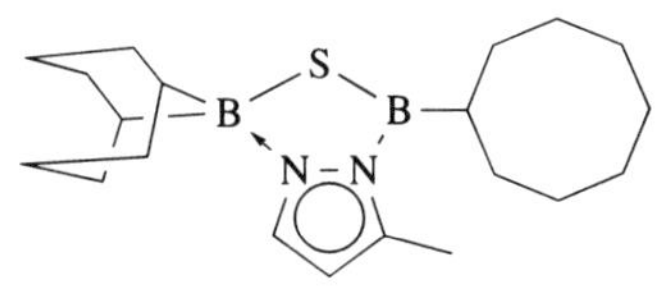

(81)

4.3.6.2 Borabicyclononanes (excluding 9-BBN) and borabicyclodecanes

The first bicyclic chiral hydroborating agent (**82**) (X = H) was prepared in 1982 (Equation (15)). It has been used in a range of asymmetric hydroborations of alkenes.[143]

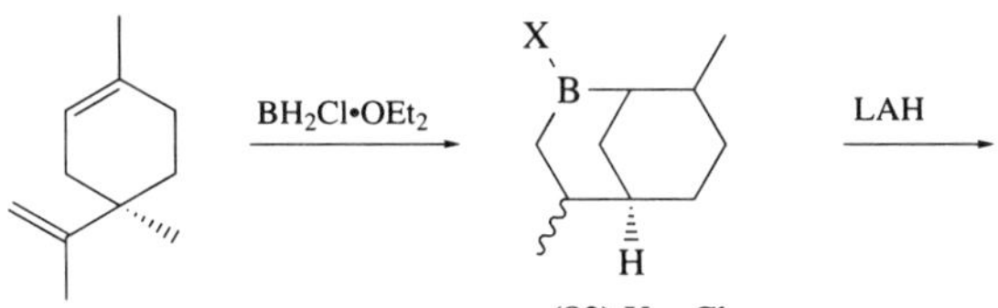

(**82**) X = H dimeric (15)

(**82**) X = Cl

The bromination and Matteson–Pasto rearrangements of some 3-borabicyclo[3.3.1]nonanes have been studied, (e.g., Equation (16)).[144] Reactions of R^3Li (R^3 = Me or Bu^n) with (**83**) (R^1 = Me; R^2 = Me, CH_2OMe, Ph, or HR^2 = $=CH_2$) give lithium salts of anions (**84**) (same R^2 and R^3). The reaction of acetyl chloride with (**84**) causes opening of the boron-containing ring.[145] Ring openings involving related borabicyclodecanes have also been studied.[146] For R^1 = MeO and R^2 = Me, the reaction of (**83**) with CH_2XLi (X = Cl or Br) at low temperature leads to (**85**) (same R^1 and R^2).[147] The dimeric 3-borabicyclo[3.3.1]nonane (**83**) (R^1 = H; R^2 = Me) reacts with alkynes ($R^3{}_2C_2$ with R^3 = Me or Pr^n) to give (**83**) (R^1 = CR^3CHR^3; R^2 = Me). In the presence of alcohols and catalytic amounts of HBr or HCl, these products rearrange to *B*-alkoxyborabicyclodecanes.[148] A route to (**83**) (R^1 = OEt; R^2H = $=CH_2$) from 1-boraadamantane and $HC(OEt)_3$ has been described. The reaction conditions (20–70 °C) are critical; at 130–140 °C, (**83**) (R^1 = OEt; R^2 = Me) is formed in addition to diethyl-1-adamantylboronate.[149] 1-Boraadamantane is also the precursor to (**83**) with R^1 = OMe and R^2 = CH_2I. This compound reacts with lithium azide in methanol and, after oxidation (H_2O_2/NaOH), (**86**) is formed; (**86**) can be converted to 1-azaadamantane by treating with $SOCl_2$ and NaOH.[150]

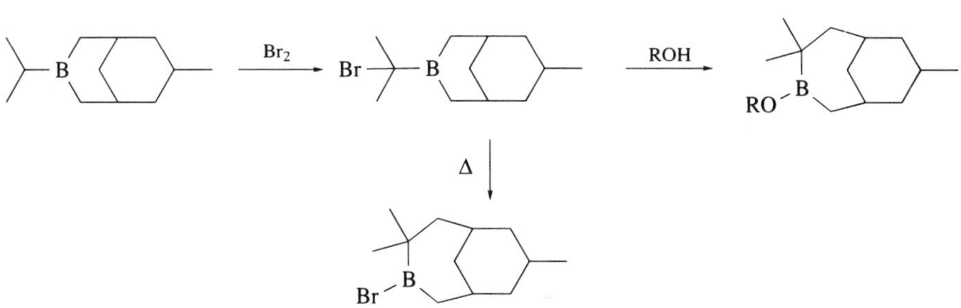

(16)

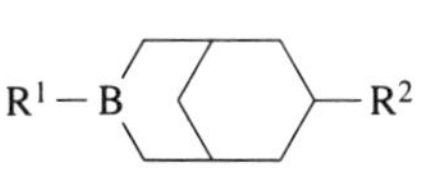

(83)

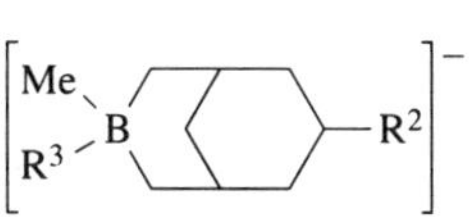

(84)

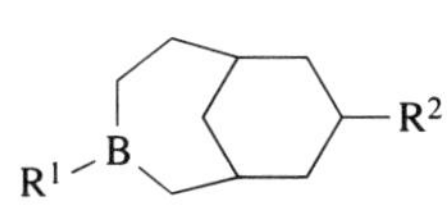

(85)

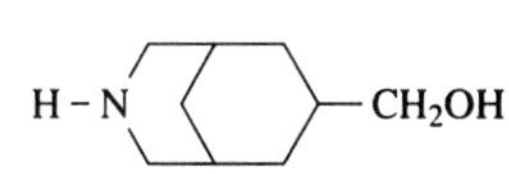

(86)

4.3.6.3 Miscellaneous simple borabicycloalkanes

Most bicyclic systems, relevant to organic synthesis, have been excluded from this discussion (see Section 4.1), but several simple systems are highlighted. Compounds (**87**) have been prepared by the reaction of appropriate 1,5,9-decatrienes with $BH_3{\cdot}L$ (e.g., L = Et_3N) in diglyme.[151] A similar approach is seen in the formation of (**88**).[152] The reaction of 1,2-$(CH{=}CH_2)_2C_4H_2$ with BH_2EtPPh_3 in the presence of methyl or benzyl iodide leads to (**89**) (compare with Equation (4)).[83]

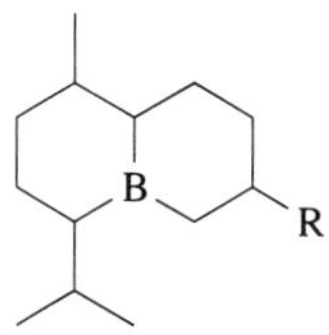

(87) R = H, Me

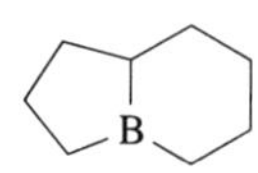

(88)

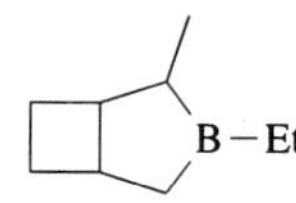

(89)

Although reactions of a series of chloro(alkylamino)boranes with allenyllithium lead to (acyclic) diallenylboranes, an attempt to prepare a *B-t*-butyl derivative yielded the bicyclic compound **(90)**; this rearranges above 20 °C via a 1,5-hydrogen migration to a carbaborane cluster. The crystal structure of **(90)** has been determined.[153,154] Related to **(90)** is **(91)** which is produced in the reaction of $\{CH(BClNPr^i_2)_2\}_2$ with Na/K alloy in benzene at reflux. The structure of **(91)** has been established crystallographically.[155]

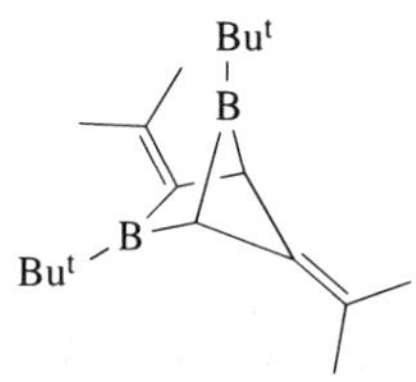

(90)

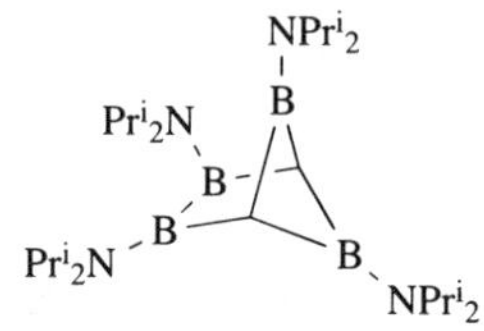

(91)

4.3.7 Boratricycloalkanes and Larger Systems

Following a similar ring-closure methodology as that used for the preparations of **(88)** and **(89)** is the reaction of monoborane with 1,5,9-cyclododecatriene to give **(92)**.[156] Trapping methylborylene (i.e., the carbene-like MeB:) with cyclohexene provides a method of forming 2-methyl-2-boratricyclo[7.4.0.0^{3,8}]tridecane, **(93)**. Methylborylene may be generated from Na/K alloy with $MeBBr_2$ but the tendency for explosion is high. A more reliable method is to treat the intercalate C_8K with $MeBBr_2$. The formation of **(93)** is stereochemically selective. A second isomer **(94)** can be prepared from $\Delta^{1,1'}$-dicyclohexene and $MeBH_2$ in THF.[157]

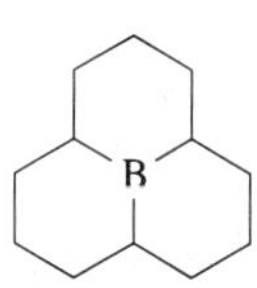

(92)

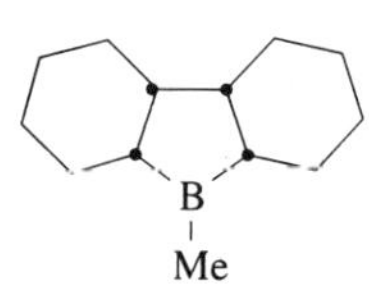

(93)

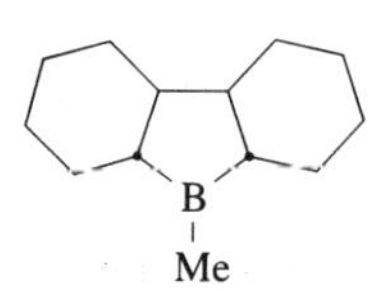

(94)

The stabilizing influence of *B*-dialkylamino substituents is reflected in a range of novel multicyclic systems. The preparation (from an appropriate 2-norbornene and $Pr^i_2NBCl_2$) and spectroscopic characterization of compounds **(95)** have been described. For $Z^1 = Z^2 = 2H$, the crystal structure of **(95)** has been elucidated.[89] Related work incorporating an unsaturated ring into the system has also been studied; the product is **(96)** and the conformation of the tetracyclic ring system has been established by x-ray diffraction.[158] When 2,5-norbornadiene is treated with Na/K alloy and $Pr^i_2NBCl_2$, compound **(97)** is formed. Crystallographic data confirm the conformation of the molecule; B–B distances are in the range 0.1715(5)–0.1723(5) nm.[159] Multiple additions of $\{Pr^i_2NB\}$ units to aromatic rings[10] occur, and these reactions have yielded a range of novel multicyclic boron–carbon systems; the methodology is analogous to that shown in Equation (1).[60,61] Reactions between PhMe or $1,3\text{-}Me_2C_6H_4$ (5 equiv.), $Pr^i_2NBF_2$ (1 equiv.) and Na/K alloy have yielded **(98)** and **(99)**, respectively. The nature of these products was revealed by x-ray analysis since structural assignments could not be made unambiguously from NMR spectroscopic data.[160] In a separate work by the same authors, products **(100)** were obtained from reactions of PhR (R = Me, Bu^t or Pr^i) with $Pr^i_2NBF_2$ and Na/K alloy. Five $\{Pr^i_2NB\}$ units add and insert into the aromatic ring to give a product which consists of a seven-membered unsaturated, and five- and six-membered saturated rings. The structure of **(100)** is confirmed by x-ray diffraction methods for R = Bu^t.[161] The tricyclic system **(101)** can also be isolated from the reaction of PhPr^i, $Pr^i_2NBF_2$ and Na/K alloy.[159] Product **(101)** (ignoring substituent R) is an isomer of **(98)**.

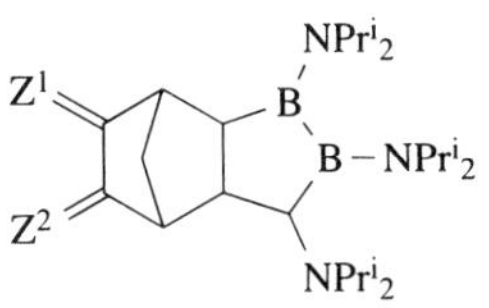

(95)
$Z^1 = Z^2 = 2H$
$Z^1 = CH_2$; $Z^2 = 2H$
$Z^1 = 2H$; $Z^2 = CHMe_2$

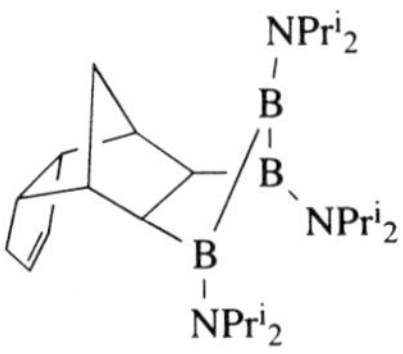

(96)

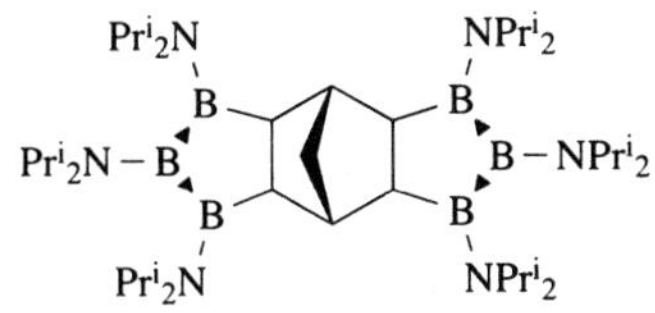

(97)

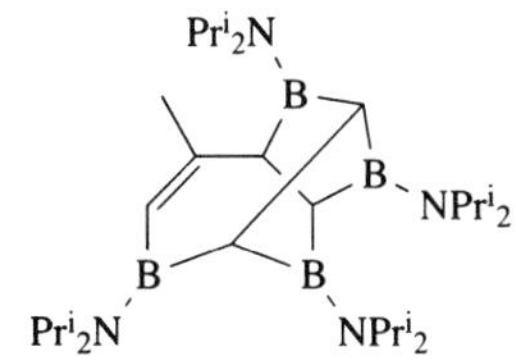

(98)

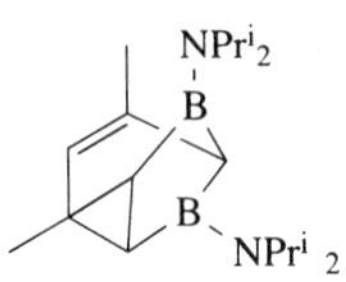

(99)

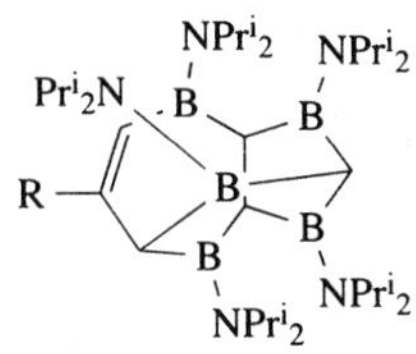

(100) R = Me, Pr^i, Bu^t

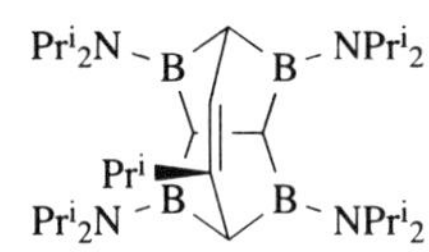

(101)

The dimerization of the diboracyclohexene (**102**), in the presence of $BF_3{\cdot}OEt_2$ as catalyst, leads to the centrosymmetric molecule (**103**), the structure of which has been elucidated (B–N = 0.1398(2) nm and 0.1401(2) nm).[162] The rearrangement of (**104**) to (**105**) is induced thermally or photochemically; (**105**) possesses an 'S_4N_4-like' cage with two long C–C distances of 0.167(1) nm being the counterparts to the long, transannular S···S interactions in S_4N_4.[163]

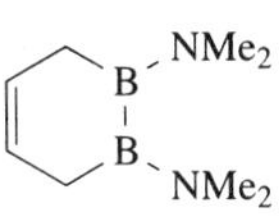

(102)

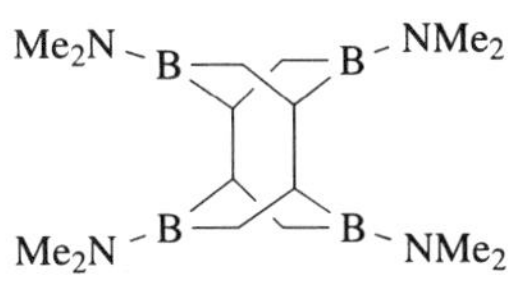

(103)

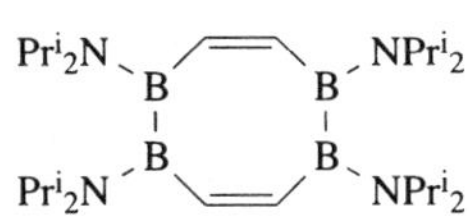

(104)

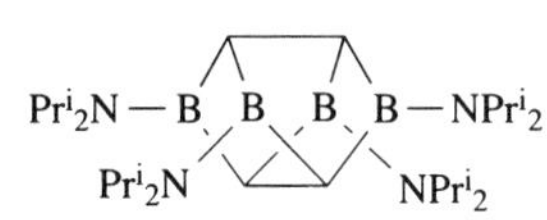

(105)

4.3.8 Boraadamantanes

A timely review by Mikhailov[21] complements the discussion by Morris[1] of the chemistry of 1-boraadamantane (**106**) and its derivatives prior to 1982. Phase transitions exhibited by 1-boraadamantane have been studied by using Raman spectroscopy over the temperature range 88–296 K; at 183 K, a transition from a plastic to ordered crystalline state occurs.[164]

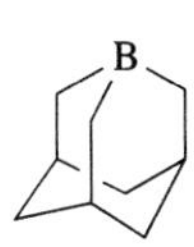

(106)

Several preparative routes to 1-boraadamantane have been developed and are summarized in routes 1,[165] 2[165] and 3[166] in Scheme 1. The 3-allyl-derivative, the precursor in route 3, isomerizes at ~130 °C; hydroboration with Et_2BH in THF of an equilibrium mixture of the two isomers leads to a mixture of (**106**) and its 4-chloro-derivative.[166] The reaction of $(CH_2{=}CHCH_2)_3B$, $HC{\equiv}C(CH_2)_3X$ (X = Cl or Br) and MeOH, and with $BH_3{\cdot}THF$ leads to (**107**).[167] The family of 1-boraadamantanes (**108**) are prepared via cyclization of appropriate alkenes in the presence of $BH_3{\cdot}L$.[168] The designed synthesis of 2-trimethylsilylmethyl-1-boraadamantane, isolated as a THF or py adduct, occurs via a series of steps, the first of which is shown in Equation (17). The precursor is prepared from triallylborane,[169] as is the starting material for 2-(2-thienyl)-1-boraadamantane, isolated as its pyridine adduct (Equation (18)).[170] Equation (19) illustrates a method to synthesize a 1-bora-2-oxaadamantane; the product is formed in 37% yield.[171]

The Lewis acidity of the boron atom in 1-boraadamantane prompts the study of a range of adducts. Complexes of type (**106**)·L^1 (e.g., $L^1 = PrNH_2$, piperidine, isoquinoline) are readily prepared by treating

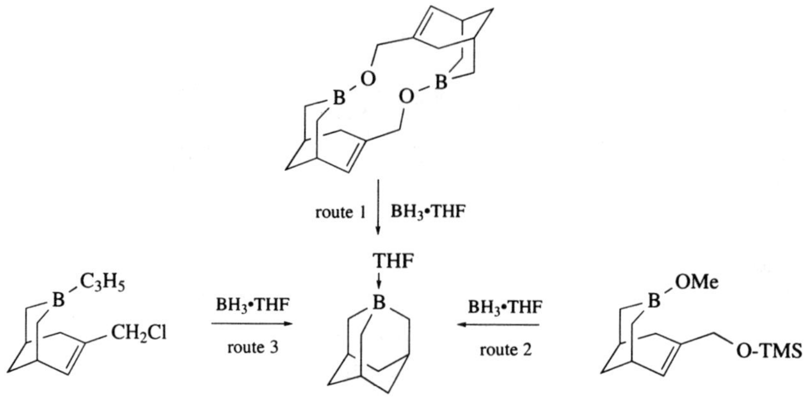

Scheme 1

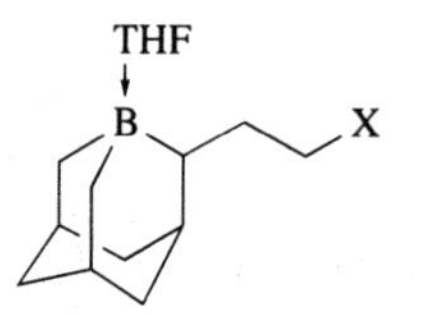

(107)
X = Cl or Br

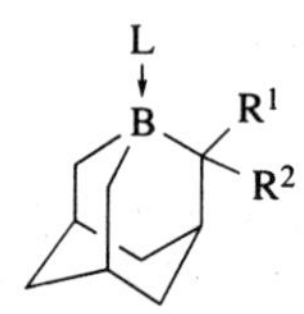

(108)
L = THF or py
R^1 = H; R^2 = Me
R^1 = R^2 = Me
R^1–R^2 = $(CH_2)_5$

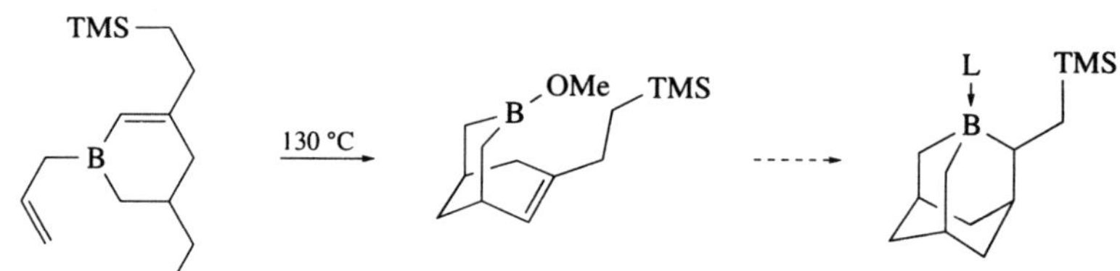

L = THF or py (17)

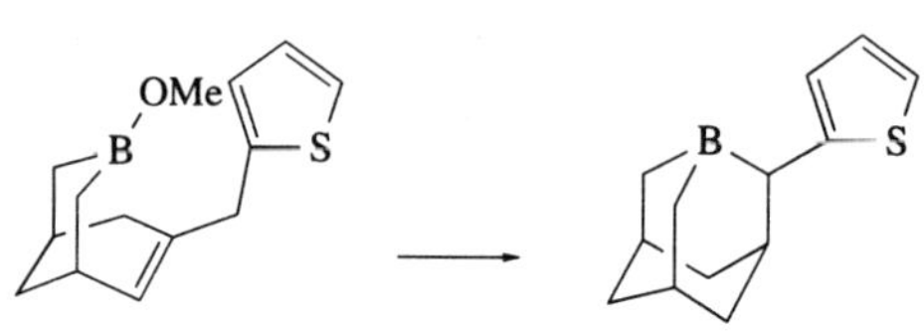

(18)

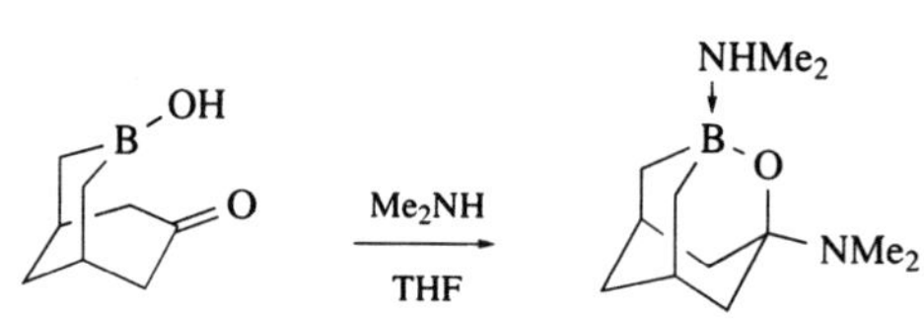

(19)

(106)·L^2 (L^2 = THF, Et_2O or Et_3N) with the appropriate base. Nineteen such complexes have been the subject of antiviral activity tests.[172] Coordination to the boron atom by nitrogen-donor ligands is also observed in a series of complexes in which the base is a pendant substituent to the adamantane cage. Equation (20) illustrates the synthesis of one such compound; related complexes have been prepared and characterized by ^{1}H, ^{11}B and ^{13}C NMR spectroscopy.[173,174] Complex (**109**), like (**69**), is a two-step redox system, the radical anions of which exhibit high thermodynamic stability.[127] Chemical ionization mass spectral data have been discussed for both intra- and intermolecular complexes.[175]

The reactivity of 1-boraadamantanes tends to be dominated by ring opening to give boracyclononanes. For example, treatment with methyllithium or trimethylamine and methanol causes (**107**) to open up.[167] 1-Boraadamantane reacts with Et_2NCl, NH_2OSO_3H, or 2,4-$(NO_2)_2C_6H_3ONH_2$ to yield boracyclononanes in which the presence of an amino substituent leads to intramolecular N → B coordination.[176] With the aldehydes RCHO (R = Ph, 4-$MeOC_6H_4$, 3,4-$(MeO)_2C_6H_3$, Pr^i) or with Me_2CO,

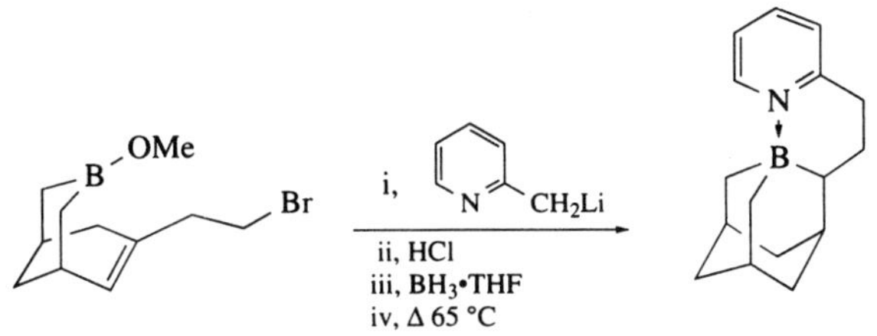

(20)

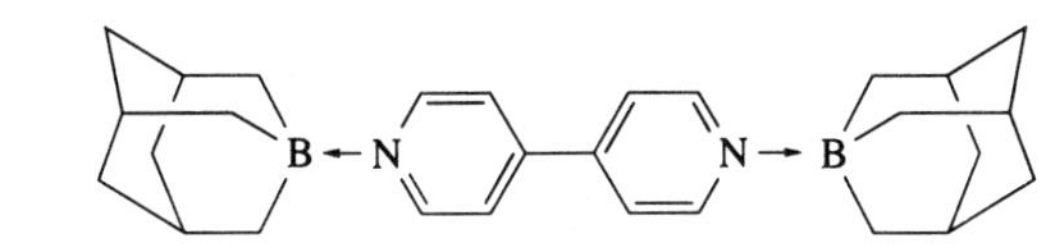

(109)

1-boraadamantane gives thermally unstable *O*-bonded adducts. These, when heated in hexane or benzene, react according to Equation (21). Cleavage of the dimer occurs in methanol for R = 4-$MeOC_6H_4$ and 3,4-$(MeO)_2C_6H_3$.[177] Lithium methoxide reacts with (**106**)·THF to give Li[(**110**)], the anion of which undergoes ring opening when it is treated with iodine at −30 °C. Further reaction with I_2 leads to the monocyclic product (**111**).[178] Benzonitrile and 4-methylbenzonitrile react with 1-boraadamantane at room temperature to give (**106**)·L (L = ArCN). Cage opening occurs when (**106**)·L is heated at 200 °C and the dimer (**112**) is produced.[179] Adduct formation also occurs with ylides (Ph_3P=CHR) and the products rearrange with loss of triphenylphosphine on heating at 132 °C.[180] Several papers describe methods of converting 1-boraadamantane to 1-azaadamantane.[150,181,182] Sources of nitrogen include LiN_3 and $EtNCl_2$.

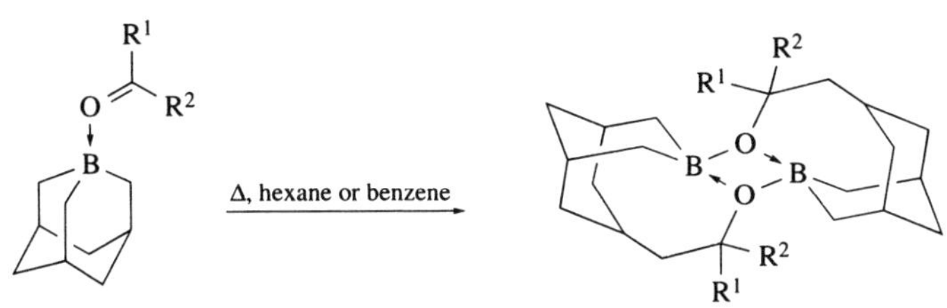

(21)

R^1 = Ph, 4-$MeOC_6H_4$, 3,4-$(MeO)_2C_6H_3$, Pr^i; R^2 = H or $R^1 = R^2$ = Me

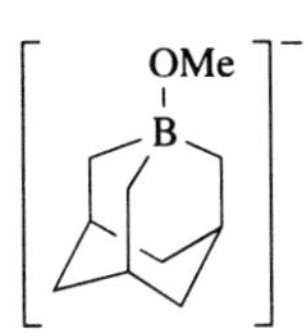

(110)

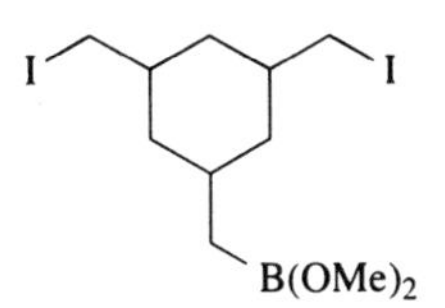

(111)

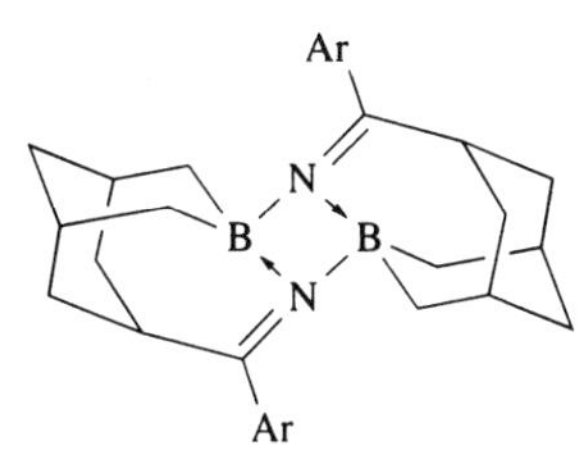

(112) Ar = Ph or 4-MeC_6H_4

The bromination of 2-Pr^i-2-boraadamantane occurs via both radical and electrophilic mechanisms. For the former route, the product is 2-(2-bromo-2-propyl)-2-boraadamantane, whereas electrophilic substitution yields 2-bromo-2-boraadamantane.[183] The reaction of the carbaborane cluster *closo*-1,5-$(CMe)_2(BEt)_3$ with potassium and iodine yields the dimer (**113**) for which a classical structure has been crystallographically confirmed.[184] The distances and angles within the C_4B_6 cage of (**113**) are similar to those previously determined for $C_4H_4B_6Me_6$. The hexachloro derivative $C_4H_4B_6Cl_6$ is prepared by pyrolysing (450 °C) $HC(BCl_2)_3$, Cl_2BCH=$CHBCl_2$ or $Cl_2BCH_2BCl_2$. The crystal structure of $C_4H_4B_6Cl_6$ has been elucidated; the B–Cl distances lie in the range 0.1755(4)–0.1769(4) nm.[185] The reaction of carbon disulfide with Na[B_3H_8] yields the interesting anion (**114**), structurally characterized as the tetraphenylphosphonium salt.[186]

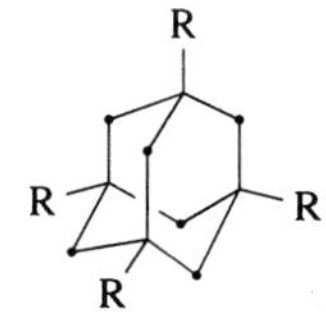

(113)
• = BEt; R = Me
• = BCl; R = H

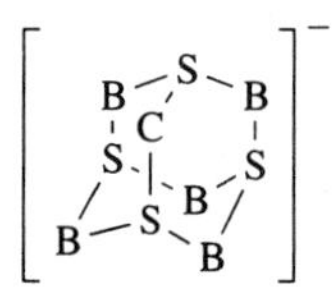

(114)
C = CH_2; B = BH_2

4.4 DERIVATIVES OF CYCLOALKENES AND BORA AROMATICS

4.4.1 Three-membered Rings

During the period 1982–1992, the chemistry of the borirene ring has expanded significantly.[14] The photolysis of $B(SiPh_3)_3$ at −196 °C generates the species Ph_3SiB: which is trapped by TMS-C≡C-TMS at room temperature to yield (**115**).[187] A second route to the C_2B–ring is the reaction of RC≡CSnMe$_3$ (R = Bu^t or SnMe$_3$) with Bu^{t}ClB–BClBu^t to yield (**116**). For R = Bu^t, (**116**) forms an adduct with 4-Bu^{t}py via a coordinate bond to the substituent boron atom. The compounds have been characterized by NMR spectroscopy.[188] Ring contraction is also a viable method (Equation (22)) of generating a borirene; members of the series of compounds (**117**) have been characterized spectroscopically.[189] Crystallographic confirmation of the three-membered ring structure has now been achieved by Eisch *et al.*; $(CMes)_2(BMes)$ may be prepared via the irradiation (λ = 300 nm) of Mes$_2$BC≡CMes. The heterocycle was found to contain a twofold axis but the carbon and boron atom sites could not be distinguished.[190] In an effort to improve upon the information available, Eisch *et al.* prepared an asymmetrical ring via the photorearrangement of Mes$_2$BC≡C(2,6-Me$_2$C$_6$H$_4$). This effectively labels the boron and carbon atoms in the borirene ring; an x-ray structure determination revealed ring distances of B–C = 0.1450(10) nm and 0.1464(11) nm and C–C = 0.1380(9) nm.[191]

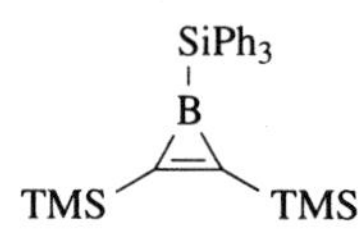

(115)

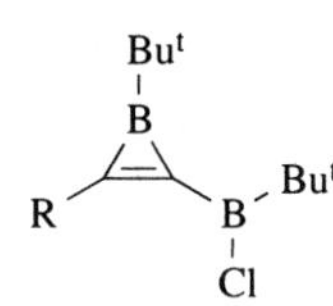

(116) R = Bu^t, SnMe$_3$

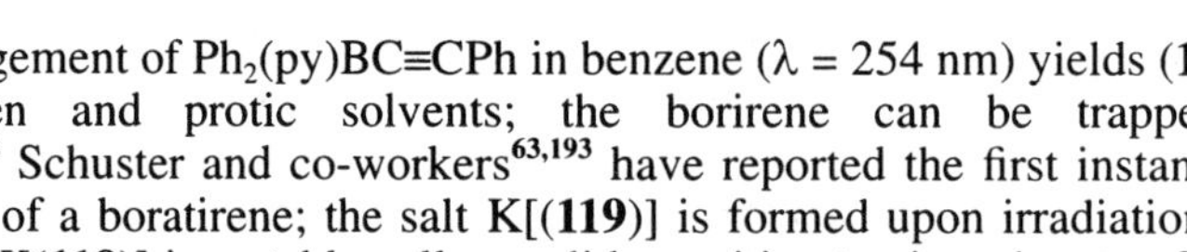

(117) (22)

R^1 =	Pr^n	Bu^n	Ph	Pr^n	Et	Pr^n
R^2 =	H	H	H	Me	Et	Pr^n

The photorearrangement of Ph$_2$(py)BC≡CPh in benzene (λ = 254 nm) yields (**118**) which is extremely sensitive to oxygen and protic solvents; the borirene can be trapped by reaction with diphenylacetylene.[192] Schuster and co-workers[63,193] have reported the first instance of the isolation and full characterization of a boratirene; the salt K[(**119**)] is formed upon irradiation (λ = 254 nm) in THF of K[Ph$_3$BC≡CPh]. K[(**119**)] is a stable yellow solid, sensitive to air and water. Some reactions of (**119**) are shown in Scheme 2.

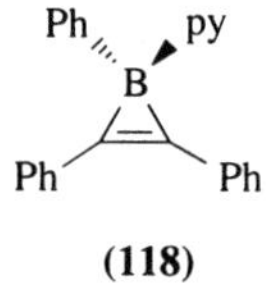

(118)

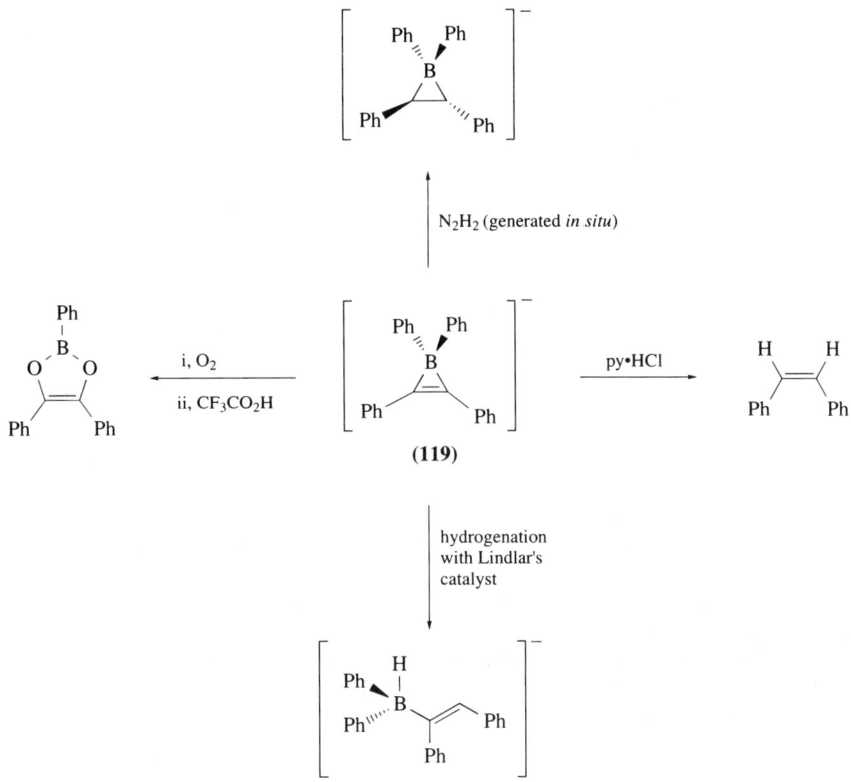

(119)

Scheme 2

The reaction of $(TMS)_2C{=}C(BBu^tCl)_2$ with four equivalents of Na/K alloy in THF yields the salt K_2[(**120**)] as shown in Equation (23). Oxidation of (**120**) yields the appropriately substituted derivative of (**11**) and (**31**), while treatment with Bu^t(TMS)NH leads to the formation of K[(**121**)]. Anions (**120**) and (**121**) have both been characterized by multinuclear NMR spectroscopy. Protonation of (**121**) results in the formation of a B–H–B bridge.[194]

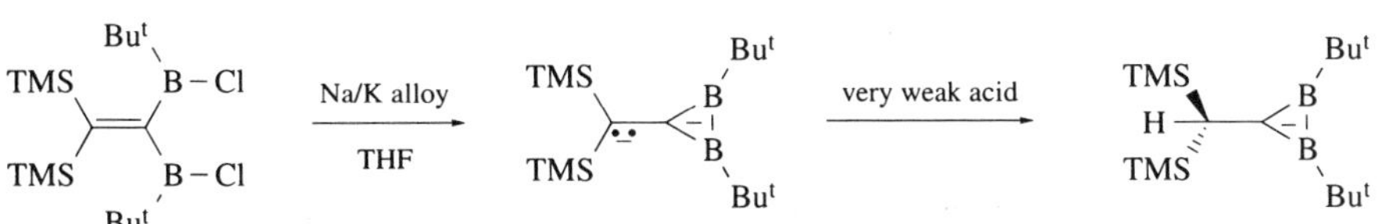

(120) (121) (23)

4.4.2 Four-membered Rings

The syntheses, characterization and chemistry of four-membered C_nB_{4-n}(n = 2 or 3) heterocycles have been significantly developed in the period from the mid-1980s to the early 1990s. Berndt and co-workers[195] have made use of the unsaturation of the exocyclic C=B bond in their unusual three-membered ring[11,35] (see Section 4.3.1) to expand the system as shown in Equation (24). A common structure for the products is proposed from NMR spectral data.[195] A similar strategy is used in the formation of the 1,2-dihydroborete (**122**). Here, the nature of the product has been confirmed by x-ray crystallography; the four-membered ring is folded (internal angle = 148.8°) along the B···C diagonal.[196] A series of compounds related to (**122**) has been prepared via [2 + 2]-cycloaddition reactions starting from 9-fluorenylidene(tetramethylpiperidino)borane and alkynes. Again, crystallographic data confirm a folded 1,2-dihydroborete geometry.[197] Similarly, $R_2C{=}BDur$ (R = TMS or $SnMe_3$) reacts with TMS-C≡C-TMS to yield 1,2-dihydroboretes. The cyclization can be reversed on heating.[198]

Compound (**123**), described as a 'mesityl-bridged methyleneborane',[199] is formed in 86% yield by oxidizing $Li_2[MesBu^tB{=}C{=}BBu^tMes]$ with tin(II) chloride. The crystal structure of (**123**) has been determined and reveals long B–C(bridge) bonds (0.1769(8) nm and 0.1748(9) nm) compared to the other two endocyclic B–C bonds (0.1448(8) nm and 0.1462(8) nm). The ring is planar.

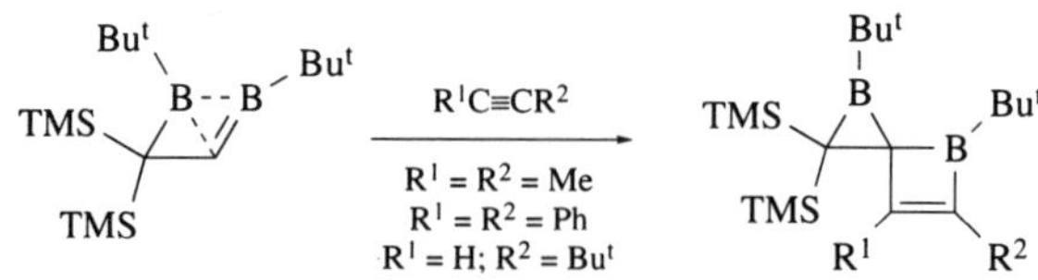

(24)

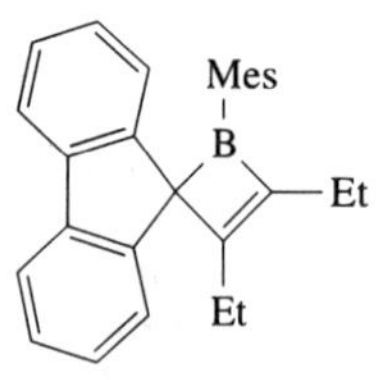

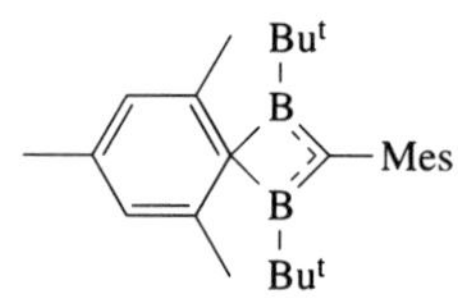

(122) **(123)**

The first derivative of 1,2-dihydro-1,2-diborete (**124**) was isolated and structurally characterized by Siebert and co-workers.[200] Compound (**124**) is prepared as colourless crystals (55% yield) from the reaction of $Pr^i_2NB(Cl)CH{=}CHB(Cl)NPr^i_2$ with Na/K alloy in benzene at reflux. Proton, ^{13}C and ^{11}B NMR spectral data support the structure which is confirmed by x-ray crystallography; distances within the planar ring are B–B = 0.175(1) nm, B–C = 0.158(1) nm and C–C = 0.131(1) nm. Thermal isomerism of (**124**) to the folded 1,3-dihydro-1,3-diborete occurs; this is expected on the basis of the calculated relative stabilities of the two isomers of the model diborete $C_2B_2H_4$ (see Section 4.2). Dimerization of (**124**) occurs in the presence of palladium phthalocyanine. The product (**104**) rearranges to give the cage molecule (**105**) (see Section 4.3.7).[163]

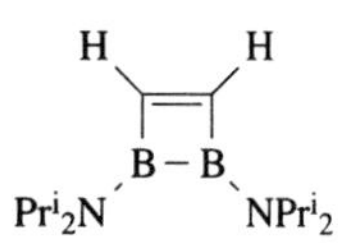

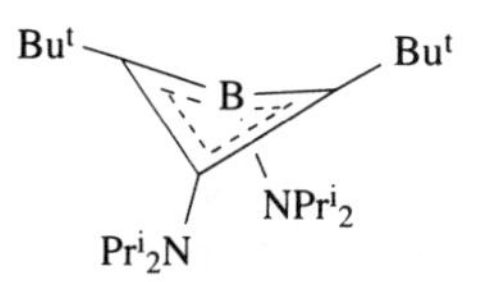

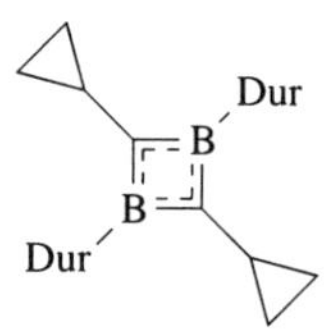

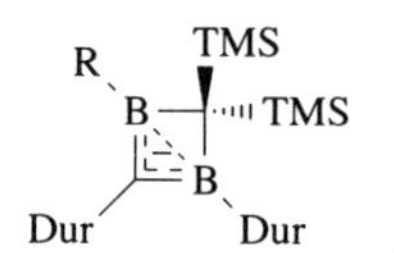

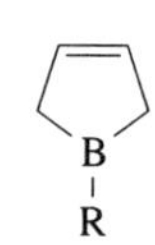

(124) **(125)** **(126)** **(127)** R = Ph or Dur **(128)**

1,3-Dihydro-1,3-diboretes can be prepared by several routes, for example, the reactions of $Bu^tB(Cl)CR{=}CRB(Cl)Bu^t$ (R = Me or Et) with Na/K alloy in pentane,[201] or of RC≡CR (R = Et or TMS) and Pr^iNBCl_2 with Na/K alloy.[202] The first 1,3-dihydro-1,3-diborete to be structurally characterized by x-ray diffraction was (**125**) which was prepared in 65% yield by treating $Pr^i_2NB(Cl)C(Bu^t){=}C(Bu^t)B(Cl)NPr^i_2$ with Na/K alloy in benzene (80 °C). Spectroscopic data (IR and NMR) are consistent with the solid-state structure for (**125**). The four-membered ring is folded (internal angle = 128°) and the endocyclic B–C distances lie in the range 0.1501–0.1506(5) nm. The transannular C···C distance is 0.181 nm.[203] The crystal structure (107 K) of the related derivative in which Bu^t groups are replaced by hydrogen atoms, reveals an internal angle of 132.4° for the diborete ring and equal B–C distances.[204] The derivative (**126**) is prepared by treating DurB(Cl)CR=CRB(Cl)Dur ($R = CH(c\text{-}C_3H_6)$) with Boganović magnesium. In its ^{13}C NMR spectrum, (**126**) exhibits a signal at δ 186.7 corresponding to strongly deshielded ring carbon atoms. The crystal structure of (**126**) has been determined; the ring is folded along the C···C diagonal.[205]

The reduction of the 1,3-dihydro-1,3-diborete $\{Bu^tBC(TMS)\}_2$ by lithium in toluene yields yellow crystals of the lithium salt (crystallographically characterized) of the corresponding dianion. This is reoxidized by $NiCl_2(PPh_3)_2$. The dianion protonates easily and reacts with MeI to give the corresponding *C*-methylated 1,3-diboretane.[206] A simple diboriranide (**127**) has been prepared and isolated as its lithium salt. Li[(**127**)] is synthesized from the bis(trimethylsilyl) analogue of (**31**) by treatment with RLi (R = Ph or Dur) followed, for R = Ph, by heating in toluene (110 °C). Results of a crystal structure analysis of Li[(**127**)] confirm that the anion possesses a folded (along B···B with internal angle = 153°) cyclic geometry.[207]

4.4.3 Five-membered Rings

Preparative routes to the 2,5-dihydroboroles (**128**) include the reaction of $MgC_4H_6{\cdot}THF$ with RBX_2 (X = Cl or Br) to yield (**128**) with various R groups.[208,209] The catalytic isomerization of (**128**) (e.g.,

R = Ph or NPr^i_2) occurs with $[(C_2H_4)_2RhCl_2]_2$ as the catalyst precursor.[209,210] Ethene reacts with $Pr^i_2NBCl_2$ and Na/K alloy to yield (**128**) with R = NPr^i_2; an analogous synthesis starting from but-2-ene is also successful, and the product here has been characterized by x-ray diffraction.[202] An attempt to prepare the acyclic compound $\{Me_3Sn(Me)C{=}C(Et)\}_2BEt$ gave, instead, (**129**). Variable-temperature multinuclear NMR spectroscopy shows that (**129**) is fluxional in solution at room temperature; the trimethylstannyl groups migrate around the ring.[211] The reaction of $Et_2NB(C{\equiv}CSnMe_3)_2$ with triethylborane gives the precursor shown in Equation (25). After heating this species at 60 °C, a mixture of the two isomers (**130**) and (**131**) is formed and on further heating, (**130**) converts to (**131**).[212] A pentasubstituted borole has been prepared as shown in Scheme 3. The direct reaction of the dilithio precursor with phenylboron dichloride has the problem that it generates, by virtue of the medium, the Et_2O adduct. In addition, $PhBCl_2$ may promote isomerization of the dilithio starting material. On the other hand, while the use of the stannyl intermediate allows a clean conversion to the final borole, the formation of the intermediate in the first place is not straightforward.[213]

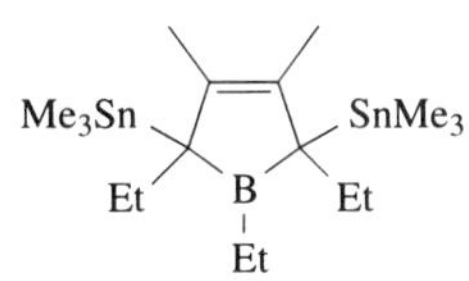

(**129**)

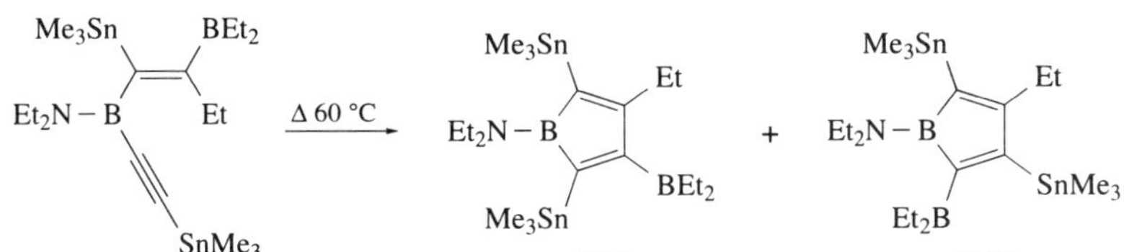

(**130**) (**131**) (25)

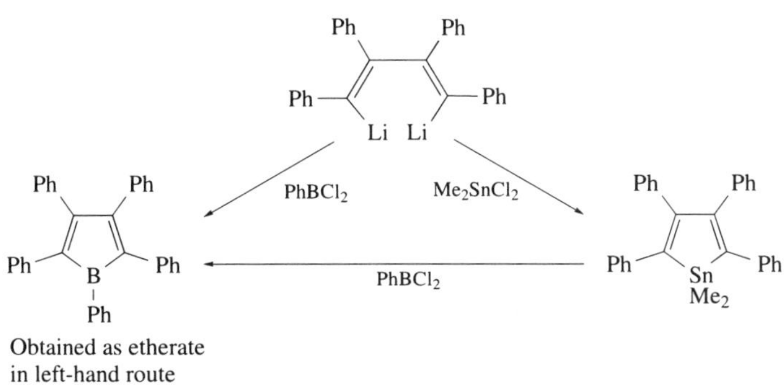

Scheme 3

Reactions of derivatives of (**128**) with alkynes have been studied and the products (**132**) have been spectroscopically characterized.[214] The reactivity of borole (**128**) (R = Ph) has been investigated. For example, it reacts with trimethylamine in ether to give the adduct (**133**), and with phenyllithium in ether, anion (**134**) is formed and is isolated as the $[Me_4N]^+$ salt.[208] The derivatives (**128**) (R^1 = dialkylamino) are quaternized by R^2Li unless steric hinderance at the boron centre can mitigate against the process. Such is the case for the sterically demanding LiTMP. Thus, (**128**) ($R^1 = NMe_2$ or NEt_2) reacts with LiTMP (R^2T) to give Li_2[(**135**)]; however, (**128**) ($R^1 = NPr^i_2$) does not react with LiTMP at 20 °C; the dilithium salt can be formed by using Bu^tLi as the base. The salts Li_2[(**135**)] form crystalline derivatives with TMEDA and the structure of Li_2[(**135**)]·TMEDA (R = Et) has been elucidated; it is a centrosymmetric dimer.[215]

Five-membered rings incorporating two boron atoms have been prepared by several methods. But-2-yne and hex-3-yne react with $I_2BCH_2BI_2$ (Equation (26)) and the diiodo derivative (**136**) can be converted to the dichloro, diethoxy and bis(methylthio) derivatives by reaction with ICl, Et_2O or $(MeS)_2$, respectively.[216] A range of compounds related to (**136**) is similarly derived from the reactions of $RCH(BI_2)_2$ (R = Me, Et, CH_2Bu^t). Treatment of the diiodo products with trimethyl- or triethylaluminum gives the corresponding di-*B*-alkyl derivatives.[217] Compounds (**137**) are formed in the reactions of $(Me_3Sn)_2C{=}CMe_2$ with $R^1(Cl)BC(Et){=}C(Et)B(Cl)R^1$ (R^1 = Me or Cl).[218] When $Cl_2BC(R^2){=}C(R^2)BCl_2$ is the precursor, *B,B'*-dichloro-derivatives related to (**137**) are formed. The chloro groups can be replaced

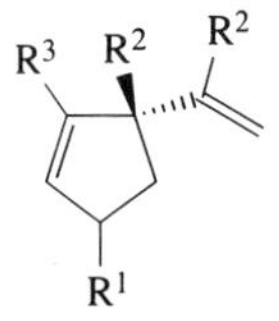

(132)
R^1 = Ph, Cy
R^2 = H, Me
R^3 = e.g. Bu, Bu^t, Ph, CH_2Cl

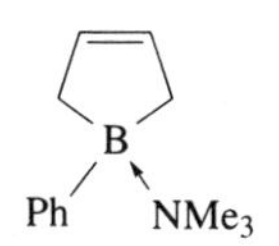

(133)

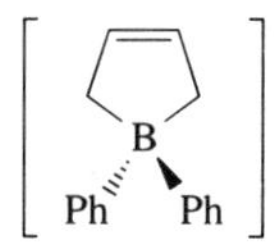

(134)

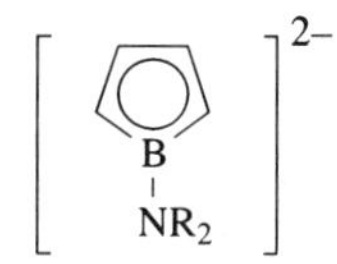

(135) R = Me, Et, Pr^i

by methyl or dialkylamino substituents.[219] On heating at 120 °C, (**137**) rearranges into a carbaborane cluster, the identity of which has not been established.[218] The reduction with potassium and subsequent methylation of a series of 1,3-diboroles have been reported.[220]

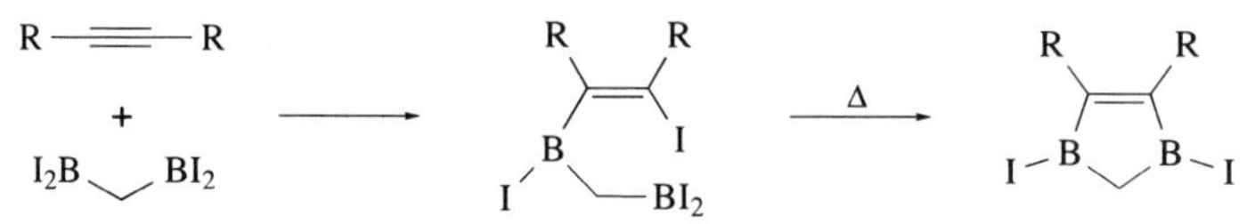

R = Me or Et

(136)

(26)

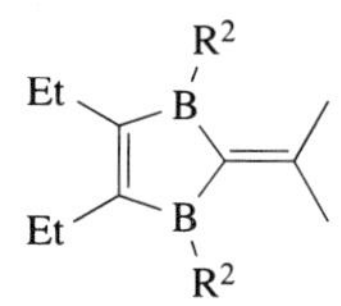

(137) R^2 = Cl or Me

4.4.4 Six-membered Rings

The use of stannyl precursors to form compounds (**138**) is well exemplified.[221,222] For RR = CH_2CH_2, (**138**) is an analogue of the phenonium ion and the unusual ^{1}H NMR spectral shifts observed for the boron-containing system are similar to those noted in the phenonium ion. This would suggest charge delocalization over the cyclopropyl ring, but the results of theoretical studies (STO–3G basis set) give no indication for such conjugation.[221] On heating (**138**) (R = H), an isomer equilibrium between the 1,4- and 1,3-dienes is established. When a mixture of the isomers is reacted with hexafluoro-2-butyne, only the 1,3-isomer undergoes a Diels–Alder cycloaddition; the products are only stable for X = OMe or NMe_2.[222] Ring opening of (**139**) occurs upon treatment with NaOMe and iodine.[223]

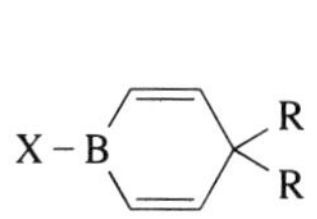

(138)
R = Me; X = Me
RR = CH_2CH_2; X = Me
R = H; X = Br, OMe, Cl, OH, F, NMe_2

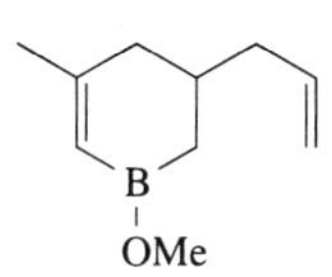

(139)
e.g. X = $MeOCH_2$, C_5H_{11}, Ph

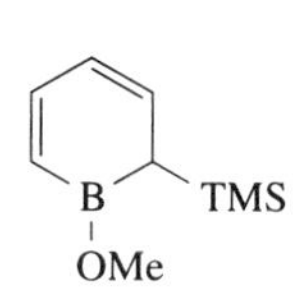

(140)

Borabenzene is an illusive species but has been trapped in the form of adducts. The pyridine adduct $C_5H_5B{\cdot}py$ has been prepared by the elimination of TMS-OMe from (**140**). ^{1}H NMR spectral and x-ray diffraction data support the aromatic character of the C_5B–ring in the product. The B–N distance is 0.1558(3) nm and the borabenzene and pyridine rings are mutually twisted by 43.3°. Similarly, the deep red pyridine adduct of 2-boranaphthalene has been prepared and characterized. The B–N distance (0.1515(15) nm) is slightly shorter than in the borabenzene adduct and the angle between the planes of the 2-boranaphthalene and pyridine units is only 8.1° suggesting some degree of conjugation.[224] The flash thermolysis of (**140**) in the presence of nitrogen yields a species proposed as $C_5H_5B{\cdot}N_2$. An absorption at 2197.6 cm^{-1} in the IR spectrum is assigned to ν(N≡N). Theory (MNDO) predicts that the reactions of borabenzene with N_2, CO and py are exothermic with enthalpies of reaction of −105.5 kJ mol^{-1}, −308.1 kJ mol^{-1} and −165.8 kJ mol^{-1}, respectively.[225]

A report of the formation of (**141**) ($R^1 = Bu^n$ and $R^2 = Me$) by the reaction of $Bu^nC{\equiv}CBu^n$ with $MeBBr_2$ and the intercalation compound C_8K has appeared,[226] but attempts to repeat this preparation have not been successful.[227] The initial work of van der Kerk *et al.* describes (**141**) ($R^1 = Bu^n$ and $R^2 = Me$) as an air- and moisture-sensitive viscous yellow liquid. In later work of Schlögl and Wrackmeyer, carbaborane clusters are obtained instead of the monocyclic product. The reduction of (**141**) ($R^1 = H$ and $R^2H = FeCp_2$) to the corresponding dianion occurs upon treatment with sodium amalgam.[228]

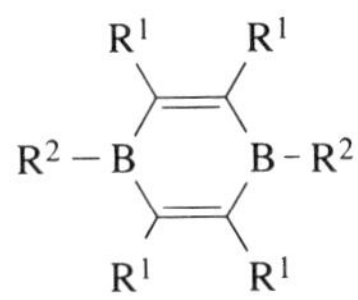

(141)

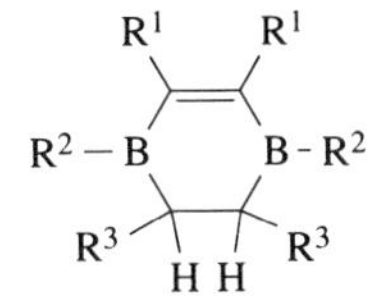

(142)

R^1 = Me, Et; R^2 = I, Me, NMe_2, OEt

R^3 = H, Me

The diboracyclohexenes (**142**) are prepared either from 1,2-bis(dichloroaluminyl)ethane, (via boron-for-aluminium displacement),[229,230] or from B_2Cl_4 and alkenes.[229] The reactivity and electrochemistry of these compounds have been studied.[229] Methylation of (**142**) ($R^1 = Me$, $R^2 = I$, $R^3 = H$) occurs at boron and the product has been characterized by multinuclear NMR spectroscopy; ring contraction via intramolecular isomerization occurs on heating (**142**) ($R^1 = R^2 = Me$, $R^3 = H$) at 160 °C.[230] Equation (27) again illustrates the use of stannyl precursors for the preparation of boron-containing rings; the products (**143**) rearrange between −30 °C and 25 °C to *nido*-tetracarbahexaboranes.[231]

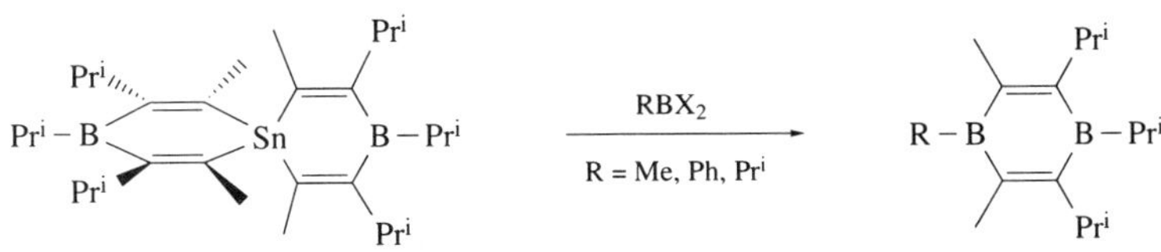

(143) (27)

The chemistry of 1,2-dibora-3,5-cyclohexadiene and 1,2-diborabenzene has been developed. $(Me_2NBCl)_2$ reacts with $C_4H_6Mg(THF)$ to yield the bis(dimethylamino) derivative of 1,2-dibora-4-cyclohexene (**102**). Treatment of this at 0 °C with lithium 2,2,6,6-tetramethylpiperidide leads to Li_2[(**144**)]. The crystal structure of $[Li(TMEDA)]_2$[(**144**)] has been determined; the C_4B_2-ring is planar and interacts with both lithium centres, one above and below the ring. There is good evidence for (**144**) possessing a delocalized 6π-electron system.[232] The cyclodimerization of (**102**) was described in Section 4.3.7[162]

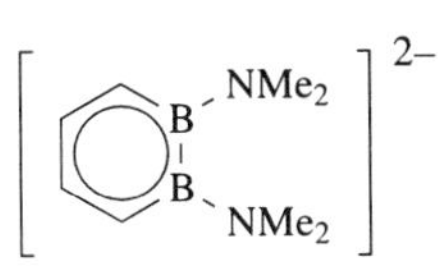

(144)

4.4.5 Seven- and Eight-membered Rings

Even though substituted derivatives have been known previously, the first synthesis of 1*H*-borepin (**145**) (R = H) was only reported in 1992.[233] It is produced when the *B*-chloro derivative is reduced with Bu_3SnH; evidence for (**145**) (R = H) comes from ^1H, ^{13}C and ^{11}B NMR spectroscopic data (^{11}B δ = +48.0, *J*(BH) = 99 Hz). The compound is moisture, oxygen and heat sensitive and has not been isolated. Derivative (**145**) (R = Me) has been prepared via boron-for-tin replacement (BMe for $SnBu^n{}_2$); the reaction is quite vigorous. The product is relatively thermally stable and undergoes no change when heated in CCl_4 at 160 °C for 1.5 h. In $CDCl_3$, however, some decomposition to benzene occurs and this is attributed to the presence of hydrogen ions.[234] Boron-for-tin exchange is also the basis of the preparation of (**146**)[235] and (**147**)[236] (see Section 4.4.6).

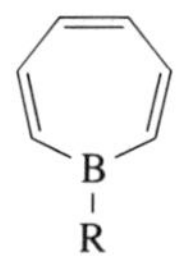

(**145**) R = H or Me

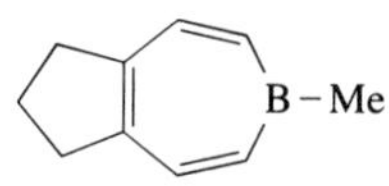

(**146**)

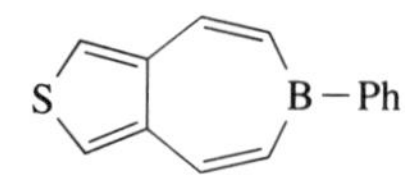

(**147**)

Heptaphenylborepin and 1-phenyl-2,3,4,5,6,7-hexa-*p*-tolylborepin have previously been prepared, each from an appropriate borole and alkyne. In a more recent report, Eisch *et al.* give a full account of the generation and thermal transformations of substituted borepin systems (see Section 4.4.6).[237]

The tetraborocine (**104**) is prepared via the dimerization of (**124**) in the presence of hydrogen and Lindlar catalyst; (**104**) has been characterized by ^{11}B and ^{1}H NMR spectroscopies and its crystal structure has been determined. The eight-membered ring is nonplanar and contains a twofold axis; B–B distances are 0.1718(2) nm.[238] The rearrangement of (**104**) into a cage structure[163] was discussed in Section 4.3.7

4.4.6 Bicyclic, Tricyclic and Larger Systems including Boron adjacent to an Aromatic Ring

A range of molecules involving unsaturated boron-containing rings has been reported in the ten years since Morris' survey of these systems.[1] One simple system is (**148**); an expansion of this borabicyclononene occurs according to Equation (28).[147] The dehydroboration of systems related to (**148**) has been studied.[239]

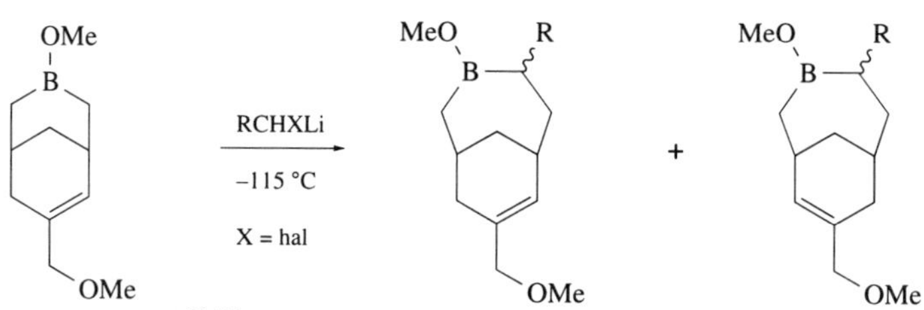

(**148**)

(28)

The use of $Pr^i_2NBX_2$ (X = F or Cl) as a source of {Pr^i_2NB}-units was highlighted in Section 4.3.7.[10,159–61] Its use is exemplified further here. Cyclooctatetraene reacts with Na/K alloy and $Pr^i_2NBX_2$ (X = F or Cl) to give (**149**); the nature of the ring system has been confirmed by x-ray crystallography and the results are consistent with multinuclear NMR spectral data.[240] 1,4-Disubstituted aromatics react with $Pr^i_2NBF_2$ and Na/K alloy to yield bicyclic products as shown in Equation (29). Structural confirmation is available for two of the derivatives and for a related dibutylamino derivative,[241] and the same framework is observed in a benzo-derivative.[242] When the starting materials are 1,4-$Pr^i_2C_6H_4$, $Pr^i_2NBF_2$ and Na/K, the product is (**150**), the crystal structure of which has been elucidated. A similar reaction occurs with benzene.[241] In contrast, (**151**) has been isolated from the reaction of one equivalent of benzene with two equivalents of $Pr^i_2NBCl_2$ (with Na/K) in DME. The spirocyclic structure is confirmed by x-ray analysis and is consistent with NMR spectral data.[243] The reaction between 1,2-$Me_2C_6H_4$ and $Pr^i_2NBF_2$ with Na/K yields (**152**) as well as an acyclic product. Structure (**152**) has been characterized by spectroscopic and x-ray diffraction methods.[244]

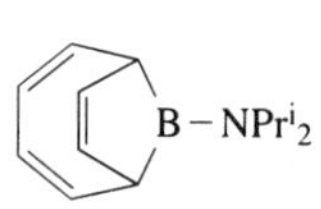

(**149**)

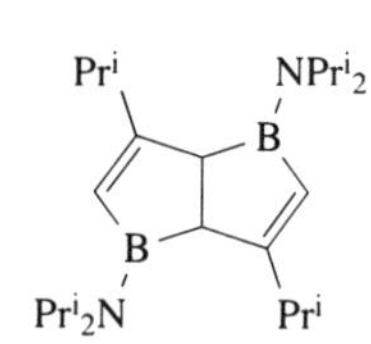

(**150**)

The reaction of $Li_2[(\mathbf{135})]$ (R = Pr^i) with $Pr^i_2NBCl_2$ in THF (−78 °C) leads to (**153**); this is the first 2,6-diborabicyclo[3.1.0]hex-3-ene derivative. Variable-temperature ^{1}H NMR spectroscopic studies

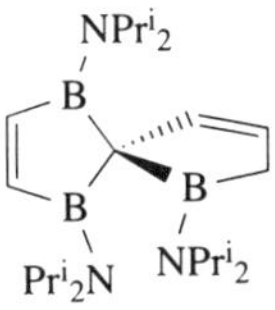

(151)

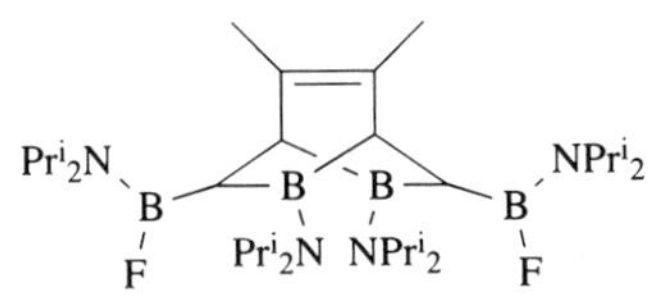

(152)

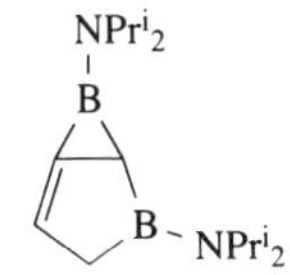

(153)

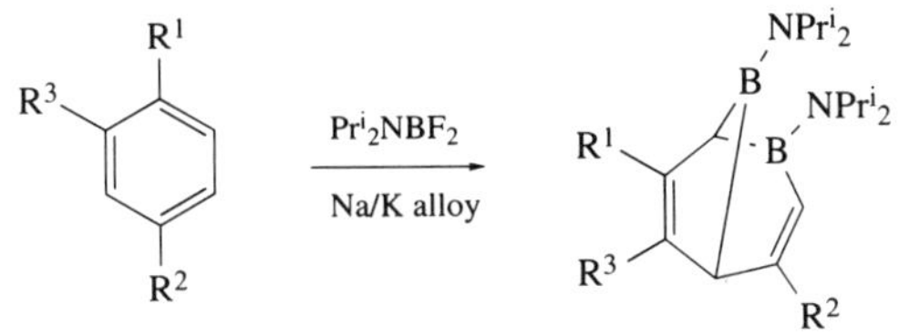

(29)

$R^1 = R^2 = Me$; $R^3 = H$ or Me
$R^1 = TMS$; $R^2 = NMe_2$; $R^3 = H$

illustrate that in solution (**153**) undergoes a tautomeric rearrangement ($\Delta G^{\ddagger} = 45 \pm 3$ kJ mol^{-1}).[245] Li$_2$[(**135**)] (R = Pr^i) is oxidized by tin(II) chloride to yield the corresponding neutral five-membered ring which dimerizes to give (**154**).[246] A derivative of (**154**) is observed as the ultimate product when Pr$^i_2$NB(CH=C=CMe$_2$)$_2$ is heated in toluene, the intermediate being the borole.[247] Similarly, when 1,2-{B(Pr$^i_2$N)CH=C=CMe$_2$}$_2$C$_6$H$_4$ is heated in mesitylene (165 °C), cyclization occurs. The product undergoes a photochemical 1,5-hydrogen shift to give (**155**); both of the cyclization products have been structurally characterized.[247]

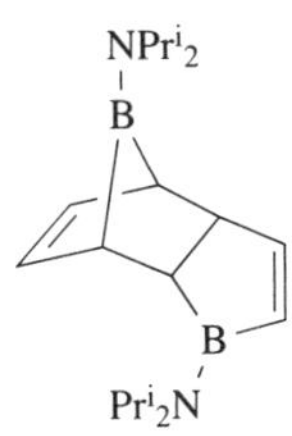

(154)

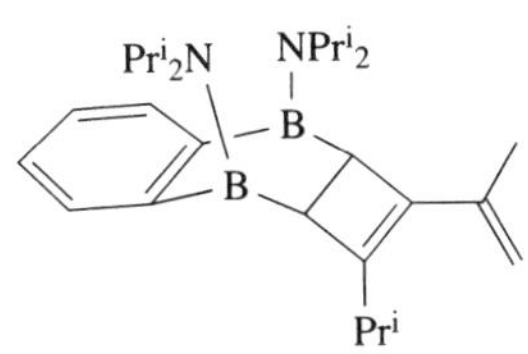

(155)

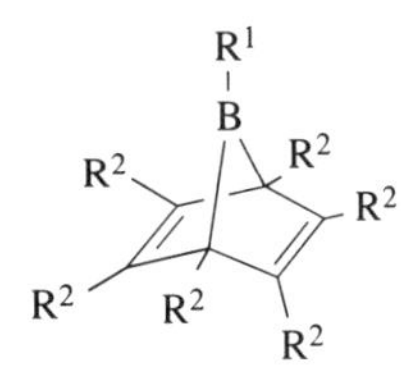

(156)

$R^2 = Ph$; $R^1 = Ph$, Me, Cl
$R^2 = 4\text{-MeC}_6H_4$; $R^1 = Ph$

Bicyclic borepin derivatives were mentioned in Section 4.4.5. The UV spectrum of (**146**) exhibits absorptions at $\lambda = 223$ nm and 315 nm, indicating that there is extensive conjugation between the boron atom and the hexatriene system.[235] Diels–Alder adducts (**156**) (formed from reactions of boroles and alkynes) are chemically degraded to benzene derivatives. Thermal rearrangement of such adducts occurs upon heating above 100 °C; a colour change to fluorescent green accompanies the isomerization. The structure of product (**157**) has been established by x-ray crystallography. The addition of NH$_3$ or py to (**157**) and its congeners causes a dramatic loss of colour, due to a hypsochromatic shift from 415 nm to 330 nm.[237]

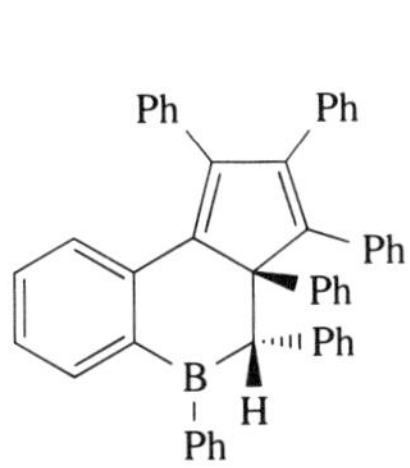

(157)

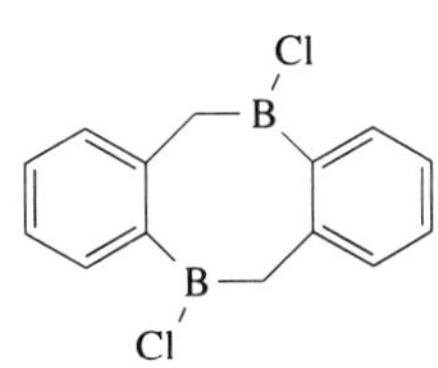

(158)

Compound (**157**) provides a link to a discussion of ring systems in which the ring–boron atom lies in a site adjacent to an aromatic ring. This location gives the possibility of interactions between π-electrons of the aromatic ring and the available $2p$ AO on the boron atom. Routes to such systems vary. The preparation of (**29**) was detailed in Section 4.3.1.[65] Ring closures are achieved by flash vacuum pyrolysis, for example, 1-(BCl$_2$)-2-EtC$_6$H$_4$ (700 °C),[247,248] 1-(BCl$_2$)-2-Pr^{n}C$_6$H$_4$ (700 °C)[247] and

1-(CH_2BCl_2)-2-MeC_6H_4 (750 °C).[247] When $PhCH_2BCl_2$ is heated at 950 °C, the dimer (**158**) is produced.[247] Dehydrogenation and dimerization are observed when the bicyclic product from the ring closure of 1-(BCl_2)-2-EtC_6H_4 is heated further (900 °C). It is proposed that the intermediate is an unsaturated bicyclic system.[248] The acyclic compounds 1-BRCl-2-(CH_2Br)C_6H_4 (R = Me, Bu^i, NPr^i_2) are the first of their type. For R = NPr^i_2, cyclization occurs as shown in Scheme 4. Ring opening can again be induced, for example, by treatment with BCl_3 as illustrated.[249]

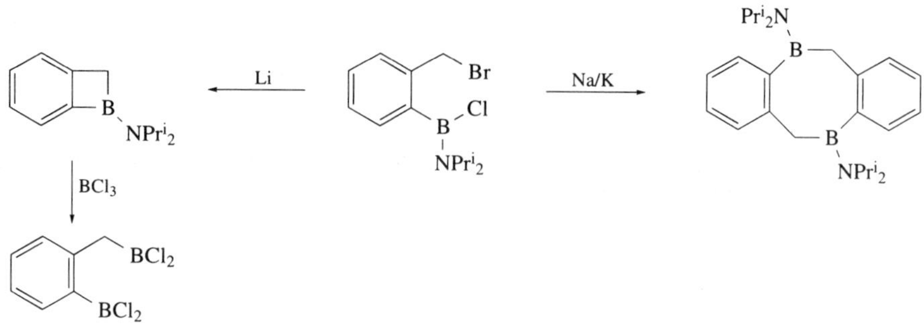

Scheme 4

The elimination of R_3SiX or R_3SnX (X = halogen; R = alkyl) is a route used to facilitate ring closure. An example is the interaction of two molecules of 1-(BClY)-2-(TMS)C_6H_4 (Y = Cl or Br).[250] The reaction of 1,2-$(SnMe_3)_2C_6H_4$ with $Cl_2BC(Et)$=$C(Et)BCl_2$ leads to (**159**) and, similarly, that of 1,2-$(BCl_2)_2C_6H_4$ with $(Me_3Sn)_2C$=CMe_2 gives (**160**). The chloro products can be derivatized (e.g., methylated with $AlMe_3$) and the new boracycles act as π-ligands to transition metals.[251] A method comparable with that used for the synthesis of (**160**) is used to prepare compound (**161**) (R = Cl). The crystal structure of (**161**) (R = NPr^i_2) has been elucidated; the seven-membered ring is folded. Related monocyclic systems have also been prepared.[252]

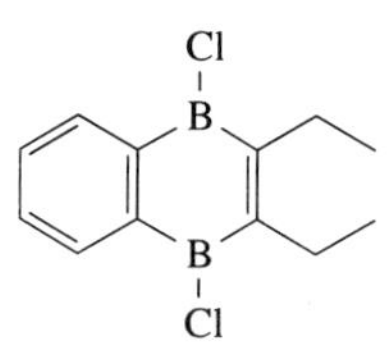

(159)

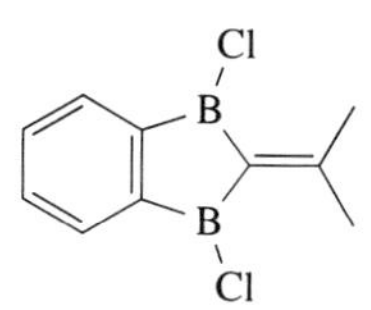

(160)

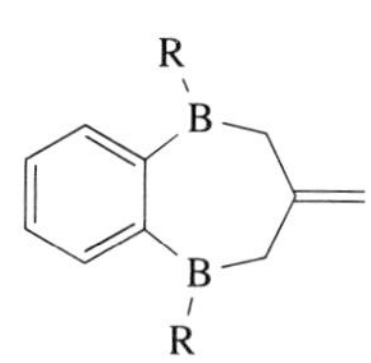

(161) R = Cl or NPr^i_2

An efficient synthesis of *B*-halo-dibenzoboroles is summarized in Equation (30). For X = Cl, reactions of (**162**) with MeOH in the presence of 2,6-Me_2py, with $Pb(SMe)_2$, and with $NHEt_2$ yield (**162**) with X = OMe, SMe and NEt_2 respectively.[253] Compound (**162**) (X = Br) is also accessed by the exchange of a {BBr} for {$SiMe_2$} unit.[254] Adducts (**162**)·L (X = Cl; L = py or methyl-substituted pyridines) have been prepared and the crystal structure of (**162**)·py (X = Cl) has been determined; the boron atom is sited tetrahedrally. Salt (instead of adduct) formation is also observed. In the reaction of (**162**)·L (L = acridine; X = Cl) with $AlCl_3$, chloride abstraction to yield $[AlCl_4]^-$ competes with base abstraction.[125]

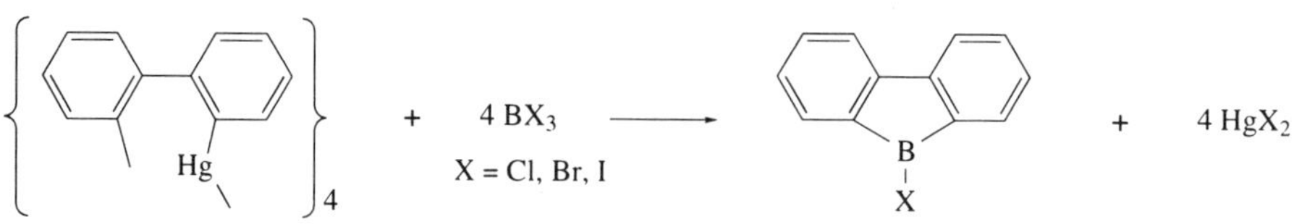

(30)

(162)

The electronic spectroscopic properties of several derivatives of dibenzoborole and of (**163**) have been studied.[255,256] The irradiation of (**163**) (R = Mes; X = C=N_2) results in loss of dinitrogen and the formation of a carbene. Reactions of this species have been investigated.[257]

When 2,4,6-$Bu^t_3C_6H_2B(OMe)_2$ is treated with LAH in the presence of TMEDA, [Li(TMEDA)][2,4,6-$Bu^t_3C_6H_2BH_3$] is formed. Subsequent hydrolysis in strongly acidic solution results in ring closure involving one of the *t*-butyl substituents.[258] An interesting cyclization is brought about in the three-step synthesis of compounds (**164**) from 1,3-$(CH_2Br)_2C_6H_3Br$. The resolution of the complexes represents the

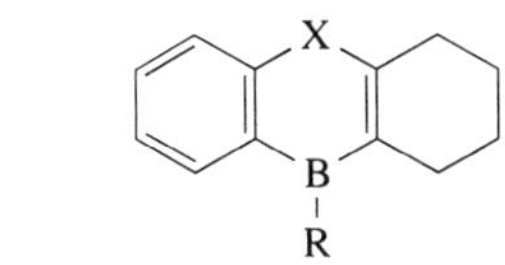

(163)
R = various e.g. X = CH_2, O, NMe

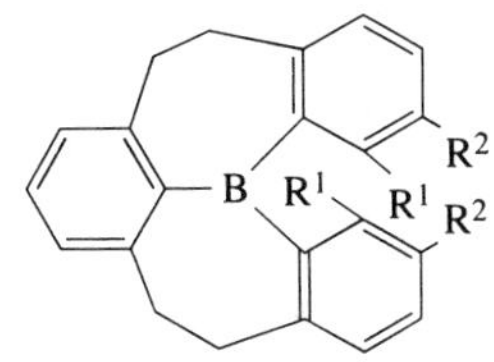

(164)
$R^1 = R^2 = H$
R^1 = Me; R^2 = H
R^1R^2 = CH=CHCH=CH

first example of optical resolution of triarylboranes. The crystal structure of (**164**) (R^1 = Me, R^2 = H) has been determined; B–C distances are in the range 0.1550–0.1574 nm and the environment about the boron atom is close to trigonal planar.[259]

4.5 RINGS CONTAINING BORON AND SILICON ATOMS

All ring sizes are discussed together in this section, since the number of compounds is relatively small.

Boron-containing products have been isolated during an investigation of the hydroboration of dialkenylsilanes. Treatment of $Me_2Si(CMe{=}CH_2)_2$ with $BH_3{\cdot}SMe_2$/MeOH or thexylborane leads to (**165**) with R = OMe or CMe_2CHMe_2, respectively. The latter complex isomerizes at 180 °C.[260]

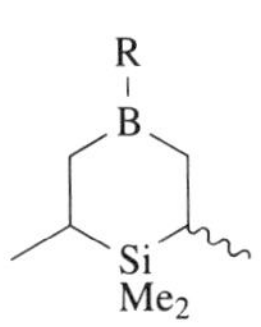

(**165**) R = OMe or CMe_2CHMe_2

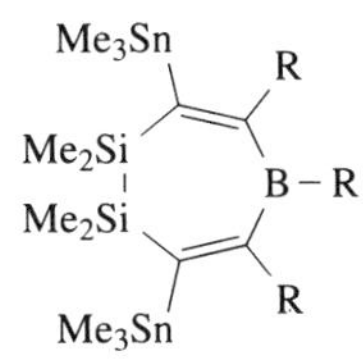

(**166**) R = Me, Et, Pr^i

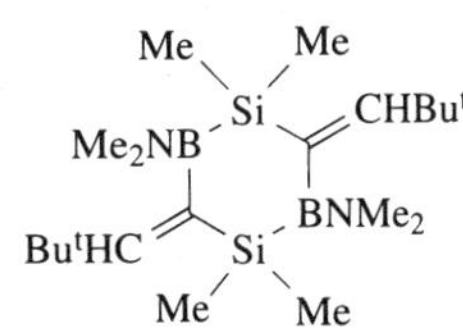

(**167**)

When BR_3 (R = Me, Et, Pr^i) reacts with $(Me_3SnC{\equiv}CSiMe_2)_2$, cyclization occurs to give compounds (**166**) which have been spectroscopically characterized. For R = Pr^i, the reaction proceeds slowly and intermediate complexes can be observed.[261] Similar reactions occur between $(HC{\equiv}CSiMe_2)_2$ and BR_3 (R = Me or Et) but for R = Pr^i, a five-membered ring is formed with the boron atom incorporated into an exo-cyclic substituent.[262] The regiospecific hydroboration of $Me_2ClSiC{\equiv}CBu^t$ with $HBCl_2$ leads to (*E*)-Bu^{t}CH=C($SiMe_2Cl$)BCl_2; this reacts with Me_3SiNMe_2 to give (*E*)-Bu^{t}CH=C($SiMe_2Cl$)B(Cl)NMe_2, which undergoes ring closure upon treatment with Na/K alloy. The six-membered ring (**167**) is obtained, although both three-membered and other isomers of the six-membered ring might be expected. Compound (**167**) has been spectroscopically and crystallographically characterized; the ring has a chair conformation.[263] (*E*)-Bu^{t}CH=C($SiMe_2Cl$)B(Cl)NMe_2 is also the precursor to the five-membered ring (**168**) which is nonplanar and exhibits a strained C–Si–C angle of 94.9(1)°.[264]

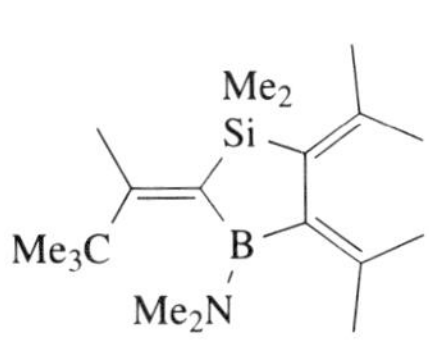

(168)

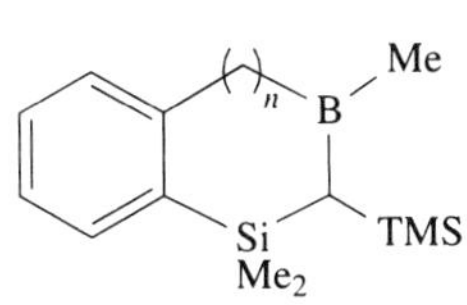

(**169**) *n* = 0 or 1

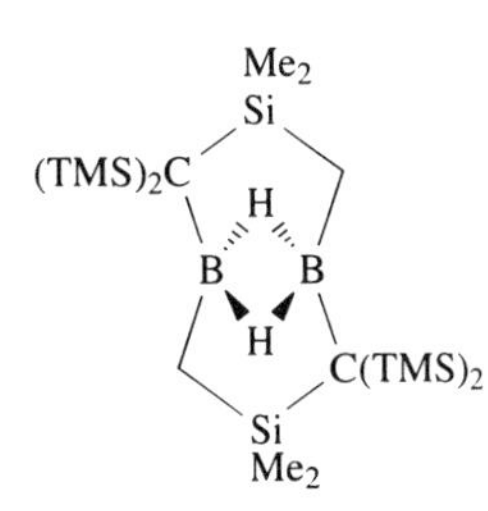

(170)

Heating RB(OMe)$C(TMS)_3$ (R = Ph or $PhCH_2$) at 560 °C leads to ring closure and the formation of (**169**) with *n* = 0 or 1, respectively.[265] When $(TMS)_3CBCl_2$ reacts with Na/K alloy, C–H bond activation occurs and (**170**) forms. The product has been characterized by IR and NMR spectroscopies and a crystal structure determination shows that the C_4B_2 unit is planar with the two $SiMe_2$ groups lying above and below the plane, respectively. The planes of the C_4B_2 and B_2H_2 bridge units are mutually perpendicular.[266]

4.6 RINGS CONTAINING BORON AND TIN, GERMANIUM AND LEAD ATOMS

Ring formation via intramolecular alkyne donation to tin or lead substituents has been observed by Wrackmeyer and co-workers as a result of the organoboration of bis(trimethylsilylethynyl)dialkylstannanes and plumbanes. Closed five-membered rings can be obtained from the π-stabilized species.[267–9] Earlier work by Wrackmeyer and co-workers reports the formation of (**171**) as one of the products in the reaction of $Me_2Sn(C{\equiv}CPh)_2$ with BR_3 (R = Me, Et, or Pr^i).[270] When $Me_2(NEt_2)SnC{\equiv}CMe$ (which exhibits a strongly basic amino group) reacts with *B*-R-9-BBN (R = Me or Et), the products are (**172**); the cyclic structure is stabilized by intramolecular N → B coordination. Elimination of diethylamine occurs when compound (**172**) is heated at 80 °C, and the cyclic products have been spectroscopically characterized.[271]

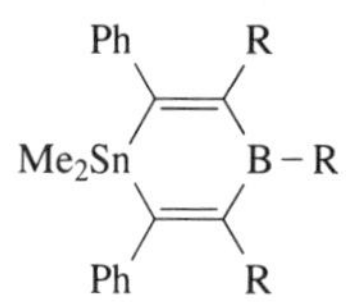

(**171**) R = Me, Et, Pr^i

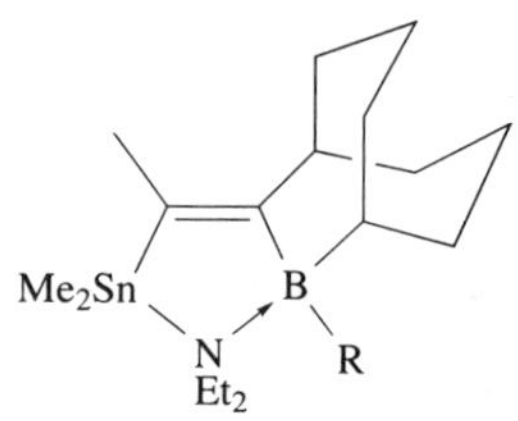

(**172**) R = Me, Et

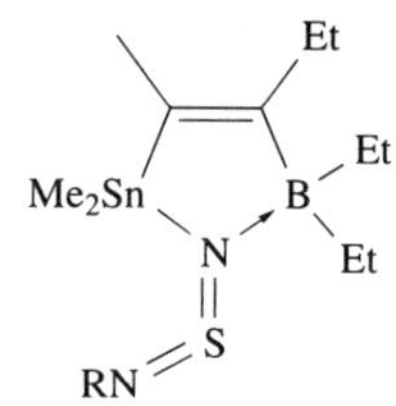

(**173**)

A range of heterocyclic molecules have been isolated in the reactions of $(Me_2ClSn)MeC{=}CEt(BEt_2)$ with the sulfur diimide salts, K[RNSN] (R = Bu^t, TMS and $Bu^t{}_2P$) and $K_2[NSN]$. The crystal structures of two products (**173**) (R = Bu^t and TMS) have been determined. The B–N distances in the two compounds are 0.1636(2) and 0.1628(4) nm, respectively.[272]

A novel $BCGe_3$ ring is formed by the addition of three Me_2Ge units to 9-fluorenylidene(tetramethyl-piperidino)borane. The product (**174**) has been characterized by multinuclear NMR spectroscopy and by x-ray diffraction. The five-membered ring has an envelope conformation and distances within the ring are B–C = 0.1615(15) nm, C–Ge = 0.2076(9) nm, B–Ge = 0.2165(10) nm and Ge–Ge = 0.2402(2) and 0.2435(2) nm.[273]

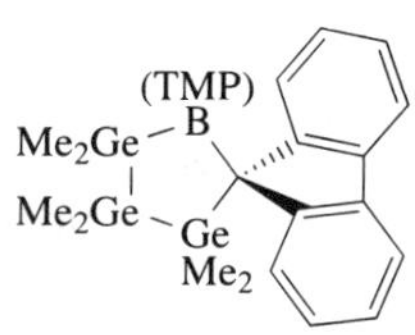

(**174**)

4.7 RINGS CONTAINING BORON AND NITROGEN ATOMS

Reviews covering aspects of cyclic compounds containing boron and nitrogen atoms are listed in Table 1. The chemistry of borazines is beyond the scope of this review. References concerning this field are contained within reviews by Maringgele[27c] and Paetzold.[12] In addition, the condensation of borazines is important;[274,275] work by Nöth and co-workers describes the synthesis of a B_6N_7 phenalene, a fragment of a layer with a graphite-like structure.[276]

4.7.1 Three-membered Rings

Three-membered rings fall mainly into the two categories of B_2N and BN_2 systems. One BNC ring has been prepared by the addition of an $\{NR^1\}$ unit (from R^1N_3 with R^1 = Ph or TMS) to $(TMP)B{=}CR^2{}_2$.[277]

The BN_2 framework is assembled in the reactions of $\{(TMP)ClB\}_2N(Pr^i)NH(Pr^i)$ with Bu^tLi,[278] (TMS)RNB(F)N(TBDMS)NH(TBDMS) (R = TMS or Bu^t) with Bu^tLi,[279] and $\{(Bu^tMe_2Si)NLi\}_2$ with $(TMS)_2NBF_2$.[280] For the product (**175**), NMR spectroscopic data suggest that the isopropyl groups lie out of the plane of the ring; this contrasts with the results for the B_2N ring described below.[278] On the other hand, multinuclear NMR spectral data for (**176**) suggest equivalent TBDMS groups lying in the ring plane[279] and this has been confirmed by x-ray diffraction.[280] The *exo*-cyclic B–N bond is shorter (0.1406(3) nm) than the *endo*-cyclic bonds (0.1426(3) nm).

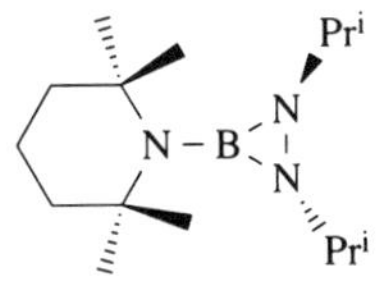

(175)

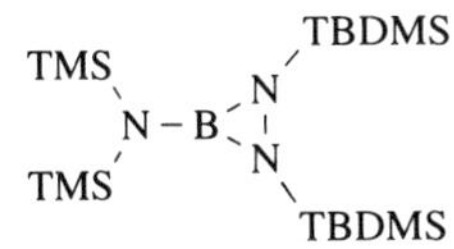

(176)

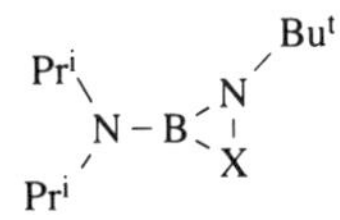

(177)
X = B(NPr^i_2), $SiMe_2$, SiMeEt, or P(NPr^i_2)

The B_2N heterocycle {(TMP)B}$_2NBu^t$ can be prepared by the dehalogenation of {(TMP)ClB}$_2NBu^t$.[278,281] NMR spectral data indicate that the substituents lie in the plane of the ring[278] and are consistent with there being no rotation about the *exo*-cyclic B–N bonds.[281] A crystal structure of {(TMP)B}$_2NBu^t$, however, confirms that in the solid state, only one tetramethylpiperidino substituent lies in plane.[281] $(Bu^tB)_2NBu^t$ is similarly prepared by reacting $(Bu^tClB)_2NBu^t$ (synthesized from $Bu^tB{\equiv}NBu^t$) with Na/K alloy.[282] Related reactions (see also Section 4.7.3) lead to compounds (**177**), which have been characterized in solution by NMR spectroscopy. The crystal structure of $(Pr^i_2NB)_2NBu^t$ has been elucidated.[283]

The reaction of $BH_3{\cdot}THF$ with the heterocycle $(Bu^tB)_2NBu^t$ was expected[284] to give an adduct, but a bicyclic product (**178**), was obtained. The structure of (**178**) (considered to be a *nido*-cluster) has been confirmed crystallographically and, in solution, NMR spectral data indicate the retention of a rigid structure. The azadiboriridine $(Bu^tB)(Pr^iB)NBu^t$ undergoes a similar addition reaction. The results of *ab initio* calculations on model systems are consistent with the experimental observations for (**178**) and its congener.[284] Each three-membered ring in $(Bu^tB)_2NBu^t$ and $(Bu^tB)(Pr^iB)NBu^t$ is expanded upon reaction with hex-3-yne. The products are (**179**) and they are reduced by the action of lithium in THF. Ring expansion of $(Bu^tB)_2NBu^t$ with $EtB{\equiv}NBu^t$ leads to (**180**) which has been structurally characterized. B–N distances within the ring range from 0.1404(2) nm to 0.1491(2) nm and the B–B bond length is 0.1789(2) nm.[285]

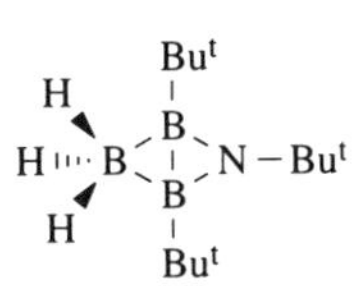

(178)

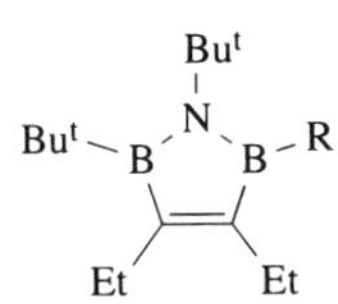

(179) R = Bu^t or Pr^i

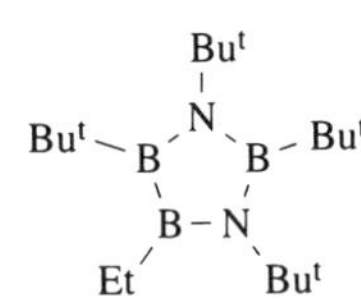

(180)

4.7.2 Four-membered Rings

The chemistry of the B_2N_2–ring is now quite well developed (see also Structure (**112**)). Preparative methods include the degradation of the borazine {FB–NR}$_3$ (R = Bu^t or TMS) by Bu^tLi to give {Bu^tBNR}$_2$, but here, the ring size of the product is dependent upon the steric requirements of the substituents. For R = Pr^i, the first product is a trimer but on heating to 200 °C, this degrades to a dimer.[286] Ring contraction is also observed when (**181**) is treated with sodium to give the product {R^2BNR^1}$_2$ (R^1, R^2 and R^3 as in (**181**)). Related reactions in which small sulfur-containing rings are produced have been explored.[287] The elimination of TMS-F from TMS-N(R)BBu^tF (R = 2,6-$Pr^i_2C_6H_3$) on heating is accompanied by the formation of {$RNBBu^t$}$_2$.[288] The formation of a BN_2Si ring has also been observed.[288] Related to this is the incorporation of other group 14 elements in the reactions of ECl_4 with PhB(NRLi)$_2$ (R = Pr^i or Bu^t for E = Ge, and R = Bu^t for E = Sn). These proceed with loss of lithium chloride to give compounds (**182**).[289] Nöth and co-workers have shown that the Sn_2N_2 ring of {Me_2SnNBu^t}$_2$ can be transformed into a B_2N_2 ring by treating with RBX_2 (R = Me, Et, Pr^i, Ph or Cl; X = Cl or Br). An alternative route to {$RBNBu^t$}$_2$ (R = Me) is the reaction of $Bu^tN(SnMe_2Br)_2$ with $MeBBr_2$. The reactions have been monitored by NMR spectroscopy. For R = Me, {$RBNBu^t$}$_2$ dimerizes and the nature of the tetramer has been confirmed by x-ray crystallography.[290] In contrast to the dehalogenation of {$(Bu^t_2N)ClB$}$_2NBu^t$ by alkali metal which leads to a three-membered heterocycle, dehalogenation by Li_2EPh (E = P or As) yields a four-membered ring in which the group 15 element is incorporated.[283]

Iminoboranes $R^1B{\equiv}NR^2$ are suitable precursors to B_2N_2 systems.[12] The cyclic dimer {(TMP)$BNBu^t$}$_2$ has been the subject of a single crystal x-ray diffraction study. The four-membered ring is planar and

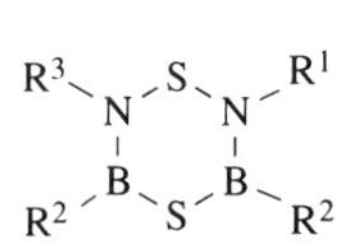

(181)
$R^1 = R^3 = Bu^t$; $R^2 = Me$
$R^1 = Bu^t$, Ph, TMS; $R^2 = NEt_2$; $R^3 = TMS$

(182)
E = Ge; R = Bu^t or Pr^i
E = Sn; R = Bu^t

diamond-shaped with the piperidino groups lying orthogonal to the B_2N_2 plane. The *endo*-cyclic B–N distances are equivalent within experimental error (0.1467 nm).[291] A series of reactions between (TMP)B≡NBu^t and boron halides lead to cationic ring systems.[292] The iminoborane $Pr^iB{\equiv}NBu^t$ is metastable at −78 °C and trimerizes readily. Interestingly, the trimer exhibits a Dewar benzene-like structure with the central B–N bond being significantly longer than the two outer edges in keeping with a localized bonding picture[293] (see also the work of Meller and co-workers).[286] Similarly, compounds $RB{\equiv}NBu^t$ (R = Me, Et and Pr^n) trimerize (50 °C) but this time form the expected borazines. Under suitable conditions (−10 to −20 °C), dimers are formed. For R = Bu, the dimer is kinetically stable with respect to the trimer but the latter is produced when $BuB{\equiv}NBu^t$ and $\{BuBNBu^t\}_2$ react together.[294] For R = Bu^t, both the monomer and dimer have been studied in detail and this includes IR and photoelectron spectral analyses and crystallographic characterization.[295] In $Bu^tB{\equiv}NBu^t$, the CBNC framework is linear and the B–N bond length is 0.1258(4) nm. The structure of the dimer[295] compares favourably with that of the piperidino-derivative described above.[291] The treatment of $RB{\equiv}NBu^t$ (R = Me, Et, Bu) with methyl lithium in the presence of TMEDA gives compounds (**183**) in which the methyl group has been transferred to one boron atom and the four-membered ring is nonplanar.[296] Related B_2N_2 cyclic systems containing one three- and one four-coordinate boron atom have been studied by Nöth and co-workers.[297]

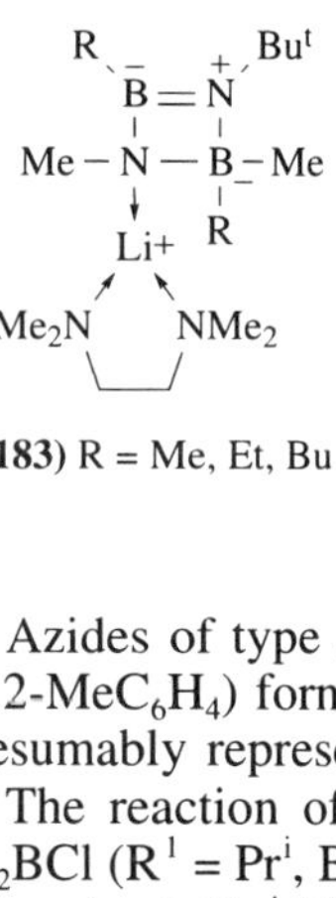

(183) R = Me, Et, Bu

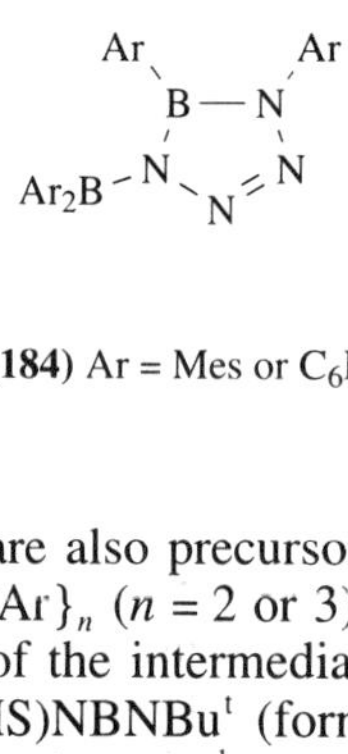

(184) Ar = Mes or C_6F_5

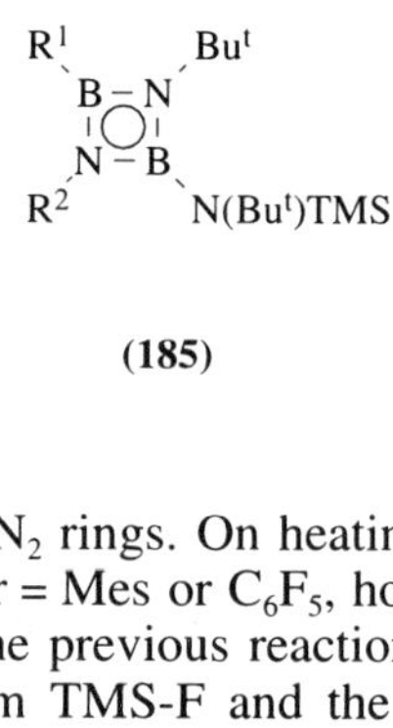

(185)

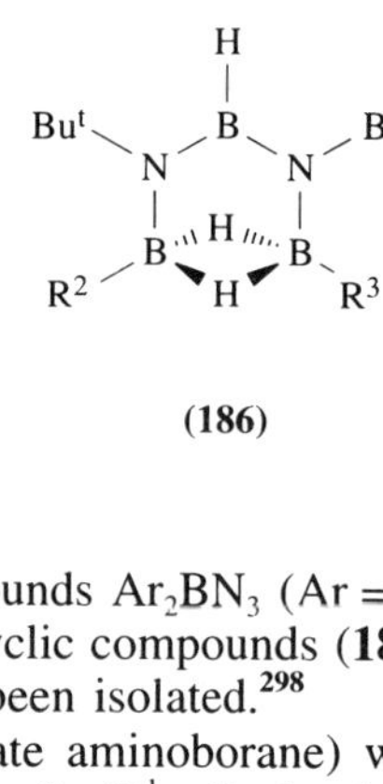

(186)

Azides of type Ar_2BN_3 are also precursors to B_2N_2 rings. On heating, compounds Ar_2BN_3 (Ar = Ph or 2-MeC_6H_4) form $\{ArBNAr\}_n$ (n = 2 or 3). For Ar = Mes or C_6F_5, however, cyclic compounds (**184**), presumably representative of the intermediates in the previous reactions, have been isolated.[298]

The reaction of Bu^t(TMS)$NBNBu^t$ (formed from TMS-F and the appropriate aminoborane) with $R^1{}_2BCl$ ($R^1 = Pr^i$, Bu, Bu^i) leads to $\{R^1BNBu^t\}_2$. The crystal structure of this dimer for R^1 = Bu has been determined. Bu^t(TMS)$NBNBu^t$ reacts with the iminoboranes $R^2N{\equiv}BR^1$ ($R^2 = Bu^t$ and R^1 = N(Bu)TMS; $R^2 = R^1 = Pr^i$) to yield (**185**), and with aldehydes to give BNCO-ring systems.[299]

The reactivity of $\{R^1BNBu^t\}_2$ (R^1 = Me, Et, Bu^n, Bu^t) with Lewis acids, protic agents, alkynes and ketones,[300] azides RN_3 (R = $PhCH_2$, Pr, Bu^i, Ph) and the nitrone PhHCN(Me)O[301] have been studied by Paetzold and co-workers. With BH_3, the rings $\{R^1BNBu^t\}_2$ (R^1 = Et, Pr, Bu^n, Bu^t, N(TMS)Bu^t) react to give the diazapentaboranes (**186**) in which R^2 and R^3 = H or R^1. The structure of (**186**) ($R^2 = Bu^n$ and R^3 = H) has been determined.[302] Ring expansion to borazines occurs when the iminoboranes $R^2N{\equiv}BR^1$ react with $\{R^2NBR^1\}_2$ (R^1 = Me, Et, Pr, Pr^i; $R^2 = Pr^i$, Bu^t).[303]

A B_2CN ring is accessed from MeB=C(TMS)$_2$. The [2 + 2]-cycloadditions of RB=C(TMS)$_2$ (R = Me or Bu^t) lead to (**187**). For R = Bu^t, the monomeric starting material can be stored but for R = Me, dimerization is facile. The reaction of RB=C(TMS)$_2$ with Ph_2CO (R = Me or Bu^t) gives a BC_2O-ring and that of MeB=C(TMS)$_2$ with $Bu^tN{\equiv}BBu^t$ gives (**188**).[304]

4.7.3 Five-membered Rings

This section is arranged according to the numbers of boron and nitrogen atoms in the five-membered ring and with respect to the degree of ring saturation.

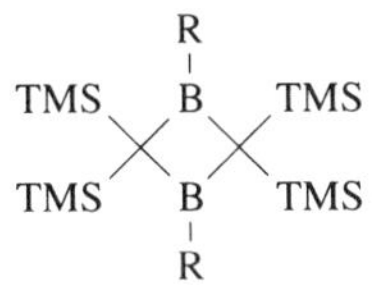

(187) R = Me or Bu^t

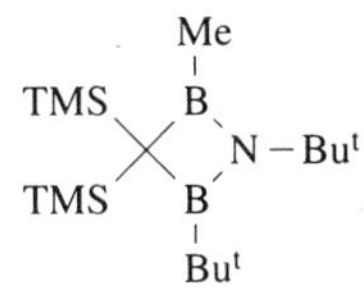

(188)

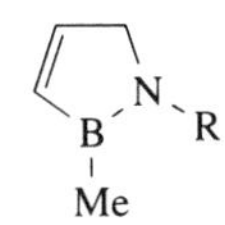

(189) R = TMS, Bu^t

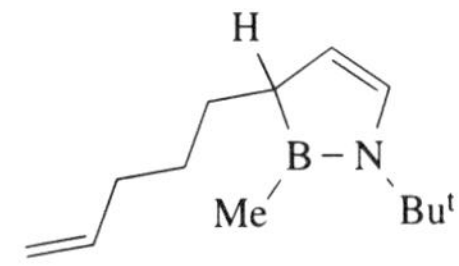

(190)

The unsaturated ring (**189**) (R = TMS) is deprotonated by Li(TMP) in THF (−78 °C) to generate a cyclic anion that can be used to form metal sandwich complexes. MNDO calculations have been used to investigate both (**189**) and its conjugate base.[305] For R = Bu^t, the lithium salt of the anion of (**189**) reacts with CH_2=$CH(CH_2)_3Br$ to give (**190**). Although (**190**) is isolable by distillation and can be characterized by spectroscopic methods, it undergoes a reversible rearrangement via B–C bond coupling to give a tricyclic system.[306] In a related system, the elimination of LiCl drives the coupling (by $SiCl_4$) of two azaborole rings to give (**191**) which has been crystallographically characterized.[307]

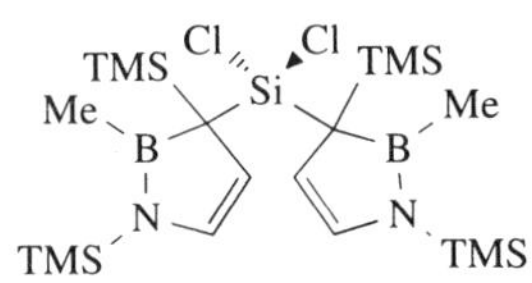

(191)

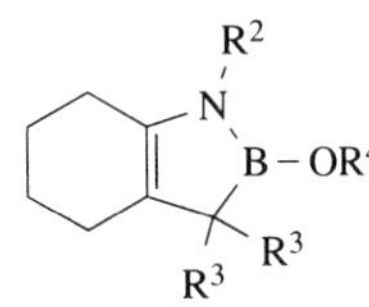

(192)

Ring closure occurs upon the carbonylation of $R^1N(R^2)BR^3_2$ (R^1 = c-C_6H_9; R^2 = Ph or Bu^i; R^3 = Bu or Pr) and treatment with R^4OH (R^4 = Hex^n or Bu^n) to give compounds of type (**192**).[308] Electron impact mass spectrometric data for a range of compounds related to (**192**) have been detailed[309] as have data for a series of similar saturated systems.[310]

The C_2BN_2 heterocycle (**193**) is prepared from Pr^{i}N=C=NPr$^i_2$, $Pr^i_2NBF_2$ and Na/K alloy. The solvent is critical; in hexane, (**193**) predominates but in DME, four products are obtained. The ring in (**193**) is disrupted upon alkaline hydrolysis.[311] The bulky Cy substituents help to stabilize (**194**); it is stable towards air and water. A crystal structure determination reveals that in (**194**), the two five-membered rings are mutually twisted by 80.7° and the B–B distance is 0.1721(9) nm. By comparing PES data for (**194**) and $B_2(NMe_2)_4$, it is proposed that (**194**) adopts a staggered conformation in the gas phase.[312] The reaction of the chloro-derivative (**195**) (X = Cl) with $(Bu^tO)_2SiBr_2$ and potassium leads to a mixture of the products (**196**). The structures are proposed on the basis of spectroscopic data.[313] The transamination of tris(dimethylamino)borane with α,ω-dimethylalkylenediamines has been studied; the bridged species (**197**) reacts with BCl_3 to give (**195**) (X = Cl). Heterocycle (**195**) (X = F) is formed when (**195**) (X = SMe) is treated with SbF_3. The fluoro-compound dimerizes in an unusual manner to (**198**) and this compound is fluxional in solution.[314] The reaction of (**195**) (X = OSO_2CF_3) with phenanthridine leads to the formation of the salt [(**199**)][SO_3CF_3]; this result has been confirmed by x-ray crystallography. Such a salt cannot be stabilized by pyridine.[315] Adducts form between (**195**) (X = Me, Ph, Cl or Br) and R_nBY_{3-n} (Y = halide; n = 0–2). A single crystal x-ray diffraction study for the adduct with X = Y = Cl illustrates that there are two enantiomers in the unit cell; each C_2N_2B ring has an envelope conformation, and there is evidence for strong π-bonding within the ring between the boron atom and the nitrogen atom not involved in coordination to the boron halide.[316]

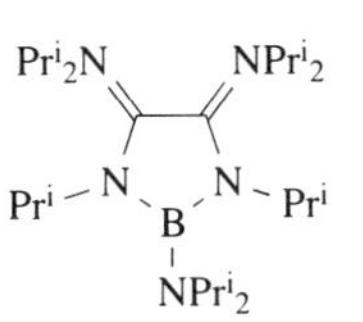

(193)

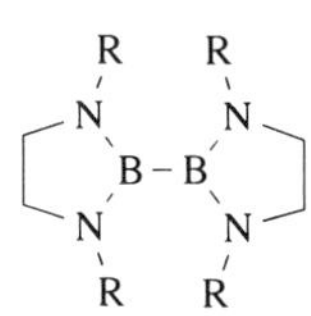

(194) R = Cy

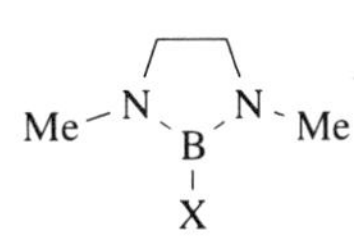

(195)

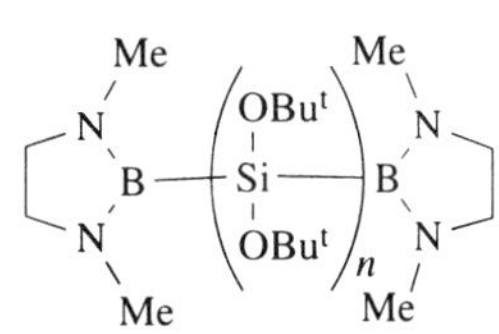

(196) n = 1 or 2

A family of substituted 1,3,2-diazaborolidines (saturated heterocycles) has been prepared. Methods used include the reaction of $MeB(NMe_2)_2$ with $R^1HN(CH_2)_2NHR^2$ (R^1 = H; R^2 = Pr^i, Bu^t, TMS), the reaction of $MeBBr_2$ and $R^1HN(CH_2)_2NHR^2$ (R^1 = H; R^2 = Pr^i, Bu^t, TMS) in the presence of trialkylamine, and the reaction between $MeBBr_2$ and $LiR^1N(CH_2)_2NR^2Li$ (R^1 = H and R^2 = Pr^i, Bu^t, TMS; R^1 = Me, Et and R^2 = Me; R^1 = R^2 = Et or Pr^i). Several products have been crystallographically characterized. Dehydrogenation of the 1,3,2-diazaborolidines leads to the corresponding 2,3-dihydro-1H-1,3,2-diazaboroles.[317] Potassium salts may be prepared by reaction of (**200**) (X = TMS) with KOBu^t;

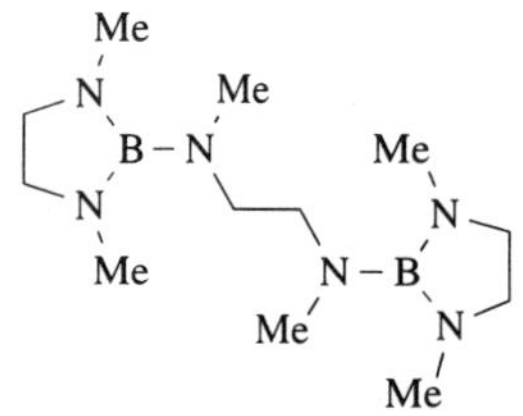

(197)

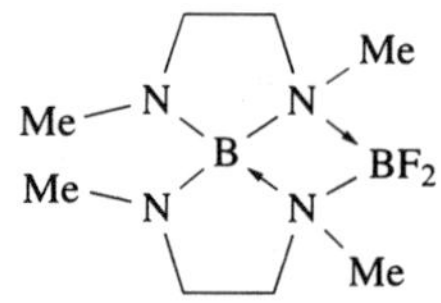

(198)

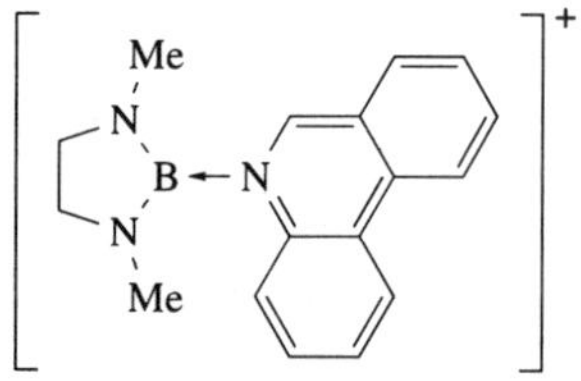

(199)

the products have been characterized by ^{11}B and ^{1}H NMR spectroscopy and, for (**200**) (R = Bu^t; X = K), by single crystal x-ray diffraction. Interestingly, the crystals contain both $KOBu^t$ and $K_2O_2SiMe_2$. In the structure, 12 anions from (**200**) associate with $[Bu^tO]^-$ and $[Me_2SiO_2]^{2-}$ anions and the overall unit contains 42 K^+ cations. The net result is a highly symmetric ring-like assembly.[318]

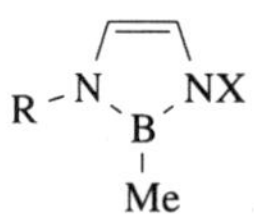

(200)

R = Me, Et, Pr^i, Bu^t

X = K or TMS

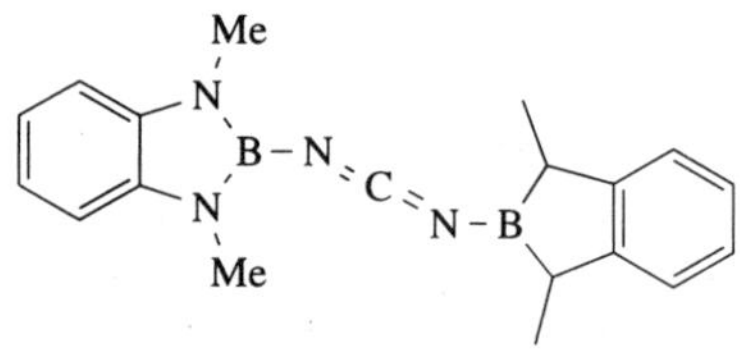

(201)

Heterocycles containing one boron and two nitrogen atoms and which are fused to aromatic rings have also been prepared.[319–22] These include (**201**) which is prepared from the corresponding *B*-chloro-derivative and TMS-N=C=NTMS.[321] Structural analysis of (**201**) reveals the presence of a nonlinear BNCNB chain ($N\hat{C}N$ = 171.9(3)°, $B\hat{N}C$ = 145.9(2)° and 163.1(3)°).[322]

The electronic structures of a group of five-membered rings with a CBN_3 framework have been studied by using x-ray electronic spectroscopy[323] and the CNDO/S method has been used to analyse the electronic spectra of compounds (**202**).[324] Mass spectral data for this family of compounds show that, for R^1 = H, fragmentation involves loss of PhCN.[325] The electrochemical properties of some of these compounds have been studied by a commutative polarographic technique.[326]

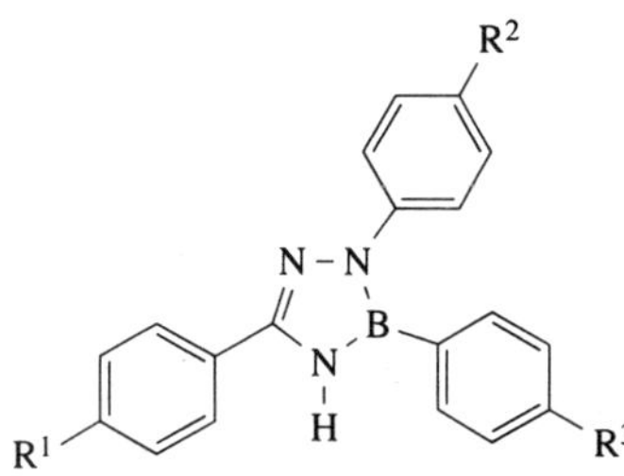

(202)

R^1 = H, Br, Me, OMe, NO_2

R^2 = H, Br, Me, NO_2

R^3 = H, Br, Me, OMe, CO_2H, CO_2Et

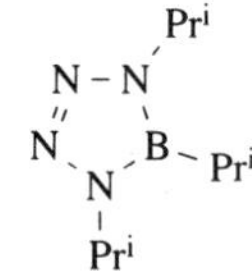

(203)

Tetraazaborolines (BN_4 rings) can be prepared by reacting azides (R^3N_3) with $R^1N(TMS)B(Cl)R^2$ under mild conditions for a wide range of substituents.[327] Iminoboranes[12] $R^1B{\equiv}NR^2$ (e.g., R^1 = Me, Et, Pr, Bu and R^2 = Bu^t) undergo [3 + 2]-cycloaddition reactions with azides (e.g., PhN_3, TMS-N_3, $PhCH_2N_3$), to yield tetraazaborolines.[12,328–30] The results of a crystal structure determination of (**203**) illustrate that there is extensive π-delocalization throughout the ring; endocyclic bond lengths are B–N = 0.1411(5) nm and 0.1422(4) nm and N–N = 0.1375(4) nm, 0.1279(4) nm and 0.1376(4) nm.[331]

Miscellaneous five-membered boron- and nitrogen-containing rings include those formed in the cyclocondensation reactions of $R(Cl)B(CH_2)_nB(Cl)R$ (R = Me or Cl; n = 2 or 3) with $(Me_3E)_2X$ (E = Si or Sn; X = NMe, O).[332] The pathway of the reaction of $CH_2\{B(NPr^i_2)Cl\}_2$ with Na/K alloy is solvent dependent. In petroleum ether, (**204**) is produced but, in THF, one amino-nitrogen atom is incorporated into the ring to give a B_2C_2N framework.[333] The elimination of LiCl from between $MeB\{N(Bu^t)Li\}_2$ and $Me_2Sn(Br)CH_2Br$ leads to (**205**), the low-temperature crystal structure of which shows that the ring is planar with B–N distances of 0.1450(7) nm and 0.1444(7) nm. The tin moiety in the ring can be exchanged for a BMe group by reacting (**205**) with $MeBBr_2$.[334] The preparation of (**206**) from MeCN, PhCH=NPh and an appropriate iminoalkylborane has been reported along with related chemistry.[335]

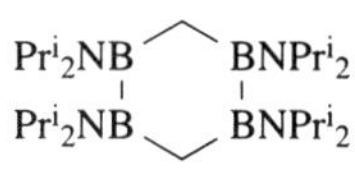

(**204**)

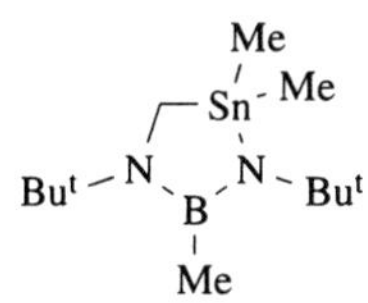

(**205**)

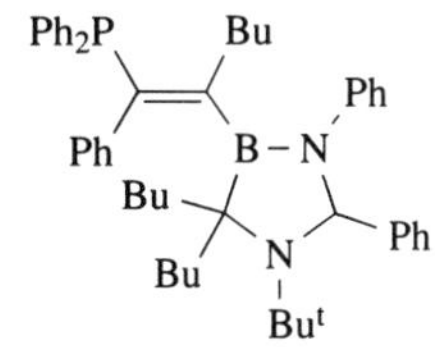

(**206**)

4.7.4 Six-membered Rings (excluding Borazines)

Six-membered rings incorporating one boron and one nitrogen atom fall into the categories of having the heteroatoms adjacent or apart. The photolysis of compounds $XB(CR{=}CHR)_2$ ($X = 2\text{-}R^1\text{-}4\text{-}R^2\text{-}6\text{-}R^3\text{-}C_6H_2NMe$) causes nonoxidative cyclization. Spectroscopic results suggest that the pathway involves an excited singlet state with conrotatory electrocyclic ring closure followed by a suprafacial 1,5-hydrogen shift.[336,337] The reaction between 4-azido-3,3'-bithienyl and $PhBCl_2$ in benzene proceeds with the evolution of dinitrogen to give (**207**) (X = Cl). After treatment with methanol, (**207**) (X = OMe) is formed. Both compounds are moisture sensitive and convert to (**207**) (X = OH) which in turn undergoes coupling via condensation to yield a B–O–B bonded species.[338,339] A related sequence of reactions starting with 3-(cyclohex-1-enyl)-4-azidothiophen yields an analogous product,[338] and other thiophene derivatives have been reported.[340] Compound (**208**) complexes with iodine and the electronic spectrum exhibits an absorption at 358 nm ($\varepsilon = 960 \pm 50$). It is concluded that the presence of the B–N bond in the aromatic system leads to a preference for a π-donor (rather than n-donor) interaction with the I_2 molecule.[341] The crystal structure of the azaboraanthracene (**209**) has been determined. The tricyclic framework is essentially planar, and the phenyl ring is twisted through 49.4° with respect to this plane. The average endocyclic B–C distance is 0.1512 nm.[342]

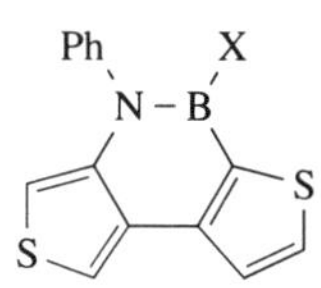

(**207**) X = Cl, OH, OMe

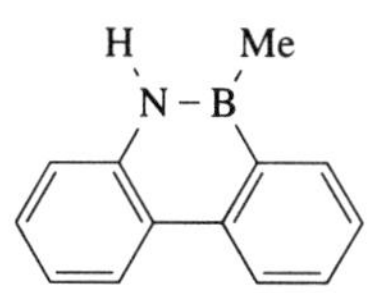

(**208**)

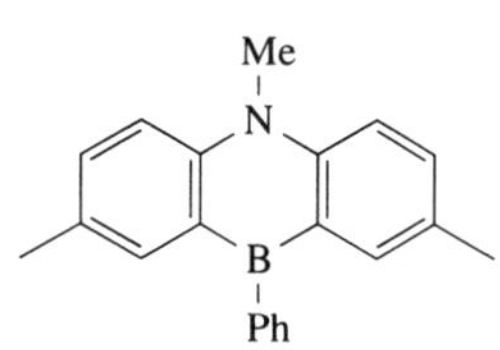

(**209**)

Compounds in the family (**210**) are prepared from BBr_3 (with $FeCl_3$ catalyst) and *N*-(organosulfonyl)hydrazones of arene (Z) aldehydes or ketones. The activities of the compounds against bacteria *in vitro* and *in vivo* have been studied.[343] The C_3BN_2 ring is degraded in sodium hydroxide solution.[344]

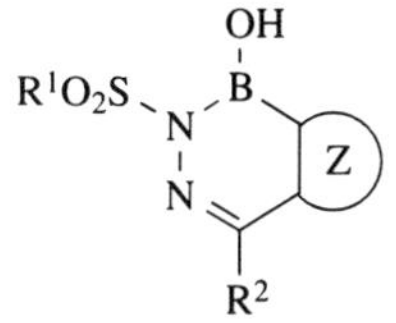

Ring Z = benzene, naphthalene, thiophene, furan, pyrrole
R^1 = various
R^2 = H, Me, Ph

(**210**)

Addition reactions of iminoboranes[12] are utilized in a variety of boron and nitrogen ring formations as described in other sections. A further example is seen in the reaction of $R^1B{\equiv}NR^2$ ($R^1 = R^2 = Pr^i$; $R^1 = Bu$, $R^2 = Bu^t$) with $B(CH_2CH{=}CH_2)_3$ which leads to $H_2C{=}CHCH_2B(R^1)N(R^2)B(CH_2CH{=}CH_2)_2$. This undergoes thermal intramolecular allyl-boration to yield a C_3BNB system.[345] The same cyclic framework is generated as HCl is eliminated during the reaction of $Pr^i_2N(Cl)B(CH_2)_3B(Cl)NPr^i_2$ with Bu^tNH_2.[346]

The syntheses of members of the group of 1,3,2-diazaboracyclohexanes (**211**) have been described. With pyrazoles, 1:1 adducts are formed which are in equilibrium with the uncomplexed species in solution.[347] The same cyclic framework appears as a polymer precursor. The reaction of (**212**) ($R^1 = R^2 = H$) with Bu^nLi and $Me_2NBR^3(Cl)$ (R^3 = Ph or NMe_2) leads to (**212**) (R^1 = H and $R^2 = BR^3(NMe_2)$). This step is repeated to give the doubly *N*-substituted compound. A crystal structure analysis of (**212**) ($R^1 = R^2 = B(NMe_2)_2$) confirms the presence of a rigid –BNBNB– backbone, making the molecule a suitable model for a linear high polymer.[348]

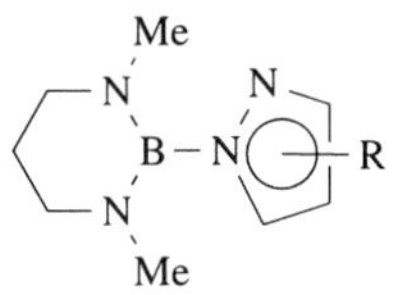

(**211**) R = various

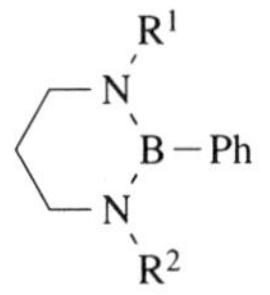

(**212**)

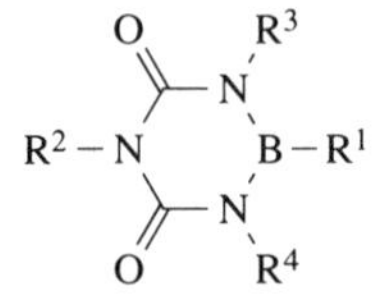

(**213**)

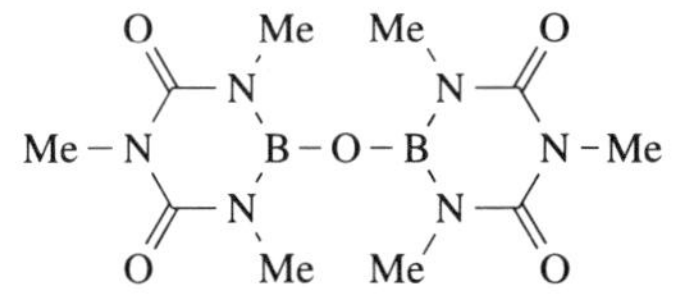

(**214**)

Six- and eight-membered cyclic products are obtained when Me_2BBr and *N,N,N'*-triorganylureas react together. Examples of the smaller of the two ring types are (**213**) ($R^1 = R^3 = R^4$ = Me and R^2 = Et; R^1 = Me and $R^2 = R^3 = R^4 = Pr^i$). Structure (**213**) ($R^1 = R^3$ = Me and $R^2 = R^4 = Bu^n$) is the product in the reaction of $Bu^nN(TMS)C(O)N(Me)C(O)N(TMS)Bu^n$ with $MeBBr_2$.[349] The range of these uracil analogues is now extensive[349–52] and includes condensed products such as (**214**) which has been isolated as an adduct with dimethylamine.[352] Related systems are the bicyclics (**215**), prepared by treating R_3B with 2-aminothiazole followed by R^2NCO in benzene. The crystal structure of the derivative with R^1 = Pr and R^2 = Ph has been elucidated.[353]

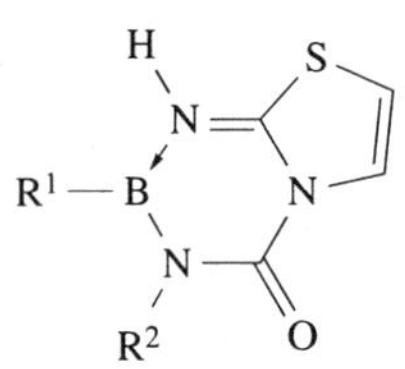

(**215**)
R^1 = Pr, Bu
R^2 = Me, Ph, 2-naphthyl

Scheme 5 illustrates the formation of a cationic C_3NBN system. The products have been spectroscopically characterized and AM1 calculations give evidence for π-interaction between the boron atom, the two adjacent nitrogen atoms and the *exo*-cyclic boron substituent, which, in the theoretical work, is fluorine.[354]

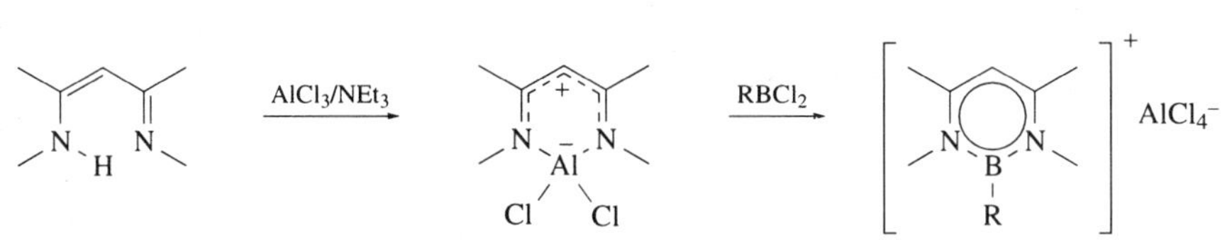

Scheme 5

Isomers of the model cyclic compound $C_2B_2N_2H_6$ have been the subjects of nonempirical HF SCF MO calculations (minimal STO-*N*G (*N* = 3, 4 and 6) basis set). Results show that the energy difference between isomers (**216**) and (**217**) is 48 kJ mol^{-1} with (**216**) being the more stable.[355]

Diazadiboranaphthalenes (**218**) (characterized by NMR spectroscopy) are produced in the reactions of BBr_3 and $2\text{-}R^2C_6H_3NH_2(R^3)$ (R^2 = Br or Me; R^3 = H, 3-Cl, or 4-Cl) followed, as appropriate, by methylation ((**218**), $R^4 = 2\text{-}R^2C_6H_3(R^3)$; R^1 = Me or Br). If BCl_3 replaces the tribromide in the reaction, the products are borazines.[356] The ^1H and ^{13}C NMR spectra of four derivatives of (**218**) have been fully assigned by use of heteronuclear shift correlation and homonuclear COSY 2D NMR spectroscopies. The ^{11}B NMR spectral chemical shifts for one compound have been reported.[357]

The bicyclic compound (**219**) ($R = SCF_3$) is formed as a stable yellow liquid from the reaction of CF_3SCl with (**219**) (R = H).[358] Studies of a macrocyclic triamine with a pendant imidazole group led to

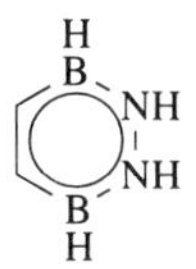

(216)

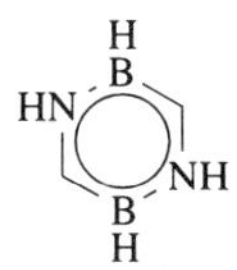

(217)

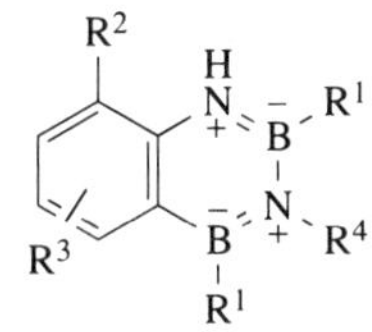

(218)

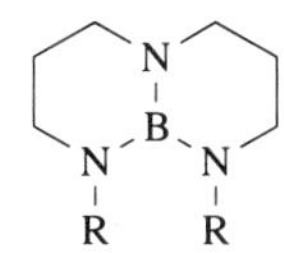

(219) R = H or SCF_3

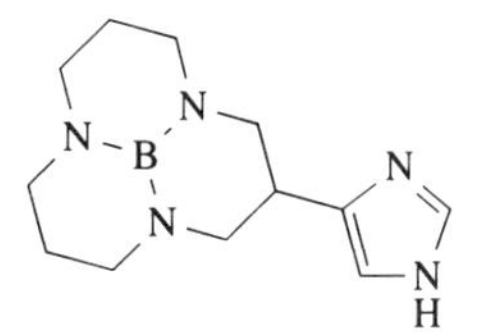

(220)

the isolation of the novel tricyclic complex (**220**). Formation of complex (**220**) occurs when the mono-oxo derivative of the corresponding macrocyclic triamine is reduced with $BH_3{\cdot}THF$ at 65 °C and the mixture is then treated with hydrochloric acid. Compound (**220**) is a colourless crystalline solid whose IR spectrum exhibits absorptions at 1471 and 1439 cm^{-1} which are assigned to ν(BN).[359]

4.7.5 Rings in which the Boron Atom is Four Coordinate

Boron- and nitrogen-containing cyclic compounds in which the boron atom is four coordinate are numerous and are not covered comprehensively in this review; compounds involving chelates such as acac have not been described fully. Compounds are organized by ring size and type.

The reactions of $(CF_3)_2BNR_2$ (R = Me or Et) with isocyanates and isothiocyanates have been studied; [2 + 2]-cycloadditions lead to four-membered chelate complexes, the reactivities of which have been examined. The structures of the two complexes (**221**) have been determined and have B–N distances of 0.1535(8) nm and 0.1635(8) nm for X = O and 0.152(1) nm and 0.161(1) nm for X = S.[360]

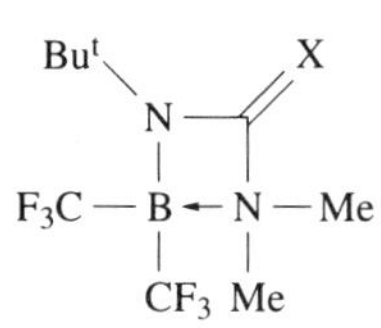

(221)
X = O or S

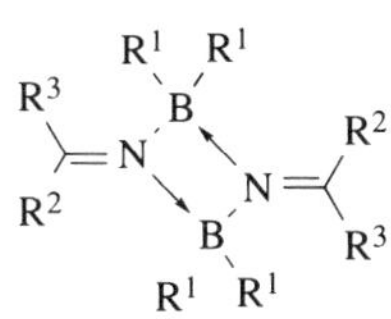

(222)

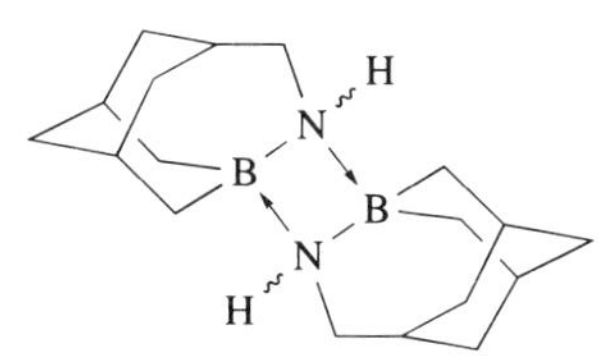

(223)

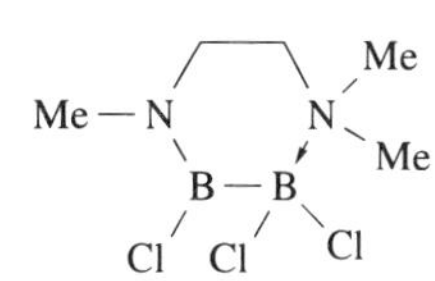

(224)

Triethylborane reacts with benzonitrile (100–270 °C) to give (**222**) with R^1 = Et, R^2 = H, R^3 = Ph, but the product decomposes under the reaction conditions.[361] A successful synthesis of another derivative (R^1 = Me, R^2 = H, R^3 = Ph) has been reported where the reactants are benzonitrile and dimethylborane. The product has been crystallographically characterized and a centrosymmetric molecule confirmed; the *endo*-cyclic angle at each boron atom is 87°.[362] An analogous system (R^1 = Pr, R^2 = SPr or SBu, R^3 = $PhCH_2$) has been prepared from $PhCH_2CN$ and Pr_2BSPr or Pr_2BSBu.[363] Further examples of (**222**) have been studied by Yalpani *et al.*[364] and the dependence of the reaction pathway upon conditions and nature of the nitrile have been emphasized.[363,364]

Trichloro- and tribromoboranes react with the iminoborane[12] (TMP)B≡NBu^t to yield four-membered rings containing BX_2 (X = Cl or Br) groups. The structure of the product for X = Br has been confirmed by x-ray diffraction.[365] The chemistry of borabicyclononanes was described in Section 4.3.6, but an example pertinent to the present section is the thermally induced (80 °C) dimerization of 3-hydroxy-7α-aminomethyl-3-borabicyclo[3.3.1]nonane to give isomers of (**223**). The isomer in which the two *N*-hydrogen atoms are *cis* to one another has been crystallographically characterized.[366]

The reactions of B_2Cl_4 with various difunctional bases leads to chelate complexes in which either one or both boron atoms are incorporated into the ring. The compounds have been spectroscopically characterized and, for (**224**), the crystal structure confirms a puckered ring (B–B = 0.1689 nm).[367] The hydroboration of 1,4-diallylpiperazine with $Et_3N{\cdot}BH_3$ gives a product in which intramolecular N → B coordination occurs to give a tricyclic structure.[368] Similar observations are reported for the product of the hydroboration of 2-allyl-1,2,3,4-tetrahydroisoquinoline.[369] The [4 + 1]-cycloadditions which occur when isonitriles react with $R^1_2BN(R^2)C(R^3){=}CHR^4$ (R^1, R^2, R^3, R^4 = alkyl and aryl) lead first to five-membered ring chelate complexes, the thermal rearrangement of which is accompanied by a change in the coordination number of the boron atom from four to three.[370]

Diboron tetrachloride reacts with alkynes (e.g., hex-3-yne) to give *cis*-bis(dichloroboryl)alkenes. Treatment of these products with $TMS(NMe_2)$ or Pr^i_2NH yields bis(chlorodialkylaminoboryl)ethenes, and in the three instances shown in Structure (**225**), intramolecular N → B donation occurs.[371] A five-membered chelate is also formed in the reaction of PhNC and $R^2NHC(Ph)=NBR^1_2$ ($R^1 = R^2 = Pr^i$; $R^1 = Pr^i$ and $R^2 = Bu^t$).[372] Coordinate bond formation between nitrogen and boron atoms which are associated with arene systems is observed in $2\text{-}(CH_2NMe_2)C_6H_4B(OH)_2$[373] and in (**226**), and in a related quinoline derivative.[374] In the presence of BCl_3, the trimethylsilyl group in complex (**226**) is replaced by a BCl_2 substituent. This product has been structurally characterized and the B–N distances within the four-membered ring are 0.1587(7) nm and 0.1574(7) nm compared to 0.1371(8) nm for the *exo*-cyclic B–N bond.[374]

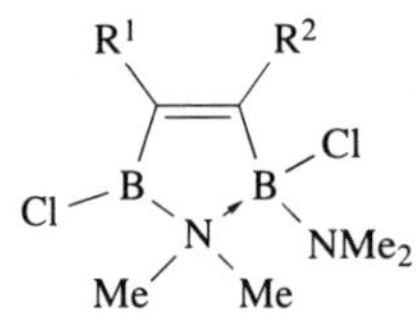

(225)
$R^1 = R^2 = Me$ or Et
$R^1 = Bu^t$, $R^2 = H$

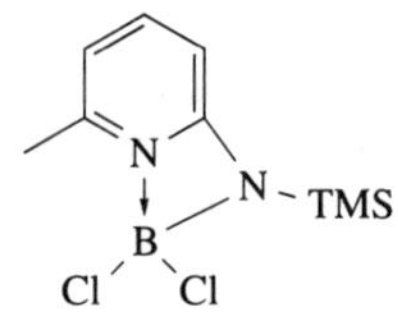

(226)

Photoelectron and IR spectroscopic studies on (**227**) and related complexes have been complemented by MNDO calculations. The results are consistent with the N → B interaction being similar in nature to that in $Me_3N{\cdot}BH_3$; in (**227**), the IR spectrum exhibits an absorption at 1039 cm^{-1} assigned to ν(BN). Previous PES results for (**228**) have been reassigned after consideration of the newly available MNDO results.[375]

(227)

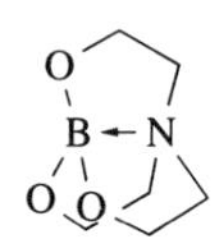

(228)

Miller *et al.* have studied a range of 1,4-diazonia-2,5-diboratacyclohexanes, (**229**).[376–9] Structure (**229**) ($R^1 = H$; $R^2 = Me$; $R^3 = CH_2Bu^t$) has been characterized by mass, IR and NMR spectroscopies and the stereochemistry of the molecule has been established; *cis*- and *trans*-diastereomers are present.[377] The reactivity of (**229**) ($R^1 = R^3 = H$; $R^2 = Me$) has been explored; reagents included in this study are Cl_2, Br_2, I_2, B_2H_6, $BH_3{\cdot}NEt_3$, BMe_3, PMe_3, $Me_2NC(O)Cl$ and Me_3NBH_2I. The 2-iodo and 2,5-diiodo-derivatives of (**229**) are useful synthons for further derivatization, and the cyclic framework is robust.[378] Adding a *B*-carboxylate functionality (accessed via a *B*-isonitrile substituent) permits some useful further chemistry to be carried out.[379]

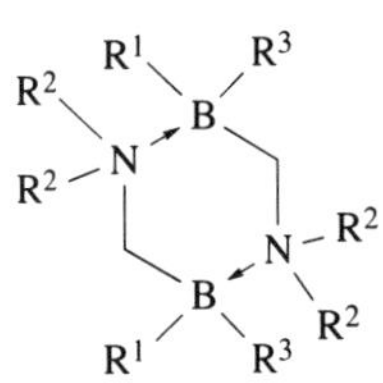

(229)

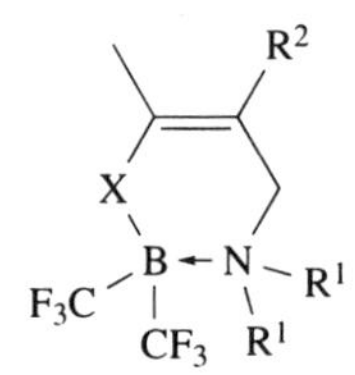

(230)
X = CH_2 or O
R^1 = Me or Et
R^2 = H or Me

The preparation and spectroscopic (IR, ^{11}B, ^{13}C and ^1H NMR) characterization of the cyclic dimer of (*N*-ethylcarbamoyl)borane have been reported.[380] The [2 + 4]-cycloaddition reactions of $(CF_3)_2B{=}NR^1_2$ (R^1 = Me or Et) and $MeC(X)CR^2{=}CH_2$ (X = CH_2 or O) lead to compounds (**230**); the structures of representative products have been confirmed by crystallography.[381] Cyclic products result when R_2BX (X = OEt and R = Ph; R = X = Et, OMe or F) reacts with $CH_2\{C(NH)OEt\}_2$. Protonation of the new compounds occurs with HBF_4.[382] Related neutral rings are formed when $BF_3{\cdot}OEt_2$ reacts with MeN=C(Me)CH=C(Me)NMeH. Again, the product is readily protonated by HBF_4. The crystal structures of both the neutral and cationic cyclic species have been determined, and results have been compared with calculated data. The experimental B–N distance in the former is 0.1524(2) nm and in the cation is 0.1549(5) nm.[383]

Heterocyclics of the series (**231**) have been prepared from PhNHC(Me)=NPh.[384,385] X-ray diffraction data for (**231**) (R^1 = Ac; R^2 = Bu) confirm that, in the solid state, centrosymmetric dimers form supported by NH···O interactions.[385] Related to these molecules are compounds (**232**) which have been structurally characterized.[386] In a similar preparative strategy, a bicyclic system related to (**232**) is formed when Bu_2BSBu is treated with MeHNC(O)NHR (R = 2-thiazolyl);[386] further members of this thiazole-derived family have also been synthesized.[387,388] Derivatives (**233**) can be prepared by heating appropriately substituted tetrazoles with nitriles. The molecular structure of (**233**) (R^1 = Pr; R^2 = 2-MeC_6H_4) has been elucidated.[389] Cycloaddition reactions between 2-($NHBR_2$)py (R = Pr^i or Bu) and PhN=C=NPh, $NCCH_2CN$ or PhNC have also been studied and give products of type (**234**).[390] The electronic structure of model compound (**234**) (R^1 = Pr; R^2 = H) has been studied (CNDO/2) and compared with that of (**235**) (X = O; R^1 = Cl; R^2 = Bu; R^3 = Ph). There is a greater π-bond order in the former compound; the strengths of the B–N interactions in (**234**) are affected by ring protonation, but in (**235**), only the B–N_{R^3} bond order is altered.[391] Compounds (**235**) (X = NR^4; R^1 = H; R^2 = Pr^i; R^3 = H; R^4 = Ph, PhCO or benzimidazol-2-yl) are formed in high yield by the [2 + 4]-cycloadditions of R^4NHC≡N and 2-($NHBPr^i_2$)py. A crystal structure determination of (**235**) (X = NR^4; R^1 = H; R^2 = Pr^i, R^3 = H; R^4 = PhCO) confirms the bicyclic geometry and also shows there to be both intra- and intermolecular hydrogen bonding. IR spectral data for solutions of this compound in $CHCl_3$ are consistent with the loss of the intermolecular interactions.[392] Boron chelates such as (**236**) exhibit luminescent properties.[393]

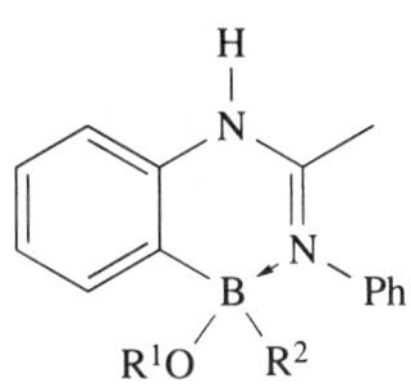

(231)

R^1 = H, Me, Ac; R^2 = OR

R^1 = H; R^2 = Cl

R^1 = Ac; R^2 = Bu

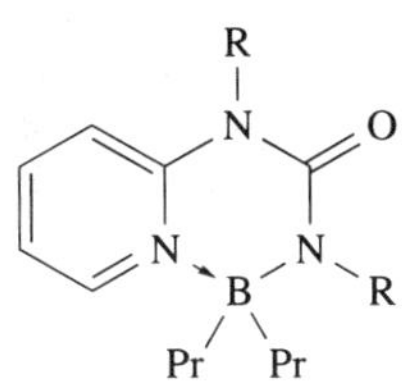

(232)

R = H or Me

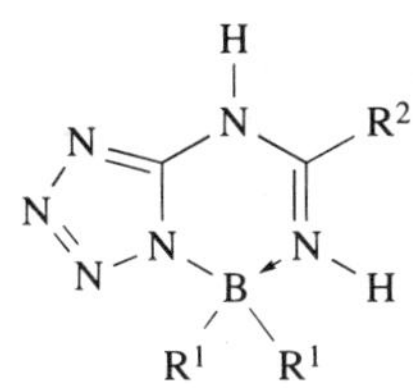

(233)

R^1 = Pr, Bu

R^2 = Me, Pr, Ph, 2-MeC_6H_4

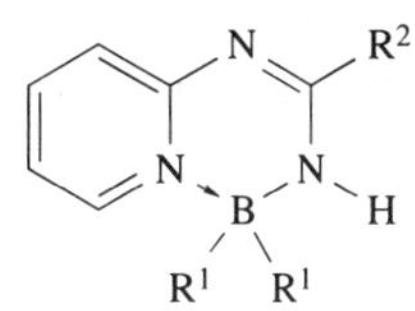

(234)

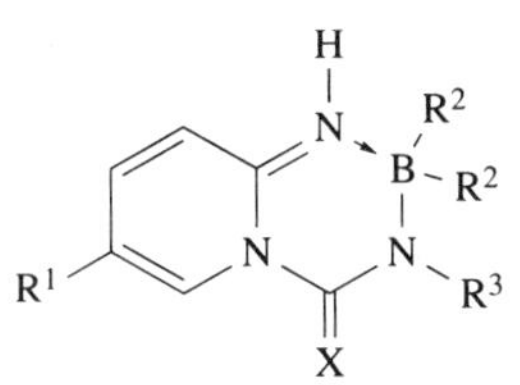

(235)

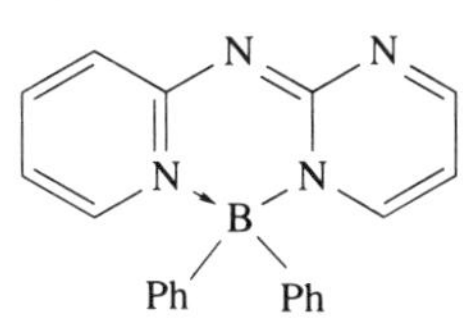

(236)

Pyrazole derivatives containing boron atoms are well established. Monomeric pyrazol-1-ylboranes form when Hpz reacts with (**195**) (X = NMe_2). However, the monomer tends to associate with a second equivalent of pyrazole to give (**237**). Association also occurs with substituted pyrazoles, for example, 3-methylpyrazole. The monomer (**195**) (X = pz) reacts with Me_2NBEt_2 to yield (**238**).[394] The reaction of Hpz with $\{Me_2NBF_2\}_2$ leads to the elimination of two equivalents of dimethylamine and the formation of (**239**) ($R^1 = R^2 = R^3 = R^4$ = F) (see below). Similar reactions using larger heterocyclic precursors (e.g., $\{ClBNBu^t\}_4$) have also been investigated but were unsuccessful; this has been attributed to steric factors.[395] On the other hand, reactions of the borazines $\{R^1BNR^2\}_3$ (R^1 = Et, Ph and R^2 = H, Me) with Hpz result in ring contraction to give heterocycles (**240**).[396] The reaction of Hpz with (**241**) is solvent dependent; in boiling ether, the five-membered ring is destroyed, but in boiling toluene, condensation products are obtained.[395]

Dimeric complexes of type (**239**) have been known for some time, and the B_2N_4 framework is relatively stable. Niedenzu and co-workers have presented ^{13}C, ^1H, ^{11}B and ^{15}N NMR spectroscopic data (not previously available) for these species (with a range of substituents) and for related monomeric compounds.[397] Bromination of (**239**) ($R^1 = R^2 = R^3 = R^4$ = H) is achieved on treatment with BBr_3 or Br_2. Mono- and di(*B,B'*)substituted products have been identified; both *cis*- and *trans*-isomers of the dibrominated compound have been characterized. The tribromo-derivative can be formed but it is in equilibrium with the di- and tetrabromo compounds. Crystal structures of (**239**) ($R^1 = R^2 = R^3 = R^4$ = H or Br) have been determined. The B_2N_4 ring exhibits a boat conformation in the unsubstituted complex

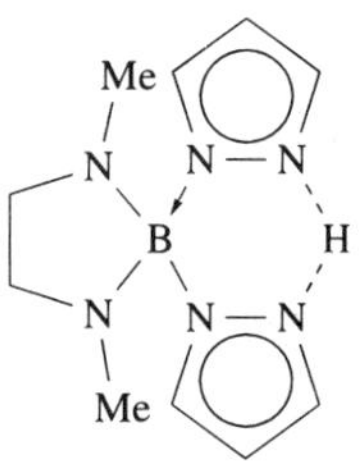

(237)

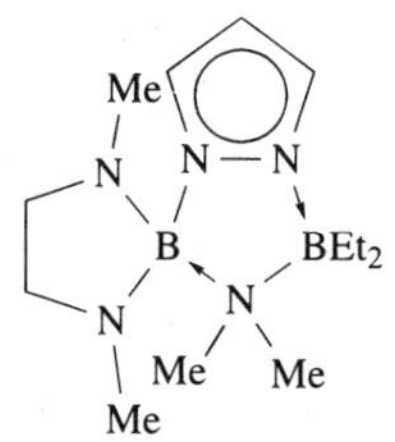

(238)

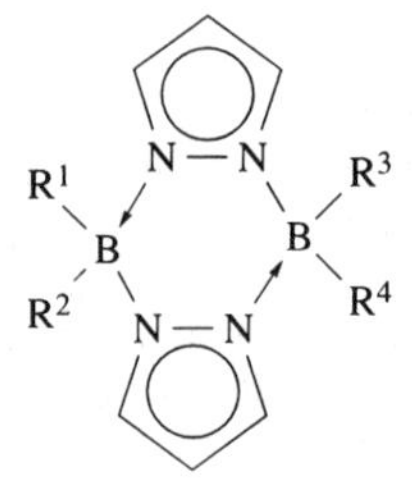

(239)

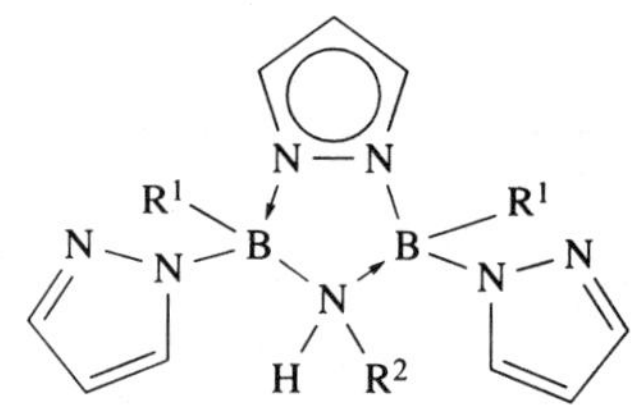

(240)

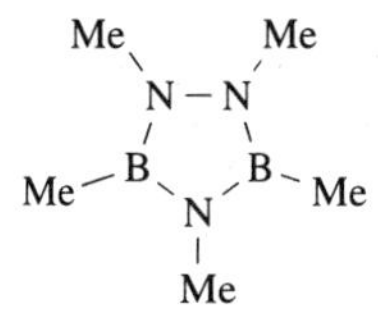

(241)

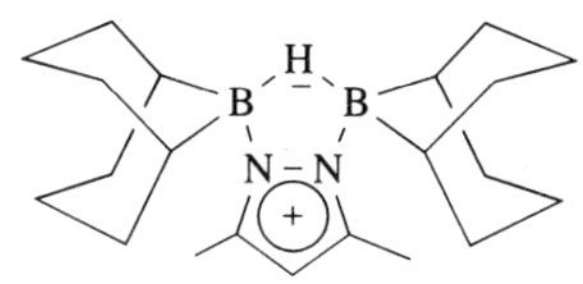

(242)

but is planar in the tetrafluoro species. This latter result mimics that for the tetrachloro derivative.[398] Fluorination of (**239**) ($R^1 = R^2 = R^3 = R^4 = H$) is achieved by using $BF_3{\cdot}OEt_2$ resulting in both tetrafluoro and di(*B,B*)fluoro derivatives. The latter undergoes chlorination to give $Cl_2B(\mu\text{-pz})_2BF_2$. The crystal structure of (**239**) ($R^1 = R^2 = R^3 = R^4 = F$) has been determined. The preparations and characterizations of $X_2B(\mu\text{-pz*})BX_2$ (Hpz* = *C*-substituted pyrazoles; X = Cl or Br) and other related derivatives have also been reported.[399] $Et(pz)B(\mu\text{-pz})_2BEt(pz)$ is formed when $EtB(NMe_2)_2$ reacts with Hpz in toluene at reflux; the reaction temperature is critical. Both *cis*- and *trans*-isomers of the product have been structurally characterized.[396]

Yalpani and co-workers have added valuable contributions to the area of pyrazolylborane chemistry. The dimer $(9\text{-BBN})_2$ reacts with Hpz to yield complexes of type (**239**) in which R^1R^2 and R^3R^4 are 9-borabicyclo[3.3.1]nonane moieties. The structure of the product has been confirmed by x-ray crystallography; the B_2N_4 ring exhibits a boat conformation, and the molecule suffers considerably from steric crowding. Analogous *C*-substituted pyrazole derivatives have been prepared, and whether or not a monomeric or dimeric structure is adopted depends on the nature of the *C*-substituent.[400] Adduct formation by monomeric *B*-pyrazolyl-9-borabicyclo[3.3.1]nonanes with excess 9-BBN or various pyrazoles (e.g., Structure (**242**)) has been studied in detail. The bridging hydrogen atom in (**242**) is asymmetically positioned, and, although this observation may well be a consequence of crystal disordering, the possibility of a weak association is consistent with mass spectral fragmentation patterns.[401]

An especially interesting derivative arises from ferrocenylpyrazole (HFcpz). The preparation of HFcpz has been described and it reacts with $Me_3N{\cdot}BH_3$ in toluene under reflux to give $H_2B(\mu\text{-Fcpz})BH_2$.[402] In the reaction of 3,5-di-*t*-butylpyrazole and triethylborane a monomeric compound results if the reaction temperature is 170 °C. However, in the range 220–230 °C, (**243**) is formed, and this is also the case for 3,5-di-*t*-butyl-4-methylpyrazole and 3,5-di-*t*-butyl-4-ethylpyrazole.[403]

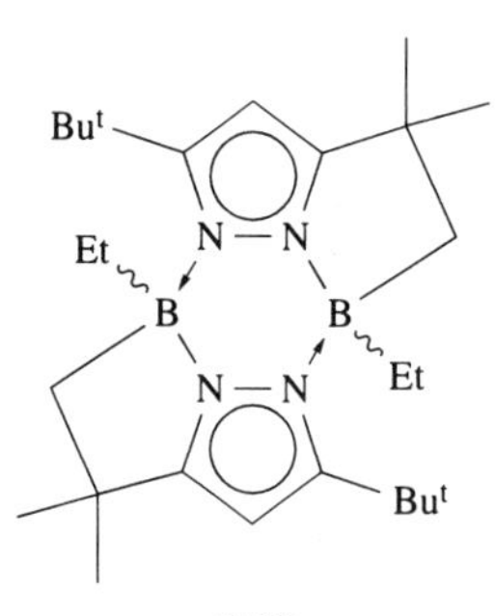

(243)

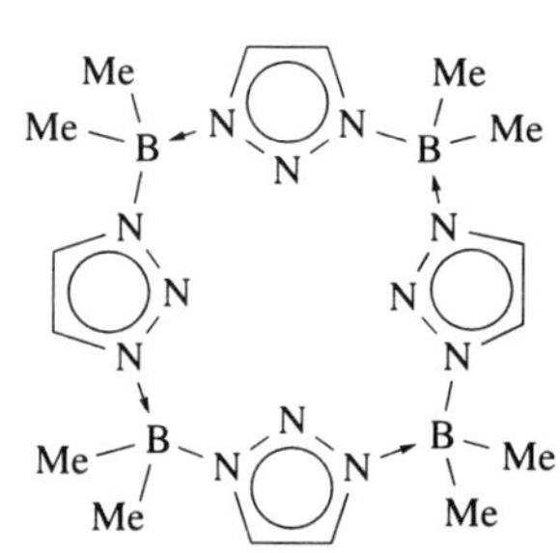

(244)

Triazole derivatives, $\{H_2BL\}_n$ have been synthesized by reacting 1,2,4-triazole, HL, with $Me_3N{\cdot}BH_3$. Mass spectral data for the products give evidence for values of $n \leq 6$. Oligomeric products also predominate in the reactions of 1,2,4-triazole (HL) with R_3B (e.g., R = Et), R_2BX (X = Cl, Br; e.g.,

R = Me, Ph) and TMS-L, or Me_2NBR_2 (e.g., R = Ph) and HL. When R_3B (e.g., R = Et) or Me_2NBPr_2 is treated with 3-NH_2-1,2,4-triazole, the formation of dimers (with structures similar to (**239**) but with the pyrazole rings replaced by triazoles) is observed.[404] An intriguing group of macrocyclic molecules are the cyclotetrakis(triazolylboranes) of which (**244**) is representative. This compound has been prepared by Niedenzu *et al.*[404,405] and the nature of the cyclic system has been confirmed crystallographically. Similarly, benzotriazole has been incorporated into a tetrameric species. The structural studies on the two macrocycles are complemented by ASED molecular orbital calculations. The oligomerization of the model H_2BL is kinetically controlled.[405]

4.8 RINGS CONTAINING BORON AND PHOSPHORUS ATOMS

A review by Power[24] covers the chemistry of rings with B_xP_y skeletons for the greater part of the period of the present survey, and readers are directed to this work. This section emphasizes systems which incorporate carbon into B_xP_y ring systems while also mentioning (in less detail) and updating the material concerning the binary boron–phosphorus heterocycles.

4.8.1 Three-membered Rings

Three-membered rings are not well represented[24] but examples include the diphosphaborirane $(Bu^tP)_2(BNPr^i{}_2)$. This system reacts with potassium or potassium naphthalide in THF with concomitant ring opening via P–P bond cleavage.[406]

4.8.2 Four-membered Rings

A series of four-membered rings with the B_2P_2 framework has been prepared and characterized[407–12] in work that is reviewed by Power.[24] Phosphorus atoms (and thus boron atoms) can be adjacent (e.g., from the [2 + 2]-cyclocondensation of KBu^tPPBu^tK and $Cl(Me_2N)BB(NMe_2)Cl)$[407] or apart[408–12] (e.g., from the reaction of $(TMP)B(Cl)PH(CEt_3)$ and $NaN(TMS)_2)$.[409] The presence of a bulky amine substituent (e.g., TMP) on the boron atom faciliates the formation of a transannular P–P bond, a reaction that is photochemically activated as shown in Equation (31) (see Section 4.8.6).[412]

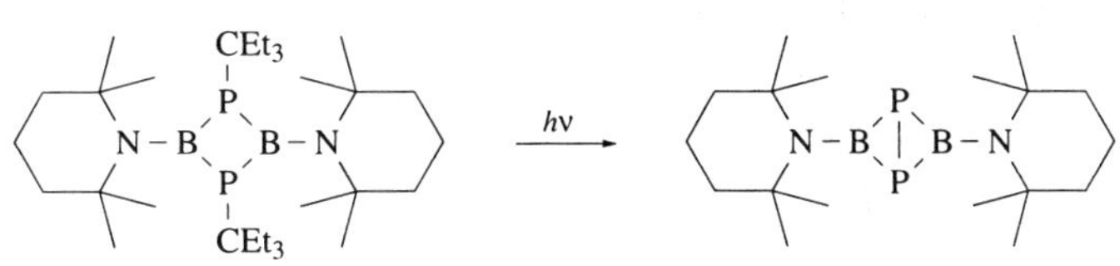

(31)

4.8.3 Five-membered Rings

The C_2B_2P ring framework dominates the field of five-membered boron and phosphorus-containing heterocycles. The elimination of lithium chloride drives the formation of (**245**) in the reactions of $PhPLi_2$ with $Cl(R^1{}_2N)BCH{=}CHB(NR^1{}_2)Cl$ (R^1 = Me, Et, Pr^i; R^2 = Ph). For $R^1 = Pr^i$ and R^2 = Ph (**245**) (a compound with a foul odour, and which decomposes in air) has been characterized by x-ray crystallography. The phenyl substituent lies perpendicular to the ring, which is itself nonplanar.[413] Related reactions of $PhPLi_2$ and $Cl(Et)BC(Et){=}C(R)SiMe_2Cl$ (e.g., R = Me) lead to phosphasilaboroles. In the solid state, dimers exist supported by P → B coordinate bonds.[414]

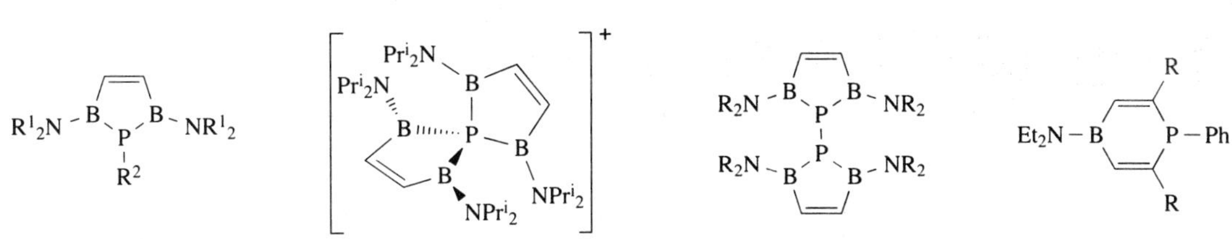

(**245**) (**246**) (**247**) R = Et or Pr^i (**248**) R = H or Me

The spirocyclic system [(**246**)]Cl has been prepared by Siebert and co-workers from the reaction of $Cl(Pr^i{}_2N)BCH{=}CHB(NPr^i{}_2)Cl$ with $P(TMS)_3$ via (**245**) ($R^1 = Pr^i$ and R^2 = TMS). The crystal structure of [(**246**)][BPh_4] has been determined; the phosphorus atom is in a distorted tetrahedral environment and P–B distances are in the range 0.1957–0.1961(5) nm.[415] Related to this work is the preparation of (**247**) (R = Et or Pr^i). These species are produced via (**245**) (R^1 = Et or Pr^i; R^2 = TMS or H). For R^2 = H, ring coupling is achieved by treatment with $Bu^t{}_2Hg$, while for R^2 = TMS, coupling occurs upon reaction with C_2Cl_6. The coupled ring structure has been confirmed crystallographically for $R^1 = Pr^i$; the P–P distance is 0.2192(1) nm, and the average B–P distance is 0.1934(3) nm.[416] Further investigations of the C_2B_2P cyclic system have been presented.[417]

4.8.4 Six-membered Rings

The chemistry of the B_3P_3 ring has been developed by Power and co-workers.[24,418–20] The reaction of $MesBBr_2$ and (c-C_6H_{11})PHLi leads to {(c-C_6H_{11})PBMes}$_3$, the first structurally characterized analogue of a borazine.[418] Spectroscopic and structural data support the presence of significant aromatic character, although by increasing the bulk of the substituents, there is a tendency to form the antiaromatic B_2P_2 system instead.

The hydrophosphination (with $PhPH_2$) of $Et_2NB(C{\equiv}CR)_2$ (R = H or Me) leads to (**248**). Analogous hydroarsination reactions can be carried out. The reactivity of compound (**248**) (R = Me) has been examined. Attack at phosphorus occurs with S_8 or Se_8 (oxidation) and with alkali metals, M, (to give P^- M^+). In the latter case, the alkali metal salts react with alkyl bromides to yield new *P*-alkyl derivatives of (**248**). Compound (**248**) (R = Me) reacts with methanol and with *t*-butyl lithium to give the *B*-methoxy and *B*-*t*-butyl derivatives, respectively.[421] In a cyclocondensation which resembles that described in Section 4.8.3 for the preparation of a four-membered ring,[407] KBu^tPPBu^tK reacts with $Cl(R_2N)BCH{=}CHB(NR_2)Cl$ (R = Me or Pr^i) to yield compounds (**249**). A crystal structure determination for (**249**) (R = Pr^i) has confirmed a nonplanar ring geometry, with one *t*-butyl group above and one below the ring. Pertinent distances are B–N = 0.1415(3) nm and B–P = 0.1957(3) nm.[422]

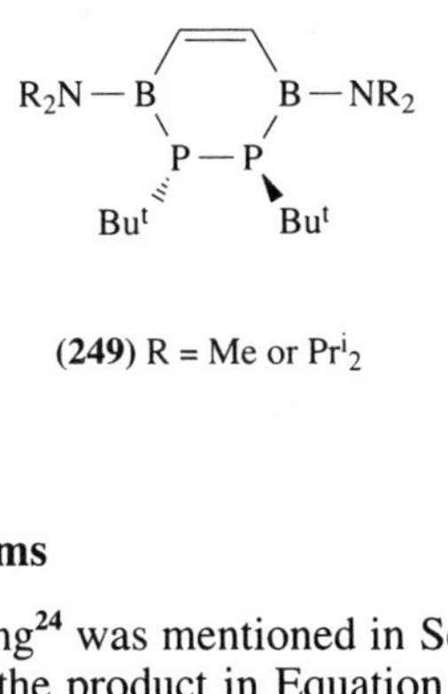

(**249**) R = Me or $Pr^i{}_2$

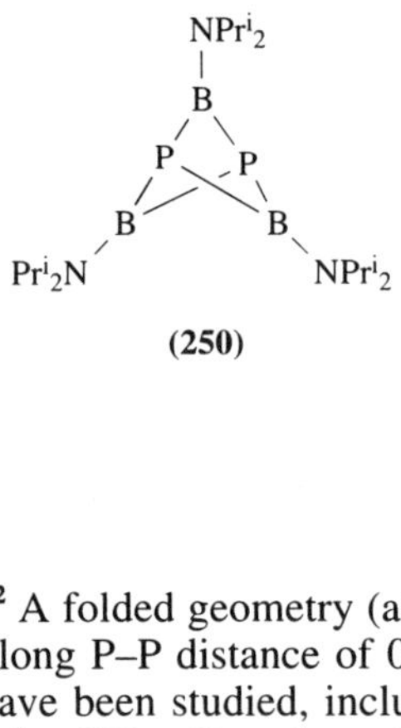

(**250**)

4.8.5 Bicyclic Systems

A bicyclic B_2P_2 ring[24] was mentioned in Section 4.8.3.[412] A folded geometry (along the P–P axis) has been established for the product in Equation (31), but the long P–P distance of 0.2349(2) nm indicates a degree of ring strain.[412] Some reactions of this bicycle have been studied, including coordination via one or two phosphorus atoms to transition metal centres and to BBr_3. Reaction with $BH_3{\cdot}THF$ results in cleavage of the system and the formation of (TMP)BH_2. Related reactions include ring expansions (e.g., with Li(TMEDA), and Et_2NBCl_2) and atom substitution (sulfur for phosphorus with the concomitant formation of a B_2S_2 monocycle).[423]

The bicyclic compound (**250**) is described as a cage by the authors.[424] It arises from the reaction of $Cl(Pr^i{}_2N)BP(TMS)_2$ with $(Pr^i{}_2N)BCl_2$, and the crystal structure has been determined.[24,424]

4.8.6 Rings in which the Boron Atom is Four Coordinate

Reactions of the previously known heterocycle (**251**) with aldehydes have been studied. Ring expansion occurs as the aldehyde inserts between the boron and phosphorus atoms. The products are stable under an inert atmosphere, although IR spectral data are consistent with dissociation of the new six-membered rings on standing in dilute solution.[425]

The reaction of $BHBr_2{\cdot}SMe_2$ with $MeC\{CH_2PMe_2\}_3$ produces a bicyclic dicationic adduct, isolated as the dibromide salt. The crystal structure of the complex has been determined; in solution, NMR

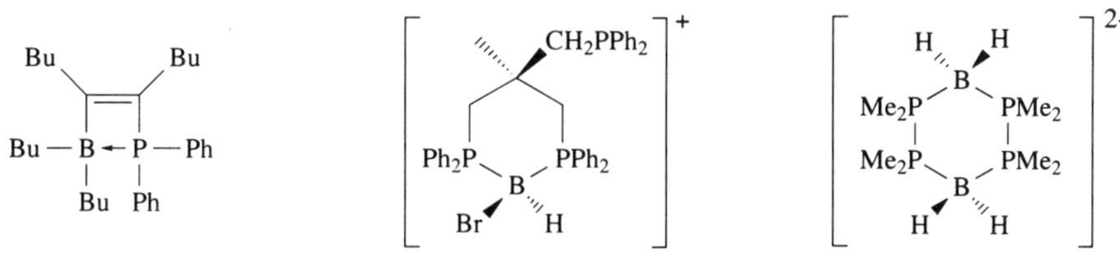

(251) (252) (253)

spectroscopic data are consistent with the retention of the C_{3v} symmetry observed in the solid state. Similarly, $MeC\{CH_2PPh_2\}_3$ reacts with $BHBr_2 \cdot SMe_2$, but now the chelate (**252**) is formed with one phosphine group in a 'dangling' environment. Cation (**252**) is isolated as the bromide.[426] Adducts (1:2) are formed between $BH_{3-x}Br_x \cdot SMe_2$ and Me_2PPMe_2; cyclic and acyclic products are obtained. One example is the reaction of $BH_2Br \cdot SMe_2$ and Me_2PPMe_2 which yields [(**253**)]Br_2, the structure of which has been confirmed by x-ray analysis. The dication exhibits a chair conformation with B–P distances of 0.1924(4) nm and 0.1918(4) nm and P–P bond lengths of 0.2190(1) nm.[427] Catecholborane and dicyclohexylborane form 1:1 adducts with PhP=C(Bu^t)(O-TMS) but spontaneous β-elimination of trimethylsiloxycatecholborane or trimethylsiloxydicyclohexylborane, respectively, leads to cyclic products.[428]

4.9 RINGS CONTAINING BORON AND OXYGEN ATOMS

A full discussion of cyclic systems involving boron and oxygen atoms should include cyclotriboroxane (boroxin) and its derivatives, and borates, but these classes of compound are beyond the scope of this review. An article addressing the structures of borates and polyborates is available.[25] This present survey focuses on relatively simple systems, most including carbon as well as boron and oxygen atoms.

A variety of heterocycles including borolanes, borinanes and boroxins have been the subjects of an ^{17}O NMR spectroscopic study; results for 33 compounds are tabulated, and the chemical shift values suggest that the magnetic screening of the oxygen atom depends on the degree of B–O π-bonding.[429]

4.9.1 Three-membered Rings

The first oxadiboririane (**254**), was reported in 1992 by Paetzold and co-workers. It is produced according to Equation (32)). The structure has been crystallographically confirmed, but suffers from a disorder involving the oxygen atom. Pertinent distances are B–O = 0.1545(5) nm and 0.1510(6) nm and B–B = 0.1601(7) nm; the latter is remarkably short.[430]

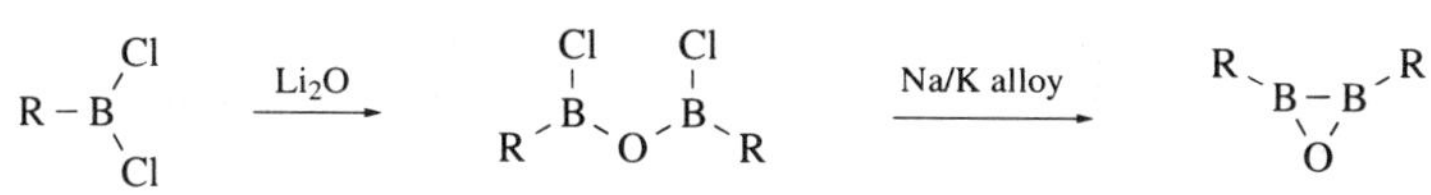

(254) (32)

4.9.2 Four-membered Rings

Four membered boron–oxygen rings are represented by B_2O_2, CBO_2 and C_2BO systems. The reaction of RLi (R = 2,4,6-$Bu^t_3C_6H_2$) with $B(OMe)_3$ in hexane yields $RB(OMe)_2$, which hydrolyses (65 °C and 0.05 torr) to give (**255**) (R = 2,4,6-$Bu^t_3C_6H_2$). In its Raman spectrum, absorptions at 927 cm^{-1} and 944 cm^{-1} are assigned to the symmetric ring stretching mode. Photolysis of (**255**) (R = 2,4,6-$Bu^t_3C_6H_2$) in the presence of trapping reagents (e.g., 2,2,4,4-tetramethyl-2,4-disila-1-oxacyclopentane) leads to products that are interpreted in terms of the reaction of an intermediate oxoborane RB=O.[431] Compound (**255**) (R = TMP) and the corresponding diselenadiboretane have been isolated. The crystal structure of (**255**) (R = TMP) confirms a planar B_2O_2 ring with B–O distances of 0.1417(4) nm and 0.1412(4) nm.[432] The synthesis of the dithiocarbazate complex (**256**) is accomplished by the reaction of $PhB(OH)_2$ with $B(OCPr^i)_3$ in toluene followed by treatment with the appropriate dithiocarbazate ligand. Compound (**256**) has been spectroscopically (IR, ^1H and ^{11}B NMR) characterized; in solution, there is evidence for the existence of a tautomeric equilibrium involving the ligand. A cyclic $AlBO_2$ system related to (**256**) has also been prepared;[433] see also Equation (21).

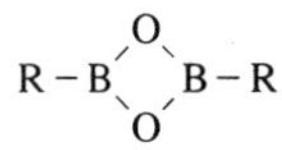

(255)
R = 2, 4, 6-$Bu^t{}_3C_6H_2$ or TMP

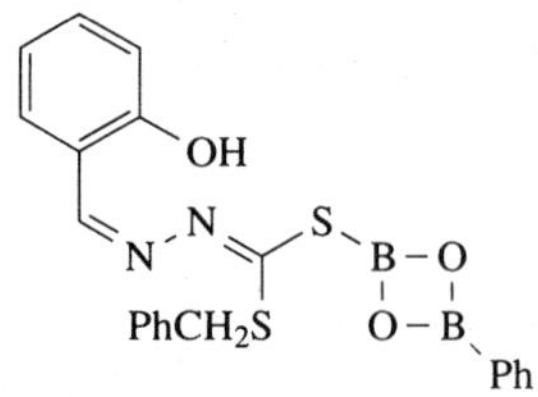

(256)

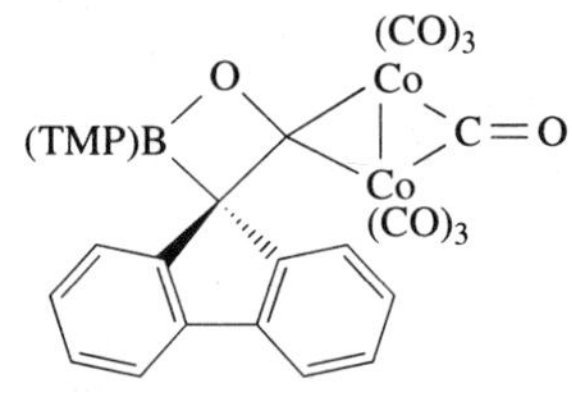

(257)

Among the reactions of $(TMS)(Bu^t)NB{=}CHBu^t$ that have been reported[434] are those with PhCHO and Ph_2CO. These give the [2 + 2]-cycloaddition products C_2BO rings, which are colourless and crystalline.[434] The formation of a 1,2-oxaboretane is shown in Equation (33).[435] The reaction of the precursor with elemental sulfur is described in Section 4.10.2. When cobalt octacarbonyl reacts with (9-fluorenylidene)(TMP)borane under ambient conditions, a [2 + 2]-cycloaddition occurs involving one carbonyl ligand. The heterocycle (**257**) has been isolated from the reaction, and the crystal structure of this product has been determined.[436]

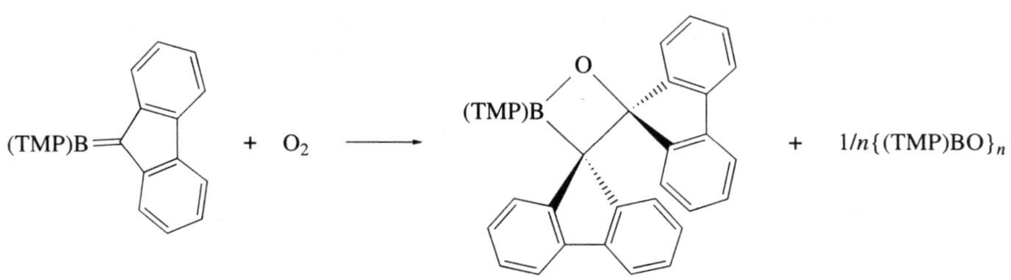

(33)

4.9.3 Five-membered Rings

4.9.3.1 Monocyclic five-membered ring systems

As part of a study of the stereoselective synthesis of α,β-dihydroxyketones, 1,3,2-dioxaboroles have been prepared; MeC(O)CH(OH)Me reacts with $PhB(OH)_2$ in toluene under reflux to yield, specifically, (**258**) (R^1 = Ph and R^2 = Me). This heterocycle reacts readily with aldehydes (PhCHO, CEt_2HCHO, n-$C_7H_{15}CHO$).[437] Boroles with only a boron substituent have been less accessible than their 1,3,2-substituted analogues. Wulff *et al.* have produced a series of compounds (**258**) (R^2 = H and R^1 = c-C_6H_{11}, OMe, Ph, 2-MeC_6H_4 or 2,6-$Me_2C_6H_2$). Reactions of these heterocycles with aldehydes have been studied and it has been found that by binding the 1,3,2-dioxaboroles to a cross-linked polymer, complicating reactions can be avoided. The applications of this polymer-bound system in organic synthesis (e.g., lengthening of sugar chains) are discussed.[438] The oxygen-linked molecule (**259**) is prepared via the reaction between benzoin and boric acid.[439]

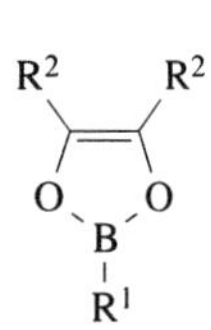

(258)
R^1 = Ph; R^2 = Me
R^1 = c-C_6H_{11}, MeO, Ph, 2-MeC_6H_4,
2,6-$Me_2C_6H_3$; R^2 = H

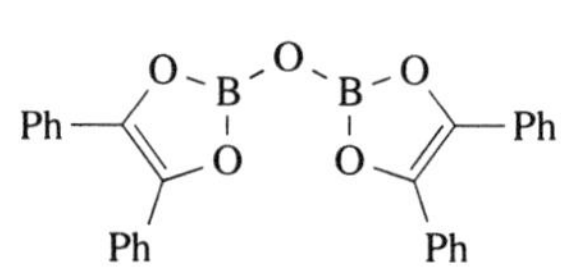

(259)

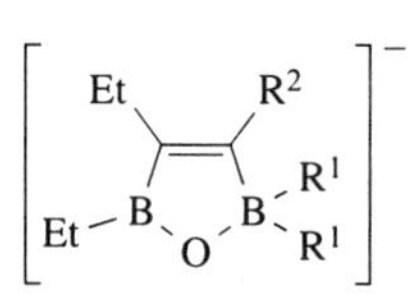

(260)
R^1 = Et, Ph; R^2 = Me, Et or Ph
$R^1{}_2$ = 9-BBN; R^2 = Me or Et

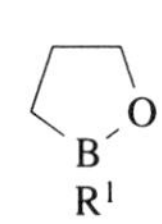

(261)

The reactions of *cis*-$Et_2BC(Et)$=$C(R^2)BR^1_2$ (R^1 = Et, Ph and R^2 = Me, Et or Ph; or R^1_2 = 9-BBN and R^2 = Me, Et) with NaOH or KOH in toluene yields an intermediate anion that undergoes loss of ethane on heating above 100 °C to give the heterocyclic anion, (**260**).[440]

Saturated rings are represented by C_3BO and C_2BO_2 systems. The derivative (**261**) with a *B*-allyloxy substituent reacts with R^1OH (R^1 = $Me(CH_2)_4$, *c*-C_6H_{11}, *c*-C_8H_{15}, $NH_2CH_2CH_2$) to yield (**261**). Linkage of two oxaborolane rings occurs when (**261**) (R^1 = allyloxy) reacts with $HOCHR^3CHR^3OH$ (R^3 = H or Me). Derivatives (**261**) (R^1 = Cl, Bu and allyl) have also been prepared from the same *B*-allyloxy precursor;[441] related chemistry has been reported.[442] Raman spectroscopic data for (**261**) (R^1 = $OCH_2C{\equiv}CH$, $N(CH_2CH{=}CH_2)_2$, $CH_2CH{=}CH_2$, $OCH_2CH{=}CH_2$, OCH_2CH_2OMe, OCH_2CH_2Cl or OCH_2CH_2Br) and for a series of bis(oxaborolanes) have been detailed, and are analysed in terms of the molecular structure.[443] The same authors have shown, via Raman spectroscopy, that for (**262**) (n = 0; R^1 = H, Me, CH_2Cl; R^2 = H), tautomerism between (**262a**) and (**262b**) occurs. When n = 0, and R^1 = R^2 = Me, or when n = 1–3, Structure (**262a**) is favoured even at elevated temperatures.[444]

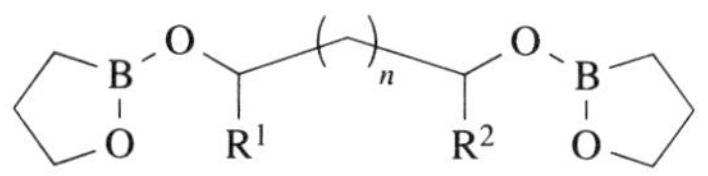

(262a)

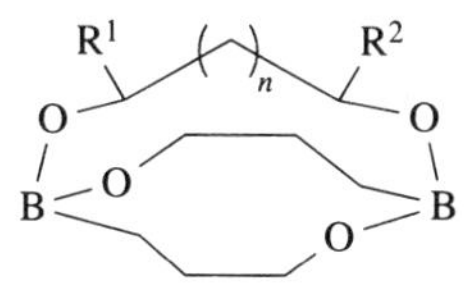

(262b)

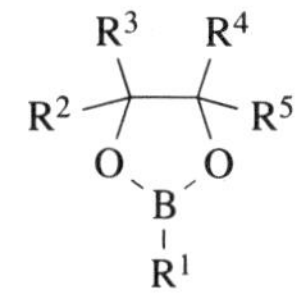

(263)

The C_2BO_2 ring is exemplified by a range of systems. The syntheses of (**263**) (R^1 = $PhSCH_2$; R^2 = R^3 = R^4 = R^5 = H or Me) from $PhSCH_2B(OH)_2$ have been described. The tetramethyl-substituted compound deprotonates at the *B*-methylene group when treated with $LiNPr^i_2$.[445] Related systems have also been reported.[446,447] The γ-methoxycrotylboronates (**263**) (R^1 = $MeOC(Me){=}CHCH_2$ and R^2 = R^3 = R^4 = R^5 = Me) have been synthesized in 85–90% isomeric purity. Reactions of (*E*)-(**263**) with several aldehydes give mainly *syn*-homoallylic alcohols.[448] In related work, the hydroboration of $R^6R^7CHC{\equiv}CH$ (R^6 = $PhCH_2O$, PhH_2S, TBDMS-O, PhS and R^7 = H, or R^6 = PhS, TMS-O, PhSe and R^7 = Me) with cyclohexylborane followed by oxidation with Me_3NO and *trans*-esterification with pinacol leads to (**263**) (R^1 = $CH{=}CH(CHR^6R^7)$ and R^2 = R^3 = R^4 = R^5 = Me).[449] The dioxaborolanes (**263**) (R^1 = $(CH_2)_4Me$ and R^2 = R^3 = R^4 = R^5 = H or Me or R^2 = R^3 = R^4 = H and R^5 = Me) are prepared with $Me(CH_2)_4BCl_2$ as the source of boron.[450] Compounds (**263**) (R^1 = substituted cyclopropyl groups; R^2 = R^3 = R^4 = R^5 = Me) are prepared with high stereoselectivity by the cycloaddition of carbenes to alkenylboronic acid esters.[451]

B-Aryl substituted derivatives of (**263**) include those with R^1 = 2,5-$(NHAc)_2C_6H_3$ and R^2 = R^3 = R^4 = R^5 = H,[452] and with R^1 = 4-$(CH_2)_nNCS$ and R^2 = R^3 = R^4 = R^5 = Me.[453] In the former, the O–B interaction stabilizes the system towards hydrolysis which suggests that there is a potential application in carbohydrate synthesis. Interest in the second derivative[453] arises from application in labelling monoclonal antibodies with radioisotopes of technetium and rhenium. The polymerization of divinylbenzene with (**263**) (R^1 = 4-$(CH{=}CH_2)C_6H_4$ and R^2 = R^3 = R^4 = R^5 = H) yields a polymer-supported borolane; hydrolysis of this product cleaves the heterocycle leaving a polymer-supported $B(OH)_2$ group.[454]

An improved synthesis of (**264**) (R = Pr^i) has been developed. Factors which influence the enantio- and diastereoselectivity of the reactions of (**264**) (R = Pr^i) with chiral and achiral aldehydes have been studied. The reactions are stereochemically sensitive to both temperature and solvent.[455] Related chemistry of (**264**) (R = Et or Pr^i) has also been described.[456]

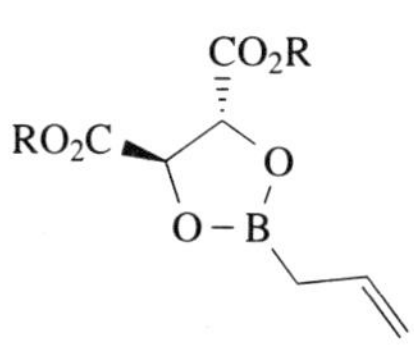

(264) R = Et or Pr^i

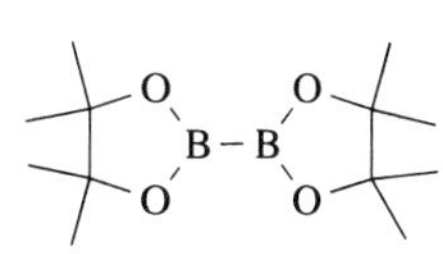

(265)

Several studies have addressed the C_2BO_2 ring structure. Solution properties of 2-hydroxy-1,3,2-dioxaborolane have been investigated by use of ^{13}C NMR spectroscopy. The heterocycle adopts a half-chair conformation.[457] Conformational studies of (**264**) (R = Pr^i) and some related derivatives have been carried out both in solution and the solid state.[458] The crystal structure of (**265**) has been determined and the results show that the two heterocycles are almost coplanar. The B–B distance is 0.1711(6) nm.[459]

4.9.3.2 Five-membered heterocycles adjacent to aromatic rings

This section concentrates on systems of type (**266**). For $R^1 = H$ (catecholborane), the vapour phase far IR spectrum of (**266**) has been analysed and results compared with related systems.[460] Preparative methods to this group of compounds include the reaction of $B(OH)_3$ or B_2O_3 with 1,2-$(OH)_2C_6H_4$ and R^2OH (e.g., $R^2 = Ph$, 1-naphthyl, 8-quinolyl, 4-MeC_6H_4, 4-BzC_6H_4, 4-$NH_2C_6H_4$, 1-benzimidazolyl-methyl, 4-$C_6H_4NHC_6H_4$-4) to yield (**266**) with $R^1 = OR^2$.[461–3] The products are light and heat stabilizers for PVC. In the case of derivative (**267**), significant fungicidal activity is noted.[463] A comparison of ^{13}C NMR spectral data between (**266**) ($R^1 = 4\text{-}OC_6H_4Me$) and related nonboron containing compounds allows the effect of the presence of the boron atom to be assessed.[464]

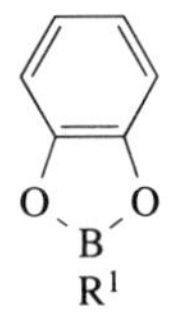

(266)

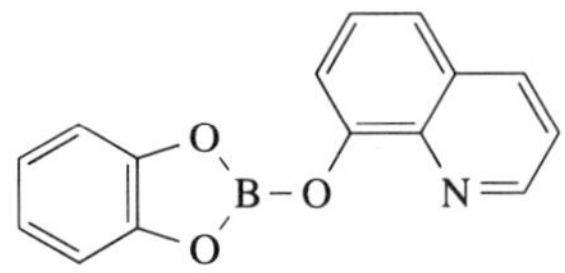

(267)

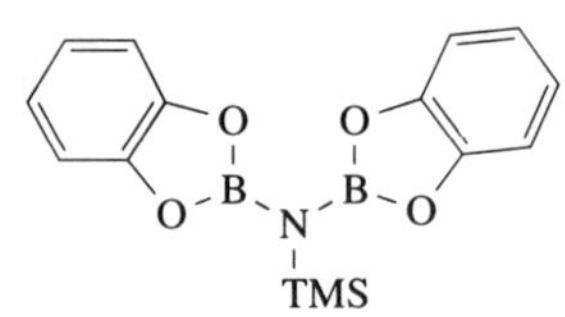

(268)

Compound (**266**) ($R^1 = Cl$) is used as a precursor to the derivatives (**266**) ($R^1 = N(TMS)$) and (**268**). The former heterocycle is the only product when (**266**) ($R^1 = Cl$) and $N(TMS)_3$ react in a 1:1 ratio. Increasing the amount of (**266**) ($R^1 = Cl$) favours (**268**) as the product. The new heterocycles have been characterized spectroscopically and the crystal structure of (**268**) has been determined. The nitrogen atom is in a trigonal planar environment, but the two benzodioxaborole rings are not coplanar with respect to one another; twist angles of 3.9° and 16.2° are observed for two crystallographically independent molecules.[465] Compound (**266**) ($R^1 = C(TMS)_3$) has been prepared and characterized. Unlike simple alkyl derivatives, this compound is not reduced by LAH. However, hydrolysis does occur, yielding $(TMS)_3CB(OH)_2$.[466] Eaborn *et al.* have also synthesized (**266**) ($R^1 = C(SiMe_2Ph)_3$) by reacting catechol with $(THF)_3Li(\mu\text{-}H)_3BC(SiMe_2Ph)_3$.[467]

Reactions of heterocycles of type (**266**) include that of the derivative with $R^1 = H$ with alkynes, $R^2C{\equiv}CH$ (R^2 = various). The products here are (**266**) ($R^1 = CH{=}CHR^2$) hydrolysis of which cleaves the dioxaborole ring.[468] For $R^1 = OPr^i$, the reactions of (**266**) with semicarbazones have been investigated.[469] Palladium catalysed cross-coupling of (*E*)-1-alkenyl-1,3,2-benzodioxaboroles with allylic phenoxides provides a route to 1,4-alkadienes. For example, when (**266**) ($R^1 = CH{=}CHBu$) and Et_3N are treated with aqueous NaOH and H_2O_2 in the presence of $Pd(PPh_3)_4$, a 65% yield of (*E,E*)-$BuCH{=}CHCH_2CH{=}CHPh$ is obtained.[470]

A high-yield synthesis to (**266**) ($R^1 = H$) is the reaction of (**269**) with sodium borohydride in diglyme. Other products of the reaction are Na[(**270**)] and B_2H_6. Problems concerning the separation of catecholatoborane from the solvent are overcome by the addition of a small quantity of lithium chloride to the reaction mixture and using Et_2O as solvent.[471] Intercalation by the host molecule (**271**) has been studied.[472,473] Examples of guest molecules are $PhCH_2NH_2$ and alcohols, and the crystal structure of (**271**)·MeOH·$PhCH_2NH_2$ has been determined. Here, $[PhCH_2NH_3]^+$ interacts with the macrocycle while $[MeO]^-$ forms an $O \rightarrow B$ coordinate bond to give a four-coordinate boron atom.[472]

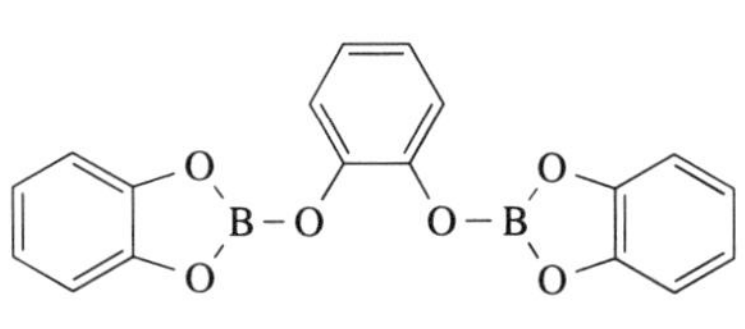

(269)

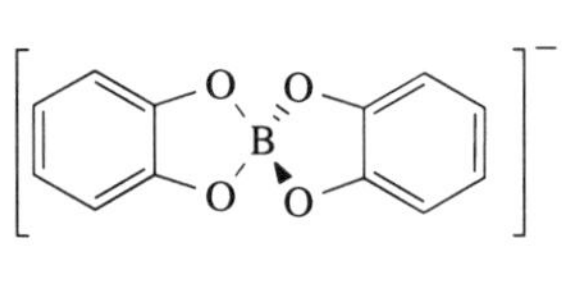

(270)

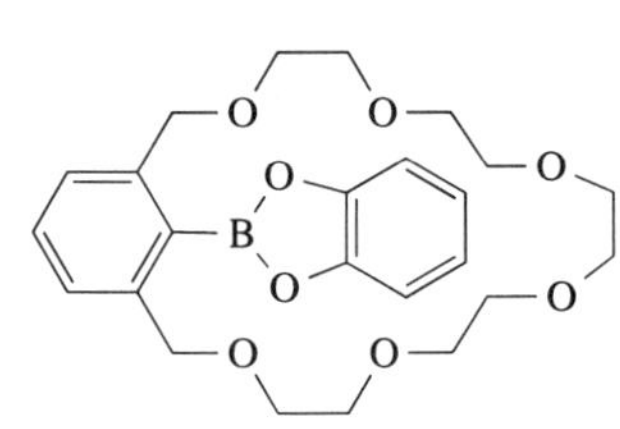

(271)

4.9.4 Six-membered Rings

Among six-membered boron and oxygen-containing heterocycles (boroxins and borates excluded), the C_3BO_2 framework is probably the most commonly represented. Physical properties reported for rings of type (**272**) include mass spectrometric measurements ($R^1 = Ph$; $R^2 = Me$ or CH_2NMe_2; $R^3 = CH_2OH$, CH_2NH_2, CH_2NHMe or CH_2NMe_2),[474] analyses of IR and Raman spectra,[475] and ^1H NMR spectral data for (**272**) (R^1 = Et, Pr^n, Pr^i, Bu^i; R^2 = H; R^3 = H, Pr^i, Bu^i).[476] NMR spectroscopic results for (**272**) (e.g., R^1 = OH, Et, Pr^i, Bu^i, OPr; $R^2 = R^3$ = Me, Et, $PhCH_2$, Ph, PhO, $MeCO_2$, $EtCO_2$) suggest that, when $R^2 = R^3$, the heterocycle interconverts between two energetically equivalent forms but, when $R^2 \neq R^3$, one form is favoured.[477] The conformations of a range of compounds of type (**272**) (R^1 = Pr^i or Bu^i; R^2 = H; R^3 = Me, Pr^i, Bu^i, Bu^t, hexyl, heptyl, Ph, $PhCH_2$, CH_2=$CHCH_2$, NO_2) have been studied by ^1H NMR spectroscopy. Most of the heterocycles adopt a conformation such that the R^3 substituent is in an equatorial position but the NO_2 group prefers an axial site.[478]

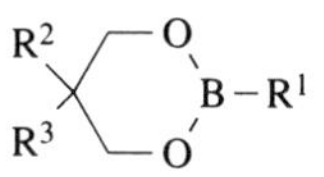

(272)

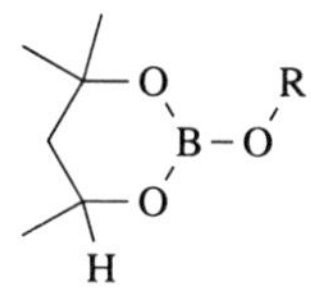

(273)

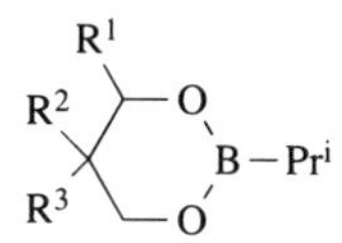

(274)
R^1, R^2 = H or Me
R^3 = H, Me, Pr^i, Bu^t, $PhCH_2$ or NO_2

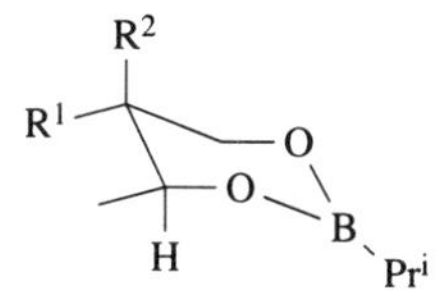

(275) R^1 = H and R^2 = Me
(276) R^1 = Me and R^2 = H

Crystal structure determinations of six-membered rings have been carried out for (**272**) with R^1 = Ph, R^2 = Me and R^3 = 4-$NO_2C_6H_4CH$=NO,[479] and with R^1 = Ph, R^2 = Me, and R^3 = NO_2,[480] and for (**273**) (R = $SiPh_3$).[481] The first two structures, determined by Kliegel *et al.*, should be compared with related molecules described in a series of articles by Kliegel *et al.* published in *Can. J. Chem.* over the period of this review. The monocyclic, rather than bicyclic, structure of (**272**) (R^1 = Ph; R^2 = Me; R^3 = 4-$NO_2C_6H_4CH$=NO) is attributed to the fact that the nitrone group exhibits low basicity and does not form an intramolecular O → B interaction.[479] Similarly, with R^3 = NO_2, no intramolecular O → B bond is formed.[480] In the solid state, heterocycle (**273**) (R = $SiPh_3$) adopts a conformation that is between a half-chair and a 'sofa'.[481]

The relative basicities of compounds (**274**) have been measured by use of IR spectroscopy. The results, which indicate that the basicity of a dioxaborinane is significantly less than that of a corresponding 1,3-dioxane, are in keeping with electron densities calculated by molecular orbital methods.[482] The potential uses of organoboron compounds as lubricating oil additives and pesticides may run into problems in terms of the hydrolytic stabilities of the materials. In attempts to improve the latter, studies have shown that heterocycles of type (**273**) (R = H) possess quite high hydrolytic stability (see also Sections 4.9.6 and 4.11.4).[483]

In a series of papers, Kuznetsov and Gren have discussed the syntheses, isomerization and reactions of a group of six-membered 1,3,2-dioxaborinanes.[481–95] When a 56:44 mixture of *cis*-(**275**) and *trans*-(**276**) is heated at 150–170 °C in the presence of 3–10% $ZnCl_2$, *cis*-(**275**) irreversibly converts into *trans*-(**276**). The rate of the conversion (which takes between 7 and 50 hours) depends on the concentration of the $ZnCl_2$; the presence of the catalyst is essential.[485] The isomers can be distinguished by characteristic IR spectral absorptions in the 400–650 cm^{-1} region, and similarly for 4,6-dimethyl derivatives. The *cis*-isomers also exhibit weak absorptions at 1060 cm^{-1} and 1080 cm^{-1} and these are not present in spectra of the *trans*-isomers.[486] ^1H and ^{13}C NMR spectroscopic data for the isomers have also been detailed.[487] The free energy difference between *cis*-(**275**) and *trans*-(**276**) is ≥10.5 kJ mol^{-1}.[488]

Some preparative methods are as follows. Along with a group of five-membered heterocycles (see Section 4.9.3.1), the six-membered ring (**272**) (R^1 = $PhSCH_2$; $R^2 = R^3$ = Me) has been prepared from $PhSCH_2B(OH)_2$.[445] The reaction of $(Pr^iO)_2BCHCl_2$ with $CH_2OHCMe_2CH_2OH$ leads to (**272**) (R^1 = $CHCl_2$; $R^2 = R^3$ = Me). This product is reduced to the R^1 = CH_2Cl derivative by the action of Bu_3SnH; conversion to the analogous iodo-derivative is then readily achieved by treatment with sodium iodide. It is also possible to reduce $(Pr^iO)_2BCHCl_2$ prior to cyclization.[496] The reaction of boric acid and 1,3-propanediol in aqueous DMSO leads to (**272**) (R^1 = OH or $OCH_2CH_2CH_2OH$; $R^2 = R^3$ = H); when two equivalents of diol are used, a bis(heterocyclic) product can be obtained, the two rings being connected by a {$BOCH_2CH_2CH_2OB$} chain.[497] The linked system (**277**) ($R^1 = R^2$ = H) is more acidic than Ph_3CH and is deprotonated (at the methylene bridge) by Li(TMP) in the presence of TMEDA in THF. The lithium salt so formed reacts with alkyl halides to yield (**277**) (R^1 = H; R^2 = alkyl). Bridge deprotonation can occur again and gives a route to (**277**) (R^1, R^2 = alkyl).[498] On the other hand, (**272**) (R^1 = various; $R^2 = R^3$ = H) reacts directly with organolithium compounds, LiR^4, to give Li[(**278**)].[499]

Properties of compounds of type (**279**) have been studied.[500] They exhibit antimicrobial activity *in vitro* and the boron atom appears to be essential for activity; electron withdrawing *B*-substituents are effective.

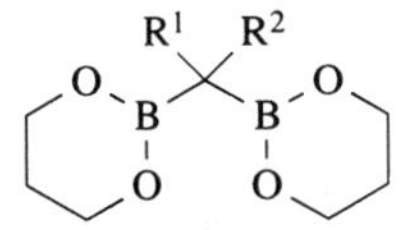

(277)
R^1 and R^2 = H or aikyl

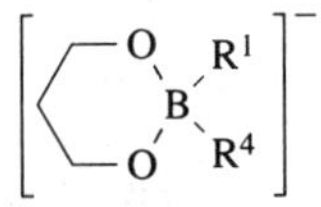

(278)

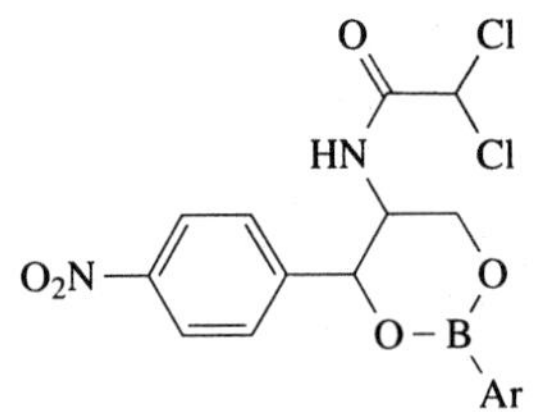

(279)
Ar = Ph or substituted Ph

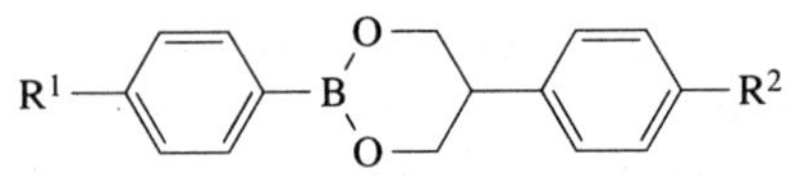

(280)
R^1 = CN and R^2 = Bu^n
R^1 = CN or CO_2H; R^2 = $C_6H_{13}{}^n$, $C_4H_9{}^nO$, $C_8H_{17}{}^nO$

A range of derivatives of six-membered 1,3,2-dioxaborinanes has been involved in studies relating to liquid crystalline materials.[501–10] This work has developed significantly since 1989. The preparation of molecules (**280**) (R^1 = CN or CO_2H; R^2 = n-C_6H_{13}, n-C_4H_9O, n-$C_8H_{17}O$) has been described; the cyano derivatives are accessed via their carboxylate analogues. The new compounds exhibit liquid crystalline properties and the transition temperatures are given in Table 5.[502] Related work examines molecules (**280**) with a wider range of R^1 groups; mesomorphic phases are exhibited.[504] Polymers with organic backbones and derivatized 1,3,2-dioxaborinanes in side-chain sites also exhibit mesomorphic phases.[505] Structural and NMR spectroscopic data for compound (**280**) (R^1 = CN; R^2 = Bu^n) show that the two phenyl rings are mutually perpendicular; the butyl chain is extended in a *trans*-conformation providing the molecule with an almost linear geometry.[507] Compounds (**281**) have been prepared and at low temperatures, they exhibit nematic phases and a high positive dielectric anisotropy.[510]

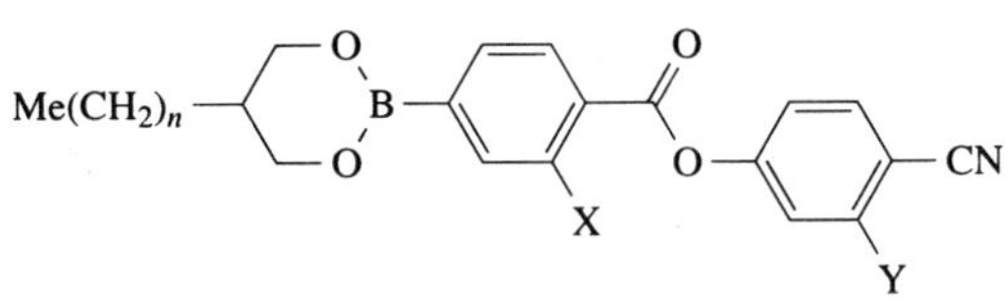

(281)
n = 2–6; X and Y = H, F, Cl

Table 5 Transition temperatures for liquid crystalline materials based on (**280**).

R^1	R^2	$T(C\text{-}meso)^d$ (°C)	$T(meso\text{-}I)^e$ (°C)
CO_2H	$C_6H_{13}{}^n$	200(n)[b]	260(d)[c]
CO_2H	$C_4H_9{}^nO$	250(n)	280(d)
CO_2H	$C_8H_{17}{}^nO$	202(n)	273(d)
CN	$C_6H_{13}{}^n$	98(n)	122.4
CN	$C_4H_9{}^nO$	117(n)	166.0
CN	$C_8H_{17}{}^nO$	95[a]	144.8

[a] T(C-smectic). [b] n = nematic. [c] d = decomposition. [d] C = crystal. [e] I = isotropic.

Heterocycles of type (**272**) in which R^1 is an inorganic moiety have been prepared and characterized. The condensation of dithiophosphates with (**272**) (R^1 = Cl; R^2 = R^3 = H) gives, for example, compounds (**272**) (R^1 = SP(O)$(OEt)_2$; R^2 = R^3 = H) and (**282**). The products (characterized by IR and ^{11}B, ^1H, ^{13}C and ^{31}P NMR spectroscopies) are volatile and hydrolyse readily. Reactions with organotin oxides have been studied.[511] Heterocycles (**272**) (R^1 = Cl; R^2, R^3 = H, Me) react with sodium salts of *O*-alkyl phosphonic acids to yield (**272**) (R^1 = OP(O)H(OR^4); R^4 = Et, Pr, Bu; R^2 = R^3 = H, Me). The structures of the products have been proposed on the basis of multinuclear NMR spectroscopic results.[512]

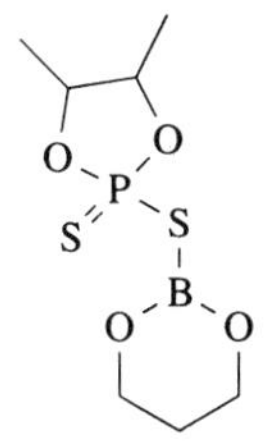

(282)

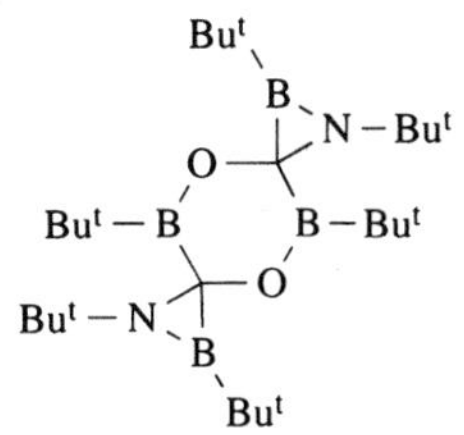

(283)

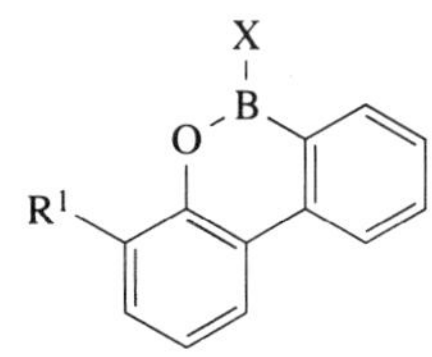

(284)

Tri-*t*-butylazadiboriridine (NB_2 ring) reacts with CO to yield (**283**). Results of a crystal structure determination have confirmed that there are significant π contributions (not indicated in the diagram) both to the B–O interactions (0.136 nm) and to the B–N bonds (0.133 nm). The formation of (**283**) is unexpected and a pathway involving an intermediate four-membered ring is proposed.[513] Metallation with BuLi or sodium of the phenols 2,6-$Ph_2C_6H_3OH$ and 2-PhC_6H_4OH (R^2OH) followed by reaction with BY_3(Y = Cl or Br) leads to products (**284**) (R^1 = H or Ph; X = Cl, Br or OR^2) as well as related B–O–B bridged compounds. X-ray diffraction results confirm the structures of the bridged derivatives. The reaction of $LiN(TMS)_2$ with (**284**) (R^1 = Ph; X = Cl) results in the formation of LiCl and (**284**) (R^1 = Ph; X = $N(TMS)_2$).[514]

4.9.5 Eight-membered Rings

Solid state and solution properties of systems of type $Ph_2P(O)OBEt_2$ have been studied. A monomer ⇌ dimer equilibrium exists in solution and the dimer persists in the solid state. The $P_2O_4B_2$ ring of $\{Ph_2P(O)OBEt_2\}_2$ exhibits a boat conformation. Similarly, in the solid state *B*-$Ph_2P(O)O$-9-BBN occurs as a dimer, exhibiting an eight-membered heterocycle with a half-chair conformation.[515] Eight-membered ring systems have also been studied in conjunction with the search for organoboron compounds exhibiting high hydrolytic stabilities (see Section 4.9.4).[483]

4.9.6 Bicyclic Systems (excluding Five-membered Heterocycles adjacent to Aromatic Rings)

A range of derivatives with the frameworks shown in (**285**) and (**286**) has been synthesized; these have applications in colour photographic materials.[516] The preparation of (**287**) from $(2\text{-}LiC_6H_4)_2O$ has been detailed.[517] Heterocycle (**288**) (R = OH) can be derived from 1,8-dilithionaphthalene. Treatment of this product with methanol in benzene yields (**288**) (R = OMe). This, in turn, reacts with PCl_5 giving (**288**) (R = Cl) and with MeMgBr to give (**288**) (R = Me).[518]

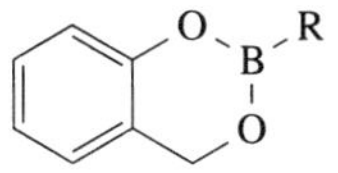

(**285**) R = alkyl or aryl

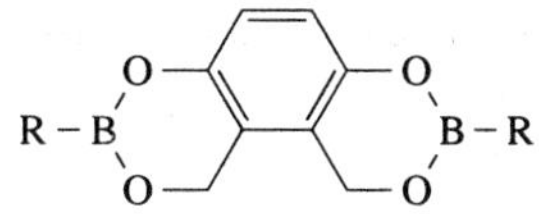

(**286**) R = alkyl or aryl

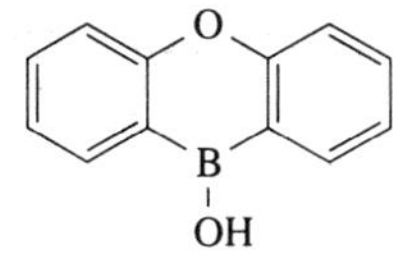

(**287**)

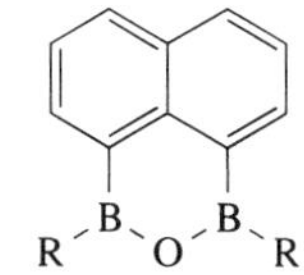

(**288**) R = OMe, Cl, Me, OH

Yalpani *et al.* have prepared compounds (**289**) according to Equation (34). The products are relatively stable, withstanding falsh photolysis at 550 °C. Despite possessing five-membered C_2O_2B units, compounds (**289**) do not behave in an analogous fashion to the saturated systems described in Section 4.9.3.1. Adducts are formed with a variety of amines.[519] Interestingly, when the precursor in Equation (34) is treated with $Bu^tCO_2BEt_2$, the reaction takes a different course and (**290**) is formed.[520] This compound has been characterized by IR and multinuclear NMR spectroscopies,[520] and exhibits thermochromic properties[520–3] being colourless at room temperature and yellow at low temperatures. The bis(*B*-phenyl) analogue of (**290**) is nonthermochromic. Results of crystallographic investigations of the two forms (20 °C and −30 °C) of (**290**) and of the bis(*B*-phenyl) derivative (20 °C) are informative. In both room temperature structures, the boron atoms are sp^2 hybridized. On cooling (**290**), the ethyl groups bend out of the plane of the bicyclic ring system and this change is concomitant with the formation of intermolecular O → B bond formation. This influences the spacing between layers of flat molecules; effectively, a reversible polymerization process occurs in the solid state. Over a temperature range of $-73\ ^\circ C < T < +40\ ^\circ C$ significant changes to the crystal cell volume take place.[521] A range of other derivatives has been studied with an emphasis on the mechanisms of phase transitions and hysteresis.[522,523] Exchange reactions between $\{EtB^{18}O\}_3$ and the bis(*B*-propyl) derivative of (**290**), and between $\{EtB^{18}O\}_3$ and (**289**) (R = Ph) have been compared; labelled oxygen is incorporated in the latter case but not in the former case.[524]

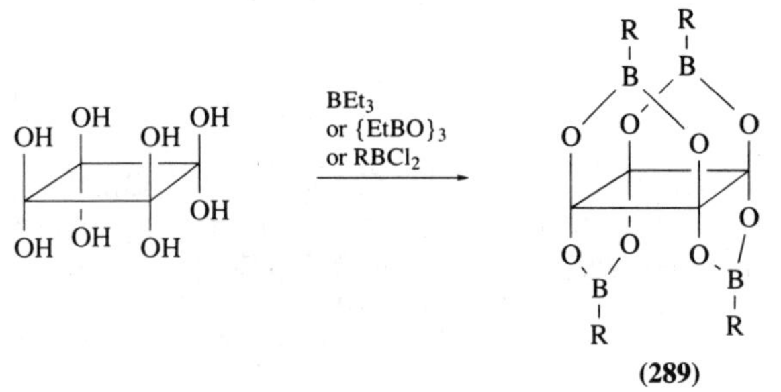

(34)

(**289**)
R = Et, Pr, Pr^i, Ph

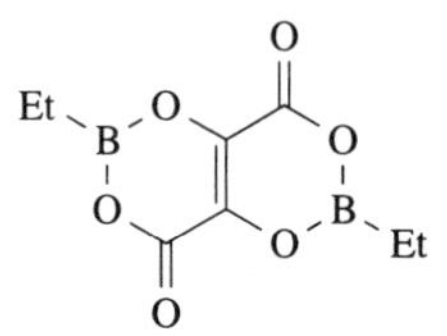

(**290**)

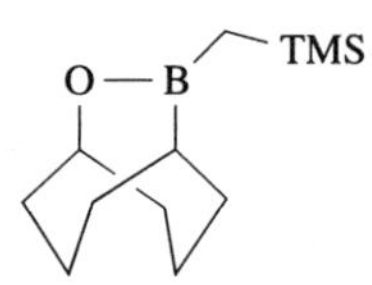

(**291**)

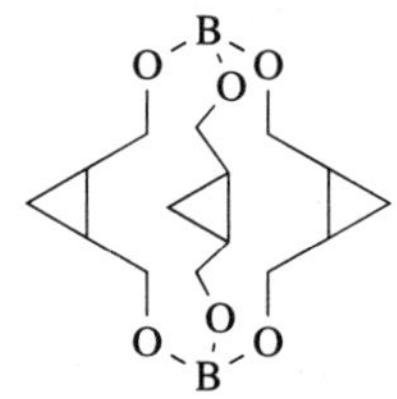

(**292**)

Miscellaneous compounds which come into this section are a derivative of (*S*)-pinandiol,[525] and heterocycles (**291**)[526] and (**292**).[527] The palladium catalysed cross-coupling of vinyl, alkynyl and aryl bromides with (**291**) give silylmethylated products with full retention of configuration in excellent yields.[526] The optically active borate (**292**) is synthesized from *trans*-1,2-cyclopropanedicarboxylic acid and $BH_3{\cdot}THF$. The novel nature of (**292**) is revealed in a crystal structure determination; each boron atom is in a trigonal planar environment. Infrared, Raman, and vibrational circular dichroism spectroscopic properties have been detailed.[527]

4.9.7 Rings in which the Boron Atom is Four Coordinate

Heterocycles in this section are organized according to ring size. In some cases the cyclic structure comes about because of intramolecular O → B coordinate bond formation, while in other systems, the four-coordinate boron environment is a result of adduct formation.

In the solid state, Me_2BOBMe_2 exhibits a dimeric structure containing a four-membered B_2O_2 central ring. The BC_2 frameworks of the terminal BMe_2 units are coplanar with the ring plane while the methyl groups attached directly to the ring lie above and below the heterocycle.[528]

The reaction of $LiC(R^1)$=$C(R^2)C(R^3)(O)$ (e.g., $R^1 = R^3$ = OMe and R^2 = H, or $R^1 = R^2$ = OMe and R^3 = NHCMePh) with Et_3B leads to the replacement of the lithium by an Et_2B substituent. An intramolecular O → B interaction causes cyclization as confirmed for $R^1 = R^2$ = OMe and R^3 = NHCMePh; the B–O distance is 0.1595(11) nm.[529] Similarly, in (**293**), a coordinate bond stabilizes a cyclic structure as confirmed crystallographically.[530] The salts M[(**294**)] (M = Na or K; e.g., $R^1_2 = Et_2$, R^2 = Me, Et, Ph or $R^1_2 = C_8H_{14}$, R^2 = Me, Et) are formed when $R^1_2BC(R^2)$=$C(Et)BEt_2$ reacts with MOH. Ethane is eliminated from some of the anions when they are heated above 100 °C.[531]

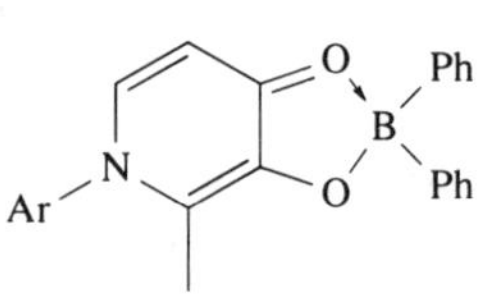

(**293**) Ar = 4-MeC_6H_4

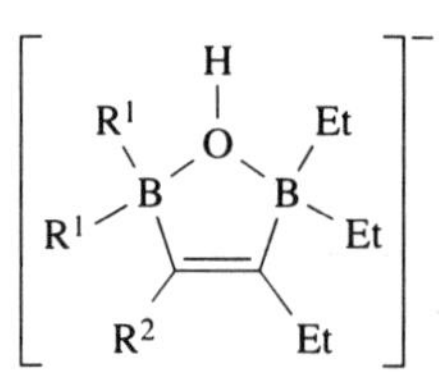

(**294**)

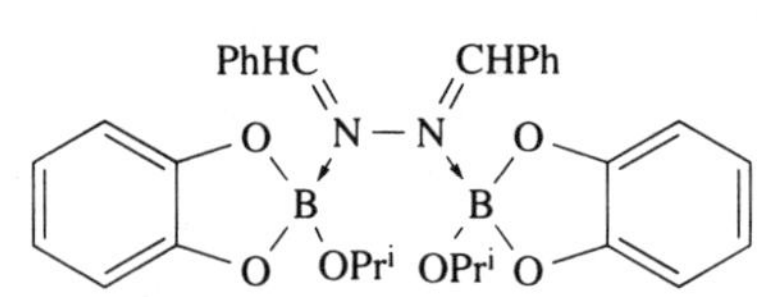

(**295**)

Adducts (**266**)·L (R^1 = OBu and L = 2-, 3- or 4-$AcC_6H_4NH_2$ or 4-$AcC_6H_4NMe_2$) and (**285**)·L (R = OBu or OPr^i and L as before) have been prepared and characterized by physicochemical methods. Spectroscopic data are consistent with the presence of a tetrahedrally sited boron atom in each compound.[532] The salt Na[(**270**)] is also formed during the high-yield synthesis to catecholatoborane (**266**); see Section 4.9.3.2.[471] Anion (**270**) has been independently prepared from boric acid and 1,2-$(OH)_2C_6H_4$.[463] The coordination of azomethine nitrogen to boron is spectroscopically evidenced in a series of complexes, for example, (**295**), which are prepared by the direct combination of the base and the heterocycles.[533]

Six-membered heterocycles incorporating a four-coordinate boron atom include dialkylborylacetylacetonate complexes. Although they are thermally stable, complexes in which the boron substituents are, for example, *c*-C_6H_{11} or Bu^n, undergo alkyl migration when photolysed to generate products with a three-coordinate boron centre.[534] A range of β-ketonate complexes has been studied.[535–41] The reaction of the trimer $\{F_2BOMe\}_3$ with acid anhydrides (e.g., MeC(O)OC(O)Me) yield derivatives of type (**296**). The crystal structure of (**296**) has been determined and two conformational isomers are present in the lattice.[542] The heterocycle (**297**) has been prepared,[543] and the reaction of oxirane and $Ph_2BOCH_2CH_2NEt_2$ yields an addition product with an eight-membered cyclic structure.[544]

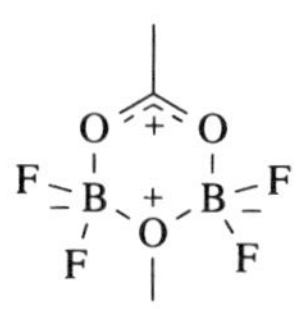

(**296**)

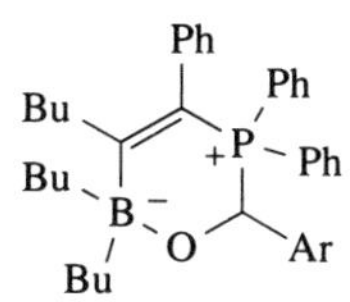

(**297**) Ar = 2-HOC_6H_4

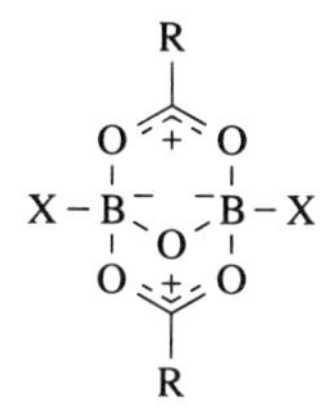

(**298**)

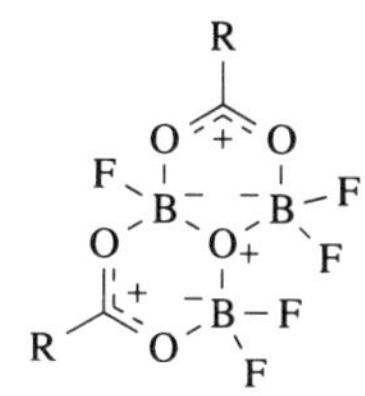

(**299**) R = Me, Et or Pr

Compounds of type (**298**) have been synthesized and characterized.[545–7] The derivative with X = Et and R = Bu^t may be used to introduce *O*-ethylboranediyl groups into polyhydroxy compounds.[545] Reactions between carboxylic acid anhydrides and $\{F_2BOR\}_3$ lead to (**298**) (X = F; R = Me, Et and Pr). The crystal structure of the derivative with R = Me has been elucidated.[546] These compounds react with boron trifluoride to give insertion products (**299**) and for R = Et, the compound has been characterized crystallographically, and by solid state and solution NMR spectroscopy.[547]

4.10 RINGS CONTAINING BORON AND SULFUR OR SELENIUM ATOMS

4.10.1 Four-membered Rings

Four-membered heterocycles exhibiting a B_2S_2-framework and structurally analogous to (**255**) are represented by $\{(Pr^i_2N)BS\}_2$,[548,549] $\{(Me_2N)BS\}_2$,[549] and $\{(TMP)BS\}_2$.[549] The selenium-containing heterocycle $\{(TMP)BSe\}_2$ has been structurally characterized.[432]

The reactions of *B*-Cl-9-BBN with LiSR, or $\{9\text{-BBN}\}_2$ with HSR (R = Me, Et, Pr, Bu^t or Ph) yield *B*-RS-9-BBN. This is monomeric in solution at room temperature but is dimeric in the solid state. X-ray crystal data show that the dimer has C_{2v} symmetry and thus both methyl groups are on the same side of the B_2S_2 ring; B–S distances are 0.1990(2) nm and 0.1998(2) nm.[550]

4.10.2 Five-membered Rings

Unsaturated sulfur-containing heterocycles of type (**300**) (E = S) have been prepared from (**301**) (E = S) by reactions with alkynes; reactions of the unsaturated heterocycles have been studied as discussed below. Compound (**300**) (E = Se; $R^1 = R^2$ = Et; X = 2,6-$Me_2C_6H_3NH$) has been prepared and characterized by ^1H, ^{13}C, ^{14}N, ^{11}B and ^{77}Se NMR spectroscopy and x-ray diffraction.[551] The chemistry of the family of compounds (**300**) with E = Se and S is quite developed. For (**300**) (E = Se; $R^1 = R^2$ = Et, X = Me or Bu), reactions with isocyanates cause ring expansion, and similarly with $OCN(CH_2)_6NCO$, (**302**) is formed.[552] When (**300**) (E = Se; $R^1 = R^2$ = Et; X = Me or Et) reacts with $Bu^tN{=}S{=}NBu^t$, one selenium atom is replaced by a $\{NBu^t\}$ unit.[553] Parallel reactions occur when (**300**) (E = S; $R^1 = R^2$ = X = Me, or $R^1 = R^2$ = Et and X = Me) reacts with $Bu^tN{=}S{=}NBu^t$.[554] Insertion occurs with $^iPrN{=}C{=}N^iPr$.[555] Exchange of the *B*-substituent occurs when (**300**) (E = Se; $R^1 = R^2$ = Et; X = I) reacts with alcohols, thiols, Me_2S_2, Ph_2S_2, $KOBu^t$ and Et_2O.[556] Reactions of TMS-Y (Y = CN, NCO, or NSO) with (**300**) (E = S; $R^1 = R^2$ = Me, Et or Pr; X = Br) result in exchange of X for Y. The cyano-derivative reacts with sulfur, retaining the heterocyclic framework and giving the derivative with X = NCS.[554]

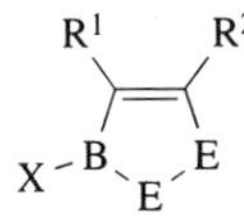

(**300**)
E = S or Se

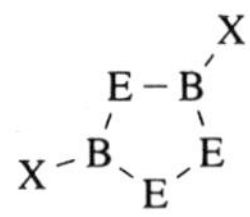

(**301**)
E = S or Se

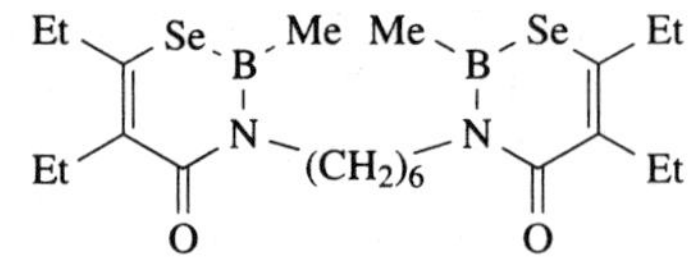

(**302**)

Saturated heterocycles (**303**) (R = alkyl) find use in the synthesis of RCO_2H.[557] For R = Cl, (**303**) (R = Cl) is a precursor to a range of derivatives, for example, with R = Cp*.[82] Heating (**303**) (R = Cl) in toluene with an equivalent of $N(TMS)_3$ gives (**303**) (R = $N(TMS)_2$), but a trimethylamino-bridged bis(heterocyclic) system forms in low yield in 2:1 and 3:1 reactions. The products have been spectroscopically characterized.[465] Treatment of (**303**) (R = Cl) with potassium eliminates KCl and causes ring coupling. The crystal structure of product (**304**) has been determined; the two five-membered rings are in a twist conformation. The B–S bonds (in the range 0.1787(2)–0.1812(2) nm) are regarded as possessing single bond character.[558] Complexes of (**303**) (R = OPr^i) with semicarbazones and thiosemicarbazones have been prepared.[559]

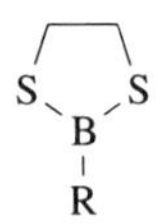

(**303**)

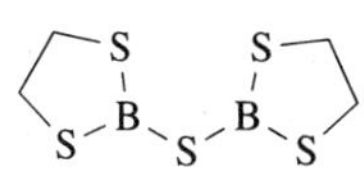

(**304**)

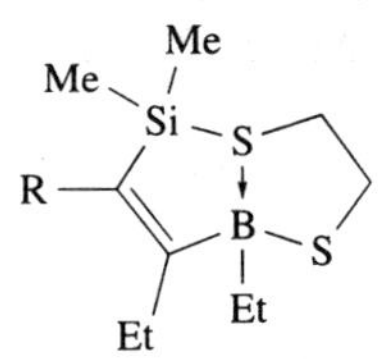

(**305**)
R = Me or CMe=CH_2

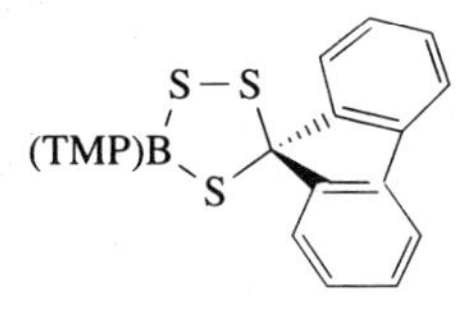

(**306**)

Ring closure of (EtClB)C(Et)=C(R)($SiMe_2Cl$) by treatment with MeSLi provides the precursor to (**305**). The latter is formed after reaction with $LiSCH_2CH_2SLi$. Compound (**305**), and related complexes, have been characterized by multinuclear NMR spectroscopy and mass spectrometry, and the structure has been crystallographically confirmed for the derivative formed from 2-$LiSC_6H_4SLi$.[560] A not dissimilar complex has been structurally characterized by Furusaki *et al.*[561]

Elemental sulfur reacts with the boron-containing precursor in Equation (33) to give (**306**).[435] The B_2S_3 ring system (**301**) (E = S) has received much attention. For X = Br, reactions of (**301**) (E = S) with alkynes, $R^1C{\equiv}CR^2$, yield (**300**) (X = Br; E = S). The chemistry of these unsaturated systems has been investigated; for example, the reaction of 2,6-$Me_2C_6H_3NHLi$ with (**300**) (E = S; X = Br; $R^1 = R^2$ = Et) gives (**300**) (X = 2,6-$Me_2C_6H_3NH$). The crystal structure of this derivative has been determined[562] (compare with the analogous selenium heterocycle).[551] The reaction of $MeBBr_2$ and H_2S_x to yield (**301**) (E = S; X = Me) provides a suitable method of preparing the ^{10}B labelled compound.[563] The selenium analogues (**301**) (E = Se; X = I or Ph) have been studied by solid-state ^{10}B and ^{11}B NMR spectroscopy and comparison of the data with those for B_3S_3 rings suggests that the presence of a chalcogen–chalcogen bond increases the donation of π-electrons into the empty 2*p* orbital of the boron atom.[564] Diborane reacts with H_2S_2, H_2S_3 or crude sulfane oil to give (**301**) (E = S; X = H). This compound reacts with alkynes, $R^2C{\equiv}CR^1$ (where R^1 is less sterically demanding than R^2) to give (**301**) (E = S; X = C(R^1)=CHR^2).[565] The bromide groups of (**301**) (E = S; X = Br) may be replaced by alkyl, amino and isothiocyanato groups (Y) via reactions with LiY.[566] With R^2NSO, (**301**) (E = S; X = Br or Cl) reacts to give (**307**) (R^2 = TMS or C_6F_5 for X = Br; R^2 = TMS for X = Cl). The bromo substituents in (**307**) can be replaced in a similar manner to those in (**301**).[566] Ring cleavage occurs when (**301**) (E = S; e.g., X = Me, Et, Ph, NMe_2) is treated with TMS-N(R^2)P=NBu^t (R^2 = TMS or Bu^t) giving products which are derivatives of a {SBNP} heterocycle.[567] Adduct formation between (**195**) (X = Me) and (**301**) (E = S; X = Br) occurs and leads ultimately to exchange products. The crystal structure of the adduct, supported by an S → B and two N → B interactions, has been determined.[568] The mechanism of exchange between (**301**) (E = S; X = Me) and cyclic $(MeN)_3(BMe)_2$ has been investigated.[569] In reactions of (**301**) (E = S; X = Me, NEt_2, $N(TMS)_2$) with sulfur diimides, six-membered $N_2S_2B_2$ heterocycles are produced.[570,571] Product characterization by ^1H, ^{13}C, ^{11}B, ^{14}N and ^{29}Si NMR spectroscopy is supported by crystallographic data for two derivatives. Each six-membered ring is nonplanar, being folded along the transannular S···S vector.[571]

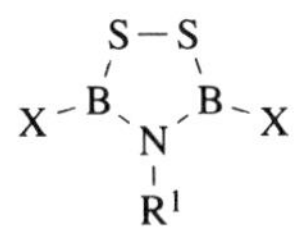

(307)
X = Cl or Br
R^1 = TMS or C_6F_5

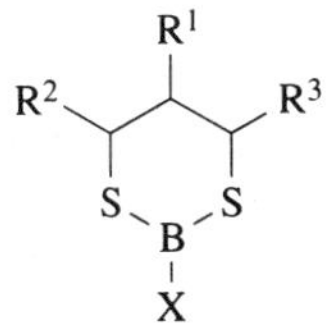

(308)
X = Cl, Br, Ph, Pr^i

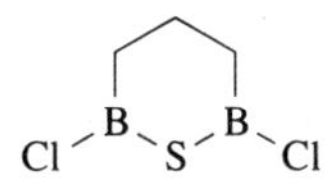

(309)

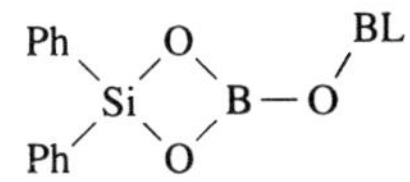

(310)

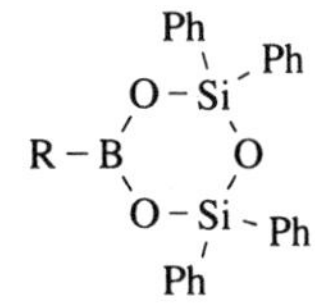

(311)
R = Ph or Mes

4.10.3 Six-membered Rings

Absorptions in the IR and Raman spectra of the dithiaboracyclohexanes (**308**) (X = Cl, Br or Ph; $R^1 = R^2 = R^3$ = H) have been assigned on the assumption that the molecules possess C_s symmetry.[572] An NMR spectroscopic study of heterocycles in the same family (X = Pr^i; $R^1 = R^2$ = H and R^3 = alkyl, or R^1 = alkyl and $R^2 = R^3$ = H) shows that the ring conformation is predominantly that of a 'couch' with alkyl groups taking equatorial sites. For $R^1 = R^2$ = Me and R^3 = H, or for $R^2 = R^3$ = Me and R^1 = H, (X = Pr^i in both cases), (**308**) exhibits a mixture of *trans*- and *cis*-isomers; the couch conformation is again preferred except in the case of *trans*-(**308**) ($R^1 = R^3$ = Me and R^1 = H) which adopts a twist-tub.[573] Borinanes of this type (X = Pr^i; R^1 = Me or Pr^i; $R^2 = R^3$ = H) react with 1,3-dioxanes with removal of the BX unit.[574]

Heterocycle (**309**) is prepared from $Cl_2B(CH_2)_3BCl_2$ and $S(TMS)_2$.[346] A $C_2B_2S_2$ ring results from the thermal decomposition of $[Me_3N{\cdot}BH_2CH_2SMe_2]I$ which is formed from $LiCH_2SMe$ and $BH_3{\cdot}NMe_3$, followed by treatment with Me_3NHCl and MeI. The crystal structure of the product, $\{BH_2CH_2SMe\}_2$, has been elucidated; the ring, which is supported by two S → B coordinate bonds, adopts a chair conformation with the methyl substituents in equatorial sites.[575] Isomerization of 1,4,2,3,-dithiaborinane and related derivatives to five-membered C_2S_2B-systems has been observed; the B–B bond is retained. The isomerization competes with redox processes.[576]

4.11 RINGS CONTAINING BORON AND MORE THAN ONE OTHER HETEROATOM

4.11.1 Boron with Silicon or Tin and Nitrogen

A series of four-membered rings with a BN_2E framework (E = Si or P) has been prepared (see Section 4.11.3).[577] Six-membered heterocycles with SiN_3B_2 and Si_2N_3B frameworks have been structurally characterized. Both rings are near planar. In solution, ^1H, ^{11}B, ^{27}Si and ^{13}C NMR spectral data are consistent with the solid-state structures.[578] Heterocycles possessing C_2SnNB frameworks, for example, (**172**), (**173**), (**182**) and (**205**) have been discussed in Sections 4.6 and 4.7.

4.11.2 Boron with Silicon and Oxygen

The preparation of a series of Schiff base (L) complexes (**310**) has been reported.[579] The condensation of $\{Ph_2Si(OH)\}_2O$ with $RB(OH)_2$ (R = Ph or Mes) leads to (**311**).[580] Eight-membered heterocycles (**312**) can be prepared from $Ph_2Si(OH)_2$ and $PhB(OH)_2$,[580] or from $Bu^t{}_2Si(OH)_2$ and $PhBCl_2$.[581] Both six- and eight-membered heterocycles are planar.[580,581]

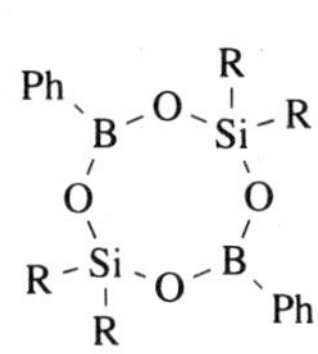

(312) R = Ph or Bu^t

4.11.3 Boron with Nitrogen and Phosphorus, Arsenic or Antimony

Three-membered BNP heterocycles have been prepared and characterized by the groups of Nöth[582] and Paetzold.[583,584] For example, the reaction of $R^1P(TMS)Cl$ with $Bu^tB{\equiv}N^tBu$ yields (**313**) ($R^1 = Bu^t$, Pr^i_2N, $(TMS)_2CH$ or Cp*; $R^2 = R^3 = Bu^t$).[583] Instead of a [2 + 2]-cycloaddition, the reaction of $Bu^tP{=}NBu^t$ and $Bu^tB{\equiv}NBu^t$ leads to BN_2P and BN_2P_2 heterocycles. The crystal structure of the former, (**314**), has been determined.[584] When (**313**) ($R^1 = Pr^i$; R^2 = TMP; $R^3 = Bu^t$) is treated with sulfur or selenium, chalcogen insertion with concomitant oxidation of the phosphorus atom occurs.[582] Four-membered BN_2P heterocycles (**315**) (E = PMe, P(O)Me, P(S)Me; R^2 = Me, Et or Ph; R^3 = Me or Bu^t; and E = P(S)Me; R^2 = F, Cl, Br; $R^3 = Bu^t$) have been prepared; for example, heating MeP(S){N-(Me)$BMe_2\}_2$ eliminates trimethylborane.[577] When $(TMP)B{\equiv}NR^1$ ($R^1 = Bu^t$ or $C_6H_3Pr^i_2$) reacts with EX_3(E = P, As or Sb; X = F, Cl or Br), the compounds $(TMP)B(X)NR^1(EX_2)$ forms. Crystal structure determinations of the products for X = Cl, E = As or Sb, and $R^1 = Bu^t$ show there to be ring formation via N_{TMP}–E interaction (N–As = 0.1869(4) nm and N–Sb = 0.2071(3) nm), but this is not apparent in solution from NMR spectral data.[585] Heterocycle (**316**) is prepared by the dehalogenation of a diborylamine precursor; the ring is nonplanar, being folded along the B···B axis (internal angle = 160°). Selenium-for-phosphorus exchange occurs when (**316**) reacts with elemental selenium.[586] Reactions of $(TMS)_2NP{=}N$-TMS with $ClXBCH_2BXCl$ (X = Cl, Me, Ph or Mes) yield five-membered PNBCB heterocycles. NMR spectral data suggest that the conformation of the ring depends on the steric requirements of X; a planar ring is proposed for X = Cl and this is confirmed by an x-ray diffraction analysis.[587]

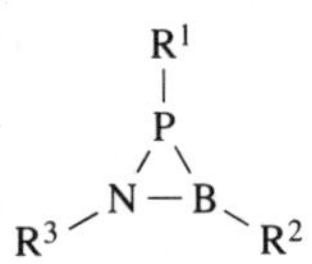

(313)

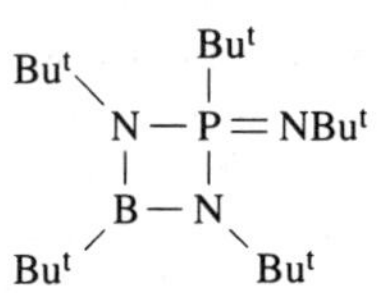

(314)

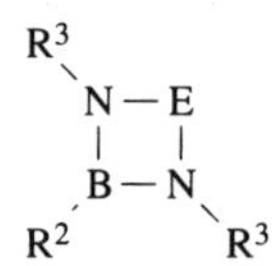

(315)
E = PR^1, $P(O)R^1$, $P(S)R^1$

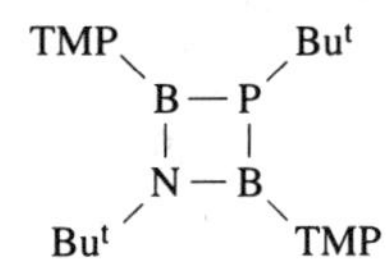

(316)

4.11.4 Boron with Nitrogen and Oxygen

Additions of CO_2, COS, CS_2 and CSe_2 to $(TMP)B{\equiv}NBu^t$ lead to four-membered heterocycles. Thermal or photolytic decomposition of these products leads to $\{(TMP)B\}_2E_2$ (E = O, S or Se).[588] Related work reports the construction of five-membered rings from the addition of CO_2, COS and CS_2 to the {RBN(Me)NMe} unit (R = Me or Ph) (originating from tetraazaborinanes).[589] Amino acids react with the iminoborane $(TMP)B{\equiv}NBu^t$ to give both acyclic and five-membered cyclic products; the latter are preferred when the substituent on the α-carbon atom is sterically demanding.[590] Heterocycles (**317**) have been prepared and characterized by spectroscopic[591] and crystallographic[592] methods. For $R^1 = NEt_2$, the solution NMR spectral data suggest that there is restricted rotation about the B–N bond.[591]

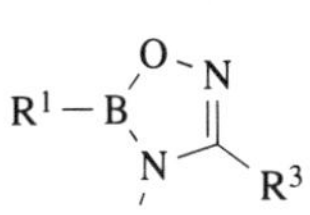

(317)
R^1 = Ph; R^2 = H, Me, Et, Pr, Bu; R^3 = Ph
$R^1 = NEt_2$; R^2 = H or Pr; R^3 = Ph
R^1 = Ph; R^2 = H; R^3 = 2-X-4-YC_6H_3 where
X = H and Y = Me or Br; X = Me or Cl and R^2 = H

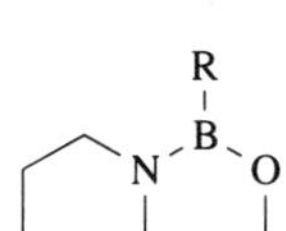

(318)
R = Ph, NMe_2

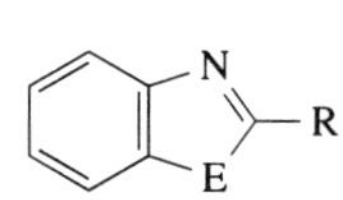

(319)
E = O or S; R = H or Me

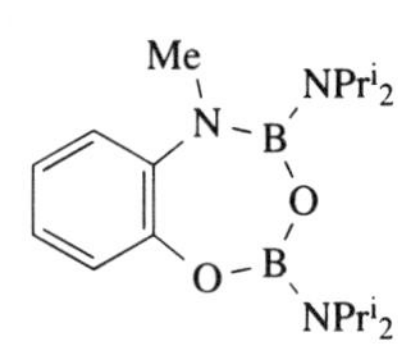

(320)

A range of saturated heterocycles incorporating boron, nitrogen, oxygen and carbon atoms is known. The large group which exhibit N → B coordinate bonds have generally been omitted from this review, but interested readers should, in particular, follow up the work of Kliegel and co-workers published mainly in *Can. J. Chem.* Heterocycles (**318**) are prepared from 2-(2-hydroxyethyl)piperidine with $RB(NMe_2)_2$ (R = Ph or NMe_2). The strategy can be applied to synthesize similar compounds.[593] Related bicyclic systems are catalysts for the enantioselective reduction of acyl dithianes[594] or ketones.[595] The reaction of (**319**) (E = O) with Na/K alloy and $Pr^i_2NBCl_2$ leads to acyclic (and lower yield cyclic) products in addition to (**320**) when R = Me. Parallel chemistry does not occur when E = S (see Section 4.11.7).[596] The elimination of LiCl leads to ring closure in the reaction of BCl_3 with 2,6-$(C(CF_3)_2OLi)_2$-4-R-py (R = various). The proximity of the pyridine nitrogen atom to the boron facilitates transannular N → B coordinate bond formation.[597]

Extensive work by Kliegel and co-workers concerning heterocycles which exhibit N → B coordinate bonds has already been mentioned. A series of papers[598] from this group addresses the reactions of $ArB(OH)_2$ (Ar = Mes or 4-$MeOC_6H_4$) and $RN(OH)CH_2N(OH)R$ (R = CMe_2CN or $CHPhCH_2NO_2$). The choice of R is made with the intention of lowering the basicity of the nitrogen atom to which R is attached. The aryl groups, particularly the mesitylene group, are sterically demanding. In these cases, the products are CN_2O_2B-monocycles. The crystal structure of the product for Ar = Ph and R = CMe_2CN confirms the formation of the eight-membered heterocycle (**321**); although this has a boat–boat conformation, the transannular N···B distances of 0.2755(3) nm and 0.2475(3) nm are rather long to be in keeping with significant bonding interactions.

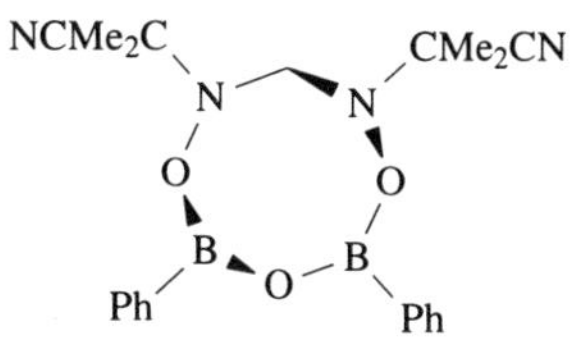

(**321**)

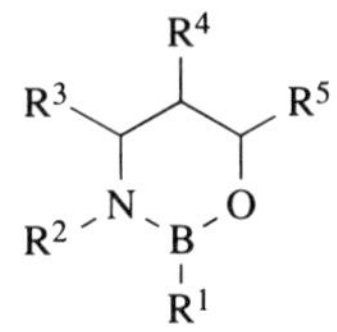

(**322**)

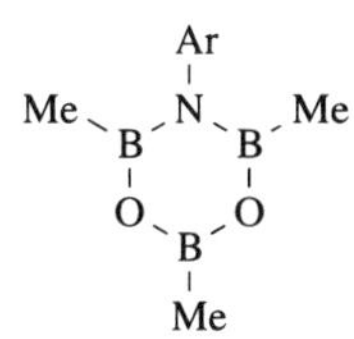

(**323**) Ar = 4-FC_6H_4

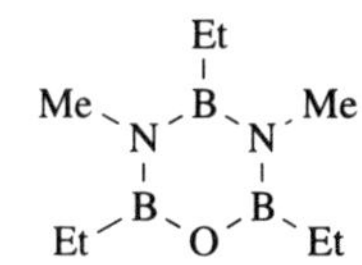

(**324**)

Six-membered heterocycles, including substituted derivatives of type (**322**) have received much synthetic and spectroscopic attention;[599–609] the work is related to studies by the same authors[484–95] and is discussed in Section 4.9.4. One synthetic method is the reaction of $R^1B(OCH_2CHMe_2)_2$ (R^1 = Pr, Pr^i, Bu) with $NH_2CHMeCHMeCH_2OH$ which is stereoselective.[601] The IR spectra of *cis*- and *trans*-stereoisomers of (**322**) (R^1 = CH_2CHMe_2; R^2 = H; R^3 = Me or H; R^4 = Me; R^5 = H or Me) exhibit strong absorptions around 3450, 1603, 1489, 1312, 1208 and 550 cm^{-1} which are assigned to ν(NH), δ(NH), ν(BH), ν(BO) and out-of-plane skeletal vibrations of the CB(O)(N) unit.[603]

Noncarbon containing six-membered rings include (**323**), the crystal structure of which has been determined.[610] At elevated temperatures, $\{(Me_2N)EtB\}_2O$ rearranges to give an equilibrium mixture of $\{EtBO\}_3$ and $(Me_2N)_2BEt$. $\{(Me_2N)EtB\}_2O$ reacts with $(MeHN)_2BEt$ to give (**324**). Bicyclic systems having a framework similar to (**298**) but with two oxygen atoms replaced by NR groups have also been prepared.[611] The reaction of $\{(Me_2N)PhB\}_2O$ with Hpz in toluene under reflux gives an unexpected product in which a $PhBO_2$ unit bridges between the two boron atoms of a PhB(μ-pz)$_2$BPh framework.[612] This, and the analogous tris(*B*-ethyl) derivative, are thermally stable and do not hydrolyse at room temperature. Dissolution in protic solvents causes structural degradation, and with $SOCl_2$, the bridging RBO_2 unit is replaced by two terminal B–Cl groups. The crystal structure of EtB(μ-pz)$_2$(μ-O_2BEt)BEt has been determined.[613] Shore and co-workers have isolated the unexpected compound (**325**) from the reaction of Me_2NBEt_2 and 2,2'-biimidazole. The crystal structure reveals four-coordinate boron atoms and a B–O–B bridge with a large angle of 151°. A possible pathway for the formation of (**325**) has been discussed.[614]

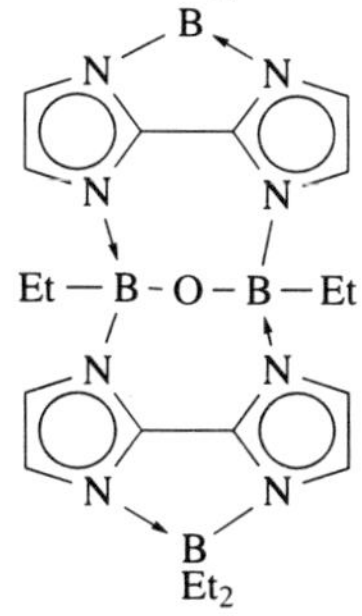

(**325**)

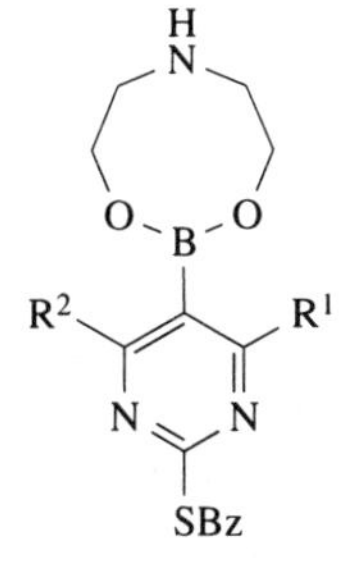

(**326**)
R^1 = H or Pr;
R^2 = OMe or SBz

Studies focusing attention on the hydrolytic stabilities of cyclic borates (see Section 4.9.4) have included eight-membered NC_4O_2B rings and related bicyclic systems.[483] The same framework appears in (**326**), that is, thiouracil derivatives which are potentially useful for boron neutron capture therapy (BNCT).[615] The reaction of $N(2\text{-}OHC_6H_4)_3$ with $BH_3{\cdot}THF$ leads to ring closure via dihydrogen elimination. The heterocyclic cage-like product has been characterized by multinuclear NMR spectroscopy, and it is proposed that an N → B interaction is present. The compound is active towards both Lewis bases (L) and acids (E) forming L → B and N → E interactions, respectively.[616]

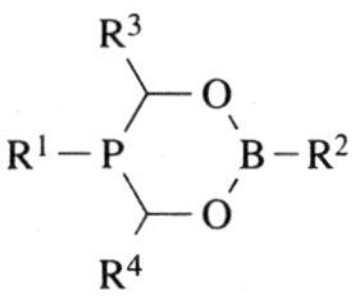

(327)

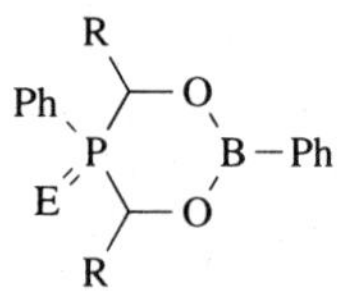

(328)
E = S or Se; R = Me, Ph, Pr^i

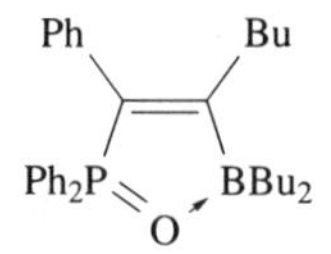

(329)

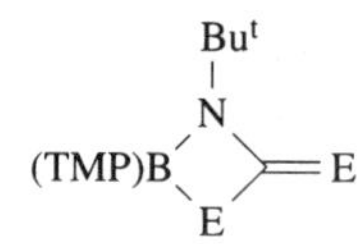

(330)
E = S or Se

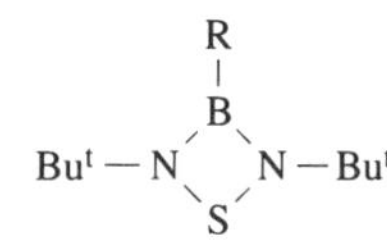

(331)
R = Ph or C_6F_5

4.11.5 Boron with Phosphorus and Oxygen

Heterocycles in this section that have received most attention are the six-membered 1,3,2,5-dioxaboraphosphorinanes (**327**);[617–28] an eight-membered heterocycle was described in Section 4.9.5. Compound (**327**) ($R^1 = R^2 = Ph$ and $R^3 = R^4$ = H, Me, Pr^i or Ph) can be prepared from $PhP(CHR^3/R^4OH)_2$ and $\{PhBO\}_3$. Conformational isomerism among these systems has been studied by ^{31}P NMR spectroscopy.[617] The addition of $PhP(CH_2OH)_2$ to Bu_3B and PhNC leads to (**327**) (R^1 = Ph; $R^2 = CBu_3$; $R^3 = R^4$ = H)[620,621] and treatment with phenylamine cleaves the ring[621] forming an adduct with copper(I) iodide in pyridine.[621] Electrochemical studies of a series of 1,3,2,5-dioxaboraphosphorinanes show that they exhibit irreversible reductions in the region −0.6 V to −2.8 V. Products identified after controlled-potential electrolysis indicate that P–C and B–O bond cleavages occur in the electrogenerated monoanion and dianion, respectively.[622] The crystal structure of (**327**) ($R^1 = R^2$ = Ph; $R^3 = R^4$ = H) has been elucidated.[623] Reactions of 1,3,2,5-dioxaboraphosphorinanes that have been studied include those with chloral,[626] $4\text{-}NO_2C_6H_4CHO$[626] and BH_3THF.[627] The ^1H NMR spectra of a series of oxidized products (**328**) and their pyridine adducts have been recorded and the values of $^2J(PH)$ coupling constants have been used to investigate structural variation.[628]

Heterocycle (**329**), (XO), is prepared by the aerial oxidation of the four-membered ring X which has a direct P → B interaction. The O → B bond in (**329**) is strong; the electronic structure of the heterocycle has been analysed.[629]

4.11.6 Boron with Nitrogen, Phosphorus and Oxygen

The iminoborane[12] (TMP)B≡NBu^t inserts into the five-membered heterocycle OCH_2CH_2OPX (X = F or Cl) via cleavage of one P–O bond. An analogous reaction occurs with the benzo-derivatives, this time with X = F, Cl, Br or Me.[630]

4.11.7 Boron with Nitrogen and Sulfur or Selenium

Four membered rings with the BNCE framework (E = S and Se) are accessed from the iminoborane[12] (TMP)B≡NBu^t and CS_2 or CSe_2 (see also Section 4.11.4). The crystal structure of (**330**) has been elucidated; the B–S and B–N distances are 0.1894(3) nm and 0.1452(3) nm, respectively. Under thermal or photolytic conditions, these heterocycles decompose to give Bu^{t}C=N=C=E and cyclic $\{(TMP)BE\}_2$.[588] The planar nature of the BSBN ring in $(Pr^i_2NB)_2S(NPh)$ (prepared by ring degradation, see below) has been confirmed by an x-ray diffraction study; the angle B–S–B is 70°.[631] The SN_2B ring skeleton can be assembled by reacting $RB(SMe)_2$ (R = Me, Et, Ph, C_6F_5) with sulfur diimides. NMR spectral data for a series of these products are complemented by crystallographic data for heterocycles (**331**).[632,633] An oxidized derivative of (**331**) (R = Ph) containing an {S=NBu^t} unit is formed via the reaction of $PhB(NBu^tLi)_2$ with $Bu^tN{=}SCl_2$.[634]

Some rings containing boron, sulfur and nitrogen atoms and derived from five-membered boron and sulfur heterocycles were included in Section 4.10.2. In Section 4.11.4, the reaction of (**319**) (E = O) with

Na/K alloy and $Pr^i_2NBCl_2$ was described; for (**319**) (E = S), an analogous reaction using $R^2_2NBX_2$ (e.g., R^2 = Et or Pr^i; X = F or Cl) yields (**332**) as a major product.[597] A five-membered BNNCS-ring results from the [2 + 3]-cycloaddition of CS_2 to the {RBN(Me)NMe} unit (R = Me or Ph) of tetraazadiborinanes. The crystal structure of the product for R = Me has been determined and bond distances are consistent with aromatic character.[590] The bromination of (**307**) (R^1 = X = Me) is achieved by using BBr_3; initially only one *B*–Me group is exchanged but further treatment with BBr_3 yields (**307**) (X = Br; R^1 = Me). Results of labelling experiments using $^{10}BBr_3$ show that the ^{10}B label becomes incorporated into the heterocycle (**333**) during exchange of X = Br for Me.[635] Degradation from a five- to four-membered ring occurs when (**307**) (X = NR^2_2; R^2 and R^1 = alkyl or TMS) is treated with excess sodium.[631,636] The products have been spectroscopically characterized and the ring structure is confirmed by the crystal structure determination of $(Pr^i_2NB)_2S(NPh)$ (see above). The reaction of Hpz with (**307**) (R^1 = Ph; X = Me) proceeds via loss of phenylamine and the formation of MeB(μ-pz)$_2$(μ-S_2)BMe. This should be compared with similar pyrazole derivatives described in Section 4.11.4. The crystal structure of MeB(μ-pz)$_2$(μ-S_2)BMe has been determined; the S–S bridge is of length 0.2105(1) nm.[637]

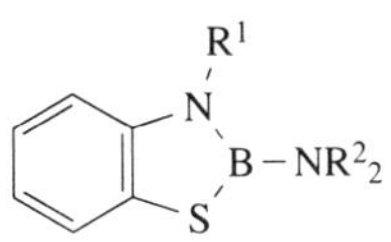

(**332**)
R^1 = H or Me
e.g., R^2 = Et or Pr^i

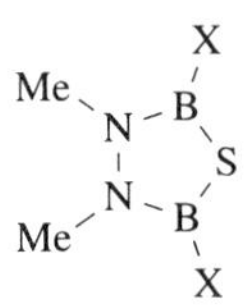

(**333**)
X = Me or Br

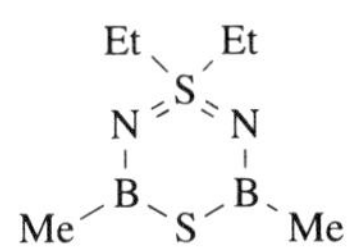

(**334**)

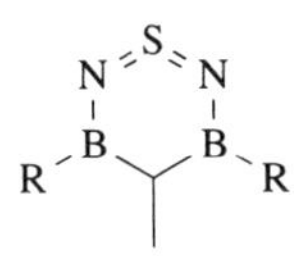

(**335**)
R = $N(TMS)_2$

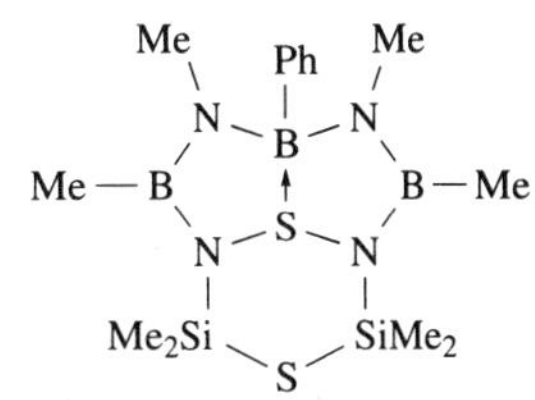

(**336**)

Ring expansion to a six-membered heterocycle occurs when (**307**) (X = Me; R^1 = Ph or Bu) or (**301**) (E = Se; X = Me) react with sulfur diimides, such as TMS-N=S=N-TMS. The products have B_2N_3S and N_2B_2SSe frameworks, respectively.[638] Related expansions are observed in the reaction of (**301**) (E = S; X = Me) with $R_2S(N\text{-}TMS)_2$ (R = Me, Et or Pr^i).[639–41] The lower oxidation state sulfur in the $N_2B_2S_2$ heterocycles so formed can be exchanged for an NH group[639,640] or oxygen atom[639] by treatment with ammonia or water, respectively. All products in this series have been characterized by multinuclear NMR spectroscopy with supporting crystallographic data for example, of derivative (**334**).[641] Ring expansion is also the method used to prepare (**335**).[634] More exotic assemblies can be accessed as shown in the synthesis of (**336**) from the addition of MeHNB(Ph)BMeH to a six-membered framework. However, in relation to the starting material, the product comes as a surprise, since the original six-membered ring possessed an $S_2B_2N_2$ rather than $S_2Si_2N_2$ frame.[642] Further tricyclic arrays have been prepared and characterized by Nöth and co-workers.[643]

The reaction of $(Me_2N)_2BCl$ with $O_2S\{NMe(TMS)\}_2$ yields an eight-membered $S_2N_4B_2$ heterocycle via the elimination of TMS-Cl. Related systems have also been studied.[644]

4.11.8 Boron with Oxygen and Sulfur

Heterocycles containing boron, oxygen and sulfur atoms are not well exemplified. The elimination of two equivalents of HCl when BCl_3 and 2-SH-4-OMeC_6H_3OH react results in cyclization to give a C_2SOB skeleton. The remaining B–Cl group can be further utilized to build up linked systems of type (**337**). The new products have been characterized by spectroscopic techniques.[645] Substituted 1,3,2-oxathiaborinanes (**338**) have been studied in solution by NMR spectroscopy with an emphasis on ring conformation,[646] and the work has been extended to other derivatives.[647]

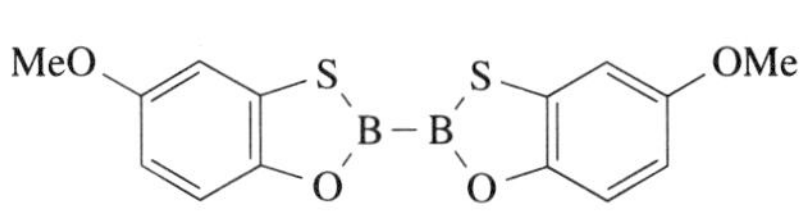

(**337**)

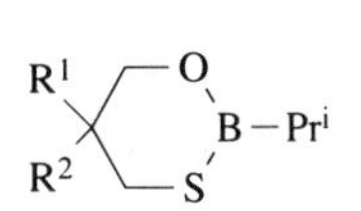

(**338**)
R^1 = R^2 = H or Me
R^1 = H; R^2 = Me or Pr^i

ACKNOWLEDGEMENT

The author wishes to acknowledge that part of this chapter was prepared at the University Chemical Laboratory, Cambridge, UK.

4.12 REFERENCES

1. J. H. Morris in 'COMC-I', vol. 1, p. 311.
2. (a) K. Smith and A. Pelter, in 'Comprehensive Organic Synthesis', eds. B. M. Trost and I. Fleming, Pergamon, Oxford, 1991, vol. 8, p. 703; (b) A. Pelter and K. Smith, in 'Comprehensive Organic Synthesis', eds. B. M. Trost and I. Fleming, Pergamon, Oxford, 1991, vol. 8, p. 487.
3. (a) G. W. Kabalka, *J. Organomet. Chem.*, 1983, **245**, 1; 1984, **274**, 1; 1986, **298**, 1; 1987, **318**, 1; (b) G. W. Kabalka and L. H. M. Guindi, *J. Organomet. Chem.*, 1988, **337**, 169; 1989, **360**, 1; 1990, **392**, 1; 1991, **404**, 1.
4. (a) A. Pelter, K. Smith and H. C. Brown, 'Borane Reagents', Academic Press, London, 1988; (b) H. C. Brown and P. V. Ramachandran, *Pure Appl. Chem.*, 1991, **63**, 307.
5. (a) D. S. Matteson, *Acc. Chem. Res.*, 1988, **21**, 294; (b) D. S. Matteson, *Tetrahedron*, 1989, **45**, 1859; (c) D. S. Matteson, *Chem. Rev.*, 1989, **89**, 1535; (d) D. S. Matteson, *Pure Appl. Chem.*, 1991, **63**, 339.
6. L. Deloux and M. Srebnik, *Chem. Rev.*, 1993, **93**, 763.
7. S. E. Thomas, 'Organic Synthesis: The Roles of Boron and Silicon', Oxford University Press, Oxford, 1991.
8. A. Pelter, K. Smith and K. Jones, *Pure Appl. Chem.*, 1991, **63**, 403.
9. K. Niedenzu, *J. Organomet. Chem.*, 1984, **261**, 1.
10. A. Meller, *Pure Appl. Chem.*, 1991, **63**, 395.
11. A. Berndt, *Angew. Chem., Int. Ed. Engl.*, 1993, **32**, 985.
12. P. Paetzold, *Adv. Inorg. Chem.*, 1987, **31**, 123.
13. H. Grützmacher, *Angew. Chem., Int. Ed. Engl.*, 1992, **31**, 1329.
14. J. J. Eisch, B. Shafii and M. P. Boleslawski, *Pure Appl. Chem.*, 1991, **63**, 365.
15. D. E. Kaufmann and W. Schacht, *Pure Appl. Chem.*, 1991, **63**, 383.
16. J. R. Bowser and T. P. Fehlner, in 'Rings, Clusters and Polymers of Main Group and Transition Elements', ed. H. W. Roesky, Elsevier, Amsterdam, 1989, p. 1.
17. D. B. Sowerby in 'The Chemistry of Inorganic Homo- and Heterocycles', eds. I. Haiduc and D. B. Sowerby, Academic Press, London, 1987, p. 103.
18. I. Haiduc in 'The Chemistry of Inorganic Homo- and Heterocycles', eds. I. Haiduc and D. B. Sowerby, Academic Press, London, 1987, p. 109.
19. W. Siebert in 'The Chemistry of Inorganic Homo- and Heterocycles', eds. I. Haiduc and D. B. Sowerby, Academic Press, London, 1987, p. 143.
20. P. Kölle and H. Nöth, *Chem. Rev.*, 1985, **85**, 399.
21. B. M. Mikhailov, *Pure Appl. Chem.*, 1983, **55**, 1439.
22. R. Köster and M. Yalpani, *Pure Appl. Chem.*, 1991, **63**, 387.
23. J. H. Morris, H. J. Gysling and D. Reed, *Chem. Rev.*, 1985, **85**, 51.
24. P. P. Power, *Angew. Chem., Int. Ed. Engl.*, 1990, **29**, 449.
25. G. Heller, *Topics in Current Chemistry*, 1986, **131**, 39.
26. K. Niedenzu and S. Trofimenko, *Topics in Current Chemistry*, 1986, **131**, 1.
27. (a) P. Paetzold, *Pure Appl. Chem.*, 1991, **63**, 345; (b) H. Nöth, G. Geisberger, G. Linti, D. Loderer, W. Rattay and E. Salzbrenner, *Pure Appl. Chem.*, 1991, **63**, 351; (c) W. Maringgele, in 'The Chemistry of Inorganic Homo- and Heterocycles', eds. I. Haiduc and D. B. Sowerby, Academic Press, London, 1987, p. 17.
28. (a) S. Y. Shaw, D. A. DuBois and R. H. Neilson, *ACS Symp. Ser.*, 1988, **360** (*Inorg. Organomet. Polym.*), 385; (b) K. J. L. Paciorek, W. Krone-Schmidt, D. H. Harris, R. H. Kratzer and K. J. Wynne, *ibid*, 392; (c) R. T. Paine and C. K. Narula, *Chem. Rev.*, 1990, **90**, 73; (d) L. G. Sneddon, M. G. L. Mirabelli, A. T. Lynch, P. J. Fazen, K. Su and J. S. Beck, *Pure Appl. Chem.*, 1991, **63**, 407.
29. T. Clark, D. Wilhelm and P. von R. Schleyer, *Tetrahedron Lett.*, 1982, **23**, 3547.
30. K. Jug, *J. Org. Chem.*, 1984, **49**, 4475.
31. P. H. M. Budzelaar and P. von R. Schleyer, *J. Am. Chem. Soc.*, 1986, **108**, 3967.
32. P. H. M. Budzelaar, P. von R. Schleyer and K. Krogh-Jespersen, *Angew. Chem., Int. Ed. Engl.*, 1984, **23**, 825.
33. P. H. M. Budzelaar, K. Krogh-Jespersen, T. Clark and P. von R. Schleyer, *J. Am. Chem. Soc.*, 1985, **107**, 2773.
34. G. Frenking and H. F. Schaefer III, *Chem. Phys. Lett.*, 1984, **109**, 521.
35. H. Klusik and A. Berndt, *Angew. Chem., Int. Ed., Engl.*, 1983, **22**, 877.
36. P. H. M. Budzelaar, S. M. van der Kerk, K. Krogh-Jespersen and P. von R. Schleyer, *J. Am. Chem. Soc.*, 1986, **108**, 3960.
37. G. Frenking, *Chem. Phys. Lett.*, 1984, **111**, 529.
38. P. H. M. Budzelaar, A. J. Kos, T. Clark and P. von R. Schleyer, *Organometallics*, 1985, **4**, 429.
39. P. H. M. Budzelaar, E. Kraka, D. Cremer and P. von R. Schleyer, *J. Am. Chem. Soc.*, 1986, **108**, 561.
40. P. von R. Schleyer, P. H. M. Budzelaar, D. Cremer and E. Kraka, *Angew. Chem., Int. Ed. Engl.*, 1984, **23**, 374.
41. D. Cremer, J. Gauss, P. von R. Schleyer and P. H. M. Budzelaar, *Angew. Chem. Int. Ed. Engl.*, 1984, **23**, 370.
42. M. Bühl, P. von R. Schleyer, M. A. Ibrahim and T. Clark, *J. Am. Chem. Soc.*, 1991, **113**, 2466.
43. N. Tyutyulkov and V. Monev, *Dokl. Bolg. Akad. Nauk.*, 1985, **38**, 1513 (*Chem. Abstr.*, 1986, **104**, 206 363v).
44. R. M. Minyaev, G. V. Orlova and V. I. Minkin, *Zh. Org. Khim.*, 1989, **25**, 2033.
45. G. Raabe, E. Heyne, W. Schleker and J. Fleischhauer, *Z. Naturforsch., Teil A*, 1984, **39**, 678.
46. G. Raabe, W. Schleker, E. Heyne and J. Fleischhauer, *Z. Naturforsch., Teil A*, 1987, **42**, 352.
47. J. Cioslowski and P. J. Hay, *J. Am. Chem. Soc.*, 1990, **112**, 1707.
48. J. M. Schulman, R. L. Disch and M. L. Sabio, *J. Am. Chem. Soc.*, 1982, **104**, 3785.
49. R. L. Disch, M. L. Sabio and J. M. Schulman, *Tetrahedron Lett.*, 1983, **24**, 1863.

50. J. M. Schulman, R. L. Disch and M. L. Sabio, *J. Am. Chem. Soc.*, 1984, **106**, 7696.
51. J. M. Schulman and R. L. Disch, *Organometallics*, 1989, **8**, 733.
52. S. T. Massey and R. W. Zoellner, *Int. J. Quantum Chem.*, 1991, **39**, 787.
53. B. Maouche, J. Gayoso and O. Ouamerali, *Rev. Roum. Chim.*, 1984, **29**, 613 (*Chem. Abstr.*, 1985, **102**, 113 558f).
54. P. von R. Schleyer and M. Bühl, *Angew. Chem. Int. Ed. Engl.*, 1990, **29**, 304.
55. A. L. Companion, F. Liu and K. Niedenzu, *Inorg. Chem.*, 1985, **24**, 1738.
56. W. N. Lipscomb and L. Massa, *Inorg. Chem.*, 1992, **31**, 2297.
57. J. R. Bowser, D. A. Jelski and T. F. George, *Inorg. Chem.*, 1992, **31**, 154.
58. T. Fukunaga, H. E. Simmons, J. J. Wendoloski and M. D. Gordon, *J. Am. Chem. Soc.*, 1983, **105**, 2729.
59. V. P. Vorob'eva, E. A. Miroshnichenko, I. I. Solomennikova, G. I. Zelehan, E. Lukevics and Yu. A. Lebedev, *Izv. Akad. Nauk SSSR, Ser. Khim.*, 1984, 1493.
60. A. Meller, D. Bromm, W. Maringgele, D. Böhler and G. Elter, *J. Organomet. Chem.*, 1988, **347**, 11.
61. A. Meller, W. Maringgele, G. Elter, D. Bromm, M. Noltemeyer and G. M. Sheldrick, *Chem. Ber.*, 1987, **120**, 1437.
62. M. Kropp, K. Bhamidapaty and G. B. Schuster, *J. Am. Chem. Soc.*, 1988, **110**, 6252.
63. M. A. Kropp, M. Baillargeon, K. M. Park, K. Bhamidapaty and G. B. Schuster, *J. Am. Chem. Soc.*, 1991, **113**, 2155.
64. J. D. Wilkey and G. B. Schuster, *J. Am. Chem. Soc.*, 1988, **110**, 7569; 1991, **113**, 2149.
65. H. Michel *et al.*, *Angew. Chem., Int. Ed. Engl.*, 1992, **31**, 607.
66. A. Tapper, T. Schmitz and P. Paetzold, *Chem. Ber.*, 1989, **122**, 595.
67. C. Wieczorek, J. Allwohn, G. Schmidt-Lukasch, R. Hunold, W. Massa and A. Berndt, *Angew. Chem., Int. Ed. Engl.*, 1990, **29**, 398.
68. H. Meyer, G. Schmidt-Lukasch, G. Baum, W. Massa and A. Berndt, *Z. Naturforsch., Teil B*, 1988, **43**, 801.
69. P. Willershausen *et al.*, *Angew. Chem., Int. Ed. Engl.*, 1992, **31**, 1384.
70. A. Höfner, B. Ziegler, R. Hunold, P. Willershausen, W. Massa and A. Berndt, *Angew. Chem., Int. Ed. Engl.*, 1991, **30**, 594.
71. M. Baudler, K. Rockstein and W. Oehlert, *Chem. Ber.*, 1991, **124**, 1149.
72. T. Mennekes, P. Paetzold, R. Boese and D. Bläser, *Angew. Chem., Int. Ed. Engl.*, 1991, **30**, 173.
73. R. Wehrmann, H. Klusik and A. Berndt, *Angew. Chem., Int. Ed. Engl.*, 1984, **23**, 826.
74. H. Meyer, G. Baum, W. Massa, S. Berger and A. Berndt, *Angew. Chem., Int. Ed. Engl.*, 1987, **26**, 546.
75. H. Meyer, G. Baum, W. Massa and A. Berndt, *Angew. Chem., Int. Ed. Engl.*, 1987, **26**, 798.
76. R. Hunold *et al.*, *Angew. Chem., Int. Ed. Engl.*, 1989, **28**, 781.
77. M. Pilz, M. Stadler, R. Hunold, J. Allwohn, W. Massa and A. Berndt, *Angew. Chem., Int. Ed. Engl.*, 1989, **28**, 784.
78. P. Hornbach, M. Hildenbrand, H. Pritzkow and W. Siebert, *Angew. Chem., Int. Ed. Engl.*, 1986, **25**, 1112.
79. G. Karger, P. Hornbach, A. Krämer, H. Pritzkow and W. Siebert, *Chem. Ber.*, 1989, **122**, 1881.
80. A. Krämer, J. K. Uhm, S. E. Garner, H. Pritzkow and W. Siebert, *Z. Naturforsch., Teil B*, 1990, **45**, 1019.
81. B. Glaser, E. Hanecker, H. Nöth and H. Wagner, *Chem. Ber.*, 1987, **120**, 659.
82. P. Jutzi, B. Krato, M. Hursthouse and A. J. Howes, *Chem. Ber.*, 1987, **120**, 565.
83. H. J. Bestman, K. Sühs and T. Röder, *Angew. Chem., Int. Ed. Engl.*, 1981, **20**, 1038.
84. G. E. Herberich, U. Eigendorf and C. Ganter, *J. Organomet. Chem.*, 1991, **402**, C17.
85. U. S. Racherla and G. G. Pai, *Heterocycles*, 1982, **18**, 285.
86. G. E. Herberich, C. Ganter, L. Wesemann and R. Boese, *Angew. Chem., Int. Ed. Engl.*, 1990, **29**, 912.
87. G. E. Herberich, C. Ganter and L. Wesemann, *Chem. Ber.*, 1990, **123**, 49.
88. M. Enders, A. Krämer, H. Pritzkow and W. Siebert, *Angew. Chem., Int. Ed. Engl.*, 1991, **30**, 84.
89. D. Bromm, U. Seebold, M. Noltemeyer and A. Meller, *Chem. Ber.*, 1991, **124**, 2645.
90. H. C. Brown and G. G. Pai, *Heterocycles*, 1982, **17**, 77.
91. H. C. Brown and G. G. Pai, *J. Organomet. Chem.*, 1983, **250**, 13.
92. B. Singaram, T. E. Cole and H. C. Brown, *Organometallics*, 1984, **3**, 1520.
93. G. E. Herberich, U. Englert, C. Ganter and L. Wesemann, *Chem. Ber.*, 1992, **125**, 23.
94. A. Maercker, W. Brieden, T. Schmidt and H. D. Lutz, *Angew. Chem., Int. Ed. Engl.*, 1989, **28**, 477.
95. H. Fisch, H. Pritzkow and W. Siebert, *Angew. Chem., Int. Ed. Engl.*, 1984, **23**, 608.
96. D. Basavaiah, *Heterocycles*, 1982, **18**, 153.
97. D. Basavaiah and H. C. Brown, *J. Org. Chem.*, 1982, **47**, 1792.
98. H. C. Brown, A. S. Phadke and M. V. Rangaishenvi, *Heteroatom. Chem.*, 1990, **1**, 83.
99. H. C. Brown, A. S. Phadke and M. V. Rangaishenvi, *J. Am. Chem. Soc.*, 1988, **110**, 6263.
100. M. Yalpani, R. Boese and R. Köster, *Chem. Ber.*, 1987, **120**, 607.
101. W. Biffar and H. Nöth, *Chem. Ber.*, 1982, **115**, 934.
102. R. Köster and G. Siedel, *Inorg. Synth.*, 1983, **22**, 198.
103. J. S. Cha, M. S. Yoon, K. W. Lee and J. C. Lee, *Heterocycles*, 1988, **27**, 1455.
104. R. Köster and G. Siedel, *Chem. Ber.*, 1992, **125**, 617.
105. C. G. Whiteley, *S. Afr. J. Chem.*, 1982, **35**, 9.
106. W. Kaim and A. Schulz, *Angew. Chem., Int. Ed. Engl.*, 1984, **23**, 615.
107. W. Biffar and H. Nöth, *Z. Naturforsch., Teil B*, 1981, **36**, 1509.
108. R. Köster, W. Schüssler and R. Boese, *Chem. Ber.*, 1990, **123**, 1945.
109. P. Idelmann, G. Müller, W. R. Scheidt, W. Schüssler and R. Köster, *Angew. Chem., Int. Ed. Engl.*, 1984, **23**, 153.
110. W. Kliegel and D. Nanninga, *Chem. Ber.*, 1983, **116**, 2616.
111. P. Galow, A. Sebald and B. Wrackmeyer, *J. Organomet. Chem.*, 1983, **259**, 253.
112. D. P. Workman and S. G. Shore, in 'Electron Deficient Boron and Carbon Clusters', eds. G. A. Olah, K. Wade and R. E. Williams, Wiley, New York, 1991, chap. 10, p. 237 and references therein.
113. R. G. Bulgakov, V. P. Vlad, B. A. Tishin, G. L. Maistrenko, G. A. Tolstikov and V. P. Kazakov, *Izv. Akad. Nauk SSSR, Ser. Khim.*, 1987, 456.
114. I. Jabre, M. Saquet and A. Thuillier, *Phosphorus, Sulfur Silicon Relat. Elem.*, 1991, **56**, 283.
115. S. Toyota and M. Oki, *Bull. Chem. Soc. Jpn.*, 1992, **65**, 2857.
116. R. Contreras and B. Wrackmeyer, *Spectrochim. Acta, Part A*, 1982, **38**, 941.
117. B. M. Mikhailov, T. A. Shchegoleva, E. M. Shashkova and V. G. Kiselev, *Izv. Akad. Nauk SSSR, Ser. Khim.*, 1985, 2575.
118. B. Wrackmeyer and C. Bihlmayer, *J. Chem. Soc., Chem. Commun.*, 1981, 1093.

119. B. Wrackmeyer, C. Bihlmayer and M. Schilling, *Chem. Ber.*, 1983, **116**, 3182.
120. C. Bihlmayer and B. Wrackmeyer, *Z. Naturforsch., Teil B*, 1981, **36**, 1265.
121. B. Wrackmeyer and S. T. Abu-Orabi, *Chem. Ber.*, 1987, **120**, 1603.
122. C. Bihlmayer, S. T. Abu-Orabi and B. Wrackmeyer, *J. Organomet. Chem.*, 1987, **321**, 25.
123. B. Wrackmeyer, *J. Organomet. Chem.*, 1989, **364**, 331.
124. H. C. Brown, J. Chandrasekharan and K. K. Wang, *J. Org. Chem.*, 1983, **48**, 3689.
125. C. K. Narula and H. Nöth, *Inorg. Chem.*, 1985, **24**, 2532.
126. S. Hünig and I. Wehner, *Heterocycles*, 1989, **28**, 359.
127. S. Hünig and I. Wehner, *Liebigs Ann. Chem.*, 1989, 299.
128. M. Yalpani, R. Köster, R. Boese and W. A. Brett, *Angew. Chem., Int. Ed. Engl.*, 1990, **29**, 302.
129. L. Komorowski, A. Meller and K. Niedenzu, *Inorg. Chem.*, 1990, **29**, 538.
130. M. Yalpani, R. Boese and R. Köster, *Chem. Ber.*, 1988, **121**, 287.
131. U. Brand, S. Hönig, K. Peters and H. G. von Schering, *Chem. Ber.*, 1991, **124**, 1187.
132. J. A. Babon, B. P. Roberts and A. C. H. Tsang, *J. Chem. Soc., Chem. Commun.*, 1985, 955.
133. J. A. Babon, B. P. Roberts and A. C. H. Tsang, *J. Chem. Res. (S)*, 1986, 334.
134. R. Köster and M. Yalpani, *Angew. Chem., Int. Ed. Engl.*, 1985, **24**, 572.
135. R. Köster, Y.-H. Tsay, C. Krüger and J. Serwatowski, *Chem. Ber.*, 1986, **119**, 1174.
136. R. Köster and G. Seidel, *Z. Naturforsch., Teil B*, 1988, **43**, 687.
137. R. Köster, G. Seidel and M. Yalpani, *Chem. Ber.*, 1989, **122**, 1815.
138. M. Yalpani, J. Serwatowski and R. Köster, *Chem. Ber.*, 1989, **122**, 3.
139. M. Yalpani, R. Köster, R. Boese and M. Sulkowski, *Chem. Ber.*, 1989, **122**, 9.
140. M. Yalpani, R. Köster and R. Boese, *Chem. Ber.*, 1992, **125**, 309.
141. M. Yalpani, R. Köster and R. Boese, *Chem. Ber.*, 1989, **122**, 19.
142. M. Yalpani, R. Boese and R. Köster, *Chem. Ber.*, 1989, **122**, 1231.
143. P. K. Jadhav and S. U. Kulkarni, *Heterocycles*, 1982, **18**, 169.
144. L. S. Vasilyev, V. V. Veselovskii, M. I. Struchkova and B. M. Mikhailov, *J. Organomet. Chem.*, 1982, **226**, 115.
145. M. E. Gurskii, S. V. Baranin, A. S. Shashkov, A. I. Lutsenko and B. M. Mikhailov, *J. Organomet. Chem.*, 1983, **246**, 129.
146. M. E. Gurskii, S. V. Baranin, A. I. Lutsenko and B. M. Mikhailov, *J. Organomet. Chem.*, 1984, **270**, 17.
147. M. E. Gurskii, S. V. Baranin and B. M. Mikhailov, *J. Organomet. Chem.*, 1984, **270**, 9.
148. L. S. Vasilyev, O. D. Smirnova, M. I. Struchkova and B. M. Mikhailov, *J. Organomet. Chem.*, 1984, **275**, 19.
149. M. E. Gurskii, T. V. Potapova, B. M. Mikhailov, A. V. Ignatenko and Yu. N. Bubnov, *Metalloorg. Khim.*, 1990, **3**, 1195.
150. Yu. N. Bubnov, M. E. Gurskii and D. G. Pershin, *Izv. Akad. Nauk SSSR, Ser. Khim.*, 1990, 952.
151. C. F. Reichert, W. E. Pye and T. A. Bryson, *Tetrahedron*, 1981, **37**, 2441.
152. H. C. Brown, E. Negishi and W. C. Dickason, *J. Org. Chem.*, 1985, **50**, 520.
153. M. Enders, H. Pritzkow and W. Siebert, *Chem. Ber.*, 1992, **125**, 1981.
154. M. Enders, H. Pritzkow and W. Siebert, *Angew. Chem., Int. Ed. Engl.*, 1992, **31**, 606.
155. A. Krämer, H. Pritzkow and W. Siebert, *Angew. Chem., Int. Ed. Engl.*, 1990, **29**, 292.
156. E. Negishi and H. C. Brown, *Org. Synth.*, 1983, **61**, 103.
157. S. M. van der Kerk, J. C. Roos-Venekamp, A. J. M. van Beijnen and G. J. M. van der Kerk, *Polyhedron*, 1983, **2**, 1337.
158. A. Meller, D. Bromm, W. Maringgele, A. Heine, D. Stalke and G. M. Sheldrick, *Chem. Ber.*, 1990, **123**, 293.
159. W. Maringgele, *et al.*, *Chem. Ber.*, 1992, **125**, 1559.
160. A. Meller, U. Seebold, W. Maringgele, M. Noltemeyer and G. M. Sheldrick, *J. Am. Chem. Soc.*, 1989, **111**, 8299.
161. W. Maringgele, *et al.*, *Organometallics*, 1991, **10**, 2097.
162. G. Herberich, M. Hostalek, N. Klaff and E. Raabe, *Chem. Ber.*, 1990, **123**, 45.
163. A. Krämer, H. Pritzkow and W. Siebert, *Z. Naturforsch., Teil B*, 1989, **44**, 96.
164. S. S. Bukalov, L. A. Leites, Yu. N. Bubnov, M. E. Gurskii and T. V. Potapova, *Izv. Akad. Nauk SSSR, Ser. Khim.*, 1989, 483.
165. B. M. Mikhailov and T. K. Baryshnikova, *J. Organomet. Chem.*, 1981, **219**, 295.
166. B. M. Mikhailov and K. L. Cherkasova, *J. Organomet. Chem.*, 1983, **246**, 9.
167. M. E. Gurskii, T. V. Potapova, K. L. Cherkasova and Yu. N. Bubnov, *Izv. Akad. Nauk SSSR, Ser. Khim.*, 1988, 415.
168. Yu. N. Bubnov and A. I. Grandberg, *Izv. Akad. Nauk SSSR, Ser. Khim.*, 1986, 1451.
169. B. M. Mikhailov and M. Yu. Etinger, *Izv. Akad. Nauk SSSR, Ser. Khim.*, 1984, 1197.
170. B. M. Mikhailov and K. L. Cherkasova, *Izv. Akad. Nauk SSSR, Ser. Khim.*, 1984, 1837.
171. Yu. N. Bubnov, M. E. Gurskii and A. V. Geiderikh, *Metalloorg. Khim.*, 1988, **1**, 1433.
172. N. A. Lagutkin, *et al.*, *Khim.-Farm. Zh.*, 1983, **17**, 1077 (*Chem. Abstr.*, 1984, **100**, 156 672q).
173. Yu. N. Bubnov, M. E. Gurskii and T. V. Potapova, *Izv. Akad. Nauk SSSR, Ser. Khim.*, 1987, 1195.
174. Yu. N. Bubnov, T. V. Potapova and M. E. Gurskii, *J. Organomet. Chem.*, 1991, **412**, 311.
175. V. I. Kadentsev, *et al.*, *Izv. Akad. Nauk SSSR, Ser. Khim.*, 1988, 2282.
176. B. M. Mikhailov, E. A. Shagova and M. Yu. Etinger, *J. Organomet. Chem.*, 1981, **220**, 1.
177. B. M. Mikhailov, T. K. Baryshnikova and A. S. Shashkov, *J. Organomet. Chem.*, 1981, **219**, 301.
178. B. M. Mikhailov, M. E. Gurskii and D. G. Pershin, *J. Organomet. Chem.*, 1983, **246**, 19.
179. B. M. Mikhailov and T. K. Baryshnikova, *J. Organomet. Chem.*, 1984, **260**, 25.
180. M. E. Gurskii, D. G. Pershin and B. M. Mikhailov, *J. Organomet. Chem.*, 1984, **260**, 17.
181. B. M. Mikhailov and E. A. Shagova, *J. Organomet. Chem.*, 1983, **258**, 131.
182. Yu. N. Bubnov, M. E. Gurskii and D. G. Pershin, *J. Organomet. Chem.*, 1991, **412**, 1.
183. B. M. Mikhailov, T. A. Shchegoleva, E. M. Shashkova and V. G. Kiselev, *J. Organomet. Chem.*, 1983, **250**, 23.
184. R. Köster, G. Siedel and B. Wrackmeyer, *Angew. Chem., Int. Ed. Engl.*, 1985, **24**, 326.
185. W. Haubold, W. Keller and G. Sawitzki, *J. Organomet. Chem.*, 1989, **367**, 19.
186. K. Wolfer, H.-D. Hausen and H. Binder, *Z. Naturforsch., Teil B*, 1985, **40**, 235.
187. B. Pachaly and R. West, *Angew. Chem., Int. Ed. Engl.*, 1984, **23**, 454.
188. C. Pues and A. Berndt, *Angew. Chem., Int. Ed. Engl.*, 1984, **23**, 306.
189. C. Habben and A. Meller, *Chem. Ber.*, 1984, **117**, 2531.
190. J. J. Eisch, B. Shafii and A. L. Rheingold, *J. Am. Chem. Soc.*, 1987, **109**, 2526.

191. J. J. Eisch, B. Shafii, J. D. Odom and A. L. Rheingold, *J. Am. Chem. Soc.*, 1990, **112**, 1847.
192. J. J. Eisch, F. Shen and K. Tamao, *Heterocycles*, 1982, **18**, 245.
193. M. A. Kropp and G. B. Schuster, *J. Am. Chem. Soc.*, 1989, **111**, 2316.
194. R. Wehrmann, H. Meyer and A. Berndt, *Angew. Chem., Int. Ed. Engl.*, 1985, **24**, 788.
195. R. Wehrmann, H. Klusik and A. Berndt, *Angew. Chem., Int. Ed. Engl.*, 1984, **23**, 369.
196. C. Pues, G. Baum, W. Massa and A. Berndt, *Z. Naturforsch., Teil B*, 1988, **43**, 275.
197. B. Glaser, E. P. Mayer, H. Nöth, W. Rattay and U. Wietelmann, *Z. Naturforsch., Teil B*, 1988, **43**, 449.
198. M. Pilz, H. Michel and A. Berndt, *Angew. Chem. Int. Ed. Engl.*, 1990, **29**, 401.
199. M. Pilz, J. Allwohn, W. Massa and A. Berndt, *Angew. Chem., Int. Ed. Engl.*, 1990, **29**, 399.
200. M. Hildenbrand, H. Pritzkow and W. Siebert, *Angew. Chem., Int. Ed. Engl.*, 1985, **24**, 759.
201. R. Wehrmann, C. Pues, H. Klusik and A. Berndt, *Angew. Chem., Int. Ed. Engl.*, 1984, **23**, 372.
202. D. Bromm, D. Stalke, A. Heine, A. Meller and G. M. Sheldrick, *J. Organomet. Chem.*, 1990, **386**, 1.
203. M. Hildenbrand, H. Pritzkow, U. Zenneck and W. Siebert, *Angew. Chem., Int. Ed. Engl.*, 1984, **23**, 371.
204. H. Irngartinger, J. Hauck and W. Siebert, *Z. Naturforsch., Teil B*, 1991, **46**, 1621.
205. M. Pilz, J. Allwohn, M. Böhl, P. von R. Schleyer and A. Berndt, *Z. Naturforsch., Teil B*, 1991, **46**, 1085.
206. G. Schmidt, G. Baum, W. Massa and A. Berndt, *Angew. Chem., Int. Ed. Engl.*, 1986, **25**, 1111.
207. P. Willershausen *et al.*, *Angew. Chem., Int. Ed. Engl.*, 1992, **31**, 1238.
208. G. E. Herberich, B. Hessner and D. Söhnen, *J. Organomet. Chem.*, 1982, **233**, C35.
209. G. E. Herberich, *et al.*, *Chem. Ber.*, 1986, **119**, 420.
210. G. E. Herberich, B. Hessner and D. Söhnen, *J. Organomet. Chem.*, 1983, **256**, C23.
211. B. Wrackmeyer, *Organometallics*, 1984, **3**, 1.
212. A. Sebald and B. Wrackmeyer, *J. Organomet. Chem.*, 1986, **307**, 157.
213. J. J. Eisch, J. E. Galle and S. Kozima, *J. Am. Chem. Soc.*, 1986, **108**, 379.
214. G. E. Herberich and S. Wang, *Chem. Ber.*, 1990, **123**, 1625.
215. G. E. Herberich, M. Hostalek, R. Laven and R. Boese, *Angew. Chem., Int. Ed. Engl.*, 1990, **29**, 317 and reference 1 therein.
216. W. Siebert, U. Ender and W. Herter, *Z. Naturforsch., Teil B*, 1985, **40**, 326.
217. G. Knörzer and W. Siebert, *Z. Naturforsch., Teil B*, 1990, **45**, 15.
218. V. Schäfer, H. Pritzkow and W. Siebert, *Chem. Ber.*, 1989, **122**, 401.
219. K. Stumpf, W. Siebert, R. Köster and G. Siedel, *Z. Naturforsch., Teil B*, 1987, **42**, 186.
220. V. Schäfer, H. Pritzkow and W. Siebert, *Angew. Chem., Int. Ed. Engl.*, 1988, **27**, 299.
221. A. J. Ashe, S. T. Abu-Orabi, O. Eisenstein and H. F. Sandford, *J. Org. Chem.*, 1983, **48**, 901.
222. G. Maier, J. Henkelmann and H. P. Reisenauer, *Angew. Chem., Int. Ed. Engl.*, 1985, **24**, 1065.
223. B. M. Mikhailov and L. I. Lavrinovich, *J. Organomet. Chem.*, 1984, **264**, 289; **268**, 5.
224. R. Boese, *et al.*, *Chem. Ber.*, 1985, **118**, 1644.
225. G. Maier, H. P. Reisenauer, J. Henkelmann and C. Kliche, *Angew. Chem., Int. Ed. Engl.*, 1988, **27**, 295.
226. S. M. van der Kerk, P. H. M. Budzelaar, A. L. M. van Eekeren and G. J. M. van der Kerk, *Polyhedron*, 1984, **3**, 271.
227. R. Schlögl and B. Wrackmeyer, *Polyhedron*, 1985, **4**, 885.
228. G. Herberich and B. Hessner, *Chem. Ber.*, 1982, **115**, 3115.
229. J. K. Uhm, D. Hu, U. Zenneck, H. Pritzkow and W. Siebert, *Taehan Hwahakhoe Chi*, 1990, **34**, 490 (*Chem. Abstr.*, 1991, **114**, 42 853h).
230. J. K. Uhm, H. Roemich, H. Wadepohl and W. Siebert, *Z. Naturforsch., Teil B*, 1988, **43**, 306.
231. B. Wrackmeyer and G. Kehr, *Polyhedron*, 1991, **10**, 1497.
232. G. Herberich, B. Hessner and M. Hostalek, *Angew. Chem., Int. Ed. Engl.*, 1986, **25**, 642.
233. A. J. Ashe, J. W. Kampf, Y. Nakadaira and J. M. Pace, *Angew. Chem., Int. Ed. Engl.*, 1992, **31**, 1255.
234. Y. Nakadaira, R. Sato and H. Sakurai, *Chem. Lett.*, 1987, 1451.
235. A. J. Ashe and F. J. Drone, *J. Am. Chem. Soc.*, 1987, **109**, 1879.
236. Y. Sugihara, T. Yagi, I. Murata and A. Imamura, *J. Am. Chem. Soc.*, 1992, **114**, 1479.
237. J. J. Eisch, J. E. Galle, B. Shafii and A. L. Rheingold, *Organometallics*, 1990, **9**, 2342.
238. A. Krämer, H. Pritzkow and W. Siebert, *Angew. Chem., Int. Ed. Engl.*, 1988, **27**, 926.
239. Yu. N. Bubnov, M. E. Gurskii and A. I. Grandberg, *Izv. Akad. Nauk SSSR, Ser. Khim.*, 1987, 2625.
240. W. Maringgele, D. Stalke, A. Heine, A. Meller and G. M. Sheldrick, *Chem. Ber.*, 1990, **123**, 489.
241. A. Meller, *et al.*, *Chem. Ber.*, 1991, **124**, 1907.
242. M. Noltemeyer, F. Pauer, D. Bromm and A. Meller, *Acta Crystallogr., Sect. C*, 1990, **46**, 1981.
243. A. Meller, D. Bromm, W. Maringgele, A. Heine, D. Stalke and G. M. Sheldrick, *J. Chem. Soc., Chem. Commun.*, 1990, 741.
244. A. Meller, U. Seebold, W. Maringgele and M. Noltemeyer, *Chem. Ber.*, 1990, **123**, 967.
245. G. E. Herberich, H. Ohst and H. Mayer, *Angew. Chem., Int. Ed. Engl.*, 1984, **23**, 969.
246. G. E. Herberich and H. Ohst, *Chem. Ber.*, 1985, **118**, 4303.
247. W. Schacht and D. Kaufmann, *Chem. Ber.*, 1987, **120**, 1331.
248. W. Schacht and D. Kaufmann, *Angew. Chem., Int. Ed. Engl.*, 1987, **26**, 665.
249. W. Schacht and D. Kaufmann, *J. Organomet. Chem.*, 1988, **339**, 33.
250. W. Schacht and D. Kaufmann, *J. Organomet. Chem.*, 1987, **331**, 139.
251. A. Fessenbecker, H. Schulz, H. Pritzkow and W. Siebert, *Chem. Ber.*, 1990, **123**, 2273.
252. G. Knörzer, H. Seyffer, H. Pritzkow and W. Siebert, *Z. Naturforsch., Teil B*, 1990, **45**, 985.
253. C. K. Narula and H. Nöth, *J. Organomet. Chem.*, 1985, **281**, 131.
254. U. Gross and D. Kaufmann, *Chem. Ber.*, 1987, **120**, 991.
255. V. A. Kuznetsov, A. Z. Gamzatov, S. Yu. Barabonin, V. O. Reikhsfel'd and S. V. Nesterova, *Zh. Obshch. Khim.*, 1990, **60**, 2505.
256. S. V. Nesterova, A. Z. Gamzatov, V. A. Kuznetsov and V. O. Reikhsfel'd, *Zh. Obshch. Khim.*, 1988, **58**, 1296.
257. S. C. Lapin, B.-E. Brauer and G. B. Schuster, *J. Am. Chem. Soc.*, 1984, **106**, 2092.
258. M. Groteklaes and P. Paetzold, *Chem. Ber.*, 1988, **121**, 809.
259. K. Okada, H. Inokawa and M. Oda, *Tetrahedron Lett.*, 1991, **32**, 6363.
260. J. A. Soderquist and A. Hassner, *J. Org. Chem.*, 1983, **48**, 1801.

261. A. Sebald, P. Seiberlich and B. Wrackmeyer, *J. Organomet. Chem.*, 1986, **303**, 73.
262. B. Wrackmeyer, *J. Chem. Soc., Chem. Commun.*, 1988, 1624.
263. W. W. C. Wong-Chi-Man, H. Pritzkow and W. Siebert, *Z. Naturforsch., Teil B*, 1990, **45**, 1597.
264. W. W. C. Wong-Chi-Man, H. Pritzkow and W. Siebert, *Z. Naturforsch., Teil B*, 1991, **46**, 1127.
265. P. Paetzold, T. Schmitz, A. Tapper and R. Ziembinski, *Chem. Ber.*, 1990, **123**, 747.
266. T. Mennekes, P. Paetzold and R. Boese, *Angew. Chem., Int. Ed. Engl.*, 1990, **29**, 899.
267. B. Wrackmeyer and K. Horchler, *J. Organomet. Chem.*, 1990, **399**, 1.
268. B. Wrackmeyer and K. Horchler, *Z. Naturforsch., Teil B*, 1990, **45**, 437.
269. B. Wrackmeyer, G. Kehr, A. Sebald and J. Kümmerlen, *Chem. Ber.*, 1992, **125**, 1597.
270. S. Kerschl and B. Wrackmeyer, *Z. Naturforsch., Teil B*, 1984, **39**, 1037.
271. S. Kerschl and B. Wrackmeyer, *Chem. Ber.*, 1988, **121**, 1451.
272. B. Wrackmeyer, S. M. Frank, M. Herberhold, H. Borrmann and A. Simon, *Chem. Ber.*, 1991, **124**, 691.
273. E. P. Mayer, H. Nöth, W. Rattay and U. Wietelmann, *Chem. Ber.*, 1992, **125**, 401.
274. J. Bai, K. Niedenzu, J. Serwatowska and J. Serwatowski, *Inorg. Chem.*, 1991, **30**, 4631.
275. J. Bai, K. Niedenzu, J. Serwatowska and J. Serwatowski, *Inorg. Chem.*, 1992, **31**, 228.
276. T. Gasparis-Ebeling and H. Nöth, *Angew. Chem., Int. Ed. Engl.*, 1984, **23**, 303.
277. B. Glaser and H. Nöth, *Angew. Chem., Int. Ed. Engl.*, 1985, **24**, 416.
278. F. Dirschl, H. Nöth and W. Wagner, *J. Chem. Soc., Chem. Commun.*, 1984, 1533.
279. U. Klingebiel, *Angew. Chem., Int. Ed. Engl.*, 1984, **23**, 815.
280. R. Boese and U. Klingebiel, *J. Organomet. Chem.*, 1986, **306**, 295.
281. F. Dirschl, E. Hanecker, H. Nöth, W. Rattay and W. Wagner, *Z. Naturforsch., Teil B*, 1986, **41**, 32.
282. R. Boese, B. Kröckert and P. Paetzold, *Chem. Ber.*, 1987, **120**, 1913.
283. K.-H. van Bonn, P. Schreyer, P. Paetzold and R. Boese, *Chem. Ber.*, 1988, **121**, 1045.
284. P. Paetzold, B. Redenz-Stormanns, R. Boese, M. Bühl and P. von R Schleyer, *Angew. Chem., Int. Ed. Engl.*, 1990, **29**, 1059.
285. P. Paetzold, B. Redenz-Stormanns and R. Boese, *Chem. Ber.*, 1991, **124**, 2435.
286. H.-A. Steuer, A. Meller and G. Elter, *J. Organomet. Chem.*, 1985, **295**, 1.
287. C. Habben, A. Meller and S. Pusch, *Z. Naturforsch., Teil B*, 1988, **43**, 959.
288. G. Elter, A. Meller and W. Luthin, *Z. Anorg. Allg. Chem.*, 1988, **560**, 18.
289. D. Fest, C. D. Habben, A. Meller, G. M. Sheldrick, D. Stalke and F. Pauer, *Chem. Ber.*, 1990, **123**, 703.
290. T. Franz, E. Hanecker, H. Nöth, W. Stöcker, W. Storch and G. Winter, *Chem. Ber.*, 1986, **119**, 900.
291. H. Nöth, M. Schwartz and S. Weber, *Chem. Ber.*, 1985, **118**, 4716.
292. H. Nöth and S. Weber, *Angew. Chem., Int. Ed. Engl.*, 1984, **23**, 994.
293. P. Paetzold, C. von Plotho, G. Schmid and R. Boese, *Z. Naturforsch., Teil B*, 1984, **39**, 1069.
294. K. Delpy, H.-U. Meier, P. Paetzold and C. von Plotho, *Z. Naturforsch., Teil B*, 1984, **39**, 1696.
295. P. Paetzold *et al.*, *Chem. Ber.*, 1984, **117**, 1089.
296. P. Paetzold, C. Pelzer and R. Boese, *Chem. Ber.*, 1988, **121**, 51.
297. H. Nöth, P. Otto and W. Storch, *Chem. Ber.*, 1986, **119**, 2517.
298. P. Paetzold and R. Truppat, *Chem. Ber.*, 1983, **116**, 1531.
299. P. Paetzold, E. Schröder, G. Schmid and R. Boese, *Chem. Ber.*, 1985, **118**, 3205.
300. P. Schreyer, P. Paetzold and R. Boese, *Chem. Ber.*, 1988, **121**, 195.
301. B. Thiele, P. Schreyer, U. Englert, P. Paetzold, R. Boese and B. Wrackmeyer, *Chem. Ber.*, 1991, **124**, 2209.
302. S. Küpper, U. Englert and P. Paetzold, *Heteroatom. Chem.*, 1990, **1**, 479.
303. P. Paetzold, J. Kiesgen, K. Krahe, H. U. Meier and R. Boese, *Z. Naturforsch., Teil B*, 1991, **46**, 853.
304. R. Boese, P. Paetzold, A. Tapper and R. Ziembinski, *Chem. Ber.*, 1989, **122**, 1057.
305. S. Amirkhalili, R. Boese, U. Höhner, D. Kampmann, G. Schmid and P. Rademacher, *Chem. Ber.*, 1982, **115**, 732.
306. G. Schmid, W. Meyer-Zaika, R. Boese and N. Augart, *Angew. Chem., Int. Ed. Engl.*, 1988, **27**, 952.
307. G. Schmid, D. Zaika, J. Lehr, N. Augart and R. Boese, *Chem. Ber.*, 1988, **121**, 1873.
308. V. A. Dorokhov and K. L. Cherkasova, *Izv. Akad. Nauk SSSR, Ser. Khim.*, 1985, 1933.
309. A. S. Bleshinskaya, O. S. Chizhov, B. M. Zolotarev, B. M. Mikhailov and V. A. Dorokhov, *Izv. Akad. Nauk SSSR, Ser. Khim.*, 1989, 593.
310. A. S. Bleshinskaya, O. G. Boldyreva, B. M. Zolotarev, V. A. Dorokhov, O. S. Chizhov and B. M. Mikhailov, *Izv. Akad. Nauk SSSR, Ser. Khim.*, 1989, 589.
311. W. Maringgele, D. Stalke, A. Heine and A. Meller, *Eur. J. Solid State Inorg. Chem.*, 1991, **28**, 867.
312. G. Ferguson, M. Parvez, R. P. Brint, D. C. M. Power, T. R. Spalding and D. R. Lloyd, *J. Chem. Soc., Dalton Trans.*, 1986, 2283.
313. J. Pfeiffer, W. Maringgele and A. Meller, *Z. Anorg. Allg. Chem.*, 1984, **511**, 185.
314. K. Anton, H. Fußstetter and H. Nöth, *Chem. Ber.*, 1984, **117**, 2542.
315. C. K. Narula and H. Nöth, *J. Chem. Soc., Chem. Commun.*, 1984, 1023.
316. K. Anton, H. Nöth and H. Pommerening, *Chem. Ber.*, 1984, **117**, 2479.
317. G. Schmid, M. Polk and R. Boese, *Inorg. Chem.*, 1990, **29**, 4421.
318. G. Schmid, J. Lehr, M. Polk and R. Boese, *Angew. Chem., Int. Ed. Engl.*, 1991, **30**, 1015.
319. F. Dallacker, E. M. Both-Pollmann and W. Muellners, *Chem.-Ztg.*, 1984, **108**, 287.
320. V. V. Kanabur and S. H. Dandegaonker, *Curr. Sci.*, 1985, **54**, 1199.
321. W. Einholz and W. Haubold, *Z. Naturforsch., Teil B*, 1986, **41**, 1367.
322. G. Sawitzki, W. Einholz and W. Haubold, *Z. Naturforsch., Teil B*, 1988, **43**, 1179.
323. B. E. Zaitsev, Z. V. Bezuglaya, G. V. Avramenko, T. M. Ivanova, V. N. Listyutenko and B. I. Stepanov, *Zh. Obshch. Khim.*, 1985, **55**, 1398.
324. Z. V. Bezuglaya, G. V. Avramenko, V. N. Listyutenko and B. E. Zaitsev, *Zh. Obshch. Khim.*, 1988, **58**, 2683.
325. Z. V. Bezuglaya, G. V. Avramenko and B. I. Stepanov, *Zh. Vses. Khim. O-va. im. D. I. Mendeleeva*, 1988, **33**, 355 (*Chem. Abstr.*, 1989, **110**, 135 287j).
326. V. A. Shapovalov, Yu. M. Kargin, B. I. Stepanov, Z. V. Bezuglaya, V. D. Bezuglyi and G. V. Avramenko, *Zh. Obshch. Khim.*, 1984, **54**, 148.
327. F. Truchet, B. Carboni, M. Vaultier and R. Carrié, *J. Organomet. Chem.*, 1987, **327**, 1.

328. P. Paetzold and C. von Plotho, *Chem. Ber.*, 1982, **115**, 2819.
329. K. H. van Bonn, T. von Bennigsen-Mackiewicz, J. Kiesgen, C. von Plotho and P. Paetzold, *Z. Naturforsch., Teil B*, 1988, **43**, 61.
330. P. Paetzold *et al.*, *Chem. Ber.*, 1988, **121**, 61.
331. W. A. Brett, P. Rademacher and R. Boese, *Acta Crystallogr., Sect. C*, 1990, **46**, 689.
332. W. Haubold, A. Gemmler and U. Kraatz, *Z. Anorg. Allg. Chem.*, 1983, **507**, 222.
333. H. Fisch, H. Pritzkow and W. Siebert, *Z. Naturforsch., Teil B*, 1988, **43**, 658.
334. G. Schmid, D. Vehreschild-Yzermann and R. Boese, *J. Organomet. Chem.*, 1987, **326**, 307.
335. A. S. Balueva, A. A. Karasik, G. N. Nikonov and B. A. Arbuzov, *Izv. Akad. Nauk SSSR, Ser. Khim.*, 1990, 2147.
336. T. J. Sobieralski and K. G. Hancock, *J. Am. Chem. Soc.*, 1982, **104**, 7533.
337. T. J. Sobieralski and K. G. Hancock, *J. Am. Chem. Soc.*, 1982, **104**, 7542.
338. R. Leardini and P. Zanirato, *J. Chem. Soc., Chem. Commun.*, 1983, 396.
339. S. Gronowitz, I. Ander and P. Zanirato, *Chem. Scr.*, 1983, **22**, 55.
340. P. Zanirato, *J. Organomet. Chem.*, 1985, **293**, 285.
341. J. E. Frey, G. M. Marchand and R. S. Bolton, *Inorg. Chem.*, 1982, **21**, 3239.
342. V. K. Bel'skii, S. V. Nesterova and V. O. Reikhsfel'd, *Zh. Strukt. Khim.*, 1987, **28**, 186.
343. M. A. Grassberger, F. Furnowsky and J. Hildebrandt, *J. Med. Chem.*, 1984, **27**, 947.
344. M. A. Grassberger, *Liebigs. Ann. Chem.*, 1985, 683.
345. P. Paetzold, J. Kiesgen, C. von Plotho and H. Schwan, *Z. Naturforsch., Teil B*, 1988, **43**, 1209.
346. G. Knörzer, H. Seyffer and W. Siebert, *Z. Naturforsch., Teil B*, 1990, **45**, 1136.
347. F. Alam and K. Niedenzu, *J. Organomet. Chem.*, 1983, **243**, 19.
348. S. Y. Shaw, D. A. DuBois, W. H. Watson and R. H. Neilson, *Inorg. Chem.*, 1988, **27**, 974.
349. W. Maringgele, *Chem. Ber.*, 1982, **115**, 3271.
350. J. Bielawski, K. Niedenzu and J. S. Stewart, *Z. Naturforsch., Teil B*, 1985, **40**, 389.
351. L. Komorowski and K. Niedenzu, *Inorg. Chem.*, 1989, **28**, 804.
352. L. Komorowski and K. Niedenzu, *Z. Naturforsch., Teil B*, 1989, **44**, 1421.
353. V. A. Dorokhov, O. G. Boldyreva, Z. A. Starikova, I. A. Teslya and B. M. Mikhailov, *Izv. Akad. Nauk SSSR, Ser. Khim.*, 1985, 2570.
354. N. Kuhn, A. Kuhn, J. Lewandowski and M. Speis, *Chem. Ber.*, 1991, **124**, 2197.
355. I. Motoc and O. E. Polansky, *Z. Naturforsch., Teil B*, 1984, **39**, 1053.
356. S. Allaoud, M. E. Mouhtadi and B. Frange, *Nouv. J. Chim.*, 1985, **9**, 499.
357. S. Allaoud, H. Bitar, B. Frange and R. Faure, *Magn. Reson. Chem.*, 1989, **27**, 809.
358. O. D. Gupta, W. A. Kamil and J. M. Shreeve, *Inorg. Chem.*, 1985, **24**, 2126.
359. E. Kimura, Y. Kurogi, M. Shionoya and M. Shiro, *Inorg. Chem.*, 1991, **30**, 4524.
360. A. Ansorge *et al.*, *J. Organomet. Chem.*, 1991, **407**, 283.
361. L. N. Mikhailova, B. I. Tikhomirov and Zh. A. Matveeva, *Vestn. Leningr. Univ., Fiz., Khim.*, 1982, 82 (*Chem. Abstr.*, 1982, **96**, 217 909v).
362. J. R. Jennings, *et al.*, *J. Organomet. Chem.*, 1983, **249**, C1.
363. V. A. Dorokhov and O. G. Boldyreva, *Izv. Akad. Nauk SSSR, Ser. Khim.*, 1986, 2089.
364. M. Yalpani, R. Köster and R. Boese, *Chem. Ber.*, 1992, **125**, 15.
365. H. Nöth and S. Weber, *Z. Naturforsch., Teil B*, 1983, **38**, 1460.
366. M. E. Gurskii, D. G. Pershin, Yu. N. Bubnov, A. V. Polyakov and A. I. Yanovskii, *Metalloorg. Khim.*, 1989, **2**, 1071.
367. W. Haubold, J. Hrebicek and G. Sawitzki, *Z. Naturforsch., Teil B*, 1984, **39**, 1027.
368. S. Kafka and M. Ferles, *Collect. Czech. Chem. Commun.*, 1985, **50**, 2275.
369. M. Ferles and S. Kafka, *Collect. Czech. Chem. Commun.*, 1983, **48**, 1068.
370. V. Dorokhov, O. Boldyreva, A. Shaskov and B. Mikhailov, *Heterocycles*, 1982, **18**, 87.
371. W. Siebert, M. Hildenbrand, P. Hornbach, G. Karger and H. Pritzkow, *Z. Naturforsch., Teil B*, 1989, **44**, 1179.
372. V. A. Dorokhov, L. G. Vorontsova, M. G. Kurella, O. S. Chizhov and B. M. Mikhailov, *Izv. Akad. Nauk SSSR, Ser. Khim.*, 1982, 2370.
373. M. Lauer and G. Wulff, *J. Organomet. Chem.*, 1983, **256**, 1.
374. L. M. Engelhardt, G. E. Jacobsen, P. C. Junk, C. L. Raston and A. H. White, *J. Chem. Soc., Chem. Commun.*, 1990, 89.
375. K.-J. Lee, P. D. Livant, M. L. McKee and S. D. Worley, *J. Am. Chem. Soc.*, 1985, **107**, 5901.
376. N. E. Miller, *J. Organomet. Chem.*, 1984, **269**, 123.
377. N. E. Miller, *Inorg. Chem.*, 1988, **27**, 2196.
378. N. E. Miller and D. L. Reznick, *J. Organomet. Chem.*, 1988, **349**, 11.
379. N. E. Miller, *Inorg. Chem.*, 1991, **30**, 2228.
380. M. R. M. D. Charandabi, D. A. Feakes, M. L. Ettel and K. W. Morse, *Inorg. Chem.*, 1991, **30**, 2433.
381. A. Ansorge *et al.*, *J. Organomet. Chem.*, 1990, **396**, 253.
382. B. Krug and K. Hartke, *Arch. Pharm. (Weinheim, Ger.)*, 1984, **317**, 883 (*Chem. Abstr.*, 1985, **102**, 6599y).
383. N. Kuhn, A. Kuhn, M. Speis, D. Bläser and R. Boese, *Chem. Ber.*, 1990, **123**, 1301.
384. L. G. Vorontsova, O. S. Chizhov, O. G. Boldyreva, V. A. Dorokhov and B. M. Mikhailov, *Izv. Akad. Nauk SSSR, Ser. Khim.*, 1985, 329.
385. O. G. Boldyreva, V. A. Dorokhov and B. M. Mikhailov, *Izv. Akad. Nauk SSSR, Ser. Khim.*, 1985, 428.
386. I. A. Teslya, Z. A. Starikova, O. G. Boldyreva and V. A. Dorokhov, *Koord. Khim.*, 1987, **13**, 1572.
387. V. A. Dorokhov, O. G. Boldyreva and M. N. Bochkareva, *Izv. Akad. Nauk SSSR, Ser. Khim.*, 1987, 2822.
388. V. A. Dorokhov, O. G. Boldyreva, B. M. Mikhailov, Z. A. Starikova and I. A. Teslya, *Izv. Akad. Nauk SSSR, Ser. Khim.*, 1985, 431.
389. V. A. Dorokhov, A. R. Amamchyan, Z. A. Starikova and I. A. Teslya, *Izv. Akad. Nauk SSSR, Ser. Khim.*, 1987, 2049.
390. V. A. Dorokhov and E. A. Shagova, *Izv. Akad. Nauk SSSR, Ser. Khim.*, 1987, 1139.
391. I. A. Teslya, I. A. Misurkin and Z. A. Starikova, *Zh. Obshch. Khim.*, 1989, **59**, 2282.
392. V. A. Dorokhov, M. F. Gordeev, Z. K. Dem'yanets, M. N. Bochkareva, L. G. Vorontsova and M. G. Kurella, *Izv. Akad. Nauk SSSR, Ser. Khim.*, 1990, 407.

393. G. M. Kaplan, A. N. Frolov, N. I. Rtishchev, I. M. Sosonkin, A. V. El'tsov and T. K. Ponomareva, *Zh. Obshch. Khim.*, 1991, **61**, 1810.
394. F. Alam and K. Niedenzu, *J. Organomet. Chem.*, 1982, **240**, 107.
395. C. Habben, L. Komorowski, W. Maringgele, A. Meller and K. Niedenzu, *Inorg. Chem.*, 1989, **28**, 2659.
396. J. Bielawski, M. K. Das, E. Hanecker, K. Niedenzu and H. Nöth, *Inorg. Chem.*, 1986, **25**, 4623.
397. W. J. Layton, K. Niedenzu and S. L. Smith, *Z. Anorg. Allg. Chem.*, 1982, **495**, 52.
398. E. Hanecker, T. G. Hodgkins, K. Niedenzu and H. Nöth, *Inorg. Chem.*, 1985, **24**, 459.
399. C. M. Clarke, *et al.*, *Inorg. Chem.*, 1987, **26**, 2310.
400. M. Yalpani, R. Boese and R. Köster, *Chem. Ber.*, 1990, **123**, 1275.
401. M. Yalpani, R. Boese and R. Köster, *Chem. Ber.*, 1990, **123**, 1285.
402. M. Yalpani, R. Köster and R. Boese, *Chem. Ber.*, 1991, **124**, 1699.
403. K. Niedenzu, J. Serwatowski and S. Trofimenko, *Inorg. Chem.*, 1991, **30**, 524.
404. K. Niedenzu and K. R. Woodrum, *Inorg. Chem.*, 1989, **28**, 4022.
405. C. P. Brock, A. L. Companion, L. D. Kock and K. Niedenzu, *Inorg. Chem.*, 1991, **30**, 784.
406. M. Baudler and G. Kupprat, *Z. Anorg. Allg. Chem.*, 1986, **533**, 146.
407. M. Baudler and M. Hintze, *Z. Anorg. Allg. Chem.*, 1985, **522**, 184.
408. A. M. Arif, A. H. Cowley, M. Pakulski and J. M. Power, *J. Chem. Soc., Chem. Commun.*, 1986, 889.
409. P. Kölle, H. Nöth and R. T. Paine, *Chem. Ber.*, 1986, **119**, 2681.
410. J. Escudie, C. Couret, M. Lazraq and B. Garrigues, *Synth. React. Inorg. Metal-Org. Chem.*, 1987, **17**, 379.
411. P. Kölle, G. Linti, H. Nöth, G. L. Wood, C. K. Narula and R. T. Paine, *Chem. Ber.*, 1988, **121**, 871.
412. P. Kölle, G. Linti, H. Nöth and K. Polborn, *J. Organomet. Chem.*, 1988, **355**, 7.
413. M. Drieß, H. Pritzkow and W. Siebert, *Angew. Chem., Int. Ed. Engl.*, 1987, **26**, 781.
414. R. Köster, G. Seidel, G. Müller, R. Boese and B. Wrackmeyer, *Chem. Ber.*, 1988, **121**, 1381.
415. M. Drieß, P. Frankhauser, H. Pritzkow and W. Siebert, *Angew. Chem., Int. Ed. Engl.*, 1989, **28**, 217.
416. M. Drieß, P. Frankhauser, H. Pritzkow and W. Siebert, *Chem. Ber.*, 1991, **124**, 1497.
417. P. Frankhauser, M. Drieß, H. Pritzkow and W. Siebert, *Chem. Ber.*, 1992, **125**, 1341.
418. H. V. R. Dias and P. P. Power, *Angew. Chem., Int. Ed. Engl.*, 1987, **26**, 1270.
419. H. V. R. Dias and P. P. Power, *J. Am. Chem. Soc.*, 1989, **111**, 144.
420. H. V. R. Dias, D. C. Pestana, M. Petrie and P. P. Power, *Phosphorus, Sulfur Silicon Relat. Elem.*, 1990, **51–52**, 87.
421. H.-O. Berger and H. Nöth, *J. Organomet. Chem.*, 1983, **250**, 33.
422. M. Drieß, H. Pritzkow and W. Siebert, *Chem. Ber.*, 1989, **122**, 467.
423. G. Linti and H. Nöth, *Z. Anorg. Allg. Chem.*, 1991, **593**, 124.
424. G. L. Wood, E. N. Duesler, C. K. Narula, R. T. Paine and H. Nöth, *J. Chem. Soc., Chem. Commun.*, 1987, 496.
425. A. S. Balueva and O. A. Erastov, *Izv. Akad. Nauk SSSR, Ser. Khim.*, 1988, 163.
426. H. Schmidbaur, T. Wimmer, G. Reber and G. Müller, *Angew. Chem., Int. Ed. Engl.*, 1988, **27**, 1071.
427. H. Schmidbaur, T. Wimmer, A. Grohmann, O. Steigelmann and G. Müller, *Chem. Ber.*, 1989, **122**, 1607.
428. A. S. Ionkin, S. N. Ignat'eva, V. M. Nekhoroshkov, Yu. Ya. Efremov and B. A. Arbuzov, *Phosphorus, Sulfur Silicon Relat. Elem.*, 1990, **53**, 1.
429. B. Wrackmeyer and R. Köster, *Chem. Ber.*, 1982, **115**, 2022.
430. P. Paetzold, L. Géret-Baumgarten and R. Boese, *Angew. Chem., Int. Ed. Engl.*, 1992, **31**, 1040.
431. B. Pachaly and R. West, *J. Am. Chem. Soc.*, 1985, **107**, 2987.
432. E. Hanecker, H. Nöth and U. Wietelmann, *Chem. Ber.*, 1986, **119**, 1904.
433. V. Chaturvedi and J. P. Tandon, *Main Group Metal Chem.*, 1990, **13**, 259.
434. I. Manners and P. Paetzold, *J. Chem. Soc., Chem. Commun.*, 1988, 183.
435. E. P. Mayer and H. Nöth, *Chem. Ber.*, 1990, **123**, 963.
436. S. Channareddy, G. Linti and H. Nöth, *Z. Naturforsch., Teil B*, 1990, **45**, 823.
437. T. Mukaiyama and M. Yamaguchi, *Chem. Lett.*, 1982, 509.
438. G. Wulff and A. Hansen, *Angew. Chem., Int. Ed. Engl.*, 1986, **25**, 560.
439. G. D. Schaumberg, K. M. Battarai and D. P. Gautam, *J. Nepal Chem. Soc.*, 1983, **3**, 11 (*Chem. Abstr.*, 1985, **103**, 142 055r).
440. R. Köster and G. Seidel, *Chem. Ber.*, 1992, **125**, 627.
441. W. Zhou, G. Zhang and H. Ding, *Youji Huaxue*, 1982, 19 (*Chem. Abstr.*, 1982, **96**, 181 336q).
442. W. Zhou, S. Liang and W. Lo, *Synthesis*, 1990, 685.
443. Z. Zhu, Y. Fang, D. He, W. Zhou, G. Zhang and H. Ding, *Huaxue Xuebao*, 1985, **43**, 640 (*Chem. Abstr.*, 1986, **105**, 24 299s).
444. Z. Zhu, W. Zhou, Y. Fang, D. He, G. Zhang and H. Ding, *Huaxue Xuebao*, 1985, **43**, 1212 (*Chem. Abstr.*, 1986, **105**, 133 941r).
445. D. S. Matteson and K. H. Arne, *Organometallics*, 1982, **1**, 280.
446. D. S. Matteson and D. Majumdar, *Organometallics*, 1983, **2**, 230.
447. D. J. S. Tsai and D. S. Matteson, *Organometallics*, 1983, **2**, 236.
448. R. W. Hoffmann and R. Metternich, *Liebigs Ann. Chem.*, 1985, 2390.
449. R. W. Hoffmann and S. Dresely, *Synthesis*, 1988, 103.
450. V. V. Kuznetsov, V. A. Bukin and A. I. Gren, *Izv. Vyssh. Uchbn. ZavEd. Khim. Khim. Teknol.*, 1990, **33**, 34 (*Chem. Abstr.*, 1990, **113**, 191 434d).
451. P. Fontani, B. Carboni, M. Vaultier and G. Mass, *Synthesis*, 1991, 605.
452. S. X. Cai and J. F. W. Keana, *Bioconjugate Chem.*, 1991, **2**, 317.
453. K. Ramalingam and D. P. Nowotnik, *Org. Prep. Proced. Int.*, 1991, **23**, 729 (*Chem. Abstr.*, 1992, **116**, 59 438r).
454. A. Sarhan, M. M. Ali and M. Y. Abd El-Al, *J. Prakt. Chem.*, 1988, **330**, 71.
455. W. R. Roush, L. K. Hoong, M. A. J. Palmer and J. C. Park, *J. Org. Chem.*, 1990, **55**, 4109.
456. W. R. Roush, L. K. Hoong, M. A. J. Palmer, J. A. Straub and A. D. Palkowitz, *J. Org. Chem.*, 1990, **55**, 4117.
457. K. Rossi and K. Pihlaja, *Acta Chem. Scand.*, 1985, **39B**, 671.
458. W. R. Roush, A. M. Ratz and J. A. Jablonowski, *J. Org. Chem.*, 1992, **57**, 2047.
459. H. Nöth, *Z. Naturforsch., Teil B*, 1984, **39**, 1463.
460. T. L. Smithson, R. A. Shaw and H. Wieser, *J. Chem. Phys.*, 1984, **81**, 4281.

461. V. I. Grachek, S. F. Naumova, G. R. Motol'ko, A. V. Smolyakov, N. N. Bukanova and N. S. Kozlov, *Dokl. Akad. Nauk SSSR*, 1985, **29**, 536.
462. V. I. Grachek, G. R. Motol'ko, S. F. Naumova and N. S. Kozlov, *Dokl. Akad. Nauk SSSR*, 1988, **32**, 235.
463. V. I. Grachek, G. R. Motol'ko and S. F. Naumova, *Vestsi. Akad. Navuk B. SSR, Ser. Khim. Navuk*, 1990, 52 (*Chem. Abstr.*, 1990, **113**, 115 375v).
464. A. F. Abramov, V. I. Grachek, G. R. Motol'ko and S. F. Naumova, *Zh. Obsch. Khim.*, 1988, **58**, 1571.
465. J. Fr. Janik, C. K. Narula, E. G. Gulliver, E. N. Duesler and R. T. Paine, *Inorg. Chem.*, 1988, **27**, 1222.
466. C. Eaborn, M. N. El-Kheli, N. Retta and J. D. Smith, *J. Organomet. Chem.*, 1983, **249**, 23.
467. C. Eaborn, M. N. El-Kheli, P. B. Hitchcock and J. D. Smith, *J. Chem. Soc., Chem. Commun.*, 1984, 1673.
468. G. W. Kabalka, K. A. R. Sastry and V. Somayaji, *Heterocycles*, 1982, **18**, 157.
469. L. Bhal, R. V. Singh and J. P. Tandon, *Acta Chim. Hung.*, 1984, **115**, 251.
470. F. Sasaya and N. Miyaura, *Bull. Korean Chem. Soc.*, 1987, **8**, 329.
471. D. Männig and H. Nöth, *J. Chem. Soc., Dalton Trans.*, 1985, 1689.
472. M. T. Reetz, C. M. Niemeyer and K. Harms, *Angew. Chem., Int. Ed. Engl.*, 1991, **30**, 1474.
473. M. T. Reetz, C. M. Niemeyer, M. Hermes and R. Goddard, *Angew. Chem., Int. Ed. Engl.*, 1992, **31**, 1017.
474. J. Bartulin, M. Zarraga, H. Zunza and H. Mansilla, *Bol. Soc. Chil. Quim.*, 1982, **27**, 230 (*Chem. Abstr.*, 1982, **97**, 82 279n).
475. V. V. Kuznetsov, L. I. Alekseenko, A. I. Staikov and A. I. Gren, *Ukr. Khim. Zh.*, 1988, **54**, 1315 (*Chem. Abstr.*, 1989, **110**, 201 795b).
476. V. V. Kuznetsov, K. S. Zakharov and A. I. Gren, *Teor. Eksp. Khim.*, 1984, **20**, 742 (*Chem. Abstr.*, 1985, **102**, 166 803p).
477. A. I. Gren, V. V. Kuznetsov and K. S. Zakharov, *Khim. Geterotsikl. Soedin*, 1986, 558 (*Chem. Abstr.*, 1986, **105**, 97533y).
478. V. V. Kuznetsov, K. S. Zakharov, I. V. Petrovskii and A. I. Gren, *Khim. Geterotsikl. Soedin*, 1990, 1107 (*Chem. Abstr.*, 1991, **114**, 42 851f).
479. W. Kliegel, L. Preu, S. J. Rettig and J. Trotter, *Can. J. Chem.*, 1985, **63**, 509.
480. W. Kliegel, L. Preu, S. J. Rettig and J. Trotter, *Can. J. Chem.*, 1986, **64**, 1855.
481. G. Ferguson, A. J. Lough, J. P. Sheehan and T. R. Spalding, *Acta Crystallogr., Sect. C*, 1990, **46**, 1252.
482. V. V. Kuznetsov, A. I. Gren, L. I. Alekseenko and E. D. Novikova, *Ukr. Khim. Zh.*, 1987, **53**, 535 (*Chem. Abstr.*, 1988, **108**, 75 455j).
483. A. Dicko, C. Coste and J. Bastide, *Bull. Soc. Chim. Fr.*, 1982, II-153.
484. V. V. Kuznetsov and A. I. Gren, *Dopov. Akad. Nauk Ukr. RSR, Ser. B*, 1983, 33 (*Chem. Abstr.*, 1983, **98**, 126 185f).
485. V. V. Kuznetsov and A. I. Gren, *Khim. Geterotsikl. Soedin*, 1983, 993 (*Chem. Abstr.*, 1983, **99**, 195 042q).
486. V. V. Kuznetsov and A. I. Gren, *Dopov. Akad. Nauk Ukr. RSR, Ser. B*, 1984, 37 (*Chem. Abstr.*, 1985, **102**, 6592r).
487. V. V. Kuznetsov and A. I. Gren, *Zh. Obshch. Khim.*, 1984, **54**, 2263.
488. V. V. Kuznetsov and A. I. Gren, *Ukr. Khim. Zh.*, 1985, **51**, 535 (*Chem. Abstr.*, 1986, **104**, 148 953z).
489. V. V. Kuznetsov and A. I. Gren, *Zh. Obshch. Khim.*, 1986, **56**, 613.
490. V. V. Kuznetsov and A. I. Gren, *Zh. Org. Khim.*, 1983, **19**, 1987.
491. V. V. Kuznetsov and A. I. Gren, *Zh. Obshch. Khim.*, 1985, **55**, 1651.
492. V. V. Kuznetsov and A. I. Gren, *Zh. Org. Khim.*, 1986, **22**, 2237.
493. V. V. Kuznetsov and A. I. Gren, *Khim. Geterotsikl. Soedin*, 1989, 994 (*Chem. Abstr.*, 1990, **112**, 179 062c).
494. V. V. Kuznetsov and A. I. Gren, *Zh. Org. Khim.*, 1985, **21**, 2016.
495. V. V. Kuznetsov and A. I. Gren, *Ukr. Khim. Zh.*, 1986, **52**, 882 (*Chem. Abstr.*, 1987, **107**, 176 080t).
496. P. G. M. Wuts and P. A. Thompson, *J. Organomet. Chem.*, 1982, **234**, 137.
497. I. A. Krivenko, A. Kamars and E. M. Shvarts, *Latv. Kim. Z.*, 1991, 233 (*Chem. Abstr.*, 1991, **115**, 92338k).
498. D. S. Matteson and R. J. Moody, *Organometallics*, 1982, **1**, 20.
499. H. C. Brown, T. E. Cole and M. Srebnik, *Organometallics*, 1985, **4**, 1788.
500. S. I. M. Mubarak, J. B. Stanford and J. K. Sugden, *Drug Dev. Ind. Pharm.*, 1984, **10**, 1131; 1617.
501. K. Seto and S. Takahashi, *Mol. Cryst. Liq. Cryst., Lett. Sect.*, 1985, **2**, 197.
502. K. Seto, S. Takahashi and T. Tahara, *J. Chem. Soc., Chem. Commun.*, 1985, 122.
503. B. Hahn and V. Percec, *Macromolecules*, 1987, **20**, 2961.
504. H. Matsubara, K. Seto, T. Tahara and S. Takahashi, *Bull. Chem. Soc. Jpn.*, 1989, **62**, 3896.
505. H. Matsubara, K. Seto, H. Tabuchi, H. Imazaki and S. Takahashi, *Chem. Lett.*, 1989, 1519.
506. H. Matsubara, S. Takahashi, K. Seto and H. Imazaki, *Mol. Cryst. Liq. Cryst.*, 1990, **180B**, 337.
507. H. Matsubara *et al.*, *Bull. Chem. Soc. Jpn.*, 1991, **64**, 2103.
508. V. S. Bezborodov, *Zh. Org. Khim.*, 1989, **25**, 2168.
509. V. S. Bezborodov and V. I. Lapanik, *Liq. Cryst.*, 1991, **10**, 803.
510. V. S. Bezborodov, M. F. Grebenkin and V. I. Lapanik, *Zh. Org. Khim.*, 1991, **27**, 385.
511. O. P. Singh, R. K. Mehrotra and G. Srivastrava, *Phosphorus, Sulfur Silicon Relat. Elem.*, 1991, **60**, 147.
512. O. P. Singh, R. K. Mehrotra and G. Srivastrava, *Synth. React. Inorg. Met.-Org. Chem.*, 1991, **21**, 717.
513. P. Paetzold, B. Redenz-Stormanns and R. Boese, *Angew. Chem., Int. Ed. Engl.*, 1990, **29**, 900.
514. W. Maringgele, A. Meller, M. Noltemeyer and G. M. Sheldrick, *Z. Anorg. Allg. Chem.*, 1986, **536**, 24.
515. L. Synoradzki, R. Mynott, J. Anbei, C. Krüger, Y.-H. Tsay and R. Köster, *Chem. Ber.*, 1984, **117**, 2863.
516. C. van de Sande and A. Verhecken, *Res. Discl.*, 1984, **243**, 305.
517. N. L. Kostenko, S. V. Nesterova, S. V. Toldov, N. K. Skvortov and V. O. Reikhsfel'd, *Zh. Obsch. Khim.*, 1987, **57**, 716.
518. H. E. Katz, *J. Org. Chem.*, 1985, **50**, 2575.
519. M. Yalpani, R. Köster and G. Wilke, *Chem. Ber.*, 1983, **116**, 1336.
520. M. Yalpani and R. Köster, *Chem. Ber.*, 1983, **116**, 3332.
521. M. Yalpani, R. Boese and D. Bläser, *Chem. Ber.*, 1983, **116**, 3338.
522. M. Yalpani, W. R. Scheidt and K. Seevogel, *J. Am. Chem. Soc.*, 1985, **107**, 1684.
523. M. Yalpani and W. E. Klotzbücher, *Z. Naturforsch., Teil B*, 1985, **40**, 222.
524. M. Yalpani, W. Schmöller and D. Henneberg, *Z. Naturforsch., Teil B*, 1984, **39**, 1241.
525. D. S. Matteson and E. C. Beedle, *Heteroatom Chem.*, 1990, **1**, 135.
526. J. A. Soderquist, B. Santiago and I. Rivera, *Tetrahedron Lett.*, 1990, **31**, 4981.
527. V. J. Heintz, W. A. Freeman and T. A. Keiderling, *Inorg. Chem.*, 1983, **22**, 2319.
528. H. Borrmann, A. Simon and H. Vahrenkamp, *Angew. Chem., Int. Ed. Engl.*, 1989, **28**, 180.

529. N. C. Barua, K. Evertz, G. Huttner and R. R. Schmidt, *Chem. Ber.*, 1987, **120**, 213.
530. C. Orvig, S. J. Rettig, J. Trotter and Z. Zhang, *Can. J. Chem.*, 1990, **68**, 1803.
531. R. Köster and G. Seidel, *Chem. Ber.*, 1992, **125**, 627.
532. A. S. Kothiwal, A. Singh, A. K. Rai and R. C. Mehrotra, *Indian J. Chem., Sect. A*, 1985, **24**, 828.
533. V. Chaturvedi, L. Chaturvedi and J. P. Tandon, *Main Group Metal Chem.*, 1990, **13**, 253.
534. K. Okada, Y. Hosoda and M. Oda, *J. Am. Chem. Soc.*, 1986, **108**, 321.
535. K. Hartke, B. Krug and R. Hoffmann, *Liebigs Ann. Chem.*, 1984, 370.
536. C. K. Narula and H. Nöth, *Z. Naturforsch., Teil B*, 1983, **38**, 1161.
537. M. V. Koz'menko, M. V. Mamaenko and O. A. Korotkikh, *Khim. Vys. Energ.*, 1990, **24**, 243 (*Chem. Abstr.*, 1990, **113**, 106 148w).
538. R. Boese, R. Köster and M. Yalpani, *Chem. Ber.*, 1985, **118**, 670.
539. V. S. Markin, P. I. Abramenko and I. I. Boiko, *Zh. Vses. Khim. O-va. im D. I. Mendeleeva*, 1984, **29**, 586.
540. S. P. Starkov, A. I. Panasenko, N. L. Polyanskaya and M. N. Volkotrub, *Izv. Vyssh. Uchebn. Zaved., Khim. Khim. Tekhnol.*, 1988, **31**, 32.
541. R. C. F. Jones, M. J. Begley, G. E. Peterson and S. Sumaria, *J. Chem. Soc., Perkin Trans. I*, 1990, 1959.
542. H. Binder, W. Matheis, H.-J. Deiseroth and H. Fu-Son, *Z. Naturforsch., Teil B*, 1983, **38**, 554.
543. A. S. Balueva and O. A. Erastov, *Izv. Akad. Nauk SSSR, Ser. Khim.*, 1987, 1199.
544. W. Kliegel, S. J. Rettig and J. Trotter, *Can. J. Chem.*, 1988, **66**, 1091.
545. R. Köster and W. Fenzl, *Inorg. Synth.*, 1983, **22**, 196.
546. H. Binder, W. Matheis, H.-J. Deiseroth and H. Fu-Son, *Z. Naturforsch., Teil B*, 1984, **39**, 1717.
547. H. Binder, W. Matheis, G. Heckmann, H. J. Deiseroth and F. Han, *Z. Naturforsch., Teil B*, 1985, **40**, 934.
548. W. Maringgele and A. Meller, *Z. Anorg. Allg. Chem.*, 1989, **572**, 140.
549. H. Nöth and W. Rattay, *J. Organomet. Chem.*, 1986, **312**, 139.
550. R. Köster, G. Seidel and R. Boese, *Chem. Ber.*, 1988, **121**, 1137.
551. C. Habben, A. Meller, M. Noltemeyer and G. M. Sheldrick, *Z. Naturforsch., Teil B*, 1986, **41**, 1093.
552. C. Habben and A. Meller, *Chem. Ber.*, 1986, **119**, 1189.
553. C. D. Habben, *Chem. Ber.*, 1988, **121**, 1967.
554. C. D. Habben, *Z. Anorg. Allg. Chem.*, 1989, **573**, 199.
555. C. D. Habben, *Z. Naturforsch., Teil B*, 1991, **46**, 1267.
556. C. D. Habben and S. Pusch, *Z. Anorg. Allg. Chem.*, 1991, **593**, 217.
557. H. C. Brown and T. Imai, *J. Org. Chem.*, 1984, **49**, 892.
558. H. Nöth, H. Pommerening and R. Staudigl, *Z. Naturforsch., Teil B*, 1988, **43**, 501.
559. L. Bahl, R. V. Singh and J. P. Tandon, *Indian J. Chem., Sect A*, 1983, **22**, 951.
560. R. Köster, G. Seidel and R. Boese, *Chem. Ber.*, 1990, **123**, 2109.
561. A. Furusaki, Z. Weike and A. Suzuki, *Bull. Chem. Soc. Jpn.*, 1986, **59**, 313.
562. M. Noltemeyer, G. M. Sheldrick, C. Habben and A. Meller, *Z. Naturforsch., Teil B*, 1983, **38**, 1182.
563. H. Nöth and R. Staudigl, *Inorg. Synth.*, 1983, **22**, 218.
564. R. Conrady-Pigorsch, W. Müller-Warmuth, G. Schwetlik, M. Wienkenhöver and B. Krebs, *Ber. Bunsen-Ges. Phys. Chem.*, 1991, **95**, 453 (*Chem. Abstr.*, 1991, **115**, 63 130f).
565. H. Binder, W. Ehmann and K. Dermentzis, *Z. Naturforsch., Teil B*, 1991, **46**, 581.
566. A. Meller and C. Habben, *Monatsh. Chem.*, 1982, **113**, 139.
567. D. Fest, C. D. Habben and A. Meller, *Chem. Ber.*, 1989, **122**, 861.
568. H. Nöth and R. Staudigl, *Chem. Ber.*, 1982, **115**, 813.
569. H. Nöth and R. Staudigl, *Chem. Ber.*, 1982, **115**, 1555.
570. C. Habben and A. Meller, *Z. Naturforsch., Teil B*, 1984, **39**, 1022.
571. C. Habben, A. Meller, M. Noltemeyer and G. M. Sheldrick, *J. Organomet. Chem.*, 1985, **288**, 1.
572. G. Davidson and K. P. Ewer, *Spectrochim. Acta, Part A*, 1986, **42**, 913.
573. V. A. Bacherikov, A. I. Gren, E. B. Topchaya, V. Ya. Gorbatyuk and O. S. Timofeev, *Zh. Obshch. Khim.*, 1990, **60**, 119.
574. A. I. Gren, V. A. Bacherikov and A. M. Turyanskaya, *Zh. Org. Khim.*, 1986, **22**, 2621.
575. H. Nöth and D. Sedlak, *Chem. Ber.*, 1983, **116**, 1479.
576. H. Nöth and H. Pommerening, *Chem. Ber.*, 1986, **119**, 2261.
577. W. Jacksties, H. Nöth and W. Storch, *Chem. Ber.*, 1985, **118**, 2030.
578. E. Hanecker and H. Nöth, *Z. Naturforsch., Teil B*, 1985, **40**, 717.
579. V. Chaturvedi and J. P. Tandon, *Indian J. Chem., Sect. A*, 1985, **24**, 641.
580. B. J. Brisdon, M. F. Mahon, K. C. Molloy and P. J. Schofield, *J. Organomet. Chem.*, 1992, **436**, 11.
581. A. Mazzah, A. Haoudi-Mazzah, M. Noltemeyer and H. W. Roesky, *Z. Anorg. Allg. Chem.*, 1991, **604**, 93.
582. P. Kölle, H. Nöth and R. T. Paine, *Chem. Ber.*, 1989, **122**, 423.
583. R. Streubel, E. Niecke and P. Paetzold, *Chem. Ber.*, 1991, **124**, 765.
584. D. Gudat, E. Niecke, M. Nieger and P. Paetzold, *Chem. Ber.*, 1988, **121**, 565.
585. A. Brandl and H. Nöth, *Chem. Ber.*, 1988, **121**, 1321.
586. P. Kölle, H. Nöth, R. T. Paine and W. Rattay, *Z. Naturforsch., Teil B*, 1988, **43**, 1439.
587. B. Ederer, H. Ederle and H. Nöth, *Chem. Ber.*, 1992, **125**, 2213.
588. D. Männig, C. K. Narula, H. Nöth and U. Wietelmann, *Chem. Ber.*, 1985, **118**, 3748.
589. F. Kumpfmüller, D. Nölle, H. Nöth, H. Pommerening and R. Staudigl, *Chem. Ber.*, 1985, **118**, 483.
590. G. Geisberger and H. Nöth, *Chem. Ber.*, 1990, **123**, 953.
591. R. H. Cragg and T. J. Miller, *J. Chem. Soc., Dalton Trans.*, 1982, 907.
592. M. Moehrle, K. Bluhme-Hensen, B. Middelhauve, D. Mootz and H. Wunderlich, *Z. Naturforsch., Teil B*, 1991, **46**, 1219.
593. R. H. Cragg and T. J. Miller, *J. Organomet. Chem.*, 1985, **294**, 1.
594. M. P. DeNinno, R. J. Perner and L. Lijewski, *Tetrahedron Lett.*, 1990, **31**, 7415.
595. A. V. R. Rao, M. K. Gurjar, P. A. Sharma and V. Kaiwar, *Tetrahedron Lett.*, 1990, **31**, 2341.
596. I. Uzarewicz, W. Maringgele and A. Meller, *Z. Anorg. Allg. Chem.*, 1990, **590**, 151.
597. D. Y. Lee and J. C. Martin, *J. Am. Chem. Soc.*, 1984, **106**, 5745.
598. W. Kliegel, G. Lubkowitz, S. J. Rettig and J. Trotter, *Can. J. Chem.*, 1991, **69**, 234; 1217; 1227.

599. A. R. Kalyuskii, V. V. Kuznetsov and A. I. Gren, *Khim. Geterotsikl. Soedin.*, 1987, 712 (*Chem. Abstr.*, 1988, **108**, 94 612z).
600. A. R. Kalyuskii, S. E. Zhuchenko, V. V. Kuznetsov and A. I. Gren, *Zh. Org. Khim.*, 1989, **25**, 2202.
601. A. R. Kalyuskii, V. V. Kuznetsov and A. I. Gren, *Zh. Org. Khim.*, 1989, **25**, 2021.
602. A. R. Kalyuskii, V. V. Kuznetsov and A. I. Gren, *Zh. Obshch. Khim.*, 1989, **59**, 1438.
603. A. R. Kalyuskii, V. V. Kuznetsov, S. P. Krasnoshchekaya and A. I. Gren, *Dokl. Akad. Nauk SSSR, Ser. B*, 1989, 41.
604. A. R. Kalyuskii, V. V. Kuznetsov, Yu. E. Brusilovskii, M. G. Glukhova, A. I. Gren and V. Ya. Gorbatyuk, *Zh. Org. Khim.*, 1990, **26**, 2498.
605. A. R. Kalyuskii, V. V. Kuznetsov and A. I. Gren, *Khim. Geterotsikl. Soedin.*, 1990, 854 (*Chem. Abstr.*, 1991, **114**, 102 139s).
606. A. R. Kalyuskii, V. V. Kuznetsov, O. S. Timofeev and A. I. Gren, *Zh. Obshch. Khim.*, 1990, **60**, 2093.
607. A. R. Kalyuskii, V. V. Kuznetsov, Yu. E. Shapiro and A. I. Gren, *Zh. Fiz. Khim.*, 1990, **64**, 2535.
608. A. R. Kalyuskii, V. V. Kuznetsov and A. I. Gren, *Khim. Geterotsikl. Soedin.*, 1991, 1262 (*Chem. Abstr.*, 1992, **116**, 214 572a).
609. A. R. Kalyuskii, V. V. Kuznetsov and A. I. Gren, *Zh. Obshch. Khim.*, 1991, **61**, 1351.
610. A. Meller, C. Habben, M. Noltemeyer and G. M. Sheldrick, *Z. Naturforsch., Teil B*, 1982, **37**, 1504.
611. M. Komorowska, K. Niedenzu and W. Weber, *Inorg. Chem.*, 1990, **29**, 289.
612. J. Bielawski and K. Niedenzu, *Inorg. Chem.*, 1986, **25**, 1771.
613. L.-Y. Hsu, J. F. Martiategui, K. Niedenzu and S. G. Shore, *Inorg. Chem.*, 1987, **26**, 143.
614. K. Niedenzu, H. Deng, D. Knoeppel, J. Krause and S. G. Shore, *Inorg. Chem.*, 1992, **31**, 3162.
615. W. Tjarks and D. Gabel, *J. Med. Chem.*, 1991, **34**, 315.
616. M. A. Paz-Sandoval, C. Fernández-Vincent, G. Uribe, R. Contreras and A. Klaébé, *Polyhedron*, 1988, **7**, 679.
617. B. A. Arbuzov, O. A. Erastov, G. N. Nikonov, I. P. Romoanova, R. P. Arshinova and R. A. Kadyrov, *Izv. Akad. Nauk SSSR, Ser. Khim.*, 1983, 1374; 1846.
618. S. N. Ignat'eva, G. N. Nikonov, O. A. Erastov and B. A. Arbuzov, *Izv. Akad. Nauk SSSR, Ser. Khim.*, 1985, 1102.
619. B. A. Arbuzov, G. N. Nikonov and O. A. Erastov, *Izv. Akad. Nauk SSSR, Ser. Khim.*, 1986, 643.
620. A. A. Karasik, O. A. Erastov and B. A. Arbuzov, *Izv. Akad. Nauk SSSR, Ser. Khim.*, 1988, 2409.
621. A. A. Karasik, G. N. Nikonov, B. A. Arbuzov, R. A. Musin and Yu. Ya. Efremov, *Izv. Akad. Nauk SSSR, Ser. Khim.*, 1991, 719.
622. E. V. Troepol'skaya, G. N. Nikonov and A. A. Karasik, *Izv. Akad. Nauk SSSR, Ser. Khim.*, 1991, 358.
623. I. A. Litvinov, O. N. Kataeva, G. N. Nikonov, O. A. Erastov, V. A. Naumov and B. A. Arbuzov, *Izv. Akad. Nauk SSSR, Ser. Khim.*, 1986, 2715.
624. G. N. Nikonov, I. A. Litvinov, S. N. Ignat'eva, O. A. Erastov and B. A. Arbuzov, *Izv. Akad. Nauk SSSR, Ser. Khim.*, 1988, 159.
625. G. N. Nikonov *et al.*, *Izv. Akad. Nauk SSSR, Ser. Khim.*, 1988, 155.
626. G. N. Nikonov, S. N. Ignat'eva, A. S. Balueva, O. A. Erastov and B. A. Arbuzov, *Izv. Akad. Nauk SSSR, Ser. Khim.*, 1988, 2412.
627. B. A. Arbuzov, G. N. Nikonov, A. A. Karasik and E. V. Malova, *Izv. Akad. Nauk SSSR, Ser. Khim.*, 1990, 1120.
628. B. A. Arbuzov, O. A. Erastov, G. N. Nikonov and A. S. Ionkin, *Izv. Akad. Nauk SSSR, Ser. Khim.*, 1984, 2501.
629. L. V. Ermolaeva, A. S. Balueva, O. A. Erastov, N. N. Sarvaarova, S. G. Vul'fson and B. A. Arbuzov, *Zh. Obshch. Khim.*, 1990, **60**, 755.
630. A. Brandl and H. Nöth, *Chem. Ber.*, 1990, **123**, 53.
631. C. D. Habben and M. Noltemeyer, *Z. Naturforsch., Teil B*, 1989, **44**, 853.
632. C. D. Habben, A. Heine, G. M. Sheldrick, D. Stalke, M. Bühl and P. von R. Schleyer, *Chem. Ber.*, 1991, **124**, 47.
633. C. D. Habben, R. Herbst-Irmer and M. Noltemeyer, *Z. Naturforsch., Teil B*, 1991, **46**, 625.
634. C. D. Habben, *Z. Anorg. Allg. Chem.*, 1991, **606**, 229.
635. H. Nöth and R. Staudigl, *Chem. Ber.*, 1982, **115**, 3011.
636. C. D. Habben, *Chem. Ber.*, 1987, **120**, 999.
637. M. K. Das, K. Niedenzu and H. Nöth, 1988, **27**, 1112.
638. C. Habben and A. Meller, *Chem. Ber.*, 1986, **119**, 9.
639. D. Fest, C. Habben and A. Meller, *Chem. Ber.*, 1986, **119**, 3121.
640. C. Habben, A. Meller, M. Noltemeyer and G. M. Sheldrick, *Z. Naturforsch., Teil B*, 1986, **41**, 799.
641. C. D. Habben and M. Noltemeyer, *Z. Naturforsch., Teil B*, 1988, **43**, 1683.
642. C. Habben, A. Meller, M. Noltemeyer and G. M. Sheldrick, *Angew. Chem., Int. Ed. Engl.*, 1986, **25**, 741.
643. A. Kendrick, H. Nöth, B. Stalla and W. Storch, *J. Chem. Soc., Dalton Trans.*, 1989, 1311.
644. W. Einholz, G. Frey and W. Haubold, *Z. Naturforsch., Teil B*, 1989, **44**, 47.
645. V. K. Andrä and A. Straube, *Z. Anorg. Allg. Chem.*, 1982, **490**, 191.
646. A. I. Gren, V. A. Bacherikov and K. S. Zakharov, *Zh. Obshch. Khim.*, 1987, **57**, 2016.
647. V. A. Bacherikov, A. I. Gren, V. Ya. Gorbatyuk and O. S. Timofeev, *Khim. Geterosikl. Soedin.*, 1990, 242 (*Chem. Abstr.*, 1990, **113**, 97 653c).

5
Boron Rings Ligated to Metals

GERHARD E. HERBERICH
Technische Hochschule Aachen, Germany

5.1 GENERAL ASPECTS 197
5.1.1 Metals and Ligands 197
5.1.2 Tripledecker Complexes 198
5.1.3 Quadrupledecker Complexes and More Extended Systems 199
5.2 SYSTEMS WITH BORON–CARBON RING LIGANDS 201
5.2.1 Complexes of Borirenes 201
5.2.2 Complexes of Five-membered Rings 202
5.2.2.1 Complexes of C_4B rings 202
5.2.2.2 Complexes of C_3B_2 rings 204
5.2.2.3 Complexes of C_2B_3 rings 207
5.2.2.4 Complexes of diborafulvene systems 208
5.2.3 Complexes of Six-membered Rings 209
5.2.3.1 Complexes of C_5B rings 209
5.2.3.2 Complexes of C_4B_2 rings 209
5.2.4 Complexes of Borepines 210
5.3 SYSTEMS WITH INORGANIC RING LIGANDS 211
5.3.1 Complexes of Boron–Nitrogen Ligands 211
5.3.2 Complexes of Boron–Phosphorus Ligands 211
5.4 SYSTEMS WITH BORON–CARBON RING LIGANDS WITH ADDITIONAL HETEROATOMS 212
5.4.1 Complexes of C_3BN Rings 212
5.4.2 Complexes of 1,2,5-Heterosilaborolenes 213
5.4.3 Complexes of 1,2,5-Phosphadiborolenes 214
5.4.4 Complexes of 1,2,5-Thiadiborolenes 214
5.5 REFERENCES 214

5.1 GENERAL ASPECTS

During the 1980s the field covered in this chapter developed at a rapid pace and by the mid-1990s it had lost its exotic character. The specific synthetic methods used have become more rational and systematic. Review articles about the whole field are not available but several specialized topics have been reviewed and these will be quoted in the appropriate context.

5.1.1 Metals and Ligands

In principle, all metals that possess the ability of polyhapto-bonding to π-electron systems can also form complexes with boron ring systems. In practice the most commonly used metals are those of the triads from manganese to nickel. There are also a few compounds of earlier *d*-metals (Ti, V, Nb, Ta, Cr, Mo, W) and of *p*-metals (Tl, Sn), but *f*-metals are conspicuously absent.

The area of the boron–carbon ligands (Section 5.2) is the most developed field, especially in the domain of five- and six-membered rings. Many new ring systems have become accessible including also a number of benzo-anellated systems. The most important advances are: (i) the new and very broad development of the complexes of the 1*H*-borole ligand (**1**);[1] (ii) an enormous growth of the chemistry of the 1,3-diborolenyl ligand (**2**)[2] and of the 1,2,3,-triboracyclopentenyl ligand (**3**);[3] and (iii) the solution of the 1*H*-borepine (**4**) problem with the development of an efficient and well-documented approach to this ring system.[4] Two notably missing ring ligands are the C_3BH_4 (boracyclobutenyl (**5**)) and the $C_3B_2H_5$ ring system (**6**) with adjacent boron atoms.

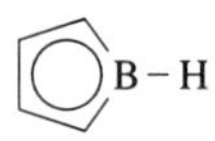

(1) π^4

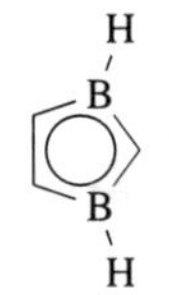

(2) π^3

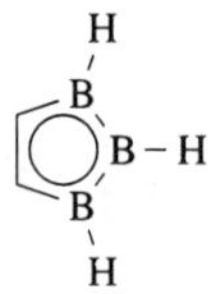

(3) π^2

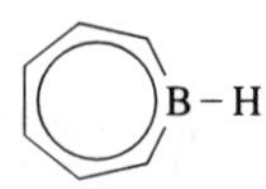

(4) π^6

(5) π^3

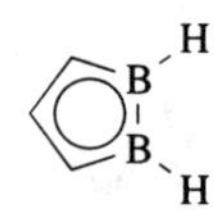

(6) π^3

Complexes of new inorganic ligands (e.g., $N_2B_2H_4$ (**7**),[5] $P_2B_2H_4$ (**8**)[6] and $P_3B_3H_6$ (**9**)[6]) (Section 5.3) and of several new mixed boron–carbon–heteroatom ligands (e.g., cyclic azaboradienes (**10**)) (Section 5.4) have been described. In this area the new chemistry of the 1,2-azaborolinyl ligand (**11**) is of particular importance.[7] The selection of known inorganic and mixed ligands in the mid-1990s is still rather erratic. By way of interpolation we may expect that all conceivable π-electron systems with skeleton atoms of intermediate electronegativity (B, C, P, As, Sb, S, Se and to some extent N) will be candidates for future developments.

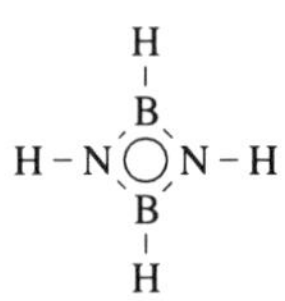

(7) π^4

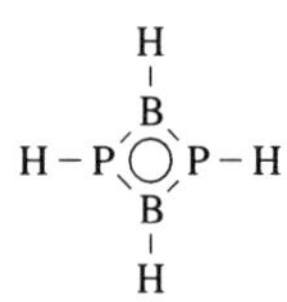

(8) π^4

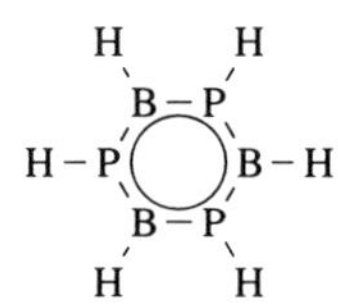

(9) π^6

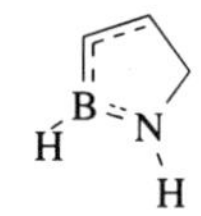

(10) π^4

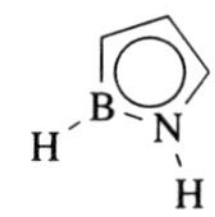

(11) π^5

5.1.2 Tripledecker Complexes

The characteristic structural feature of tripledecker complexes is the cyclic π-electron system which is bifacially bonded to two metal centres thereby building up a *closo*–polyhedron. About half of the ring ligands that are known to act as central or bridging ligands in tripledecker complexes contain boron atoms. The eight pertinent ligands comprise the five-membered rings C_4BH_5 (**1**),[1] 1,3-$C_3B_2H_5$ (**2**),[2] 1,2,3-$C_2B_3H_5$ (**3**),[3] 1,2,4-$C_2B_3H_5$ (**12**), and $C_2B_2SH_4$ (**13**), and the six-membered rings C_5BH_6 (**14**), 1,2-$C_4B_2H_6$ (**15**) and 1,4-$C_4B_2H_6$ (**16**). Many earlier results have been brought together in a wider context in a review on the chemistry of the 1,3-diborolenyl complexes.[2] The more recent additions to the list of central ligands are the rings (**14**)[8] and (**15**).[9]

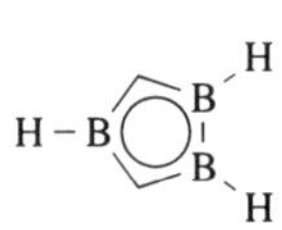

(12) π^2

(13) π^4

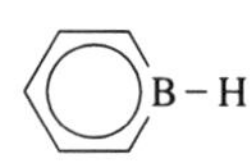

(14) π^5

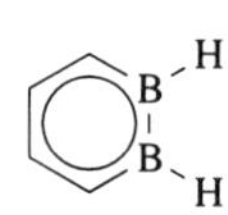

(15) π^4

(16) π^4

The bonding situation in tripledecker complexes has been discussed in several papers.[10] Wade's rules predict maximal stability for an electron count of 30.[11] Hoffmann's classical work on the basis of extended Hückel MO calculations has shown that tripledecker complexes possess two closed-shell configurations, one at lower energy with an electron count of 30 and one at higher energy with 34 electrons.[12] The extent of *d*-orbital splitting will depend on the nature of the metals and ligands present. The presence of strong ligands (e.g., of CO ligands) and/or of heavier *d*-metals will cause large *d*-orbital splittings with the consequence of a pronounced preference for complexes with an electron count of 30. Alternatively, when the metal–ligand interactions are comparatively weak (e.g., for cyclopentadienyl and benzene ligands) the magic number 30 will have little meaning.

Empirically the electron count of the known tripledecker complexes varies from 26 to 34, but an overwhelming majority of complexes with central ligands (**1**)–(**3**) or (**12**)–(**16**) possess 30 valence electrons. Thus, it is empirically justified to relate to a '30e rule'. The family of sandwich tripledecker complexes (**17**) with 29–34 valence electrons stands for the exceptions to the rule,[13] while in the vanadium complex (**18**) the carbonyl ligands reinforce the 30e rule.[14] We note in passing that very few 34-electron tripledecker complexes (about five types) are known: an occasionally quoted '30/34e rule' is therefore unjustified.

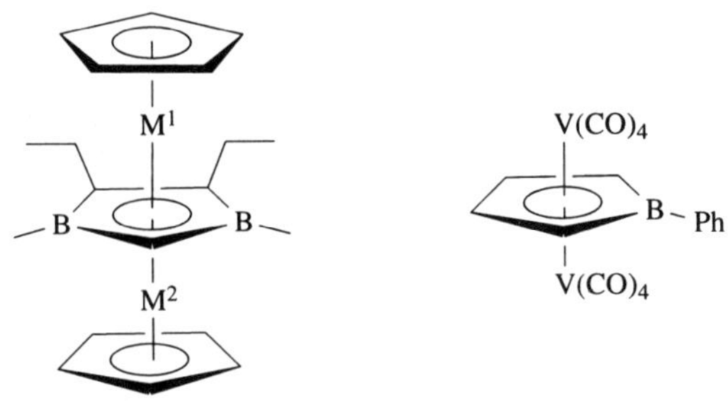

(**17**) M^1 = Fe, Co, Ni; M^2 = Co, Ni (**18**)

The popularity of the 1976 Hoffmann paper[12] is certainly due to the simplicity of the methods and arguments used and to the empirical success of the 30e rule. However, it has been emphasized that more sophisticated bonding considerations are needed for the less symmetric heterobimetallic tripledecker complexes of type (**17**) and for any quantitative prediction of spectral properties. For the bonding description in complexes of type (**17**) Hartree–Fock INDO SCF calculations were sufficient in some cases, in others correlation effects played a large role.[13]

In common understanding, tripledecker complexes possess a *closo*-structure. Recently, closely related examples of dinuclear 34 electron complexes have been found which have doubly-opened or slipped tripledecker structures. The ground-state structure of complex (**19**) shows a bridging diborafulvene ligand which is ligated to one Rh(cod) fragment in a trimethylenemethane-like fashion and to the other one as a 1,2-diboracyclopentenyl ring.[15] At room temperature, complex (**19**) is fluxional and the two Rh(cod) fragments interchange their role at high rate with respect to the NMR spectroscopic timescale.[15] A similar type of structure is found in the rigid complex (**20**) (and several related compounds); the bridging diborafulvene ligand is again ligated to one $Fe(CO)_3$ fragment in a trimethylenemethane-like coordination mode and to the second $Fe(CO)_3$ fragment as a diborolenyl ligand.[16–17]

(**19**) (**20**)

5.1.3 Quadrupledecker Complexes and More Extended Systems

In principle, there are two ways to build more extended structures: the construction of larger stacks with bifacially bound bridging ligands, and the construction of multiunit systems with σ-bond linkages between the various units.

The first approach follows the lines established for the synthesis of tripledecker structures.[2] The first quadrupledecker or tetradecker complex 'MnFeMn' (**21**) was described in 1978 and contains a bridging 1,2,5-thiadiborolene ligand.[18] The majority of known examples are based on bridging 1,3-diborolenyl (**2**) and 1,2,3-triboracyclopentenyl ligands (**3**).[3] There is also one representative 'CoRuCo' (**22**) with bridging 1*H*-borole ligands[19] and one 'NiNiNi' (**23**) with bridging 1,4-diboracyclohexadiene ligands.[20] Quadrupledecker complexes with two different bridging ligands such as the complex 'CoCoRh' (**24**) have also been made.[21] The present confines of this area seem rather accidential and certainly reflect the predilections of the research groups involved.

A further extension of the same architectural motifs has lead to the first pentadecker complex (**25**) in 1982,[22] the first hexadecker complex (**26**) in 1985,[23] and to the synthesis of two multilayer sandwich

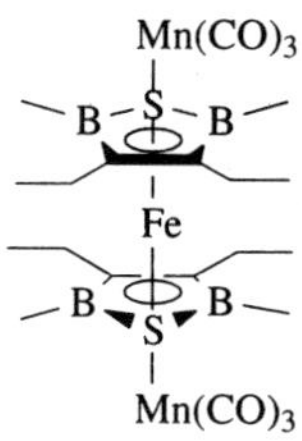

(21)

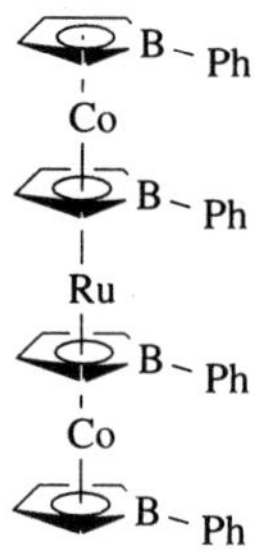

(22)

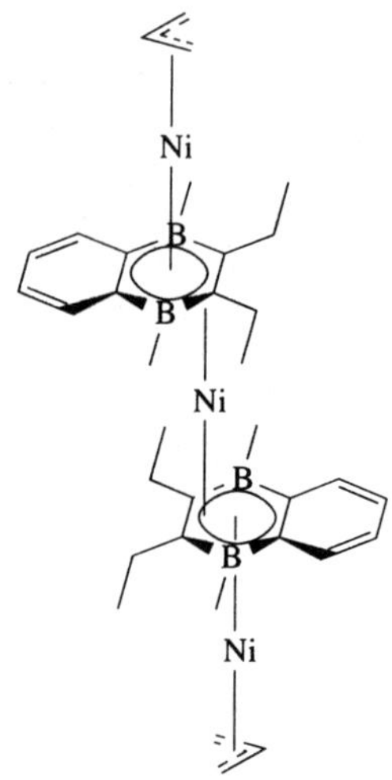

(23)

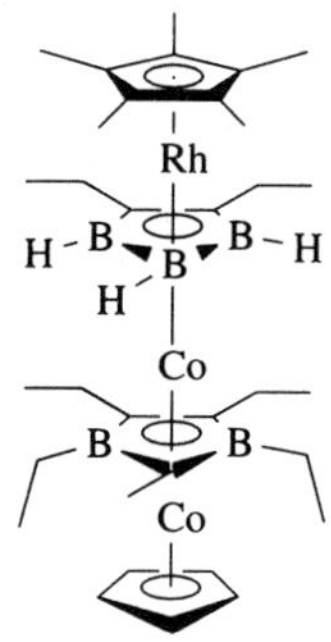

(24)

compounds, the air-sensitive nickel-based polymer (**27**)[24] and the robust rhodium-based polymer (**28**).[25] These materials are the first representatives of a new type of one-dimensional semiconductor.

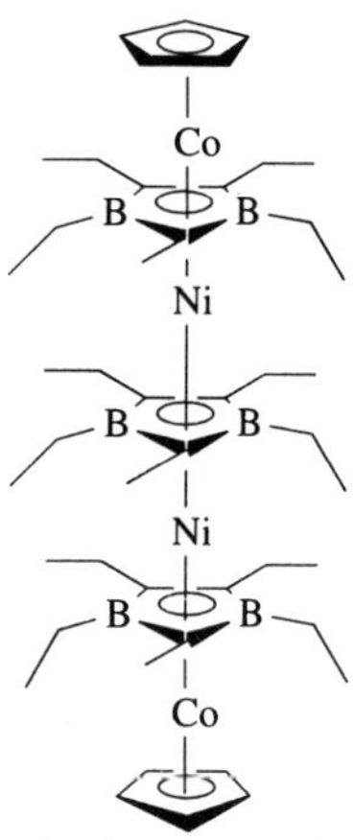

(25)

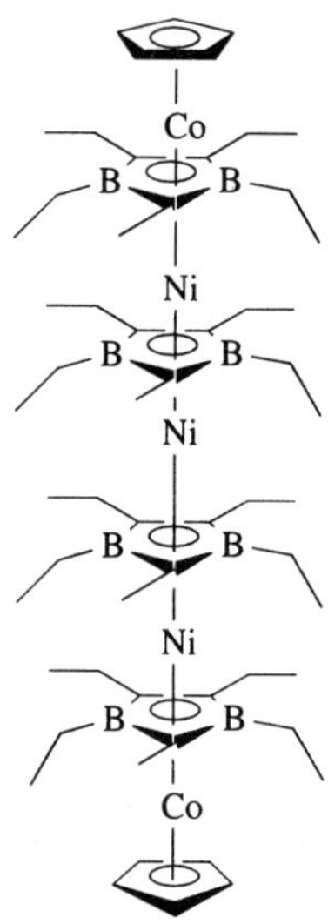

(26)

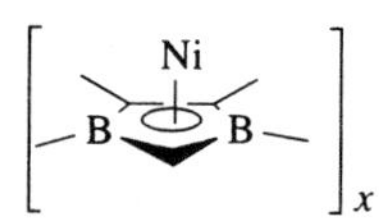

(27)

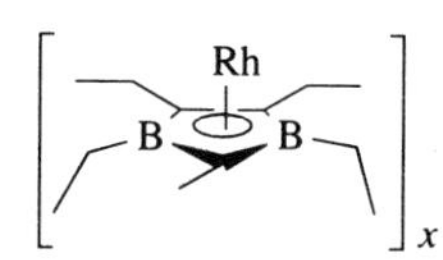

(28)

The second approach to extended structures uses σ-linkages to connect sandwich, tripledecker and quadrupledecker complexes to larger units. The prototypal structure here is provided by biferrocenyl with a σ-linkage between two ferrocene units. Suitable linking groups are, for example, biphenyl (**29**), pentafulvalene dianion (**30**) and the mixed linkage group benzyltetramethyl-cyclopentadienide (**31**). The combination of the two approaches gives at hand a molecular kit which allows to construct and to synthesize designed structures of considerable complexity and fine-tuned properties. This development has been reviewed in detail.[3]

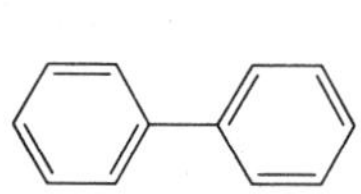

(29)

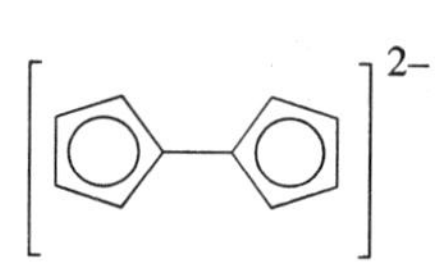

(30)

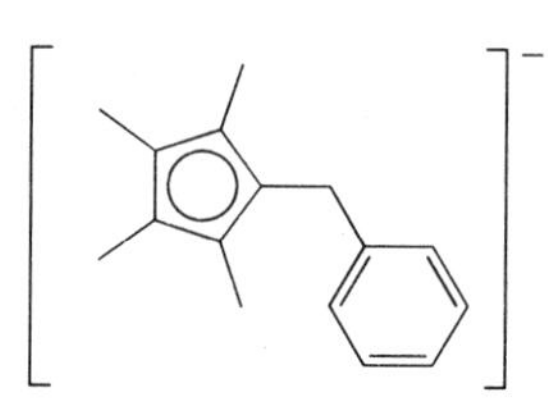

(31)

The electron count for oligodecker structures is even less strict than for tripledecker systems. A '12*n* + 6' rule where *n* is the number of metal centres has been put forward as each added stack adds 12 electrons to the electron count.[3,25] Thus, a quadrupledecker complex would require 42 electrons as realized in the diamagnetic complexes 'MnFeMn' (**21**) or 'CoRuCo' (**22**). However, the family of quadrupledecker complexes $[M\cdot(\mathbf{32})_2]$ (**33**) (M = Cr, Mn, Fe, Co, Ni, Cu, Zn) with 40–46 electrons establishes an isolobal analogy to the 3*d*-metal metallocenes. With the exception of the zinc compound these quadrupledecker complexes are paramagnetic.[2,26] In terms of ligand field theory, the anion (**32**) is a weaker ligand than Cp^- which allows high-spin behaviour in the case of manganocene only. Thus, the '12*n* + 6' rule has no meaning here. The bonding situation in the series of quadrupledecker complexes (**33**) has also been analysed by extended Hückel methods.[10]

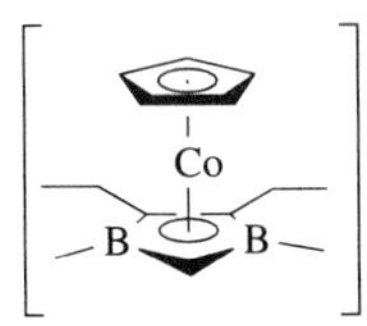

(**32**)

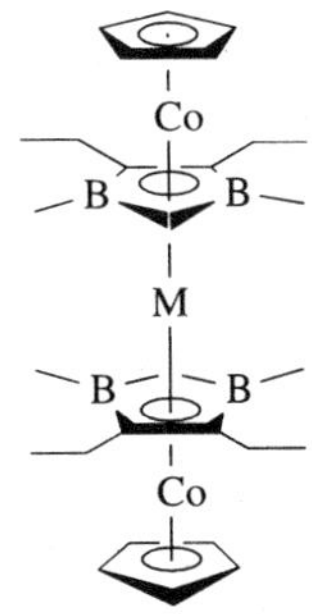

(**33**)

5.2 SYSTEMS WITH BORON–CARBON RING LIGANDS

The basic structural motif of this chapter, the coordination of a simple boraethene $CH_2{=}BH$ (**34**) or borataethene $[CH_2{=}BH_2]^-$ (**35**) to a transition metal, is still waiting to be realized. However, complexes derived from the closely related 1,2-azaboraallene $CH_2{=}B{=}NH_2$ (**36**) have been described.

$H_2C{=}BH$ (**34**) $[H_2C{=}BH_2]^-$ (**35**) $H_2C{=}B{=}NH_2$ (**36**)

The 9-fluorenylidene (2,2,6,6-tetramethylpiperidino)borane (**37**) is the first kinetically stabilized 1,2-azaboraallene. In spite of the steric crowding encountered in (**37**), this compound can directly be added to 16 electron fragments; for instance, addition to $Fe(CO)_4$ gives the η-(C,B) complex (**38**).[27–28] Analogous complexes with the 16 electron fragments $Cr(CO)_2(C_6H_6)$, $Mn(CO)_2(C_5H_4Me)$, $Fe(CO)_3L$ (L = e.g., $P(OMe)_3$, $SbPh_3$) and Co(CO)Cp have also been described.[29] Generally the observed metal–boron bond lengths are short (e.g., Fe–B 0.2125(5) nm for (**38**)) and the B–N distances indicate strong B–N π-bonding (B–N 0.1377(6) nm for (**38**)).[27,29]

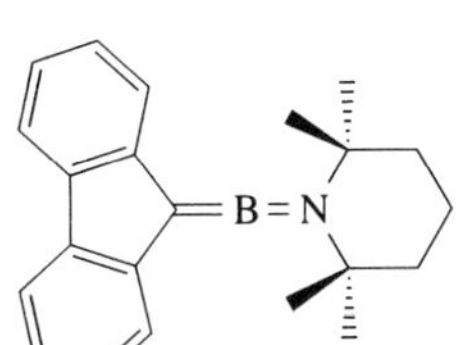

(**37**)

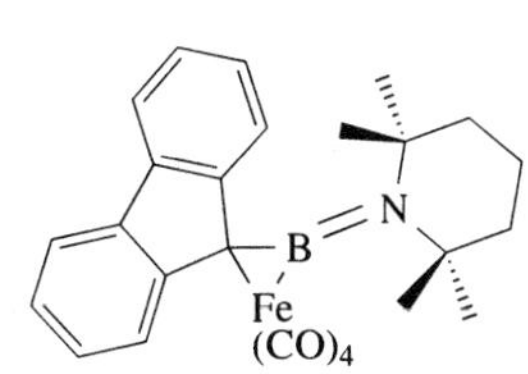

(**38**)

5.2.1 Complexes of Borirenes

The first borirene complex $[(2\text{-}MeC_2BH_2)Fe_3(CO)_9]$ (**39**) was found as an unexpected by-product from the reaction of $[Fe(py)_6][Fe_4(CO)_{13}]$ with $BH_3{\cdot}SMe_2$.[30] It can be described as 2-methylborirene facially bonded to a triangular Fe_3 face of a $[Fe(CO)_3]_3$ fragment or alternatively as a *closo* cluster electronically analogous to $B_6H_6^{2-}$.

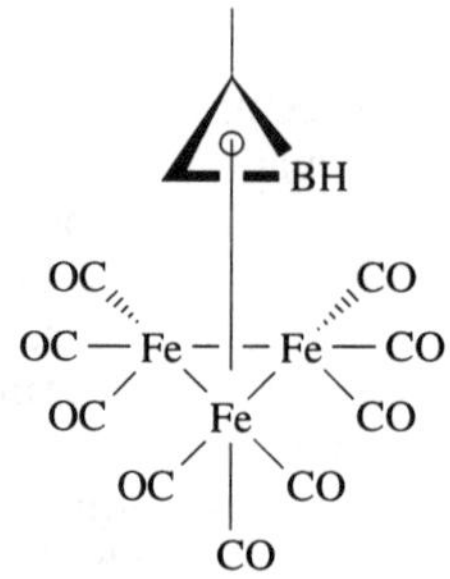

(39)

5.2.2 Complexes of Five-membered Rings

5.2.2.1 Complexes of C_4B rings

Although the first complex of 1*H*-borole, the tripledecker complex [{μ-(2-Et-C_4H_3BPh)}-{$Mn(CO)_3$}$_2$],[31] was discovered as early as 1976, an efficient and versatile entry into this field was only developed in the years after 1982. Meanwhile, complexes of all metals from the vanadium group to the nickel group (with the exception of technetium) have been described and a review is available.[1]

The starting point is the simple synthesis of 2,5-dihydro-1*H*-boroles (the '3-borolenes') (**40**) from magnesium–butadiene and organoboron dihalides (Scheme 1).[32] Catalytic isomerization of the 3-borolenes gives the less reactive 2-borolenes (**41**).[32] The 1-(dialkylamino) derivatives of the borolenes can be lithiated to give the dilithio derivatives Li_2·(**42**) (Scheme 2).[33–34]

(40) **(41)**

R = Me, Ph, NMe_2, NEt_2; X = Cl, Br

Scheme 1

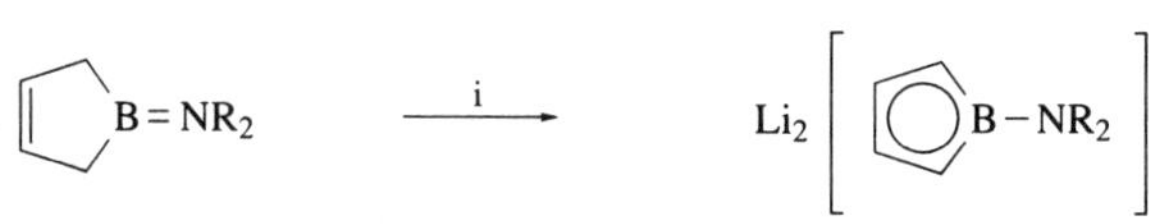

Li_2•(**42**)

i, LiTMP for R = Me, Et; Bu^{t}Li for R = Pr^i
(**a**) R = Me; (**b**) R = Et; (**c**) R = Pr^i

Scheme 2

Complexes of 1*H*-boroles can be made by essentially three methods: via the dianions $(\mathbf{42})^{2-}$, from borolenes by dehydrogenating complex formation and via borole–ammonia adducts. In older work, Li_2·(**42c**)[33] was treated with metal halides, in some cases in the presence of CO, to give diisopropylamino compounds as, for example, [$Cr(CO)_2L_2$], [$Mn(CO)L_2$], [$Fe(CO)_3L$], [$Fe(CO)L_2$], [Co_2L_3] and NiL_2 with L = $C_4H_4BNPr^i_2$ (**42c**).[35] These compounds are inert with respect to substitution at boron and form ammonio derivatives on protonation (Scheme 3).[36]

When the less sterically hindered dimethylamino and diethylamino derivatives [Li_2(TMEDA)·(**42a**)] and [Li_2(TMEDA)·(**42b**)][34] are used, subsequent substitution allows modification of the substituent at boron by standard methods. This was demonstrated for the nickel complexes [$Ni(C_4H_4BR)_2$] (e.g., R = H, F, Cl, Br, I, OMe, NEt_2, Me, Ph).[37]

Dehydrogenating complex formation takes place when borolenes are treated with suitable substrates; in the course of these reactions the borolene is in part dehydrogenated to give the borole–metal complex and in part consumed by the transferred hydrogen. Suitable substrates are (i) noble-metal compounds with hydrogen transferring ability (e.g., [{$RuCl_2(C_6Me_6)$}$_2$], [$RuCl_2(PPh_3)_3$], [$OsCl_2(PPh_3)_3$], [{$RhCl(C_2H_4)_2$}$_2$], [$RhCl(PPh_3)_3$]);[38] (ii) the metal carbonyls [$V(CO)_6$],[14] [$Mn_2(CO)_{10}$],[39] $Fe(CO)_5$ under

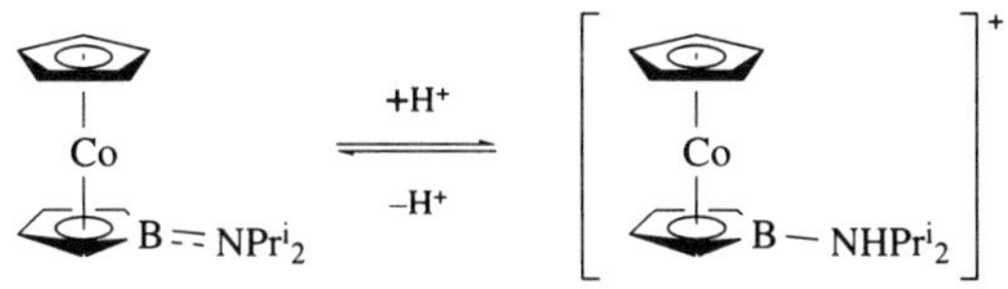

(43) **(44)**

Scheme 3

irradiation,[39] $Ru_3(CO)_{12}$,[38] $Os_3(CO)_{12}$[38] and $Co_2(CO)_8$;[39] and (iii) $NiCp_2$.[40] Although, in most cases, the products are straightforward mononuclear 18 electron complexes or 34 electron metal–metal bonded dimers or 30- electron tripledecker complexes, the system $NiCp_2/C_4H_6BPh$ is somewhat unusual (Scheme 4). In this case the reaction gives a primary product, the blue compound (**45**) with a boracyclopentenyl ligand and then the 34-electron tripledecker complex (**46**); at rather drastic temperatures, decomposition of (**46**) produces the bisligand–nickel sandwich complex (**47**).

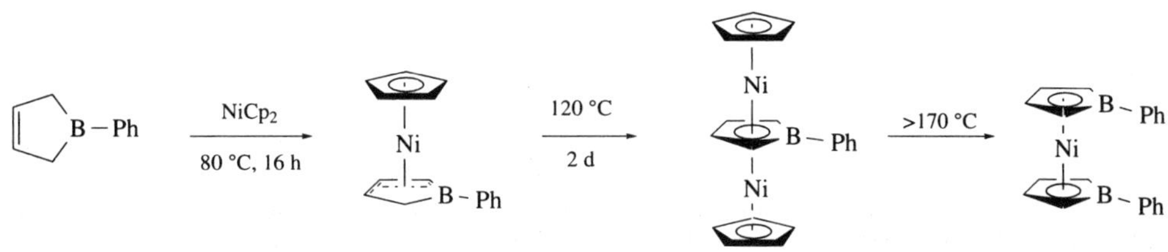

(45) **(46)** **(47)**

Scheme 4

Borole–ammonia adducts (**48**) have been obtained in solution from an ammonia degradation of $[CoI(CO)_2L]$ ($L = C_4H_4BR$, R = Me, Ph).[41] These adducts may serve as a source of the borole ligand when mild reaction conditions are feasible, and complexes of type $[M(CO)_4L]$ (M = Cr, Mo, W) (**49**),[41] [M(cod)L] (**50**) and $[ML_2]$ (**51**) (M = Ni, Pd, Pt)[42] have been obtained in this way.

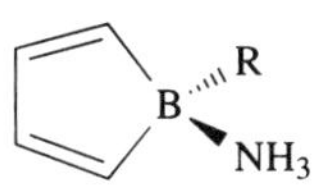

(48) R = Me, Ph

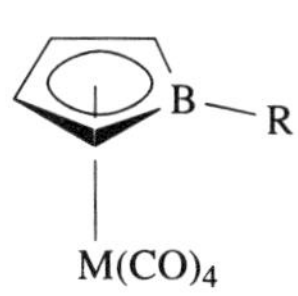

(49) M = Cr, Mo, W

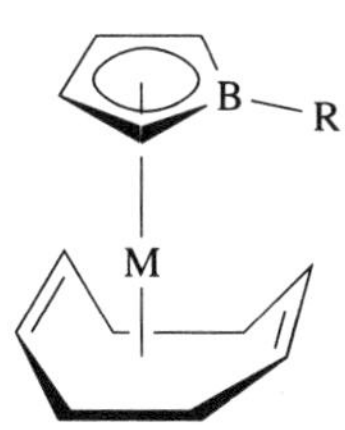

(50) M = Ni, Pd, Pt

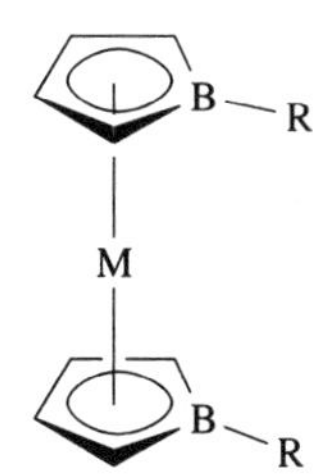

(51) M = Ni, Pd, Pt

Although the above syntheses add the metal to a preformed C–B skeleton one case has been found where the boron is added to an organometallic substrate. Heating the cobaltole (**52**) with $BH_3 \cdot THF$ (or better with $BHCl_2 \cdot SMe_2$) gives *inter alia* the borole–cobalt complex (**53**).[43]

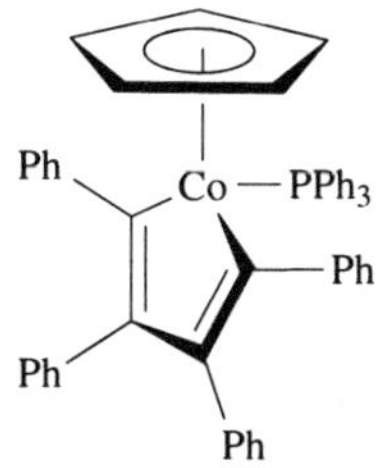

(52)

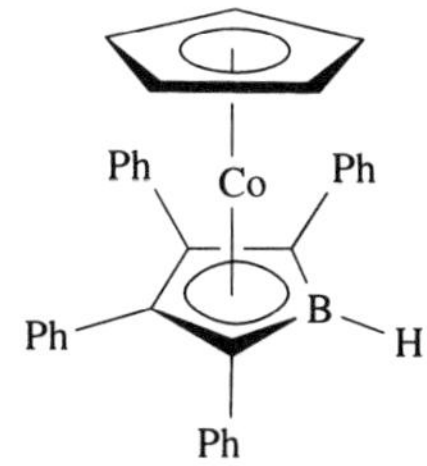

(53)

On the basis of qualitative MO considerations the borole ligand can be compared with diene ligands and with the cyclopentadienyl ring. It possesses a low-lying LUMO making it a stronger acceptor than a diene and the interaction with the metal is more covalent than in the cyclopentadienyl case, in agreement with spectroscopic evidence.[19,39]

Structurally the borole metal interaction shows a characteristic slip distortion; the projection of the metal onto the (best) plane through the ring carbons does not coincide with the geometric centre of the ring but is shifted away from the boron. The ligand is also slightly folded along C-2/5 with the boron bending away from the metal. For instance, for [Fe(CO)$_3$(C$_4$H$_4$BPh)] the slip amounts to 0.0069 nm and the folding angle is 6.1°.[39] Heterosubstituents at boron which allow π-interactions with the boron increase the energy of the borole LUMO; they stabilize the free 1*H*-borole and destabilize its complexes. This effect is particularly large for dialkylamino substituents. Thus, the complexes *cis*-Cr(CO)$_2$(C$_4$H$_4$BNPr$^i_2$)$_2$ and Mn(CO)(C$_4$H$_4$BNPr$^i_2$)$_2$ show atypically large slip distortions (0.0195 nm and 0.0136 nm) and folding angles (14.4° and 16.4°) and only weak direct metal–boron interactions; the aminoborole ligand has become diene-like.[44]

Formally replacing a carbon atom in a Cp ring with a boron atom causes increased barriers to internal rotation and pronounced preferential conformations. For some bisborole–nickel complexes restricted rotation could be observed by NMR spectroscopic methods; for example, the Gibbs activation energy, $\Delta G^{\ddagger}(170\ \mathrm{K}) = 32 \pm 4\ \mathrm{kJ\ mol^{-1}}$ for [Ni(C$_4$H$_4$BPh)$_2$].[42] Again amino substituents render the borole ring more diene-like and the barrier to internal rotation is increased; for example, $\Delta G^{\ddagger}(230\ \mathrm{K}) = 50 \pm 3\ \mathrm{kJ\ mol^{-1}}$ for [Ni(C$_4$H$_4$BNMe$_2$)$_2$].[37]

In general, borole–metal complexes are thermally much more robust than comparable diene complexes; this is illustrated by the relatively high decomposition temperature of 250 °C of the palladium(0) complex [Pd(C$_4$H$_4$BPh)$_2$].[42] Their redox potentials show drastic anodic shifts when compared with isoelectronic cyclopentadienyl complexes or even when compared with cyclobutadiene complexes of the same type; for instance, the complex [Ni(C$_4$H$_4$BMe)$_2$] (isoelectronic with dimethylferrocene) shows a reversible reduction at −1.49 V and an irreversible oxidation at 1.39 V vs. SCE in CH$_2$Cl$_2$.[37] Borole ligands also have a remarkable ability to stabilize anionic complexes as, for example, (**54**) and (**55**);[45] the anions (**55**) are stable even in neutral aqueous solution.[19]

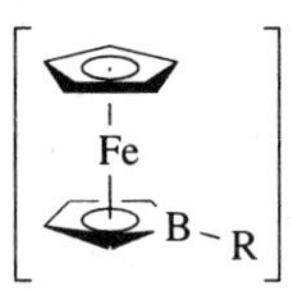

(54) R = Me, Ph

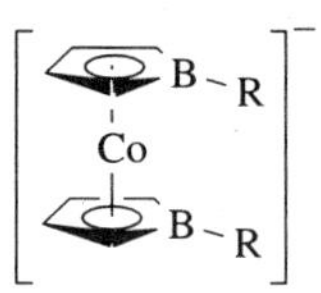

(55)

Electrophilic attack usually takes place at the α-position with respect to the boron. Thus, the sandwich complex [Co(C$_4$H$_4$BPh)Cp] (**56**) undergoes fast, fully regioselective proton–deuterium exchange with CF$_3$CO$_2$D/CDCl$_3$ at ambient temperature to give the 2,5-dideutero derivative, and a Friedel–Crafts acetylation with MeCOCl/SnCl$_4$/CH$_2$Cl$_2$ to produce the 2-acetyl derivative.[19] The isoelectronic 'borataferrocene' ion (**54**) (R = Ph) is protonated at the iron centre to give an iron hydride (**57**) with a weak Fe···H···B three-centre two-electron (3c–2e) interaction and with MeI/NaH alkylation takes place, again in the α-position, to give the 2,5-dimethyl derivative of (**57**).[45]

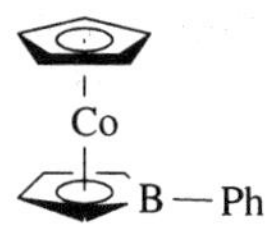

(56)

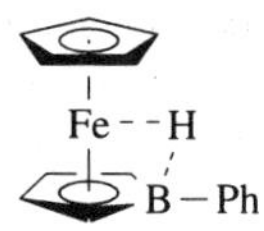

(57)

The anions (**54**) and (**55**) (and other similar anions) readily undergo stacking reactions with a large variety of metalloelectrophiles (e.g., [Cr(CO)$_3$(NCMe)$_3$], [{RhCl(cod)}$_2$]) to give tripledecker complexes.[19,46] The quadrupledecker complex (**22**) has also been made in this way.[19] When the stacking reaction can be combined with a nucleophilic degradation of different regiochemistry the result may be a borole transfer reaction. This aspect of tripledecker reactivity has been discussed in a conference report;[47] the transfer of a borole ligand from (**54**) via the tripledecker complex 'ReFe' (**58**) to an otherwise inaccessible (borole)rhenate (**59**) illustrates the principle (Scheme 5).[48]

5.2.2.2 Complexes of C$_3$B$_2$ rings

The chemistry of the 1,3-C$_3$B$_2$ ring systems with nonadjacent boron atoms is particularly rich, whereas the 1,2-C$_3$B$_2$ ring with adjacent boron atoms still presents largely unsolved synthetic problems. The attempted preparation of a 1,2-diborolenyl complex from a 4-methylene-1,2-diborolane (**60**) and

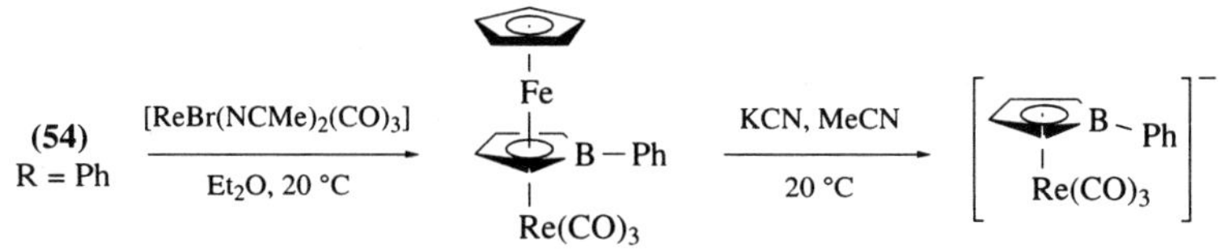

(58) (59)

Scheme 5

$[CpCo(C_2H_4)_2]$ resulted in cleavage of the B–B bond and gave a fluxional boratabutadiene complex (**61**); the corresponding 3-methylene-1,2-diborolane reacted analogously.[49]

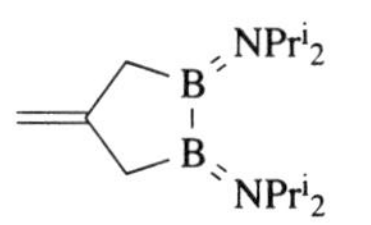

(**60**)

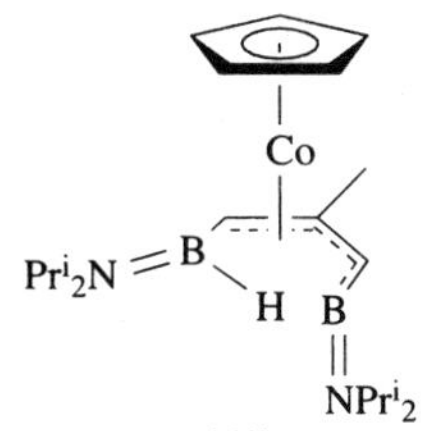

(**61**)

A review on complexes of 2,3-dihydro-1*H*-1,3-diboroles (1,3-diborolenes) (**62**) and 2,3-dihydro-1*H*-1,3-diborolyl (1,3-diborolenyl) ligands (**63**) acting as ligands to Fe, Co, Ni, Rh and Pt is available,[2] and more detailed accounts are concentrated in two important papers.[13,50] In earlier work the 1,3-diborolene (**62a**) had been used as starting material. In thermal reactions one hydrogen atom of (**62a**) is lost and complexes of the 1,3-diborolenyl (**63a**) were obtained. This three electron ligand may be bonded monofacially in sandwich and half-sandwich complexes or, more frequently, acts as a bifacially bonded, bridging or central ligand in tripledecker complexes.

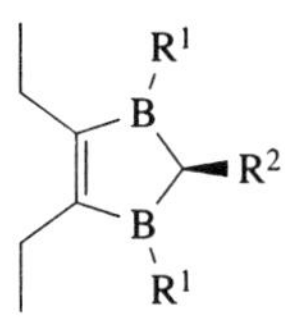

(**62**)

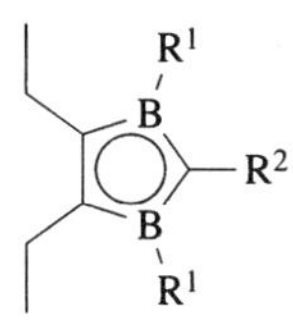

(**63**)

(a) R^1 = Et, R^2 = Me
(b) R^1 = Me, R^2 = H

Under mild reaction conditions, earlier stages of the complex formation may be observed. When the 1,3-diborolene (**62b**) is treated with Jonas' complex $[CpCo(C_2H_4)_2]$, the red 1,3-diborolene complex (**64b**) with a nonclassical hydridic hydrogen atom is obtained.[50] The 2-methyl derivative (**62a**) possesses lower symmetry and the CoCp fragment may in principle add to one of the two diastereotopic faces of the ring. Experimentally two primary products are found. The predominant complex, the red isomer (**64a**) possesses a 2-methyl group in the pseudoequatorial *endo*-position and a hydridic hydrogen atom attached to the *exo*-face of the ligand.[2,50–51] There is a minor, less stable, yellow isomer (**65a**), which quickly rearranges to the more stable isomer (**64a**); here the hydridic hydrogen atom of the yellow isomer is presumably located at the *endo*-side of the ring and directly bonded to cobalt.[51]

The complexes (**64**) possess unusual properties. The axial hydridic hydrogen atoms are highly shielded (for (**64a**): $\delta(^1H) = -8.62$q, in C_6D_6) and C-2 shows reduced coupling (for (**64a**): $^3J(^{13}C–^1H)$ = 81 Hz). Thus, this hydrogen atom is bonded to C-2 and is at the same time engaged in 3c–2e bonding. X-ray structural work located the hydridic H atom on the *exo*-side of the ligand; however, it could not distinguish between a structure (**64a**) with the hydridic H atom in the molecular mirror plane and a structure (**64b**) with statistical disorder of two C–B bridge-protonated enantiomers. Both structures imply three-centre bonding ((**64a**): 3c–2e H···C···Co; (**64b**): 3c–2e C···H···B). Theoretical analysis by means of EH and INDO MO methods reveals two very close energy minima corresponding to the two structures discussed ((**64b**) is found 0.39 eV above (**64a**)).[50] In the light of the usual uncertainties of calculation without geometry optimization, the question, which of the two isomeric structures represents the ground state structure of (**64a,b**) seems not to be finally settled and so formula (**64**) is to be understood as provisional.

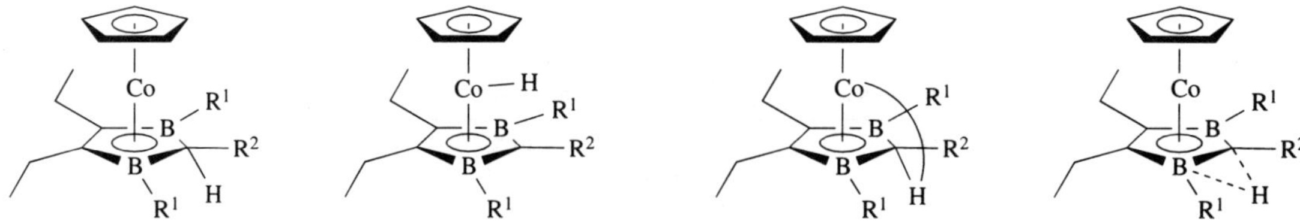

(64) **(65)** **(64a)** **(64b)**

(a) R^1 = Et, R^2 = Me; (b) R^1 = Me, R^2 = H

In case of the closely related toluene–iron complex [Fe(PhMe)·(**62a**)] the structural problem could be solved by the combined application of x-ray and neutron diffraction work. The experimental result so obtained corresponds to Structure (**64b**) discussed above. The hydridic hydrogen occupies one of the two C–B bridge positions on the *exo*-side of the 1,3-diborolene ligand. The resulting two enantiomers are disordered in the crystal giving an effective structure with one-half of a hydrogen atom above each of the two C–B bonds of the molecule.[2]

The reaction of the 1,3-diborolene (**62a**) with [Pt(cod)$_2$] provides a second example for a facial differentiation in the course of the reaction. Two products (**66**) and (**67**) were isolated from the reaction mixture.[52] The minor product (**66**) is a 16 electron complex with an η^4-bonded ligand (**62a**); the 2-methylmethylene group of the ligand is bent away from the metal with a folding angle of 17.1°. The major product (**67**) is a 16 electron complex with a η^5-bonded ligand (**63a**), this time folded towards the metal with a folding angle of 28°, and a 1,4,5-η^3-(4-cyclooctenyl) ligand. Addition of a Pt(cod) fragment to the more hindered face of (**62a**) gives the observed product (**66**), while addition to the less hindered face places the hydrogen atom 2-H into the vicinity of the coordinatively unsaturated platinum which mediates a transannular hydrogen shift to the cod counter-ligand to give the observed product (**67**) (Scheme 6).[52]

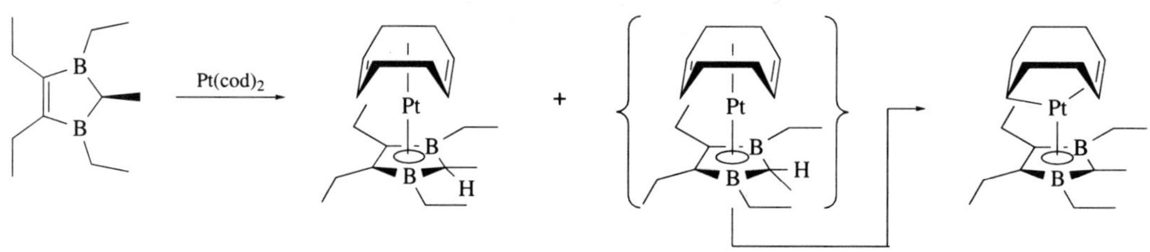

(62a) **(66)** **(67)**

Scheme 6

The axial hydrogen of the cobalt sandwich complexes (**64a,b**) is markedly acidic. These complexes (**64a,b**) react with elemental potassium in THF to give H_2 and the potassium salts K[(**68a,b**)].[50] The 18-electron anions (**68a,b**)$^-$ and Cp^- are isolobal, and consequently the reactivity of these anions is in many respects analogous to that of Cp^-. They undergo protonation, deuteration and methylation at C-2 to give products of type (**64**).[50] Treatment with [AuCl(PPh$_3$)] produces gold derivatives (**69a,b**). In the structure of (**69b**) the Au(PPh$_3$) group is found above C-2 of the C_3B_2-ring with a slight inward tilt (7.9°). This corresponds to a strong Au–C-2 bond (0.218(1) nm) and two weak Au–B interactions (0.249(1) nm and 0.253(1) nm) and the connectivity is intermediate between η^1- and η^3-bonding.[53] Structures with η^5-bonded metals were found in the thallium tripledecker complex [Tl{$Me_2C_2(BMe)_2CH$}CoCp] (**70**),[54] in the bent tin quadrupeldecker complex [Sn·(**68b**)$_2$],[55] and in the quadrupledecked metallocene analogues (**33**) mentioned earlier.

As an alternative synthetic method, mostly entropy-driven thermal reactions were used to synthesize a wide variety of tripledecker complexes,[13] and later of more extended structures. The reaction of the 32-electron tripledecker complex (**71**) with [Fe$_2$(CO)$_9$] may serve as a more recent example (Scheme 7).[56] Upon heating in toluene two parallel reactions take place. Replacement of the NiCp fragment with a Fe(CO)$_3$ fragment gives the labile 31-electron tripledecker complex (**72**) and insertion of a Fe(CO)$_3$ fragment into the Ni–Cp bond produces the trinuclear complex (**73**) (subsequently (**73**) undergoes a slow dismutation).

A practical synthesis of the 2,3-dihydro-1*H*-benzo[d]-1,3-diboroles (**74**) has been developed.[57] They readily form complexes of the related benzoborolyl ligand (**75**) as, for example, the 31-electron tripledecker complex [(**75b**)·(CoCp)$_2$][57] and quadrupledecker complexes such as the diamagnetic 46-electron complex [Ni{(**75b**)·(NiCp)}$_2$].[58]

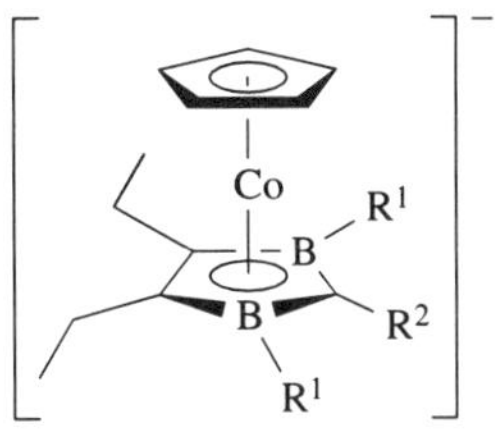

(68)

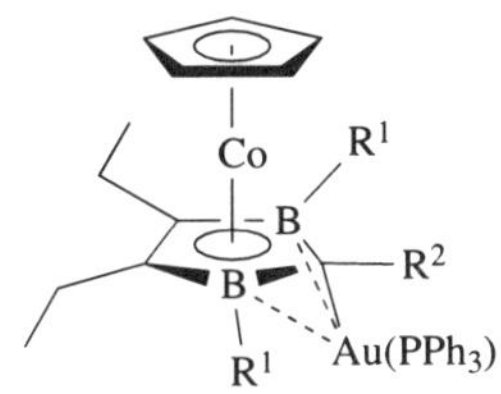

(69)

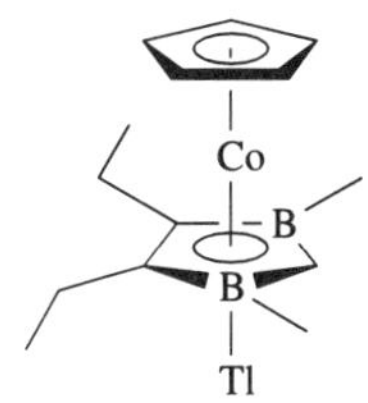

(70)

(a) R^1 = Et, R^2 = Me; (b) R^1 = Me, R^2 = H

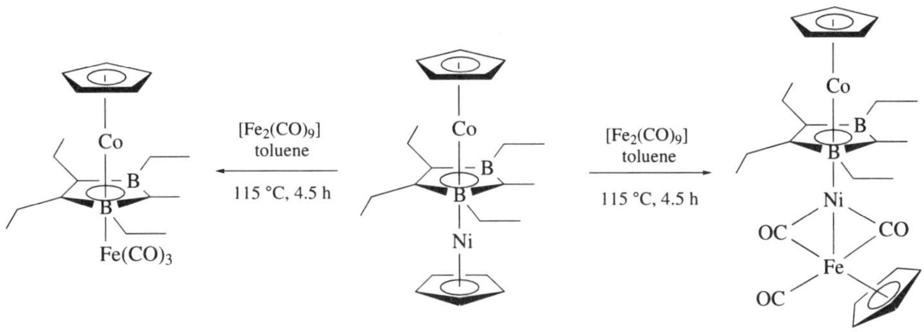

(72) **(71)** **(73)**

Scheme 7

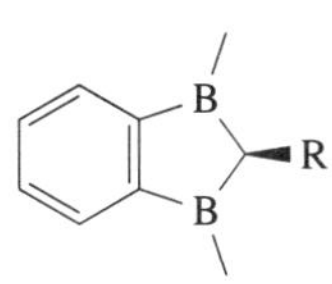

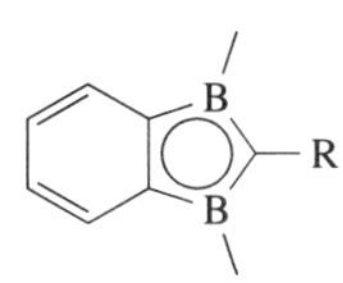

(74) **(75)**

(a) R = Me; (b) R = H

5.2.2.3 Complexes of C_2B_3 rings

The two isomers (**3**) and (**12**) of the five-membered ring system $C_2B_3H_5$ have been known since the early 1970s. Although the chemistry of the 1,2,4-isomer (**12**) has remained very limited, the study of complexes of the 1,2,3-isomer (**3**) has seen much progress in recent years.[3]

The synthetic approach to complexes of the 1,2,3-$C_2B_3H_5$ ligand has to be rather indirect as the highly charged anions $[C_2B_3H_5]^{4-}$ are unknown. The general approach is illustrated in Schemes 8 and 9. Pentaborane(9) (readily available in the USA)[3] reacts with alkynes in the absence of a solvent to give the *nido*-carbaboranes *nido*-2,3-$R_2C_2B_4H_6$ (**76**) (R = e.g., Et, $PhCH_2$). These can be deprotonated to produce monoanions $(\mathbf{77})^-$ or dianions $(\mathbf{78})^{2-}$ (Scheme 8). These anions, either alone or in combination with carbanion reagents such as NaCp, LiC_5Me_5 or K_2cot, react with suitable halides to produce sandwich-like *closo*-metallacarbaboranes $[(R_2C_2B_4H_4)ML]$ (**79**) where ML is usually a 14-electron fragment. Details of this chemistry can be found elsewhere in this work.

The apex of the *closo*-metallacarbaboranes $[(R_2C_2B_4H_4)ML]$ (**79**) can readily be removed by base in the presence of a proton donor; TMEDA in the presence of a small quantity of water is both efficient and convenient. This decapitation provides *nido*-metallacarbaboranes $[(R_2C_2B_3H_5)ML]$ (**80**) which in turn can be deprotonated to give monoanions $(\mathbf{81})^-$ or dianions $(\mathbf{82})^{2-}$; these are now complexes of the 1,2,3-$C_2B_3H_5$ ring (**3**) (Scheme 9).[3] This reaction sequence has mainly been realized and developed for ML = Fe(1,3,5-C_8H_{10})[59] and (indirectly) Fe(arene),[59] Ru(arene),[60] $CoCp$[60] and $CoCp^*$,[60–61] and $RhCp^*$.[62]

The process of electrophilic stacking can now be applied to the anions of type $(\mathbf{81})^-$ and $(\mathbf{82})^{2-}$ to produce tripledecker complexes with bridging $R_2C_2B_3H_3$ ligands. Further variability is achieved by means of mixed-ligand syntheses with carbanion reagents or with the *nido*-carbaborane anions $(\mathbf{77})^-$ or dianions $(\mathbf{78})^{2-}$. In the latter case the process of decapitation and electrophilic stacking can be repeated.

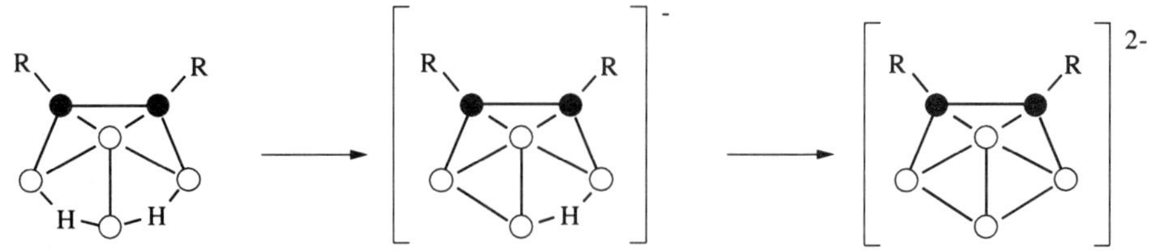

(76) **(77)$^-$** **(78)$^{2-}$**

○ = BH, ● = C

Scheme 8

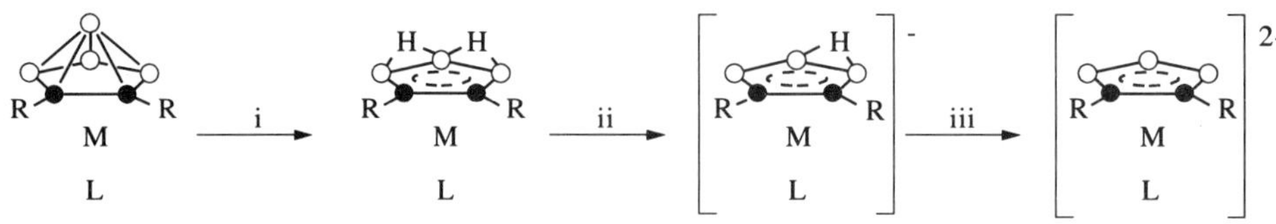

(79) **(80)** **(81)$^-$** **(82)$^{2-}$**

i, TMEDA (H_2O), 20 °C; ii, NaH, THF; iii, LiBu, THF, 0 °C

Scheme 9

Many of the new tripledecker and quadrupledecker complexes with 1,2,3-$C_2B_3H_5$ (**3**) and likewise 1,3-$C_3B_2H_5$ ligands (**2**) are chemically remarkably robust. They usually show several, often reversible redox processes with robust mixed-valence intermediates. This fosters the idea that extended structures (cf. Section 5.1.3) of oligodecker units interconnected by σ-linkages could be valuable new materials, possibly with interesting conducting properties.[3] The methodology to build such systems is at hand.

5.2.2.4 Complexes of diborafulvene systems

Complexes of two isomeric diborafulvenes have been made, and the doubly opened tripledecker complexes (**19**) and (**20**) (Section 5.1.2) are especially interesting representatives. The dihydrodiborafulvene ligand (**83**) forms complexes [$Cr(CO)_4$·(**83**)],[63] [$Mo(CO)_4$·(**83**)] (**84**),[63] [$Fe(CO)_3$·(**83**)][64] and [Ni·($\mathbf{83}$)$_2$].[63] The *exo*-methylene of the ligand in the molybdenum complex (**84**) is bent towards the metal, thereby coming into closer contact with the metal. Furthermore, the *exo*-methylene group seems to be polar and certainly highly polarizable. The chromium complex readily adds PEt_3 to give a zwitterionic adduct [$Cr(CO)_4${η-$Et_2C_2(BMe)_2C$–CMe_2PEt_3}] with a phosphoniomethyl substituent and an anionic chromium centre. Similarly the benzofulvene (**85**) forms a complex (**86**) with one $Fe(CO)_3$ group[65] and the previously mentioned complex (**20**) with two $Fe(CO)_3$ groups.[16] Complex (**86**) shows extremely unsymmetric bonding of the iron atom to the *exo*-methylene group (Fe–C-2 0.2121(5) nm, Fe–C-6 0.2817(5) nm).[65] The bonding situation in (**86**), (**20**) and related compounds has been analysed in detail.[16–17]

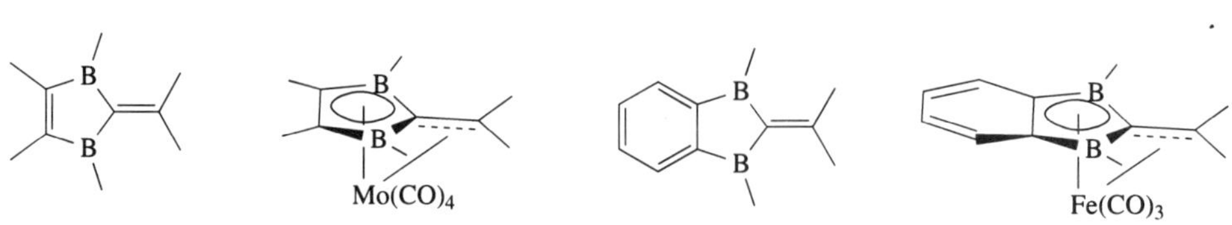

(83) **(84)** **(85)** **(86)**

5.2.3 Complexes of Six-membered Rings

5.2.3.1 Complexes of C_5B rings

Complexes with borabenzene and boratabenzene ring ligands have been reviewed in detail.[66] It has been shown that the 19 electron complex [CpFe(C_6Me_6)] reacts with organoboron dihalides to give a new class of 19 electron complexes [Fe(C_5H_5BR)(C_6Me_6)] (R = Me, Ph, NEt_2, OH) (**87**); products from the corresponding ring expansion of the arene ring could not be detected.[67] The chemistry of the complexes (**87**) resembles that of cobaltocene closely.[67]

The attack of metalloelectrophiles on the sandwich complex [(Cp*Ru(C_5H_5BMe)] has been studied. With the ruthenium-electrophile obtained from [(Cp*RuCl)$_4$] and $AgCF_3SO_3$ in acetone the first tripledecker complex with a bridging borabenzene ligand (**88**)·$CF_3SO_3^-$ has been obtained and structurally characterized.[8] With the rhodium-electrophile [Cp*Rh($OCMe_2$)$_3$]$^{2+}$ a labile tripledecker dication (**89**) is formed which is prone to nucleophilic degradation; the resulting boratabenzene transfer reaction is from ruthenium to rhodium and gives the rhodicenium analogue [Cp*Rh(C_5H_5BMe)]$^+$.[8]

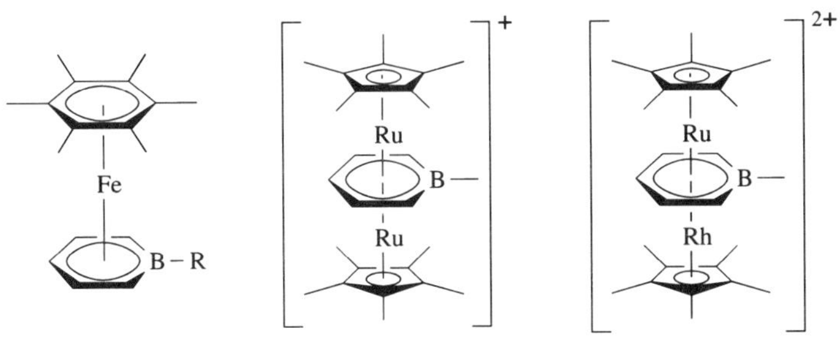

(**87**) (**88**) (**89**)

5.2.3.2 Complexes of C_4B_2 rings

Complexes of 1,2-dibora-3,5-cyclohexadienes have been obtained from the 1,2-diboratabenzene salt [Li(TMEDA)]$_2$[$C_4H_4B_2(NMe_2)_2$] and transition metal halides.[9] Examples are the rhodium sandwich complex (**90**),[68] the rhodium tripledecker complex (**91**)[9] and the compounds [(C_6Me_6)RuL], [NiL_2], [Fe(CO)L_2] and [$Co_2(CO)_4L_2$] with L = $C_4H_4B_2(NMe_2)_2$.[68] The dimethylamino substituent can be substituted by some Lewis-acidic reagents. Thus, the rhodium complex (**90**) reacts with BCl_3 to produce the dichloro derivative (**92**), while more basic reagents such as LAH and Al_2Me_6 induce a remarkable ring contraction to give borole complexes.[68]

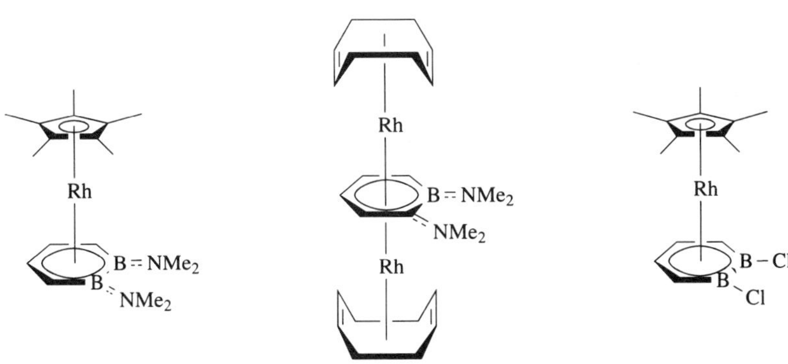

(**90**) (**91**) (**92**)

Only one attempt to synthesize complexes with a 1,3-diborabenzene ring system has been reported so far. Metal-assisted ring opening of the bicyclic $C_4B_2H_6$-isomer (**93**) produced an oily $Fe(CO)_3$ complex (**94a**) and after a substitution step a crystalline PMe_3 derivative (**94b**).[69] The structure of (**94b**) shows gross distortions of the ring. The two boranediyl groups are bent away from the metal resulting in rather long Fe–B separations (0.2457(3) nm) and the bonding is close to an $\eta^{3:1}$-allyl-yl situation.[69]

A new approach to the chemistry of 1,4-diborabenzene ligands has been developed. The 1,4-dibora-2-cyclohexenes (**95**) may undergo dehydrogenative complex formation when treated with hydrogen

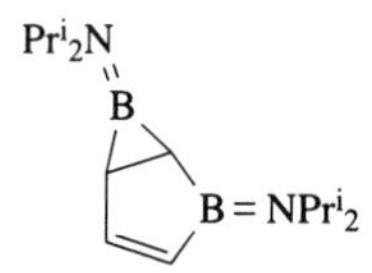

(93)

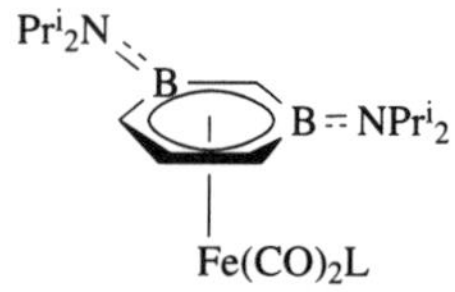

(94)
(a) L = CO; (b) L = PMe$_3$

transferring substrates. Condensation of Fe(g.) into toluene and subsequent addition of (**95b**) results in the formation of the sandwich complex (**96**) and heating (**95a**) with [{Ni(CO)Cp}$_2$] produces the 'NiNi' tripledecker complex (**97**), one of those rare examples of a 34-electron tripledecker sandwich complex.[70]

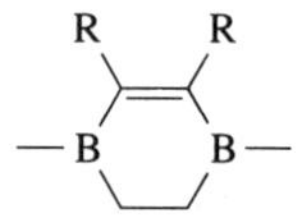

(95)
(**a**) R = Me
(**b**) R = Et

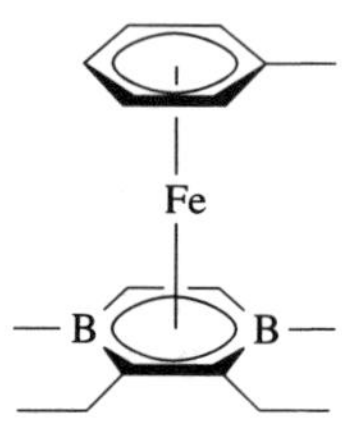

(96)

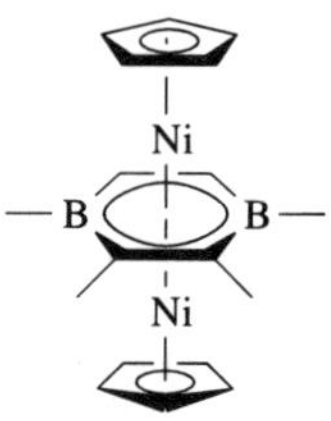

(97)

1,4-Diboranaphthalene (**98**) is a stable heterocycle which may act as a four-electron ligand. For instance, (**98**) forms complexes [Fe(CO)$_3$·(**98**)],[65] [(**98**)·(FeCp)$_2$],[71] [Ni(cod)·(**98**)][20] and [Ni·(**98**)$_2$].[20] The metal in these complexes is more tightly bonded to the ethene moiety than to the benzene moiety of the ligand as seen from marked slip distortions (cf. Ni–C-2 0.2183(3) nm and Ni–C-4a 0.2341(3) nm for Ni(cod)·(**98**)).[20] A different bonding mode is found in the diamagnetic 44 electron tetradecker complex (**23**) where the central nickel atom is only tetrahapto-ligated to the two bridging ligands (**98**) that is, in a pseudo-square-planar situation.[20] 9,10-Dihydro-9,10-diboraanthracene (**99**) reacts with [Fe(CO)$_3$(C$_8$H$_{14}$)$_2$] to form a complex [Fe(CO)$_3$·{η^6-(**99**)}] and polynuclear products [{μ,η^4,η^6-(**99**)}{Fe(CO)$_3$}$_2$] and [{μ,η^4,η^4,η^6-(**99**)}{Fe(CO)$_3$}$_3$] with one or two additional η^4-Fe(CO)$_3$ fragments *anti* to the η^6-Fe(CO)$_3$ group of the inner C$_4$B$_2$ ring.[72]

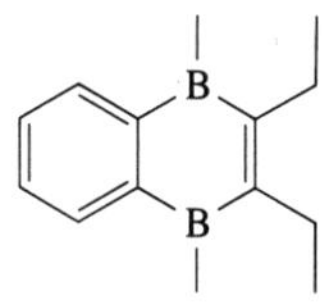

(98)

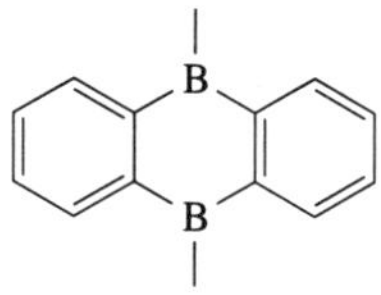

(99)

5.2.4 Complexes of Borepines

1*H*-Borepine is one of the fundamental aromatic heterocycles. An overview of earlier borepine chemistry is available.[73] Simple carbon-unsubstituted 1*H*-borepines with various substituents at boron (e.g., H, Me, Ph, NMe$_2$, OMe, Cl) have become available.[4] The first complex of a borepine was the tricarbonylchromium complex (**100**) with a fused cyclopenteno ring.[74] More recently tricarbonylmolybdenum complexes (**101**) of carbon-unsubstituted 1*H*-borepines have been described. The compounds (**101b**)[75] and (**101c**)[76] were obtained from the free borepines by treatment with [Mo(CO)$_3$(py)$_3$]/BF$_3$·OEt$_2$, while the fully unsubstituted (**101a**) was made from (**101b**) and Li[BHEt$_3$].[75] The structures of (**101b**)[75] and (**101c**)[76] prove the η^7-bonding mode of the borepine ligand. A similar benzoborepine complex (**102**) has also been made.[77]

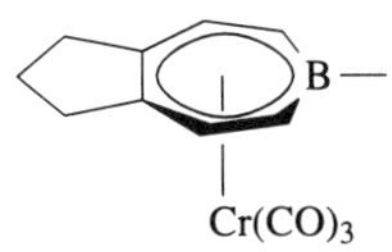

(100)

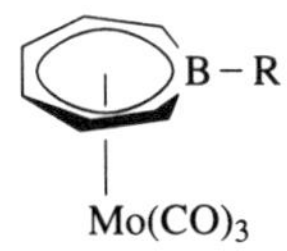

(101)
(a) R = H
(b) R = Cl
(c) R = Ph

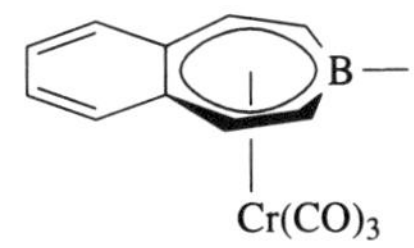

(102)

5.3 SYSTEMS WITH INORGANIC RING LIGANDS

5.3.1 Complexes of Boron–Nitrogen Ligands

The notion that BN and CC units are isoelectronic has driven much of the quest for new ligand systems. The valence electrons in a BN unit are more localized than in a CC unit. Therefore, the ensuing complexes of boron–nitrogen ligands are less stable and therefore less varied and numerous than the complexes of their carbon–carbon analogues. Three new classes of complexes have been described.[78]

Iminoboranes are able to form transition metal complexes as side-on coordinated ligands and as bridging ligands such as acetylenes. Thus, the bis(*t*-butyl)iminoborane (**103a**) reacts with [Cp_2NbH_3] to give complex (**104**) with the boron in the central position of the bent sandwich moiety and with a weak 3c–2e Nb–H→B interaction.[79] With [$Co_2(CO)_8$], a dinuclear complex (**105**) with a tetrahedrane-type structure is formed.[80]

RB≡NBu^t

(103)
(a) R = Bu^t
(b) R = Bu

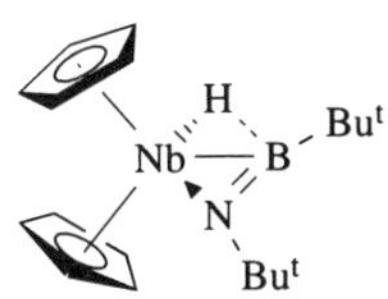

(104)

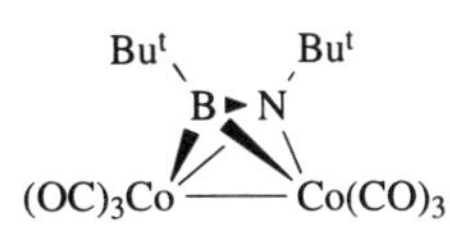

(105)

1,3,2,4-Diazadiboretidines (**106**) are heteroanalogues of cyclobutadiene. They form η^4-complexes [$ML(R^1{}_2B_2N_2R^2{}_2)$] with ML = $Cr(CO)_4$, $Mo(CO)_4$, $W(CO)_4$, $Fe(CO)_3$, CoCp and R^1 = Pr, Pr^i, Bu, Bu^t.[5] Thermally iminoboranes undergo cyclotrimerization. A catalytic cyclodimerization was observed when (**103b**) or Pr^{i}B≡NPr^i were treated with the manganese complex [$Mn(CO)_2(THF)(C_5H_4Me)$];[81] with [$Cr(CO)_5(THF)$] in THF the iminoborane (**103b**) gave the diazadiboretidine complex (**107**) in a stoichiometric cyclodimerization together with the borazine (BuBNBu^t)$_3$.[5]

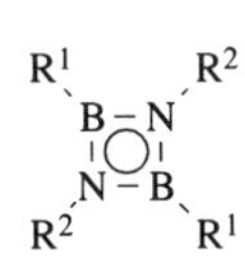

(106)

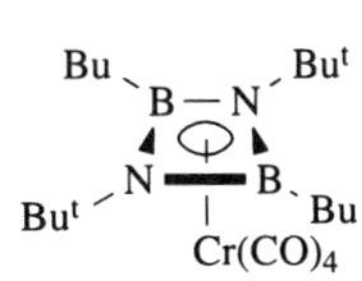

(107)

5.3.2 Complexes of Boron–Phosphorus Ligands

The first complexes of boron–phosphorus ring systems have been described. Both the 1,3,2,4-diphosphadiboretane (**108**) and the 1,3,5,2,4,6-triphosphatriborinane (**109**) may act as ligands to transition metals, and complexes [$Cr(CO)_4$·(**108**)] (**110**) and [$Cr(CO)_3$·(**109**)] (**111**) were made by conventional methods.[6] Irradiation of (**108**) in the presence of [$Fe(CO)_5$] results in stepwise complex formation, first producing an η^3-(*B, P, B*)-bonded intermediate [$Fe(CO)_4$·(**108**)] and then the final product [$Fe(CO)_3$·(**108**)].[6]

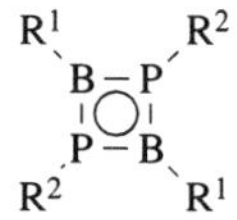

(108)

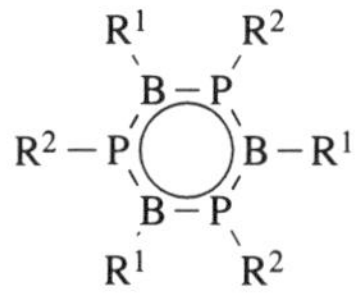

(109)

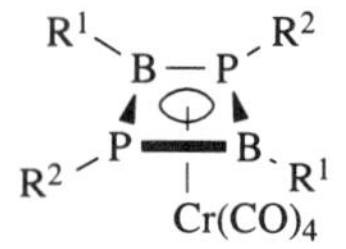

(110)

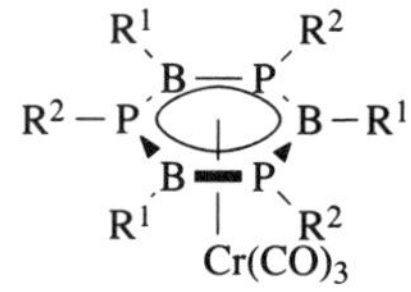

(111)

R^1 = 2,4,6-$Me_3C_6H_2$, R^2 = Bu^t

5.4 SYSTEMS WITH BORON–CARBON RING LIGANDS WITH ADDITIONAL HETEROATOMS

5.4.1 Complexes of C_3BN Rings

Formal replacement of a CC group in cyclopentadiene with a BN group produces the 2,5-dihydro-1*H*-1,2-azaboroles or 1,2-azaborolines (**112**). Lithiation of these (usually with LiTMP, lithium 2,2,6,6-tetramethylpiperide, in THF) provides 2,5-dihydro-1*H*-1,2-azaborolides or 1,2-azaborolinides $(\mathbf{113})^-$.[82] Complexes of (**112**) and many more of the ligand (**113**) have been made, mainly in the mid-1980s. This chemistry has been reviewed.[7]

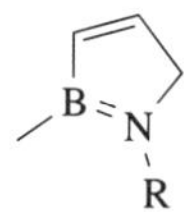

(112)

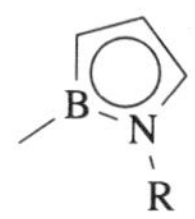

(113)

(**a**) R = Bu^t; (**b**) R = TMS

Complexes of (**113**) can be made by one of two methods: from the 1,2-azaborolines (**112**) via dehydrogenating complex formation or, more generally, by treating suitable halides with salts Li·(**113**). For instance, gentle heating of $[Co_2(CO)_8]$ with (**112a**) produces a deep red dicarbonyl $[Co(CO)_2\cdot(\mathbf{113a})]$.[83] With the milder reagent $[Co(C_2H_4)_2Cp]$ the expected primary product of such a reaction, the azaboroline complexes (**114**) can be isolated in small yields; these in turn slowly loose hydrogen to give the paramagnetic azaborolinyl complexes (**115**).[84] The lithium salt Li·(**113a**) reacts with the chromium group hexacarbonyls to give lithium salts of the complex anions (**116**).[85]

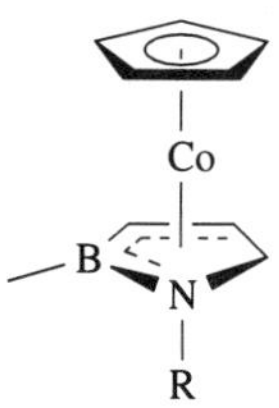

(114)

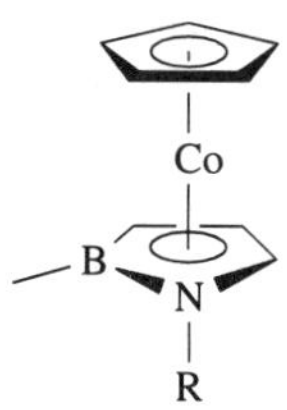

(115)

R = Bu^t or TMS

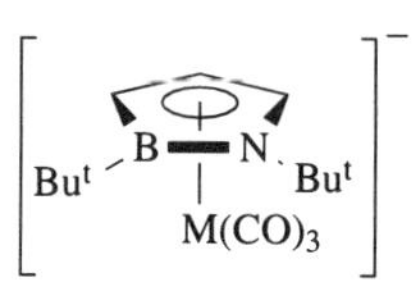

(116)

M = Cr, Mo, W

Due to the prochiral nature of the 1,2-azaborolinyl ligands (**113**) syntheses of metallocene analogues will give rise to mixtures of diastereomers.[7] In several cases pairs of *rac*- and *meso*-isomers have been structurally characterized. The racemic cobaltocene analogue *rac*-$[Co\cdot(\mathbf{113b})_2]$ shows a pronounced slip distortion (0.02 nm) with the cobalt atom shifted away from the NB-region of the ring[86] which seems to be caused by the excess electron. In case of the diamagnetic (!) nickelocene analogue *rac*-$[Ni\cdot(\mathbf{113a})_2]$, the slip distortion is even more pronounced (0.0445 nm) and the structure approaches that of a 16-electron bis(allyl)nickel derivative.[87]

The beryllocene analogue $[Be\cdot(\mathbf{113a})_2]$ (**117**) again exists as two, slowly interconverting diastereomers. For the *rac*-form, a low-temperature structure determination at −150 °C revealed an η^1,η^5-bonding situation in the crystal, while in solution the diastereomers are fluxional, each one showing only one type of azaborolinyl ring.[88] The closely related mixed compound $[CpBe\cdot(\mathbf{113a})]$ (**118**) show η^5-bonding for the Cp ring and η^1-bonding for the azaborolinyl ligand in toluene and in THF at −50 °C; at elevated temperatures THF solutions of (**118**) show a pronounced electric conductivity and their ^{1}H NMR spectra indicate extensive dissociation.[88]

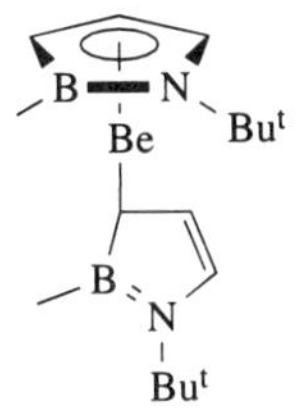

(117)

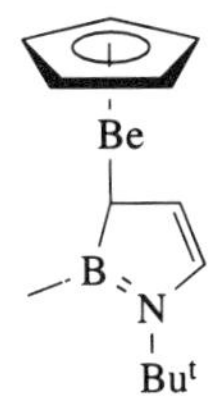

(118)

In an exploratory study, the enantiomerically pure half-sandwich complex [Fe(CO)$_2$I·(**113a**)] was made. Starting from *rac*-Na[Fe(CO)$_2$·(**113a**)] the diastereomeric menthyl ethers [Fe(CO)$_2$(CH$_2$OR)·(**113a**)] (R = (1*R*,3*R*,4*S*)-menthyl) were made and separated by HPLC chromatography. Oxidative degradation with iodine removed the chiral auxiliary and produced enantiomerically pure iodides.[89]

5.4.2 Complexes of 1,2,5-Heterosilaborolenes

Rings of type (**119**) may be considered as heterocyclic analogues of butadiene. In the cases of X = S, Se, NR^1 and PR^1- these heterocycles may act as η^4-4, electron ligands, although complexes of the 1,2,5-oxasilaborolenes could not be obtained.[90]

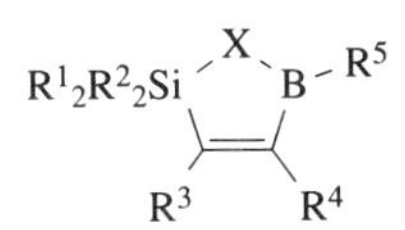

(119)
X = O, S, Se, NR^2, PR^2

The chemistry of the 1,2,5-azasilaborolenes (**120**) is most widely developed. The complexes [Cr(CO)$_4$·(**120a**)], [Mo(CO)$_4$·(**120a**)], [Fe(CO)$_3$·(**120a**)], [CpCo·(**120a**)], [CpRh·(**120a**)], the dimer [{RhCl·(**120a**)}$_2$], [IrCl(C$_2$H$_4$)·(**120a**)], *meso*-[Ni·(**120a**)$_2$] and many closely related compounds have been made.[91–92] The unsaturated side chain of the isopropenyl derivative (**120b**) can give rise to regioisomers. Structure determinations of (**120c**) with a complex substituent Ph·Cr(CO)$_3$ at the nitrogen atom (N–B 0.1447(4) nm and 0.142.7(6) nm, B–C 0.1586(5) nm and 0.1580(7) nm, C–C 0.1337(4) nm and 0.1323(7) nm and of its CoCp complex [CoCp·(**120c**)] (**121**) (N–B 0.1533(6) nm, B–C 0.1527(7) nm, C–C 0.1417(7) nm demonstrate the structural changes of the azaboradiene system brought about by complex formation with the CoCp fragment.[92]

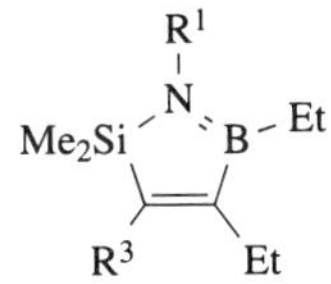

(120)

(a) R^1 = R^3 = Me
(b) R^1 = Me, R^3 = CMe=CH$_2$
(c) R^1 = Ph•Cr(CO)$_3$, R^3 = Me

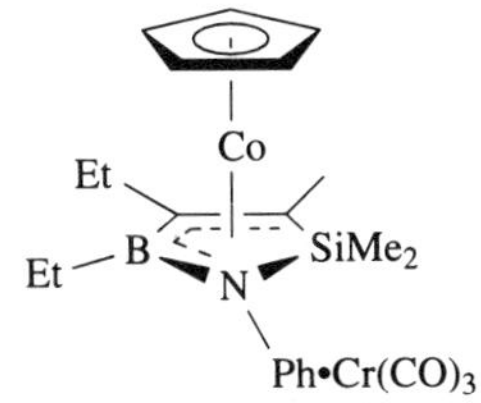

(121)

Similar complexes of 1,2,5-phosphasilaborolenes,[93] 1,2,5-thiasilaborolenes[94] and 1,2,5-selenasilaborolenes[95] have been described. In each case representative x-ray structural work has proved the η^4-bonding mode of the 1,2-heteroborabutadiene ring moiety. The complexes of the 1,2,5-phosphasilaborolenes are thermally most robust. In this case intriguing rearrangement reactions have been observed at temperatures above 150 °C. These involve an interchange of the substituents at boron and phosphorus and to a minor extent also an interchange of the two carbon-bonded substituents.[93] Mechanistic information about these rearrangement reactions is not available.

5.4.3 Complexes of 1,2,5-Phosphadiborolenes

Recent work has shown that the 1,2,5-phosphadiborolene ring system may also act as a four electron ligand. 2,5-Dihydro-1-phenyl-1*H*–1,2,5-phosphadiborole (**122**) reacts with $[Fe_2(CO)_9]$ to give a tricarbonyliron derivative (**123**).[96] In (**123**) the C_2B_2P ring skeleton is folded along the line B1,B3 by 29.6° and the bonding situation is close to η^3-bonding with very long Fe–B separations (average 0.2562 nm).[96] This structure again demonstrates the perturbing and destabilizing influence of an exocyclic *B*-amino substituent (cf. structure (**94b**))

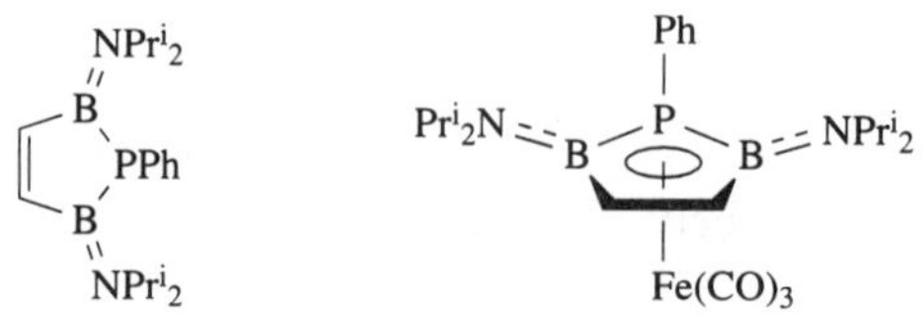

(122) **(123)**

5.4.4 Complexes of 1,2,5-Thiadiborolenes

Two important extensions to the older work on complexes of the 1,2,5-thiadiborolene ring system have appeared. Treatment of the 2,5-dihydro-1,2,5-thiadiborole (**124**) with $[NiCp_2]$ or $[\{Ni(CO)Cp\}_2]$ gives the green 19 electron sandwich complex [NiCp·(**124**)] (**125**); the conceivable 34-electron tripledecker complex [(**124**)·$(NiCp)_2$] was not found.[97] A new class of complexes [Fe(arene)·(**124**)] was made accessible by means of cocondensation of Fe(g.) with (**124**) and arenes; the benzene complex (**126**) was structurally characterized.[98]

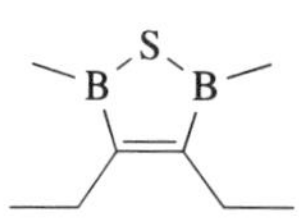

(124)

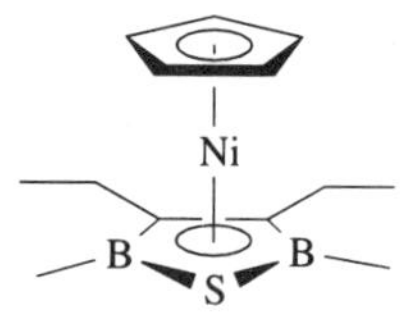

(125)

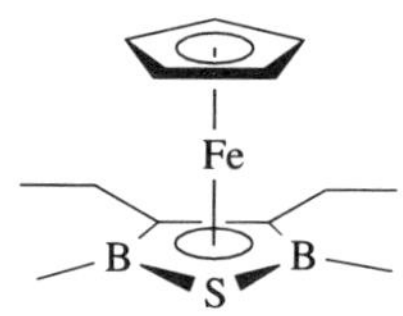

(126)

5.5 REFERENCES

1. G. E. Herberich, *Chem. Rev.*, manuscript in preparation.
2. W. Siebert, *Angew. Chem.*, 1985, **97**, 924; *Angew. Chem., Int. Ed. Engl.*, 1985, **24**, 943.
3. R. N. Grimes, *Chem. Rev.*, 1992, **92**, 251.
4. A. J. Ashe, III, W. Klein and R. Rousseau, *Organometallics*, 1993, **12**, 3225.
5. P. Paetzold, K. Delpy and R. Boese, *Z. Naturforsch. Teil B*, 1988, **43**, 839.
6. B. Kaufmann, H. Nöth, R. T. Paine, K. Polborn and M. Thomann, *Angew. Chem.*, 1993, **105**, 1534; *Angew. Chem., Int. Ed. Engl.*, 1993, **32**, 1446.
7. G. Schmid, *Comments Inorg. Chem.*, 1985, **4**, 17.
8. G. E. Herberich, U. Englert and D. Pubanz, *J. Organomet. Chem.*, 1993, **459**, 1.
9. G. E. Herberich, B. Hessner and M. Hostalek, *Angew. Chem.*, 1986, **98**, 637; *Angew. Chem., Int. Ed. Engl.*, 1986, **25**, 642.
10. E. D. Jemmis and A. C. Reddy, *J. Am. Chem. Soc.*, 1990, **112**, 722, and references therein.
11. K. Wade, *Adv. Inorg. Chem. Radiochem.*, 1976, **18**, 1.
12. J. W. Lauher, M. Elian, R. H. Summerville and R. Hoffmann, *J. Am. Chem. Soc.*, 1976, **98**, 3219.
13. J. Edwin, *et al. J. Am. Chem. Soc.*, 1983, **105**, 2582.
14. G. E. Herberich, I. Hausmann and N. Klaff, *Angew. Chem.*, 1989, **101**, 328; *Angew. Chem., Int. Ed. Engl.*, 1989, **28**, 319.
15. G. E. Herberich, C. Ganter, L. Wesemann and R. Boese, *J. Organomet. Chem.*, 1990, **394**, C1.
16. A. Fessenbecker, M. Enders, H. Pritzkow, W. Siebert, I. Hyla-Kryspin and R. Gleiter, *Chem. Ber.*, 1991, **124**, 1505.
17. Z. Nagy-Magos, A. Fessenbecker, H. Pritzkow, W. Siebert, I. Hyla-Kryspin and R. Gleiter, *Chem. Ber.*, 1991, **124**, 2685.
18. W. Siebert, C. Böhle, C. Krüger and Y.-H. Tsay, *Angew. Chem.*, 1978, **90**, 558; *Angew. Chem., Int. Ed. Engl.*, 1978, **17**, 527.
19. G. E. Herberich, B. Hessner and R. Saive, *J. Organomet. Chem.*, 1987, **319**, 9.
20. H. Schulz, H. Pritzkow and W. Siebert, *Chem. Ber.*, 1992, **125**, 993.
21. A. Fessenbecker, *et al.*, *Inorg. Chem.*, 1990, **29**, 5164.
22. M. W. Whiteley, H. Pritzkow, U. Zenneck and W. Siebert, *Angew. Chem.*, 1982, **94**, 464; *Angew. Chem., Int. Ed. Engl.*, 1982, **21**, 453.
23. T. Kuhlmann and W. Siebert, *Z. Naturforsch., Teil B*, 1985, **40**, 167.
24. T. Kuhlmann, S. Roth, J. Rozière and W. Siebert, *Angew. Chem.*, 1986, **98**, 87; *Angew. Chem., Int. Ed. Engl.*, 1986, **25**, 105.
25. W. Siebert, *Pure Appl. Chem.*, 1988, **60**, 1345.

26. W. Siebert, J. Edwin, H. Wadepohl and H. Pritzkow, *Angew. Chem.*, 1982, **94**, 148; *Angew. Chem., Int. Ed. Engl.*, 1982, **21**, 149.
27. S. Helm and H. Nöth, *Angew. Chem.*, 1988, **100**, 1378; *Angew. Chem., Int. Ed. Engl.*, 1988, **27**, 1331.
28. S. Channareddy, G. Linti and H. Nöth, *Angew. Chem.*, 1990, **102**, 222; *Angew. Chem., Int. Ed. Engl.*, 1990, **29**, 199.
29. S. W. Helm, G. Linti, H. Nöth, S. Channareddy and P. Hofmann, *Chem. Ber.*, 1992, **125**, 73.
30. X. Meng, T. P. Fehlner and A. L. Rheingold, *Organometallics*, 1990, **9**, 534.
31. G. E. Herberich, J. Hengesbach, G. Huttner, A. Frank and U. Schubert, *J. Organomet. Chem.*, 1983, **246**, 141.
32. G. E. Herberich, *et al.*, *Chem. Ber.*, 1986, **119**, 420.
33. G. E. Herberich and H. Ohst, *Z. Naturforsch. Teil B*, 1983, **38**, 1388.
34. G. E. Herberich, M. Hostalek, R. Laven and R. Boese, *Angew. Chem.*, 1990, **102**, 330; *Angew. Chem., Int. Ed. Engl.*, 1990, **29**, 317.
35. G. E. Herberich and H. Ohst, *Chem. Ber.*, 1985, **118**, 4303.
36. G. E. Herberich, M. Negele and H. Ohst, *Chem. Ber.*, 1991, **124**, 25.
37. G. E. Herberich, U. Englert, M. Hostalek and R. Laven, *Chem. Ber.*, 1991, **124**, 17.
38. G. E. Herberich, W. Boveleth, B. Hessner, M. Hostalek, D. P. J. Köffer and M. Negele, *J. Organomet. Chem.*, 1987, **319**, 311.
39. G. E. Herberich, W. Boveleth, B. Hessner, D. P. J. Köffer, M. Negele and R. Saive, *J. Organomet. Chem.*, 1986, **308**, 153.
40. G. E. Herberich, I. Hausmann, B. Hessner and M. Negele, *J. Organomet. Chem.*, 1989, **362**, 259.
41. G. E. Herberich, B. Hessner, M. Negele and J. A. K. Howard, *J. Organomet. Chem.*, 1987, **336**, 29.
42. G. E. Herberich and M. Negele, *J. Organomet. Chem.*, 1988, **350**, 81.
43. F. -E. Hong, C. W. Eigenbrot and T. P. Fehlner, *J. Am. Chem. Soc.*, 1989, **111**, 949.
44. G. E. Herberich, B. Hessner, H. Ohst and I. A. Raap, *J. Organomet. Chem.*, 1988, **348**, 305.
45. G. E. Herberich, *et al.*, *Organometallics*, 1994, **13**, 619.
46. G. E. Herberich, D. P. J. Köffer and K. M. Peters, *Chem. Ber.*, 1991, **124**, 1947.
47. G. E. Herberich, *et al.*, *J. Organomet. Chem.*, 1989, **372**, 53.
48. G. E. Herberich, B. J. Dunne and B. Hessner, *Angew. Chem.*, 1989, **101**, 798; *Angew. Chem., Int. Ed. Engl.*, 1989, **28**, 737.
49. G. Gabbert, W. Weinmann, H. Pritzkow and W. Siebert, *Angew. Chem.*, 1992, **104**, 1670; *Angew. Chem., Int. Ed. Engl.*, 1992, **31**, 1603.
50. J. Edwin, *et al. Organometallics*, 1983, **2**, 1666.
51. K. Stumpf, W. Siebert, R. Köster and G. Seidel, *Z. Naturforsch. Teil B*, 1987, **42**, 186.
52. H. Wadepohl, H. Römich, H. Pritzkow and W. Siebert, *Chem. Ber.*, 1988, **121**, 2103.
53. K. Geilich, K. Stumpf, H. Pritzkow and W. Siebert, *Chem. Ber.*, 1987, **120**, 911.
54. K. Stumpf, H. Pritzkow and W. Siebert, *Angew. Chem.*, 1985, **97**, 64; *Angew. Chem., Int. Ed. Engl.*, 1985, **24**, 71.
55. H. Wadepohl, H. Pritzkow and W. Siebert, *Organometallics*, 1983, **2**, 1899.
56. J. Edwin, M. W. Whiteley, W. Herter and W. Siebert, *J. Organomet. Chem.*, 1990, **394**, 329.
57. H. Schulz, T. Deforth and W. Siebert, *Z. Naturforsch. Teil B*, 1993, **48**, 68.
58. W. Siebert, *et al.*, *Chem. Ber.*, 1993, **126**, 1587.
59. R. N. Grimes, *Pure Appl. Chem.*, 1987, **59**, 847.
60. J. H. Davis Jr., E. Sinn and R. N. Grimes, *J. Am. Chem. Soc.*, 1989, **111**, 4776.
61. M. A. Benvenuto and R. N. Grimes, *Inorg. Chem.*, 1991, **30**, 2836.
62. A. Fessenbecker, *et al.*, *Inorg. Chem.*, 1990, **29**, 5157.
63. V. Schäfer, H. Pritzkow and W. Siebert, *Chem. Ber.*, 1989, **122**, 401.
64. G. Brodt and W. Siebert, *Chem. Ber.*, 1989, **122**, 633.
65. A. Fessenbecker, H. Schulz, H. Pritzkow and W. Siebert, *Chem. Ber.*, 1990, **123**, 2273.
66. G. E. Herberich and H. Ohst, *Adv. Organomet. Chem.*, 1986, **25**, 199.
67. G. E. Herberich, W. Klein and T. P. Spaniol, *Organometallics*, 1993, **12**, 2660.
68. G. E. Herberich, B. Hessner and M. Hostalek, *J. Organomet. Chem.*, 1988, **355**, 473.
69. G. E. Herberich, B. Hessner, N. Klaff and H. Ohst, *J. Organomet. Chem.*, 1989, **375**, 161.
70. K.-F. Wörner, J.-K. Uhm, H. Pritzkow and W. Siebert, *Chem. Ber.*, 1990, **123**, 1239.
71. H. Schulz, H. Pritzkow and W. Siebert, *Chem. Ber.*, 1992, **125**, 987.
72. H. Schulz, H. Pritzkow and W. Siebert, *Chem. Ber* 1991, **124**, 2203.
73. A. J. Ashe, III, F. J. Drone, C. M. Kausch, J. Kroker and S. M. Al-Taweel, *Pure Appl. Chem.*, 1990, **62**, 513.
74. A. J. Ashe, III and F. J. Drone, *J. Am. Chem. Soc.*, 1987, **109**, 1879 and *J. Am. Chem. Soc.*, 1988, **110**, 6599.
75. A. J. Ashe, III, J. W. Kampf, W. Klein, R. Rousseau, *Angew. Chem.*, 1993, **105**, 1112; *Angew. Chem., Int. Ed. Engl.*, 1993, **32**, 1065.
76. A. J. Ashe, III, J. W. Kampf, Y. Nakadaira and J. M. Pace, *Angew. Chem.*, 1992, **104**, 1267; *Angew. Chem., Int. Ed. Engl.*, 1992, **31**, 1255.
77. A. J. Ashe, III, J. W. Kampf, C. M. Kausch, H. Konishi, M. O. Kristen and J. Kroker, *Organometallics*, 1990, **9**, 2944.
78. P. Paetzold, *Pure Appl. Chem.*, 1991, **63**, 345.
79. E. Bulak, G. E. Herberich, I. Manners, H. Mayer and P. Paetzold, *Angew. Chem.*, 1988, **100**, 964; *Angew. Chem., Int. Ed. Engl.*, 1988, **27**, 958.
80. P. Paetzold and K. Delpy, *Chem. Ber.*, 1985, **118**, 2552.
81. K. Delpy, D. Schmitz and P. Paetzold, *Chem. Ber.*, 1983, **116**, 2994.
82. G. Schmid, D. Zaika, J. Lehr, N. Augart and R. Boese, *Chem. Ber.*, 1988, **121**, 1873.
83. G. Schmid and F. Schmidt, *Chem. Ber.*, 1986, **119**, 1766.
84. G. Schmid, U. Höhner and D. Kampmann, *Z. Naturforsch., Teil B*, 1983, **38**, 1094.
85. G. Schmid, F. Schmidt and R. Boese, *Chem. Ber.*, 1985, **118**, 1949.
86. G. Schmid, S. Amirkhalili, U. Höhner, D. Kampmann and R. Boese, *Chem. Ber.*, 1982, **115**, 3830.
87. G. Schmid, D. Kampmann, U. Höhner, D. Bläser and R. Boese, *Chem. Ber.*, 1984, **117**, 1052.
88. G. Schmid, O. Boltsch and R. Boese, *Organometallics*, 1987, **6**, 435.
89. G. Schmid and T. Rohling, *J. Organomet. Chem.*, 1989, **375**, 21.
90. R. Köster, G. Seidel, R. Boese and B. Wrackmeyer, *Chem. Ber.*, 1988, **121**, 597.
91. R. Köster, G. Seidel, B. Wrackmeyer and D. Schlosser, *Chem. Ber.*, 1989, **122**, 2055.

92. R. Köster, G. Seidel, C. Krüger, G. Müller, A. Jiang and R. Boese, *Chem. Ber.*, 1989, **122**, 2075.
93. R. Köster, G. Seidel, R. Boese and B. Wrackmeyer, *Chem. Ber.*, 1988, **121**, 1941.
94. R. Köster, G. Seidel, R. Boese and B. Wrackmeyer, *Chem. Ber.*, 1988, **121**, 709.
95. R. Köster, G. Seidel, R. Boese and B. Wrackmeyer, *Chem. Ber.*, 1988, **121**, 1955.
96. P. Frankhauser, M. Driess, H. Pritzkow and W. Siebert, *Chem. Ber.*, 1992, **125**, 1341.
97. W. Siebert, M. El-Essawi, R. Full and J. Heck, *Z. Naturforsch. Teil B*, 1985, **40**, 458.
98. U. Zenneck, L. Suber, H. Pritzkow and W. Siebert, *Chem. Ber.*, 1986, **119**, 971.

6
Polyhedral Carbaboranes

T. ONAK
California State University, Los Angeles, CA, USA

6.1 INTRODUCTION 217

6.2 CARBABORANES 218

6.2.1 Literature Reviews 218
6.2.2 Structural Interrelationships and Accompanying Theory 218
6.2.3 Physical Properties 220
6.2.3.1 Diffraction studies 220
6.2.3.2 Spectroscopic properties 222
6.2.3.3 Mass spectrometry 226
6.2.3.4 Additional physical properties 226
6.2.4 Syntheses 227
6.2.4.1 Carbaboranes from polyboranes and alkynes 227
6.2.4.2 Cage growth as a result of carbon insertion reactions into carbaboranes 228
6.2.4.3 Cage growth as a result of boron insertion reactions 228
6.2.4.4 Cage fusion reactions 229
6.2.4.5 Dehydrogenation reactions 229
6.2.4.6 Selective cage degradation 229
6.2.4.7 Coupled carbaboranes 231
6.2.4.8 Additional methods of preparation 231
6.2.5 Chemical Properties 232
6.2.5.1 Skeletal rearrangements, cage fluxional behavior, and isomer stabilities 232
6.2.5.2 Hydrogen acidity studies 234
6.2.5.3 Reduction reactions and related chemistry 235
6.2.5.4 Reactions of carbaboranes with Lewis bases 235
6.2.5.5 Derivative chemistry 236
6.2.5.6 Polymer applications 244
6.2.5.7 Other applications 245

6.3 ORGANOPOLYBORON HYDRIDES 245

6.3.1 Physical Properties 245
6.3.2 Preparation of Organopolyboranes 245
6.3.3 Reactions of Organopolyboranes 246

6.4 REFERENCES 247

6.1 INTRODUCTION

This chapter treats the subject of organoboranes in which carbon is either an integral part of a polyhedral cage (carbaboranes) or is directly attached, as a member of a substituent group, exopolyhedral to a polyboron hydride (organopolyboron hydrides). Work dated earlier than 1981 is found in *COMC-I*.[1] The present review updates the studies published through to 1992 and an early portion of 1993.

6.2 CARBABORANES

6.2.1 Literature Reviews

Review articles on the carbaboranes since the mid-1980s include various annual surveys,[2–7] as well as other series published less frequently.[8–12] Structural parameters for a number of carbaboranes are available in two chapter reviews;[13–14] and additional structural patterns among carbaborane structures have been put forward.[15–18] Certain theoretical features of carbaboranes have been discussed and, in some cases, in a fashion that intercorrelates these polyhedral systems to other cage compounds.[19–24]

Other reviews and accounts covering specialized aspects of carbaborane chemistry include: carbaboranes viewed in terms of "hypercoordinate" carbon compounds,[25–6] analogies between carbaboranes and carbocations,[15,27] electron distributions in boranes and carbaboranes,[28] electron deficiency aspects of the small carbaboranes,[29] structural comparisons between carbaboranes and the higher borides,[30] small carbaboranes,[31] syntheses and reactions of nine- and ten-atom carbaboranes,[32] monocarbaborane anions $[CB_8H_{10}]^-$, $[CB_{10}H_{11}]^-$, $[CB_{11}H_{12}]^-$ and their derivatives,[33] carbaboranes other than $C_2B_{10}H_{12}$,[34] $C_2B_{10}H_{12}$ and their derivatives,[35] boron-substituted derivatives of $C_2B_{10}H_{12}$,[36] chemistry of cage-carbon bound (*exo*-polyhedral) nitrogen derivatives of $C_2B_{10}H_{12}$,[37] ^{11}B NMR spectroscopy,[38] vibrational spectroscopy,[39] synthetic strategies,[40–1] *C*-trimethylsilyl-substituted carbaboranes,[42] $C_2B_{10}H_{12}$ compounds containing boron–element bonds,[43] carboranyl derivatives of nontransition metals,[44–6] cyclometallated carbaboranes containing a boron–transition metal bond,[47] transition metal promoted reactions of carbaboranes,[48–9] boron cluster compounds in coordination chemistry,[50] potential applications,[51] and carbaborane containing polymers.[52–61]

Valiant attempts have been made to systematize the nomenclature for carbaborane related compounds;[62] however, most of the pertinent literature is devoid of the more novel and cumbersome aspects of these proposals.

6.2.2 Structural Interrelationships and Accompanying Theory

A recent review[15] has put into perspective most of the prevailing structural paradigms for the carbaboranes. Theoretical calculations on carbaborane clusters are expectedly abundant and include the following studies. Molecular orbital calculations on $C_2B_2H_4$ indicate that a tetrahedrane (carbaborane) geometry for this molecule is less stable than a 1,3-diboretane structure.[63–4] *Nido*-2,3,4,5-$C_4B_2H_6$ is more stable than *nido*-1,2,3,5-$C_4B_2H_6$ by 168.2 kJ mol^{-1} as estimated at the [6-31G*//3-21G + MP3/6-31G//3-21G − 6-31G//3-21G] level.[65] The electronic structures of 1,5-$C_2B_3H_5$ and 1,5-$C_2B_3H_3$, studied at the 6-31G* *ab initio* level of theory, include a three-center two-electron C–B–C bond. The strength of this three-center bond is low and its C–C bond is long. Analysis of the bonding in these carbaboranes along with that in 1,3-diborabicyclo[1.1.1]pentane shows that the strength of these three-center bonds is directly related to the nature of the bridging group. It is also proposed that 1,3-diboretene, the carborane analogue of bicyclo[1.1.0]butane, possesses a C–B–C bond.[66] *Ab initio* calculations at the 6-31G* level of theory confirm that 1,5-$C_2B_3H_5$ and 1,6-$C_2B_4H_6$ are more stable than other isomers with similar cage geometry.[67] A six-electron rule for explaining observed cage isomer stabilities of *closo*-$C_2B_3H_5$, *closo*-$C_2B_5H_7$, and *closo*-$C_2B_6H_8$ isomers has been developed.[68] *Ab initio* calculations at the STO-3G level indicate the following order of stabilities: for *closo*-X,Y-$C_2B_3H_5$, X,Y = 1,5 > X,Y = 1,2 > 2,3; for *closo*-X,Y-$C_2B_4H_6$, X,Y = 1,6 > 1,2; for *closo*-X,Y-$C_2B_5H_7$, X,Y = 2,4 > 2,3 > 1,2 > 1,7; for *closo*-X,Y-$C_2B_6H_8$, X,Y = 1,7 > 1,6 > 1,2 > 1,3 > 1,5 > 3,4; for *closo*-X,Y-$C_2B_7H_9$, X,Y = 4,5 > 3,4 > 1,4 > 1,3 > 1,8; for *closo*-X,Y-$C_2B_8H_{10}$, X,Y = 1,10 > 1,6 > 2,7 > 1,2 > 2,4 > 2,6 > 2,3; for *closo*-X,Y-$C_2B_9H_{11}$, X,Y = 2,3 > 2,10 > 2,9 > 2,6; and for *closo*-X,Y-$C_2B_{10}H_{12}$, X,Y = 1,12 > 1,7 > 1,2.[69–70] From *ab initio* MO calculations on a series of *closo*-carbaboranes $C_2B_{n-2}H_n$ $(5 \leq n \leq 12)$ it was determined that the external B–H bonds have the homopolar bonding character of a typical two-center two-electron bond.[71]

A barrier of ca. 42 kJ mol^{-1} is predicted from *ab initio* MO calculations for the exchange of terminal hydrogens between 1,5-$C_2B_3H_5$ and B_2H_6. At the [MP2/6-31G*](+ZPC) level it is calculated that the reaction of 1,5-$C_2B_3H_5$ and BH_3 to form 1,6-$C_2B_4H_6$ and H_2 is 139.3 kJ mol^{-1} exothermic while the reaction to form $C_2B_4H_8$ is 138.9 kJ mol^{-1} exothermic.[72] The carbon-capped square pyramid of *nido*-1,2-$C_2B_3H_7$ is only slightly more stable than an envelope-like classical 3,4,5-triboracyclo-1,2-pentene-$\mu_{3,4}$-H-$\mu_{4,5}$-H structure at the HF/3-21G level.[73] Geometry optimized STO-3G calculations on the nearly octahedral *closo*-1,2-$C_2B_4H_6$, and $[1,2\text{-}C_2B_4H_7]^+$ ion (with a bridging hydrogen on a B-3–B-4–B-6 face), indicate that the proton affinity of 1,2-$C_2B_4H_6$ is 810.4 kJ mol^{-1}, which is in reasonable agreement with the experimentally determined value.[74] The results of calculations using the (9*s*,5*p*) basis set of

Huzinaga are used along with the Laplacian of the charge density to predict a greater susceptibility of 1,6-$C_2B_4H_6$, compared with 1,5-$C_2B_3H_5$, toward nucleophilic attack; the reverse order is predicted for susceptibility toward electrophilic attack.[75] *Ab initio* calculations at the MP2/6-31G*//3-21G level of theory for 1-CH-1,2-$C_2B_3H_4$, a singlet carbene attached to the cage, causes a complete disruption of the cage structure. By contrast, 1-CH-1,2-$C_2B_4H_5$, is an energy minimum. On both potential energy surfaces, the global minimum appears to be an expanded cage isostructural and isoelectronic with the next higher carbaborane.[76] The order of cage isomer stabilities within each of the $C_2B_4H_6$ and $C_2B_5H_7$ systems does not change in proceeding from the 6-31G* to the MP2/6-31G* levels of theory.[68]

Ab initio derived energies at various levels of theory are reported for 1,2- and 1,6-$C_2B_4H_6$, as well as for 2,3-$C_2B_5H_7$.[77] An *ab initio* MO calculational approach to the rearrangement of *closo*-$C_2B_4H_6$ isomers shows that a diamond–square–diamond process may be important and that two benzvalene-like structures, both +167 kJ mol^{-1} above 1,6-$C_2B_4H_6$, could be involved; however, one such process is blocked by a HOMO–LUMO crossing. The 1,2-$C_2B_4H_6$ has a predicted barrier to formation of the more stable 1,6-$C_2B_4H_6$ of about 163.7 kJ mol^{-1}.[78] The Roby projection-density method of population analysis involving 4-31G calculations has been applied to 1,2-$C_2B_4H_6$, 1,6-$C_2B_4H_6$, and 2,3-$C_2B_4H_8$.[79] The most stable structure of 1-(SiH_2)-2,3-$C_2B_4H_6$ is predicted to have an SiH_2 group centered over the pentagonal face of a pyramid in such a manner that two terminal Si–H bonds lie in the symmetry plane.[77] *Ab initio* calculations carried out on the nonclassical and classical forms of $C_2B_4H_8$ at the MP2/6-31G* + ZPC// 3-21G level of theory indicates that the classical form, which is a six-membered ring in a chair conformation, is 236.5 kJ mol^{-1} higher in energy than the *nido*-2,3-$C_2B_4H_8$ configuration.[80] Extended Hückel MO calculations have been made on $C_2B_4H_8$, and the results compared to those of related cage species.[81] A theoretical study of reaction of H_2 with CB_5H_5 to form CB_5H_7 has been carried out at the 4-31G and 3-21G* levels of theory.[82] *Ab initio* MO calculations with full geometry optimization yield an octahedral structure for *closo*-1-CB_5H_7 that is in good agreement with the experimentally determined structure.[83] The *ab initio* STO-3G calculated energy of $[CB_5H_6]^-$ ion is reported.[74] and a MNDO computational comparison between $[CB_5H_6]^-$ and several other geometrically related polyboranes has been made.[84] From *ab initio* considerations at the MP2/6-31G*//3-21G level of theory, corrected for zero-point energy differences, the bridging hydrogen of *closo*-1-CB_5H_7 migrates from one B–B–B trigonal face to another with a barrier of 62 kJ mol^{-1}.[85–6] In accordance with experimental data, calculated results at the MP2/6-31G*//3-21G level of theory indicate a 6.3 kJ mol^{-1} lowering of the activation energy for bridging hydrogen tautomerization of CB_5H_7 when the terminal hydrogen attached to the cage carbon is replaced by a methyl group. At the same level of theory the calculated equilibrium ratio 2.2:1 of 2-Me-*closo*-1-CB_5H_6/4-Me-*closo*-1-CB_5H_6 is in reasonable agreement with the experimentally observed level of 2.8:1 at 25 °C.[87] The calculated (MP2/6-31G*//3-21G) relative stabilities of the four X-Me-*nido*-2-CB_5H_8 isomers are $X = 3 > 4 > 1 > 2$, which is consistent with experimental observations.[87] Experimentally determined stabilities among the three *B*-Cl-*closo*-2,4-$C_2B_5H_6$ derivatives correlate well with results obtained from the application of the *ab initio* MO method.[88] The LUMO of 1,6-$C_2B_6H_8$ is concentrated predominantly on the symmetry related and four-connected boron atoms B-4 and B-8 (29% on each), and to a lesser extent on the carbon atoms C-1 and C-6 (11% on each); this molecular orbital lies at −8.5 eV.[89] An examination of structures generated during the *ab initio* MO conversion of 3,5-$C_2B_6H_8$ to the 1,7-$C_2B_6H_8$ isomer indicates that the computational rearrangement follows the diamond–square–diamond pathway.[70] The geometry-optimized total energies and bond distances for the *closo*-carbaboranes 3,5-$C_2B_6H_8$, 1,7-$C_2B_7H_9$, and 1,2-$C_2B_7H_9$ calculated by the *ab initio* SCF MO method using the STO-3G basis set are reported. Relative energies are compared with those of the other carbaborane isomers in the eight- and nine-(skeletal)-atom classes.[90] The effect of the carbon atoms in the capping positions of $C_2B_7H_9$, when compared with $[B_9H_9]^{2-}$, is to drain electronic charge from the prism ends of the tricapped trigonal prismatic cluster; this results in the lengthening of the triangle edges of the prism.[91] The STO-3G Mulliken atomic charges and overlap populations of 4-CB_8H_{14} have been calculated.[92–3] An expanded cage of formula $C_3B_{10}H_{12}$ is only 14.2 kJ mol^{-1} higher (at the MP2/6-31G*//3-21G + ZPC level of theory) than the carboranylcarbene 1-CH-*closo*-1,2-$C_2B_{10}H_{11}$ which suggests that a mechanism suggested earlier for the degenerate rearrangement of the carboranylcarbene is possible. Calculations are also reported for the dianion [*nido*-$C_2B_{10}H_{12}]^{2-}$, which is related to the 13-vertex *closo* cage by removal of a C^{2+} vertex.[76,94] The vibrational spectra for $[CB_{11}H_{12}]^-$ with C_{5v} symmetry and for 1,12-$C_2B_{10}H_{12}$ with D_{5d} symmetry are calculated at the 3-21G level of theory; for 1,12-$C_2B_{10}H_{12}$ the results compare favorably with experimental infrared and Raman spectra.[95]

Optimized geometry parameters for $[CB_3H_4]^{3-}$, $[CB_4H_5]^{3-}$, and $[CB_5H_6]^{3-}$ have been derived from MNDO molecular orbital calculations.[96] The calculated proton affinity for 1,5-$C_2B_3H_5$ is 603.6 kJ mol^{-1}, as determined by the MNDO method, and a carbon protonated structure is reported.[97] Both MNDO and *ab initio* (employing the 3-21G basis set) studies on the protonation of 1,6-$C_2B_4H_6$ predict B–B edge protonation; the MNDO method predicts B–B–B face protonation of 2,4-$C_2B_5H_7$ and 1,12-$C_2B_{10}H_{12}$; the MNDO calculated proton affinity for 2-CB_5H_9 is 692.8 kJ mol^{-1}.[97–8] Isomer stabilities determined by the MNDO method have been correlated to experimental results for the following compounds: *B*-X-*closo*-2,4-$C_2B_5H_6$ (X = Me, Cl, Br, I, Me_3N^+), *B,B'*-X_2-*closo*-2,4-$C_2B_5H_5$ (X = Me, Cl, Br, I),[88,99–102] *B*-Me-*B*-Cl-*closo*-2,4-$C_2B_5H_5$,[100–1] *B*-Cl-*B'*,*B''*-Et_2-*closo*-2,4-$C_2B_5H_4$, and *B,B'*-(*closo*-2,4-$C_2B_5H_6)_2$.[88] MNDO calculated dipole moments for all of the *B*-X-*closo*-2,4-$C_2B_5H_6$ (X = Me, Cl) isomers have been correlated to isomer volatility trends.[101] MNDO MO calculations on $C_2B_3H_5$, $C_2B_4H_6$, $C_2B_5H_7$, $C_2B_6H_8$, $C_2B_7H_9$, $C_2B_8H_{10}$, and 1,12-$C_2B_{10}H_{12}$, are used to assess the accuracy of Stone's theory of polyborane electronic structure and bonding.[103] Tensor surface harmonic theory is used to describe the structure and bonding both in $[C_2B_7H_9]^{2-}$ and $[C_2B_9H_{11}]^{2-}$.[104] In the course of parametrization of boron for AM1 calculations, various properties of the following carbaboranes were included: CB_5H_7, CB_5H_9, 2,3,4,5-$C_4B_2H_6$, 1,5-$C_2B_3H_5$, 1,2-$C_2B_4H_6$, and 1,6-$C_2B_4H_6$.[105] AM1 calculations were performed on a number of neutral and anionic derivatives of $[7\text{-}CB_{10}H_{13}]^-$ and $[1\text{-}CB_{11}H_{12}]^-$, and heats of formation, ionization potentials, bond indices, and atomic charges, were derived.[106]

Hartree–Fock calculations on 1,2-$C_2B_{10}H_{12}$ and some related systems indicate that, in good agreement with experimental data, three of the four types of boron sites in the carbaborane are associated with nearly identical boron (1*s*) core excitation energies of 101.9 eV; the fourth site yields an excitation energy of 190.9 eV.[107] Upon considering the Mulliken net atomic charge populations calculated from extended Hückel wave functions for a polyhedral $[B_nH_n]^{2-}$ and $[C_{n-1}N]^-$ reference compounds, along with the assumption that carbons of the related isoelectronic carbaborane $C_2B_mH_{m+2}$ prefer to be at sites where electron density is greatest in the reference compound $[C_{n-1}N]^-$, the order of stabilities within cage $C_2B_mH_{m+2}$ isomer sets are predicted.[108] The full icosahedral chirality polynomial has been simplified for chiral monosubstituted carbaboranes of the general formula 1,X-$C_2B_{10}H_{11}$-Y-R (X = 2,7; Y = 1–12).[109]

Additional theoretical calculations have been carried out on *closo*-1,5-$C_2B_3H_5$,[110] 1,2-$C_2B_4H_6$,[111] *closo*-1,6-$C_2B_4H_6$,[110,112–13] 2,4-$C_2B_5H_7$,[113] $C_4B_6H_{10}$,[114] 4,6-$C_2B_7H_{13}$,[93] $[C_2B_8H_{10}]^{2-}$,[114] $C_4B_8H_{12}$,[114] $[C_2B_9H_{11}]^{2-}$,[115] 1,2-$C_2B_{10}H_{12}$,[116] 1,2-$C_2B_{10}H_{12}$ with a *C*-attached allyl anion,[3] 1,7-$C_2B_{10}H_{12}$,[116] and 1,12-$C_2B_{10}H_{12}$.[117]

6.2.3 Physical Properties

6.2.3.1 Diffraction studies

The structures of numerous carbaboranes have been determined by x-ray diffraction techniques, as listed in Table 1. The x-ray determined structure of the anion, $[1,2\text{-}C_2B_{10}H_{10}\text{-}1\text{-}C_6H_5\text{-}2\text{-}O]^-$, consists of a *nido*-shaped $[C_6H_5CB_{10}H_{10}]^-$ residue capped by a pentuply-bridging carbonyl group with a C–O distance of 0.1245(3) nm.[118] From x-ray data the *exo*-carbon–carbon distance in $[(Ph_3PMe)_2]_2[(C_2B_{10}H_{11})_2]$ is found to be 0.1377 nm which is indicative of substantial double-bond character. The corresponding distance in $[(Ph_3PMe)_2]_2$[μ-9,10-CH-(μ-9',10'-CH-*nido*-7'-$CB_{10}H_{11}$)-*nido*-7-$CB_{10}H_{11}$] is 0.1515 nm.[119–20] From crystal structure studies of several salts of $[CB_{11}H_{12}]^-$ it is suggested that this carbaborane anion is a poorer σ-donor than most anionic spectator ions in common use.[121–4] An x-ray diffraction study of $Bz_4C_4B_8H_8$, discloses that the C_4B_8 cage is locked into an open-cage geometry (**4**) as a consequence of severe intramolecular crowding of the benzyl groups.[125]

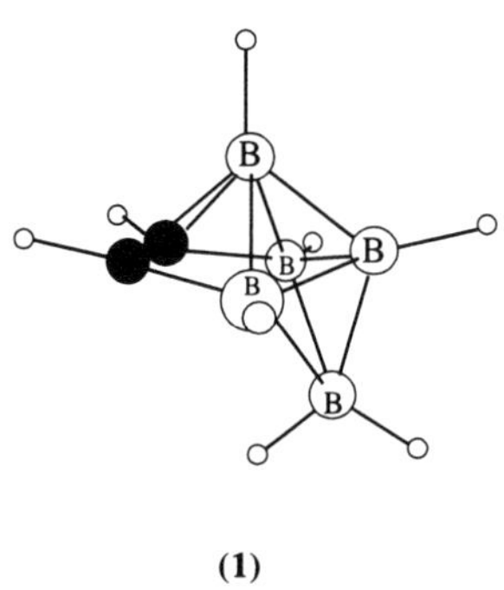

(1)

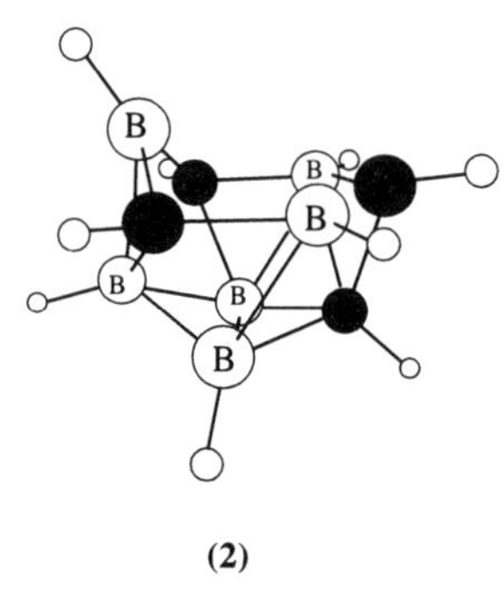

(2)

Application of electron diffraction techniques to the pentagonal pyramidal *nido*-2,3-$C_2B_4H_7$-2-TMS indicates that the cage bond lengths are: 0.146 nm for C–C, 0.154 nm for C-2–B-6 and C-3–B-4,

Table 1 Structure studies on carbaboranes and carbaborane derivatives using x-ray diffraction.

Compound	*Ref.*
2,3,4-$((Me_2CH)_2N)_3$-1,5-$C_2B_3H_2$	126
$(THF{\cdot}Na^+)_2$[2,3-$(TMS)_2$-2,3-$C_2B_4H_5^-]_2$	127
{$Na^+{}_2(THF)_4$[2,4-$(TMS)_2$-2,4-$C_2B_4H_4]^{2-}\}_2$	128
$[Me_4N]$[*nido*-3,4-$C_2B_5H_6$-3,4-Et_2] (**1**)	129
exo-6-$(Me_3P^+CH_2)$-3,4-Et_2-*nido*-3,4-$C_2B_5H_5^-$	130
$(C_2H_5)_4$-2,6,8,10-$C_4B_6Et_6$ (**2**)	131
Bu_4N[*nido*-4,5-$C_2B_6H_9$]	132
arachno-6-$(NCCH_2)$-5,6,7-$C_3B_7H_{12}$	133
$Me_4C_4B_7H_8$-11-Br	134
$(R)_4C_4B_8H_8$ R = Me, Et, (**3**) (**4**)	135
$Bz_4C_4B_8H_8$	125
1,10-(1′-$SiMe_2$-2′-Me-*closo*-1′,2′-$C_2B_{10}H_{10})_2$-*closo*-1,10-$C_2B_8H_8$	136
arachno-6,9-$C_2B_8H_{13}$-6-TMS	137
$[Cp^*{}_2Ir_2Cl_3]^+$[*closo*-1-$CB_9H_{10}]^-$	138
Na[7,8-(1,13-dithia-4,7,10-trioxatridecane-1,13-diyl)-7,8-$C_2B_9H_{10}$]	139
nido-9-PPh_3-7,8-$C_2B_9H_{11}$	140
$[Me_4N]$[7,8-(3′,6′,9′-trioxaundecane-1′,11′-dithiolato-S,S′)-7,8-$C_2B_9H_{10}$]	141
$[C_6H_5CH_2NMe_3]$[7-(CH_2OMe)-7,8-*nido*-$C_2B_9H_{11}$]	142
9-Ph_2PH-7,8-$C_2B_9H_{11}$ and 10-Ph_2PH-7,8-$C_2B_9H_{11}$	143
10-[2-(2-dimethylsulfonioethoxy)ethoxy]-*nido*-7,8-$C_2B_9H_{11}$	144
$[H(DMSO)_2][C_2B_9H_{12}]$	145
Cs[7,8-$C_2B_9H_{12}$]	146
nido-2,7-$C_2B_9H_{10}$-11-Bz-2,7-Me_2	147
7,8-$C_2B_9H_{11}$-5-$SOMe_2$	148, 149
$[Me_4N]$[7,8-$C_2B_9H_{11}$-7-Ph]	3
Cs[7,8-$C_2B_9H_{10}$-9-I-7-C_6H_4-4′-NCS]	150
$\mu_{10,11}$-H-9-Me_2S-*nido*-7,8-$C_2B_9H_{10}$	151
7,9-$C_2B_9H_{11}$-8-SMe_2	152
$[Me_3NH]_2$[7,8-$C_2B_9H_{10}$-7,8-$(S)_2]_2$	153
$[Me_3NH]$[7,8-$C_2B_9H_{10}$-7,8-$(SH)_2$]	153
[7,8-(ethane-1′,2′-dithiolato-S,S′)dicarba-*nido*-undecaborato]-bis(triphenylphosphine)rhodium(I)	154
nido-8-Ph_3P-7-$CB_{10}H_{12}$	155
6-Me_2S-2-$[(TMS)_2CH]CB_{10}H_9$	156
6-Me_2S-4-HO-2-$[(TMS)_2CH]CB_{10}H_8$	156
cobaltocenium salt of 2-$[(TMS)_2CH]$-*closo*-$CB_{10}H_{10}^-$	156
9-$(C_6H_5)_3P$-7-$[(TMS)_2CH]$-*nido*-$CB_{10}H_{11}$	157
$[(Ph_3PMe)_2]_2$[μ-9,10-CH-(μ-9′,10′-CH-*nido*-7′-$CB_{10}H_{11}$)-*nido*-7-$CB_{10}H_{11}$]	119, 120
8,9,10,12-F_4-1,2-$C_2B_{10}H_8$	35
[1,2-$C_2B_{10}H_{11}$-9-IPh]I	158
1,12-$C_2B_{10}H_{11}$-2-$PO(OMe)_2$	159, 160
1,2-$C_2B_{10}H_{10}$-1,2-(-$CH_2P(Me)CH_2$-)	161
9-Me-1,2-$C_2B_{10}H_{11}$	162
Li[*closo*-1,2-$C_2B_{10}H_{10}$-1-Me]·$[MeN(CH_2CH_2NMe_2)_2]$	163
1,2-$C_2B_{10}H_{10}$-1,2-[-$CH_2SiMe_2OSiMe_2CH_2$-$]_2$	2c
1,2-$C_2B_{10}H_{10}$-1-Ph-2-HgX (X = Br, I)	3
$TlLCl_2$ (HL = 1,7-$C_2B_{10}H_{10}$-1,7-Me_2)	4c
$Ru(CO)_2$(1-Ph_2PCH_2-1,2-$C_2B_{10}H_{10}$)(1-Ph_2PCH_2-1,2-$C_2B_{10}H_{11}$)(Cl)	164
9-Cl-1,2-CH_2CH_2-*closo*-1,2-$C_2B_{10}H_{10}$	165
1-CH_2OMe-1,2-*closo*-$C_2B_{10}H_{11}$	142
9-Ph-1,7-$C_2B_{10}H_{11}$	166
1,2-$C_2B_{10}H_{11}$-1-TMS_3	2c
1,2-$C_2B_{10}H_{11}$-1-$GeMe_3$	2c
1,7-$C_2B_{10}H_9$-1-CH_2NMe_2-7-Ph-4-$Re(CO)_4$	167
1,7-$C_2B_{10}H_{10}$-1,7-$[CH_2SiMe_2OH]_2$	168
1,7-$C_2B_{10}H_9$-1,7-Me_2-9-Tl(bis-trifluoroacetato)(bipyridyl)	3
1,2-$C_2B_{10}H_2$-3,4,5,6,7,8,9,10,11,12-$Cl_{10}{\cdot}Me_2SO$	4a
1,2-$C_2B_{10}H_{10}$-1,2-[Ph-4′-OC_6H_4-4″-C(O)Me$]_2$	169, 170
1,2-$C_2B_{10}H_{11}$-9-$SPPh_2$	171
1,2-(1,10-dithia-4,7-dioxadecane-1,10-diyl)-1,2-$C_2B_{10}H_{10}$	139
1,2-(3′-oxapentane-1′,5′-dithiolato-SS′)-1′,2′-$C_2B_{10}H_{10}$	172
9-azido-1,7-$C_2B_{10}H_{11}$	5c
1-(1′-*closo*-2′-Me-1′,2′-$C_2B_{10}H_{10}$)-2,3-$Cl_2C_6H_3$	173
1,4-[1,2-$C_2B_{10}H_{10}$-1,4-diyl]-2-butanone	174
$C_2B_{10}H_{11}{\cdot}CH_2C_{27}H_{45}O$ (i.e., methylcholesterol)	175
8,9,10,12-Cl_4-1,2-$(HOCH_2)_2$-$C_2B_{10}H_6$	176
$[PPh_3Me]$[*nido*-7,9-Me_2-7,9-$C_2B_{10}H_{11}$]	177
8-(1,2-$C_2B_{10}H_{11}$-3-)-5,6-$C_2B_8H_{11}$	178
$[Ph_3P]_3Rh^+$[*nido*-7,8-$C_2B_9H_{11}R]^-$ (R = 1′-(*closo*-1′,2′-$C_2B_{10}H_{11}$))	179

Table 1 (continued)

Compound	*Ref.*
$(1,7\text{-}C_2B_{10}H_{11}\text{-}9\text{-})_2O$	180
$[(1,7\text{-}C_2B_{10}H_{11}\text{-}9\text{-})_2Br]BF_4$	181
1,4-bis(1′-*closo*-2′-Me-1′,2′-$C_2B_{10}H_{10})_2C_6H_4$	173
$[(TMS\text{-}O)_2(1,2\text{-}C_2B_{10}H_{11}\text{-}1\text{-}CH_2\text{-})Si]_2O$	168
Me_4N^+[*nido*-$(C_6H_5)C_2B_{10}H_{11}$-*closo*-$PhC_2B_{10}H_{10}]^-$	182
$(1,2\text{-}C_2B_{10}H_{11}\text{-}9\text{-}S\text{-})_2CH_2$	183
$Cl_2Si(1,2\text{-}C_2B_{10}H_{10})_2SiCl_2$	184
1,2-(1′,2′-$C_2B_{10}H_{10}$-1′-$SiMe_2CMe_3$-2′-CH_2-$)_2$-C_6H_4	185
$[(Ph_3PMe)_2]_2[(C_2B_{10}H_{11})_2]$, [(15-crown-5)$_3Na_2][(C_2B_{10}H_{11})_2]$	119, 120
1,3,5-[4′-(1,2-$C_2B_{10}H_{11}$-1-)biphenyl-4-yl]$_3$benzene	4e
1,3,5-(1,2-$C_2B_{10}H_{11}$-1-CH_2Ph-4′-$)_3$-C_6H_3	186
Li[(-1-Hg-1,2-$C_2B_{10}H_{10}$-2-$)_4$Cl]	187
(1,2-$C_2B_{10}H_{10}$-2-Me-1-$)_2Mg\cdot 2C_4H_8O_2$	188
(-1-Hg-1,2-$C_2B_{10}H_{10}$-2-$)_4\cdot(THF)_4\cdot(H_2O)_2$	189
1-Me_2N-2-Me_2S-*closo*-1-$CB_{11}H_{10}$	190
Me_4N[1-Me_2N-1-*closo*-1-$CB_{11}H_{11}$]	191
$Me_3NCB_{11}H_{11}$	192
1-$CB_{11}H_{11}$-1-NMe_3	193
$Ag(CB_{11}H_{12})\cdot IrCl(CO)(PPh_3)_2\cdot 0.5PhF$	194

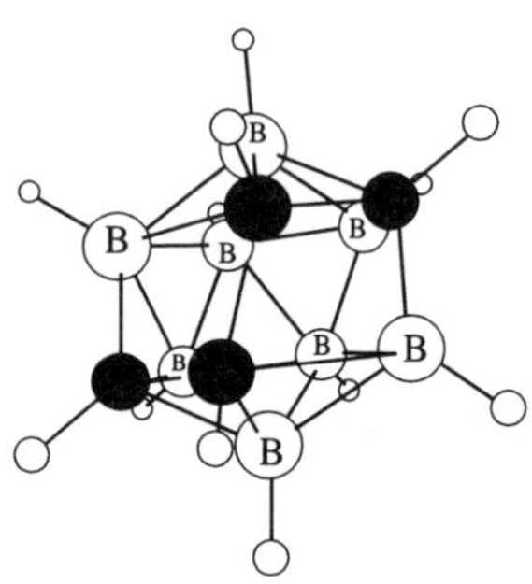

(3)

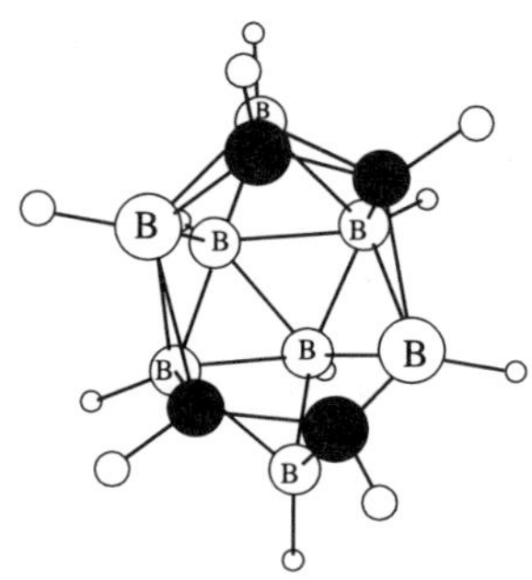

(4)

0.176 nm for B-4–B-5, 0.183 nm for C-2–B-1 and C-3–B-1, 0.176 nm and 0.169 nm for the two B-1–B-(basal) bonds.[195] Other carbaborane structures studied by electron diffraction include the (near) icosahedral cage derivatives 1,12-$C_2B_{10}H_{10}$-1,12-$(Me)_2$, 1,12-$C_2B_{10}H_{10}$-1,12-I_2,[3] 1,1'-bis(*p*-carbaborane), 1-(1'-1',12'-$C_2B_{10}H_{11}$)-1,12-$C_2B_{10}H_{11}$,[196] (1,12-$C_2B_{10}H_{11}$-1-$)_2$Hg,[197] and (1,12-$C_2B_{10}H_{11}$-2-$)_2$Hg.[198]

6.2.3.2 Spectroscopic properties

(i) Nuclear magnetic resonance spectroscopy

NMR spectroscopy has been employed for the identification and structure elucidation of nearly every compound reported in this chapter; and for these data the reader is advised to consult the original references cited herein (mostly in the synthetic portions of this chapter). In this section the referenced studies are those largely dealing with correlative work in the field as well as with experimental NMR spectroscopic highlights. Some NMR calculational methods (e.g., IGLO (individual gauge for localized orbitals), GIAO (gauge independent atomic orbitals)[199–204] have been applied to *ab initio* optimized structures and are now at the confidence level that rivals diffraction methods for structure determinations; the best NMR correlations among the carbaboranes are found with the ^{11}B NMR spectroscopic data. Empirical correlations of ^{11}B NMR spectroscopic chemical shift patterns have also been employed in the predictions of carbaborane structures.[205–6]

The ^{11}B NMR spectral chemical shifts of the carbaboranes 1,5-$C_2B_3H_5$, 1-CB_5H_7, 1,6-$C_2B_4H_6$, 1,2-$C_2B_4H_6$, 2,4-$C_2B_5H_7$, 1,2-$C_2B_3H_7$, 2-CB_5H_9, 2,3-$C_2B_4H_8$, 2,3,4-$C_3B_3H_7$, and 2,3,4,5-$C_4B_2H_6$ have been calculated using the IGLO method. The IGLO derived ^{11}B chemical shifts, calculated using the geometries at a correlated level (MP2/6-31G*), agree better with the experimentally obtained shifts than those IGLO derived shifts calculated from experimental geometries.[199] By using the *ab initio*/IGLO/

NMR method,[199] the structure of *closo*-$C_2B_6H_8$ was confirmed as having the same bisdisphenoidal framework in the fluid phase as in the crystalline phase, and the [*closo*-$CB_7H_8]^-$ is predicted to have a fluxional *closo* geometry in solution.[201] The *ab initio*/IGLO/NMR method, when applied to *nido*-2,6-$C_2B_6H_{10}$, strongly suggests that the experimental ^{11}B NMR chemical shifts are compatible with a static structure.[202] The NMR spectroscopic data for *nido*-4,5-$C_2B_6H_{10}$ support a gross cage geometry similar to that confirmed for the anion, but indicate that the two bridging hydrogens of the compound are tautomerizing on the NMR spectroscopic timescale. IGLO ^{11}B chemical shift calculations using the 3-21G optimized geometry for *nido*-4,5-$C_2B_6H_{10}$ correlate well with experimentally obtained values.[132] Good correlations are also obtained between the experimental ^{11}B chemical shifts and IGLO shift calculations made on the 6-31G* optimized geometries of both the [*nido*-3,5-$C_2B_6H_9]^-$ ion and 6-Me_3N-*nido*-3,5-$C_2B_6H_8$ adduct (**5**).[203] Other *ab initio*/IGLO/NMR studies have been carried out on the [*closo*-1-$CB_9H_{10}]^-$ ion[200] and on several geometries of formula $C_4B_6H_{10}$.[131] Utilizing Fenske–Hall wave functions and eigenvalues combined with the Ramsey sum of states approximation, the sign and magnitude of the paramagnetic contribution to the shielding correlates reasonably well with the observed ^{11}B chemical shift differences between boron resonances within 2-X-*closo*-1,6-$C_2B_4H_5$ (X = H, F, Cl, I, Me) compounds.[207]

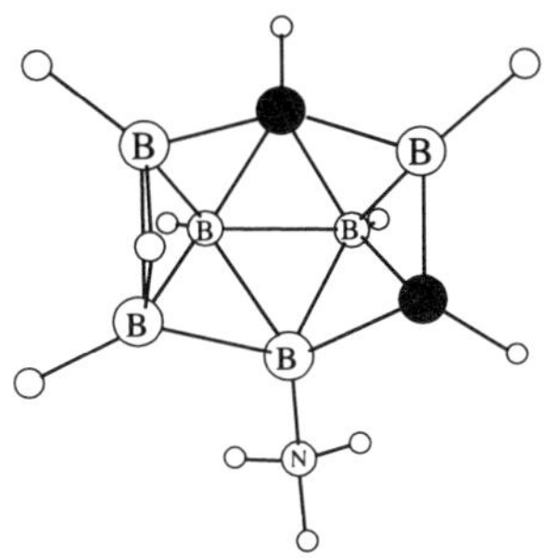

(5)

The ^{11}B and ^{13}C NMR spectra of $Et_4C_4B_6Et_6$ are consistent with a skeletal structure analogous to that of $B_{10}H_{14}$, and there is some evidence of cage fluxional behavior at room temperature.[208] *Arachno*-4,5-$C_2B_7H_{13}$ (**6**), from NMR spectral analyses, may be the correct formula for a compound previously thought to be *nido*-2,6-$C_2B_7H_{11}$.[209–10] A revised structure for the [6,9-$C_2B_8H_{10}]^{2-}$ anion (**7**) is suggested on the basis of NMR spectroscopic data.[211]

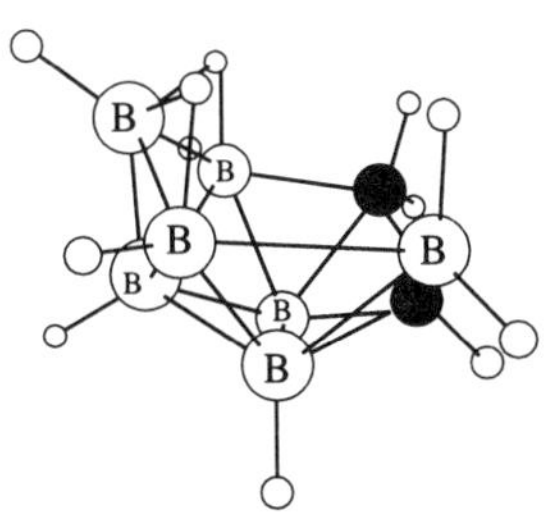

(6)

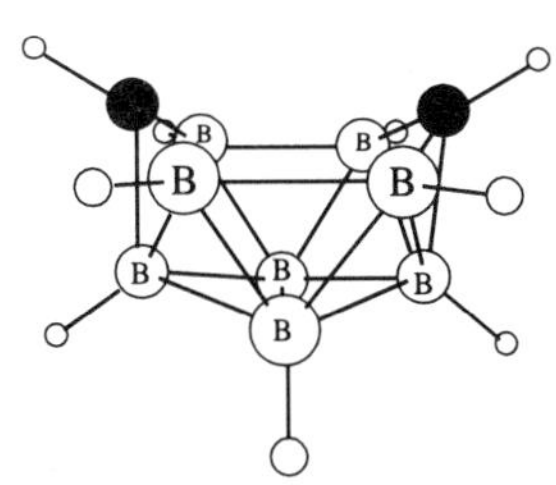

(7)

For the boron atoms in known *closo*-$C_2B_nH_{n+2}$ (n = 3–10) compounds a cage "umbrella" angle, derived from known structural data, is used in an empirical relationship (which includes a dependency on the number of adjacent cage carbon atoms) to successfully predict $^1J(^{11}B^1H)$ values.[212] Measurements of high-resolution ^{11}B NMR spectra of *nido*-2,3-$C_2B_4H_6$-2,3-$(C_2H_5)_2$ in a wide variety of polar solvents reveal solvent-induced shifts ranging from −0.4 to +0.5 ppm compared with those measured in hexane.[213] ^{11}B NMR aromatic-solvent-induced shift (ASIS) effects, observed for *closo*-$C_2B_5H_7$ and two of its B-chloro derivatives, are correlated to atom charge densities;[214] and ^1H ASIS effects for a number of closo-$C_2B_nH_{n+2}$ (n = 3–10) compounds are correlated to PRDDO (partial retention diatomic differential overlap) MO derived hydrogen charge densities.[215]

Substituent effects on the ^{11}B NMR chemical shifts have been correlated among the known F, Cl, Br, and I derivatives of 2,4-$C_2B_5H_7$; two trends are summarized as follows: (a) downfield shifts are observed for boron atoms directly attached to the halogen, with greater shifts observed for boron atoms with smaller halogens; and (b) significant upfield shifts are noted for boron atoms antipodal (i.e., 7-position) to substituted 1-positions, with greater antipodal-boron upfield shifts found when smaller halogens are attached to B-1.[216]

In agreement with predictions from MO calculations $^1J(^{11}B^{11}B)$ for *closo*-pentaalkyl-1,5-dicarbapentaboranes(5) is <10 Hz.[217] The exopolyhedral $^1J(^{11}B^{11}B)$ for 2:2'-[1,5-$C_2B_3H_4]_2$ is found to be 137 Hz which corresponds to ~39% *s*-character for the coupled boron atoms.[218] The exopolyhedral $J(^{11}B^{11}B)$ values for $[2,4\text{-}C_2B_5H_6]_2$ are found to be 119–124 Hz for the 1:3'-isomer, 151 Hz for 3:3'-isomer, and ≥100 Hz for 3:5'-isomer. An intracage coupling constant of 9.5 Hz for B-1–B-5 of 3:3'-$[C_2B_5H_6]_2$ is found.[219] A coupling constant of 9.5 Hz for $^1J(^{11}\text{B-1}^{11}\text{B-5})$ of 2,4-$C_2B_5H_7$ is measured.[220] Resolution enhanced proton decoupled ^{11}B NMR spectra of 5-X-2,4-$C_2B_5H_6$ (X = Cl, Br) exhibit partially relaxed 1:1:1:1 quartets for B-5 and B-6 in each compound with $^1J(^{11}\text{B-5},^{11}\text{B-6})$ values that are in the vicinity of 30 Hz. In addition, $^1J(^{11}\text{B-5}, ^{11}\text{B-6})$ values of 36 Hz for 1,5-Cl_2-2,4-$C_2B_5H_5$, 42 Hz for 3,5-Cl_2-2,4-$C_2B_5H_5$, 25 Hz for $[5\text{-}Me_3N\text{-}6\text{-}Me\text{-}2,4\text{-}C_2B_5H_5]^+$, and 27 Hz for 5-Me-6-Cl-2,4-$C_2B_5H_5$ are found.[218] Two-dimensional ^{11}B–^{11}B NMR spectroscopy confirms many of the B–B atom connectivities in $(Me)_4C_4B_7H_9$, 1,2- and 1,7-$C_2B_{10}H_{12}$.[220] A two-dimensional ^{11}B–^{11}B NMR analysis of 7-CB_8H_{12} suggests that the hydrogen that bridges B-6 and B-8 is involved in a tautomerism that involves terminal hydrogen character on each of these two boron sites.[221] Boron atoms of *arachno*-4-CB_8H_{14}, connected by unsymmetrical bridges, have been shown to give rise to observable coupling in the ^{11}B COSY spectrum.[222] Coupling constants, $J(^{13}\text{C-6–}^{11}\text{B-7})$ ~42 Hz and $J(^{13}\text{C-6–}^{13}\text{C-Me}) = 39$ Hz, have been measured for 6-Me-6-^{13}C-5,6,9-$C_3B_7H_{10}$ (**8**).[223] $J(^{199}Hg–^{11}B)$ values are ca. 1170–1366 Hz for various 9-RHg derivatives of 1,2- and 1,7-$C_2B_{10}H_{12}$.[4a] Values of J(BP) ~300–330 Hz are found for *B*-PO$(OMe)_2$ derivatives of 1,7-$C_2B_{10}H_{12}$ and 1,12-$C_2B_{10}H_{12}$.[159–60]

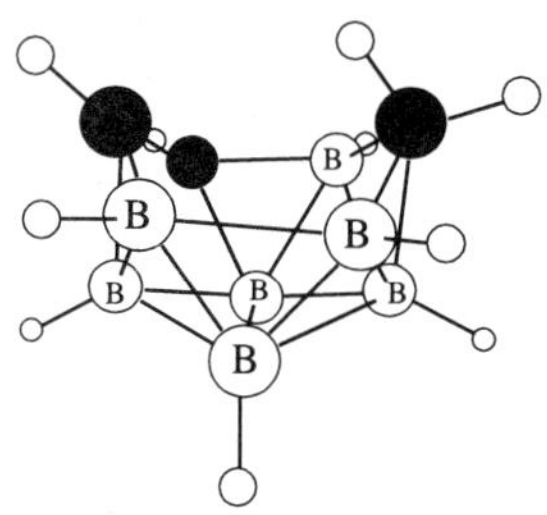

(**8**)

Proton NMR second moments and spin–lattice relaxation times have been measured for 1,2-$C_2B_{10}H_{12}$ over the temperature range 320–77 K.[224] Microdetermination of boron in 1,2-$C_2B_{10}H_{12}$ and 1,7-$C_2B_{10}H_{11}$-9-Br has been carried out by Fourier-transform NMR spectroscopy.[225] ^{11}B and ^{13}C MAS solid-state NMR of 1,2-, 1,7-, and 1,12-$C_2B_{10}H_{12}$ are comparable with those of solution spectra. Substitution by a phenyl group results in extensive broadening of both ^{11}B and ^{13}C signals of the derivative 1-Ph-1,2-$C_2B_{10}H_{11}$.[226] Molecular motion in crystalline phases of 1,2-$C_2B_{10}H_{12}$,[227] 1,7-$C_2B_{10}H_{12}$,[228] and 1,12-$C_2B_{10}H_{12}$[39] have been studied by NMR.

The $[7,8\text{-}C_2B_9H_{12}]^-$ ion has been shown, via NMR spectroscopy of the $(PhCH_2)Me_3N^+$ salt and an x-ray crystallographic study of $[H(DMSO)_2][C_2B_9H_{12}]$, to contain an *endo*-H atom bonded to B-10 and not, as previously thought, a B–H–B bridge. The presence of an *endo*-H atom in $[7,8\text{-}C_2B_9H_{12}]^-$ is rationalized by consideration of the form of the HOMO of $[7,8\text{-}C_2B_9H_{11}]^{2-}$. Analysis of the results of EHMO (extended Hückel molecular orbital) calculations confirms that the *endo*-H atom bonds to the cage via a four-center two-electron bond, and allows the relative magnitudes of *endo* and *exo* B–H coupling constants to be understood as well as the observation of weak coupling between the *endo*-H atom and B-9 and B-11.[145,229] The ^{11}B and ^1H NMR spectroscopic properties of *closo*-1,2-$C_2B_{10}H_{12}$ and *nido*-7,8-$C_2B_9H_{13}$ are correlated to bonding trends.[229]

Additional NMR spectroscopic data are available for *nido*-2,3-$C_2B_4H_6$-2,3-$(C_2H_5)_2$,[230] *nido*-hexaalkyl-2,3,4,5-tetracarbahexaboranes(6),[231] and 4,6-$C_2B_7H_{13}$ (**9**).[93]

(ii) Vibrational spectroscopy

A survey published in 1992[39] of vibrational spectroscopy of carbaboranes and related compounds has outlined most of the important aspects of this subject. The mean ν(BH) frequencies for *closo*-carbaboranes lie in the range 2600–2660 cm^{-1} with little or no correlation to cage size or boron

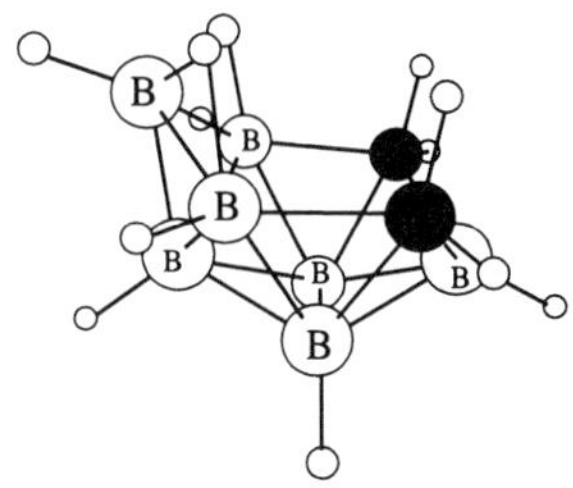

(9)

coordination number. The position of the ν(CH) band depends on the coordination number of the carbon atom, the highest frequency, 3160 cm^{-1}, observed for the smallest *closo* carbaborane, 1,5-$C_2B_3H_5$, with its four-coordinated carbon atoms. When the coordination number increases by "one," the ν(CH) frequency decreases by ~50 cm^{-1}.[39]

The strong B–H stretching absorption bands for the coupled compounds 1:2'-[1,5-$C_2B_3H_4$]$_2$, 2:2',1':2"-[1,5-$C_2B_3H_4$][1',5'-$C_2B_3H_3$][1",5"-$C_2B_3H_4$], and 2:2',3':1"-[1,5-$C_2B_3H_4$][1',5'-$C_2B_3H_3$][1",5"-$C_2B_3H_4$] appear at 2620 cm^{-1}, 2615 cm^{-1}, and 2602 cm^{-1}, respectively.[232] From the IR and Raman spectra of *closo*-1,6-$C_2B_4H_6$ the following assignments have been made: (frequencies in cm^{-1}) 3118 and 3107 (CH stretching); 2667, 2664, and 2647 (BH stretching); 1177, 1096, 799, 775, 474, and 447 (BC stretching); and 1261, 1159, and 986 (BB stretching).[233]

The temperature dependence of the Raman spectrum of the plastic solid carbaborane 1,10-$C_2B_8H_{10}$, studied in the range 15–470 K, shows that the compound has only one plastic phase, probably with anisotropic reorientations, and a phase transition to an ordered crystal takes place at 188 K.[39] The Raman spectrum of two isomers of Cs[$C_2B_9H_{12}$] shows an assigned bridging BHB band at ca. 2080 cm^{-1} for [7,8-$C_2B_9H_{12}$]$^-$ and bands at 2125 cm,$^{-1}$, 1912 cm^{-1}, 1840 cm^{-1}, and 1770 cm^{-1} for [7,9-$C_2B_9H_{12}$]$^-$;[234] problems with the vibrational assignment in these isomers are discussed.[39] The zwitterions 5(6)-Me_2SO-7,8-$C_2B_9H_{11}$ and 6-C_5H_5N-7,8-$C_2B_9H_{11}$ both show extra hydrogen vibrations as weak broad bands in the region around 2000 cm^{-1}, characteristic of B–H–B bridges, as confirmed by a study of the D-analogues.[39] IR spectra are available for *nido*-2,7-$C_2B_9H_{12}$-11-Me and its bridged deuterated analogue, prepared by treating *nido*-2,7-$C_2B_9H_{12}$-11-Me with NaOD/D_2O followed by D_2SO_4.[39]

The IR and Raman C–H stretching bands in each of the *closo*-$C_2B_{10}H_{12}$ isomers nearly coincide, indicating that the two C–H stretching coordinates in each isomer do not couple even when the two carbon atoms are adjacent, as in the 1,2-$C_2B_{10}H_{12}$ case.[39] The Raman spectrum of 1,2-$C_2B_{10}H_{12}$ displays frequency bands at (in cm^{-1}) 475, 584, 716, 773, 918, 986, 1140, 2629, 3066 (C–H stretching).[235–6] Variable temperature vibrational studies of $C_2B_{10}H_{12}$ isomers in the solid phase have been conducted.[3,39,237–8] Vibrational spectroscopy has revealed unusual splitting of the C–H vibration in the carbon-bonded CH_2(1,2-$C_2B_{10}H_{11}$)$_2$.[4a] From IR spectra of 1,2-$C_2B_{10}H_{11}$-X-C≡CH (X = 1,3,9) it is concluded that the ethynyl C–H bond has greater proton-donating capacity than the carbaborane C–H bond. The carboranyl group increases the proton-donating capacity of acetylenic C–H bonds in comparison with C_4H_9C≡CH; this effect decreases in the series $X = 1 > 3 > 9$.[5c,239] Analyses of the vibrational spectra of *o*- and *m*-carbaboranes containing C-, B-3-, or B-9-Me, -vinyl, or -ethynyl groups indicated that the substituent vibrational frequencies for carbon-substituted carbaboranes were higher than those for the boron-substituted compounds.[5c] Solid complexes of boron 1,X-$C_2B_{10}H_2$-*B*-Cl_{10} (X = 1,7) with some oxygen- and nitrogen-containing bases (e.g., dioxane, dimethylformamide, pyridine, R_3N) were investigated by IR and Raman spectroscopy. Complexes containing C–H···O bonds are characterized by a relative C–H stretching frequency shift of up to 12%. C–H···N bonds, with Me_3N, for example, are stronger with a relative shift of approximately 18%. Et_3N complexes form a $NH^+\cdots C^-$ proton transfer hydrogen bond while pyridine can give either C–H···N or $C^-\cdots{}^+HN$ hydrogen-bonded adducts depending on the solvent and temperature.[4a,39] An IR study of 1,12-$C_2B_{10}H_{11}$-2-PO(OMe)$_2$ supports the existance of a CH···OP hydrogen bond.[39]

Other carbaboranes which have been the subject of vibrational studies include: alkyl derivatives of *nido*-CB_5H_9,[240] organic derivatives of *nido*-2,3-$C_2B_4H_8$,[125,241–5] boron substituted alkenyl derivatives of 2,4-$C_2B_5H_7$,[246] 5-[HO(CF_3)$_2$C]-2,4-$C_2B_5H_6$,[247] 5,6-$C_2B_6H_{12}$,[248] *nido*-2,6-$C_2B_7H_{11}$,[249] $Me_4C_4B_7H_8$-11-Br, as well as the two isomers of $Me_4C_4B_7H_9$,[134] *arachno*-6,9-$C_2B_8H_{13}$-6-TMS,[137] alkyl derivatives of 2,3,7,8-$C_4B_8H_{12}$ (**3**)(**4**),[243–4,250] salts of [CB_9H_{10}]$^-$,[251–2] [$CB_{10}H_{11}$]$^-$,[253] $CB_{10}H_{12}$ derivatives,[156] 1,2-$C_2B_{10}H_{11}$-9-Me,[254] substituted alkyl derivatives of 1,2-$C_2B_{10}H_{12}$,[255] bis-9-*m*-carboranyl ether,[39] 3- and 9-F-1,2-$C_2B_{10}H_{11}$,[39] 1,2-$C_2B_{10}H_{11}$-1-CH_2PPh_2,[256] Ni(X)(1,2-$C_2B_{10}H_{10}$-1-P[NMe_2]$_2$-2-PPh_2) (X = Cl, Br, I, SCN),[257] the boron-bonded derivatives TlL_2Cl (HL = 1,2- and 1,7-$C_2B_{10}H_{12}$) and TlL_2Br (HL = 1,2- and 1,7-

$C_2B_{10}H_{12}$), and carbon-bonded TIL_2Cl (HL = 1,2-$C_2B_{10}H_{12}$),[39] mercury derivatives of 1,7- and 1,12-$C_2B_{10}H_{12}$,[39] tin derivatives of 1,2- and 1,7-$C_2B_{10}H_{12}$,[258] (1,7-$C_2B_{10}H_{11}$-9-)$_2$,[259] M[$CB_{11}H_{12}$] salts,[260] and [$ML_4(H_2O)_2$][$CB_{11}H_{12}$]$_2$·H_2O (M = Co, Ni; L = *m*-nitrobenzoylhydrazine).[251]

6.2.3.3 Mass spectrometry

Electron resonance capture mass spectra of the radical anions of 1,2-$C_2B_{10}H_{12}$ and 1,2-$C_2B_{10}H_{11}$-1-Me in the gas phase confirm the existence of long-lived negative ions in these carbaboranes.[261] An estimation of charge distributions in 1,2- and 1,7-$C_2B_{10}H_{12}$ have been derived from a photoionization mass spectrometric study of various vinyl and aryl derivatives.[5a] Adiabatic ionization energies of various boron-halo derivatives of 1,2- and 1,7-$C_2B_{10}H_{12}$, determined by photoionization mass spectrometry, indicated that there is conjugation between nonbonding orbitals of the halogens and the carbaborane electronic system. A relation between ionization energy and halogen electronegativity was obtained for monohalo derivatives, and the electron density distribution in the 9, 8, and 4 positions of the 1,2-$C_2B_{10}H_{12}$ icosahedron was determined.[262] Mass spectroscopic data have also been reported for alkyl derivatives of *nido*-2,3-$C_2B_4H_8$,[125,242–3,245] for $R_2C_4B_8H_{10}$ (R = indenylmethyl, fluorenylmethyl) compounds,[243] for $Me_4C_4B_7H_8$-11-Br, for the two isomers of $Me_4C_4B_7H_9$,[126] for 1,2-, 1,7-, and 1,12-$C_2B_{10}H_{12}$,[4d] for 1,2-$C_2B_{10}H_{10}$-1,2-(CH_2O-)$_2$ spiro derivatives of cycloperchlorotriphosphazene ($Cl_6P_3N_3$),[5a] for 1,X-$C_2B_{10}H_{10}$-1-R-9-HgX (X = 2, 7, R = H, Me; X = Cl, Br, I), for (1,X-$C_2B_{10}H_{10}$-1-R-B-)$_2$Hg (X = 2, 7, R = H, Me; X = 12, B = 2, R = H), and for 1,12-$C_2B_{10}H_{11}$-2-HgCl.[263]

6.2.3.4 Additional physical properties

From dipole moment determinations on seven polyfluoro derivatives of 1,2- and 1,7-$C_2B_{10}H_{12}$ it is suggested that decreased dipole moments of the 1,7-$C_2B_{10}H_{12}$ derivatives are due to interactions of the orbitals of the unshared pair of electrons on fluorine with *p*-orbitals of the carbaborane.[264]

The standard enthalpies of combustion, formation, and vaporization of 1,6-$C_2B_4H_6$ and 2,4-$C_2B_5H_7$ have been measured.[4c] Thermochemical characteristics for 1-Me, 9-Me,[162] 1-Me_3C,[2c] and 9-hexyl[265] derivatives of 1,2-$C_2B_{10}H_{12}$ have been determined. The experimental boron nuclear quadrupole couplings for the three $C_2B_{10}H_{12}$ isomers are in good agreement with *ab initio* calculations of the electric field gradient tensors.[266–7] Measurements on the various boron environments of 1,2-$C_2B_{10}H_{12}$ point to ^{10}B–^{11}B quadrupole coupling constant ratios of ~2.[268–70] The ^{127}I NQR spectra of [1,2-$C_2B_{10}H_{11}$-9-IC_6H_5]BF_4, [1,7-$C_2B_{10}H_{11}$-9-IC_6H_5]BF_4, and [1,12-$C_2B_{10}H_{11}$-2-IC_6H_5]BF_4 show quadrupole coupling constants (e^2Qq_{zz}, in MHz) of 1554.37, 1600.61, and 1598.54, respectively.[4c]

Short-lived absorption spectra of a methanolic solution of 1,12-$C_2B_{10}H_{12}$, under the action of a pulse of accelerated electrons, have been recorded over the whole spectral range from the UV to the IR spectra.[271] The UV spectra of 1-C_6H_5-1,12-, 2-C_6H_5-1,7-, 9-C_6H_5-1,2-, and 9-C_6H_5-1,7-$C_2B_{10}H_{11}$ show there to be little or no conjugation between the *closo*-carboranyl groups and the benzene ring. In the spectra of 3-Ph-7,8-, 5-Ph-7,8-, 5-Ph-7,8-, 1-Ph-7,9-, and 7-Ph-7,9-derivatives of the [$C_2B_9H_{12}$]$^-$ ion, the vibrational structure of the electronic band remains, its maximum being situated on the shoulder of a stronger absorption from the *nido*-polyhedron.[5c] UV spectra have also been reported for alkyl derivatives of *nido*-$C_2B_4H_8$,[245] various salts of the [$CB_{10}H_{11}$]$^-$ ion,[253] 1,2- and 1,7-$C_2B_{10}H_{12}$,[271] and M[$CB_{11}H_{12}$].[260]

The first two band systems in the UV photoelectron spectrum of 1,12-$C_2B_{10}H_{11}$ are assigned to cluster π-bonding orbitals and σ-B–H(C–H)-bonding orbitals, respectively. The interaction between the *exo*-substituent and cluster π-orbitals (in 1-SH-1,12-$C_2B_{10}H_{11}$, 2-SH-1,12-$C_2B_{10}H_{11}$, 1-X-1,2-$C_2B_{10}H_{11}$ (X = SH, Ph), 3-Ph-1,2-$C_2B_{10}H_{11}$, 4-X-1,2-$C_2B_{10}H_{11}$, 8-X-1,2-$C_2B_{10}H_{11}$, 9-X-1,2-$C_2B_{10}H_{11}$, 9-X-1,7-$C_2B_{10}H_{11}$ (X = SH, MeS) and 1-SH-1,7-$C_2B_{10}H_{11}$) is weak. The electron density at particular vertices in 1,2-$C_2B_{10}H_{12}$ decreases in the order $9 \geq 8 > 4 > 3 > 1$.[272] The He^I PES results for 5-X-2,4-$C_2B_5H_6$ and 5,6-X_2-2,4-$C_2B_5H_5$ (X = Cl, Br, I) indicate that, after inductive effects are taken into account, π-type interactions between cluster surface (π-*endo*) orbitals and filled halogen *p*-orbitals account for the observed substituent effects on the photoelectron spectra of the substituted carbaborane.[102] The He^I photoelectron spectra of 1,10-$C_2B_8H_{10}$ and 1,12-$C_2B_{10}H_{12}$ have been reported.[273] The He^I and He^{II} photoelectron spectra of 1,2- and 1,7-$C_2B_{10}H_{12}$ are assigned in terms of correspondence of bands to B–H bonding or cluster-localized orbitals. The experimental assignments agree with MNDO calculations on the molecules, which allowed assignment of the cluster orbitals and ionizations in terms of Stone's (1980, 1981) theory of *closo*-boranes.[274–5] He^I photoelectron spectra of 1,X–$C_2B_{10}H_{11}$-9-Y (X = 2,7; Y = Cl, Br, I) support the contention that there is conjugative interaction of cluster π-type orbitals with the halogen nonbonding orbitals in the case of Y = Cl.[275]

The pulse radiolysis of 1,2-, 1,7-, and 1,12-$C_2B_{10}H_{12}$ give short-lived absorption spectra of di- and monoanions.[3] Angle-resolved photoemission provides evidence that initial adsorption of 2,3-Et_2-*nido*-2,3-$C_2B_4H_6$ on silicon surfaces at 90 K occurs with loss of ethyl groups; assignment of the photoemission features, and the molecular orientation of the cage, is aided by MNDO calculations.[276] Inner shell electron energy-loss spectroscopy of 2,3-Et_2-*nido*-2,3-$C_2B_4H_6$ and *closo*-1,2-$C_2B_{10}H_{12}$ have been correlated with MNDO calculational studies.[277]

The zero-field splitting parameters of the triplet ESR spectra of the carboranylcarbene 1,2-$C_2B_{10}H_{11}$-1-CH and its 2-Me-derivative indicate little delocalization of the π-spin density onto the carboranyl moiety.[278] The ESR spectra of carboranyl-substituted arenecyclopentadienyl complexes of iron are interpreted in terms of steric interactions.[279] Photolysis of 1,12-$C_2B_{10}H_{11}Tl(O_2CCF_3)_2$, 1,2-$C_2B_{10}H_{11}InCl_2$(THF), or 1,2-$C_2B_{10}H_{11}Ga(O_2CCF_3)_2$ in PhMe containing 3,5- or 3,6-di-*t*-butyl-*o*-benzoquinone gives complex carbaborane radicals in the ESR. Photolysis of 1,7-$C_2B_{10}H_{11}$ TlXR (X = R = Cl; X = Br, R = Ph, Et) with these benzoquinones gave radicals with the B–Tl bond still intact.[35] Other ESR studies have focused on the decomposition of the phosphorus–mercury bond in BHgP(O)$(OR)_2C_2B_{10}H_{11}$ by UV light,[35] on ketone adducts of borocentered 1,2-$C_2B_{10}H_{12}$, 1,7-$C_2B_{10}H_{12}$, and 1,12-$C_2B_{10}H_{12}$ radicals,[4c] and on the interaction of metal-centered free radicals with a-diketones containing 1-methyl-*o*-and *m*-carboranyl groups.[280] The boron K-edge spectra of 1,2-$C_2B_{10}H_{12}$ was investigated to characterize the sensitivity of such spectra to the local environment.[281]

The microwave structure of 5-F-*closo*-2,4-$C_2B_5H_6$ indicates a slightly shortened B–F bond length, thus suggesting some B–F double bond character.[282] The redox behavior of 1,2-, 1,7-, and 1,12-$C_2B_{10}H_{12}$ is significantly changed by incorporating tin and mercury substituents on a boron atom.[5a] From trends in the values of the substituent constants σ_I and σ_R (determined by the ^{19}F NMR Taft method) for the 9-*m*-carboranylhalonium groups $[9\text{-}m\text{-}C_2B_{10}H_{11}X]^+$ (X = Cl, Br, I), it is proposed that regioselectivity and regiospecificity in nucleophilic substitution reactions of these systems are determined by steric, not electronic, factors.[283]

6.2.4 Syntheses

6.2.4.1 Carbaboranes from polyboranes and alkynes

A *hypho*-$C_3B_4H_{12}$ has been synthesized by a "hot–cold" reaction of B_4H_{10} with propyne.[284] It is suggested, from the results of a theoretical study (STO-3G level) on the reactions of B_4H_{10} with ethene and ethyne to form hydrogen and the *closo*-carbaboranes 1,2- and 1,6-$C_2B_4H_6$, that the driving force for these reactions (via *arachno* and *nido* precursors) is provided by a progressive increase in the B–C bonding accompanied by transfer of electronic charge to the carbon atoms.[285]

The *nido*-2,3-$C_2B_4H_6$-2,3-R^1,R^2 species (e.g., R^1 and/or R^2 = $Me(CH_2)_n$, $Ph(CH_2)_n$, indenylmethyl, fluorenylmethyl) are obtained from the reaction of corresponding substituted acetylenes with B_5H_9 in the presence of Et_3N;[125,241–4,286–7] the diphenyl compound cannot be prepared in this manner, (i.e., with Et_3N) and is obtained instead in a thermal reaction of B_5H_9 with diphenylacetylene in the absence of amine.[244] Open-chain dialkynes EtC≡C$(CH_2)_n$C≡CEt (*n* = 4, 6), the trialkyne, HC≡C$(CH_2)_5$C≡C$(CH_2)_5$C≡CH, and certain cyclic dialkynes when treated with B_5H_9 and Et_3N give the corresponding *nido*-carbaborane oligomers.[245] 2,3-$C_2B_4H_6$-2-TMS-3-R (R = TMS, Me, H) is prepared from B_5H_9 and TMS-C≡CR.[195]

Closo-1,2-$C_2B_8H_8$-1,2-Me_2 is formed from the reaction of 1,2'-$(B_5H_8)_2$ with 2-butyne catalyzed by 2,6-dimethylpyridine.[288] The yield of *nido*-5,6-$C_2B_8H_{10}$-5,6-Me_2 from $B_9H_{12}\cdot OEt_2$ and MeC≡CMe is maximized when the reactants are employed in a 4:1 ratio of butyne:$B_9H_{13}\cdot OEt_2$. This implies that hydroboration may be directly or indirectly responsible for removing one boron atom from the nonaborane cage as indicated in Equation (1). In a similar fashion the parent 5,6-$C_2B_8H_{12}$ is prepared from acetylene and $B_9H_{13}\cdot OEt_2$.[289]

$$B_9H_{13}\bullet OEt_2 + 4\,MeC{\equiv}CMe \longrightarrow 5,6\text{-}C_2B_8H_{10}\text{-}5,6\text{-}Me_2 + B(MeC{=}CHMe)_3 + Et_2O \quad (1)$$

The Me_2S-promoted reaction of $B_{10}H_{14}$ with TMS-C≡C-TMS produces 9-(Me_2S)-7-$[(TMS)_2CH]CB_{10}H_{11}$. The cluster geometry of the product is based on an icosahedron missing one vertex, with an exopolyhedral $(TMS)_2CH$ group bonded to the cage carbon.[156] Acetylenic cyclotriphosphazenes react with $(MeCN)_2B_{10}H_{12}$ to give phosphazene derivatives of 1,2-$C_2B_{10}H_{12}$.[290] (1,2-$C_2B_{10}H_{11}$-1-CH_2O-$)_2$P(R)X (R = Me, X = S; R = Me, Ph, X = O) is formed from the reaction of $B_{10}H_{14}$ with $(HC{\equiv}CCH_2O)_2$P(R)X in the presence of $PhNMe_2$. Similarly, (1,2-$C_2B_{10}H_{11}$-1-CH_2O-$)_3$PO is obtained from $B_{10}H_{14}$ and $(HC{\equiv}CCH_2O)_3$PO.[291] 1,2-(1',2'-$C_2B_{10}H_{10}$-1'-$MeCO_2CH_2$-2'-$)_2$-C_6H_4, is formed from $B_{10}H_{14}$/1,2-($MeCO_2CH_2C{\equiv}C$-)$_2$-C_6H_4/C_5H_5N.[4b] A series of ether carbaboranes 1-R^1-2-R^2-1,2-*closo*-$C_2B_{10}H_{10}$ ($R^1 = R^2 = CH_2OMe$; $R^1 = CH_2OMe$, $R^2 = H$; $R^1 = (CH_2)_2OMe$, $R^2 = H$; $R^1 = (CH_2)_3OMe$, $R^2 = H$) are prepared from $B_{10}H_{14}$ and corresponding acetylenic ethers in the presence of either Me_2S or MeCN.[142] Additional carbaboranes that have been prepared from $(MeCN)_2B_{10}H_{12}$ and the appropriate substituted alkyne are: (1,2-$C_2B_{10}H_{11}$-1-$CH_2CH_2CH_2$-)$_2$,[292] (1,2-$C_2B_{10}H_{11}$-1-)$_2CH_2$,[293] (1,2-$C_2B_{10}H_{11}$-2)-uridine compounds,[294–5] nitroimidazole-linked,[296] pyrimidine-linked,[297] and steroidal-linked[298] carbaboranes.

6.2.4.2 Cage growth as a result of carbon insertion reactions into carbaboranes

The reaction of *nido*-2,3-$C_2B_4H_6$-2,3-Et_2 with 2-butyne in the presence of NaH and $NiCl_2$ results in a two-carbon insertion into the cage system to produce the tetracarbon carbaborane *nido*-4,5,7,8-$C_4B_4H_4$-4,5-Me_2-7,8-Et_2.[299] The reaction of *arachno*-4,5-$C_2B_7H_{13}$ (**6**) with acetylenes in hexane at 120 °C gives rise to two representatives of the 10-vertex family of tricarbaboranes, *arachno*-μ-6,9-RCH-9-R-5,6,9-$C_3B_7H_{11}$ (R = H (**10**), Me) and *nido*-5,6,10-Me_3-5,6,10-$C_3B_7H_8$,[300] as well as the tetracarbaboranes *nido*-5,6,8,9-$C_4B_6H_{10}$ and *nido*-2,7,8,11-$C_4B_7H_{11}$.[301] The [*arachno*-6,9-$C_2B_7H_{12}$]$^-$ anion reacts with cyanoacetylene to give the polyhedral *arachno*-tricarbon carbaborane, *arachno*-6-($NCCH_2$)-5,6,7-$C_3B_7H_{12}$.[133] The reaction of MeCN with the [*arachno*-$C_2B_7H_{12}$]$^-$ ion results in a carbon insertion product, [*nido*-$MeC_3B_7H_9$]$^-$; acidification of the latter ion gives *nido*-6-Me-5,6,9-$C_3B_7H_{10}$ (**8**).[302–3] When $Me^{13}CN$ is used in this sequence 6-Me-6-^{13}C-5,6,9-$C_3B_7H_{10}$ is obtained.[223]

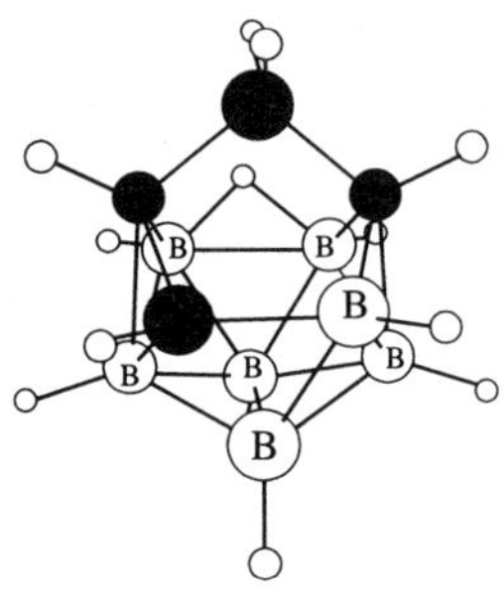

(10)

6.2.4.3 Cage growth as a result of boron insertion reactions

The reaction of 1,5-$C_2B_3H_5$ and BH_3 to give 1,6-$C_2B_4H_6$ and H_2 is predicted to be 139.4 kJ mol^{-1} exothermic at the [MP2/6-31G*](+ZPC) *ab initio* level of theory; the formation of *nido*-2,3-$C_2B_4H_8$ from the same two reagents is calculated to be 138.9 kJ mol^{-1} exothermic.[72] Reaction of 2,3-Et_2-*nido*-2,3-$C_2B_4H_6$ with $Et_3N{\cdot}BH_3$ at 140 °C gives 2,3-Et_2-*closo*-2,3-$C_2B_5H_5$;[304] when carried out at 170 °C the reaction also produces 1,7-Et_2-*closo*-1,7-$C_2B_6H_6$, and 1,7-Et_2-*closo*-1,7-$C_2B_7H_7$. 2,3-Et_2-*closo*-2,3-$C_2B_5H_5$ is also prepared by the reaction of Li[*nido*-2,3-Et_2-*nido*-2,3-$C_2B_4H_5$] with $Me_2S{\cdot}BH_3$ or by the reaction of NaLi[2,3-Et_2-*nido*-2,3-$C_2B_4H_4$] with $Me_2S{\cdot}BHBr_2$. Reaction of NaLi[2,3-$Et_2C_2B_4H_4$] with $PhBCl_2$ or $MeBBr_2$ affords the mono-B-substituted compounds *closo*-1-Ph-2,3-Et_2-*closo*-2,3-$C_2B_5H_4$, and 1-Me-2,3-Et_2-*closo*-2,3-$C_2B_5H_4$, respectively.[305]

The *arachno*-carbaborane 5,6-$C_2B_6H_{12}$ (**11**) is prepared from the $PtBr_2$ catalyzed reaction of B_2H_6 with 1,5-$C_2B_3H_5$.[248] *Arachno*-6,9-$C_2B_8H_{13}$-6-TMS (**12**) is prepared by the cage expansion reaction of Na[2,3-$C_2B_4H_6$-2,3-$(TMS)_2$] with B_5H_9.[137]

Treatment of 7-$CB_{10}H_{12}$-7-NMe_3 with Me_3NBH_3 at 180–200 °C proceeds by splitting off one methyl group and inserting one boron vertex to obtain 1-$CB_{11}H_{11}$-1-$NHMe_2$.[306] Thermal decomposition of [$BH_2(NMe_3)_2$][*nido*-7-$CB_{10}H_{13}$] at 270 °C produces 7-Me_3N-*closo*-1-$CB_{11}H_{11}$ along with the 2- and 12-Me_3N-1-$CB_{11}H_{11}$ isomers as side products.[307] Both *nido*-7-Me_3N-7-$CB_{10}H_{12}$ and its 4-Cl derivative undergo cage expansion reactions with $Et_3N{\cdot}BH_3$ to give [Et_3NMe][1-Me_2N-1-$CB_{11}H_{11}$].[5a,106] Reactions of $(C_2H_5)_3N{\cdot}^{10}BH_3$ with both *nido*-[7-$CB_{10}H_{13}$]$^-$ and 7-Me_3N-*nido*-7-$CB_{10}H_{12}$ give, respectively, *closo*-[1-$CB_{11}H_{12}$]$^-$ and [1-Me_2N-1-$CB_{11}H_{11}$]$^-$ labeled by ^{10}B exclusively at the B-2 site. Apparent cage rearrangement is observed in the reaction of 7-Me_3N-8-$C_6H_5CH_2$-*nido*-7-$CB_{10}H_{11}$ with Et_3NBH_3 to produce only the 1-Me_3N-7-$C_6H_5CH_2$-1-$CB_{11}H_{11}$ *closo* isomer.[307] A cage capping reaction to produce [*closo*-$CB_{11}H_{12}$]$^-$ is found upon reacting *nido*-8-Me_2S-7-$CB_{10}H_{12}$ with $Bu_3P{\cdot}BH_3$.[155] Similarly, S_2(7-

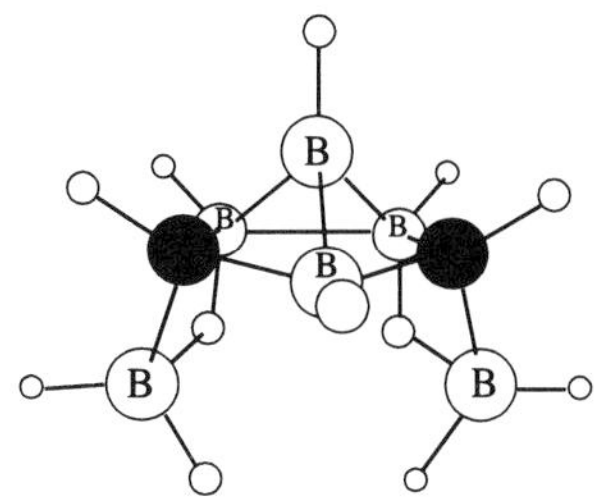

(11)

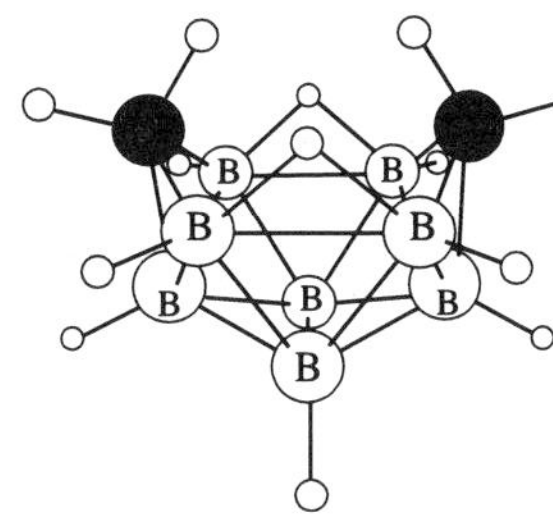

(12)

Me_3N-*nido*-7-$CB_{10}H_{11})_2$ undergoes boron insertion with Et_3NBH_3 to give [Et_3NMe][1-Me_2N-2-HS-*closo*-1-$CB_{11}H_{10}$].[190]

6.2.4.4 Cage fusion reactions

Dimerization of *closo*-1,5-C_2B_3-1,5-Me_2-2,3,4-Et_3 occurs upon treatment with potassium followed by I_2; an adamantane-like structure is proposed for the product which in turn rearranges at 160 °C to a $Me_4C_4B_6(C_2H_5)_6$ carbaborane containing an icosahedral framework minus two adjacent vertices.[131,208,308] The *nido*-1,2,8,10-$C_4B_7H_{11}$ is produced from the gas-phase pyrolysis of 1,5-$C_2B_3H_5$ at 400 °C.[232]

The carbaborane fusion products $R_4C_4B_8H_8$ (R = alkyl, phenyl) are prepared from the treatment of the bridge-deprotonated anions of the respective *nido*-2,3-$C_2B_4H_6$-2,3-R_2 with $FeCl_2/O_2$. The $R_4C_4B_8H_8$ compounds exist in solution as either open-cage (**4**) or closed-cage (**3**) isomers or a mixture of the two.[125,244,250,309] Disproportionation of $(R_2C_2B_4H_4)_2FeH_2$ to $R_4C_4B_8H_8$ and $(R_2C_2B_4H_4)_2Fe_2(THF)_2$ is catalyzed by traces of $FeCl_3$ in THF. At $FeCl_3/(R_2C_2B_4H_4)_2FeH_2$ molar ratios of 2.0 or greater, in THF, $R_4C_4B_8H_8$ is quantitatively produced.[250,309] The fusion of two C_2B_4 cages to give $R^1{}_2R^2{}_2C_4B_8H_8$ (R^1 = Et, Bu^n, CH_2Ph; R^2 = Me, Et, Bu^n, CH_2Ph) has been accomplished by reacting $Na^+[R^2{}_2C_2B_4H_5]^-$ salts with $(R^1{}_2C_2B_4H_4)Fe(\eta^6$-$C_8H_{10})$ complexes at 150 °C in the absence of solvent.[310] Boron-substituted *nido*-2,3-$C_2B_4H_5$-2,3-$R^1{}_2$-4-R^2 ($R^1 = R^2$ = Et, $PhCH_2$) compounds are bridge-deprotonated by NaH, complexed to Fe^{2+}, and oxidatively fused to produce the hexasubstituted $R^1{}_4R^2{}_2C_4B_8H_6$ carbaboranes.[311] The monosubstituted [*nido*-2,3-$C_2B_4H_6$-2-R]$^-$ anions (R = indenylmethyl, fluorenylmethyl) undergo slow Fe^{2+}-promoted fusion to form $R_2C_4B_8H_{10}$. However, the corresponding disubstituted $[R_2C_2B_4H_5]^-$ anions do not fuse, which is attributed to steric hindrance.[243] The oligomeric $C_2B_4H_8$ derivatives produced from the reaction of polyalkynes and B_5H_9 in the presence of Et_3N are converted to products incorporating C_4B_8 cages by oxidative fusion methods.[245] Mixed-ligand fusion of the anions of *nido*-2,3-$(PhCH_2)_2C_2B_4H_6$ and $[(OC)_3Cr](PhCH_2)_2C_2B_4H_6$ in the presence of $FeCl_2$ gives $(PhCH_2)_4C_4B_8H_8$.[312] *Nido*-2,3-$C_2B_4H_6$-2,3-[TMS]$_2$ undergoes thermolytic fusion at 210 °C to give *nido*-2,3,7,8-$C_4B_8H_{10}$-2,7-[TMS]$_2$.[313]

6.2.4.5 Dehydrogenation reactions

Thermal decomposition of Na[*arachno*-$C_4B_7H_{12}$] at 413 K gives *nido*-7,8,9,10-$C_4B_7H_{11}$.[301] 4-CB_8H_{14} is dehydrogenated at 623 K to give 4-(or 7-)CB_8H_{12}.[92–3] 8-Me_3N-9-Me-*arachno*-6-CB_9H_{12} is converted to 8-Me_3N-9-Me-*nido*-6-CB_9H_{10} by the use of K_2CO_3/Me_2CO.[314] Upon thermolysis *nido*-8-Me_2S-7-$CB_{10}H_{12}$ undergoes a cage closure reaction to give *closo*-4-Me_2S-2-$CB_{10}H_{10}$.[155]

6.2.4.6 Selective cage degradation

The quantitative degradation of *closo*-2,4-$C_2B_5H_7$ to [*nido*-2,4-$C_2B_4H_7$]$^-$ is accomplished with the use of Li[N(R)$_2$], R = alkyl; a short lived intermediate in the reaction is proposed to be [$\mu_{5,6}$-(R_2NBH)-*nido*-2,4-$C_2B_4H_6$]$^-$ (**13**).[315] The structural assignment for the intermediate is supported by *ab initio*/IGLO/NMR calculations.[204] The fluoride ion in aprotic solvents is effective in promoting a quantitative "debor" reaction on the *closo*-carbaboranes 2,4-$C_2B_5H_7$, 1,2-$C_2B_{10}H_{12}$, and 1,7-$C_2B_{10}H_{12}$; the obtained *nido*-carbaborane anions are [2,4-$C_2B_4H_7$]$^-$, [7,8-$C_2B_9H_{12}$]$^-$, and [7,9-$C_2B_9H_{12}$]$^-$, respectively.[316]

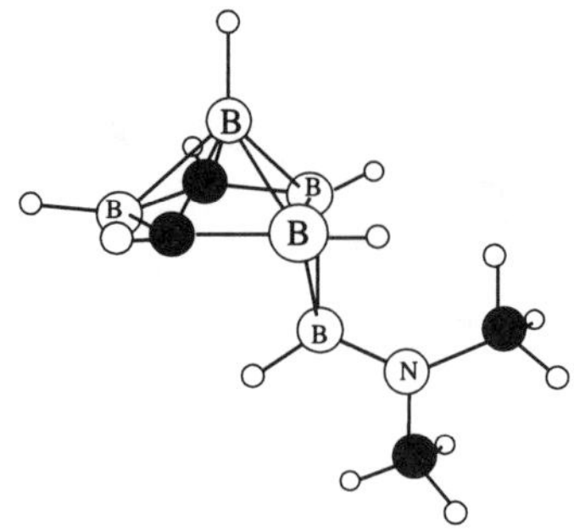

(13)

Arachno-$[4,5\text{-}C_2B_6H_{11}]^-$ (**14**) with C_s symmetry is produced via selective degradation of *closo*-1,2-$C_2B_8H_{10}$.[317] Reaction of *arachno*-4,5-$C_2B_7H_{13}$ (**6**) carbaborane with H_2O/NaCN produces the *hypho*-$[C_2B_6H_{13}]^-$ anion (**15**).[318] The acetonitrile-induced degradation of [*arachno*-4,5-$C_2B_7H_{12}]^-$ gives rise to [*nido*-4,5-$C_2B_6H_9]^-$, which in turn is converted to *nido*-4,5-$C_2B_6H_{10}$ by protonation of the ion with HCl.[132]

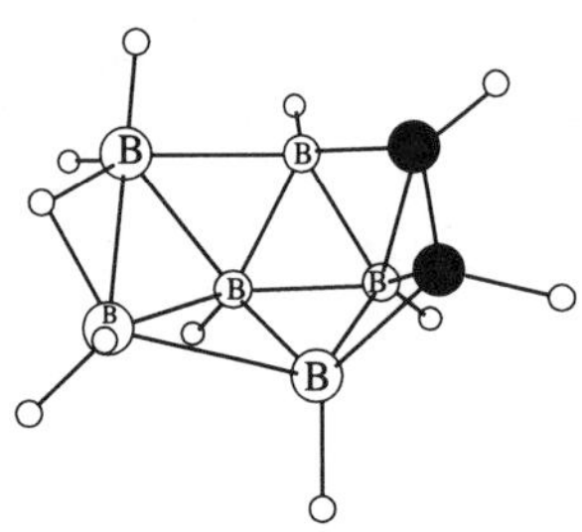

(14)

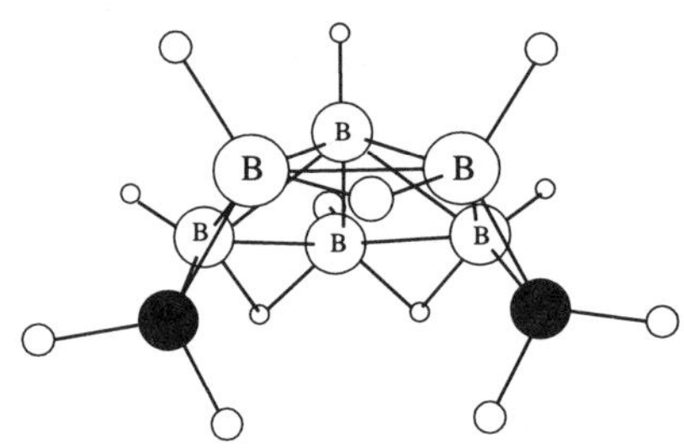

(15)

Reaction of 6-Me_3N-*arachno*-4-CB_8H_{12} with NaH/THF affords the *closo*-$[1\text{-}CB_7H_8]^-$ ion.[314] Degradation of 6-Me_3N-6-CB_9H_{11} with KOH/MeOH yields $\mu_{3,4}$-Me_3N-CB_5H_{11}.[92] The reaction of Na[*nido*-6-CB_9H_{12}] with $FeCl_3$/HCl/H_2O yields *arachno*-4-CB_8H_{14}.[92–3] Treatment of K[7,8-$C_2B_9H_{12}$] with H_2CO/HCl/H_2O gives rise to *nido*-2,6-$C_2B_7H_{11}$.[249] Degradation of $[7,9\text{-}C_2B_9H_{12}]^-$ with $NaNO_2$/HCl affords *arachno*-6,8-$C_2B_7H_{13}$.[319]

The 9-Me-10-HO-*nido*-$[7\text{-}CB_{10}H_{11}]^-$ ion is formed from the action of K_2CO_3/H_2O on the reactive isomer of the $[C_2B_{10}H_{13}]^-$ anion; treatment of 9-Me-10-HO-*nido*-$[7\text{-}CB_{10}H_{11}]^-$ with HCl/H_2O produces *arachno*-4-CB_7H_{13} (**16**).[314] Treatment of the reactive isomer of the $[C_2B_{10}H_{13}]^-$ anion with Me_3N/H_2O/THF gives 8-Me_3N-9-Me-*arachno*-6-CB_9H_{12} and 6-Me_3N-*arachno*-4-CB_8H_{12}.[314] The reaction of $Na_2[C_2B_{10}H_{12}]$ with $Tl(O_2CMe)$ leads to degradation of the C_2B_{10} cage, and affords $Tl_3(C_2B_9H_{12})O$.[320] A rapid high yield conversion of 1,12-$C_2B_{10}H_{12}$ to [K(18-crown-6)][*nido*-2,9-$C_2B_9H_{12}$] in the presence of KOH/18-crown-6 ether is the best method to prepare the [*nido*-2,9-$C_2B_9H_{12}]^-$ ion.[321] Treating 9-F-, 9,12-F_2-, 8,9,12-F_3- (**A**), or 8,9,10,12-F_4 derivatives of 1,2-$C_2B_{10}H_{12}$ with NaOH in MeOH gives 6-F-, 1,6-F_2-, 1,5,6-F_3- (**B**) and 5,6,10-F_3- (**C**), or 1,5,6,10-F_4- derivatives of $[7,8\text{-}C_2B_9H_{12}]^-$ anions, respectively. Regioselectivity of this debor reaction is noted in the reaction of (**A**), which gives two isomeric products, (**B**) and (**C**), in a 1:2 ratio.[5c] Refluxing $[NH_4]_2[1,2\text{-}C_2B_{10}H_{10}\text{-}1,2\text{-}(S)_2]$ in EtOH containing 1,2-Cl_2-C_6H_4/NaI, followed by the addition of aqueous Me_4NCl, yields $[Me_4N][7,8\text{-}C_2B_9H_{10}\text{-}7,8\text{-}(SH)_2]$.[153] Similar chemistry utilizing $BrCH_2CH_2Br$ instead of 1,2-Cl_2-C_6H_4/NaI as a reagent results in the formation of $[Me_4N]_2[(CH_2S)_2C_2B_9H_{10}]_2$.[322] A mechanism that explains the degradation pathways of the *C,C'*-dithiolato and monothiolato derivatives of $C_2B_{10}H_{12}$ to their respective $[C_2B_9H_{12}]^-$ derivatives is proposed.[141] 7,9-$C_2B_9H_{11}$-B-NC_5H_5, is produced from the action of pyridine on $[1,7\text{-}C_2B_{10}H_{11}\text{-}9\text{-}Br^+Ph][BF_4]^-$.[4e] Treating 3-$NH_2$-1,2-$C_2B_{10}H_{11}$ with KOH/EtOH (or with piperidine) gives $[3\text{-}NH_2\text{-}7,8\text{-}C_2B_9H_{11}]^-$ and the parent $[7,8\text{-}C_2B_9H_{12}]^-$.[323] Action of KOH on $RCB_{10}H_{10}CN{=}NCB_{10}H_{10}CR$ (R = H, Me, Ph) produces both *closo–nido*-$[RCB_{10}H_{10}CN{=}NCB_9H_{10}CR]^-$ and *nido–nido*-$[RCB_9H_{10}CN\text{–}NCB_9H_{10}CR]^{2-}$ ions. The *closo–closo* compound $RCB_{10}H_{10}CNHNHCB_{10}H_{10}CR$ can be converted into the corresponding *closo–nido*-$[RCB_{10}H_{10}CNHNHCB_9H_{10}CR]^-$ ions in the presence of piperidine base.[324] 1-Methyl-1-[(1,2-$C_2B_{10}H_{11}$-1-)methylene]-3,3,5,5-tetrachlorocyclotriphosphazene is converted by a base such as $C_5H_{10}NH$ to the respective [*nido*-$C_2B_9H_{12}]^-$ derivative ion.[290] Action of KOH/EtOH on (1,2-

$C_2B_{10}H_{11}$-1-$CH_2CH_2CH_2)_2$ yields the [7-(1',2'-$C_2B_{10}H_{10}$-1'-$(CH_2)_6$)-7,8-$C_2B_9H_{11}]^-$ ion.[292] Partial cage degradation of 1-R^1-2-R^2-1,2-*closo*-$C_2B_{10}H_{10}$ ($R^1 = R^2 = CH_2OMe$; $R^1 = (CH_2)_nOMe$, $R^2 = H$) affords the respective derivatives of [7,8-*nido*-$C_2B_9H_{12}]^-$ and [7,8-*nido*-$C_2B_9H_{11}]^{2-}$.[142] Treatment of 1,2-$C_2B_{10}H_{11}$-1-C_6H_4-4'-NH_2 with KOH/EtOH gives [7,8-$C_2B_9H_{11}$-7-C_6H_4-4'-$NH_2]^-$.[150]

The controlled degradation of $Me_4C_4B_8H_8$ in 95% EtOH in air gives $Me_4C_4B_7H_9$.[134]

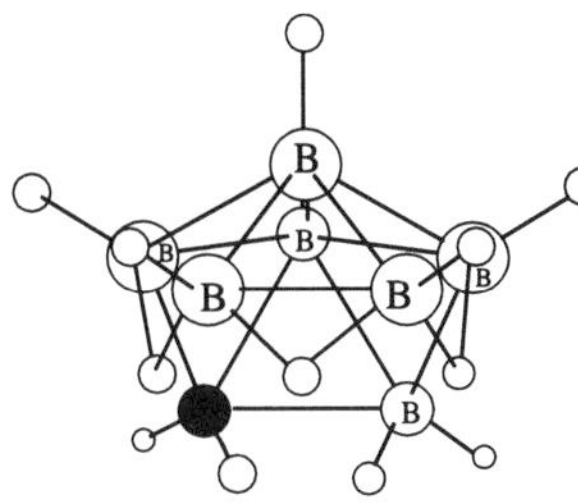

(16)

6.2.4.7 Coupled carbaboranes

The B–B coupled carbaborane (1,5-$C_2B_3H_4$-2-$)_2$ is formed from 1,5-$C_2B_3H_5$ in the presence of $PtBr_2$.[325] Similarly, (1,6-$C_2B_4H_5$-2$)_2$ is formed from 1,6-$C_2B_4H_6$ in the presence of $PtBr_2$.[325] 1:2'-[1,5-$C_2B_3H_4]_2$, 2:2',1':2"-[1,5-$C_2B_3H_4$][1',5'-$C_2B_3H_3$][1",5"-$C_2B_3H_4$], and 2:2',3':1"-[1,5-$C_2B_3H_4$][1',5'-$C_2B_3H_3$][1",5"-$C_2B_3H_4$] are produced from the gas-phase pyrolysis of 1,5-$C_2B_3H_5$ at 400 °C.[232] The mixed coupled carbaborane 2':2-[1',5'-$C_2B_3H_4$][1,6-$C_2B_4H_5$] is formed from the copyrolysis of 1,5-$C_2B_3H_5$ and 1,6-$C_2B_4H_6$ at 400 °C.[232]

The coupled compound 3-(8'-*nido*-5',6'-$C_2B_8H_{11}$)-1,2-$C_2B_{10}H_{11}$ is prepared by heating *nido*-7,8-$C_2B_9H_{13}$ at 80 °C.[326] When 1,10-$(SiMe_2Cl)_2$-*closo*-1,10-$C_2B_8H_8$ is reacted with two equivalents of 1-Me-2-MgBr-*closo*-1,2-$C_2B_{10}H_{10}$, 1,10-(1'-$SiMe_2$-2'-Me-*closo*-1',2'-$C_2B_{10}H_{10})_2$-*closo*-1,10-$C_2B_8H_8$ is formed.[136] Reduction of 1,1'-(*closo*-1',2'-$C_2B_{10}H_{11}$)-*closo*-1,2-$C_2B_{10}H_{11}$ with two equivalents of sodium or sodium naphthalide yields $Na_2[C_2B_{10}H_{11}]_2$; further reaction with another two equivalents of either of the two reducing agents gives rise to a species which can be diprotonated to give [μ-9,10-CH-(μ-9',10'-CH-*nido*-7'-$CB_{10}H_{11}$)-*nido*-7-$CB_{10}H_{11}]^{2-}$.[119–20] The coupled carbaborane (1,7-$C_2B_{10}H_{11}$-9-$)_2$ is formed on catalytic decomposition of (1,7-$C_2B_{10}H_{11}$-9-$)_2$Hg in the presence of $(Ph_3P)_3RhCl$ in hexamethylphosphoramide.[259] Li[*nido*-(Ph)$C_2B_{10}H_{11}$-*closo*-(Ph)$C_2B_{10}H_{10}$] is prepared by treating 1,2-$C_2B_{10}H_{10}$-1-C_6H_5-2-Li with DMF or DMSO.[177] The action of sodium in MeCN solution on 1-R-1,2-$C_2B_{10}H_{11}$ (R = Me, Ph) gives the coupled anion [7-R-12-(1'-R-1',2'-$C_2B_{10}H_{10}$-2'-)-7,8-$C_2B_9H_{10}]^-$.[327]

6.2.4.8 Additional methods of preparation

Peralkylated derivatives of *nido*-2,3,4,5-$C_4B_2H_6$ are obtained from the rearrangement of peralkylated 1,4-dibora-2,5-cyclohexadienes.[328] *Nido*-2,3,4,5-$C_4B_2H_4$-1-Me-2-$NCHMe_2$ is prepared by treating the lithium salt of the borole dianion, $Li_2[C_4H_4BNCHMe_2]$ with $MeBBr_2$.[329] Reaction of $(Cl_2B)_2CHCH(BCl_2)_2$ with $(Me_2CH)_2NH$, and subsequently the NaK_8 alloy, gives rise to a bicyclic isomer of the *closo*-carbaborane $(CH)_2(BR)_4$ (R = $N(CHMe_2)_2$).[330] 1,2-$(TMS)_2$-*closo*-1,2-$C_2B_4H_4$ is produced from the reaction of 2,3-$[TMS]_2$-*closo*-1,2,3-$SnC_2B_4H_4$ with $PtCl_4$ or $PtCl_2$.[331]

At 110 °C the (metallavinyl)carbaborane complex *nido*-4,5-μ-[CpFe(CO)(PPh_3)(η^1:μ-*trans*-MeC=CMe)]-2,3-$Et_2C_2B_4H_5$ converts to 4,5-Me_2-*nido*-7,8-$Et_2C_4B_4H_4$. The latter compound is also obtained from the reaction of *nido*-2,3-$Et_2C_2B_4H_6$ with 2-butyne catalyzed by various ruthenium carbonyl complexes.[332]

Flash thermolysis (355 °C) of 2-R-B_5H_8 (R = ethenyl, 1-*trans*-1-propenyl, *cis*-2-but-2-enyl) yields alkyl derivatives of CB_5H_9.[240]

6.2.5 Chemical Properties

6.2.5.1 Skeletal rearrangements, cage fluxional behavior, and isomer stabilities

Theoretical studies on cage rearrangements have supported DSD (diamond–square–diamond) mechanisms for some structural systems, whereas other studies point the way to either modified DSD mechanisms, or still other mechanistic routes. A molecular orbital energy level correlation for the DSD rearrangement process as applied to the trigonal-bipyramidal $C_2B_3H_5$ shows that such a mechanism is blocked by the crossing of filled and vacant molecular orbitals, indicating that such a process faces a very high activation barrier.[333–4] Isomerization barriers have been calculated for rearrangements on the $C_2B_3H_5$ potential energy surface. Polarization and electron correlation effects have been estimated based on geometries calculated at the HF/3-21G level. The DSD mechanism relating the 1,5- and 2,3-isomers involves a HOMO–LUMO crossing that is ca. 335 kJ mol^{-1} above the energy of the 1,5-isomer. An alternative mechanism for cage rearrangement is the rotation of two equatorial atoms about an axis in the equatorial plane. No HOMO–LUMO crossing is encountered; however, the barrier is approximately the same as with the DSD mechanism.[335]

It is suggested that the rearrangement mechanism for the conversion of 1,2-$C_2B_4H_6$ to 1,6-$C_2B_4H_6$, as investigated by *ab initio* calculations, proceeds in two steps with a classical benzvalene-like intermediate. The first step, rate-determining, is a modified DSD rearrangement where the symmetry of the transition state is reduced from C_{2v} to C_2, thus changing the step from orbitally forbidden (C_{2v} symmetry) to orbitally allowed (C_2 symmetry). The second step involves a concerted DSD rearrangement the net result of which is a local bond rotation. The activation barrier is calculated to be 187.1 kJ mol^{-1} (MP4/6-31G* + ZPC//6-31G*), which is in good agreement with an experimental estimate (176–188 kJ mol^{-1}). A third calculated transition state provides a low-energy pathway for the interchange of two sets of equivalent borons in the intermediate.[78,336]

At least three processes have been proposed for intramolecular isomerizations of the *closo*-carbaboranes, $C_2B_{n-2}H_n$, $n = 5–12$ that preserve the polyhedral framework while redistributing the carbon atoms to new positions. The three processes are the DSD mechanism, the triangular face rotation mechanism, and the single-edge cleavage process. Reaction sequences for carbaborane rearrangements are proposed in relation to *ab initio* SCF MO calculated energies and orbital symmetry considerations of individual isomers and to pertinent experimental observations.[334,337–40] Tensor-surface-harmonic theoretical studies of polyborane rearrangements appear to have some bearing on certain carbaborane rearrangements.[20–1]

Thermolysis of 2,3-Et_2-*closo*-2,3-$C_2B_5H_5$ at 320 °C in the gas phase results in an isomerization to 2,4-Et_2-*closo*-2,4-$C_2B_5H_5$.[305] From *ab initio* considerations it is predicted that a rearrangement of 2,3-$C_2B_5H_7$ to 2,4-$C_2B_5H_7$ could occur via one square-capped prism transition state to form the 1,2-$C_2B_5H_7$ isomer (which is 173.3 kJ mol^{-1} less stable than the 2,4-$C_2B_5H_7$ isomer), and finally over a second square-capped prism transition state to form the 2,4-$C_2B_5H_7$ isomer.[78] It is estimated that the degenerate DSD rearrangement of 2,4-$C_2B_5H_7$ which involves a square-capped prism as the transition state involves a barrier of 254 kJ mol^{-1}.[78] The thermal (295 °C) rearrangement rate patterns of 5- and 3-R-*closo*-2,4-$C_2B_5H_6$ (R = Me) to an equilibrium mixture of the three *B*-Me-*closo*-2,4-$C_2B_5H_6$ isomers best correlate with a DSD framework reorganization that move substituent-bearing boron atoms to new positions, while the two-cage carbon atoms essentially remain at nonadjacent equatorial sites in the pentagonal bipyramid.[341] The same mechanistic model applies to the thermal rearrangement results where R = Et, Cl, Br, I for the mono-boron-substituted compound, and similarly for 5,6-R_2-*closo*-2,4-$C_2B_5H_5$,[99–100,102,216,342] 5-Me-6-Cl-*closo*-2,4-$C_2B_5H_5$,[100] and 3-Cl-5,6-$(C_2H_5)_2$-*closo*-2,4-$C_2B_5H_4$.[342] In each case the experimental rearrangements have led to equilibrium quantities of isomeric boron-substituted derivatives of 2,4-$C_2B_5H_7$. The isomer sequence observed in the rearrangement of 3-Cl-5,6-$(C_2H_5)_2$-*closo*-2,4-$C_2B_5H_4$ rules out a 1,2-substituent-shift mechanism.[342] The first isomer produced during the course of the 5-Me-6-Cl-2,4-$C_2B_5H_5$ 295 °C rearrangement is 1-Me-5-Cl-2,4-$C_2B_5H_5$, followed by the appearance of 3-Me-5-Cl-2,4-$C_2B_5H_5$;[100] the order of appearance of these and the other five isomers rules out a triangle-face-rotation (TFR) mechanism in which cage-carbon atoms stay in low-coordination nonadjacent positions, and is consistent with a DSD mechanism.[100] Derived rearrangement rate constant comparisons[99–100,216] among related mono- and disubstituted compounds, 5-X-*closo*-2,4-$C_2B_5H_6$, 5,6-X_2-*closo*-2,4-$C_2B_5H_5$ (X = Me, Cl), and 5-Me-6-Cl-*closo*-2,4-$C_2B_5H_5$ indicate that (i) the presence of a second substituent accelerates the apparent translocation of the first substituent (i.e., the nature of the "stationary" substituent is more critical in affecting the rearrangement rate than that of the "migrating" substituent), and (ii) the effect of chlorine in this regard is greater than that of the methyl group. Chlorine with its unshared pairs of electrons may well "back-bond" to the cage boron(s), thereby facilitating the formation of a partially "open" intermediate (or transition state) as proposed in the course of a DSD rearrangement. When the stationary substituent is the methyl group, the significantly lower rate constants associated with the migrating group (in the disubstituted systems) are attributed to the more moderate electron donating nature, and/or potential hyperconjugative character, of the methyl group. The activation energies for the interconversion of *B*-Cl-2,4-$C_2B_5H_6$ isomers are estimated to be in the vicinity of 180–200 kJ mol^{-1}.[100]

Experimentally determined stabilities[99–100,102,216,341–2] among numerous *closo*-2,4-$C_2B_5H_7$ derivatives (*B*-halo, *B*-alkyl, *B* Me_3N^+-, *B*,*B*'-dihalo, *B*,*B*'-dialkyl, *B*-Me *B*'-Cl-, *B*-Cl-*B*',*B*"-Et_2-) in various isomer sets correlate very well to results obtained from the application of the MNDO semiempirical MO method[88,101] and, in three isomer sets (the *B*-Cl,[88] *B*-Me, *B*,*B*'-Me_2[340]), by application of the *ab initio* SCF method. Among the monosubstituted compounds *B*-X-2,4-$C_2B_5H_6$ (X = Me, Cl, Br) it is established that the 3-X-isomer is the most stable in each of these three-isomer sets,[99–100,102,216,341] whereas when X = I the 5-isomer is the most stable.[216] The 1-X-isomer is the least stable in all cases except where X = Me. A substituent-positional "additivity" effect has been successfully used for the prediction of *B*-substituted dicarbaheptaborane isomer stabilities.[99]

A consideration of rearrangement pathways between *closo*-$C_2B_6H_8$ isomers via a DSD mechanism, combined with *ab initio* SCF MO results, lead to the expectation that the isolation of the 1,2-isomer should be feasible.[334] *Ab initio* SCF MO optimization of 3,5-$C_2B_6H_8$ in one study has been unsuccessful and results in the production of the 1,7-$C_2B_6H_8$ isomer instead; examination of calculated structures generated during the optimization procedure indicates that the computational rearrangement follows the DSD pathway.[70]

It has been assessed by MO considerations that among two plausible DSD rearrangement mechanisms for $C_2B_7H_9$ the "single" DSD mechanism is forbidden by the principle of conservation of orbital symmetry, whereas a "double" DSD process is allowed. By means of the double DSD mechanism the interconversion of all isomers is possible through the closed cycle (1,7 = 3,4 = 1,8 = 4,5 = 1,2 = 1,4 = 1,7 = etc.).[334,343] An MO consideration of the DSD rearrangement process for the *closo*-$C_2B_8H_{10}$ system indicates that the 2,7-isomer is an intermediate between two known isomers (1,10- and 1,6-) of lower energy, while the 1,2-isomer is stabilized behind a barrier presented by a higher energy (2,3-) isomer.[70,334] Possible rearrangment pathways between *closo*-$C_2B_9H_{11}$ isomers via a DSD mechanism have been considered.[334] From SCF MO calculations it is predicted that 2,9- and 2,8-$C_2B_9H_{11}$ positional isomers should be metastable, blocked from rearranging to more stable structures by higher energy isomers that intervene along the reaction path. The geometry-optimized total energies for 17 positional isomers of $C_2B_9H_{11}$ are reported.[69–70,344] When an attempt was made to obtain an *ab initio* MO energy minimum for 2,6-$C_2B_9H_{11}$, this framework reorganized via a DSD process to the 2,3-isomer.[70] Protonating Cs[*nido*-7,9-$C_2B_9H_9$-8-R^1-7,9-R^2_2] (R^1 = Me, $PhCH_2$; R^2 = H, Me) gives rearranged compounds *nido*-2,7-$C_2B_9H_{10}$-11-R^1-2,7-R^2_2 and Cs[*nido*-2,7-$C_2B_9H_{10}$-9-R^1-2,7-R^2_2].[147]

Thermolysis of 9-Me_2S-7-[$(TMS)_2CH$]$CB_{10}H_{11}$ in refluxing toluene yields 6-Me_2S-2-[$(TMS)_2CH$]$CB_{10}H_9$ as the major product, while solid-state thermolysis of 9-Me_2S-7-[$(TMS)_2CH$]$CB_{10}H_{11}$ gives 6-Me_2S-2-[$(TMS)_2CH$]$CB_{10}H_9$.[156]

The isomerization of 1,2-$C_2B_{10}H_{12}$ to 1,7-$C_2B_{10}H_{12}$, which proceeds at normal pressures at 400 °C, is completely suppressed at pressures >1000 MPa.[345] Thermal isomerizations of the labeled *closo* carbaboranes 3-^{10}B-1,2-$C_2{''}B_9H_{12}$ and 2-^{10}B-1,7-$C_2{''}B_9H_{12}$ followed by ^{10}B and ^{11}B NMR analyses shows that (i) 3-^{10}B-1,2-$C_2{''}B_9H_{12}$ undergoes rearrangements that completely scramble the enriched boron atom over the entire cage, (ii) these rearrangements occur at a rate faster than the conversion of 1,2-$C_2B_{10}H_{12}$ to 1,7-$C_2B_{10}H_{12}$, (iii) the 1,7-$C_2B_{10}H_{12}$ that is formed does not reform 1,2-$C_2B_{10}H_{12}$ at a detectable rate, and (iv) at temperatures in which 1,2-$C_2B_{10}H_{12}$ forms 1,7-$C_2B_{10}H_{12}$ at a significant rate, the latter undergoes no further rearrangements. A mechanism that explains the movement of ^{10}B at 350 °C in 1,2-$C_2B_{10}H_{12}$ involves a 12-vertex *nido* intermediate. A number of previously considered mechanisms, including simple exchange between two boron sites, TFR in an icosahedron, DSD twist, and rotation of pentagonal pyramids, are extensions of the *nido* intermediate mechanism. The thermal rearrangement results for 3-F-3-^{10}B-1,2-$C_2{''}B_9H_{11}$ demonstrate that independent intramolecular fluorine migration does not occur during rearrangement.[346] A study of various $[B_{12}H_{12}]^{2-}$ structures provides indirect support for TFR as a mechanism that can best account for the observed isomerizations of the related isoelectronic carbaboranes 1,2-, 1,7-, and 1,12-$C_2B_{10}H_{12}$.[347] From a thermal rearrangement study of a 1,2-$C_2B_{10}H_{12}$ derivative having a three-carbon bridge spanning adjacent cage positions, it is rationalized that an "extended" triangle rotation mechanism can best account for the observed results in this derivative.[348] Isomerization of 1,2-$C_2B_{10}H_{11}$-3-R (R = CH=CH_2, C≡CH) in the gas phase at 560–580 °C gives a mixture of 1,7-$C_2B_{10}H_{11}$-2-R and 1,7-$C_2B_{10}H_{11}$-4-R.[349]

A 1,2,4,5-tetraalkyl derivative of *nido*-$C_4B_2H_6$ is proposed to be a product of the thermal rearrangement of a 2,5-diborabicyclo[2.1.1]hexane derivative.[350] Fluxionality in a highly substituted *nido*-2,3,4,5-$C_4B_2H_6$ has been observed when attachments are mesityl groups.[351] Upon treating $Li_2[C_4H_4BNCHMe_2]$ with $Me_2CHNBCl_2$ a bis-$NCHMe_2$ substituted C_4B_2 species is produced that does not have *nido*-cage geometry but instead is a fluxional 2,6-diborabicyclo[3.1.0]hex-3-ene derivative.[329] Et_4-1,2,3,9-$C_4B_6Et_6$ with a skeletal structure analogous to that of $B_{10}H_{14}$, is prepared from the 160 °C isomerization of a compound with the same formula but with an adamantane structure having skeletal carbons at the four high coordination (bridgehead) sites.[308] Et_4-1,2,3,9-$C_4B_6Et_6$ and the isomeric Et_4-2,6,8,10-$C_4B_6Et_6$ (**2**) both have a dynamic structure as deduced by NMR spectroscopic studies.[131,208,308]

The reversible cage rearrangements of tetra-*C*-alkyltetracarbadodecaboranes, $R_4C_4B_8H_8$ (R = Me, Et, Pr), as examined by NMR spectroscopy, indicate that each compound exists in solution (in a variety of solvents) as a mixture of two cage isomers, designated **A** and **B**, with K_{eq} values ([**B**]/[**A**] in toluene at 25 °C) of 0.56, 2.2, and 1.9 for R = Me, Et, and Pr, respectively. In the solid state, $R_4C_4B_8H_8$ (R = Me) exists as isomer **A**, (**3**); when R = Et the compound crystallizes as isomer **B**, which x-ray crystallography establishes as an open framework (**4**) comprised of two pyramidal C_2B_4 units joined at their basal B–B edges. When R = C_3H_7 the compound contains both **A** and **B** isomers in the solid state. All three $R_4C_4B_8H_8$ (R = Me, Et, Pr), when placed in solution, undergo changes in the [**B**]:[**A**] ratio until equilibrium is reached; however, the process is much more rapid in the case where R = Me (minutes at ambient temperature) than in the higher homologues, which require hours. In each case, removal of solvent causes the compound to revert to its original solid-state cage isomer. From variable-temperature ^{11}B NMR measurements the values of ΔH and ΔS for the **A** → **B** conversion, which involves cleavage of a framework C–C bond, are very small; the ΔH values range from 6.2 kJ mol^{-1} for R = Me to 9.8 kJ mol^{-1} for R = Pr, the slight increase attributed to an apparent steric effect.[135]

6.2.5.2 Hydrogen acidity studies

The order of hydrogen acidities in the carbaboranes is normally $H\mu > HC > HB$. For those carbaboranes with no bridging hydrogens the removal of cage CH hydrogen with metal alkyls is a common method to provide a carbaborane anion to be used in reactions that usually result in substitution at the cage carbon.

Bridge deprotonation of $C_2B_4H_8$ occurs with the action of $Ph_3P{=}CH_2$; the latter reagent does not deprotonate the *C,C'*-dimethyl derivative of $C_2B_4H_8$.[352] A study of bridging hydrogen deprotonation reactions of *nido*-2,3-$C_2B_4H_6$-2-$R^1$3-R^2 with NaH or KH (in THF) showed that the reaction rate is pseudo-first order and independent of the amount of metal hydride present, and is governed by substituent steric effects. 2,3-$C_2B_4H_6$-2,3-Ph_2 exhibits very fast reaction rates at the higher temperatures measured (> −40 °C) and anomalously high activation energy (50.2 kJ mol^{-1}) and low entropy of activation (20 eu). In this instance, electron withdrawal by the phenyl groups via resonance, combined with pronounced steric effects by the phenyl groups, are proposed to account for the observed behavior.[243–5,353] Treatment of 2,3-$(TMS)_2$-2,3-$C_2B_4H_6$ with NaH in THF produces the ion cluster $(THF{\cdot}Na^+)_2[2,3\text{-}(TMS)_2\text{-}2,3\text{-}C_2B_4H_5^-]_2$.[127]

Lithiation of *closo*-2,4-$C_2B_5H_7$ with LiC_4H_9 followed by Cl-TMS (or $ClSiMe_2H$) gives rise to 2,4-$C_2B_5H_5$-2,4-$(TMS)_2$ (or 2,4-$C_2B_5H_5$-2,4-$(SiMe_2H)_2$); $Li_2[2,4\text{-}C_2B_5H_5]$ reacts with deuterium oxide to give the *C,C'*-dideuterio derivative of 2,4-$C_2B_5H_7$.[354] Deprotonation of $Me_4C_4B_7H_9$ with NaH in THF generates the salt $Na[Me_4C_4B_7H_8]$ from which the CH proton on the bridging CHMe group has been removed. Protonation of $Na[Me_4C_4B_7H_8]$ gives a new isomer of $Me_4C_4B_7H_9$.[354] 4-CB_8H_{14} is deprotonated with K_2CO_3/H_2O to yield an anion, $[4\text{-}CB_8H_{13}]^-$,[92–3] in which all three B–H(bridging) hydrogen atoms and the *endo*-C–H hydrogen atom are involved in an exchange process.[222] Treating $[Me_3NH][7,8\text{-}C_2B_9H_{12}]$ with NaH affords $[7,8\text{-}C_2B_9H_{11}]^{2-}$.[355]

The acid dissociation constants of $C_2B_9H_{13}$, as potentiometrically and spectrophotometrically determined, are $pK_a(1) = 2.98$ and $pK_a(2) = 14.25$.[356] A pK_a of 6.18 has been measured for 3-(8'-*nido*-5',6'-$C_2B_8H_{11}$)-1,2-$C_2B_{10}H_{11}$.[326] The pK_a values, in 50% EtOH, for 1,2-$C_2B_{10}H_{11}$-X-SH are 3.30 for X = 1, 9.32 for X = 8, 10.08 for X = 9; for 1,7-$C_2B_{10}H_{11}$-X-SH, 5.30 for X = 1, 7.38 for X = 4, 9.45 for X = 9; and for 1,12-$C_2B_{10}H_{11}$-X-SH, 5.85 for X = 1, 7.75 for X = 2.[357] The acidities for 1,2-$C_2B_{10}H_{11}$-X-CO_2H follow the order X = 1 > 3 > 4 > 8 > 9; for 1,7-$C_2B_{10}H_{11}$-X-CO_2H, X = 1 > 2 > 5 > 4 > 9; and for 1,12-$C_2B_{10}H_{11}$-X-CO_2H, X = 1 > 2. The acidities for 1,Y-$C_2B_{10}H_{11}$-1-CO_2H follow the order X = 1 > 7 > 12.[357]

6.2.5.3 Reduction reactions and related chemistry

Upon treatment of *closo*-1,5-C_2B_3-1,5-Me_2-2,3,4-$(C_2H_5)_3$ with potassium metal the ^{11}B NMR resonances attributed to $[closo\text{-}1,5\text{-}C_2B_3\text{-}1,5\text{-}Me_2\text{-}2,3,4\text{-}Et_3]^{2-}$ are observed.[308] The two-electron reduction of 1,2-$(TMS)_2$-*closo*-1,2-$C_2B_4H_4$ (with adjacent carbon atoms) with sodium naphthalenide in THF produces the disodium salt of the dianion $[2,4\text{-}(TMS)_2\text{-}nido\text{-}2,4\text{-}C_2B_4H_4]^{2-}$ with carbon atoms apart.[128] Et_{10}-*nido*-2,6,8,10-C_4B_6 (**2**) reacts with alkali metals (K, Cs) to form the $[C_4B_6Et_{10}]^{2-}$ dianion, which is oxidized back to the Et_{10}-*nido*-2,6,8,10-C_4B_6 with I_2, $FeCl_2$, or $HgCl_2$.[131] Reduction of *nido*-5,6-$C_2B_8H_{12}$ with $NaBH_4$ in EtOH/NaOH gives rise to *arachno*-6,9-$C_2B_8H_{14}$ (**12**). The latter carbaborane is also produced by the action of Na/Hg in ethanol on 5,6-$C_2B_8H_{12}$.[358–9]

Reduction of the $[nido\text{-}6\text{-}CB_9H_{12}]^-$ ion with sodium metal in refluxing liquid ammonia gives $[arachno\text{-}6\text{-}CB_9H_{14}]^-$,[360a] Treatment of 6-$Me_3N$-6-$CB_9H_{11}$ with sodium in liquid NH_3 affords $[nido\text{-}6\text{-}CB_9H_{12}]^-$ [92] Na/liquid NH_3 also reduces 1-$CB_{11}H_{11}$-1-NMe_3 to the parent $[1\text{-}CB_{11}H_{12}]^-$ anion.[306]

Reduction of 1,1'-(*closo*-1',2'-$C_2B_{10}H_{11}$)-*closo*-1,2-$C_2B_{10}H_{11}$ with two equivalents of sodium, or sodium naphthalide, yields $Na_2[C_2B_{10}H_{11}]_2$; further reaction with another two equivalents of either of the two reducing agents gives rise to a species which can be diprotonated to give $[\mu\text{-}9,10\text{-}CH\text{-}(\mu\text{-}9',10'\text{-}CH\text{-}nido\text{-}7'\text{-}CB_{10}H_{11})\text{-}nido\text{-}7\text{-}CB_{10}H_{11}]^{2-}$.[112,120] Cathodic C–H bond cleavage is observed, and *o*-carboranyl anions are formed when 1,2-$C_2B_{10}H_{12}$ and some of its alkyl and aryl derivatives are electrochemically reduced on transition metal electrodes in aprotic solvents and in the presence of etraalkylammonium salts.[35]

6.2.5.4 Reactions of carbaboranes with Lewis bases

(i) With hydride ion or fluoride ion

Reaction of *closo*-2,3-$C_2B_5H_5$-2,3-Et_2 with $Li[B(Et_3)H]$ affords the seven-vertex *nido*-carbaborane anion $[nido\text{-}3,4\text{-}C_2B_5H_6\text{-}3,4\text{-}Et_2]^-$ (**1**) which has a structure based on a dodecahedron missing one five-connected vertex.[129,305] The reaction of *closo*-$C_2B_6H_8$ with LiH in THF produces the $[nido\text{-}1,3\text{-}C_2B_6H_9]^-$ ion, the C_1 skeletal structure of which can be considered as a fragment of an icosahedron in which the carbons are in 1,3-positions relative to one another around a six-atom open face; a bridging hydrogen is located between two adjacent low-coordination boron atoms along the open face.[203]

Upon reaction with hydride ion, both *nido*-8-Me_2S-7-$CB_{10}H_{12}$[155] and *nido*-7-Me_3N-7-$CB_{10}H_{12}$[106] undergo a cage closure reaction to produce $[closo\text{-}CB_{10}H_{11}]^-$. Reaction of 9-$Me_2S$-7-$[(TMS)_2CH]$-$CB_{10}H_{11}$ with NaH, or with $LiBH(C_2H_5)_3$, results in proton loss, Me_2S dissociation, and cage closure to yield the $2\text{-}[(TMS)_2CH]\text{-}closo\text{-}CB_{10}H_{10}]^-$ anion.[156]

Fluoride ion, in aprotic solvents, attaches to one of the four equivalent borons in *closo*-1,6-$C_2B_4H_6$ and converts to the $[5\text{-}F\text{-}nido\text{-}2,4\text{-}C_2B_4H_6]^-$ ion.[316] Reaction of $Bu_4N^+F^-$/THF with $C_2B_7H_9$ produces the $[arachno\text{-}2,6\text{-}C_2B_6H_{11}]^-$ ion, the structure of which has been elucidated by the *ab initio*/IGLO/NMR method.[360b] For other reactions with fluoride ion see (Sections 6.2.4.6 and 6.2.5.5(*i*)(*a*)).

(ii) With nitrogen bases

An ammonia complex of *closo*-1,5-$C_2B_3H_5$ was sought by *ab initio* calculations at the HF/3-21G level but the starting geometry optimized to separate moieties.[73] Whereas the parent *closo*-2,4-$C_2B_5H_7$ does not react with Me_3N, both 3-Cl-2,4-$C_2B_5H_6$ and 5-Cl-2,4-$C_2B_5H_6$ form 1:1 adducts with this amine. The adducts, $Me_3N{\cdot}Z\text{-}ClC_2B_5H_6$ (Z = 3,5), react with BCl_3 to yield the quaternary ammonium salts, $[Z\text{-}Me_3N\text{-}closo\text{-}2,4\text{-}C_2B_5H_6][BCl_4]$. The reaction of Me_3N with 1-Cl-2,4-$C_2B_5H_6$ proceeds with more difficulty than with the other two *B*-Cl isomers, but once formed, the 1:1 adduct combines with BCl_3 to give the rearrangement product $[3\text{-}Me_3N\text{-}closo\text{-}2,4\text{-}C_2B_5H_6][BCl_4]$. When 5-Cl-6-Me-$C_2B_5H_5$ is employed as the starting carbaborane in the Me_3N/BCl_3 reaction sequence, $[5\text{-}Me_3N\text{-}6\text{-}Me\text{-}closo\text{-}2,4\text{-}C_2B_5H_5][BCl_4]$ is produced. The reaction of 5-Cl-2,4-$C_2B_5H_6$ with Me_3P produces the $Me_3P{\cdot}5\text{-}ClC_2B_5H_6$ adduct which in turn is converted to $[5\text{-}Me_3P\text{-}closo\text{-}2,4\text{-}C_2B_5H_6][BCl_4]$ on the addition of BCl_3.[361]

Trimethylamine reacts with *closo*-$C_2B_6H_8$ to give a weakly bound adduct and eventually forms a more strongly bound $Me_3N{\cdot}C_2B_6H_8$ adduct (**5**). The cage carbon and boron atoms of (**5**) are arranged in a fashion which can be considered as an icosahedral fragment in which the carbon atoms are in 1,3-positions relative to one another around a six-atom open face; the lone bridging hydrogen atom is located between two adjacent low-coordination boron atoms (5,6) along the open face.[203]

A compound with *nido*-7,9-$C_2B_9H_{11}$-10- units attached to each of the two nitrogen atoms of 4-carbomethoxypyrazole is prepared from *closo*-1,8-$C_2B_9H_{11}$ and the anion of 4-carbomethoxypyrazole.[362]

6.2.5.5 Derivative chemistry

(i) Cage boron attached species

(a) Derivatives containing B_{cage}–halogen bonds. Fluoride ion, in aprotic solvents, reacts with *closo*-1,6-$C_2B_4H_6$ to give the [5-F-*nido*-2,4-$C_2B_4H_6$]$^-$ ion.[316] *Closo*-$C_2B_4Cl_4H_2$ is reported as a product from the action of HCl on 2,3-$C_2B_4H_6$-2,3-$(TMS)_2$; however, a very low-field ^{11}B NMR spectral chemical shift, $\delta = 62.76$ ppm,[363] for this purported tetrachloro derivative of $C_2B_4H_6$ is suggestive of a classical environment for the boron atoms. The relative reactivity of 2-Cl-1,6-$C_2B_4H_5$ is found to be significantly greater than that of the parent 1,6-$C_2B_4H_6$ toward bromination with $Br_2/AlCl_3$ to give the respective *B*-brominated derivatives. In the case of the 2-chloro-derivative the *trans* (4)-position is attacked. Both 1,6-$C_2B_4H_6$ and 1,10-$C_2B_8H_{10}$ are found to be less reactive than 2,4-$C_2B_5H_7$ toward *B*-chlorination with $Cl_2/AlCl_3$. The relative reactivities of the three *B*-methyl- or three *B*-chloro-isomers (*B* = 1, 3, 5) of 2,4-$C_2B_5H_7$ toward chlorination ($Cl_2/AlCl_3$) do not appear to be much affected by the position of the substituent; in all of these reactions it is the 6-position (for 5-Me-2,4-$C_2B_5H_6$) and the 5,6-positions (for the 1- and the 3-Me-2,4-$C_2B_5H_6$ isomers) that are primarily involved in further substitution. The reactivity of 2,4-$C_2B_5H_7$ toward $Cl_2/AlCl_3$ is found to be considerably less than that of benzene toward the same reagents.[364]

Halogen exchange is observed between *B*-halo derivatives of *closo*-2,4-$C_2B_5H_7$ and tetraalkylammonium halides, and is possible only when the "reagent" halide ion is smaller than the "leaving" halide. For example, both 3- and 5-I-$C_2B_5H_6$ react with bromide ion to give quantitative amounts of the respective 3- and 5-Br-$C_2B_5H_6$ isomers. Similarly, 5-Br-$C_2B_5H_6$ is converted to 5-Cl-$C_2B_5H_6$ in the presence of Cl^-. This same halide ion reacts with the carbaborane adduct Me_3N·5,6-$Br_2C_2B_5H_5$ to give both 5-Cl-6-Br-$C_2B_5H_5$ and 5,6-Cl_2-$C_2B_5H_5$. Fluoride ion reacts with 5-Br-$C_2B_5H_6$ or Me_3N·5-Cl-$C_2B_5H_6$ to give 5-F-$C_2B_5H_6$. Two other fluorine substituted carbaboranes, 3-F-$C_2B_5H_6$ and 3,5-F_2-$C_2B_5H_5$, are produced from fluoride ion and 3-I- and 3,5-I_2- derivatives of $C_2B_5H_7$, respectively. The rate of substitution as the halide ion is varied, is in correspondence with the expected nucleophilicity trend $F^- > Cl^- > Br^-$ in nonaqueous solvents.[365–6] The 1:1 amine-bromocarbaborane adduct Me_3N·5-Br-$C_2B_5H_6$ reacts with CH_2Cl_2 to form 5-Cl-2,4-$C_2B_5H_6$, via a presumed [5-Me_3N-2,4-$C_2B_5H_6$]$^+$ intermediate.[367] Reaction of I_2 with *closo*-2,4-$C_2B_5H_7$ in the presence of $AlCl_3$ gives initial halogen attack at the 5- and 6-cage positions, whereas the same reaction in the absence of a Lewis acid affords 3-I-2,4-$C_2B_5H_6$ as the initial product, followed by formation of 3,5-I_2-2,4-$C_2B_5H_5$. With bromine, however, the mono- and dibromocarbaborane product(s) are the same with or without the presence of the Lewis acid, producing the 5-Br- and 5,6-Br_2- derivatives.[102,261] The action of ICl or IBr, in the presence of $AlCl_3$, on *closo*-2,4-$C_2B_5H_7$ produces predominately *B*-iodo products. The same reagents, without the presence of a Lewis acid, produce *B*-chloro and *B*-bromo products, respectively. The reaction of Et_2NSF_3 with 5-Br-*closo*-2,4-$C_2B_5H_6$, or with Me_3N·5-Br-2,4-$C_2B_5H_6$, yields 5-F-*closo*-2,4-$C_2B_5H_6$.[368] Halogenation of *closo*-2,4-$C_2B_5H_7$ with BX_3 (X = Cl, Br, I) at elevated temperature yields 3-X-2,4-$C_2B_5H_6$ as the principal product, the rate of halogenation increasing with the size of the halogen on BX_3. The order of reactivity between some $C_2B_5H_7$ compounds and BCl_3 is 5,6-Et_2-$C_2B_5H_5$ > 2,4-$C_2B_5H_7$ > 5-Cl-$C_2B_5H_6$. *Closo*-2,4-$C_2B_5H_7$ reacts with Me_3CCl to give 3-Cl-2,4-$C_2B_5H_6$ at 295 °C, but the rate of formation of 3-Cl-$C_2B_5H_6$ from these reagents is less facile than from the reaction of 2,4-$C_2B_5H_7$ and BCl_3. 3-Br-2,4-$C_2B_5H_6$ is formed from the reaction of 2,4-$C_2B_5H_7$ and $BrCH_2CH_2Br$ at temperatures higher than 300 °C.[369] 5-Cl-2,4-$C_2B_5H_6$ reacts with $NaBH_4$ to give the parent 2,4-$C_2B_5H_7$.[361] Halogenation of 2,3-Et_2-*closo*-2,3-$C_2B_5H_5$ with elemental bromine or with $Br_2/AlBr_3$ results in the formation of 5-Br-2,3-Et_2-*closo*-2,3-$C_2B_5H_4$.[305] $C_2B_5X_7$ (X = Cl, Br) and $C_2B_7Cl_9$ (presumed to be *carbaborane* derivatives) are obtained from the co-pyrolysis of B_2X_4 and CX_4 or C_2X_4.[370]

The AlX_3-catalyzed (X = Cl, Br, I) halogenation of *arachno*-4,5-$C_2B_7H_{13}$ (**6**) with anhydrous HX produces a series of 6-substituted derivatives, 6-X-4,5-$C_2B_7H_{12}$. The same compounds, along with 6,8-I_2-4,5-$C_2B_7H_{11}$, are obtained in noncatalyzed reactions using elemental halogens.[371] Direct $AlCl_3$ catalyzed halogenation of *arachno*-4,6-$C_2B_7H_{13}$ (**9**) with CCl_4, Br_2, or I_2 gives rise to monosubstituted 4,6-$C_2B_7H_{12}$-3-X (X = Cl, Br, I), 4,6-$C_2B_7H_{12}$-5-X (X = Br, I), and disubstituted 4,6-$C_2B_7H_{11}$-3,5-X_2 (X = Cl, Br); uncatalyzed iodination also produces the 3,5-I_2- derivative.[372] Electrophilic bromination of *arachno*-$Me_4C_4B_7H_9$ produces $Me_4C_4B_7H_8$-11-Br.[134]

Addition of HX (X = F, Cl, Br, I) to *nido*-[6,9-$C_2B_8H_{10}$]$^{2-}$ (**7**) gives *arachno*-5-X-6,9-$C_2B_8H_{13}$ (**12**). The mechanism probably involves stereoselective addition of HX to the B-5–C-6 bond followed by addition of two further hydrogen atoms, one at the bridging site and the other at the C-9 atom.[358–9,373] Electrophilic halogenation of 6,9-$C_2B_8H_{14}$ (**12**) occurs at the B-1 site to give a series of 1-X-6,9-$C_2B_8H_{13}$ (X = Cl, Br, I) derivatives.[358–9]

Me_4N[7,8-$C_2B_9H_{11}$-7-C_6H_4-4'-NCS], upon treatment with sodium *N*-chloro-*p*-toluenesulfonamide and NaI in a phosphate buffer, gives Me_4N[7,8-$C_2B_9H_{10}$-9-I-7-C_6H_4-4'-NCS].[150] Reaction of $Li_2[C_2B_9H_{11}]$ with $Me_2S{\cdot}BBr_3$ leads to 3-Br-1,2-$C_2B_{10}H_{11}$.[374]

Nido-7-Me_3N-7-$CB_{10}H_{12}$, when treated electrochemically with HCl/H_2O, results in chlorine substitution at the 4-position, whereas the action of HCl/$AlCl_3$ produces a product with chlorine substitution at the 9-position.[106] Reactions of 1,2- and 1,7-$C_2B_{10}H_{12}$ with SbF_5 results in replacement of hydrogen by fluorine at the boron atoms, to give in succession 9-F-, 9,12-F_2-, 8,9,12-F_3-, and 8,9,10,12-F_4 derivatives of 1,2-$C_2B_{10}H_{12}$; and 9-F-, and 9,10-F_2- derivatives of 1,7-$C_2B_{10}H_{12}$.[4d,5a,35] [9-F-1,7-$C_2B_{10}H_{10}$]$^+$[SbF_6]$^-$ has also been obtained from the action of SbF_5 on 1,7-$C_2B_{10}H_{12}$.[5a] Reaction of 1,2- and 1,7-$C_2B_{10}H_{12}$ with SbF_5/SO_2 and perfluorotoluene gives 9-fluoro-12-(2,3,4,5,6-pentafluorobenzyl)-1,2-$C_2B_{10}H_{11}$ and 9-fluoro-10-(2,3,4,5,6-pentafluorobenzyl)-1,7-$C_2B_{10}H_{10}$, respectively.[35] 1,12-$C_2B_{10}H_{11}$-2-F is produced from the action of F^- on [1,12-$C_2B_{10}H_{11}$-2-IC_6H_5]$^+$.[375] Chlorination of 1,2-(-CH_2CH_2-)-*closo*-1,2-$C_2B_{10}H_{10}$, containing a four-membered exocyclic unit, with CCl_4/$AlCl_3$ produces the 9-Cl-derivative.[165] Chlorination of 9-I-1,2-$C_2B_{10}H_{11}$ leads to 9-Cl_2I-1,2-$C_2B_{10}H_{11}$.[35] 1,7-$C_2B_{10}H_{11}$-9-$Br^+C_6H_5$[BF_4]$^-$ and (1,7-$C_2B_{10}H_{11}$-9-)$_2Br^+$[BF_4]$^-$ are obtained by treating a mixture of C_6H_6/CH_2Cl_2/MeCN/BrF_3 with (1,7-$C_2B_{10}H_{11}$-9-)$_2$Hg followed by $BF_3{\cdot}OEt_2$.[376–7] [1,12-$C_2B_{10}H_{11}$-1-IC_6H_5]BF_4 is prepared by the reaction of 1,12-$C_2B_{10}H_{11}$-2-HgI with C_6H_6/(MeCO)$_2$O/H_2SO_4/$K_2S_2O_8$.[375,378] The reactions of phenyl(*B*-carboranyl)iodonium salts (derived from all three isomers of $C_2B_{10}H_{12}$) with various nucleophiles, L, can result in nucleophilic substitution (L = F^-, Br^-) or free radical reactions (L = OH^-, pyridine, Ph_3P). Products of both types of reactions are observed when L = NC^-.[4c] Reactions of [9-phenyliodo-1,7-$C_2B_{10}H_{11}$][BF_4] with species such as halides, pseudohalides, amido-, and benzoyl-groups give 9-substituted products.[4a,379] Several pathways are observed for reactions of [1,7-$C_2B_{10}H_{11}$-9-ClC_6H_5]$^+$[BF_4]$^-$ with nucleophiles N_3^-, NO_2^-, OH^-, F^-, C_5H_5N, and PPh_3, (i) nucleophilic substitution, (ii) one-electron reduction, and (iii) degradation of the carbaborane icosahedron. Degradation occurs with increasing ease in the series of ions [1,7-$C_2B_{10}H_{11}$-9-X-Ph]$^+$, X = I, Br, Cl; one-electron reduction reactions lead to only halogen–phenyl bond cleavage regardless of the halogen atom of the halonium cation, and the regioselectivity of S_N reactions changes dramatically. Nucleophilic substitution in the iodonium cation occurs only at a carbaborane boron atom. For the bromonium and the chloronium cations, it takes place both at the carbaborane boron and at the phenyl carbon.[380] The preparation of 3-I-1,2-$C_2B_{10}H_{11}$ is possible from the action of BI_3 on $Li_2[C_2B_9H_{11}]$.[381] 9,12-I_2-1,2-$C_2B_{10}H_{10}$ is produced from I_2 and 1,2-$C_2B_{10}H_{12}$.[381] Reactions of 1,2-$C_2B_{10}H_{12}$ with I_2 in CCl_4 and $CHCl_3$ in the presence of $AlCl_3$ gives mainly 1,2-$C_2B_{10}H_{11}$-9-I in addition to the 9-Cl, 9,12-Cl_2, 9-Cl, 12-I, and 9,12-I_2 derivatives; similar bromination gives 1,2-$C_2B_{10}H_{11}$-9-Br. Iodination of 1,7-$C_2B_{10}H_{12}$ with I_2 in CCl_4 and $CHCl_3$ in the presence of $AlCl_3$ gives mainly 1,7-$C_2B_{10}H_{11}$-9-I in addition to the 9-Cl, 9,10-Cl_2, 9-Cl-10-I, and 9,10-I_2 derivatives; similar bromination gives mainly 1,2-$C_2B_{10}H_{11}$-9-Br and 1,2-$C_2B_{10}H_{11}$-9-Cl.[35] Heating 9-I-1,2-$C_2B_{10}H_{11}$ at 390–400 °C gives a mixture of *B*-I-1,2-$C_2B_{10}H_{11}$ (*B* = 3, 4, 8, 9); further heating at 550–570 °C gives a mixture of *B*-I-1,7-$C_2B_{10}H_{11}$ (*B* = 4, 5, 9) isomers.[382]

Direct halogenation of [1-$CB_{11}H_{12}$]$^-$ affords [7,8,9,10,12-Cl_5-1-$CB_{11}H_7$]$^-$, [7,8,9,10,11,12-X_6-1-$CB_{11}H_6$]$^-$ (X = Cl, Br), [12-I-$CB_{11}H_{11}$]$^-$, and [7,12-I_2-1-$CB_{11}H_{10}$]$^-$.[383] *Closo*-1-Me_3N-1-$CB_{11}H_{11}$, when treated electrochemically with HCl/$MeCO_2H$, results in chlorine substitution at the 12 position.[106]

(b) Derivatives containing B_{cage}–chalcogen bonds. Photolysis of the small carbaborane 2,4-$C_2B_5H_7$ with hexafluoroacetone results in the formation of 5-[HOC(CF_3)$_2$C(CF_3)$_2$O-2,4-$C_2B_5H_6$.[247] The action of sodium amalgam in ethanol on 5,6-$C_2B_8H_{12}$ affords 5-C_2H_5O-6,9-$C_2B_8H_{13}$; 5,5'-O-(6,9-$C_2B_8H_{13}$)$_2$ is obtained from the reaction of HF/H_2O with *nido*-[6,9-$C_2B_8H_{10}$]$^{2-}$ (**7**).[358–9,373] Two compounds, 5,5'-X-(*arachno*-6,9-$C_2B_8H_{13}$)$_2$, X = O, OS(O_2)O, are obtained from the action of H_2SO_4 on Na_2[*nido*-6,9-$C_2B_8H_{10}$] (**7**).[384] The *nido*-[8-HO-7-$CB_{10}H_{12}$]$^-$ anion is obtained by treating 7-Me_3N-*nido*-7-$CB_{10}H_{12}$ with NaH followed by hydrolysis.[106,385] 1,7-$C_2B_{10}H_{11}$-9-OH is formed from the reaction of [(1,7-$C_2B_{10}H_{11}$-9)-(aryl)X]$^+$[BF_4]$^-$ (X = Br, I) with NO_2^-/H_2O.[376,379] Similarly, 1,12-$C_2B_{10}H_{11}$-2-OH is produced from [1,12-$C_2B_{10}H_{11}$-1-IC_6H_5]BF_4 and aqueous $NaNO_2$.[4b] (1,7-$C_2B_{10}H_{11}$-9-)$_2$O is obtained from the alkaline hydrolysis of (1,7-$C_2B_{10}H_{11}$-9-)(C_6H_5)$Br^+BF_4^-$.[35]

Reaction of 4,6-$C_2B_7H_{13}$ (**9**) with elemental sulfur at 130 °C produces 4,6-$C_2B_7H_{12}$-3-SH and 4,6-$C_2B_7H_{12}$-5-SH.[372] 7,8-$C_2B_9H_{11}$-9-SMe_2 is afforded from the reaction of K[7,8-$C_2B_9H_{12}$] with Me_2SO and aqueous H_2SO_4.[386] The reaction of the [$CB_{10}H_{13}$]$^-$ anion with Me_2S/H_2SO_4 results in hydrogen loss and formation of *nido*-8-Me_2S-7-$CB_{10}H_{12}$.[155] Friedel–Crafts substitution of 7-Me_3N-*nido*-7-$CB_{10}H_{12}$ with S_2Cl_2 produces 4(6)-HS-7-Me_3N-*nido*-7-$CB_{10}H_{11}$ and 4,4'-, 6-6'-, and 4,6'-S_2(7-Me_3N-*nido*-7-

$CB_{10}H_{11})_2$.[190] (1,X-$C_2B_{10}H_{11}$-9-)$_2S_2$ (X = 2, 7) is obtained from the reaction of 1,X-$C_2B_{10}H_{12}$ with $S_2Cl_2/AlCl_3$.[35] (1,2-$C_2B_{10}H_{11}$-9-S-)$_2$POMe is formed from the action of $MePOCl_2$ on 1,2-$C_2B_{10}H_{11}$-9-SH in the presence of NEt_3.[35] 9-[SP(O)(OR)$_2$]-1,2-$C_2B_{10}H_{11}$ (R = Me, Et) is prepared by the reaction of ClP(OEt)$_2$ with 9-HS-1,2-$C_2B_{10}H_{11}$ in the presence of Et_3N.[35] 12-L-1-$CB_{11}H_{11}$ (L = SMe_2, $MeSCH_2SMe$) are reported.[383,387]

The 9-[SeP(O)(OR)$_2$]-1,2-$C_2B_{10}H_{11}$ (R = Me, Et) compounds are prepared by refluxing (RO)$_3$P with di-(1,2-$C_2B_{10}H_{11}$-9-)$_2Se_2$ in toluene.[35] The syntheses of other carbaborane derivatives with boron–selenium bonds have been reported.[3,4e,35,388] (1,X-$C_2B_{10}H_{11}$-9-)$_2Te_2$ (X = 2, 7) is formed from the reaction of 1,X-$C_2B_{10}H_{11}$-9-$TeCl_3$ with $Na_2S/EtOH/H_2O$.[4b] (1,X-$C_2B_{10}H_{11}$-9-)$_2Te_2$ (X = 2, 7) is formed from the reaction of 1,X-$C_2B_{10}H_{11}$-9-$TeCl_3$ with $Na_2S/EtOH/H_2O$; (1,2-$C_2B_{10}H_{11}$-9-)$_2Te_2$ is also produced from 1,2-$C_2B_{10}H_{11}$-9-$TeCl_3$ and $N_2H_4{\cdot}H_2O$.[4e]

(c) Derivatives containing B_{cage}–N and B_{cage}–P bonds. 2,3,4-Tris(diisopropylamino)-1,5-dicarba-*closo*-pentaborane(5) is prepared from a retro Diels–Alder reaction involving a 1,2,3-tris-(diisopropylamino)triborolane.[126] [*B*-Me_3N-*closo*-2,4-$C_2B_5H_6$]$^+$ ions (*B* = 3, 5) are formed from the action of Me_3N on *B*-Cl-2,4-$C_2B_5H_6$ followed by addition of BCl_3. An analogous 5-Me_3P derivative is formed in a similar fashion.[361] A series of charge-compensated boron-substituted carbaborane cage derivatives, *nido*-9-L-7,8-$C_2B_9H_{11}$ and *nido*-10-L-7,8-$C_2B_9H_{11}$ (L = NEt_3, $PhNMe_2$, DMF, $MeCONMe_2$, 4-(methoxycarbonyl)pyridine), are prepared by the ferric chloride-promoted oxidative coupling reaction of K[*nido*-7,8-$C_2B_9H_{12}$] with the Lewis base, L. Another series of neutral boron substituted carbaborane cage derivatives, *nido*-10-L-7,9-$C_2B_9H_{11}$, are synthesized directly from the reaction of closo-1,8-$C_2B_9H_{11}$ with the nitrogen-containing ligand, L.[389] Reactions of [(1,7-$C_2B_{10}H_{11}$-9-)(aryl)X]$^+$[BF_4]$^-$ (X = Br, I) with NO_2^-, N_3^-, and PPh_3 lead to the substituted carbaboranes 1,7-$C_2B_{10}H_{11}$-9-X (X = NO_2, N_3, PPh_3).[376,379] 1,7-$C_2B_{10}H_{11}$-9-Br$^+$Ph[BF_4]$^-$ reacts with pyridine, and subsequently with $MeCO_2H/H_2O$, to give *B*-pyridinium-7,9-$C_2B_9H_{11}$.[376] 1,7-$C_2B_{10}H_{11}$-2-NH_2, and 1,7-$C_2B_{10}H_{10}$-2-NH_2-6-Me are formed from the reaction of 1,2-$C_2B_{10}H_{11}$-3-NH_2 with Na/NH_3 and subsequently with CH_2I_2, CHI_3, and CI_4; in the case of CI_4 a small amount of 1,7-$C_2B_{10}H_{10}$-*B*-NH_2-*B*-I is also formed.[3] (1,2-$C_2B_{10}H_{11}$-3)-NHCONH-(1,2-$C_2B_{10}H_{11}$-3-) is obtained from 1,2-$C_2B_{10}H_{11}$-3-$NH_2/KMnO_4/Me_2CO$.[4b] Some acylation and alkylation reactions of the amine group in 9-NH_2-1,7-$C_2B_{10}H_{11}$, and oxidation of NH_2 in 3-NH_2-1,2-$C_2B_{10}H_{11}$, have been studied.[4b] Reaction of (1,2-$C_2B_{10}H_{11}$-3-NH_2-1-CH_2)$_2$ with maleic and acetic anhydrides yields (1,2-$C_2B_{10}H_{11}$-3-N(-COCH=CHCO-)-1-CH_2)$_2$.[2c] *Nido*-7-Me_3N-7-$CB_{10}H_{12}$, when treated with Tl($OCOCF_3$)$_3$/NEt_3/THF, affords 8-Et_3N-7-Me_3N-7-$CB_{10}H_{10}$.[106] Electrophilic *B*-substitution of *closo*-[1-$CB_{11}H_{12}$]$^-$ with H_2NOSO_3H provides 7-H_3N-*closo*-1-$CB_{11}H_{11}$.[307]

6-Me_3P-3,4-Et_2-*nido*-3,4-$C_2B_5H_5$ is the result of partial cage opening of 2,3-Et_2-*closo*-2,3-$C_2B_5H_5$ with Me_3P.[305] Reaction of Ph_2PCl with K[7,8-$C_2B_9H_{12}$] gives 9-Ph_2PH-7,8-$C_2B_9H_{11}$. A similar reaction with the dianion Na_2[7,8-$C_2B_9H_{11}$] gives 10-Ph_2PH-7,8-$C_2B_9H_{11}$.[143] *Nido*-9-PPh_3-7,8-$C_2B_9H_{11}$ is formed from the reaction of Tl[*closo*-3,1,2-$TlC_2B_9H_{11}$] with PPh_3 in the presence of AgBr.[140] Reaction of *nido*-8-Me_2S-7-$CB_{10}H_{12}$ with Ph_3P yields *nido*-8-Ph_3P-7-$CB_{10}H_{12}$.[155] The phosphonate ester 1,7-$C_2B_{10}H_{11}$-9-PO(OMe)$_2$ is prepared from (1,7-$C_2B_{10}H_{11}$-9-)$_2$Hg and (MeO)$_3$P irradiated with UV light.[35] 1,12-$C_2B_{10}H_{11}$-2-PO(OMe)$_2$ is formed in a similar fashion.[159–60]

(d) Derivatives containing exo-cage B–C bonds. *B*-Alkenyl derivatives of 1,5-$C_2B_3H_5$, 1,6-$C_2B_4H_6$, and 2,3-$C_2B_4H_8$ are prepared from the (RC_2R)$Co_2(CO)_6$ (R = H, Me, Et) catalyzed reactions of alkynes with the parent carbaborane.[246] The formation of the *nido*-2,3-$C_2B_4H_5$-2,3-R^1_2-4-R^2 products (R^2 = Me, $MeCH_2$, $PhCH_2$, Ph(CH_2)$_3$, $MeC_6H_4CH_2$) from reactions of [*nido*-2,3-$C_2B_4H_5$-2,3-R^1_2]$^-$ ions (R^1 = Et, $PhCH_2$) with organic halides (R^2X) are proposed to occur via *nido*-2,3-$C_2B_4H_5$-2,3-R^1_2-$\mu_{4,5}$-R^2 bridged intermediates, which rapidly rearrange to the 4-substituted products.[311] Reaction of 2,4-$C_2B_5H_7$ with BPh_3 at 160–165 °C produces 3-Ph-2,4-$C_2B_5H_6$ and some 5-Ph-2,4-$C_2B_5H_6$.[369] 2,3-Et_2-*closo*-2,3-$C_2B_5H_5$ reacts with $Me_3P{=}CH_2$ to give the seven-vertex zwitterion, *exo*-6-($Me_3P^+CH_2$)-3,4-Et_2-*nido*-3,4-$C_2B_5H_5^-$.[130] Boron substituted alkenyl derivatives of 2,4-$C_2B_5H_7$ are prepared from the (MeC_2Me)$Co_2(CO)_6$ catalyzed reaction of 2-butyne with the parent carbaborane.[246] Photolysis of the small carbaborane 2,4-$C_2B_5H_7$ with hexafluoroacetone results in the formation of 5-[HO(CF_3)$_2$C]-2,4-$C_2B_5H_6$.[247] The reactivity of 2,4-$C_2B_5H_7$ toward alkylation with $MeCl/AlCl_3$ is considerably less than that of benzene, slightly less than that of PhCl, and greater than that of 1,6-$C_2B_4H_6$. The relative reactivities of the substituted carbaborane systems 5-X-2,4-$C_2B_5H_6$ toward $RCl/AlCl_3$ (R = alkyl), to give 5-X-6-R-2,4-$C_2B_5H_5$, are X = Me > H > Cl ~I.[361,364]

[*Nido*-7,9-$C_2B_9H_9$-7,9-Me_2-8-CH_2Ph]$^-$ is obtained from [*nido*-7,9-$C_2B_9H_9$-7,9-Me_2]$^{2-}$ and $ClCH_2Ph$. Protonation of [*nido*-7,9-$C_2B_9H_9$-7,9-Me_2-8-CH_2Ph]$^-$ with $MeCO_2H/H_2O$ produces the neutral *nido*-2,7-$C_2B_9H_{10}$-2,7-Me_2-11-CH_2Ph. Upon treating the latter compound with $CsCl/NH_3$ the reverse rearrangement occurs and [*nido*-7,9-$C_2B_9H_9$-7,9-Me_2-8-CH_2Ph]$^-$ is formed.[390] *Nido*-2,7-$C_2B_9H_{12}$-11-Me is produced from the addition of Na/NH_3 to K[*nido*-7,8-$C_2B_9H_{12}$] followed by addition of MeI. Its bridged deuterated analogue is prepared by treating *nido*-2,7-$C_2B_9H_{12}$-11-Me with $NaOD/D_2O$ followed by D_2SO_4.[254] Reaction of $PhMe_2SiCl$ with $[C_2B_9H_{11}]^{2-}$ in THF at reduced temperatures gives both the *endo*- and *exo*-(10-phenyldimethylsilyl)-7,8-dicarbaundecaborate anion.[391]

The compounds *B*-OC-7-Me_3N-7-$CB_{10}H_{10}$ and *B*-HO_2C-7-Me_3N-7-$CB_{10}H_{10}$ were obtained from the treatment of *nido*-7-Me_3N-7-$CB_{10}H_{12}$ with NaH followed by the addition of oxalyl chloride.[106] Organic groups can be attached to B-9 of both 1,2- and 1,7-$C_2B_{10}H_{12}$ by reactions of Grignard reagents with 9-halocarbaboranes in the presence of palladium and nickel catalysts.[35,162,265] 9-Alkyl-, 9-aryl-, and 9-ethynyl-1,2- or -1,7-dicarba-*closo*-dodecaboranes were prepared via the cross-coupling reaction between alkyl-, aryl-, or ethynylzinc compounds and 9-I-1,2- or 9-I-1,7-1,2-$C_2B_{10}H_{11}$ in the presence of homogeneous palladium catalysts.[392] The $(Ph_3P)_2PdCl_2$-catalyzed coupling reaction of 3-I-1,2-$C_2B_{10}H_{11}$ and RMgBr is used to prepare 3-alkyl-1,2-$C_2B_{10}H_{11}$ from 3-I-1,2-$C_2B_{10}H_{11}$, and 9,12-dialkyl-1,2-$C_2B_{10}H_{10}$ from 9,12-I_2-1,2-$C_2B_{10}H_{10}$ and RMgBr.[381] The isomeric boron substituted carbaboranes *B*-Ph-1,7-$C_2B_{10}H_{11}$ (B = 2, 4, 5, 9) are the products of thermal (630–750 °C) condensation of 1,7-$C_2B_{10}H_{12}$ and benzene.[4c] 9-R-1,7-$C_2B_{10}H_{11}$ (R = Ph, C_6F_5, 1,4-$Me_2C_6H_3$) are prepared by UV photolysis of (1,7-$C_2B_{10}H_{11})_2$Hg or 1,7-$C_2B_{10}H_9Tl(O_2CCF_3)_2$ in RH.[35] Alkylations of 1,X-$C_2B_{10}H_{12}$ by either Me_2CHY or PrY (X = 2,7; Y = Cl, Br) in the presence of $AlCl_3$ give similar results, a mixture of *B*-Me_2CH-1,X-$C_2B_{10}H_{11}$ (B = 4, 8, 9) and *B*-$MeCH_2CH_2$-1,X-$C_2B_{10}H_{11}$ (B = 8, 9).[382,393] Treatment of 1,2-(or 1,7-)-$C_2B_{10}H_{11}$-9-CH_2CH_2Me or 1,2-(or 1,7-)$C_2B_{10}H_{11}$-9-$CHMe_3$ with HCl/$AlCl_3$ produces a mixture consisting of 1,X-$C_2B_{10}H_{12}$ and 1,X-$C_2B_{10}H_{11}$-*B*-C_3H_7 (where X = 2,7 and C_3H_7 is a mixture of CH_2CH_2Me and $CHMe_2$), and 1,2-(or 1,7-)-$C_2B_{10}H_{10}$-*B*,*B*'-$(C_3H_7)_2$.[393–4] In the presence of BF_3, 1,2-$C_2B_{10}H_{12}$ undergoes a Friedel–Crafts type reaction with the cluster complex $[(\mu\text{-H})_3Os_3(CO)_9(\mu_3\text{-}C)]_3(O_3B_3O_3)$ to form the boron-substituted complex 9-$[(\mu\text{-H})_3Os_3(CO)_9(\mu_3\text{-C})]$-(1,2-$C_2B_{10}H_{11}$).[395] Other *B*-alkyl derivatives of 1,2-$C_2B_{10}H_{12}$ are obtained by alkylation under electrophilic conditions.[2c] Treatment of 9-CH_2=$CHCH_2$-1,2-(or 1,7-)-$C_2B_{10}H_{11}$ with Br_2 in $CHCl_3$ gives 9-$CH_2BrCHBrCH_2$-1,2-(or 1,7-)$C_2B_{10}H_{11}$, which when treated with $NaNH_2/NH_3$ followed by acid workup yields propargylcarbaboranes 9-HC≡CCH_2-1,2-(or 1,7-)-$C_2B_{10}H_{11}$.[396] Electrophilic reactions of 1,2- and 1,7-$C_2B_{10}H_{11}$-9-C_6H_5 with a variety of reagents (e.g., MeCOCl/$AlCl_3$, $Br_2/AlCl_3$) give both *meta*- and *para*-phenyl substitution products.[169,182] Chromium tricarbonyl complexes of phenyl derivatives of 1,2-$C_2B_{10}H_{12}$ have been synthesized.[4a] 1,7-$C_2B_{10}H_{11}$-9-CN is the product from the reaction of $[1,7\text{-}C_2B_{10}H_{11}\text{-}9\text{-}Br^+C_6H_5][BF_4]^-$ with KCN·18-crown-6. Similarly, 1,12-$C_2B_{10}H_{11}$-2-CN is produced from $[1,12\text{-}C_2B_{10}H_{11}\text{-}2\text{-}IC_6H_5]^+$ and KCN·18-crown-6.[4c] 1,2-$C_2B_{10}H_{12}$ reacts with carboethoxycarbene, :$CHCO_2$Et, to form all four possible products of B–H insertion, 1,2-$C_2B_{10}H_{11}$-X-CH_2CO_2Et (X = 3, 4, 8, and 9).[397] The generation of a *B*-substituted carbaboranylcarbene is accomplished by irradiation of 1,2-Me_2-3-N_2CH-*closo*-1,2-$C_2B_{10}H_9$; subsequent reactions in the presence of various alkenes are characterized by unusual triplet reactivity.[398] Treating 1,X-$C_2B_{10}H_{11}$-9-COCl (X = 2, 7) with $C_6H_6/AlCl_3$ gives the respective ketones 1,X-$C_2B_{10}H_{11}$-9-COPh which, when treated with PhMgBr or LAH, convert to the carbinols 1,X-$C_2B_{10}H_{11}$-9-C(OH)Ph_2 and 1,X-$C_2B_{10}H_{11}$-9-CH(OH)Ph. Protonation of the carbinols with CF_3CO_2H/CH_2Cl_2 give carbocations in which the cationic center is thought to be directly connected to a six-coordinate B atom.[5a] 1,2- and 1,7-$C_2B_{10}H_{11}$-9-C(O)Re$(CO)_5$ are prepared by treating 1,2- or 1,7-$C_2B_{10}H_{11}$-9-COCl with NaRe$(CO)_5$.[5b]

(e) Other cage-boron attached derivatives. A *closo*-carbaborane containing a trigonal boron σ-bonded to a cage carbaborane, 2-Br_2B-*closo*-1,6-$C_2B_4H_5$, has been isolated from the reaction of *closo*-1,6-$C_2B_4H_6$ and BBr_3.[369] The coupled B_2H_6-polyhedral cage species 2:1',2'-[1,6-$C_2B_4H_5$][B_2H_5] (**17**) is obtained from the $PtBr_2$ catalyzed reaction between 1,6-$C_2B_4H_6$ and B_2H_6.[248] Mercuration of the anions [*nido*-7-R-7,8-$C_2B_9H_{11}]^-$ (R = H, Me, Me_2CH) with $HgCl_2$/LiBr/KOH/H_2O gives $[10,10'\text{-}(7\text{-R-}7,8\text{-}C_2B_9H_{10})_2Hg]^{2-}$.[399] Cs[1,2-$C_2B_{10}H_{11}$-1-$BH_3$] is obtained from the action of $(C_2H_5)_3NBH_3$ on Cs[7,8-$C_2B_9H_{12}$]. The same reagent, $(C_2H_5)_3NBH_3$, reacts with the 1,2- and 1,7-$C_2B_{10}H_{12}$ isomers as well as with Cs[7,9-$C_2B_9H_{12}$] to give 1,2-$C_2B_{10}H_{12}$-1-$BH_2N(C_2H_5)_3$ in the first instance, and 1,7-$C_2B_{10}H_{11}$-1-$BH_2N(C_2H_5)_3$ in the latter two instances. Reaction of Cs[1,2-$C_2B_{10}H_{11}$-1-BH_3] with aqueous H^+ produces 1,2-$C_2B_{10}H_{12}$.[400] 9-Fluorosulfato-1,2- and 1,7-$C_2B_{10}H_{11}$ are prepared by reaction of $HOSO_2F$ and SbF_5 with 1,2- and 1,7-$C_2B_{10}H_{12}$.[401] Electrophilic deuteration of *closo*-$[1\text{-}CB_{11}H_{12}]^-$ in DCl/D_2O confirms the expected order of reactivity on individual skeletal atoms, decreasing in the series B-12 > B-7–11 > B-2–6 > C-1.[307]

(ii) Cage carbon attached species

(a) Derivatives containing C_{cage}–chalcogen bonds. Deprotonation of 1,2-$C_2B_{10}H_{10}$-1-C_6H_5-2-OH by *N*,*N*,*N*',*N*'-tetramethylnaphthalene-1,8-diamine gives a salt whose anion, $[1,2\text{-}C_2B_{10}H_{10}\text{-}1\text{-}C_6H_5\text{-}2\text{-}O]^-$ effectively consists of a *nido* shaped $[C_6H_5CB_{10}H_{10}]^-$ residue capped by a pentuply-bridging carbonyl

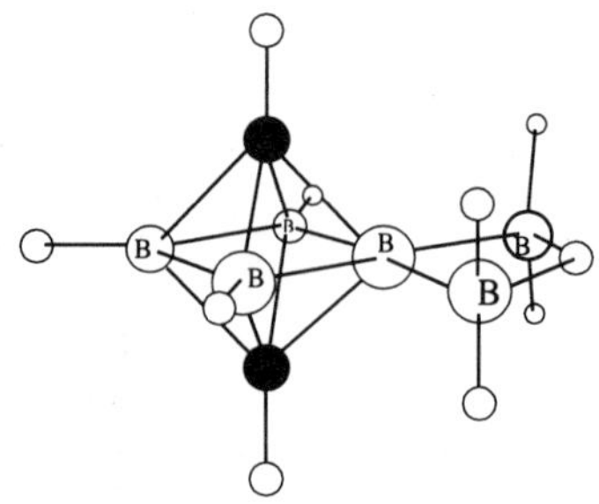

(17)

group with a C–O distance of 0.1245(3) nm and an IR frequency of 1460 cm^{-1}.[118] $[1\text{-HO-1-CB}_{11}H_{11}]^-$ is produced from the reaction of $1\text{-H}_3\text{N-1-CB}_{11}H_{11}$ with $NaNO_2$, then with Zn/HCl.[383]

$[Me_3NH]_2[7,8\text{-}C_2B_9H_{10}\text{-}7,8\text{-}(S)_2]_2$ is produced from the reaction of $[Me_3NH][7,8\text{-}C_2B_9H_{10}\text{-}7,8\text{-}(SH)_2]$ with $NaHCO_3/H_2O/NaI/I_2$.[153] Reaction of the anion 7,8-(ethane-1',2'-dithiolato-S,S')dicarba-*nido*-undecaborate with $RhCl(PPh_3)_3$ gives [7,8-(ethane-1',2'-dithiolato-S,S')dicarba-*nido*-undecaborato]-bis(triphenylphosphine)rhodium.[154] The reaction of $1,2\text{-}C_2B_{10}H_{10}\text{-}1,2\text{-}(SH)_2$, under basic conditions, with organic dihalogenated compounds results in the formation of macrocycles incorporating 7,8-dicarba-*nido*-undecaborate or 1,2-dicarba-*closo*-dodecaborane units.[139,141,172,322] Various palladium complexes of $1,2\text{-}C_2B_{10}H_{10}\text{-}1,2\text{-}(SH)_2$ were studied.[402] Treatment of 9-SCN derivatives of both 1,2- and $1,7\text{-}C_2B_{10}H_{12}$ with $K_2[C_8H_8]$ proceeds with SC cleavage to give $C_2B_{10}H_{11}\text{-9-SH}$ and $(C_2B_{10}H_{11}\text{-9-S-})_2$ coupled products.[4a] Treatment of $(1,2\text{-}C_2B_{10}H_{10}\text{-1-Li-2-})_2SiMe_2$ with SCl_2 produces the cyclic–SiCCSCC– in which the two pairs of adjacent carbons are each part of a C_2B_{10} carbaborane cage.[4e] The synthesis of *S*-$(1,2\text{-}C_2B_{10}H_{11}\text{-1-yl})$thiophosphates and thiophosphonates[403] and *S*-$(1,2\text{-}C_2B_{10}H_{11}\text{-9-yl})$diphenylthiophosphinites[171] have been reported. *C*-substituted derivatives of the general formula $[1\text{-RS–CB}_{11}H_{11}]^-$ (e.g., R = H, Me) are reported.[383,387]

(b) Derivatives containing C_{cage}–N, C_{cage}–P, and C_{cage}–As bonds. $6\text{-Me}_2\text{C=NH-}nido\text{-6-CB}_9H_{11}$ reacts with $(MeO)_2SO_2$ to form $6\text{-Me}_3\text{N-6-CB}_9H_{11}$, and reacts with KOH and subsequently with HCl to form $6\text{-H}_3\text{N-6-CB}_9H_{11}$. The latter compound reacts with $NaNO_2$, Me_2S, and dilute HCl to give $6\text{-Me}_2\text{S-6-CB}_9H_{11}$. The reduction of $6\text{-H}_3\text{N-6-CB}_9H_{11}$ with Na/NH_3 produces the parent $[6\text{-CB}_9H_{12}]^-$ ion.[404] The reactivity of the ammine group in the $7\text{-H}_3\text{N-7-CB}_{10}H_{12}$ carbaborane was examined and a series of 7-L-$7\text{-CB}_{10}H_{12}$ derivatives (L = $Me_2C{=}NH$, $(CH_2)_5C{=}NH$, $PhCH_2(PhCH{=})N$, $HO_2CCH_2NH_2$, $MeCONH_2$, $(CH_2)_6N_4$, Me_2S) were characterized.[405] Methylation of $1\text{-CB}_{11}H_{11}\text{-1-NHMe}_2$ produces $1\text{-CB}_{11}H_{11}\text{-1-NMe}_3$ which can be reduced to the parent $[1\text{-CB}_{11}H_{12}]^-$ anion with Na/liquid NH_3.[306] Zwitterionic 1-L-$1\text{-CB}_{11}H_{11}$ (L = NH_3, $NHMe_2$, NMe_3, NH(Et)Me, NMe=C(OH)Ph, SMe_2) are reported. Reaction of $7\text{-H}_3\text{N-7-CB}_{11}H_{12}$ with $NaBH_4$ and $Et_3N{\cdot}BH_3$, followed by hydrolysis, gives $1\text{-H}_3\text{N-1-CB}_{11}H_{11}$.[383,387] Treatment of $1\text{-H}_3\text{N-1-CB}_{11}H_{11}$ with $(MeO)_2SO_2$ in base gives $1\text{-Me}_3\text{N-1-CB}_{11}H_{11}$.[383] $1\text{-NH}_2\text{-1,X-}C_2B_{10}H_{11}$ (X = 1, 7) is obtained from hydrogenation of the corresponding $1\text{-PhN=N-1,X-}C_2B_{10}H_{11}$.[35] Formation of $[1,2\text{-}C_2B_{10}H_{10}\text{-1-}(1',2'\text{-}C_2B_{10}H_{11}\text{-1'-})\text{-2-NH}]_2$ is accomplished by treatment of $1,2\text{-}C_2B_{10}H_{10}\text{-1-}(1',2'\text{-}C_2B_{10}H_{11}\text{-1'-})\text{-2-NH}_2$ with Na/NH_3 followed by $KMnO_4$.[292,406] Oxidation of the $[HCB_{10}H_{10}CCB_{10}H_{10}CNH_2]^{4-}$ anion by $K[MnO_4]$ in liquid NH_3 gives the hydrazo compound $HCB_{10}H_{10}CCB_{10}H_{10}CNHNHCB_{10}H_{10}CCB_{10}H_{10}CH$ exclusively. The 1,1'-azo-1,2-carbaboranes $RCB_{10}H_{10}CN{=}NCB_{10}H_{10}CR$ (R = H, Me, Ph), are formed from the oxidation of $[1,2\text{-}C_2B_{10}H_{10}\text{-1-R-2-NH}_2]^{2-}$ anions by $K[MnO_4]$ in liquid NH_3. $RCB_{10}H_{10}CN{=}NCB_{10}H_{10}CR$ is reduced to $RCB_{10}H_{10}CNHNHCB_{10}H_{10}CR$, and degraded by KOH to *closo–nido*-$[RCB_{10}H_{10}CN{=}NCB_9H_{10}CR]^-$ and *nido–nido*-$[RCB_9H_{10}CN{=}NCB_9H_{10}CR]^{2-}$.[406] $R^1CB_{10}H_{10}CN{=}NCB_{10}H_{10}CR^1$ (R^1 = H, Me, Ph) reacts with R^2Li (R^2 = Me, Et, Ph, 4-ClC_6H_4, $4\text{-Me}_2NC_6H_4$) to produce the *N*-monosubstituted derivatives $R^1CB_{10}H_{10}CNHNR^2CB_{10}H_{10}CR^1$, along with small amounts of the hydrazocarbaboranes $R^1CB_{10}H_{10}CNHNHCB_{10}H_{10}CR^1$.[324] Treatment of $1,2\text{-}C_2B_{10}H_{10}\text{-1-Li-2-}(1',2'\text{-}C_2B_{10}H_{11}\text{-1'-})$ with PhN_3 produces $1,2\text{-}C_2B_{10}H_{10}\text{-1-N}_3\text{(H)Ph-2-}(1',2'\text{-}C_2B_{10}H_{11}\text{-1'-})$; the latter compound reacts with $MeCO_2H$ to give $1,2\text{-}C_2B_{10}H_{10}\text{-1-NH}_2\text{-2-}(1',2'\text{-}C_2B_{10}H_{11}\text{-1'-})$. $1,2\text{-}C_2B_{10}H_{10}\text{-1-NH}_2\text{-2-}(1',2'\text{-}C_2B_{10}H_{11}\text{-1'-})$ and NOCl react to form $1,2\text{-}C_2B_{10}H_{10}\text{-1-Cl-2-}(1',2'\text{-}C_2B_{10}H_{11}\text{-1'-})$.[292]

The phosphinocarbaboranes, $1,2\text{-}C_2B_{10}H_{10}\text{-}1,2\text{-}(PR_2)_2$ (R = Ph, NMe_2) react with $[N(C_2H_5)_4][AuBr_2]$ to give $1,2\text{-}C_2B_{10}H_{10}\text{-}1,2\text{-}(PR_2AuBr)_2$. Similar products are obtained with the mixed substituted phosphino carbaboranes, $1,2\text{-}C_2B_{10}H_{10}\text{-1-PR}^1{}_2\text{-2-PR}^2{}_2$ (R^1 = Ph, R^2 = NMe_2 or H).[407] $1,2\text{-}C_2B_{10}H_{10}\text{-}1,2\text{-}[PPh_2]_2L$ reacts with $Rh_2Cl_2(CO)_4$ and $RhCl(PPh_3)_3$ or $RhCl(CO)(PPh_3)_2$ to give RhCl(CO)L and $RhCl(PPh_3)L$, respectively; alkenes are catalytically hydrogenated by H_2 in the presence of $RhCl(PPh_3)L$ at 60 atm and catalytically unselectively hydroformylated by CO and H_2 in the presence of

$RhCl(CO)L$.[408] $(1,2\text{-}C_2B_{10}H_{11}\text{-}1\text{-}P(C_6H_5)_2)_2RhCl(CO)$ is formed from $1,2\text{-}C_2B_{10}H_{11}\text{-}1\text{-}P(C_6H_5)_2$ and $[Rh(cod)Cl]_2$.[409] Complexes of the type *cis*- and *trans*-$Ni(X)(1,2\text{-}C_2B_{10}H_{10}\text{-}1\text{-}P[NMe_2]_2\text{-}2\text{-}PPh_2)$ (X = Cl, Br, I, SCN)[257] and $[1,2\text{-}C_2B_{10}H_{11}\text{-}1\text{-}P(C_6H_5)_2]_2MX_2$ (M = Co, Ni; X = Cl, Br, SCN)[410] have been prepared.

$(1,7\text{-}C_2B_{10}H_{10}\text{-}1\text{-}Me\text{-}7\text{-})_2AsCl$, formed from $1,7\text{-}C_2B_{10}H_{10}\text{-}1\text{-}CHMe_2\text{-}7\text{-}Li$ and $AsCl_3$, reacts with Cl_2/H_2O to form $(1,7\text{-}C_2B_{10}H_{10}\text{-}1\text{-}Me\text{-}7\text{-})_2AsO_2H$. Similar chemistry occurs with the analogous derivatives of the 1,2- isomer of $C_2B_{10}H_{12}$.[3]

(c) Derivatives containing C_{cage}–C, Si, Sn, Ge, Pb bonds. Reaction of $1,2\text{-}C_2B_8H_{10}$ with LiC_4H_9 and subsequently with MeI produces $1,2\text{-}C_2B_8H_8\text{-}1,2\text{-}Me_2$; the results of mono *C*-methylation of this same carbaborane system suggest that the 2-position is more prone toward alkylation by this substitution sequence than the 1- position of this cage.[411] $Me_4N[7,8\text{-}C_2B_9H_{11}\text{-}7\text{-}C_6H_4\text{-}4'\text{-}NH_2]$, reacts with Cl_2CS to form $Me_4N[7,8\text{-}C_6H_4\text{-}4'\text{-}NCS]$.[150] 1-Methyl-1-[(carboranyl)methylene]-3,3,5,5-tetrachlorocyclotriphosphazene anion (carboranyl = $C_2B_9H_{11}$-) reacts with $Rh[PPh_3]_3Cl$ to form a metallocarboranylphosphazene derivative. The corresponding dianion reacts with $M(CO)_6$ (M = Mo, W) to yield bismetallacarboranyl phosphazene derivatives.[290,412]

C-Alkylation of $1,2\text{-}C_2B_{10}H_{12}$ by alkyl halides is effected in the presence of solid KOH, THF, or Me_2SO, and dibenzo-18-crown-6.[35] Mechanistic studies have been made on the reactions of $Li[1,2\text{-}C_2B_{10}H_{11}\text{-}1\text{-}CHMe_2]$ with EtBr and EtI to produce $1,2\text{-}C_2B_{10}H_{10}\text{-}1\text{-}C_2H_5\text{-}2\text{-}CHMe_2$.[2c] $1,2\text{-}CH_2CH_2$-*closo*-$1,2\text{-}C_2B_{10}H_{10}$, with a four-membered exocyclic unit, is formed from the reaction of $1\text{-}XCH_2CH_2\text{-}1,2\text{-}C_2B_{10}H_{11}$ (X = Br, TsO) with Bu^nLi.[165] High-pressure thermolysis of $1,2\text{-}C_2B_{10}H_{11}\text{-}1\text{-}CHMe{=}CH_2$ affords $1,2\text{-}C_2B_{10}H_{11}\text{-}2\text{-}CH(Me)CH_2CH_2C({=}CH_2)\text{-}(1,2\text{-}C_2B_{10}H_{11}\text{-}2)$.[345] Reaction of $1,2\text{-}C_2B_{10}H_{10}\text{-}1,2\text{-}(CH_2CH{=}CH_2)_2$ with $WCl_6/SnMe_4$ or $WOCl_4/SnMe_4$ produces the dihydrobenzocarbaborane $1,2\text{-}C_2B_{10}H_{10}\text{-}(\text{-}1\text{-}CH_2CH{=}CHCH_2\text{-}2\text{-})$.[4b] $1,2\text{-}C_2B_{10}H_{11}\text{-}1\text{-}CH_2CH{=}CH_2$, prepared from $1,2\text{-}C_2B_{10}H_{12}$, NaH, $CH_2{=}CHCH_2Br$, and NaI in $MeOCH_2CH_2OMe$, rearranges to $1,2\text{-}C_2B_{10}H_{11}\text{-}1\text{-}CH{=}CHMe$ upon heating with $NaOCMe_3$ in $HOCMe_3$.[413] The reaction of $1,2\text{-}C_2B_{10}H_{10}\text{-}1\text{-}R\text{-}2\text{-}CF{=}CFCF_3$ (R = Me, Ph) with $Me_3COOCMe_3$ under UV light leads to a carbaborane radical species.[4e] Mono- and bifunctional isopropenylphenyl derivatives of $1,2\text{-}C_2B_{10}H_{12}$ are synthesized by the Wittig reaction from the corresponding acetophenyl derivatives and $CH_2{=}PPh_3$.[414] $(1,X\text{-}C_2B_{10}H_{11}\text{-}1\text{-}C{\equiv}C)_2M[PPh_2]_2$ (X = 2, 7, M = Pd, Pt) is produced from the action of $Cl_2M[PPh_2]_2$ on $1,2\text{-}C_2B_{10}H_{11}\text{-}1\text{-}C{\equiv}CLi$.[3] $1,2\text{-}C_2B_{10}H\text{-}B\text{-}Cl_{10}\text{-}1\text{-}R$ (R = $CH_2CH_2CO_2H$, $CH_2CHOHCH_2OCH_2CH{=}CH_2$) compounds are prepared from the action of $1,2\text{-}C_2B_{10}H_2\text{-}B\text{-}Cl_{10}$ with LiC_4H_9 followed by the addition of the bromo or epoxy derivatives of R.[415]

$1,7\text{-}C_2B_{10}H_{11}\text{-}7\text{-}C_6H_5$ is formed from the reductive-elimination reaction of the six-coordinate Ir^{III} complex $(1,7\text{-}C_2B_{10}H_{11}\text{-}7\text{-}C_6H_5\text{-}1\text{-})Ir(H)(Cl)(CO)[PPh_3]_2$.[416] $1,2\text{-}C_2B_{10}H_{11}\text{-}1\text{-}Ph$, containing tritium in the 2-position, is prepared by treating the corresponding $1,2\text{-}C_2B_{10}H_{10}\text{-}1\text{-}Ph\text{-}2\text{-}Li$ with tritiated water.[417] Mononitration of the phenyl ring of $1,2\text{-}C_2B_{10}H_{11}\text{-}1\text{-}Ph$ produces a 4% *ortho*-, 26% *meta*-, and 70% *para*-isomer mixture.[418]

Treating $1\text{-}HOCH_2\text{-}1,7\text{-}C_2B_{10}H_{11}$ with $SOCl_2$ in $CHCl_3$ gives bis(1-*m*-carboranylmethyl)sulfite. Formation of $1\text{-}ClCH_2\text{-}1,7\text{-}C_2B_{10}H_{11}$ during the reaction of $1\text{-}HOCH_2\text{-}1,7\text{-}C_2B_{10}H_{11}$ with $SOCl_2$ proceeds through an intermediate bis(1-*o*-carboranylmethyl)sulfite.[419] Other derivatives of $1,2\text{-}C_2B_{10}H_{12}$ with hydroxymethyl attachments, including $1\text{-}HOCH_2\text{-}2\text{-}HOCH_2CH(OH)CH_2\text{-}1,2\text{-}C_2B_{10}H_{10}$,[420] have been reported.[4b] The alkoxymethyl derivative $1\text{-}ROCH_2\text{-}1,X\text{-}C_2B_{10}H_{11}$ (R = H, $CH_2{=}CHCH_2$, X = 1, 7) has been studied.[35] $1,2\text{-}C_2B_{10}H_{10}\text{-}1,2\text{-}(CH_2OH)_2$ reacts with hexachlorocyclophosphazine to form carboranylmethylphosphazine derivatives.[4e] *o*-Carboranylene–cyclotriphosphazene was isolated from the reaction mixture of bis(hydroxymethyl)-*o*-carbaborane with hexachlorocyclotriphosphazene.[421] The reaction of $1,2\text{-}C_2B_{10}H_{10}\text{-}1,2\text{-}(CH_2OH)_2$ with PCl_5 results in a spiro phosphorus derivative, $[1,2\text{-}C_2B_{10}H_{10}\text{-}1,2\text{-}(CH_2O)_2]_2PCl$ and $1,2\text{-}C_2B_{10}H_{10}\text{-}1,2\text{-}(CH_2)_2O$.[422] Hydrolysis of $1,2\text{-}C_2B_{10}H_{10}\text{-}1,2\text{-}(CH_2OPClOCH_2)$, prepared by cyclizing $1,2\text{-}C_2B_{10}H_{10}\text{-}1,2\text{-}(CH_2OH)_2$ with PCl_3, yields $1,2\text{-}C_2B_{10}H_{10}\text{-}1,2\text{-}(CH_2OP(H)(O)OCH_2)$.[255] Compounds related to this, $(1,2\text{-}C_2B_{10}H_{11}\text{-}1\text{-}CH_2O)_3PO$,[3] $1\text{-}R^1R^2P(O)OCH_2\text{-}1,2\text{-}C_2B_{10}H_{11}$,[423] and $(1,7\text{-}C_2B_{10}H_{11}\text{-}1\text{-}CH_2O)_3PO$ have also been studied.[424] Complexes of $1,2\text{-}C_2B_{10}H_{11}\text{-}1\text{-}CH_2PPh_2$ with Pd, Cu, Os, Ru,[5b] Ir, and Rh[4a] have been characterized. $1,2\text{-}C_2B_{10}H_{10}\text{-}1,2\text{-}(CH_2Cl)_2$ reacts with magnesium to form $1,2\text{-}C_2B_{10}H_{10}\text{-}1,2\text{-}(CH_2CH_2)$ containing a four-membered C_4 ring.[2c]

Free-radical initiated reactions of $1,2\text{-}C_2B_{10}H_{12}$ with PhCN and naphthalene give cyanophenyl and naphthyl derivatives of this carbaborane.[4c] The kinetics of acylation of aniline derivatives with $1,2\text{-}C_2B_{10}H_{10}\text{-}1\text{-}Ph\text{-}2\text{-}COCl$ have been studied.[4d] Refluxing $(1,2\text{-}C_2B_{10}H_{10}\text{-}1\text{-}CO_2)_2NiL$ (L = 2,2'-bipyridyl) in $1,2\text{-}Me_2\text{-}C_6H_4$ affords $(1,2\text{-}C_2B_{10}H_{11}\text{-}1)_2NiL$ and $1,2\text{-}C_2B_{10}H_{12}$.[425] Additional studies on carboranylcarboxyl compounds have been reported.[4e,35] Initial reaction rates are reported for the condensation reactions of $1,7\text{-}C_2B_{10}H_{10}\text{-}1,7\text{-}(CHO)_2$ with various aniline derivatives.[426] The treatment of $Li[1,2\text{-}C_2B_{10}H_{10}\text{-}1\text{-}R]$ (R = Me, $CHMe_2$, Ph) with $ClCH_2(C_2H_3S)$, where C_2H_3S contains a C_2S three-membered ring, produces $1,2\text{-}C_2B_{10}H_{10}\text{-}1\text{-}R\text{-}2\text{-}CH_2(C_2H_3S)$;[3] reactions of $1,2\text{-}C_2B_{10}H_{10}\text{-}1\text{-}R\text{-}2\text{-}CH_2\text{-}CHSCH_2$ (R = $CHMe_2$, Ph) variously with amines, aldehydes, ketones, SO_2Cl_2, H_2O_2, Cl_2, Br_2, I_2, and KSH, have centered on the chemistry of the epithio group.[3,4b,427–8]

$1,2\text{-}C_2B_{10}H_{10}\text{-}1,2\text{-}(CH_2MgBr)_2$, prepared from $1,2\text{-}C_2B_{10}H_{10}\text{-}1,2\text{-}(CH_2Br)_2$ and magnesium, reacts with $(Ph_3)_2PMCl_2$ (M = Pt, Pd, Ni) to give $1,2\text{-}C_2B_{10}H_{10}\text{-}1,2\text{-}(CH_2\text{-})_2M[PPh_3]_2$ containing a five-membered C_4M ring. Analogous treatment of $1,2\text{-}C_2B_{10}H_{10}\text{-}1,2\text{-}(CH_2MgBr)_2$ with Cp_2MCl_2 (M = Ti Zr) yields $1,2\text{-}C_2B_{10}H_{10}\text{-}1,2\text{-}(CH_2\text{-})_2MCp_2$. Reaction of $1,2\text{-}C_2B_{10}H_{10}\text{-}1,2\text{-}(CH_2MgBr)_2$ with MCl_4 (M = Si, Sn, Ge) provides the spiroheterocyclic compounds $[1,2\text{-}C_2B_{10}H_{10}\text{-}1,2\text{-}(\text{-}CH_2\text{-})_2]_2M$.[35] Other related heterocyclic rings incorporating phosphorus and arsenic have been studied.[161] In addition, five-membered heterocycles containing two carbon atoms of the dicarbadodecaborane unit, two methylene groups, and Ni, Pd, Pt, Ti, Zr, Si, Ge, or Sn atoms, as well as six-membered heterocycles containing four carbon atoms of two carbaborane units, a Si atom and Ni, Co, Pd, Cu, or S atoms, have been prepared.[35] Stable carboranylmethyl complexes of Fe and Ni, $(1,2\text{-}C_2B_{10}H_{10}\text{-}1\text{-}Me\text{-}2\text{-}CH_2\text{-})Fe(CO)_2Cp$ and $(1,2\text{-}C_2B_{10}H_{10}\text{-}1\text{-}Me\text{-}2\text{-}CH_2\text{-})Ni(PPh_3)Cp$, are obtained by reaction of the corresponding carboranylmethylmagnesium chlorides with halogen complexes of iron and nickel. $(1,2\text{-}C_2B_{10}H_{10}\text{-}1\text{-}Me\text{-}2\text{-}CH_2\text{-})Fe(CO)_2Cp$, when treated with Br_2 or $HgCl_2$, undergoes rearrangement with migration of the carboranylmethyl group from the iron atom to the carbonyl group. Similar chemistry occurs with the 1,7-isomer of the carbaborane.[2c,429] Reaction of $CoCl_2$·(bipyridine) with $1,2\text{-}C_2B_{10}H_{10}\text{-}1\text{-}CO_2Li\text{-}2\text{-}Li$ produces $1,2\text{-}C_2B_{10}H_{10}\text{-}1,2\text{-}(CO_2Co)$ containing a five-membered C_3OCo ring.[4a] $[(1,7\text{-}C_2B_{10}H_{11}\text{-}1\text{-})C(OH)(Me)C_5H_4]_2Fe$ is prepared from the reaction of $Li_2[1,7\text{-}C_2B_{10}H_{10}]$ with $(MeCOC_5H_4)FeCp$; dehydration of $[(1,7\text{-}C_2B_{10}H_{11}\text{-}1\text{-})C(OH)(Me)C_5H_4]_2Fe$ yields $[(1,7\text{-}C_2B_{10}H_{11}\text{-}1\text{-})C(=CH_2)C_5H_4]_2Fe$.[430] Treatment of $[1,2\text{-}C_2B_{10}H_{10}\text{-}1\text{-}R^1\text{-}2\text{-}C_5H_2O\text{-}2',6'\text{-}R^2{}_2][ClO_4]$ (where $R^1 = CMe_3$, Ph, and R^2 = Me, Ph) with NH_3, $MeNH_2$, $PhNH_2$, N_2H_4, or $PhNNH_2$ yields a variety of nitrogen heterocycles with a carboranyl substituent.[431] $Li[1\text{-}Me\text{-}1,2\text{-}C_2B_{10}H_{10}]$ reacts with $(\eta^6\text{-}o\text{-}C_6H_4X_2)Cr(CO)_3$ to give (when X = Cl) 1-(1'-*closo*-2'-Me-1',2'-$C_2B_{10}H_{10}$)-2,3-$Cl_2C_6H_3$ or (when X = F) 1-(1-*closo*-2'-Me-1',2'-$C_2B_{10}H_{10}$)-2-FC_6H_4. Reaction of two equivalents of $Li[1\text{-}Me\text{-}1,2\text{-}C_2B_{10}H_{10}]$ with $(\eta^6\text{-}p\text{-}C_6H_4F_2)Cr(CO)_3$ results in the displacement of both fluorine atoms on the arene ring to yield 1,4-bis(1'-*closo*-2'-Me-1',2'-$C_2B_{10}H_{10})_2C_6H_4$.[173]

Glycosylcarbaborane derivatives were prepared from lithiated 1,2-dicarba-*closo*-dodecaboranes and aldehydic sugars.[432] An optimum procedure for the esterification of $1,2\text{-}C_2B_{10}H_{11}$-1-carboxylic acid esters with long chain unsaturated fatty alcohols is based on reaction of the carbaborane acid chloride and alcohol in the presence of 4-dimethylaminopyridine in CH_2Cl_2.[433] The acylation of aniline by *m*- and *p*-carbaboranecarboxylic acid chlorides in several nonaqueous solvents have been studied.[5a,434] The reactivities of 1,7-bis[(4-aminophenyl)amido]- and 1,7-bis[(4-aminophenyl)carboxy]-*m*-carbaborane with benzoyl chloride were determined.[435] Allyl and propynyl esters of 1,7- and 1,12-$C_2B_{10}H_{12}$ dicarboxylic acids have been synthesized.[4a]

Treatment of $2,3\text{-}C_2B_4H_6\text{-}2,3\text{-}[TMS]_2$ with HCl produces both $2,3\text{-}C_2B_4H_7$-2-TMS and $2,3\text{-}C_2B_4H_8$.[363,436] *Nido*-$2,3\text{-}C_2B_4H_7$-2-TMS, also obtained from the action of $NaHF_2$ on $2,3\text{-}C_2B_4H_6\text{-}2,3\text{-}[TMS]_2$,[436] reduces aldehydes and ketones but not styrene.[437] $Li_2[2,4\text{-}C_2B_5H_5]$ reacts with $ClSiMe_2H$ to give $2,4\text{-}C_2B_5H_5\text{-}2,4\text{-}[SiMe_2H]_2$.[354] The biscarbaborane, $Me_2Si(1,2\text{-}C_2B_{10}H_{10})_2M$(2,2'-bipyridyl), with a MC_2SiC_2 (M = Ni, Co, Pd, Cu) six-membered ring is prepared from MCl_2(bpy) and $Li_2[(1,2\text{-}C_2B_{10}H_{10})_2SiMe_2]$.[35] $(1,2\text{-}C_2B_{10}H_{10}\text{-}1\text{-}Me\text{-}X\text{-})_2SiC_3H_6$ (X = 2, 7) is afforded from the reaction of $Li[1,X\text{-}C_2B_{10}H_{10}\text{-}1\text{-}Me]$ and $Cl_2SiC_3H_6$ (where SiC_3H_6 is a silacyclobutane ring).[4b] $[1,2\text{-}C_2B_{10}H_{10}\text{-}1\text{-}Me\text{-}2\text{-}CH_2SiMe_2\text{-}]_2O$ is formed by the action of LiC_4H_9 on $1,2\text{-}C_2B_{10}H_{10}\text{-}1\text{-}Me\text{-}2\text{-}CH_2SiMe_2OMe$.[438] $(1,X\text{-}C_2B_{10}H_{10}\text{-}1\text{-}Me\text{-}2\text{-}CH_2SiMe_2\text{-})_2O$, (X = 2, 7) is prepared from the action of water on $1,X\text{-}C_2B_{10}H_{10}\text{-}1\text{-}Me\text{-}2\text{-}CH_2SiMe_2OLi$.[4c] $1,2\text{-}C_2B_{10}H_{10}\text{-}1\text{-}Me\text{-}2\text{-}CH_2SiMe_2(1',2'\text{-}C_2B_{10}H_{10}\text{-}1'\text{-}Me\text{-}2'\text{-})$ is produced from $1,2\text{-}C_2B_{10}H_{10}\text{-}1\text{-}Me\text{-}2\text{-}Li/1,2\text{-}C_2B_{10}H_{10}\text{-}1\text{-}Me\text{-}2\text{-}CH_2SiMe_2Cl$.[438] $ClCH_2SiMe_2O[SiMe_2CH_2CB_{10}H_{10}CH_2SiMe_2O]_2SiMe_2CH_2Cl$, containing the 1,7-isomer of the $CB_{10}H_{10}C$ units, is obtained from the action of $Li_2[1,7\text{-}C_2B_{10}H_{10}]$ on $[ClCH_2SiMe_2]_2O$.[4c] The reaction of *t*-butyldimethylsilyl chloride with the C-lithiated derivative of $1,2\text{-}C_2B_{10}H_{12}$ yields the carboranylsilane derivative $1\text{-}Me_2SiCMe_3\text{-}1,2\text{-}C_2B_{10}H_{11}$. The latter compound serves as a synthon to other C(cage) monosubstituted *closo*-$1,2\text{-}C_2B_{10}H_{12}$ derivatives.[185,439]

Studies on other carboranyl derivatives include the following molecules: $(1,2\text{-}C_2B_{10}H_{10}\text{-}1\text{-}Me\text{-}2\text{-}CH_2)_2CH_2$,[292] 1-Me-2-phenylalkyl-$1,2\text{-}C_2B_{10}H_{10}$[440] 1-ethynyl derivatives of $1,12\text{-}C_2B_{10}H_{12}$,[35] compounds based on 1-cyclopentadiene-$1,2\text{-}C_2B_{10}H_{11}$,[4a] $1,2\text{-}R_2\text{-}1,2\text{-}C_2B_{10}H_{10}$ (R = 4'-Ph-C_6H_4-, 4''-$MeOC_6H_4O$-4'-C_6H_4-, 4''-$MeOC_6H_4$-4'-C_6H_4-, 4''-$MeOC_6H_4$-4'-C_6H_4),[441] $1\text{-}CF_3COCO\text{-}2\text{-}Me\text{-}1,2\text{-}C_2B_{10}H_{10}$,[442] $1\text{-}O_2NCH_2CH(R)\text{-}1,2\text{-}C_2B_{10}H_{11}$ (R = aryl),[35] 1-(ω-aminoalkyl)-$1,2\text{-}C_2B_{10}H_{11}$,[443–4] $(1,7\text{-}C_2B_{10}H_{10}\text{-}1\text{-}Me\text{-}7\text{-}CH_2)_2Hg$,[2c] $(1,7\text{-}C_2B_{10}H_{10}\text{-}1\text{-}Me\text{-}7\text{-}CH_2)_3As$,[424] 1-HON=C(Cl)-$1,7\text{-}C_2B_{10}H_{11}$,[35] 1-ONC-$1,7\text{-}C_2B_{10}H_{11}$,[35] 1-TMS-P=C(O-TMS)-$1,7\text{-}C_2B_{10}H_{11}$,[35] and $1,7\text{-}[HOSiMe_2CH_2]_2\text{-}1,7\text{-}C_2B_{10}H_{10}$.[2c] Additional carbon attached derivatives are cited in the *Gmelin series*[8–9] and in other studies.[2–5,445]

(iii) B_{cage} and C_{cage} attached metal derivatives

The preparation of $(1,2\text{-}C_2B_{10}H_{10}\text{-}2\text{-Me-}1\text{-})_2Mg\cdot 2C_4H_8O_2$ is effected by a dioxane-induced disproportionation of $1,2\text{-}C_2B_{10}H_{10}\text{-}2\text{-Me-}1\text{-MgBr}$.[188] Reaction of $C_2B_{10}H_2Cl_{10}$ with R_3M (R = Me, Me_2CHCH_2; M = Al, In) gives $R_2MC_2B_{10}HCl_{10}$.[35] $[(1,Y\text{-}C_2B_{10}H_{11}\text{-}1\text{-R-Y-})_2AlH_2]M$ (R = H, Me, Ph, $CH_2{=}C(Me)$-, M = Li, Na; Y = 2, 7, 12) are prepared by treating $1,Y\text{-}C_2B_{10}H_{11}\text{-}1\text{-R}$ with $M[AlH_4]$ in tetrahydrofuran.[5b] $1,2\text{-}C_2B_{10}H_{11}\text{-}9\text{-}Ga(O_2CCF_3)_2$ is formed from the action of $Ga(O_2CCF_3)_3$ on $1,2\text{-}C_2B_{10}H_{11}\text{-}Hg(O_2CCF_3)$.[35] $1,2\text{-}C_2B_{10}H_{11}\text{-}1\text{-Ph}$ reacts with $MeGaCl_2$ to afford $1,2\text{-}C_2B_{10}H_{10}\text{-}1\text{-}C_6H_5\text{-}2\text{-}Ga(Me)Cl$; $(1,2\text{-}C_2B_{10}H_{10}\text{-}1\text{-Ph-}2\text{-})_2GaCl$ is formed from the action of $1,2\text{-}C_2B_{10}H_{10}\text{-}1\text{-}C_6H_5\text{-}2\text{-Li}$ and $GaCl_3$. $1,2\text{-}C_2B_{10}H_{11}\text{-}1\text{-Ph}$ reacts with Me_3Ga in the presence of $[Me_2NCH_2]_2$ to give $1,2\text{-}C_2B_{10}H_{10}\text{-}1\text{-Ph-}2\text{-}GaMe_2\cdot[Me_2NCH_2]_2$. In a similar fashion the aluminum, indium, and thallium derivatives, $1,2\text{-}C_2B_{10}H_{10}\text{-}1\text{-Ph-}2\text{-}MMe_2\cdot[Me_2NCH_2]_2$, M = Al, In, Tl, are prepared from Me_3M and the phenylcarbaborane.[446] The rate constants for B-thallation of 1,2- and $1,7\text{-}C_2B_{10}H_{12}$, and various derivatives thereof, by $Tl(O_2CCF_3)_3$ have been reported.[2c,447] $1,12\text{-}C_2B_{10}H_{11}\text{-}2\text{-}Tl(OCOCF_3)_2$ is prepared from the action of $Tl(OCOCF_3)_3/CF_3CO_2H$ on $1,12\text{-}C_2B_{10}H_9\text{-}2\text{-}HgOCOCF_3$.[4c] Refluxing $1,2\text{-}C_2B_{10}H_{11}\text{-}9\text{-}Tl(O_2CCF_3)_2$ with mercury in DMF yields $1,2\text{-}C_2B_{10}H_{11}\text{-}9\text{-}HgO_2CCF_3$ and $(1,2\text{-}C_2B_{10}H_{11}\text{-}9\text{-})_2Hg$.[4b] $1,7\text{-}C_2B_{10}H_9\text{-}1\text{-}R^1\text{-}7\text{-}R^2\text{-}9\text{-}TlSn(O_2CCH_2COMe)_2Br_2$ ($R^1 = R^2$ = H, Me) is obtained by treatment of $1,7\text{-}C_2B_{10}H_9\text{-}1\text{-}R^1\text{-}7\text{-}R^2\text{-}9\text{-}TlBr_2$ with $Sn(O_2CCH_2COMe)_2$.[448] Other studies of thallium derivatives of $C_2B_{10}H_{12}$ have been reported.[2c,449] Manganese-bound carbaboranes have been studied.[3]

Reaction of $Li[1,2\text{-}C_2B_{10}H_{11}]$ with $CoCl_2\cdot L$ (L = 2,2'-bipyridyl) yields $(1,2\text{-}C_2B_{10}H_{11}\text{-}1\text{-})_2Co\cdot L$. The dilithio salt $Li_2[1,2\text{-}C_2B_{10}H_{10}]$ reacts with $CoCl_2\cdot L$ to produce $(1,2\text{-}C_2B_{10}H_{10}\text{-}1,2\text{-})Co\cdot L$ which is proposed to have a C_2Co three-membered ring.[450] The preparation of $(1,2\text{-}C_2B_{10}H_{11}\text{-}1\text{-})NiCp\cdot PPh_3$ has been reported.[451] $(1,2\text{-}C_2B_{10}H_{11}\text{-}1\text{-})_2M\cdot L$ (M = Ni, Pd; L = 2,2'-bipyridyl) is formed from the action of $MX_2\cdot L$ (X = halogen) on $1,2\text{-}C_2B_{10}H_{11}\text{-}1\text{-Li}$.[4a] Metal complexes with a metal–boron bond are prepared by cyclopalladation of $1,Y\text{-}C_2B_{10}H_{10}\text{-}1\text{-}R^1\text{-}2\text{-}CH_2NR^2_2$ (Y = 2, 7; R^1 = H, Ph; R^2 = Me, Et) with $(PhCN)_2PdCl_2$.[4b] Reaction of $1,2\text{-}C_2B_{10}H_{11}\text{-}1\text{-}CH_2NEt_2$ with $MeRe(CO)_5$ at 100 °C gives $1,2\text{-}C_2B_{10}H_{10}\text{-}1\text{-}CH_2NEt_2\text{-}B\text{-}Re(CO)_4$.[35] The addition reactions of *trans*-$[Ir\text{-}(1,Y\text{-}C_2B_{10}H_{10}\text{-}2\text{-R-}1\text{-})(CO)L^1L^2]$, where Y = 1, 7, $L^1 = L^2 = PPh_3$, $P(Me)Ph_2$, and R = H, Me, Ph, with HCl, HBr, Cl_2, and Br_2 have been studied both in the solid state and solution using a variety of solvents. The HX adducts of general formula $[Ir(H)(X)(1,Y\text{-}C_2B_{10}H_{11}\text{-}1)(CO)L^1L^2]$ (X = Cl, Br, I) reductively eliminate the carbaborane molecule $1,Y\text{-}C_2B_{10}H_{12}$.[452] Reactions of $[Ir(H)_2(\sigma\text{-}1,7\text{-}C_2B_{10}H_{10}\text{-}7\text{-Ph})(CO)(RCN)(PPh_3)]$ (R = Me, Ph) with alkenes and alkynes, and subsequent reductive-elimination of the alkane or alkene molecule from the obtained products, have been examined.[453] $1,7\text{-}C_2B_{10}H_{10}\text{-}1\text{-Ph-}7\text{-}Ir(CO)(PhCN)[PPh_3]$ is found to be an effective catalyst for the homogeneous hydrogenation of terminal alkenes and alkynes.[454] Still other iridium bound carbaboranes have been reported.[455]

$(1,Y\text{-}C_2B_{10}H_{11}\text{-}9)_2Hg$ (X = 2,7) compounds are obtained from the electrochemical symmetrization of $1,Y\text{-}C_2B_{10}H_{11}\text{-}9\text{-HgX}$ (X = Cl, Br, CF_3CO_2) in acetone or acetonitrile as solvent, utilizing platinum or Cu–Hg as the cathode, and magnesium or platinum as the anode.[4c] $(1,Y\text{-}C_2B_{10}H_{11}\text{-}9\text{-})_2Hg$ (Y = 2,7) is formed from the reaction of $1,Y\text{-}C_2B_{10}H_{11}\text{-}9\text{-HgCl}$ with $Na_2SnO_2/NaOH$.[35] $(1,2\text{-}C_2B_{10}H_{10}\text{-}1\text{-R-}2\text{-})_2Hg$ is prepared by the action of $HgCl_2/LiBr$, in aqueous DMF, on $1,2\text{-}C_2B_{10}H_{11}\text{-}1\text{-R}$ (R = Me, Ph, CH_2Cl).[35] The photoreactions of $(1,Y\text{-}C_2B_{10}H_{11}\text{-}9\text{-})_2Hg$ (Y = 7,12) with $(MeO)_3P$ give dimethyl B-carboranylphosphonates.[4d] The radical products in the reaction of $(1,2\text{-}C_2B_{10}H_{11}\text{-})_2Hg$ (with 3,6-di-*t*-butyl-1,2-benzoquinone) have been studied by ESR.[4c,35] *m*-Carbaboran-9-yl radicals, formed in the decomposition of $(1,7\text{-}C_2B_{10}H_{11}\text{-}9\text{-})_2Hg$, add to the double bond in tetrachloroethylene, trichloroethylene, and 1-hexene.[35] The electroreduction mechanism of $1,2\text{-}C_2B_{10}H_{11}\text{-}9\text{-}HgP(O)(OR)_2$ differs from the reduction mechanism of the corresponding *m*-carbaborane derivative in that in the first case, a cleavage of the B–Hg bond occurs and, in the second case, a cleavage of the Hg–P bond occurs.[35] Reactions of the B-bonded carbaborane mercury derivatives with $(C_6F_5)_3GeGe(Et)_3$ affords compounds with B–Hg–Ge or Ge–Hg–B–B–Hg–Ge chains.[4b] Treatment of $(1,2\text{-}C_2B_{10}H_{11}\text{-}9\text{-})_2Hg$ with naphthyllithium gives $[(1,2\text{-}C_2B_{10}H_{11}\text{-}9)_2Hg]^{4-}$. Reaction of $(1,Y\text{-}C_2B_{10}H_{11}\text{-}9)_2Hg$ (Y = 2,7) with LiC_4H_9 and subsequently with TMS-Cl gives $[1,2\text{-}C_2B_{10}H_9\text{-}1\text{-TMS-}9]_2Hg$.[3] $(1,2\text{-}C_2B_{10}H_{11}\text{-}9)HgSn(\text{acetylacetonate})_2\text{-}(\text{-}9\text{-}1,2\text{-}C_2B_{10}H_{11})$ is formed from the action of $Sn(\text{acetylacetonate})_2$ on $(1,2\text{-}C_2B_{10}H_{11}\text{-}9)_2Hg$.[456] The reaction of HgI_2 with *closo*-$1,2\text{-}Li_2\text{-}1,2\text{-}C_2B_{10}H_{10}$ yields the cyclic tetramer, $Li_2[(1\text{-Hg-}1,2\text{-}C_2B_{10}H_{10}\text{-}2)_4I_2]$, which has been studied in terms of host–guest chemistry.[187] Treatment of $Li_2[(1\text{-Hg-}1,2\text{-}C_2B_{10}H_{10}\text{-}2)_4I_2]$ with AgOAc gives the neutral tetramer $(1\text{-Hg-}1,2\text{-}C_2B_{10}H_{10}\text{-}2)_4$.[189] Reactions of B-bonded mercury derivatives of $C_2B_{10}H_{12}$ with tin(II) acetylacetonate affords carbaborane compounds with *exo*-cage B–Hg–Sn–B bonds.[4d] Other mercury derivatives have been prepared and the chemistry of these compounds to produce Se, Te, Tl, and Sn derivatives of $C_2B_{10}H_{12}$ have been studied.[3,4a] Mercury reagents of $C_2B_{10}H_{12}$ have been used to prepare B- and C-attached carbaborane derivatives of the lanthanides Sm, Eu, Tm, Yb.[4b,35,457] $(1,2\text{-}C_2B_{10}H_{10}\text{-}1\text{-}C_6H_5\text{-}2)_3M$ (M = La, Tm, Yb) is formed from $1,2\text{-}C_2B_{10}H_{10}\text{-}1\text{-}C_6H_5\text{-}2\text{-}Li/MCl_3$.[35]

(iv) Miscellaneous chemistry

Deuteration of 4,6-$C_2B_7H_{13}$ (**9**) in a $DCl/AlCl_3/CS_2$ system affords the 3,5-dideuterio derivative.[372] Conversion of the racemic 5,6-$C_2B_8H_{12}$ to its levorotatory enantiomer is accomplished with the use of (+)-*N*-methylcamphidine.[458–9] The [*closo*-1-CB_9H_{10}]$^-$ anion is formed from the action of [6-NMe_3-*nido*-6-CB_9H_{11}] with piperidine in THF.[131] 9-L-6-CB_9H_{13} (L = Me_2S, MeCN, PPh_3) is formed from the reaction of [Me_4N][*nido*-6-CB_9H_{12}] with HCl/L.[92] 1,2-$C_2B_{10}H_{12}$ is produced from the thermal decomposition of (1,2-$C_2B_{10}H_{10}$-1-CO_2)$_2$Ni(2,2'-bipyridyl).[425] 1,2-Dehydro-*o*-carbaborane, a carbaborane analogue of benzyne, has been generated through loss of bromide ion from the [2-Br-1,2-$C_2B_{10}H_{10}$]$^-$ anion. This intermediate undergoes the ene reaction as well as [2 + 2] and [2 + 4] cycloadditions.[460–3] Li[*closo*-1,2-$C_2B_{10}H_{10}$-1-Me]·[MeN($CH_2CH_2NMe_2$)$_2$] is prepared by successive treatments of 1-Me-*closo*-1,2-$C_2B_{10}H_{11}$ with LiC_4H_9 and MeN($CH_2CH_2NMe_2$)$_2$.[163]

6.2.5.6 Polymer applications

The early carbaborane incorporated polymers were mostly based on chemistry that mimicked silicone polymer work.[8–9,52,55,58–9] High thermal stabilities, as well as other desirable properties, of many such polymers have logically led to other studies in which the $C_2B_{10}H_{12}$ cage is incorporated into polyamides,[4–5,464] polyamine–polyimides,[465] polyurethanes, polyazomethines,[5,466] polyphosphazenes,[4c,5c,52,435] polyesters,[3,4a] cellulose esters,[4b] polymethacrylates,[4c–d,414,467] polyarylenes,[2c,4a,4e,468–9] epoxy resins,[470–4] and polyorganosilazanes.[475–6]

Carbaborane exocyclic vinyl silanes of the type, 1,2-$C_2B_{10}H_{10}$-1,2-(CH_2OSi(R)(CH=CH_2)OCH_2) (R = Me, OMe, Cl) have been homopolymerized and copolymerized with methyl methacrylate.[3] The hydrolytic polycondensation of $C_2B_{10}H_{11}$-$CH_2SiR^1R^2Cl$ (R^1, R^2 = Me, Cl) occurs with the formation of oligomers with a cyclic or silsesquioxane structure.[2c] The stabilizing effect of carbaborane groups in siloxanes and silsesquioxanes have been related to the formation of additional cross-links by carbaborane groups during thermal oxidation at 400 °C.[52] Carbaborane siloxane oligomers, in which the carbaborane is either 1,2- or 1,7-$C_2B_{10}H_x$, are prepared from the hydrolysis of $C_2B_{10}H_{11}$-1-CH_2Si(Me)(OMe)$_2$ and $C_2B_{10}H_{11}$-1-CH_2SiCl_3.[438] Certain carbaborane–silicone polymers, with C_2B_{10} carbaborane frameworks, have been prepared for use in connection with specialty chromatographic columns.[477–80] The Raman spectra of silicone polymers incorporating 1,7-$C_2B_{10}H_{12}$ cages have been studied,[236] and other polymers based on the incorporation of the $C_2B_{10}H_{12}$ into organosiloxanes have been reported.[4a–b] Pyrolysis of a 1,7-C_2B_{10}-containing siloxane polymer produces the parent 1,7-$C_2B_{10}H_{12}$,[481] and pyrolysis of carbaborane–siloxane polymers based on $C_2B_{10}H_{10}$ units can produce ceramic composites.[4a,482–3] Highly cross-linked 1,7-bis(hydroxydimethylsilyl)-*m*-carbaborane-triacetoxymethylsilane copolymer rubber coatings have been prepared that show resistance to attack by atomic oxygen.[484]

1,2-$C_2B_{10}H_{10}$-1,2-(CH_2OH)$_2$ reacts with organospirobicyclosilazasiloxanes to give siloxane polymers.[485] Polymers obtained from carbaboranediols have been characterized by gel-permeation chromatography and ^{1}H NMR spectroscopy.[486] Thermal oxidation of resite prepared from the 1,2-$C_2B_{10}H_{10}$-1,2-(C_6H_4-4'-OH)$_2$/HCHO copolymer proceeds with formation of water-soluble boron containing compounds.[4d] Copolymerization of epoxy oligomers based on 1,2-$C_2B_{10}H_{10}$-1,2-(C_6H_4-4-OH)$_2$ with 4,4'-diphenylmethane diisocyanate produces materials with better cure rates when the relative amounts of the carbaborane moiety are higher.[487] The effect of B_2O_3 on the thermal oxidative stabilization of carbaborane polymers was studied in tests with 1,2-$C_2B_{10}H_{10}$-1,2-(C_6H_4-4'-OH)$_2$/HCHO and 1,2-$C_2B_{10}H_{10}$-1,2-(C_6H_4-4'-OH)$_2$/C_6H_5OH/HCHO copolymers.[4d] Polymers derived from an *N,N'*-bis(hydroxymethyl)aminophenyl derivative of 1,2-$C_2B_{10}H_{12}$ have been reported.[488]

NMR spectroscopic relaxation processes were investigated in polyesters prepared from carboxyphenyl derivatives of 1,2- and 1,7-$C_2B_{10}H_{12}$.[489] Carbaborane-containing poly(arylacetylenes) in which the carbaborane fragment is 1,2-$C_2B_{10}H_{10}$-1,2 are obtained by treating poly(*p*-diethynylbenzene) with decaborane in organic amide solvents or in toluene in the presence of $PhNMe_2$.[4b,4d] A copolymer of carboranylstyrene and acryloferrocene has been reported,[4d] and the thermal stablity of the product peroxide-initiated polymerization of a *C,C'*-bis(acryloyloxymethyl) derivative of 1,2-$C_2B_{10}H_{12}$ has been studied.[490] Polycyclocondensation of bis[4-(4'-acetylphenoxy)phenyl]-*C,C'*-*o*-carbaborane with various monoacetylarylenes is effected at 400 °C.[491] Polyaryl systems containing icosahedral 1,2-$C_2B_{10}H_{10}$ fragments in the main chain are prepared by triflic acid-promoted polycondensation of aromatic dicarboxylic acids with aromatic diethers.[492] The temperature dependency of the dielectric constant of polyarylates containing 1,2-$C_2B_{10}H_{10}$-1,2-units in the chain has been examined.[2c] The spin–spin and spin–lattice relaxation times for polyarylates consisting of dihydroxydiphenyl ether/1,7-$C_2B_{10}H_{10}$-1,7-

copolymers, both in solution and in the solid state, have been studied over a range of temperatures.[493] Oligomeric products are formed from radical addition of 1,2- and 1,7-$C_2B_{10}H_{12}$ to 1,2-dichlorethylene.[35]

A polymer with repeating units of $SC_2B_{10}H_{10}SCH_2C_6H_4CH_2$, where the $C_2B_{10}H_{10}$ unit contains adjacent carbons in the cage and where the benzene ring is *ortho* substituted, is prepared by the action of $K_2[1,2\text{-}C_2B_{10}H_{10}\text{-}1,2\text{-}(S)_2]$ with $C_6H_4\text{-}1,2\text{-}(CH_2Br)_2$.[322] Polymerization of the carbaborane epithiopropane, 1,2-$C_2B_{10}H_{10}$-1-Ph-2-CH_2CHCH_2S by $BF_3{\cdot}OEt_2$ gives a carbaborane substituted polythioethylene.[4b] *m*-Carboranylene-bridged poly(2,3-dihydrobenzothiazoles) were synthesized by polycondensation of 1,7-dicarbadodecaborane(12)-1,7-dicarboxaldehyde with 4,4'-diamino-[1,1'-biphenyl]3,3'-dithiol.[5c]

$[Et_4N]OH$ degrades the carbaborane cages in polyamines which contain 1,7-$C_2B_{10}H_x$ moieties in the polymeric chain; the product is a polyamine which contains $C_2B_9H_x$ units.[4b] Polyamides containing 1,7-$C_2B_{10}H_x$ groups were converted by treatment with KOH/EtOH to polymers in which the carbaborane moiety was degraded to C_2B_9 cages.[494] Branched $-[-OC-CB_{10}H_{10}C-CONH-C_6H_4-C_6H_4-NH-]_m$ copolymers containing $[Et_4N][C_2B_9H_{10}]$ units in place of some of the 1,7-$C_2B_{10}H_{10}$ units are useful as heat stabilizers for polyamides.[4c]

m-Carbaboranedicarboxylic acid chloride reacts with 1,3-dianilino-4,6-diaminobenzene to give a polyamide which in turn can be cyclized to give a carbaborane containing polybenzimidazole.[5c] A synthetic study of carbaborane-containing poly(naphthoylenebenzimidazole) has been conducted.[5c] *m*-Carbaborane-containing polymeric Schiff bases are prepared by polycondensation of *m*-carboranylenediamine, 1,7-bis(4-aminophenylcarboxy)-1,7-$C_2B_{10}H_{10}$, or 1,7-bis(4-aminophenylamido)-1,7-$C_2B_{10}H_{10}$ with dialdehydes.[5a,495]

6.2.5.7 Other applications

$Me_4N[7,8\text{-}C_2B_9H_{11}\text{-}7\text{-}C_6H_4\text{-}4'\text{-}NH_2]$,[150] 1,2-$C_2B_{10}H_{11}$-1-$C_6H_4$-4'-$N_2^+$ ion with tritium enrichment in the 2-position,[418] several glycosyl carbaborane derivatives,[496] and tritiated derivatives of phenyl isothiocyanate derivatives[497] of 1,2-$C_2B_{10}H_{12}$ and of 7,8-$[C_2B_9H_{12}]^-$ have been synthesized and used in connection with tumor localization of boron-10-labeled antibodies. The preparations of carbaboranyl amino acids and peptides with *closo*-1,2-$C_2B_{10}H_x$ and/or $C_2B_9H_x$ attached cage units,[498–500] porphyrins with attached carbaborane moieties,[501–4] carbaboranyl nucleoside[505] and nucleotide[506] derivatives, and *o*-carboranylmethyl ethers of steroids[298] are described for use in boron neutron capture therapy of tumors. Treatment of 1,2-$C_2B_{10}H_{12}$ with cyclodextrins gives rise to inclusion complexes,[507] and polyols of a cascade type have been employed as a water-solubilizing element of carbaborane derivatives for boron neutron capture therapy.[508]

6.3 ORGANOPOLYBORON HYDRIDES

6.3.1 Physical Properties

Equilibration studies on 2- and 3-MeB_5H_{10} indicates that the latter isomer is very slightly more stable than the former in $CDCl_3$ solution.[509] IR spectral absorptions for alkenyl derivatives of B_5H_9 are reported.[240] An electron diffraction study of tetraborane(8) carbonyl in the gas phase has resulted in the structure determination of an *endo–exo* isomeric mixture; the B–C bond distance is determined to be 0.1517 nm.[510–11] Cyclic voltammetry of *arachno*-$[B_{10}H_{13}(CN)]^{2-}$ shows four oxidation waves.[512] The cyano group in $[B_{10}H_{13}(CN)]^{2-}$ anion occupies an *endo*-position with respect to the *arachno*-B_{10} cage in both solution (^{11}B NMR spectrum of cesium salt) and the solid (x-ray crystal structure of Me_3NPh^+ salt) states.[513] The He^I and He^{II} photoelectron spectra of 1,12-$(OC)_2$-$B_{12}H_{10}$ are assigned in terms of correspondence of bands to BH bonding or cluster-localized orbitals; the experimental assignments agree with MNDO calculations on the molecules, which allow assignment of the cluster orbitals and ionizations in terms of Stone's (1980, 1981) theory of *closo*-boranes.[274]

6.3.2 Preparation of Organopolyboranes

Tetrakis(*t*-butyl)tetraborane(4), $Bu^t_4B_4$, is prepared from the reaction of B_4Cl_4 with Bu^tLi; use of EtLi as the alkyl source gives rise to EtB_4Cl_3, and $Et_2B_4Cl_2$.[514] Tetra-*t*-butyltetraboratetrahedrane, $R_4B_4(R = Bu^t)$, is prepared by reduction of Bu^tBF_2 with Na–K alloy. X-ray analysis of $Bu^t_4B_4$ gives an average B–B bond distance of 0.171 nm and a B–C bond distance of 0.157 nm.[515]

2-Aryl derivatives of pentaborane(9), 2-*R*–B_5H_8 (R = Ph, tolyl, xylyl, $C_6H_2Me_3$), are prepared in low yield by $AlCl_3$-catalyzed action of 2-ClB_5H_8 on alkylbenzenes.[516] Early transition-metal halides $HfCl_4$ and $ZrCl_4$ have been studied as catalysts in the formation of 1-alkylpentaborane(9) compounds from alkyl halides and pentaborane(9).[517] In the presence of BF_3, the cluster complex $[(\mu\text{-H})_3Os_3(CO)_9\text{-}(\mu_3\text{-C})]_3(O_3B_3O_3)$ undergoes a Friedel–Crafts type reaction with B_5H_9 to form the boron-substituted complex $(\mu\text{-H})_3Os_3(CO)_9(\mu_3\text{-C})(B_5H_8)$. Substitution occurs at the apical site of the B_5 square pyramid of the pentaborane(9) derivative, the site expected to be most susceptible to electrophilic attack.[395,518] The reactions of pentaborane(9) with either ethyne, propyne, or 2-butyne, in the presence of catalytic amounts of $Ir(CO)C(PPh_3)_2$ give the corresponding alkenylpentaboranes 2-$(CHR^1{=}CR^2)B_5H_8$ (R^1, R^2 = H or Me). The reaction corresponds to a BH addition to the alkyne in a *cis*-fashion with the reaction of propyne occurring in a predominately anti-Markownikoff manner. Similarly, complexes $(R^1C{\equiv}R^2)Co_2(CO)_6$ (R^1, R^2 = H, Me) also activate alkynes for BH addition with pentaborane(9), but, in contrast with the results obtained with $Ir(CO)Cl(PPh_3)_2$, internal alkynes are more reactive than terminal alkynes and the reaction with propyne gives predominately Markownikoff addition.[240] B_5H_9 is trisalkylated with $LiEt_3BH$ to give 2,3,4-$Et_3B_5H_6$, whereas use of $Li(EtCHMe)_3BH$ results in formation of the monoalkyl product, 2-*s*-butylpentaborane(9). A mechanism to the first named product is proposed involving an intermediate, $[Et_3B_6H_{10}]^-$, followed by ethyl group migration to adjacent boron atoms; subsequent expulsion of $[BH_4]^-$ would then result in 2,3,4-$Et_3B_5H_6$.[519] The gas-phase photolysis reaction at 300 nm of pentaborane(9) and hexafluoroacetone produces both the basal and apical perfluoropropanol-substituted pentaborane compounds, RB_5H_8 (R = $C(CF_3)_2OH$).[247] The reaction of B_5H_9 with NaCN occurs in a 1:1 molar ratio at −30 °C to +10 °C to yield $Na[B_5H_9CN]$, which can be isolated in the form of a dioxanate; from ^{11}B NMR spectroscopic data, a structure related to that of B_5H_{11} is suggested for the product.[520] Synthesis 1-X-μ-$(Me_2NCH_2)B_5H_7$ (X = H, Et, Br), in which a C–N group replaces a bridging hydrogen of B_5H_9, is prepared from the reaction of $[Me_2N{=}CH_2]I$ with salts of the $[B_5H_8]^-$, $[1\text{-}EtB_5H_7]^-$ and $[1\text{-}BrB_5H_7]^-$ anions.[521]

2-Me-B_6H_9 is obtained from the treatment of *arachno*-4-CB_7H_{13} (**16**) with HCl/H_2O, and is also produced from the reaction of the reactive isomer of the $[C_2B_{10}H_{13}]^-$ anion with HCl/H_2O.[314] Formation of octabromomethylnonaborane, MeB_9Br_8, as well as other alkylated nonaboranes (EtB_9Br_8, $Me_2B_9Br_7$, and $MeEtB_9Br_7$), are obtained from the thermal decomposition of $(Et_3NH)_2B_{10}Br_{10}$ at temperatures ~430 °C.[522] The equatorially monosubstituted species [*closo*-2-OC-$B_{10}H_9]^-$ is prepared from the reaction of [*closo*-$B_{10}H_{10}]^{2-}$ with oxalyl chloride; an x-ray structure determination of two salts of this carbonylpolyborane anion shows that the C–O bond distance is 0.113 nm and the B–C bond distance is 0.148 nm. [*closo*-2-OC-$B_{10}H_9]^-$ reacts with H_2O to form [*closo*-2-$HO_2CB_{10}H_9]^{2-}$ and with amines and alcohols to form amides and esters, respectively.[523]

6.3.3 Reactions of Organopolyboranes

The conversion of $[1\text{-}MeB_5H_7]^-$ to 2,2-$(1\text{-}MeB_5H_7)_2$ is effected with $FeCl_2/FeCl_3$.[524] Hydrogen–deuterium exchange reactions of both 1- and 2-MeB_5H_8 with DBF_2 lead to the conclusion that the ratio of rates of exchange of an apical:basal hydrogen atom in the pentaborane system is ~6:1 (counting the terminal hydrogen atoms only), respectively.[525] At 1000 °C, in the presence of argon, poly(2-vinyl-pentaborane(9)) forms B_4C;[526] in the presence of NH_3 at ~1400 °C, this substituted pentaborane forms amorphous BN.[527] The reaction of the $[B_5H_8]^-$ anion with *B*-chloro-9-borabicyclo[3.2.1]nonane (*B*-Cl-9-BBN) produces a boron-bridge substituted *nido*-pentaborane, (μ-9-BBN)B_5H_8. The solid-state structure of (μ-9-BBN)B_5H_8 reveals that the 9-BBN fragment retains its characteristic chair–chair conformation. In solution at room temperature the 9-BBN fragment rotates relative to the pentaborane framework. From a dynamic ^{13}C NMR spectroscopic experiment the free energy of activation for the rotation has been found to be 60.6 ± 0.4 kJ mol^{-1}. An x-ray structure shows that the bridging boron is 0.187–0.192 nm from the basal B-2 and B-3. The B–C bond distance is 0.156 nm. In Et_2O solution (μ-9-BBN)B_5H_8 slowly converts to a *nido*-hexaborane derivative, 4,5-(cycloocta-1,5-diyl)B_6H_8, in which the bridging boron atom has moved into, and expanded, the cluster. The result is the conversion of the [3.3.1]ring system of 9-BBN to a [3.3.2]bicycle in which two adjacent basal boron atoms of a *nido*-hexaborane cluster provide the two-atom bridge. The solid-state x-ray structure of 4,5-(cycloocta-1,5-diyl)B_6H_8 reveals a chair–boat conformation for the [3.3.2] bicyclic ring system. In addition, two independent molecules are observed that differ primarily in the placement of the hydrogen atoms bridging the basal edges of the hexaborane fragment. The B–C bond distances are in the range 0.158–0.160 nm.[528]

3-MeB_6H_{11}, selectively ^{10}B-enriched in the B-1, B-2, and B-6 positions, reacts with Me_2O to produce 2-MeB_5H_8, which is ^{10}B-enriched at B-4 and, to a lesser extent, at B-3,5. In the presence of 2,6-lutidine the ^{10}B label in 2-MeB_5H_8 equilibrates into all boron positions except the methyl substituted B-2.[529] *B*-propyl derivatives of $B_{10}H_{14}$ have been used to prepare *B*-propyl derivatives of 1,2-$C_2B_{10}H_{12}$.[530]

6.4 REFERENCES

1. T. Onak in 'COMC-I' vol. 1, p. 411.
2. (a) J. B. Leach, *Specialist Periodical Reports: Organomet. Chem.*, 1981, **9**, 55; (b) *ibid.*, 1982, **10**, 48; (c) *ibid.*, 1983, **11**, 61.
3. J. H. Morris, *Specialist Periodical Reports: Organomet. Chem.*, 1984, **12**, 47.
4. (a) T. R. Spalding, *Specialist Periodical Reports: Organomet. Chem.*, 1985, **13**, 38; (b) *ibid.*, 1986, **14**, 40; (c) *ibid.*, 1987, **15**, 39; (d) *ibid.*, 1987, **16**, 39; (e) *ibid.*, 1989, **17**, 35.
5. (a) C. E. Housecroft, *Specialist Periodical Reports: Organomet. Chem.*, 1989, **18**, 37; (b) *ibid.*, 1990, **19**, 36; (c) *ibid.*, 1991, **20**, 34.
6. D. S. Matteson, *J. Organomet. Chem.*, 1981, **207**, 13.
7. F. A. Hart, A. G. Massey, P. G. Harrison and J. H. Holloway, *Ann. Rep. Prog. Chem., Sect. A*, 1980/81, **77**, 3; A. J. Welch *Ann. Rep. Prog. Chem., Sect. A A*, 1981/2, **78**, 19; 1982/3 **79**, 19; 1983/4, **80**, 19.
8. T. Onak, 'Gmelin Handbook of Inorganic Chemistry', Boron Compounds, 2nd supplement, vol. 2, Springer, Berlin, 1982, p. 223.
9. T. Onak, 'Gmelin Handbook of Inorganic Chemistry', Boron Compounds, 3rd supplement, vol. 4, Springer, Berlin, 1988, p. 153.
10. M. Grassberger and R. Köster, 'Houben-Weyl Methoden der Organischen Chemie', vol. XIII/3c, Organoborverbindungen III, Georg Thieme Verlag, Stuttgart, 1984, p. 92.
11. G. Wrackmeyer and R. Köster, 'Houben-Weyl Methoden der Organischen Chemie', vol. XIII/3c, Organoborverbindungen III, Georg Thieme Verlag, Stuttgart, 1984, p. 377.
12. T. Onak in 'Inorganic Reactions and Methods', eds. J. J. Zuckerman and A. P. Hagen, VCH, New York, 1990, vol. 10, p. 118.
13. R. A. Beaudet, in 'Advances in Boron and the Boranes', eds. J. F. Liebman, A. Greenberg and R. E. Williams, VCH, New York, 1988, p. 417.
14. V. S. Mastryukov, O. V. Dorofeeva and L. V. Vilkov, *Usp. Khim.*, 1980, **49**, 2377; *Russ. Chem. Rev.* 1980, **49**, 1181.
15. R. E. Williams, *Chem. Rev.*, 1992, **92**, 177.
16. R. E. Williams, in 'Electron Deficient Boron and Carbon Clusters', eds. G. A. Olah, K. Wade and R. E. Williams, Wiley, New York, 1991, p. 11.
17. M. E. O'Neill and K. Wade, *Polyhedron*, 1984, **3**, 199.
18. L. V. Vilkov, *Stud. Phys. Theor. Chem.*, 1981, **13**, 117 (*Chem. Abstr.*, 1982, **96**, 198 671t).
19. D. M. P. Mingos, *Inorg. Chem.*, 1985, **24**, 114.
20. D. J. Wales and A. J. Stone, *Inorg. Chem.*, 1987, **26**, 3845.
21. D. J. Wales and D. M. P. Mingos, L. Zhenyang, *Inorg. Chem.*, 1989, **28**, 2754.
22. M. C. Favas and D. L. Kepert, *Prog. Inorg. Chem.*, 1981, **28**, 309.
23. A. Tang, *Youji Huaxue*, 1984, 227.
24. Q. Li and A. Tang, *Gaodeng Xuexiao Huaxue Xuebao*, 1991, **12**, 69 (*Chem. Abstr.*, 1991, **115**, 114 576r).
25. G. A. Olah and G. K. S. Prakash, *Chem. Bri.*, 1983, **19**, 916.
26. G. A. Olah, G. K. S. Prakash, R. E. Williams, L. D. Field and K. Wade, 'Hypercarbon Chemistry', Wiley, New York, 1987.
27. R. E. Williams, G. K. S. Prakash, L. D. Field and G. A. Olah, in 'Advances in Boron and the Boranes', ed. J. F. Liebman, A. Greenberg and R. E. Williams, VCH, New York, 1988, p. 191.
28. K. Wade, in 'Electron Deficient Boron and Carbon Clusters', eds. G. A. Olah, K. Wade and R. E. Williams, Wiley, New York, 1991, p. 95.
29. T. Onak and K. Fuller, in 'Electron Deficient Boron and Carbon Clusters', eds. G. A. Olah, K. Wade and R. E. Williams, Wiley, New York, 1991, p. 183.
30. W. N. Lipscomb, *J. Less-Common Met.*, 1981, **82**, 1.
31. T. Onak, in 'Advances in Boron and the Boranes', eds. J. F. Liebman, A. Greenberg and R. E. Williams, VCH, New York, 1988, p. 125.
32. B. Stibr, J. Plesek and S. Hermanek, in 'Advances in Boron and the Boranes', eds. J. F. Liebman, A. Greenberg and R. E. Williams, VCH, New York, 1988, p. 35.
33. I. B. Sivaev, A. Kayumov, A. B. Yakushev, K. A. Solntsev and N. T. Kuznetsov, *Koord. Khim.*, 1989, **15**, 1466.
34. B. Stibr, *Chem. Rev.*, 1992, **92**, 225.
35. V. I. Bregadze, *Chem. Rev.*, 1992, **92**, 209.
36. V. N. Kalinin, *Usp. Khim.*, 1980, **49**, 2188; *Russ. Chem. Rev.*, 1980, **49**, 1084 (*Chem. Abstr.*, 1981, **94**, 139 864u).
37. I. B. Sivaev, K. A. Solntsev and N. T. Kuznetsov, *Koord. Khim.*, 1992, **18**, 227.
38. (a) A. R. Siedle, *Ann. Rep. NMR Spectrosc.*, 1988, **20**, 205; (b) A. R. Siedle, *Ann. Rep. NMR Spectrosc.*, 1982, **12**, 177.
39. L. A. Leites, *Chem. Rev.*, 1992, **92**, 279.
40. R. N. Grimes, in 'Advances in Boron and the Boranes', eds. J. F. Liebman, A. Greenberg and R. E. Williams, VCH, New York, 1988, p. 235.
41. R. N. Grimes, *Adv. Inorg. Chem. Radiochem.*, 1983, **26**, 55.
42. N. S. Hosmane and J. A. Maguire, in 'Electron Deficient Boron and Carbon Clusters', eds. G. A. Olah, K. Wade and R. E. Williams, Wiley, New York, 1991, p. 215.
43. V. V. Grushin, V. I. Bregadze and V. N. Kalinin, *J. Organomet. Chem. Libr.*, 1988, **20**, 1 (*Chem. Abstr.*, 1988, **109**, 170 491x).
44. V. I. Bregadze, V. T. Kampel, A. Y. Usyatinskii and N. N. Godovikov, *Pure Appl. Chem.*, 1991, **63**, 357.

45. V. I. Bregadze, V. T. Kampel, A. Y. Usyatinskii and N. N. Godovikov, *Pure Appl. Chem.*, 1991, **63**, 835.
46. V. I. Bregadze, *Proc. Indian Acad. Sci., Sect. A*, 1989, **55**, 414.
47. V. N. Kalinin, A. V. Usatov and L. I. Zakharkin, *Metalloorg. Khim.*, 1989, **2**, 54.
48. L. G. Sneddon, *Pure Appl. Chem.*, 1987, **59**, 837.
49. E. W. Corcoran and L. G. Sneddon, in 'Advances in Boron and the Boranes', eds. J. F. Liebman, A. Greenberg and R. E. Williams, VCH, New York, 1988, p. 71.
50. N. T. Kuznetsov and K. A. Solntsev, *Koord. Khim.*, 1991, **17**, 1157.
51. J. Plesek, *Chem. Rev.*, 1992, **92**, 269.
52. T. Onak, in 'Inorganic Reactions and Methods', eds. J. J. Zuckerman and A. P. Hagen, VCH, New York, 1990, vol. 17, p. 97.
53. V. V. Korshak, S. A. Pavlova, P. N. Gribkova and T. N. Balykova, *Acta Polym.*, 1981, **32**, 61.
54. E. N. Peters, *Ind. Eng. Chem., Prod. Res. Dev.*, 1984, **23**, 28.
55. N. I. Bekasova, *Usp. Khim.* 1984, **53**, 107.
56. H. R. Allcock, *ACS Symp. Ser.*, 1983, **232**, 49.
57. E. Y. Sorokina and V. G. Berezkin, *Neftekhimiya*, 1986, **26**, 464.
58. O. A. Mel'nik, A. A. Sakharova and T. M. Frunze, *Usp. Khim.*, 1988, **57**, 1529; *Russ. Chem. Rev.*, 1988, **57**, 875.
59. N. I. Bekasova and N. G. Komarova, *Usp. Khim.*, 1992, **61**, 647.
60. E. Bortel, *Polimery (Warsaw)*, 1990, **35**, 145.
61. V. V. Korshak, N. I. Bekasova and L. G. Komarova, *Plast. Massy*, 1989, 5.
62. J. B. Casey, W. J. Evans and W. H. Powell, *Inorg. Chem.*, 1981, **20**, 1333; 1981, **20**, 3556; 1983, **22**, 2228; 1983, **22**, 2236; 1984, **23**, 4132.
63. K. Krogh-Jespersen, D. Cremer, J. D. Dill, J. A. Pople and P. v. R. Schleyer, *J. Am. Chem. Soc.*, 1981, **103**, 2589.
64. P. H. M. Budzelaar, K. Krogh-Jespersen, T. Clark and P. v. R. Schleyer, *J. Am. Chem. Soc.*, 1985, **107**, 2773.
65. P. H. M. Budzelaar, S. M. Van der Kerk, K. Krogh-Jespersen and P. v. R. Schleyer, *J. Am. Chem. Soc.*, 1986, **108**, 3960.
66. L. H. Reed and L. C. Allen, *Int. J. Quantum Chem., Symp.*, 1991, **25**, 489.
67. M. L. McKee, *J. Phys. Chem.*, 1991, **95**, 9273.
68. E. D. Jemmis, *J. Am. Chem. Soc.*, 1982, **104**, 7017.
69. J. J. Ott and B. M. Gimarc, *J. Comput. Chem.*, 1986, **7**, 673.
70. B. M. Gimarc and J. J. Ott, *J. Am. Chem. Soc.*, 1987, **109**, 1388.
71. K. Takano, M. Izuho and H. Hosoya, *J. Phys. Chem.*, 1992, **96**, 6962.
72. M. L. McKee, *J. Am. Chem. Soc.*, 1988, **110**, 4208.
73. M. L. McKee, *Inorg. Chem.* 1988, **27**, 4241.
74. T. Whelan and P. Brint, *J. Chem. Soc., Faraday Trans. 2*, 1985, 267.
75. R. F. W. Bader and D. A. Legare, *Can. J. Chem.*, 1992, **70**, 657.
76. M. L. McKee, *J. Am. Chem. Soc.*, 1991, **113**, 9448.
77. M. L. McKee, *J. Phys. Chem.*, 1992, **96**, 1679.
78. M. L. McKee, *J. Am. Chem. Soc.*, 1988, **110**, 5317.
79. D. W. J. Cruickshank, A. Chablo, M. Eisenstein and P. L. Reidy, *Acta Chem. Scand. A*, 1988, **42**, 530.
80. M. L. McKee, *J. Phys. Chem.*, 1989, **93**, 1265.
81. D. N. Cox, D. M. P. Mingos and R. Hoffmann, *J. Chem. Soc., Dalton Trans.*, 1981, 1788.
82. A. M. Mebel and O. P. Charkin, *Zh. Neorg. Khim.*, 1991, **36**, 2880.
83. P. Brint, E. F. Healy, T. R. Spalding and T. Whelan, *J. Chem. Soc., Dalton Trans.*, 1981, 2515.
84. P. Brint and B. Sangchakr, *J. Chem. Soc., Dalton Trans.*, 1988, 105.
85. M. L. McKee, *J. Phys. Chem.*, 1989, **93**, 3426.
86. A. M. Mebel and O. P. Charkin, *Zh. Neorg. Khim.*, 1990, **35**, 312; *Russ. J. Inorg. Chem.*, 1990, **35**, 174.
87. T. Onak, J. Tseng, D. Tran, M. Correa, S. Herrera and J. Arias, *Inorg. Chem.*, 1992, **31**, 2161.
88. T. Onak, E. O'Gorman, T. Banuelos, C. Alfonso and M. Yu, *Inorg. Chem.*, 1990, **29**, 335.
89. D. M. P. Mingos and A. J. Welch, *J. Chem. Soc., Dalton Trans.*, 1980, 1674.
90. B. M. Gimarc, B. M. Dai and J. J. Ott, *J. Comput. Chem.*, 1989, **10**, 14.
91. M. E. O'Neill and K. Wade, *Polyhedron*, 1983, **2**, 963.
92. K. Base, B. Stibr, J. Dolansky and J. Duben, *Collect. Czech. Chem. Commun.*, 1981, **46**, 2345.
93. J. Dolansky, S. Hermanek and R. Zahradnik, *Collect. Czech. Chem. Commun.*, 1981, **46**, 2479.
94. M. L. McKee, *J. Am. Chem. Soc.*, 1992, **114**, 5856.
95. T. A. Green, A. C. Switendick and D. Emin, *J. Chem. Phys.*, 1988, **89**, 6815.
96. E. D. Jemmis and P. N. V. Pavankumar, *Proc. Indian Acad. Sci. Chem. Sci.*, 1984, **93**, 479.
97. R. L. DeKock and C. P. Jasperse, *Inorg. Chem.* 1983, **22**, 3843.
98. G. F. Mitchell and A. J. Welch, *J. Chem. Soc., Dalton Trans.*, 1987, 1017.
99. Z. J. Abdou *et al.*, *Inorg. Chem.*, 1985, **24**, 2363.
100. Z. J. Abdou, G. Abdou, T. Onak and S. Lee, *Inorg. Chem.*, 1986, **25**, 2678.
101. E. O'Gorman, T. Banuelos and T. Onak, *Inorg. Chem.*, 1988, **27**, 912.
102. G. A. Beltram, C. Jasperse, M. A. Cavanaugh and T. P. Fehlner, *Inorg. Chem.*, 1990, **29**, 329.
103. P. Brint, J. P. Cronin, E. Seward and T. Whelan, *J. Chem. Soc., Dalton Trans.*, 1983, 975.
104. A. J. Stone and M. J. Alderton, *Inorg. Chem.*, 1982, **21**, 2297.
105. M. J. S. Dewar, C. Jie and E. G. Zoebisch, *Organometallics*, 1988, **7**, 513.
106. S. A. Khan, J. H. Morris and S. Siddiqui, *J. Chem. Soc., Dalton Trans.*, 1990, 2053.
107. T. A. Green, *AIP Conf. Proc.*, 1991, **231**, 42.
108. J. J. Ott and B. M. Gimarc, *J. Am. Chem. Soc.* 1986, **108**, 4303.
109. R. B. King, *J. Math. Chem.*, 1987, **1**, 45.
110. S. G. Semenov, *Teor. Eksp. Khim.*, 1987, **23**, 451; *Theor. Exp. Chem. (USSR)*, 1987, **23**, 422.
111. P. G. Perkins and J. J. P. Stewart, *J. Chem. Soc., Faraday Trans.*, **2**, 1982, 285.
112. S. G. Semenov, *Zh. Strukt. Khim.*, 1981, **22**, 164; *J. Struct. Chem. (USSR)*, 1981, **22**, 776.
113. R. Mageswaran and N. J. Fitzpatrick, *J. Natl. Sci. Counc. Sri Lanka*, 1987, **15**, 47 (*Chem. Abstr.*, 1990, **112**, 7670h).
114. S. Lee, *Inorg. Chem.*, 1992, **31**, 3063.

115. A. Goursot, E. Penigault, H. Chermette and J. G. Fripiat, *Can. J. Chem.*, 1986, **64**, 1752.
116. D. D. Sung, J. D. Lee and S. K. Choi, *Bull. Korean Chem. Soc.*, 1987, **8**, 63.
117. C. L. Beckel, G. E. Obarski, M. Z. Fuka and J. D. Fritts, *Mater. Res. Soc. Symp. Proc.*, 1987, 95.
118. D. A. Brown, W. Clegg, H. M. Colquhoun, J. A. Daniels, I. R. Stephenson and K. Wade, *J. Chem. Soc., Chem. Commun.*, 1987, 889.
119. T. D. Getman, C. B. Knobler and M. F. Hawthorne, *Inorg. Chem.*, 1992, **31**, 101.
120. T. D. Getman, C. B. Knobler and M. F. Hawthorne, *J. Am. Chem. Soc.*, 1990, **112**, 4593.
121. K. Shelly, C. A. Reed, Y. J. Lee and W. R. Scheidt, *J. Am. Chem. Soc.*, 1986, **108**, 3117.
122. G. P. Gupta, G. Lang, Y. J. Lee, W. R. Scheidt, K. Shelly and C. A. Reed, *Inorg. Chem.*, 1987, **26**, 3022.
123. K. Shelly, D. C. Finster, Y. J. Lee, W. R. Scheidt and C. A. Reed, *J. Am. Chem. Soc.* 1985, **107**, 5955.
124. D. J. Liston, Y. J. Lee, W. R. Scheidt and C. A. Reed, *J. Am. Chem. Soc.*, 1989, **111**, 6643.
125. J. T. Spencer, M. R. Pourian, R. J. Butcher, E. Sinn and R. N. Grimes, *Organometallics*, 1987, **6**, 335.
126. D. Bromm, U. Seebold, M. Noltemeyer and A. Meller, *Chem. Ber.*, 1991, **124**, 2645.
127. N. S. Hosmane, U. Siriwardane, G. Zhang, H. Zhu and J. A. Maguire, *J. Chem. Soc., Chem. Commun.*, 1989, 1128.
128. N. S. Hosmane *et al.*, *Inorg. Chem.*, 1991, **30**, 3793.
129. J. S. Beck, W. Quintana and L. G. Sneddon, *Organometallics*, 1988, **7**, 1015.
130. K. Su, P. J. Fazen, P. J. Carroll and L. G. Sneddon, *Organometallics*, 1992, **11**, 2715.
131. R. Köster *et al.*, *Chem. Ber.*, 1991, **124**, 2715; *Chem. Ber.*, 1992, **125**, 663.
132. S. O. Kang, J. W. Bausch, P. J. Carroll and L. G. Sneddon, *J. Am. Chem. Soc.*, 1992, **114**, 6248.
133. K. Su, B. Barnum, P. J. Carroll and L. G. Sneddon, *J. Am. Chem. Soc.*, 1992, **114**, 2730.
134. D. C. Finster and R. N. Grimes, *J. Am. Chem. Soc.*, 1981, **103**, 2675.
135. T. L. Venable, R. B. Maynard and R. N. Grimes, *J. Am. Chem. Soc.*, 1984, **106**, 6187.
136. T. D. Getman, P. M. Garrett, C. B. Knobler, M. F. Hawthorne, K. Thorne and J. D. MacKenzie, *Organometallics*, 1992, **11**, 2723.
137. J. R. Wermer, N. S. Hosmane, J. J. Alexander, U. Siriwardane and S. G. Shore, *Inorg. Chem.*, 1986, **25**, 4351.
138. K. Nestor, B. Stibr, J. D. Kennedy, M. Thornton-Pett and T. Jelinek, *Collect. Czech. Chem. Commun.*, 1992, **57**, 1262.
139. F. Teixidor, C. Vinas, J. Rius, C. Miravitlles and J. Casabo, *Inorg. Chem.*, 1990, **29**, 149.
140. K. M. Kim, Y. Do, C. B. Knobler and M. F. Hawthorne, *Bull. Korean Chem. Soc.*, 1989, **10**, 321.
141. F. Teixidor *et al.*, *Inorg. Chim. Acta*, 1990, **176**, 287.
142. K. F. Shaw and A. J. Welch, *Polyhedron*, 1992, **11**, 157.
143. L. I. Zakharkin *et al.*, *Metalloorg. Khim.*, 1989, **2**, 1274.
144. V. Subrtova, V. Petricek and L. Hummel, *Acta Crystallogr., Sect. C: Cryst. Struct. Commun.*, 1989, **C45**, 1964.
145. J. Buchanan, E. J. M. Hamilton, D. Reed and A. J. Welch, *J. Chem. Soc., Dalton Trans.*, 1990, 677.
146. J. Rius, A. Romerosa, F. Teixidor, J. Casabo and C. Miravitlles, *Inorg. Chem.*, 1991, **30**, 1376.
147. L. I. Zakharkin, G. G. Zhigareva, V. A. Antonovich, A. I. Yanovskii and Y. T. Struchkov, *Zh. Obshch. Khim.*, 1986, **56**, 2066; *J. Gen. Chem. (USSR)*, 1986, **56**, 1823.
148. V. V. Grushin, T. P. Tolstaya, A. I. Yanovskii and Y. T. Struchkov, *Dokl. Akad. Nauk SSSR*, 1984, **276**, 1136; *Dokl. Chem. Proc. Acad. Sci. USSR*, 1984, **276**, 205.
149. V. V. Grushin, T. P. Tolstaya, A. I. Yanovskii and Y. T. Struchkov, *Izv. Akad. Nauk SSSR, Ser. Khim.*, 1984, 855.
150. E. A. Mizusawa, M. R. Thompson and M. F. Hawthorne, *Inorg. Chem.*, 1985, **24**, 1911.
151. J. Cowie, E. J. M. Hamilton, J. C. V. Laurie and A. J. Welch, *Acta Crystallogr., Sect. C: Cryst. Struct. Commun.*, 1988, **44**, 1648.
152. V. Subrtova, C. Novak, A. Linek and J. Hasek, *Acta Crystallogr. C*, 1984, **40**, 1955.
153. C. Vinas, W. M. Butler, F. Teixidor and R. W. Rudolph, *Inorg. Chem.*, 1986, **25**, 4369.
154. F. Teixidor *et al.*, *Inorg. Chim. Acta*, 1990, **176**, 61.
155. W. Quintana and L. G. Sneddon, *Inorg. Chem.*, 1990, **29**, 3242.
156. R. L. Ernest, W. Quintana, R. Rosen, P. J. Carroll and L. G. Sneddon, *Organometallics*, 1987, **6**, 80.
157. W. Quintana, R. L. Ernest, P. J. Carroll and L. G. Sneddon, *Organometallics*, 1988, **7**, 166.
158. V. M. Ionov, M. Y. Subbotin, V. V. Grushin, T. P. Tolstaya, I. N. Lisichkina and L. A. Aslanov, *Zh. Strukt. Khim.*, 1983, **24**, 139; *J. Struct. Chem. (USSR)*, 1983, **24**, 638.
159. A. I. Yanovskii, Y. T. Struchkov, V. T. Kampel, V. I. Bregadze, N. N. Godovikov and M. I. Kabachnik, *Izv. Akad. Nauk SSSR, Ser. Khim.*, 1986, 2299.
160. V. I. Bregadze *et al.*, *Dokl. Akad. Nauk SSSR*, 1985, **285**, 1127; *Dokl. Chem. Proc. Acad. Sci. USSR*, 1985, **285**, 364.
161. L. I. Zakharkin, M. G. Meiramov, V. A. Antonovich, A. V. Kazantsev, A. I. Yanovskii and Y. T. Struchkov, *Zh. Obshch. Khim.*, 1983, **53**, 90; *J. Gen. Chem. (USSR)*, 1983, **53**, 73.
162. G. L. Gal'chenko, N. B. Tamm, V. K. Pavlovich, V. A. Ol'shevskaya and L. I. Zakharkin, *Metalloorg. Khim.*, 1990, **3**, 414.
163. W. Clegg, D. A. Brown, S. J. Bryan and K. Wade, *Polyhedron*, 1984, **3**, 307.
164. A. V. Polyakov, A. I. Yanovskii, Y. T. Struchkov, V. N. Kalinin, A. V. Usatov and L. I. Zakharkin, *Koord. Khim.*, 1988, **14**, 1278; *Soviet J. Coord. Chem.*, 1988, **14**, 718.
165. R. P. L'Esperance, Z. H. Li, D. Van Engen and M. Jones, *Inorg. Chem.*, 1989, **28**, 1823.
166. Yu. A. Kabachii, A. I. Yanovskii, P. M. Valetskii, Y. T. Struchkov, S. V. Vinogradova and V. V. Korshak, *Dokl. Akad. Nauk SSSR*, 1985, **280**, 1390.
167. A. I. Yanovskii, Y. T. Struchkov, V. N. Kalinin, A. V. Usatov and L. I. Zakharkin, *Koord. Khim.*, 1982, **8**, 1700.
168. A. I. Yanovskii, *et al.*, *Zh. Strukt. Khim.*, 1982, **23**, 88.
169. L. I. Zakharkin, V. A. Ol'shevskaya and V. A. Antonovich, *Zh. Org. Khim.*, 1987, **23**, 1691; *Russ. J. Org. Chem.*, 1987, **23**, 1513.
170. S. V. Lindeman, I. A. Khotina, M. M. Teplyakov, Y. Struchkov and V. V. Korshak, *Makromol. Chem.*, 1988, **189**, 471.
171. V. P. Balema, E. G. Rys, N. N. Godovikov, A. V. Polyakov and Y. T. Struchkov, *Izv. Akad. Nauk SSSR, Ser. Khim.*, 1992, 180.
172. F. Teixidor *et al.*, *J. Chem. Soc., Dalton Trans.*, 1990, 525.
173. T. J. Henly, C. B. Knobler and M. F. Hawthorne, *Organometallics*, 1992, **11**, 2313.
174. S. H. Wu, D. Van Engen and M. Jones, *Acta Crystallogr., Sect. C: Cryst. Struct. Commun.*, 1990, **C46**, 656.
175. V. Subrtova, V. Petricek and K. Maly, *Collect. Czech. Chem. Commun.*, 1991, **56**, 1983.

176. G. M. Rusakova, A. I. Gusev, B. P. Parfenov, S. Ya. Pechurina and I. V. Chesnokova, *Zh. Strukt. Khim.*, 1990, **31**, 147.
177. T. D. Getman, C. B. Knobler and M. F. Hawthorne, *Inorg. Chem.*, 1990, **29**, 158.
178. V. Subrtova, A. Linek and J. Hasek, *Acta Crystallogr. B*, 1982, **38**, 3147.
179. C. B. Knobler *et al.*, *J. Am. Chem. Soc.*, 1984, **106**, 2990.
180. V. V. Grushin, A. V. Polyakov, A. I. Yanovskii, Y. T. Struchkov, S. S. Bukalov and L. A. Leites, *Metalloorg. Khim.*, 1991, **4**, 828.
181. A. I. Yanovskii, Y. T. Struchkov, V. V. Grushin, T. P. Tolstaya and I. I. Demkina, *Zh. Struct. Khim.*, 1988, **29**, 89; *J. Struct. Chem. (USSR)*, 1988, **29**, 422.
182. L. I. Zakharkin, G. G. Zhigareva, A. V. Polyakov, A. I. Yanovskii and Y. T. Struchkov, *Izv. Akad. Nauk SSSR, Ser. Khim.*, 1987, 872.
183. C. Novak, V. Subrtova, A. Linek and J. Hasek, *Acta Crystallogr., Cryst. Struct. Commun.*, 1983, **C39**, 1393.
184. R. R. Ryan and R. Schaeffer, *Cryst. Struct. Commun.*, 1981, **10**, 133.
185. F. A. Gomez, S. E. Johnson and M. F. Hawthorne, *J. Am. Chem. Soc.*, 1991, **113**, 5915.
186. S. V. Lindeman *et al.*, *Makromol. Chem.*, 1984, **185**, 417.
187. X. Yang, C. C. Knobler and M. F. Hawthorne, *Angew. Chem.*, 1991, **103**, 1519; *Angew. Chem., Int. Ed. Engl.*, 1991, **30**, 1507.
188. W. Clegg, D. A. Brown, S. J. Bryan and K. Wade, *J. Organomet. Chem.*, 1987, **325**, 39.
189. X. Yang, S. E. Johnson, S. I. Khan and M. F. Hawthorne, *Angew. Chem.*, 1992, **104**, 886; *Angew. Chem., Int. Ed. Engl.*, 1992, **31**, 893.
190. S. A. Khan, J. H. Morris, M. Harman and M. B. Hursthouse, *J. Chem. Soc., Dalton Trans.*, 1992, 119.
191. C. Novak, V. Subrtova, V. Petricek, L. Hummel and J. Hasek, *Collect. Czech. Chem. Commun.*, 1990, **55**, 653.
192. A. B. Yakushev, I. B. Sivaev, K. A. Solntsev, I. Y. Kuznetsov, L. A. Butman and N. T. Kuznetsov, *Zh. Neorg. Khim.*, 1988, **33**, 1398.
193. K. Maly, V. Subrtova and V. Petricek, *Acta Crystallogr., C*, 1987, **43**, 593.
194. D. J. Liston, C. A. Reed, C. W. Eigenbrot and W. R. Scheidt, *Inorg. Chem.*, 1987, **26**, 2739.
195. N. S. Hosmane, N. N. Sirmokadam and M. N. Mollenhauer, *J. Organomet. Chem.*, 1985, **279**, 359.
196. O. V. Dorofeeva, V. S. Mastryukov, L. V. Vilkov, A. V. Golubinskii and A. I. Kovredov, *Zh. Strukt. Khim.*, 1988, **29**, 137.
197. V. S. Mastryukov, A. V. Polyakov, A. V. Golubinskii, L. V. Vilkov, V. T. Kampel and V. I. Bregadze, *Metalloorg. Khim.*, 1989, **2**, 684.
198. V. S. Mastryukov *et al.*, *Metalloorg. Khim.*, 1991, **4**, 132.
199. M. Bühl and P. v. R. Schleyer, *J. Am. Chem. Soc.*, 1992, **114**, 477.
200. M. Bühl, P. V. R. Schleyer, Z. Havlas, D. Hnyk and S. Hermanek, *Inorg. Chem.*, 1991, **30**, 3107.
201. J. W. Bausch, G. K. S. Prakash and R. E. Williams, *Inorg. Chem.*, 1992, **31**, 3763.
202. J. W. Bausch, G. K. S. Prakash, M. Bühl, P. v. R. Schleyer and R. E. Williams, *Inorg. Chem.*, 1992, **31**, 3060.
203. T. Onak *et al.*, *Inorg. Chem.*, 1992, **31**, 3910.
204. T. Onak, *et al.*, *Inorg. Chem.*, 1993, **32**, 487.
205. S. Hermanek, *Chem. Rev.*, 1992, **92**, 325.
206. F. Teixidor, C. Vinas and R. W. Rudolph, *Inorg. Chem.*, 1986, **25**, 3339.
207. T. P. Fehlner, P. T. Czech and R. F. Fenske, *Inorg. Chem.*, 1990, **29**, 3103.
208. R. Köster, G. Seidel and B. Wrackmeyer, *Angew. Chem.*, 1984, **96**, 520; *Angew. Chem., Int. Ed. Engl.*, 1984, **23**, 512.
209. S. Hermanek, T. Jelinek, J. Plesek, B. Stibr and J. Fusek, *J. Chem. Soc., Chem. Commun.*, 1987, 927.
210. S. Hermanek, T. Jelinek, J. Plesek, B. Stibr and J. Fusek, *Collect. Czech. Chem. Commun.*, 1988, **53**, 2742.
211. B. Stibr, Z. Janousek, K. Base, S. Hermanek, J. Plesek and I. A. Zakharova, *Collect. Czech. Chem. Commun.*, 1984, **49**, 1891.
212. W. Jarvis, Z. J. Abdou and T. Onak, *Polyhedron*, 1983, **2**, 1067.
213. T. L. Venable, C. T. Brewer and R. N. Grimes, *Inorg. Chem.*, 1985, **24**, 4751.
214. T. Onak, Z. J. Abdou, G. Abdou and S. Herrera, *Heteroatom Chem.*, 1992, **3**, 219.
215. W. Jarvis, W. Inman, B. Powell, E. W. DiStefano and T. Onak, *J. Magn. Reson.*, 1981, **43**, 302.
216. B. Ng, T. Onak, T. Banuelos, F. Gomez and E. W. DiStefano, *Inorg. Chem.*, 1985, **24**, 4091.
217. R. Köster and B. Wrackmeyer, *Z. Naturforsch., Teil B*, 1981, **36**, 704.
218. W. Nam, M. Soltis, C. Gordon, S. Lee and T. Onak, *J. Magn. Reson.*, 1984, **59**, 399.
219. J. A. Anderson, R. J. Astheimer, J. D. Odom and L. G. Sneddon, *J. Am. Chem. Soc.*, 1984, **106**, 2275.
220. T. L. Venable, W. C. Hutton and R. N. Grimes, *J. Am. Chem. Soc.*, 1982, **104**, 4716; *J. Am. Chem. Soc.*, 1984, **106**, 29.
221. S. Hermanek, J. Fusek, B. Stibr, J. Plesek and T. Jelinek, *Polyhedron*, 1986, **5**, 1873.
222. O. W. Howarth, M. J. Jasztal, J. G. Taylor and M. G. H. Wallbridge, *Polyhedron* 1985, **4**, 1461.
223. C. A. Plumb and L. G. Sneddon, *Organometallics*, 1992, **11**, 1681.
224. E. C. Reynhardt, A. Watton and H. E. Petch, *J. Magn. Reson.*, 1982, **46**, 453.
225. F. Kasler and M. Tierney, *Mikrochim. Acta*, 1981, **2**, 301.
226. R. K. Harris, J. Bowles, I. R. Stephenson and E. H. Wong, *Spectrochim. Acta, Part A*, 1988, **44**, 273.
227. E. C. Reynhardt, *J. Magn. Reson.*, 1986, **69**, 337.
228. E. C. Reynhardt and S. Froneman, *Mol. Phys.*, 1991, **74**, 61.
229. X. L. R. Fontaine *et al.*, *J. Chem. Soc., Dalton Trans.*, 1990, 681.
230. J. T. Spencer and R. N. Grimes, *Organometallics*, 1987, **6**, 323.
231. B. Wrackmeyer, *Z. Naturforsch., Teil B*, 1982, **37**, 412.
232. R. J. Astheimer and L. G. Sneddon, *Inorg. Chem.*, 1983, **22**, 1928.
233. J. Bragin, D. S. Urevig and M. Diem, *J. Raman Spectrosc.*, 1982, **12**, 86.
234. L. A. Leites, S. S. Bukalov, L. E. Vinogradova, V. N. Kalinin, N. I. Kobel'kova and L. I. Zakharkin, *Izv. Akad. Nauk SSSR, Ser. Khim.*, 1984, 954.
235. J. A. Shelnutt, B. Morosin, D. Emin, A. Mullendore, G. Slack and C. Wood, *Am. Inst. Phys. Conf. Proc.*, 1986, **140**, 312.
236. J. A. Shelnutt, *Microbeam Anal., Proc.*, 1986, **21**, 35.
237. S. S. Bukalov and L. A. Leites, *Opt. Spektrosk.*, 1984, **56**, 10.
238. R. M. Paroli, N. T. Kawai, G. Lord, I. S. Butler and D. F. R. Gilson, *Inorg. Chem.*, 1989, **28**, 1819.

239. L. E. Vinogradova, L. A. Leites, A. I. Kovredov, V. A. Ol'shevskaya and L. I. Zakharkin, *Izv. Akad. Nauk SSSR, Ser. Khim.*, 1982, 1663.
240. R. Wilczynski and L. G. Sneddon, *Inorg. Chem.*, 1981, **20**, 3955.
241. R. B. Maynard, L. Borodinsky and R. N. Grimes, *Inorg. Synth.*, 1983, **22**, 211.
242. J. T. Spencer and R. N. Grimes, *Organometallics*, 1987, **6**, 328.
243. M. E. Fessler, J. T. Spencer, J. F. Lomax and R. N. Grimes, *Inorg. Chem.*, 1988, **27**, 3069.
244. H. A. Boyter and R. N. Grimes, *Inorg. Chem.*, 1988, **27**, 3075.
245. H. A. Boyter and R. N. Grimes, *Inorg. Chem.*, 1988, **27**, 3080.
246. R. Wilczynski and L. G. Sneddon, *Inorg. Chem.*, 1982, **21**, 506.
247. R. J. Astheimer and L. G. Sneddon, *Inorg. Chem.*, 1984, **23**, 3207.
248. E. W. Corcoran and L. G. Sneddon, *J. Am. Chem. Soc.*, 1985, **107**, 7446.
249. B. Stibr, J. Plesek and S. Hermanek, *Inorg. Synth.*, 1983, **22**, 237.
250. R. B. Maynard and R. N. Grimes, *Inorg. Synth.*, 1983, **22**, 215.
251. A. G. Kayumov, K. A. Solntsev, L. V. Goeva, N. T. Kuznetsov and O. G. Ellert, *Zh. Neorg. Khim.*, 1988, **33**, 1771; *Russ. J. Inorg. Chem.*, 1988, **33**, 1007.
252. S. F. Myasoedov, T. V. Tsimerinova, K. A. Solntsev and N. T. Kuznetsov, *Koord. Khim.*, 1984, **10**, 1174.
253. S. F. Myasoedov, A. L. Popov, K. A. Solntsev and N. T. Kuznetsov, *Zh. Neorg. Khim.*, 1985, **30**, 98; *Russ. J. Inorg. Chem.*, 1985, **30**, 54.
254. L. A. Leites, S. S. Bukalov, L. E. Vinogradova, S. P. Knyazev and Y. A. Strelenko, *Izv. Akad. Nauk SSSR, Ser. Khim.*, 1986, 1801.
255. E. E. Nifant'ev, T. Y. Prokof'eva, R. K. Magdeeva, L. S. Sklyarskii and K. D. Galstyan, *Zh. Obshch. Khim.*, 1982, **52**, 217; *J. Gen. Chem. (USSR)*, 1982, **52**, 197.
256. V. N. Kalinin, A. V. Usatov and L. I. Zakharkin, *Zh. Obshch. Khim.*, 1987, **57**, 2508.
257. J. G. Contreras, J. Pena and L. M. S. Trivino, *Inorg. Chim. Acta*, 1984, **87**, 181.
258. L. A. Leites *et al.*, *Metalloorg. Khim.*, 1989, **2**, 654.
259. L. I. Zakharkin, N. Y. Krainova, I. V. Pisareva and V. A. Antonovich, *Synth. React. Inorg. Metal-Org. Chem.*, 1985, **15**, 655.
260. S. F. Myasoedov, T. V. Tsimerinova, K. A. Solntsev and N. T. Kuznetsov, *Zh. Neorg. Khim.* 1984, **29**, 1421; *Russ. J. Inorg. Chem.*, 1984, **29**, 817.
261. V. A. Mazunov, Y. S. Nekrasov, V. I. Khvostenko and V. I. Stanko, *Izv. Akad. Nauk SSSR, Ser. Khim.*, 1983, 223; *Bull. Acad. Sci. USSR Div. Chem. Sci.*, 1983, 197.
262. V. N. Pustobaev, V. M. Orlov, V. A. Ol'shevskaya and L. I. Zakharkin, *Metalloorg. Khim.*, 1991, **4**, 646.
263. N. I. Vasyukova, Y. S. Nekrasov, V. T. Kampel and V. I. Bregadze, *Izv. Akad. Nauk SSSR, Ser. Khim.*, 1987, 567; *Bull. Acad. Sci. USSR Div. Chem. Sci.*, 1987, 517.
264. G. N. Kartsev, S. M. Ignat'eva, V. N. Lebedev and L. I. Zakharkin, *Metalloorg. Khim.*, 1989, **2**, 1398.
265. G. L. Gal'chenko, N. B. Tamm, V. K. Pavlovich, V. A. Ol'shevskaya and L. I. Zakharkin, *Metalloorg. Khim.*, 1990, **3**, 259.
266. A. Lötz and J. Voitländer, *J. Chem. Phys.*, 1991, **95**, 3208.
267. Y. L. Pascal and O. Convert, *Magn. Reson. Chem.*, 1991, **29**, 308.
268. B. H. Ruessink and C. MacLean, *J. Chem. Phys.*, 1986, **85**, 3217.
269. A. Lötz and J. Voitländer, *J. Chem. Phys.*, 1986, **85**, 3136.
270. A. J. Leffler, *J. Chem. Phys.*, 1984, **81**, 2574.
271. G. M. Zimina, A. V. Vannikov and V. I. Stanko, *Khim. Vys. Energ.*, 1982, **16**, 335.
272. T. Vondrak, J. Plesek, S. Hermanek, B. Stibr, *Polyhedron*, 1989, **8**, 805.
273. T. P. Fehlner, M. Wu, B. J. Meneghelli and R. W. Rudolph, *Inorg. Chem.*, 1980, **19**, 49.
274. P. Brint, B. Sangchakr, M. McGrath, T. R. Spalding and R. J. Suffolk, *Inorg. Chem.*, 1990, **29**, 47.
275. T. Vondrak, *Polyhedron*, 1987, **6**, 1559.
276. S. Lee, D. Li, P. A. Dowben, F. K. Perkins, M. Onellion and J. T. Spencer, *J. Am. Chem. Soc.*, 1991, **113**, 8444.
277. S. Lee, P. A. Dowben, A. T. Wen, A. P. Hitchcock, J. A. Glass and J. T. Spencer, *J. Vac. Sci. Technol., A*, 1992, **10**, 881.
278. R. S. Hutton, H. D. Roth and S. Chari, *J. Phys. Chem.*, 1981, **85**, 753.
279. S. P. Solodovnikov, L. M. Bronshtein, L. S. Shilovtseva, Y. A. Kabachii and P. M. Valetskii, *Metalloorg. Khim.*, 1988, **1**, 856.
280. B. L. Tumanskii *et al.*, *Izv. Akad. Nauk SSSR, Ser. Khim.*, 1989, 1139.
281. F. J. Esposto, *et al.*, *J. Vac. Sci. Technol., A*, 1991, **9**, 1663.
282. S. S. Durso, R. A. Sheeks, E. V. O'Gorman and R. A. Beaudet, *Inorg. Chem.*, 1989, **28**, 1615.
283. V. V. Grushin, I. I. Demkina, T. P. Tolstaya, M. V. Galakhov and V. I. Bakhmutov, *Metalloorg. Khim.*, 1989, **2**, 727.
284. R. Greatrex, N. N. Greenwood and M. J. Kirk, *J. Chem. Soc., Chem. Commun.*, 1991, 1510.
285. R. L. DeKock, T. P. Fehlner, C. E. Housecroft, T. V. Lubben and K. Wade, *Inorg. Chem.*, 1982, **21**, 25.
286. R. G. Swisher, E. Sinn and R. N. Grimes, *Organometallics*, 1985, **4**, 890.
287. S. M. Cendrowski-Guillaume and J. T. Spencer, *Organometallics*, 1992, **11**, 969.
288. J. J. Briguglio, P. J. Carroll, E. W. Corcoran and L. G. Sneddon, *Inorg. Chem.*, 1986, **25**, 4618.
289. S. H. Lawrence, J. R. Wermer, S. K. Boocock, M. A. Banks, P. C. Keller and S. G. Shore, *Inorg. Chem.*, 1986, **25**, 367.
290. H. R. Allcock, A. G. Scopelianos, R. R. Whittle and N. M. Tollefson, *J. Am. Chem. Soc.*, 1983, **105**, 1316.
291. E. G. Rys, N. N. Godovikov and M. I. Kabachnik, *Izv. Akad. Nauk SSSR, Ser. Khim.*, 1983, 2640; *Bull. Acad. Sci. USSR Div. Chem. Sci.*, 1983, 2372.
292. T. Totani, K. Aono, K. Yamamoto and K. Tawara, *J. Med. Chem.*, 1981, **24** 1492.
293. L. A. Leites, L. E. Vinogradova, A. I. Kovredov, Z. S. Shaugumbekova and L. I. Zakharkin, *Izv. Akad. Nauk SSSR, Ser. Khim.*, 1982, 2170.
294. Y. Yamamoto *et al.*, *J. Chem. Soc., Chem. Commun.*, 1992, 157.
295. A. K. M. Anisuzzaman, F. Alam and A. H. Soloway, *Polyhedron*, 1990, **9**, 891.
296. M. Scobie and M. D. Threadgill, *J. Chem. Soc., Chem. Commun.*, 1992, 939.
297. R. C. Reynolds, T. W. Trask and W. D. Sedwick, *J. Org. Chem.*, 1991, **56**, 2391.
298. L. Schneiderova *et al.*, *Collect. Czech. Chem. Commun.*, 1992, **57**, 463.
299. M. G. L. Mirabelli and L. G. Sneddon, *Organometallics*, 1986, **5**, 1510.

300. B. Stibr *et al.*, *J. Chem. Soc., Chem. Commun.*, 1987, 1106.
301. B. Stibr, T. Jelinek, E. Drdakova, S. Hermanek and J. Plesek, *Polyhedron*, 1988, **7**, 669.
302. S. O. Kang, G. T. Furst and L. G. Sneddon, *Inorg. Chem.*, 1989, **28**, 2339.
303. C. A. Plumb, P. J. Carroll and L. G. Sneddon, *Organometallics*, 1992, **11**, 1665.
304. J. S. Beck, A. P. Kahn and L. G. Sneddon, *Organometallics*, 1986, **5**, 2552.
305. J. S. Beck and L. G. Sneddon, *Inorg. Chem.*, 1990, **29**, 295.
306. J. Plesek, T. Jelinek, E. Drdakova, S. Hermanek and B. Stibr, *Collect. Czech. Chem. Commun.*, 1984, **49**, 1559.
307. T. Jelinek, J. Plesek, F. Mares, S. Hermanek and B. Stibr, *Polyhedron*, 1987, **6**, 1981.
308. R. Köster, G. Seidel and B. Wrackmeyer, *Angew. Chem.*, 1985, **97**, 317.
309. R. B. Maynard and R. N. Grimes, *J. Am. Chem. Soc.*, 1982, **104**, 5983.
310. K. K. Fonda and R. N. Grimes, *Polyhedron*, 1990, **9**, 949.
311. J. H. Davis and R. N. Grimes, *Inorg. Chem.*, 1988, **27**, 4213.
312. T. Whelan, J. T. Spencer, M. R. Pourian and R. N. Grimes, *Inorg. Chem.*, 1987, **26**, 3116.
313. N. S. Hosmane, M. Dehghan and S. Davies, *J. Am. Chem. Soc.*, 1984, **106**, 6435.
314. J. Plesek, T. Jelinek, B. Stibr and S. Hermanek, *J. Chem. Soc., Chem. Commun.*, 1988, 348.
315. Z. J. Abdou, F. Gomez, G. Abdou and T. Onak, *Inorg. Chem.*, 1988, **27**, 3679.
316. H. Tomita, H. Luu and T. Onak, *Inorg. Chem.*, 1991, **30**, 812.
317. T. Jelinek, B. Stibr, S. Hermanek and J. Plesek, *J. Chem. Soc., Chem. Commun.*, 1989, 804.
318. T. Jelinek, J. Plesek, S. Hermanek and B. Stibr, *Main Group Metal Chem.*, 1987, **10**, 397.
319. K. Base, S. Hermanek and F. Hanousek, *J. Chem. Soc., Chem. Commun.*, 1984, 299.
320. N. W. Alcock, J. G. Taylor and M. G. H. Wallbridge, *J. Chem. Soc., Dalton Trans.*, 1987, 1805.
321. D. C. Busby and M. F. Hawthorne, *Inorg. Chem.*, 1982, **21**, 4101.
322. F. Teixidor and R. W. Rudolph, *J. Organomet. Chem.*, 1983, **241**, 301.
323. R. J. Zaldivar, R. W. Kobayashi, G. S. Rellick and J. M. Yang, *Carbon*, 1991, **29**, 1145.
324. K. Aono and T. Totani, *J. Chem. Soc., Dalton Trans.*, 1981, 1196.
325. E. W. Corcoran and L. G. Sneddon, *J. Am. Chem. Soc.*, 1984, **106**, 7793.
326. Z. Janousek, J. Plesek, B. Stibr and S. Hermanek, *Collect. Czech. Chem. Commun.*, 1983, **48**, 228.
327. L. I. Zakharkin, V. A. Ol'shevskaya, V. A. Antonovich, A. V. Lebedev and A. V. Bukhtiarov, *Metalloorg. Khim.*, 1990, **3**, 20.
328. B. Wrackmeyer and G. Kehr, *Polyhedron*, 1991, **10**, 1497.
329. G. E. Herberich, H. Ohst and H. Mayer, *Angew. Chem.*, 1984, **96**, 975.
330. A. Kraemer, H. Pritzkow and W. Siebert, *Angew. Chem.*, 1990, **102**, 333.
331. N. S. Hosmane *et al.*, *Inorg. Chem.*, 1990, **29**, 2698.
332. M. G. L. Mirabelli, P. J. Carroll and L. G. Sneddon, *J. Am. Chem. Soc.*, 1989, **111**, 592.
333. B. M. Gimarc and J. J. Ott, *Inorg. Chem.* 1986, **25**, 83.
334. B. M. Gimarc and J. J. Ott, *Stud. Phys. Theor. Chem.*, 1987, **51**, 285.
335. M. L. McKee, *Theor. Chem.*, 1988, **45**, 191.
336. M. L. McKee, *J. Am. Chem. Soc.*, 1992, **114**, 879.
337. B. M. Gimarc and J. J. Ott, *Main Group Met. Chem.*, 1989, **12**, 77.
338. B. M. Gimarc and J. J. Ott, *J. Math. Chem.*, 1990, **5**, 359.
339. D. M. P. Mingos and D. J. Wales, in 'Electron Deficient Boron and Carbon Clusters', eds. G. A. Olah, K. Wade and R. E. Williams, Wiley, New York, 1991, p. 143.
340. J. J. Ott, C. A. Brown and B. M. Gimarc, *Inorg. Chem.*, 1989, **28**, 4269.
341. B. Oh and T. Onak, *Inorg. Chem.*, 1982, **21**, 3150.
342. W. Nam, Z. J. Abdou, H. Lee, T. Banuelos and T. Onak, *Inorg. Chem.*, 1989, **28**, 669.
343. B. M. Gimarc and J. J. Ott, *Inorg. Chem.*, 1986, **25**, 2708.
344. B. M. Gimarc, B. Dai, D. S. Warren and J. J. Ott, *J. Am. Chem. Soc.*, 1990, **112**, 2597.
345. V. V. Gavrilov *et al.*, *Izv. Akad. Nauk SSSR, Ser. Khim.*, 1986, 2367.
346. G. M. Edvenson and D. F. Gaines, *Inorg. Chem.*, 1990, **29**, 1210.
347. B. M. Gimarc, D. S. Warren, J. J. Ott and C. Brown, *Inorg. Chem.*, 1991, **30**, 1598.
348. S. H. Wu and M. Jones, *J. Am. Chem. Soc.*, 1989, **111**, 5373.
349. V. N. Kalinin, N. I. Kobel'kova and L. I. Zakharkin, *Izv. Akad. Nauk SSSR, Ser. Khim.*, 1982, 1661; *Bull. Acad. Sci. USSR Div. Chem. Sci.*, 1982, 1479.
350. M. Enders, H. Pritzkow and W. Siebert, *Angew. Chem.*, 1992, **104**, 628.
351. H. Michel *et al.*, *Angew. Chem., Int. Ed. Engl.*, 1992, **31**, 607.
352. L. Barton, P. K. Rush, T. Zhu, P. Nevels and M. H. Owens, *Inorg. Chem.* 1989, **28**, 381.
353. M. E. Fessler, T. Whelan, J. T. Spencer and R. N. Grimes, *J. Am. Chem. Soc.*, 1987, **109**, 7416.
354. A. F. Zhigach, A. B. Petrunin and D. B. Bekker, *Zh. Obshch. Khim.*, 1980, **50**, 2727.
355. P. E. Behnken and M. F. Hawthorne, *Inorg. Chem.*, 1984, **23**, 3420.
356. N. A. Truba and B. I. Nabivanets, *Zh. Obshch. Khim.*, 1979, **49**, 1333.
357. L. I. Zakharkin and V. A. Ol'shevskaya, *Zh. Obshch. Khim.*, 1987, **57**, 368.
358. B. Stibr, Z. Janousek, J. Plesek, T. Jelinek and S. Hermanek, *Collect. Czech. Chem. Commun.*, 1987, **52**, 103.
359. Z. Janousek, J. Plesek, S. Hermanek and B. Stibr, *Polyhedron*, 1985, **4**, 1797.
360. (a) B. Stibr, T. Jelinek, J. Plesek and S. Hermanek, *J. Chem. Soc. Chem. Commun.*, 1987, 963; (b) T. Onak, D. Tran, J. Tseng, M. Diaz, J. Arias and S. Herrera, *J. Am. Chem. Soc.*, 1993, **115**, 9210.
361. G. Siwapinyoyos and T. Onak, *Inorg. Chem.*, 1982, **21**, 156.
362. M. F. Hawthorne *et al.*, *J. Am. Chem. Soc.*, 1990, **112**, 5365.
363. N. S. Hosmane, M. S. Islam and E. G. Burns, *Inorg. Chem.*, 1987, **26**, 3236.
364. W. Nam and T. Onak, *Inorg. Chem.*, 1987, **26**, 1581.
365. B. Ng, T. Onak and K. Fuller, *Inorg. Chem.*, 1985, **24**, 4371.
366. B. Ng and T. Onak, *J. Fluorine Chem.*, 1985, **27**, 119.
367. K. Fuller and T. Onak, *J. Organomet. Chem.*, 1983, **249**, C6.
368. F. A. Gomez, T. Onak, J. Arias and C. Alfonso, *Main Group Met. Chem.*, 1990, **13**, 237.

369. W. Nam and T. Onak, *Inorg. Chem.*, 1987, **26**, 48.
370. W. Haubold and W. J. Keller, *J. Organomet. Chem.*, 1989, **361**, C54.
371. T. Jelinek, B. Stibr, F. Mares, J. Plesek and S. Hermanek, *Polyhedron*, 1987, **6**, 1737.
372. T. Jelinek, S. Hermanek, B. Stibr and J. Plesek, *Polyhedron*, 1986, **5**, 1303.
373. B. Stibr, Z. Janousek, J. Plesek, T. Jelinek and S. Hermanek, *J. Chem. Soc. Chem. Commun.*, 1985, 1365.
374. J. Li and M. Jones, *Inorg. Chem.*, 1990, **29**, 4162.
375. V. V. Grushin, T. P. Tolstaya and I. N. Lisichkina, *Izv. Akad. Nauk SSSR, Ser. Khim.*, 1983, 2165; *Bull. Acad. Sci. USSR Div. Chem. Sci.*, 1983, 1957.
376. V. V. Grushin, I. I. Demkina and T. P. Tolstaya, *Izv. Akad. Nauk SSSR, Ser. Khim.*, 1985, 2111; *Bull. Acad. Sci. USSR Div. Chem. Sci.*, 1985, 1950.
377. V. V. Grushin, T. P. Tolstaya and I. N. Lisichkina, *Izv. Akad. Nauk SSSR, Ser. Khim.*, 1982, 2412; *Bull. Acad. Sci. USSR Div. Chem. Sci.*, 1982, 2127.
378. V. V. Grushin, I. I. Demkina and T. P. Tolstaya, *Inorg. Chem.*, 1991, **30**, 4860.
379. V. V. Grushin, I. I. Demkina and T. P. Tolstaya, *J. Chem. Soc., Perkin Trans. 2*, 1992, 505.
380. V. V. Grushin, I. I. Demkina and T. P. Tolstaya, *Inorg. Chem.*, 1991, **30**, 1760.
381. J. Li, C. F. Logan and M. Jones, *Inorg. Chem.*, 1991, **30**, 4866.
382. L. I. Zakharkin, A. I. Kovredov, V. A. Ol'shevskaya and S. V. Vitt, *Zh. Obshch. Khim.*, 1985, **55**, 1019; *J. Gen. Chem. (USSR)*, 1985, **55**, 908.
383. T. Jelinek, J. Plesek, S. Hermanek and B. Stibr, *Collect. Czech. Chem. Commun.*, 1986, **51**, 819.
384. B. Stibr, J. Plesek, T. Jelinek, S. Hermanek, K. A. Solntsev and N. T. Kuznetsov, *Collect. Czech. Chem. Commun.*, 1987, **52**, 957.
385. S. A. Khan, *J. Chem. Soc. Pak.*, 1991, **13**, 38.
386. J. Plesek, Z. Janousek and S. Hermanek, *Inorg. Synth.*, 1983, **22**, 239.
387. J. Plesek, T. Jelinek and B. Stibr, *Polyhedron*, 1984, **3**, 1351.
388. L. I. Zakharkin, N. Y. Krainova, G. G. Zhigareva and I. V. Pisareva, *Izv. Akad. Nauk SSSR, Ser. Khim.*, 1982, 1650; *Bull. Acad. Sci. USSR Div. Chem. Sci.*, 1982, 1468.
389. H. C. Kang, S. S. Lee, C. B. Knobler and M. F. Hawthorne, *Inorg. Chem.*, 1991, **30**, 2024.
390. L. I. Zakharkin, G. G. Zhigareva, V. A. Antonovich, A. I. Yanovskii and Y. T. Struchkov, *Izv. Akad. Nauk SSSR, Ser. Khim.*, 1984, 1672.
391. L. I. Zakharkin, G. G. Zhigareva, V. A. Ol'shevskaya and V. A. Antonovich, *Metalloorg. Khim.*, 1990, **3**, 556.
392. L. I. Zakharkin and V. A. Ol'shevskaya, *Synth. React. Inorg. Metal-Org. Chem.*, 1991, **21**, 1041.
393. L. I. Zakharkin, A. I. Kovredov, V. A. Ol'shevskaya and S. V. Vitt, *Izv. Akad. Nauk SSSR, Ser. Khim.*, 1983, 1680; *Bull. Acad. Sci. USSR Div. Chem. Sci.*, 1983, 1528.
394. L. I. Zakharkin, A. I. Kovredov, V. A. Ol'shevskaya and S. V. Vitt, *Zh. Obshch. Khim.*, 1985, **55** 2501; *J. Gen. Chem. (USSR)*, 1985, **55**, 2220.
395. J. R. Wermer, D. Y. Jan, T. D. Getman, E. Moher and S. G. Shore, *Inorg. Chem.*, 1988, **27**, 4274.
396. A. I. Kovredov, Z. S. Shaugumbekova and L. I. Zakharkin, *Metalloorg. Khim.*, 1988, **1**, 179.
397. G. Zheng and M. Jones, *J. Am. Chem. Soc.*, 1983, **105**, 6487.
398. J. Li and M. Jones, *J. Am. Chem. Soc.*, 1992, **114**, 1094.
399. L. I. Zakharkin and V. A. Ol'shevskaya, *Zh. Obshch. Khim.*, 1992, **62**, 138.
400. J. Plesek, T. Jelinek, S. Hermanek and B. Stibr, *Collect. Czech. Chem. Commun.*, 1986, **51**, 81.
401. V. N. Lebedev, M. V. Troitskaya and L. I. Zakharkin, *Metalloorg. Khim.*, 1991, **4**, 954.
404. T. Jelinek, B. Stibr, J. Plesek and S. Hermanek, *J. Organomet. Chem.*, 1986, **307**, C13.
405. T. Jelinek, J. Plesek, S. Hermanek and B. Stibr, *Collect. Czech. Chem. Commun.*, 1985, **50**, 1376.
406. K. Aono and T. Totani, *J. Chem. Soc., Dalton Trans.*, 1981, 1190.
407. S. Al-Baker, W. E. Hill and C. A. McAuliffe, *J. Chem. Soc., Dalton Trans.*, 1985, 1387.
408. F. A. Hart and D. W. Owen, *Inorg. Chim. Acta*, 1985, **103**, L1.
409. G. Dyer, R. M. Wharf and W. E. Hill, *Inorg. Chim. Acta*, 1987, **133**, 137.
410. J. P. Pena, L. M. S. Trivino and J. G. Contreras, *Bol. Soc. Chil. Quim.*, 1982, **27**, 40 (*Chem. Abstr.*, 1982, **97**, 216 409r).
411. E. H. Wong and G. T. Nalband, *Inorg. Chim. Acta*, 1981, **53**, L139.
412. H. R. Allcock, *Am. Chem. Soc., Div. Pet. Chem.*, 1982, **27**, 652.
413. J. Plesek, B. Stibr, E. Drdakova, Z. Plzak and S. Hermanek, *Chem. Ind. (London)*, 1982, 778.
414. M. M. Teplyakov, I. A. Khotina, A. A. Sakharova, O. A. Mel'nik, V. S. Papkov and Y. P. Kvachev, *Makromol. Chem.*, 1992, **193**, 351.
415. D. Gabel and R. Walczyna, *Z. Naturforsch., Teil C*, 1982, **37**, 1038.
416. M. Basato, F. Morandini, B. Longato and S. Bresadola, *Inorg. Chem.*, 1984, **23**, 649.
417. E. Mizusawa, H. L. Dahlman, S. J. Bennett, D. M. Goldenberg and M. F. Hawthorne, *Proc. Natl. Acad. Sci. USA*, 1982, **79**, 3011.
418. D. M. Goldenberg, R. M. Sharkey, F. J. Primus, E. Mizusawa and M. F. Hawthorne, *Proc. Natl. Acad. Sci. USA*, 1984, **81**, 560.
419. L. I. Zakharkin, M. G. Meiramov, K. M. Uyzbaev and N. F. Shemyakin, *Zh. Obshch. Khim.*, 1991, **61**, 1349.
420. H. Nemoto, F. G. Rong, Y. and J. Yamamoto, *Org. Chem.*, 1990, **55**, 6065.
421. M. P. Prigozhina, P. V. Petrovskii, L. G. Komarova and A. L. Rusanov, *Izv. Akad. Nauk SSSR, Ser. Khim.*, 1992, 177.
422. A. V. Kazantsev, M. G. Meiramov, A. I. Kovredov and L. I. Zakharkin, *Izv. Akad. Nauk SSSR, Ser. Khim.*, 1982, 1603; *Bull. Acad. Sci. USSR Div. Chem. Sci.*, 1982, 1425.
423. M. I. Kabachnik, L. S. Zakharov, E. L. Gefter, G. N. Molchanova and P. V. Petrovskii, *Izv. Akad. Nauk SSSR, Ser. Khim.*, 1990, 1933.
424. L. I. Zakharkin, A. I. Kovredov, A. V. Kazantsev and M. G. Meiramov, *Zh. Obshch. Khim.*, 1981, **51**, 357; *J. Gen. Chem. (USSR)*, 1981, **51**, 289.
425. V. P. Prokopovich, A. A. Erdman, N. A. Maier and Y. A. Ol'dekop, *Dokl. Akad. Nauk SSSR*, 1984, **278**, 1141; *Dokl. Chem. Proc. Acad. Sci. USSR*, 1984, **278**, 360.
426. G. Rabilloud and B. Sillion, *J. Chem. Res. (S)*, 1981, 264.

427. A. V. Kazantsev, T. V. Shustova and L. I. Zakharkin, *Izv. Akad. Nauk SSSR, Ser. Khim.*, 1982, 2134; *Bull. Acad. Sci. USSR Div. Chem. Sci.*, 1982, 1884.
428. L. I. Zakharkin, T. V. Shustova and A. V. Kazantsev, *Zh. Obshch. Khim.*, 1981, **51**, 1071; *J. Gen. Chem. (USSR)*, 1981, **51**, 893.
429. A. I. Kovredov, M. G. Meiramov, A. V. Kazantsev and L. I. Zakharkin, *Zh. Obshch. Khim.*, 1981, **51**, 854; *J. Gen. Chem. (USSR)*, 1981, **51**, 708.
430. S. L. Sosin, V. P. Alekseeva, M. D. Litvinova, T. M. Ezhova and Y. V. Shevchenko, *Dokl. Akad. Nauk SSSR*, 1983, **271**, 871.
431. O. V. Drygina, G. N. Dorofeenko and O. Y. Okhlobystin, *Khim. Geterotsikl. Soedin.*, 1981, 454.
432. J. L. Maurer, F. Berchier, A. J. Serino, C. B. Knobler and M. F. Hawthorne, *J. Org. Chem.*, 1990, **55**, 838.
433. S. B. Kahl, *Tetrahedron Lett.*, 1990, **31**, 1517.
434. E. A. Chizhova, N. S. Ivanova, S. M. Tanevska-Osinska and N. I. Bekasova, *Izv. Vyssh. Uchebn. Zaved., Khim. Khim. Tekhnol.*, 1992, **35**, 64.
435. N. I. Bekasova, L. G. Komarova, N. G. Komarova, N. S. Ivanova, E. A. Chizhova and O. K. Shebanova, *Zh. Obshch. Khim.*, 1991, **61**, 2224.
436. N. S. Hosmane, N. N. Maldar, S. B. Potts, D. W. H. Rankin and H. E. Robertson, *Inorg. Chem.*, 1986, **25**, 1561.
437. U. Von Arx, P. R. Pradhan and R. Keese, *Chimia*, 1986, **40**, 13.
438. V. N. Kalinin, B. A. Izmailov, A. A. Kazantsev, V. D. Myakushev, A. A. Zhdanov and L. I. Zakharkin, *J. Organomet. Chem.*, 1981, **216**, 295.
439. F. A. Gomez and M. F. Hawthorne, *J. Org. Chem.*, 1992, **57**, 1384.
440. L. I. Zakharkin, V. A. Ol'shevskaya and T. Y. Poroshina, *Zh. Org. Khim.*, 1989, **25**, 2152.
441. M. M. Teplyakov, I. A. Khotina, T. L. Gelashvili and V. V. Korshak, *Dokl. Akad. Nauk SSSR*, 1983, **271**, 874; *Dokl. Chem. Proc. Acad. Sci. USSR*, 1983, **271**, 252.
442. Y. V. Zeifman, S. A. Postovoi, V. N. Lebedev and L. I. Zakharkin, *Zh. Obshch. Khim.*, 1988, **58**, 936.
443. J. G. Wilson A. K. M. Anisuzzaman, F. Alam and A. H. Soloway, *Inorg. Chem.*, 1992, **31**, 1955.
444. J. Malmquist and S. Sjoeberg, *Inorg. Chem.*, 1992, **31**, 2534.
445. S.-H. Wu and M. Jones, *Inorg. Chem.*, 1988, **27**, 2005.; *ibid.*, 1986, **25**, 4802.
446. V. I. Bregadze, A. Y. Usyatinskii, V. T. Kampel, L. M. Golubinskaya and N. N. Godovikov, *Izv. Akad. Nauk SSSR, Ser. Khim.*, 1985, 1212.
447. A. Y. Usyatinskii, V. I. Bregadze, T. M. Shcherbina and N. N. Godovikov, *Izv. Akad. Nauk SSSR, Ser. Khim.*, 1981, 1428.
448. V. I. Bregadze, A. Y. Usyatinskii, G. Z. Suleimanov and N. N. Godovikov, *Izv. Akad. Nauk SSSR, Ser. Khim.*, 1981, 1927.
449. A. Y. Usyatinskii, Z. V. Todres, T. M. Shcherbina, V. I. Bregadze and N. N. Godovikov, *Izv. Akad. Nauk SSSR, Ser. Khim.*, 1983, 1640.
450. Y. A. Ol'dekop, N. A. Maier, A. A. Erdman and V. P. Prokopovich, *Zh. Obshch. Khim.*, 1982, **52**, 2256; *J. Gen. Chem. (USSR)*, 1982, **52**, 2008.
451. Y. A. Ol'dekop, N. A. Maier, A. A. Erdman and V. P. Prokopovich, *Vestsi Akad. Navuk BSSR, Ser. Khim. Navuk*, 1983, 114.
452. B. Longato, F. Morandini and S. Bresadola, *Inorg. Chim. Acta*, 1980, **39**, 27.
453. B. Longato and S. Bresadola, *Inorg. Chem.*, 1982, **21**, 168.
454. F. Morandini, B. Longato and S. Bresadola, *J. Organomet. Chem.*, 1982, **239**, 377.
455. M. Basato, B. Longato, F. Morandini and S. Bresadola, *Inorg. Chem.*, 1984, **23**, 3972.
456. G. Z. Suleimanov, V. I. Bregadze, V. T. Kampel, M. V. Petriashvili, N. N. Godovikov and V. I. Sokolov, *Izv. Akad. Nauk SSSR, Ser. Khim.*, 1982, 1606.
457. G. Z. Suleimanov *et al.*, *Zh. Org. Khim.*, 1983, **19**, 2258.
458. B. Stibr, J. Plesek and A. Zobacova, *Polyhedron*, 1982, **1**, 824.
459. B. Stibr, S. Hermanek, Z. Janousek, Z. Plzak, J. Dolansky and J. Plesek, *Polyhedron*, 1982, **1**, 822.
460. H. L. Gingrich, T. Ghosh, Q. Huang and M. Jones, *J. Am. Chem. Soc.*, 1990, **112**, 4082.
461. T. Ghosh, H. L. Gingrich, C. K. Kam, E. C. Mobraaten and M. Jones, *J. Am. Chem. Soc.*, 1991, **113**, 1313.
462. Q. Huang, H. L. Gingrich and M. Jones, *Inorg. Chem.*, 1991, **30**, 3254.
463. H. L. Gingrich, Q. Huang, A. L. Morales and M. Jones, *J. Org. Chem.*, 1992, **57**, 3803.
464. T. M. Babchinitser, L. G. Komarova, Y. A. Gladkova, N. I. Bekasova, S. A. Pavlova and V. V. Korshak, *Vysokomol. Soedin. A*, 1989, **31**, 91.
465. N. I. Bekasova, N. G. Komarova, L. G. Komarova and V. A. Sergeev, *Vysokomol. Soedin. B*, 1991, **33**, 2159.
466. V. A. Sergeev, N. I. Bekasova, M. A. Surikova, N. G. Komarova and S. A. Arnautov, *Plast. Massy*, 1991, 28.
467. A. A. Sakharova, O. A. Mel'nik, T. M. Frunze and V. N. Kalinin, *Vysokomol. Soedin. B*, 1985, **27**, 883.
468. V. V. Korshak, M. M. Teplyakov, T. L. Gelashvili, V. N. Kalinin and L. I. Zakharkin *Vysokomol. Soedin. A*, 1980, **22**, 262.
469. L. M. Bronshtein *et al.*, *Vysokomol. Soedin. A*, 1990, **32**, 1331.
470. E. F. Kolpikova *et al.*, *Plast. Massy*, 1990, 46.
471. A. I. Kalachev, L. I. Komarova, N. I. Kuryakova, P. M. Veletskii and S. V. Vinogradova, *Plast. Massy*, 1990, 10.
472. A. I. Kalachev, L. I. Komarova, N. I. Kuryakova and S. V. Vinogradova, *Plast. Massy*, 1989, 49.
473. A. K. Bonetskaya *et al.*, *Vysokomol. Soedin. B*, 1991, **33**, 551.
474. A. K. Bonetskaya *et al.*, *Vysokomol. Soedin. B*, 1990, **32**, 190.
475. V. N. Kalinin, B. A. Izmailov, A. A. Kazantsev, A. A. Zhdanov and L. I. Zakharkin, *Zh. Obshch. Khim.*, 1981, **51**, 859; *J. Gen. Chem. (USSR)*, 1981, **51**, 713.
476. M. M. Teplyakov, I. A. Khotina, A. V. Dmitrenko, V. V. Korshak, B. A. Ismailov and V. D. Myakushev, *Makromol. Chem.*, 1990, **191**, 809.
477. A. I. Mikhal'skii, D. D. Novikov, N. A. Tetenok and I. I. Miroshnikova, *Zh. Fiz. Khim.*, 1991, **65**, 2750.
478. G. M. Janini and N. T. Filfil, *J. Chromatogr.*, 1989, **469**, 43.
479. N. S. Nikitina, Z. A. Klimak and N. A. Nikulicheva, *Neftepererab. Neftekhim. (Moscow)*, 1985, 29.
480. J. Kusmierz, E. Malinski, W. Czerwiec and J. Szafranek, *J. Chromatogr.*, 1985, **331**, 219.
481. L. G. Sarto, S. A. Estes, P. C. Uden, S. Siggia and R. M. Barnes, *Anal. Lett.*, 1981, **14**, 205.
482. J. Jamet, J. R. Spann, R. W. Rice, D. Lewis and W. S. Coblenz, *Ceram. Eng. Sci. Proc.* 1984, **5**, 677.
483. W. S. Coblenz, G. H. Wiseman, P. B. Davis and R. W. Rice, *Mater. Sci. Res.*, 1984, **17**, 271.
484. J. Kulig, G. Jefferis and M. Litt, *Polym. Mater. Sci. Eng.*, 1989, **61**, 219.

485. G. V. Kotrelev and E. A. Zhdanova, *Vysokomol. Soedin. B*, 1981, **23**, 289.
486. L. A. Nock, J. M. Goldwasser and H. G. Adolph, *J. Polym. Sci., Part A: Polym. Chem.*, 1991, **29**, 1133.
487. A. K. Bonetskaya *et al.*, *Plast. Massy*, 1988, 10.
488. V. A. Sergeev, V. K. Shitikov, G. N. Koloskova, V. I. Pakhomov and N. V. Batenina, *Plast. Massy*, 1980, 24.
489. R. M. Aliguliev, Y. V. Zelenev, G. M. Aliev, D. M. Khiteeva and V. S. Matochkin, *Acta Polym.*, 1980, **31**, 400.
490. N. I. Makarevich, N. I. Sushko, A. I. Ivanov, L. S. Lyapina and B. P. Parfenov, *Vestsi Akad. Navuk B. SSR, Ser. Khim. Navuk*, 1984, 63.
491. V. V. Korshak, M. M. Teplyakov, I. A. Khotina and A. I. Kovalev, *Acta Polym.*, 1988, **39**, 440.
492. H. M. Colquhoun, J. A. Daniels, I. R. Stephenson and K. Wade, *Polym. Commun.*, 1991, **32**, 272.
493. A. I. Tregub, K. S. Issajew, A. S. Achrijew, P. I. Iwanow, W. S. Matotschkin and J. V. Selenew, *Plaste Kautsch.*, 1981, **28**, 137.
494. N. I. Bekasova, L. G. Komarova, G. A. Kats and V. V. Korshak, *Acta Polym.*, 1986, **37**, 438.
495. V. V. Korshak, N. I. Bekasova, N. G. Komarova, V. V. Vagin and A. I. Solomatina, *Vysokomol. Soedin. B*, 1989, **31**, 54.
496. J. L. Maurer, A. J. Serino and M. F. Hawthorne, *Organometallics*, 1988, **7**, 2519.
497. A. Varadarajan, R. M. Sharkey, D. M. Goldenberg and M. F. Hawthorne, *Bioconjugate Chem.*, 1991, **2**, 102.
498. A. Varadarajan and M. F. Hawthorne, *Bioconjugate Chem.*, 1991, **2**, 242.
499. J. L. Fauchere, O. Leukart, A. Eberle and R. Schwyzer, *Helv. Chim. Acta*, 1979, **62**, 1385.
500. R. J. Paxton, B. G. Beatty, A. Varadarajan and M. F. Hawthorne, *Bioconjugate Chem.*, 1992, **3**, 241.
501. M. Miura, D. Gabel, G. Oenbrink and R. G. Fairchild, *Tetrahedron Lett.*, 1990, **31**, 2247.
502. R. C. Haushalter, W. M. Butler and R. W. Rudolph, *J. Am. Chem. Soc.* 1981, **103**, 2620.
503. S. B. Kahl *et al.*, *Proc. Natl. Acad. Sci. USA.*, 1990, **87**, 7265.
504. S. B. Kahl and M. S. Koo, *J. Chem. Soc., Chem. Commun.*, 1990, 1769.
505. Y. Yamamoto, *Pure Appl. Chem.*, 1991, **63**, 423.
506. A. H. Soloway, A. K. M. Anisuzzaman, F. Alam, R. F. Barth and L. Liu, *Pure Appl. Chem.*, 1991, **63**, 411.
507. A. Harada and S. Takahashi, *J. Chem. Soc., Chem. Commun.*, 1988, 1352.
508. H. Nemoto, J. G. Wilson, H. Nakamura and Y. Yamamoto, *J. Org. Chem.*, 1992, **57**, 435.
509. J. H. Osborne, R. C. P. Hill and D. M. Ritter, *Inorg. Chem.*, 1986, **25**, 372.
510. S. J. Cranson, P. M. Davies, R. Greatrex, D. W. H. Rankin and H. E. Robertson, *J. Chem. Soc., Dalton Trans.*, 1990, 101.
511. N. N. Greenwood, in 'Electron Deficient Boron and Carbon Clusters', eds. G. A. Olah, K. Wade and R. E. Williams, Wiley, New York, 1991, p. 165.
512. M. Arunchaiya, P. C. Choi and J. H. Morris, *J. Chem. Res. (S)*, 1985, 216.
513. K. Base, N. W. Alcock, O. W. Howarth, H. R. Powell, A. T. Harrison and M. G. H. Wallbridge, *J. Chem. Soc., Chem. Commun.*, 1988, 341.
514. T. Davan and J. A. Morrison, *J. Chem. Soc., Chem. Commun.*, 1981, 250.
515. T. Mennekes, P. Paetzold, R. Boese and D. Blaeser, *Angew. Chem.*, 1991, **103**, 199; *Angew. Chem., Int. Ed. Engl.*, 1991, **30**, 173.
516. J. A. Heppert and D. F. Gaines, *Inorg. Chem.*, 1982, **21**, 4117.
517. D. F. Gaines and L. K. Lee, *Organometallics*, 1990, **9**, 1748.
518. D. P. Workman and S. G. Shore, in 'Electron Deficient Boron and Carbon Clusters', eds. G. A. Olah, K. Wade and R. E. Williams, Wiley, New York, 1991, p. 237.
519. D. F. Gaines and G. A. Steehler, *J. Chem. Soc., Chem. Commun.*, 1984, 1127.
520. J. G. Taylor and M. G. H. Wallbridge, *Polyhedron*, 1985, **4**, 321.
521. D. F. Gaines and D. E. Coons, *Inorg. Chem.*, 1986, **25**, 364.
522. D. Saulys and J. A. Morrison, *Inorg. Chem.*, 1980, **19**, 3057.
523. K. Shelly, C. B. Knobler and M. F. Hawthorne, *Inorg. Chem.*, 1992, **31**, 2889.
524. C. T. Brewer and R. N. Grimes, *J. Am. Chem. Soc.*, 1985, **107**, 3552.
525. G. A. Kline and R. F. Porter, *Inorg. Chem.*, 1981, **20**, 292.
526. M. G. L. Mirabelli and L. G. Sneddon, *J. Am. Chem. Soc.*, 1988, **110**, 3305.
527. M. G. L. Mirabelli and L. G. Sneddon, *Inorg. Chem.*, 1988, **27**, 3271.
528. G. M. Edvenson, D. F. Gaines, H. A. Harris and C. F. Campana, *Organometallics*, 1990, **9**, 401.
529. D. F. Gaines and D. E. Coons, *J. Am. Chem. Soc.*, 1985, **107**, 3266.
530. L. I. Zakharkin, V. A. Ol'shevskaya and A. I. Kovredov, *Zh. Obshch. Khim.*, 1985, **55**, 949.

7
Main-group Heteroboranes

LEE J. TODD
Indiana University, Bloomington, IN, USA

7.1 INTRODUCTION 257

7.2 DERIVATIVES CONTAINING GROUP 13 ELEMENTS (Al, Ga, In, Tl) 258
- *7.2.1 Alumina- and Gallaboranes* 258
- *7.2.2 Alumina- Galla-, Inda-, and Thallacarbaboranes* 258

7.3 DERIVATIVES CONTAINING GROUP 14 ELEMENTS (Si, Ge, Sn, Pb) 259
- *7.3.1 Silaboranes and Silacarbaboranes* 259
- *7.3.2 Borane Derivatives Containing Germanium, Tin, or Lead* 259
- *7.3.3 Carbaborane Derivatives Containing Germanium, Tin, or Lead* 260

7.4 DERIVATIVES CONTAINING NITROGEN 260
- *7.4.1 Azaboranes* 260
- *7.4.2 Diazaboranes* 261
- *7.4.3 Azacarbaboranes* 262
- *7.4.4 Transition Metal Derivatives of Azaboranes* 262

7.5 DERIVATIVES CONTAINING PHOSPHORUS 263
- *7.5.1 Phosphaboranes* 263
- *7.5.2 Diphosphaboranes* 264
- *7.5.3 Phosphacarbaboranes* 265
- *7.5.4 Transition Metal Complexes of Phosphaboranes* 265

7.6 DERIVATIVES CONTAINING ARSENIC 265
- *7.6.1 Arsaboranes* 265
- *7.6.2 Diarsaboranes* 266
- *7.6.3 Metal Complexes of Arsa- and Diarsaboranes* 266

7.7 STIBA- AND BISMABORANES AND THEIR DERIVATIVES 267

7.8 OXABORANES AND OXAMETALLABORANES 267

7.9 DERIVATIVES CONTAINING SULFUR 268
- *7.9.1 Thiaboranes and Their Metal Derivatives* 268
- *7.9.2 Dithiaboranes and Their Metal Derivatives* 269
- *7.9.3 Thiacarbaboranes* 269

7.10 SELENA- AND TELLURABORANES AND THEIR METAL DERIVATIVES 270

7.11 REFERENCES 271

7.1 INTRODUCTION

This chapter surveys not only heterocarbaboranes, which was the only subject in Chapter 5.6 of *COMC-I* (1982), but also heteroboranes and their metal complexes. Most of the molecules covered here have a cage structure containing several BH units and one or more main-group heteroatoms. Compounds with heterocyclic structures or bicyclic structures are generally not covered (see Chapter 4, this volume).

Some of the earlier heteroborane chemistry up to 1981 was covered in a review entitled *Metallaboron Cage Compounds of the Main Group Metals*.[1] More recent partial reviews of heteroborane chemistry can be found in a volume published in honor of Burg.[2] In addition, a survey of the ^{11}B NMR spectroscopic data of main-group heteroboranes was included in a chapter on NMR spectroscopy by Hermanek.[3] Another review of ^{11}B NMR spectroscopy by Siedle includes much useful information about heteroboranes.[4]

The available space for this review precludes the incorporation of all new material for this review period. I hope that the material presented here gives an overview of the research activity in this area for this period. In some cases, the reader is referred to a review for further pertinent data in a specific area. In most cases, schematic structures will be used instead of the more space-consuming ORTEP plots from the x-ray crystal structure determinations.

7.2 DERIVATIVES CONTAINING GROUP 13 ELEMENTS (Al, Ga, In, Tl)

7.2.1 Alumina- and Gallaboranes

The first *closo*-aluminaborane was obtained in good yield by reaction of $Na_2B_{11}H_{13}$ in liquid $AlMe_3$ with one equivalent of glyme at 100 °C to form $Na_2[B_{11}H_{11}AlMe]$ (**1**) and methane.[5] Compound (**1**) reacts with water or methanol producing $Na_2[B_{11}H_{13}]$. A single-crystal x-ray structure determination of $(AsPh_4)_2[B_{11}H_{11}AlMe]$ showed the *closo* icosahedral structure of this compound.[5]

The novel mixed tetraborane(10) derivative, $H_2GaB_3H_8$ (**2**) was prepared by reaction of $[H_2GaCl]_2$ with $(Bu^n{}_4N)[B_3H_8]$ at 243 K.[6] Electron diffraction measurements are consistent with (**2**) having a tetraborane structure with the gallium atom in a wing (B_2) position. Compound (**2**) decomposes at room temperature giving metallic gallium, H_2, and B_4H_{10} from the condensed phase or a solid (GaB_2), H_2 and B_2H_6 from the gas phase.

7.2.2 Alumina- Galla-, Inda-, and Thallacarbaboranes

The acid–base chemistry of group 13 metallacarbaboranes was first recognized by the formation of oxygen base complexes of $[3\text{-Et-}1,2(R)_2\text{-}3,2,1\text{-AlC}_2B_9H_9]$ (R = Me or H).[7–8] The *closo*-gallacarbaborane, $[1\text{-Bu}^t\text{-}1\text{-Ga-}2,3\text{-}(TMS)_2\text{-}2,3\text{-}C_2B_4H_4]$ formed a bridged complex with 0.5 equiv. 2,2'-bipyrimidine[9] and a 1:1 bidentate complex with 1 equiv. of this base.[10] An x-ray structure of $[1\text{-}(2,2'\text{-bipy(imidine)-}1\text{-Bu}^t\text{-}1\text{-Ga-}2,3(TMS)_2\text{-}2,3\text{-}C_2B_4H_4]$ shows that the gallium atom is considerably displaced away from the carbon atoms in the C_2B_3 face. The molecule looks very much like a *nido* structure. However, the steric demands of the *exo* cage groups confuse the structural issue of the cage.

The synthesis and structure of the indacarbaborane, *closo*-$[1\text{-}(Me_2CH)\text{-}1\text{-In-}2,3\text{-}(TMS)_2\text{-}2,3\text{-}C_2B_4H_{14}]$ have been reported.[11] The solid-state structure consists of a dimeric cluster of two opposed InC_2B_4 cages bonded together by B–H–In bridge units.

The x-ray structure of PPN[*closo*-3,1,2-$TlC_2B_9H_{11}$] shows a similar bonded dimer to the indacarbaborane.[12] The structure study shows separated PPN^+ cations and a dimeric anion. The $B_9C_2H_{11}$ ligand is bonded in a pentahapto manner to one thallium atom and, at the same time, bridge bonded (M–H–B) via two BH units on the open face to the second thallium atom. Numerous examples of B–H–M interactions have been cited in metallaborane and metallacarbaborane chemistry.[13] Interesting ^{205}Tl–^{11}B spin coupling to some, but not all, ^{11}B signals of this thallacarbaborane are observed. In another report, the x-ray structure of Tl[*closo*-3,1,2-$Tl(CMe)_2B_9H_9$] showed the presence of isolated Tl^+ and $[Tl(CMe)_2B_9H_9]^-$ ions.[14] The anion has a slipped icosahedral structure with the thallium bonded primarily to the three boron atoms of the dicarbollide ligand. The role of the cation appears to be important for the overall intermolecular arrangement in these different salts.

Treatment of *nido*-2,3-$Et_2C_2B_4H_6$ with $Et_3N{\cdot}AlH_3$ at 0 °C produced 1 equiv. of H_2 and inserted a $Et_3N{\cdot}AlH_2$ unit in place of a bridge hydrogen in the starting carbaborane.[15] At 50 °C, another equivalent of H_2 was lost, forming the seven-vertex molecule, *nido*-$[6\text{-}Et_3N\text{-}6\text{-AlH-}3,4\text{-}(EtC)_2B_4H_4]$ (**3**). Compound (**3**) reacted with another equivalent of 2,3-$Et_2C_2B_4H_6$ at 70 °C to form the bis(carboranyl)aluminum molecule (**4**) illustrated in Scheme 1.

Heating of *closo*-$[3\text{-Et-}3,1,2\text{-AlC}_2B_9H_{11}]$ in an aromatic solvent at 90 °C formed the dimeric species, *commo*-$[3,3'\text{-Al}\{exo\text{-}[3,9\text{-}(\mu\text{-H})_2\text{-AlEt}_2\text{-}3,1,2\text{-AlC}_2B_9H_9](3',1',2'\text{AlC}_2B_9H_{11})\}]$ (**5**). Compound (**5**) can be regarded as a zwitterion composed of a *commo*-$[3,3'\text{-Al}(3,1,2\text{-AlC}_2B_9H_{11})_2]^-$ anion (**6**) complexed via B–H–Al bonds with a $AlEt_2{}^+$ cation.[16] The anion (**6**) has also been obtained by the reaction illustrated

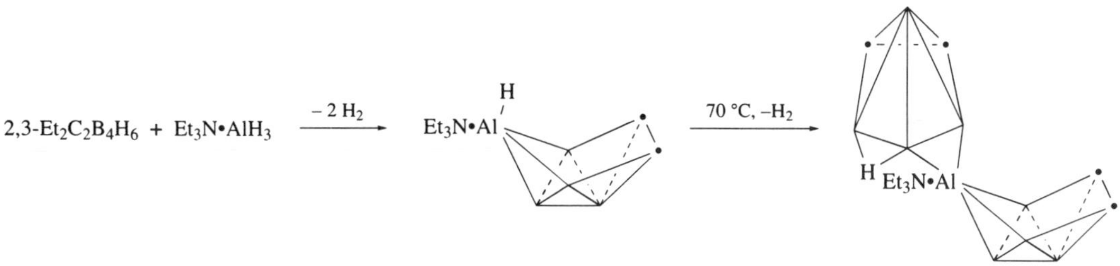

(3) • = C–Et (4)

Scheme 1

in Equation (1). The corresponding gallium anion, *commo*-[3,3'-Ga(3,1,2-GaC$_2$B$_9$H$_{11}$)$_2$]$^-$ was prepared by similar chemistry to Equation (1). The [B$_9$C$_2$H$_{11}$]$^{2-}$ ion, when bound to metals in an η^5 bonding mode, is a formal six-electron donor. In anion (**6**), two dicarbollide ligands are bound to aluminum(III) in η^5-fashion, giving a formal 12 electrons to the central aluminum atom. The presently favored description has four of these electrons residing in two nonbonding molecular orbitals associated with the dicarbollide ligands.[17]

$$\text{Tl}[\textit{closo}\text{-3,1,2-TlC}_2\text{B}_9\text{H}_{11}] + 4\ \text{AlMe}_2\text{Cl} \xrightarrow{\text{toluene}} \text{Tl}[\textit{commo}\text{-Al(7,8-C}_2\text{B}_9\text{H}_{11})_2] + 2\ \text{Tl}[\text{AlMe}_3\text{Cl}] + \text{Tl}[\text{AlMe}_2\text{Cl}_2] \quad (1)$$

The syntheses and structural characterization of the σ-bonded bridging species *nido*-[μ-6,9-AlEt(OEt$_2$)-6,9-C$_2$B$_8$H$_{10}$], [Al(η-6,9-C$_2$B$_8$H$_{10}$)$_2$]$^-$, and [Al(η-2,7-C$_2$B$_6$H$_8$)$_2$$^-$ have been reported.[18–19]

7.3 DERIVATIVES CONTAINING GROUP 14 ELEMENTS (Si, Ge, Sn, Pb)

7.3.1 Silaboranes and Silacarbaboranes

A benzene solution of B$_{10}$H$_{14}$ and Me(H)Si[NMe$_2$]$_2$ was heated at reflux to form 6,9(Me$_2$NH)$_2$B$_{10}$H$_{12}$ and a 15% yield of 1,2-dimethyl-1,2-disila-*closo*-dodecarborane(12) (**7**), the first disilaborane.[20–1] An x-ray structure determination indicates that (**7**) has a structure very similar to *ortho*-carbaborane. Compound (**7**) is air sensitive. It melts at 201–203 °C (sealed capillary) and decomposes with gas evolution above 230 °C. It reacts with, and is degraded by, alcoholic KOH in THF at room temperature within minutes.

Reaction of silicon tetrachloride with 2 equiv. of Li$_2$[7,8-C$_2$B$_9$H$_{11}$] in refluxing benzene formed *commo*-[3,3'-Si(3,1,2-SiC$_2$B$_9$H$_{11}$)$_2$] (**8**), the first sandwich complex containing silicon.[22–3] A related silacarbaborane sandwich compound, [2,2',3,3'-(TMS)$_4$-*commo*-1,1'-Si(1,2,3-Si(1,2,3-SiC$_2$B$_4$H$_4$)$_2$], was prepared by similar chemistry subsequently.[24] At 450 °C in a flow-through reactor, (**8**) decomposed giving a silicon mirror and forming *closo*-1,8-C$_2$B$_9$H$_{11}$. Compound (**1**) hydrolyzes to form the [7,8-C$_2$B$_9$H$_{12}$]$^-$ ion. Pyridine reacted with (**8**) to form [(10-*exo*-η^1-7,8-C$_2$B$_9$H$_{11}$)(10-*endo*-η^1-7,8-C$_2$B$_9$H$_{11}$)Si(C$_5$H$_5$N)$_2$] whose structure was determined by an x-ray diffraction study.[23] Reaction of (**8**) with Me$_3$P formed a different type of product, as illustrated in Figure 1.[23] This structure resulted from nucleophilic attack of one B–H vertex by trimethylphosphine to form a (Me$_3$P)$_2$BH-B(cage) boronium function single-bonded to a *nido*-B$_8$C$_2$Si cage. The other B$_9$C$_2$H$_{11}$ ligand attached to silicon is strongly slipped away from the carbon atoms.

7.3.2 Borane Derivatives Containing Germanium, Tin, or Lead

The incorporation of heavier group 14 elements into boron hydride cage structures has not been investigated very extensively. The *nido* 11-vertex derivatives 7-Me$_2$MB$_{10}$H$_{12}$ (M = Ge or Sn) have been prepared by reaction of NaB$_{10}$H$_{13}$ with Me$_2$MCl$_2$.[25] The salt (Ph$_4$As)$_2$B$_{10}$H$_{12}$ reacted with SnCl$_2$ and Me$_2$SnCl$_2$ to give (Ph$_4$As)$_2$B$_{10}$H$_{12}$SnCl$_2$ and (Ph$_4$As)$_2$B$_{10}$H$_{12}$SnMe$_2$Cl$_2$, respectively.[26] The ^{119}Sn Mössbauer isomer shifts suggest that these dianions contain tin(II) and tin(IV), respectively. The structures of these salts will be of interest.

The dianions, [B$_{11}$H$_{11}$E]$^{2-}$ (E = Ge or Sn) have been prepared in good yield starting with [B$_{11}$H$_{14}$]$^-$, ECl$_2$ and Bu^{n}Li.[27] The [B$_{11}$H$_{11}$Pb]$^{2-}$ ion was made from [B$_{11}$H$_{14}$]$^-$ and PbCl$_2$ in aqueous sodium

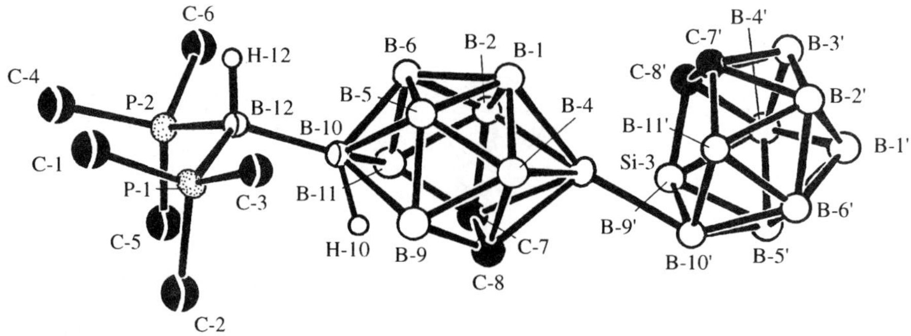

Figure 1 The structure of the product obtained by reaction of compound (**8**) with 2 equiv. of Me_3P.

hydroxide.[27] Treatment of the respective dianion with methyl iodide in THF formed *closo*-$[B_{11}H_{11}ECH_3]^{3-}$ (E = Ge or Sn). A single-crystal x-ray structure determination of [1-Me-1-$SnB_{11}H_{11}$]$PPh_3Me{\cdot}CH_2Cl_2$ demonstrated the *closo* icosahedral structure of the monoanion.

7.3.3 Carbaborane Derivatives Containing Germanium, Tin, or Lead

Carbaborane derivatives containing germanium, tin, or lead were reviewed in 1988.[28] The crystal structures of *closo*-[1-Sn-2(TMS)-3-Me-2,3-$C_2B_4H_4$] and its 1:1 complex with 2,2'-bipyridine have been determined.[29–30] There is a small displacement of the tin atom away from the carbon atoms in the C_2B_3 face in the parent stannacarbaborane. In the bipyridine complex, the tin atom displacement is much larger. The tin–bipyridine unit can thus be considered as η^3-bonded to the three boron atoms of the open face.

Reaction of $GeCl_4$ with Li[2,3-$(TMS)_2$-$C_2B_4H_5$] formed both the germanium(II) derivative, *closo*-[1-Ge-2,3-$(TMS)_2$-2,3-$C_2B_4H_4$] and the interesting germanium(IV) compound, *commo*-[$\{(TMS)_2C_2B_4H_4\}_2$Ge].[31]

Reaction of *closo*-[1-Sn-2-(TMS)-3-Me-2,3-$C_2B_4H_4$] with $TiCl_4$ followed a redox pathway giving [1,1'-Sn{2-(TMS)-3-Me-2,3-$C_2B_4H_4$}$_2$] and $TiCl_3$.[32] The tin atom in this sandwich product is in a formal +4 oxidation state. Interestingly, the C_2B_4 ligands are not parallel to one another in this compound as found in the corresponding silicon and germanium *commo*-metallacarbaborane derivatives, but are bent.

Twelve vertex heterocarbaboranes have also yielded interesting results. The aluminacarbaborane, [2,3-Me_2-1-Et-1,2,3-$AlC_2B_9H_9$] when reacted with Me_2SnCl_2 formed *nido*-[$Me_2Sn\{(MeC)_2B_9H_9\}$].[33] The Me_2Sn group probably is η^3-bonded to the three boron atoms in the open face of the $(MeC)_2B_9H_9$ unit.

Treatment of the salt $(Ph_3PMe)[TlB_9H_9(CMe)_2]$ with $GeCl_2$ or $SnCl_2$ in a 1:1 ratio leads to the ionic compounds $(Ph_3PMe)[ClE(B_9H_9C_2Me_2)]$, E = Ge or Sn.[34] An x-ray structure study of the germanium derivative shows that it is a *nido* 12-vertex cage anion in which the germanium atom is η^3-bonded to the three boron atoms in the open face of the $[7,8\text{-}C_2Me_2B_9H_9]^{2-}$ ligand. With a 2:1 ratio of the salt and ECl_2 reagents, the compounds $(Ph_3PMe)_2[E(B_9H_9C_2Me_2)_2]$ were formed. The structure of these dianions will be of interest.

7.4 DERIVATIVES CONTAINING NITROGEN

7.4.1 Azaboranes

The early syntheses and chemistry of $[B_9H_{12}NH]^-$, $B_9H_{11}NH$, and $B_8H_{12}NH$ have been reviewed.[35–6] The $B_9H_{11}NH$ molecule is readily attacked by water and converted to $B_8H_{12}NH$. The boron atom removed in this reaction can be stabilized by complexation with various neutral Lewis bases. The x-ray structure of one of these complexes, *arachno*-[9-{(C_6H_{11})NC}-6-NHB_9H_{11},] has been reported.[37] Slow passage of *nido*-$B_9H_{11}NH$ through an evacuated hot tube (400 °C) formed *closo*-1-NHB_9H_9.[37] This proved for the first time that a very electronegative nitrogen atom could be incorporated with fourfold connectivity into an electron-deficient *closo* cage structure.

An attempt to insert a boron atom into $B_9H_{11}NH$ by reaction with $Et_3N{\cdot}BH_3$ formed $Et_3N{\cdot}BH_2NB_9H_{11}$ instead.[38]

A new access to azaboranes was achieved by reaction of $B_{10}H_{12}(SMe_2)_2$ with excess HN_3 to form $B_{10}H_{12}(N_3)NH_2$, a product with a bridging NH_2 group.[39] Thermolysis of this azide derivative in boiling xylene formed a 5% yield of *nido*-7-$NHB_{10}H_{12}$ (**9**). Treatment of (**9**) with an excess of $Et_3N{\cdot}BH_3$ at 140 °C formed $(Et_3NH)[NB_{11}H_{11}]$ (**10**) in good yield.[40] Reaction of (**10**) with HBF_4 afforded sublimable *closo*-$HNB_{11}H_{11}$ (**11**). The gas-phase electron diffraction data and theoretical calculations suggest a distorted icosahedral structure for this parent azaborane. The main distortions are short N–B distances (0.1716(9) nm) and expansion of the pentagonal belt adjacent to nitrogen (0.1825(6) nm).[41] Compound (**10**) was treated with MeO_3SCF_3 to form $MeNB_{11}H_{11}$.[42]

A new method for preparing azaboranes involves thermal dehydrogenation of *arachno*-$B_9H_{13}(NH_2R)$ to form *nido*-$B_9H_{11}NR$.[43] The *nido* compound when treated with excess $Me_2S{\cdot}BH_3$ at 170 °C in decalin formed *closo*-$B_{11}H_{11}NR$.

The *closo*-$MeNB_{11}H_{11}$ can be opened up to a *nido* 12-vertex structure by reaction with methanol.[44] The $[N(PPh_3)_2][MeNB_{11}H_{11}(OMe)]$ (**12**) salt has the *nido* structure shown in Figure 2. Attack of $MeNB_{11}H_{11}$ with Me^-, Bu^tO^-, or F^- forms similar *nido* compounds.

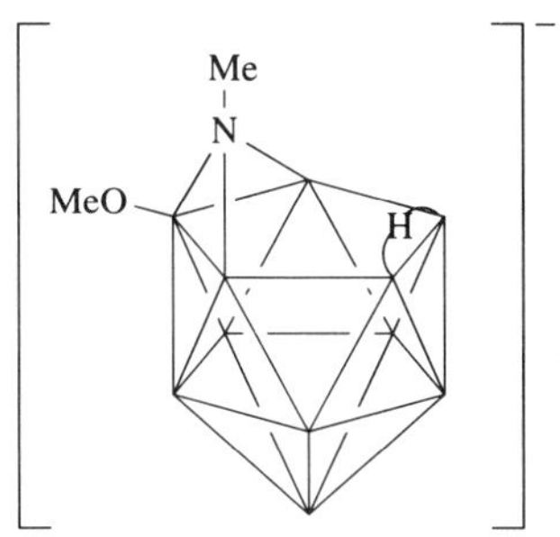

Figure 2 Schematic structural representation of the anion of (**12**).

7.4.2 Diazaboranes

The first cluster diazaboranes reported were the peralkylated *nido*-type molecules with the general formula $(Bu^t)_2N_2B_4R^1{}_2R^2{}_2$ (where $R^1{:}R^2$ = Me:Me, $Pr^i{:}Pr^i$, and $Bu^t{:}Pr^i$).[45] This type of molecule was prepared in a two-step sequence from an iminoborane as illustrated in Equation (2). The x-ray structure of the diazaborane was determined. The N_2B_4 skeleton is a *nido*-fragment of a pentagonal bipyramid with a base vertex missing as illustrated in Figure 3. The synthesis of $(TMS)_2N_2B_4$ has been reported.[46]

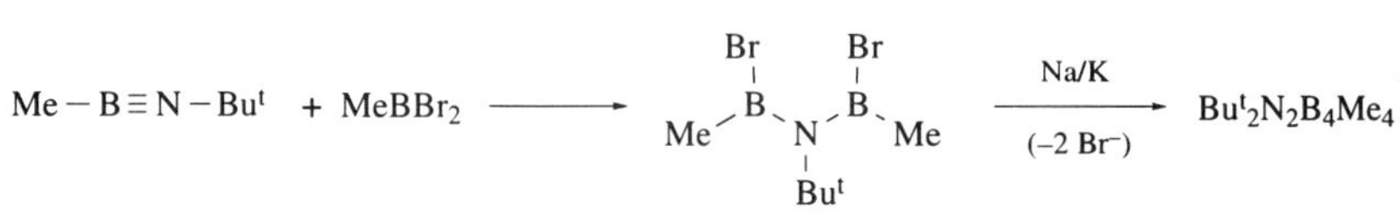

(**13**) (2)

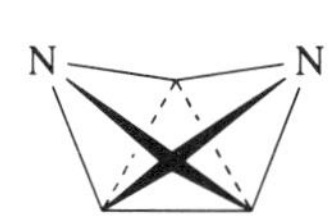

Figure 3 Schematic structural representation of the *nido*-N_2B_4 skeleton.

When (**13**) in the reaction sequence in Equation (2) has the structure (**13a**), dechlorination with Na/K formed (**14**), an azadiboriridine. Compound (**14**) reacted with $THF{\cdot}BH_3$ to form the unusual acid–base adduct illustrated in Figure 4.[47] The NB_3 skeleton can be regarded as a *nido*-fragment of a trigonal bipyramid. This compound should be in the azaborane section (Section 7.7.4.1). However, due to its mode of formation, it is cited here.

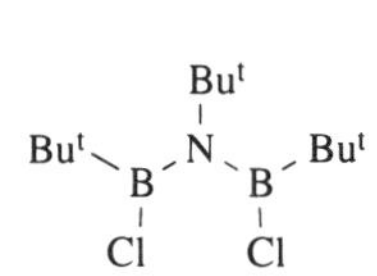

(**13a**)

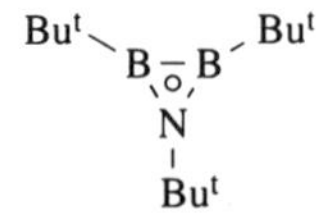

(14)

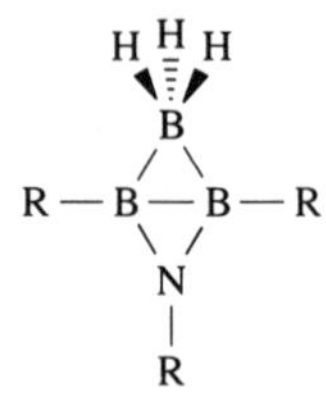

Figure 4 Schematic structural representation of the NB_3 skeleton.

Reaction of *arachno*-9(MeCN)-6-NB_9H_{12} or *nido*-6-NB_9H_{11} with sodium nitrite in THF solution followed by acidification (CF_3CO_2H) formed *arachno*-6,9-$N_2B_8H_{12}$ (**15**) in low yield.[48] The new compound was partially characterized by NMR (^1H and ^{11}B) spectroscopy and mass spectrometry. The 2:4:2 pattern of doublets in the ^{11}B NMR spectrum of (**15**) as well as the pronounced bridge hydrogen coupling on the area four resonance is characteristic of other known *arachno*-6,9-$E_2B_8H_{10}$ species (i.e., $Se_2B_8H_{10}$[49–50] and $(H_2C)_2B_8H_{10}$[51]).

7.4.3 Azacarbaboranes

Reaction of *closo*-1,2-$C_2B_8H_{10}$ with Bu^tNH_2 in THF at room temperature formed mainly *anti*-11-Bu^t-*arachno*-5,10,11-$C_2NB_8H_{12}$ (see Figure 5 for the proposed structure).[52] A small amount of the corresponding *syn* isomer was also formed. This is an interesting new type of cage expansion process converting a 10-vertex *closo* system to an 11-vertex *arachno* molecule.

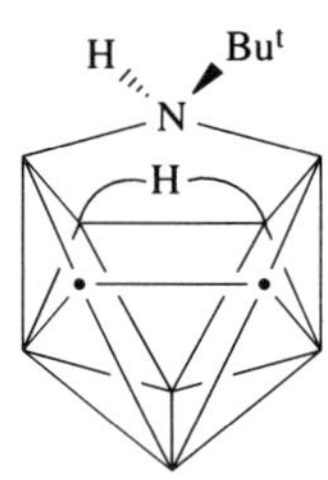

Figure 5 Proposed structure of *anti*-11-Bu^t-*arachno*-5,10,11-$C_2NB_8H_{12}$, where • = CH.

Treatment of *arachno*-4-CB_8H_{14} with sodium nitrite in benzene–water followed by acidification gave *arachno*-6,9-CNB_8H_{13}.[53] This is a $[B_{10}H_{14}]^{2-}$-type structure in which the 6- and 9-BH_2^- units are replaced with one CH_2 and one NH unit.

7.4.4 Transition Metal Derivatives of Azaboranes

Deprotonation of 7-$HNB_{10}H_{12}$ with $LiBHEt_3$ and then reaction with $(Ph_3P)_3RhCl$ formed *closo*-$[\{(Ph_3P)_2RhH\}HNB_{10}H_{10}]$, the first icosahedral metallazaborane.[54] The orange crystalline solid can be handled briefly in air. An x-ray structure analysis of the complex confirms a distorted $RhNB_{10}$ icosahedron. The small size of the nitrogen atom leads to relatively short B–N distances (0.163–0.1728 nm) in the cage.

Reaction of *nido*-6-HNB_9H_{11} and $[(\eta\text{-Cp*})IrCl_2]_2$ in the presence of *N*,*N*,*N*',*N*'-tetramethylnaphthalene-1,8-diamine gave *nido*-[9-Cl-8-(η-Cp*)-8,7-$IrNHB_9H_{10}$] (**16**) in moderate yield.[55] The structure of this open-faced, 11-vertex azametallaborane is illustrated in Figure 6. A sulfur-containing analogue has also been reported.[56] The mechanism of formation of this *nido* azametallaborane is more complex than just halide displacement from the starting $[(\eta\text{-Cp*})IrCl_2]_2$ by the incipient $[B_9H_{10}NH]^-$. The regioselective attachment of the chlorine substituent to B-9 to give a 40% yield of this specific substituted product is of interest.

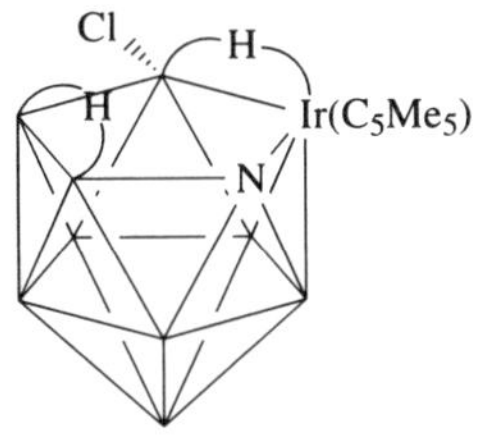

Figure 6 Schematic structural representation of compound (**16**).

The reaction of *nido*-6-HNB_9H_{11} with $[(\eta\text{-}MeC_6H_4\text{-}4\text{-}Pri)RuCl_2]_2$ results in the formation of *closo*-[1-(η-MeC_6H_4-4-$^iPr^i$)-1,2-$RuNHB_9H_9$] in low yield.[57] Preliminary NMR spectroscopic results suggest that a variety of other azaruthenaboranes are also formed in this reaction. Treatment of 6-HNB_9H_{11} and cyclopentadiene with excess sodium hydride in THF followed by reaction with $CoCl_2$ formed *closo*-[1-(Cp)-1,2-$CoNHB_9H_9$] (**17**) in 45% yield.[58] Single-crystal x-ray analysis gave the structure illustrated in Figure 7. Compound (**17**) could be deprotonated at nitrogen and then reacted with methyl iodide to form *closo*-[1-(C_5H_5)-1,2-$CoN(Me)B_9H_9$] in good yield.

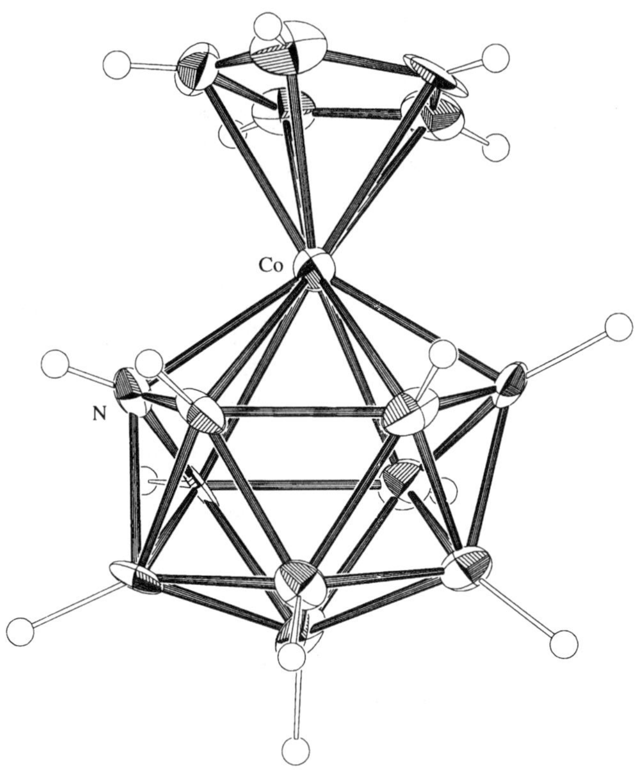

Figure 7 Structure of [1-(η-Cp)-1,2-$CoNHB_9H_9$] (**17**).

7.5 DERIVATIVES CONTAINING PHOSPHORUS

7.5.1 Phosphaboranes

The compound 5,6-$\mu(PPh_2)B_{10}H_{13}$ (which was reported in 1962)[59] can be deprotonated with aqueous sodium hydroxide to give the $[6,9\text{-}\mu(PPh_2)B_{10}H_{12}]^-$ ion.[60] The structure was confirmed by a single-crystal x-ray diffraction analysis of $(Ph_3PMe)[B_{10}H_{12}PPh_2]$.[61] The structure resembles that of $[B_{10}H_{14}]^{2-}$, but it has a bridging $-PPh_2-$ group between the B-6 and B-9 positions instead of *endo*-terminal hydrogen atoms.

In 1991, a review of bridged and cage-inserted phosphaboranes including molecular orbital calculations appeared.[62] Geometrical isomers of the phosphorus-bridged pentaboranes of the general formula $R^1R^2PB_5H_8$ were reported previously.[63–4] More recently, the bridged derivatives

$(TMS)_2C{=}PB_5H_8$ and $(TMS)_2CHP(Cl)B_5H_8$ were obtained from $[B_5H_8]^-$ and the appropriate phosphorus chloride.[65] The ^{11}B NMR spectra of the two types of derivatives are quite different. The authors speculate that while $(TMS)_2CHP(Cl)B_5H_8$ is a μ_2-phosphino-bridged pentaborane (having a three-electron-donating unit), the $(TMS)_2C{=}PB_5H_8$ derivative may have a different type of bonding (μ_3-phosphorus, with bonding to the apex boron). An x-ray structure determination of the μ_3-compound is definitely needed to resolve this question. Similar compounds were obtained by a hydroboration reaction of B_5H_9 with a phosphaalkene as illustrated in Equation (3).[66–7] The product in Equation (3) reacts with water and eliminates H_2 to give TMS-O(But)C=PB_5H_8, which is similar to the second product reported by Gaines.[65] In the early 1990s, $(\mu_2\text{-}Ph_2P)B_5H_8$ was prepared.[68] An x-ray structural determination of a $Fe(CO)_2(Cp)$-substituted derivative showed that the μ_2-diphenylphosphino group used the lone pair of electrons on phosphorus in the μ_2-phosphino-bridged three-electron-donating unit.[68]

TMS-P=C(Bu^t)O-TMS + B_5H_9 ⟶ 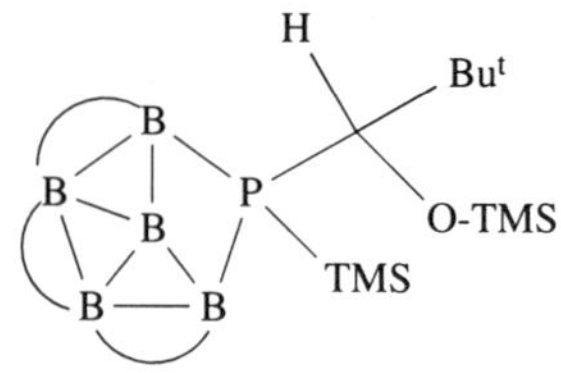 (3)

The synthesis of phosphorus-containing boranes has been slow relative to other heteroboranes. This is due, in part, to the complexity of the product mixtures and also to the chemical instability of some phosphaboranes. The first polyhedral phosphaborane was obtained by reaction of $[B_{11}H_{13}]^{2-}$ with $PhPCl_2$ to give *closo*-$PhPB_{11}H_{11}$ in low yield after an extensive purification procedure.[69] This was followed several years later by a general synthesis of *nido*-7-$RPB_{10}H_{12}$ (R = Me, Et, Pr, Ph) and their metal complexes, $[Co(B_{10}H_{10}PR)(Cp)]$.[70] The $[7\text{-}RPB_{10}H_{11}]^-$ derivatives were prepared by reacting $B_{10}H_{14}$ and $RPCl_2$ in the presence of NaH or triethylamine. Protonation then gave 7-$RPB_{10}H_{12}$. More recently, phosphaboranes have been obtained from the reaction of $B_{10}H_{14}$ with PCl_3 and triethylamine.[71] The derivative *closo*-6-Et_3N-2-PB_9H_8 was isolated from this type of reaction. An x-ray structure study of this compound as an enantiomeric mixture has been reported.[71] The parent anion, *closo*-$[2\text{-}PB_9H_9]^-$, was also isolated from this reaction mixture.[72] The PCl_3 appears to function as a hydride acceptor, converting $[2\text{-}PB_9H_9]^-$ to 6-Et_3N-2-PB_9H_8 in the presence of triethylamine.

A series of new icosahedral monophosphaborane derivatives has been reported. Treatment of $[B_{11}H_{13}]^{2-}$ with $MePCl_2$ in THF formed a low yield of the air-sensitive *closo*-$MePB_{11}H_{11}$.[73] An x-ray study of this compound has been reported.

Treatment of $(Me_3NH)[B_{11}H_{14}]$ with 4 equiv. of Bu^{n}Li and then PCl_3 in THF formed a 29% yield of *closo*-2-Me_3N-1-$PB_{11}H_{10}$.[72] This compound was fully characterized, including an x-ray study. A second compound believed to be *closo*-7-Me_3N-1-$PB_{11}H_{10}$ was isolated from this reaction but not fully characterized. When the trimethylamine was removed in the reaction above before the PCl_3 was added, the parent molecular ion $[PB_{11}H_{11}]^-$ was isolated.[72] As suggested above, the PCl_3 appears to act as a hydride abstractor and in the presence of Me_3N, the Me_3N–$PB_{11}H_{10}$ isomers are formed.

7.5.2 Diphosphaboranes

Thermolysis of a 2:1 mixture of $(Pr^i{}_2N)B(Cl)P(TMS)_2$ and $Pr^i{}_2NBCl_2$ leads to formation of the trigonal-bipyramidal cage compound, $P_2(Pr^i{}_2NB)_3$. The phosphorus atoms are in axial positions.[74] A more controlled synthesis of P_2B_3 cages was developed from reactions of 1,3,2,4-diphosphadiboretanes $(HPBNR_2)_2$ with Bu^{n}Li, followed by addition of R_2NBCl_2 and then Bu^{t}Li.[75] The B–N distance of $P_2(Pr^i{}_2NB)_3$ of 0.134(2) nm is consistent with considerable B–N π-bonding in this type of cage structure. This suggests that there may be little B–B bonding in this type of cluster.

A diphosphaborane was obtained in low yield by the reaction of B_2Cl_4 with PCl_3 at 300 °C to form *closo*-1,2-$P_2B_4Cl_4$.[76] Subsequently, the reaction of $B_{10}H_{14}$ with PCl_3 and triethylamine formed 1,2-$P_2B_{10}H_{10}$ in 12% yield.[71] The thermal conversion of the 1,2-isomer to 1,7-$P_2B_{10}H_{10}$ occurred at 560–590 °C in a sealed tube.[71] Degradation of 1,2-$P_2B_{10}H_{10}$ with aqueous sodium hydroxide gave $[7\text{-}PB_{10}H_{12}]^-$ exclusively. This anionic phosphaborane was fully characterized by this method of synthesis. In contrast, treatment of 1,2-$P_2B_{10}H_{10}$ with dry, oxygen-free piperidine–chloroform gave *nido*-$[7,8\text{-}P_2B_9H_{10}]^-$ in good yield. The heteroboranes 1,2-$PAsB_{10}H_{10}$ and 1,2-$PSbB_{10}H_{10}$ were obtained in low yield by reaction of $B_{10}H_{14}$ with triethylamine and mixtures of PCl_3 and $AsCl_3$ or PCl_3 and SbI_3, respectively.[72] The last compound in this series, 1,2-$PBiB_{10}H_{10}$, was prepared by reaction of $[7\text{-}PB_{10}H_{12}]^-$ with triethylamine and $BiCl_3$.[77]

7.5.3 Phosphacarbaboranes

The reaction between $Na(THF)Li[2,3\text{-}(TMS)_2C_2B_4H_4]$ and 2,4,6-$Bu^t_3C_6H_2PCl_2$ formed air-sensitive 1-(2,4,6-$Bu^t_3C_6H_2$)-1-P-2,3-$(TMS)_2$-2,3-$C_2B_4H_4$.[78] Based on NMR spectroscopic data, a *closo*-geometry was proposed for this compound.

Recently, it was reported that $NaLi[2,3\text{-}R^i_2\text{-}2,3\text{-}C_2B_4H_4]$ (R = Et or Bz) and R^2PCl_2 (R^2 = Me, Bu^t, Ph) react to form $R^2P(RC)_2B_4H_4$.[79] However, in this case (via an *ab initio*/IGLO/NMR study), it is proposed that these phosphadicarbaboranes have a seven-vertex *nido* configuration. Resolution of the structures of these phosphadicarbaboranes await x-ray structure determinations.

Phosphorus trichloride reacted with $Li_2[7,8\text{-}(MeC)_2B_9H_9]$ to form $ClP(MeC)_2B_9H_9$.[80] This phosphacarbaborane, when treated with Pr^iMgCl, gave $Pr^iP(MeC)_2B_9H_9$ and, when treated with $AlCl_3$, gave $ClP(MeC)_2B_9H_9{\cdot}AlCl_3$, respectively. This last reaction shows that the lone pair of electrons on phosphorus is not strongly involved in bonding to the cage framework and is available for adduct formation with $AlCl_3$. The x-ray structure determination of the corresponding arsenic derivative, $ClAs(MeC)_2B_9H_9$, shows that the ClAs unit is bonded mainly to the three boron atoms in the open face but not to the two carbon atoms (see Figure 8). This molecule can be viewed as a *nido*-$(MeC)_2B_9H_9$ 11-vertex structure with a μ_3-bridging ClAs (or ClP) two-electron-donor bridging unit.

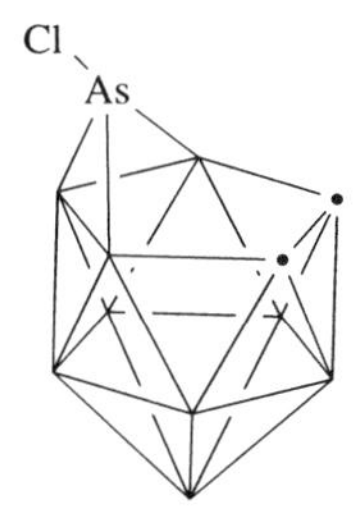

Figure 8 Schematic structural representation of $[AsCl\{(MeC)_2B_9H_9\}]$, where • = CMe.

7.5.4 Transition Metal Complexes of Phosphaboranes

Metallaphosphaboranes such as $[2\text{-}(\eta\text{-}Cp)\text{-}1\text{-}R\text{-}2,1\text{-}CoPB_{10}H_{10}]$ (R = Me, Et, Pr, Ph) and σ-bonded complexes have been known for some time.[70,81a] More recently, it was found that $[7\text{-}PhPB_{10}H_{11}]^-$ reacted with *cis*-$[PtCl_2(PMe_2Ph)_2]$ to form the *closo* species $[1\text{-}Ph\text{-}2(PMe_2Ph)_2\text{-}2,1\text{-}PtPB_{10}H_{10}]$.[81b] The x-ray structure determination shows that the platinum and phosphorus atoms occupy adjacent positions in the distorted icosahedron. In contrast, treatment of $(NEt_4)[PhPB_{10}H_{11}]$ with $[RhCl_2(\eta\text{-}Cp^*)]_2$ in refluxing CH_2Cl_2 gave the tentatively-identified compound [1-Ph-2,3-(η-Cp*)-*closo*-2,3,1-$Rh_2PB_9H_9$], along with a number of other products.[81b]

7.6 DERIVATIVES CONTAINING ARSENIC

7.6.1 Arsaboranes

The first reported synthesis of arsaboranes was by Little and co-workers.[82] This report indicated that reaction of $B_{10}H_{14}$ with $AsCl_3$ in diethyl ether with excess sodium hydride or sodium borohydride as base and reducing agent formed $[7\text{-}B_{10}H_{12}As]^-$ in 34% yield. Unfortunately, this product had to be purified by column chromatography. A second, synthetic route seemed to be preferred because the product was obtained in pure form without chromatography but in only 15% yield.[82] This alternative procedure employed THF as solvent, triethylamine as the base, and zinc powder as the reducing agent. While there is more than one possible multistep pathway by which this chemical transformation could occur, the overall balanced reduction is that shown in Equation (4).

$$B_{10}H_{14} + AsI_3 + 2\,Et_3N + 2\,e^- \longrightarrow Et_3NH[B_{10}H_{12}As] + Et_3NHI + 2\,I^- \quad (4)$$

More recently, it was found that zinc powder is not an effective reducing agent in this reaction.[83] The best yield of $(NMe_4)[B_{10}H_{12}As]$ (50% based on starting $B_{10}H_{14}$) was obtained using AsI_3 as the arsenic source and $NaBH_4$ as the reducing agent.[83]

The chemistry of the $[7\text{-}AsB_{10}H_{12}]^-$ ion has not been very extensive. Treatment of this arsaborane anion with triethylamine borane in diglyme at 160 °C formed the icosahedral anion $[B_{11}H_{11}As]^-$ in 52% yield.[82] Heating solid $Cs[B_{10}H_{12}As]$ at 375 °C also formed $[B_{11}H_{11}As]^-$ but in lower yield.[82] In 1987, it was reported that $(HNEt_3)[B_{11}H_{14}]$ reacted with As_2O_3 in aqueous KOH to form $[B_{11}H_{11}As]^-$ in 48% yield.[83] The $[B_{11}H_{14}]^-$ ion is readily available in good yield by the reaction of sodium borohydride with $BF_3{\bullet}O(C_2H_5)_2$ in diglyme at 105 °C.[84]

Methyl iodide in THF reacted with $(Me_4N)[7\text{-}B_{10}H_{12}As]$ to form sublimable $7\text{-}B_{10}H_{12}AsMe$ in 76% yield. This compound can be deprotonated in aqueous ammonia to form $[7\text{-}B_{10}H_{11}AsMe]^-$.[82] More vigorous treatment of $7\text{-}B_{10}H_{12}AsMe$ with sodium in liquid ammonia led to demethylation and formation of $[7\text{-}B_{10}H_{12}As]^-$.

The organoarsenic compound $PhAsCl_2$ reacted with $B_{10}H_{14}$ in excess triethylamine at reflux to form the $[PhAsB_{10}H_{11}]^-$ ion in low yield.[82] This crude product was passed through a strongly acidic ion exchange resin to form sublimable $PhAsB_{10}H_{12}$. Treatment of the neutral product with aqueous ammonia and then tetramethylammonium chloride gave the fully characterized salt $(NMe_4)[PhAsB_{10}H_{11}]$.[82]

7.6.2 Diarsaboranes

The reaction of $B_{10}H_{14}$ with 2 equiv. of $AsCl_3$ in excess triethylamine formed $1,2\text{-}B_{10}H_{10}As_2$ in moderate yield.[83] The use of $AsCl_3$ gave a much better yield of the icosahedral diarsa compound than when $AsBr_3$ or AsI_3 were used. Treatment of $1,2\text{-}B_{10}H_{10}As_2$ with excess piperidine at room temperature for several hours removed a boron atom adjacent to both arsenic atoms to form the piperidinium salt of the $[7,8\text{-}B_9H_{10}As_2]^-$ ion. Reaction of this anion with methyl iodide in THF at room temperature formed $7,8\text{-}B_9H_{10}As_2Me$ in which the methyl group is bonded to one of the two equivalent arsenic atoms.[85]

7.6.3 Metal Complexes of Arsa- and Diarsaboranes

A growing number of transition metal derivatives of the arsa- and diarsaboranes have been reported (see Table 1). The two iron-containing compounds are η^1 complexes with an arsenic–iron single bond joining a $Fe(CO)_2(\eta\text{-}Cp)$ unit to an arsa- or diarsaborane.[81] The remaining compounds employ the arsaboranes as η^5-ligands. Some of the first complexes reported (i.e., $[M(B_{10}H_{10}As)_2]^{n-}$ ($M = Co$, $n = 3$; $M = Ni$, $n = 2$) and $[Co(7,8\text{-}As_2B_9H_9)_2]^-$) were prepared in strong aqueous base.[86] More gentle anhydrous methods using amines as base have been employed more recently to form new metal complexes.[83]

Table 1 Transition metal derivatives of the arsa- and diarsaboranes.

Compound	*Ref.*	*Compound*	*Ref.*
$(NMe_4)_3[Co(B_{10}H_{10}As)_2]$[a]	87	$[RhH(B_9H_9As_2)(PPh_3)_2]$[b]	92
$(NMe_4)_2[Ni(B_{10}H_{10}As)_2]$[a]	87	$[Rh(B_9H_9As_2)(Cp)]$[b]	92
$(NMe_4)[Co(B_{10}H_{10}As)(Cp)]$[a]	87	$[Ni(B_9H_9As_2)(PPhMe_2)_2]$[b]	89
$[Fe(AsB_{10}H_{12})(CO)_2(Cp)]$[a]	81	$[Pd(B_9H_9As_2)(PPhMe_2)_2]$[b]	88,89
$(NMe_4)[Co(B_9H_9As_2)_2]$[b]	87	$[Pt(B_9H_9As_2)(PPhMe_2)_2]$[b]	88
$[Co(B_9H_9As_2)(Cp)]$[b]	87	$[Pd(B_9H_9As_2)(PPh_3)_2]$[b]	88
$[Ni(B_9H_9As_2)(Ph_2PCH_2CH_2PPh_2)]$[b]	83	$[\{(Cp)Co\}_2(B_8H_8As_2)]$	83
$[Fe(As_2B_9H_{10})(CO)_2(C_2H_5)]$[b]	81		

[a] Derived from the $[7\text{-}B_{10}H_{12}As]^-$ ion. [b] Derived from the $[7,8\text{-}B_9H_{10}As_2]^-$ ion.

One interesting example is the synthesis of $[\{Co(Cp)\}_2B_8H_8As_2]$ (**18**) by the prolonged reaction of $[Co(Cp)(7,8\text{-}As_2B_9H_9)]$ (**19**) with piperidine in the presence of $CoCl_2$ and C_5H_6.[83] The reaction is believed to involve removal of the boron atom adjacent to both arsenic atoms of (**19**) and subsequent insertion of a CoCp group to form (**18**). The proposed schematic cluster structure of (**18**) is given in Figure 9.

A series of $[M(Me_2PhP)_2(7,8\text{-}As_2B_9H_9)]$ (M = Ni, Pd, Pt) complexes has been prepared and fully characterized.[87–8] In the nickel and palladium reactions, a second product with phosphine substitution at a boron atom of the diarsaborane ligand was observed, as illustrated in Figure 10.[87–8] Formally this is an interchange of phosphine and hydrido ligands between metal and diarsaborane ligand, respectively, followed by a chlorine (from solvent) for hydrogen interchange. This type of reaction had been observed previously in metallacarbaboranes[89–90] and other cage systems.

Figure 9 Schematic cluster structure of $[\{Co(Cp)\}_2B_8H_8As_2]$ (**18**), where M = Co(η-Cp).

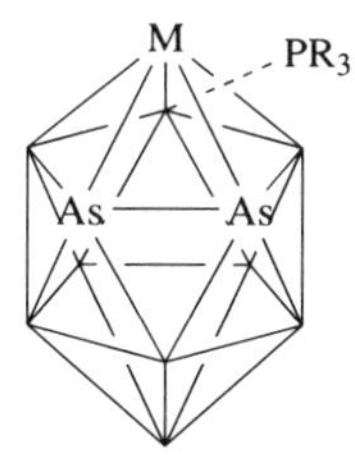

Figure 10 Schematic cluster structure of phosphine-substituted metalladiarsaborane complex, where M = $PdCl(PR_3)$.

7.7 STIBA- AND BISMABORANES AND THEIR DERIVATIVES

Treatment of $B_{10}H_{14}$ with excess $SbCl_3$ and triethylamine in THF solution formed *closo*-1,2-$Sb_2B_{10}H_{10}$ (**20**) in low yield.[85] Compound (**20**) could be purified by sublimation at 100 °C and only showed signs of thermal decomposition at its melting point, 296 °C. Reaction of (**20**) with piperidine at 50 °C formed a salt believed to be $(Me_4N)[7,8\text{-}Sb_2B_9H_{10}]$ (**21**) via ^{11}B NMR spectroscopy. The salt readily decomposed and was not fully characterized. The 11-vertex *nido* anion (**21**), when treated with $CoCl_2$ and cracked cyclopentadiene in excess piperidine, gave *closo*-[1-Cp-1,2,3-$CoSb_2B_9H_9$].[85] More recently, (**21**) was treated with Bu^{n}Li and then $[PdCl_2(Me_2PhP)_2]$ to form *closo*-[1,1-$(Me_2PhP)_2$-1,2,3-$PdSb_2B_9H_9$].[88]

The icosahedral anion, $[B_{11}H_{11}Sb]^-$ was formed when $[B_{11}H_{14}]^-$ was reacted with $SbCl_3$ and triethylamine in THF.[85]

Despite the expectations of thermal instability, the syntheses and characterization of several icosahedral bismaboranes has been accomplished. Reaction of $BiCl_3$, excess triethylamine and $B_{10}H_{14}$ at room temperature gave 1,2-$Bi_2B_{10}H_{10}$ (**22**) in low yield.[91] When neat piperidine reacted with (**22**) at room temperature for several hours, the major product was $[B_{10}H_{10}]^{2-}$. No ^{11}B NMR spectral evidence was found for the presence of $[7,8\text{-}B_9H_{10}B_2]^-$ or $[B_{10}H_{12}Bi]^-$. The mixed heteroboranes 1,2-$EBiB_{10}H_{10}$ (E = P, As, or Sb) have also been prepared.[91] Deprotonation of $[B_{11}H_{14}]^-$ with Bu^{n}Li and then reaction with $BiCl_3$ formed $[B_{11}H_{11}Bi]^-$ in 25% yield.[91]

7.8 OXABORANES AND OXAMETALLABORANES

The first compound reported that incorporated oxygen into the cage framework was the *nido* 10-vertex metal complex, $[(\eta\text{-}1,3,5\text{-}Me_3C_6H_3)FeOB_8H_{10}]$.[92] An x-ray structure determination showed that the complex had a decaborane(14) framework with the iron atom in the five-coordinate 2-position and the oxygen atom in the 6-position on the open face. The next example reported was the 12-vertex *nido* species, $[(\eta\text{-}Cp^*)Rh(Cl)OB_{10}H_9(PMe_2Ph)]$.[93–4] A closely related third oxametallaborane $[(\eta\text{-}Cp^*)Rh(NEt_3)OB_{10}H_{10}]$ was reported in 1990.[95] The schematic structures of these oxametallaboranes are given in Figure 11.

During the reaction of $[B_{11}H_{14}]^-$ with M_2O_3 (M = As, Sb, Bi) in aqueous sodium hydroxide which gave $[B_{11}H_{11}M]^-$ as the main product, a side product now formulated as *nido*-$[B_{11}H_{12}O]^-$ was obtained.[96–7] While ^{11}B and ^1H NMR spectroscopic data are consistent with the proposed formulation of this first oxaborane, an x-ray study will be needed to confirm the structure.

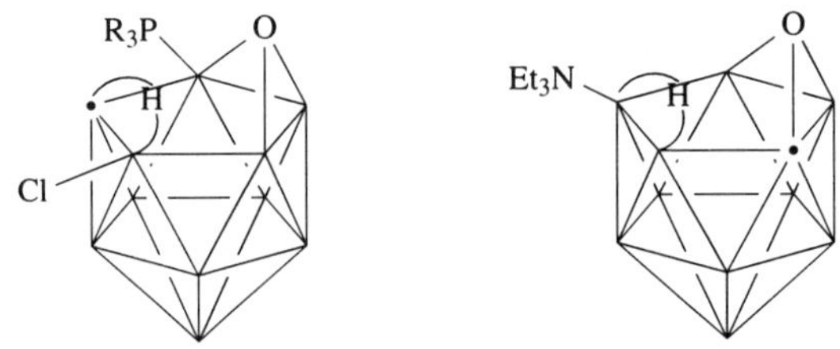

Figure 11 Schematic structures of oxarhodaboranes, where • = Rh(η-Cp).

7.9 DERIVATIVES CONTAINING SULFUR

7.9.1 Thiaboranes and Their Metal Derivatives

Reaction of thermally generated iron atoms with toluene and then *nido*-7-$SB_{10}H_{12}$ at low temperature formed *closo*-[2-(η-C_6H_5Me)-2,1-$FeSB_{10}H_{10}$].[98] A single-crystal x-ray study confirmed the distorted icosahedral structure for this complex.[98] Similarly, reaction of iron vapor with toluene and then *nido*-6-SB_9H_{11} produced *nido*-8-(η-PhMe)-8,7-$FeSB_9H_{11}$.[98] An x-ray structure study confirmed that this complex has an 11-vertex *nido* geometry with the iron and sulfur atoms in adjacent positions in the open face of the cage structure.[98]

The ready availability of *nido*-6-SB_9H_{11} and *arachno*-[6-SB_9H_{12}]$^-$ has set the stage for extensive investigations of many ruthenium-, rhodium-, and iridium-containing complexes. Treatment of Cs[6-SB_9H_{12}] with [RhCl(PPh_3)$_3$] in ethanol formed [(Ph_3P)$_2$RhSB_9H_{10}] (**23**) in good yield and an x-ray study yielded the schematic cluster structure shown in Figure 12.[99]

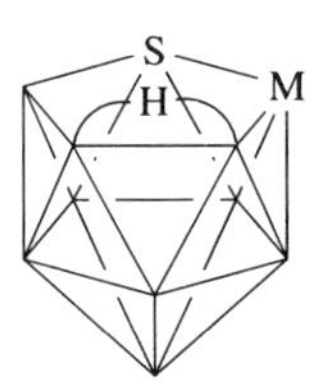

Figure 12 Schematic cluster structure of [(Ph_3P)$_2$RhSB_9H_{10}] (**23**), where M = Rh(PPh_3)$_2$.

Compound (**23**) appears to have a *nido* structure in the solid state and as such, would be formally a 16-electron complex at rhodium. The compound is fluxional in solution as indicated by variable-temperature ^{11}B NMR spectroscopy. Compound (**23**) readily reacts with more [RhCl(PPh_3)$_3$] to give the unusual 12-vertex *closo* complex, [2,3-(PPh_3)$_2$-3-Cl-μ-2:3-(Cl)-2-$Ph_2PC_6H_4$-2,3,1-$Rh_2SB_9H_8$] (**24**). This complex has a unique bridging chlorine atom (three-electron donor) between the two adjacent rhodium atoms. There is also a carbon–boron bond between a phenyl group of one triphenylphosphine ligand and the unique boron atom adjacent to both rhodium atoms.

Reaction of Cs[6-SB_9H_{12}] with [(η-Cp*)$RhCl_2$]$_2$ formed 1-*S*-2,3-[(η-Cp*)Rh]$_2$-7-ClB_9H_8 (**25**) and a number of other unidentified colored products.[100]

An x-ray structural determination of (**25**) illustrated by the schematic structure of Figure 13 shows this complex to be a *closo* 12-vertex cluster with the chlorine atom bonded to the boron atom adjacent to both rhodium atoms. A similar type of hydrogen atom substitution on a boron atom (B–C bond) was seen in (**24**).

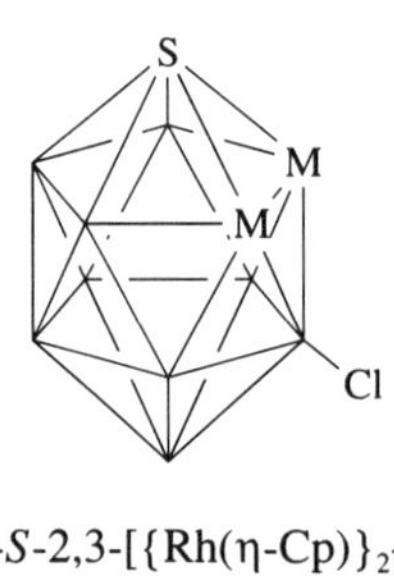

Figure 13 Schematic structure of 1-*S*-2,3-[{Rh(η-Cp)}$_2$-7-ClB_9H_8] (**25**), where M = Rh(η-Cp).

Reaction of $[IrCl_2(\eta\text{-}Cp^*)]_2$ with $Cs[6\text{-}SB_9H_{12}]$ gave a different isolated set of products than was observed for the corresponding rhodium reagent, namely *nido*-$[(\eta\text{-}Cp^*)IrSB_9H_{10}Cl]$ (two isomers; see Figure 14) and *closo*-$[(\eta\text{-}Cp^*)IrSB_9H_9]$.[56]

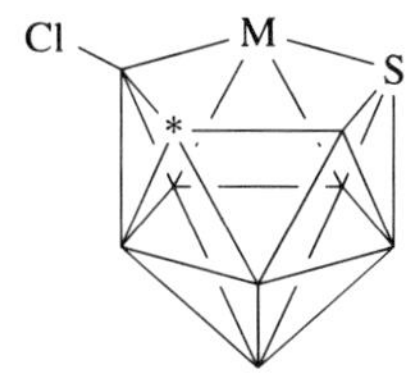

Figure 14 Schematic structure of $[(\eta\text{-}Cp^*)IrSB_9H_{10}Cl]$, where M = Ir($\eta$-Cp*), and * = position of chlorine substitution in the second isomer.

Similar chemistry was observed during the reaction of $[RuCl_2(\eta\text{-}p\text{-cymene})]_2$ with $Cs[6\text{-}SB_9H_{12}]$ and *N,N,N'*, *N'*-tetramethylnaphthalene-1,8-diamine (proton sponge) forming *nido*-$[(\eta\text{-}p\text{-cymene})RuSB_9H_{10}Cl]$ with the configuration illustrated in Figure 14 as well as *closo*-$[\{\eta\text{-}p\text{-cymene})Ru\}_2S\text{-}B_9H_9]$ and a monochloro-substituted derivative with the configuration of Figure 13.[101]

7.9.2 Dithiaboranes and Their Metal Derivatives

The synthesis of the dithiaborane, *arachno*-6,8-$S_2B_7H_9$ (**26**), in good yield from *arachno*-$[6\text{-}SB_9H_{12}]^-$ and $KHSO_3$ in aqueous acid was the starting point for much more dithiaborane chemistry.[102]

Compound (**26**) was deprotonated with NaH/THF to give *arachno*-$[S_2B_7H_8]^-$, which was reacted with LiCp* and $CoCl_2$ to form the 11-vertex cluster *nido*-$[8,10\text{-}(\eta\text{-}Cp^*)_2Co_2\text{-}7,9\text{-}S_2B_7H_7]$ (**27**) and the open triple-decker complex, *nido*-$[4,6\text{-}(Cp^*Co)_2\text{-}3,5\text{-}S_2B_2H_2]$ (**28**).[103] Small amounts of several other dithiacobaltaborane clusters were isolated as well. The proposed structure of (**28**) is illustrated in Figure 15. Compound (**27**) was found to isomerize at 250 °C and 300 °C to two other complexes in which one and then both of the CoCp* units are proposed to have moved out of the open face of the 11-vertex *nido* cage.

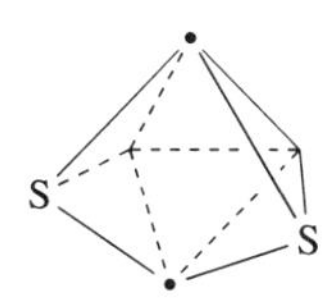

Figure 15 Proposed structure of $[4,6\text{-}(CoCp^*)_2\text{-}3,5\text{-}S_2B_2H_2]$ (**28**), where • = CoCp*.

Reaction of *arachno*-$[S_2B_7H_8]^-$ with refluxing acetonitrile, followed by protonation with aqueous HCl, gave *hypho*-5-Me-5,11,7,14-$CNS_2B_7H_9$.[104] A single-crystal x-ray structural study revealed that the compound has the *hypho* structure shown in Figure 16. The C-5–N-11 bond distance (0.1298(5) nm) is within the normal C=N double-bond range. Thus, the molecule can be viewed as a nine-vertex *hypho*-thiaborane with a bridging imine group.

Reaction of *arachno*-$[6,8\text{-}S_2B_7H_8]^-$ with acetone lead to removal of one boron atom and formation of *hypho*-$[S_2B_6H_9]^-$ (**29**) in good yield.[105] Treatment of (**29**) with excess methyl iodide formed $(MeS)_2B_6H_8$ (**30**). A single-crystal x-ray determination of (**30**) revealed an eight-vertex *hypho* structure that could be derived from a *nido*-$S_2B_9H_9$ structure by removal of three boron atoms as illustrated in Figure 17. Reaction of (**29**) with $BH_3\cdot THF$ formed *hypho*-$[S_2B_7H_{10}]^-$ (**31**), whose proposed structure is illustrated in Figure 17(d). The anion, *arachno*-$[S_2B_7H_8]^-$, was also reduced by $BH_3\cdot THF$ to (**31**).

7.9.3 Thiacarbaboranes

Reaction of *arachno*-4-CB_8H_{14} with excess S_8 and triethylamine formed *arachno*-CSB_8H_{12}.[106] This is a $[B_{10}H_{14}]^{2-}$ -type structure in which the 6 and 9 BH_2^- cage units are replaced with isoelectronic CH_2 and S units.

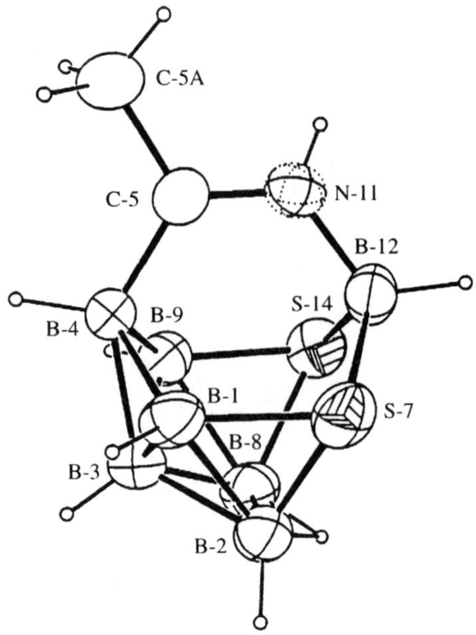

Figure 16 ORTEP drawing of the molecular structure of *hypho*-5-Me-5,11,7,14-$CNS_2B_7H_9$.

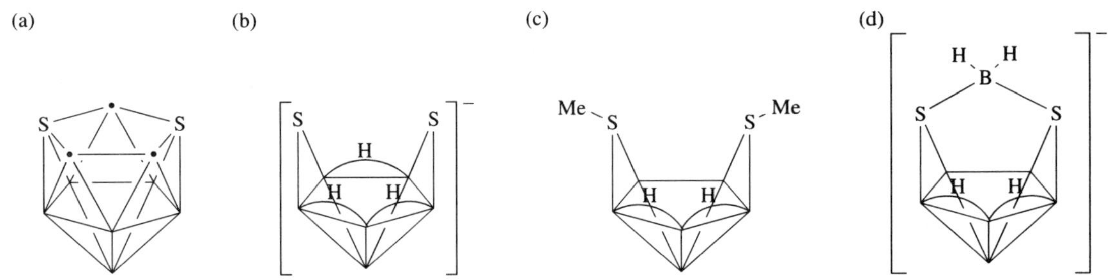

Figure 17 Schematic structures of (a) *nido*-$S_2B_9H_9$ (where • = boron atom to be removed); (b) *hypho*-$[S_2B_6H_9]^-$ (**29**); (c) *hypho*-$(MeS)_2B_6H_8$ (**30**); and (d) *hypho*-$[S_2B_7H_{10}]^-$ (**31**).

7.10 SELENA- AND TELLURABORANES AND THEIR METAL DERIVATIVES

The diselenaborane, $B_9H_9Se_2$ (**32**) was obtained as an unexpected minor product during the reaction of $B_{10}H_{14}$ with polyselenide ion in aqueous base, which generated $[B_{10}H_{11}Se]^-$ as the major product.[107] Subsequent study proved that polyselenide ion reacted with $[B_9H_{14}]^-$ (which could be generated from $B_{10}H_{14}$) to form $B_9H_9Se_2$ in good yield.[108–9] Compound (**32**) was determined to be *nido*-7,9-$Se_2B_9H_9$ by ^{11}B NMR spectroscopy at 192.48 MHz.[110] Compound (**32**) reacted with strong base to give an anionic intermediate that reacted further with cyclopentadiene monomer, $CoCl_2$ and triethylamine to give complexes such as [(η-Cp)Co($B_9H_9Se_2$)] and [{(η-Cp)Co}$_2$(B_9H_9Se)].[111]

In related chemistry, $B_9H_{13}{\cdot}SMe_2$ was found to react with polyselenide ion to form both Na[$B_8H_9Se_2$] and Na[$B_9H_{12}Se$].[108] Acidification of the products gave *arachno*-6,9-$B_8H_{10}Se_2$ (**33**) in low yield.[109] The complex, *nido*-[(η-Cp)Co($B_8H_8Se_2$)] was formed by reaction of (**33**) with $CoCl_2$ cyclopentadiene monomer and triethylamine.[108]

The synthesis of the telluraborane, $B_{10}H_{12}Te$ and its iron and cobalt complexes has been reviewed previously.[112] Reaction of [$PdCl_2(PR_3)_2$] (R_3 = Ph_3, $MePh_2$, or Me_2Ph) with $[7\text{-}EB_{10}H_{11}]^-$ (E = Se or Te) in THF at room temperature formed mainly *closo*[2,2-PR_3-2,1-$PdEB_{10}H_{10}$].[113]

Similarly, the *closo* compounds [2,2-$(PR_3)_2$-1,2-$SePtB_{10}H_{10}$] (R_3 = Et_3, $Bu^n{}_3$, or Me_2Ph) were obtained in moderate yield from the reaction between *cis*-[$PtCl_2(PR_3)_2$] and $[7\text{-}SeB_{10}H_{11}]^-$ in THF.[114] In one of the reactions above, a second complex, *arachno*-[9,9-$(PMe_2Ph)_2$-6,9-$SePtB_8H_{10}$] (**34**) was isolated. X-ray crystallographic characterization of this complex is illustrated in the schematic representation of Figure 18. Thiaplatinaborane molecules of this type have been characterized in the past.

In refluxing toluene [$PdX^1{}_2(PPh_3)_2$] (X^1 = Cl or I) and $[7\text{-}TeB_{10}H_{11}]^-$ afforded mainly *closo*-[2-X^1-2-PPh_3-2,1-$PdTeB_{10}H_9(PPh_3)$] (**35**). The treatment of (**35**) (X^1 = I) with mercury(II) salts in THF formed

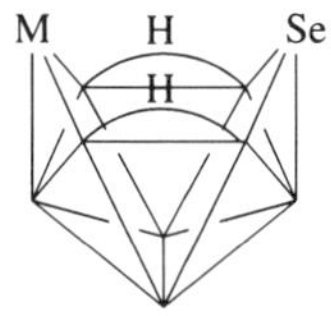

Figure 18 Schematic structure of compound (**34**), where M = $Pt(PMe_2Ph)_2$.

closo-[2-X^2-2-PPh_3-2,1-$PdTeB_{10}H_9(PPh_3)$] (**36**) (X^2 = Cl, Br, CN, SCN, or O_2CMe). A single-crystal x-ray structure of (**36**) (X^2 = O_2CMe) is illustrated in Figure 19.[115]

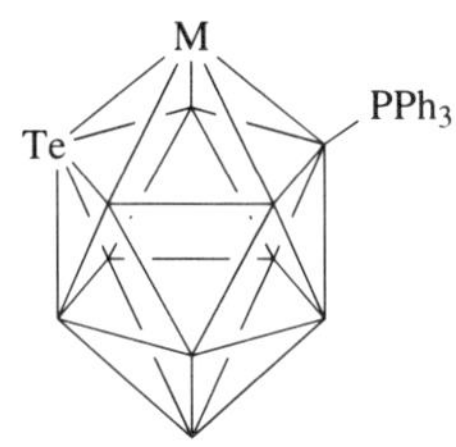

Figure 19 Schematic configuration of compound (**36**), where M = $Pd(PPh_3)O_2CMe$.

Treatment of (**35**) (X = I) with $AgBF_4$ in toluene and then isolation of the product in moist air gave the metal-centered cationic complex *closo*-[2(H_2O)-2-PPh_3-2,1-$PdTeB_{10}H_9(PPh_3)$][BF_4].[116] This compound was characterized by x-ray crystallography and has a configuration similar to that of (**36**) (Figure 19, where M = $Pd(H_2O)PPh_3$). The water molecule of this complex can be displaced by a variety of neutral Lewis bases to form *closo*-[2-L-2(PPh_3)-2,1-$PdTeB_{10}H_9(PPh_3)_2$][BF_4] (L = CO, CNR, NCMe, RNH_2, THF, PMe_2Ph).

Reaction of [$PtCl_2(PR_3)_2$] (R_3 = Et_3, $Bu^n{}_3$, Me_2Ph) with [7-$TeB_{10}H_{11}$]$^-$ in refluxing THF gave *closo*-[2,2(PR_3)$_2$-1,2-$TePtB_{10}H_{10}$].[117] In a different type of reaction *closo*-$SeB_{11}H_{11}$ with $Pt(PPh_3)_4$ in refluxing ethanol gave [$PtH(PPh_3)_3$][$SeB_{10}H_{11}$] (minor product) and *closo*-[2,2-(PPh_3)$_2$-1,2-$SePtB_{10}H_{10}$] (**37**).[115] The x-ray structure determination of (**37**) showed that the platinum and selenium atoms were in adjacent positions in the cage structure.

7.11 REFERENCES

1. L. J. Todd, in 'Metal Interactions with Boron Clusters', ed. R. N. Grimes, Plenum, New York, 1982, chap. 4.
2. J. F. Liebman, A. Greenberg and R. E. Williams (eds.) 'Advances in Boron and the Boranes', VCH, New York, 1988.
3. S. Hermanek, *Chem. Rev.*, 1992, **92**, 325.
4. A. R. Siedle, *Annu. Rep. NMR Spectrosc.*, 1988, **20**, 205.
5. T. D. Getman and S. G. Shore, *Inorg. Chem.*, 1988, **27**, 3439.
6. C. R. Pulham, A. J. Downs, D. W. H. Rankin and H. E. Robertson, *J. Chem. Soc., Dalton Trans.*, 1992, 1509.
7. B. M. Mikhailov and T. V. Potapova, *Izv. Akad. Nauk. SSSR*, 1968, **5**, 1153.
8. P. Jutzi and P. Galow, *J. Organomet. Chem.*, 1987, **319**, 139.
9. N. S. Hosmane, K.-J. Lu, H. Zhang, L. Jia, A. H. Cowley and M. A. Mardones, *Organometallics*, 1991, **10**, 963.
10. N. S. Hosmane, H. Zhang, K.-J. Lu, J. A. Maguire, A. H. Cowley and M. A. Mardones, *Struct. Chem.*, 1992, **3**, 183.
11. N. S. Hosmane, K.-J. Lu, H. Zhang, A. H. Cowley and M. A. Mardones, *Organometallics*, 1991, **10**, 392.
12. M. J. Manning, C. B. Knobler, M. F. Hawthorne and Y. Do, *Inorg. Chem.*, 1991, **30**, 3589.
13. R. N. Grimes, in 'Metal Interactions with Boron Clusters', ed. R. N. Grimes, Plenum, New York, 1982, chap 7.
14. P. Jutzi, D. Wegener and M. B. Hursthouse, *Chem. Ber.*, 1991, **124**, 295.
15. J. S. Beck and L. G. Sneddon, *J. Am. Chem. Soc.*, 1988, **110**, 3467.
16. D. M. Schubert, M. A. Bandman, W. S. Rees, Jr., C. B. Knobler, P. Lu, W. Nam and M. F. Hawthorne, *Organometallics*, 1990, **9**, 2046.
17. W. S. Rees, Jr., D. M. Schubert, C. B. Knobler and M. F. Hawthorne, *J. Am. Chem. Soc.*, 1986, **108**, 5367.
18. D. M. Schubert, C. B. Knobler, W. S. Rees, Jr. and M. F. Hawthorne, *Organometallics*, 1987, **6**, 201; 203.
19. D. M. Schubert, C. B. Knobler and M. F. Hawthorne, *Organometallics*, 1987, **6**, 1353.
20. D. Seyferth, K. Büchner, W. S. Rees, Jr. and W. M. Davis, *Angew. Chem., Int. Ed. Engl.*, 1990, 918.
21. D. Seyferth *et al.*, *J. Am. Chem. Soc.*, 1993, **115**, 3586.
22. W. S. Rees, Jr., D. M. Schubert, C. B. Knobler and M. F. Hawthorne, *J. Am. Chem. Soc.*, 1986, **108**, 5369.
23. D. M. Schubert, W. S. Rees, Jr., C. B. Knobler and M. F. Hawthorne, *Organometallics*, 1990, **9**, 2938.
24. U. Siriwardane, M. S. Islam, T. A. West, N. S. Hosmane, J. A. Maguire and A. H. Cowley, *J. Am. Chem. Soc.*, 1987, **109**, 4600.
25. T. J. Dupont, R. E. Loffredo, R. C. Haltiwanger, C. A. Turner and A. D. Norman, *Inorg. Chem.*, 1978, **17**, 2062.
26. N. N. Greenwood and B. Youll, *J. Chem. Soc., Dalton Trans.*, 1975, 158.
27. R. W. Chapman, J. G. Kester, K. Folting, W. E. Streib and L. J. Todd, *Inorg. Chem.*, 1992, **31**, 979.

28. N. S. Hosmane and J. A. Maguire, in 'Advances in Boron and the Boranes', eds. J. F. Liebman, A. Greenberg and R. E. Williams, VCH, New York, 1988, pp. 303–28.
29. N. S. Hosmane, P. de Meester, N. N. Maldar, S. B. Potts, S. S. C. Chu and R. H. Herber, *Organometallics*, 1986, **5**, 772.
30. A. H. Cowley, P. Galow, N. S. Hosmane, P. Jutzi and N. C. Norman, *J. Chem. Soc., Chem. Commun.*, 1984, 1564.
31. N. S. Hosmane, P. de Meester, U. Siriwardane, M. S. Islam and S. S. C. Chu, *J. Am. Chem. Soc.*, 1986, **108**, 6050.
32. L. Jia, H. Zhang and N. S. Hosmane, *Organometallics*, 1992, **11**, 2957.
33. P. Jutzi and P. Galow, *J. Organomet. Chem.*, 1987, **319**, 139.
34. P. Jutzi, D. Wegener, H.-G. Stammler, A. Karaulov and M. B. Hursthouse, *Inorg. Chim. Acta*, 1992, **198–200**, 369.
35. L. J. Todd, A. Arafat, J. Baer and J. C. Huffman, in 'Advances in Boron and the Boranes', eds., J. F. Liebman, A. Greenberg and R. E. Williams, VCH, New York, 1988, p. 287.
36. K. Base, *Collect. Czech. Chem. Commun.*, 1983, **48**, 2593.
37. A. Arafat, J. Baer, J. C. Huffman and L. J. Todd, *Inorg. Chem.*, 1986, **25**, 3757.
38. J. Müller, P. Paetzold, U. Englert and J. Runsink, *Chem. Ber.*, 1992, **125**, 97.
39. J. Müller, P. Paetzold and R. Boese, *Heteroatom. Chem.*, 1990, **1**, 461.
40. J. Müller, J. Runsink and P. Paetzold, *Angew. Chem., Int. Ed. Engl.*, 1991, **30**, 175.
41. D. Hnyk *et al.*, *Inorg. Chem.*, 1993, **32**, 2442.
42. F. Meyer, J. Müller, P. Paetzold and R. Boese, *Angew. Chem., Int. Ed. Engl.*, 1992, **31**, 1227.
43. F. Meyer, J. Müller, M. U. Schmidt and P. Paetzold, *Inorg. Chem.*, 1993, **32**, 5053.
44. F. Meyer, J. Müller, P. Paetzold and R. Boese, *Angew. Chem., Int. Ed. Engl.*, 1992, **31**, 1227.
45. R. Boese, B. Krockert and P. Paetzold, *Chem. Ber.*, 1987, **120**, 1913.
46. P. Paetzold and B. Redenz-Stormanns, *Inorg. Synth.*, 1992, **29**, 54.
47. P. Paetzold, B. Redenz-Stormanns, R. Boese, M. Bühl and P. v. R. Schleyer, *Angew. Chem., Int. Ed. Engl.*, 1990, **29**, 1059.
48. B. Stibr, J. D. Kennedy and T. Jelinek, *J. Chem. Soc., Chem. Commun.*, 1990, 1309.
49. G. D. Friesen, R. L. Kump and L. J. Todd, *Inorg. Synth.*, 1992, **29**, 105.
50. G. D. Friesen, A. Barriola and L. J. Todd, *Chem. Ind. (London)*, 1978, 631.
51. B. Stibr, J. Plesek and S. Hermanek, *Collect. Czech. Chem. Commun.*, 1974, **39**, 1805.
52. Z. Janousek, J. Fusek and B. Stibr, *J. Chem. Soc., Dalton Trans.*, 1992, 2649.
53. J. Holub, T. Jelinek, J. Plesek, B. Stibr, S. Hermanek and J. D. Kennedy, *J. Chem. Soc., Chem. Commun.*, 1991, 1389.
54. H. P. Hansen, J. Müller, U. Englert and P. Paetzold, *Angew. Chem., Int. Ed. Engl.*, 1991, **30**, 1377.
55. K. Nestor, X. L. R. Fontaine, J. D. Kennedy, B. Stibr, K. Base and M. Thornton-Pett, *Collect. Czech. Chem. Commun.*, 1991, **56**, 1607.
56. K. Nestor, X. L. R. Fontaine, N. N. Greenwood, J. D. Kennedy and M. Thornton-Pett, *J. Chem. Soc., Dalton Trans.*, 1991, 2657.
57. K. Base *et al.*, *J. Chem. Soc., Chem. Commun.*, 1988, 1240.
58. J. G. Kester, J. C. Huffman and L. J. Todd, *Inorg. Chem.*, 1988, **27**, 4528.
59. E. L. Muetterties and V. D. Aftandilian, *Inorg. Chem.*, 1962, **1**, 731.
60. M. A. Beckett and J. D. Kennedy, *J. Chem. Soc., Chem. Commun.*, 1983, 575.
61. M. Thornton-Pett, M. A. Beckett and J. D. Kennedy, *J. Chem. Soc., Dalton Trans.*, 1986, 303.
62. J. A. Glass, Jr., T. A. Whelan and J. T. Spencer, *Organometallics*, 1991, **10**, 1148.
63. A. B. Burg and H. Heinen, *Inorg. Chem.*, 1968, **7**, 1021.
64. I. B. Mishra and A. B. Burg, *Inorg. Chem.*, 1972, **11**, 664.
65. D. E. Coons and D. F. Gaines, *Inorg. Chem.*, 1987, **26**, 1985.
66. R. W. Miller, K. J. Donaghy and J. T. Spencer, *Organometallics*, 1991, **10**, 1161.
67. R. W. Miller, K. J. Donaghy and J. T. Spencer, *Phosphorus Sulfur*, 1991, **57**, 287.
68. B. H. Goodreau, R. L. Ostrander and J. T. Spencer, *Inorg. Chem.*, 1991, **30**, 2066.
69. J. L. Little, J. T. Moran and L. J. Todd, *J. Am. Chem. Soc.*, 1967, **89**, 5495.
70. J. L. Little, *Inorg. Chem.*, 1976, **15**, 114.
71. J. L. Little, J. G. Kester, J. C. Huffman and L. J. Todd, *Inorg. Chem.*, 1989, **28**, 1087.
72. J. L. Little, M. A. Whitesell, R. W. Chapman, J. G. Kester, J. C. Huffman and L. J. Todd, *Inorg. Chem.*, 1993, **32**, 3369.
73. T. D. Getman, H.-B. Deng, L. Y. Hsu and S. G. Shore, *Inorg. Chem.*, 1989, **28**, 3612.
74. G. L. Wood, E. N. Duesler, C. K. Narula, R. T. Paine and H. Nöth, *J. Chem. Soc., Chem. Commun.*, 1987, 496.
75. D. Dou, G. L. Wood, E. Duesler, R. T. Paine and H. Nöth, *Inorg. Chem.*, 1992, **31**, 3756.
76. W. Haubold, W. Keller and G. Sawitzki, *Angew. Chem., Int. Ed. Engl.*, 1988, **27**, 925.
77. J. L. Little, M. A. Whitesell, J. G. Kester, K. Folting and L. J. Todd, *Inorg. Chem.*, 1990, **29**, 804.
78. N. S. Hosmane, K. -J. Lu, A. H. Cowley and M. A. Mardones, *Inorg. Chem.*, 1991, **30**, 1325.
79. B. A. Barnum, W. Keller, J. W. Bausch and L. G. Sneddon, in 'Abstracts Boron USA Workshop, 1992', Washington State University, Washington, DC, 1992 No. 61.
80. P. Jutzi, D. Wegener and M. Hursthouse, *J. Organomet. Chem.*, 1991, **418** 277.
81. (a) T. Yamamoto and L. J. Todd, *J. Organomet. Chem.*, 1974, **67**, 75; (b) S. R. Bunkhall, X. L. R. Fontaine, N. N. Greenwood, J. D. Kennedy and M. Thornton-Pett, *J. Chem. Soc., Dalton Trans.*, 1990, 73.
82. J. L. Little, S. S. Pao and K. K. Sugathan, *Inorg. Chem.*, 1974, **13**, 1752.
83. T. P. Hanusa, N. R. de Parisi, J. G. Kester, A. Arafat and L. J. Todd, *Inorg. Chem.*, 1987, **26**, 4100.
84. G. B. Dunks, K. Barker, E. Hedaya, C. Hefner, K. Palmer-Ordonez and P. Remec, *Inorg. Chem.*, 1981, **20**, 1692.
85. J. L. Little, *Inorg. Chem.*, 1979, **18**, 1598.
86. J. L. Little and S. S. Pao, *Inorg. Chem.*, 1978, **17**, 584.
87. M. McGrath, T. R. Spalding, X. L. R. Fontaine, J. D. Kennedy and M. Thornton-Pett, *J. Chem. Soc., Dalton Trans.*, 1991, 3223.
88. S. A. Jasper, Jr., S. Roach, J. N. Stipp, J. C. Huffman and L. J. Todd, *Inorg. Chem.*, 1993, **32**, 3072.
89. R. E. King, III, S. B. Miller, C. B. Knobler and M. F. Hawthorne, *Inorg. Chem.*, 1983, **22**, 3548.
90. X. L. R. Fontaine, J. D. Kennedy, M. McGrath and T. R. Spalding, *Mag. Res. Chem.*, 1991, **29**, 711.
91. J. L. Little, M. A. Whitesell, J. G. Kester, K. Folting and L. J. Todd, *Inorg. Chem.*, 1990, **29**, 804.
92. R. P. Micciche, J. J. Briguglio and L. G. Sneddon, *Inorg. Chem.*, 1984, **23**, 3992.

93. X. L. R. Fontaine, H. Fowkes, N. N. Greenwood, J. D. Kennedy and M. Thornton-Pett, *J. Chem. Soc., Chem. Commun.*, 1985, 1722.
94. X. L. R. Fontaine, H. Fowkes, N. N. Greenwood, J. D. Kennedy and M. Thornton-Pett, *J. Chem. Soc., Dalton Trans.*, 1987, 2417.
95. E. J. Ditzel *et al.*, *J. Chem. Soc., Chem. Commun.*, 1990, 1692.
96. A. Ouassas, C. R'Kha, H. Mongeot and B. Frange, *Inorg. Chim. Acta*, 1991, **180**, 257.
97. A. Ouassas, B. Fenet, H. Mongeot and B. Frange, in 'Current Topics in the Chemistry of Boron', Royal Society of Chemistry, Cambridge, 1993.
98. S. O. Kang, P. J. Carroll and L. G. Sneddon, *Organometallics*, 1988, **7**, 772.
99. G. Ferguson *et al.*, *J. Chem. Soc., Chem. Commun.*, 1990, 891.
100. G. Ferguson, M. F. Hawthorne, B. Kaitner and F. J. Lalor, *Acta Crystallogr., Part C*. 1984, **40**, 1707.
101. M. Bown, X. L. R. Fontaine, N. N. Greenwood and J. D. Kennedy, *Z. Anorg. Allg. Chem.*, 1991, **602**, 17.
102. J. Plesek, S. Hermanek and Z. Janousek, *Collect. Czech. Chem. Commun.*, 1977, **42**, 785.
103. S. O. Kang and L. G. Sneddon, *Inorg. Chem.*, 1988, **27**, 3769.
104. S. O. Kang, G. T. Furst and L. G. Sneddon, *Inorg. Chem.*, 1989, **28**, 2339.
105. S. O. Kang and L. G. Sneddon, *J. Am. Chem. Soc.*, 1989, **111**, 3281.
106. J. Holub, T. Jelinek, J. Plesek, B. Stibr, S. Hermanek and J. D. Kennedy, *J. Chem. Soc., Chem. Commun.*, 1991, 1389.
107. J. L. Little, G. D. Friesen and L. J. Todd, *Inorg. Chem.*, 1977, **16**, 869.
108. G. D. Friesen, A. Barriola and L. J. Todd, *Chem. Ind. (London)*, 1978, 631.
109. (a) G. D. Friesen, T. P. Hanusa and L. J. Todd, *Inorg. Synth.*, 1992, **29**, 103; (b) G. D. Friesen, R. L. Kump and L. J. Todd, *Inorg. Synth.*, 1992, **29**, 105.
110. O. N. Dhubhghaill, D. Reed and T. R. Spalding, *Polyhedron*, 1993, **12**, 1977.
111. G. D. Friesen, A. Barriola, P. Daluga, P. Ragatz, J. C. Huffman and L. J. Todd, *Inorg. Chem.*, 1980, **19**, 458.
112. L. J. Todd, in 'Metal Interactions with Boron Clusters', ed. R. N. Grimes, Plenum, New York, 1982, chap. 4.
113. G. Ferguson, J. F. Gallagher, M. McGrath, J. P. Sheehan, T. R. Spalding and J. D. Kennedy, *J. Chem. Soc., Dalton Trans.*, 1993, 27.
114. Faridoon *et al.*, *J. Chem. Soc., Dalton Trans.*, 1989, 1657.
115. G. Ferguson, M. Parvez, J. A. MacCurtain, O. N. Dhubhghaill, T. R. Spalding and D. Reed, *J. Chem. Soc., Dalton Trans.*, 1987, 699.
116. J. P. Sheehan, T. R. Spalding, G. Ferguson, J. F. Gallagher, B. Kaitner and J. D. Kennedy, *J. Chem. Soc., Dalton Trans.*, 1993, 35.
117. G. Ferguson, J. D. Kennedy, X. L. R. Fontaine, Faridoon and T. R. Spalding, *J. Chem. Soc., Dalton Trans.*, 1988, 2555.

8
Metallaboranes

LAWRENCE BARTON and DILEEP K. SRIVASTAVA
University of Missouri–St. Louis, MO, USA

8.1 INTRODUCTION 277

8.2 UNITS CONTAINING ONE BORON ATOM 278

8.2.1 Tetrahydroborate Derivatives 278
8.2.1.1 Introduction 278
8.2.1.2 Group 3 metals 278
8.2.1.3 Group 4 metals 278
8.2.1.4 Group 5 metals 280
8.2.1.5 Group 6 and 7 metals 283
8.2.1.6 Group 8 and 9 metals 284
8.2.1.7 Group 11 metals 286
8.2.1.8 Lanthanide metals 286
8.2.1.9 Actinide metals 286
8.2.2 Trihydroborate Derivatives 287
8.2.2.1 Trihydroborate compounds of the transition metals 287
8.2.2.2 Trihydroborate compounds of the lanthanide metals 288
8.2.2.3 Trihydroborate compounds of the actinide metals 290
8.2.3 Dihydroborate Derivatives 290
8.2.4 Monohydroborate Derivatives 292
8.2.5 Boryl Derivatives of the Transition Metals 292
8.2.6 Metal-rich Metallaboranes 293
8.2.6.1 Introduction 293
8.2.6.2 Trimetal clusters 296
8.2.6.3 Tetrametal clusters 298
8.2.6.4 Clusters containing more than four metal atoms 303

8.3 UNITS CONTAINING TWO BORON ATOMS 304

8.3.1 Heptahydrodiborate Species 304
8.3.2 Hexahydrodiborate Derivatives 306
8.3.2.1 Group 5 metals 306
8.3.2.2 Group 8 metals 306
8.3.3 Pentahydrodiborate Derivatives 307
8.3.3.1 Group 6 metals 308
8.3.3.2 Group 8 metals 308
8.3.3.3 Group 9 metals 309
8.3.3.4 Group 10 metals 310
8.3.4 Diborane(4) Derivatives 311
8.3.5 Diborane(2) Derivatives 312

8.4 UNITS CONTAINING THREE BORON ATOMS 313

8.4.1 Introduction 313
8.4.2 Octahydrotriborate Derivatives 315
8.4.2.1 Group 5 metals 315
8.4.2.2 Group 6 metals 315
8.4.2.3 Group 8 metals 315
8.4.2.4 Group 9 and 10 metals 316
8.4.2.5 Lanthanides 316
8.4.3 Heptahydrotriborate Derivatives 316
8.4.3.1 Group 8 metals 316
8.4.3.2 Group 9 metals 316

8.4.3.3 Group 10 metals 316
8.4.3.4 Group 11 metals 317
8.4.4 Hexahydrotriborate Derivatives 317
8.4.4.1 Group 8 metals 317
8.4.5 Trihydrotriborate Derivatives 318

8.5 UNITS CONTAINING FOUR BORON ATOMS 319
8.5.1 Nonahydrotetraborate Derivatives 319
8.5.1.1 Group 5 metals 319
8.5.1.2 Group 8 metals 319
8.5.2 Octahydrotetraborate Derivatives 319
8.5.2.1 Group 8 metals 319
8.5.2.2 Group 9 metals 320
8.5.3 Heptahydrotetraborate Species 321
8.5.4 Hexahydrotetraborate Species 321
8.5.4.1 Group 8 derivatives 321
8.5.4.2 Group 9 derivatives 321
8.5.5 Tetrahydrotetraborate Species 322
8.5.6 Dihydrotetraborate Complexes 322

8.6 UNITS CONTAINING FIVE BORON ATOMS 323
8.6.1 Decahydropentaborate Derivatives 323
8.6.1.1 Group 7 metals 323
8.6.1.2 Group 9 metals 323
8.6.2 Nonahydropentaborate Derivatives 323
8.6.2.1 Group 8 metals 323
8.6.2.2 Group 9 metals 324
8.6.3 Octahydropentaborate Derivatives 326
8.6.3.1 Group 8 metals 326
8.6.3.2 Group 9 metals 326
8.6.3.3 Group 11 metals 326
8.6.3.4 Group 12 metals 326
8.6.4 Heptahydropentaborate Derivatives 327
8.6.4.1 Group 8 metals 327
8.6.4.2 Group 9 metals 327
8.6.5 Pentahydropentaborate Derivatives 327

8.7 UNITS CONTAINING SIX BORON ATOMS 327
8.7.1 Nonahydrohexaborate Derivatives 327
8.7.1.1 Group 8 metals 327
8.7.1.2 Group 10 metals 327
8.7.2 Hexahydrohexaborate Derivatives 328
8.7.2.1 Group 8 metals 328
8.7.2.2 Group 10 metals 328

8.8 UNITS CONTAINING SEVEN BORON ATOMS 329

8.9 UNITS CONTAINING EIGHT BORON ATOMS 331
8.9.1 Introductory Remarks 331
8.9.2 Tetradecahydrooctaborate Derivatives 331
8.9.2.1 Group 8 metals 332
8.9.2.2 Group 9 metals 332
8.9.3 Dodecahydrooctaborate Derivatives 333
8.9.3.1 Group 7 metals 333
8.9.3.2 Group 8 metals 334
8.9.3.3 Group 9 metals 335
8.9.3.4 Group 10 metals 336
8.9.3.5 Group 11 metals 337
8.9.4 Octahydrooctaborate Derivatives 337
8.9.5 Conjoined Octaborate Derivatives 337

8.10 UNITS CONTAINING NINE BORON ATOMS 337
8.10.1 Introductory Remarks 337
8.10.2 Pentadecahydrononaborate Derivatives 338
8.10.2.1 Group 8 metals 338
8.10.2.2 Group 9 metals 338
8.10.2.3 Group 11 metals 339
8.10.3 Tridecahydrononaborate Derivatives 340
8.10.4 Undecahydrononaborate Derivatives 344
8.10.4.1 Group 8 metals 344
8.10.4.2 Group 9 metals 347
8.10.4.3 Group 10 metals 349

8.11 UNITS CONTAINING TEN BORON ATOMS 349

8.11.1 Hexadecahydrodecaborate Derivatives 349
8.11.1.1 Lanthanides metals 349
8.11.2 Tetradecahydrodecaborate Derivatives 350
8.11.2.1 Group 8 metals 350
8.11.2.2 Group 9 metals 351
8.11.2.3 Group 10 metals 351
8.11.2.4 Group 11 metals 352
8.11.2.5 Group 12 metals 352
8.11.3 Dodecahydrodecaborate Derivatives 352
8.11.3.1 Phosphine-ligated systems 355
8.11.3.2 η^5*- or* η^6*-Arene-ligated systems* 357
8.11.4 Miscellaneous Metallaboranes Containing 10 Boron Atoms 358
8.11.4.1 Group 8 metals 358
8.11.4.2 Group 9 metals 361
8.11.4.3 Group 10 metals 362
8.11.4.4 Group 11 metals 362

8.12 UNITS CONTAINING 11 BORON ATOMS 363

8.13 UNITS CONTAINING 12 OR MORE BORON ATOMS 363
8.13.1 Species Containing 12 Boron Atoms 363
8.13.1.1 Group 8 metals 363
8.13.1.2 Group 10 metals 363
8.13.2 Species Containing 13 Boron Atoms 363
8.13.3 Species Containing 14 Boron Atoms 364
8.13.4 Species Containing 16 Boron Atoms 364
8.13.5 Species Containing 17 Boron Atoms 364
8.13.6 Species Containing 18 Boron Atoms 365
8.13.7 Species Containing 20 Boron Atoms 365

8.14 REFERENCES 366

8.1 INTRODUCTION

There have been several substantial contributions to metallaborane chemistry since the publication of *COMC-I* and the chapter on organometallic compounds with bonds between a transition metal and boron by Gilbert, Boocock, and Shore.[1] Several reviews have appeared on metallaboranes, the most comprehensive was that by Kennedy which appeared in two parts.[2] Three books have been published which contain much material on the chemistry of metallaboranes. One contains contributions by several major researchers in the field,[3] another was published as a result of a conference held in honor of the 80th birthday of Anton Burg and contains contributions from several major researchers in the field in the USA,[4] and the third is a monograph containing much recent information on metallaboranes, aimed primarily at undergraduates.[5] Other contributions include a survey of the analogy between metallaboranes and metal–hydrocarbon clusters,[6] reviews on some of the earlier work on the rapidly expanding field of metal-rich borane clusters and boride-like species,[7–8] and more personalized accounts of specific author's contributions to the field.[9–17] Finally, detailed information on metallaboranes may be found in the *Gmelin Handbook on the Boron Hydrogen and Noble Gas System*, three volumes of which have been written since 1982.[18–19]

Notable advances in metallaborane chemistry include an extension of the range of bonding in the tetrahydroborate moiety, for example, in $\{(\eta^5\text{-}C_5H_3Bu^t_2)_2Ce[\mu:\eta^4\text{-}(\mu_3\text{-}H)_2B(\mu_2\text{-}H)_2]\}_2$, there are two tetrahydroborate groups bridging two cerium atoms such that each BH_4 contains two μ_2 and two μ_3 hydrogen atoms.[20] A new major area of progress in monoborane derivatives involves the discovery of the metal-rich metallaboranes[7–8] and species containing encapsulated boron atoms such that they resemble metal borides.[21–4]

Another area that has attracted much attention is the discovery of systems whose structures and skeletal electron counts do not conform to the polyhedral skeletal electron pair theory (PSEPT). Typically the clusters have fewer electrons than are required by the PSEPT.[25] A controversy has arisen concerning whether the electron deficiency should be considered on the metal center or associated with delocalized cluster bonding. The two points of view term the species hyper- (*closo*, *nido*, *arachno*)[26–7] or iso- (*closo*, etc.),[11] respectively. The former view emphasizes the need for further theoretical study,[25] whereas the latter view suggests more experimental study;[28] the latter view also suggests that the notion of metal vertices contributing two electrons and only three orbitals to cluster bonding is too restrictive. Theoretical work seems to contest the idea that the metal could contribute four orbitals to cluster bonding.[29] The iso- (or hyper-) clusters are related to the "normal" species by simple diamond–square–diamond (DSD) rearrangements.[30] One explanation for the existence of *nido*-species with four-membered open faces and *closo*-species with novel geometries is the availability of low-energy rearrangements via DSD mechanisms, some of which are used to explain spectroscopic fluxional processes implied from NMR spectroscopic measurements. Some of these ideas are mentioned in the following chapters but a detailed presentation is beyond the scope of this treatment.

8.2 UNITS CONTAINING ONE BORON ATOM

8.2.1 Tetrahydroborate Derivatives

8.2.1.1 Introduction

In this section systems which do not contain any organic groups are not covered. Thus, unlike *COMC-I*,[1] all tetrahydroborate complexes are not all included, only those containing organic groups either as ligands, as part of the modification of the BH_4 moiety or in some case as gegenions, are included. Since there is such a large number of species, many of them are presented only in tables. The tables include compound listings with data on physical appearance, melting point, available spectroscopic and related data, and a brief indication of the preparation. For the latter the methods for tetrahydroborate complexes are generalized in the following reactions. (A) Reaction between a transition metal halide (usually a chloride) and an alkali metal tetrahydroborate (usually $NaBH_4$). (B) Reaction between a transition metal alkoxide or similar complex with BH_3·THF or B_2H_6. (C) Reaction of a binary transition metal compound with a mixture of $NaBH_4$ and ligand in EtOH or another solvent. (D) Reaction of a transition metal complex of the ligand with $NaBH_4$ in EtOH or another solvent. (E) Reaction of a transition metal tetrahydroborate salt with ligand in a solvent. (F) Reaction of a transition metal hydride with BH_3·THF or B_2H_6 in ether. The methods are shown by the letters A–F in the tables.

For tetrahydroborate complexes, the BH_4 moiety functions as a mono-, bi- or tridentate ligand and many examples are given in *COMC-I*.[1] The number of bridging hydrogen atoms in the BH_4 complex is governed by the number of other ligands on the metal atom, and also by the size of the metal atoms. Thus, zirconium has a greater propensity for the formation of η^3-BH_4 complexes than does titanium. The structures are not illustrated herein unless there are unusual or interesting features which distinguish them, nor are any structural parameters included.

8.2.1.2 Group 3 metals

Tetrahydroborate derivatives of group 3 metals are few in number and they are listed in Table 1. The species $Cp_2Sc(BH_4)$ (**1**) contains a triply bridging BH_4 moiety[31] as do the related $C_5H_3(TMS)_2$ derivatives $[C_5H_3(TMS)_2]_2Sc[BH_4]$ (**2**) and $[C_5H_3(TMS)_2]_2Sc[BH_4]$ (**3**).[32] These three systems are fluxional, undergoing rapid bridge–terminal hydrogen atom exchange.

Table 1 Tetrahydroborate compounds of group 3 metals.

Compound	*No.*	*Preparation method*	*Color*	*Melting point (°C)*	*Physical data*	*Ref.*
$Cp_2Sc(BH_4)$	(**1**)	D	White	Sublimes 120–130 (10^{-3} torr)	^{11}B, ^{1}H, ^{54}Sc NMR; IR	31
$[C_5H_3(TMS)_2]_2Sc[BH_4]$	(**2**)	D	White	82–84	^{11}B, ^{1}H, NMR; IR; crystal structure	32
$[C_5H_3(TMS)_2]_2Y[BH_4]$	(**3**)	D	White	130–128	^{11}B, ^{1}H, NMR; IR	32

8.2.1.3 Group 4 metals

There is a large number of group 4 metal tetrahydroborate complexes just as there are many simple binary tetrahydroborates. Many of them were prepared as possible precursors to high-tech materials such as TiB_2 and ZrB_2. A listing is given in Table 2.

Table 2 Tetrahydroborate compounds of group 4 metals.

Compound	*No.*	*Preparation method*	*Color/form*	*Melting point* (°C)	*Physical data*	*Ref.*
$[CH_2(C_5H_4)_2Ti(BH_4)]$	**(4)**		Black crystals	sublimes 70–80 °C/ 10^{-2} torr	IR, ESR	33
$[(Salen)Ti[BH_4]_2 \cdot THF]$	**(5)**	D	Orange crystals		IR	34
$[Ti(BH_4)_3(PMe_3)_2]$	**(6)**	E	Blue crystals	104 decomposes	^1H NMR, IR, ESR	35, 36
$[Ti(BH_4)_3(PEt_3)_2]$	**(7)**	D	Blue needles	75 decomposes	^1H NMR, IR, ESR	36
$[Ti(BH_4)_3(PMe_2Ph)_2)]$	**(8)**	D	Blue prisms	75 decomposes	^1H NMR, IR, ESR	36
$[Ti(BH_4)_3(POMe_3)_2]$	**(9)**	D	Blue prisms		IR	36
$[Ti(BH_4)_3(dmpe)]$	**(10)**	D	Pale-blue solid	190 decomposes	IR	36
$[Ti(BH_4)_3(DME)]$	**(11)**	A	Light-blue solid		IR, dipole moment	37
$[Ti(BH_4)_2(dmpe)_2]$	**(12)**	D	Red-orange prisms	195 decomposes	^1H NMR; IR; dipole moment	38
$[Ti(BD_4)_2(dmpe)_2]$	**(13)**	D		IR	38	
$[(Bisphenoxo)Ti(BH_4)_2]$	**(15)**	D	Red crystals		^1H, ^{11}B NMR; IR; crystal structure	40
$[(Bisphenoxo)Zr(THF)(BH_4)_2]$	**(16)**	D	White crystals		^1H, ^{11}B NMR; IR; crystal structure	40
$[(Cp_2Zr(\eta^2\text{-}BH_4)(AlH_2\{BH_4\}_2)]$	**(17)**	$Cp_2Zr(H)AlH_4 + B_2H_6$				41
$[Cp_2ZrCl(BH_4)]$	**(18)**	$Cp_2ZrCl_2 + [B_3H_8]^-$				42
$[Cp_2Zr(BH_4)_2]$	**(19)**	F		Sublimes 100 °C/ 10^{-3} torr	IR; ^1H, ^{11}B NMR; MS	43
$[CpZr(BH_4)_3]$	**(20)**	D	Colorless crystals	Sublimes 35–40 °C/ 0.1 torr	IR; Raman; ^1H, ^{11}B, electron diffraction structure	44
$[Zr_2H_3(BH_4)_5(PMe_3)_2]$	**(21)**	A	Amber crystals	103 decomposes	^1H, ^{11}B, ^{31}P NMR; IR; crystal structure	45
$[Zr_2H_4(BH_4)_4(dmpe)_2]$	**(22)**	(**21**)+L	Colorless crystals	185 decomposes	^1H, ^{11}B, ^{31}P NMR; IR; crystal structure	46
$[Zr_3H_6(BH_4)_6(PMe_3)_4]$	**(23)**	D	Colorless crystals	140 decomposes	^1H, ^{11}B, ^{31}P NMR; IR; crystal structure	46
$[ZrH(BH_4)_3(dmpe)]$	**(24)**	A	Off-white platelets	86 decomposes	^{11}B, ^1H, ^{31}P NMR; IR	45
$[ZrH_2(BH_4)_2(dmpe)_2]$	**(25)**	A	White solid	141–42	^{11}B, ^1H, ^{31}P NMR; IR	45
$[Hf_2H_3(BH_4)_5(PMe_3)_2]$	**(26)**	E	Colorless prisms	135 decomposes	^1H, ^{11}B, ^{31}P NMR; IR; crystal structure	45
$[HfH(BH_4)_3(dmpe)]$	**(27)**	E	Off-white blocks	198 decomposes	^1H, ^{11}B, ^{31}P NMR; IR	45
$[HfH_2(BH_4)_2(dmpe)_2]$	**(28)**	E	White	88	^1H, ^{11}B, ^{31}P NMR; IR	45
$[Hf_3H_6(BH_4)_6(PMe_3)_4]$	**(29)**	E	Off-white microcrystals	125–127 decomposes	^1H, ^{11}B, ^{31}P NMR; IR	46
$[Hf_2H_4(BH_4)_4(dmpe)_2]$	**(30)**	(**26**) + L	Colorless crystals	200 decomposes	^1H, ^{11}B, ^{31}P NMR; IR	46
$[Hf(BH_4)_3(N\{SiMe_2CH_2PMe_2\}_2)]$	**(31)**	D	Colorless crystals	119 decomposes	^{11}B, ^1H, ^{31}P NMR; IR	47, 48
$[Hf(BH_4)_3[N(SiMe_2CH_2PPr^i)_2]$	**(32)**	E	White crystals		^1H, ^{31}P NMR, IR	48
$[Hf\{N(SiMe_2CH_2PMe_2)_2\}]_2(\mu\text{-}H)_3(BH_4)_3$	**(33)**	E	Off-white crystals	134–136	^{11}B, ^1H, ^{31}P NMR; IR; crystal structure	47, 48
$[Hf\{N(SiMe_2CH_2PMe_2)_2\}]_2(\mu\text{-}H)_4(BH_4)_2$	**(34)**	E	Yellow flakes	146	^1H, ^{31}P NMR; IR	48
$Hf(BH_4)_2CH_2TMS[NSiMe_2CH_2PMe_2)_2]$	**(35)**	E	Pale-yellow crystals		^1H, ^{31}P NMR	48

The compound $CH_2(C_5H_4)_2Ti(BH_4)$ (**4**) contains bidentate BH_4 groups[33] as does (Salen)Ti$[BH_4]_2\cdot$THF (**5**)[34] but the latter contains what is effectively a seven-coordinate titanium atom and also BNH_2 groups bonded to the titanium through a bridging hydrogen atom. As seen in Structure (**5**), there are two electron deficient B–H–Ti–N four-membered rings. $[Ti(BH_4)_3(PMe_3)_2]$ (**6**)[35–6] also contains a bidentate BH_4 group but the other two are quite unusual. The orientation around the titanium is approximately trigonal bipyramidal with the phosphine ligands occupying axial positions and the three BH_4 groups in equatorial positions. The bidentate group sits on a mirror plane with the μ hydrogen atoms on the equatorial plane of the complex. The other two BH_4 groups bond "side" on with only one μ hydrogen atom and a direct Ti–B bond. The resulting system is a type of Ti···H···B agostic interaction. The "normal" BH_4 Ti–B distance is 0.24 nm, whereas the agostic Ti···B distance is 0.227 nm. $[Ti(BH_4)_3(DME)]$ (**11**) also has a distorted trigonal bipyramidal structure with two bidentate BH_4 groups in equatorial positions and another in an axial position. The DME ligand chelates both axial and equatorial positions.[37] The more crowded system, $[Ti(BH_4)_2(dmpe)_2]$ (**12**), exhibits eight-coordinate geometry, if the μ hydrogen atoms are included, with the dmpe ligands and the metal forming a square plane with two bidentate BH_4 groups in axial positions. This contrasts with the related vanadium complex, $V(\eta^1\text{-}BH_4)_2(dmpe)_2$ (**14**),[38–9] which has a similar structure but the BH_4 groups are unidentate.

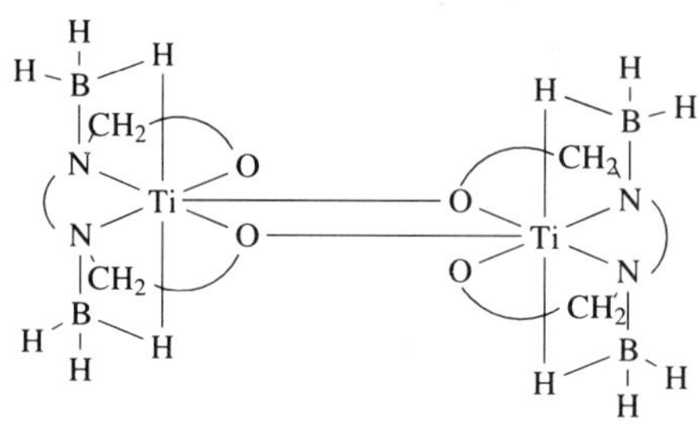

(5)

The compound (bisphenoxo)$Ti(BH_4)_2$ (**15**) contains two tridentate BH_4 ligands which, with the two oxygen atoms, represents tetrahedral coordination around the titanium atom but the related species (bisphenoxo)Zr(THF)$(BH_4)_2$ (**16**) contains one bidentate and one tridentate BH_4 group.[40] This complex contains a THF group so the approximate coordination around zirconium is a distorted trigonal bipyramid with the two BH_4 groups and one of the chelating oxygen atoms in the equatorial region and the other chelating oxygenation, and the oxygen atom from the THF, in axial positions.

Several tetrahydroborate complexes of zirconium are listed in Table 2. The zirconium complexes tend to favor triply bridging BH_4 groups but the mixed species are also quite common. $CpZr(BH_4)_3$ (**20**) has three triply bridging BH_4 groups,[44] but the binuclear polyhydride complex $Zr_2H_3(BH_4)_5(PMe_3)_2$ (**21**)[45] contains both double and triple bridged BH_4 groups. The related species and $Zr_2H_4(BH_4)_4(dmpe)_2$ (**22**) $Zr_3H_6(BH_4)_6(PMe_3)_4$ (**23**),[46] $[ZrH(BH_4)_3(dmpe)]$ (**24**)[45] and $[ZrH_2(BH_4)_2(dmpe)_2]$ (**25**)[45] contain both η^2- and η^3-BH_4 groups as well as bridging hydrogen atoms and they are fluxional. Members of a similar series of hafnium polyhydride tetrahydroborates (**26**)–(**30**) are also fluxional. A related series of mixed hafnium hydride–tetrahydroborate complexes provides a new feature.[47–8] These systems contain the chelating ligand $[N(SiMe_2CH_2PMe_2)_2]$ and the binuclear species contain two hafnium atoms bridged by hydrogen atoms. They are formed by the treatment of (**31**) with base. The base removes BH_3 from the BH_4 groups to afford hydrido-species, according to, for example, (Equation (1)). The structure of product (**33**) is unique in that it contains all three types of BH_4 groups, η^1, η^2 and η^3.[47–8]

$$2Hf[BH_4]_3[N(SiMe_2CH_2PMe_2)_2]\ (\mathbf{31}) + 3PMe_3 \longrightarrow [HfN(SiMe_2CH_2PMe_2)_2](BH_4)_3(\mu\text{-}H)_3\ (\mathbf{33}) + 3H_3B{\bullet}PMe_3 \quad (1)$$

8.2.1.4 Group 5 metals

The group 5 derivatives are listed in Table 3. $V_2Zn_2H_4[BH_4]_2(PMePh_2)_4$ (**36**)[49] and $[V(\mu\text{-}Cl)(\mu\text{-}dppm)(BH_4)]_2$ (**37**)[50] are binuclear systems. $Ta(BH_4)H_2(PMe_3)_4$ (**45**)[52] is a mixed hydrido-tetrahydroborato complex with the tantalum atom essentially seven coordinate with pentagonal bipyramidal geometry. Two phosphine ligands occupy axial positions and the bidentate BH_4 group has its μ hydrogen atoms in a plane perpendicular to the equatorial plane.

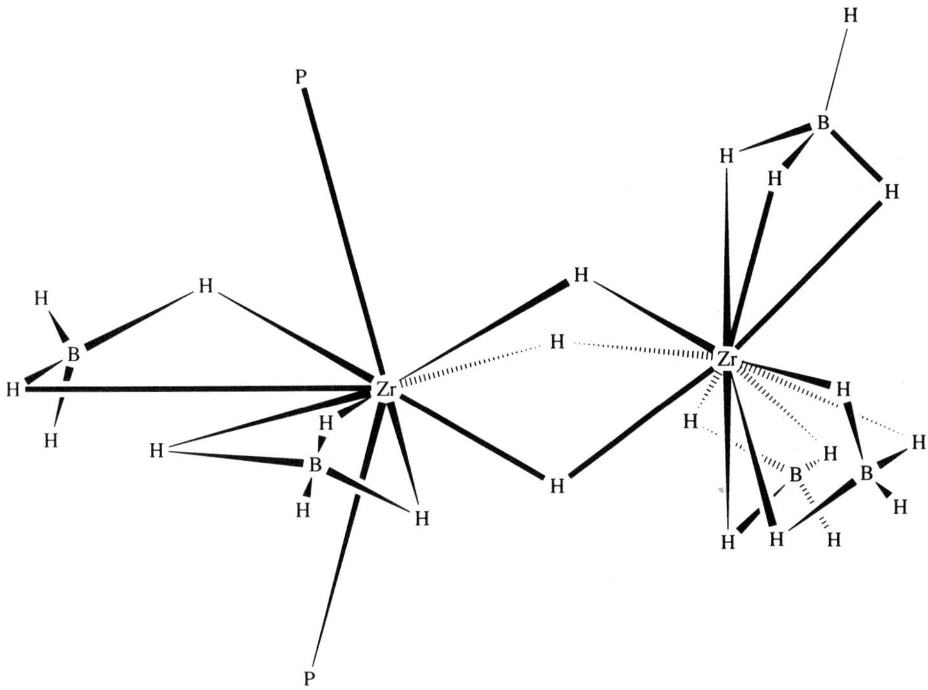
H
B
H
H
H
H
Zr
Zr
P
P
H
H
H
B
H
H
H
B
H
H
H
H
H
B
H
H
H
H
H
H
H

(21)

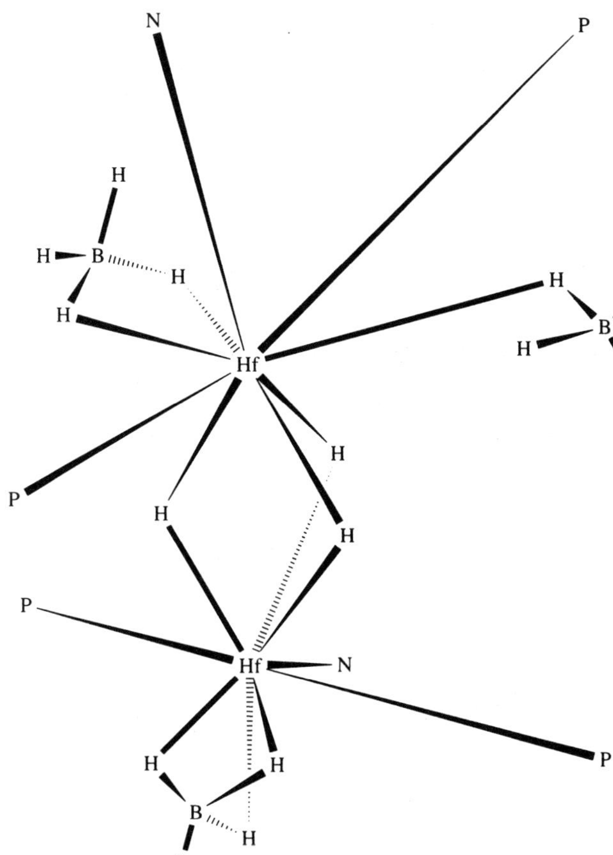
N
P
H
H
B
H
H
Hf
H
B
H
H
H
P
H
H
H
P
Hf
N
P
H
H
B
H
H

(33)

Table 3 Tetrahydroborate compounds of group 5 metals.

Compound	*No.*	*Preparation method*	*Color/form*	*Melting point* (°C)	*Physical data*	*Ref.*
$[V_2Zn_2H_4(BH_4)_2(PMePh_2)_4]$	**(36)**	E	Dark-red crystals		^{11}B, ^{31}P NMR; IR; crystal structure	49
$[V(\mu\text{-}Cl)(\mu\text{-dppm})(BH_4)]_2$	**(37)**	C	Dark-red crystals		ESR; crystal structure	50
$[Na(dme)][V(BH_4)_4]$	**(38)**	C	Purple powder	51–52	IR	39, 51
$[Li(Et_2O)][V(BH_4)_4]$	**(39)**	A	Purple oil		ESR	39, 51
$[V(BH_4)_3(PMe_3)_2]$	**(40)**	E	Bright-green prisms	94 decomposes	^{1}H NMR; magnetic moment, IR; crystal structure	39, 51
$[V(BH_4)_2(dmpe)_2]$	**(14)**	D	Gray crystals	140 decomposes	^{1}H NMR; magnetic moment, IR; crystal structure	38, 39
$[V(BH_4)H(PMe_3)_2]_2$	**(41)**	(**40**) + PMe_3	Lavender solid	101 decomposes	^{1}H NMR; magnetic moment, IR	51
$[V(BH_4)_2(PMe_3)_2]_2\mu\text{-}O$	**(42)**	(**40**) + H_2O	Purple needles	120 decomposes	^{1}H NMR; IR; crystal structure	51
$[V(BH_4)_3(THF)_3]$	**(43)**	D	Purple crystals	140 decomposes	^{1}H NMR; magnetic moment; IR	51
$[V(BH_4)Cl(THF)_2]_2$	**(44)**	D	Purple prisms	100 decomposes	^{1}H NMR; magnetic moment; IR; ESR	51, 39
$[Ta(BH_4)H_2(PMe_3)_4]$	**(45)**	D	Emerald green	Sublimes, 25 °C/10^{-5} torr	^{1}H, NMR; crystal structure	52
$[TaCpCp^*(BH_4)]$	**(46)**	E	Green crystals		^{1}H, ^{11}B NMR	53

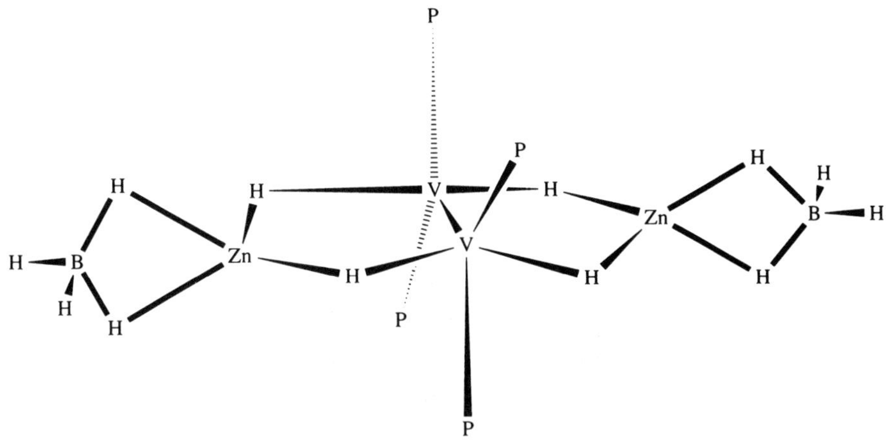

(36)

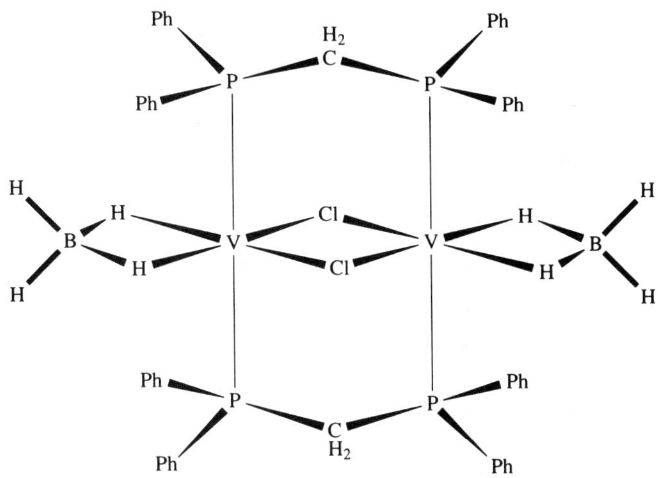

(37)

A series of vanadium BH_4 complexes, analogous to the titanium complexes described above, is also listed in Table 3. $V(\eta^2\text{-}BH_4)_3(PMe_3)_2$ (**40**)[39,51] has approximate trigonal bipyramidal geometry with the phosphine ligands in axial positions, like (**6**), but the three BH_4 groups are all bidentate. As described above $V(\eta^1\text{-}BH_4)_2(dmpe)_2$ (**14**)[38–9] contains unidentate BH_4 groups but is otherwise quite similar to (**12**). $[V(BH_4)\mu\text{-}H(PMe_3)_2]_2$ (**41**) is a derivative of (**40**) with $[V(BH_4)(PMe_3)_2]$ groups bridged at the vanadium atoms through two hydrogen atoms[51] and $[\{V(BH_4)_2(PMe_3)_2\}_2\mu\text{-}O]$ (**42**) contains two $[V(BH_4)_2(PMe_3)_2]$ groups bridged at the vanadium atoms by a single oxygen atom.[51]

8.2.1.5 Group 6 and 7 metals

Tetrahydroborate complexes of groups 6 and 7 are given in Table 4. The species represented by the general formula, $[PPN][(CO)_5M(\mu\text{-}H)(\mu\text{-}BH_3)]$ (M = Cr, W, Mo) (**47**), appear to be η^1-complexes,[54] whereas $[PPN][(\mu\text{-}H)_2BH_2Cr(CO)_4]$ (**48**) is an η^2-complex.[55] Compound (**47**) is expected to be octahedral and (**48**) adopts a distorted octahedral shape with the two μ hydrogen atoms occupying two of the octahedral coordination sites. (**48**) is converted into (**47**) (M = Cr) by treatment with CO and in turn the latter is converted to the hydrido complex by treatment with piperidine, according to Equations (2) and (3).

$$[(\mu\text{-}H)_2BH_2Cr(CO)_4]^- \ \mathbf{(48)} \ + \ CO \longrightarrow [(CO)_5Cr\text{-}\mu\text{-}H\text{-}\mu\text{-}BH_3]^- \ \mathbf{(47)} \qquad (2)$$

$$[(CO)_5Cr\text{-}\mu\text{-}H\text{-}\mu\text{-}BH_3]^- \ \mathbf{(47)} \ + \ \text{piperidine} \longrightarrow [(CO)_5\mu H]^- \ + \ \text{pip}\bullet BH_3 \qquad (3)$$

Table 4 Tetrahydroborate compounds of group 6 and 7 metals.

Compound	*No.*	*Preparation method*	*Color/ form*	*Melting point (°C)*	*Physical data*	*Ref.*
$PPN[(CO)_5MBH_4]$, (M = Cr, W, Mo)	**(47)**	$LM(CO)_5$ + $PPNBH_4$	Yellow solid		^1H, ^{13}C NMR; IR	54
$PPN[(\mu\text{-}H)_2BH_2Cr(CO)_4]$	**(48)**	D	Red crystals		^1H, ^{13}C NMR; IR; crystal structure	53
$Cr(\eta^2\text{-}BH_4)H(dmpe)_2$	**(49)**	D	Red prisms	120 (180)[56] decomposes	^1H NMR; IR; magnetic moment crystal structure[38]	38, 56
$[W(\eta^2\text{-}BH_4)H_3(PMe_3)_3]$	**(50)**	D	Orange crystals	180 decomposes	^1H NMR; IR	56
trans-$W(\eta^2\text{-}BH_4)(CO)(NO)(PMe_3)_2$	**(51)**	F	Purple crystals		^1H, ^{13}C NMR; IR; crystal structure.	57
$Mn(BH_4)_2(THF)_2$	**(52)**	C	Yellow oil		IR; thermal analysis	58
$Mn(BH_4)_2(THF)_3$	**(53)**	C	Flesh colored crystals		IR; thermal analysis, crystal structure.	58
$Mn_2(\mu\text{-}H)(\mu\text{-}BH_4)(CO)_6\text{-}(\mu\text{-}dppm)$	**(54)**	F			^1H, ^{31}P NMR; IR; crystal structure	59
$Re(\eta^1,\eta^2\text{-}BH_4)_2H(PMe_2Ph)_3$	**(55)**	D	Bright orange crystals	84 decomposes	^1H, ^{31}P NMR; IR	56

The complex $[Cr(\eta^2\text{-}BH_4)H(dmpe)_2]$ (**49**), is paramagnetic. IR specta suggest unidentate ligation of the BH_4 group, and ^1H NMR spectral data are consistent with *trans*-octahedral geometry.[38,56] The rhenium complex $[Re(\eta^1,\eta^2\text{-}BH_4)_2H(PMe_2Ph)_3]$ (**55**) appears to contain both η^1-, and η^2-BH_4 groups in a species which has seven-coordinate pentagonal pyramidal geometry. The unidentate group occupies an axial position and one of the phosphines occupies the other.[56] The species (**50**)–(**53**)[56–8] all possess bidentate BH_4 groups but $[Mn_2(\mu\text{-}H)(\mu\text{-}BH_4)(CO)_6(\mu\text{-}dppm)]$ (**54**)[59] contains a unique bridging BH_4 group.

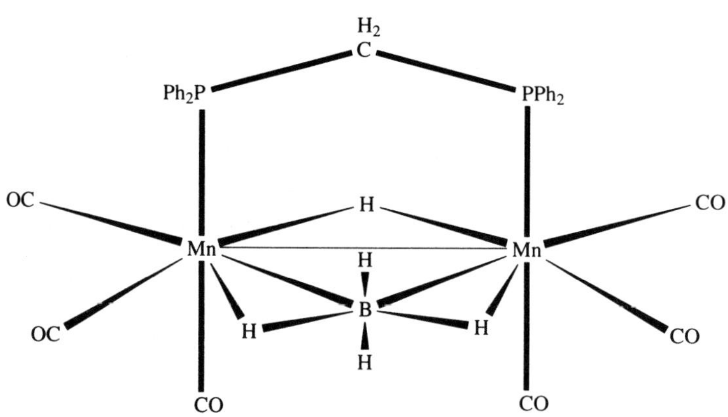

(54)

8.2.1.6 Group 8 and 9 metals

A list of tetrahydroborate compounds of group 8 and 9 metals is given in Table 5. Not included in this listing are the metal-rich species which are described separately in Section 8.2.6. The group 8 complexes $[triphosFe(H)(BH_4)]$ (**56**)[60], $[(dmpe)_2Fe(H)(BH_4)]$ (**57**),[61–2] $[(ttp)Ru(H)(BH_4)]$ (**58**),[63] and $[(PP_3)Ru(H)(BH_4)]$ (**59**)[64] all contain both single hydride ligands and a BH_4 ligand in effectively distorted octahedral complexes. In (**56**) the BH_4 ligand is bidentate, rigid at low temperature but at higher temperatures all the metal-bound hydrogen atoms exchange. Both (**57**) and (**59**) contain a unidentate BH_4 ligand so the only exchange mechanism inferred from NMR spectral data is hydrogen atom exchange within the BH_4 group. Compound (**58**) contains a bidentate BH_4 group and in this case too, terminal–bridging hydrogen exchange occurs within the BH_4 group but does not involve the unique Fe–H bond. The triphos analogue of (**58**), $[\{(triphos)(H)Ru\}_2(\mu\eta^2\text{-}BH_4)][BPh_4]$ (**60**)[65] obtained by treatment of $[(triphos)(H)RuBH_4]$ with Me_2O and isolation as the BPh_4 salt, contains a bridging η^2-BH_4 group. $Os[BH_4]H_3[P(c\text{-}C_5H_9)_3]_2$ (**61**)[66] contains a bidentate BH_4 group in a pentagonal bipyramidal complex with the two phosphine ligands in axial positions. NMR spectra suggest that the bridging hydrogen atoms exchange with the Os–H hydrogens without involvement of the terminal hydrogen atoms on boron.

Table 5 Tetrahydroborate compounds of group 8 and 9 metals.

Compound	*No.*	*Preparation method*	*Color/ form*	*Melting point* (°C)	*Physical data*	*Ref.*
triphosFe(H)[BH_4]	(**56**)	D	Bright-red crystals		^{1}H NMR; IR; magnetic moment, crystal structure	60
$(dmpe)_2$Fe(H)[BH_4][a]	(**57**)	D	Yellow solid		^{1}H, ^{11}B, ^{31}P NMR; IR	61, 62
(ttp)Ru(H)(BH_4)[b]	(**58**)	D	Pale-yellow crystals	180	^{1}H, NMR, IR; crystal structure	63
[(PP_3)Ru(H)(BH_4)][c]	(**59**)	D	Pale-cream crystals		^{1}H, ^{31}P NMR; IR	64
[(triphos)(H)(Ru]$_2$-(μη^2-BH_4)] [BPh_4]	(**60**)	D	Pale yellow crystals		^{1}H, ^{31}P NMR; IR; crystal structure	65
Os[BH_4]H_3[P(c-C_5H_9)$_3$]$_2$	(**61**)	F	Yellow crystals		^{1}H NMR; IR; crystal structure	66
(terpy)Co[BH_4]	(**62**)	D	Gray crystals structure		^{1}H NMR; IR; MS; crystal structure	67
(dpppe)$_2$Co$_2$[BH_4]$_2$·0.5C_6H_6[d]	(**63**)	C	Dark-green crystals		IR; crystal structure	68
[Cp*IrH]$_2$(μ-H)(μ-BH_4)	(**64**)	D	Yellow-orange crystals		^{1}H, ^{13}C, ^{11}B; NMR; IR; crystal structure	69

[a] dmpe = [(Me_2P)CH_2]$_2$. [b] ttp = [PhP(C_3H_6P(Ph)$_2$)$_2$]. [c] PP_3 = P($CH_2CH_2PPh_2$)$_3$. [d] dpppe = [CH_2($C_2H_4PPh_2$)$_2$].

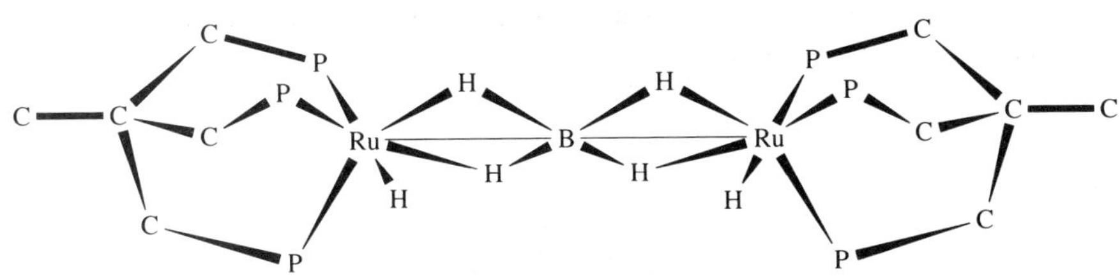

(**60**)

The group 9 species listed in Table 5 provide three different types of BH_4 ligand. [(terpy)Co(BH_4)] (**62**)[67] (terpy, 2,2',2''-terpyridyl) contains a η^2-BH_4 group, whereas (dpppe)$_2$Co$_2$[BH_4]$_2$·0.5C_6H_6(**63**)[68] is a binuclear species with a pair of bridging BH_4 groups each containing a pair of Co–H–B bonds, a μ$_3$-H atom and a terminal hydrogen atom. The iridium complex [Cp*IrH]$_2$(μ-H)(μ-BH_4) (**64**)[69] is also binuclear, the η^2-BH_4 group bridges the two iridium atoms both of which also bear hydrogen atoms. In this species the hydrogen atoms in the BH_4 group appear not to exchange.

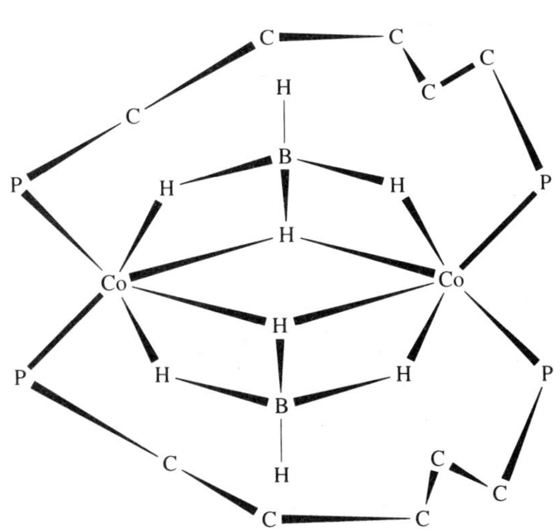

(**63**)

8.2.1.7 Group 11 metals

The best known tetrahydroborate complexes and those for which criteria for the identification of the attachment modes were developed are the Cu^{I} species. Whether the BH_4 group bonds to the metal with η^1- or η^2-coordination is determined either by steric effects or the tendency of the copper to achieve tetrahedral geometry. Several new systems have been prepared and examples of both modes of attachment are observed. These are listed in Table 6. The mixed ligand species $[Ph_3PCu(phen)(BH_4)]$ **(65)**[70–1] and $[Ph_3PCu\{(2,9\text{-}Me_2)(4,7\text{-}Ph_2)phen\}(BH_4)]$ **(66)**[69,71] contain η^2-BH_4 groups as do bis(phosphine) complexes such as **(72)**.[75] On the other hand, the species (triphos)$CuBH_4$ **(67)**,[72] $(np_3)CuBH_4$ **(70)**[73] and $(PHPh_2)_3CuBH_4$ **(72)**[75–6] are unambiguously η^2, but there is some question about the mode of attachment of the BH_4 group in $(etp_3)CuBH_4$ **(71)** and $(PCy_3)_2CuBH_4$ **(68)**. $[(Cu(PPh_3)_2)_2BH_4][ClO_4]$ **(69)**[74] contains a bidentate BH_4 unit which bridges two copper atoms such that there are only bridging hydrogen atoms. In **(69)** the H–B–H angles are smaller than in other $CuBH_4$ with η^2-BH_4 ligands and the P–Cu–P angle is also smaller.

Table 6 Tetrahydroborate compounds of group 11 metals.

Compound	*No.*	*Preparation method*	*Color/ form*	*Melting point (°C)*	*Physical data*	*Ref.*
α-$[Ph_3PCu(phen)[BH_4]$	**(65)**	E	Yellow crystals		IR; x-ray and neutron crystal structure	70, 71
β-$[Ph_3PCu(phen)[BH_4]$	**(65)**	E	Red crystals		IR; x-ray and neutron crystal structure	70, 71
$Ph_3PCu[(2,9\text{-}Me_2)\text{-}(4,7\text{-}Ph_2)phen][BH_4]$	**(66)**	E	Orange–yellow crystals		IR; x-ray and neutron crystal structure	71
(triphos)$CuBH_4$	**(67)**	D	White solid		IR; crystal structure	72
$(PCy)_2Cu(BH_4)$	**(68)**	D	White solid		^1H, ^{31}P, ^{13}C NMR; IR	73
$[(Cu(PPh_3)_2)_2BH_4][ClO_4]$	**(69)**	L_2CuBH_4 + $[ClO_4]$			crystal structure	74
$(np_3)Cu(BH_4)$[a]	**(70)**	D	White solid		^1H, ^{31}P, ^{13}C NMR; IR	73
$(etp_3)Cu(BH_4)$[b]	**(71)**	D	White solid		^1H, ^{31}P, ^{13}C NMR; IR	73
$(PHPh_2)_3CuBH_4$	**(72)**	D	Oil		IR	75
$C_2H_4[N\{C_2H_4PPh_2\}_2CuBH_4]_2$	**(73)**	D	White crystals		^1H, ^{13}C, ^{31}P, ^{11}B NMR; IR; crystal structure	77

[a] (np_3) = tris{2-(diphenylphosphino)ethylamine}. [b] (etp_3) = 1,1,1-tris{(diphenylphosphino)-ethyl)ethane}.

8.2.1.8 Lanthanide metals

The data on tetrahydroborate complexes of the lanthanides are given in Table 7. Essentially all the data for **(74)**–**(80)**[79–81] are tabulated. The species **(80)**–**(84)**[32] contain the ligand $[C_5H_3(TMS)_2]$ which facilitates solubility in hydrocarbon solvents. These latter complexes, on the basis of IR data, are presumed to contain tridentate BH_4 ligands. The species $[(\eta^5\text{-}C_5H_3Bu^t_2)_2Ce\{\mu{:}\eta^4(\mu_3\text{-}H)_2BH_2\}]_2$ **(106)**[20] contains a unique pair of η^4-BH_4 groups bridging two Cp_2Ce units in which each BH_4 group contains two μ_3- and two μ_2 hydrogen atoms. The tetrahedral arrangement around the boron atoms is quite distorted and the Ce···B distance is larger than in related complexes.

8.2.1.9 Actinide metals

Several species of tetrahydroborate compounds of actinides are listed in Table 8. Due to the size of the actinide elements, most BH_4 complexes are tridentate. Although hydrogen atoms are difficult to locate in complexes containing very heavy atoms such as uranium or thorium, well-established criteria based on the metal–boron distance[92] and on IR spectra[32] allow the assignment of the bonding mode in cases where the hydrogen atoms were not located. All the systems listed in Table 8 contain η^3-BH_4 groups exclusively except for $[(dmpe)_2U(BH_4)_3]$ **(112)**[86] which contains two η^3- and one η^2-BH_4 group. Structure **(112)** represents pentagonal bipyramidal coordination of the nonhydrogen atoms around uranium. The axial BH_4 group is bidentate [U–B(2) = 0.284(3) nm] and the equatorial BH_4 groups are tridentate [U–B(1) = 0.268(4) nm]. The related system $[U(BH_4)_3(2\text{-}Ph_2PPy)_2]$ **(111)** contains three η^2-BH_4 groups, two axial and one equatorial, resulting in a very distorted pentagonal pyramidal arrangement around the uranium atom.[85]

Table 7 Tetrahydroborate compounds of the lanthanide metals.

Compound	*No.*	*Preparation method*	*Color/ form*	*Melting point* (°C)	*Physical data*	*Ref.*
$Gd(BH_4)_3{\cdot}DME$,	(**74**)	Sublimation of (**90**)	White		IR; thermal studies	78
$Tb(BH_4)_3{\cdot}DME$,	(**75**)	Sublimation of (**91**)	White		IR; thermal studies	78
$Dy(BH_4)_3{\cdot}DME$,	(**76**)	Sublimation of (**92**)	White		IR; thermal studies	78
$Ho(BH_4)_3{\cdot}DME$,	(**77**)	Sublimation of (**93**)	Pale yellow		IR; thermal studies	78
$Tm(BH_4)_3{\cdot}DME$,	(**78**)	Sublimation of (**95**)	Pink		IR; thermal studies	78
$Lu(BH_4)_3{\cdot}DME$,	(**79**)	Sublimation of (**97**)	Pale green		IR; thermal studies	78
$Er(BH_4)_3{\cdot}DME$,	(**80**)	Sublimation of (**94**)	White		IR; thermal studies	78
$[NBu^n{}_4][La(BH_4)_4{\cdot}THF]$	(**81**)	E	White		^{11}B NMR; IR	79
$NaLa[BH_4]_4{\cdot}DME$,	(**82**)	C	White		IR; thermal studies	80
$NaCe[BH_4]_4{}^4DME$,	(**83**)	C	White		IR; thermal studies	80
$[NBu^n{}_4][Pr(BH_4)_4{\cdot}THF]$	(**84**)	E			^{11}B NMR; IR	79
$[NBu^n{}_4][Nd(BH_4)_4{\cdot}THF]$	(**85**)	E			^{11}B NMR; IR	79
$NaPr[BH_4]_4{\cdot}4DME$,	(**86**)	C	Green		IR; thermal studies	80
$NaNd[BH_4]_4{\cdot}4DME$,	(**87**)	C	Lilac		IR; thermal studies	80
$NaSm[BH_4]_4{\cdot}4DME$,	(**88**)	C	White		IR; thermal studies	80
$NaEu[BH_4]_4{\cdot}4DME$,	(**89**)	C	Yellow		IR; thermal studies	80
$NaGd[BH_4]_4{\cdot}4DME$,	(**90**)	C	White		IR; thermal studies	81
$NaTb[BH_4]_4{\cdot}4DME$,	(**91**)	C	White		IR; thermal studies	81
$NaDy[BH_4]_4{\cdot}3DME$	(**92**)	C	Pale yellow		IR; thermal studies	81
$NaHo[BH_4]_4{\cdot}3DME$,	(**93**)	C	Yellow		IR; thermal studies	81
$NaEr[BH_4]_4{\cdot}3DME$	(**94**)	C	Pink		IR; thermal studies	81
$NaTm[BH_4]_4{\cdot}3DME$	(**95**)	C	Pale green		IR; thermal studies	81
$NaYb[BH_4]_4{\cdot}3DME$	(**96**)	C	White		IR; thermal studies	81
$NaLu[BH_4]_4{\cdot}3DME$	(**97**)	C	White		IR; thermal studies	81
$[NBu^n{}_4][Nd(BH_4)_5]$	(**98**)	E			^{11}B NMR; IR	79
$[NBu^n{}_4][Pr(BH_4)_5]$	(**99**)	E			^{11}B NMR; IR	79
$[NBu^n{}_4][Nd(BH_4)_5]$	(**100**)	E			^{11}B NMR; IR	79
$[C_5H_3(TMS)_2]_2La[BH_4]$	(**101**)	D	White	93–95	^{11}B, ^1H NMR; IR	32
$[C_5H_3(TMS)_2]_2Pr[BH_4]$	(**102**)	D	Green	105–107	^{11}B, ^1H NMR; IR	32
$[C_5H_3(TMS)_2]_2Nd[BH_4]$	(**103**)	D	Blue violet	113–115	^{11}B, ^1H NMR; IR	32
$[C_5H_3(TMS)_2]_2Sm[BH_4]$	(**104**)	D	Yellow	115–117	^{11}B, ^1H NMR; IR	32
$[C_5H_3(TMS)_2]_2Yb[BH_4]$	(**105**)	D	Maroon	130–132	^{11}B, ^1H NMR; IR	32
$\{(\eta^5\text{-}C_5H_3Bu^t{}_2)_2Ce[\mu{:}\eta^4\text{-}(\mu_3\text{-}H)_2BH_2]\}_2$	(**106**)	D	Orange-yellow crystals		Crystal structure	20

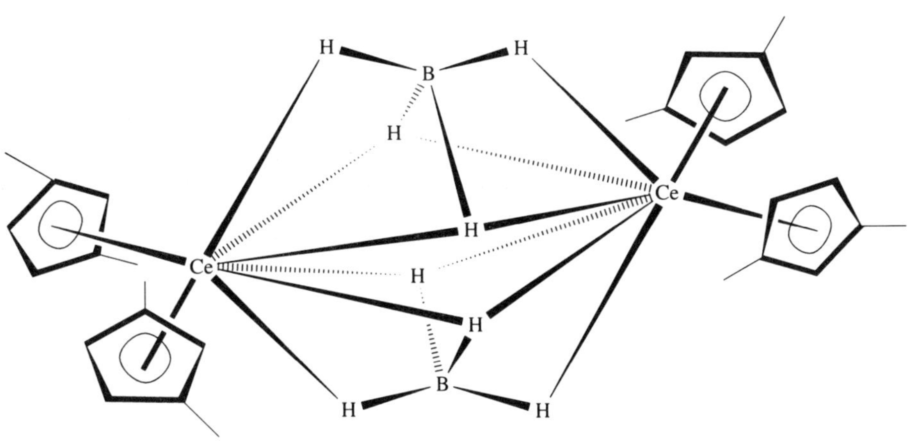

(106)

8.2.2 Trihydroborate Derivatives

8.2.2.1 Trihydroborate compounds of the transition metals

The trihydroborate complexes listed in Table 9 provide a range of different examples. $[Zr(BH_3Me)_4]$ (**132**)[93–5] contains four $MeBH_3$ ligands each of which bond through three μ hydrogen atoms to the

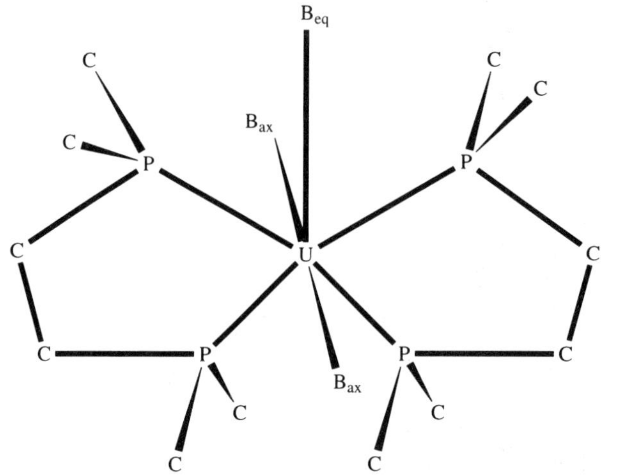

(112)

zirconium atom in a system in which the four boron atoms form a tetrahedron around the zirconium. [$Cp_2Zr(H)BH_3Me$] (**133**),[96] on the other hand, contains a bidentate BH_3Me group with the two Cp groups, the hydride ion and the boron atom forming a distorted tetrahedron around the zirconium. $CpMo(CO)_2[P(BH_3)(Ph)\{N(TMS)_2\}]$ (**134**)[97] is a molybdenaphosphenium complex containing a borane group bonded to the phosphorus atom and also μ-H-bonded to the molybdenum atom forming an unusual $\overline{MoPB(H_2)\text{-}(\mu\text{-}H)}$ four-membered ring. $[Au_2(dppm)_2(S_2CNEt_2)][BH_3CN]$ (**135**)[98] is an unbridged "A-frame" with each BH_3CN group bonding to a single gold atom through a single μ hydrogen atom. Thus, the gold coordination is T-shaped, as illustrated in Structure (**135**).

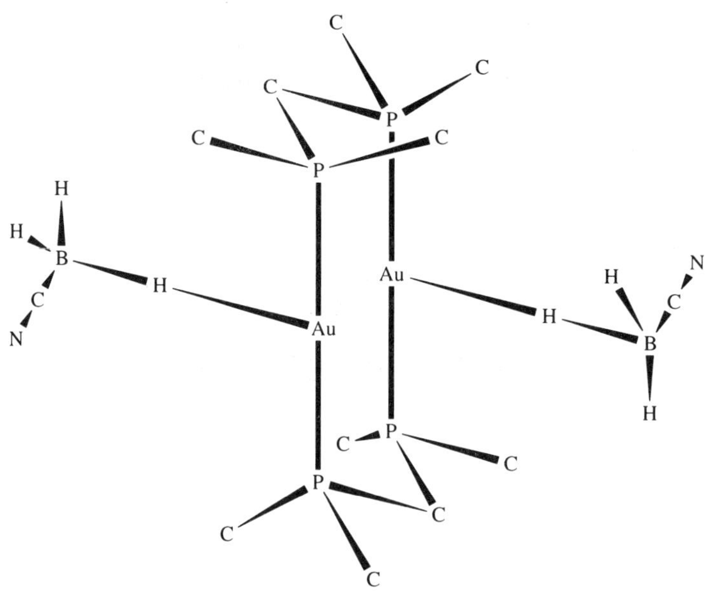

(135)

8.2.2.2 Trihydroborate compounds of the lanthanide metals

Several lanthanide complexes of the [BH_3Me] anion are listed in Table 10. IR data for (**136**)–(**145**) and x-ray structural data, for $Ho[BH_3Me]_3{\cdot}2Py$ (**140**) and $Yb[BH_3Me]_3{\cdot}2OEt_2$ (**141**)[99] confirm that they contain η^3-[BH_3Me] groups.

Table 8 Tetrahydroborate compounds of the actinide metals.

Compound	*No.*	*Preparation method*	*Color/form*	*Melting point (°C)*	*Physical data*	*Ref.*
$(C_9H_7)_3Th(BH_4)$	**(107)**	E	Yellow solid	Sublimes 130 °C/10^{-3} torr	^1H, ^{13}C NMR; IR; MS; magnetic susceptibility	82
$(C_9H_7)_2Th(BH_4)_2$	**(108)**	E	Red-brown solid	Sublimes	^1H, ^{13}C NMR; IR; MS; magnetic susceptibility	82
$Cp_2Th(BH_4)_2$	**(109)**	E	White solid	175 °C/10^{-4} torr	^1H, ^{13}C, ^{11}B NMR; IR; MS; magnetic susceptibility	82, 83
$U(BH_4)_3\cdot 3THF$	**(110)**	F	Dark-brown crystals		^1H, ^{11}B NMR; IR; crystal structure	84
$U[BH_4]_3(2\text{-}Ph_2PPy)_2$	**(111)**	E	Purple crystals		^1H NMR; IR; crystal structure	85
$(dmpe)_2U[BH_4]_3$	**(112)**	E	Red–olive crystals		IR; crystal structure	86
$(PPh_3O)_2U(BH_4)_4$	**(113)**	E	Green crystals		Crystal structure	87
$CpU(BH_4)_3$	**(114)**	E	Orange-yellow solid	Sublimes 20 °C/10^{-2} torr	^1H, ^{11}B, ^{13}C NMR; IR; crystal structure	83, 88, 89
$Cp_2U(BH_4)_2$	**(115)**	E	Red solid	Sublimes 80 °C/10^{-2} torr	^1H, ^{11}B, ^{13}C NMR; IR	83, 88, 89
$Cp_2U(BH_4)_2[Na(18\text{-crown-}6)]$	**(116)**	E	Red crystals		^1H NMR; IR	89
$Na[Cp_2U(BH_4)_2]$	**(117)**	**(128)** + Cp^-			^1H NMR	90
$Na[Cp_3U(BH_4)]$	**(118)**	**(128)** + Cp^-			^1H NMR	90
$(C_5H_4Me)_2U(BH_4)_2$	**(119)**	C	Brown-yellow solid	Sublimes 55 °C/10^{-2} torr	^1H, ^{11}B, ^{13}C NMR; IR	83
$(\eta\text{-}2,4\text{-}Me_2C_5H_5)U(BH_4)_3$	**(120)**	C	Green crystals	Sublimes 60 °C/10^{-2} torr	^1H NMR; IR; crystal structure	89
$(\eta\text{-}2,4\text{-}Me_2C_5H_5)_2U(BH_4)_2$	**(121)**	C	Red microcrystals		^1H NMR; IR	89
$(\eta\text{-}2,4\text{-}Me_2C_5H_5)_2U(BH_4)_2[K(18\text{-crown-}6)]$	**(122)**	**(128)** + $Me_2C_5H_3^-$	Red crystals		^1H NMR; IR	89
$(C_5H_4SiMe_3)_2U(BH_4)_2$	**(123)**	C	Red oil		^1H, ^{11}B, ^{13}C NMR; IR	83
$(\eta^5\text{-}C_5Me_4P)_2U(BH_4)_2$	**(124)**	E	Brown crystals		^1H, ^{31}P NMR; crystal structure	91
$(\eta^5\text{-}C_5Me_4P)_2U(BH_4)(THF)$	**(125)**	**(124)** + THF			^1H, ^{31}P NMR	91
$(\eta^5\text{-}C_5Me_4P)_2U(BH_4)(PPh_3O)_2$	**(126)**	**(124)** + Ph_3PO			^1H, ^{31}P NMR	91
$[(\eta^5\text{-}C_5Me_4P)_2U(BH_4)]_2$	**(127)**	**(124)** + Na/Hg			^1H, ^{31}P NMR	91
$(\eta^6\text{-}C_6Me_6)U(BH_4)_3$	**(128)**	E	Red crystals		^1H NMR; IR	90
$(\eta^6\text{-}C_6H_3Me_3)U(BH_4)_3$	**(129)**	$U(BH_4)_4 + C_6H_3Me_3$			^1H NMR; crystal structure	90
$(\eta\text{-}C_9H_7)_2U(BH_4)_2$	**(130)**	E	Red-brown solid	Sublimes	^1H, ^{13}C NMR; IR; MS; magnetic susceptibility	82
$(\eta\text{-}C_9H_7)_3U(BH_4)$	**(131)**	E			^1H NMR	82

Table 9 Trihydroborate compounds of the transition metals.

Compound	No.	Preparation method	Color/ form	Melting point (°C)	Physical data	Ref.
$[Zr(BH_3Me)_4]$	**(132)**	A	White crystals	90–92	^{1}H NMR; IR; ESR; crystal structure	93–5
$[Cp_2Zr(H)BH_3Me]$	**(133)**	D	White needles	Sublimes	^{1}H, ^{11}B NMR; IR; crystal structure	96
$CpMo(CO)_2[P(BH_3)(Ph)$-{$N(TMS)_2$}]	**(134)**	B	Pale yellow-green solid		^{1}H; ^{31}P, ^{11}B, ^{13}C NMR; IR; crystal structure	97
$[Au_2(dppm)_2(S_2CNEt_2)]$ $[BH_3CN]_2$	**(135)**	C[a]	White solid	202 decomposes	^{1}H NMR; IR; crystal structure	98

[a] $[RBH_3]$ instead of $[BH_4]$.

Table 10 Trihydroborate compounds of the lanthanide metals.

Compound	No.	Preparation method	Color/ form	Melting point (°C)	Physical data	Ref.
$Ho[BH_3Me]_3{\cdot}OEt_2$	**(136)**	A			^{1}H NMR; IR	99
$Ho[BH_3Me]_3{\cdot}THF$	**(137)**	**(114)** + THF			^{1}H NMR; IR	99
$Ho[BH_3Me]_3{\cdot}2THF$	**(138)**	**(114)** + THF	Pale-orange crystals	133–134	^{1}H NMR; IR	99
$Ho[BH_3Me]_3{\cdot}py$	**(139)**	**(114)** + Py	Pale-orange crystals	115–119	^{1}H NMR; IR	99
$Ho[BH_3Me]_3{\cdot}2py$	**(140)**	**(114)** + Py	Pale-orange crystals	132–134	^{1}H NMR; IR; crystal structure	99
$Yb[BH_3Me]_3{\cdot}OEt_2$	**(141)**	A	White crystals	59–60	^{1}H NMR; IR; crystal structure	99
$Yb[BH_3Me]_3{\cdot}THF$	**(142)**	**(119)** + THF			^{1}H NMR; IR	99
$Yb[BH_3Me]_3{\cdot}2THF$	**(143)**	**(119)** + THF			^{1}H NMR; IR	99
$Lu[BH_3Me]_3{\cdot}OEt_2$	**(144)**	A			^{1}H NMR; IR	99
$Lu[BH_3Me]_3{\cdot}THF$	**(145)**	**(122)** + THF	White solid	110–112	^{1}H NMR; IR	99

8.2.2.3 Trihydroborate compounds of the actinide metals

Several actinide complexes of the $[BH_3Me]$ anion also exist and they are listed in Table 11. The species $M[BH_3Me]_4$, where M = Th, U, Np, and Zr, are all similar in structure, each containing η^3-$[BH_3Me]$ groups, with 12 bridging hydrogen atoms surrounding the metal atom. The thorium and uranium species, **(146)** and **(149)**, crystallize differently from the zirconium and neptunium species **(132)** and **(154)** in that the latter consist of body-centered packing of the molecules, whereas the former belong to the triclinic space groups.[93–4,101–2] The adducts $Th[BH_3Me]_2{\cdot}Et_2O$ **(147)**[100] and $[Th(BH_3Me)_4{\cdot}THF]_2$ **(148)**[100] are dimeric species and contain pairs of bridging $[BH_3Me]$ groups. The hydrogen atoms in each of the $[BH_3Me]$ groups connect the two metal atoms such that one is triply bridging, and of the other two, one bonds to one metal atom and the other to the second metal atom reminiscent of the bonding in **(63)**.[68] The bridging hydrogen atoms in **(147)** are *cis* to each other whereas in **(148)** they are *trans* across a center of symmetry. The related systems $U(BH_3Me)_4{\cdot}2THF$ **(150)**[103] and $[U(BH_3Me)_4{\cdot}THT]_2$ **(151)**[103] are different (THT = tetrahydrothiophene). They both contain η^3-$[BH_3Me]$ groups, the former is monomeric with *trans*-THF ligands in an octahedral system, whereas the latter dimeric species is bridged by the sulfur atoms of the THT ligands such that the USUS plane is perpendicular to the plane of the THT ligands and the coordination around uranium is distorted octahedral, as it is in the chelated species $[(dmpe)U(BH_3Me)_4]$ **(152)**.[104] The more crowded $(dmpe)_2U[MeBH_3]_4$ **(153)**[104] contains $[BH_3Me]$ groups which are bonded differently. The molecule adopts a distorted pentagonal bipyramidal geometry with both dmpe ligands and a bidentate $[BH_3Me]$ group in equatorial positions and two tridentate $[BH_3Me]$ groups in axial positions.

8.2.3 Dihydroborate Derivatives

A series of systems with the general formula $M[BR_2H_2]Cl$ where R = succinimidyl (suc)[108] or phthalidimyl (phthal)[109] are given in Table 12 along with some physical data. The tetranuclear system $[(Cp_2TiH_2)_2AlH_2BH_2]$ **(155)**[107] has been structurally characterized, and consists of two (CpTi) groups bonded each by two μ-H atoms to an aluminum atom which is in turn bonded to a bidentate BH_4 group which is quite distorted from tetrahedral geometry. The Al···B distance (0.227 nm) is unusually long. Two cobalt-substituted borane complexes are described in Table 12. The unstable species $(CO)_4CoBH_2{\cdot}THF$ **(163)**[110] is prepared in the reaction between $Co_2(CO)_8$ and $BH_3{\cdot}THF$ at −15 °C. As

Table 11 Trihydroborate compounds of the actinide metals.

Compound	*No.*	*Preparation method*	*Color/ form*	*Melting point (°C)*	*Physical data*	*Ref.*
$Th[BH_3Me]_4$	(**146**)	A	White crystals	62–63	^1H NMR; IR; crystal structure	93
$[Th(BH_3Me)_4]_2 \cdot Et_2O$,	(**147**)	A in Et_2O	White crystals	92–93	^1H, ^{11}B NMR; IR; crystal structure	100
$[Th(BH_3Me)_4 \cdot THF]_2$	(**148**)	A in THF	White crystals	112–114	^1H, ^{11}B NMR; IR; crystal structure	100
$U[BH_3Me]_4$	(**149**)	A	Violet crystals	72–74	^1H NMR; IR; crystal structure	93, 94, 101, 102
$U(BH_3Me)_4 \cdot 2THF$	(**150**)	A in THF	Green crystals	95–98	^1H NMR; IR; crystal structure	103
$[U(BH_3Me)_4 \cdot THT]_2$	(**151**)	A in THT	Dark-green crystals	92–94	^1H NMR; IR; crystal structure	103
$(dmpe)U[BH_3Me]_4$	(**152**)	E	Green crystals	131–132	^1H NMR; IR; crystal structure	104
$(dmpe)_2U[BH_3Me]_3$	(**153**)	E	Brown–black needles	130–132	^1H NMR; IR; crystal structure	104
$Np[BH_3Me]_4$	(**154**)	A	Dark green crystals	62–63	^1H NMR; IR; UV–VIS; crystal structure	93, 94, 105, 106

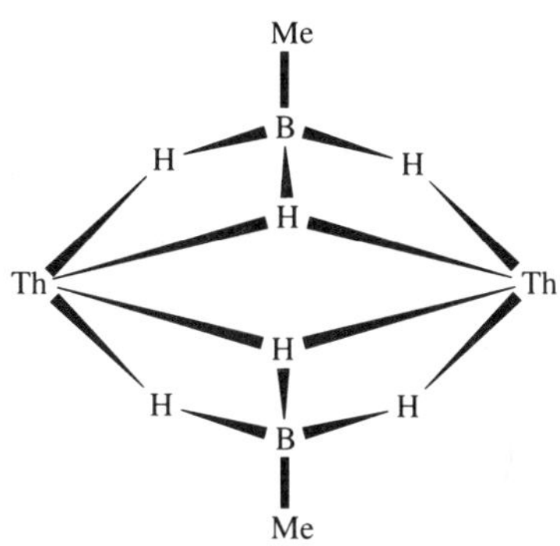

(**147**)

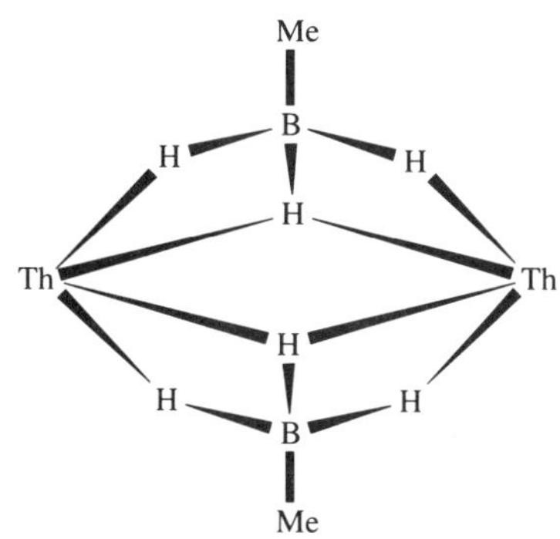

(**148**)

seen in Structure (**163**), it is a THF adduct of the substituted borane $(CO)_4CoBH_2$. Although the species decomposes rapidly, it is characterized by IR and NMR spectral data. A formal derivative of (**163**) is available from the reaction between $CoCl_2$ and $NaBH_4$ in the presence of CO and dppm.[111] The product of this reaction, $[(CO)_2(\eta^1\text{-dppm})Co(\mu\text{-dppm})BH_2]$ (**164**), for which a crystal structure determination confirms the structure, is derived in a formal sense from (**163**) by the replacement of two CO ligands on the Co by two dppm ligands, one monodentate and one bidentate, with the phosphorus of the monodentate group replacing the THF, as indicated in (**164**).

Table 12 Dihydroborate compounds of the transition metals.

Compound	*No.*	*Preparation method*	*Color/ form*	*Melting point (°C)*	*Physical data*	*Ref.*
$[(Cp_2TiH_2)_2AlH_2BH_2]$	(**155**)	D	Dark-brown needles		Crystal structure	107
$Cr[B(suc)_2H_2]_2Cl$	(**156**)	A	Black	360	IR;UV–VIS; magnetic moment	108
$Mn[B(suc)_2H_2]_2Cl$	(**157**)	A	Black	360	IR; UV–VIS; magnetic moment	108
$Mn[(phthal)_2BH_2]Cl$	(**158**)	A	Yellow-brown crystals	260 decomposes	IR; UV–VIS; magnetic moment	109
$Fe[(phthal)_2BH_2]Cl$	(**159**)	A	Dark-brown crystals	275 decomposes	IR; UV–VIS; magnetic moment	109
$Fe[B(suc)_2H_2]_2Cl$	(**160**)	A	Black	360	IR; UV–VIS; magnetic moment	108
$Co[B(suc)_2H_2]_2Cl$	(**161**)	A	Black		IR; UV–VIS; magnetic moment	108
$Co[(phthal)_2BH_2]Cl$	(**162**)	A	Yellow crystals	265 decomposes	IR; UV–VIS; magnetic moment	109
$(CO)_4CoBH_2 \cdot THF$	(**163**)	B	Red in solution		IR; ^1H, ^{11}B NMR; MS	110
$[(CO)_2(\eta^1\text{-dppm})Co\text{-}(\mu\text{-dppm})BH_2]$	(**164**)	C	Yellow crystals		IR; ^{31}P, ^{11}B NMR; crystal structure	111
$Ni[B(suc)_2H_2]_2Cl$	(**165**)	A	Black		IR; UV–VIS; magnetic moment	108
$Ni[(phthal)_2BH_2]Cl$	(**166**)	A	Light green	268	IR; UV–VIS; magnetic moment	109
$Cu[B(suc)_2H_2]_2Cl$	(**167**)	A	Crystals black		IR; UV–VIS; magnetic moment	108
$Cu[(phthal)_2BH_2]Cl$	(**168**)	A	Dark-blue crystals	270	IR; UV–VIS; magnetic moment	109

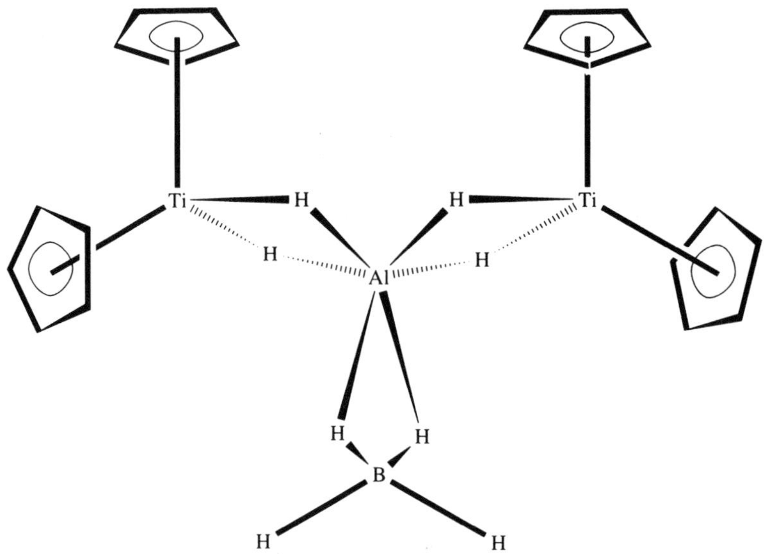

(155)

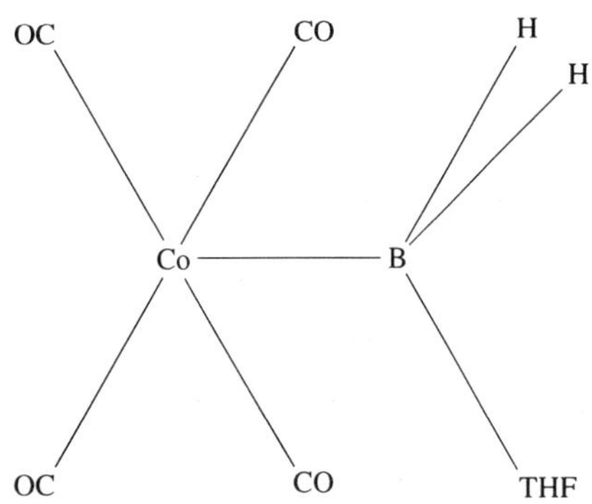

(163)

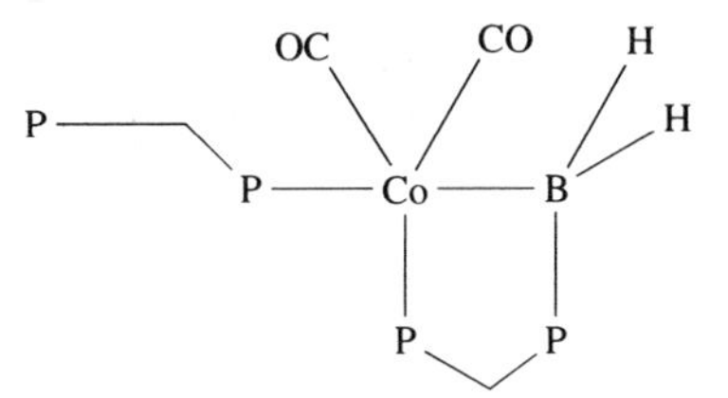

(164)

8.2.4 Monohydroborate Derivatives

Table 13 lists several species with the general formula $M[BHR_3]$ for R = succinimidyl,[113] phthalimidyl[114] and imidazolyl[116] along with physical data.

$[Cp_2\overline{ZrOCH_2B(C_8H_{14})}H]_2$ (**169**)[112] is the unusual product of the insertion of the borane, 9-BBN, into the metallaoxirane Z–C bond of the (η-formaldehyde)zirconocene dimer indicated in Scheme 1. Another insertion of a borane into an organometallic compound occurs in the preparation of $[W_2\{\mu\text{-}MeCB(H)Et\}(CO)_4Cp_2]$ (**172**)[115] from the reaction between $[CpW(\equiv CMe)(CO)_2]$ and $BH_3{\cdot}THF$ in THF. The structure (**172**) of this unusual species, must involve cleavage of the C≡W bond and incorporation of all three H atoms of the BH_3 group into the final product. 9-BBN does not afford a similar product.

8.2.5 Boryl Derivatives of the Transition Metals

Table 14 lists the boryl complexes of the transition metals. A novel series of complexes of the boraalkene 9-fluorenylidene(tetramethylpiperidino)borane with the isolobal fragments $Fe(CO)_3L$, $CpCo(CO)$, $Cp^*Mn(CO)_2$ and $\eta\text{-}(C_6H_6)Cr(CO)_2$[117–19] has been reported and examples are illustrated in Scheme 2. In these systems, the metal–($R_2NB{=}CH_2$) bonding appears to resemble conventional metal–alkene coordination and accounts for the stability of these unusual systems.[119] The other boryl systems listed in Table 14 are more conventional complexes of the transition metals. Thus $[RhHCl(BO_2C_6H_4)(PPr^i_3)_2]$ (**204**)[120] and $[RhHCl(BO_2C_6H_4)(PPh_3)_2]$ (**205**)[121] are trigonal bipyramidal

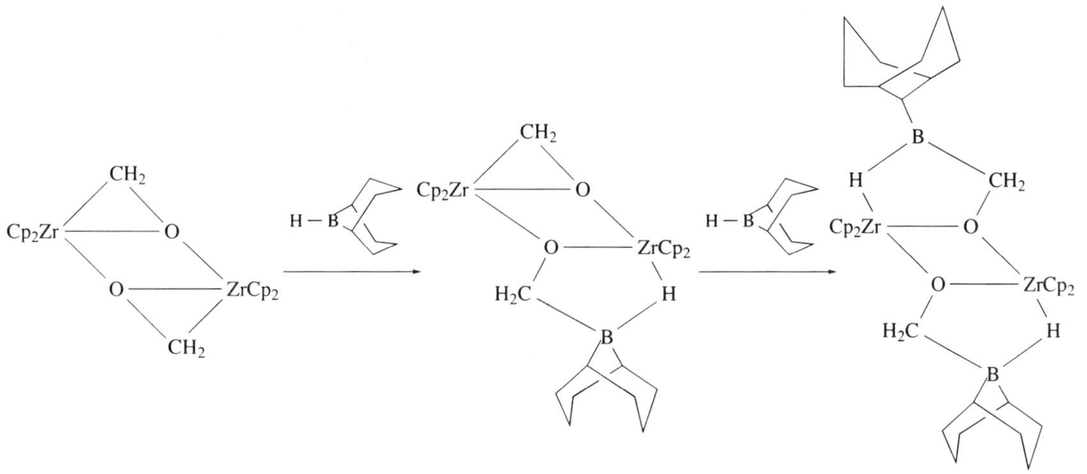

(169)

Scheme 1

Table 13 Monohydroborate compounds of the transition metals.

Compound	*No.*	*Preparation method*	*Color/ form*	*Melting point (°C)*	*Physical data*	*Ref.*
$[Cp_2\overline{ZrOCH_2B(C_8H_{14})H}]_2$	**(169)**	a		206 decomposes	^1H, ^{13}C NMR; crystal structure	112
$Cr[B(suc)_3H]_3$	**(170)**	A	Black		^1H NMR; IR; magnetic moment; UV–VIS	113
$Cr[(phthal)_3BH]_3$	**(171)**	A	Light green	340 decomposes	^1H NMR; IR; magnetic moment; UV–VIS	114
$[W_2\{\mu\text{-MeCB(H)Et}\}\text{-}(CO)_4Cp_2]$	**(172)**	B	Red–orange crystals		^1H, ^{11}B, ^{13}C NMR; IR; crystal structure	115
$Mn[(imidazolyl)_3BH]_2$	**(173)**	A	Grey		IR; magnetic moment; UV–VIS	116
$Mn[B(suc)_3H]_2$	**(174)**	A	Black		^1H NMR; IR; magnetic moment; UV–VIS	113
$Mn[(phthal)_3BH]_2$	**(175)**	A	Yellow	335 decomposes	^1H NMR; IR; magnetic moment; UV–VIS	114
$Fe[B(suc)_3H]_3$	**(176)**	A	Black		^1H NMR; IR; magnetic moment; UV–VIS	113
$Fe[(imidazolyl)_3BH]_3$	**(177)**	A	Gold		IR; magnetic moment; UV–VIS	116
$Fe[(phthal)_3BH]_3$	**(178)**	A	Brown	320 decomposes	^1H NMR; IR; magnetic moment; UV–VIS	114
$Co[(imidazolyl)_3BH]_2$	**(179)**	A	Purple		IR; magnetic moment; US–VIS	116
$Co[B(suc)_3H]_2$	**(180)**	A	Black		^1H NMR; IR; magnetic moment; UV–VIS	113
$Ni[(imidazolyl)_3BH]_2$	**(181)**	A	Black		IR; magnetic moment; UV–VIS	116
$Ni[B(suc)_3H]_2$	**(182)**	A	Black		^1H NMR; IR; magnetic moment; UV–VIS	113
$Ni[(phthal)_3BH]_2$	**(183)**	A	Dark green	305 decomposes	^1H NMR; IR; magnetic moment; UV–VIS	114
$Cu[(imidazolyl)_3BH]_2$	**(184)**	A	Blue		IR; magnetic moment; UV–VIS	116
$Cu[B(suc)_3H]_2$	**(185)**	A	Black		^1H NMR; IR; magnetic moment; UV–VIS	113
$Cu[(phthal)_3BH]_2$	**(186)**	A	Green	320 decomposes	^1H NMR; IR; magnetic moment; UV–VIS	114
$Zn[B(suc)_3H]_2$	**(187)**	A	Black		^1H NMR; IR; magnetic moment; UV–VIS	113
$Cd[B(suc)_3H]_2$	**(188)**	A	Black		^1H NMR; IR; magnetic moment; UV–VIS	113
$Hg[B(suc)_3H]_2$	**(189)**	A	Black		^1H NMR; IR; magnetic moment; UV–VIS	113

[a] Metal complex + 9-BBN

group 9 metal complexes with the boryl moiety in an equatorial position and *mer*-$Ir(H)(BO_2C_6H_4)(Cl)(PMe_3)_3$ **(207)**[123] and *fac*-$[IrH_2(PMe_3)_3(BC_8H_{14})$ **(208)**[124] are octahedral. The structures of (**204**) and **207**) are typical of these species. The species (**204**)–(**208**) catalyze the hydroboration of alkenes.[120–4]

8.2.6 Metal-rich Metallaboranes

8.2.6.1 Introduction

In this section, metal-rich species are defined as those in which the number of metal atoms is greater than the number of boron atoms. A listing of such systems is given in Table 15. This is a very new area of research and all the publications have appeared since the early 1980s. The area has been reviewed

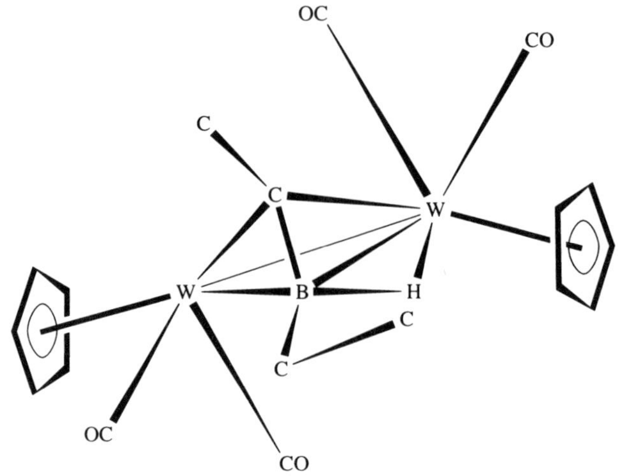

(172)

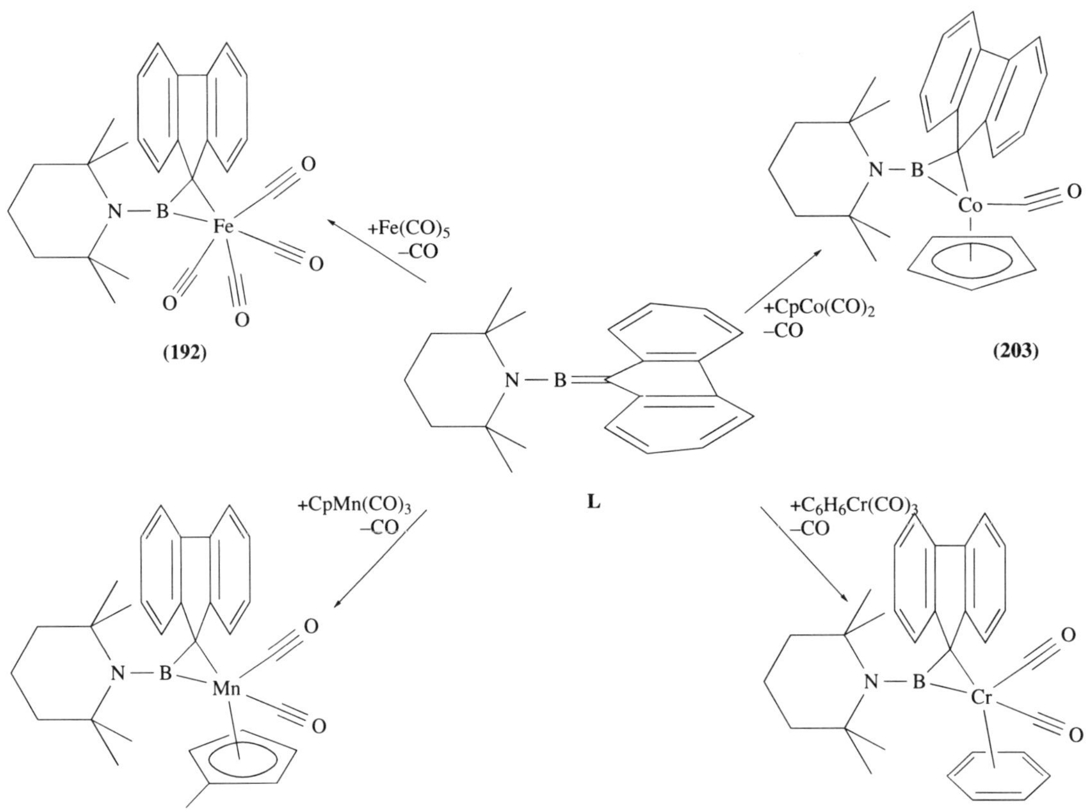

(192) (203) L (191) (190)

Scheme 2

very recently[7] and earlier reviews have also appeared.[8,156] The two prototypical systems $HFe_4(BH_2)(CO)_{12}$ (**218**)[21,130] and $HFe_3(CO)_9BH_4$ (**209**)[125–6] contain partially encapsulated boron atoms and are analogues of well-known organometallic clusters which contain carbidic carbon atoms. Thus, $HFe_4(BH_2)(CO)_{12}$ is a structural and isoelectronic analogue of the so-called Tachikawa cluster $HFe_4(CH)(CO)_{12}$[157–9] and $HFe_3(CO)_9BH_4$ is an analogue of the μ_3-alkylidyne triosmium complex $H_3Os_3(\mu_3\text{-}CH)(CO)_9$.[160–1] The full range of systems includes species such as $[HRu_6(CO)_{17}B]$,[23,142] $[H_2Fe_3Rh_3(CO)_{15}B]$,[24] $[Fe_4(CO)_{12}B(AuPPh_3)_3]$,[22] and $Ru_9(CO)_{23}B_2$[142] which contain completely encapsulated boridic boron atoms. This section is arranged according to the number of metal atoms in the cluster.

Table 14 Diboryl compounds of the transition metals.

Compound	*No.*	*Preparation method*	*Color/ form*	*Melting point (°C)*	*Physical data*	*Ref.*
$(\eta^6\text{-}C_6H_6)(CO)_2[\eta^2\text{-}CR_2(tmp)B]Cr$[a,b]	**(190)**	$(C_6H_6)CrCO_3 + L + h\nu$[c]	Orange	114	^1H, ^{11}B, ^{13}C NMR; IR	117
$(CO)_2[\eta^2\text{-}CR_2(tmp)B](Cp^*)Mn$	**(191)**	$Cp^*Mn(CO)_3 + L + h\nu$	Yellow	131–134 decomposes	^1H, ^{11}B, ^{13}C NMR; IR; crystal structure	117
$(CO)_4[\eta^2\text{-}CR_2(tmp)B]Fe$	**(192)**	$L + Fe_2(CO)_9$	Light brown	136–140 decomposes	^1H, ^{11}B, ^{13}C NMR; IR; MS; crystal structure	117, 118
$(CO)_3[\eta^2\text{-}CR_2(tmp)B]Fe$	**(193)**	$L + Fe(CO)_5 + h\nu$	Dark red	160–165 decomposes	^1H, ^{11}B, ^{13}C NMR; IR; MS; crystal structure	119
$(CO)_3PPh_3[\eta^2\text{-}CR_2(tmp)B]Fe$	**(194)**	(**193**) + PPh_3	Yellow	200–205 decomposes	^1H, ^{11}B, ^{31}P, ^{13}C NMR; IR; MS; crystal structure	119
$(CO)_3PCy_3[\eta^2\text{-}CR_2(tmp)B]Fe$	**(195)**	$L + Fe(CO)_5 + h\nu + PR_3$	Yellow orange	185 decomposes	^1H, ^{11}B, $^{13}C^{31}P$ NMR; IR; MS	119
$(CO)_3P(OMe)_3[\eta^2\text{-}CR_2(tmp)B]Fe$	**(196)**	$L + Fe(CO)_5 + h\nu + PR_3$	Orange	175 decomposes	^1H, ^{11}B, $^{13}C^{31}P$ NMR; IR; MS	119
$(CO)_3(PMe_3)[\eta^2\text{-}CR_2(tmp)B]Fe$	**(197)**	$L + Fe(CO)_5 + h\nu + PR_3$	Orange	168–170 decomposes	^1H, ^{11}B, ^{13}C NMR; IR; crystal structure	117
$(CO)_3(Me_2PhP)[\eta^2\text{-}CR_2(tmp)B]Fe$	**(198)**	$L + Fe(CO)_5 + h\nu + PR_3$	Yellow	154 decomposes	^1H, ^{11}B, ^{13}C NMR; IR	117
$(CO)_3(MePh_2P)[\eta^2\text{-}CR_2(tmp)B]Fe$	**(199)**	$L + Fe(CO)_5 + h\nu + Pr_3$	Orange	178–180 decomposes	^1H, ^{11}B, ^{13}C NMR; IR	117
$(CO)_3(PCl_3)[\eta^2\text{-}CR_2(tmp)B]Fe$	**(200)**	$L + Fe(CO)_5 + h\nu + PR_3$	Orange	124	^1H, ^{11}B, ^{13}C NMR; IR	117
$(CO)_3(Cl_2PhP)[\eta^2\text{-}CR_2(tmp)B]Fe$	**(201)**	$L + Fe(CO)_5 + h\nu + PR_3$	Orange–yellow	155 decomposes	^1H, ^{11}B, ^{13}C NMR; IR	117
$(CO)_3SbPh_3[\eta\text{-}CR_2(tmp)B]Fe$	**(202)**	(**193**) + $SbPh_3$	Yellow–orange	136–140 decomposes	^1H, ^{11}B, ^{13}C NMR; IR; MS	119
$(CO)(\eta^5\text{-}Cp)[\eta^2\text{-}CR_2(tmp)B]Co$	**(203)**	$L + CpCo(CO)2 + h\nu$	Red brown	115 decomposes	^1H, ^{11}B, ^{13}C NMR; IR; crystal structure	117
$[RhHCl(BO_2C_6H_4)(PPr^i_3)_2]$[d]	**(204)**	$HB + [RhCl(N_2)(PPri_3)_2]$[e]	Golden crystals		^1H, ^{11}B, ^{31}P NMR; crystal structure	120
$RhHCl(BO_2C_6H_4)(PPh_3)_2$	**(205)**	$HB + [RhCl(PPh_3)_3$		78 decomposes	^1H, ^{11}B, NMR: IR	121, 122
mer-$IrPMe_3(BO_2C_6H_4)(Cl)\{C(H)C(H)Bu^t\}$	**(206)**	(**207**) + HC≡CtBu			^1H, ^{11}B, ^{31}P NMR	123
mer-$Ir(H)(BO)_2C_6H_4)(Cl)(PMe_3)_3$	**(207)**	$HB + Ir(COE)(PMe_3)_3Cl$	Red crystals crystals		^1H, ^{11}B, ^{31}P, ^{13}C NMR; crystal structure	123
fac-$[IrH_2(PMe_3)_3(BC_8H_{14})$[f]	**(208)**	9-BBN + $IrH(PMe_3)_3$	White crystals		^1H, ^{11}B ^{31}P; crystal structure	124

[a] CR_2 = 9-fluorenylidene. [b] (tmp)B = (2,2,6,6-tetramethylpiperidino)borane. [c] L = 9-fluorenylidene(2,2,6,6-tetramethylpiperidino)borane. [d] $BO_2C_6H_4$ = 1,2-phenylenedioxy)boryl. [e] HB = $HBO_2C_6H_4$. [f] BC_8H_{14} = borabicyclo[3.3.1]nonyl.

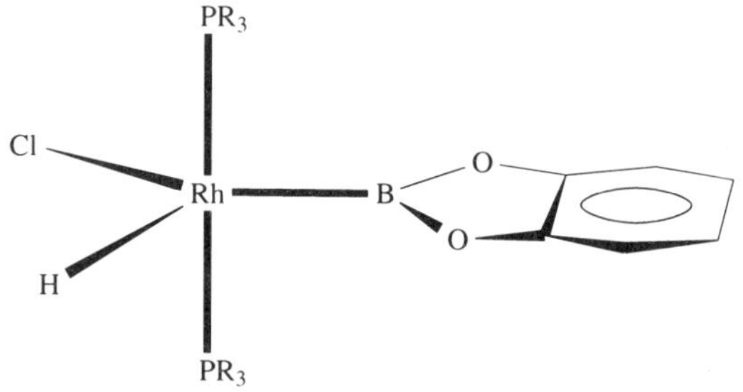

(204)

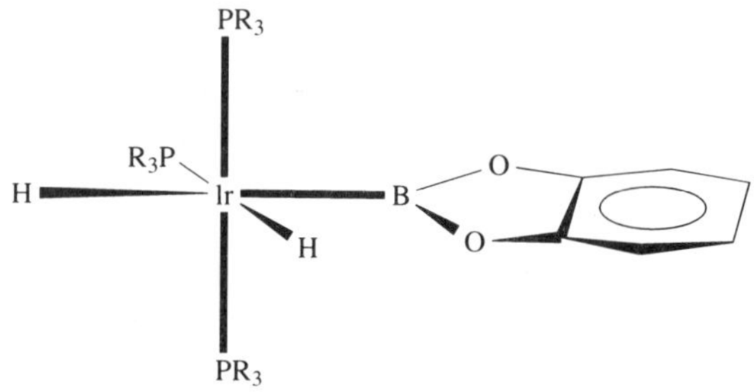

(207)

8.2.6.2 Trimetal clusters

The triiron complex $(\mu\text{-H})Fe_3(CO)_9(\mu_3\text{-}BH_4)$ (**209**)[125–6] is available in low yield by protonation of the product of the reaction between Na[Fe(CO)$_4$C(O)Me] and $BH_3\cdot$THF, and in much higher yields from the reaction between $Fe(CO)_3(\eta^2$-*cis*-cyclooctene) and $BH_3\cdot L$ (L = THF or Me_2S).[162] The structure of (**209**) is given in Scheme 3 and it consists of a triangular array of $Fe(CO)_3$ units capped by a boron atom with three Fe–B bridging hydrogen atoms and one of the Fe–Fe axes also bridged by a hydrogen. (**209**) is readily deprotonated using [PPN]Cl to afford its conjugate base $[(H)Fe_3(CO)_9(\mu_3\text{-}BH_3)]^-$ (**210**)[126–7] which is also shown in Scheme 3. The species (**210**) reacts with Lewis bases to afford products which are the consequence of H_2 replacement rather than replacement of CO. Thus, when (**210**) is treated with CO the species [PPN][$(\mu\text{-CO})Fe_3(CO)_9BH_2$] (**214**),[127,129] which contains a bridging CO and a single Fe–H–B bridge, is formed. Treatment with PMe_2Ph results in the formation of (**215**) in which the phosphine has replaced H_2 and a CO is displaced into a bridging position. If $Fe(CO)_5$ is added to the reaction mixture used for the formation of (**209**), another species, $[(\mu\text{-H})(\mu\text{-CO})Fe_3(CO)_9BH_2]$ (**213**),[127] is isolated. This latter species is the conjugate acid of (**214**) and deprotonation of (**213**) with [PPN]Cl affords (**214**). The species (**213**) is isoelectronic with the triosmium cluster $[(\mu\text{-H})_3Os_3(CO)_9BCO]$ (**253**),[149–50] but the structure is quite different and (**213**) exhibits no tendency to rearrange. The *B*-methyl derivative of (**209**), $[(\mu\text{-H})Fe_3(CO)_9(BH_3Me)]$ (**211**),[126] is available if Li[BEt$_3$H] is included in the reaction mixture used for the preparation of (**209**) and it undergoes similar chemistry to that exhibited by (**209**). An important reaction of $[(H)Fe_3(CO)_9(\mu_3\text{-}BH_3)]^-$ (**210**) is the addition of another $Fe(CO)_3$ moiety in a cluster expansion reaction to afford $[HFe_4(CO)_{12}BH]^-$ (**219**),[131] which is described in Section 8.2.6.3.

The ruthenium analogue of (**209**), $[Ru_3(CO)_9BH_5]$ (**234**),[141] formed in the reaction between $Ru_3(CO)_{12}$ and $BH_3\cdot$THF followed by treatment with Li[BEt$_3$H], exists as two isomers. Their structures are given in Scheme 4 and they exist in approximately equal amounts at room temperature. At higher temperatures NMR spectra suggest that they are interconverting. One isomer is the structural analogue of (**209**) and the other has an additional Fe–H–Fe bridge bond and only two B–H–Fe bonds. They are readily deprotonated using [PPN]Cl to afford the anion $[Ru_3(CO)_9BH_4]^-$ (**235**)[142] which is structurally identical to the iron analogue (**210**). This is illustrated in Scheme 4 along with some chemistry which is described in the next section.

A unique triosmium cluster, $[(\mu\text{-H})_3Os_3(CO)_9BCO]$ (**253**),[149–50] which contains a μ_3-BCO unit is formed in the hydroboration of $(\mu\text{-H})_2Os_3(CO)_{10}$ by B_2H_6 in the presence of NEt_3. The structure of (**253**) is given in Scheme 5 along with some chemistry. A mechanism for the formation of (**253**), which requires insertion of boron into an Os–CO bond, is described.[149–50] The CO may be replaced by PMe_3 to afford (**254**)[149–50] and it is reduced by $BH_3\cdot$THF to afford $[(\mu\text{-H})_3Os_3(CO)_9BCH_2]$ (**255**),[151] which a structural and electronic analogue of the triosmium vinylidene cluster $Os_3H_2(\mu\text{-C=CH}_2)(CO)_9$.[163] This and several other reactions of (**253**) are summarized in Scheme 5. They include activation of the BCO unit with the Lewis acids BCl_3, BBr_3, *B*-Cl-9-BBN, and $PhBCl_2$ to afford species containing the CBX_2 (X = Cl, Br) grouping via a unique intramolecular interchange of the boron and carbon atoms.[152,154] Not shown in Scheme 5 is the reaction of (**253**) with catechol–borane to afford the species $[(\mu\text{-H})_3Os_3(CO)_9B(\mu_3\text{-COBO}_2C_6H_4)]$ (**260**)[150] and other interchange reactions which afford derivatives of the methylidyne cluster $H_3Os_3(\mu_3\text{-CH})(CO)_9$.[154] If the hydroboration of $(\mu\text{-H})_2Os_3(CO)_{10}$ is carried out in CH_2Cl_2 in the presence of excess THF, the boroxine-supported triosmium methylidyne carbonyl cluster $[(\mu\text{-H})_2Os_3(CO)_9(\mu_3\text{-C-})]_3[B_3O_3]$ (**261**)[150] is formed. (**261**) contains a boroxine ring and to each boron atom is bonded a (**253**) cluster through the oxygen of the carbonyl. This species may be converted to σ-bonded polyhedral borane derivatives of the triosmium methylidyne cluster by treatment with the borane in the presence of Lewis acids. Thus, $[(\mu\text{-H})_2Os_3(CO)_9(\mu_3\text{-C-1-}B_5H_8)]$ (**262**)[153] is formed on

Table 15 Metal-rich hydroborate compounds of the transition metals.

Compound	*No.*	*Preparation method*	*Color/form*	*Physical data*	*Ref.*
$(\mu\text{-H})Fe_3(CO)_9(\mu_3\text{-}BH_4)$	**(209)**	B	Dark-red crystals	^1H, ^{11}B NMR; IR; crystal structure	125, 126
$PPN[(H)Fe_3(CO)_9(BH_3)]$	**(210)**	**(209)** + PPNCl	Red-orange	^1H, ^{11}B NMR; IR; MS; Mössbauer	127, 126
$[(\mu\text{-H})Fe_3(CO)_9(BH_3Me)]$	**(211)**	a	Orange	^1H, ^{11}B NMR; IR; MS; Mössbauer	126
$PPN[(\mu\text{-H})Fe_3(CO)_9(BH_2Me)]$	**(212)**	**(211)** + PPNCl	Red	^1H, ^{11}B NMR; IR	126
$[(\mu\text{-H})(\mu\text{-CO})Fe_3(CO)_9BH_2]$	**(213)**	B	Brown	^1H, ^{11}B NMR; IR; MS	127
$PPN[(\mu\text{-CO})Fe_3(CO)_9BH_2]$	**(214)**	**(210)** + CO	Brown	^1H, ^{11}B NMR; IR	128, 129
$PPN[Fe_3(CO)_8(\mu\text{-CO})\text{-}(PhMe_2P)BH_2]$	**(215)**	**(210)** + $PhMe_2P$	Dark brown	^1H, ^{31}P, ^{11}B NMR; IR	129
$[(\mu\text{-H})Fe_3(CO)_9(\mu\text{-CO})BHMe]$	**(216)**	a	Brown	^1H, ^{11}B NMR; MS	129
$PPN[\mu\text{-CO})Fe_3(CO)_9BHMe]$	**(217)**	(**216** + PPNCl	Brown	^1H, ^{11}B NMR	129
$[HFe_4(CO)_{12}BH_2]$	**(218)**	**(270)** + $Fe_2(CO)_9$	Brown	^1H, ^{11}B NMR; crystal structure	21, 130
$PPN[HFe_4(CO)_{12}BH]$	**(219)**	**(210)** + $Fe_2(CO)_9$	Brown	^1H, ^{11}B NMR; IR	131, 132
$[PPN][HFe_4(CO)_{12-n}\text{-}(PPhMe_2)_nBH]$ $(n = 1,2)$	**(220)**	**(219)** + $PPhMe_2$		^1H, ^{31}P, ^{11}B NMR; IR	133
$HFe_4(CO)_{12}(AuPEt_3)BH$	**(221)**	**(219)** + LAuCl	Dark green	^1H, ^{31}P, ^{11}B NMR; IR;	134
$HFe_4(CO)_{12}\{AuP(2\text{-}MeC_6H_4)_3BH$	**(222)**	**(219)** + LAuCl	Light green	^1H, ^{31}P, ^{11}B NMR; IR; crystal structure	135
$HFe_4(CO)_{12}(AuPPh_3)BH$	**(223)**	**(219)** + LAuCl	Dark green brown	^1H, ^{31}P, ^{11}B NMR; IR	134
$Fe_4(CO)_{12}(AuPPh_3)_2BH$	**(224)**	**(219)** + 2LAuCl	Green brown	^1H, ^{31}P, ^{11}B NMR; IR; crystal structure	136
$Fe_4(CO)_{12}(AuPPh_3)(AuPR_3)BH$	**(225)**	**(219)** + LAuCl/ L′AuCl	Dark green	^1H, ^{31}P, ^{11}B NMR	134
$Fe_4(CO)_{12}BHAu_2\{AsPh_3\}_2$	**(226)**	**(219)** + LAuCl	Dark green	^1H, ^{11}B NMR; IR; crystal structure	137
$[Fe_4(CO)_{12}B\{AuPPh_3\}_3]$	**(227)**	**(219)** + $[(LAuCl)_3O]^+$		^1H, ^{11}B, ^{13}C NMR; IR; crystal structure	22
$[AuL_2][\{HFe_4(CO)_{12}BH\}_2Au]$ $(L = PMe_3, PEt_3, PPhMe_2, PPh_2Me)$	**(228)**	**(219)** + LAuCl	Dark green	^1H, ^{11}B, ^{31}P NMR; IR; MS; crystal structure	135, 138
$[(\mu\text{-H})Fe_4(CO)_{12}CBH_2)]$	**(229)**	G	Deep brown	^1H, ^{11}B, ^{13}C NMR; IR; crystal structure	139, 140
$[HFe_4(CO)_{12}C]BHBr$	**(230)**	G with $BH_2Br{\cdot}SH_2$	Black needles	^1H, ^{11}B, ^{13}C NMR; IR; crystal structure	140
$PPN[\textit{trans}\text{-}Fe_4Rh_2(CO)_{16}B]$	**(231)**	**(219)** + $[Rh(CO)_2Cl]_2$	Dark brown	^1H, ^{11}B NMR; IR	24
$PPN[\textit{cis}\text{-}Fe_4Rh_2(CO)_{16}B]$	**(232)**	**(219)** + $[Rh(CO)_2Cl]_2$	Dark brown	^1H, ^{11}B NMR; IR; MS; crystal structure	24
mer-$[H_2Fe_3Rh_3(CO)_{15}B]$	**(233)**	**(231)** + H^+	Brown	^1H, ^{11}B NMR; IR; crystal structure	24
$[Ru_3(CO)_9BH_5]$	**(234)**	G		^1H, ^{11}B NMR; IR	141
$PPN[Ru_3(CO)_9BH_4]$	**(235)**	**(234)** + PPNCl	Yellow	^1H, ^{11}B IR; MS	142
$HRu_4(CO)_{12}BH_2$	**(236)**	G	Dark-yellow crystals	^1H, ^{11}B NMR; IR; crystal structure	23, 143, 144
$[PPN][HRu_4(CO)_{12}BH]$	**(237)**	**(236)** + KH	Brown solid	^1H, ^{11}B NMR; IR	143, 144
$HRu_4(CO)_{12}Au(PPh_3)BH$	**(238)**	**(237)** + LAuCl	Orange–brown solid	^1H, ^{11}B NMR; IR; MS	143
$HRu_4(CO)_{12}\{Au(PPh_3)\}_2BH$	**(239)**	**(237)** + LAuCl	Red solid	^1H, ^{11}B NMR; IR; MS; crystal structure	143
$HRu_4(CO)_{12}BHAu\{P(2\text{-}Me\text{-}C_6H_4)_3\}$	**(240)**	**(237)** + LAuCl	Dark-orange crystals	^1H, ^{11}B, ^{31}P NMR; IR; MS	135
$Ru_4(CO)_{12}BHAu_2\{P(2\text{-}Me\text{-}C_6H_4)_3\}_2$	**(241)**	**(237)** + LAuCl	Bright orange	^1H, ^{31}P, ^{11}B NMR; IR; MS	135
$PPN[Ru_4(CO)_{12}BHAuP\text{-}(2\text{-}MeC_6H_4)_3]$	**(242)**	**(240)** + NEt_3	Bright orange	^1H, ^{31}P, ^{11}B NMR; IR; MS	135
$PPN[\{HRu_4(CO)_{12}BH\}_2Au]$	**(243)**	**(237)** + ClAu(dppm)AuCl	Bright-yellow crystals	^1H, ^{31}P, ^{11}B NMR; IR; MS; crystal structure	135
$[H(CpM)\{Ru_3(CO)_{11}\}BH]$ $(M = W, Mo)$	**(244)**	$[CpM(CO)_3]$ + **(235)**	Orange	^1H, ^{11}B NMR; IR; MS; crystal structure	145
$[HFeRu_3(CO)_{12}BH_2]$	**(245)**	$Fe(CO)_5$ + **(191a)**	Yellow	^1H, ^{11}B NMR	142
$PPN[HFeRu_3(CO)_{12}BH]$	**(246)**	**(245)** + PPNCl	Red–orange	^1H, ^{11}B NMR	142
$[\{Ru_4H(CO)_{12}B\}_2Cu_4(\mu\text{-Cl})]$	**(247)**	**(237)** + $[CuL_4]^+$	Red solid	^{11}B, ^1H NMR; MS; IR; crystal structure	146
$PPN[\{HRu_4(CO)_{12}BH\}_2Ag]$	**(248)**	**(237)** + $[AgL_4]^+$	Yellow	^1H NMR; IR; MS	146
$[\{Ru_4H(CO)_{12}B\}_2Ag_4(\mu\text{-Cl})]$	**(249)**	**(237)** + $[AgL_4]^+$	Red solid	^{11}B, ^1H NMR; IR; MS	146
$[HRu_6(CO)_{17}B]$	**(250)**	$Ru_3(CO)_{12}$ + $BH_3{\cdot}THF$	Red solid	^{11}B, ^1H NMR; IR; MS; crystal structure	23, 142, 147
$[PPN][Ru_6(H)_2(CO)_{18}B]$	**(251)**	Anion of **(283)** + $Ru_3(CO)_{10}(MeCN)_2$	Orange solid	^1H, ^{11}B NMR; IR; crystal structure	148
$Ru_9(CO)_{23}B_2$	**(252)**	**(46)** + $h\nu$	MS		142

Table 15 (continued)

Compound	*No.*	*Preparation method*	*Color/form*	*Physical data*	*Ref.*
$[(\mu\text{-H})_3(CO)_9Os_3BCO]$	(**253**)	F	Bright yellow	^1H, ^{13}C, ^{11}B NMR; IR; crystal structure	149, 150
$[(\mu\text{-H})_3(CO)_9Os_3BPMe_3]$	(**254**)	(**253**) + PMe_3	Yellow solid	^1H, ^{11}B, ^{31}P NMR, IR	149, 150
$(\mu\text{-H})_3(CO)_9Os_3BCH_2$	(**255**)	(**253**) + BH_3	Light yellow	^1H, ^{11}B, IR; MS; crystal structure	151
$(\mu\text{-H})_3(CO)_9Os_3BCX_2(X = Cl, Br)$	(**256**)	(**253**) + BX_3	White solid	^1H, ^{11}B, ^{13}C NMR; IR; MS; crystal structure	152, 154
$(\mu\text{-H})_3(CO)_9Os_3B[\mu_3\text{-}\eta^2\text{-}C(OBC_8H_{14})B(Cl)]$	(**257**)	(**253**) + B-Cl-9-BBN	Pale yellow	^1H, ^{11}B, ^{13}C NMR; IR; MS; crystal structure	154
$(\mu\text{-H})_3(CO)_9Os_3CBX_2L$ ($L = NMe_3, PMe_3, PPh_3$)	(**258**)	(**256**) + L	Light yellow	^1H, ^{11}B NMR; IR; MS	154
$(\mu\text{-H})_3(CO)_9Os_3B[\mu_3\text{-}\eta^2\text{-}C\{OB(Ph)Cl\}B(Cl)]$	(**259**)	(**253**) + $PhBCl_2$	White solid	^1H, ^{11}B, ^{13}C NMR; IR; MS	154
$[(\mu\text{-H})_3(CO)_9Os_3B(\mu_3\text{-}COBO_2C_6H_4)]$	(**260**)	(**253**) + $HBO_2C_6H_4$	Cream solid	^1H, ^{11}B, ^{13}C NMR; IR; MS	150
$[(\mu\text{-H})_2Os_3(CO)_9(\mu_3\text{-}C\text{-})]_3[B_3O_3O_3]$	(**261**)	F/THF	Off-white	^1H, ^{11}B, ^{13}C NMR; IR; crystal structure	150
$[(\mu\text{-H})_2Os_3(CO)_9(\mu_3\text{-}C\text{-}1\text{-}B_5H_8)]$	(**262**)	(**261**) + $B_5H_9/AlCl_3$	White	^1H, ^{11}B NMR, MS	153
$[(CpCo)_3PBPh_2]$	(**263**)	B	Brown crystals	^{11}B NMR; MS; crystal structure	155

[a] $Li[BHEt_3] + Fe(CO)_5 + BH_3{\cdot}THF$.

treatment with B_5H_9 in the presence of $AlCl_3$ and carboranyl derivatives also containing σ-bonds between the methylidyne carbon atom and boron are similarly available.[153]

The last member of this series of trimetallaborane clusters to be described in this section is $[(CpCo)_3PBPh_2]$ (**263**).[155] It is obtained in the reaction between $CpCo(PPh_3)_2$ and $BH_3{\cdot}THF$ at −78 °C. The structure of (**263**) must involve the migration of a Ph group from phosphorus to boron.

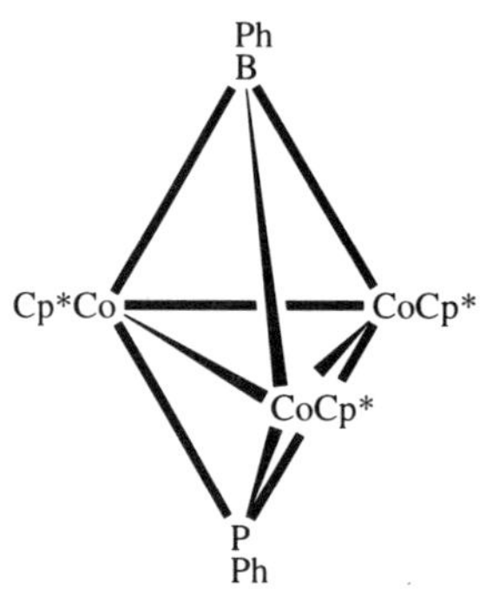

(**263**)

8.2.6.3 Tetrametal clusters

The discovery of the tetrametal "butterfly" cluster $HFe_4(CO)_{12}BH_2$ (**218**)[21,130] stimulated much work.[156] This species is the product of the reaction between $B_2H_6Fe_2(CO)_6$, which is described later (see Section 8.3.2.2), and $Fe_2(CO)_9$. It consists of a butterfly of four $Fe(CO)_3$ units with a BH_2 bridging the "wingtips" and a μ-hydrogen atom between the other two iron atoms, as indicated in Scheme 6. Deprotonation of (**218**) with [PPN]Cl affords $[PPN][HFe_4(CO)_{12}BH]$ (**219**)[131–2] in which one of the Fe–H–B atoms has been removed. In fact all three protons may be removed from (**218**) to form the anion $[Fe_4(CO)_{12}B]^{3-}$ in a process in which NMR spectroscopic data suggest the order of proton removal is: loss of Fe–H–B followed by Fe–H–Fe, and then Fe–H–B.[164] An alternative route to the anion $[HFe_4(CO)_{12}BH]^-$ (**219**) involves a cluster expansion resulting from the treatment of (**210**) with $Fe_2(CO)_9$. This reaction, in which H_2 and $Fe(CO)_5$ are formed in a process where the two skeletal electrons provided by a $Fe(CO)_3$ unit replace those of the leaving H_2, is also shown in Scheme 6. Treatment of (**219**) with phosphine results in the successive replacement of one CO on each of the wingtip iron atoms. The ultimate product, $[PPN][Fe_4(CO)_{10}(PPhMe_2)_2BH_2]$ (**220**),[132–3] which contains two moles of base, has undergone a hydrogen atom migration from the Fe–H–Fe position to the vacant wingtip Fe–B position. This is shown in Scheme 6.

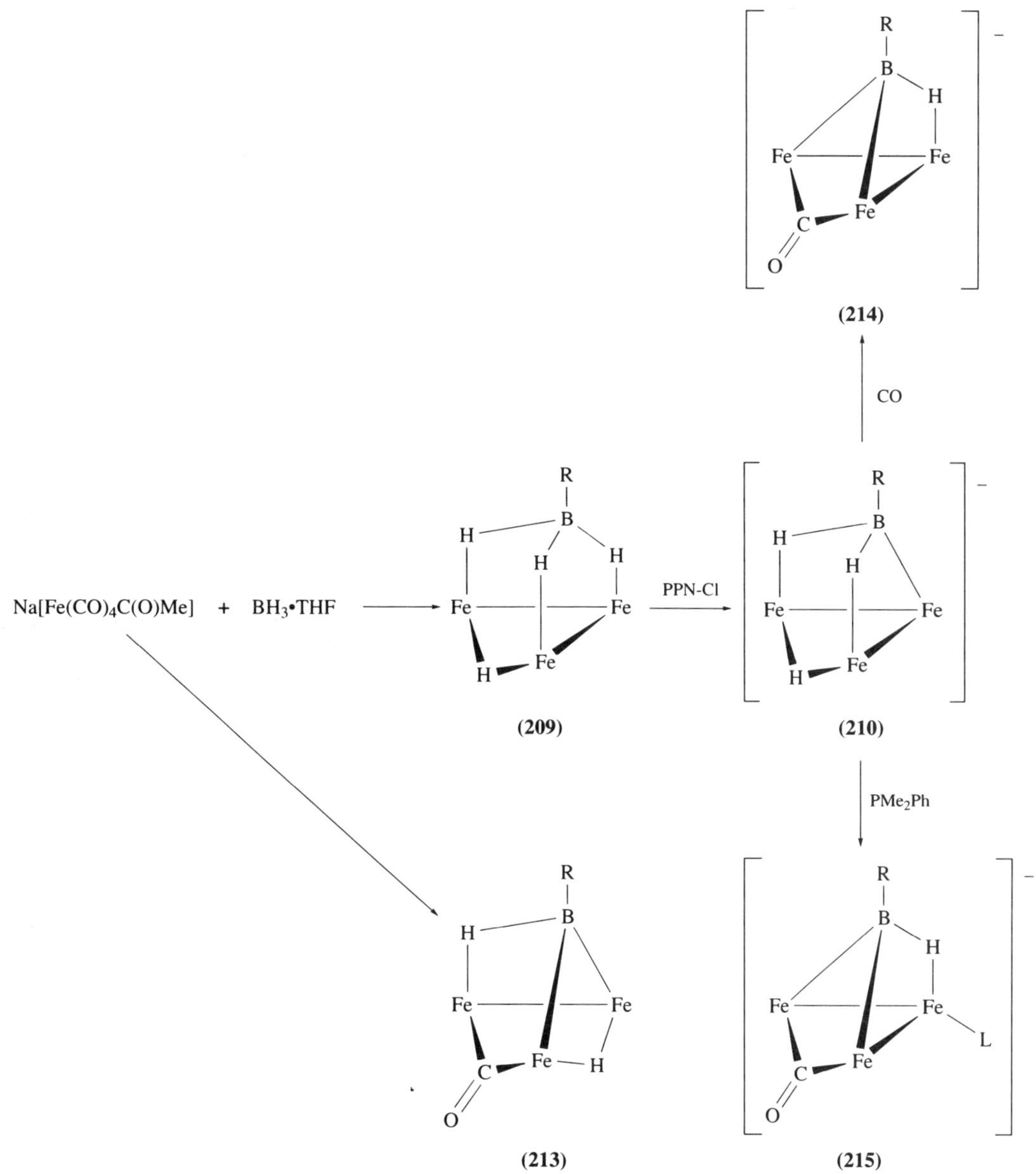

Fe = $Fe(CO)_3$
FeL = $Fe(CO)_2PMe_2Ph$
R = H or Me

Scheme 3

A series of clusters in which the hydrogen atoms in (**218**) are replaced by gold(I) phosphine electrophiles has been prepared. Some of this chemistry is summarized in Scheme 7 and many of the new compounds are included in Table 15. The species $HFe_4(CO)_{12}(AuPPh_3)BH$ (**223**)[134] is obtained by treatment of (**219**) with an equimolar quantity of Ph_3PAuCl. The resulting structure, which is given in Scheme 7, is one in which a R_3PAu group replaces one of the Fe–H–B hydrogen atoms in (**218**).[135] Addition of a second mole of Ph_3PAuCl to (**223**) or addition of two moles of Ph_3PAuCl to (**219**) affords $Fe_4(CO)_{12}(AuPPh_3)_2BH$ (**224**)[136] and over longer periods a third species $[Fe_4(CO)_{12}B(AuPPh_3)_3]$ (**227**)[22] is formed. Products with several phosphine ligands are available including those with mixed systems[136–8] and the related species $[Fe_4(CO)_{12}B(AuAsPh_3)_2H]$ (**226**)[137] has been prepared. $Fe_4(CO)_{12}(AuPPh_3)_2BH$ (**224**), which contains two $AuPR_3$ units, exists as two isomers, as indicated in Scheme 7. Phosphines with small organyl groups tend to favor the symmetrical system (**224a**), which has the two $AuPR_3$ units bridging the Fe–B positions, with the hydrogen atom in the Fe–H–Fe position, but bulkier phosphines favor (**224b**) in which both gold atoms are adjacent and the μ-hydrogen atom occupies one of the original Fe–H–B sites.[165–7] The species in which all three hydrogen atoms have been replaced by $AuPR_3$

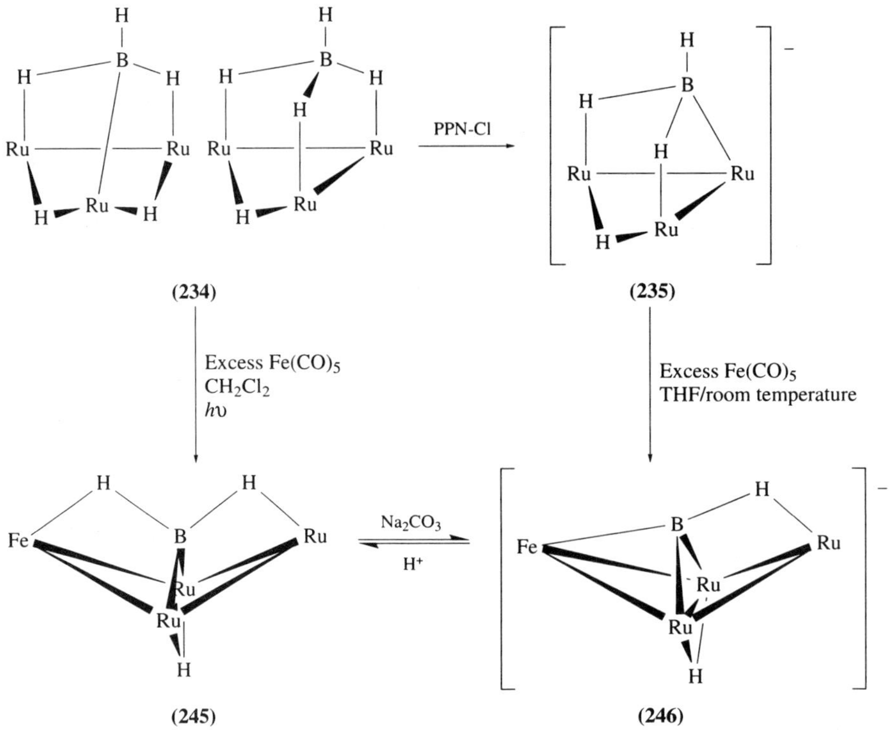

Scheme 4

groups, shown in Scheme 7 as (**227**), contains a boron atom which is completely encapsulated by the iron and gold atoms and is thus a metallaboride. This species is more easily prepared in higher yields by treatment of (**219**) with $[(R_3PAu)_3O][BF_4]$.[22] The metallaboride nature of (**227**) is indicated by its relative stability. (**227**) is air-stable, whereas the precursor (**218**) is quite air-sensitive. Isolation of (**227**) is complicated by the formation of the complex species $[\{HFe_4(CO)_{12}BH\}_2Au][\{Au(PR_3)_2\}BH]$ (**228**),[138,135] when phosphines such as PMe_2Ph or $PMePh_2$ are used. This latter contains a gold atom fused between two "face-to-face" $[HFe_4(CO)_{12}BH]$ units as indicated in (**228**).

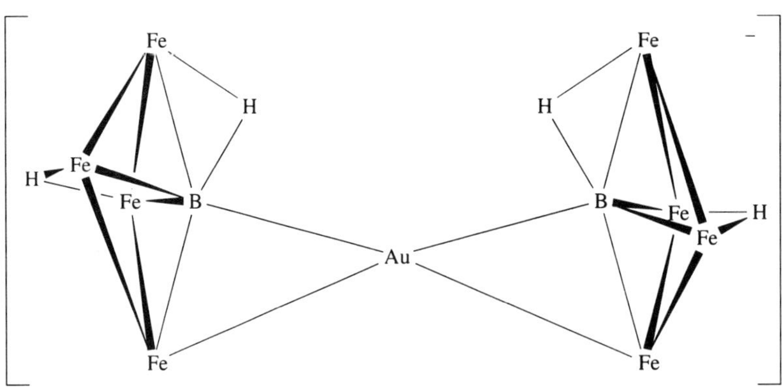

(228)

A trace product in the preparation of (**209**) from THF and $Fe_2(CO)_9$ is the carbido cluster $[(\mu\text{-}H)Fe_4(CO)_{12}CBH_2)]$ (**229**).[139–40] (**229**) is best described as a borane substituted by a carbido cluster. Derivatives with the general formula $[(\mu\text{-}H)Fe_4(CO)_{12}CBXY)]$ (**230**) (X = Y = Cl, Br; X = H, Y = Cl, Br, OH) are also available.[140]

An analogous series of Ru_4 clusters exists.[23,143–4] These are listed in Table 15 and include $HRu_4(CO)_{12}BH_2$ (**236**), $[PPN][HRu_4(CO)_{12}BH]$ (**233**) and a series of $AuPR_3$ derivatives (**238**)–(**243**) similar to those described above.[143,135] Mixed metal butterfly clusters are available from the mixture of

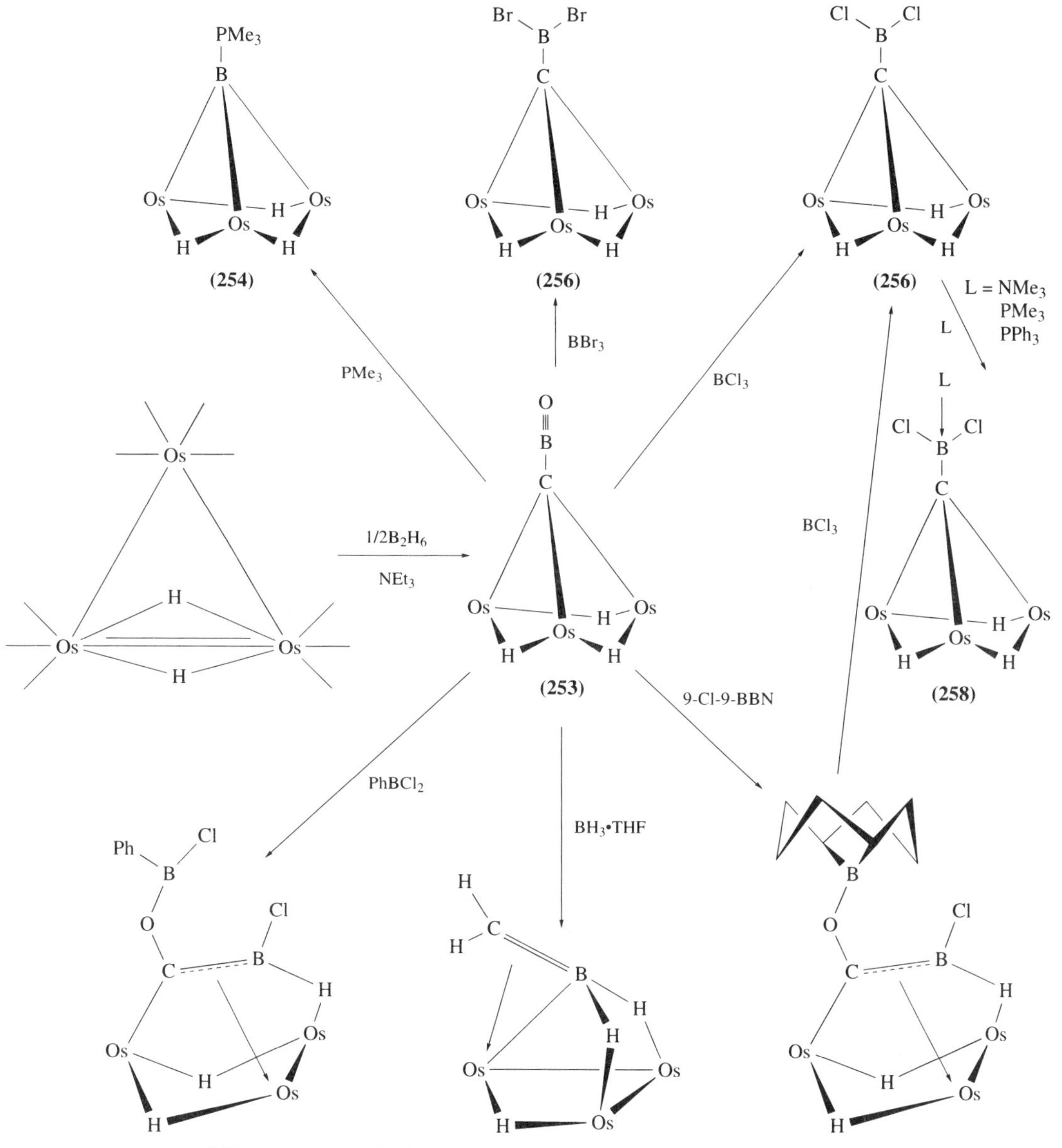

Scheme 5

isomers of [$Ru_3(CO)_9BH_5$] (**234**).[142–3] This chemistry is summarized in Scheme 4. Photolysis of (**234**) in the presence of $Fe(CO)_5$ affords [$HFeRu_3(CO)_{12}BH_2$] (**245**), whereas treatment of the conjugate base of (**234**), [$Ru_3(CO)_9BH_4$]$^-$, with $Fe(CO)_5$ affords PPN[$HFeRu_3(CO)_{12}BH$] (**246**).[142] The related species [H(CpM)$Ru_3(CO)_{11}$)BH] (**244**), where M = W or Mo, is available from similar cluster expansion reactions of (**234**) with {CpM$(CO)_3$}$_2$.[145]

The reactivity of $HRu_4(CO)_{12}BH_2$ (**236**) and [H(CpW)$Ru_3(CO)_{11}$)BH] (**244**) towards alkynes has been studied.[168–70] Reaction of diphenyacetylene with (**236**) under photolysis conditions results in insertion of the alkyne into the cluster such that B–H bond activation, B–C bond formation, and Ru_{hinge}–$Ru_{wingtip}$ bond cleavage occurs to afford [$HRu_4(CO)_{12}$B(H)C(Ph)CHPh] (**264**).[168–9] On the other hand [H(CpW)$Ru_3(CO)_{11}$)BH] (**244**) reacts with diphenylacetylene to afford a product [H(CpW)$Ru_3(CO)_{11}$BC(Ph)CPhH] (**265**), in which the butterfly skeleton is retained and the alkyne interacts with the boron atom and one ruthenium atom only. Reaction of (**236**) with 1-phenyl-1-propyne affords low yields of [$HRu_4(CO)_{12}$B(H)C(Ph)CMeH], presumed to have a structure similar to that of (**264**) and 1-phenyl-2,3-dimethylazulene, a formal dimer of the initial alkyne.[170]

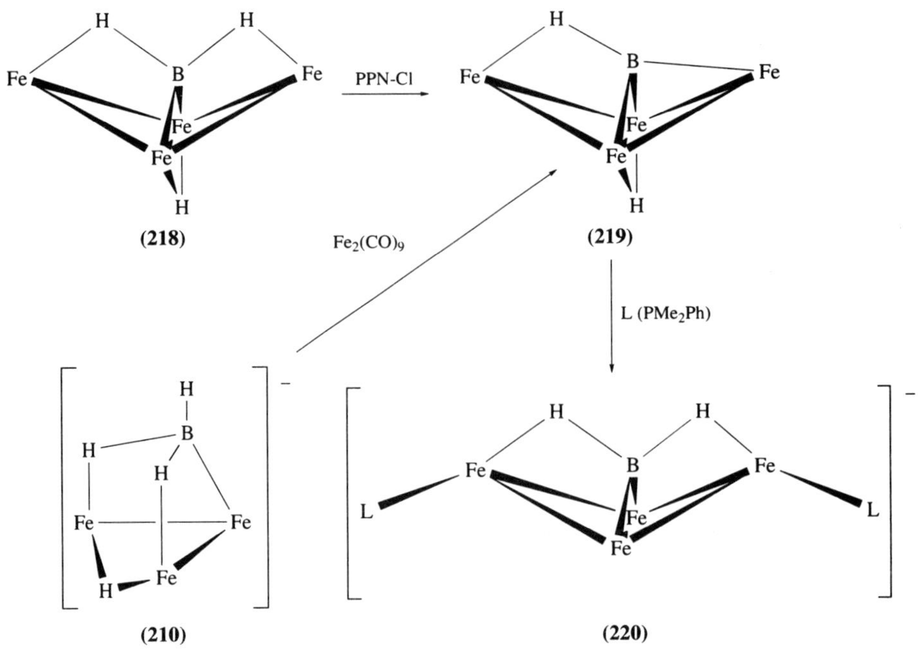

Fe = $Fe(CO)_3$, FeL = $Fe(CO)_2PMe_2Ph$

Scheme 6

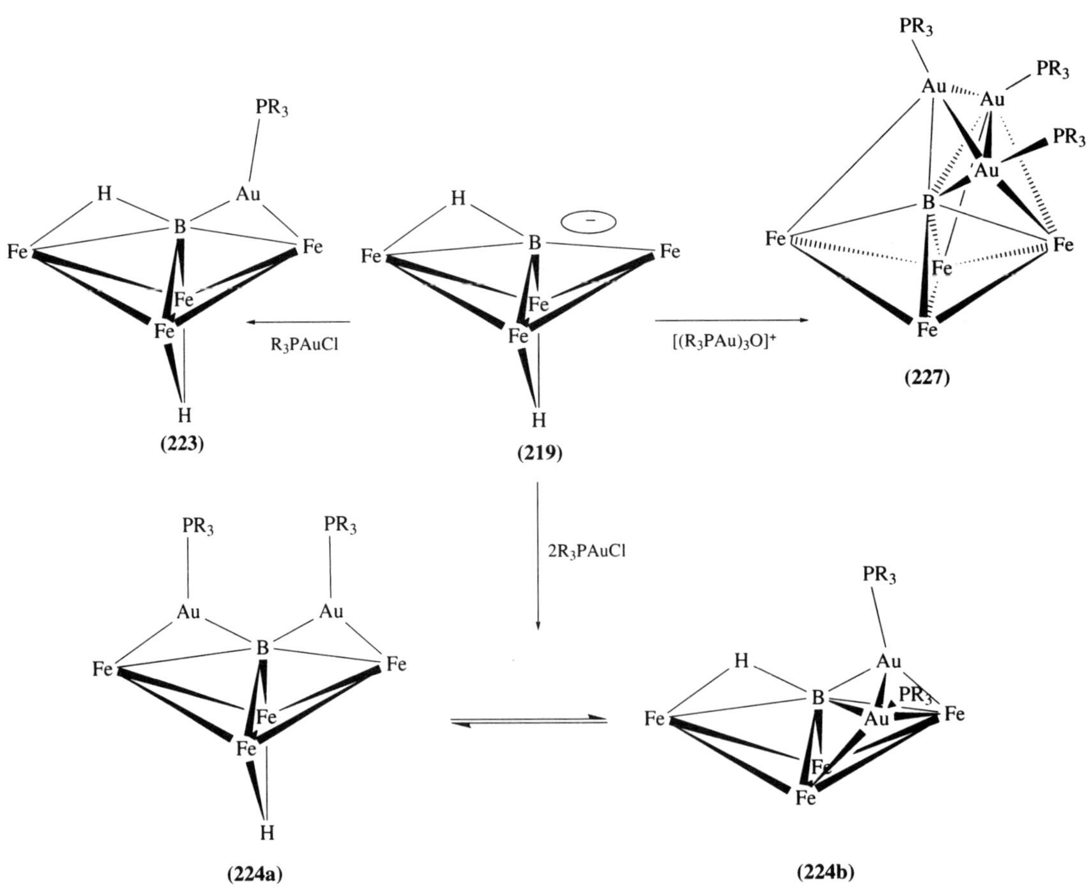

Scheme 7

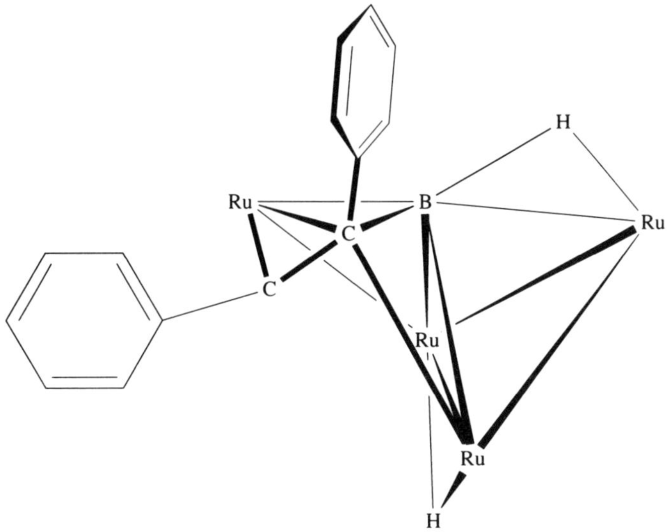

(264)

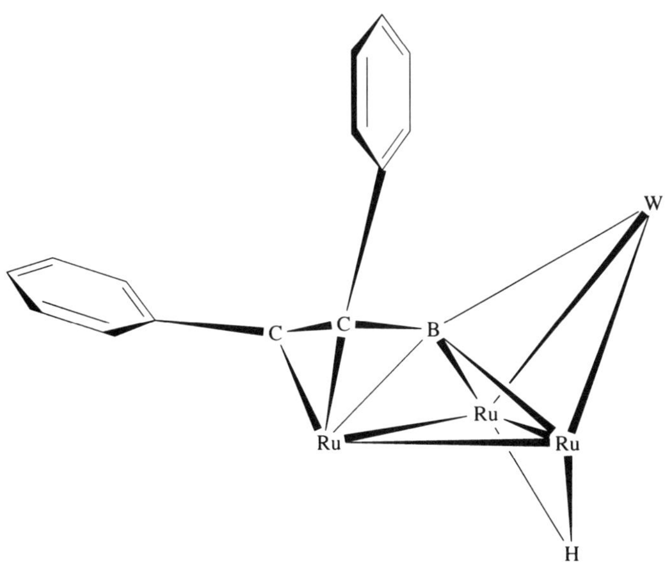

(265)

Ru = Ru(CO)$_3$
W = CpW(CO)$_2$

8.2.6.4 Clusters containing more than four metal atoms

The members of this group are analogues of the well-known octahedral carbidic clusters such as $[Ru_6C(CO)_{17}]$[171] which contain a carbon atom at the center of an octahedron of ruthenium atoms. A boridic cluster isoelectronic with this latter, the 86 electron *closo*-system, $[HRu_6B(CO)_{17}]$ (**250**),[23,142] is obtained in 10% yield by stirring $Ru_3(CO)_{12}$ and $BH_3{\cdot}THF$ in toluene at 75 °C for 5 h, and independently, the conjugate base of (**250**) from the reaction between $[Ru_3(CO)_{11}(NCMe)]$ and $BH_3{\cdot}NMe_3$ at room temperature.[147] An isomer of (**250**) is obtained in the photolysis or thermal decomposition of $[Ru_3(CO)_9BH_5]$ (**234**).[140] Another product of the thermal decomposition of (**234**) is a species tentatively identified as $Ru_9(CO)_{23}B_2$ (**252**) whose proposed structure is a pair of face-fused Ru_6 octahedra each containing an interstitial boron atom.[142] A completely different hexaruthenium boride cluster, the 90 electron *arachno*-trigonal pyramidal $[PPN][Ru_6(H_2)(CO)_{18}B]$ (**251**),[148] is obtained from the reaction between $[Ru_3(CO)_{10}(NCMe)_2]$ and $[PPN][Ru_3(CO)_9B_2H_5]$ (see Section 8.3.3.2). In structure

(**251**) two bridging hydrogen atoms are located. Mixed metal *closo*-octahedral metallaborane clusters are also available. [PPN][$Fe_4Rh_2(CO)_{16}B$],[24] the first *closo*-octahedral metallaborane to be observed, is prepared in the reaction between [PPN][$HFe_4(CO)_{12}BH$] and [$Rh(CO)_2Cl]_2$. The initially formed species in this reaction is the *cis*-isomer (**232**) which rearranges to afford the *trans*-isomer (**231**). Protonation of (**231**) affords [$H_2Fe_3Rh_3(CO)_{15}B$] (**233**) which is formed exclusively as the *mer*-isomer.[24]

The species [PPN][{$HRu_4(CO)_{12}B$}$_2Cu_4(\mu$-Cl)][Cl] (**247**)[146] is obtained as the major product from the reaction between [PPN][$HRu_4(CO)_{12}BH$] and [$Cu(NCMe)_4$][BF_4]. The structure (**247**), consists of two $HRu_4(CO)_{12}B$ butterfly subclusters fused via a tetrahedral Cu_4 unit, one edge of which is bridged by a chlorine ligand. The hydrogen atoms were not located in the structure determination but NMR spectroscopic data indicate that they are not associated with the boron atom. When silver is used instead of copper, the silver analogue of (**228**), that is [PPN][{$HRu_4(CO)_{12}BH$}$_2$Ag] (**248**), is formed along with the silver analogue of (**247**), that is [PPN][{$HRu_4(CO)_{12}B$}$_2Ag_4(\mu$-Cl)][Cl] (**249**).[146]

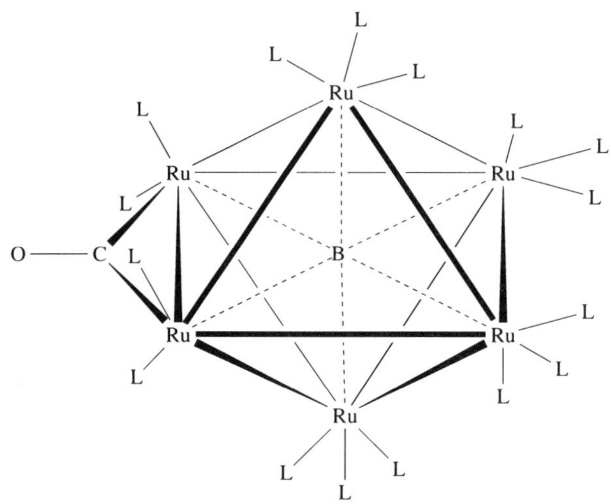

(250)

L = $CO_{terminal}$

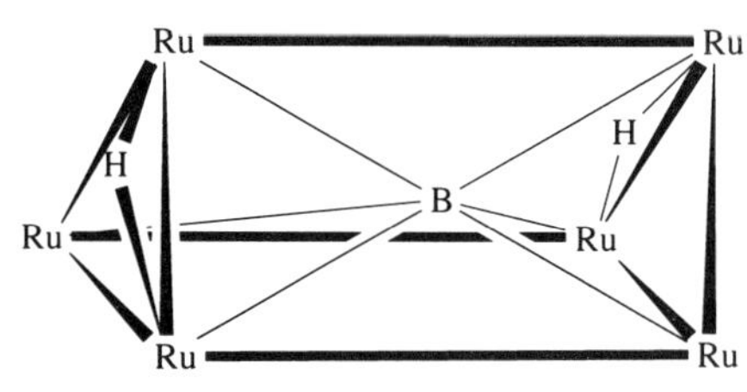

(251)

Ru = $Ru(CO)_3$

8.3 UNITS CONTAINING TWO BORON ATOMS

8.3.1 Heptahydrodiborate Species

[$Cp^*Ru(PMe_3)(\eta^2$-B_2H_7)] (**266**),[172] the first example of an η^2-B_2H_7 bonding geometry in a three vertex metallaborane species, is obtained as a minor product from the reaction between [$Cp^*Ru(PMe_3)Cl_2$] and excess of $NaBH_4$. The crystal structure (**266**), reveals a {Ru}$(H_\mu)_2(B_2H_5)$ central unit.

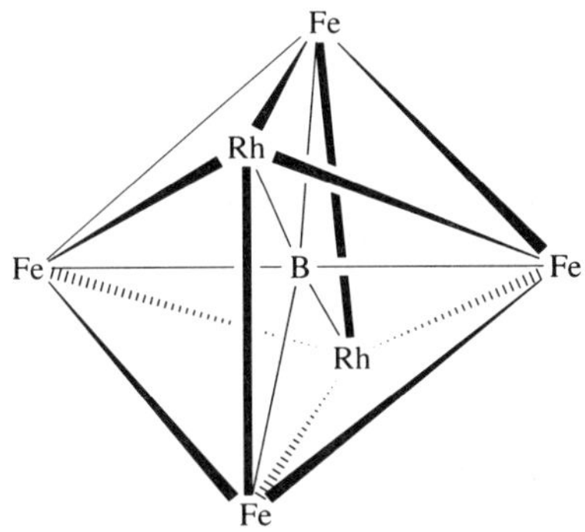

(231)

Fe = Fe(CO)$_3$
Rh = Rh(CO)$_2$

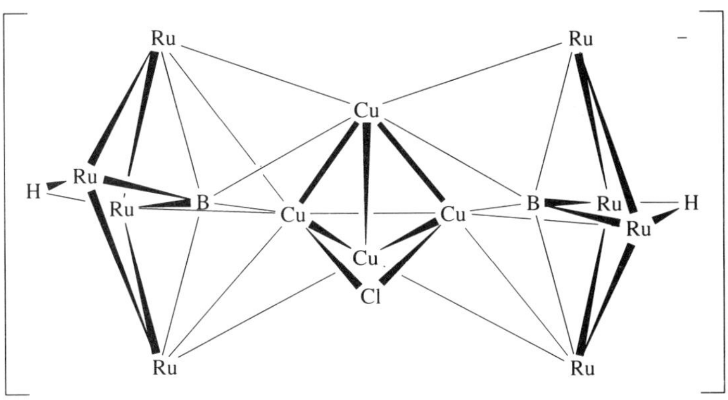

(247)

Ru = Ru(CO)$_3$

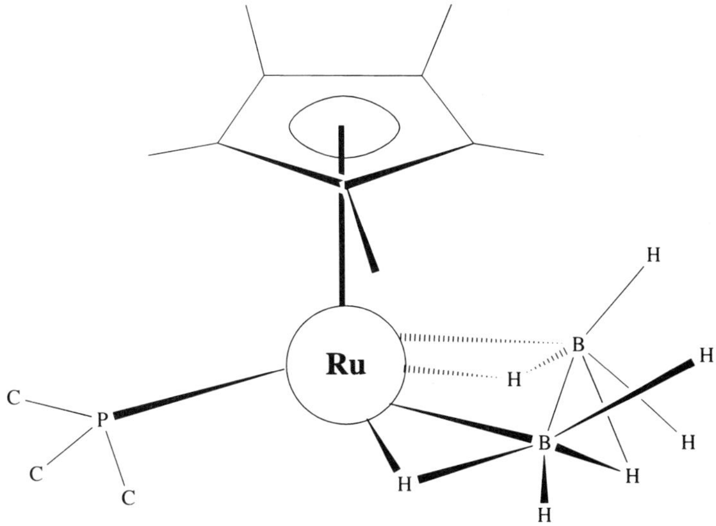

(266)

8.3.2 Hexahydrodiborate Derivatives

8.3.2.1 Group 5 metals

$[Cp^*_2Ta_2(\mu\text{-}X)_2(B_2H_6)]$ (X = Br, Cl) (**267**)[173–5] is prepared as a blue crystalline solid in the reaction between $Cp^*_2Ta_2(\mu\text{-}X)_4$ and two equivalents of $LiBH_4$ in ether. The reactions are given in Scheme 8. NMR spectra suggest that the species contains inequivalent BH_t groups and inequivalent boron atoms resulting in a $(H_\mu)H_tB\text{–}H\text{–}BH_t(H_\mu)_2$ group. A crystal structure determination confirms that the unsymmetric $(H_\mu)H_tB\text{–}H\text{–}BH_t(H_\mu)_2$ grouping exists in the major solution species but NMR spectra indicate the presence of a mixture of species. The related system, $[Cp_2Ta_2(B_2H_6)_2]$ (**268**),[173–5] is formed as a violet solid by addition of four equivalents of $LiBH_4$ to $Cp^*_2Ta_2(\mu\text{-}Cl)_4$ or two equivalents of $LiBH_4$ to $(C_5Me_5)_2Ta_2(\mu\text{-}Br)_2(B_2H_6)$. The NMR spectra suggest a mirror plane with symmetric bridging $(H_\mu)_2H_tB\text{–}H\text{–}BH_t(H_\mu)_2$ groups as indicated in Scheme 8.

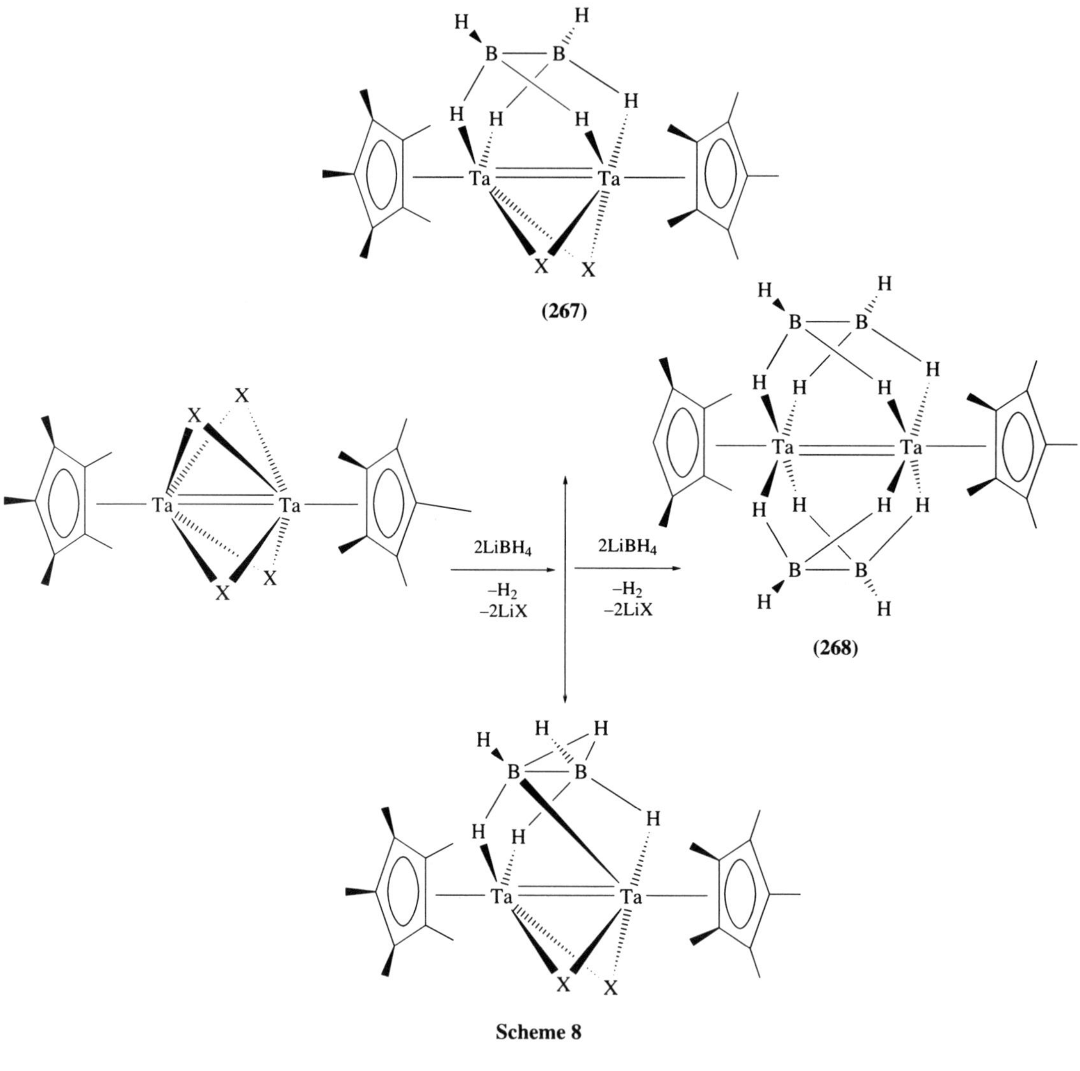

Scheme 8

8.3.2.2 Group 8 metals

The anion $[B_2H_5Fe_2(CO)_6]^-$ (**269**),[176] the conjugate base of $[B_2H_6Fe_2(CO)_6]$ (**270**) which was described in *COMC-I*, is available as orange crystals as either the Ph_4As^+ or the Et_3NH^+ salt. Spectral properties for $[AsPh_4][B_2H_5Fe_2(CO)_6]$ are compared with those for $B_2H_6Fe_2(CO)_6$, and the structures of the acid–base pair, identified on the basis of high-field NMR and Mössbauer spectra, are given in Structures (**269**) and (**270**). Both species exhibit fluxional processes involving terminal and bridging hydrogens. A second proton is removed from $B_2H_6Fe(CO)_6$ by slow treatment with Bu^nLi at low temperature to afford $Li_2[B_2H_4Fe_2(CO)_6]$ (**271**).[177] Treatment of (**271**) with an excess of $Fe(CO)_4Br_2$ affords the coupled cage species *conjuncto*-$B_4H_8Fe_4(CO)_{12}$ (**272**) in what is apparently an oxidative coupling reaction. The x-ray crystal structure, confirmed by a low-temperature determination, is given in Structure (**272**).

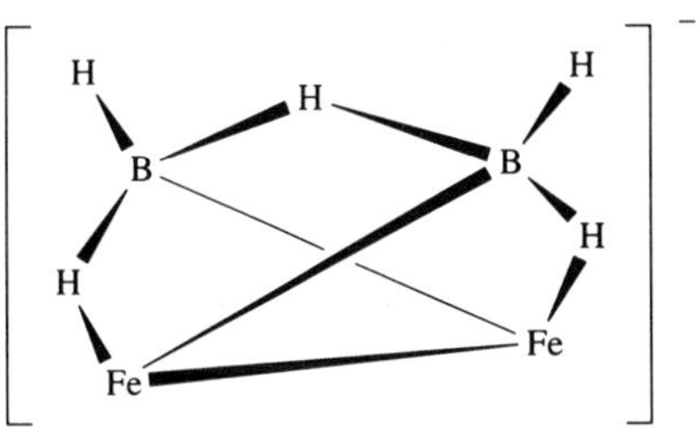

(269)

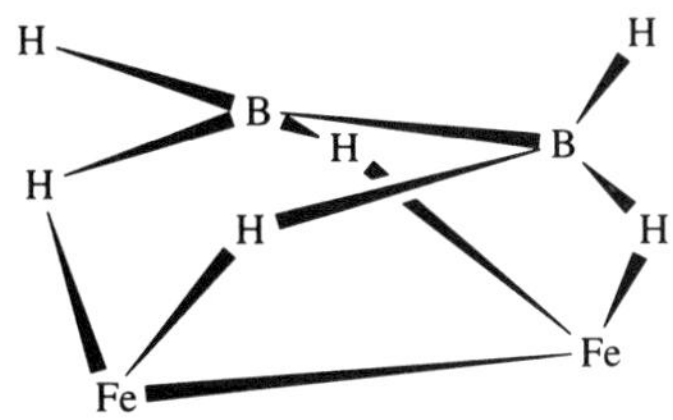

(270)

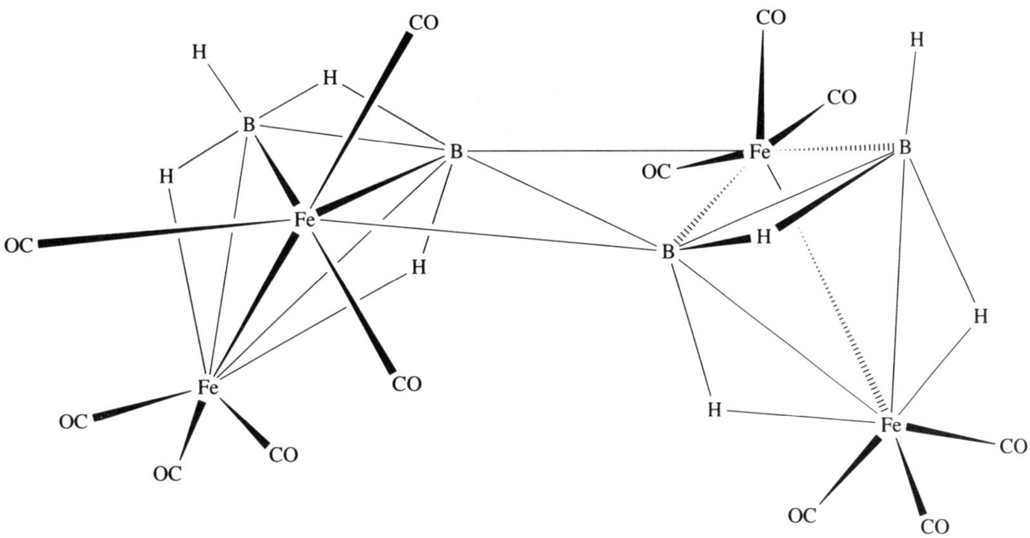

(272)

8.3.3 Pentahydrodiborate Derivatives

The pentahydrodiborate ligand coordinated to a transition metal is the most commonly encountered example of a metalladerivative of diborane(6), and some data are given in Table 16.

Table 16 Spectral data for (η^2-B_2H_5) complexes.

Compound	*No.*	*Color/ form*	*Melting point (°C)*	*Spectral data*	*Ref.*
$[Fe(CO)_4(\eta^2\text{-}B_2H_5)]^-$	(**274**)	Brown solid	50 °C decomposes	^{11}B, ^1H, ^{13}C NMR; IR, Mössbauer	178
$[Ru(CO)_4(\eta^2\text{-}B_2H_5)]^-$	(**275**)			^{11}B, ^1H, ^{13}C NMR; IR	178
$[Os(CO)_4(\eta^2\text{-}B_2H_5)]^-$	(**276**)			^{11}B, ^1H, ^{13}C NMR; IR	178
$CpFe(CO)_2(\eta^2\text{-}B_2H_5)$	(**277**)	Orange–yellow crystals	Sublimes	^{11}B, ^1H, ^{13}C NMR; IR, Mössbauer crystal structure	178
$CpRu(CO)_2(\eta^2\text{-}B_2H_5)$	(**278**)	Yellow solid	Decomposes	^{11}B, ^1H, ^{13}C NMR; IR	178
$(Cp)_2Mo(H)(\eta^2\text{-}B_2H_5)$	(**273**)	Yellow solid		^{11}B, ^1H NMR; crystal structure	172
$(CpCo)_2(\mu\text{-}PPh_2)B_2H_5$	(**284**)	Dark red crystals		Crystal structure	182

8.3.3.1 Group 6 metals

$[Cp_2MoH(\eta\text{-}B_2H_5)]$ (**273**)[172] is prepared as a yellow solid in the photolysis of a mixture of $[Cp_2MoH_2]$ and excess $BH_3{\cdot}THF$, or by treatment of $[Cp_2MoCl_2]$ with $Na[B_3H_8]$. A crystal structure determination reveals the geometry given in Structure (**273**). It consists of a Mo–B_2 triangle with a H_μ bridging the two BH_2 groups and a unique hydrogen atom bonded to molybdenum which is not bridging to a boron atom.

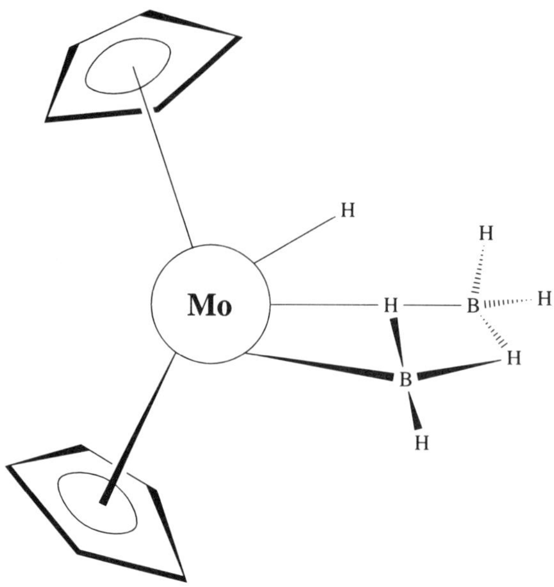

(273)

8.3.3.2 Group 8 metals

The series of species $K[M(CO)_4(\eta^2\text{-}B_2H_5)]$ for M = Fe (**274**), Ru (**275**) and Os (**276**) and also $[CpM'(CO)_2(\eta^2\text{-}B_2H_5)]$, for M' = Fe (**277**), Ru (**278**), described as the iron species in *COMC-I*,[178] has been extended and data for all the known members of the class are given in Table 16. Although the structures of the $K[M(CO)_4(\eta^2\text{-}B_2H_5)]$ species, assumed from spectral data are analogues of B_2H_6, they are also analogues of metal–alkene complexes. Mössbauer spectra for the iron species provide support for the existence of a three-center two-electron bond involving dsp^3 type bonding and they mitigate against the iron being six-coordinate. Relative stabilities of the species decrease in the order Fe≥ Os ≫ Ru. An x-ray structure determination of (**277**), obtained at −174 °C, reveals that the species also has the structure of B_2H_6 in which one of the bridging protons has been replaced by $CpFe(CO)_2$. The bonding between the B_2H_5 group and the iron atom is considered to be similar to that for $K[M(CO)_4(\eta^2\text{-}B_2H_5)]$, but in this case the orbital from iron which interacts in a three-center two-electron bond is considered to be d^2sp^3 hybridized. Treatment of (**276**) with HCl gives $H(CO)_4Os(B_2H_5)$ (**279**)[179] according to Equation (4). NMR spectra for (**279**) suggest a structure with the hydrogen atom residing in a position *cis* to the B_2H_5 ligand. Treatment of (**276**) with $[Me_3O][BF_4]$ or $(PPh_3)AuCl$ affords $[Me_3O(CO)_4Os(B_2H_5)]$ (**280**) and $(PPh_3)Au(CO)_4Os(B_2H_5)$ (**281**), respectively.

$$K[(CO)_4Os(B_2H_5)] + HCl \longrightarrow H(CO)_4Os(B_2H_5) + KCl \qquad (4)$$

$[HRu_3(CO)_9B_2H_5]$ (**282**)[180] and $[HRu_3(CO)_8(PPh_3)B_2H_5]$ (**283**)[181] are unique examples of $\eta^2\text{-}B_2H_5$ coordination. The former (**282**) is prepared as a minor product in the reaction between $Ru_3(CO)_{12}$, $BH_3{\cdot}THF$, and $Li[BEt_3H]$ and in higher yield from the reaction between $[B_3H_8]^-$ and $Ru_3(CO)_{12-x}(MeCN)_x$. Spectral data suggest the structure is similar to that given in Structure (**283**). If (**282**) is photolyzed in the presence of PPh_3, the monosubstituted derivative, (**283**) is obtained as a yellow crystalline solid. An x-ray structure determination for (**283**) confirms that both species are analogues of B_5H_9 in which three BH units have been subrogated by three isolobal $Ru(CO)_2L$ (L = CO, PPh_3) groups. An alternative description of the structure of (**282**) and (**283**) is that they are analogues of an alkene coordinated to a Ru_3 cluster. Deprotonation of (**282**) to $[Ru_3(CO)_9B_2H_5]^-$ removes the Ru–H–Ru bridging proton but the resulting anion shows no fluxionality of the bridging hydrogen atoms, similar to the related species, $[Fe_2(CO)_6B_3H_6]^-$ (**307**) and in contrast to the parent borane anion $B_5H_8^-$.

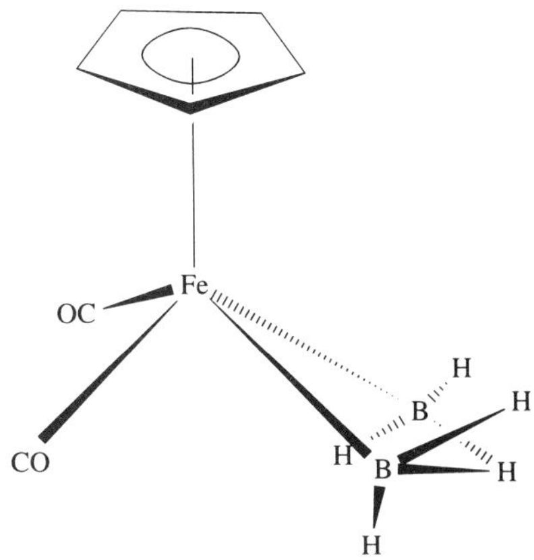

(274)–(276)

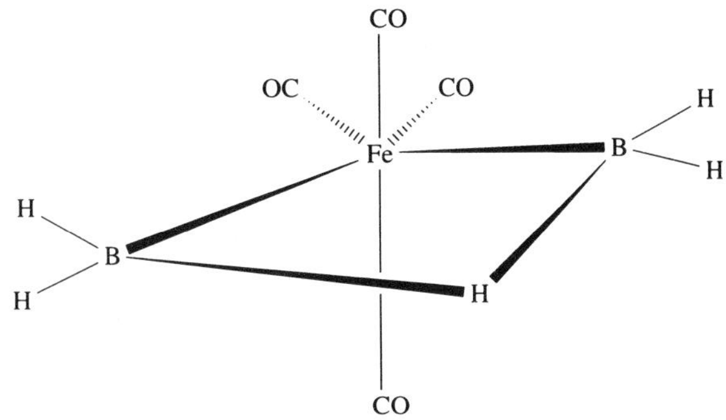

(277)–(278)

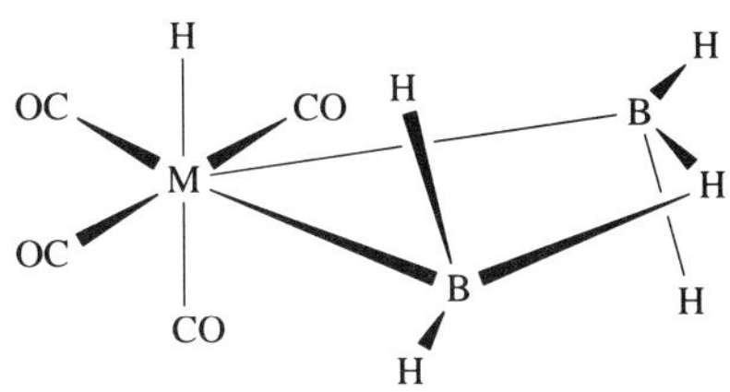

(279)

M = Os, Ru

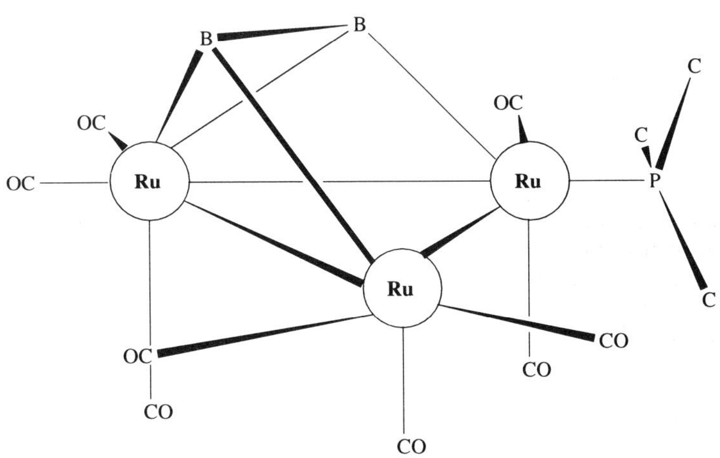

(283)

8.3.3.3 Group 9 metals

$Cp_2Co_2(\mu\text{-}PPh_2)B_2H_5$ **(284)**[182] is formed as dark-red air-stable crystals as one of the products of the reaction between $BH_3{\cdot}THF$ and $CpCo(PPh_3)_2$. A crystal structure determination reveals that the species contains a PPh_2 bridged $(CpCo)_2$ fragment which is also asymmetrically bridged by a B_2H_5 ligand. One boron atom, B-1, is bonded to both cobalt atoms while the other, B-2 is bonded to only one cobalt atom. One Co–B axis is bridged by a hydrogen atom, whereas the other is bridged by a BH_3 moiety forming a B–H–B bond. This form of the B_2H_5 ligand differs from the others and it may be described as an analogue of a C_2H_3 ligand with a σ–π bonding mode.

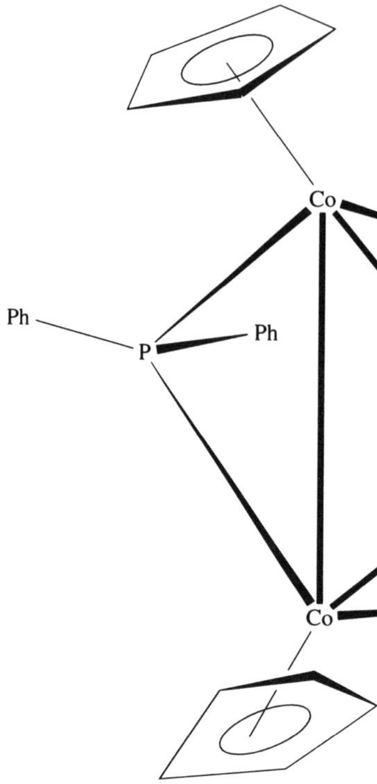

(284)

8.3.3.4 Group 10 metals

The B_2H_5 moiety is also found in the species $[Pt_2(PMe_2Ph)_2(B_2H_5)(B_6H_9)]$ (**285**),[183] a complex metallaborane which contains a η^3-B_2H_5 ligand as well as a η^3-B_6H_9 ligand. The system is described later but the B_2H_5 portion of the molecule is seen in Structure (**285**). It resembles that in $[Cp_2Co_2(\mu\text{-}PPh_2)B_2H_5]$ (**284**).

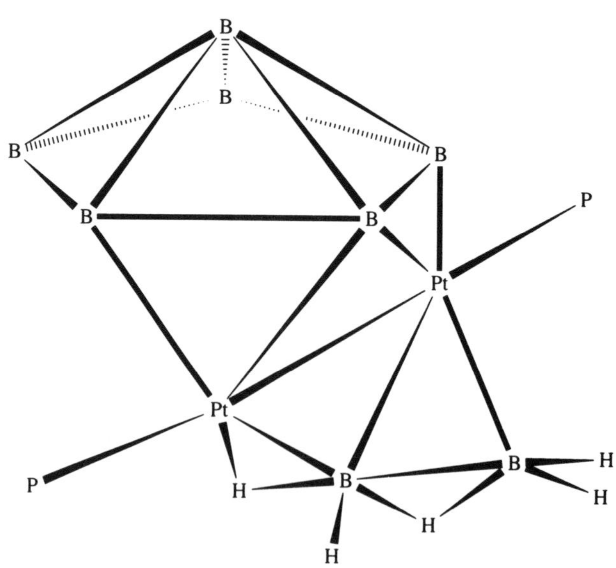

(285)

8.3.4 Diborane(4) Derivatives

$[ZnCl_2(B_2H_4{\cdot}2PMe_3)]$ (**286**)[184] is formed as a stable solid in the reaction between $B_2H_4{\cdot}2PMe_3$ and $ZnCl_2$ in CH_2Cl_2 at ambient temperature. Structure (**286**) involves simple coordination of the diborane(4) adduct to the metal to afford a $Zn(HB)_2$ system reminiscent of the $[B_3H_8]^-$ anion.[185] The related species $[Ni(CO)_2\{B_2H_4{\cdot}2PMe_3\}]$ (**287**),[186] $[Cr(CO)_4(B_2H_4{\cdot}2PMe_3)]$ (**288**),[187] and $[Cu(B_2H_4{\cdot}2PMe_3)_2]X$,(X = Cl, I) (**289, 290**)[188] have also been prepared. The structures of these species are similar to (**286**). In (**287**) the nickel atom is tetrahedrally coordinated to two CO ligands and the B_2H_4 moiety, and in (**288**) the geometry around the chromium atom is approximately octahedral with the metal bonding to four CO atoms and the B_2H_4 moiety. The $[Cu\{B_2H_4{\cdot}2PMe_3\}_2]$ cation in (**290**) involves a Cu^I ion coordinated to two diborane(4) adducts to afford a tetrahedral complex. The yellow solid (**288**) is the final product in the photolytic reaction between $Cr(CO)_6$ and $B_2H_4{\cdot}2PMe_3$. The initially formed species is $[Cr(CO)_5(B_2H_4{\cdot}2PMe_3)]$ (**292**)[187] for which an x-ray structure determination indicates monodentate coordination of the diborane(4) moiety through a single Cr–H–B bridge bond. The diborane(4) ligand adopts an eclipsed conformation in (**292**), as it does in its other metal complexes, in contrast with the free borane which adopts the staggered conformation.

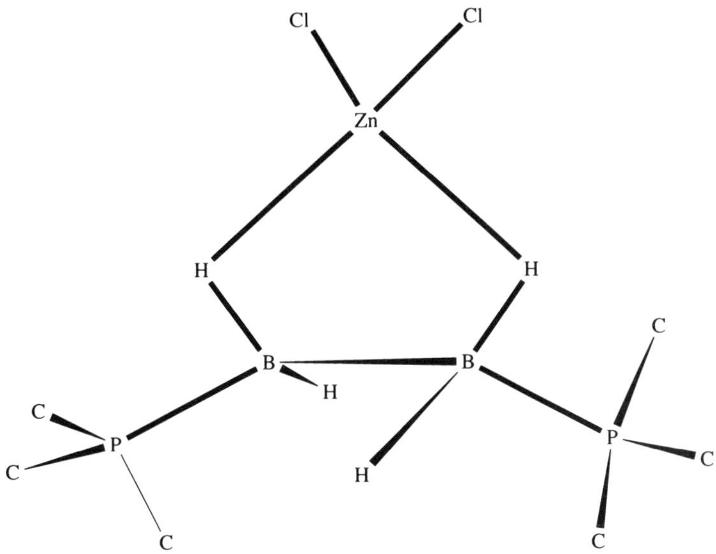

(286)

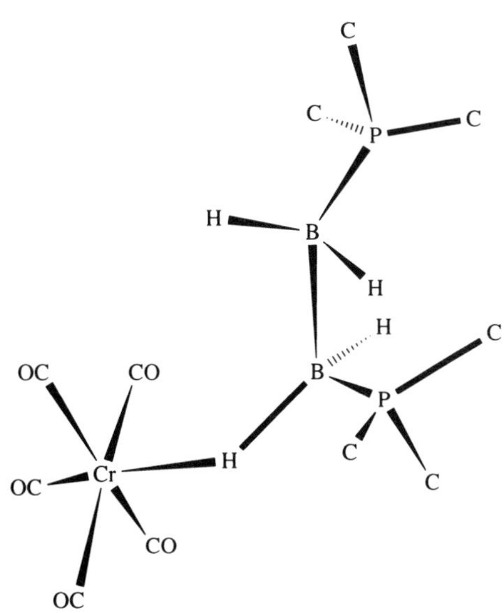

(292)

8.3.5 Diborane(2) Derivatives

$[(\mu_3\text{-H})_2[CpCo]_4B_2H_2]$ (**293**)[189] is obtained as ruby-red, air-stable crystals by stirring excess $BH_3{\cdot}THF$ and $CpCo(PPh_3)(C_2Et_2)$. The structure is that of a *closo*-system with adjacent boron atoms. 2-Ph-1,3,6,7,2-$[CpCo]_4PB_2H_2$ (**294**)[190] is obtained as green crystals by varying the conditions used for the reaction described for (**293**) in order to reduce the borane:cobalt ratio to 2.5:1 by adding $BH_3{\cdot}THF$ slowly over a 1 h period. The structure (**294**) indicates that a PhP group has inserted into the $[CpCo]_4B_2H_2$ cluster but the results suggest that under BH_3 deficient conditions, a $CpCo(PPh_3)$ fragment has incorporated into the cluster building process followed by reductive elimination of C_6H_6 to yield the PPh fragment-containing pentagonal bipyramidal structure.

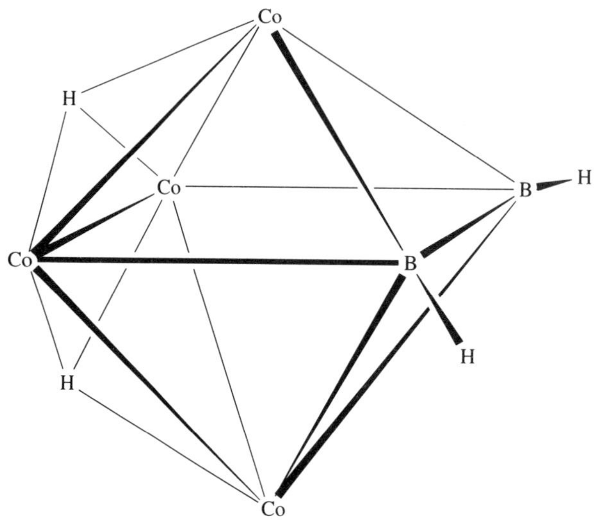

(293)

Co = CpCo

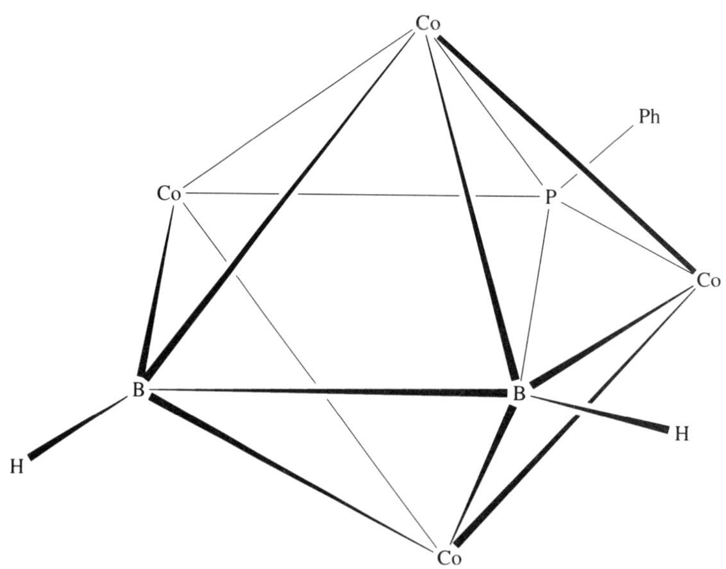

(294)

Co = CpCo

8.4 UNITS CONTAINING THREE BORON ATOMS

8.4.1 Introduction

There are several complexes of triboron ligands with metal centers and their classification, in terms of structure and bonding, has been addressed in the literature.[2a,191–2] The 2-metalla-*arachno*-tetraboranes involve replacement of B-2, that is, a BH_2 group in B_4H_{10} by an isolobal metal group such that the borane functions as a bidentate ligand and formally donates two electron pairs to the metal ML_n^+ moiety via two M–H–B bridge bonds. This is exemplified in [$NbCp_2B_3H_8$] (**295**).[193] The molecule retains the butterfly structure and the metal occupies one of the wingtip positions. Other examples of this type of structure are found for [$WH_3(PMe)_3B_3H_8$] (**296**),[194] [$(CO)H(PPh_3)_2OsB_3H_8$] (**297**),[195] [$RuH(B_3H_8)(CO)(PPh_3)_2$] (**298**),[196–9] and [$(PPh_3)_2H_2IrB_3H_8$] (**299**)[195] which are described below. The other commonly observed structure is that in which the metal moiety replaces B-1 in B_4H_{10} and the triborane is effectively a tricoordinate ligand in which there are no M–H–B bridge bonds, as indicated for the *arachno*-iridatetraborane, [$(IrB_3H_7)(CO)H(PPh_3)_2$] (**300**).[200] The three boron atoms and the metal atom are almost planar with the central boron atom out of the plane such that the dihedral butterfly angle is 127°. The first example of this class to be observed was $(PMe_2Ph)_2PtB_3H_7$ (**301**)[201] which is described in *COMC-I* but the structure, which suggested that the Pt–B_3H_7 interaction was asymmetric, was disordered. The platinum complex (**301**) was referred to as a borallyl complex and was considered to involve donation of two pairs of electrons analogous to conventional allyl complexes. The structure of (**301**) and a series of palladium and platinum analogues of the type [(L–L)MB_3H_7] (**302**)–(**306**)[202–4] suggest that, in a formal sense, this class of species involves the donation of three pairs of electrons to the metal by the B_3H_7 ligand. The 2-metallatetraboranes typically have a B–B–B angle of ~60°, whereas in the 1-metallatetraboranes the B–B–B angle is between 112° and 120°. The 2-metallatetraboranes are obtained for the earlier transition metals, whereas the 1-metallatetraboranes are found for the later transition metals. Another mode of coordination of the B_3H_7 moiety involves the dimetallapentaborane, [$(CO)_6Fe_2B_3H_7$] (**307**),[205] which is related to the species (**282**) [$HRu_3(CO)_9B_2H_5$] described above. Compound (**307**) has a B–B–B angle of 90° and the Fe–B_3 bonding may be considered to be intermediate between that in (**300**) and (**297**). An additional example of triborane coordination is found in [$CpFe(CO)B_3H_7(PPh_2)$](**308**)[206] and this is described below.

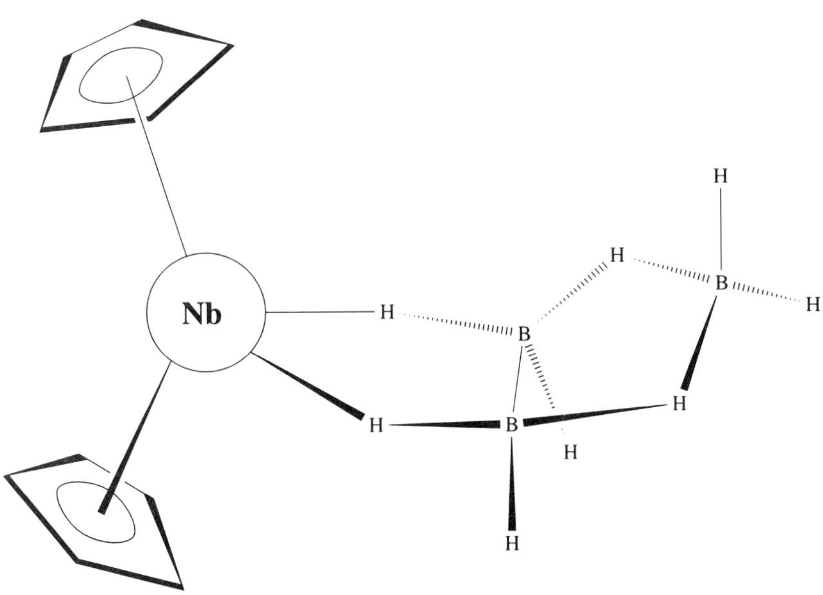

(295)

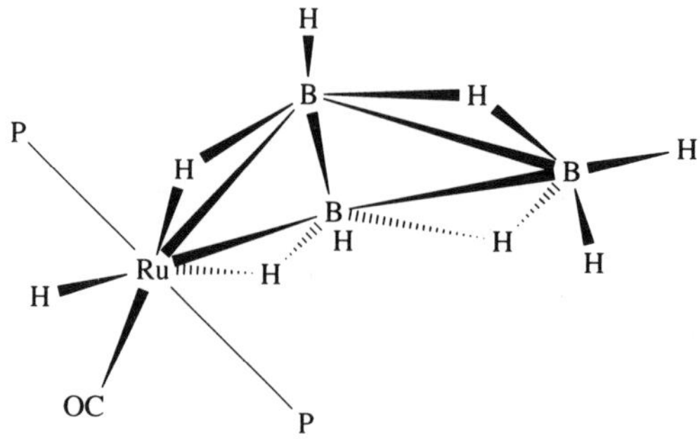
H
B
H
P
H
B
H
B
H
H
Ru
H
H
H
OC
P

(298)

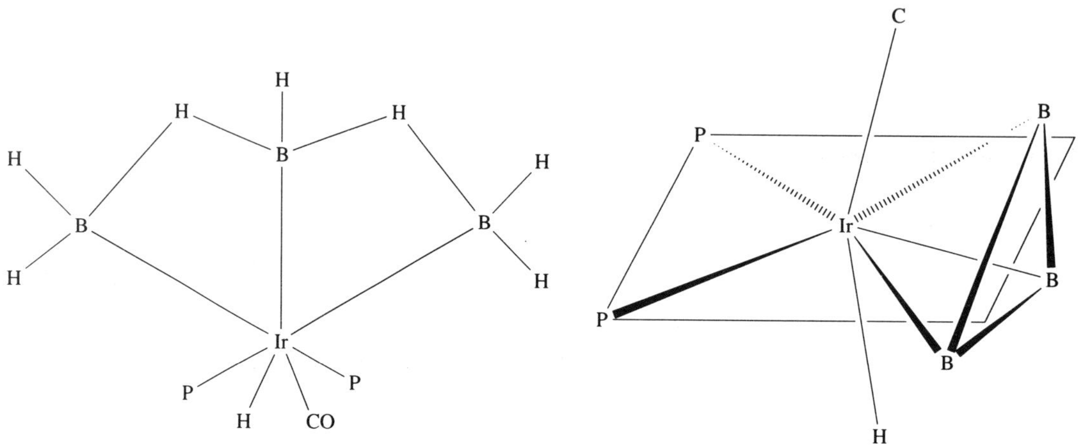
H
H
H
B
H
H
B
B
H
H
H
Ir
P
P
H
CO
C
P
B
Ir
B
P
B
H

(300)

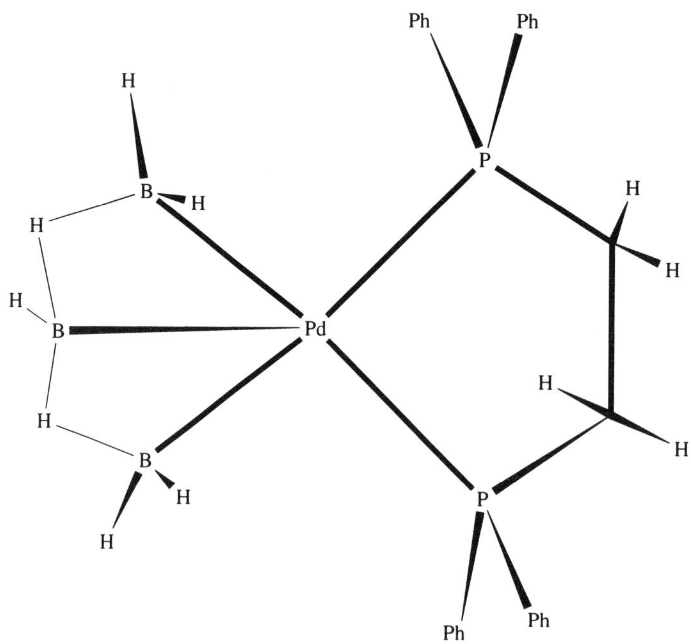
Ph
Ph
H
P
B
H
H
H
H
H
B
Pd
H
H
H
B
H
P
H
Ph
Ph

(303)

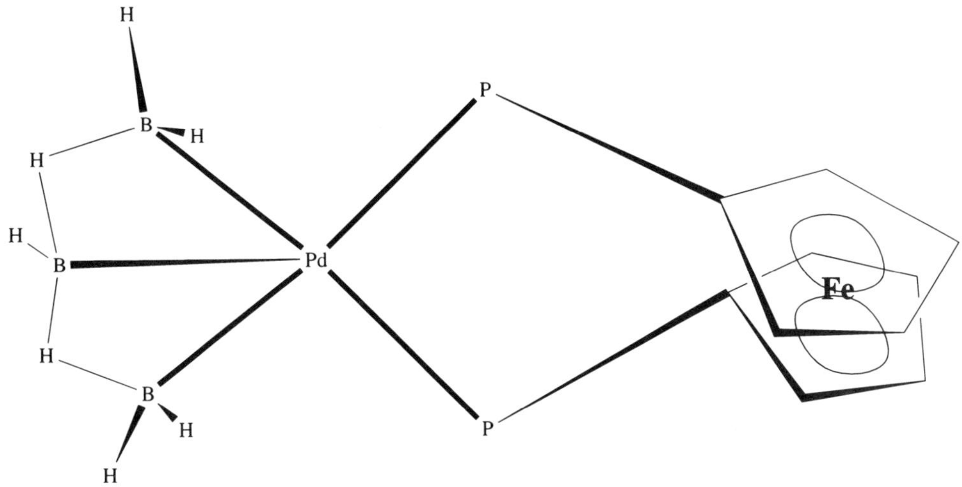

(306)

8.4.2 Octahydrotriborate Derivatives

8.4.2.1 Group 5 metals

$[NbCp_2B_3H_8]$ (**295**)[193] is obtained from the reaction between B_5H_9 and Cp_2NbH_3 at 0 °C and isolated from a mixture of $[Cp_2NbB_4H_9]$ (**309**) and $[Cp_2NbBH_4]$ by fractional crystallization. Spectral data are consistent with the solid-state structure. The butterfly geometry (**295**) for one of the two independent molecules in the asymmetric unit, can be considered to be an analogue of the *arachno*-B_4H_{10} cluster.

8.4.2.2 Group 6 metals

$[WH_3(PMe)_3B_3H_8]$ (**296**)[194] is prepared in ~90% yield from the reaction of $WH_6(PMe)_3$ with $BH_3{\cdot}THF$ at room temperature, according to Equation (5). The reaction proceeds via a purple intermediate which reacts with excess $BH_3{\cdot}THF$ to give a yellow–brown product. NMR spectral data suggest a fluxional process involving the hydrogen atoms on tungsten. The crystal structure indicates that (**296**) is another analogue of an *arachno*-B_4H_{10} cluster, in which tungsten is bonded through W–H–B bridge bonds to each of the two boron atoms on the edge of the B_3 triangle, similar to that illustrated for (**295**).

$$WH_6(PMe)_3 + 3BH_3{\bullet}THF \longrightarrow [WH_3(PMe)_3B_3H_8]\ \mathbf{(296)} + 2H_2 + 3THF \qquad (5)$$

8.4.2.3 Group 8 metals

$[(CO)H(PPh_3)_2OsB_3H_8]$ (**297**),[195] the first air-stable osmaborane, is prepared by the reaction between $Tl[B_3H_8]$ and $Os(CO)ClH(PPh_3)_3$ in CH_2Cl_2 at ambient temperature, as a colorless solid in 65% yield. NMR spectral data suggest a structure similar to that given for (**298**), described below, and to that of other B_3 metallaboranes in which the $[B_3H_8]^-$ group functions as a bidentate base. The metal atom is an 18-electron system and it subrogates B-2 in B_4H_{10}. Reaction between $[Ru(CO)ClH(PPh_3)_3]$ or $[RuH(CO)(NCMe)_2(PPh_3)_2]ClO_4$ and $[B_3H_8]^-$ salts, in CH_2Cl_2 at ambient temperature, affords the colorless, air-stable, crystalline *arachno*-2-ruthenatetraborane, $[RuH(B_3H_8)(CO)(PPh_3)_2]$ (**298**),[196–9] in 87–93% yield. The crystal structure, depicted in (**298**), reveals distorted octahedral geometry around the ruthenium atom, due primarily to inequivalent steric constraints associated with *exo*- and *endo*-sites for phosphine coordination to the RuB_3 butterfly. The spectral properties suggest that the structure is retained in solution and the molecule is fluxional above room temperature. Reaction of (**298**) with *N*-halosuccinimides produces stable halometallatetraboranes $[RuX(B_3H_8)(CO)(PPh_3)_2]$ (X = Cl, Br, I) The NMR spectral data suggested a typical *trans*-P–Ru–P arrangement for all the three complexes including less favored iodide species.

8.4.2.4 Group 9 and 10 metals

$[(PPh_3)_2H_2IrB_3H_8]$ (**299**),[195] is prepared similarly to the preparation of the osmaborane (**297**) from $IrCl(PPh_3)_3$ and $Tl[B_3H_8]$. The proposed structure is similar to that for the osmium complex except that there are two hydrogen atoms on the iridium atom. $[CpNi(PPh_3)][B_3H_8]$ (**310**)[207] is available from the reaction between $(C_5H_5)Ni(PPh_3)Cl$ and $[Et_4N][B_3H_8]$ and PPh_3.

8.4.2.5 Lanthanides

$(THF)_xYb[B_3H_8]_2$ (**311**)[208] is formed from the reduction of $BH_3{\cdot}THF$ in the ytterbium amalgam according to Equation (6). It is air and moisture sensitive and yield data are unavailable due to decomposition in the absence of solvent.

$$2Yb\text{–}Hg + 8BH_3{\bullet}THF \longrightarrow Yb[BH_4]_2 + (THF)_xYb[B_3H_8]_2\ (\mathbf{311}) + THF \qquad (6)$$

8.4.3 Heptahydrotriborate Derivatives

8.4.3.1 Group 8 metals

The crystal structure of $[Fe_2(CO)_6B_3H_7]$ (**307**)[205] consists of a B_3H_7 moiety attached to one iron of a bent "sawhorse" $Fe_2(CO)_6$ fragment through two Fe–H–B bridge hydrogen bonds and to the other iron through multicenter iron–boron bonding. The molecule has approximate C_s symmetry, one iron being six-coordinate with distorted octahedral geometry while the other iron is effectively seven coordinate. The simplest description of the molecular structure is as a derivative of B_5H_9 in which one apical and one basal BH group are replaced by $Fe(CO)_3$ groups. $[CpFe(CO)B_3H_7(PPh_2)]$ (**308**)[206], is prepared, in 28% yield as a dark-orange air-stable solid, by the reaction of $[CpFe(CO)B_4H_6(PPh_2)]$ with silica gel. The formation of (**308**) takes place as a result of loss of a vertex boron atom as outlined in Scheme 9. The proposed structure of (**308**), consistent with spectroscopic data, is depicted in Scheme 9 and it represents the first structurally characterized metal triborane complex exhibiting a 1103 styx structure. It consists of a substituted *arachno*-B_4H_{10} parent structure in which one vertex has been replaced by a $[(PPh_2)FeCpCO]$ fragment and one of the expected M–H–B bridging hydrogens has been replaced by a bridging PPh_2 group. An alternative description for (**308**) is based on a metallated $B_3H_8^-$ structure in which the iron fragment coordinates to one boron atom of the $B_3H_8^-$ ligand through a M–H–B interaction and the phosphorus atom bridges between the iron and the adjacent boron atom in much the same fashion as a bridging M–H–B hydrogen atom.

8.4.3.2 Group 9 metals

$[(IrB_3H_7)(CO)H(PPh_3)_2]$ (**300**),[200] for which the preparation from $Tl[B_3H_8]$ and *trans*-[Ir-(CO)Cl$(PPh_3)_2$] is discussed in *COMC-I*, has been structurally characterized. The structure of the coordination sphere of the iridium and the metallaborane cluster is given in Structure (**300**). The IrB_3H_7 moiety is a four vertex *arachno*-cluster with the expected butterfly structure. The iridium atom is on a "hinge" position. The authors suggest that the species is best described as a tridentate *hypho*-$[B_3H_7]^{4-}$ ligand bonding to a $[Ir(V)]^{4+}$ center as in Structure (**300**). This structure contrasts with that of (**299**) described in Section 8.4.2.4.

8.4.3.3 Group 10 metals

$(PMe_2Ph)_2PtB_3H_7$ (**301**), which has been described in *COMC-I*, is prepared chromatographically in 8% yield as one of several products in a reaction of 4-(SMe_2)-7-(OMe)-B_9H_{10} with $[PtCl_2(PMe_2Ph)_2]$ and proton sponge.[183]

A number of palladium metallaboranes with the formula $(Ph_2P(CH_2)_nPPh_2)Pd(B_3H_7)$, $n = 1$ (**303**), $n = 2$ (**303**), $n = 3$ (**304**), $n = 4$ (**305**), which are analogues of (**301**), have been prepared by the reaction of $(Ph_2P(CH_2)_nPPh_2)PdCl_2$ with $B_3H_8^-$ in 2–35% yield.[202] The yield of these compounds depends upon the organic chain length of the phosphino ligand and is maximum when $n = 2$. No reaction is observed

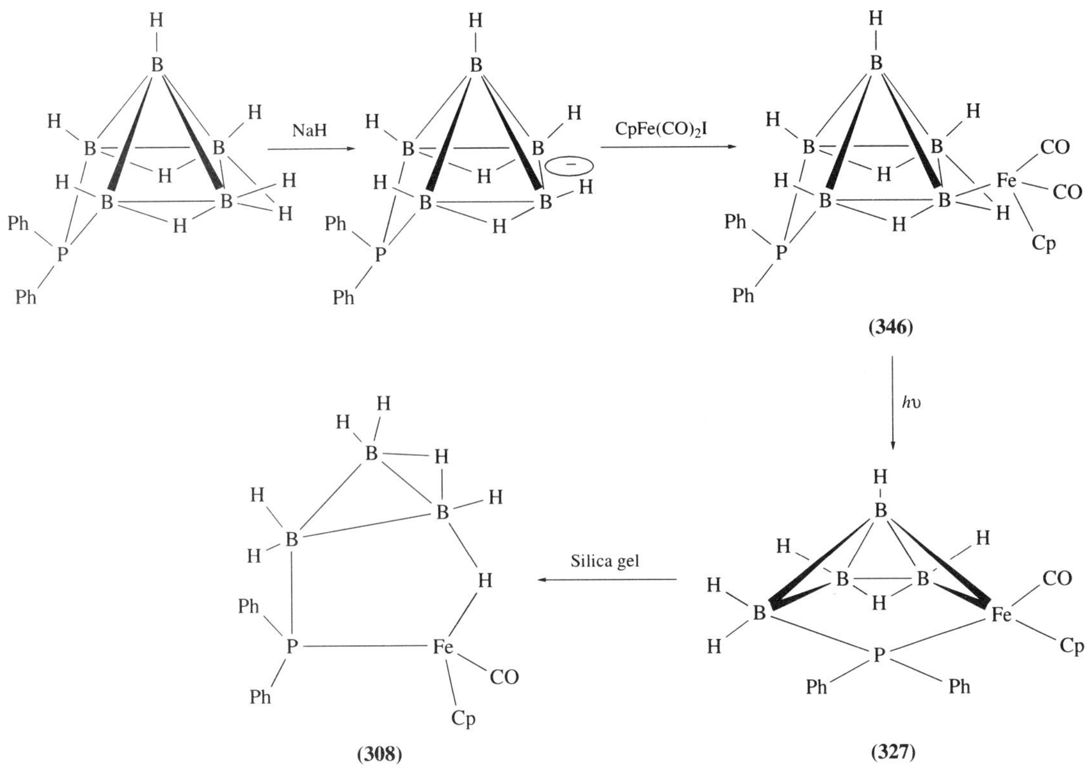

Scheme 9

with longer organic chain length (n = 5, 6). The structure of (**303**), is that of a palladium(II) atom in a pseudo-square-planar environment with the $\{B_3H_7\}$ ligand adopting an η^3-bonding mode as in $(Me_2PhP)_2PtB_3H_7$, (**301**) and (dppf)PdB_3H_7 (**306**) which is described below. The two phosphorus donor atoms occupy two coordination sites which are mutually *cis* and the "π-allyl" B_3H_7 unit occupies the remaining two sites. The $[Ph_2P(CH_2)_nPPh_2]Pd$ unit acts as a three-electron fragment occupying one "hinge" site of an *arachno*-PtB_3 butterfly cluster. A series of related platinatetraboranes also exists for n = 2, 3, 4.[203] A related species 1-$[\{Ph_2P(C_5H_4)_2Fe\}PtB_3H_7]$ (**306**),[204] is obtained from the reaction between $[B_3H_8]^-$ and $[Fe(\eta^5\text{-}C_5H_4PPh_2)_2]PdCl_2$.

8.4.3.4 Group 11 metals

$[(PPh_3)_2Cu(H_3BCNB_3H_7)]$ (**312**),[209] may be prepared by anodic dissolution of copper in acetonitrite solution of $[N(Bu^n)_4][H_3BCNB_3H_7]$. The process is carried out in a divided cell, supporting electrolyte $[N(Bu^n)_4][BF_4]$ is in the cathode compartment, and the borane anion salt is used as its own supporting electrolyte in the anode compartments.

8.4.4 Hexahydrotriborate Derivatives

8.4.4.1 Group 8 metals

PPN$[\{Fe(CO)_3\}_2B_3H_6]$ (**313**)[210] is prepared by deprotonating its conjugate acid $[\{Fe(CO)_3\}_2B_3H_7]$ (**307**) with [PPN]Cl in MeOH. The deprotonation is easily reversed by treatment with H_3PO_4 suggesting the reversible transformations shown in Scheme 10. NMR spectral data of (**313**) are consistent with the structure shown in Scheme 10. The data suggest that the species (**313**) and its conjugate acid are not fluxional at room temperature. This behavior is ascribed to the influence of the metal atom on the B–H–Fe bonds. The spectral data also suggest that a Fe–H–B proton is removed on deprotonation in preference to a B–H–B proton. MO calculations suggest that the observed structure is stabilized by shifting excess negative charge from iron to the CO ligands.

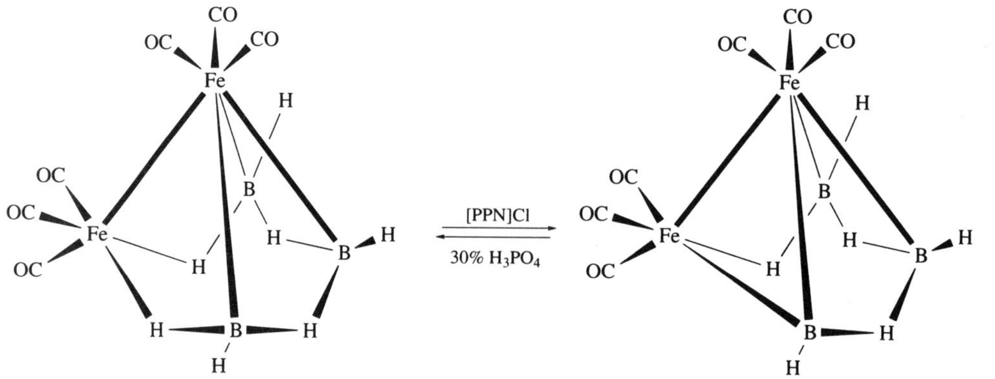

(307) (313)

Scheme 10

8.4.5 Trihydrotriborate Derivatives

The structure of $(CpCo)_3(CO)B_3H_3$ (**314**),[211] which was described previously in *COMC-I*, has been redetermined. Structure (**314**) consists of an octahedron formed by three cobalt atoms and three boron atoms and the triangle formed by the three cobalt atoms is capped by a carbonyl group.

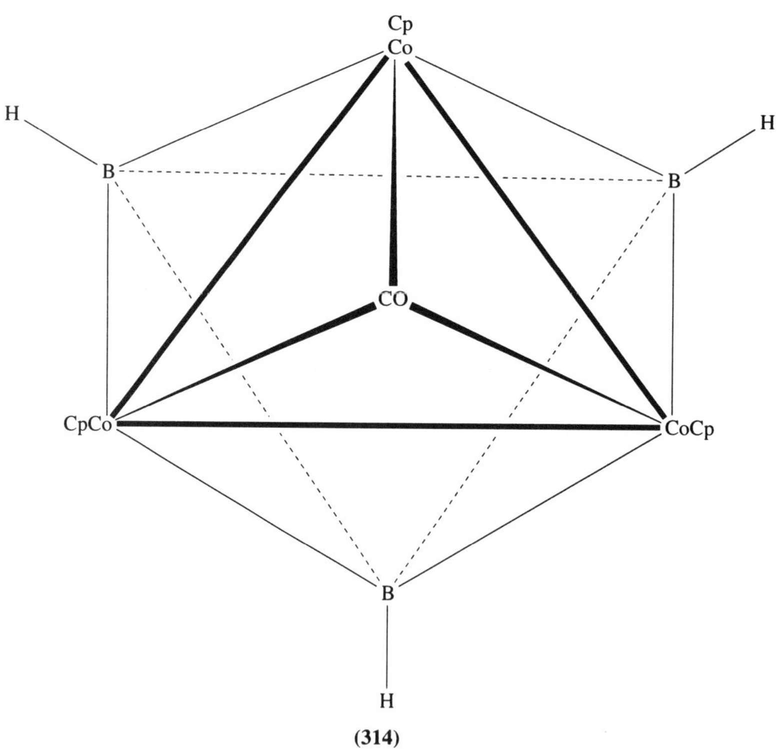

(314)

8.5 UNITS CONTAINING FOUR BORON ATOMS

8.5.1 Nonahydrotetraborate Derivatives

8.5.1.1 Group 5 metals

[$Cp_2NbB_4H_9$] (**309**)[193] is prepared in the reaction between B_5H_9 and Cp_2NbH_3 at 0 °C. NMR spectra suggest the structure given in (**309**); that is the niobium moiety subrogates the 2-position in *arachno*-B_5H_{11}. Thus, this system is the first example of a $[B_4H_9]^-$ ligand coordinating to a metal employing η^2-coordination. Other examples of $[B_4H_9]^-$ coordination, including those described below, all involve η^4-coordination in systems in which the metal subrogates the 1-position in B_5H_{11}.

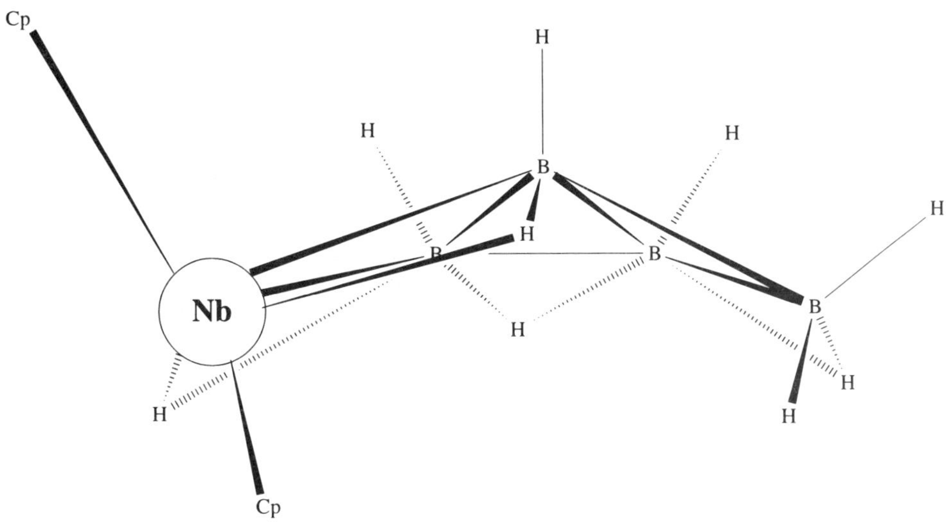

(309)

8.5.1.2 Group 8 metals

[1-Cp*$RuHB_4H_9$] (**315**)[212] is formed as one of the products in the reaction between [(η-C_6Me_6)$RuCl_2$]$_2$ with K[B_4H_9] in CH_2Cl_2/THF at −35 °C. Spectral, data are consistent with the structures of (**315**) and (**316**). [1,1,1-(CO)(PMe_3)$_2$Ir(η^4-B_4H_9)] (**316**)[213] is formed in low yield as a pale-yellow air-stable diamagnetic solid, in the reaction between [NEt_4][B_9H_{12}] and [Ir(CO)Cl(PMe_3)$_2$Cl] in CH_2Cl_2. The species is related to [1,1,1-(CO)(PMe_2Ph)$_2$Ir(η^4-B_4H_9)], described in *COMC-I*, and spectral data are consistent with the structures of (**315**) and (**316**). [2,3-{Fe(PMe_3)$_2$}$_2$(μ-H)B_4H_9] (**317**)[214] is an unusual example of coordination of the $[B_4H_9]^-$ anion. It is formed in low yield as a black crystalline solid (dark red–brown in solution) from the reaction between B_5H_9 and [Fe(PMe_3)$_3$(η^2-CH_2PMe_2)H]. The structure is that of a *nido*-pentaborane(9) analogue with the iron atoms occupying adjacent basal positions with a unique BH_3 group capping the Fe_2B face, as indicated in Structure (**317**). Thus it is a *nido*-six-vertex cluster with the monocapped square pyramidal structure containing seven skeletal electron pairs, and the structure is apparently favored over the expected *closo*-octahedron.

8.5.2 Octahydrotetraborate Derivatives

8.5.2.1 Group 8 metals

[2-{Fe(PMe_3)$_3$}B_4H_8], (**318**)[214] is the major product obtained in reaction in which (**317**) was isolated. It is an orange crystalline solid, unstable in air but stable in the solid state in the absence of air for 2–3 days. The structure, suggested by NMR spectra, is one in which the Fe(PMe_3)$_3$ moiety subrogates a BH group in B_5H_9 in a basal position. The structure given in (**318**) is generic for the series of *nido*-2-metallaboranes including (**317**)–(**320**) and others previously reported may be described as a complex

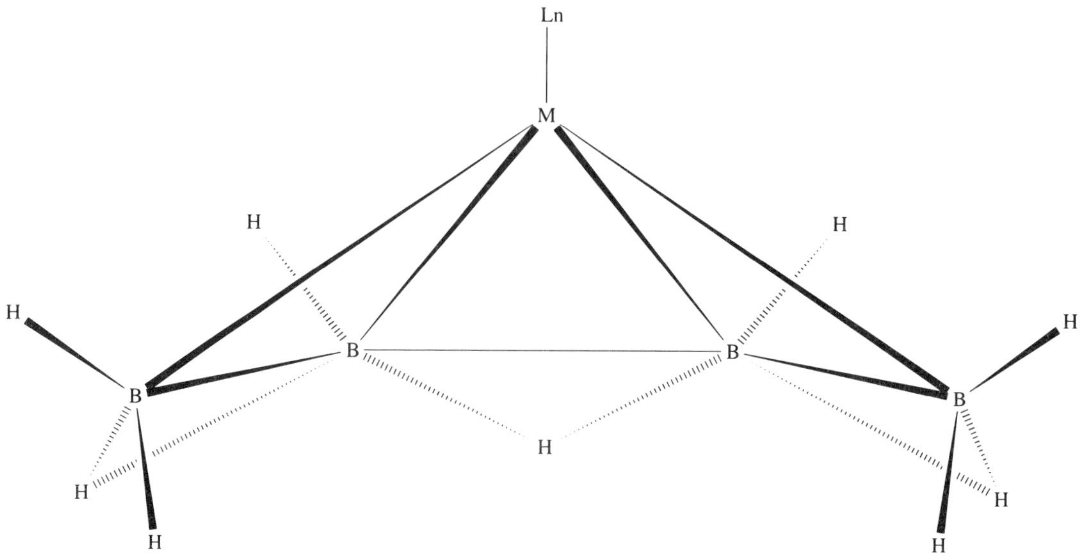

(315), (316)

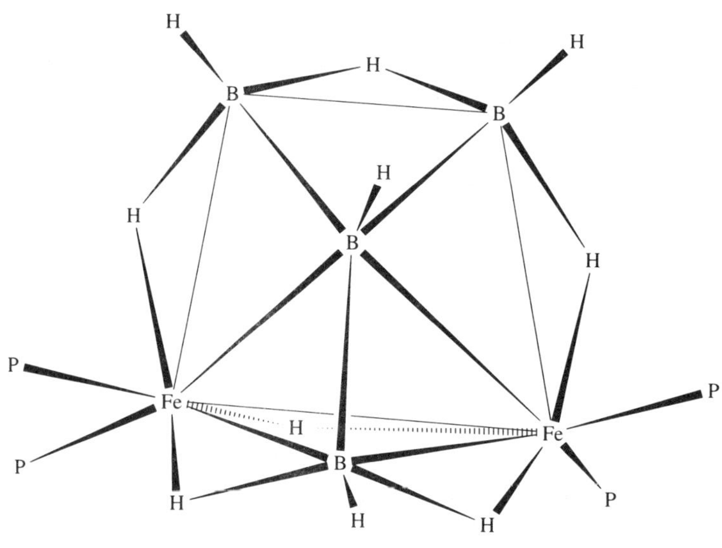

(317)

between a tridentate $[B_4H_8]^{2-}$ ligand with a d^6 metal center. $[(CO)(PPh_3)_2RuB_4H_8]$ **(318)**[196] is available as a light-yellow solid in low yield from the reaction between $[B_5H_8]^-$ and $[Ru(CO)ClH(PPh_3)_3]$ in THF/CH_2Cl_2 at low temperature. NMR spectra are consistent with the structure similar to that in Structure **(318)**. $[(OsB_4H_8)(CO)(PPh_3)_2]$ **(319)**[195] also belongs to this class of systems as does $[2(\eta\text{-}C_6Me_6)RuB_4H_8]$ **(320)**,[212] prepared from the reaction between $[(\eta\text{-}C_6Me_6)RuCl_2]_2$ and $K[B_6H_{11}]$ in CH_2Cl_2/THF at −25 °C. A 1-isomer of this class of *nido*-metallapentaboranes, $[1\text{-}\{Fe(CO)_3\}B_4H_8]$ **(321)**, which is described in *COMC-I*,[215–16] has been studied in an effort to understand the bonding in such systems.

8.5.2.2 Group 9 metals

The species related to the series **(317)**–**(320)**, $[2\text{-}Cp^*CoB_4H_8]$ **(322)**, is prepared in reactions analogous to the preparation of $[2\text{-}CpCoB_4H_8]$ **(323)**.[217] The structure of $[1\text{-}CpCoB_4H_8]$ **(324)**[218] has been confirmed as a *nido*-1-metallapentaborane with the CpCo group replacing the BH group at the 1-position. The related $[1\text{-}Cp^*CoB_4H_8]$ **(325)**[218] is available in higher yields. Yellow crystals of **(325)** are obtained from slow thermolysis of the red **(321)** in a controlled process. The two isomers of $CpCoB_4H_8$, 2-$CpCoB_4H_8$ **(323)** and 1-$CpCoB_4H_8$ **(324)**, have been used in transition metal-catalyzed cage fusion reactions to afford dicobaltadecaboranes by deprotonation with KH in THF followed by stirring at room

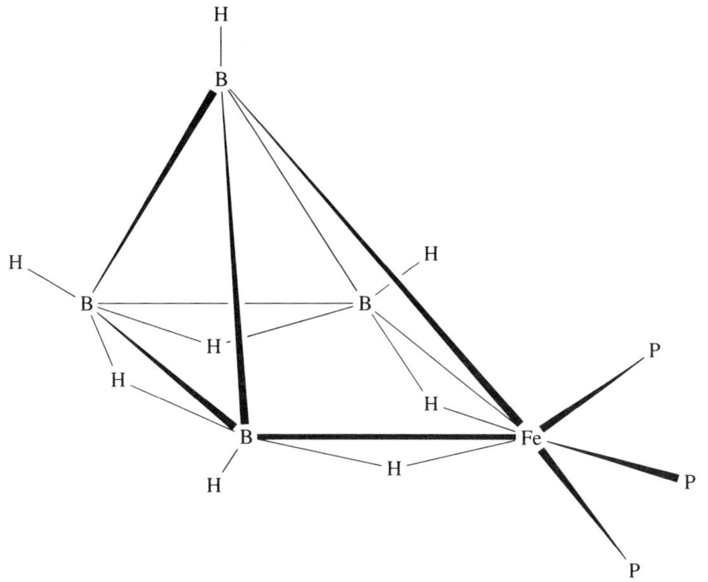

(318)

temperature in the presence of a slight excess of $RuCl_3$. The products are the isomers of *nido*-$Cp_2Co_2B_8H_{12}$ (see Section 8.9.3.3), and the actual isomers of the latter obtained depend on the individual isomers of $[CpCoB_4H_7]^-$ used in the reactions.[219–20] (**324**) was used in a seminal experiment to demonstrate the power of 2D $^{11}B–^{11}B$ NMR spectroscopy.[221]

8.5.3 Heptahydrotetraborate Species

$[CpMo(C_5H_4)(B_4H_7)]$ (**326**)[222] is formed if the photolysis of a mixture of $[Cp_2MoH_2]$ and excess $BH_3{\cdot}THF$, in the preparation of $[MoCp_2H(\eta^2\text{-}B_2H_5)]$ (**172**), is prolonged or if $[MoCp_2H(\eta^2\text{-}B_2H_5)]$ is thermolyzed. A single crystal structure determination along with detailed NMR studies indicate the structure illustrated in (**326**), which is correctly identified as $[CpMo(\eta^5{:}\eta^1\text{-}C_5H_4)B_4H_7]$ and exists as an *arachno*-B_5H_{11} analogue. A CpMo moiety occupies the 2-position in the parent borane and a Cp group σ-bonded to the boron in the 5-position also bonds in an η^5-mode to the molybdenum atom, affording a system with a skeletal electron count which conforms to the *arachno*-designation. Extended thermolysis converts (**326**) into the carbaborane $[CpMo(\eta^3{:}\eta^2\text{-}C_3H_3)C_2B_9H_9]$, a species in which the Cp ring has been partially incorporated into the borane cage.

8.5.4 Hexahydrotetraborate Species

8.5.4.1 Group 8 derivatives

$[CpFe(CO)B_4H_6(PPh_2)]$ (**327**)[206] is also a species based on the structure of *arachno*-B_5H_{11} and it is comparable with (**326**). It is isolated as a yellow air-stable solid in the photolysis of $[CpFe(CO)_2B_5H_7(\mu\text{-}PPh_2)]$ (**346**), which is described in Section 8.6.3.4. NMR spectra suggest the structure indicated in Scheme 9. The B_5H_{11} skeleton has the B-2 atom replaced by a $CpFe(CO)(\mu\text{-}PPh_2)$ fragment and an *endo*-terminal hydrogen atom on the opposite basal boron has been replaced by a terminally bonded PPh_2 group which is itself bonded to the iron center. In addition the μ hydrogen atom between B-2 and B-3 is now terminally bonded to B-2.

8.5.4.2 Group 9 derivatives

1,2-$(Cp^*{}_2Co)_2B_4H_6$ (**328**)[217,219] is one of several products obtained as a violet solid in the reaction between LiCp, $[B_5H_8]^-$, and $CoCl_2$. NMR spectra suggest a *closo*-octahedral structure with two sets of equivalent boron atoms, and two hydrogen atoms capping the faces containing the adjacent cobalt atoms.

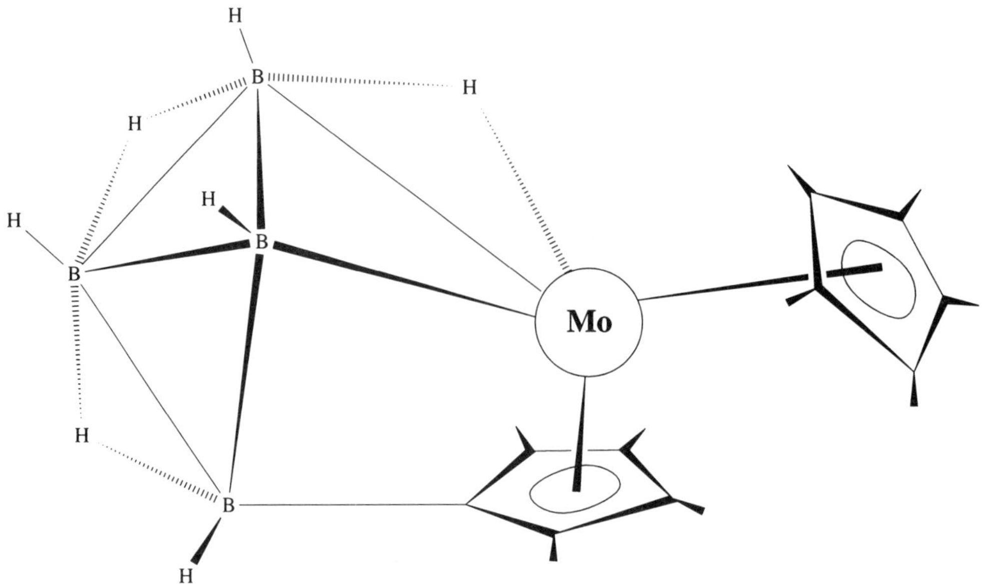

(326)

8.5.5 Tetrahydrotetraborate Species

$[1,2,3\text{-}(Cp^*Co)_3B_4H_4]$ (**329**)[217,223] is also formed in the reaction between LiCp*, $[B_5H_8]^-$, and $CoCl_2$. It is a yellow crystalline solid and an x-ray structure determination indicates that it consists of a Co_3B_3 octahedron whose Co_3 face is capped by a BH group.

8.5.6 Dihydrotetraborate Complexes

$[(CO)_3(PPh_3)(Ph_2\overline{PC_6H_4)_2(Ir_2}B_4H_2)]$ (**330**)[224] is isolated from the degradation of $[B_{10}H_{10}]^{2-}$ with *trans*-$[Ir(CO)Cl(PPh_3)_2]$ in methanol. It is isolated as yellow crystals and a structure determination reveals that the species adopts the $[B_6H_6]^{2-}$ structure with adjacent iridium atoms subrogating BH groups in the 1- and 2-positions. The molecule also exhibits double *ortho*-boronation involving equivalent phenyl groups on each PPh_3 group and the 3- and 5-positions as indicated in Structure (**330**).

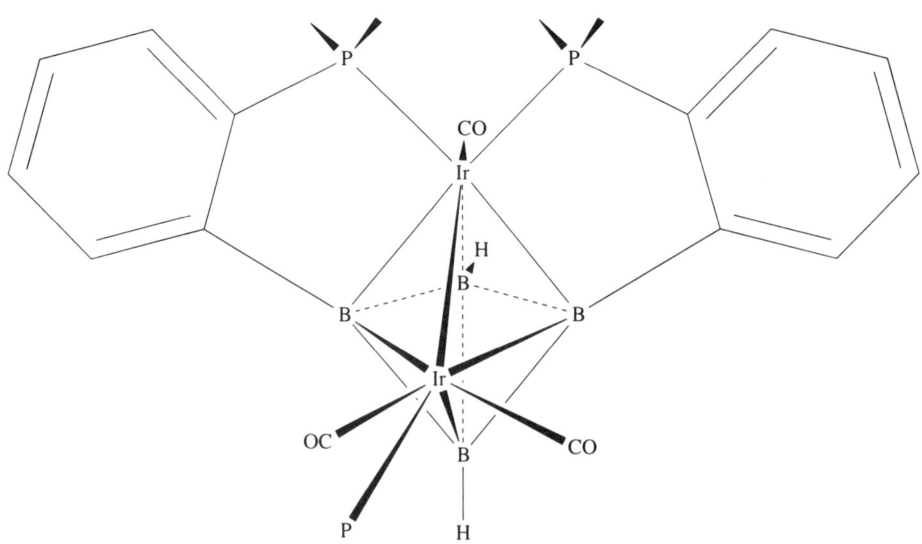

(330)

8.6 UNITS CONTAINING FIVE BORON ATOMS

8.6.1 Decahydropentaborate Derivatives

8.6.1.1 Group 7 metals

2,2,2-$(CO)_3$-2-MnB_5H_{10} (**331**)[225] described previously in *COMC-I*, is now available in higher yields from B_5H_9, $Mn_2(CO)_{10}$, and H_2, by performing the reaction in a stainless steel cylinder under 100 atm H_2, at 120 °C for 14 h in the presence of 5% Ru/C as catalyst. The related species, 2,2,2-$(CO)_3$-2-ReB_5H_{10} (**332**), is obtained similarly in 7–12% yields.

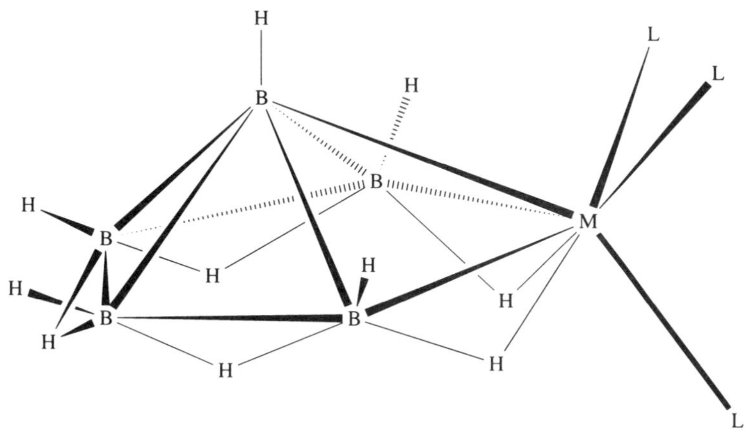

(332)

8.6.1.2 Group 9 metals

[2,3-$EtC_2B_4H_4$]-2-Co[B_5H_{10}] (**333**)[226] is formed in the reaction between Na[B_5H_8], $CoCl_2$ and Na[2,3-$EtC_2B_4H_4$] for several hours at −78 °C in THF. (**333**) is one of the products isolated and the proposed structure is that of a *commo*-system with a *nido*-2-cobaltahexaborane cluster sharing a common vertex with a *closo*-2-cobalta-3,4-dicarbaheptaborane moiety.

8.6.2 Nonahydropentaborate Derivatives

8.6.2.1 Group 8 metals

[4-$(CO)_3FeB_5H_9$] (**334**)[227] has been studied theoretically using self-consistent charge calculations. The most stable configuration is the [4-$(CO)_3FeB_5H_9$] isomer, which is the assumed structure of the species based on data for derivatives. The HOMO for [4-$(CO)_3FeB_5H_9$] is directly comparable with that of the parent borane B_6H_{10}. [4,4,4-(CO)$(PPh_3)_2$-*nido*-4-RuB_5H_9] (**335**)[196] is formed as a by-product in low yield in the synthesis of [CO$(PPh_3)_2RuB_4H_8$] (**318**) which is described in Section 8.5.2.1. It could not be isolated pure and it decomposes to give [RuCO$(PPh_3)_2B_4H_8$]. NMR spectra are consistent with a structure similar to that shown in Structure (**332**), except that the B-2–B-6 position does not contain a bridging hydrogen atom. It is formally best described as a *nido*-ruthenahexaborane in which a tridentate *arachno*-$[B_5H_9]^{2-}$ ligand coordinates to a $[Ru(CO)(PPh_3)_2]^{2+}$ moiety which does not contribute skeletal electrons to the B_5 cluster. The structure differs from that of the first metallahexaborane discovered, 4-$(CO)_3FeB_5H_9$, (**334**) in that the metal atom sits in a position which bridges to two boron atoms through H-bridges and is distal from the basal B–B bond, whereas in the latter, the iron atom is located in a position adjacent to the basal B–B bond.

[$(Ph_3P)_2$(CO)OsB_5H_9], (**336**)[195] is isoelectronic with (**335**) and it is obtained in high yield, as a pale yellow crystalline solid, from the reaction between Li[B_5H_8] and OsHCl(CO)$[PPh_3]_2$ in THF/CH_2Cl_2. NMR spectral data are consistent with a structure similar to that for (**335**). Deprotonation of (**336**) with LiMe or NaH, followed by treatment with [$PtCl_2(PMe_2Ph)_2$] affords the *nido*-osmaplatinaheptaborane,

$[(CO)(PPh_3)_2\overline{HOs(PMe_2Ph)ClPt}B_5H_7]$ (**337**)[228–30] as a lemon-yellow crystalline solid in 35% yield according to Equations (7) and (8). The reaction is considered to proceed via the replacement of a Cl^- by the osmaborane anion, thus adding a second metal atom to the cluster. An x-ray structure determination reveals of resulting species has the structure given in Structure (**337**). The initially formed cluster, (**a**) in Scheme 11 presumably flexes (**b**) to accommodate the second metal center during the course of its transition to the *nido* seven-vertex cluster (**c**). During this process, a bridging atom rearrangement takes place about the open face. The presence of the Os–H–Pt bridging hydrogen atom is apparent from NMR spectra. The structure conforms geometrically to the *nido* seven-vertex class although the electron count, of 16 skeletal electron pairs, would suggest that it ought to be a *closo*-system. The influence of the platinum moiety which prefers square planar coordination, and thus provides a two orbital contribution rather than a conical three orbital contribution, is ascribed as the reason for this apparent "violation" of the polyhedral skeleton electron pair theory (PSEPT) rules. Attempts to introduce a second platinum atom into the cluster by treatment with $[PtCl_2(PMe_2Ph)_2]$ and excess NaH results in the formation of the novel Structure (**338**).[229] The reaction proceeds such that PPh_3 is exchanged for PMe_2Ph on the Pt, HCl is eliminated and one of the phenyl groups migrates from a PPh_3 group to give a PPh_2 group and a boron-bonded phenyl group. The overall stoichiometry is given in Equation (9) and the resulting structure involves a PPh_2 group bridging two metal atoms as indicated in Structure (**338**). $[4-(\eta^6-C_6Me_6Ru)B_5H_9]$ (**339**)[212] an analogue of (**335**) and (**336**) has been reported as one of the products of the reaction between $[(\eta^6-Cp^*)RuCl_2]_2$ and $[B_6H_{11}]^-$ but details have not appeared.

$$[(Ph_3P)_2(CO)OsB_5H_9]\ (\mathbf{336}) + LiMe \longrightarrow MeH + Li^+[(Ph_3P)_2(CO)OsB_5H_8]^- \quad (7)$$

$$[(Ph_3P)_2(CO)OsB_5H_9]^- + [PtCl_2(PMe_2Ph)_2] \longrightarrow [(CO)(PPh_3)_2\overline{HOs(PMe_2Ph)ClPt}B_5H_7]\ (\mathbf{337}) + Cl^- + PMe_2Ph \quad (8)$$

$$[(CO)(PPh_3)_2\overline{HOs(PMe_2Ph)ClPt}B_5H_7]\ (\mathbf{337}) + NaH \longrightarrow [(CO)(PPh_3)\overline{(PPh_2)Os(PPh_3)Pt}B_5H_7Ph]\ (\mathbf{338}) + H_2 + NaCl + PMe_2Ph \quad (9)$$

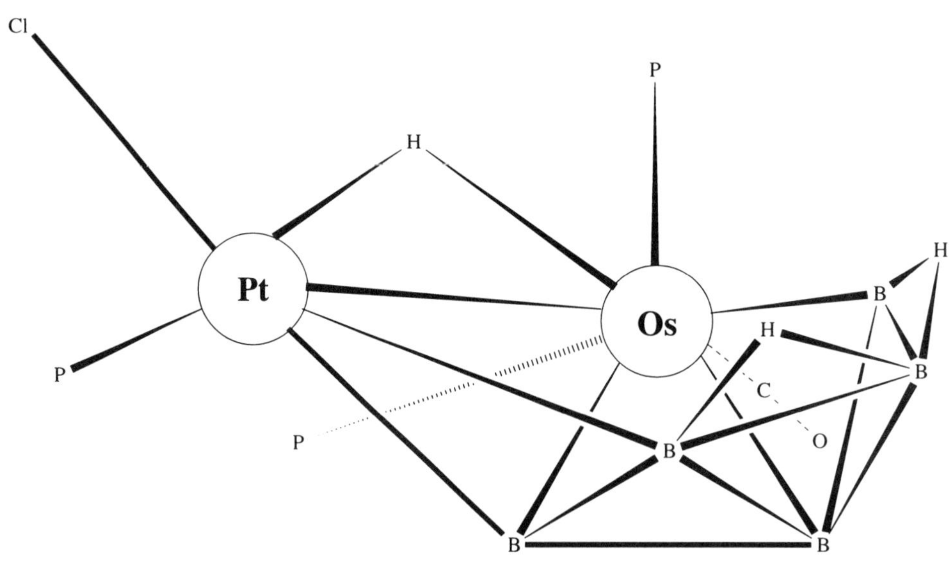

(**337**)

8.6.2.2 Group 9 metals

$[Cp^*_2Co_2B_5H_9]$ (**340**)[217] is formed in very low yield in the reaction between LiCp*, $CoCl_2$ and $Na[B_5H_8]$ at –76 °C. NMR spectral data suggest that Structure (**340**) is a *nido*-seven-vertex cluster, the first uncomplexed example of such a system.

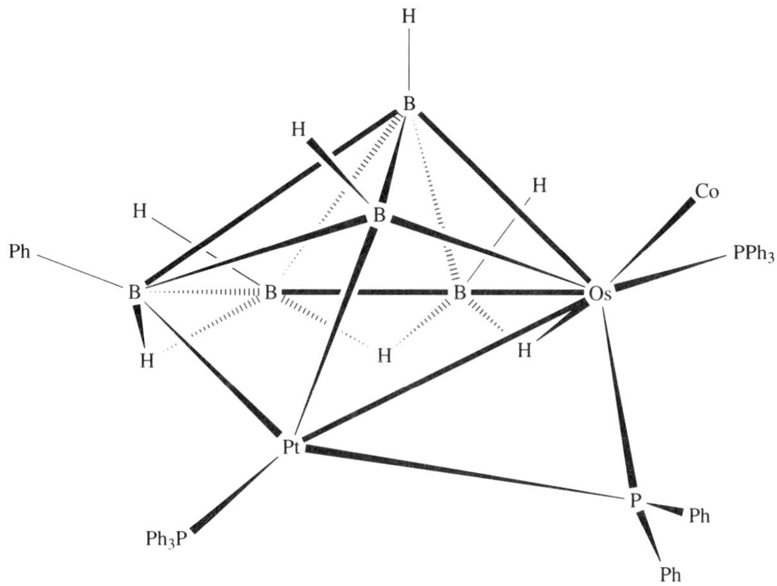

(338)

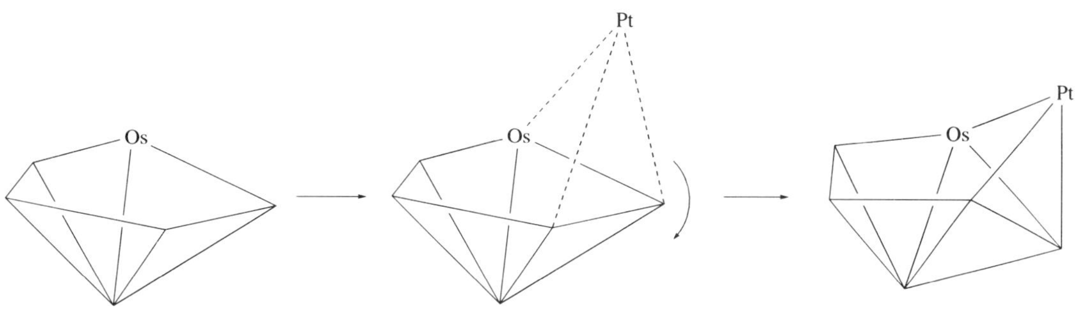

Scheme 11

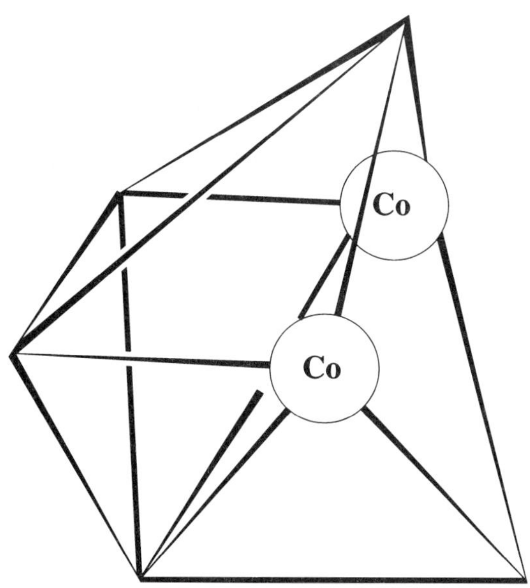

(340)

Co = Cp*Co

8.6.3 Octahydropentaborate Derivatives

8.6.3.1 Group 8 metals

2-[$CpFe(CO)_2$]B_5H_8 (**341**)[231] is obtained as a yellow oil from the reaction between $CpFe(CO)_2I$ and Li[B_5H_8] in Et_2O. Spectral data indicate that the iron moiety is σ-bonded to a basal boron atom. Compound (**341**) is relatively thermally stable, but only in the absence of air.

8.6.3.2 Group 9 metals

2-[$Co(CO)_4$]B_5H_8 (**342**)[231] is isolated in moderate yield as a yellow–brown liquid from the pyrolysis of $Co_2(CO)_8$ in liquid B_5H_9, and in lower yields from the photolysis of the same reaction mixture, and also from the reaction between Na[$Co(CO)_4$] and 2-ClB_5H_8. It is thermally unstable and spectral data support the formulation as the 2-substituted B_5H_9 derivative.

8.6.3.3 Group 11 metals

Cu[Ph_2PCH_2]$_2B_5H_8$ (**343**)[232] is available from the reaction between K[B_5H_8] and Cu[Ph_2PCH_2]$_2$I in Me_2O/CH_2Cl_2 at −35 °C. NMR spectroscopic data are consistent with the structure of (**343**). Two resonances are observed for the basal boron atoms and this is ascribed to the absence of fluxionality in the molecule which would require a basal B–B bond. [Au(PPh_3)B_5H_8] (**344**)[233] has been prepared from the reaction between LiB_5H_8 and [Au(PPh_3)NO_3] in ether and characterized by a crystal structure determination. The pale-red solid is a *nido*-pentaborane cluster in which a Au(PPh_3) group has replaced a bridging hydrogen atom and is thus similar to the proposed structure for (**343**).

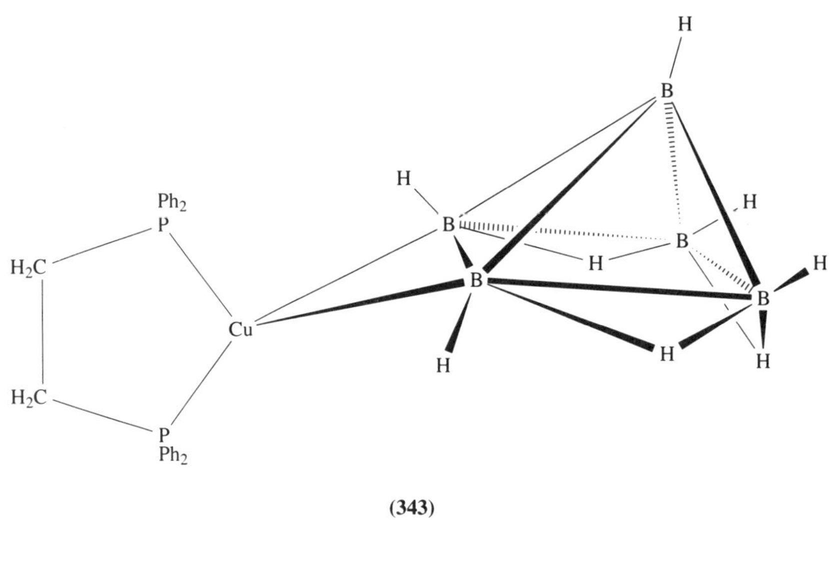

(343)

8.6.3.4 Group 12 metals

[2,3-μ-(MeZn)B_5H_8] (**345**)[234] is formed in the reaction between [2,3-μ-(CpBe)B_5H_8] and Me_2Zn in benzene over an extended period. NMR spectroscopic data suggest that the reaction indicated in Equation (10) takes place.

$$[2,3\text{-}\mu\text{-}(CpBe)B_5H_8] + Me_2Zn \longrightarrow [2,3\text{-}\mu\text{-}(MeZn)B_5H_8] + CpBeMe \qquad (10)$$

8.6.4 Heptahydropentaborate Derivatives

8.6.4.1 Group 8 metals

$[Cp(CO)_2FeB_5H_7(\mu\text{-}PPh_2)]$ (**346**)[235] is obtained, as a yellow air-stable solid, from the deprotonation of $[(\mu\text{-}PPh_2)B_5H_8]$ with NaH followed by reaction with $Fe(Cp)(CO)_2I$. The resulting species is a rare example of a *nido*-pentaborane(9) cluster with substitution of bridging and terminal hydrogen atoms with different substituents. The structure, indicated by spectral data and confirmed by an x-ray study is given in Scheme 9.

8.6.4.2 Group 9 metals

$1,2\text{-}(C_5Me_5)_2Co_2B_5H_7$ (**347**)[217] is formed in very low yield as a burgundy crystalline solid in the reaction between LiCp*, $CoCl_2$, and $Na[B_5H_8]$ at −76 °C. NMR spectral data suggest a structure that is a *closo*-seven-vertex pentagonal bipyramidal cluster with a Cp*Co group in both axial and equatorial positions. The extra two hydrogen atoms, required for the correct skeletal electron count, are located in face capping positions adjacent to the two cobalt atoms.

8.6.5 Pentahydropentaborate Derivatives

$[B_5H_5Fe(CO)_3]^{2-}$ (**348**)[216,236] has not been observed but calculations have been performed on this and related systems to assess the effect of the orbitals on the $Fe(CO)_3$ moiety on the position of bridging and terminal hydrogen atoms in the cluster.

8.7 UNITS CONTAINING SIX BORON ATOMS

8.7.1 Nonahydrohexaborate Derivatives

8.7.1.1 Group 8 metals

$Cu(PPh_3)_2B_6H_9Fe(CO)_4$ (**349**)[237] is prepared as a pale-yellow powder in the reaction between $K[Fe(CO)_4B_6H_9]$ and $[(PPh_3)_2CuBr{\cdot}0.5C_6H_6]$ in CH_2Cl_2 at −35 °C. It is air and light sensitive but is stable in the dark for several days at ambient temperature and for weeks under vacuum. Spectral data suggest the structure given in (**349**); the Cu^I moiety has inserted into a basal B–B bond in the $Fe(CO)_4B_6H_9$ moiety generated by the removal of a bridging hydrogen atom. Thus, (**349**) is an *arachno*-heterobimetallaoctaborane.

8.7.1.2 Group 10 metals

There are several organoplatinaboranes which contain the hexaboranyl moiety, although in all of the known systems except $PtCl_2(B_6H_{10})_2$, which was described in *COMC-I*, they are actually quite complex systems. $[PMe_2Ph]_2Pt_2[\eta\text{-}B_6H_9]_2$ (**350**)[238] contains two bidentate *nido*-$[\eta\text{-}B_6H_9]$ ligands bridged by the $[(PMe_2Ph)Pt\text{–}Pt(PMe_2Ph)]^{2+}$ moiety as seen in Structure (**350**). A related system, $[Pt_2(\mu\text{-}\eta^3\text{-}B_6H_9)(\mu\text{-}\{\eta^3\text{-}B_6H_8\text{-}\eta^2\text{-}[PtH\text{-}cis\text{-}(PMe_2Ph)_2]\}(PMe_2Ph)_2]$, (**351**),[238] has an additional platinum atom coordinated to the other side of one of the B_6H_9 moieties. $[Pt_2(PMe_2Ph)_2(\eta^3\text{-}B_2H_5)(\eta^3\text{-}B_6H_9)]$ (**285**)[183,239] also contains the B_6H_9 moiety bonded to two platinum atoms, as indicated in (**285**). $[(PMe_2Ph)_4Pt_3B_{14}H_{16}]$ (**352**),[239] where the hydrogen atoms on the boron cage are omitted for clarity as are the two PMe_2Ph groups on Pt(2) and the single PMe_2Ph groups on Pt(7) and Pt(10), has two *nido*-B_6H_9 moieties bridged by two boron atoms and three platinum atoms and $(PMe_2Ph)_2Pt_2B_{16}H_{15}(C_6H_4Me)(PMe_2Ph)$ (**353**)[240] consists of a *closo*-$[Pt_2B_{10}]$ unit and a *nido*-$[Pt_2B_6]$ unit joined with the Pt_2B faces in common, as seen in Structure (**353**) where the hydrogen atoms and the organyl groups on the phosphorus atoms are omitted for clarity. Also containing a $[B_6H_9]$ unit is the species $(PMe_2Ph)PtB_{16}H_{18}(PMe_2Ph)$ (**354**)[239] which contains a *nido*-11-vertex PtB_{10} cluster linked to a smaller eight-vertex PtB_7 cluster which contains the *nido*-B_6H_9 subunit as seen in Structure (**354**). Compounds (**352**)–(**354**) are described later in Sections 8.13.3 and 8.13.4.

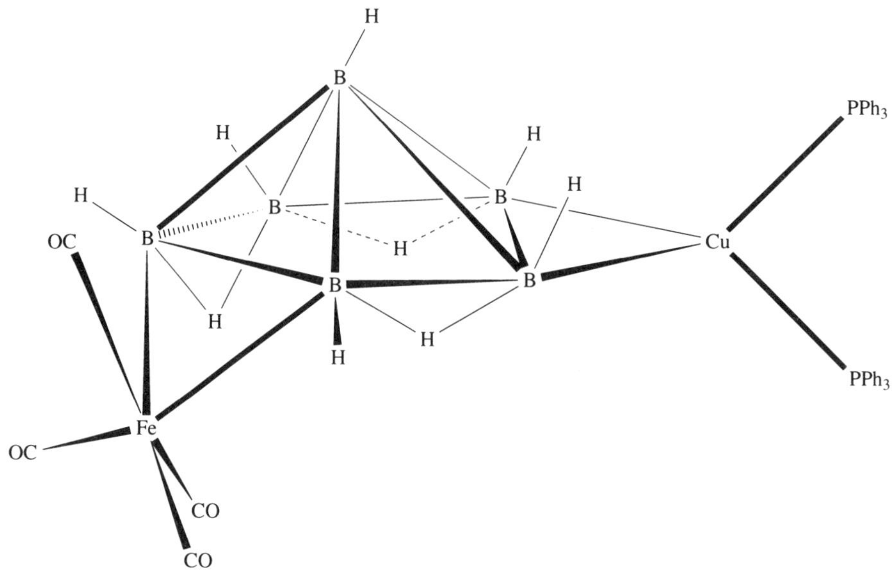

(349)

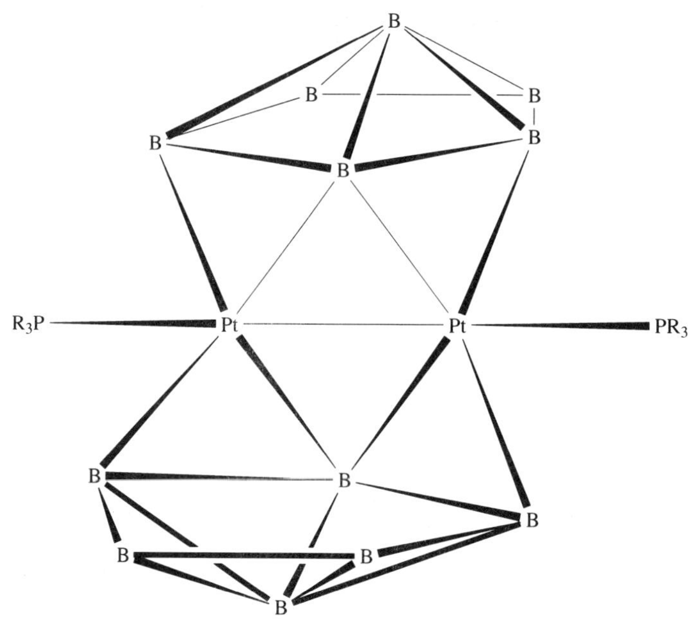

(350)

8.7.2 Hexahydrohexaborate Derivatives

8.7.2.1 Group 8 metals

$[B_6H_6Fe(CO)_3]^{2-}$ (**355**)[216] has been studied using MO calculations. The analogue of $[B_7H_7]^{2-}$ is predicted to be stable and would exist with longer than normal Fe–B-(2–6) distances and the ring hydrogen atoms in the B_6 plane.

8.7.2.2 Group 10 metals

$[Bu_4N][CpNiB_6H_6]$ (**356**)[241] is prepared from the reaction between $[Bu_4N]_2[B_6H_6]$ and nicklelocene in MeCN as pale-brown air-stable crystals. Spectral properties suggest structure (**356**), an analogue of $[B_7H_7]^{2-}$.

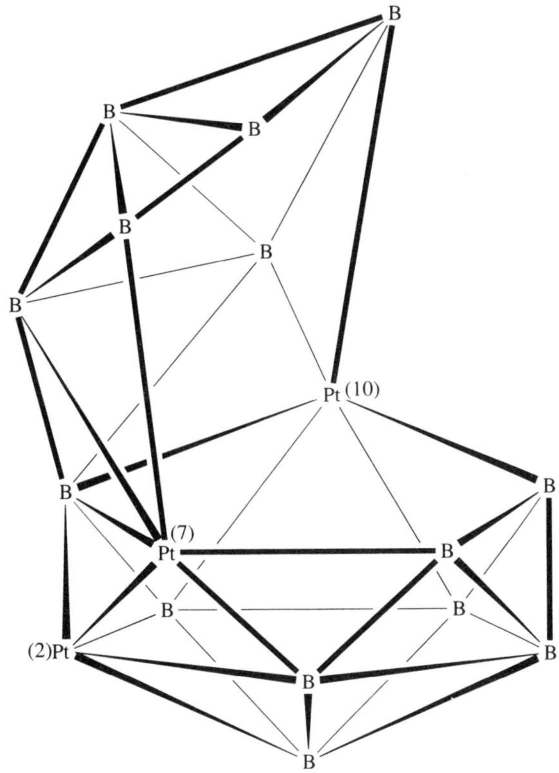

(352)

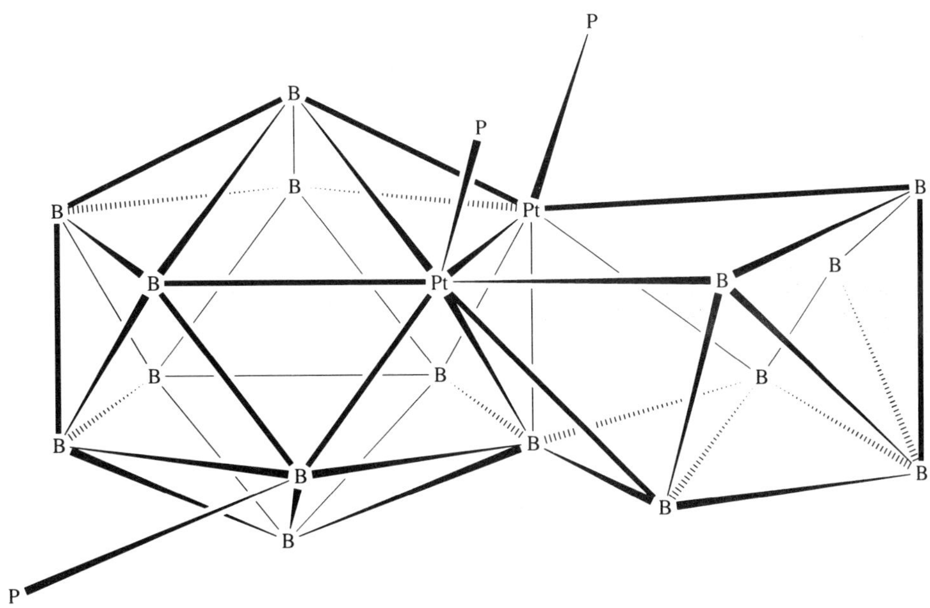

(353)

8.8 UNITS CONTAINING SEVEN BORON ATOMS

$[Bu_4N][3,3,3,3\text{-}(CO)_4\text{-}arachno\text{-}WB_7H_{12}]$ (**357**)[242] was the first eight-vertex contiguous metallaborane to be prepared. It is obtained from the reaction between $[W(CO)_3(MeCN)_3]$ and $[Bu_4N][arachno\text{-}B_9H_{14}]$

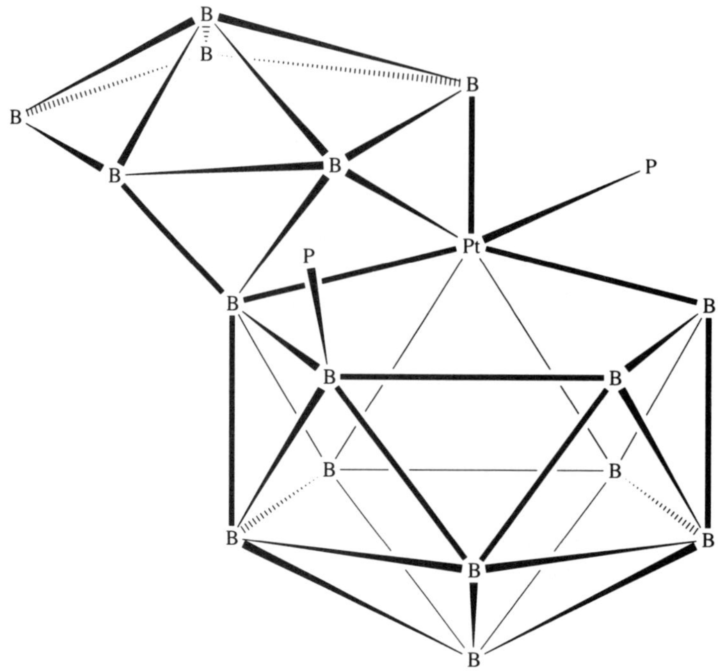

(354)

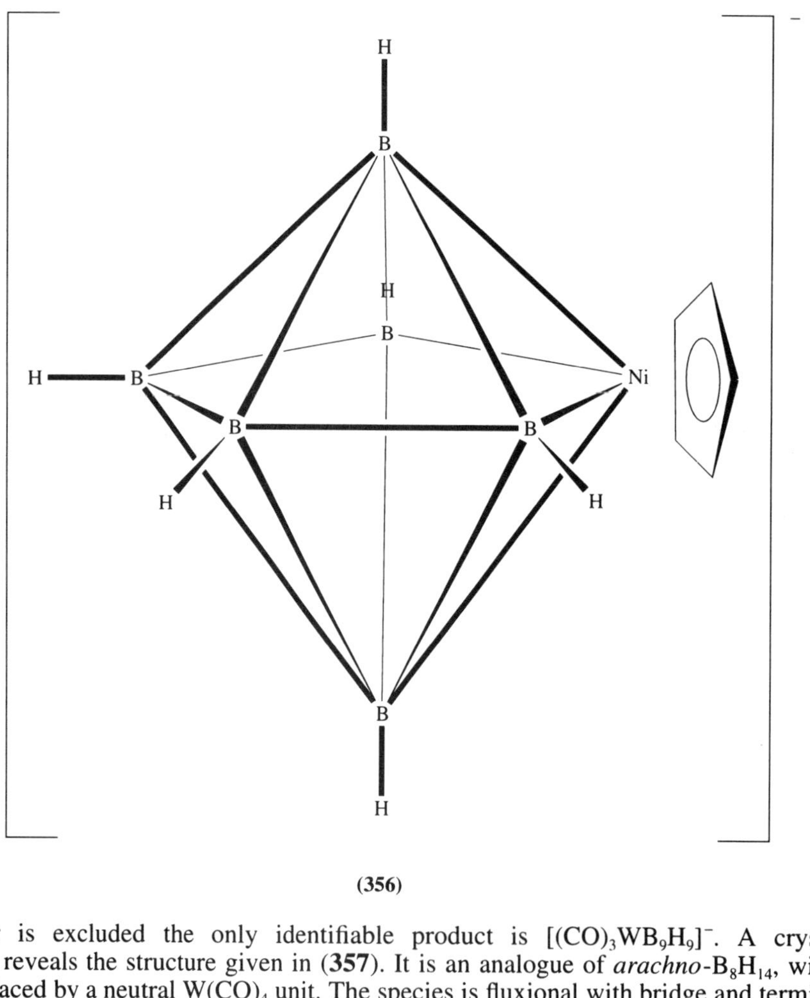

(356)

in air. If air is excluded the only identifiable product is $[(CO)_3WB_9H_9]^-$. A crystal structure determination reveals the structure given in (**357**). It is an analogue of *arachno*-B_8H_{14}, with formally a BH_2^+ unit replaced by a neutral $W(CO)_4$ unit. The species is fluxional with bridge and terminal hydrogen atom-exchange at positions H-8 and H-$\mu_{3,8}$.

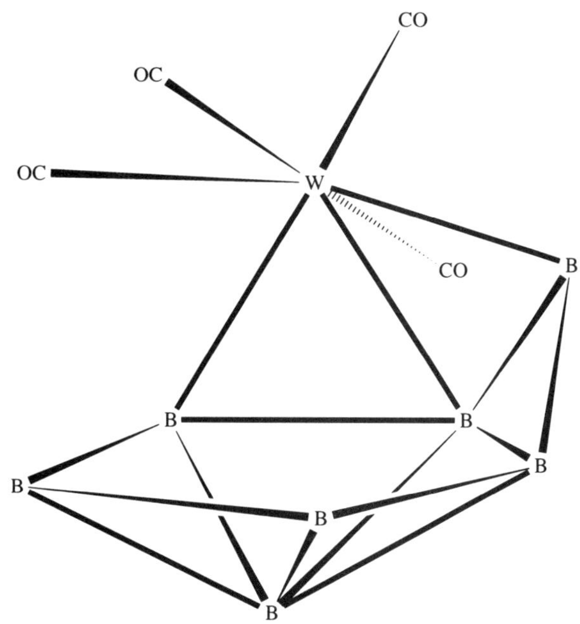

(357)

8.9 UNITS CONTAINING EIGHT BORON ATOMS

8.9.1 Introductory Remarks

The number of systems which appear not to obey the PSEPT is now quite large and there is a brief discussion of this in the introduction. An example of this is found among the recently discovered metallanonaboranes. Thus, *nido*-[$(PMe_3)_2(CO)Ir(B_8H_{10}Cl)$] (**358**) loses four electrons to afford $H(PMe_3)_2Ir(B_8H_7Cl)$ (**359**), as indicated in Equation (11).[243] (**359**) is two electrons short of the predicted $2n + 2$ electron count and adopts a geometry different from that predicted for *closo*-systems by the PSEPT. The structure, confirmed by a crystal structure determination, is given in (**359**). The structure is that of an idealized tricapped trigonal prism with one difference. The iridium atom caps one of the rectangular faces and boron atoms cap the other two but the borane to metal coordination appears to be an η^6 mode with the six boron atoms adopting an inverted chair conformation. The resulting cluster thus deviates from the polyhedral skeletal electron pair theory both in terms of structure and the fact that it would appear to contain four skeletal electrons fewer than the *nido*-cluster (**358**). Further discussion of these subtleties is beyond the scope of this chapter but some aspects are presented in Section 8.10.4.2. The more typical metallanonaboranes derive from classical boranes by subrogation of a boron vertex with a metal-containing group.

$$\textit{nido}\text{-}(PMe_3)_2(CO)Ir(B_8H_{10}Cl)\ \mathbf{(358)} \xrightarrow[\text{-}H_2,\ CO]{135\ ^\circ C} H(PMe_3)_2Ir(B_8H_7Cl)\ \mathbf{(359)} \qquad (11)$$

8.9.2 Tetradecahydrooctaborate Derivatives

These species are derivatives containing formally a ML^+ grouping coordinating to a $[B_8H_{14}]^-$ ion. Alternatively the systems are considered as formal *arachno*-metallanonaboranes and are thus formally derived from *arachno*-iso-B_9H_{15}, with a metal atom subrogating a BH group. Thus, we have included in this section all the species which are classified, according to their skeletal electron count, as *arachno*-clusters.

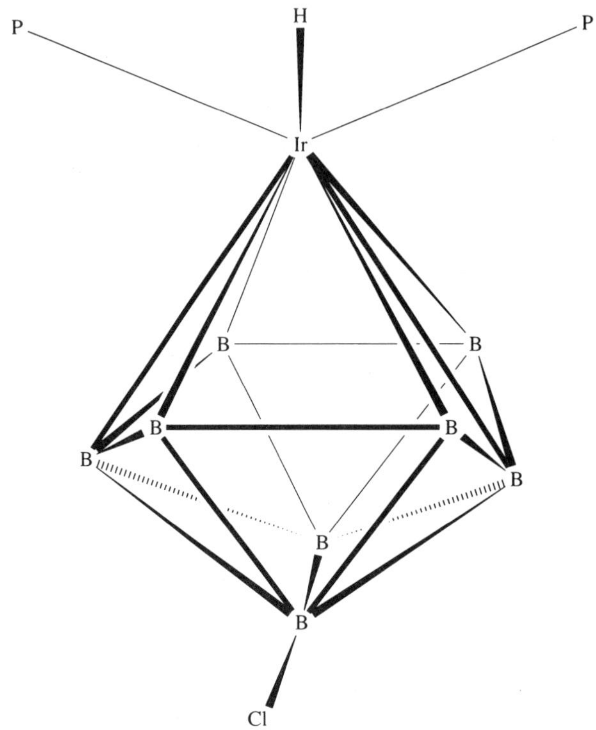

(359)

8.9.2.1 Group 8 metals

[1-(η^6-C_6H_6)-*n*-*arachno*-1-RuB_8H_{14}] (**360**)[244] is prepared in low yield, as a yellow air-stable crystalline solid, from the reaction of K[B_6H_{11}] with [(η^6-C_6H_6)$RuCl_2$]$_2$ at −25 °C. Compound (**360**) is an analogue of *n*-B_9H_{15} rather than of iso-B_9H_{15}, as has been the case for known *arachno*-metallanonaboranes, as indicated in Structure (**360**) which also gives the idealized cluster geometry.

8.9.2.2 Group 9 metals

[2,5-(μ-PPh_2)-1-($PHPh_2$)-2-(Cp*Rh)-B_8H_9] (**361**)[245] is prepared as a yellow air-stable solid from the reaction between [6-Cp*-*nido*-6-RhB_9H_{13}] and $PHPh_2$ in CH_2Cl_2. The *arachno*-system retains much *nido*-character presumably due to the unusual bridging PPh_2 group which could serve as a pseudo-B–B bond in the 2,5 position. There is a B–B bond in the analogous position in the *nido*-structure. A simplified representation of the structure and that of the *nido*-nonaborane are given in Structure (**361**). Compound (**361**) is very similar in structure to a *nido*-B_9-cluster and spectral data for (**361**) resemble those for *nido*-B_9H_{13}, but the PPh_2 moiety provides two electron pairs to the cluster to render it as an *arachno*-system. *Sym*-[4,4,4,4-(Me_3P)$_2$(CO)(H)IrB_8H_{12}] (**362**), and *asym*-[4,4,4,4-(Me_3P)$_2$(CO)-(H)IrB_8H_{12}] (**363**)[246] are obtained as colorless solids, in low yields, from the ambient temperature reaction between *trans*-[(Me_3P)$_2$Ir(CO)Cl] and [NEt_4][B_9H_{12}] in CH_2Cl_2. NMR spectroscopic and x-ray data for (**364**) suggest the structures given (**362**)–(**364**). The colorless solid, [(Me_3P)$_2$(CO)(H)IrB_8H_{11}Cl] (**364**),[245–6] is formed in low-yield as a by-product in the preparation of (**362**) and (**363**) described above. The structures represented as (**362**)–(**364**) all involve the replacement of the BH group in position 4 of [B_9H_{13}L] by the iridium moiety. The iridium moiety is isoelectronic with BH_2 so one of the hydrogen atoms at position 7 in [B_9H_{14}]$^-$ is not present in this series of iridanonaboranes. Derivative chemistry of these systems is available. Thus if (**363**) is treated with KH and then [$PtCl_2$(PMe_3)$_2$], the colorless solid bimetalladecaborane, [(Me_3P)$_2$(CO)(H)IrB_8H_{10}Pt(PMe_3)$_2$] (**365**),[247] is obtained. Structure (**365**) indicates a decaborane(16) cage in which the iridium and platinum moieties subrogate BH_2^- groups at the 9-, and 6-positions, respectively. Thermolysis of (**363**) affords a *nido*-iridanonaborane (**366**) according to Equation (12).[243,246–7] [4-(Cp*)-4-(*endo*-MeCN)-*arachno*-4-IrB_8H_{12}], (**367**)[248] is obtained as pale-yellow crystals among the products of the reaction between the *nido*-

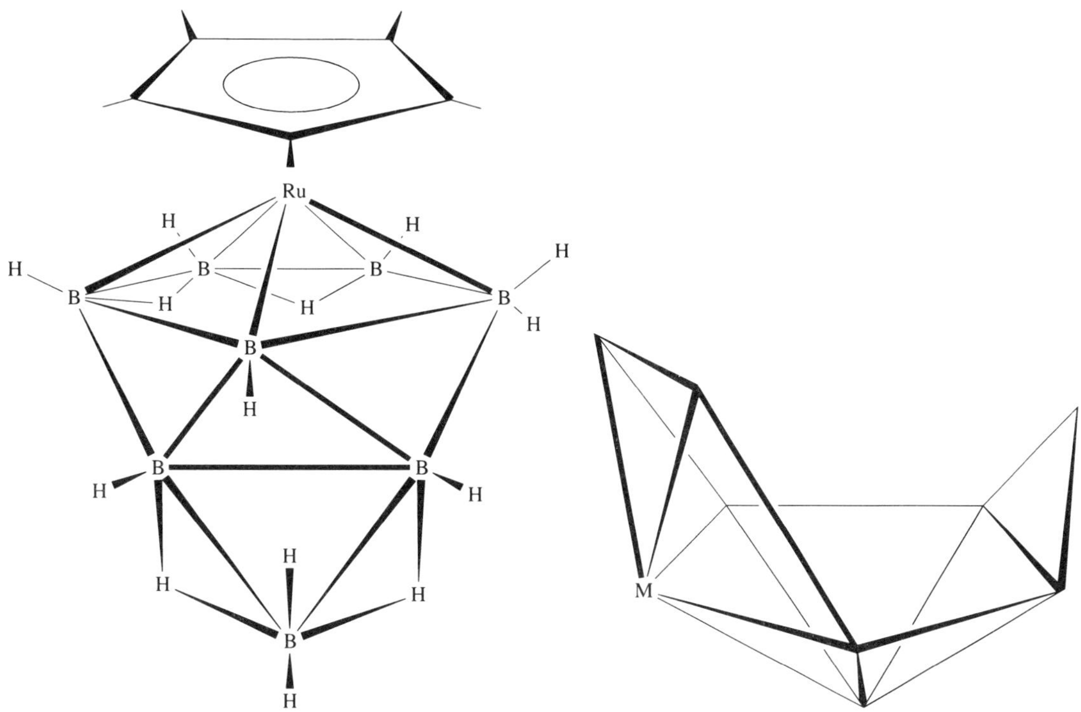

(a) **(360)** **(b)**

iridadecaborane [$Cp^{*}IrB_9H_{13}$] with MeCN. The MeNC ligand occupies an *endo*-position on the iridium atom in a cluster which contains four Hμ atoms symmetrically disposed in the open face of the formal *arachno*-nonaborane cluster.

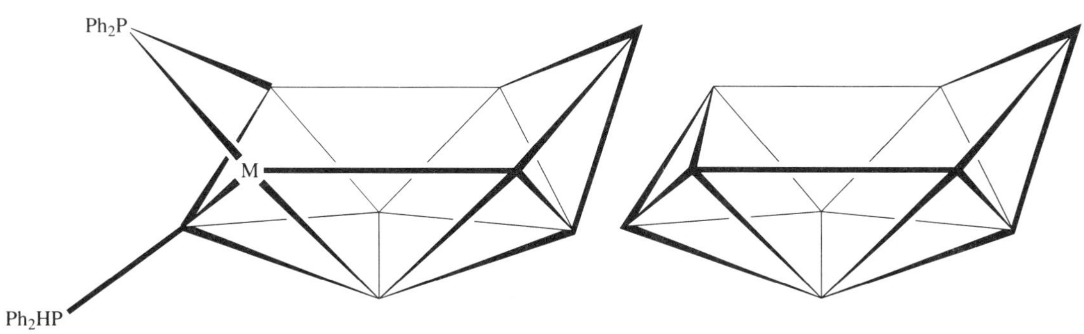

(a) **(361)** **(b)**

$$[(Me_3P)_2(CO)(H)IrB_8H_{12}]\ \mathbf{(363)} \xrightarrow{\Delta} nido\text{-}2,2,2\text{-}[(Me_3P)_2(CO)\text{-}2\text{-}IrB_8H_{11}]\ \mathbf{(366)} + H_2 \qquad (12)$$

8.9.3 Dodecahydrooctaborate Derivatives

8.9.3.1 Group 7 metals

[2,2,2-$(PMe_2Ph)_3$-2,2-H_2-*nido*-2-ReB_8H_{11}] (**368**)[249] is prepared by thermolysis of the 9-OEt derivative of [6,6,6-$(PMe_2Ph)_3$-6-H-*nido*-6-ReB_9H_{13}] (**369**) in *sym*-$C_2D_2Cl_4$ at 100 °C. The structure of the yellow air-stable crystalline solid is given in Scheme 12 which indicates that formation takes place by degradation of the nine-boron precursor involving the elimination of one boron atom.

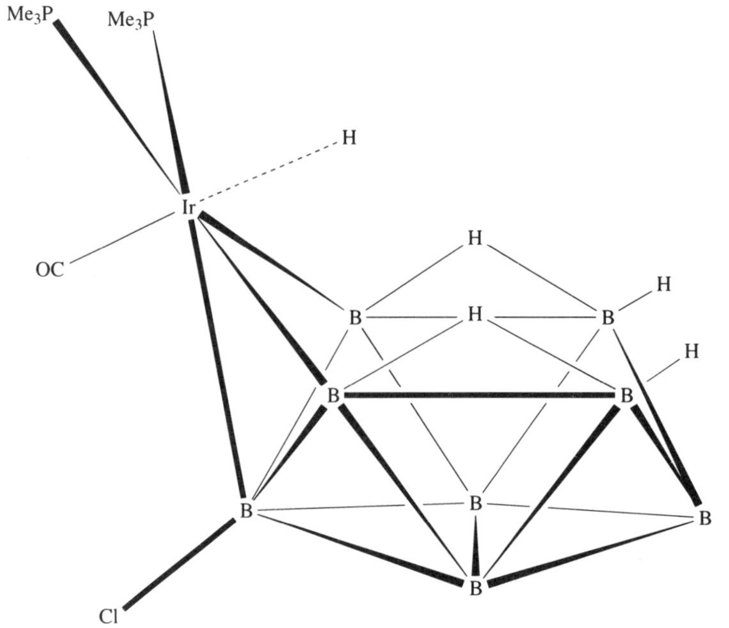

(362)–(364)

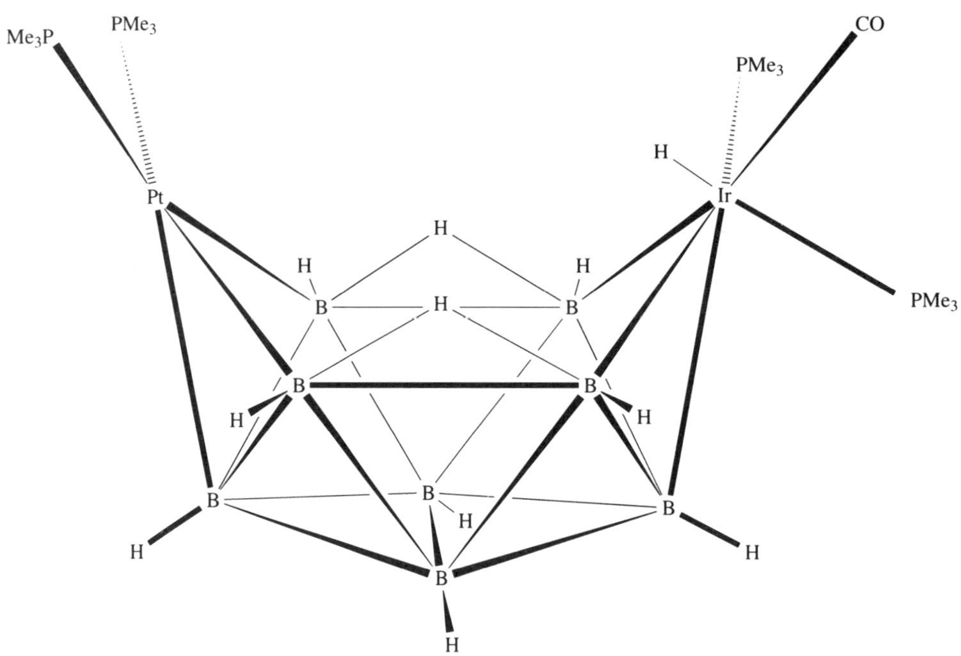

(365)

8.9.3.2 Group 8 metals

nido-2-[η-Mes-2-Fe-6-OB_8H_{10}] (**370**),[250] is obtained as a trace by-product in the reaction of iron vapor with $C_6Me_3H_3$ and $B_{10}H_{14}$, presumably from air oxidation during workup of other products. The yellow crystalline solid has a true *nido*-structure, with the iron moiety occupying the 2-position in $B_{10}H_{14}$ and the oxygen atom occupying the 6-position.

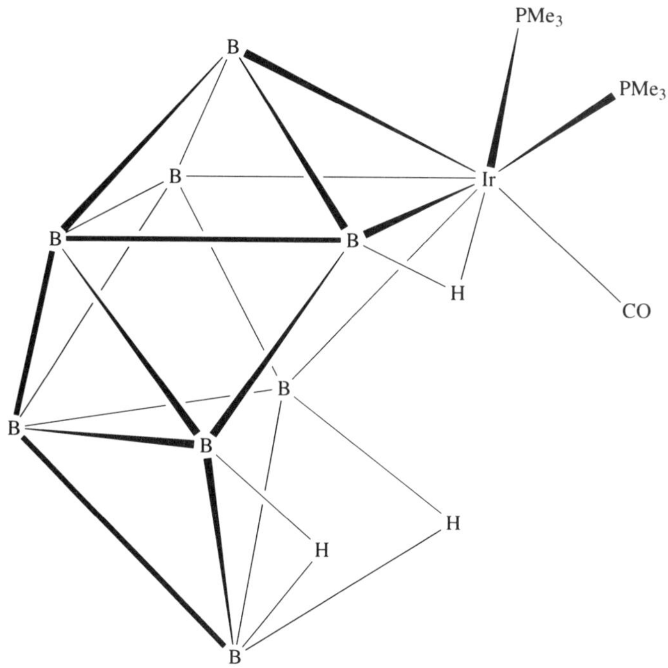

(366)

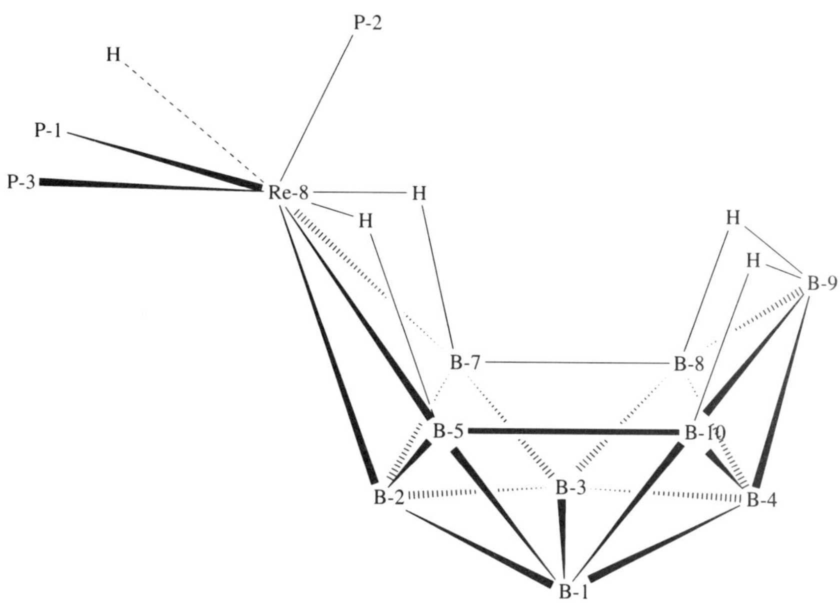

(369)

8.9.3.3 Group 9 metals

[6,9-$(Cp^*Co)_2B_8H_{12}$] (**371**) and [5,7-$(Cp^*Co)_2B_8H_{12}$] (**372**)[217,251] are available in the reaction between LiCp*, $CoCl_2$, and $[B_9H_{14}]^-$ as green and dark-brown crystals, respectively. The related 5,9-isomer, [5,9-$(Cp^*Co)_2B_8H_{12}$] (**373**) is obtained as a light-brown crystalline solid from the reaction between $[B_5H_8]^-$, LiCp*, and $CoCl_2$. The systems are true formal analogues of $B_{10}H_{14}$ with the Cp*Co groups subrogating the 6,9-, 5,7-, and 5,9-positions in $B_{10}H_{14}$, respectively. The species 5,8- (**374**), 1,5- (**375**), 1,7- (**376**), and 2,4-$Cp_2Co_2B_8H_{12}$ (**377**),[220] a series of isomers of [$(CpCo)_2B_8H_{10}$], are formed in the metal-promoted fusion of isomers of $[CpCoB_4H_7]^-$ as indicated in Scheme 13. Total yields ranged from 23% when starting with 2-$[CpCoB_4H_7]^-$ to 57% when starting with 1-$[CpCoB_4H_7]^-$ and the individual isomer

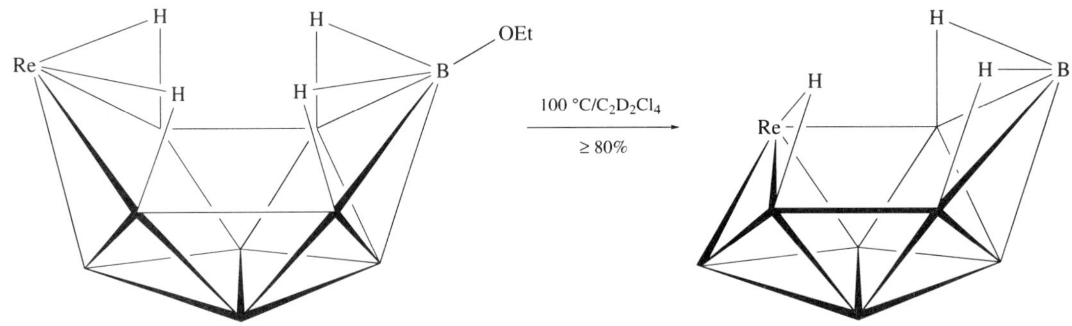

(369) **(368)**

Re = Re(PMe$_2$Ph)$_3$H$_2$

Scheme 12

mixtures formed depends on the starting materials. Previously the 5,7- and the 6,9-isomers were known (see *COMC-I*).

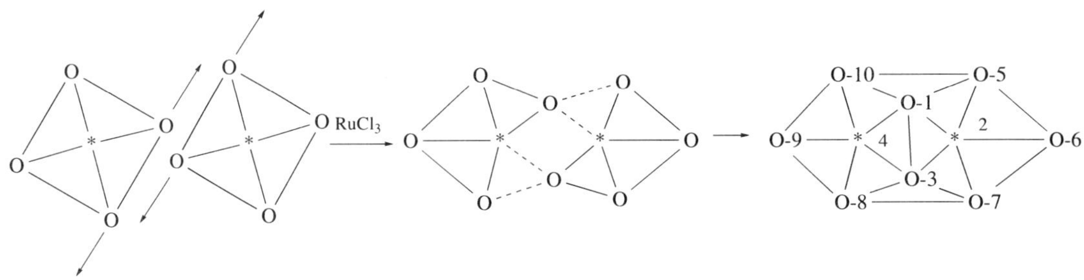

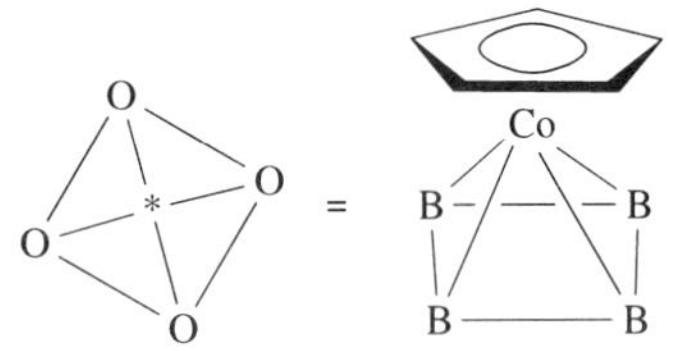

Scheme 13

[2-(Cp*)-*nido*-2-RhB$_8$H$_{10}$-5-(PMe$_2$Ph)] (**378**),[252] an analogue of [PMe$_2$Ph)$_3$H$_2$ReB$_8$H$_{11}$ (**368**) described above, is prepared as an orange solid one of three major products in the reaction between [6-(Cp*)-*nido*-6-RhB$_9$H$_{13}$] with excess PMe$_2$Ph in CH$_2$Cl$_2$ solution at ambient temperature. The formation of *nido*-rhodanonaborane is considered to be analogous to the formation of the related *nido*-rhenanonaborane.

nido-2,2,2-[(Me$_3$P)$_2$(CO)-2-IrB$_8$H$_{11}$] (**366**) is formed as described above in the thermolysis of *asym*-[(Me$_3$P)$_2$(CO)(H)IrB$_8$H$_{12}$] (**363**).[243,246–7] The pale-yellow crystalline solid is a *nido*-system consisting of a [B$_9$H$_{12}$]$^-$ cage with the iridium moiety subrogating the 2-position, as indicated in (**366**). The related species *nido*-2,2,2-[(Me$_3$P)$_3$(CO)-2-IrB$_8$H$_{11}$] (**379**), and *nido*-2,2,2-[(Me$_3$P)$_2$(CO)-2-IrB$_8$H$_{10}$Cl] (**380**) are obtained from thermolysis of the corresponding *arachno*-iridanonaboranes.

8.9.3.4 Group 10 metals

[7,7-(dppe)NiB$_8$H$_{12}$] (**381**)[238] tentatively assigned by its ^{11}B NMR spectrum, is prepared in low yield from the reaction of NiCl$_2$, dppe, and [NMe$_{14}$][B$_9$H$_{14}$] in CH$_2$Cl$_2$. [4,4-(PMe$_2$Ph)$_2$-*arachno*-4-PtB$_8$H$_{12}$] (**382**),[239–40,253–4] a colorless solid initially prepared from the reaction between [B$_9$H$_{14}$]$^-$ and *cis*-PtCl$_2$(PMe$_2$Ph)$_2$, is also available from the reactions between *cis*-PtCl$_2$(PMe$_2$Ph)$_2$ and 6,6'-(B$_{10}$H$_{13}$)$_2$O, *cis*-PtCl$_2$(PMe$_2$Ph)$_2$ and B$_9$H$_{13}$·SMe$_2$, PtCl$_2$(PMe$_2$Ph)$_2$ and B$_{10}$H$_{12}$·SMe$_2$, PtCl$_2$(PMe$_2$Ph)$_2$ and [B$_{10}$H$_{11}$Ph]$^-$. Its structure resembles that of the [B$_9$H$_{14}$]$^-$ anion and the *arachno*-metallanonaboranes described previously. The PtB$_8$ cage contains a six-atom open face characteristic of an *arachno*-nine-vertex cluster and different from the *nido*-nine-vertex clusters which contain five- or four-atom faces. The electron count for this system would suggest a *nido*-system, however, the L$_2$Pt group behaves like the BH$_2^-$ group it appears to subrogate and interacts in a η^3 mode with the B$_8$ cage. Deprotonation of (**382**) and treatment of the resulting anion with *cis*-$PtCl_2(PMe_2Ph)_2$ or $[PdCl_2(PMe_2Ph)_2]$ affords the dimetalla-*arachno*-decaboranes(16), $[(PhMe_2P)_4Pt_2B_8H_{10}]$ (**383**) and $[(PhMe_2P)_4PtPdB_8H_{10}]$ (**384**), respectively. The generic structure for these species is similar to that for (**365**). Thermolysis of (**382**) under varying conditions affords a range of novel *conjuncto*-macropolyhedral species which are described later. Derivatives of $(PhMe_2P)_2PtB_8H_{12}$ are available[183] and they include [2-Cl-4,4-$(PhMe_2P)_2$-*arachno*-4-PtB_8H_{11}] (**385**), and [8-OMe-4,4-$(PhMe_2P)_2$-*arachno*-4-PtB_8H_{11}] (**386**). $(PPh_3)_2PtB_8H_{12}$ (**387**), $[(PPh_3)_2PdB_8H_{12}]$ (**388**), and $[(Me_2PhP)_2PdB_8H_{12}]$ (**389**) are similarly available in reactions analogous to those for the preparation of (**382**).

8.9.3.5 Group 11 metals

$[S_2CNEt_2AuB_8H_{12}]$ (**390**) and $[(S_2CNEt_2)_2Au_2B_8H_{10}]$ (**391**)[255] are prepared in low yield from the reaction between $[B_9H_{12}]^-$ and $AuBr_2[S_2CNEt_2]$ in CH_2Cl_2 at 25 °C as a very pale-yellow solid and a bright-yellow crystalline solid, respectively. The structures of the species are analogous to the *arachno*-mono- and diplatinaboranes described in the previous section. Thus, they each contain hydrogen bridges symmetrically disposed in the positions related to 5,10 and 7,8 in decaborane(10). $[(H_{11}C_6)_3PAu]_2B_8H_{10}$, (**392**)[256] is formed from the reaction between $[AuMe(PCy_3)]$ and $B_{10}H_{12}(PPh_3)_2$. The yellow crystalline solid resembles (**391**) with one major difference. The Au···Au distance in (**392**) is 0.065 nm shorter than that in (**391**) and although the electron count fits the *arachno*-designation it is clear that (**392**) fits neither the standard description of a *nido*- nor an *arachno*-cluster. The authors relate the structure to an analogue of B_8H_{12}, in that both the latter and (**392**) may be considered to be derived from a common $[B_8H_{10}]^{2-}$ core interacting with two protons or two isolobal $[(H_{11}C_6)_3PAu]^+$ fragments.

8.9.4 Octahydrooctaborate Derivatives

[1,1,1-$[(Me_3P)_2$(H)-1-IrB_8H_8] is formed as a red solid in low yield from the thermolysis of (**366**).[11–13,243] Thermolysis of the chloro-substituted *nido*-iridanonaborane affords 1,1,1-$[(Me_3P)_2(H)IrB_8H_7Cl]$ (**359**) which is described in Section 8.9.1

8.9.5 Conjoined Octaborate Derivatives

Conjuncto-$[Pt_2(PMe_2Ph)_2(\eta^3\text{-}B_2H_5)(\eta^3\text{-}B_6H_9)]$ (**285**),[183,257] a yellow crystalline solid, is prepared as a minor product in the reaction between either 1-Cl-4-(SMe_2)-B_9H_{12} or 4-(SMe_2)-7-(OMe)-B_9H_{12} and $[PtCl_2(PMe_2Ph)_2]$ in the presence of "proton sponge." In the structure, illustrated in (**285**), contains a B_6H_9 cluster and a B_2H_5 cluster bridged by a $[Pt_2(PMe_2Ph)_2]$ moiety. The B_6H_9 subcluster bridges the Pt_2 unit exactly in the same way as it does in $[Pt_2(PMe_2Ph)_2(\eta^3\text{-}B_6H_9)_2]$ (**350**), and the B_2H_5 moiety resembles that in (**284**).

8.10 UNITS CONTAINING NINE BORON ATOMS

8.10.1 Introductory Remarks

There are many metallaboranes containing nine boron atoms. Clusters belonging to the *arachno*-, *nido*-, and *closo*-systems represent species in which metal-containing groups subrogate one or more boron vertices in $[B_{10}H_{14}]^{2-}$, $B_{10}H_{14}$, and $[B_{10}H_{10}]^{2-}$, respectively. One approach is to regard the metal moiety as a unipositive ion which interacts with the anions obtained by the removal of a proton from the neutral parent borane cluster. Although this is clearly a major oversimplification, it provides a convenient mode for the organization of this section. Thus, the *arachno*-, *nido*-, and *closo*-metalladecaboranes are considered as complexes of the units $[B_9H_{15}]$, $[B_9H_{13}]$, and $[B_9H_{11}]$ in this treatment.

8.10.2 Pentadecahydrononaborate Derivatives

8.10.2.1 Group 8 metals

[μ-6,9-(NMe_2)-10-(PMe_2Ph)-5-(η^6-C_6Me_6)-*arachno*-5-RuB_9H_{10}] (**394**)[258] is a derivative of [6-η^6-C_6Me_6-*nido*-6-RuB_9H_{13}] (**393**). Treatment of (**393**) with MeCN affords an unstable yellow solid adduct (**395**) which is stabilized as (**394**) by addition of PMe_2Ph. Compound (**394**) is an orange air-stable crystalline solid. The structures of (**393**)–(**395**) are given in Scheme 14 which suggests the mode of formation of (**394**) from (**393**). [(η-C_6Me_6)Ru(NMe_2CH_2)B_9H_{11}] (**395**) is an unstable species containing a –$NMeCH_2$– moiety bridging the 6- and 9-positions of the essentially *nido*-framework. Decomposition of (**395**) affords the 10-vertex so-called *isocloso*-species [1-(η^6-C_6Me_6)-*isocloso*-1-RuB_9H_9] (**396**) described below. Reaction between MeCN and the 8-OMe derivative of (**393**), that is [6-(η^6-C_6Me_6)-8-(OMe)-*nido*-6-RuB_9H_{12}] (**397**) traps an unusual 12- vertex aza-monocarboranyl cluster containing four-, five-, and six-membered open faces,[259] [5-(η^6-C_6Me_6)-7-(OMe)-*arachno*-5-RuN(Me)C(H)B_9H_{11}].

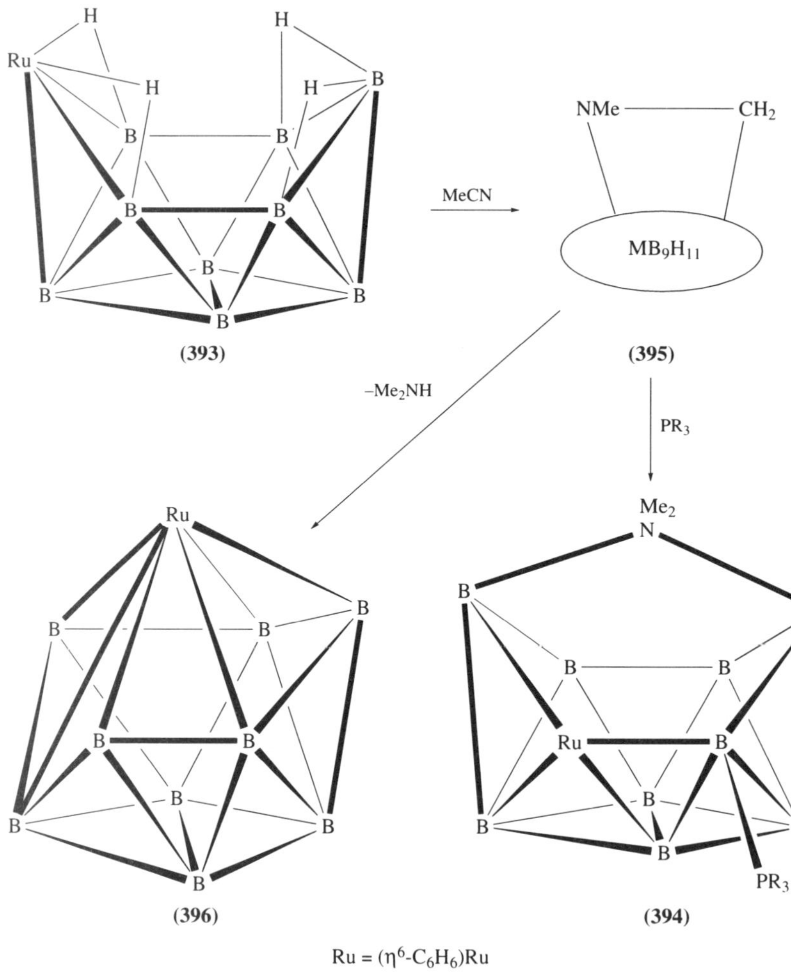

Scheme 14

8.10.2.2 Group 9 metals

Chemistry related to that described in the previous section exists for the Cp*Rh system but in this case the intermediate [6-Cp*-6,9-$(MeCN)_2$-*arachno*-6-(RhB_9H_{11})] (**398**) is obtained.[260] Reaction of [6-(Cp*)-*nido*-6-(RhB_9H_{13})] (**399**) with MeCN in toluene affords a yellow air-stable crystalline product, identified by NMR spectroscopy as (**398**). It is obtained in good yield according to Equation (13), in a reaction reminiscent of the reaction of bases with $B_{10}H_{14}$. The proposed structure is similar to that given in (**398**), determined for the related species [6,9-(*p*-MeC_6H_4NC)$_2$(Cp*RhB_9H_{11}) (**400**), which is obtained as a yellow crystalline solid by treatment of (**399**) with p-MeC_6H_4NC in toluene.[260] If (**398**) is heated in boiling toluene, carbon incorporation occurs to yield a dicarbon metallacarbaborane [1-(Cp*)-2-Me-*closo*-1,2,3-$RhC_2B_8H_9$].[259] Related chemistry for the iridaborane exists.[248] Treatment of [6-(Cp*)-*nido*-6-(IrB_9H_{13})] (**401**) with MeCN affords, in addition to [4-(Cp*)-4-(*endo*-MeCN)-*arachno*-4-IrB_8H_{12}] (**367**), a pale-yellow solid identified as [6-(Cp*)-6-(*endo*-MeCN)-*arachno*-6-IrB_9H_{11}-9-(*exo*-MeCN)] (**402**), in 61% yield. NMR spectra suggest a structure analogous to that for (**400**) in which the MeCN ligand is in an *exo*-position and the other is in an *endo*-position.

$$[6\text{-}(Cp^*RhB_9H_{13})]\ \mathbf{(399)} + 2MeCN \longrightarrow [(MeCN)_2(Cp^*RhB_9H_{11})]\ \mathbf{(398)} + H_2 \qquad (13)$$

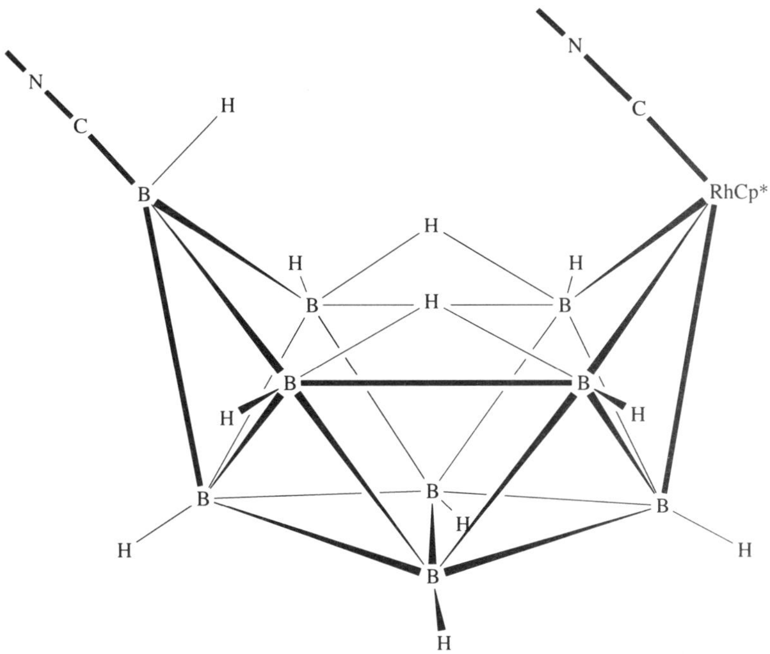

(398)

8.10.2.3 Group 11 metals

A series of metallaboranes, $Cu(PPh_3)_2(B_9H_{13}X)$ (**403**), where X = H, NCS, NCSe, $NCBPH_3$, $NCBH_3$ and $NC(BH_2)NCBH_3$, have been prepared and characterized spectroscopically.[261] They are obtained in good yield by allowing $Cu(PPh_3)_2(BH_4)$ and the $[N(PPh_3)_2]^+$ salt of the borane ion to react in CH_2Cl_2. NMR spectra suggest that the structures of the first four species are fluxional at room temperature resulting in a plane of symmetry with the copper moiety bridging to two of the three *endo*-hydrogens as indicated in Structure (**403**). The last two compounds are expected to involve the copper atom coordinating the terminal BH_3 group via a pair of Cu–H–B bridge bonds.

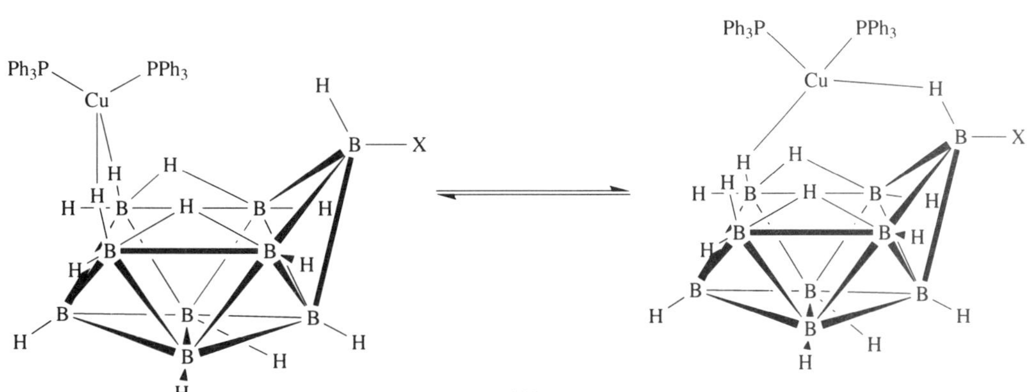

(403)

8.10.3 Tridecahydrononaborate Derivatives

These systems are formal derivatives of $B_{10}H_{14}$ and most of them are quite similar in structure so that *nido*-metalladecaboranes containing metals from different groups in the periodic table are treated in the same section. There is a series of *nido*-6-phosphino-6-metalladecaboranes which includes species containing the metals Mo (**404**), (**405**),[262] W (**406**),[263] (**407**),[262] Re (**369**),[263–4] Ru (**408**)–(**411**),[196] Os (**412**),[263] and Ir (**413**),[265] (**414**).[266] They are quite similar to each other and the structure of $[(PMe_2Ph)_3HReB_9H_{13}]$ (**369**), given in Section 8.9.3.1, is representative of the series. Table 17 lists the species and their preparation, and a summary of some bond distances in the series is given in Table 18. The metal–boron distances decrease across the periodic table, W > Re > Os > Ir, and Mo > Ru, but other trends are less obvious. Cage distortion in any regular mode is not clear from an examination of the data listed in Table 18. The effect of substitution of different metals for boron at the 6-position in the $B_{10}H_{14}$ cluster has little effect on the NMR spectra and this emphasizes the compatibility of these metal centers in the polyhedral boron cluster.

Table 17 Preparation and properties of 6-(phosphino)-6-metalladecaboranes.

Species	*No.*	*Preparation*	*Appearance/properties*	*Yield*	*Ref.*
$[(CO)_2(PPh_3)_3MoB_9H_{13}]$	(**404**)	$[Mo(CO)_3Cl_2(PPh_3)_2] + [B_9H_{14}]^-$	Bright-purple solid	4%	262
$[(PMe_2Ph)_3H_2WB_9H_{13}]$	(**406**)	$[WH_6(PMe_2Ph)_3] + [B_9H_{14}]^-$	Blue air-stable crystals	21%	263
$[(CO)_3(PMe_2Ph)WB_9H_{13}]$	(**407**)	$W(CO)_6 + Cl_2 + [B_9H_{14}]^- + PMe_2Ph$	Purple air-stable solid	10%	262
$[(PMe_2Ph)_3HReB_9H_{13}]$	(**369**)	mer-$[ReCl_3(PMe_2Ph)_3] + [B_9H_{14}]^-$	Purple crystals	40%	263, 264
$[(PMe_2Ph)_3RuB_9H_{13}]$	(**408**)	$[RuCl_3(PMe_2Ph)_3] + [B_9H_{14}]^-$ in refluxing EtOH	Red solid	35%	196
$[(CO)(PPh_3)_2RuB_9H_{13}]$	(**409**)	$[Ru(CO)ClH(PPh_3)_3] + [B_9H_{14}]^-$ in CH_2Cl_2	Bright-orange solid	35%	196
$[(CO)(PPh_3)_2RuB_9H_{11}(PPh_3)]$	(**410**)	$[Ru(CO)ClH(PPh_3)_3] + [B_9H_{14}]^-$ in CH_2Cl_2	Light-yellow solid	5%	196
$[(PPh_3)_2HRuB_9H_{12}(PPh_3)]$	(**411**)	$[RuCl_2(PPh_3)_4] + [B_9H_{14}]^-$ in CH_2Cl_2	Burgundy-red solid (trace)		196
$[(PMe_2Ph)_3OsB_9H_{13}]$	(**412**)	$[OsCl_3(PMe_2Ph)_3] + [B_9H_{14}]^-$ in refluxing EtOH	Orange solid	32%	196
$[(PPh_3)_2HIrB_9H_{13}]$	(**413**)	$IrCl(PPh_3)_3 + [B_9H_{14}]^-$ in CH_2Cl_2	Bright-yellow solid	80%	265
$[(\eta^2{:}\eta^2\text{-cod})ClIrB_9H_{13}$	(**414**)	$[Ir(cod)Cl_2] + [B_9H_{14}]^-$	Red–orange solid	5%	266

Table 18 Comparative structural data for 6-(phosphino)-6-metalladecaboranes.249,263

		Bond distance (nm)				
Species	*No.*	M–B-2	M–B-5	M–B-7	B-1–B-3	B-5–B-10
$[(CO)_2(PMe_2Ph)_2Mo\text{-}B_9H_{12}Cl]$	(**405**)	0.2397	0.2442	0.2465	0.1771	0.2041
$[(PPh_3)_2HRuB_9H_{12}(PPh_3)]$	(**411**)	0.2300	0.2311	0.2263	0.1764	0.2011
$[(PMe_2Ph)_3H_2WB_9H_{13}]$	(**407**)	0.2432	0.2418	0.2472	0.1751	1.2053
$[(PMe_2Ph)_3HReB_9H_{13}]$	(**369**)	0.2378	0.2366	0.2386	0.1805	0.2094
$[(PMe_2Ph)_3OsB_9H_{13}]$	(**412**)	0.2337	0.2344	0.2243	0.1771	0.2016
$[(PPh_3)_2HIrB_9H_{13}]$	(**413**)	0.2269	0.2289	0.2281	0.1788	0.1977

In addition to the series of 6-phosphino-6-metalla-*nido*-decaboranes, there is another series of analogous metalladecaboranes. This one includes the η^6-hexamethyl-benzenemetalla-*nido*-decaboranes[252,260,269–71] and the η^5-pentamethyl-cyclopentadienylmetalla-*nido*-decaboranes.[245,260,269–73] Structures of the 1, 2, and 5 isomers are given in Figure 1 and that for the 6-isomer is given in Scheme 14. Preparative and other data are given in Table 19 for the 6-isomers. $[6\text{-}\eta^6\text{-}C_6Me_6\text{-}nido\text{-}6\text{-}RuB_9H_{13}]$ (**393**)[212,268] is typical of these species. The $[(\eta^6\text{-}C_6Me_6)]$ group replaces a BH vertex at the 6-position in $B_{10}H_{14}$ and thus acts as a conventional two-electron three orbital donor to the metallaborane cluster. The proposed structure is consistent with the polyhedral skeletal electron pair theory and the metal atom is an 18 electron d^6 Ru^{II} complex. Compound (**393**) is also accessible from the reaction between $[(\eta^6\text{-}C_6Me_6)RuCl_2]_2$ and $[N^nBu_4][B_9H_{12}]$ in CH_2Cl_2. Reactions of (**393**) with bases proceed analogously to those of the isoelectronic $B_{10}H_{14}$ with bases except that harsher conditions are required and *nido*-products are observed.[268] Thus, similar to the reaction of *nido*-$B_{10}H_{14}$ with phosphines to afford *arachno*-$B_{10}H_{12}$ diadducts, $[6\text{-}\eta^6\text{-}C_6Me_6\text{-}nido\text{-}6\text{-}RuB_9H_{13}]$ (**393**) reacts with PMe_2Ph in refluxing CH_2Cl_2 to afford $[5\text{-}\eta^6\text{-}C_6Me_6\text{-}nido\text{-}5\text{-}RuB_9H_{11}\text{-}7\text{-}(PMe_2Ph)]$ (**420**), a yellow crystalline solid in 77% yield according to Equation (14). The related species $[5\text{-}\eta^6\text{-}C_6Me_6\text{-}nido\text{-}5\text{-}RuB_9H_{11}\text{-}7\text{-}(PPh_3)]$ (**421**) forms in 54% yield as a yellow crystalline solid when (**393**) reacts with PPh_3.[268] Reaction between (**393**) and SMe_2 in refluxing toluene affords the 10-vertex *isocloso*-species $[1\text{-}\eta^6\text{-}C_6Me_6\text{-}isocloso\text{-}RuB_9H_9]$ (**396**) which is described below.[269]

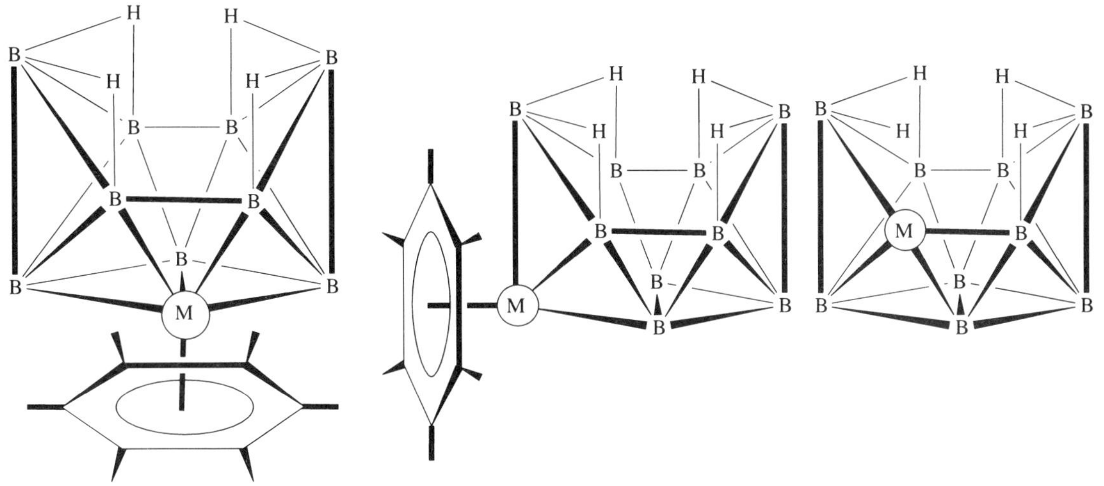

(422a) **(422b)** **(422c)**

Figure 1 Structures of three of the four isomers of [n-η-C_6Me-*nido*-n-MB_9H_{13}]. (a) n = 1; (b) n = 2; and (c) n = 5. For (c) the aromatic ring is omitted for clarity.

Table 19 Preparation and properties of 6-(η^n-C_nMe_n)-*nido*-6-MB_9H_{13} species.

Species	*No.*	*Reaction*	*Yield (%)*	*Appearance*	*Ref.*
6-(η^6-C_6Me_6)-6-RuB_9H_{13}	**(393)**	$[B_9H_{14}]^-$ + [{Ru(η^6-C_6Me_6)Cl_2}$_2$]	80	Orange–yellow solid	212, 268
[6-(η^6-C_6Me_6)-8-(OEt)-6-RuB_9H_{12}]	**(415)**	$[TeB_{10}H_{11}]^-$ + [{Ru(η^6-C_6Me_6)Cl_2}$_2$]	38	Orange solid	274
[5-Cl-6-(η^6-C_6Me_6)-6-RuB_9H_{12}]	**(416)**	$[B_9H_{12}]^-$ + [{Ru(η^6-C_6Me_6)Cl_2}$_2$]	12		262
6-(η^6-C_6Me_6)-6-OsB_9H_{13}	**(417)**	$[B_9H_{14}]^-$ + [{Os(η^6-C_6Me_6)Cl_2}$_2$]	13	Orange crystals	268
6-(Cp*)-6-CoB_9H_{13}	**(418)**	Li[Cp*] + $CoCl_2$ + $[B_9H_{14}]^-$	25	Burgundy solid	217, 251
6-(Cp*)-6-RhB_9H_{13}	**(398)**	$[B_9H_{14}]^-$ + [Rh(Cp*Cl_2)$_2$]	93	Blood-orange solid	271
6-(Cp*)-6-IrB_9H_{13}	**(419)**	$[B_9H_{14}]^-$ + [Ir(Cp*Cl_2)$_2$	55	Orange solid	270

[6-η^6-C_6Me_6-6-RuB_9H_{13}] **(393)** + PMe_2Ph ⟶ [5-η^6-C_6Me_6-5-RuB_9H_{11}-7-(PMe_2Ph)] **(420)** + H_2 (14)

The parent 5-(η^6-C_6Me_6)-*nido*-5-metalladecaboranes are also known. For example [5-(η^6-C_6Me_6)-*nido*-5-RuB_9H_{13}] **(422c)**[270] is obtained as a yellow crystalline solid in 18% yield from the reaction between $[B_{10}H_{13}]^-$, prepared *in situ* from proton sponge in CH_2Cl_2 and $B_{10}H_{14}$, and [{Ru(η^6-C_6Me_6)Cl_2}$_2$] in CH_2Cl_2. NMR spectra of the species are consistent with the structure given in Figure 1(c). The related 5-(η^6-C_6Me_6)-*nido*-5-OsB_9H_{13} **(423)**, 5-(Cp*)-*nido*-5-RhB_9H_{13} **(424)**, and 5-(Cp*)-*nido*-5-IrB_9H_{13} **(425)** are similarly prepared. The details are given in Table 20 along with some data on the by-products formed. The iron species, 5-(η^6-$C_6Me_3H_3$)-*nido*-5-FeB_9H_{13} **(426)** and 5-(η^6-C_6H_5Me)-*nido*-5-FeB_9H_{13} **(427)**, also listed in Table 20, are prepared in the reaction between ligand, $B_{10}H_{14}$, and iron atoms in a metal atom reactor.[250]

The 1- and 2-isomers of the ruthenium complex, shown in Figure 1(a) and 1(b), respectively, [1-η^6-C_6Me_6-*nido*-1-RuB_9H_{13}] **(422a)** and [2-η^6-C_6Me_6-*nido*-1-RuB_9H_{13}] **(422b)**,[267] are unexpected products of the reaction between *arachno*-$[B_6H_{11}]^-$ and [(η^6-C_6Me_6)$RuCl_2$]$_2$. They are both yellow air-stable solids and are isolated in 9% and 5% yields, respectively. The structure of **(422b)** is quite similar to that of the parent $B_{10}H_{14}$ but for the 1-isomer, **(422a)** the B-5···B-10 axis is nonbonding and thus the cluster is quite open and resembles, somewhat, an *arachno*-system with a seven-membered face. The B-5–B-10 distance is 0.2497 nm compared with 0.2 nm for $B_{10}H_{14}$. The structures of the 5- and 6-isomers have dimensions quite similar to those for $B_{10}H_{14}$ except for the bonds adjacent to the metal atom.

Reaction chemistry of the (η^n-arene)-*nido*-metalladecaboranes, described above, leads to novel systems. [6-(Cp*)-*nido*-6-RhB_9H_{13}] **(424)** reacts with SMe_2 to afford several products including [Cp*RhB_9(SMe_2)$H_{10}RhB_9H_7$(SMe_2)$_2$ **(433)**.[269] A structure determination identifies the latter as a *conjuncto*-species involving a unique *isonido*-type 11-vertex cluster {7',9'-Rh_2B_9} and a closo-10-vertex {2-RhB_9} cluster joined at a common rhodium atom. Reaction of **(424)** with excess of PMe_2Ph in CH_2Cl_2 affords [2-(Cp*)-*closo*-2-RhB_9H_7-3,10-(PMe_2Ph)$_2$]**(434)** which is described in Section 8.10.4.2, and the two species [5-(Cp*)-*nido*-5-RhB_9H_{11}-7-(PMe_2Ph)] **(435)**, and [2-(Cp*)-*nido*-2-RhB_8H_{10}-5-

Table 20 Preparation and properties of 5-(η^n-C_nMe_n)-*nido*-5-MB_9H_{13} species.

Species	*No.*	*Reaction*	*Yield (%)*	*Appearance*	*Ref.*
5-(η^6-$C_6Me_3H_3$)-5-FeB_9H_{13}	**(426)**	$B_{10}H_{14}$ + Fe(g) + $C_6Me_3H_3$	<1	Reddish purple	250
5-(η^6-C_6H_5Me)-5-FeB_9H_{13}	**(427)**	$B_{10}H_{14}$ + Fe(g) + PhMe	<1	Pinkish purple	250
5-(η^6-C_6Me_6)-5-RuB_9H_{13}	**(422)**	$[B_{10}H_{13}]^-$ + [{Ru(η^6-C_6Me_6)Cl_2}$_2$]	19	Yellow crystals	270
[5-(η^6-C_6Me_6)-6-OMe-5-RuB_9H_{12}]	**(428)**	[{Ru(η^6-C_6Me_6)Cl_2}$_2$] + $[B_{10}H_{13}]^-$	32	Yellow crystalline air-stable solid	275
5-(η^6-C_6Me_6)-5-OsB_9H_{13}	**(423)**	$[B_{10}H_{13}]^-$ + [{Os(η^6-C_6Me_6)Cl_2}$_2$]	22	Yellow air-stable solid	270
[6-(OH)-5-(η^6-C_6Me_6)-5-OsB_9H_{12}]	**(429)**	Formed in prep. of **(423)**	6		270
5-(Cp*)-5-RhB_9H_{13}	**(424)**	$[B_{10}H_{13}]^-$ + [$(RhCp^*Cl_2)_2$]	21	Yellow crystals	270
[9,9′-{5-(Cp*)-5-RhB_9H_{12}}$_2$O]	**(430)**	Formed in prep. of **(424)**	27		270
[9-(OH)-5-(Cp*)-5-RhB_9H_{12}]	**(431)**	Formed in prep. of **(424)**	21		270
5-(Cp*)-5-IrB_9H_{13}	**(425)**	$[B_9H_{14}]^-$ + [$(IrCp^*Cl_2)_2$]	6	Yellow solid	270

(PMe_2Ph)] **(378)**[252] which have been described in Section 8.9.3.3. **(435)**[252] is obtained in 35% yield as a red–orange air-stable microcrystalline solid whose structure is similar to that shown in Figure 1(c), with the phosphine bonded at the 7-position and only three μ-hydrogen atoms in the open face. The position corresponding to the 6,7-position of $B_{10}H_{14}$ does not contain a bridging hydrogen atom.

An unusual reaction in which a *nido*-rhodadecaborane is formed involves the air oxidation of [*nido*-(Cp*)$RhB_{10}H_{13}$Cl] **(436)**. Chromatography on silica in air leads to several products including [μ-9,9′-O-{5-(Cp*)-*nido*-5-RhB_9H_{12}}$_2$] **(430)**.[276] It is formed in 28% yield and a crystal structure determination confirms the species as a pair of (*nido*-rhodadecaboranyl) cluster bridged at the 9-position by an oxygen atom. [6-(Cp*)-6-IrB_9H_{13}] **(419)** reacts with $(Cp^*RhCl_2)_2$ in the presence of proton sponge in CH_2Cl_2 to afford the bimetallaborane [1,4-$(Cp^*)_2$-μ-1,4-H-*isonido*-1,4-$RhIrB_9H_{10}$] **(437)**,[277] an open-faced *nido*-like cluster with an electron count suggestive of a *closo*-system. The formation of the species is visualized in Scheme 15. Structure **(a)** represents the starting iridium complex, **(b)** indicates a rearranged version of the latter, and **(c)** shows the insertion of the rhodium complex into the cage. Although not observed in the x-ray structure, mass spectral data suggest the presence of two additional hydrogen atoms and NMR data suggest the presence bridge hydrogens between the two metal atoms and also between the pair of boron across the open face from the metal atoms.

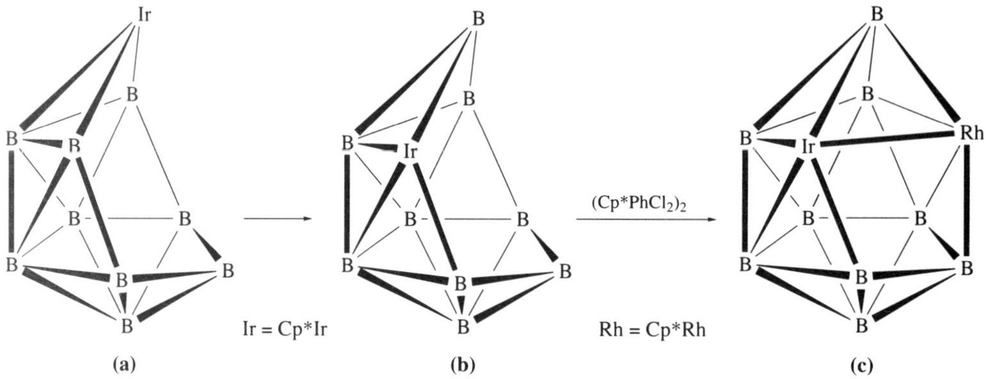

(437)

Scheme 15

Among the 6-metalladecaboranes listed in Table 17, the irida-species are perhaps the most interesting. [6-H-6,6-$(PPh_3)_2$-*nido*-6-IrB_9H_{13}] **(413)**[265–6] is prepared in low yield from excess $[B_9H_{14}]^-$ and *trans*-Ir(CO)Cl$(PPh_3)_2$ and in high yield from $IrCl(PPh_3)_3$. The PMe_3 derivative $HIrB_9H_{13}(PMe_3)_2$ **(438)** and the cyclooctadiene derivative 6-(η^2:η^2-C_8H_{12})-6-Cl-*nido*-6-IrB_9H_{13} **(414)** are known and some details are given in Table 17. Pyrolysis of **(413)** and **(438)** affords novel *closo*-iridadecaboranes and these are described in the next section. If these latter syntheses are carried out using salts of the $[B_9H_{12}]^-$ ion instead of $[B_9H_{14}]^-$, mixtures of products which contain novel *ortho*-boronated species are obtained. Thus reaction of $[B_9H_{12}]^-$ with [$IrCl(PPh_3)_3$] in CH_2Cl_2 at 25 °C affords a 80% yield of the bright-yellow solid [6,6,6,-(PPh_3)(Ph_2P-*o*-C_6H_4)H-*nido*-6-IrB_9H_{12}-5] **(439)**,[265,278] which is identified by NMR spectra as the *ortho*-boronation product. It exists in two isomers *sym*-(**440**) and *asym*-(**440**), the former converts to the latter on thermolysis at 65 °C. Their preparation and structures are given in Scheme 16. Another product of the reaction between $[B_9H_{12}]^-$ and $[IrCl(PPh_3)_3]$ is an isomer of (**439**), $[5\text{-H-5-(PPh}_3)\text{-5-(PPh}_2\text{-}o\text{-}C_6H_4\text{-}nido\text{-5-IrB}_9H_{12}\text{-2}]$ (**440**),[265] obtained in up to 20% yield as a yellow solid. The structure of (**440**) has been confirmed by a crystal structure determination. Also formed in the above reaction in 5% yield is the species $[5\text{-H-5,7,-(PPh}_3)_2\text{-5-(PPh}_2\text{-}o\text{-}C_6H_4\text{-}nido\text{-5-IrB}_9H_{10}\text{-2}]$ (**441**)[265] for which a crystal structure determination indicates that the additional phosphine is on B-7 in the open face which has one μ-hydrogen atom less than (**440**).

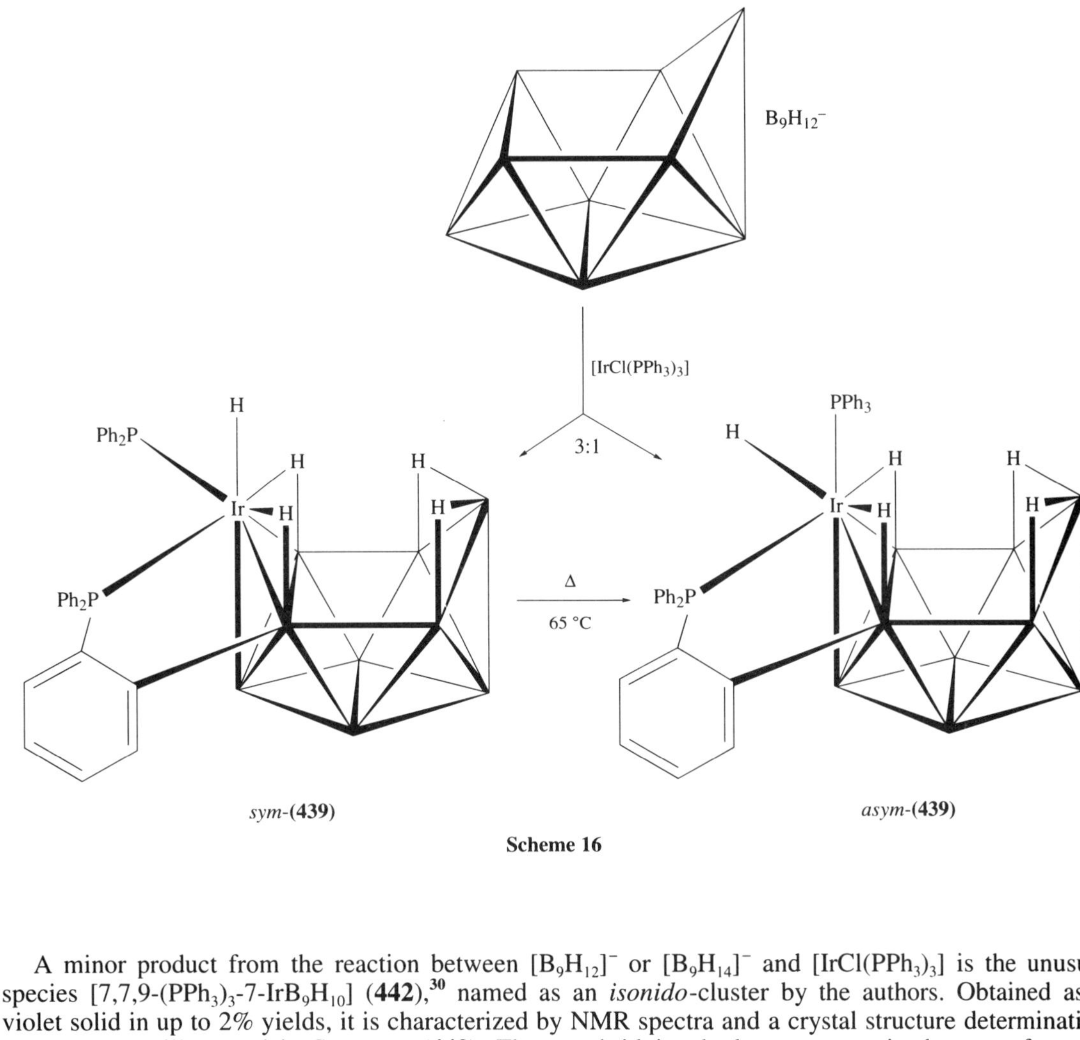

sym-(**439**) *asym*-(**439**)

Scheme 16

A minor product from the reaction between $[B_9H_{12}]^-$ or $[B_9H_{14}]^-$ and $[IrCl(PPh_3)_3]$ is the unusual species $[7,7,9\text{-(PPh}_3)_3\text{-7-IrB}_9H_{10}]$ (**442**),[30] named as an *isonido*-cluster by the authors. Obtained as a violet solid in up to 2% yields, it is characterized by NMR spectra and a crystal structure determination as the system illustrated in Structure (**442**). The two bridging hydrogen atoms in the open face are assigned on the basis of NMR spectroscopy and they are fluxional. Although beyond the scope of this treatment structure (**442**) may be considered as being derived from the *closo*-11-vertex octadecahedron by the removal of a vertex of four connectivity. An extension of this point is to be found below in Section 8.10.3.2, where *isocloso*-metalladecaboranes are described.

Treatment of the orthoboronated *nido*-6-iridadecaborane, *asym*-(**439**), with KH followed by addition of *cis*-$[PtCl_2(PMe_3)_2]$ affords the bimetallaundecaborane (**443**)[279] as an orange-red air-stable solid in 50% yield. A crystal structure determination reveals the structure shown in Scheme 17.

Thermolysis of 1,1,2,2-$C_2H_4Cl_2$ solutions of $[5\text{-H-5,7,-(PPh}_3)\text{-5-(PPh}_2\text{-}o\text{-}C_6H_4\text{-}nido\text{-5-IrB}_9H_{12}\text{-2}]$ (**441**) result in the loss of H_2 gas and the formation of the several *isonido*-species including $[7,9,\text{-(PPh}_3)_2\text{-7-(}o\text{-Ph}_2PC_6H_4)\text{-}isonido\text{-7-IrB}_9H_9\text{-10}]$ (**444**), an *ortho*-boronated analogue of (**442**).[30] Another of these *isonido*-systems $[3\text{-(OEt)-4,9-(PPh}_3)\text{-7,7,-Cl-(}o\text{-Ph}_2PC_6H_4)\text{-}isonido\text{-7-IrB}_9H_5\text{-10}]\cdot CH_2Cl_2$ (**445**) is available from the thermolysis of (**441**). Its structure has been determined and it conforms to Structure (**444**).[280]

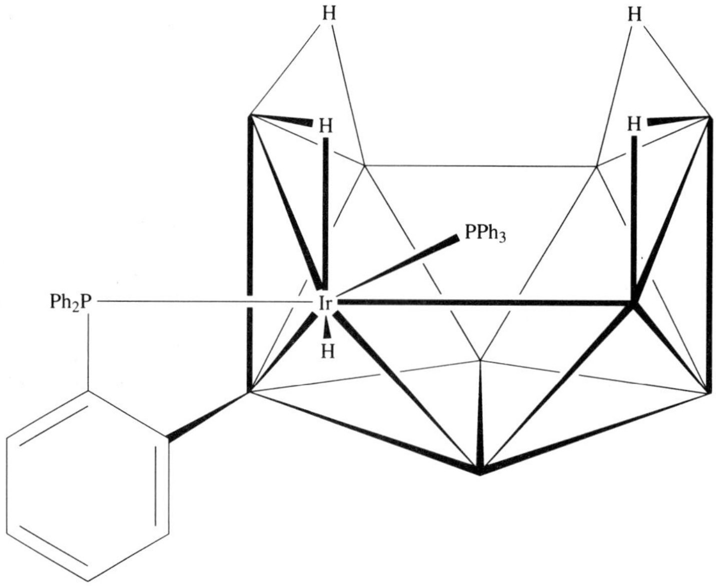

(440)

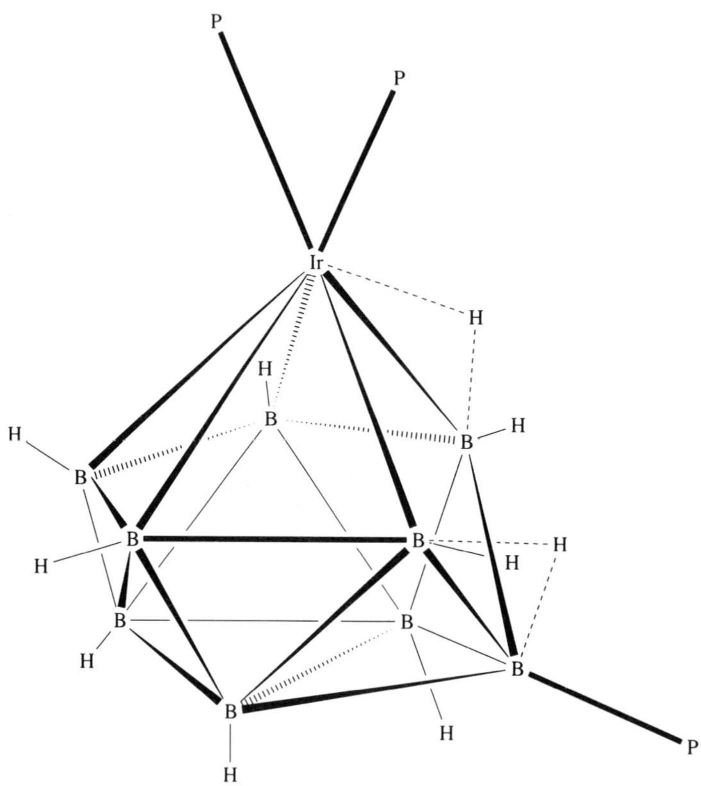

(442)

8.10.4 Undecahydrononaborate Derivatives

8.10.4.1 Group 8 metals

[1-(η^6-C_6Me_6)-*isocloso*-1-RuB_9H_9] (**396**)[269] is prepared in 80% yield from the reaction between SMe_2 and [6-(η^6-C_6Me_6)-*nido*-RuB_9H_{13}] (**393**) in refluxing toluene. The species is identified by the NMR

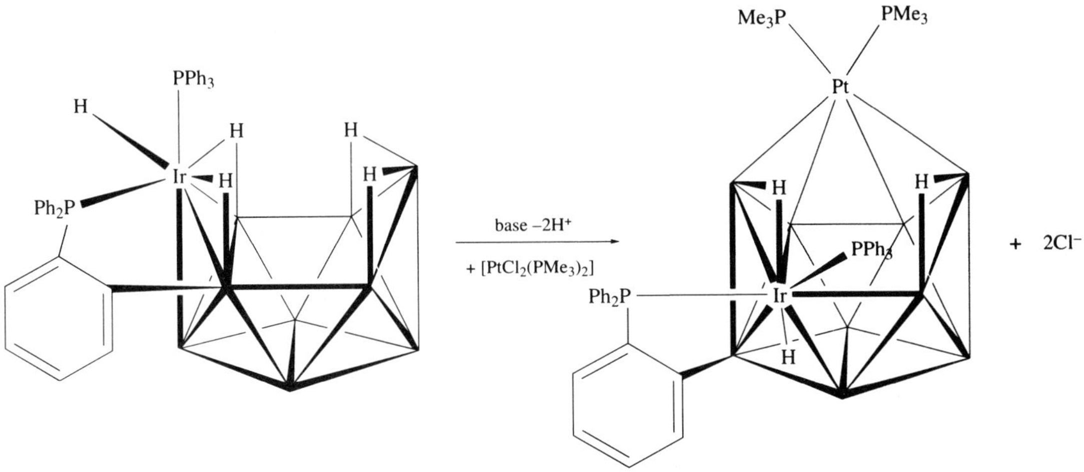

asym-(**439**) (**443**)

Scheme 17

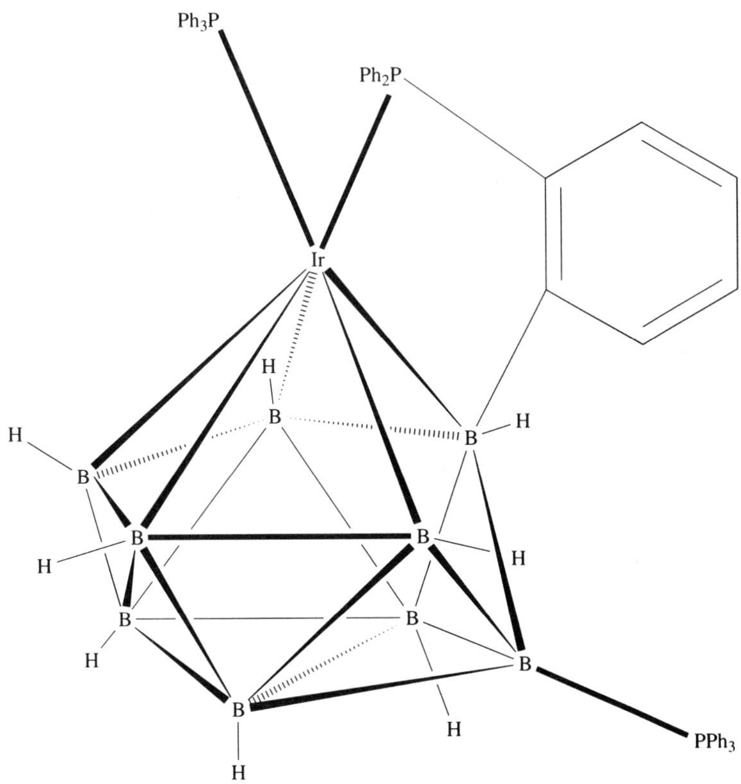

(**444**)

spectrum which is quite reminiscent of that of the *ortho*-boronated *isocloso* iridaborane, [$(PPh_3)(Ph_2P$-o-$C_6H_4)HIrB_9H_8$] (**446**) which was discovered earlier and is described below. The proposed structure, which was shown in Scheme 14, suggests there are three sets of boron atoms and this was confirmed by an x-ray study of the derivative [1,1,1-$(PPh_3)HCl$-1-RuB_9H_7-3,5-$(PPh_3)_2$] (**447**).[281] The species exhibits a 1:3:3:3:3 $RuB_3B_3B_3$ cluster structure which has C_{3v} symmetry with three "belts,". of three boron atoms. The geometry is clearly a closed polyhedron, and the electron count is two electrons short of the correct $2n + 2$, so the authors designate the species as an *isocloso* cluster. Compound (**396**) is also formed in the reaction between (**393**) and MeCN.[258] The reaction, which forms (**395**) as an intermediate and was mentioned in Section 8.10.2.1 proceeds according to Equation (15), and thus serves to reduce MeCN to MeNHMe during the oxidation of (**393**) to the *isocloso* product (**396**). Reaction between

$[B_9H_{12}]^-$ and $[\{(\eta^6\text{-}C_6Me_6)RuCl_2\}_2]$ in CH_2Cl_2 affords the 5-Cl derivative of (**393**) along with traces of (**393**) and (**396**).[282] The 2-ethoxy-derivative of (**396**), $[1\text{-}(\eta^6\text{-}C_6Me_6)$-*isocloso*-$1\text{-}RuB_9H_8\text{-}2\text{-}(OEt)]$ (**448**)[197] is prepared by the reaction between $[6\text{-}(\eta^6\text{-}C_6Me_6)ClRuB_3H_8]$ and $[NHEt_3][B_{10}H_{10}]$ as a yellow air-stable solid in 2% yield. $[1,1,1\text{-}(PPh_3)HCl\text{-}1\text{-}RuB_9H_7\text{-}3,5\text{-}(PPh_3)_2]$ (**447**),[281] is prepared in 39% yield as an orange air-stable solid in the reaction between $[NEt_4][B_9H_{14}]$ and $[RuCl_2(PPh_3)_3]$ in CH_2Cl_2 as indicated in Equation (16), but since phosphine-boranes such as $BH_3{\cdot}PPh_3$ and $B_9H_{13}{\cdot}PPh_3$ are formed in the reaction, the process is clearly much more complex. $[1\text{-}(\eta^6\text{-}MeC_6H_4\text{-}4\text{-}Pr^i)$-*isocloso*-$1\text{-}RuB_9H_9]$ (**449**)[282] is available in 2% yield as a yellow air-stable solid from the reaction between $[B_9H_{12}]^-$ and $[\{(\eta^6\text{-}MeC_6H_4\text{-}4\text{-}Pr^i)RuCl_2\}_2]$.

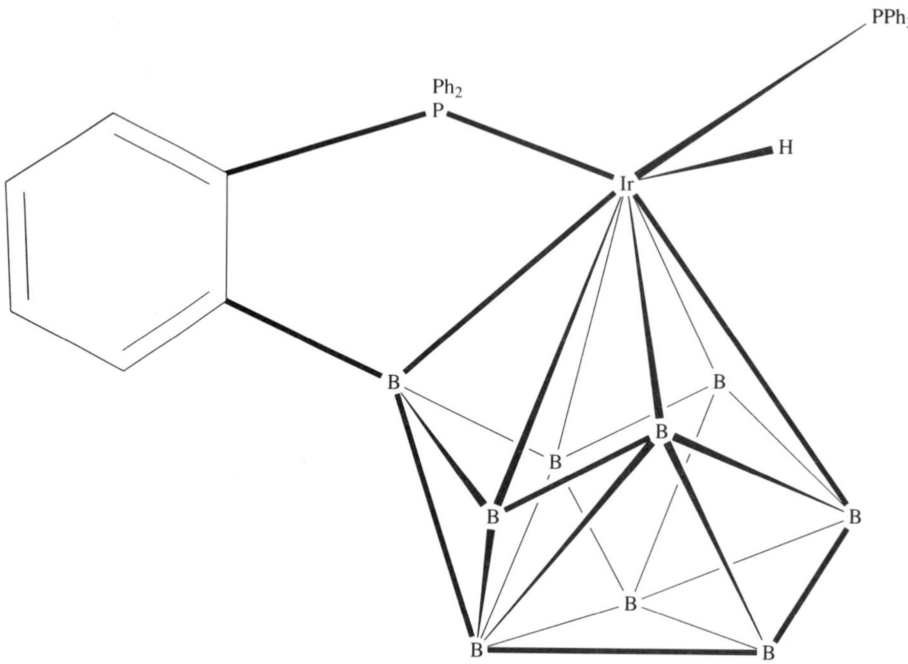

(**446**)

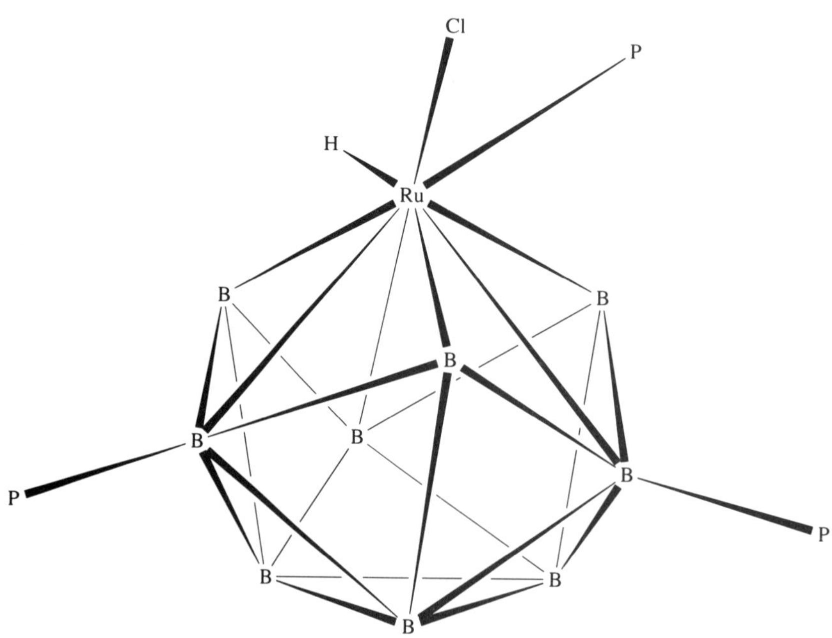

(**447**)

$MeCN + [(C_6Me_6)RuB_9H_{13}]$ **(393)** ⟶ $MeNHMe + [(\eta^6\text{-}C_6Me_6)RuB_9H_9]$ **(396)** (15)

$[B_9H_{14}]^- + [RuCl_2(PPh_3)_3]$ ⟶ $[(PPh_3)HClRuB_9H_7(PPh_3)_2]$ **(447)** $+ 3H_2 + Cl^-$ (16)

8.10.4.2 Group 9 metals

$[2\text{-}(\eta^5\text{-}C_5Me_5)$-*closo*-$2\text{-}RhB_9H_7\text{-}3,10\text{-}(PMe_2Ph)_2]$ (**434**)[252] is prepared in 25% yield in the reaction between $[6\text{-}(Cp^*)\text{-}6\text{-}RhB_9H_{13}]$ (**399**) and PMe_2Ph in refluxing toluene, as a yellow air-stable solid. The structure, illustrated in Structure (**450**) with the *exo* hydrogens and organyl groups of phosphorus atoms omitted for clarity, and confirmed by an x-ray diffraction study, is the expected bicapped square antiprism with the Cp*Rh moiety replacing atom B-2. The $B(PMe_2Ph)$ groupings at positions 2, and 10, serve to replace BH^- moieties in $[B_{10}H_{10}]^{2-}$, so since Cp*Rh is isolobal with BH, the species has a skeletal electron count consistent with the PSEPT. The bond distances are quite normal except that the Rh–B-1 value is 0.016–0.022 nm shorter than the others. The major product of this reaction is [5-(Cp*)-*nido*-$5\text{-}RhB_9H_{11}\text{-}7\text{-}(PMe_2Ph)]$ (**450**).

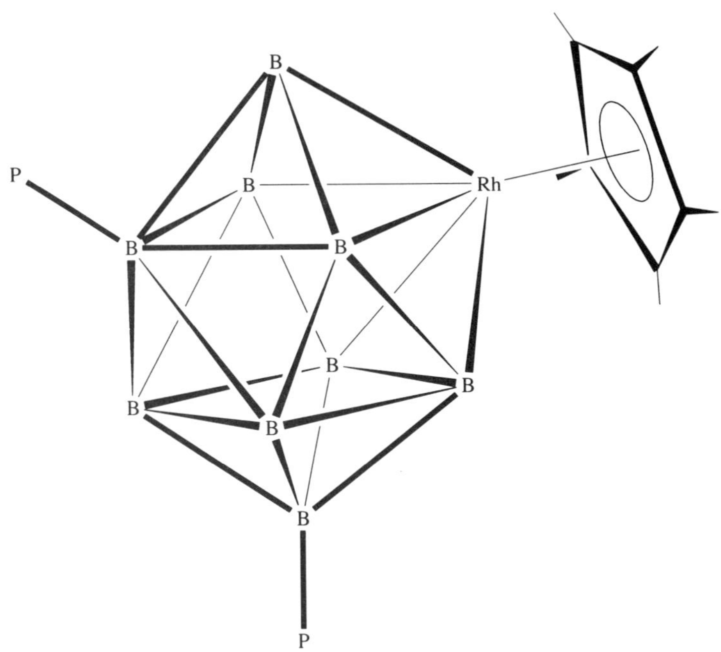

(450)

The formation of $[1\text{-}(\eta^6\text{-}C_6Me_6)$-*isocloso*-$1\text{-}RuB_9H_9]$ (**396**)[269] by the reaction between SMe_2 and $[6\text{-}(\eta^6\text{-}C_6Me_6)$-*nido*-$RuB_9H_{13}]$ (**393**) is not applicable to the rhodium system. Thus reaction between SMe_2 and [6-(Cp*)-*nido*-$RhB_9H_{13}]$ (**399**) gives a variety of products one of which is the unusual *conjuncto*-species $[(\eta^5\text{-}C_5Me_5)RhB_9(SMe_2)\text{-}H_{10}RhB_9H_7(SMe_2)_2]$ (**433**)[269] described in Section 8.10.3.

$[10\text{-}(PPh_3)\text{-}2,2,2\text{-}(PH_3)_2(Ph_2P\text{-}o\text{-}C_6H_4)H$-*closo*-$2\text{-}IrB_9H_7\text{-}1]$ (**451**)[283] is formed by the passage of ethyne through a refluxing solution of (**439**). It is isolated as a pale-yellow solid in 47% yield. Structure (**451**) and it is a conventional *closo*-cluster with two inequivalent PH_3 groups. The cluster is quite distorted and it appears to resemble an "*isonido*" cluster but the presence of the large iridium atom may simply distort it. The role of the ethyne in the reaction is not understood but it may play a role in the reduction of the PPh_3 groups to PH_3.

Pyrolysis of $[6\text{-}H\text{-}6,6\text{-}(PPh_3)_2$-*nido*-$6\text{-}IrB_9H_{13}]$ (**413**) or *asym*-(**439**), $[6,6,6,\text{-}(PPh_3)(Ph_2P\text{-}o\text{-}C_6H_4)H$-*nido*-$6\text{-}IrB_9H_{12}\text{-}5]$,[278,284] affords $[1,1,1\text{-}H\{PPh_3\}(Ph_2P\text{-}o\text{-}C_6H_4)$-*isocloso*-$1\text{-}IrB_9H_8\text{-}2]$ (**446**) as orange–yellow crystals in quantitative yield. Identified by NMR spectra and a crystal structure determination, (**446**) was the prototype *isocloso*-cluster. It is assumed that the pyrolyses proceed with loss of H_2 and rearrangement involving orthoboronation. Similarly, pyrolysis of $[6\text{-}H\text{-}6,6\text{-}(PMe_3)_2$-*nido*-$6\text{-}IrB_9H_{13}]$ (**413**) affords $[1,1,1\text{-}H(PMe_3)_2$-*isocloso*-$1\text{-}IrB_9H_9]$(**449**).[278,284]

The relationship between *isonido*-clusters, for example (**442**) and (**444**), to the normal species which conform to the PSEPT, and to *isocloso*-clusters, such as (**396**) and (**446**), is explained in terms of a DSD

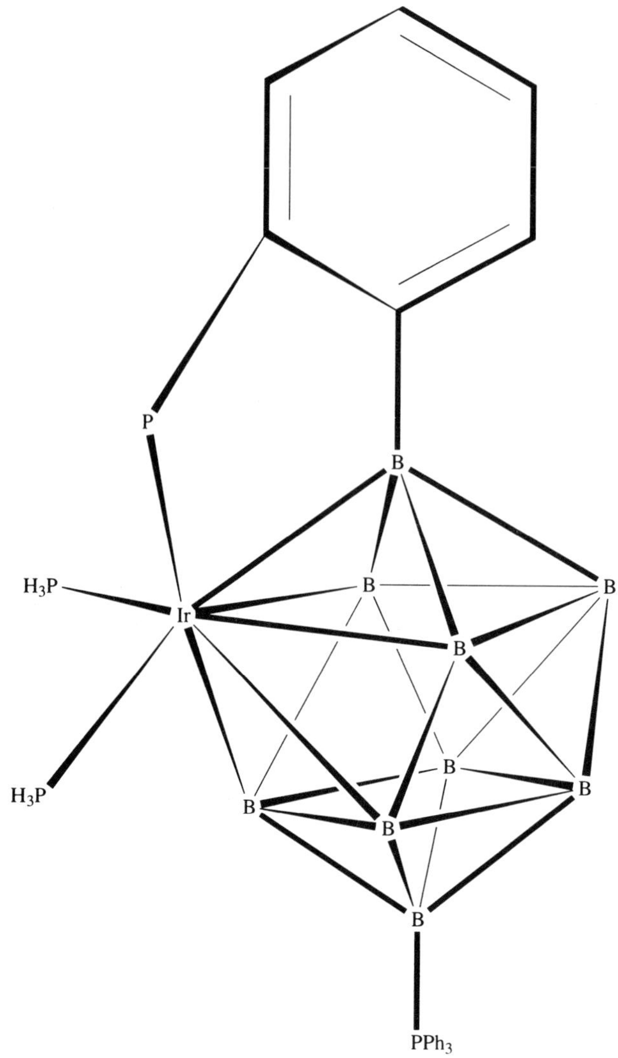

(451)

rearrangement of the normal systems.[30] This is illustrated in Scheme 18. As indicated in the scheme, a DSD rearrangement of the "normal" metalladecaborane affords the *isocloso*-cluster which exhibits the characteristic trigonal symmetry involving a 1:3:3:3 arrangement of metal moiety and boron atoms, respectively, as observed for the species (**396**), (**446**), and others. The intermediate (**b**) has the characteristic "square" face observed for *isonido*-clusters (**442**) and (**444**), etc., described above.

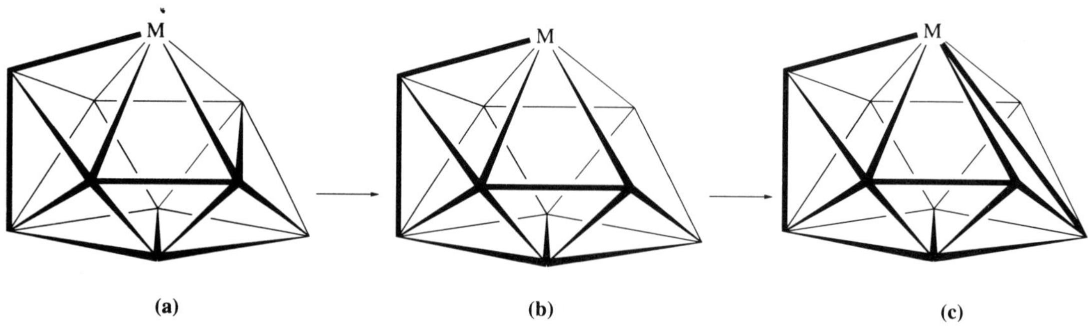

Scheme 18

8.10.4.3 Group 10 metals

[1,1-$(PMe_2Ph)_2$-2,4-Cl_2-*closo*-1-NiB_9H_7] (**452**)[238] is a minor product from the reaction between *cis*-[$NiCl_2(PMe_2Ph)_2$] and $(B_{10}H_{13})_2O$ in the presence of proton sponge in MeOH. It is obtained as a red crystalline solid in < 1% yield. It is identified by a crystal structure and, as seen in Structure (**452**), it is clearly a conventional *closo*-cluster with a BH^{2-} group replaced by the $Ni(PMe_2Ph)_2$ moiety, thus rendering the nickel as Ni^{IV}. The authors suggest that there may be significant differences between this system and others in which the NiL_2 group replaces a BH moiety. The system is actually two electrons short for a "normal" *closo*-cluster. Bond distances are quite normal and only the B–Ni distances deviate noticeably from the from the corresponding distances in $[B_{10}H_{10}]^{2-}$. [*Closo*-$(PMe_2Ph)_3Pt_2B_9H_8(OMe)$], (**453**)[183] is prepared as a green crystalline solid in the reaction between 4-(SMe_2)-7-(OMe)-B_9H_{12}, proton sponge, and [$PtCl_2(PMe_2Ph)_2$]. Structural data are unavailable but NMR spectra indicate that the platinum atoms are adjacent.

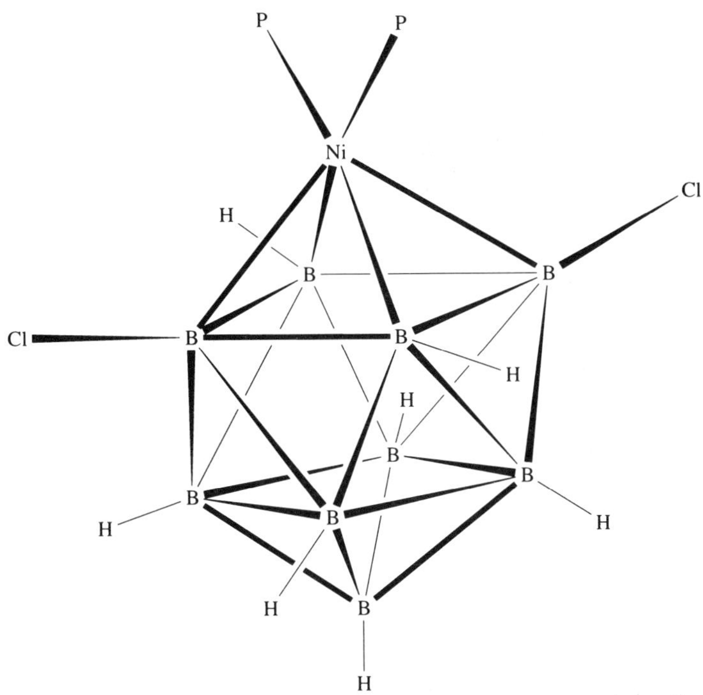

(452)

8.11 UNITS CONTAINING TEN BORON ATOMS

8.11.1 Hexadecahydrodecaborate Derivatives

8.11.1.1 Lanthanides metals

$[(MeCN)_6Yb][(\mu\text{-}H_2)B_{10}H_{12}]\cdot 2MeCN$ (**454**) is obtained by treatment of $B_{10}H_{14}$ with ytterbium metal in liquid NH_3 at –40 °C, followed by removal of NH_3 and addition of MeCN.[285] A crystal structure determination indicates that the $B_{10}H_{14}$ unit coordinates to ytterbium through two B–H–Yb bridges from adjacent BH_2-6 and BH-5 terminal *exo*-positions on the boron cage as seen in Structure (**454**). NMR spectra suggest a fluxional process in which coordination to positions (5) and (7) exchanges via a singly bridged B-6–H-6–Yb intermediate. Europium and samarium derivatives are also available from the reduction of $B_{10}H_{14}$ with metal in liquid NH_3 or with rare-earth amalgams. Some data are given in Table 21.

Table 21 Lanthanide hexadecahydrodecaborate derivatives.285

Species	*No.*	*Color*	*Physical data*
$[(DME)_xSm][B_{10}H_{15}]$	**(455)**	Red–green	^{11}B NMR and IR spectra
$[(DME)_x(MeCN)_ySm][B_{10}H_{14}]$	**(456)**	Green–black	^{11}B NMR and IR spectra
$[(MeCN)_xEu][B_{10}H_{14}]$	**(457)**	Bright yellow	^{11}B NMR and IR spectra
$[(NH_{3x}Eu][B_{10}H_{14}]$	**(458)**	Pale green	^{11}B NMR and IR spectra
$[(MeCN)_xEu][B_{10}H_{13}][B_{10}H_{15}]$	**(459)**	Yellow–white	IR spectra
$[(MeCN)_6Yb][(\mu\text{-}H_2)\text{-}B_{10}H_{12}]\cdot 2MeCN$	**(454)**	Yellow–orange	X-ray structure, NMR, IR
$[(NH_3)_3Yb][B_{10}H_{14}]$	**(460)**	Lavender solid	IR spectra
$[(MeCN)_3Yb][B_{10}H_{14}]$	**(461)**	Bright yellow	^{11}B NMR and IR spectra
$[(MeCN)_xYb][B_{10}H_{13}][B_{10}H_{15}]$	**(462)**	Pea green	^{11}B NMR and IR spectra

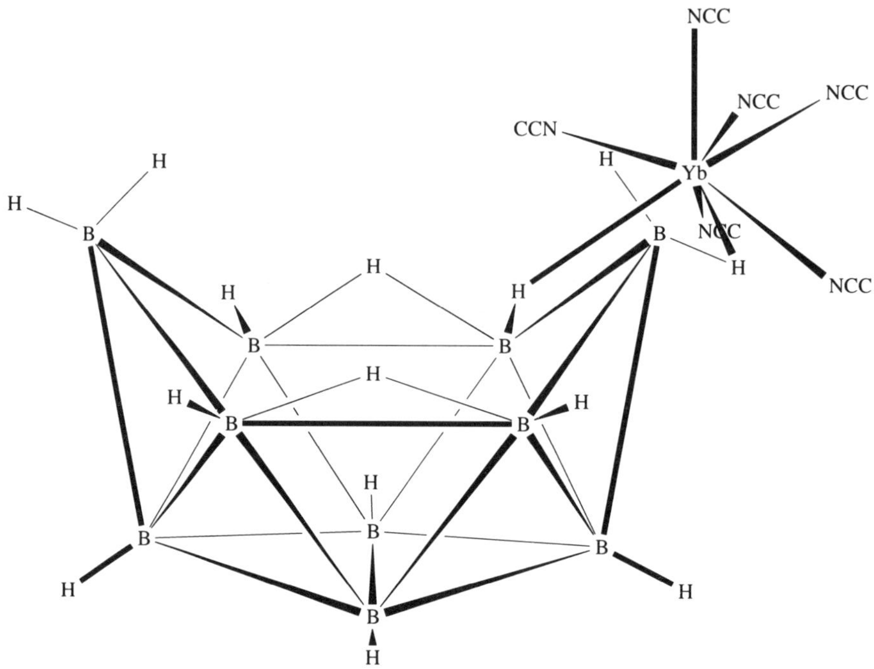

(454)

8.11.2 Tetradecahydrodecaborate Derivatives

8.11.2.1 Group 8 metals

$[7\text{-}(\eta^6\text{-}C_6Me_6)\text{-}nido\text{-}7\text{-}RuB_{10}H_{13}]^-$ **(463)**[286] is prepared in the reaction between $[\{Ru(\eta^6\text{-}C_6Me_6)Cl_2\}_2]$ and $[B_{10}H_{14}]^{2-}$ in MeCN at 25 °C, as a yellow air-stable solid in 86% yield. The structure of the species, as determined by x-ray diffraction, is given in Structure (**463a**) and the idealized structure is shown diagrammatically in Structure (**463b**). The metal replaces a boron atom in the open face of an 11-vertex *nido*-cluster. There are three bridging hydrogen atoms in the open face of the cluster, in contrast with the unsubstituted cluster, $[B_{11}H_{14}]^-$ which contains only two and an *endo*-hydrogen on the plane of symmetry.[287] This is unusual for a metallaundecaborane and the only other reported examples are the rhodium species described below. The other known metallaundecaboranes involve structures in which the metal moiety sits on one side of the open face of $B_{10}H_{14}$ with two bridging hydrogen atoms at the B-8–B-9 and the B-10–B-11 positions, as indicated in Structure (**463c**). An example of the latter is the series of *nido*-platinaundecaboranes described in *COMC-I* and in Section 8.11.2.3. The anion (**463**) has approximate mirror plane symmetry although there is no crystallographically imposed mirror plane. The B_4Ru face is almost flat and the B–B distances in the face are longer than normal and resemble those in the anion $[B_{10}H_{14}]^{2-}$. The three bridging hydrogen atoms are observed at a single resonance in the NMR indicating fluctionality and the motion is not arrested at temperatures down to −50 °C. This latter observation is more typical of *arachno*-boranes and it reflects retention of *arachno*-character from the precursor borane $[B_{10}H_{14}]^{2-}$.

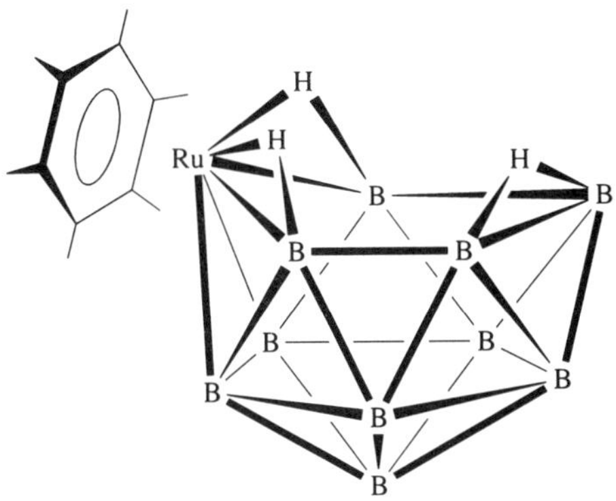

(463a)

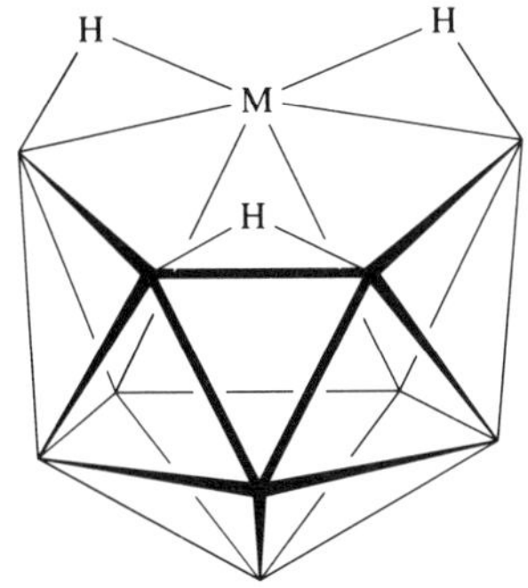

(463b)

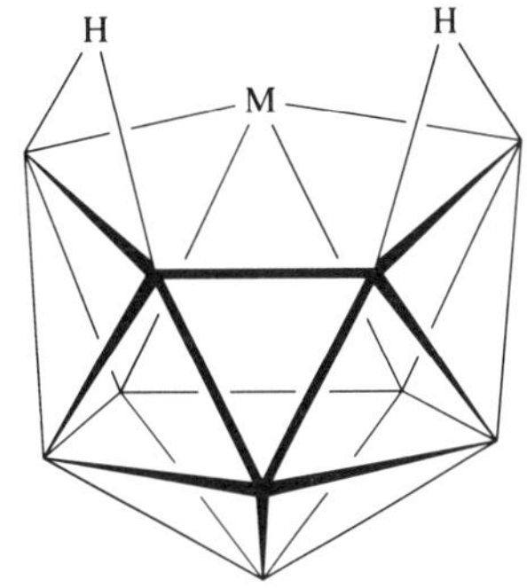

(463c)

8.11.2.2 Group 9 metals

[7-(Cp*)-*nido*-7-$RhB_{10}H_{11}$-8-Cl-11-(PMe_2Ph)]·CH_2Cl_2 (**464**)[288–9] has the triple hydrogen-bridged structure observed for the related *nido*-ruthenaundecaborane (**463**) described above. It is prepared from [(Cp*)$RhCl_2]_2$ and *arachno*-[$B_{10}H_{12}(PMe_2Ph)_2$] as a yellow air-stable solid in 21% yield. Reaction of (**464**) with excess PMe_2Ph in CH_2Cl_2 affords [7-(Cp*)-*nido*-7,12-$RhOB_{10}H_9$-8-Cl-11-(PMe_2Ph)] (**465**)[289–90] which contains an oxygen atom coordinated to the three boron atoms in the open face. (**465**) is presumably obtained by reaction with traces of adventitious H_2O according to Equation (17). [*nido*-(Cp*)$RhB_{10}H_{13}Cl$] (**466**),[276] a structurally characterized quadruply hydrogen-bridged 11-vertex species, is obtained in the reaction between $B_{10}H_{14}$ and [{Rh(Cp*)$Cl_2\}_2$] in the presence of proton sponge, but details have not appeared yet.[291]

$$[\text{7-(Cp*)-}\textit{nido}\text{-7-RhB}_{10}\text{H}_{11}\text{-8-Cl-11-(PMe}_2\text{Ph)}] + \text{H}_2\text{O} \xrightarrow{25\ ^\circ\text{C}/2\ \text{d}} [\text{7-(Cp*)-}\textit{nido}\text{-7,12-RhOB}_{10}\text{H}_9\text{-8-Cl-11-(PMe}_2\text{Ph)}] + 2\text{H}_2 \qquad (17)$$

8.11.2.3 Group 10 metals

7,7-[$PMe_2Ph]_2$-*nido*-7-$PtB_{10}H_{12}$ (**467**) initially described in *COMC-I*, is now available in much higher yields from the reaction between $B_{10}H_{14}$ and *cis*-$PtCl_2[PMe_2Ph]_2$ in the presence of proton sponge.[292] Up to 57% yields of the bright-yellow crystalline solid, which melts with decomposition at 198 °C, are obtained. The structure of (**467**) resembles the structure shown as Structure (**463c**), with two μ-hydrogen atoms in the open face. These latter indicate that it is a derivative of *nido*-decaborane in that the two bridging hydrogen atoms from $B_{10}H_{14}$ have been replaced by two three-center two-electron bonds from platinum to the cage. On the other hand, the overall skeletal electron pair count appears to be short of two electrons for a *nido*-system. One approach to account for this has been made by Welch *et al.*[293] using EHMO–FMO calculations but its presentation is beyond the scope of this treatment. As part of this ongoing study, a series of known systems [$PhCH_2NMe_3$][7,7,7-$(CO)_3$-7-Co-*nido*-$B_{10}H_{12}$] (**468**),[294] [$PhCH_2NMe_3]_2[(B_{10}H_{12})_2Pd]$·MeCN (**469**),[295] and [$PhCH_2NMe_3]_2[(B_{10}H_{12})_2Pt]$ (**470**)[296] have been resynthesized and crystal structures determined for each. Compound (**468**) is described as a true *nido*-icosahedral species with a structure analogous to that in Structure (**463c**). The $Co(CO)_3$ unit conforms to an effective mirror plane through the cage with one CO ligand situated in an *endo*-position. [$PhCH_2NMe_3]_2[(B_{10}H_{12})_2Pt]$ (**470**) has a structure similar to that of the other M[$B_{10}H_{12}$] systems described in this section except that there is some asymmetry of the polyhedron which are ascribed to crystal packing effects. [$PhCH_2NMe_3]_2[(B_{10}H_{12})_2Pd]$·MeCN (**469**) contains a center of inversion and the solvent molecule appears to remove the distortion.

[$(PMe_2Ph)_2PtB_{10}H_{12}$] (**467**) reacts with PMe_2Ph in refluxing toluene to afford an 11-vertex *nido*-platinaundecaborane, [$(PMe_2Ph)_4Pt_2B_{10}H_9Cl$][297] which is described in Section 8.11.4.3. Substituted derivatives of (**467**) are available in reaction of the metal reagent with SMe_2·$B_{10}H_{12}$.[292,298] Thus, 8-Cl-7,7-[$PMe_2Ph]_2$-*nido*-7-$PtB_{10}H_{11}$ (**471**) and 7-Cl-7-[$PMe_2Ph]_2$-8-SMe_2-*nido*-7-$PtB_{10}H_{11}$ (**472**) are available from the reactions of $PtCl_2(PMe_2Ph)_2$ or [$PtCl_2PMe_2Ph]_2$, with SMe_2·$B_{10}H_{12}$, as golden-yellow plates

(18%) and a cream colored solid (14%), respectively. [μ-2,7-(SCSNEt$_2$)-7-(PMe$_2$Ph)-*nido*-7-PtB$_{10}$H$_{11}$] (**473**) is another example of the class of *nido*-metallaundecaboranes with only two bridging hydrogen atoms in the open face.[299] It is prepared as yellow-orange crystals in 52% yield from the reaction between [(PMe$_2$Ph)$_2$PtB$_{10}$H$_{12}$] and AuBr$_2$(CSCSNEt$_2$)] in CH$_2$Cl$_2$ under dry N$_2$. The structure is similar to that shown in Structure (**463c**) except that there is an exopolyhedral *N,N*-diethyldithiocarbamate ligand bridging the platinum atom and the adjacent unique B-2 atom to form a Pt–B–S–C–S five-membered ring. The platinum–borane bonding is twisted relative to other known platinaboranes and this is presumably due to the influence of the five-membered ring. NMR spectral data are consistent with the observed structure.

8.11.2.4 Group 11 metals

[(H$_{12}$B$_{10}$Au)(AuPEt$_3$)$_4$(AuB$_{10}$H$_{12}$)], (**474**),[300–1] the final product in the preparation of a series of novel auraboranes, is available from the reaction between B$_{10}$H$_{14}$ with R$_3$PAuMe (R = Et, Ph) in CH$_2$Cl$_2$, as indicated in Scheme 19. The intermediate in this process, [B$_{10}$H$_{12}$AuPR$_3$]$_2$ (**475**) which is shown in Scheme 19, involves two *nido*-icosahedral 7-AuB$_{10}$H$_{12}$ cages linked by an Au–Au' bond. The reaction may be arrested at the initial stage, that is, [R$_3$PAu(B$_{10}$H$_{13}$)] (**476**), if bulky phosphines are used. Alternatively, the [5,6-μ-(AuPR$_3$)-*nido*-B$_{10}$H$_{13}$] (**476**) may be prepared from [Et$_3$NH][B$_{10}$H$_{13}$] and [AuCl{P(C$_6$H$_{11}$)$_3$}] in CH$_2$Cl$_2$ as colorless crystals in 51% yield. Similarly, [5,6-μ-(AuP{C$_6$H$_4$Me-2}$_3$)-*nido*-B$_{10}$H$_{13}$] is obtained as very pale-yellow air-stable crystals in 65% yield. The structures of both the (**476**) species, R = c-C$_6$H$_{11}$ or C$_6$H$_4$Me-2, are *nido*-B$_{10}$H$_{14}$ cages with a R$_3$PAu$^+$ moiety replacing a bridging proton, formally at the 5,6 position as indicated in Scheme 19.[301]

Other products are available from the above reaction.[302] If a solution of B$_{10}$H$_{14}$ in CH$_2$Cl$_2$ is added slowly to a slight excess of R$_3$PAuMe (R = cycl or *o*-tolyl), in addition to the [5,6-μ-(AuPR)$_3$-*nido*-B$_{10}$H$_{13}$], small amounts of the yellow anions [(B$_{10}$H$_{12}$)Au(B$_{10}$H$_{12}$)]$^-$ (**477**) and [(B$_{10}$H$_{12}$)Au(B$_{10}$H$_{13}$)]$^{2-}$ (**478**) are obtained as their [(R$_3$P)$_2$Au]$^+$ salts. The structures of the two species and [(R$_3$P)$_2$Au][(B$_{10}$H$_{12}$)Au(B$_{10}$H$_{12}$)] (**477**) and [(R$_3$P)$_2$Au]$_2$[(B$_{10}$H$_{12}$)Au(B$_{10}$H$_{13}$)] (**478**) are known and are given in Structures (**477**) and (**478**), respectively. The anion in [(R$_3$P)$_2$Au][(B$_{10}$H$_{12}$)Au(B$_{10}$H$_{12}$)] (**477**) contains two *nido*-cage conjoined at the common 13th vertex which is formally Au^{3+} and is isostructural with [M(B$_{10}$H$_{12}$)$^-$] (M = Ni, Pd, Pt) which is described in *COMC-I*. The anion [(B$_{10}$H$_{12}$)Au(B$_{10}$H$_{13}$)]$^{2-}$ (**478**) has no precedent in borane chemistry. It consists of [AuB$_{10}$H$_{12}$] and [AuB$_{10}$H$_{13}$] cages fused about a common gold vertex. The upper half in Structure (**478**) has *nido*-icosahedral geometry with an AuB$_4$ open face. This is analogous to the subunit found in [(H$_{12}$B$_{10}$Au)(AuPR$_3$)$_4$(AuB$_{10}$H$_{12}$)], illustrated in Scheme 19 and appears to be thermodynamically stable. The [AuB$_{10}$H$_{13}$] fragment is an *arachno*-cluster with a six-membered open face. It is the first such metallaborane cluster and may be considered to arise from removal of two vertices of a 13-vertex *closo*-cluster. The three bridging hydrogen atoms are assumed from the stoichiometry and the ^{1}H NMR spectrum. [5,6-μ-(AuPR$_3$)-*nido*-B$_{10}$H$_{13}$] (**476**) may be deprotonated with NEt$_3$ or KOH to afford a salt of [*nido*-{AuP(C$_5$H$_{11}$)$_3$}-B$_{10}$H$_{12}$]$^-$ (**479**).[303] The [HNEt]$^+$ salt is isolated as red crystals and is completely characterized; the structure is illustrated in Structure (**479**). Deprotonation removes the Hμ$_{9,10}$ atom and presumably the gold–phosphine moiety slips from μ to μ$_4$ on the decaboranyl framework. The deprotonation is reversible, treatment of (**479**) with H$^+$ affords (**476**).

8.11.2.5 Group 12 metals

The species [NMe$_4$]$_2$[ZnB$_{10}$H$_{12}$] (**480**)[304] is obtained from the reaction between a slurry of the metal with B$_{10}$H$_{14}$. The metal is prepared by reduction of ZnCl$_2$ with potassium in glyme at 50 °C. (**480**) is obtained as a yellow crystalline solid in 24% yield. Similarly, [CoB$_{10}$H$_{10}$]$^{2-}$ and [NiB$_{10}$H$_{12}$]$^{2-}$ salts are prepared in lower yields.

8.11.3 Dodecahydrodecaborate Derivatives

The systems described in the following section are obtained by the capping of a C_{2v} open 10-vertex *nido*-decaboranyl-type cluster with a ligated metal moiety. This gives the gross structure expected for a *closo*-undecaborane. The structural data for the species, however, indicate that simple *nido*-character is not present for the borane cage since the B-6–B-7, B-3–B-4 distances are much shorter than the

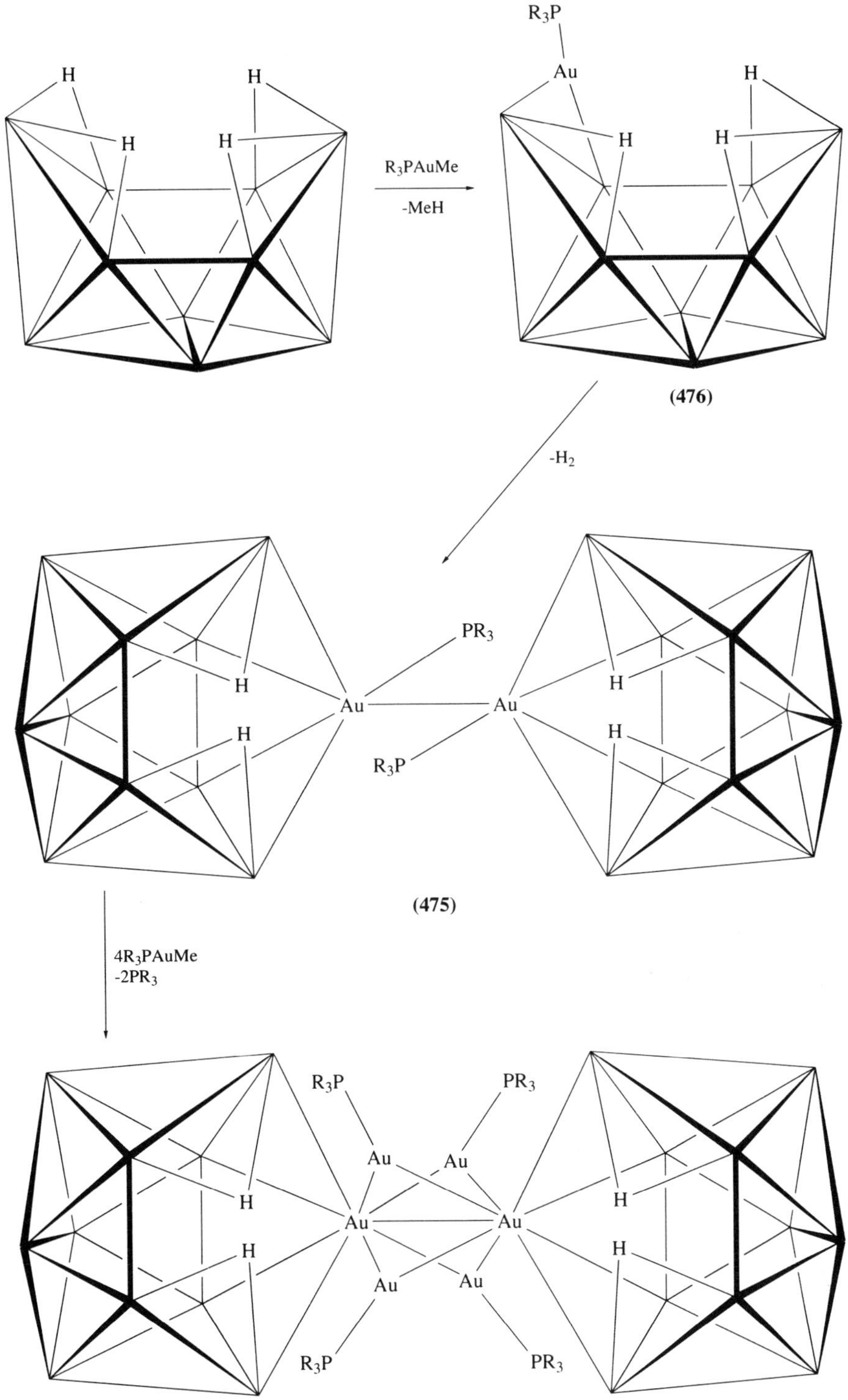

Scheme 19

corresponding distances in $B_{10}H_{14}$. Also, the NMR data differ widely from those for $B_{10}H_{14}$. The electron count for the species is two short for an 11-vertex *closo*-system but the structure appears to fit in well with that predicted by the PSEPT for a 11-vertex *closo*-cluster. For [1,1-$(PPh_3)_2$-2,5-$(OEt)_2$-*isocloso*-1-$RuB_{10}H_8$] (**481**), the authors suggest that the neutral $Ru(PPh_3)_2$ center contributes four electrons to the bonding in the RuB_{10} cluster which thus achieves the 24-electron 11-vertex *closo*-count. The authors describe this species and the others described in this section as *isocloso* and assign the metal

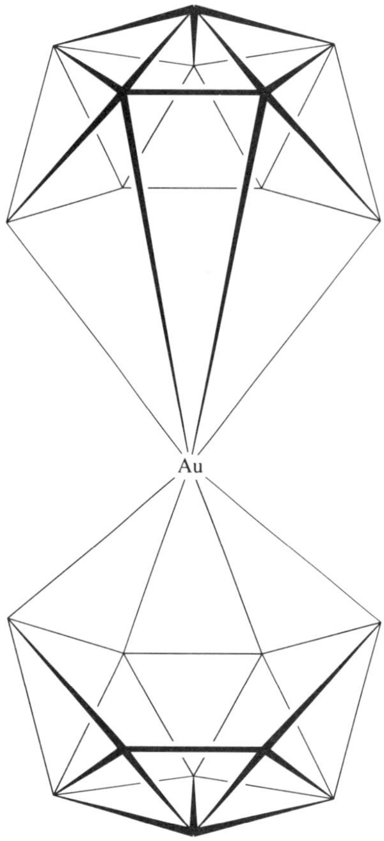

(477)

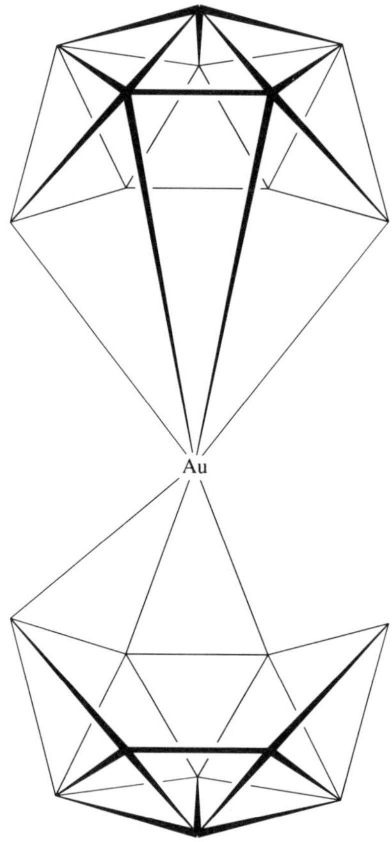

(478)

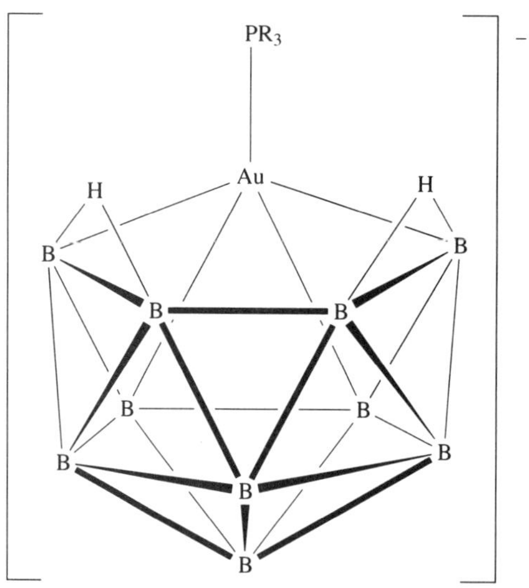

(479)

contribution as four orbitals rather than the three predicted by the classical PSEPT theory. These systems are related to the *isocloso*-metalladecaboranes described in Section 8.10.4.3 in that they are derived from the "normal". *closo*-11-vertex system by a DSD rearrangement. It is well known that the $[B_{11}H_{11}]^{2-}$ anion is fluxional and that a possible mechanism for this takes place via a series of DSD rearrangements which generate identical structures with the borons scrambled.[305] Thus, for the 11-vertex systems the "normal" *closo*- and *isocloso*-systems are identical.

In the treatment of these *isocloso*-metallaundecaboranes, the general structures are all similar and thus the examples are not presented according to their position in the periodic table but according to the type of ligand on the metal.

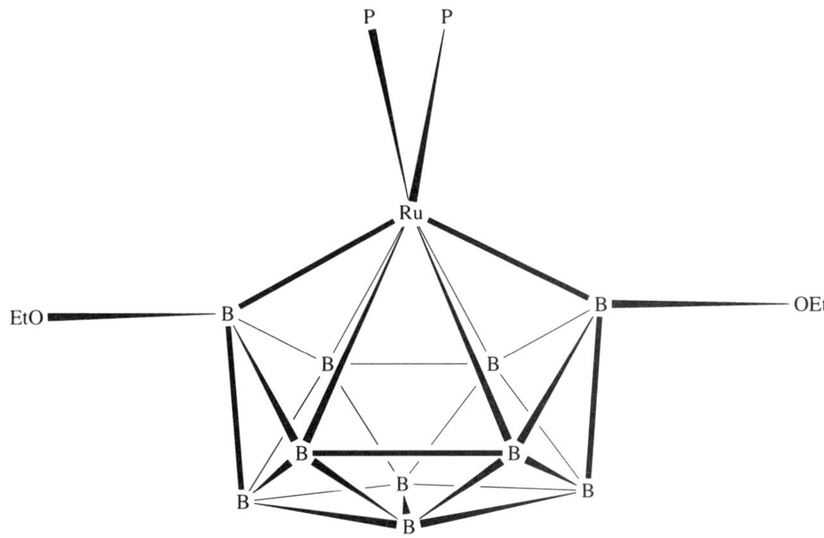

(481)

8.11.3.1 Phosphine-ligated systems

Table 22 lists the preparation and some properties of systems containing the metals ruthenium,[281,306–9] osmium,[307–8] and rhodium.[310] The basic structure of these systems is illustrated by that for 1,1-$(PPh_3)_2$-2,5-$(OEt)_2$-*isocloso*-1-$RuB_{10}H_8$] in Structure (**481**).[281,306] The PMe_2Ph species (**482**) is also available as is the osmium analogue (**483**) but attempts to prepare the osmium analogue of (**481**) by carrying out the reaction in refluxing ethanol results in the formation of [2,5-$(OEt)_2$-1-$(PPh_3)(Ph_2PC_6H_4)$-*closo*-$OsB_{10}H_7$-3] (**484**), an orthoboronated system which is shown in Structure (**484**). [1,1-$(PPh_3)_2$-*closo*-1-$RuB_{10}H_8$-2,5-$(OEt)_2$] (**481**) or [1,1-$(PMe_2Ph)_2$-*closo*-1-$OsB_{10}H_8$-2,5-$(OEt)_2$] (**482**) react with $RuCl_2(PPh_2Me)_3$ or $OsCl_2(PPh_2Me)_3$ in refluxing EtOH to afford a series of species with the formula [$(PR_3)_2$ClMClM'$(PPh_3)B_{10}H_8(OEt)_2$] where M/M' are Ru/Ru (**486**), Ru/Os (**486**), Os/Ru (**487**) and Os/Os (**488**).[308] Both reagents lose a mole of phosphine in the reaction and the second metal sits on the Ru–B-3–B-4 face bonded through Ru–Cl–Ru, Ru–H–B and Ru–H–B bridges, respectively.

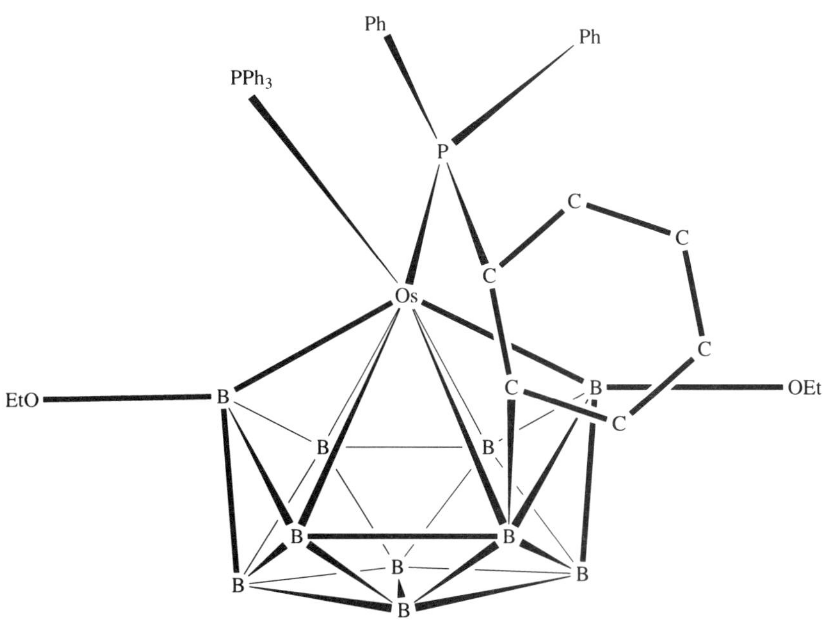

(484)

Table 22 Preparation and properties of phosphine-ligated 1-metalla-*isocloso*-undecaboranes.

Species	*No.*	*Reaction*	*Yield* (%)	*Appearance/structural data*	*Ref.*
$[1,1\text{-}(PPh_3)_2\text{-}2,5\text{-}(OEt)_2\text{-}RuB_{10}H_8]$	(**481**)	$[RuCl_2(PPh_3)_3] + [B_{10}H_{10}]^{2-}$ under N_2 in EtOH	80	Red crystalline solid, NMR, x-ray	281, 306
$[1,1\text{-}(PMe_2Ph)_2\text{-}2,5\text{-}(OEt)_2\text{-}RuB_{10}H_8]$	(**482**)	$[RuCl_2(PMe_2Ph)_3] + [B_{10}H_{10}]^{2-}$	52	Orange solid, NMR	307
$[1,1\text{-}(PMe_2Ph)_2\text{-}1\text{-}OsB_{10}H_8\text{-}2,5\text{-}(OEt)_2]$	(**483**)	$[OsCl_2(PMe_2Ph)_3] + [B_{10}H_{10}]^{2-}$	42	Yellow air-stable crystals, NMR	307
$[(Ph_2PC_6H_4)$-*closo*-$OsB_{10}H_7\text{-}3\ 2,5\text{-}(OEt)_2\text{-}1\text{-}(PPh_3)]$	(**484**)	$[OsCl_2(PPh_3)_3] + [B_{10}H_{10}]^{2-}$ in refluxing EtOH under N_2	39	Amber-brown crystals, NMR, x-ray	307
$[(PPh_3)_2ClRuClRu'(PPh_3)B_{10}H_8(OEt)_2]$	(**485**)	(**481**) + $RuCl_2(PPh_2Me)_3$	80	Red–brown solid, NMR, x-ray	308
$[(PPh_3)_2ClRuClOs'(PPh_3)B_{10}H_8(OEt)_2]$	(**486**)	(**481**) + $OsCl_2(PPh_2Me)_3$	80	Olive–brown solid, NMR	308
$[(PPh_3)_2ClOsClRu'(PPh_3)B_{10}H_8(OEt)_2]$	(**487**)	(**482**) + $RuCl_2(PPh_2Me)_3$	80	Olive–brown solid, NMR	308
$[(PPh_3)_2ClOsClOs'(PPh_3)B_{10}H_8(OEt)_2]$	(**488**)	(**482**) + $OsCl_2(PPh_2Me)_3$	80	Red–brown solid, NMR	308
$[\{1\text{-}(PPh_3)_2Ru\}\text{-}(\mu\text{-}MeCO_2)_3\text{-}(\mu\text{-}H)_2\text{-}1\text{-}RuB_{10}H_7]$	(**489**)	$RuCl_2(PPh_3)_3] + [B_{10}H_{10}]^{2-}$ in Me_2CO	1.5	Red air-stable crystals, NMR, x-ray	309
$[(PPh_3)(MeCO_2)_2\text{-}1\text{-}RuB_{10}H_7\text{-}3\text{-}OH]$	(**490**)	$RuCl_2(PPh_3)_3] + [B_{10}H_{10}]^{2-}$ in Me_2CO	3.5	Orange–yellow crystals, NMR, x-ray	309
$[1,1\text{-}(PMe_2Ph)_2\text{-}1,2\text{-}\mu\text{-}H\text{-}2,5\text{-}(OMe)_2\text{-}1\text{-}RhB_{10}H_8]$	(**491**)	$[RhCl_3(PMe_2Ph)_3] + [B_{10}H_{10}]$ in refluxing MeOH	5	Air-stable bright orange solid, NMR, x-ray	310
$[1,1\text{-}(PMe_2Ph)_2\text{-}1,2\text{-}\mu\text{-}H\text{-}2\text{-}Cl\text{-}5\text{-}OMe\text{-}1\text{-}RhB_{10}H_8]$	(**492**)	$[RhCl_3(PMe_2Ph)_3] + [B_{10}H_{10}]$ in refluxing MeOH	2	Orange solid, NMR, x-ray	310

If the reaction between $[B_{10}H_{10}]^{2-}$ and $[RuCl_2(PPh_3)_3]$ is carried out in the presence of acetone in refluxing THF, two new products are obtained.[309] One, [{1-$(PPh_3)_2$Ru}-(μ-$MeCO_2)_3$-(μ-H)$_2$-*closo*-1-$RuB_{10}H_7$] (**489**),[309] is obtained as a red air-stable crystalline solid in 1.5% yield. NMR data and an x-ray structure determination indicated in Structure (**489**). A second ruthenium atom is linked to the cage by four different kinds of bridge; two hydrogen bridges, one from boron and one from ruthenium, and two *O,O'*-acetate bridges, one from boron and one from ruthenium, as indicated in Structure (**489**). A related species, [$(PPh_3)(MeCO_2)_2$-*closo*-1-$RuB_{10}H_7$-3-OH] (**490**),[309] is a second product from the reaction between $[RuCl_2(PPh_3)_2]$ and $[B_{10}H_{10}]^{2-}$ in acetone. The orange–yellow crystalline monoruthenaborane is obtained in 3.5% yield and characterized by NMR spectroscopy and an x-ray structural study. The species also exhibits *exo*-polyhedral borane-to-metal acetate bridging, and also a boron-hydroxy group as indicated in Structure (**490**), with terminal hydrogen atoms on boron and some carbon and hydrogen atoms omitted for clarity. [1,1-$(PMe_2Ph)_2$-1,2-μ-H-2,5-$(OMe)_2$-*isocloso*-1-$RhB_{10}H_8$] (**491**)[310] is related to (**481**) and (**482**) and it exists as two rotamers which differ by the orientation of the OMe groups and rotation of the P–C and Rh–P bonds. The RhB_{10} cluster conforms to idealized C_s symmetry due to the presence of a bridging hydrogen atom between Rh and B-2. The unique hydrogen atom is located in the structure and the Rh–P-1–P-2 bonding plane tilts away slightly from the unique hydrogen atom and in addition Rh–B-2 is longer than Rh–B-5. The 5-methoxy-derivative [1,1-$(PMe_2Ph)_2$-1,2-μ-H-2-Cl-5-OMe-*isocloso*-1-$RhB_{10}H_8$] (**492**)[310] is also isolated in this product mixture in 2% yield and is also identified by a crystal structure determination and NMR spectroscopic data.

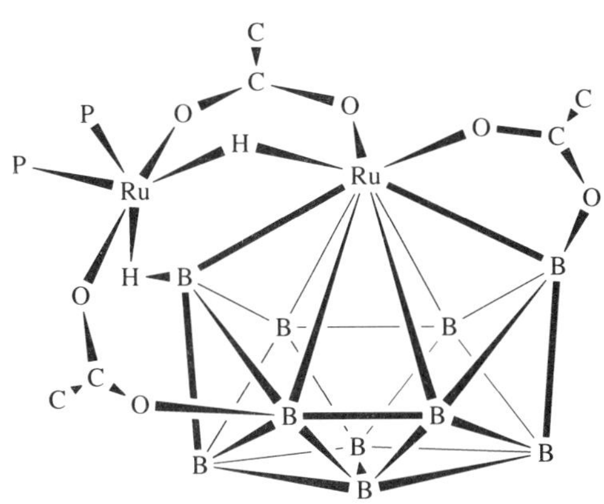

(489)

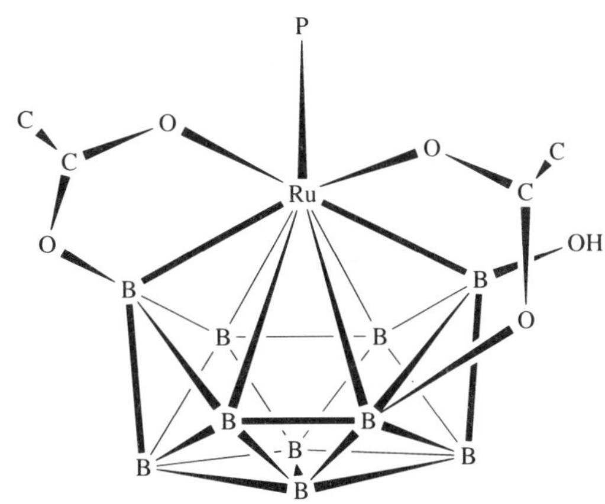

(490)

8.11.3.2 *η*⁵- or *η*⁶-Arene-ligated systems

There is another class of *isocloso*-undecaboranes which are similar to those described in the previous paragraphs. This class involves incorporation of an η^5- or η^6-aromatic hydrocarbon–metal group into the 1-position of an octadecahedral cluster such that the *closo*-cluster again requires that the metal moiety provides four orbitals and four electrons to what was initially a *nido*-type B_{10} cluster. Some of these systems, which have the general structure given in Structure (**499**), for 1-(1-Me-4-$Pr^iC_6H_4$)$RuB_{10}H_{10}$, with the terminal hydrogen atoms omitted for clarity, are listed in Table 23. They are prepared by the reaction of the metal reagent with $[B_{10}H_{10}]^{2-}$ salts[275–6,197,311–12] or by the reaction of ligand, $B_{10}H_{14}$, and metal atoms in a metal atom vapor reactor.[250] [{$(\eta^6$-$C_6Me_6)_2Ru_2H_4$}-*closo*-1-$RuB_{10}H_8(OEt)_2$] (**504**)[197] is obtained, as a dark-red crystalline solid in 32% yield, from the reaction between *closo*-$[B_{10}H_{10}]^{2-}$, [Ru(η^6-C_6Me_6)Cl_2] and Tl$[B_3H_8]$. (**504**) was characterized by a crystal structure determination, as indicated in (**504**) where the (η^6-C_6Me_6) moieties on Ru-2 and Ru-3 and the hydrogen associated with the boron cage are omitted for clarity. It is an 11-vertex *closo*-type 1-metallaundecaborane cluster with an ethoxy substituent at the B-2,3 positions. The cage is fused at the 1-position to a triruthenium subcluster. Presumably the $[B_3H_8]^-$ moiety serves as a reducing agent in the process to form the $H_2Ru_3H_2$ subcluster. The neutral [$(\eta^6$-$C_6Me_6)_2Ru_2H_4$] can be regarded as a bidentate ligand to the Ru^I center at the 1-position of the octadecahedron. The triangular Ru_3 unit is essentially an iscosceles triangle. Ru-2 and Ru-3 are linked by two mutually *trans*-Ru–H–Ru bridging atoms with one of the bridges bent in an *endo*-mode over the plane and the other *exo* to it. In addition each of these ruthenium atoms is bonded to Ru-1 by a Ru–H–Ru bridge, the bridges being out of the Ru_3 plane and on the opposite side to the

$\mu_{2,3}$ hydrogen (*endo*). (**506**) reacts with CO in CH_2Cl_2 to give a 94% yield of the air-stable burgundy-red complex, $[\{(\eta_6\text{-}C_6Me_6)_2Ru_2H_2(CO)_2\}\text{-}RuB_{10}H_8(OEt)_2]$ (**505**),[314] for which a crystal structure determination indicates that the major metallaundecaborane cluster is unchanged in the reaction but that the two Ru-2–H–Ru-3 bridges have been replaced by Ru-2–CO–Ru-3 bridges, as indicated in Structure (**505**). The two CO ligands are equidistant from Ru-2 and Ru-3 and are almost perpendicular to the Ru_3 plane.

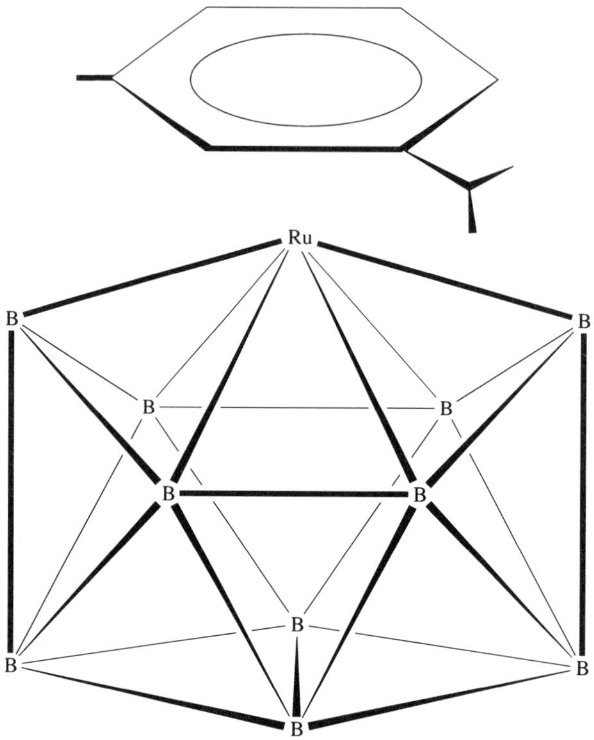

(499)

$[1\text{-}Cp^*)\text{-}1\text{-}RhB_{10}H_{10}]$ (**501**) and $[1\text{-}(Cp^*)\text{-}1\text{-}RhB_{10}H_9\text{-}2\text{-}(OMe)]$ (**502**)[312] are listed in Table 23 and their molecular structures are similar to the other species in Table 23 and to that found in (**504**). They warrant little further discussion except that they also appear to violate the PSEPT in that if the (Cp*Rh) moiety contributes three orbitals and two electrons to the open $B_{10}H_{10}$ fragment the resulting system is two electrons fewer than required for an idealized *closo*-undecaborane. The system may be considered to contain a Rh^V such that the (Cp*Rh) moiety contributes four orbitals and four electrons to the cluster rendering what these authors have previously described as an *isocloso*-undecaborane. The derivative $[1\text{-}(Cp^*)\text{-}4\text{-}(NEt_2)\text{-}1\text{-}RhB_{10}H_9]$ (**503**)[313] is formed in 66% yield, as a dark-purple solid, from the reaction between $[1\text{-}(Cp^*)\text{-}1\text{-}RhB_{10}H_{10}]$ (**501**) and Et_2NH in CH_2Cl_2 at room temperature. A crystal structure determination reveals that the cluster is a derivative of (**501**) with a planar NEt_2 group at the 4-position. The B–N distance is much shorter than expected for a single B–N bond which suggests π-bonding between the nitrogen atom and B-4. NMR data indicate a barrier to rotation about the B–N bond of 53.8 kJ mol^{-1} at –100 °C.

8.11.4 Miscellaneous Metallaboranes Containing 10 Boron Atoms

8.11.4.1 Group 8 metals

$4\text{-}[CpFe(CO)_2\text{-}2{:}2'\text{-}(B_5H_7)(B_5H_8)]$ (**506**),[315] a unique triple cluster, is prepared from $Li[2{:}2'\text{-}(B_5H_7)(B_5H_8)]$ and $CpFe(CO)_2Cl$ as red solid. NMR spectra suggest Structure (**506**).

Table 23 Preparation and properties of 1-(η^n-C_nH_n)-*isocloso*-$MB_{10}H_{10}$ systems.

Species	*No.*	*Reaction*	*Yield (%)*	*Physical properties*	*Ref.*
[1-(η^6-$C_6Me_3H_3$)$FeB_{10}H_{10}$]	(**493**)	Fe(g) + mesitylene + $B_{10}H_{14}$		Air-sensitive yellow solid, NMR and IR	250
[1-(η^6-C_6H_5Me)$FeB_{10}H_{10}$]	(**494**)	Fe(g) + $C_6H_6CH_3$ + $B_{10}H_{14}$		Air-stable yellow solid, NMR and IR	250
[1-(η^6-C_6Me_6)$RuB_{10}H_{10}$]	(**495**)	[Ru(η^6-(C_6Me_6)Cl_2] + $[B_{10}H_{10}]^{2-}$ in CH_2Cl_2	80	Air-stable yellow crystals, NMR and IR	275
[2-Cl-1-(η^6-C_6Me_6)$RuB_{10}H_9$]	(**496**)	In preparation of (**259**) (trace)		Yellow air-stable solid, NMR and IR	275
[1-(η^6-C_6Me_6)-2-(OMe)$RuB_{10}H_9$]	(**497**)	In preparation of (**259**) in MeOH	52	Yellow crystals, NMR and IR	275
[1-(η^6-C_6Me_6)-4-(OMe)$RuB_{10}H_9$]	(**498**)	In preparation of (**259**) in MeOH	trace	Yellow crystals, NMR and IR	275
[1-(η^6-pcym)$RuB_{10}H_{10}$][a]	(**499**)	[Ru(η^6-pcym)Cl_2] + $[B_{10}H_{10}]^{2-}$ in CH_2Cl_2	87	Air-stable yellow orange solid, x-ray structure	311
[1-(η^6-C_6Me_6)-2-(OEt)$RuB_{10}H_9$]	(**502**)	[Ru(η^6-C_6Me_6)ClB_3H_8] + $[B_{10}H_{10}]^{2-}$ in EtOH		Air-stable yellow solid, NMR and IR	197
[1-Cp*$RhB_{10}H_{10}$]	(**501**)	[RhCp*Cl_2] + $[B_{10}H_{10}]^{2-}$ in CH_2Cl_2	45	Air-stable orange crystals, NMR and IR	312
[1-Cp*$RhB_{10}H_9$-2-(OMe)]	(**502**)	[RhCp*Cl_2] + $[B_{10}H_{10}]^{2-}$ in MeOH	38	Air-stable orange crystals, x-ray structure	312
[1-Cp*-4-(NEt_2)$RhB_{10}H_9$]	(**503**)	(**501**) + Et_2NH in CH_2Cl_2	66	Dark-purple air-stable solid, x-ray structure	313

[a] pcym = 1-Me-4-$Pr^iC_6H_4$.

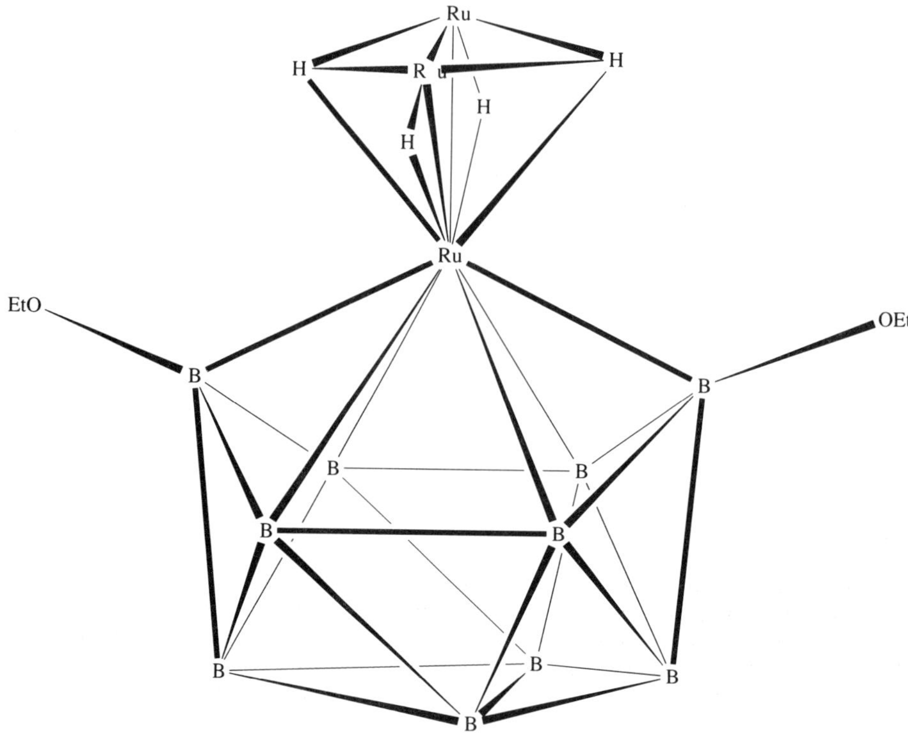

(504)

Ru = (η^6-C_6Me_6)Ru

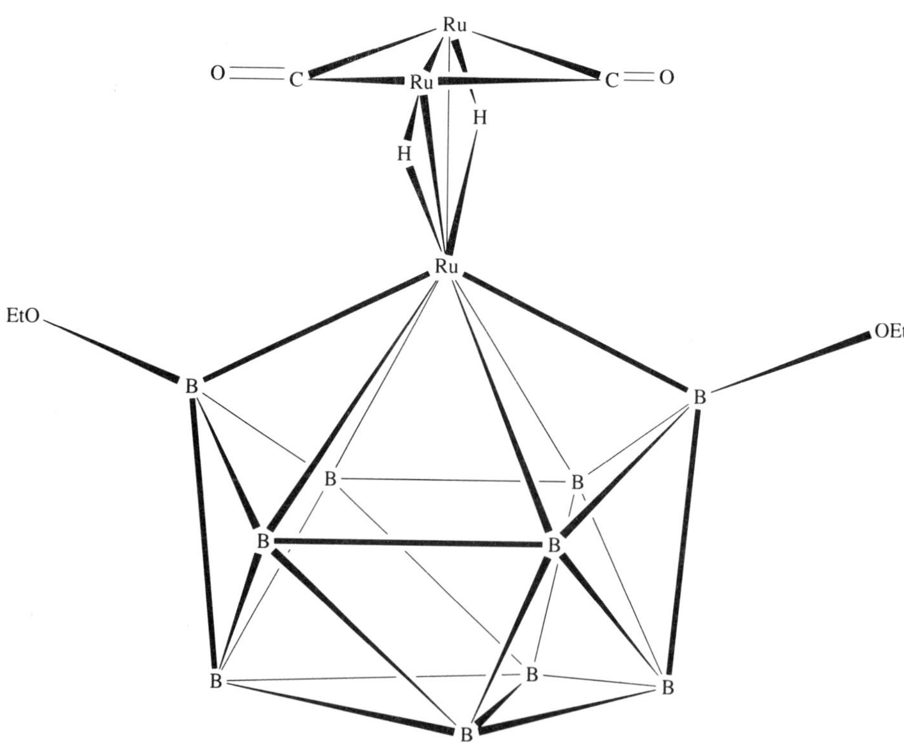

(505)

Ru = (η^6-C_6Me_6)Ru

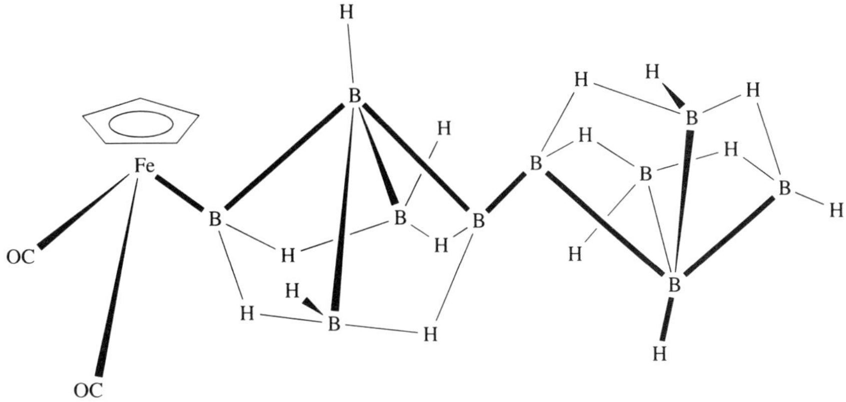

(506)

8.11.4.2 Group 9 metals

A unique system in which a Co–Co bond exists is [*closo*-1,1,2,2-$(CO)_4$-1,2-μ-CO-4,11-$(SEt_2)_2$-1,2-$Co_2B_{10}H_8$] (**507**).[316] It is prepared as orange needles in 22% yield from the reaction between *arachno*-$B_{10}H_{12}(SEt)_2$ and $Co_2(CO)_8$, analogous to the manner in which alkynes insert into decaboranes to afford 1,2-$C_2B_{10}H_{12}$. The 4,7-$(SEt_2)_2$-isomer, [*closo*-1,1,2,2-$(CO)_4$-1,2-μ-CO-4,7-$(SEt_2)_2$-1,2-$Co_2B_{10}H_8$] (**508**) is obtained as orange plates in 22% yield. The two species are also available from the reaction between *arachno*-$B_{10}H_{12}(SEt)_2$ and $Co_4(CO)_{12}$ in combined yield of 48%. The reactions proceed as indicated in Equations (18) and 19. Crystal structures of the two species characterize them as analogues of 1,2-$C_2B_{10}H_{12}$ in which the $Co_2(CO)_5$ group containing a bridging CO has replaced the C_2H_2 group in the carbaborane as indicated in Structure (**507**) with the terminal hydrogen atoms omitted for clarity.

$$\textit{arachno}\text{-}B_{10}H_{12}(SEt_2) + Co_2(CO)_8 \longrightarrow \textit{closo}\text{-}B_{10}H_8(SEt_2)_2Co_2(CO)_5 + 3CO + 2H_2 \quad (18)$$

$$2\ \textit{arachno}\text{-}B_{10}H_{12}(SEt_2) + Co_4(CO)_{12} \longrightarrow 2\ \textit{closo}\text{-}B_{10}H_8(SEt_2)_2Co_2(CO)_5 + 2CO + 4H_2 \quad (19)$$

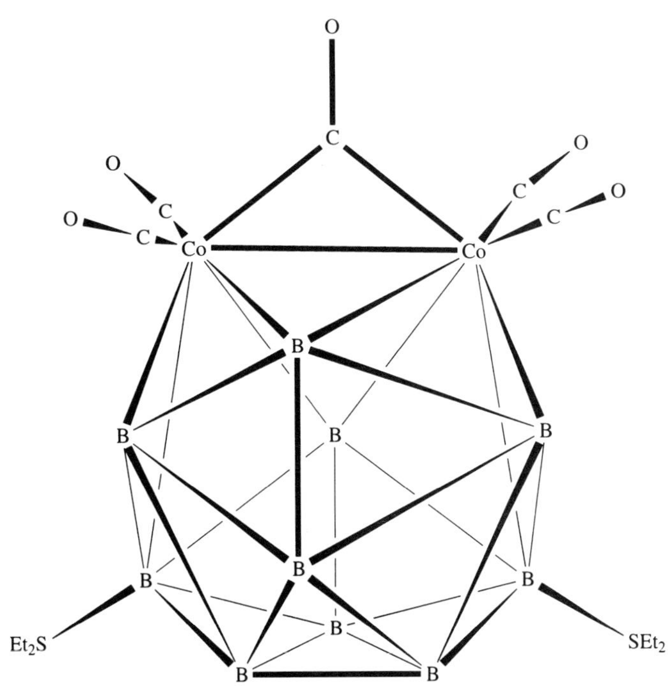

(507)

8.11.4.3 Group 10 metals

[1-Cl-1,2,2,4-$(PMe_2Ph)_4$-*closo*-1,2-$Pt_2B_{10}H_9$] (**509**)[297] is obtained in an unusual cluster disproportionation reaction of 7-platinaundecaborane [$(PMe_2Ph)_2PtB_{10}H_{12}$)] (**467**) with [$PMe_2Ph$] as indicated in Equation 20. The yields of the 10- and 12-vertex species are both about 12%. (**509**) is obtained as a by-product resulting from reactions with traces of chlorine impurities in the phosphine. A crystal structure determination of (**509**) indicates that the diplatinadodecaborane has a *closo*-structure.

$$2[(PMe_2Ph)_2PtB_{10}H_{12})] \ (\mathbf{467}) \ + \ 2[PMe_2Ph] \longrightarrow [(PMe_2Ph)_4Pt_2(B_{10}H_{10})] \ + \ [(B_{10}H_8)(PMe_2Ph)_2] \quad (20)$$

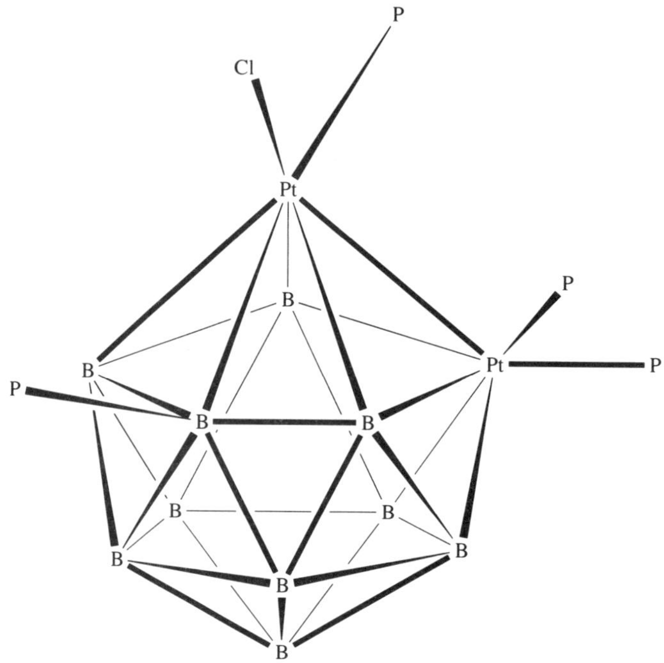

(509)

8.11.4.4 Group 11 metals

$[Et_3NH][Cu(1\text{-}B_{10}H_9N_2)_2]$ (**510**)[317] is an unusual system in which two $[1\text{-}B_{10}H_9N_2]^-$ clusters are linked by a copper atom. It is isolated as pinkish plates in 14% yield from the reaction of $[Et_3NH][1\text{-}B_{10}H_9N_2]$ with CuCl in refluxing $MeCN/H_2O$. The structure, confirmed by an x-ray structure determination, is illustrated in (**510**) with terminal hydrogen atoms omitted, for clarity. The copper atom is tetrahedrally bonded to the two $[1\text{-}B_{10}H_9N_2]$ polyhedral units through apical and equatorial B–H vertices derived from each polyhedron.

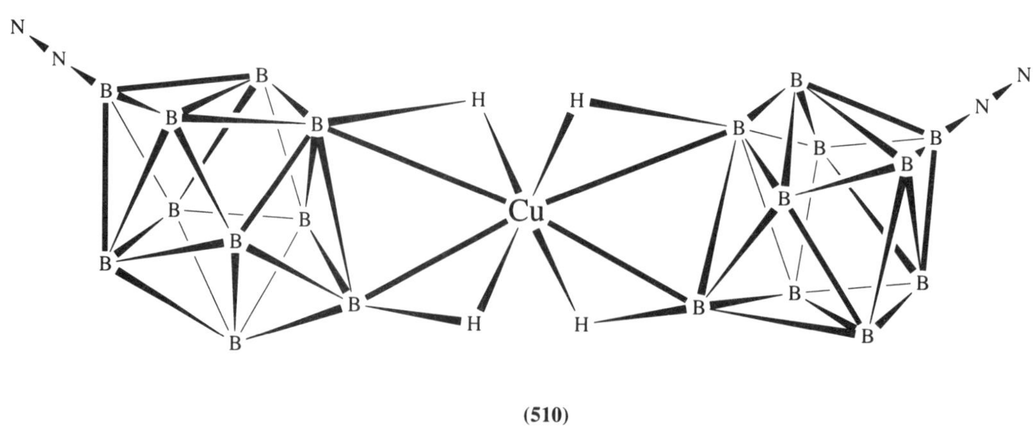

(510)

8.12 UNITS CONTAINING 11 BORON ATOMS

There are very few metallaboranes containing 11 boron atoms which fit into the format of this treatment, although the carbaborane analogue of $[B_{11}H_{11}]^{2-}$ is found in a large number of metallaboranes and several main group $REB_{11}H_{11}$ species are known (for example, E = Al, N, P) and covered elsewhere in this volume.

$[Re(CO)_5]B_{11}H_{14}$, **(511)**[318] is prepared by allowing $[Re(CO)_5Br]$ and $Na[B_{11}H_{13}]$ to react in THF at 25 °C. It is extracted with CH_2Cl_2 and obtained as a light brown hygroscopic powder in 20% yield. It contains the $[B_{11}H_{14}]^-$ anion, as indicated by NMR and IR spectra, and attempts to convert it or the related species (**512**) to a metallaborane have been unsuccessful. A second material isolated from the same product mixture, by crystallization with THF, is $Na[Re(CO)_5]B_{11}H_{14}$, (**512**). It is converted to the Cs^+ salt in 35% yield and identified by NMR and IR spectra. $[Rh(CO)(PPh_3)_2]B_{11}H_{14}$, **(513)**[318] is prepared from *trans*-$Rh(CO)(PPh_3)_2Br$ and $Na_2[B_{11}H_{13}]$. The initial product is $Cs[Rh(CO)(PPh_3)_2]B_{11}H_{13}$ (**515**) which is protonated to form (**513**). NMR spectral data are available for (**513**) and IR data for both (**513**) and (**514**).

$[B_{11}H_{11}Fe(CO)_3]^{2-}$ **(515)**[216] has been studied using MO calculations with a view to probing structure. (**515**) is considered to be stable with quite short Fe–B bond distances. If a second metal fragment is introduced, it appears not to have a preferred location, the energy differences in the various possibilities being quite small.

8.13 UNITS CONTAINING 12 OR MORE BORON ATOMS

8.13.1 Species Containing 12 Boron Atoms

8.13.1.1 Group 8 metals

$[(PMe_2Ph)_3RuB_{12}H_{12}]$ **(515)**[319] is prepared in the reaction between $Na_2[B_{12}H_{12}]$ and $[RuCl_3(PMe_2Ph)_3]$ in refluxing ethanol. It is isolated as a sparingly soluble lemon-yellow solid in 10% yield and identified by its NMR spectra and by comparison with the 7-triethylamino derivative for which a crystal structure is available. The assumed structure is given in Structure (**516**) ((**a**) along the threefold axis with the hydrogen atoms and organyl groups on phosphorus omitted for clarity, (b) a view perpendicular to the threefold axis showing the bridging hydrogen atoms) and NMR spectral L data in CD_2Cl_2 at 21 °C are available. The $Ru(PMe_2Ph)_3$ group coordinates to the $B_{12}H_{12}$ cluster through a triangular group of three adjacent terminal hydrogen atoms. This type of metal coordination to borane through three bridging hydrogen atoms is unique for a *closo*-cluster but it is observed for other species such as $[(CO)_3MnB_3H_8]$ (see *COMC-I*). The species may also be considered as a 13-vertex capped icosahedral cluster with the neutral $Ru(PMe_2Ph)_3$ group capping the triangular face. It thus contains 13 skeletal electron pairs and conforms to the PSEPT for a capped cluster. The chloro derivative bis(triphenyl-phosphine)ruthenium(II), $[(PPh_3)_2ClRuB_{12}H_{11}(NEt_3)]$ **(517)**[319] is prepared in the reaction between $[NHEt_3]_2[B_{12}H_{12}]$ and $[RuCl_2(PPh_3)_3]$ in refluxing EtOH as a bright-orange solid in 30% yield. Single crystal x-ray analysis confirms that the structure is the same as that in (**516**) except that one of the ligands on ruthenium is chlorine and the 7-position has a NEt_3 group bonded through nitrogen to B-7.

8.13.1.2 Group 10 metals

$[Pt_2(\mu\text{-}\eta^3\text{-}B_6H_9)_2(PMe_2Ph)_2]$, **(350)**[238–9] reported previously in *COMC-I*, and in Section 8.7.1.2, is available in alternative low-yield routes and a complete structural determination is available. It is prepared in 3.6% yield in the reaction between 6,6'-$O(B_{10}H_{13})_2$ and *cis*-$[PtCl_2(PMe_2Ph)_2]$[238] and in 4% yield from the pyrolysis of $[(PMe_2Ph)_2PtB_8H_{12}]$.[239] Yellow crystalline needles suitable for x-ray analysis are obtained. The general features of the structure have been described previously, except for bond length and angle data; however, these are now available.

8.13.2 Species Containing 13 Boron Atoms

$[B_{13}H_{13}Fe(CO)_3]^{2-}$ **(518)**[216] has been probed theoretically. The species is predicted to be stable and the Fe–B distances are predicted to be shorter than expected from data for analogous systems.

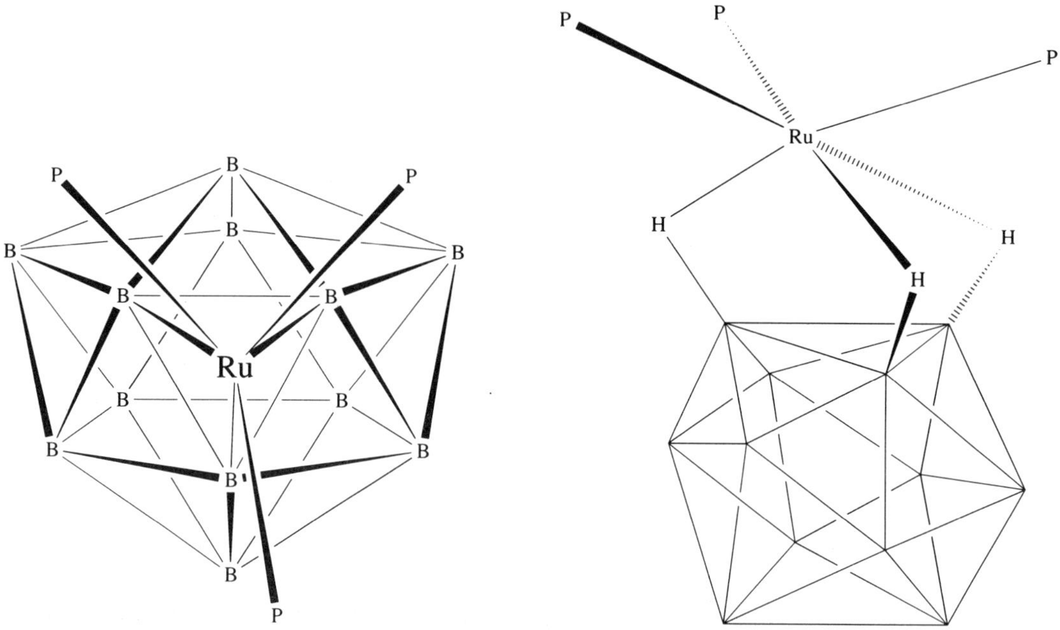

(a) (b)

(516)

8.13.3 Species Containing 14 Boron Atoms

$[(PMe_2Ph)_4Pt_3B_{14}H_{16}]$ (**352**)[239,320] is prepared in the thermolysis of $[(PMe_2Ph)_2PtB_8H_{12}]$ (**382**) in refluxing toluene under N_2. It is obtained as green crystals in 6% yield. The structure of $[(PMe_2Ph)_4Pt_3B_{14}H_{16}]$ is given in (**352**) (see Section 8.7.1.2.) The species may be considered as a *nido*-type 11-vertex 2,7,10-Pt_3B_8 cluster, slightly distorted from the idealized $[B_{11}H_{14}]^-$ structure, fused to an *arachno*-type 9-vertex 6',8'-Pt_2B_7 cluster. The latter has its Pt-6'–B-7'–Pt-8' atoms in common with the former and the distortions are ascribed to the accommodation of the large platinum atoms. The result is a 17-vertex triangular-faced polyhedron with a seven-membered PtB_2PtB_3 open face. The hydrogen atoms are not located in the structure determination but $^1H\{^{11}B\}$ NMR spectra suggest that each boron atom except B-11 has a terminal hydrogen atom bonded to it and that there are three bridging hydrogen atoms.

8.13.4 Species Containing 16 Boron Atoms

$[(PMe_2Ph)PtB_{16}H_{18}(PMe_2Ph)]$, (**354**)[239,321] is isolated as a red solid in 2% yield as another of the products from the pyrolysis of $[(PMe_2Ph)_2PtB_8H_{12}]$ (**382**) in refluxing toluene. The structure of (**354**), confirmed by a crystal structure determination, is given in Section 8.7.1.2. It is a *conjuncto*-species consisting of a *nido*-11-vertex PtB_{10} cluster, based on an icosahedron with a vertex missing, and an *arachno*-8-vertex cluster. The two clusters are fused across their open faces with Pt-7–B-8 as a common edge. $[(PMe_2Ph)_2Pt_2B_{16}H_{15}(C_6H_4Me)(PMe_2Ph)]$ (**353**)[240] is yet another species formed in the pyrolysis of $[(PMe_2Ph)_2PtB_8H_{12}]$ (**382**) in refluxing toluene. The formation of this dark-orange air-stable solid which is isolated in <1% yield is favored by shorter reaction times and it may be a precursor to the monoplatina-species described previously. A preliminary x-ray structure determination reveals a *conjuncto*-system consisting of a *closo*-$[Pt_2B_{10}]$ subcluster and a *nido*-$[Pt_2B_7]$ subcluster fused via a common Pt_2B closed triangular face as indicated in (**353**) (see Section 8.7.1.2).

8.13.5 Species Containing 17 Boron Atoms

Two metallaboranes containing 17 boron atoms have been prepared. These unusual species have structures which are based on the well-characterized *iso*-$B_{18}H_{22}$. One, iso-8-$CpCoB_{17}H_{21}$ (**519**),[322] exists as an iso-$B_{18}H_{22}$ cluster with the CpCo group replacing a BH vertex in the 8-position. The other, [(Cp*Rh)$_2$B$_{17}$H$_{19}$] (**520**),[323] involves one rhodium moiety adding to one side of iso-$B_{18}H_{22}$ while the other rhodium moiety adds to the other side with concomitant loss of one BH moiety. [iso-8-CpCoB$_{17}$H$_{21}$] (**519**) is prepared as a red solid in 2.2% yield from the reaction between cobalt vapor, *n*-$B_{18}H_{22}$ and cyclopentadiene in a metal atom reactor. The ^{11}B NMR spectrum and IR spectrum are available. It has an asymmetric fused cage structure based on iso-$B_{18}H_{22}$ and not *n*-$B_{18}H_{22}$. (**519**) forms by degradative insertion of the cobalt moiety and isomerization from an *n*-to an iso-B_{18} structure. [(Cp*Rh)$_2$B$_{17}$H$_{19}$], (**520**) is prepared in the reaction between [NEt$_4$]$_2$[*n*-$B_{18}H_{20}$] and [Cp*RhCl$_2$]$_2$ in CH_2Cl_2 at 21 °C in 14% yield. A preliminary crystal structure determination reveals Structure (**520**). The structure of [(Cp*Rh)$_2$B$_{17}$H$_{19}$] is that of a 19-vertex cluster comprising a *nido*-11-vertex RhB$_{10}$ subcluster and a *nido*-10-vertex RhB$_9$ subcluster joined at the B-10–B-11 [=B-8'–B-4'] edge. Rhodium-7' has added to the edge of the *n*-$B_{18}H_{22}$ with the elimination of an opposing BH vertex. The second rhodium atom adds to the other open face as shown in (**520**) and Rh-7 bonds to B-12,4' through a bridging hydrogen atom, indicated by a broken line.

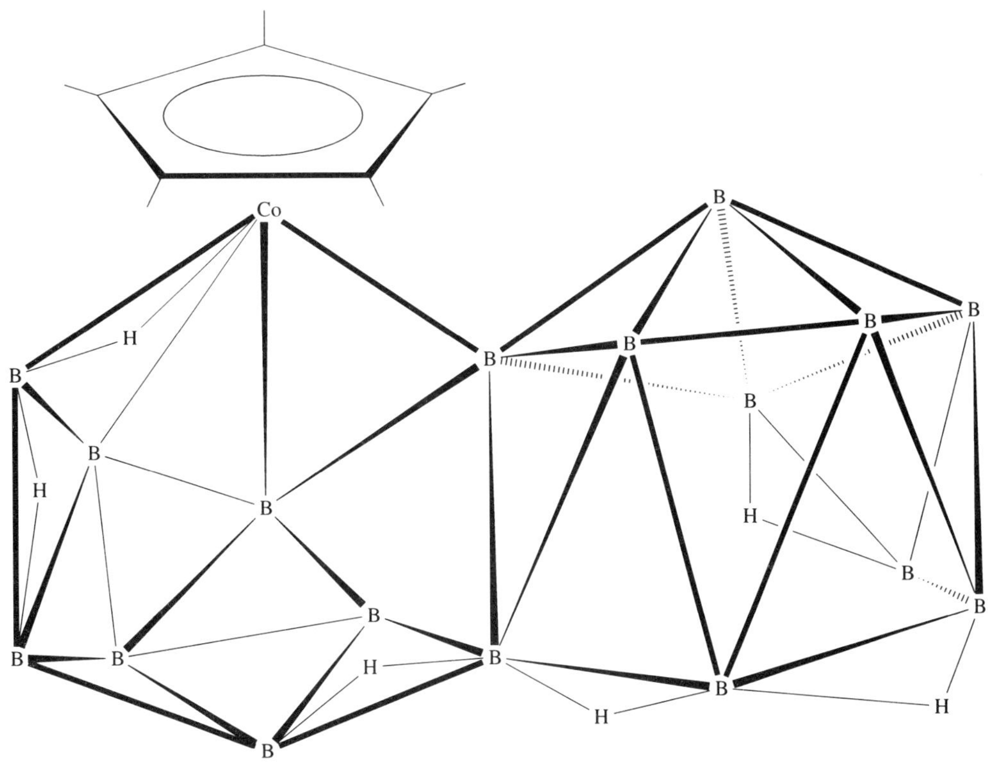

(519)

8.13.6 Species Containing 18 Boron Atoms

[PMe$_2$Ph]Pt-η^4-*n*-$B_{18}H_{20}$] (**521**)[324] and the isomer [PMe$_2$Ph]Pt-η^4-*i*-$B_{18}H_{20}$][324] are available from the reaction between *cis*-[PtCl$_2$(PMe$_2$Ph)]$_2$ and an equimolar mixture of the two isomers in the presence of base. The products are yellow air-stable crystalline solids formed by the replacement of two protons in one of the two open faces in each of the isomers by a L_2Pt^{2+} moiety. The species are analogous to the species L_2Pt-η-$B_{10}H_{12}$ (**509**) described earlier.

8.13.7 Species Containing 20 Boron Atoms

[CpFeC$_5$H$_4$CH$_2$N(Me)$_2$C$_2$H$_5$]$_2$[B$_{20}$H$_{18}$] (**522**),[325] has been prepared as yellow cubic crystals in quantitative yield which allowed an accurate structure determination.

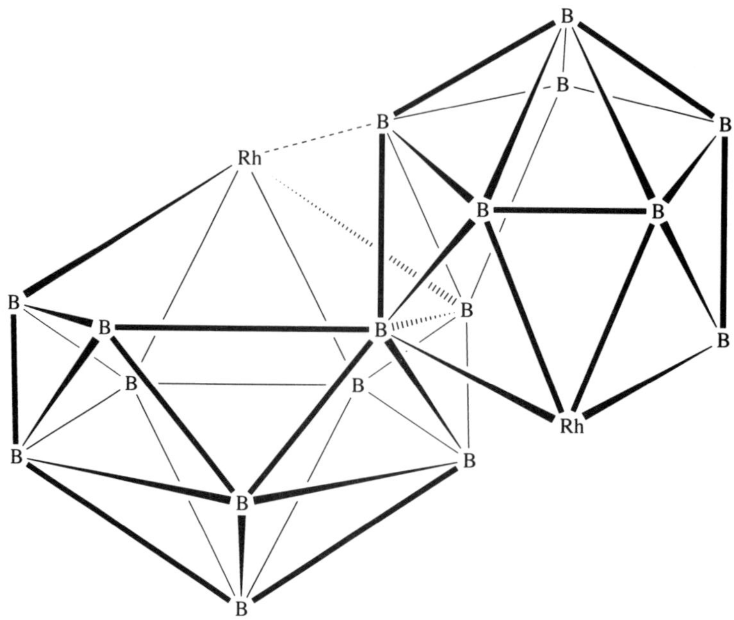

(520)

Rh = Cp*Rh

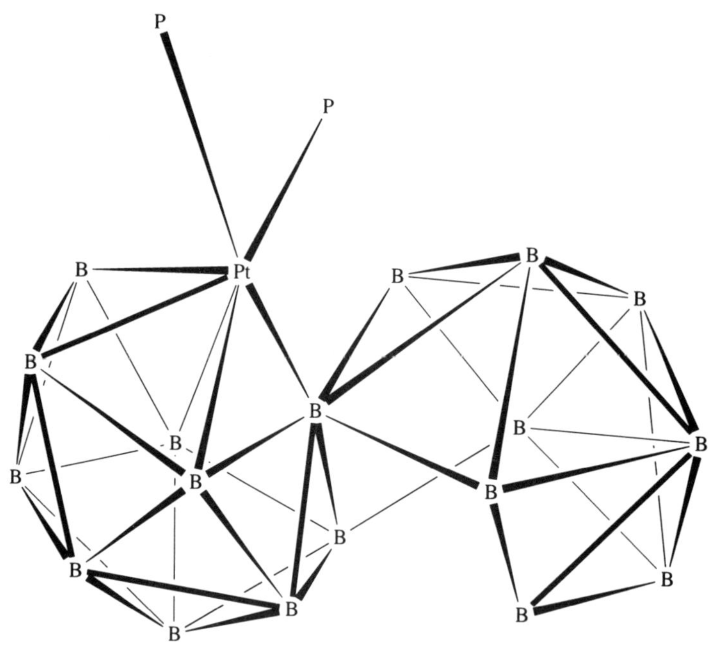

(521)

8.14 REFERENCES

1. K. B. Gilbert, S. K. Boocock and S. G. Shore, in 'COMC-I', vol. 6, pp. 879–945.
2. J. D. Kennedy, *Prog. Inorg. Chem.*, 1984, **32**, 519; J. D. Kennedy, *Prog. Inorg. Chem.*, 1986, **34**, 211.
3. R. N. Grimes 'Metal Interactions with Boron Clusters', (ed.) Plenum, New York, 1982; and articles included therein.
4. J. F. Liebman, A. Greenberg and R. E. Williams, (eds.) 'Advances in Boron and the Boranes', VCH, Weinheim, 1988.
5. C. E. Housecroft, 'Boranes and Metalloboranes: Structure, Bonding and Reactivity', 2nd edn., Ellis Horwood, Chichester, 1994.
6. C. E. Housecroft and T. P. Fehlner, *Adv. Organomet. Chem.*, 1982, **21**, 57.
7. C. E. Housecroft, *Adv. Organomet. Chem.*, 1991, **33**, 1.

8. T. P. Fehlner, *New. J. Chem.*, 1988, **12**, 307.
9. N. N. Greenwood, in 'Rings Cages and Clusters of the Main Group Elements', ACS Symposium Series No. 232, ed. A. H. Cowley, American Chemical Society, Washington, DC, 1983, pp. 125–138.
10. R. N. Grimes, *Acc. Chem. Res.*, 1983, **16**, 22.
11. N. N. Greenwood, *Pure Appl. Chem.*, 1983, **55**, 77.
12. N. N. Greenwood, *Pure Appl. Chem.*, 1983, **55**, 1415.
13. N. N. Greenwood, in 'Inorganic Chemistry Toward the 21st Century', ed. ACS Symposium Series No. 211, M. H. Chisholm, American Chemical Society, Washington, DC, 1983, p. 333.
14. N. N. Greenwood, *Chem. Soc. Rev.*, 1984, **13**, 353.
15. N. N. Greenwood and J. D. Kennedy, *Pure Appl. Chem.* 1991, **63**, 317.
16. D. P. Workman and S. G. Shore, in 'Electron Deficient Boron and Carbon Clusters', eds. G. A. Olah, K. Wade and R. E. Williams, Wiley, New York, 1990, pp. 237–260.
17. T. P. Fehlner, in 'Electron Deficient Boron and Carbon Clusters', eds. G. A. Olah, K. Wade and R. E. Williams, Wiley, New York, 1990, pp. 287–307.
18. L. Barton, 'Gmelin Handbook of Inorganic Chemistry', Boron Compounds, 3rd suppl., vol. 1, Springer, Berlin, 1987, pp. 2–241.
19 (a) L. Barton, 'Gmelin Handbook of Inorganic Chemistry, Boron Compounds', 4th suppl., vol. 1a, Springer, Berlin, 1994, pp. 9–155; (b) vol. 1b, 1995 (in press).
20. E. B. Lobkovsky, Y. K. Gun'ko, B. M. Bulychev, V. K. Belsky, G. L. Soloveichik and M. Y. Antipin, *J. Organomet. Chem.*, 1991, **406**, 343.
21. K. S. Wong, W. R. Scheidt and T. P. Fehlner, *J. Am. Chem. Soc.*, 1982, **104**, 1111.
22. K. S. Harpp, C. E. Housecroft, A. L. Rheingold and M. S. Shongwe, *J. Chem. Soc., Chem. Commun.*, 1988, 965.
23. F-E. Hong, T. J. Coffy, D. A. McCarthy and S. G. Shore, *Inorg. Chem.*, 1989, **28** 3284.
24. A. K. Bandyopadhyay, R. Khattar, J. Puga, T. P. Fehlner and A. L. Rheingold, *Inorg. Chem.*, 1992, **31**, 465.
25. R. L. Johnston, D. M. P. Mingos and P. Sherwood, *New J. Chem.*, 1991, **15**, 831.
26. C. W. Jung, R. T. Baker and M. F. Hawthorne, *J. Am. Chem. Soc.*, 1981, **103**, 810.
27. R. T. Baker, *Inorg. Chem.*, 1986, **25**, 109.
28. J. D. Kennedy, *Inorg. Chem.*, 1986, **25**, 111.
29. R. L. Johnston and D. M. P. Mingos, *Inorg. Chem.*, 1986, **25**, 3321.
30. J. Bould, J. D. Kennedy, and M. Thornton-Pett, *J. Chem. Soc., Dalton Trans.*, 1992, 563.
31. M. Mancini *et al.*, *Inorg. Chem.* 1984, **23**, 1072.
32. M. F. Lappert, A. Singh, J. L. Atwood and W. E. Hunter, *J. Chem. Soc., Chem. Commun.*, 1983, 206.
33. A. I. Sizov and B. M. Bulychev, *Zh. Neorg. Khim.*, 1981, **26**, 2652; *Russ. J. Inorg. Chem.*, 1981, **26**, 1421.
34. G. Dell'Amico, F. Marchetti and C. Floriani, *J. Chem. Soc., Dalton Trans.*, 1982, 2197.
35. J. A. Jensen and G. S. Girolami, *J. Chem. Soc., Chem. Commun.*, 1986, 1160.
36. J. A. Jensen, S. R. Wilson and G. S. Girolami, *J. Am. Chem. Soc.*, 1988, **110**, 4977.
37. J. A. Jensen, J. E. Gozum, D. M. Pollina and G. S. Girolami, *J. Am. Chem. Soc.*, 1988, **110**, 1643.
38. J. A. Jensen and G. S. Girolami, *Inorg. Chem.*, 1989, **28**, 2107.
39. J. A. Jensen and G. S. Girolami, *J. Am. Chem. Soc.*, 1988, **110**, 4450.
40. F. Corazza, C. Floriani, A. Chiesi-Villa and C. Guastini. *Inorg. Chem.*, 1991, **30**, 145.
41. V. D. Sasnovskaya, L. V. Titov and V. Ya. Rosolovskii, *Izv. Akad. Nauk SSSR, Ser. Khim.*, 1984, 1408 (*Chem. Abstr.*, 1984, **101**, 171 415).
42. V. D. Sasnovskaya and L. V. Titov, *Izv. Akad. Nauk SSSR, Ser. Khim.*, 1984, 417 (*Chem. Abstr.*, 1984, **100**, 192 021).
43. D. Männig and H. Nöth, *J. Organomet. Chem.*, 1984, **275**, 169.
44. A. G. Csaszar, L. Hedberg, K. Hedberg, R. C. Burns, A. T. Wen and M. J. McGlinchey, *Inorg. Chem.*, 1991, **30**, 1371.
45. J. E. Gozum and G. S. Girolami, *J. Am. Chem. Soc.*, 1991, **113**, 3829.
46. J. E. Gozum, S. R. Wilson and G. S. Girolami, *J. Am. Chem. Soc.*, 1992, **114**, 9483.
47. M. D. Fryzuk and H. D. Williams, *Organometallics*, 1983, **2**, 162.
48. M. D. Fryzuk, S. J. Rettig, A. Westerhaus and H. D. Williams *Inorg. Chem.*, 1985, **24**, 4316.
49. R. L. Bansemer, J. C. Huffman and K. G. Caulton, *J. Am. Chem. Soc.*, 1983, **105**, 6163.
50. F. A. Cotton, S. A. Duraj and W. J. Roth, *Inorg. Chem*, 1984, **23**, 4113.
51. J. A. Jensen and G. S. Girolami, *Inorg. Chem.*, 1989, **28**, 2114.
52. M. L. Luetkens, J. C. Huffman and A. P. Sattelberger, *J. Am. Chem. Soc.*, 1985, **107**, 3361.
53. M. L. H. Green and L-L. Wong, *J. Chem. Soc., Chem. Commun.*, 1989, 571.
54. M. Y. Darensbourg and S. Slater, *J. Am. Chem. Soc.*, 1981, **103**, 5914.
55. M. Y. Darensbourg, R. Bau, M. W. Marks, R. R. Burch, Jr., J. C. Deaton and S. Slater, *J. Am. Chem. Soc.*, 1982, **104**, 6961.
56. A. R. Barron, J. E. Salt, G. Wilkinson, M. Motevalli and M. B. Hursthouse, *Polyhedron*, 1986, **5**, 1833.
57. A. A. H. van der Zeijden, V. Shklover and H. Berke, *Inorg. Chem.*, 1991, **30**, 4393.
58. V. D. Makhaev, A. P. Borisov, T. P. Gnilomedova, E. B. Lobkovskii and A. N. Chekhlov, *Izv. Akad. Nauk SSSR, Ser. Khim.*, 1987, 1712; *Bull. Acad. Sci. USSR*, 1987, **36**, 1582.
59. R. Carreno, V. Riera, M. A. Ruiz, Y. Jeannin and M. Philoche-Levisalles, *J. Chem. Soc., Chem. Commun.*, 1990, 15.
60. C. A. Ghilardi, P. Innocenti, S. Midollini and A. Orlandini, *J. Chem. Soc., Dalton Trans.*, 1985, 605.
61. M. V. Baker and L. D. Field, *J. Chem. Soc., Chem. Commun.*, 1984, 996.
62. R. Bau, H. S. H. Yuan, M. V. Baker and L. D. Field, *Inorg. Chim. Acta*, 1986, **114**, L2.
63. J. B. Letts, T. J. Mazanec and D. W. Meek, *J. Am. Chem. Soc.*, 1982, **104**, 3898.
64. C. Bianchini, P. J. Perez, M. Peruzzini, F. Zanobini and A. Vacca, *Inorg. Chem.*, 1991, **30**, 279.
65. L. F. Rhodes, L. M. Venanzi, C. Sorato and A. Albinati, *Inorg. Chem.*, 1986, **25**, 3335.
66. P. W. Frost, J. A. K. Howard and J. L. Spencer, *J. Chem. Soc., Chem. Commun.*, 1984, 1362.
67. E. J. Corey, N. J. Cooper, W. M. Canning, W. N. Lipscomb and T. F. Koetzle, *Inorg. Chem.*, 1982, **21** 192.
68. D. G. Holah, A. N. Hughes, S. Maciaszek, V. R. Magnuson and K. O. Parker, *Inorg. Chem.*, 1985, **24**, 3956.
69. T. M. Gilbert, F. J. Hollander and R. G. Bergman, *J. Am. Chem. Soc.*, 1985, **107**, 3508.
70. B. E. Green, C. H. L. Kennard, G. Smith, B. D. James and A. H. White, *Cryst. Struct. Commun.*, 1981, **10**, 1245.
71. B. E. Green *et al.*, *Inorg. Chim. Acta.*, 1984, **83**, 177.

72. C. A. Ghilardi, S. Midollini and A. Orlandini, *Inorg. Chem.*, 1982, **21**, 4096.
73. C. Bianchini, C. A. Ghilardi, A. Meli, S. Midollini and A. Orlandini, *Inorg. Chem.*, 1985, **24**, 924.
74. B. E. Green, C. H. L. Kennard, G. Smith, B. D. James, P. C. Healy and A. H. White, *Inorg. Chim. Acta*, 1984, **81**, 147.
75. V. D. Makhaev, A. P. Borisov, E. B. Lobkovskii, V. B. Polyakova and K. N. Semenenko, *Izv. Akad. Nauk SSSR, Ser. Khim.*, 1985, 1881; *Bull. Acad. Sci. USSR*, 1985, **36**, 1731.
76. J. C. Bommer and K. W. Morse, *Inorg. Chem.*, 1983, **22**, 592.
77. M. M. T. Khan, P. Paul and K. Venkatasubramanian, *Polyhedron*, 1991, **10**, 1827.
78. V. D. Makhaev, A. P. Borisov and K. N. Semenenko, *Zh. Neorg. Khim.*, 1986, **31**, 1586; *Russ. J. Inorg. Chem.*, 1986, **31**, 908.
79. L. V. Titov, L. A. Gavrilova, U. Mirsaidov, T. G. Vershinnikova, G. N. Baiko and V. Ya. Rosolovskii, *Zh. Neorg. Khim.*, 1982, **27**, 1953; *Russ. J. Inorg. Chem.*, 1982, **27**, 1104.
80. V. D. Makhaev, A. P. Borisov, B. P. Tarasov and K. N. Semenenko, *Zh. Neorg. Khim.*, 1981, **26**, 2645; *Russ. J. Inorg. Chem.*, 1981, **26**, 1417.
81. V. D. Makhaev, A. P. Borisov, B. P. Tarasov, and K. N. Semenenko, *Zh. Neorg. Khim.*, 1983, **28**, 340; *Russ. J. Inorg. Chem.*, 1983, **28**, 186.
82. S. Bettonville and J. Goffart, *J. Organomet. Chem.*, 1988, **356**, 297.
83. P. Zanella *et al.*, *Inorg. Chim. Acta*, 1988, **144**, 129.
84. D. Männig and H. Nöth, *Z. Anorg. Allg. Chem.*, 1986, **543**, 66.
85. H. J. Wasserman, D. C. Moody, R. T. Paine, R. R. Ryan and K. V. Salazar, *J. Chem Soc., Chem. Commun.*, 1984, 533.
86. H. J. Wasserman, D. C. Moody and R. R. Ryan, *J. Chem Soc., Chem. Commun.*, 1984, 532.
87. P. Charpin, M. Nierlich, G. Chevrier, D. Vigner, M. Lance and D. Baudry, *Acta Crystallogr.*, 1987, **C43**, 1255.
88. D. Baudry *et al.*, *J. Chem. Soc., Chem. Commun.*, 1985, 1553.
89. D. Baudry *et al.*, *J. Organomet. Chem.*, 1989, **371**, 163.
90. D. Baudry *et al.*, *J. Organomet. Chem.*, 1989, **371**, 155.
91. D. Baudry, M. Ephritikhine, F. Nief, L. Ricard and F. Mathey, *Angew. Chem., Int. Ed. Engl.*, 1990, **29**, 1485.
92. N. Edelstein, *Inorg. Chem.*, 1981, **20**, 297.
93. R. Shinomoto, E. Gamp, N. M. Edelstein and D. H. Templeton, *Inorg. Chem.*, 1983, **22**, 2351.
94. K. Rajnak, E. Gamp, R. Banks, R. Shinomoto and N. Edelstein, *Inorg. Chim. Acta*, 1984, **95**, 29.
95. K. Rajnak, R. H. Banks, E. Gamp and N. Edelstein, *J. Chem. Phys.*, 1984, **80**, 5951.
96. W. K. Kot, N. M. Edelstein and A. Zalkin, *Inorg. Chem.*, 1987, **26**, 1339.
97. W. F. McNamara, E. N. Duesler, R. T. Paine, J. V. Ortiz, P. Kölle and H. Nöth, *Organometallics*, 1986, **5**, 380.
98. M. N. I. Khan, C. King, D. D. Heinrich, J. P. Fackler, Jr. and L. C. Porter, *Inorg. Chem.*, 1989, **28**, 2150.
99. R. Shinomoto, A. Zalkin and N. M. Edelstein, *Inorg. Chim. Acta*, 1987, **139**, 97.
100. R. Shinomoto, J. G. Brennan, N. M. Edelstein and A. Zalkin, *Inorg. Chem.*, 1985, **24**, 2896.
101. K. Rajnak, E. Gamp, R. Shinomoto and N. Edelstein, *J. Chem. Phys.*, 1984, **80**, 5942.
102. E. Gamp, R. Shinomoto, N. Edelstein and B. R. McGarvey, *Inorg. Chem.*, 1987, **26**, 2177.
103. R. Shinomoto, A. Zalkin, and N. M. Edelstein, *Inorg. Chim. Acta*, 1987, **139**, 91.
104. J. Brennan, R. Shinomoto, A. Zalkin and N. Edelstein, *Inorg. Chem.*, 1984, **23**, 4143.
105. K. Rajnak, R. H. Banks, E. Gamp and N. Edelstein, *J. Chem. Phys.*, 1984, **80**, 5951.
106. E. Gamp and N. Edelstein, *J. Chem. Phys.*, 1984, **80**, 5963.
107. A. I. Sizov, I. V. Molodnitskaya, B. M. Bulychev, V. K. Bel'skii and G. L. Soloveichik, *J. Organomet. Chem.*, 1988, **344**, 185.
108. S. A. A. Zaidi, M. Jaria, S. Khan and Z. A. Siddiqi, *Indian J. Chem., Sect. A*, 1985, **24**, 314.
109. S. A. A. Zaidi, M. Jaria, R. Kureshy, M. Yamin and Z. A. Siddiqi, *Bull. Soc. Chim. Fr.*, 1985, 177.
110. J. D. Basil, A. A. Aradi, N. K. Bhattacharyya, N. P. Rath, C. Eigenbrot and T. P. Fehlner, *Inorg. Chem.*, 1990, **29**, 1260.
111. D. J. Elliot *et al.*, *Inorg. Chem.*, 1990, **29**, 5014.
112. G. Erker, U. Hoffmann, R. Zwettler and C. Krüger, *J. Organomet. Chem.*, 1989, **367**, C15.
113. S. A. A. Zaidi, M. Jaria and Z. A. Siddiqi, *Bull. Soc. Chim. Fr.*, 1987, 599.
114. S. A. A. Zaidi, M. Jaria and Z. A. Siddiqi, *Synth. React. Inorg. Metal-Org. Chem.*, 1990, **16**, 1067.
115. G. A. Carriedo, G. P. Elliot, J. A. K. Howard, D. B. Lewis and F. G. A. Stone, *J. Chem. Soc., Chem. Commun.*, 1984, 1585.
116. S. A. A. Zaidi, T. A. Khan, S. R. A. Zaidi and Z. A. Siddiqi, *Polyhedron*, 1985, **4**, 1163.
117. S. Helm, G. Linti, H. Nöth, S. Channareddy and P. Hofmann, *Chem. Ber.*, 1992, **125**, 73.
118. S. Helm and H. Nöth, *Angew. Chem., Int. Ed. Engl.*, 1988, **27**, 1331.
119. S. Channareddy, G. Linti and H. Nöth, *Angew. Chem., Int. Ed. Engl.*, 1990, **29**, 199.
120. S. A. Westcott, N. J. Taylor, T. B. Marder, R. T. Baker, N. J. Jones and J. C. Calabrese, *J. Chem. Soc., Chem. Commun.*, 1991, 304.
121. K. Burgess, W. A. van der Donk, S. A. Westcott, T. B. Marder, R. T. Baker and J. C. Calabrese, *J. Am. Chem. Soc.*, 1992, **114**, 9350.
122. D. Männig and H. Nöth, *Angew. Chem., Int. Ed. Engl.*, 1985, **24**, 878.
123. J. R. Knorr and J. S. Merola, *Organometallics*, 1990, **9**, 3008.
124. R. T. Baker, *et al.*, *J. Am. Chem. Soc.*, 1990, **112**, 9399.
125. J. C. Vites, C. Eigenbrot and T. P. Fehlner, *J. Am. Chem. Soc.*, 1984, **106**, 4633.
126. J. C. Vites, C. E. Housecroft, C. Eigenbrot, M. L. Buhl, G. J. Long and T. P. Fehlner, *J. Am. Chem. Soc.*, 1986, **108**, 3304.
127. J. C. Vites, C. E. Housecroft, G. B. Jacobsen and T. P. Fehlner, *Organometallics*, 1984, **3**, 1591.
128. C. E. Housecroft and T. P. Fehlner, *Inorg. Chem.*, 1986, **25**, 404.
129. C. E. Housecroft and T. P. Fehlner, *J. Am. Chem. Soc.*, 1986, **108**, 4867,
130. T. P. Fehlner, C. E. Housecroft, W. R. Scheidt and K. S. Wong, *Organometallics*, 1983, **2**, 825.
131. C. E. Housecroft and T. P. Fehlner, *Organometallics*, 1986, **5**, 379.
132. C. E. Housecroft, M. L. Buhl, G. J. Long and T. P. Fehlner, *J. Am. Chem. Soc.*, 1987, **109**, 3323.
133. C. E. Housecroft and T. P. Fehlner, *Organometallics*, 1986, **5**, 1279.
134. K. S. Harpp and C. E. Housecroft, *J. Organomet. Chem.*, 1988, **340**, 389.
135. S. M. Draper, C. E. Housecroft, J. E. Rees, M. S. Shongwe, B. S. Haggerty and A. L. Rheingold, *Organometallics*, 1992, **11**, 2356.

136. C. E. Housecroft and A. L. Rheingold, *J. Am. Chem. Soc.*, 1986, **108**, 6420.
137. C. E. Housecroft, M. S. Shongwe, A. L. Rheingold and P. Zanello, *J. Organomet. Chem.*, 1991, **408**, 7.
138. C. E. Housecroft, A. L. Rheingold and M. S. Shongwe, *J. Chem. Soc., Chem. Commun.*, 1988, 1630.
139. X. Meng, N. P. Rath and T. P. Fehlner, *J. Am. Chem. Soc.*, 1989, **111**, 3422.
140. X. Meng, N. P. Rath, T. P. Fehlner and A. L. Rheingold, *Organometallics*, 1991, **10**, 1986.
141. A. K. Chipperfield and C. E. Housecroft, *J. Organomet. Chem.*, 1988, **349**, C17.
142. S. M. Draper, C. E. Housecroft, A. K. Keep, D. M. Matthews, A. L. Rheingold and X. Song, *J. Organomet. Chem.* 1992, **423**, 241.
143. A. K. Chipperfield, C. E. Housecroft and A. L. Rheingold, *Organometallics*, 1990, **9**, 681.
144. F-E. Hong, D. A. McCarthy, J. P. White, III, C. E. Cottrell and S. G. Shore, *Inorg. Chem.*, 1990, **29**, 2874
145. C. E. Housecroft, D. M. Matthews, A. L. Rheingold and X. Song, *J. Chem. Soc., Dalton Trans.*, 1992, 2855.
146. S. M. Draper, A. D. Hattersley, C. E. Housecroft and A. L. Rheingold, *J. Chem. Soc., Chem. Commun.*, 1992, 1365.
147. A. K. Chipperfield, C. E. Housecroft and P. R. Raithby, *Organometallics*, 1990, **9**, 479.
148. C. E. Housecroft, D. M. Matthews, A. L. Rheingold and X. Song, *J. Chem. Soc., Chem. Commun.*, 1992, 842.
149. S. G. Shore, D.-Y. Jan, L.-Y. Hsu and W.-L Hsu, *J. Am. Chem. Soc.*, 1983, **105**, 5923.
150. D.-Y. Jan, D. P. Workman, L-Y. Hsu, J. A. Krause and S. G. Shore, *Inorg. Chem.*, 1992, **31**, 5123.
151. D-Y. Jan and S. G. Shore, *Organometallics*, 1987, **6** 428.
152. D-Y. Jan, L-Y. Hsu, D. P. Workman and S. G. Shore, *Organometallics*, 1987, **6** 1984.
153. J. R. Wermer, D-Y Jan, T. D. Getman, E. Moher and S. G. Shore, *Inorg. Chem.*, 1988, **27**, 4274.
154. D. P. Workman, D-Y Jan and S. G. Shore, *Inorg. Chem.*, 1990, **29**, 3518.
155. J. Feilong, T. P. Fehlner and A. L. Rheingold, *Angew. Chem., Int. Ed. Engl.*, 1988, **27** 424.
156. C. E. Housecroft, *Polyhedron*, 1987, **6**, 1935.
157. M. Tachikawa and E. L. Muetterties, *J. Am. Chem. Soc.*, 1980, **102**, 4541.
158. J. S. Bradley, *Phil. Trans. R. Soc., London*, Ser. A, 1982, **308**, 103.
159. E. M. Holt, K. H. Whitmire and D. F. Shriver, *J. Organomet. Chem.*, 1981, **213**, 125.
160. R. B. Calvert and J. R. Shapley, *J. Am. Chem. Soc.*, 1977, **99**, 5225.
161. A. J. Deeming, *Adv. Organomet. Chem.*, 1986, **26**, 1.
162. X. Meng, A. K. Bandyopadhyay, T. P. Fehlner and F.-W. Grevels, *J. Organomet. Chem.*, 1990, **394**, 15
163. A. J. Deeming and M. Underhill, *J. Chem. Soc., Chem. Commun.*, 1973, 277.
164. N. P. Rath and T. P. Fehlner, *J. Am. Chem. Soc.*, 1987, **109**, 5273.
165. C. E. Housecroft and A. L. Rheingold, *Organometallics*, 1987, **6**, 1332.
166. C. E. Housecroft, M. S. Shongwe and A. L. Rheingold, *Organometallics*, 1988, **7**, 1885.
167. C. E. Housecroft, M. S. Shongwe and A. L. Rheingold, *Organometallics*, 1989, **8**, 2651.
168. A. K. Chipperfield, B. S. Haggerty, C. E. Housecroft and A. L. Rheingold, *J. Chem. Soc., Chem. Commun.*, 1990, 1174.
169. C. E. Housecroft *et al.*, *Organometallics*, 1992, **11**, 4048.
170. S. M. Draper, C. E. Housecroft, A. K. Keep, D. M. Matthews, B. S. Haggerty and A. L. Rheingold, *J. Organomet. Chem.* 1991, **410**, C44.
171. A. Sirigu, M. Bianchi and E. Benedetti, *J. Chem. Soc., Chem Commun.*, 1969, 596.
172. P. D. Grebenik *et al.*, *J. Chem. Soc., Chem. Commun.*, 1988, 799.
173. L. Messerle, *Chem. Rev.*, 1988, **88**, 1229.
174. C. Ting and L. Messerle, *Inorg. Chem.*, 1989, **28** 171.
175. C. Ting and L. Messerle, *J. Am. Chem. Soc.*, 1989, **111**, 3449.
176. G. B. Jacobsen *et al.*, *Inorg. Chem.*, 1987, **26**, 4040.
177. C. S. Jun, X. Meng, K. J. Haller and T. P. Fehlner. *J. Am. Chem. Soc.*, 1991, **113**, 3603.
178. T. J. Coffy, G. Medford, J. Plotkin, G. J. Long, J. C. Huffman and S. G. Shore, *Organometallics*, 1989, **8**, 2404.
179. T. J. Coffy and S. G. Shore, *J. Organomet. Chem.*, 1990, **394**, C27.
180. A. K. Chipperfield, C. E. Housecroft and D. M. Matthews. *J. Organomet. Chem.*, 1990, **384**, C38.
181. C. E. Housecroft, D. M. Matthews and A. L. Rheingold, *J. Chem. Soc., Chem. Commun.*, 1992, 323.
182. J. Feilong, T. P. Fehlner and A. L. Rheingold, *J. Organomet. Chem.*, 1988, **348**, C22.
183. R. Ahmad, J. E. Crook, N. N. Greenwood and J. D. Kennedy, *J. Chem. Soc., Dalton Trans.*, 1986, 2433.
184. S. A. Snow, M. Shimoi, C. D. Ostler, B. K. Thompson, G. Kodama and R. W. Parry, *Inorg. Chem.*, 1984, **23**, 511.
185. G. F. Mitchell and A. J. Welch, *J. Chem. Soc., Dalton Trans.*, 1987, 1017.
186. S. A. Snow and G. Kodama, *Inorg. Chem.*, 1985, **24**, 795.
187. M. Shimoi, K. Katoh and H. Ogino, *J. Chem. Soc., Chem. Commun.*, 1990, 811.
188. M. Shimoi, K. Katoh, H. Tobita and H. Ogino, *Inorg. Chem.*, 1990, **29**, 814.
189. J. Feilong, T. P. Fehlner and A. L. Rheingold, *J. Am. Chem. Soc.*, 1987, **109**, 1860.
190. J. Feilong, T. P. Fehlner and A. L. Rheingold, *J. Chem. Soc., Chem. Commun.*, 1987, 1395.
191. C. E. Housecroft and T. P. Fehlner, *Inorg. Chem.*, 1982, **21**, 1739.
192. J. Bould, N. N. Greenwood, J. D. Kennedy and W. S. McDonald, *J. Chem. Soc., Dalton Trans.*, 1985, 1843.
193. P. D. Grebenik, J. B. Leach, J. M. Pounds, M. L. H. Green and P. Mountford, *J. Organomet. Chem.*, 1990, **382**, C1.
194. P. D. Grebenik, J. B. Leach, M. L. H. Green and N. M. Walker, *J. Organomet. Chem.*, 1988, **345**, C31.
195. J. Bould, N. N. Greenwood and J. D. Kennedy, *J. Organomet. Chem.*, 1983, **249**, 11.
196. N. N. Greenwood, J. D. Kennedy, M. Thornton-Pett and J. D. Woollins, *J. Chem. Soc., Dalton Trans.*, 1985, 2397.
197. M. Bown, X. L. R. Fontaine, N. N. Greenwood, P. MacKinnon, J. D. Kennedy and M. Thornton-Pett, *J. Chem. Soc., Dalton Trans.*, 1987, 2781.
198. I. D. Burns, A. F. Hill, A. R. Thompsett, N. W. Alcock and K. S. Claire, *J. Organomet. Chem.*, 1992, **425**, C8.
199. N. W. Alcock, I. D. Burns, K. S. Claire and A. F. Hill, *Inorg. Chem.*, 1992, **31**, 4606.
200. J. Bould, N. N. Greenwood, J. D. Kennedy and W. S. McDonald, *J. Chem. Soc., Dalton Trans.*, 1985, 1843.
201. L. J. Guggenberger, A. R. Kane and E. L. Muetterties, *J. Am. Chem. Soc.*, 1972, **94**, 5665.
202. C. E. Housecroft, B. A. M. Shaykh, A. L. Rheingold and B. S. Haggerty, *Inorg. Chem.*, 1991, **30**, 125.
203. C. E. Housecroft, S. M. Owen, P. R. Raithby and B. A. M. Shaykh, *Organometallics*, 1990, **9**, 1617.
204. B. S. Haggerty, C. E. Housecroft, A. L. Rheingold and B. A. M. Shaykh, *J. Chem. Soc., Dalton Trans.* 1991, 2175.
205. K. J. Haller, E. L. Andersen and T. P. Fehlner, *Inorg. Chem.*, 1981, **20**, 309.

206. B. H. Goodreau, L. R. Orlando and J. T. Spencer, *J. Am. Chem. Soc.*, 1992, **114**, 3827.
207. L. Zhang, Y. Huang, C. Xinlan and J. He, 'Gaodeng Xuexaio Huaxue Xuebao', 1985, **6**, 872 (*Chem. Abstr.*, 1986, **105**, 191 365)
208. T. G. Hill, R. A. Godfroid, J. P. White, III, S. G. Shore, *Inorg. Chem.*, 1991, **30**, 2952.
209. G. B. Jacobsen, J. H. Morris and D. Reed, *J. Chem. Res.(S)*, 1982, 319.
210. C. E. Housecroft, *Inorg. Chem.*, 1986, **25**, 3108.
211. J. M. Gromek, and J. Donohue, *Cryst. Struct. Commun.* 1981, **10**, 849.
212. M. Bown, N. N. Greenwood and J. D. Kennedy, *J. Organomet. Chem.* 1986, **309**, C67.
213. J. Bould, N. N. Greenwood and J. D. Kennedy, *J. Chem. Soc., Dalton Trans.*, 1982, 481.
214. P. D. Grebenik, M. L. H. Green, M. A. Kelland, J. B. Leach and P. Mountford, *J. Chem. Soc., Chem. Commun.*, 1990, 1234.
215. R. L. DeKock and T. P. Fehlner, *Polyhedron*, 1982, **1**, 521.
216. E. D. Jemmis, P. N. V. P. Kumar and G. N. Sastry, *Polyhedron*, 1990, **9**, 2359.
217. T. L. Venable and R. N. Grimes, *Inorg. Chem.*, 1982, **21**, 887.
218. T. L. Venable, E. Sinn and R. N. Grimes, *J. Chem. Soc., Dalton Trans.*, 1984, 2275.
219. T. L. Venable, C. T. Brewer and R. N. Grimes, *Inorg. Chem.*, 1985, **24**, 4751.
220. C. T. Brewer and R. N. Grimes, *J. Am. Chem. Soc*, 1985, **107**, 3552.
221. T. L. Venable, W. C. Hutton and R. N. Grimes, *J. Am. Chem. Soc.*, 1984, **106**, 29.
222. P. D. Grebenik, M. L. H. Green, M. A. Kelland, J. B. Leach and P. Mountford, *J. Chem. Soc., Chem. Commun.*, 1989, 1397.
223. T. L. Venable, E. Sinn and R. N. Grimes, *Inorg. Chem.*, 1982, **21**, 904.
224. J. E. Crook, N. N. Greenwood, J. D. Kennedy and W. S. McDonald, *J. Chem. Soc., Chem. Commun.*, 1982, 383.
225. D. E. Coons and D. F. Gaines, *Inorg. Chem.*, 1985, **24**, 3774.
226. L. Borodinsky and R. N. Grimes, *Inorg. Chem.*, 1982, **21**, 1921.
227. P. Brint, W. K. Pelin and T. R. Spalding, *J. Chem. Soc., Dalton Trans.*, 1981, 546.
228. J. Bould, J. E. Crook, N. N. Greenwood and J. D. Kennedy, *J. Chem. Soc., Chem. Commun.*, 1983, 951.
229. J. Bould, J. E. Crook, N. N. Greenwood and J. D. Kennedy, *J. Chem. Soc., Dalton Trans.*, 1991, 185.
230. J. D. Kennedy, *Main Group Metal Chemistry*, 1989, **12**, 149.
231. M. B. Fischer, D. F. Gaines and J. A. Ulman, *J. Organomet. Chem.* 1982, **231**, 55.
232. P. K. Rush and L. Barton, *Polyhedron*, 1985, **4**, 1741.
233. N. W. Alcock, L. Parkhill and M. G. H. Wallbridge, *Acta Crystallogr.*, 1985, **C41**, 716.
234. D. F. Gaines, K. M. Coleson and J. C. Calabrese, *Inorg. Chem.*, 1981, **20**, 2185.
235. B. H. Goodreau, R. L. Ostrander and J. T. Spencer, *Inorg. Chem.*, 1991, **30**, 2066.
236. W. K. Pelin, T. R. Spalding and R. P. Brint, *J. Chem. Res. (S)* 1982, 120.
237. L. Barton and D. K. Srivastava, *Organometallics*, 1991, **10**, 2982.
238. N. N. Greenwood, M. J. Hails, J. D. Kennedy and W. S. McDonald, *J. Chem. Soc., Dalton Trans.*, 1985, 953.
239. M. A. Beckett, J. E. Crook, N. N. Greenwood and J. D. Kennedy, *J. Chem. Soc., Dalton Trans.*, 1986, 1879.
240. M. A. Beckett, N. N. Greenwood, J. D. Kennedy, P. A. Salter and M. Thornton-Pett, *J. Chem. Soc., Chem. Commun.*, 1986, 556.
241. D. M. Vinitskii, *et al.*, *Russ. J. Inorg. Chem.*, 1984, **29**, 984.
242. X. L. R. Fontaine, N. N. Greenwood, J. D. Kennedy, I. Macpherson and M. Thornton-Pett, *J. Chem. Soc., Chem. Commun.*, 1987, 476.
243. J. Bould, J. E. Crook, N. N. Greenwood, J. D. Kennedy and W. S. McDonald, *J. Chem. Soc., Chem. Commun.*, 1982, 346.
244. M. Bown, X. L. R. Fontaine, N. N. Greenwood, J. D. Kennedy and M. Thornton-Pett, *J. Organomet. Chem.*, 1986, **315**, C1.
245. X. L. R. Fontaine, N. N. Greenwood, J. D. Kennedy, P. MacKinnon and M. Thornton-Pett, *J. Chem. Soc., Dalton Trans.*, 1988, 2059.
246. J. Bould, J. E. Crook, N. N. Greenwood and J. D. Kennedy, *J. Chem. Soc., Dalton Trans.*, 1984, 1903.
247. J. Bould, N. N. Greenwood and J. D. Kennedy, *J. Chem. Soc., Dalton Trans.*, 1984, 2477.
248. K. Nestor, X. L. R. Fontaine, N. N. Greenwood, J. D. Kennedy and M. Thornton-Pett, *J. Chem. Soc., Dalton Trans.*, 1989, 1465.
249. M. A. Beckett, M. Bown, X. L. R. Fontaine, N. N. Greenwood, J. D. Kennedy and M. Thornton-Pett, *J. Chem. Soc., Dalton Trans.*, 1988, 1969.
250. R. P. Micciche, J. J. Briguglio and L. G. Sneddon, *Inorg. Chem.*, 1984, **23**, 3992.
251. T. L. Venable, E. Sinn and R. N. Grimes, *Inorg. Chem.*, 1982, **21**, 895.
252. X. L. R. Fontaine, H. Fowkes, N. N. Greenwood, J. D. Kennedy and M. Thornton-Pett, *J. Chem. Soc., Dalton Trans.*, 1987, 1431.
253. S. K. Boocock, N. N. Greenwood, M. J. Hails, J. D. Kennedy and W. S. McDonald, *J. Chem. Soc., Dalton Trans.*, 1981, 1415.
254. S. R. Bunkhall, X. L. R. Fontaine, N. N. Greenwood, J. D. Kennedy and M. Thornton-Pett, *J. Chem. Soc., Dalton Trans.*, 1990, 73.
255. M. A. Beckett, J. E. Crook, N. N. Greenwood and J. D. Kennedy, *J. Chem. Soc., Dalton Trans.*, 1984, 1427.
256. A. J. Wynd and A. J. Welch, *J. Chem. Soc., Dalton Trans.*, 1990, 2803.
257. R. Ahmad, J. E. Crook, N. N. Greenwood, J. D. Kennedy and W. S. McDonald, *J. Chem. Soc., Chem. Commun.*, 1982, 1019.
258. E. J. Ditzel, X. L. R. Fontaine, N. N. Greenwood, J. D. Kennedy, Z. Sisan, M. Thornton-Pett. *J. Chem. Soc., Chem. Commun.*, 1989, 1762.
259. E. J. Ditzel *et al.*, *J. Chem. Soc., Chem. Commun.*, 1990, 1741.
260. X. L. R. Fontaine, N. N. Greenwood, J. D. Kennedy, P. MacKinnon and M. Thornton-Pett, *J. Chem. Soc., Dalton Trans.*, 1988, 2809.
261. D. G. Meina and J. H. Morris, *J. Chem. Soc., Dalton Trans.*, 1985, 1903.
262. N. N. Greenwood, J. D. Kennedy, I. MacPherson and M. Thornton-Pett, *Z. Anorg. Allg. Chem.*, 1986, **540**, 45.
263. M. A. Beckett, N. N. Greenwood, J. D. Kennedy and M. Thornton-Pett, *J. Chem. Soc., Dalton Trans.*, 1986, 795.
264. M. A. Beckett, N. N. Greenwood, J. D. Kennedy and M. Thornton-Pett, *J. Chem. Soc., Dalton Trans.*, 1985, 1119.
265. J. Bould, J. E. Crook, N. N. Greenwood, J. D. Kennedy and M. Thornton-Pett, *J. Chem. Soc., Dalton Trans.*, 1990, 1441.

266. S. K. Boocock, J. Bould, N. N. Greenwood, J. D. Kennedy and W. S. McDonald, *J. Chem. Soc., Dalton Trans.*, 1982, 713.
267. M. Bown, X. L. R. Fontaine, N. N. Greenwood, J. D. Kennedy and P. MacKinnon, *J. Chem. Soc., Chem. Commun.*, 1987, 817.
268. M. Bown, X. L. R. Fontaine, N. N. Greenwood and J. D. Kennedy, *J. Organomet. Chem.*, 1987, **325**, 233.
269. E. J. Ditzel, X. L. R. Fontaine, N. N. Greenwood, J. D. Kennedy and M. Thornton-Pett, *J. Chem. Soc., Chem. Commun.*, 1989, 1262.
270. M. Bown *et al.*, *J. Chem. Soc., Dalton Trans.*, 1988, 2597.
271. X. L. R. Fontaine, H. Fowkes, N. N. Greenwood, J. D. Kennedy and M. Thornton-Pett, *J. Chem. Soc., Dalton Trans.*, 1986, 547.
272. X. L. R. Fontaine and J. D. Kennedy *J. Chem. Soc., Chem. Commun.*, 1986, 779.
273. K. Nestor, X. L. R. Fontaine, N. N. Greenwood, J. D. Kennedy and M. Thornton-Pett, *J. Chem. Soc., Chem. Commun.*, 1989, 455.
274. Faridoon, M. Magrath, T. R. Spalding, J. D. Kennedy, X. L. R. Fontaine and M. R. Thornton-Pett, *J. Chem. Soc., Dalton Trans.*, 1990, 1819.
275. M. Bown, X. L. R. Fontaine, N. N. Greenwood, J. D. Kennedy and M. Thornton-Pett, *J. Chem. Soc., Dalton Trans.*, 1988, 925.
276. E. J. Ditzel *et al.*, *J. Chem. Soc., Chem. Commun.*, 1990, 1692.
277. K. Nestor, X. L. R. Fontaine, N. N. Greenwood, J. D. Kennedy and M. Thornton-Pett, *J. Chem. Soc., Chem. Commun.*, 1989, 455.
278. J. Bould, N. N. Greenwood, J. D. Kennedy and W. S. McDonald, *J. Chem. Soc., Chem. Commun.*, 1982, 465.
279. J. Bould, J. E. Crook, N. N. Greenwood, J. D. Kennedy and W. S. McDonald, *J. Chem. Soc., Chem. Commun.*, 1983, 949.
280. J. Bould, P. Brint, J. D. Kennedy, and M. Thornton-Pett, *Acta Crystallogr., Sect. C*, 1990, **46**, 1010.
281. J. E. Crook, M. Elrington, N. N. Greenwood, J. D. Kennedy, M. Thornton-Pett and J. D. Woollins, *J. Chem. Soc., Dalton Trans.*, 1985, 2407.
282. M. Bown, X. L. R. Fontaine, N. N. Greenwood and J. D. Kennedy, *Z. Anorg. Allg. Chem.*, 1991, **598**, 45.
283. J. Bould, P. Brint, X. L. R. Fontaine, J. D. Kennedy and M. Thornton-Pett, *J. Chem. Soc., Chem. Commun.*, 1989, 1763.
284. J. Bould, N. N. Greenwood and J. D. Kennedy, *J. Chem. Soc., Dalton Trans.*, 1990, 1451.
285. J. P. White, H-B. Deng and S. G. Shore, *J. Am. Chem. Soc.*, 1989, **111**, 8946.
286. M. Bown, X. L. R. Fontaine, N. N. Greenwood, J. D. Kennedy and M. Thornton-Pett, *J. Chem. Soc., Dalton Trans.*, 1987, 1169.
287. T. D. Getman, J. A. Krause and S. G. Shore, *Inorg. Chem.*, 1988, **27**, 2398.
288. X. L. R. Fontaine, H. Fowkes, N. N. Greenwood, J. D. Kennedy and M. Thornton-Pett, *J. Chem. Soc., Chem. Commun.*, 1985, 1165.
289. X. L. R. Fontaine, H. Fowkes, N. N. Greenwood J. D. Kennedy and M. Thornton-Pett, *J. Chem. Soc., Dalton Trans.*, 1987, 2417.
290. X. L. R. Fontaine, H. Fowkes, N. N. Greenwood, J. D. Kennedy and M. Thornton-Pett, *J. Chem. Soc., Chem. Commun.*, 1985, 1722.
291 J. D. Kennedy, personal communication. Cited in Ref. 276 as crystallographically characterized and 'to be published'.
292. S. K. Boocock, N. N. Greenwood, J. D. Kennedy, W. S. McDonald and J. Staves, *J. Chem. Soc., Dalton Trans.*, 1981, 2573.
293. S. A. Macgregor *et al.*, *J. Chem. Soc., Dalton Trans.*, 1991, 3317.
294. S. A. Macgregor, L. J. Yellowlees and A. J. Welch, *Acta Crystallogr., Sect. C*, 1990, **46**, 551.
295. S. A. Macgregor, J. A. Scanlan, L. J. Yellowlees, and A. J. Welch, *Acta. Crystallogr., Sect. C*, 1991, **47**, 513.
296. S. A. Macgregor, L. J. Yellowlees and A. J. Welch, *Acta. Crystallogr., Sect. C*, 1990, **46**, 1399.
297. Y. M. Cheek, J. D. Kennedy and M. Thornton-Pett, *Inorg. Chim. Acta*, 1985, **99**, L43.
298. J. E. Crook, N. N. Greenwood, J. D. Kennedy and W. S. McDonald, *J. Chem. Soc., Dalton Trans.*, 1984, 2487.
299. M. A. Beckett, N. N. Greenwood, J. D. Kennedy and M. Thornton-Pett, *Polyhedron*, 1985, **4**, 505.
300. A. J. Wynd, S. E. Robins, D. A. Welch and A. J. Welch, *J. Chem. Soc., Chem. Commun.*, 1985, 819.
301. A. J. Wynd, A. J. McLennan, D. Reed and A. J. Welch, *J. Chem. Soc., Dalton Trans.*, 1987, 2761.
302. A. J. Wynd and A. J. Welch, *J. Chem. Soc., Chem. Commun.*, 1987, 1174.
303. A. J. Wynd, A. J. Welch and R. V. Parish, *J. Chem. Soc., Dalton Trans.*, 1990, 2185.
304. D. F. Gaines and G. A. Steehler, *J. Chem. Soc., Chem. Commun.*, 1982, 122.
305. D. A. Kleier, D. A. Dixon and W. N. Lipscomb, *Inorg. Chem.*, 1978, **17**, 166.
306. J. E. Crook, M. Elrington, N. N. Greenwood, J. D. Kennedy and J. D. Wollins, *Polyhedron*, 1984, **3**, 901.
307. M. Elrington, N. N. Greenwood, J. D. Kennedy and M. Thornton-Pett, *J. Chem. Soc., Dalton Trans.*, 1986, 2277.
308. M. Elrington, N. N. Greenwood, J. D. Kennedy and M. Thornton-Pett, *J. Chem. Soc., Chem. Commun.*, 1984, 1398.
309. X. L. R. Fontaine, N. N. Greenwood, J. D. Kennedy, M. Thornton-Pett and Z. Peiju, *J. Chem. Soc., Chem. Commun.*, 1987, 1717.
310. H. Fowkes, N. N. Greenwood, J. D. Kennedy and M. Thornton-Pett, *J. Chem. Soc., Dalton Trans.*, 1986, 517.
311. M. Bown, X. L. R. Fontaine, N. N. Greenwood, J. D. Kennedy and M. Thornton-Pett, *J. Chem. Soc., Dalton Trans.*, 1990, 3039.
312. X. L. R. Fontaine, H. Fowkes, N. N. Greenwood, J. D. Kennedy and M. Thornton-Pett, *J. Chem. Soc., Dalton Trans.*, 1987, 2417.
313. E. J. Ditzel, X. L. R. Fontaine, N. N. Greenwood, J. D. Kennedy and M. Thornton-Pett, *J. Chem. Soc., Chem. Commun.*, 1989, 1115.
314. M. Bown, J. M. Waters, X. L. R. Fontaine and J. D. Kennedy, *J. Organonmet. Chem.*, 1990, **390**, 379.
315. A. M. Barriola, *Acta Cient. Venezolana*, 1981, **32**, 72.
316. D. M. Schubert, C. B. Knobler, P. A. Wegner and M. F. Hawthorne, *J. Am. Chem. Soc.*, 1988, **110**, 5219.
317. L-L. Ng, B. K. Ng, K. Shelly, C. B. Knobler and M. F. Hawthorne. *Inorg. Chem.*, 1991, **30**, 4278.
318. H. Mongeot, A. Tangi and B. Bonnetot, *Inorg. Chim. Acta*, 1989, **161**, 143.
319. M. Elrington, N. N. Greenwood, J. D. Kennedy and M. Thornton-Pett, *J. Chem. Soc., Dalton Trans.*, 1987, 451.
320. M. A. Beckett, J. E. Crook, N. N. Greenwood and J. D. Kennedy, *J. Chem. Soc., Chem. Commun.*, 1983, 1228.
321. M. A. Beckett, J. E. Crook, N. N. Greenwood, J. D. Kennedy and W. S. McDonald, *J. Chem. Soc., Chem. Commun.*, 1982, 552.

322. S. O. Kang and L. G. Sneddon, *Inorg. Chem.*, 1988, **27**, 587.
323. X. L. R. Fontaine, N. N. Greenwood, J. D. Kennedy, P. I. MacKinnon and M. Thornton-Pett, *J. Chem. Soc., Chem. Commun.*, 1986, 1111.
324. Y. M. Cheek, N. N. Greenwood, J. D. Kennedy and W. S. McDonald, *J. Chem. Soc., Chem Commun.*, 1982, 80.
325. U. Siriwardane, S. S. C. Chu, N. S. Hosmane, H. Zhu and G. Zhang, *Acta Crystallogr., Sect. C*, 1989, **45**, 333.

9
Transition Metal Metallacarbaboranes

RUSSELL N. GRIMES
University of Virginia, Charlottesville, VA, USA

9.1 INTRODUCTION 373
9.1.1 Scope of Field 373
9.1.2 Trends Since 1982 374
9.2 GENERAL FEATURES OF STRUCTURE, BONDING, AND REACTIVITY 374
9.2.1 Carbaboranes as Ligands: Comparison with Hydrocarbons 374
9.2.2 Metallacarbaboranes as Clusters and Sandwich Complexes 375
9.2.3 General Properties of Metallacarbaboranes 376
9.3 METHODS OF SYNTHESIS 376
9.4 REACTIONS OF METALLACARBABORANES 377
9.5 MONOCARBON METALLACARBABORANES 377
9.6 DICARBON METALLACARBABORANES 381
9.6.1 Small (6- and 7- vertex) Cages 381
9.6.2 Intermediate (8- to 11-vertex) Cages 394
9.6.2.1 Synthesis 394
9.6.2.2 Structures and bonding 398
9.6.2.3 Wedged metallacarbaboranes 398
9.6.3 Icosahedral (12-vertex) Cages 399
9.6.3.1 Synthesis 411
9.6.3.2 Cage rearrangement and modification 414
9.6.3.3 Spectroscopic and other physical properties 414
9.6.3.4 Applications 415
9.6.4 13-vertex Cages 417
9.7 TRICARBON METALLACARBABORANES 419
9.8 TETRACARBON METALLACARBABORANES 421
9.9 REFERENCES 425

9.1 INTRODUCTION

9.1.1 Scope of Field

The growth of metallacarbaborane chemistry following *COMC-I* has been spectacular. In that 1982 publication, the chapter dealing with this area[1a] covered metallaboranes *and* metallacarbaboranes, and included both main group and transition metals. Reflecting the expansion of the field since that time, the present survey is concerned solely with transition metal metallacarbaboranes, with separate coverage of metallacarboranes, *exo*-polyhedral metallated carbaboranes, and main group metallacarbaboranes in other chapters of this work. A number of reviews on various aspects of metallacarbaborane chemistry have appeared,[1b–v] including an excellent recent treatment of transition metal and lanthanide complexes by Saxena and Hosmane[1b] and a wide-ranging discussion of structure and bonding by Yanovskii.[1u] The March/April 1992 "all-boron" issue of *Chemical Reviews* contains several articles that relate directly or indirectly to metallacarbaboranes, including discussions of cage structure and bonding,[2] practical applications,[3] and multidecker sandwich chemistry.[1c]

This chapter is concerned with carbaboranes (borane clusters containing skeletal carbon atoms) in which one or more transition metal, lanthanide element, or actinide element atoms are present in the cage framework, except for the special case of metal sandwich complexes of planar organoboron ligands. The latter class of compounds, which can be viewed as metal–carbon–boron clusters and thus fall under the formal definition of metallacarbaboranes, are treated in Chapter 5, this volume, and will not be covered here. The large family of *exo*-metal carbaborane derivatives, in which metal atoms reside only outside the cage framework, are included in Chapter 6, this volume. There are inevitable areas of overlap with other chapters in this volume, including "hybrid" metal sandwich complexes that incorporate both carbaborane and organoborane ligands, and metallacarbaboranes that contain transition metal and main group heteroatoms in the same polyhedral framework (a relatively rare group). Species of these types are included here.

9.1.2 Trends Since 1982

Since the mid-1980s, carbaborane ligands have proved to be synthetically valuable tools in transition metal organometallic chemistry, assuming the role of hydrocarbon groups, such as $[\eta\text{-}C_5H_5]^-$ (Cp), $[\eta\text{-}C_5Me_5]^-$ (Cp*), and arenes, in metal sandwich complexes. As carbaboranes have become more accessible, their extraordinary versatility, tailorability, and stabilizing power has drawn increasing attention from organometallic chemists. At present, the most extensively utilized carbaborane substrates in synthesis are *C*- and *B*-substituted derivatives of *nido*-2,3-$[C_2B_4H_6]^{2-}$, *nido*-$[C_2B_9H_{11}]^{2-}$ ("dicarbollide"), and to a lesser extent *nido*-$[C_2B_{10}H_{12}]^{2-}$, the last two also finding extensive use as parent species. Metal coordination to the open faces of these ligands generates *closo*-polyhedral MC_2B_n clusters having respectively 7, 12, and 13 vertices; such complexes are synthons for a variety of other compound types, including polymetallated clusters, multidecker sandwiches, and linked-cage systems.

These carbaborane ligands, and most others, have planar or nearly planar binding faces and are isolobal and isoelectronic analogues of Cp, Cp*, and arenes; this is an important advantage in their systematic exploitation in organometallic synthesis. Metallacarbaborane chemistry since 1982 has been largely centered on complexes of C_2B_n ligands in which n is 3, 4, 9, or 10, and the vast majority of complexes reported in this period are in this group. Much of the interest in these compounds has focused on three areas: development of metallacarbaborane catalysts and reagents, exploration of new cluster systems, and the designed synthesis of multicluster–multisandwich oligomers and polymers. Work in the first two categories was well under way at the time of the earlier review, and has advanced considerably since then; the last-named area has emerged almost entirely since the mid-1980s.

Important progress has been achieved as well on other aspects of metallacarbaborane chemistry, including monocarbon, tricarbon, and tetracarbon systems. Investigations of these areas, discussed below, have uncovered novel boron cluster chemistry and opened possible new roles for metallacarbaboranes in planned synthesis.

9.2 GENERAL FEATURES OF STRUCTURE, BONDING, AND REACTIVITY

9.2.1 Carbaboranes as Ligands: Comparison with Hydrocarbons

The properties of carbaborane ligands and their binding to metal centers have been treated extensively in the 1982 chapter and in numerous subsequent reviews.[1] Typically, an n-membered open face on the carbaborane unit—for example, a planar C_2B_3 ring—supplies n atomic orbitals that interact with metal orbitals (usually three) of appropriate symmetry to form a molecular orbital set that contains three bonding MOs. The six electrons required for a filled-shell configuration are normally provided by the ligand anion, for example, $[C_2B_4H_6]^{2-}$ or $[C_2B_9H_{11}]^{2-}$ (Figure 1). While the ligand can also be represented as a neutral four-electron donor (e.g., $C_2B_4H_6$), ionic descriptions involving anionic ligands and metal cations are conventional (and more realistic in terms of the actual synthesis of most metallacarbaboranes from anionic carbaboranes).

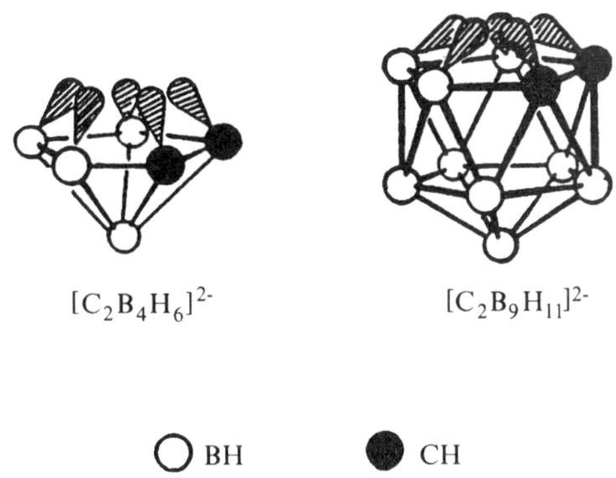

Figure 1 Structures of $[C_2B_4H_6]^{2-}$ and $[C_2B_9H_{11}]^{2-}$ ligands showing atomic orbitals available for metal sandwich bonding.

As this picture indicates, carbaborane ligands are isolectronic and isolobal with common hydrocarbon ligands such as Cp, Cp*, and arenes, and MO descriptions of metal–carbaborane interactions are closely analogous to those of metallocenes and arene–metal sandwiches. Quantitative molecular orbital treatments of structure and bonding in metallacarbaboranes have been published in recent years,[4] and advances in cluster electron-counting theory have appeared.[5] In general, owing to the low electronegativity of boron relative to carbon, face-bonding (i.e., η^5 or η^6) interactions of carbaborane units with metal ions are more covalent, hence stronger, than are metal–hydrocarbon sandwich bonds; this is a major reason for the typically higher (often much higher) chemical and thermal stabilities of metallacarbaboranes compared to their metallocene and metal–arene cousins. The formal dinegative charge possessed by dicarbon carbaborane ligands is a further advantage, since high metal oxidation states are readily stabilized. For example, robust complexes of cobalt(IV) and nickel(IV) are well known in metallacarbaborane chemistry.

Carbaborane ligands can be used exclusively, as in *commo*-metallacarbaboranes such as $[Co(C_2B_9H_{11})_2]^-$, but are more commonly employed in combination with other groups to create mixed-ligand complexes, for example, 1,2,3-Cp*Co($Et_2C_2B_4H_4$). Mixed-ligand species account for most of the advances in metallacarbaborane chemistry since the mid-1980s and now constitute a large, new subfield of organometallic chemistry that has significantly broadened the horizons of that discipline.

9.2.2 Metallacarbaboranes as Clusters and Sandwich Complexes

Throughout their history metallacarbaboranes have been described both as polyhedral metal–boron–carbon clusters and as metal–carbaborane complexes. This "duality" is more than a semantic distinction, in that the synthesis and chemical properties of metallacarbaboranes relate closely both to boron clusters and to transition metal sandwiches. An illustrative example is afforded by indenylferracarbaboranes of the type 1,2,3-(C_9H_7)Fe($Et_2C_2B_4H_4$), which undergo both C_5-ring to C_6-ring metal migration (transition metal organometallic chemistry) and expansion of the FeC_2B_4 cage via insertion of nickel (typical of polyhedral boranes).[6]

Both models are firmly grounded in theory: metallacarbaboranes are formally derived from carbaboranes via replacement of one or more BH vertices with isoelectronic metal–ligand units (e.g., CpCo), and generally are amenable to treatment by the polyhedral skeletal electron-counting theory,[4e,4m,7] but are equally well treated as metal–ligand complexes. However, in this author's opinion there are, or ought to be, some commonsense limits to the use of the term "carbaborane ligand." Such a designation is clearly warranted for carbaborane anions that are known to exist in both metal-complexed and uncomplexed form, for example, $[Et_2C_2B_4H_4]^{2-}$ and $[C_2B_9H_{11}]^{2-}$. It is also useful to treat as ligands certain other commonly appearing carbaborane moieties such as $[Et_2C_2B_3H_5]^{2-}$ and $[Et_2C_2B_3H_3]^{4-}$ that are found only as metal complexes, but are close steric and isoelectronic analogues of hydrocarbon ligands (in this case Cp and Cp*). For metallacarbaborane cages in which the nonmetal portion (the boron–carbon skeleton) has no independent existence and no special structural significance *per se*, referring to such fragments as "ligands" is not particularly useful.

A frequent source of confusion is the label "electron-deficient" which is still, unfortunately, often misapplied to boranes, carbaboranes, and their metal derivatives. Since virtually all isolable boron-containing clusters are stable, filled-shell electronic systems with low electron affinity, the continued application of this term to such compounds is inappropriate, outmoded, and can be seriously misleading (as attested by flagrant errors in some textbooks, e.g., "... boranes are reactive because they are electron-deficient"). The molecule BH_3, which has a vacant bonding orbital, *is* electron-deficient; its dimer, B_2H_6, and its isoelectronic analogue, C_2H_4, have no vacant bonding orbitals and are in no meaningful sense deficient in electrons, nor are most of the thousands of carbaboranes and other boron cluster compounds. "Electron-deficiency" was originally applied to boranes at a time when their structures and bonding were not understood and the reactivity of the lower boron hydrides was misinterpreted. Through tradition and inertia, the misuse of this term continues, but it should be understood that in this context it is strictly jargon signifying nonclassical bonding, and conveys no truth related to actual electronic properties.

9.2.3 General Properties of Metallacarbaboranes

With few exceptions, metallacarbaboranes are stable, colored crystalline solids that are soluble in organic solvents, survive exposure to air indefinitely, and melt well above room temperature. Many of them undergo reversible chemical or electrochemical oxidation and reduction. For the more extensively studied classes of metallacarbaboranes, notably the *nido*-$M(C_2B_3)$, *closo*-$M(C_2B_4)$, and *closo*-$M(C_2B_9)$ systems, methods have been developed for the controlled introduction of substituent groups at cage boron and carbon locations, and to a lesser extent, for replacement or modification of metal-bound ligands.[1a–c] Where decomposition or instability is encountered, it is usually associated with the *exo*-polyhedral (noncarbaborane) ligands or substituents, rather than with the metallacarbaborane cage. The robustness of these compounds under varying conditions provides a basis for extensive development of this chemistry and for their exploitation as reagents and as building blocks for the construction of large systems.[1c] Indeed, few other classes of organometallics are as well suited for such purposes as the metallacarbaboranes, a point to be elaborated in subsequent sections.

9.3 METHODS OF SYNTHESIS

Earlier reviews[1a–c] described a number of methods for the insertion of metals into carbaborane substrates, only one of which has been extensively used in recent years, namely, the reaction of metal cations or cationic metal–ligand groups with carbaborane anions (e.g., Equations (1)–(4)). Other approaches, such as the reaction of metal reagents with neutral carbaboranes, insertion of hydrocarbons into metallaboranes, and metal replacement and transfer, are less controllable and frequently require special conditions or less accessible reagents. Serendipitous findings, such as unexpected carbon insertion into metallaborane cages, have also produced metallacarbaboranes. While these reactions do not qualify as general synthetic routes, they often provide fascinating new insights into cluster chemistry and a number of them are described throughout this chapter.

$$2\,[C_2B_9H_{12}]^- + CoCl_2 \xrightarrow[\text{aq. NaOH}]{\text{air}} [Co(C_2B_9H_{11})_2]^- + Co^0 \quad (1)$$

$$[2,3\text{-}Et_2C_2B_4H_5]^- + [C_5Me_5]^- + CoCl_2 \xrightarrow[\text{THF}]{\text{air}} 1,2,3\text{-}Cp^*Co(Et_2C_2B_4H_4) + H_2 \quad (2)$$

$$[2,3\text{-}Et_2C_2B_4H_4]^{2-} + 0.5\,[(\text{arene})RuCl_2]_2 \longrightarrow 1,2,3\text{-}(\text{arene})Ru(Et_2C_2B_4H_4) \quad (3)$$

$$Tl^+[TlC_2B_9H_{11}]^{2-} + [(C_6Me_6)_2Co]^+ \longrightarrow \textit{endo}\text{-}(C_6Me_6H)Co(C_2B_9H_{11}) \quad (4)$$

Metal cation–carbaborane anion reactions have been employed to prepare a wide range of metallacarbaborane products, often (but not always) in good to excellent yields. This approach emphasizes the organometallic nature of these species and, for the most part, draws on standard organometallic metal reagents and procedures. Little is known of the detailed mechanism of metal complexation at the carbaborane face, but in general metallation evidently occurs initially at an electron-rich B–B edge, followed by binding of the metal to the remainder of the face. In support of this idea, it is well established that the presence of *exo*-polyhedral boron-attached substituents on C_2B_3 open faces (especially in small carbaborane anions[1c]) can profoundly influence the metal attack via electronic effects on the carbaborane framework. Specific examples of such effects and their application in synthesis are seen in multidecker sandwich formation, discussed in Section 9.6.1

Arene–metal–carbaborane sandwiches can often be synthesized from a carbaborane anion, a metal cation, and the arene mono- or dianion, if the latter species (for example, naphthaleneide or fluoreneide) can be generated in solution.[8] Alternatively, (arene)Fe($R^1R^2C_2B_4H_4$) sandwiches are preparable via thermal displacement of the cyclooctatriene ligand from 1,2,3-(η^6-C_8H_{10})Fe($R^1R^1C_2B_4H_4$) complexes with the desired arene, as shown in Equation (5).[1c,1h,9] Similarly, photolysis of 3,1,2-$(CO)_3Fe(C_2B_9H_{11})$ in benzene affords $(C_6H_6)Fe(C_2B_9H_{11})$.[10] For the synthesis of arene–metallacarbaboranes of second- and third-row transition metals, (arene)metal dihalides can be employed to insert the "preassembled" arene–metal unit into the cage,[11] as in Equations (3) and (4).

$$(\eta^6\text{-}C_8H_{10})Fe(Et_2C_2B_4H_4) + \text{naphthalene} \longrightarrow (\eta^6\text{-}C_{10}H_8)Fe(Et_2C_2B_4H_4) + C_8H_{10} \quad (5)$$

Metal vapor reactions with carbaboranes were studied in detail by Sneddon and co-workers and led to the discovery of several new metallacarbaborane cage systems and isomers.[12] This approach has the capability of generating novel species that do not form under more conventional conditions; however, it is applicable to only a limited range of metals and allows no control over the products formed, which are usually obtained as mixtures with low individual yields.

Metal-promoted oxidative fusion[1a,1c,1h,1q,13] can generate tetracarbon dimetallacarbaboranes from small metallacarbaborane substrates at ambient temperatures (e.g., Equation (6)). Although reactions of this kind are well established in boron cluster chemistry and have been demonstrated with binary boron hydrides, carbaboranes, and their metal complexes, fusion has not been observed in metallacarbaboranes other than MC_2B_3 species. A more general route to tetracarbon metallacarbaborane clusters,[14] discussed further below, is via metal insertion into tetracarbon carbaborane anions such as $[Et_4C_4B_8H_8]^{2-}$.

$$2\,[Cp^*Co(Et_2C_2B_3H_4)]^- \xrightarrow[\text{THF}]{[O]} Cp^*_2Co_2(Et_4C_4B_6H_6) + H_2 \quad (6)$$

9.4 REACTIONS OF METALLACARBABORANES

Chemical modification of metallacarbaborane complexes can occur via attack of a reagent species on the carbaborane ligand, the external ligand(s), or the metal center. Intramolecular rearrangement can also occur, usually requiring an elevated temperature although there are notable exceptions such as oxidative fusion reactions and "*exo–endo*" metal migration in certain dicarbollide complexes. In general, reactions of the *exo*-polyhedral (nonboron) ligands are characteristic of the groups involved, but they are often modified by the presence of the carbaborane cage. Processes occurring at the metal are common in complexes having labile ligands (especially important, of course, in catalytically active species), but can also involve metal protonation and deprotonation, and the above-mentioned oxidative fusion reaction, in which the metal is expelled from the cage entirely. Reactions on the carbaborane framework itself can take place at B–H or C–H vertices, and are often important synthetic routes to substituted derivatives. Removal of B–H vertices via strong bases (or in some cases acids) is widely employed to create open faces that, in turn, can be coordinated to a new metal center or modified via the introduction of substituents.

A type of reaction that is characteristic of boron clusters is reductive cage opening and its inverse, oxidative cage closure. These processes, studied extensively by Hawthorne and co-workers in the early days of metallacarbaborane chemistry and extended to smaller polyhedra by Grimes *et al.*, can be exploited to increase or decrease the cluster size (polyhedral expansion and contraction), but have received less attention in recent years.

These reaction types were reviewed earlier[1a] and a general treatment will not be given here. However, some of them have found extensive application in the tailored synthesis of metallacarbaborane-based organometallic systems and are discussed in later sections of this chapter.

9.5 MONOCARBON METALLACARBABORANES

Relatively few monocarbon complexes are known (Table 1), prepared in most cases from neutral or CB_nH_m carbaboranes or their bridge-deprotonated anions via treatment with metal reagents (Equations (7)–(9)).[15,16,28] The crystallographically determined structure of the $PtCB_7$ cluster[16] is shown in Figure 2.

Similar reactions of these and other monocarbon carbaboranes, for example, *arachno*-$[CB_8H_{13}]^-$ and $[CB_9H_{12}]^-$, have given the 10- and 11-vertex products listed in Table 1. In some cases,[17,19] cage

Table 1 Monocarbon metallacarbaboranes (since 1982).

Compound[a]	*Data*[b]	*References*
9-Vertex polyhedra, 1 metal atom[c]		
Iridium		
$(CO)(Ph_3P)_2Ir(CB_7H_8)$	X, B, H	15
Platinum		
arachno-6,4-$(Ph_3P)_2Pt(CB_7H_{11})$	X, B(2D), H(2D), P	16
10-Vertex polyhedra, 1 metal atom		
Cobalt		
$[2,1\text{-}CpCo(CB_8H_9)]^-$	B, H	17,19
$[2,1\text{-}CpCo(CB_8H_9)]^-$	X	18
Nickel		
6,1-$CpNi(CB_8H_9)$	B, H	17,19
10,1-$CpNi(CB_8H_9)$	B, H	17,19
Ruthenium		
arachno-9,6-$CpH(PPh_3)Ru(CB_8H_{12})$	X, B, H, P	20
nido-9,6-$(CO)_3Ru(CB_8H_{10})PPh_3$	X, B, H, P	20
Osmium		
nido-9,6-$(CO)_3Os(CB_8H_{10})PPh_3$	X, B, H, P	20
Iridium		
$(Ph_3P)HIr(MeCO_2)(CB_8H_7)PPh_3$	X	21
2,10-$(Ph_3P)_2HIr(CB_8H_8)$-1-PPh_3	X, B, H	22
nido-9,6-$(Ph_3P)_2HIr(CB_8H_{10})$	X	23
arachno-9,6-$(CO)(Ph_3P)_2HIr(CB_8H_{12})$ (symmetric and asymmetric isomers)	X(asym), B, H(2D), P	24
Platinum		
nido-9,6-$(Ph_3P)_2Pt(CB_8H_{12})$	X	25,26
nido-9,6-$[(o\text{-}C_6H_4)Ph_2P]_2Pt(CB_8H_{10})$	X	27
arachno-9,6-$(PMe_2Ph)_2Pt(CB_8H_{12})$	X	28
10-Vertex polyhedra, 2 metal atoms		
Nickel		
$Cp_2Ni_2(CB_7H_8)$	S	19
6,7,1-$Cp_2Ni_2(CB_7H_8)$	X	29
11-Vertex polyhedra, 1 metal atom		
Iron		
$[2,1\text{-}CpFe(CB_9H_{10})]^-$	X	30
Cobalt		
2,1-$CpCo[(H_3N)(CB_9H_9)]$	B, H	31
$\{2,1\text{-}CpCo[(H_2N)CB_9H_9]\}^-$	B, H	31
2,1-$CpCo[(NMe_2H)CB_9H_9]$	B, H	31
$CpCo[(H_3N)CB_9H_9]$	B	32
$\{CpCo[(H_2N)CB_9H_9]\}^-$	B	32
$CpCo[(Me_2HN)CB_9H_9]$	B	32
Rhodium		
2,1-$Cp^*Rh[(NHEt)CB_9H_8]$-7-CNEt	X, B, H	33
nido-8,7,9-$(Ph_3P)_2(H)Rh(SCB_8H_{10})$	X, B, P	34
2,1,3-$(Ph_3P)_2(H)Rh(SCB_8H_9)$	X, B, P	34
Platinum		
nido-8,7-$(PMe_2Ph)_2Pt(CB_9H_{11})$	X, B(2D), H(2D), P, Pt	28
12-Vertex polyhedra, 1 metal atom		
Cobalt		
$[2,1\text{-}CpCo(CB_{10}H_{11})]^-$	X	30
2,1-$CpCo\{[(TMS)_2CH]CB_{10}H_9\}$-*n*-$SMe_2$ (*n* = 7, 12)	B, H, I	35
2,1-$CpCo\{[(TMSi)_2CH]CB_{10}H_9\}(C_5H_4)CoCp$	B, H	35
2,1-$(MeC_6H_5)Co\{[(TMS)_2CH]CB_{10}H_{10}\}$	X, B, H, I	35
Nickel		
2,1,7-$(Ph_3P)BrNi(Me\text{-}PCB_9H_{10})$		36
2,1-$CpNi\{[(TMS)_2CH]CB_{10}H_{10}\}$	X, B, H, I	35
Molybdenum		
$[2,1\text{-}(CO)_3(Ph_3P)Mo(HO)(CB_{10}H_{10})]^-$	B(2D), H, I	37
Rhodium		
2,1-$(Ph_3P)_2HRh(CB_{10}H_{10})$-1-$NH_2$	H, P, I	38
2,1-$Br(Ph_3P)Rh[(NHR^1R^2)CB_{10}H_{10}]$ (R^1 = H, $CH_2CH{=}CHMe$; R^2 = $CH_2CH{=}CHMe$)	X, H, P, I	39
$\{[2,1\text{-}(Ph_3P)(\mu\text{-}NH_2)Rh(CB_{10}H_{10})]_2(\mu\text{-}H)\}^-$	X, H, P	38
$\{[2,1\text{-}(Ph_3P)(\mu\text{-}NH_2)RH(CB_{10}H_{10})]_2(\mu\text{-}Br)\}^-$	X, H, P, I	39
2,1,7-$(Ph_3P)_2Rh(Me\text{-}PCB_9H_{10})$		36
Ruthenium		
arachno-$(C_6Me_6)Ru(Me\text{-}NCB_9H_{12})$	X, B, H	33

[a] Clusters are *closo* except when otherwise designated; metals are listed in order of increasing atomic number. Cp = η-C_5H_5, Cp* = η-C_5Me_5.
[b] X = x-ray diffraction, B = ^{11}B NMR, H = ^1H NMR, P = ^{31}P NMR, 2D = two-dimensional (COSY) NMR, I = infrared, S = mass spectral data only.
[c] Metal atoms in metallacarbaborane framework.

$$[CB_7H_8]^- + Ir(CO)Cl(PPh_3)_2 \xrightarrow{87\%} closo\text{-}(CO)(PPh_3)_2Ir(CB_7H_8) \quad (7)$$

$$nido\text{-}1\text{-}CB_8H_{12} + Pt(PPh_3)_4 \xrightarrow{26\%} arachno\text{-}6,4\text{-}(PPh_3)_2Pt(CB_7H_{11}) \quad (8)$$

$$[CB_9H_{14}]^- + PtCl_2(PMe_2Ph)_2 \xrightarrow{46\text{–}64\%} nido\text{-}8,7\text{-}(PMe_2Ph)_2Pt(CB_9H_{11}) + Cl^- + H_2 + HCl \quad (9)$$

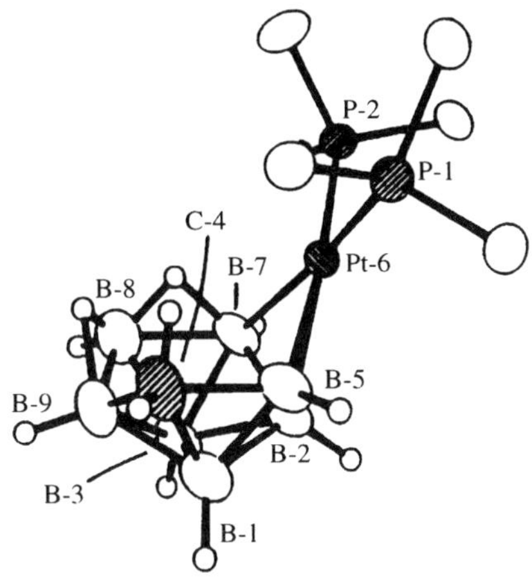

Figure 2 Structure of *arachno*-6,4-$(Ph_3P)_2PtCB_7H_{11}$ (reproduced by permission of the Royal Society of Chemistry from *J. Chem. Soc., Dalton Trans.*, 1993, 1269).

rearrangement has been observed following metal insertion; thermal isomerization has also been used to generate new isomers.[17,27] As a class these compounds are structurally well characterized, most of them conforming to the geometries predicated from skeletal electron-count theory. The decaboranelike structure of the *nido*-ruthenacarbaborane in Figure 3 is typical.[20]

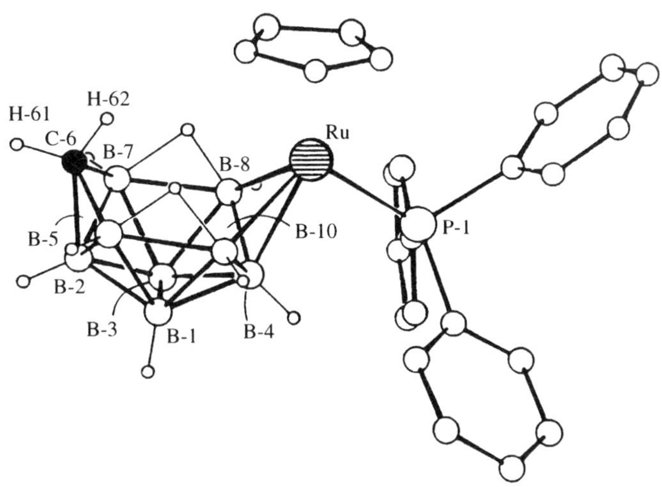

Figure 3 Structure of *arachno*-9,6-$CpH(PPh_3)Ru(CB_8H_{12})$ (hydrogen on ruthenium not shown) (reproduced by permission of the Royal Society of Chemistry from *J. Chem. Soc., Dalton Trans.*, 1987, 2793).

With one exception, all of these complexes have a single metal atom in the cage framework; the sole example of a dimetallic species is 6,7,1-$Cp_2Ni_2(CB_7H_8)$, prepared from $Na^+[CB_8H_{13}]^-$ and nickelocene,[19] and structurally defined as a *closo* 10-vertex polyhedron with the nickels occupying adjacent vertices.[29]

Monocarbon metallacarbaboranes have been synthesized via carbon insertion into metallaborane clusters (Equation (10)).[33] Also unusual is the preparation of thiametallacarbaboranes from a thiacarbaborane (Equation (11)).[34]

Insertion of metals into *nido* 11-vertex $CB_{10}H_{11}$ carbaboranes offers a straightforward route to *closo* 12-vertex MCB_{10} clusters, and a number of these have been prepared since 1982 (Table 1). The methods employed include reactions of neutral carbaboranes with $CpCo(CO)_2$, $CpNi(CO)_2$, or cobalt vapor.[35] The *nido*-monocarbaborane $H_3NCB_{10}H_{12}$ on treatment with $RhCl(PPh_3)_3$ in the presence of $Bu^t_4N^+OH^-$ gave

$$RNC{\equiv}C + nido\text{-}6\text{-}Cp^{*}Rh(B_9H_{13}) \longrightarrow 6,9\text{-}(RNC)_2\text{-}6\text{-}CpRh(B_9H_{11}) \xrightarrow[\text{toluene}]{\text{heat}} 1,2\text{-}Cp^{*}Rh[(NHR)CB_9H_9\text{-}7\text{-}CNEt] \quad (10)$$

$$CSB_8H_{12} + RhCl(PPh_3)_3 \longrightarrow nido\text{-}8,7,9\text{-}(PPh_3)_2(H)Rh(SCB_8H_{10}) + 2,1,3\text{-}(PPh_3)_2(H)Rh(SCB_8H_9) \quad (11)$$

the dimeric NH_2-bridged anion[38] $[\{2,1\text{-}(Ph_3P)(\mu\text{-}NH_2)Rh(CB_{10}H_{10})\}_2(\mu\text{-}H)]^-$. Reactions of this ion with alkenyl halides gave monomeric complexes characterized as $2,1\text{-}Br(Ph_3P)Rh[(NHR^1R^2)CB_{10}H_{10}]$ (R^1 = H, $CH_2CH{=}CHMe$; $R^2 = CH_2CH{=}CHMe$) via Br-bridged dimeric intermediates, one of whose structures is shown in Figure 4.[39] The monomeric rhodium complexes have 16-electron metal centers and are related to catalytically active dicarbon hydridorhodacarbaborane clusters discussed in Section 9.6.3.4.

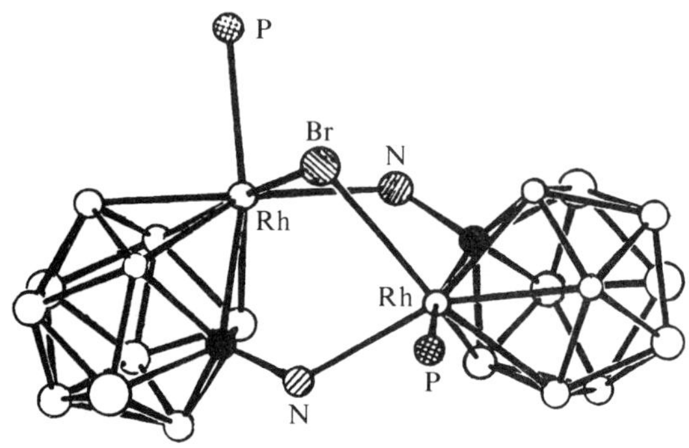

Figure 4 Structure of $[2,1\text{-}(Ph_3P)(\mu\text{-}NH_2)Rh(CB_{10}H_{10})]_2(\mu\text{-}Br)$. Cage carbon atoms are solid black (reprinted with permission from *Inorg. Chem.*, 1993, **32**, 3393; copyright 1993 American Chemical Society).

In a very different approach, *arachno*-$[B_{10}H_{14}]^{2-}$ was combined with $Mo(CO)_3(MeCN)_3$ and PPh_3 to give *closo*-$[2,1\text{-}(CO)_3(Ph_3P)Mo(HO)(CB_{10}H_{10})]^-$ in 50% yield, the product probably forming via intramolecular CO transfer from the metal reagent.[37]

Several 12-vertex monocarbon heterometallacarbaboranes that incorporate nitrogen or phosphorus in the cluster framework have been described (Equations (12) and (13)).[33,36] The molecular structure of the unusual 12-vertex *arachno*-ruthenacarbaazaborane is depicted in Figure 5.

$$MeN{\equiv}C + nido\text{-}6\text{-}Cp^{*}Rh(B_9H_{12}\text{-}OMe) \longrightarrow 6,9\text{-}(MeNC)_2\text{-}6\text{-}CpRh(B_9H_{11}) \longrightarrow arachno\text{-}(C_6Me_6)Ru(Me\text{-}NCB_9H_{12}) \quad (12)$$

$$2,1,7\text{-}(PPh_3)_2Rh(MePCB_9H_{10}) \xleftarrow[\text{THF}]{(PPh_3)_3RhCl} [1,7\text{-}PCB_9H_{10}Me]^- \xrightarrow[\text{THF}]{(PPh_3)_3NiBr_2} 2,1,7\text{-}(PPh_3)BrNi(MePCB_9H_{10}) \quad (13)$$

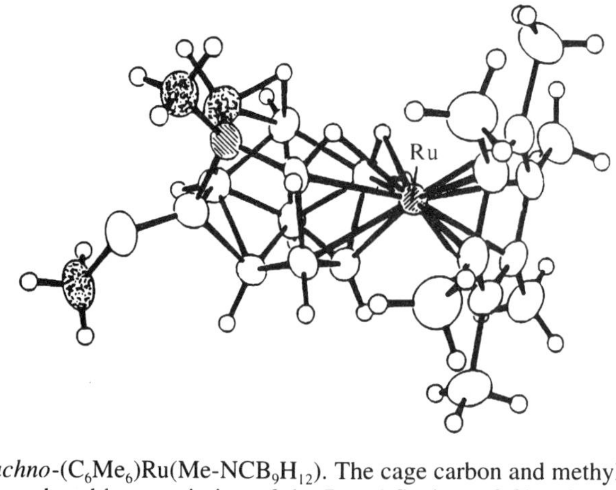

Figure 5 Structure of *arachno*-$(C_6Me_6)Ru(Me\text{-}NCB_9H_{12})$. The cage carbon and methyl atoms are speckled and the cage nitrogen is shaded (reproduced by permission of the Royal Society of Chemistry from *J. Chem. Soc., Chem. Commun.*, 1990, 1741).

9.6 DICARBON METALLACARBABORANES

9.6.1 Small (6- and 7- vertex) Cages

The chemistry of *closo*-MC_2B_4 and $M^1C_2B_3M^2$ clusters, prepared from the exceedingly versatile *nido*-2,3-$[R^1R^2C_2B_4H_5]^-$ and *nido*-2,3-$[R^1R^2C_2B_4H_4]^{2-}$ anions (Section 9.3), has developed substantially since 1982 and now forms a large subfield of organometallic sandwich chemistry.[1b–d] Some appreciation of the growth of this area can be gained from the listing of characterized compounds in Table 2. The development of small metallacarbaborane chemistry[1b,c,1h,1t] has recently been accelerated by improvements in the synthesis of *nido*-2,3-$R^1R^2C_2B_4H_6$ carbaboranes, which can now be prepared on a mole scale[96b] under mild conditions[96a] from B_5H_9 and alkynes. As shown schematically in Figure 6, mixed-ligand *closo*-sandwich complexes of the type 1,2,3-$LM(R^1R^2C_2B_4H_6)$, in which L is an organic ligand and M^1 is usually Fe, Co, or Ru, can be converted to open-faced (*nido*) species via base-promoted removal of the apex BH ("decapitation") followed by bridge deprotonation and introduction of a second metal–ligand unit (M^2 = Fe, Co, Ru, Rh, or Os) to create triple-decker sandwiches.[11c,d,57] Alternatively, coordination of two double-decker units to a third metal forms tetradecker stacks[89,91] that are slightly bent in the middle (8–15°), as shown by x-ray diffraction studies on several complexes. Figure 7(a) depicts the structure of a CoNiCo diacetyl sandwich.[89] In the mid-1990s, this synthetic approach was extended to the preparation of a five-cobalt hexadecker sandwich (Figure 7(b)).[95] Multidecker stacks of this type are of interest as synthons for multi-sandwich systems, as discussed below.

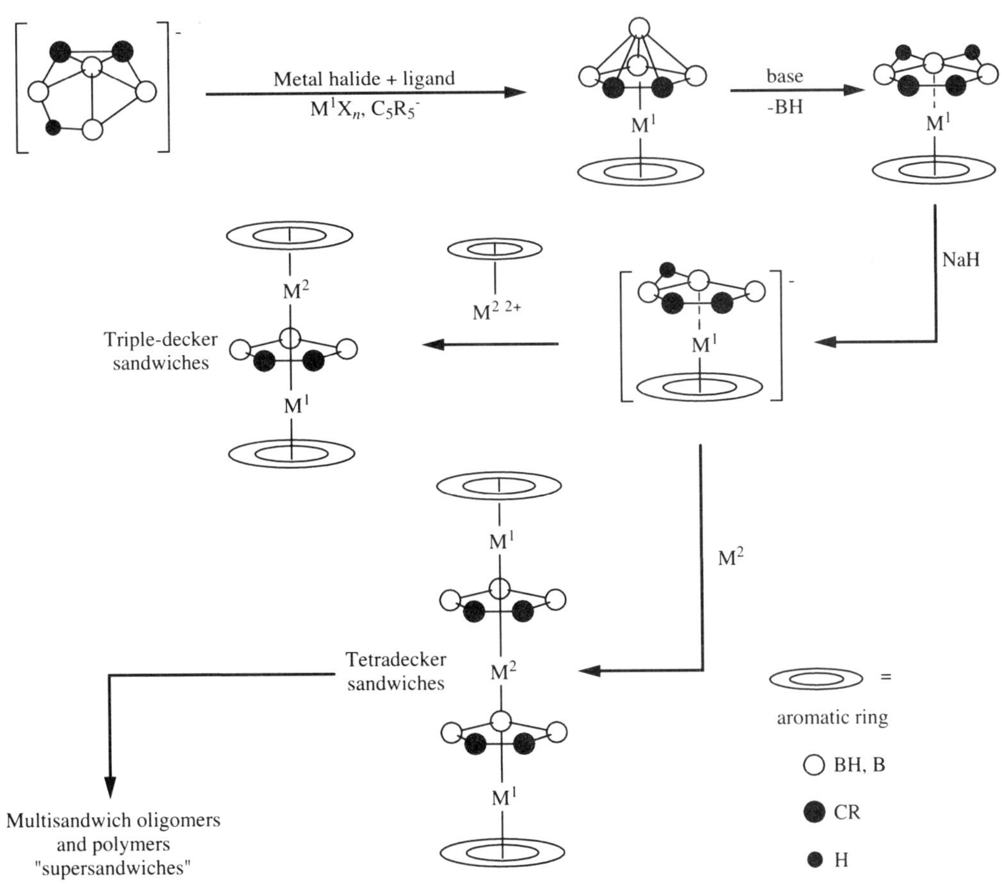

Figure 6 Synthetic routes to metallacarbaborane sandwich complexes.

"Hybrid" triple- and tetradecker complexes that combine carbaborane (C_2B_3, C_2B_4) and diborole (C_3B_2) ligands have been prepared via stacking reactions employing the 1,2,3,5-$[CpCo(Et_2MeC_3B_2Et_2)]^-$ substrate (Equation (14)).[58,88]

Metal complexation reactions of *nido*-2,3-$LM(R^1R^2C_2B_3H_4)^-$ anions are sensitive to the presence of substituents on the C_2B_3 open face (especially on boron), and methods for introducing organic and inorganic groups at specific boron positions have been reported.[43,44,51,52] Varying the R^1 and R^2 groups

Table 2 Six- and seven-vertex dicarbon metallacarbaboranes (since 1982).

Compound[a]	*Data*[b]	*References*
6-Vertex polyhedra, 1 metal atom[c,d]		
Iron		
2,1,6-$(CO)_3Fe(C_2B_3H_5)$	E	40
nido-$(C_6H_6)Fe(Et_2C_2B_3H_5)$	B, H, I	41
nido-$(C_6Me_6)Fe(Et_2C_2B_3H_5)$	B, I	41
nido-$(C_6H_5\text{-})Fe[\text{-}(CH_2)_3C_2B_3H_6]$	B, H, I	9b
nido-$(MeC_6H_5)Fe(Et_2C_2B_3H_5)$	B	12a
nido-$(C_6Me_6)Fe(Et_2C_2B_3H_5)$	B(2D)	42
nido-$(C_6Me_6)Fe(Et_2C_2B_3H_3\text{-4-I-5-Me})$	B, H, I	43
nido-$(C_6Me_6)Fe(Et_2C_2B_3H_{5-n}R_n)$ (R_n = 5-Me; 4,5-Me_2; 4,5,6-Me_3; 5-CH_2Ph)	B, H, I	43
nido-$(C_6Me_6)Fe(Et_2C_2B_3H_4\text{-5-}CH_2C_6H_4CH_2Br)$	B, H, I	44
p-[*nido*-$(C_6Me_6)Fe(Et_2C_2B_3H_4\text{-5-}CH_2)]_2C_6H_4$	B, H, I	44
nido-$(C_6Me_6)Fe(Et_2C_2B_3H_4)\text{-5-}(CH_2C_6H_5)Fe(Et_2C_2B_4H_4)$	B, H, I	44
nido-$(C_6Me_6)Fe(Et_2C_2B_3H_4)\text{-5-}(CH_2C_6H_5)Fe(Et_2C_2B_3H_5)$	B, H, I	44
p-[*nido*-$(C_6Me_6)Fe(Et_2C_2B_3H_4\text{-5-}CH_2)]_2(C_6H_4)Fe(Et_2C_2B_4H_4)$	B, H, I	44
nido-$(\eta^6\text{-}C_8H_{10})Fe(Me_2C_2B_3H_5)$ (C_8H_{10} = cyclooctatriene)	B, H	9a
nido-$(\eta^6\text{-}C_9H_8)Fe(Et_2C_2B_3H_5)$ (C_9H_8 = indene)	B, H, C, I, E	6
nido-$(Me_4C_4S)Fe(Et_2C_2B_3H_5)$	B, H, I, U	45
nido-$(Et_4C_4B_8H_7)FeH(Et_2C_2B_3H_4)$ (*B–B*)	B, I	14c
Cobalt		
nido, closo-$Cp_2Co^+[(Et_2C_2B_3H_5)Co(Et_2C_2B_4H_4)]^-$	B, H, I	46
nido, closo-$Cp_2Co^+[(Et_2C_2B_3H_4\text{-5-Me})Co(Et_2C_2B_4H_4)]^-$	B, H, I	46
nido-$Cp^*Co(C_2B_3H_7)$	B, H, I	47
nido-$Cp^*Co(C_2B_3H_6Cl)$	H, I	47
nido-$Cp^*Co(ClC_2B_3H_6)$	B, H, I	47
nido-$Cp^*Co(Me_2C_2B_3H_5)$	H, I	48
nido-$Cp^*Co(Me_2C_2B_3H_4\text{-5-Me})$	B, H, I	48
nido-$Cp^*Co(Me_2C_2B_3H_3\text{-4,6-}Me_2)$	H, I	48
nido-$Cp^*Co(Me_2C_2B_3H_2\text{-4,5,6-}Me_3)$	X, B, H, I	48
nido-$Cp^*Co(Me_2C_2B_3H_4\text{-5-Cl})$	H, I	48
nido-$Cp^*Co(Me_2C_2B_3H_3\text{-4-Me-5-Cl})$	H, I	48
nido-$Cp^*Co(Me_2C_2B_3H_2\text{-4,6-}Me_2\text{-5-Cl})$	B, H, I	48
nido-$Cp^*Co(Me_2C_2B_3H_4\text{-}\mu\text{-HgCl})$	B(2D)	42
nido-$(5\text{-}B_9H_{12}C_4H_8O)Co(Me_2C_2B_3H_5)$	X	49
nido-$CpCo(Et_2C_2B_3H_5)$	B, H, I, U	11c
nido-$Cp^*Co(Et_2C_2B_3H_5)$	B, H, I, U	11c
nido-$[Cp^*Co(Et_2C_2B_3H_3)]_2$ (bent isomer)	X, B, H, U, I	13b
nido-$[Cp^*Co(Et_2C_2B_3H_3)]_2$ (planar isomer)	X, B, H	13b
nido-$[Cp^*Co(Et_2C_2B_3H_4)]_2$ (*B–B*)	B, H	13b
nido-$Cp^*Co(Et_2C_2B_3H_4\text{-5-R})$ (R = Me, Et, $CH_2C_6H_4CH_2Br$, $CH_2C_6H_4Me$)	B, H, I	43
nido-$Cp^*Co(Et_2C_2B_3H_{5-n}R_n)$ (R_n = 4,5-Me_2; 4,5,6-Me_3)	B, H, I	43
nido-$Cp^*Co(Et_2C_2B_3H_3\text{-4-I-5-R})$ (R = Me, Et)	B, H, I	43
nido-$Cp^*Co[(Et_2C_2B_3H_4\text{-5-}(C{=}CH_2)O(C{=}O)Me]$	X, B, H, C, I	50
nido-$Cp^*Co[(Et_2C_2B_3H_4\text{-5-}(C{=}O)Me]$	B, H, C, I	50
nido-$Cp^*Co(Et_2C_2B_3H_4\text{-5-R})$ (R = $CH_2C_6H_5$, CH_2CF_3, $CH_2C_6F_5$, $CH_2C_2H_5$, $CHMe_2$, CH_2CN, $CH_2CH_2NH_2$, CH_2OMe, $CH_2C(O)Me$, CH_2CCH, CH_2CCMe, CH_2CCCH_2Cl, $CH_2CH{=}CH_2$, $CH_2C(O)OMe$, $CH_2C(O)OH$, $CH_2OC(O)Me$, CH_2OH, C_4H_8O)	B, H, C, I, U	51
nido-$Cp^*Co(Et_2C_2B_3H_4\text{-5-R})$ (R = $C({=}CH_2)OC(O)Me$, $C(O)Me$, $C(O)CH_2C(O)OMe$, Cl, Br, I, CN)	B, H, C, I, U	52
nido-$Cp^*Co(Et_2C_2B_3H_4\text{-5-}C({=}CH_2)OC(O)Me$	X	52
nido-$Cp^*Co(Et_2C_2B_3H_4\text{-4,6-R})$ (R = $C(O)CF_3$, Cl, Br, I, CN)	B, H, C, I, U	52
nido-$Cp^*Co(Et_2C_2B_3H_2\text{-}Cl_3)$	B, H, C, I, U	52
nido-$Cp^*Co(Et_2C_2B_3H_3\text{-5-Me-4(6)-Cl})$	B, H, C, I, U	52
nido-$Cp^*Co(Et_2C_2B_3H_2\text{-5-Me-4,6-}Cl_2)$	B, H, C, I, U	52
nido-$Cp^*Co(Et_2C_2B_3H_3\text{-4,6-}X_2)$ (X = Cl, Br, I)	B, H, C, I, U	52
nido-$Cp^*Co(Et_2C_2B_3H_3\text{-5-}C_4H_8O)$	U	52
nido-$Cp^*Co(Et_2C_2B_3H_3\text{-4,6-}NO_2)$	H	52
nido-$Cp^*Co(Et_2C_2B_3H_3\text{-4,6-}CHOHCMe_3)$	B, H, I, U	52
nido-$Cp^*Co(Et_2C_2B_3H_3\text{-4,6-R})$ (R = $(CH_2)_4OC(O)CMe_3$, $(CH_2)_4OH$)	B, H, I, U	52
nido-$Cp^*Co(Et_2C_2B_3H_3C(O)CMe_3)$	B, H, I, U	52
nido-$(5\text{-}B_9H_{12}\text{-}C_4H_8O)Co(Et_2C_2B_3H_5)$	B, H, I	53
nido-$(6\text{-}B_9H_{12}\text{-}C_4H_8O)Co(Et_2C_2B_3H_5)$	B, H, I	53
nido,closo-$(Et_2C_2B_4H_4)(Et_2C_2B_3H_5)Co_2(C_5H_4)_2$ ($(C_5H_4)_2$ = fulvalene)	B, H, I, U	11d
nido-$[(Et_2C_2B_3H_5)Co(C_5H_4)]_2$ ($(C_5H_4)_2$ = fulvalene)	B, H, I, U	11d
nido-$[(Et_2C_2B_3H_5)Co(C_5Me_4)]_2(C_6H_4)$	X, B, H, I, U	11d
nido-$[(Et_2C_2B_3H_5)Co(C_5Me_4)]_2(C_6H_4)_2$	B, H, I, U	11d
p-[*nido*-$Cp^*Co(Et_2C_2B_3H_4\text{-5-}CH_2)]_2C_6H_4$	B, H, I	44

Table 2 (continued)

Compound[a]	*Data*[b]	*References*
o-[*nido*-Cp*Co($Et_2C_2B_3H_4$-5-CH_2)]$_2C_6H_4$	B, H, I	44
nido-($PhCH_2C_5Me_4$)Co($Et_2C_2B_3H_4$-5-R) (R = H, CH_2Ph)	B, H, I	54
nido,closo-($PhCH_2C_5Me_4$)Co($Et_2C_2B_3H_4$-5-CH_2Ph)Fe($Et_2C_2B_4H_4$)	B, H, I	54
nido,closo-($Et_2C_2B_4H_4$)Fe($PhCH_2C_5Me_4$)Co($Et_2C_2B_3H_4$-5-CH_2Ph)	B, H, I	54
nido-($Et_2C_2B_3H_3$-4-R^1-5-R^2)Co(C_5Me_4)C_6H_4(C_5Me_4)Co($Et_2C_2B_3H_3$-4-R^3-5-R^4) (R^1,R^2,R^3,R^4 = H, Cl, Br, Me)	B, H	55
nido,closo-M[($Et_2C_2B_3H$-4-R^1-5-R^2)Co(C_5Me_4)C_6H_4(C_5Me_4) Co($Et_2C_2B_3H_4R^3$)]$_2$ (R^1,R^2,R^3 = H, Cl, Me; M = Co, CoH, Ni)	B, H, U, FAB	55
nido,closo-Co[($Et_2C_2B_3H$-4,5-Cl_2)Co(C_5Me_4)C_6H_4(C_5Me_4)Co($Et_2C_2B_3H_5$)]$_2$	X	55
nido,closo-[($Et_2C_2B_3H_4Cl$)Co(C_5Me_4)]$_2$[(C_5Me_4)Co($Et_2C_2B_3H_2Cl$) Co($Et_2C_2B_3H_2Cl$)Co(C_5Me_4)]$_n$(C_6H_4)$_4$ (*n* = 3, 5)	FAB	55
nido[($Et_2C_2B_3H_4$-5-Me)Co(C_5H_4)]$_2$	B, H	55
nido-[($Et_2C_2B_3H_3$-4,6-Br_2)Co(C_5H_4)]$_2$	B, H	55
nido-[($Et_2C_2B_3H_2$-4,6-Br_2-5-Me)Co(C_5H_4)]$_2$	X, H	55
nido,closo-[($Et_2C_2B_3H_4$-5-Me)Co(C_5H_4)$_2$Co($Et_2C_2B_3H_2$-5-Me)]$_2$Co	B, H, U, FAB	55
nido,closo-[($Et_2C_2B_3H_4$-5-Me)Co(C_5H_4)$_2$$Co_3$($Et_2C_2B_3H_2$-5-Me)$_2$($C_5H_4$)]$_2$	H, U, FAB	559
nido-($NC_4Me_2R_2$)Co($Et_2C_2B_3H_5$) (R = Me, H)	B, H, I	56
nido-($NC_4Me_2R_2$)Co($Et_2C_2B_3H_5$) (R = Me, H)	B, H, I, U	57
nido-(PC_4Me_4)Co($Et_2C_2B_3H_5$)	B, H, P, I, U	57
nido-($R_2C_2B_3H_5$)Co($Et_2MeC_3B_2Et_2$)CoCp (R = Et, CH_2Ph)	B, H, I, E	58
p-nido-{Cp*Co[(TMS)$C_2B_3H_5$]CH_2}$_2C_6H_4$	B, H, I	47
nido-Cp*Co[(TMS)$_2C_2B_3H_5$]	H, I	47
nido-Cp*Co[(TMS)$_2C_2B_3H_4Cl$]	H, I	47
nido-Cp*Co[(TMS)$C_2B_3H_6$]	H, I	47
nido-Cp*Co[(TMS)$C_2B_3H_5Cl$]	H, I	47
nido-(C_5Ph_5)Co[(TMS)$_2C_2B_3H_5$]	H, I	47
Ruthenium		
nido-(MeC_6H_4-$CHMe_2$)Ru($Et_2C_2B_3H_5$)	B, H, I, U	11c
p-[*nido*-($MeC_6H_4CHMe_2$)Ru($Et_2C_2B_3H_4$-5-CH_2)]$_2C_6H_4$	B, H, I	44
p-[*nido*-($MeC_6H_4CHMe_2$)Ru($Et_2C_2B_3H_4$-5-CH_2)]$_2$(C_6H_4)Fe($Et_2C_2B_4H_4$)	B, H, I	44
nido-($MeC_6H_4CHMe_2$)Ru($Et_2C_2B_3H_{5-n}R_n$) (R_n = 5-Me; 4,5-Me_2; 4,5,6-Me_3)	B, H, I	43
Rhodium		
nido-Cp*Rh($Et_2C_2B_3H_5$)	B, H, I, E	58
6-Vertex polyhedra, 3 metal atoms		
$(CO)_9Fe_3MeHC_2BH$	X, B, H, C, I	59
7-Vertex polyhedra, 1 metal atom		
Titanium		
(η-C_8H_8)Ti($Et_2C_2B_4H_4$)	X, B, H, I	60,61
(η-C_8H_8)Ti($Et_2C_2B_4H_4$)	H, detailed synthesis	62
(η-C_8H_8)Ti($Et_2C_2B_4H_4$)	B(solvent shifts)	63
(η-C_8H_8)Ti($Et_2C_2B_4H_3$-5-I	B, I	60
(η-C_8H_8)Ti($Et_2C_2B_4H_2$-4,5-I_2	B, I	60
Vanadium		
(η-C_8H_8)V($Et_2C_2B_4H_4$)	X, M, g, I	60,61
Chromium		
(η-C_7H_7)Cr($Et_2C_2B_4H_4$)	X, I	60
Cr[(TMS)$_2C_2B_4H_4$]$_2$	X, B, H, C, I	64
Cr[(TMS)$_2C_2B_4H_4$]$_2^-$	X*, M	64
Cr[(TMS)$_2C_2B_4H_4$]$_2$	X, B, H, C, I, M	4k
Cr[(TMS)$RC_2B_4H_4$]$_2^-$ (R = TMS, Me, H)	X(TMS, H), M(TMS),I	4k
Manganese		
Mn_3[(TMS)$_2C_2B_4H_4$]$_4^{3-}$	X, I, M, g*	65
Iron		
Cp_2Co^+[HFe($Et_2C_2B_4H_4$)$_2$]$^-$	B, H	46
Cp_2Co^+[Fe($Et_2C_2B_4H_4$)$_2$]$^-$	H, U	46
$H_2Fe(Et_2C_2B_4H_4)_2$	B, detailed synthesis	66
$H_2Fe(Me_2C_2B_4H_4)_2$	m	67
$H_2Fe[(PhCH_2)_2C_2B_4H_4]_2$	B, H, I	68
H_2Fe($Et_2C_2B_4H_3$-4-Et)$_2$	B, H, I	69
$H_2Fe[Pr^i_2C_2B_4H_4]_2$	B	70
$Fe_2[PhCH_2)_2C_2B_4H_5]_2$	B, H, I	9b
$(Ph_2PCH_2)_2ClFe(Et_2C_2B_4H_4)$	X, M, I, U, g	71
$CpFe(C_2B_4H_6)$	E	40
Cp*FeH($Et_2C_2B_4H_4$)	B, H, E, U	72
Cp*Fe($Et_2C_2B_4H_4$)	B, H, E, U, g, C	72
Cp*Fe($Et_2C_2B_4H_4$)$^-$	B, H, E, C	72

Table 2 (continued)

Compound[a]	*Data*[b]	*References*
$[(Et_2C_2B_4H_4)FeH(C_5Me_4)]_2C_6H_4$	X, B, H, E, U	72
$(Et_2C_2B_4H_4)FeH(C_5Me_4)C_6H_4(C_5Me_4)Fe(Et_2C_2B_4H_4)$	B, H, M, g, E	72
$[(Et_2C_2B_4H_4)Fe(C_5Me_4)]_2C_6H_4$	B, H, M, g, E	72
$LFe(Et_2C_2B_4H_4)$ ($L = C_6Me_6, C_6Me_5H, C_6Me_4H_2, C_6Me_3H_3$, 1,4-$C_6Me_2H_4$, C_6MeH_5, C_6H_6)	E	73
$(C_6H_6)Fe(Bu_2C_2B_4H_4)$	E	73
$(C_6Me_6)Fe(Et_2C_2B_4H_4)^+$	g	73
$(C_6Me_6)Fe(Et_2C_2B_4H_4)^-$	g	73
$(C_6Me_3H_3)Fe(Et_2C_2B_4H_4)^-$	g	73
$(C_6H_6)Fe(Et_2C_2B_4H_4)$	B, detailed synthesis	62
$(C_6H_6)Fe(Me_2C_2B_4H_4)$	B, H, C, I	74
$(C_6H_6)Fe(Me_2C_2B_4H_4)$	X, B, H, C	9a
$[(C_6H_6)Fe(Et_2C_2B_4H_3]_2CHMeCH_2$	X, B, H, I	41
$(C_6H_6)Fe(Et_2C_2B_4H_4)$	E	75
$(C_6H_6)Fe(Et_2C_2B_4H_4)$	B(solvent shifts)	63
$(C_6H_5Me)Fe(Et_2C_2B_4H_4)$	X, B, H, C, I	12a
$(C_6H_5C_6H_5)Fe(Et_2C_2B_4H_4)$	X, B, H, C, I	41
$(C_6H_5C_6H_5)[Fe(Et_2C_2B_4H_4]_2$	H, I	41
$(C_6H_5)Fe[(CH_2)_3C_2B_4H_5]$	X, B, H, I	9b
$(C_6H_5C_6H_5)[Fe(Et_2C_2B_4H_4)]_2$	E	75
$(p\text{-}MeC_6H_4Me)Fe(C_2B_4H_4)$	B, I	76
p-[*nido*-$(MeC_6H_4CHMe_2)Ru(Et_2C_2B_3H_4\text{-}5\text{-}CH_2)]_2(C_6H_4)Fe(Et_2C_2B_4H_4)$	B, H, I	44
$(C_6Me_3H_3)Fe(Me_2C_2B_4H_4)$	X, B, H, C, I	9a
$(C_6Me_6)Fe(Me_2C_2B_4H_4)$	X, B, H, C, I	9a
$(C_6Me_6)Fe(Me_2C_2B_4H_4)$	B, H, I	12b
$(C_6Me_6)Fe(Me_2C_2B_4H_3)(C_2Me_2H_3)$ (2 isomers)	B	12b
5-[*nido*-$(Et_2C_2B_4H_4)Fe(C_6H_5CH_2)](Et_2C_2B_3H_4)Fe(C_6Me_6)$	B, H, I	44
p-[*nido*-$(C_6Me_6)Fe(Et_2C_2B_3H_4\text{-}5\text{-}CH_2)]_2(C_6H_4)Fe(Et_2C_2B_4H_4)$	B, H, I	44
$(Et_2C_2B_4H_4)Fe(PhCH_2C_5Me_4)Co(Et_2C_2B_4H_4)$	B, H, I	54
$(PhCH_2C_5Me_4)Co(Et_2C_2B_3H_4\text{-}5\text{-}CH_2Ph)Fe(Et_2C_2B_4H_4)$	B, H, I	54
$(Et_2C_2B_4H_4)Fe(PhCH_2C_5Me_4)Co(Et_2C_2B_3H_4\text{-}5\text{-}CH_2Ph)$	B, H, I	54
$[(Et_2C_2B_4H_4)Fe(PhCH_2C_5Me_4)Co]_2(Et_2C_2B_3H_3)$	B, H, I	54
$(Et_2C_2B_4H_4)Fe(PhCH_2C_5Me_4)Co(Et_2C_2B_3H_3)Co(C_5Me_4CH_2Ph)$	B, H, I	54
$(\eta^6\text{-}C_8H_{10})Fe(Me_2C_2B_4H_4)$ (C_8H_{10} = cyclooctatriene)	m	67
$(\eta^6\text{-}C_8H_{10})Fe(Me_2C_2B_4H_4)$	X, B, H, C, I	74
$(\eta^6\text{-}C_8H_{10})Fe(Et_2C_2B_4H_4)$	m	77
$(\eta^6\text{-}C_8H_{10})FePh(CH_2)_3(C_2B_4H_5)$	B, H, I	9b
$(\eta^6\text{-}C_8H_{10})FePh(CH_2)_2(C_2B_4H_5)$	B, H, I	9b
$(\eta^6\text{-}C_8H_{10})Fe[(CO)_3CrPhCH_2](PhCH_2)C_2B_4H_4)$	B, H, I	78
$(\eta^6\text{-}C_8H_{10})Fe[(CO)_3CrPhCH_2]_2C_2B_4H_4)$	B, H, I	78
$(\eta^6\text{-}C_8H_{10})Fe(Et_2C_2B_4H_4)$	B, detailed synthesis	62
$(\eta^6\text{-}C_8H_{10})Fe(Et_2C_2B_4H_4)$	B(solvent shifts)	63
$(\eta^6\text{-}C_9H_8)Fe(Et_2C_2B_4H_4)$ (C_9H_8 = indene)	X, B, H, C, I, E	6
$[(\eta^5\text{-}C_9H_7)Fe(Et_2C_2B_4H_4)]_2$	B, H, I	6
$(\eta^5\text{-}C_9H_7)FeH(Et_2C_2B_4H_4)$	B, H, C, I, E	6
$(\eta^5\text{-}C_9H_7)Fe(Et_2C_2B_4H_4)$	H, I, g, E	6
$(\eta^6\text{-}C_{10}H_{12})Fe(Et_2C_2B_4H_4)$ ($C_{10}H_{12}$ = tetrahydronaphthalene)	X, B, H, I	9c
$(\eta^6\text{-}C_{10}H_8)Fe(Et_2C_2B_4H_4)$ ($C_{10}H_8$ = naphthalene)	X, B, H, I*	9c
$(\eta^6\text{-}C_{13}H_{10})Fe[(PhCH_2)_2C_2B_4H_4]$ ($C_{13}H_{10}$ = fluorene)	B, H, I	78
$(\eta^6\text{-}C_{14}H_{10})Fe(Et_2C_2B_4H_4)$ (unsymmetrical) ($C_{14}H_{10}$ = phenanthrene)	X, B, H, I	9c
$(\eta^6\text{-}C_{14}H_{10})Fe(Et_2C_2B_4H_4)$ (symmetrical)	B, H, I	9c
$(\eta^6\text{-}C_{14}H_{12})Fe](PhCH_2)_2C_2B_4H_4]$ ($C_{14}H_{12}$ = 9,10-dihydroanthracene)	B, H, I	78
$(\eta^6\text{-}C_{14}H_{12})Fe_2[(PhCH_2)_2C_2B_4H_4]_2$	B, H, I	78
$(C_{16}H_{18})Fe(Me_2C_2B_4H_4)$	B, H, C, I	74
$(C_{16}H_{18})Fe[(PhCH_2)_2C_2B_4H_4]$	B, H, I	78
$(\eta^6\text{-}C_{16}H_{16})Fe[(PhCH_2)_2C_2B_4H_4]$ ($C_{16}H_{16}$ = [2.2]paracyclophane)	B, H, I	78
$(C_{16}H_{18})Fe(Et_2C_2B_4H_4)$	m	77
$(Et_4C_4B_8H_7)FeH_2(Et_2C_2B_4H_3)$ (*B–B*) (isomer mixture)	B	14c
$(Et_4C_4B_8H_8OC_4H_8)Fe(Et_2C_2B_4H_4)$	B, I	14c
$CpCo(Et_2MeC_3B_2Et_2)Fe(Et_2C_2B_4H_4)$	U, I, g, E	79
$CpCo(Et_2MeC_3B_2Et_2)Fe(Et_2C_2B_4H_4)^-$	H*, E	79
$CpCo(Et_2MeC_3B_2Et_2)Fe(Et_2C_2B_4H_4)^{2-}$	H*, g, E	79
$CpCo(Et_2MeC_3B_2Et_2)Fe[(PhCH_2)_2C_2B_4H_4]$	H, U, I, g, E	79
$CpCo(Et_2MeC_3B_2Et_2)Fe[(PhCH_2)_2C_2B_4H_4]^-$	H, E	79
$CpCo(Et_2MeC_3B_2Et_2)Fe[(PhCH_2)_2C_2B_4H_4]^{2-}$	H, g, E	79
$(Me_4C_4S)Fe(Et_2C_2B_4H_4)$	B, H, I, U, E	45
$(HNMe_2C_8H_4)Fe(Et_2C_2B_4H_4)$	B, H, I, U	45
$(MeNMe_2C_8H_4)Fe(Et_2C_2B_4H_4)$	B, H, I, U	45

Table 2 (continued)

Compound[a]	*Data*[b]	*References*
Cobalt		
$Cp_2Co^+[Co(Et_2C_2B_4H_4)_2]^-$	X, B, H, U	46
nido,closo-$Cp_2Co^+[(Et_2C_2B_3H_5)Co(Et_2C_2B_4H_4)]^-$	B, H, I	46
nido,closo-$Cp_2Co^+[(Et_2C_2B_3H_4\text{-5-Me})Co(Et_2C_2B_4H_4)]^-$	B, H, I	46
$(Ph_2PCH_2)_2XCo(Et_2C_2B_4H_4)$ (X = Cl, CN, I)	X, B, H, P, I, U	71
$(Ph_2PCH_2)_2(Cl)Co(Ph_2C_2B_4H_4)$	B, H, C, U, I	80
$CpCo(C_2B_4H_6)$	E	40
$CpCo(Me_2C_2B_4H_4)$	E	40
$CpCo(Et_2C_2B_4H_4)$	B, H, I, U	11c
$CpCo[(TMS)_2C_2B_4H_3]\text{-}\mu(5)\text{-}B_2H_5$	X, B, H, I	12c
$CpCo[(TMS)_2C_2B_4H_4]$	B, H	12c
$CpCo(Et_2MeC_3B_2Et_2)Co(R_2C_2B_4H_4)$ (R = Et, CH_2Ph)	B, H, I, E	58
1,2,4-$CpCo(C_2B_4H_6)$	E	40
nido-$CpCo(Me_2C_2B_4H_3)\text{-}Me_2C_2B_4H_5$ (2 isomers)	X	81
$[(Et_2C_2B_4H_4)Co(C_5H_4)]_2$ ($(C_5H_4)_2$ = fulvalene)	X, B, H, I, U	11d
nido,closo-$(Et_2C_2B_4H_4)(Et_2C_2B_3H_5)Co_2(C_5H_4)_2$ ($(C_5H_4)_2$ = fulvalene)	B, H, I, U	11d
$CH_2[(C_5H_4)Co(Et_2C_2B_4H_4)]_2$	X, B, H, I	82
$Cp^*Co(C_2B_4H_6)$	B, H, I	47
$Cp^*Co(Et_2C_2B_4H_4)$	B, H, I, U	11c
$Cp^*Co(EtC_2B_4H_5)$	B, H, I	47
$Cp^*Co[(TMS)_2C_2B_4H_4]$	B, H, I	47
$Cp^*Co[(TMS)C_2B_4H_5]$	B, H, I	47
$Cp^*Co[(PhCH_2)EtC_2B_4H_4]$	H, I	47
$[Cp^*Co(C_2B_4H_5)]_2CH_2$	B, H, I	47
p-$[Cp^*Co(C_2B_4H_5)CH_2]_2C_6H_4$	B, H, I	47
p-$\{Cp^*Co[(TMS)C_2B_4H_4]CH_2\}_2C_6H_4$	B, H, I	47
m-$\{Cp^*Co[(TMS)C_2B_4H_4]CH_2\}_2C_6H_4$	B, H, I	47
o-$\{Cp^*Co[(TMS)C_2B_4H_4]CH_2\}_2C_6H_4$	B, H, I	47
$Cp^*Co(Me_2C_2B_4H_4)$	B, H, I	48
$[(Et_2C_2B_4H_4)Co(C_5Me_4)]_2(C_6H_4)$	B, H, I, U	11d
$[(Et_2C_2B_4H_4)Co(C_5Me_4)]_2(C_6H_4)_2$	B, H, I, U	11d
$(Et_2C_2B_4H_4)Co(C_5Me_4C_6H_4C_6H_5)$	B, H, I, U	11d
$(Et_2C_2B_4H_4)Fe(PhCH_2C_5Me_4)Co(Et_2C_2B_4H_4)$	B, H, I	54
$(PhCH_2C_5Me_4)Co(Et_2C_2B_4H_4)$	B, H, I	54
$(C_5Ph_5)Co(C_2B_4H_6)$	H, I	47
$(C_5Ph_5)Co[(TMS)_2C_2B_4H_4]$	B, H, I	47
$(PC_4Me_4)Co(Et_2C_2B_4H_4)$	B, H, P, I, U	57
$(2\text{-}B_5H_{10})Co(Et_2C_2B_4H_4)$	B, H, I	53
$(5\text{-}B_9H_{12}C_4H_8O)Co(Et_2C_2B_4H_4)$	B, H, I	53
$(5\text{-}B_9H_{12}C_4H_8O)Co(Et_2C_2B_4H_4)$	X	49
$(Et_4C_4B_8H_7OC_4H_8)Co(Et_2C_2B_4H_4)$	X, B, H, I	14c
$(Et_4C_4B_8H_7OC_4H_8)Co(Et_2C_2B_4H_3)$ (*B–B*)	B, H, I	14c
Nickel		
$(Ph_3P)BrNi(Et_2C_2B_4H_3\text{-5-}PPh_3)$	X, B, I, U	71
1,2,4-$Ni[(TMS)_2C_2B_4H_4]_2$	X, B, H, C, I	100
1,2,4-$(Me_2NC_2H_4NMe_2)Ni[(TMS)_2C_2B_4H_4]$	X, B, H, C, I	100
Yttrium		
$(C_4H_8O)(Cl)Y[(TMS)_2C_2B_4H_4]_2(C_4H_8O)Li^-$	X, B, H, C, I	83
Zirconium		
$(C_4H_8O)ClZr[(TMS)_2C_2B_4H_4]_2^-Li(C_4H_8O)_2^+$	X, B, H, C, I	84
Ruthenium		
$(C_6H_6)Ru(Et_2C_2B_4H_4)$	E	75
p-[*nido*-$(MeC_6H_4CHMe_2)Ru(Et_2C_2B_3H_4\text{-5-}CH_2)]_2(C_6H_4)Fe(Et_2C_2B_4H_4)$	B, H, I	44
$LRu(Et_2C_2B_4H_4)$ (L = 1,4-$CHMe_2C_6H_4Me$, C_6H_6)	E	73
$(C_6H_6)Ru(Et_2C_2B_4H_4)$	B, H, I, U	11c
$(C_6Me_6)Ru(Et_2C_2B_4H_4)$	B, H, I, U	11c
$(MeC_6H_4CHMe_2)Ru(Et_2C_2B_4H_4)$	B, H, I, U	11c
$(MeC_6H_4CHMe_2)Ru[(PhCH_2)_2C_2B_4H_4]$	B, H, I, U	11c
$(C_{10}H_{14})Ru[(TMS)_2C_2B_4H_4]$	B, H, I	47
Rhodium		
$(Et_2C_2B_4H_4)Rh(Et_2MeC_3B_2Et_2)CoCp$	X, B, H, I	58
$Cp^*Rh(Et_2C_2B_4H_4)$	B, H, I, E	58
Hafnium		
$Cl(C_4H_8O)Hf[Me(TMS)C_2B_4H_4]_2\cdot Li(C_4H_8O)$	X	85
Osmium		
$(CO)_3Os[(TMS)_2C_2B_4H_4)]$	B, H, C, I, Si	86
Gadolinium		
μ,*closo*-$[Gd(TMS)_2C_2B_4H_4]_3[\mu_2\text{-Li-}(TMS)_2)_2C_2B_4H_4]_3(\mu_3\text{-OMe})\text{-}[\mu_2\text{-Li}(C_4H_8O)]_3(\mu_3\text{-O})$	X	87

Table 2 (continued)

Compound[a]	*Data*[b]	*References*
7-Vertex polyhedra, 2 metal atoms		
Iron–cobalt		
$Cp_2CoFe(Et_2C_2B_3H_3)$	U, I	11c
Dicobalt		
$Cp^*Co(Et_2C_2B_3H_3)CoCp$	B, H	13b
$Cp^*_2Co_2(Et_2C_2B_3H_3)$	E	57
$Cp^*Co(Et_2C_2B_3H_2R)Co(Et_2MeC_3B_2Et_2)CoCp$ (R = H, Cl)	B, H, I, E[H]	88
$Cp^*_2Co_2(Me_2C_2B_3$-4,6-Me_2-5-Cl)	B, H, I	48
$Cp^*_2Co_2(Me_2C_2B_3Me_3)$	H, I	48
$Cp^*_2Co_2[(MeSi)_2C_2B_3H_3]$	B, I	48
$Cp^*_2Co_2(C_2B_3H_5)$	B, H, I	48
$[Cp^*Co(Et_2C_2B_3H_2$-5-Cl$)]_2Co$	X, g, U	89
$[Cp^*Co(Et_2C_2B_3H_2$-5-Br$)]_2Co$	U	89
$[Cp^*Co(Et_2C_2B_3H_2$-5-Me$)]_2Co$	U	89
$[Cp^*Co(Et_2C_2B_3H_2$-5-$CH_2CCMe)]_2Co$	U	89
$[Cp^*Co(Et_2C_2B_3H_3)]_2Co$	U	89
$[(PhCH_2C_5Me_4)Co]_2(Et_2C_2B_3H_3)$	B, H, I	54
$[(Et_2C_2B_4H_4)Fe(PhCH_2C_5Me_4)Co]_2(Et_2C_2B_3H_3)$	B, H, I	54
$(Et_2C_2B_4H_4)Fe(PhCH_2C_5Me_4)Co(Et_2C_2B_3H_3)Co(C_5Me_5CH_2Ph)$	B, H, I	54
p-$\{Cp^*_2Co_2[(Me_3Si)C_2B_3HCl_2]CH_2\}_2C_6H_4$	B, H, I	47
nido,closo-$M[(Et_2C_2B_3H$-4-R^1-5-$R^2)Co(C_5Me_4)C_6H_4(C_5Me_4)$-$Co(Et_2C_2B_3H_4$-$R^3)]_2$ (R^1, R^2, R^3 = H, Cl, Me; M = Co, CoH)	B, H, U, FAB	55
nido,closo-$Co[(Et_2C_2B_3H$-4,5-$Cl_2)Co(C_5Me_4)C_6H_4(C_5Me_4)Co(Et_2C_2B_3H_5)]_2$	X	55
nido,closo-$[(Et_2C_2B_3H_4Cl)Co(C_5Me_4)]_2[(C_5Me_4)Co(Et_2C_2B_3H_2Cl)$-$Co(Et_2C_2B_3H_2Cl)Co(C_5Me_4)]_n(C_6H_4)_4$ (*n* = 3, 5)	FAB	55
$[(Et_2C_2B_3H$-4,5-$Cl_2)_2Co_3(C_5Me_4)_2]_2$	X*	55
$[Cp^*Co(Et_2C_2B_3H_2$-5-Me$)CoH(Et_2C_2B_3H_3)]_2Co$	X, U, FAB	95
nido,closo-$[(Et_2C_2B_3H_4$-5-Me$)Co(C_5H_4)_2Co(Et_2C_2B_3H_2$-5-Me$)]_2Co$	B, H, U, FAB	55
nido,closo-$[(Et_2C_2B_3H_4$-5-Me$)Co(C_5H_4)_2Co_3(Et_2C_2B_3H_2$-5-Me$)_2(C_5H_4)]_2$	H, U, FAB	55
$[Cp^*Co(Et_2C_2B_3H_2$-5-X$)]_2M$ (X = Cl, Br, I, Me, C(O)Me, $MeCCCH_2$; M = Co, Ni)	E, g(Ni, Cl)	102a
$Cp^*Co(Et_2C_2B_3H_2$-5-Cl$)Ru(Et_2C_2B_3H$-4,5-$Cl_2)CoCp^*$	E, g	102a
$[Cp^*Co(Et_2C_2B_3H$-4,5-$Cl_2)]_2Ni$	B, H, C, E	102a
$[Cp^*Co(Et_2C_2B_3FCl_2)]_2Ni$	H, F, E	102a
$[(Et_2C_2B_3H_4R)CoL]_2(C_6H_4)LCo(Et_2C_2B_3H_2R)M(Et_2C_2B_3H_2R)CoL]_n(C_6H_4)$ (*n* = 1, 3, 5; L = C_5Me_4; M = Co, Ni; R = Cl, Me)	E	102b
$(NC_4Me_2R_2)Co(Et_2C_2B_3H_3)CoCp^*$ (R = Me, H)	B, H, I	56
$(NC_4Me_2R_2)_2Co_2(Et_2C_2B_3H_3)$ (R = Me, H)	B, H, I	56
$(NC_4Me_4)_2Co_2(Et_2C_2B_3H_3)$	X	90
$(NC_4Me_2R_2)Co(Et_2C_2B_3H_3)CoCp^*$ (R = Me, H)	B, H, I, U, E	57
$(NC_4Me_2R_2)_2Co_2(Et_2C_2B_3H_3)$	X, B, H, I, U, E*	57
$[(NC_4Me_4)Co(Et_2C_2B_3H_3)CoCp^*]^+I^-$	B, H, I, U	57
$(NC_4Me_4)Co(Et_2C_2B_3H_2I)CoCp^*$	B, H, I, U	57
$(NC_4Me_4)Co(Et_2C_2B_3H_2Cl)CoCp^*$	B, H, I, U	57
$(NC_4Me_4)Co(Et_2C_2B_3H_2Br)CoCp^*$	B, H, I, U	57
$(NC_4Me_4)_2Co_2(Et_2C_2B_3H_2I)$	B, H, I, U	57
$(H_3BNC_4Me_4)Co(Et_2C_2B_3H_3)CoCp^*$	B, H, I, U	57
$(H_3BNC_4Me_4)_2Co_2(Et_2C_2B_3H_3)$	B, H, I, U	57
$(PC_4Me_4)Co(Et_2C_2B_3H_3)CoCp^*$	B, H, P, I, U, E	57
$(PC_4Me_4)_2Co_2(Et_2C_2B_3H_3)$	B, H, P, I, U, E	57
$(PC_4Me_4)Co(Et_2C_2B_3H_3)Co(NC_4Me_4)$	X, B, H, P, I, U, E	57
Cobalt–Nickel		
$[Cp^*_2Co_2(Me_2C_2B_3H_2$-5-Me$)]_2Ni$	B, H, I	48
$[Cp^*Co(Et_2C_2B_3H_2$-5-$CH_2CCMe)]_2Ni$	B, H	89
$[Cp^*Co(Et_2C_2B_3H_2$-5-Cl$)]_2Ni$	X, B, H, C, U	89
$[Cp^*Co(Et_2C_2B_3H_2$-5-Br$)]_2Ni$	B, H, C	89
$[Cp^*Co(Et_2C_2B_3H_2$-5-Me$)]_2Ni$	B, H, C, U	89
$[Cp^*Co(Et_2C_2B_3H_2$-5-C(O)Me$)]_2Ni$	X, B, H, C, U	89
$[Cp^*Co(Et_2C_2B_3H_2$-5-C(O)Me$)]_2Ni$	X, B, H, C, I	91
nido,closo-$Ni[(Et_2C_2B_3H$-4-R^1-5-$R^2)Co(C_5Me_4)C_6H_4(C_5Me_4)$-$Co(Et_2C_2B_3H_4$-$R^3)]_2$ (R^1, R^2, R^3 = H, Cl, Me)	B, H, U, FAB	55
Cobalt–Ruthenium		
$Cp^*Co(Et_2C_2B_3H_3)Ru(CO)_3$	B, H, I, U	11c
$Cp^*Co(Et_2C_2B_3H_2$-5-Cl$)Ru(Et_2C_2B_3H$-4,5-$Cl_2)CoCp^*$	X, B, H, U	89
$[CpCo(Et_2C_2B_3H_3)Ru(MeC_6H_4$-$CHMe_2)]^n$ (*n* = +, 0, −, 2−)	E, U, g(+, −)	92
$(CO)_2Ph(MeO)C{=}Ru(Et_2C_2B_3H_2Cl)CoCp^*$	X, B, H, C, I, U, E	93
$(MeC_6H_4CHMe_2)Ru(Et_2C_2B_3H_3)Co(Et_2MeC_3B_2Et_2)CoCp$	B, H, I, E	88
$[(MeC_6H_4CHMe_2)Ru(Et_2C_2B_3H_3)Co(C_5H_4)]_2$	B, H, I, U	11d

Table 2 (continued)

Compound[a]	*Data*[b]	*References*
$[(MeC_6H_4CHMe_2)Ru(Et_2C_2B_3H_3)Co(C_5Me_4)]_2(C_6H_4)_2$	B, H, I, U	11d
$[(MeC_6H_4CHMe_2)Ru(Et_2C_2B_3H_3)Co(C_5Me_4)]_2(C_6H_4)$	X, B, H, I, U	11d
$CpCo(Et_2C_2B_3H_3)Ru(MeC_6H_4\text{-}CHMe_2)$	X, B, H, I, U	11c
$CpCo(Et_2C_2B_3H_2Cl)Ru(MeC_6H_4\text{-}CHMe_2)$	B, H, I, U	11c
$(NC_4Me_2R_2)Co(Et_2C_2B_3H_3)Ru(p\text{-}CHMe_2C_6H_4Me)$ (R = Me, H)	B, H, I	56
$(NC_4Me_2R_2)Co(Et_2C_2B_3H_3)Ru(p\text{-}CHMe_2C_6H_4Me)$ (R = Me, H)	B, H, I, U, E	57
$(PC_4Me_4)Co(Et_2C_2B_3H_3)Ru(MeC_6H_4CHMe_2)$	B, H, P, I, U, E	57
Cobalt–Rhodium		
$Cp^*Co(Et_2C_2B_3H_2\text{-}4\text{-}R)RhCp^*$ (R = H, Cl)	B, H, I, E[H]	58
$Cp^*Rh(Et_2C_2B_3H_2\text{-}R)Co(Et_2MeC_3B_2Et_2)CoCp$ (R = H, Cl)	X[H], B, H, I, E[H]	88
Cobalt–osmium		
$(MeC_6H_4CHMe_2)Os(Cp)Co(Et_2C_2B_3H_3)$	B, H, I, U	11c
$(MeC_6H_4CHMe_2)Os(Cp)Co(Et_2C_2B_3H_2Cl)$	B, H, I, U	11c
Diruthenium		
$(MeC_6H_4CHMe_2)_2Ru_2(Et_2C_2B_3H_3)$	X, B, H, I, U	11c
$[(MeC_6H_4CHMe_2)_2Ru_2(Et_2C_2B_3H_3)]^n$ (n = 2+, +, 0, −)	X, B, H, I, U	92
Ruthenium–Rhodium		
$(MeC_6H_4CHMe_2)Ru(Et_2C_2B_3H_3)RhCp^*$	B, H, I, E	58
Ruthenium–osmium		
$(MeC_6H_4CHMe_2)Ru(Et_2C_2B_3H_3)Os(MeC_6H_4CHMe_2)$	X, B, H, I, U	11c
$(MeC_6H_4CHMe_2)Ru(Et_2C_2B_3H_2Cl)Os(MeC_6H_4CHMe_2)$	B, H, I, U	11c
Dirhodium		
$Cp^*_2Rh_2(Et_2C_2B_3H_2\text{-}4\text{-}R)$ (R = H, Cl)	B, H, I, E[H]	58
7(8)-Vertex wedged polyhedra		
B-wedged-$CpFe_2(Me_2C_2B_4H_4)_2$	B, I	94
B-wedged-$CpCoFe[(PhCH_2)_2C_2B_4H_4]_2$	B, H, I	68
B-wedged-$(Et_3P)_2CoFe(Me_2C_2B_4H_4)_2$	X, B, H, P, I	94
B-wedged-$(Et_3P)_2PtFe(Me_2C_2B_4H_4)_2$	X*, B, H, P, I	94
Fe-wedged-$(C_4H_8O)_2Fe_2(Et_2C_2B_4H_4)_2$	H, I, U	13a
Fe-wedged-$C_2H_4(MeO)_2Fe_2(Me_2C_2B_4H_4)_2$	X, m, M, g	67
Fe-wedged-$(C_4H_8O)_2Fe_2(Et_2C_2B_4H_4)_2$	m	67
Fe-wedged-$(C_4H_8O)_2Fe_2(Me_2C_2B_4H_4)_2$	m	67

[a] Clusters are *closo* except when otherwise designated; metals are in order of increasing atomic number. Cp = η-C_5H_5, Cp* = η-C_5Me_5. [b] X = x-ray diffraction, B = ^{11}B NMR, H = ^{1}H NMR, C = ^{13}C NMR, P = ^{31}P NMR, 2D = two-dimensional (COSY) NMR, E = electrochemical data, M = magnetic susceptibility data, I = infrared data, U = UV–visible data, g = ESR (EPR) data, m = Mössbauer spectra, FAB = fast atom bombardment mass spectral data, S = mass spectral data only. Asterisks indicate that only partial data are given. [c] Metal atoms in metallacarbaborane framework. [d] All *nido*-MC_2B_3, *closo*-MC_2B_4, and *closo*-$M^1M^2C_2B_3$ clusters have adjacent skeletal carbon atoms (2,3) unless otherwise indicated.

attached to cage *carbon* atoms presents a different problem. To a limited extent this can be done by first preparing the desired *nido*-2,3-$R^1R^2C_2B_4H_6$ carborane via reaction of the $R^1C{\equiv}CR^2$ alkyne with B_5H_9,[80] but a more versatile approach[47,48] utilizes *C,C'*-disilylated metallacarbaboranes from which the SiH_3 or TMS groups are displaced *catalytically* by treatment with $Bu_4N^+F^-$. As noted in Section 9.3 and Equation (5), (arene)$Fe(R^1R^2C_2B_4H_4)$ sandwiches can be prepared from arene anions or via thermal displacement of the cyclooctatriene ligand from $(\eta^6\text{-}C_8H_{10})Fe(R^1R^2C_2B_4H_4)$ complexes. The reaction is quite general and has been used to prepare a variety of multimetal–multilevel sandwich compounds.[44,54]

These synthetic methods have been exploited in the designed synthesis of multisandwich and linked-cluster assemblies, such as the "staircase" oligomers depicted in Figure 8, that are possible precursors to new types of low-dimensional conducting or semiconducting chain polymers.[55] In related work, a CH_2-bridged biscobaltacarbaborane, $CH_2[(C_5H_4)Co(Et_2C_2B_4H_4)]_2$, has been prepared from the $[CH_2(C_5H_5)_2]^{2-}$ ion.[82]

In addition to their role as synthetic building blocks, *nido*-$[R^1R^2C_2B_4H_4]^{2-}$ ligands have the ability to stabilize transition metal organometallic systems that are otherwise unstable or unknown. For example, arene sandwich complexes of iron(III) and ruthenium(III) are virtually nonexistent, but long-lived $[(arene)M^{III}(Et_2C_2B_4H_4)]^+$ cations in which M is Fe or Ru are reversibly generated on electrochemical oxidation of the M^{II} neutral complexes.[73] Similarly, the octahapto complexes[60] 1,2,3-$(\eta\text{-}C_8H_8)M^{IV}$-$(Et_2C_2B_4H_4)$ (M = V, Ti) are air-stable solids that contrast with the general instability of titanium complexes of planar $[C_8H_8]^{2-}$ (and the nonexistence of known vanadium counterparts). The robustness of small metallacarbaborane complexes allows them to undergo a considerable range of chemical and electrochemical reactions without degradation of the cage. The rich chemistry of indenylferracarboranes, referred to earlier in Section 9.2.2, is illustrative; another example is afforded by 1,2,3-$Cp^*Fe^{II}H(Et_2C_2B_4H_4)$ and its bimetallic counterpart 1,2,3-$[(C_5Me_4)Fe^{II}H(Et_2C_2B_4H_4)]_2C_6H_4$, which

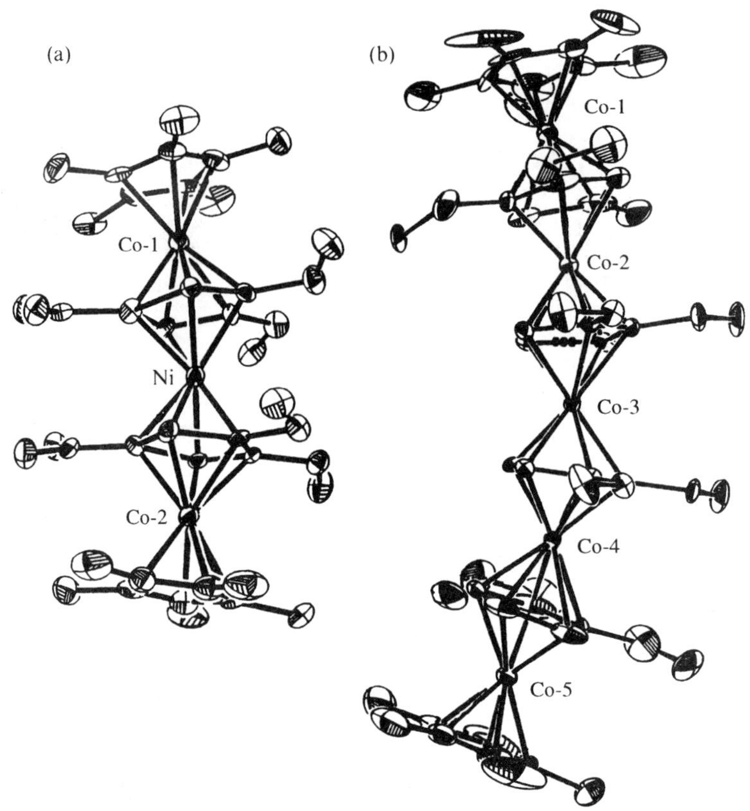

Figure 7 Structures of C_2B_3-bridged tetradecker and hexadecker sandwiches: (a) [Cp*Co($Et_2C_2B_3H_2$-5-C(O)-Me)]$_2$Ni and (b) [Cp*Co($Et_2C_2B_3H_2$-5-Me)CoH($Et_2C_2B_3H_3$)]$_2$Co ((a) reprinted with permission from *Inorg. Chem.*, 1992, **31**, 5202; copyright 1992 American Chemical Society: (b) reproduced with permission from *J. Am. Chem. Soc.*, 1994, **116**, 2687; copyright 1994 American Chemical Society).

$$[CpCo(Et_2MeC_3B_2Et_2)]^- + [Et_2C_2B_4H_5]^- \xrightarrow[THF]{CoCl_2} CpCo(Et_2MeC_3B_2Et_2)Co(Et_2C_2B_4H_4) \xrightarrow{TMEDA}$$

$$nido\ CpCo(Et_2MeC_3B_2Et_2)Co(Et_2C_2B_3H_5) \xrightarrow{2\ LiHBEt_3} \xrightarrow{(Cp^*RhCl_2)_2} CpCo(Et_2MeC_3B_2Et_2)Co(Et_2C_2B_3H_3)RhCp^* \quad (14)$$

undergo reversible redox processes to generate, in the former case, an Fe^{III} species, and in the latter case, Fe^{II}/Fe^{III} and Fe^{III}/Fe^{III} systems (Figure 9).[72] The crystal structure of [1,2,3-$(C_5Me_4)Fe^{II}H(Et_2C_2B_4H_4)]_2C_6H_4$ (Figure 10)[72] established that the metal-bound protons are in fact capping FeB_2 faces, as had long been conjectured for similar $H_nMC_2B_4$ clusters.[1a]

Hydrocarbon–metal–carbaborane complexes of the types just discussed can be regarded as bridges between boron cluster and transition metal organometallic chemistry. In contrast, systems in which hydrocarbon ligands are absent, and the metal bonds exclusively to small carbaborane units (often together with solvent or other weakly bound ligands), have generated some truly novel stereochemistry. Indeed, it is in this area that one most clearly sees the distinctive behavior of the small metallacarbaboranes that contrasts with that of the dicarbollyl ($[C_2B_9H_{11}]^{2-}$) complexes. The special character of small metallacarbaborane chemistry was evident as early as 1974, when the bis(carboranyl) complexes $(2,3\text{-}R_2C_2B_4H_4)_2Fe^{II}H_2$ and $(2,3\text{-}R_2C_2B_4H_4)_2Co^{III}H$ were shown to undergo quantitative face-to-face oxidative fusion of the carbaborane ligands to form $R_4C_4B_8H_8$ cages[1a,1c,1h,1q]—a reaction that was subsequently extended to other small borane and heteroborane clusters[1q,13] but, significantly, is *not* exhibited by larger units (edge coupling[97] rather than cage fusion is observed). Recently, the work of Hosmane and co-workers has generated some remarkable multicage systems, an example of which is the trinuclear gadolinium complex[87] shown in Figure 11. Other important structurally characterized compounds from that laboratory, discussed in a 1993 review,[1b] include bis(carboranyl) complexes of yttrium,[83] zirconium,[84] and hafnium[85] (all bent sandwiches); chromium sandwiches containing chromium(III) and chromium(IV) respectively;[4k,64] and a manganacarbaborane trimer that incorporates four $[(TMS)_2C_2B_4H_4]^{2-}$ ligands, three manganese atoms, and three lithium atoms (Figure 12).[65]

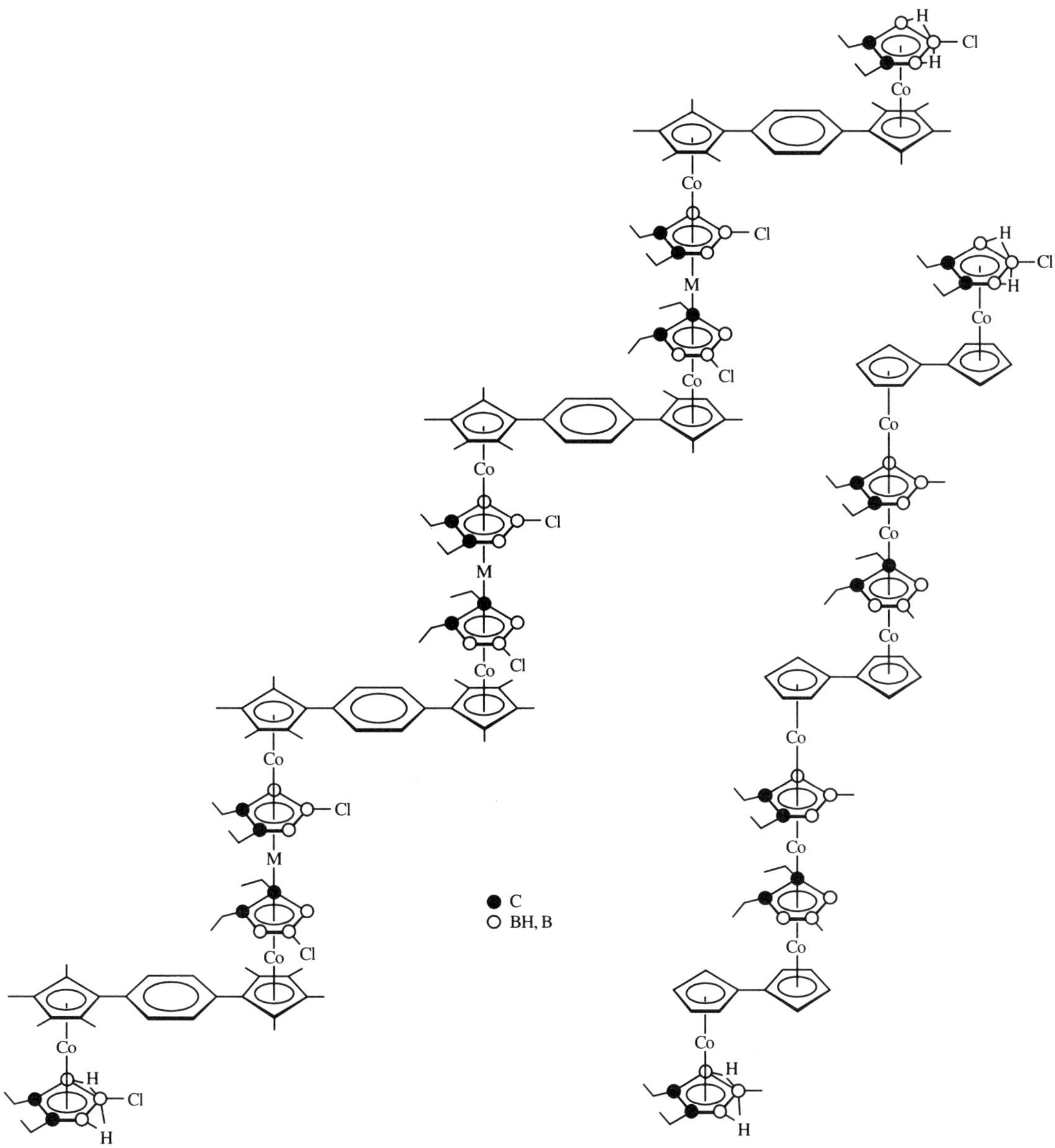

Figure 8 Proposed structures of cobaltacarbaborane staircase oligomers (reprinted with permission from *J. Am. Chem. Soc.*, 1993, **115**, 6143; copyright 1993 American Chemical Society).

These structurally complex systems involve not only intramolecular metal–carbaborane binding but, in some cases, metal–metal, and other types of interactions as well. Efforts to gain a detailed understanding of their formation and electronic structures certainly present a challenge to currently available quantitative methods.[1b] A useful contribution[4h] was made by applying the Fenske–Hall quantum method to the seven-vertex *closo* systems (bipy)Sn($Me_2C_2B_4H_4$) and $(CO)_3Fe(Me_2C_2B_4H_4)$, which revealed that the observed slippage of the tin atom toward the middle boron (B5) is caused by an antibonding interaction between a lone pair on the metal and the cage carbons; in the ferracarbaborane, the corresponding interaction involves the t_{2g} set on iron and is weak, and hence slippage does not occur and the metal is centered over the carbaborane face. Self-consistent charge and configuration calculations on the parent ferracarbaborane cluster indicated that the binding of the $Fe(CO)_3$ group to the carbaborane ligand is weak in comparison to that in $C_2B_5H_7$ carbaboranes.[4g] However, it is notable that the tin atom in 1,2,3-Sn[$(TMS)_2C_2B_4H_4$] is replaced by an $Os(CO)_3$ group on treatment with $Os_3(CO)_{12}$ at 150 °C.[86] Remarkably, the osmacarbaborane product is a colorless liquid.

Although most metallacarbaborane clusters of the MC_2B_4 and $M^1C_2B_3M^2$ types have adjacent (2,3) cage carbon atoms, isomeric species having nonadjacent (2,4) carbons can be prepared via thermal rearrangement of 2,3-isomers, or by insertion of metals into *closo*-$R_2C_2B_4H_4$ carbaboranes or

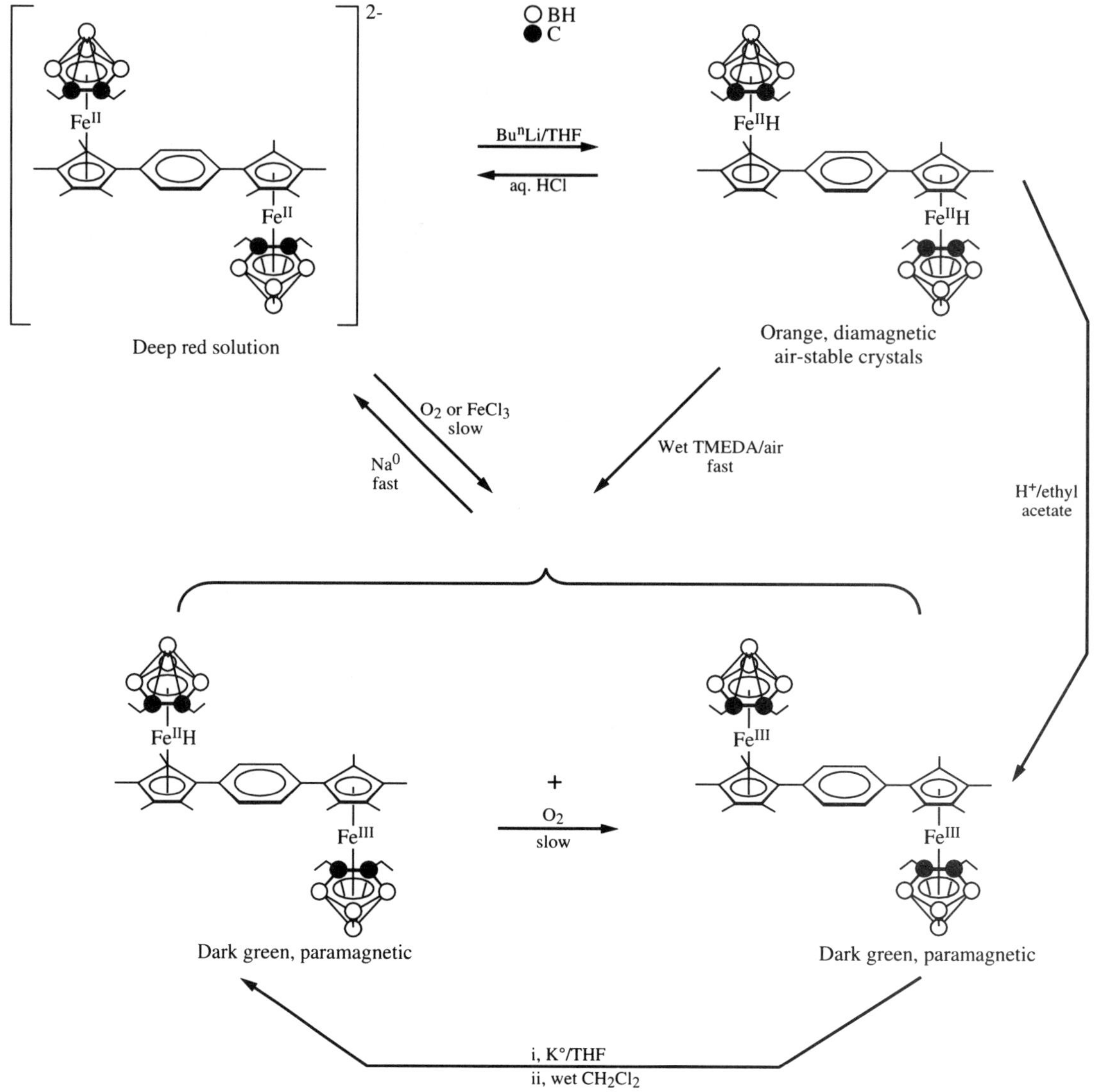

Figure 9 Redox chemistry of a phenylene-linked bis(ferracarbaborane) complex (reprinted with permission from *J. Am. Chem. Soc.*, 1992, **114**, 5214; copyright 1992 American Chemical Society).

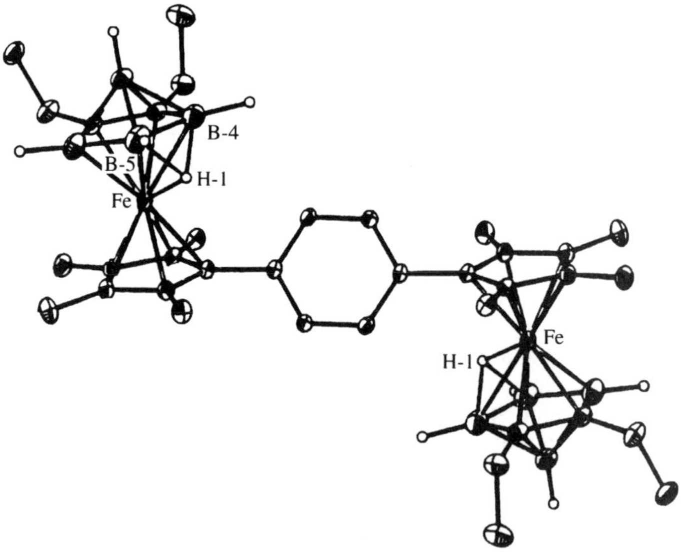

Figure 10 Structure of $[(Et_2C_2B_4H_4)FeH(C_5Me_4)]_2C_6H_4$ showing face-capping hydrogens (H1) (reprinted with permission from *J. Am. Chem. Soc.*, 1992, **114**, 5214; copyright 1992 American Chemical Society).

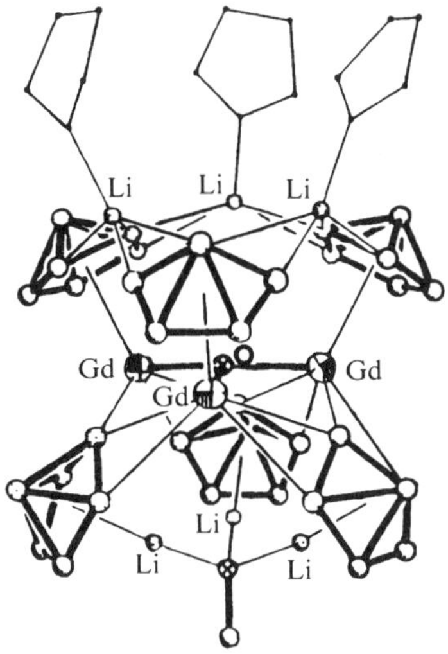

Figure 11 Structure of μ,*closo*-[Gd(TMS)$_2$C$_2$B$_4$H$_4$]$_3$ [μ$_2$-Li-(TMS)$_2$C$_2$B$_4$H$_4$]$_3$(μ$_3$-OMe)[μ$_2$-Li(C$_4$H$_8$O)]$_3$(μ$_3$-O) (reproduced by permission of VCH from *Angew. Chem., Int. Ed. Engl.*, 1992, **31**, 432).

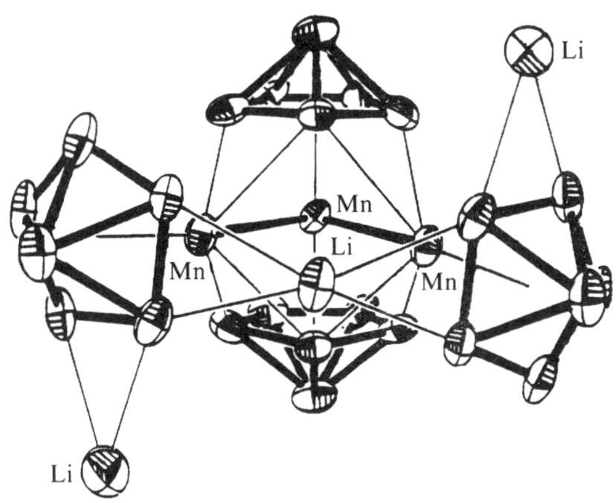

Figure 12 Structure of Li$_3${Mn$_3$[(TMS)$_2$C$_2$B$_4$H$_4$]$_4$} (reprinted with permission from *J. Am. Chem. Soc.*, 1991, **113**, 8531; copyright 1991 American Chemical Society).

$[R_2C_2B_nH_n]^{2-}$ ($n = 4$ or 5) anions obtained via reduction of small *closo*-carbaboranes with alkali metals.[1a,98] Recently, Hosmane reopened this area with the serendipitous discovery that the reaction of $NiCl_2$ with the *nido*-2,3-[(TMS)$_2$C$_2$B$_4$H$_4$]$^{2-}$ dianion gave *closo*-2,3-(TMS)$_2$C$_2$B$_4$H$_4$ and nickel(0) via polyhedral contraction (see above); reduction of the *closo* species with sodium naphthaleneide gave the *nido*-2,4-[(TMS)$_2$C$_2$B$_4$H$_4$]$^{2-}$ dianion.[99] Treatment of dilithio derivatives of the 2,4-dianion with $NiCl_2$ gave *closo*-nickelacarbaboranes with nonadjacent cage carbons, as shown in Figure 13.[100]

The binding capabilities of small *nido*-carbaborane ligands are not limited to η^5-coordination of the C_2B_3 face to metals, as underlined by a number of recent findings. Suitably ligated electron rich metals can form stable B–B-bridged (η^2) complexes, as in μ-4,5-L$_2$Cu-(2,3-R$_2$C$_2$B$_4$H$_5$) (L = Ph$_3$P, or L$_2$ = dppe; R = H, Me),[101a] which are analogous to earlier metal-bridged derivatives.[1a] The nickel species μ-(dppe)NiCl-(2,3-C$_2$B$_4$H$_7$) is unstable above −80 °C, losing HCl to form *closo*-1,2,3-(dppe)Ni(C$_2$B$_4$H$_6$).[101b] Edge bonding on the C_2B_3 open face can also take place between two carbaborane ligands, as in the bridge-coupled complexes shown in Figure 14 that involve three-center, two-electron intercage linkages.[13b,81] It will be noted that the molecules in Figures 14(b) and 14(c) are isomers having, respectively, bent and coplanar $(C_2B_3)_2$ ring moieties.[13b] This type of dimer represents an intermediate stage between single σ-bond B–B linkage and full fusion to form a single polyhedral cage. The latter situation has been encountered with small cobaltacarbaboranes, as described below in Section 9.8.

The synthetic versatility, tailorability, and stabilizing power of metal complexes of the *nido*-2,3-[R^{1}R^{2}C$_2$B$_4$H$_4$]$^{2-}$ ligand encourages the extension of this chemistry into new areas, especially those having potential applications.[1c,3] Synthetic approaches to linked-sandwich assemblies such as those described earlier may eventually lead to new types of low-dimensional electrical conductors.[1c] In this

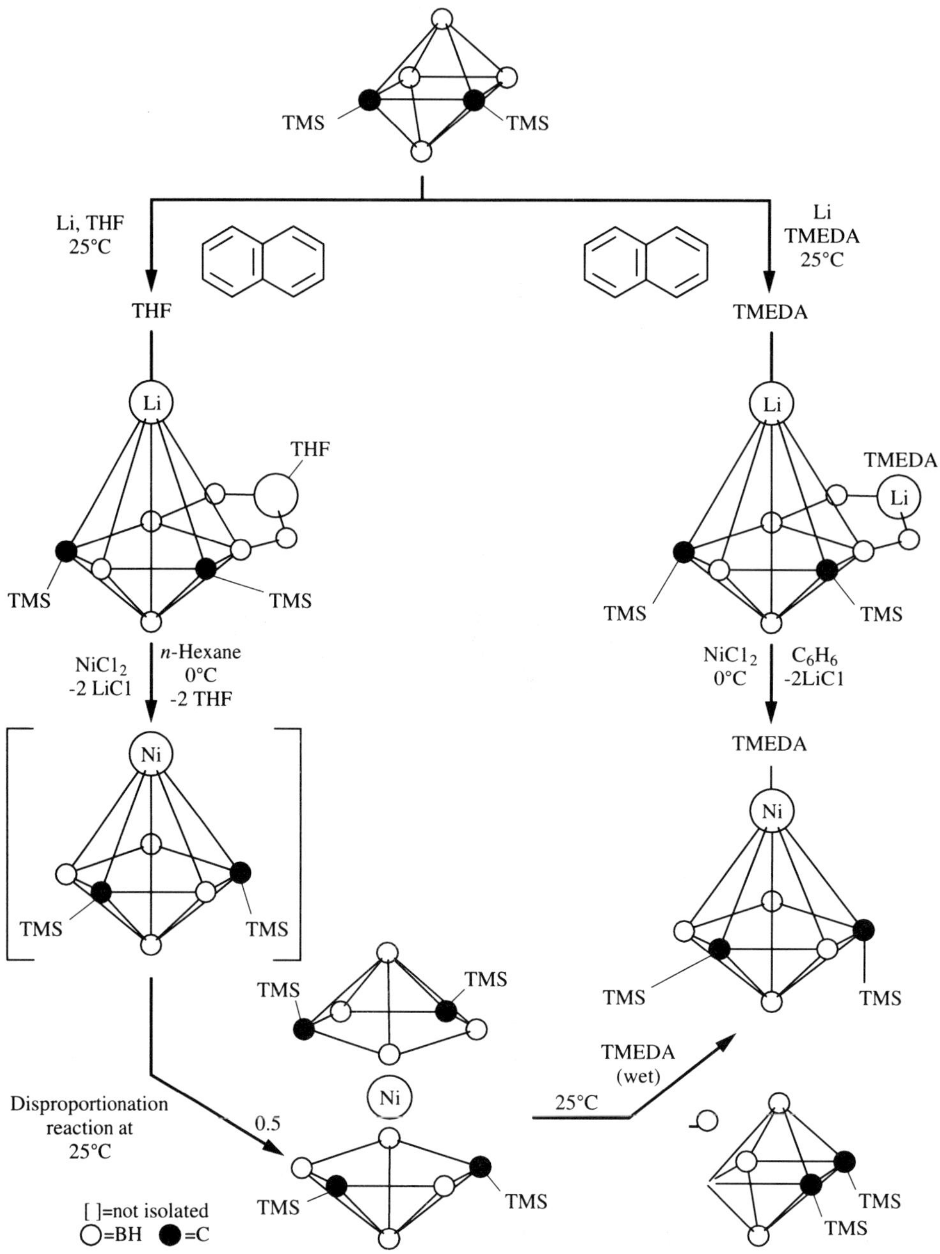

Figure 13 Synthetic route to metal complexes of the *nido*-2,4-$[(Me_3Si)_2C_2B_4H_4]^{2-}$ ligand (reprinted with permission from *Organometallics*, 1993, **12**, 3933; copyright 1993 American Chemical Society).

connection, the degree of electron delocalization in carbaborane-bridged multidecker stacks is of central interest, and recent spectroelectrochemical work on paramagnetic diruthenium and cobalt–ruthenium metallacarbaborane triple-decker cations has established that these species are completely delocalized (Class III) mixed-valent systems.[92] Electrochemical studies of a series of tetradecker sandwiches of the formula $[Cp^*Co(Et_2C_2B_3H_2\text{-}5\text{-}R)]_2M$ where M is Co^{IV}, Ni^{IV}, or Ru^{IV} have provided strong evidence that these, too, are highly delocalized.[102] The electronic spectra of these intensely colored, nearly black, complexes can be correlated, to a degree, with energy level calculations on an idealized $CpCoCpCoCpCoCp^+$ tetradecker system,[4i] and support a model involving charge-transfer excitation from bonding MOs on the $Cp^*CoC_2B_3$ end units to empty antibonding orbitals on M.[89]

New materials exhibiting useful electronic, magnetic, and/or optical properties might also, in principle, be obtained from salts of metallacarbaborane anions and suitable organic or organometallic cations. A step in this direction was the synthesis of several mixed-sandwich cobaltocenium metallacarbaboraneide salts in reactions as depicted in Figure 15, in which the cation and anion are

(a)

(b)

(c)

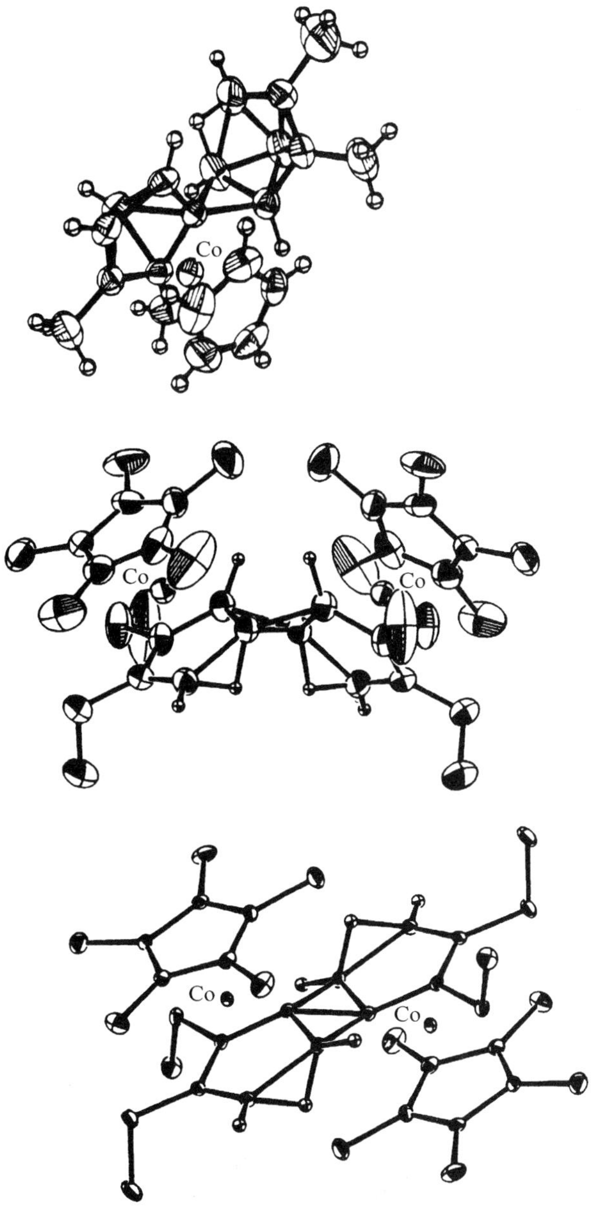

Figure 14 Structures of (a) *nido*-$CpCo(Me_2C_2B_4H_3)Me_2C_2B_4H_5$, (b) *nido*-$[Cp^*Co(Et_2C_2B_3H_3)]_2$ (bent isomer), and (c) *nido*-$[Cp^*Co(Et_2C_2B_3H_3)]_2$ ("planar" isomer) ((a) reprinted with permission from *Inorg. Chem.*, 1982, **21**, 1328; (b), (c) from *Inorg. Chem.*, 1993, **32**, 2156; copyright 1982:1993 American Chemical Society).

sterically similar and are paired in the solid state.[46] The salt in which M = Fe is paramagnetic and has an unusual solid-state structure that is under investigation at the time of writing.

In a different direction, small metallacarbaborane complexes are also being explored as reagents for organic synthesis. The first Fischer carbene incorporating a carbaborane ligand has been prepared and structurally characterized[93] and was found to undergo the Dötz benzannulation reaction, as shown in Figure 16. Related complexes of tantalum and zirconium show catalytic activity in alkene polymerization,[103] as do 12-vertex metallacarbaboranes of these elements, discussed in Section 9.6.3.4.

A recently prepared triiron cluster that qualifies as a metallacarbaborane, $(CO)_9Fe_3MeHC_2BH$, has a *closo*-octahedral structure[59] with an Fe_3 face, and was prepared from $[Fe(py)_6][Fe_4(CO)_{13}]$ and $BH_2Br{\cdot}SMe_2$ in toluene at 75 °C. Although its synthetic origin is quite distinct, this six-vertex cage—which has seven skeletal electron pairs and therefore obeys the skeletal electron-counting rules—emphasizes the close relationship between the polyhedral borane and metal cluster families, a topic of increasing importance in theory and experiment.[4m]

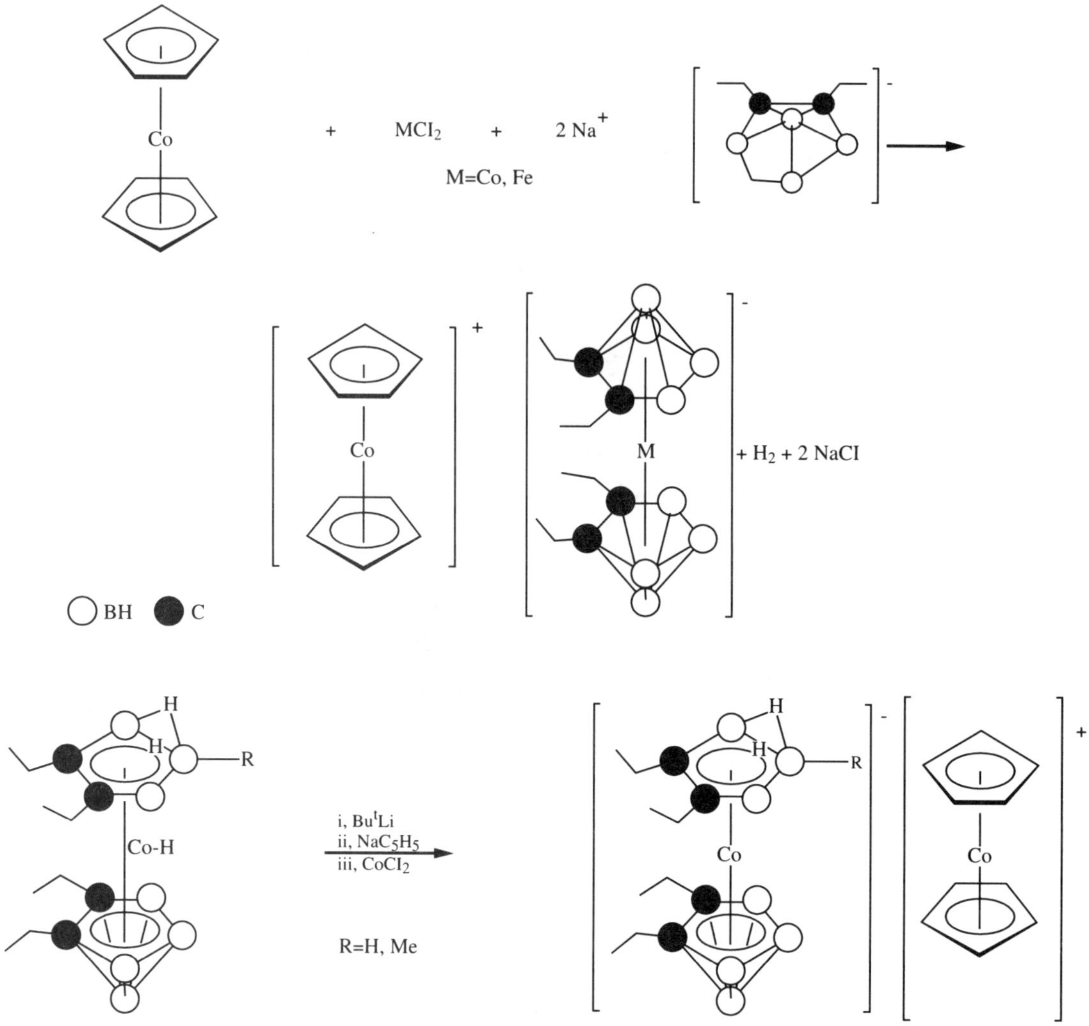

Figure 15 Synthesis of cobaltocenium metallacarboranide mixed-sandwich salts (reprinted with permission from *Inorg. Chem.*, 1993, **32**, 3188; copyright 1993 American Chemical Society).

9.6.2 Intermediate (8- to 11-vertex) Cages

Metallacarbaborane polyhedra having more than seven and fewer than 12 cage atoms are few in number, as the listing in Table 3 indicates. Unlike the chemistry of seven- and 12-vertex cages, for which extensive synthetic methodologies have been developed as described in the preceding and following sections, general routes to the intermediate species are not currently available. Primarily, this is due to the lesser accessibility of the midsized precursor carbaboranes; also, since these carbaboranes lack planar, well-defined bonding faces (such as those found in the *nido*-C_2B_4 and -C_2B_9 ligands) to which metals coordinate preferentially, metal attack may occur at more than one cage site. Intermediate metallacarbaboranes have usually been obtained in relatively uncontrolled, nonspecific reactions, often together with other products. Consequently, the directed synthesis of specific complexes in this class is rarely possible, and studies tend to focus on individual compounds as they are found. Nonetheless, much intriguing stereochemistry has been uncovered in this area, and it continues to attract the interest of researchers.

9.6.2.1 Synthesis

MC_2B_n clusters in which $n = 5$–8 have been prepared via direct metal insertion into neutral carbaboranes using metal vapor (Equation (15))[12a,12c,d] or organometallic reagents (Equations

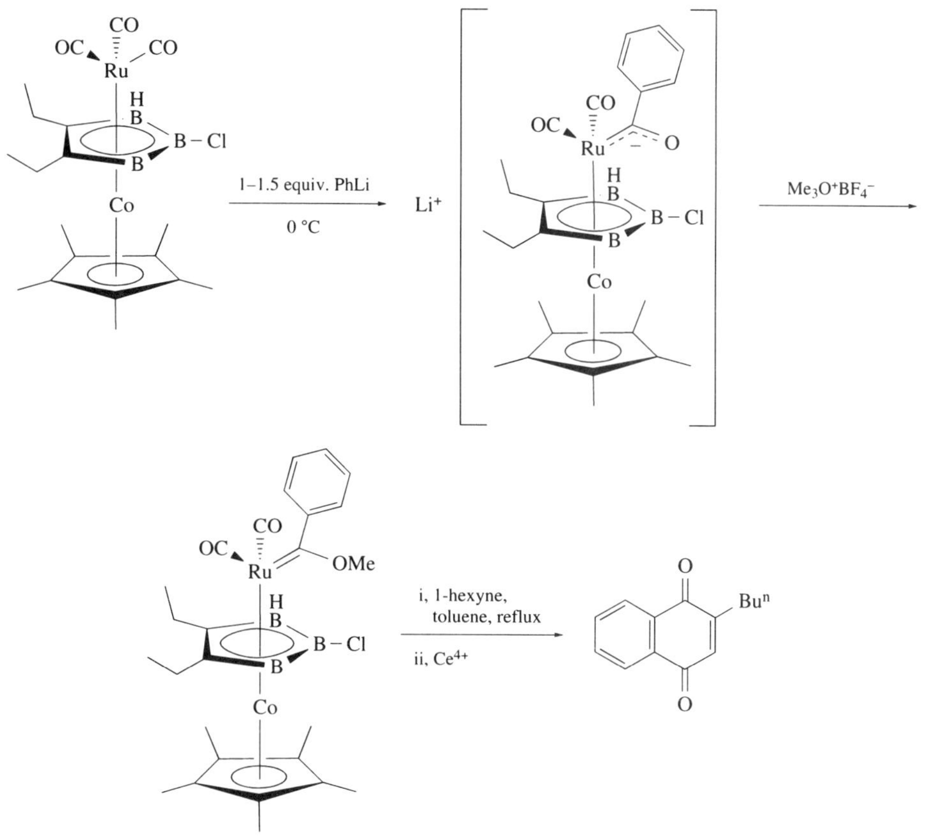

Figure 16 Synthesis and reactivity of $(CO)_2Ph(MeO)C{=}Ru(Et_2C_2B_3H_2Cl)CoCp^*$, a dimetallacarbaborane Fischer carbene (reprinted with permission from *J. Am. Chem. Soc.*, 1992, **114**, 8733; copyright 1992, American Chemical Society).

(16)–(18)).[11e,104,106,109,111,116] Direct metal insertion into *closo*-carbaboranes, as in Equations (17) and (18), a method originally explored during the 1970s in the laboratories of Grimes[1a,98b] and Stone,[1a,119] is "roulette-wheel" chemistry allowing no control over the product structures, but it has generated some remarkable cage geometries. The structure of the dicobalt product in Equation (17), which contains an *exo*-polyhedral metal center, is shown in Figure 17 with a proposed bonding scheme that satisfies the electronic requirements for a *closo*-cage system as well as 18-electron valence shells for both metals.[104] The monoplatinum cluster generated in Equation (18), whose 18 skeletal electrons predict a *closo*-dodecahedral geometry, has a four-membered Pt–B–B–C open face[106] and is one of many examples of distorted metallacarbaborane polyhedra that are said to "violate" Wade's rules (a subject discussed below). The diplatinum species obtained in the same reaction is, as expected, a tricapped trigonal prism, albeit with several unusually long bonds; a Pt–Pt bond forms one of the capping edges.

$$\textit{nido}\text{-2,6-}C_2B_7H_{11} \xrightarrow[\text{Cp, PhMe, } C_6Me_3H_3,\ C_2Me_2]{\text{Co, Fe, or Ni}} LM(C_2B_7H_n) \qquad (15)$$

$$LM = CpCo,\ (PhMe)Fe,\ (C_6Me_3H_3)Fe,\ (\eta^3\text{-}C_4Me_4H)Ni;\ n = 9\text{–}13$$

$$\textit{arachno}\text{-4,5-}C_2B_7H_{13} + Pt(PPh_3)_4 \xrightarrow[80\%]{C_6H_6,\ RT} \textit{arachno}\text{-9,5,6-}(PPh_3)Pt(C_2B_7H_{11}) \qquad (16)$$

$$\textit{closo}\text{-2,4-}Me_2C_2B_5H_5 + Co(PEt_3)_4 \xrightarrow[70\%]{C_6H_6,\ RT} \mu(4,8)\text{-}[(PEt_3)_3CoH_2]\text{-4,1,7-}(PEt_3)Co(Me_2C_2B_5H_4) \qquad (17)$$

$$\textit{closo}\text{-2,4-}R_2C_2B_5H_5 + Pt_2(\mu\text{-}C_8H_{12})(PEt_3)_4 \xrightarrow[R = H,\ Me]{} \text{4,1,7-}(PEt_3)_2Pt(R_2C_2B_5H_5) + (PEt_3)_4Pt_2(R_2C_2B_5H_5) \qquad (18)$$

The tried-and-tested approach of metal insertion into open-cage carbaborane anions (see above) generates complexes in the 8- to 11-vertex range (Equation (19)).[108] Intermediate metallacarbaboranes

Table 3 8- to 11-vertex dicarbon metallacarbaboranes (since 1982).

Compound[a]	*Data*[b]	*References*
8-Vertex polyhedra, 1 metal atom[c]		
Iron		
4,1,7-$(Bu^tNC)_3Fe(Me_2C_2B_5H_5)$	X, I	104
Cobalt		
3,1,2-$CpCo(Me_2C_2B_5H_5)$	X	105
5,1,8-$CpCo[(TMS)_2C_2B_5H_5]$	B, H	12c
$\mu[(Et_3P)_2CoH]$-$\mu(Ph_3P)$-$\mu(H)$-4,1,7-$(Et_3P)Co(Me_2C_2B_5H_4)$	X, B, H, P, I	104
Platinum		
nido-4,1,7-$(Et_3P)_2Pt(C_2B_5H_7)$	B, P, I	106
nido-4,1,7-$(Et_3P)_2Pt(Me_2C_2B_5H_5)$	X, B, P, I	106
8-Vertex polyhedra, 2 metal atoms		
Iron–nickel		
$(\eta^5$-$C_9H_7)Fe(Et_2C_2B_4H_4)NiCp^*$	X, B, H, C, I	6
$(\eta^5$-$C_9H_7)Fe(Et_2C_2B_4H_4)Ni(Et_2MeC_3B_2Et_2)CoCp$	B, H, I, E	6
$Cp^*Fe(Et_2C_2B_4H_4)NiCp^*$	X, B, H, I	6
$Cp^*Fe(Et_2C_2B_4H_4)Ni(Et_2HC_3B_2Me_2)CoCp$	B, H, I	6
9-Vertex polyhedra, 1 metal atom		
Iron		
$(MeC_6H_5)Fe(Et_2C_2B_6H_6)$	S	12a
Cobalt		
1,4,5-$CpCo[(TMS)_2C_2B_6H_6]$	B, H, I	12c
1,4,6-$CpCo[(TMS)_2C_2B_6H_6]$	B, H, I	12c
9-Vertex polyhedra, 2 metal atoms		
Dicobalt		
(2,4-$C_2B_5H_6$)-1,8,5,6-$Cp_2Co_2(C_2B_5H_6)$	X	107
Diplatinum		
1,6,4,5-$(Et_3P)_4Pt_2(Me_2C_2B_5H_5)$	X, B, P, I	106
1,6,4,5-$(Et_3P)_4Pt_2(C_2B_5H_7)$	B, P, I	106
10-Vertex polyhedra, 1 metal atom		
Iron		
$(MeC_6H_5)Fe(Et_2C_2B_7H_7)$	S	12a
Cobalt		
(6-B_9H_{12}-OC_4H_8)-6,1,2-$Co(Et_2C_2B_7H_7)$	B, H, I	53
(6-B_9H_{12}-OC_4H_8)-6,1,2-$Co(Et_2C_2B_7H_7)$	X	49
2,1,4-$CpCo(C_2B_7H_9)$	B, H, I	12d
arachno-4,2,3-$CpCo(C_2B_7H_{13})$	B, H, I	12d
Ruthenium		
nido-2,5,6-$(C_6H_6)Ru(C_2B_7H_{11})$	X, B, H	11b
2,1,6-$[C_6Me_6]Ru(C_2B_7H_9)$	X, B, H	108
Rhodium		
2,1,6-$Cp^*Rh(C_2B_7H_9)$	B, H	108
Platinum		
arachno-9,5,6-$(Ph_3P)_2Pt(C_2B_7H_{11})$	X, B, H(2D)	109
11-Vertex polyhedra, 1 metal atom		
Cobalt		
1,2,3-$CpCo(C_2B_8H_9$-10-Br)	X, B	110
Nickel		
nido-7,8,9-$CpNi(C_2B_8H_{11})$	X, I	111
nido-7,8,9-$CpNi(C_2B_8H_{10}$-μ-$AuPPh_3)$	X, B, I	111
Ruthenium		
1,2,4-$(C_6H_6)Ru(C_2B_8H_{10})$	B, H	11b
1,2,4-$(MeC_6H_4Pr^i)Ru(C_2B_8H_8)$	X, B(2D*)	112
1,2,4-$(C_6Me_6)Ru(C_2B_8H_8)$	B(2D*)	112
1,2,4-$(MeC_6H_4CHMe_2)Ru(Me_2C_2B_8H_8)$	X, B, H	11e
1,2,3-$(MeC_6H_4CHMe_2)Ru(Me_2C_2B_8H_8)$	X, B, H	11e
1,2,3-$(C_6Me_6)Ru(MeC_2B_8H_9)$	B, H	11e
1,2,4-$(C_6Me_6)Ru(Me_2C_2B_8H_8)$	B, H	11e
1,2,4-$(C_6Me_6)Ru(MeC_2B_8H_8$-3-X) (X = H, Cl)	B, H	11e
Rhodium		
1,2,3-$Cp^*Rh(MeC_2B_8H_9)$	X, B, H	33
nido-$(Ph_3P)_2Rh(C_2B_8H_{11})$	X	113
nido-2,8,10-$Cp^*RhMe_2C_2B_8H_8)$	X, B(2D*), H(2D*)	114
Osmium		
nido-2,8,10-$(C_6Me_6)Os(Me_2C_2B_8H_8)$	X, B(2D*), H(2D*)	114

Table 3 (continued)

Compound[a]	*Data*[b]	*References*
Iridium		
1,2,4-$(Ph_3P)_2(H)Ir(C_2B_8H_{10})$ (found *nido*)	X, B, H	115
Platinum		
nido-9,7,8-$(Et_3P)_2HPt(C_2B_8H_{11})$	X, P, B, H, I	116
nido-9,7,8-$(Et_3P)HPt(C_2B_8H_9$-10-$PEt_3)$	X, B, H, P, I	116
1,2,3-$(Me_2PhP)_2Pt(C_2B_8H_9$-2-Me)	X, B, H	117
1,2,3-$(Me_2PhP)_2Pt(C_2B_8H_{10})$	X, B, H	117
1,2,3-$(Me_2PhP)_2Pt(C_2B_8H_9$-2-Ph)	B, H	117
1,2,3-$[P(OMe)_3]_2Pt(C_2B_8H_{10})$	X, B, H	118
1,2,3-$[P(OMe)_3]_2Pt(C_2B_8H_9$-2-Ph)	X, B, H	118

[a] Clusters are *closo* except when otherwise designated. Cp = η-C_5H_5, Cp* = η-C_5Me_5. [b] X = x-ray diffraction, B = ^{11}B NMR, H = ^{1}H NMR, C = ^{13}C NMR, P = ^{31}P NMR, 2D = two-dimensional (COSY) NMR, I = infrared data, E = electrochemical data, S = mass spectral data only. Asterisks indicate that only partial data are given. [c] Metal atoms in metallacarbaborane framework.

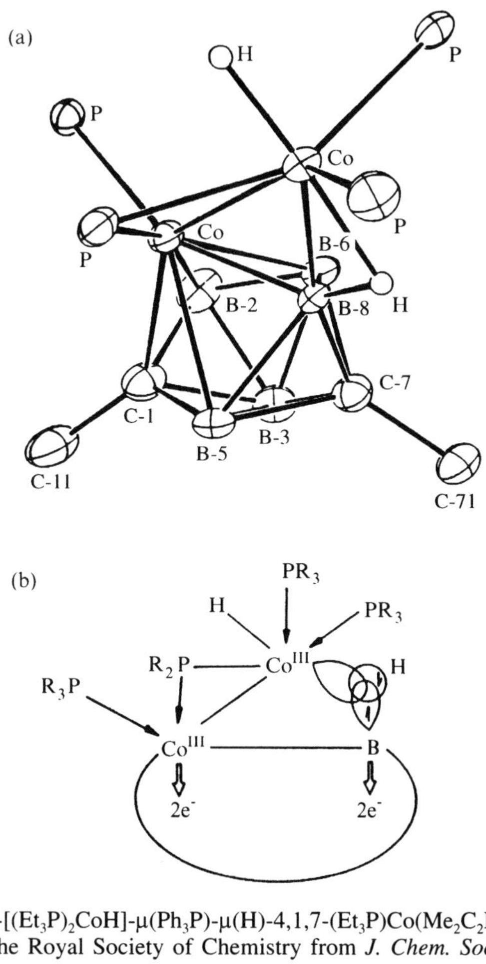

Figure 17 (a) Structure of μ-$[(Et_3P)_2CoH]$-$\mu(Ph_3P)$-$\mu(H)$-4,1,7-$(Et_3P)Co(Me_2C_2B_5H_4)$ and (b) bonding diagram (reproduced by permission of the Royal Society of Chemistry from *J. Chem. Soc., Chem Commun.*, 1982, 688).

have also been obtained via degradation of icosahedral clusters (Equation (20))[116] and by insertion of metals into seven-vertex metallacarbaboranes, in the latter case giving heterobimetallic products (Equation (21)).[6] A very different metallacarbaborane synthesis[33] involved the reaction of the rhodaborane *nido*-6-$Cp^*RhB_9H_{13}$ with EtN≡C to form the adduct *arachno*-6,9-$(EtNC)_2$-$Cp^*RhB_9H_{11}$; in boiling toluene the adduct underwent carbon incorporation to form one- and two-carbon 11-vertex *closo*-metallacarbaboranes.

$$arachno\text{-}4,6\text{-}[C_2B_7H_{12}]^- + [(C_6Me_6)RuCl_2]_2 \xrightarrow[60\%]{CHCl_3} closo\text{-}2,1,6\text{-}(C_6Me_6)Ru(C_2B_7H_9) \quad (19)$$

$$closo\text{-}3,1,2\text{-}(C_6H_6)Ru(C_2B_9H_{11}) + KOH \longrightarrow closo\text{-}1,2,4\text{-}(C_6H_6)Ru(C_2B_8H_{10}) + nido\text{-}2,5,6\text{-}(C_6H_6)Ru(C_2B_7H_{11}) \quad (20)$$

$$closo\text{-}3,1,2\text{-}Cp^*Fe(Et_2C_2B_4H_4) \xrightarrow[\text{ii, }(Cp^*NiBr)_2]{\text{i, K}} closo\text{-}Cp^*_2NiFe(Me_2C_2B_4H_4) \quad (21)$$

9.6.2.2 Structures and bonding

X-ray crystallography has established the structures of a number of 8- to 11-vertex metallacarbaboranes (Table 3), most of which conform to the skeletal electron-counting rules. For example, the eight-vertex $NiFeC_2B_4$ polyhedron formed in Equation (21) has a *closo*-dodecahedral geometry as expected for an 18 skeletal electron system ($n + 1$ pairs), while 24-electron 10-vertex cages such as $(C_6H_6)Ru(C_2B_7H_{11})$ (Equation (20)) have a *nido* geometry analogous to $B_{10}H_{14}$, an isoelectronic borane. As noted above, however, many metallaboranes and metallacarbaboranes exhibit structural "distortions" (i.e., deviations from what we naively anticipate) that usually consist of either (i) unexpectedly long, nonbonding atom–atom distances that create open faces on the polyhedral framework or (ii) n-vertex *closo*-cage geometries that, while still deltahedral (all faces triangular) nevertheless differ from the structure usually found for that particular number of vertices. Interpretation of these distortions has engendered some lively controversy, centered largely on the issue of the nature and extent of electronic perturbation of the cluster by the metal centers. For example, *hyper-closo* species,[120] which have only n skeletal electron pairs (e.g., the 10-vertex, 20-electron diferracarbaborane $Cp_2Fe_2(C_2B_6H_8)$),[121] normally adopt capped polyhedral geometry since this satisfies the requirement for a filled-shell electronic system (the "capping principle"[4m]). An alternative view, espoused by Kennedy,[122] holds that such departures from "normal" geometry arise from a four-orbital contribution from the metal to the cluster, as distinct from the usual three-orbital contribution, and he has labeled such species "*iso-closo,*" extending the same idea to "*iso-nido*" and "*iso-arachno*" categories. This proposal has not been widely adopted, perhaps because it seems to run counter to Occam's Razor. In any event, Johnston *et al.*[4j,123] have examined the controversy from a theoretical viewpoint and concluded that the *hyper-closo* description is reasonable, and more generally applicable than the *iso-closo* model, although the latter may be useful in certain cases.

On the broader question of how the presence of metals influences the structures of polyhedral borane cages, a caveat is necessary. Metallaborane and metallacarbaborane systems that *appear* incompatible with the skeletal electron-counting rules may in fact be kinetically stabilized isolable intermediates that reflect mainly the mode of synthesis—that is, the metal is trapped at or near the point of attack on the borane skeleton, and a low-energy pathway for rearrangement is not available. Cases in point are the family of 14-vertex $Cp_2Fe_2(Me_4C_4B_8H_8)$ isomers discussed elsewhere[1a,124] and a recently described 12-vertex *klado*-$Co_2C_4B_6$ cluster, described in Section 9.8, that might appear to be an especially severe "violation" of Wade's rules but in fact is a kinetically trapped species. Caution seems well advised in drawing conclusions about the effects of metals on skeletal bonding based on the structures of clusters that formed under low-energy conditions, particularly where more than one cage metal vertex is present. Moreover, cage distortions can also be produced by steric interactions—as, for example, when bulky substituents are involved[223]—and by placement of electron-donor *exo*-polyhedral ligands on the metal, as in "*nido*"-3,1,2-$(NMe_2CH_2)_2Pd(C_2B_9H_{11})$, which exhibits a slipped geometry.[239]

9.6.2.3 Wedged metallacarbaboranes

commo-Bis(carboranyl) clusters having two seven-vertex cages joined at a common metal center with an additional atom (metal or boron) "wedged" between the polyhedra form a structurally distinct class. These complexes, of which the examples prepared since 1982 are listed at the end of Table 2, can also be viewed as edge-fused eight-vertex *commo* systems. In the majority of cases the wedging unit is a BH group that was displaced from its original location in one of the C_2B_3 rings by an incoming metal–ligand moiety,[1a,68,94] but transition metal or main group atoms can also occupy the wedge (Figure 18). The diiron complex depicted in Figure 18(b) is an intermediate in the conversion of $FeH_2(R_2C_2B_4H_4)_2$ to $R_4C_4B_8H_8$ via oxidative fusion.[67] This compound, an extremely oxygen-sensitive purple solid, has a low-spin iron(II) center at the *commo* vertex and a high-spin quasi-tetrahedral iron(II) in the wedging position, so that the molecule is paramagnetic with four unpaired electrons. Iron-wedged complexes of this type form from $FeH_2(R_2C_2B_4H_4)_2$ on standing in solution, and form more rapidly in the presence of a trace of $FeCl_3$; exposure to oxidants causes instant fusion to $R_4C_4B_8H_8$ with expulsion of the metals.[67]

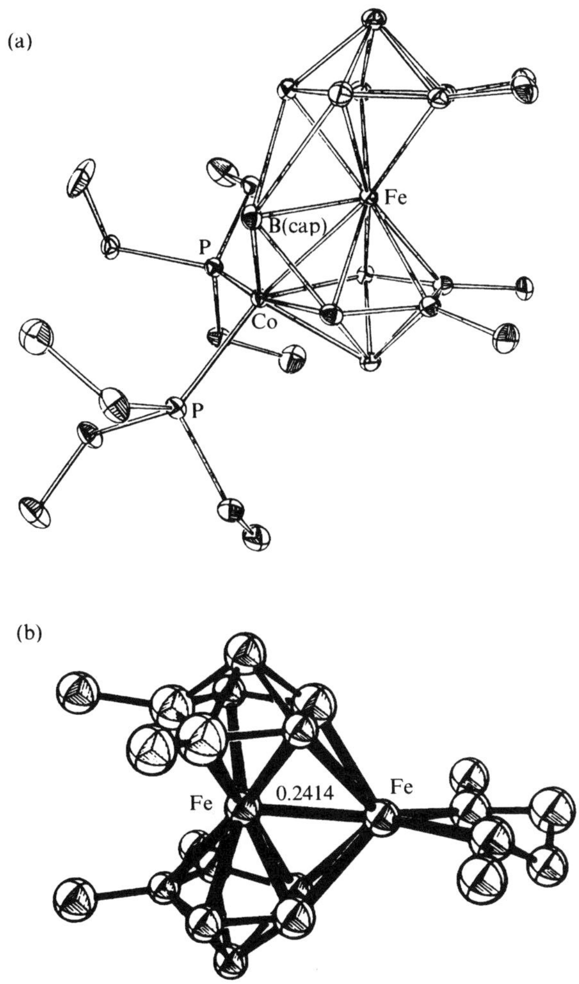

Figure 18 Structures of (a) B-wedged-$(Et_3P)_2CoFe(Me_2C_2B_4H_4)_2$ and (b) Fe-wedged-$C_2H_4(MeO)_2$-$Fe_2(Me_2C_2B_4H_4)_2$ ((a) reproduced by permission of the Royal Society of Chemistry from *J. Chem. Soc., Dalton Trans.*, 1982, 1679; (b) reprinted with permission from *J. Am. Chem. Soc.*, 1982, **104**, 5987; copyright 1982 American Chemical Society).

One might expect to find analogous wedged complexes in the larger metallacarborane families, but so far none have been reported (icosahedral *commo*-metallacarbaboranes having *exo*-polyhedral interligand bridging groups are known, as described below).

9.6.3 Icosahedral (12-vertex) Cages

Metal sandwich complexes of the 7,8- and 7,9-$[R^1R^2C_2B_9H_9]^{2-}$ (dicarbollide) ligands constitute the most widely studied class of metallacarbaboranes by far, reflecting primarily the commercial availability of the $R^1R^2C_2B_{10}H_{10}$ carbaboranes. While much of the work in this area emanates from the laboratories of Hawthorne and of Stone, these complexes have attracted attention from an impressive array of other groups as well. The level of activity is evident from Table 4, which lists the considerable number of new compounds that have been characterized since 1982. In this period, substantial progress has been made in the development of synthetic routes to specific compound classes, elucidation of molecular structures, detailed understanding of electronic structure and bonding, and application of 12-vertex metallacarbaboranes in several areas. Each of these topics is briefly surveyed below, with the reader referred to specialized reviews and research papers for detailed discussion.

Table 4 12-Vertex dicarbon metallacarbaboranes (since 1982).

Compound[a]	*Data*[b]	*References*
One metal atom[c]		
Scandium		
3,1,2-Cp*[(TMS)$_2$CH]Sc(C$_2$B$_9$H$_{11}$)$^-$Li$^+$	X	125
3,1,2-Cp*[(TMS)$_2$CH]Sc(C$_2$B$_9$H$_{11}$)$^-$Li$^+$	B, H, C	126
{3,1,2-Cp*[(TMS)$_2$CH]Sc(C$_2$B$_9$H$_{11}$)}$_2$Li·Li(OC$_4$H$_8$)$_3$	X	126
[3,1,2-Cp*(H)Sc(C$_2$B$_9$H$_{11}$)]$_2^{2-}$	X, H, C	126
3,1,2-(OC$_4$H$_8$)$_3$Cp*Sc(C$_2$B$_9$H$_{11}$)	B, H, C	126
Chromium		
3,1,2-Cr(C$_2$B$_9$H$_{10}$)$_2$-μ-(CH$_2$)$_4^-$	B, H, C, I	127
3,1,2-(Ph$_3$Sn)(CO)$_3$Cr(C$_2$B$_9$H$_{11}$)	B, Sn, I	128
3,1,2-L$_2$Pt=Cr(CO)$_3$(Me$_2$C$_2$B$_9$H$_9$) (L = Ph$_3$P, Et$_3$P)	B, H, C, P	129
Manganese		
3,1,2-(CO)$_3$Mn(RC$_6$H$_4$C$_2$B$_9$H$_{10}$)$^-$ (R = H, *m*-F, *p*-F)	F, I	130
3,1,2-(CO)$_3$Mn(C$_2$B$_9$H$_{10}$-4-SMe$_2$)	X, B, H	131
3,1,2-(CO)$_3$Mn(C$_2$B$_9$H$_{10}$-9-SMe$_2$)		132
3,1,2-(CO)$_3$Mn(C$_2$B$_9$H$_{11}$)·CD$^-$ (CD = α-, or β-cyclodextrin)	H, CD	133
Iron		
3,1,2-Fe(C$_2$B$_9$H$_{11}$)$_2^{2-}$	X	134
3,1,2-Fe(C$_2$B$_9$H$_{10}$-8-NEt$_3$)$_2$	X, B, H, I	134
3,1,2-Fe(C$_2$B$_9$H$_{10}$-4-C$_5$H$_4$NCOMe)$_2$	B, H, I	134
2,1,7-Fe(C$_2$B$_9$H$_{10}$-11-C$_5$H$_5$N)$_2$	B, H, I	134
3,1,2-Fe(C$_2$B$_9$H$_{11}$)$_2^{2-}$	(electrochem. synth.)	135
3,1,2-Fe(C$_2$B$_9$H$_{11}$)$^{2-}$	(electrochem. prep.)	136
3,1,2-Fe(R$_2$C$_2$B$_9$H$_9$)$_2^-$	r	137
3,1,2-[Fe(C$_2$B$_9$H$_{11}$)$_2$]$_2$Fe^{II}·4L (L = bipy, phen)		138
3,1,2-Fe(C$_2$B$_9$H$_{11}$)$_2^-$Na$^+$·2L (L = bipy, phen)		138
3,1,2-Fe(C$_2$B$_9$H$_{10}$-8-Cl)$_2$	X	139
3,1,2-Fe(C$_2$B$_9$H$_{10}$)$_2$-μ-8,8-MeO	X	140
2,1,7-Fe(C$_2$B$_9$H$_{10}$)$_2$-μ-N$_2$C$_3$H$_3$	X, E, B, H	141
3,1,2-CpFe(C$_2$B$_9$H$_{11}$)	E	40
3,1,2-CpFe(C$_2$B$_9$H$_{10}$-9-Br)		142
3,1,2-CpFe(C$_2$B$_9$H$_{10}$-9-HgCl)	B, I	142
3,1,2-CpFe(CH$_2$OHC$_2$B$_9$H$_{10}$)	I	143
3,1,2-CpFe(CHOC$_2$B$_9$H$_{10}$)	I	143
3,1,2-CpFe(CH$_2$COOHC$_2$B$_9$H$_{10}$)	I	143
3,1,2-CpFe(CH$_2$OAcC$_2$B$_9$H$_{10}$)	I	143
3,1,2-CpFe(CH$_2$OEtC$_2$B$_9$H$_{10}$)	I	143
3,1,2-CpFe(CH$_2$OCH$_2$CH$_2$OMeC$_2$B$_9$H$_{10}$)	I	143
3,1,2-CpFe[MeCH(OH)C$_2$B$_9$H$_{10}$]	I	143
3,1,2-CpFe[CH(OH)PhC$_2$B$_9$H$_{10}$]	I	143
3,1,2-CpFe(CH$_2$OHC$_2$B$_9$H$_{10}$)$^-$	H, I	143
3,1,2-CpFe(CHOEt$_2$C$_2$B$_9$H$_{10}$)$^-$	H, I	143
3,1,2-CpFe[MeCH(OH)C$_2$B$_9$H$_{10}$]$^-$	H, I	143
3,1,2-CpFe(CH$_2$ClC$_2$B$_9$H$_{10}$)	I	143
3,1,2-CpFe(COOHC$_2$B$_9$H$_{10}$)	I	143
3,1,2-CpFe[CH(OEt)$_2$C$_2$B$_9$H$_{10}$]	X, I	143
3,1,2-CpFe[C(O)MeC$_2$B$_9$H$_{10}$]	I	143
3,1,2-CpFe(MeC$_2$B$_9$H$_{10}$)	I	143
3,1,2-CpFe[C(O)CH$_2$DC$_2$B$_9$H$_{10}$]	S	143
3,1,2-CpFe(CH$_2$C$_2$B$_9$H$_{10}$) [zwitterion]	H, I	144
3,1,2-CpFe(CHMeC$_2$B$_9$H$_{10}$) [zwitterion]	H, I	144
3,1,2-CpFe(CHPhC$_2$B$_9$H$_{10}$) [zwitterion]	X, I	144
3,1,2-CpFe(CMe$_2$C$_2$B$_9$H$_{10}$) [zwitterion]	I	144
3,1,2-CpFe(C[R]=CH$_2$C$_2$B$_9$H$_{10}$)$^-$ [zwitterion]	H, I	144
3,1,2-CpFe(C$_3$H$_3$RC$_2$B$_9$H$_{10}$) (R = H, Ph)	H, C, I	145
3,1,2-CpFe(PhCH$_2$CH=CHC$_2$B$_9$H$_{10}$)	I	145
3,1,2-CpFe(MeCH=CHC$_2$B$_9$H$_{10}$)$^-$	H, I	145
3,1,2-CpFe(CH$_2$=CHCH$_2$C$_2$B$_9$H$_{10}$)$^-$	H, I	145
3,1,2-CpFe(HCCC$_2$B$_9$H$_{10}$)	I	146
3,1,2-CpFe(MeCOC$_2$B$_9$H$_{10}$)	I	146
3,1,2-CpFe[CH$_2$=C(OMe)C$_2$B$_9$H$_{10}$]	I	146
3,1,2-CpFe(HCCC$_2$B$_9$H$_{10}$)$^-$	I	146
3,1,2-CpFe(H$_2$C=CC$_2$B$_9$H$_{10}$)	H, I	146
3,1,2-CpFe[C(Me)ClC$_2$B$_9$H$_{10}$]	H, I	146
3,1,2-CpFe[C(Me)BrC$_2$B$_9$H$_{10}$]	H, I	146
3,1,2-CpFe(C$_2$B$_9$H$_9$-8-R^1-9-R^2-12-R^3) (R^1, R^2, R^3 = H, OH, Cl, Br)	S(detailed)	147
3,1,2-CpFe(*m*,*p*-FC$_6$H$_4$C$_2$B$_9$H$_{10}$)$^-$	F(Taft)	148
3,1-[(3,1,2-C$_5$H$_4$)Fe(C$_2$B$_9$H$_{11}$)]$_2$C(Me)$_2$CH$_2$C(Me)SMe$_2$	I	149

Table 4 (continued)

Compound[a]	*Data*[b]	*References*
3,1-$[(3,1,2\text{-}C_5H_4)Fe(C_2B_9H_{11})]_2C(Me)_2CH{=}CMe$	I	149
3,1,2-$[CH_2{=}C(Me)C_5H_4]Fe(C_2B_9H_{11})$	I	149
3,1,2-$[CH_2{=}C(Me)C_5H_4]Fe(C_2B_9H_{11})^-$	H, I	149
3,1,2-$[Me_2(O)Me_2CC_5H_4]Fe(C_2B_9H_{11})^-$	H, I	149
3,1,2-$[Me_2(S)Me_2CC_5H_4]Fe(C_2B_9H_{11})^-$	H, I	149
3,1,2-$Fe(C_2B_9H_{10})_2$-μ-$(CH_2)_4^-$	X, B, H, C, I	127
3,1,2-$Fe(C_2B_9H_{10})_2$-μ-$TosN(CH_2CH_2)_2^-$	B, I	127
3,1,2-$(C_6H_6)Fe(C_2B_9H_{11})$	X, B, H, I	10
2,1,7-$RFe(Me_2C_2B_9H_9)$ (R = C_6H_6,1,3-$C_6H_4Me_2$, $C_{10}H_8$)	B, H, I	150
2,1,7-$(MeC_6H_5)Fe(Me_2C_2B_9H_9)$	X, B, H, I	150
2,1,7-$[(MeO)_3P]_3Fe(Me_2C_2B_9H_9)$	P	150
2,1,7-$(CO)_3Fe(Me_2C_2B_9H_9)$	I	150
3,1,2-$(MeC_6H_5)Fe(C_2B_9H_{11})$	X, B, H	151
3,1,2-(1,4-$Me_2C_6H_4)Fe(C_2B_9H_{11})$	X, B, H	151
3,1,2-(1,3,5-$C_6Me_3H_3)Fe(C_2B_9H_{11})$	X, B, H, I	11a
3,1,2-$(C_6Me_6)Fe(C_2B_9H_{11})$	B, H, I	11a
3,1,2-$(CO)L^1L^2Fe(C_2B_9H_{11})$ (L^1, L^2 = CO, PPh_3, MeCN, $P(OMe)_3$)	B, H	152
3,1,2-$(CO)_2(PPh_3)Fe(C_2B_9H_{11})$	X	152
3,1,2-$(CO)_2[P(OMe)_3]Fe(C_2B_9H_{11})$	X	152
3,1,2-$(CO)(MeCN)(PPh_3)Fe(C_2B_9H_{11})$	X	152
3,1,2-$(CO)_2Fe(C_2B_9H_{11})^{2-}$	B	153
3,1,2-$(CO)_2LFe(C_2B_9H_{11})^-$ (L = Me, $PhCH_2$, Ph_3Sn, COMe)	X[Ph_3Sn], B, H	153
3,1,2-$(CO)(\eta^3\text{-}CH_2CRCH_2)Fe(C_2B_9H_{11})^-$ (R = H, Me)	X[R = H], B, H	153
3,1,2-$(CO)(COMe)(PMe_3)Fe(C_2B_9H_{11})^-$	X, B, H	153
Cobalt		
3,1,2-$CpCo(C_2B_9H_{10})$-4-Me_2S	(separation of enantiomers)	154
3,1,2-$CpCo(C_2B_9H_{11})$	(detailed synthesis)	155
3,1,2-$CpCo(C_2B_9H_{11})$	X	156
3,1,2-$CpCo(C_2B_9H_{11})$	isomerization	157
3,1,2-$CpCo(C_2B_9H_{11})$	E	40
3,1,2-$Co(C_2B_9H_{11})_2^-$	E	40
2,1,7-$Co(C_2B_9H_{11})_2^-$	E	40
2,1,12-$CpCo(C_2B_9H_{11})$	B, I	158
3,1,2-$CpCo(R(C_2B_9H_{10})$ (R = H, Me, Ph, $HOCH_2$, $ClCH_2$, HOOC)	S	159
3,1,2-$Cp^*Co(C_2B_9H_{11})$	B(2D),H(2D)	160
3,1,2-(*endo*-$C_6Me_6H)Co(C_2B_9H_{11})$	B, H, I	11b
3,1,2-$(\eta^5\text{-}C_{13}H_9)Co(C_2B_9H_{11})$ ($C_{13}H_9$ = fluorenyl)	X	161
3,1,2-$L_4[Co(C_2B_9H_{11})_2]_2$ (L = bipy, phen)		162
3,1,2-$[Co(C_2B_9H_{11})_2]Na{\cdot}2L$ (L = bipy, phen)		162
3,1,2-$Co(C_2B_9H_{11})_2^-$	X	163
3,1,2-$Co(C_2B_9H_{11})_2^-$	B, H, I	164
3,1,2-$Co(C_2B_9H_{11})_2^-$	B, H, I, U	165
3,1,2-$Co(C_2B_9H_{11})_2^-$	M	166
3,1,2-$Co(C_2B_9H_{11})_2^-$	electrochem. prep.	167
3,1,2-$Co(C_2B_9H_{11})_2^-$	dielec. permittivity	168
3,1,2-$(C_2B_9H_{11})Co(C_2B_9H_{10})$-8-$X)^-$ (X = Cl, I)	B, H, U, I	165
3,1,2-$(C_2B_9H_{11})Co(C_2B_9H_{10})$-8-$Br)^-$	B, H, U, I, E	165
3,1,2-$Co(C_2B_9H_{10})$-8-$Cl)_2$	X	139
3,1,2-$Co(C_2B_9H_{10})$-8-$X)_2^-$ (X = Cl, Br, I)	B, H, U, I	165
3,1,2-$Co(C_2B_9H_{10})$-8-$I)_2^-$	X	169
3,1,2-$(C_2B_9H_{10})$-8-$Cl)Co(C_2B_9H_9$-8,9-$Cl_2)^-$	B, H, U, I	165
3,1,2-$Co(C_2B_9H_{10})$-9-$Cl)_2^-$	B, H, U, I	165
3,1,2-$Co(C_2B_9H_9$-8,9-$Cl_2)_2^-$	B, H, U, I	165
3,1,2-$Co(C_2B_9H_9$-9,12-$Cl_2)_2^-$	B, H, U, I	165
3,1,2-$Co(C_2B_9H_8$-8,9,12-$X_3)_2^-$ (X = Cl, I)	B, H, U, I	165
3,1,2-$(C_2B_9H_{11})Co(C_2B_9H_{10})$-8-$I)^-$	X	170
3,1,2-$CpCo(MeEt_2C_3B_2Et_2)Co(C_2B_9H_{11})$	X, B, H, E	171
2,1,7-$Co(C_2B_9H_{10})_2$-μ-6,6′-ER (E = O, S, N)	(separation of enantiomers)	172
3,1,2-$Co(C_2B_9H_{10})_2$-μ-8,8-O^-	X	30
3,1,2-$Co(C_2B_9H_{10})_2$-μ-$TosN(CH_2CH_2)_2^-$	B, I	127
3,1,2-$Co(C_2B_9H_{10})_2$-μ-$(CH_2)_n^-$ (*n* = 3–5)	X(*n* = 3, 4), B, H, C, I	127
2,1,7-$Co(C_2B_9H_{10})_2$-μ-$N_2C_3H_3$	X, E, B, H	141
2,1,7-$Co(C_2B_9H_{10})_2$-μ-$N_2C_3H_2$-CO_2H	X, B, H	141
2,1,7-$^{57}Co(C_2B_9H_{10})_2$-μ-$N_2C_3H_2$-CO_2H		141
3,1,2-$Co(C_2B_9H_{10})_2$-μ-1,2-$C_6H_4^-$		173
3,1,2-$Co(C_2B_9H_{10})_2$-μ-$SCH_2C(O)OMe$	X	174
3,1,2-$Co(C_2B_9H_{10})_2$-μ-I	B, H	175
3,1,2-$(C_2B_9H_{10}$-8-$L)Co(1,2\text{-}(C_2B_9H_{10})$-8-I) (L = Me_2S, C_5H_5N, Me_3N, H_3N)	B, H	175

Table 4 (continued)

Compound[a]	*Data*[b]	*References*
3,1,2-$Co(C_2B_9H_{10})C_5H_5NCOOMe)_2Cl$	B, H, I	134
2,1,7-$Co(C_2B_9H_{10})_2$-μ-$[N_2C_3H_2(C(O)OMe)]^-$	X	176
3,1,2-$Co(R_2C_2B_9H_9)_2^-$	r	137
3,1,2-(η^5-C_9H_7)$Co(C_2B_9H_{11})$ (C_9H_7 = indenyl)	X, B, H, U, E	156
3,1,2-(η^5-C_9H_7)$Co(R^1R^2C_2B_9H_9)$ (R^1R^2 = Ph, H; CH_2OMe, H; CH_2OMe, CH_2OMe)	X, B, H, I	177
3,1,2-$Co(C_2B_9H_8Br_3)_2^-$	E	178
3,1,2-$Co(C_2B_9H_8$-8,9,12-$Br_3)_2^-$	X	179
3,1,2-$Co_2(C_2B_9H_{11})_2(C_2B_8H_{10})^{2-}$	M	166
3,1,2-$Co_3(C_2B_9H_{11})_2(C_2B_8H_{10})_2^{3-}$	M	166
3,1,2-$Co_4(C_2B_9H_{11})_2(C_2B_8H_{10})_3^{4-}$	M	166
3,1,2-$Co_2(C_2B_9H_{11})_2(C_2B_8H_{10})^{2-}$	B, H, I	164
3,1,2-$Co_3(C_2B_9H_{11})_2(C_2B_8H_{10})_2^{3-}$	B, H, I	164
3,1,2-$Co_4(C_2B_9H_{11})_2(C_2B_8H_{10})_3^{4-}$	B, H, I	164
Nickel		
3,1,2-$(Ph_3P)(CO)Ni(C_2B_9H_{11})$	B, H, P	180
3,1,2-$(Ph_3P)(NC_5H_5)Ni(C_2B_9H_{11})$	B, H, P	180
2,1,7-$(Ph_3P)_2Ni(C_2B_9H_{11})$	B, H, P	180
2,1,12-$(Ph_3P)_2Ni(C_2B_9H_{11})$	B, H, P	180
3,1,2-$L_2Ni(C_2B_9H_{11})$ (L = Ph_3P, (p-$MeC_6H_4)_3P$, Et_3P, Ph_2MeP, $PhMe_2P$)	B, H, P	180
3,1,2-$[Ph_2P(CH_2)_3PPh_2]_2Ni(C_2B_9H_{11})$	H, P	180
3,1,2-$(Ph_3P)HNiC_2B_9H_{10}$-8-PPh_3	B, H, P	180
3,1,2-$(Ph_3P)(Cl)NiC_2B_9H_{10}$-8-$PPh_3$	B, H, P	180
3,1,2-[(p-$MeC_6H_4)_3P]HNiC_2B_9H_{10}$-8-P($p$-$MeC_6H_4)_3$	B, H, P	180
3,1,2-$Ni(C_2B_9H_{10}$-8-$PPh_3)_2$-μ-$(CO)_2$	X, B, H, P	180
3,1,2-(2,2-$NC_5H_4)_2)Ni(C_2B_9H_{11})$	H, C, I	181
3,1,2-(bipy)$Ni(C_2B_9H_{11})$	B, H, I	182
3,1,2-(bipy)$Ni(C_2B_9H_{11})$	I	183
2,1,7-(bipy)$Ni(C_2B_9H_{11})$	I	183
3,1,2-(bipy)$Ni(C_2B_9H_{11})$	X	184
2,1,7-(bipy)$Ni(C_2B_9H_{11})$	X	184
$Ni^{2+}[Ni(1,2$-$(C_2B_9H_{11})_2^-]_2\cdot 4L$ (L = bipy, pyridine)	I	182
$Na^+[Ni(1,2$-$(C_2B_9H_{11})_2^-]\cdot 2$bipy	I	182
3,1,2-$CpCo(MeEt_2C_3B_2Et_2)Ni(C_2B_9H_{11})$	X, H, E	171
3,1,2-Ph_2B(pyrazolyl)$_2Ni(C_2B_9H_{11})^-$	X, B, H, I, U	185
3,1,2-$[Ni(C_2B_9H_{11})_2]_2Ni\cdot 4L$ (L = phen)		186
3,1,2-$Ni(C_2B_9H_{11})_2\cdot 2L$ (L = phen)		186
2,1,7-$Ni(C_2B_9H_{11})_2\cdot 2L$ (L = phen)		186
3,1,2-$Ni(C_2B_9H_{10}$-8-$OMe)_2$	X	187
2,1,7-$Ni(C_2B_9H_{10})_2$-μ- $N_2C_3H_3$	X(meso, (±)), E, B, H	141
3,1,2-$Ni(C_2B_9H_{10})_2$-μ-$(CH_2)_4^-$	X, B, H, C, I	127
3,1,2-$Ni(R_2C_2B_9H_9)_2^-$	r	137
Copper		
3,1,2-$(Ph_3P)Cu(C_2B_9H_{11})^-$	X, B, H, P, I	188
3,1,2-$(Ph_3P)CuC_2B_9H_{10}$-4-$NC_5H_4CO_2Me$	X, B, P, H, I	188
3,1,2-$(Ph_3P)CuC_2B_9H_9$-*exo*-4,8-(μ-H)$_2Cu(PPh_3)$	X, B, P, H, I	188
nido-Cu_3(μ-H)$_3$(7,8-$C_2B_9H_9$-9-$NC_5H_4CO_2Me)_3$	X, B, H, I	188
3,1,2-$(Ph_3P)CuC_2B_9H_{10}$-4-SMe_2	X, B, H	189
2,1,7-$Cu(C_2B_9H_{10})_2$-μ-$N_2C_3H_3$	X, E, B, H	141
3,1,2-$(Ph_3P)CuC_2B_9H_9$-μ-$CuH_2(Ph_3P)$	X, B, H, P, I	190
Zirconium		
3,1,2-Cp*(MeC=CMe_2)$Zr(C_2B_9H_{11})$	X	191
3,1,2-(Cp*$Zr(C_2B_9H_{11})_2$(μ-CH_2)	X	191
3,1,2-(Et_8porphyrin)$Zr(C_2B_9H_{11})$	X, B, H, C, I, U	192
Molybdenum (no exo-*polyhedral metals)*		
3,1,2-$(Ph_3Sn)(CO)_3Mo(C_2B_9H_{11})$	B, Sn, I	128
3,1,2-$(CO)_2$(μ-$MeC_6H_4C)Mo(C_2B_9H_{11})^-$	B, H, C	193
3,1,2-$(CO)_3MoC_2B_9H_{10}$-1-CH_2-[*cyclo*-$N_3P_3(C_5H_{10}N)_4Me$]	P, I	194
3,1,2-CpMo(μ-$C_3H_3)C_2B_9H_9$	X, B, H, C	195
3,1,2-$(L)(CO)_3Mo(Me_2C_2B_9H_8CH_2C_6H_4Me)$ (L = PMe_3, CO)	B, H, C, P	196
3,1,2-$[(MeO)_3P](CO)(MeC_6H_4C)Mo(Me_2C_2B_9H_9)^-$	B, H, C, P	196
3,1,2-(μ-$MeC_6H_4C)(CO)_3Mo(Me_2C_2B_9H_9)^-$	B, H, C, P	196
3,1,2-μ-MeC_6H_4C=$Mo(CO)_2(PMe_3)(Me_2C_2B_9H_9)^-$	H, C	197
3,1,2-$(CO)_2(C_3H_5)Mo(R_2C_2B_9H_9)$ (R = H, Me)	B, H, C, P, I	198
3,1,2-$(CO)_2L^1L^2Mo(Me_2C_2B_9H_9)$ (L^1, L^2 = CO, Ph_3P)	B, H, C, P, I	198
3,1,2-$(CO)_2(C_4H_6)Mo(C_2B_9H_{11})$	B, H, C, P, I	198
3,1,2-(CO)L(RC≡CR)$Mo(R_2C_2B_9H_9)$ (L = Me_2C_2, Ph_2C_2, Ph_3P, PMe_2Ph; R = Me, Ph)	B, H, C, P, I	198

Table 4 (continued)

Compound[a]	*Data*[b]	*References*
Molybdenum (cage), iron (exo-*polyhedral*)		
3,1,2-[μ-$(CO)_6Fe_2$]$(CO)_2$(μ-MeC_6H_4C)(μ-H)Mo($Me_2C_2B_9H_7$)	B, H, C, P	196
3,1,2-[μ-$(CO)_6Fe_2$]$(CO)_2$(μ-MeC_6H_4C)Mo$(Me_2C_2B_9H_7)^-$	X, B, H, C, P	196
3,1,2-[μ-$(CO)_6Fe_2$]$(CO)_2$(μ-MeC_6H_4C)Mo$(Me_2C_2B_9H_7)^-$	X, B, H, C, P	196
Molybdenum (cage), cobalt (exo-*polyhedral*)		
(η-C_4Me_4)(CO)Co(μ-MeC_6H_4C)$(CO)_2$Mo($Me_2C_2B_9H_9$)	B*, H, C, I*	317
Molybdenum (cage), copper (exo-*polyhedral*)		
3,1,2-[μ-Cu(H)(CO)]$_2$-[3,1,2-(CO)Mo$C_2B_9H_{10}]_2^{2-}$	X, B, H, I	199
Molybdenum (cage), molybdenum (exo-*polyhedral*)		
3,1,2-(η-C_7H_7)Mo(μ-MeC_6H_4C)(CO)(L)Mo($Me_2C_2B_9H_9$) (L = CO, PMe_3)	B, H, C, P	193
3,1,2-($η^5$-C_9H_7)Mo[P$(OMe)_3$](μ-MeC_6H_4C)$(CO)_2$Mo($Me_2C_2B_9H_9$) (C_9H_7 = indenyl)	B, H, C, P	193
3,1,2-μ-RC-CpLMo-$(CO)_2$Mo($Me_2C_2B_9H_8$-$CH_2C_6H_4Me$) (R = Me, C_6H_4Me; L = CO, PMe_3)	B, H, C	200
3,1,2-$(CO)_7Mo_2$(μ-MeC_6H_4C)$(CO)_2$Mo$(R_2C_2B_9H_9)^-$ (R = Me, H)	B, H, C	201
[3,1,2-(CO)Mo$C_2B_9H_{10}$-1-CH_2PMeN[Mo$(CO)_4$]P$(C_5H_{10}N)_2$N]$_n^{2-}$	P, I	194
Molybdenum (cage), rhodium (exo-*polyhedral*)		
3,1,2-(μ-CO)[$(Ph_3)_2$Rh][P$(OMe)_3$](MeC_6H_4C)Mo($Me_2C_2B_9H_9$)	B, H, C, P	196
Molybdenum (cage), tungsten (exo-*polyhedral*)		
3,1,2-$(CO)_2$[(C_6H_4MeC=W(CO)Cp]-Mo($Me_2C_2B_9H_9$)	B, H, C, P, I	198
3,1,2-μ-RC(CpLW)$(CO)_2$Mo($Me_2C_2B_9H_8CH_2C_6H_4Me$) (R = Me, C_6H_4Me; L = CO, PMe_3)	B, H, C	200
Molybdenum (cage), platinum (exo-*polyhedral*)		
3,1,2-L_2Pt=Mo$(CO)_3$($Me_2C_2B_9H_9$) (L = Ph_3P, Et_3P)	B, H, C, P	129
3,1,2-(PMe_2Ph)LPt(μ-CC_6H_4Me)$(CO)_2$Mo($Me_2C_2B_9H_8$) (L = CO, PMe_2Ph)	B, H, C, P	202
3,1,2-$(Et_3P)_2$Pt$(CO)_2$Mo($Me_2C_2B_9H_8CH_2C_6H_4Me$)	I	202
3,1,2-$(Et_3P)_2$PtH$(CO)_2$LMo($Me_2C_2B_9H_7CH_2C_6H_4Me$) (L = CO, PEt_3, dppm, Me_3)	B, H, C, P	202
3,1,2-$(Et_3P)_2$Pt=Mo$(CO)_2$H($Me_2C_2B_9H_7CH_2C_6H_4Me$)	B, H, C, P	202
Molybdenum (cage), gold (exo-*polyhedral*)		
3,1,2-(μ-Ph_3PAu)[P$(OMe)_3$](CO)(MeC_6H_4C)Mo($Me_2C_2B_9H_9$)	B, H, C, P	196
Ruthenium		
3,1,2-(C_6H_6)Ru($C_2B_9H_{11}$)	B, H, I	11a
3,1,2-(C_6H_6)Ru($C_2B_9H_{11}$)	X	150
3,1,2-(C_6H_6)Ru($C_2B_9H_{11}$)	B, H, E	157
3,1,2-(C_6Me_6)Ru($C_2B_9H_{11}$)	B(2D), H(2D)	203
3,1,2-(C_6Me_6)Ru($C_2B_9H_{11}$)	X	204
3,1,2-(S_2CH)(Ph_3P)Ru($C_2B_9H_{11}$)	X	205
3,1,2-$(CO)_3$Ru($C_2B_9H_{11}$)	I	150
3,1,2-$(CO)_3$Ru($C_2B_9H_{11}$)	B, I	206
3,1,2-$(CO)_2$HRu$(C_2B_9H_{11})^-$	B, I	206
3,1,2-$(CO)_2$(COOMe)Ru$(C_2B_9H_{11})^-$	B, H, I	206
3,1,2-$(CO)_2$(COMe)Ru$(C_2B_9H_{11})^-$	B, H, I	206
3,1,2-$(CO)_2$($OCOCF_3$)Ru$(C_2B_9H_{11})^-$	B, H, I	206
3,1,2-[(CO)(μ-CO)Ru$(C_2B_9H_{11})]_2^{2-}$	B, H, I	206
Rhodium (no exo-*polyhedral metals)*		
3,1,2-(Ph_3P)(η-C_3H_5)Rh($C_2B_9H_{11}$)	X, B, P	207
3,1,2-(Ph_3P)($η^3$-C_4H_7)Rh($C_2B_9H_{11}$)	H, P	207
3,1,2-(Ph_3P)(MeCOMe)Rh($C_2B_9H_{11}$)	H, P	207
2,1,7-(Ph_3P)(MeCOMe)Rh($C_2B_9H_{11}$)	H, P	207
[18-crown-6][3,1,2-(Ph_3P)($PhCH_2$)BrRh$(C_2B_9H_{11})]^-$	H, P	207
3,1,2-(Ph_3P)(CO)(COMe)Rh($C_2B_9H_{11}$)	H, P	207
2,1,7-(Ph_3P)(CO)(COMe)Rh($C_2B_9H_{11}$)	H, P	207
2,1,7-(Ph_3P)(CO)($COCH_2Ph$)Rh($C_2B_9H_{11}$)	H, P	207
3,1,2-(Ph_3P)(CO)MeRh($C_2B_9H_{11}$)	H	207
3,1,2-CpRh($C_2B_9H_{11}$)	B, H	157
2,1,7-(CpRh($C_2B_9H_{11}$)	B, H	157
3,1,2-(Cp*Rh($C_2B_9H_{11}$)	X, H(2D), B	160
2,1,7-L^1L^2(Cl)Rh($C_2B_9H_{11}$) (L^1 = Ph_3P; L^2 = CO, C_5H_5N, MeCN, Ph_3P; L^1L^2 = $(Ph_2PCH_2)_2$)	P	208
3,1,2-L_2XRh($R^1R^2C_2B_9H_9$) (X = H, D, Cl; L = Ph_3P, PEt_3, $(Ph_2PCH_2)_2$, $PhMe_2P$; R^1, R^2 = H, D, Me, Bu, Ph)	P, H, I	208
3,1,2-[CH_2CH_2C(O)OBu](Ph_3P)Rh($MeC_2B_9H_{10}$)	X, B, H, P	208
2,1,12-$(Ph_3P)_2$HRh($C_2B_9H_{11}$)	P, H, I	208
d-3,1,2-$(Ph_3P)_2$HRh($MeC_2B_9H_{10}$)	P, o	208
3,1,2-[CH_2CH(Ph)C(O)OEt](Ph_3P)Rh($MeC_2B_9H_{10}$)	H, P	208
2,1,7-L_2HRh($RC_2B_9H_{10}$) (L = Ph_3P, PEt_3, $PhMe_2P$; R = H,Me,Ph)	P, H, I	208

Table 4 (continued)

Compound[a]	*Data*[b]	*References*
3,1,2-[$CH_2CH(Me)C(O)OMe$]((Ph_3P))Rh($MeRC_2B_9H_9$) (R = H,Me)	H, P	208
2,1,7-(Ph_3P)(Cl)Rh($R^1R^2C_2B_9H_9$) (R^1 = H, Me; R^2 = Ph, Me)	H, P	208
2,1,7-(Ph_3P)(Cl)Rh($C_2B_9H_{11}$)	X, P	208
2,1,7-($C_7H_7CH_2$)Rh($C_2B_9H_{11}$) ($C_7H_7CH_2$ = 2-methylenenorbornadienyl)	H, C	209
3,1,2-($C_7H_7CH_2$)Rh($R^1R^2C_2B_9H_9$) (R^1 = H, Me, Ph; R^2 = H, Me)	X[Me_2], H, C	209
3,1,2-(C_9H_7)Rh($PhC_2B_9H_{10}$)	X*	210
3,1,2-(C_9H_7)Rh[$(MeOCH_2)_2C_2B_9H_9$]	B(2D), H, I	210
3,1,2-$(Ph_3P)_2$(Cl)Rh($C_2B_9H_{11}$)		211
3,1,2-$(Ph_3P)_2$BrRh($C_2B_9H_{11}$)	H, P, I	211
3,1,2-(NO_3)$(Ph_3P)_2$Rh($C_2B_9H_{11}$)	P, I	211
3,1,2-$(Ph_3P)_2$(HSO_4)Rh($C_2B_9H_{11}$)	B, P, I	211
3,1,2-$(Ph_3P)_2$DRh($C_2B_9H_{11}$)	I	211
3,1,2-(NO_3)$(PhMe_2P)_2$Rh($C_2B_9H_{11}$)	B, P, I	211
3,1,2-(NO_3)(Ph_3P)(CO)Rh($C_2B_9H_{11}$)	P, I	211
3,1,2-(NO_3)(Ph_3P)Rh($C_2B_9H_{11}$)	B, P, I	211
3,1,2-(Ph_3P)[C(Ph)C(Ph_3P)CHC(Ph)]Rh($C_2B_9H_{11}$)	X, H, P, I	211
[3,1,2-(Ph_3P)(μ-CN)Rh($C_2B_9H_{11}$)]$_4$	X, B, H, P, I	211
3,1,2-(Ph_3P)(CO)(Cl)Rh($C_2B_9H_{11}$)	P, I	211
3,1,2-$(NCMe)_2$(Ph_3P)Rh($C_2B_9H_{11}$)$^+$$HSO_4^-$	B, P, I	211
3,1,2-$(Ph_3P)_2$Me(COO)Rh($C_2B_9H_{11}$)	I*	212
2,1,7-$(Ph_3P)_2$Me(COO)Rh($C_2B_9H_{11}$)	B, H, P	212
2,1,12-$(Ph_3P)_2$Me(COO)Rh($C_2B_9H_{11}$)	B, H, P	212
3,1,2-(Ph_3P)Me(COO)Rh($C_2B_9H_{11}$)	B, H, P	212
3,1,2-(Ph_3P)(Ph_2P-*o*-C_6H_4)Rh($C_2B_9H_{11}$)	X, B, H, P*	212
2,1,7-(Ph_3P)[MeCHO(O)CMe]Rh($C_2B_9H_{11}$)	B, H, P	212
2,1,12-$(Ph_3P)_2$(H)Rh($C_2B_9H_{11}$)	H, P, I	158
3,1,2-$(Ph_3P)_2$Rh($C_2B_9H_{11}$)$^-$	X, B, P, I	213
2,1,7-$(Ph_3P)_2$Rh($C_2B_9H_{11}$)$^-$	X, B, P, I	213
3,1,2-(L)(Ph_3P)Rh($C_2B_9H_{11}$)$^-$ (L = CO, C_2H_4)	B, P, I	213
2,1,7-(L)(Ph_3P)Rh($C_2B_9H_{11}$)$^-$(L = CO, C_2H_4)	B, P, I	213
2,1,12-(L)(Ph_3P)Rh($C_2B_9H_{11}$)$^-$ (L = CO, Ph_3P)	B, P, I	213
3,1,2-$Ph_2P(C_6H_5)$(Ph_3P)Rh($C_2B_9H_{11}$)	X	214
3,1,2-(NO)(Ph_3P)Rh($Me_2C_2B_9H_9$)	B, H, C, P, I*	215
2,1,7-(Ph_3P)[η^2-C(*m*-FC_6H_4)NOC(=O)]Rh($C_2B_9H_{11}$)$^-$	X, H, P, I	216
3,1,2-(Ph_3P)[η^2-C(*m*-FC_6H_4)NOC(=O)]Rh($C_2B_9H_{11}$)$^-$	H, P, I	216
3,1,2-(Ph_3P)[η^2-C(*m*-C_6H_5)NOC(=O)]Rh($C_2B_9H_{11}$)$^-$	H, P, I	216
[3,1,2-(R_3P)Rh($C_2B_9H_{11}$)]$_2$ (R = Et, Me_2Ph)	B, H, P	217
μ-(Ph_3P)(CO)Rh-3,1,2-(Ph_3P)(CO)Rh($C_2B_9H_{11}$)	X, B, H, C, P	218
3,1,2-(Ph_3P)Br_2Rh($C_2B_9H_{11}$)$^-$$Ph_3PH^+$ (ion pair)	X, B, H, P, I	219
3,1,2-(Ph_3P)Cl_2Rh($C_2B_9H_{11}$)$^-$	B, H, P, I	219
3,1,2-(Ph_3P)I_2Rh($C_2B_9H_{11}$)$^-$	B, H, P, I	219
3,1,2-(Ph_3P)Br_2Rh($C_2B_9H_{11}$)$^-$$K^+$ (18-crown-6)	B, H, P, I	219
3,1,2-HB(pyrazolyl)$_3$Rh($C_2B_9H_{11}$)	X, B, H, I, U	185
2,1,7-HB(pyrazolyl)$_3$Rh($C_2B_9H_{11}$)	X, B, H, I, U	185
[(*t*-BuC)$_2P_2$]Rh($C_2B_9H_{11}$)	X, B, H, C, P	220
3,1,2-$(Ph_3P)_2$HRh($C_2B_9H_{10}$)-1-CH_2-[*cyclo*-$N_3P_3(C_5H_{10}N)_4$Me]	H, P, I	194
[3,1,2-$(Ph_3P)_2$HRh($C_2B_9H_{10}$)-1-CH_2PMeNP$(C_5H_{10}N)_2$N]$_n$	P, I	194
3,1,2-H$(PPh_3)_2$Rh{[PhNHC(O)]$_2C_2B_9H_{10}$}	I, (catal. hydrosilyl.)	221
3,1,2-{H$(PPh_3)_2$Rh[HNPhPhNHC(O)](CO)$C_2B_9H_{10}$}$_n$	I, (catal. hydrosily.)	221
2,1,8-$(Et_3P)_2$HRh($C_2B_9H_{10}$)-1-$C_2B_{10}H_{11}$	H, P, I	222
3,1,2-$(Ph_3P)_2$HRh($C_2B_9H_9$)-μ-$CH_2C_6H_4CH_2$	H, P, I	222
3,1,2-$(Et_3P)_2$HRh($C_2B_9H_9$)-μ-$CH_2C_6H_4CH_2$	H, B, P, I	222
[2,1,8-$(Et_3P)_2$HRh($MePhC_2B_9H_9$)]$_2$	H, B, P, I	222
3,1,2-$(Et_3P)_2$HRh($Me_2C_2B_9H_9$)	H, B, P, I	222
3,1,2-$(Et_3P)_2$HRh($C_2B_9H_9$)-μ-$(CH_2)_3$	H, B, P, I	222
pseudo-nido-3,1,2-Cp*Rh($Ph_2C_2B_9H_9$)	X, B, H, I	223
3,1,2-(EtC_5H_4)Rh($Me_2C_2B_9H_9$)	B, H	224
3,1,2-(C_7H_8)Rh($Me_2C_2B_9H_9$)$^-$ (C_7H_8 = norbornadiene)	H	224
3,1,2-(C_7H_9)Rh($Me_2C_2B_9H_9$)	X, H	224
3,1,2-(η^5-C_8H_{11})Rh($Me_2C_2B_9H_9$)	X, H, C	225
3,1,2-(η^3-C_8H_{13})(CO)Rh($Me_2C_2B_9H_9$)	H, C	225
3,1,2-$(Et_2P)_2$HRh($Me_2C_2B_9H_9$)	H, C	225
3,1,2-(Me_2P)ClRh($Me_2C_2B_9H_9$)	X, H, C	225
μ-HC(C_6H_4Me)-3,1,2-(C_8H_{12})Rh($Me_2C_2B_9H_9$)	B, H	226
μ-HC(C_6H_4Me)-3,1,2-(Ph_3P)(CO)Rh($Me_2C_2B_9H_9$)	B, H	226
(L = CO, PMe_2Ph; R = Me, H)	X(L = CO, R = Me), B, H, C, P	226
3,1,2-(η^3-C_8H_{13})Rh($Me_2C_2B_9H_9$)	X, H	227

Table 4 (continued)

Compound[a]	*Data*[b]	*References*
3,1,2-(η^4-C_8H_{12})Rh($R^1R^2C_2B_9H_9$)$^-$ (R^1, R^2 = Me, Me; H, H; H, Ph)	H	227
2,1,7-$(Ph_3P)_2$(H)Rh($MePhC_2B_9H_9$)	X, B, H, P, I	228
3,1,2-$(Ph_3P)_2$HRh($C_2B_9H_9$)-μ-$CH_2C_6H_4CH_2$	X	229
2,1,8-$(Et_3P)_2$HRh($PhMeC_2B_9H_9$)	X	229
3,1,2-$(CO)_2$Rh($Me_2C_2B_9H_9$)$^-$	H, C	230
3,1,2-(Ph_3P)(CO)Rh($Me_2C_2B_9H_9$)$^-$	H, C, P	231
3,1,2-(Ph_3P)(CO)HRh($Me_2C_2B_9H_9$)	H, C, P	231
μ(B,Rh)–(R^1HC)-3,1,2-L(CO)Rh($R^2{}_2C_2B_9H_9$) (R^1 = C_6H_4Me, $C_6H_3Me_2$; L = Ph_3C, PO; R^2 = H, Me)	X[Ph_3C, C_6H_4Me, H], B, H, C, P	232
[3,1,2-L(CO)Rh($R_2C_2B_9H_8$)-4-CH_2R]$^-$ (L = Ph_3C, PO; R = H, Me)	B, H, C, P	232
Rhodium (cage), cobalt (exo-*polyhedral*)		
[$(Bu^tC)_2P_2Co(CO)_2(C_4Me_4)$]Rh($C_2B_9H_{11}$)	X, B, H, C, P	220
μ-(Me_4C_4)$(CO)_2$Co-3,1,2-(Ph_3P)(CO)Rh($C_2B_9H_{11}$)	X, B, H, C, P	218
Rhodium (cage), manganese (exo-*polyhedral*)		
3,1,2-[μ-Mn$(CO)_4$]L(CO)Rh($Me_2C_2B_9H_9$) (L = Ph_3P, CO)	B, H, C, P, I*	215
Rhodium (cage), copper (exo-*polyhedral*)		
3,1,2-(Ph_3P)Cu(L)(CO)Rh($Me_2C_2B_9H_9$) (L = CO, Ph_3P, Me_3P)	H, C, P	231
3,1,2-(Ph_3PCu)(C_8H_{12})Rh($Me_2C_2B_9H_9$)	X,B,H,C,P,I*	215
Rhodium (cage), rhodium (exo-*polyhedral*)		
3,1,2-($C_2B_9H_{11}$)Rh($C_2B_9H_{10}$)-8-(C_9H_6)Rh(C_9H_7)	X, B(2D), H, I	210
3,1,2-$(Et_3P)_2$(H)Rh($C_2B_9H_{10}$)Rh[(CODH)$C_2B_9H_{10}$] (*C–C*)	X, B, H, P	217
[3,1,2-(Et_3P)Rh($C_2B_9H_{10}$)]$_2$ (*C–C*; *Rh–Rh*)	X, P	217
[3,1,2-(Et_3P)Rh($C_2B_9H_{10}$)]((μ-H)$_2$; *Rh–Rh*; *C–C*)	X, P	233
[3,1,2-(Ph_3P)Rh($C_2B_9H_{11}$)]$_2$(*Rh–Rh*)	X, P	217
[3,1,2-(Ph_3P)Rh($PhC_2B_9H_{10}$)]$_2$(*Rh–Rh*)	X, B, H, P	217
(CO)Rh-3,1,2-$(CO)_2$HRh($Me_2C_2B_9H_9$)	H, C, P	231
3,1,2-$(CO)_3Rh_2(Me_2C_2B_9H_9)_2$$^-$	X, B, H, C, P	231
Rhodium (cage), tungsten (exo-*polyhedral*)		
Cp$(CO)_2$W(C_6H_4MeC)[3,1,2-$(CO)_2$Rh($Me_2C_2B_9H_9$)]$_2Au_2$	X, H, C	230
3,1,2-Cp$(CO)_2$W(C_6H_4MeCAu)(CO)LRh($R_2C_2B_9H_9$) (L = CO, Ph_3P; R = H, Me)	X(L = PPh_3, R = H), H, C	230
2,1,7-$(CO)_2$CpW(μ-CC_6H_4Me)Au(CO)(Ph_3P)Rh($C_2B_9H_{11}$)	H, C, P	234
Rhodium (cage), rhenium (exo-*polyhedral*)		
μ-(C_5H_5R)$(CO)_2$Re-3,1,2-(CO)LRh($Me_2C_2B_9H_9$) (L = CO, PMe_2Ph; R = Me, H)	X(L = CO, R = Me), B, H, C, P	226
Rhodium (cage), iridium (exo-*polyhedral*)		
μ-(Ph_3P)$(CO)_2$(H)Ir-3,1,2-(Ph_3P)(CO)Rh($C_2B_9H_{10}$)	X, B, H, C, P	218
μ-(Ph_3P)$(CO)_2$Ir-3,1,2-(Ph_3P)(CO)Rh($C_2B_9H_{10}$)$^-$	B, H, C, P	218
Rhodium (cage), platinum (exo-*polyhedral*)		
3,1,2-$(Et_3P)_2$Pt(μ-H)(μ-CO)-(Ph_3P)Rh($C_2B_9H_{11}$)	X, B, H, C, P, Pt	235
3,1,2-(R_2HC_2)(Et_3P)Pt(CO)(Ph_3P)Rh($C_2B_9H_{11}$) (R = Ph, C_6H_4Me)	X(C_6H_4Me), B, H, C, P, Pt	235
μ-RLPt-2,1,7-(Ph_3P)(CO)Rh($C_2B_9H_{11}$) (R = H, Me, Ph; L = PEt_3, PMe_2Ph)	B, H, P	236
μ-RLPt-3,1,2-(Ph_3P)(CO)Rh($C_2B_9H_{11}$) (R = Me, Ph; L = PEt_3, PPh_3, PMe_2Ph)	B, H, C, P	236
μ-L^1L^2Pt-2,1,7-(Ph_3P)(CO)Rh($C_2B_9H_{10}$) (L^1 = PEt_3, PPh_3, PMe_2Ph; L^2 = PEt_3, CO, PMe_2Ph)	B, H, C, P	236
$(Et_3P)_2$Pt(μ-H)(μ-CO)-2,1,7-(Ph_3P)Rh($C_2B_9H_{11}$)	X, B, H, C, P	236
μ-(PhC_2Ph)$(Et_3P)_2$Pt-2,1,7-(Ph_3P)(CO)Rh($C_2B_9H_{10}$)	B, H, C, P	236
μ-L_2Pt-3,1,2-(Ph_3P)(CO)Rh($C_2B_9H_{10}$) (L = PEt_3, PMe_2Ph)	B, H, C, P	236
Rhodium (cage), gold (exo-*polyhedral*)		
3,1,2-$(Ph_3P)_2$(Ph_3PAu)Rh($C_2B_9H_{11}$)	H, B, C, P	237
3,1,2-$(Ph_3P)_2$(Ph_3PAu)Rh($Me_2C_2B_9H_9$)	H, B, C, P	237
3,1,2-(Ph_3P)(H)$(Ph_3PAu)_2$Rh($Me_2C_2B_9H_9$)	H, B, C, P	237
2,1,7-(Ph_3P)Au(CO)(Ph_3P)Rh($C_2B_9H_{11}$)	H, C, P	234
2,1,7-$(CH_2)_n$[Ph_2PAu(CO)(Ph_3P)Rh($C_2B_9H_{11}$)]$_2$ (*n* = 2–6)	H, C, P	234
2,1,7-(CH=CH)[Ph_2PAu(CO)(Ph_3P)Rh($C_2B_9H_{11}$)]$_2$ (2 isomers)	H, C, P	234
3,1,2-(CH=CH)[Ph_2PAu(CO)(Ph_3P)Rh($C_2B_9H_{11}$)]$_2$ (2 isomers)	H, C, P	234
2,1,7-$(CO)_2$CpW(μ-CC_6H_4Me)Au(CO)(Ph_3P)Rh($C_2B_9H_{11}$)	H, C, P	234
[$(Bu^tC)_2P_2$Au(PPh_3)]Rh($C_2B_9H_{11}$)	X, B, H, C, P	220
3,1,2-$(Ph_3P)_2$(Ph_3PAu)Rh($Me_2C_2B_9H_9$)	H, P	238
3,1,2-(Ph_3P)(H)$(Ph_3PAu)_2$Rh($Me_2C_2B_9H_9$)	X, H, P	238
3,1,2-[(Ph_3P)Au](CO)LRh($R_2C_2B_9H_9$) (L = CO, Ph_3P; R = H, Me)	H, C, P	230
3,1,2-Cp$(CO)_2$W(C_6H_4MeCAu)(CO)LRh($R_2C_2B_9H_9$) (L = CO, Ph_3 P; R = H, Me)	X(L = PPh_3, R = H), H, C	230
Cp$(CO)_2$W(C_6H_4MeC)[3,1,2-$(CO)_2$Rh($Me_2C_2B_9H_9$)]$_2Au_2$	X, H, C	230
3,1,2-(Ph_3P)Au(Ph_3P)(CO)Rh($Me_2C_2B_9H_9$)	H, C, P	231

Table 4 (continued)

Compound[a]	*Data*[b]	*References*
Palladium		
3,1,2-$L_2Pd(C_2B_9H_{11})$ (L_2 = $(NMe_2)_2C_2H_4$, $(Ph_2PCH_2)_2$, 1,5-C_8H_{12}, C_4Ph_4)	B	239
3,1,2-$L_2Pd(C_2B_9H_{11})$ (L = $(MeO)_3P$, Bu^tNC, NH_3)	B	239
3,1,2-$(C_8H_{12})Pd(C_2B_9H_{11})$	X	240
'*nido*'-3,1,2-$(NMe_2CH_2)_2Pd(C_2B_9H_{11})$	X, B	239
3,1,2-$(Me_3P)_2Pd(C_2B_9H_{11})$	X, B	239
Samarium		
3,1,2-$(C_4H_8O)_4Sm(C_2B_9H_{11})$	X, B, H, I, M	241
3,1,2-$(C_4H_8O)_2Sm(C_2B_9H_{11})_2^-$	X, M, B, I	241
Gadolinium		
3,1,2-$Gd(C_2B_9H_{11})_2^-$	B, g	242
Ytterbium		
3,1,2-$(C_4H_8O)_4Yb(C_2B_9H_{11})$	B, H, I, M	241
3,1,2-$(HCONMe_2)_4Yb(C_2B_9H_{11})$	X	241
3,1,2-$(C_4H_8O)_2Yb(C_2B_9H_{11})_2^-$	X, M, B, I	241
Hafnium		
$Cp^*(MeC{=}CMe_2)Hf(C_2B_9H_{11})$		191
$(Cp^*Hf(C_2B_9H_{11})_2(\mu\text{-}CH_2)$		191
Tantalum		
3,1,2-$Cl_3Ta(C_2B_9H_{11})$	X, B, H, C(2D)	243
3,1,2-$(MeC_5H_4)Cl_2Ta(C_2B_9H_{11})$	X, B, H, C	243
3,1,2-$CpCl_2Ta(C_2B_9H_{11})$	B, H, C	243
3,1,2-$Me_3Ta(C_2B_9H_{11})$	B, H, C	243
3,1,2-$(MeC_5H_4)Me_2Ta(C_2B_9H_{11})$	B, H, C	243
Tungsten (no exo-*polyhedral metals)*		
2,1,8-$(CO)_4W(Me_2C_2B_9H_8\text{-}11\text{-}CH_2R)$ (R = *p*-C_6H_4Me, Me)	B, H, C	244
2,1,8-$(CO)_3(X)W(Me_2C_2B_9H_8\text{-}11\text{-}CH_2R)^-$ (X = Cl, I; R = *p*-C_6H_4Me, Me)	X(Cl, C_6H_4Me), B, H, C	244
2,1,8-$(CO)_2(RC{=}CPh)W(Me_2C_2B_9H_8\text{-}11\text{-}CH_2C_6H_4Me)$ (R = Ph, Me)	X(Me), B, H, C	244
2,1,7-$(CO)_3(X)W(Me_2C_2B_9H_9)^-$ (X = Cl, I)	X(I), B, H, C	244
2,1,8-$(CO)_3(R)W(Me_2C_2B_9H_8\text{-}11\text{-}CH_2C_6H_4Me)^-$ (R = I, Cl)	X[Cl], B, H, C, I	245
2,1,8-$(CO)_2(PhC{=}CR)W(Me_2C_2B_9H_8\text{-}11\text{-}CH_2C_6H_4Me)$ (R = Me, Ph)	X[Me], B, H, C, I	245
3,1,2-$(Ph_3Sn)(CO)_3W(C_2B_9H_{11})$	X, B, Sn, I	128
3,1,2-$(Ph_3P)(CO)_3W(Me_2C_2B_9H_8\text{-}4\text{-}CH_2C_6H_4Me)$	B, H, C, P	246
3,1,2-$L_2(CO)_2W(Me_2C_2B_9H_8\text{-}4\text{-}CH_2R)$ (R = C_6H_4Me, Me; L = CO, PhC_2Ph, PPh_3, $CNBu^t$)	X(C_6H_4Me,PPh_3), B, H, C, P	246
3,1,2-$(PhC{=}CPh)_2(CO)W(Me_2C_2B_9H_8\text{-}4\text{-}CH_2R)$ (R = C_6H_4Me, Me)	B, H, C, P	246
3,1,2-$(PhC{=}CPh)(PMe_3)(CO)W(Me_2C_2B_9H_8\text{-}4\text{-}CH_2C_6H_4Me)$	B, H, C, P	246
3,1,2-$(CO)_3(Ph_3P)W(Me_2C_2B_9H_8CH_2R)$ (R = C_6H_4Me,C_6H_4OMe)	X(C_6H_4OMe), B, H, C, P	247
2,1,8-$(CO)_3(I)W(Me_2C_2B_9H_8CH_2R)^-$ (R = C_6H_4Me,C_6H_4OMe)	B, H, C, P	247
3,1,2-$(CO)_2L_2W(Me_2C_2B_9H_8CH_2R)$ (L = CO, Ph_3P, $PHPh_2$, CNBu, PhC_2Ph; R = C_6H_4Me, C_6H_4OMe)	B, H, C, P	247
3,1,2-$(CO)_2(Ph_2PCH_2PPh_2)W(Me_2C_2B_9H_8C_6H_4OMe)$	B, H, C, P	247
3,1,2-$(CO)_2(Ph_2PCH_2PPh_2CHR)W(Me_2C_2B_9H_9)$ (R = C_6H_4Me, C_6H_4OMe)	P	247
3,1,2-$(CO)(PhC{=}CPh)_2W(Me_2C_2B_9H_8CH_2R)$ (R = C_6H_4Me,C_6H_4OMe)	B, H, C, P	247
3,1,2-$(CO)_3(Ph_3P)W(Me_2C_2B_9H_8C_2R)$ (R = $C_6H_4CH_2OMe$, C_6H_4Me, C_6H_4OMe)	B, H, C, P	248
3,1,2-$(CO)_2(RC)W(Me_2C_2B_9H_9)^-$ (R = Me,MeC_6H_4,$MeOC_6H_4$,$MeOCH_2C_6H_4$)	B, H, C, P	248
3,1,2-$(CO)_2L_2W[Me_2C_2B_9H_7\text{-}\mu\text{-}(CH_2)_2C_6H_4]$ (L = CO, Ph_2PH)	X(Ph_2PH), B, H, C, P	248
3,1,2-$(CO)_2(Ph_2PCH_2PPh_2)W[Me_2C_2B_9H_7\text{-}\mu\text{-}(CH_2)_2C_6H_4]$	B, H, C, P	248
3,1,2-$(CO)(PhC{=}CPh)_2W[Me_2C_2B_9H_7\text{-}\mu\text{-}(CH_2)_2C_6H_4]$	B, H, C, P	248
3,1,2-$(CO)(PhC{=}CPh)_2W[Me_2C_2B_9H_8CH_2R$ (R = C_6H_4Me,C_6H_4OMe)	B, H, C, P	248
2,1,8-$(CO)_3(I)W[Me_2C_2B_9H_7\text{-}\mu\text{-}(CH_2)_2C_6H_4]^-$	X, B, H, C, P	248
2,1,8-$(CO)_2(Bu^tC{=}CH)W[Me_2C_2B_9H_7\text{-}\mu\text{-}(CH_2)_2C_6H_4]^-$	B, H, C, P	248
2,1,8-$(CO)(Bu^tC{=}CH)_2W[Me_2C_2B_9H_7\text{-}\mu\text{-}(CH_2)_2C_6H_4]^-$	B, H, C, P	248
3,1,2-$(CO)_3(WC_2B_9H_{10})$-1-CH_2[*cyclo*-$N_3P_3(C_5H_{10}N)_4Me$]	P, I	194
3,1,2-$(CO)_2(CC_6H_4Me)W(Me_2C_2B_9H_9)^-$	H, C, P, I	249
3,1,2-$(Ph_3P)_2Rh(CO)_2(\mu\text{-}CC_6H_4Me)W(Me_2C_2B_9H_9)$	X, H, C, P, I	249
3,1,2-$(Ph_3P)_2Rh(CO)_2(\mu\text{-}CC_6H_4Me)W(Me_2C_2B_9H_8\text{-}4\text{-}C_7H_9)$	X, C, P, I	249
3,1,2-$(Ph_3P)Au(\mu\text{-}CC_6H_4Me)(CO)_2W(Me_2C_2B_9H_9)$	X, C, P, I	249
3,1,2-$(MeC_6H_4C)(CO)_2W(C_2B_9H_{11})^-$	H, C, I	250
3,1,2-$(MeC)(CO)_2W(Me_2C_2B_9H_9)^-)$	H, C, I	250
3,1,2-$(MeC_6H_4C)(CO)_2W(Me_2C_2B_9H_9)^-$	H, C, P, I	250
3,1,2-$(\mu\text{-}CR)(CO)_2W(Me_2C_2B_9H_9)^-$ (R = Ph,*p*-MeC_6H_4)	X[R = *p*-MeC_6H_4], H, C, P	251

Table 4 (continued)

Compound[a]	*Data*[b]	*References*
3,1,2-(μ-C-*p*-MeC$_6$H$_4$)(CO)$_2$W(Me$_2$C$_2$B$_9$H$_9$)$^-$	X[amended]	252
3,1,2-(CO)$_2$(μ-H)[μ-C(C$_6$H$_4$Me)]W(Me$_2$C$_2$B$_9$H$_9$)	H, C	253
3,1,2-(PMe$_3$)(CO)$_2$[μ-C(C$_6$H$_4$Me)]W(Me$_2$C$_2$B$_9$H$_9$)$^-$	H, C, P	253
3,1,2-(PMe$_3$)(CO)[ROC=C(C$_6$H$_4$Me)]W(Me$_2$C$_2$B$_9$H$_9$) (R = H,Me)	H, C, P	253
3,1,2-(CO)$_2$(μ-BuC$_2$C)W(Me$_2$C$_2$B$_9$H$_9$)$^-$	B, H, C	193
3,1,2-(CO)$_2$(μ-MeC$_6$H$_4$C)W(Me$_2$C$_2$B$_9$H$_9$)$^-$	B, H, C	193
3,1,2-(CO)$_2$(R^{1}C)W(R^{2}MeC$_2$B$_9$H$_9$)$^-$ (R^1 = C$_6$H$_4$Me, Me; R^2 = H, Me)	H, C	254
3,1,2-(CO)$_2$(CH[R^1]PPh$_2$[CH$_2$]$_n$PPh$_2$)W(R^{2}MeC$_2$B$_9$H$_9$) (R^1 = C$_6$H$_4$Me,Me; R^2 = H, Me; n = 1, 2)	X(C$_6$H$_4$Me, Me), B, H, C, P	254
3,1,2-(CO)$_2$(R$_2$PCH$_2$CH$_2$PR$_2$)W(Me$_2$C$_2$B$_9$H$_8$-4-Et) (R = Ph, Me)	B, H, C, P	254
3,1,2-(CO)$_2$ButC$_2$-C)W(Me$_2$C$_2$B$_9$H$_9$)$^-$	H, C	255
2,1,7-(CO)$_2$(ButC$_2$-C)W(Me$_2$C$_2$B$_9$H$_9$)$^-$	H, C	255
3,1,2-R(CO)$_2$W(Me$_2$C$_2$B$_9$H$_9$)$^-$ (R = *o,p*-MeC$_6$H$_4$C, Me$_2$C$_6$H$_3$C)	H, C, P	256
3,1,2-μ-MeC$_6$H$_4$C=W(CO)$_2$(PMe$_3$)(Me$_2$C$_2$B$_9$H$_9$)$^-$	H, C	197
3,1,2-(CO)$_2$(C$_3$H$_5$)W(R$_2$C$_2$B$_9$H$_9$) (R = H, Me)	B, H, C, P, I*	198
3,1,2-(CO)$_2$L^{1}L^{2}W(Me$_2$C$_2$B$_9$H$_9$) (L^1, L^2 = CO, Ph$_3$P)	B, H, C, P, I*	198
3,1,2-(CO)L(RC=CR)W(R$_2$C$_2$B$_9$H$_9$) (L = Me$_2$C$_2$, Ph$_2$C$_2$, Ph$_3$P, PMe$_2$Ph; R = Me, Ph)	B, H, C, P, I*	198
Tungsten (cage), iron (exo-*polyhedral*)		
3,1,2-(CO)$_3$Fe(μ-CO)(μ-CHR)(CO)$_2$W(Me$_2$C$_2$B$_9$H$_8$)$^-$ (R = Ph,*p*-MeC$_6$H$_4$)	X[R = *p*-MeC$_6$H$_4$], B, H, C	257
3,1,2-(CO)$_6$Fe$_2$(μ-AuPPh$_3$)(CMe)(CO)$_2$W(Me$_2$C$_2$B$_9$H$_7$)	B, H, C	257
3,1,2-(CO)$_6$Fe$_2$(CR)(CO)$_2$W(Me$_2$C$_2$B$_9$H$_7$)$^-$ (R = Me, Ph, *p*-MeC$_6$H$_4$)	X[R = Ph], B, H, C	257
3,1,2-[μ-(CO)$_3$Fe](μ-RC)(μ-CO)(CO)W(Me$_2$C$_2$B$_9$H$_9$)$^-$ (R = C$_6$H$_3$Me$_2$, C$_6$H$_4$Me)	X(C$_6$H$_3$Me$_2$), C	258
3,1,2-[μ-(CO)$_3$Fe](μ-RCH)(μ-CO)(CO)W(Me$_2$C$_2$B$_9$H$_8$)$^-$ (R = *o,p*-C$_6$H$_4$Me)	X, C	258
3,1,2-[μ-(CO)$_3$Fe)$_2$(μ-RC)(CO)$_2$W(Me$_2$C$_2$B$_9$H$_9$)$^-$ (R = *o,p*-C$_6$H$_4$Me)	C	258
3,1,2-[μ-(CO)$_6$Fe$_2$](CO)$_2$(μ-MeC$_6$H$_4$C)W(Me$_2$C$_2$B$_9$H$_7$)$^-$	B, H, C, P	196
3,1,2-(CO)$_6$Fe$_2$(μ-CH=C=CBu)(CO)$_2$(μ-CO)W(Me$_2$C$_2$B$_9$H$_8$)$^-$	X, B, H, C	259
3,1,2-(CO)$_6$Fe$_2$(μ-CH=C=CBu)(CO)$_2$(μ-COH)W(Me$_2$C$_2$B$_9$H$_8$)	X, B, H, C	259
3,1,2-(μ-R)(CO)$_2$[μ-(CO)$_6$Fe$_2$]W(Me$_2$C$_2$B$_9$H$_9$)$^-$ (R = *o,p*-MeC$_6$H$_4$C)	B, H, C	256
3,1,2-(μ-CO)(μ-R)(CO)$_2$[μ-(CO)$_3$Fe]W(Me$_2$C$_2$B$_9$H$_9$)$^-$ (R = *o,p*-MeC$_6$H$_4$C)	B, H, C	256
3,1,2-(μ-CO)(μ-R)(CO)[μ-(CO)$_2$(L)Fe]W(Me$_2$C$_2$B$_9$H$_9$)$^-$ (R = *o*-MeC$_6$H$_4$C, Me$_2$C$_6$H$_3$C; L = CO, PMe$_2$Ph)	H, C, P	256
3,1,2-(μ-CO)(μ-Me$_2$C$_6$H$_3$C)(CO)[μ-(CO)$_2$(CO)Fe]W(Me$_2$C$_2$B$_9$H$_9$)$^-$	X	256
Tungsten (cage), cobalt (exo-*polyhedral*)		
(η-C$_4$Me$_4$)(CO)Co(μ-RC)(CO)$_2$W(Me$_2$C$_2$B$_9$H$_9$) (R = MeC$_6$H$_4$,Me)	B*, H, C, I*	317
3,1,2-[μ-(CO)$_4$Co$_2$)(μ-PhC)(CO)$_2$W(Me$_2$C$_2$B$_9$H$_9$)$^-$	B, H, C	258
3,1,2-[μ-(CO)$_6$Co$_2$)(μ-PhC)(CO)$_2$W(Me$_2$C$_2$B$_9$H$_9$)$^-$	X, H	258
(μ-CR)(CO)$_8$Co$_2$W(Me$_2$C$_2$B$_9$H$_9$)$^-$ (R = Me, Ph,*m,p*-MeC$_6$H$_4$)	X[Ph], B, H, C	260
(μ-CR)(CO)$_6$Co$_2$W(Me$_2$C$_2$B$_9$H$_9$)$^-$ (2 B–H–Co) (R = Me, Ph,*m,p*-MeC$_6$H$_4$)	C, B, H	260
[Cp*Co(CO)$_2$CpW(MeC$_6$H$_4$C)]Au[MeC=](CO)$_2$W(MeC$_2$B$_9$H$_9$)	H, C	261
[Cp(CO)$_2$M=CR^1]Au[(R^{2}C)Cp*Co(CO)$_2$]W(Me$_2$C$_2$B$_9$H$_9$) (M = W, Mo; R^1, R^2 = Me, C$_6$H$_4$Me)	X[Mo, C$_6$H$_4$Me], H, C	261
Tungsten (cage), copper (exo-*polyhedral*)		
μ-CuCl-μ-RC(CO)$_2$W(Me$_2$C$_2$B$_9$H$_9$)$^-$ (R = C$_6$H$_4$Me,Bu^{t}C$_2$)	H, C	262
μ-(CuCl)$_2$-μ-RC(CO)$_2$W(Me$_2$C$_2$B$_9$H$_9$)$^-$ (R = C$_6$H$_4$Me,Bu^{t}C$_2$)	H, C	262
μ-CNR^{2}Cu-μ-R^{1}C-(CO)$_2$W(Me$_2$C$_2$B$_9$H$_9$) (R^1 = C$_6$H$_4$Me,Bu^{t}C$_2$; R^2 = C$_6$H$_3$Me$_2$, Bu^t)	H, C	262
Cu[μ-RC(CO)$_2$W(Me$_2$C$_2$B$_9$H$_9$)]$_2$$^-$ (R = C$_6$H$_4$Me,Bu^{t}C$_2$)	X[ButC$_2$], H, C	262
Tungsten (cage), molybdenum (exo-*polyhedral*)		
3,1,2-(C$_9$H$_7$)Mo(CO)(μ-BuC$_2$C)(CO)$_2$W(Me$_2$C$_2$B$_9$H$_9$)	B, H, C	193
3,1,2-(C$_7$H$_7$)Mo(μ-RC)(CO)$_2$W(Me$_2$C$_2$B$_9$H$_9$) (R = MeC$_6$H$_4$, BuC$_2$C)	B, H, C	193
3,1,2-(C$_7$H$_7$)Mo(μ-MeC$_6$H$_4$C)(CO)(L)W(Me$_2$C$_2$B$_9$H$_9$) (L = CO, PMe$_3$)	B, H, C, P	193
3,1,2-(C$_7$H$_7$)Mo(μ-MeC$_6$H$_4$C)(CO)$_2$W(MeC$_2$B$_9$H$_{10}$)	B, H, C	193
3,1,2-[MeC$_6$H$_4$CH=C$_3$(Bu^t)CpMo(CO)$_2$](CO)$_2$W(C$_2$B$_9$H$_{11}$)	X, H, C	255
3,1,2-[MeC$_6$H$_4$CH=C$_3$(Bu^t)H]CpMo(CO)$_2$W(C$_2$B$_9$H$_{11}$)	H, C	255
[Cp(CO)$_2$Mo=CR^1]Au[(R^{2}C)Cp*Co(CO)$_2$]W(Me$_2$C$_2$B$_9$H$_9$) (R^1,R^2 = Me, C$_6$H$_4$Me)	X[C$_6$H$_4$Me], H, C	261
3,1,2-(CO)$_2$[(C$_9$H$_7$)(CO)(μ-CC$_6$H$_4$Me)Mo]W(Me$_2$C$_2$B$_9$H$_9$) (C$_9$H$_7$ = indenyl)	X, B, H, C, I	263
3,1,2-(CO)[(C$_9$H$_7$)(μ-CO)(μ-CC$_6$H$_4$Me)(Me$_2$PCH$_2$PMe$_2$)Mo]W(Me$_2$C$_2$B$_9$H$_9$)	P, C, I	263
nido[slipped]-3,1,2-(CO)(Et$_2$C$_2$)[(CO)$_2$(C$_9$H$_7$)(μ-CHC$_6$H$_4$Me)Mo]-W(η^3-Me$_2$C$_2$B$_9$H$_8$)	X, B, H, C, I	263
3,1,2-(CO)$_2$[(C$_9$H$_7$)(CO)(μ-CC$_6$H$_4$Me)Mo]W(C$_2$B$_9$H$_{11}$)	B, H, C, I	264
3,1,2-(CO)$_2$[(C$_9$H$_7$)(CO)(μ-CC$_6$H$_4$Me)Mo]W(Me$_2$C$_2$B$_9$H$_9$)	X, B, H, C, I	264
2,1,7-(CO)$_2$[(C$_9$H$_7$)(CO)(μ-CC$_6$H$_4$Me)Mo]W(Me$_2$C$_2$B$_9$H$_9$)	X, B, H, C, I	264

Table 4 (continued)

Compound[a]	*Data*[b]	*References*
3,1,2-$(CO)_2[(C_9H_7)(CO)(\mu\text{-}CHC_6H_4Me)Mo]W(Me_2C_2B_9H_9)^-$	B, H, C, I	264
3,1,2-$(CO)_2[(C_9H_7)(PMe_3)(CO)(\mu\text{-}CC_6H_4Me)Mo]W(Me_2C_2B_9H_9)$	B, H, C, P, I	264
3,1,2-$(CO)(PMe_3)[(C_9H_7)(CO)(\mu\text{-}CC_6H_4Me)Mo]W(Me_2C_2B_9H_{11})$	B, H, C, P, I	264
3,1,2-$(CO)[(C_9H_7)(CO)(\mu\text{-}CC_6H_4Me)(Me_2PCH_2PMe_2)Mo]W(Me_2C_2B_9H_9)$	B, H, C, P, I	264
3,1,2-$(CO)_2[(C_2H_7)(L)(\mu\text{-}CC_6H_4Me)Mo]W(Me_2C_2B_9H_9)$ [L = $N_2C(C_6H_4Me\text{-}4)_2$, N_2CPh_2; C_9H_7 = indenyl]	B, H, C	265
3,1,2-$(CO)_7Mo_2(\mu\text{-}MeC_6H_4C)(CO)_2W(R_2C_2B_9H_9)^-$ (R = Me, H)	B, H, C	201
3,1,2-$(CO)_6(L)Mo_2(\mu\text{-}BuC_2C)(CO)_2W(Me_2C_2B_9H_9)^-$ [L = CO, $(Ph_2P)_2CH_2$]	B, H, C, P	201
3,1,2-$(CO)_6(Me_3P)Mo_2(\mu\text{-}MeC_6H_4C)(CO)_2W(Me_2C_2B_9H_9)^-$	X, B, H, C	201
3,1,2-$[Cp(CO)_2Mo{=}CR^1]Au[R^2C{=}](CO)_2W(Me_2C_2B_9H_9)$ (R^1, R^2 = Me, C_6H_4Me)	H, C, P*	266
3,1,2-$[Cp(CO)_2Mo{=}CR^1]Au[L_2Pt(R^2C)(CO)_2]W(Me_2C_2B_9H_9)$ (R^1, R^2 = Me, C_6H_4Me; L_2 = C_8H_{12}, $2PMe_2Ph$)	H, C, P*	266
3,1,2-$(C_9H_7)(PHPh_2)Mo(\mu\text{-}MeC_6H_4C)(CO)_2W(Me_2C_2B_9H_9)$	B, H, C	267
3,1,2-$(C_7H_7)Mo(\mu\text{-}HOCCC_6H_4Me)(\mu\text{-}Ph_2P)(CO)W(RC_2B_9H_{10})$ (R = H, Me)	X(H), B, H, C, P	267
3,1,2-$(C_9H_7)Mo(CO)_2(\mu\text{-}Ph_2P)(CO)W(Me_2C_2B_9H_8\text{-}4\text{-}CH_2C_6H_4Me)$	X, B, H, C, P	267
3,1,2-$(C_7H_7)Mo(\mu\text{-}MeCCC_6H_4Me)(\mu\text{-}CH_2)(\mu\text{-}O)(OEt)W(Me_2C_2B_9H_9)$	X, B, H, C	268
3,1,2-$\mu\text{-}R^1C\text{-}(C_7H_7)Mo(CO)(R^1CCR^2)\text{-}W(Me_2C_2B_9H_9)$ (R^1 = C_6H_4Me,$ButC_2$; R^2 = Me, Et)	H, C	269
3,1,2-$\mu\text{-}Me_4C_4(C_7H_7)Mo(MeC_6H_4CH)W(Me_2C_2B_9H_8)$	X, B, H, C	269
3,1,2-$\mu\text{-}C_6H_4Me(C_7H_7)Mo(CO)(RCCR)W(Me_2C_2B_9H_{10})$ (R = Me, Ph)	H, C	269
3,1,2-$\mu\text{-}RCCpLMo(CO)_2W(Me_2C_2B_9H_8CH_2C_6H_4Me)$ (R = Me, C_6H_4Me; L = CO, PMe_3)	B, H, C	200
2,1,7-$\mu\text{-}C_6H_4Me(CO)CpMo(CO)_2W(Me_2C_2B_9H_9)$	B, H, C	200
3,1,2-$(CO)_2[CpMo(CO)](\mu\text{-}CR^1)W(Me_2C_2B_9H_8CH_2R^2)$ (R^1 = Me, C_6H_4Me; R^2 = Me, C_6H_4Me)	B, H, C	270
3,1,2-$(CO)_2(L)[CpMo(CO)](\mu\text{-}CR)W(Me_2C_2B_9H_8CH_2C_6H_4Me)$ (R = Me, C_6H_4Me; L = CO,PMe_3)	B, H, C, P	270
Tungsten (cage), ruthenium (exo-*polyhedral*)		
3,1,2-$CpRu(\mu\text{-}H)(CO)(\mu\text{-}CC_6H_4Me)(CO)_2W(Me_2C_2B_9H_8)$	X, B, H, C, I	271
3,1,2-$CpRu(\mu\text{-}H)(CO)(\mu\text{-}CMe)(CO)_2W(Me_2C_2B_9H_8)$	B, H, C, I	271
3,1,2-$CpRu(CO)(\mu\text{-}CC_6H_4Me)(CO)_2W(Me_2C_2B_9H_8)$	X, B, H, C	271
3,1,2-$CpRu(CO)(\mu\text{-}CMe)(CO)_2W(Me_2C_2B_9H_8)$	B, H, C	271
3,1,2-$CpRu(CO)(\mu\text{-}CHC_6H_4Me)(CO)_2W(Me_2C_2B_9H_8)$	H*	271
3,1,2-$CpRu(CO)(\mu\text{-}CHC_6H_4Me)(CO)_2(Me_3P)W(Me_2C_2B_9H_8)$	B, H, C, P, I	271
3,1,2-$(\mu\text{-}R)(CO)_2[Cp(L)Ru]W(Me_2C_2B_9H_9)$ (R = $Me_2C_6H_3C$; L = CO, PMe_2Ph)	B(2D), H, C, P	256
Tungsten (cage), rhodium (exo-*polyhedral*)		
3,1,2-$(Ph_3P)_2Rh(CO)_2(\mu\text{-}CC_6H_4Me)W(Me_2C_2B_9H_9)$	X, H, C, P, I	250
3,1,2-$(Ph_3P)_2Rh(CO)_2(\mu\text{-}CC_6H_4Me)W(C_2B_9H_{11})$	H, C, P, I	250
3,1,2-$(Et_3P)_2Rh(CO)_2(\mu\text{-}CC_6H_4Me)W(Me_2C_2B_9H_9)$	H, C, P, I	250
3,1,2 $(Ph_3P)_2Rh(CO)_2(\mu\text{-}CC_6H_4Mc)W(Me_2C_2B_9H_8)\text{-}4\text{-}C_7H_9)$	X, H, C, P, I	250
3,1,2-$(Ph_3P)_2Rh(CO)_2(\mu\text{-}CC_6H_4Me)W(C_2B_9H_{10}C_7H_9)$	H, C, P, I	250
3,1,2-$(Ph_2PCH_2)_2Rh(CO)_2(\mu\text{-}CC_6H_4Me)W(Me_2C_2B_9H_9)$	H, C, P, I	250
3,1,2-$(\mu\text{-}CO)[(Ph_3)_2Rh](CO)(MeC_6H_4C)W(Me_2C_2B_9H_9)$	B, H, C, P	196
Tungsten (cage), tungsten (exo-*polyhedral*)		
3,1,2-$Cp(CO)W(MeC_6H_4)C(CO)_2W(R_2C_2B_9H_9)$ (R = H, Me)	B, H, C	272
3,1,2-$W(CO)_2(\mu\text{-}H)[\mu\text{-}C_2(C_6H_4Me)_2](CO)_2W(C_2B_9H_{11})_2^-$	H, C	253
3,1,2-$[MeC_6H_4CH{=}C_3(Bu^t)CpW(CO)_2](CO)_2W(C_2B_9H_{11})$	H, C	255
3,1,2-$[MeC_6H_4CH\text{–}C_3(Bu^t)H]CpW(CO)_2(CO)_2W(C_2B_9H_{11})$	X, H, C	255
3,1,2-$(CO)_2W(Et_2C_2B_9H_9)(\mu\text{-}RC)(CO)W(Et_2C_2B_9H_8CH_2R)^-$ (R = C_6H_4Me, Me) (*W–W*)	B, H, C, I	273
closo,hypercloso-3,1,2-$(CO)_2W(Et_2C_2B_9H_9)(\mu\text{-}RC)W(Et_2C_2B_9H_8CH_2R)^-$ (R = C_6H_4Me, Me) (*W–W*)	X(C_6H_4Me), B, H, C, I	273
3,1,2-$(CO)_2W(Et_2C_2B_9H_9)(\mu\text{-}RC)(Bu^tNC)W(Et_2C_2B_9H_8CH_2R)^-$ (R = C_6H_4Me, Me) (*W–W*)	B, H, C, I	273
3,1,2-$(CO)_2W(Et_2C_2B_9H_9)(\mu\text{-}MeC_6H_4C)(Bu^t(H)NC)W(Et_2C_2B_9H_8\text{-}CH_2C_6H_4Me)$ (*W–W*)	B, H, C, I	273
3,1,2-$[Cp^*Co(CO)_2CpW(MeC_6H_4C)]Au[MeC{=}](CO)_2W(Me_2C_2B_9H_9)$	H, C	261
3,1,2-$[Cp(CO)_2W{=}CR^2]Au[(R^1C)Cp^*Co(CO)_2]W(Me_2C_2B_9H_9)$ (R^1, R^2 = Me, C_6H_4Me)	H, C	261
$[3,1,2\text{-}(CO)W(C_2B_9H_{10})\text{-}1\text{-}CH_2PMeN[W(CO)_4]P(C_5H_{10}N)_2N]_n^{2-}$	P, I	194
$(CO)_2[(C_9H_7)(CO)(\mu\text{-}CC_6H_4Me)W]\text{-}W(Me_2C_2B_9H_9)$ (C_9H_7 = indenyl)	X, B, H, C, I	263
3,1,2-$(CO)_2[(C_9H_7)(CO)(\mu\text{-}CC_6H_4Me)W]W(Me_2C_2B_9H_9)$	X, B, H, C, I	264
2,1,7-$(CO)_2[(C_9H_7)(CO)(\mu\text{-}CC_6H_4Me)W]W(Me_2C_2B_9H_9)$	X, B, H, C, I	264
3,1,2-$(CO)_2[(C_9H_7)L(\mu\text{-}CC_6H_4Me)W]W(Me_2C_2B_9H_9)$ (L = PMe_3, $N_2(C_6H_4Me)_2$)	B, H, C, P, I	264
3,1,2-$(CO)_2[(C_9H_7)(PMe_3)(CO)(\mu\text{-}CC_6H_4Me)W]W(Me_2C_2B_9H_9)$	B, H, C, P, I	264

Table 4 (continued)

Compound[a]	*Data*[b]	*References*
3,1,2-$(CO)_2[(C_9H_7)(L)(\mu\text{-}CC_6H_4Me)W]W(Me_2C_2B_9H_9)$ (L = $N_2C(C_6H_4Me\text{-}4)_2$, N_2CPh_2, PMe_3, O, NC_6H_4Me-4; C_9H_7 = indenyl)	B, H, C	265
3,1,2-$(CO)_2[(C_9H_7)[N_2C(C_6H_4Me\text{-}4)_2](\mu\text{-}CC_6H_4Me)W]W(Me_2C_2B_9H_9)$	X	265
3,1,2-$(CO)_2[(C_9H_7)(O)(\mu\text{-}CC_6H_4Me)W]W(Me_2C_2B_9H_9)$	X	265
3,1,2-$\mu\text{-}C_6H_4Me\text{-}(CO)CpW(CO)_2W(Me_2C_2B_9H_9)$	B, H, C	200
2,1,7-$\mu\text{-}C_6H_4Me(CO)CpW(CO)_2W(Me_2C_2B_9H_9)$	X, B, H, C	200
3,1,2-$\mu\text{-}RCCpLW(CO)_2W(Me_2C_2B_9H_8CH_2C_6H_4Me)$ (R = Me, C_6H_4Me; L = CO, PMe_3)	X[Me,CO], B, H, C	200
3,1,2-$(CO)_2[CpW(CO)](\mu\text{-}CR^1)W(Me_2C_2B_9H_8CH_2R^2)$ (R^1 = Me, C_6H_4Me; R^2 = Me, C_6H_4Me)	B, H, C	270
3,1,2-$(CO)_2(L)[CpW(CO)](\mu\text{-}CR)W(Me_2C_2B_9H_8CH_2C_6H_4Me)$ (R = Me, C_6H_4Me; L = CO, PMe_3)	B, H, C, P	270
3,1,2-$(CO)_2[CpW(CO)PMe_3](\mu\text{-}CMe)W(Me_2C_2B_9H_8Et)$	B, H, C, P	270
$(CO)_2W[CpW(CO)L](\mu\text{-}CC_6H_4Me)W(Me_2C_2B_9H_8CH_2C_6H_4Me)$ (L = CO, PMe_3)	X(PMe_3), B, H, C, P	270
3,1,2-$\mu\text{-}Cp(CO)W{=}CR^1(CO)_2W(Me_2C_2B_9H_8CH_2R^2)$ (R^1 = $CC(Bu^t)$, Me, $CC(C_6H_4Me)$; R^2 = C_6H_4Me, Me)	B, H, C	274
3,1,2-$[Cp(CO)_2W{=}CR^1]Au[R^2C{=}](CO)_2W(Me_2C_2B_9H_9)$ (R^1, R^2 = Me, C_6H_4Me)	H, C, P*	266
3,1,2-$[Cp(CO)_2W{=}CR^1]Au[L_2Pt(R^2C)(CO)_2]W(Me_2C_2B_9H_9)$ (R^1, R^2 = Me, C_6H_4Me; L_2 = C_8H_{12}, $2PMe_2Ph$)	H, C, P*	266
Tungsten (cage), iridium (exo-*polyhedral*)		
3,1,2-$[\mu\text{-}(PEt_3)_2Ir](\mu\text{-}RC)(\mu\text{-}CO)(CO)W(Me_2C_2B_9H_9)$ (R = *p*-C_6H_4Me)	B, H, P	258
3,1,2-$[\mu\text{-}(PEt_3)_2Ir](RC)(CO)_2(H)W(Me_2C_2B_9H_8)$ (R = *p*-C_6H_4Me)	X, B, H, P	258
3,1,2-$[\mu\text{-}(CO)(H)Ir](Ph_2P_2C_2H_4)](\mu\text{-}RC)(CO)_2W(Me_2C_2B_9H_8)$ (R = *p*-C_6H_4Me)	B, H, C, P	258
3,1,2-$[\mu\text{-}(CO)(H)Ir][(N_2[C_5H_4]_2)](\mu\text{-}RC)(CO)_2W(Me_2C_2B_9H_8)$ (R = *p*-C_6H_4Me)	B, H, C	258
3,1,2-$L^1L^2Ir(\mu\text{-}MeC_6H_4C)(CO)_2W(Me_2C_2B_9H_8\text{-}\mu\text{-}H)$ (L^1, L^2 = Ph_3P, Et_3P, $(PhO)_3P$, $(MeO)_3P$)	X[L^1, L^2 = Et_3P], B, H, C, P	275
3,1,2-$L_nIr(\mu\text{-}MeC_6H_4C)(CO)_2W(Me_2C_2B_9H_8\text{-}\mu\text{-})$ (L_n = $2P(OMe)_3$ + H + CO; $3PMe_3$ + H; (dppe)(CO) + H; CO + bipy + H)	X[L^1, L^2 = Et_3P], B, H, C, P	275
3,1,2-$L_3(H)Ir(\mu\text{-}MeC_6H_4C)(CO)_2W(Me_2C_2B_9H_8\text{-}\mu\text{-})$ (L = Me_3P, $(MeO)_3P$)	X, B, H, C, P	275
Tungsten (cage), platinum (exo-*polyhedral*)		
3,1,2-$[\mu\text{-}(CC_6H_3Me_2)(Et_3P)(H)Pt(CO)_2W(Me_2C_2B_9H_9)$ (*Pt–H–B*)	B, H, C, P	276
3,1,2-$[\mu\text{-}(CC_6H_3Me_2)(Et_3P)Pt(CO)_2W(Me_2C_2B_9H_8)$ (*Pt–B*) (2 isomers)	X[1 isomer], B, H, C, P	276
3,1,2-$[\mu\text{-}(CC_6H_3Me_2)(Et_3P)(L)Pt(CO)_2W(Me_2C_2B_9H_8)$ (*Pt–B*) (L = CO, $CNBu^t$)	X[CO], B, H, C, P	276
3,1,2-$\mu\text{-}(PhMe_2P)_2Pt\text{-}\mu\text{-}MeC_6H_4C(CO)_2W(Me_2C_2B_9H_9)$ (2 isomers)	X, B, H, C, P	277
2,1,7-$\mu\text{-}(PhMe_2P)_2Pt\text{-}\mu\text{-}MeC_6H_4C(CO)_2W(Me_2C_2B_9H_9)$ (2 isomers)	X, B, H, C, P	277
3,1,2-$Pt[\mu\text{-}MeC_6H_4C(CO)_2W(Me_2C_2B_9H_9)]_2$	B, H, C, P	277
3,1,2-$L_2Pt{=}W(CO)_3(Me_2C_2B_9H_9)$ (L = Ph_3P, Et_3P)	B, H, C, P	129
3,1,2-$(\mu\text{-}CR)(C_8H_{12})Pt(Ph_3P)Au(CO)_2W(Me_2C_2B_9H_9)^-$ (R = Ph, *p*-MeC_6H_4)	C(Ph), Pt(Ph), H(Ph)	251
3,1,2-$(\mu\text{-}CR)(CO)_2(C_8H_{12})PtW(Me_2C_2B_9H_9)^-$ (R = Ph, *p*-MeC_6H_4)	H, C, P	251
3,1,2-$(\mu\text{-}CR^1)(CO)_2(PR^2_3)Pt\text{-}W(Me_2C_2B_9H_9)^-$[($R^1$ = Ph, *p*-MeC_6H_4; R^2 = $PMePh_2$, PMe_2Ph)]	C(Ph), Pt(Ph), H, P	251
3,1,2-$(Et_3P)_2Pt(B)(\mu\text{-}H)(CO)_2(PMe_3)W(Me_2C_2B_9H_7\text{-}p\text{-}CH_2C_6H_4Me)$	X, B, H, C, P, I	278
3,1,2-$(Et_3P)_2Pt(B)(\mu\text{-}H)(CO)_2(L)W(Me_2C_2B_9H_7\text{-}p\text{-}CH_2C_6H_4Me)$ (L = CO, $CNBu^t$, PPh_2H)	B, H, C, P, I	278
3,1,2-$(Et_3P)_2Pt(B)(\mu\text{-}H)(CO)_3W(Me_2C_2B_9H_7R$ (R = H, Et)	X[Et], B, H, C, P, I	278
3,1,2-$(Et_3P)_2Pt(B)(\mu\text{-}H)(CO)_3W(Me_2C_2B_9H_8)$	X, B, H, C, P, I	278
3,1,2-$(Et_3P)_2Pt(\mu\text{-}H)_2(CO)_2(L)W(Me_2C_2B_9H_7\text{-}p\text{-}CH_2C_6H_4Me^+$ (L = CO, $CNBu^t$, PPh_2H)	B, H, C, P, I	278
3,1,2-'*hypercloso*'-$(Et_3P)_2(\mu\text{-}CO)_2Pt{=}W(Me_2C_2B_9H_8R)$ (R = H, Et)	B, H, C, P, I	278
3,1,2-'*hypercloso*'-$(Et_3P)_2(\mu\text{-}CO)_2Pt{=}W(Me_2C_2B_9H_8\text{-}p\text{-}CH_2C_6H_4Me)$	X, B, H, C, P, I	278
3,1,2-$(\mu\text{-}CO)(\mu\text{-}R)(CO)PtL_2W(Me_2C_2B_9H_9)^-$ (R = *o,p*-MeC_6H_4C, $Me_2C_6H_3C$; L = C_8H_{12}, $(PMePh_2)_2$)	H, C, P, Pt	256
3,1,2-$(\mu\text{-}CO)(\mu\text{-}R)(CO)(\mu\text{-}Ph_3PAu)PtLW(Me_2C_2B_9H_9)$ (R = $Me_2C_6H_3C$; L = C_8H_{12})	H, C, P, Pt	256
3,1,2-$(CH_2)_n[Ph_2PAu(\mu\text{-}RC)(C_8H_{12})Pt(CO)_2W(Me_2C_2B_9H_9)]_2$ (*n* = 2–6; R = *p*-MeC_6H_4)	B, H, C, P	279
3,1,2-$[Cp(CO)_2Mo{=}CR^1]Au[L_2Pt(R^2C)(CO)_2]W(Me_2C_2B_9H_9)$ (R^1, R^2 = Me, C_6H_4Me; L_2 = C_8H_{12}, $2PMe_2Ph$)	H, C, P*	266
3,1,2-$[Cp(CO)_2W{=}CR^1]Au[L_2Pt(R^2C)(CO)_2]W(Me_2C_2B_9H_9)$ (R^1, R^2 = Me, C_6H_4Me; L_2 = C_8H_{12}, $2PMe_2Ph$)	H, C, P*	266

Table 4 (continued)

Compound[a]	Data[b]	References
Tungsten (cage), gold (exo-*polyhedral*)		
3,1,2-$(Ph_3P)Au(\mu\text{-}CC_6H_4Me)(CO)_2W(Me_2C_2B_9H_9)$	X, C, P, I	249
3,1,2-$(\mu\text{-}Ph_3PAuMeC_6H_4C)(CO)_2W(Me_2C_2B_9H_9)$	X, H, C, P, I	250
3,1,2-$(\mu\text{-}Ph_3PAuMeC_6H_4C)(CO)_2W(C_2B_9H_{11})$	H, C, P, I	250
3,1,2-$(\mu\text{-}Ph_3PAuMeC)(CO)_2W(Me_2C_2B_9H_9)$	H, C, P, I	250
3,1,2-$\mu,\mu'\text{-}(MeC_6H_4C)_2Au[(CO)_2W(Me_2C_2B_9H_9)]_2^-$	H, C, I	250
3,1,2-$(\mu\text{-}CR)(C_8H_{12})Pt(Ph_3P)Au(CO)_2W(Me_2C_2B_9H_9)^-$ (R = Ph, *p*-MeC_6H_4)	C(Ph), Pt(Ph), H(Ph)	251
3,1,2-$(\mu\text{-}Ph_3PAu)R(CO)_2W(Me_2C_2B_9H_9)$ (R = *o,p*-MeC_6H_4C, $Me_2C_6H_3C$)	H, C	256
3,1,2-$(\mu\text{-}Ph_3PAu)(CO)_2(MeC_6H_4C)W(Me_2C_2B_9H_9)$	B, H, C, P	196
3,1,2-$[Cp^*Co(CO)_2CpW(MeC_6H_4C)]Au[MeC{=}](CO)_2W(Me_2C_2B_9H_9)$	H, C	261
3,1,2-$[Cp(CO)_2M{=}CR^2]Au[(R^2C)Cp^*Co(CO)_2]W(Me_2C_2B_9H_9)$ (M = W, Mo; R^1, R^2 = Me, C_6H_4Me)	X[Mo, C_6H_4Me], H, C	261
3,1,2-$(Ph_3P)Au(\mu\text{-}RC)(CO)_2W(Me_2C_2B_9H_9)$ (R = BuC_2, MeC_6H_4)	B, H, C	201
3,1,2-$L^1{}_2Pt(L^2)Au(\mu\text{-}RC)\text{-}(CO)_2W(Me_2C_2B_9H_9)$ (R = Bu^tC_2, MeC_6H_4; $L^1{}_2$ = $(PhMe_2P)_2$, C_8H_{12}; L^2 = Ph_3P, Me_2PhP)	B, H, C, P	201
3,1,2-$\mu\text{-}AuCl\text{-}\mu\text{-}RC(CO)_2W(Me_2C_2B_9H_9)^-$ (R = C_6H_4Me,Bu^tC_2)	H, C	262
3,1,2-$Au[\mu\text{-}RC(CO)_2W(Me_2C_2B_9H_9)]_2^-$ (R = C_6H_4Me, Bu^tC_2)	H, C	262
3,1,2-$[Cp(CO)_2Mo{=}CR^1]Au[R^2C{=}](CO)_2W(Me_2C_2B_9H_9)$ (R^1, R^2 = Me, C_6H_4Me)	H, C, P*	266
3,1,2-$[Cp(CO)_2W{=}CR^1]Au[R^2C{=}](CO)_2W(Me_2C_2B_9H_9)$ (R^1, R^2 = Me, C_6H_4Me)	H, C, P*	266
3,1,2-$[Cp(CO)_2Mo{=}CR^1]Au[L_2Pt(R^2C)(CO)_2]W(Me_2C_2B_9H_9)$ (R^1, R^2 = Me, C_6H_4Me; L_2 = C_8H_{12}, $2PMe_2Ph$)	H, C, P*	266
3,1,2-$[Cp(CO)_2W{=}CR^1]Au[L_2Pt(R^2C)(CO)_2]W(Me_2C_2B_9H_9)$ (R^1, R^2 = Me, C_6H_4Me; L_2 = C_8H_{12}, $2PMe_2Ph$)	H, C, P*	266
3,1,2-$(CH_2)_n[Ph_2PAu(\mu\text{-}RC)(CO)_2W(Me_2C_2B_9H_9)]_2$ (*n* = 2–6; R = *p*-MeC_6H_4)	X[*n* = 4], B, H, C, P	279
3,1,2-$(CH{=}CH)[Ph_2PAu(\mu\text{-}RC)(CO)_2W(Me_2C_2B_9H_9)]_2$ (2 isomers)	B, H, C, P	279
3,1,2-$(CH_2)_n[Ph_2PAu(\mu\text{-}RC)(C_8H_{12})Pt(CO)_2W(Me_2C_2B_9H_9)]_2$ (*n* = 2–6; R = *p*-MeC_6H_4)	B, H, C, P	279
3,1,2-$(CO)_6Fe_2(\mu\text{-}AuPPh_3)(CMe)(CO)_2W(Me_2C_2B_9H_7)$	B, H, C	257
Rhenium		
3,1,2-$(CO)_3Re(C_2B_9H_{10}\text{-}9\text{-}SMe_2)$		132
3,1,2-$(CO)_3Re(C_2B_9H_{11})\cdot CD^-$ (CD = α or β-cyclodextrin)	X, H, CD	133
3,1,2-$(CO)_3Re[(CHR)C_2B_9H_{10}]$ (R = H, Me)		280
3,1,2-$(CO)_3Re(RC_2B_9H_{10})$ (R = CH_2OH, $CH{=}CH_2$, $CMe{=}CH_2$)		280
3,1,2-$(CO)_3Re(C_2B_9H_{10}\text{-}1\text{-}C_6H_4R^-)$ (R = H, *m*-F, *p*-F)	F, I	130
Osmium		
3,1,2-$(C_6H_6)Os(C_2B_9H_{11})$	B, H	11b
Iridium		
3,1,2-$Cp^*Ir(C_2B_9H_{11})$	X, B, H(2D)	160
2,1,8-$(Ph_3P)_2HIr(C_2B_9H_{10}\text{-}8\text{-}Ph)$	X, B, H, P, I	281
2,1,8-$(Ph_3P)_2HIr(C_2B_9H_{10}\text{-}8\text{-}Me)$	B, H, P, I	281
2,1,8-$(Ph_3P)_2HIr(C_2B_9H_{10})\text{-}8\text{-}1,2(C_2B_9H_{11})$	B, H, P, I	281
3,1,2-$(CO)(Ph_3)Ir(C_2B_9H_{11})^-$	B, P, I	213
3,1,2-$(Ph_3PAu)(CO)LIr(Me_2C_2B_9H_9)$ (L = PPh_3, CO)	P, I	238
3,1,2-$(Ph_3P)_4Au_4(CO)Ir(CO)_2Ir(Me_2C_2B_9H_9)$	B, P, I	238
3,1,2-$(Ph_3P)[\eta^2\text{-}C(p\text{-}ClC_6H_4)NOC({=}O)]Ir(C_2B_9H_{11})^-$	H, P, I	216
3,1,2-$(Ph_3PAu)(CO)LIr(Me_2C_2B_9H_9)$ (L = Ph_3P, CO)	X(Ph_3P), B, H, C, P	282
3,1,2-$(Ph_3P)_4Au_4(CO)Ir(CO)_2Ir(Me_2C_2B_9H_8)^+(Me_2C_2B_9H_9)^-$	B, H, C, P	282
3,1,2-$(Ph_3PAu)(Ph_3P)_2Ir(C_2B_9H_{11})$	B, H, C, P	282
nido-3,1,2-$(CO)_2Ir(Me_2C_2B_9H_9^-$	B, H, C, P	282
nido-3,1,2-$(Ph_3PAu)(Ph_3P)_2(H)Ir(Me_2C_2B_9H_8)$	X, B, H, C, P	282
Platinum		
3,1,2-$Pt(C_2B_9H_{11})_2$	X, B, H	283
3,1,2-$(Ph_2PCH_2)Pt(C_2B_9H_{11})$	B	239
3,1,2-$(PMe_2Ph)_2Pt(PhC_2B_9H_{10})$	X, P	284
3,1,11-$(PMe_2Ph)_2Pt(PhC_2B_9H_9\text{-}11\text{-}R)$ (R = H, Ph)	X, P	284
3,1,11-$(PMe_2Ph)_2Pt(C_2B_9H_{10}\text{-}11\text{-}Ph)$	X, P	284
3,1,2-$(Ph_2PCH_2)_2Pt(C_2B_9H_{11})$	X, B	285
Gold		
nido-7,8-$(C_2B_9H_{11})\text{-}10\text{-}Au(PPh_3)^-$	X, B, H, P	286
Mercury		
nido-10,7,8-$(Ph_3P)Hg(Ph_2C_2B_9H_9)$	X	287
Samarium		
3,1,2-$(C_4B_8O)_4Sm(C_2B_9H_{11})$	B, H, I	288
3,1,2-$(C_4H_8O)_4Sm(C_2B_9H_{11})$	X, B, H, I, M	241

Table 4 (continued)

Compound[a]	*Data*[b]	*References*
$3,1,2\text{-}(C_4H_8O)_2Sm(C_2B_9H_{11})_2^-$	H	288
$3,1,2\text{-}(C_2H_8O)_2Sm(C_2B_9H_{11})_2^-$	X, M, B, I	241
Gadolinium		
$3,1,2\text{-}Gd(C_2B_9H_{11})_2^-$	B, g	242
Ytterbium		
$3,1,2\text{-}(C_4H_8O)_4Yb(C_2B_9H_{11})$	B, H, I	288
$3,1,2\text{-}(HCONMe_2)_4Yb(C_2B_9H_{11})$	X, B, H, I	288
$3,1,2\text{-}(C_4H_8O)_4Yb(C_2B_9H_{11})$	B, H, I, M	241
$3,1,2\text{-}(HCONMe_2)_4Yb(C_2B_9H_{11})$	X	241
$3,1,2\text{-}(C_4H_8O)_2Yb(C_2B_9H_{11})_2^-$	M, B, I	241
2 Metal atoms[c]		
Chromium		
$2,3,1,7\text{-}Cp_2Cr_2(C_2B_8H_{10})$	X, I	111
Cobalt		
$2,4,1,7\text{-}Cp_2Co_2(C_2B_8H_{10})$		157
$3,6,1,2\text{-}Co_2(C_2B_9H_{11})_2(C_2B_8H_{10})^{2-}$	B, H, I	164
$3,6,1,2\text{-}Co_3(C_2B_9H_{11})_2(C_2B_8H_{10})_2^{3-}$	B, H, I	164
$3,6,1,2\text{-}Co_4(C_2B_9H_{11})_2(C_2B_8H_{10})_3^{4-}$	B, H, I	164
$3,6,1,2\text{-}Co_2(C_2B_9H_{11})_2(C_2B_8H_{10})^{2-}$	M	166
$3,6,1,2\text{-}Co_3(C_2B_9H_{11})_2(C_2B_8H_{10})_2^{3-}$	M	166
$3,6,1,2\text{-}Co_4(C_2B_9H_{11})_2(C_2B_8H_{10})_3^{4-}$	M	166

[a] Clusters are *closo* expect when otherwise designated; metals are in order of increasing atomic number. Cp = η-C_5H_5, Cp* = η-C_5Me_5. [b] X = x-ray diffraction, B = ^{11}B NMR, H = ^{1}H NMR, C = ^{13}C NMR, P = ^{31}P NMR, F = ^{19}F NMR, CD = circular dichroism, o = optical rotation, D = two-dimensional (COSY) NMR, E = electrochemical data, M = magnetic susceptibility data, I = infrared data, U = UV–visible data, g = ESR (EPR) data, m = Mössbauer spectra, FAB = fast atom bombardment mass spectral data, R = Raman data, S = mass spectral data only. Asterisks indicate that only partial data are given. [c] Metal atoms in metallacarbaborane framework.

9.6.3.1 Synthesis

The dicarbollide ligands have assumed a significant role in organometallic chemistry and can stabilize a seemingly unlimited variety of monometallic and polymetallic cluster systems. This is well illustrated by the work of Stone (reviewed elsewhere[1m,n]) in the areas of alkylidyne complexes[197,202,246–8,253–5,259,270,272,274] and metallacarbaboranes having bridging carbene or carbyne ligands.[193,196,200,201,226,230,232,249–52,256,257,260–2,264–7,269,271,275–9] The alkylidyne work centers on the ability of the $[\eta^5\text{-}R^1R^2C_2B_9H_9]^{2-}$ unit to replace its isoelectronic analogue $[\eta\text{-}C_5R_5]^-$ (R = H or Me) in species such as $W(\equiv CR)(CO)_2(\eta\text{-}C_5R_5)$. Protonation of such carbaborane–alkylidyne complexes of tungsten or molybdenum induces migration of the CR unit to a boron atom on the carbaborane ligand;[246] when the protonation is conducted in the presence of other reagents, dimetallic compounds are produced via alkylidyne–ligand coupling.[255] For example, protonation of a 1:1 mixture of $[NEt_4]^+[W(\equiv CR)(CO)_2(C_2B_9H_{11})]^-$ and $Mo(\equiv CC_6H_4Me)(CO)_2(\eta\text{-}C_5H_5)$ with $HBF_4\cdot Et_2O$ gave the molybdenum–tungsten species shown in Figure 19.[255] A plethora of different products is generated in such reactions by variation of the added substrate, the acid employed for protonation, and the metallacarbaborane salt. The stereochemistry and other aspects of this very complex chemistry have been discussed in several articles.[1b,1m,n]

An even more extensive area explored by the Stone group is that of polymetallic complexes with bridging carbene or carbyne ligands, on which well over a hundred papers have been published (see reviews cited earlier). This chemistry effectively bridges (literally and figuratively) the separately huge fields of metal cluster and metallacarbaborane chemistry. Although there appears to be little synthetic control over the nature or structures of the products, an impressive wealth of structurally novel species have been obtained via reactions of alkylidyne metallacarbaborane anions of the above type with low-valent metal species, for example, $AuCl(PPh_3)$. Figure 20 depicts several examples of such products.[279]

Metallacarbaboranes featuring metal–metal bonds have received much attention. Reactions of $[Rh(CO)L(R_2C_2B_9H_9)]^-$ anions (L = CO or PPh_3, R = H or Me) with chlorogold reagents afford products having direct Au–Rh bonds; alternatively, salts of [10-*endo*-$[Au\text{-}PPh_3]$-*nido*-$Me_2C_2B_9H_9]^-$ can be treated with rhodium complexes to give products in which the carbaborane ligand is transferred from gold to rhodium.[230,231,234,237,238,282] Similar chemistry generates species having Fe–W, Co–W, Ir–W, Rh–Cu, Rh–Mn, Rh–Pt, Rh–Ir, Mo–W, Mo–Pt, and other metal–metal bonds.[129,215,218,235,236,258,263,268] Somewhat different heterodinuclear complexes of the type $[(CO)_3(SnPh_3)M(C_2B_9H_{11}]^-$ having Sn–M bonds

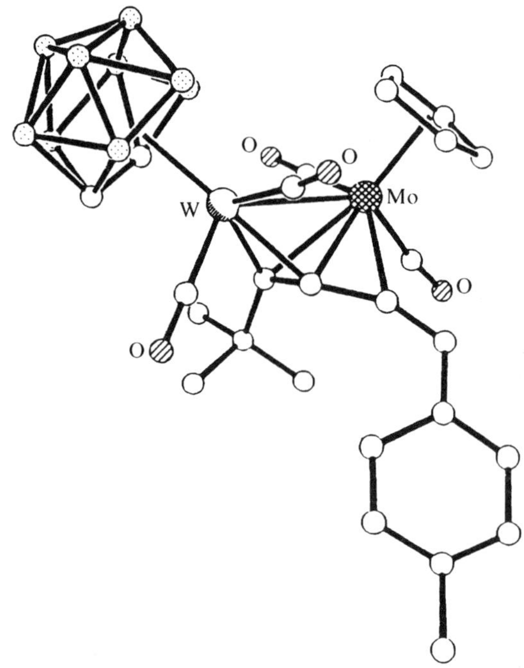

Figure 19 Structure of 3,1,2-[$MeC_6H_4CH{=}C_3(Bu^t)CpMo(CO)_2$]$(CO)_2W(C_2B_9H_{11})$ (reproduced by permission of the Royal Society of Chemistry from *J. Chem. Soc., Dalton Trans.*, 1992, 2685).

(M = Cr, Mo, W) have also been prepared,[128] as have several novel copper species having Cu–Cu bonds, including a "pinwheel" cluster incorporating a Cu_3 triangle (Figure 21).[188,190,199]

Straightforward syntheses of other classes of monometallic (ligand)$_x$M($R^1R^2C_2B_9H_9$) organometallacarbaboranes in which the ligand is arene,[10,11a,b,150,151,157] allyl,[198] and diphosphacyclobutadiene[220] have been described. Hawthorne and co-workers have prepared metallacarbaborane–pyrazolylborate[185] and –porphyrin[192] complexes, thereby linking carbaborane chemistry to these two large fields. Indeed, the versatility of the 7,8- and 7,9-[$C_2B_9H_{11}$]$^{2-}$ anions and their *B*- and *C*-substituted derivatives has allowed their introduction into virtually every area of organotransition metal chemistry. In many cases the carbaborane ligands serve as surrogates for Cp, Cp*, or arene units, as in indenyl– and fluorenyl–metal–carbaborane sandwiches,[156,161,177,210] Fe^{II} and Ru^{III} complexes,[152,153,203,204] complexes of Rh or Ir,[160,205,209,225] bis(carboranyl) zwitterions[132] of the type $(CO)_3M(C_2B_9H_{10}\text{-}9\text{-}SMe_2)$, and mixed-ligand diborole–dicarbollyl sandwiches, 3,1,2-$CpCo(MeEt_2C_3B_2Et_2)M(C_2B_9H_{11})$ (M = Co, Ni).[171] However, the special electronic character, reactivity, and steric properties of these ligands allow a much richer metal chemistry than that of their hydrocarbon counterparts, often stabilizing novel structures. In addition to the examples already mentioned, Chetcuti *et al.* have synthesized new kinds of metallacycles via 1,3-dipolar cycloaddition involving low-valent metal carbonyls,[216] while Allcock *et al.* have prepared carbaborane-stabilized cyclic and polymeric phosphazene complexes of Rh, Mo, and W.[194]

Dicarbollide complexes of certain elements are of special interest, especially those related to catalysis (see above), for example, rhodium, iridium, and the early transition metals. The first 12-vertex metallacarbaboranes of the lanthanide elements have been prepared, including species of the type $L_nM(C_2B_9H_{11})$ where L is THF or DMF and M is Sm or Yb.[241,288] Bercaw and co-workers[125,126] prepared the first scandium metallacarbaboranes, including a dimeric species containing Sc–H bonds; however, owing to a strong reciprocal B–H dative bonding interaction with the electron-poor scandium atom, this species proved too unreactive to catalyze alkene polymerization.

Coordination of dicarbollide ligands to very electron-rich metals such as copper, mercury, or gold invariably produces open-cage structures in which the metal is strongly bound to only one or two atoms on the C_2B_3 face (representative examples can be found in references 286, 287, 289, and 290). Complexes of this class are usually described as *exo*-metallated carbaborane derivatives and are covered by Chapter 6, this volume.

The utility of the dicarbollide ligands can be enhanced by introducing substituents such as amines or alkyl sulfides that reduce the 2– charge to 1– or 0. Such "charge-compensated" carbaborane units[131,134,189] allow the synthesis of neutral complexes having low oxidation state metal centers, as in 3,1,2-$Fe^{II}(C_2B_9H_{10}NEt_3)_2$.

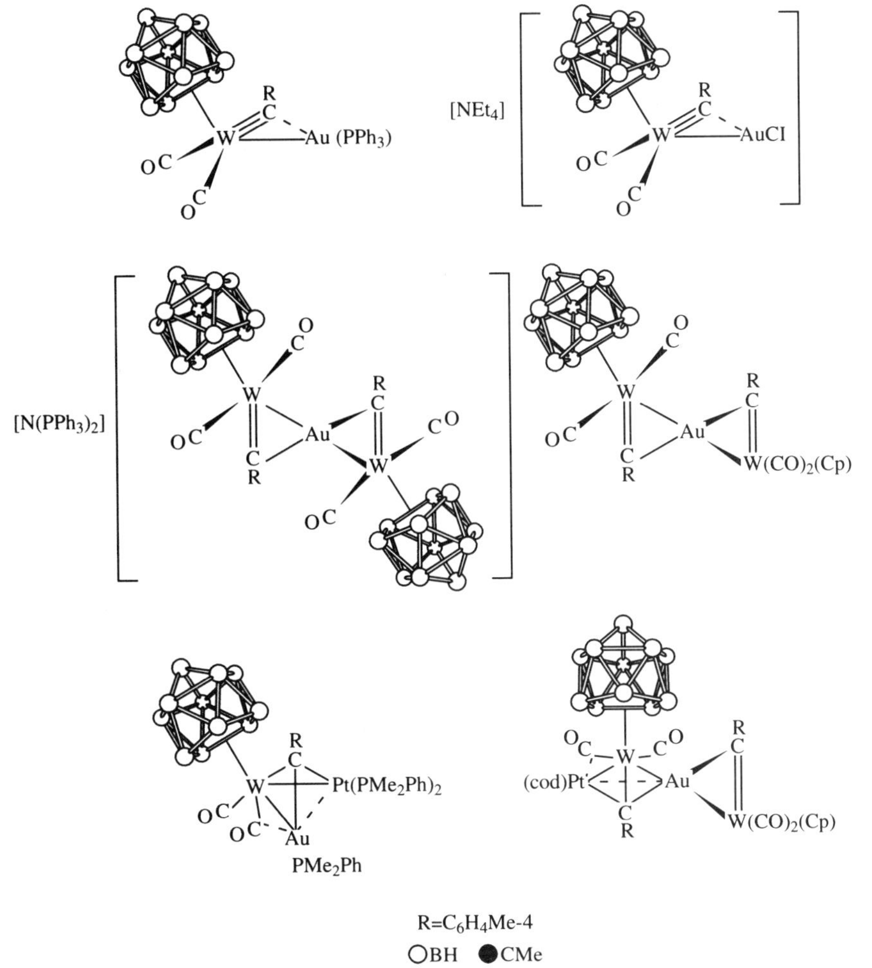

Figure 20 Examples of polynuclear metal complexes having *exo*-polyhedral carbene or carbyne bridges (reproduced by permission of the Royal Society of Chemistry from *J. Chem. Soc., Dalton Trans.*, 1992, 2495).

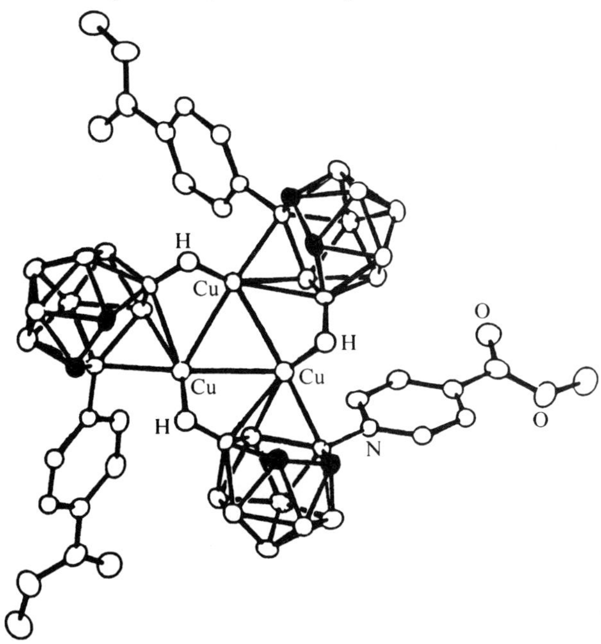

Figure 21 Structure of *nido*-$Cu_3(\mu\text{-}H)_3(7,8\text{-}C_2B_9H_9\text{-}9\text{-}NC_5H_4COOMe)_3$ (reprinted with permission from *Inorg. Chem.*, 1988, **27**, 1716; copyright 1988 American Chemical Society).

While most dicarbollide metal complexes are prepared from the $[R^1R^2C_2B_9H_{10}]^-$ or $[R^1R^2C_2B_9H_9]^{2-}$ anions and metal cations, alternative approaches have been used. The group at Minsk has synthesized $[M(C_2B_9H_{11})_2]^-$ sandwiches (M = Fe, Ni, Co) via electrolysis of solutions of carbaborane anions using metal electrodes.[135,136,167] The same workers obtained isomeric 3,1,2- and 2,1,7-nickelacarboranes via thermal decomposition of σ-*o*-carboranyl nickel derivatives.[181,182,291] A most unusual synthesis of an icosahedral metallacarbaborane was the formation of 3,1,2-CpCoMo[(μ-C_3H_3)$C_2B_9H_9$] from *arachno*-Cp(C_5H_4)MoB_4H_7 in very low yield, in a process where two carbon atoms of a Cp ring are incorporated into the cage.[195]

Cage-substituted derivatives of 12-vertex complexes continue to be characterized in large numbers (Table 4), often for specific applications, as discussed below. Czech workers have extended their pioneering studies on intramolecularly bridged bis(dicarbollyl) compounds via the synthesis of the iodinium-bridged complex[175] 3,1,2-Co$(C_2B_9H_{10})_2$-μ-8,8'-I, the corresponding bromonium-bridged complex (not isolated), and the related species [2,1,7-Co$(C_2B_9H_{10})_2$-μ-6,6'-ER]$^-$ in which the bridging group E is O, S, or N, and R is a substituent that varies in size and electronic properties.[140,172,174] Enantiomers of these chiral complexes have been separated by liquid chromatography on a β-cyclodextrin column; CD spectra of the enantiomers eluted first or last show the same absolute configuration regardless of E and R.[172] Optical resolution was similarly achieved on the asymmetrically substituted derivative 3,1,2-CpCo($C_2B_9H_{10}$-4-Me_2S).[154]

9.6.3.2 *Cage rearrangement and modification*

Thermal isomerization of metallacarbaborane cages, a well-known phenomenon,[1a] has been further explored in recent years. Rearrangements entailing separation of the cage carbon atoms can occur at relatively low energy, as shown in the conversions of 3,1,2- to 3,1,11-PtC_2B_9 icosahedra[284] at 25–55 °C and of several alkylidyne–tungstacarboranes at temperatures as low as −78 °C;[244,245,273] in one case, reversibility was observed.[244] (Reversible cage isomerizations have been long known in 6- to 9-vertex cobaltacarbaboranes.[98d]) The bis(phosphino)hydridoiridacarbaboranes 3,1,2-$(Ph_3P)_2$HIr(R$C_2B_9H_{10}$) (R = H, Ph) were readily converted to the corresponding 2,1,8-isomers on reflux in benzene.[281] Erdman and co-workers have investigated the synthesis and thermal isomerization of icosahedral nickelacarbaboranes of the type LNi($C_2B_9H_{11}$) where L is bipy or phen.[183,184,186] Cage rearrangements are also induced by reduction over sodium amalgam, occuring at room temperature in ruthenium and rhodium complexes and at 85 °C in cobaltacarbaboranes.[157] A *closo* to *hypercloso* transformation was effected by protonation of a tungsten–alkylidyne complex.[273]

Thermolysis of bis(triarylphosphine)nickel dicarbollyl complexes leads not to cage isomerization but, rather, to migration of phosphine from the metal to the cage. In the presence of CO in benzene at 80 °C, $(Ph_3P)_2$Ni($C_2B_9H_{11}$) was converted to a dimer containing a direct Ni–Ni bond.[180]

9.6.3.3 *Spectroscopic and other physical properties*

Volkov *et al.*[166a] have examined the Raman spectra of Fe^{III}, Co^{III}, Ni^{III}, and Sn^{II} dicarbollides, and have investigated the diamagnetism of multicage Co^{III} complexes of the type $[Co_n(C_2B_9H_{11})_2(C_2B_8H_{10})_{n-1}]^{n-}$ (n = 1–4), confirming that all the metal centers are indeed low-spin d^6. Phase transitions in solid 3,1,2- and 2,1,7-Co$(C_2B_9H_{11})_2$ have also been examined.[166b]

The electrical conductivity of cobalt complexes of chlorinated dicarbollides has been studied and the dissociation constants of Cs^+, Sr^{2+}, and Ce^{3+} salts determined as part of a study of these ions as extraction agents (see below).[168] The stoichiometry of complex formation in these systems has also been investigated[292] via IR, NMR, and x-ray photoelectron spectroscopies, which demonstrated that the cobalt centers and B–Cl groups are involved in the complexation of alkali and alkaline earth metal ions (an area of significant practical application, as discussed in the next section). Following earlier electrochemical studies of icosahedral metallacarbaboranes by Geiger and Brennan,[40] Russian workers have explored the relation between redox potentials of these species and the ionization potentials of the corresponding metals.[293] Matel *et al.* have investigated the radiolysis of bis(dicarbollyl)cobalt complexes in aqueous media, in which interest centered on the oxidation of the complexes by OH radicals, forming mono- and dihydroxy derivatives.[294,295]

9.6.3.4 Applications

In recent years major efforts have been directed to utilizing the distinctive structural and electronic attributes of metallacarbaboranes in practical ways, described in a review by Plesek.[3] Progress in the more important areas will be summarized here.

(i) Catalysis

Dicarbollyl complexes of Ru, Rh, and Ir have been studied since the mid-1970s, primarily by Hawthorne and his students,[1g] as catalyst precursors in the hydrogenation, hydrosilylation, and hydroformylation of alkenes. Investigations since the mid-1980s have focused on the synthesis, stereochemistry, and reactivity of these species and the mechanisms of the catalytic processes.[206–8,211–14,217,219,222,224,227–9,233,296–301] The most intensively studied catalyst precursors have been the *closo*-rhodacarbaborane 3,1,2-$(Ph_3P)_2HRh^{III}(C_2B_9H_{11})$ and its 2,1,7- and 2,1,12-cage isomers, which, for example, promote the rapid hydrogenation of isopropenyl acetate to acetic acid and propene; use of D_2 generates isotopically pure $MeCO_2D$.[300] These icosahedral complexes are in equilibrium in solution with *exo-nido* rhodium(I) tautomers whose metal centers are outside the cage and bonded to it via a pair of B–H–Rh bridges (Figure 22(a)). The rhodium(I) centers in the *exo-nido* species migrate on the surface of the carbaborane cage via rapid formation and breaking of Rh–H–B bonds. Extensive studies have shown that the rhodium(I) center in the *exo-nido* tautomer slowly undergoes regiospecific oxidative addition to B–H to form a B–Rh^{III}–H array which is catalytically active.[300] A new mechanism has been proposed for hydrogenation catalysis that supersedes earlier schemes;[298,299] a simplified version is presented in Figure 22(b).[300] It is now known that the *exo-nido* tautomers are the sources of the catalytically active species in a range of reactions, including hydrogenation and isomerization of alkenes and H–D exchange. On the other hand, in rhodacarbaborane-catalyzed reactions of alkenyl acetates with triethylsilane, *closo*-3,1,2-$(Ph_3P)_2HRh(C_2B_9H_{11})$ was found to be far superior to *exo-nido*-$(Ph_3P)_2$Rh-7,8-μ-$(CH_2)_3$-7,8-$C_2B_9H_{10}$ as a catalyst precursor, readily generating trialkylsilyl acetates.[301]

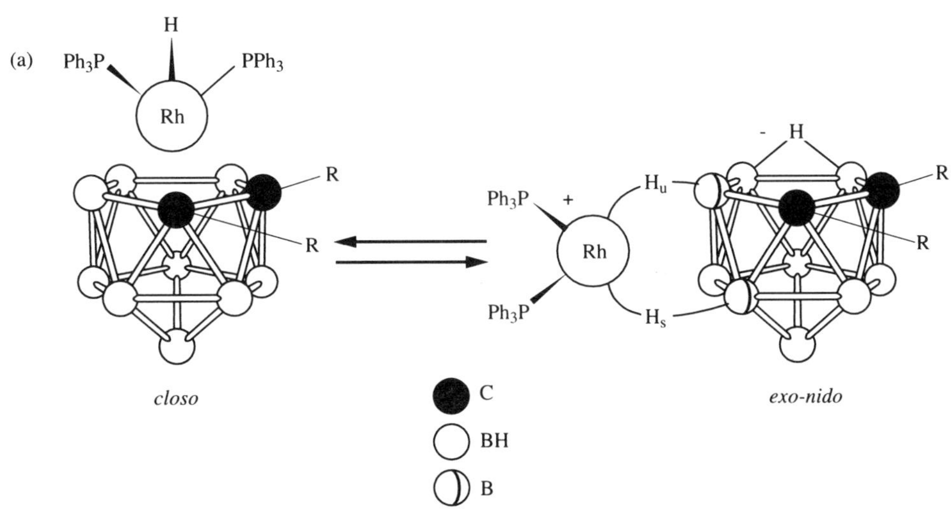

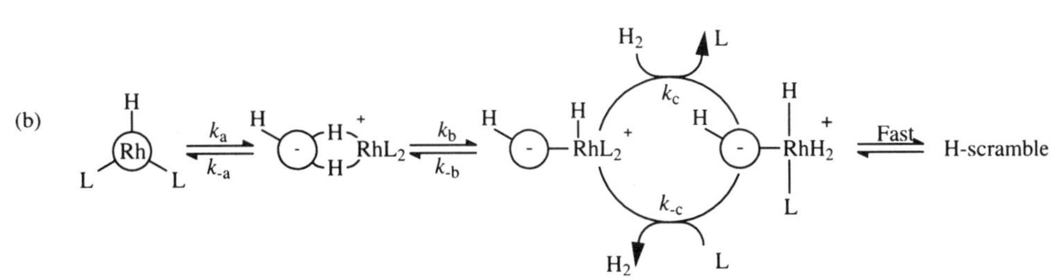

Figure 22 (a) Equilibrium involving *closo* and *exo-nido* rhodacarbaborane tautomers. (b) Proposed mechanism for hydrogenation catalysis ((a) reprinted with permission from *Organometallics*, 1990, **9**, 2327; (b) from *J. Am. Chem. Soc.*, 1989, **111**, 7475; copyright 1990/1989 American Chemical Society).

Similar work involving dicarbollyl rhodium complexes in catalysis has been reported, in considerably less detail, by Russian investigators. The hydrogenation and isomerization of alkenes is promoted by copolymers containing bound –[3,1,2-$(Ph_3P)_2HRh(C_2B_9H_{10})$-1-$CMeCH_2$]– units,[302] while polymeric 2,1,7-rhodacarbaboranes catalyze the hydrosilylation of 1-hexene.[221] Monomeric 3,1,2-$(Ph_3P)_2HRh(C_2B_9H_{11})$ is reported to be an effective catalyst for the alcoholysis of hydriosilanes to silyl ethers.[303] Belgian and Russian workers have examined the use of rhodacarbaborane catalysts in the stereoselective hydrogenation of methacycline to doxycycline (a tetracycline antibiotic employed in chemotherapy)[304] and the same groups have found that the ruthenacarbaboranes 3,1,2-$(Ph_3P)_2ClHRu(R_2C_2B_9H_9)$ and *exo-nido*-$[(Ph_3P)_2ClRu^+(R_2C_2B_9H_{10})]^-$ (R = H, Me) catalyze alkene cyclopropanation reactions.[305] In an indirect application, $[M(C_2B_9H_{11})_2]^-$ complexes have been employed as "noncoordinating" anions with $[Cp_2ZrMe]^+$ in catalysis of alkene polymerization.[306] This exploits the extremely low basicity of the metallacarbaborane anions, a property explored also by Buchwald *et al.*[307]

Dicarbollyl complexes of the early transition metals are catalytically active in alkene polymerization reactions, as shown by Jordan.[191,308] Species of the type $Cp^*RM(C_2B_9H_{11})$ (M = Zr, Hf; R = alkyl) are neutral analogues of Cp_2MR^+ cations that are the presumed active species in metallocene-based Ziegler catalysts. Accordingly, the metallacarbaboranes show typical "electrophilic metal alkyl" behavior, in that they bind Lewis bases, are reactive toward insertion and intramolecular C–H activation, exhibit agostic M–H–C and M–H–B interactions, and promote ethylene polymerization and propene oligomerization.[308] Tantalum(V) complexes such as $(\eta\text{-}C_5H_4Me)Me_2Ta(C_2B_9H_{11})$ and $Me_3Ta(C_2B_9H_{11})$[243] may exhibit reactivity similar to the corresponding cationic $Cp^*R_2ML_n{}^+$ and $Cp_2RML_n{}^+$ species.

(ii) Stabilization of organic groups

Zakharkin and co-workers have studied extensively the electronic properties of metallacarbaboranes such as 3,1,2-$CpFe^{II}(C_2B_9H_{10})$ and 3,1,2-$(CO)_3M(C_2B_9H_{10})$ (M = Mn, Re) as substituents on organic molecules.[130,143–6,148,149,280] These moieties stabilize carbocations very effectively, indeed more strongly than does the ferrocenyl group. Detailed studies of their electronic effects via the Taft method[130,148] revealed that while $CpFe^{II}(C_2B_9H_{10})$ is a better electron donor than $(CO)_3Mn(C_2B_9H_{10})$ toward substrates bound to the cage carbon, the latter group is a more effective stabilizing agent; hence there appears to be no correlation between carbocation stabilization and electron donation, contrary to earlier ideas.[130]

(iii) Metal extraction and trapping

A distinctive characteristic of most carbaborane and metallacarbaborane metal salts is their solubility in organic solvents. This property has been exploited in the solvent extraction of alkali and alkaline earth metals, Pd^{2+}, Pb^{2+}, Tl^+, lanthanides, actinides, and others, and finds application in the analysis of milk and other biological samples as well as the recovery of radionuclides from spent nuclear fuels.[3,178,309] The most extensively used reagents are the $[Co(C_2B_9H_{11})_2]^-$ anion and its *B*-hexachloro derivative, using nitrobenzene as the extracting solvent. Heavy alkali metal cations, especially Sr^{2+} and Ba^{2+}, are easily separated from other metals and from each other using the cobalt reagent with appropriate additives.[310–14] Large-scale recovery of ^{137}Cs and ^{90}Sr is accomplished very effectively with the $[Co(C_2B_9H_8Cl_3)_2]^-$ anion which is is stable in 3 M nitric acid, a reagent employed in the process.[3]

(iv) Biological and medical applications

A particularly imaginative and potentially important application of the metal-coordinating properties and other attributes of carbaborane anions is the development of "Venus flytrap" bridged-dicarbollide reagents for the capture of radioactive metal ions and their subsequent use in tumor imaging and therapy.[127,141,173,176,315] The coordinating ligands consist of two 11-vertex *nido*-carbaborane units connected by a linking polymethylene or other carbon chain,[127] or by a pyrazole ring.[141,176] Figure 23(a) depicts a pyrazole-bridged monoanion, obtained as a pair of diastereomers, which on removal of both B–H–B protons acquires a trinegative charge and readily coordinates to trivalent transition metal ions (Figure 23(b)). A derivative having a carbomethoxy group on the pyrazole ring, and incorporating the radioisotope $^{57}Co^{3+}$, has been esterified and subsequently attached to an anticarcinogenic monoclonal antibody with full retention of immunological activity and excellent tumor localization in mice.[315,316] The highly stable metallacarbaborane clusters are not metabolized in the body, affording a major advantage over conventional radiopharmaceutical carries such as amino carboxylates, and may find use in the delivery of other radioisotopes such as ^{66}Ni, ^{67}Cu, and ^{105}Rh in diagnosis and therapy.[141,315]

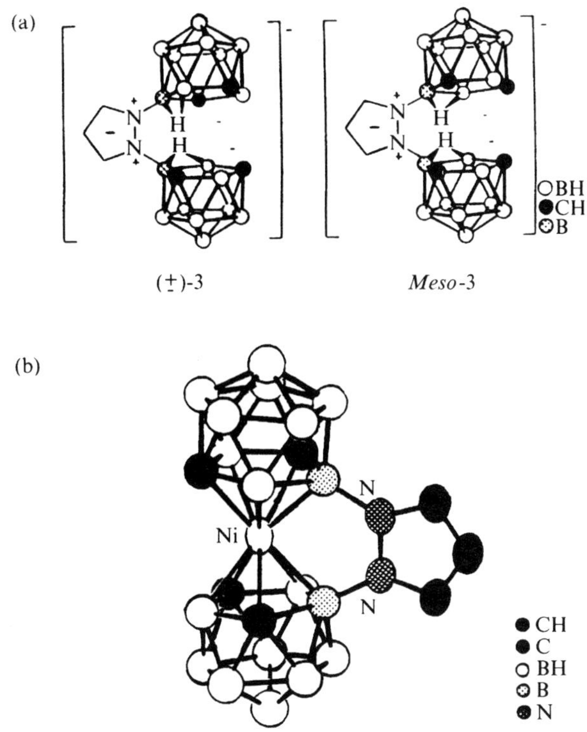

Figure 23 Structures of (a) pyrazole-bridged "Venus flytrap" monoanions and (b) nickel(III) sandwich complex 2,1,7-$Ni(C_2B_9H_{10})_2$-μ-$N_2C_3H_3$ (reprinted with permission from *J. Am. Chem. Soc.*, 1992, **114**, 9003; copyright 1992 American Chemical Society).

Inclusion complexes in which the mangana- and rhenacarbaboranes 3,1,2-$(CO)_4M(C_2B_9H_{11})$ are bound into α- and β-cyclodextrin have been prepared and structurally characterized,[133] and are of interest as enzyme models and for their potential application in catalysis; very strong binding was observed in aqueous solution with β-cyclodextrin as host. Finally, the gadolinium complex[242] 3,1,2-$[Gd(C_2B_9H_{11})_2]^-$ is worthy of mention given the high thermal neutron capture cross-section of the ^{155}Gd and ^{157}Gd isotopes (higher than that of ^{10}B, which is the basis of boron neutron capture therapy). In principle, such compounds could find application in gadolinium-based tumor treatment.

9.6.4 13-vertex Cages

The coordination of *nido*-$[C_2B_{10}H_{12}]^{2-}$ ligands or their derivatives to metal ions, obtained via reductive cage opening of icosahedral carboranes, form supraicosahedral 13-vertex *closo*-MC_2B_{10} clusters. As demonstrated in extensive earlier work,[1a] these complexes often exhibit chemistry distinct from that of their 12-vertex counterparts, for example, undergoing facile cage rearrangement and degradation. Since 1982 a number of studies involving 13-vertex systems have been published by Stone and his associates,[129,193,200,244,317–22] generally extending the main themes of their work on icosahedral MC_2B_9 alkylidene/alkylidyne-bridged systems. The characterized compounds from this and other laboratories are listed in Table 5.

The stereochemistry of heterobimetallic and -trimetallic cluster systems incorporating MC_2B_{10} cages, prepared from metal–alkylidyne complexes of molybdenum and tungsten,[317–21] is similar in many respects to that of analogous icosahedral systems (see above and earlier reviews).[1b,1m] It is found that the carbaborane ligand can adopt either a "spectator" role (a term not without ambiguities), or can form B–M or B–H–M bonds to the metal.[317] In a departure from 12-vertex chemistry, protonation of the 13-vertex alkylidyne metallacarbaboranes often leads to facile loss of the protonated boron vertex, generating an icosahedral MC_2B_9 cluster.[244,319] In one of the more intriguing findings, protonation of the alkylidyne clusters $[(CO)_2(RC)W(Me_2C_2B_{10}H_{10})]^-$ (R = Me or C_6H_4Me) gave dimeric products featuring

Table 5 13-vertex dicarbon metallacarbaboranes (since 1982).

Compound	*Data*[a]	*References*
Cobalt (cage), cobalt (exo-*polyhedral)*		
1,2,4-$(Et_3P)Co(MeC_2B_{10}H_{10})$-μ-$Co(Et_3P)_2$-μ-$(H)_2$	X	318
Molybdenum (cage), cobalt (exo-*polyhedral)*		
(μ-$MeC_6H_4C)(CO)_2Cp^*Co{=}Mo(Me_2C_2B_{10}H_{10})^-$	H, C, I*	317
(μ-$MeC_6H_4C)(CO)_2Cp^*Co{=}Mo(Me_2C_2B_{10}H_9)$(*Co–B*)	B, H, C*, I*	317
$(Ph_3P)Au$(μ-$MeC_6H_4C)(CO)_2Cp^*Co{=}Mo(Me_2C_2B_{10}H_{10})$	H, C, I*	317
(μ-$MeC_6H_4C)(CO)_2$(η-$C_4Me_4)Co{=}Mo(Me_2C_2B_{10}H_{10})$	X, B, H, C*, I*	317
Molybdenum (cage), platinum (exo-*polyhedral)*		
$L^1L^2Pt{=}Mo(CO)_3(Me_2C_2B_{10}H_{10})$ ($L^1 = Ph_3P, Et_3P$; $L^2 = Ph_3P, CO$)	X[Ph_3P, CO], B, C, H, P, I*	129
$(Ph_3P)(CO)Pt{=}Mo(CO)_3(Me_2C_2B_{10}H_9)^-$ (*Pt–B*)	B, H, C, P, I*	129
Molybdenum (cage), gold (exo-*polyhedral)*		
$(Ph_3P)Au$(μ-$MeC_6H_4C)(CO)_2Cp^*Co{=}Mo(Me_2C_2B_{10}H_{10})$	H, C, I*	317
Palladium		
$(Ph_2PCH_2)_2Pd(C_2B_{10}H_{12})$	X, B	285
Tungsten (no exo-*polyhedral metals)*		
1,2,4-$(CO)_3(R^1C)W(R^2{}_2C_2B_{10}H_{10})^-$ ($R^1 = MeC_6H_4$, Me; R^2 = H, Me)	H, C, I*	319
1,2,3-$(RC)(CO)_2W(Me_2C_2B_{10}H_{10})^-$ ($R = MeC_6H_4, Me_2C_6H_3$)	X[$Me_2C_6H_3C$], H, C, I*	320
1,2,3-$(CO)_2(RC)W(Me_2C_2B_{10}H_{10})^-$ ($R = MeC_6H_4, Me_2C_6H_3$)	X[$Me_2C_6H_3$], H, C, I*	321
1,2,4-$(CO)_4W(Me_2C_2B_{10}H_9CH_2C_6H_4Me)$	B, H, C, P, I*	319
1,2,4-$(CO)_2(PhC{=}CPh)W(Me_2C_2B_{10}H_9CH_2R)$ ($R = C_6H_4Me$, Me)	B, H, C, P, I*	319
1,2,4-$(CO)_2L_2W(R_2C_2B_{10}H_{10})$ (L = CO, PPh_3; R = H, Me)	B, H, C, P, I*	319
1,2,4-$(CO)_2(PhC{=}CPh)W(Me_2C_2B_{10}H_{10})$	B, H, C, P, I*	319
1,2,4-$(CO)_4$(μ-$R^1C)W_2(R^1{}_2C_2B_{10}H_{10})(C_2B_{10}H_9$-$CH_2$-$R^2)$ (R^1 = H, Me; $R^2 = Me, C_6H_4Me$)	X[Me,Me],B, H, C, P, I*	319
Tungsten (cage), iron (exo-*polyhedral)*		
$[Fe_2(CO)_6(CO)_2W(Me_2C_2B_{10}H_{10})$-μ-$CC_6H_4Me]^-$	B, H, C, I*	320
$[Fe(CO)_2(CO)_2W(Me_2C_2B_{10}H_{10})$-μ-$CC_6H_3Me_2]^-$	H, C, I*	321
$[Fe(CO)_2(CO)_2W(Me_2C_2B_{10}H_{10})$-μ-$CC_6H_3Me_2]^-$	X, B, H, C, I*	320
Tungsten (cage), cobalt (exo-*polyhedral)*		
(μ-$MeC_6H_4C)(CO)_2Cp^*Co{=}W(Me_2C_2B_{10}H_{10})^-$	H, C, I*	317
(μ-$MeC_6H_4C)(CO)_2Cp^*Co{=}W(Me_2C_2B_{10}H_9)$ (*Co–B*)	B, H, C*, I*	317
$(Ph_3P)Au$(μ-$MeC_6H_4C)(CO)_2Cp^*Co{=}W(Me_2C_2B_{10}H_{10})$	X, H, C, I*	317
(μ-$MeC_6H_4C)(CO)_2$(η-$C_4Me_4)Co{=}W(Me_2C_2B_{10}H_{10})$	B, H, C*, I*	317
Tungsten (cage), molydenum (exo-*polyhedral)*		
μ-C_6H_4Me-$LCpMo$-$(CO)_2W(Me_2C_2B_{10}H_{10})$ (L = CO, PMe_3)	X[PMe_3], B, H, C, I*	200
$(C_7H_7)Mo$(μ-$MeC_6H_4C)(CO)_2W(MeC_2B_{10}H_{10})$	B, H, C, I*	193
Tungsten (cage), rhodium (exo-*polyhedral)*		
(μ-CO)$[(Ph_3P)_2Rh]$(μ-R)$(CO)W(Me_2C_2B_{10}H_{10})$ ($R = MeC_6H_4C, Me_2C_6H_3C$)	B, H, C, P, I*	320
μ-$(Ph_3P)_2Rh$-μ-$RC(CO)_2W(Me_2C_2B_{10}H_{10})$ ($R = MeC_6H_4, Me_2C_6H_3$)	H, P, I*	321
Tungsten (cage), tungsten (exo-*polyhedral)*		
μ-$C_6H_4MeLCpW(CO)_2W(Me_2C_2B_{10}H_{10})$ (L = CO, PMe_3)	B, H, C, I*	200
Tungsten (cage), platinum (exo-*polyhedral)*		
$L^1L^2Pt{=}W(CO)_3(Me_2C_2B_{10}H_{10})$ ($L^1 = Ph_3P, Et_3P$; $L^2 = Ph_3P$, CO)	B, C, H, P, I*	129
$(Ph_3P)(CO)Pt{=}W(CO)_3(Me_2C_2B_{10}H_9)^-$ (*Pt–B*)	B, H, C, P, I*	129
Tungsten (cage), gold (exo-*polyhedral)*		
(μ-Ph_3PAu)(μ-$RC)(CO)_2W(Me_2C_2B_{10}H_{10})$ ($R = MeC_6H_4, Me_2C_6H_3$)	H, C, I*	321
(μ-$Ph_3PAu)(R)(CO)_2W(Me_2C_2B_{10}H_{10})$ ($R = MeC_6H_4C, Me_2C_6H_3C$)	H, C, P, I*	320
$(Ph_3P)Au$(μ-$MeC_6H_4C)(CO)_2Cp^*Co{=}W(Me_2C_2B_{10}H_{10})$	X, H, C, I*	317
Iridium		
$(Ph_3P)_2(H)Ir(C_2B_{10}H_{11}OMe)$	X, B*	285
Gold		
nido-$(Ph_3P)Au(7,9$-$Me_2C_2B_{10}H_{10})^-$	X, B, H, C, P	322
Samarium		
$[(C_4H_8O)_3Sm(C_2B_{10}H_{12})]_\infty$	M, B, H, I	323
Europium		
1,2,4-$(C_4H_8O)_2Eu(C_2B_{10}H_{12})_2{}^{2-}$	X, B	324
1,2,4-$(C_4H_8O)_2Eu(C_2B_{10}H_{12})_2{}^{2-}$	X, B, H, I	323
1,2,4-$[L_3Eu(C_2B_{10}H_{12})]_n$ ($L = OC_4H_8$, MeCN)	X[MeCN], B, H, I	323
Ytterbium		
$(C_4H_8O)_4Yb(C_2B_{10}H_{12})$	B*, H, M, I	323

[a] X = x-ray diffraction, B = ^{11}B NMR, H = ^1H NMR, C = ^{13}C NMR, P = ^{31}P NMR, M = magnetic susceptibility, I = infrared. Asterisks designate that only partial data are given.

a B–W–W triangular link and a B–H–W bridge (Figure 24).[319] The propensity of some 13-vertex metallacarbaboranes to undergo low-energy rearrangement involving separation of the cage carbon atoms,[244,321] in some cases reversibly,[244] is a notable feature of these systems, reminiscent of previously observed reversible isomerizations in small metallacarbaboranes.[98d] A rare example of an open 13-vertex cage is afforded by the gold complex[322] *nido*-$(Ph_3P)Au(7,9\text{-}Me_2C_2B_{10}H_{10})^-$ in which the metal is coordinated to just three borons on the six-membered open face of the carbaborane ligand.

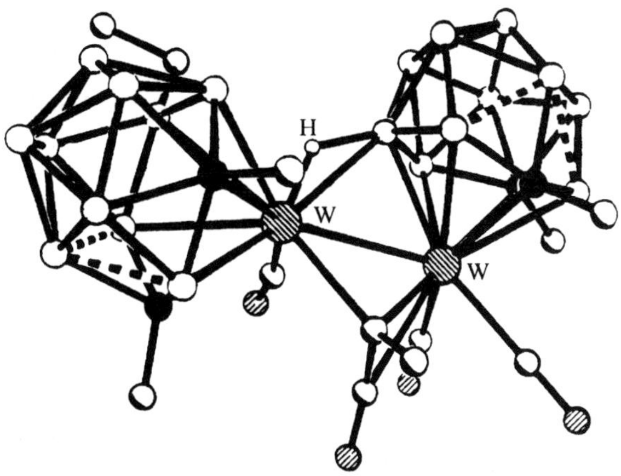

Figure 24 Structure of 1,2,4-$(CO)_4(\mu\text{-}MeC)W_2(Me_2C_2B_{10}H_{10})(C_2B_{10}H_9Et)$. Framework carbon atoms are solid black (reprinted with permission from *Organometallics*, 1993, **12**, 1131; copyright 1993 American Chemical Society).

In a different application of 13-vertex chemistry,[323,324] lanthanacarbaboranes have been prepared via reactions of $LnI_2(THF)_2$ salts (Ln = Sm, Eu, or Yb) with *nido*-7,9-$[C_2B_{10}H_{12}]^{2-}$. Replacement of the THF ligands in the europium complex with MeCN generates a novel polymeric solid, $[Eu(C_2B_{10}H_{12})(MeCN)_3]_\infty$ whose structure is shown in Figure 25.[323]

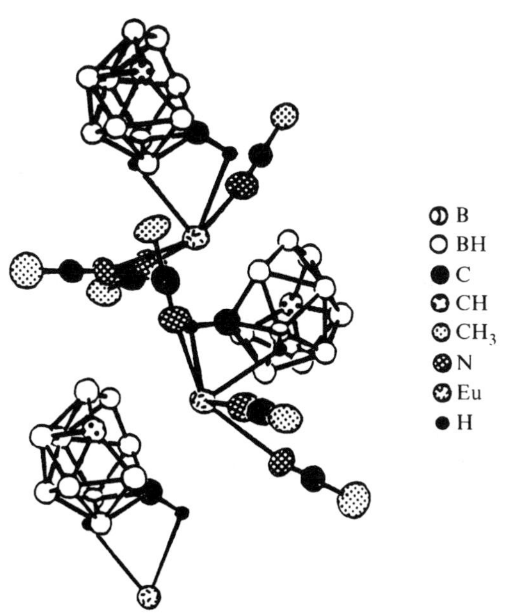

Figure 25 Structure of 1,2,4-$[(MeCN)_3Eu(C_2B_{10}H_{12})]_n$ showing three repeating units of the spiral polymeric chain (reprinted with permission from *Inorg. Chem.*, 1992, **31**, 268; copyright 1992 American Chemical Society).

9.7 TRICARBON METALLACARBABORANES

The few metal derivatives of tricarbon carbaboranes reported since the earlier review[1a] are listed in Table 6. Metal sandwich complexes incorporating planar C_3B_2 (diborole) ring ligands, of which a considerable number have been characterized by Siebert and his co-workers, can be viewed as tricarbon carbaboranes but are discussed separately in Chapter 5, this volume. However, a recent outgrowth of this chemistry has been the synthesis of boron-capped diborole complexes which contain seven-vertex pentagonal bipyramidal MC_3B_3 clusters[325–7] and are metallacarbaboranes by any definition. The species shown in Figure 26 were obtained from diborolyl starting materials, in a mechanistically undefined process in which capping of the C_3B_2 rings occurs via transfer of B–Me units.[325,326] The 2,3,5-tricarbon carbaborane ligands, which have not been isolated in uncomplexed form, are isomeric with the previously prepared 2,3,4-$[Me_nC_3B_3H_{6-n}]^-$ ions ($n = 1, 2$)[328] (having three adjacent carbons in the basal ring) which form stable managanacarbaboranes of the type 1,2,3,4-$(CO)_3Mn(Me_nC_3B_3H_{6-n})$.[329] They are also isolobal with $C_5H_5^-$, and electrochemical and NMR spectroscopic studies of the 2,3,5-tricarbon metallacarbaboranes by Siebert *et al.*[327] indicate that the metal–ligand bonding is very similar to that in metallocenes.

Table 6 Tricarbon metallacarbaboranes (since 1982).

Compound[a]	*Data*[b]	*References*
7-Vertex polyhedra, 1 metal atom		
Nickel		
1,2,3,5-$Ni(HEt_2C_3B_3Me_3)_2$	X, M, E, g	326
1,2,3,5-$(HEt_2C_3B_3Me_3)_2Ni_2(HEt_2C_3B_2Me_2)$	X, B, H	325
1,2,3,5-$(HEt_2C_3B_3Me_3)_2Ni_2(HEt_2C_3B_2Me_2)$	E, g, H	327
1,2,3,5-$(HMe_2C_3B_3Me_3)_2Ni_2(HMe_2C_3B_2Me_2)$	B, H	325
1,2,3,5-$(HEt_2C_3B_3Me_3)Ni(HEt_2C_3B_2Me_2)$	X, E	327
1,2,3,5-$(HEt_2C_3B_3Me_3)Ni(HEt_2C_3B_2Me_2)$	B, H	325
1,2,3,5-$(C_8H_{12})Ni(HEt_2C_3B_3Me_3)^+$	X, E	327
1,2,3,5-$CpNi(HEt_2C_3B_2Me_2)Ni(HEt_2C_3B_3Me_3)$	E, g, H	327
1,2,3,5-$[(HMe_2C_3B_3Me_3)Ni(HMe_2C_3B_2Me_2)]_2Ni$	H	325
11-Vertex polyhedra, 1 metal atom		
Manganese		
1,2,3,4-$(CO)_3Mn(2\text{-}MeC_3B_7H_9)$	B, H, C, I	330
Iron		
1,2,3,4-$CpFe(n\text{-}MeC_3B_7H_9)$ ($n = 2, 4$)	X, B(2d), H, I, u	330
1,2,3,5-$Fe(n\text{-}MeC_3B_7H_9)_2$ ($n = 2, 5$)	X, B, H, I, u	331
(5-Me-2,3,5-$MeC_3B_7H_9$)Fe(4-Me-2,3,4-$C_3B_7H_9$)	X, B, H, I, u	331
Fe(10-Me-2,3,10-$C_3B_7H_9)_2$	X, B, H, I, u	331
1,2,3,4-CpFe(n-Me-2,3,4-$C_3B_7H_9$) ($n = 2, 4$; ^{13}C labeled)	H, C	332
Fe(2-Me-2,3,5-$C_3B_7H_9)_2$ (^{13}C labeled)	H, C	332
Fe(10-Me-2,3,10-$C_3B_7H_9)_2$ (^{13}C labeled)	H, C	332
(5-Me-2,3,5-$C_3B_7H_9$)Fe(4-Me-2,3,4-$C_3B_7H_9$) (^{13}C labeled)	H, C	332
Cobalt		
Co(2-Me-2,3,5-$C_3B_7H_9)_2$	X, I, u	331
(5-Me-2,3,5-$C_3B_7H_9$)Co(2-Me-2,3,5-$C_3B_7H_9$)	X, I, u	331
Nickel		
nido-9,7,8,10-$CpNi(8\text{-}MeC_3B_7H_9)$	B, H, I	330
nido-11,7,8,10-(η^3-Me_4C_4H)$NiC_3B_7H_7$	X, B, H, I	12d

[a] Clusters are *closo* except when otherwise designated; metals are in order of increasing atomic number. Cp = η-C_5H_5, Cp* = η-C_5Me_5. [b] X = x-ray diffraction, B = ^{11}B NMR, H = ^1H NMR, C = ^{13}C NMR, M = magnetic susceptibility, U = UV–visible data, I = infrared data, g = ESR (EPR) data. Asterisks designate that only partial data are given.

Sneddon's group[330–2] has pioneered the development of large tricarbon metallacarbaborane chemistry, initially isolating an 11-vertex $NiC_3B_7H_7$ cluster from the reaction of nickel vapor and *arachno*-$C_2B_7H_{13}$ with 2-butyne.[12d] Subsequently, a more controllable approach was developed based on the *nido*-[6-Me-5,6,9-$C_3B_7H_9]^-$ ion, an isoelectronic analogue of the decaborane anion $[B_{10}H_{13}]^-$ that is proposed to have a similar basketlike structure. The mononegative carbaborane ligand can function as a surrogate for $C_5H_5^-$, combining with divalent metal ions to form neutral $M(C_3B_7)_2$ or $CpM(C_3B_7)$ complexes. This fact, together with the nonplanar six-membered open face on the ligand, leads to a stereochemistry for these compounds that is quite distinct, for a given metal ion, from that of most dicarbon metallacarbaboranes. For example, the tricarbon carbaborane analogue of a bis(dicarbollyl) dianion $[M^{II}(C_2B_9H_{11})_2]^{2-}$ is neutral $M^{II}(MeC_3B_7H_9)_2$. Similarly, neutral $CpM^{II}(MeC_3B_7H_9)$ sandwiches are analogous to $[CpM^{II}(C_2B_9H_{11})]^-$ monoanions. Species of both types have been prepared via reaction of the carbaborane anion with first-row transition metal ions, in several cases generating isomers that differ in the location of the C–Me unit; thermolysis of these products generates additional isomers.[331] Studies employing ^{13}C labeling demonstrated that the isomerism occurs via cage rearrangement, with no

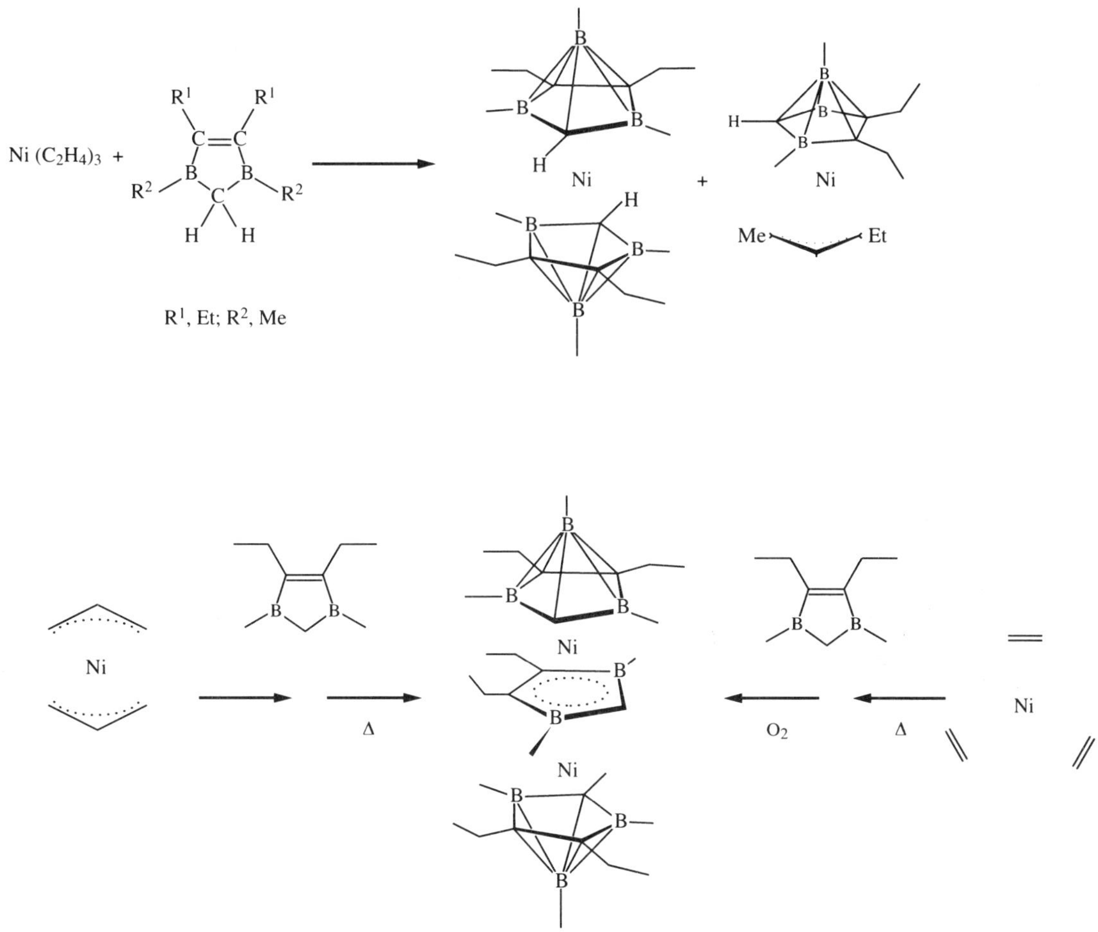

Figure 26 Synthetic routes to 1,2,3,5-NiC_3B_3 cluster compounds (reproduced by permission of VCH from *Agnew. Chem., Int. Ed. Engl.*, 1986, **25**, 1099 and reprinted with permission from *Organometallics*, 1988, **7**, 2316; copyright 1988 American Chemical Society).

cleavage of the methyl–cage carbon bond.[332] Typical of the crystallographically characterized clusters are those shown in Figure 27.[330,331]

Remarkably, although the bis(carboranyl)cobalt isomers (see Figure 27(b)) are analogues of cobaltocene, they are stable to air and common oxidizing agents, and the Co^{II}–Co^{II} couple is 1.8 V more positive than that of cobaltocene.[331] This observation clearly underlines the ability of the tricarbon carbaborane ligand to stabilize +2 metal oxidation states.

9.8 TETRACARBON METALLACARBABORANES

Metallacarbaboranes having four cage carbon atoms have been obtained via several routes, including metal vapor synthesis,[12a] oxidative fusion of dicarbon metallacarbaboranes,[1a,13b,41,48,57] metal insertion into tetracarbon carbaboranes,[1a,14a–c] and, in one case, boron insertion into a metallacycle.[333] (Curiously, while the tetracarbon *nido*-carbaborane $C_4B_2H_6$ and its derivatives have been known since the mid-1960s,[334] no metallacarbaborane complexes of this ligand have been reported.) Table 7 lists the compounds characterized since 1982. As in the case of the diborole–metal complexes mentioned earlier, there is a large class of sandwich compounds bridged by planar borole (C_4B) rings, developed primarily by Herberich. Although these complexes could be classed as tetracarbon metallacarbaboranes, they are included in the treatment of organoboron ring systems in Chapter 5, this volume. The one exception listed in Table 7 is the *nido*-CoC_4B cluster prepared by Fehlner *et al.*[333] from the metallacycle $Cp(PPh_3)CoC_4Ph_4$ and BH_3·THF, and several B-subsituted derivatives.

The most controllable general route to tetracarbon metallacarbaboranes is based on the insertion of metal ions into *arachno*-$[R_4C_4B_8H_8]^{2-}$ ions (R = alkyl), developed since the mid-1970s.[1a,1q,14a–c] These

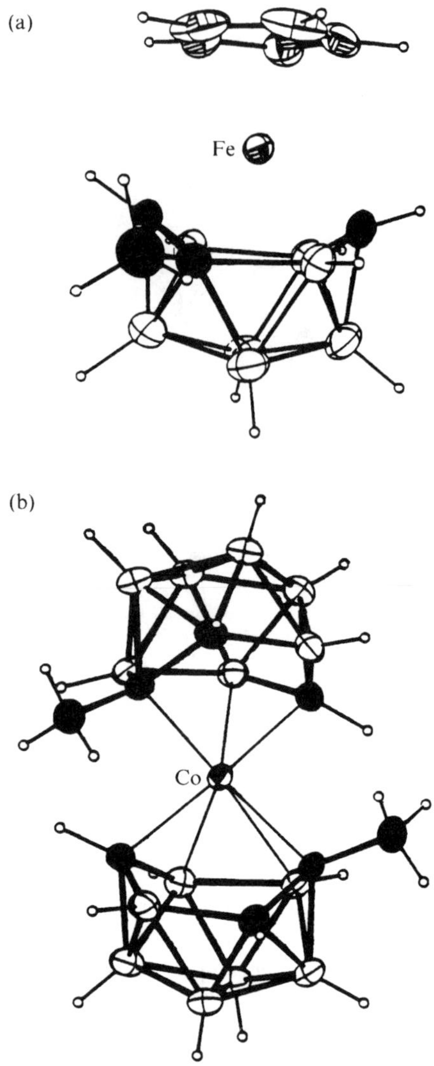

Figure 27 Structures of (a) 1,2,3,4-$CpFe(4\text{-}MeC_3B_7H_9)$ and (b) $Co(2\text{-}Me\text{-}2,3,5\text{-}C_3B_7H_9)_2$. Cage carbon and methyl carbon atoms are solid black ((a) reprinted with permission from *Organometallics*, 1992, **11**, 1665; (b) from *Organometallics*, 1992, **11**, 1672; copyright 1992 American Chemical Society).

syntheses generate 12- to 14-vertex mono- and dimetallacarbaboranes, including mono- and bis(carboranyl) species. In the latter category are both homoligand $[(M_n(R_4C_4B_8H_8)_2]$ (n = 1, 2) and heteroligand $(R^1_4C_4B_8H_8)M(R^2_2C_2B_4H_4)$ complexes of Cr, Fe, and Co.[14a–c] The stereochemistry is often complex, generating structures having intramolecular interligand B–B or B–X–B bonding (Figure 28(a)).[14b] The proposed structure of an unusual biscobalt species having edge-fused 14-vertex polyhedra is presented in Figure 28(b) with its correlated spectroscopy (COSY) ^{11}B–^{11}B 2D-NMR spectrum.[14b]

Tetracarbon metallacarbaboranes can also form via oxidative fusion of dicarbon *nido*-metallacarbaborane substrates (usually MC_2B_4 cages), as mentioned at several points in this review (see Sections 9.3, 9.6.1, 9.6.2.3, and Equation (6). In work on the synthesis of tetradecker sandwich complexes via the coordination of two $[Cp^*Co(Et_2C_2B_3H_2\text{-}5\text{-}R)]^{2-}$ substrate ions to M^{2+}, an approach which works well for M = Co, Ni, and Ru[55,89] (Section 9.6.1), it was found that when M = Fe, the only isolable products were the fused tetracarbon species $Cp^*_2Co_2(Et_4C_4B_6H_4R_2)$ where R is H, Me, Et, or Cl.[13b] X-ray crystallography on the parent and dichloro products revealed a novel cage geometry, shown in Figure 29, which is a fragment of a T_d 16-vertex *closo*-polyhedron with four missing vertices and hence belongs to the "*klado*" structural class.[13b] Although this cage geometry has not been observed previously, it has been postulated as an intermediate species.[335] In terms of skeletal electron-counting theory,[4e,4m,7] this cluster is *apparently* in clear violation; as a 28-electron, 12-vertex cage it would be expected to adopt a *nido* structure corresponding to a 13-vertex polyhedron with one vertex removed. This is, in fact, the geometry found earlier[335] for an isomeric cluster, $Cp_2Co_2(C_4B_6H_{10})$. Consequently, it

Table 7 Tetracarbon metallacarbaboranes (since 1982).

Compound[a,b]	*Data*[c]	*References*
6-Vertex polyhedra, one metal atom		
Cobalt		
nido-1-$CpCoPh_4C_4BH$	X, B, H, I, U	333
nido-1-$CpCoPh_4C_4BOH$	B, H, I, U	333
nido-1-$CpCoPh_4C_4BR$ (R = Br, Ph, Bu)	B, H	333
nido-1-$CpCoPh_4C_4BR$ (R = OMe, $OSiMe_3$)	H	333
8-Vertex polyhedra, one metal atom		
Iron		
arachno,nido-1,4,5,7,8-$(C_6Me_6)Fe(Me_4C_4B_3H_3)$	X, B, H, I	12b
arachno,nido-1,4,5,7,8-$(C_6H_5Me)Fe(Me_4C_4B_3H_3)$	B, H, I	12b
9-Vertex polyhedra, one metal atom		
Cobalt		
nido-$(Et_4C_4B_4H_6)CoH(Et_4C_4B_8H_8)$	B, H, I	14b
10-Vertex polyhedra, one metal atom		
Iron		
nido-2,6,7,9,10-$(C_6H_5Me)Fe(Me_4C_4B_5H_5)$	X, B, H, I	12b
11-Vertex polyhedra, one metal atom		
Iron		
nido-2,7,8,9,10-$(C_6H_6)Fe(Et_4C_4B_6H_6)$	X*, B(2D), H, I	41
nido?-$(C_6Me_6)Fe(Me_4C_4B_6H_6)$	S	12b
12-Vertex polyhedra, one metal atom		
Chromium		
nido-$CpCr(Et_4C_4B_7H_7)$(green isomer)	X	14a
nido?-$CpCr(Et_4C_4B_7H_7)$(yellow isomer)	S	14a
Iron		
nido?-$(MeC_6H_5)Fe(Et_4C_4B_7H_7)$	S	12a
Cobalt		
nido-$(NC_4Me_4)Co(Et_4C_4B_7H_7)$	B, H, I	57
12-Vertex polyhedra, two metal atoms		
Cobalt		
$Cp^*_2Co_2(C_4B_6H_8Cl_2)$	B, H, I	48
nido-$Cp^*_2Co_2(Et_4C_4B_6H_6)$	B, H, I, U	57
nido-$(NC_4Me_4)_2Co_2(Et_4C_4B_6H_6)$	B, H, I, U	57
klado-$Cp^*_2Co_2(Et_4C_4B_6H_4X_2)$ (X = H, Me, Et, Cl)	X(H, Cl), B, H, U(Cl), I(Cl)	13b
13-Vertex polyhedra, one metal atom		
Chromium		
nido-$CpCr(Et_4C_4B_8H_8)$	X, I	14a
Iron		
nido-$(Et_4C_4B_8H_7)FeH(Et_2C_2B_3H_4)$ (*B–B*)	B, I	14c
$(Et_4C_4B_8H_7)FeH_2(Et_2C_2B_4H_3)$ (*B–B*) (isomer mixture)	B	14c
$(Et_4C_4B_8H_8OC_4H_8)Fe(Et_2C_2B_4H_4)$	B, I	14c
Cobalt		
nido-$(Et_4C_4B_8H_8)CoH(Et_4C_4B_8H_8OC_4H_8)$	B(2D), H, I	14c
nido-$CoH(Et_4C_4B_8H_7)_2$ (*B–B*)	B, H, I	14c
$(Et_4C_4B_8H_7OC_4H_8)Co(Et_2C_2B_4H_4)$	X, B, H, I	14c
$(Et_4C_4B_8H_7OC_4H_8)Co(Et_2C_2B_4H_3)$ (*B–B*)	B, H, I	14c
nido-$(Et_4C_4B_8H_8)Co(Et_4C_4B_8H_7OC_4H_8)$	I	14b
nido-$CoH(Et_4C_4B_8H_7)_2$ (*B–B*)	B(2D), I	14b
nido-$(Et_4C_4B_4H_6)CoH(Et_4C_4B_8H_8)$	B, H, I	14b
arachno-$CoH(Et_4C_4B_8H_7)_2(OCMe_2)_2$	X, B, H, I	14b
nido-$Co(Et_4C_4B_8H_7)_2(OH)$ (2 isomers)	B, I	14b
nido-$CoH(Et_4C_4B_8H_6OH)_2$	B, I	14b
14-Vertex polyhedra, two metal atoms		
$Co_2(Et_4C_4B_8H_8)_2$	B(2D), H, I	14b
nido-$Co_2(Et_4C_4B_8H_7)_2$ (*B–B*) (3 isomers)	B, H, I	14

[a] Clusters are *closo* except when otherwise designated; metals are in order of increasing atomic number. Cp = η-C_5H_5, Cp* = η-C_5Me_5. [b] X = x-ray diffraction, B = ^{11}B NMR, H = ^1H NMR, C = ^{13}C NMR, 2D = two-dimensional (COSY) NMR, U = UV–visible data, I = infrared data, S = mass spectral data only. Asterisks designate that only partial data are given.

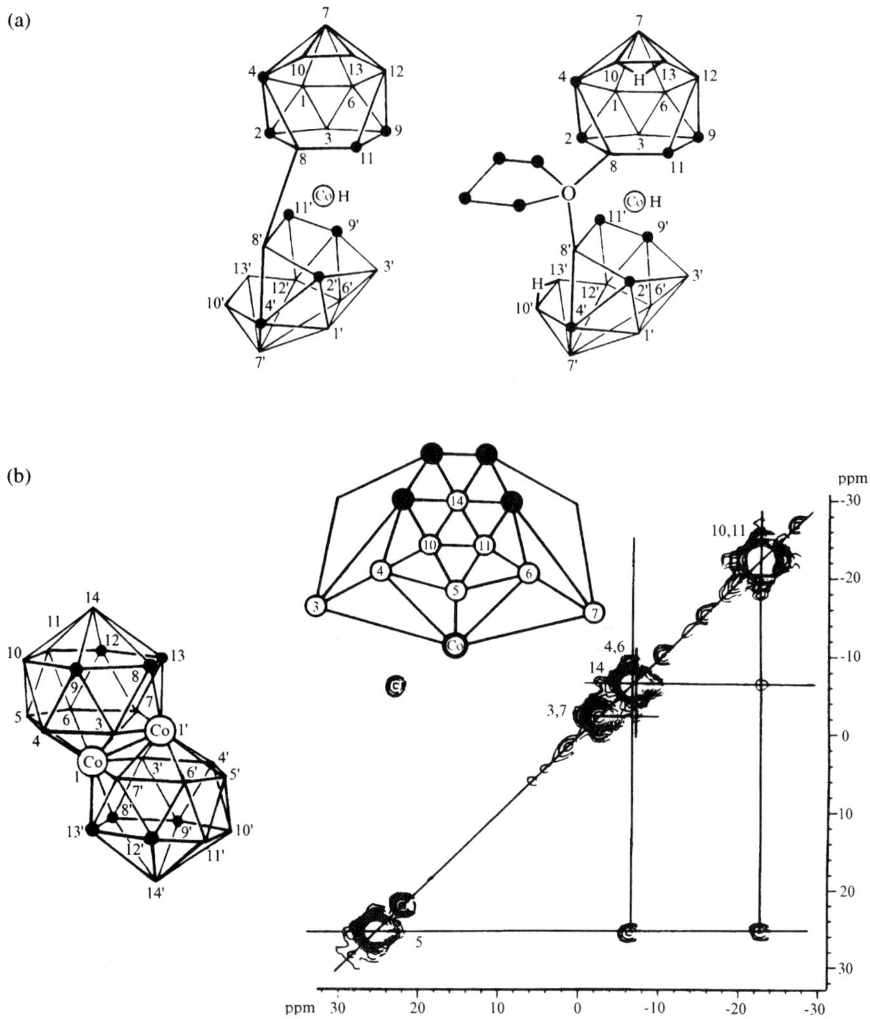

Figure 28 (a) Proposed structures of *nido*-$CoH(Et_4C_4B_8H_7)_2$ and *nido*-$(Et_4C_4B_8H_8)CoH(Et_4C_4B_8H_8\text{-}OC_4H_8)$ based on ^{11}B-^{11}B COSY 2D-NMR spectra. (b) Proposed structure of $Co_2(Et_4C_4B_8H_8)_2$ and its ^{11}B–^{11}B COSY 2D-NMR spectrum ((a) reprinted with permission from *Inorg. Chem.*, 1985, **24**, 834; (b) from *Inorg. Chem.*, 1985, **24**, 826; copyright 1985 American Chemical Society).

has been suggested[13b] that the *klado*-metallacarbaborane cage is a kinetically stabilized species and does not represent a thermodynamically preferred cage structure. This finding underlines the limitations inherent in applying qualitative electron-counting arguments to the prediction and rationalization of cluster structures as discussed above.

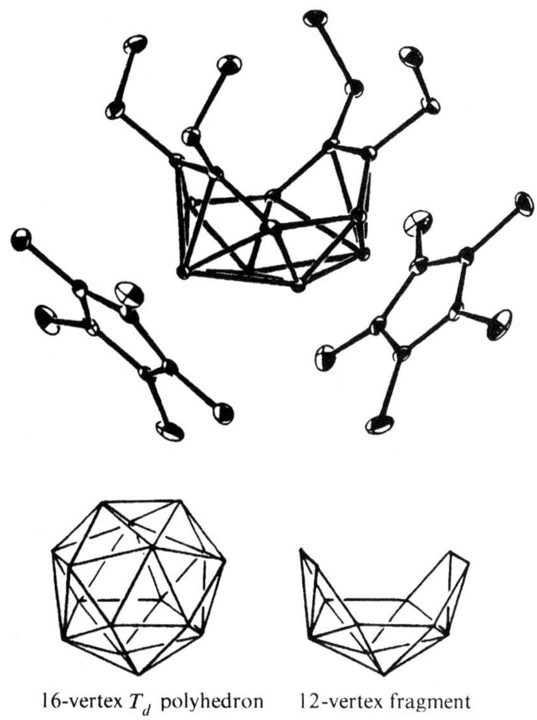

Figure 29 Structure of *klado*-$Cp^*{}_2Co_2(Et_4C_4B_6H_6)$ (reprinted with permission from *Inorg. Chem.*, 1993, **32**, 2156; copyright 1993 American Chemical Society).

9.9 REFERENCES

1. (a) R. N. Grimes in 'COMC-I', Vol. 1 Chap. 5.5, p.459; (b) A. K. Saxena and N. S. Hosmane, *Chem. Rev.*, 1993, **93**, 1081; (c) R. N. Grimes, *Chem. Rev.*, 1992, **92**, 251; (d) R. N. Grimes, *Pure Appl. Chem*, 1991, **63**, 369; (e) D. M. Schubert, M. J. Manning and M. F. Hawthorne, *Phosphorus, Sulfur, Silicon Relat. Elem.*, 1989, **41**, 253; (f) E. W. Corcoran Jr. and L. G. Sneddon, in 'Advances in Boron and the Boranes', (Mol. Struct. Energ. Vol. 5), eds. J. F. Liebman, A. Greenberg and R. E. Williams, VCH, New York, 1988, Chapter 4, pp. 71–89; (g) M. F. Hawthorne, in 'Advances in Boron and the Boranes', (Mol. Struct. Energ. Vol. 5), eds. J. F. Liebman, A. Greenberg and R. E. Williams, VCH, New York, 1988, chap. 10, pp. 225–233; (h) R. N. Grimes, in 'Advances in Boron and the Boranes' (Mol. Struct. Energ. Vol. 5), ed. J. F. Liebman, A. Greenberg and R. E. Williams, VCH, New York, 1988, Chapter 11, pp. 235–263; (i) L. G. Sneddon, *Pure Appl. Chem.*, 1987, **59**, 837; (j) R. N. Grimes, *Pure Appl. Chem.*, 1987, **59**, 847; (k) R. N. Grimes, *Am. Institute Physics Conf. Proc.*, 1986, **140**, 31; (m) F. G. A. Stone, *Adv. Organomet. Chem.*, 1990, **31**, 53; (n) F. G. A. Stone, *Pure Appl. Chem.*, 1986, **58**, 529; (o) L. I. Zakharkin, V. V. Kobak and G. G. Zhigareva, *Russ. Chem. Rev.*, 1986, **55**, 531 (translated from *Uspekhi Khimii*, 1986, **55**, 974); (p) J. H. Morris, H. J. Gysling and D. Reed, *Chem. Rev.*, 1985, **85**, 51; (q) R. N. Grimes, *Adv. Inorg. Chem. Radiochem.*, 1983, **26**, 55; (r) N. N. Greenwood, *Chem. Soc. Rev.*, 1984, **13**, 353; (s) R. N. Grimes (ed.), 'Metal Interactions with Boron Clusters', Plenum, New York, 1982; (t) R. N. Grimes, in 'Electron-Deficient Boron and Carbon Clusters', ed. G. A. Olah, K. Wade and R. E. Williams, New York, 1991, Chapter 11, pp. 261–285; (u) A. I. Yanovskii, *Russ. Chem. Rev.*, 1985, **54**, 515 (translated from 'Uspekhi Khimii', 1985, **54**, 881); (v) R. N. Grimes, *New Scientist*, 1991, **132**, 45.
2. R. E. Williams, *Chem. Rev.*, 1992, **92**, 177.
3. J. Plesek, *Chem. Rev.*, 1992, **92**, 269.
4. (a) M. J. Calhorda, D. M. P. Mingos and A. J. Welch, *J. Organomet. Chem.*, 1982, **228**, 309; (b) W. K. Pelin, T. R. Spalding and R. P. Brint, *J. Chem. Res.(S)*, 1982, 120 (*Chem. Abstr.*, 1982, **97**; 115 593r); (c) M. J. Calhorda and D. M. P. Mingos, *J. Organomet. Chem.*, 1982, **229**, 229; (d) E. D. Jemmis, *J. Am. Chem. Soc.*, 1982, **104**, 7017; (e) M. E. O'Neill and K. Wade, in 'Metal Interactions with Boron Clusters', ed. R. N. Grimes, Plenum, New York, 1982, Chapter 1, pp. 1–41; (f) Q.-S. Li, H.-T. Yu and A.-C. Tang, *Theor. Chim. Acta*, 1986, **70**, 379; (g) R. Mageswaran and N. J. Fitzpatrick, *J. Natl. Sci. Counc. Sri Lanka*, 1987, **15**, 47 (*Chem. Abstr.*, 1990, **112**, 7670h); (h) R. D. Barreto, T. P. Fehlner and N. S. Hosmane, *Inorg. Chem.*, 1988, **27**, 453; (i) E. D. Jemmis and A. C. Reddy, *J. Am. Chem. Soc.*, 1990, **112**, 722; (j) R. L. Johnston, D. M. P. Mingos and P. Sherwood, *Nouv. J. Chim.*, 1991, **15**, 831; (k) A. R. Oki *et al.*, *Organometallics*, 1992, **11**, 4202; (m) D. M. P. Mingos and D. J. Wales, 'Introduction to Cluster Chemistry', Prentice Hall, Englewood Cliffs, NJ, 1990.
5. (a) A. J. Stone, *Polyhedron*, 1984, **3**, 1299; (b) B. K. Teo, *Inorg. Chem.*, 1985, **24**, 4209; (c) B. K. Teo and N. J. A. Sloane, *Inorg. Chem.*, 1985, **24**, 4545; (d) D. G. Evans and D. M. P. Mingos, *Organometallics*, 1983, **2**, 435.
6. A. Fessenbecker *et al.*, *J. Am. Chem. Soc.*, 1991, **113**, 3061.
7. (a) K. Wade, *Adv. Inorg. Chem. Radiochem.*, 1976, **18**, 1; (b) D. M. P. Mingos, *Acc. Chem. Res.*, 1984, **17**, 311.
8. J. T. Spencer and R. N. Grimes, *Organometallics*, 1987, **6**, 323.
9. (a) R. G. Swisher, E. Sinn and R. N. Grimes, *Organometallics*, 1983, **2**, 506; (b) *ibid.*, 1985, **4**, 890; (c) *ibid.*, 1985, **4**, 896.
10. S. S. Lee, C. B. Knobler and M. F. Hawthorne, *J. Organomet. Chem.*, 1990, **394**, 29.

11. (a) T. P. Hanusa, J. C. Huffman and L. J. Todd, *Polyhedron*, 1982, **1**, 77; (b) T. P. Hanusa, J. C. Huffman, T. L. Curtis and L. J. Todd, *Inorg. Chem.*, 1985, **24**, 787; (c) J. H. Davis, Jr., E. Sinn and R. N. Grimes, *J. Am. Chem. Soc.*, 1989, **111**, 4776; (d) *ibid.*, 1989, **111**, 4784; (e) M. Bown, X. L. R. Fontaine, N. N. Greenwood, J. D. Kennedy and M. Thornton-Pett, *J. Chem. Soc., Dalton Trans.*, 1990, 3039.
12. (a) R. P. Micciche and L. G. Sneddon, *Organometallics*, 1983, **2**, 674; (b) R. P. Micciche J. J. Briguglio and L. G. Sneddon, *Organometallics*, 1984, **3**, 1396; (c) J. J. Briguglio and L. G. Sneddon, *Organometallics*, 1985, **4**, 721; (d) J. J. Briguglio and L. G. Sneddon, *Organometallics*, 1986, **5**, 327.
13. (a) R. B. Maynard and R. N. Grimes, *J. Am. Chem. Soc.*, 1982, **104**, 5983; (b) K. W. Piepgrass *et al.*, *Inorg. Chem.*, 1993, **32**, 2156.
14. (a) R. B. Maynard, Z.-T. Wang, E. Sinn and R. N. Grimes, *Inorg. Chem.*, 1983, **22**, 873; (b) Z.-T. Wang, E. Sinn and R. N. Grimes, *Inorg. Chem.*, 1985, **24**, 826; (c) Z.-T. Wang, E. Sinn and R. N. Grimes, *Inorg. Chem.*, 1985, **24**, 834.
15. B. Stibr *et al.*, *Collect. Czech. Chem. Commun.*, 1992, **57**, 1439.
16. J. H. Jones, B. Stibr, J. D. Kennedy, A. D. Lawrence and M. Thornton-Pett, *J. Chem. Soc., Dalton Trans.*, 1993, 1269.
17. B. Stibr *et al.*, *Collect. Czech. Chem. Commun.*, 1984, **49**, 1660.
18. K. A. Solntsev *et al.*, *Koord. Khim.*, 1983, **9**, 993 (*Chem. Abstr.*, 1983, **99**, 167 334t).
19. B. Stibr *et al.*, *Polyhedron*, 1982, **1**, 833.
20. N. W. Alcock, M. J. Jasztal and M. G. H. Wallbridge, *J. Chem. Soc., Dalton Trans.*, 1987, 2793.
21. J. E. Crook, N. N. Greenwood, J. D. Kennedy and W. S. McDonald, *J. Chem. Soc., Chem. Commun.*, 1983, 83.
22. N. W. Alcock, J. G. Taylor and M. G. H. Wallbridge, *J. Chem. Soc., Chem. Commun.*, 1983, 1168.
23. N. W. Alcock, J. G. Taylor and M. G. H. Wallbridge, *Acta Crystallogr. Sect. C*, 1988, **44**, 2079.
24. X. L. R. Fontaine *et al.*, *J. Chem. Soc., Dalton Trans.*, 1990, 2887.
25. G. A. Kukina, V. S. Sergienko, M. A. Porai-Koshits, K. Base and I. A. Zakharova, *Izv. Akad. Nauk SSSR, Ser. Khim.*, 1981, 2838.
26. G. A. Kukina, M. A. Porai-Koshits, V. S. Sergienko, Yu. V. Zefirov and G. G. Sadikov, *Koord. Khim.*, 1985, **11**, 385 (*Chem. Abstr.*, 1985, **103**, 123 679f).
27. G. A. Kukina, V. S. Sergienko and M. A. Porai-Koshits, *Koord. Khim.*, 1985, **11**, 400 (*Chem. Abstr.*, 1985, **103**, 71 480n).
28. B. Stibr, T. Jelinek, J. D. Kennedy, X. L. R. Fontaine and M. Thornton-Pett, *J. Chem. Soc., Dalton Trans.*, 1993, 1261.
29. K. A. Solntsev *et al.*, *Koord. Khim.*, 1984, **10**, 1132 (*Chem. Abstr.*, 1984, **101**, 211 437q).
30. V. Subrtova and J. Plesek, *Cesk. Cas. Fyz.*, 1982, **32**, 73 (*Chem. Abstr.*, 1982, **96**, 208 716c).
31. T. Jelinek, B. Stibr, J. Plesek and S. Hermanek, *J. Organomet. Chem.*, 1986, **307**, C13.
32. J. G. Kester, J. C. Huffman and L. J. Todd, *Inorg. Chem.*, 1988, **27**, 4528.
33. E. J. Ditzel *et al.*, *J. Chem. Soc., Chem. Commun.*, 1990, 1741.
34. K. Nestor, J. D. Kennedy, M. Thornton-Pett, J. Holub and B. Stibr, *Inorg. Chem.*, 1992, **31**, 3339.
35. W. Quintana, R. L. Ernest, P. J. Carroll and L. G. Sneddon, *Organometallics*, 1988, **7**, 166.
36. L. I. Zakharkin and G. G. Zhigareva, *Zh. Obshch. Khim.*, 1982, **52**, 2802.
37. X. L. R. Fontaine, N. N. Greenwood, J. D. Kennedy, P. I. MacKinnon and I. Macpherson, *J. Chem. Soc., Dalton Trans.*, 1987, 2385.
38. J. A. Walker, C. A. O'Con, L. Zheng, C. B. Knobler and M. F. Hawthorne, *J. Chem. Soc., Chem. Commun.*, 1983, 803.
39. I. T. Chizhevsky *et al.*, *Inorg. Chem.*, 1993, **32**, 3393.
40. W. E. Geiger and D. E. Brennan, *Inorg. Chem.*, 1982, **21**, 1963.
41. R. G. Swisher, R. J. Butcher, E. Sinn and R. N. Grimes, *Organometallics*, 1985, **4**, 882.
42. T. L. Venable, W. C. Hutton and R. N. Grimes, *J. Am. Chem. Soc.*, 1984, **106**, 29.
43. J. H. Davis, Jr., M. D. Attwood and R. N. Grimes, *Organometallics*, 1990, **9**, 1171.
44. M. D. Attwood, J. H. Davis, Jr. and R. N. Grimes, *Organometallics*, 1990, **9**, 1177.
45. K. J. Chase and R. N. Grimes, *Inorg. Chem.*, 1991, **30**, 3957.
46. X. Meng, S. Waterworth, M. Sabat and R. N. Grimes, *Inorg. Chem.*, 1993, **32**, 3188.
47. M. A. Benvenuto and R. N. Grimes, *Inorg. Chem.*, 1992, **31**, 3897.
48. M. A. Benvenuto, M. Sabat and R. N. Grimes, *Inorg. Chem.*, 1992, **31**, 3904.
49. L. Borodinsky, E. Sinn and R. N. Grimes, *Inorg. Chem.*, 1982, **21**, 1928.
50. K. W. Piepgrass, J. H. Davis, Jr., M. Sabat and R. N. Grimes, *J. Am. Chem. Soc.*, 1991, **113**, 680.
51. K. W. Piepgrass and R. N. Grimes, *Organometallics*, 1992, **11**, 2397.
52. K. W. Piepgrass, K. E. Stockman, M. Sabat and R. N. Grimes, *Organometallics*, 1992, **11**, 2404.
53. L. Borodinsky and R. N. Grimes, *Inorg. Chem.*, 1982, **21**, 1921.
54. J. H. Davis, Jr., M. A. Benvenuto and R. N. Grimes, *Inorg. Chem.*, 1991, **30**, 1765.
55. X. Meng, M. Sabat and R. N. Grimes, *J. Am. Chem. Soc.*, 1993, **115**, 6143.
56. K. J. Chase and R. N. Grimes, *Organometallics*, 1989, **8**, 2492.
57. K. J. Chase, R. F. Bryan, M. K. Woode and R. N. Grimes, *Organometallics*, 1991, **10**, 2631.
58. A. Fessenbecker *et al.*, *Inorg. Chem.*, 1990, **29**, 5157.
59. X. Meng, T. P. Fehlner and A. L. Rheingold, *Organometallics*, 1990, **9**, 534.
60. R. G. Swisher, E. Sinn and R. N. Grimes, *Organometallics*, 1984, **3**, 599.
61. R. G. Swisher, E. Sinn, G. A. Brewer and R. N. Grimes, *J. Am. Chem. Soc.*, 1983, **105**, 2079.
62. R. G. Swisher and R. N. Grimes, *Organometallic Syntheses*, 1986, **3**, 104.
63. T. L. Venable, C. T. Brewer and R. N. Grimes, *Inorg. Chem.*, 1985, **24**, 4751.
64. A. R. Oki *et al.*, *Organometallics*, 1991, **10**, 2996.
65. A. R. Oki, H. Zhang, N. S. Hosmane, H. Ro and W. E. Hatfield, *J. Am. Chem. Soc.*, 1991, **113**, 8531.
66. R. B. Maynard and R. N. Grimes, *Inorg. Synth.*, 1983, **22**, 215.
67. R. N. Grimes, R. B. Maynard, E. Sinn, G. A. Brewer and G. J. Long, *J. Am. Chem. Soc.*, 1982, **104**, 5987.
68. J. T. Spencer, M. R. Pourian, R. J. Butcher, E. Sinn and R. N. Grimes, *Organometallics*, 1987, **6**, 335.
69. J. H. Davis, Jr. and R. N. Grimes, *Inorg. Chem.*, 1988, **27**, 4213.
70. T. L. Venable, R. B. Maynard and R. N. Grimes, *J. Am. Chem. Soc.*, 1984, **106**, 6187.
71. H. A. Boyter, Jr., R. G. Swisher, E. Sinn and R. N. Grimes, *Inorg. Chem.*, 1985, **24**, 3810.
72. M. Stephan *et al.*, *J. Am. Chem. Soc.*, 1992, **114**, 5214.
73. J. W. Merkert, W. E. Geiger, M. D. Attwood and R. N. Grimes, *Organometallics*, 1991, **10**, 3545.

74. R. B. Maynard, R. G. Swisher and R. N. Grimes, *Organometallics*, 1983, **2**, 500.
75. J. W. Merkert, W. E. Geiger, J. H. Davis, Jr., M. D. Attwood and R. N. Grimes, *Organometallics*, 1989, **8**, 1580.
76. R. N. Grimes, in 'VI International Meeting on Boron Chemistry, Torun, Poland, July 1990', Abstract SA-6.
77. R. N. Grimes, in 'VIII International Meeting on Boron Chemistry, Knoxville, TN, July 1993', Abstract SA-1.
78. J. T. Spencer and R. N. Grimes, *Organometallics*, 1987, **6**, 328.
79. M. D. Attwood *et al.*, *Organometallics*, 1989, **8**, 1300.
80. H. A. Boyter, Jr. and R. N. Grimes, *Inorg. Chem.*, 1988, **27**, 3075.
81. A. J. Borelli, Jr., J. S. Plotkin and L. G. Sneddon, *Inorg. Chem.*, 1982, **21**, 1328.
82. C. A. Plumb, P. J. Carroll and L. G. Sneddon, *Inorg. Chem.*, 1991, **30**, 4678.
83. A. R. Oki, H. Zhang and N. S. Hosmane, *Organometallics*, 1991, **10**, 3964.
84. U. Siriwardane, H. Zhang and N. S. Hosmane, *J. Am. Chem. Soc.*, 1990, **112**, 9637.
85. H. Zhang, L. Jia and N. S. Hosmane, *Acta Crystallogr., Sect. C*, 1993, **49**, 453.
86. N. S. Hosmane and N. N. Sirmokadam, *Organometallics*, 1984, **3**, 1119.
87. A. R. Oki, H. Zhang and N. S. Hosmane, *Angew. Chem., Int. Ed. Engl.*, 1992, **31**, 432.
88. A. Fessenbecker *et al.*, *Inorg. Chem.*, 1990, **29**, 5164.
89. K. W. Piepgrass, X. Meng, M. Hölscher, M. Sabat and R. N. Grimes, *Inorg. Chem.*, 1992, **31**, 5202.
90. R. J. Butcher, W. L. Darby and E. Sinn, *Inorg. Chim. Acta*, 1993, **203**, 51.
91. K. W. Piepgrass, J. H. Davis, Jr., M. Sabat and R. N. Grimes, *J. Am. Chem. Soc.*, 1991, **113**, 681.
92. J. W. Merkert, J. H. Davis, Jr., W. E. Geiger and R. N. Grimes, *J. Am. Chem. Soc.*, 1992, **114**, 9846.
93. K. E. Stockman, M. Sabat, M. G. Finn and R. N. Grimes, *J. Am. Chem. Soc.*, 1992, **114**, 8733.
94. G. K. Barker, M. P. Garcia, M. Green, F. G. A. Stone and A. J. Welch, *J. Chem. Soc., Dalton Trans.*, 1982, 1679.
95. X. Wang, M. Sabat and R. N. Grimes, *J. Am. Chem. Soc.*, 1994, **116**, 2687.
96. (a) S. M. Cendrowski-Guillaume and J. T. Spencer, *Organometallics*, 1992, **11**, 969; (b) K. E. Stockman, B. Gangnus, X. Meng and R. N. Grimes, manuscript in preparation.
97. D. F. Gaines, C. K. Nelson and G. A. Steehler, *J. Am. Chem. Soc.*, 1984, **106**, 7266.
98. (a) V. R. Miller and R. N. Grimes, *J. Am. Chem. Soc.*, 1973, **95**, 2830; (b) V. R. Miller, L. G. Sneddon, D. C. Beer and R. N. Grimes, *J. Am. Chem. Soc.*, 1974, **96**, 3090; (c) L. G. Sneddon, D. C. Beer and R. N. Grimes, *J. Am. Chem. Soc.*, 1973, **95**, 6623; (d) V. R. Miller and R. N. Grimes, *J. Am. Chem. Soc.*, 1975, **97**, 4213.
99. N. S. Hosmane *et al.*, *Inorg. Chem.*, 1991, **30**, 3793.
100. H. Zhang *et al.*, *Organometallics*, 1993, **12**, 3933.
101. (a) L. Barton and P. K. Rush, *Inorg. Chem.*, 1986, **25**, 91; (b) L. Barton and P. K. Rush, *Inorg. Chem.*, 1985, **24**, 3413.
102. (a) J. R. Pipal and R. N. Grimes, *Organometallics*, 1993, **12**, 4452; (b) J. R. Pipal and R. N. Grimes, *Organometallics*, 1993, **12**, 4459.
103. K. E. Stockman, M. Sabat, M. G. Finn and R. N. Grimes, in 'VIII International Meeting on Boron Chemistry', Knoxville, Tennessee, July 1993, Abstract PN-16.
104. G. K. Barker *et al.*, *J. Chem. Soc., Chem. Commun.*, 1982, 688.
105. G. J. Zimmerman and L. G. Sneddon, *Acta Crystallogr., Sect. C*, 1983, **39**, 856.
106. G. K. Barker, M. P. Garcia, M. Green, F. G. A. Stone and A. J. Welch, *J. Chem. Soc., Chem. Commun.*, 1982, 46.
107. R. P. Micciche, J. S. Plotkin and L. G. Sneddon, *Inorg. Chem.*, 1983, **22**, 1765.
108. M. Bown *et al.*, *J. Chem. Soc., Chem. Commun.*, 1988, 974.
109. B. Stibr *et al.*, *Collect. Czech. Chem. Commun.*, 1991, **56**, 646.
110. L. I. Zakharkin, L. N. Kanakhina, A. I. Yanovskii and V. A. Antonovich, *Koord. Khim.*, 1981, **7**, 1692 (*Chem. Abstr.*, 1982, **96**, 104 331g).
111. G. K. Barker *et al.*, *Chem. Soc., Chem. Commun.*, 1983, 277.
112. M. Bown, X. L. R. Fontaine, N. N. Greenwood, J. D. Kennedy and M. Thornton-Pett, *Organometallics*, 1987, **6**, 2254.
113. P. Lu, C. B. Knobler and M. F. Hawthorne, *Acta Crystallogr., Sect. C*, 1984, **40**, 1704.
114. M. Bown, X. L. R. Fontaine, N. N. Greenwood, J. D. Kennedy and M. Thornton-Pett, *J. Chem. Soc., Chem. Commun.*, 1987, 1650.
115. K. Nestor *et al.*, *Inorg. Chem.*, 1989, **28**, 2219.
116. G. K. Barker, M. Green, F. G. A. Stone, W. C. Wolsey and A. J. Welch, *J. Chem. Soc., Dalton Trans.*, 1983, 2063.
117. J. D. Kennedy, M. Thornton-Pett, B. Stibr and T. Jelinek, *Inorg. Chem.*, 1991, **30**, 4481.
118. J. D. Kennedy, K. Nestor, B. Stibr, M. Thornton-Pett and G. S. A. Zammit, *J. Organomet. Chem.*, 1992, **437**, C1.
119. M. Green, J. L. Spencer and F. G. A. Stone, *J. Chem. Soc., Dalton Trans.*, 1979, 1679.
120. R. T. Baker, *Inorg. Chem.*, 1986, **25**, 109.
121. K. P. Callahan, W. J. Evans, F. Y. Lo, C. E. Strouse and M. F. Hawthorne, *J. Am. Chem. Soc.*, 1975, **97**, 296.
122. (a) J. D. Kennedy, *Inorg. Chem.*, 1986, **25**, 111; (b) J. D. Kennedy, *Progr. Inorg. Chem.*, 1984, **32**, 519.
123. R. L. Johnston and D. M. P. Mingos, *Inorg. Chem.*, 1986, **25**, 3321.
124. W. M. Maxwell, R. Weiss, E. Sinn and R. N. Grimes, *J. Am. Chem. Soc.*, 1977, **99**, 4016.
125. R. E. Marsh, W. P. Schaefer, G. C. Bazan and J. E. Bercaw, *Acta Crystallogr., Sect. C*, 1992, **48**, 1416.
126. G. C. Bazan, W. P. Schaefer and J. E. Bercaw, *Organometallics*, 1993, **12**, 2126.
127. F. A. Gomez, S. E. Johnson, C. B. Knobler and M. F. Hawthorne, *Inorg. Chem.*, 1992, **31**, 3558.
128. J. Kim, Y. Do, Y. S. Sohn, C. B. Knobler and M. F. Hawthorne, *J. Organomet. Chem.*, 1991, **418**, C1.
129. S. J. Dossett, D. F. Mullica, E. L. Sappenfield, F. G. A. Stone and M. J. Went, *J. Chem. Soc., Dalton Trans.*, 1993, 281.
130. L. I. Zakharkin and V. A. Ol'shevskaya, *Izv. Akad. Nauk SSSR, Ser. Khim.*, 1987, 2296.
131. J. Cowie, E. J. M. Hamilton, J. C. V. Laurie and A. J. Welch, *J. Organomet. Chem.*, 1990, **394**, 1.
132. L. I. Zakharkin, V. A. Ol'shevskaya and T. Yu. Poroshina, *Metalloorg. Khim.*, 1989, **2**, 363, (*Chem. Abstr.*, 1990, **112**, 139 305c).
133. P. A. Chetcuti, P. Moser and G. Rihs, *Organometallics*, 1991, **10**, 2895.
134. H. C. Kang, S. S. Lee, C. B. Knobler and M. F. Hawthorne, *Inorg. Chem.*, 1991, **30**, 2024.
135. V. L. Shirokii, A. A. Erdman, Z. P. Zubreichuk, N. A. Maier and Yu. A. Ol'dekop, *Zh. Obshch. Khim.*, 1983, **53**, 951.
136. A. A. Erdman, V. L. Shirokii, Z. P. Zubreichuk, N. A. Maier and Yu. A. Ol'dekop, *Vestsi. Ak. Navuk, BSSR, Ser. Khim. Navuk*, 1984, 74 (*Chem. Abstr.*, 1984, **101**, 171 445g).
137. K. V. Kradenov, S. G. Vasil'eva, V. V. Volkov and B. A. Kolesov, *Izv. Sib. Otd. Akad. Nauk SSSR, Ser. Khim.*, 1990, **2**, 23.

138. A. A. Erdman, Z. P. Zubreichuk, N. A. Maier and Yu. A. Ol'dekop, *Vestsi Akad. Navuk BSSR, Ser. Khim. Navuk*, 1989, 109, (*Chem. Abstr.*, 1990, **112**, 139 422p).
139. N. I. Kirillova *et al.*, *Metalloorg. Khim.*, 1989, **2**, 859 (*Chem. Abstr.*, 1990, **112**, 98 798v).
140. V. Subrtova, K. Maly, V. Petricek and A. Linek, *Acta Crystallogr., Sect. B*, 1982, **38**, 2028.
141. A. Varadarajan *et al.*, *J. Am. Chem. Soc.*, 1992, **114**, 9003.
142. L. I. Zakharkin, V. V. Kobak and V. A. Antonovich, *Zh. Obshch. Khim.*, 1983, **53**, 2153.
143. L. I. Zakharkin, V. V. Kobak, A. I. Yanovskii and Yu. T. Struchkov, *J. Organomet. Chem.*, 1982, **228**, 119.
144. L. I. Zakharkin, V. V. Kobak, A. I. Yanovskii and Yu. T. Struchkov, *Tetrahedron*, 1982, **38**, 3515.
145. L. I. Zakharkin, V. V. Kobak, A. I. Yanovskii and V. A. Antonovich, *Izv. Akad. Nauk SSSR, Ser. Khim.*, 1983, 2809.
146. L. I. Zakharkin and V. V. Kobak, *J. Organomet. Chem.*, 1984, **270**, 229.
147. N. I. Vasyukova, D. B. Zagorevskii, V. V. Kobak, A. I. Kovredov and L. I. Zakharkin, *Izv. Akad. Nauk SSSR, Ser. Khim.*, 1984, 2246.
148. L. I. Zakharkin and V. V. Kobak, *Izv. Akad. Nauk SSSR, Ser. Khim.*, 1985, **6**, 1449.
149. L. I. Zakharkin and V. V. Kobak, *Izv. Akad. Nauk SSSR, Ser. Khim.*, 1983, 2611.
150. M. P. Garcia *et al.*, *J. Chem. Soc., Dalton Trans.*, 1985, 2343.
151. H. C. Kang, C. B. Knobler and M. F. Hawthorne, *Inorg. Chem.*, 1987, **26**, 3409.
152. S. S. Lee, C. B. Knobler and M. F. Hawthorne, *Organometallics*, 1991, **10**, 670.
153. S. S. Lee, C. B. Knobler and M. F. Hawthorne, *Organometallics*, 1991, **10**, 1054.
154. J. Plesek, B. Grüner and P. Malon, *J. Chromatogr.*, 1992, **626**, 197.
155. J. Plesek, S. Hermanek and B. Stibr, *Inorg. Synth.* 1983, **22**, 235.
156. D. E. Smith and A. J. Welch, *Organometallics*, 1986, **5**, 760.
157. T. P. Hanusa and L. J. Todd, *Polyhedron*, 1985, **4**, 2063.
158. D. C. Busby and M. F. Hawthorne, *Inorg. Chem.*, 1982, **21**, 4101.
159. N. I. Vasyukova, Yu. S. Nekrasov, V. A. Ol'shevskaya and L. I. Zakharkin, *Metalloorg. Khim.*, 1988, **1**, 65.
160. X. L. R. Fontaine *et al.*, *J. Chem. Soc., Dalton Trans.*, 1990, 681.
161. Z. G. Lewis and A. J. Welch, *Acta Crystallogr., Sect. C*, 1992, **48**, 53.
162. A. A. Erdman *et al.*, *Vestsi Akad. Navuk BSSR, Ser. Khim. Navuk*, 1986, 66 (*Chem. Abstr.*, 1987, **107**, 236 950p).
163. L. Borodinsky, E. Sinn and R. N. Grimes, *Inorg. Chem.*, 1982, **21**, 1686.
164. S. Ya. Dvurechenskaya, V. V. Volkov, S. V. Tkachev and L. S. Den'kina, *Russ. J. Inorg. Chem. (Zh. Neorg. Khim.)*, 1982, **27**, 1740.
165. L. Matel, F. Macasek, P. Rajec, S. Hermanek and J. Plesek, *Polyhedron*, 1982, **1**, 511.
166. (a) V. V. Volkov, V. N. Ikorskii and S. Ya. Dvurechenskaya, *Izv. Akad. Nauk SSSR, Ser. Khim.*, 1983, 252; (b) V. V. Volkov, G. S. Yurjev, S. G. Vasilyeva, K. G. Myakishev and G. S. Voronina, *Sibirskii Khim. Zh.*, 1991, **1**, 109.
167. A. A. Erdman, Z. P. Zubreichuk, V. L. Shirokii, N. A. Maier and Yu. A. Ol'dekop, *Vestsi. Ak. Navuk, BSSR, Ser. Khim. Navuk*, 1984, 86 (*Chem. Abstr.*, 1984, **101**, 192 180g).
168. V. V. Romanovskii, M. A. Afonin and V. N. Romanovskii, *Soviet Radiochem.*, 1990, 521.
169. P. Sivy *et al.*, *Acta Crystallogr., Sect. C*, 1986, **42**, 28.
170. P. Sivy *et al.*, *Acta Crystallogr., Sect. C*, 1986, **42**, 30.
171. J. M. Forward, D. M. P. Mingos, W. Siebert, J. Hauss and H. R. Powell, *J. Chem. Soc., Dalton Trans.*, 1993, 1783.
172. J. Plesek, B. Grüner, T. Vanek and H. Votavova, *J. Chromatogr.*, 1993, **633**, 73.
173. K. Shelly, C. B. Knobler and M. F. Hawthorne, *New. J. Chem.*, 1988, **12**, 317.
174. I. Cisarova and V. Petricek, *Acta Crystallogr., Sect. C*, 1986, **42**, 663.
175. J. Plesek, B. Stibr and S. Hermanek, *Collect. Czech. Chem. Commun.*, 1984, **49**, 1492.
176. M. F. Hawthorne *et al.*, *J. Am. Chem. Soc.*, 1990, **112**, 5365.
177. Z. G. Lewis, D. Reed and A. J. Welch, *J. Chem. Soc., Dalton Trans.*, 1992, 731.
178. E. Makrlik and P. Vanura, *Talanta*, 1985, **32**, 423.
179. P. Sivy *et al.*, *Acta Crystallogr.*, 1986, **C42**, 24.
180. R. E. King, III, S. B. Miller, C. B. Knobler and M. F. Hawthorne, *Inorg. Chem.*, 1983, **22**, 3548.
181. V. P. Prokopovich, A. A. Erdman, N. A. Maier and Yu. A. Ol'dekop, *Dokl. Akad. Nauk SSSR*, 1984, **278**, 1141.
182. N. A. Maier, A. A. Erdman, Z. P. Zubreichuk, V. P. Prokopovich and Yu. A. Ol'dekop, *J. Organomet. Chem.*, 1985, **292**, 297.
183. A. A. Erdman, Z. P. Zubreichuk, N. A. Maier and Yu. A. Ol'dekop, *Zh. Obshch. Khim.*, 1988, **58**, 334.
184. A. A. Erdman *et al.*, *Koord. Khim.*, 1989, **15**, 122 (*Chem. Abstr.*, 1990, **112**, 77 477z).
185. D. M. Schubert, C. B. Knobler, S. Trofimenko and M. F. Hawthorne, *Inorg. Chem.*, 1990, **29**, 2364.
186. A. A. Erdman, Z. P. Zubreichuk, N. A. Maier, I. A. Shingel and Yu. A. Ol'dekop, *Vestsi Akad. Navuk BSSR, Ser. Khim. Navuk*, 1989, 41 (*Chem. Abstr.*, 1990, **112**, 56 223y).
187. V. Petricek, K. Maly, A. Petrina, K. Base and A. Linek, *Z. Krystallogr.*, 1984, **166**, 1 (*Chem. Abstr.*, 1984, **101**, 31 453h).
188. H. C. Kang, Y. Do, C. B. Knobler and M. F. Hawthorne, *Inorg. Chem.*, 1988, **27**, 1716.
189. E. J. M. Hamilton and A. J. Welch, *Polyhedron*, 1991, **10**, 471.
190. Y. Do, H. C. Kang, C. B. Knobler and M. F. Hawthorne, *Inorg. Chem.*, 1987, **26**, 2348.
191. D. J. Crowther, N. C. Baenziger and R. F. Jordan, *J. Am. Chem. Soc.*, 1991, **113**, 1455.
192. J. Arnold, S. E. Johnson, C. B. Knobler and M. F. Hawthorne, *J. Am. Chem. Soc.*, 1992, **114**, 3996.
193. S. J. Dossett, I. J. Hart and F. G. A. Stone, *J. Chem. Soc., Dalton Trans.*, 1990, 3481.
194. H. R. Allcock, A. G. Scopelianos, R. R. Whittle and N. M. Tollefson, *J. Am. Chem. Soc.*, 1983, **105**, 1316.
195. P. D. Grebenik, M. L. H. Green, M. A. Kelland, J. B. Leach and P. Mountford, *J. Chem. Soc., Chem. Commun.*, 1989, 1397.
196. D. D. Devore, C. Emmerich, J. A. K. Howard and F. G. A. Stone, *J. Chem. Soc., Dalton Trans.*, 1989, 797.
197. P. Dahlke, J. C. Jeffery, M. D. Mortimer and F. G. A. Stone, *Polyhedron*, 1992, **11**, 1587.
198. S. J. Dossett, S. Li and F. G. A. Stone, *J. Chem. Soc., Dalton Trans.*, 1993, 1585.
199. Y. Do, C. B. Knobler and M. F. Hawthorne, *J. Am. Chem. Soc.*, 1987, **109**, 1853.
200. S. A. Brew, N. Carr, M. D. Mortimer and F. G. A. Stone, *J. Chem. Soc., Dalton Trans.*, 1991, 811.
201. S. J. Dossett, I. J. Hart, M. U. Pilotti and F. G. A. Stone, *J. Chem. Soc., Dalton Trans.*, 1990, 3489.
202. J. C. Jeffery, M. D. Mortimer and F. G. A. Stone, *Inorg. Chim. Acta*, 1992, **198–200**, 593.
203. M. Bown *et al.*, *Mag. Reson. Chem.*, 1989, **27**, 947.

204. M. Bown *et al.*, *Acta Crystallogr., Sect. C*, 1990, **46**, 995.
205. G. Ferguson *et al.*, *Acta Crystallogr., Sect. C*, 1990, **46**, 1402.
206. P. E. Behnken and M. F. Hawthorne, *Inorg. Chem.*, 1984, **23**, 3420.
207. J. A. Walker, L. Zheng, C. B. Knobler, J. Soto and M. F. Hawthorne, *Inorg. Chem.*, 1987, **26**, 1608.
208. R. T. Baker *et al.*, *J. Am. Chem. Soc.*, 1984, **106**, 2965.
209. L. I. Zakharkin *et al.*, *J. Organomet. Chem.*, 1988, **358**, 449.
210. Z. G. Lewis and A. J. Welch, *J. Organomet. Chem.*, 1992, **438**, 353.
211. W. C. Kalb *et al.*, *Inorg. Chem.*, 1982, **21**, 4027.
212. R. E. King, III, D. C. Busby and M. F. Hawthorne, *J. Organomet. Chem.*, 1985, **279**, 103.
213. J. A. Walker, C. B. Knobler and M. F. Hawthorne, *Inorg. Chem.*, 1985, **24**, 2688.
214. C. B. Knobler, R. E. King, III and M. F. Hawthorne, *Acta Crystallogr., Sect. C*, 1986, **42**, 159.
215. J. C. Jeffery, F. G. A. Stone and I. Topaloglu, *J. Organomet. Chem.*, 1993, **451**, 205.
216. P. A. Chetcuti, J. A. Walker, C. B. Knobler and M. F. Hawthorne, *Organometallics*, 1988, **7**, 641.
217. P. E. Behnken *et al.*, *J. Am. Chem. Soc.*, 1985, **107**, 932.
218. J. R. Fernandez, G. F. Helm, J. A. K. Howard, M. U. Pilotti and F. G. A. Stone, *J. Chem. Soc., Dalton Trans.*, 1990, 1747.
219. L. Zheng, R. T. Baker, C. B. Knobler, J. A. Walker and M. F. Hawthorne, *Inorg. Chem.*, 1983, **22**, 3350.
220. H. F. Dare, J. A. K. Howard, M. U. Pilotti, F. G. A. Stone and J. Szameitat, *J. Chem. Soc., Dalton Trans.*, 1990, 2263.
221. G. A. Kats, L. G. Komarova and A. L. Rusanov, *Izv. Akad. Nauk SSSR, Ser. Khim.*, 1991, 960 (English, edn. p. 855).
222. J. A. Long, T. B. Marder, P. E. Behnken and M. F. Hawthorne, *J. Am. Chem. Soc.*, 1984, **106**, 2979.
223. Z. G. Lewis and A. J. Welch, *J. Organomet. Chem.*, 1992, **430**, C45.
224. D. M. Speckman, C. B. Knobler and M. F. Hawthorne, *Organometallics*, 1985, **4**, 1692.
225. J. C. Jeffery, F. G. A. Stone and I. Topaloglu, *Polyhedron*, 1993, **12**, 319.
226. M. U. Pilotti, F. G. A. Stone and I. Topaloglu, *J. Chem. Soc., Dalton Trans.*, 1991, 1621.
227. D. M. Speckman, C. B. Knobler and M. F. Hawthorne, *Organometallics*, 1985, **4**, 426.
228. J. D. Hewes, M. Thompson and M. F. Hawthorne, *Organometallics*, 1985, **4**, 13.
229. C. B. Knobler *et al.*, *J. Am. Chem. Soc.*, 1984, **106**, 2990.
230. N. Carr *et al.*, *J. Chem. Soc., Dalton Trans.*, 1990, 2253.
231. M. U. Pilotti, I. Topaloglu and F. G. A. Stone, *J. Chem. Soc., Dalton Trans.*, 1991, 1355.
232. M. U. Pilotti and F. G. A. Stone, *J. Chem. Soc., Dalton Trans.*, 1990, 2625.
233. P. E. Behnken, C. B. Knobler and M. F. Hawthorne, *Angew. Chem., Int. Ed. Engl.*, 1983, **22**, 722.
234. J. E. Goldberg and F. G. A. Stone, *Polyhedron*, 1992, **11**, 2841.
235. J. E. Goldberg, J. A. K. Howard, H. Müller, M. U. Pilotti and F. G. A. Stone, *J. Chem. Soc., Dalton Trans.*, 1990, 3055.
236. J. E. Goldberg, D. F. Mullica, E. L. Sappenfield and F. G. A. Stone, *J. Chem. Soc., Dalton Trans.*, 1992, 2693.
237. J. C. Jeffery, P. A. Jelliss and F. G. A. Stone, *J. Chem. Soc., Dalton Trans.*, 1993, 1073.
238. J. A. K. Howard, J. C. Jeffery, P. A. Jelliss, T. Sommerfeld and F. G. A. Stone, *J. Chem. Soc., Chem. Commun.*, 1991, 1664.
239. H. M. Colquhoun, T. J. Greenhough and M. G. H. Wallbridge, *J. Chem. Soc., Dalton Trans.*, 1985, 761.
240. D. E. Smith and A. J. Welch, *Acta Crystallogr., Sect. C*, 1986, **42**, 1717.
241. M. J. Manning, C. B. Knobler, R. Khattar and M. F. Hawthorne, *Inorg. Chem.*, 1991, **30**, 2009.
242. V. N. Lebedev, N. F. Shemyakin, S. P. Soldovnikov and L. I. Zakharkin, *Metalloorg. Khim.*, 1988, **1**, 718 (*Chem. Abstr.*, 1989, **111**, 134 368v).
243. R. Uhrhammer, D. J. Crowther, J. D. Olson, D. C. Swenson and R. F. Jordan, *Organometallics*, 1992, **11**, 3098.
244. S. A. Brew, N. Carr, J. C. Jeffery, M. U. Pilotti and F. G. A. Stone, *J. Am. Chem. Soc.*, 1992, **114**, 2203.
245. S. A. Brew, J. C. Jeffery, M. U. Pilotti and F. G. A. Stone, *J. Am. Chem. Soc.*, 1990, **112**, 6148.
246. S. A. Brew, D. D. Devore, P. D. Jenkins, M. U. Pilotti and F. G. A. Stone, *J. Chem. Soc., Dalton Trans.*, 1992, 393.
247. J. C. Jeffery, S. Li, D. W. I. Sams and F. G. A. Stone, *J. Chem. Soc., Dalton Trans.*, 1992, 877.
248. J. C. Jeffery, S. Li and F. G. A. Stone, *Organometallics*, 1992, **11**, 1902.
249. M. Green, J. A. K. Howard, A. P. James, C. M. Nunn and F. G. A. Stone, *J. Chem. Soc., Chem. Commun.*, 1984, 1113.
250. M. Green, J. A. K. Howard, A. P. James, C. M. Nunn and F. G. A. Stone, *J. Chem. Soc., Dalton Trans.*, 1987, 61.
251. F.-E. Baumann, J. A. K. Howard, O. Johnson and F. G. A. Stone, *J. Chem. Soc., Dalton Trans.*, 1987, 2661.
252. O. Johnson, J. A. K. Howard, M. Kapan and G. M. Reisner, *J. Chem. Soc., Dalton Trans.*, 1988, 2903.
253. A. P. James and F. G. A. Stone, *J. Organomet. Chem.*, 1986, **310**, 47.
254. S. A. Brew, P. D. Jenkins, J. C. Jeffery and F. G. A. Stone, *J. Chem. Soc., Dalton Trans.*, 1992, 401.
255. G. C. Bruce, D. F. Mullica, E. L. Sappenfield and F. G. A. Stone, *J. Chem. Soc., Dalton Trans.*, 1992, 2685.
256. F.-E. Baumann, J. A. K. Howard, R. J. Musgrove, P. Sherwood and F. G. A. Stone, *J. Chem. Soc., Dalton Trans.*, 1988, 1879.
257. F.-E. Baumann, J. A. K. Howard, O. Johnson and F. G. A. Stone, *J. Chem. Soc., Dalton Trans.*, 1987, 2917.
258. F.-E. Baumann, *et al.*, *J. Chem. Soc., Chem. Commun.*, 1987, 1881.
259. N. Carr, S. J. Dossett and F. G. A. Stone, *J. Organomet. Chem.*, 1991, **413**, 223.
260. F.-E. Baumann, J. A. K. Howard, R. J. Musgrove, P. Sherwood and F. G. A. Stone, *J. Chem. Soc., Chem. Commun.*, 1988, 1891.
261. N. Carr, M. C. Gimeno and F. G. A. Stone, *J. Chem. Soc., Dalton Trans.*, 1990, 2247.
262. J.-L. Cabioch, S. J. Dossett, I. J. Hart, M. U. Pilotti and F. G. A. Stone, *J. Chem. Soc., Dalton Trans.*, 1991, 519.
263. M. Green *et al.*, *J. Chem. Soc., Chem. Commun.*, 1985, 1778.
264. M. Green *et al.*, *J. Chem. Soc., Dalton Trans.*, 1987, 81.
265. J. A. K. Howard, A. P. James, A. N. de M. Jelfs, C. M. Nunn and F. G. A. Stone, *J. Chem. Soc., Dalton Trans.*, 1987, 1221.
266. M. C. Gimeno and F. G. A. Stone, *J. Chem. Soc., Dalton Trans.*, 1990, 2239.
267. S. A. Brew, S. J. Dossett, J. C. Jeffery and F. G. A. Stone, *J. Chem. Soc., Dalton Trans.*, 1990, 3709.
268. S. J. Dossett, M. U. Pilotti and F. G. A. Stone, *Polyhedron*, 1990, **9**, 2953.
269. S. J. Dossett, I. J. Hart, M. U. Pilotti and F. G. A. Stone, *J. Chem. Soc., Dalton Trans.*, 1991, 511.
270. S. A. Brew and F. G. A. Stone, *J. Chem. Soc., Dalton Trans.*, 1992, 867.
271. M. Green, J. A. K. Howard, A. N. de M. Jelfs, O. Johnson and F. G. A. Stone, *J. Chem. Soc., Dalton Trans.*, 1987, 73.
272. S. A. Brew, J. C. Jeffery, M. D. Mortimer and F. G. A. Stone, *J. Chem. Soc., Dalton Trans.*, 1992, 1365.
273. N. Carr, D. F. Mullica, E. L. Sappenfield and F. G. A. Stone, *Organometallics*, 1992, **11**, 3697.

274. G. C. Bruce and F. G. A. Stone, *Polyhedron*, 1992, **11**, 1607.
275. J. C. Jeffery, M. A. Ruiz, P. Sherwood and F. G. A. Stone, *J. Chem. Soc., Dalton Trans.*, 1989, 1845.
276. D. D. Devore, J. A. K. Howard, J. C. Jeffery, M. U. Pilotti and F. G. A. Stone, *J. Chem. Soc., Dalton Trans.*, 1989, 303.
277. N. Carr, M. C. Gimeno and F. G. A. Stone, *J. Chem. Soc., Dalton Trans.*, 1990, 2617.
278. M. J. Attfield, J. A. K. Howard, A. N. de M. Jelfs, C. M. Nunn and F. G. A. Stone, *J. Chem. Soc., Dalton Trans.*, 1987, 2219.
279. J. E. Goldberg, D. F. Mullica, E. L. Sappenfield and F. G. A. Stone, *J. Chem. Soc., Dalton Trans.*, 1992, 2495.
280. L. I. Zakharkin, V. V. Kobak, N. Yu. Krainova and V. A. Antonovich, *Izv. Akad. Nauk SSSR, Ser. Khim.*, 1984, 2157.
281. J. A. Doi, E. A. Mizusawa, C. B. Knobler and M. F. Hawthorne, *Inorg. Chem.*, 1984, **23**, 1482.
282. J. C. Jeffery, P. A. Jelliss and F. G. A. Stone, *J. Chem. Soc., Dalton Trans.*, 1993, 1083.
283. E. A. Chernyshev *et al.*, *Zh. Obshch. Khim.*, 1983, **53**, 1433.
284. D. R. Baghurst *et al.*, *J. Organomet. Chem.*, 1993, **447**, C14.
285. N. W. Alcock, J. G. Taylor and M. G. H. Wallbridge, *J. Chem. Soc., Dalton Trans.*, 1987, 1805.
286. E. J. M. Hamilton and A. J. Welch, *Polyhedron*, 1990, **9**, 2407.
287. Z. G. Lewis and A. J. Welch, *Acta Crystallogr., Sect. C*, 1993, **49**, 715.
288. M. J. Manning, C. B. Knobler and M. F. Hawthorne, *J. Am. Chem. Soc.*, 1988, **110**, 4458.
289. L. I. Zakharkin, V. A. Ol'shevskaya, V. N. Lebedev and P. V. Petrovskii, *Zh. Obshch. Khim.*, 1992, **62**, 132 (English edn. p. 109).
290. L. I. Zakharkin and V. A. Ol'shevskaya, *Zh. Obshch. Khim.*, 1992, **62**, 138 (English edn. p. 114).
291. V. P. Prokopovich, A. A. Erdman, N. A. Maier and Yu. A. Ol'dekop, *Vestsi Akad. Nauk BSSR, Ser. Khim. Navuk*, 1987, **5**, 37.
292. M. A. Afonin *et al.*, *Radiokhimiya*, 1989, **31**, 79 (*Chem. Abstr.*, 1990, **112**, 158 302r).
293. V. V. Strelets and S. V. Kukharenko, *Metalloorg. Khim.*, 1988, **1**, 692.
294. L. Matel, L. Vasekova and P. Racay, *J. Radioanal. Nucl. Chem.*, 1988, **126**, 375.
295. L. Matel and L. Vasekova, *J. Radioanal. Nucl. Chem.*, 1989, **136**, 17.
296. J. D. Hewes, C. W. Kreimendahl, T. B. Marder and M. F. Hawthorne, *J. Am. Chem. Soc.*, 1984, **106**, 5757.
297. J. A. Long, T. B. Marder and M. F. Hawthorne, *J. Am. Chem. Soc.*, 1984, **106**, 3004.
298. P. E. Behnken *et al.*, *J. Am. Chem. Soc.*, 1984, **106**, 3011.
299. P. E. Behnken *et al.*, *J. Am. Chem. Soc.*, 1984, **106**, 7444.
300. J. A. Belmont *et al.*, *J. Am. Chem. Soc.*, 1989, **111**, 7475.
301. H. C. Kang and M. F. Hawthorne, *Organometallics*, 1990, **9**, 2327.
302. V. N. Kalinin *et al.*, *Izv. Akad. Nauk SSSR, Ser. Khim.*, 1984, 2151; *ibid.*, 1985, 2442.
303. L. I. Zakharkin and G. G. Zhigareva, *Izv. Akad. Nauk SSSR, Ser. Khim.*, 1992, **41**, 1284 (English).
304. B. Pirotte *et al.*, *Tetrahedron Lett.*, 1993, **34**, 1471.
305. A. Demonceau *et al.*, *Tetrahedron Lett.*, 1992, **33**, 2009.
306. G. G. Hlatky, R. R. Eckman and H. W. Turner, *Organometallics*, 1992, **11**, 1413.
307. R. B. Grossman, R. A. Doyle and S. L. Buchwald, *Organometallics*, 1991, **10**, 1501.
308. D. J. Crowther and R. F. Jordan, *Makromol. Chem., Macromol. Symp.*, 1993, **66**, 121.
309. J. Plesek *et. al.*, *Collect. Czech. Chem. Commun.*, 1984, **49**, 2776.
310. V. Koprda, V. Scasnar and P. Galan, *J. Radioanal. Chem.*, 1983, **80**, 55.
311. E. Makrlik and P. Vanura, *J. Radioanal. Nucl. Chem.*, 1985, **92**, 71.
312. E. Makrlik and P. Vanura, *J. Radioanal. Nucl. Chem.*, 1985, **96**, 381, 451.
313. P. Vanura and E. Makrlik, *Collect. Czech. Chem. Commun.*, 1985, **50**, 581.
314. I. M. Ivanov, V. V. Volkov and N. K. Kalish, *Sibirs. Khim. Zh.*, 1991, 78.
315. R. J. Paxton *et al.*, *Proc. Natl. Acad. Sci. USA*, 1991, **88**, 3387.
316. B. G. Beatty *et al.*, *J. Nucl. Med.*, 1993, **34**, 1294.
317. N. Carr, J. R. Fernandez and F. G. A. Stone, *Organometallics*, 1991, **10**, 2718.
318. G. K. Barker, M. P. Garcia, M. Green, F. G. A. Stone and A. J. Welch, *J. Chem. Soc., Chem. Commun.*, 1983, 137.
319. N. Carr, D. F. Mullica, E. L. Sappenfield and F. G. A. Stone, *Organometallics*, 1993, **12**, 1131.
320. S. J. Crennell, D. D. Devore, S. J. B. Henderson, J. A. K. Howard and F. G. A. Stone, *J. Chem. Soc., Dalton Trans.*, 1989, 1363.
321. D. D. Devore, S. J. B. Henderson, J. A. K. Howard and F. G. A. Stone, *J. Organomet. Chem.*, 1988, **358**, C6.
322. J. C. Jeffery, P. A. Jelliss and F. G. A. Stone, *Inorg. Chem.*, 1993, **32**, 3382.
323. R. Khattar, M. J. Manning, C. B. Knobler, S. E. Johnson and M. F. Hawthorne, *Inorg. Chem.*, 1992, **31**, 268.
324. R. Khattar, C. B. Knobler, S. E. Johnson and M. F. Hawthorne, *Inorg. Chem.*, 1991, **30**, 1970.
325. T. Kuhlmann, H. Pritzkow, U. Zenneck and W. Siebert, *Angew. Chem., Int. Ed. Engl.*, 1984, **23**, 965.
326. J. Zwecker, H. Pritzkow, U. Zenneck and W. Siebert, *Angew. Chem., Int. Ed. Engl.*, 1986, **25**, 1099.
327. J. Zwecker, T. Kuhlmann, H. Pritzkow, W. Siebert and U. Zenneck, *Organometallics*, 1988, **7**, 2316.
328. D. A. Franz, J. W. Howard and R. N. Grimes, *J. Am. Chem. Soc.*, 1969, **91**, 4010.
329. J. W. Howard and R. N. Grimes, *Inorg. Chem.*, 1972, **11**, 263.
330. C. A. Plumb, P. J. Carroll and L. G. Sneddon, *Organometallics*, 1992, **11**, 1665.
331. C. A. Plumb, P. J. Carroll and L. G. Sneddon, *Organometallics*, 1992, **11**, 1672.
332. C. A. Plumb and L. G. Sneddon, *Organometallics*, 1992, **11**, 1681.
333. F.-E. Hong, C. W. Eigenbrot and T. P. Fehlner, *J. Am. Chem. Soc.*, 1989, **111**, 949.
334. (a) P. Binger, *Tetrahedron Lett.*, 1966, 2675; *Angew. Chem., Int. Ed. Engl.*, 1968, **7**, 286; (b) T. P. Onak and G. T. F. Wong, *J. Am. Chem. Soc.*, 1970, **92**, 5226; (c) V. R. Miller and R. N. Grimes, *Inorg. Chem.*, 1972, **11**, 862.
335. K.-S. Wong, J. R. Bowser, J. R. Pipal and R. N. Grimes, *J. Am. Chem. Soc.*, 1978, **100**, 5045.

10
Aluminum

JOHN J. EISCH
State University of New York at Binghamton, NY, USA

10.1 INTRODUCTION 432
10.1.1 Status of the Recent Technical Literature 432
10.1.2 Emerging Role of Organoaluminum Compounds in Modern Chemistry 433
10.1.3 Vantage Point of this Treatment 433

10.2 OVERVIEW OF RECENT DEVELOPMENTS 433
10.2.1 Structural Studies of Coordination to Aluminum(III) Centers 433
10.2.2 Steric Factors in the Structure and Reactivity of Aluminum(III) Compounds 434
10.2.3 Generation and Isolation of Aluminum(II) and Aluminum(I) Intermediates 435
10.2.4 Experimental Detection and Computational Clairvoyance of Reactive Intermediates 436
10.2.5 Newly Elucidated Reaction Pathways 436
10.2.6 Selective Organoaluminum Reagents in Organic Synthesis 438
10.2.7 Synergistic Reactions with Nonmain Group Metal Partners 439
10.2.8 Developments in Research Techniques 440
10.2.9 Reviews of Organoaluminum Chemistry 441

10.3 STRUCTURE AND BONDING 441
10.3.1 Subvalent Aluminum Intermediates 441
10.3.1.1 Spectroscopic detection and ab initio calculations 441
10.3.1.2 Chemical trapping of transient intermediates 443
10.3.1.3 Structural verification of labile compounds 445
10.3.2 The Question of π Bonding in Monomeric Tricoordinate Organoaluminum Compounds and in Associated Aluminum Organyls 447
10.3.3 Oligomers of Mixed Organoaluminums, R_nAlE_{3-n}, and Attendant Equilibria 449
10.3.4 The Structural Enigma of Aluminoxanes 450
10.3.5 Unusual and Variegated Coordination in Organoaluminum Complexes 454

10.4 PREPARATION 455
10.4.1 Organization of this Survey 455
10.4.2 Aluminum–Carbon Bonds 456
10.4.2.1 General considerations 456
10.4.2.2 Hydroalumination 458
10.4.2.3 Metathesis between metal alkyls and aluminum compounds 460
10.4.2.4 Displacement reactions with aluminum metal and a less active metal alkyl 461
10.4.3 Aluminum–Hydrogen Bonds 461
10.4.4 Aluminum–Nonmetal Bonds 462
10.4.5 Aluminum–Metal Bonds 462
10.4.6 Interconversions of Aluminum–Carbon Bonds 463
10.4.6.1 Preparative scope 463
10.4.6.2 Alkene exchange with aluminum alkyls 464
10.4.6.3 Carbalumination 464
10.4.6.4 Alumination of carbon Brønsted acids 466

10.5 CHEMICAL REACTIONS OF ALUMINUM–CARBON BONDS WITH NONTRANSITION METAL SUBSTRATES 467
10.5.1 Scope and Relation of Available Reviews 467
10.5.2 Homolysis vs. Heterolysis as the Prevailing Reaction Mechanism 467
10.5.3 Substitution Reactions of Carbon–Heteroatom Bonds 468
10.5.4 Addition Reactions to Rings or π Bonds 469

10.5.4.1 Epoxides 469
10.5.4.2 Carbon–carbon double bonded substrates 470
10.5.4.3 Carbonyl and related substrates 472
10.5.4.4 Related A≌B substrates 475
10.5.5 Rearrangements Induced by R_3Al 476

10.6 INTERACTIONS OF ORGANOALUMINUM REAGENTS AND TRANSITION METAL COMPOUNDS 478
10.6.1 Background, Significance, and Scope of Such Interactions 478
10.6.1.1 Historical development 478
10.6.1.2 Scope of relevant transition metal compounds 479
10.6.1.3 Survey of the reactions of transition metal salts with organoaluminum compounds 480
10.6.2 Mechanisms of Important Reactions Promoted by Transition Metal–Aluminum Alkyl Combinations 482
10.6.2.1 Ziegler–Natta catalysis 482
10.6.2.2 Ziegler nickel effect 483
10.6.2.3 Alkene and alkyne oligomerization and hydrogenation processes 484

10.7 FUNCTIONAL GROUP TRANSFORMATIONS IN ORGANIC SYNTHESIS 485
10.7.1 Scope and Organization 485
10.7.2 Generation of the Required Organoaluminum Reagents 486
10.7.3 Suitable Organic Substrates 487
10.7.3.1 Reactivity trends of substrates towards organoaluminum reagents 487
10.7.3.2 Epoxides 487
10.7.3.3 Ethers 491
10.7.3.4 Halides 491
10.7.3.5 Carbonyl and related derivatives 492
10.7.3.6 α,β-Unsaturated carbonyl derivatives 494
10.7.3.7 Akenic and alkynic derivatives 495

10.8 SUMMATION AND PROSPECT 496

10.9 REFERENCES 497

10.1 INTRODUCTION

10.1.1 Status of the Recent Technical Literature

Keeping abreast of new findings in any branch of science is a Sisyphean task (see *COMC-I* for a survey of the literature up to 1980[1]). A recent STN search of "all organoaluminum compounds since 1979" produced more than 1000 citations of new scientific publications.[2] This number does not include the many patents that have been issued since 1979 on the preparation of organoaluminum compounds and their applications in industry as polymerization cocatalysts, ceramic precursors, specialty chemicals in electronic device fabrication, and selective reagents in organic synthesis, to name the most prominent uses. In preparing this chapter an analysis of this patent literature and the integration of such findings into the present treatment has been forgone. In defense of this conscious omission it is earnestly pleaded that such digestion of the patent literature would have turned an already onerous task into a Herculean challenge. On a more rational plane, it can be argued that any really novel, fundamental aspect of organoaluminum chemistry, which underlies a patented invention, will sooner or later appear in a scientific publication and hence be encompassed by a literature search such as that employed for this chapter.

Even with this self-imposed limitation on the scientific literature consulted, it has been necessary to sift and glean from these 1000 citations those that represent truly novel findings or reach significant conclusions. This admittedly subjective judgment has reduced the number cited and analyzed in this chapter to about 300. In some cases, this has led to a detailed treatment of one article, out of a group of several contemporaneous articles on that topic, and a simple reference to the remaining articles. In so doing, I am not making any judgments on originality or scientific priority, but rather responding to that publication which seems most clearly or persuasively written. It should be stressed that what is most evident from this assessment of the literature is the widespread and high level of experimental and theoretical organoaluminum research being conducted throughout the world.

10.1.2 Emerging Role of Organoaluminum Compounds in Modern Chemistry

At the conclusion of the previous chapter on aluminum, a vista was offered of how organoaluminum chemistry might develop during the remainder of the twentieth century. It was suggested that the significance of this field might lie in "novel stoichiometric procedures for forming carbon–carbon bonds, the controlled oligomerization or polymerization of hydrocarbons, and the efficient conversion of carbon monoxide into useful fuels and chemicals."[1] These predictions have been largely borne out by pervasive developments in industry and in academia. The lure of a potentially low-cost class of organometallics has attracted the interest of especially the polymer industry, where an oligomeric methylaluminoxane (MAO) has become the cocatalyst of choice in Ziegler–Natta stereoregular alkene polymerizations.[3] Moreover, an array of aluminum hydrides and mixed alkyls of the type, R_nAlE_{3-n}, has come to occupy a valued position in organic synthesis for the selective formation of C–E bonds, where E = C, H, O, N, or X.[4] Further, aluminum alkyls have proved to be unique precursors for the production of semiconductor materials and of tailor-made ceramics.[5] Thus, such organoaluminum compounds are no longer overshadowed by their lithium or magnesium counterparts in either utility or versatility.

Despite these predictable developments in organoaluminum chemistry, other recent lines of research and discovery have burgeoned in a most surprising manner. In a little over a decade subvalent organoaluminum compounds have been generated on surfaces or in rigid matrices and a number of such compounds have been isolated, either as such or as adducts with unsaturated hydrocarbons. A celebrated instance of such a compound's isolation and full structural characterization is the synthesis of the tetrameric pentamethylcyclopentadienylaluminum(I) by the Schnöckel group.[6] These exciting experimental achievements have engendered a lively branch of computational chemistry dealing with the *ab initio* calculations of aluminum atom–hydrocarbon interactions. The theoretical exploration of aluminum–alkyne adduct configurations is most illustrative.[7,8] On the other extreme, namely higher coordination numbers for aluminum centers, are the unexpected findings that organoaluminum compounds can easily attain covalent ligation numbers of five,[9] six,[10] and even seven.[11] These advances in molecular structure show how circumscribed the knowledge of carbon–aluminum bonding was in the late 1970s. Finally, these achievements support the contention that organoaluminum chemistry will also play a prominent role in future studies of chemical bonding and reaction mechanisms.

10.1.3 Vantage Point of this Treatment

The views of recent advances in organoaluminum chemistry given in this chapter are predicated on the systematic exposition of the principles of structure and bonding, dynamic equilibria, preparation and chemical reactions, and synthetic applications of organoaluminum compounds, as presented in the previous chapter.[1] The treatment of newer developments follows substantially the same format, except that the chemical interaction of transition metal compounds with organoaluminum reagents has grown so much in importance that it merits separate consideration. On the other hand, Section 10.7 will be restricted in coverage to a general treatment of methodology and to salient advances in the use of organoaluminum reagents for specific organic substrates. Since this topic receives detailed and extensive consideration in Chapter 6, Volume 11, excessive overlap is thereby be minimized. Before proceeding with such a sequential coverage, however, it may be helpful to present an overview of research highlights in each of these areas.

10.2 OVERVIEW OF RECENT DEVELOPMENTS

10.2.1 Structural Studies of Coordination to Aluminum(III) Centers

Diverse x-ray crystallographic studies have explored the steric limitations of Lewis base interactions with aluminum(III) centers. Bulky ligands, such as those in (**1**), favor the formation of monomeric tricoordinate derivatives,[12] rather than the ubiquitous bridging dimer encountered even with cyclopropyl bridges, as in (**2**).[13] By the use of chelating or crown multidentate Lewis bases, coordination numbers of five,[9] six,[10] and even seven[11] have been achieved, as exemplified by the Schiff-base complex (**3**), the aminoalkoxyl derivative $[\{N(CH_2CH_2O)_3\}_2Al_2Me_3]$ (**4**) (in which N-1 is joined to O-2, O-3, and O-4 by ethene bridges, and N-2 is similarly bonded to other oxygen atoms), and the crown ether cation (**5**), respectively.

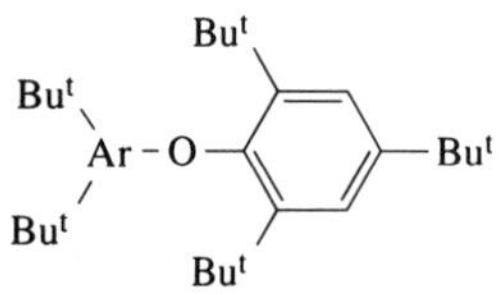

(1)

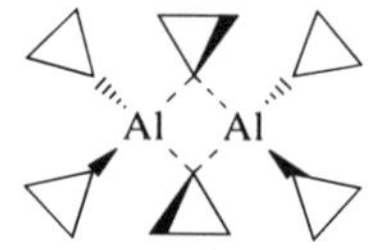

(2)

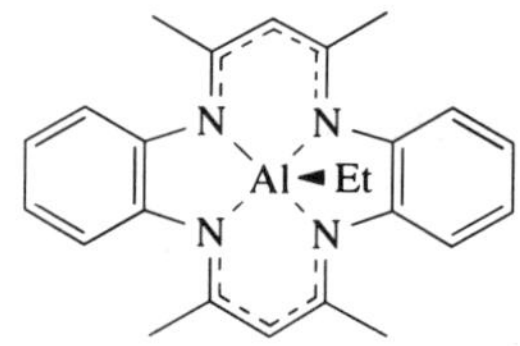

(3)

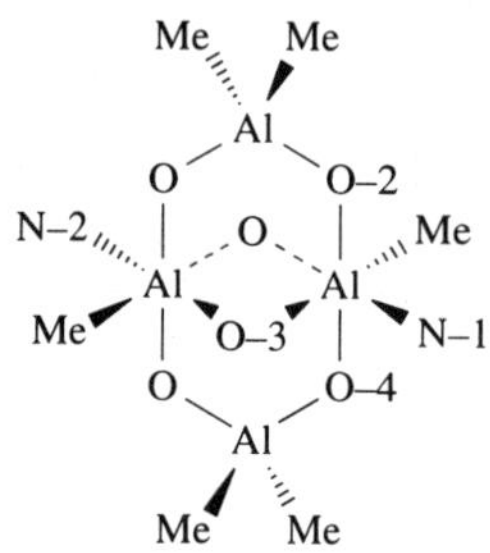

(4)

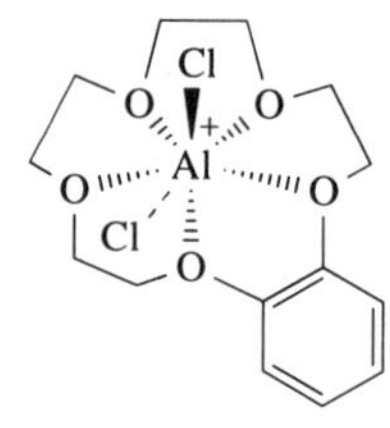

(5)

In addition to σ-bonding contributions to the foregoing coordination numbers of aluminum(III), the relatively short Al–O bonds in tricoordinate (**1**) and in tetracoordinate (**6**) have been assessed as to whether they would support the existence of a π-bonding component in such Al–O interactions (**7**) and (**8**). Surprisingly, such π-bonding is assigned a relatively minor role by one research group for tricoordinate (**1**),[12] while another group espouses π-bonding for tetracoordinate (**6**).[14,15]

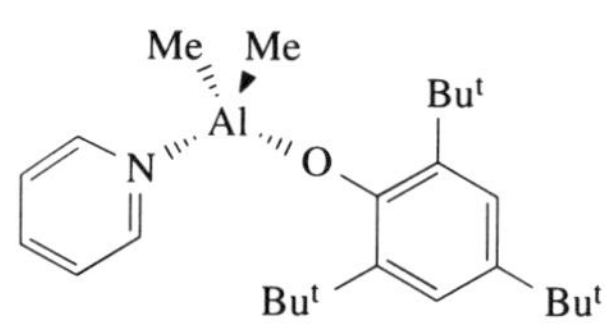

(6)

(1) ⟷ $Bu^t_2\overset{-}{Al}{=}\overset{+}{O}{-}Ar$
(7)

(6) ⟷ $\overset{+}{Py}Me_2\overset{2-}{Al}{=}\overset{+}{O}{-}Ar$
(8)

10.2.2 Steric Factors in the Structure and Reactivity of Aluminum(III) Compounds

The increase in covalent radius in going from tricoordinate boron to tricoordinate aluminum, 0.081 nm to 0.125 nm, means that steric inhibition of R_3Al dimerization (Equation (1)) sets in only with much more bulky groups. Accordingly, trimethylaluminum is firmly dimeric in the solid state or in solution, while trimethylborane shows no sign of association. On the other hand, the bridging character of the triphenylaluminum dimer is completely disrupted by the *o*-methyl groups in trimesitylaluminum (**9**), which is monomeric both in solution and in the solid state.[16] Similar steric hindrance seems responsible for the monomeric character of trineopentylaluminum (**10**) and of tribenzylaluminum (**11**) in solution,[17] although the latter shows an unusually short separation between the aluminum center of one monomer and the *ortho* carbon of a second unit in the solid state.[18] The same steric factors that interfere with dimerization also are operative in coordination. For example, (**9**) forms a weak complex with THF and the Al–O bond resulting from such coordination is distinctly longer (0.004 nm) than that in less hindered etherates.[19,20]

As already mentioned, organoaluminum aryloxides with bulky *ortho* substituents, such as (**1**), also tend to exist as monomers,[14,15] and the isolation of the discrete Cl_2Al^+ can be largely ascribed to the steric screening provided by the coordinated benzo-15-crown-5 ligand (**5**) which serves to insulate this aluminum-centered cation from the $[AlCl_3Et]^-$ anion.[11] Similarly, the remarkable resistance to hydrolysis

(1)

M = B, Al
R = H, organyl

of ethylaluminum tetradentate Schiff base complexes such as (**3**) may partly arise from the difficulty with which water molecules can gain access to the aluminum center. As the ligands are made more rigid and the aluminum thus becomes less accessible, the half-life of hydrolysis increases.[9]

The interaction of ketones with aluminum alkyls leading to carbalumination of the C=O bond has long been viewed as involving prior Lewis complexation.[21–3] By use of diaryloxy(methyl)aluminum (**12**), such an intermediate complex (**13**) with benzophenone has been isolated (Equation (2)) and its structure determined by x-ray diffraction (XRD).[15] The methyl group shows no tendency at 20 °C to migrate from aluminum to the carbonyl carbon; this finding is consonant with the work of Ashby who suggests that two Me_3Al monomers are involved in such transition states for the carbalumination of ketones. Indeed, when (**13**) is treated with MeLi in ether, smooth addition to the carbonyl group occurs with the formation of (**15**), most likely by way of the Ashby transition state (**14**) (Equation (3)).

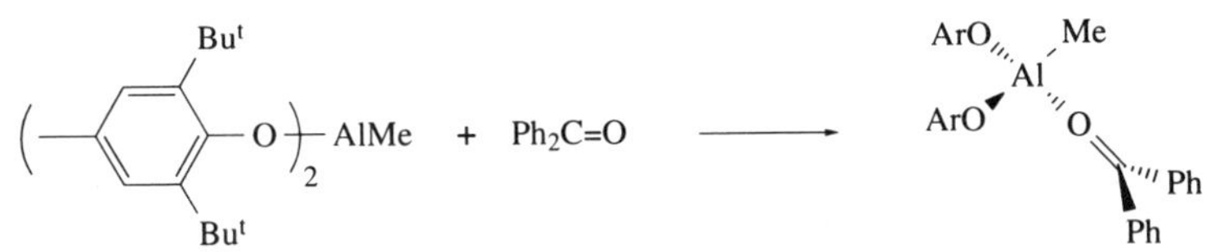

(2)

(**12**) Ar = 2,6-di-Bu^t-4-methylphenyl (**13**)

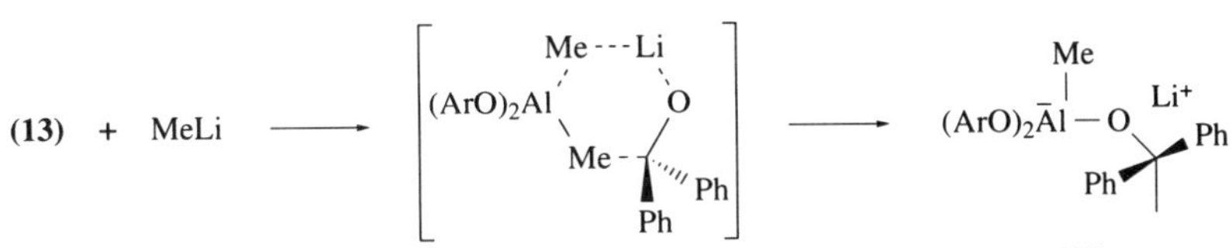

(3)

(**14**) (**15**)

10.2.3 Generation and Isolation of Aluminum(II) and Aluminum(I) Intermediates

Low-valent aluminum compounds have been known for some time: in 1948 Klemm *et al.* generated AlCl from aluminum and chlorine at 1000 °C;[24,25] in 1968 it was reported that the photolysis of dimeric Ph_3Al produced mostly biphenyl, arylaluminum hydrides, and aluminum metal, observations consistent with the intermediacy of phenylaluminum(I);[26] and in 1988 Olah *et al.* published a detailed reinvestigation[27] of the Hall–Nash reaction[28,29] (Equation (4)) and concluded that there was strong evidence for the intermediacy of aluminum(II) halides in such transformations.

$$H_2C{=}CH_2 \;+\; Al \;+\; AlX_3 \longrightarrow (MeCH_2)_nAlX_{3-n} \qquad (4)$$

However, since the late 1980s impressive strides have been made in providing stable prototypes of subvalent organoaluminum compounds for fundamental studies. Uhl and co-workers first cleared the scientific stage by demonstrating that the reported synthesis of tetraisobutyldialuminum from diisobutylaluminum chloride ((**16**), R = Bu^i) and potassium[30] was not verifiable (Equation (5), (**17**)),[31] and then conducted a successful synthesis of such a formally aluminum(II) structure by the dehalogenative coupling of bis[bis(trimethylsilyl)methyl]aluminum chloride (Equation (5), (**16**) and (**17**), R = $(TMS)_2CH$).[32] The structure of the resulting tetrakis[bis(trimethyl)methyl]dialuminum, which proved to be surprisingly stable (decomposing at 219 °C), was fully characterized by XRD and appropriate spectroscopic data and hence represents the first clear-cut example of an organoaluminum(II) type.

$$\underset{(\mathbf{16})}{2\,R_2AlCl} \;+\; 2\,K \longrightarrow \underset{(\mathbf{17})}{R_2Al{-}AlR_2} \qquad (5)$$

In similar manner, Schnöckel and co-workers have produced the first unambiguous example of a relatively stable organoaluminum(I) compound by treating the metastable AlCl in ethereal solution with bis(pentamethylcyclopentadienyl)magnesium (Equation (6)).[6] The resulting η-pentamethylcyclopentadienylaluminum(I) (**18**), which is stable at 25 °C, was shown by XRD to be tetrameric and to be composed of a core tetrahedron of aluminum atoms, to whose individual apices the Cp* groups are bonded in a pentahapto manner.[33]

$$4\ AlCl{\bullet}(Et_2O)_x\ +\ 2\ Cp_2{}^*Mg \xrightarrow[-\ 2\ MgCl_2]{} \underset{\textbf{(18)}}{(Cp^*Al)_4} \qquad (6)$$

10.2.4 Experimental Detection and Computational Clairvoyance of Reactive Intermediates

Various advanced spectroscopic and magnetic instrumental techniques, combined with modern surface science and matrix isolation procedures, have during the 1980s produced a wealth of experimental clues as to the formation and persistence of adducts between aluminum atoms and hydrocarbons. Since the structural data obtainable from individual experiments are tantalizingly suggestive rather than definitive, there has been a clear and urgent need for a theoretical treatment of the various ways in which the available atomic aluminum orbitals can interact with those of the hydrocarbon substrate, be it methane,[34] ethyne,[7,8] ethene,[35] or benzene.[36] Such *ab initio*, state-of-the-art quantum mechanical calculations on various geometrical arrangements of aluminum–hydrocarbon interactions have sought to identify structures of minimum energy that would accord with the limited experimental data now available. Perhaps more important, however, is the heuristic and predictive value of such computations. Not only do such calculations encourage experimentalists to seek further particular data to verify or reject the various theoretical structural models, but also they may well guide the experimentalist in the choice of the individual organic substrate with which the bonding sought would be most favorable. A bright future for this field is assured, if both experimentalists and theoreticians pay close attention to each other's insights.

10.2.5 Newly Elucidated Reaction Pathways

The dominant reaction pathway that emerged from the detailed mechanistic studies of Eisch's and Ashby's group for the carbalumination and hydroalumination[4] of unsaturated organic substrates, C⩵E (E = C, O, N) has been that monomeric, tricoordinate $R^2{}_2Al–R^1$ (**19**) (R^1 = alkyl, 1-alkenyl, 1-alkynyl, aryl or H) attacks C⩵E in an electrophilic fashion in the rate-limiting step (Equation (7)).[1] In transition state (**20**), involving the carbalumination of ketones, it is likely that two units of $R^2{}_2Al–R^1$ participate.[23] In most cases of either carbalumination[37,38] or hydroalumination[4] the product (**21**) associates into dimers. Since such additions are strongly retarded by Lewis bases ((R_2O or R_3N), and as there are no indications that free radicals are involved under *uncatalyzed* conditions, these reactions are properly considered as electrophilic additions, Ad_E2.[39]

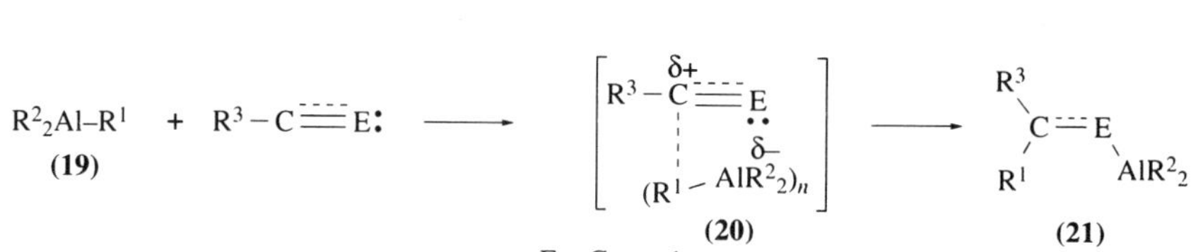

E = C, *n* = 1
E = O, *n* = 2

(7)

On the other hand, there is growing evidence that many reactions, involving not only aluminum metal but also aluminum alkyls or hydrides, occur via free-radical intermediates. Not surprisingly, free-radical activity has been noted in the decomposition of organoaluminum peroxides,[40] the irradiation of aluminum alkyls,[41] and the interaction of aluminum alkyls with strong electron acceptors like quinones.[42] Possibly unexpected, however, is the generation of radical anions (**23**) when sterically hindered aryl ketones[43] and azaaromatic heterocycles,[44] $R^1{}_2C$=E ((**22**) E = N, O) are treated with aluminum alkyls and hydrides, respectively (Equation (8), (**19**) R^1 = R, H).

Furthermore, the recent synthesis of stable aluminum(II) and aluminum(I) complexes, (**17**) and (**18**), makes the proposal of such subvalent intermediates in certain reactions more credible. These reactions would include the photolysis of dimeric Ph_3Al (**24**), which produces, presumably via (**25**), biphenyl

$$\underset{\mathbf{(19)}}{R^2{}_2Al{-}R^1} + \underset{\mathbf{(22)}}{R^1{}_2C{=}E} \longrightarrow \underset{\mathbf{(23)}}{[R^1{}_2\dot{C}{-}\ddot{E}]^-} + [R^2{}_2\dot{Al}R^1]^+ \qquad (8)$$

directly and an aluminum intermediate proposed to be PhAl (**26**), since (**26**) could be captured by diphenylacetylene (**27**) to produce (**28**). The structure of (**28**) was deduced by deuteriodealumination to yield (**29**). The importance of dimeric (**24**) for such photolysis is seen in the failure of monomeric etherate (**30**) to undergo a similar reaction (Scheme 1).[26,45]

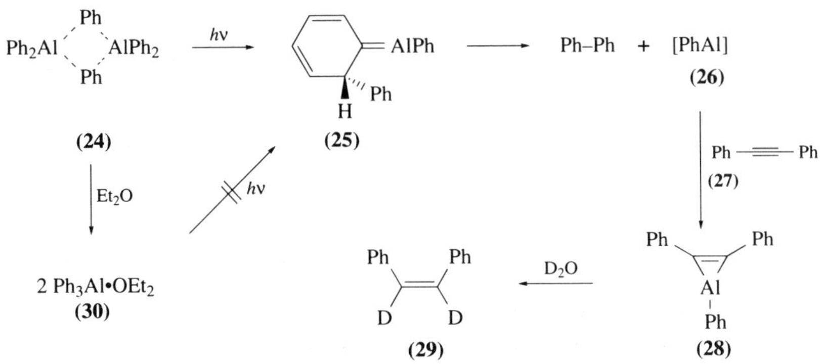

Scheme 1

In a similar vein, the interaction of aluminum and AlX_3 at moderate temperatures to form aluminum(II) halides gains in credence from the findings of two groups that the kinetically controlled product of the Hall–Nash reaction (Equation (4)) is actually 1,2-bis(dihaloalumino)ethane (**32**).[27,46] From this it can be reasonably proposed that paramagnetic AlX_2 (**31a**) or diamagnetic Al_2X_4 (**31b**) is generated and that these add in a vicinal fashion to ethene. Olah group studied the Al–AlX_3 system intensively and obtained spectroscopic evidence for the formation of (**31b**) in the condensed state and the existence of (**31a**) as a pyridine complex. Furthermore, (**31b**) was found to react with cyclohexene and the product to yield (**33**) upon deuteriolysis (Scheme 2).[27] It is unfortunate that the stereochemistry of such AlX_2 additions is apparently unknown. A differentiation between *syn* and *anti* addition should be readily apparent with an alkyne (**34**): were only (**31b**) serving as the addend, then only (**35**) would be expected.[47] However, if (**31a**) is the reactive intermediate, then a significant proportion of (**36**) should be formed.

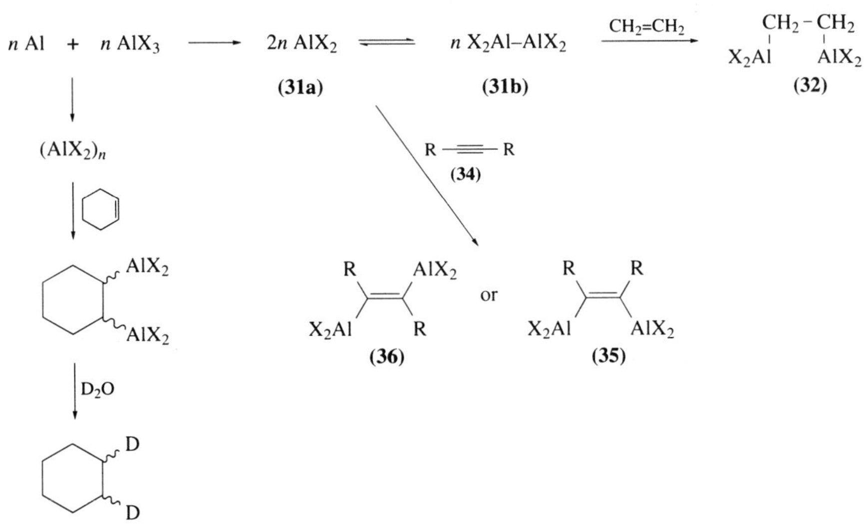

Scheme 2

10.2.6 Selective Organoaluminum Reagents in Organic Synthesis

The differences in the reactivity of the C–Al bonds in triorganylaluminums is at once a limitation as well as an advantage for the synthetic applications of these reagents. That only one C–Al bond of $R^2{}_2Al–R^1$ reacts readily with an oxygen-containing substrate (Equation (9)) is surely a wasteful drawback. But if the reactivity of the R^1–Al bond is sufficiently greater than that of the other two R^2–Al linkages, then a great and useful selectivity becomes available. Just such a $R^2{}_2Al–R^1$ system is readily prepared by the *syn* hydroalumination of alkynes by dialkylaluminum hydrides and the *anti* hydroalumination of alkynes by lithium trialkylaluminum hydrides (Equations (10) and (11)). The synthetic utility of such Al–H additions has recently received careful evaluation and has been found to be the most versatile method for the synthesis of stereoregular alkenes, dienes, and their functionalized derivatives.[4] Derivatives are readily obtained by the selective reactions of adducts (**38**) and (**39**) because of the heightened reactivity of the vinylic–aluminum bond towards a variety of electrophiles E (E = CO_2, H_2CO, X^+, H^+, etc.), which readily cleave the C–Al bond with retention of configuration ((**40**), Equation (12)).[48,49]

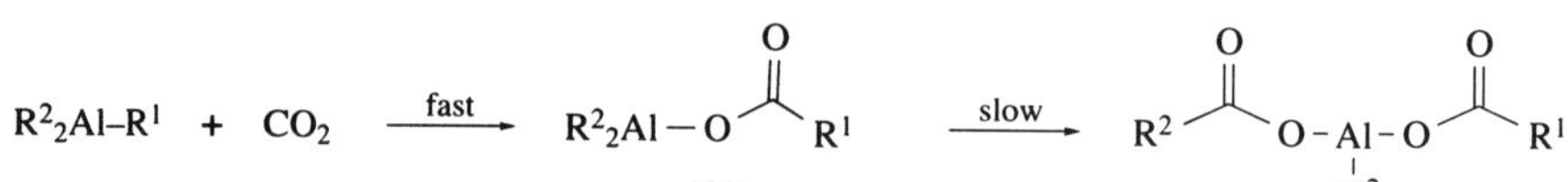

(**37**) (9)

(**38**) (10)

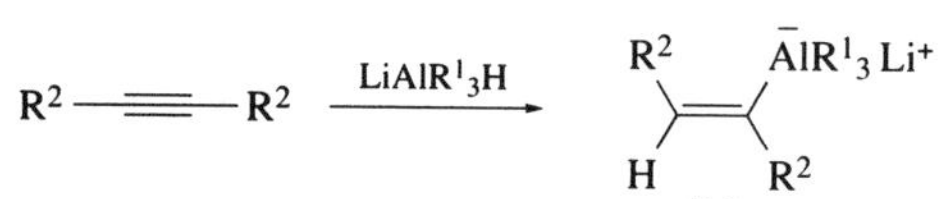

(**39**) (11)

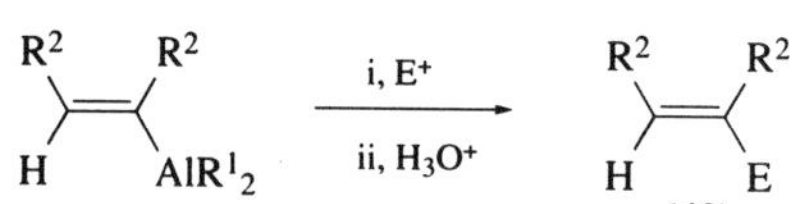

(**40**) (12)

The marked drop in the reactivity of the remaining C–Al bonds in such oxyaluminum derivatives as (**37**) (Equation (9)) is undoubtedly due in large part to the stability of Al–O bonds and to the tendency of such linkages to promote intermolecular aggregation. Witness the firm oligomers formed by aluminum alkoxides. These reactivity and structural consequences of Al–O bonds are all manifestations of the great oxophilic character of aluminum. It is this same oxophilicity that provides the great driving force for the facile, selective reaction of specific organoaluminum reagents with oxygen-containing substrates. Given herewith is an illustrative spectrum of such high-yielding transformations; in all cases at least one Al–O bond is formed and in some reactions an energy bonus in the form of an aluminoxane grouping, Al–O–Al, is gained (Equations (13)–(19)).[50–5]

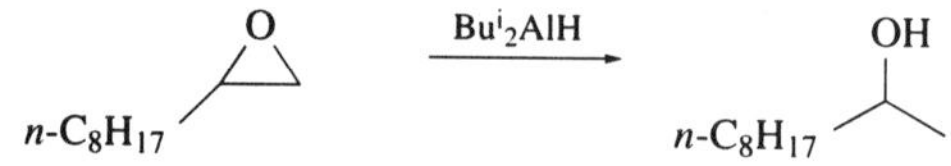

(13)

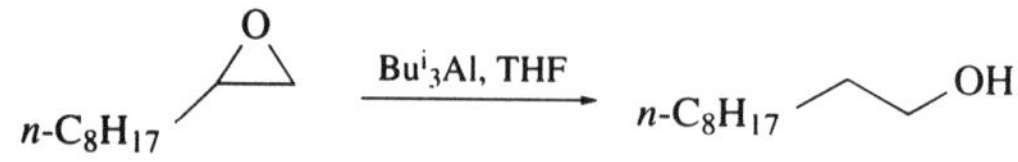

(14)

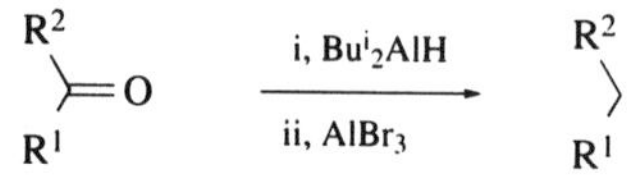

(15)

Renewed interest has emerged for the use of organoaluminum halides or alkoxides in various Lewis acid-catalyzed transformations. The advantage in substituting R_nAlX_{3-n} for AlX_3 is twofold: the organic

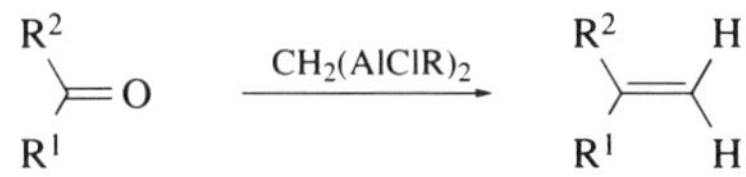

(16)

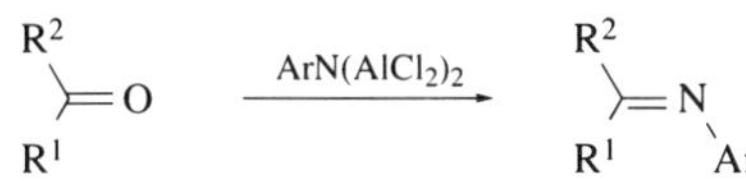

(17)

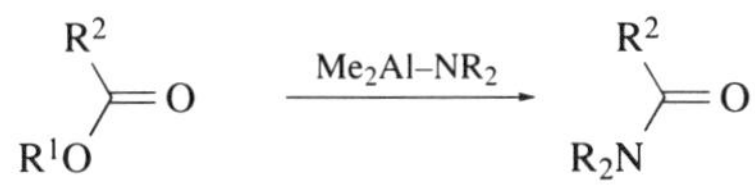

(18)

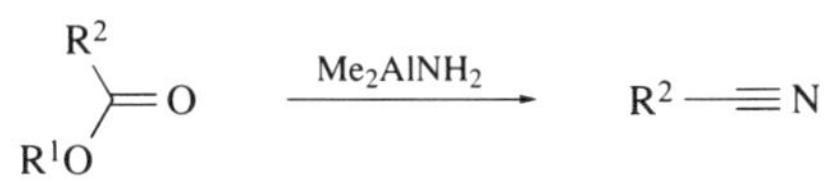

(19)

Lewis acids are freely soluble in hydrocarbons and halocarbons; and it is possible to vary the Lewis acidity by changing the number of X ligands since the acidity decreases in the series, $AlX_3 > RAlX_2 > R_2AlX > R_3Al$. The following are several interesting Lewis acid-catalyzed reactions effected by R_nAlX_{3-n} compounds (Equations (20)–(23)).[56–9]

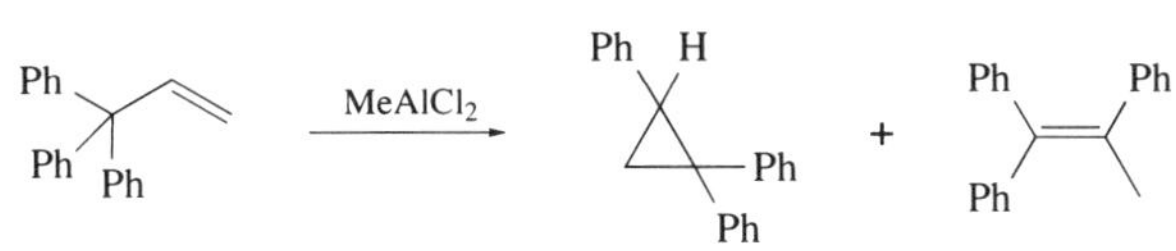

(20)

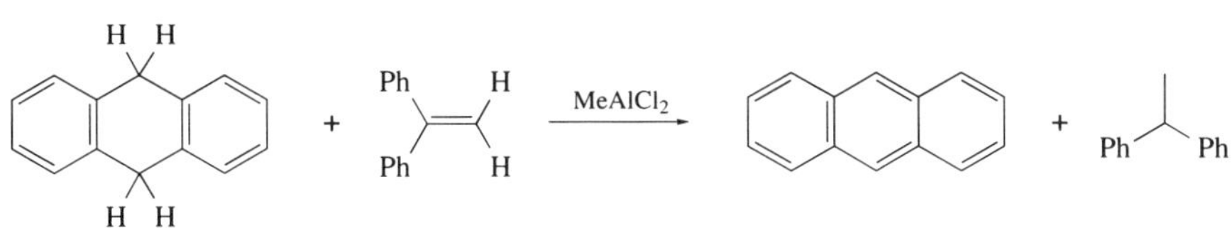

(21)

(22)

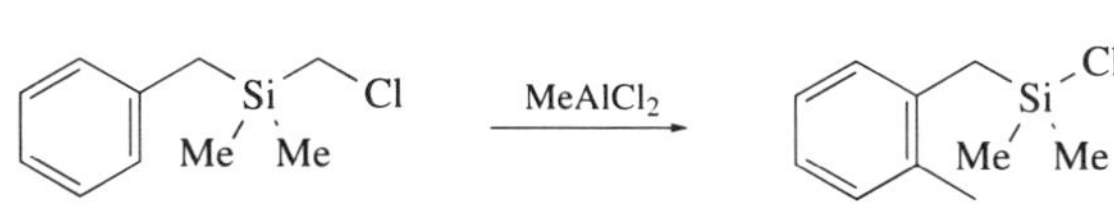

(23)

10.2.7 Synergistic Reactions with Nonmain Group Metal Partners

Since the 1950s, the interactions of organoaluminum compounds with a variety of transition metal, lanthanide, and even actinide salts have aroused the keen interest of both academic and industrial chemists. Such heightened interest had its origin in some fundamental studies of Ziegler in Germany, which were motivated by sheer curiosity but which led to a completely unexpected finding of monumental significance. (The instructive story of how this discovery came to light is related elsewhere.[60]) The pioneering research of the Ziegler group in the early 1950s, coupled with the subsequent, perceptive studies of the Natta group, led to the discovery of titanium and zirconium combinations with aluminum alkyls that polymerize ethene at low pressure (Ziegler) and can polymerize α-alkenes with both regio- and stereoselectivity (Natta). That this discovery of a new, highly advantageous method for alkene polymerization has revolutionized the chemical industry worldwide is evidenced by the thousands of patents that have been issued on such so-called Ziegler–Natta catalysis. The culminating tribute to the influence of the researches of Karl Ziegler and of Giulio Natta came in 1963, when they were jointly awarded the Nobel Prize for the work they had begun a scant 10 years before.[61]

By exploring a spectrum of other transition metal salt–aluminum alkyl combinations, chemists have identified catalysts or reagents for other valuable organic transformations, a sampling of which are the following: Wilke's cyclooligomerization of butadiene;[62] the stereoselective *cis*-1,4-polymerization of 1,3-dienes;[63] Wilke's nickel(0)-catalyzed regioselective dimerization of propene;[64] Tebbe's titanium-based methylenating agent for carbonyl substrates;[65] Kaminsky's highly active methylaluminoxane-promoted metallocene catalysts for polymerizing ethene;[3] and Ewen's[66] and Kaminsky's[67] bridged metallocene–methylaluminoxane (MAO) catalysts for polymerizing propene stereoselectively in homogeneous media, that is, without the aid of a heterogeneous phase.

These and many other catalyst combinations have largely been uncovered by empirical searches and only since the early 1970s has the molecular basis for this bimetallic catalysis begun to be understood. With this fundamental insight into catalyst structure comes the expectation that future soluble or specifically anchored catalysts can be designed to achieve greater loco-, regio-, and stereoselective reaction with a given organic substrate, such as a 1,3-diene. Illustrative of this important trend is the intensive research being conducted on soluble stereoregular propene polymerization catalysts based on MAO combinations with bridged and substituted zirconocene derivatives (**41**). The presence and location of substituents can foster either the syndiospecific[68] or the isospecific polymerization of propene from suitable derivatives of (**41**) (Scheme 3).[69,70]

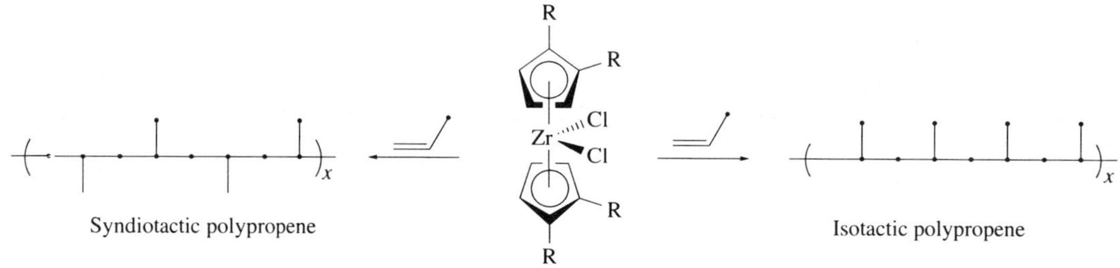

(**41**)
R groups can be bridges between Cp rings

+

MAO

Scheme 3

10.2.8 Developments in Research Techniques

The widespread application of x-ray diffraction for organoaluminum structure determination has yielded much detailed information on the coordination of ligands to the aluminum ((**3**)–(**6**)) and the precise nature of autoassociation in oligomers of the R_3Al or R_nAlE_{3-n} type, such as (**2**). The most illuminating XRD structures recently obtained are those of a Breslow–Natta soluble polymerization catalyst for ethene (**42**) and the insertion product of (**42**) with the ethene surrogate, trimethyl(phenylethynyl)silane. This adduct (**43**) can be viewed as an analogue of the first ethene insertion product of (**42**), which cannot react further because of the steric shielding of the σ Ti–C bond.[71]

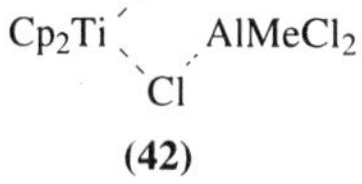

(**42**)

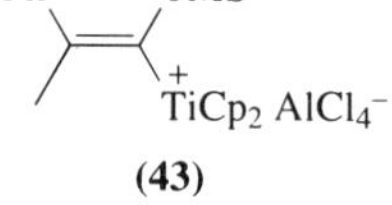

(**43**)

Aluminum-27 NMR spectroscopy has also now come into its own and sufficient empirical data on relative chemical shifts have been gathered that it is possible to reach some reliable conclusions about the coordination sphere around the aluminum center.[72,73] However, the natural signal breadth, especially at low temperatures, tends to limit the value of measurements in some situations.

As previously mentioned, extensive studies of aluminum atom interactions with organic substrates in frozen matrices, and as adsorbed on surfaces, have been carried out and a wealth of IR, UV, MS, and ESR data now are available to theoreticians, who can now apply a "reality test" to their calculated molecular models. The feasibility of such experimental measurements would seem to assure much more future study of such aluminum reactions and the relevant transient intermediates by advanced spectroscopic methods. Typifying this trend is the use of neutralization–reionization mass spectrometry (NRMS) to identify the formation of $AlCH_2^+$, $AlMe\cdot^+$, and $Al(Me)_2^+$ fragments from Me_3Al in the gas phase of a mass spectrometer.[74]

10.2.9 Reviews of Organoaluminum Chemistry

In addition to the basic chapter in *COMC-I* and the reviews cited therein,[1] aspects of organoaluminum chemistry have been assessed annually to 1981 in the *Annual Surveys* volumes of the *Journal of Organometallic Chemistry*.[75] Since then, focused reviews of selective topics have appeared: the hydroalumination of unsaturated hydrocarbons and the chemical reactions of the resulting organoalanes in 1991;[4] the preparation and reactions of unsaturated organoalanes in 1984;[76] and the synthesis, structure, and chemical reactions of aluminoxanes in 1990.[77]

10.3 STRUCTURE AND BONDING

10.3.1 Subvalent Aluminum Intermediates

Organoaluminum compounds of lower coordination or having fewer than six valence electrons vary tremendously in stability, from the spectroscopically detectable but transient existence of MeAl in the gas phase[74] to the stable, isolable tetramer of Cp*Al.[6] This wide variation in kinetic stability illustrates a lesson of considerable applicability in preparative chemistry: the kinetic stabilization and isolation of compounds of subvalent elements are generally fostered by employing space-filling ligands to screen off the ligating center from intermolecular modes of decomposition. Besides the recent dramatic successes in generating phosphinidene (Ar_2P) and silene (Ar_2Si) derivatives and isolating their π-bonded dimers,[78] the tetramer of Bu^tB[79] (**44**) and the dimesitylene complex of $GaCl{\cdot}GaCl_3$[80] (**45**) further illustrate the value of this approach. The various efforts in forming univalent compounds of group 13 have recently been surveyed.[81]

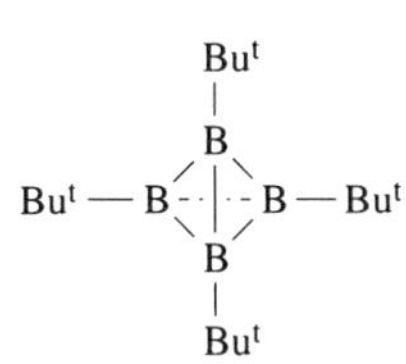

(44)

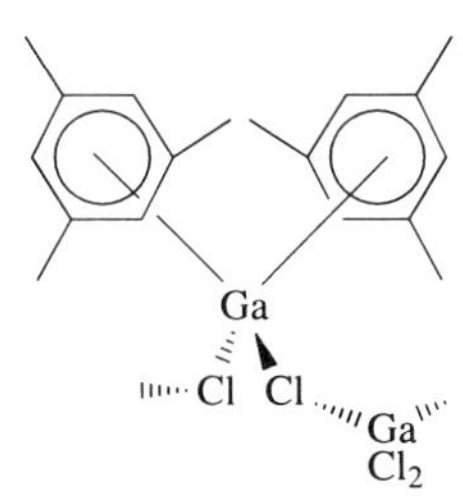

(45)

10.3.1.1 Spectroscopic detection and ab initio calculations

The pioneering publication in the area of spectroscopic detection was the ESR study of Kasai and McLeod in 1975 of aluminum atom–ethene complexes in a neon matrix at 4 K.[82] Subsequent ESR measurements at various temperatures,[35,83,84] IR data in frozen argon,[85] and finally recent laser studies,[86] have unambiguously pointed to the formation of symmetrical π complex (**46**) having a lower value of 66.9 kJ mol^{-1} as the binding energy of Al^0 to the π bond. By NRMS the existence of (**46**) as a perfectly stable molecule in the gas phase has been confirmed.[74] Contemporaneous *ab initio* calculations, employing simultaneous geometry optimization and proper use of electron correlation, has finally provided a satisfactory description of the π bonding in (**46**).[87] Generated from Et_3Al via $AlC_2H_4^+$ (**47**) by NRMS, (**46**) was shown to have a binding energy exceeding 46 kJ mol^{-1}. Both (**47**) and its isomer (**48**), H–C̈–Al–Me$^+$, are cogenerated in the MS ionization chamber but do not interconvert, even though (**48**) lies less than 20.9 kJ mol^{-1} above (**47**) (Equation (24)).

By applying the technique of NRMS to ions generated from Me_3Al, the formation of cations $AlCH_2^+$, $AlMe^+$, and $Al(Me)_2^+$ and their neutralized counterparts, (**49**), (**50**), and (**51**), has been established.[74] Furthermore, evidence suggests that (**49**), $\bullet Al{=}CH_2$, may well have a strong Al–C bond, as the calculated value of 322 kJ mol^{-1} proposes.[88]

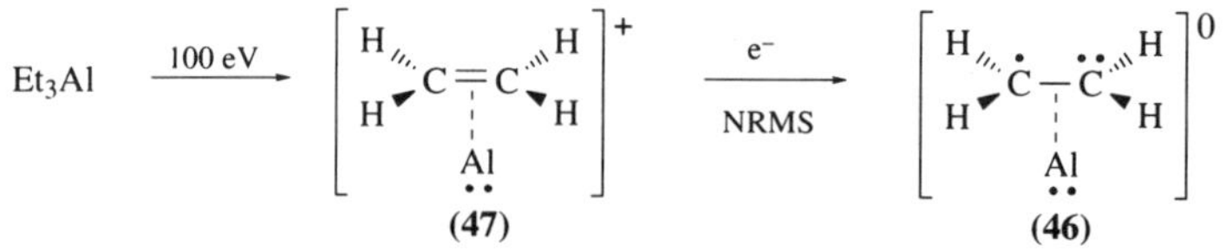

(24)

Further ESR studies by the Kasai group[89] reported in 1977 on the interaction of aluminum atoms with ethyne in matrices led to the suggestion that a σ complex (**52**), rather than a π complex (compare (**46**)), is formed (Equation (25)), since the ^{27}Al hyperfine structure showed two nonequivalent protons.[89]

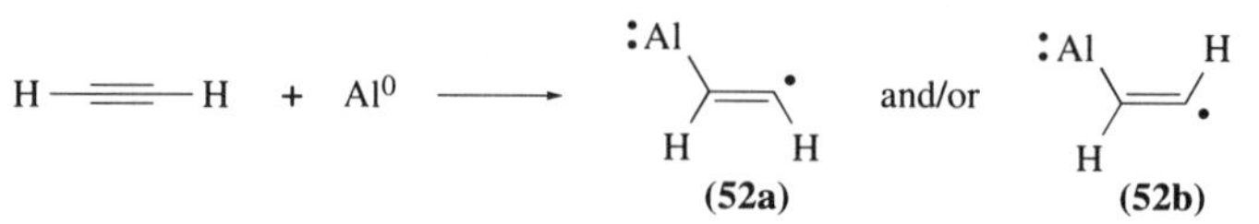

(25)

Theoreticians have taken on the challenge of rationalizing these data and the results are thus far inconclusive.[90] One group has calculated the binding energies of eight different isomeric electronic states and has concluded that the σ complexes (**52a**) and (**52b**), as well as the π complex are bound by only 71–79.5 kJ mol^{-1}, while isomers Al–Ċ=CH$_2$ and H–ȦlC≡CH are bound by 113–117.2 kJ mol^{-1}.[8] In another calculation, electron correlation was taken explicity into account and the symmetrical π complex was found to be the most stable isomer with the *trans* σ complex (**52b**) comparable in energy.[7] Since the π complex has not been observed experimentally, it is difficult to find much merit in the latter calculation. However, because the inclusion of electron correlation should make such *ab initio* calculations more rigorous, the net effect is to make the reader wary of the state of the art in quantum mechanical calculations.

Two further experimental and computational studies of aluminum–unsaturated system interactions are those involving carbon monoxide[91] and benzene.[36] To understand the observed formation of $Al(CO)_2$ and $Al(C_6H_6)$ in frozen matrices, the geometries and binding energies of various isomeric arrangements were calculated by *ab initio* methods. With due allowances for the vagaries of such calculations, one group found only a weak binding energy for AlCO (8.4–16.7 kJ mol^{-1}), a complex not found in such matrices, but estimated a binding energy of about 58.6 kJ mol^{-1} for $Al(CO)_2$, a complex whose existence has been established.[91]

Another group has optimized five isomeric structures for $Al(C_6H_6)$ and has concluded that either (**53**) or (**54**) is the most likely structure of the complex. In both, the aluminum center interacts with *para*-carbon centers, but in (**54**) the aluminum is tetrahapto.[36]

Insertion of Al^0 into σ carbon–hydrogen bonds in frozen matrices or on inert surfaces has been established experimentally.[92,93] Photochemically excited (^2D) Al^0 reacted with methane to form Me–Ȧ–H, (**55**) which also underwent reductive elimination on further irradiation. By isotopic substitution, (**55**) was shown to have a bent geometry (Equation (26)).[34]

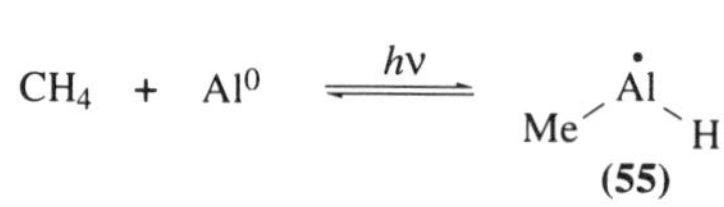

(26)

With propene, ground-state (^2P) Al^0 inserted into the allylic C–H bond to yield η^3-allylaluminum hydride (**56**), as demonstrated by ESR spectroscopy (Equation (27)).[94]

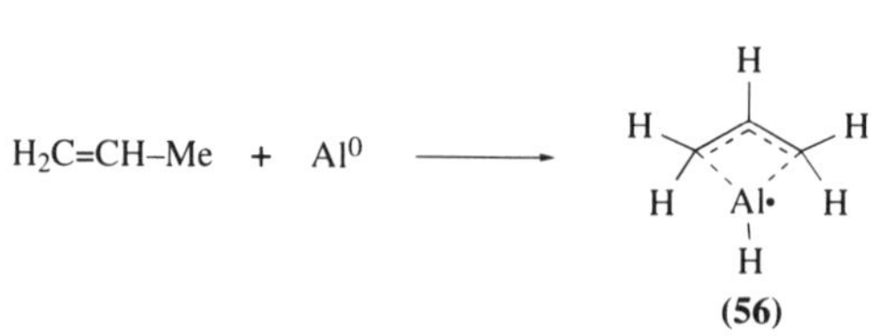

(27)

In the reverse of the reactions in Equations (26) and (27), the reductive elimination from $Me_2Al{-}O^-$ (**57**) in the gas phase via collision-induced decomposition (CID) has yielded the aluminum(I) base $Al{-}O^-$ (**58**). The gas-phase basicities of (**57**) and (**58**) were found to be very similar. The general MS technique, employing a flowing afterglow–selected ion flow tube setup (FA–SIFT), permitted the study of the reaction of Me_3Al with OH^- to generate (**57**), (**58**), and (**59**) and to monitor their decompositions (Equation (28)).[95] The stability of the methylene(dimethyl)aluminate anion (**59**) is considerable, since it comprises 15% of the reaction products.

$$\underset{\textbf{(59)}}{[Me_2Al{=}CH_2]^-} \xleftarrow[-\,Me_3Al,\ -\,H_2O]{} (Me_3Al)_2 + OH^- \xrightarrow[-\,Me_3Al,\ -\,MeH]{} \underset{\textbf{(57)}}{Me_2AlO^-} \longrightarrow \underset{\textbf{(58)}}{AlO^-} + 2\,Me\bullet \qquad (28)$$

10.3.1.2 Chemical trapping of transient intermediates

The isolation and chemical scavenging of subvalent aluminum compounds on a preparatively useful scale have been resoundingly successful since the early 1970s. Even prior to 1970, there were isolated reports on the intermediacy of Al^{I} and Al^{II} compounds in thermal or photochemical processes of Al^{III} systems. The high-temperature generation of AlCl and the photolysis of Ph_3Al to PhAl, which were cited in Section 10.2.3, are cases in point. The reactions of diorganylaluminum halides with alkali metals give further indications of the transitory formation of $R_2Al\bullet$. First, the interaction of Ph_2AlCl with sodium yields Al^0 and phenylsodium (Equation (29)).[96] Second, the reaction of Et_2AlCl and Li^0 in the presence of pyridine produces the persistent radical (**60**), which was characterized by electrical conductivity and ESR spectroscopy (Equation (30)).[97,98]

$$(Ph_2AlCl)_2 \xrightarrow[-\,NaCl]{Na^0} [Ph_2Al\bullet] \xrightarrow{Na^0} PhNa + Al^0 + (Ph_3Al)_2 \qquad (29)$$

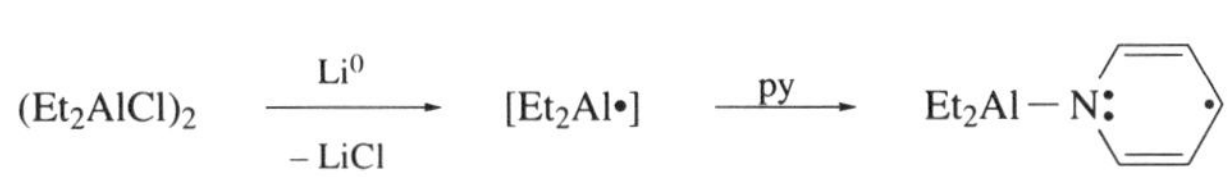

(30)

(**60**)

In the tradition of the latter work, Kaim isolated and characterized azaaromatic radicals and radical anions from the individual reactions of R_3Al and R_2AlCl with alkali metals in the presence of 2,2'-bipyridine, 4,4'-bipyridine, and 1,4-diazines.[99–101] Although such radicals are undoubtedly derived from $R^2{}_2AlR^1$, the unpaired spin is largely on the azaaromatic π systems and not on the aluminum center. The paramagnetic products formed from the interaction of R_3Al with hindered ketones,[43] quinones,[42,102] and azabisisobutyronitriles[103] lead to the formation of complex (**61**) and then to similar radicals complexed with R_2Al (**62**) and, most likely, alkyl radicals (Equation (31)). It is thus unlikely that free $R_2Al\bullet$ intermediates are involved.

$$R_3Al + {:}E{=}F \longrightarrow \underset{\textbf{(61)}}{[R_3Al{:}E{=}F]} \xrightarrow{-\,R\bullet} \underset{\textbf{(62)}}{R_2Al{-}\dot{E}{-}F} \qquad (31)$$

$E = O, N$

$F = CR, CR_2, NR$

Real progress in this area quickened when it was found that AlX could be trapped in a noble gas matrix and then used as a feasible reagent.[25,104] Furthermore, although it is not certain what the degree of association for $AlCl_2$ is, the reagent can be generated easily *in situ* from Al^0 and $AlCl_3$ below 100 °C and efficiently trapped by ethene to produce $Cl_2AlCH_2{-}CH_2AlCl_2$ ((**32**) X = Cl).[27,46] Schnöckel's group used AlCl to prepare tetrameric Cp*Al (**18**)[6] and also studied its reactions with alkynes and 1,3-dienes. The reaction of AlCl with 2-butyne leads to a dimer, (**67**), of substituted 1,4-dialuminacyclohexadiene (**66**), which can be viewed as arising from a carbene-like addition of AlCl (**63**) to the alkyne, polar opening of (**64**) to zwitterionic (**65**), and dimerization to (**66**), which should be a reactive antiaromatic ring because of its four-electron π system.[105] This instability of (**66**) would be a further driving force for forming (**67**), which can be viewed as a mutual π complexation of the aluminum centers and π bonds in two units of (**66**) (Scheme 4).[106] From NMR spectroscopy it has been further concluded that (**67**) is stable in benzene solution but dissociates into the dietherate of (**66**) with the addition of Et_2O or THF. Furthermore, the observations that (**67**) can be vaporized at 140 °C and its parent ion detected by mass spectrometry are further convincing indications of its relative stability. Since no ion corresponding to

(**66**) is detectable, the dimerization energy due to π complexation of aluminum in an η^2-manner in (**67**) must be at least 129.7 kJ mol^{-1}.[107] From the remarkable thermal decomposition of (**67**) to yield hexamethylbenzene as the only organic compound, it appears that (**67**) collapses directly into hexamethylbenzene (**63**), and (**64**). The last fragment could readily regenerate (**67**) by the suggested sequence in Scheme 4.

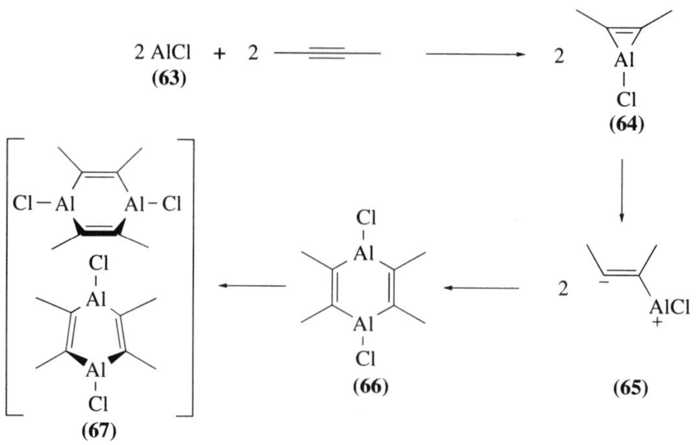

Scheme 4

Large-scale SCF calculations have been carried out on (**66**) and (**67**) and a Davidson–Roby population analysis then applied to the resulting SCF wavefunctions to obtain an estimate of the dimerization energy of (**66**) → (**67**). The resulting value of 108.8 kJ mol^{-1} is in reasonable agreement with the experimental minimum. Pertinent to the suggestion of the intermediacy of (**64**) and (**65**) in the interaction of AlCl with 2-butyne are the recent *ab initio* calculations of the reactions of HAl with ethyne and ethene.[108] For such interactions the carbenoid-like addition to the acetylenic π bond was found to yield aluminirene (**68**) with a dissociation energy of 115.9 kJ mol^{-1}. Alternative insertion into a C–H bond (**69**) or hydroalumination (**70**) leads to a ethynylalane or vinylaline, which are more stable than (**68**) by 60.3 kJ mol^{-1} and 82.9 kJ mol^{-1}, respectively. Therefore, (**69**) and (**70**) would be favored by thermodynamic factors but (**68**) might possess kinetic stability, once formed. Analogous trends were suggested for the aluminirane system and its open-chain isomers.[108]

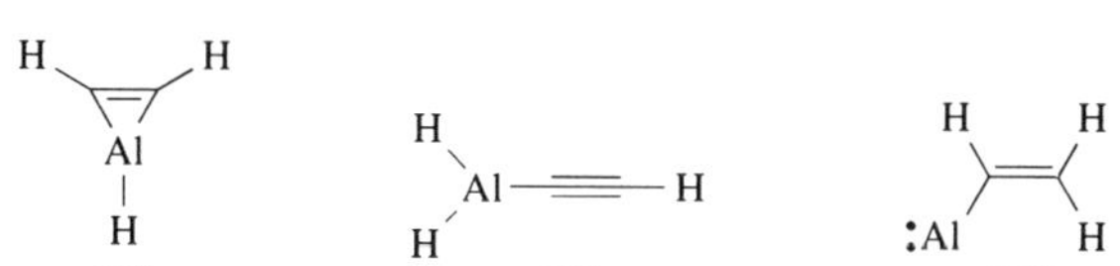

(**68**) (**69**) (**70**)

Similar to the reaction with 2-butyne, reaction of AlCl with 2,3-dimethyl-1,3-butadiene could be viewed as yielding 1-chloro-3,4-dimethyl-2,5-dihydroaluminole (**71**) and the subsequent hexamerization of (**71**) via intraunit C–Al and Al–Cl interactions to yield (**72**) (Scheme 5).[109] The allylic nature of the C–Al bonds in (**71**) should render them especially prone to exchange with other aluminum centers, so that the more stable seven-membered ring units with tetracoordinate aluminum centers can be formed as in (**72**), which possesses a noteworthy structure of almost perfect D_{3h} symmetry.

The interception of subvalent aluminum halides by ethene lies at the heart of recent studies that show that the first-formed product from ethene, $AlCl_3$, and Al^0 is the vicinal isomer (**73**), which rearranges on heating into (**74**) (Equation (32)).[27,46] The Olah group has investigated the reaction of $AlCl_3$, and Al^0 in hydrocarbon slurry at 120 °C or under high vacuum at 580 °C. They have examined the resulting powders by IR and powder XRD and have attempted to assign bands to $(AlCl_2)_n$ (n = 1, 2,...), which are not explained by AlCl or $AlCl_3$. Since no sample of supposed $AlCl_2$ or $AlBr_2$ was ever isolated in the pure state, no reliable analytical data, such as aluminium and halide percentages and aluminum:halide ratios, seem to have been obtained. Furthermore, AlX and AlX_3 are admittedly present in such mixtures and any AlX_2 would be expected to disproportionate into AlX_3 and AlX or Al^0, even at room temperature. Accordingly, even if the spectral deductions based on IR spectroscopic or ESR evidence for the presence of AlX_2 were correct, it would be difficult to attribute the reactions of such "AlX_2" powders unambiguously to $AlCl_2$ (Equation (33)) when it is known that AlX and AlX_3 are also present. The insightful work of the Schnöckel group on the addition of AlCl to alkynes and dienes (Schemes 4 and 5) suggest the more attractive pathway to (**73**) could be given by Equation (34).

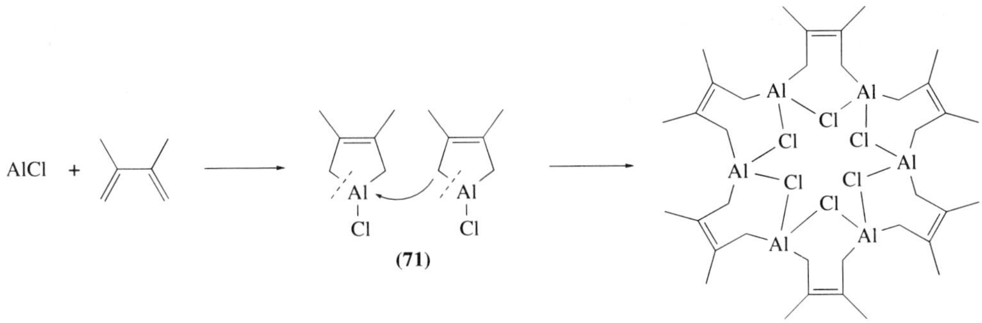

(71) (72)

Scheme 5

(73) (74) (32)

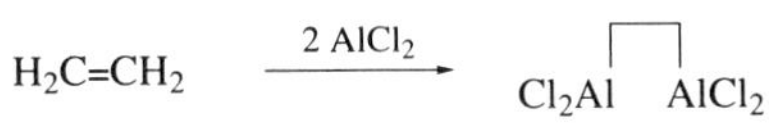

(73) (33)

(73) (34)

Not until analytically pure and well-characterized samples of AlX_2 are available can it be hoped to demonstrate clearly that (**73**) can arise from aluminum dichloride. The same ambiguity beclouds all the reported chemical reactions of alkenes and of alkyl or aryl halides with the supposed AlX_2 generated from AlX_3 and aluminum.[27] From the relatively greater kinetic stability of AlCl, especially as a solvated etherate,[25] its interaction with an organic halide would seem to be especially straightforward, involving a precedented oxidative addition reaction (Equation (35)), a reaction well known with thallium(I) halides.[110]

$$AlCl + R\text{–}X \longrightarrow R\text{–}Al(X)Cl \qquad (35)$$

10.3.1.3 Structural verification of labile compounds

The first organoaluminum(I) compound stable under normal conditions, whose structure can be determined by XRD, is tetrameric η-pentamethylcyclopentadienyl-aluminum ((**18**), Equations (6) and (36)).[33] The average C–Al distance is 0.2334 nm, very close to the C–Mg distance of 0.2348 nm of decamethylmagnesocene. This pentahapto bonding and short C–Al separation appears to shield the aluminum nucleus greatly. At a chemical shift of $\delta = -80.8$ ppm, this ^{27}Al signal is the highest known for a stable system. However, a toluene solution of (**18**) above 30 °C exhibits a new signal at −149.5 ppm, which agrees closely with the calculated chemical shift of −143 ppm for the monomer (Equation (36)).

$$(Cp^*Al)_4 \rightleftharpoons 4\,Cp^*Al \quad \Delta H^o = 150.7\ \text{kJ mol}^{-1} \qquad (36)$$

The simple tetramer of cyclopentadienylaluminum can be similarly generated (see Equation (6)) but tends to disproportionate to Cp_3Al and Al^0 above −60 °C. At −80 °C this compound displays an ^{27}Al NMR signal at −111 ppm, which corresponds well with the signal calculated for the tetramer (−105 ppm), rather than that for the monomer (−172 ppm).

In this context it appears fitting to mention the recent synthesis of the first organoidal–aluminum(I) compound, tri-*t*-butylsilylaluminum (**75**) prepared as shown in Equation (37). This compound has not yet yielded to XRD structure determination, but there is some MS evidence that it also exists as a tetramer. However, calculations cannot account for the observed ^{27}Al NMR shift of 64.5 ppm.[111]

$$AlCl + Bu^t_3SiNa \xrightarrow[-\ NaCl]{PhMe-Et_2O} \underset{(\mathbf{75})}{(Bu^t_3SiAl)_x} \quad (37)$$

The first synthesis of diorganoaluminum(II) compounds (**76**) and (**77**) as dimers with Al–Al bonds was reported in the late 1970s to have resulted when diisobutylaluminum chloride or trisobutylaluminum were treated with potassium[30,112] (Equations (38) and (39)).

$$2\ Bu^i_2Al{-}Cl \xrightarrow[-\ KCl]{K} \underset{(\mathbf{76})}{Bu^i_2Al{-}AlBu^i_2} \quad (38)$$

$$2\ Bu^i_3Al + 2\ K \longrightarrow 2\ K^+\ \underset{(\mathbf{77})}{(Bu^i_3Al{-}AlBu^i_3)^{2-}} \quad (39)$$

These claims by the Hoberg group that compounds (**76**) and (**77**) are the principal reaction products could not be corroborated by the newer investigations of the Uhl group. Uhl and co-workers have shown that the reagents in Equation (39) lead chiefly to potassium tetraisobutylaluminate (**78**), which was formed under a variety of experimental conditions and whose nature was unambiguously established by XRD (Equation (40)).[113] No evidence was obtained that any other stable organic product, such as (**77**), was also formed. For the reaction of Bu^i_2AlCl with potassium, instead of isolating the tetraisobutyldialuminum ((**76**) in Equation (38)) that Hoberg and Krause proposed as the main product,[112] Uhl and co-workers isolated dipotassium dodecaiisobutyldodecaaluminate (**79**), whose XRD analysis showed that the twelve Bu^i–Al units were arranged as an Al_{12} icosahedron (Equation (41)).[114] Units of (**76**) are possibly generated as transient intermediates but they clearly do not persist in the reaction. It is appealing to speculate, in light of the relative stability of aluminum(I) intermediates, that any (**76**) formed would dissociate and disproportionate (Equation (42)) and that the Bu^i–Al units would cluster together and be reduced (Equation (43)).

$$4\ Bu^i_3Al + 3\ K \longrightarrow 3\ K^+\ \underset{(\mathbf{78})}{(AlBu^i_4)^-} + 4\ Al \quad (40)$$

$$24\ Bu^i_2Al{-}Cl + 26\ K \xrightarrow{-KCl} 2\ K^+\ \underset{(\mathbf{79})}{(Al{-}Bu^i)_{12}^{2-}} + 12\ Bu^i_3Al \quad (41)$$

$$Bu^i_2Al{-}AlBu^i_2 \longrightarrow 2\ Bu^i_2Al\bullet \longrightarrow Bu^i_3Al + Bu^iAl \quad (42)$$

$$12\ Bu^iAl \longrightarrow (Bu^iAl)_{12} \xrightarrow{2\ K^0} 2\ K^+\ \underset{(\mathbf{79})}{(Al{-}Bu^i)_{12}^{2}} \quad (43)$$

Besides setting aside the claims of the Hoberg group, as summarized in Equations (38) and (39), Uhl has achieved the first genuine synthesis of a tetraorganodialuminum(4) by employing the space-filling bis(trimethylsilyl)methyl group as R in R_2AlCl (Equation (44)).[32] Dialuminum compound (**80**) was isolated in 59% yield together with 15% of tris[bis(trimethylsilyl)methyl]aluminum (**81**) and about 15% of an unknown product. The x-ray crystal analysis of (**80**) established its structure with an Al–Al bond length of 0.266 nm. The compound is remarkably stable: only above 219 °C does it decompose into (**81**) and Al^0.

$$2\ [(TMS)_2CH]_2AlCl + 2\ K \xrightarrow{-2\ KCl} \underset{(\mathbf{80})}{[(TMS)_2CH]_2Al{-}Al[CH(TMS)_2]} + \underset{(\mathbf{81})}{[(TMS)_2CH]_3Al} \quad (44)$$

Most recently, the tetraaryl analogue of (**80**) has also been synthesized ($R_2Al{-}AlR_2$, where R = 2,4,6-tri-*n*-propylphenyl) and its x-ray structural analysis performed.[115c] Moreover, both tetraorganodialuminum compounds have been reduced by alkali metals to the corresponding radical anions, $[R_2Al{\dot{-}}Al_2]^-$ (**80a**), whose XRD structure shows a significant shortening of the Al–Al bond over that in the neutral precursor. Such shortening is in accord with the π-bonding character of the Al–Al interaction.[115a]

The concatenating ability of aluminum was first suggested in 1966 by Schram to account for an unusual product (**82**) arising from the interaction of trimethylaluminum with tetrakis-

(dimethylamino)diborane (Equation (45)).[116] Spectroscopic data and isopiestic molecular weight support for the molecular formula and the conventional valency of aluminum suggest the presence of three Al–Al bonds, possibly among four aluminum centers in a trigonal pyramid (**82a**), with three Me_2N groups (one bonded to Al-1 and two bridging Al-2–Al-3 and Al-3–Al-4, respectively).

$$(Me_2N)_2B\text{–}B(NMe_2)_2 \;+\; Me_3Al \longrightarrow \underset{\textbf{(82)}}{Al_4B(NMe_2)_3(Me)_6} \;+\; \text{known compounds} \qquad (45)$$

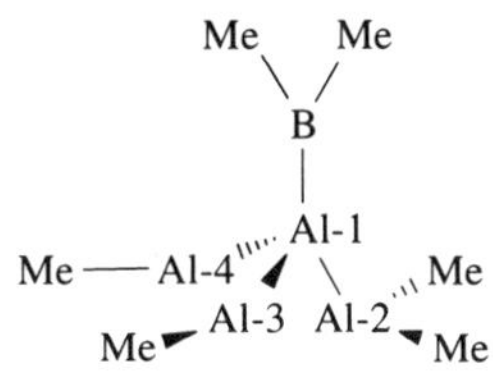

(82a)

In more recent work, Schram[116b] reinvestigated the reaction of diisobutylaluminum hydride with potassium that Zakharkin and Gavrilenko had proposed as proceeding as depicted in Equation (46).[117] This conclusion draws for its analogy on the original interpretation of Hoberg and Krause for the $Bu^i_3Al\text{–}K^0$ reaction (Equation (39)). As might be expected from Uhl's work (Equation (40)), Schram's reexamination and complete analysis of the golden-brown reaction product (**84**) shows no evidence for an Al–Al bond as shown in (**83**). Were (**83**) the correct structure, then hydrolysis should have yielded three moles of H_2, rather than the two moles observed experimentally. Experimental analyses for potassium, aluminum, isobutyl, and hydride, together with isopiestic MW and multinuclear NMR data that support equivalent magnetic environments for the potassium, aluminum, and isobutyl centers, are consistent with (**84**) existing as a symmetrical dimer of $K(AlBu^i_3H)$ that does not have an Al–Al bond. One possible structure is shown in (**84a**) (Equation (47)).

$$[(Me_2CHCH_2)_2AlH]_3 \;+\; 2\,K \longrightarrow \underset{\textbf{(83)}}{K_2[HBu^i_2Al\text{–}AlBu^i_2H]} \qquad (46)$$

$$[(Me_2CHCH_2)_2AlH] \;+\; 2\,K \longrightarrow \underset{\textbf{(84)}}{[K(AlBu^i_2H]_2} \;+\; Al \;+\; KH \quad \text{or} \quad 2\,K^+ \qquad (47)$$

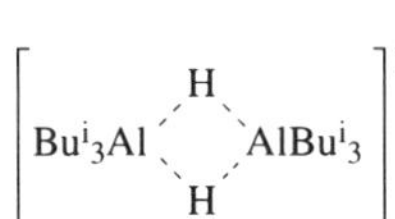

(84a)

10.3.2 The Question of π Bonding in Monomeric Tricoordinate Organoaluminum Compounds and in Associated Aluminum Organyls

The availability of an unshared electron pair or π bonding on an atom adjacent to a tri- or tetracoordinate aluminum center raises the possibility of either $p_\pi\text{–}3p_\pi$ overlap (**85a**) or $p_\pi\text{–}3d_\pi$ overlap (**85b**) with the aluminum (Scheme 6). Experimental support for such π bonding would be expected to arise as shortening of the E–Al separation, compared with a situation where π bonding would be less likely. Thus, a comparison of the terminal Al–C bonds in dimeric trimethylaluminum (**86**)[118] and dimeric triphenylaluminum (**87**)[119] finds the bond distances, at 0.197 ± 0.001 nm, almost undistinguishable. Thus, shortening due to π bonding (**87a**) is not in evidence.

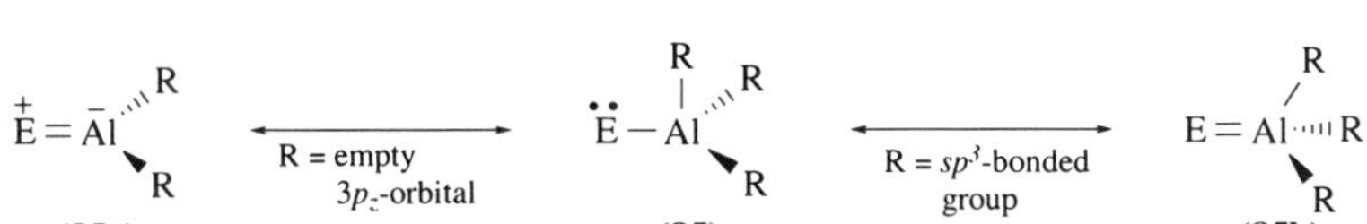

Scheme 6

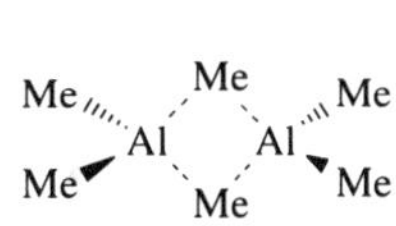

(86)

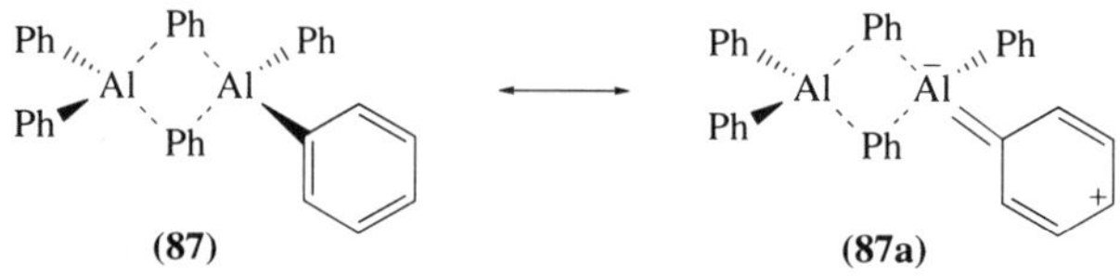

(87) (87a)

The C–Al separation in monomeric Me_3Al is 0.1957 nm,[120] while the C–Al separation in monomeric trimesitylaluminum (**88**) (where the π bonding suggested by (**87a**) could be operative) is actually, at 0.1995 nm, longer.[16] This finding not only rules out π bonding in (**88**) but supports the conclusion that nonbonded repulsions are more important in determining the C–Al separations. The terminal C–Al bond length in dimeric tri-*o*-tolylaluminum, 0.1997 nm,[121] compared with the analogous C–Al bond in (**87**), 0.1958 nm, further corroborates this view.

The question of a π bonding contribution to the O–Al bond has been addressed by the XRD structure analysis of monomeric aryloxyaluminum compounds. For the tricoordinate situation, the average O–Al separation in di-*t*-butylaluminum 2,4,6-tri-*t*-butylphenoxide (**1**) has been shown to be 0.1709 nm. This value is actually slightly longer than that predicted from an ionic model, which leads to an estimate of 0.166–0.168 nm. Thus, there is no compelling reason to invoke π bonding.[12] In the structure of the pyridine complex of dimethylaluminum 2,6-di-*t*-butyl-4-methylphenoxide (**6**), Barron has proposed π-O–Al bonding because the Al–O–C angles (141–156°) are considered to be unusually large for a main-group alkoxide and the O–Al distances (0.170–0.174 nm) to be unusually short.[14] However, the specific dimensions of the alkoxides that support this conclusion are not given. Furthermore, the O–Al separation in the tricoordinate dimethylaluminum 2,6-di-*t*-butyl-4-methylphenoxide (**6**) is reported as 0.1686 nm, a normal separation not requiring π bonding according to Power.[12] That coordination of pyridine would lengthen the O–A bond by 0.005 nm would appear normal, without invoking π bonding in either (**1**) or (**6**).

On the other hand, there is ample evidence of interunit or intermolecular π bonding in associated triorganylaluminum compounds. Such π bonding is evident in the dimeric structure of diphenyl(phenylethynyl)aluminum (**89**), where the positioning of the inclined alkynyl bridges can be viewed as interunit π complexation of the π cloud with the aluminum center.[122] However, it is worth recalling that even the dimerization of Me_3Al can be interpreted in terms of a π interaction of the bridging methyl groups with the σ- and η-bonded $Me_2Al–AlMe_2$ framework (**90**).

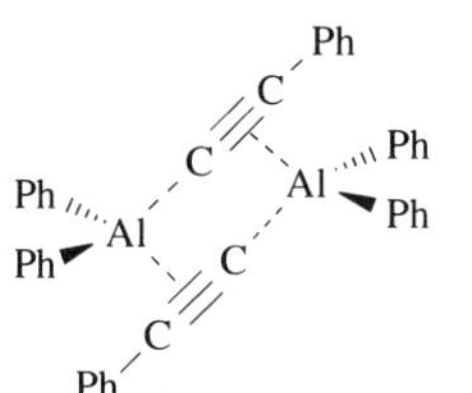

(89)

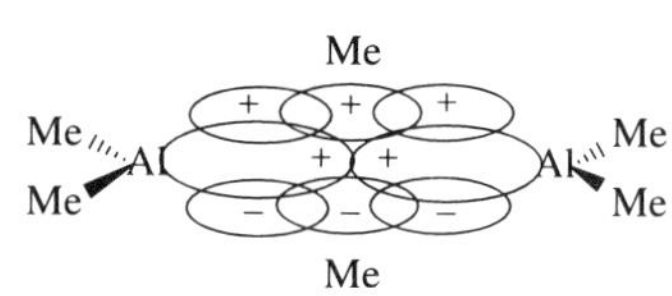

(90)

The monomeric nature of trimesitylaluminum,[16] compared with the firmly dimeric triphenylaluminum, shows how the *ortho* methyl groups can interfere decisively with aryl bridging through nonbonded atom repulsions. Similar steric hindrance prevents the dimerization of tribenzylaluminum in solution but, in the solid state, the aluminum center of one Bz_3Al unit is directly above an *ortho* C–H bond of another Bz_3Al unit and only 0.2453 nm away (**91**), a separation only 0.025 nm greater than in other bridged bonds.[18]

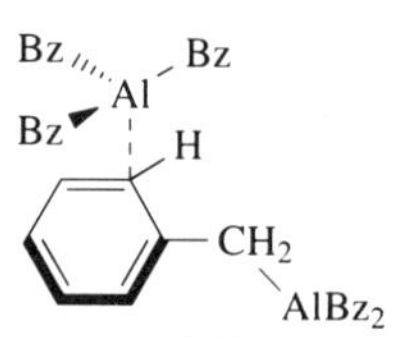

(91)

Since the ring bonding in cyclopropane is best interpreted as being some blend of σ and π bonding, it may cause no surprise to learn that the structure of dimeric tricyclopropylaluminum is comparable in stability to the dimers of Me_3Al and Ph_3Al and has similar C–Al separations for the bridging (0.2074 nm) and terminal (0.1944 nm) cyclopropyl groups. The stabilization of the cyclopropyl bridge by use of *p* orbitals located on the bridging α carbon atom and the nonbonding aluminum orbitals (**92**) offers an explanation that can be supported by CNDO II calculations.[13]

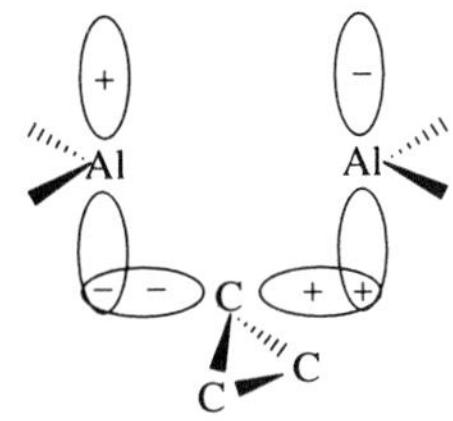

(92)

Finally, such interunit π interactions can lead to polymeric chains, as in the molecular structure of dimethyl(cyclopentadienyl)aluminum, where C-1 and C-3 of the bridging cyclopentadienyl ring are attached to the repeating Me_2Al unit (**93**). Attachment at these carbon centers may be determined by steric factors, because a change in hapticity to η^2 or η^5 should not be difficult.[123]

(93)

10.3.3 Oligomers of Mixed Organoaluminums, R_nAlE_{3-n}, and Attendant Equilibria

The interaction of the empty $3p_z$ orbital in tricoordinate triorganylaluminums with π electrons of an organic ligand is the basis for the autoassociation of $R^2{}_nAlR^1{}_{3-n}$ units, as discussed in the preceding section. If one of the ligands in R_nAlE_{3-n} (**94**) has an unshared electron pair, then the stage is set for a strong interunit interaction of the Lewis acidic aluminum center with the Lewis basic ligand E, either in an intraunit chelating fashion (**95**), for example, E = $(CH_2)_2$–O–R^2 or $(CH_2)_3$-$NR^1{}_2$) or in an interunit, oligomerizing monomer (**96**).

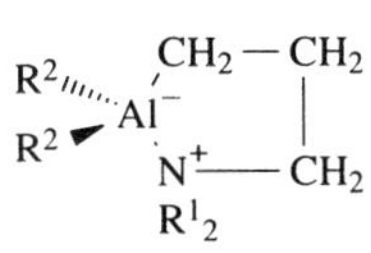

(95)

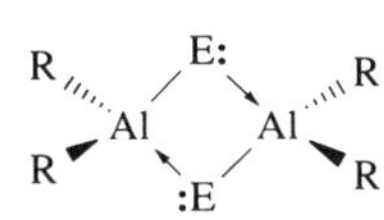

(96)

Equilibria among monomer, dimer, trimer, and higher oligomers of (**94**) are determined by the Lewis basicity of the bridging group in associated structures such as (**96**), with the strength of the bridging decreasing in the order, E = NR_2, OR, X. The steric requirements of R can interfere altogether with even a dimerizing association. Thus, although dialkylaluminum phenoxides exist as dimers with bridging phenoxyl groups ((**96**) E = OPh),[124] the corresponding dialkylaluminum 2,6-di-*t*-butyl-4-methylphenoxide (**97**) is monomeric.[12,15,125] Although (**97**) is sterically hindered from dimerizing, it can still act as a Lewis acid toward phosphines and amines, such as Me_3P and pyridines. Even ketones such as benzophenone can form a 1:1 complex with methylaluminum bis(2,6-di-*t*-butyl-4-methylphenoxide) ((**13**), Equation (2)). The XRD structure of this benzophenone adduct with (**97**) reveals a distorted tetrahedral array about the aluminum atom (**98**). From the observed aryl oxide Al–O distances (0.1721–0.1733 nm) and the angular disposition of the ketonic and aryloxy groups about aluminum, it is proposed that π bonding between the aryloxy lone pairs and the ketonic Al–O σ antibonding orbital determine the rotational orientation of the aryloxy groups. However, as discussed in the previous section, an Al–O separation of 0.170–0.174 nm for tetracoordinate aluminum is not unusually short and hence does not compel the invocation of π bonding to account for it. As to the spatial rotation of the aryloxy groups in (**98**), it is probable that such a conformation could result simply from a minimization of nonbonded interactions, without the involvement of π bonding, a factor of unproved significance.

A wide variety of mixed organoaluminum complexes have been prepared by allowing a proton-bearing ligand, EH, to react with an aluminum alkyl. In several cases the resulting R_nAlE_{3-n} has been shown to form a dimer with unusual coordination. Thus the reaction given in Equation (48) yields a complex containing two five-coordinate aluminum centers in square-pyramidal environments (**99**).[126] Similarly, the reaction of Me_3Al with 1,3-diphenyltriazene leads to a hexacoordinate complex (**100**) with an unusually long Al–N bond (0.2155 nm) *trans* to the methyl group (Equation (49)).[127] Further, the reaction of Et_3Al with tetradentate Schiff bases, such as *N,N*'-bis(salicylaldehydo)ethenediamine (salen)

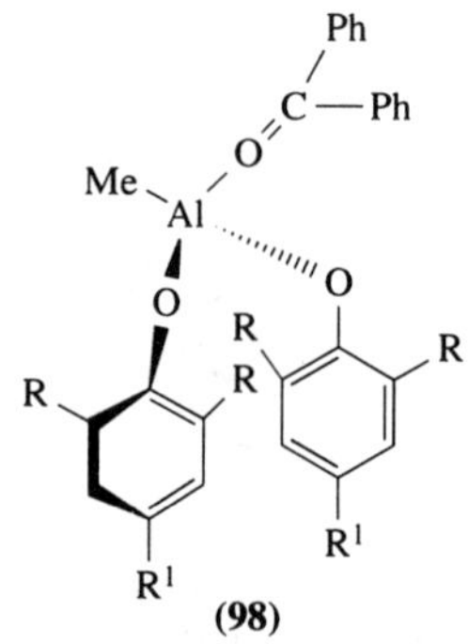

(98)

(**101**), yields monoethylaluminum derivatives (**102**) constrained by a pentacoordinate, square-pyramidal geometry (Equation (50)).[9] Finally, as already mentioned and cited in (**4**), the reaction of Me_3Al with $N(CH_2CH_2OH)_3$ provided a complex containing an Al_4O_6 cluster and representing the first structurally characterized octahedral aluminum alkyl.[10]

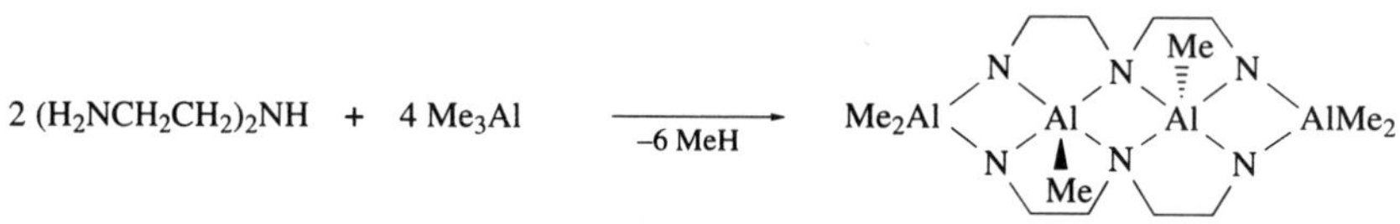

(99) (48)

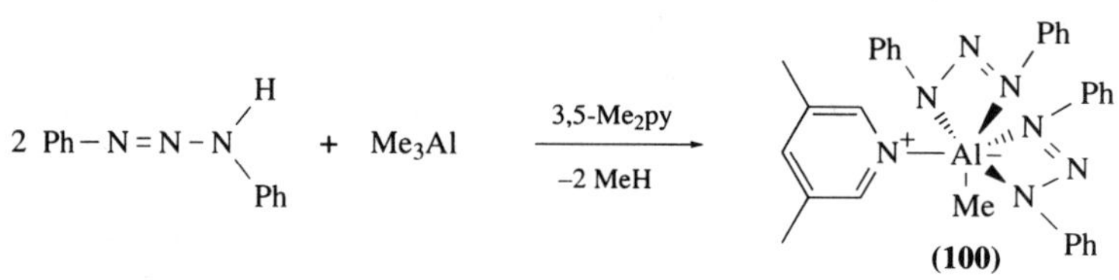

(100) (49)

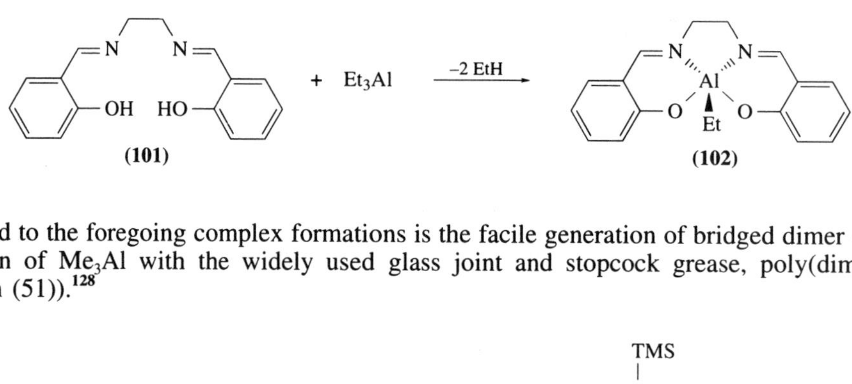

(101) **(102)** (50)

Related to the foregoing complex formations is the facile generation of bridged dimer (**103**) from the interaction of Me_3Al with the widely used glass joint and stopcock grease, poly(dimethyl)siloxane (Equation (51)).[128]

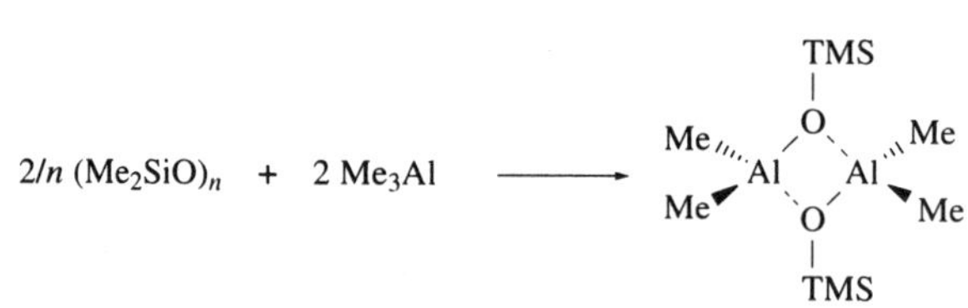

(103) (51)

10.3.4 The Structural Enigma of Aluminoxanes

Perhaps the most difficult structural question in mixed organoaluminums is posed by those products arising from the partial hydrolysis of aluminum alkyls. Known generically as aluminoxanes or alumoxanes, these inadequately characterized mixtures of components having Al–O–Al linkages have been found to be superb cocatalysts for the transition metal-catalyzed polymerization of alkenes.[3] Just what the role of aluminoxanes is in such so-called Ziegler–Natta catalysis is not yet clear and it is for an understanding of this catalysis that the question of aluminoxane structure takes on such urgency.

If exactly two alkyl groups per aluminum are hydrolyzed, a chain of alternating oxygen and alkylaluminum units ((**104**), Equation (52)) should result with terminal R_2Al groups. However, because of the Lewis acidity of the aluminum centers and the Lewis basicity of the oxygen sites, the possibility of interchain complexation, cyclization to cycloaluminoxanes (**105**) and cross-linking of chains via junctions such as (**106**) adds enormously to the structural possibilities and hence the complexities of a given aluminoxane sample. The considerable amount of physical and chemical investigation published in the open literature and aimed at elucidating the structural units in aluminoxanes has been reviewed recently by Pasynkiewicz, who is forced to conclude that insufficient structural information has been gathered to enable an understanding of the role of aluminoxanes in Ziegler–Natta catalysis.[77]

$$n\,R_3Al \;+\; (n-2)H_2O \;\longrightarrow\; R_2Al\!\left(O-\underset{}{\overset{R}{Al}}-O-\overset{R}{Al}-O\right)_{(n/2-1)}\!AlR_2 \qquad (52)$$

(104)

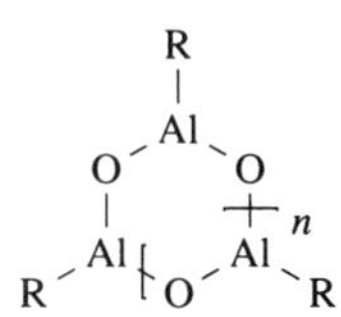

(105)

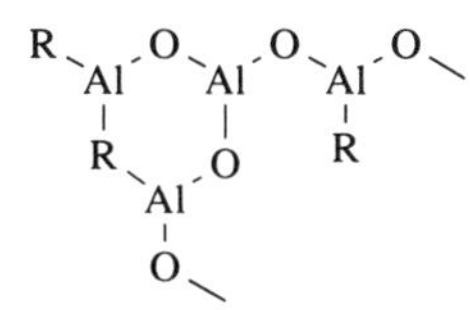

(106)

A large part of the structural challenge presented by aluminoxanes is that such idealized chains, rings, or branches as depicted in (**104**)–(**106**) are metastable intermediates, which are thermodynamically unstable ultimately with respect to alumina and the volatile aluminum alkyl (Equation (53)).[129] In addition, at much lower temperatures the aluminum alkyls can cause such poly(alkylaluminoxanes) as (**104**) or (**105**) to be in equilibrium with tetraalkyldialuminoxanes (**107**), which are also associated (to at least a dimer and possibly a trimer) and can exist as isomeric arrangements of alkylaluminoxanes complexed with trialkylaluminum units (e.g., (**108**), Scheme 7). In the presence of electron donors, such as nitriles, amines, or ethers, aluminoxanes probably exist as monomers complexed in a 1:1 ratio with these Lewis bases.[130,131]

$$3\,(RAlO)_n \xrightarrow{\Delta} n\,R_3Al\uparrow \;+\; n\,Al_2O_3 \qquad (53)$$

$$(\mathbf{104}) \text{ or } (\mathbf{105}) + R_3Al \rightleftharpoons n\,(R_2Al\text{–}O\text{–}AlR_2)_x \rightleftharpoons$$

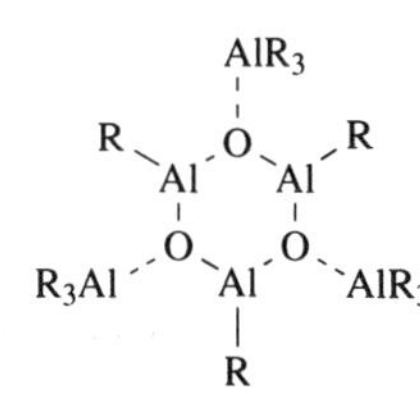

(107) **(108)**

Scheme 7

This rich variety of isomeric complexes, either in ready equilibrium with each other or kinetically controlled in their formation from R_3Al and water sources at various temperatures and in donor solvents, is the reason for the large number of empirical procedures and rather arbitrary experimental descriptions for producing poly(alkylaluminoxanes) suitable as Ziegler–Natta cocatalysts in alkene polymerization. The same complicated array of isomeric arrangements helps explain why the catalytic efficacy of aluminoxanes is dependent not only on rigid adherence to a given preparative procedure but also on a deterioration in the catalytic activity ("aging") upon storage, unless appropriately stabilized. Due to the commercial importance of alkylaluminoxanes such details on the production and stabilization of the catalytic activity of aluminoxanes are largely the subject of patents or the matter of trade secrets.

At least in broad outline, however, the methods for the protolytic formation of aluminoxanes are well known. The most obvious method, discovered to their amazement (and probable trepidation) by the pioneering colleagues of Ziegler, is the partial hydrolysis of aluminum alkyls. To convert this reaction from a life-threatening pyrotechnic spectacle into a tranquil preparation of aluminoxanes, controlled amounts of water must be introduced into a solution of R_3Al at low temperatures.[132] Although an ether may be used to moderate the vigor of hydrolysis, it should not complex with the resulting aluminoxanes. Diisopropyl ether has been recommended.[133] Another method of slowing the rates of hydrolysis of an aluminum alkyl is to introduce the water as an insoluble phase, such as employing the metal salt hydrates of aluminum sulfate or copper sulfate.[134]

Other nonprotonolytic preparations of aluminoxanes are summarized in Equations (54)–(59) and Scheme 8.[53,135–40] The methods depicted in Equations (58) and (59) do not share the drawback of the others which do not produce aluminoxanes suitable for catalytic applications, since a potentially deactivating Lewis base is involved.

$$3\,Et_3Al + CO_2 \longrightarrow Et_2Al\text{–}O\text{–}AlEt_2 + Et_3COAlEt_2 \qquad (54)$$

$$RC(=O)NR^1_2 + 2\,Bu^i_2AlH \longrightarrow Bu^i_2Al\text{–}O\text{–}AlBu^i_2 + R^2CH_2NR^1_2 \qquad (55)$$

$$Ph_2C{=}O + PhN(AlCl_2)_2 \longrightarrow Ph_2C{=}NPh + Cl_2Al\text{–}O\text{–}AlCl_2 \qquad (56)$$

$$PbO + Me_2AlCl \longrightarrow \text{mixture} \xrightarrow{Et_2O} Me_2Al\text{–}O\text{–}AlMeCl{\bullet}Et_2O \qquad (57)$$

$$LiOAlR_2 + R_2AlCl \xrightarrow{-\,LiCl} R_2Al\text{–}O\text{–}AlR_2 \qquad (58)$$

$$R_2B\text{–}O\text{–}BR_2 + 2\,Et_3Al \longrightarrow Et_2Al\text{–}O\text{–}AlEt_2 + 2\,R_2BEt \qquad (59)$$

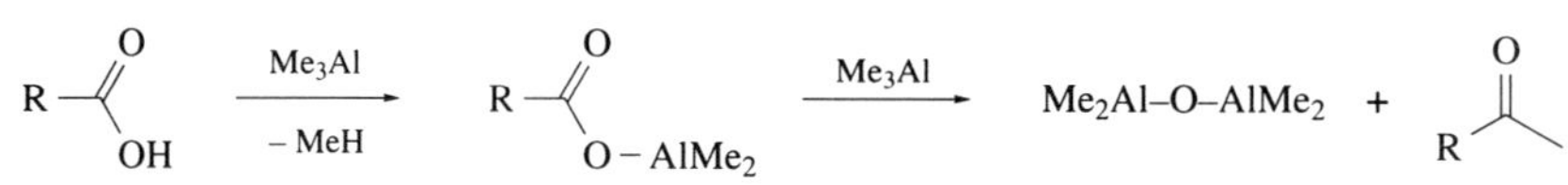

Scheme 8

As should be clear from previous comments, the structural detail for aluminoxanes in solution has not been arrived at unambiguously. Local structural subunits, such as those depicted in Structures (**104**)–(**108**), undoubtedly seem to play a role among the many structural variants in complex equilibrium with each other. Perhaps this conclusion is all that physicochemical studies will ever be able to establish. The "structure of aluminoxanes" is perhaps akin to the equally complex questions from the natural realm, which seek to define the "structure of lignin" or the "structure of coal." If so, then it is of dubious import when, from such complex metastable equilibria, occasionally a well-definable complex precipitates and is amenable to XRD structure analysis. What the demonstrable structure of such an aluminoxane component conveys about the structural nature of aluminoxanes in solution is unclear. With this serious caveat in mind, the nature of some of these crystalline complexes is now considered. The discussion will be restricted to aluminoxane complexes free of other main group or transition metal ions and it will be seen that no reliable XRD structure has been obtained, excepting the recent work on *t*-butylaluminoxanes, which is discussed later.[141]

Since $(Et_2Al)_2O$ was found to have a molecular weight corresponding to a tetramer at a given concentration, further ^1H, ^{27}Al, and ^{17}O NMR studies are consistent with the tricoordination of oxygen and the tetracoordination of aluminum as proposed in (**109**).[142] Some equally reasonable but similarly hypothetical, dynamic structures have been proposed for aluminoxanes having the empirical formulas of $Me_2Br_2Al_2O$[143] and $Bu^i_3ClAl_2O$,[144] but even granted that the proposed structures accurately render the bonding for these dialuminoxanes *in isolation*, they really provide no insight into the bonding and complex equilibria in alkylaluminoxane mixtures.

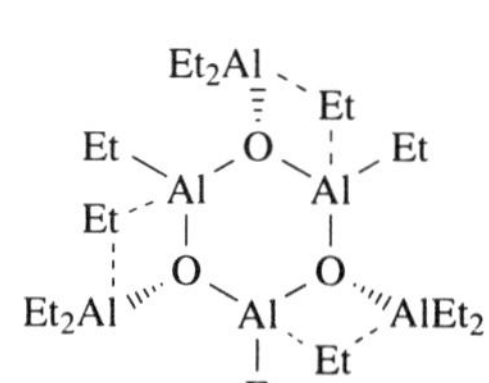

(109)

At this point it should be remembered that the burning curiosity concerning aluminoxane structure in reference to catalytic activity really arises for methylaluminoxanes, rather than for alkylaluminoxanes in general. The cocatalytic activity of methylaluminoxanes is generally greatly superior to that of ethyl-, propyl-, or butylaluminoxanes. From this sharply gradated catalytic activity it is abundantly clear that a Me–Al bond is much more efficacious than anything larger. It is equally clear that a methyl group in R_3Al has a profound influence on structure because Me_3Al exists as a stable dimer while Pr_3Al and $Bu^i{}_3Al$ tend to be monomers. The smaller size and superior bridging capability of the methyl group in methylaluminoxanes should cause both the structure and any catalytic activity of $(MeAlO)_n$ (**110**) to be quite different to that of higher alkylaluminoxanes.

The recent results of the Barron group on the structures of oligomers of tetra-*t*-butyldialuminoxane (**113**) and *t*-butylaluminoxanes (**114**) are probably best viewed in this light.[141] The isolation and crystal structure determination of a number of hydrolysis products of tri-*t*-butylaluminum (**111**) and their Lewis base adducts have provided fundamental insight into the bonding possibilities of $Bu^t{}_2AlO$ and Bu^tAlO units in the resulting oligomers (Scheme 9, (**112**)–(**115**)). However, it is highly questionable whether such structural insights are readily transferable to interpreting the structure of methylaluminoxanes, given the markedly different steric requirements of Me vs. Bu^t, as well as the superior bridging tendency of the methyl group in electron-deficient Al···R···Al bonding.

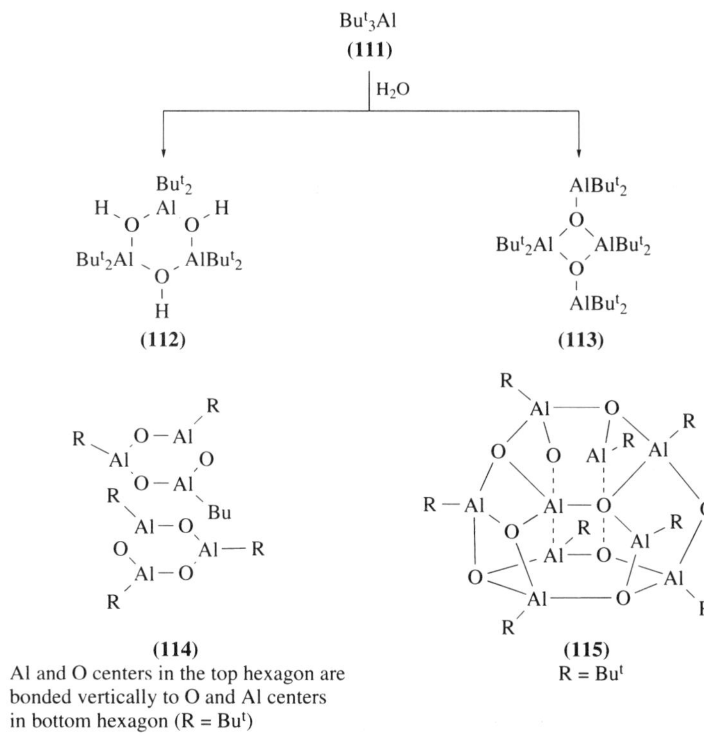

Scheme 9

Regardless of what ultimate relevance the foregoing *t*-butylaluminoxane structures prove to have for solving the structural enigma of methylaluminoxanes, there is the further question of cocatalytic activity of alkylaluminoxanes in the Ziegler–Natta stereoregular polymerization of α-alkenes.[3] Normally the alkylaluminoxanes catalyst is added in great excess (up to 1000 equiv.) over the transition metal component and the methylaluminoxane is markedly more active than the corresponding ethyl or higher alkylaluminoxanes or even ordinary alkyls such as Me_3Al or Et_2AlCl. To explain the considerable cocatalytic action of excess methylaluminoxane in Ziegler–Natta catalyst systems, it has been suggested that $(MeAlO)_n$ (**110**) functions as a most efficient scavenger of protic or oxidizing impurities that otherwise would deactivate the transition metal active site. However, this theory is discredited by the failure of other R_nAlE_{3-n} components to exhibit a comparable cocatalytic activity, even though they are also effective scavengers of moisture and oxygen.

Accordingly, the explanation for the catalytic activity of MAOs has been sought through identifying special structural features in MAO that could cause such cocatalysis. Moreover, since a huge excess of MAO is required per equivalent of transition metal component, such a MAO structural unit might constitute a small percentage of the various components in an MAO sample. An illustrative example of this approach in trying to understand the catalytic activity of MAO is a recent investigation that concludes that the active units in MAO are the terminal Me_2Al groups of the aluminoxanes chains, $\text{-(}Al(MeE)\text{–}O\text{)-}_n$ (compare (**104**) in Equation (52)).[145] However, what the electronic basis would be for the activity of such a unit remains unclear. At present the structure(s) and the cocatalytic activity of methylaluminoxanes remain unsolved problems and represent significant challenges for future research.

10.3.5 Unusual and Variegated Coordination in Organoaluminum Complexes

In Section 10.2 it was mentioned that the first authenticated example of a six-coordinate organoaluminum compound (**4**) was synthesized in 1988 from the reaction of Me_3Al with $N(CH_2CH_2OH)_3$.[10] Four years prior to this finding, the seven-coordinate complex of the Cl_2Al^+ cation with benzo-15-crown-5 (**5**) with a $(AlEtCl_3)^-$ anion resulted from the reaction of $EtAlCl_2$ with the crown ether.[11] The noteworthy disproportionation of the $EtAlCl_2$ occurring in this reaction is mirrored in the reaction of Et_2AlCl with the macrocyclic tetradentate amine, 1,4,8,11-tetrazacyclotetradecane (**116**). The resulting cage complex (**117**) is generated by the *in situ* disproportionation of Et_2AlCl into Et_3Al and $EtAlCl_2$ (Equation (60)).[146]

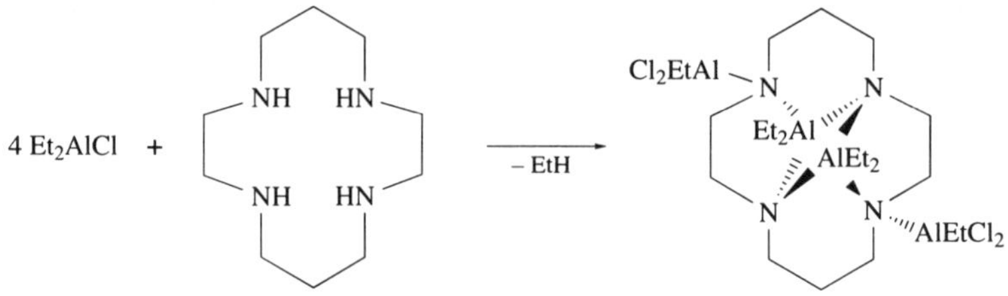

(116) **(117)** (60)

As might be expected, crown ethers and thiacrown ethers form well-defined complexes with Me_3Al units, but no such transannular pentacoordinate bridging of Me_3Al or Me_2Al fragments arises as in (**117**); the Me_3Al unit remains strictly exocyclic, as in (**118**) and (**119**).[147,148]

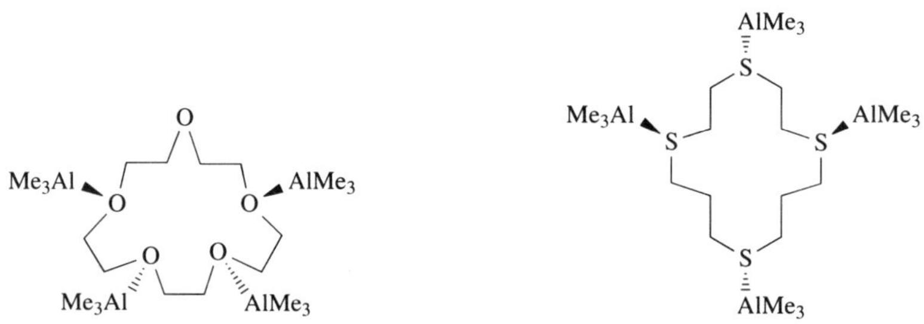

(118) **(119)**

Possibly the richest variations in coordination arise from the interactions of Me_3Al with primary amines. Stepwise interaction of Me_3Al with NH_3 could be viewed as proceeding through intermediates (**120**)–(**122**) upon thermolysis with the ultimate formation of the diamondlike lattice of aluminum nitride (Scheme 10).[149,150] Intermediates (**121**) and (**122**) could oligomerize to cyclic or cage structures or form open-chain polymers prior to the final elimination of methane and the formation of AlN. Since the early 1960s such discrete oligomers have been intercepted by employing both various R_3Al compounds and primary amines having bulky organic groups. Thus, the reaction in Equation (61) yielded the cubane-like tetramer of PhAl=NPh (**123**).[151] In such interactions the first elimination of RH yields (**121**), as a dimer (**124**), which can be isolated in many cases and structurally characterized.[152,153]

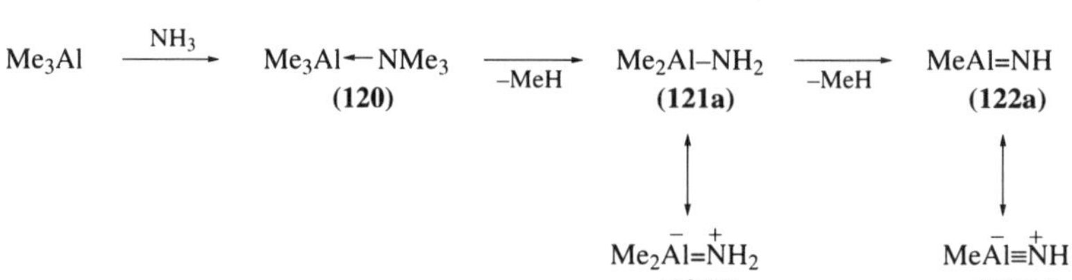

Scheme 10

Depending on the nature of R and R^1, (**124**) can be further transformed into other oligomers:[150] if R^2 = Me and R^1 = mesityl, a cubane-like tetramer is formed, see (**123**); if R^2 = Me and R^1 = 2,6-Pr^i_2-C_6H_3, an aluminazene trimer results (**125**); and if R^2 = Me and R^1 = Ph, the hexagonal cage hexamer (**126**) is obtained.[154] Since the Al_3N_3 array in (**125**) is planar, it is of interest to inquire whether the ring attains any stabilization through π-electron delocalization. Although the Al–N bond length is shorter (0.178 nm) than other Al–N bonds in polyiminoalanes (0.189–0.196 nm), calculations at the SCF level with the basis set 6-31G* indicate little or no delocalization for ((**125**) R^2, R^1 = H).[155]

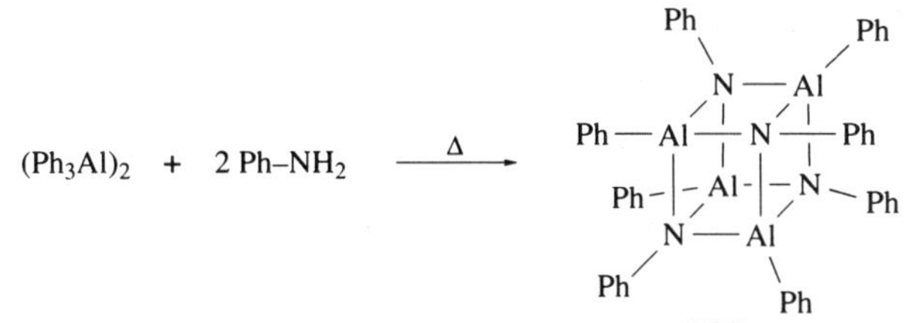

(123)

(61)

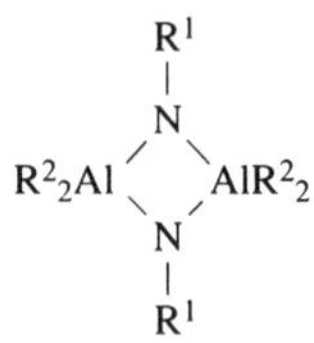

(124)

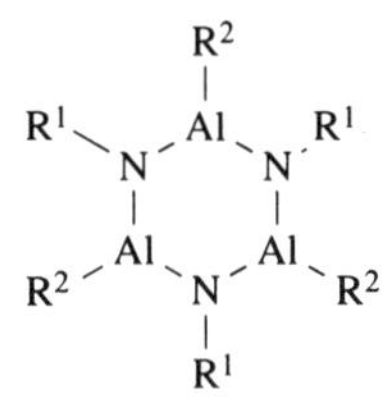

(125)

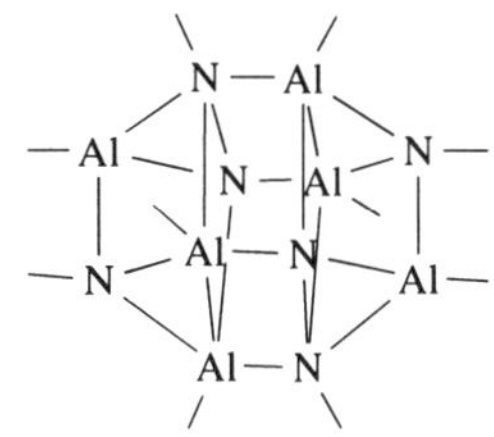

(126)

Finally, the unusual pentacoordination thought to be involved in alkali metal aluminates deserves mention. An ^{27}Al NMR study has shown that alkali metal aluminum hydrides and tetraorganylaluminates undergo ready group exchange (Equation (62)).[156] As a likely intermediate for such exchange, pentacoordinate-bridged aluminum centers have an obvious appeal (**127**). Support for this proposal comes from the dimeric character of $NaAlEt_3H$ and $KAlBu^i_3H$ in alkane solution. Furthermore, the low-energy Al–H band at 1461 cm^{-1} in the IR spectrum would be consistent with Al···H···Al bonding.[157,158] A suggested structure for such a dimer is (**128**).[116]

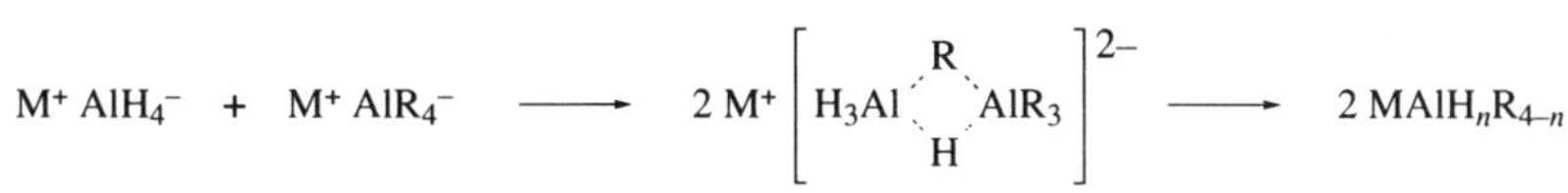

(127)

(62)

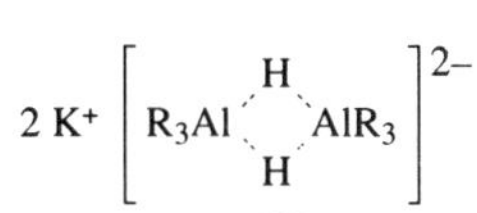

(128)

10.4 PREPARATION

10.4.1 Organization of this Survey

The original treatment of the available preparative methods for organoaluminum compounds outlined the synthesis of various compound types, ranging from symmetrical and unsymmetrical triorganylaluminums, R_3Al and $R^2{}_nAlR^1{}_{3-n}$, through hydrides, R_nAlH_{3-n} and $M^1AlR_nH_{4-n}$, to mixed derivatives, R_nAlE_{m-n}, where E = X, O, S, N, and so on.[1] Since these various preparations evolved from over 120 years of research, it is not to be expected that many novel preparative approaches would have been discovered since the early 1980s.

Therefore, to lend a different and possibly useful viewpoint to this survey of preparative methods, the present treatment is organized according to the covalent bonds made between aluminum and a second atom. This organization will permit an appreciation, in a systematic manner, of how Al–C, Al–H, Al–E (E = X, O, C, or N) and even aluminum–metal bonds can be fashioned. This survey is followed by an analysis of the practical ways by which one type of Al–C bond can be transformed into another kind of Al–C bond. Stress has also been laid on some known, but less common preparative methods that have received less attention than they appear to warrant. Organized in this manner, the present survey of preparative methods for organoaluminum derivatives should prove to be usefully complementary to and not duplicative of the original survey in *COMC-I*.

10.4.2 Aluminum–Carbon Bonds

10.4.2.1 General considerations

The most general and hence most useful methods of forming Al–C bonds are (i) the addition of Al–H bonds to carbon–carbon unsaturation (hydroalumination); (ii) the interaction of an Al–X, Al–O, or Al–N bond with an organometallic derivative of a more active or even a less active metal in a metathesis reaction; and (iii) the interaction of aluminum metal with an organometallic derivative of a less reactive metal (typically metals of group 12, such as mercury, or higher). Each of these methods deserves individual appraisal.

Other reported methods for making Al–C bonds either have serious limitations in our present state of knowledge or involve the generation of reactive intermediates whose utility remains to be corroborated by future research. These limited or undeveloped routes to Al–C bonds will be briefly discussed as a group in the following paragraphs.

The first, rather classical method is the interaction of aluminum metal with an organic halide. The interaction of aluminum with ethyl iodide to produce a 1:1 mixture of $EtAlI_2$ and Et_2AlI, as reported by Hallwachs and Schafarik in 1859, represents one of the first successful formations of an Al–C bond (Equation (63)).[159] Not only does the method have the drawback of producing a mixture of organoaluminum derivatives, (**129**) and (**130**), but these products are strong Lewis acids, which can attack the RX to unleash carbenium-ion decompositions producing alkenes and oligomers, as well as HX that can cause violent protodemetallation of (**129**) and (**130**). This undesirable decomposition can sometimes be minimized by use of ethers as the reaction medium, but the products, (**129**) and (**130**), are then obtained as stable monoetherates. In the absence of ethers the reaction has proved a manageable way to produce the so-called sesquihalide mixture of (**129**) and (**130**), if R = Me, Et, and Ph. The chemist should not lose sight of the potential for the reaction in Equation (63) to proceed in an *explosive manner*.[160] It is possible that future research on the use of aluminum powders of optimal particle size, appropriate promoters, sonication, and/or high-speed stirring rates may extend the utility and applicability of this method. An example underlining the promise of this approach is the smooth, high-yielding preparation of bis(dichloroalumino)methane (**131**) according to Equation (64) by using some CH_2Br_2 as a promoter and removing fine particles of aluminum from the aluminum powder employed.[52]

$$2\,Al + 3\,RX \longrightarrow \underset{\mathbf{(129)}}{RAlX_2} + \underset{\mathbf{(130)}}{R_2AlX} \qquad (63)$$

$$3\,CH_2Cl_2 + 4\,Al \xrightarrow{CH_2Br_2} \underset{\mathbf{(131)}}{CH_2(AlCl_2)_2} + 2/n\,(\!-\underset{\displaystyle Cl}{\underset{|}{Al}}-CH_2\!)_n \qquad (64)$$

A second, quite limited method for producing Al–C bonds is the interaction of an aluminum hydride, R_2AlH (especially as the R_3N complex) or LAH, with a carbon acid of sufficient acidity. The latter requirement limits the reactive C–H bonds to those of terminal alkynes (Equation (65)),[161] cyclopentadienes (Equation (66)),[162] and highly acidic sp^2 C–H bonds adjacent to heteroatoms (Equation (67)).[163] It should be noted that the tetracoordinate aluminum derivative is obtained in all cases (**132**)–(**134**). If the tricoordinate aluminum (R_3Al) is sought, the method is of little value.

$$Bu^i{}_2AlH{\bullet}NEt_3 + Bu{-}C{\equiv}CH \xrightarrow[-H_2]{} \underset{\mathbf{(132)}}{Bu{-}C{\equiv}C{-}AlBu^i{}_2{\bullet}NEt_3} \qquad (65)$$

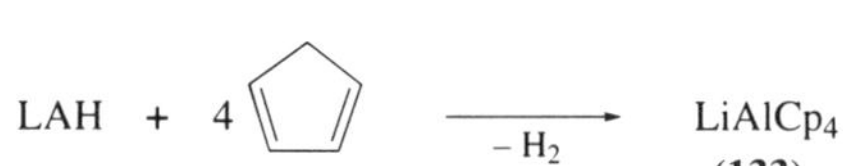

(133) (66)

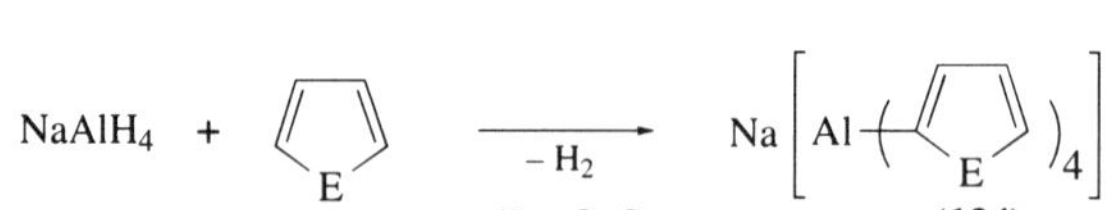

E = O, S **(134)** (67)

A third, potentially general route to unusual Al–C bonds is the addition of subvalent aluminum halides to carbon–carbon unsaturation. As discussed in Section 10.3.1.2, compelling evidence has been presented that AlCl can add readily both to alkynes and 1,3-dienes (Schemes 4 and 5). Furthermore, evidence that $AlCl_2$ or an oligomer can add to alkenes has been published (Equation (32)), but such an interpretation should, at this stage of knowledge, be considered ambiguous. The alternative interpretation of the reaction, namely the addition of AlCl and then $AlCl_3$ to the alkene (Equation (34)), remains a distinct possibility.

Related to such thermally induced vicinal dialuminations of C=C and C≡C bonds by AlCl and/or $AlCl_3$ are the photochemically promoted formation of adducts (**135**) and (**136**) (Equation (68)).[164,165]

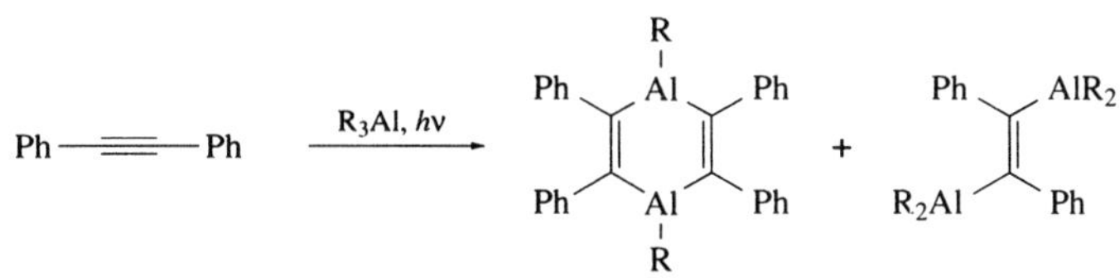

(**135**) (**136**)

R = Et, Me_2CH, Me_2CHCH_2, Bu^t

(68)

A fourth, inherently interesting way of making Al–C bonds is by the addition of aluminum(III) halides to carbon–carbon unsaturation. Such haloaluminations have not been authenticated, as have the analogous haloborations (Equation (69)).[166,167] However, judging by the formation of (**138**) from 2-butyne and $AlCl_3$, there is reason to believe that a chloroalumination adduct (**137a**) or its zwitterionic counterpart (**137b**) could be involved (Scheme 11).[168] Likewise, the interesting rearrangements of 3,3,3-triphenylpropene catalyzed by $MeAlCl_2$ can be rationalized as proceeding through adducts (**139a**) or (**139b**) to produce isomers (**140**) and (**141**) (Scheme 12).[56] From these examples it might be suggested that, as a way of making Al–C bonds, haloalumination may prove of greatest value in making such bonds in transitory intermediates formed in a variety of alkene and alkyne oligomerization processes.

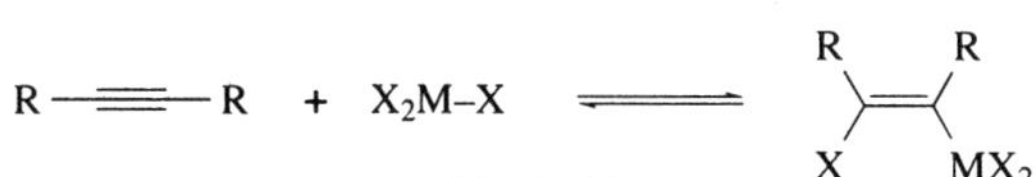

M = B, Al

(69)

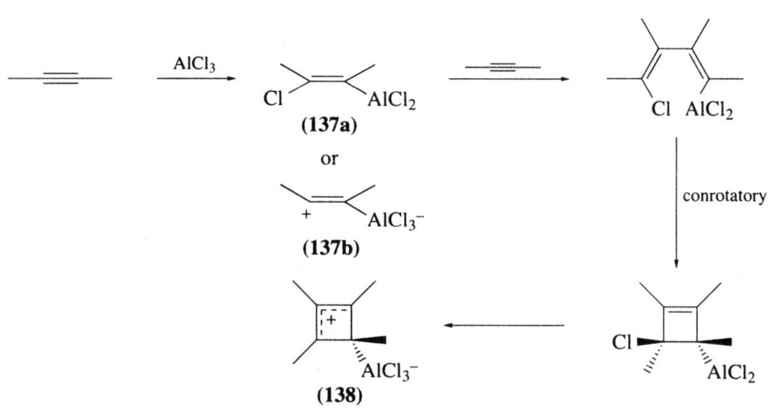

Scheme 11

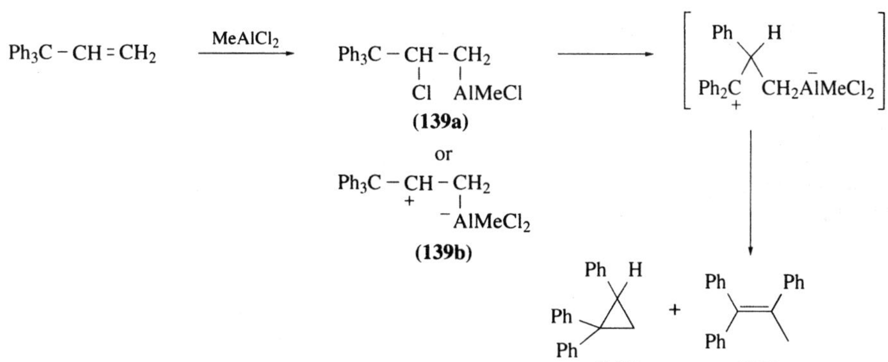

Scheme 12

10.4.2.2 Hydroalumination

The addition of aluminum hydrides to alkenic and alkynic carbon–carbon unsaturation, namely the hydroalumination reaction, is undoubtedly the most broadly useful method of forming Al–C bonds. Since this reaction and its applicability to organic synthesis were the subject of an extensive review in 1990,[4] only the most salient features of this reaction and its value for preparing specific organoaluminum compounds are discussed here.

Hydroalumination is the molecular basis for the large-scale, industrial synthesis of symmetrical triorganylaluminum (**142**) from the corresponding α-alkene, powdered and activated aluminum, and dihydrogen.[169] If the alkene bears bulky substituents at the β position, as with isobutene,[169] 2,4,4-trimethyl-1-pentene, 4-vinylcyclohexene, dipentene, or trimethyl(vinyl)silane,[170] a one-step, so-called direct synthesis gives acceptable or even good yields of (**142**) (Equation (70)).

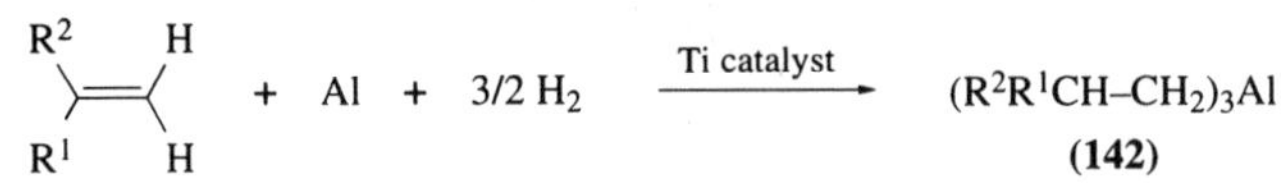

(70)

For straight-chain α-alkenes, the aluminum hydride is prepared in the presence of the desired R_3Al in order to trap and stabilize the AlH_3 by formation of R_2AlH (Equation (71)). Note that no alkene is present at this stage in case it inserts into the C–Al bond to yield chain-growth (i.e., Aufbau) products (Equation (72)). In the second stage and at a lower temperature the resulting R_2AlH is allowed to hydroaluminate the alkene with the formation of (**142**) (Equation (73)). With alkenes not bearing any heteroatoms, the hydroalumination displays a marked regioselectivity and produces over 90% of the primary alkyl–aluminum bond, RCH_2CH_2–Al.[171] The presence of a trialkylsilyl group on the β carbon, however, favors the reverse regioselectivity (Equation (74)).[172] Further effects of substituents on both the regiochemistry and stereochemistry of such additions are summarized later.

$$2\,(RCH_2CH_2)_3Al \;+\; Al \;+\; 3/2\,H_2 \xrightarrow{\Delta} 3\,(RCH_2CH_2)_2AlH \qquad (71)$$

$$(RCH_2CH_2)_3Al \;+\; 3x\,H_2C{=}CH_2 \longrightarrow [RCH_2CH_2(CH_2CH_2)_x]_3Al \qquad (72)$$

$$(RCH_2CH_2)_2AlH \;+\; RCH{=}CH_2 \longrightarrow \underset{\textbf{(142)}}{(RCH_2CH_2)_3Al} \qquad (73)$$

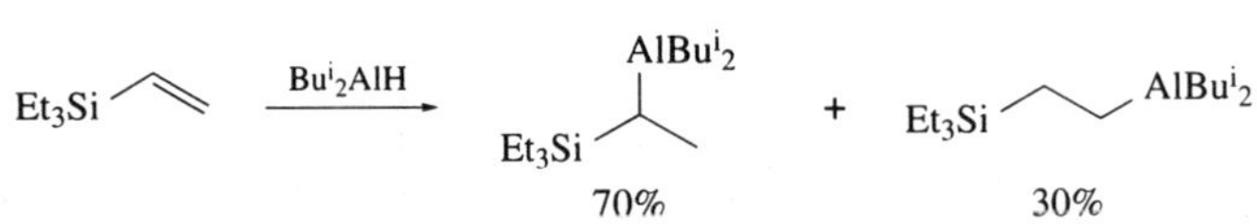

(74)

Hydroalumination is well suited for preparing a wide variety of unsymmetrical aluminum alkyls of the type $R^2{}_2AlR^1$ (**143**); the simple addition of the commercially available Et_2AlH or $Bu^i{}_2AlH$ to a wide variety of terminal or strained C=C bonds provides high-yielding routes to a variety of unsymmetrical alkyls. Illustrative are the adducts produced from 1-decene (**144**)[173], norbornene (**145**),[174] camphene (**146**),[171] acenaphthylene (**147**),[175] 2-deuterio-3,3-dimethylindene (**148**),[176] and 2,6-dimethyl-2,7-octadiene (**149**).[177] The following observations are noteworthy: (i) isomer (**144**) is accompanied by less than 5% of the 2-decyl isomer; (ii) the formation of (**145**) and (**146**) results from exclusively *exo* attack on the norbornene skeleton; (iii) (**147**) is the only organoaluminum to result from acenaphthylene and R_2AlH, even though reaction of this adduct with H_2O or with $R_2C{=}O$ forms products of attack of these reagents at the *ortho* carbon (indicated by *); (iv) (**148**) is the exclusively formed adduct, resulting from a regiospecific, *syn* stereoselective hydroalumination in ether; (v) (**149**) results from a highly regioselective, locoselective addition; and (vi) these $R^2{}_2AlR^1$ compounds maintain their integrity if formed and stored at moderate temperatures (<50 °C); at higher temperatures, disproportionation tends to take place (Equation (75)).

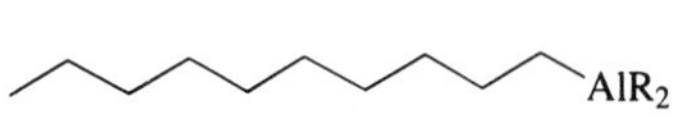

(144)

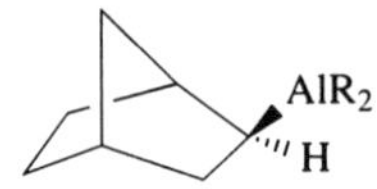

(145)

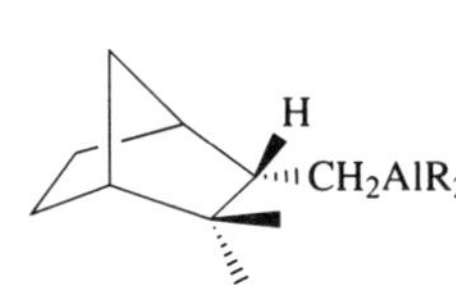

(146)

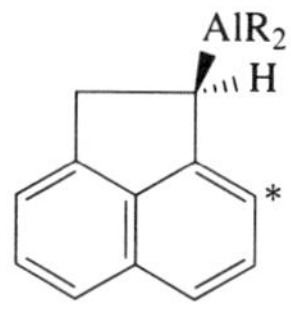

(147)

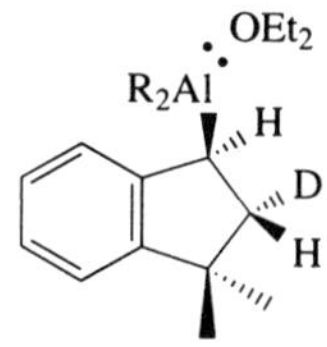

(148)

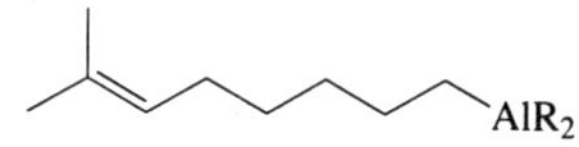

(149)

$$R^2{}_2AlR^1 \rightleftharpoons R^2AlR^1{}_2 + R^2{}_3Al + R^1{}_3Al \quad (75)$$

Perhaps the most interesting type of unsymmetrical triorganylaluminums available through hydroalumination are the 1-alkenyl(dialkyl)aluminums (**150**) available by the addition of R_2AlH to alkynes. With unsymmetrical alkynes, the variations in regioselectivity and stereoselectivity, which are caused by the steric and electronic effects of substituents, can produce significant differences in the proportions of the four possible products (Equation (76)).

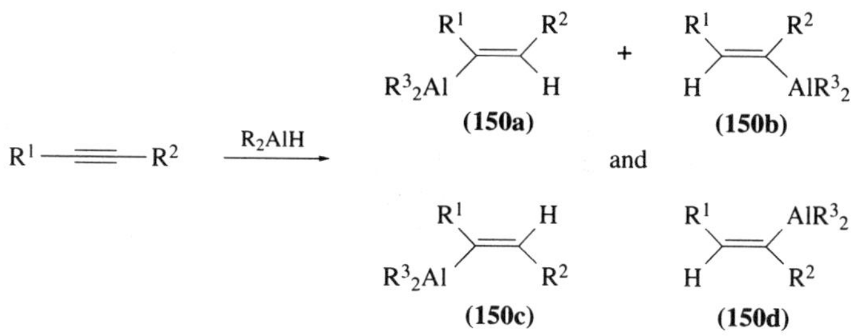

(76)

The original review of hydroalumination should be consulted for the details of how a broad spectrum of substituents on the alkyne influences the proportions of adducts (**150a**)–(**150d**) that are produced (R^1,R^2 = alkyl, aryl, H, R_3Si, R_3Ge, R_3Sn, R_2N, R_2P, RO, RS, RSO_2 and Li, R_2Al).[4] However, the following stereochemical trends should prove most useful in attempting selective formation of one of these adducts. The *syn* course of hydroalumination ((**150a**) and (**150b**)) is generally the kinetically controlled mode of addition. When R^1, R^2 = H, alkyl, or aryl, such *syn* adducts are relatively stable and are converted to the *anti* adducts only at higher temperatures and by way of the bis-hydroalumination adducts (e.g., (**151**)) in an addition–elimination sequence (Scheme 13). When R^1 or R^2 = TMS or Me_3Ge, the resulting *syn* adducts isomerize promptly into the *anti* adducts (above 0 °C), in the absence of ethers or amines. In the presence of ethers or amines, the *syn* adducts are stable at moderate temperatures, to suggest that the free $3p_z$ orbital on aluminum in (**153**) lowers the C=C bond rotational barrier (Scheme 14). The *syn* or *anti* mode of hydroalumination with alkyne substituents as Me_2P, Me_2N, EtS, EtO, or $PhSO_2$ would seem to be explicable in terms of how such substituents promote or retard one of the two isomerization pathways exemplified in Schemes 13 and 14.

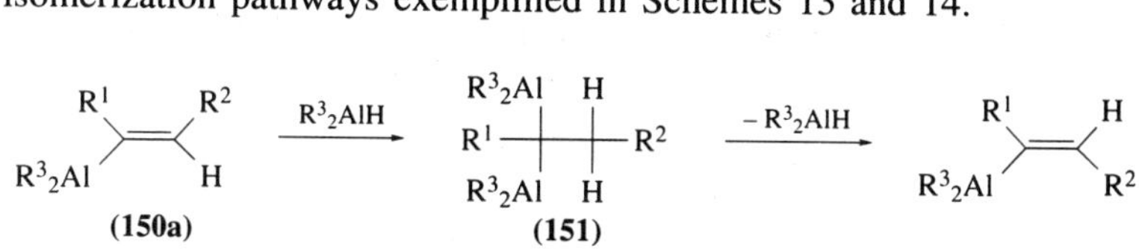

Scheme 13

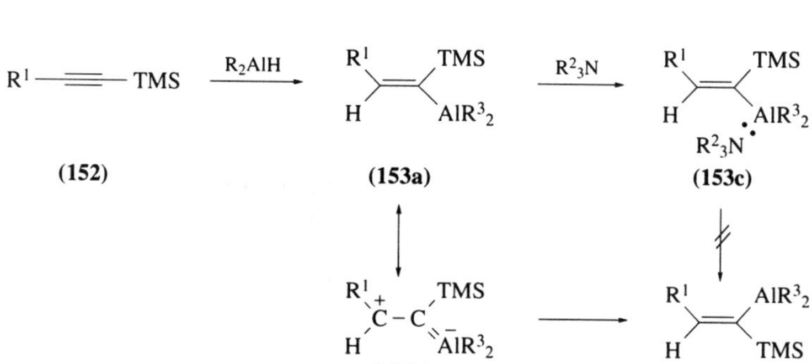

Scheme 14

In a similar manner, certain regiochemical regularities caused by substituents are noteworthy. (i) With alkyl(aryl)acetylenes, R^1–C≡C–Ph, the Al–H addition favors isomer (**156**) over (**157**) both for polar reasons and from steric considerations (Equation (77)). (ii) With ynamines and alkynic ethers ((**158**) E = OR or NR_2) the preponderant adduct is (**160**), suggesting an electrophilic addition fostered by the heteroatom E (**159**) and subsequent *syn–anti* isomerization (Equation (78)). (iii) Although RS and R_2P substituents might have been expected to exert an influence similar to RO and R_2N groups, they do not and the addition proceeds in strictly a *syn* fashion and the preponderant regioisomer formed has the R_2Al adjacent to the heteroatom (**162**):(**163**) = 85:15 (Equation (79)). (iv) As with 1-alkenes, 1-alkynes give almost exclusively the 1-alkenyl regioisomer (**164**), but small to moderate proportions of the 1-alkynyl derivative (**165**) are formed by competing metallation (Equation (80)).

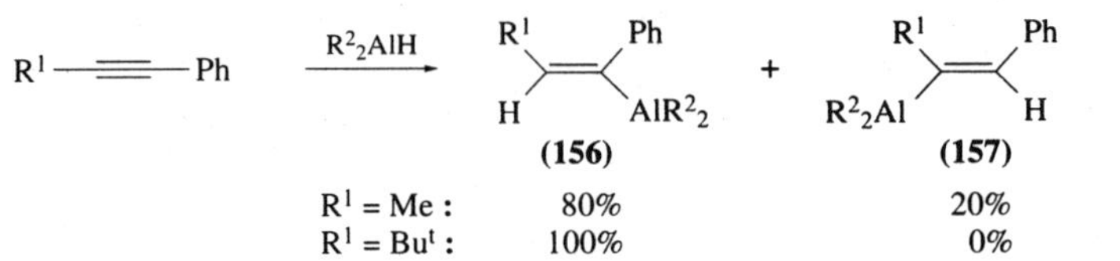

(77)

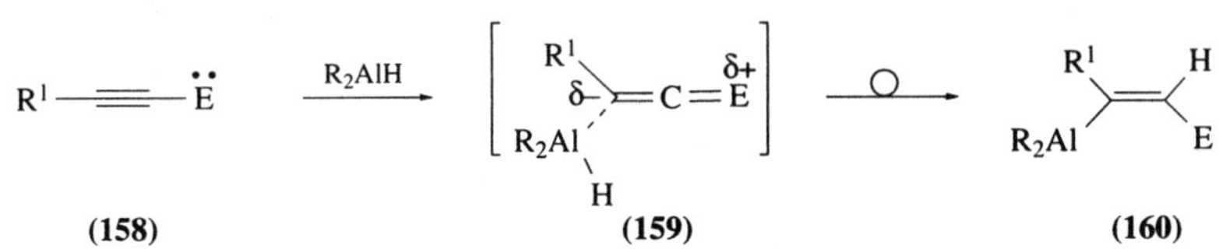

(78)

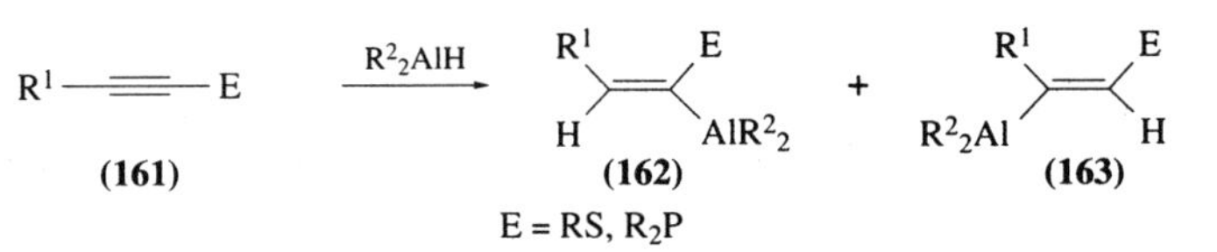

(79)

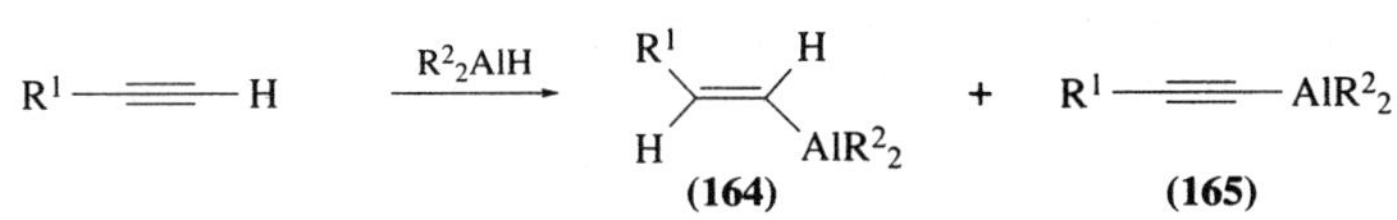

(80)

In this connection it should be mentioned that lithium aluminum hydrides, such as LAH or $LiAlR_3H$, have shown some utility in the hydroalumination of alkenes and alkynes, especially in the presence of titanium-salt promoters. However, these reagents are more prone to the metallation side reaction shown in the formation of (**165**) (see Equations (66) and (67)).

10.4.2.3 Metathesis between metal alkyls and aluminum compounds

The interaction of an aluminum–nonmetal bond with a metal alkyl is a well-established, broadly applicable way of forming Al–C bonds. In the case where the metal alkyl is of a more reactive metal than aluminum, such as those of groups 1 and 2, this preparative method has made use of Grignard and oganoalkali reagents to convert $AlCl_3$ into triorganylaluminums. If the metal alkyl reagent contains ether, the resulting aluminum product will be formed as the etherate, which may prove most difficult to disrupt without profound decomposition (Equation (81)).[178]

$$3\ Bu^tMgCl\ +\ AlCl_3\ \xrightarrow[-\ MgCl_2]{Et_2O}\ Bu^t_3Al\bullet OEt_2\ \xrightarrow{\Delta}\ \text{decomposes} \qquad (81)$$

If ether-free RMgCl is produced or hydrocarbon solutions of RLi are employed, the unsolvated R_3Al is then accessible.[178] In quite a similar manner $R^2_2AlR^1$ derivatives can be prepared from R^2_2AlCl or R^1AlCl_2 derivatives. Although such unsymmetrical types can be prepared at moderate temperatures, they are prone to disproportionation or decomposition especially in the unsolvated form (Equation (75)). Hence, higher temperatures and certainly vacuum distillation should be avoided in isolating or purifying $R^2_2AlR^1$.

Although the preparative scope and limitations of employing such active metal alkyls are rather well understood, this is not the case of metathesis with less active alkyls. In these cases, equilibria are often set up in such exchanges, whereby only a part of the organic groups is transferred to aluminum and the equilibrium must be driven by product removal to complete such transfer (Equation (82)).[179] Such equilibria leading to a new Al–C bond formation should prove significant both for novel preparative routes to organoaluminum derivatives (Equation (83))[180] and for an understanding of the cocatalytic role of the aluminum component in Ziegler–Natta polymerization (Equation (84)).[181]

$$Ph_3B + Et_3Al \xrightleftharpoons{\Delta} Ph_3Al + \underset{\text{(removed by vacuum)}}{Et_3B\uparrow} \tag{82}$$

(83)

$$Cp_2Ti(Me)(Cl) + AlCl_3 \rightleftharpoons Cp_2Ti(Cl)(Cl) + MeAlCl_2 \tag{84}$$

10.4.2.4 Displacement reactions with aluminum metal and a less active metal alkyl

The reaction of aluminum turnings or granules with mercury alkyls or aryls dates from 1856 and is thus the oldest method for preparing unsolvated trialkyl- and triarylaluminums[182] (Equation (85)). The method has even proved serviceable for the preparation of R_3Al compounds that might be expected to undergo facile rearrangements: triisopropylaluminum can be prepared free of any formation of *n*-propylaluminum groups, conceivably arising from a thermal dehydroalumination and rehydroalumination with the expected regioselectivity (see Equation (73)).

$$2\,Al + 3\,R_2Hg \xrightarrow{\Delta} 2\,R_3Al + 3\,Hg \tag{85}$$

The recent synthesis of relatively stable organoaluminum(I) derivatives (Equation (6)) should reawaken interest in the possible mechanism of such displacement reactions. Since all groups are unlikely to transfer from mercury to aluminum in one step, aluminum(I) intermediates (**166**) are likely to be formed and might be detected or trapped in some manner (Scheme 15).

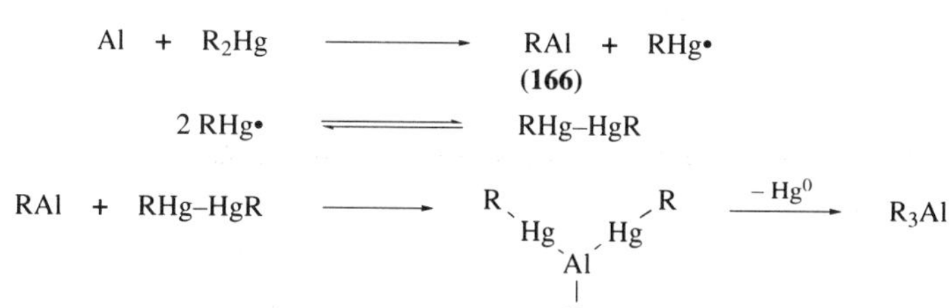

Scheme 15

10.4.3 Aluminum–Hydrogen Bonds

Possibly the most generally applicable and reliable route to a dialkylaluminum hydride is the metathesis reaction given in Equation (86). This method is readily adaptable to producing pure deuterides (**167**) free of D–H exchange products from the alkyl groups (Equation (86)).[183]

$$R_2AlCl + LiD \xrightarrow[-\,LiCl]{} \underset{\mathbf{(167)}}{R_2Al{-}D} \tag{86}$$

A second method of some generality is the redistribution reaction between R_3Al and aluminum hydride, wherein the latter component is generated *in situ* from activated aluminum powder and dihydrogen or is added as the separately prepared $AlH_3{\cdot}OEt_2$ (Equation (87)).[184]

$$2\,R_3Al \;+\; AlH_3 \longrightarrow 3\,R_2AlH \qquad (87)$$
R = alkyl, aryl, 1,2-disubstituted vinyl

A third route to such hydrides is the thermal, selective dehydroalumination of aluminum alkyls having substituents (R^1) at the β position (Equation (88)).[169] A variant of this method, applicable even when $R^1 = H$, is the heating of the aluminum alkyl with dihydrogen under pressure, which reduces (**168**) under catalysis by titanium or chromium salts and drives the equilibrium to the right. This procedure is particularly applicable in industrial practice.[185]

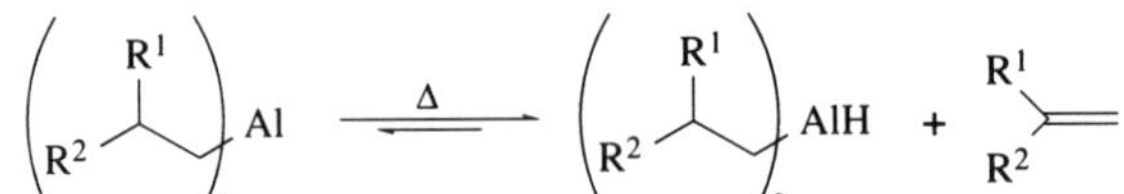

(**168**)

R^1 = alkyl, aryl (88)

10.4.4 Aluminum–Nonmetal Bonds

The most versatile route to establishing bonds between aluminum and halogen, oxygen, sulfur, or nitrogen (E) is the protodealumination reaction of an Al–C bond with appropriate amounts of the corresponding Brønsted acid, $H\text{-}ER^1_x$ (Equation (89)). The ease and vigor of reaction depend greatly on the kinetic acidity of $H\text{-}ER^1_x$: hydrogen halides or alcohols cause rapid reaction at room temperature, while primary or secondary amines require heating. A wealth of derivatives, often as oligomeric complexes, can be generated in this manner. Many of the novel structures exhibited by such mixed organoaluminum derivatives, $R^2_{3-n}Al(ER^1_x)_n$, are surveyed in Section 10.3.3.

$$R^2_3Al \;+\; n\,H\text{-}ER^1_x \longrightarrow R^2_{3-n}Al(ER^1_x)_n \;+\; n\,R^2H \qquad (89)$$

Although a number of specialized and useful ways of making Al–E bonds are available, the only other versatile route to Al–O and Al–X bonds is the redistribution reaction with the salts, $Al(OR^1)_3$[160,186] and AlX_3 (Equations (90) and (91)).

$$2\,Et_3Al \;+\; Al(OEt)_3 \longrightarrow 3\,Et_2AlOEt \qquad (90)$$

$$Me_3Al \;+\; 2\,AlCl_3 \longrightarrow 3\,MeAlCl_2 \qquad (91)$$

Particular Al–O bond formations of great industrial importance arise from the complete oxidation of aluminum alkyls to alkoxides (Equation (92)) and their partial hydrolysis to alkylaluminoxanes (Equation (93)). The former reaction is the basis for the Ziegler process for producing higher straight-chain alcohols from ethene (see Equation (72)); the latter reaction, for the case of R = Me, represents the route to MAO, a most important cocatalyst in Ziegler–Natta polymerizations (Section 10.3.4).

$$R_3Al \xrightarrow{O_2} Al(OR)_3 \qquad (92)$$

$$R_3Al \xrightarrow{H_2O} -\!\!\left[\underset{}{\overset{R}{\overset{|}{Al}}}-O\right]\!\!-_x \qquad (93)$$

10.4.5 Aluminum–Metal Bonds

Of far-reaching importance to organoaluminum chemistry have been both the unambiguous synthesis of an Al–Al bond by Uhl (Equation (5))[32] and the cogent evidence adduced by Olah group suggesting that $AlCl_2$ can be formed from aluminum and $AlCl_3$ at high temperatures and that it may be associated as shown in Scheme 2.[27] The unexpected kinetic stability of tetrakis(dimethylsilylmethyl)dialuminum (**17**), undergoing decomposition only at >219 °C, suggests that previous attempts to dehalogenate R_2AlCl (where $R = Bu^i$) failed because $R_2Al\bullet$ radicals underwent alternative reactions other than coupling (Scheme 16) more readily and not because the Al–Al bond is inherently too weak.

A potentially general route to aluminum–metalloid bonds, such as Al–Si, Al–Ge, Al–Sn, Al–Pb, or aluminum–transition metal (M_t) bonds, is the interaction of an organoaluminum halide with either

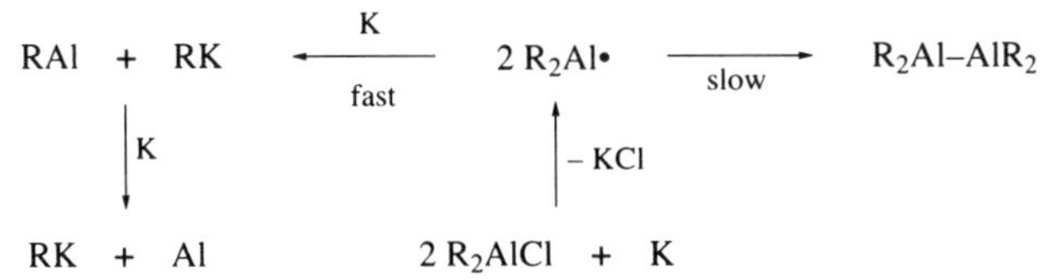

Scheme 16

$R^1{}_3MNa$ or $NaM_t(CO)_n$ (Equations (94) and (95)). Illustrative of this approach is the preparation of $Et_2Al{-}Sn{-}Bu^n{}_3$.[187]

$$R_2AlCl \xrightarrow[-\,NaCl]{NaM_t(CO)_n} R_2Al{-}M_t(CO)_n \qquad (94)$$

$$R^2{}_2AlCl \xrightarrow[-\,NaCl]{R^1{}_3MNa} R^2{}_2Al{-}MR^1{}_3 \qquad (95)$$

10.4.6 Interconversions of Aluminum–Carbon Bonds

10.4.6.1 Preparative scope

The relatively inexpensive, commercially available triethylaluminum and triisobutylaluminum serve as versatile starting materials for preparing a wide variety of other alkyl, vinylic, aryl, and acetylenic aluminum derivatives by way of: (i) exchange reactions between R_3Al and a metal organyl derivative (Section 10.4.2.3); (ii) displacement reactions between R_3Al and alkene; (iii) addition of Al–C bonds of R_3Al to alkenes or alkynes (carbalumination); and (iv) the alumination of carbon Brønsted acids by R_3Al.

Of limited generality for making new C–Al bonds is the dehalogenation of organylaluminum sesquihalides (Equation (96)), which are produced from the reaction of organic halides with aluminum (Equation (97)). However, the method is still especially important as an efficient route to Me_3Al.[188] Likewise, the thermal disproportionation of methylaluminum sesquiiodide is of interest for the preparation of this important aluminum alkyl (Equation (97)).[189]

$$MeAlCl_2 + Me_2AlCl \xrightarrow[-NaCl]{Na} Me_3Al + Al \qquad (96)$$

$$3\,MeI + 2\,Al \longrightarrow MeAlI_2 + Me_2AlI \xrightarrow{\Delta} Me_3Al\uparrow + AlI_3 \qquad (97)$$

Interconversion of Al–C bonds by isomerization is not in general a practical preparative method as much as it is an inconvenience or a frustration for the chemist seeking to obtain one pure organoaluminum isomer. Some of commonly encountered isomerizations are exemplified in Equations (98)–(102) and Scheme 17.[178,190–4]

$$Bu^t{}_3Al \xrightarrow{\Delta} (Me_2CH{-}CH_2)_3Al \qquad (98)$$

(99)

(100)

(101)

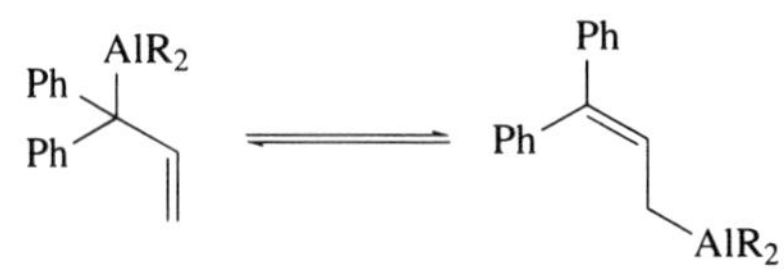

(102)

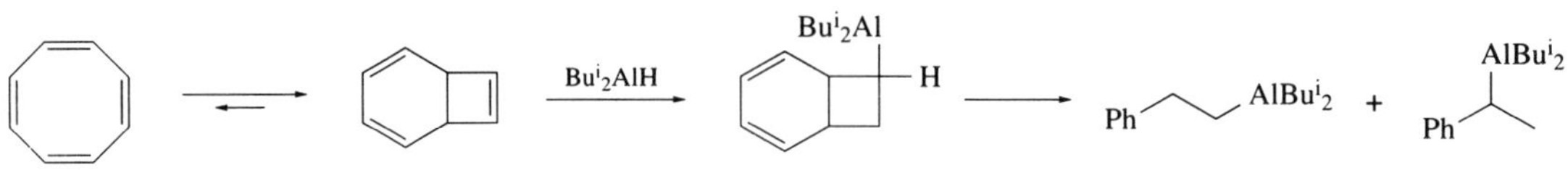

Scheme 17

10.4.6.2 Alkene exchange with aluminum alkyls

Especially with Bu^i_3Al, heating with or without a nickel catalyst sets up an equilibrium with the products of dehydroalumination (Equation (103)). An excess of a more volatile alkene ($CH_2{=}CH_2$, Equation (104)) or a slight excess of a less volatile α-alkene ($CH_2{=}CHC_6H_{13}$, Equation (105)) permits the displacement of isobutene and the preparation of a different R_3Al. In this situation Bu^i_3Al serves as a convenient source of AlH_3.

$$Bu^i_3Al \overset{\Delta}{\rightleftharpoons} Bu^i_2AlH + H_2C{=}CMe_2 \tag{103}$$

$$Bu^i_3Al \xrightarrow[-\,H_2C=CMe_2]{H_2C=CH_2} Et_3Al \tag{104}$$

$$Bu^i_3Al \xrightarrow[-\,H_2C=CMe_2]{H_2C=CHR} (RCH_2CH_2)_3Al \tag{105}$$

10.4.6.3 Carbalumination

The insertion of ethene into aluminum–ethyl bonds is the basis of the original Ziegler Aufbau or chain growth process for preparing mixtures of higher straight-chain α-alkenes or alcohols (C_4–C_{30}) (Scheme 8).[169]

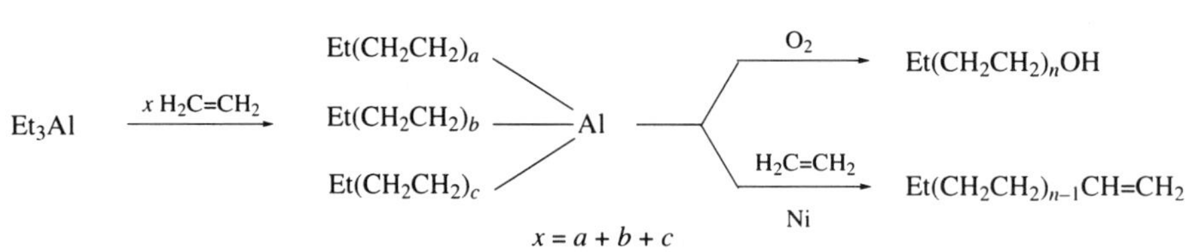

Scheme 18

Insertions of α-alkenes into Al–C bonds proceeds much more slowly and hence a faster, competing dehydroalumination of the adduct usually limits the insertion of one or two units. The catalytic dimerization of α-alkenes by Bu^i_3Al utilizes this difference in insertion and dehydroalumination rates (Scheme 9).

Similarly, ethyne inserts readily into an aluminum–alkyl bond but here the adduct is less reactive to further insertions.[195] This reaction has been developed as a way of synthesizing (*Z*)-1-alkenylaluminum reagents that yield a variety of isomerically pure (*Z*)-alkenes (Scheme 20).[196]

Attempted carbalumination of terminal alkynes leads to moderate or large amounts of the undesired metallated derivative $R^1{-}C{\equiv}C{-}AlR^2_2$. To avoid this side reaction, the silyl derivative, for example, (**169**), is commonly employed. To accelerate the carbometallation, which generally occurs in a *syn*

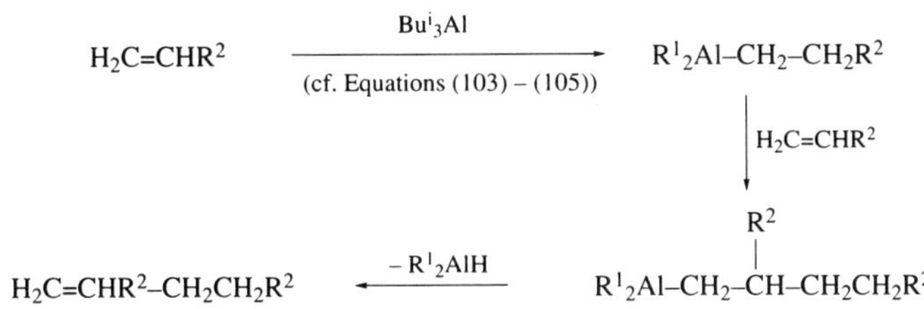

Scheme 19

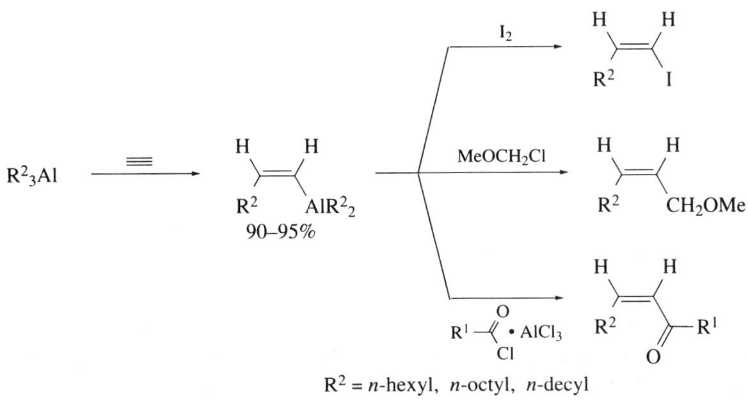

Scheme 20

manner, stoichiometric complexes of aluminum reagent either with a dialkylmagnesium (Equation (106))[197] or with a titanocene or zirconocene component (Equation (107)) can be employed.[198] In both cases the metal in the adduct (M) need not be aluminum.

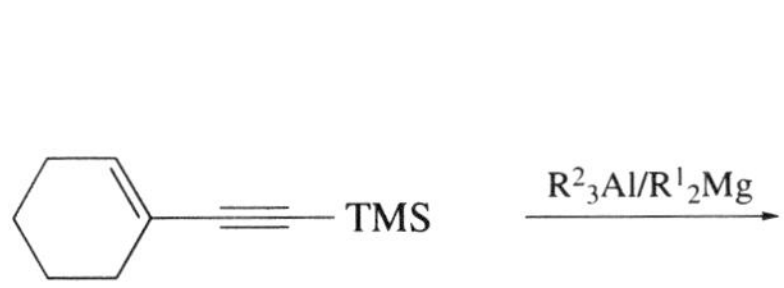

(169) (106)

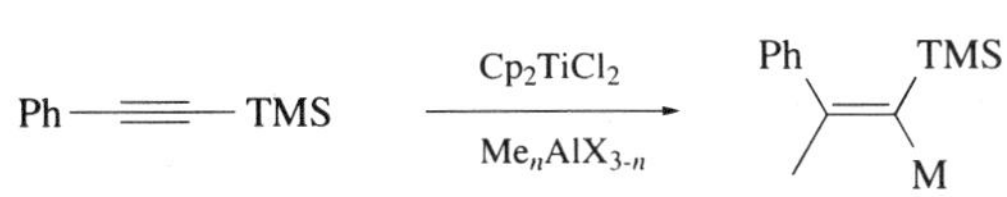

(107)

As is evident from Equations (101) and (108), such carbaluminations can occur intramolecularily and do involve *syn* addition to the π bond.[199]

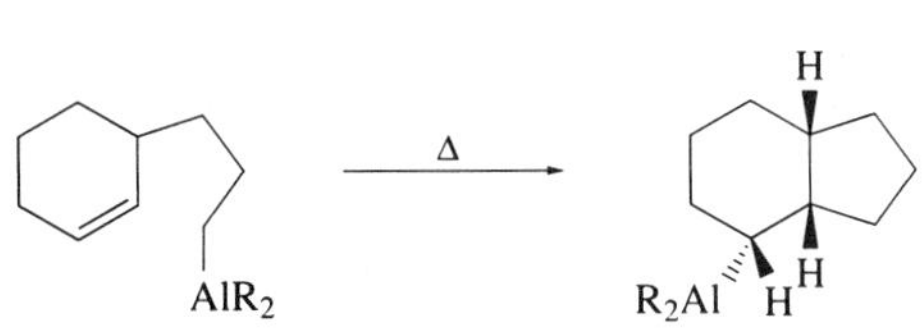

(108)

Two special, interrelated examples of carbalumination involve the insertion of a carbene or carbenoid intermediate into a Al–C bond, resulting in a homologization of the carbon chain. In one case, diazomethane acts as a source of methylene (Equation (109)).[200] The reaction is a methylene analogue of the Ziegler ethene–Aufbau process (see Equation (72) and Scheme 18).

Although methylene insertions occur in a statistical manner with aluminum–alkyl bonds, the behavior of (*Z*)-1-butenyl(diethyl)aluminum (**170**) (Scheme 21) reveals a locoselective insertion into the aluminum–vinylic bond. The allylic aluminum product (**171**) probably exists as equilibrating isomers, similar to the main product (**173**) resulting from the hydroalumination of 1,1-diphenyl-1,3-butadiene (**172**) (Equation 110).[193] In a similar vien, diphenyldiazomethane readily reacts with an aluminum–aryl bond, but the insertion occurs at only one Al–C bond (Equation (111)).[201]

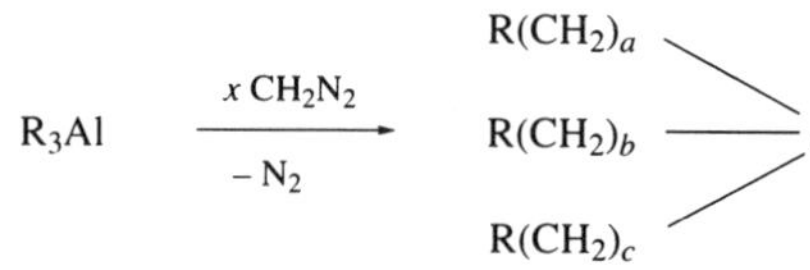

$x = a + b + c$ (109)

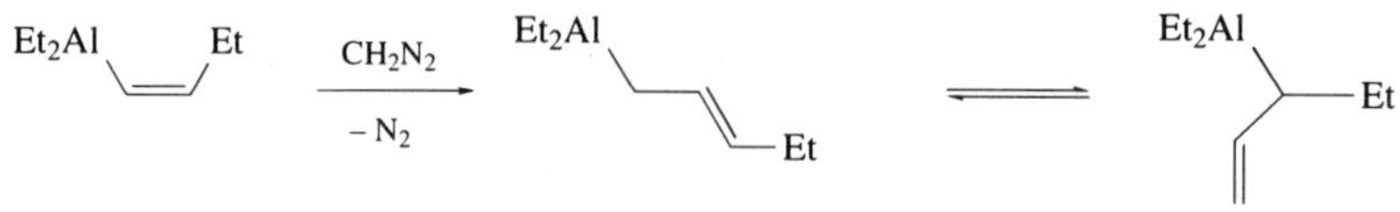

(170) **(171a)** **(171b)**

Scheme 21

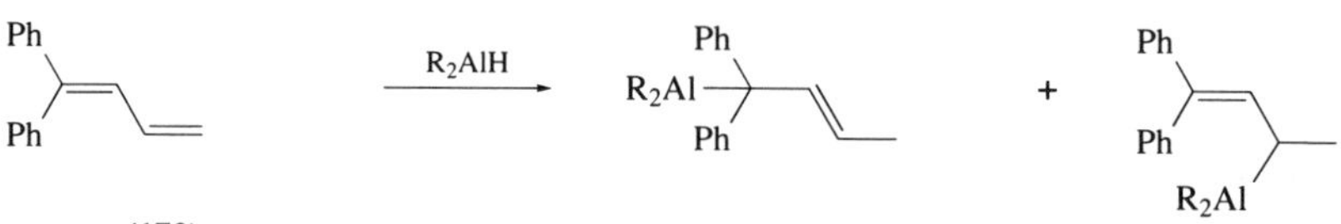

(172) **(173a)** **(173b)** (110)

$$Ph_3Al \xrightarrow[-N_2]{Ph_2CN_2} Ph_2Al\text{–}CPh_3 \qquad (111)$$

The other interesting carbalumination involving a carbenoid precursor occurs with α-halomethylaluminum derivatives (**174**), themselves formed by way of a diazomethane insertion into Al–X bonds (Scheme 22). Derivatives (**174**) can in turn carbaluminate alkenes and lead, via γ elimination, to cyclopropanes.[200]

$$R_2Al\text{–}I \xrightarrow[-N_2]{CH_2N_2} \underset{\mathbf{(174)}}{R_2Al\text{–}CH_2\text{–}I} \xrightarrow{H_2C{=}CH_2} [R_2Al\text{–}CH_2\text{–}CH_2\text{–}CH_2I] \xrightarrow[-R_2AlI]{} \triangledown$$

Scheme 22

10.4.6.4 Alumination of carbon Brønsted acids

Although the metallation of carbon acids is a useful approach to preparing an array of organomagnesium and a large number of organolithium compounds, this approach has only limited scope for organoaluminum derivatives. Terminal alkynes[202] and substituted cyclopentadienes[203] undergo alumination with Ph_3Al at moderate temperatures (Equations (112) and (113)).

$$Ph_3Al \xrightarrow[-C_6H_6]{Ph\text{—}\equiv} (Ph_2Al\text{—}\equiv\text{—}Ph)_2 \qquad (112)$$

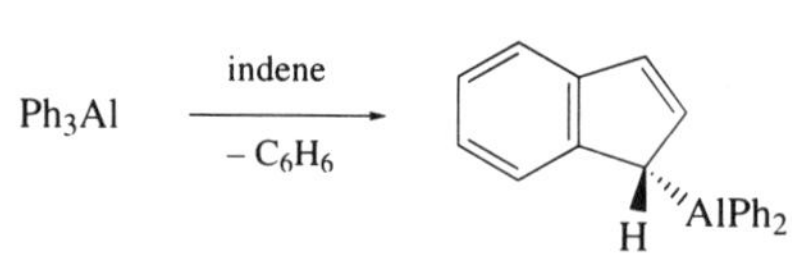

(113)

The decomposition of Ph_3Al over 250 °C appears to involve alumination at the *meta* position, since iodinolysis leads to *m*-diodobenzene (Equation (114)).[204] As might be expected, such aluminations occur more readily when an intramolecular pathway is available (Equation (115)).[205]

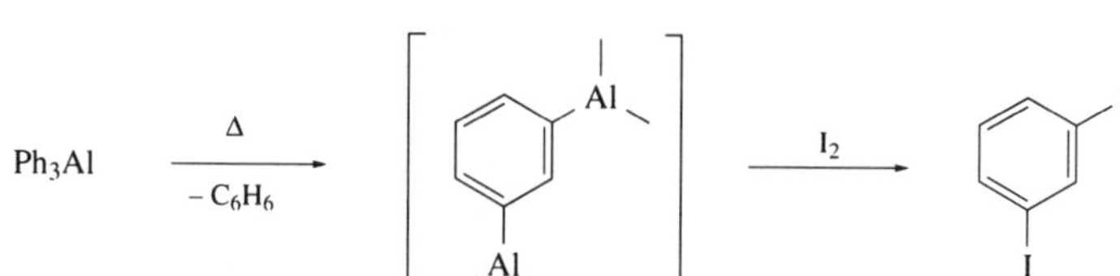

(114)

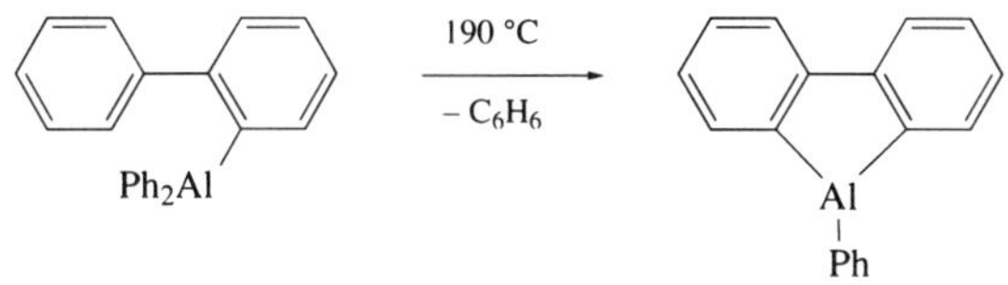

(115)

The alumination of arenes can be achieved in high conversion under milder conditions by employing the R_3Al metallating agent in its "ate" complex form, $Na^+ \; AlR_4^-$. However, the resulting aluminum aryl is also necessarily then produced as an ate complex (**175**). As in lithiation reactions promoted by KO–Bu^t, this alumination is also catalyzed by NaOR (Equation (116)).[206] The triarylaluminum could probably be obtained from the product by careful addition of 0.33 equivalent of finely powdered $AlCl_3$ to a benzene suspension of (**175**).

$$NaAlEt_4 + 4\,C_6H_6 \xrightarrow{NaOR} \underset{\textbf{(175)}}{NaAlPh_4} + 4\,C_2H_6 \qquad (116)$$

10.5 CHEMICAL REACTIONS OF ALUMINUM–CARBON BONDS WITH NONTRANSITION METAL SUBSTRATES

10.5.1 Scope and Relation of Available Reviews

In this section the reactions of aluminum–carbon bonds with substrates not involving a transition metal component are surveyed. Due to their complex course and because of their importance in stereoselective synthesis and polymerization, the reactions of organoaluminum compounds with transition metal substrates are discussed separately in Section 10.6. Moreover, the discussion here is limited to specific reactions of Al–C bonds with nontransition metal substrates. For a discussion of the reactions of organoaluminum reagents involving Al–H, Al–X, Al–O, or Al–N bonds or involving Al–C bonds for an overall stereoselective transformation of the carbon skeleton, Section 10.7 deals with those conversions most valuable in organic synthesis.

Since the late 1960s a number of detailed or specific reviews of organoaluminum reactions have appeared. There are the authoritative treatments of Lehmkuhl *et al.*[207] and of Mole and Jeffrey[208] in 1970 and the chapter in *COMC-I*.[1] A discussion of the reactions of unsaturated organoaluminum compounds by Zweifel and Miller appeared in 1984[76] and a survey of the reactions of the hydroalumination adducts of alkenes and alkynes by Eisch in 1990.[4] Rather narrowly focused reviews of the reactions of unsaturated organoaluminum halides,[209] of diisobutylaluminum hydride,[210] and of the sterically hindered organoaluminum aryloxides and amides developed by the Yamamoto group[211–13] have also been published.

10.5.2 Homolysis vs. Heterolysis as the Prevailing Reaction Mechanism

The extensive variety and broad scope of previous mechanistic studies of organoaluminum compounds support the conclusion that polar reaction pathways involving the heterolysis of the Al–C bond are most often observed. Reactions are thus initiated by electrophilic attack of the Lewis-acidic aluminum reagents, R_3Al (**175**), on an unshared electron pair (**176**), or on the π cloud (**177**) to effect the substitution or addition process (Scheme 23). The evidence drawn from reaction rates, reaction order, solvent effects, and both regio- and stereochemistry are evaluated and discussed in detail in *COMC-I*.[1]

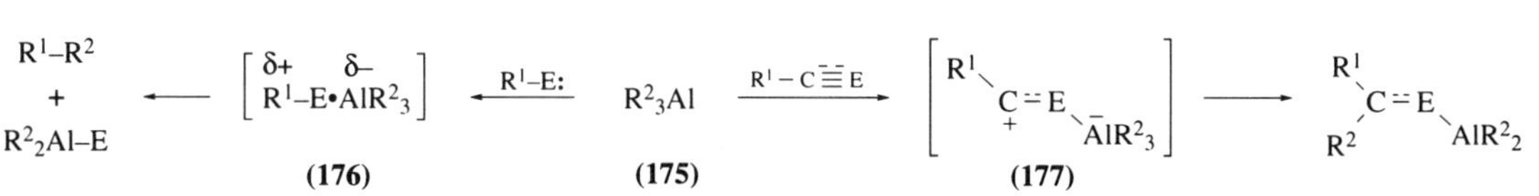

Scheme 23

Homolytic cleavages of the Al–C bond, although observed in only a minority of reactions, are by no means negligible and often lead to a variety of unusual and unexpected products. Reactions conducted with free-radical substrates, such as dioxygen, or under photolytic or extreme thermolytic conditions naturally tend to follow Al–C bond homolysis pathways and generate free-radical intermediates. The important Ziegler ethene growth process for producing higher straight-chain alcohols by final air oxidation of the aluminum alkyls is a case in point (Equation (92)). Likewise, the photochemical dialumination of alkynes by R_3Al (Equation (68)) undoubtedly involves Al–C bond homolysis. The catalytic action of Bu^i_3Al on the reaction of alkyl iodides at 450–600 K with an aluminum-coated silicon surface (Equation (117))[214] and the reaction of 1,3- or 1,4-dihaloalkanes with an aluminum surface (Equation (118))[215] seem to proceed via radical intermediates. A similar conclusion can be reached for the photochemically promoted polymerization of methyl methacrylate by methylaluminum tetraphenylporphyrin.[216]

$$RI + Al{\bullet}Si \xrightarrow{Bu^i_3Al} R_nSiH_{4-n} \qquad (117)$$

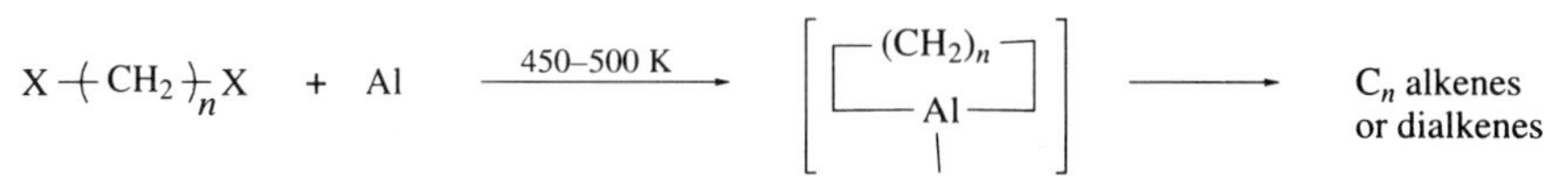
(118)

However, there is also considerable evidence of such free-radical processes occurring under milder, less favorable conditions. Not surprisingly, the reaction of Et_3Al with *t*-butyl peresters occurs by a combination of heterolytic and homolytic paths, while Ph_3Al gave a clean heterolytic reaction (Equation (119)).[217] The reaction of aluminum alkyls with chlorinated hydrocarbons, a process of great importance to industrial safety, has been shown to produce alkyl radicals and to be capable of explosive rapidity.[218] The latter hazard is suggestive of an uncontrolled, free-radical chain propagation. The formation of singlet radical pairs, $[R{\cdot}/{\cdot}CCl_3]^s$, has been established for the interaction of R_3Al with CCl_4 in cyclohexane by CIDNP measurements.[219]

$$Ph_3Al + PhCO_3Bu^t \longrightarrow Ph_2Al{-}O_2CPh + PhOBu^t \qquad (119)$$

Another common mode of radical formation from R_3Al is a single-electron transfer with an electron-accepting organic substrate (EA in Equation (120)). As the EA substrate, quinones,[42,102] tetracyanoethene,[220] and azaaromatic heterocycles[44,99,100] have proved most suitable. With this solid evidence for the occurrence of radicals in R_3Al reactions, a growing importance for homolytic processes in future mechanistic studies can be predicted. Yet, in the following discussion of substitution, addition, and rearrangement reactions of R_3Al, a heterolytic interpretation of the reaction course will be found more widely applicable.

$$R_3Al + EA \longrightarrow R_2Al^{+}{\bullet}EA^{-} + R{\bullet} \qquad (120)$$

10.5.3 Substitution Reactions of Carbon–Heteroatom Bonds

The principal substitution reaction of recent interest is the interaction of a suitable C–O bond with R_3Al; the driving force for this process is the presence of a relatively labile C–O bond, as in an acetal, and the formation of the strong Al–O bond (Scheme 24).[221] Complexation (**178**) and incipient carbenium-ion formation (**179**) seem to be important steps in these cleavages. The recent examples given in Equations (121)–(130), involving orthoformates, acetals, unsaturated ethers, and other derivatives, proceed in good yield in many cases and show that the reaction has considerable utility.[222–30]

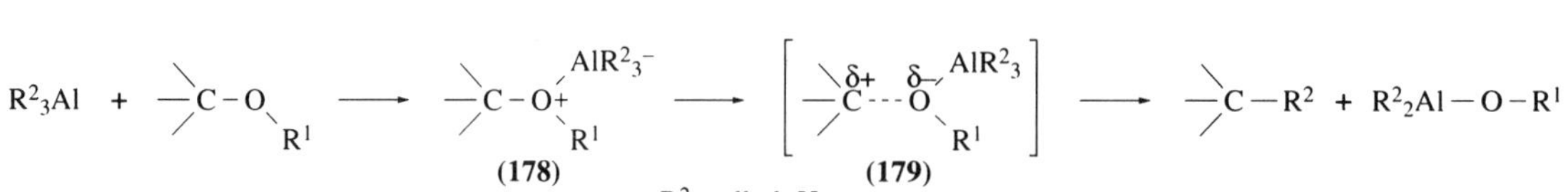
(**178**) (**179**)

R^2 = alkyl, H

Scheme 24

$$HC(OEt)_3 \xrightarrow{Bu^i_2AlH} CH_2(OEt)_2 \qquad (121)$$

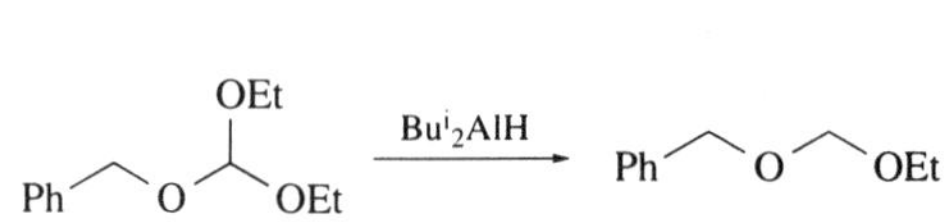
(122)

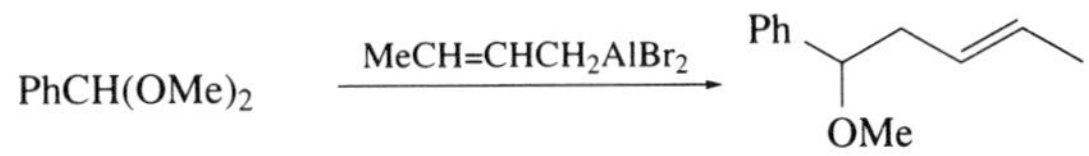
(123)

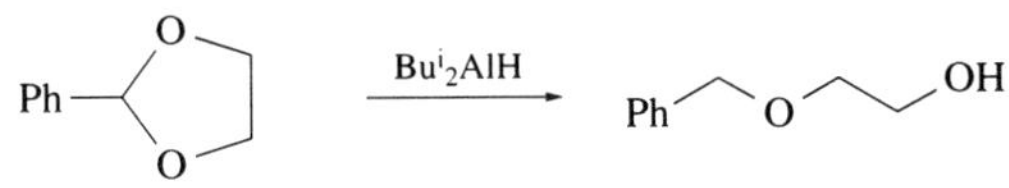
(124)

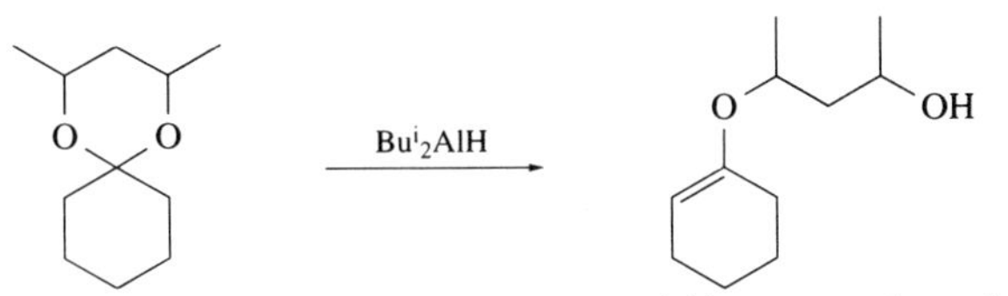
useful in asymmetric synthesis
(125)

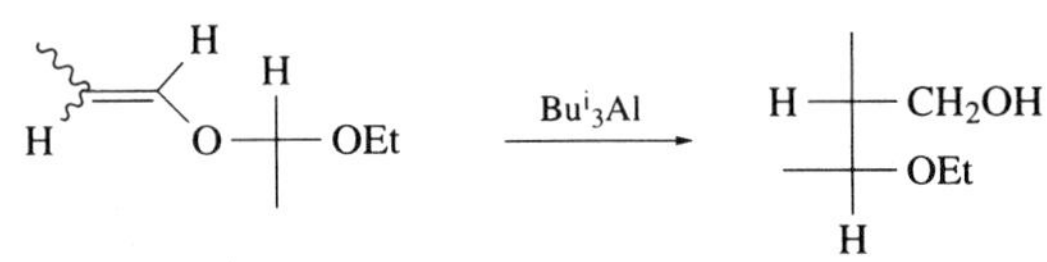
reductive rearrangement
(126)

$$PhCH_2OCH_2CH{=}CH_2 \xrightarrow{Bu^i_3Al/AlX_3} PhCH_2OH \qquad (127)$$

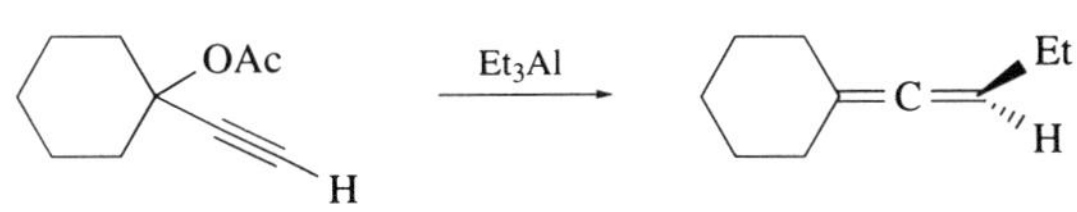
(128)

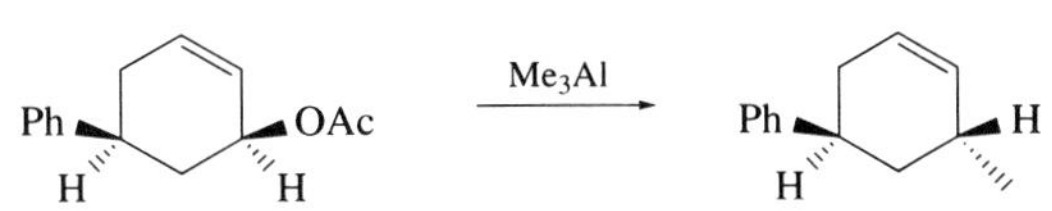
(129)

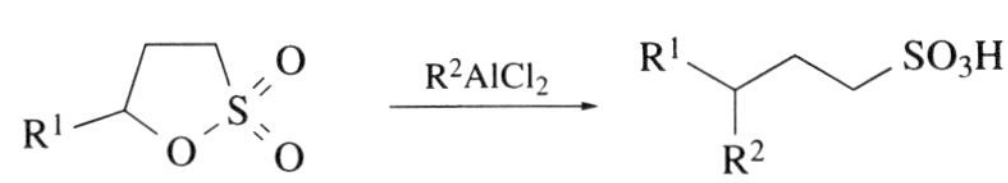
(130)

10.5.4 Addition Reactions to Rings or π Bonds

10.5.4.1 Epoxides

Most related to the foregoing cleavages are the ring opening additions to epoxides. Such reactions are capable of proceeding with high stereoselectivity. Optically active (**180**), for example, gives 94% of (**181**) with Me_3Al (Equation (131)).[231]

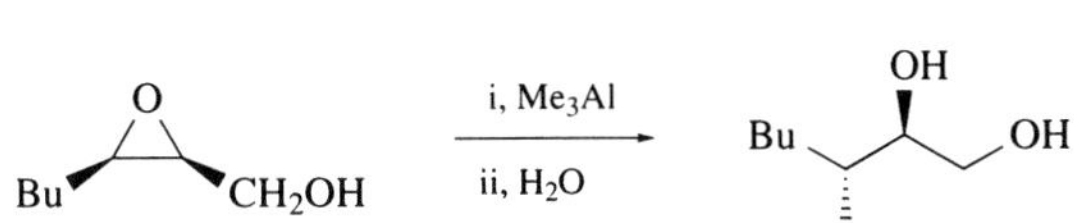
(**180**) (**181**)
inversion of configuration at C-3
(131)

Regioselectivity can also be achieved by use of lithium tetraorganylaluminates, such as $LiAlBuMe_3$, as shown in Equation (132). The product ratio of (**182**):(**183**) was found to be 65:1.[232]

A peculiar cleavage of an α-silyl epoxide, possibly related to some carbenoid cleavages of such systems by RLi and previously reported by Eisch and Galle,[233] is depicted in Scheme 25.[234]

Such epoxide openings can be of great value for coupling rings and of interest to carbohydrate chemists, as illustrated in Equation (133).[235]

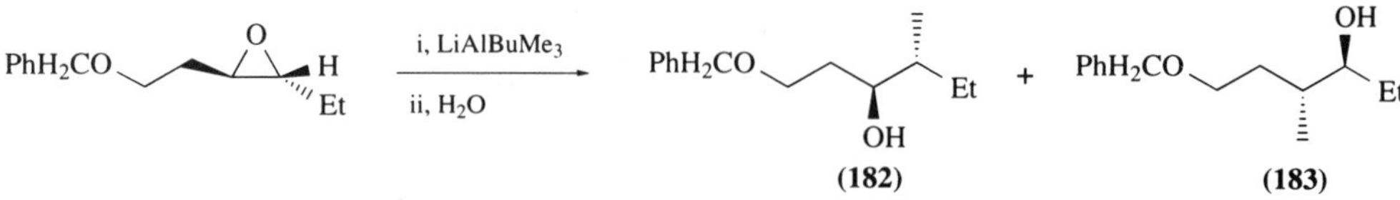

(132)

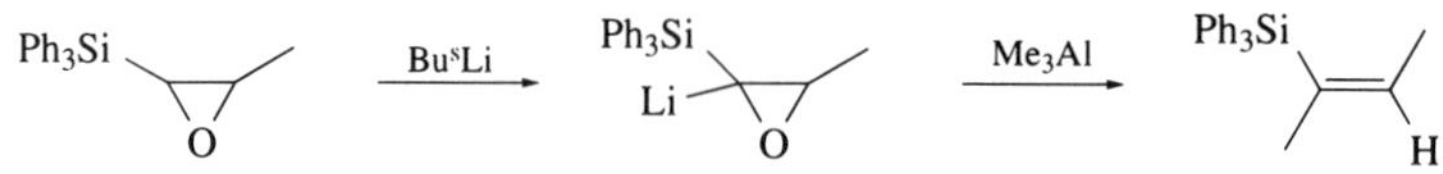

Scheme 25

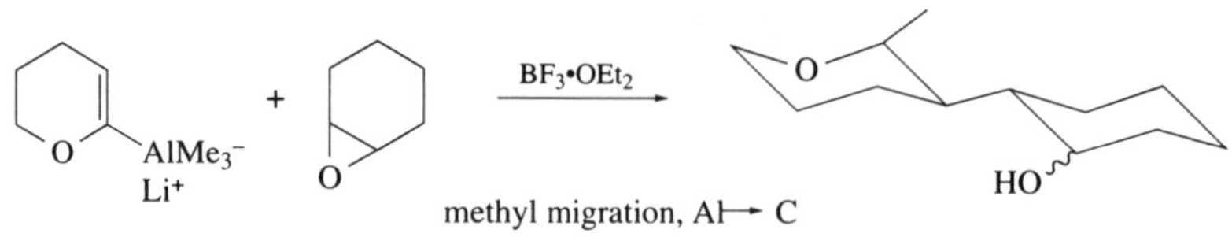

(133)

10.5.4.2 Carbon–carbon double bonded substrates

The additions of Al–C and Al–H bonds to carbon–carbon unsaturation (carbalumination and hydroalumination) are considered in Section 10.4. It remains here only to discuss some recent studies of the mechanisms of such additions. For example, the Ziegler–Rienäcker cyclization[192] of certain α,ω-alkadienes by R_2AlH has been reexamined with 1,5-hexadienes (**184**) and found to involve mono- and dihydroaluminated adducts ((**185**) and (**186**) in Scheme 26).[236] It was found that either etherate formation (**187**) or diadduct formation (**186**) retarded the rate of cyclization. These findings support the view that intramolecular carbocyclization occurs by the tricoordinate aluminum in (**185**) making an electrophilic attack on the distal C=C bond to yield (**188**).[37,38]

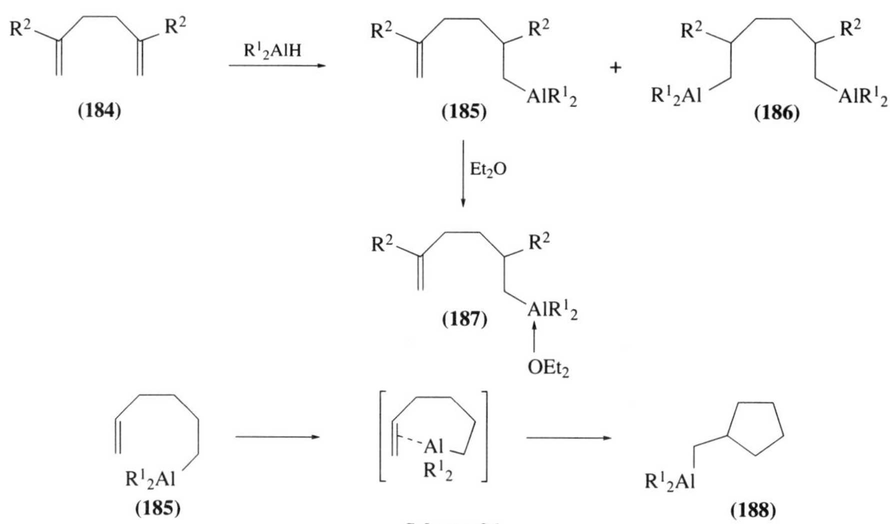

Scheme 26

The kinetics, regiochemistry, and stereochemistry of the hydroalumination of alkenes have been determined for cycloalkenes such as (**189**) and (**190**). The *syn* adducts (**191**)[176] and (**192**)[175] were initially formed but tended to undergo configurational inversion below 100 °C unless the aluminum center was coordinated with ethers or amines (e.g., (**191**) ⇌ (**193**)). Although such adducts were shown by NMR spectroscopy to have only benzylic–aluminum bonds, they gave varying amounts of rearranged products, (e.g., (**195**) and (**196**)) upon treatment with electrophiles (Scheme 27).[175]

Various novel aspects of the hydralumination reaction have been reported. In the reduction of allylic alcohols, such as (**197**), cyclic intermediates of the type (**198**) have been shown to be formed in significant amounts (Scheme 28).[237] The hydroalumination of 1-trimethylsilyl-1,3-alkadynes, such as (**199**), with $LiAlR_2H_2$ shows a locoselectivity for the C-3–C-4 C≡C bond and effects reduction in a *trans* fashion (Equation (134)).[238] Such a reduction thus takes a course completely opposite to that expected with R_2AlH, where the C-1–C-2 C≡C bond is reduced in a *cis* fashion.[239]

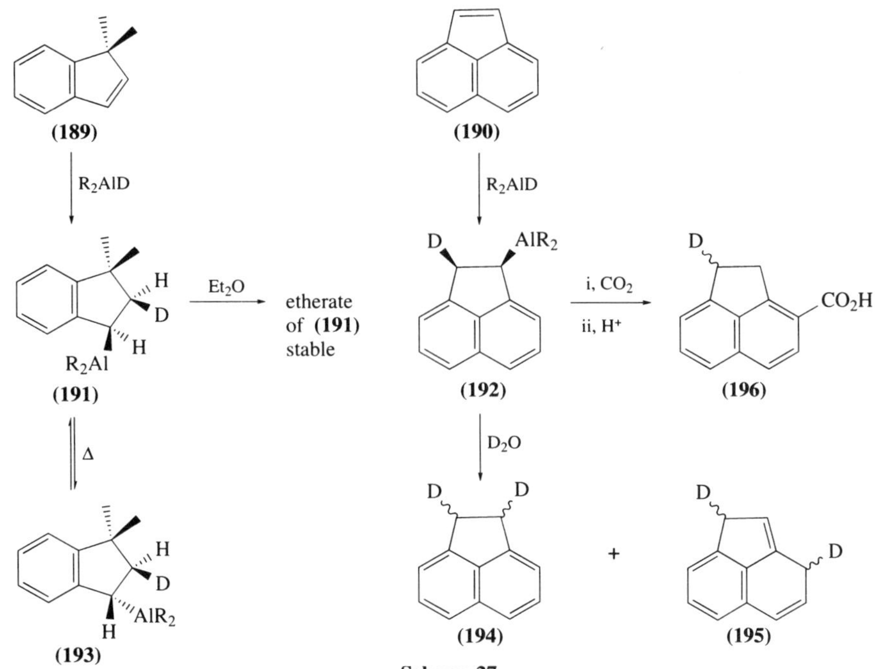

Scheme 27

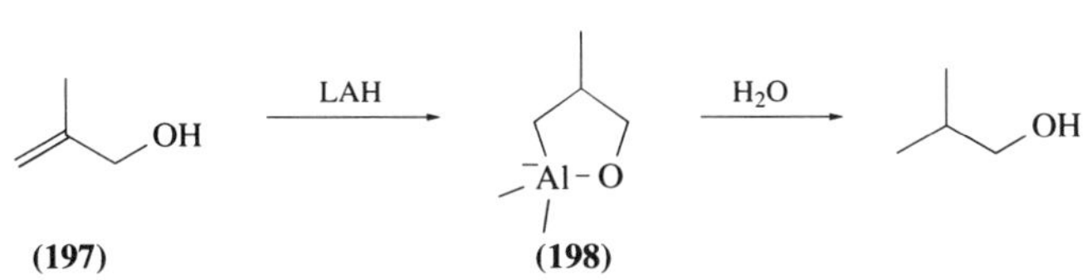

Scheme 28

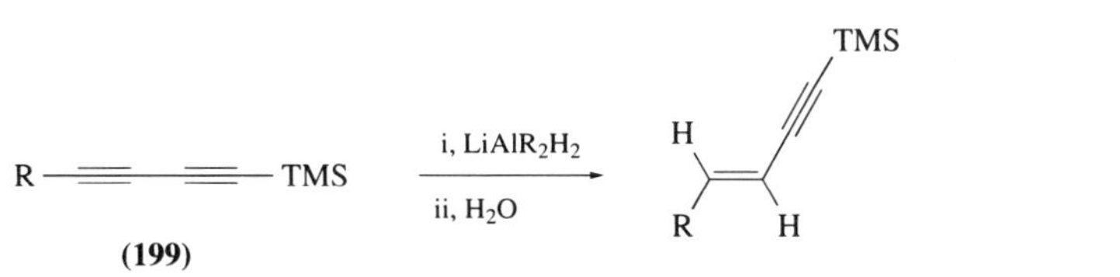

(134)

The hydralumination of vinylsilanes (**200**) with R_2AlH has been known to give a 70:30 ratio of (**201**) and (**202**) at 75 °C.[172] When the reaction is conducted with Bu^i_3Al at 150 °C, the regioselectivity is reversed and (**202**) is now formed in 74% yield.[240] The product ratio at 75 °C was evidently kinetically controlled and at 150 °C thermodynamically controlled (Equation (135)).

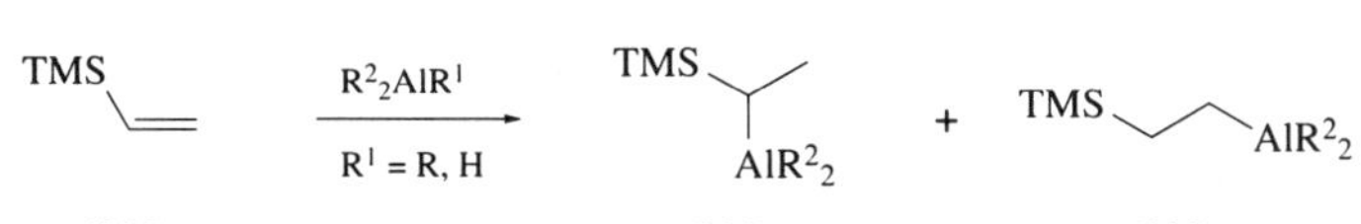

(135)

Finally, the hydroalumination of 1-chloro-1-alkynes (**203**) proceeds in a highly regio- and stereoselective manner to yield adducts (**204**), which constitute a novel class of carbenoid synthons (Equation (136)) for the further synthesis of trisubstituted alkenes of controlled stereochemistry.[241]

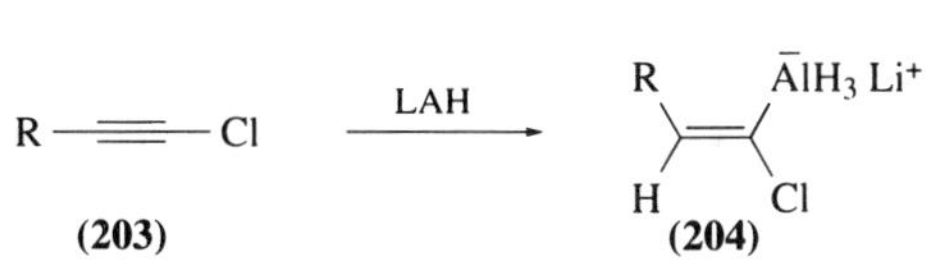

(136)

10.5.4.3 Carbonyl and related substrates

The reaction of organoaluminum reagents with carbonyl substrates has continued to receive much attention since the early 1970s with no sign of abatement. The pronounced oxophilicity of tricoordinate aluminum for the carbonyl group ensures that reaction will occur. However, this reactivity often leads to a variety of desired and undesired reactions, such as principally alkylation (**205**) and reduction (**206**), as well as deoxygenation (**207**), enolate salt formation (**208**), and aldol condensation (**209**), which are exemplified in Scheme 29 with cyclohexanone.[242] The desideratum then for novel aluminum reagents in this area is a high selectivity for each of these individual chemical reactions.

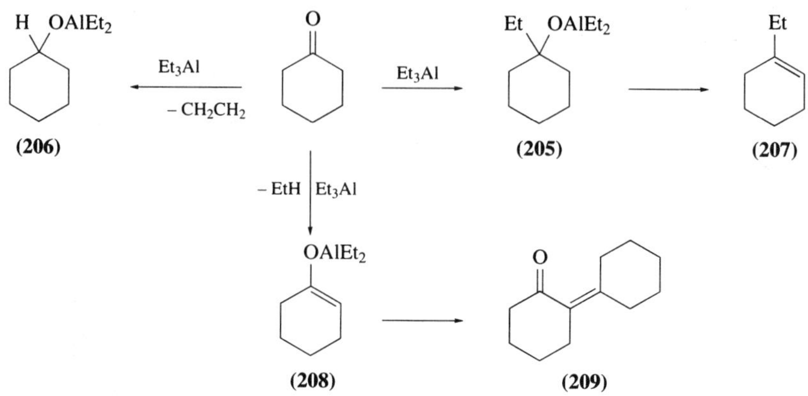

Scheme 29

A number of recent studies have established that the first step in the interaction of R_3Al with the carbonyl group is complex formation. The crystal structure of such a complex with benzophenone (**13**) has proved this view decisively.[15] Further, there is NMR spectral evidence that a carbonyl group may well form a chelating complex with the dialuminomethane (**210**) as a prelude to addition (Scheme 30).[243] Moreover, Me_3Al forms a zwitterionic complex with H_2CO (**211**) that can react with alkenes to produce both an adduct and a homoallylic alcohol (Equation (137)).[244] The process, suitably modified, seems to be of considerable synthetic potential (see Section 10.7.3.5).

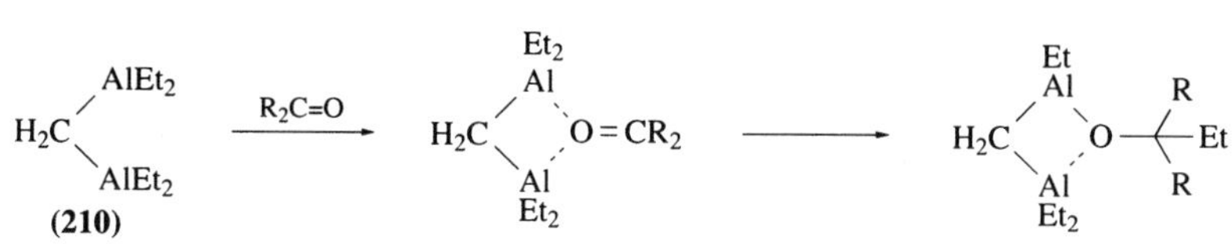

Scheme 30

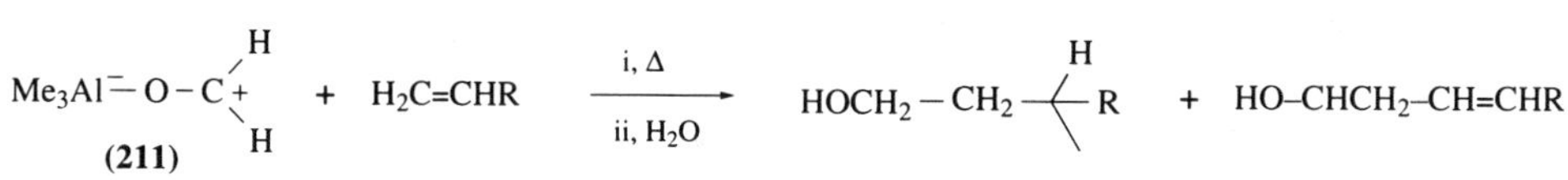

(137)

Pertinent to such complexation and the nature of the transition state for the addition of R_3Al to carbonyl substrates is the behavior of (**191**) (as its etherate) and undeuteriated (**192**). The question at issue has been whether a four-membered (**212**) or a six-membered (**213**) transition state is involved. Considerations of strain and conservation of orbital overlap would tend to favor (**213**) and kinetic measurements are consistent with transition state (**213**). Stereochemical arguments further bolster the case for (**213**): were a four-membered transition state accessible for the concerted addition, then the stereochemically defined (**191**) should add to (**214**) with retention of configuration (**216**). Instead, the addition leads to complete epimerization at C-1 (1:1 mixture of (**216**) and (**217**), Scheme 31).[176] Apparently, the initial organoaluminum–ketone complex (cf. (**211**)) dissociates into an ion pair (**215**), which recombines randomly to form adducts (**216**) and (**217**) and no four-membered transition state plays a role.

The behavior of (**192**) towards ketones bolsters the conclusion that a six-membered variant of (**213**) is the lower-energy transition state. At low temperatures reaction of (**192**) with ketones leads only to adduct (**219**), attainable via transition state (**218**), and none of adduct (**220**), attainable through a four-membered transition state such as (**212**). On warming to 25 °C (**219**) rearranges to (**220**), presumably through dissociation to an ion pair (cf. (**215**)) and its recombination to the thermodynamically more stable adduct (**220**) (Scheme 32).[175]

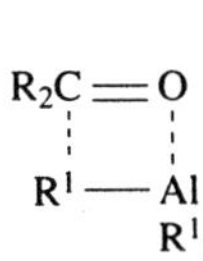

(212)

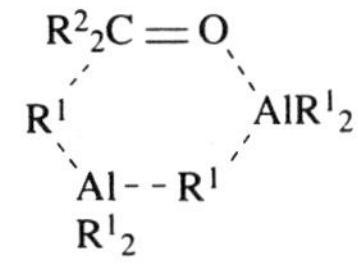

(213)

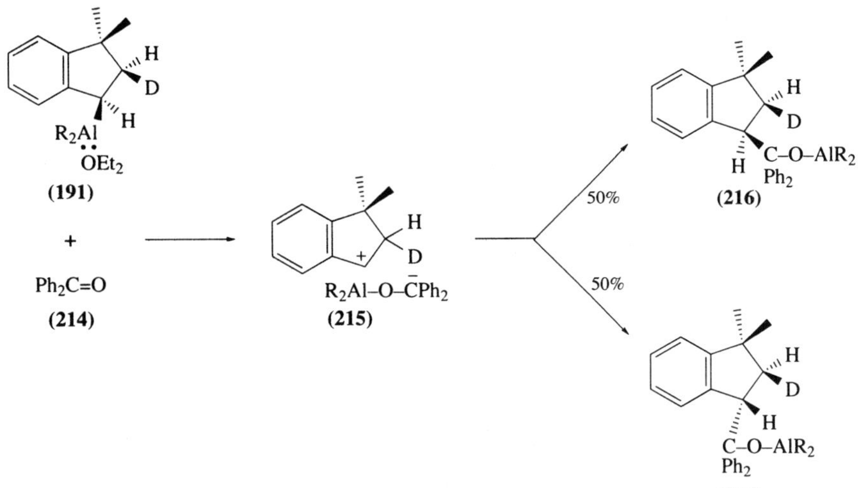

Scheme 31

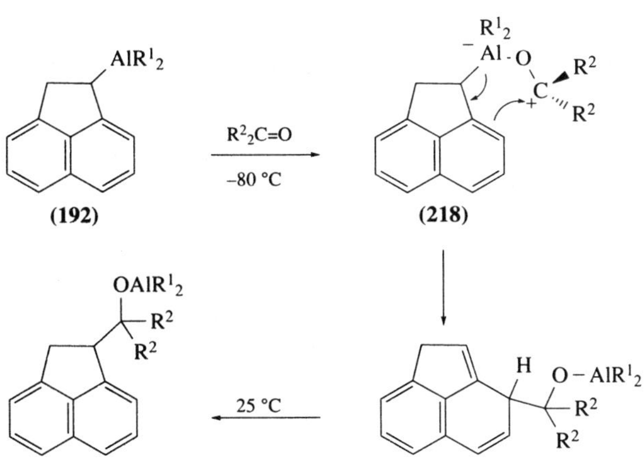

Scheme 32

A similar, interesting variation in the attack of electrophiles (E^+), such as carbonyl compounds and allylic halides, on allylic lithium, boron and aluminum substrates ((**221**) M = Li, AlR_3^-, BR_3^-) has been reported, with the aluminates and borates generally giving more of the proximal attack ((**222**), Equation (138)).[245]

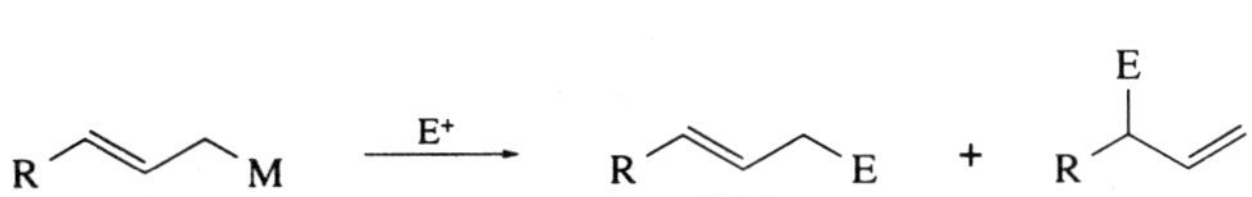

(221) **(222)** (138)

R = RO, RS, RSe, R_3Si

Several groups have developed selective ketone syntheses from such carbonyl addition reactions. Thus lithium tetraorganylaluminates react with acyl chlorides to produce unsymmetrical ketones in 75–90% yield (Equation (139)).[246] Alternatively, mixtures of $R^3{}_3Al$ and R_2AlOR^2 can be substituted for $LiAlR_4$ (Equation (140)).[247,248] A further variant makes use of the reaction of 8-(acyloxy)quinolines with $R^2{}_2AlCl$ or R^2AlCl_2 at low temperatures in CH_2Cl_2 (Equation (141)).[249,250]

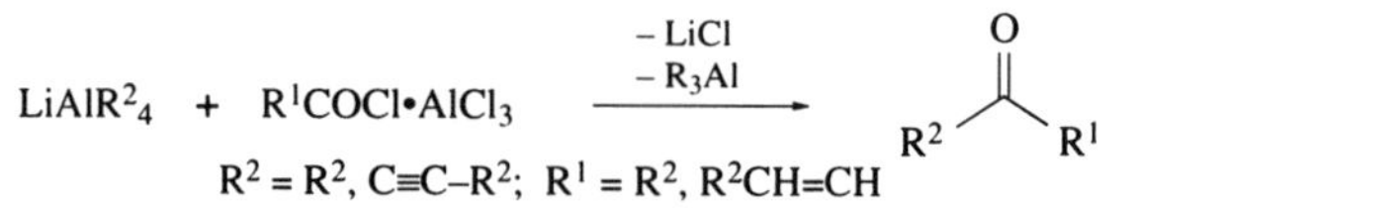

(139)

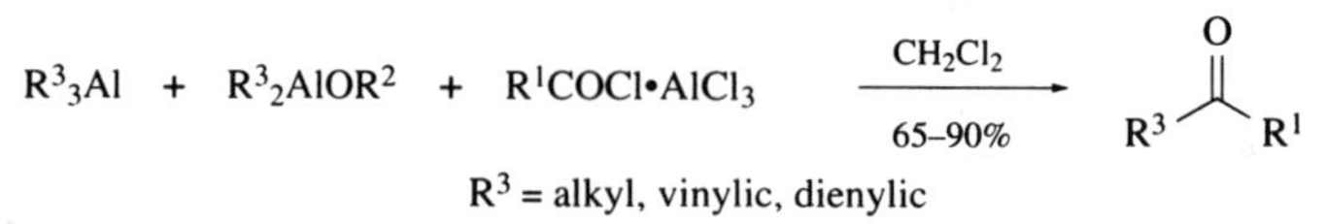

(140)

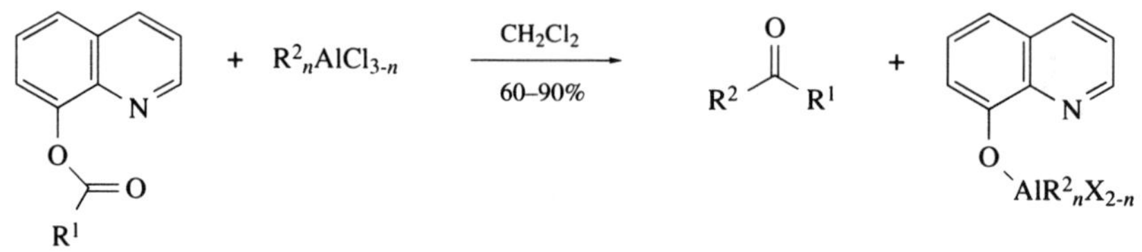

(141)

Selective addition of one type of Al–C bond is often of pivotal value in synthesis. One such group is the 1-alkynyl synthon, which can be selectively added by $R_2Al–C≡C–R$ reagents (Equation (142)).[251,252] Another is the (*Z*)-γ-methoxyallyl moiety delivered by (**224**), which in turn is produced from (**223**) (Scheme 33).[253]

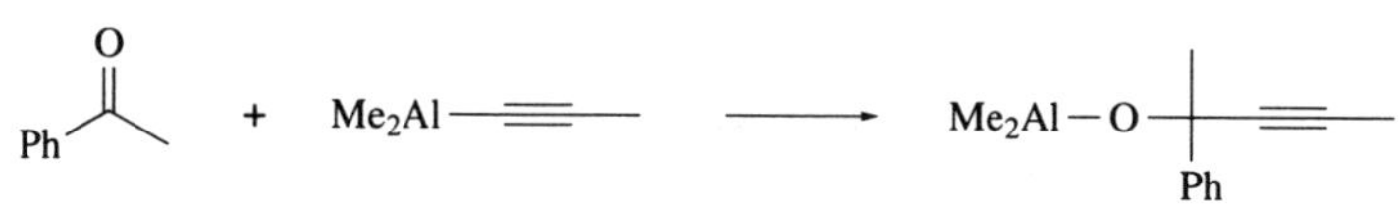

(142)

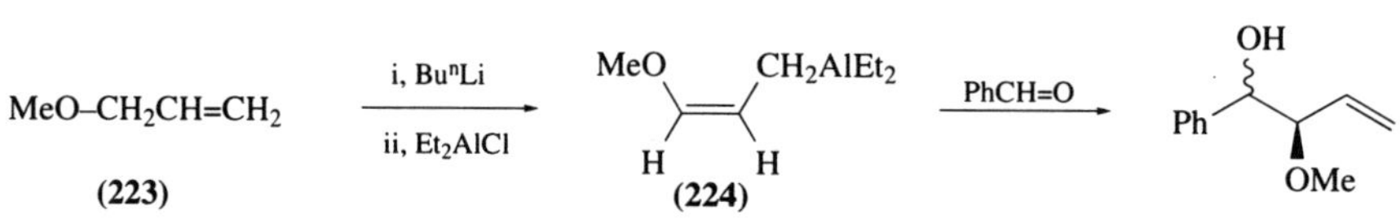

Scheme 33

An aluminum alkylenating agent for ketones has recently been introduced. Equivalent to the Wittig reagent, geminal dialuminomethanes of the type $CH_2(AlR^2R^1)_2$ ((**225**) R^2, R^1 = Cl, Et or Me) are able to convert ketones directly into the corresponding methylene compound. As is evident from Scheme 34, the reaction has considerable scope.[52] Where the appropriate geminal dialuminoalkane is accessible, alkylenation can be achieved with higher carbon units, but the reaction is suitable only with diaryl ketones (Equation 143); with aliphatic ketones aldol condensation and reduction are the dominant reactions (Equation 144).[52]

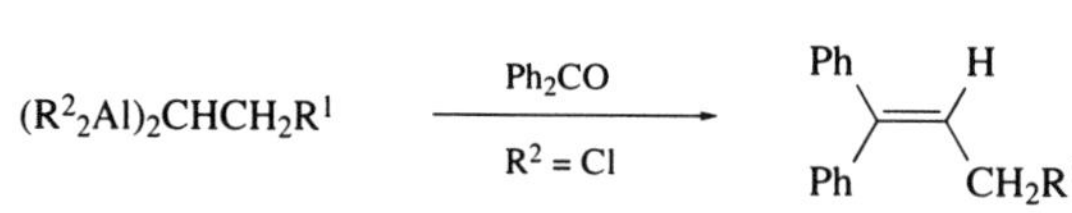

(143)

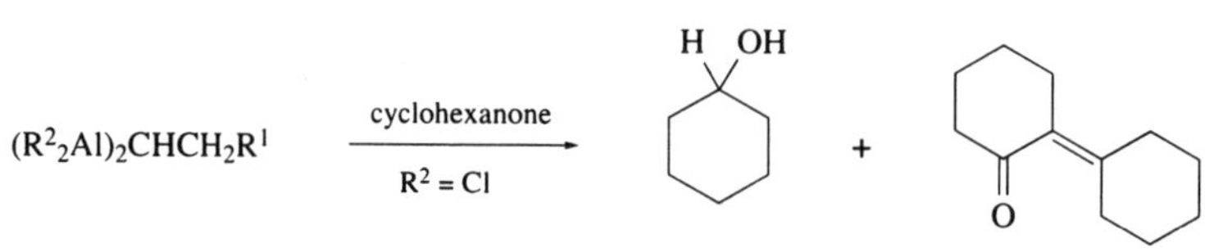

(144)

With α,β-unsaturated ketones the possibility of the 1,4-addition of R_3Al also exists, and such conjugate addition can become the principal mode of reaction. For example, a key step in the synthesis of a sex pheromone is the conjugate addition shown in Equation (145)).[254] Addition of Me_2AlI or Ph_2AlI to enones, such as (**226**), occurs principally in a 1,4-manner (Equation (146)).[255] Similarly, a combination of the bulky $MeAl(OAr)_2$ with RLi or RMgX (**227**) leads to an unusual 1,4-addition of the resulting tetraorganoaluminate complex to α,β-unsaturated ketones such as (**226**).[256] Furthermore, the same complex effects equatorial alkylation of cyclic ketones (**228**) (Equation (147)).[257]

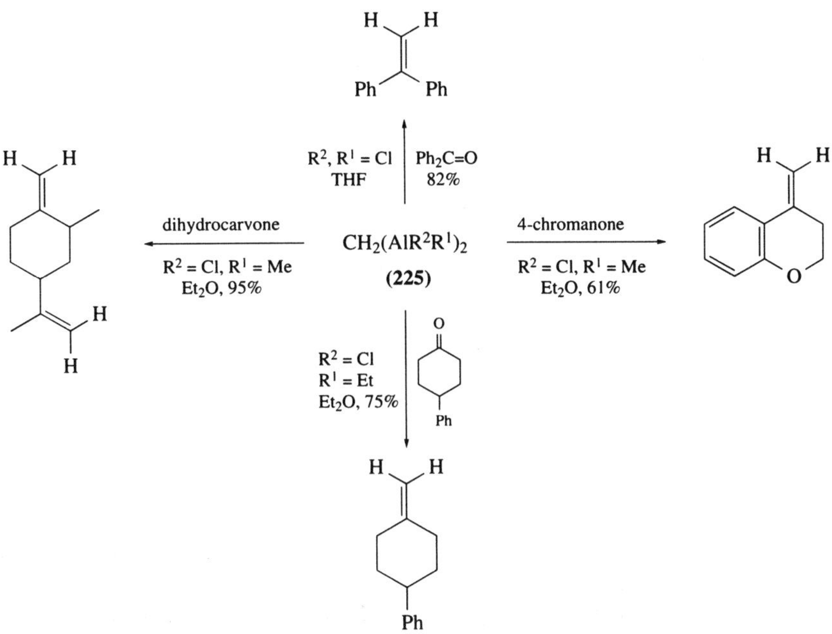

Scheme 34

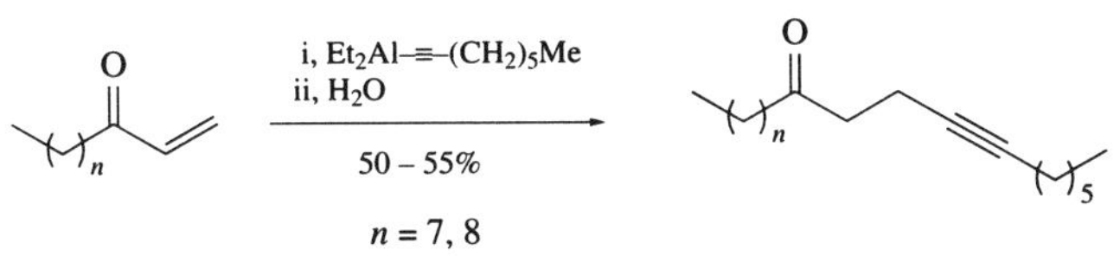

(145)

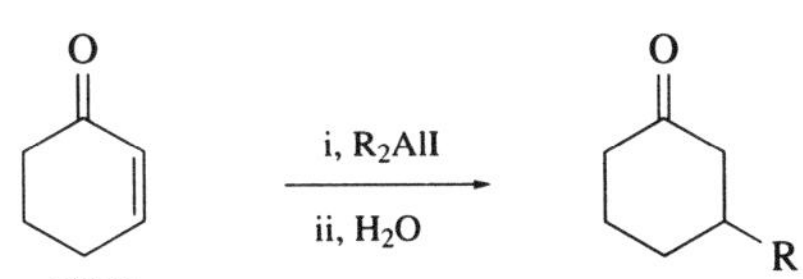

(226)

(146)

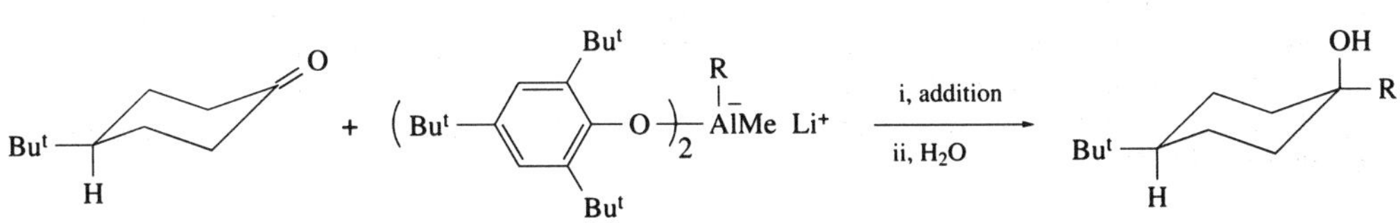

(228) **(227)**

(147)

10.5.4.4 Related A≡B substrates

Other unsaturated substrates add R_3Al or R_2AlH, but often the resulting adducts yield a variety of products. Thus, benzonitrile with tetraalkyldialuminoxanes can give simple alkylation (Equation (148)), but other $(R_2Al)_2O$ give principally alkylated dimers and trimers (Equation (149)).[258] In a similar manner, α-imino ketones (**229**) react with R_3Al to give chelated adducts (**230**) that react further to form dimers (Scheme 35).[259] As with RLi and RMgX, nitrosobenzene reacts with R_3Al to yield several dimeric products (Equation (150)).[260] Despite such reduction of the NO group, α-nitro alkenes undergo smooth, selective conjugate addition of 1-alkenylaluminum reagents (Scheme 36); a 96:4 *cis*:*trans* mixture is formed in >90% yield.[261]

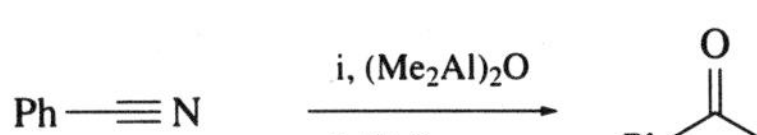

(148)

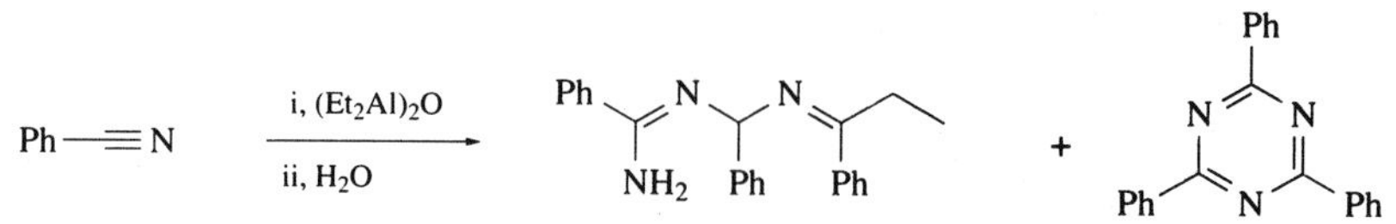
(149)

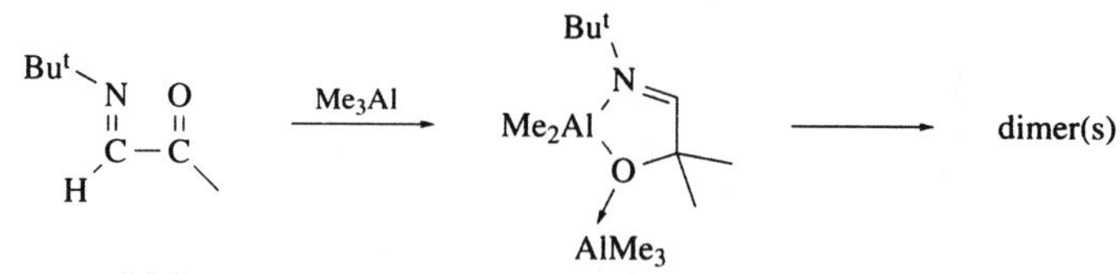
(229) (230)

Scheme 35

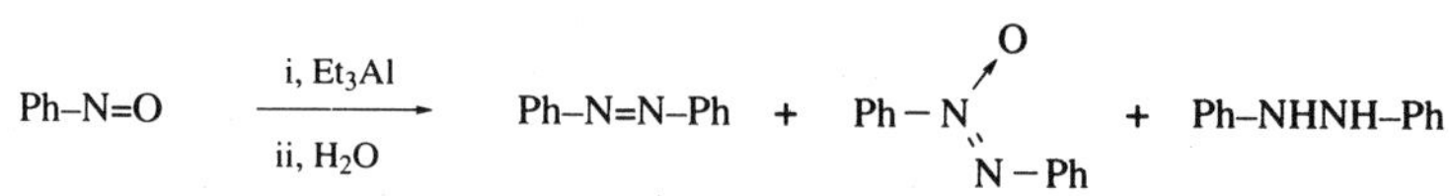
(150)

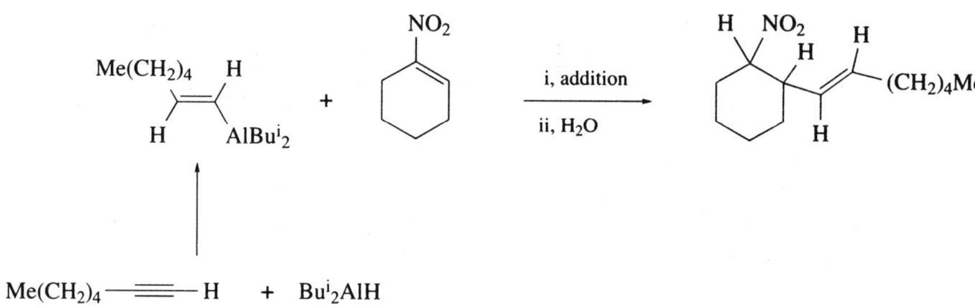
Scheme 36

An interesting variant of adding Al–C bonds to unsaturated substrates is the addition of aluminum–carbon congener bonds to C=O linkages. Two such examples are depicted in Equations (151) and (152).[262,187] Although such additions are curiosities at the present, they may become significant in the future.

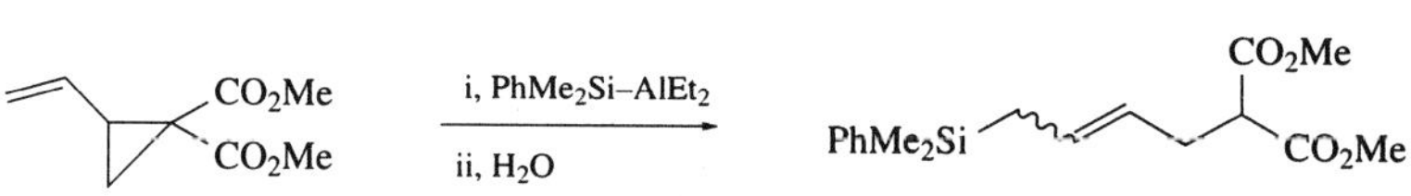
(151)

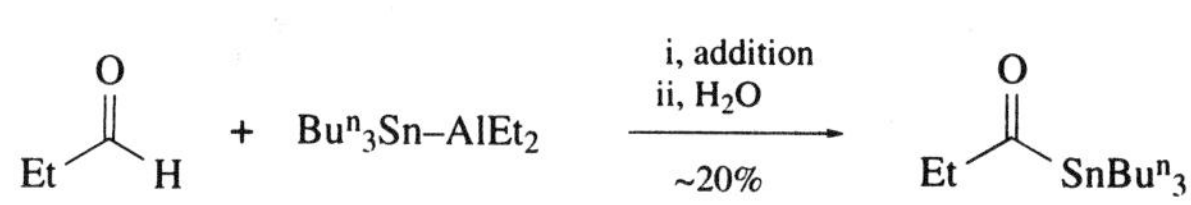
(152)

10.5.5 Rearrangements Induced by R_3Al

Skeletal or C=C bond rearrangements of organoaluminum compounds have already been shown to occur readily during their preparation and reactions, as documented by Equations (98)–(102) and by Schemes 17, 26, and 27. Furthermore, complex rearrangements of interest are: (i) the alkylation of homoallylic esters, such as steroidal tosylate (**231**) (Equation (153)),[263] (ii) the reduction of dihydrofurans (**232**) by Bu^i_3Al (Equation (154)),[264] (iii) the stereospecific 1,2-aryl shift upon treating ketal (**233**) with $LiAlMe_3Bu^n$ (Equation (155));[265] and (iv) the carbocyclization of epoxides, such as (**234**) and (**235**), via an intramolecular ene reaction by methyl bis(4-bromo-2,6-di-*t*-butyl-phenoxide (**236**) (Equations (156) and (157)).[266]

Finally, both the Beckmann rearrangement (Equation (158))[267,268] and the Beckmann fragmentation (Equation (159))[269] have been found to be induced by R_3Al and to be accompanied by an alkylation step. In both these reactions the oxophilic, Lewis-acidic action of R_3Al clearly promotes heterolysis of the N–O bond.

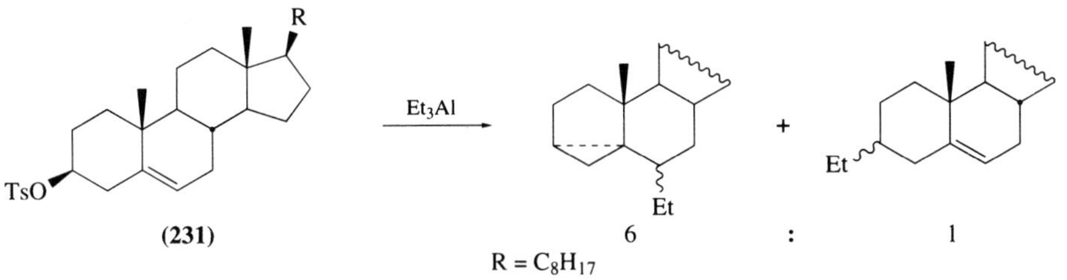

(153)

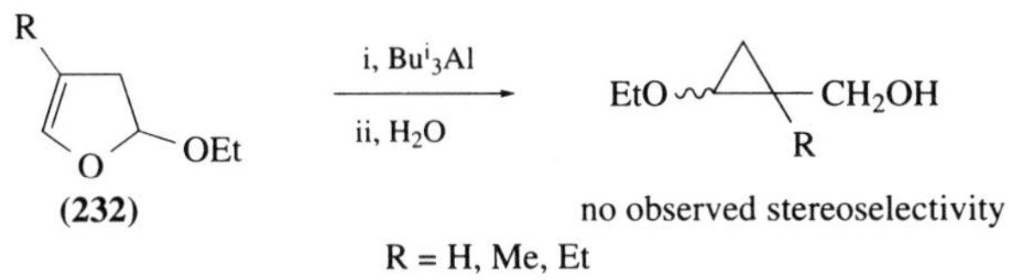

(154)

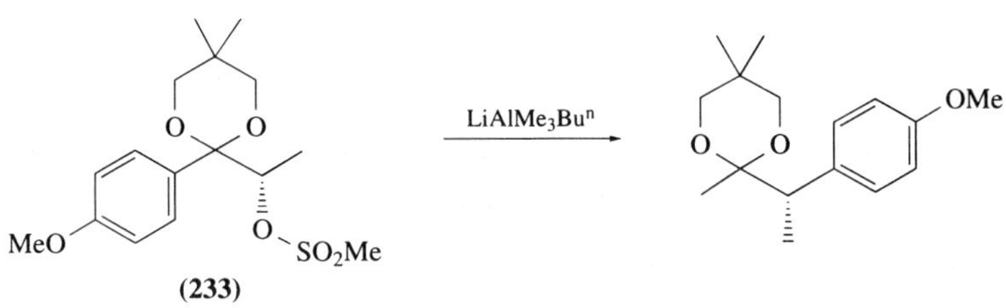

(155)

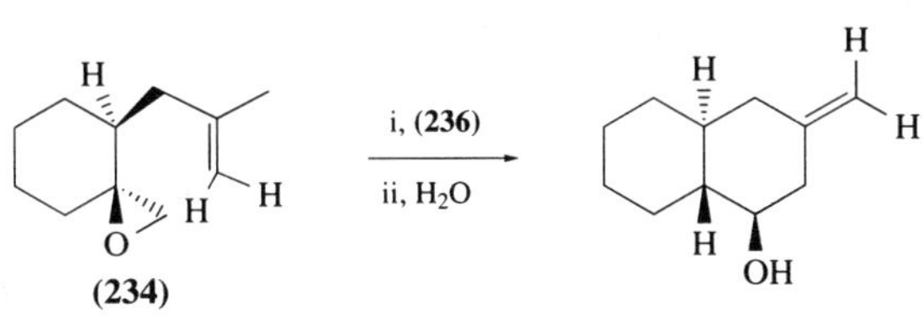

(156)

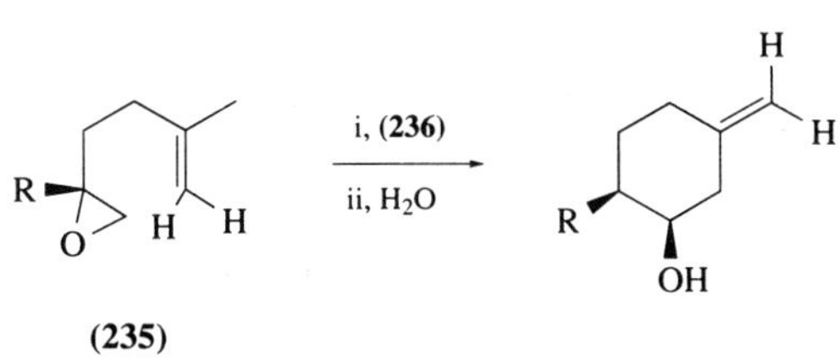

(157)

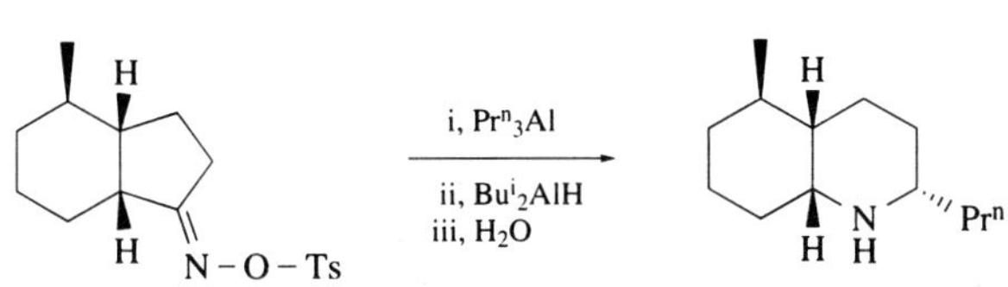

(158)

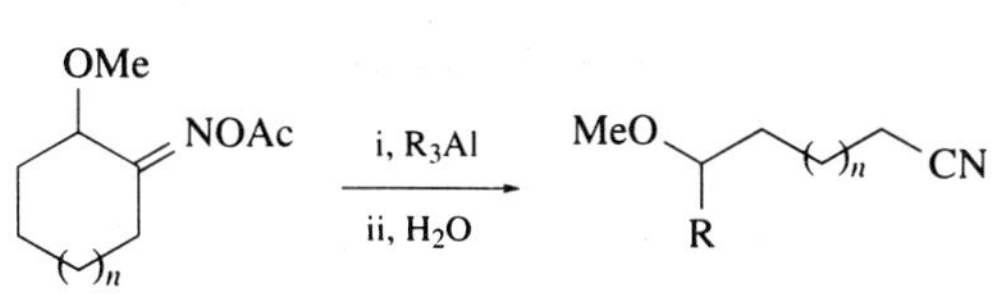

(159)

10.6 INTERACTIONS OF ORGANOALUMINUM REAGENTS AND TRANSITION METAL COMPOUNDS

10.6.1 Background, Significance, and Scope of Such Interactions

10.6.1.1 Historical development

One of the pivotal discoveries, responsible for the astonishing growth since the 1950s of organometallic chemistry in general and of organoaluminum chemistry in particular, was the serendipitous observation of an unexpected reaction between an aluminum alkyl and a transition metal salt. In 1953 Ziegler and Holzkamp reported that the course of the recently discovered "Ziegler alkene growth process" (Equation (160)) was drastically changed by the presence of traces of nickel salts (Scheme 37). In this situation, only one ethene unit was inserted to form aluminum–butyl units **(237)** and thereafter AlH_3 was transferred to a new ethene unit with the liberation of 1-butene. The "alkene growth process" was turned thereby, in an unforeseen and disconcerting way, into an "alkene displacement process" or a catalytic dimerization of ethene to 1-butene. This surprising phenomenon has come to be known as the "Ziegler nickel effect."[270–3]

$$Et_3Al \xrightarrow{x\ H_2C{=}CH_2} [Et(CH_2CH_2)_a][Et(CH_2CH_2)_b][Et(CH_2CH_2)_c]Al \qquad (160)$$

$$x = a + b + c = 1\text{–}15\ H_2C{=}CH_2 \text{ units}$$

$$Et_3Al \xrightarrow{3\ H_2C{=}CH_2} \underset{\mathbf{(237)}}{Bu^n{}_3Al} \xrightarrow[3\ H_2C{=}CH_2]{\text{Ni salts, } -Et_3Al} 3\ EtCH{=}CH_2$$

Scheme 37

Since this profound effect of nickel on an organoaluminum reaction was both exciting and mysterious, Ziegler and his group set out both to understand the molecular basis of the nickel effect and to survey what effects other transition metal salts might exert on organoaluminum reactions. Research on the mechanistic aspects has been underway since the early 1950s and a fundamental insight is now beginning to emerge. However, extraordinary success was enjoyed almost immediately by Ziegler's group when they carried out an empirical evaluation of the effect of transition metal salts on the ethene–triethylaluminum reaction (Equation (160)). With $TiCl_4$ or $ZrCl_4$ present in the reaction mixture, they found that the alkene displacement process was not promoted but to their further astonishment (and undoubted delight), the alkene growth process was extraordinarily accelerated. Under these experimental conditions, however, not tens of ethene units were inserted into the metal–carbon bond, as with Et_3Al alone, but tens of thousands (Equation (161)).[60]

$$Et_3Al \xrightarrow[TiCl_4]{x\ H_2C{=}CH_2} [Et(CH_2CH_2)_x]_nM \qquad (161)$$

$$x = 10^5 \text{ (av.)}$$
$$M = Ti, Al$$

The resulting hydrocarbon, after hydrolysis to remove metallic salts, proved to be linear high-density polyethene having an of average molecular weight of 3×10^6. This unexpected production of polyethene of high density and melting point under such mild experimental conditions unleashed a torrent of research on whether such combinations of aluminum alkyls and transition metal salts would also polymerize α-alkenes and 1,3-dienes. The subsequent discoveries by Natta and his group that propene could be polymerized either randomly to atactic polypropene or stereoselectively to either isotactic or syndiotactic polypropene by means of various titanium(IV), titanium(III), or vanadium(IV) salts in combination with aluminum alkyls brought the interest of the international chemical industry to fever pitch. Such Ziegler–Natta catalysis, as it has come to be termed, has been found to be applicable to both the polymerization and oligomerization of diverse vinyl monomers, alkynes, and 1,3-dienes. It can be appreciated that all these discoveries in the area of Ziegler–Natta catalysis were made by trial-and-error experimentation, without an understanding on a molecular basis of how the transition metal and the aluminum components interact to bring about the catalytic effect. As with the Ziegler nickel effect, research on the fundamental basis of Ziegler–Natta catalysis has been under way since the early 1950s but has only led to reliable mechanistic insights since the mid-1980s.

10.6.1.2 Scope of relevant transition metal compounds

From the foregoing background, it is understandable why the interactions of aluminum alkyls with the salts of both early transition metals (titanium, zirconium, and hafnium as well as those of groups 3 and 5) and late transition metals (nickel, palladium, and platinum, as well as those of groups 8, 9, and 11) have been the subject of much research. An electronic rationale for the particular suitability of early transition metal salts in Ziegler–Natta catalysis would be that these metals as neutral (**238**)[274,300] or cationic centers (**239**) would be stronger Lewis acids and thus complex firmly with alkenes (**240**).[71,275] Furthermore, such a metal center would be a more reactive electrophile and thereby promote the rearrangement of (**240**) into (**241**). The role of the aluminum component in this phase of the reaction is then that of a Lewis acid which removes either neutral donor molecules •D (THF, OEt_2, etc.) or an anion A^- (X^-, R^-, RO^-) and opens up a coordination site on the transition metal M_t for the alkene ((**242**) and (**243**), Scheme 38). It is clear, therefore, why the $R^2{}_nAlR^1{}_{3-n}$ is usually an essential component in the active Ziegler–Natta catalyst.

$Cp_2Ti{-}R$ (**238**) $\quad$ $Cp_2\overset{+}{M}{-}R$ (**239**)

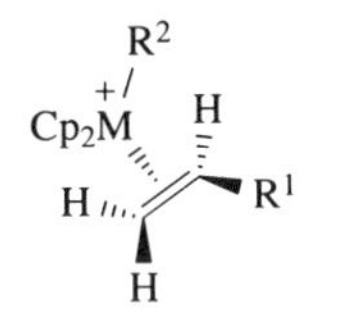

(**240**) M = Ti, Zr $\quad$ $Cp_2\overset{+}{M}{-}CH_2CHR^2R^1$ (**241**)

$$\underset{(242)}{R^2{-}M_tD_{n-1}} \xleftarrow{R^2{-}M_tD_n} R^2{}_nAlR^1{}_{3-n} \xrightarrow{R^2{-}M_tA_n} \underset{(243)}{R^2{-}\overset{+}{M_t}A_{n-1}}$$

R^2, R^1 = alkyl, X, OR

Scheme 38

The catalytic activity of late transition metal salts in the Ziegler nickel effect and in other alkene and alkyne oligomerization reactions is now known to be due to the formation of subvalent transition metal complexes, such as Ni^0, Pd^0, Co^0, Rh^1, and Fe^0. The role of the aluminum component in these processes is to reduce the transition metal salt to the necessary subvalent state by successive alkylation and M_t–C bond homolysis (Equation (162)).[272] Once the subvalent complex is produced (e.g., (**244**)), it is capable of producing its catalytic action in the absence of an aluminum component. The disproportionation of 1,4-cyclohexadiene ((**245**), Equation (163)) is such an example.[273] Nevertheless, such reactions are often cocatalyzed by organoaluminum Lewis acids, such as R_nAlX_{3-n}.[276]

$$Ni(acac)_2 + 2\,Et_3Al \xrightarrow[-\,Et_2Al\,acac]{} [Et_2Ni] \xrightarrow[-\,2\,Et\bullet]{2\,cod} \underset{(244)}{(cod)_2Ni} \qquad (162)$$

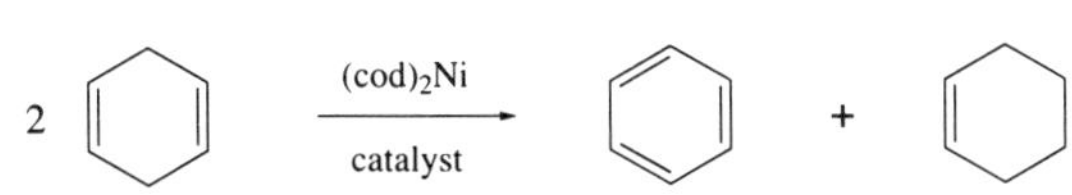

(**245**) $\qquad$ (163)

The preponderance of research on aluminum alkyl–transition metal salt interactions has been done with compounds of titanium, zirconium, vanadium, nickel, and cobalt, although a few studies involving hafnium, tantalum, iron, palladium, and copper have appeared recently. The next section discusses the scope and variety of the observed reactions between R_3Al and such transition metal salts.

10.6.1.3 Survey of the reactions of transition metal salts with organoaluminum compounds

The most general reaction observed with almost all transition metal salts is a metal–metal exchange leading to a transition metal–carbon bond (Equation (164)). How kinetically stable the M_t–C bond is in (**246**) is strongly dependent on the nature of the ligands (L), the oxidation state n, the charge of M_t, and the presence of complexed $R^1{}_2AlE$. Without such stabilizing features, most M_t–C bonds undergo facile homolysis by thermal or photochemical promotion (see Equation (162)). The formation of methyltitanocene chloride (**247**) from titanocene dichloride is illustrative of such metal–metal exchanges: the equilibrium lies further to the right as n in Me_nAlCl_{3-n} is successively 1, 2, and 3 (Equation (165)).[181] The Ti–C bond is stabilized against homolysis by the donor Cp ligands and by the complexation between (**247**) and (**248**) which lends some cationic character to the titanium center.

$$L_xM_t^{n}\text{–E} + R^1{}_2\text{Al–}R^2 \xrightarrow[E\,=\,X,\ OR,\ acac]{} \underset{(\mathbf{246})}{L_xM_t^{n-z}\text{–}R^2} + R^1{}_2\text{Al–E} \qquad (164)$$

$$Cp_2TiCl_2 + Me_nAlCl_{3-n} \rightleftharpoons \underset{(\mathbf{247})}{Cp_2TiMeCl} + \underset{(\mathbf{248})}{Me_{n-1}AlCl_{4-n}} \qquad (165)$$

This latter factor represents the second important reaction between $L_xM_t^{n}$–E and $R^2{}_2Al$–R^1, namely a Lewis base–Lewis acid heterolysis of the M_t–E bond (Scheme 39). Such heterolysis can be partial, leading to a bridging or contact ion pair (**249**); or such heterolysis can be complete, producing a solvent-separated ion pair (**250**). Examples of both types of ion pairs, established by NMR spectroscopy or by XRD, are offered in Equations (166)–(169).[71,277,278] Structure (**251**) is further informative in showing how remarkably the positive charge stabilizes the Ti–C bond against homolysis.

$$L_xM_t\text{–E} + R^2{}_2\text{Al–}R^1 \rightleftharpoons \underset{(\mathbf{249})}{\overset{\delta+}{L_xM_t}\cdots\cdots E\cdots\cdots \overset{\delta-}{AlR^2{}_2R^1}} \rightleftharpoons \underset{(\mathbf{250})}{L_xM_t^{+} \parallel {}^{-}AlER^2{}_2R^1}$$

Scheme 39

(166)

(167)

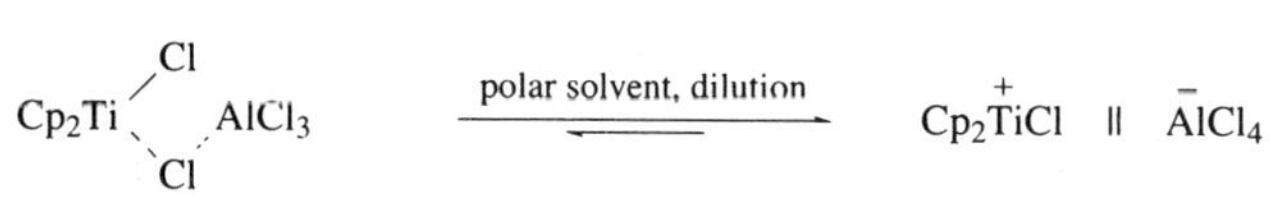

(168)

(169)

(**251**)

The third way in which M_t–E and $R^2{}_2Al$–R^1 can interact is by reduction, which is favored for those cases where (**246**) is too unstable and undergoes rapid homolysis (Equations (162) and (170)).[279] It is noteworthy that the usually labile R_2Ni intermediates (see Equation (162)) are greatly stabilized by chelating ligands (Equation (171)).[280]

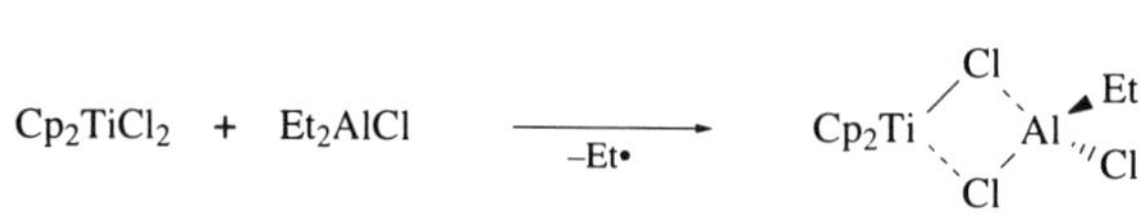

(170)

A fourth course of reaction is an alumination of the methylene carbon adjacent to a positively polarized metal (**252**), which situation appears to be the decisive factor in the formation of the Tebbe reagent (**253**) from Cp_2TiCl_2 and Me_3Al (Equation (172)).[281,282] Related to this interaction but instead involving alumination of the carbon β to the metal is the reaction of zirconocene dichloride with Et_3Al.

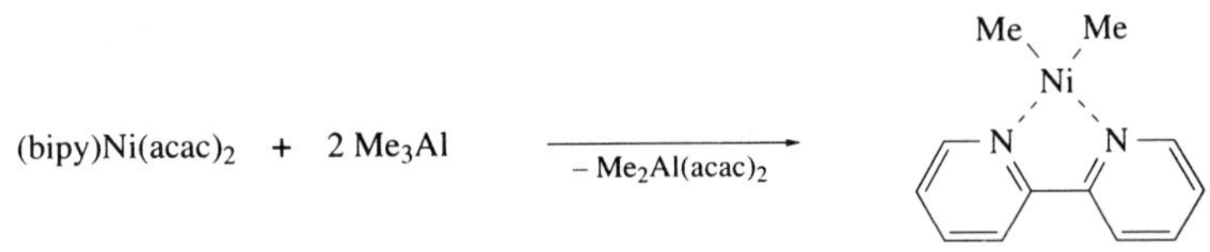

(171)

The remarkable products (**255**) and (**256**) have been fully characterized by XRD and are presumed to form by way of intermediate (**254**) (Equation (173)).[283–5] It is likely that β alumination in (**254**) is fostered by an agostic interaction between $Zr^{\delta+}$ and the β hydrogen.

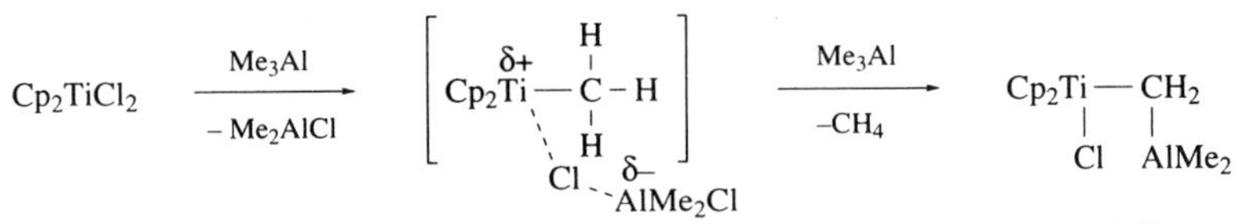

(**252**) (**253**) (172)

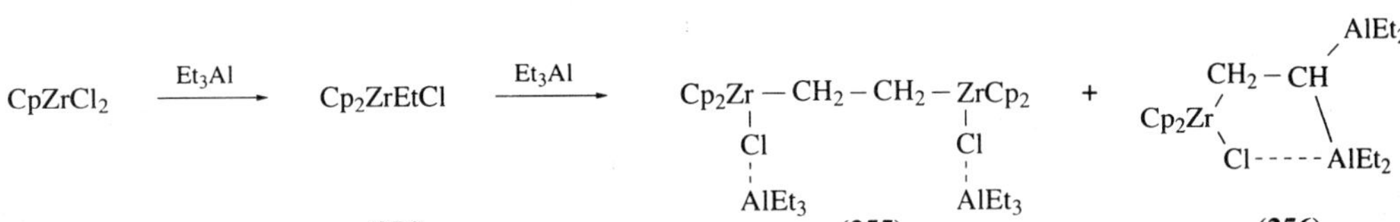

(**254**) (**255**) (**256**) (173)

A fifth type of reaction, namely the metal-catalyzed elimination of aluminum hydride from aluminum alkyls of the structure $(R^2R^1CH–CH_2)_3Al$, may also result from an agostic interaction of M_t and the β hydrogen of an alkyl chain. As illustrated for the Ziegler nickel effect, there are two modes by which such an agostic bonding might be exerted: (i) complexation of the aluminum alkyl with a nickel(0) center which act as a Lewis base ((**257**), Scheme 40); and (ii) oxidative addition of the R_3Al to nickel(0) (**258**), followed by a Ni–H bond elimination and a reductive elimination (Scheme 41). In the latter pathway all reactions are considered to be reversible. Which mode of reaction is the principal path has not yet been established but there is now no doubt that nickel(0) is the active oxidation state.[272,273]

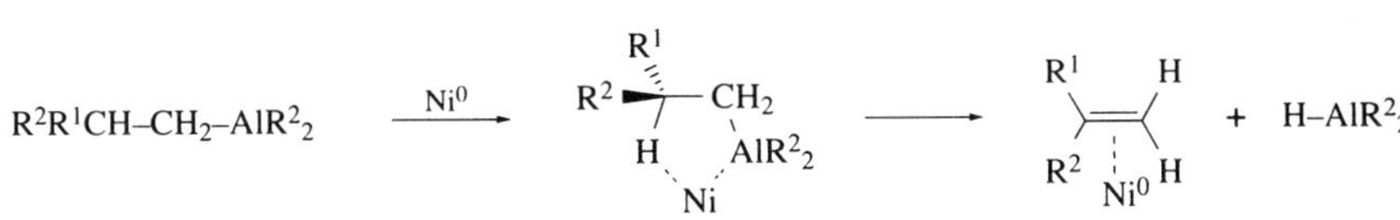

(**257**)

Scheme 40

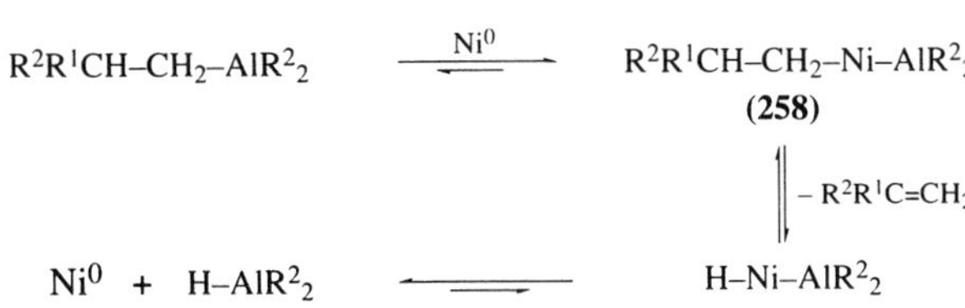

Scheme 41

A sixth reaction occurs between Al–H bonds and a subvalent M_t complex. The reaction can be complex but it leads to dihydrogen evolution and the formation of M_t–Al bonds. Two examples of this process are given in Equations (174) and (175).[273,286]

$$(H_2C{=}CH_2)_3Ni \xrightarrow[<-70\,°C]{NaAlHR_3} Na[R_3AlH{\bullet}Ni(H_2C{=}CH_2)_2] \xrightarrow{CO} 2\,Na^+\,[Ni_6(CO)_{12}]^{2-} \qquad (174)$$

$$(cod)Ni(bipy) \xrightarrow[-H_2]{LAH} [Li(THF)_nAlH_2{\bullet}Ni{\bullet}bipy]_x \qquad (175)$$

A seventh reaction is the cleavage of M_t–M_t bonds by $R^2{}_2Al–R^1$ reagents. Cleavage of (**259**) by Et_2AlCl gave the paramagnetic nickel(I) compound (**260**), together with its complexes with Et_2AlCl (Equation (176)).[287]

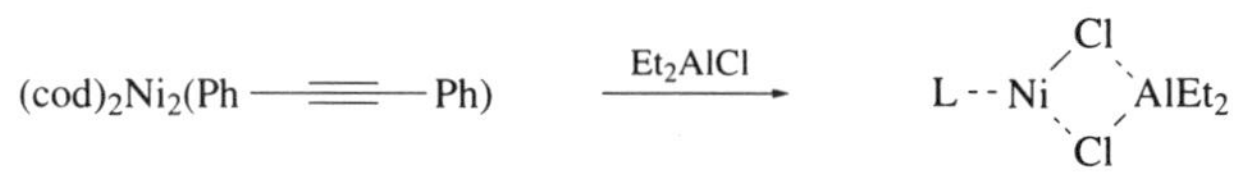

(259) (260) (176)

An eighth reaction variant is the addition of an alkyl or hydride group from the aluminum alkyl to a metal cationic complex onto a site other than the metal (Equation 177).[288]

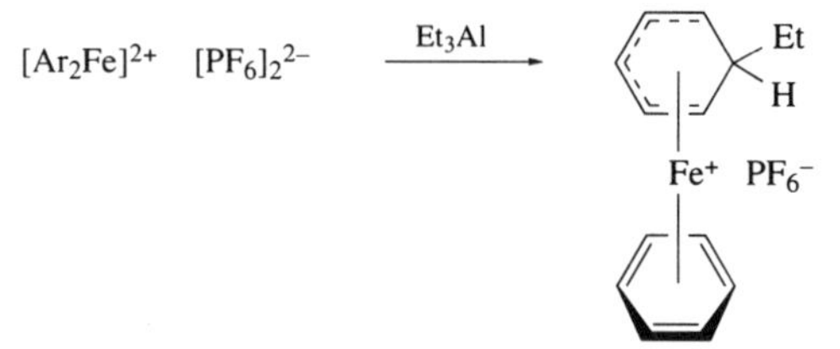

(177)

A ninth reaction exists in two forms, an aluminum alkyl-assisted, transition metal carbometallation and a transition metal-assisted carbalumination. The former process is the key chain propagation step in Ziegler–Natta catalysis (Equation (178)),[289–92] while the latter is the basis for the zirconocene-catalyzed carbalumination of alkynes (Equation 179).[293]

$$Cp_2TiMeCl{\bullet}MeAlCl_2 \xrightarrow{x\ H_2C{=}CH_2} Cp_2Ti[(CH_2CH_2)_x]MeCl{\bullet}MeAlCl_2 \qquad (178)$$

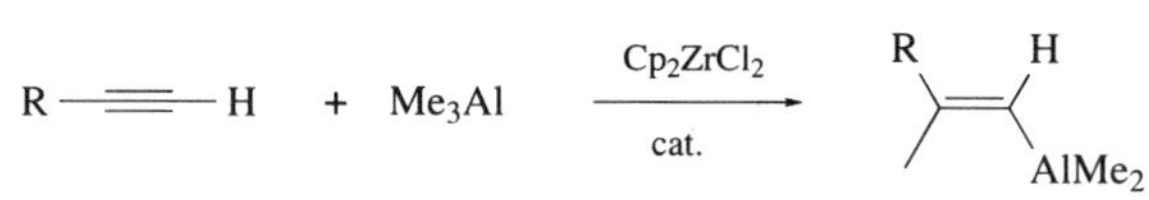

(179)

A tenth mode of reaction is the cleavage of a metallocene-1,3-alkadiene complex (**261**), in equilibrium with metallocycle (**262**), with an aluminum halide to produce (**263**) (Scheme 42).[294]

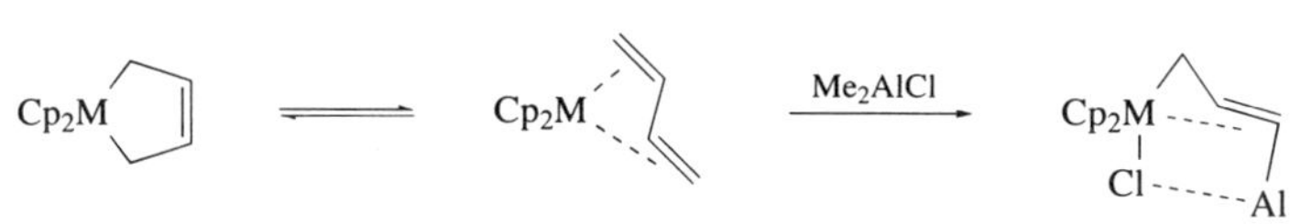

(**262**) M = Zr, Hf (**261**, *trans*) (**263**) Me$_2$

Scheme 42

The richness of the modes of interaction possible between aluminum alkyls and transition compounds assures that future research in this area will remain strong and that novel aspects of catalyzed C–C and C–H bond formation will continue to emerge.

10.6.2 Mechanisms of Important Reactions Promoted by Transition Metal–Aluminum Alkyl Combinations

10.6.2.1 Ziegler–Natta catalysis

Despite the years of industrial research, from the early 1950s, and the thousands of barely distinguishable patents on Ziegler–Natta catalysts,[61] only since the early 1980s have significant strides been made in establishing the mechanisms and stereochemical basis of such polymerization. This recent progress has largely issued from investigations of the behavior of soluble metallocene dihalides or dialkyls towards organoaluminum compounds in homogeneous media.[278,289–92] The bis(η-cyclopentadienyl)titanium(IV) and bis(η-cyclopentadienyl)zirconium(IV) derivatives (**264**), as well as the bridged chiral *rac*-bis(η^5-indenyl) derivatives of titanium[66] and zirconium[295,296] (**265**), and the prochiral, bridged cyclopentadienyl–fluorenyl derivatives (**266**)[297,298] have most often been employed. The aluminum component has often been R_nAlCl_{3-n}, where R = Me or Et and n = 1, 2, or 3. Most recently, a mixture of oligomers of the empirical formula $(MeAlO)_n$, methylaluminoxane (MAO), has been the cocatalyst of choice.[3] It is now generally accepted that R_nAlCl_{3-n} or MAO serve at least three different functions when used to activate (**264**)–(**266**) where R^1, R^2 = Cl. These are: (i) as an alkylating agent by metal–metal exchange (Equation (164)) to produce a M_t–C bond; (ii) as a Lewis acid to abstract a chloride and thus to generate a cationic vacancy on the metal center (Scheme 39); and (iii) by virtue of its size, as a regulator of the equilibrium between contact and solvent-separated ion pairs (Scheme 43) compare (**249**) and (**250**) in Scheme 39.[299]

$Cp_2M^2R^2R^1$
(264)
M^2 = Ti, Zr, Hf
M^1 = H_2C, Me_2Si
R^2, R^1 = Me, Cl

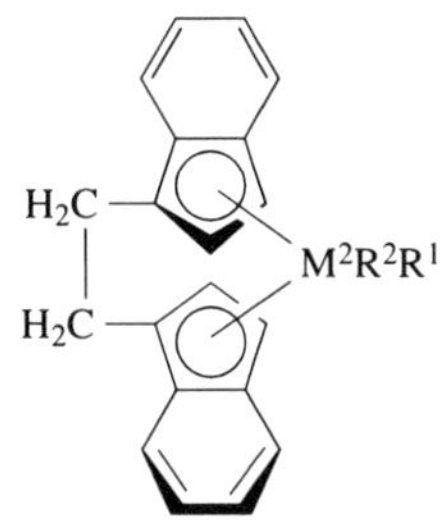

(265)

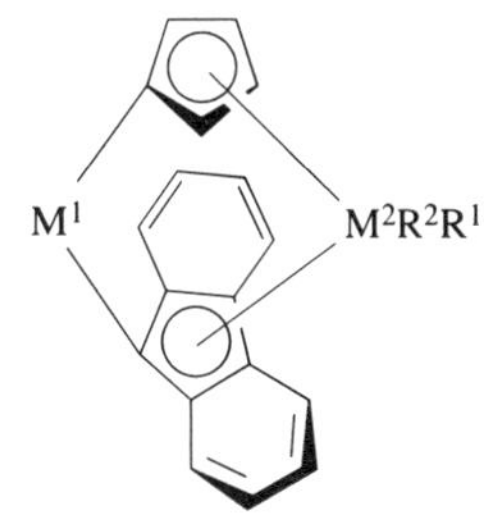

(266)

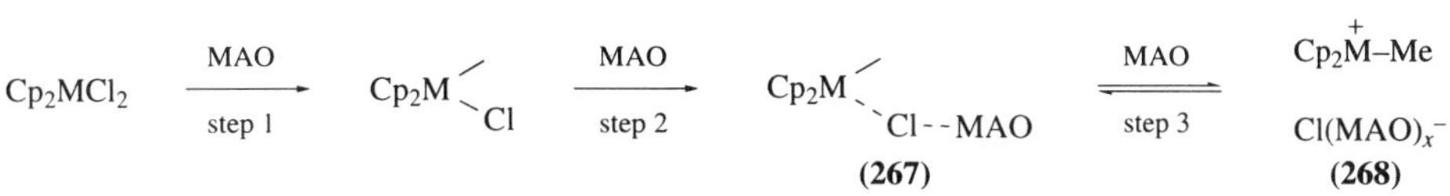

Scheme 43

Evidence has now been adduced that solvent-separated ion pairs (**268**) are more reactive that contact ion pairs (**267**) in the polymerization of both ethene[278] and propene[298] in homogeneous media. It is not yet certain what the relative role of such ion pairs is in the stereospecific polymerization of α-alkenes, but a recent study indicates that solvent-separated ion pairs are less syndiospecific.[298] From these observations on the cocatalytic effects of MAO emerges the hypothesis that the superior cocatalytic action of MAO should be ascribed to its poly-Lewis-acidic character that sequesters the anion (E = Cl, R, OR in Scheme 39) and thereby promotes solvent-separated ion pair formation. Unfortunately, the structural units in MAO are poorly defined, whether they be ring or linear chelating subunits, so this suggestion that MAO may function as a crown-Lewis acid cannot yet be corroborated.[77]

The study of homogeneous catalysts for the polymerization of ethene has also been conducted with η-allyltitanocene (**269**).[300] This compound itself is inactive and it shows a remarkable specificity as to which alkylaluminum component is a suitable cocatalyst. Neither Me_3Al nor $MeAlCl_2$ produces an ethene polymerization catalyst. Their inactivity can be explained by the inability of Me_3Al to disrupt the η^3-coordination of the allyl group in (**269**) and by the complete transfer of the allyl group from (**269**) to form (**268**) by the action of $MeAlCl_2$ (Scheme 44). On the other hand, Me_2AlCl does activate (**269**) for the polymerization of ethene. The efficacy of Me_2AlCl is suggested to arise from its changing the η^3-bonding of the allyl group in (**269**) to η^1-bonding in (**271**), while maintaining the requisite Ti–C bond (contrast this reaction with the formation of (**270**)). Furthermore, the Ti···Cl···Al bridge in (**271a**) could be broken to permit an open coordination site for ethene, as in (**271b**). It is noteworthy that addition of one equivalent of an ether completely deactivates the catalyst, probably because of the formation of (**272**).

In comparison with the polymerization catalyst system Cp_2TiCl_2–$MeAlCl_2$, the allyltitanocene–Me_2AlCl combination is distinctly less active; this difference may reflect the higher electrophilicity of a titanocene(IV) cation (Cp_2TiR^+) for ethene compared with that of a neutral titanocene(III) system (**271b**).[301]

10.6.2.2 Ziegler nickel effect

Through the efforts of the Wilke group in Mülheim[272] and of Eisch's group,[273] the molecular basis of the Ziegler nickel effect has been gradually, although not completely, elucidated. It is now clear that the active oxidation state is nickel(0), rather than nickel(I) or nickel(II). Addition of nickel(II) salts to an excess of an aluminum alkyl leads to a rapid and complete reduction to nickel(0). Furthermore, the nickel effect can be readily induced by the direct addition of a nickel(0) complex, such as $(cod)_2Ni$, to R_3Al. In Wilke's interpretation, the elimination of an Al–H bond and its transfer to another alkene occur while both the R_3Al and alkene remain coordinated to the same Ni^0 center ((**273a**) and (**273b**) in Equation (180)). In Eisch's view such a concerted transfer mediated by a Ni^0 atom should encounter

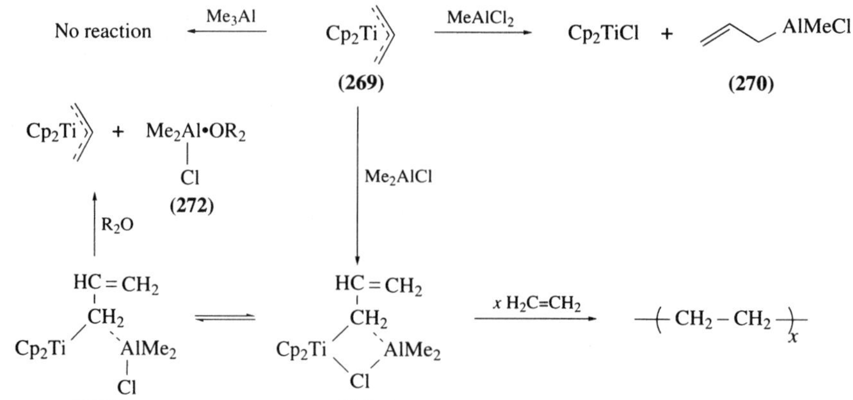

Scheme 44

considerable steric hindrance, as in the case of the tris(norbornene)nickel complex (**274**), which undergoes smooth Al–H transfer to the *exo* face of the π bond (Scheme 45).[273] Putative intermediate (**275**) would seem to offer no easy access of the β hydrogen and AlR_2 groups to the termini of the C=C bond.

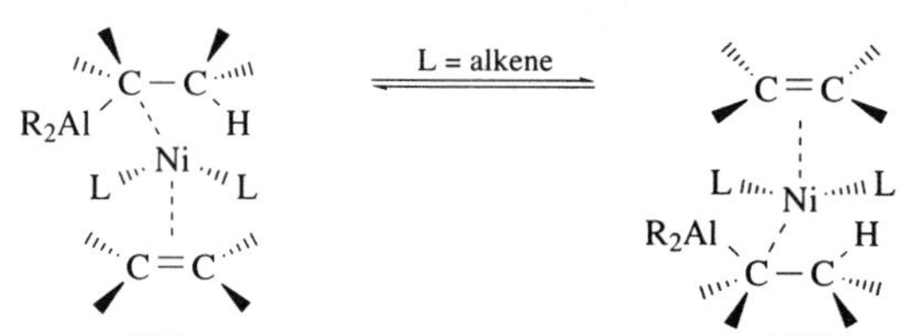

(180)

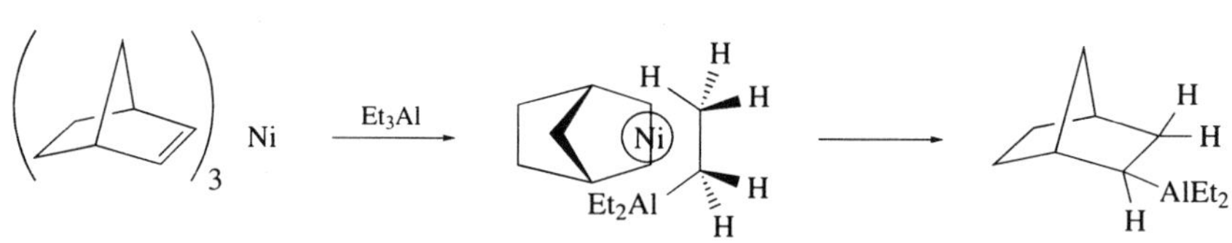

Scheme 45

For this and other reasons, the nickel(0)-induced elimination of R_2AlH from R_3Al in a separate, initial step to form R_2Al–Ni–H (**276**) is preferred. The hydrometallation of another alkene by (**276**) to yield (**277**), followed by the elimination of nickel(0) would complete the transfer (**278**). The observation that nickel(0) promotes the hydroalumination of alkenes[270] and alkynes[271] can be readily accommodated into this scheme by including an alternative route to (**276**) from nickel(0) and R_2AlH (Scheme 46). For various other experimental evidence concerning the deactivation and reactivation of the nickel catalyst, as well as support for the intermediacy of Ni–Al bonds, the original publications should be consulted.[272,273]

10.6.2.3 Alkene and alkyne oligomerization and hydrogenation processes

A huge number of significant industrial processes for the homo- and heterooligomerization of alkenes, alkynes, and 1,3-dienes have been developed whose catalyst is generated from the interaction of various transition metal salts and aluminum alkyls. Similar catalyst combinations have been found effective for the hydrogenation of the same unsaturated substrates. In may of these processes the aluminum alkyl serves principally as a reductant for the transition metal, as is depicted in Equations (162) and (170); the subvalent M_t then effects the subsequent oligomerizations by a sequence of oxidative additions and reductive eliminations (Schemes 47 and 48)[302,303] or the hydrogenation via a metal hydride (see (**278**) in Scheme 49). In some cases, the R_nAlE_{3-n} present may further catalyze the process by producing the catalyst in cationic form (cf. Scheme 39). This latter effect is operative in the Wilke dimerization of propene by allylnickel halide–phosphine–R_nAlX_{3-n} combinations (Scheme 49). The actual catalyst appears to be a phosphine-coordinated cationic nickel hydride (**278**).[64]

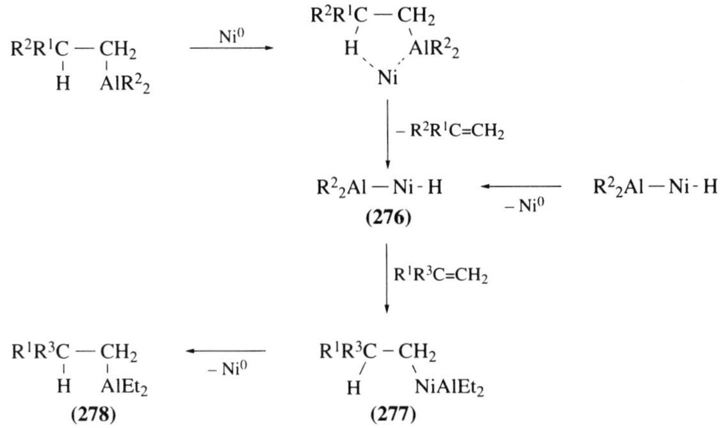

Scheme 46

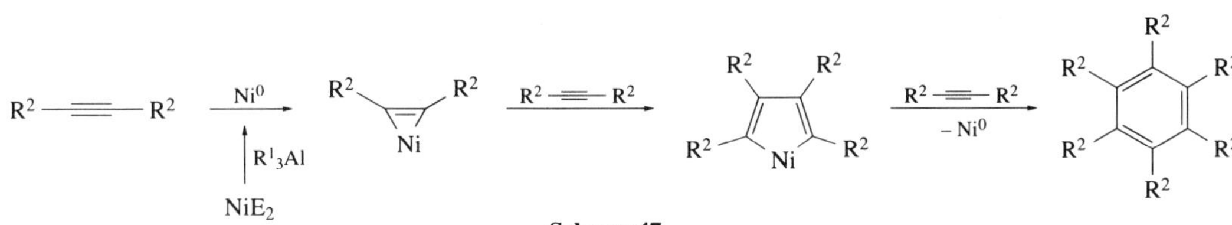

Scheme 47

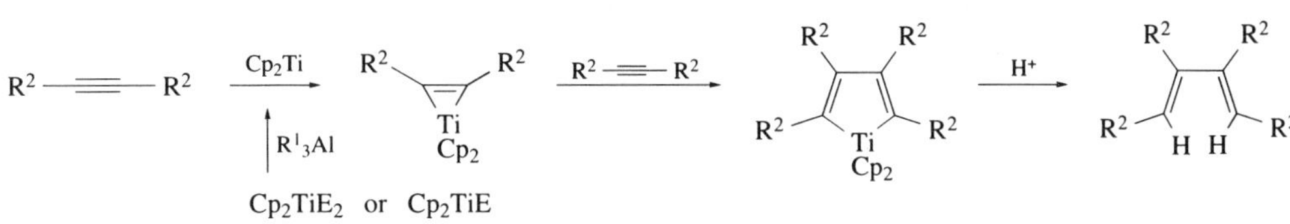

Scheme 48

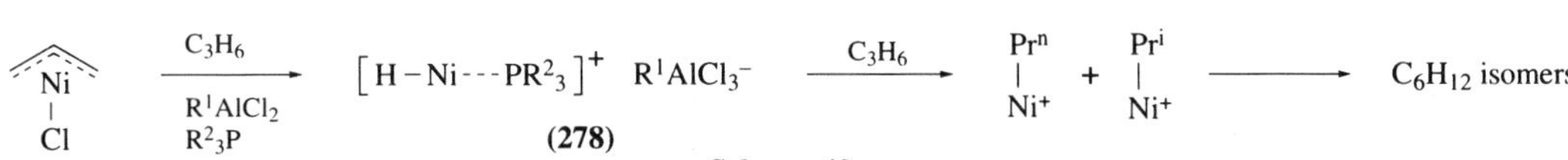

Scheme 49

10.7 FUNCTIONAL GROUP TRANSFORMATIONS IN ORGANIC SYNTHESIS

10.7.1 Scope and Organization

The principal reactions of the aluminum–carbon bond are described in Sections 10.4.5, 10.5, and 10.6. Here the focus is placed on the organic substrates towards which organoaluminum reagents are particularly reactive and on how certain functional groups in such substrates can be advantageously transformed. Where such transformations can be effected in a highly selective manner, the resulting reaction can prove of great value in organic synthesis. For example, the isomerization of an unsymmetrical epoxide (**279**) selectively into one allylic alcohol (**280**) or its reductive cleavage preferentially into one alcohol (**281**) is of enormous interest to the synthetic organic chemist (Scheme 50). Therefore, the transformations discussed in this section will be classified by the type of organic substrate. A more comprehensive treatment of the applications of organoaluminum reagents in organic synthesis is given in Chapter 6, Volume 11. In this chapter a systematic discussion is given on how carbon–element bonds can be formed from R_2AlE in a selective, efficient manner. In this way, the two treatments of R_2AlE reagents will complement each other effectively.

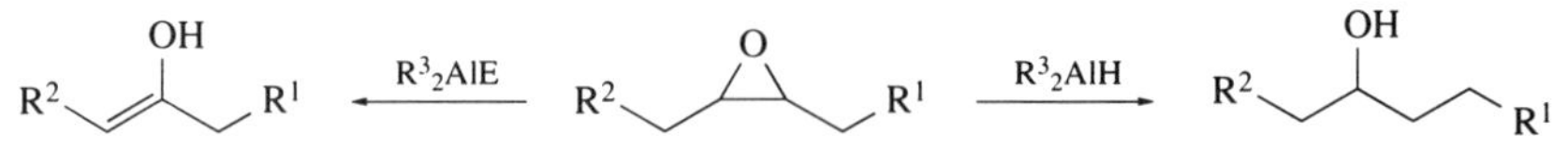

(280) (279) (281)

Scheme 50

10.7.2 Generation of the Required Organoaluminum Reagents

In most reactions of triorganylaluminum reagents only one Al–C bond is readily available for reaction. The requisite trialkylaluminum is commercially available for the lower homologues Me_3Al, Et_3Al, $Pr^n{}_3Al$, $Bu^n{}_3Al$, $Bu^i{}_3Al$, $Hex^n{}_3Al$, $Oct^n{}_3Al$, and $Dec^n{}_3Al$.[304] Other members can be readily prepared by the alkene displacement reaction (Equation (181)).[169]

$$3\ R\text{–}CH{=}CH_2 + Bu^i{}_3Al \rightleftharpoons (RCH_2CH_2)_3Al + 3\ Me_2C{=}CH_2 \qquad (181)$$

Vinylic aluminum derivatives, often of a defined stereochemistry, can be obtained by the regio- and stereoselective hydroalumination of a wide variety of acetylenic derivatives (e.g., Scheme 51;[239] cf. Section 10.4.2.2).[4] It is especially advantageous that the aluminum–vinylic bond reacts preferentially over the aluminum–alkyl bonds and that it usually reacts with retention of configuration.[76]

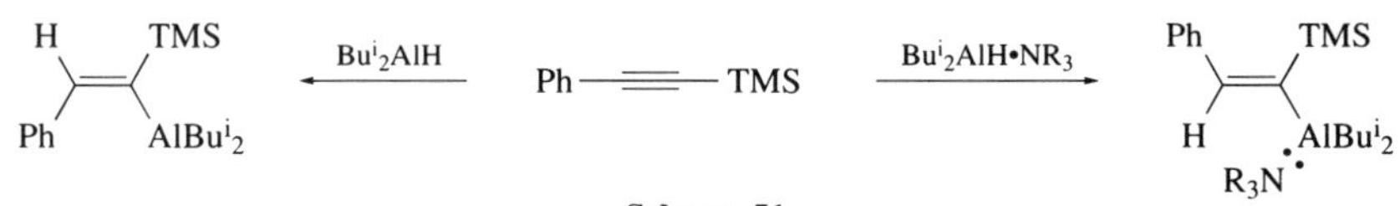

Scheme 51

Acetylenic aluminum reagents are in some cases directly accessible by the alumination of 1-alkynes[202] or in all cases from the alkynyl lithium derivative (Scheme 52).[195]

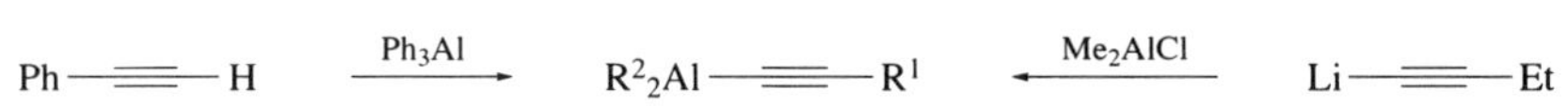

$R^2, R^1 = Ph$
$R^2 = Me;\ R^1 = Et$

Scheme 52

The synthesis of representative organoaluminum derivatives is described in considerable experimental detail in *Organometallic Syntheses*.[304,305]

Fortunate for organic synthesis is the finding that the vinylic or acetylenic group (R^1) in $R^2{}_2AlR^1$ is more reactive than the alkyl groups (R).

Such hydride sources as $Bu^i{}_2AH$ or $Bu^i{}_3Al$ (Equations (104) and (105)) are commercially available and they can be combined with MeLi to produce the complex hydride sources as $LiAlMeBu^i{}_2H$ and $LiAlMeBu^i{}_3$.

A variety of reagents containing aluminum–heteroatom bonds are readily produced by protodealumination of R_3Al or metal salt metatheses with R_2AlCl (Section 10.4.4). Some valuable reagents of particular utility can be made by these methods (Equations (182)–(185)).[53,306–8]

$$Me_3Al + H\text{–}NHMe \xrightarrow{-\,CH_4} Me_2Al\text{–}NHMe \qquad (182)$$

$$2\ EtAlCl_2 + H_2NPh \xrightarrow{-\,C_2H_6} (Cl_2Al)_2NPh \qquad (183)$$

$$Et_3Al + H\text{–}C{\equiv}N \xrightarrow{-\,C_2H_6} Et_2Al\text{–}CN \qquad (184)$$

$$Et_2AlCl + NaN_3 \xrightarrow{-\,NaCl} Et_2Al\text{–}N_3 \qquad (185)$$

10.7.3 Suitable Organic Substrates

10.7.3.1 Reactivity trends of substrates towards organoaluminum reagents

Since $R^2{}_2Al–R^1$ reagents (R^1 = alkyl, vinylic, acetylenic, H, X, OR, NR_2, etc.) are generally Lewis acidic, substrates bearing unshared electrons pairs or π electrons should be especially prone to attack. Among substrates having C–O, C–N, and C–X bonds $R^2{}_2Al–R^1$ will complex at the heteroatom E but heterolysis of the C–E bond is neither rapid nor clean as to its formation of products (Equation (186)) in the case of simple halides, amines, or ethers. Only when the C–E bond is in a strained ring, as with epoxides or aziridines, or where there are geminal C–E bonds, as with acetals or ketals, do such cleavages occur smoothly and rapidly. With C–X bonds the aluminum reagent may have to be activated by employing it as an ate complex ($LiAlR^2{}_3R^1$) or the cleavage may require catalysis by a transition metal, such as nickel, cobalt, or palladium.

(186)

The generally most suitable substrates for $R^2{}_2Al–R^1$ are unsaturated substrates containing C=C, C≡C, C=O, or C=C–C=O linkages. Addition of $R^2{}_2Al–R^1$ to these systems will generally be retarded in donor solvents where the electrophilic character of $R^2{}_2Al–R^1$ is sharply depressed by complex formation. Moreover, if R^2 has β hydrogens, hydroalumination of the C=E linkage will complete with Al–R^1 bond addition (Equation (187)). With multiply unsaturated systems, such as (R–C≡C–R, mono vs. biadditions (hydroaluminations) can compete, whereas with α,β-unsaturated carbonyls, 1,2 vs. 1,4 addition becomes a possibility (Equation (188)).

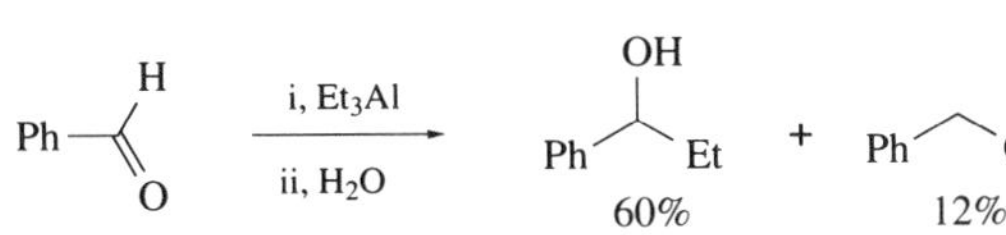

(187)

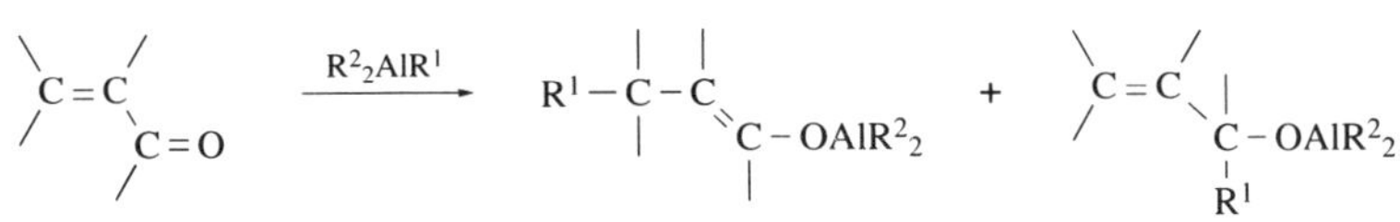

(188)

10.7.3.2 Epoxides

The cleavage of epoxides by organoaluminum reagents ($R^2{}_2Al–R^1$, where R^1 = H, X, NR_2, Me, or enolate salt) can be achieved with great regioselectivity and stereoselectivity and is thus a transformation of great importance in synthesis. The reductive cleavage of terminal epoxides with aluminum hydride sources illustrates the synthetic potential.[50] The reduction of a typical alkyl-substituted epoxide (**282**) with $Bu^i{}_2AlH$ in heptane gives 95% of exclusively the secondary alcohol (**283**); however, use of $Bu^i{}_2AlH$ in THF provides only the primary alcohol (**284**) in 93% yield (Scheme 53). With a phenyl-substituted epoxide the reduction with $Bu^i{}_3Al$ in THF produces an 85% yield of a 93:7 mixture of (**284**) and (**283**), whereas LAH gives a 97:3 mixture of (**283**) and (**284**). The silyl-substituted epoxide forms a 90% yield of exclusively the primary alcohol (**284**) with $Bu^i{}_2AlH$ in heptane, while $Bu^i{}_3Al$ in THF yields 75% of a 65:35 mixture of (**283**) and (**284**).

These empirical observations can be interpreted to mean that $Bu^i{}_2AlH$, which is actually present as a trimer in hydrocarbon media, but as a Lewis complex of the monomer in donor media, acts as a nucleophilic hydride source whose site of attack is governed by steric factors (**285**). With $Bu^i{}_3Al$, complexation with the epoxide probably precedes isobutene elimination and the generation of the Al–H bond. The transition state for ring opening may be attained only with considerable stretching of a C–O bond and the development of carbenium-ion character (**286**), hence favoring primary alcohol formation.

The behavior of the silyl-substituted epoxide toward $Bu^i{}_2AlH$ and $Bu^i{}_3Al$ is the reverse of that shown by the alkyl and aryl-substituted epoxides is not surprising. Electronic effects operative in the ring-opening reactions of α,β-epoxyalkylsilanes can rationally account for this behavior. Nucleophilic attack of LAH or $LiAlD_4$ on such epoxides is known to favor attachment of the hydride ion to the carbon α to silicon and the exclusive formation of the primary alcohol (Equation (189)).

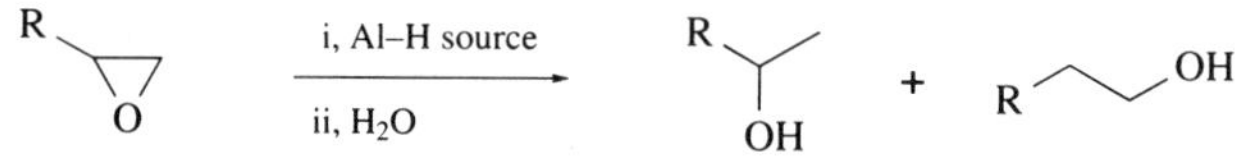

(282) (283) (284)

R	*Al–H*	*Yield of* (**283**) (%)	*Yield of* (**284**) (%)
C_8H_{17}	Bu^i_2AlH	100	0
	Bu^i_3Al	0	100
C_6H_5	Bu^i_3Al	7	93
	LAH	97	3
Ph_3Si	Bu^i_2AlH	0	100
	Bu^i_3Al	65	35

Scheme 53

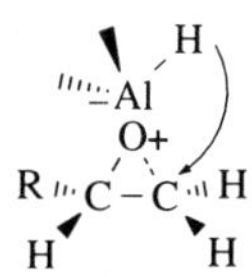

(285)

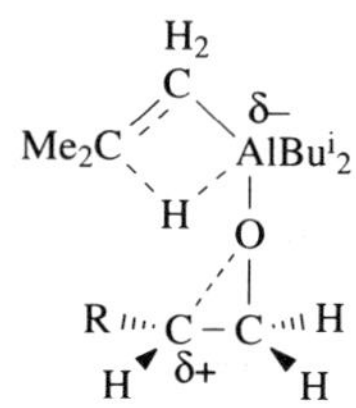

(286)

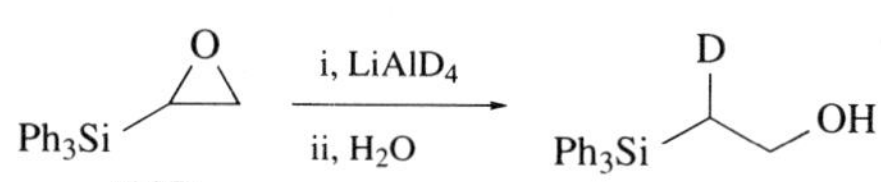

(287) (189)

This regiochemistry is proposed to result via a transition state stabilized by partial pentacoordination of the hydride ion with silicon.[309] In the present study it could accordingly be proposed that the analogous reduction of (**287**) with Bu^i_2AlH would be determined by a transition state such as (**288**). Conversely, the tendency of Bu^i_3Al in THF to favor formation of the secondary alcohol (**283**) would be consistent with a transition state, as in (**286**), again resembling a structure leading to the more stable carbenium ion (**289a**). In this situation, however, the more stable carbenium ion site is the primary carbon, because the positive charge can be stabilized by carbon–silicon σ-bond hyperconjugation ((**289b**), Equation (190)).[310]

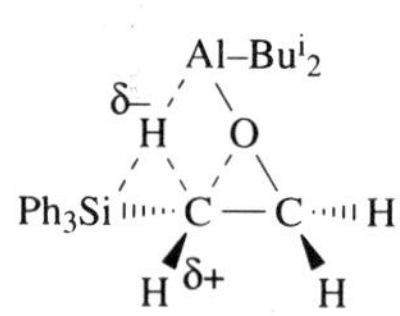

(288)

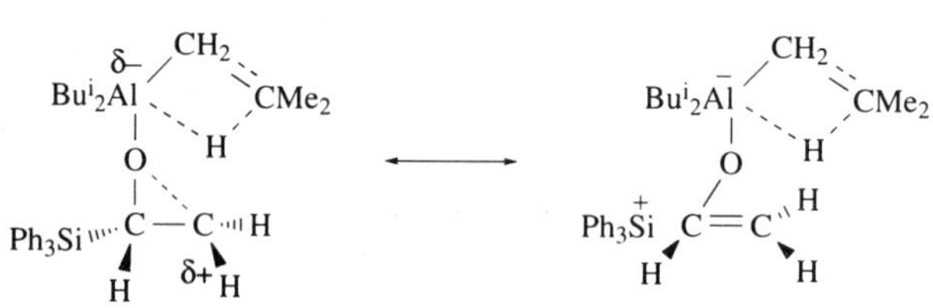

(289a) **(289b)** (190)

In keeping with these views, the regio- and stereoselectivity of reducing (**290**) to (**292**) by diisobutylaluminum 2,6-di-*t*-butylphenoxides can be ascribed to a carbenium-ion-like transition state (**291**) in which the hydride is delivered to the less hindered *exo* side (Scheme 54).[311]

The selective halogenative cleavage of epoxides (**293**) provides an appealing route to the isomeric halohydrins, (**294**) and (**295**), themselves valuable precursors via an oxidation step to α-halocarbonyl

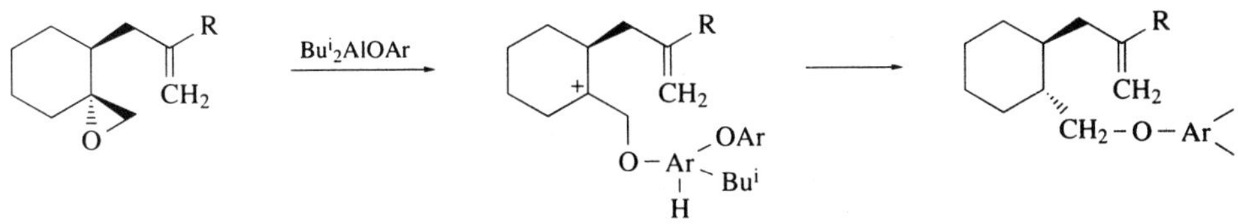

(290) (291) (292)

Scheme 54

derivatives.[312] Such cleavages can be effected under remarkably mild conditions in excellent yields by HX and various metal halides. With most aluminum halides a moderate to good selectivity favoring the secondary alcohol (**294**) is observed (Scheme 55). Only when AlI_3 or the $Bu^i{}_2AlCl$ is employed can the secondary alcohol (**294**) be exclusively formed. No modified aluminum reagent studied, however, was able to produce the primary alcohol (**295**) in high selectivity. Therefore, it is noteworthy that the titanium reagents, $TiBr_4$ and $TiBr_4{\cdot}Ti(NEt)_4$, are capable of causing the alternative, halogenative cleavages, each in a highly selective manner (Scheme 55).[312]

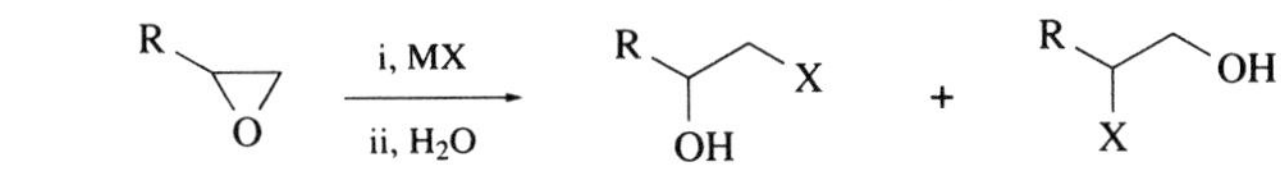

(**293**) R = C_8H_{17} (**294**) (**295**)

MX	*Yield of* (**294**) (%)	*Yield of* (**295**) (%)
$AlCl_3$	60–80	40–20
AlI_3	100	0
$Bu^i{}_2AlCl{\bullet}NEt_3$	100	0
$TiBr_4$	9	91
$TiBr_4{\bullet}Ti(NEt_2)_4$	94	6

Scheme 55

Both the regio- and stereoselective cleavage of epoxide (**296**) is attained by Et_2AlCl; a 100% yield of a 92:8 mixture of chlorohydrins (**297**) and (**298**) can be produced (Equation (191)).[313]

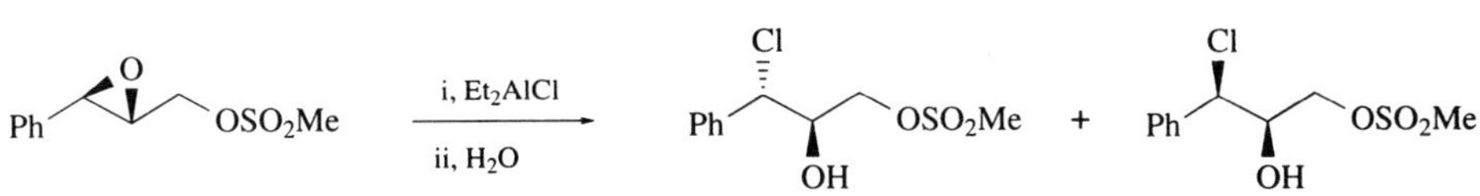

(**296**) (**297**) (**298**) (191)

The two regiochemical modes of cleaving the foregoing epoxides by metal halides can be viewed as occurring via two limiting mechanistic pathways, either electrophilic attack by MX_n, giving a complex (**299**) which can lead to the more stable carbenium ion-like transition state (**300**) (Equation (192)) or a nucleophilic by halide ion on the epoxide–metal halide complex (**299**), giving the more stable transition state (**301**). These mechanistic extremes closely resemble S_N1 and S_N2 models for aliphatic nucleophilic displacement. The regioselectivity resulting from the cleavage of epoxides by metal salts ought therefore to be determined by the relative energies of transition states (**300**) and (**301**), that is, whether electrophilic attack or nucleophilic attack is product determining (Equation (192)). It is clear why AlI_3 would favor the formation of halohydrin (**294**), for nucleophilic attack by I^- should be faster via transition state (**301**). Similarly, when R = phenyl in (**299**), the carbenium-ion transition state (**300**) would be lower in energy.

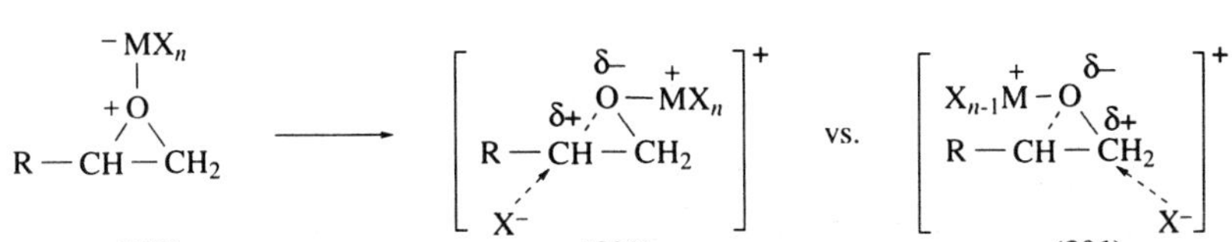

(**299**) (**300**) (**301**) (192)

The stereoselective *trans* opening of an epoxide by a chiral aluminum amide has been clearly demonstrated by the reaction of epoxides (**302**) with aluminum amide (**303**) (Equation (193)).[314] Also, the alkylative cleavage of optically active epoxide (**304**) proceeds regioselectively to produce (**306**); undoubtedly the alkoxyaluminum intermediate (**305**) is decisive on the course of addition (Scheme 56).[315] Furthermore, the inversion of configuration in such ring openings is shown by the reaction of (**307**) with Et_3Al (Equation (194)).[316]

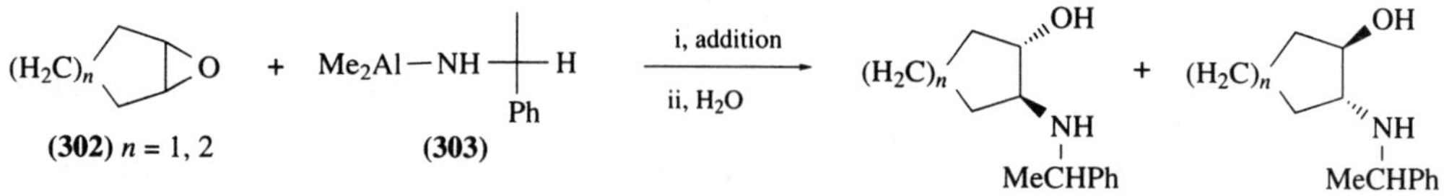

(**302**) n = 1, 2 (**303**) (193)

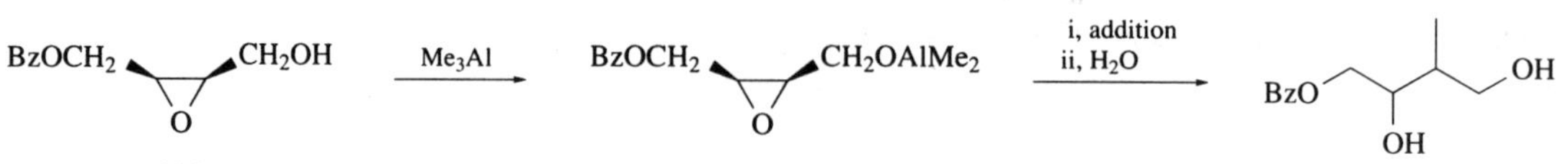

(**304**) (**305**) (**306**)

Scheme 56

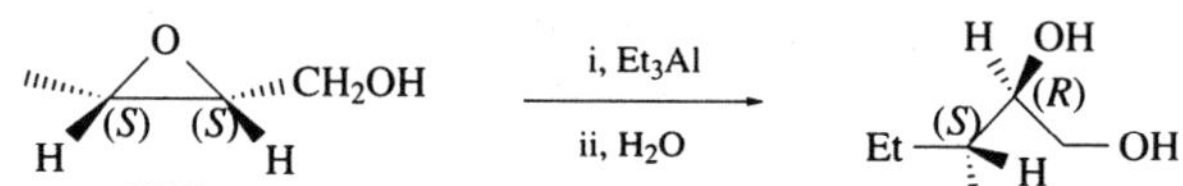

(**307**) (194)

The opening of 1,2-epoxyalkanes (**308**) by (*E*) aluminum enolates (**309**) has been shown to proceed with pronounced diastereoselectivity to produce *syn*:*anti* ratios of (**311**) ranging from 84:16 to 95:5. The dominant *syn* diastereomer is that predicted to be favored from transition state configuration (**310**) (Scheme 57).[317]

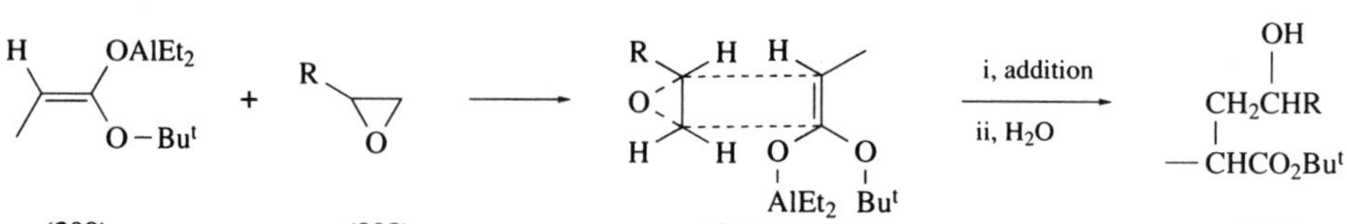

(**309**) (**308**) (**310**) (**311**)

Scheme 57

Finally, the isomerizing cleavage of certain epoxides by $R^2{}_2Al–R^1$ reagents should be mentioned. In the early 1960s, Kirchhof showed that cyclic epoxides, such as 1,2-epoxycyclodecane, can be isomerized by $Bu^i{}_2AlH$ at elevated temperatures into the corresponding allylic alcohol (Equation 195).[318] The same reaction with $Bu^i{}_3Al$ has received considerable attention for converting terpenes to perfume bases.[319] Subsequently, Yamamoto's group claims to have achieved such isomerizations at 0 °C in benzene and in a regioselective manner by employing diethylaluminum 2,2,6,6-tetramethylpiperidide (DATMP) (Equation 196), which was prepared *in situ* from Et_2AlCl and LiTMP.[320] Since, however, LiCl remained in the reaction mixture, there arises the question of whether LiCl may play a role in the isomerizing action of DATMP. It has been observed, for example, that DATMP containing LiCl reacts with a terminal alkyl-substituted epoxide to give a chlorohydrin, but that LiCl-free DATMP gives the expected amine cleavage product (Scheme 58).[312] Clearly the role of LiCl and the actual mechanism of such isomerizations merit further study.

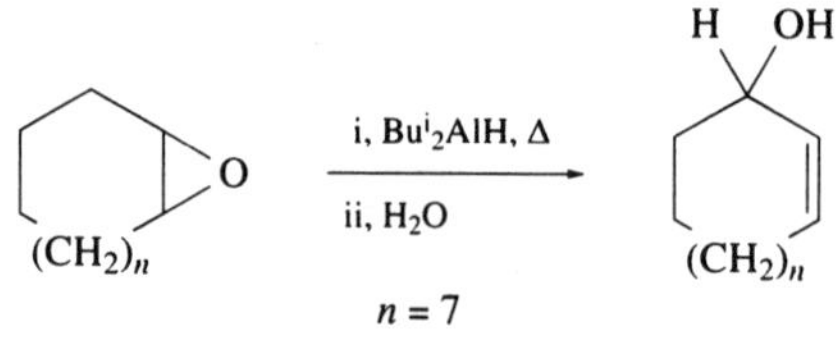

(195)

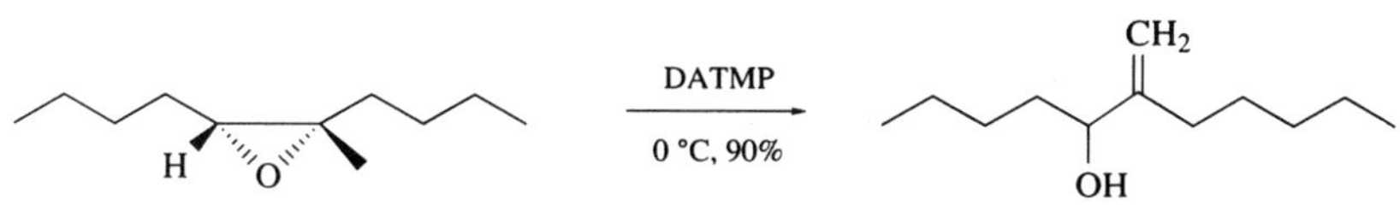

(196)

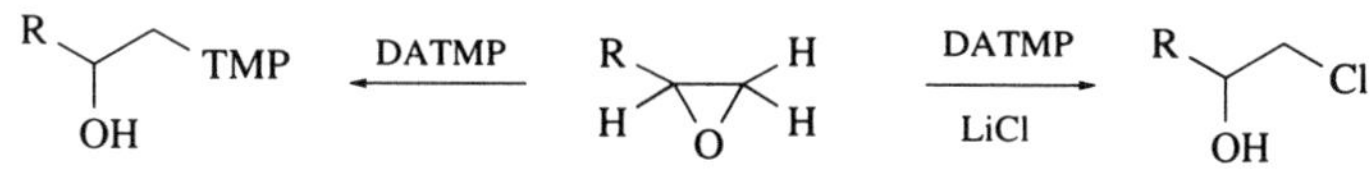

Scheme 58

10.7.3.3 Ethers

Aluminum halides of the formula R_nAlX_{3-n} have been shown to readily cleave ethers of the aryloxy,[321] tetrahydropyranyl,[322] pyranyl,[323] and acetal[324] types under mild conditions and thus can be used, among other purposes, to selectively unmask hydroxyl groups protected by a tetrahydropyranyl derivative. This cleavage occurs without disturbing the *t*-butyldimethylsilyl ether (Equation (197)).[325]

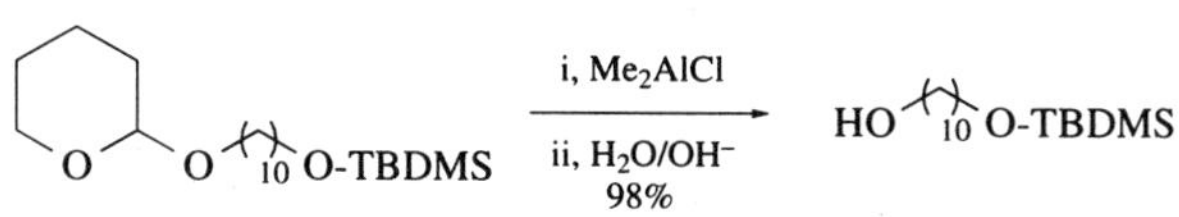

(197)

In some cases with R_2AlH or R_3Al the cleavages can be accompanied by regiospecific alkylation or reduction. Illustrations of some preparatively useful cleavages are given in Equations (198)–(201).[323,324,326,327]

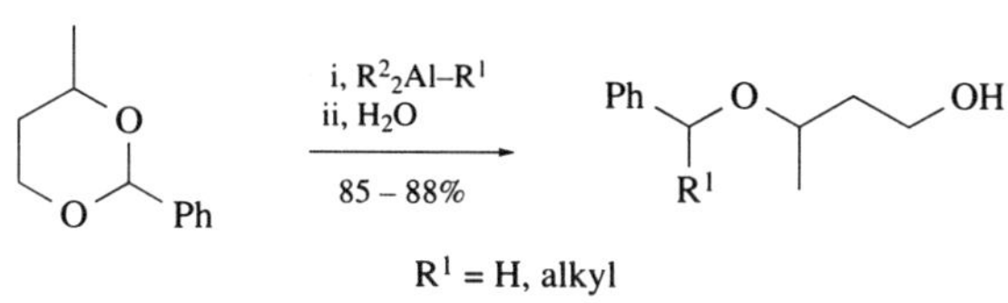

(198)

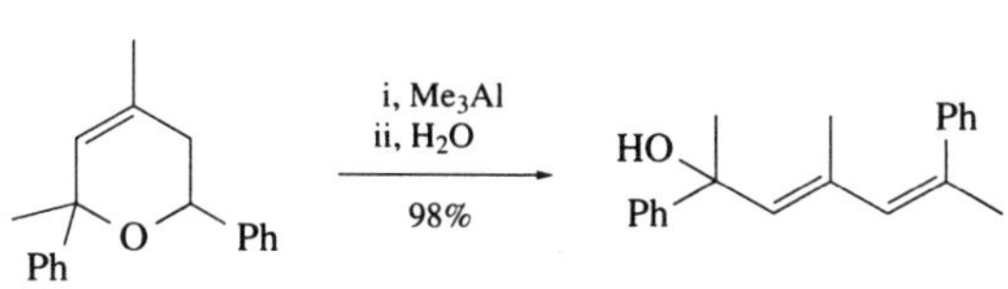

(199)

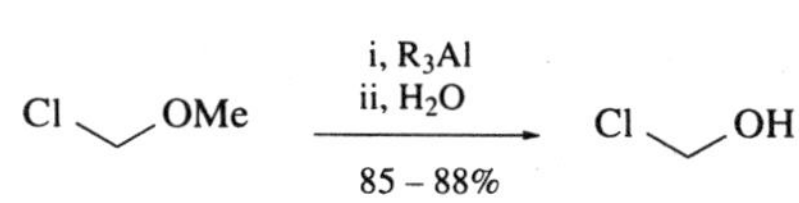

(200)

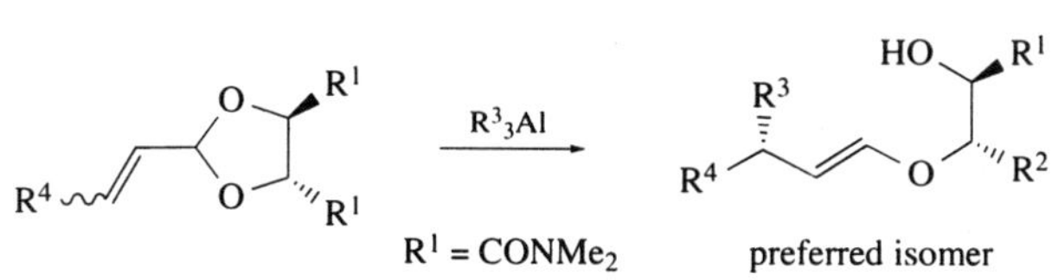

(201)

10.7.3.4 Halides

Alkyl halides undergo C–C bond coupling reactions only slowly with triorganoaluminum reagents (e.g., (**312**)), but the lithium aluminate complex couples smoothly and stereoselectively (**313**) when R^1 is *n*-alkyl or allylic (Scheme 59).[328,329]

With R_3Al and allenic bromides a smooth coupling without rearrangement can be achieved (Equation (202)).[330] Alternatively, the lithium allenyl aluminate (**315**) can be coupled with allylic halides efficiently (Equation (203)).[331] In a related coupling reaction, the $PhSO_2$ group serves as surrogate leaving group instead of halide (Equation (204)).[332]

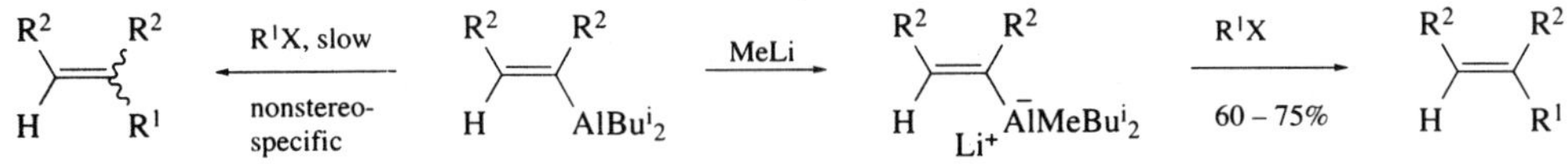

(312) (313) (314)

Scheme 59

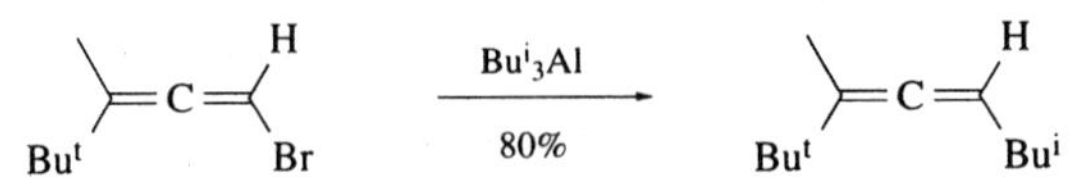

(202)

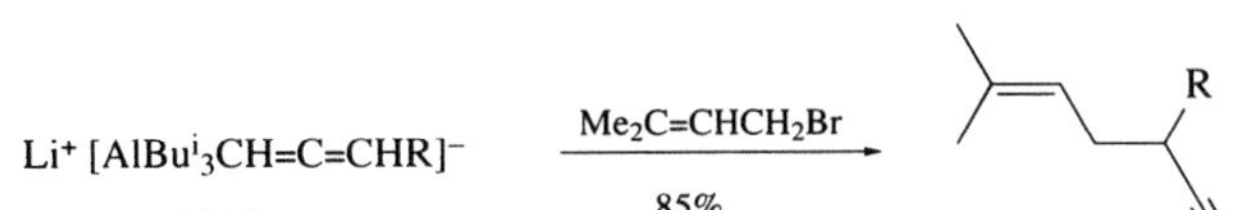

(315)

(203)

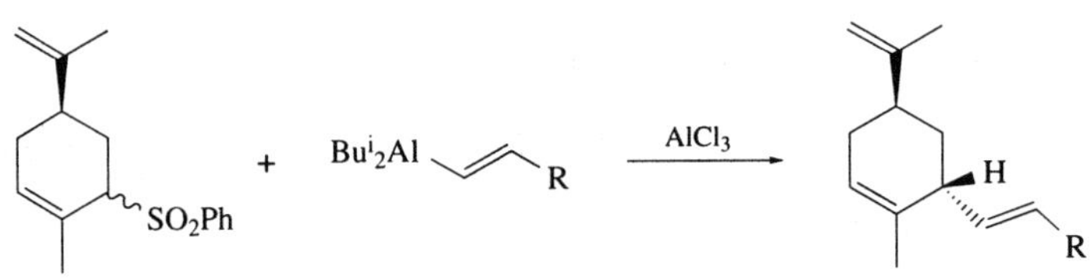

(204)

As a final example, consider a most interesting and mysterious cyclizing and coupling reaction catalyzed by a transition metal (Equation (205)).[333] It can be hoped that many other unusual couplings remain to be discovered.

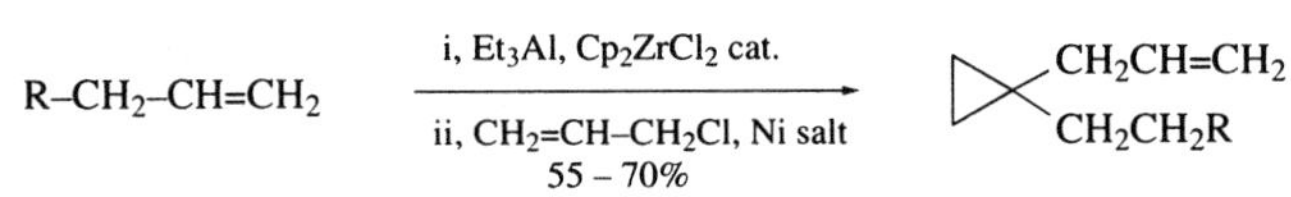

(205)

10.7.3.5 Carbonyl and related derivatives

The addition of an aluminum–alkyl bond to a C=O linkage is generally not a clean reaction (Scheme 29) and offers no advantage over the alternative additions of RLi or RMgX to effect such conversions. The special advantages of employing $R^2{}_2AlR^1$ for such additions are discussed in Section 10.5.4.3. Discussed here are unusual variants of R_3Al–carbonyl interactions and the addition of aluminum–heteroatom bonds.

Two unusual reactions of R_3Al reagents with carbonyl substrates are the intramolecular ene reaction given in Equation (22), promoted by methylaluminum bis(4-bromo-2,6-di-*t*-butylphenoxide) (mabr),[57] and the use of a formaldehyde complex with methylaluminum bis(2,6-diphenylphenoxide) (maph) as a monomeric, selective source of $H_2C{=}O$ in a number of organometallic and enolate condensation reactions (Scheme 60).[334] This latter reagent represents an improvement of Snider's original concept of employing the $H_2C{=}O{\cdot}Me_3Al$ complex for electrophilic attack on alkenes (Equation (137)).

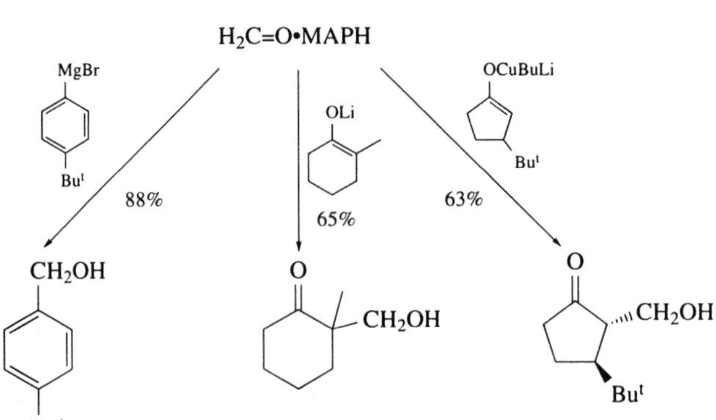

Scheme 60

The addition of hydrogen and deoxygenation of either ketonic carbonyls by organoaluminum Lewis acids is a process that has been found applicable to either alkyl- or aryl-substituted ketones and their corresponding secondary alcohols, as exemplified by the reduction of benzophenone, acetophenone, and 5-nonanone (Equations (206)–(208)).[51] To avoid aldol condensation in the reduction of ketones, it is advantageous to reduce the ketone first with Bu^i_2AlH to produce the aluminum secondary alcoholate before introducing the strong Lewis acid ($AlBr_3$, Equation (208) and Scheme 61). When the secondary alcohol is the starting point, the same alcoholate can be formed by adding Bu^i_3Al first (Scheme 61). With dialkyl ketones or their alcohols, the resulting alkane contains a considerable amount of alkene, but this can readily be reduced by a terminal hydrogenation.

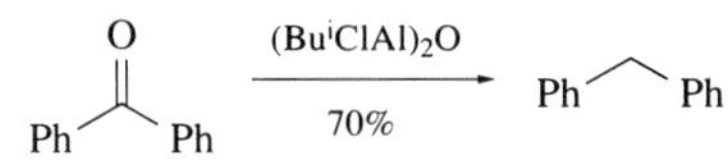

(206)

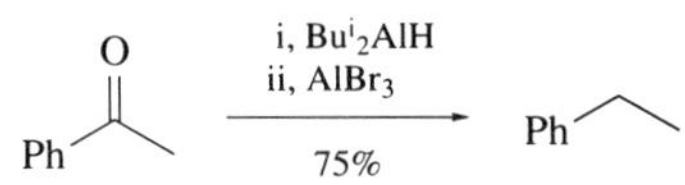

(207)

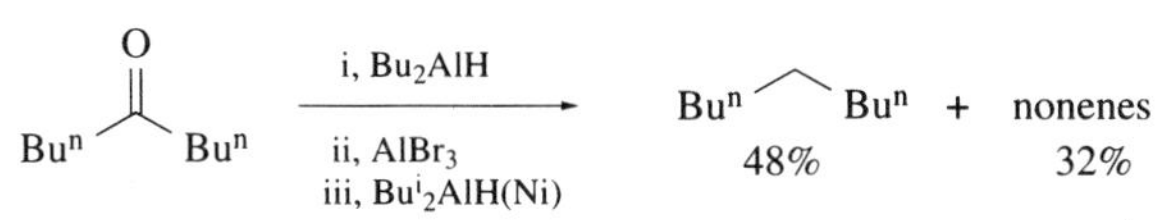

(208)

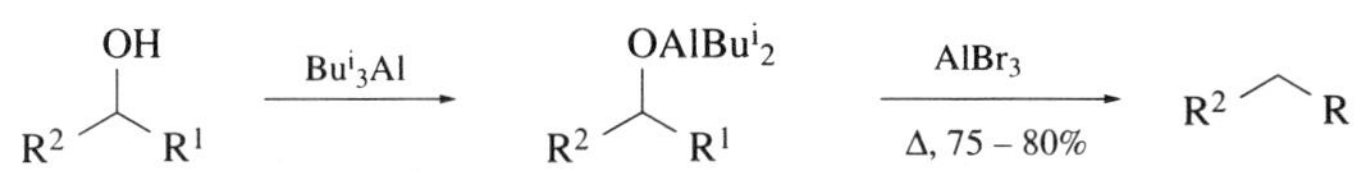

R^2, R^1 = Ph, Ph; Ph, Me; Bu^n, Bu^n

Scheme 61

In an extension of a known reaction, further examples of reducing CO–NR_2 linkages by Bu^i_3Al into CH_2–NR_2 groups have been reported.[335]

The addition of suitable aluminum amides or imides to carbonyl or cyano groups yields nitrogen derivatives under mild conditions in an efficient manner. For example, bis(dichloroaluminum) phenylimide (**316**) is a highly effective iminating agent for aldehydes and ketones (Equation (209)); its reactivity is ascribed to the formation of the highly stable dialuminoxane by product (**317**).[53]

(316) **(317)**

(209)

The superior selectivity of (**316**) as an iminating agent, compared with the traditional Reddelien agent, $PhNH_2{\cdot}ZnCl_2{\cdot}HCl$ (**318**) is seen in the response of chalcone (**319**) (Scheme 62). Iminating agent (**316**) achieves the desired reaction in high yield; agent (**318**) gives only the product of conjugate addition.

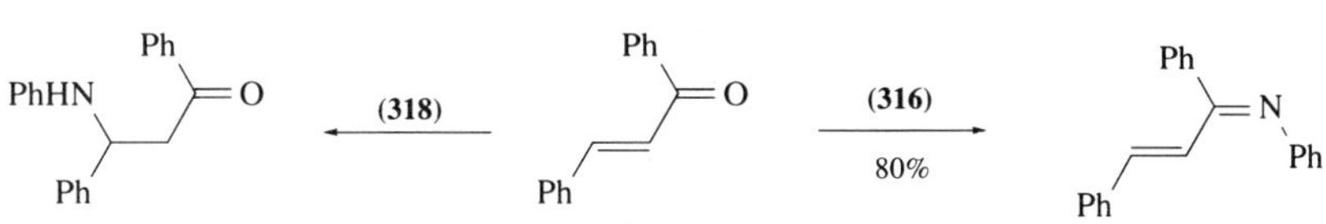

(319)

Scheme 62

The foregoing reagent can also be used to convert acyl chlorides into *N*-phenylamides. However, a more general amidating reagent, applicable both to esters and nitriles, is generated from NH_3, RNH_2 or R_2NH, and Me_3Al. In a one-pot reaction with esters the corresponding primary, secondary, or tertiary amide is formed in 60–70% yield (Equation (210)).[54,55] With nitrile substrates amidines are formed equally well (Equation (211)). Here again the driving force of these reactions is the forming of an Al–O bond. The same principle is undoubtedly operative in the conversion of esters into selenoesters by Me_2Al–SeMe (Equation (212)).[336]

$$R^3\text{–}CO_2R^1 + Me_3Al + R^2{}_2NH \longrightarrow R^3\text{–}CONR^2{}_2 + CH_4 + Me_2AlOR^1 \quad R^3 = \text{alkyl, H} \tag{210}$$

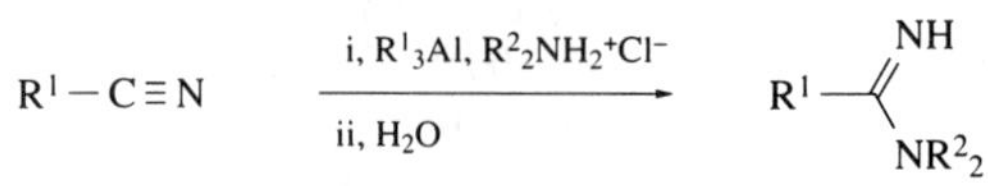

(211)

(212)

In a similar vein to such amidations is the reaction of γ-butyrolactone with aniline (**320**) in the presence of Me_2AlCl (Equation (213)).[337]

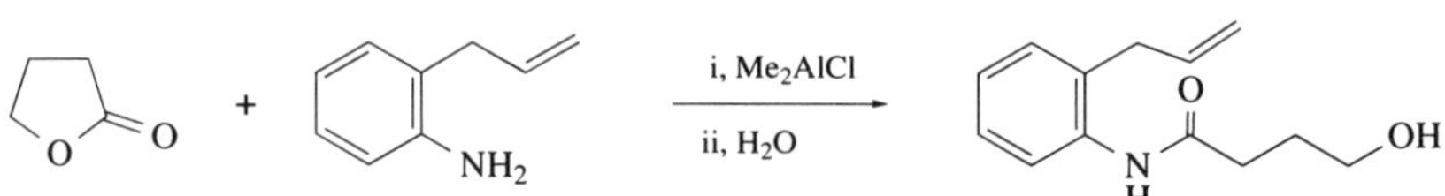

(320)

(213)

10.7.3.6 α,β-Unsaturated carbonyl derivatives

Organoaluminum compounds have long been known to undergo 1,4 or conjugate addition to these systems. Chalcone, for example (**319**), adds Ph_3Al in this manner almost quantitatively. Thus, it is not surprising that Al–Et, Al–CN, Al–N_3, Al–TePh, and Al–H bonds give large amounts of such conjugate addition products (Equations (214)–(218)).[338–42]

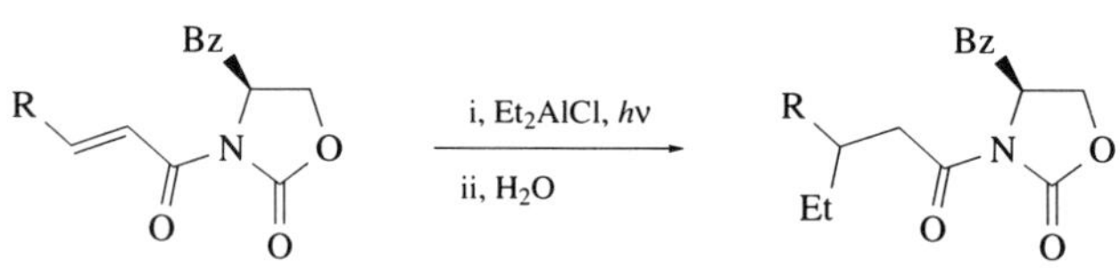

(214)

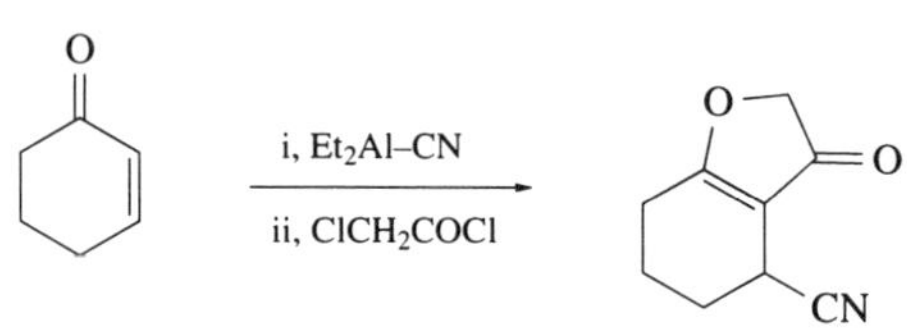

(215)

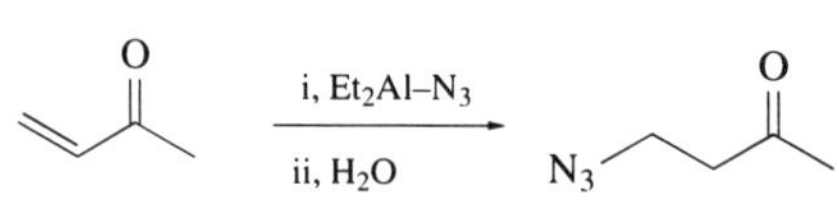

(216)

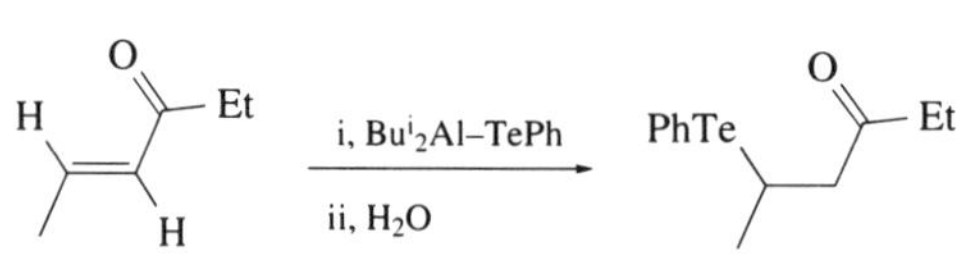

(217)

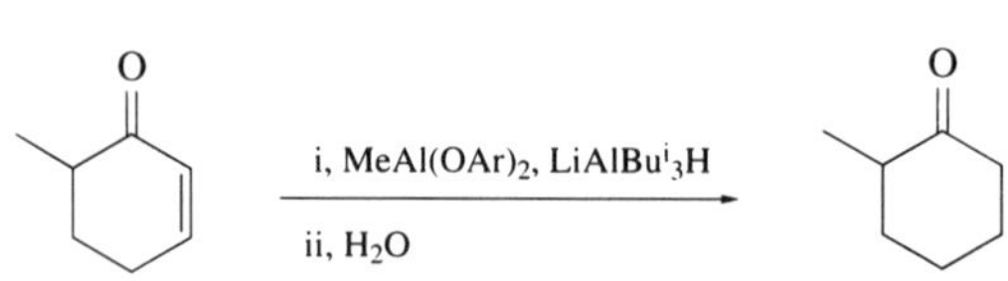

(218)

These conjugate additions can probably occur, in individual cases, by polar or radical mechanisms but with initial complexation of the aluminum reagent $R^2{}_2Al\text{–}R^1$ at the carbonyl carbon. Through a six-membered transition state (**321**) the group R^1 is proximal to the β carbon and thus it is in a position to attack, regardless of whether R^1 has anionic or radical character. Alternatively, two units of $R^2{}_2Al\text{–}R^1$ associated through an Al–R^2–Al bridge may be involved (**322**).

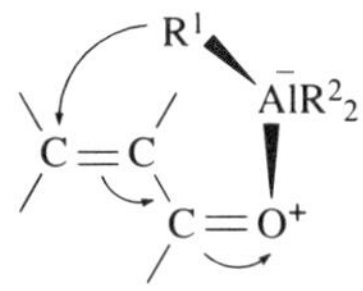

(321)

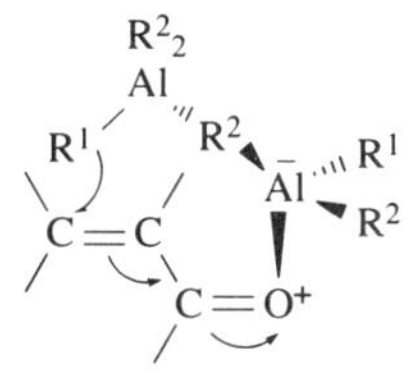

(322)

10.7.3.7 Akenic and alkynic derivatives

Although organoaluminum compounds first attracted attention through Ziegler's discovery of the facile addition of triethylaluminum to ethene (cf. Section 10.6.1.1 and Equation (72)), the number of R_3Al compounds and the number of alkenes and alkynes undergoing such additions cleanly and in high yield is really quite limited. For example, even Me_3Al reacts too slowly with ethene and β-branched R_3Al, such as $Bu^i{}_3Al$, tend to transfer an Al–H bond to the alkene rather than an Al–R bond. Higher terminal alkenes than ethene undergo such carbaluminations so slowly ((**323**), step 1 in Scheme 63) that Al–H bond eliminations occur (step 2). The resulting R_2AlH can cause reduction and dimerization of the alkene (step 3) (cf. Scheme 19).

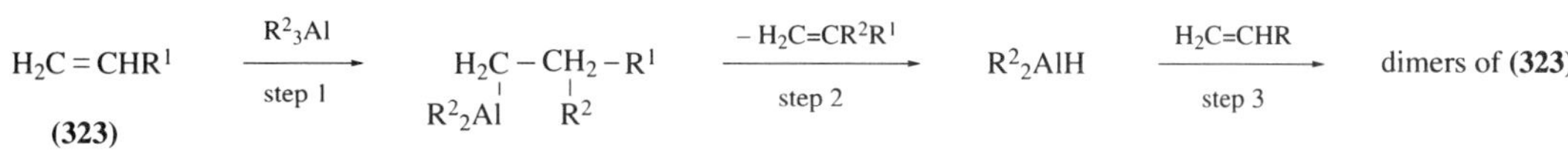

Scheme 63

Likewise, of the terminal alkynes only acetylene undergoes a clean, stereoselective carbalumination with trialkylaluminum derivatives (cf. Scheme 20). With monosubstituted alkenes, alumination competes strongly with carbalumination. With Ph_3Al, such alumination constitutes a useful method for preparing alkynylaluminum derivative (Scheme 52).[202]

Unsaturated hydrocarbons with internal C=C or C≡C linkages undergo carbalumination even more slowly than their terminally unsaturated counterparts and generally the reactivity of the unsaturated linkage falls sharply in the order (**324**) > (**325**) > (**326**).[4] Thus, the substrates (**324**)–(**326**), where R = Ph, react with Ph_3Al under the following conditions: (i) (**324**) cleanly in benzene at 80 °C;[37,38] (ii) (**325**) at 200 °C;[343] (iii) (**326**) no reaction at 200 °C. With strained counterparts of (**325**), such as benzonorbornadiene or norbornadiene (**327**), carbalumination occurs readily below 100 °C.[344] (**327**) undergoes a biscarbalumination with high regioselectivity to produce (**328**) (Equation (219)).[203]

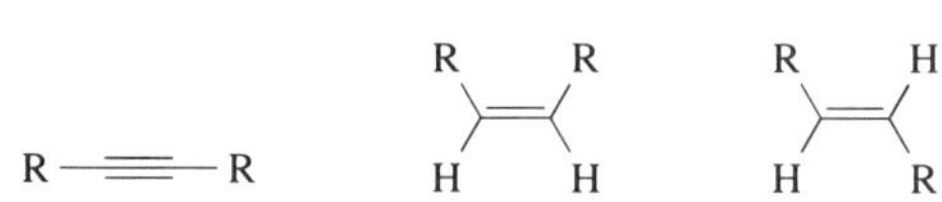

(324) **(325)** **(326)**

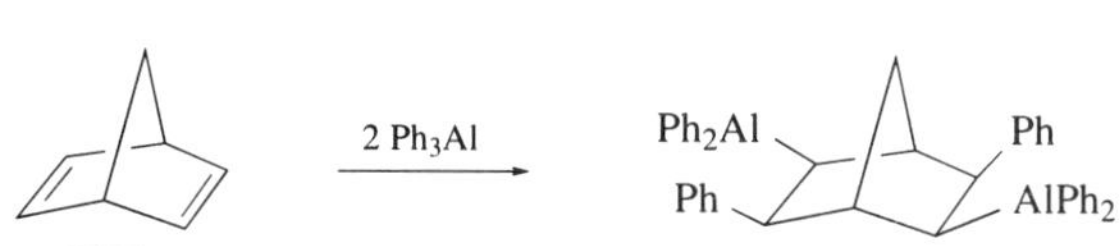

(219)

When such carbaluminations of (**324**)–(**326**) are attempted with aluminum alkyls having β C–H bonds, such as $Bu^i{}_3Al$, then hydroalumination competes strongly (Scheme 64).[195]

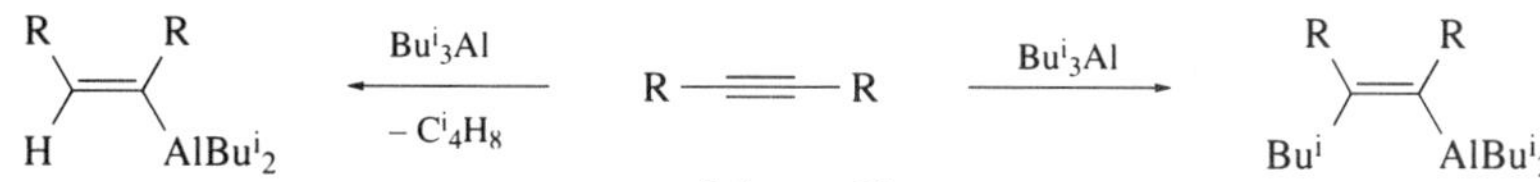

Scheme 64

The foregoing discussion serves to highlight both the remaining problems and the unrealized potential of the Ziegler carbalumination reaction in organic synthesis. The ability of unsolvated R_3Al agents to form C–C bonds, often in a regio- and stereoselective manner, with unactivated alkenes and alkynes would be of enormous value in the controlled synthesis of carbon skeletons, if the reaction were general for R_3Al types and proceeded cleanly. It is this property that is their chief advantage over organolithium or organomagnesium reagents. What is necessary, therefore, to increase the scope and feasibility of carbalumination is the discovery of ways to increase the rate of carbalumination and suppress the elimination of $(RCH_2CH_2)_2AlH$ from $(RCH_2CH_2)_3Al$ agents.

Partial solutions to this overriding problem have been reported. The smooth regioselective and *syn* addition of Me_3Al to terminal alkynes has been known to be catalyzed by Cp_2ZrCl_2 (cf. Equation (179)). The similarly selective addition of a methyl or ethyl group to trialkylsilyl-masked terminal alkynes (**329**) can be achieved by stoichiometric 1:1 ratios of $RAlCl_2$ and Cp_2TiCl_2 (Equation (220)).[198] In this reaction the carbometallating agent is the ion-pair complex, $Cp_2TiR^+\ AlCl_4^-$. In another situation, the zirconium-catalyzed addition of methallyl(dimethyl)aluminum (**330**) to 5-decyne proceeds smoothly to give *syn* adduct (**331**) (Equation (221)).[345] Likewise, the aluminocuprate (**332**) reagent resulting from the interaction of CuBr with R_3Al reacts with terminal alkynes to give approximately a 2:1 ratio of (**333**) and (**334**) in 80% yield (Equation (222)).[346,347]

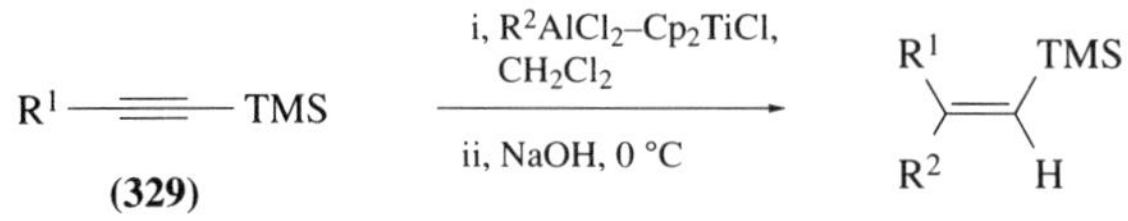

(329) (220)

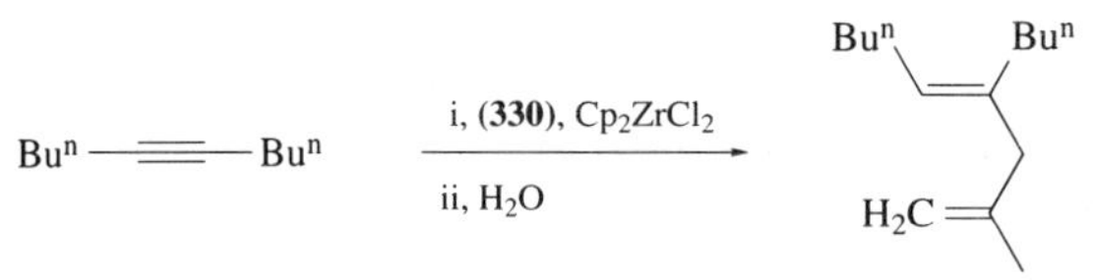

(331) (221)

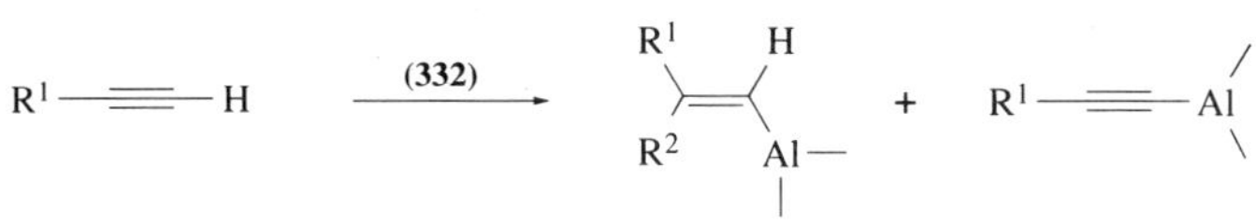

(333) (334) (222)

Finally, a related carbalumination promoted by an electrophile, $Me_2Si^+(SMe)$, is shown in Equation (223).[348] From these examples, it is clear that no completely general method of promoted carbalumination has yet been found. However, the addition of alkyl, allylic, and 1-alkynyl groups to carbon–carbon unsaturation can now be achieved in a number of synthetically useful situations. Such findings will surely encourage further research on this important aspect of organic synthesis.

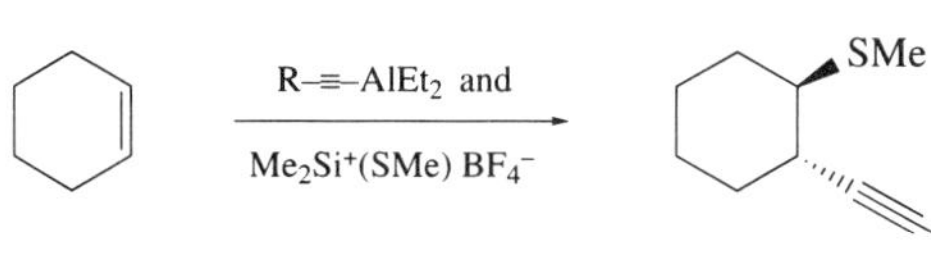

(223)

In contrast with the current limited utility of the Ziegler carbalumination reaction in organic synthesis, the hydroalumination of alkenes and alkynes by neutral, hydrocarbon-soluble R_2AlH reagents is already a well-developed, versatile method for the regioselective and stereoselective synthesis of functionalized hydrocarbon derivatives. Examples of the synthetic applications of the hydroalumination reaction are given throughout this chapter (especially Section 10.4.2.2) and a detailed discussion has appeared in the series, *Comprehensive Organic Synthesis*.[4]

10.8 SUMMATION AND PROSPECT

In this review of just 15 years of organoaluminum research it is remarkable to perceive the unforeseen growth and ramifications that this supposedly mature field[1] has undergone. In the area of structure and bonding alone the nature of organoaluminum chemistry has been dramatically transformed. The isolation and full x-ray structural characterization of the tetrameric η-pentamethylcyclopentadienylaluminum(I)[23] and the monomeric tetraorganyldialuminum systems[32,115] clearly demonstrate that organoaluminum(I) and organoaluminum(II) compounds are feasible intermediates in the reactions of aluminum metal with organic substrates and in the photochemical and free-radical reactions of organoaluminum(III) reactants. Although the scope of σ bonding to an aluminum center has thus been clarified, the nature and extent of the corresponding π bonding remain to be established experimentally, especially in monomeric AlE_3 units. Such π bonding in Al–O bonds has received both an affirmative[14] and a negative[12] judgment.

Keeping pace with the establishment of oxidation states of 3, 2, and 1 for isolated organoaluminum compounds has been the detection of unstable organoaluminum intermediates of hydrocarbons through powerful and sensitive spectroscopic instrumentation. The availability of such extraordinary experimental data has encouraged theoreticians to carry out *ab initio* calculations, in order to choose among possible configurations. This welcome symbiosis among spectroscopists and theoreticians bodes well for the future of insights into structure and bonding.

The obvious Lewis acidity in tricoordinate organoaluminum(III) compounds has received much attention, both through structural studies aimed at establishing higher coordination numbers for aluminum(III) and through an interest in R_3Al reaction mechanisms for stoichiometric and catalytic processes. In the former structural studies, a variety of Lewis-basic ligands have been found to yield a rich array of R_3Al complexes exhibiting coordination numbers of four naturally, five,[9] six[10] and even seven.[11] In the latter investigations the many reactions of R_3Al with organic or with transition metal substrates have been shown to stem from the availability of a tricoordinate R_3Al Lewis acid, which either itself initiates reaction by an electrophilic attack on the substrate (carbalumination or hydroalumination of a C$\equiv$E linkage) or generates a nonaluminum electrophile that completes the reaction (Ziegler–Natta polymerization). The same Lewis acidity of covalently bonded R_3Al units is also responsible for the numerous thermal rearrangements to which such compounds are prone (Section 10.5.5). Such rearrangements contribute both to the joys and the frustrations of chemists, who strive to apply organoaluminum reagents in organic synthesis.

Despite and sometimes because of such rearrangements, organoaluminum compounds have now won a secure and highly valued position in organic synthetic methodology for the functionalization of carbon–carbon unsaturation and the transformation of carbonyl and reactive ether linkages into other functional groups. Such reactions can often be achieved with high loco-, regio-, and stereoselectivity and in a manner clearly superior to that attainable with the corresponding organomagnesium or organolithium reagents. In addition to their special role in stoichiometric organometallic reactions, organoaluminum compounds are the preferred cocatalysts in a variety of important industrial processes catalyzed by transition metals, such as Ziegler–Natta alkene polymerization,[61] the Wilke alkene dimerization[64] and the Ziegler nickel effect.[270] The combined ability of aluminum alkyls, such as polymeric MAO, to function as an alkylating agent, a reductant, and a Lewis acid, forms the basis of their extraordinary importance in the chemical industry.

In light of the astonishing, largely unforeseen developments in organoaluminum chemistry over the recent past, one is hesitant to play the role of prophet. However, with the large number of fruitful lines of research that have recently emerged, in structure and bonding, in synthetic methodology, and in industrial catalysis, it is no reckless prophecy to expect significant, exciting discoveries in organoaluminum chemistry over the far horizon.

10.9 REFERENCES

1. J. J. Eisch, in 'COMC-I', vol. 1, p. 555.
2. *STN International*, American Chemical Society, Washington, DC, 1993.
3. H. Sinn, W. Kaminsky, H.-J. Vollmer and R. Woldt, *Angew. Chem., Int. Ed. Engl.*, 1980, **19**, 390.
4. J. J. Eisch, in 'Comprehensive Organic Synthesis', eds. B. M. Trost and I. Fleming, Pergamon, Oxford, 1992, vol. 8, p. 733.
5. R. M. Laine (ed.) 'Inorganic and Organometallic Polymers with Special Properties', Kluwer, Dordrecht, 1992, pp. 435.
6. C. Dohmeier, C. Robl, M. Tacke and H. Schnöckel, *Angew. Chem.*, 1991, **103**, 594; *Angew. Chem., Int. Ed. Engl.* 1991, **30**, 564.
7. J. S. Tse, *J. Am. Chem. Soc.*, 1990, **112**, 5060.
8. Y. Xie, B. F. Yates and H. F. Schaefer, III, *J. Am. Chem. Soc.*, 1990, **112**, 517.
9. S. J. Dzugan and V. L. Goedken, *Inorg. Chem.*, 1986, **25**, 2858.
10. M. D. Healy and A. R. Barron, *J. Am. Chem. Soc.*, 1989, **111**, 398.
11. S. G. Bott, H. Elgamal and J. L. Atwood, *J. Am. Chem. Soc.*, 1985, **107**, 1796.
12. M. A. Petrie, M. M. Olmstead and P. P. Power, *J. Am. Chem. Soc.*, 1991, **113**, 8704.
13. W. H. Ilsley, M. D. Glick, J. P. Oliver and J. W. Moore, *Inorg. Chem.*, 1980, **19**, 3572.
14. M. D. Healy, J. W. Ziller and A. R. Barron, *J. Am. Chem. Soc.*, 1990, **112**, 2949.
15. M. B. Power, S. G. Bott, J. L. Atwood and A. R. Barron, *J. Am. Chem. Soc.*, 1990, **112**, 3446.
16. J. J. Jerius, J. M. Hahn, A. F. M. M. Rahman, O. Mols, W. H. Ilsley and J. P. Oliver, *Organometallics*, 1986, **5**, 1812.
17. O. T. Beachley, Jr. and L. Victoriano, *Organometallics*, 1987, **7**, 63.
18. A. F. M. M. Rahman, K. F. Siddiqui and J. P. Oliver, *Organometallics*, 1982, **1**, 881.
19. V. Srini, J. De Mel and J. P. Oliver, *Organometallics*, 1989, **8**, 827.

20. A. Haaland, S. Samdal and O. Stokkeland, *J. Organomet. Chem.*, 1977, **134**, 165.
21. G. Wittig, *Angew. Chem.*, 1958, **70**, 65.
22. J. Laemmle, E. C. Ashby and P. V. Roling, *J. Org. Chem.*, 1973, **38**, 2526.
23. H. M. Neumann, J. Laemmle and E. C. Ashby, *J. Am. Chem. Soc.*, 1973, **95**, 2597.
24. W. Klemm, E. Voss and K. Geiersberger, *Z. Anorg. Allg. Chem.*, 1948, **256**, 15.
25. M. Tacke and H. Schnöckel, *Inorg. Chem.*, 1989, **28**, 2895.
26. J. J. Eisch and J. L. Considine, *J. Am. Chem. Soc.*, 1968, 90, 6257.
27. G. A. Olah *et al.*, *J. Am. Chem. Soc.*, 1988, **110**, 3231.
28. F. C. Hall and A. W. Nash, *J. Inst. Pt. Technol.*, 1937, **23**, 679.
29. F. C. Hall and A. W. Nash, *J. Inst. Pt. Technol.*, 1938, **24**, 471.
30. H. Hoberg and S. Krause, *Angew. Chem.*, 1976, **88**, 760.
31. W. Uhl, *Z. Anorg. Allg. Chem.*, 1989, **570**, 37.
32. W. Uhl, *Z. Naturforsch., Teil B*, 1988, **43**, 1113.
33. J. Gauss, U. Schneider, R. Ahlrichs, C. Dohmeier and H. Schnöckel, *J. Am. Chem. Soc.*, 1993, **115**, 2402.
34. J. M. Parinis and G. A. Ozin, *J. Phys. Chem.*, 1989, **93**, 1204.
35. J. H. B. Chenier, J. A. Howard and B. Mile, *J. Am. Chem. Soc.*, 1987, **109**, 4109.
36. S. J. Silva and J. D. Head, *J. Am. Chem. Soc.*, 1992, **114**, 6479.
37. J. J. Eisch and C. K. Hordis, *J. Am. Chem. Soc.*, 1971, **93**, 2974.
38. J. J. Eisch and C. K. Hordis, *J. Am. Chem. Soc.*, 1971, **93**, 4496.
39. J. March, 'Advanced Organic Chemistry', 4th edn., Wiley, New York, 1992, p. 734.
40. V. A. Dodonov, L. P. Stepovik, S. M. Sofronova and E. N. Kosolapkina, *Zh. Obshch. Khim.*, 1990, **60**, 1130.
41. J. L. Brum, S. Deshmukh and B. Koplitz, *Chem. Phys. Lett.*, 1990, **165**, 413.
42. A. Alberola, A. M. Nogal Gonzalez, J. A. Martinez de Ilarduya and F. J. Pulido, *An. Quim., Ser. C.*, 1982, **78**, 166.
43. E. C. Ashby and A. B. Goel, *J. Organomet. Chem.*, 1981, **221**, C15.
44. W. Kaim, *Z. Naturforsch., Teil B*, 1982, **37B**, 783.
45. J. L. Considine, Ph.D. Thesis, The Catholic University of America, Washington, 1972.
46. H. Martin and H. Bretinger, *Z. Naturforsch., Teil B*, 1985, **40**, 182.
47. T. D. Coyle and J. J. Ritter, in 'Advances in Organometallic Chemistry', eds. F. G. A. Stone and R. West, Academic Press, New York, 1972, vol. 10, p. 237.
48. J. J. Eisch and M. W. Foxton, *J. Organomet. Chem.*, 1968, **11**, P7.
49. G. Zweifel and R. B. Steele, *J. Am. Chem. Soc.*, 1967, **89**, 5085.
50. J. J. Eisch, Z. R. Liu and M. Singh, *J. Org. Chem.*, 1992, **57**, 1618.
51. J. J. Eisch Z. R. Liu and M. P. Boleslawski, *J. Org. Chem.*, 1992, **57**, 2143.
52. A. M. Piotrowski, D. B. Malpass, M. P. Boleslawski and J. J. Eisch, *J. Org. Chem.*, 1988, **53**, 2829.
53. J. J. Eisch and R. Sanchez, *J. Org. Chem.*, 1986, **51**, 1848.
54. J. I. Levin, E. Turos and S. M. Weinreb, *Synth. Commun.*, 1982, **12**, 989.
55. J. L. Wood, N. A. Khatri and S. M. Weinreb, *Tetrahedron Lett.*, 1979, (51), 4907.
56. J. J. Eisch, S. R. Sexsmith and M. Singh, *Energy Fuels*, 1989, **3**, 761.
57. K. Maruoka, T. Ooi and H. Yamamoto, *J. Am. Chem. Soc.*, 1990, **112**, 9011.
58. K. Maruoka, N. Murase, T. Ooi and H. Yamamoto, *Synlett* 1991, 857.
59. J. J. Eisch and C. S. Chiu, *Heteroat. Chem.*, 1994, **5**, 265.
60. J. J. Eisch, *J. Chem. Educ.*, 1983, **60**, 1009.
61. J. Boor, Jr., 'Ziegler-Natta Catalysts and Polymerizations', Academic Press, New York, 1979.
62. G. Wilke, *Angew. Chem.*, 1963, **75**, 10.
63. B. A. Dolgoplosk, E. N. Kropacheva, E. N. Kennikova, E. I. Kuznetsova and K. G. Goldova, *Dokl. Akad. Nauk SSSR*, 1960, **135**, 847.
64. U. Birkenstock, H. Bönnemann, B. Bogdanovic, D. Walter and G. Wilke, in 'Advances in Chemistry Series', American Chemical Society, Washington, DC, 1968, vol. 70, p. 250.
65. F. N. Tebbe, G. W. Parshall and G. S. Reddy, *J. Am. Chem. Soc.*, 1978, **100**, 3611.
66. J. A. Ewen, *J. Am. Chem. Soc.*, 1984, **106**, 6355.
67. W. Kaminsky, K. Külper, H. H. Brintzinger and F. R. W. P. Wild, *Angew. Chem.*, 1985, **97**, 507.
68. J. A. Ewen, R. L. Jones, A. Razavi and J. D. Ferrara, *J. Am. Chem. Soc.*, 1988, **110**, 6255.
69. W. Kaminsky, A. Ahlers and N. Müller-Lindenhof, *Angew. Chem.*, 1989, **101**, 1304.
70. W. Spaleck, M. Antberg, V. Dolle, R. Klein, J. Rohrmann and A. Winter, *New. J. Chem.*, 1990, **14**, 499.
71. J. J. Eisch, A. M. Piotrowski, S. K. Brownstein, E. J. Gabe and F. L. Lee, *J. Am. Chem. Soc.*, 1985, **107**, 7219.
72. R. Benn, A. Rufinska, H. Lehmkuhl, E. Janssen and C. Krüger, *Angew. Chem.*, 1983, **95**, 808.
73. R. Benn and A. Rufinska, *Angew. Chem., Int. Ed. Engl.*, 1986, **25**, 861.
74. R. Srinivas, D. Süzle and H. Schwarz, *J. Am. Chem. Soc.*, 1990, **112**, 8334.
75. J. D. Smith, *J. Organomet. Chem.*, 1983, **257**, 17.
76. G. Zweifel and J. A. Miller, in 'Organic Reactions', ed. W. G. Dauben, Wiley, New York, 1984, vol. 32, chap. 2, p. 375.
77. S. Pasynkiewicz, *Polyhedron*, 1990, **9**, 429.
78. M. Yoshifuyi and A. G. Brook, in 'Abstracts of PSIBLOCS Euchem Conference', Paris-Palaiseau, France, August 22–26, 1988.
79. T. Mennekes, P. Paetzold, R. Boese and D. Bläser, *Angew. Chem.*, 1991, **103**, 199.
80. H. Schmidbaur, U. Thewalt and T. Zafiropoulos, *Chem. Ber.*, 1984, **117**, 3381.
81. P. Paetzold, *Angew. Chem.*, 1991, **103**, 559.
82. P. H. Kasai and D. McLeod, Jr., *J. Am. Chem. Soc.*, 1975, **97**, 5609.
83. P. H. Kasai, *J. Am. Chem. Soc.*, 1982, **104**, 1165.
84. J. A. Howard, B. Mile, J. S. Tse and H. Morris, *Faraday Trans.*, 1987, **83**, 3701.
85. L. Manceron and L. Andrews, *J. Phys. Chem.*, 1989, **93**, 2934.
86. L. Manceron and L. Andrews, *J. Phys. Chem.*, 1990, **94**, 3513.
87. Y. Xie, B. F. Yates, Y. Yamaguchi and H. F. Schaefer, III, *J. Am. Chem. Soc.*, 1989, **111**, 6163.
88. D. J. Fox, D. Ray, P. C. Rubesin and H. F. Schaefer, III, *J. Chem. Phys.*, 1980, **73**, 3246.

89. P. H. Kasai, D. McLeod and T. Watanabe, *J. Am. Chem. Soc.*, 1977, **99**, 3521.
90. S. A. Mitchell, B. Simard, D. M. Rayner and P. A. Hackett, *J. Phys. Chem.*, 1988, **92**, 1655.
91. S. Sakai and S. Inagaki, *J. Am. Chem. Soc.*, 1990, **112**, 7961.
92. J. H. B. Chenier, C. A. Hampson, J. A. Howard, B. Mile and R. J. Sutcliffe, *J. Phys. Chem.*, 1986, **90**, 1524.
93. J. H. B. Chenier, C. A. Hampson, J. A. Howard and B. Mile, *J. Chem. Soc., Chem. Commun.*, 1986, 730.
94. M. Histed, J. A. Howard, H. Morris and B. Mile, *J. Am. Chem. Soc.*, 1988, **110**, 5290.
95. R. Damrauer, M. Krempp, M. W. Schmidt and M. S. Gordon, *J. Am. Chem. Soc.*, 1991, **113**, 2393.
96. W. C. Kaska, Ph.D. Thesis, University of Michigan, 1963.
97. R. Köster, G. Benedikt and H. W. Schrötter, *Angew. Chem.*, 1964, **76**, 649.
98. H. Lehmkuhl, G. Fuchs and R. Köster, *Tetrahedron Lett.*, 1965, 2511.
99. W. Kaim, *J. Organomet. Chem.*, 1980, **201**, C5.
100. W. Kaim, *J. Organomet. Chem.*, 1981, **215**, 325.
101. W. Kaim, *J. Organomet. Chem.*, 1981, **215**, 337.
102. Z. Florjanczyk and E. Szymanska-Zachara, *J. Organomet. Chem.*, 1983, **259**, 127.
103. N. S. Enikolopyan, F. S. D'yachkovskii, Z. M. Dzabieva, P. E. Matkovskii and A. F. Shestakov, *Dokl. Akad. Nauk SSSR*, 1981, **261**, 1145.
104. H. Schnöckel, *Z. Naturforsch., Teil B*, 1976, **31**, 1291.
105. J. J. Eisch, J. E. Galle and S. Kozima, *J. Am. Chem. Soc.*, 1986, **108**, 379.
106. H. Schnöckel, M. Leimkühler, R. Lotz and R. Mattes, *Angew. Chem.*, 1986, **98**, 929.
107. R. Ahlrichs, M. Häser, H. Schnöckel and M. Tacke, *Chem. Phys. Lett.*, 1989, **154**, 104.
108. Y. Xie and H. F. Schaefer, III, *J. Am. Chem. Soc.*, 1990, **112**, 5393.
109. C. Dohmeier, R. Mattes and H. Schnöckel, *J. Chem. Soc., Chem. Commun.*, 1990, 358.
110. H. Kurosawa and R. Okawara, *Organomet. Chem. Rev. A*, 1970, **6**, 65.
111. N. Wiberg, in 'Frontiers in Organosilicon Chemistry', eds. A. Bassindale and P. P. Gaspar, The Royal Society of Chemistry, Cambridge, 1991, p. 263.
112. H. Hoberg and S. Krause, *Angew. Chem.*, 1978, **90**, 1013.
113. W. Uhl and J. Wagner, *J. Organomet. Chem.*, 1992, **427**, 151.
114. W. Hiller, K. W. Klinkhammer, W. Uhl and J. Wagner, *Angew. Chem.*, 1991, **103**, 182.
115. (a) W. Uhl, A. Vester, W. Kaim and J. Poppe, *J. Organomet. Chem.*, 1993, **454**, 9; (b) C. Pluta, K.-R. Pörschke, C. Krüger and K. Hildebrandt, *Angew. Chem., Int. Ed. Engl.*, 1993, **32**, 388; (c) R. J. Wehmschulte, K. Ruhlandt-Senge, M. M. Olmstead, H. Hope, B. E. Sturgeon and P. P. Power, *Inorg. Chem.*, 1993, **32**, 2983.
116. (a) E. P. Schram, *Inorg. Chem.*, 1966, **5**, 1291; (b) E. P. Schram, *Inorg. Chim. Acta*, 1987, **134**, 117.
117. L. I. Zakharkin and V. V. Gavrilenko, *Zh. Obshch. Khim.*, 1962, **32**, 689.
118. R. G. Vranka and E. L. Amma, *J. Am. Chem. Soc.*, 1967, **89**, 3121.
119. J. F. Malone and W. S. McDonald, *J. Chem. Soc.*, 1972, 2646.
120. A. Almenningen, S. Halvorsen and A. Haaland, *Acta Chem. Scand.*, 1971, **25**, 1937.
121. M. Barber, D. Liptak and J. P. Oliver, *Organometallics*, 1982, **1**, 1307.
122. G. D. Stucky, A. M. McPherson, W. E. Rhine, J. J. Eisch and J. L. Considine, *J. Am. Chem. Soc.*, 1974, **96**, 1941.
123. B. Teclé, P. W. R. Corfield and J. P. Oliver, *Inorg. Chem.*, 1982, **21**, 458.
124. T. Mole, *Aust. J. Chem.*, 1966, **19**, 373.
125. M. D. Healy, D. A. Wierda and A. R. Barron, *Organometallics*, 1988, **7**, 2543.
126. G. H. Robinson and S. A. Sangokoya, *J. Am. Chem. Soc.*, 1987, **109**, 6852.
127. J. T. Leman and A. R. Barron, *Organometallics*, 1989, **8**, 1828.
128. A. W. Apblett and A. R. Barron, *Organometallics*, 1990, **9**, 2137.
129. L. Siergiejczyk and L. Synoradzki, *J. Organomet. Chem.*, 1986, **311**, 253.
130. M. Boleslawski, J. Serwatowski and S. Pasynkiewicz, *J. Organomet. Chem.*, 1978, **161**, 279.
131. A. Storr, K. Jones and A. W. Laubengayer, *J. Am. Chem. Soc.*, 1968, **90**, 3173.
132. S. I. Ishida, *J. Polym. Sci.*, 1962, **62**, 1.
133. T. Saegusa, Y. Fuji, H. Fuji and J. Furukawa, *Makromol. Chem.*, 1962, **55**, 232.
134. G. A. Razuvaev, Yu. A. Sangalov, Yu. Ya. Nel'Kenbaum and K. S. Minsker, *Izv. Akad. Nauk SSSR. Ser. Khim.*, 1975, 2547.
135. K. Ziegler, F. Krupp, K. Weyer and W. Larbig, *Liebigs Ann. Chem.*, 1960, **629**, 251.
136. L. I. Zakharkin and I. M. Khorlina, *Izv. Akad. Nauk SSSR, Ser. Khim.*, 1959, 2146.
137. D. W. Harney, A. Meisters and T. Mole, *Aust. J. Chem.*, 1974, **27**, 1639.
138. M. Boleslawski, S. Pasynkiewicz, K. Jaworski and A. Sadownik, *J. Organomet. Chem.*, 1975, **97**, 15.
139. N. Ueyama, T. Araki and H. Tani, *Inorg. Chem.*, **12**, 2218.
140. J. Serwatowski, Ph.D. Thesis, Wydawnictwa Politechniki, Warsaw, 1989.
141. M. R. Mason, J. M. Smith, S. G. Bott and A. R. Barron, *J. Am. Chem. Soc.*, 1993, **115**, 4971.
142. M. Boleslawski, S. Pasynkiewicz, A. Kunicki and J. Serwatowski, *J. Organomet. Chem.*, 1976, **116**, 285.
143. L. Siergiejczyk, M. Boleslawski and L. Synoradzki, *Polimery-Tworzywa Wielkoczasteczkowe*, 1986, 397.
144. L. Siergiejczyk and M. Boleslawski, *Polimery-Tworzyma Wielkoczasteczkowe*, 1986, 395.
145. R. Steiger, Ph.D. Thesis, University of Hamburg, 1990.
146. G. H. Robinson and S. A. Sangokoya, *Organometallics*, 1988, **7**, 1453.
147. J. L. Atwood, D. C. Hrncir, R. Shakir, M. S. Dalton, R. D. Priester and R. D. Rogers, *Organometallics*, 1982, **1**, 1021.
148. G. H. Robinson and S. A. Sangokoya, *J. Am. Chem. Soc.*, 1988, **110**, 1494.
149. G. Bähr, 'FIAT Review of German Science, 1939–1946. Inorganic Chemistry', Dieterich'sche Verlagsbuchhandlung, Wiesbaden, 1948, Part II, p. 155.
150. K. M. Waggoner and P. P. Power, *J. Am. Chem. Soc.*, 1991, **113**, 3385.
151. T. R. R. McDonald and W. S. McDonald, *Acta Crystallogr., Sect. B*, 1972, **28**, 1619.
152. J. C. Bartke, A. Haaland and D. P. Novak, *Acta Chem. Scand.*, 1975, **29A**, 273.
153. H. Hess, A. Hinderer and S. Steinhauser, *Z. Anorg. Chem.*, 1970, **377**, 1.
154. A.-A. I. Al-Wassil, P. B. Hitchcock, S. Sarisaban, J. D. Smith and C. L. Wilson, *J. Chem. Soc., Dalton Trans.*, 1985, 1929.
155. W. H. Fink and J. C. Richards, *J. Am. Chem. Soc.*, 1991, **113**, 3393.
156. V. V. Gavrilenko, M. I. Vinnikova, V. A. Antonovich and L. I. Zakharkin, *Izv. Akad. Nauk SSSR, Ser. Khim.*, 1982, 2367.

157. J. J. Eisch and S. G. Rhee, *J. Organomet. Chem.*, 1972, **42**, C73.
158. O. T. Beachley, Jr., C. Tessier-Youngs, R. G. Simmons and R. B. Hallock, *Inorg. Chem.*, 1982, **21**, 1970.
159. W. Hallwachs and A. Schafarik, *Liebigs Ann. Chem.*, 1859, **109**, 207.
160. A v. Grosse and J. M. Mavity, *J. Org. Chem.*, 1940, **5**, 106.
161. P. Binger, *Angew. Chem.*, 1963, **75**, 686.
162. L. I. Zakharkin and V. V. Gavrilenko, *Zh. Obshch. Khim.*, 1963, **33**, 3112.
163. L. I. Zakharkin and V. V. Gavrilenko, *Izv. Akad. Nauk SSSR, Ser. Khim.*, 1963, 1882.
164. H. Hoberg and F. Aznar, *J. Organomet. Chem.*, 1979, **164**, C13.
165. H. Hoberg and F. Aznar, *Angew. Chem.*, 1980, **92**, 132.
166. M. F. Lappert and B. Prokai, *J. Organomet. Chem.*, 1964, **1**, 384.
167. J. J. Eisch and L. J. Gonsior, *J. Organomet. Chem.*, 1967, **8**, 53.
168. C. Krüger, P. J. Roberts, Y. H. Tsay and J. B. Koster, *J. Organomet. Chem.*, 1974, **78**, 69.
169. K. Ziegler, H.-G. Gellert, H. Lehmkuhl, W. Pfohl and K. Zosel, *Liebigs Ann. Chem.*, 1960, **629**, 1.
170. A. V. Kuchin *et al.*, *Zh. Obshch. Khim.*, 1980, **50**, 911.
171. K. Ziegler, F. Krupp and K. Zosel, *Liebigs Ann. Chem.*, 1960, **629**, 241.
172. J. J. Eisch and G. R. Husk, *J. Org. Chem.*, 1964, **29**, 254.
173. K. Ziegler, H.-G. Gellert, H. Martin, K. Nagel and J. Schneider, *Liebigs Ann. Chem.*, 1954, **589**, 91.
174. R. Schimpf and P. Heimbach, *Chem. Ber.*, 1970, **103**, 2122.
175. J. J. Eisch and K. C. Fichter, *J. Org. Chem.*, 1984, **49**, 4631.
176. J. J. Eisch and K. C. Fichter, *J. Organomet. Chem.*, 1983, **250**, 63.
177. R. Rienäcker and G. Ohloff, *Angew. Chem.*, 1961, **73**, 240.
178. H. Lehmkuhl, *Liebigs Ann. Chem.*, 1968, **719**, 40.
179. R. Köster and G. Bruno, *Liebigs Ann. Chem.*, 1960, **629**, 89.
180. J. J. Eisch and K. Zerche-Mackenzie, Invention disclosure, State University of New York, Research Foundation, 1994.
181. J. J. Eisch and K. R. Caldwell, in 'Homogeneous Transition Metal Catalyzed Reactions', eds. W. R. Moser and D. W. Slocum, Advances in Chemistry Series No. 230, American Chemical Society, Washington, DC, 1992, p. 575.
182. G. B. Buckton and W. Odling, *Liebigs Ann. Chem. (Suppl.)*, 1865, **4**, 109.
183. J. J. Eisch, 'Organometallic Syntheses', Academic Press, New York, 1981, vol. 2, p. 136.
184. J. J. Eisch and S. G. Rhee, *J. Organomet. Chem.*, 1975, **86**, 143.
185. H. Lehmkuhl and K. Ziegler, in 'Methoden der Organischen Chemie', ed. E. Mueller, Thieme, Stuttgart, 1970, vol. XIII/4, p. 35.
186. T. Mole, *Aust. J. Chem.*, 166, **19**, 373.
187. M. Kosugi, H. Naka, H. Sano and T. Migita, *Bull. Chem. Soc. Jpn.*, 1987, **60**, 3462.
188. H. Lehmkuhl and K. Ziegler, in 'Methoden der Organischen Chemie', ed. E. Mueller, Thieme, Stuttgart, 1970, vol. XIII/4, p. 52.
189. K. S. Pitzer and H. S. Gutowsky, *J. Am. Chem. Soc.*, 1946, **68**, 2204.
190. J. J. Eisch and K. C. Fichter, *J. Am. Chem. Soc.*, 1974, **96**, 6815.
191. J. J. Eisch and S. G. Rhee, *J. Am. Chem. Soc.*, 1974, **97**, 4673.
192. R. Rienäcker and G. F. Goethel, *Angew. Chem.*, 1967, **79**, 862.
193. J. J. Eisch and G. R. Husk, *J. Organomet. Chem.*, 1974, **64**, 41.
194. A. V. Kuchin, A. Yu. Spivak, V. P. Yur'ev and G. A. Tolstikov, *Izv. Akad. Nau, SSSR, Ser. Khim.*, 1979, 473.
195. G. Wilke and H. Müller, *Liebigs Ann. Chem.*, 1960, **629**, 222.
196. N. I. Andreeva, A. V. Kuchin and G. A. Tolstikov, *Zh. Obshch. Khim.*, 1985, **55**, 1316.
197. H. Hayami, K. Oshima and H. Nozaki, *Tetrahedron Lett.*, 1984, **25**, 4433.
198. J. J. Eisch, R. J. Manfre and D. A. Komar, *J. Organomet. Chem.*, 1978, **159**, C13.
199. J. J. Eisch and G. R. Husk, *J. Org. Chem.*, 1966, **31**, 3419.
200. H. Hoberg, *Liebigs Ann. Chem.*, 1966, **695**, 1.
201. H. Hoberg and E. Ziegler, *Angew. Chem.*, 1967, **79**, 411.
202. J. J. Eisch and W. C. Kaska, *J. Organomet. Chem.*, 1964, **2**, 184.
203. J. J. Eisch and S. J. Y. Liu, *J. Organomet. Chem.*, 1970, **21**, 285.
204. J. J. Eisch and W. C. Kaska, *J. Am. Chem. Soc.*, 1962, **84**, 1501.
205. J. J. Eisch and W. C. Kaska, *J. Am. Chem. Soc.*, 1966, **88**, 2976.
206. H. Lehmkuhl and R. Schäfer, *Liebigs Ann. Chem.*, 1967, **705**, 32.
207. H. Lehmkuhl, K. Ziegler and H. G. Gellert, 'Methoden der Organischen Chemie', ed. E. Mueller, Thieme, Stuttgart, 1970, vol. XIII/4, p. 314.
208. T. Mole and E. A. Jeffrey, 'Organoaluminum Compounds', Elsevier, Amsterdam, 1972, p. 465.
209. F. Barbot, *Bull. Soc. Chim. Fr.*, 1984, (1–2, pt. 2), 83.
210. N. M. Yoon and Y. S. Gyoung, *J. Org. Chem.*, 1985, **50**, 2443.
211. H. Yamamoto and K. Maruoka, *Pure Appl. Chem.*, 1983, **55**, 1853.
212. H. Yamamoto and K. Maruoka, *Farumashia*, 1984, **20**, 766.
213. H. Yamamoto and K. Maruoka, *Pure Appl. Chem.*, 1988, **60**, 21.
214. B. E. Bent, C.-T. Kao, B. R. Zegarski, L. H. Dubois and R. G. Nuzzo, *J. Am. Chem. Soc.*, 1991, **113**, 9112.
215. B. E. Bent, R. G. Nuzzo, B. R. Zegarski and L. H. Dubois, *J. Am. Chem. Soc.*, 1991, **113**, 1143.
216. M. Kuroki, T. Watanabe, T. Aida and S. Inoue, *J. Am. Chem. Soc.*, 1991, **113**, 5903.
217. V. A. Dodonov, S. M. Sofronova and L. P. Stepovik, *Zh. Obshch. Khim.*, 1986, **56**, 1566.
218. W. H. Thomas, *Ind. Eng. Chem., Prod. Res. Dev.*, 1982, **21**, 120.
219. I. Kh. Teregulov, R. A. Sadykov and A. V. Kuchin, *Metalloorg. Khim.*, 1991, **4**, 141.
220. Z. Florjanczyk and U. Iwaniak, *J. Organomet. Chem.*, 1983, **252**, 275.
221. A. A. Volkov, S. S. Zlotskii, E. Kh. Kravets and D. L. Rakhmankulov, *Zh. Obshch. Khim.*, 1986, **56**, 2711.
222. A. A. Volkov, S. S. Zlotskii, E. Kh. Kravets and D. L. Rakhmankulov, *Zh. Vses. Khim. Ova.*, 1987, **32**, 345.
223. F. Barbot and P. Miginiac, *J. Organomet. Chem.*, 1979, **170**, 1.
224. A. A. Volkov, S. S. Zlotskii, E. Kh. Kravets and D. L. Rakhmankulov, *Izv. Vyssh. Uchebn. Zaved., Khim. Khim. Tekhnol.*, 1987, **30**, 110.

225. Y. Naruse and H. Yamamoto, *Tetrahedron Lett.*, 1986, **27**, 1363.
226. R. Menicagli, C. Malanga, M. Dell'Innocenti and L. Lardicci, *J. Org. Chem.*, 1987, **52**, 5700.
227. A. Alberola, B. Bolado, A. M. Gonzalez Nogal and M. R. Poblet, *An. Quim., Ser. C*, 1980, **76**, 5.
228. G. A. Tolstikov, T. Yu. Romanova and A. V. Kuchin, *Izv. Akad. Nauk SSSR, Ser. Khim.*, 1981, 1438.
229. C. Gallina, *Tetrahedron Lett.*, 1985, **26**, 519.
230. G. Sonnek and G. Mueller, *Z. Chem.*, 1982, **22**, 102.
231. T. Suzuki, H. Saimoto, H. Tomioka, K. Oshima and H. Nozaki, *Tetrahedron Lett.*, 1982, **23**, 3597.
232. L. A. Flippin, P. A. Brown and K. Jalali-Araghi, *J. Org. Chem.*, 1989, **54**, 3588.
233. J. J. Eisch and J. E. Galle, *J. Organomet. Chem.*, 1988, **341**, 293.
234. M. Taniguchi, K. Oshima and K. Utimoto, *Tetrahedron Lett.*, 1991, **32**, 2783.
235. A. Alexakis, J. Hanaizi, D. Jachiet, J. F. Normant and L. Toupet, *Tetrahedron Lett.*, 1990, **31**, 1271.
236. V. I. Smetanyuk, G. F. Brodovskaya, A. A. Antonov, M. V. Ul'yanova and N. S. Nametkin, *Neftekhimiya*, 1983, **23**, 784.
237. M. Vincens, R. Fadel and M. Vidal, *C.R. Acad. Sci., Ser. 2*, 1987, **305**, 85.
238. J. A. Miller and G. Zweifel, *J. Am. Chem. Soc.*, 1983, **105**, 1383.
239. J. J. Eisch and M. W. Foxton, *J. Org. Chem.*, 1971, **36**, 3520.
240. A. V. Kuchin, L. I. Akhmetov, V. P. Yur'ev and G. A. Tolstikov, *Zh. Obshch. Khim.*, 1981, **51**, 1855.
241. G. Zweifel, W. Lewis and H. P. On, *J. Am. Chem. Soc.*, 1979, **101**, 5101.
242. Y. Baba, *Bull. Chem. Soc. Jpn.*, 1968, **41**, 2173.
243. E. C. Ashby and R. S. Smith, *J. Organomet. Chem.*, 1982, **225**, 71.
244. B. B. Snider, R. Cordova and R. T. Price, *J. Org. Chem.*, 1982, **47**, 3643.
245. Y. Yamamoto, H. Yatagai, Y. Saito and K. Maruyama, *J. Org. Chem.*, 1984, **49**, 1096.
246. G. A. Tolstikov, F. Kh. Valitov and A. V. Kuchin, *Zh. Obshch. Khim.*, 1983, **53**, 2284.
247. G. A. Tolstikov, F. Kh. Valitov and A. V. Kuchin, *Zh. Obshch. Khim.*, 1980, **50**, 2715.
248. G. A. Tolstikov, F. Kh. Valitov and A. V. Kuchin, *Zh. Obshch. Khim.*, 1981, **51**, 1599.
249. G. A. Tolstikov, F. Kh. Valitov and A. V. Kuchin, *Zh. Obshch. Khim.*, 1981, **51**, 1682.
250. G. A. Tolstikov, F. Kh. Valitov and A. V. Kuchin, *Zh. Obshch. Khim.*, 1982, **52**, 1328.
251. K. B. Starowieyski and A. Becalska, *J. Organomet. Chem.*, 1984, **269**, 223.
252. K. B. Starowieyski, A. Becalska and A. Okninski, *J. Organomet. Chem.*, 1985, **293**, 7.
253. M. Koreeda and Y. Tanaka, *J. Chem. Soc., Chem. Commun.*, 1982, 845.
254. H. J. Bestmann and M. Schmidt, *Tetrahedron Lett.*, 1985, **26**, 6171.
255. E. C. Ashby and S. A. Noding, *J. Org. Chem.*, 1979, **44**, 4792.
256. K. Maruoka, K. Nonoshita and H. Yamamoto, *Tetrahedron Lett.*, 1987, **28**, 5723.
257. K. Maruoka, T. Itoh, M. Sakurai, K. Nonshita and H. Yamamoto, *J. Am. Chem. Soc.*, 1988, **110**, 3588.
258. A. Piotrowski, A. Kunicki and S. Pasynkiewicz, *J. Organomet. Chem.*, 1980, **201**, 105.
259. M. R. P. Van Vliet, G. Van Koten, M. S. De Keijser and K. Vrieze, *Organometallics*, 1987, **6**, 1652.
260. A. Alberola and R. Pedrosa, *An. Quim., Ser. C*, 1980, **76**, 21.
261. A. Pecunioso and R. Menicagli, *Tetrahedron*, 1987, **43**, 5411.
262. K. Fugami, K. Oshima, K. Utimoto and H. Nozaki, *Bull. Chem. Soc. Jpn.*, 1987, **60**, 2509.
263. G. A. Tolstikov, A. Yu. Spivak and A. V. Kuchin, *Izv. Akad. Nauk SSSR, Ser. Khim.*, 1983, 1452.
264. R. Menicagli, C. Malanga, M. Guidi and L. Lardicci, *Tetrahedron*, 1987, **43**, 171.
265. Y. Honda, E. Morita and G. Tsuchihashi, *Chem. Lett.*, 1986, 277.
266. K. Maruoka, N. Murase, T. Ooi and H. Yamamoto, *Synlett*, 1991, 857.
267. K. Hattori, Y. Matsumura, T. Miyazaki, K. Maruoka and H. Yamamoto, *J. Am. Chem. Soc.*, 1981, **103**, 7368.
268. K. Maruoka, T. Miyazaki, M. Ando, Y. Matsumura, S. Sakane, K. Hattori and H. Yamamoto, *J. Am. Chem. Soc.*, 1983, **105**, 2831.
269. H. Fujioka, T. Yamanaka, K. Takuma, M. Miyazaki and Y. Kita, *J. Chem. Soc., Chem. Commun.*, 1991, 533.
270. K. Ziegler, E. Holzkamp, H. Breil and H. Martin, *Angew. Chem.*, 1955, **67**, 541.
271. J. J. Eisch and M. W. Foxton, *J. Organomet. Chem.*, 1968, **12**, P33.
272. K. Fischer, K. Jonas, P. Misbach, R. Stabba and G. Wilke, *Angew. Chem.*, 1973, **85**, 1002.
273. J. J. Eisch, S. R. Sexsmith and K. C. Fichter, *J. Organomet. Chem.*, 1990, **382**, 273.
274. J. J. Eisch, *J. Am. Chem. Soc.*, 1962, **84**, 3830.
275. R. F. Jordan, C. S. Bajgur, R. Willett and B. Scott, *J. Am. Chem. Soc.*, 1986, **108**, 7410.
276. S. R. Sexsmith, Ph.D. Thesis, State University of New York at Binghamton.
277. X. Yang, C. L. Stern and T. J. Marks, *J. Am. Chem. Soc.*, 1991, **113**, 3623.
278. J. J. Eisch, S. I. Pombrik and G. X. Zheng, *Organometallics*, 1993, **12**, 3856.
279. J. N. Hay and R. M. S. Obaid, *Eur. Polym. J.*, 1978, **14**, 965.
280. G. Wilke, *Pure Appl. Chem.*, 1968, **17**, 179.
281. T. S. Chou, S. B. Huang and W. H. Hsu, *J. Chim. Chem. Soc. (Taipei)*, 1982, **30**, 277.
282. L. F. Cannizzo and R. H. Grubbs, *J. Org. Chem.*, 1985, **50**, 2386.
283. J. Kopf, W. Kaminsky and H. J. Vollmer, *Cryst. Struct. Commun.*, 1980, **9**, 197.
284. J. Kopf, H. J. Vollmer and W. Kaminsky, *Cryst. Struct. Commun.*, 1980, **9**, 271.
285. W. Kaminsky, J. Kopf, H. Sinn and H. J. Vollmer, *Angew. Chem., Int. Ed. Engl.*, 1976, **15**, 629.
286. W. Kleimann, K. R. Poerschke and G. Wilke, *Chem. Ber.*, 1985, **118**, 323.
287. V. M. Akhmedov, E. G. Ismailov, A. G. Piraliev, Ya. A. Abbasov and A. G. Guseinov, *Koord. Khim.*, 1986, **12**, 1575.
288. J. L. Atwood, S. D. Chirstie, M. D. Clerk, D. A. Osmond, K. C. Struge and M. J. Zaworotko, *Organometallics*, 1992, **11**, 337.
289. W. P. Long, *J. Am. Chem. Soc.*, 1959, **81**, 5312.
290. J. J. Eisch, M. P. Boleslawski and A. M. Piotrowski, in 'Transition Metals and Organometallics as Catalysts for Olefin Polymeriation', eds. W. Kaminsky and H. Sinn, Springer, Berlin, 1988, pp. 371–388.
291. S. K. Brownstein, E. J. Gabe, N. F. Han, F. L. Lee, Y. LePage, A. M. Piotrowski and J. J. Eisch, *J. Chem. Res. (S)*, 1992, 214.
292. J. J. Eisch, K. R. Caldwell, S. Werner and C. Krüger, *Organometallics*, 1991, **10**, 3417.
293. T. Yoshida and E. Negishi, *J. Am. Chem. Soc.*, 1981, **103**, 4985.

294. G. Erker, R. Noe, C. Krüger and S. Werner, *Organometallics*, 1992, **11**, 4174.
295. W. Kaminsky, K. Külper and S. Niedoba, *Makromol. Chem., Macromol. Sym.*, 1986, **3**, 377.
296. W. Herrmann, J. Rohrmann, E. Herdtweck, W. Spaleck and A. Winter, *Angew. Chem., Int. Ed. Engl.*, 1989, **28**, 1511.
297. J. A. Ewen, L. Haspeslagh, J. L. Atwood and H. Zhang, *J. Am. Chem. Soc.*, 1987, **109**, 6544.
298. N. Herfert and G. Fink, *Makromol. Chem.*, 1992, **193**, 773.
299. J. J. Eisch, S. I. Pombrik, S. Gürtzgen, R. Rieger and W. Uzick, in 'Proceedings of the Symposium on Catalyst Design for Tailor-Made Polyolefins', Hokuriku, Japan, March 10–12, Elsevier, Tokyo, 1994.
300. J. J. Eisch and M. P. Boleslawski, *J. Organomet. Chem.*, 1987, **334**, C1.
301. S. C. Wu, M.Sc. Thesis, State University of New York at Binghamton, 1993.
302. J. J. Eisch, J. E. Galle, A. A. Aradi and M. P. Boleslawski, *J. Organomet. Chem.*, 1986, **312**, 399.
303. A. Furlani, G. Moretti and A. Guerrieri, *J. Polym. Sci., B*, 1967, **5**, 523.
304. J. J. Eisch and R. B. King (eds.), 'Organometallics Syntheses', Academic Press and Elsevier, 1981, vol. 2.
305. J. J. Eisch and R. B. King (eds.), 'Organometallic Syntheses', Academic Press and Elsevier, 1986, vol. 3 and 1988 vol. 4.
306. N. Davidson and H. C. Brown, *J. Am. Chem. Soc.*, 1942, **64**, 316.
307. W. Nagata and M. Yoshioka, *Tetrahedron Lett.*, 1966, 1913.
308. M. I. Prince and K. Weiss, *J. Organomet. Chem.*, 1966, **5**, 584.
309. J. J. Eisch and C. S. Chiu, *J. Organomet. Chem.*, 1988, **358**, C1.
310. W. Hanstein, H. J. Berwin and T. G. Traylor, *J. Am. Chem. Soc.*, 1970, **92**, 829.
311. K. Maruoka, S. Saito, T. Ooi and H. Yamamoto, *Synlett*, 1991, 255.
312. J. J. Eisch, Z. R. Liu, X. Ma and G.-X. Zheng, *J. Org. Chem.*, 1992, **57**, 5140.
313. L. X. Gao, H. Saitoh, F. Feng and A. Murai, *Chem. Lett.*, 1991, 1787.
314. L. E. Overman and S. Sugai, *J. Org. Chem.*, 1985, **50**, 4154.
315. W. R. Roush, M. A. Adam and S. M. Peseckis, *Tetrahedron Lett.*, 1983, **24**, 1377.
316. M. Honda, T. Igarashi, N. Marubayashi and T. Komori, *Liebigs Ann. Chem.*, 1991, 595.
317. T. J. Sturm, A. E. Marolewski, D. S. Rezenka and S. K. Taylor, *J. Org. Chem.*, 1989, **54**, 2039.
318. W. Kirchhof, *Chem. Ber.*, 1960, **93**, 2712.
319. P. Teisseire, M. Plattier, W. Wojnarowski and G. Ourisson, *Bull. Soc. Chim. Fr.*, 1966, 2749.
320. A. Yasuda, S. Tanaka, K. Oshima, H. Yamamoto and H. Nozaki, *J. Am. Chem. Soc.*, 1974, **96**, 6513.
321. K. B. Starowieyski and Z. Rzepkowska, *J. Organomet. Chem.*, 1987, **322**, 309.
322. D. S. Brown, S. V. Ley and M. Bruno, *Heterocycles*, 1989, **28**, 773.
323. A. Alberola, C. Andres, A. Gonzalez Ortega, R. Pedrosa and M. Vicente, *J. Chem. Soc., Perkin Trans.*, 1987, **1**, 2125.
324. A. A. Volkov, S. S. Zlotskii, E. Kh. Kravets, L. V. Spirikhin and D. L. Rakhmankulov, *Khim. Geterotsikl. Soedin.*, 1985, (8), 1036.
325. Y. Ogawa and M. Shibasaki, *Tetrahedron Lett.*, 1984, **25**, 663.
326. A. V. Kuchin, S. A. Markova and G. A. Tolstikov, *Izv. Akad. Nauk SSSR, Ser. Khim.*, 1988, 425.
327. K. Maruoka, S. Nakai, M. Sakurai and H. Yamamoto, *Synthesis*, 1986, 130.
328. J. J. Eisch and G. A. Damasevitz, *J. Org. Chem.*, 1976, **41**, 2214.
329. K. Uchida, K. Utimoto and H. Nozaki, *J. Org. Chem.*, 1976, **41**, 2215.
330. A. M. Caporusso, L. Lardicci and F. DaSettimo, *Gazz. Chim. Ital.*, 1986, **116**, 599.
331. G. Hahn and G. Zweifel, *Synthesis*, 1983, 883.
332. B. M. Trost and M. R. Ghadiri, *J. Am. Chem. Soc.*, 1986, **108**, 1098.
333. U. M. Dzhemilev, A. G. Ibragimov, A. P. Zolotarev, R. R. Muslukov and G. A. Tolstikov, *Izv. Akad. Nauk SSSR, Ser. Khim.*, 1990, 1190.
334. K. Maruoka, A. B. Concepcion, N. Hirayama and H. Yamamoto, *J. Am. Chem. Soc.*, 1990, **112**, 7422.
335. A. Alberola, A. M. Gonzalez Nogal and M. D. Guerra Suarez, *An. Quim. Ser. C*, 1980, **76**, 28.
336. A. P. Kozikowski and A. Ames, *Tetrahedron*, 1985, **41**, 4821.
337. A. G. M. Barrett and D. Dhanak, *Tetrahedron Lett.*, 1987, **28**, 3327.
338. K. Rueck and H. Kunz, *Angew. Chem.*, 1991, **103**, 712.
339. S. Bernasconi, G. Jommi, S. Montanari and M. Sisti, *Synthesis*, 1987, 1126.
340. B. Y. Chung, Y. S. Park, I. S. Cho and B. C. Hyun, *Bull. Korean Chem. Soc.*, 1988, **9**, 269.
341. K. Sasaki, Y. Aso, T. Otsubo and F. Ogura, *Chem. Lett.*, 1989, 607.
342. K. Nonoshita, K. Maruoka and H. Yamamoto, *Bull. Chem. Soc. Jpn.*, 1988, **61**, 2241.
343. J. J. Eisch, N. E. Burlinson and M. Boleslawski, *J. Organomet. Chem.*, 1976, **111**, 137.
344. J. J. Eisch and N. E. Burlinson, *J. Am. Chem. Soc.*, 1976, **98**, 753.
345. Y. Wei, *Youji Huaxue*, 1988, **8**, 343.
346. N. I. Andreeva, A. V. Kuchin and G. A. Tolstikov, *Zh. Obshch. Khim.*, 1985, **55**, 693.
347. A. V. Kuchin, G. A. Tolstikov and N. I. Andreeva, *Izv. Akad. Nauk SSSR, Ser. Khim.*, 1987, 397.
348. B. M. Trost and S. J. Martin, *J. Am. Chem. Soc.*, 1984, **106**, 4263.

11
Gallium, Indium and Thallium, Excluding Transition Metal Derivatives

MICHAEL A. PAVER, CHRISTOPHER A. RUSSELL
and DOMINIC S. WRIGHT
University of Cambridge, UK

11.1 INTRODUCTION AND SCOPE	503
11.2 ALKYL AND ARYL DERIVATIVES	504
11.2.1 Syntheses and Structures of R_3M, Ar_3M and Lewis-base Adducts	504
11.2.2 Organohalogen and Pseudohalogen Derivatives	511
11.2.3 Cationic and Anionic Derivatives	514
11.3 ORGANOMETALLIC HYDRIDES	514
11.4 ORGANOMETALLICS CONTAINING HETEROATOMIC GROUPS	517
11.4.1 Introduction	517
11.4.2 Syntheses and Structure of Group 13–15 Complexes Containing Ga, In and Tl Bonded to N, P, As and Sb	518
11.4.3 Syntheses and Structure of Group 13–16 Complexes Containing Ga, In and Tl Bonded to O, S, Se and Te	524
11.5 LOW OXIDATION STATE COMPLEXES	528
11.5.1 Cyclopentadienyl Compounds	529
11.5.2 Carbaborane Compounds	532
11.5.3 Arene Complexes	534
11.6 METAL–METAL BONDED COMPLEXES	537
11.7 REFERENCES	540

11.1 INTRODUCTION AND SCOPE

The years from 1982 to 1992 have seen an explosion of interest in group 13 (Al, Ga, In, Tl) compounds and particularly in the organometallic derivatives of the heavier metals (M = Ga, In and Tl). Many exciting advances have been made particularly concerning the structures of these organometallic complexes in the solid state and the observation of a variety of new classes of compounds, for example, π arene complexes and metal–metal bonding. Undoubtedly, one of the major driving forces in recent years in this area has been the increased application of these metals in the electronics industry, for example, in semiconductors such as GaAs, GaN and InP.[1] As a consequence, a principal focus of much of this work has been the development of new volatile organometallic precursors for the epitaxial growth of thin films for electronic device applications (e.g., metal organic chemical vapour deposition, MOCVD), which comprise, in molecular form, a well-defined stoichiometry; for example, heteroatomic derivatives containing combinations of mainly gallium and indium bonded to nitrogen, phosphorus and arsenic (= E) in molecular ring or cluster arrangements $(ME)_n$. Cowley and Jones, and Wells have reviewed the structures and chemistry of group 13 metal complexes in which two-electron bonds to group 15 elements are present.[1] The thirst for new materials has been followed by the application of new synthetic strategies to these species and by the use of new heteroatom-containing ligands in order to effect changes in their molecular structure, for example, their aggregation into rings or clusters and to promote desirable physical properties (particularly high volatility). The most prolific workers involved in this area include Barron (Harvard), Beachley Jr. (Buffalo, NY), Cowley (Austin, Texas), Jones (Austin, Texas), Power (Davis, California), Theopold (Delaware), and Wells (Duke, N. Carolina).

A further flourishing area of research has been the investigation of low-valent (+1 oxidation state) compounds generally containing π or sterically bulky ligands bonded to gallium, indium or thallium. While cyclopentadienide derivatives continue to feature heavily in the chemistry of the low-valent metal organometallics of the heavier group 13 elements and GaCp* has recently been synthesized,[2] perhaps the most major advance made in the period 1982–1992 has been the investigation of low-valent complexes containing π-bonded arene ligands in the solid state. Although it had been known since 1881 that gallium(I) dihalides (actually mixed valance salts $Ga^+{\cdot}GaX_4^-$) are soluble in benzene, it had not previously proved possible to thoroughly investigate the natures of the complexes which are formed. In his pioneering work, Schmidbaur (Munich) showed that the species are π-arene complexes of gallium(I).[3] Since then a host of arene ligands have been employed and some extremely elegant work on such gallium(I) and indium(I) complexes has been published. Schmidbaur has written an extensive review on this area[3] and, in addition, a thorough review by Jutzi is also a valuable guide to the structural diversity of these species and to their comparison with the structural and bonding patterns in other main group metal π complexes.[4] A variety of novel metallacarbaborane sandwich and half-sandwich complexes, in which the carbaborane anions are isoelectronic with cyclopentadienide, have been prepared and structurally elucidated, notably by Hawthorne (California, LA).

Dramatic advances have been made also in the investigation of metal–metal bonded (M–M) complexes in recent years by Uhl (Stuttgart) and others for aluminum,[5] gallium[6] and indium.[7] Here, the metals are in either the +3 or +1 oxidation states and the metal–metal bonds are stabilized by sterically bulky organic groups. The reactivity and theoretical nature of these bonds represents a new and exciting area for future investigations.

Overall, the progress made during this period on gallium, indium and thallium organometallic chemistry owes much to the development and availability of x-ray crystallography, particularly low-temperature methods, and by advances in the handling and synthesis of these commonly extremely air- and moisture-sensitive and thermally unstable compounds. The structural elucidation of these complexes in the solid state is now commonplace and perhaps this single feature constitutes the most major difference between the flavour of the review in *COMC-I* and this review.

This chapter will concern all of the above areas of research with particular stress on the key structural (solid-state and gas-phase) and synthetic advances made in the period from 1982 to 1992. In addition, the application of the variety of spectroscopic methods (NMR, IR and others) will be surveyed. In view of the potential applications of many of the materials discussed, the use of these as precursors for electronic device applications is also mentioned where appropriate. The chapter will be confined to organometallic species in the strict sense of compounds containing C–M bonds to gallium, indium or thallium. However, where particular comparisons can be made with the structures or behaviours of related 'inorganic' derivatives (not necessarily containing C–M bonds), brief mention of these species will be given. Complexes containing gallium, indium or thallium bonded to transition metals are discussed elsewhere in this volume (see Chapter 12, this volume) and will only be mentioned briefly here.

11.2 ALKYL AND ARYL DERIVATIVES

11.2.1 Syntheses and Structures of R_3M, Ar_3M and Lewis-base Adducts

The trialkyls of group 13 metals and their adducts with amines and phosphines have been successfully used for the MOCVD of 13–15 semiconductors.[8,9] Recent interest has concentrated on the adducts of such trialkyl/triaryl species with suitable Lewis bases. The advantages of such derivatives over the parent species are the reduced fire and toxicity hazards, consequently easing handling and purification procedures.

The majority of these involve the co-ordination of a nonprotic donor to the group 13 metal centre, for example, $(Me_3Ga)_2{\cdot}(dppe)$ (**1**) and $(Me_3In)_2{\cdot}(dppe)$ (**2**),[10a] $Me_3InN(C_2H_4)_3N$ (**3**),[11] $Ph_3Ga{\cdot}P(TMS_3)_3$

(**4**),[12] [(TMS-CH_2)$_3$Ga]$_2$·$Me_2NC_2H_4NMe_2$ (**5**),[13] Me_3Ga·N(C_2H_4)$_3$N·$GaMe_3$ (**6**),[14] (urotropine)·*x*$GaMe_3$ (urotropine = $C_6H_{12}N_4$, $x = 1$ (**7**) or 2 (**8**)),[15] (TMS-CH_2)$_3$In·As(TMS)$_3$ (**9**), Ph_3Ga·As(TMS)$_3$ (**10**)[16] and (Me_3Ga)$_2$(dibenzo-18-crown-6) (**11**).[17] Thus, there are many examples of R_3M species acting as acceptors of one lone pair. However, only rarely are adducts formed where two or more Lewis base donor atoms are present per indium atom. Recently, the adduct Me_3In·(Pr^iNCH_2)$_3$ (**12**)[18] (where (Pr^iNCH_2)$_3$ = *N,N',N''*-triisopropyl-1,3,5-triazacyclohexane) has been synthesized; it exhibits trimethylindium accepting three lone pairs. The solid-state structure of (**12**) shows the indium atom to reside above the six-membered ring, with the isopropyl groups in an equatorial position and the nitrogen atoms bonded to indium are staggered with respect to the indium–methyl bonds. Interestingly, the angles between the indium–carbon bonds give a shallow InC_3 pyramid (average C–In–C 115.4°), which is typical for adducts where trimethylindium has accepted only one electron pair. The In–C bonds, however, show a slight elongation (In–C-1 0.2223 nm and In–C-2 0.2247 nm) compared with previous reported adducts (ranging from 0.2147 nm to 0.2186 nm). The In–N bond lengths in (**12**) are all equal and are the longest yet to be reported, 0.2776 nm (cf. 0.238 nm in (**20**)[11]). The equality of these bond lengths suggests that the interactions of the indium with each nitrogen atom are similar; their unprecedented long bond lengths suggesting a very much weaker interaction than in the normal two-electron adduct bonds.

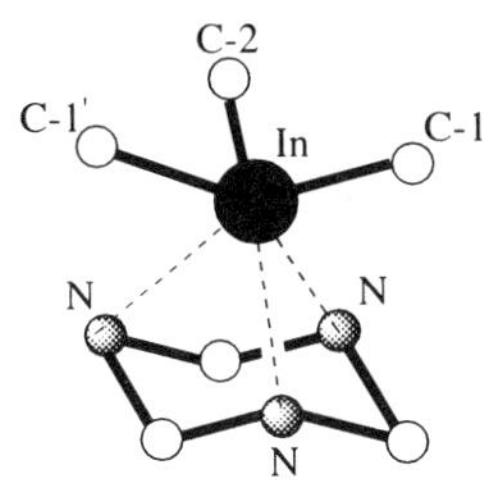

(**12**)

Among the other types of adducts are those involving protic donors, where metallation of the donor has not occurred, for example, Me_3Ga·NH_2Bu^t (**13**) and Me_3In·NH_2Bu^t (**14**),[19] (Me_3CCH_2)$_3$Ga·$PHPh_2$ (**15**),[20] Bu^t_3Ga·NH_2Ph (**16**),[21] Me_3Ga·$NHCy_2$ (**17**) and Me_3Ga·NH(CHMe)(CH_2)$_3$CHMe (**18**),[22] HcyclenAs·$GaMe_3$ (**19**) (Hcyclen = HN[(CH_2)$_2$N]$_3CH_2CH_2$)[23] and $Me_3InNHMe(CH_2)_2NHMeInMe_3$ (**20**).[11]

The growing importance of group 13 derivatives in the production of thin films and epitactic layers has prompted a search for new precursors. Owing to the limited chemical and thermal stability of such precursors, difficulties are often encountered in the epitaxial process, especially in evaporation or in the formation of alkoxy impurities due to traces of oxygen. By the application of intermolecular adducts, for example, Me_3Ga·NMe_3 or Me_3In·PEt_3, as source materials, such side reactions are drastically reduced. However, even these precursors of significantly enhanced thermal stability cause evaporation problems, their vapour pressures being significantly reduced.[24] Schumann *et al.* have used the concept of intramolecular coordination to yield a series of monomeric, volatile compounds which are liquids at room temperature or very low melting solids. The intramolecularly base-stabilized complex Ga[(CH_2)$_3$]$_3$N (**21**), was obtained by the reaction of $GaCl_3$ with three equivalents of N[(CH_2)$_3$MgCl]$_3$[25] (Scheme 1). Complex (**21**) is a colourless solid melting at 54 °C and is only slowly attacked by atmospheric oxygen. An x-ray study of (**21**) shows the gallium atom to be surrounded trigonally monopyramidally by three CH_2 groups and the amine nitrogen. The GaC_3 moiety is trigonal planar while the electron deficiency of the gallium centre is balanced by the intramolecular coordination of the nitrogen atom to the apex of the pyramid (Ga–N 0.2095 nm, Ga–C 0.1998 nm, C–Ga–C 120°, C–Ga–N 89.2°), the Ga–N distance being slightly longer than the sum of the covalent radii of gallium and nitrogen (0.195 nm) due to the strain in the molecular framework.

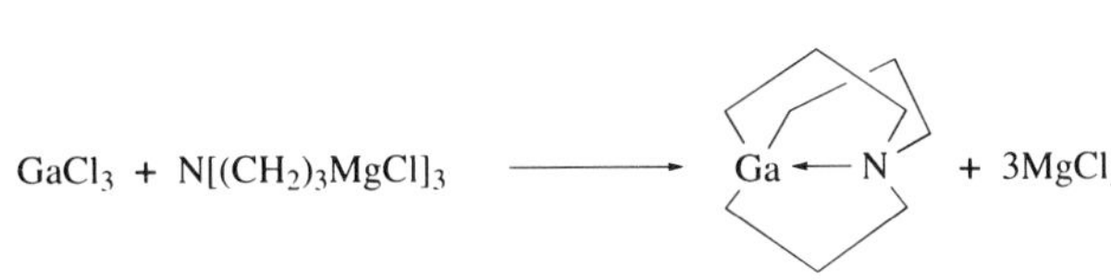

(**21**)

Scheme 1

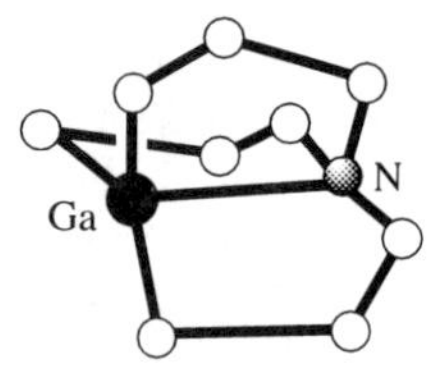

(21)

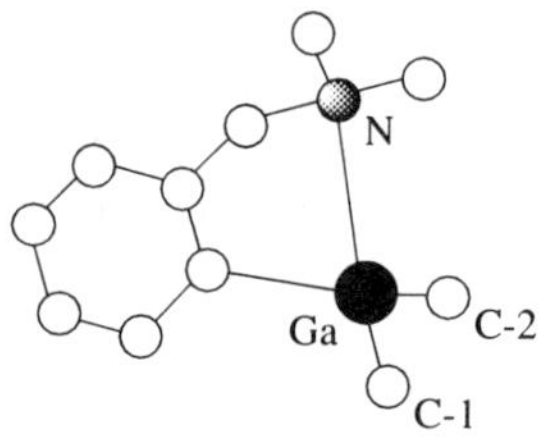

(22)

Such a trigonal planar GaC_3 moiety is remarkable, since in all previous triorganogallium–amine adducts, the gallium has assumed a tetrahedral geometry.[10–23] Other similar base-stabilized intramolecular complexes have been structurally characterized by Schumann and co-workers. The reaction between Me_2GaCl and *o*-[(dimethylamino)methyl]phenyllithium leads to the formation of [*o*-{(dimethylamino)methyl}phenyl]dimethylgallium (**22**), the analogous indium complex, [*o*-{(diethylamino)methyl}phenyl]dimethylindium (**23**), having also been synthesized and structurally characterized.[26]

Both of these compounds are almost air stable and show only slow decomposition in contact with moisture. Their vapour pressures are similar to that of $Me_3Ga{\cdot}NMe_3$,[24] this only being suitable for low-pressure metal organic vapour phase epitaxy (MOVPE). The crystal structures of (**22**) and (**23**) show the metals to be in distorted tetrahedral environments, being bonded to three carbon atoms and the nitrogen of the amino group. In (**22**) the GaC_3 moiety is close to planarity with the gallium atom pulled out of the C-1–C-2–C-3 plane by 0.0284 nm, while the indium in (**23**) resides 0.0244 nm above this plane. The tetrahedral distortion showing nearly 'normal' angles (C-1–Ga–N 104.7°, C-2–Ga–N 105.1°) and also a significantly different angle (C-3–Ga–N 84.4°) is caused by the transannular Ga–N interaction. This also manifests itself in the C-8–C-3–Ga angle which is 110.7° (the expected angle would be 120° for sp^2 hybridized C-3). As with (**21**) the Ga–N contact is longer (0.213 nm) than the sum of the covalent radii of gallium and nitrogen (0.195 nm). Similar effects are seen in the indium analogue (**23**).

Among other trialkyl or triaryl group 13 derivatives are those containing no donor atoms. The period 1982–1992 has seen the structural characterization of several new derivatives. Tris(benzyl)indium, $(PhCH_2)_3In$ (**24**) has been synthesized by the reaction of $PhCH_2MgCl$ with $InCl_3$ in toluene.[27] An x-ray crystallographic study shows (**24**) to have an almost trigonal planar GaC_3 centre, the sum of the C–In–C angles being 356.3° with the indium atom lying 24.6 pm above the C_3 plane. Salient structural parameters are: average In–C 0.2195 nm, average In–C–C_{Ph} 0.1150 nm. The solid-state structure exhibits close In···C intermolecular contacts between the *ortho* and *meta* carbons of one phenyl group and the indium of a neighbouring molecule (0.300 nm and 0.308 nm, respectively; cf. sum of van der Waals radii of carbon and indium, 0.360 nm). This compares with the recently redetermined crystal structure of Me_3In (**25**), which comprises molecules bound into cyclic tetramers by intermolecular In···C contacts of 0.3083 nm.[28,29] In these tetramers each indium is 0.441 nm out of the plane of the other three, and each is linked by four longer In···C intermolecular contacts of 0.3558 nm to four others, to give an infinite three-dimensional structure. The effects of the solid-state packing on the internal geometry can be seen in the In–C bond distances: the carbon which participates in the closer In···C contact has a bond length of 0.2179 nm, whereas the nonparticipating carbons have an average length of 0.2129 nm. Although the individual C–In–C angles differ significantly from 120°, their sum is 359.94° indicating an essentially planar InC_3 skeleton.

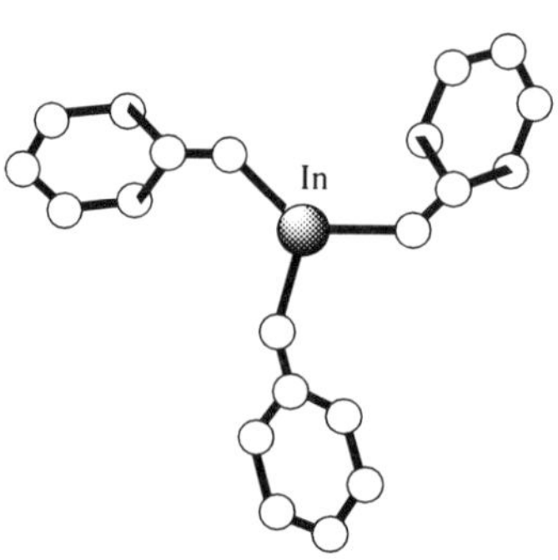

(24)

The structures of trimesitylgallium, $(Mes)_3Ga$ (**26**) and the indium analogue, $(Mes)_3In$ (**27**) have been determined.[30,31] In terms of gallium chemistry, (**26**) represents one of only a few gallium compounds containing bulky ligands to be fully characterized, and indeed the first containing bulky aromatic substituents. It can be prepared in high yield from the reaction of the mesityl Grignard reagent in ether and $GaCl_3$.

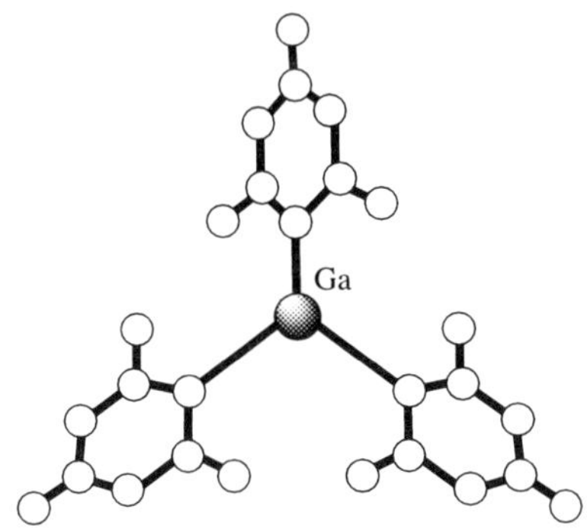

(26)

It appears that $(Mes)_3Ga$ is an exceedingly weak Lewis acid, probably due to the steric bulk of the mesityl substituents, neither Et_2O nor THF forming adducts which are stable to dissociation at room temperature. The crystal structure of (**26**) consists of discrete molecular units, there being no abnormally close intermolecular contacts. The Ga–C(aromatic) distances are equivalent, 0.1968 nm, the gallium atom lying 6.3 pm above the C_3 plane with C–Ga–C = 119.93°, so that the coordination geometry about gallium is close to the ideal trigonal planar. The aromatic rings take up a 'propeller-like' arrangement and make a dihedral angle of 55.9° with the coordination plane. In many ways the properties of the indium analogue, $(Mes)_3In$ (**27**), echo those of (**26**). The In–C distances vary from 0.2163(5) nm to 0.2170(5) nm, the geometry about indium being planar. However, whereas the mesityl groups in (**26**) are arranged in a propeller-like fashion with each ring related by a crystallographic C_3 axis, the three rings in (**27**) are crystallographically inequivalent with each ring angled differently with respect to the InC_3 core plane. It is as a consequence of this conformation that there is a wide range of intramolecular C···C contacts between adjacent *ortho* methyl groups, several of these being within the sum of the van der Waals radii (such an effect is not observed in (**26**)). Thus, it is clear that these unfavourable distortions away from C_3 symmetry must result from the increase in In–C distance (average 0.2166 nm) relative to that observed in $(Mes)_3Ga$, 0.1968 nm or $(Mes)_3Al$, 0.1995 nm, allowing the rings to adopt a new conformation. The driving force for this could be either steric (crystal packing) or electronic. The low-temperature ^{13}C NMR spectrum exhibits two resonances for the *ortho* methyl groups, the coupling constant, $J(^{13}C-^{1}H)$, for the more downfield resonance (102.0 Hz) being in the region previously observed for species containing C–H···M interactions.[32] These two lines broaden and merge into a singlet at room temperature. Additionally, the band at 2710 cm^{-1} in the IR spectrum may be associated with an agostic C–H bond. Thus, spectroscopic evidence supports a C–H···In interaction in solution.

Cowley, Jones and co-workers have recently successfully synthesized and structurally characterized the trifluoromethyl analogue of (**27**), $[2,4,6\text{-}(CF_3)_3C_6H_2]_3In$ (**28**).[33] Again, the InC_3 geometry is planar, the In–C bond length of 0.2189(9) nm being somewhat longer than in (**27**) (0.2168 nm) presumably due to the larger steric demands of the fluorinated ligand. Structure (**28**) adopts a 'propeller-type' conformation, in which the twist angles of the aryl rings with respect to the InC_3 plane are very similar (49.4°, 49.50° and 54.8°). This contrast to $(Mes)_3In$ may result from packing; however, the six (trigonal prismatic) intramolecular In···F contacts from the o-CF_3 groups (average distance 0.276 nm, sum of van der Waals radii, 0.340 nm) may also play a role in the conformational preference.

Following the synthesis of $(Mes)_3Ga$, the structures of two complexes containing π-bound $Mo(CO)_3$ units to the aryl rings have been reported by Beachley Jr. in order to probe the way in which the chemical properties of the group 13 moiety were affected by the electron-withdrawing $Mo(CO)_3$ unit. Reaction of $(Mes)_3Ga$ with $Mo(CO)_6$ in DME resulted in the formation of $[(CO)_3Mo(\eta\text{-Mes})]Ga(Mes)_2$ (**29**) and $[(CO)_3Mo(\eta\text{-Mes})]_2Ga(Mes)$ (**30**).[34]

These represent the first examples of arene–group 6 metal tricarbonyl complexes which have gallium-containing substituents bonded to an arene ring. Depending on the stoichiometry employed, the monomolybdenum or the dimolybdenum species could be isolated as the major products. However, the trimolybdenum species was never observed even when an excess of $Mo(CO)_6$ was reacted. The crystal structures of (**29**) and (**30**) consist of discrete molecules and, as in the case of $(Mes)_3Ga$, there are no

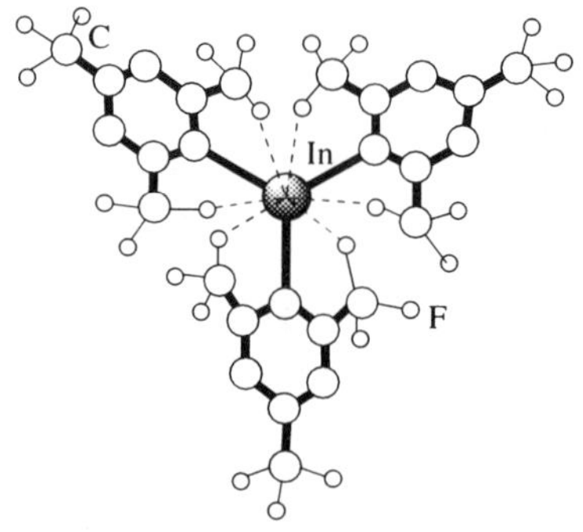

(28)

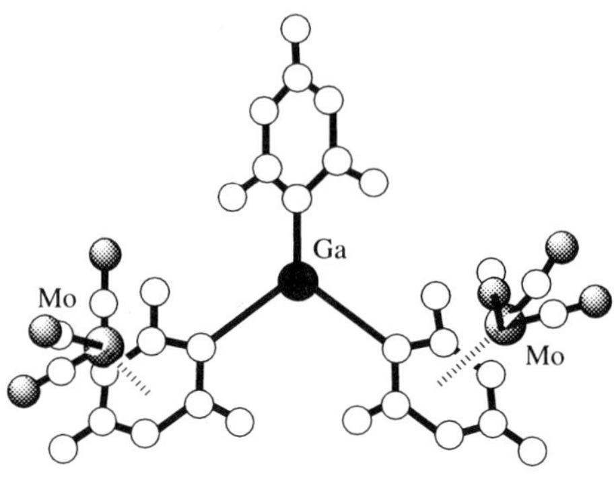

(30)

abnormally close intermolecular interactions. The geometry around the central gallium differs in the two complexes. In the case of (**29**), the gallium atom is distorted from trigonal planar and lies 1.1 pm above the plane of the three bonded carbon atoms. Additionally, two of the C–Ga–C angles deviate considerably from the ideal angle of 120° (127.2° and 109.3°), the remaining angle being 123.5°. The geometry of (**30**) is closer to being trigonal planar. As with other triarylgallium compounds, the aromatic rings in both species are twisted from coplanarity and exhibit a propeller-like arrangement. The effect of the coordinated molybdenum atoms in (**29**) and (**30**) is to slightly lengthen the Ga–C bonds, for example, 0.1955(14) nm for the mesityl group not bound to molybdenum in (**30**) and 0.1991(10) nm for those bound to molybdenum. An unusual feature of the chemistry of these two species is their apparent stability in air for short periods, the ^{1}H NMR spectra of samples exposed for 1 h being identical to fresh samples. This suggests that the $Mo(CO)_3$ moiety hinders the approach of oxygen and water to the gallium centre. When (**29**) and (**30**) are reacted with anhydrous HCl, the gallium–mesityl carbon bond to the uncomplexed mesityl ring was preferentially cleaved, as shown by monitoring the reaction by ^{1}H NMR spectroscopy. This may be related to steric effects. Similarly, the reaction between $(Mes)Mo(CO)_3$ and $(Mes)GaCl_2$ results in the formation of $[(CO)_3Mo(Mes)]GaCl_2$, indicating that $(Mes)GaCl_2$ is apparently a better electron-donating ligand to $Mo(CO)_3$ than mesitylene.

An alternative class of potential precursors to semiconductors are compounds of the type $R^1{}_2MCH_2ER^2{}_2$ (M = group 13 element, E = group 15 element). This class of compound has a methylene group that bridges the group 13 and 15 elements in order to fix the group 13:15 ratio at 1, as required. Thus, such compounds incorporate Lewis acid and base sites without the possibility of π-bonding in the monomeric species between the vacant orbital on the group 13 atom and the electron pair on the group 15 atom. Beachley, Churchill and co-workers have reported the synthesis and characterization of $R_2MCH_2PPh_2$ (R = CH_2CMe_3, CH_2TMS; M = Ga, In).[35] They can be prepared from the reaction between R_2MCl and $LiCH_2PPh_2$ in good yield. An x-ray crystallographic study on the indium complex $[(Me_3CCH_2)_2InCH_2PPh_2]_2$ (**31**) reveals a dimeric, six-membered ring with a 'chair' conformation. The In–P···In'–P' moiety is planar with the two methylene carbons lying above and below this plane. Relevant bond lengths and angles are: average $In–C_{neopentyl}$ 0.2222 nm, average In–CH_2 0.2258 nm, average In–P 0.2699 nm (cf. 0.2755 nm in $(Me_3In)_2$·(dppe)[10b]). Angles about the indium atoms deviate substantially from the ideal tetrahedral value, the largest being that between the bulky neopentyl ligands, C–In–C 122.1°. Cryoscopic studies in benzene suggest that the structure is retained in solution (i.e., remains dimeric), the ^{31}P and ^{1}H NMR spectra also imply no observable changes in the degree of association with concentration. It is noteworthy that since compounds such as (**31**) may be viewed as simple adducts, their structure is retained in solution, especially when compared with species such as $(Me_3CCH_2)_3Ga{\cdot}PHPh_2$ which are significantly dissociated in benzene.[20]

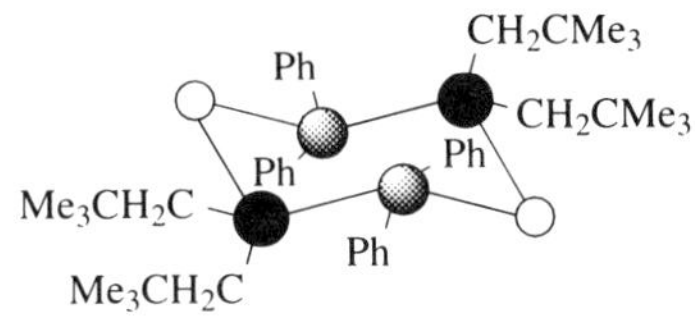

(31)

While the potential utility of organo–group 13 compounds as precursors to semiconducting materials is principally responsible for the renaissance in this area, the literature reveals a paucity of studies concerning ferrocenyl derivatives of organogallium species. Recently, the first example of a ferrocenylgallane, [(η-Cp)Fe(η-$CpGaMe_2$)]$_2$ (**32**), has been structurally characterized.[36]

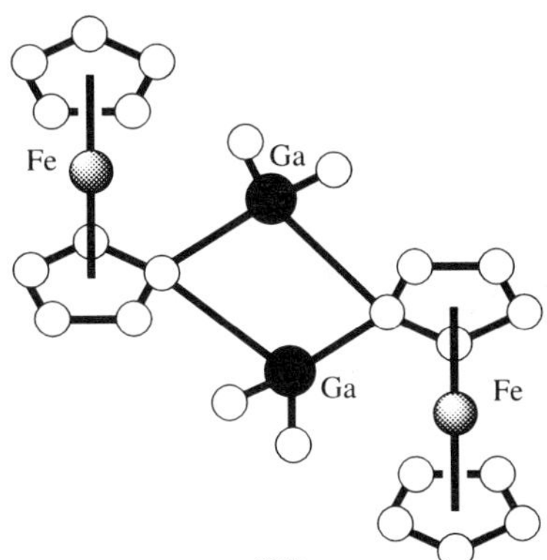

(32)

The structure consists of two ferrocenyl units bridged by two dimethylgallium units, via carbon atoms of substituted cyclopentadienyl groups, resulting in an asymmetric Ga_2C_2 four-membered ring. The Ga_2C_2 ring of (**32**) contains two short Ga–C bonds of 0.1992 nm (comparable with those found in trimethlygallium adducts) and two long Ga–C bonds of 0.2587 nm. Hence, (**32**) may be viewed as a weakly bound dimer of ferrocenyldimethylgallane, (η-Cp)Fe(η-$CpGaMe_2$). Bond angles within the ring are Ga–C–Ga, 80.7° and C–Ga–C, 99.3°, with a Ga···Ga distance of 0.2999 nm, indicating no significant metal–metal interaction. In the structure of (**32**) the long Ga–C bond distance (0.2587 nm) is among the longest gallium–carbon distances reported.

Work by Erker *et al.* demonstrates the application of trialkylgallium derivatives to form complexes exhibiting thermodynamically stable planar tetracoordinate carbon, electronically stabilized by the combined action of the alkylgallium moiety and a bent metallocene unit. Such compounds exhibit a $R_2Ga(\mu\text{-}R^1,\mu\text{-}R^2)$ unit in which the gallium centre is connected to the adjacent early transition metal centre by forming two stable σ-hydrocarbyl bridges.[37] The role of the trimethylgallium species is twofold: initially to act as a Lewis acid to abstract the Lewis basic trimethylphosphane ligand, and subsequently to act as a coupling component to react with the *in situ*-generated reactive (η^2-cyclohexyne)$ZrCp_2$ intermediate (Scheme 2). An x-ray crystallographic study of the products of such reactions was undertaken, namely $Cp_2Zr(\mu\text{-}\eta^1{:}\eta^2\text{-}C_6H_8)(\mu\text{-}CH_3)GaMe_2$ (**33**) and $Cp_2Zr(\mu\text{-}\eta^1{:}\eta^2\text{-}C_6H_4)(\mu\text{-}CH_3)GaMe_2$ (**34**).

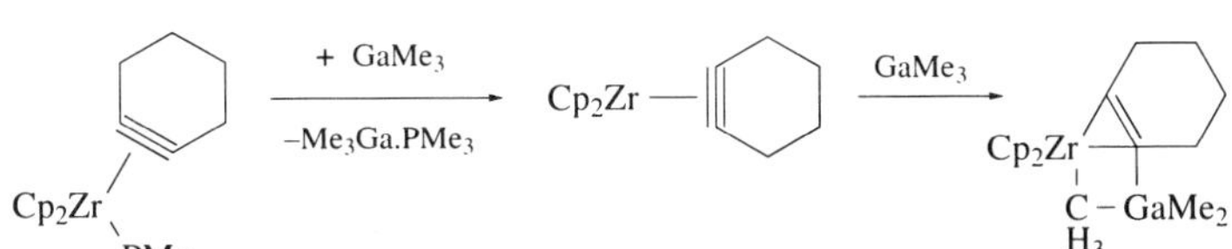

Scheme 2

The most remarkable feature of (**33**) is that the gallium centre is connected to zirconium by means of two different σ-hydrocarbyl bridges. One of them contains the cyclo-C_6H_8 system, which is η^1-bonded to gallium and η^2-coordinated to zirconium. Carbon atom C-2 is planar tetracoordinate, having

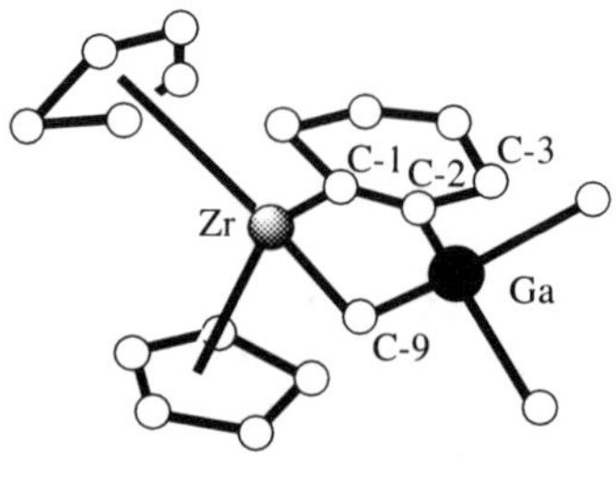

(33)

associated bond lengths of C-2–C-3 0.1572 nm, C-2–C-1 0.1314 nm (i.e., in the range of a C=C bond) and Zr–C-2 0.2423 nm. The Ga–C-2 length of 0.2096 nm is slightly longer than usually observed for Ga–C σ-bonds, whereas the terminal Ga–C bonds to the methyl groups are as expected, with an average length of 0.1974 nm. The bonding geometry at C-2 is as expected for a typical electronically stabilized planar tetracoordinate carbon derivative, exhibiting a three-centre two-electron metal–carbon–metal σ-interaction (C-1–C-2–C-3 100.3°, C-3–C-2–Ga 79.3°, Ga–C-2–Zr 61.1°). The coordination environment about the gallium is pseudotetrahedral, featuring two σ-bridging groups connecting it with zirconium. One bridge involves an sp^2 orbital at C-2 in a three-centre two-electron bonding situation, while the gallium and zirconium atoms are further connected by means of a μ-Me_3 group. The Zr–C-9–Ga angle is 74.6°, while there is a long Ga–C-9 bond of 0.2142 nm. The properties of (**34**) are largely analogous. Both complexes react with alkenes to yield organozirconium species and trimethylgallium.

Studies of aluminum alkyl-based liquid clathrates have been widely studied but gallium derivatives have been somewhat neglected. The reaction between rubidium acetate and trimethylgallium (1:2 ratio) yields Rb[Me_3GaOC(Me)OGaMe_3] (**35**).[38] The complex consists of two tetrahedrally coordinated gallium atoms lying 0.078 nm and 0.013 nm out of the acetate plane (average Ga–O 0.2069 nm, Ga–C 0.200 nm). The closest contacts of the oxygen atoms to the rubidium cation are 0.2861 nm and 0.3021 nm.

A final class of R_3M complexes are those involving crown ethers and macrocyclic amines and mixed nitrogen–oxygen donors, with interest centring around liquid clathrates. An example of this is the reaction between Me_3Ga and diaza-18-crown-6 (~5:1), leading to the formation of [Me_2Ga][$C_{12}H_{25}N_2O_4$][GaMe_3]$_2$ (**36**).[39] This represents the first organogallium–hetero crown-ether complex to be structurally characterized. Unusually, the gallium moiety within the cavity is five-coordinate. There are two types of organogallium units present: two Me_3Ga groups are terminally bound tetrahedrally to nitrogen, whereas the central gallium is bound in a distorted trigonal pyramidal manner (being bound to two oxygens, a nitrogen and two methyl groups). The terminal Ga–N bond lengths of 0.2164 nm are significantly longer than that of the gallium in the cavity, Ga–N 0.2024 nm, whereas the Ga–O distances are significantly longer than usual. It is also worthy of note that the terminal Me_3Ga units are bound exclusively to nitrogen, even though there are twice as many oxygen atoms as nitrogen atoms within the aza crown, particularly considering the great affinity of gallium for oxygen.

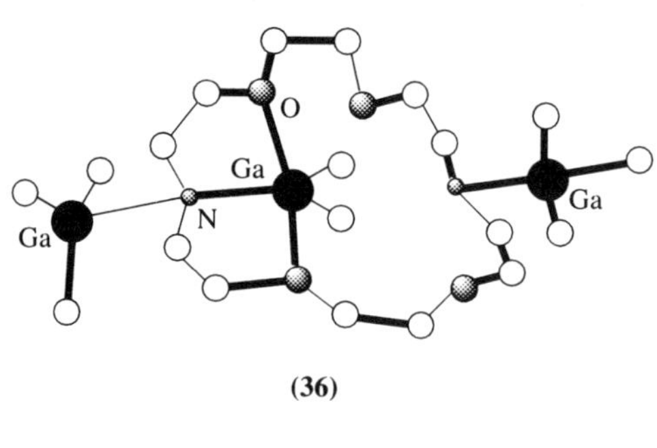

(36)

The same workers have also synthesized and structurally characterized the gallium species (Me_3Ga)$_4$(Me_4[14]aneN_4) (**37**) and (MeGa)$_2$[14]aneN_4(GaMe_3)$_2$ (**38**), from Me_4[14]aneN_4 and [14]aneN_4, respectively.[40] The structure of (**37**) is that of a simple adduct, the four Me_3Ga groups being bound terminally to the four nitrogen atoms within the macrocycle (average Ga–N 0.2192 nm). However, [14]aneN_4 possesses acidic hydrogen atoms and is thus amenable to condensation reactions. The structure of (**38**) reveals that the macrocycle has experienced severe distortion.

There are two types of methylgallium moieties, condensation having occurred with only two of the four reacting trimethylgallium molecules. Hence, two are bound as simple adducts (average Ga–N

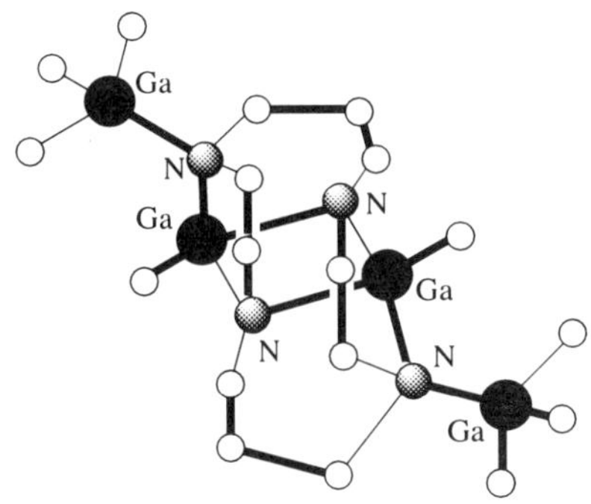

(38)

0.1995 nm and Ga–N 0.2126 nm), whereas the two condensed fragments form a Ga_2N_2 ring residing in the central cavity of the macrocycle. The Ga_2N_2 ring is planar and slightly asymmetric (average Ga–N 0.2008 nm), with a Ga–N–Ga angle of 90.2° and a N–Ga–N angle of 89.8°. Related structures to (**37**) are $(GaMe_3)_2$[14]aneN_4 (**39**) and $(AlMe_3)_2$[14]aneN_4 (**40**).[41] Reaction of (**40**) with two moles of trimethylgallium leads to the two Me_3Ga units forming adducts with two of the nitrogens from the aza crown, while the two Me_3Al units originally present in (**40**) participate in condensation with the aza hydrogen atoms of the crown yielding $(MeAl)_2$[14]ane$N_4(GaMe_3)_2$ (**41**).[41] This is isostructural with the gallium analogue (**38**), possessing an Al_2N_2 ring (Al–N 0.1943 nm).

11.2.2 Organohalogen and Pseudohalogen Derivatives

Triorganometal (R_3M, R = alkyl or aryl) and mixed organometallic halogen (R_2MX and RMX_2, R = alkyl or aryl, X = halogen or pseudohalogen) derivatives have continued to be an active area of research in the period from 1982 to 1992 mainly due to their use as precursors for the synthesis of device materials. Central research interests have involved the continued investigation of structural trends in the solid state and the modification of these structures by the employment of particular organic groups attached to the metals (e.g., bulky substituents and those containing pendant intramolecular donor centres). The latter has been aimed both at the tuning of these species as reagents for MOCVD applications (e.g., volatility) and at increasing the stability of these often highly air- and moisture-sensitive compounds. Group 13 adducts with various neutral heteroatomic Lewis bases (e.g., 1,2-bis(diphenylphosphido)ethane[42]) have also been employed as direct replacements for the group 13 trialkyls conventionally used in the MOCVD process. The adducts (e.g., $Me_3In{\cdot}PEt_3$[43]) are more stable than their components and are therefore easier to handle and store.[9]

Oligomerization through M–X–M bridges (which increases the coordination sphere of the metal centres) into dimers and polymers is the norm for the organohalogen derivatives in the solid state. Monomeric mixed organohalogen complexes are formed only where functionalized organic groups, in which heteroatomic and/or pendant heteroatomic groups are present, satisfy the coordination sphere of the metal, or where intermolecular Lewis base donation occurs.

A range of coordination geometries is observed for monomeric organohalogen complexes in the solid state in which inter- or intramolecular Lewis base solvation of the metal centres is present. These range from four-coordinate tetrahedral adducts to irregular five- or six-coordinate complexes. Simple adducts, such as [$MeInCl_2$. NH_2Bu^t] (**42**),[44] [$EtInX_2{\cdot}Me_2NCH_2CH_2NMe_2$] (X = Br, I) (**43**),[45] [$Cy_2GaBr{\cdot}NH_2Ph$] (**44**),[46] [$Ph_2(Cl)Ga{\cdot}P(TMS)_3$] (**45**),[12] [$(Me_3CCH_2)_2Ga(Cl){\cdot}As(TMS)_3$] (**46**),[47] [$(Ph_2)PCH_2P(Ph_2){\cdot}Ga(Cl_2)Me$] (**47**),[48] have been prepared and structurally characterized in the solid state. Generally, four-coordinate geometries occur here at the metal centres, an exception being (**43**) where five-coordinate indium is present. Structures (**42**) and (**44**) are of interest in regard to the presence of potentially reactive acid N–H protons on their primary amine Lewis base donors. As such, both can be regarded as trapped intermediates for the reaction of primary amines with organogallium species. The solid-state structures of (**42**) and (**44**) make an interesting contast. Whereas in (**44**) short intramolecular (0.248 nm) N–H···Br–Ga interactions are present, in (**42**) association of the monomer units occurs via intermolecular N–H···Cl–In interactions into a polymer.

The structure and solution dynamics of complex (**47**) is also worthy of note. Only one of the PPh_2 donor centres is attached to the gallium atom in the solid state. However, in solution a dynamic process in which the gallium centre flips between the two PPh_2 groups is implicated. The reactivity of the ligand $(Ph_2)PCH_2P(Ph_2)$ towards $M(Me)_3$ (M = Al, Ga) is very different to that of early main group metal

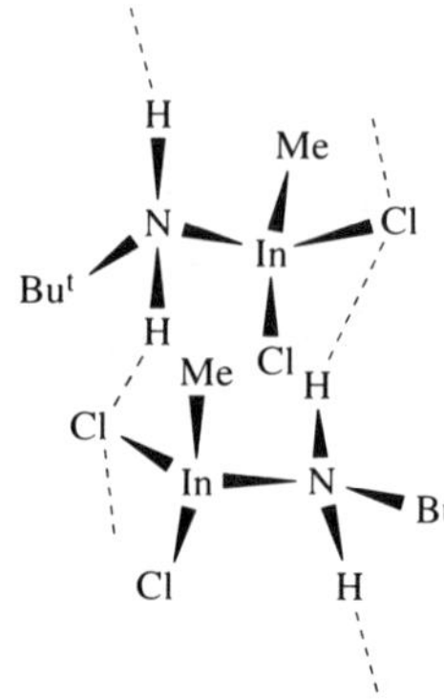

(42)

organometallics which deprotonate the ligand at the methylene group,[49] and is also at variance with the reactivity of $(Ph_2)PNHP(Ph_2)$, which is easily deprotonated by $M(Me)_3$.[50]

The base-stabilized complexes $[Ga(Cl)\{2\text{-}C(TMS)_2C_5H_4N\}_2]$ (**48**),[51] $[Ga(Cl_2)\{2,6\text{-}(N(Me_2)CH_2)_2C_6H_3\}]$ (**49**),[52] $[Tl(Cl_2)\{2\text{-}N(Me_2)CH_2C_6H_3\}]_2$ (**50**),[53] $[InBr\{MeN(CH_2)_3\}_2]_\infty$ (**51**),[54] $[In(Cl)(Me)\{2,6\text{-}(N(Et_2)CH_2)_2C_6H_3\}]$ (**52**),[55] have all been characterized in the solid state and exhibit five- or six-coordination of their metal centres. Complexes (**48**), (**49**) and (**52**) are all monomers in the solid state. However, whereas (**49**) and (**52**) have pseudotrigonal bipyramidal metal centres, (**48**) has a square-based pyramidal geometry. The reduction of the latter with Na/K in THF gives a monomeric radical $[Ga\{2\text{-}C(TMS)_2C_5H_4N\}_2]$• (**53**) in solution. In the absence of solvent, reversible reductive coupling of the ligands to substituted 4,4'-dipyridyl ligands occurs as is shown by the isolation of the 4,4'-dipyridyl ligand after hydrolysis and oxidation.

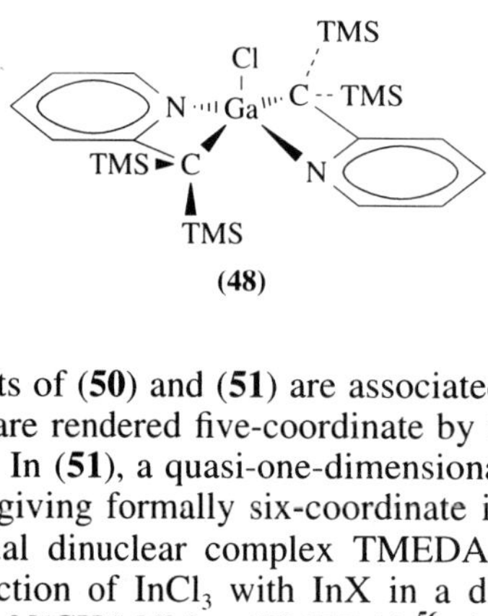

(48)

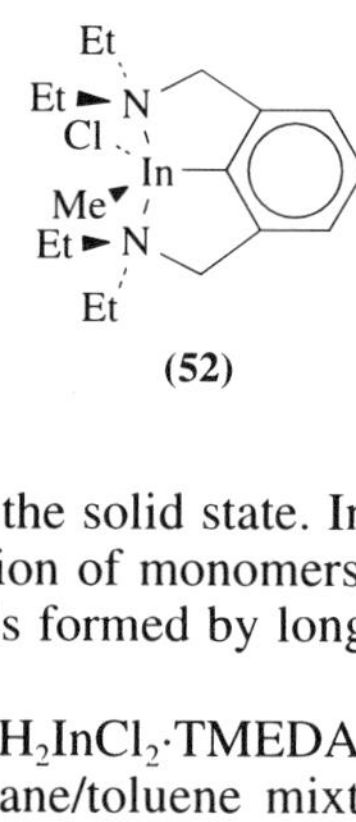

(52)

The monomer units of (**50**) and (**51**) are associated further in the solid state. In the structure of (**50**), the thallium centres are rendered five-coordinate by the association of monomers into dimeric units by $Tl(\mu\text{-}Cl)_2Tl$ bridging. In (**51**), a quasi-one-dimensional polymer is formed by long-range intermolecular In···Br interactions, giving formally six-coordinate indium.

The highly unusual dinuclear complex $TMEDA\cdot(Cl)(X)InCH_2InCl_2\cdot TMEDA$ (X = Cl, Br) (**54**) is produced by the reaction of $InCl_3$ with InX in a dichloromethane/toluene mixture together with the bidentate ligand $Me_2N(CH_2)_2NMe_2$ (TMEDA).[56] A tentative mechanism for the formation of (**54**) involves the initial formation of $ClCH_2In(Cl)(X)$ (**55**) by the interaction of CH_2Cl_2 with InX, followed by its reaction with the In–In-bonded dimer $(Cl)(X)In\text{–}InCl_2$ (**56**) (Scheme 3).

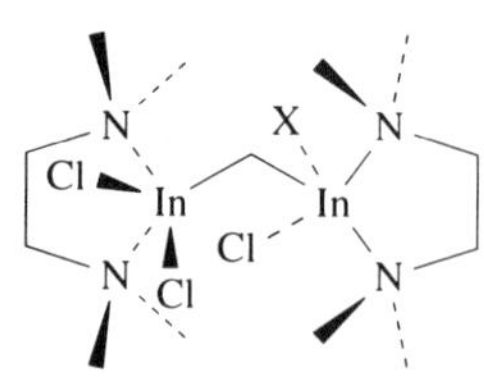

(54)

As pointed out earlier, generally dimeric or polymeric complexes occur even where bulky organic substituents are attached to the metals by association of monomer unit via halogen bridges. This is illustrated by the host of derivatives of the general formulae $R_nMX_{(3-n)}$ (X = halogen, n = 1 or 2) which fall into this structural class. These include the complexes $[(2,4,6\text{-}Me_3C_6H_2)_2In(\mu\text{-}Cl)]_2$ (**57**),[31] *trans*-$[(Cp^*)(Cl)Ga(\mu\text{-}Cl)]_2$ (**58**),[57] $[Cp^*_2Ga(\mu\text{-}Cl)]_2$ (**59**),[57] $[Bu^t_2Ga(\mu\text{-}I)]_2$ (**60**),[58] $[(2,4,6\text{-}Me_3C_6H_2)_2In(\mu\text{-}I)]_2$ (**61**),[59] *trans*-$[\{(TMS)_2CH\}(Pr^i)In(\mu\text{-}Cl)]_2$ (**62**),[60] *trans*-$[(Bu^t)(Cl)Ga(\mu\text{-}Cl)]_2$ (**63**),[61] $[(C_6H_{11})_2Ga(\mu\text{-}Cl)]_2$ (**64**),[61] *trans*-$[\{H_2B(pz)_2\}(Me)In(\mu\text{-}Cl)]_2$ (pz = pyrazolyl, $(CH)_3N_2$) (**65**),[62] $[(2,4,6\text{-}Me_3C_6H_2)In(\mu\text{-}I)_2]_\infty$

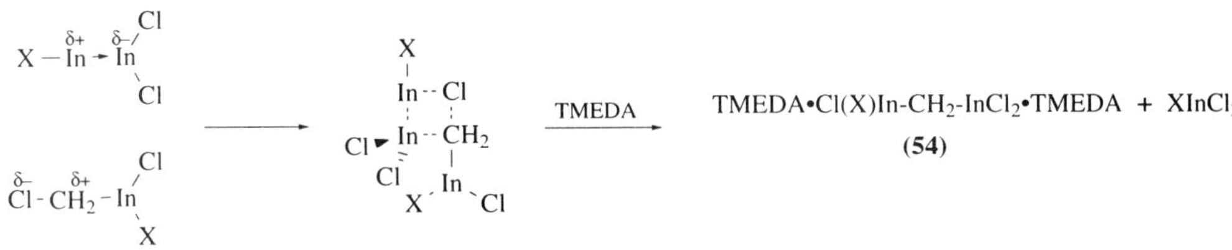

Scheme 3

(66),[59] $[Pr^i_2In(\mu\text{-}Cl)]_\infty$ **(67)**,[63] $[(2,4,6\text{-}Me_3C_6H_2)Ga(Cl)(\mu\text{-}Cl)]_\infty$ **(68)**,[64] $[$*trans*$\text{-}\{(CH_2CMe_3)(CH_2TMS)In(\mu\text{-}Cl)\}_2]_\infty$ **(69)**.[65]

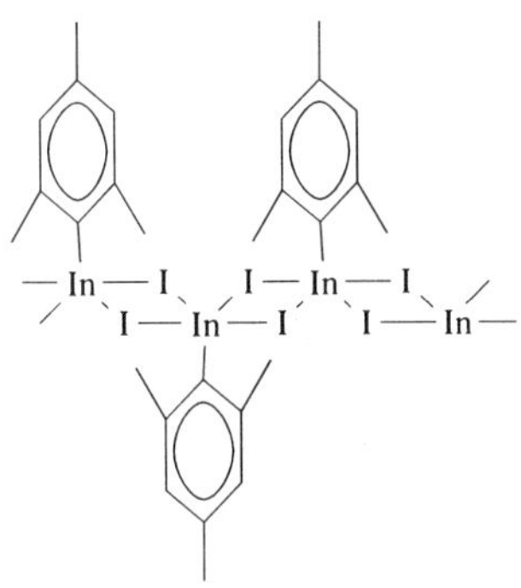

(66)

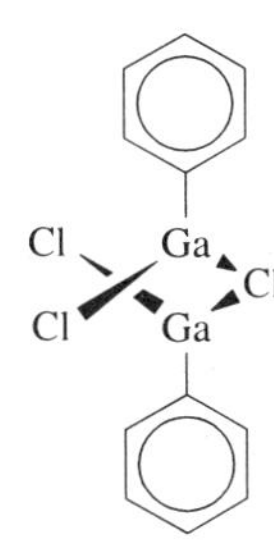

(68)

Dimeric complexes **(57)**–**(65)** almost always have planar $M(\mu\text{-}X)_2M$ rings in the solid state and it is noteworthy that, where there is the possibility of isomerism, the *trans* isomers always occur in the solid state. An exception to this is that of **(61)** in which a 'butterfly-shaped' $(InI)_2$ core is present with a fold angle of 23.2° from planar. This is in contrast with the planar core geometry of the chlorine analogue **(57)**. The latter stresses the subtle influence of steric bulk of the organic groups and the strength of bridging on the nature of the dimeric cores (Section 11.3 gives further examples of this effect). The nature of these dimeric species in the solid state can often change when in solution. Thus, **(61)** is, in contast with its dimeric bromine analogue, largely monomeric in benzene and CH_2Cl_2 solutions. The degree of association in solution presumably reflects the effectiveness of the halogen bridging between the metal centres.

An examination of the structures of the polymeric complexes listed earlier, **(66)**–**(69)**, illustrates the effect of the steric requirements of the organic groups attached to the metals on the extent of association in the solid state. The ladder polymer structures of **(67)** and **(69)** can be seen as coming about by the lateral association of dimeric units. This is exemplified particularly in the structure of **(69)** where dimer units are only loosely associated by long-range In···Cl interactions (0.3523(3) nm) along the *a*-axis. The structures of polymeric **(66)** and of the related complex **(68)** make an interesting comparison. Whereas **(66)** has a structure in which the indium centres are linked by two μ-I bridges, in complex **(68)** the gallium centres form a polymer strand in which only one μ-Cl bridge is present, the other chlorine atom on each planar gallium being terminally attached. Undoubtedly the steric influence of the very bulky mesityl groups in these complexes, together with the smaller size of gallium compared with indium, explains this structural difference.

The polymeric complexes $[Me_2InCN]_\infty$ **(70)**[66] and $[Me_2GaN_3]_\infty$ **(71)**[67] (see Section 11.4.2) have similar polymeric spiral structures in the solid state. The latter are related to the structures of $[Ph_2SbNCS]_\infty$ **(72)**[68] and $[Me_3SnCN]_\infty$ **(73)**.[69]

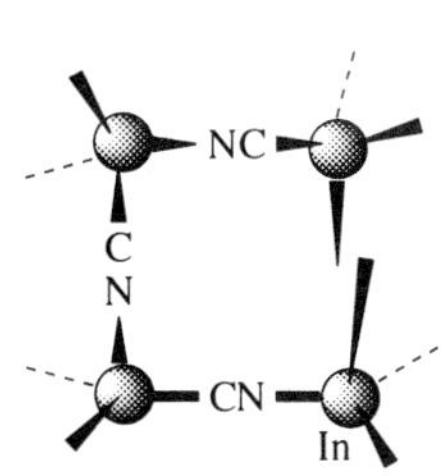

(70)

11.2.3 Cationic and Anionic Derivatives

Anionic organometallic derivatives have been known for many years, as would be expected for complexes of type MR_3 having a vacant orbital. The chapter by Tuck in *COMC-I* provides an excellent review of the area up to 1982. However, this chapter has dealt mainly with x-ray structural studies of new compounds and as such there are relatively few new compounds to report. Many of the compounds applicable to this section have been previously discussed in other sections, notably in organohalogen derivatives.

Synthetic methods of preparing these compounds still centre largely on the addition of tetraalkyl ammonium halide salts to R_3M, or the addition of metal hydrides/alkyls/aryls to R_3M. One new potential synthetic route has been discovered by Uhl (Stuttgart) using the mixed valence salts $M[MX_4]$. It is found during the course of the coupling reactions of these mixed valence salts with bulky lithium alkyls, that not only are the metal–metal-bonded species $R_2M–MR_2$ (M = Ga, In) produced, but also metal anions are formed as by-products. For instance, mixing In_2Br_4 and $LiCH(TMS)_2$ yields $[(TMEDA)Li(\mu\text{-}Br)_2In(CH(TMS)_2)_2]$ (**74**) as a by-product, whereas using Ga_2Br_4 with the same lithium alkyl yields the corresponding lithium tetraalkylgallate as a by-product.[70,71] The exact nature of the products is highly dependent on the solvent system employed.

Several anionic complexes such as $[K(15\text{-crown-}5)_2][Pr^i_2InCl_2]$ (**75**),[63] $[Tl(Mes)_2][TlCl_3(Mes)]$ (**76**),[72] $[Br_3InCH_2PPh_3]$ (**77**)[73] and $[Br_3InCH_2TMEDA]$ (**78**)[73] have also been prepared. Compound (**76**) is particularly notable for the fact that it is an aryl derivative which has previously proved difficult to synthesize by traditional approaches. It is notable that instead of using alkali metal aryls as had previously been used, (**76**) is synthesized from the silver aryl compound.

Two further structures of anionic complexes have been reported after a reexamination of the interactions of alkali metal hydrides with $M(CH_2TMS)_3$ (M = Ga, In). Previous work had suggested that reductive elimination occurs.[74,75] However, mixing $Ga(CH_2TMS)_3$ with KH produces $KGa(CH_2TMS)_3H$ (**79**), whereas mixing $In(CH_2TMS)_3$ with KH yields the tetraalkyl indate $KIn(CH_2TMS)_4$ (**80**).[76] The structures of (**79**) and (**80**) have, as expected, tetrahedral anions.

Despite the well-established routes to anionic complexes, their chemistry still remains relatively uninvestigated.

Like their anionic counterparts, cations of M^{III}, particularly those of thallium have been known for many years. In polar solvents, TlR_2X dissociates to form the essentially linear cation TlR_2^+.[77] The first crystal structure containing the Me_2Tl^+ cation followed in 1967.[78] This section deals exclusively with this type of diorganothallium cation.

Since 1982 there have been many further examples of crystal structures involving the essentially linear $TlMe_2^+$ cation, principally possessing different donor ligands attached to thallium. These include $[Me_2Tl\cdot\text{dicyclohexano-18-crown-6}]^+$ (**81**),[79] $[Me_2Tl\cdot\text{tridecalino-18-crown-6}]^+$ (**82**),[80] $[Me_2Tl\cdot\text{dinaphthaleno-18-crown-6}]^+$ (**83**),[81] $[Me_2Tl\cdot\{(py)_2CH_2\}NO_3]_2$ (**84**)[82] and $[Me_2Tl\cdot\{(terpyridyl)(H_2O)\}NO_3]$ (**85**).[82]

Several different cations involving alternative alkyl/aryl groups have also been reported. $[Tl(Mes)_2][TlCl_3(Mes)]$ (**76**) contains the first diaryl thallium cation which is essentially linear (C–Tl–C = 173.1°).[72] The diacetonyl complex $[Tl\{CH_2C(O)Me\}_2(\mu\text{-}CF_3SO_3)(bipy)]_2$ (**86**) has also been reported with a slightly bent TlR_2^+ fragment (C–Tl–C = 160.8°).[83] This compound is important due to the role of thallium in the organic chemistry of ketones.

Several structural studies have been published on the dialkyl indium analogues. Dimethyl indium(III) complexes were isolated from the reaction of dimethyl indium nitrate with a variety of neutral nitrogen-donor ligands in aqueous methanol.[84] The geometry of the $InMe_2$ fragment is significantly distorted away from linear as a result of interactions with the nitrogen-donor ligands.

A further diorganoindium compound, $[Pr^i_2In(THF)_2][BF_4]$ (**87**) has recently been prepared.[85] The C–In–C angle is 160.3°, with the structure consisting of $[Pr^i_2In(THF)_2]^+$ units linked by BF_4^- anions to form an infinite polymer.

Only very recently has the structural characterization of a dialkylgallium cation been reported. Reaction of Bu^tNH_2 with Me_2GaBr in ether yields $[Me_2Ga(Bu^tNH_2)_2]Br$ (**88**).[86] The gallium cation adopts a highly distorted tetrahedral geometry, resulting in a C–Ga–C angle of 121.5°.

11.3 ORGANOMETALLIC HYDRIDES

The extensive covalent chemistry of boron hydrides finds no parallel with the heavier elements of group 13. Organometallic hydrides of gallium, indium and thallium are characterized by extreme air and moisture sensitivity as well as thermal instability. Thus, their characterization is in its infancy, there being only a handful of structurally authenticated examples.

Synthetic strategies are very much limited to the substitution of halides by powerful hydride donors, such as $LiAlH_4$, or to the addition of H^- by alkali metal hydrides, for example, MH (M = Li, Na, K). The structures are seldom monomeric, the existence of hydride bridges between metal centres typifying most of this class of organometallic compounds.

Various amine adducts of gallane, GaH_3, have been structurally characterized in the gas phase by electron diffraction, the simplest being trimethylamine–gallane, $Me_3N{\cdot}GaH_3$ (Ga–H, 0.1497 nm) (**89**).[87] Other monomeric adducts include, $(GaH_3)_2$(TMEDA), (**90**) $(GaH_3)_2$(dmpe) (**91**) and GaH_3(quinuclidine) (quinuclidine = $N(CH_2CH_2)_3CH$) (**92**).[88,89] The structure of gaseous dimethylamidogallane has also been determined by electron diffraction, the major species being shown not to be the monomer Me_2NGaH_2 as suggested previously, but the dimer $(Me_2NGaH_2)_2$ (**93**) with a cyclic Ga_2N_2 skeleton (Ga–N 0.2027 nm, Ga–H 0.1487 nm, Ga–N–Ga 90.6°).[90] Thus, for many years the chemistry of gallium hydrides appeared to be dominated by complexes in which the gallane moiety was stabilized through coordination to a suitable donor. The free hydrides thus appeared to hold a shadowy existence, the gas-phase structure of gallane having only recently been elucidated via synthesis from $(H_2GaCl)_2$ and $LiGaH_4$. Electron diffraction studies have shown it to have a diborane-like structure, $H_2Ga(\mu\text{-}H_2)_2GaH_2$ (**94**),[91] the salient structural parameters being Ga–H_{br} 0.171 nm, Ga–H_t 0.152 nm and Ga–H_{br}–Ga 97.9 °. It undergoes analogous reactions to diborane, giving adducts with PH_3 and Me_3N through symmetric bridge cleavage, and ionic $H_2Ga(NH_3)_2{}^+GaH_4{}^-$ with ammonia through asymmetric cleavage.

The structures of gaseous dimethylgallane (**95**), synthesized by the reaction between Me_3Ga and $NaGaH_4$, and dimethylgallium chloride (**96**) have only recently been elucidated by electron diffraction.[92] These results indicate that the predominant vapour species at low pressures and temperatures of 290–350 K are dimeric molecules, again with diborane-like structures, $Me_2Ga(\mu\text{-}X)_2GaMe_2$ (X = H or Cl), with the heavy atom skeleton conforming to D_{2h} symmetry. Dimethylgallane thus represents the first organometallic gallium hydride containing a $Ga(\mu\text{-}H)_2Ga$ bridging unit to be structurally characterized. It is worthy of note that the Ga···Ga distance in the hydride bridged dimer $(Me_2GaH)_2$ is markedly shorter than in the chlorine-bridged analogue (0.261 nm and 0.330 nm, respectively), implying some sort of direct interaction between the two group 13 metals. This is to be compared with twice the covalent radius of tetrahedrally coordinated gallium, 0.252 nm,[93] and also the Ga–Ga distances found within various forms of elemental gallium, ranging from 0.247 nm to 0.307 nm.[94] Thus, it is not clear whether this difference is a manifestation of steric effects or merely a reflection of the different donor capacity of the bridging atom.

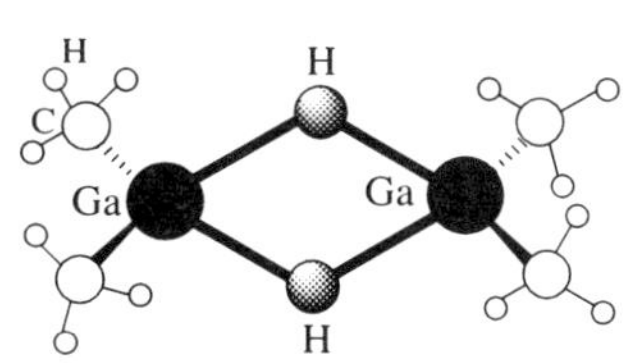

(**95**)

Previously the only known indium hydrides which were stable above room temperature were the solid $(InH)_x$, of unknown structure,[95] the trialkylindates $M[InR_3H]$ (with M = Na or K; R = Me, Et[96] or TMS-CH_2[76]) and sodium diethylindate, $Na(InEt_2H_2)$.[96] Recently, Eaborn, Smith and co-workers have managed to stabilize and structurally characterize an alkyltrihydroindate in the solid state, by utilizing the bulky tris(trimethylsilyl)methyl group (Tsi). The reaction between $[Li(THF)_3(\mu\text{-}Cl)In(Tsi)Cl_2]$ and an excess of $LiAlH_4$ at low temperature yielded crystals of $(TMS)_3CIn(H)(\mu\text{-}H)Li(THF)_2(\mu\text{-}H)In(\mu\text{-}H)$-$(H)C(TMS)_3$ (**97**).[97] Although no hydrogen atoms could be located in the crystal structure, their presence could be inferred from the observation of In–H stretches in the IR spectrum (1635, 1660, 1695, 1725 cm^{-1}). Quadrupolar interactions with ^{115}In ($I = \frac{9}{2}$) or possible intramolecular exchange made signals from In–H protons weak, broad and difficult to observe. However, a broad signal at δ 4.7 ppm from In–H–Li protons in (**97**) was detected by the $^6Li\{^1H\}$ nuclear Overhauser effect, suggesting that the hydrogen bridges between lithium and indium persist in toluene or THF. The In–C bonds are inclined at ~34° on either side of the plane defined by the lithium and indium atoms, so that if each indium atom bears a terminal hydrogen, leaving three for the bridging positions, coordination of both lithium and indium is roughly tetrahedral. Thus, the inferred structure of (**97**) fits both crystallographic and spectroscopic data. Structure (**97**) may either be regarded as electron deficient with electron density from In–H bonds delocalized onto the lithium or as a tightly interlocked ion pair $[Li(THF)_2][(Tsi)_2In_2H_5]$ in which the diindate ion has displaced THF from the coordination sphere of the lithium with a gain in electrostatic energy as cation and anion come closer together. Realistically, it is probable that both descriptions are required for a full account of the bonding.

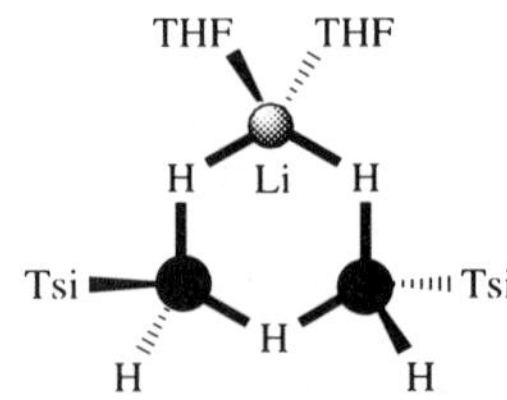

(97)

Recent progress has been made towards the nature of the reactions that occur between $Ga(CH_2TMS)_3$ and $In(CH_2TMS)_3$ with alkali metal hydrides (Beachley *et al.*). Previous reports suggested that reductive elimination reactions occurred through the thermal decomposition of the transient hydride to yield $KGa(CH_2TMS)_2$ and $SiMe_4$,[74,75] but a reinvestigation has shown this not to be the case. The product from the reaction between $Ga(CH_2TMS)_3$ and KH has been identified as $KGaH(CH_2TMS)_3$ (**79**) by an x-ray crystallographic study and spectroscopic data.[11] This complex is thermally stable up to 200 °C. The structure consists of discrete K^+ and $[GaH(CH_2TMS)_3]^-$ ions, two of each occupying the asymmetric unit. Each gallium(III) atom is in a tetrahedral coordination environment, average Ga–C 0.203 nm and C–Ga–C 111.2°, two independent Ga–H distances of 0.136 nm and 0.149 nm. The primary products of the analogous indium reaction appear to be $KInH(CH_2TMS)_3$ (**98**) and $KIn(CH_2TMS)_4$, with no evidence being found to enforce the previous claim of these being necessary intermediates in the reductive elimination pathway to low oxidation state organoindium(I) compounds, $MIn(CH_2TMS)_2$. Although $KInH(CH_2TMS)_3$ is thermally degraded at room temperature, exhaustive tests on the decomposition products under numerous reaction conditions suggest more than just a simple reductive elimination reaction. The major products include indium metal and Me_4Si, there being no evidence for the formation of $MIn(CH_2TMS)_2$. Thus, no stoichiometric reductive elimination occurs, $KGaH(CH_2TMS)_3$ being thermally stable under normal synthetic conditions, while $KInH(CH_2TMS)_3$ decomposes at room temperature to indium metal and not to organoindium(I) compounds. This difference in thermal stability between the gallium and indium derivatives can be related to the ease of the dissociative reactions of the four-coordinate ions and their relative metal–carbon bond strengths.

Full characterizations of compounds containing indium–hydrogen bonds have been hampered by the inherent instability of such a system. However, studies of the reaction between $KInH(CH_2CMe_3)_3$ and $In(CH_2CMe_3)_3$ have led to full characterization (x-ray crystallographic, physical and spectroscopic) of the hydride, $K[H\{In(CH_2CMe_3)_3\}_2]$ (**99**).[98] Reaction between $In(CH_2CMe_3)_3$ and KH initially yields the simple Lewis acid–base adduct, $KHIn(CH_2CMe_3)_3$; however, this readily reacts with more $In(CH_2CMe_3)_3$ to form (**99**). In an excess of KH this then further reacts to produce $KHIn(CH_2CMe_3)_3$ as the final product. The solid-state structure of (**99**) is best viewed as an ordered array of K^+ cations and $[(Me_3CCH_2)_3In\text{–}H\text{–}In(CH_2CMe_3)_3]^-$ anions in a 1:1 ratio, all of which are bisected by a crystallographic twofold axis. The $[(Me_3CCH_2)_3In\text{–}H\text{–}In(CH_2CMe_3)_3]^-$ anions have precise C_2 symmetry. As with other related electron-deficient systems, the bridging hydride ligands are involved in two-electron three-centre bonds. Observed In–H distances range from 0.193 nm to 0.195 nm in the asymmetric unit, while the In–H–In angles vary from 151° to 161°. The In–(μ-H) distance is, as expected, substantially longer (by ~0.02 nm) than the predicted terminal In–H distance of ~0.175 nm. The geometry about each indium atom is that of a slightly flattened tetrahedron formed by the three α-carbon atoms of the neopentyl groups and the bridging hydride ligand. The potassium counterions have their closest interactions with the hydride ligands (average K···H 0.278 nm). Cryoscopic molecular weight determinations in benzene reveal the presence of ion pairs with the simplest formula of the compound. Thus, $K[H\{In(CH_2CMe_3)_3\}_2]$ can be envisaged to exist in solution as species related to the units identified in the solid state and to have a hydride ion serving as a Lewis base to two $In(CH_2CMe_3)_3$ molecules as Lewis acids.

Solution studies on $KHIn(CH_2CMe_3)_3$ reveal the presence of dimeric units, $[K\{HIn(CH_2CMe_3)_3\}]_2$, the calculated molecular weight not varying significantly with concentration as with (**99**), thus implying neither associative nor dissociative equilibria are significant. Two possible structures have been proposed, involving on the one hand a five-coordinate indium species with bridging hydride ions, $(Me_3CCH_2)_3In(\mu\text{-}H\cdots K)_2In(CH_2CMe_3)_3$, or alternatively tetrahedral $[In(CH_2CMe_3)_3]^-$ anions with bridging potassium ions, $(Me_3CCH_2)_3InH(\mu\text{-}K)_2HIn(CH_2CMe_3)_3$. ^1H NMR spectroscopic studies on mixtures of $K[HIn(CH_2CMe_3)_3]$ and $K[H\{In(CH_2CMe_3)_3\}_2]$ exhibit only one set of resonances at chemical shifts which are intermediate between the two compounds, indicating the rapid exchange of neopentyl and hydride ligands between the two compounds. It is worthy of note that these two compounds are inert towards decomposition reactions over long periods of time, in contrast with the previously discussed $KHIn(CH_2TMS)_3$.

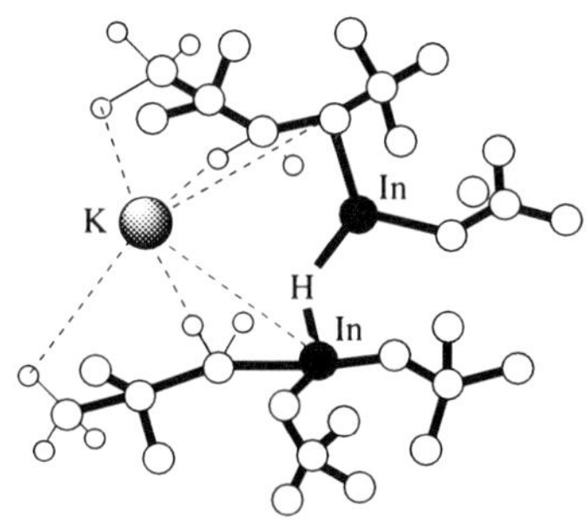

(99)

Another class of hydride derivatives, especially for the heavier members of group 13 (In, Tl), are the carbaborane complexes. The solid-state structure of [*closo*-3,1,2-$TlC_2B_9H_{11}$]$^-$ (**100**) exhibits well-separated PPN^+ cations and discrete 'dimeric' anions.[99] In the structure of (**100**) the carbaborane ligand is bonded in a pentahapto fashion with one of the thallium centres, while at the same time this ligand is approximately in a 'bridging' position between the two thallium ions. This bridged array is stabilized by a set of metal–hydrogen–boron (M–H–B) interactions between the terminal hydrogens of one carbaborane ligand and the thallium which is 'capping' the other carbaborane ligand. Although the distances involved here are rather long (Tl–H 0.262 nm) and no evidence for coupling to ^{205}Tl was observed in the ^{1}H NMR spectrum, they compare well with previously reported metal–hydrogen separations,[100] especially bearing in mind the relative size of the thallium(I) ion (the ionic radius of the thallous ion has been estimated to be between 0.15 nm and 0.17 nm, depending on coordination number).

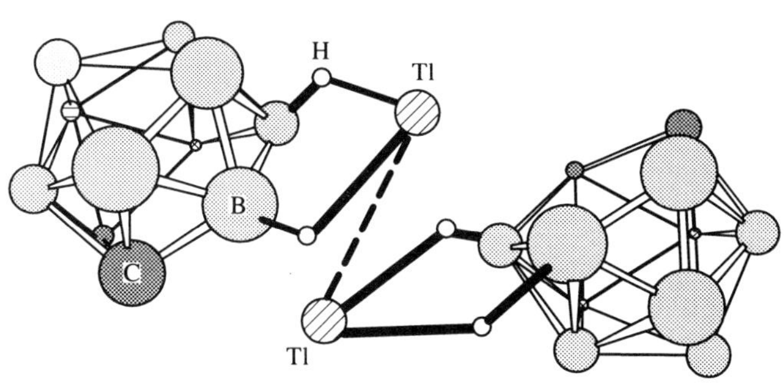

(100)

11.4 ORGANOMETALLICS CONTAINING HETEROATOMIC GROUPS

11.4.1 Introduction

As pointed out in Section 11.1, complexes containing two-electron two-centre bonds between gallium, indium and thallium (= M) and various heteroatoms (group 15, E = N, P; group 16, E = O, S), semimetals (group 15, E = As; group 16, E = Se) and some metals (group 15, E = Sb; group 16, E = Te), have been increasingly studied since the mid-1980s. This interest has been motivated by the potential uses of the indium and gallium complexes, in particular, as single-source precursors for MOCVD applications and by the desire to synthesize materials which contain a well-defined stoichiometry of the metals to the heteroatoms, semimetals or metals prior to deposition.[1] These complexes can then in principle, be degraded by thermal or chemical means to give the desired semiconducting phases (e.g., GaN, GaAs and InP).

There are a number of advantages in using such single-source precursors for the deposition of thin films over conventional methods, for example the MOCVD synthesis of GaAs from AsH_3 and $GaMe_3$ (Equation (1)) at elevated temperatures (600–700 °C).[101] These include the lower toxicity of the solid single-source precursors, their ease of purification over gaseous reactants and their lower sensitivity to oxygen and/or moisture. Additionally, many of these materials decompose under less extreme conditions than conventionally employed.[1]

$$AsH_3 + GaMe_3 \xrightarrow{600\text{–}700\ °C} GaAs + 4MeH \qquad (1)$$

Much of this section will concentrate on the increasing variety of complexes which have been characterized in the solid state and on the structural and bonding trends gleaned from such investigations. The thermal and chemical degradation of some of these materials into electronic materials will also be mentioned.

11.4.2 Syntheses and Structure of Group 13–15 Complexes Containing Ga, In and Tl Bonded to N, P, As and Sb

Most work since the early 1980s has concerned the investigation of metal phosphide and arsenide complexes due to the potential application of these complexes in the synthesis of semiconducting phases such as GaAs, MP (M = Ga and In). Cowley and Jones, and later Wells, have reviewed the syntheses and structures of complexes containing two-electron bonds between gallium or indium and nitrogen, phosphorus, arsenic and antimony, and their applications.[1]

As a whole the structures of these complexes are molecular rings or clusters in the solid state, containing $(ME)_n$ cores in which the heteroatoms bridge the metal centres. They have been most commonly prepared by the metallation of the E–H bonds (E = N, P, As, Sb) in organic acids (R^1R^2EH; R^1, R^2 = alkyl, aryl, or H) with trivalent gallium or indium organometallics (R^3_3 M) (Equation (2)) in hydrocarbon (toluene, hexane, etc.) and/or aprotic donor solvents (Et_2O, THF).

$$R^1R^2EH + R^3_3M \longrightarrow [R^3_2MER^1R^2]_n + R^3H \qquad (2)$$

An alternative route to heteroatomic complexes is the transmetallation of gallium, indium and thallium halide salts (MX_3, X = halogen) with early main group reagents (e.g., R^1R^2ELi; Equation (3)). This may produce mixed ligand complexes containing one or two halides attached to the group 13 metals per formula unit depending on the stoichiometries of the reactions. Metallation of mixed organometallic halide complexes (mainly of the monohalides, R_2MX) has also been employed in the synthesis of a range of heteroatomic complexes (Equation (4)).

$$MX_3 + 3R^1R^2ELi \longrightarrow (R^1R^2E)_3M + 3LiX \qquad (3)$$

$$R^1_2MX + R^2R^3ELi \longrightarrow [R^1_2MER^2R^3]_n + LiX \qquad (4)$$

Further synthetic approaches used since the early 1980s have been silyl halide elimination (Equation (5)) and equilibration (Equation (6)). The former has been found to be most useful where bulky substituents are present, and for antimony derivatives.

$$R^1_2E\text{-TMS} + MX_3 \longrightarrow [X_2MER_2]_n + X\text{-TMS} \qquad (5)$$

$$M(ER^1_2)_3 + R^2_3M \longrightarrow [R^1_2EMR^2_2]_n \qquad (6)$$

Surprisingly, before 1986 there had been no solid-state structure work on gallium or indium phosphides, arsenides and stibanes. This situation was despite the fact that Coates and co-workers had shown as early as 1963 that Me_3Ga and Me_3In react thermally with secondary phosphines and arsines (R_2EH, R = alkyl or aryl, E = P or As) to give materials of composition $(Me_2MER_2)_2$.[102] However, this situation has changed drastically and there are now a large range of monomeric, dimeric and trimeric $(EM)_n$ ring complexes, and a number of cage compounds containing combinations of gallium, indium and thallium with various nitrogen, phosphorus, arsenic and antimony heteroatom groups which have been characterized in the solid state.

Examples of monomeric organometallic complexes that have been structurally characterized are $[(Mes)_2As]_3Ga$ (**101**),[103] $Cp^*_2GaAs(TMS)_2$ (**102**),[104] $Bu^tGaAsBu^t_2$ (**103**)[105] and $In(PBu^t_2)_3$ (**104**).[106] In all these complexes the gallium centres are approximately trigonal planar, and the arsenic and phosphorus centres are distinctly pyramidal. In (**101**) the gallium atom is ~0.015 nm out of the As_3 mean plane and there is considerable distortion in the attachment of the $As(Mes)_2$ groups to gallium which assume a 'propeller' shape about the metal centre.[103] The orientation of the $As(TMS)_2$ group to the Cp^*_2Ga fragment of (**101**), perpendicular to the line of the C–Ga–C, argues against any π-bonding between gallium and arsenic. Such multiple bonding also does not seem likely in (**102**)–(**104**). In all the complexes (**101**)–(**104**) the Ga–As and In–P bond lengths are close to the sum of the covalent radii of the atoms involved (average 0.2492 nm in (**101**),[103] 0.2433(4) nm in (**102**)[104] and 0.2446(3) nm in (**103**);[105] cf. sum of the covalent radii of gallium and arsenic, 0.246 nm,[93] average 0.2562 nm in (**104**);[106] cf. sum of covalent radii of indium and phosphorus, 0.251 nm[8]. The slow decomposition of (**102**) to GaAs upon addition of Bu^tOH in pentane has been followed by UV–visible spectroscopy.[104] The results indicate that the process occurs with the intermediate formation of small clusters $(GaAs)_n$ which aggregate into amorphous $(GaAs)_\infty$. Monomeric, solvated GaAs species are thought to be involved in this aggregation process.

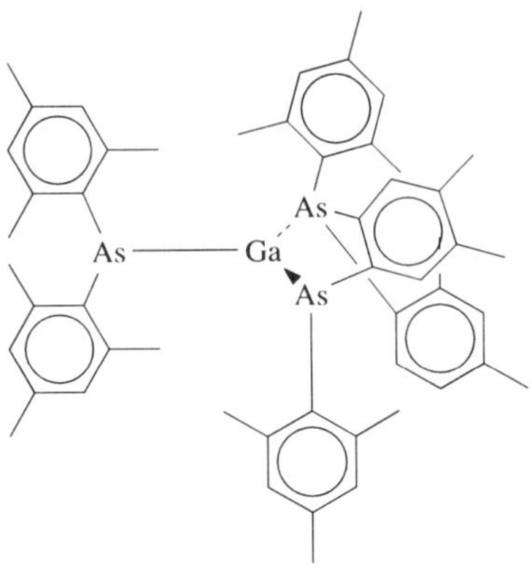

(101)

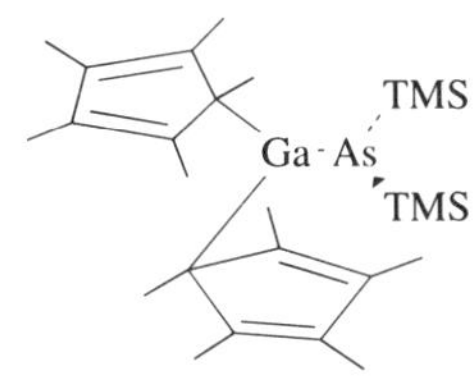

(102)

Several monomeric phosphinogallanes in which gallium is three-coordinate have been prepared using bulky substituents on $P[Bu^t_2GaPRSiPh_3]$ ($R = 2,4,6\text{-}C_6H_2Bu^t_3$ (**105**),[107] $2,4,6\text{-}C_6H_2Pr^i_3$ (**106**), $2,4,6\text{-}C_6H_2Me_3$ (**107**) and TMS (**108**)[108]) by the transmetallation reactions of the asymmetric lithium phosphides ($LiPRSiPh_3$) with Bu^t_2GaCl. The presence of only one Ga–P bond in these 'ethylene analogues' allows the investigation of potential $p\pi$–$p\pi$ multiple bonding to be more readily assessed since the effects of any π-bonding will not, as in $M[N(TMS)_2]_3$ (M = Al (**109**)[109,110] and Ga(**110**)[110]), be divided between several M–E bonds. These species are the first well-characterized examples of unassociated monophosphinogallanes. It appears in general for species of the formulae $R^1_2ME(R^2)R$ (M = Ga and Al; E = P or As[109]) that the strength of multiple bonding in the Ga(Al)–P and Ga(Al)–P bonds is dependent mainly on the inversion barrier at the phosphorus or arsenic centres (~0–151 kJ mol^{-1} for phosphanes and up to 188 kJ mol^{-1} for arsanes). Multiple bonding should be encouraged by the presence of bulky electropositive substituents which can lower the inversion barrier (cf. B–P and B–As compounds[111,112]). X-ray structural and NMR spectroscopic (^1H, ^{31}P) evidence support there being weak π-interactions in certain of these monophosphinogallanes. This is borne out by the correllation between the Ga–P bond distances in these species and the twist angles between the lone pair of phosphorus and the empty p-orbital of gallium observed in their solid-state structures. For $R = 2,4,6\text{-}C_6H_2Bu^t_3$ (**105**) and $2,4,6\text{-}C_6H_2Pr^i_3$ (**106**), in which the gallium is in an approximately trigonal planar geometry and where the twist angles are below 4°, the Ga–P distances are the shortest (0.2295(1) nm in (**105**) and 0.2296(1) nm in (**106**)) and at the lower end of the range for single Ga–P bond lengths (0.234 nm in $Ga[PH(2,4,6\text{-}C_6H_2Bu^t_3)]_3$ (**111**) and 0.232 nm in $Bu^tGa[PH(2,4,6\text{-}C_6H_2Bu^t_3)]_2$ (**112**)[108]). In contrast, for R = TMS (**108**) the Ga–P distance increases (0.2358(4) nm) and the twist angle is far greater than in (**105**) and (**106**) (76.5°). It is interesting to note that the weak π-bonding observed in (**105**) and (**106**) is disrupted in $Bu^t_2GaP(Ph)B(2,4,6\text{-}C_6H_2Me_3)$ (**113**) where more effective B–P $p\pi$–$p\pi$ bonding competes with the Ga–P π-bonding, and both the Ga–P distance and the twist angle of the phosphorus lone pair (with respect to the gallium lone pair) are intermediate between the values observed for (**105**) and (**106**), and (**108**) (Ga–P 0.2319(1) nm, twist angle 56.4°). Variable-temperature ^1H NMR spectroscopic studies on (**105**) show that there is a 53.2 kJ mol^{-1} barrier to rotation of the Ga–P bond (~35.5–37.7 kJ mol^{-1} in (**106**) and an insignificant barrier in (**107**)). This value compares with 44.4 kJ mol^{-1} for rotation of the Ga–P bonds in the 'allyl cation' analogue $(2,4,6\text{-}Me_3C_6H_2)P[Ga(2,4,6\text{-}Pr^i_3C_6H_2)_2]_2$ (**114**) in which the gallium centre is also planar.[113] It should also be noted that in $Bu^t_2GaAs[CH(TMS)_3](SiPh_3)$ (**115**) there is no evidence for π-bonding (the arsenic centre being pyramidal and the twist angle being 70.6°). The geometry at arsenic and the Ga–As bond (0.2459 nm) in the latter are similar to those in other known monomeric Ga–As-bonded complexes (**101**), (**102**) and (**103**).[103–5]

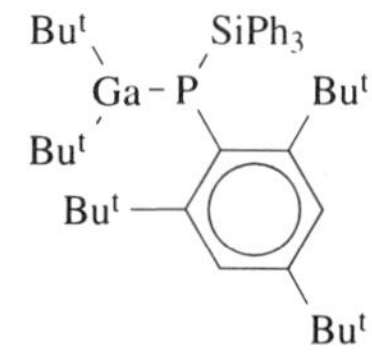

(105)

In a similar vein, Ga–N π-bonding has been investigated in $Bu^t_2GaN(R)SiPh_3$ (R = Bu^t (**116**) and 1-adamantyl (**117**)), $(2,4,6\text{-}Pr^i_3C_6H_2)_2GaNH(2,6\text{-}Pr^i_2C_6H_3)$ (**118**) and in $(2,4,6\text{-}Pr^i_3C_6H_2)GaNPh_2$ (**119**).[114] Somewhat surprisingly, on the basis of the results for the Ga–P-bonded systems above and bearing in mind the fact that the *p*-orbital on gallium and the lone pair orbital on nitrogen should be better matched than those on gallium and phosphorus, there is no clear evidence for Ga–N π-bonding in these species.

In addition to the neutral monomers, mononuclear anion complexes are also worthy of mention. Several of these have been characterized in the solid state in recent years, for example, in $Li(THF)_4^+\cdot[M(EPh_2)_4]^-$ (M = Ga, E = P (**120**); M = Ga, E = As (**121**); M = In, E = P (**122**),[115] and in $[V_2(\mu\text{-}Cl)_3(THF)_6]^+\cdot[Ga(PBu^t_2)_2Cl_2]^-$ (**123**).[1] All have very distorted tetrahedral gallium or indium centres and possess M–E bonds whose lengths are between those found in dimeric complexes containing $(ME)_2$ cores (see later) and in GaP, GaAs and InP.

Among the first gallium and arsenic complexes to be structurally characterized were the dimeric complexes $[Ga(\mu\text{-}EBu^t_2)R_2]_2$ (E = P, As, R = Me (**124**); E = P, As; R = Bu^n (**125**)) prepared by the *in situ* transmetallation reactions of $GaCl_3$ with one equivalent of $LiEBu^t_2$ and two equivalents of RLi in THF or toluene. All these species have central planar four-membered $(ME)_2$ cores in which the gallium centres are bridged by phosphorus and arsenic.[116] The Ga–P and Ga–As bond lengths in these complexes are slightly greater than expected for covalent bonds (Ga–P average 0.2475 nm and Ga–As average 0.2551 nm; cf. sum of covalent radii of Ga–P and Ga–As 0.234 nm and 0.246 nm, respectively). The planar core geometries observed in these complexes contrast with the puckered core geometry found in the bulky thiolate complex $[I_2Ga(\mu\text{-}SPr^i)]_2$ (**126**) in which a fold angle of 43.3(2)° is observed on the Ga···Ga diagonal.[117]

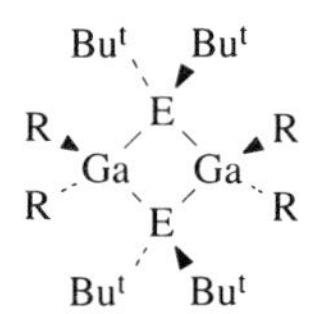

(124) and **(125)**

Dimeric structures similar to those above and of general formulae $[R^1R^2MER^3R^4]_2$ are now common for gallium and indium amides, phosphides and arsenides, and a few stibanes are also known. Phosphides and arsenides are $[Ph_2Ga\{\mu\text{-}As(CH_2TMS)_2\}]_2$ (**127**),[118] $[(TMS\text{-}CH_2)_2In(\mu\text{-}PPh_2)]_2$ (**128**),[119] $[(Me_3CCH_2)_2Ga(\mu\text{-}PPh_2)]_2$ (**129**),[119] $[Et_2M(\mu\text{-}EBu^t_2)]_2$ (M = In, E = P (**130**);[106] M = Ga, E = As (**131**)[1]), $[(TMS\text{-}CH_2)_2In\{\mu\text{-}E(TMS)_2\}_2]_2$ (E = P (**132**),[120] E = As (**133**)[121]), $[(TMS\text{-}CH_2)_2Ga(\mu\text{-}AsBu^t_2)]_2$ (**134**),[1] $[\{(TMS\text{-}CH_2)_2As\}_2Ga\{\mu\text{-}As(CH_2TMS)_2\}]_2$ (**135**),[122] *trans*-$[Bu^t(TMS\text{-}CC)Ga(\mu\text{-}PEt_2)]_2$ (**136**),[123] $[(Me_3CCH_2)_2GaAs(TMS)_2]_2$ (**46**),[47] *trans*-$[Bu^t_2Ga\{\mu\text{-}PH(C_5H_9)\}]_2$ (**137**)[124] and *trans*-$[Bu^t_2Ga(\mu\text{-}EH(Bu^t)]_2$ (E = P (**138**), E = As (**139**));[125] amides are $[Me_2In(\mu\text{-}NR^1R^2)]_2$ (R^1R^2 N = Et_2N (**140**),[126] Pr^i_2N (**141**),[126] $MeN(CH_2)_2N$ (**142**),[127] $(TMS)_2N$ (**143**),[126] $(Me_3Ge)_2N$ (**144**)[128]) and *trans*-$[(PhMe_2CCH_2)_2Ga\{\mu\text{-}N(H)(Pr)\}]_2$ (**145**);[129] mixed halogen complexes are $[Cl_2Ga\{\mu\text{-}P(TMS)_2\}]_2$ (**146**),[130] *trans*-$[(Bu^t_2Sb)(Cl)In(\mu\text{-}SbBu^t_2)]_2$ (**147**),[131] *trans*-$[Br(TMS\text{-}CH_2)_2AsGa(\mu\text{-}As(CH_2TMS)_2]_2$ (**148**),[132] *trans*-$[(Cp^*)(Cl)In\{\mu\text{-}P(TMS)_2\}]_2$ (**149**),[133] *trans*-$[(2,4,6\text{-}Bu^t_3C_6H_2PH)Bu^tGa(\mu\text{-}Cl)]_2$ (**150**),[134] *trans*-$[(Pr^i)(Cl)In\{\mu\text{-}NH(Bu^t)\}]_2$ (**151**)[135] and *trans*-$[(2,4,6\text{-}Me_3C_6H_2)\{(TMS)_2N\}In(\mu\text{-}Cl)]_2$ (**152**).[136] Almost all of these complexes have four-membered $(ME)_2$ core structures which are generally planar, exceptions are the halide-bridged dimers (**150**)[134] and (**152**)[135] where evidently the bulky heteroatomic groups are unable to bridged effectively.

The $(ME)_2$ cores of the heteroatom-bridged dimers (E = N, P, As, Sb) are characteristically asymmetric in terms of their M–E bond lengths, having rhombic cores with two longer and two shorter bonds. In the case of complexes containing both *endo*- and *exo*-cyclic bonds, M–E (bonds (**135**), (**147**) and (**148**)) the *endo*-cyclic bonds are always longer than the *exo*-cyclic bonds.[1] An exception to these general trends is that of $[(TMS\text{-}CH_2)_2AsGa\{\mu\text{-}As(CH_2TMS)_2\}]_2$ (**135**) which is a loose dimer in which the $(GaAs)_2$ core is, similar to $[I_2Ga(\mu\text{-}SPr^i)]_2$ (**126**), puckered (dihedral angle between the core atoms 13.6° in (**133**), cf. 36.7° in (**126**)).[117]

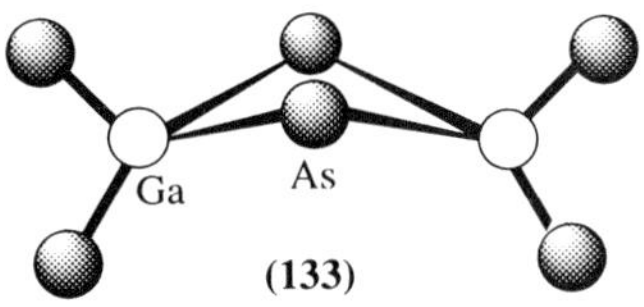

(133)

A further interesting point is that of potential *cis–trans* isomerism resulting from the optional orientations within the bridging primary amide, phosphide or arsenide ligands (in the case of complexes (**137**)–(**139**), (**145**) and (**150**)) or in the orientation of the terminal groups (in the case of (**136**) and (**147**)–(**152**)). In all cases the *trans* isomers are observed in the solid state, presumably due to steric reasons. However, these complexes are commonly involved in *cis–trans* equilibria in solution. For example, ^{31}P NMR spectroscopic studies in benzene at +35 °C on complex (**138**) indicate that *cis* and *trans* isomers are present in a ratio of 2.5:1, and for complex (**148**) ^{13}C NMR spectroscopic studies in toluene (21–87 °C) show that both an interchange of the terminal and bridging $As(CH_2TMS)_2$ groups ($\Delta H_{act} = \sim 78.3$ kJ mol^{-1}) and *cis–trans* isomerism $\Delta H_{act} = \sim 72.8$ kJ mol^{-1} are occurring. Complexes (**136**) and (**149**)–(**152**) are of particular interest since they contain three different groups attached to the group 13 metals.

Dimeric complexes containing both a bridging heteroatomic group (ER_2) and a bridging halogen have been prepared by the silyl elimination reactions of R_2MX with $E(TMS)_3$ (E = P, As) in a 2:1 molar ratio (Equation (7)); examples are $[R_2M\{\mu\text{-}E(TMS)_2\}(\mu\text{-}X)MR_2]$ (R = TMS-CH_2, Me_3CCH_2, Ph; M = Ga, In; E = P, As) (**153**).[12,120,121,137,138]

$$R_2MX + E(TMS)_3 \longrightarrow [R_2M(\mu\text{-E-TMS})(\mu\text{-}X)MR_2] \quad (7)$$

Whether dimeric or trimeric $(ME)_n$ rings are observed for the amides, phosphides or arsenides is dependent, at least in the solid state, on the steric bulk and requirements of the organic groups involved. Thus, increasing the steric bulk of the substituents results, as seen above, in dimeric complexes. Trimeric complexes $(R_2MER^1_2)_3$ exhibit a variety of apparently energetically close $(ME)_3$ ring conformations in the solid state (planar, boat, twist and irregular boat). The conformations of these rings is influenced by the steric bulk of the substituents and possibly also by the inversion energies for the endocyclic E centres. Additionally, potential dative E–M π-bonding may play a role.[1]

The trimeric complexes $[R_2Ga(\mu\text{-}EH_2)]_3$ have been prepared by the reactions of EH_3 (−196 °C) with GaR_3 (E = N, R = Me (**154**);[139] E = P, R = Bu^t (**155**);[140] E = As, R = Bu^t (**156**);[140] E = N, R = Bu^t (**157**)[141]). These are important in assessing the mechanism of formation of GaN, GaP and GaAs from these precursors in their vapour-phase deposition. The reaction of $GaMe_3$ with NH_3 initially produces the $Me_3Ga{\cdot}NH_3$ (**158**) adduct before ultimate decomposition into GaN by successive loss of MeH. A similar reaction has been postulated in the gas-phase synthesis of GaAs from Me_3Ga and AsH_3. Heating the adduct (**158**) to 120 °C for 1 h under a pressure of 450 torr of dry nitrogen gives the trimeric $[Me_2Ga(\mu\text{-}NH_2)]_3$ (**154**) complex in high yield[139] and not, as previously supposed, a dimeric adduct.[142] The complex has an irregular boat-shaped core geometry in which four of the ring atoms are almost coplanar. The other two atoms are above the ring plane. The increase in the lowest energy Ga–N stretching infrared mode at 77 K from 357 cm^{-1} in the adduct (**158**) to 422 cm^{-1} in complex (**154**) denotes, at least in part, the increase in Ga–N bonding. The nature of the $(GaN)_3$ ring found in (**154**) is similar to that present in the aluminum analogue $[Me_2Al(\mu\text{-}NH_2)]_3$ (**159**)[143] but contrasts with the essentially planar core geometry observed for $[Bu^t_2Ga(\mu\text{-}NH_2)]_3$ (**157**) in the solid state.[140] The differences in the core geometries observed for $[Bu^t_2Ga(\mu\text{-}NH_2)]_3$ (**157**) and $[Me_2Ga(\mu\text{-}NH_2)]_3$ (**154**) stresses the subtle influence of steric factors on the observed core geometries of these trimeric complexes. Complex (**157**) is prepared in a similar way to (**154**), or by the transmetallation reaction Bu^t_2GaCl with $NaNH_2$ in almost quantitative yield.[141]

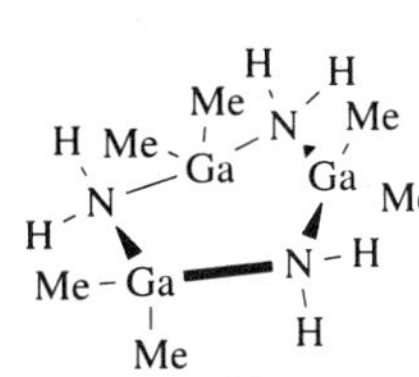

(154)

The condensation of PH_3 or AsH_3 with Bu^t_3Ga (neat or in benzene) at −196 °C followed by heating to +25 °C gives the $[Bu^t_2Ga(\mu\text{-}EH_2)]_3$ complexes (E = P and As).[140] The complex $[Bu^t_2Ga(\mu\text{-}PH_2)]_3$ (**155**) has been characterized in the solid state and, like the NH_2 analogue (**157**), has a planar core geometry reflecting the steric influence of the organic groups involved. $[Bu^t_2Ga(\mu\text{-}OH)]_3$ (**160**), which was originally identified as an impurity present in the synthesis of complex (**157**) and which can be prepared by the addition of a slight excess of H_2O to Bu^t_3Ga (90%), also has a planar $(GaO)_3$ ring structure.[140,141]

Structurally characterized trimeric complexes of gallium and indium are far rarer than the dimeric complexes mentioned earlier. Other structurally characterized examples include $[Me_2In(\mu\text{-}AsMe_2)]_3$ (**161**) (planar and puckered conformers),[144] $[Me_2In(\mu\text{-}SbBu^t_2)]_3$ (**162**) (twist boat conformer),[1] $[Me_2Ga(\mu\text{-}AsPr^i_2)]_3$ (**163**) (boat conformer),[1] $[Br_2Ga(\mu\text{-}As(CH_2TMS)_2)]_3$ (**164**) (twist boat conformer),[1,131] $[Cl_2Ga(\mu\text{-}SbBu^t_2)]_3$ (**165**) (irregular boat conformer),[144] $[Me_2Ga(\mu\text{-}SbBu^t_2)]_3$ (**166**) (twist boat conformer),[1] and $[(Me_3CCH_2)ClGa(\mu\text{-}PPh_2)]_3$ (**167**) (axial, axial, equatorial boat conformer)[145] and $[Me_2Ga(\mu\text{-}PPr^i_2)]_3$ (**168**) (distorted boat conformer).[146] Complex (**161**) is particularly worthy of note since two independent conformers are present in the same asymmetric unit. In the planar conformer, the deviation from the least-squares $(InAs)_3$ plane is from 0.0097(4) nm to −0.0139(3) nm. In the puckered conformer the three indium atoms and one arsenic atom are almost coplanar, and the remaining two arsenic centres are placed above and below this plane. The mixed halogen organophosphide (**167**) is also of interest in that it is the only trimeric complex so far characterized in the solid state in which three different groups are attached to gallium. The bulky neopentyl groups adopt an axial, axial, equatorial conformation with respect to the boat-shaped $(GaP)_3$ core.

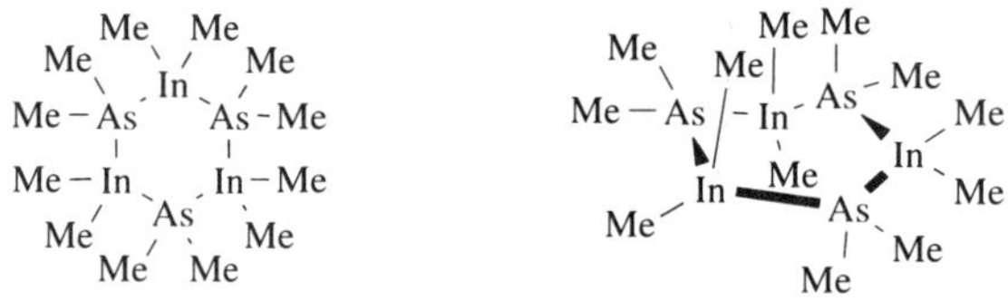

(161)

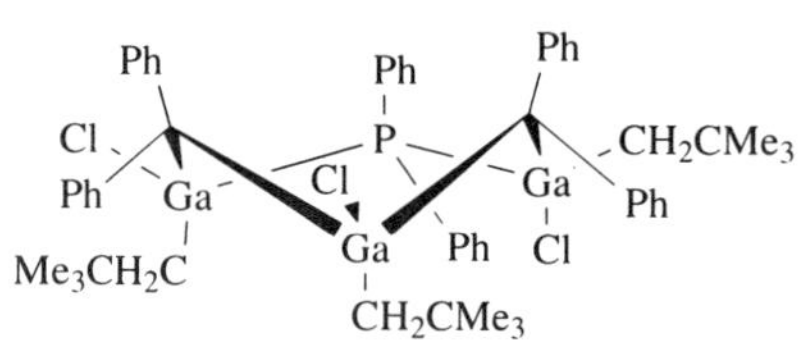

(167)

In order to investigate the possibility of π-delocalization and aromatic stabilization, $[(2,4,6\text{-}Ph_3C_6H_2)GaP(cyclo\text{-}C_6H_{11})]_3$ (**169**) has been prepared by the reaction of $Li_2P(cyclo\text{-}C_6H_{11})$ with $(2,4,6\text{-}Ph_3C_6H_{11})GaCl_2$.[147] The complex is the first example of a ring system containing heavy *p*-block elements both of which are three-coordinate. The fact that the $(GaP)_3$ core geometry is puckered suggests that little or no aromatic stabilization occurs here. However, it is interesting to note that the Ga–P bonds in the core of (**169**) are the shortest yet observed (~0.228–0.234 nm). The shortness of the Ga–P bonds in (**169**) reflects the fact that both gallium and phosphorus are only three-coordinate rather than being attributable to significant Ga–P localized π-bonding. This can be surmised from the observation of only slightly longer Ga–P bonds in the dimeric Lewis base-stabilized complex $[\{(2,6\text{-}CH_2NMe_2)_2\}C_6H_3GaP{\cdot}SiPh_3]_2$ (**170**) in which four-coordinate gallium and three coordinate phosphorus are present.[52]

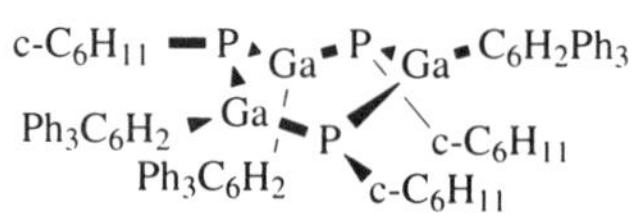

(169)

Cage compounds of gallium and indium with group 15 elements are a rarity. There are only five that have been characterized in the solid state: $[(PhAsH)R_2Ga(PhAs)_6(RGa)_4]$ (R = TMS-CH_2) (**171**),[148] $(RGaPR)_2GaCl_2R$ (R = Bu^t) (**172**),[149] $[Ga_4R^1_3(PR^2)_4PHR^2]$ (R^1 = 2,4,6-$Pr^i_3C_6H_2$, R^2 = 1-adamantyl) (**173**),[150] $Ga_2P_4Bu^t_6$ (**174**),[151] $[Bu^tGa(\mu\text{-}PBu^t)_2(GaBu^t_2)_2]$ (**175**),[152] $[Pr^iInPSiPh_3]_4$ (**176**)[153] and $[Bu^tMPSiPh_3]$ (**177**) (M = Ga and In).[154] Structure (**171**) was prepared by the reaction of $(TMS\text{-}CH_2)_3Ga$

with $PhAsH_2$ together with the previously known compound $(PhAs)_6$. Although the mechanism involved in the formation of (**171**) is not clear, the same authors have shown that diarsines are by-products of the reaction of hindered gallanes with secondary arsines.[148] Thus, it is not altogether surprising that (**171**) contains both Ga–As and As–As bonds in its As_7Ga_5 core. The location of the lone hydrogen of the neutral-bonded PhAsH fragment of the core is suggested by the larger mean bond angles at the unique arsenic (average 113.3°) compared with those at the other arsenic centres in the core.[148]

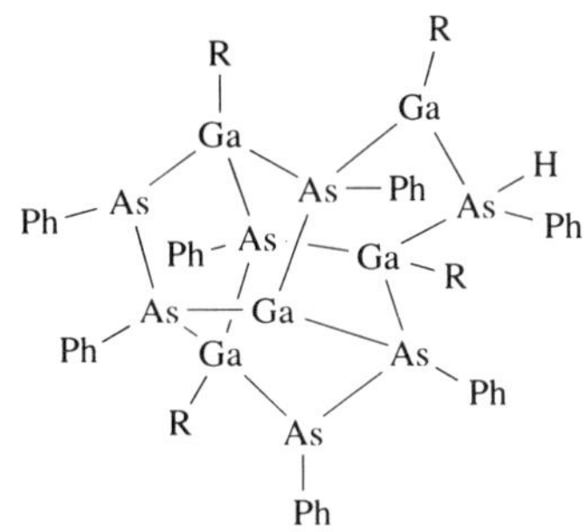

(171)

In (**172**), which was prepared by the reaction of $K_2[Bu^tPPBu^t]$ with Bu^tGaCl_2, the Bu^tGaCl_2 unit behaves simultaneously as a Lewis acid and as a Lewis base.[149] The Ga_3P_4 core is best considered as resulting from the cocomplexation of a boat-shaped Ga_2P_4 dimer with Bu^tGaCl_2 (Scheme 4). However, there is no direct evidence for the formation of the dimeric Ga_2P_4 units prior to the formation of (**172**).

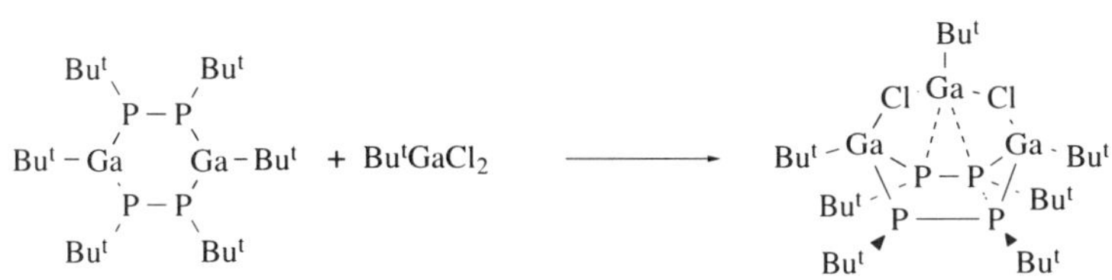

Scheme 4

A Ga_4P_5 core is observed in the solid-state structure of (**173**) which consists of a six-membered $(GaP)_3$ ring fused with two four-membered $(GaP)_2$ rings.[150] The complex is prepared by the reaction of Li_2P(1-adamantyl) with $Cl_2Ga(2,4,6\text{-}Pr^i{}_3C_6H_2)_2$ (1:1 equivalents). The location of the acidic hydrogen atom, on P-1, is indicated by the longer Ga–P distances compared with those in the remainder of the core.

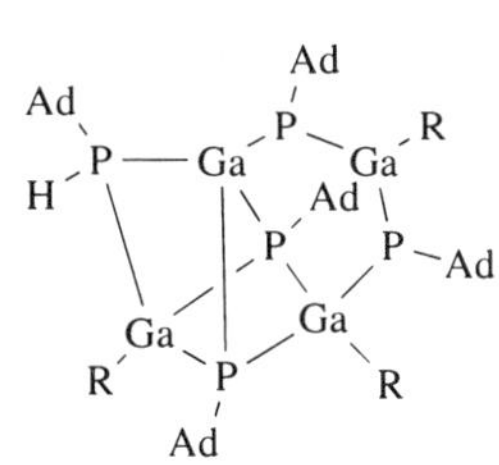

(173)

The reaction of $(Bu^t)_3Ga$ with white phosphorus results in the formation of complex (**174**).[151] The complex, which has a GaP_4 cage structure, can be regarded as forming by the insertion of a gallium atom into a P_4 unit of white phosphorus. The mechanism of the reaction producing (**174**) presumably involves the addition of a Ga–C bond across one of the P–P bonds of the P_4 unit, as occurs in the reactions of alkyllithiums with white phosphorus.[155]

The compound $[Bu^tGa(\mu\text{-}PBu^t)_2(GaBu^t{}_2)_2]$ (**175**), which is prepared by the reaction of Li_2PBu^t with $Bu^t{}_2GaCl$ (2:1 equivalents, respectively) in Et_2O, has a distorted trigonal bipyramidal Ga_3P_2 core structure in which both three- and four-coordinate gallium are present.[152] Interestingly, the 1:1 reaction of $Bu^t{}_2GaCl$ and Li_2PBu^t produces $[(Bu^tGa)(Bu^t{}_2Ga)(\mu\text{-}PBu^t)_2Li{\cdot}2Et_2O]$ (**178**), thus shedding light on the mechanism of formation of (**175**) and of other cage and cluster compounds such as the GaAs complex (**171**) and the GaP complex (**173**) which are structurally related. Compound (**176**), prepared by

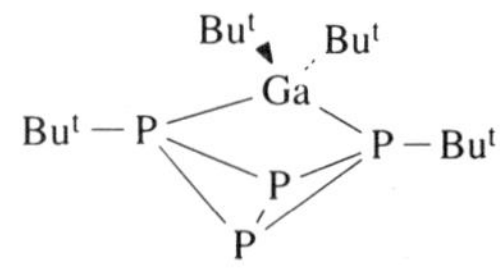

(174)

the reaction of Li_2PSiPh_3 with Pr^iInI_2 (1:1 equivalents), has a slightly distorted cubane structure in the solid state.[153] Polycrystalline films of InP (15.0–20.0 nm thick) are deposited using (**176**) as a precursor at 500 °C. The complex, although photolytically unstable, is surprisingly stable to alcoholysis by EtOH or Bu^tOH. $[Bu^tMPSiPh_3]$ (**177**) (M = Ga and In) have also been prepared and structurally characterized. These have similar cubane structures to (**176**).[154,155]

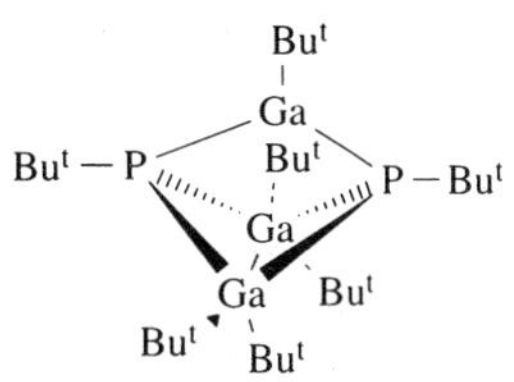

(175)

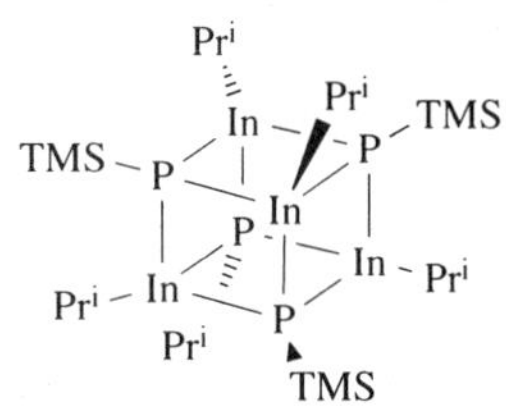

(176)

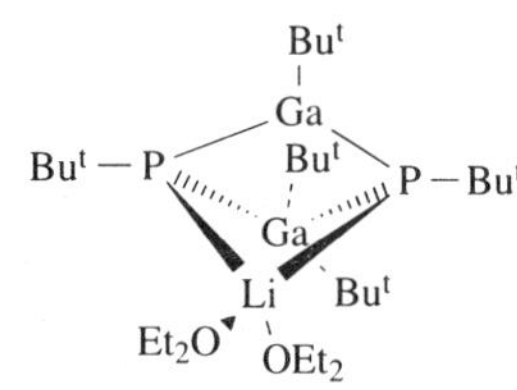

(178)

There are a few polymeric structures worthy of note. The structure of $[Me_2GaN_3]_\infty$ (**71**), prepared by the low-temperature reaction of Me_2GaCl with NaN_3, is that of a helical chain polymer (along the *a*-axis) in the solid state.[67] The complex can be efficiently thermally decomposed under relatively mild conditions (400 °C, 10 torr) into GaN.

An unusual method of oligomerization is seen in $[Me_2Si(NBu^t)_2InMe]_2$ (**179**) which has a ladder structure in the solid state consisting of a central $(InN)_2$ ring dimer.[156]

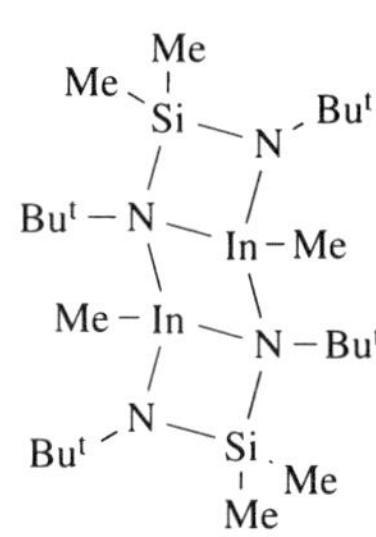

(179)

11.4.3 Syntheses and Structure of Group 13–16 Complexes Containing Ga, In and Tl Bonded to O, S, Se and Te

Somewhat suprisingly, bearing in mind the potential applications of single-source group 13–16 complexes in the deposition of binary metal oxide and chalcogenide phases which have useful electronic and optical properties, there are comparatively few examples of well-characterized organometallic complexes containing gallium, indium and thallium bonded to group 16 elements (E = O, S, Se, Te), especially where simple (unfunctionalized) organic groups (RE^-, R = alkyl or aryl) are present. This section concentrates particularly on the structurally characterized complexes containing such unfunctionalized organic groups, excluding the numerous inorganic derivatives, for example, $TlGaSe_2$,[62] Ga_4E_{10} (E = S, Se), $Tl_2Te_2^{\ 2-}$[157,158] and $[In_2Se_2]^{4-}$.[159] However, as in Section 11.4.2, it is valuable to discuss coordination (or organometallic) compounds which do not necessarily have Ga–, In– or Tl–C bonds, but which are structurally related to heteroatomic derivatives which do. The syntheses of these derivatives are similar to those discussed for the group 13–15 complexes and have generally involved metallation or transmetallation procedures (see Section 11.4.2). A selection of these synthetic routes, together with further unusual strategies, will be discussed in connection with individual compounds rather than in a separate section.

Neutral monomeric complexes, particularly those containing three-coordinate gallium, indium or thallium, have rarely been characterized in the solid state. Especially for the largest metals (indium and thallium), the relatively high Lewis-base acidities tend to result in oligomerization or polymerization except where Lewis bases are present or where intramolecular donation of pendant heteroatomic groups to the metal can occur, thus saturating their coordination spheres, or where very bulky organic groups are attached to the metals. Monomeric complexes that have been structurally characterized possess a range of coordination numbers (three to six) for the group 13 metals and include $E^I(\mu\text{-}OBu^t)_3Sn$ (**180**) (three-coordinate E = In and Tl),[160,161] $[2\text{-}CH{:}NOH\text{-}C_6H_3O]_2GaMe$ (**181**) (five-coordinate gallium via intramolecular donation by oxygen and nitrogen),[162] $[2\text{-}NH_2C_6H_4OGaMe_2]$ (**182**) (four-coordinate gallium via intramolecular donation by oxygen and nitrogen),[163] $[NH_2CH_2CO_2GaMe_2]$ (**183**) (four-coordinate gallium via intramolecular donation by oxygen and nitrogen),[164] $C_5H_3N(2\text{-}CH_2NH_2)(3\text{-}O)GaMe_2$ (**184**) (four-coordinate gallium via intramolecular donation by oxygen and nitrogen),[165] $[(THF)Me_2TlS_2CPPh_2]$ (**185**) (five-coordinate thallium),[166] $[Et_2O{\cdot}In\{S(2,4,6\text{-}(CF_3)_3)C_6H_2\}_3]$ (**186**) (four-coordinate indium),[167] $In(SePh)_3{\cdot}2(NC_5H_5)$ (**187**) (five-coordinate indium),[168] $GaBu^n[S(2,4,6\text{-}Bu^t_3C_6H_2)]_2$ (**188**) (three-coordinate gallium),[169] $M[E(2,4,6\text{-}Bu^t_3C_6H_2)]_3$ (**189**) (M = Ga, In; E = S, Se; (three-coordinate gallium and indium)[169–71] and $[(TMS)_2CH]_2GaTeSi(TMS)_3$ (**190**) (three-coordinate gallium).[172] In addition, a number of mononuclear anionic derivatives, in which the metals are commonly four-coordinate, have been prepared and structurally characterized, for example, in $[PPh_4]^+{\cdot}[TlBr(SPh)_3]^-$ (**191**),[173] $[R^1_4N]^+{\cdot}[Ga(SR^2)_4]^-$ $R^1 = Pr^n$, R^2 = Et (**192**); R^1 = Et, R^2 = Ph (**193**)).[174]

$M[E(2,4,6\text{-}Bu^t_3C_6H_2)]_3$ (**189**) (M = Ga, In; E = S, Se)[169–71] derivatives are all very structurally similar and contain almost completely trigonal planar gallium and indium centres. In contrast with the indium derivatives, the gallium complexes have relatively close Ga···H contacts with the Bu^t groups of the organic ligand which are around the sum of the van der Waals radii. This may reflect the larger Lewis acidity of gallium compared with indium, although it is not clear whether these close contacts are real interactions. The In–S and In–Se bonds in the sulfur and selenium derivatives (In–S average 0.2398(6) nm, In–Se average 0.2506(3) nm) are the shortest yet observed. However, there does not appear to be any structural indication of M–E π-bonding in any of these complexes, judging by the orientation of the $2,4,6\text{-}Bu^t_3C_6H_2$ groups. Rather, the short bond lengths are probably a consequence of the low coordination number of indium in these complexes.

In the compound $[(TMS)_2CH]_2GaTeSi(TMS)_3$ (**190**),[172] which can be prepared by the unusual reaction of the Ga–Ga bonded dimer $[\{(TMS)_2CH\}_2Ga]_2$ (**194**) with the Te–Te-bonded dimer $[(TMS)_3SiTe]_2$ (**195**) at 100 °C in toluene (Scheme 5) or by the more conventional transmetallation reaction of $(TMS)_3SiTeLi$ with $[(TMS)_2CH]_2GaCl$, there is a ~50 kJ mol^{-1} barrier to rotation of the Ga–Te bond. This value is similar to that observed for Ga–E (E = N, P) bonded monomers in which only one such bond is present and may imply an element of π-bonding is present here.[114]

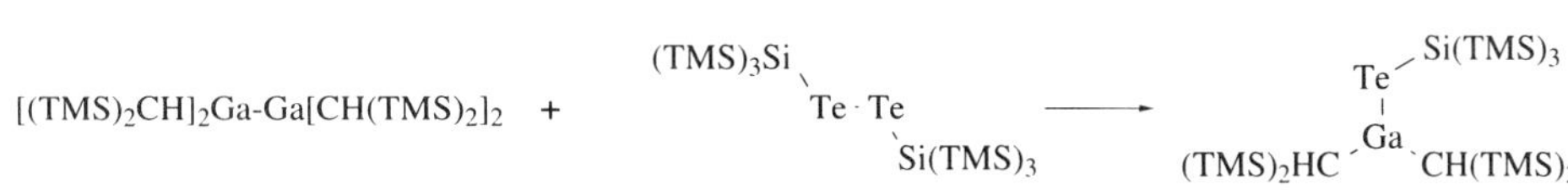

(**194**) (**195**) (**190**)

Scheme 5

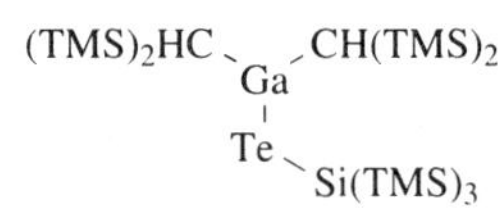

(**190**)

Dimeric species, comprising a central $(ME)_2$ ring (M = Ga, In, Tl; E = O, S, Se, Te), are also becoming increasingly common. Dimeric complexes containing simple monofunctional organic groups which have been characterized in the solid state include: the alkoxides $[Bu^t_2Ga(\mu\text{-}OBu^t)]_2$ (**196**),[61b] $[Bu^t_2In(\mu\text{-}OEt)]_2$ (**197**),[175] $[Cp_2Ga(\mu\text{-}OEt)]_2$ (**198**),[176] *trans*-$[E^I\{\mu\text{-}O[2,4,6\text{-}(CF_3)_3C_6H_2]\}]_2$ (E = In (**199**),[177] Tl (**200**),[178]), *trans*-$[Bu^t_2In(\mu\text{-}OOBu^t)]_2$ (**201**)[179]; the sulfides $[Et_4N]^+{\cdot}[(PhS)_2Ga(\mu\text{-}S)]_2^-$ (**202**),[174] *trans*-$[Ph_2Ga(\mu\text{-}SEt)]_2$ (**203**),[180] *trans*-$[Tl^I\{\mu\text{-}SSi(OBu^t)_3\}]_2$ (**204**),[181] *trans*-$[Bu^t_2Ga(\mu\text{-}SH)]_2$ (**205**),[182a] *trans*-$[I_2Ga(\mu\text{-}SPr^i)]_2$ (**126**);[117] and the selenides and tellurides *trans*-$[(Me_3CCH_2)_2MEPh]_2$ (M = Ga, E = Te (**206**);[183] M = In, E = Se (**207**)[184]).

These species bear a close relationship to the dimeric gallium, indium and thallium amide, phosphide and arsenide complexes mentioned earlier (Section 11.4.2). However, there is a large variation in the conformations of the $(ME)_2$ rings where E = S, Se or Te. Thus, whereas all the complexes (**196**)–(**205**) have planar cores, complexes (**126**) (fold angle about Ga···Ga 143.3°),[117] (**206**) (fold angle about

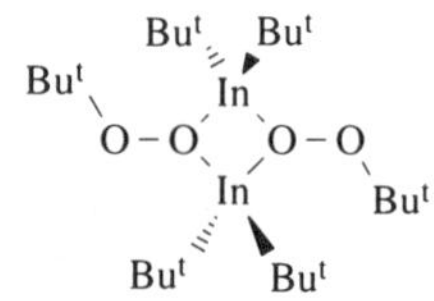

(201)

Ga···Ga 150.2°)[183] and (**207**) (fold angle about In···In 149.4°)[184] have butterfly-shaped $(ME)_2$ cores. The folding of these rings has been attributed to the steric interaction of the organic groups attached to In and Ga, and to S, Se and Te. It can be seen that for more bulky bridging groups a nonplanar core geometry becomes possible in order to relieve steric repulsion. As expected, the Ga–Te bonds in (**206**)[183] (average 0.276 nm, cf. sum of covalent radii 0.262 nm[93]) and the In–Se bonds in (**207**)[184] (0.2744 nm, cf. sum of covalent radii 0.258 nm[93]) are longer than in monomeric species in which terminal, rather than bridging, metal contacts are made (Ga–Te 0.2535 nm in $[(TMS)_2CH]_2GaTeSi(TMS)_3$ (**190**),[172] and In–Se 0.2506 nm in $In[E(2,4,6\text{-}Bu^t_3C_6H_2)]_3$ (**189**) (E = Se)[170]).

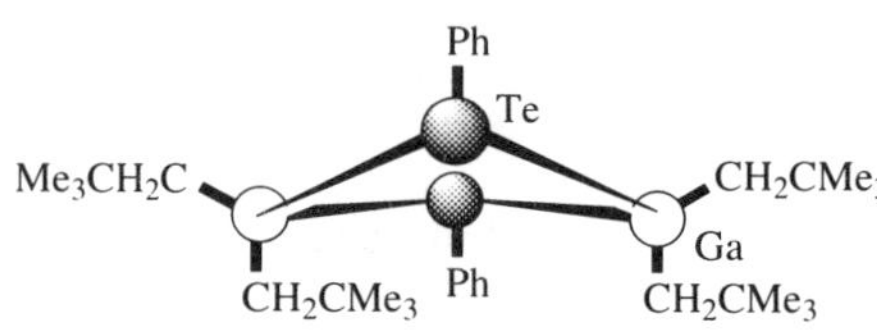

(206)

The dimeric structure of *trans*-$[Bu^t_2Ga(\mu\text{-}SH)]_2$ (**205**),[182a] in which the S–H atoms assume a *trans* conformation with respect to the core and which is prepared by the reaction of SH_2 with Ph_3Ga, can be compared to those of its N, P, As and O analogues (complexes (**154**)–(**157**), and (**158**) mentioned in Section 11.4.2[139–41]) which are all trimeric with $(GaE)_3$ ring cores. The unusual anion of $[Et_4N]^+\cdot[(PhS)_2Ga(\mu\text{-}S)]^-$ (**202**),[174] in which μ-S centres bridge the gallium atoms together, and the two-coodinate indium(I) and thallium(I) metal centres present in *trans*-$[E\{\mu\text{-}O[2,4,6\text{-}(CF_3)_3C_6H_2]\}]_2$ (E = In (**199**),[177] Tl (**200**)[178]) and *trans*-$[Tl(\mu\text{-}SSi(OBu^t)_3]_2$ (**204**)[181] are also worthy of note.

The peroxide complex *trans*-$[Bu^t_2In(\mu\text{-}OOBu^t)]_2$ (**201**),[179] formed by the careful oxidation of Bu^t_3In with oxygen or pyridine *N*-oxide, is the first example of such a species characterized in the solid state for any group 13 metal.

In addition to dimers containing monofunctional organic groups, dimers can also occur where multifunctional bridging groups are present, for example, in the alkoxides $[Me_2Ga\{(\mu\text{-}1\text{-}O)(2\text{-}O)C_7H_5\}]_2$ (**208**),[185] $[Me_2Ga\{(\mu\text{-}2\text{-}OCH_2)(py)\}]_2$ (**209**),[186] $[Me_2Ga\{\mu\text{-}9\text{-}O(1\text{-}quinoline)\}]_2$ (**210**),[165] $[Me_2Ga\{(\mu\text{-}1\text{-}O)(2\text{-}OMe)C_6H_4\}]_2$ (**211**),[187] $[Mc_2In\{\mu\text{-}O(2\text{-}CH{:}O)C_6H_4\}]_2$ (**212**),[188] $[Me_2In(\mu\text{-}OC(Ph)(2\text{-}PyrN)_2]_2$ (**213**),[84] and others.[189] Although the $(ME)_2$ dimer cores are retained, the coordination numbers of the group 13 metals is increased to five or even six by the coordination of pendant heteroatomic groups.

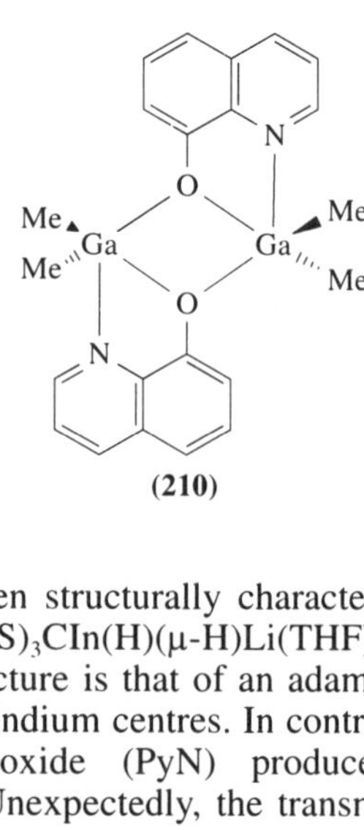

(210)

Only three alkoxide cages have been structurally characterized $[\{O(TMS)_3CIn\}_4(OH)_6]$ (**214**)[190] is prepared by the hydrolysis of $[(TMS)_3CIn(H)(\mu\text{-}H)Li(THF)_2(\mu\text{-}H)In(\mu\text{-}H)(H)C(TMS)_3]$ (**97**)[97] with aqueous methanol. The solid-state structure is that of an adamantane In_4O_7 core in which a central μ_4-O atom is attached equivalently to the indium centres. In contrast with the previous structure, oxidation of $[Me_2InPPh_2]_2$ with pyridine *N*-oxide (PyN) produces the tetranuclear cubane structure $[InMe(OH)(O_2PPh_2)_4]\cdot 4PyN$ (**215**).[191] Unexpectedly, the transmetallation reaction of partially hydrated

$InCl_3$ produces the pentanuclear cage complex $[(InOPr^i)_5(\mu_2\text{-}OPr^i)(\mu_3\text{-}OPr^i)(\mu_5\text{-}O)]$ (**216**) as the only isoluble product.[192] The structure is that of a square-based pyramidal array of indium atoms linked together by μ_3 and μ_2 alkoxide bridges and a central square-based pyramidal μ_5-O atom.

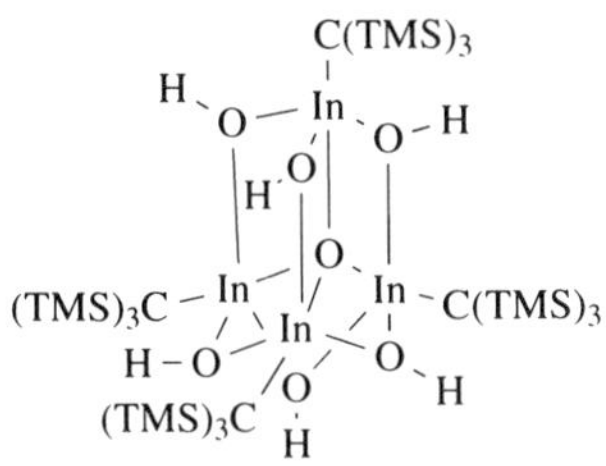

(**214**)

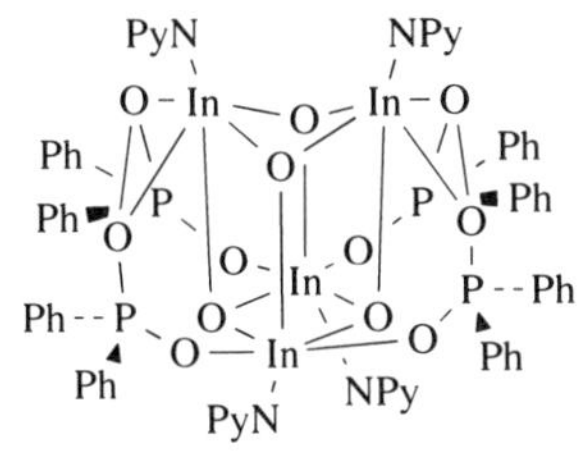

(**215**)

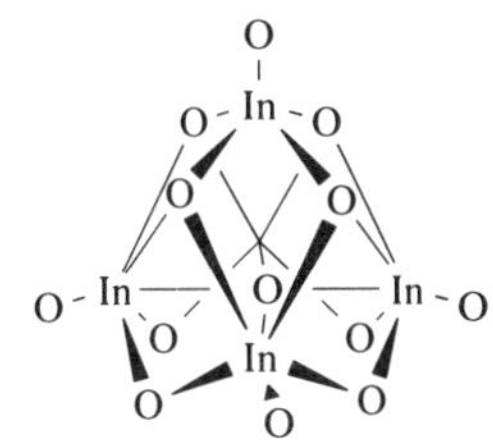

(**216**)

The interaction of sulfur,[182,193,194] selenium[193,194] and tellurium[193,194] with $GaBu^t_3$ has been examined and a range of remarkable cage complexes has been characterized. The reaction of S_8 with $GaBu^t_3$ leads to the formation of a mixture of $[(Bu^t)_2Ga(\mu\text{-}SSBu^t)]_2$ (**217**), $[(Bu^t)_2Ga(\mu\text{-}SSBu^t)(\mu\text{-}SBu^t)Ga(Bu^t)_2]_2$ (**218**) in a 1:3 molar ratio together with trace amounts of $[(Bu^t)_2Ga(\mu\text{-}SBu^t)]_2$ (**219**).[194] In the presence of S_8, thermolysis of (**217**)–(**219**) gives a range of di-*t*-butyl-polysulfides $Bu^t\text{-}S(S)_nS\text{-}Bu^t$ (n = 1–4) and the cubane $[Bu^tGa(\mu\text{-}S)]_4$ (**220**).[194] The latter can also be prepared by the thermolysis of *trans*-$[Bu^t_2Ga(\mu\text{-}SH)]_2$ (**205**).[182,194]

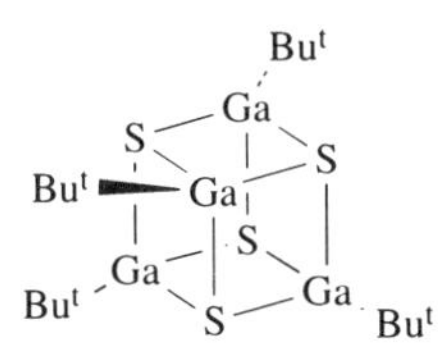

(**220**)

A similar reaction of $GaBu^t_3$ with Se_8 gives $[Bu^tGa(\mu\text{-}Se)]_4$ (**221**) directly, which has a similar cubane structure to (**220**) in the solid state.[194] In contrast with the reactions with S_8 and Se_8, the reaction of $GaBu^t_3$ with metallic selenium and tellurium gives $[(Bu^t)_2Ga(\mu\text{-}EBu^t)]_2$ (**222**) (E = Se, Te).[194]

$[Bu^tGa(\mu\text{-}S)]_4$ (**220**) has an extremely interesting thermochemistry. Repeated sublimation at atmospheric pressure produces the octamer $[Bu^tGa(\mu\text{-}S)]_8$ (**223**) in low yield[193] which is thought to have a drum structure similar to those observed for tin oxides.[195] Thermolysis of (**220**) in hexane produces the heptamer $[Bu^tGa(\mu\text{-}S)]_7$ (**224**) whose crystal structure has been determined,[193] while thermolysis of the tetramer (**220**) in pentane gives the octamer (**223**). Dissolution of $[Bu^tGa(\mu\text{-}S)]_x$ (x = 4, 7, 8) in pyridine produces the trimeric donor complex $[Bu^tGa(\mu\text{-}S)py]_3$ (**225**).[193] The latter is the only such trimeric chalcogenide complex to have been structurally characterized.

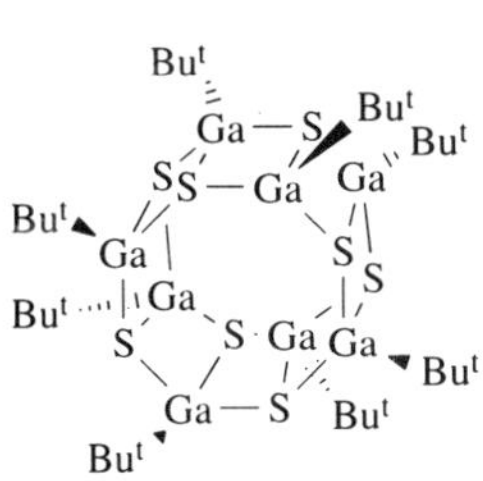

(**223**)

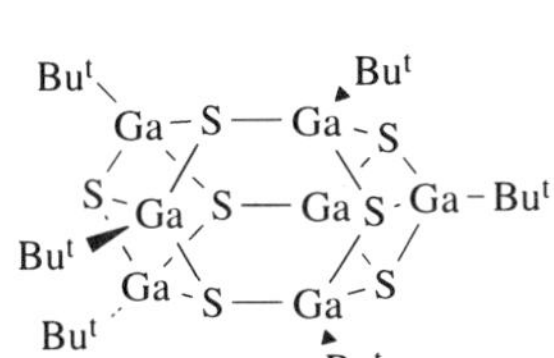

(**224**)

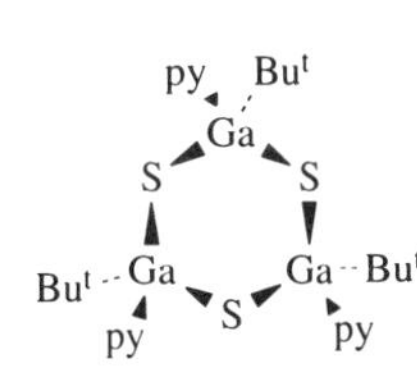

(**225**)

Although thermolysis of the trimer (**225**) gives the tetramer (**220**), thermolysis in the solid state gives the hexamer $[Bu^tGa(\mu\text{-}S)]_6$ (**226**) whose structure in the solid state has been determined.[193] Addition of pyridine to the hexamer (**226**) regenerates the pyridine-solvated trimer (**225**) and sublimation of the hexamer (**226**) or the trimer (**225**) gives the tetramer (**220**).[193]

Thermolysis of $[Bu^t_2Ga(\mu\text{-}OH)]_3$ (**160**) (described in Section 11.5.2) produces $[Bu^t_2Ga(\mu\text{-}O)]_9$ (**227**) whose structure is probably related to the heptamer (**224**).[193] This nonamer is the first example of a gallium oxide cage complex.

A very different tetrameric cage structure to that of the cubanes (**220**) and (**221**) occurs in the solid state for $Ga_4I_4(\mu\text{-}SMe)S_2$ (**228**).[196] Here an adamantane-type cage (similar to $Ga_4S_{10}^{2-}$, $In_4S_{10}^{2-}$ and

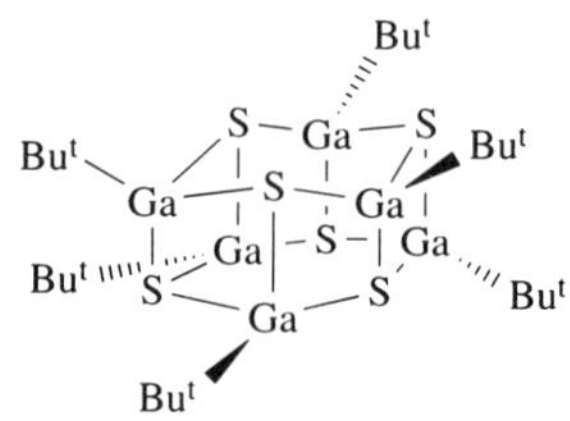

(226)

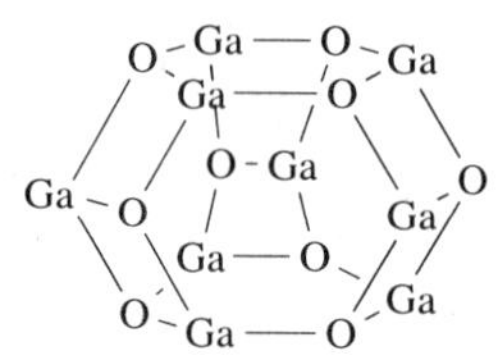

(227)

$In_4Se_{10}^{2-}$ [197]) is observed in which there is a Ga_4S_6 core with the iodine atoms being terminally attached to gallium. Highly complicated cage structures are observed in the solid state for Tl^ISPh (**229**), which exists as $[Tl_5(SPh)_6]^-\cdot[Tl_7(SPh)_6]^+$, and for Tl^ISBu^t (**230**), which exists as $[Tl_8(SBu^t)_8]$.[198]

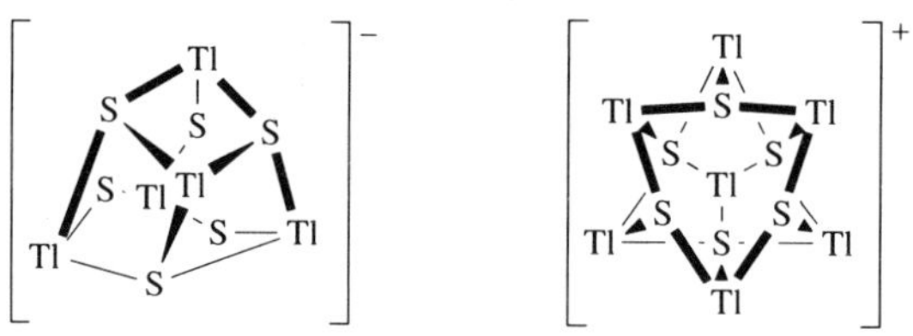

(229)

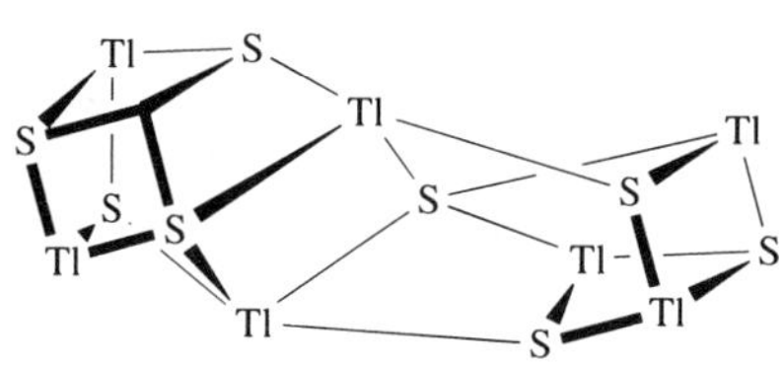

(230)

In contrast with the cage structures mentioned above, Tl^ISCH_2Ph (**231**)[198] and $[2,4,6\text{-}(CF_3)_3C_6H_2S]Tl^I$ (**232**)[199] form infinite polymeric ladders in the solid state which are constructed from $(TlS)_2$ dimer units sharing edges. The structure of $[In(SePh)_3]_\infty$ (**233**) is also built up, from its basic $\cdots In(\mu\text{-}SePh)_3In\cdots$ units, into an infinite polymer.[168]

11.5 LOW OXIDATION STATE COMPLEXES

There have been many significant and exciting advances in the low-valent chemistry of gallium, indium and thallium since the mid-1980s. Whereas the +1 oxidation state in organometallic compounds has been known for many years for thallium and indium, it was virtually unknown for gallium. Pioneering work by Schmidbaur on mixed-valence arene complexes, and by Schnöckel on the isolation and characterization of the isolated gallium(I) species has changed this situation dramatically. There has also been a wealth of new data on indium and thallium, both in the new area of arene complexes, and in relatively 'old' areas such as cyclopentadienyl derivatives. Several carbaborane derivatives have also been prepared to complement the cyclopentadienyl work. In addition, there have been several reviews on the area. Schmidbaur has reviewed 'Arene Complexes of Univalent Gallium, Indium and Thallium',[3] Jutzi has a more general review of 'π-Bonding to Main Group Elements',[4] and Paetzold has a short review on 'Monovalency of the Third Main-Group Elements—News from EX-Type Compounds', which covers both organometallic and low-valent halide systems.[200]

11.5.1 Cyclopentadienyl Compounds

Gallium(II) chloride has proved an ideal stable starting compound for the investigation of low-valent gallium chemistry; it is a mixed valent species, $Ga^+GaCl_4^-$ in the crystalline solid and in the melt. The investigation of pure gallium(I) compounds as opposed to these mixed species has proved a somewhat harder task. Pioneering work by Schnöckel has made this possible with the preparation and characterization of GaCl.[201] This is prepared by passing HCl gas over gallium metal at 1200 K and a pressure of less than 0.2 mbar, and condensing the GaCl product in a toluene/ether matrix at −196 °C (Equation (8)). The product remains stable up to 0 °C. Layering this red metastable solution with pentane at −30 °C precipitates yellow air- and temperature-sensitive crystals. An x-ray crystallographic investigation shows it to be a molecular compound $[\{(Et_2O)_2ClGa\}Ga\{GaCl_2(Et_2O)\}_3]$ (**234**)[202] (see Section 11.6). The variety of oxidation states between 0 and +2 that can be formally assigned to each gallium atom implies a disproportionation reaction from GaCl to form this product, or a comproportionation ($Ga^{III} + Ga^{I}$) arising from small amounts of $GaCl_3$ formed in the original reaction. The preparation of GaCl may prove a convenient starting point for the investigation of low-valent gallium chemistry, with both the gallium(I) products and any disproportionation products potentially yielding new and exciting results.

$$Ga(l.) + HCl(g.) \xrightarrow[96\text{–}99\%]{1200\ K,\ 0.2\ mbar} GaCl(g.) + 1/2H_2(g.) \qquad (8)$$

Following the synthesis and crystal structure determination of $[Al_4Cp^*_4]$, GaCp was the only missing member in the series of cyclopentadienyl compounds of the monovalent group 13 compounds. To demonstrate the synthetic utility of GaCl, Schnöckel and co-workers used it as a starting point to prepare and characterize GaCp (**235**).[2] Addition of biscyclopentadienyl magnesium to the metastable solution of GaCl in toluene/ether at −78 °C, followed by recondensation at ~10^{-2} mbar produces an air-sensitive solution of GaCp. The product was characterized by ^{71}Ga NMR spectroscopy, with a shift of $\delta = -714$ (relative to $[Ga(H_2O)_6]$ external standard) being very similar to those reported for gallium(I) in mixed-valence species, and mass spectrometry which indicated that the species was monomeric. Carbon-13 and ^1H NMR spectroscopies show only one signal each, showing η^5 coordination of the Cp or a rapid fluxional process. MP2 *ab initio* calculations support the conclusions drawn from the experiments that GaCp is monomeric with η^5 coordination of the Cp in a C_{5v} structure.

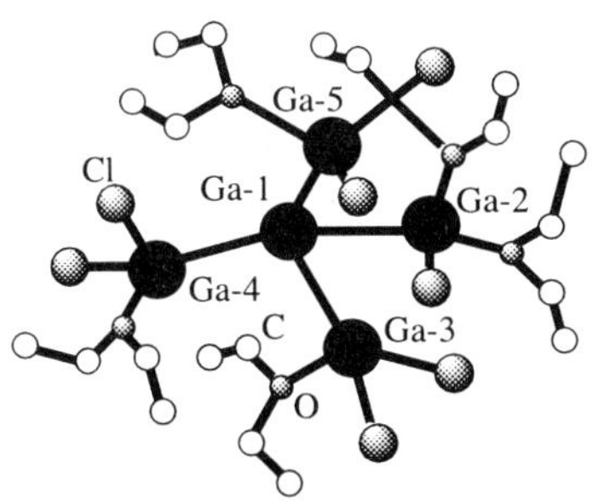

(**234**)

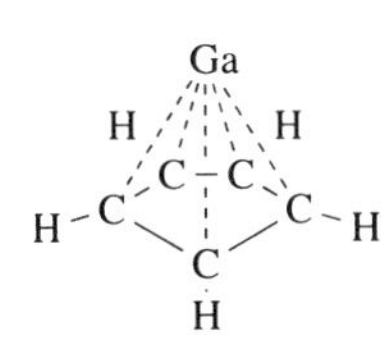

(**235**)

A small number of cyclopentadienyl gallium(III) compounds are also known. There has been much debate over the hapticity of this type of compound which was discussed in *COMC-I*. A new compound, tris(cyclopentadienyl)gallium, $[Ga(\eta^1\text{-}Cp)_3]$ (**236**) has now been prepared and characterized.[203] It shows that the transition from η^5 coordination in GaCp through η^2 or η^3 bonding in Me_2GaCp to monohapto σ-bonded Cp in $GaCp_3$ is complete. The compound is prepared by addition of lithium cyclopentadienide to gallium(III) chloride. It is a white pentane-soluble material which sublimes at 40–42 °C but decomposes at 45 °C. An x-ray crystallographic investigation shows discrete $GaCp_3$ monomers. The Cp rings are σ-coordinated in the trigonal planar molecule. Molecular weight measurements in hydrocarbons suggest that the compound is monomeric with σ-coordination of the Cp rings. The ^1H NMR spectrum shows only a single signal, consistent with the expected fluxional nature of the molecule. The low-temperature ^1H NMR spectrum has not been investigated.

There have also been many investigations of cyclopentadienyl derivatives of indium and thallium. The solid-state structures of the parent compounds, InCp and TlCp, are both well-known polymeric zig-zag chains, $[(\mu\text{-}\eta\text{-}Cp)E]_\infty$, in which the metal centres behave as acceptors to the bridging Cp ligands.[204] In the gas phase, both compounds are monomeric with η^5 coordination of the metal to the Cp ring.[205]

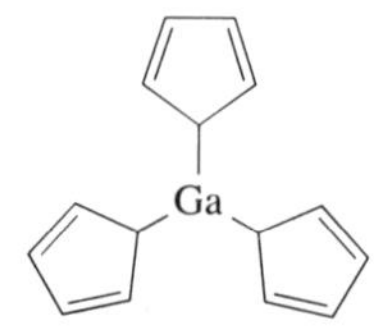

(236)

The original x-ray diffraction studies were, however, semiquantitative which led Beachley and co-workers to reinvestigate the structure of InCp and further to study other InCp derivatives.

The reinvestigation of CpIn (**237**) produced a more detailed analysis than was previously possible.[206] In the infinite zig-zag chains each indium interacts in an η^5 fashion with two Cp rings ($Cp_{centroid}$–In–$Cp_{centroid}$ 128.02°, In–$Cp_{centroid}$ 0.2726 nm). Likewise, each Cp ring interacts with two indium atoms (In–$Cp_{centroid}$–In 176.99°). In addition the short In–In distances may indicate weak In–In interactions between individual chains, with each indium interacting with two other indium atoms.

The structure of InC_5H_4Me (**238**) is very similar to that of InCp but with certain minor differences.[206] The angle at indium is slightly more open at 130.66° and the In–$Cp_{centroid}$ distance reduced to 0.2609 nm. The angle at Cp is closer to linear at 179.74°. More significantly, interchain interactions operate through an inversion centre such that each indium interacts with only one other indium atom. Furthermore, the gas-phase structure of $In(C_5H_4Me)$ was investigated by electron diffraction. Like InCp, it consists of discrete monomeric units with the indium situated above the ring centroid. The indium–ring centroid distance of 0.2309 nm is again smaller than that observed in InCp (cf. 0.2332 nm). This suggests that the electron-donating methyl group makes C_5H_4Me a better ligand than C_5H_5 towards indium(I). The decreased antibonding s–a_1 (i.e., π) interaction between the metal and the ring is believed to be responsible.[207] Cryoscopic measurements in cyclohexane show that InC_5H_4Me exhibits a degree of association between 1.49 and 1.62, consistent with a monomer–dimer equilibrium.

Two other infinite chains of InCp derivatives, $In(C_5H_4TMS)$ (**239**) and $In(C_5H_4CMe_3)$ (**240**) have also been prepared.[208,209] These two compounds are very similar in structure but, unlike InCp and $In(C_5H_4Me)$, they show no In–In interstrand interactions. Table 1 summarizes the main bond lengths and angles in all these infinite structures.

Table 1 Structural data for infinite cyclopentadienyl derivatives of indium(I).

Derivative	*In–Cp$_c$^{1}* (nm)	*In–Cp$_c$^{2}* (nm)	*In–Cp$_c$–In* (°)	*Cp$_c$^{1}–In–Cp$_c$^{2}* (°)
In (Cp)	0.2726	0.2687	176.99	128.02
In (C_5H_4Me)	0.2609	0.2771	179.74	130.66
In (C_5H_4TMS)	0.2609	0.2822	175.94	131.78
In ($C_5H_4CMe_3$)	0.2530	0.285	173.4	128.2

In complete contrast, the x-ray structure of InCp* (**241**) consists of a hexameric structure.[7] The In–In distances are at very similar distances to those in InCp and $In(C_5H_4Me)$ (~0.295 nm) even though the structures are extremely different. This, together with the volatility of $In(C_5Me_5)$ suggests that the octahedral cluster has only marginal stability.

Concerning smaller oligomers, a 'quasidimeric' species, $InC_5(Bz)_5$ (**242**) has been reported.[210]

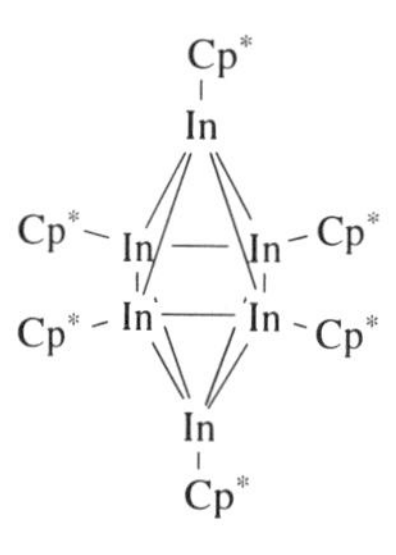

(241)

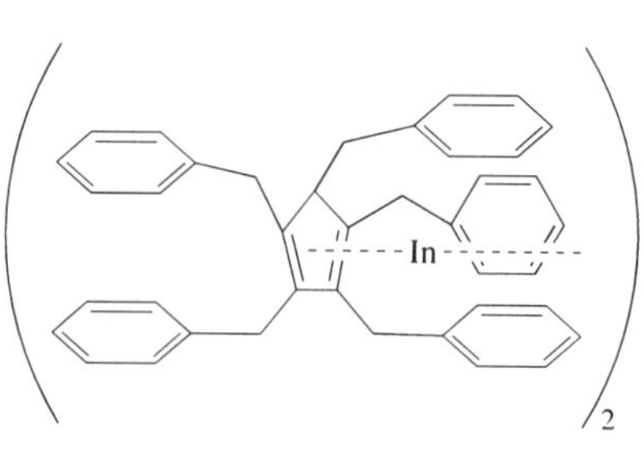

(242)

The quasidimeric label arises because two single molecules associate via a centre of symmetry between the two metal atoms to give the impression of a dimer. Three of the five benzyl groups of each

Cp face towards the metal atom. The In–In distance of 0.3631 nm is far shorter than those reported in all other InCp compounds. However, this is not thought to arise from metal–metal bonding, but rather from the packing effects of the organic envelope around each metal. It is suggested that similar packing of organic ligands gives rise to the close In–In interactions observed in other compounds. However, theoretical work by Janiak and Hoffmann has suggested that a modicum of metal–metal bonding is possible in such species.[211]

One compound that makes no claims as to any intramolecular interactions is the monomeric In(η-$C_5H_4PPh_2$) (**243**). The smallest In–In distance is considerably longer than in the other reported compounds, at 0.5929 nm, and there is no interaction with the lone pair on the phosphorus atom.

All of these indium cyclopentadienyl derivatives are exceptionally air- and water-sensitive. Being the only stable indium(I) organometallics, they hold great potential for exploring indium(I) chemistry.

Thallium cyclopentadienyl derivatives are, on the whole, more stable and accessible than their indium counterparts. Several thallium analogues of the indium compounds described earlier have been prepared. These include Tl(C_5H_4TMS) (**244**)[212] and Tl[$C_5(Bz)_5$] (**245**).[213] These compounds form exact structural analogues of the indium derivatives in that the former is an infinite polymer and the latter is a quasidimer. The thallium analogue of In(C_5Me_5) has also been prepared but, unlike the hexameric indium counterpart, TlCp* (**246**) crystallizes as an infinite polymer.[214] Two further infinite polymeric derivatives, Tl[$C_5Me_4(SiMe_2Ph)$] (**247**) and Tl[$C_5Me_4(SiMe_2(Bz))$] (**248**) have also been structurally characterized, but their structures are unremarkable.[215]

A new structural type in this field has also been observed with the hexameric 'doughnut' molecule $[Tl\{C_5H_3(TMS)_2\text{-}1,3\}]_6$ (**249**).[212] This hexameric structure contrasts with the polymeric [Tl(C_5H_4TMS)] and the monomeric $[Tl\{C_5H_2(TMS)_3\text{-}1,2,4\}]$ (by cryoscopic measurements in benzene). This shows that the degree of polymerization decreases with increasing TMS substitution in the C_5 ring. It is argued that this demonstrates the general principle that lipophilicity and volatility (or indeed covalent character as judged by a decreasing tendency for molecular aggregation) for Cp complexes increases with increasing TMS substitution.[216] A counterargument suggests that the zig-zag chains of $[CpTl]_\infty$ adopt a *trans* orientation, and that increasing the steric bulk on one side of the cyclopentadienyl moiety will first produce a *cis* orientated molecule, as in the doughnut $[Tl\{C_5H_3(TMS)_2\text{-}1,3\}]_6$, and then monomers when the steric bulk is too extreme to allow any oligomeric aggregation.[200]

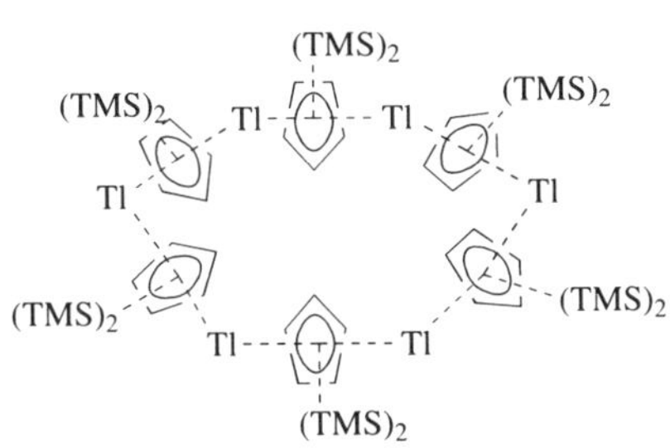

(**249**)

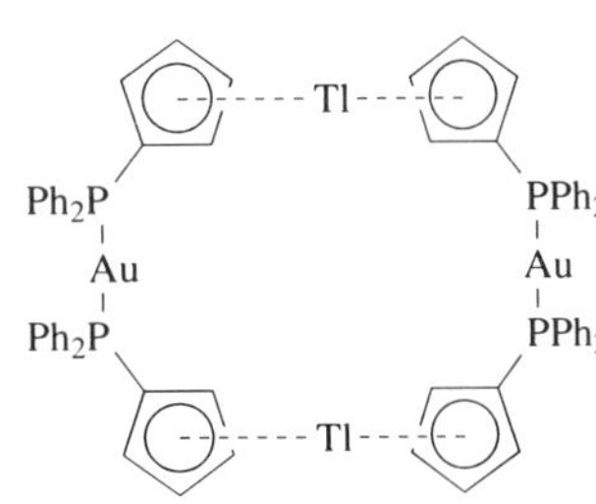

(**250**)

A mixed metal compound containing thallium cyclopentadienyl derivatives has also been prepared. It is a mixed thallium–gold cluster $[TlAu(C_5H_4PPh_2)_2]_2$ (**250**),[217] prepared by the reaction of $TlC_5H_4PPh_2$ with AuCl(SMe_2) in benzene. Each gold is coordinated to two phosphino groups, and each thallium is η^5 bound to two cyclopentadienyl fragments. The authors interpret this as two gold diphosphide units held together by thallium atoms, although recent work suggests that it may be best viewed as two $TlCp_2$ units held together by the gold–phosphine linkage. There remains some ambiguity on this point.[218]

Previously, $[CpTl]_\infty$ had been used mainly as a cyclopentadienyl transfer agent for use with transition metal ions, but very little is known about its chemistry owing to its low solubility in organic solvents. However, the reaction of bis(cyclopentadienyl) magnesium with cyclopentadienyl thallium leads to nucleophilic addition of Cp^- to CpTl. Addition of pentamethyldiethylenetriamine (PMDETA) leads to crystals of the ion-separated $[Cp_2Tl]^-[CpMgPMDETA]^+$ (**251**).[218] Similarly, reaction of CpTl with CpLi and PMDETA (1:2:1) results in the ion-contacted [CpTl(μ-Cp)Li·PMDETA] (**252**). A crystallographic investigation of these complexes revealed that both are monomeric with bent Cp_2Tl^- fragments (Cp–Tl–Cp 156.7° in (**251**) and 153.3° in (**252**), respectively). The Cp_2Tl^- fragment is thus isoelectronic (14 electrons$^-$) and isostructural with stannocene.

Ab initio MO calculations show that the bent C_s geometry of Cp_2Tl^- is only marginally more stable than the linear D_{5d} conformer (by 3.5 kJ mol^{-1}). This implies the thallocene anion is highly flexible and that the thallium possesses a largely spherical lone pair of electrons. In addition, one of the Cp rings is more weakly attached to the thallium atom than the other.

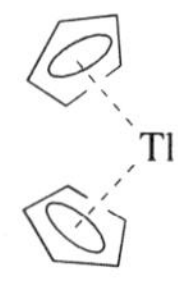

(251)

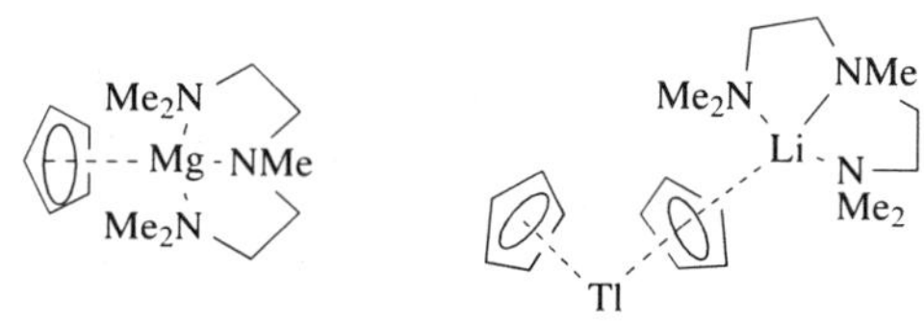

(252)

11.5.2 Carbaborane Compounds

Low-valent carbaborane compounds of thallium have been known for many years.[219] Similarly, many aluminacarbaboranes have been prepared and characterized. However, very little structural data were available on galla- and indacarbaboranes. The formative steps in this field of study have now been made and this work is described below.

The carbaboranes of the lighter elements are with the metal in the higher +3 oxidation state. However, because of the isolobal analogy between cyclopentadienyl and carbaboranes, and because the compounds have π-bound carbaboranes, these compounds are considered here. They had previously been characterized on the basis of NMR spectroscopic data alone, save for *closo*-3-Me-MeCH$_2$-3,1,2-GaC$_2$B$_9$H$_{11}$ (**253**).[220] Only recently have more unequivocal x-ray stucture determinations of these types of compounds been made.

The first to be considered is a compound that does not contain a carbaborane cage, but a π complex between a thallium atom and a C$_3$B$_2$ ring akin to the cyclopentadienyl compounds described above. The workers in Heidelberg have shown that the reaction of (η-Cp)(η-C$_2$B$_2$C)cobalt (C$_2$B$_2$C = 1,3,4,5-tetramethyl-2,3-dihydro-1,3 1-diborole) with CpTl in THF yields Tl(μ-C$_2$B$_2$C)Co(η-Cp) (**254**).[221] This orange–brown compound is air-sensitive, soluble in nonpolar solvents and sublimable in a high vacuum. An x-ray structure analysis on (**254**) shows that the thallium does not lie over the centroid of the C$_2$B$_2$C ring, but is displaced by 0.024 nm towards the carbon bearing the highest positive charge in the C$_2$B$_2$C ring. Units of (**254**) are monomeric, and it is suggested that the stronger acceptor functions of the C$_2$B$_2$C ring prevent dissociation into ions in the solid state.

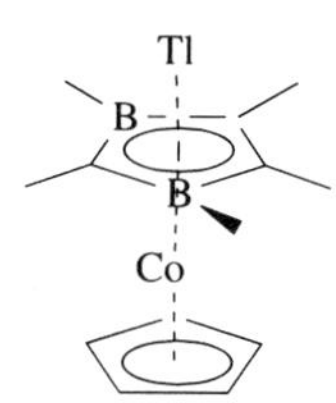

(254)

Although the synthesis and spectroscopic data on some indium carbaboranes has been reported for many years,[222,223] no structural data were available until 1991. Workers in Dallas and at Austin reacted [Na$^+$Li$^+${2,3-(TMS)$_2$C$_2$B$_4$H$_4$}$^{2-}$] with (Me)$_2$CHInI$_2$ producing the *closo*-indacarbaborane (Me)$_2$CHIn[2,3-(TMS)$_2$-2,3-C$_2$B$_4$H$_4$] (**255**) as white air-sensitive crystals.[224] The crystal structure of (**255**) shows it to be a dimeric cluster. However, as with (**242**),[210] it is more of a 'pseudodimer' with monomeric units linked by a crystallographic centre of inversion, rather than a dimer linked by strong metal–metal bonds. The In–B bond lengths of (**255**) fall into two distinct categories with six intracage distances averaging 0.239 nm, and four intercage distances averaging 0.304 nm. The indium is coordinated in a pentahapto fashion to the C$_2$B$_3$ face, but is significantly slipped towards η^3 coordination. The ^{11}B NMR spectrum shows broad, ill-defined resonances that do not permit a distinction between η^5 and η^3 coordination in solution.

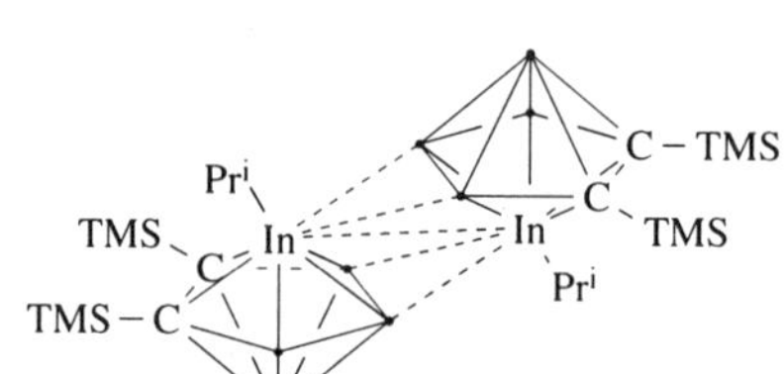

(255)

Several carbaboranes of gallium have been prepared which complement the relative wealth of information on aluminum–carbaborane complexes. Much of the early work on these systems made structural assignments on the basis of NMR spectroscopy alone, but now several x-ray stucture determinations are available. One technique used in the synthesis of these compounds has been the use of other group 13 carbaboranes as transmetallation reagents. This technique has been used previously to prepare other metallacarbaborane compounds and more recently to prepare aluminacarbaboranes.[225,226] The approach has now been extended to prepare a compound which yielded the first x-ray structure determination of a gallacarbaborane sandwich.[227,228]

Reaction of excess $GaCl_3$ with Tl[*closo*-3,1,2-$TlC_2B_9H_{11}$] produces [Tl{Ga($C_2B_9H_{11}$)$_2$}] (**256**) in quantitative yield. The gallacarbaborane cage is slightly bent, with an angle between the normals to the open faces of the carbaborane cages of 174.5°. The gallium atom is noticeably slipped towards the three boron atoms on each face of the carbaborane cage which suggests η^3 bonding. This contrasts with the analogous aluminum compound which demonstrates η^5 bonding to each carbaborane cage. It has been previously suggested that the slippage to η^3 coordination occurs as a result of back-bonding from the filled *d*-orbitals into the nonbonding MOs of the carbaborane ligand.[223,229] More recent calculations can model the slippage using only gallium *s*- and *p*-orbitals interacting with the carbaborane molecular orbitals.[230] Another interesting structural feature of (**256**) is the eclipsed nature of the carbaborane cages. This has been explained by nonbonded interactions between the carbaborane hydrogen atoms and the thallium cation. Each thallium cation in the unit cell is surrounded by five gallacarbaborane ions, resulting in six cage hydrogens that approach closely (<0.32 nm) to the thallium atom. As a result of ligand rotation the cages become eclipsed, and the sandwich must distort to the slightly bent structure in order to maximize gallium–cage orbital overlap.

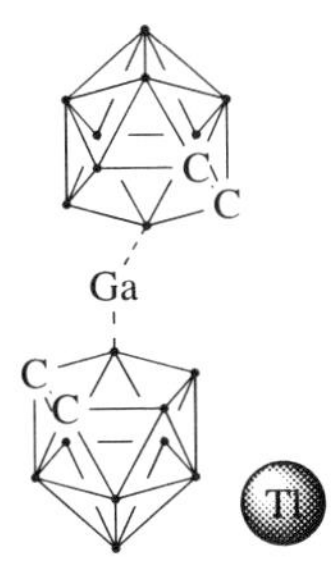

(**256**)

Two further gallacarbaborane cages, that of the half-sandwich *closo*-1-Bu^t-1-Ga-2,3-$(TMS)_2$-2,3-$C_2B_4H_4$ (**257**), and the adduct it forms with 2,2'-bipyrimidine, [1,1'-(2,2'-$C_8H_6N_4$){1-Bu^t-1-Ga-2,3-$(TMS)_2$-2,3-$C_2B_4H_4$}$_2$] (**258**) have also been characterized by x-ray crystallography.[231]

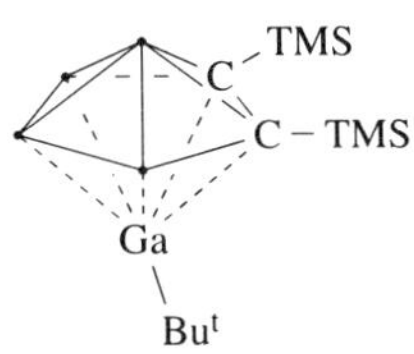

(**257**)

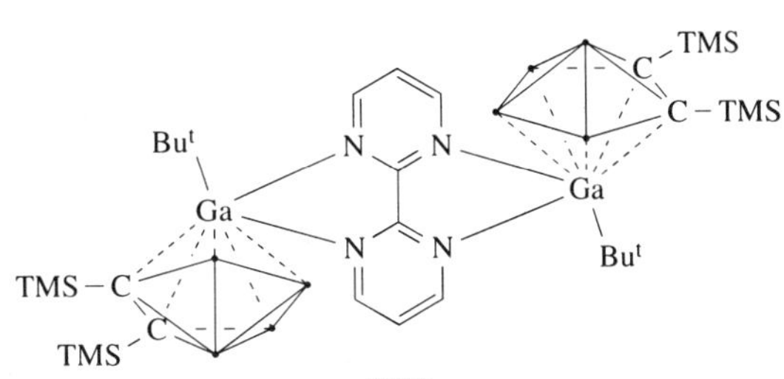

(**258**)

Compund (**257**) is formed by the reaction of [GaBu^{t}Cl$_2$]$_2$ with *nido*-carbaborane dianionic salts Na^+Li^+2,3-$(TMS)_2$-2,3-$C_2B_4H_4$ in 1:2 molar ratio. The x-ray structure reveals a gallacarbaborane of *closo* geometry in which the gallium atom occupies an apical vertex in a distorted pentagonal bipyramid. The carbaborane is approximately η^3-coordinated to the gallium. Interestingly, the tertiary butyl group of the gallium makes a pronounced tilt towards the cage carbons on the C_2B_3 face ($C_2B_{3\ centroid}$–Ga–Bu^t = 24°). This has been observed previously in 1-Me-1-Ga-2,3-$C_2B_4H_6$,[223] and has been explained by extended Hückel calculations which show that bending the gallium–alkyl bond increases the bonding between the gallium and the carbaborane cage.[230]

Reaction of (**257**) with 2,2'-bipyrimidine instantaneouslyproduces (**258**). In this structure, the bipyrimidine bridges the two gallacarbaboranes. The binding of the strong nitrogen donors to the gallium in each cage induces η^1 coordination of the gallium to the carbaborane, with the gallium bound to the unique boron on the C_2B_3 face. Consequently, the tertiary butyl group is even more tilted ($C_2B_{3\,centroid}$–Ga–Bu^t = 56°). These findings are consistent with what is observed when Lewis bases coordinate to aluminacarbaboranes, stannacarbaboranes, germanacarbaboranes and plumbacarbaboranes.[227,232–5]

11.5.3 Arene Complexes

Arene complexes of transition metals have been widely investigated following the discovery of bis(benzene) chromium in 1955.[236] However, data on main group metal arene complexes have been sparse, despite the fact that it has been known since 1881 that gallium(II) chloride is soluble in aromatic solvents. Arene complexes of the alkali and alkaline earth metals have been investigated and found to be η^2- or η^3-coordinated.[3] Work on group 14 arene complexes started much earlier and has shown that tin(II) and lead(II) form 1:1 arene complexes with η^6-arene coordination.[237,238] However, little data were available on group 13 complexes despite the fact that it had long been known that the mixed-valent metal salts of group 13 are soluble in aromatic solvents which opposes conventional solvation theories. Cryoscopic molecular weight determinations indicate strong concentration dependence for the degree of association, with dimers prevailing between 0.05 M and 0.1 M, and monomers only being reached at infinite dilution,[239,240] The nonconducting nature of the solutions further pointed to tight ion-pair formation. Since the low-valent group 14 metal ions are isoelectronic with the low-valent group 13 (i.e., $(n-1)d^{10}\ ns^2$), predictions drawn from this link were the only indication of the structure of the corresponding group 13 compounds. However, group 13 arene complexes have recently been isolated and unambiguously identified in their own right.

The arene complexes of group 13 are important for several reasons, as summarized by Schmidbaur.[3]

(i) They are powerful reducing agents.
(ii) The arenes should be activated to nucleophilic attack.
(iii) The arene solubility of gallium(I) and indium(I) may provide a route to ultrahigh purity metals for use in semiconductors.
(iv) These salts may provide a convenient room-temperature synthetic route into investigating low-valent group 13 chemistry for those elements whose monovalent compounds are otherwise difficult to synthesize.

When a hot anhydrous saturated benzene solution of $Ga(GaCl_4)$ is allowed to cool, colourless crystals are obtained. An x-ray diffraction study shows that the crystals consist of dimeric units, $[\{(C_6H_6)_2Ga\}^+(GaCl_4)^-]_2\cdot 3C_6H_6$ (**259**).

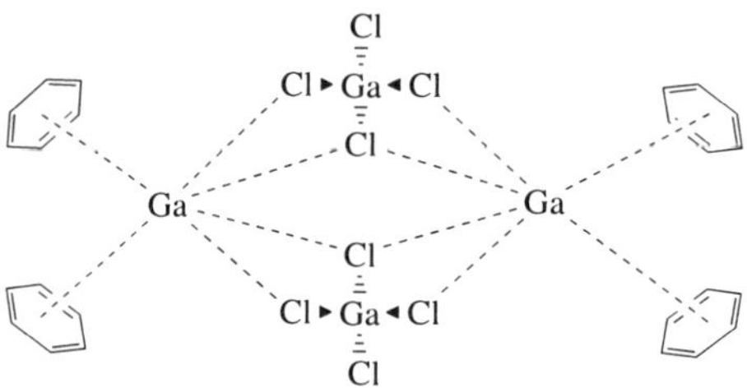

(259)

The two gallium(I) centres form two bent sandwiches with η^6-coordinated benzenes, and an interplane angle of 124.4°. The gallium(I)–benzene$_{centroid}$ distances differ markedly (0.2764 nm and 0.2929 nm, respectively). Four chlorine atoms from two different $GaCl_4^-$ anions surround the gallium(I) centre irregularly. The Ga^{III}–Cl distances range from 0.2152 nm to 0.2183 nm, whereas the Ga^I–Cl distances are much larger, ranging from 0.3097 nm to 0.3716 nm, indicating only weak interactions between the gallium(I) centre and the $GaCl_4^-$ unit.

Following this study, the product formed by dissolving $Ga[GaCl_4]$ in mesitylene was investigated.[241] This compound was shown to be $[(C_6H_3Me_3)_2Ga]^+[GaCl_4]^-$ (**260**) and is so stable that the mesitylene adducts are retained even under vacuum at +20 °C. The x-ray structural study again shows the bent gallium(I) arene sandwich is present (the interplane angle is now increased to 140.3°, presumably for steric reasons). The mesitylenes remain coordinated in an η^6 fashion to the gallium(I) centre. However, the $GaCl_4^-$ unit coordinates weakly through only one chlorine atom to the gallium(I) centre as a result of steric hinderance from the mesitylene groups. A second chlorine of the $GaCl_4^-$ coordinates to a different gallium(I) centre. Hence the units are linked into a polymeric chain.

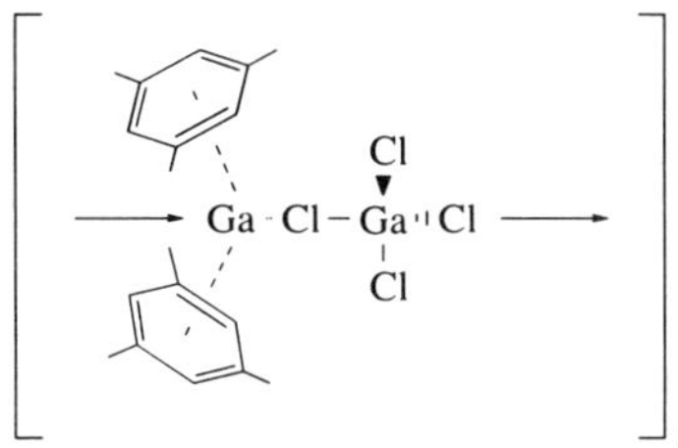

(260)

It is noteworthy that when $Ga[GaCl_4]$ is dissolved in a mixed benzene/mesitylene solution, only the mesitylene is taken up by the gallium(I) centre. Hence, it appears that methyl substitution of the benzene strongly favours gallium(I) complexation.

However, if $Ga[GaCl_4]$ is dissolved in toluene, followed by addition of durene (durene = 1,2,4,5-tetramethylbenzene) at high concentration (1.4 M) then the microcrystalline product, not structurally characterized, is $[(durene)Ga]^+[GaCl_4]^-$ (**261**).[242] With lower concentrations of durene (0.34 M), crystals suitable for x-ray diffraction are obtained. The product is $[(\eta\text{-}C_6H_5Me)(\eta\text{-}C_6H_2Me_4)Ga]^+[GaCl_4]^-$ (**262**).[242] Its structure is essentially the same as the benzene adduct (**259**).

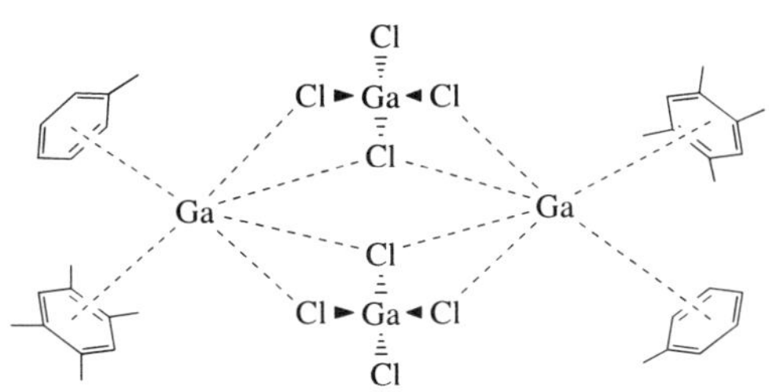

(262)

The interplane angle is 122°, very similar to that in (**259**). The durenes are arranged in a *trans* conformation. The most interesting point about the molecule is that the gallium(I)–durene$_{centroid}$ distance is considerably less, at 0.2642 nm, than the gallium(I)–toluene$_{centroid}$ distance of 0.3039 nm. This reinforces the proposition that the more heavily substituted arenes are favoured in gallium(I) complexation. In (**262**), in contrast with (**259**), steric requirements prevent the coordination of a second durene molecule. There is, however, still sufficient room for a toluene molecule to coordinate. One must also recall that even the two gallium(I)–arene distances for the bis(benzene) complex (**259**) differ markedly.[243]

If the reaction is performed using hexamethyl benzene/toluene solution, then the steric bulk becomes so extreme that only one hexamethylbenzene can coordinate to the gallium(I) centre. The product, $[(\eta\text{-}C_6Me_6)Ga]^+[GaCl_4]^-$ (**263**) is the first, and at the time of writing the only monoarene complex of gallium(I) structurally characterized.[244]

There are four formula units in the unit cell. One of the gallium(I) ions is unique, and posseses a metal–ring distance of only 0.243 nm. The other three have metal–ring distances of 0.251 nm. The overall structure may be described as consisting of four $GaCl_4^-$ anions surrounded by four $[Ga(C_6Me_6)]^+$ cations, that is, an inorganic core surrounded by a hydrocarbon envelope.

On changing the anion to $GaBr_4^-$ then slight changes are observed in the structure upon dissolution in toluene/hexamethylbenzene. The product $[(\eta\text{-}C_6Me_6)Ga]^+[GaBr_4]^-$ (**264**) forms a complicated array similar to (**263**).[245] It differs in that the coordination sphere around each gallium(I) is completed by five bromine atoms, as opposed to six or seven chlorine atoms that surround each gallium(I) in (**263**).

The arene complexes of the heavier group 13 metals have proved more difficult to obtain than the gallium analogues. However, by utilizing the fact that a higher degree of arene substitution produces more stable complexes, and that mesitylene allows bis substitution at the metal, then indium and thallium analogues have been prepared. At the time of writing there is only one example of a simple arene complex, that of the bis(mesitylene) compound, $[(C_6H_3Me_3)_2In]^+[InBr_4]^-$ (**265**).[246] Like the gallium analogue, it forms a chain polymer, although the larger ionic radii for indium and bromine mean that only three In(I)–Br contacts are made as opposed to the four in (**260**). The interplane angle is 132.7°, slightly more compressed than in (**260**) as the larger ionic radii of indium causes less steric congestion.

For thallium, it is found that thallium(I) tetrahalometalates(III) are soluble in benzene, mesitylene or hexamethylbenzene. The mesitylene complex with the tetrabromogallate anion was the first of these arene complexes to be structurally characterized.[247] The product, however, is not simply the thallium analogue of (**265**), but a more complex mixture of both the mono and bis(arene) compounds having the overall formula, $[(\eta\text{-}C_6H_3Me_3)_6Tl_4(GaBr_4)_4]$ (**266**). Two of the thallium atoms are monocoordinated and the other two are biscoordinated. The thallium–arene distance in the monoarene cations is 0.294 nm, as opposed to 0.300 nm and 0.302 nm in the bis(arene) cations. The interplane angle in the bis(arene) cations is 119.5°, even more compressed than in (**265**) for the same reason that (**260**) is more compressed than (**265**), (i.e., the larger ionic radii of thallium causes less steric congestion). Hence the observation of the monoarene species is all the more surprising. The reason probably lies in the complex network existing between the thallium–arene cations and the $GaBr_4^-$ anions. These nonbonding Tl···Br contacts form the basis of the network. Since the number of these contacts varies, some Tl^+ cations can coordinate one mesitylene and others two mesitylenes, hence the unusual stoichiometry.

Noirot *et al.*, by using less-coordinating anions have isolated less complex structures incorporating $Tl(Mes)_2$ units.[248] The first compound is dimeric $[(\eta\text{-}C_6H_3Me_3)_2TlOTeF_5]_2$ (**267**), prepared by dissolving $TlOTeF_5$ in mesitylene. The second is $[(\eta\text{-}C_6H_3Me_3)_2Tl]^+[B(OTeF_5)_4]^-$ (**268**), prepared by mixing $TlOTeF_5$ with $B(OTeF_5)_3$ in mesitylene.

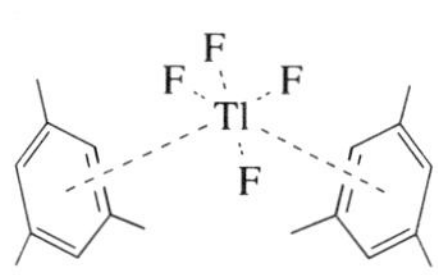

(267)

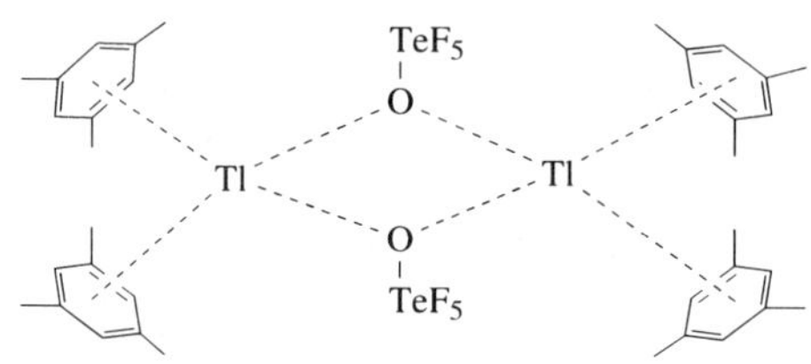

(268)

Dimeric (**267**) consists of a Tl_2O_2 core. The mesitylenes are η^6-coordinated with interplane angles of 118° and 121°, respectively, (each bis(arene) thallium unit is crystallographically independent) and $Tl\text{–}C_{arene}$ distances averaging 0.336 nm. This is very similar to those in (**266**). For (**268**), the bis(arene) thallium units are connected by weak Tl···F interactions from two different $B(OTeF_5)_4$ units to form a polymeric structure. The mesitylenes are again η^6-coordinated, the interplane angle being 129°, but the average $Tl\text{–}C_{arene}$ distance of 0.317 nm is now much shorter than in (**266**) or (**267**). This demonstrates the very poor coordinating power of the $B(OTeF_5)_4$ anion, leading to stronger thallium–mesitylene bonds.

Schmidbaur has extended the work beyond simple arenes to investigate cyclophanes, thus using the chelate effect to stabilize the compounds. Once again the gallium systems have proved the easiest to investigate. Addition of Ga_2Cl_4 or Ga_2Br_4 to [2,2]paracyclophane yields microcrystalline materials that are much less sensitive to air and moisture than the nonbridged counterparts. Preliminary tests suggested that this would have a double-bridged sandwich structure akin to the chromium(0) complex.[249,250] However, the x-ray structure determination of the Ga_2Br_4 complex found that instead of the gallium(I) being in between the cyclophane interstices, it becomes attached to the outside of the bridged arenes, and coordinates to two cyclophane units to form an infinite polymer (**269a**).[251]

The benzene rings of each cyclophane unit remain parallel to each other, with the angle at gallium(I) between each cyclophane being 123.3°. Further, each gallium(I) is coordinated to two $GaBr_4$ units. These further coordinate to other gallium(I) centres, hence linking the structure into a three-dimensional lattice.

On extending this work to study the reaction of [2,2]paracyclophane with $In[InBr_4]$ in mesitylene and $Tl[GaBr_4]$ in toluene, results analogous to $Ga[GaBr_4]$ were found.[252] The angle at the metal is more open, at 118.8° (In) and 115.4° (Tl), as was found in the unbridged mesitylene examples described above.

On using still larger paracyclophanes such as [3,3,3]paracyclophane, still more complicated polymeric arrays might have been expected. However, somewhat surprisingly, a macrocyclic tris(arene) complex $[\{(\eta\text{-}[2,2,2]paracyclophane)Ga\}GaBr_4]$ (**269b**) is observed, with the gallium(I) situated in the middle of the cyclophane cavity.[253] Gallium(I) is approximately equidistant from the centre of each benzene ring, but lies 0.43 nm above the plane of three rings due to the close proximity of the bromine atom of a $GaBr_4^-$ unit. This type of η^{18} coordination is unprecedented in the area of metal–π-complex chemistry.

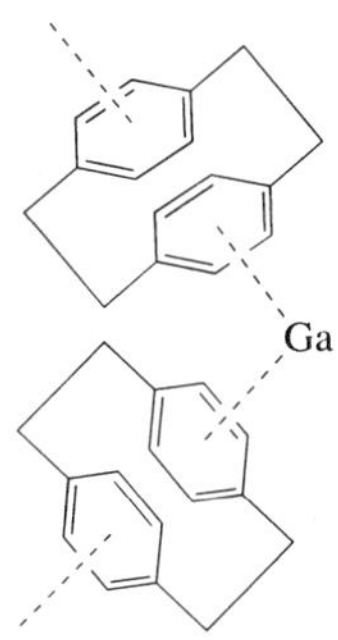

(269a)

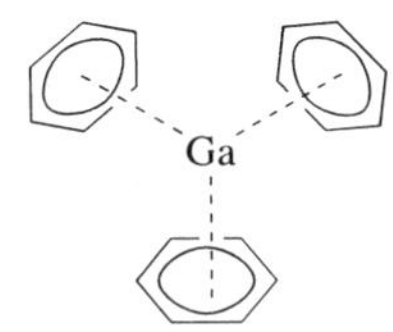

(269b)

11.6 METAL–METAL BONDED COMPLEXES

Examples of complexes containing metal–metal bonds between group 13 elements are few and far between. Typically, bulky substituents are required in order to stabilize such derivatives, which are otherwise prone to disproportionation reactions into elemental metal and trivalent metal derivatives. Although the occurrence of metal–metal contacts is not a new feature in group 13 chemistry, examples are rare, many being rather dubious contacts often as a result of crystal packing instead of fully-fledged metal–metal bonds.

Until recently the only examples of such metal–metal bonds were to be found in species such as $Br_2Ga–GaBr_2·2dioxan$[254] or in indium sulfides and selenides. Uhl and co-workers have synthesized the gallium organometallic, $[\{(TMS)_2HC\}_2Ga]_2$ (**270**),[255] by the reaction of bis(trimethylsilyl)methyllithium with $Br_2Ga–GaBr_2·2dioxan$. This molecule is isotypic with the aluminum analogue, $[\{(TMS)_2HC\}_2Al]_2$, and as such exemplifies the occurrence of a Ga–Ga bond.[256] Cryoscopic molecular mass measurements in benzene indicate (**270**) to be a monomer in solution. An x-ray crystallographic study shows the molecule to be almost planar, the C–Ga–Ga–C dihedral angle being only 8° from the ideal 180° and the gallium atoms lying 0.0046 nm and 0.0015 nm out of the plane defined by the gallium atom and the two associated carbons; other pertinent data are Ga–Ga 0.2541 nm, average Ga–C 0.1992 nm, average Ga–Ga–C 122.0°. In a similar manner Uhl and coworkers synthesized the indium analogue, $[\{(TMS)_2HC\}_2In]_2$ (**271**),[71] in a similar reaction utilizing $In_2Br_4·2TMEDA$. The In–In bond length of 0.2828 nm compares with the other homologues, Ga–Ga 0.2541 nm and Al–Al 0.2660 nm. Thus, this distance is similar to that observed in the indium halides, $In_2Br_3I·2TMEDA$ 0.2775 nm and In_2Br_3 0.2688 nm.[257,258] As with the aluminum and gallium homologues, the molecular structure of (**271**) is almost planar, although the deviations from planarity are more prominent than for the others, the indium atoms lying 0.0039 nm and 0.0042 nm out of the plane. This compares with a distance of 0.0191 nm in the indium(III) complex, $[(TMS)_2HC]_3In$.[259]

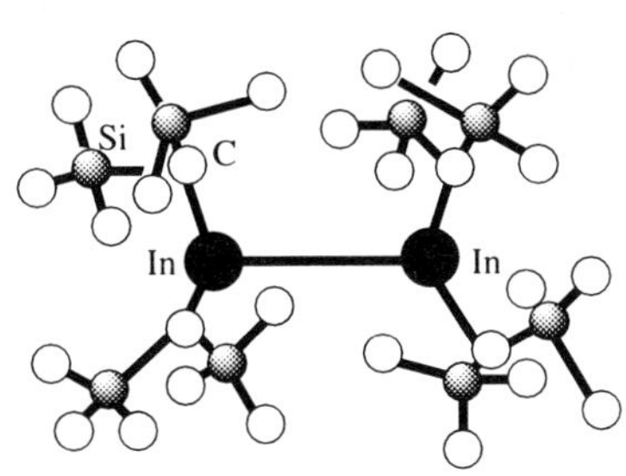

(271)

In addition to the gallium–gallium bonded species $[\{(TMS)_2HC\}_2Ga]_2$, further work using this sterically demanding ligand has been forthcoming with stabilized organogallium metal–metal bonded species. Noticeably less-bulky substituents tend to lead to disproportionation, yielding exclusively elemental gallium and gallium(III) derivatives. Reaction of $Ga_2Br_4·2dioxane$ at −60 °C with ether-free tris(trimethylsilyl)methyllithium in pentane resulted in the isolation of a dark red crystalline compound, $[GaC(TMS)_3]_4$ (**272**).[6] The gallium atoms exhibit the unusual +1 oxidation state which is formed by disproportionation. Unusually, for a low oxidation-state gallium organometallic, (**272**) is remarkably thermally stable, decomposing only above 255 °C and can be stored in air for several months without change. Depending on the method of characterization, several degrees of oligomerization have been determined: in the mass spectrum only the monomer is detected with no heavier fragments, such as those expected for oligomeric units, being observed. Cryoscopic molecular mass determinations in benzene over a range of concentrations indicate degrees of aggregation ranging from a trimer (0.023 mol l^{-1}) to a monomer as dissociation increases on dilution (0.0014 mol l^{-1}). UV–visible spectra indicate the presence of metal–metal bonds (252, 294 and 435 nm), the high wavelength absorption only falling off at very high dilution, indicating monomerization with increasing dilution (cf. 360 nm for $[\{(TMS)_2HC\}_2Ga]_2$). The crystal structure of (**272**) reveals an almost undistorted tetrahedral Ga_4 skeleton, with an average Ga–Ga distance of 0.2688 nm. This is, as expected, 0.015 nm longer than in the tetraalkyldigallium derivative $[\{(TMS)_2HC\}_2Ga]_2$ with an isolated Ga–Ga distance of 0.2541 nm.[255] This lengthening on increasing oligomerization compares with that observed in the aluminum cluster $Al_4Cp^*_4$ (Al–Al 0.2769 nm,[5] compared with the dialuminum derivative $[\{(TMS)_2HC\}_2Al]_2$,[256] Al–Al 0.2660 nm). In addition, the Ga–C bond lengths (average 0.208 nm) are significantly longer than in (**270**), average 0.1994 nm, being comparable with those obtained for bonds to bridging carbon atoms.

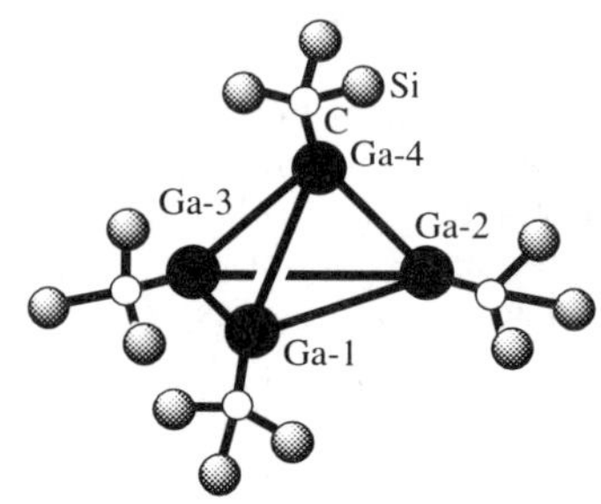

(272)

A very recent application of the 2,4,6-$(CF_3)_3C_6H_2$ ligand, to stabilize gallium(II) and indium(II) derivatives with metal–metal bonds, has been reported. The reaction between equimolar quantities of InCl and 2,4,6-$(CF_3)_3C_6H_2Li$ in ether yields the indium(II) dimer, $[\{2,4,6\text{-}(CF_3)_3C_6H_2\}_2In]_2$ (**273**).[33] Here, the implied disproportionation, $2In^I \rightarrow In^0 + In^{II}$ is unusual. However, traces of the corresponding [2,4,6-$(CF_3)_3C_6H_2]_3In$ species could also be detected in the reaction mixture; this was more satisfactorily prepared from InI_3.

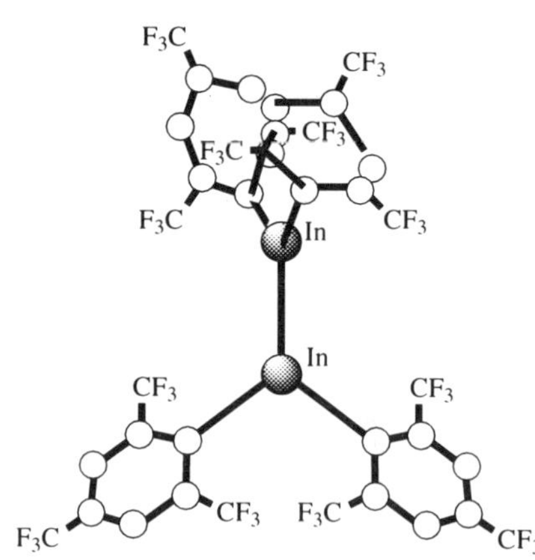

(273)

The crystal structure of (**273**) features an In–In bond length of 0.2744(2) nm, consistent with a bond order of unity.[71] Both indium centres adopt trigonal planar geometries, with the C–In–C angles approximately 3° smaller than the C–In–In angles. The dihedral angle between the C–In–C planes is 94.1(5)°, presumably as a consequence of minimizing steric repulsions between *o*-CF_3 groups and maximizing intramolecular In···F interactions. As in the case of B_2R_4 derivatives, there is no evidence to support the existence of any π-type interaction between the group 13 metals.[260] There are, however, eight intramolecular In···F contacts (one for each *o*-CF_3 group) which fall in the range 0.2801(10)–0.2957(13) nm and are, thus, shorter than the sum of the van der Waals radii[261] of 0.340 nm (however, the average In···F distance of 0.2856(13) nm is considerably longer than the sum of the covalent radii,[261] 0.216 nm, and significantly longer than those in InF_3,[262] 0.206 nm). Arguably, however, it may be the presence of these interactions which gives rise to the stability of this species. It has been suggested that the high nucleophilicity and low basicity of this ligand minimize reduction of the metal centre.[263] Coupled with electrostatic repulsions between CF_3 groups, oligomerization is inhibited and stability is enhanced.

Since the early 1980s there has been a renaissance of group 13 elements in low oxidation states, in particular, halide complexes exhibiting metal–metal interactions or true electron-pair bonds. By cocondensing GaCl and a mixture of toluene/diethylether (4:1) at −196 °C, GaCl, which is metastable at room temperature, was obtained in solution. An x-ray crystallographic study showed the structure to be $[(Et_2O)_2ClGa]Ga[GaCl_2(Et_2O)]_3$ (**234**) (see Section 11.5.1).[202]

The central atom Ga-1 is tetrahedrally surrounded by another four gallium atoms, Ga-2–Ga-5. Each of these is likewise tetrahedrally coordinated; two chlorine atoms and one ether molecule bind to Ga-3, Ga-4 and Ga-5, while Ga-2 is bound to only one chlorine atom and two ether molecules. Consequently, oxidation states of 0 for Ga-1, +1 for Ga-2 and +2 for Ga-3–Ga-5 can be inferred. Despite this the Ga–Ga distances vary only slightly, between 0.2417(2) nm and 0.2450(2) nm; the longest distance observed is between Ga-1 and Ga-2. This ties in with the assignment of oxidation states, bearing in mind the relative sizes of the gallium atoms (in gallium metal a short bond of 0.244 nm is observed in addition to six longer bonds from 0.270 nm to 0.279 nm).[264] The Ga–Cl bond lengths fall within the expected range, 0.2208–0.2224 nm (cf. Ga–Cl 0.217 nm in $Ga_2Cl_4{\cdot}dioxan_2$).[254] Angles within the Ga_5 tetrahedron can be explained by repulsive electrostatic interactions. Thus, angles that involve two gallium(II) ions are wider than the tetrahedral angle, while angles engaging gallium(I) are smaller, resulting from less repulsion. The preparation of (**234**) from GaCl also yields small amounts of $GaCl_3$, hence implying disproportionation reactions ($2Ga^{I} \rightarrow Ga^{0} + Ga^{II}$) and possibly also comproportionation reactions ($Ga^{III} + Ga^{I} \rightarrow 2Ga^{II}$).

Among other group 13 derivatives are the cyclopentadienide complexes which have been reported to exhibit metal–metal interactions. Thus, pentabenzylcyclopentadienylindium(I), $(Bz)_5C_5In$ (**242**),[210] exists as two monomeric species associated to a quasidimer by In···In contacts in the solid state. The distance of 0.3631 nm between the two indium atoms of a dimeric unit is like that between the two thallium atoms (0.3632 nm) in the dimeric unit of the analogous thallium compound.[213] However, this In–In distance is significantly shorter than the interchain In···In distances of 0.3986 nm in $(CpIn)_\infty$ (**237**) and $(MeC_5H_4In)_\infty$ (**238**),[206] or the intraclusteral In···In distances of 0.3963 nm and 0.3943 nm in $(Cp^*In)_6$ (**241**),[265,7] which are interpreted in terms of weak In···In interactions. However, in this case, since the metal–metal distances are the same in the indium and thallium derivatives, it has been suggested that the packing effect of the ligand enforces dimerization of these molecules by organic–envelope interaction rather than by metal–metal interaction. The compound InCp* (**241**) crystallizes such that the individual InCp* units are arranged about points of crystallographic S_6 symmetry, whereby the indium atoms are on the interior and the Cp* units are on the exterior of hexameric 'clusters'.[265,7]

The work initiated by Uhl and co-workers, namely the synthesis of $R_2M{-}MR_2$ derivatives (M = Al, Ga, Tl; R = $CH(TMS)_2$),[255,256,71] has been extended via one-electron reductions of such systems. Thus, Power and co-workers effected the lithium reduction of $Trip_2Al{-}AlTrip_2$ (Trip = 2,4,6-$Pr^i{}_3C_6H_2$) to $[Li(TMEDA)_2]^+[Trip_2Al{\dot{-}}AlTrip_2]^{\bullet-}$ (**274**) or $[Li(12\text{-crown-}4)_2]^+[Trip_2Al{\dot{-}}AlTrip_2]^{\bullet-}$ (**275**).[266] In doing so, the aluminum–aluminum bond length decreases from 0.2647 nm to 0.2470 nm and the torsion angles between the planes of the aluminum atoms decreases from 44.8° to 1.4°. Solutions of complex (**274**) exhibit strong EPR signals, there being an 11 line pattern consistent with the equal coupling between the unpaired electron and the two aluminum nuclei (^{27}Al, I = $\frac{5}{2}$). Spectroscopic and structural data support the existence of a one-electron π bond between the aluminum atoms. This work has also been undertaken for the gallium analogue, $Trip_2Ga{-}GaTrip_2$, to yield $[Li(12\text{-crown-}4)_2]^+[Trip_2Ga{\dot{-}}GaTrip_2]^{\bullet-}$ (**276**).[266] This species has a gallium–gallium bond length of 0.2343 nm, which is substantially shorter (by 0.017 nm) than in the neutral precursor, 0.2515 nm (Table 2). Again, these data combined with a low coupling constant between the unpaired electron and gallium nuclei, are consistent with the formation of a one-electron Ga–Ga π bond resulting from the overlap of two gallium 4*p* orbitals.

The above data show that in both cases the one-electron reduction process results in a M–M shortening of ~0.017 nm, as well as a large decrease in the torsion angles between the planes at the metals (and a slight lengthening in the M–C distances). It is surprising that the addition of just one electron leads to such a dramatic change, especially bearing in mind the only minor changes in σ-hybridization contributing to the shortening ($sp^3 \rightarrow sp^2$). Such changes are substantially greater (by ~0.01 nm) than that observed for the two-electron reduction of tetraorganodiboron species to give $(R_2BBR_2)^{2-}$ ions.[267] It has been argued that one possible explanation for this is the charge separations across the M–C and M–M bonds in (**274**)–(**276**), which result in partial positive and negative charges on the M and C atoms, respectively. Thus, in the neutral precursors, this ionic character is expected to

Table 2 Comparison of the bond distances and angles surrounding the metals in $Trip_2M{-}MTrip_2$ and the anion $[Trip_2M{\doteq}MTrip_2]^{\bullet -}$ (M = Al, Ga).

Parameter	*Trip₂M–MTrip₂* M = Al	M = Ga	*[Trip₂M⋯MTrip₂]•⁻* M = Al(**274**)	M = Ga
M–M (nm)	0.2647(3)	0.2515(3)	0.2470(2)	0.2343(2)
M–C (nm)	0.1996(3)	0.2008(7)	0.2021(1)	0.2038(10)
C–M–C (°)	11.37(2)	11.22(4)	10.73(9) 11.68(7)	11.15(5) 11.74(6)
Twist angle between the planes at the metals (°)	44.8	43.8	1.4	15.5

increase the M–M bond lengths and decrease the M–C bond lengths relative to what is predicted from the sum of their radii. Addition of an electron tends to cancel these ionic interactions by placing negative charge density in a π orbital, thus increasing the M–C bond length and dramatically shortening the M–M bond. Similar observations have also been reported by Pörschke and co-workers,[268] following the one-electron reduction of the aluminum–aluminum-bonded dimer, $[\{(TMS)_2HC\}_2Al]_2$, to $[Li(TMEDA)_2]^+$-$[Al_2\{CH(TMS)_2\}_4]^{\bullet -}$.

11.7 REFERENCES

1. (a) A. H. Cowley and R. A. Jones, *Angew. Chem.*, 1989, **101**, 1235; *Angew. Chem., Int. Ed. Engl.*, 1989, **28**, 1208; (b) R. L. Wells, *Coord. Chem. Rev.*, 1992, 273.
2. D. Loos, H. Schnöckel, J. Gauss, U. Schneider, *Angew. Chem.*, 1992, **104**, 1376; *Angew. Chem., Int. Ed. Engl.*, 1992, **31**, 1362.
3. H. Schmidbaur, *Angew. Chem.*, 1985, **97**, 893; *Angew Chem., Int. Ed. Engl.*, 1985, **24**, 893.
4. P. Jutzi, *Adv. Organomet. Chem.*, 1986, **26**, 217.
5. C. Dohmeier, C. Robl, M. Tacke and H. Schnöckel, *Angew. Chem.*, 1991, **103**, 594; *Angew. Chem. Int. Ed. Engl.*, 1991, **30**, 564.
6. W. Hiller, M. Layh, W. Schwarz and W. Uhl, *Angew. Chem.*, 1992, **104**, 1378; *Angew. Chem. Int. Ed. Engl.*, 1992, **31**, 1364.
7. O. T. Beachley, Jr., M. R. Churchill, J. C. Fettinger, J. C. Pazik and L. Victoriano, *J. Am. Chem. Soc.*, 1986, **108**, 4666.
8. H. M. Manasevit and W. I. Simpson, *J. Electrochem. Soc.*, 1969, **116**, 1725; *J. Electrochem. Soc.*, 1973, **120**, 135.
9. R. H. Moss, *J. Cryst. Growth*, 1984, **68**, 78.
10. (a) A. Boardman, R. W. H. Small and I. J. Worrall, *Inorg. Chim. Acta.*, 1986, **119**, L13; (b) D. C. Bradley, *et al.*, *Polyhedron*, 1988, **14**, 1289.
11. D. C. Bradley, H. Dawes, D. M. Frigo, M. B. Hursthouse and B. Hussain, *J. Organomet. Chem.*, 1987, **325**, 55.
12. R. L. Wells, S. R. Aubuchon, M. F. Self, J. P. Jasinski, R. C. Woudenberg and R. J. Butcher, *Organometallics*, 1992, **11**, 3370.
13. R. B. Hallock, W. E. Hunter, J. L. Atwood and O. T. Beachley, Jr., *Organometallics*, 1985, **4**, 547.
14. A. M. Bradford, D. C. Bradley, M. B. Hursthouse and M. Motevalli, *Organometallics*, 1992, **11**, 111.
15. H. Krause, K. Sille, H. D. Hausen and J. Weidlein, *J. Organomet. Chem.*, 1982, **235**, 253.
16. R. L. Wells, A. T. McPhail, L. J. Jones, III, M. F. Self and R. J. Butcher, *Organometallics*, 1992, **11**, 2694.
17. G. H. Robinson, W. E. Hunter, S. G. Bott and J. L. Atwood, *J. Organomet. Chem.*, 1987, **326**, 9.
18. D. C. Bradley, D. M. Frigo, I. S. Harding, M. B. Hursthouse and M. Motevalli, *J. Chem. Soc., Chem. Commun.*, 1992, 577.
19. D. A. Atwood, R. A. Jones, A. H. Cowley, J. L. Atwood and S. G. Bott, *J. Organomet. Chem.*, 1992, **434**, 143.
20. M. A. Banks, O. T. Beachley, Jr. and J. D. Maloney, *Polyhedron*, 1990, **9**, 335.
21. D. A. Atwood, R. A. Jones, A. H. Cowley, J. L. Atwood and S. G. Bott, *Polyhedron*, 1991, **10**, 1897.
22. D. C. Bradley, H. M. Dawes, M. B. Hursthouse, L. M. Smith and M. Thornton-Pett, *Polyhedron*, 1990, **9**, 343.
23. D. V. Khasnis, H. Zhang and M. Lattman, *Organometallics*, 1992, **11**, 3748.
24. (a) G. Wiberg, T. Johannsen and O. Stecher, *Z. Anorg. Allg. Chem.*, 1943, **251**, 114; (b) G. E. Coates and R. A. Whitcombe, *J. Chem. Soc.*, 1956, 3351.
25. (a) H. Schumann, U. Hartmann, A. Dietrich and J. Pickardt, *Angew. Chem.*, 1988, **100**, 1119; *Angew. Chem., Int. Ed. Engl.*, 1988, **27**, 1077; (b) H. Schumann *et al.*, *Chem. Ber.*, 1991, **124**, 1113.
26. H. Schumann *et al.*, *Chem. Ber.*, 1990, **123**, 2093.
27. B. Neumüller, *Z. Anorg. Allg. Chem.*, 1991, **592**, 42.
28. E. L. Amma and R. E. Rundle, *J. Am. Chem. Soc.*, 1958, **80**, 4141.
29. A. J. Blake and S. Cradock, *J. Chem. Soc., Dalton Trans.*, 1990, 2393.
30. O. T. Beachley, Jr., M. R. Churchill, J. C. Pazik and J. W. Ziller, *Organometallics*, 1986, **5**, 1814.
31. J. T. Leman and A. R. Barron, *Organometallics*, 1989, **8**, 2214.
32. M. Brookhart and M. L. H. Green, *J. Organomet. Chem.*, 1983, **250**, 395.
33. R. D. Schluter, A. H. Cowley, D. A. Atwood, R. A. Jones, M. R. Bond and C. J. Carrano, *J. Am. Chem. Soc.*, 1993, **115**, 2070.
34. O. T. Beachley, Jr., T. L. Royster, Jr., W. J. Youngs, E. A. Zarate and C. A. Tessier-Youngs, *Organometallics*, 1989, **8**, 1679.
35. O. T. Beachley, Jr., M. A. Banks, Jr., M. R. Churchill, W. G. Feighery and J. C. Fettinger, *Organometallics*, 1991, **10**, 3036.
36. B. Lee, W. T. Pennington, J. A. Laske and G. H. Robinson, *Organometallics*, 1990, **9**, 2864.
37. G. Erker, M. Albrecht, C. Krüger and S. Werner, *J. Am. Chem. Soc.*, 1992, **114**, 8531.

38. M. J. Zaworotko, R. D. Rogers and J. L. Atwood, *Organometallics*, 1982, **1**, 1179.
39. B. Lee, W. T. Pennington and G. H. Robinson, *Organometallics*, 1990, **9**, 1709.
40. B. Lee, W. T. Pennington, G. H. Robinson and R. D. Rogers, *J. Organomet. Chem.*, 1990, **396**, 269.
41. G. H. Robinson, W. T. Pennington, B. Lee, M. F. Self and D. C. Hrncir, *Inorg. Chem.*, 1991, **30**, 809.
42. D. C. Bradley, M. M. Faktor, E. A. D. White, D. M. Young, *Chemitronics*, 1988, **3**, 50; A. H. Moore, M. D. Scott, J. J. Davis, D. C. Bradley, M. M. Faktor and H. Chudzynska, *J. Cryst. Growth*, 1988, **77**, 19.
43. D. C. Bradley, M. M. Faktor, M. Scott and E. A. D. White, *J. Cryst. Growth*, 1986, **75**, 101.
44. M. Veith and O. Recktenwald, *J. Organomet. Chem.*, 1984, **264**, 19.
45. M. A. Khan, C. Peppe and D. G. Tuck, *J. Organomet. Chem.*, 1985, **280**, 17.
46. D. A. Atwood, A. H. Cowley and R. A. Jones, *J. Organomet. Chem.*, 1992, **430**, C29.
47. R. L. Wells, A. T. McPhail, J. W. Pasterczyk and J. W. Alvanipour, *Organometallics*, 1992, **11**, 226.
48. H. Schmidbaur, S. Lauteschläger and G. Müller, *J. Organomet. Chem.*, 1985, **281**, 25.
49. K. Issleib and H.-P. Abicht, *J. Prakt. Chem.*, 1970, **312**, 456; K. Issleib and H.-P. Abicht and H. Winkelmann, *Z. Anorg. Allg. Chem.*, 1972, **388**, 89.
50. H. Schmidbaur, S. Lauteschläger and B. Milewski-Mahrla, *Chem. Ber.*, 1983, **116**, 1403.
51. U. Kynast, B. W. Skelton, A. H. White, M. J. Henderson and C. L. Raston, *J. Organomet. Chem.*, 1990, **384**, C1.
52. A. H. Cowley, R. A. Jones, M. A. Mardones, J. Ruiz, J. L. Atwood and S. G. Bott, *Angew. Chem.*, 1990, **102**, 1169; *Angew. Chem., Int. Ed. Engl.*, 1990, **29**, 1150.
53. J. Vincente, J. A. Abad, J. F. Gutierrez-Jugo and P. G. Jones, *J. Chem. Soc., Dalton Trans.*, 1989, 2241.
54. H. Schumann, F. H. Görlitz, T. D. Seuss and W. Wassermann, *Chem. Ber.*, 1992, **125**, 3.
55. H. Schumann, W. Wassermann and A. Dietrich, *J. Organomet. Chem.*, 1989, **365**, 11.
56. M. A. Khan, C. Peppe and D. G. Tuck, *Organometallics*, 1985, **5**, 525.
57. O. T. Beachley, Jr., R. B. Hallock, H. M. Zhang and J. L. Atwood, *Organometallics*, 1985, **4**, 1675.
58. G. G. Hoffmann, D. Hellert and M. Warren, *J. Organomet. Chem.*, 1993, **444**, 21.
59. J. T. Leman, J. W. Ziller and A. R. Barron, *Organometallics*, 1991, **10**, 1766.
60. B. Neumüller, *Z. Naturforsch., Teil B*, 1991, **46**, 1539.
61. D. A. Atwood, A. H. Cowley, R. A. Jones, M. A. Mardones, J. L. Atwood and S. G. Bott, *J. Coord. Chem.*, 1992, **25**, 233; M. B. Power, W. M. Cleaver, A. W. Apblett, A. R. Barron and J. W. Ziller, *Polyhedron*, 1992, **11**, 477.
62. D. L. Reger, S. J. Knox, A. L. Rheingold and B. S. Haggerty, *Organometallics*, 1990, **9**, 2581.
63. B. Neumüller, F. Gahlmann, M. Schäfer and S. Magull, *J. Organomet. Chem.*, 1992, **440**, 263.
64. O. T. Beachley, Jr., M. R. Churchill, J. C. Pazik and J. W. Ziller, *Organometallics*, 1987, **6**, 2088.
65. O. T. Beachley, Jr., J. D. Maloney, M. R. Churchill and C. H. Lake, *Organometallics*, 1991, **10**, 3568.
66. J. Blank, H.-D. Hausen and J. Weidlein, *J. Organomet. Chem.*, 1993, **444**, C4.
67. D. A. Atwood, R. A. Jones, A. H. Cowley, J. L. Atwood and S. G. Bott, *J. Organomet. Chem.*, 1990, **394**, C6.
68. G. E. Forster, G. Southerington, M. J. Begley and D. B. Sowerby, *J. Chem. Soc., Chem. Commun.*, 1991, 54.
69. E. O. Schlemper and D. Britton, *Inorg. Chem.*, 1966, **5**, 507.
70. W. Uhl, K. Klinkhammer, M. Layh and W. Massa, *Chem. Ber.*, 1991, **124**, 279.
71. W. Uhl, M. Layh and W. Hiller, *J. Organomet. Chem.*, 1989, **368**, 139.
72. A. Laguna, E. Fernández, A. Mendía, M. E. Ruiz-Romero and P. G. Jones, *J. Organomet. Chem.*, 1989, **365**, 201.
73. T. A. Annan, D. G. Tuck, M. A. Khan and C. Peppe, *Organometallics*, 1991, **10**, 2159.
74. O. T. Beachley, Jr. and R. N. Rusinko, *Inorg. Chem.*, 1981, **20**, 1367.
75. O. T. Beachley, Jr. and R. G. Simmons, *Inorg. Chem.*, 1980, **19**, 3042.
76. R. B. Hallock *et al.*, *Inorg. Chem.*, 1983, **22**, 3683.
77. G. D. Shier and R. S. Drago, *J. Organomet. Chem.*, 1966, **5**, 330.
78. T. L. Blundell and H. M. Powell, *J. Chem. Soc., Chem. Commun.*, 1967, 54.
79. D. L. Hughes and M. R. Truter, *J. Chem. Soc., Chem. Commun.*, 1982, 727.
80. K. Kobiro, M. Takahashi, Y. Odaira, Y. Kawasaki, Y. Kai and N. Kasai, *J. Chem. Soc., Dalton Trans.*, 1986, 2613.
81. K. Kobiro, S. Takada, Y. Odaira and Y. Kawasaki, *J. Chem. Soc., Dalton Trans.*, 1986, 1767.
82. A. J. Canty, K. Mills, B. W. Skelton and A. H. White, *J. Chem. Soc., Dalton Trans.*, 1986, 939.
83. J. Vincente, J. Abad, G. Cara and P. G. Jones, *Angew. Chem.*, 1990, **102**, 1184; *Angew. Chem., Int. Ed. Engl.*, 1990, **29**, 1125.
84. A. J. Canty, L. A. Titcombe, B. W. Skelton and A. H. White, *J. Chem. Soc., Dalton Trans.*, 1988, 35.
85. B. Neumüller and F. Gahlmann, *J. Organomet. Chem.*, 1991, **414**, 271.
86. D. A. Atwood, R. A. Jones, A. H. Cowley, S. G. Bott and J. L. Atwood, *J. Organomet. Chem.*, 1992, **425**, C1.
87. (a) P. L. Baxter, A. J. Downs and D. W. H. Rankin, *J. Chem. Soc., Dalton Trans.*, 1984, 1755; (b) N. N. Greenwood, A. Storr and M. G. H. Wallbridge, *Inorg. Chem.*, 1963, **2**, 1036.
88. D. O'Hare, J. S. Foord, T. C. M. Page and T. J. Whitaker, *J. Chem. Soc., Chem. Commun.*, 1991, 1445.
89. J. L. Atwood, S. G. Bott, F. M. Elms, C. Jones and C. L. Raston, *Inorg. Chem.*, 1991, **30**, 3792.
90. P. L. Baxter, A. J. Downs, D. W. H. Rankin and H. E. Robertson, *J. Chem. Soc., Dalton Trans.*, 1985, 807.
91. C. R. Pulham, A. J. Downs, M. J. Goode, D. W. H. Rankin and H. E. Robertson, *J. Am. Chem. Soc.*, 1991, **113**, 5149.
92. (a) P. L. Baxter, A. J. Downs, M. J. Goode, D. W. H. Rankin and H. E. Robertson, *J. Chem. Soc., Chem. Commun.*, 1986, 805; (b) P. L. Baxter, A. J. Downs, M. J. Goode, D. W. H. Rankin and H. E. Robertson, *J. Chem. Soc., Dalton Trans.*, 1990, 2873.
93. L. Pauling, 'The Nature of the Chemical Bond', 3rd edn., Cornell University Press, Ithaca, NY, 1960, p. 120.
94. A. F. Wells, 'Structual Inorganic Chemistry', 5th edn., Clarendon Press, Oxford, 1984, p. 1278.
95. E. Wiberg and E. Amberger, 'Hydrides of the Elements of Main Groups I–IV', Elsevier, Amsterdam, 1971, p. 454.
96. V. V. Gavrilenko, V. S. Kolesov and L. I. Zakharkin, *Zh. Obshch. Khim.*, 1977, **47**, 964; 1979, **49**, 1845.
97. A. G. Avent, C. Eaborn, P. B. Hitchcock, J. D. Smith and A. C. Sullivan, *J. Chem. Soc., Chem. Commun.*, 1986, 988.
98. O. T. Beachley, Jr., Sun-Hua L. Chao, M. R. Churchill and R. F. See, *Organometallics*, 1992, **11**, 1486.
99. M. J. Manning, C. B. Knobler, M. F. Hawthorne and Y. Do, *Inorg. Chem.*, 1991, **30**, 3589.
100. R. N. Grimes, in 'Metal Interactions with Boron Clusters', ed. R. N. Grimes, Plenum, New York, 1982, Chap. 7.
101. (a) H. M. Manasevit, *Appl. Phys. Lett.*, 1968, **12**, 156; (b) T. Nakanis, *J. Cryst. Growth*, 1984, **68**, 156.

102. (a) G. E. Coates and J. Graham, *J. Chem. Soc.*, 1963, 233; (b) O. T. Beachley, Jr. and G. E. Coates, *J. Chem. Soc.*, 1965, 3241.
103. C. G. Pitt, K. T. Higa, A. T. McPhail, and R. L. Wells, *Inorg. Chem.*, 1986, **25**, 2483.
104. E. K. Byrne, L. Parkanyi, K. H. Theopold, *Science*, 1988, **241**, 332.
105. K. T. Higa and C. George, *Organometallics*, 1990, **9**, 275.
106. N. W. Alcock, I. A. Degnan, M. G. H. Wallbridge, H. R. Powell. M. McPartlin and G. M. Sheldrick, *J. Organomet. Chem.*, 1989, **361**, C33.
107. M. A. Petrie, K. Ruhlandte-Senge and P. P. Power, *Inorg. Chem.*, 1992, **31**, 4038.
108. M. A. Petrie and P. P. Power, *J. Chem. Soc., Dalton Trans.*, 1993, 1737.
109. H. Bürger, J. Cictron, U. Goetze, U. Wannagat and W. J. Wismar, *J. Organomet. Chem.*, 1971, **33**, 1.
110. G. M. Sheldrick and W. S. Sheldrick, *J. Chem. Soc., A*, 1969, 2279.
111. M. A. Petrie, S. C. Shoner, H. V. R. Dias and P. P. Power, *Angew. Chem.*, 1990, **102**, 1061; *Angew. Chem., Int. Ed. Engl.*, 1990, **29**, 1033.
112. D. C. Pestana and P. P. Power, *J. Am. Chem. Soc.*, 1991, **113**, 8426.
113. M. A. Petrie and P. P. Power, *Inorg. Chem.*, 1993, **32**, 1309.
114. K. M. Waggoner, K. Ruhlandte-Senge, R. J. Wehmschulte, X. He, M. M. Olmstead and P. P. Power, *Inorg. Chem.*, 1993, **32**, 2557.
115. C. J. Carrano, A. H. Cowley, D. M. Giolando, R. A. Jones, C. M. Nunn and J. M. Power, *Inorg. Chem.*, 1988, **27**, 2709.
116. A. M. Arif *et al.*, *J. Chem. Soc., Chem. Commun.*, 1986, 1543.
117. C. J. Hoffmann and C. Burschka, *Angew. Chem.*, 1985, **97**, 965; *Angew. Chem., Int. Ed. Engl.*, 1985, **24**, 970.
118. R. L. Wells, A. P. Purdy, A. T. McPhail and C. G. Pitt, *J. Organomet. Chem.*, 1986, **308**, 281.
119. O. T. Beachley, Jr., J. P. Kopasz, H. Zhang, W. E. Hunter and J. L. Atwood, *J. Organomet. Chem.*, 1987, **325**, 69.
120. R. L. Wells, A. T. McPhail and M. F. Self, *Organometallics*, 1992, **11**, 221.
121. R. L. Wells, L. J. Jones, A. T. McPhail and A. Avanipour, *Organometallics*, 1991, **10**, 2345.
122. R. L. Wells, A. P. Purdy, K. T. Higa, A. T. McPhail and C. G. Pitt, *J. Organomet. Chem.*, 1987, **325**, C7.
123. K. E. Lee, K. T. Higa and R. A. Nissan, *Organometallics*, 1992, **11**, 2816.
124. D. E. Heaton, R. A. Jones, K. B. Kidd, A. H. Cowley and C. M. Nunn, *Polyhedron*, 1988, **7**, 1901.
125. D. A. Atwood, A. H. Cowley, P. R. Harris, R. A. Jones, S. U. Koschmieder and C. M. Nunn, *J. Organomet. Chem.*, 1993, **449**, 61.
126. K. A. Aitchison *et al.*, *J. Organomet. Chem.*, 1989, **366**, 11.
127. A. M. Arif, D. C. Bradley, H. Dawes, D. M. Frigo, M. B. Hursthouse and B. Hussain, *J. Chem. Soc., Dalton Trans*, 1987, 2159.
128. M. Trapp, H.-D. Hausen, G. Weckler and J. Weidlein, *J. Organomet. Chem.*, 1993, **450**, 53.
129. O. T. Beachley, Jr., M. J. Noble, M. R. Churchill and C. H. Lake, *Organometallics*, 1992, **11**, 1051.
130. R. L. Wells, M. F. Self, A. T. McPhail, S. R. Aubuchon, R. C. Woudenberg and J. C. Jasinski, *Organometallics*, 1993, **12**, 2832.
131. A. R. Barron, A. H. Cowley, R. A. Jones, C. M. Nunn, D. L. Westmoreland, *Polyhedron*, 1988, **7**, 77.
132. A. P. Purdy, R. L. Wells, A. T. McPhail and C. G. Pitt, *Organometallics*, 1987, **6**, 2099.
133. K. H. Theopold and T. Douglas, *Inorg. Chem.*, 1991, **30**, 594.
134. A. H. Cowley, R. A. Jones, M. Mardones, J. L. Atwood and S. G. Bott, *Heteroat. Chem.*, 1991, **2**, 11.
135. B. Neumüller, *Z. Naturforsch., Teil B*, 1990, **45**, 1559.
136. B. Neumüller, *Z. Naturforsch., Teil B*, 1991, **46**, 753.
137. R. L. Wells, J. W. Pasterczyk, A. T. McPhail, J. D. Johansen and A. Alvanipour, *J. Organomet. Chem.*, 1991, **407**, 17.
138. (a) R. L. Wells, W. K. Holley, S. Shafieezad, A. T. McPhail and C. G. Pitt, *Phosphorus, Sulphur, Silicon Relat. Elem.*, 1989, **41**, 15; (b) W. K. Holley, R. L. Wells, S. Shafieezad, A. T. McPhail and C. G. Pitt, *J. Organomet. Chem.*, 1990, **381**, 15.
139. M. J. Almond, M. G. B. Drew, C. E. Jenkins and D. A. Rice, *J. Chem. Soc., Dalton Trans.*, 1992, 5.
140. A. H. Cowley, P. R. Harris, R. A. Jones and C. M. Nunn, *Organometallics*, 1991, **10**, 652.
141. D. A. Atwood *et al.*, *Organometallics*, 1993, **12**, 24.
142. G. E. Coates, *J. Chem. Soc.*, 1951, 2003.
143. L. V. Interrante, G. A. Sigel, M. Garbauskas, C. Hejna and G. A. Slack, *Inorg. Chem.*, 1989, **28**, 252.
144. A. H. Cowley, R. A. Jones, K. B. Kidd, C. M. Nunn and D. L. Westmoreland, *J. Organomet. Chem.*, 1988, **341**, C1.
145. O. T. Beachley, Jr., J. D. Maloney and R. D. Rodgers, *J. Organomet. Chem.*, 1993, **449**, 69.
146. A. H. Cowley, R. A. Jones, M. A. Mardones and C. M. Nunn, *Organometallics*, 1991, **10**, 1635.
147. H. Hope, D. C. Pestana and P. P. Power, *Angew. Chem*, 1990, **103**, 726; *Angew. Chem., Int. Ed. Engl.*, 1991, **30**, 691.
148. R. L. Wells, A. P. Purdy, A. T. McPhail and C. G. Pitt, *J. Chem. Soc., Chem. Commun.*, 1986, 487.
149. A. H. Cowley, R. A. Jones, M. A. Mardones, J. L. Atwood and S. G. Bott, *Angew. Chem.*, 1991, **103**, 1163; *Angew. Chem., Int. Ed. Engl.*, 1991, **30**, 1141.
150. K. M. Waggoner, S. Parkin, D. C. Pestana, H. Hope and P. P. Power, *J. Am. Chem. Soc.*, 1991, **113**, 3597.
151. M. B. Power and A. R. Barron, *Angew. Chem.*, 1991, **103**, 1403; *Angew. Chem., Int. Ed. Engl.*, 1991, **30**, 1353.
152. M. P. Petrie and P. P. Power, *Organometallics*, 1993, **12**, 1592.
153. D. A. Atwood, A. H. Cowley, R. A. Jones and M. A. Mardones, *J. Organomet. Chem.*, 1993, **449**, C1.
154. A. H. Cowley, R. A. Jones, M. A. Mardones, J. L. Atwood and S. G. Bott, *Angew. Chem.*, 1990; *Angew. Chem., Int. Ed. Engl.*, 1990, **29**, 1409.
155. R. Riedel, H.-D. Hausen and E. Fluck, *Angew. Chem.*, 1985, **97**, 1050; *Angew. Chem., Int. Ed. Engl.*, 1985, **24**, 1056.
156. M. Vieth, H. Lange, O. Recktenwald, W. Frank, *J. Organomet. Chem.*, 1985, **294**, 273.
157. D. Müller and H. Hahn, *Z. Anorg. Allg. Chem.*, 1978, **438**, 258.
158. R. C. Burns, J. D. Corbett, *J. Am. Chem. Soc.*, 1981, **103**, 2627.
159. M. G. Kanatzidis and S. Dhingra, *Inorg. Chem.*, 1989, **28**, 2024.
160. M. Veith and K. Kunze, *Angew. Chem.*, 1991, **103**, 92; *Angew. Chem., Int. Ed. Engl.*, 1991, **30**, 95.
161. M. Veith and R. Rosler, *Angew. Chem.*, 1982, **94**, 867; *Angew. Chem., Int. Ed. Engl.*, 1982, **21**, 858.
162. S. J. Rettig, A. Storr and J. Trotter, *Acta Crystallogr., Part C*, 1992, **48**, 1587.
163. G. D. Gracey, S. J. Rettig, A. Storr, J. Trotter, *Can. J. Chem.*, 1987, **65**, 2469.
164. S. J. Rettig, A. Storr and J. Trotter, *Can. J. Chem.*, 1987, **65**, 1349.

165. E. C. Onyiriuka, S. J. Rettig, A. Storr and J. Trotter, *Can. J. Chem.*, 1987, **65**, 782.
166. E. M. Vaquez-Lopez, A. Sauchez, J. Casas, J. Sardo and E. E. Castellano, *J. Organomet. Chem.*, 1992, **438**, 29.
167. N. Bertel, M. Noltemeyer and H. W. Roesky, *Z. Anorg. Allg. Chem.*, 1990, **588**, **102**.
168. T. A. Annan, R. Kumar, H. E. Mabrouk, D. G. Tuck and R. K. Chadha, *Polyhedron*, 1989, **8**, 865.
169. K. Ruhlandte-Senge and P. P. Power, *Inorg. Chem.*, 1991, **30**, 2633.
170. K. Ruhlandte-Senge and P. P. Power, *Inorg. Chem.*, 1993, **32**, 3478.
171. K. Ruhlandte-Senge and P. P. Power, *Inorg. Chem.*, 1991, **30**, 3683.
172. W. Uhl, M. Layh, G. Becker, K. Klinkhammer and T. Hildenbrand, *Chem. Ber.*, 1992, **125**, 1547.
173. R. K. Chadha, P. C. Hayes, H. E. Mabrouk and D. G. Tuck, *Can. J. Chem.*, 1987, **65**, 804.
174. L. E. Maelia and S. A. Koch, *Inorg. Chem.*, 1986, **25**, 1896.
175. D. C. Bradley, D. M. Frigo, M. B. Hursthouse and B. Hussain, *Organometallics*, 1988, **7**, 1112.
176. A. H. Cowley, S. K. Mehrotra, J. L. Atwood and W. E. Hunter, *Organometallics*, 1985, **4**, 1115.
177. M. Scholz, M. Noltemeyer and W. H. Roesky, *Angew. Chem.*, 1989, **101**, 1419; *Angew. Chem., Int. Ed. Engl.*, 1989, **28**, 1383.
178. H. W. Roesky, M. Scholz, M. Noltemeyer and F. T. Edelmann, *Inorg. Chem.*, 1989, **28**, 3829.
179. W. M. Cleaver and A. R. Barron, *J. Am. Chem. Soc.*, 1989, **111**, 8966.
180. G. G. Hoffmann and C. J. Burschka, *J. Organomet. Chem.*, 1984, **267**, 229.
181. W. Wojnowski, K. Peters, E.-M. Peters and H. G. v. Schnering, *Z. Anorg. Allg. Chem.*, 1985, **531**, 147.
182. (a) M. B. Power and A. R. Barron, *J. Chem. Soc., Chem. Commun.*, 1991, 1315; (b) A. H. Cowley, R. A. Jones, P. R. Harris, D. A. Atwood, L. Contreras and C. Burek, *Angew. Chem.*, 1991, **103**, 1164; *Angew. Chem., Int. Ed. Engl.*, 1991, **30**, 1143.
183. M. A. Banks, O. T. Beachley, Jr., H. J. Gysling and H. R. Luss, *Organometallics*, 1990, **9**, 1979.
184. O. T. Beachley, Jr., J. C. Lee, H. J. Gysling, S.-H. L. Chao, M. R. Churchill and C. H. Lake, *Organometallics*, 1992, **11**, 3144.
185. S. J. Rettig, A. Storr and J. Trotter, *Can. J. Chem.*, 1984, **62**, 1705.
186. S. J. Rettig, A. Storr, J. Trotter and K. Uhrich, *Can. J. Chem.*, 1984, **62**, 2783.
187. D. G. Hendershot, M. Barber, R. Kumar and J. P. Oliver, *Organometallics*, 1991, **10**, 3302.
188. N. W. Alcock, I. A. Degnan, S. M. Roe and M. G. H. Wallbridge, *J. Organomet. Chem.*, 1991, **414**, 285.
189. A. J. Canty, L. A. Titcombe, B. W. Skelton and A. H. White, *Inorg. Chim. Acta*, 1986, **117**, L35.
190. S. S. Al-Juaid *et al.*, *J. Chem. Soc., Chem. Commun.*, 1986, 908.
191. A. M. Arif and A. R. Barron, *Polyhedron*, 1988, **7**, 2091.
192. (a) D. C. Bradley, H. Chudzynska, D. M. Frigo, M. B. Hursthouse and M. A. Mazid, *J. Chem. Soc., Chem. Commun.*, 1988, 1258. (b) D. C. Bradley, H. Chudzynska, D. M. Frigo, M. E. Hammond, M. B. Hursthouse and M. A. Mazid, *Polyhedron*, 1990, **9**, 719.
193. M. B. Power, J. W. Ziller and A. R. Barron, *Organometallics*, 1992, **11**, 2783.
194. M. B. Power, J. W. Ziller, A. N. Tyler and A. R. Barron, *Organometallics*, 1992, **11**, 1055.
195. R. R. Holmes, *Acc. Chem. Res.*, 1989, **22**, 190.
196. A. Boardman, S. E. Jeffs, R. W. H. Small and I. J. Worrall, *Inorg. Chim. Acta*, 1984, **83**, L39.
197. B. Krebs, D. Voelker and K. Stiller, *Inorg. Chim. Acta*, 1982, **65**, L101.
198. B. Krebs and A. Brömmelhaus, *Angew. Chem.*, 1989, **101**, 1726; *Angew. Chem., Int. Ed. Engl.*, 1989, **28**, 1682.
199. D. Labahn. E. Pohl, R. Herbst-Irmer, D. Stalke, H. W. Roesky and G. M. Sheldrick, *Chem. Ber.*, 1991, **124**, 1127.
200. P. Paetzold, *Angew. Chem.*, 1991, **103**, 559; *Angew. Chem., Int. Ed. Engl.*, 1991, **30**, 544.
201. M. Tacke, H. Kreienkamp, L. Plaggenborg and H. Schnöckel, *Z. Anorg. Allg. Chem.*, 1991, **604**, 35.
202. D. Loos, H. Schnöckel and D. Fenske, *Angew. Chem.*, 1993, **105**, 1124; *Angew. Chem., Int. Ed. Engl.*, 1993, **32**, 1059.
203. O. T. Beachley, Jr., T. D. Getman, R. U. Kriss, R. B. Hallock, W. E. Hunter and J. L. Atwood, *Organometallics*, 1985, **4**, 751.
204. E. Frasson F. Menegus and C. Pannatoni, *Nature*, 1963, **199**, 1087.
205. S. Shibata, L. S. Bartell and R. M. Gavin Jr., *J. Chem. Phys.*, 1964, **41**, 717.
206. O. T. Beachley, Jr., J. C. Pazik, T. E. Glassman, M. R. Churchill, J. C. Fettinger and R. Blom, *Organometallics*, 1988, **7**, 1051.
207. P. Jutzi, F. Kohl, P. Hofmann, C. Krüger and Y. H. Tsay, *Chem. Ber.*, 1980, **113**, 757.
208. O. T. Beachley, Jr., J. F. Lees, T. E. Glassman, M. R. Churchill and L. A. Buttry, *Organometallics*, 1990, **9**, 2488.
209. O. T. Beachley, Jr., J. F. Less and R. D. Rogers, *J. Organomet. Chem.*, 1991, **418**, 165.
210. H. Schumann, C. Janiak, F. Görlitz, J. Loebel and A. Dietrich, *J. Organomet. Chem.*, 1989, **363**, 243.
211. C. Janiak and R. Hoffmann, *J. Am. Chem. Soc.*, 1990, **112**, 5924.
212. S. Harvey, C. L. Raston, B. W. Skelton, A. H. White, M. F. Lappert and G. Srivastava, *J. Organomet. Chem.*, 1987, **328**, C1.
213. H. Schumann, C. Janiak, J. Pickardt and U. Börner, *Angew. Chem.*, 1987, **99**, 788; *Angew. Chem., Int. Ed. Engl.*, 1987, **26**, 789.
214. H. Werner, H. Otto and H. J. Kraus, *J. Organomet. Chem.*, 1986, **315**, C57.
215. H. Schumann, H. Kucht, A. Dietrich and L. Esser, *Chem. Ber.*, 1990, **123**, 1811.
216. M. F. Lappert, A. Singh, J. L. Atwood and W. E. Hunter, *J. Chem. Soc., Chem. Commun.*, 1981, 1190.
217. G. K Anderson and N. P. Rath, *J. Organomet. Chem.*, 1991, **414**, 129.
218. D. R. Armstrong *et al.*, *Angew. Chem.*, 1993, **105**, 1807; *Angew. Chem., Int. Ed. Engl.*, 1993, **32**, 1774.
219. J. L. Spencer, M. Green and F. G. A. Stone, *J. Chem. Soc., Chem. Commun.*, 1972, 1178.
220. D. A. T. Young, R. J. Wiersema and M. F. Hawthorne, *J. Am. Chem. Soc.*, 1971, **93**, 5687.
221. K. Stumpf, H. Pritzkow and W. Siebert, *Angew. Chem.*, 1985, **97**, 64; *Angew. Chem., Int. Ed. Engl.*, 1985, **24**, 71.
222. N. N. Greenwood, B. S. Thomas and D. W. White, *J. Chem. Soc., Dalton Trans.*, 1975, 299.
223. R. N. Grimes, W. J. Rademaker, M. L. Denniston, R. F. Bryan and P. T. Greene, *J. Am. Chem. Soc.*, 1972, **94**, 1865.
224. N. S. Hosmane, K. Lu, H. Zhang, A. H. Cowley and M. A. Mardones, *Organometallics*, 1991, **10**, 392.
225. P. Jutzi and P. Galow, *J. Organomet. Chem.*, 1987, **319**, 139.
226. M. A. Bandman, C. B. Knobler and M. F. Hawthorne, *Inorg. Chem.*, 1988, **27**, 2399.
227. D. M. Schubert *et al.*, *Organometallics*, 1990, **9**, 2046.

228. M. A. Bandman, C. B. Knobler and M. F. Hawthorne, *Inorg. Chem.*, 1989, **28**, 1204.
229. L. F. Warren, Jr. and M. F. Hawthorne, *J. Am. Chem. Soc.*, 1968, **90**, 4823.
230. E. Canadell, O. Eisenstein and J. Rubio, *Organometallics*, 1984, **3**, 759.
231. N. S. Hosmane, K. Lu, H. Zhang, L. Jia, A. H. Cowley and M. A. Mardones, *Organometallics*, 1991, **10**, 963.
232. N. S. Hosmane, P. de Meester, N. N. Maldar, S. B. Potts, S. S. C. Chu and R. H. Herber, *Organometallics*, 1986, **5**, 772.
233. N. S. Hosmane, M. S. Islam, B. S. Penkston, U. Siriwardane, J. J. Banewicz and J. A. Maguire, *Organometallics*, 1988, **7**, 2340.
234. N. S. Hosmane *et al.*, *Organometallics*, 1989, **8**, 1769.
235. N. S. Hosmane, K.-J Ku, H. Zhu, U. Siriwardane, M. S. Shet and J. A. Maguire, *Organometallics*, 1990, **9**, 808.
236. E. O. Fischer and W. Hafner, *Z. Naturforsch., Teil B*, 1955, **10**, 665.
237. M. S. Weininger, P. F. Rodesiler and E. L. Amma, *Inorg. Chem.*, 1979, **18**, 751.
238. R. E. Rundle and J. D. Corbett, *J. Am. Chem. Soc.*, 1957, **79**, 757.
239. J. D. Corbett and A. Heshaft, *J. Am. Chem. Soc.*, 1958, **80**, 1530.
240. R. K. McMullan and J. D. Corbett, *J. Am. Chem. Soc.*, 1958, **80**, 4761.
241. H. Schmidbaur, U. Thewalt and T. Zafiropoulos, *Chem. Ber.*, 1984, **117**, 3381.
242. H. Schmidbaur, R. Novak, B. Huber and G. Müller, *Polyhedron*, 1990, **9**, 283.
243. H. Schmidbaur, U. Thewalt and T. Zafiropoulos, *Organometallics*, 1983, **2**, 1550.
244. H. Schmidbaur, U. Thewalt and T. Zafiropoulos, *Z. Naturforsch., Teil B*, 1984, **39**, 1642.
245. H. Schmidbaur, U. Thewalt and T. Zafiropoulos, *Angew. Chem.*, 1984, **96**, 60; *Angew. Chem., Int. Ed. Engl.*, 1984, **23**, 76.
246. J. Ebenhöch, G. Müller, J. Riede and H. Schmidbaur, *Angew. Chem.*, 1984, **96**, 367; *Angew. Chem., Int. Ed. Engl.*, 1984, **23**, 396.
247. H. Schmidbaur, W. Bublak, J. Riede and G. Müller, *Angew. Chem.*, 1984, **97**, 402; *Angew. Chem., Int. Ed. Engl.*, 1985, **24**, 415.
248. M. D. Noirot, O. P. Anderson and S. H. Strauss, *Inorg. Chem.*, 1987, **26**, 2216.
249. A. R. Koray, M. L. Ziegler, N. E. Blank and M. W. Haenel, *Tetrahedron Lett.*, 1979, 2465.
250. C. Elschenbroich, L. Möckel and U. Zenneck, *Angew. Chem.*, 1978, **90**, 560; *Angew. Chem., Int. Ed. Engl.*, 1978, **17**, 531.
251. H. Schmidbaur, W. Bublak, B. Huber and G. Müller, *Organometallics*, 1986, **5**, 1647.
252. H. Schmidbaur, W. Bublak, B. Huber, J. Hofmann and G. Müller, *Chem. Ber.*, 1989, **122**, 265.
253. H. Schmidbaur, R. Hager, B. Huber and G. Müller, *Angew. Chem.*, 1987, **99**, 354; *Angew. Chem., Int. Ed. Engl.*, 1987, **26**, 338.
254. R. W. H. Small and I. J. Worrall, *Acta Crystallogr., Part B*, 1982, **38**, 250.
255. W. Uhl, M. Layh and T. Hildenbrand, *J. Organomet. Chem.*, 1989, **364**, 289.
256. W. Uhl, *Z. Naturfosch., Teil B*, 1988, **43**, 1113.
257. M. A. Khan, C. Peppe and D. G. Tuck, *Can. J. Chem.*, 1984, **62**, 601.
258. G. Meyer and T. Staffel, *Z. Anorg. Allg. Chem.*, 1987, **552**, 113.
259. A. J. Carty, M. J. S. Gynane, M. F. Lappert, S. J. Miles, A. Singh and N. J. Taylor, *Inorg. Chem.*, 1980, **19**, 3637.
260. A. Moezzi, M. M. Olmstead, R. A. Bartlett and P. P. Power, *Organometallics*, 1992, **11**, 2383.
261. W. W. Porterfield, 'Inorganic Chemistry. A Unified Approach', Addison Wesley, Reading, MA, 1983, p. 168.
262. S. Schneider and R. Hoppe, *Z. Anorg. Allg. Chem.*, 1970, **376**, 277.
263. M. B. Power, J. B. Wilking and A. R. Barron, *Chemtracts: Inorg. Chem.*, 1991, **3**, 146.
264. A. J. Bradley, *Z. Kyrstallogr.*, 1935, **91**, 302.
265. O. T. Beachley, Jr. *et al.*, *Organometallics*, 1989, **8**, 346.
266. R. J. Wehmschulte, K. Ruhlandt-Senge, M. M. Olmstead, H. Hope, B. E. Sturgeon and P. P. Power, *Inorg. Chem.*, 1993, **32**, 2983.
267. (a) H. Klusik and A. Berndt, *Angew. Chem.*, 1981, **93**, 903; *Angew. Chem. Int. Ed. Engl.*, 1981, **20**, 870; (b) A. Moezzi, M. M. Olmstead and P. P. Power, *J. Am. Chem. Soc.*, 1992, **114**, 2715; (c) A. Moezzi, R. A. Bartlett and P. P. Power, *Angew. Chem*, 1992, **104**, 1075; *Angew. Chem. Int. Ed. Engl.*, 1992, **31**, 1082.
268. C. Pluta, K. R. Pörschke, C. Krüger and K. Hildenbrand, *Angew. Chem.*, 1993, **105**, 451; *Angew. Chem. Int. Ed. Engl.*, 1993, **32**, 388.

12
Transition Metal Complexes of Aluminum, Gallium, Indium and Thallium

CLAIRE J. CARMALT and NICHOLAS C. NORMAN
University of Newcastle upon Tyne, UK
and
LUCY M. CLARKSON
E. I. DuPont de Nemours & Co. Inc., Deepwater, NJ, USA

12.1 INTRODUCTION 546

12.2 ALUMINUM 546

12.2.1 Aluminum–(μ-H)–Metal Bonds 546
12.2.1.1 Group 4 metals 546
12.2.1.2 Group 6 metals 548
12.2.1.3 Group 7 metals 550
12.2.1.4 Group 8 metals 550
12.2.1.5 Group 10 metals 551
12.2.1.6 f-Block metals 551
12.2.2 Aluminum–(μ-Alkyl)–Metal Bonds 552
12.2.2.1 Group 6 metals 552
12.2.3 Direct Aluminum–Metal Bonds 552
12.2.3.1 Group 9 metals 552

12.3 GALLIUM 553

12.3.1 Direct Gallium–Metal Bonds 553
12.3.1.1 Group 6 metals 553
12.3.1.2 Group 7 metals 553
12.3.1.3 Group 8 metals 555
12.3.1.4 Group 9 metals 555
12.3.1.5 Group 10 metals 557

12.4 INDIUM 557

12.4.1 Direct Indium–Metal Bonds 557
12.4.1.1 Group 6 metals 557
12.4.1.2 Group 7 metals 560
12.4.1.3 Group 8 metals 562
12.4.1.4 Group 9 metals 564
12.4.1.5 Group 10 metals 566
12.4.1.6 Metalloporphyrin complexes 566

12.5 THALLIUM 567

12.5.1 Direct Thallium–Metal Bonds 567
12.5.1.1 Group 6 metals 567
12.5.1.2 Group 7 metals 567

12.5.1.3 Group 8 metals 567
12.5.1.4 Group 9 metals 569
12.5.1.5 Group 10 metals 569
12.5.1.6 Group 11 metals 569
12.5.1.7 Metalloporphyrin complexes 569

12.6 REFERENCES 570

12.1 INTRODUCTION

This chapter deals with compounds containing organotransition metal fragments and either aluminum, gallium, indium or thallium. The coverage, in terms of compound types, is much the same as that in the chapter of the same title in *COMC-I* by Boocock and Shore,[1] but in order to avoid duplication, we have covered the literature only from 1980 onwards. This chapter should therefore be seen as an update of the first edition and in order to facilitate use of both volumes, we have adopted similar section headings and classified the compounds accordingly. Structural diagrams have been redrawn for this review from crystallographic coordinates. A number of reviews have appeared since 1980 which deal, sometimes in part, with compounds relevant to this topic.[2–6]

12.2 ALUMINUM

12.2.1 Aluminum–(μ-H)–Metal Bonds

12.2.1.1 Group 4 metals

An extensive range of biscyclopentadienyltitanium(III) aluminohydride complexes has been synthesized by Bulychev[5] according to Equations (1)–(3).

$$[TiClCp_2] + LAH \longrightarrow [TiCp_2\{(\mu\text{-}H)_2AlH_2\}] + LiCl \quad (1)$$

$$[Ti(BH_4)Cp_2] + LAH \longrightarrow [TiCp_2\{(\mu\text{-}H)_2AlH_2\}] + LiBH_4 \quad (2)$$

$$[Ti(hal)Cp_2] + AlH_{3-x}(hal)_x \longrightarrow [TiCp_2\{(\mu\text{-}H)_2AlH_{2-x}X_x\}] \quad (3)$$

Particular emphasis was given to the properties of the aluminohydride species as alkene isomerization catalysts and mechanistic studies on complexes of the form $[TiCp_2\{(\mu\text{-}H)_2AlH_{2-x}X_x\}]$ (**1**) and $[Ti(\eta\text{-}C_5H_4(CH_2)_nC_5H_4)\{(\mu\text{-}H)_2AlH_{2-x}X_x\}]$[7] (**2**) (X = halide) established that the alkene coordinates to the aluminum centre prior to the hydride migration and isomerization steps.

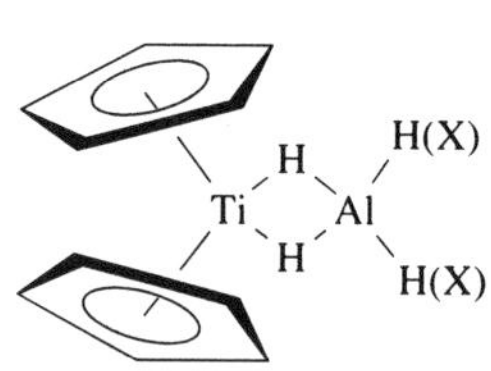

(1)

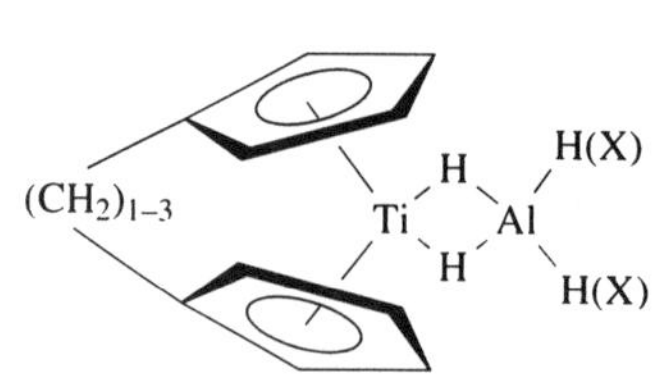

(2)

In addition to the mechanistic and reactivity studies, x-ray crystal structure determinations have been carried out on the following compounds: $[\{TiCp_2\}_2\{\mu\text{-}AlH_4(N(Me)_2CH_2CH_2N(Me)_2)AlH_4\}]$ (**3a**);[8] $[TiCp_2\{(\mu\text{-}H)_2AlCl_2(Et_2O)\}]$ (**3b**);[9] $[\{TiCp_2(\mu\text{-}H)_2Al(\mu\text{-}H)(\eta^1{:}\eta^5\text{-}C_5H_4)TiCp(\mu\text{-}H)\}_2]$ (**4**);[10] $[\{TiCp^*_2\}_2\text{-}\{(\mu\text{-}H)_2AlH(\mu\text{-}H)_2AlH(\mu\text{-}H)_2\}]$ (**5a**);[11] $[\{TiCp_2\}_2\{(\mu\text{-}H)_2AlH(\mu\text{-}OEt)_2AlH(\mu\text{-}H)_2\}]$ (**5b**);[12] $[\{TiCp_2\}_2\text{-}\{(\mu\text{-}H)_2AlH(\mu\text{-}NEt_2)_2AlH(\mu\text{-}H)_2\}]$ (**5c**);[12] $[\{TiCp_2\}_2\{(\mu\text{-}H)_2AlH(\mu\text{-}OMe)_2AlH(\mu\text{-}H)_2\}]$ (**5d**);[13] $[\{TiCp^*_2\}_2\{(\mu\text{-}H)_2AlH(\mu\text{-}H)_2\}]$ (**6a**);[11] $[\{TiCp_2\}_2\{(\mu\text{-}H)_2AlCl(\mu\text{-}H)_2\}]$ (**6b**);[14] $[\{TiCp_2\}_2\text{-}\{(\mu\text{-}H)_2Al(BH_4)(\mu\text{-}H)_2\}]$ (**6c**)[15] $[\{TiCp_2\}_2\{(\mu\text{-}H)_2AlMe(\mu\text{-}H)_2\}]$ (**6d**);[16] $[\{(TiCp(\eta\text{-}C_5H_4CH_2)\{(\mu\text{-}H)_2Al\})_2(\mu_3\text{-}O)\}_2]$ (**7**),[17] and $[\{TiCp_2(\mu\text{-}H)_2Al(\mu\text{-}H)(\eta^1{:}\eta^5\text{-}C_5H_4)TiCp(\mu\text{-}H)\}_2]$ (**8**).[16] Compounds (**3**), (**5**) and (**6**) are shown diagrammatically and illustrate the presence of the $Ti\text{–}(\mu\text{-}H)_2\text{–}Al$ groups specifically the formally $[AlH_5]^{2-}$ and $[Al_2H_8]^{2-}$ groups in the dinuclear systems; compounds (**6d**) and (**5c**) are also shown in Figures 1 and 2, respectively.

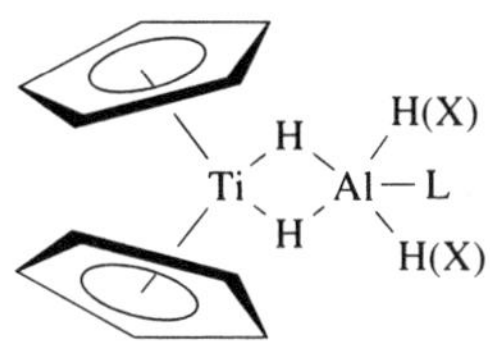

(3)
(a) X = H, L = 1/2 TMEDA; **(b)** X = Cl, L = Et_2O

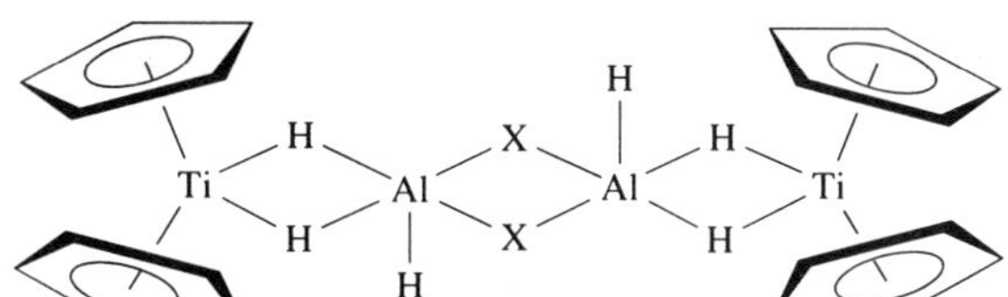

(5)
(a) X = HCp*; **(b)** X = OEt; **(c)** X = NEt_2; **(d)** X = OMe

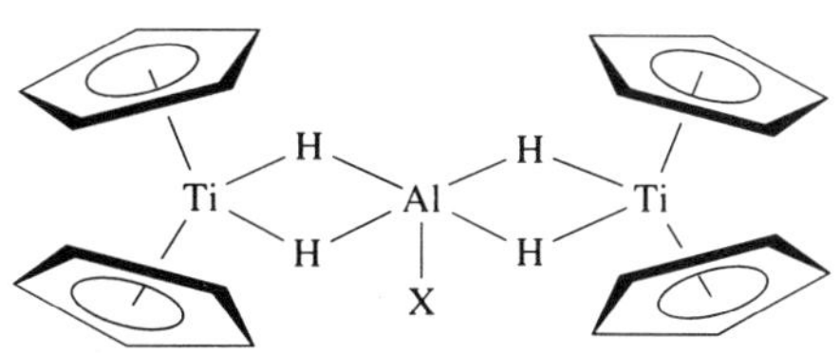

(6)
(a) X = HCp*; **(b)** X = Cl; **(c)** X = BH_4; **(d)** X = Me

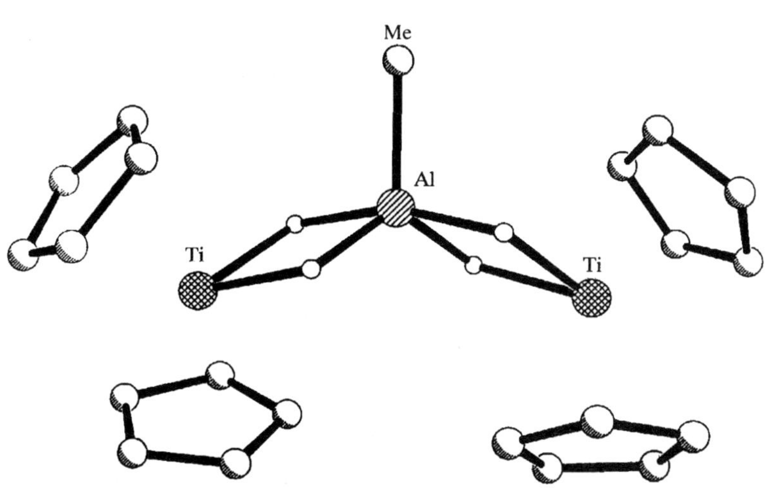

Figure 1 Molecular structure of (**6d**).

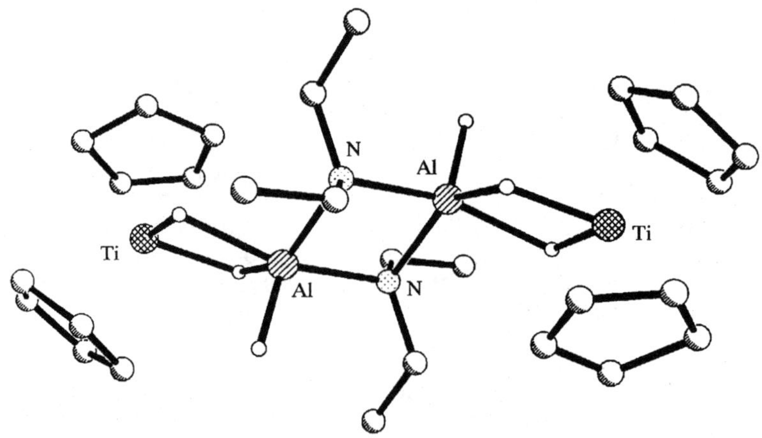

Figure 2 Molecular structure of (**5c**).

A range of alkylated titanocene derivatives of the form $[Ti(Cp-xMe_x)_2(AlH_4)]$ analogous to (**1**) prepared from $[TiClCp_2]$ and LAH has also been reported by Mach *et al.*[18] together with studies on their reactivity towards butadiene which yields 1-methyl-η-allyltitanocenes.

In addition to the biscyclopentadienyltitanium(III) complexes, the triscyclopentadienylzirconium(IV) complex $[ZrCp_3(AlHEt_3)]$ (**9**)[19] has also been described. Compound (**9**) was prepared from the reaction between $[ZrHCp_3]$ and $AlEt_3$ and in solution an equilibrium between the reactants and (**9**) exists, but it is (**9**) which crystallizes from solution; its structure was established by x-ray crystallography.

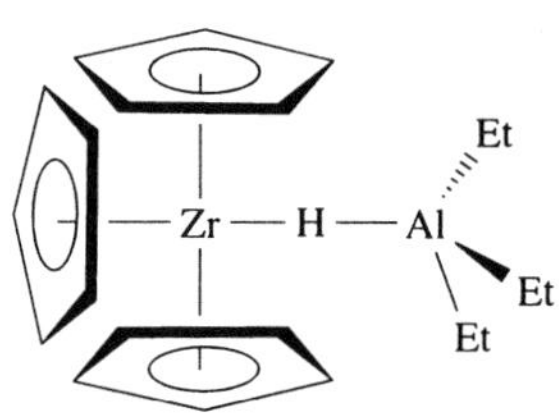

(**9**)

12.2.1.2 Group 6 metals

The reaction between $[WCl_4(PMe_3)_3]$ and LAH in diethyl ether followed by addition of TMEDA $[Me_2N(CH_2)_2NMe_2]$ affords the tungsten(IV) complex $[\{WH_3(PMe_3)_3\}_2\{(\mu\text{-}H)_2AlH(\mu\text{-}H)_2\}]$ (**10**),[20,21] which contains a bridging $[AlH_5]^{2-}$ group analogous to that found in (**6a**); the structure is shown in Figure 3. If the reaction is carried out without TMEDA, the tungsten(IV) compound $[\{WH_3(PMe_3)_3\}_2\{(\mu\text{-}H)_2AlH(\mu\text{-}H)_2AlH(\mu\text{-}\text{-})_2\}]$ (**11a**)[21] is formed, which contains an $[Al_2H_8]^{2-}$ bridge as found in (**5a**). The same reaction in refluxing THF also without TMEDA, results in a similar compound (**11b**), a complex analogous to (**11a**) but in which the two aluminum atoms are bridged by OBu^n groups (derived from the ring opening of THF) rather than hydride.

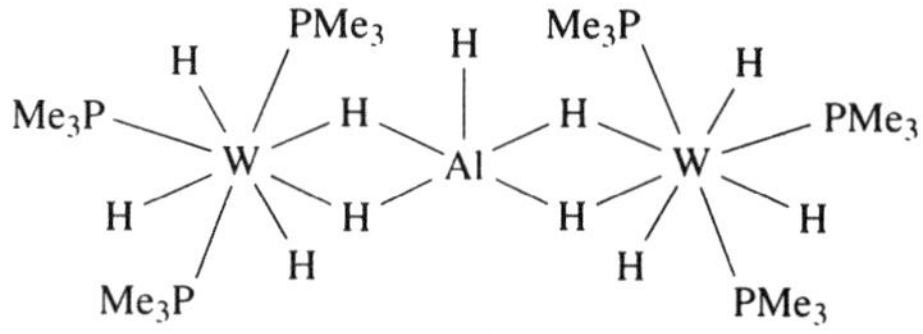

(**10**)

The metal(II) complexes $[\{MX(L)_4\}_2\{(\mu\text{-}H)_2AlH(\mu\text{-}H)_2AlH(\mu\text{-}H)_2\}]$ (**12a**)–(**12d**)[22] have been prepared from the reactions between the corresponding metal dichloride complexes $[MCl_2(L)_4]$ and LAH in Et_2O or THF and are presumed to have structures analogous to (**11a**) with a bridging $[Al_2H_8]^{2-}$ unit, on the basis of NMR spectroscopic data.

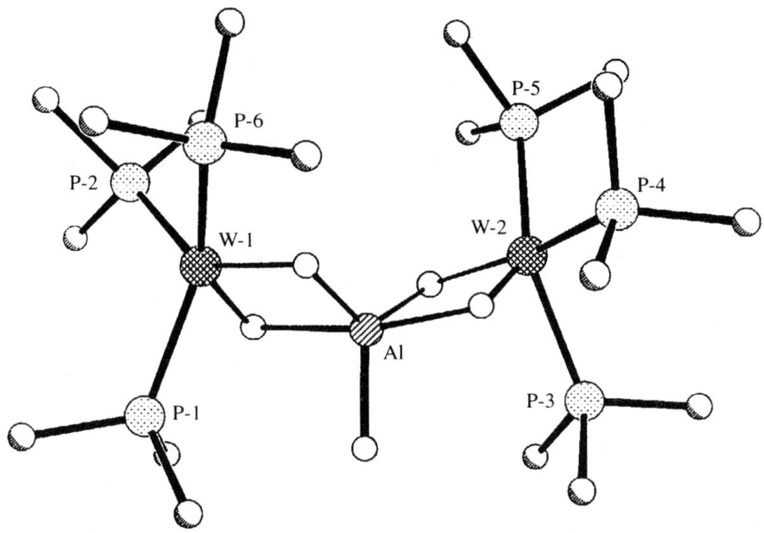

Figure 3 Molecular structure of (**10**); hyrdride ligands to tungsten are not shown.

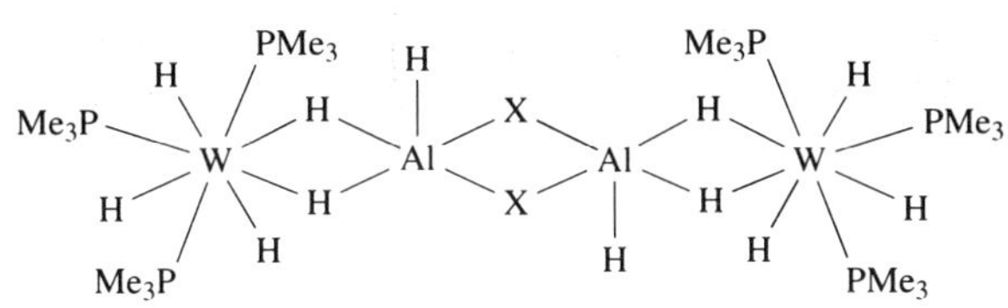

(11)
(**a**) X = H; (**b**) X = OBu^n

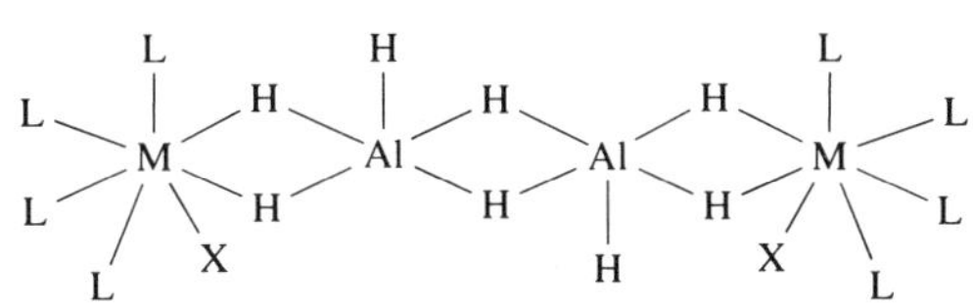

(12)
(**a**) M = Cr, 2L = dmpe, X = H; (**b**) M = Mo, L = PMe_3, X = H;
(**c**) M = W, L = PMe_3, X = H; (**d**) M = W, L = PMe_3, X = Cl

The reactivity of (**11a**) towards a number of nitrogen bases has also been explored.[23] With RCN (R = Me or Et) the complexes $[\{WH_3(PMe_3)_3\}_2\{(\mu\text{-H})_2AlH(\mu\text{-N=CHR})_2AlH(\mu\text{-H})_2\}]$ (**13**) ((**13a**), R = Me; (**13b**), R = Et) are formed, (**13b**) having been characterized by x-ray crystallography. The structure is very similar to that of (**11a**) but the central two hydrides which bridge between the aluminum centres have been replaced by μ-N=CHEt groups resulting from addition of an Al–H bond across the C≡N bond; prior coordination of the RCN unit to the aluminum centre is presumed. The reaction of (**11a**) with $[NHMe_3]Cl$ affords the monotungsten(IV) complex $[WH_2(PMe_3)_3\{(\mu\text{-H})_3AlCl_2(NMe_3)\}]$ (**14**), the structure of which was also determined by x-ray crystallography, revealing the presence of three hydrogen bridges between the tungsten and aluminum atoms.

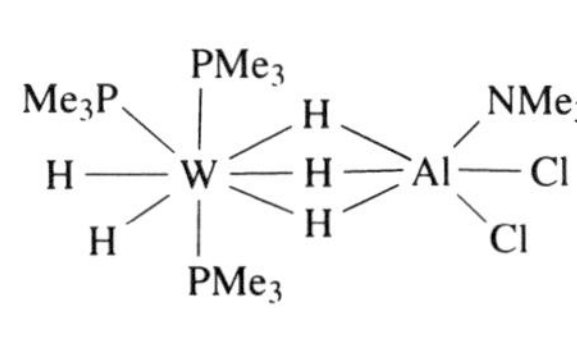

(14)

Another tungsten(IV) complex, $[WCp_2\{(\mu\text{-H})_2AlMe_3\}]$ (**15**), is formed in the reaction between the tungstocene dihydride $[WH_2Cp_2]$ and Al_2Me_6 in toluene[24] and the crystallographically determined structure is shown. The dideutero derivative of (**15**), $[WCp_2\{(\mu\text{-D})_2AlMe_3\}]$, undergoes H–D scrambling between the W–$(\mu\text{-H})_2$–Al and C_5H_5 C–H positions, although no deuterium incorporation into the Me C–H bonds is observed.

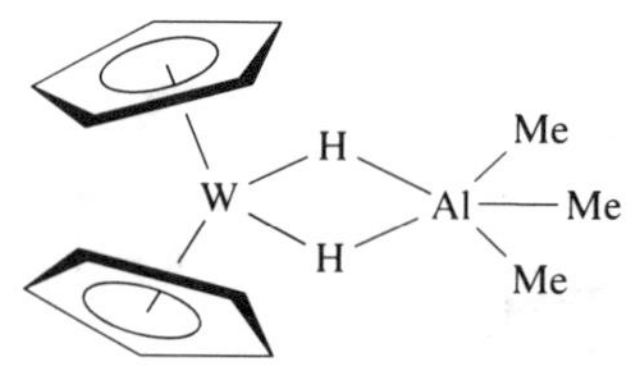

(15)

12.2.1.3 Group 7 metals

The reaction between [$MnBr_2$(dmpe)$_2$] and LAH in toluene affords the manganese(I) complex [{Mn(dmpe)$_2$}$_2${(μ-H)$_2$AlH(μ-H)$_2$AlH(μ-H)$_2$}] (**16**),[25] which is analogous to the chromium complex (**12a**) with the exception that there are no terminal hydrides bonded to the manganese centres. An x-ray crystallographic study revealed the presence of a $[Al_2H_8]^{2-}$ bridging unit as found in (**5a**), (**11a**) and (**12**).

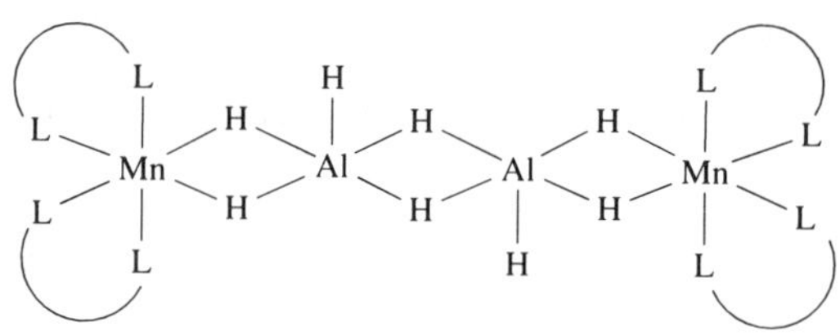

(16)

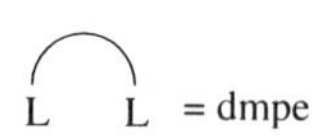

The rhenium(III) complexes [{ReH_2(PMe$_2$Ph)$_3$}$_2${(μ-H)$_2$AlH(μ-H)$_2$AlH(μ-H)$_2$}] (**17**), analogous to the tungsten complex (**11a**), and [{ReH_2(PMe$_2$Ph)$_3$}$_2${(μ-H)$_2$AlH(μ-OBu^n)$_2$AlH(μ-H)$_2$}] (**18**), analogous to (**11b**), have been prepared from the reaction between LAH and [$ReCl_3$(PMe$_2$Ph)$_3$] in THF,[21] the latter being formed on heating in this solvent. NMR spectroscopic data are consistent with the structures of (**17**) and (**18**) being similar to those of (**11a**) and (**11b**), respectively, with the exception that the rhenium centres are bonded to two terminal hydrides as opposed to three in the case of the tungsten compounds. The relationship between the rhenium complexes (**17**) and (**18**) and the manganese complex (**16**) is that one phosphorus centre in (**16**) is replaced by two hydrides in (**17**) and (**18**).

Other examples of hydridorhenium aluminates are formed in the reactions between Al_2Me_6 and [ReH_7P_2] and [ReH_5P_3] (where P = PMe$_2$Ph or PMePh$_2$)[26] in benzene with the elimination of methane. In the former case the compounds formed are [ReH_4(P)$_2${(μ-H)$_2$AlMe$_2$}] ((**19**), P = PMe$_2$Ph; (**20**), P = PMePh$_2$) and in the latter case, the compounds isolated are [ReH_2(P)$_3${(μ-H)$_2$AlMe$_2$}] ((**21**), P = PMe$_2$Ph; (**22**), P = PMePh$_2$). All complexes exhibit fluxional behaviour involving exchange of terminal and bridging hydrides and the structures of (**20**) and (**22**) have been determined by x-ray crystallography and are also shown diagramatically. Complex (**20**) has two hydrides bridging between the rhenium and aluminum centres whereas in (**22**) a third hydride is close enough to the aluminum atom to be described as bridging.

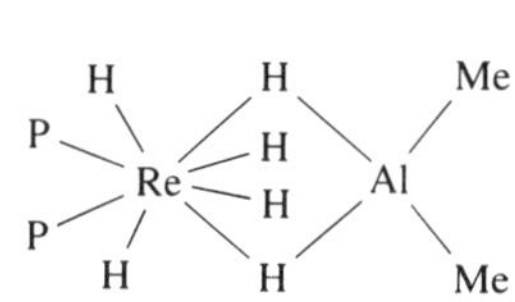

(20)
P = PMePh$_2$

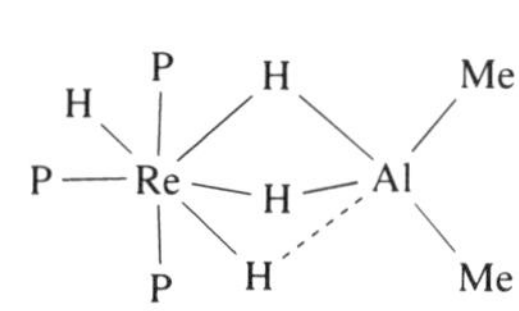

(22)
P = PMePh$_2$

12.2.1.4 Group 8 metals

The reaction between [$RuCl_2$(P)$_4$] and LAH in Et_2O affords the ruthenium(II) complexes [{RuH(P)$_3$}$_2${(μ-H)$_2$AlH(μ-H)$_2$AlH(μ-H)$_2$}] ((**23a**), P = PMe$_3$; (**23b**), P = PEtPh$_2$; (**23c**), P = PPh$_3$),[27] which spectroscopic data indicate have structures with $[Al_2H_8]^{2-}$ bridges analogous to (**5a**), (**11a**), (**12**), (**16**) and (**17**) and in which the three phosphines are in a *fac* configuration. The reaction between (**23a**) and TMEDA affords the monoaluminum ruthenium(II) species $[\{RuH(PMe_3)_3\}_2\{(\mu\text{-}H)_2AlH(\mu\text{-}H)_2\}]$ analogous to the tungsten complex (**10**). The osmium analogues of (**23a**) and (**23c**) were prepared similarly from $[OsCl_2(P)_4]$ (P = PMe_3, PPh_3).[27]

The cyclopentadienyl complex [$RuCl_2$(P)Cp*] also reacts with LAH in Et_2O to give the complexes [Ru(P)Cp*{AlH_4}] (**24**) (P = PMe_3, PEt_3, PPr^i_3, $PMePh_2$, PPh_3),[28] which undergo alcoholysis reactions to afford [RuH_3(P)Cp*] and $Al(OR)_3$.

12.2.1.5 Group 10 metals

The reaction between [Ni(cdt)] (cdt = cyclododecatriene) and hydridotrialkylaluminates [Q][$AlHR_3$] (R = Me, Et, Bu^i) affords the ionic complexes [Q][Ni(cdt){(μ-H)AlR_3}] (**25**) in which the nickel and aluminum centres are linked by a single μ-H bridge.[29]

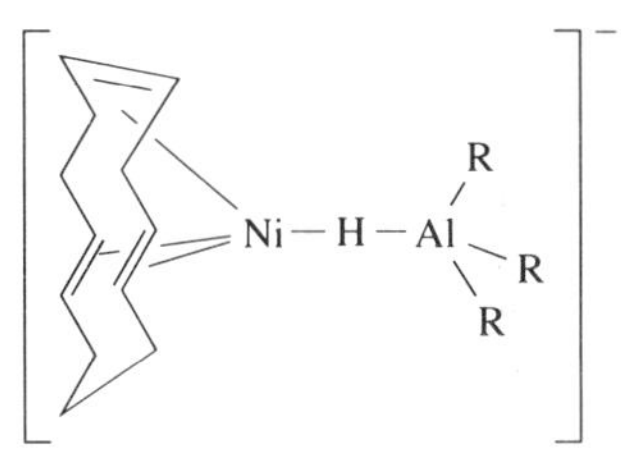

(**25**)

Similar reactivity is observed in the reactions between $[Ni(C_2H_4)_3]$ and hydridotrialkylaluminates in Et_2O–THF or Et_2O–amine mixtures, resulting in the temperature-sensitive species [Q][$Ni(C_2H_4)_2$-{(μ-H)AlR_3}] (**26**).[30] The gallium analogues have also been prepared, these being the only examples of M–H–Ga interactions. AlR_3 is liberated on treatment with CO to form $[NiH(CO)_3]^-$ as the initial product. In the reaction between [Ni(cdt)] and either $LiAlH_2Bu^i_2$ or LAH in Et_2O or Et_2O–TMEDA, the temperature-sensitive compounds [Li(TMEDA)$_2$][{Ni(cdt)}$_2${(μ-H)$_2AlBu^i_2$}]·2Et_2O(**27**), [Ni(cdt)-{AlH_4}]·2Et_2O, and [Li(TMEDA)$_2$][Ni(cdt){AlH_4}] are produced.[31]

A number of studies have also been carried out on aluminum hydride transfer to unsaturated hydrocarbons resulting from reactions between organoaluminum hydrides and nickel(0) compounds. The basic types of reactions observed are nickel-catalyzed alkene formation from AlR_3, nickel-catalyzed hydroalumination from $AlHR_2$ and unsaturated hydrocarbons, and the stoichiometric reaction of nickel(0) compounds with LAH or $AlHR_2$ to form intermediates of the form [Ni(L)AlH_2Li].[32]

12.2.1.6 f-Block metals

The lanthanide complexes [$MClCp_2$] (M = Lu, Y) react with LAH in benzene–Et_2O in the presence of the Lewis bases NEt_3 and THF to give the dimeric complexes [(MCp_2-{(μ-H)$_2AlH_2$(L)})$_2$] ((**28a**), M = Y, L = NEt_3; (**28b**), M = Y, L = THF; (**28c**), M = Lu, L = NEt_3; (**28d**), M = Lu, L = THF)[33] the structures of which have been established by x-ray crystallography and which are illustrated diagramatically. The yttrium and lutetium centres are linked by two μ-H bridges, each of which is also part of a bridge to an aluminum atom; each aluminum is also linked to one M atom by a μ-H bridge. In the particular case of (**28c**), prolonged exposure of the crystal to x rays (i.e., the crystal on which the original data were obtained) resulted in a solid-state transformation to an isomeric species (**29**) made up of two identical monomers with *no* hydride bridging between the lutetium centres. The two monomers are also illustrated in an orientation which reflects this transformation due, in the solid state, to movement of the hydrogen atoms with the heavy atoms remaining in much the same positions.

A related samarium(III) complex, [{Sm(η-$C_5H_4Bu^t$)$_2${(μ-H)$_3$AlH(THF)}$_2$] (**30**), was obtained from the reaction between the samarium(II) species Na[Sm(η-$C_5H_4Bu^t$)$_3$]·THF and two equivalents of AlH_3 in THF, the samarium being oxidized in this process.[34] The structure illustrated was determined by x-ray crystallography; it differs from that of (**28c**) in that both samarium atoms have a hydride bridge to each aluminum atom.

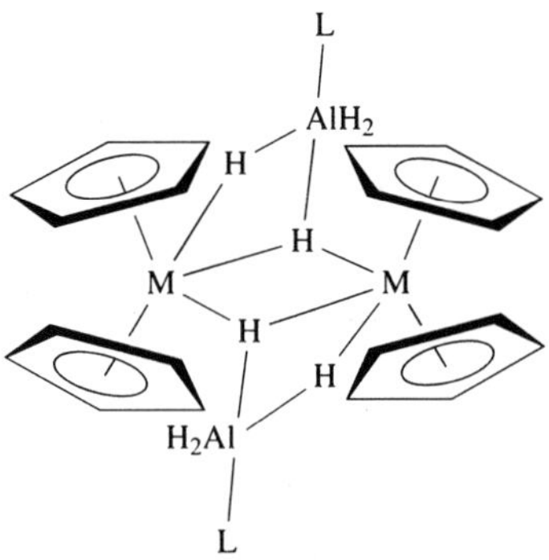

(28)
(a) M = Y, L = NEt_3; **(b)** M = Y, L = THF
(c) M = Lu, L = NEt_3; **(d)** M = Lu, L = THF

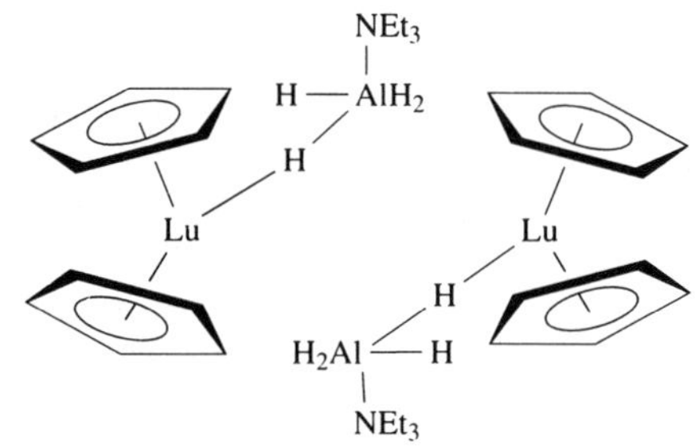

(29)

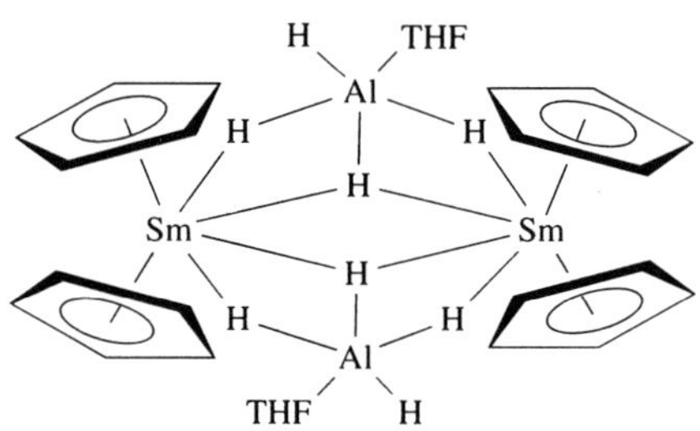

(30)

12.2.2 Aluminum–(μ-Alkyl)–Metal Bonds

12.2.2.1 Group 6 metals

A structure is proposed for the complex $[Mo(NO)_2(OPr^iAlCl_2Et)_2]$ (**31**)[35] in which the ethyl groups bonded to the aluminum centres interact with the molybdenum centre as shown. Compound (**31**) was prepared from the reaction between $[\{Mo(NO)_2(OPr^i)_2\}_n]$ and $EtAlCl_2$.

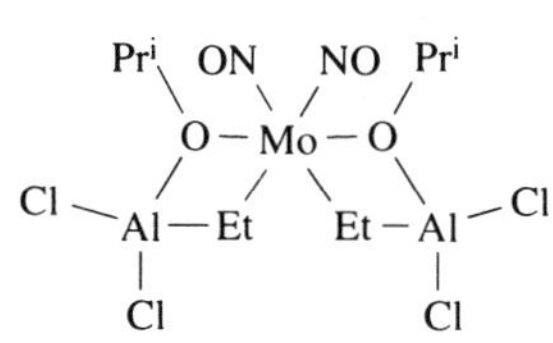

(31)

12.2.3 Direct Aluminum–Metal Bonds

12.2.3.1 Group 9 metals

Reactions between a series of cyclopentadienylrhodium complexes and trialkylaluminum compounds afford the adducts $[Rh(L)_2(AlR_3)Cp]$ ((**32a**), L = PMe_3, PMe_2Ph, PEt_3, R = Me; (**32b**), L = PMe_3, R = Et; L = PMe_3 and C_2H_4, R = Me) together with the complex $[Rh(PMe_3)_2(Al_2Me_4Cl_2)Cp]$ (**33**) when the aluminum reagent is $Al_2Me_4Cl_2$.[36] Complex (**33**) was characterized by x-ray crystallography and contains an unsupported Rh–Al bond (0.254 8(1) nm). The structure is shown in Figure 4 and has been described, in view of the coordination geometry around the rhodium-bound aluminum, as a closely interacting $[Rh(PMe_3)_2(Al_2Me_2)Cp]^+$ $[AlMe_2Cl_2]^-$ ion pair; the Rh–Al bond is very labile in solution.

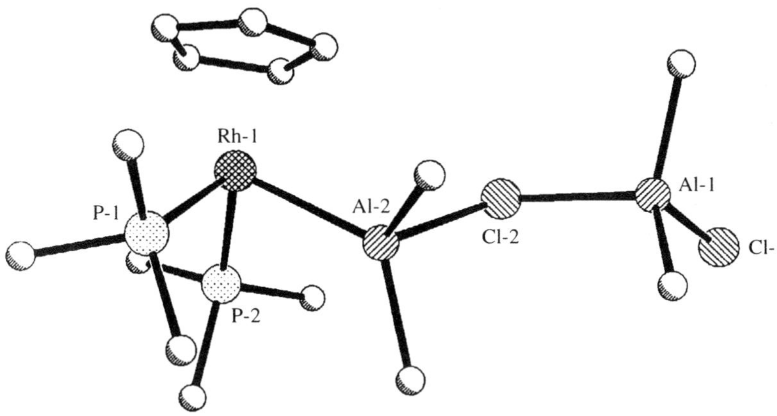

Figure 4 Molecular structure of (**33**).

12.3 GALLIUM

12.3.1 Direct Gallium–Metal Bonds

12.3.1.1 Group 6 metals

Ultrasonic irradiation of gallium metal with diiodine affords a species which has been formulated as gallium monoiodide, GaI. Although this is not a homogeneous material, it reacts as GaI and a number of compounds with transition metal–gallium bonds have been made as a result of a formal insertion of this unit into metal–metal or metal–halogen bonds. Thus, the reaction between $[MoI(CO)_3(\eta\text{-}C_5H_4Me)]$ and GaI followed by recrystallization of the product from Et_2O gave the complex $[GaI_2(Et_2O)\{Mo(CO)_3(\eta\text{-}C_5H_4Me)\}]$ (**34**),[37] the structure of which was established by x-ray crystallography. An analogous reaction between GaI and $[MoI(CO)_2(\eta\text{-}C_7H_7)]$ gave a similar complex, $[GaI_2(THF)\{Mo(CO)_2(\eta\text{-}C_7H_7)\}]$ (**35**), also from insertion of GaI into the Mo–I bond, whilst the reaction between GaI and $[Mo_2(CO)_6(\eta\text{-}C_5H_4Me)_2]$ resulted in formal insertion into the Mo–Mo bond and formation of $[GaI\{Mo(CO)_3(\eta\text{-}C_5H_4Me)\}_2]$ (**36**).

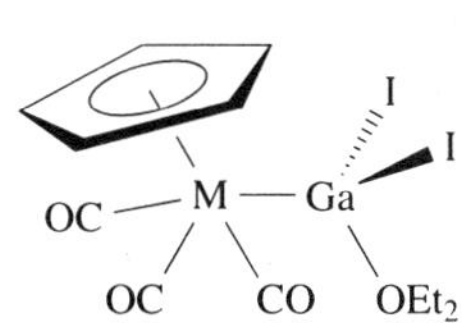

(34)

The complex $[Ga\{Mo(CO)_3Cp\}_3]$ (**37**)[38] has been prepared from the reaction between $GaCl_3$ and $3K[Mo(CO)_3Cp]$ in THF. Solution infrared spectroscopic studies indicate that partial dissociation of $[Mo(CO)_3Cp]^-$ anions occurs in THF and MeCN but not in toluene or CH_2Cl_2. Similar properties are observed for the related indium complexes which are discussed in Section 12.4.1.1.

12.3.1.2 Group 7 metals

A great deal has been published on the indium and thallium complexes $[E\{Mn(CO)_5\}_3]$ (E = In, Tl),[1] but much less was known about the gallium derivative. One report states that the reaction between $GaCl_3$ and $3Na[Mn(CO)_5]$ in THF affords the anionic complex $Na[GaCl\{Mn(CO)_5\}_3]$ (**38**)[39] rather than the neutral species $[Ga\{Mn(CO)_5\}_3]$ (**39**) (in fact, three species are obtainable from this reaction with the general formula $[GaCl_{4-n}\{Mn(CO)_5\}_n]$ (n = 1, 2, 3) depending on the amount of $Na[Mn(CO)_5]$ present). A subsequent report, however, showed that an orange material could be extracted in Et_2O from the above reaction with an infrared spectrum and analytical data consistent with the neutral complex (**39**),[38] although the crystals obtained were not suitable for x-ray crystallography and hence a definitive structural characterization. There are no crystal structures of $[E\{M(CO)_5\}_3]$ compounds for E = Ga, In or Tl and M = Mn or Re.

A more extensive chemistry exists for rhenium derivatives. The reaction between GaI_3 and $[Re_2(CO)_{10}]$ in the presence of gallium metal in xylene at 413 K for 4 d afforded yellow crystals of the complex $[Re_2(CO)_8\{\mu\text{-}GaRe(CO)_5\}_2]$ (**40**),[40] the structure of which is shown in Figure 5, consisting of an $Re_2(CO)_8$ unit in which the formal Re–Re bond is bridged by two $GaRe(CO)_5$ fragments.

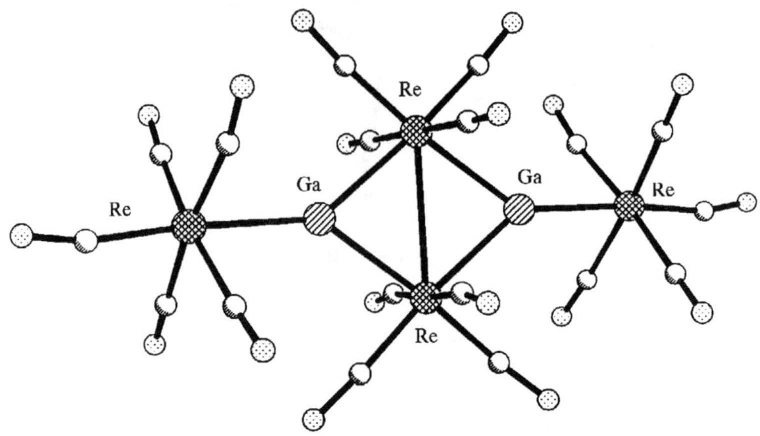

Figure 5 Molecular structure of (**40**).

An analogous reaction between GaI_3 and *trans*-$[Re_2(CO)_8(PPh_3)_2]$ in boiling xylene afforded yellow crystals of the trirhenium complex $[Re_2(CO)_4(PPh_3)_2(\mu\text{-}I)_2\{GaRe(CO)_4(PPh_3)\}]$ (**41a**)[41] the structure of which is shown diagrammatically, and in which the central Re–Re bond is bridged by two iodides and one $GaRe(CO)_4(PPh_3)$ fragment; the bromo derivative (**41b**) was prepared similarly from $GaBr_3$ in boiling mesitylene.[42]

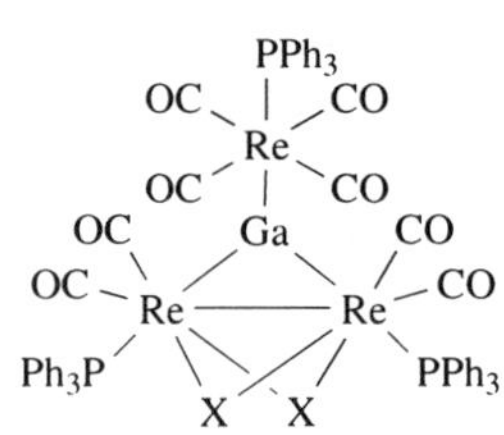

(41)
(a) X = I; **(b)** X = Br

If $GaCl_3$ is reacted with *trans*-$[Re_2(CO)_8(PPh_3)_2]$ under similar conditions, the tetrarhenium cluster $[Re_3(CO)_6(PPh_3)_3(\mu_2\text{-}Cl)_3\{\mu_3\text{-}GaRe(CO)_4(PPh_3)\}]$ (**42**)[42] is formed and a further example of a gallium–rhenium complex was obtained from the reaction between (**41a**) and CO, namely $[Re_2(CO)_6(PPh_3)_2(\mu\text{-}I)(\mu\text{-}GaI_2)]$ (**43**).[42] Both complexes were characterized crystallographically.

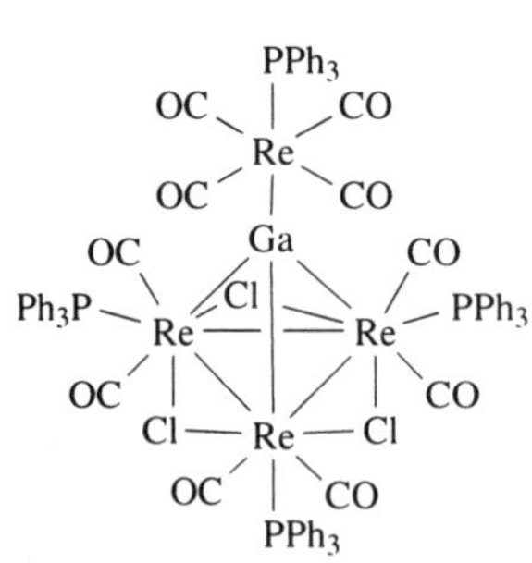

(42)

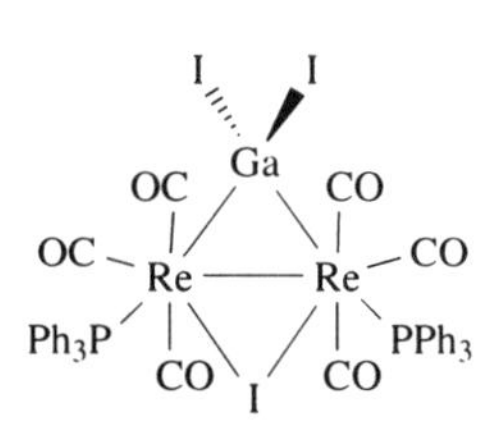

(43)

In another reaction, prolonged heating of gallium trihalides with $[Re_2(CO)_{10}]$ at 300 °C gave (**40**) together with the tetrahedral cluster $[Re_4(CO)_{12}\{\mu_3\text{-}GaRe(CO)_5\}_4]$ (**44**),[43] the structure of which is assumed to be similar to that of the indium derivative,[1] which consists of a central Re_4 tetrahedron with each rhenium atom bonded to three terminal carbonyl ligands and with each face of the tetrahedron bridged by an In(Ga)Re$(CO)_5$ fragment.

12.3.1.3 Group 8 metals

The reaction between GaI and $[FeI(CO)_2Cp]$ affords the Fe–Ga complex $[GaI_2(Et_2O)_n\{Fe(CO)_2Cp\}]$ (**45**)[37] as a result of insertion of GaI into the Fe–I bond in a similar fashion to that observed for (**34**). A subsequent reaction between (**45**) and two equivalents of $Na[C_5H_4Me]$ followed by reaction with pyridine led to substitution of the iodides by C_5H_4Me and the formation of a pyridine adduct $[Ga(\eta^1\text{-}C_5H_4Me)_2(py)\{Fe(CO)_2Cp\}]$ (**46**),[37] which was characterized by x-ray crystallography, showing the methylcyclopentadienyl groups to be η^1-coordinated to gallium.

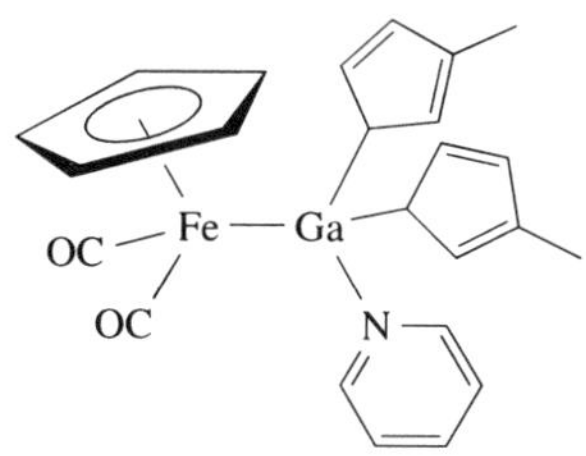

(**46**)

Treatment of $GaCl_3$ with three equivalents of $K[Fe(CO)_2Cp]$ affords the triiron–gallium complex $[Ga\{Fe(CO)_2Cp\}_3]$ (**47**),[38] which contains a trigonal planar gallium atom bonded to three $Fe(CO)_2Cp$ fragments by unsupported Ga–Fe bonds; the structure is shown in Figure 6. A similar reaction in which $GaCl_3$ reacts with two equivalents of $K[Fe(CO)_2Cp]$ afforded the diiron complex $[GaCl\{Fe(CO)_2Cp\}_2]$ (**48**).[38]

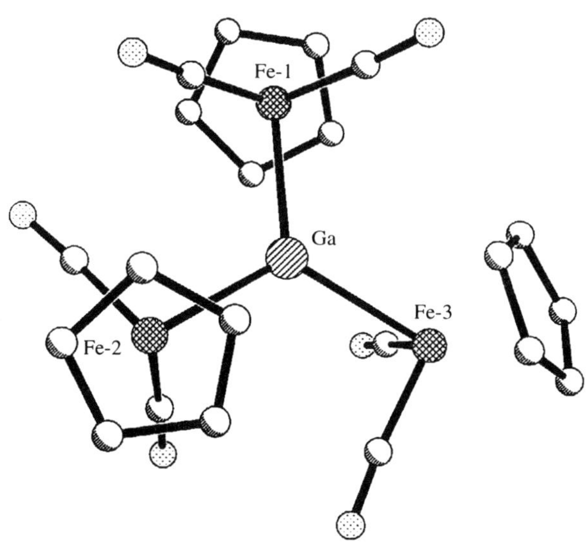

Figure 6 Molecular structure of (**47**).

The crystal structure of the vinylgallium–iron complex $[Fe_2(CO)_8\{\mu\text{-}Ga(C_2H_3)(THF)\}_2]$ (**49**) has been described[44] in which two $Fe(CO)_4$ groups are bridged by two vinyl gallium fragments. A THF ligand is also coordinated to each gallium.

12.3.1.4 Group 9 metals

Of all transition metal–gallium complexes, those involving bonds to cobalt are the most numerous in terms of both the compounds themselves and the applications to which use they have been put. The reaction between $GaCl_3$ and two or three equivalents of $K[Co(CO)_3(PPh_3)]$ affords the complexes

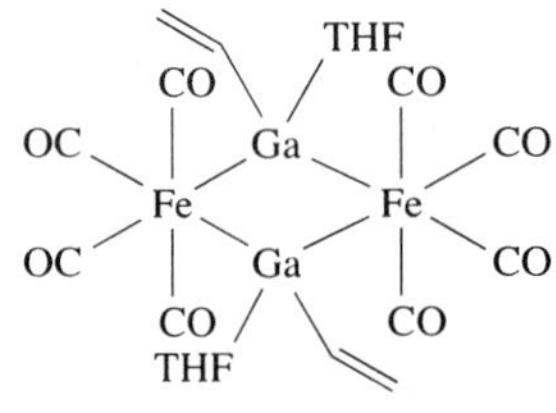

(49)

$[GaCl\{Co(CO)_3(PPh_3)\}_2]$ (**50**) and $[Ga\{Co(CO)_3(PPh_3)\}_3]$ (**51**),[38] respectively, which have been characterized spectroscopically. Furthermore, (**50**) can be prepared from (**51**) and $GaCl_3$ according to the Equation (4) reaction.

$$2[Ga\{Co(CO)_3(PPh_3)\}_3] + GaCl_3 \longrightarrow 3[GaCl\{Co(CO)_3(PPh_3)\}_2] \quad (4)$$

The organo complex $[Ga(CH_2CMe_3)_2(THF)\{Co(CO)_4\}]$ (**52**)[45,46] has been prepared from the reaction between $GaCl(CH_2CMe_3)_2$ and $Na[Co(CO)_4]$ in THF and a subsequent report[47] described the crystal structure of (**52**) together with the synthesis of the related complex $[Ga(CH_2\text{-}TMS)_2(THF)\{Co(CO)_4\}]$ (**53**). A number of reactions of (**52**) and (**53**) were also described and these are illustrated in Scheme 1 (dmpm = 1,2-bis(dimethylphosphino)methane).

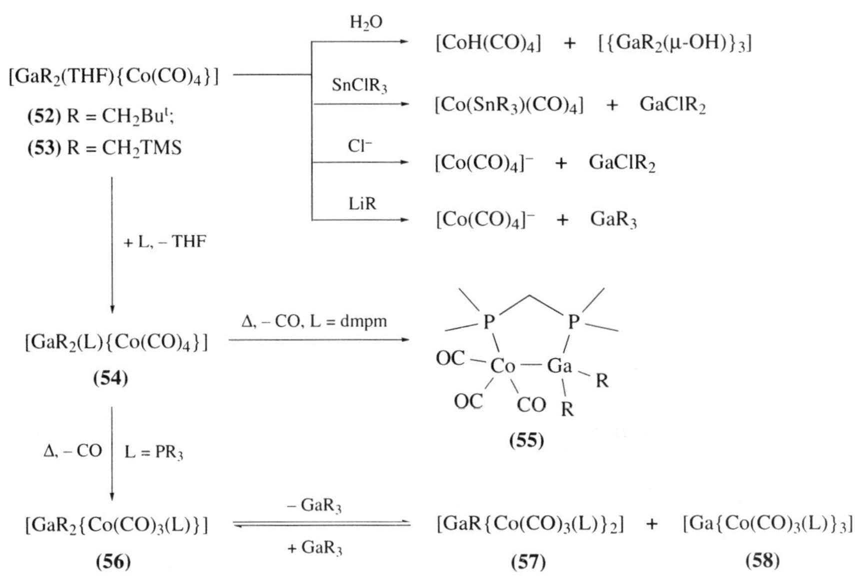

Scheme 1

In addition to the organo derivatives mentioned above, a number of derivatives containing organonitrogen side groups have also been synthesized in which the nitrogen atom is coordinated to the gallium centre.[48] These complexes are shown in Scheme 2.

The structure of (**59c**) was established by x-ray crystallography.[48] A further report by Fischer *et al.*[46] described in addition to (**52**) and (**60a**), the synthesis of the complexes $[GaH_2(NMe_3)\{Co(CO)_4\}]$ (**61**) and $[GaH_2(NMe_3)\{Mn(CO)_5\}]$ (**62**) from GaH_2Cl and the corresponding metal carbonylate anion in the presence of NMe_3, together with the x-ray crystal structure of (**60a**). The gallium complex $[GaEt(CH_2CH_2CH_2NEt_2)\{Fe(CO)_2Cp\}]$ (**63**) and the aluminum complexes $[Al(CH_2CMe_3)(CH_2CH_2CH_2NEt_2)\{ML_n\}]$ (**64a**), $ML_n = Fe(CO)_2Cp$; (**64b**), $ML_n = Co(CO)_3(PMe_3)$; (**64c**) $ML_n = Ni(CO)Cp$) were also described.[46]

In addition to the synthetic and structural studies carried out, a number of the cobalt–gallium species have been investigated as vapour-phase or solid-phase precursors to binary intermetallic phases such as β-CoGa.[47] Similar work has also been done on a number of the indium complexes. These types of intermetallic phase have important applications in semiconductor technology.

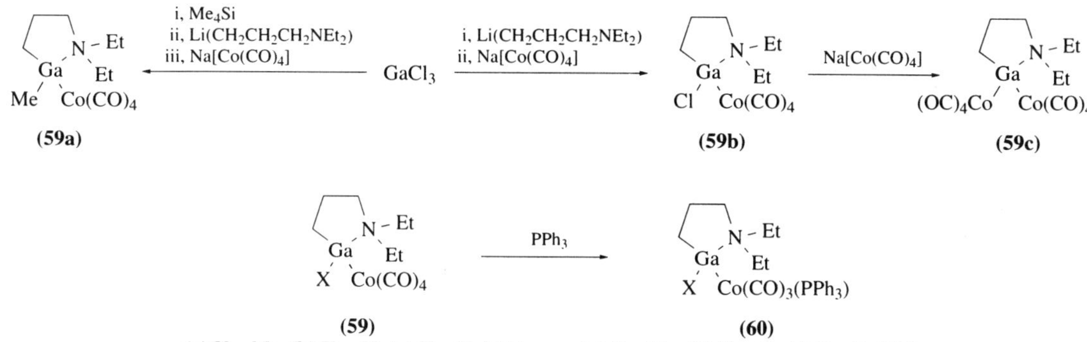

(a) X = Me; (b) X = Cl; (c) X = Co(CO)4 (a) X = Me; (b) X = Cl; (c) X = Co(CO)4

Scheme 2

12.3.1.5 Group 10 metals

Two examples of complexes containing bonds to a group 10 metal have been described. The reaction between K[Ni(CO)Cp] and $GaCl(CH_2CMe_3)_2$ affords the complex $[Ga(CH_2CMe_3)_2(THF)\{Ni(CO)Cp\}]$ (**65**),[50] which was structurally characterized. The platinum complex $[Pt(P\text{–}P)(CH_2CMe_3)\text{-}\{Ga(CH_2CMe_3)_2\}]$ (**66**)[51] ((P–P) = $(c\text{-}C_6H_{11})_2PCH_2CH_2P(c\text{-}C_6H_{11})_2$) was prepared from the reaction between $Ga(CH_2CMe_3)_3$ and $[Pt(P\text{–}P)(CH_2CMe_3)H]$ by elimination of CMe_4.

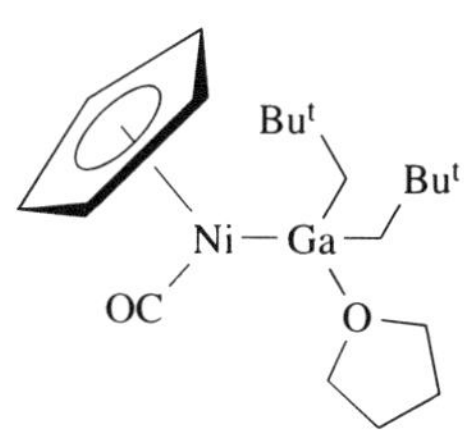

(65)

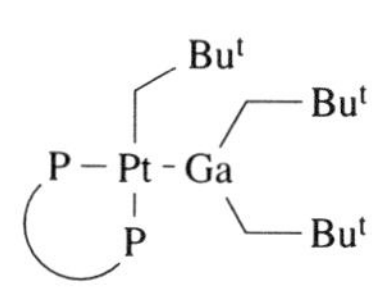

(66)

12.4 INDIUM

12.4.1 Direct Indium–Metal Bonds

12.4.1.1 Group 6 metals

The reaction between $InCl_3$ and $3Na[Mo(CO)_3Cp]$ in THF affords the trimolybdenum–indium complex $[In\{Mo(CO)_3Cp\}_3]$ (**67**),[52] which was characterized crystallographically. A view of (**67**) is shown in Figure 7 and details of the structure will be referred to later in this section.

With two equivalents of $Na[Mo(CO)_3Cp]$, the product isolated was $[Na(THF)_2][InCl_2\{Mo(CO)_3Cp\}_2]$ (**68**), whereas with one equivalent, no chloride substitution was observed and the adduct complex $[Na(THF)_2][InCl_3\{Mo(CO)_3Cp\}]$ (**69**) was formed. Both complexes were characterized by x-ray crystallography and exist in the solid state as centrosymmetric dimers as shown. The tungsten analgoues were prepared in a similar manner and characterized by spectroscopic and analytical methods.

The anion $[InCl_3\{Mo(CO)_3Cp\}]^-$ in (**69**) has also been reported in association with the cation $[Mo_3(CO)_3Cp_3(\mu\text{-}CO)_3(\mu_3\text{-}O)]^+$ derived from the reaction between $InCl_3$ and $[Mo_2(CO)_6Cp_2]$ in diglyme, in which the neutral cluster $[Mo_3(CO)_4Cp_3(\mu\text{-}Cl)(\mu_3\text{-}O)]$ is also produced.[53]

Additional studies on (**67**) have also been reported, including the crystal structure of a THF solvate which has a molecular structure essentially identical with that originally described,[52,54] together with the chromium analogue $[In\{Cr(CO)_3Cp\}_3]$ (**70**). All three structures (the two for (**67**) and that for (**70**)) are similar in that the orientations of the $M(CO)_3Cp$ fragments are the same (see Figure 7) and are such that the symmetry plane of the $M(CO)_3Cp$ fragment is coplanar with the M_3E plane. Extended Hückel molecular orbital calculations indicate that there is little in the way of M–In π-bonding and that the observed conformations about the M–In bonds are largely dictated by steric, as opposed to electronic,

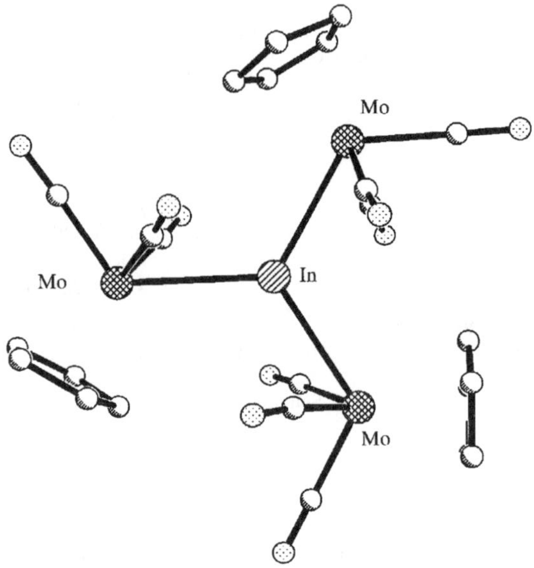

Figure 7 Molecular structure of (**67**).

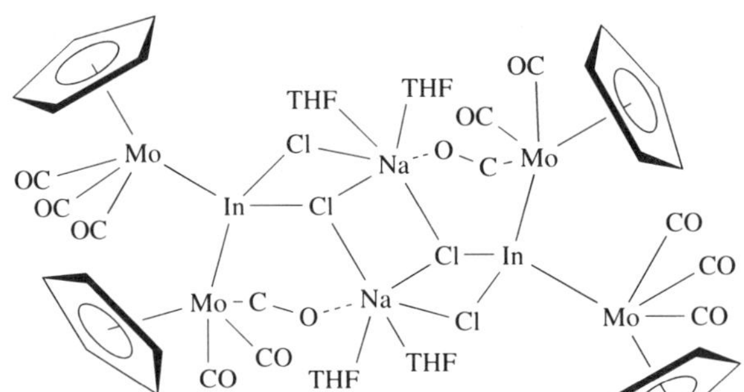

(**68**)

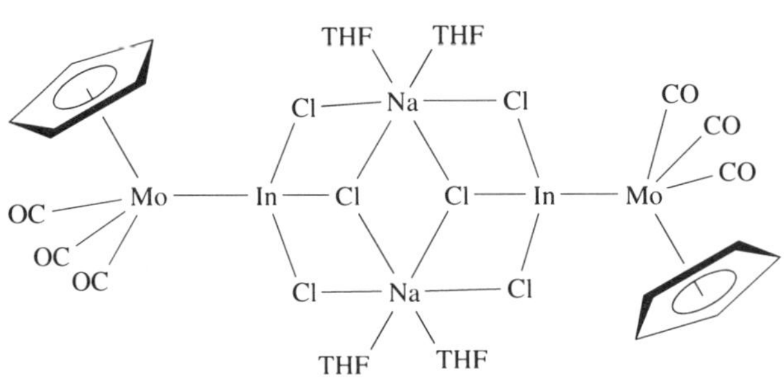

(**69**)

factors. In addition to the structural studies, the solution-state properties of (**67**) and (**70**) were also considered. Infrared spectroscopic data revealed that in noncoordinating solvents such as CH_2Cl_2 and toluene, the complexes are undissociated, whereas in MeCN and acetone extensive dissociation into what is probably $[In(solv)_x\{M(CO)_3Cp\}_2]^+$ $[Mo(CO)_3Cp]^-$ occurs. In THF, no dissociation is observed in CaF_2 infrared cells, but in KBr cells a reaction takes place resulting in $[InBr\{M(CO)_3Cp\}_2]$ and $K[Mo(CO)_3Cp]$.[52,54]

Synthetic and solution infrared spectroscopic studies have also been described for a range of compounds with the general formula $[InX(L)\{M(CO)_3Cp\}_2]$, where X = Cl, I; L = THF, py; M = Cr, Mo, W.[55] The chromium complex $[Na(THF)_2][InCl_2\{Cr(CO)_3Cp\}_2]$ (**71**) was prepared by the same method as (**68**) and in both cases only one chloride is substituted with a dichloro-anionic species resulting. The neutral monochloro-complexes $[InCl\{M(CO)_3Cp\}_2]$ ((**72a**), M = Cr; (**72b**), M = Mo) were obtained from the reactions between the dimer $[M_2(CO)_6Cp_2]$ (M = Cr, Mo) and indium monochloride (InCl) in toluene, this being a formal inserion of InCl into the M–M bond in a similar fashion to that observed with GaI in the synthesis of (**36**). No structural data are available for these species, although they are unlikely to be monomeric and probably exist as chloro-bridged dimers or polymers. Treatment with a Lewis base ligand (L), however, resulted in a series of monomeric complexes, three of which were structurally characterized, namely $[InCl(THF)\{Mo(CO)_3Cp\}_2]$ (**73a**), $[InCl(py)\{Cr(CO)_3Cp\}_2]$ (**73b**) and $[NMe_4][InCl_2\{Mo(CO)_3Cp\}_2]$ (**73c**), in which, for (**73c**), the ligand is chloride, resulting in an anionic species similar to (**68**) but in which the cation is $[NMe_4]^+$. Compound (**73a**) is shown in Figure 8. Solution infrared spectroscopic studies indicate that in MeCN solution, some dissociation of $[M(CO)_3Cp]^-$ occurs as seen in the trimetal complexes.

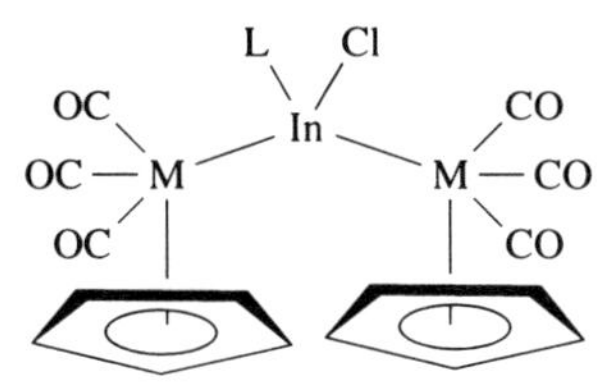

(73)
(**a**) M = Mo, L = THF; (**b**) M = Cr, L = py; (**c**) M = Mo, L = Cl⁻

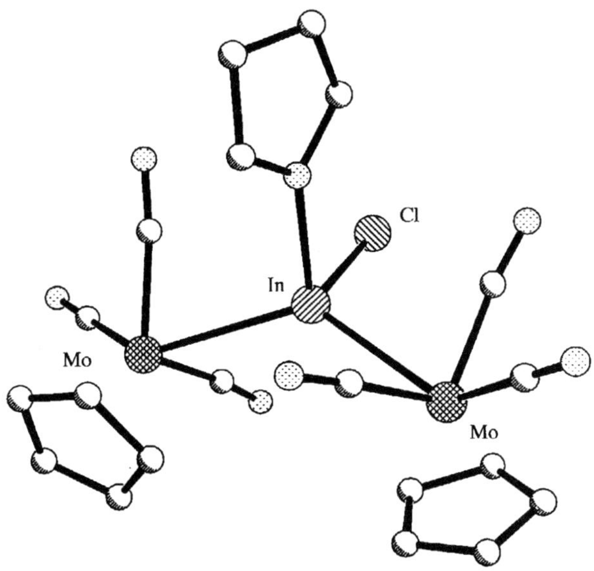

Figure 8 Molecular structure of (**73a**).

The reactivity of the complexes $[InBr\{Mo(CO)_3Cp\}_2]$ and $[InCl_2\{Mo(CO)_3Cp\}]$ towards the reagents $E(TMS)_2$ (E = chalcogen) in THF has been studied. In the former case, reaction with $S(TMS)_2$ affords the cubic cluster $[\{In_4S_4\{Mo(CO)_3Cp\}_4]$ (**74**) whereas the reaction of the latter with $Te(TMS)_2$ gave the tellurium analogue $[\{In_4Te_4\{Mo(CO)_3Cp\}_4]$ (**75**).[56] A view of (**75**) is shown in Figure 9.

Some photochemical studies involving molybdenum and indium have also been described. Photolysis of $[Mo_2(CO)_6Cp_2]$ and InR_3 (R = Et, Bu^t, neopentyl) efficiently forms the complexes $[InR_2\text{-}\{Mo(CO)_3Cp\}]$ (**76**).[57] These reactions are believed to occur via a bimolecular radical substitution process which demonstrates a substantial strength for the Mo–In bond in these complexes and consistent with this result is the fact that (**67**) is formed (reversibly) in a photochemical reaction between $[Mo_2(CO)_6Cp_2]$ and indium metal.

In all of the above complexes, the indium is present in the 3+ oxidation state, or at least in a trivalent form. Examples of what can be thought of as complexes of indium(I) and a transition metal fragment are formed in the reactions between InX_3 (X = Cl, Br, I) and $Na_2[M_2(CO)_{10}]$ (M = Cr, Mo, W) in THF.[58] The crystal structure of $[\{InBr(THF)\{Cr(CO)_5\}_x]$ (**77**) has been determined, and the formula shown illustrates the ladder polymeric structure, the monomeric unit of which is a THF-solvated InBr group acting as a two-electron donor ligand to a $Cr(CO)_5$ fragment.

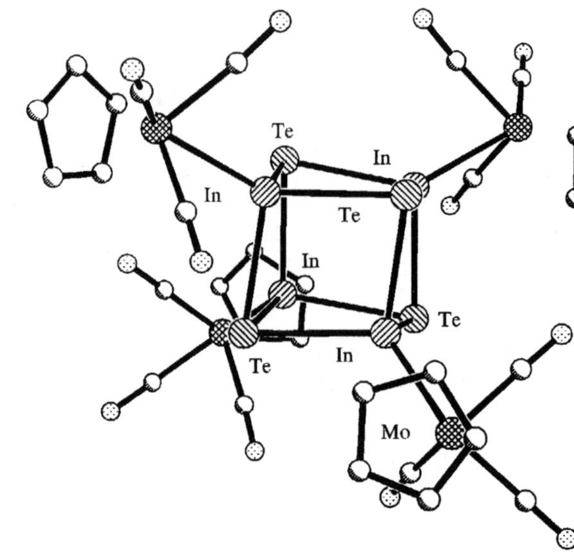

Figure 9 Molecular structure of (**75**).

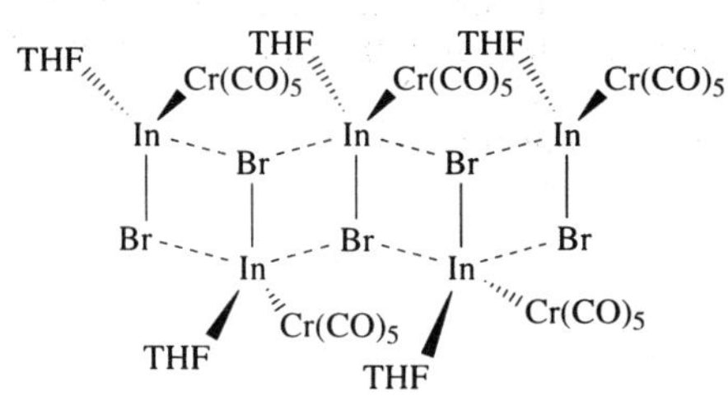

(77)

Another example of indium(I) coordinated to a transition metal is found in $[(CO)_5MoSn\{\mu\text{-}OBu^t\}_3InMo(CO)_5]$ (**78**).[59] The chromium analogue is also known, together with a complex that contains only one $Mo(CO)_5$ fragment, which is observed to be bonded to the In^I centre rather than the Sn^{II} centre, indicating that the former is the better donor in this situation.

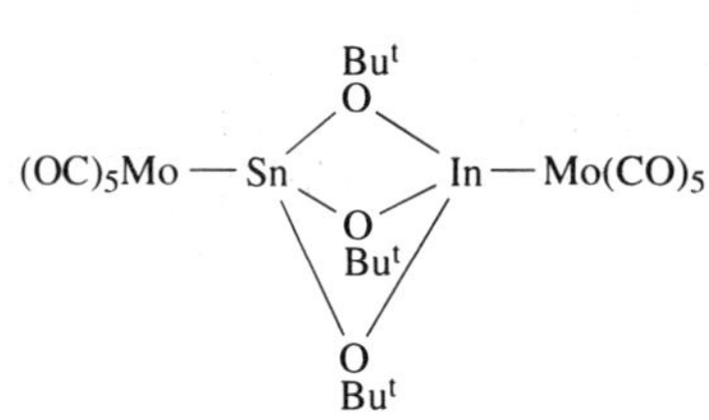

(78)

12.4.1.2 Group 7 metals

The indium complexes $[InCl_{4-n}\{Mn(CO)_5\}_n]$ (n = 1, 2, 3) analogous to the gallium species described in Section 12.3.1.2 are formed in the reactions between $InCl_3$ and one, two or three equivalents of $Na[Mn(CO)_5]$.[39]

The reaction between InI_3 and $[Re_2(CO)_{10}]$ in a pressure-balanced Schlenk tube at 473 K affords two complexes, $[In\{\mu\text{-}Re_2(CO)_6(\mu\text{-}I)_2\}\{\mu\text{-}Re_2(CO)_8(\mu\text{-}I)\}]$ (**79**) and $[In\{\mu\text{-}I_2Re(CO)_4\}\{Re(CO)_5\}_2]$ (**80**).[60]

The reaction between $[Re_2(CO)_8(\mu\text{-}PPh_2)_2]$ and InX_3 (X = Cl, Br, I) in xylene at 220 °C affords a number of products, most of which are rhenium phosphido-containing clusters, but one of the isolated products was the cluster $[Re_2(CO)_4\{InI_2(H_2O)\}_2(\mu\text{-}PPh_2)_2]$ (**81**), which was characterized by x-ray crystallography and contains a phosphido-bridged dirhenium unit in which each rhenium is bonded to an InI_2 group with the indium also coordinated by a water molecule.[61]

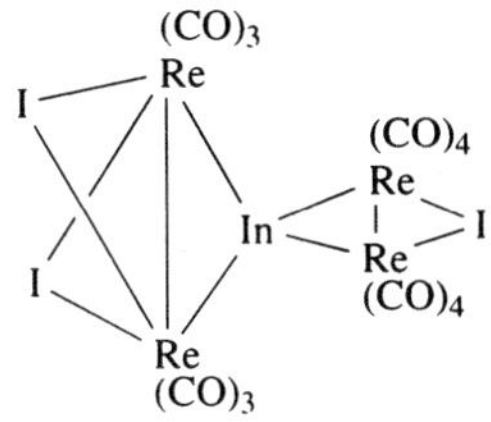

(79)

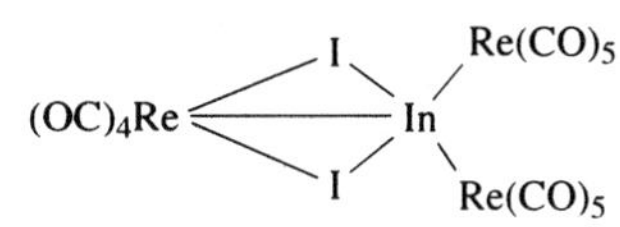

(80)

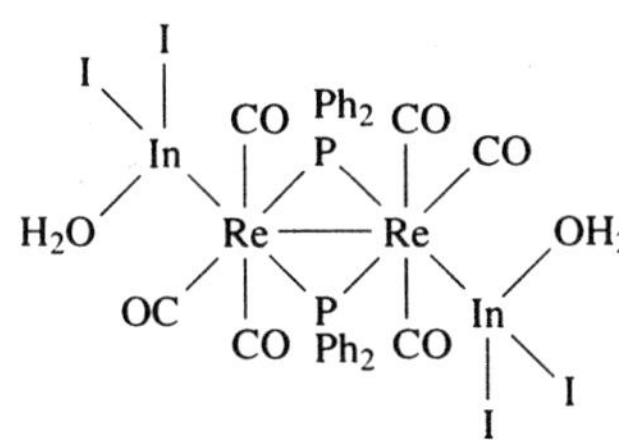

(81)

The reaction of the manganese carbonylate anion complex $[AsPh_4]_2[Mn_3(CO)_{10}(\mu\text{-}CO)_2(\mu\text{-}H)]$ with $InCl_3$ affords two manganese–indium clusters with the formulas $[AsPh_4][In\{Mn_3(CO)_{12}(\mu\text{-}H)\}_2]$ (**82**) and $[AsPh_4][InCl\{Mn(CO)_5\}\{Mn_3(CO)_{12}(\mu\text{-}H)\}]$ (**83**), both of which were characterized by x-ray crystallography.[62] In (**82**), the indium atom bridges between the edges of two orthogonal Mn_3 triangles and is in a distorted tetrahedral environment.

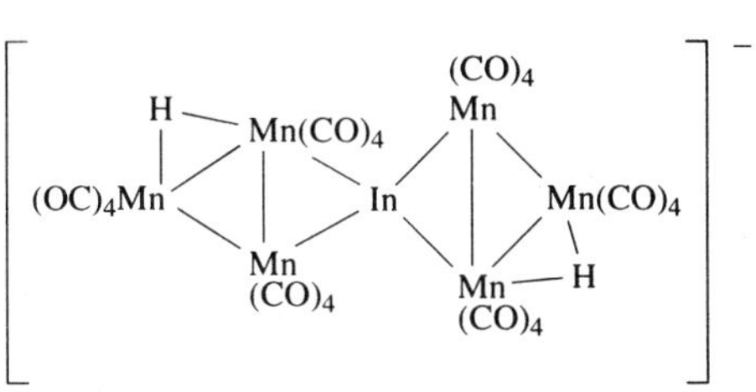

(82)

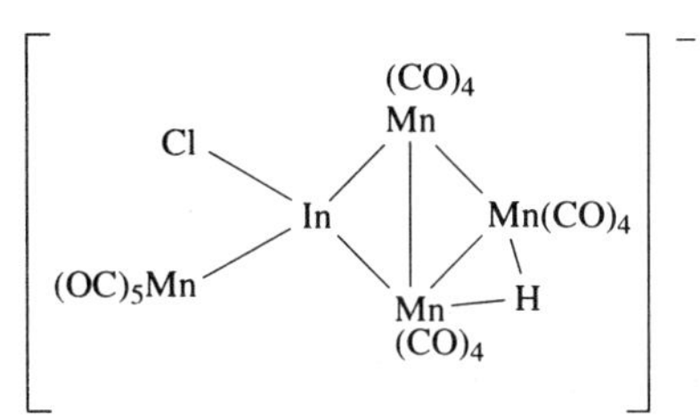

(83)

If, however, the completely deprotonated anion as its K^+ salt $K_3[Mn_3(CO)_{12}]$ is reacted with $InCl_3$ followed by $[AsPh_4]Cl$, two different products are formed, one of which contains indium in an unusual pentagonal planar coordination environment,[63] $[AsPh_4]_2[Mn_2(CO)_8(\mu\text{-}InCl\{Mn(CO)_5\})_2]$ (**84**) and $[AsPh_4]_2[In\{Mn(CO)_4\}_5]$ (**85**).

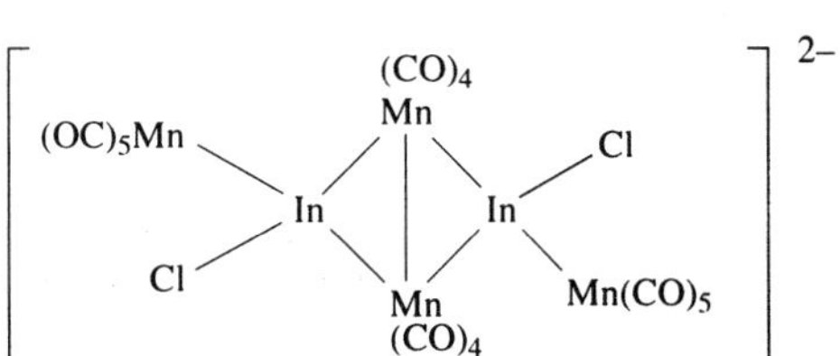

(84)

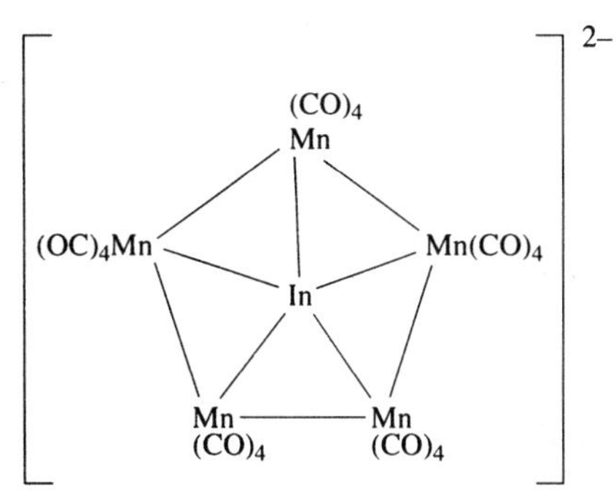

(85)

The complex $[In(CH_2CH_2CH_2NMe_2)_2\{Mn(CO)_5\}]$ (**86**) has been prepared from the reaction between $InBr(CH_2CH_2CH_2NMe_2)_2$ and $Na[Mn(CO)_5]$;[46] other transition metal derivatives of this indium–ligand fragment are discussed later.

12.4.1.3 Group 8 metals

The reaction between $InCl_3$ and three equivalents of $Na[Fe(CO)_2(\eta\text{-}C_5H_4R)]$ affords the triiron complexes $[In\{Fe(CO)_2(\eta\text{-}C_5H_4R)\}_3]$ (**87**) (R = H, Me), spectroscopic data for which are analogous to the gallium complex (**47**).[64] Subsequent studies revealed that the diiron complex $[InCl\{Fe(CO)_2\text{-}Cp\}_2]$ (**88**) could be synthesized either from the reaction between $InCl_3$ and two equivalents of $Na[Fe(CO)_2Cp]$ or from the reaction between $[Fe_2(CO)_4Cp_2]$ and InCl.[65] Complex (**88**) was characterized by x-ray crystallography and shown to be a chloro-bridged dimer, as shown in Figure 10.

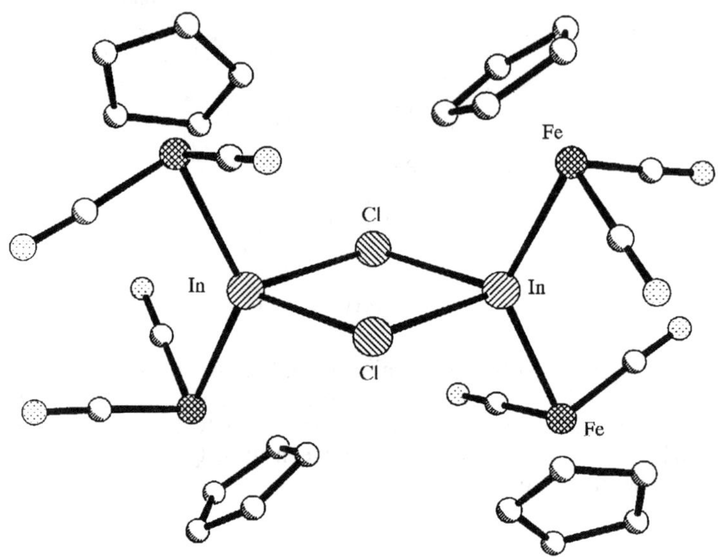

Figure 10 Molecular structure of (**88**).

The iodo analogue $[InI\{Fe(CO)_2Cp\}_2]$ (**89**) was prepared similarly and adducts of both (**88**) and (**89**) with halide anions and neutral two-electron donors were prepared, $[InX_2\{Fe(CO)_2Cp\}_2]^-$ ((**90a**), X = Cl; (**90b**), X = I) and $[InX(L)\{Fe(CO)_2Cp\}_2]$ ((**91a**), X = Cl, L = py; (**91b**), X = Cl, I, L = PMe_2Ph). Compound (**91b**) where X = Cl was structurally characterized by x-ray crystallography and contains a distorted tetrahedral indium centre bonded to two irons, one chlorine, and the phosphorus of a phosphine ligand.

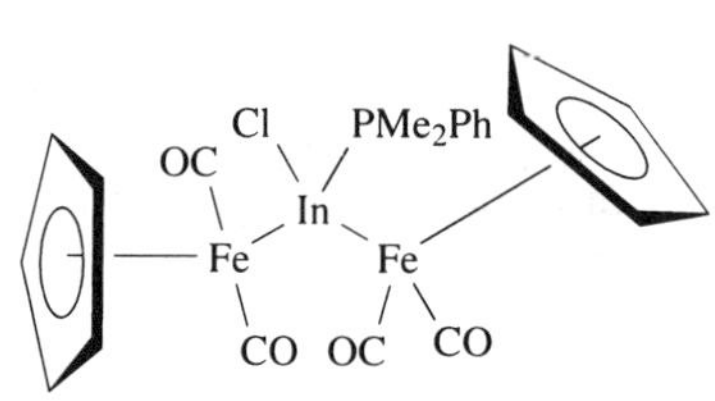

(**91b**)

The ruthenium analogues $[In\{Ru(CO)_2Cp\}_3]$ (**92**), $[InCl\{Ru(CO)_2Cp\}_2]$ (**93**) and $[N(PPh_3)_2][InCl_2\{Ru(CO)_2Cp\}_2]$ (**94**) have also been characterized.[66] Compounds (**92**) and (**93**) were structurally characterized; (**93**) is isomorphous with (**88**) whilst (**92**) is similar to the gallium analogue (**47**) but differs in the conformations of the $M(CO)_2Cp$ fragments about the E–M bonds.

The reaction between $[InBr\{Fe(CO)_2Cp\}_2]$ and $Se(TMS)_2$ affords the cubic cluster $[In_4Se_4\text{-}\{Fe(CO)_2Cp\}_4]$ (**95**) and the reaction between $[InCl_2\{Fe(CO)_2Cp\}]$, reported by Clarkson *et al.*,[65] with $S(TMS)_2$ affords $[In_4S_4\{Fe(CO)_2Cp\}_4]$ (**96**).[56] The structure of (**95**) was established by x-ray crystallography and reveals a cubic structure very similar to (**75**) shown in Figure 9.

The organoindium complex $[In(CH_2CH_2CH_2NMe_2)_2\{Fe(CO)_2Cp\}]$ (**97**) has been prepared from the reaction between $InBr(CH_2CH_2CH_2NMe_2)_2$ and $Na[Fe(CO)_2Cp]$ and characterized by x-ray crystallography.[46] The indium centre is five-coordinate, being bonded to an iron and two carbons in a trigonal planar arrangement, with the nitrogen atoms above and below this plane giving an overall geometry close to trigonal bipyramidal. The manganese derivative (**86**) is presumed to have a similar structure, as is a cobalt example mentioned later.

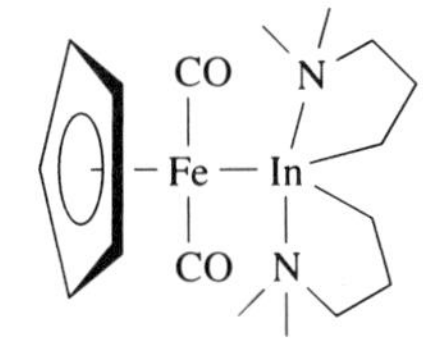

(97)

A range of indium–iron carbonyl clusters has been reported. The complex $[NEt_4]_2[In_2Fe_6(CO)_{24}]$ (**98**) has been synthesized from a 1:1 reaction between $InCl_3$ and $Fe(CO)_5$–KOH–MeOH solution and also from the reaction between $InCl_3$ and $[NEt_4]_2[Fe_2(CO)_8]$ in MeCN. The structure, represented below, comprises a central In_2Fe_2 rhombus in which each indium atom also bridges across an $Fe_2(CO)_8$ unit.[67]

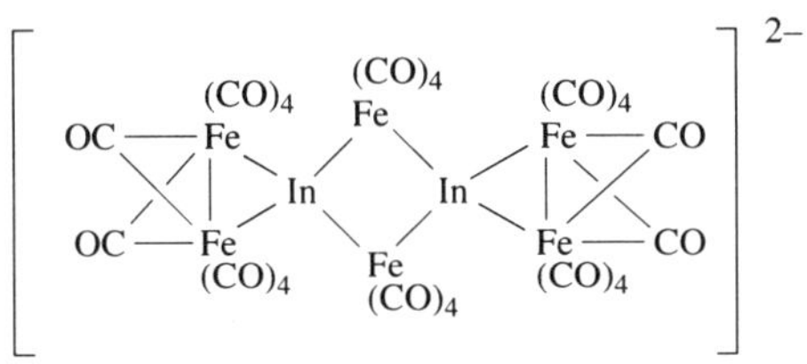

(98)

A fragment of (**98**), $[InFe_2(CO)_8]^-$ stabilized by bipy (2,2'-bipyridyl), $[In(bipy)\{Fe(CO)_4\}_2]$ (**99**)[67] has been obtained from the addition of bipy to an orange solution obtained from the reaction between $InCl_3$ and methanolic $Fe(CO)_5$–KOH in a 1:2 ratio and believed to contain $[NEt_4]_2[In_2Fe_4(CO)_{16}]$, the thallium analogue of which is discussed later.

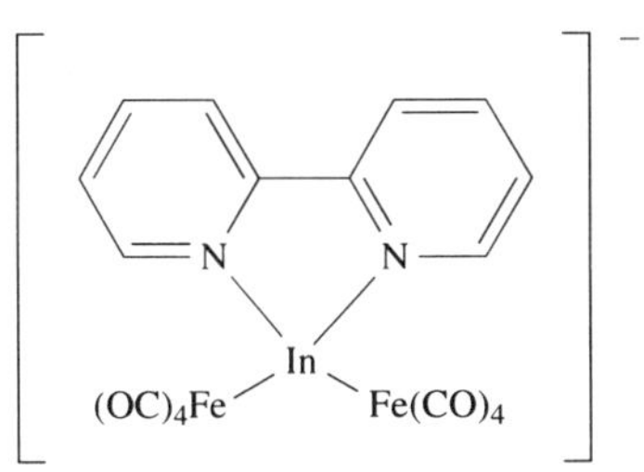

(99)

A compound similar to (**98**), namely $K[In\{Fe_2(CO)_8\}_2]$ (**100**), has been prepared from the reaction between $Fe_3(CO)_{12}$ and activated indium powder in the presence of potassium chloride. The structure was established by x-ray crystallography.[68] We can view this as a spirocyclic compound with indium(III) bonded to two $[Fe_2(CO)_8]^{2-}$ units.

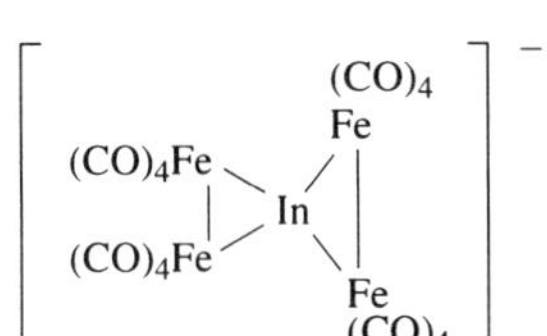

(100)

The complex $[Fe_2(CO)_8\{\mu\text{-InCl(THF)}\}_2]$ (**101**) is formed in the reaction between either $Na_2[Fe(CO)_4]$ or $Na_2[Fe_2(CO)_8]$ and $InCl_3$ or from the reaction between $Fe_3(CO)_{12}$ and InCl in THF. Complex (**101**) forms an adduct with bromide $[Fe_2(CO)_8\{\mu\text{-InClBr}\}_2]^{2-}$ (**102**), which has been structurally characterized as its $[NBu^n_4]^+$ salt.[68]

The reaction between $InBr_3$ and 3.5 equivalents of $Na_2[Fe(CO)_4]$ in THF affords the triiron complex $[In\{Fe(CO)_4\}_3]^{3-}$, which was isolated and characterized as its $[NMe_3(CH_2Ph)]^+$ salt, $[NMe_3(CH_2Ph)]_3$-$[In\{Fe(CO)_4\}_3]$ (**103**).[69] Oxidation of (**103**) with $AgBF_4$ affords the anions $[In_2Fe_6(CO)_{24}]^{4-}$ and

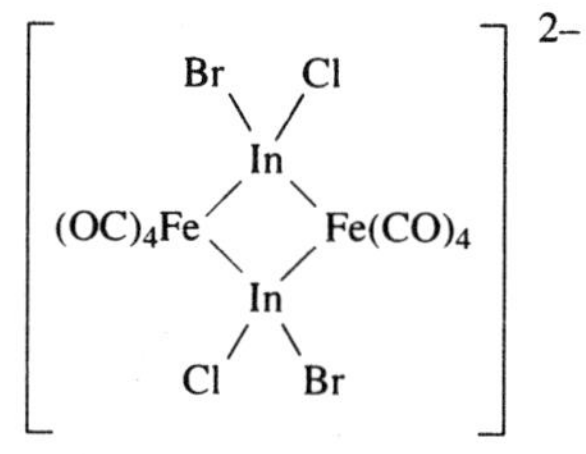

(102)

$[In_2Fe_6(CO)_{24}]^{2-}$; the latter is the same as in (**98**). With $InBr_3$, (**103**) gives the complexes $[InBr_{3-x}\{Fe(CO)_4\}_x]^{x-}$ ($x = 1, 2$); when $x = 1$, the anion is dimeric, that is, $[Fe_2(CO)_8\{\mu\text{-}InBr_2\}_2]^{2-}$ (**104**), and related to (**102**); it has also been structurally characterized.[69]

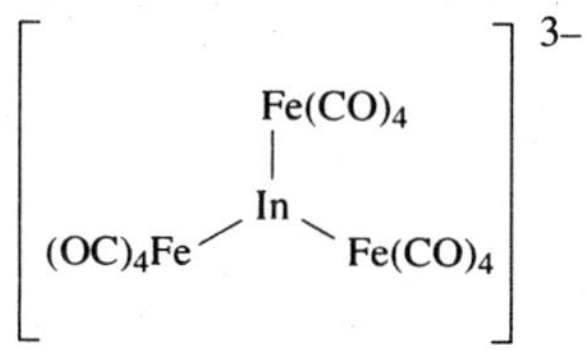

(103)

$Li[InCl_3\{C(TMS)_3\}]$ reacts with $Na_2[Fe(CO)_4]$ to give the complex $[In_2\{C(TMS)_3\}_2(\mu\text{-}Cl)_2\{\mu\text{-}Fe(CO)_4\}]$ (**105**), which has the structure shown with a diindium unit bridged by two chlorides and an $Fe(CO)_4$ group.[70]

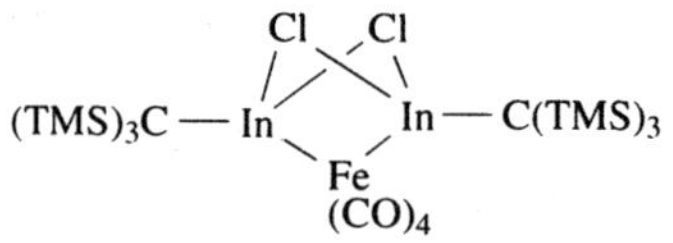

(105)

The reaction between $K[Fe(CO)_3\{Si(OMe)_3\}(PPh_2CH_2C(O)Ph)]$ and $InCl_3$ affords the complex $[Fe(CO)_3\{Si(OMe)_3\}\{InCl_2\}(PPh_2CH_2C(O)Ph)]$ (**106**), in which the phosphino ketone ligand $PPh_2CH_2C(O)Ph$ bridges between the iron and indium centres. If the analogous phosphino cyanide $PPh_2CH_2CH_2CN$ is used, a similar complex (**107**) is formed which is apparently dimeric with (μ-Cl) bridges between the indium centres, although a mononuclear species (**108**) is formed when picoline is added.[71]

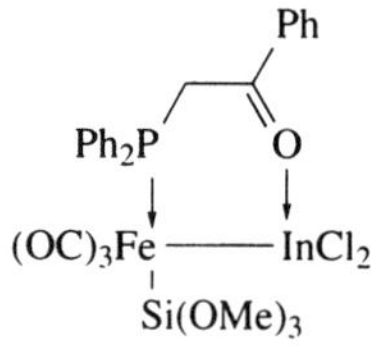

(106)

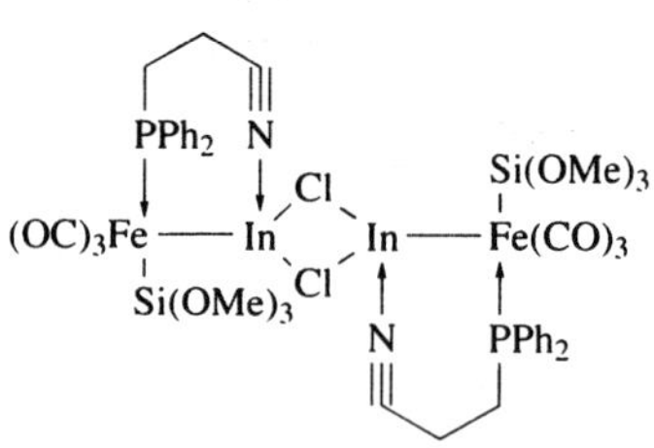

(107)

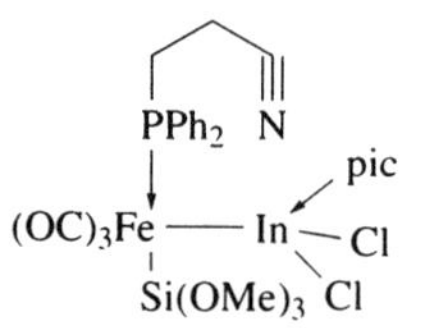

(108)

12.4.1.4 Group 9 metals

The complex $[In(CH_2CH_2CH_2NMe_2)_2\{Co(CO)_4\}]$ (**109**) is analogous to the iron complex (**97**) and has been described[46] together with the dicobalt complex $[In(CH_2\text{-}TMS)(THF)\{Co(CO)_4\}_2]$ (**110**)[46] obtained from the reaction between $[InCl(CH_2\text{-}TMS)_2]$ and $Na[Co(CO)_4]$. In this latter case, a redistribution reaction has occurred involving the expected monocobalt–indium species probably according to Equation (5). Compound (**110**) was structurally characterized and the Co_2InC unit is almost planar.

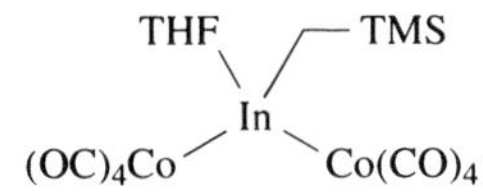

(110)

$$2\,[InCl(CH_2\text{-}TMS)_2] + 2\,Na[Co(CO)_4] \longrightarrow 2\,[In(CH_2\text{-}TMS)_2\{Co(CO)_4\}] + 2\,NaCl \longrightarrow [In(CH_2\text{-}TMS)\{Co(CO)_4\}_2] + [In(CH_2\text{-}TMS)_3] \quad (5)$$

The reaction between $InCl_3$ and one equivalent of $K[Co(CO)_3(PPh_3)]$ affords the complex $[InCl_2(PPh_3)\{Co(CO)_4\}]$ (**111**),[72] which was characterized by x-ray crystallography. In solution the initial product is probably $[InCl_2\{Co(CO)_3(PPh_3)\}]$, from which (**111**) is formed (in lower yield) as a result of phosphine migration from cobalt to indium. $InCl_3$ and two equivalents of $K[Co(CO)_3(PPh_3)]$ affords two complexes. The first was isolated and structurally characterized as $[Co(CO)_3(PPh_3)_2][InCl_2\{Co(CO)_4\}_2]$ (**112**), which again indicates phosphine lability, whereas the second complex was identified in solution by infrared spectroscopy as probably $[InCl\{Co(CO)_3(PPh_3)\}_2]$ (**113**).[72] The anion in (**112**) has also been characterized as its PPN salt ($PPN = N(PPh_3)_2$), $[PPN][InCl_2\{Co(CO)_4\}_2]$.[72] The reaction between $[Co_2(CO)_6(PPh_3)_2]$ and InCl also affords a dicobalt–indium complex $[InCl(PPh_3)\{Co(CO)_3(PPh_3)\}_2]$ (**114**),[72] resulting from a formal insertion reaction of InCl into the Co–Co bond.

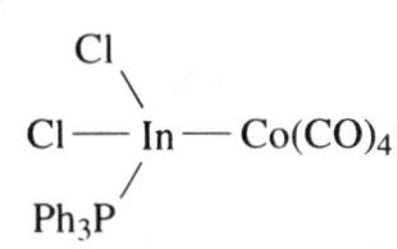

(111)

Indium trichloride and three equivalents of $K[Co(CO)_3(PPh_3)]$ afford the tricobalt complex $[In\{Co(CO)_3(PPh_3)\}_3]$ (**115**).

The complex $[NEt_4][InI\{Co(CO)_4\}_3]$ (**116**) has been isolated from the reaction between $K[Co(CO)_4]$ and $[In\{Co(CO)_4\}_3]$ followed by addition of $[NEt_4]I$.[73] The structure is shown in Figure 11.

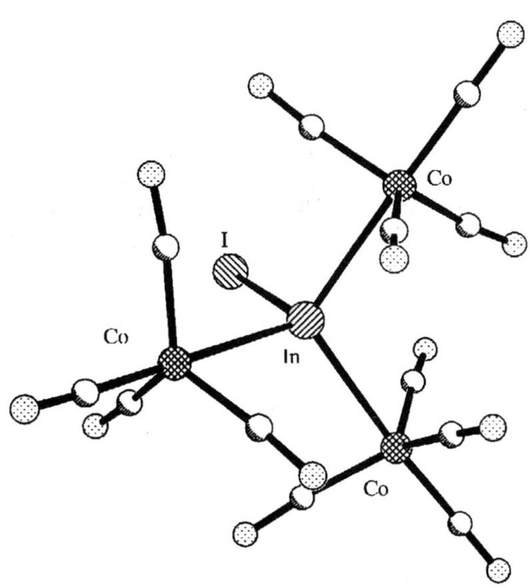

Figure 11 Molecular structure of (**116**).

Little work has been done with rhodium but one example of an iridium–indium bond is known. The reaction between $[IrMe(PMe_3)_4]$ and either $InBrMe_2$ or $InMe_3$ affords the complexes $[IrMe_2(InMeBr)(PMe_3)_3]$ (**117**) and $[IrMe_2(InMe_2)(PMe_3)_3]$ (**118a**), both of which exist as *fac* and *mer* isomers as shown for (**117**).[74]

The reaction between $[IrMe(PMe_3)_4]$ and $InEt_3$ proceeds by a more complicated route illustrated in Scheme 3, but resulting in the complex $[IrH(Et)(InEt_2)(PMe_3)_3]$ (**118b**) which has been characterized crystallographically.

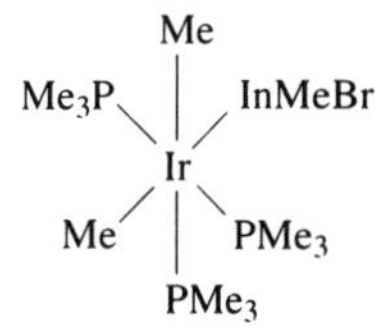

(117a)

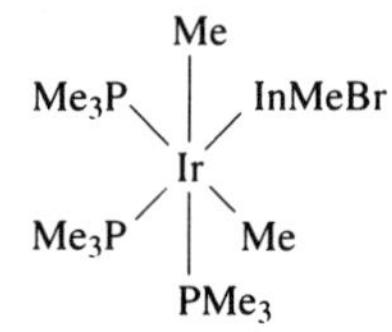

(117b)

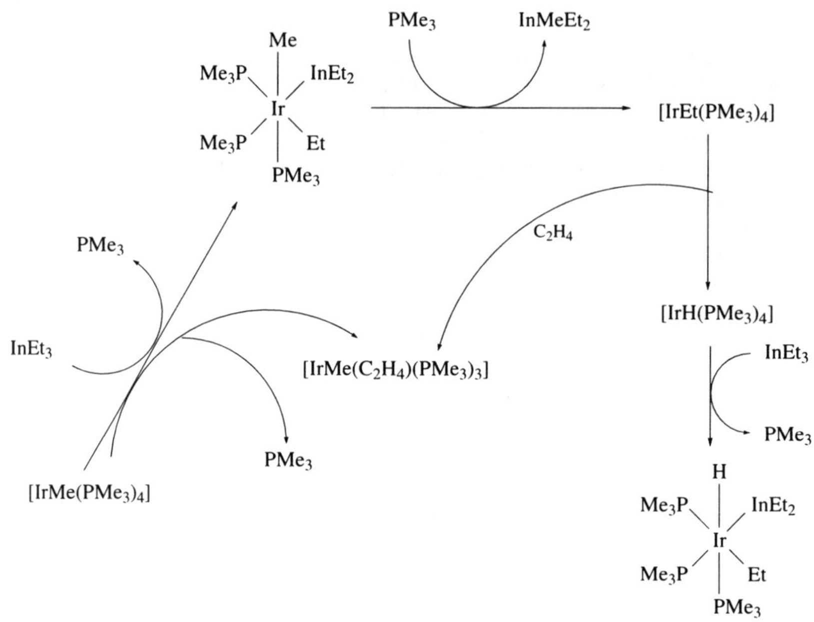

(118)

Scheme 3

12.4.1.5 Group 10 metals

The nickel–indium complex $[In(CH_2CH_2CH_2NMe_2)_2\{Ni(CO)Cp\}]$ (**119**) analogous to (**97**), has been described,[46,50] being prepared by the reaction between K[Ni(CO)Cp] and $In\text{-}Br(CH_2CH_2CH_2NMe_2)_2$.

The reaction between the platinum complex $[PtH(R)\{Cy_2PCH_2CH_2PCy_2\}]$ and InR_3 ($R = CH_2CMe_3$ or CH_2-TMS) affords the platinum–indium species $[PtR(InR_2)\{Cy_2PCH_2CH_2PCy_2\}]$ (**120**) (Cy = cyclohexyl); for $R = CH_2$-TMS the compound has been characterized by x-ray crystallography and the PtP_2In plane is found to be almost perpendicular to the C_2InPt plane.[45]

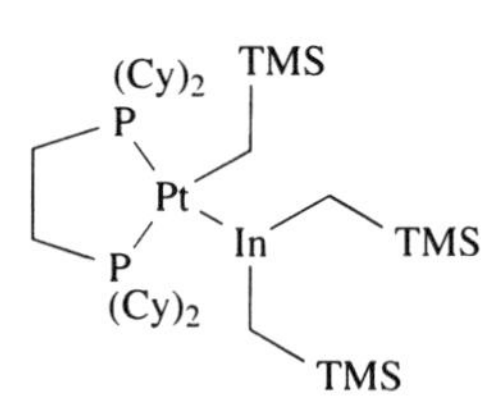

(120)

12.4.1.6 Metalloporphyrin complexes

A number of metalloporphyrin complexes in which a group 13 metal is bonded within the porphyrin ring and also bonded to an organotransition metal fragment are known.

Octaethylporphyrin (H_2oep) and tetraphenylporphyrin (H_2tpp) complexes have been the most widely studied. Examples include (oep)In–ML_n, where $ML_n = Co(CO)_4$,[75–7] $Mn(CO)_5$,[75–7] $Mo(CO)_3Cp$,[75–7] $W(CO)_3Cp$,[75–7] $Cr(CO)_3Cp$,[76,77] $Re(CO)_5$,[78] or $\{(tpp)In\}_2Fe(CO)_4$;[75] (tpp)In–ML_n, where ML_n =

$Mo(CO)_3Cp$,[77,79,80] $W(CO)_3Cp$,[77,79] $Mn(CO)_5$,[77,81] $Co(CO)_4$,[77,81] $Cr(CO)_3Cp$,[77] or $Re(CO)_5$[78] (the aluminum and gallium analogues of the $Re(CO)_5$ complexes have also been prepared[78]); and (oep)Rh–In(oep);[82] for reviews, see refs. 83 and 84. X-ray structures have been described for (oep)In–$Mn(CO)_5$,[76] (oep)In–$Mo(CO)_3Cp$[80] and (oep)Rh–In(oep).[82]

12.5 THALLIUM

12.5.1 Direct Thallium–Metal Bonds

12.5.1.1 Group 6 metals

The thallium complexes $[Tl\{Cr(CO)_3Cp\}_3]$ (**121**) and $[Tl\{Mo(CO)_3Cp\}_3]$ (**122**) have been prepared and structurally characterized, and are isomorphous with (**70**) and one form of (**67**) (the other form of (**67**) which is solvated has the same molecular structure).[54] The structure of (**122**) has been described previously[85] but in this case, the $Mo(CO)_3Cp$ fragments adopt different conformations about the Mo–Tl bonds. Compound (**122**) can be prepared from the reaction between $TlCl_3$ and three equivalents of $K[Mo(CO)_3Cp]$, but is also formed by disproportionation of the thallium(I) compound $Tl[Mo(CO)_3Cp]$ in THF solution. Compound (**121**) is formed in high yield from the reaction between thallium metal and $[Cr_2(CO)_6Cp_2]$.

A range of compounds of the form $[Q][Tl(2,4,6\text{-}F_3C_6H_2)_3(ML_n)]$ (**122**) where Q = NBu^n_4 or PPN and ML_n = $Co(CO)_4$, $Mn(CO)_5$, $Mo(CO)_3Cp$ or $W(CO)_3Cp$, have been prepared from the corresponding ammonium or PPN salts of the ML_n anions and $[Tl(2,4,6\text{-}F_3C_6H_2)_3(diox)]$ or $[Tl(2,4,6\text{-}F_3C_6H_2)_3]$, spectroscopic data for which are presented.[86]

12.5.1.2 Group 7 metals

The reaction between TlCl and $Na[Re(CO)_5]$ affords the unstable thallium(I) compound $Tl[Re(CO)_5]$, which disproportionates into thallium metal and the thallium(III) species $[Tl\{Re(CO)_5\}_3]$ (**123**), which itself undergoes a thermal rearrangement to the complex $[Re_2(CO)_8\{\mu\text{-}TlRe(CO)_5\}_2]$ (**124**), which is analogous to (**40**).[39,40]

12.5.1.3 Group 8 metals

The most extensive series of compounds which fall within this section is a range of thallium–iron carbonylate anions that are mostly prepared from reactions between thallium(I) or thallium(III) salts and methanolic $Fe(CO)_5$–KOH.

The reaction between $K[FeH(CO)_4]$ and a variety of thallium salts followed by the addition of $[NEt_4]Cl$ or $[NEt_4]Br$ affords the hexanuclear complex $[NEt_4]_2[\{Fe(CO)_4\}_2\{\mu\text{-}TlFe(CO)_4\}_2]$ (**125**), which consists of a central Fe_2Tl_2 rhombus with exocyclic $Fe(CO)_4$ groups bonded to each thallium atom.[87] The orange indium compound from which (**99**) was prepared is probably similar.[67]

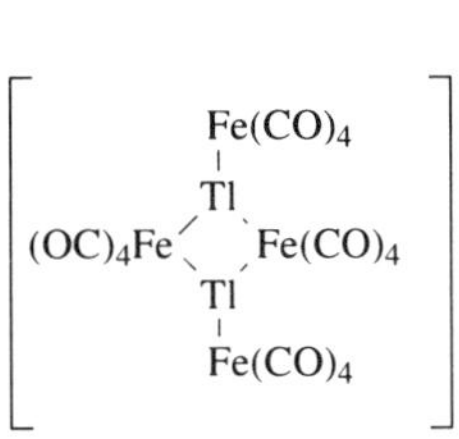

(125)

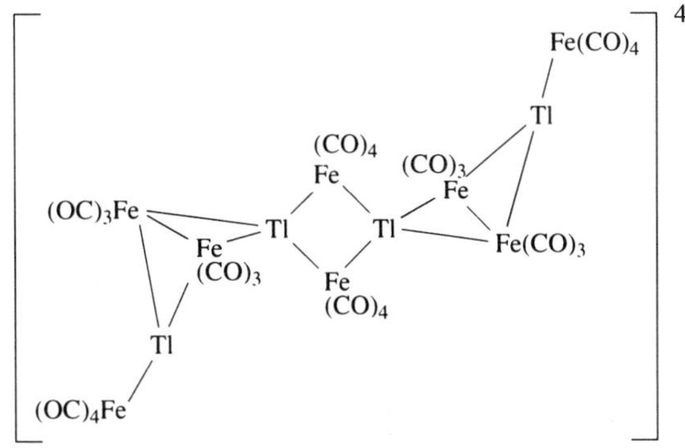

(126)

Solutions of (**125**) tend to lose CO, resulting in the larger clusters $[NEt_4]_4[Tl_4Fe_8(CO)_{30}]$ (**126**) and $[NEt_4]_6[Tl_6Fe_{10}(CO)_{36}]$ (**127**) both of which have been characterized by x-ray crystallography.[87,88]

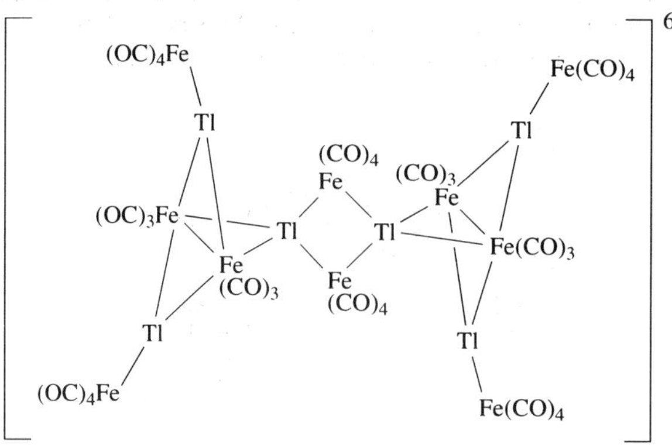

(127)

An additional anionic cluster $[NEt_4]_2[Tl_2Fe_6(CO)_{24}]$ (**128**), isomorphous with (**98**), is formed by oxidation of (**125**)[89] and reactions of (**125**) and various bidentate nitrogen ligands afford the complexes $[Tl(N\text{–}N)\{Fe(CO)_4\}_2]$ (**129**).[90]

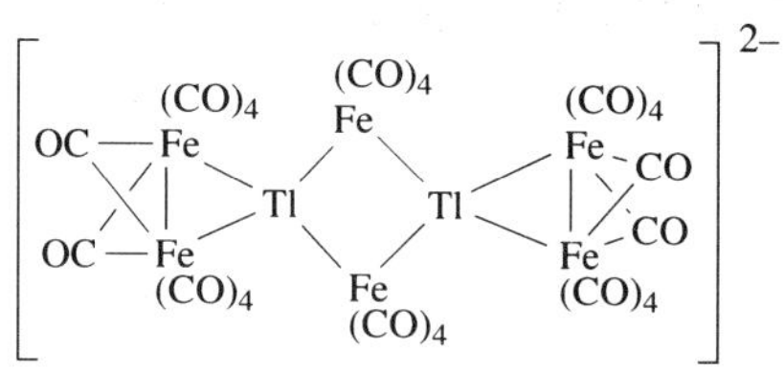

(128)

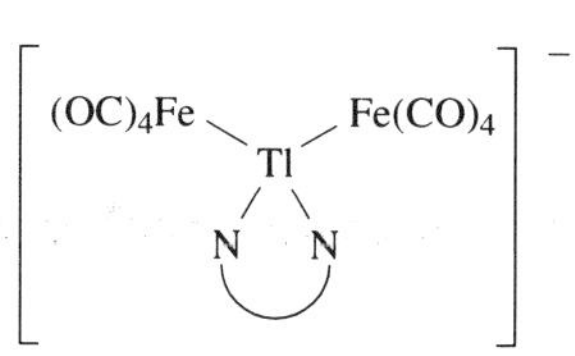

(129)

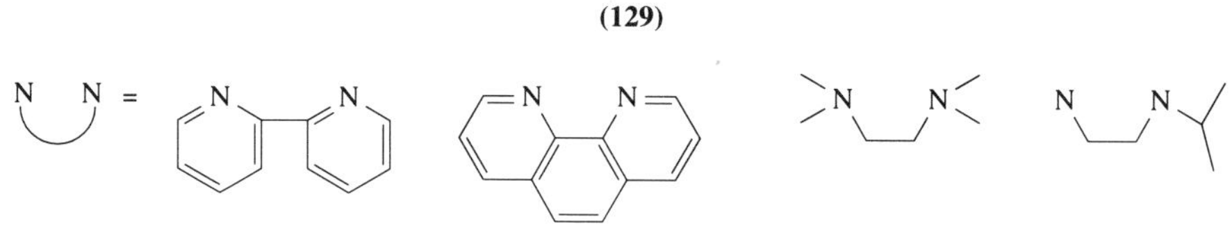

Thallium–ruthenium bonding is found in the cluster $[AsPh_4][Tl\{Ru_6C(CO)_{16}\}_2]$ (**130**),[91] in which a thallium atom is found in a slightly twisted square planar environment between edges of two $Ru_6C(CO)_{16}$ octahedra.

The thallium(I) complex $Tl[Fe(CO)_3(NO)]$ (**131**) has also been described and structurally characterized but does not contain a Tl–Fe bond; the structure is ionic with separated Tl^+ and $[Fe(CO)_3(NO)]^-$ ions[92] and is very similar to that of the isoelectronic $Tl[Co(CO)_4]$.[1] A range of thallium(I) transition metal carbonylate compounds of the form $Tl[ML_n]$ have been described and these are listed in ref. 1. The only structural data available are for the two complexes mentioned above, which show no Tl–M bonding, but some solution infrared studies have provided results consistent with the presence of such bonds in solution in some cases. Solution and solid-state extended x-ray absorption fine structure (EXAFS) spectra on $Tl[Co(CO)_4]$ and $Tl[Mo(CO)_3Cp]$, however, gave no evidence of such bonds.[93]

12.5.1.4 Group 9 metals

The crystal structure of the cationic complex $[Ir_2Cl_2(CO)_2\{\mu\text{-}P(Ph)_2CH_2As(Ph)CH_2P(Ph)_2\}_2(\mu\text{-}Tl)]^+$ (**132**)[94] has been described in which a Tl^+ cation bridges between two iridium centres and this constitutes a genuine example of a transition metal to thallium(I) covalent bond. This interpretation (as opposed to an ionic formulation) is supported by solution spectroscopic studies.

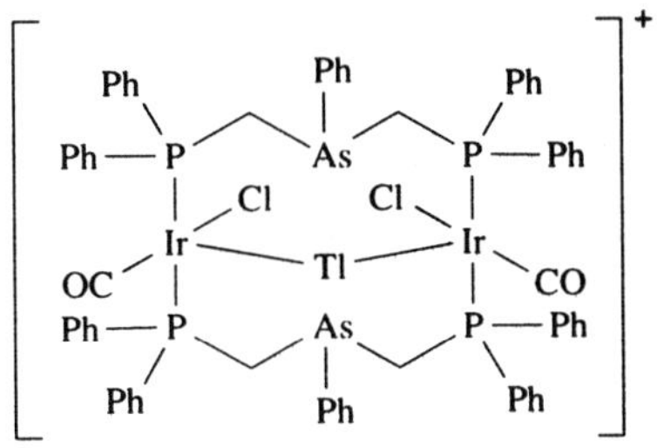

(132)

12.5.1.5 Group 10 metals

The reaction between $[Rh_2Cl_2(\eta\text{-}C_8H_{12})_2]$ and $[Pt_3(CO)_3(PCy_3)_3]$ in the presence of $TlPF_6$ led to the isolation of the tetrahedral cluster $[TlPt_3(CO)_3(PCy_3)_3][RhCl_2(\eta\text{-}C_8H_{12})]$ (**133**), which was characterized by x-ray crystallography.[95] One way of viewing the cation $[TlPt_3(CO)_3(PCy_3)_3]^+$ is as a Tl^{3+} cation closely associated with a $[Pt_3(CO)_3(PCy_3)_3]^{2-}$ dianion.

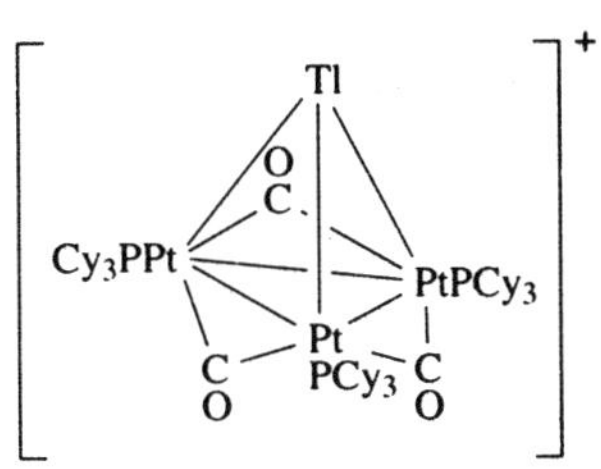

(133)

The structure of $[Tl_2Pt(CN)_4)]$ (**134**) has been described[96] in which Tl^I–metal bonds (platinum in this case) are also present.

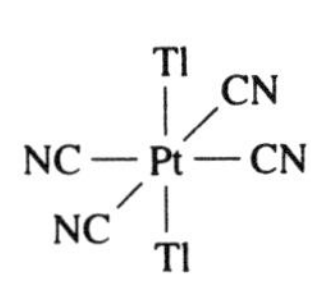

(134)

12.5.1.6 Group 11 metals

A gold–thallium(I) interaction is present in $[\{AuTl(Ph_2PCH_2S)\}_2]$ (**135**).[97]

12.5.1.7 Metalloporphyrin complexes

Octaethylporphyrin (oep) and tetraphenylporphyrin (tpp) complexes have also been reported for thallium. Examples include (oep)Tl–ML_n, where ML_n = $Mo(CO)_3Cp$,[77,98,99] $Cr(CO)_3Cp$,[77,99] $W(CO)_3Cp$,[77,99] $Mn(CO)_5$,[77,99] $Co(CO)_4$,[77,99] or $Re(CO)_5$,[78] and (tpp)In–ML_n, where ML_n = $Mo(CO)_3Cp$,[77] $Cr(CO)_3Cp$,[77] $W(CO)_3Cp$,[77] $Mn(CO)_5$,[77] $Co(CO)_4$,[77] or $Re(CO)_5$.[78]

X-ray structures have been described for (oep)Tl–$Mn(CO)_5$[99] and (oep)Tl–$Mo(CO)_3Cp$.[98]

12.6 REFERENCES

1. S. K. Boocock and S. G. Shore, in 'COMC-I', vol. 6, p. 947.
2. N. A. Compton, R. J. Errington and N. C. Norman, *Adv. Organomet. Chem.*, 1990, **31**, 91.
3. K. H. Whitmire, *J. Coord. Chem., Sect. B*, 1988, **17**, 95.
4. A. R. Barron and G. Wilkinson, *Polyhedron*, 1986, **5**, 1897.
5. B. M. Bulychev, *Polyhedron*, 1990, **9**, 387.
6. T. P. Fehlner (ed.), 'Inorganometallic Chemistry', Plenum, New York, 1992.
7. B. M. Bulychev, E. V. Evdokimova, A. I. Sizov and G. L. Soloveichik, *J. Organomet. Chem.*, 1982, **239**, 313.
8. E. B. Lobkovskii, G. L. Soloveichik, A. I. Sizov, B. M. Bulychev, A. I. Gusev and N. I. Kirillova, *J. Organomet. Chem.*, 1984, **265**, 167.
9. E. B. Lobkovskii, G. L. Soloveichik, B. M. Bulychev, R. G. Gerr and Y. T. Struchkov, *J. Organomet. Chem.*, 1984, **270**, 45.
10. E. B. Lobkovskii, G. L. Soloveichik, A. I. Sizov and B. M. Bulychev, *J. Organomet. Chem.*, 1985, **280**, 53.
11. V. K. Bel'skii, A. I. Sizov, B. M. Bulychev and G. L. Soloveichik, *J. Organomet. Chem.*, 1985, **280**, 67.
12. T. Yu. Sokolova, A. I. Sizov, B. M. Bulychev, E. A. Rozova, V. K. Bel'skii and G. L. Soloveichik, *J. Organomet. Chem.*, 1990, **288**, 11.
13. V. K. Bel'skii, A. I. Sizov and G. L. Soloveichik, *Koord. Khim.*, 1985, **11**, 1003.
14. A. I. Sizov *et al.*, *J. Organomet. Chem.*, 1987, **335**, 323.
15. A. I. Sizov, I. V. Molodnitskaya, B. M. Bulychev, V. K. Bel'skii and G. L. Soloveichik, *J. Organomet. Chem.*, 1988, **344**, 185.
16. A. I. Sizov, I. V. Molodnitskaya, B. M. Bulychev, E. V. Evdokimova, V. K. Bel'skii and G. L. Soloveichik, *J. Organomet. Chem.*, 1988, **344**, 293.
17. E. B. Lobkovskii, A. I. Sizov, B. M. Bulychev, I. V. Sokolova and G. L. Soloveichik, *J. Organomet. Chem.*, 1987, **319**, 69.
18. K. Mach, H. Antropiusova, V. Varga and V. Hanus, *J. Organomet. Chem.*, 1988, **358**, 123.
19. J. Kopf, H.-J. Vollmer and W. Kaminsky, *Cryst. Struct. Commun.*, 1980, **9**, 985.
20. A. R. Barron, M. B. Hursthouse, M. Motevalli and G. Wilkinson, *J. Chem. Soc., Chem. Commun.*, 1985, 664.
21. A. R. Barron, D. Lyons, G. Wilkinson, M. Motevalli, A. J. Howes and M. B. Hursthouse, *J. Chem. Soc., Dalton Trans.*, 1986, 279.
22. A. R. Barron, J. E. Salt and G. Wilkinson, *J. Chem. Soc., Dalton Trans.*, 1986, 1329.
23. A. R. Barron, G. Wilkinson, M. Motevalli and M. B. Hursthouse, *J. Chem. Soc., Dalton Trans.*, 1987, 837.
24. J. W. Bruno, J. C. Huffman and K. G. Caulton, *J. Am. Chem. Soc.*, 1984, **106**, 444.
25. G. S. Girolami, G. Wilkinson, M. Thornton-Pett and M. B. Hursthouse, *J. Am. Chem. Soc.*, 1983, **105**, 6752; G. S. Girolami *et al.*, *J. Chem. Soc., Dalton Trans.*, 1985, 921.
26. W. A. Skupinski, J. C. Huffman, J. W. Bruno and K. G. Caulton, *J. Am. Chem. Soc.*, 1984, **106**, 8128.
27. A. R. Barron and G. Wilkinson, *J. Chem. Soc., Dalton Trans.*, 1986, 287.
28. D. H. Lee, H. Suzuki and Y. Moro-oka, *J. Organomet. Chem.*, 1987, **330**, C20.
29. K. R. Pörschke and G. Wilke, *Chem. Ber.*, 1985, **118**, 313.
30. W. Kleimann, K. R. Pörschke and G. Wilke, *Chem. Ber.*, 1985, **118**, 323.
31. K. R. Pörschke and G. Wilke, *J. Organomet. Chem.*, 1988, **358**, 519; G. Wilke, *Angew. Chem. Int. Ed. Engl.*, 1988, **27**, 185.
32. J. J. Eisch, S. R. Sexsmith and K. C. Fichter, *J. Organomet. Chem.*, 1990, **382**, 273.
33. V. K. Bel'skii, A. B. Erofeev, B. M. Bulychev and G. L. Soloveichik, *J. Organomet. Chem.*, 1984, **265**, 123; V. K. Bel'skii, B. M. Bulychev, A. B. Erofeev and G. L. Soloveichik, *ibid.*, 1984, **268**, 107; S. Y. Knjazhanski, B. M. Bulychev, V. K. Bel'skii and G. L. Soloveichik, *ibid.*, 1987, **327**, 173.
34. Yu. K. Gun'ko, B. M. Bulychev, A. I. Sizov, V. K. Bel'skii and G. L. Soloveichik, *J. Organomet. Chem.*, 1990, **390**, 153.
35. A. Keller, *J. Organomet. Chem.*, 1990, **385**, 285.
36. J. M. Mayer and J. C. Calabrese, *Organometallics*, 1984, **3**, 1292.
37. M. L. H. Green, P. Mountford, G. J. Smout and S. R. Speel, *Polyhedron*, 1990, **9**, 2763.
38. R. M. Campbell, L. M. Clarkson, W. Clegg, D. C. R. Hockless, N. L. Pickett and N. C. Norman, *Chem. Ber.*, 1992, **125**, 55.
39. H.-J. Haupt, F. Neumann and B. Schwab, *Z. Anorg. Allg. Chem.*, 1982, **485**, 234.
40. H.-J. Haupt, U. Flörke and H. Preut, *Acta Crystallogr. Sect. C*, 1986, **42**, 665.
41. U. Flörke, P. Balsaa and H.-J. Haupt, *Acta Crystallogr., Sect. C*, 1986, **42**, 275.
42. H.-J. Haupt, P. Balsaa and U. Flörke, *Z. Anorg. Allg. Chem.*, 1988, **557**, 69.
43. H.-J. Haupt, P. Balsaa and B. Schwab, *Z. Anorg. Allg. Chem.*, 1985, **521**, 15.
44. J. C. Vanderhooft, R. D. Ernst, F. W. Cagle, R. J. Neustadt and T. H. Cymbaluk, *Inorg. Chem.*, 1982, **21**, 1876.
45. R. A. Fischer and J. Behm, *J. Organomet. Chem.*, 1991, **413**, C10.
46. R. A. Fischer, J. Behm, Th. Priermeier and W. Scherer, *Angew. Chem., Int. Ed. Engl.*, 1993, **32**, 746.
47. R. A. Fischer and J. Behm, *Chem. Ber.*, 1992, **125**, 37.
48. R. A. Fischer, J. Behm and Th. Priermeier, *J. Organomet. Chem.*, 1992, **429**, 275.
49. Y.-J. Chen, H. D. Kaesz, Y. K. Kim, H.-J. Müller, R. S. Williams and Z. Xue, *Appl. Phys. Lett.*, 1989, **55**, 2760; H. D. Kaesz *et al.*, *New J. Chem.*, 1990, **14**, 527; F. Maury, L. Brandt and H. D. Kaesz, *J. Organomet. Chem.*, 1993, **449**, 159; R. Fischer, W. Scherer and M. Kleine, *Angew. Chem., Int. Ed. Engl.*, 1993, **32**, 748.
50. R. A. Fischer, J. Behm, E. Herdtweck and C. Kronseder, *J. Organomet. Chem.*, 1992, **437**, C29.
51. R. A. Fischer, H. D. Kaesz, S. I. Khan and H.-J. Müller, *Inorg. Chem.*, 1990, **29**, 1601.
52. L. M. Clarkson, W. Clegg, N. C. Norman, A. J. Tucker and P. M. Webster, *Inorg. Chem.*, 1988, **27**, 2653.
53. B. Nuber, W. Schatz and M. L. Ziegler, *Z. Naturforsch., Teil B*, 1990, **45**, 508.
54. L. M. Clarkson, W. Clegg, D. C. R. Hockless, N. C. Norman and T. B. Marder, *J. Chem. Soc., Dalton Trans.*, 1991, 2229.
55. L. M. Clarkson *et al.*, *J. Chem. Soc., Dalton Trans.*, 1991, 2241.
56. K. Merzweiler, F. Rudolph and L. Brands, *Z. Naturforsch., Teil B*, 1992, **47**, 470.
57. D. L. Thorn, *J. Organomet. Chem.*, 1991, **405**, 161.
58. H. Behrens, M. Moll, E. Sixtus and G. Thiele, *Z. Naturforsch., Teil B*, 1977, **32**, 1109.
59. M. Veith and K. Kunze, *Angew. Chem., Int. Ed. Engl.*, 1991, **30**, 95.
60. H.-J. Haupt and U. Flörke, *Acta Crystallogr., Sect. C*, 1989, **45**, 1718.

61. H.-J. Haupt, M. Woyciechowski and U. Flörke, *Z. Anorg. Allg. Chem.*, 1991, **592**, 153.
62. W. Schatz, H.-P. Neumann, B. Nuber, B. Kanellakopulos and M. L. Ziegler, *Chem. Ber.*, 1991, **124**, 453.
63. M. Schollenberger, B. Nuber and M. L. Ziegler, *Angew. Chem., Int. Ed. Engl.*, 1992, **31**, 350.
64. N. C. Norman and P. M. Webster, *Z. Naturforsch., Teil B*, 1989, **44**, 91.
65. L. M. Clarkson, N. C. Norman and L. J. Farrugia, *Organometallics*, 1991, **10**, 1286.
66. J. C. Calabrese, L. M. Clarkson, T. B. Marder, N. C. Norman and N. J. Taylor, *J. Chem. Soc., Dalton Trans.*, 1992, 3525.
67. J. M. Cassidy and K. H. Whitmire, *Acta Crystallogr., Sect. C*, 1990, **46**, 1781.
68. C.-C. Lin, G. Kong, H. Cho and B. R. Whittlesey, *Inorg. Chem.*, 1993, **32**, 2705.
69. V. G. Albano, M. Cané, M. C. Iapalucci, G. Longoni and M. Monari, *J. Organomet. Chem.*, 1991, **407**, C9.
70. J. L. Atwood *et al.*, *J. Chem. Soc., Dalton Trans.*, 1987, 747.
71. F. Balegroune *et al.*, *J. Cluster Sci.*, 1992, **3**, 275.
72. L. M. Clarkson, K. McCrudden, N. C. Norman and L. J. Farrugia, *Polyhedron*, 1990, **9**, 2533.
73. L. M. Clarkson, L. J. Farrugia and N. C. Norman, *Acta Crystallogr., Sect. C*, 1991, **47**, 2525.
74. D. L. Thorn and R. L. Harlow, *J. Am. Chem. Soc.*, 1989, **111**, 2575.
75. P. Cocolios, C. Moise and R. Guilard, *J. Organomet. Chem.*, 1982, **228**, C43.
76. R. Guilard *et al.*, *Inorg. Chem.*, 1987, **26**, 2467.
77. R. Guilard, A. Tabard, A. Zrineh and M. Ferhat, *J. Organomet. Chem.*, 1990, **389**, 315.
78. R. Guilard *et al.*, *J. Organomet. Chem.*, 1991, **401**, 227.
79. P. Cocolios, D. Chang, O. Vittori, R. Guilard, C. Moise and K. M. Kadish, *J. Am. Chem. Soc.*, 1984, **106**, 5724.
80. C. Lecomte, A. Habbou, P. Mitaine, P. Richard and R. Guilard, *Acta Crystallogr., Sect. C*, 1989, **45**, 1226.
81. S. Onaka *et al.*, *Inorg. Chem.*, 1985, **24**, 1070.
82. N. L. Jones, P. J. Carroll and B. B. Wayland, *Organometallics*, 1986, **5**, 33.
83. R. Guilard and K. M. Kadish, *Comments Inorg. Chem.*, 1988, **7**, 287.
84. R. Guilard, C. Lecomte and K. M. Kadish, *Struct. Bonding*, 1987, **64**, 205.
85. J. Rajaram and J. A. Ibers, *Inorg. Chem.*, 1973, **12**, 1313.
86. R. Uson, A. Laguna, J. A. Abad and E. de Jesus, *J. Chem. Soc., Dalton Trans.*, 1983, 1127.
87. K. H. Whitmire, J. M. Cassidy, A. L. Rheingold and R. R. Ryan, *Inorg. Chem.*, 1988, **27**, 1347.
88. K. H. Whitmire, R. R. Ryan, H. J. Wasserman, T. A. Albright and S.-K. Kang, *J. Am. Chem. Soc.*, 1986, **108**, 6831.
89. J. M. Cassidy and K. H. Whitmire, *Inorg. Chem.*, 1989, **28**, 1432.
90. J. M. Cassidy and K. H. Whitmire, *Inorg. Chem.*, 1989, **28**, 1435.
91. G. B. Ansell, M. A. Modrick and J. S. Bradley, *Acta Crystallogr., Sect. C*, 1984, **40**, 1315.
92. L. M. Clarkson, W. Clegg, D. C. R. Hockless and N. C. Norman, *Acta Crystallogr., Sect. C*, 1992, **48**, 236.
93. L. M. Clarkson, Ph.D. Thesis, University of Newcastle, 1991.
94. A. L. Balch, J. K. Nagle, M. M. Olmstead and P. E. Reedy, Jr., *J. Am. Chem. Soc.*, 1987, **109**, 4123.
95. O. J. Ezomo, D. M. P. Mingos and I. D. Williams, *J. Chem. Soc., Chem. Commun.*, 1987, 924.
96. J. K. Nagle, A. L. Balch and M. M. Olmstead, *J. Am. Chem. Soc.*, 1988, **110**, 319.
97. S. Wang, J. P. Fackler, Jr., C. King and J. C. Wang, *J. Am. Chem. Soc.*, 1988, **110**, 3308.
98. P. Richard, A. Zrineh, R. Guilard, A. Habbou and C. Lecomte, *Acta Crystallogr., Sect. C*, 1989, **45**, 1224.
99. R. Guilard *et al.*, *Inorg. Chem.*, 1988, **27**, 697.

Author Index

This Author Index comprises an alphabetical listing of the names of the authors cited in the text and the references listed at the end of each chapter in this volume.

Each entry consists of the author's name, followed by a list of numbers, for example

Templeton, J. L., 366, 385[233] (350, 366), 387[370] ((363))

For each name, the page numbers for the citation in the reference list are given, followed by the reference number in superscript and the page number(s) in parentheses of where that reference is cited in the text. Where a name is referred to in text only, the page number of the citation appears with no superscript number. References cited both in the text and in the tables are included.

Although much effort has gone into eliminating inaccuracies resulting from the use of different combinations of initials by the same author, the use by some journals of only one initial, and different spellings of the same name as a result of the transliteration processes, the accuracy of some entries may have been affected by these factors.

Aaresild, E., 121[40] (60)
Abad, J. A., 541[53] (512), 541[83] (514), 571[86] (567)
Abbasov, Ya. A., 501[287] (481)
Abd El-Al, M. Y., 192[454] (173)
Abdou, G., 248[100] (220, 232, 233), 250[214] (223), 252[315] (229)
Abdou, Z. J., 248[99] (220, 232, 233), 248[100] (220, 232, 233), 250[212] (223), 250[214] (223), 252[315] (229), 252[342] (232, 233)
Abicht, H.-P., 122[128] (70, 71, 76), 541[49] (512)
Abramenko, P. I., 194[539] (179)
Abramov, A. F., 193[464] (174)
Abu-Orabi, S. T., 188[121] (140), 188[122] (140), 189[221] (152)
Achrijew, A. S., 255[493] (245)
Adam, M. A., 502[315] (490)
Adolph, H. G., 255[486] (244)
Afonin, M. A., 428[168] (400, 414), 430[292] (414)
Aftandilian, V. D., 272[59] (263)
Agafonov, N. E., 124[207] (83)
Agami, C., 32[104f] (13)
Ahlers, A., 498[69] (440)
Ahlrichs, R., 498[33] (436, 445), 499[107] (444)
Ahmad, R., 369[183] (310, 316, 327, 337, 349), 370[257] (337)
Aida, T., 500[216] (468)
Ainslie, R. D., 122[113] (69, 86, 101)
Aitchison, K. A., 542[126] (520)
Akhmedov, V. M., 501[287] (481)
Akhmetov, L. I., 501[240] (471)
Akhtar, J., 55[36] (48)
Akkerman, O. S., 120[2] (58, 72, 74, 87–9), 120[16] (59, 71), 121[60] (63), 121[61] (63, 74, 77, 90), 121[78] (65, 74, 77, 90), 122[137a] (71, 83, 84), 122[137b] (71, 83), 122[138a] (71), 122[138b] (71), 122[140] (71, 85), 122[141a] (71, 83), 122[141b] (71, 83), 122[142] (71), 122[143a] (71, 85, 92), 122[143b] (71, 76, 85, 92), 123[146] (71, 76, 83, 85, 96), 123[156] (73), 124[238] (85), 124[239] (85), 125[257] (88, 89, 93), 125[262] (90, 96), 125[270] (94), 125[278] (96, 102)
Alam, F., 191[347] (164), 192[394] (167), 251[295] (228), 254[443] (242), 255[506] (245)
Alami, M., 124[219] (83)
Al-Baker, S., 253[407] (240)
Albano, V. G., 571[69] (563, 564)
Alberola, A., 498[42] (436, 443, 468), 501[227] (468), 501[260] (475), 502[323] (491), 502[335] (493)
Albinati, A., 367[65] (284, 285)
Albrecht, M., 540[37] (509)
Albright, T. A., 571[88] (568)
Alcock, N. W., 252[320] (230), 255[513] (245), 369[198] (313, 315), 369[199] (313, 315), 370[233] (326), 426[20] (378, 379), 426[22] (378), 426[23] (378), 430[285] (400, 418), 542[106] (518–20), 543[188] (526)
Alderton, M. J., 248[104] (220)
Alekseenko, L. I., 193[475] (174), 193[482] (175)
Alekseeva, V. P., 254[430] (242)
Alexakis, A., 122[99] (67, 68, 102), 501[235] (469)
Alexander, J. J., 249[137] (221, 225, 228)
Alexandratos, S., 34[219b] (30)
Al-Fekri, D. M., 122[124] (70)
Alfonso, C., 248[88] (219, 220, 233), 252[368] (236)
Ali, M. M., 192[454] (173)
Aliev, G. M., 255[489] (244)
Aliguliev, R. M., 255[489] (244)
Al-Juaid, S. S., 123[151] (72, 85, 87, 88, 92), 543[190] (526)
Allaoud, S., 191[356] (164), 191[357] (164)
Allcock, H. R., 124[224] (83), 248[56] (218), 251[290] (227, 230, 241), 253[412] (241), 412, 428[194] (400, 412)
Allen, L. C., 248[66] (218)
Allen, P. E. M., 123[165] (76)
Allwohn, J., 187[67] (134), 187[77] (135), 189[199] (149), 189[205] (150)
Almenningen, A., 499[120] (448)
Almlöf, J., 125[271] (94, 101)
Almond, M. J., 542[139] (521, 526)
Alonso, T., 125[294] (98–100)
Alsdorf, H., 32[84f] (11), 32[85c] (11), 32[85d] (11)
Al-Taweel, S. M., 215[73] (210)
Alvanipour, A., 542[137] (521)
Alvanipour, J. W., 541[47] (511, 520)
Al-Wassil, A.-A. I., 499[154] (454)
Amamchyan, A. R., 191[389] (167)
Amano, T., 122[115] (69)
Amberger, E., 541[95] (515)
Ames, A., 502[336] (493)
Amirkhalili, S., 190[305] (161), 215[86] (212)
Amma, E. L., 499[118] (447), 540[28] (506), 544[237] (534)
Amstutz, R., 32[87] (12), 32[91] (12), 32[96] (12), 32[114] (15), 32[115] (15)
Anand, N., 122[131] (70)
Anbei, J., 193[515] (177)
Ander, I., 191[339] (163)
Andersen, E. L., 369[205] (313, 316)
Andersen, R. A., 55[34] (46), 55[40] (48), 55[48] (52), 124[247] (86, 94, 101), 125[266] (93, 112, 114), 125[269] (94, 112, 113), 126[368] (113), 126[369] (113, 114)
Anderson, G. K, 543[217] (531)
Anderson, J. A., 250[219] (224)
Anderson, O. P., 544[248] (536)
Anderson, R. A., 34[205] (26)
Ando, M., 501[268] (476)
Andrä, V. K., 195[645] (185)
Andreeva, N. I., 500[196] (464), 502[346] (496), 502[347] (496)
Andres, C., 502[323] (491)
Andrews, L., 498[85] (441), 498[86] (441)
Andrews, P. C., 31[56] (10)
Angermund, K., 125[260] (90, 93),

125[300] (99)
Anikina, E. V., 124[223] (83)
Anisuzzaman, A. K. M., 251[295] (228), 254[443] (242), 255[506] (245)
Annan, T. A., 541[73] (514), 543[168] (525, 528)
Ansell, G. B., 571[91] (568)
Ansorge, A., 191[360] (165), 191[381] (166)
Antberg, M., 498[70] (440)
Anthonsen, T., 124[204] (82)
Antipin, M. Y., 367[20] (277, 286, 287)
Anton, K., 190[314] (161), 190[316] (161)
Anton, U., 124[211b] (83)
Antonov, A. A., 501[236] (470)
Antonovich, V. A., 249[147] (221, 233), 249[161] (221, 242), 249[169] (221, 239), 251[259] (226, 231), 252[327] (231), 253[390] (238), 253[391] (239), 427[110] (396), 428[142] (400), 428[145] (400, 416), 430[280] (400, 416), 499[156] (455)
Antropiusova, H., 570[18] (548)
Aoki, Y., 124[219] (83)
Aono, K., 251[292] (228, 231, 240, 242), 252[324] (230, 240), 253[406] (240)
Apblett, A. W., 499[128] (450), 541[61] (512)
Appenrodt, J., 30[1a] (1)
Aradi, A. A., 368[110] (290, 291), 502[302] (484)
Arafat, A., 272[35] (260), 272[37] (260), 272[83] (265, 266)
Araki, T., 499[139] (452)
Arbuzov, B. A., 191[335] (163), 192[428] (171), 195[617] (184), 195[618] (184), 195[619] (184), 195[620] (184), 195[621] (184), 195[623] (184), 195[624] (184), 195[626] (184), 195[627] (184), 195[628] (184), 195[629] (184)
Arest-Yakubovich, A. A., 126[352] (109, 119), 126[357a] (111, 119), 126[357b] (111), 126[358] (111), 126[359] (111)
Arias, J., 248[87] (219), 252[360b] (235), 252[368] (236)
Arif, A. M., 32[78] (11), 192[408] (169), 542[116] (520), 542[127] (520), 543[191] (526)
Arkhireeva, T. M., 55[49] (52)
Armstrong, D. R., 30[14] (2), 31[30a] (6, 8), 31[30b] (6, 8), 31[37] (6), 31[38a] (6), 31[67] (10), 32[113] (15), 32[122] (15), 33[127] (15), 33[129] (15), 33[131] (17), 543[218] (531)
Arnautov, S. A., 254[466] (244)
Arne, K. H., 192[445] (173, 175)
Arnold, J., 32[110] (15), 32[116] (15), 428[192] (400, 412)
Aroney, M. J., 54[27] (45), 54[28] (45), 55[33] (46)
Arshinova, R. P., 195[617] (184)
Arunchaiya, M., 255[512] (245)
Ashby, E. C., 121[45] (61), 122[113] (69, 86, 101), 122[124] (70), 123[186a] (80, 81), 123[186b] (80, 81), 123[192a] (80, 81), 125[305] (101), 125[307] (101), 435, 436, 498[22] (435), 498[23] (435, 436, 496), 498[43] (436, 443), 501[243] (472), 501[255] (474)
Ashe, A. J., III, 189[221] (152), 189[233] (153), 189[235] (154, 155), 214[4] (198, 210), 215[73] (210), 215[74] (210), 215[75] (210), 215[76] (210), 215[77] (210)
Ashworth, P. A., 124[217] (83)
Aslam, M., 32[112] (15)
Aslanov, L. A., 249[158] (221)
Aso, Y., 123[178] (78), 502[341] (494)
Astheimer, R. J., 250[219] (224), 250[232] (225, 229, 231), 251[247] (225, 237, 238, 246)
Atoji, M., 33[179] (20), 33[180b] (20)
Attfield, M. J., 430[278] (400, 411)
Attwood, M. D., 426[43] (381, 382), 426[44] (381, 382, 387), 426[73] (382, 387), 427[75] (382), 427[79] (382)
Atwood, D. A., 540[19] (505, 506), 540[21] (505, 506), 540[33] (507, 538), 541[46] (511), 541[61] (512), 541[67] (513, 524), 541[86] (514), 542[125] (520), 542[141] (521, 522, 526), 542[153] (522, 524), 543[182b] (527)
Atwood, J. L., 31[29] (6, 8), 31[31] (6), 31[35] (6), 31[43] (8), 31[51] (9), 32[90] (12), 54[24] (42, 43), 367[32] (278, 286, 287), 497[11] (433, 434, 454, 497), 497[15] (434, 435, 449, 472), 499[147] (454), 501[288] (482), 502[297] (482), 540[13] (505, 506), 540[17] (505, 506), 540[19] (505, 506), 540[21] (505, 506), 541[38] (510), 541[52] (512, 522), 541[57] (512), 541[61] (512), 541[67] (513, 524), 541[86] (514), 541[89] (515), 542[119] (520), 542[134] (520), 542[149] (522, 523), 542[154] (522, 524), 543[176] (525), 543[203] (529), 543[216] (531), 571[70] (564)
Aubuchon, S. R., 540[12] (504, 506, 511, 521), 542[130] (520)
Augart, N., 190[306] (161), 190[307] (161), 215[82] (212)
Aultman, J. F., 120[11] (58)
Avanipour, A., 542[121] (520, 521)
Avent, A. G., 541[97] (515, 526)
Avramenko, G. V., 190[323] (162), 190[324] (162), 190[325] (162), 190[326] (162)
Aznar, F., 500[164] (457), 500[165] (457)
Azoulay, R., 55[59] (54)

Baba, Y., 501[242] (472)
Babchinitser, T. M., 254[464] (244)
Babon, J. A., 188[132] (141), 188[133] (141, 142)
Babudri, F., 124[215] (83)
Bacherikov, V. A., 194[573] (181), 194[574] (181), 195[646] (185), 195[647] (185)
Badaev, F. Z., 121[30] (59, 62), 121[56] (62)
Bader, R. F. W., 248[75] (219)
Baenziger, N. C., 428[191] (400, 416)
Baer, J., 272[35] (260), 272[37] (260)
Baghurst, D. R., 430[284] (400, 414)
Bahl, L., 194[559] (180)
Bähr, G., 499[149] (454)
Bai, J., 190[274] (158), 190[275] (158)
Baidakova, Z. M., 126[358] (111), 126[359] (111)
Baiko, G. N., 368[79] (286, 287)
Baillargeon, M., 187[63] (133, 148)
Bajgur, C. S., 501[275] (479)
Baker, E. B., 34[218a] (30)
Baker, K. V., 120[18] (59, 77)
Baker, M. V., 367[61] (284, 285), 367[62] (284, 285)
Baker, R. T., 367[26] (277), 367[27] (277), 368[120] (292, 293, 295), 368[121] (292, 293, 295), 368[124] (293, 295), 427[120] (398), 429[208] (400, 415), 429[219] (400, 415)
Bakhmutov, V. I., 251[283] (227)
Balch, A. L., 571[94] (569), 571[96] (569)
Balegroune, F., 571[71] (564)
Balema, V. P., 249[171] (221, 240)
Balsaa, P., 570[41] (554), 570[42] (554), 570[43] (554)
Balueva, A. S., 191[335] (163), 192[425] (170), 194[543] (179), 195[626] (184), 195[629] (184)
Bal'yan, Kh. V., 122[111] (68)
Balykova, T. N., 248[53] (218)
Bandman, M. A., 271[16] (258), 543[226] (533), 544[228] (533)
Bandyopadhyay, A. K., 367[24] (277, 294, 297, 304), 369[162] (296)
Banewicz, J. J., 544[233] (534)
Banister, A. J., 32[111] (15), 32[113] (15)
Banks, M. A., Jr., 251[289] (227), 540[20] (505, 506, 509), 540[35] (508), 543[183] (525, 526)
Banks, R. H., 368[94] (287, 290, 291), 368[95] (287, 290), 368[105] (291)
Bansemer, R. L., 367[49] (280, 282)
Banuelos, T., 248[88] (219, 220, 233), 248[101] (220, 233), 250[216] (224, 232, 233), 252[342] (232, 233)
Barabonin, S. Yu., 189[255] (156)
Baranin, S. V., 188[145] (143), 188[146] (143), 188[147] (143, 154)
Barber, M., 499[121] (448), 543[187] (526)
Barbot, F., 122[119] (70), 500[209] (467), 500[223] (468)
Barker, G. K., 427[94] (382, 398), 427[104] (394–6), 427[106] (394–6), 427[111] (394, 396, 400), 427[116] (394, 396, 397), 430[318] (417, 418)
Barker, K., 272[84] (266)
Barnes, R. M., 254[481] (244)
Barnett, N. D. R., 31[63] (10)
Barnum, B. A., 249[133] (221, 228), 272[79] (265)
Barr, D., 30[13] (2), 31[18] (3), 31[19] (3), 31[20] (3), 31[21] (3), 31[23] (4), 31[30a] (6, 8), 31[30b] (6, 8), 31[37] (6), 31[44] (8, 9), 33[132] (17)
Barreto, R. D., 425[4h] (375, 389)
Barrett, A. G. M., 502[337] (494)
Barriola, A. M., 272[50] (262), 273[108] (270), 273[111] (270), 371[315] (358)
Barron, A. R., 367[56] (284), 448, 453, 497[10] (433, 450, 454, 497), 497[14] (434, 448, 497), 497[15] (434, 435, 449, 472), 499[125] (449), 499[127] (449), 499[128] (450), 499[141] (452, 453), 504, 540[31] (507, 512), 541[59] (512), 541[61] (512), 542[131] (520, 522), 542[151] (522, 523), 543[179] (525,

526), 543[182a] (525–7), 543[191] (526), 543[193] (527), 543[194] (527), 544[263] (539), 570[4] (546), 570[20] (548), 570[21] (548, 550), 570[22] (548), 570[23] (549), 570[27] (550, 551)
Bartell, L. S., 543[205] (529)
Barth, R. F., 255[506] (245)
Bartke, J. C., 499[152] (454)
Bartlett, R. A., 31[42] (8, 9), 31[47] (9), 31[48] (9), 31[52] (9), 31[54] (9, 11), 31[55] (9), 31[75] (10), 32[112] (15), 33[141] (19), 544[260] (538), 544[267c] (539)
Bartmann, E., 121[28] (59, 60, 100, 102), 125[310a] (101), 125[310b] (102)
Bartoli, G., 124[209] (83)
Barton, L., 252[352] (234), 367[18] (277), 367[19a] (277), 370[232] (326), 370[237] (327), 427[101a] (391), 427[101b] (391)
Bartulin, J., 193[474] (174)
Barua, N. C., 194[529] (178)
Baryshnikova, T. K., 188[165] (145), 188[177] (147), 188[179] (147)
Basato, M., 253[416] (241), 254[455] (243)
Basavaiah, D., 187[96] (137), 187[97] (137)
Base, K., 248[92] (219, 229, 230, 234, 235, 244), 250[211] (223), 252[319] (230), 255[513] (245), 272[36] (260), 272[55] (262), 272[57] (263), 426[25] (378), 428[187] (400)
Basha, F. Z., 122[134] (70)
Basil, J. D., 368[110] (290, 291)
Basova, R. V., 126[357a] (111, 119), 126[358] (111), 126[359] (111)
Basset, J.-M., 34[203] (26)
Bassindale, A., 499[111] (445)
Bastide, J., 193[483] (175, 177, 184)
Batalina, I. N., 122[107] (67, 87, 95)
Batenina, N. V., 255[488] (244)
Batlaw, R., 121[50a] (61), 121[50b] (61)
Battarai, K. M., 192[439] (172)
Bau, R., 367[55] (283), 367[62] (284, 285)
Baudler, M., 187[71] (134), 192[406] (169), 192[407] (169, 170)
Baudry, D., 368[87] (289), 368[88] (289), 368[89] (289), 368[90] (289), 368[91] (289)
Bauer, P., 121[33] (60)
Bauer, W., 32[88] (12), 32[104i] (13), 33[153] (19)
Baum, G., 187[68] (134), 187[74] (135), 187[75] (135), 189[196] (149), 189[206] (150)
Baumann, F.-E., 429[251] (400, 411), 429[256] (400, 411), 429[257] (400, 411), 429[258] (400, 411), 429[260] (400, 411)
Bausch, J. W., 249[132] (221, 223, 230), 250[201] (222, 223), 250[202] (222, 223), 272[79] (265)
Baxter, P. L., 541[87a] (515), 541[90] (515), 541[92a] (515), 541[92b] (515)
Bazan, G. C., 427[125] (400, 412), 427[126] (400, 412)
Beachley, O. T., Jr., 497[17] (434), 500[158] (455), 504, 507, 508, 516, 530, 540[7] (504, 530, 539), 540[13] (505, 506), 540[20] (505, 506, 509), 540[30] (507), 540[34] (507), 540[35] (508), 541[57] (512), 541[64] (513), 541[65] (513), 541[74] (514, 516), 541[75] (514, 516), 541[98] (516), 542[102b] (518), 542[119] (520), 542[129] (520), 542[145] (522), 543[183] (525, 526), 543[184] (525, 526), 543[203] (529), 543[206] (530, 539), 543[208] (530), 543[209] (530), 544[265] (539)
Beattie, J. K., 45, 54[25] (44, 47), 54[26] (45), 54[29] (45), 54[30] (45), 54[31] (46)
Beatty, B. G., 255[500] (245), 430[316] (416)
Beaudet, R. A., 247[13] (218), 251[282] (227)
Becalska, A., 501[251] (474), 501[252] (474)
Beck, J. S., 186[28] (130), 249[129] (221, 235), 252[304] (228), 252[305] (228, 232, 235, 236, 238), 271[15] (258)
Beckel, C. L., 249[117] (220)
Becker, G., 32[76] (11), 32[80] (11), 32[81] (11), 32[117] (15), 32[118] (15), 32[120d] (15), 543[172] (525, 526)
Beckett, M. A., 272[60] (263), 272[61] (263), 370[239] (327, 336, 363, 364), 370[240] (327, 336, 364), 370[249] (333), 370[255] (337), 370[263] (340), 370[264] (340), 371[299] (352), 371[320] (364), 371[321] (364)
Beedle, E. C., 193[525] (178)
Beer, D. C., 427[98b] (391, 395), 427[98c] (391)
Begley, M. J., 194[541] (179), 541[68] (513)
Behm, J., 570[45] (556, 566), 570[46] (556, 561, 562, 564, 566), 570[47] (556), 570[48] (556), 570[50] (557, 566)
Behnken, P. E., 252[355] (234), 429[206] (400, 415), 429[217] (400, 415), 429[222] (400, 415), 429[233] (400, 415), 430[298] (415), 430[299] (415)
Behrens, H., 570[58] (559)
Behrens, U., 34[187] (23), 125[264] (91)
Behrooz, M., 124[210] (83)
Bekasova, N. I., 248[55] (218, 244), 248[59] (218, 244), 248[61] (218), 254[434] (242), 254[435] (242, 244), 254[464] (244), 254[465] (244), 254[466] (244), 255[494] (245), 255[495] (245)
Bekker, D. B., 252[354] (234, 242)
Bell, N. A., 54[3] (36, 38, 39, 42, 44, 45, 47–53), 54[16] (40), 55[37] (48), 55[43] (50), 55[45] (50, 51), 55[46] (52, 53), 55[55] (53)
Bellenbaum, M., 122[91] (66)
Belmont, J. A., 430[300] (415)
Belozerov, A. I., 123[176b] (78)
Belsky, V. K., 191[342] (163), 367[20] (277, 286, 287), 368[107] (290, 291), 570[11] (547), 570[12] (547), 570[13] (547), 570[15] (547), 570[16] (547), 570[33] (551), 570[34] (551)
Beltram, G. A., 248[102] (220, 226, 232, 233, 236)
Bencivenni, L., 55[42] (50)
Benedetti, E., 369[171] (303)
Benedikt, G., 499[97] (443)
Benkeser, R. A., 123[196] (82)
Benn, R., 123[164] (76, 77, 93), 124[228] (84), 124[248] (86, 93), 125[265] (93, 101, 102), 125[301] (99, 100), 126[336] (107), 498[72] (440), 498[73] (440)
Bennett, S. G., 124[211a] (83)
Bennett, S. J., 253[417] (241)
Beno, M. A., 33[169] (19)
Bent, B. E., 500[214] (468), 500[215] (468)
Benvenuto, M. A., 215[61] (207), 426[47] (382, 387), 426[48] (382, 387, 421, 423), 426[54] (382, 387)
Bercaw, J. E., 412, 427[125] (400, 412), 427[126] (400, 412)
Berchier, F., 254[432] (242)
Berezkin, V. G., 248[57] (218)
Bergbreiter, D. E., 123[179] (78), 126[317] (103)
Berger, H.-O., 192[421] (170)
Berger, S., 187[74] (135)
Bergman, R. G., 367[69] (285, 286)
BergStresser, G. L., 125[256] (88, 89, 93)
Berke, C. M., 54[8] (37)
Berke, H., 367[57] (284)
Bernardon, C., 123[196] (82)
Bernasconi, S., 502[339] (494)
Bernath, P. F., 120[7] (58)
Berndt, A., 130, 134, 149, 186[11] (130, 131, 134, 149), 186[35] (131, 134, 149), 187[67] (134), 187[68] (134), 187[70] (134), 187[73] (135), 187[74] (135), 187[75] (135), 187[77] (135), 188[188] (148), 189[194] (149), 189[195] (149), 189[196] (149), 189[198] (149), 189[199] (149), 189[201] (150), 189[205] (150), 189[206] (150), 544[267a] (539)
Berner, H. J., 126[320] (104)
Bernstein, M. P., 31[64] (10)
Bersohn, M., 122[130] (70)
Bertel, N., 543[167] (525)
Bertozzi, S., 122[93] (67)
Berwin, H. J., 502[310] (488)
Bestman, H. J., 187[83] (135, 138, 143)
Bestmann, H. J., 501[254] (474)
Bettonville, S., 368[82] (289)
Bezborodov, V. S., 193[508] (176), 193[509] (176), 193[510] (176)
Bezuglaya, Z. V., 190[323] (162), 190[324] (162), 190[325] (162), 190[326] (162)
Bezuglyi, V. D., 190[326] (162)
Bhagi, A. K., 125[312] (102)
Bhal, L., 193[469] (174)
Bhamidapaty, K., 187[62] (133), 187[63] (133, 148)
Bhattacharyya, N. K., 368[110] (290, 291)
Bianchi, M., 369[171] (303)
Bianchini, C., 367[64] (284, 285), 368[73] (286)
Bickelhaupt, F., 71, 120[2] (58, 72, 74, 87–9), 120[5] (58, 64, 74, 93, 96, 102), 120[16] (59, 71), 121[60] (63), 121[61] (63, 74, 77, 90), 121[78] (65, 74, 77, 90), 121[84] (66, 71, 76, 92), 122[135] (71, 76, 83, 84,

92, 96), 122[136] (71, 83, 84), 122[137a] (71, 83, 84), 122[137b] (71, 83), 122[138a] (71), 122[138b] (71), 122[139] (71), 122[140] (71, 85), 122[141a] (71, 83), 122[141b] (71, 83), 122[142] (71), 122[143a] (71, 85, 92), 122[143b] (71, 76, 85, 92), 123[146] (71, 76, 83, 85, 96), 123[156] (73), 124[238] (85), 124[239] (85), 125[257] (88, 89, 93), 125[262] (90, 96), 125[270] (94), 125[278] (96, 102)
Bielawski, J., 191[350] (164), 192[396] (167, 168), 195[612] (183)
Biffar, W., 187[101] (138), 187[107] (139)
Bihlmayer, C., 187[118] (140), 188[119] (140), 188[120] (140), 188[122] (140)
Bindal, R. D., 122[131] (70)
Binder, H., 188[186] (147), 194[542] (179), 194[546] (179), 194[547] (179), 194[565] (180)
Binger, P., 430[334a] (421), 500[161] (456)
Birchall, T., 33[124] (15)
Birkenstock, U., 498[64] (440, 484, 497)
Bishop, P. M., 120[21b] (59, 80)
Bitar, H., 191[357] (164)
Blaeser, D., 255[515] (245)
Blake, A. J., 540[29] (506)
Blank, J., 541[66] (513)
Blank, N. E., 544[249] (536)
Bläser, D., 125[314] (103), 187[72] (134), 191[383] (166), 193[521] (177, 178), 215[87] (212), 498[79] (441)
Bleshinskaya, A. S., 190[309] (161), 190[310] (161)
Block, E., 32[112] (15)
Blom, R., 124[247] (86, 94, 101), 125[269] (94, 112, 113), 126[368] (113), 126[369] (113, 114), 543[206] (530, 539)
Blomberg, C., 121[84] (66, 71, 76, 92)
Bluhme-Hensen, K., 194[592] (182)
Blundell, T. L., 541[78] (514)
Boardman, A., 540[10a] (504, 506), 543[196] (527)
Boche, G., 2, 11, 14, 30[10] (2, 11, 14, 17, 30), 31[38b] (6), 32[105] (13), 33[133] (17), 33[134] (17), 33[138] (19), 33[139] (19), 33[146] (19), 33[167] (19, 24), 123[158] (74)
Bochkareva, M. N., 191[387] (167), 191[392] (167)
Boersma, J., 33[153] (19)
Boese, R., 31[45] (9), 31[49] (9), 55[35] (47, 48), 125[314] (103), 187[72] (134), 187[86] (136), 187[100] (138), 187[108] (139), 188[128] (141), 188[130] (141), 188[139] (142), 188[140] (142), 188[141] (142), 188[142] (142), 189[215] (151), 189[224] (152), 190[266] (157), 190[280] (158), 190[282] (159), 190[283] (159), 190[284] (159), 190[285] (159), 190[293] (160), 190[296] (160), 190[299] (160), 190[300] (160), 190[301] (160), 190[303] (160), 190[304] (160), 190[305] (161), 190[306] (161), 190[307] (161), 190[317] (161), 190[318] (162), 191[331] (162), 191[334] (163), 191[364] (165), 191[383] (166), 192[400] (168), 192[401] (168), 192[402] (168), 192[414] (169), 192[430] (171), 193[513] (177), 193[521] (177, 178), 194[538] (179), 194[550] (179), 194[560] (180), 214[5] (198, 211), 214[15] (199), 215[34] (202), 215[82] (212), 215[85] (212), 215[86] (212), 215[87] (212), 215[88] (212), 215[90] (213), 216[92] (213), 216[93] (213), 216[94] (213), 216[95] (213), 255[515] (245), 272[39] (261), 272[42] (261), 272[44] (261), 272[45] (261), 272[47] (261), 498[79] (441)
Bogdanovic, B., 120[6] (58, 59, 99, 100, 107), 120[24] (59, 99), 121[26a] (59, 70, 100), 121[26b] (59, 99, 100), 121[29] (59, 100), 124[249] (86, 90, 93–5), 124[250a] (86, 100), 124[250b] (86, 100), 125[287] (98, 100), 125[297] (99, 100), 125[299] (99), 125[302] (100), 125[303] (100), 498[64] (440, 484, 497)
Böhl, M., 189[205] (150)
Böhland, H., 126[320] (104)
Böhle, C., 214[18] (199)
Böhler, D., 187[60] (133, 144)
Boiko, I. I., 194[539] (179)
Bolado, B., 501[227] (468)
Boldyreva, O. G., 190[310] (161), 191[353] (164), 191[363] (165), 191[370] (165), 191[384] (167), 191[385] (167), 191[386] (167), 191[387] (167), 191[388] (167)
Boleslawski, M. P., 186[14] (130, 148), 498[51] (438, 493), 498[52] (438, 456, 474), 499[130] (451), 499[138] (452), 499[142] (452), 499[143] (452), 499[144] (452), 501[290] (482), 502[300] (479, 483), 502[302] (484), 502[343] (495)
Bolton, R. S., 191[341] (163)
Boltsch, O., 55[35] (47, 48), 215[88] (212)
Bommer, J. C., 368[76] (286)
Bonasia, J., 32[110] (15), 32[116] (15)
Boncella, J. M., 125[269] (94, 112, 113), 126[368] (113)
Bond, M. R., 540[33] (507, 538)
Bonetskaya, A. K., 254[473] (244), 254[474] (244), 255[487] (244)
Bonini, B. F., 124[205a] (82)
Bönnemann, H., 121[29] (59, 100), 127[381] (117), 498[64] (440, 484, 497)
Bonnetot, B., 371[318] (363)
Boocock, S. K., 251[289] (227), 277, 366[1] (277, 278), 370[253] (336), 371[266] (340, 342), 371[292] (351), 546, 570[1] (546, 553, 554, 568)
Boor, J., Jr., 498[61] (439, 482, 497)
Bopegedera, A. M. R. P., 120[7] (58)
Borelli, A. J., Jr., 427[81] (382, 391)
Borisov, A. P., 367[58] (284), 368[75] (286), 368[78] (287), 368[80] (286, 287), 368[81] (286, 287)
Börner, U., 543[213] (531, 539)
Borodinsky, L., 251[241] (225, 227), 370[226] (323), 426[49] (382, 396), 426[53] (382, 396), 428[163] (400)
Borrmann, H., 190[272] (158), 193[528] (178)
Borst, J. B., 32[121b] (15)
Bortel, E., 248[60] (218)
Both-Pollmann, E. M., 190[319] (162)
Bott, S. G., 31[43] (8), 54[24] (42, 43), 497[11] (433, 434, 454, 497), 497[15] (434, 435, 449, 472), 499[141] (452, 453), 540[17] (505, 506), 540[19] (505, 506), 540[21] (505, 506), 541[52] (512, 522), 541[61] (512), 541[67] (513, 524), 541[86] (514), 541[89] (515), 542[134] (520), 542[149] (522, 523), 542[154] (522, 524)
Böttcher, B., 31[31] (6)
Böttcher, P., 32[117] (15)
Bottka, N., 55[60] (54)
Boudin, A., 121[37] (60), 124[221] (83)
Boudjouk, P., 123[145] (71)
Bould, J., 367[30] (278, 343, 348), 369[192] (313), 369[195] (313, 315, 316, 320, 323), 369[200] (313, 316), 370[213] (319), 370[228] (324), 370[229] (324), 370[243] (331, 332, 336, 337), 370[246] (332, 336), 370[247] (332, 336), 370[265] (340, 342, 343), 371[266] (340, 342), 371[278] (342, 347), 371[279] (343), 371[280] (343), 371[283] (347), 371[284] (347)
Boveleth, W., 215[38] (202, 203), 215[39] (202–4)
Bowers, J. R., Jr., 123[186b] (80, 81)
Bowles, J., 250[226] (224)
Bowman, R. P., 121[44] (60, 79)
Bown, M., 273[101] (269), 369[197] (313, 315, 346, 357, 359), 370[212] (319, 320, 324, 340, 341), 370[244] (332), 370[249] (333), 371[267] (341), 371[268] (340, 341), 371[270] (340–2), 371[275] (342, 357, 359), 371[282] (346), 371[286] (350), 371[311] (357, 359), 371[314] (358), 426[11e] (377, 394, 396), 427[108] (395, 396), 427[112] (396), 427[114] (396), 428[203] (400, 412), 429[204] (400, 412)
Bowser, J. R., 186[16] (130), 187[57] (132), 430[335] (422)
Boyter, H. A., Jr., 251[244] (225, 227, 229, 234), 251[245] (225–7, 229, 234), 426[71] (382), 427[80] (382, 387)
Bradford, A. M., 540[14] (505, 506)
Bradley, A. J., 544[264] (539)
Bradley, D., 32[82a] (11)
Bradley, D. C., 540[10b] (506, 508), 540[11] (504–6, 516), 540[14] (505, 506), 540[18] (505, 506), 540[22] (505, 506), 541[42] (511), 541[43] (511), 542[127] (520), 543[175] (525), 543[192a] (527), 543[192b] (527)
Bradley, J. S., 369[158] (294), 571[91] (568)
Bragin, J., 250[233] (225)
Brand, U., 188[131] (141)
Brandl, A., 194[585] (182), 195[630] (184)
Brands, L., 570[56] (559, 562)
Brandsma, L., 31[28] (5), 31[60] (10), 33[153] (19), 121[79] (65, 71)
Brandt, L., 570[49]
Brauer, B.-E., 189[257] (156)
Brauer, D. J., 34[204] (26, 27)
Braun, S., 33[135] (17)

Brazier, C. R., 120[7] (58)
Bregadze, V. I., 247[35] (218, 221, 227, 235, 237–43, 245), 247[43] (218), 247[44] (218), 248[45] (218), 248[46] (218), 249[159] (221, 224, 238), 249[160] (221, 224, 238), 250[197] (222), 251[263] (226), 254[446] (243), 254[447] (243), 254[448] (243), 254[449] (243), 254[456] (243)
Bregar, Z., 123[174] (78)
Breil, H., 501[270] (478, 484, 497)
Bremer, M., 31[46] (9), 31[59] (10), 31[60] (10)
Brennan, D. E., 414, 426[40] (382, 400, 414)
Brennan, J. G., 368[100] (290, 291), 368[104] (290, 291)
Bresadola, S., 253[416] (241), 254[452] (243), 254[453] (243), 254[454] (243), 254[455] (243)
Bretinger, H., 498[46] (437, 443, 444)
Brett, W. A., 188[128] (141), 191[331] (162)
Brew, S. A., 428[200] (400, 411, 417, 418), 429[244] (400, 414, 417, 419), 429[245] (400, 414), 429[246] (400, 411), 429[254] (400, 411), 429[267] (400, 411), 429[270] (400, 411), 429[272] (400, 411)
Brewer, C. T., 250[213] (223), 255[524] (246), 370[219] (321), 370[220] (321, 335), 426[63] (382)
Brewer, G. A., 426[61] (382), 426[67] (382, 398)
Brieden, W., 187[94] (137)
Briguglio, J. J., 251[288] (227), 272[92] (267), 370[250] (20, 334, 341, 342, 357, 359), 426[12b] (377, 382, 423), 426[12c] (377, 382, 394, 396), 426[12d] (377, 396, 420)
Brinkmann, R., 121[29] (59, 100)
Brint, P., 190[312] (161), 248[74] (218, 219), 248[83] (219), 248[84] (219), 248[103] (220), 251[274] (226, 245), 370[227] (323), 370[236] (327), 371[280] (343), 371[283] (347), 425[4b] (375)
Brintzinger, H. H., 498[67] (440)
Brisdon, B. J., 194[580] (181)
Britton, D., 541[69] (513)
Brock, C. P., 192[405] (169)
Brock, T. H., 124[228] (84)
Brodovskaya, G. F., 501[236] (470)
Brodt, G., 215[64] (208)
Bromm, D., 187[60] (133, 144), 187[61] (133, 144), 187[89] (136, 144), 188[158] (144), 189[202] (150, 151), 189[242] (154), 189[243] (154), 249[126] (221, 226, 238)
Brömmelhaus, A., 543[198] (528)
Bronshtein, L. M., 251[279] (227), 254[469] (244)
Brook, A. G., 498[78] (441)
Brooker, S., 32[93] (12)
Brookhart, M., 540[32] (507)
Brooks, J. J., 33[125] (15), 33[140] (19)
Brooks, W. M., 125[295] (98, 100)
Broquet, C., 122[110] (68)
Brown, C. A., 252[340] (232, 233), 252[347] (233)
Brown, D. A., 125[253] (88), 249[118] (220, 240), 249[163] (221, 244), 250[188] (221, 243)
Brown, D. S., 502[322] (491)
Brown, D. W., 55[56] (54)
Brown, H. C., 124[220] (83), 136, 137, 186[4] (130, 138), 187[90] (136), 187[91] (136), 187[92] (136), 187[97] (137), 187[98] (137), 187[99] (137), 188[124] (140), 188[152] (143), 188[156] (144), 193[499] (175), 194[557] (180), 502[306] (486)
Brown, J. M., 120[18] (59, 77)
Brown, P. A., 501[232] (469)
Brown, T. L., 34[213b] (29), 34[218c] (30), 34[218d] (30)
Brownstein, S. K., 498[71] (440, 479, 480), 501[291] (482)
Bruce, G. C., 429[255] (400, 411), 430[274] (400, 411)
Bruin, J. W., 122[137b] (71, 83)
Brum, J. L., 498[41] (436)
Bruno, G., 500[179] (461)
Bruno, J. W., 570[24] (549), 570[26] (550)
Bruno, M., 502[322] (491)
Brusilovskii, Yu. E., 195[604] (183)
Bryan, R. F., 426[57] (381, 382, 421, 423), 543[223] (532, 533)
Bryan, S. J., 125[253] (88), 249[163] (221, 244), 250[188] (221, 243)
Bryson, T. A., 188[151] (143)
Bsata, M., 33[164] (19)
Bublak, W., 544[247] (536), 544[251] (536), 544[252] (536)
Bubnov, Yu. N., 188[149] (143), 188[150] (143, 147), 188[164] (145), 188[167] (145, 146), 188[168] (145), 188[171] (145), 188[173] (146), 188[174] (146), 188[182] (147), 189[239] (154), 191[366] (165)
Buchanan, J., 249[145] (221, 224)
Buchmeiser, W., 33[164] (19)
Büchner, K., 271[20] (259)
Büchner, W., 32[84d] (11)
Buchwald, S. L., 416, 430[307] (416)
Buckton, G. B., 500[182] (461)
Budzelaar, P. H. M., 131, 186[31] (131), 186[32] (131, 134), 186[33] (131), 186[36] (131), 186[38] (131), 186[39] (131), 186[40] (131), 186[41] (131), 189[226] (153), 248[64] (218), 248[65] (218)
Bühl, M., 186[42] (131), 187[54] (132), 190[284] (159), 195[632] (184), 250[199] (222, 223), 250[200] (222, 223), 250[202] (222, 223), 272[47] (261)
Buhl, M. L., 368[126] (294, 296, 297), 368[132] (297, 298)
Bukalov, S. S., 188[164] (145), 250[180] (221), 250[234] (225), 250[237] (225), 251[254] (225, 239)
Bukanova, N. N., 193[461] (174)
Bukhtiarov, A. V., 252[327] (231)
Bukin, V. A., 192[450] (173)
Bulak, E., 215[79] (211)
Bulgakov, R. G., 187[113] (139)
Bulychev, B. M., 55[49] (52), 367[20] (277, 286, 287), 367[33] (279, 280), 368[107] (290, 291), 546, 570[5] (546), 570[7], 570[8] (547), 570[9] (547), 570[10] (547), 570[11] (547), 570[12] (547), 570[15] (547), 570[16] (547), 570[17] (547), 570[33] (551), 570[34] (551)
Bunkhall, S. R., 272[81b] (265, 266), 370[254] (336)
Burch, R. R., Jr., 367[55] (283)
Burek, C., 543[182b] (527)
Burg, A. B., 258, 272[63] (263), 272[64] (263), 277
Bürger, H., 542[109] (519)
Burgess, K., 368[121] (292, 293, 295)
Burkey, D. J., 126[348] (109, 112–14)
Burlinson, N. E., 502[343] (495), 502[344] (495)
Burnashova, T. D., 123[147] (71)
Burns, C. J., 55[34] (46), 55[40] (48), 125[266] (93, 112, 114), 125[269] (94, 112, 113), 126[368] (113), 126[369] (113, 114)
Burns, E. G., 252[363] (236, 242)
Burns, I. D., 369[198] (313, 315), 369[199] (313, 315)
Burns, R. C., 367[44] (279, 280), 542[158] (524)
Burns, T. P., 120[22] (59, 86, 101), 120[23] (59), 122[127] (70)
Burschka, C. J., 542[117] (520, 525), 543[180] (525)
Burton, D. J., 121[68] (64)
Busby, D. C., 252[321] (230), 428[158] (400), 429[212] (400, 415)
Buschmann, E., 124[205b] (82)
Bushby, R., 34[220a] (30)
Butcher, R. J., 249[125] (220, 221, 225–7, 229), 426[41] (382, 421, 423), 426[68] (382, 398), 427[90] (382), 540[12] (504, 506, 511, 521), 540[16] (505, 506)
Butin, K. P., 123[176b] (78)
Butler, I. S., 250[238] (225)
Butler, S., 122[90] (66)
Butler, W., 32[120a] (15)
Butler, W. M., 249[153] (221, 230, 240), 255[502] (245)
Butman, L. A., 250[192] (221)
Buttrus, N. H., 126[325] (105)
Buttry, L. A., 543[208] (530)
Byrne, E. K., 542[104] (518, 519)

Cabioch, J.-L., 429[262] (400, 411)
Cagle, F. W., 570[44] (555)
Cahiez, G., 124[214] (83), 124[219] (83)
Cai, S. X., 192[452] (173)
Calabrese, J. C., 55[41] (49, 50), 368[120] (292, 293, 295), 368[121] (292, 293, 295), 370[234] (326), 570[36] (552), 571[66] (562)
Caldwell, K. R., 500[181] (461, 480), 501[292] (482)
Calhorda, M. J., 425[4a] (375), 425[4c] (375)
Callahan, K. P., 427[121] (398)
Calvert, R. B., 369[160] (294)
Campana, C. F., 255[528] (246)
Campbell, G. L., 121[32] (60)
Campbell, R. M., 570[38] (553, 555, 556)
Canadell, E., 54[17] (40), 544[230] (533)
Cané, M., 571[69] (563, 564)
Canning, W. M., 367[67] (285)
Cannizzo, L. F., 501[282] (480)
Canty, A. J., 541[82] (514), 541[84] (514, 526), 543[189] (526)
Caporusso, A. M., 502[330] (491)

Cara, G., 541[83] (514)
Carboni, B., 190[327] (162), 192[451] (173)
Carloni, P., 123[176a] (78)
Carpenter, G. B., 32[100] (12), 32[101] (12)
Carr, N., 428[200] (400, 411, 417, 418), 429[230] (400, 411), 429[244] (400, 414, 417, 419), 429[259] (400, 411), 429[261] (400, 411), 429[273] (400, 414), 430[277] (400, 411), 430[317] (400, 417, 418), 430[319] (417–19)
Carrano, C. J., 540[33] (507, 538), 542[115] (520)
Carreno, R., 367[59] (284)
Carrié, R., 190[327] (162)
Carriedo, G. A., 368[115] (292, 293)
Carroll, P. J., 249[130] (221, 238), 249[132] (221, 223, 230), 249[133] (221, 228), 249[156] (221, 225, 227, 233, 235), 249[157] (221), 251[288] (227), 252[303] (228), 252[332] (231), 273[98] (268), 426[35] (378, 379), 427[82] (382, 387), 430[330] (420, 421), 430[331] (420, 421), 571[82] (567)
Carty, A. J., 544[259] (537)
Casabo, J., 249[139] (221, 240), 249[146] (221)
Casas, J., 543[166] (525)
Casey, J. B., 248[62] (218)
Cassidy, J. M., 571[67] (563, 567), 571[87] (567, 568), 571[89] (568), 571[90] (568)
Castellano, E. E., 543[166] (525)
Castle, P. L., 124[218] (83)
Caulton, K. G., 32[82b] (11), 32[95] (12), 34[202] (26), 367[49] (280, 282), 570[24] (549), 570[26] (550)
Cavanaugh, M. A., 248[102] (220, 226, 232, 233, 236)
Cederbaum, F. E., 124[227] (84)
Cendrowski-Guillaume, S. M., 251[287] (227), 427[96a] (381)
Cerveau, G., 121[37] (60), 124[221] (83)
Cetinkaya, B., 31[41] (8), 32[90] (12)
Cha, J. S., 187[103] (138)
Chablo, A., 248[79] (219)
Chadha, R. K., 543[168] (525, 528), 543[173] (525)
Chan, L.-H., 31[15] (3)
Chandrasekharan, J., 188[124] (140)
Chang, D., 571[79] (566, 567)
Channareddy, S., 192[436] (172), 215[28] (201), 215[29] (201), 368[117] (292, 295), 368[119] (292, 295)
Chao, M. R., 541[98] (516)
Chao, S.-H. L., 543[184] (525, 526)
Chapman, R. W., 271[27] (259), 272[72] (264)
Charandabi, M. R. M. D., 191[380] (166)
Chari, S., 251[278] (227)
Charkin, O. P., 248[82] (219), 248[86] (219)
Charpin, P., 368[87] (289)
Chase, K. J., 426[45] (382), 426[56] (382), 426[57] (381, 382, 421, 423)
Chattopadhyay, S., 121[75a] (65, 70)
Chaturvedi, L., 194[533] (179)
Chaturvedi, V., 192[433] (172), 194[533] (179), 194[579] (181)
Cheek, Y. M., 371[297] (351, 362), 372[324] (365)
Cheetham, A. K., 33[174] (20)
Chekhlov, A. N., 367[58] (284)
Chekulaeva, L. A., 126[333] (106)
Chen, G. J., 122[122] (70)
Chen, H., 31[42] (8, 9), 31[48] (9), 31[52] (9)
Chen, J., 122[123] (70)
Chen, L. S., 122[122] (70)
Chen, Q., 122[123] (70)
Chen, T., 126[344] (108, 109, 117, 120)
Chen, Y.-J., 570[49]
Cheng, W. L., 124[217] (83)
Chenier, J. H. B., 498[35] (436, 441), 499[92] (442), 499[93] (442)
Cherkasov, L. N., 127[386] (119)
Cherkasov, V. N., 127[386] (119)
Cherkasova, K. L., 188[166] (145), 188[167] (145, 146), 188[170] (145), 190[308] (161)
Chermette, H., 249[115] (220)
Chernyshev, E. A., 430[283] (400)
Chesnokova, I. V., 250[176] (221)
Chetcuti, P. A., 412, 427[133] (400, 417), 429[216] (400, 412)
Chevrier, G., 368[87] (289)
Chiesi-Villa, A., 367[40] (279, 280)
Chipperfield, A. K., 369[141] (296, 297), 369[143] (297, 300, 301), 369[147] (297, 303), 369[168] (301), 369[180] (308)
Chirstie, S. D., 501[288] (482)
Chisholm, M. H., 367[13] (277, 337)
Chiu, C. S., 498[59] (439), 502[309] (488)
Chizhevsky, I. T., 426[39] (378, 380)
Chizhov, O. S., 190[309] (161), 190[310] (161), 191[372] (166), 191[384] (167)
Chizhova, E. A., 254[434] (242), 254[435] (242, 244)
Cho, H., 571[68] (563)
Cho, I. S., 502[340] (494)
Choi, P. C., 255[512] (245)
Choi, S. K., 249[116] (220)
Chou, T. S., 501[281] (480)
Christophersen, M. N., 32[97] (12)
Chu, S. S. C., 272[29] (260), 272[31] (260), 372[325] (365), 544[232] (534)
Chudzynska, H., 541[42] (511), 543[192a] (527), 543[192b] (527)
Chuit, C., 121[37] (60), 124[221] (83)
Chung, B. Y., 502[340] (494)
Churchill, M. R., 508, 540[7] (504, 530, 539), 540[30] (507), 540[35] (508), 541[64] (513), 541[65] (513), 542[129] (520), 543[184] (525, 526), 543[206] (530, 539), 543[208] (530)
Churchill, R. F., 541[98] (516)
Chwojnowski, A., 54[15] (39), 126[335] (107)
Cictron, J., 542[109] (519)
Cioslowski, J., 132, 186[47] (132)
Cisarova, I., 428[174] (400, 414)
Claire, K. S., 369[198] (313, 315), 369[199] (313, 315)
Clark, T., 125[306] (101), 186[29] (131), 186[33] (131), 186[38] (131), 186[42] (131), 248[64] (218)
Clarke, C. M., 192[399] (168)
Clarkson, L. M., 562, 570[38] (553, 555, 556), 570[52] (557, 558), 570[54] (557, 558, 567), 570[55] (558), 571[65] (562), 571[66] (562), 571[72] (565), 571[73] (565), 571[92] (568), 571[93] (568)
Cleaver, W. M., 541[61] (512), 543[179] (525, 526)
Clegg, W., 30[12] (2), 30[13] (2), 31[18] (3), 31[19] (3), 31[20] (3), 31[21] (3), 31[22] (4), 31[23] (4), 31[30a] (6, 8), 31[37] (6), 31[44] (8, 9), 31[56] (10), 31[57] (10), 31[63] (10), 31[66] (10), 31[67] (10), 32[111] (15), 32[113] (15), 125[253] (88), 126[316] (103), 249[118] (220, 240), 249[163] (221, 244), 250[188] (221, 243), 570[38] (553, 555, 556), 570[52] (557, 558), 570[54] (557, 558, 567), 571[92] (568)
Clerk, M. D., 501[288] (482)
Cloke, F. G. N., 126[350] (109, 110)
Clough, S., 120[19] (59)
Cloux, R., 125[274] (95)
Coan, P. S., 32[95] (12)
Coates, G. E., 35, 42, 54[16] (40), 55[43] (50), 55[45] (50, 51), 518, 540[24b] (505, 506), 542[102a] (518), 542[102b] (518), 542[142] (521)
Coblenz, W. S., 254[482] (244), 254[483] (244)
Cockcroft, J. K., 33[154b] (19, 20), 33[172a] (20), 33[172b] (20), 33[177] (20)
Cocolios, P., 571[75] (566), 571[79] (566, 567)
Coffy, T. J., 367[23] (277, 294, 297, 300, 303), 369[178] (307, 308), 369[179] (308)
Cole, T. E., 124[220] (83), 187[92] (136), 193[499] (175)
Coleson, K. M., 54[10] (38, 50), 55[41] (49, 50), 370[234] (326)
Collins, J. B., 33[182] (21)
Collum, D. B., 2, 31[61] (10), 31[62] (10)
Colquhoun, H. M., 249[118] (220, 240), 255[492] (244), 429[239] (398, 400)
Companion, A. L., 187[55] (132), 192[405] (169)
Compton, N. A., 570[2] (546)
Concepcion, A. B., 502[334] (492)
Concepcion, R., 127[382] (117)
Condroski, K., 123[184] (79)
Conn, R. S. E., 122[108] (68)
Conrady-Pigorsch, R., 194[564] (180)
Considine, J. L., 498[26] (435, 437), 498[45] (437), 499[122] (448)
Consiglio, G., 124[217] (83)
Contreras, J. G., 251[257] (225, 241), 253[410] (241)
Contreras, L., 543[182b] (527)
Contreras, R., 187[116] (139), 195[616] (184)
Convert, O., 251[267] (226)
Coons, D. E., 255[521] (246), 255[529] (247), 272[65] (264), 370[225] (323)
Cooper, M. K., 54[27] (45), 54[28] (45), 55[33] (46)
Cooper, N. J., 124[229] (84), 367[67] (285)
Corazza, F., 367[40] (279, 280)

Corbelin, S., 33[172a] (20), 33[172b] (20), 34[184] (23, 24), 34[185] (23, 25), 34[191] (23)
Corbett, J. D., 542[158] (524), 544[238] (534), 544[239] (534), 544[240] (534)
Corcoran, E. W., Jr., 248[49] (218), 251[248] (225, 228, 239), 251[288] (227), 252[325] (231), 425[1f] (374)
Cordova, R., 501[244] (472)
Corey, E. J., 367[67] (285)
Corfield, P. W. R., 499[123] (449)
Correa, M., 248[87] (219)
Corriu, R. J. P., 121[37] (60), 124[221] (83)
Corsi, J., 34[199] (26)
Coste, C., 193[483] (175, 177, 184)
Cotton, F. A., 34[217] (30), 367[50] (280, 282)
Cottrell, C. E., 369[144] (297, 300)
Couret, C., 192[410] (169)
Cowie, J., 249[151] (221), 427[131] (400, 412)
Cowley, A. H., 55[39] (48), 192[408] (169), 271[9] (258), 271[10] (258), 271[11] (258), 271[24] (259), 272[30] (260), 272[78] (265), 367[9] (277), 504, 507, 518, 540[1a] (503, 504, 517, 518, 520–2), 540[19] (505, 506), 540[21] (505, 506), 540[33] (507, 538), 541[46] (511), 541[52] (512, 522), 541[61] (512), 541[67] (513, 524), 541[86] (514), 542[115] (520), 542[124] (520), 542[125] (520), 542[131] (520, 522), 542[134] (520), 542[140] (521, 522, 526), 542[144] (522), 542[146] (522), 542[149] (522, 523), 542[153] (522, 524), 542[154] (522, 524), 543[176] (525), 543[182b] (527), 543[224] (532), 544[231] (533)
Cox, D. E., 33[183] (21)
Cox, D. N., 248[81] (219)
Coyle, T. D., 498[47] (437)
Cradock, S., 540[29] (506)
Cragg, R. H., 194[591] (182), 194[593] (182)
Cramer, R. E., 123[161] (74, 109)
Cranson, S. J., 255[510] (245)
Cremer, D., 186[39] (131), 186[40] (131), 186[41] (131), 248[63] (218)
Crennell, S. J., 430[320] (417, 418)
Crimmins, M. T., 124[212] (83)
Cronin, J. P., 248[103] (220)
Crook, J. E., 369[183] (310, 316, 327, 337, 349), 370[224] (322), 370[228] (324), 370[229] (324), 370[239] (327, 336, 363, 364), 370[243] (331, 332, 336, 337), 370[246] (332, 336), 370[255] (337), 370[257] (337), 370[265] (340, 342, 343), 371[279] (343), 371[281] (345, 346, 355, 356), 371[298] (351), 371[306] (355, 356), 371[320] (364), 371[321] (364), 426[21] (378)
Crowther, D. J., 428[191] (400, 416), 429[243] (400, 416), 430[308] (416)
Cruickshank, D. W. J., 248[79] (219)
Csaszar, A. G., 367[44] (279, 280)
Cunkle, G. T., 121[81] (65, 72)
Curtis, T. L., 426[11b] (377, 396, 400)
Cymbaluk, T. H., 570[44] (555)
Czech, P. T., 250[207] (223)
Czerwiec, W., 254[480] (244)

Dagonneau, M., 123[188] (80, 81)
Dahlhaus, J., 126[366] (112, 114)
Dahlke, P., 428[197] (400, 411)
Dahlman, H. L., 253[417] (241)
Dai, B. M., 248[90] (219), 252[344] (233)
Dais, H. V. R., 32[120e] (15)
Dallacker, F., 190[319] (162)
Dalpozzo, R., 124[209] (83)
Dalton, M. S., 499[147] (454)
Daluga, P., 273[111] (270)
Damasevitz, G. A., 502[328] (491)
Damrauer, R., 499[95] (443)
Dandegaonker, S. H., 190[320] (162)
Dang, H., 121[43] (60)
Daniels, J. A., 249[118] (220, 240), 255[492] (244)
Dankwardt, J. W., 121[75a] (65, 70)
Darby, W. L., 427[90] (382)
Dare, H. F., 429[220] (400, 412)
Darensbourg, M. Y., 367[54] (4, 283), 367[55] (283)
Darling, G. D., 123[149] (72)
Das, M. K., 192[396] (167, 168), 195[637] (185)
DaSettimo, F., 502[330] (491)
Dauben, W. G., 498[76] (441, 467, 486)
Davan, T., 255[514] (245)
Davidson, G., 194[572] (181)
Davidson, M. G., 32[122] (15), 33[126] (15)
Davidson, N., 502[306] (486)
Davies, C. E. H., 126[347] (109, 110)
Davies, J. E., 32[85b] (11)
Davies, P. M., 255[510] (245)
Davies, S., 252[313] (229)
Davis, J. H., Jr, 215[60] (207), 252[311] (229, 238), 426[11c] (377, 381, 382), 426[43] (381, 382), 426[44] (381, 382, 387), 426[50] (382), 426[54] (382, 387), 426[69] (382), 427[75] (382), 427[91] (381, 382), 427[92] (382, 392)
Davis, J. J., 541[42] (511)
Davis, P. B., 254[483] (244)
Davis, S. R., 121[59] (62, 78)
Davis, W. M., 271[20] (259)
Dawes, H., 540[11] (504–6, 516), 542[127] (520)
Dawes, H. M., 540[22] (505, 506)
de Boer, H. J. R., 121[60] (63), 122[141b] (71, 83), 123[146] (71, 76, 83, 85, 96), 124[238] (85)
de Jesus, E., 571[86] (567)
de Kanter, F. J. J., 33[156] (19)
De Keijser, M. S., 501[259] (475)
de Meester, P., 272[29] (260), 272[31] (260), 544[232] (534)
De Mel, J., 497[19] (434)
de Parisi, N. R., 272[83] (265, 266)
Deaton, J. C., 367[55] (283)
DeBernardis, J. F., 122[134] (70)
Dedek, V., 122[120] (70)
Dedopoulou, D., 124[212] (83)
Deeming, A. J., 369[161] (294), 369[163] (296)
Deforth, T., 215[57] (206)
Defrees, D., 125[293] (98)
Degnan, I. A., 542[106] (518–20), 543[188] (526)
Dehghan, M., 252[313] (229)
Deiseroth, H.-J., 194[542] (179), 194[546] (179), 194[547] (179)
DeKock, R. L., 248[97] (219, 220), 251[285] (227), 370[215] (320)
Dell'Amico, G., 367[34] (279, 280)
Dell'Innocenti, M., 501[226] (468)
Deloux, L., 186[6] (130)
Delpy, K., 190[294] (160), 214[5] (198, 211), 215[80] (211), 215[81] (211)
Delrosario, J. D., 124[203] (82)
Demkina, I. I., 250[181] (221), 251[283] (227), 253[376] (237, 238), 253[378] (237), 253[379] (237, 238), 253[380] (237)
Demonceau, A., 430[305] (416)
Dem'yanets, Z. K., 191[392] (167)
Deng, H.-B., 195[614] (183), 272[73] (264), 371[285] (349)
DeNinno, M. P., 194[594] (182)
Den'kina, L. S., 428[164] (400)
Denniston, M. L., 543[223] (532, 533)
Dermentzis, K., 194[565] (180)
Deschler, U., 127[377] (116)
Deshmukh, S., 498[41] (436)
Desorcie, J. L., 124[224] (83)
DeStephano, J. P., 125[275] (95, 106)
Dettole, A., 124[215] (83)
Deutch, J. E., 121[38] (60), 121[49] (61)
Devore, D. D., 428[196] (400, 411), 429[246] (400, 411), 430[276] (400, 411), 430[320] (417, 418), 430[321] (417–19)
Dewar, M. J. S., 248[105] (220)
DeYoung, J., 123[173] (78)
Dhanak, D., 502[337] (494)
Dhingra, S., 542[159] (524)
Dhubhghaill, O. N., 273[110] (270), 273[115] (271)
Dias, H. V. R., 31[42] (8, 9), 31[52] (9), 31[54] (9, 11), 33[141] (19), 192[418] (170), 192[419] (170), 192[420] (170), 542[111] (519)
Diaz, M., 252[360b] (235)
Dickason, W. C., 188[152] (143)
Dickerson, R. E., 34[213b] (29)
Dicko, A., 193[483] (175, 177, 184)
Diem, M., 250[233] (225)
Dietrich, A., 540[25a] (505), 541[55] (512), 543[210] (530, 532, 539), 543[215] (531)
Dietrich, H., 33[145] (19), 33[155b] (19), 34[190] (23), 34[192] (23, 24)
Dill, J. D., 248[63] (218)
Ding, H., 192[441] (173), 192[443] (173), 192[444] (173)
Dirschl, F., 190[278] (158, 159), 190[281] (159)
Disch, R. L., 186[48] (132), 186[49] (132), 187[50] (132), 187[51] (132)
DiStefano, E. W., 250[215] (223), 250[216] (224, 232, 233)
Ditzel, E. J., 273[95] (267), 370[258] (338, 345), 370[259] (338, 339), 371[269] (340, 341, 344, 347), 371[276] (342, 351, 357), 371[313] (358, 359), 426[33] (378–80, 396, 397)
Dixon, D. A., 371[305] (354)
Dixon, N. J., 124[217] (83)
Dmitrenko, A. V., 254[476] (244)
Do, Y., 249[140] (221, 238), 271[12] (258), 427[128] (400, 412), 428[188] (400, 412), 428[190] (400, 412), 428[199] (400, 412), 541[99] (517)

Dobbousi, B. O., 32[116] (15)
Dodonov, V. A., 498[40] (436), 500[217] (468)
Dohmeier, C., 123[160] (74, 109), 497[6] (433, 436, 441, 443), 498[33] (436, 445), 499[109] (444), 540[5] (504, 538)
Doi, J. A., 430[281] (400, 414)
Dolansky, J., 248[92] (219, 229, 230, 234, 235, 244), 248[93] (219, 220, 224, 229, 230, 234), 254[459] (244)
Dolg, M., 126[373] (114)
Dolgoplosk, B. A., 498[63] (440)
Dolle, V., 498[70] (440)
Domalski, M. S., 122[94] (67), 122[95] (67, 102)
Donaghy, K. J., 272[66] (264), 272[67] (264)
Donndelinger, P., 123[180] (79)
Donohue, J., 370[211] (318)
Doran, M. A., 34[218a] (30)
Dorofeenko, G. N., 254[431] (242)
Dorofeeva, O. V., 247[14] (218), 250[196] (222)
Dorokhov, V. A., 190[308] (161), 190[309] (161), 190[310] (161), 191[353] (164), 191[363] (165), 191[370] (165), 191[372] (166), 191[384] (167), 191[385] (167), 191[386] (167), 191[387] (167), 191[388] (167), 191[389] (167), 191[390] (167), 191[392] (167)
Dossett, S. J., 427[129] (400, 411, 417, 418), 428[193] (400, 411, 417, 418), 428[198] (400, 412), 428[201] (400, 411), 429[259] (400, 411), 429[262] (400, 411), 429[267] (400, 411), 429[268] (400, 411), 429[269] (400, 411)
Dou, D., 272[75] (264)
Douglas, T., 542[133] (520)
Dowben, P. A., 251[276] (227), 251[277] (227)
Downs, A. J., 271[6] (258), 541[87a] (515), 541[90] (515), 541[91] (515), 541[92a] (515), 541[92b] (515)
Doyle, R. A., 430[307] (416)
Dräger, M., 124[222] (83)
Drago, R. S., 541[77] (514)
Drake, S. R., 126[360] (111, 112, 115)
Draper, S. M., 368[135] (297, 299, 300), 369[142] (294, 296, 297, 301, 303), 369[146] (297, 304), 369[170] (301)
Drdakova, E., 252[301] (228, 229), 252[306] (228, 235, 240), 253[413] (241)
Dresely, S., 192[449] (173)
Drew, M. G. B., 542[139] (521, 526)
Driess, M., 32[76] (11), 192[413] (169), 192[415] (170), 192[416] (170), 192[417] (170), 192[422] (170), 216[96] (214)
Drone, F. J., 189[235] (154, 155), 215[73] (210), 215[74] (210)
Drygina, O. V., 254[431] (242)
Du, C.-J. F., 121[69] (65, 83)
du Mont, W.-W., 32[109] (15)
Duben, J., 248[92] (219, 229, 230, 234, 235, 244)
DuBois, D. A., 186[28] (130), 191[348] (164)
Dubois, J. E., 121[33] (60)
Dubois, L. H., 121[31] (59, 62), 500[214] (468), 500[215] (468)
Duboudin, J.-G., 125[311] (102)
Duda, B., 123[162] (75)
Duesler, E. N., 192[424] (170), 193[465] (174, 180), 272[74] (264), 272[75] (264), 368[97] (288, 290)
Duff, A. W., 124[244] (86, 95, 96)
Dugrand, L., 55[59] (54)
Dunitz, J. D., 32[86] (12), 32[87] (12), 32[91] (12), 32[96] (12), 32[114] (15), 32[115] (15)
Dunken, H., 32[84b] (11)
Dunks, G. B., 272[84] (266)
Dunne, B. J., 215[48] (204)
Dupont, T. J., 271[25] (259)
Duraj, S. A., 367[50] (280, 282)
Durani, S., 122[131] (70)
Durso, S. S., 251[282] (227)
Durst, T., 32[104g] (13)
Dvurechenskaya, S. Ya., 428[164] (400), 428[166a] (400, 414)
D'yachkovskii, F. S., 499[103] (443)
Dyer, G., 253[409] (241)
Dykstra, C. E., 121[58] (62, 78)
Dykstra, R. R., 32[121b] (15)
Dzabieva, Z. M., 499[103] (443)
Dzhemilev, U. M., 87, 122[98] (67, 97, 98), 122[102a] (67, 87, 95), 122[102b] (67, 87, 95, 103), 122[107] (67, 87, 95), 123[169] (77, 92), 124[242] (86, 98), 125[251] (87, 95), 125[252] (87), 126[318] (103), 502[333] (492)
Dzugan, S. J., 497[9] (433, 435, 450, 497)

Eaborn, C. A., 33[171] (19), 123[151] (72, 85, 87, 88, 92), 174, 193[466] (174), 193[467] (174), 515, 541[97] (515, 526)
Eaton, P. E., 121[67] (64, 70, 72, 103), 121[81] (65, 72)
Ebel, H. F., 33[137c] (19, 30)
Ebenhöch, J., 544[246] (535)
Eberle, A., 255[499] (245)
Ebcrson, L., 123[176a] (78)
Eckert, T. S., 120[14] (59)
Eckman, R. R., 430[306] (416)
Eddy, M. M., 33[183] (21)
Edelmann, F. T., 32[93] (12), 543[178] (525, 526)
Edelstein, N. M., 368[92] (286), 368[93] (287, 290, 291), 368[94] (287, 290, 291), 368[95] (287, 290), 368[96] (288, 290), 368[99] (288, 290), 368[100] (290, 291), 368[101] (290, 291), 368[102] (290, 291), 368[103] (290, 291), 368[104] (290, 291), 368[105] (291), 368[106] (291)
Ederer, B., 194[587] (182)
Ederle, H., 194[587] (182)
Edvenson, G. M., 252[346] (233), 255[528] (246)
Edwards, A. J., 32[77] (11), 33[128] (15)
Edwin, J., 214[13] (199, 205, 206), 215[26] (201), 215[50] (205, 206), 215[56] (206)
Efremov, Yu. Ya., 192[428] (171), 195[621] (184)
Ehmann, W., 194[565] (180)
Eigenbrot, C. W., 215[43] (203), 250[194] (221), 368[110] (290, 291), 368[125] (294, 296, 297), 368[126] (294, 296, 297), 430[333] (421, 423)
Eigendorf, U., 187[84] (135, 137)
Einholz, W., 190[321] (162), 190[322] (162), 195[644] (185)
Eisch, J. J., 120[16] (59, 71), 120[24] (59, 99), 122[141b] (71, 83), 124[210] (83), 124[238] (85), 124[239] (85), 124[240] (85, 86, 96), 124[243a] (86, 95), 124[243b] (86, 95), 124[250b] (86, 100), 125[309] (101), 148, 154, 186[14] (130, 148), 188[190] (148), 189[191] (148), 189[192] (148), 189[213] (151), 189[237] (154, 155), 436, 467, 469, 483, 497[1] (432, 433, 436, 441, 455, 467, 496), 497[4] (433, 436, 438, 441, 458, 459, 467, 486, 495, 496), 498[26] (435, 437), 498[37] (436, 470, 495), 498[38] (436, 470, 495), 498[48] (438), 498[50] (438, 487), 498[51] (438, 493), 498[52] (438, 456, 474), 498[53] (438, 452, 486, 493), 498[56] (439, 457), 498[59] (439), 498[60] (439, 478), 498[71] (440, 479, 480), 499[105] (443), 499[122] (448), 500[157] (455), 500[167] (457), 500[172] (458, 471), 500[175] (458, 470, 472), 500[176] (458, 470, 472), 500[180] (461), 500[181] (461, 480), 500[183] (461), 500[184] (461), 500[190] (463), 500[191] (463), 500[193] (463, 465), 500[198] (465, 496), 500[199] (465), 500[202] (466, 486, 495), 500[203] (466, 495), 500[204] (466), 500[205] (466), 501[233] (469), 501[239] (470, 486), 501[271] (478, 484), 501[273] (478, 479, 481, 483, 484), 501[274] (479), 501[278] (480, 482, 483), 501[290] (482), 501[291] (482), 501[292] (482), 502[299] (483), 502[300] (479, 483), 502[302] (484), 502[304] (486), 502[305] (486), 502[309] (488), 502[312] (488–90), 502[328] (491), 502[343] (495), 502[344] (495), 570[32] (551)
Eisenstein, M., 248[79] (219)
Eisenstein, O., 54[17] (40), 189[221] (152), 544[230] (533)
El-Essawi, M., 216[97] (214)
Elgamal, H., 497[11] (433, 434, 454, 497)
Elian, M., 214[12] (198, 199)
Eliel, E. L., 125[274] (95)
El-Kheli, M. N., 193[466] (174), 193[467] (174)
Ellert, O. G., 251[251] (225, 226)
Elliot, D. J., 368[111] (291)
Elliot, G. P., 368[115] (292, 293)
Ellis, A. M., 120[8] (58)
Ellrich, K., 126[321] (104)
Elms, F. M., 541[89] (515)
Elrington, M., 371[281] (345, 346, 355, 356), 371[306] (355, 356), 371[307] (355, 356), 371[308] (355, 356), 371[319] (363)
Elschenbroich, C., 544[250] (536)
Elter, G., 187[60] (133, 144), 187[61] (133, 144), 190[286] (159, 160), 190[288] (159)
El'tsov, A. V., 192[393] (167)

Emanuelson, T., 127[382] (117)
Emin, D., 248[95] (219), 250[235] (225)
Emmerich, C., 428[196] (400, 411)
Ender, U., 189[216] (151)
Enders, M., 187[88] (136), 188[153] (144), 188[154] (144), 214[16] (199, 208), 252[350] (233)
Engelen, B., 33[164] (19)
Engelhardt, L. M., 31[39] (8), 31[40] (8), 31[65] (10), 33[142] (19), 121[25b] (59, 60, 70, 100), 123[144] (71, 83, 84), 123[155] (73, 99, 111), 126[315] (103), 126[363] (112–14), 191[374] (166)
Englert, U., 187[93] (137), 190[301] (160), 190[302] (160), 214[8] (198, 209), 215[37] (202, 204), 272[38] (260), 272[54] (262)
Enikolopyan, N. S., 499[103] (443)
Ent, H., 122[142] (71)
Ephritikhine, M., 368[91] (289)
Erastov, O. A., 192[425] (170), 194[543] (179), 195[617] (184), 195[618] (184), 195[619] (184), 195[620] (184), 195[623] (184), 195[624] (184), 195[626] (184), 195[628] (184), 195[629] (184)
Erdik, E., 122[109] (68, 83)
Erdman, A. A., 253[425] (241, 244), 254[450] (243), 254[451] (243), 414, 427[135] (400, 414), 427[136] (400, 414), 428[138] (400), 428[162] (400), 428[167] (400, 414), 428[181] (400, 414), 428[182] (400, 414), 428[183] (400, 414), 428[184] (400, 414), 428[186] (400, 414), 430[291] (414)
Erickson, W. F., 122[94] (67)
Ericson, A., 123[163b] (75)
Erker, G., 125[291] (98), 368[112] (292, 293), 502[294] (482), 509, 540[37] (509)
Ermolaeva, L. V., 195[629] (184)
Ernest, R. L., 249[156] (221, 225, 227, 233, 235), 249[157] (221), 426[35] (378, 379)
Ernst, R. D., 570[44] (555)
Erofeev, A. B., 570[33] (551)
Errington, R. J., 570[2] (546)
Escudie, J., 192[410] (169)
Espe, M. P., 127[382] (117)
Esposto, F. J., 251[281] (227)
Esser, L., 543[215] (531)
Estes, S. A., 254[481] (244)
Etinger, M. Yu., 188[169] (145), 188[176] (146)
Ettel, M. L., 191[380] (166)
Evans, D. G., 425[5d] (375)
Evans, W. J., 248[62] (218), 427[121] (398)
Evdokimova, E. V., 570[7], 570[16] (547)
Evers, E. A. I. M., 125[270] (94)
Evertz, K., 194[529] (178)
Evstafeeva, N. E., 55[53] (53)
Ewen, J. A., 440, 498[66] (440, 482), 498[68] (440), 502[297] (482)
Ewer, K. P., 194[572] (181)
Ezhova, T. M., 254[430] (242)
Ezomo, O. J., 571[95] (569)

Fabicon, R. M., 126[329] (106)
Fabre, J.-L., 122[97] (67, 83)
Fackler, J. P., Jr., 368[98] (288, 290), 571[97] (569)
Fadel, R., 501[237] (470)
Faegri, K., Jr., 125[271] (94, 101), 126[369] (113, 114)
Fagan, P. J., 121[63] (63, 84)
Fairchild, R. G., 255[501] (245)
Faktor, M. M., 541[42] (511), 541[43] (511)
Falorni, M., 55[50] (53), 55[51] (53), 55[52] (53), 124[202] (82)
Fang, Y., 192[443] (173), 192[444] (173)
Faridoon, 273[114] (270), 273[117] (271), 371[274] (341)
Farkas, J., Jr., 121[35] (60), 125[277] (96, 105)
Farley, P. S., 121[48] (61)
Farrugia, L. J., 571[65] (562), 571[72] (565), 571[73] (565)
Fatti, G., 122[93] (67)
Fauchere, J. L., 255[499] (245)
Faure, R., 191[357] (164)
Favas, M. C., 247[22] (218)
Fazen, P. J., 186[28] (130), 249[130] (221, 238)
Feakes, D. A., 191[380] (166)
Fedorenko, T. V., 121[34] (60)
Fehlner, T. P., 186[16] (130), 215[30] (201), 215[43] (203), 248[102] (220, 226, 232, 233, 236), 250[207] (223), 251[273] (226), 251[285] (227), 366[6] (277), 367[8] (277, 294), 367[17] (277), 367[21] (277, 294, 297, 298), 367[24] (277, 294, 297, 304), 368[110] (290, 291), 368[125] (294, 296, 297), 368[126] (294, 296, 297), 368[127] (296, 297), 368[128] (297), 368[129] (296, 297), 368[130] (294, 297, 298), 368[131] (296–8), 368[132] (297, 298), 368[133] (297, 298), 369[139] (297, 300), 369[140] (297, 300, 303), 369[155] (297, 298), 369[162] (296), 369[164] (298), 369[177] (307), 369[182] (307, 309), 369[189] (312), 369[190] (312), 369[191] (313), 369[205] (313, 316), 370[215] (320), 421, 425[4h] (375, 389), 426[59] (382, 393), 430[333] (421, 423), 570[6] (546)
Fehr, C., 126[328] (106)
Feighery, W. G., 540[35] (508)
Feilong, J., 369[155] (297, 298), 369[182] (307, 309), 369[189] (312), 369[190] (312)
Fenet, B., 273[97] (267)
Feng, F., 502[313] (489)
Feng, X., 31[47] (9)
Fenske, D., 123[160] (74, 109), 543[202] (529, 539)
Fenske, R. F., 250[207] (223)
Fenzl, W., 194[545] (179)
Ferguson, G., 190[312] (161), 193[481] (175), 273[99] (268), 273[100] (268), 273[113] (270), 273[115] (271), 273[116] (271), 273[117] (271), 429[205] (400, 412)
Ferhat, M., 571[77] (566, 567, 569)
Ferles, M., 191[368] (165), 191[369] (165)
Fernández, E., 541[72] (514)
Fernandez, J. R., 429[218] (400, 411), 430[317] (400, 417, 418)
Fernández-Vincent, C., 195[616] (184)
Ferrara, J. D., 498[68] (440)
Fessenbecker, A., 189[251] (156), 214[16] (199, 208), 214[17] (199, 208), 214[21] (199), 215[62] (207), 215[65] (208, 210), 425[6] (375, 382, 396, 397), 426[58] (381, 382), 427[88] (381, 382)
Fessler, M. E., 251[243] (225–7, 229, 234), 252[353] (234)
Fest, D., 190[289] (159), 194[567] (180), 195[639] (185)
Fettinger, J. C., 540[7] (504, 530, 539), 540[35] (508), 543[206] (530, 539)
Fiandanese, V., 124[215] (83)
Fichter, K. C., 500[175] (458, 470, 472), 500[176] (458, 470, 472), 500[190] (463), 501[273] (478, 479, 481, 483, 484), 570[32] (551)
Field, L. D., 54[30] (45), 247[26] (218), 247[27] (218), 367[61] (284, 285), 367[62] (284, 285)
Fieser, M., 31[24] (5)
Filfil, N. T., 254[478] (244)
Fink, G., 502[298] (482, 483)
Fink, W. H., 499[155] (454)
Finn, M. G., 427[93] (382, 393), 427[103] (393)
Finster, D. C., 249[123] (220), 249[134] (221, 225, 231, 236)
Fisch, H., 187[95] (137), 191[333] (163)
Fischer, E. O., 544[236] (534)
Fischer, K., 501[272] (478, 479, 481, 483, 484)
Fischer, M. B., 370[231] (326)
Fischer, R. A., 556, 570[45] (556, 566), 570[46] (556, 561, 562, 564, 566), 570[47] (556), 570[48] (556), 570[49], 570[50] (557, 566), 570[51] (557)
Fitch, A. N., 33[172a] (20), 33[172b] (20)
Fitzpatrick, N. J., 248[113] (220), 425[4g] (375, 389)
Fjeldberg, T., 31[32] (6, 9)
Fjellvåg, H., 125[292] (98)
Fleischhauer, J., 186[45] (131), 186[46] (131, 132)
Fleming, I., 186[2] (130, 138), 497[4] (433, 436, 438, 441, 458, 459, 467, 486, 495, 496)
Flippin, L. A., 501[232] (469)
Flock, R., 31[46] (9)
Flood, L., 124[208a] (83)
Floriani, C., 367[34] (279, 280), 367[40] (279, 280)
Florjanczyk, Z., 499[102] (443, 468), 500[220] (468)
Flörke, U., 570[40] (554, 567), 570[41] (554), 570[42] (554), 570[60] (560), 571[61] (560)
Fluck, E., 542[155] (523, 524)
Folting, K., 271[27] (259), 272[77] (264), 272[91] (267)
Fomina, T. N., 122[111] (68)
Fonda, K. K., 252[310] (229)
Fontaine, X. L. R., 250[229] (224), 272[55] (262), 272[56] (262, 269), 272[81b] (265, 266), 272[87] (266), 272[90] (266), 273[93] (267), 273[94] (267), 273[101] (269), 273[117] (271), 369[197] (313, 315, 346, 357, 359), 370[242] (329), 370[244] (332), 370[245] (332, 340), 370[248] (332, 339), 370[249] (333), 370[252]

(336, 340, 342, 347), 370[254] (336), 370[258] (338, 345), 370[260] (338–40), 371[267] (341), 371[268] (340, 341), 371[269] (340, 341, 344, 347), 371[271] (340, 341), 371[272] (340), 371[273] (340), 371[274] (341), 371[275] (342, 357, 359), 371[277] (342), 371[282] (346), 371[283] (347), 371[286] (350), 371[288] (351), 371[289] (351), 371[290] (351), 371[309] (355–7), 371[311] (357, 359), 371[312] (357–9), 371[313] (358, 359), 371[314] (358), 372[323] (365), 426[11e] (377, 394, 396), 426[24] (378), 426[28] (377, 378), 426[37] (378, 380), 427[112] (396), 427[114] (396), 428[160] (400, 412)
Fontani, P., 192[451] (173)
Fontanille, M., 125[267] (93)
Foord, J. S., 541[88] (515)
Foppl, H., 33[181] (20)
Forschner, T. C., 124[229] (84)
Forster, G. E., 541[68] (513)
Forward, J. M., 428[171] (400, 412)
Fowkes, H., 273[93] (267), 273[94] (267), 370[252] (336, 340, 342, 347), 371[271] (340, 341), 371[288] (351), 371[289] (351), 371[290] (351), 371[310] (355–7), 371[312] (357–9)
Fox, D. J., 498[88] (441)
Foxton, M. W., 498[48] (438), 501[239] (470, 486), 501[271] (478, 484)
François, B., 126[354] (110, 111), 126[355] (111), 126[356] (111), 127[376] (116)
Frange, B., 191[356] (164), 191[357] (164), 273[96] (267), 273[97] (267)
Frank, A., 215[31] (202)
Frank, S. M., 190[272] (158)
Frank, W., 542[156] (524)
Frankhauser, P., 192[415] (170), 192[416] (170), 192[417] (170), 216[96] (214)
Franz, D. A., 430[328] (420)
Franz, T., 190[290] (159)
Fraser, R. R., 32[104h] (13)
Frasson, E., 543[204] (529)
Fréchet, J. M. J., 123[149] (72)
Freeman, P. K., 125[298] (99)
Freeman, W. A., 193[527] (178)
Freijee, F. J. M., 122[143a] (71, 85, 92), 122[143b] (71, 76, 85, 92)
Frenking, G., 33[139] (19), 54[13] (39), 131, 186[34] (131, 134), 186[37] (131)
Frey, G., 195[644] (185)
Frey, J. E., 191[341] (163)
Friesen, G. D., 272[49] (262), 272[50] (262), 273[107] (270), 273[108] (270), 273[109a] (270), 273[109b] (270), 273[111] (270)
Frigo, D. M., 540[11] (504–6, 516), 540[18] (505, 506), 542[127] (520), 543[175] (525), 543[192a] (527), 543[192b] (527)
Fripiat, J. G., 249[115] (220)
Fritts, J. D., 249[117] (220)
Frolov, A. N., 192[393] (167)
Froneman, S., 250[228] (224)
Frost, P. W., 367[66] (284, 285)
Frunze, T. M., 248[58] (218, 244), 254[467] (244)
Frye, S. V., 125[274] (95)
Fryzuk, M. D., 367[47] (279, 280), 367[48] (279, 280)
Fuchs, G., 499[98] (443)
Fuentes, G. R., 32[95] (12)
Fugami, K., 501[262] (476)
Fuji, H., 499[133] (451)
Fuji, Y., 499[133] (451)
Fujihara, H., 121[71] (65), 121[82] (66)
Fujikura, S., 123[182] (79), 123[183] (79)
Fujioka, H., 501[269] (476)
Fujita, M., 122[125] (70)
Fuka, M. Z., 249[117] (220)
Fukunaga, T., 187[58] (132)
Full, R., 216[97] (214)
Fuller, D. J., 31[62] (10)
Fuller, K., 247[29] (218), 252[365] (236), 252[367] (236)
Furlani, A., 502[303] (484)
Furnowsky, F., 191[343] (163)
Furst, G. T., 252[302] (228), 273[104] (269)
Fürstner, A., 59, 120[17] (59)
Furukawa, J., 499[133] (451)
Furukawa, N., 121[71] (65), 121[82] (66)
Furusaki, A., 180, 194[561] (180)
Fusek, J., 250[209] (223), 250[210] (223), 250[221] (224), 272[52] (262)
Fu-Son, H., 194[542] (179), 194[546] (179)
Fußstetter, H., 190[314] (161)

Gabbert, G., 215[49] (205)
Gabe, E. J., 498[71] (440, 479, 480), 501[291] (482)
Gabel, D., 195[615] (184), 253[415] (241), 255[501] (245)
Gagas, D., 123[184] (79)
Gahlmann, F., 541[63] (513, 514), 541[85] (514)
Gaimaldinov, R. G., 125[252] (87)
Gaines, D. F., 54[10] (38, 50), 54[23] (42), 55[41] (49, 50), 252[346] (233), 255[516] (246), 255[517] (246), 255[519] (246), 255[521] (246), 255[528] (246), 255[529] (247), 272[65] (264), 370[225] (323), 370[231] (326), 370[234] (326), 371[304] (352), 427[97] (388)
Gais, H.-J., 33[135] (17)
Galakhov, M. V., 251[283] (227)
Galan, P., 430[310] (416)
Gal'chenko, G. L., 249[162] (221, 226, 239), 251[265] (226, 239)
Galiano-Roth, A. S., 31[62] (10)
Galindo, J., 126[328] (106)
Gallagher, J. F., 273[113] (270), 273[116] (271)
Gallagher, M. J., 121[25c] (59, 60, 70, 100, 102)
Galle, J. E., 124[210] (83), 189[213] (151), 189[237] (154, 155), 469, 499[105] (443), 501[233] (469), 502[302] (484)
Gallina, C., 501[229] (468)
Gallucci, J. C., 34[197] (22)
Galow, P., 187[111] (139), 271[8] (258), 272[30] (260), 272[33] (260), 543[225] (533)
Galstyan, K. D., 251[255] (225, 241)
Gamp, E., 368[93] (287, 290, 291), 368[94] (287, 290, 291), 368[95] (287, 290), 368[101] (290, 291), 368[102] (290, 291), 368[105] (291), 368[106] (291)
Gamzatov, A. Z., 189[255] (156), 189[256] (156)
Gangnus, B., 427[96b] (381)
Ganter, C., 187[84] (135, 137), 187[86] (136), 187[87] (136, 137), 187[93] (137), 214[15] (199)
Gao, L. X., 502[313] (489)
Garbauskas, M., 542[143] (521)
Garcia, M. P., 427[94] (382, 398), 427[106] (394–6), 428[150] (400, 412), 430[318] (417, 418)
Gardiner, M. G., 124[245] (86, 92, 93, 95, 112–14), 127[385] (118)
Garner, S. E., 187[80] (135)
Garrett, P. M., 249[136] (221, 231)
Garrigues, B., 192[410] (169)
Garst, J. F., 61, 121[46] (61, 62), 121[49] (61), 121[50a] (61), 121[50b] (61)
Gaskill, D. K., 55[60] (54)
Gaspar, P. P., 499[111] (445)
Gasparis-Ebeling, T., 190[276] (158)
Gauss, J., 186[41] (131), 498[33] (436, 445), 540[2] (504, 529)
Gautam, D. P., 192[439] (172)
Gautheron, B., 34[197] (22)
Gavin, R. M., Jr., 543[205] (529)
Gavrilenko, V. V., 126[333] (106), 447, 499[117] (447), 499[156] (455), 500[162] (456), 500[163] (456), 541[96] (515)
Gavrilov, V. V., 252[345] (233, 241)
Gavrilova, L. A., 368[79] (286, 287)
Gayoso, J., 187[53] (132)
Gefter, E. L., 253[423] (241)
Geiderikh, A. V., 188[171] (145)
Geiersberger, K., 498[24] (435)
Geiger, W. E., 414, 426[40] (382, 400, 414), 426[73] (382, 387), 427[75] (382), 427[92] (382, 392)
Geilich, K., 215[53] (206)
Geisberger, G., 186[27] (130), 194[590] (182, 185)
Geissler, M., 32[92] (12), 33[160] (19, 21), 125[263] (91, 105)
Gelashvili, T. L., 254[441] (242), 254[468] (244)
Gellert, H.-G., 500[169] (458, 462, 464, 486), 500[173] (458), 500[207] (467)
Gemmler, A., 191[332] (162)
George, B., 120[19] (59)
George, C., 542[105] (518, 519)
George, T. F., 187[57] (132)
Géret-Baumgarten, L., 192[430] (171)
Gerr, R. G., 570[9] (547)
Gershkokhen, S. L., 126[332] (106), 126[334] (106)
Gerteis, R. L., 34[213b] (29)
Getman, T. D., 249[119] (220, 221, 231), 249[120] (220, 221, 231, 235), 249[136] (221, 231), 250[177] (221, 231), 253[395] (239, 246), 271[5] (258), 272[73] (264), 369[153] (296–8), 371[287] (350), 543[203] (529)
Geurink, P. J. A., 33[156] (19)
Ghadiri, M. R., 502[332] (491)
Ghilardi, C. A., 367[60] (284, 285), 368[72] (286), 368[73] (286)
Ghosh, T., 254[460] (244), 254[461] (244)
Giacomelli, G., 55[50] (53), 55[51] (53), 55[52] (53), 124[202] (82)
Gigli, R., 55[42] (50)

Gilbert, K. B., 277, 366[1] (277, 278)
Gilbert, T. M., 367[69] (285, 286)
Gilchrist, J. H., 31[61] (10), 31[62] (10)
Gilje, J. W., 123[161] (74, 109)
Gill, G. B., 31[28] (5)
Gill, W. R., 32[111] (15), 32[113] (15)
Gilson, D. F. R., 250[238] (225)
Gimarc, B. M., 248[69] (218, 233), 248[70] (218, 219, 233), 248[90] (219), 248[108] (220), 252[333] (232), 252[334] (232, 233), 252[337] (232), 252[338] (232), 252[340] (232, 233), 252[343] (233), 252[344] (233), 252[347] (233)
Gimeno, M. C., 429[261] (400, 411), 429[266] (400, 411), 430[277] (400, 411)
Gindelberger, D. E., 32[116] (15)
Gingrich, H. L., 254[460] (244), 254[461] (244), 254[462] (244), 254[463] (244)
Giolando, D. M., 542[115] (520)
Girolami, G. S., 367[35] (279, 280), 367[36] (279, 280), 367[37] (279, 280), 367[38] (279, 280, 282–4), 367[39] (280, 282, 283), 367[45] (279, 280), 367[46] (279, 280), 367[51] (282, 283), 570[25] (550)
Gladkova, Y. A., 254[464] (244)
Glaser, B., 187[81] (135), 189[197] (149), 190[277] (158)
Glass, J. A., Jr., 251[277] (227), 272[62] (263)
Glassman, T. E., 543[206] (530, 539), 543[208] (530)
Glegg, W., 32[99] (12)
Gleiter, R., 214[16] (199, 208), 214[17] (199, 208)
Glick, M. D., 32[120a] (15), 32[120b] (15), 497[13] (433, 448)
Glukhova, M. G., 195[604] (183)
Gnilomedova, T. P., 367[58] (284)
Goddard, R., 34[201] (26), 55[36] (48), 193[473] (174)
Godfroid, R. A., 370[208] (316)
Godovikov, N. N., 247[44] (218), 248[45] (218), 249[159] (221, 224, 238), 249[171] (221, 240), 251[291] (227), 254[446] (243), 254[447] (243), 254[448] (243), 254[449] (243), 254[456] (243)
Goedken, V. L., 121[47c] (61, 62, 77), 497[9] (433, 435, 450, 497)
Goel, A. B., 123[192a] (80, 81), 125[305] (101), 125[307] (101), 498[43] (436, 443)
Goethel, G. F., 500[192] (463, 470)
Goetze, U., 542[109] (519)
Goeva, L. V., 251[251] (225, 226)
Goffart, J., 368[82] (289)
Goldberg, J. E., 429[234] (400, 411), 429[235] (400, 411), 429[236] (400, 411), 430[279] (400, 411)
Goldenberg, D. M., 253[417] (241), 253[418] (241, 245), 255[497] (245)
Goldman, E., 120[19] (59)
Goldova, K. G., 498[63] (440)
Goldwasser, J. M., 255[486] (244)
Golubinskaya, L. M., 254[446] (243)
Golubinskii, A. V., 250[196] (222), 250[197] (222)
Gomez, F. A., 250[185] (221, 242), 250[216] (224, 232, 233), 252[315] (229), 252[368] (236), 254[439] (242), 427[127] (400, 416)
Gonsior, L. J., 500[167] (457)
Gonzalez Nogal, A. M., 501[227] (468), 502[335] (493)
Gonzalez Ortega, A., 502[323] (491)
Goode, M. J., 541[91] (515), 541[92a] (515), 541[92b] (515)
Goodreau, B. H., 272[68] (264), 370[206] (313, 316, 321), 370[235] (327)
Gorbatyuk, V. Ya., 194[573] (181), 195[604] (183), 195[647] (185)
Gordeev, M. F., 191[392] (167)
Gordon, C., 250[218] (224)
Gordon, M. D., 187[58] (132)
Gordon, M. S., 499[95] (443)
Görlitz, F. H., 541[54] (512), 543[210] (530, 532, 539)
Götz, N., 124[205b] (82)
Goursot, A., 249[115] (220)
Gozum, J. E., 367[37] (279, 280), 367[45] (279, 280), 367[46] (279, 280)
Graalmann, O., 32[99] (12)
Gracey, G. D., 542[163] (525)
Grachek, V. I., 193[461] (174), 193[462] (174), 193[463] (174, 179), 193[464] (174)
Graf, W., 54[5] (36)
Graham, J., 542[102a] (518)
Grandberg, A. I., 188[168] (145), 189[239] (154)
Grassberger, M. A., 191[343] (163), 191[344] (163), 247[10] (218)
Greatrex, R., 251[284] (227), 255[510] (245)
Grebenik, P. D., 369[172] (304, 307, 308), 369[193] (313, 315, 319), 369[194] (313, 315), 370[214] (319), 370[222] (321), 428[195] (400, 414)
Grebenkin, M. F., 193[510] (176)
Greci, L., 123[176a] (78)
Green, B. E., 367[70] (286), 367[71] (286), 368[74] (286)
Green, M. A., 34[202] (26)
Green, M. L. H., 367[53] (282, 284), 369[193] (313, 315, 319), 369[194] (313, 315), 370[214] (319), 370[222] (321), 427[94] (382, 398), 427[106] (394–6), 427[116] (394, 396, 397), 427[119] (395), 428[195] (400, 414), 429[249] (400, 411), 429[250] (400, 411), 429[263] (400, 411), 429[264] (400, 411), 429[271] (400, 411), 430[318] (417, 418), 540[32] (507), 543[219] (532), 570[37] (553, 555)
Green, T. A., 248[95] (219), 248[107] (220)
Greenberg, A., 247[13] (218), 247[27] (218), 247[31] (218), 247[32] (218), 247[40] (218), 248[49] (218), 271[2] (258), 272[28] (260), 272[35] (260), 366[4] (277), 425[1f] (374), 425[1g] (374, 415), 425[1h] (374, 377, 381, 388)
Greene, P. T., 543[223] (532, 533)
Greenhough, T. J., 429[239] (398, 400)
Greenwood, N. N., 251[284] (227), 255[511] (245), 271[26] (259), 272[56] (262, 269), 272[81b] (265, 266), 273[93] (267), 273[94] (267), 273[101] (269), 367[9] (277), 367[11] (277, 337), 367[12] (277, 337), 367[13] (277, 337), 367[14] (277), 367[15] (277), 369[183] (310, 316, 327, 337, 349), 369[192] (313), 369[195] (313, 315, 316, 320, 323), 369[196] (313, 315, 320, 323, 340), 369[197] (313, 315, 346, 357, 359), 369[200] (313, 316), 370[212] (319, 320, 324, 340, 341), 370[213] (319), 370[224] (322), 370[228] (324), 370[229] (324), 370[238] (327, 336, 349, 363), 370[239] (327, 336, 363, 364), 370[240] (327, 336, 364), 370[242] (329), 370[243] (331, 332, 336, 337), 370[244] (332), 370[245] (332, 340), 370[246] (332, 336), 370[247] (332, 336), 370[248] (332, 339), 370[249] (333), 370[252] (336, 340, 342, 347), 370[253] (336), 370[254] (336), 370[255] (337), 370[257] (337), 370[258] (338, 345), 370[260] (338–40), 370[262] (340, 341), 370[263] (340), 370[264] (340), 370[265] (340, 342, 343), 371[266] (340, 342), 371[267] (341), 371[268] (340, 341), 371[269] (340, 341, 344, 347), 371[271] (340, 341), 371[273] (340), 371[275] (342, 357, 359), 371[277] (342), 371[278] (342, 347), 371[279] (343), 371[281] (345, 346, 355, 356), 371[282] (346), 371[284] (347), 371[286] (350), 371[288] (351), 371[289] (351), 371[290] (351), 371[292] (351), 371[298] (351), 371[299] (352), 371[306] (355, 356), 371[307] (355, 356), 371[308] (355, 356), 371[309] (355–7), 371[310] (355–7), 371[311] (357, 359), 371[312] (357–9), 371[313] (358, 359), 371[319] (363), 371[320] (364), 371[321] (364), 372[323] (365), 372[324] (365), 425[1r] (374), 426[11e] (377, 394, 396), 426[21] (378), 426[37] (378, 380), 427[112] (396), 427[114] (396), 541[87b] (515), 543[222] (532)
Gregory, C. D., 34[207] (26)
Gregory, K., 2, 30[4] (2, 30), 31[59] (10), 31[60] (10)
Gregory, P., 34[192] (23, 24)
Greiser, T., 32[85a] (11), 34[210] (29), 126[323] (105)
Gren, A. I., 175, 192[450] (173), 193[475] (174), 193[476] (175), 193[477] (175), 193[478] (175), 193[482] (175), 193[484] (175, 183), 193[485] (175, 183), 193[486] (175, 183), 193[487] (175, 183), 193[488] (175, 183), 193[489] (175, 183), 193[490] (175, 183), 193[491] (175, 183), 193[492] (175, 183), 193[493] (175, 183), 193[494] (175, 183), 193[495] (175, 183), 194[573] (181), 194[574] (181), 195[599] (183), 195[600] (183), 195[601] (183), 195[602] (183), 195[603] (183), 195[604] (183), 195[605] (183), 195[606] (183), 195[607] (183), 195[608] (183), 195[609] (183), 195[646] (185), 195[647] (185)
Grevels, F.-W., 369[162] (296)

Grey, R. A., 34[199] (26)
Gribanova, E. V., 124[242] (86, 98)
Gribkova, P. N., 248[53] (218)
Grimes, R. N., 214[3] (198–201, 207, 208), 215[59] (207), 215[60] (207), 215[61] (207), 247[40] (218), 247[41] (218), 249[125] (220, 221, 225–7, 229), 249[134] (221, 225, 231, 236), 249[135] (221, 234), 250[213] (223), 250[220] (224), 250[230] (224), 251[241] (225, 227), 251[242] (225–7), 251[243] (225–7, 229, 234), 251[244] (225, 227, 229, 234), 251[245] (225–7, 229, 234), 251[250] (225, 229), 251[286] (227), 252[309] (229), 252[310] (229), 252[311] (229, 238), 252[312] (229), 252[353] (234), 255[524] (246), 271[1] (258), 271[13] (258), 273[112] (270), 366[3] (277), 367[10] (277), 370[217] (320–2, 324, 327, 335, 341), 370[218] (320), 370[219] (321), 370[220] (321, 335), 370[221] (321), 370[223] (322), 370[226] (323), 370[251] (335, 341), 377, 395, 425[1a] (373, 374, 376, 377, 388, 391, 395, 398, 414, 417, 419, 421), 425[1c] (374, 376, 377, 381, 388, 391), 425[1d] (374, 381), 425[1h] (374, 377, 381, 388), 425[1j] (374), 425[1k] (374), 425[1q] (374, 377, 388, 421), 425[1s] (374), 425[1t] (374, 381), 425[1v] (374), 425[4e] (375, 422), 425[8] (377), 425[9a] (377, 382), 426[11c] (377, 381, 382), 426[13a] (377, 382, 388), 426[14a] (377, 421–3), 426[14b] (377, 421–3), 426[14c] (377, 382, 421–3), 426[41] (382, 421, 423), 426[42] (382), 426[43] (381, 382), 426[44] (381, 382, 387), 426[45] (382), 426[46] (382, 393), 426[47] (382, 387), 426[48] (382, 387, 421, 423), 426[49] (382, 396), 426[50] (382), 426[51] (381, 382), 426[52] (381, 382), 426[53] (382, 396), 426[54] (382, 387), 426[55] (382, 387, 422), 426[56] (382), 426[57] (381, 382, 421, 423), 426[60] (382, 387), 426[61] (382), 426[62] (382), 426[63] (382), 426[66] (382), 426[67] (382, 398), 426[68] (382, 398), 426[69] (382), 426[70] (382), 426[71] (382), 426[73] (382, 387), 427[74] (382), 427[75] (382), 427[76] (382), 427[77] (382), 427[78] (382), 427[80] (382, 387), 427[89] (381, 382, 392, 422), 427[91] (381, 382), 427[92] (382, 392), 427[93] (382, 393), 427[95] (381, 382), 427[96b] (381), 427[98a] (391), 427[98b] (391, 395), 427[98c] (391), 427[98d] (391, 414, 419), 427[102a] (382, 392), 427[102b] (382, 392), 427[103] (393), 427[124] (398), 428[163] (400), 430[328] (420), 430[329] (420), 430[334c] (421), 430[335] (422), 541[100] (517), 543[223] (532, 533)
Grohmann, A., 192[427] (171)
Gromek, J. M., 121[74] (65, 74, 82), 370[211] (318)
Gronowitz, S., 191[339] (163)
Gropen, O., 125[293] (98)
Gross, J., 34[203] (26)
Gross, U., 189[254] (156)
Grosse, A v., 500[160] (456, 462)
Grossi, L., 124[209] (83)
Grossman, R. B., 430[307] (416)
Groteklaes, M., 189[258] (156)
Grove, D. M., 32[94] (12, 13)
Grubbs, R. H., 501[282] (480)
Grüner, B., 428[154] (400, 414), 428[172] (400, 414)
Grüning, R. J., 31[31] (6), 31[35] (6)
Grushin, V. V., 247[43] (218), 249[148] (221), 249[149] (221), 249[158] (221), 250[180] (221), 250[181] (221), 251[283] (227), 253[375] (237), 253[376] (237, 238), 253[377] (237), 253[378] (237), 253[379] (237, 238), 253[380] (237)
Gruter, G.-J. M., 125[278] (96, 102)
Grützmacher, H.-F., 32[85d] (11), 32[103] (12), 186[13] (130, 131)
Guastini., C., 367[40] (279, 280)
Gudat, D., 194[584] (182)
Guerra Suarez, M. D., 502[335] (493)
Guerrieri, A., 502[303] (484)
Guggenberger, L. J., 369[201] (313)
Guidi, M., 501[264] (476)
Guilard, R., 571[75] (566), 571[76] (566, 567), 571[77] (566, 567, 569), 571[78] (566, 567, 569), 571[79] (566, 567), 571[80] (566, 567), 571[83], 571[84], 571[98] (569), 571[99] (569)
Guindi, L. H. M., 186[3] (130, 138)
Gulliver, E. G., 193[465] (174, 180)
Gümrükcü, I., 32[90] (12)
Gun'ko, Yu. K., 367[20] (277, 286, 287), 570[34] (551)
Günther, H., 33[135] (17)
Gupta, G. P., 249[122] (220)
Gupta, O. D., 191[358] (164)
Gurjar, M. K., 194[595] (182)
Gurskii, M. E., 188[145] (143), 188[146] (143), 188[147] (143, 154), 188[149] (143), 188[150] (143, 147), 188[164] (145), 188[167] (145, 146), 188[171] (145), 188[173] (146), 188[174] (146), 188[178] (147), 188[180] (147), 188[182] (147), 189[239] (154), 191[366] (165)
Gürtzgen, S., 502[299] (483)
Guseinov, A. G., 501[287] (481)
Gusev, A. I., 250[176] (221), 570[8] (547)
Gutierrez-Jugo, J. F., 541[53] (512)
Gutowsky, H. S., 500[189] (463)
Gynane, M. J. S., 544[259] (537)
Gyoung, Y. S., 500[210] (467)
Gysling, H. J., 186[23] (130), 425[1p] (374), 543[183] (525, 526), 543[184] (525, 526)

Haaland, A., 124[247] (86, 94, 101), 125[293] (98), 126[368] (113), 498[20] (434), 499[120] (448), 499[152] (454)
Haase, M., 32[99] (12)
Habaue, S., 122[118] (70, 77), 126[345] (108, 109)
Habben, C. D., 188[189] (148), 190[287] (159), 190[289] (159), 192[395] (167), 194[551] (179, 180), 194[552] (179), 194[553] (179), 194[554] (179, 180), 194[555] (179), 194[556] (180), 194[562] (180), 194[566] (180), 194[567] (180), 194[570] (180), 194[571] (180), 195[610] (183), 195[631] (184, 185), 195[632] (184), 195[633] (184), 195[634] (184, 185), 195[636] (185), 195[638] (185), 195[639] (185), 195[640] (185), 195[641] (185), 195[642] (185)
Habbou, A., 571[80] (566, 567), 571[98] (569)
Hackett, P. A., 499[90] (442)
Haenel, M. W., 544[249] (536)
Hafner, W., 544[236] (534)
Hagen, A. P., 247[12] (218), 248[52] (218, 244)
Hager, R., 544[253] (536)
Haggerty, B. S., 368[135] (297, 299, 300), 369[168] (301), 369[170] (301), 369[202] (313, 316), 369[204] (313, 317), 541[62] (512, 524)
Hagias, S., 123[165] (76)
Hahn, B., 193[503] (176)
Hahn, G., 502[331] (491)
Hahn, H., 542[157] (524)
Hahn, J. M., 497[16] (434, 448)
Haiduc, I., 186[17] (130), 186[18] (130), 186[19] (130), 186[27] (130)
Hails, M. J., 370[238] (327, 336, 349, 363), 370[253] (336)
Hall, F. C., 498[28] (435), 498[29] (435)
Hall, P. L., 31[61] (10)
Hallade, M. W., 123[181b] (79)
Hallberg, A., 123[163a] (75)
Haller, K. J., 369[177] (307), 369[205] (313, 316)
Hallock, R. B., 500[158] (455), 540[13] (505, 506), 541[57] (512), 541[76] (514, 515), 543[203] (529)
Hallwachs, W., 456, 500[159] (456)
Haltiwanger, R. C., 271[25] (259)
Halvorsen, S., 499[120] (448)
Haly, B. D., 124[220] (83)
Hambley, T. W., 54[25] (44, 47)
Hamdouchi, C., 121[47c] (61, 62, 77), 121[53] (61), 126[343] (108, 109, 117)
Hamilton, E. J. M., 249[145] (221, 224), 249[151] (221), 427[131] (400, 412), 428[189] (400, 412), 430[286] (400)
Hammel, A., 126[362] (112–14, 117, 118)
Hammerschmidt, W. W., 121[39] (60)
Hammond, M. E., 543[192b] (527)
Hampel, B., 34[195] (22)
Hampel, F., 126[375] (115)
Hampson, C. A., 499[92] (442), 499[93] (442)
Han, F., 194[547] (179)
Han, N. F., 501[291] (482)
Han, R., 55[44] (50–2), 126[322] (104)
Hanafusa, T., 123[191] (80, 81)
Hanaizi, J., 501[235] (469)
Hanawalt, E. M., 126[326] (105)
Hanay, W., 126[320] (104)
Hancock, K. G., 191[336] (163), 191[337] (163)
Hanecker, E., 187[81] (135), 190[281] (159), 190[290] (159), 192[396] (167, 168), 192[398] (168), 192[432] (171, 179), 194[578] (181)

Hanousek, F., 252[319] (230)
Hansen, A., 192[438] (172)
Hansen, E., 121[41] (60)
Hansen, H. P., 272[54] (262)
Hansen, J., 121[74] (65, 74, 82)
Hanson, G. R., 127[385] (118)
Hanstein, W., 502[310] (488)
Hanus, V., 570[18] (548)
Hanusa, T. P., 120[3] (58, 107, 110, 112, 114, 119), 120[4] (58, 107, 109, 112, 113, 115), 126[341] (107, 112), 126[347] (109, 110), 126[348] (109, 112–14), 126[349] (91, 109, 110, 112, 115, 118, 119), 126[361] (112), 126[365] (112–14), 126[367] (112, 113), 126[370] (113, 114), 126[371] (113, 115, 120), 126[372] (113, 114), 127[387] (119), 272[83] (265, 266), 273[109a] (270), 426[11a] (377, 400, 412), 426[11b] (377, 396, 400), 428[157] (400, 412, 414)
Haoudi-Mazzah, A., 194[581] (181)
Harada, A., 255[507] (245)
Harder, C. L., 121[44] (60, 79)
Harder, S., 32[106] (13), 33[153] (19), 34[196] (22)
Harding, I. S., 540[18] (505, 506)
Harlow, R. L., 571[74] (565)
Harm, K., 31[38b] (6)
Harman, M., 250[190] (221, 229, 237)
Harmat, N., 125[313] (102)
Harms, K., 32[105] (13), 33[133] (17), 33[134] (17), 33[138] (19), 33[139] (19), 123[158] (74), 193[472] (174)
Harney, D. W., 499[137] (452)
Harpp, K. S., 367[22] (277, 294, 297, 299, 300), 368[134] (297, 299)
Harris, D. H., 186[28] (130)
Harris, H. A., 54[23] (42), 255[528] (246)
Harris, P. R., 542[125] (520), 542[140] (521, 522, 526), 543[182b] (527)
Harris, R. K., 250[226] (224)
Harrison, A. T., 31[62] (10), 255[513] (245)
Harrison, P. G., 32[119c] (14), 247[7] (218)
Harrison, W. H., 33[183] (21)
Hart, F. A., 247[7] (218), 253[408] (240)
Hart, H., 121[69] (65, 83)
Hart, I. J., 428[193] (400, 411, 417, 418), 428[201] (400, 411), 429[262] (400, 411), 429[269] (400, 411)
Hartke, K., 191[382] (166), 194[535] (179)
Hartley, F. R., 31[28] (5)
Hartmann, H.-M., 32[120d] (15)
Hartmann, U., 540[25a] (505)
Harvey, S., 121[25a] (59, 60, 63, 70, 71, 100), 121[25c] (59, 60, 70, 100, 102), 121[27] (59, 72, 99, 100), 123[155] (73, 99, 111), 125[294] (98–100), 543[212] (531)
Hase, T. A., 123[177] (78)
Hasek, J., 249[152] (221), 250[178] (221), 250[183] (221), 250[191] (221)
Häser, M., 499[107] (444)
Hasler, P., 121[39] (60)
Haspeslagh, L., 502[297] (482)
Hassner, A., 189[260] (157)
Hatada, K., 123[150] (72)
Hatajima, T., 124[212] (83)
Hatfield, W. E., 426[65] (382, 388)
Hattersley, A. D., 369[146] (297, 304)
Hattori, K., 501[267] (476), 501[268] (476)
Haubold, W., 188[185] (147), 190[321] (162), 190[322] (162), 191[332] (162), 191[367] (165), 195[644] (185), 253[370] (236), 272[76] (264)
Hauck, J., 189[204] (150)
Haupt, H.-J., 570[39] (553, 560, 567), 570[40] (554, 567), 570[41] (554), 570[42] (554), 570[43] (554), 570[60] (560), 571[61] (560)
Hauschild, K., 122[91] (66)
Hausen, H.-D., 31[33] (6), 188[186] (147), 540[15] (505, 506), 541[66] (513), 542[128] (520), 542[155] (523, 524)
Haushalter, R. C., 255[502] (245)
Hausmann, I., 214[14] (199, 202), 215[40] (203)
Hauss, J., 428[171] (400, 412)
Havlas, Z., 250[200] (222, 223)
Hawkins, S. M., 31[43] (8)
Hawthorne, M. F., 127[378a] (116), 127[378b] (116), 249[119] (220, 221, 231), 249[120] (220, 221, 231, 235), 249[136] (221, 231), 249[140] (221, 238), 249[150] (221, 231, 237, 241, 245), 249[173] (221, 242), 250[177] (221, 231), 250[185] (221, 242), 250[187] (221, 243), 250[189] (221, 243), 252[321] (230), 252[355] (234), 252[362] (236), 253[389] (238), 253[417] (241), 253[418] (241, 245), 254[432] (242), 254[439] (242), 255[496] (245), 255[497] (245), 255[498] (245), 255[500] (245), 255[523] (246), 271[12] (258), 271[16] (258), 271[17] (259), 271[18] (259), 271[19] (259), 271[22] (259), 271[23] (259), 272[89] (266), 273[100] (268), 367[26] (277), 371[316] (361), 371[317] (362), 377, 399, 412, 415, 425[1e] (374), 425[1g] (374, 415), 425[10] (377, 400, 412), 426[38] (378, 380), 427[113] (396), 427[121] (398), 427[127] (400, 416), 427[128] (400, 412), 427[134] (400, 412), 428[151] (400, 412), 428[152] (400, 412), 428[153] (400, 412), 428[158] (400), 428[173] (400, 416), 428[176] (400, 416), 428[180] (400, 414), 428[185] (400, 412), 428[188] (400, 412), 428[190] (400, 412), 428[192] (400, 412), 428[199] (400, 412), 429[206] (400, 415), 429[207] (400, 415), 429[212] (400, 415), 429[213] (400, 415), 429[214] (400, 415), 429[216] (400, 412), 429[219] (400, 415), 429[222] (400, 415), 429[224] (400, 415), 429[227] (400, 415), 429[228] (400, 415), 429[233] (400, 415), 429[241] (400, 412), 430[281] (400, 414), 430[288] (400, 412), 430[296] (415), 430[297] (415), 430[301] (415), 430[323] (418, 419), 430[324] (418, 419), 504, 541[99] (517), 543[220] (532), 543[226] (533), 544[228] (533), 544[229] (533)
Hay, J. N., 501[279] (480)
Hay, P. J., 186[47] (132)
Hayami, H., 500[197] (465)
Hayashi, N., 123[175] (78)
Hayashi, T., 124[216] (83)
Hayes, P. C., 121[36] (60), 543[173] (525)
He, D., 192[443] (173), 192[444] (173)
He, J., 370[207] (316)
He, X., 542[114] (520, 525)
Head, J. D., 498[36] (436, 442)
Healy, E. F., 248[83] (219)
Healy, M. D., 497[10] (433, 450, 454, 497), 497[14] (434, 448, 497), 499[125] (449)
Healy, P. C., 368[74] (286)
Heathcock, C. H., 31[27] (5)
Heaton, D. E., 542[124] (520)
Heck, J., 216[97] (214)
Heckmann, G., 194[547] (179)
Hedaya, E., 272[84] (266)
Hedberg, K., 367[44] (279, 280)
Hedberg, L., 367[44] (279, 280)
Hefner, C., 272[84] (266)
Heimbach, P., 500[174] (458)
Heine, A., 188[158] (144), 189[202] (150, 151), 189[240] (154), 189[243] (154), 190[311] (161), 195[632] (184)
Heinen, H., 272[63] (263)
Heinoja, K., 120[20b] (59, 60), 121[41] (60)
Heinrich, D. D., 368[98] (288, 290)
Heintz, V. J., 193[527] (178)
Heinz, K. J., 124[211a] (83)
Hejna, C., 542[143] (521)
Helary, G., 125[267] (93)
Heller, G., 186[25] (130, 171)
Hellert, D., 541[58] (512)
Helm, G. F., 429[218] (400, 411)
Helm, S. W., 215[27] (201), 215[29] (201), 368[117] (292, 295), 368[118] (292, 295)
Hemer, I., 122[120] (70)
Hencken, G., 33[154a] (19, 20), 34[213c] (29)
Hendershot, D. G., 543[187] (526)
Henderson, K. W., 126[316] (103)
Henderson, M. J., 124[246] (86, 90), 541[51] (512)
Henderson, S. J. B., 430[320] (417, 418), 430[321] (417–19)
Hengesbach, J., 215[31] (202)
Hengge, E., 32[120d] (15)
Henkelmann, J., 189[222] (152), 189[225] (153)
Henly, T. J., 249[173] (221, 242)
Henneberg, D., 193[524] (178)
Heppert, J. A., 255[516] (246)
Herber, R. H., 272[29] (260), 544[232] (534)
Herberhold, M., 190[272] (158)
Herberich, G. E., 125[288] (98), 135, 187[84] (135, 137), 187[86] (136), 187[87] (136, 137), 187[93] (137), 188[162] (145, 153), 189[208] (150, 151), 189[209] (150, 151), 189[210] (151), 189[214] (151), 189[215] (151), 189[228] (153), 189[232] (153), 189[245] (155), 189[246] (155), 214[1] (198, 202), 214[8] (198, 209), 214[9] (198, 209), 214[14] (199, 202), 214[15] (199), 214[19] (199, 203, 204), 215[31] (202), 215[32] (202), 215[33] (202), 215[34] (202), 215[35] (202), 215[36] (202), 215[37] (202, 204), 215[38]

(202, 203), 215[39] (202–4), 215[40] (203), 215[41] (203), 215[42] (203, 204), 215[44] (204), 215[45] (204), 215[46] (204), 215[47] (204), 215[48] (204), 215[66] (209), 215[67] (209), 215[68] (209), 215[69] (209), 215[79] (211), 252[329] (231, 234), 421
Herbst-Irmer, R., 195[633] (184), 543[199] (528)
Herdtweck, E., 502[296] (482), 570[50] (557, 566)
Herfert, N., 502[298] (482, 483)
Hermanek, S., 247[32] (218), 248[93] (219, 220, 224, 229, 230, 234), 250[200] (222, 223), 250[205] (222), 250[209] (223), 250[210] (223), 250[211] (223), 250[221] (224), 251[249] (225, 230), 251[272] (226), 252[301] (228, 229), 252[306] (228, 235, 240), 252[307] (228, 238, 239), 252[314] (229, 230, 246), 252[317] (230), 252[318] (230), 252[319] (230), 252[326] (231, 234), 252[358] (235, 237), 252[359] (235, 237), 252[360a] (235), 253[371] (236), 253[372] (236, 237, 244), 253[373] (237), 253[383] (237, 238, 240), 253[384] (237), 253[386] (237), 253[400] (239), 253[404] (240), 253[405] (240), 253[413] (241), 254[459] (244), 258, 271[3] (258), 272[51] (262), 272[53] (262), 273[102] (269), 273[106] (269), 426[31] (378), 428[155] (400), 428[165] (400), 428[175] (400, 414)
Hermes, M., 193[473] (174)
Herrera, S., 248[87] (219), 250[214] (223), 252[360b] (235)
Herrmann, W., 502[296] (482)
Herter, W., 189[216] (151), 215[56] (206)
Heshaft, A., 544[239] (534)
Heslop, J. A., 55[43] (50)
Hess, H., 499[153] (454)
Hessner, B., 189[208] (150, 151), 189[210] (151), 189[228] (153), 189[232] (153), 214[9] (198, 209), 214[19] (199, 203, 204), 215[38] (202, 203), 215[39] (202–4), 215[40] (203), 215[41] (203), 215[44] (204), 215[48] (204), 215[68] (209), 215[69] (209)
Hewes, J. D., 429[228] (400, 415), 430[296] (415)
Hey, E., 31[70] (10), 31[71] (10), 31[74] (10)
Hey-Hawkins, E., 31[73] (10)
Heyn, A. S., 122[94] (67)
Heyne, E., 186[45] (131), 186[46] (131, 132)
Higa, K. T., 542[103] (518, 519), 542[105] (518, 519), 542[122] (520), 542[123] (520)
Hildebrandt, J., 191[343] (163)
Hildebrandt, K., 499[115b] (496)
Hildenbrand, K., 544[268] (540)
Hildenbrand, M., 187[78] (135), 189[200] (150), 189[203] (150), 191[371] (165)
Hildenbrand, T., 543[172] (525, 526), 544[255] (537–9)
Hill, A. F., 369[198] (313, 315), 369[199] (313, 315)
Hill, C. L., 121[42] (60)
Hill, E. A., 121[44] (60, 79), 121[65] (63), 123[180] (79), 123[181a] (79), 123[181b] (79), 123[184] (79)
Hill, R. C. P., 255[509] (245)
Hill, T. G., 370[208] (316)
Hill, W. E., 253[407] (240), 253[409] (241)
Hillenbrand, D. F., 54[10] (38, 50)
Hiller, W., 33[149] (19), 499[114] (446), 540[6] (504, 537), 541[71] (514, 537–9)
Hinderer, A., 499[153] (454)
Hintze, M., 192[407] (169, 170)
Hiraga, Y., 127[383] (118)
Hirashima, T., 122[101] (67)
Hirayama, N., 502[334] (492)
Histed, M., 499[94] (442)
Hitchcock, A. P., 251[277] (227)
Hitchcock, P. B., 31[32] (6, 9), 31[41] (8), 31[69] (10), 31[70] (10), 32[79] (11), 32[123] (15), 33[171] (19), 123[151] (72, 85, 87, 88, 92), 123[152] (72, 73, 86–8, 92), 124[244] (86, 95, 96), 126[350] (109, 110), 193[467] (174), 499[154] (454), 541[97] (515, 526)
Hlatky, G. G., 430[306] (416)
Hnyk, D., 250[200] (222, 223), 272[41] (261)
Hoberg, H., 446, 447, 498[30] (435, 446), 499[112] (446), 500[164] (457), 500[165] (457), 500[200] (465, 466), 500[201] (465)
Hockless, D. C. R., 570[38] (553, 555, 556), 570[54] (557, 558, 567), 571[92] (568)
Hodgkins, T. G., 192[398] (168)
Hodgson, K. O., 34[209] (26)
Hodgson, S. M., 31[37] (6)
Hoffer, R. K., 126[330] (106)
Hoffman, D., 34[194] (23)
Hoffman, K., 32[103] (12), 34[212] (29), 34[214] (29)
Hoffmann, C. J., 542[117] (520, 525)
Hoffmann, G. G., 541[58] (512), 543[180] (525)
Hoffmann, R., 194[535] (179), 198, 199, 214[12] (198, 199), 248[81] (219), 531, 543[211] (531)
Hoffmann, R. W., 192[448] (173), 192[449] (173)
Hoffmann, U., 368[112] (292, 293)
Hofmann, F. R., 126[320] (104)
Hofmann, J., 544[252] (536)
Hofmann, P., 215[29] (201), 368[117] (292, 295), 543[207] (530)
Höfner, A., 187[70] (134)
Hogenbirk, M., 122[138b] (71), 123[156] (73)
Hogerheide, M. P., 32[94] (12, 13)
Höhner, U., 190[305] (161), 215[84] (212), 215[86] (212), 215[87] (212)
Holah, D. G., 367[68] (285, 290)
Hollander, F. J., 367[69] (285, 286)
Holley, W. K., 542[138a] (521), 542[138b] (521)
Holloway, J. H., 247[7] (218)
Hollstein, W., 33[133] (17)
Holm, T., 81, 121[64] (63, 78), 123[170] (78), 123[187] (80, 81), 123[190] (80, 81), 123[192b] (80, 81), 123[197] (82), 123[198a] (82), 124[208a] (83)
Holmes, A. B., 121[62] (63)
Holmes, R. R., 543[195] (527)
Hölscher, M., 427[89] (381, 382, 392, 422)
Holt, E. M., 369[159] (294)
Holt, T., 124[208b] (83)
Holtkamp, H. C., 121[84] (66, 71, 76, 92)
Holtz, J., 30[1c] (1, 18)
Holub, J., 272[53] (262), 273[106] (269), 426[34] (378, 379)
Holzkamp, E., 478, 501[270] (478, 484, 497)
Hommes, H., 121[79] (65, 71)
Honda, M., 502[316] (490)
Honda, Y., 501[265] (476)
Hong, F.-E., 215[43] (203), 367[23] (277, 294, 297, 300, 303), 369[144] (297, 300), 430[333] (421, 423)
Hönig, S., 188[131] (141)
Hoong, L. K., 192[455] (173), 192[456] (173)
Hope, H., 31[54] (9, 11), 31[68] (10), 32[89] (12), 33[158] (19, 29), 33[169] (19), 499[115c] (446, 496), 542[147] (522), 542[150] (522, 523), 544[266] (539)
Hoppe, R., 544[262] (539)
Horak, M., 120[15] (59, 60), 120[20a] (59, 60), 121[40] (60), 121[41] (60)
Horchler, K., 190[267] (158), 190[268] (158)
Hordis, C. K., 498[37] (436, 470, 495), 498[38] (436, 470, 495)
Hornbach, P., 187[78] (135), 187[79] (135), 191[371] (165)
Hosmane, N. S., 247[42] (218), 249[127] (221, 234), 249[128] (221, 235), 249[137] (221, 225, 228), 250[195] (222, 227), 252[313] (229), 252[331] (231), 252[363] (236, 242), 254[436] (242), 271[9] (258), 271[10] (258), 271[11] (258), 271[24] (259), 272[28] (260), 272[29] (260), 272[30] (260), 272[31] (260), 272[32] (260), 272[78] (265), 372[325] (365), 374, 388, 391, 425[1b] (374, 376, 381, 388, 389, 411, 417), 425[4h] (375, 389), 426[65] (382, 388), 427[83] (382, 388), 427[84] (382, 388), 427[85] (382, 388), 427[86] (382, 389), 427[87] (382, 388), 427[99] (391), 543[224] (532), 544[231] (533), 544[232] (534), 544[233] (534), 544[234] (534), 544[235] (534)
Hosoda, Y., 194[534] (179)
Hosoya, H., 248[71] (218)
Hostalek, M., 188[162] (145, 153), 189[215] (151), 189[232] (153), 214[9] (198, 209), 215[34] (202), 215[37] (202, 204), 215[38] (202, 203), 215[68] (209)
Houri, A. F., 122[104] (67)
Housecroft, C. E., 247[5a] (218, 226–8, 237, 239, 242, 244, 245), 251[285] (227), 366[5] (277), 366[6] (277), 366[7] (277, 294), 367[22] (277, 294, 297, 299, 300), 368[126] (294, 296, 297), 368[127] (296, 297), 368[128] (297), 368[129] (296, 297), 368[130] (294, 297, 298), 368[131] (296–8), 368[132] (297, 298), 368[133] (297, 298), 368[134] (297,

299), 368[135] (297, 299, 300), 369[136] (297, 299), 369[137] (297, 299), 369[138] (297, 299, 300), 369[141] (296, 297), 369[142] (294, 296, 297, 301, 303), 369[143] (297, 300, 301), 369[145] (297, 301), 369[146] (297, 304), 369[147] (297, 303), 369[148] (297, 303), 369[156] (294, 298), 369[165] (299), 369[166] (299), 369[167] (299), 369[168] (301), 369[169] (301), 369[170] (301), 369[180] (308), 369[181] (308), 369[191] (313), 369[202] (313, 316), 369[203] (313, 317), 369[204] (313, 317), 370[210] (317)
Hoveyda, A. H., 122[104] (67)
Howard, J. A., 498[35] (436, 441), 498[84] (441), 499[92] (442), 499[93] (442), 499[94] (442)
Howard, J. A. K., 123[152] (72, 73, 86–8, 92), 215[41] (203), 367[66] (284, 285), 368[115] (292, 293), 428[196] (400, 411), 429[218] (400, 411), 429[220] (400, 412), 429[235] (400, 411), 429[238] (400, 411), 429[249] (400, 411), 429[250] (400, 411), 429[251] (400, 411), 429[252] (400, 411), 429[256] (400, 411), 429[257] (400, 411), 429[260] (400, 411), 429[265] (400, 411), 429[271] (400, 411), 430[276] (400, 411), 430[278] (400, 411), 430[320] (417, 418), 430[321] (417–19)
Howard, J. W., 430[328] (420), 430[329] (420)
Howarth, O. W., 250[222] (224, 234), 255[513] (245)
Howes, A. J., 187[82] (135, 139, 180), 570[21] (548, 550)
Hrebicek, J., 191[367] (165)
Hrncir, D. C., 499[147] (454), 541[41] (511)
Hsieh, K., 123[184] (79)
Hsu, L.-Y., 195[613] (183), 272[73] (264), 369[149] (296, 297), 369[150] (296, 297), 369[152] (296, 297)
Hsu, W. H., 501[281] (480)
Hsu, W.-L, 369[149] (296, 297)
Hu, D., 189[229] (153)
Huang, Q., 254[460] (244), 254[462] (244), 254[463] (244)
Huang, S. B., 501[281] (480)
Huang, Y., 370[207] (316)
Huber, B., 126[364] (112–14), 544[242] (535), 544[251] (536), 544[252] (536), 544[253] (536)
Hubert-Pfalzgraf, L. G., 32[82b] (11)
Hues, S. M., 55[60] (54)
Huffman, J. C., 34[202] (26), 126[347] (109, 110), 126[349] (91, 109, 110, 112, 115, 118, 119), 126[367] (112, 113), 126[370] (113, 114), 126[372] (113, 114), 272[35] (260), 272[37] (260), 272[58] (263), 272[71] (264), 272[72] (264), 272[88] (266, 267), 273[111] (270), 367[49] (280, 282), 367[52] (280, 282), 369[178] (307, 308), 426[11a] (377, 400, 412), 426[11b] (377, 396, 400), 426[32] (378), 570[24] (549), 570[26] (550)
Hughes, A. N., 367[68] (285, 290)
Hughes, D. L., 541[79] (514)
Hughes, N., 120[18] (59, 77)
Hummel, L., 249[144] (221), 250[191] (221)
Hünig, S., 188[126] (140), 188[127] (140, 146)
Hunold, R., 187[67] (134), 187[70] (134), 187[76] (135), 187[77] (135)
Hunter, W. E., 367[32] (278, 286, 287), 540[13] (505, 506), 540[17] (505, 506), 542[119] (520), 543[176] (525), 543[203] (529), 543[216] (531)
Hursthouse, M. B., 187[82] (135, 139, 180), 250[190] (221, 229, 237), 271[14] (258), 272[34] (260), 272[80] (265), 367[56] (284), 540[11] (504–6, 516), 540[14] (505, 506), 540[18] (505, 506), 540[22] (505, 506), 542[127] (520), 543[175] (525), 543[192a] (527), 543[192b] (527), 570[20] (548), 570[21] (548, 550), 570[23] (549), 570[25] (550)
Husk, G. R., 500[172] (458, 471), 500[193] (463, 465), 500[199] (465)
Hussain, B., 540[11] (504–6, 516), 542[127] (520), 543[175] (525)
Hutchings, D. S., 127[379] (117)
Hutchinson, L. L., 125[298] (99)
Huttner, G., 194[529] (178), 215[31] (202)
Hutton, R. S., 251[278] (227)
Hutton, W. C., 250[220] (224), 370[221] (321), 426[42] (382)
Hvoslef, J., 32[89] (12)
Hyla-Kryspin, I., 214[16] (199, 208), 214[17] (199, 208)
Hyun, B. C., 502[340] (494)

Iapalucci, M. C., 571[69] (563, 564)
Ibers, J. A., 571[85] (567)
Ibragimov, A. A., 123[169] (77, 92)
Ibragimov, A. G., 122[98] (67, 97, 98), 123[169] (77, 92), 124[242] (86, 98), 502[333] (492)
Ibrahim, M. A., 186[42] (131)
Ichikawa, K., 121[72] (65)
Idelmann, P., 187[109] (139)
Ievins, A., 34[212] (29)
Igarashi, T., 502[316] (490)
Ignatenko, A. V., 188[149] (143)
Ignat'eva, S. M., 251[264] (226)
Ignat'eva, S. N., 192[428] (171), 195[618] (184), 195[624] (184), 195[626] (184)
Iida, H., 122[125] (70)
Ikorskii, V. N., 428[166a] (400, 414)
Ilsley, W. H., 497[13] (433, 448), 497[16] (434, 448)
Imai, T., 194[557] (180)
Imamoto, T., 124[212] (83)
Imamura, A., 189[236] (154)
Imazaki, H., 193[505] (176), 193[506] (176)
Imi, K., 123[182] (79)
Imizu, Y., 120[10] (58, 59, 62, 78, 108)
Inagaki, S., 499[91] (442)
Indolese, A., 124[217] (83)
Inman, W., 250[215] (223)
Innocenti, P., 367[60] (284, 285)
Inokawa, H., 189[259] (157)
Inoue, M., 123[183] (79)
Inoue, S., 500[216] (468)
Interrante, L. V., 542[143] (521)
Ionkin, A. S., 192[428] (171), 195[628] (184)
Ionov, V. M., 249[158] (221)
Irngartinger, H., 189[204] (150)
Ishchenko, V. M., 55[49] (52)
Ishida, N., 122[133] (70)
Ishida, S. I., 499[132] (451)
Ishihara, K., 122[129] (70, 82)
Ishikawa, N., 122[90] (66)
Ishino, Y., 122[101] (67)
Islam, M. S., 252[363] (236, 242), 271[24] (259), 272[31] (260), 544[233] (534)
Isley, W. H., 32[120b] (15), 32[120c] (15)
Ismailov, B. A., 254[476] (244)
Ismailov, E. G., 501[287] (481)
Issajew, K. S., 255[493] (245)
Issleib, K., 541[49] (512)
Ito, Y., 124[216] (83)
Itoh, T., 501[257] (474)
Itsuno, S., 123[149] (72)
Ivanov, A. I., 255[490] (244)
Ivanov, I. M., 430[314] (416)
Ivanova, N. S., 254[434] (242), 254[435] (242, 244)
Ivanova, T. M., 190[323] (162)
Ivanova, Z. G., 123[147] (71)
Iwaniak, U., 500[220] (468)
Iwanow, P. I., 255[493] (245)
Izmailov, B. A., 254[438] (242, 244), 254[475] (244)
Izuho, M., 248[71] (218)

Jablonowski, J. A., 192[458] (173)
Jabre, I., 187[114] (139)
Jachiet, D., 501[235] (469)
Jackman, 2
Jacksties, W., 194[577] (181, 182)
Jacobsen, G. B., 368[127] (296, 297), 369[176] (306), 370[209] (317)
Jacobsen, G. E., 31[65] (10), 191[374] (166)
Jadhav, P. K., 188[143] (143)
Jalali-Araghi, K., 501[232] (469)
Jalander, L., 123[199] (82)
James, A. P., 429[249] (400, 411), 429[250] (400, 411), 429[253] (400, 411), 429[265] (400, 411)
James, B. D., 367[70] (286), 368[74] (286)
James, L. M., 122[126] (70)
Jamet, J., 254[482] (244)
Jan, D.-Y., 253[395] (239, 246), 369[149] (296, 297), 369[150] (296, 297), 369[151] (296, 297), 369[152] (296, 297), 369[153] (296–8), 369[154] (296, 297)
Janas, Z., 124[226] (84)
Janiak, C., 531, 543[210] (530, 532, 539), 543[211] (531), 543[213] (531, 539)
Janik, J. Fr., 193[465] (174, 180)
Janini, G. M., 254[478] (244)
Janke, N., 121[26a] (59, 70, 100), 121[26b] (59, 99, 100), 125[287] (98, 100), 125[299] (99), 125[302] (100)
Janousek, Z., 250[211] (223), 252[326] (231, 234), 252[358] (235, 237), 252[359] (235, 237), 253[373] (237), 253[386] (237), 254[459] (244), 272[52] (262), 273[102] (269)
Janssen, E., 498[72] (440)
Jaria, M., 368[108] (290, 291), 368[109] (290, 291), 368[113] (292, 293), 368[114] (292, 293)

Jarvis, W., 250[212] (223), 250[215] (223)
Jasien, P. G., 121[58] (62, 78)
Jasinski, J. C., 542[130] (520)
Jasinski, J. P., 540[12] (504, 506, 511, 521)
Jasper, S. A., Jr., 272[88] (266, 267)
Jasperse, C. P., 248[97] (219, 220), 248[102] (220, 226, 232, 233, 236)
Jastrzebski, J. T. B. H., 32[97] (12), 32[102] (12)
Jasztal, M. J., 250[222] (224, 234), 426[20] (378, 379)
Jaworski, K., 499[138] (452)
Jeannin, Y., 367[59] (284)
Jefferis, G., 254[484] (244)
Jeffery, J. C., 428[197] (400, 411), 428[202] (400, 411), 429[215] (400, 411), 429[225] (400, 412), 429[237] (400, 411), 429[238] (400, 411), 429[244] (400, 414, 417, 419), 429[245] (400, 414), 429[247] (400, 411), 429[248] (400, 411), 429[254] (400, 411), 429[267] (400, 411), 429[272] (400, 411), 430[275] (400, 411), 430[276] (400, 411), 430[282] (400, 411), 430[322] (417–19)
Jeffrey, E. A., 467, 500[208] (467)
Jeffs, S. E., 543[196] (527)
Jelfs, A. N. de M., 429[265] (400, 411), 429[271] (400, 411), 430[278] (400, 411)
Jelinek, T., 249[138] (221), 250[209] (223), 250[210] (223), 250[221] (224), 252[301] (228, 229), 252[306] (228, 235, 240), 252[307] (228, 238, 239), 252[314] (229, 230, 246), 252[317] (230), 252[318] (230), 252[358] (235, 237), 252[360a] (235), 253[371] (236), 253[372] (236, 237, 244), 253[373] (237), 253[383] (237, 238, 240), 253[384] (237), 253[387] (238, 240), 253[400] (239), 253[404] (240), 253[405] (240), 272[48] (262), 272[53] (262), 273[106] (269), 426[28] (377, 378), 426[31] (378), 427[117] (396)
Jelliss, P. A., 429[237] (400, 411), 429[238] (400, 411), 430[282] (400, 411), 430[322] (417–19)
Jelski, D. A., 187[57] (132)
Jemmis, E. D., 54[19] (40), 55[38] (48–50), 214[10] (198, 201), 248[68] (218, 219), 248[96] (219), 370[216] (320, 327, 328, 363), 425[4d] (375), 425[4i] (375, 392)
Jenkins, C. E., 542[139] (521, 526)
Jenkins, P. D., 429[246] (400, 411), 429[254] (400, 411)
Jennings, J. R., 120[12] (58), 191[362] (165)
Jensen, J. A., 367[35] (279, 280), 367[36] (279, 280), 367[37] (279, 280), 367[38] (279, 280, 282–4), 367[39] (280, 282, 283), 367[51] (282, 283)
Jerius, J. J., 497[16] (434, 448)
Jia, L., 271[9] (258), 272[32] (260), 427[85] (382, 388), 544[231] (533)
Jiang, A., 216[92] (213)
Jie, C., 248[105] (220)
Johannsen, T., 540[24a] (505, 506)
Johansen, J. D., 542[137] (521)
Johnson, O., 429[251] (400, 411), 429[252] (400, 411), 429[257] (400, 411), 429[271] (400, 411)
Johnson, S. E., 250[185] (221, 242), 250[189] (221, 243), 427[127] (400, 416), 428[192] (400, 412), 430[323] (418, 419), 430[324] (418, 419)
Johnston, R. L., 367[25] (277), 367[29] (277), 398, 425[4j] (375, 398), 427[123] (398)
Jolly, B. S., 31[40] (8), 126[315] (103)
Jolly, P. W., 124[228] (84)
Jommi, G., 502[339] (494)
Jonas, K., 501[272] (478, 479, 481, 483, 484)
Jones, C., 541[89] (515)
Jones, J. H., 426[16] (377, 378)
Jones, K., 186[8] (130), 499[131] (451)
Jones, L. A., 121[32] (60)
Jones, L. J., III, 540[16] (505, 506), 542[121] (520, 521)
Jones, M., 249[165] (221, 237, 241), 249[174] (221), 252[348] (233), 253[374] (237), 253[381] (237, 239), 253[397] (239), 253[398] (239), 254[445] (242), 254[460] (244), 254[461] (244), 254[462] (244), 254[463] (244)
Jones, N. J., 368[120] (292, 293, 295)
Jones, N. L., 571[82] (567)
Jones, P. G., 541[53] (512), 541[72] (514), 541[83] (514)
Jones, R. A., 31[72] (10), 32[78] (11), 54[24] (42, 43), 504, 507, 518, 540[1a] (503, 504, 517, 518, 520–2), 540[19] (505, 506), 540[21] (505, 506), 540[33] (507, 538), 541[46] (511), 541[52] (512, 522), 541[61] (512), 541[67] (513, 524), 541[86] (514), 542[115] (520), 542[124] (520), 542[125] (520), 542[131] (520, 522), 542[134] (520), 542[140] (521, 522, 526), 542[144] (522), 542[146] (522), 542[149] (522, 523), 542[153] (522, 524), 542[154] (522, 524), 543[182b] (527)
Jones, R. C. F., 194[541] (179)
Jones, R. L., 498[68] (440)
Jones, R. V. H., 122[106] (67, 71, 87, 95)
Jones, S. R., 126[341] (107, 112)
Jordan, K. D., 123[171] (78)
Jordan, R. F., 416, 428[191] (400, 416), 429[243] (400, 416), 430[308] (416), 501[275] (479)
Joshi, N. N., 124[220] (83)
Jousseaume, B., 125[311] (102)
Jug, K., 186[30] (131)
Julia, M., 122[97] (67, 83)
Jun, C. S., 369[177] (307)
Jung, C. W., 367[26] (277)
Junk, P. C., 31[40] (8), 121[25a] (59, 60, 63, 70, 71, 100), 125[294] (98–100), 126[315] (103), 126[363] (112–14), 127[379] (117), 127[385] (118), 191[374] (166)
Jurkschat, K., 122[128] (70, 71, 76)
Jutzi, P., 34[195] (22), 123[159] (74, 86, 92, 93), 126[364] (112–14), 126[366] (112, 114), 187[82] (135, 139, 180), 271[8] (258), 271[14] (258), 272[30] (260), 272[33] (260), 272[34] (260), 272[80] (265), 504, 528, 540[4] (504, 528), 543[207] (530), 543[225] (533)
Juza, R., 33[180a] (20)

Kabachii, Y. A., 249[166] (221), 251[279] (227)
Kabachnik, M. I., 249[159] (221, 224, 238), 251[291] (227), 253[423] (241)
Kabalka, G. W., 186[3] (130, 138), 193[468] (174)
Kadentsev, V. I., 188[175] (146)
Kadish, K. M., 571[79] (566, 567), 571[83], 571[84]
Kadota, I., 124[219] (83)
Kadyrov, R. A., 195[617] (184)
Kaesz, H. D., 570[49], 570[51] (557)
Kafka, S., 191[368] (165), 191[369] (165)
Kageyama, H., 123[153] (73, 83), 123[154] (73)
Kahl, S. B., 254[433] (242), 255[503] (245), 255[504] (245)
Kahn, A. P., 252[304] (228)
Kahn, B. E., 120[22] (59, 86, 101)
Kai, Y., 54[11] (38), 54[12] (38), 123[153] (73, 83), 123[154] (73), 125[286] (97), 541[80] (514)
Kaim, W., 54[20] (40, 41), 54[21] (40, 41), 54[22] (40, 41), 125[273] (95, 103), 126[319] (103), 187[106] (139), 443, 498[44] (436, 468), 499[99] (443, 468), 499[100] (443, 468), 499[101] (443), 499[115a] (446, 496)
Kaino, M., 122[129] (70, 82)
Kaiser, S. L., 34[202] (26)
Kaitner, B., 273[100] (268), 273[116] (271)
Kaiwar, V., 194[595] (182)
Kajihara, Y., 125[281] (97, 98)
Kalachev, A. I., 254[471] (244), 254[472] (244)
Kalb, W. C., 429[211] (400, 415)
Kalinin, V. N., 247[36] (218), 247[43] (218), 248[47] (218), 249[164] (221), 249[167] (221), 250[234] (225), 251[256] (225), 252[349] (233), 254[438] (242, 244), 254[467] (244), 254[468] (244), 254[475] (244), 430[302] (416)
Kalish, N. K., 430[314] (416)
Kalyuskii, A. R., 195[599] (183), 195[600] (183), 195[601] (183), 195[602] (183), 195[603] (183), 195[604] (183), 195[605] (183), 195[606] (183), 195[607] (183), 195[608] (183), 195[609] (183)
Kam, C. K., 254[461] (244)
Kamars, A., 193[497] (175)
Kambe, N., 121[83] (66, 109)
Kamienski, C. W., 124[240] (85, 86, 96)
Kamil, W. A., 191[358] (164)
Kaminsky, W., 440, 497[3] (433, 440, 450, 453, 482), 498[67] (440), 498[69] (440), 501[283] (481), 501[284] (481), 501[285] (481), 501[290] (482), 502[295] (482), 570[19] (548)
Kampel, V. T., 247[44] (218), 248[45] (218), 249[159] (221, 224, 238), 250[197] (222), 251[263] (226), 254[446] (243), 254[456] (243)
Kampf, J. W., 189[233] (153), 215[75] (210), 215[76] (210), 215[77] (210)
Kampmann, D., 190[305] (161), 215[84]

(212), 215[86] (212), 215[87] (212)
Kanabur, V. V., 190[320] (162)
Kanakhina, L. N., 427[110] (396)
Kanatani, R., 122[92] (67)
Kanatzidis, M. G., 542[159] (524)
Kanda, T., 121[83] (66, 109)
Kane, A. R., 369[201] (313)
Kane, D. J., 32[77] (11)
Kanellakopulos, B., 571[62] (561)
Kaneti, J., 123[172] (78)
Kang, H. C., 253[389] (238), 427[134] (400, 412), 428[151] (400, 412), 428[188] (400, 412), 428[190] (400, 412), 430[301] (415)
Kang, J., 122[96] (67)
Kang, S. O., 249[132] (221, 223, 230), 252[302] (228), 273[98] (268), 273[103] (269), 273[104] (269), 273[105] (269), 372[322] (364)
Kang, S.-K., 571[88] (568)
Kanters, J. A., 33[153] (19)
Kao, C.-T., 500[214] (468)
Kapan, M., 429[252] (400, 411)
Kapfer, C. A., 123[145] (71)
Kapil, R. S., 122[131] (70)
Kaplan, G. M., 192[393] (167)
Kapoor, P. N., 125[312] (102)
Kapoor, R. N., 125[312] (102)
Karasik, A. A., 191[335] (163), 195[620] (184), 195[621] (184), 195[622] (184), 195[627] (184)
Karaulov, A., 272[34] (260)
Karen, P., 125[292] (98)
Karger, G., 187[79] (135), 191[371] (165)
Kargin, Yu. M., 190[326] (162)
Karle, I. L., 33[168] (19)
Karras, M., 122[108] (68)
Kartsev, G. N., 251[264] (226)
Kasai, N., 54[11] (38), 54[12] (38), 123[153] (73, 83), 123[154] (73), 541[80] (514)
Kasai, P. H., 441, 442, 498[82] (441), 498[83] (441), 499[89] (442)
Kaska, W. C., 499[96] (443), 500[202] (466, 486, 495), 500[204] (466), 500[205] (466)
Kasler, F., 250[225] (224)
Kataeva, O. N., 195[623] (184)
Katagiri, T., 123[193] (80, 81), 123[198b] (82)
Katoh, K., 369[187] (311), 369[188] (311)
Kats, G. A., 255[494] (245), 429[221] (400, 416)
Katz, H. E., 193[518] (177)
Kaubi, K., 120[20a] (59, 60)
Kaufmann, B., 214[6] (198, 211)
Kaufmann, D. E., 186[15] (130), 189[247] (155, 156), 189[248] (155, 156), 189[249] (156), 189[250] (156), 189[254] (156)
Kaufmann, E., 34[221] (30)
Kaupp, M., 126[353] (110), 126[373] (114)
Kausch, C. M., 215[73] (210), 215[77] (210)
Kawa, H., 33[183] (21), 123[173] (78)
Kawai, N. T., 250[238] (225)
Kawaichi, N., 124[235] (85)
Kawasaki, Y., 541[80] (514), 541[81] (514)
Kayumov, A. G., 247[33] (218), 251[251] (225, 226)
Kazakov, V. P., 187[113] (139)
Kazantsev, A. A., 254[438] (242, 244), 254[475] (244)
Kazantsev, A. V., 249[161] (221, 242), 253[422] (241), 253[424] (241, 242), 254[427] (241), 254[428] (241), 254[429] (242)
Kean, D. M., 121[32] (60)
Keana, J. F. W., 192[452] (173)
Keep, A. K., 369[142] (294, 296, 297, 301, 303), 369[170] (301)
Keese, R., 254[437] (242)
Kehr, G., 189[231] (153), 190[269] (158), 252[328] (231)
Keiderling, T. A., 193[527] (178)
Kelland, M. A., 370[214] (319), 370[222] (321), 428[195] (400, 414)
Keller, A., 570[35] (552)
Keller, P. C., 251[289] (227)
Keller, W., 188[185] (147), 272[76] (264), 272[79] (265)
Keller, W. J., 253[370] (236)
Kendrick, A., 195[643] (185)
Kennard, C. H. L., 124[245] (86, 92, 93, 95, 112–14), 367[70] (286), 368[74] (286)
Kennedy, J. D., 249[138] (221), 272[48] (262), 272[53] (262), 272[55] (262), 272[56] (262, 269), 272[60] (263), 272[61] (263), 272[81b] (265, 266), 272[87] (266), 272[90] (266), 273[93] (267), 273[94] (267), 273[101] (269), 273[106] (269), 273[113] (270), 273[116] (271), 273[117] (271), 277, 366[2] (277, 313), 367[15] (277), 367[28] (277), 367[30] (278, 343, 348), 369[183] (310, 316, 327, 337, 349), 369[192] (313), 369[195] (313, 315, 316, 320, 323), 369[196] (313, 315, 320, 323, 340), 369[197] (313, 315, 346, 357, 359), 369[200] (313, 316), 370[212] (319, 320, 324, 340, 341), 370[213] (319), 370[224] (322), 370[228] (324), 370[229] (324), 370[230] (324), 370[238] (327, 336, 349, 363), 370[239] (327, 336, 363, 364), 370[240] (327, 336, 364), 370[242] (329), 370[243] (331, 332, 336, 337), 370[244] (332), 370[245] (332, 340), 370[246] (332, 336), 370[247] (332, 336), 370[248] (332, 339), 370[249] (333), 370[252] (336, 340, 342, 347), 370[253] (336), 370[254] (336), 370[255] (337), 370[257] (337), 370[258] (338, 345), 370[260] (338–40), 370[262] (340, 341), 370[263] (340), 370[264] (340), 370[265] (340, 342, 343), 371[266] (340, 342), 371[267] (341), 371[268] (340, 341), 371[269] (340, 341, 344, 347), 371[271] (340, 341), 371[272] (340), 371[273] (340), 371[274] (341), 371[275] (342, 357, 359), 371[277] (342), 371[278] (342, 347), 371[279] (343), 371[280] (343), 371[281] (345, 346, 355, 356), 371[282] (346), 371[283] (347), 371[284] (347), 371[286] (350), 371[288] (351), 371[289] (351), 371[290] (351), 371[291] (351), 371[292] (351), 371[297] (351, 362), 371[298] (351), 371[299] (352), 371[306] (355, 356), 371[307] (355, 356), 371[308] (355, 356), 371[309] (355–7), 371[310] (355–7), 371[311] (357, 359), 371[312] (357–9), 371[313] (358, 359), 371[314] (358), 371[319] (363), 371[320] (364), 371[321] (364), 372[323] (365), 372[324] (365), 398, 426[11e] (377, 394, 396), 426[16] (377, 378), 426[21] (378), 426[28] (377, 378), 426[34] (378, 379), 426[37] (378, 380), 427[112] (396), 427[114] (396), 427[117] (396), 427[118] (396), 427[122a] (398), 427[122b] (398)
Kennikova, E. N., 498[63] (440)
Kepert, D. L., 247[22] (218)
Kerschl, S., 190[270] (158), 190[271] (158)
Kester, J. G., 271[27] (259), 272[58] (263), 272[71] (264), 272[72] (264), 272[77] (264), 272[83] (265, 266), 272[91] (267), 426[32] (378)
Khalilov, L. M., 123[169] (77, 92), 124[242] (86, 98), 125[251] (87, 95)
Khan, M. A., 541[45] (511), 541[56] (512), 541[73] (514), 544[257] (537)
Khan, M. M. T., 368[77] (286)
Khan, M. N. I., 368[98] (288, 290)
Khan, S., 368[108] (290, 291)
Khan, S. A., 248[106] (220, 228, 235, 237–9), 250[190] (221, 229, 237), 253[385] (237)
Khan, S. I., 250[189] (221, 243), 570[51] (557)
Khan, T. A., 368[116] (292, 293)
Khasnis, D. V., 540[23] (505, 506)
Khatri, N. A., 498[55] (438, 493)
Khattar, R., 127[378a] (116), 127[378b] (116), 367[24] (277, 294, 297, 304), 429[241] (400, 412), 430[323] (418, 419), 430[324] (418, 419)
Khiteeva, D. M., 255[489] (244)
Khorlina, I. M., 499[136] (452)
Khotina, I. A., 249[170] (221), 253[414] (241, 244), 254[441] (242), 254[476] (244), 255[491] (244)
Khvostenko, V. I., 251[261] (226, 236)
Kidd, K. B., 32[78] (11), 542[124] (520), 542[144] (522)
Kiesgen, J., 190[303] (160), 191[329] (162), 191[345] (163)
Kilaas, L., 124[204] (82)
Kiljunen, H., 123[177] (78)
Kim, J., 427[128] (400, 412)
Kim, K. M., 249[140] (221, 238)
Kim, Y. J., 31[62] (10)
Kim, Y. K., 570[49]
Kimura, E., 191[359] (165)
King, B. A., 125[276] (96)
King, C., 368[98] (288, 290), 571[97] (569)
King, R. B., 120[16] (59, 71), 120[24] (59, 99), 122[141b] (71, 83), 124[238] (85), 124[239] (85), 124[240] (85, 86, 96), 124[243b] (86, 95), 124[250b] (86, 100), 125[309] (101), 248[109] (220), 502[304] (486), 502[305] (486)
King, R. E., III, 272[89] (266), 428[180] (400, 414), 429[212] (400, 415), 429[214] (400, 415)
Kinsley, S. A., 33[147] (19, 25, 26), 127[380] (117, 119)
Kinzelmann, H.-G., 121[26a] (59, 70,

100), 121[26b] (59, 99, 100), 125[287] (98, 100)
Kirby, S. L., 121[32] (60)
Kirchhof, W., 490, 502[318] (490)
Kirillova, N. I., 428[139] (400), 570[8] (547)
Kirk, M. J., 251[284] (227)
Kirmse, W., 121[54b] (61)
Kiselev, V. G., 187[117] (139), 188[183] (147)
Kita, Y., 501[269] (476)
Kitamura, M., 123[200] (82, 106)
Kitayama, T., 123[150] (72)
Kjonaas, R. A., 126[330] (106)
Klabunde, K. J., 120[10] (58, 59, 62, 78, 108), 124[233] (84)
Klaébé, A., 195[616] (184)
Klaff, N., 188[162] (145, 153), 214[14] (199, 202), 215[69] (209)
Kleier, D. A., 371[305] (354)
Kleimann, W., 501[286] (481), 570[30] (551)
Klein, H.-F., 34[200] (26), 34[203] (26), 126[321] (104)
Klein, R., 498[70] (440)
Klein, W., 214[4] (198, 210), 215[67] (209), 215[75] (210)
Kleine, M., 570[49]
Klemm, W., 435, 498[24] (435)
Kliche, C., 189[225] (153)
Kliegel, W., 175, 182, 183, 187[110] (139), 193[479] (175), 193[480] (175), 194[544] (179), 194[598] (183)
Klimak, Z. A., 254[479] (244)
Kline, G. A., 255[525] (246)
Klingebiel, U., 31[45] (9), 32[99] (12), 190[279] (158), 190[280] (158)
Klinkhammer, K., 541[70] (514), 543[172] (525, 526)
Klinkhammer, K. W., 32[117] (15), 32[118] (15), 499[114] (446)
Klotzbücher, W. E., 193[523] (177, 178)
Klumpp, G. W., 33[156] (19), 122[138b] (71), 122[139] (71)
Klusener, P. A. A., 31[60] (10)
Klusik, H., 134, 186[35] (131, 134, 149), 187[73] (135), 189[195] (149), 189[201] (150), 544[267a] (539)
Knight, K. S., 122[105a] (67, 71, 95), 122[105b] (67, 71)
Knjazhanski, S. Y., 570[33] (551)
Knobler, C. B., 127[378a] (116), 127[378b] (116), 249[119] (220, 221, 231), 249[120] (220, 221, 231, 235), 249[136] (221, 231), 249[140] (221, 238), 249[173] (221, 242), 250[177] (221, 231), 250[179] (221), 253[389] (238), 254[432] (242), 255[523] (246), 271[12] (258), 271[16] (258), 271[17] (259), 271[18] (259), 271[19] (259), 271[22] (259), 271[23] (259), 272[89] (266), 371[316] (361), 371[317] (362), 425[10] (377, 400, 412), 426[38] (378, 380), 427[113] (396), 427[127] (400, 416), 427[128] (400, 412), 427[134] (400, 412), 428[151] (400, 412), 428[152] (400, 412), 428[153] (400, 412), 428[173] (400, 416), 428[180] (400, 414), 428[185] (400, 412), 428[188] (400, 412), 428[190] (400, 412), 428[192] (400, 412), 428[199] (400, 412), 429[207] (400, 415), 429[213] (400, 415), 429[214] (400, 415), 429[216] (400, 412), 429[219] (400, 415), 429[224] (400, 415), 429[227] (400, 415), 429[229] (400, 415), 429[233] (400, 415), 429[241] (400, 412), 430[281] (400, 414), 430[288] (400, 412), 430[323] (418, 419), 430[324] (418, 419), 541[99] (517), 543[226] (533), 544[228] (533)
Knobler, C. C., 250[187] (221, 243)
Knochel, P., 121[76] (65, 77), 126[331] (106)
Knoeppel, D., 195[614] (183)
Knorr, J. R., 368[123] (293, 295)
Knörzer, G., 189[217] (151), 189[252] (156), 191[346] (163, 181)
Knox, S. J., 541[62] (512, 524)
Knyazev, S. P., 251[254] (225, 239)
Kobak, V. V., 425[10] (374), 428[142] (400), 428[143] (400, 416), 428[144] (400, 416), 428[145] (400, 416), 428[146] (400, 416), 428[147] (400), 428[148] (400, 416), 428[149] (400, 416), 430[280] (400, 416)
Kobayashi, R. W., 252[323] (230)
Kobayashi, Y., 122[116] (69)
Kobel'kova, N. I., 250[234] (225), 252[349] (233)
Kobiro, K., 541[80] (514), 541[81] (514)
Koch, S. A., 543[174] (525, 526)
Koch, W., 33[139] (19)
Kocienski, P. J., 124[217] (83)
Kock, L. D., 192[405] (169)
Kodama, G., 369[184] (311), 369[186] (311)
Kodama, K., 55[58] (54)
Koetzle, T. F., 367[67] (285)
Köffer, D. P. J., 215[38] (202, 203), 215[39] (202–4), 215[46] (204)
Kohl, F., 543[207] (530)
Kolesov, B. A., 427[137] (400)
Kolesov, V. S., 541[96] (515)
Kölle, P., 186[20] (130), 192[409] (169), 192[411] (169), 192[412] (169, 170), 194[582] (182), 194[586] (182), 368[97] (288, 290)
Koloskova, G. N., 255[488] (244)
Kolpikova, E. F., 254[470] (244)
Komar, D. A., 500[198] (465, 496)
Komarova, L. G., 248[61] (218), 253[421] (241), 254[435] (242, 244), 254[464] (244), 254[465] (244), 255[494] (245), 429[221] (400, 416)
Komarova, L. I., 254[471] (244), 254[472] (244)
Komarova, N. G., 248[59] (218, 244), 254[435] (242, 244), 254[465] (244), 254[466] (244), 255[495] (245)
Komori, T., 502[316] (490)
Komorowska, M., 195[611] (183)
Komorowski, L., 188[129] (141), 191[351] (164), 191[352] (164), 192[395] (167)
Kong, G., 571[68] (563)
Kong, S., 121[43] (60)
König, H., 126[321] (104)
Konishi, H., 215[77] (210)
Koo, M. S., 255[504] (245)
Kopasz, J. P., 542[119] (520)
Kopf, J., 31[58] (10), 31[59] (10), 32[85b] (11), 33[143] (19), 33[160] (19, 21), 34[184] (23, 24), 34[185] (23, 25), 34[191] (23), 34[208] (26), 34[210] (29), 125[263] (91, 105), 126[323] (105), 501[283] (481), 501[284] (481), 501[285] (481), 570[19] (548)
Koplitz, B., 498[41] (436)
Koppert, S., 126[321] (104)
Koppetsch, G., 124[249] (86, 90, 93–5), 124[250b] (86, 100)
Koprda, V., 430[310] (416)
Koray, A. R., 544[249] (536)
Koreeda, M., 501[253] (474)
Korneev, N. N., 126[332] (106), 126[334] (106)
Korotkikh, O. A., 194[537] (179)
Korshak, V. V., 248[53] (218), 248[61] (218), 249[166] (221), 249[170] (221), 254[441] (242), 254[464] (244), 254[468] (244), 254[476] (244), 255[491] (244), 255[494] (245), 255[495] (245)
Kos, A. J., 123[172] (78), 186[38] (131)
Koschmieder, S. U., 54[24] (42, 43), 542[125] (520)
Kosolapkina, E. N., 498[40] (436)
Kosolapov, A. M., 121[57] (62, 108)
Kostenko, N. L., 193[517] (177)
Köster, H., 33[162] (19), 33[166] (19, 24), 33[176] (20)
Koster, J. B., 500[168] (457)
Köster, K., 34[193] (23)
Köster, R., 186[22] (130, 138, 141), 187[100] (138), 187[102] (138), 187[104] (138), 187[108] (139), 187[109] (139), 188[128] (141), 188[130] (141), 188[134] (141), 188[135] (142), 188[136] (142), 188[137] (142), 188[138] (142), 188[139] (142), 188[140] (142), 188[141] (142), 188[142] (142), 188[184] (147), 189[219] (152), 191[364] (165), 192[400] (168), 192[401] (168), 192[402] (168), 192[414] (169), 192[429] (171), 192[440] (173), 193[515] (177), 193[519] (177), 193[520] (177), 194[531] (178), 194[538] (179), 194[545] (179), 194[550] (179), 194[560] (180), 215[51] (205), 215[90] (213), 215[91] (213), 216[92] (213), 216[93] (213), 216[94] (213), 216[95] (213), 247[10] (218), 247[11] (218), 249[131] (221, 223, 229, 234, 235, 244), 250[208] (223, 229, 234), 250[217] (224), 252[308] (229, 234, 235), 499[97] (443), 499[98] (443), 500[179] (461)
Kosugi, M., 500[187] (463, 476)
Kot, W. K., 368[96] (288, 290)
Kothiwal, A. S., 194[532] (179)
Kotrelev, G. V., 255[485] (244)
Kotsuki, H., 124[219] (83)
Kottke, T., 32[93] (12), 33[152] (19, 20)
Kovalev, A. I., 255[491] (244)
Kovredov, A. I., 250[196] (222), 251[239] (225), 251[293] (228), 253[382] (237, 239), 253[393] (239), 253[394] (239), 253[396] (239), 253[422] (241), 253[424] (241, 242), 254[429] (242), 255[530] (247), 428[147] (400)
Kozikowski, A. P., 502[336] (493)

Kozima, S., 189[213] (151), 499[105] (443)
Kozlov, N. S., 193[461] (174), 193[462] (174)
Koz'menko, M. V., 194[537] (179)
Kraatz, U., 191[332] (162)
Kradenov, K. V., 427[137] (400)
Kraemer, A., 252[330] (231)
Krahe, K., 190[303] (160)
Krainova, N. Y., 251[259] (226, 231), 253[388] (238), 430[280] (400, 416)
Krajenbrink, F. G., 55[56] (54), 55[57] (54)
Kraka, E., 186[39] (131), 186[40] (131)
Kramarz, W., 125[272b] (95)
Krämer, A., 187[79] (135), 187[80] (135), 187[88] (136), 188[155] (144), 188[163] (145, 150, 154), 189[238] (154)
Krasnoshchekaya, S. P., 195[603] (183)
Krato, B., 187[82] (135, 139, 180)
Kratzer, R. H., 186[28] (130)
Kraus, H. J., 543[214] (531)
Krause, H., 540[15] (505, 506)
Krause, J., 32[84b] (11), 195[614] (183)
Krause, J. A., 369[150] (296, 297), 371[287] (350)
Krause, S., 446, 447, 498[30] (435, 446), 499[112] (446)
Kravets, E. Kh., 500[221] (468), 500[222] (468), 500[224] (468), 502[324] (491)
Krebs, B., 194[564] (180), 543[197] (527), 543[198] (528)
Kreienkamp, H., 543[201] (529)
Kreimendahl, C. W., 430[296] (415)
Krempp, M., 499[95] (443)
Kriss, R. U., 543[203] (529)
Kristal'nyi, E. V., 126[357a] (111, 119), 126[358] (111), 126[359] (111)
Kristen, M. O., 126[366] (112, 114), 215[77] (210)
Krivenko, I. A., 193[497] (175)
Kröckert, B., 190[282] (159)
Krockert, B., 272[45] (261)
Krogh-Jespersen, K., 186[32] (131, 134), 186[33] (131), 186[36] (131), 248[63] (218), 248[64] (218), 248[65] (218)
Kroker, J., 215[73] (210), 215[77] (210)
Krone-Schmidt, W., 186[28] (130)
Kronseder, C., 570[50] (557, 566)
Kropacheva, E. N., 498[63] (440)
Kropp, M. A., 187[62] (133), 187[63] (133, 148), 189[193] (148)
Krug, B., 191[382] (166), 194[535] (179)
Krüger, C., 34[201] (26), 34[204] (26, 27), 123[159] (74, 86, 92, 93), 124[249] (86, 90, 93–5), 125[265] (93, 101, 102), 125[291] (98), 125[299] (99), 125[300] (99), 125[302] (100), 126[336] (107), 188[135] (142), 193[515] (177), 214[18] (199), 216[92] (213), 368[112] (292, 293), 498[72] (440), 499[115b] (496), 500[168] (457), 501[292] (482), 502[294] (482), 540[37] (509), 543[207] (530), 544[268] (540)
Krupp, F., 499[135] (452), 500[171] (458)
Ku, K.-J, 544[235] (534)
Kubiniok, S., 32[109] (15)
Kuchin, A. V., 500[170] (458), 500[194] (463), 500[196] (464), 500[219] (468), 501[228] (468), 501[240] (471), 501[246] (473), 501[247] (473), 501[248] (473), 501[249] (473), 501[250] (473), 501[263] (476), 502[326] (491), 502[346] (496), 502[347] (496)
Kucht, H., 543[215] (531)
Kuhlmann, T., 214[23] (199), 214[24] (200), 430[325] (420), 430[327] (420)
Kuhn, A., 191[354] (164), 191[383] (166)
Kuhn, N., 125[314] (103), 191[354] (164), 191[383] (166)
Kühr, H., 32[85c] (11), 32[85d] (11)
Kukharenko, S. V., 430[293] (414)
Kukina, G. A., 426[25] (378), 426[26] (378), 426[27] (378, 379)
Kukovinets, A. G., 125[251] (87, 95)
Kulig, J., 254[484] (244)
Kulkarni, S. U., 188[143] (143)
Külper, K., 498[67] (440), 502[295] (482)
Kumada, M., 122[92] (67), 124[216] (83)
Kumar, P. N. V. P., 54[19] (40), 370[216] (320, 327, 328, 363)
Kumar, R., 543[168] (525, 528), 543[187] (526)
Kümmerlen, J., 190[269] (158)
Kump, R. L., 272[49] (262), 273[109b] (270)
Kumpfmüller, F., 194[589] (182)
Kündig, E. P., 120[21a] (59, 79), 120[21b] (59, 80)
Kunicki, A., 499[142] (452), 501[258] (475)
Kunz, H., 502[338] (494)
Kunze, K., 542[160] (525), 570[59] (560)
Küpper, S., 190[302] (160)
Kupprat, G., 192[406] (169)
Kurella, M. G., 191[372] (166), 191[392] (167)
Kureshy, R., 368[109] (290, 291)
Kurogi, Y., 191[359] (165)
Kuroki, M., 500[216] (468)
Kurosawa, H., 499[110] (445)
Kuryakova, N. I., 254[471] (244), 254[472] (244)
Kusabayashi, S., 124[210] (83)
Kushlan, D. M., 125[268] (93)
Kusmierz, J., 254[480] (244)
Kuwajima, I., 122[132] (70), 124[219] (83)
Kuznetsov, I. Y., 250[192] (221)
Kuznetsov, N. T., 247[33] (218), 247[37] (218), 248[50] (218), 250[192] (221), 251[251] (225, 226), 251[252] (225), 251[253] (225, 226), 251[260] (226), 253[384] (237)
Kuznetsov, V. A., 189[255] (156), 189[256] (156)
Kuznetsov, V. V., 175, 192[450] (173), 193[475] (174), 193[476] (175), 193[477] (175), 193[478] (175), 193[482] (175), 193[484] (175, 183), 193[485] (175, 183), 193[486] (175, 183), 193[487] (175, 183), 193[488] (175, 183), 193[489] (175, 183), 193[490] (175, 183), 193[491] (175, 183), 193[492] (175, 183), 193[493] (175, 183), 193[494] (175, 183), 193[495] (175, 183), 195[599] (183), 195[600] (183), 195[601] (183), 195[602] (183), 195[603] (183), 195[604] (183), 195[605] (183), 195[606] (183), 195[607] (183), 195[608] (183), 195[609] (183)
Kuznetsova, E. I., 498[63] (440)
Kvachev, Y. P., 253[414] (241, 244)
Kynast, U., 541[51] (512)

Labahn., D., 543[199] (528)
Laboue, B., 124[214] (83)
Laemmle, J., 498[22] (435), 498[23] (435, 436, 496)
Lagow, R., 33[183] (21)
Lagow, R. J., 123[173] (78)
Laguna, A., 541[72] (514), 571[86] (567)
Lagutkin, N. A., 188[172] (146)
Lai, Y.-H., 59, 120[13] (59)
Laine, R. M., 497[5] (433)
Lake, C. H., 541[65] (513), 542[129] (520), 543[184] (525, 526)
Lalonde, J. J., 126[317] (103)
Lalor, F. J., 273[100] (268)
Lambert, C., 34[222] (30)
Lambertsen, T., 33[154b] (19, 20), 33[177] (20)
Lance, M., 368[87] (289)
Lang, G., 249[122] (220)
Lange, H., 542[156] (524)
Lange, L., 32[109] (15)
Langlotz, I., 31[38b] (6)
Lapanik, V. I., 193[509] (176), 193[510] (176)
Lapin, S. C., 189[257] (156)
Lapkin, I. I., 55[53] (53), 55[54] (53)
Lappert, M. F., 2, 31[29] (6, 8), 31[32] (6, 9), 31[41] (8), 31[43] (8), 31[51] (9), 31[69] (10), 31[70] (10), 32[79] (11), 32[90] (12), 32[123] (15), 33[142] (19), 123[152] (72, 73, 86–8, 92), 123[167] (76, 84, 85, 88, 98), 124[225] (83, 84), 124[234] (84, 85), 124[244] (86, 95, 96), 124[246] (86, 90), 126[350] (109, 110), 367[32] (278, 286, 287), 500[166] (457), 543[212] (531), 543[216] (531), 544[259] (537)
Larbig, W., 499[135] (452)
Lardicci, L., 55[50] (53), 55[51] (53), 55[52] (53), 124[202] (82), 501[226] (468), 501[264] (476), 502[330] (491)
Lartiges, S., 32[117] (15)
Lashley, L. K., 124[211a] (83)
Laske, J. A., 540[36] (509)
Lattman, M., 55[39] (48), 540[23] (505, 506)
Laube, T., 32[86] (12), 32[87] (12), 32[88] (12), 32[91] (12), 32[115] (15)
Laubengayer, A. W., 499[131] (451)
Lauer, M., 191[373] (166)
Lauher, J. W., 214[12] (198, 199)
Laurie, J. C. V., 249[151] (221), 427[131] (400, 412)
Lauteschläger, S., 541[48] (511), 541[50] (512)
Laven, R., 189[215] (151), 215[34] (202), 215[37] (202, 204)
Lavrinovich, L. I., 189[223] (152)
Lawler, R. G., 121[73] (65)
Lawless, G. A., 32[123] (15), 126[350] (109, 110)
Lawrence, A. D., 426[16] (377, 378)
Lawrence, K. E., 121[50a] (61), 121[50b] (61)

Lawrence, L. M., 121[42] (60)
Lawrence, S. H., 251[289] (227)
Layh, M., 33[149] (19), 540[6] (504, 537), 541[70] (514), 541[71] (514, 537–9), 543[172] (525, 526), 544[255] (537–9)
Layton, W. J., 192[397] (167)
Lazraq, M., 192[410] (169)
Lazzaroni, R., 122[93] (67)
Leach, J. B., 247[2a] (218, 242), 369[193] (313, 315, 319), 369[194] (313, 315), 370[214] (319), 370[222] (321), 428[195] (400, 414)
Leardini, R., 191[338] (163)
Lebedev, A. V., 252[327] (231)
Lebedev, V. N., 251[264] (226), 253[401] (239), 254[442] (242), 429[242] (400, 417), 430[289]
Lebedev, Yu. A., 187[59] (133)
Lecomte, C., 571[80] (566, 567), 571[84], 571[98] (569)
Lee, B., 540[36] (509), 541[39] (510), 541[40] (510), 541[41] (511)
Lee, C.-H., 121[67] (64, 70, 72, 103)
Lee, D. H., 570[28] (551)
Lee, D. Y., 194[597] (183, 185)
Lee, F. L., 498[71] (440, 479, 480), 501[291] (482)
Lee, H., 252[342] (232, 233)
Lee, J. C., 187[103] (138), 543[184] (525, 526)
Lee, J. D., 249[116] (220)
Lee, K., 125[281] (97, 98)
Lee, K. E., 542[123] (520)
Lee, K. S., 33[157] (19)
Lee, K. W., 187[103] (138)
Lee, K.-J., 191[375] (166)
Lee, L. K., 255[517] (246)
Lee, S., 248[100] (220, 232, 233), 248[114] (220), 250[218] (224), 251[276] (227), 251[277] (227)
Lee, S. S., 253[389] (238), 425[10] (377, 400, 412), 427[134] (400, 412), 428[152] (400, 412), 428[153] (400, 412)
Lee, Y. J., 249[121] (220), 249[122] (220), 249[123] (220), 249[124] (220)
Lees, J. F., 543[208] (530)
Lefevre-Jenot, L., 125[267] (93)
Leffers, W., 34[195] (22), 126[364] (112–14)
Leffler, A. J., 251[270] (226)
Legare, D. A., 248[75] (219)
Lehmkuhl, H., 66, 122[85] (66, 67, 79), 122[87] (66), 122[91] (66), 123[164] (76, 77, 93), 124[248] (86, 93), 125[265] (93, 101, 102), 125[300] (99), 126[336] (107), 126[337] (107), 467, 498[72] (440), 499[98] (443), 500[169] (458, 462, 464, 486), 500[178] (460, 463), 500[185] (462), 500[188] (463), 500[206] (467), 500[207] (467)
Lehr, G. F., 121[73] (65)
Lehr, J., 190[307] (161), 190[318] (162), 215[82] (212)
Leimkühler, M., 499[106] (443)
Leites, L. A., 188[164] (145), 247[39] (218, 224–6), 250[180] (221), 250[234] (225), 250[237] (225), 251[239] (225), 251[254] (225, 239), 251[258] (226), 251[293] (228)
Leman, J. T., 499[127] (449), 540[31] (507, 512), 541[59] (512)
LePage, Y., 501[291] (482)
L'Esperance, R. P., 249[165] (221, 237, 241)
Less, J. F., 543[209] (530)
Letts, J. B., 367[63] (284, 285)
Leukart, O., 255[499] (245)
Leung, W.-P., 31[51] (9), 123[144] (71, 83, 84), 123[152] (72, 73, 86–8, 92), 123[168] (76), 124[225] (83, 84)
Levin, J. I., 498[54] (438, 493)
Levine, L. J., 126[361] (112)
Lewandos, G. S., 120[19] (59)
Lewandowski, J., 191[354] (164)
Lewis, D., 254[482] (244)
Lewis, D. B., 368[115] (292, 293)
Lewis, D. P., 122[106] (67, 71, 87, 95)
Lewis, W., 501[241] (471)
Lewis, Z. G., 428[161] (400, 412), 428[177] (400, 412), 429[210] (400, 412), 429[223] (398, 400), 430[287] (400)
Ley, S. V., 502[322] (491)
Li, B., 123[181a] (79)
Li, D., 251[276] (227)
Li, J., 253[374] (237), 253[381] (237, 239), 253[398] (239)
Li, P. T.-J., 120[23] (59)
Li, Q., 247[24] (218)
Li, Q.-S., 425[4f] (375)
Li, S., 428[198] (400, 412), 429[247] (400, 411), 429[248] (400, 411)
Li, Z. H., 249[165] (221, 237, 241)
Liang, S., 192[442] (173)
Liao, S.-T., 120[24] (59, 99), 123[148] (72), 125[297] (99, 100)
Lichtmann, L. S., 55[56] (54)
Liebman, J. F., 247[13] (218), 247[27] (218), 247[31] (218), 247[32] (218), 247[40] (218), 248[49] (218), 271[2] (258), 272[28] (260), 272[35] (260), 366[4] (277), 425[1f] (374), 425[1g] (374, 415), 425[1h] (374, 377, 381, 388)
Lijewski, L., 194[594] (182)
Lim, D., 32[104a] (13)
Limberg, M., 120[15] (59, 60)
Lin, C.-C., 571[68] (563)
Lin, S.-T., 124[233] (84)
Lincoln, F. J., 125[295] (98, 100)
Lincoln, S. F., 123[165] (76)
Lindeman, S. V., 249[170] (221), 250[186] (221)
Lindner, H. J., 33[135] (17)
Lindsell, W. E., 120[1] (58–62, 64, 66, 71–5, 77–9, 83, 85, 87, 88, 91, 92, 94, 96, 97, 101, 103, 105, 109–12, 115, 119), 126[351] (109)
Linek, A., 249[152] (221), 250[178] (221), 250[183] (221), 428[140] (400, 414), 428[187] (400)
Link, D. C., 121[65] (63), 123[180] (79)
Linti, G., 186[27] (130), 192[411] (169), 192[412] (169, 170), 192[423] (170), 192[436] (172), 215[28] (201), 215[29] (201), 368[117] (292, 295), 368[119] (292, 295)
Lipscomb, W. N., 187[56] (132), 247[30] (218), 367[67] (285), 371[305] (354)
Liptak, D., 499[121] (448)
Lisichkina, I. N., 249[158] (221), 253[375] (237), 253[377] (237)
Liston, D. J., 249[124] (220), 250[194] (221)
Listyutenko, V. N., 190[323] (162), 190[324] (162)
Litt, M., 254[484] (244)
Little, J. L., 265, 272[69] (264), 272[70] (264, 265), 272[71] (264), 272[72] (264), 272[77] (264), 272[82] (265, 266), 272[85] (266, 267), 272[86] (266), 272[91] (267), 273[107] (270)
Litvinov, I. A., 195[623] (184), 195[624] (184)
Litvinova, M. D., 254[430] (242)
Liu, F., 187[55] (132)
Liu, L., 121[59] (62, 78), 255[506] (245)
Liu, P., 121[43] (60)
Liu, S. J. Y., 500[203] (466, 495)
Liu, T. C., 124[213] (83)
Liu, Y., 121[43] (60)
Liu, Z. R., 498[50] (438, 487), 498[51] (438, 493), 502[312] (488–90)
Livant, P. D., 191[375] (166)
Lloyd, D. R., 190[312] (161)
Lo, F. Y., 427[121] (398)
Lo, W., 192[442] (173)
Lobkovskii, E. B., 54[32] (46), 367[20] (277, 286, 287), 367[58] (284), 368[75] (286), 570[8] (547), 570[9] (547), 570[10] (547), 570[17] (547)
Lochmann, E., 32[104i] (13)
Lochmann, L., 32[104a] (13), 32[104b] (13), 32[105] (13)
Loderer, D., 186[27] (130)
Loebel, J., 543[210] (530, 532, 539)
Loffredo, R. E., 271[25] (259)
Logan, C. F., 253[381] (237, 239)
Lomakina, S. I., 125[252] (87)
Lomax, J. F., 251[243] (225–7, 229, 234)
Long, G. J., 368[126] (294, 296, 297), 368[132] (297, 298), 369[178] (307, 308), 426[67] (382, 398)
Long, J. A., 429[222] (400, 415), 430[297] (415)
Long, W. P., 501[289] (482)
Longato, B., 253[416] (241), 254[452] (243), 254[453] (243), 254[454] (243), 254[455] (243)
Longoni, G., 571[69] (563, 564)
Loos, D., 123[160] (74, 109), 540[2] (504, 529), 543[202] (529, 539)
Lorberth, J., 34[188] (22, 23)
Lord, G., 250[238] (225)
Lorenzen, N. P., 31[58] (10), 31[59] (10), 34[184] (23, 24), 34[191] (23), 34[206] (26)
Lötz, A., 251[266] (226), 251[269] (226)
Lotz, R., 499[106] (443)
Lough, A. J., 193[481] (175)
Lu, K., 543[224] (532), 544[231] (533)
Lu, K.-J., 271[9] (258), 271[10] (258), 271[11] (258), 272[78] (265)
Lu, P., 271[16] (258), 427[113] (396)
Lubben, T. V., 251[285] (227)
Lubkowitz, G., 194[598] (183)
Lucken, E. A. C., 33[155a] (19, 20)
Luetkens, M. L., 367[52] (280, 282)
Luh, T. Y., 124[217] (83)
Lühder, K., 124[236] (85)
Lukevics, E., 187[59] (133)
Luss, H. R., 543[183] (525, 526)
Lüthi, H. P., 125[271] (94, 101)
Luthin, W., 190[288] (159)
Lutsenko, A. I., 188[145] (143), 188[146]

(143)
Lutz, H. D., 187[94] (137)
Luu, H., 252[316] (229, 235, 236)
Lyapina, L. S., 255[490] (244)
Lynch, A. T., 186[28] (130)
Lyons, D., 570[21] (548, 550)

Ma, X., 502[312] (488–90)
Maatta, E. A., 121[63] (63, 84)
Mabrouk, H. E., 543[168] (525, 528), 543[173] (525)
Macasek, F., 428[165] (400)
McAuliffe, C. A., 253[407] (240)
McCarthy, D. A., 367[23] (277, 294, 297, 300, 303), 369[144] (297, 300)
McCormick, M. J., 126[341] (107, 112), 126[347] (109, 110), 126[361] (112)
McCrudden, K., 571[72] (565)
MacCurtain, J. A., 273[115] (271)
McDonald, T. R. R., 499[151] (454)
McDonald, W. S., 369[192] (313), 369[200] (313, 316), 370[224] (322), 370[238] (327, 336, 349, 363), 370[243] (331, 332, 336, 337), 370[253] (336), 370[257] (337), 371[266] (340, 342), 371[278] (342, 347), 371[279] (343), 371[292] (351), 371[298] (351), 371[321] (364), 372[324] (365), 426[21] (378), 499[119] (447), 499[151] (454)
McElroy, B. J., 124[240] (85, 86, 96)
McGarvey, B. R., 368[102] (290, 291)
McGeary, C. A., 123[151] (72, 85, 87, 88, 92)
McGinnity, J. J., 125[295] (98, 100)
McGlinchey, M. J., 367[44] (279, 280)
McGrath, M., 251[274] (226, 245), 272[87] (266), 272[90] (266), 273[113] (270)
MacGregor, M., 31[57] (10)
Macgregor, S. A., 371[293] (351), 371[294] (351), 371[295] (351), 371[296] (351)
Mach, K., 125[280a] (97), 125[280b] (97), 548, 570[18] (548)
Maciaszek, S., 367[68] (285, 290)
McKee, M. L., 191[375] (166), 248[67] (218), 248[72] (218, 228), 248[73] (218, 235), 248[76] (219), 248[77] (219), 248[78] (219, 232), 248[80] (219), 248[85] (219), 248[94] (219), 252[335] (232), 252[336] (232)
McKeever, L. D., 34[218a] (30)
McKelvey, J. M., 34[219b] (30)
MacKenzie, J. D., 249[136] (221, 231)
Mackey, O. N. D., 125[308] (101, 102)
MacKinnon, P., 369[197] (313, 315, 346, 357, 359), 370[245] (332, 340), 370[260] (338–40), 371[267] (341), 372[323] (365), 426[37] (378, 380)
MacLean, C., 251[268] (226)
McLennan, A. J., 371[301] (352)
McLeod, D., Jr., 441, 498[82] (441), 499[89] (442)
McMullan, R. K., 544[240] (534)
McNamara, W. F., 368[97] (288, 290)
McPartlin, M., 542[106] (518–20)
McPhail, A. T., 540[16] (505, 506), 541[47] (511, 520), 542[103] (518, 519), 542[118] (520), 542[120] (520, 521), 542[121] (520, 521), 542[122] (520), 542[130] (520), 542[132] (520), 542[137] (521), 542[138a] (521), 542[138b] (521), 542[148] (522, 523)
McPherson, A. M., 499[122] (448)
Macpherson, I., 370[242] (329), 370[262] (340, 341), 426[37] (378, 380)
Madeja, K., 124[236] (85)
Madsen, J. Ø., 123[190] (80, 81)
Maelia, L. E., 543[174] (525, 526)
Maercker, A., 33[136] (17), 33[164] (19), 187[94] (137)
Maerker, A., 125[309] (101)
Maetzke, T., 33[144] (19)
Magdeeva, R. K., 251[255] (225, 241)
Mageswaran, R., 248[113] (220), 425[4g] (375, 389)
Magnuson, V. R., 367[68] (285, 290)
Magrath, M., 371[274] (341)
Maguire, J. A., 247[42] (218), 249[127] (221, 234), 271[10] (258), 271[24] (259), 272[28] (260), 544[233] (534), 544[235] (534)
Magull, S., 541[63] (513, 514)
Mahdi, W., 33[145] (19), 34[190] (23), 34[192] (23, 24)
Mahon, M. F., 194[580] (181)
Mahrwald, R., 125[313] (102)
Maier, G., 189[222] (152), 189[225] (153)
Maier, N. A., 253[425] (241, 244), 254[450] (243), 254[451] (243), 427[135] (400, 414), 427[136] (400, 414), 428[138] (400), 428[167] (400, 414), 428[181] (400, 414), 428[182] (400, 414), 428[183] (400, 414), 428[186] (400, 414), 430[291] (414)
Mair, C., 123[165] (76)
Mair, F. S., 31[66] (10)
Maistrenko, G. L., 187[113] (139)
Majid, T. N., 121[76] (65, 77)
Majumdar, D., 192[446] (173)
Makarevich, N. I., 255[490] (244)
Makhaev, V. D., 367[58] (284), 368[75] (286), 368[78] (287), 368[80] (286, 287), 368[81] (286, 287)
Makrlik, E., 428[178] (400, 416), 430[311] (416), 430[312] (416), 430[313] (416)
Malanga, C., 501[226] (468), 501[264] (476)
Maldar, N. N., 254[436] (242), 272[29] (260), 544[232] (534)
Malhotra, K. C., 32[83] (11)
Malinski, E., 254[480] (244)
Malmquist, J., 254[444] (242)
Malon, P., 428[154] (400, 414)
Malone, J. F., 499[119] (447)
Maloney, J. D., 540[20] (505, 506, 509), 541[65] (513), 542[145] (522)
Malova, E. V., 195[627] (184)
Malpass, D. B., 498[52] (438, 456, 474)
Maly, K., 249[175] (221), 250[193] (221), 428[140] (400, 414), 428[187] (400)
Mamaenko, M. V., 194[537] (179)
Manasevit, H. M., 540[8] (504, 519), 541[101a] (517)
Manceron, L., 498[85] (441), 498[86] (441)
Mancini, M., 367[31] (278)
Manfre, R. J., 500[198] (465, 496)
Manners, I., 192[434] (172), 215[79] (211)
Männig, D., 193[471] (174, 179), 194[588] (182, 184), 367[43] (279), 368[84] (289), 368[122] (293, 295)
Manning, M. J., 271[12] (258), 425[1e] (374), 429[241] (400, 412), 430[288] (400, 412), 430[323] (418, 419), 541[99] (517)
Manriquez, J. M., 121[63] (63, 84)
Mansilla, H., 193[474] (174)
Mansour, T., 32[104h] (13)
Maouche, B., 187[53] (132)
March, J., 498[39] (436)
Marchand, G. M., 191[341] (163)
Marchese, G., 124[215] (83)
Marchetti, F., 367[34] (279, 280)
Marchioro, G., 121[81] (65, 72)
Marder, T. B., 368[120] (292, 293, 295), 368[121] (292, 293, 295), 429[222] (400, 415), 430[296] (415), 430[297] (415), 570[54] (557, 558, 567), 571[66] (562)
Mardones, M. A., 271[9] (258), 271[10] (258), 271[11] (258), 272[78] (265), 541[52] (512, 522), 541[61] (512), 542[134] (520), 542[146] (522), 542[149] (522, 523), 542[153] (522, 524), 542[154] (522, 524), 543[224] (532), 544[231] (533)
Mares, F., 252[307] (228, 238, 239), 253[371] (236)
Maringgele, W., 158, 186[27] (130), 187[60] (133, 144), 187[61] (133, 144), 188[158] (144), 188[159] (144, 154), 188[160] (144, 154), 188[161] (144, 154), 189[240] (154), 189[243] (154), 189[244] (154), 190[311] (161), 190[313] (161), 191[349] (164), 192[395] (167), 193[514] (177), 194[548] (179), 194[596] (183)
Markies, P. R., 120[2] (58, 72, 74, 87–9), 121[61] (63, 74, 77, 90), 121[66] (64, 65, 70, 86, 90, 91, 93, 96, 106, 109, 110), 121[78] (65, 74, 77, 90), 121[80] (65, 74, 77), 123[157] (73, 74, 88), 123[166] (76, 77, 85, 90), 125[254] (88, 89), 125[257] (88, 89, 93), 125[261] (90), 125[262] (90, 96)
Markin, V. S., 194[539] (179)
Markova, S. A., 502[326] (491)
Marks, M. W., 367[55] (283)
Marks, T. J., 121[63] (63, 84), 501[277] (480)
Marolewski, A. E., 502[317] (490)
Marsch, M., 31[38b] (6), 32[105] (13), 33[133] (17), 33[134] (17), 33[138] (19), 33[139] (19), 123[158] (74)
Marsh, R. E., 427[125] (400, 412)
Martiategui, J. F., 195[613] (183)
Martin, H., 498[46] (437, 443, 444), 500[173] (458), 501[270] (478, 484, 497)
Martin, J. C., 194[597] (183, 185)
Martin, R. L., 32[83] (11)
Martin, R. M., 121[81] (65, 72)
Martin, S. J., 502[348] (496)
Martin, T. R., 123[167] (76, 84, 85, 88, 98), 124[234] (84, 85)
Martinez de Ilarduya, J. A., 498[42] (436, 443, 468)
Marubayashi, N., 502[316] (490)
Maruoka, K., 498[57] (439, 492), 498[58] (439), 500[211] (467), 500[212] (467), 500[213] (467), 501[256]

(474), 501[257] (474), 501[266] (476), 501[267] (476), 501[268] (476), 502[311] (488), 502[327] (491), 502[334] (492), 502[342] (494)
Maruta, M., 122[90] (66)
Maruyama, K., 81, 123[193] (80, 81), 123[198b] (82), 124[204] (82), 501[245] (473)
Marynick, D. S., 37, 54[9] (38, 48)
Mashima, K., 125[281] (97, 98)
Masiero, S., 124[205a] (82)
Mason, M. R., 499[141] (452, 453)
Mason, S. A., 123[152] (72, 73, 86–8, 92)
Mass, G., 192[451] (173)
Massa, L., 187[56] (132)
Massa, W., 32[118] (15), 34[188] (22, 23), 123[158] (74), 187[67] (134), 187[68] (134), 187[70] (134), 187[74] (135), 187[75] (135), 187[77] (135), 189[196] (149), 189[199] (149), 189[206] (150), 541[70] (514)
Massey, A. G., 247[7] (218)
Massey, S. T., 187[52] (132)
Mastryukov, V. S., 247[14] (218), 250[196] (222), 250[197] (222), 250[198] (222)
Matel, L., 414, 428[165] (400), 430[294] (414), 430[295] (414)
Matheis, W., 194[542] (179), 194[546] (179), 194[547] (179)
Mathey, F., 368[91] (289)
Mathis, C., 126[354] (110, 111), 127[376] (116)
Matkovskii, P. E., 499[103] (443)
Matochkin, V. S., 255[489] (244)
Matotschkin, W. S., 255[493] (245)
Matsubara, H., 193[504] (176), 193[505] (176), 193[506] (176), 193[507] (176)
Matsumura, K., 121[82] (66)
Matsumura, Y., 501[267] (476), 501[268] (476)
Matsuyama, T., 123[191] (80, 81)
Mattes, R., 499[106] (443), 499[109] (444)
Matteson, D. S., 186[5] (130), 192[445] (173, 175), 192[446] (173), 192[447] (173), 193[498] (175), 193[525] (178), 247[6] (218)
Matthews, D. M., 369[142] (294, 296, 297, 301, 303), 369[145] (297, 301), 369[148] (297, 303), 369[170] (301), 369[180] (308), 369[181] (308)
Matthews, P., 54[2] (35)
Matveeva, Zh. A., 191[361] (165)
Maurer, J. L., 254[432] (242), 255[496] (245)
Maury, F., 570[49]
Mavity, J. M., 500[160] (456, 462)
Maxwell, W. M., 427[124] (398)
May, A. S., 31[39] (8)
Mayer, E. P., 189[197] (149), 190[273] (158), 192[435] (172, 180)
Mayer, H., 189[245] (155), 215[79] (211), 252[329] (231, 234)
Mayer, J. M., 570[36] (552)
Maynard, R. B., 249[135] (221, 234), 251[241] (225, 227), 251[250] (225, 229), 252[309] (229), 426[13a] (377, 382, 388), 426[14a] (377, 421–3), 426[66] (382), 426[67] (382, 398), 426[70] (382), 427[74] (382)
Mazanec, T. J., 367[63] (284, 285)
Mazid, M. A., 543[192a] (527), 543[192b] (527)
Mazunov, V. A., 251[261] (226, 236)
Mazzah, A., 194[581] (181)
Mazzanti, G., 124[205a] (82)
Mebel, A. M., 248[82] (219), 248[86] (219)
Medford, G., 369[178] (307, 308)
Meek, D. W., 367[63] (284, 285)
Meh, D., 121[44] (60, 79)
Mehler, K., 122[87] (66), 123[164] (76, 77, 93), 124[248] (86, 93), 125[265] (93, 101, 102), 125[300] (99), 126[336] (107)
Mehrotra, R. C., 194[532] (179)
Mehrotra, R. K., 193[511] (176), 193[512] (176)
Mehrotra, S. K., 543[176] (525)
Meier, H.-U., 190[294] (160), 190[303] (160)
Meina, D. G., 370[261] (339)
Meiramov, M. G., 249[161] (221, 242), 253[419] (241), 253[422] (241), 253[424] (241, 242), 254[429] (242)
Meisters, A., 499[137] (452)
Meli, A., 368[73] (286)
Meller, A., 130, 160, 186[10] (130, 133, 144, 154), 187[60] (133, 144), 187[61] (133, 144), 187[89] (136, 144), 188[129] (141), 188[158] (144), 188[160] (144, 154), 188[189] (148), 189[202] (150, 151), 189[240] (154), 189[241] (154), 189[242] (154), 189[243] (154), 189[244] (154), 190[286] (159, 160), 190[287] (159), 190[288] (159), 190[289] (159), 190[311] (161), 190[313] (161), 192[395] (167), 193[514] (177), 194[548] (179), 194[551] (179, 180), 194[552] (179), 194[562] (180), 194[566] (180), 194[567] (180), 194[570] (180), 194[571] (180), 194[596] (183), 195[610] (183), 195[638] (185), 195[639] (185), 195[640] (185), 195[642] (185), 249[126] (221, 226, 238)
Mellet, R., 55[59] (54)
Mel'nik, O. A., 248[58] (218, 244), 253[414] (241, 244), 254[467] (244)
Mendía, A., 541[72] (514)
Meneghelli, B. J., 251[273] (226)
Menegus, F., 543[204] (529)
Meng, X., 215[30] (201), 369[139] (297, 300), 369[140] (297, 300, 303), 369[162] (296), 369[177] (307), 426[46] (382, 393), 426[55] (382, 387, 422), 426[59] (382, 393), 427[89] (381, 382, 392, 422), 427[96b] (381)
Menicagli, R., 501[226] (468), 501[261] (475), 501[264] (476)
Mennekes, T., 187[72] (134), 190[266] (157), 255[515] (245), 498[79] (441)
Merkert, J. W., 426[73] (382, 387), 427[75] (382), 427[92] (382, 392)
Merola, J. S., 368[123] (293, 295)
Merzweiler, K., 570[56] (559, 562)
Messerle, L., 369[173] (306), 369[174] (306), 369[175] (306)
Metternich, R., 192[448] (173)
Meunier, P., 34[197] (22)
Meyer, F., 272[42] (261), 272[43] (261), 272[44] (261)
Meyer, G., 544[258] (537)
Meyer, H., 187[68] (134), 187[74] (135), 187[75] (135), 189[194] (149)
Meyer-Zaika, W., 190[306] (161)
Miah, M. A. J., 121[75a] (65, 70)
Micciche, R. P., 272[92] (267), 370[250] (20, 334, 341, 342, 357, 359), 426[12a] (377, 382, 394, 396, 421, 423), 426[12b] (377, 382, 423), 427[107] (396)
Michael, A., 30[1a] (1)
Micha-Screttas, M., 124[241] (86, 93), 126[327] (106)
Michel, H., 187[65] (133, 155), 189[198] (149), 252[351] (234)
Middelhauve, B., 194[592] (182)
Midollini, S., 367[60] (284, 285), 368[72] (286), 368[73] (286)
Miginiac, L., 122[100] (67, 109)
Miginiac, P., 500[223] (468)
Miginiac, Ph., 122[119] (70)
Migita, T., 500[187] (463, 476)
Mikhailov, B. M., 145, 186[21] (130, 145), 187[117] (139), 188[144] (143), 188[145] (143), 188[146] (143), 188[147] (143, 154), 188[148] (143), 188[149] (143), 188[165] (145), 188[166] (145), 188[169] (145), 188[170] (145), 188[176] (146), 188[177] (147), 188[178] (147), 188[179] (147), 188[180] (147), 188[181] (147), 188[183] (147), 189[223] (152), 190[309] (161), 190[310] (161), 191[353] (164), 191[370] (165), 191[372] (166), 191[384] (167), 191[385] (167), 191[388] (167), 271[7] (258)
Mikhailova, L. N., 191[361] (165)
Mikhal'skii, A. I., 254[477] (244)
Miki, K., 123[153] (73, 83), 123[154] (73)
Mile, B., 498[35] (436, 441), 498[84] (441), 499[92] (442), 499[93] (442), 499[94] (442)
Miles, S. J., 544[259] (537)
Miles, W. H., 124[203] (82)
Milewski-Mahrla, B., 541[50] (512)
Miller, J. A., 467, 498[76] (441, 467, 486), 501[238] (470)
Miller, N. E., 166, 191[376] (166), 191[377] (166), 191[378] (166), 191[379] (166)
Miller, R. W., 272[66] (264), 272[67] (264)
Miller, S. B., 272[89] (266), 428[180] (400, 414)
Miller, T. A., 120[8] (58)
Miller, T. J., 194[591] (182), 194[593] (182)
Miller, V. R., 427[98a] (391), 427[98b] (391, 395), 427[98d] (391, 414, 419), 430[334c] (421)
Mills, K., 541[82] (514)
Mingos, D. M. P., 247[19] (218), 247[21] (218, 232), 248[81] (219), 248[89] (219), 252[339] (232), 367[25] (277), 367[29] (277), 425[4a] (375), 425[4c] (375), 425[4j] (375, 398), 425[4m] (375, 393, 398, 422), 425[5d] (375), 425[7b] (375, 422), 427[123] (398), 428[171] (400, 412), 571[95] (569)

Minkin, V. I., 186[44] (131)
Minsker, K. S., 499[134] (452)
Minyaev, R. M., 186[44] (131)
Mirabelli, M. G. L., 186[28] (130), 251[299] (228), 252[332] (231), 255[526] (246), 255[527] (246)
Miravitlles, C., 249[139] (221, 240), 249[146] (221)
Mircea, A., 55[59] (54)
Miroshnichenko, E. A., 187[59] (133)
Miroshnikova, I. I., 254[477] (244)
Mirsaidov, U., 368[79] (286, 287)
Misbach, P., 501[272] (478, 479, 481, 483, 484)
Mishra, I. B., 272[64] (263)
Misra, M. C., 31[41] (8)
Misurkin, I. A., 191[391] (167)
Mitaine, P., 571[80] (566, 567)
Mitchell, G. F., 248[98] (220), 369[185] (311)
Mitchell, S. A., 499[90] (442)
Miura, M., 255[501] (245)
Miyanaga, S., 121[77] (65)
Miyano, N., 123[191] (80, 81)
Miyaura, N., 193[470] (174)
Miyazaki, M., 501[269] (476)
Miyazaki, T., 501[267] (476), 501[268] (476)
Miyoshi, K., 124[230] (84)
Mizuno, Y., 127[384] (118)
Mizusawa, E. A., 249[150] (221, 231, 237, 241, 245), 253[417] (241), 253[418] (241, 245), 430[281] (400, 414)
Mobraaten, E. C., 254[461] (244)
Mochida, K., 126[338] (107), 126[339] (107, 108, 118), 126[340] (107), 126[346] (108, 109), 127[383] (118), 127[384] (118)
Möckel, L., 544[250] (536)
Modrick, M. A., 571[91] (568)
Moehrle, M., 194[592] (182)
Moezzi, A., 544[260] (538), 544[267b] (539), 544[267c] (539)
Moher, E., 253[395] (239, 246), 369[153] (296–8)
Moise, C., 571[75] (566), 571[79] (566, 567)
Molchanova, G. N., 253[423] (241)
Mole, T., 467, 499[124] (449), 499[137] (452), 500[186] (462), 500[208] (467)
Moll, M., 570[58] (559)
Molle, G., 121[33] (60)
Mollenhauer, M. N., 250[195] (222, 227)
Molloy, K. C., 194[580] (181)
Molodnitskaya, I. V., 368[107] (290, 291), 570[15] (547), 570[16] (547)
Mols, O., 497[16] (434, 448)
Monari, M., 571[69] (563, 564)
Moncrieff, D., 32[122] (15)
Monev, V., 186[43] (131)
Mongeot, H., 273[96] (267), 273[97] (267), 371[318] (363)
Montanari, S., 502[339] (494)
Moody, D. C., 368[85] (286, 289), 368[86] (286, 289)
Moody, R. J., 193[498] (175)
Moon, K. B., 126[341] (107, 112)
Moore, A. H., 541[42] (511)
Moore, J. W., 497[13] (433, 448)
Mootz, D., 31[31] (6), 194[592] (182)
Morales, A. L., 254[463] (244)
Moran, J. T., 272[69] (264)
Morandini, F., 253[416] (241), 254[452] (243), 254[454] (243), 254[455] (243)
Moretti, G., 502[303] (484)
Morita, E., 501[265] (476)
Morken, J. P., 122[104] (67)
Morley, C. P., 123[159] (74, 86, 92, 93), 125[308] (101, 102)
Moro-oka, Y., 570[28] (551)
Morosin, B., 250[235] (225)
Morris, H., 498[84] (441), 499[94] (442)
Morris, J. H., 130, 145, 154, 186[1] (130, 145, 154), 186[23] (130), 247[3] (218, 220–2, 225, 227, 238, 241–4), 248[106] (220, 228, 235, 237–9), 250[190] (221, 229, 237), 255[512] (245), 370[209] (317), 370[261] (339), 425[1p] (374)
Morrison, J. A., 255[514] (245), 255[522] (246)
Morrison, J. D., 31[27] (5)
Morse, K. W., 191[380] (166), 368[76] (286)
Mortimer, M. D., 428[197] (400, 411), 428[200] (400, 411, 417, 418), 428[202] (400, 411), 429[272] (400, 411)
Moseley, P. T., 55[55] (53)
Moser, P., 427[133] (400, 417)
Moser, W. R., 500[181] (461, 480)
Moses, L. M., 121[54a] (61, 77), 122[94] (67)
Mösges, C., 126[375] (115)
Moskau, D., 33[135] (17)
Moss, R. H., 540[9] (504, 511)
Motevalli, M., 367[56] (284), 540[14] (505, 506), 540[18] (505, 506), 570[20] (548), 570[21] (548, 550), 570[23] (549)
Motoc, I., 191[355] (164)
Motol'ko, G. R., 193[461] (174), 193[462] (174), 193[463] (174, 179), 193[464] (174)
Mouhtadi, M. E., 191[356] (164)
Mountford, P., 369[193] (313, 315, 319), 370[214] (319), 370[222] (321), 428[195] (400, 414), 570[37] (553, 555)
Mubarak, S. I. M., 193[500] (175)
Mueller, E., 500[185] (462), 500[188] (463), 500[207] (467)
Mueller, G., 501[230] (468)
Muellners, W., 190[319] (162)
Muetterties, E. L., 272[59] (263), 369[157] (294), 369[201] (313)
Mukaiyama, T., 192[437] (172)
Müllen, K., 124[211b] (83)
Mullendore, A., 250[235] (225)
Müller, D., 542[157] (524)
Müller, G., 54[6] (36), 125[291] (98), 126[364] (112–14), 187[109] (139), 192[414] (169), 192[426] (171), 192[427] (171), 216[92] (213), 541[48] (511), 544[242] (535), 544[246] (535), 544[247] (536), 544[251] (536), 544[252] (536), 544[253] (536)
Müller, H., 429[235] (400, 411), 500[195] (464, 486, 495)
Müller, H.-J., 570[49], 570[51] (557)
Müller, J., 272[38] (260), 272[39] (261), 272[40] (261), 272[42] (261), 272[43] (261), 272[44] (261), 272[54] (262)
Muller, P. M., 122[106] (67, 71, 87, 95)
Müller-Lindenhof, N., 498[69] (440)
Müller-Warmuth, W., 194[564] (180)
Mullica, D. F., 427[129] (400, 411, 417, 418), 429[236] (400, 411), 429[255] (400, 411), 429[273] (400, 414), 430[279] (400, 411), 430[319] (417–19)
Mulvey, R. E., 2, 30[5] (2, 30), 30[13] (2), 31[18] (3), 31[19] (3), 31[20] (3), 31[21] (3), 31[22] (4), 31[23] (4), 31[30a] (6, 8), 31[30b] (6, 8), 31[37] (6), 31[44] (8, 9), 31[56] (10), 31[57] (10), 31[63] (10), 31[67] (10), 126[316] (103)
Münch, A., 32[80] (11)
Münch Hartmann, H., 32[120d] (15)
Mundt, O., 32[81] (11)
Murai, A., 502[313] (489)
Muraoka, K., 124[210] (83)
Murase, N., 498[58] (439), 501[266] (476)
Murata, I., 189[236] (154)
Murray, B. D., 31[54] (9, 11), 32[89] (12)
Musgrove, R. J., 429[256] (400, 411), 429[260] (400, 411)
Musin, R. A., 195[621] (184)
Muslukhov, R. R., 123[169] (77, 92), 125[252] (87)
Muslukov, R. R., 502[333] (492)
Myakishev, K. G., 428[166b] (400, 414)
Myakushev, V. D., 254[438] (242, 244), 254[476] (244)
Myasoedov, S. F., 251[252] (225), 251[253] (225, 226), 251[260] (226)
Mynott, R., 124[249] (86, 90, 93–5), 125[297] (99, 100), 125[299] (99), 193[515] (177)

Nabivanets, B. I., 252[356] (234)
Nagase, S., 123[189] (80, 81), 124[210] (83)
Nagasuna, K., 125[281] (97, 98)
Nagata, W., 502[307] (486)
Nagel, K., 500[173] (458)
Nagle, J. K., 571[94] (569), 571[96] (569)
Nagy-Magos, Z., 214[17] (199, 208)
Naka, H., 500[187] (463, 476)
Nakadaira, Y., 189[233] (153), 189[234] (154), 215[76] (210)
Nakai, S., 502[327] (491)
Nakamura, A., 54[11] (38), 54[12] (38), 121[77] (65), 125[281] (97, 98), 125[291] (98)
Nakamura, H., 255[508] (245)
Nakanis, T., 541[101b] (517)
Nakanishi, H., 123[150] (72)
Nakazono, Y., 124[230] (84)
Nakhmanovich, B. I., 126[357a] (111, 119), 126[357b] (111), 126[358] (111), 126[359] (111)
Nalband, G. T., 253[411] (241)
Nam, W., 250[218] (224), 252[342] (232, 233), 252[364] (236, 238), 253[369] (236, 238, 239), 271[16] (258)
Nametkin, N. S., 501[236] (470)
Nanninga, D., 187[110] (139)
Narske, R. N., 124[233] (84)
Narula, C. K., 186[28] (130), 188[125] (140, 156), 189[253] (156), 190[315] (161), 192[411] (169), 192[424]

(170), 193[465] (174, 180), 194[536] (179), 194[588] (182, 184), 272[74] (264)
Naruse, Y., 501[225] (468)
Nash, A. W., 498[28] (435), 498[29] (435)
Naso, F., 124[215] (83)
Natta, G., 439, 478
Naumov, V. A., 195[623] (184)
Naumova, S. F., 193[461] (174), 193[462] (174), 193[463] (174, 179), 193[464] (174)
Negele, M., 215[36] (202), 215[38] (202, 203), 215[39] (202–4), 215[40] (203), 215[41] (203), 215[42] (203, 204)
Negishi, E., 122[103a] (67, 84), 122[103b] (67), 124[227] (84), 188[152] (143), 188[156] (144), 501[293] (482)
Negoro, K., 124[235] (85)
Nehls, D., 124[236] (85)
Neilson, R. H., 186[28] (130), 191[348] (164)
Nekhoroshkov, V. M., 192[428] (171)
Nekrasov, Y. S., 251[261] (226, 236), 251[263] (226), 428[159] (400)
Nel'Kenbaum, Yu. Ya., 499[134] (452)
Nelson, C. K., 427[97] (388)
Nemoto, H., 253[420] (241), 255[508] (245)
Nesmeyanova, O. A., 122[86] (66)
Nesterova, S. V., 189[255] (156), 189[256] (156), 191[342] (163), 193[517] (177)
Nestor, K., 249[138] (221), 272[55] (262), 272[56] (262, 269), 370[248] (332, 339), 371[273] (340), 371[277] (342), 426[34] (378, 379), 427[115] (396), 427[118] (396)
Neugebauer, W., 33[160] (19, 21)
Neumann, F., 570[39] (553, 560, 567)
Neumann, H. M., 498[23] (435, 436, 496)
Neumann, H.-P., 571[62] (561)
Neumüller, B., 540[27] (506), 541[60] (512), 541[63] (513, 514), 541[85] (514), 542[135] (520), 542[136] (520)
Neustadt, R. J., 570[44] (555)
Nevels, P., 252[352] (234)
Newcomb, M., 126[317] (103)
Ng, B., 250[216] (224, 232, 233), 252[365] (236), 252[366] (236)
Ng, B. K., 371[317] (362)
Ng, L-L., 371[317] (362)
Ngoviwatchai, P., 124[210] (83)
Niecke, E., 194[583] (182), 194[584] (182)
Niedenzu, K., 130, 167, 169, 186[9] (130), 186[26] (130), 187[55] (132), 188[129] (141), 190[274] (158), 190[275] (158), 191[347] (164), 191[350] (164), 191[351] (164), 191[352] (164), 192[394] (167), 192[395] (167), 192[396] (167, 168), 192[397] (167), 192[398] (168), 192[403] (168), 192[404] (169), 192[405] (169), 195[611] (183), 195[612] (183), 195[613] (183), 195[614] (183), 195[637] (185)
Niedoba, S., 502[295] (482)
Nief, F., 368[91] (289)
Nieger, M., 194[584] (182)
Niemeyer, C. M., 193[472] (174), 193[473] (174)
Nierlich, M., 368[87] (289)
Nifant'ev, E. E., 251[255] (225, 241)
Nikitina, N. S., 254[479] (244)
Nikolaeva, S. N., 55[49] (52)
Nikonov, G. N., 191[335] (163), 195[617] (184), 195[618] (184), 195[619] (184), 195[621] (184), 195[622] (184), 195[623] (184), 195[624] (184), 195[625] (184), 195[626] (184), 195[627] (184), 195[628] (184)
Nikulicheva, N. A., 254[479] (244)
Nishi, T., 121[77] (65)
Nishiyama, H., 121[70] (65, 70)
Nissan, R. A., 542[123] (520)
Nitto, Y., 122[103a] (67, 84)
Noble, M. J., 542[129] (520)
Nock, L. A., 255[486] (244)
Noding, S. A., 501[255] (474)
Noe, R., 502[294] (482)
Nogal Gonzalez, A. M., 498[42] (436, 443, 468)
Noirot, M. D., 536, 544[248] (536)
Nojima, M., 124[210] (83)
Nölle, D., 194[589] (182)
Noltemeyer, M., 187[61] (133, 144), 187[89] (136, 144), 188[160] (144, 154), 189[242] (154), 189[244] (154), 193[514] (177), 194[551] (179, 180), 194[562] (180), 194[571] (180), 194[581] (181), 195[610] (183), 195[631] (184, 185), 195[633] (184), 195[640] (185), 195[641] (185), 195[642] (185), 249[126] (221, 226, 238), 543[167] (525), 543[177] (525, 526), 543[178] (525, 526)
Nomoto, T., 125[257] (88, 89, 93), 125[262] (90, 96)
Nomura, N., 124[215] (83)
Nonoshita, K., 501[256] (474), 502[342] (494)
Nonshita, K., 501[257] (474)
Norman, A. D., 271[25] (259)
Norman, N. C., 272[30] (260), 570[2] (546), 570[38] (553, 555, 556), 570[52] (557, 558), 570[54] (557, 558, 567), 571[64] (562), 571[65] (562), 571[66] (562), 571[72] (565), 571[73] (565), 571[92] (568)
Normant, J. F., 122[99] (67, 68, 102), 126[331] (106), 501[235] (469)
Nöth, H., 158–60, 182, 185, 186[20] (130), 186[27] (130), 187[81] (135), 187[101] (138), 187[107] (139), 188[125] (140, 156), 189[197] (149), 189[253] (156), 190[273] (158), 190[276] (158), 190[277] (158), 190[278] (158, 159), 190[281] (159), 190[290] (159), 190[291] (160), 190[292] (160), 190[297] (160), 190[314] (161), 190[315] (161), 190[316] (161), 191[365] (165), 192[396] (167, 168), 192[398] (168), 192[409] (169), 192[411] (169), 192[412] (169, 170), 192[421] (170), 192[423] (170), 192[424] (170), 192[432] (171, 179), 192[435] (172, 180), 192[436] (172), 192[459] (173), 193[471] (174, 179), 194[536] (179), 194[549] (179), 194[558] (180), 194[563] (180), 194[568] (180), 194[569] (180), 194[575] (181), 194[576] (181), 194[577] (181, 182), 194[578] (181), 194[582] (182), 194[585] (182), 194[586] (182), 194[587] (182), 194[588] (182, 184), 194[589] (182), 194[590] (182, 185), 195[630] (184), 195[635] (185), 195[637] (185), 195[643] (185), 214[6] (198, 211), 215[27] (201), 215[28] (201), 215[29] (201), 272[74] (264), 272[75] (264), 367[43] (279), 368[84] (289), 368[97] (288, 290), 368[117] (292, 295), 368[118] (292, 295), 368[119] (292, 295), 368[122] (293, 295)
Novak, C., 249[152] (221), 250[183] (221), 250[191] (221)
Novak, D. P., 499[152] (454)
Novak, R., 544[242] (535)
Novikov, D. D., 254[477] (244)
Novikova, E. D., 193[482] (175)
Nowak, M., 125[272a] (95), 125[272b] (95)
Nowell, I. W., 54[16] (40)
Nowotnik, D. P., 192[453] (173)
Noyori, R., 123[200] (82, 106)
Nozaki, H., 123[182] (79), 123[183] (79), 126[330] (106), 500[197] (465), 501[231] (469), 501[262] (476), 502[320] (490), 502[329] (491)
Nuber, B., 570[53] (557), 571[62] (561), 571[63] (561)
Nudelman, N. E. S., 31[38b] (6)
Nuffer, R., 126[355] (111)
Nugent, K. W., 45, 54[25] (44, 47), 54[26] (45), 54[29] (45), 54[30] (45), 54[31] (46)
Nunn, C. M., 429[249] (400, 411), 429[250] (400, 411), 429[265] (400, 411), 430[278] (400, 411), 542[115] (520), 542[124] (520), 542[125] (520), 542[131] (520, 522), 542[140] (521, 522, 526), 542[144] (522), 542[146] (522)
Nuzzo, R. G., 121[31] (59, 62), 500[214] (468), 500[215] (468)

Oae, S., 121[82] (66)
Obaid, R. M. S., 501[279] (480)
Obarski, G. E., 249[117] (220)
O'Brien, L. C., 120[7] (58)
O'Brien, R. A., 126[344] (108, 109, 117, 120)
Ochi, M., 124[219] (83)
Ochs, R., 30[1b] (1)
O'Con, C. A., 426[38] (378, 380)
Oda, M., 189[259] (157), 194[534] (179)
Odaira, Y., 541[80] (514), 541[81] (514)
Odling, W., 500[182] (461)
Odom, J. D., 189[191] (148), 250[219] (224)
Oehlert, W., 187[71] (134)
Oenbrink, G., 255[501] (245)
Ogata, K., 124[212] (83)
Ogawa, H., 126[346] (108, 109), 127[383] (118)
Ogawa, Y., 502[325] (491)
Ogino, H., 369[187] (311), 369[188] (311)
O'Gorman, E. V., 248[88] (219, 220, 233), 248[101] (220, 233), 251[282] (227)
Ogura, F., 123[178] (78), 502[341] (494)
Ogura, K., 122[125] (70)
Ogura, S., 126[338] (107)

Oh, B., 252[341] (232, 233)
O'Hare, D., 541[88] (515)
Ohloff, G., 500[177] (458)
Ohnuma, Y., 54[11] (38)
Ohst, H., 189[245] (155), 189[246] (155), 215[33] (202), 215[35] (202), 215[36] (202), 215[44] (204), 215[66] (209), 215[69] (209), 252[329] (231, 234)
Ohtsuka, N., 55[58] (54)
Okada, K., 189[259] (157), 194[534] (179)
Okamoto, Y., 123[153] (73, 83), 123[154] (73)
Okawara, R., 499[110] (445)
Okhlobystin, O. Y., 254[431] (242)
Oki, A. R., 425[4k] (375, 382, 388), 426[64] (382, 388), 426[65] (382, 388), 427[83] (382, 388), 427[87] (382, 388)
Oki, M., 187[115] (139)
Okninski, A., 501[252] (474)
Olah, G. A., 187[112] (139), 247[16] (218), 247[25] (218), 247[26] (218), 247[27] (218), 247[28] (218), 247[29] (218), 247[42] (218), 252[339] (232), 255[511] (245), 255[518] (246), 367[16] (277), 367[17] (277), 425[1t] (374, 381), 435, 437, 444, 462, 498[27] (435, 437, 443–5, 462)
Olbrich, F., 34[208] (26)
Ol'dekop, Yu. A., 253[425] (241, 244), 254[450] (243), 254[451] (243), 427[135] (400, 414), 427[136] (400, 414), 428[138] (400), 428[167] (400, 414), 428[181] (400, 414), 428[182] (400, 414), 428[183] (400, 414), 428[186] (400, 414), 430[291] (414)
Oliver, J. P., 32[120a] (15), 32[120b] (15), 32[120c] (15), 33[165] (19), 497[13] (433, 448), 497[16] (434, 448), 497[18] (434, 448), 497[19] (434), 499[121] (448), 499[123] (449), 543[187] (526)
Olmstead, M. M., 31[42] (8, 9), 31[47] (9), 31[48] (9), 31[52] (9), 31[54] (9, 11), 31[68] (10), 31[75] (10), 32[112] (15), 32[120e] (15), 33[159] (19), 33[169] (19), 33[170] (19), 497[12] (433, 434, 448, 449, 497), 499[115c] (446, 496), 542[114] (520, 525), 544[260] (538), 544[266] (539), 544[267b] (539), 571[94] (569), 571[96] (569)
Olrich, F., 31[58] (10)
Ol'shevskaya, V. A., 249[162] (221, 226, 239), 249[169] (221, 239), 251[239] (225), 251[262] (226), 251[265] (226, 239), 252[327] (231), 252[357] (234), 253[382] (237, 239), 253[391] (239), 253[392] (239), 253[393] (239), 253[394] (239), 253[399] (239), 254[440] (242), 255[530] (247), 427[130] (400, 416), 427[132] (400, 412), 428[159] (400), 430[289], 430[290]
Olson, J. D., 429[243] (400, 416)
Omae, I., 32[119d] (14)
On, H. P., 501[241] (471)
Onak, T., 247[1] (217), 247[8] (218, 242, 244), 247[9] (218, 242, 244), 247[12] (218), 247[29] (218), 247[31] (218), 248[52] (218, 244), 248[87] (219), 248[88] (219, 220, 233), 248[100] (220, 232, 233), 248[101] (220, 233), 250[203] (222, 223, 235), 250[204] (222, 229), 250[212] (223), 250[214] (223), 250[215] (223), 250[216] (224, 232, 233), 250[218] (224), 252[315] (229), 252[316] (229, 235, 236), 252[341] (232, 233), 252[342] (232, 233), 252[360b] (235), 252[361] (235, 236, 238), 252[364] (236, 238), 252[365] (236), 252[366] (236), 252[367] (236), 252[368] (236), 253[369] (236, 238, 239)
Onak, T. P., 430[334b] (421)
Onaka, S., 571[81] (567)
O'Neil, P. A., 31[57] (10), 31[63] (10), 31[66] (10), 126[316] (103)
O'Neill, M. E., 54[14] (39, 53), 247[17] (218), 248[91] (219), 425[4e] (375, 422)
Onellion, M., 251[276] (227)
Ong, A., 120[15] (59, 60)
Onyiriuka, E. C., 543[165] (525, 526)
Ooi, T., 498[57] (439, 492), 498[58] (439), 501[266] (476), 502[311] (488)
Oppolzer, W., 79, 120[21b] (59, 80), 120[21c] (59), 123[185] (80)
Orlandini, A., 367[60] (284, 285), 368[72] (286), 368[73] (286)
Orlando, L. R., 370[206] (313, 316, 321)
Orlov, V. M., 251[262] (226)
Orlova, G. V., 186[44] (131)
Ortiz, J. V., 126[374] (114), 368[97] (288, 290)
Orvig, C., 194[530] (178)
Osborne, J. H., 255[509] (245)
Oshima, K., 126[330] (106), 500[197] (465), 501[231] (469), 501[234] (469), 501[262] (476), 502[320] (490)
Osman, A., 121[36] (60)
Osmond, D. A., 501[288] (482)
Ostler, C. D., 369[184] (311)
Ostrander, R. L., 272[68] (264), 370[235] (327)
Oswald, J., 121[45] (61)
Otsubo, T., 123[178] (78), 502[341] (494)
Ott, J. J., 248[69] (218, 233), 248[70] (218, 219, 233), 248[90] (219), 248[108] (220), 252[333] (232), 252[334] (232, 233), 252[337] (232), 252[338] (232), 252[340] (232, 233), 252[343] (233), 252[344] (233), 252[347] (233)
Otto, H., 543[214] (531)
Otto, P., 190[297] (160)
Otway, D. J., 126[360] (111, 112, 115)
Ouamerali, O., 187[53] (132)
Ouassas, A., 273[96] (267), 273[97] (267)
Ourisson, G., 502[319] (490)
Overman, L. E., 502[314] (490)
Owen, D. W., 253[408] (240)
Owen, S. M., 369[203] (313, 317)
Owens, M. H., 252[352] (234)
Ozeki, M., 55[58] (54)
Ozin, G. A., 498[34] (436, 442)
Ozol, Ya., 34[212] (29)

Pace, J. M., 189[233] (153), 215[76] (210)
Pachaly, B., 188[187] (148), 192[431] (171)
Paciorek, K. J. L., 186[28] (130)
Paetzold, P., 31[49] (9), 130, 158, 160, 171, 182, 186[12] (130, 158, 159, 162, 163, 165, 184), 186[27] (130), 187[66] (134, 135), 187[72] (134), 189[258] (156), 190[265] (157), 190[266] (157), 190[282] (159), 190[283] (159), 190[284] (159), 190[285] (159), 190[293] (160), 190[294] (160), 190[295] (160), 190[296] (160), 190[298] (160), 190[299] (160), 190[300] (160), 190[301] (160), 190[302] (160), 190[303] (160), 190[304] (160), 191[328] (162), 191[329] (162), 191[330] (162), 191[345] (163), 192[430] (171), 192[434] (172), 193[513] (177), 194[583] (182), 194[584] (182), 214[5] (198, 211), 215[78] (211), 215[79] (211), 215[80] (211), 215[81] (211), 255[515] (245), 272[38] (260), 272[39] (261), 272[40] (261), 272[42] (261), 272[43] (261), 272[44] (261), 272[45] (261), 272[46] (261), 272[47] (261), 272[54] (262), 498[79] (441), 498[81] (441), 528, 543[200] (528, 531)
Page, T. C. M., 541[88] (515)
Pai, G. G., 187[85] (135–7), 187[90] (136), 187[91] (136)
Paine, R. T., 186[28] (130), 192[409] (169), 192[411] (169), 192[424] (170), 193[465] (174, 180), 194[582] (182), 194[586] (182), 214[6] (198, 211), 272[74] (264), 272[75] (264), 368[85] (286, 289), 368[97] (288, 290)
Pajerski, A. D., 34[210] (29), 125[255] (88), 125[256] (88, 89, 93), 125[258] (89, 93), 126[329] (106)
Pakhomov, V. I., 255[488] (244)
Pakulski, M., 192[408] (169)
Palkowitz, A. D., 192[456] (173)
Palm, V., 121[41] (60)
Palmer, M. A. J., 192[455] (173), 192[456] (173)
Palmer-Ordonez, K., 272[84] (266)
Panasenko, A. A., 123[169] (77, 92)
Panasenko, A. I., 194[540] (179)
Pankowski, J., 121[53] (61)
Pannatoni, C., 543[204] (529)
Pao, S. S., 272[82] (265, 266), 272[86] (266)
Papasergio, R. I., 121[25b] (59, 60, 70, 100), 124[246] (86, 90)
Papkov, V. S., 253[414] (241, 244)
Paquette, L. A., 34[197] (22)
Parfenov, B. P., 250[176] (221), 255[490] (244)
Parinis, J. M., 498[34] (436, 442)
Parish, R. V., 371[303] (352)
Park, J. C., 192[455] (173)
Park, K. M., 187[63] (133, 148)
Park, Y. S., 502[340] (494)
Parkanyi, L., 542[104] (518, 519)
Parker, K. O., 367[68] (285, 290)
Parkhill, L., 370[233] (326)
Parkin, G., 55[44] (50–2), 126[322] (104)
Parkin, S., 542[150] (522, 523)
Paroli, R. M., 250[238] (225)
Parry, R. W., 369[184] (311)
Parshall, G. W., 498[65] (440)
Parsons, J. D., 55[56] (54), 55[57] (54)
Parvez, M., 34[210] (29), 125[255] (88),

125[256] (88, 89, 93), 125[258] (89, 93), 190[312] (161), 273[115] (271)
Pascal, Y. L., 251[267] (226)
Pasterczyk, J. W., 541[47] (511, 520), 542[137] (521)
Pasynkiewicz, S., 451, 498[77] (441, 451, 483), 499[130] (451), 499[138] (452), 499[142] (452), 501[258] (475)
Patalinghug, W. C., 127[379] (117)
Patterman, S. P., 33[168] (19)
Pattison, I., 31[17] (3)
Pauer, F., 34[186] (22, 23), 189[242] (154), 190[289] (159)
Paul, P., 368[77] (286)
Pauling, L., 541[93] (515, 519, 526)
Pavankumar, P. N. V., 248[96] (219)
Paver, M. A., 32[77] (11)
Pavlova, S. A., 248[53] (218), 254[464] (244)
Pavlovich, V. K., 249[162] (221, 226, 239), 251[265] (226, 239)
Paxton, R. J., 255[500] (245), 430[315] (416, 417)
Pazik, J. C., 540[7] (504, 530, 539), 540[30] (507), 541[64] (513), 543[206] (530, 539)
Paz-Sandoval, M. A., 195[616] (184)
Pearson, R. G., 34[207] (26)
Pearson, W. H., 121[75b] (65, 70)
Pechurina, S. Ya., 250[176] (221)
Pecunioso, A., 501[261] (475)
Pedrosa, R., 501[260] (475), 502[323] (491)
Peiju, Z., 371[309] (355–7)
Pelin, W. K., 370[227] (323), 370[236] (327), 425[4b] (375)
Pelino, M., 55[42] (50)
Pelter, A., 186[2] (130, 138), 186[4] (130, 138), 186[8] (130)
Pelzer, C., 31[49] (9), 190[296] (160)
Pena, J. P., 251[257] (225, 241), 253[410] (241)
Pendergrass, E., 123[179] (78)
Penigault, E., 249[115] (220)
Penkston, B. S., 544[233] (534)
Pennington, W. T., 540[36] (509), 541[39] (510), 541[40] (510), 541[41] (511)
Peppe, C., 541[45] (511), 541[56] (512), 541[73] (514), 544[257] (537)
Percec, V., 193[503] (176)
Pereyre, M., 32[119a] (14)
Perez, P. J., 367[64] (284, 285)
Perkins, F. K., 251[276] (227)
Perkins, P. G., 248[111] (220)
Perner, R. J., 194[594] (182)
Perret, C., 120[21a] (59, 79), 120[21b] (59, 80)
Pershin, D. G., 188[150] (143, 147), 188[178] (147), 188[180] (147), 188[182] (147), 191[366] (165)
Persson, I., 123[163a] (75), 123[163b] (75)
Peruzzini, M., 367[64] (284, 285)
Peseckis, S. M., 502[315] (490)
Pestana, D. C., 192[420] (170), 542[112] (519), 542[147] (522), 542[150] (522, 523)
Petch, H. E., 250[224] (224)
Peters, E. N., 248[54] (218)
Peters, E.-M., 543[181] (525, 526)
Peters, K., 188[131] (141), 543[181] (525, 526)
Peters, K. M., 215[46] (204)
Peterson, G. E., 194[541] (179)
Peterson, R. A., 127[382] (117)
Peterson, S. W., 34[212] (29)
Petriashvili, M. V., 254[456] (243)
Petricek, V., 249[144] (221), 249[175] (221), 250[191] (221), 250[193] (221), 428[140] (400, 414), 428[174] (400, 414), 428[187] (400)
Petrie, M., 192[420] (170)
Petrie, M. A., 497[12] (433, 434, 448, 449, 497), 542[107] (519), 542[108] (519), 542[111] (519), 542[113] (519)
Petrie, M. P., 542[152] (522, 523)
Petrina, A., 428[187] (400)
Petrov, A. A., 122[112] (68)
Petrovskii, I. V., 193[478] (175)
Petrovskii, P. V., 253[421] (241), 253[423] (241), 430[289]
Petrunin, A. B., 252[354] (234, 242)
Pez, G. P., 34[199] (26)
Pfeiffer, J., 190[313] (161)
Pfohl, W., 500[169] (458, 462, 464, 486)
Phadke, A. S., 187[98] (137), 187[99] (137)
Philoche-Levisalles, M., 367[59] (284)
Pickard, P. L., 31[16] (3)
Pickardt, J., 540[25a] (505), 543[213] (531, 539)
Pickett, N. L., 570[38] (553, 555, 556)
Piepgrass, K. W., 426[13b] (377, 382, 388, 391, 421–4), 426[50] (382), 426[51] (381, 382), 426[52] (381, 382), 427[89] (381, 382, 392, 422), 427[91] (381, 382)
Pihlaja, K., 192[457] (173)
Pill, E., 121[40] (60)
Pillai, C. N., 122[89] (66, 67, 102)
Pilotti, M. U., 428[201] (400, 411), 429[218] (400, 411), 429[220] (400, 412), 429[226] (400, 411), 429[231] (400, 411), 429[232] (400, 411), 429[235] (400, 411), 429[244] (400, 414, 417, 419), 429[245] (400, 414), 429[246] (400, 411), 429[262] (400, 411), 429[268] (400, 411), 429[269] (400, 411), 430[276] (400, 411)
Pilz, M., 187[77] (135), 189[198] (149), 189[199] (149), 189[205] (150)
Pini, D., 122[93] (67)
Pinkus, A. G., 123[195] (82)
Piotrowski, A. M., 498[52] (438, 456, 474), 498[71] (440, 479, 480), 501[258] (475), 501[290] (482), 501[291] (482)
Pipal, J. R., 427[102a] (382, 392), 427[102b] (382, 392), 430[335] (422)
Piraliev, A. G., 501[287] (481)
Pirotte, B., 430[304] (416)
Pisareva, I. V., 251[259] (226, 231), 253[388] (238)
Pitt, C. G., 542[103] (518, 519), 542[118] (520), 542[122] (520), 542[132] (520), 542[138a] (521), 542[138b] (521), 542[148] (522, 523)
Pitzer, K. S., 500[189] (463)
Plaggenborg, L., 543[201] (529)
Plass, H., 33[178] (20), 34[211b] (29)
Plattier, M., 502[319] (490)
Plavac, N., 122[130] (70)
Plesek, J., 247[32] (218), 248[51] (218), 250[209] (223), 250[210] (223), 250[211] (223), 250[221] (224), 251[249] (225, 230), 251[272] (226), 252[301] (228, 229), 252[306] (228, 235, 240), 252[307] (228, 238, 239), 252[314] (229, 230, 246), 252[317] (230), 252[318] (230), 252[326] (231, 234), 252[358] (235, 237), 252[359] (235, 237), 252[360a] (235), 253[371] (236), 253[372] (236, 237, 244), 253[373] (237), 253[383] (237, 238, 240), 253[384] (237), 253[386] (237), 253[387] (238, 240), 253[400] (239), 253[404] (240), 253[405] (240), 253[413] (241), 254[458] (244), 254[459] (244), 272[51] (262), 272[53] (262), 273[102] (269), 273[106] (269), 415, 425[3] (374, 391, 415, 416), 426[30] (378, 400), 426[31] (378), 428[154] (400, 414), 428[155] (400), 428[165] (400), 428[172] (400, 414), 428[175] (400, 414), 430[309] (416)
Plotkin, J., 369[178] (307, 308)
Plotkin, J. S., 427[81] (382, 391), 427[107] (396)
Plumb, C. A., 250[223] (224, 228), 252[303] (228), 427[82] (382, 387), 430[330] (420, 421), 430[331] (420, 421), 430[332] (420, 421)
Pluta, C., 499[115b] (496), 544[268] (540)
Plzak, Z., 253[413] (241), 254[459] (244)
Poblet, M. R., 501[227] (468)
Poerschke, K. R., 501[286] (481)
Pohl, E., 543[199] (528)
Pohl, S., 32[109] (15), 34[195] (22)
Polansky, O. E., 191[355] (164)
Polborn, K., 192[412] (169, 170), 214[6] (198, 211)
Polk, M., 190[317] (161), 190[318] (162)
Poll, W., 32[117] (15)
Pollina, D. M., 367[37] (279, 280)
Polyakov, A. V., 191[366] (165), 249[164] (221), 249[171] (221, 240), 250[180] (221), 250[182] (221, 239), 250[197] (222)
Polyakova, V. B., 368[75] (286)
Polyanskaya, N. L., 194[540] (179)
Pombrik, S. I., 501[278] (480, 482, 483), 502[299] (483)
Pommerening, H., 190[316] (161), 194[558] (180), 194[576] (181), 194[589] (182)
Ponomareva, T. K., 192[393] (167)
Pople, J. A., 248[63] (218)
Popov, A. L., 251[253] (225, 226)
Poppe, J., 499[115a] (446, 496)
Porai-Koshits, M. A., 426[25] (378), 426[26] (378), 426[27] (378, 379)
Poroshina, T. Y., 254[440] (242)
Poroshina, T. Yu., 427[132] (400, 412)
Porrschke, K. R., 34[201] (26)
Pörschke, K. R., 540, 544[268] (540), 570[29] (551), 570[30] (551), 570[31] (551)
Pörschke, K.-R., 499[115b] (496)
Porter, L. C., 368[98] (288, 290)
Porter, R. F., 255[525] (246)
Porterfield, W. W., 544[261] (538, 539)
Posisil, J., 32[104a] (13)
Posta, A., 122[120] (70)
Postovoi, S. A., 254[442] (242)

Potapova, T. V., 188[149] (143), 188[164] (145), 188[167] (145, 146), 188[173] (146), 188[174] (146), 271[7] (258)
Potts, S. B., 254[436] (242), 272[29] (260), 544[232] (534)
Pounds, J. M., 369[193] (313, 315, 319)
Pourian, M. R., 249[125] (220, 221, 225–7, 229), 252[312] (229), 426[68] (382, 398)
Powell, B., 250[215] (223)
Powell, H. M., 541[78] (514)
Powell, H. R., 255[513] (245), 428[171] (400, 412), 542[106] (518–20)
Powell, W. H., 248[62] (218)
Power, D. C. M., 190[312] (161)
Power, J. M., 192[408] (169), 542[115] (520)
Power, M. B., 497[15] (434, 435, 449, 472), 541[61] (512), 542[151] (522, 523), 543[182a] (525–7), 543[193] (527), 543[194] (527), 544[263] (539)
Power, P. P., 2, 31[42] (8, 9), 31[47] (9), 31[48] (9), 31[50] (9), 31[52] (9), 31[53] (9), 31[54] (9, 11), 31[55] (9), 31[68] (10), 31[69] (10), 31[75] (10), 32[89] (12), 32[108] (15), 32[109] (15), 32[112] (15), 32[120e] (15), 33[141] (19), 33[158] (19, 29), 33[159] (19), 33[169] (19), 33[170] (19), 169, 170, 186[24] (130, 169, 170), 192[418] (170), 192[419] (170), 192[420] (170), 448, 497[12] (433, 434, 448, 449, 497), 499[115c] (446, 496), 499[150] (454), 504, 539, 542[107] (519), 542[108] (519), 542[111] (519), 542[112] (519), 542[113] (519), 542[114] (520, 525), 542[147] (522), 542[150] (522, 523), 542[152] (522, 523), 543[169] (525), 543[170] (525, 526), 543[171] (525), 544[260] (538), 544[266] (539), 544[267b] (539), 544[267c] (539)
Pradhan, P. R., 254[437] (242)
Prakash, G. K. S., 247[25] (218), 247[26] (218), 247[27] (218), 250[201] (222, 223), 250[202] (222, 223)
Pratten, S. J., 54[27] (45), 54[28] (45), 55[33] (46)
Preu, L., 193[479] (175), 193[480] (175)
Preut, H., 570[40] (554, 567)
Price, R. T., 501[244] (472)
Priermeier, Th., 570[46] (556, 561, 562, 564, 566), 570[48] (556)
Priester, R. D., 499[147] (454)
Prigozhina, M. P., 253[421] (241)
Primus, F. J., 253[418] (241, 245)
Prince, M. I., 502[308] (486)
Pritzkow, H., 32[76] (11), 187[78] (135), 187[79] (135), 187[80] (135), 187[88] (136), 187[95] (137), 188[153] (144), 188[154] (144), 188[155] (144), 188[163] (145, 150, 154), 189[200] (150), 189[203] (150), 189[218] (151, 152), 189[220] (152), 189[229] (153), 189[238] (154), 189[251] (156), 189[252] (156), 190[263] (157), 190[264] (157), 191[333] (163), 191[371] (165), 192[413] (169), 192[415] (170), 192[416] (170), 192[417] (170), 192[422] (170), 214[16] (199, 208), 214[17] (199, 208), 214[20] (199, 210), 214[22] (199), 215[26] (201), 215[49] (205), 215[52] (206), 215[53] (206), 215[54] (206), 215[55] (206), 215[63] (208), 215[65] (208, 210), 215[70] (210), 215[71] (210), 215[72] (210), 216[96] (214), 216[98] (214), 252[330] (231), 252[350] (233), 430[325] (420), 430[326] (420), 430[327] (420), 543[221] (532)
Prokai, B., 500[166] (457)
Prokof'eva, T. Y., 251[255] (225, 241)
Prokopovich, V. P., 253[425] (241, 244), 254[450] (243), 254[451] (243), 428[181] (400, 414), 428[182] (400, 414), 430[291] (414)
Prosenc, M. H., 34[196] (22)
Pubanz, D., 214[8] (198, 209)
Pues, C., 188[188] (148), 189[196] (149), 189[201] (150)
Puga, J., 367[24] (277, 294, 297, 304)
Pulham, C. R., 271[6] (258), 541[91] (515)
Pulido, F. J., 498[42] (436, 443, 468)
Purdy, A. P., 542[118] (520), 542[122] (520), 542[132] (520), 542[148] (522, 523)
Pusch, S., 190[287] (159), 194[556] (180)
Pustobaev, V. N., 251[262] (226)
Pye, W. E., 188[151] (143)

Quintana, W., 249[129] (221, 235), 249[155] (221, 228, 229, 235, 237, 238), 249[156] (221, 225, 227, 233, 235), 249[157] (221), 426[35] (378, 379)
Quintard, J.-P., 32[119a] (14)

Raabe, E., 188[162] (145, 153)
Raabe, G., 131, 186[45] (131), 186[46] (131, 132)
Raap, I. A., 215[44] (204)
Rabe, G., 34[186] (22, 23)
Rabilloud, G., 253[426] (241)
Racay, P., 430[294] (414)
Racherla, U. S., 187[85] (135–7)
Rachon, J., 121[52] (61)
Rademacher, P., 190[305] (161), 191[331] (162)
Rademaker, W. J., 543[223] (532, 533)
Ragatz, P., 273[111] (270)
Raghavachari, K., 34[221] (30)
Rahm, R., 32[119a] (14)
Rahman, A. F. M. M., 33[165] (19), 497[16] (434, 448), 497[18] (434, 448)
Rai, A. K., 31[70] (10), 194[532] (179)
Raithby, P. R., 32[77] (11), 32[107] (13), 369[147] (297, 303), 369[203] (313, 317)
Rajaram, J., 571[85] (567)
Rajec, P., 428[165] (400)
Rajnak, K., 368[94] (287, 290, 291), 368[95] (287, 290), 368[101] (290, 291), 368[105] (291)
Rakhmankulov, D. L., 500[221] (468), 500[222] (468), 500[224] (468), 502[324] (491)
Rakita, P. E., 120[11] (58)
Rakshit, D., 124[231] (84)
Ralston, C. L., 124[234] (84, 85)
Ramachandran, P. V., 186[4] (130, 138)
Ramalingam, K., 192[453] (173)
Rangaishenvi, M. V., 187[98] (137), 187[99] (137)
Rankin, D. W. H., 254[436] (242), 255[510] (245), 271[6] (258), 541[87a] (515), 541[90] (515), 541[91] (515), 541[92a] (515), 541[92b] (515)
Rannenberg, M., 31[33] (6)
Rao, A. V. R., 194[595] (182)
Rao, E. V. K., 55[59] (54)
Raston, C. L., 31[39] (8), 31[40] (8), 31[65] (10), 31[71] (10), 33[142] (19), 121[25a] (59, 60, 63, 70, 71, 100), 121[25b] (59, 60, 70, 100), 121[25c] (59, 60, 70, 100, 102), 121[27] (59, 72, 99, 100), 123[144] (71, 83, 84), 123[155] (73, 99, 111), 123[167] (76, 84, 85, 88, 98), 123[168] (76), 124[225] (83, 84), 124[245] (86, 92, 93, 95, 112–14), 124[246] (86, 90), 125[294] (98–100), 125[295] (98, 100), 126[315] (103), 126[363] (112–14), 127[379] (117), 127[385] (118), 191[374] (166), 541[51] (512), 541[89] (515), 543[212] (531)
Rath, N. P., 368[110] (290, 291), 369[139] (297, 300), 369[140] (297, 300, 303), 369[164] (298), 543[217] (531)
Rattay, W., 186[27] (130), 189[197] (149), 190[273] (158), 190[281] (159), 194[549] (179), 194[586] (182)
Ratz, A. M., 192[458] (173)
Ray, D., 498[88] (441)
Raymond, K. N., 34[209] (26)
Rayner, D. M., 499[90] (442)
Razavi, A., 498[68] (440)
Razuvaev, G. A., 499[134] (452)
Reber, G., 192[426] (171)
Recktenwald, O., 541[44] (511), 542[156] (524)
Reddy, A. C., 214[10] (198, 201), 425[4i] (375, 392)
Reddy, G. S., 498[65] (440)
Redenz-Stormanns, B., 190[284] (159), 190[285] (159), 193[513] (177), 272[46] (261), 272[47] (261)
Reed, A. E., 34[221] (30)
Reed, C. A., 249[121] (220), 249[122] (220), 249[123] (220), 249[124] (220), 250[194] (221)
Reed, D., 30[13] (2), 31[37] (6), 31[67] (10), 32[107] (13), 32[121a] (15), 186[23] (130), 249[145] (221, 224), 273[110] (270), 273[115] (271), 370[209] (317), 371[301] (352), 425[1p] (374), 428[177] (400, 412)
Reed, L. H., 248[66] (218)
Reedy, P. E., Jr., 571[94] (569)
Rees, J. E., 368[135] (297, 299, 300)
Rees, W. S., Jr., 271[16] (258), 271[17] (259), 271[18] (259), 271[20] (259), 271[22] (259), 271[23] (259)
Reetz, M. T., 125[313] (102), 193[472] (174), 193[473] (174)
Reger, D. L., 541[62] (512, 524)
Reich, H. J., 32[121b] (15)
Reichert, C. F., 188[151] (143)
Reidy, P. L., 248[79] (219)
Reikhsfel'd, V. O., 189[255] (156), 189[256] (156), 191[342] (163), 193[517] (177)

Reisenauer, H. P., 189[222] (152), 189[225] (153)
Reisner, G. M., 429[252] (400, 411)
Reitstoen, B., 124[204] (82)
Rellick, G. S., 252[323] (230)
Remec, P., 272[84] (266)
Retta, N., 193[466] (174)
Rettig, S. J., 193[479] (175), 193[480] (175), 194[530] (178), 194[544] (179), 194[598] (183), 367[48] (279, 280), 542[162] (525), 542[163] (525), 542[164] (525), 543[165] (525, 526), 543[185] (526), 543[186] (526)
Reye, C., 121[37] (60), 124[221] (83)
Reynhardt, E. C., 250[224] (224), 250[227] (224), 250[228] (224)
Reynolds, R. C., 251[297] (228)
Rezenka, D. S., 502[317] (490)
Reznick, D. L., 191[378] (166)
Rhee, S. G., 500[157] (455), 500[184] (461), 500[191] (463)
Rheingold, A. L., 188[190] (148), 189[191] (148), 189[237] (154, 155), 215[30] (201), 367[22] (277, 294, 297, 299, 300), 367[24] (277, 294, 297, 304), 368[135] (297, 299, 300), 369[136] (297, 299), 369[137] (297, 299), 369[138] (297, 299, 300), 369[140] (297, 300, 303), 369[142] (294, 296, 297, 301, 303), 369[143] (297, 300, 301), 369[145] (297, 301), 369[146] (297, 304), 369[148] (297, 303), 369[155] (297, 298), 369[165] (299), 369[166] (299), 369[167] (299), 369[168] (301), 369[170] (301), 369[181] (308), 369[182] (307, 309), 369[189] (312), 369[190] (312), 369[202] (313, 316), 369[204] (313, 317), 426[59] (382, 393), 541[62] (512, 524), 571[87] (567, 568)
Rhine, W. E., 33[163] (19), 34[189] (23, 25), 34[198] (24), 34[212] (29), 499[122] (448)
Rhodes, L. F., 367[65] (284, 285)
Ricard, L., 368[91] (289)
Rice, D. A., 542[139] (521, 526)
Rice, R. W., 254[482] (244), 254[483] (244)
Richard, P., 571[80] (566, 567), 571[98] (569)
Richards, J. C., 499[155] (454)
Richarz, W., 121[39] (60)
Richey, H. G., Jr., 34[210] (29), 121[35] (60), 121[54a] (61, 77), 122[88] (66, 87, 95), 122[94] (67), 122[95] (67, 102), 125[255] (88), 125[256] (88, 89, 93), 125[258] (89, 93), 125[259] (90, 93, 96), 125[268] (93), 125[275] (95, 106), 125[276] (96), 125[277] (96, 105), 125[279] (96, 102), 126[326] (105), 126[329] (106)
Richmann, P. N., 123[161] (74, 109)
Richter, W. J., 125[289] (98), 125[290] (98), 125[296] (98)
Riede, J., 126[321] (104), 544[246] (535), 544[247] (536)
Riedel, R., 542[155] (523, 524)
Rieger, R., 502[299] (483)
Rieke, R. D., 59, 61, 70, 97, 101, 120[22] (59, 86, 101), 120[23] (59), 122[127] (70), 125[282] (97, 98), 125[283] (97, 98), 125[284] (97, 98), 125[285] (97, 98), 126[342] (108, 109, 117, 120), 126[344] (108, 109, 117, 120)
Rienäcker, R., 500[177] (458), 500[192] (463, 470)
Riera, V., 367[59] (284)
Rihs, G., 427[133] (400, 417)
Riikoja, A., 121[40] (60)
Ritter, D. M., 255[509] (245)
Ritter, J. J., 498[47] (437)
Rius, J., 249[139] (221, 240), 249[146] (221)
Rivera, I., 193[526] (178)
Rivera, S. L., 124[203] (82)
Riviere, H., 122[110] (68)
R'Kha, C., 273[96] (267)
Ro, H., 426[65] (382, 388)
Roach, S., 272[88] (266, 267)
Roberts, B. P., 188[132] (141), 188[133] (141, 142)
Roberts, P. J., 34[204] (26, 27), 500[168] (457)
Robertson, F. C., 126[351] (109)
Robertson, H. E., 254[436] (242), 255[510] (245), 271[6] (258), 541[90] (515), 541[91] (515), 541[92a] (515), 541[92b] (515)
Robins, S. E., 371[300] (352)
Robinson, G. H., 499[126] (449), 499[146] (454), 499[148] (454), 540[17] (505, 506), 540[36] (509), 541[39] (510), 541[40] (510), 541[41] (511)
Robl, C., 123[160] (74, 109), 497[6] (433, 436, 441, 443), 540[5] (504, 538)
Robles, E. S. J., 120[8] (58)
Rochow, E. G., 31[15] (3)
Rockstein, K., 187[71] (134)
Rode, J., 121[54b] (61)
Rode, K., 121[54b] (61)
Röder, T., 187[83] (135, 138, 143)
Rodesiler, P. F., 544[237] (534)
Rodgers, R. D., 542[145] (522)
Rodionov, V. N., 121[34] (60)
Roe, S. M., 543[188] (526)
Roemich, H., 189[230] (153)
Roesky, H. W., 32[93] (12), 34[186] (22, 23), 186[16] (130), 194[581] (181), 543[167] (525), 543[178] (525, 526), 543[199] (528)
Roesky, W. H., 543[177] (525, 526)
Rogers, R. D., 31[29] (6, 8), 31[31] (6), 499[147] (454), 541[38] (510), 541[40] (510), 543[209] (530)
Rohde, C., 125[306] (101)
Rohling, T., 215[89] (213)
Rohrmann, J., 498[70] (440), 502[296] (482)
Roling, P. V., 498[22] (435)
Roller, S., 124[222] (83)
Romanova, T. Yu., 501[228] (468)
Romanovskii, V. N., 428[168] (400, 414)
Romanovskii, V. V., 428[168] (400, 414)
Romashchenkova, N. D., 122[112] (68)
Romerosa, A., 249[146] (221)
Römich, H., 215[52] (206)
Romoanova, I. P., 195[617] (184)
Rong, F. G., 253[420] (241)
Rookhuizen, R. B., 122[142] (71)
Roos-Venekamp, J. C., 188[157] (144)
Root, K. S., 121[38] (60), 121[42] (60)
Rosen, R., 249[156] (221, 225, 227, 233, 235)
Rosini, C., 55[51] (53)
Rosler, R., 542[161] (525)
Rosolovskii, V. Ya., 367[41] (279), 368[79] (286, 287)
Rossi, K., 192[457] (173)
Rossiter, B. E., 124[206] (82, 83)
Rössler, M., 32[81] (11)
Roth, H. D., 251[278] (227)
Roth, S., 214[24] (200)
Roth, W. J., 367[50] (280, 282)
Roush, W. R., 192[455] (173), 192[456] (173), 192[458] (173), 502[315] (490)
Rousseau, R., 214[4] (198, 210), 215[75] (210)
Rousset, C. J., 122[103a] (67, 84), 122[103b] (67)
Royo, B., 32[123] (15), 126[350] (109, 110)
Royster, T. L., Jr., 540[34] (507)
Rozière, J., 214[24] (200)
Rozova, E. A., 570[12] (547)
Rtishchev, N. I., 192[393] (167)
Rubesin, P. C., 498[88] (441)
Rubio, J., 544[230] (533)
Rudashevskaya, T. Yu., 122[86] (66)
Rudolph, F., 570[56] (559, 562)
Rudolph, R. W., 249[153] (221, 230, 240), 250[206] (222), 251[273] (226), 252[322] (230, 240, 245), 255[502] (245)
Rueck, K., 502[338] (494)
Ruessink, B. H., 251[268] (226)
Rufinska, A., 123[164] (76, 77, 93), 124[228] (84), 124[248] (86, 93), 125[265] (93, 101, 102), 126[336] (107), 498[72] (440), 498[73] (440)
Ruhlandte-Senge, K., 32[108] (15), 32[109] (15), 32[120e] (15), 499[115c] (446, 496), 542[107] (519), 542[114] (520, 525), 543[169] (525), 543[170] (525, 526), 543[171] (525), 544[266] (539)
Ruiz, J., 541[52] (512, 522)
Ruiz, M. A., 367[59] (284), 430[275] (400, 411)
Ruiz-Romero, M. E., 541[72] (514)
Rundle, R. E., 540[28] (506), 544[238] (534)
Runsink, J., 272[38] (260), 272[40] (261)
Rusakova, G. M., 250[176] (221)
Rusanov, A. L., 253[421] (241), 429[221] (400, 416)
Rush, P. K., 252[352] (234), 370[232] (326), 427[101a] (391), 427[101b] (391)
Rusinko, R. N., 541[74] (514, 516)
Russell, C. A., 32[77] (11)
Russell, G. A., 124[210] (83)
Ryan, M. T., 122[122] (70)
Ryan, R. R., 250[184] (221), 368[85] (286, 289), 368[86] (286, 289), 571[87] (567, 568), 571[88] (568)
Rys, E. G., 249[171] (221, 240), 251[291] (227)
Rzepkowska, Z., 502[321] (491)

S. Trivino, L. M., 251[257] (225, 241)
Saak, W., 32[109] (15), 34[195] (22)
Sabat, M., 426[46] (382, 393), 426[48]

(382, 387, 421, 423), 426[50] (382), 426[52] (381, 382), 426[55] (382, 387, 422), 427[89] (381, 382, 392, 422), 427[91] (381, 382), 427[93] (382, 393), 427[95] (381, 382), 427[103] (393)
Sabesan, A., 123[195] (82)
Sabio, M. L., 186[48] (132), 186[49] (132), 187[50] (132)
Saburi, M., 122[103a] (67, 84)
Sadikov, G. G., 426[26] (378)
Sadownik, A., 499[138] (452)
Sadykov, R. A., 500[219] (468)
Saegusa, T., 499[133] (451)
Saheki, Y., 124[235] (85)
Saimoto, H., 501[231] (469)
Saito, S., 502[311] (488)
Saito, Y., 501[245] (473)
Saitoh, H., 502[313] (489)
Saive, R., 214[19] (199, 203, 204), 215[39] (202–4)
Sakabe, Y., 121[72] (65)
Sakai, S., 123[171] (78), 499[91] (442)
Sakane, S., 501[268] (476)
Sakata, H., 124[219] (83)
Sakharova, A. A., 248[58] (218, 244), 253[414] (241, 244), 254[467] (244)
Sakuma, Y., 55[58] (54)
Sakurai, H., 189[234] (154)
Sakurai, M., 501[257] (474), 502[327] (491)
Salazar, K. V., 368[85] (286, 289)
Salem, G., 121[25a] (59, 60, 63, 70, 71, 100), 121[25b] (59, 60, 70, 100)
Salt, J. E., 367[56] (284), 570[22] (548)
Salter, P. A., 370[240] (327, 336, 364)
Salvino, J. M., 32[98] (12)
Salzbrenner, E., 186[27] (130)
Samdal, S., 498[20] (434)
Sams, D. W. I., 429[247] (400, 411)
Sanaeva, E. P., 123[176b] (78)
Sanchez, R., 124[243a] (86, 95), 124[243b] (86, 95), 498[53] (438, 452, 486, 493)
Sandford, H. F., 189[221] (152)
Sangalov, Yu. A., 499[134] (452)
Sangchakr, B., 248[84] (219), 251[274] (226, 245)
Sangokoya, S. A., 499[126] (449), 499[146] (454), 499[148] (454)
Sano, H., 500[187] (463, 476)
Santiago, B., 193[526] (178)
Sappenfield, E. L., 427[129] (400, 411, 417, 418), 429[236] (400, 411), 429[255] (400, 411), 429[273] (400, 414), 430[279] (400, 411), 430[319] (417–19)
Saquet, M., 187[114] (139)
Sardo, J., 543[166] (525)
Sarhan, A., 192[454] (173)
Sarisaban, S., 499[154] (454)
Sarto, L. G., 254[481] (244)
Sarv, K., 121[40] (60)
Sarvaarova, N. N., 195[629] (184)
Sasada, K., 124[235] (85)
Sasahara, T., 121[72] (65)
Sasaki, K., 502[341] (494)
Sasaya, F., 193[470] (174)
Sasnovskaya, V. D., 367[41] (279), 367[42] (279)
Sastry, G. N., 370[216] (320, 327, 328, 363)
Sastry, K. A. R., 193[468] (174)
Sato, F., 122[114] (69)
Sato, R., 189[234] (154)
Satoh, N., 124[235] (85)
Sattelberger, A. P., 367[52] (280, 282)
Sattler, E., 31[73] (10)
Sauchez, A., 543[166] (525)
Sauermann, G., 33[175a] (20), 33[175b] (20), 34[213a] (29)
Saulys, D., 255[522] (246)
Sawamura, M., 124[216] (83)
Sawitzki, G., 188[185] (147), 190[322] (162), 191[367] (165), 272[76] (264)
Saxena, A. K., 374, 425[1b] (374, 376, 381, 388, 389, 411, 417)
Scanlan, J. A., 371[295] (351)
Scasnar, V., 430[310] (416)
Schaaf, T. F., 32[120a] (15), 32[120b] (15)
Schacht, W., 186[15] (130), 189[247] (155, 156), 189[248] (155, 156), 189[249] (156), 189[250] (156)
Schade, C., 2, 11, 18, 30[8] (2, 11, 18, 30), 34[190] (23), 34[192] (23, 24)
Schaefer, H. F., III, 186[34] (131, 134), 497[8] (433, 436, 442), 498[87] (441), 498[88] (441), 499[108] (444)
Schaefer, W. P., 427[125] (400, 412), 427[126] (400, 412)
Schaeffer, R., 250[184] (221)
Schafarik, A., 456, 500[159] (456)
Schäfer, M., 541[63] (513, 514)
Schäfer, R., 500[206] (467)
Schäfer, V., 189[218] (151, 152), 189[220] (152), 215[63] (208)
Schat, G., 120[16] (59, 71), 121[84] (66, 71, 76, 92), 122[137a] (71, 83, 84), 122[137b] (71, 83), 122[138a] (71), 122[138b] (71), 122[141a] (71, 83), 122[141b] (71, 83), 122[143a] (71, 85, 92), 122[143b] (71, 76, 85, 92), 123[156] (73), 125[270] (94)
Schatz, W., 570[53] (557), 571[62] (561)
Schaumberg, G. D., 192[439] (172)
Scheidt, W. R., 187[109] (139), 193[522] (177, 178), 249[121] (220), 249[122] (220), 249[123] (220), 249[124] (220), 250[194] (221), 367[21] (277, 294, 297, 298), 368[130] (294, 297, 298)
Scherer, W., 570[46] (556, 561, 562, 564, 566), 570[49]
Schier, A., 34[215] (29)
Schilling, B. E. R., 124[247] (86, 94, 101)
Schilling, M., 188[119] (140)
Schimpf, R., 500[174] (458)
Schleker, W., 186[45] (131), 186[46] (131, 132)
Schlemper, E. O., 541[69] (513)
Schlenk, W., 1, 18, 30[1a] (1), 30[1c] (1, 18)
Schleyer, P. von R., 2, 11, 18, 30[4] (2, 30), 30[7] (2, 11, 18), 30[8] (2, 11, 18, 30), 31[46] (9), 31[59] (10), 31[60] (10), 33[153] (19), 33[160] (19, 21), 34[190] (23), 34[192] (23, 24), 34[216] (29), 34[221] (30), 34[222] (30), 55[38] (48–50), 123[172] (78), 125[306] (101), 126[353] (110), 126[373] (114), 126[375] (115), 131, 186[29] (131), 186[31] (131), 186[32] (131, 134), 186[33] (131), 186[36] (131), 186[38] (131), 186[39] (131), 186[40] (131), 186[41] (131), 186[42] (131), 187[54] (132), 189[205] (150), 190[284] (159), 195[632] (184), 248[63] (218), 248[64] (218), 248[65] (218), 250[199] (222, 223), 250[200] (222, 223), 250[202] (222, 223), 272[47] (261)
Schlichte, K., 120[24] (59, 99), 125[297] (99, 100), 125[299] (99), 125[302] (100), 125[303] (100)
Schlögl, R., 153, 189[227] (153)
Schlosser, D., 215[91] (213)
Schlosser, M., 32[104c] (13), 32[104d] (13), 32[104e] (13), 33[137a] (19, 30), 33[137b] (19, 30)
Schluter, R. D., 540[33] (507, 538)
Schmid, G., 55[35] (47, 48), 125[280b] (97), 190[293] (160), 190[299] (160), 190[305] (161), 190[306] (161), 190[307] (161), 190[317] (161), 190[318] (162), 191[334] (163), 214[7] (198, 212), 215[82] (212), 215[83] (212), 215[84] (212), 215[85] (212), 215[86] (212), 215[87] (212), 215[88] (212), 215[89] (213)
Schmidbaur, H., 54[4] (36), 54[5] (36), 54[6] (36), 55[47] (52), 127[377] (116), 192[426] (171), 192[427] (171), 498[80] (441), 504, 528, 534, 536, 540[3] (504, 528, 534), 541[48] (511), 541[50] (512), 544[241] (534), 544[242] (535), 544[243] (535), 544[244] (535), 544[245] (535), 544[246] (535), 544[247] (536), 544[251] (536), 544[252] (536), 544[253] (536)
Schmidt, F., 215[83] (212), 215[85] (212)
Schmidt, G., 189[206] (150)
Schmidt, M., 501[254] (474)
Schmidt, M. U., 272[43] (261)
Schmidt, M. W., 499[95] (443)
Schmidt, R. R., 194[529] (178)
Schmidt, T., 187[94] (137)
Schmidt-Lukasch, G., 187[67] (134), 187[68] (134)
Schmitz, D., 215[81] (211)
Schmitz, T., 187[66] (134, 135), 190[265] (157)
Schmöller, W., 193[524] (178)
Schneider, J., 500[173] (458)
Schneider, M. L., 55[45] (50, 51)
Schneider, P., 120[21c] (59)
Schneider, S., 544[262] (539)
Schneider, U., 498[33] (436, 445), 540[2] (504, 529)
Schneiderova, L., 251[298] (228, 245)
Schnering, H. G. v., 543[181] (525, 526)
Schnöckel, H., 123[160] (74, 109), 433, 436, 443, 444, 497[6] (433, 436, 441, 443), 498[25] (435, 443, 445), 498[33] (436, 445), 499[104] (443), 499[106] (443), 499[107] (444), 499[109] (444), 528, 529, 540[2] (504, 529), 540[5] (504, 538), 543[201] (529), 543[202] (529, 539)
Schock, L. E., 127[382] (117)
Schofield, P. J., 194[580] (181)
Schollenberger, M., 571[63] (561)
Scholz, M., 543[177] (525, 526), 543[178] (525, 526)
Schram, E. P., 446, 447, 499[116a]

(447, 455), 499[116b] (447, 455)
Schrank, F., 32[120d] (15)
Schreyer, P., 190[283] (159), 190[300] (160), 190[301] (160)
Schröder, E., 190[299] (160)
Schroth, G., 124[228] (84), 126[336] (107)
Schrötter, H. W., 499[97] (443)
Schubert, B., 33[151] (19, 21), 33[154b] (19, 20), 33[160] (19, 21), 33[161] (19, 21), 33[177] (20), 34[210] (29), 125[264] (91), 126[324] (105)
Schubert, D. M., 271[16] (258), 271[17] (259), 271[18] (259), 271[19] (259), 271[22] (259), 271[23] (259), 371[316] (361), 425[1e] (374), 428[185] (400, 412), 543[227] (533, 534)
Schubert, U., 34[200] (26), 34[203] (26), 34[215] (29), 215[31] (202)
Schulman, J. M., 186[48] (132), 186[49] (132), 187[50] (132), 187[51] (132)
Schulten, M., 125[314] (103)
Schultz, A. G., 124[208a] (83)
Schulz, A., 187[106] (139)
Schulz, H., 189[251] (156), 214[20] (199, 210), 215[57] (206), 215[65] (208, 210), 215[71] (210), 215[72] (210)
Schulz, M., 124[232] (84)
Schumann, H., 506, 540[25a] (505), 540[25b] (505), 540[26] (506), 541[54] (512), 541[55] (512), 543[210] (530, 532, 539), 543[213] (531, 539), 543[215] (531)
Schümann, U., 31[58] (10), 33[143] (19), 33[145] (19), 34[187] (23)
Schüssler, W., 187[108] (139), 187[109] (139)
Schuster, G. B., 148, 187[62] (133), 187[63] (133, 148), 187[64] (133), 189[193] (148), 189[257] (156)
Schuster, H. U., 33[180a] (20)
Schwab, B., 570[39] (553, 560, 567), 570[43] (554)
Schwan, H., 191[345] (163)
Schwartz, M., 190[291] (160)
Schwarz, H., 498[74] (441)
Schwarz, W., 126[362] (112–14, 117, 118), 540[6] (504, 537)
Schweizer, W. B., 32[87] (12), 32[91] (12), 32[96] (12), 32[114] (15), 32[115] (15)
Schwetlik, G., 194[564] (180)
Schwickardi, M., 124[250a] (86, 100), 124[250b] (86, 100)
Schwyzer, R., 255[499] (245)
Scobie, M., 251[296] (228)
Scopelianos, A. G., 251[290] (227, 230, 241), 428[194] (400, 412)
Scott, B., 501[275] (479)
Scott, M. D., 541[42] (511), 541[43] (511)
Scott, W. J., 124[218] (83)
Screttas, C. G., 124[241] (86, 93), 126[327] (106)
Sebald, A., 187[111] (139), 189[212] (151), 190[261] (157), 190[269] (158)
Sedlak, D., 194[575] (181)
Sedwick, W. D., 251[297] (228)
See, 541[98] (516)
Seebach, D., 2, 11, 30[9] (2, 10, 11), 32[86] (12), 32[87] (12), 32[88] (12), 32[91] (12), 32[96] (12), 32[114] (15), 32[115] (15), 121[74] (65, 74, 82), 123[201] (82)
Seebold, U., 187[89] (136, 144), 188[160] (144, 154), 189[244] (154), 249[126] (221, 226, 238)
Seemann, D., 33[144] (19)
Seetz, J. W. F. L., 122[140] (71, 85), 122[141a] (71, 83), 122[141b] (71, 83), 122[142] (71)
Seevogel, K., 121[26b] (59, 99, 100), 193[522] (177, 178)
Segal, J. A., 124[244] (86, 95, 96)
Sei, T., 54[12] (38)
Seiberlich, P., 190[261] (157)
Seidel, G., 188[136] (142), 188[137] (142), 192[414] (169), 192[440] (173), 194[531] (178), 194[550] (179), 194[560] (180), 215[51] (205), 215[90] (213), 215[91] (213), 216[92] (213), 216[93] (213), 216[94] (213), 216[95] (213), 250[208] (223, 229, 234), 252[308] (229, 234, 235)
Seiler, P., 32[114] (15), 121[74] (65, 74, 82)
Seitz, L. M., 34[218c] (30)
Seki, T., 122[103a] (67, 84)
Selenew, J. V., 255[493] (245)
Self, M. F., 540[12] (504, 506, 511, 521), 540[16] (505, 506), 541[41] (511), 542[120] (520, 521), 542[130] (520)
Semenenko, K. N., 368[75] (286), 368[78] (287), 368[80] (286, 287), 368[81] (286, 287)
Semenov, S. G., 248[110] (220), 248[112] (220, 235)
Sergeev, G. B., 121[30] (59, 62), 121[56] (62), 121[57] (62, 108)
Sergeev, V. A., 254[465] (244), 254[466] (244), 255[488] (244)
Sergienko, V. S., 426[25] (378), 426[26] (378), 426[27] (378, 379)
Serino, A. J., 254[432] (242), 255[496] (245)
Serwatowska, J., 190[274] (158), 190[275] (158)
Serwatowski, J., 188[135] (142), 188[138] (142), 190[274] (158), 190[275] (158), 192[403] (168), 499[130] (451), 499[140] (452), 499[142] (452)
Seto, K., 193[501] (176), 193[502] (176), 193[504] (176), 193[505] (176), 193[506] (176)
Setzer, W. N., 2, 11, 18, 30[7] (2, 11, 18)
Seudeal, N., 121[36] (60)
Seuss, T. D., 541[54] (512)
Seward, E., 248[103] (220)
Sexsmith, S. R., 498[56] (439, 457), 501[273] (478, 479, 481, 483, 484), 501[276] (479), 570[32] (551)
Sexton, A., 120[18] (59, 77)
Seyam, A. M., 121[63] (63, 84)
Seyferth, D., 127[377] (116), 271[20] (259), 271[21] (259)
Seyffer, H., 189[252] (156), 191[346] (163, 181)
Shafieezad, S., 542[138a] (521), 542[138b] (521)
Shafii, B., 186[14] (130, 148), 188[190] (148), 189[191] (148), 189[237] (154, 155)
Shagova, E. A., 188[176] (146), 188[181] (147), 191[390] (167)
Shakir, R. J., 31[29] (6, 8), 32[90] (12), 499[147] (454)
Shakoor, A., 125[300] (99)
Shapiro, Yu. E., 195[607] (183)
Shapley, J. R., 369[160] (294)
Shapovalov, V. A., 190[326] (162)
Sharkey, R. M., 253[418] (241, 245), 255[497] (245)
Sharma, H. K., 125[312] (102)
Sharma, P. A., 194[595] (182)
Shashkov, A. S., 188[145] (143), 188[177] (147)
Shashkova, E. M., 187[117] (139), 188[183] (147)
Shaskov, A., 191[370] (165)
Shaugumbekova, Z. S., 251[293] (228), 253[396] (239)
Shaw, K. F., 249[142] (221, 228, 231)
Shaw, R. A., 192[460] (174)
Shaw, S. Y., 186[28] (130), 191[348] (164)
Shaykh, B. A. M., 369[202] (313, 316), 369[203] (313, 317), 369[204] (313, 317)
Shchegoleva, T. A., 187[117] (139), 188[183] (147)
Shcherbakova, G. I., 126[332] (106), 126[334] (106)
Shcherbina, T. M., 254[447] (243), 254[449] (243)
Shearer, H. M. M., 30[11] (2), 30[12] (2), 54[16] (40), 55[37] (48), 55[45] (50, 51), 55[46] (52, 53), 55[55] (53)
Shebanova, O. K., 254[435] (242, 244)
Sheehan, J. P., 193[481] (175), 273[113] (270), 273[116] (271)
Sheeks, R. A., 251[282] (227)
Sheldrick, G. M., 32[93] (12), 32[99] (12), 34[186] (22, 23), 187[61] (133, 144), 188[158] (144), 188[160] (144, 154), 189[202] (150, 151), 189[240] (154), 189[243] (154), 190[289] (159), 193[514] (177), 194[551] (179, 180), 194[562] (180), 194[571] (180), 195[610] (183), 195[632] (184), 195[640] (185), 195[642] (185), 542[106] (518–20), 542[110] (519), 543[199] (528)
Sheldrick, W. S., 542[110] (519)
Shelly, K., 249[121] (220), 249[122] (220), 249[123] (220), 255[523] (246), 371[317] (362), 428[173] (400, 416)
Shelnutt, J. A., 250[235] (225), 250[236] (225, 244)
Shemyakin, N. F., 253[419] (241), 429[242] (400, 417)
Shen, F., 189[192] (148)
Sherwood, P., 367[25] (277), 425[4j] (375, 398), 429[256] (400, 411), 429[260] (400, 411), 430[275] (400, 411)
Shestakov, A. F., 499[103] (443)
Shet, M. S., 544[235] (534)
Shevchenko, Y. V., 254[430] (242)
Shibasaki, M., 502[325] (491)
Shibata, S., 543[205] (529)
Shibutani, T., 121[71] (65), 121[82] (66)
Shier, G. D., 541[77] (514)
Shilovtseva, L. S., 251[279] (227)
Shimoi, M., 369[184] (311), 369[187] (311), 369[188] (311)
Shin, S. H., 34[188] (22, 23)
Shingel, I. A., 428[186] (400, 414)
Shinomoto, R., 368[93] (287, 290, 291),

368[94] (287, 290, 291), 368[99] (288, 290), 368[100] (290, 291), 368[101] (290, 291), 368[102] (290, 291), 368[103] (290, 291), 368[104] (290, 291)
Shionoya, M., 191[359] (165)
Shiragami, H., 123[182] (79)
Shiro, M., 191[359] (165)
Shirokii, V. L., 427[135] (400, 414), 427[136] (400, 414), 428[167] (400, 414)
Shitikov, V. K., 255[488] (244)
Shklover, V., 367[57] (284)
Shlcenk, W., 30[1b] (1)
Shoner, S. C., 31[48] (9), 542[111] (519)
Shongwe, M. S., 367[22] (277, 294, 297, 299, 300), 368[135] (297, 299, 300), 369[137] (297, 299), 369[138] (297, 299, 300), 369[166] (299), 369[167] (299)
Shore, S. G., 183, 187[112] (139), 195[613] (183), 195[614] (183), 249[137] (221, 225, 228), 251[289] (227), 253[395] (239, 246), 255[518] (246), 271[5] (258), 272[73] (264), 277, 366[1] (277, 278), 367[16] (277), 367[23] (277, 294, 297, 300, 303), 369[144] (297, 300), 369[149] (296, 297), 369[150] (296, 297), 369[151] (296, 297), 369[152] (296, 297), 369[153] (296–8), 369[154] (296, 297), 369[178] (307, 308), 369[179] (308), 370[208] (316), 371[285] (349), 371[287] (350), 546, 570[1] (546, 553, 554, 568)
Shreeve, J. M., 191[358] (164)
Shriver, D. F., 369[159] (294)
Shustova, T. V., 254[427] (241), 254[428] (241)
Shvarts, E. M., 193[497] (175)
Sibi, M. P., 121[75a] (65, 70)
Sibley, K. L., 122[126] (70)
Siddiqi, Z. A., 368[108] (290, 291), 368[109] (290, 291), 368[113] (292, 293), 368[114] (292, 293), 368[116] (292, 293)
Siddiqui, K. F., 497[18] (434, 448)
Siddiqui, S., 248[106] (220, 228, 235, 237–9)
Siebert, W., 150, 170, 186[19] (130), 187[78] (135), 187[79] (135), 187[80] (135), 187[88] (136), 187[95] (137), 188[153] (144), 188[154] (144), 188[155] (144), 188[163] (145, 150, 154), 189[200] (150), 189[203] (150), 189[204] (150), 189[216] (151), 189[217] (151), 189[218] (151, 152), 189[219] (152), 189[220] (152), 189[229] (153), 189[230] (153), 189[238] (154), 189[251] (156), 189[252] (156), 190[263] (157), 190[264] (157), 191[333] (163), 191[346] (163, 181), 191[371] (165), 192[413] (169), 192[415] (170), 192[416] (170), 192[417] (170), 192[422] (170), 214[2] (198, 199, 201, 205, 206), 214[16] (199, 208), 214[17] (199, 208), 214[18] (199), 214[20] (199, 210), 214[22] (199), 214[23] (199), 214[24] (200), 214[25] (200, 201), 215[26] (201), 215[49] (205), 215[51] (205), 215[52] (206), 215[53] (206), 215[54] (206), 215[55] (206), 215[56] (206), 215[57] (206), 215[58] (206), 215[63] (208), 215[64] (208), 215[65] (208, 210), 215[70] (210), 215[71] (210), 215[72] (210), 216[96] (214), 216[97] (214), 216[98] (214), 252[330] (231), 252[350] (233), 420, 428[171] (400, 412), 430[325] (420), 430[326] (420), 430[327] (420), 543[221] (532)
Siedel, G., 187[102] (138), 187[104] (138), 188[184] (147), 189[219] (152)
Siedle, A. R., 247[38a] (218), 247[38b] (218), 258, 271[4] (258)
Siergiejczyk, L., 499[129] (451), 499[143] (452), 499[144] (452)
Sigel, G. A., 542[143] (521)
Sigel, G. E., 32[112] (15)
Siggia, S., 254[481] (244)
Sikorski, P., 124[250a] (86, 100)
Sille, K., 540[15] (505, 506)
Sillion, B., 253[426] (241)
Sillmon, R. S., 55[60] (54)
Silva, S. J., 498[36] (436, 442)
Simard, B., 499[90] (442)
Simmons, H. E., 187[58] (132)
Simmons, R. G., 500[158] (455), 541[75] (514, 516)
Simon, A., 190[272] (158), 193[528] (178)
Simon, D., 124[222] (83)
Simpson, W. I., 540[8] (504, 519)
Sinani, S. V., 55[53] (53), 55[54] (53)
Singaram, B., 187[92] (136)
Singh, A., 31[29] (6, 8), 194[532] (179), 367[32] (278, 286, 287), 543[216] (531), 544[259] (537)
Singh, M., 498[50] (438, 487), 498[56] (439, 457)
Singh, O. P., 193[511] (176), 193[512] (176)
Singh, R. V., 193[469] (174), 194[559] (180)
Sinn, E., 215[60] (207), 249[125] (220, 221, 225–7, 229), 251[286] (227), 370[218] (320), 370[223] (322), 370[251] (335, 341), 425[9a] (377, 382), 426[11c] (377, 381, 382), 426[14a] (377, 421–3), 426[14b] (377, 421–3), 426[14c] (377, 382, 421–3), 426[41] (382, 421, 423), 426[49] (382, 396), 426[60] (382, 387), 426[61] (382), 426[67] (382, 398), 426[68] (382, 398), 426[71] (382), 427[90] (382), 427[124] (398), 428[163] (400)
Sinn, H., 497[3] (433, 440, 450, 453, 482), 501[285] (481), 501[290] (482)
Sirigu, A., 369[171] (303)
Siriwardane, U., 249[127] (221, 234), 249[137] (221, 225, 228), 271[24] (259), 272[31] (260), 372[325] (365), 427[84] (382, 388), 544[233] (534), 544[235] (534)
Sirmokadam, N. N., 250[195] (222, 227), 427[86] (382, 389)
Sisan, Z., 370[258] (338, 345)
Sisti, M., 502[339] (494)
Sivaev, I. B., 247[33] (218), 247[37] (218), 250[192] (221)
Sivik, M. R., 34[197] (22)
Sivy, P., 428[169] (400), 428[170] (400), 428[179] (400)
Siwapinyoyos, G., 252[361] (235, 236, 238)
Sixtus, E., 570[58] (559)
Sizov, A. I., 367[33] (279, 280), 368[107] (290, 291), 570[7], 570[8] (547), 570[10] (547), 570[11] (547), 570[12] (547), 570[13] (547), 570[14] (547), 570[15] (547), 570[16] (547), 570[17] (547), 570[34] (551)
Sjoeberg, S., 254[444] (242)
Skalitzky, D., 123[184] (79)
Skarnulis, A. J., 120[18] (59, 77)
Skelton, B. W., 31[40] (8), 31[71] (10), 123[167] (76, 84, 85, 88, 98), 123[168] (76), 124[225] (83, 84), 125[294] (98–100), 126[315] (103), 127[385] (118), 541[51] (512), 541[82] (514), 541[84] (514, 526), 543[189] (526), 543[212] (531)
Sklyarskii, L. S., 251[255] (225, 241)
Skupinski, W. A., 570[26] (550)
Skvortov, N. K., 193[517] (177)
Slack, G., 250[235] (225)
Slack, G. A., 542[143] (521)
Slade, M. J., 31[29] (6, 8)
Slater, S., 367[54] (4, 283), 367[55] (283)
Sloane, N. J. A., 425[5c] (375)
Slocum, D. W., 500[181] (461, 480)
Small, R. W. H., 540[10a] (504, 506), 543[196] (527), 544[254] (537, 539)
Smeets, W. J. J., 120[2] (58, 72, 74, 87–9), 121[61] (63, 74, 77, 90), 121[78] (65, 74, 77, 90), 123[156] (73), 125[257] (88, 89, 93), 125[262] (90, 96)
Smetanyuk, V. I., 501[236] (470)
Smid, J., 125[267] (93), 127[376] (116)
Smidbaur, H., 34[215] (29)
Smirnov, V. V., 121[30] (59, 62), 121[57] (62, 108)
Smirnova, O. D., 188[148] (143)
Smith, C. L., 122[126] (70)
Smith, D. E., 428[156] (400, 412), 429[240] (400)
Smith, D. W., 121[49] (61)
Smith, G., 367[70] (286), 368[74] (286)
Smith, J. D., 33[171] (19), 123[151] (72, 85, 87, 88, 92), 193[466] (174), 193[467] (174), 498[75] (441), 499[154] (454), 515, 541[97] (515, 526)
Smith, J. M., 499[141] (452, 453)
Smith, K., 186[2] (130, 138), 186[4] (130, 138), 186[8] (130)
Smith, L. M., 540[22] (505, 506)
Smith, R. S., 501[243] (472)
Smith, S. J., 31[69] (10), 32[79] (11)
Smith, S. L., 192[397] (167)
Smithson, T. L., 192[460] (174)
Smolyakov, A. V., 193[461] (174)
Smout, G. J., 570[37] (553, 555)
Snaith, R., 2, 30[4] (2, 30), 30[12] (2), 30[13] (2), 31[18] (3), 31[19] (3), 31[20] (3), 31[21] (3), 31[22] (4), 31[23] (4), 31[30a] (6, 8), 31[30b] (6, 8), 31[34] (6), 31[37] (6), 31[44] (8, 9), 31[67] (10), 32[107] (13), 54[18] (40)
Sneddon, L. G., 186[28] (130), 248[48] (218), 248[49] (218), 249[129] (221, 235), 249[130] (221, 238), 249[132] (221, 223, 230), 249[133] (221, 228), 249[155] (221, 228, 229, 235, 237, 238), 249[156] (221, 225, 227,

233, 235), 249[157] (221), 250[219] (224), 250[223] (224, 228), 250[232] (225, 229, 231), 251[240] (225, 231, 245, 246), 251[246] (225, 238), 251[247] (225, 237, 238, 246), 251[248] (225, 228, 239), 251[288] (227), 251[299] (228), 252[302] (228), 252[303] (228), 252[304] (228), 252[305] (228, 232, 235, 236, 238), 252[325] (231), 252[332] (231), 255[526] (246), 255[527] (246), 271[15] (258), 272[79] (265), 272[92] (267), 273[98] (268), 273[103] (269), 273[104] (269), 273[105] (269), 370[250] (20, 334, 341, 342, 357, 359), 372[322] (364), 377, 420, 425[1f] (374), 425[1i] (374), 426[12a] (377, 382, 394, 396, 421, 423), 426[12b] (377, 382, 423), 426[12c] (377, 382, 394, 396), 426[12d] (377, 396, 420), 426[35] (378, 379), 427[81] (382, 391), 427[82] (382, 387), 427[98b] (391, 395), 427[98c] (391), 427[105] (396), 427[107] (396), 430[330] (420, 421), 430[331] (420, 421), 430[332] (420, 421)
Snider, B. B., 122[108] (68), 492, 501[244] (472)
Snieckus, V., 121[75a] (65, 70)
Snow, M. R., 54[25] (44, 47)
Snow, S. A., 369[184] (311), 369[186] (311)
Snyder, D. C., 123[196] (82)
Sobieralski, T. J., 191[336] (163), 191[337] (163)
Sobota, P., 123[162] (75), 124[226] (84), 125[272a] (95), 125[272b] (95)
Sockwell, S. C., 126[347] (109, 110), 126[349] (91, 109, 110, 112, 115, 118, 119), 126[371] (113, 115, 120)
Soderquist, J. A., 189[260] (157), 193[526] (178)
Sofronova, S. M., 498[40] (436), 500[217] (468)
Sohn, Y. S., 427[128] (400, 412)
Söhnen, D., 189[208] (150, 151), 189[210] (151)
Sokolov, V. I., 254[456] (243)
Sokolova, I. V., 570[17] (547)
Sokolova, T. Yu., 570[12] (547)
Soldovnikov, S. P., 429[242] (400, 417)
Solntsev, K. A., 247[33] (218), 247[37] (218), 248[50] (218), 250[192] (221), 251[251] (225, 226), 251[252] (225), 251[253] (225, 226), 251[260] (226), 253[384] (237), 426[18] (378), 426[29] (378, 379)
Solodovnikov, S. P., 251[279] (227)
Solomatina, A. I., 255[495] (245)
Solomennikova, I. I., 187[59] (133)
Soloveichik, G. L., 55[49] (52), 367[20] (277, 286, 287), 368[107] (290, 291), 570[7], 570[8] (547), 570[9] (547), 570[10] (547), 570[11] (547), 570[12] (547), 570[13] (547), 570[15] (547), 570[16] (547), 570[17] (547), 570[33] (551), 570[34] (551)
Soloway, A. H., 251[295] (228), 254[443] (242), 255[506] (245)
Soltis, M., 250[218] (224)
Somayaji, V., 193[468] (174)
Sommerfeld, T., 429[238] (400, 411)
Song, X., 369[142] (294, 296, 297, 301, 303), 369[145] (297, 301), 369[148] (297, 303)
Sonnek, G., 501[230] (468)
Sonnentag, H., 123[184] (79)
Sonoda, N., 121[83] (66, 109)
Sooriyakumaran, R., 123[145] (71)
Sorato, C., 367[65] (284, 285)
Sorokina, E. Y., 248[57] (218)
Sosin, S. L., 254[430] (242)
Sosonkin, I. M., 192[393] (167)
Soto, J., 429[207] (400, 415)
Soutar, I., 126[351] (109)
Southerington, G., 541[68] (513)
Sowerby, D. B., 186[17] (130), 186[18] (130), 186[19] (130), 186[27] (130), 541[68] (513)
Spalding, T. R., 190[312] (161), 193[481] (175), 247[4a] (218, 221, 224, 225, 237, 239–44), 248[83] (219), 251[274] (226, 245), 272[87] (266), 272[90] (266), 273[110] (270), 273[113] (270), 273[115] (271), 273[116] (271), 273[117] (271), 370[227] (323), 370[236] (327), 371[274] (341), 425[4b] (375)
Spaleck, W., 498[70] (440), 502[296] (482)
Spaniol, T. P., 215[67] (209)
Spann, J. R., 254[482] (244)
Speckman, D. M., 429[224] (400, 415), 429[227] (400, 415)
Speel, S. R., 570[37] (553, 555)
Speis, M., 191[354] (164), 191[383] (166)
Spek, A. L., 32[94] (12, 13), 33[156] (19), 120[2] (58, 72, 74, 87–9), 121[61] (63, 74, 77, 90), 121[78] (65, 74, 77, 90), 123[156] (73), 125[257] (88, 89, 93), 125[262] (90, 96)
Spencer, J. L., 367[66] (284, 285), 427[119] (395), 543[219] (532)
Spencer, J. T., 249[125] (220, 221, 225–7, 229), 250[230] (224), 251[242] (225–7), 251[243] (225–7, 229, 234), 251[276] (227), 251[277] (227), 251[287] (227), 252[312] (229), 252[353] (234), 272[62] (263), 272[66] (264), 272[67] (264), 272[68] (264), 370[206] (313, 316, 321), 370[235] (327), 425[8] (377), 426[68] (382, 398), 427[78] (382), 427[96a] (381)
Spirikhin, L. V., 502[324] (491)
Spivak, A. Yu., 500[194] (463), 501[263] (476)
Spliethoff, B., 121[29] (59, 100)
Sporikou, C. N., 121[62] (63)
Sprich, J. D., 120[19] (59)
Sproesser, L., 124[205b] (82)
Squiller, E. P., 34[210] (29), 125[259] (90, 93, 96), 125[279] (96, 102)
Srebnik, M., 186[6] (130), 193[499] (175)
Srini, V., 497[19] (434)
Srinivas, R., 498[74] (441)
Srivastava, D. K., 370[237] (327)
Srivastava, G., 543[212] (531)
Srivastrava, G., 193[511] (176), 193[512] (176)
Stabba, R., 501[272] (478, 479, 481, 483, 484)
Stadler, M., 187[77] (135)
Staffel, T., 544[258] (537)
Stahl, S., 124[232] (84)
Staikov, A. I., 193[475] (174)
Staley, R. H., 120[9] (58)
Stalke, D., 20, 32[121a] (15), 32[122] (15), 33[126] (15), 33[152] (19, 20), 34[186] (22, 23), 188[158] (144), 189[202] (150, 151), 189[240] (154), 189[243] (154), 190[289] (159), 190[311] (161), 195[632] (184), 543[199] (528)
Stalla, B., 195[643] (185)
Stam, C. H., 32[97] (12)
Stammler, H.-G., 272[34] (260)
Stanford, J. B., 193[500] (175)
Stanger, A., 125[304] (100)
Stanko, V. I., 251[261] (226, 236), 251[271] (226)
Stapleton, L., 120[11] (58)
Starikova, Z. A., 191[353] (164), 191[386] (167), 191[388] (167), 191[389] (167), 191[391] (167)
Starkov, S. P., 194[540] (179)
Starowieyski, K. B., 54[15] (39), 55[36] (48), 126[335] (107), 501[251] (474), 501[252] (474), 502[321] (491)
Stashina, G. A., 124[207] (83)
Staudigl, R., 194[558] (180), 194[563] (180), 194[568] (180), 194[569] (180), 194[589] (182), 195[635] (185)
Stavber, S., 123[174] (78)
Staves, J., 371[292] (351)
Stecher, O., 540[24a] (505, 506)
Steehler, G. A., 255[519] (246), 371[304] (352), 427[97] (388)
Steel, H. L., 34[220a] (30)
Steele, B. R., 126[327] (106)
Steele, R. B., 498[49] (438)
Steigelmann, O., 192[427] (171)
Steiger, R., 499[145] (453)
Steinhauser, S., 499[153] (454)
Stepanov, B. I., 190[323] (162), 190[325] (162), 190[326] (162)
Stephan, M., 426[72] (382, 388)
Stephenson, I. R., 249[118] (220, 240), 250[226] (224), 255[492] (244)
Stepovik, L. P., 498[40] (436), 500[217] (468)
Stern, C. L., 501[277] (480)
Steuer, H.-A., 190[286] (159, 160)
Stevenson, G. R., 127[382] (117)
Stewart, J. J. P., 248[111] (220)
Stewart, J. S., 191[350] (164)
Stibr, B., 247[32] (218), 247[34] (218), 248[92] (219, 229, 230, 234, 235, 244), 249[138] (221), 250[209] (223), 250[210] (223), 250[211] (223), 250[221] (224), 251[249] (225, 230), 251[272] (226), 252[300] (228), 252[301] (228, 229), 252[306] (228, 235, 240), 252[307] (228, 238, 239), 252[314] (229, 230, 246), 252[317] (230), 252[318] (230), 252[326] (231, 234), 252[358] (235, 237), 252[359] (235, 237), 252[360a] (235), 253[371] (236), 253[372] (236, 237, 244), 253[373] (237), 253[383] (237, 238, 240), 253[384] (237), 253[387] (238, 240), 253[400] (239), 253[404] (240), 253[405] (240), 253[413] (241), 254[458] (244), 254[459] (244), 272[48] (262), 272[51] (262), 272[52] (262), 272[53] (262), 272[55] (262),

273[106] (269), 426[15] (377, 378), 426[16] (377, 378), 426[17] (377–9), 426[19] (377–9), 426[28] (377, 378), 426[31] (378), 426[34] (378, 379), 427[109] (394, 396), 427[117] (396), 427[118] (396), 428[155] (400), 428[175] (400, 414)
Stiller, K., 543[197] (527)
Stipa, P., 123[176a] (78)
Stipp, J. N., 272[88] (266, 267)
Stöcker, W., 190[290] (159)
Stockman, K. E., 426[52] (381, 382), 427[93] (382, 393), 427[96b] (381), 427[103] (393)
Stokkeland, O., 498[20] (434)
Stoll, H., 126[373] (114)
Stone, A. J., 220, 226, 245, 247[20] (218, 232), 248[104] (220), 425[5a] (375)
Stone, F. G. A., 368[115] (292, 293), 395, 399, 411, 417, 425[1m] (374, 411, 417), 425[1n] (374), 427[94] (382, 398), 427[106] (394–6), 427[116] (394, 396, 397), 427[119] (395), 427[129] (400, 411, 417, 418), 428[193] (400, 411, 417, 418), 428[196] (400, 411), 428[197] (400, 411), 428[198] (400, 412), 428[200] (400, 411, 417, 418), 428[201] (400, 411), 428[202] (400, 411), 429[215] (400, 411), 429[218] (400, 411), 429[220] (400, 412), 429[225] (400, 412), 429[226] (400, 411), 429[231] (400, 411), 429[232] (400, 411), 429[234] (400, 411), 429[235] (400, 411), 429[236] (400, 411), 429[237] (400, 411), 429[238] (400, 411), 429[244] (400, 414, 417, 419), 429[245] (400, 414), 429[246] (400, 411), 429[247] (400, 411), 429[248] (400, 411), 429[249] (400, 411), 429[250] (400, 411), 429[251] (400, 411), 429[253] (400, 411), 429[254] (400, 411), 429[255] (400, 411), 429[256] (400, 411), 429[257] (400, 411), 429[259] (400, 411), 429[260] (400, 411), 429[261] (400, 411), 429[262] (400, 411), 429[265] (400, 411), 429[266] (400, 411), 429[267] (400, 411), 429[268] (400, 411), 429[269] (400, 411), 429[270] (400, 411), 429[271] (400, 411), 429[272] (400, 411), 429[273] (400, 414), 430[274] (400, 411), 430[275] (400, 411), 430[276] (400, 411), 430[277] (400, 411), 430[278] (400, 411), 430[279] (400, 411), 430[282] (400, 411), 430[317] (400, 417, 418), 430[318] (417, 418), 430[319] (417–19), 430[320] (417, 418), 430[321] (417–19), 430[322] (417–19), 498[47] (437), 543[219] (532)
Storch, W., 190[290] (159), 190[297] (160), 194[577] (181, 182), 195[643] (185)
Storr, A., 499[131] (451), 541[87b] (515), 542[162] (525), 542[163] (525), 542[164] (525), 543[165] (525, 526), 543[185] (526), 543[186] (526)
Stöver, H. D. H., 123[149] (72)
Stowell, J. C., 31[25] (5)
Strandberg, R., 123[199] (82)
Straub, J. A., 192[456] (173)
Straube, A., 195[645] (185)
Strauss, S. H., 544[248] (536)
Streib, W. E., 271[27] (259)
Streih, W. E., 32[95] (12)
Streitweiser, A., Jr., 2, 30, 32[106] (13), 33[147] (19, 25, 26), 34[219a] (30), 34[219b] (30), 54[8] (37), 127[380] (117, 119)
Strelenko, Y. A., 251[254] (225, 239)
Strelets, V. V., 430[293] (414)
Streubel, R., 194[583] (182)
Strouse, C. E., 427[121] (398)
Struchkov, Y. T., 249[147] (221, 233), 249[148] (221), 249[149] (221), 249[159] (221, 224, 238), 249[161] (221, 242), 249[164] (221), 249[166] (221), 249[167] (221), 249[170] (221), 249[171] (221, 240), 250[180] (221), 250[181] (221), 250[182] (221, 239), 253[390] (238), 570[9] (547)
Struchkov, Yu. T., 428[143] (400, 416), 428[144] (400, 416)
Struchkova, M. I., 188[144] (143), 188[148] (143)
Struge, K. C., 501[288] (482)
Strunk, S., 32[104d] (13)
Stuart, A. L., 31[72] (10)
Stucky, G. D., 33[125] (15), 33[140] (19), 33[163] (19), 33[168] (19), 33[183] (21), 34[189] (23, 25), 34[198] (24), 34[212] (29), 499[122] (448)
Stumpf, K., 189[219] (152), 215[51] (205), 215[53] (206), 215[54] (206), 543[221] (532)
Sturgeon, B. E., 499[115c] (446, 496), 544[266] (539)
Sturm, T. J., 502[317] (490)
Su, K., 186[28] (130), 249[130] (221, 238), 249[133] (221, 228)
Subbotin, M. Y., 249[158] (221)
Suber, L., 216[98] (214)
Subrtova, V., 249[144] (221), 249[152] (221), 249[175] (221), 250[178] (221), 250[183] (221), 250[191] (221), 250[193] (221), 426[30] (378, 400), 428[140] (400, 414)
Sue, R. E., 121[25c] (59, 60, 70, 100, 102), 125[295] (98, 100)
Suemitsu, R., 124[230] (84)
Suffolk, R. J., 251[274] (226, 245)
Sugai, S., 502[314] (490)
Sugathan, K. K., 272[82] (265, 266)
Sugden, J. K., 193[500] (175)
Suggs, J. W., 33[157] (19)
Sugihara, Y., 189[236] (154)
Sugino, T., 121[83] (66, 109)
Sugita, T., 121[72] (65)
Sühs, K., 187[83] (135, 138, 143)
Suleimanov, G. Z., 254[448] (243), 254[456] (243), 254[457] (243)
Sulkowski, M., 188[139] (142)
Sullivan, A. C., 33[171] (19), 541[97] (515, 526)
Sultanov, R. M., 122[107] (67, 87, 95), 125[251] (87, 95), 125[252] (87), 126[318] (103)
Sumaria, S., 194[541] (179)
Summerville, R. H., 214[12] (198, 199)
Sun, X., 121[43] (60)
Sung, D. D., 249[116] (220)
Sun-Hua, L., 541[98] (516)
Surikova, M. A., 254[466] (244)
Sushko, N. I., 255[490] (244)
Sutcliffe, R. J., 499[92] (442)
Süzle, D., 498[74] (441)
Suzuki, A., 194[561] (180)
Suzuki, H., 570[28] (551)
Suzuki, N., 122[103b] (67)
Suzuki, T., 501[231] (469)
Swenson, D. C., 429[243] (400, 416)
Swift, B. L., 121[49] (61)
Swingle, N. M., 124[206] (82, 83)
Swisher, R. G., 251[286] (227), 425[9a] (377, 382), 426[41] (382, 421, 423), 426[60] (382, 387), 426[61] (382), 426[62] (382), 426[71] (382), 427[74] (382)
Switendick, A. C., 248[95] (219)
Synoradzki, L., 193[515] (177), 499[129] (451), 499[143] (452)
Szafranek, J., 254[480] (244)
Szameitat, J., 429[220] (400, 412)
Szymanska-Zachara, E., 499[102] (443, 468)

Tabard, A., 571[77] (566, 567, 569)
Tabuchi, H., 193[505] (176)
Tachikawa, M., 369[157] (294)
Tacke, M., 497[6] (433, 436, 441, 443), 498[25] (435, 443, 445), 499[107] (444), 540[5] (504, 538), 543[201] (529)
Tahara, T., 193[502] (176), 193[504] (176)
Takada, S., 541[81] (514)
Takagi, T., 122[121] (70)
Takahashi, I., 124[227] (84)
Takahashi, K., 122[125] (70)
Takahashi, M., 541[80] (514)
Takahashi, S., 193[501] (176), 193[502] (176), 193[504] (176), 193[505] (176), 193[506] (176), 255[507] (245)
Takahashi, T., 67, 122[103a] (67, 84), 122[103b] (67)
Takanishi, K., 122[132] (70)
Takano, K., 248[71] (218)
Takeuchi, H., 127[383] (118)
Takuma, K., 501[269] (476)
Tamao, K., 122[92] (67), 122[133] (70), 189[192] (148)
Tamborski, C., 122[122] (70)
Tamm, N. B., 249[162] (221, 226, 239), 251[265] (226, 239)
Tanaka, N., 123[153] (73, 83)
Tanaka, S., 502[320] (490)
Tanaka, Y., 501[253] (474)
Tanaseichuk, B. S., 123[176b] (78)
Tandon, J. P., 192[433] (172), 193[469] (174), 194[533] (179), 194[559] (180), 194[579] (181)
Tanevska-Osinska, S. M., 254[434] (242)
Tang, A., 247[23] (218), 247[24] (218)
Tang, A.-C., 425[4f] (375)
Tang, L. C., 126[354] (110, 111), 126[356] (111)
Tangi, A., 371[318] (363)
Tani, H., 499[139] (452)
Taniguchi, M., 501[234] (469)
Tanner, P. S., 126[371] (113, 115, 120), 127[387] (119)
Tapper, A., 187[66] (134, 135), 190[265] (157), 190[304] (160)

Tarasov, B. P., 368[80] (286, 287), 368[81] (286, 287)
Tatsumi, K., 125[291] (98)
Tawara, K., 251[292] (228, 231, 240, 242)
Taylor, J. G., 250[222] (224, 234), 252[320] (230), 255[520] (246), 426[22] (378), 426[23] (378), 430[285] (400, 418)
Taylor, N. J., 368[120] (292, 293, 295), 544[259] (537), 571[66] (562)
Taylor, R. G., 124[244] (86, 95, 96)
Taylor, S. K., 124[211a] (83), 502[317] (490)
Tebbe, F. N., 440, 498[65] (440)
Teclé, B., 32[120c] (15), 33[165] (19), 499[123] (449)
Teisseire, P., 502[319] (490)
Teixidor, F., 249[139] (221, 240), 249[141] (221, 230, 240), 249[146] (221), 249[153] (221, 230, 240), 249[154] (221, 240), 249[172] (221, 240), 250[206] (222), 252[322] (230, 240, 245)
Templeton, D. H., 368[93] (287, 290, 291)
Teo, B. K., 425[5b] (375), 425[5c] (375)
Teplyakov, M. M., 249[170] (221), 253[414] (241, 244), 254[441] (242), 254[468] (244), 254[476] (244), 255[491] (244)
Teregulov, I. Kh., 500[219] (468)
Tesh, K. F., 126[365] (112–14)
Teslya, I. A., 191[353] (164), 191[386] (167), 191[388] (167), 191[389] (167), 191[391] (167)
Tessier-Youngs, C. A., 500[158] (455), 540[34] (507)
Tetenok, N. A., 254[477] (244)
Thal, A., 30[1a] (1)
Theis, M., 33[136] (17)
Theonnes, D., 33[150] (19), 33[162] (19), 34[210] (29)
Theopold, K. H., 504, 542[104] (518, 519), 542[133] (520)
Thewalt, U., 125[280b] (97), 498[80] (441), 544[241] (534), 544[243] (535), 544[244] (535), 544[245] (535)
Thiele, B., 190[301] (160)
Thiele, G., 570[58] (559)
Thoennes, D., 126[323] (105)
Thomann, M., 214[6] (198, 211)
Thomas, B. S., 543[222] (532)
Thomas, S. E., 124[231] (84), 186[7] (130, 138)
Thomas, W. H., 500[218] (468)
Thompsett, A. R., 369[198] (313, 315)
Thompson, B. K., 369[184] (311)
Thompson, M., 429[228] (400, 415)
Thompson, M. R., 249[150] (221, 231, 237, 241, 245)
Thompson, P. A., 193[496] (175)
Thompson, P. E., 55[60] (54)
Thorn, D. L., 570[57] (559), 571[74] (565)
Thorne, A. J., 31[32] (6, 9), 31[41] (8)
Thorne, K., 249[136] (221, 231)
Thornton-Pett, M., 249[138] (221), 272[55] (262), 272[56] (262, 269), 272[61] (263), 272[81b] (265, 266), 272[87] (266), 273[93] (267), 273[94] (267), 367[30] (278, 343, 348), 369[196] (313, 315, 320, 323, 340), 369[197] (313, 315, 346, 357, 359), 370[240] (327, 336, 364), 370[242] (329), 370[244] (332), 370[245] (332, 340), 370[248] (332, 339), 370[249] (333), 370[252] (336, 340, 342, 347), 370[254] (336), 370[258] (338, 345), 370[260] (338–40), 370[262] (340, 341), 370[263] (340), 370[264] (340), 370[265] (340, 342, 343), 371[269] (340, 341, 344, 347), 371[271] (340, 341), 371[273] (340), 371[274] (341), 371[275] (342, 357, 359), 371[277] (342), 371[280] (343), 371[281] (345, 346, 355, 356), 371[283] (347), 371[286] (350), 371[288] (351), 371[289] (351), 371[290] (351), 371[297] (351, 362), 371[299] (352), 371[307] (355, 356), 371[308] (355, 356), 371[309] (355–7), 371[310] (355–7), 371[311] (357, 359), 371[312] (357–9), 371[313] (358, 359), 371[319] (363), 372[323] (365), 426[11e] (377, 394, 396), 426[16] (377, 378), 426[28] (377, 378), 426[34] (378, 379), 427[112] (396), 427[114] (396), 427[117] (396), 427[118] (396), 540[22] (505, 506), 570[25] (550)
Thoumazeau, E., 125[311] (102)
Threadgill, M. D., 251[296] (228)
Thuillier, A., 187[114] (139)
Tierney, M., 250[225] (224)
Tikhomirov, B. I., 191[361] (165)
Tilley, T. D., 34[205] (26)
Timofeev, O. S., 194[573] (181), 195[606] (183), 195[647] (185)
Ting, C., 369[174] (306), 369[175] (306)
Tinga, M. A. G. M., 122[138a] (71), 124[237] (85, 91, 92, 97)
Tishin, B. A., 187[113] (139)
Titcombe, L. A., 541[84] (514, 526), 543[189] (526)
Titov, L. V., 367[41] (279), 367[42] (279), 368[79] (286, 287)
Tjarks, W., 195[615] (184)
Tkachev, S. V., 428[164] (400)
Tobita, H., 369[188] (311)
Todd, L. J., 271[1] (258), 271[27] (259), 272[35] (260), 272[37] (260), 272[49] (262), 272[50] (262), 272[58] (263), 272[69] (264), 272[71] (264), 272[72] (264), 272[77] (264), 272[81a] (265, 266), 272[83] (265, 266), 272[88] (266, 267), 272[91] (267), 273[107] (270), 273[108] (270), 273[109a] (270), 273[109b] (270), 273[111] (270), 273[112] (270), 426[11a] (377, 400, 412), 426[11b] (377, 396, 400), 426[32] (378), 428[157] (400, 412, 414)
Todres, Z. V., 254[449] (243)
Tolbot, T. L., 31[16] (3)
Toldov, S. V., 193[517] (177)
Tollefson, N. M., 251[290] (227, 230, 241), 428[194] (400, 412)
Tolstaya, T. P., 249[148] (221), 249[149] (221), 249[158] (221), 250[181] (221), 251[283] (227), 253[375] (237), 253[376] (237, 238), 253[377] (237), 253[378] (237), 253[379] (237, 238), 253[380] (237)
Tolstikov, G. A., 122[98] (67, 97, 98), 122[102a] (67, 87, 95), 123[169] (77, 92), 125[252] (87), 187[113] (139), 500[194] (463), 500[196] (464), 501[228] (468), 501[240] (471), 501[246] (473), 501[247] (473), 501[248] (473), 501[249] (473), 501[250] (473), 501[263] (476), 502[326] (491), 502[333] (492), 502[346] (496), 502[347] (496)
Tomioka, H., 501[231] (469)
Tomita, H., 252[316] (229, 235, 236)
Toogood, G. E., 31[22] (4)
Topaloglu, I., 429[215] (400, 411), 429[225] (400, 412), 429[226] (400, 411), 429[231] (400, 411)
Topchaya, E. B., 194[573] (181)
Topolski, M., 121[47a] (61, 62, 77), 121[47c] (61, 62, 77), 121[53] (61)
Totani, T., 251[292] (228, 231, 240, 242), 252[324] (230, 240), 253[406] (240)
Toupet, L., 501[235] (469)
Toyota, S., 187[115] (139)
Tran, D., 248[87] (219), 252[360b] (235)
Trapp, M., 542[128] (520)
Trask, T. W., 251[297] (228)
Traylor, T. G., 502[310] (488)
Treber, J., 121[26b] (59, 99, 100), 125[302] (100)
Tregub, A. I., 255[493] (245)
Trekoval, J., 32[104b] (13)
Trivino, L. M. S., 253[410] (241)
Troepol'skaya, E. V., 195[622] (184)
Trofimenko, S., 186[26] (130), 192[403] (168), 428[185] (400, 412)
Troitskaya, M. V., 253[401] (239)
Trost, B. M., 121[75b] (65, 70), 186[2] (130, 138), 497[4] (433, 436, 438, 441, 458, 459, 467, 486, 495, 496), 502[332] (491), 502[348] (496)
Trotter, J., 193[479] (175), 193[480] (175), 194[530] (178), 194[544] (179), 194[598] (183), 542[162] (525), 542[163] (525), 542[164] (525), 543[165] (525, 526), 543[185] (526), 543[186] (526)
Troyanov, S. I., 125[280a] (97)
Truba, N. A., 252[356] (234)
Truchet, F., 190[327] (162)
Truppat, R., 190[298] (160)
Truter, M. R., 541[79] (514)
Tsai, D. J. S., 192[447] (173)
Tsang, A. C. H., 188[132] (141), 188[133] (141, 142)
Tsay, Y.-H., 34[204] (26, 27), 125[300] (99), 188[135] (142), 193[515] (177), 214[18] (199), 500[168] (457), 543[207] (530)
Tse, J. S., 497[7] (433, 436, 442), 498[84] (441)
Tseng, J., 248[87] (219), 252[360b] (235)
Tsimerinova, T. V., 251[252] (225), 251[260] (226)
Tsuchihashi, G., 501[265] (476)
Tsukada, M., 121[72] (65)
Tuck, D. G., 121[36] (60), 514, 541[45] (511), 541[56] (512), 541[73] (514), 543[168] (525, 528), 543[173] (525), 544[257] (537)
Tucker, A. J., 570[52] (557, 558)
Tucker, C. E., 121[76] (65, 77)
Tückmantel, W., 126[330] (106)
Tumanskii, B. L., 251[280] (227)

Turner, C. A., 271[25] (259)
Turner, H. W., 430[306] (416)
Turos, E., 498[54] (438, 493)
Turyanskaya, A. M., 194[574] (181)
Tuulmets, A., 60, 120[15] (59, 60), 120[20a] (59, 60), 120[20b] (59, 60), 121[40] (60), 121[41] (60)
Twiss, J., 55[37] (48), 55[46] (52, 53)
Twiss, P., 123[144] (71, 83, 84)
Tyler, A. N., 543[194] (527)
Tyutyulkov, N., 186[43] (131)

Uccello Barretta, G., 124[202] (82)
Uchibori, Y., 123[189] (80, 81)
Uchida, K., 502[329] (491)
Uden, P. C., 254[481] (244)
Ueyama, N., 499[139] (452)
Uhl, W., 33[149] (19), 435, 446, 447, 462, 498[31] (435), 498[32] (435, 446, 462, 496), 499[113] (446), 499[114] (446), 499[115a] (446, 496), 504, 514, 537, 539, 540[6] (504, 537), 541[70] (514), 541[71] (514, 537–9), 543[172] (525, 526), 544[255] (537–9), 544[256] (537–9)
Uhm, J.-K., 187[80] (135), 189[229] (153), 189[230] (153), 215[70] (210)
Uhm, S. T., 120[23] (59)
Uhrhammer, R., 429[243] (400, 416)
Uhrich, K., 543[186] (526)
Ulman, J. A., 370[231] (326)
Ul'yanova, M. V., 501[236] (470)
Underhill, M., 369[163] (296)
Ungvary, F., 121[50a] (61), 121[50b] (61)
Uppal, J. S., 120[9] (58)
Urabe, H., 122[132] (70)
Urevig, D. S., 250[233] (225)
Uribe, G., 195[616] (184)
Usatov, A. V., 248[47] (218), 249[164] (221), 249[167] (221), 251[256] (225)
Uson, R., 571[86] (567)
Usyatinskii, A. Y., 247[44] (218), 248[45] (218), 254[446] (243), 254[447] (243), 254[448] (243), 254[449] (243)
Ute, K., 123[150] (72)
Utimoto, K., 123[182] (79), 123[183] (79), 501[234] (469), 501[262] (476), 502[329] (491)
Utko, J., 124[226] (84)
Uyzbaev, K. M., 253[419] (241)
Uzarewicz, I., 194[596] (183)
Uzick, W., 502[299] (483)

Vacca, A., 367[64] (284, 285)
Vagin, V. V., 255[495] (245)
Vahrenkamp, H., 193[528] (178)
Valetskii, P. M., 249[166] (221), 251[279] (227)
Valitov, F. Kh., 501[246] (473), 501[247] (473), 501[248] (473), 501[249] (473), 501[250] (473)
van Beijnen, A. J. M., 188[157] (144)
van Bonn, K.-H., 190[283] (159), 191[329] (162)
van de Heisteeg, B. J. J., 120[16] (59, 71), 122[137a] (71, 83, 84), 122[138a] (71), 122[141a] (71, 83), 122[141b] (71, 83)
van de Sande, C., 193[516] (177)
van der Donk, W. A., 368[121] (292, 293, 295)
van der Kerk, G. J. M., 188[157] (144), 189[226] (153)
van der Kerk, S. M., 153, 186[36] (131), 188[157] (144), 189[226] (153), 248[65] (218)
van der Mieroop, W. F., 32[102] (12)
van der Schaaf, P. A., 32[94] (12, 13)
van der Wal, G., 122[143b] (71, 76, 85, 92)
van der Zeijden, A. A. H., 367[57] (284)
van Eekeren, A. L. M., 189[226] (153)
van Eikema Hommes, N. J. R., 33[156] (19), 122[138b] (71), 122[139] (71)
Van Engen, D., 249[165] (221, 237, 241), 249[174] (221)
van Klink, G. P. M., 125[278] (96, 102)
van Koten, G., 32[94] (12, 13), 32[97] (12), 32[102] (12), 501[259] (475)
Van Vliet, M. R. P., 501[259] (475)
Vanderhooft, J. C., 570[44] (555)
Vanek, T., 428[172] (400, 414)
Vannikov, A. V., 251[271] (226)
Vanura, P., 428[178] (400, 416), 430[311] (416), 430[312] (416), 430[313] (416)
Vaquez-Lopez, E. M., 543[166] (525)
Vara Prasad, J. V. N., 122[89] (66, 67, 102)
Varadarajan, A., 255[497] (245), 255[498] (245), 255[500] (245), 428[141] (400, 416, 417)
Varga, V., 125[280a] (97), 125[280b] (97), 570[18] (548)
Vasekova, L., 430[294] (414), 430[295] (414)
Vasilyev, L. S., 188[144] (143), 188[148] (143)
Vasilyeva, S. G., 427[137] (400), 428[166b] (400, 414)
Vasumathi, N., 124[220] (83)
Vasyukova, N. I., 251[263] (226), 428[147] (400), 428[159] (400)
Vaultier, M., 190[327] (162), 192[451] (173)
Vehreschild-Yzermann, D., 191[334] (163)
Veith, M., 15, 33[130] (15), 541[44] (511), 542[160] (525), 542[161] (525), 570[59] (560)
Veletskii, P. M., 254[471] (244)
Venable, T. L., 249[135] (221, 234), 250[213] (223), 250[220] (224), 370[217] (320–2, 324, 327, 335, 341), 370[218] (320), 370[219] (321), 370[221] (321), 370[223] (322), 370[251] (335, 341), 426[42] (382), 426[63] (382), 426[70] (382)
Venanzi, L. M., 367[65] (284, 285)
Venkatasubramanian, K., 368[77] (286)
Verhecken, A., 193[516] (177)
Verkruijsse, H. D., 31[28] (5), 121[79] (65, 71)
Verpeaux, J.-N., 122[97] (67, 83)
Vershinnikova, T. G., 368[79] (286, 287)
Veselovskii, V. V., 188[144] (143)
Vester, A., 499[115a] (446, 496)
Vetrone, J., 33[124] (15)
Vicente, M., 502[323] (491)
Victoriano, L., 497[17] (434), 540[7] (504, 530, 539)
Vidal, M., 501[237] (470)
Vieth, M., 542[156] (524)
Vigner, D., 368[87] (289)
Vilkov, L. V., 247[14] (218), 247[18] (218), 250[196] (222), 250[197] (222)
Villena, A., 121[78] (65, 74, 77, 90)
Vimba, S., 34[212] (29)
Vinas, C., 249[139] (221, 240), 249[153] (221, 230, 240), 250[206] (222)
Vincens, M., 501[237] (470)
Vincente, J., 541[53] (512), 541[83] (514)
Vinitskii, D. M., 370[241] (328)
Vinnikova, M. I., 499[156] (455)
Vinogradova, L. E., 250[234] (225), 251[239] (225), 251[254] (225, 239), 251[293] (228)
Vinogradova, S. V., 249[166] (221), 254[471] (244), 254[472] (244)
Vites, J. C., 368[125] (294, 296, 297), 368[126] (294, 296, 297), 368[127] (296, 297)
Vitt, S. V., 253[382] (237, 239), 253[393] (239), 253[394] (239)
Vittori, O., 571[79] (566, 567)
Vlad, V. P., 187[113] (139)
Voelker, D., 543[197] (527)
Voitländer, J., 251[266] (226), 251[269] (226)
Volden, H. V., 124[247] (86, 94, 101), 125[269] (94, 112, 113), 126[368] (113), 126[369] (113, 114)
Volkotrub, M. N., 194[540] (179)
Volkov, A. A., 500[221] (468), 500[222] (468), 500[224] (468), 502[324] (491)
Volkov, V. V., 414, 427[137] (400), 428[164] (400), 428[166a] (400, 414), 428[166b] (400, 414), 430[314] (416)
Vollhardt, K. P. C., 125[304] (100)
Vollhardte, J., 33[135] (17)
Vollmer, H.-J., 497[3] (433, 440, 450, 453, 482), 501[283] (481), 501[284] (481), 501[285] (481), 570[19] (548)
Von Arx, U., 254[437] (242)
von Bennigsen-Mackiewicz, T., 191[329] (162)
von Plotho, C., 190[293] (160), 190[294] (160), 191[328] (162), 191[329] (162), 191[345] (163)
von Schering, H. G., 188[131] (141)
Vondrak, T., 251[272] (226), 251[275] (226)
Vorob'eva, V. P., 187[59] (133)
Voronina, G. S., 428[166b] (400, 414)
Voronkov, M. G., 123[147] (71)
Vorontsova, L. G., 191[372] (166), 191[384] (167), 191[392] (167)
Vos, M., 33[156] (19)
Voss, E., 498[24] (435)
Vostrikova, O. S., 122[102a] (67, 87, 95), 122[102b] (67, 87, 95, 103), 122[107] (67, 87, 95), 125[251] (87, 95), 126[318] (103)
Votavova, H., 428[172] (400, 414)
Vranka, R. G., 499[118] (447)
Vrieze, K., 501[259] (475)
Vul'fson, S. G., 195[629] (184)

Waack, R., 33[173] (20), 34[218a] (30)
Wade, K., 2, 30[11] (2), 30[12] (2), 30[13] (2), 31[17] (3), 31[18] (3), 31[19] (3), 31[21] (3), 31[22] (4), 31[67] (10), 54[14] (39, 53), 125[253] (88), 187[112] (139), 214[11] (198), 247[16] (218), 247[17] (218), 247[26] (218), 247[28] (218), 247[29] (218), 247[42] (218), 248[91] (219), 249[118] (220, 240), 249[163] (221, 244), 250[188] (221, 243), 251[285] (227), 252[339] (232), 255[492] (244), 255[511] (245), 255[518] (246), 367[16] (277), 367[17] (277), 425[1t] (374, 381), 425[4e] (375, 422), 425[7a] (375, 422)
Wadepohl, H., 189[230] (153), 215[26] (201), 215[52] (206), 215[55] (206)
Wadman, S. N., 124[217] (83)
Waggoner, K. M., 499[150] (454), 542[114] (520, 525), 542[150] (522, 523)
Wagner, H., 187[81] (135)
Wagner, I., 32[109] (15)
Wagner, J., 499[113] (446), 499[114] (446)
Wagner, L. J., 124[224] (83)
Wagner, R., 121[44] (60, 79)
Wagner, W., 190[278] (158, 159), 190[281] (159)
Wakamoto, K., 122[101] (67)
Wakefield, B. J., 31[26] (5, 19), 121[48] (61)
Walborsky, H. M., 61, 62, 121[47a] (61, 62, 77), 121[47b] (61, 62, 77), 121[47c] (61, 62, 77), 121[51] (61, 77), 121[52] (61), 121[53] (61), 126[343] (108, 109, 117)
Walczyna, R., 253[415] (241)
Wales, D. J., 247[20] (218, 232), 247[21] (218, 232), 252[339] (232), 425[4m] (375, 393, 398, 422)
Walker, G. T., 31[30b] (6, 8), 31[67] (10)
Walker, J. A., 426[38] (378, 380), 429[207] (400, 415), 429[213] (400, 415), 429[216] (400, 412), 429[219] (400, 415)
Walker, N., 124[205b] (82)
Walker, N. M., 369[194] (313, 315)
Wallbridge, M. G. H., 250[222] (224, 234), 252[320] (230), 255[513] (245), 255[520] (246), 370[233] (326), 426[20] (378, 379), 426[22] (378), 426[23] (378), 429[239] (398, 400), 430[285] (400, 418), 541[87b] (515), 542[106] (518–20), 543[188] (526)
Walling, C., 81, 121[55] (62), 123[194] (80, 81)
Wallis, J. M., 123[159] (74, 86, 92, 93)
Walter, D., 498[64] (440, 484, 497)
Wang, J. C., 571[97] (569)
Wang, K. K., 188[124] (140)
Wang, M., 123[148] (72)
Wang, S., 189[214] (151), 571[97] (569)
Wang, X., 427[95] (381, 382)
Wang, Z.-T., 426[14a] (377, 421–3), 426[14b] (377, 421–3), 426[14c] (377, 382, 421–3)
Wannagat, U., 542[109] (519)
Wardell, J. L., 1, 30[3] (1), 31[28] (5)
Warren, D. S., 252[344] (233), 252[347] (233)
Warren, L. F., Jr., 544[229] (533)
Warren, M., 541[58] (512)
Wasserman, H. J., 368[85] (286, 289), 368[86] (286, 289), 571[88] (568)
Wassermann, W., 541[54] (512), 541[55] (512)
Watanabe, T., 499[89] (442), 500[216] (468)
Waters, J. M., 371[314] (358)
Waterworth, S., 426[46] (382, 393)
Watkins, E. K., 122[88] (66, 87, 95)
Watson, W. H., 191[348] (164)
Watton, A., 250[224] (224)
Wayland, B. B., 571[82] (567)
Waymouth, R. M., 122[105a] (67, 71, 95), 122[105b] (67, 71)
Weber, B., 123[201] (82)
Weber, S., 190[291] (160), 190[292] (160), 191[365] (165)
Weber, W., 195[611] (183)
Webster, P. M., 570[52] (557, 558), 571[64] (562)
Wecker, U., 124[232] (84)
Weckler, G., 542[128] (520)
Weese, K. J., 31[42] (8, 9), 31[47] (9)
Wegener, D., 271[14] (258), 272[34] (260), 272[80] (265)
Wegner, P. A., 371[316] (361)
Wehmeyer, R. M., 120[22] (59, 86, 101)
Wehmschulte, R. J., 499[115c] (446, 496), 542[114] (520, 525), 544[266] (539)
Wehner, I., 188[126] (140), 188[127] (140, 146)
Wehrmann, R., 187[73] (135), 189[194] (149), 189[195] (149), 189[201] (150)
Wei, Y., 502[345] (496)
Weidenmann, A., 33[154b] (19, 20)
Weidlein, J., 31[33] (6), 126[362] (112–14, 117, 118), 540[15] (505, 506), 541[66] (513), 542[128] (520)
Weike, Z., 194[561] (180)
Weininger, M. S., 544[237] (534)
Weinmann, W., 215[49] (205)
Weinreb, S. M., 498[54] (438, 493), 498[55] (438, 493)
Weiss, E., 2, 30[6] (2), 31[58] (10), 31[59] (10), 32[84c] (11), 32[84d] (11), 32[84e] (11), 32[84f] (11), 32[85a] (11), 32[85b] (11), 32[85c] (11), 32[85d] (11), 32[92] (12), 32[103] (12), 33[143] (19), 33[145] (19), 33[150] (19), 33[151] (19, 21), 33[154a] (19, 20), 33[154b] (19, 20), 33[155a] (19, 20), 33[160] (19, 21), 33[161] (19, 21), 33[162] (19), 33[166] (19, 24), 33[172a] (20), 33[172b] (20), 33[175a] (20), 33[175b] (20), 33[176] (20), 33[177] (20), 33[178] (20), 34[184] (23, 24), 34[185] (23, 25), 34[187] (23), 34[191] (23), 34[193] (23), 34[206] (26), 34[208] (26), 34[210] (29), 34[211a] (29), 34[211b] (29), 34[212] (29), 34[213a] (29), 34[213c] (29), 34[214] (29), 54[4] (36), 54[5] (36), 125[263] (91, 105), 125[264] (91), 126[323] (105), 126[324] (105)
Weiss, K., 502[308] (486)
Weiss, R., 427[124] (398)
Welch, A. J., 247[7] (218), 248[89] (219), 248[98] (220), 249[142] (221, 228, 231), 249[145] (221, 224), 249[151] (221), 351, 369[185] (311), 370[256] (337), 371[294] (351), 371[295] (351), 371[296] (351), 371[300] (352), 371[301] (352), 371[302] (352), 371[303] (352), 425[4a] (375), 427[94] (382, 398), 427[106] (394–6), 427[116] (394, 396, 397), 427[131] (400, 412), 428[156] (400, 412), 428[161] (400, 412), 428[177] (400, 412), 428[189] (400, 412), 429[210] (400, 412), 429[223] (398, 400), 429[240] (400), 430[286] (400), 430[287] (400), 430[318] (417, 418)
Welch, D. A., 371[300] (352)
Weller, F., 31[74] (10)
Wellmar, A., 123[163a] (75), 123[163b] (75)
Wells, A. F., 541[94] (515)
Wells, R. L., 504, 518, 540[1b] (503, 504, 517, 518, 520–2), 540[12] (504, 506, 511, 521), 540[16] (505, 506), 541[47] (511, 520), 542[103] (518, 519), 542[118] (520), 542[120] (520, 521), 542[121] (520, 521), 542[122] (520), 542[130] (520), 542[132] (520), 542[137] (521), 542[138a] (521), 542[138b] (521), 542[148] (522, 523)
Wen, A. T., 251[277] (227), 367[44] (279, 280)
Wendoloski, J. J., 187[58] (132)
Went, M. J., 427[129] (400, 411, 417, 418)
Werle, V., 33[180a] (20)
Wermer, J. R., 54[23] (42), 249[137] (221, 225, 228), 251[289] (227), 253[395] (239, 246), 369[153] (296–8)
Werner, H., 124[232] (84), 543[214] (531)
Werner, S., 501[292] (482), 502[294] (482), 540[37] (509)
Wesemann, L., 187[86] (136), 187[87] (136, 137), 187[93] (137), 214[15] (199)
West, P., 33[137d] (19, 30), 33[173] (20)
West, R., 188[187] (148), 192[431] (171), 498[47] (437)
West, T. A., 271[24] (259)
Westcott, S. A., 368[120] (292, 293, 295), 368[121] (292, 293, 295)
Westeppe, U., 120[24] (59, 99), 124[250b] (86, 100), 125[287] (98, 100), 125[297] (99, 100), 125[299] (99), 125[303] (100)
Westerhaus, A., 367[48] (279, 280)
Westmoreland, D. L., 542[131] (520, 522), 542[144] (522)
Weyer, K., 499[135] (452)
Wharf, R. M., 253[409] (241)
Wheatley, P. J., 32[84a] (11)
Whelan, T., 248[74] (218, 219), 248[83] (219), 248[103] (220), 252[312] (229), 252[353] (234)
Whelan, T. A., 272[62] (263)
Whetten, A., 120[10] (58, 59, 62, 78, 108)
Whitaker, T. J., 541[88] (515)
Whitby, R. J., 122[106] (67, 71, 87, 95)
Whitcombe, R. A., 540[24b] (505, 506)
White, A. H., 31[39] (8), 31[40] (8), 31[65] (10), 31[71] (10), 33[142] (19),

121[25b] (59, 60, 70, 100), 123[144] (71, 83, 84), 123[155] (73, 99, 111), 123[167] (76, 84, 85, 88, 98), 123[168] (76), 124[225] (83, 84), 124[246] (86, 90), 125[294] (98–100), 126[315] (103), 126[363] (112–14), 127[379] (117), 127[385] (118), 191[374] (166), 367[70] (286), 368[74] (286), 541[51] (512), 541[82] (514), 541[84] (514, 526), 543[189] (526), 543[212] (531)
White, D. W., 543[222] (532)
White, E. A. D., 541[42] (511), 541[43] (511)
White, J. P., III, 369[144] (297, 300), 370[208] (316), 371[285] (349)
Whitehead, G. J., 30[11] (2), 30[12] (2)
Whiteley, C. G., 187[105] (138)
Whiteley, M. W., 214[22] (199), 215[56] (206)
Whitesell, M. A., 272[72] (264), 272[77] (264), 272[91] (267)
Whitesides, G. M., 121[38] (60), 121[42] (60), 121[49] (61)
Whiting, D. A., 31[28] (5)
Whitmire, K. H., 32[93] (12), 369[159] (294), 570[3] (546), 571[67] (563, 567), 571[87] (567, 568), 571[88] (568), 571[89] (568), 571[90] (568)
Whitten, J. P., 121[48] (61)
Whitthauer, C., 32[76] (11), 32[80] (11), 32[81] (11)
Whittle, R. R., 34[210] (29), 125[259] (90, 93, 96), 125[279] (96, 102), 251[290] (227, 230, 241), 428[194] (400, 412)
Whittlesey, B. R., 571[68] (563)
Wiberg, E., 541[95] (515)
Wiberg, G., 540[24a] (505, 506)
Wiberg, N., 499[111] (445)
Widdowson, D. A., 124[218] (83)
Wieczorek, C., 187[67] (134)
Wienkenhöver, M., 194[564] (180)
Wierda, D. A., 499[125] (449)
Wiersema, R. J., 543[220] (532)
Wieser, H., 192[460] (174)
Wietelmann, U., 189[197] (149), 190[273] (158), 192[432] (171, 179), 194[588] (182, 184)
Wilczynski, R., 251[240] (225, 231, 245, 246), 251[246] (225, 238)
Wild, F. R. W. P., 498[67] (440)
Wilhelm, D., 186[29] (131)
Wilke, G., 34[201] (26), 193[519] (177), 440, 483, 498[62] (440), 498[64] (440, 484, 497), 500[195] (464, 486, 495), 501[272] (478, 479, 481, 483, 484), 501[280] (480), 501[286] (481), 570[29] (551), 570[30] (551), 570[31] (551)
Wilkey, J. D., 187[64] (133)
Wilking, J. B., 544[263] (539)
Wilkinson, A. P., 33[174] (20)
Wilkinson, G., 34[217] (30), 367[56] (284), 570[4] (546), 570[20] (548), 570[21] (548, 550), 570[22] (548), 570[23] (549), 570[25] (550), 570[27] (550, 551)
Willard, P. G., 33[157] (19)
Willershausen, P., 187[69] (134), 187[70] (134), 189[207] (150)
Willett, R., 501[275] (479)
Williams, E. H., 123[165] (76)
Williams, H. D., 367[47] (279, 280), 367[48] (279, 280)
Williams, I. D., 571[95] (569)
Williams, J. E., 34[219b] (30)
Williams, R. A., 126[348] (109, 112–14), 126[361] (112), 126[365] (112–14), 126[367] (112, 113), 126[370] (113, 114), 126[372] (113, 114), 127[387] (119)
Williams, R. E., 187[112] (139), 247[13] (218), 247[15] (218), 247[16] (218), 247[26] (218), 247[27] (218), 247[28] (218), 247[29] (218), 247[31] (218), 247[32] (218), 247[40] (218), 247[42] (218), 248[49] (218), 250[201] (222, 223), 250[202] (222, 223), 252[339] (232), 255[511] (245), 255[518] (246), 271[2] (258), 272[28] (260), 272[35] (260), 366[4] (277), 367[16] (277), 367[17] (277), 425[1f] (374), 425[1g] (374, 415), 425[1h] (374, 377, 381, 388), 425[1i] (374, 381), 425[2] (374)
Williams, R. S., 570[49]
Williams, S., 120[19] (59)
Williamson, K. L., 120[14] (59)
Williard, P. G., 31[36] (6), 32[98] (12), 32[100] (12), 32[101] (12)
Wilson, C. L., 499[154] (454)
Wilson, J. G., 254[443] (242), 255[508] (245)
Wilson, S. R., 367[36] (279, 280), 367[46] (279, 280)
Wimmer, T., 192[426] (171), 192[427] (171)
Winkelmann, H., 541[49] (512)
Winter, A., 498[70] (440), 502[296] (482)
Winter, G., 190[290] (159)
Wischmeyer, U., 122[105b] (67, 71)
Wiseman, G. H., 254[483] (244)
Wismar, W. J., 542[109] (519)
Wittig, G., 498[21] (435)
Witty, H., 34[200] (26)
Wocadlo, S., 34[188] (22, 23)
Wojnarowski, W., 502[319] (490)
Wojnowski, W., 543[181] (525, 526)
Woldt, R., 497[3] (433, 440, 450, 453, 482)
Wolfer, K., 188[186] (147)
Wolfrum, R., 34[211a] (29), 34[213a] (29)
Wollins, J. D., 371[306] (355, 356)
Wolsey, W. C., 427[116] (394, 396, 397)
Wong, E. H., 250[226] (224), 253[411] (241)
Wong, G. T. F., 430[334b] (421)
Wong, K. S., 367[21] (277, 294, 297, 298), 368[130] (294, 297, 298)
Wong, K.-S., 430[335] (422)
Wong, L-L., 367[53] (282, 284)
Wong-Chi-Man, W. W. C., 190[263] (157), 190[264] (157)
Wood, C., 250[235] (225)
Wood, G. L., 192[411] (169), 192[424] (170), 272[74] (264), 272[75] (264)
Wood, J. L., 498[55] (438, 493)
Woode, M. K., 426[57] (381, 382, 421, 423)
Woodrum, K. R., 192[404] (169)
Woollins, J. D., 369[196] (313, 315, 320, 323, 340), 371[281] (345, 346, 355, 356)
Workman, D. P., 187[112] (139), 255[518] (246), 367[16] (277), 369[150] (296, 297), 369[152] (296, 297), 369[154] (296, 297)
Worley, S. D., 191[375] (166)
Wörner, K.-F., 215[70] (210)
Worrall, I. J., 540[10a] (504, 506), 543[196] (527), 544[254] (537, 539)
Woudenberg, R. C., 540[12] (504, 506, 511, 521), 542[130] (520)
Woyciechowski, M., 571[61] (560)
Wrackmeyer, B., 140, 153, 158, 187[111] (139), 187[116] (139), 187[118] (140), 188[119] (140), 188[120] (140), 188[121] (140), 188[122] (140), 188[123] (140), 188[184] (147), 189[211] (151), 189[212] (151), 189[227] (153), 189[231] (153), 190[261] (157), 190[262] (157), 190[267] (158), 190[268] (158), 190[269] (158), 190[270] (158), 190[271] (158), 190[272] (158), 190[301] (160), 192[414] (169), 192[429] (171), 215[90] (213), 215[91] (213), 216[93] (213), 216[94] (213), 216[95] (213), 250[208] (223, 229, 234), 250[217] (224), 250[231] (224), 252[308] (229, 234, 235), 252[328] (231)
Wrackmeyer, G., 247[11] (218)
Wright, D. S., 31[37] (6), 31[44] (8, 9), 32[77] (11), 32[107] (13), 32[121a] (15), 32[122] (15), 33[126] (15)
Wright, T. C., 31[72] (10)
Wu, M., 251[273] (226)
Wu, S. C., 502[301] (483)
Wu, S.-H., 249[174] (221), 252[348] (233), 254[445] (242)
Wu, T.-C., 126[342] (108, 109, 117, 120)
Wulff, G., 172, 191[373] (166), 192[438] (172)
Wunderlich, H., 194[592] (182)
Wuts, P. G. M., 193[496] (175)
Wyatt, B. K., 31[17] (3)
Wynd, A. J., 370[256] (337), 371[300] (352), 371[301] (352), 371[302] (352), 371[303] (352)
Wynne, K. J., 186[28] (130)

Xiao, C., 126[331] (106)
Xiaojie, X., 31[53] (9)
Xie, Y., 497[8] (433, 436, 442), 498[87] (441), 499[108] (444)
Xinlan, C., 370[207] (316)
Xiong, H., 125[282] (97, 98), 125[283] (97, 98), 125[284] (97, 98), 125[285] (97, 98), 126[342] (108, 109, 117, 120)
Xiong, Y., 121[67] (64, 70, 72, 103)
Xu, X., 31[68] (10)
Xu, Y., 123[148] (72)
Xu, Z., 122[104] (67)
Xue, Z., 570[49]

Yagi, T., 189[236] (154)
Yakushev, A. B., 247[33] (218), 250[192] (221)
Yalpani, M., 165, 168, 177, 186[22] (130, 138, 141), 187[100] (138), 188[128] (141), 188[130] (141), 188[134] (141), 188[137] (142), 188[138] (142), 188[139] (142), 188[140] (142), 188[141] (142), 188[142] (142), 191[364] (165), 192[400] (168), 192[401] (168),

192[402] (168), 193[519] (177), 193[520] (177), 193[521] (177, 178), 193[522] (177, 178), 193[523] (177, 178), 193[524] (178), 194[538] (179)
Yamaguchi, M., 192[437] (172)
Yamaguchi, Y., 498[87] (441)
Yamamoto, H., 122[118] (70, 77), 122[129] (70, 82), 124[215] (83), 126[345] (108, 109), 467, 490, 498[57] (439, 492), 498[58] (439), 500[211] (467), 500[212] (467), 500[213] (467), 501[225] (468), 501[256] (474), 501[257] (474), 501[266] (476), 501[267] (476), 501[268] (476), 502[311] (488), 502[320] (490), 502[327] (491), 502[334] (492), 502[342] (494)
Yamamoto, K., 251[292] (228, 231, 240, 242)
Yamamoto, J., 253[420] (241)
Yamamoto, T., 272[81a] (265, 266)
Yamamoto, Y., 32[119b] (14), 54[7] (36), 122[117] (69), 124[204] (82), 251[294] (228), 255[505] (245), 255[508] (245), 501[245] (473)
Yamanaka, T., 501[269] (476)
Yamanishi, T., 126[338] (107), 126[340] (107)
Yamashita, H., 123[178] (78)
Yamashita, M., 124[230] (84)
Yamataka, H., 123[191] (80, 81)
Yamauchi, M., 54[12] (38)
Yamazaki, H., 123[175] (78)
Yamin, M., 368[109] (290, 291)
Yanagisawa, A., 122[118] (70, 77), 124[215] (83), 126[345] (108, 109)
Yang, J. M., 252[323] (230)
Yang, X., 250[187] (221, 243), 250[189] (221, 243), 501[277] (480)
Yang, Z.-Y., 121[68] (64)
Yanovskii, A. I., 191[366] (165), 249[147] (221, 233), 249[148] (221), 249[149] (221), 249[159] (221, 224, 238), 249[161] (221, 242), 249[164] (221), 249[166] (221), 249[167] (221), 249[168] (221), 250[180] (221), 250[181] (221), 250[182] (221, 239), 253[390] (238), 374, 425[1u] (374), 427[110] (396), 428[143] (400, 416), 428[144] (400, 416), 428[145] (400, 416)
Yarosh, O. G., 123[147] (71)
Yasuda, A., 502[320] (490)
Yasuda, H., 54[11] (38), 54[12] (38), 121[77] (65), 125[281] (97, 98), 125[291] (98)
Yasuoka, N., 54[11] (38), 54[12] (38)
Yatagai, H., 501[245] (473)
Yates, B. F., 497[8] (433, 436, 442), 498[87] (441)
Yeh, M. C. P., 126[331] (106)
Yellowlees, L. J., 371[294] (351), 371[295] (351), 371[296] (351)
Yoon, M. S., 187[103] (138)
Yoon, N. M., 500[210] (467)
Yoshida, T., 501[293] (482)
Yoshifuyi, M., 498[78] (441)
Yoshioka, M., 502[307] (486)
Youll, B., 271[26] (259)
Young, D. A. T., 543[220] (532)
Young, D. M., 541[42] (511)
Youngs, W. J., 540[34] (507)
Yu, D., 123[148] (72)
Yu, H.-T., 425[4f] (375)
Yu, M., 248[88] (219, 220, 233)
Yu, S., 123[148] (72)
Yuan, H. S. H., 367[62] (284, 285)
Yuan, K., 124[218] (83)
Yuki, H., 123[153] (73, 83), 123[154] (73)
Yurchenko, A. G., 121[34] (60)
Yur'ev, V. P., 500[194] (463), 501[240] (471)
Yurjev, G. S., 428[166b] (400, 414)

Zaegel, F., 34[197] (22)
Zafiropoulos, T., 498[80] (441), 544[241] (534), 544[243] (535), 544[244] (535), 544[245] (535)
Zagorevskii, D. B., 428[147] (400)
Zagorskii, V. V., 121[57] (62, 108)
Zagorsky, V. V., 121[56] (62)
Zahradnik, R., 248[93] (219, 220, 224, 229, 230, 234)
Zaidi, S. A. A., 368[108] (290, 291), 368[109] (290, 291), 368[113] (292, 293), 368[114] (292, 293), 368[116] (292, 293)
Zaika, D., 190[307] (161), 215[82] (212)
Zaitsev, B. E., 190[323] (162), 190[324] (162)
Zakharkin, L. I., 124[223] (83), 126[333] (106), 248[47] (218), 249[143] (221, 238), 249[147] (221, 233), 249[161] (221, 242), 249[162] (221, 226, 239), 249[164] (221), 249[167] (221), 249[169] (221, 239), 250[182] (221, 239), 250[234] (225), 251[239] (225), 251[256] (225), 251[259] (226, 231), 251[262] (226), 251[264] (226), 251[265] (226, 239), 251[293] (228), 252[327] (231), 252[349] (233), 252[357] (234), 253[382] (237, 239), 253[388] (238), 253[390] (238), 253[391] (239), 253[392] (239), 253[393] (239), 253[394] (239), 253[396] (239), 253[399] (239), 253[401] (239), 253[419] (241), 253[422] (241), 253[424] (241, 242), 254[427] (241), 254[428] (241), 254[429] (242), 254[438] (242, 244), 254[440] (242), 254[442] (242), 254[468] (244), 254[475] (244), 255[530] (247), 416, 425[1o] (374), 426[36] (378, 380), 427[110] (396), 427[130] (400, 416), 427[132] (400, 412), 428[142] (400), 428[143] (400, 416), 428[144] (400, 416), 428[145] (400, 416), 428[146] (400, 416), 428[147] (400), 428[148] (400, 416), 428[149] (400, 416), 428[159] (400), 429[209] (400, 412), 429[242] (400, 417), 430[280] (400, 416), 430[289], 430[290], 430[303] (416), 447, 499[117] (447), 499[136] (452), 499[156] (455), 500[162] (456), 500[163] (456), 541[96] (515)
Zakharov, K. S., 193[476] (175), 193[477] (175), 193[478] (175), 195[646] (185)
Zakharov, L. S., 253[423] (241)
Zakharova, I. A., 250[211] (223), 426[25] (378)
Zaldivar, R. J., 252[323] (230)
Zalkin, A., 34[205] (26), 127[380] (117, 119), 368[96] (288, 290), 368[99] (288, 290), 368[100] (290, 291), 368[103] (290, 291), 368[104] (290, 291)
Zalkin, Z., 33[147] (19, 25, 26)
Zammit, G. S. A., 427[118] (396)
Zanella, P., 368[83] (289)
Zanello, P., 369[137] (297, 299)
Zani, P., 124[205a] (82)
Zanirato, P., 191[338] (163), 191[339] (163), 191[340] (163)
Zanobini, F., 367[64] (284, 285)
Zarate, E. A., 540[34] (507)
Zarges, W., 33[134] (17), 33[138] (19), 33[139] (19)
Zarraga, M., 193[474] (174)
Zaworotko, M. J., 501[288] (482), 541[38] (510)
Zeeh, B., 124[205b] (82)
Zefirov, Yu. V., 426[26] (378)
Zegarski, B. R., 500[214] (468), 500[215] (468)
Zeifman, Y. V., 254[442] (242)
Zelehan, G. I., 187[59] (133)
Zelenev, Y. V., 255[489] (244)
Zenneck, U., 189[203] (150), 189[229] (153), 214[22] (199), 216[98] (214), 430[325] (420), 430[326] (420), 430[327] (420), 544[250] (536)
Zerche-Mackenzie, K., 500[180] (461)
Zerger, R., 33[163] (19), 34[189] (23, 25)
Zhang, G., 192[441] (173), 192[443] (173), 192[444] (173), 249[127] (221, 234), 372[325] (365)
Zhang, H. M., 31[51] (9), 271[9] (258), 271[10] (258), 271[11] (258), 272[32] (260), 426[65] (382, 388), 427[83] (382, 388), 427[84] (382, 388), 427[85] (382, 388), 427[87] (382, 388), 427[100] (382, 391), 502[297] (482), 540[23] (505, 506), 541[57] (512), 542[119] (520), 543[224] (532), 544[231] (533)
Zhang, L., 370[207] (316)
Zhang, S., 123[148] (72)
Zhang, Y. M., 124[213] (83)
Zhang, Z., 194[530] (178)
Zhdanov, A. A., 254[438] (242, 244), 254[475] (244)
Zhdanova, E. A., 255[485] (244)
Zheng, G., 253[397] (239)
Zheng, G.-X., 501[278] (480, 482, 483), 502[312] (488–90)
Zheng, L., 426[38] (378, 380), 429[207] (400, 415), 429[219] (400, 415)
Zhenyang, L., 247[21] (218, 232)
Zhigach, A. F., 252[354] (234, 242)
Zhigareva, G. G., 249[147] (221, 233), 250[182] (221, 239), 253[388] (238), 253[390] (238), 253[391] (239), 425[1o] (374), 426[36] (378, 380), 430[303] (416)
Zhou, J. P., 121[67] (64, 70, 72, 103)
Zhou, W., 192[441] (173), 192[442] (173), 192[443] (173), 192[444] (173)
Zhu, H., 249[127] (221, 234), 372[325] (365), 544[235] (534)
Zhu, T., 252[352] (234)
Zhu, Z., 192[443] (173), 192[444] (173)
Zhuchenko, S. E., 195[600] (183)
Zhulin, V. M., 124[207] (83)
Ziegler, B., 187[70] (134)
Ziegler, E., 500[201] (465)
Ziegler, K., 439, 451, 478, 495,

499[135] (452), 500[169] (458, 462, 464, 486), 500[171] (458), 500[173] (458), 500[185] (462), 500[188] (463), 500[207] (467), 501[270] (478, 484, 497)
Ziegler, M. L., 544[249] (536), 570[53] (557), 571[62] (561), 571[63] (561)
Ziembinski, R., 190[265] (157), 190[304] (160)
Ziller, J. W., 497[14] (434, 448, 497), 540[30] (507), 541[59] (512), 541[61] (512), 541[64] (513), 543[193] (527), 543[194] (527)
Zimina, G. M., 251[271] (226)
Zimmerman, G. J., 427[105] (396)
Zimmermann, C., 121[47b] (61, 62, 77)
Zinnius, A., 31[31] (6)
Zlotskii, S. S., 500[221] (468), 500[222] (468), 500[224] (468), 502[324] (491)
Zobacova, A., 254[458] (244)
Zoebisch, E. G., 248[105] (220)
Zoellner, R. W., 187[52] (132)
Zolotarev, A. P., 502[333] (492)
Zolotarev, B. M., 190[309] (161), 190[310] (161)
Zosel, K., 500[169] (458, 462, 464, 486), 500[171] (458)
Zrineh, A., 571[77] (566, 567, 569), 571[98] (569)
Zubreichuk, Z. P., 427[135] (400, 414), 427[136] (400, 414), 428[138] (400), 428[167] (400, 414), 428[182] (400, 414), 428[183] (400, 414), 428[186] (400, 414)
Zubritskii, L. M., 122[111] (68), 122[112] (68)
Zuckerman, J. J., 247[12] (218), 248[52] (218, 244)
Zunza, H., 193[474] (174)
Zupan, M., 123[174] (78)
Zwecker, J., 430[326] (420), 430[327] (420)
Zweifel, G., 467, 498[49] (438), 498[76] (441, 467, 486), 501[238] (470), 501[241] (471), 502[331] (491)
Zwettler, R., 368[112] (292, 293)

Subject Index

JOHN NEWTON
David John (Services), Slough, UK

This Subject Index contains individual entries to the text pages of Volume 1. The index covers general types of organometallic compound, specific organometallic compounds, general and specific organic compounds where their synthesis or use involves organometallic compounds, types of reaction (insertion, oxidative addition, etc.), spectroscopic techniques (NMR, IR, etc.), and topics involving organometallic compounds.

Because authors may have approached similar topics from different viewpoints, index entries to those topics may not always appear under the same headings. Both synonyms and alternatives should therefore be considered to obtain all the entries on a particular topic. Commonly used synonyms include alkyne/acetylene, compound/complex, preparation/synthesis, etc. Entries where the oxidative state of a metal has been specified occur after all the entries for the unspecified oxidation state, and the same or similar compounds may occur under both types of heading. Thus $Cr(C_6H_6)_2$ occurs under Chromium, bis(η-benzene) and again under Chromium(0), bis(η-benzene). Similar ligands may also occur in different entries. Thus a carbene–metal complex may occur under Carbene complexes, Carbene ligands, or Carbenes, as well as under the specific metal. Individual organometallic compounds may also be listed in the Cumulative Formula Index in Volume 14.

Acetylene, diphenyl-
reactions with ruthenaboranes, 301
Actinide complexes
metallacarbaboranes, 373–424
tetrahydroborates, 286
trihydroborates, 290
1,3-Alkadiynes, 1-trimethylsilyl-
hydroalumination, 470
Alkali metal alkoxides
addition to dialkylmagnesium compounds, 95
structures, 11
Alkali metal amides
reaction with Grignard reagents, 103
Alkali metal compounds
ethyl, 19
magnesium compounds, 105, 106
methyl, 19
Alkali metal organometallics, 1–30
amido, 2–10
antimony compounds, 10, 11
arsenic compounds, 10, 11
carbon–metal bonds, 18–29
σ-complexes, 18–25
π-complexes, 18–25
enolates
structures, 11
ethynides
structures, 20
fluorenyl, 25
group 15 complexes, 2–11
group 16 compounds, 11–14
imino, 2–10
lead compounds, 14, 15
mixed-metal compounds, 25–9
main group complexes, 29
transition metal complexes, 25–7
multimetallated, 16, 17
oxygen compounds, 11–13
phosphorus compounds, 10, 11
propynides
structures, 20
selenium compounds, 13, 14
silicon compounds, 14, 15
sulfur compounds, 13, 14
tellurium compounds, 13, 14
theory, 29, 30
thiolates, 14
tin compounds, 14, 15
Alkaline earth metal compounds
bis(pentamethylcyclopentadienyl)
complexes with Lewis bases, 114
cyclooctetraene, 117
heterobimetallic, 119
zincates, 119
Alkaline metal compounds
anionic polymerization
initiation, 109
Alkene Growth Process
organoaluminum compounds, 478
Alkenes
addition compounds
magnesium, 97, 98
addition of magnesium hydride
catalysis, 86
hydrogenation
organoaluminum reagents, 484
metal complexes
metallaboranes, 308
oligomerization
organoaluminum reagents, 484
reactions with
organoaluminum compounds, 495, 496
Alkynes
addition compounds
magnesium, 97, 98
reactions with ruthenaboranes, 301
1-Alkynes, 1-chloro-
hydroalumination, 471
Allyl compounds
metallation, 24
Aluminaboranes
properties, 258
Aluminacarboranes
properties, 258, 259
Aluminadicarbadodecaborane
structure, 258
Aluminates
structures, 29
Alumination
carbon Brønsted acids, 466, 467
Aluminoxanes
structures, 450–3
Aluminum
carbon bonds, 456–61
interconversions, 463–7
in organic synthesis, 485–96
reactions with nontransition metal substrates, 467–76
chemistry, 497

hydrogen bonds, 461, 462
metal bonds, 462, 463
nonmetal bonds, 462
turnings
displacement reactions, 461
Aluminum alkyls
alkene exchange, 464
Aluminum complexes
(μ-alkyl)-metal bonds, 552
f-block metals
(μ-H)-metal bonds, 551
group 4 metals
(μ-H)-metal bonds, 546–8
group 6 metals
(μ-alkyl)-metal bonds, 552
(μ-H)-metal bonds, 548, 549
group 7 metals
(μ-H)-metal bonds, 550
group 8 metals
(μ-H)-metal bonds, 550, 551
group 9 metals
direct bonds to aluminum, 552
group 10 metals
(μ-H)-metal bonds, 551
(μ-H)-metal bonds, 546–51
metathesis
metal alkyls, 460, 461
organomagnesium, 106, 107
transition metal bonds, 545–69
Aluminum(I) complexes
generation, 435, 436
isolation, 435, 436
Aluminum(II) complexes
generation, 435, 436
isolation, 435, 436
Aluminum(III) complexes
reactivity
steric factors, 434, 435
structures, 433
steric factors, 434, 435
Aluminum, trialkyl-
rearrangements, 476
Anthracene
magnesium compounds, 99, 100
Grignard reagent preparation, 63
Anthracene, 9,10-diphenyl-
reaction with magnesium, 99
Arsaboranes
metal complexes, 266
properties, 265, 266
1-Azaadamantane
synthesis, 147
1,2-Azaboraallene
metal complexes, 201
Azaboranes
properties, 260, 261
transition metal derivatives, 262, 263
Azaboroles
synthesis, 161
1*H*-1,2-Azaboroles, 2,5-dihydro-
metal complexes, 212
1*H*-1,2-Azaborolides, 2,5-dihydro-
metal complexes, 212
1,2-Azaborolines
metal complexes, 212
1,2-Azaborolinides
metal complexes, 212
Azacarboranes
properties, 262
Azadecaboranes
properties, 261
Azadiboriridine
reactions, 159
Azadiboriridine, tri[t[butyl-
reactions, 177
Azadodecaboranes
properties, 261
1,2,5-Azasilaborolenes
metal complexes, 213

B_6N_7-phenalene
synthesis, 158
Barium, allylchloro-
reactions
with carbonyl compounds, 109
Barium, bis(α-arylalkyl)-
ionic structures, 111
Barium, bis(pentamethylcyclopentadienyl)-, 112
Barium, bis(triphenylmethyl)-
complexes, 111
formation, 111
Barium, bis(1,1,3-triphenylpropyl)-
structure, 111
Barium chlorides
allyl
preparation, 108
Barium complexes
anthracene, 117
arene radical anion complexes, 117
chemistry, 57–120
Barium, dialkynyl-
preparation, 111
Barium, dicumyl-
structure, 110
Barium, dicyclopentadienyl-, 112–16
ab initio calculations, 114
Barium, difluorenyl-
formation, 115
ion pair structures, 115
Barium, difluorenyltetrammine-
crystal structure, 115
Barium, diindenyl-
formation, 115
Barium, dimethyl-
structure, 110
Beckmann rearrangement
organoaluminum compounds, 476
Benzene, nitroso-
reaction with
trialkylaluminum, 475
Benzoborepine
metal complexes, 210, 211
1*H*-Benzo[d]-1,3-diboroles, 2,3-dihydro-
synthesis, 206
Benzodioxaborole
structure, 174
Benzonitrile
reaction with tetraalkyl-dialuminoxanes, 475
Benzotriazole
structure, 169
Beryllium
(1-*t*-butyldihydro-2-methyl-1*H*-azaborolyl)cyclopentadienyl-
preparation, 46
chemistry, 35–54
dimethyl-
band structure calculations, 40
toxicity, 35
Beryllium, acylbromo-
synthesis, 53
Beryllium, bis(1-*t*-butyldihydro-2-methyl-1*H*-azaborolyl)-
preparation, 46
Beryllium, bis(cycloheptadienyl)-
NMR, 38
Beryllium borohydride, methyl-
NMR, 38
Beryllium, bromomethyl-
reactions, 53
Beryllium chloride, tetrakis-(dimethylsulfonium)-
synthesis, 36, 37
Beryllium, chloro(cyclopentadienyl)-
reaction with potassium pentaborane, 48
structure, 48
Beryllium, chlorodicyclopentadienyl-
reactions, 46
Beryllium compounds
alkyl
optically active, synthesis, 52
cyclopentadienyl-, 44–50
NMR, 50
Beryllium, (cyclooctadienyl)-
NMR, 38
Beryllium, cyclopentadienylhalo-
bonding, 50
thermodynamic calculations, 50
Beryllium, cyclopentadienyl-(pentamethylcyclopentadienyl)-
structure, 46
Beryllium, (di-*t*-butylphosphino)(pentamethyl-cyclopentadienyl)-
synthesis, 42, 43
Beryllium, dicyclopentadienyl-
benzene complex, 45
NMR, 45
ordered structure
calculation, 44
solution
dipole moment, 45
structure, 45
structure
rationalization, 45
synthesis, 44–8
vapor phase
structure, 45
Beryllium, diethyl-
metal organic vapor phase epitaxy, 54
Beryllium, (2-dimethylamino-*N*-methylethylamido)-|hydrido-|
synthesis, 50
Beryllium hydrides, 50–2

Beryllium, (pentamethylcyclopentadienyl)methyl-
  preparation, 48
Beryllium, tetra(propyn-1-yl)-(tetramethyl)-dialumino-
  preparation, 39
Beryllium, tris(3-*t*-butylpyrazolyl)hydroboratohydrido-
  synthesis, 50
Biferrocenyl
  structure, 200
Bimetallaboranes, 323
Bismaboranes
  properties, 267
Bis(oxaborolanes)
  preparation, 173
1-Boraadamantane
  complexes, 145
2-Boraadamantane, 2-isopropyl-
  bromination, 147
Boraadamantanes
  preparation, 143, 145–7
Borabenzene
  adducts, 152
  metal complexes, 209
  preparation, 152
  structure, 131, 132
3-Borabicyclo[003.003.1]nonane
  preparation, 165
9-Borabicyclo[003.003.1]nonane
  preparation, 138–42, 168
Borabicyclodecanes
  preparation, 143
Borabicyclodecanes, *B*-alkoxy-
  preparation, 143
Borabicyclononanes
  preparation, 143
Borabicyclononene
  synthesis, 154
Boracyclanes
  preparation, 137
Boracyclopentenyl ligands
  metal complexes, 203
Boraethene
  metal complexes, 201
2-Boranaphthalene
  preparation, 152
Borane, chloro(alkylamino)-
  reactions with allenyllithium, 144
Borane, cyclotetrakis(triazolyl-
  properties, 169
Borane, 9-fluorenylidene-(2,2,6,6-tetramethylpiperidino)-
  metal complexes, 201
Borane, imino-
  addition reactions, 163
  cycloaddition reactions, 162
  reactions, 159, 160
  reactions with trihaloboranes, 165
Borane, methylene-
  mesityl-bridged, 149, 150
Borane, oxo-
  reactions, 171
Borane, pyrazolyl-
  chemistry, 168
  preparation, 167
  structure, 132
Boranes, imino-
  metal complexes, 211
Boranes, methylene-
  properties, 134
Boranes, pyrazolyl-
  preparation, 140
1-Bora-2-oxaadamantane
  synthesis, 145
Boratabenzene
  metal complexes, 209
Boratabutadiene
  metal complexes, 205
Borataethene
  metal complexes, 201
Borataferrocene
  protonation, 204
Borates
  cyclic
    stabilities, 184
  structures, 29
  tetrahydro-
    group 5 metals, metallaboranes, 280–3
Borates, dihydro-
  metallaboranes, 290, 291
Borates, hexahydro-
  group 10 metals
    metallaboranes, 328
  metallaboranes, 328
Borates, monohydro-
  metallaboranes, 292
Borates, tetrahydro-
  group 3 metals
    metallaboranes, 278
  group 4 metals
    metallaboranes, 278–80
  group 6 metals
    metallaboranes, 283, 284
  group 7 metals
    metallaboranes, 283, 284
  group 8 metals
    metallaboranes, 284, 285
  group 9 metals
    metallaboranes, 284, 285
  group 11 metals
    metallaboranes, 286
  metallaboranes, 278–86
Borates, trihydro-
  actinide complexes, 290
  lanthanide complexes, 288
  metallaboranes, 287–90
  transition metal complexes, 287, 288
Boratirene
  characterization, 148
Boratricycloalkanes
  preparation, 144
Borazines
  chemistry, 158
  phosphorus analogue, 170
  reactions, 167
  synthesis, 160
Borepane
  preparation, 137, 138
Borepin, heptaphenyl-
  synthesis, 154
Borepins
  bicyclic derivatives, 155
  geometry, 132
  metal complexes, 210
1*H*-Borepins
  metal complexes, 198
  synthesis, 153, 154
Borete, 1,2-dihydro-
  structure, 131, 132, 149
Boridic clusters
  ruthenium, 303
Borinanes
  NMR, 171
  preparation, 136
  reactions, 181
Borirenes
  anions, 133
  chemistry, 148
  metal complexes, 201
  structure, 131
1,2-Borolane, 3-methylene-
  metal complexes, 205
Borolanes
  NMR, 171
  polymer-supported, 173
Borole
  metal complexes, 202–4
Borole, 2,5-dihydro-
  preparation, 150
Borole ligands
  metal complexes, 198
2-Borolenes
  synthesis, 202
3-Borolenes
  synthesis, 202
Boroles
  ammonia adducts, 203
  metal complexes
    slip distortion, 204
  preparation, 151
1*H*-Boroles, 2,5-dihydro-
  synthesis, 202
Boron carbon rings
  metal complexes, 201–10
Boron cyclic systems
  three- or four-coordinate, 129–85
Boron germanium rings
  chemistry, 158
Boron lead rings
  chemistry, 158
Boron nitrogen antimony rings
  properties, 182
Boron nitrogen arsenic rings
  properties, 182
Boron nitrogen ligands
  metal complexes, 211
Boron nitrogen oxygen rings
  properties, 182
Boron nitrogen phosphorus oxygen rings
  properties, 184
Boron nitrogen phosphorus rings
  properties, 182
Boron nitrogen rings
  chemistry, 158–69
  properties, 181
Boron nitrogen selenium rings
  properties, 184, 185
Boron nitrogen sulfur rings
  properties, 184, 185
Boron oxygen rings
  four-coordinate boron, 178, 179
  properties, 171–9
Boron oxygen sulfur rings
  properties, 185
Boron phosphorus ligands
  metal complexes, 211
Boron phosphorus oxygen rings
  properties, 184
Boron phosphorus rings
  four-coordinate boron, 170

properties, 169–71
Boron ring compounds
four-coordinate, 165–9
Boron rings
ligated to metals, 197–214
Boron selenium rings
properties, 179–81
Boron silicon oxygen rings
properties, 181
Boron silicon rings, 157
properties, 181
Boron sulfur rings
properties, 179–81
Boron tin rings
chemistry, 158
properties, 181
Boronates, methoxycrotyl-
synthesis, 173
Boroxins
NMR, 171
Borylene, methyl-
preparation, 144
Butterfly boridic clusters, 298–301

Cadmates
structures, 29
Caesium, methyl-
preparation, 20
Calcium
co-condensation with tetrahydrofuran, 107
ultrafine particles, 107
Calcium, bis[1,3-bis(trimethylsilyl)cyclopentadienyl]-, 112
Calcium, bis[bis(trimethylsilyl)-methyl]-
crystal structure, 110
Calcium, bis(*t*-butylcyclopentadienyl)-, 112
Calcium, bis(1,3-di-*t*-butylcyclopentadienyl)-, 112
Calcium, bis(methylcyclopentadienyl)-, 112
Calcium, bis(pentamethylcyclopentadienyl)-, 112
adducts with 1,4-bis(trimethylsilyl)-1,3-butadiene, 114
adducts with transition metal complexes, 115, 120
solid-state structure, 113
Calcium, bis(tetraisopropylcyclopentadienyl)-
structure, 113
Calcium, bis(trimethylsilyl)cyclopentadienyl)-, 112
Calcium, bis(triphenylmethyl)-
formation, 111
Calcium bromide, (1-methyl-2,2-diphenylcyclopropyl)-
chiral, 108
Calcium carbaboranes
preparation, 116
Calcium complexes
anthracene, 117
arene radical anion complexes, 117
chemistry, 57–120
1,3-diene
reactions, 117
Calcium, cyclopentadienyliodo-
alkylated, 110
structure, 109
Calcium, dialkynyl-
preparation, 111
Calcium, dicyclopentadienyl-, 112–16
adducts with ethers, 114
unsolvated, 112
Calcium, dimethyl-
structure, 110
Calcium, diphenyl-
structure, 110
Calcium, (ethyltetramethylcyclopentadienyl)amide-
structure, 119
Calcium fluorides
aryl, 108
Calcium halides
aryl
preparation, co-condensation reactions, 108
solvent-free, 108
Calcium iodide, cyclopentadienyl-
preparation, 109
Calcium, (pentamethylcyclopentadienyl)[bis(trimethylsilyl)methyl]-
formation, 115
Calcium, (trimethylaluminum)-cyclopentadienyl-
formation, 119, 120
Carbaboranes
ab initio/IGLO/NMR, 222
acidity, 234
alkylation, 239
C-alkylation, 241
applications, 245
boron nuclear quadrupole couplings, 226
bridging hydrogen tautomerization, 219
cage fluxional behavior, 232–4
chemical properties, 232–45
derivatives, 236–45
B_{cage}–chalcogen bonds, 237, 238
B_{cage}–halogen bonds, 236, 237
B_{cage}–nitrogen bonds, 238
B_{cage}–phosphorus bonds, 238
C_{cage}–arsenic bonds, 240, 241
C_{cage}–chalcogen bonds, 239–42
C_{cage}–nitrogen bonds, 240, 241
C_{cage}–phosphorus bonds, 240, 241
exo-cage B–C bonds, 238, 239
diffraction, 220–2
dipole moments, 220, 226
electron diffraction, 220
electrophilic deuteration, 239
electrophilic reactions, 239
fluoride derivatives, 236
fluxionality, 233
free radical initiated reactions, 241
halogenation, 236
Hartree–Fock calculations, 220
He[1] photoelectron spectra, 226
infrared spectra, 219, 225
isomer stabilities, 220, 232–4
ligands, 374, 375
mass spectrometry, 226
metal vapor reactions, 377
microwave structure, 227
molecular orbital calculations, 218
NMR, 222–4
nucleophilic attack, 219
nucleophilic substitution, 227
partial cage opening, 238
physical properties, 220–7
polyhedral, 218–45
polymers, 244
proton affinity, 218, 220
pulse radiolysis, 227
Raman spectra, 219, 223, 225
reaction with hydrides, 235
reaction with hydrogen halides, 237
reactions with fluorides, 235
reactions with Lewis bases, 235, 236
reactions with nitrogen bases, 235, 236
rearrangement, 219
mechanisms, 233
reduction, 235
siloxane oligomers, 244
skeletal rearrangements, 232–4
spectroscopy, 222–6
spiro phosphorus derivatives, 241
structures, 218–20
synthesis, 227–31
cage fusion reactions, 229
cage growth, 228
cage growth, boron insertion reactions, 228, 229
coupled carboranes, 231
dehydrogenation, 229
from polyboranes and alkynes, 227, 228
rearrangements, 231
selective cage degradation, 229–31
tensor surface harmonic theory, 220
thermochemical characteristics, 226
thermolysis, 232
UV photoelectron spectrum, 226
UV spectroscopy, 226
vibrational spectroscopy, 224–6
Carbaboranes, glycosyl-
derivatives
B_{cage} metals, 243
C_{cage} metals, 243
deuteration, 244
levorotatory enantiomers, 244
preparation, 242
Carbaboranes, halo-
halogen exchange, 236
with boron trihalides, 236
Carbalumination, 464–6, 470
Carbenes, carboranyl-
ESR spectra, 227
B-substituted, 239
Carbidic clusters

octahedral
ruthenium, 303
Carbomagnesiation
alkenes, 87
alkynes, 67
1,3-dienes, 67
enynes, 67
unsaturated C–C bonds, 66
zirconium compounds
catalysis, 68
Carbonyl compounds
reactions with
organoaluminum compounds, 492–4
organoaluminum reagents, 472–4
α,β-unsaturated
reactions with
organoaluminum compounds, 494
o-Carborane, 1,2-dehydro-
preparation, 244
nido-Carboranes
metal complexes, 0215
Carboranes
polyhedral, 217–47
Catecholatoborane
isolation, 174
synthesis, 179
Catecholborane
properties, 174
reactions, 171
Chemisorption
methyl bromide
magnesium crystal surface, 62
Chromium
dicarbon metallacarbaboranes
7-vertex, 382
12-vertex, 400
tetracarbon metallacarbaboranes
12-vertex, 423
13-vertex, 423
Chromium complexes
indium bonds, 557
thallium bonds, 567
Cobalt
dicarbon metallacarbaboranes
6-vertex, 382
7-vertex, 382
8-vertex, 396
9-vertex, 396
10-vertex, 396
11-vertex, 396
12-vertex, 400, 418
monocarbon metallacarbaboranes, 378
11-vertex, 378
12-vertex, 378
tetracarbon metallacarbaboranes
6-vertex, 423
9-vertex, 423
12-vertex, 423
13-vertex, 423
tricarbon metallacarbaboranes
11-vertex, 420
Cobalt complexes
dihydroborates, 290
diselenaboranes, 270
gallium bonds, 555
indium bonds, 564
Cobalt nickel
dicarbon metallacarbaboranes
7-vertex, 382
Cobalt osmium
dicarbon metallacarbaboranes
7-vertex, 382
Cobalt rhodium
dicarbon metallacarbaboranes
7-vertex, 382
Cobalt ruthenium
dicarbon metallacarbaboranes
7-vertex, 382
Cobaltaazaundecaboranes
preparation, 263
Cobaltole
reaction with borane tetrahydrofuran adducts, 203
Copper
dicarbon metallacarbaboranes
12-vertex, 400
Copper clusters
metallaboranes, 304
Copper complexes
heptahydrotriborates
metallaboranes, 317
Copper(I) complexes
tetrahydroborates, 286
Cycloalkanes
B-derivatives, 133–47
Cycloalkanes, 1,2-bis(methylene)-
magnesium compounds, 97
Cycloalkenes
hydroalumination, 470
Cyclooctatetraene
magnesium compounds, 98
reaction with potassium and ytterbium, 25
Cyclotriboroxane
properties, 171

Decaborates, dodecahydro-
metallaboranes, 352–8
Decaborates, hexadecahydro-
metallaboranes, 349
Decaborates, tetradecahydro-
group 8 metals
metallaboranes, 350
group 9 metals
metallaboranes, 351
group 10 metals
metallaboranes, 351, 352
group 11 metals
metallaboranes, 352
group 12 metals
metallaboranes, 352
metallaboranes, 350–2
Diamond–square–diamond rearrangements
metallaboranes, 277, 347
Diarsaboranes
metal complexes, 266
properties, 266
Diarsadodecaboranes
properties, 266
1,3,2-Diazaboracyclohexanes
synthesis, 164
Diazaboranes
properties, 261, 262
1*H*-1,3,2-Diazaborole, 2,3-dihydro-
preparation, 161
1,3,2-Diazaborolidines
synthesis, 161
Diazadecaboranes
properties, 262
Diazadiboranaphthalenes
synthesis, 164, 165
1,3,2,4-Diazadiboretidines
metal complexes, 211
Diazapentaboranes
synthesis, 160
1,4-Diazonia-2,5-diboratacyclohexanes
synthesis, 166
Dibenzoboroles
synthesis, 156
Dibismadodecaboranes
properties, 267
9,10-Diboraanthracene, 9,10-dihydro-, 210
1,2-Diborabenzene
chemistry, 153
1,3-Diborabenzene
metal complexes, 209
1,4-Diborabenzene
metal complexes, 209
2,6-Diborabicyclo[3.1.0]hex-3-ene
synthesis, 154
1,2-Dibora-3,5-cyclohexadienes
chemistry, 153
metal complexes, 209
Diboracyclohexene
dimerization, 145
1,4-Dibora-2-cyclohexenes
dehydrogenative complex formation, 209
Diboracyclohexenes
preparation, 153
Diborafulvenes
metal complexes, 199, 208
1,4-Diboranaphthalene
metal complexes, 210
Diborane(2)
metallaboranes, 312
Diboranes(4)
metallaboranes, 311
1,2-Diboratabenzene
metal complexes, 209
Diborates, heptahydro-
metallaboranes, 304
Diborates, hexahydro-
group 5 metals
metallaboranes, 306
group 8 metals
metallaboranes, 306, 307
metallaboranes, 306, 307
Diborates, pentahydro-
group 6 metals
metallaboranes, 308
group 8 metals
metallaboranes, 308
group 9 metals
metallaboranes, 309
group 10 metals
metallaboranes, 310
metallaboranes, 307–10
1,3-Diboretane
preparation, 135
Diborete
stabilities, 150
1,2-Diborete, 1,2-dihydro-
preparation, 150
1,3-Diborete, 1,3-dihydro-
preparation, 150
Diboretene

structure, 131
Diboriranide
preparation, 150
1,2-Diborolane, 4-methylene-
metal complexes, 204, 205
1*H*-1,3-Diborole, 2,3-dihydro-
metal complexes, 205
1,3-Diborolenes
metal complexes, 205
1,3-Diborolenyl ligands
metal complexes, 198, 205
1,3-Diboroles
preparation, 152
1*H*-1,3-Diborolyl, 2,3-dihydro-
metal complexes, 205
Dicobalt
dicarbon metallacarbaboranes
7-vertex, 382
9-vertex, 396
Dienes
addition compounds
magnesium, 97, 98
Dimagnesium organometallics
equilibrium, 76
Diorganoaluminum(II) complexes
synthesis, 446
1,3,2,5-Dioxaboraphosphorinanes
properties, 184
1,3,2-Dioxaborinanes
properties, 176
synthesis, 175
Dioxaborolanes
preparation, 173
1,3,2-Dioxaboroles
preparation, 172
Diphosphaboranes
properties, 264
Diphosphaborirane
properties, 169
1,3,2,4-Diphosphadiboretane
metal complexes, 211, 212
Diphosphahexaboranes
properties, 264
Diphosphaundecaboranes
properties, 264
Diplatinum
dicarbon metallacarbaboranes
9-vertex, 396
Dirhodium
dicarbon metallacarbaboranes
7-vertex, 382
Diruthenium
dicarbon metallacarbaboranes
7-vertex, 382
Diselenadiboretane
preparation, 171
Diselenaundecaboranes
properties, 270
1,2-Disila-*closo*-dodecaborane,
1,2-dimethyl-
preparation, 259
Dithiaboracyclohexanes
spectra, 181
Dithiaboranes
properties, 269
Dithiaborinanes
isomerization, 181

Epoxides
reactions with
organoaluminum
complexes, 469, 487–90
Ethers
reactions with
organoaluminum
compounds, 491
Ethylenediamine, benzyldimethyl-
dilithiation, 17
Ethyne
hydrogenation
organoaluminum reagents,
484
oligomerization
organoaluminum reagents,
484
reactions with
organoaluminum
compounds, 495, 496
Europium
dicarbon metallacarbaboranes
12-vertex, 418

Ferracarbaboranes, indenyl-
chemistry, 387
reactions, 375

Gadolinium
dicarbon metallacarbaboranes
7-vertex, 382
12-vertex, 400
Gallaboranes
properties, 258
Gallacarboranes
properties, 258, 259
Gallane, dimethyl-
structure, 515
Gallane, phosphino-
preparation, 519
Gallates
structures, 29
Gallium carbide
structure, 506
Gallium complexes
anionic, 514
arenes, 534–6
arsenic ligands, 518–24
cage compounds, 522
carboboranes, 532–4
cationic, 514
chemistry, 503–40
cyclopentadienyl, 529–31
group 6 metals
direct bonds to gallium,
553
group 7 metals
direct bonds to gallium,
553–5
group 9 metals
direct bonds to gallium,
555, 556
group 10 metals
direct bonds to gallium,
557
halogen complexes, 511–13
hydrides, 514–17
low oxidation state, 528–36
metal–metal bonds, 537–40
metal bonds, 553–7
nitrogen ligands, 518–24
oxygen ligands, 524–8
phosphorus ligands, 518–24
pseudohalogen complexes,
511–13
selenium ligands, 524–8
sulfur ligands, 524–8
tellurium ligands, 524–8
transition metal bonds, 545–69
trialkyl, 504–11
triaryl, 504–11
Gallium hydride
amine adducts, 515
Gallium, trialkyl
complexes, 509
Gallium, trimesityl-
structure, 507
Gallium, trimethyl-
crown ether complexes, 510
Germanaboranes
properties, 259, 260
Germanacarboranes
properties, 260
Germanium, bis(bis(trimethyl-
silyl)dicarbahexa-
borane)-
preparation, 260
Gold
dicarbon metallacarbaboranes
12-vertex, 400, 418
Gold(I) complexes
clusters
metallaboranes, 299
Grignard reagents
addition to alkenes, 66
addition to alkynes
copper catalysis, 68
addition to cyclopropenes, 66
alkynes
protection, 70
analysis, 78
benzylic
addition to α,β-unsaturated
carbonyl compounds, 82
enantioselective, carbonyl
compounds, 82
reactions, 82
benzylic di-
preparation, 71
carbonation
rate constants, 78
carbonyl compounds, 80–4
chiral
reactions, 82
chiral sparteine ligands
structure, 73
five-coordinate
solid-state structure, 74
six-coordinate
solid-state structure, 74
crown ether, 77
cubane di-
preparation, 72
di-
cyclization, 76
preparation, 71, 72
1,1-di-
preparation, 71
solid-state structure, 73
1,2-di-
preparation, 71
1,3-di-
preparation, 71
electron transfer, 78
2-enyl
interconversions, 77
fluorination, 78
o-functionalized aromatic
preparation, 70
halogenated
preparation, 70

in organometallic synthesis, 83, 84
in synthesis of dimetallic organometallics, 83
in synthesis of metallacycles, 84
in synthesis of metallic organometallics, 84
inversion at α-C atoms, 77
LAXS studies, 75
matrix isolation studies, 62
methyl functionalized
  preparation, 70
^{25}Mg NMR, 76
phenols
  protection, 70
poly-
  preparation, 71
polymeric poly-
  preparation, 72
preparation
  added ligands, 60
  bromine–magnesium exchange, 65
  carbanion intermediates, 63
  commercial, 58
  direct reaction, 58–63
  exchange reactions, 63–6
  from bromoanisoles, 65
  from iodopyridines, 65
  functionalized, 70
  kinetics, 60
  magnesium activation, 59
  mass transfer, 60
  mechanism, 60, 61
  mercury compounds, 65
  metal–halogen exchange, 64
  metal–hydrogen exchange, 63
  organotellurium compounds, 66
  oscillating rates of reaction, 60
  pyridyl sulfoxides, 65
  radical anion intermediates, 61
  radical intermediates, 60
  reaction with unsaturated bonds, 66–9
  retention of stereochemistry, 61
  solvents, 60
  transmetallation, 65
  Walborsky adsorption model, 61
  zero-order kinetics, 60
reactions, 78–84
  aromatic ketones, 81
  coupling, 83
  enolization of ketones, 81
  heterolytic, 80
  SET mechanisms, 80
  transition metal compounds, 83
  α,β-unsaturated carbonyl compounds, 82
skeleton rearrangements, 79, 80
solid-state
  physical properties, 72–4
  structures, 72–8
  structures, four-coordinate, 73
  structures, intramolecularly coordinated oxygen donors, 74
solutions
  structures, 75–7
spectroscopy, 77
stereochemical changes, 77
theoretical studies, 78
thermal rearrangement, 79
thermochemistry, 78
titanium tetrachloride
  alkylation, 83
tri-
  preparation, 71
triphenylmethyl
  solid-state structure, 73
unsolvated, 62
unsolvated neopentyl
  solid-state structure, 73
zirconium tetrachloride
  alkylation, 84

Hafnium
  dicarbon metallacarbaboranes
    7-vertex, 382
    12-vertex, 400
Halides
  reactions with
    organoaluminum compounds, 491, 492
Heteroboranes
  properties, 257–71
Hexaborates, hexahydro-
  group 8 metals
    metallaboranes, 328
Hexaborates, nonahydro-
  group 8 metals
    metallaboranes, 327
  group 10 metals
    bimetallaboranes, 327
  metallaboranes, 327
Hydroalumination, 456, 458–60, 470
Hydrocarbons
  metallation
    enthalpies, 63
    magnesium amides, 103
  polylithiated, 21
Hydromagnesiation
  titanium catalysis
    addition to alkenes, 69

Indacarboranes
  properties, 258, 259
Indates
  structures, 29
Indium complexes
  anionic, 514
  arenes, 534–6
  arsenic ligands, 518–24
  cage compounds, 522
  carboboranes, 532–4
  cationic, 514
  chemistry, 503–40
  cyclopentadienyl, 529–31
  direct bonds to metals, 557–67
  group 6 metals
    direct bonds to indium, 557–60
  group 7 metals
    direct bonds to indium, 560, 561
  group 8 metals
    direct bonds to indium, 562–4
  group 9 metals
    direct bonds to indium, 564, 565
  group 10 metals
    direct bonds to indium, 566
  halogen complexes, 511–13
  hydrides, 514–17
  low oxidation state, 528–36
  metal–metal bonds, 537–40
  nitrogen ligands, 518–24
  oxygen ligands, 524–8
  phosphorus ligands, 518–24
  pseudohalogen complexes, 511–13
  selenium ligands, 524–8
  sulfur ligands, 524–8
  tellurium ligands, 524–8
  transition metal bonds, 545–69
  trialkyl, 504–11
  triaryl, 504–11
Indium, trimesityl-
  structure, 507
Indium, tris(2,4,6-tris(trifluoromethyl)-phenyl)-
  structure, 507
Iridaboranes
  isocloso
    orthoboronated, 345
Iridatetraboranes, 313
Iridium
  dicarbon metallacarbaboranes
    11-vertex, 396
    12-vertex, 400, 418
  monocarbon metallacarbaboranes, 378
    10-vertex, 378
Iridium complexes
  indium bonds, 565
  metallaboranes, 342
  thiaboranes, 268
Iron
  dicarbon metallacarbaboranes
    6-vertex, 382
    7-vertex, 382
    8-vertex, 396
    9-vertex, 396
    10-vertex, 396
    12-vertex, 400
  monocarbon metallacarbaboranes
    11-vertex, 378
  tetracarbon metallacarbaboranes
    8-vertex, 423
    10-vertex, 423
    11-vertex, 423
    13-vertex, 423
  tricarbon metallacarbaboranes
    11-vertex, 420
Iron, (benzene)dicarbaborane)-
  synthesis, 377
Iron carbonyl clusters
  indium bonds, 563
Iron–cobalt
  dicarbon metallacarbaboranes
    7-vertex, 382
Iron complexes

dodecahydrooctaborates, 334
gallium bonds, 555
indium bonds, 562
tetranuclear clusters
metallaboranes, 298–301
thallium bonds, 567, 568
trinuclear clusters
metallaboranes, 296
Iron–nickel
dicarbon metallacarbaboranes
8-vertex, 396

Ketones
aluminum alkylenation, 474

Lanthanide complexes
aluminum tetrahydride derivatives, 551
metallaboranes, 349
metallacarbaboranes, 373–424
octahydrotriborates
metallaboranes, 316
tetrahydroborates, 286
trihydroborates, 288
Lithium, allyl-
structures, 24
Lithium amides
structure, 5
Lithium, butyl-
structure, 19, 20
Lithium compounds
magnesium
preparation, 105
Lithium, diarylimido-
structure, 3
Lithium, diarylimino-
structure, 3
Lithium, indenyl-
structure, 24
Lithium magnesium compounds
alkynyl
preparation, 105
benzyl
preparation, 105
Lithium, methyl-
structure, 20
tetramer
structure, 30
Lithium organoborates
formation, 119
Lithium, *t*-butyl-
structure, 19, 20
Lithium tri-*t*-butylberyllate
synthesis, 42
Lithocene
anion
structure, 22

Magnesacycles
dimerization, 92
Magnesacyclohexane
dimerization, 92
Magnesacyclopentanes
formation, 87
Magnesates
orthomagnesiation
cubane, 64
structures, 29
Magnesiation
ethene, 95
Magnesium
activation
Grignard reagent preparation, 59
active
formation from magnesium hydride, 59
addition compounds
1,4-diphenyl-1,3-butadiene, 97, 98
unsaturated organic molecules, 96–100
co-condensation
with fluorobenzene, 59
with organic halides, 62
highly active, 100
single-crystal surface, 59
slurries
Grignard reagent preparation, 59
turnings
Grignard reagent preparation, 59
ultrasonic irradiation
Grignard reagent preparation, 59
Magnesium, alkylhalo-, 70
Magnesium, allenylchloro-
preparation, 70
Magnesium, allylchloro
structure, 74
Magnesium aluminum compounds
alkynyl, 107
Magnesium, bis(fluorenyl)-
preparation, 86
Magnesium, bis(indenyl)-
preparation, 86
Magnesium, bis[tris(trimethylsilyl)methyl]-
solid-state
structure, 87
structure, 87, 88
sublimation, 85
Magnesium, butyl(*s*-butyl)-
preparation, 85, 86
Magnesium carbide
structure, 98
Magnesium, chloro(pentamethylcyclopentadienyl)-
in organometallic synthesis, 84
Magnesium compounds
anthracene
complexes with oxygen and nitrogen ligands, 100
crystal structures, 99
ethene insertion, 100
Grignard reagent preparation, 59
in organometallic synthesis, 100
polymer supported, 99
radical anion complexes, 99
reactivity, 100
solid-state NMR, 100
chemistry, 57–120
diaryl
preparation, 86
diorganyl
addition to alkenes, 87
addition to cyclopentenes, 87
alkyl bridges, 88
3-coordinate complexes, 88
4-coordinate complexes, 88
crown ether adducts, 93
crown ether complexes, 89
cryptand adducts, 93
cryptand complexes, 89
electron transfer reactions, 95
ether cleavage reactions, 96
fluxionality, 93
gas-phase structures, 94
inner complexes, 90
intramolecular coordination, 90
intramolecularly coordinated crown ethers, 90
in organometallic synthesis,
ionic disproportionation, 89
ionization, 89, 90
molecular association, solutions, 92
NMR, 92
physical properties, 87–94
polyether complexes, 89
preparation, 85–7
properties, 85–96
reactions, 94–6
reactions, organic substrates, 95
reactions with carbon monoxide, 95
reactions with crown ethers, 95
redistribution of organic ligands, 86
rotaxane threaded complexes with crown ethers, 93
rotaxane threaded structures, 89
solid-state, structures, 87
solutions, properties, 92
solvation by ethers, 88
structures, 87–94
thermal decomposition, 94
organo-
ion pair formation, 93
organolithium
reactivity, 105
Magnesium, cyclopentadienylhalo-
structure, 74
Magnesium, cyclopentadienylhydrido-
formation, 101
Magnesium, cyclopentadienylneopentyl-
gas-phase electron diffraction, 94
Magnesium, dialkynylbis(TMEDA)-
structure, 91
Magnesium, dicyclopentadienyl-
ab initio LCAO–MO calculations, 94
adducts, 93
preparation, 86
structures, 91
Magnesium, diethylamidohydride
hydromagnesiation of alkenes, 103
Magnesium, dimethyl-
addition to alkoxyketones, 95
Magnesium, dineopentyl-
mean bond dissociation energies, 94
Magnesium, diphenyl-

activation, 86
halogen exchange of crown ether complexes, 96
intramolecular metallation of crown ether complex, 96
Magnesium, diphenylvinylene-
synthesis, 97
Magnesium hydrides
organyl
hydromagnesiation, octene, 101
Magnesium lithium bromide
tris(trimethylsilyl)methyl, 105, 106
Magnesium, tetrakis[(diphenylvinylene)tetrahydrofuran]-
structure, 91
Magnesium, tetrakis[(naphthalenediyl)tetrahydrofuran]-
structure, 91
Magnesium, tetrakis-[(phenylene)tetrahydrofuran]-
structure, 91
Manganese
dicarbon metallacarbaboranes
7-vertex, 382
12-vertex, 400
tricarbon metallacarbaboranes
11-vertex, 420
Manganese complexes
indium bonds, 560
Manganese(I) complexes
aluminum tetrahydride derivatives, 550
Mercury
dicarbon metallacarbaboranes
12-vertex, 400
Metal organic chemical vapor deposition
gallium, indium, or thallium compounds, 504
Metal organic vapor phase epitaxy
organoberyllium compounds, 53
Metal powders
reactive
magnesium reactions, 100
Metallaboranes, 275–371
arene-ligated systems, 357, 358
3 boron atoms, 313
4 boron atoms, 319–22
5 boron atoms, 323–7
6 boron atoms, 327, 328
7 boron atoms, 329, 330
8 boron atoms, 331–7
9 boron atoms, 337–49
10 boron atoms, 349–62
group 8 metals, 358
group 9 metals, 361
group 10 metals, 362
group 11 metals, 362
11 boron atoms, 363
12 boron atoms, 363
group 8 metals, 363
group 10 metals, 363
13 boron atoms, 363
14 boron atoms, 364
16 boron atoms, 364
17 boron atoms, 364, 365
18 boron atoms, 365
20 boron atoms, 365
chemistry, 275–365
metal-rich, 293–304
phosphine-ligated systems, 355–7
Metallacarbaboranes, 373–424
12-vertex
alkylidyne-bridged, 411
applications, 415–17
cage rearrangement, 414
carbene-bridged, 411
carbyne-bridged, 411
catalysis, 415
physical properties, 414
spectroscopy, 414
synthesis, 411–14
13-vertex cages, 417–19
bonding, 374–6
bridged complexes, 391
clusters, 375, 376
cobaltocenium salts, 392
diborole hybrid complexes, 381, 419
dicarbon, 381–419
8–11 vertices, 394–9
8–11 vertices, bonding, 398
8–11 vertices, structure, 398
8–11 vertices, synthesis, 394–7
hydrocarbon complexes, 388
icosahedral
12-vertex, 399–417
lanthanides, 419
in medicine,
metal–metal bonds, 411
metal extraction, 416
monocarbon, 377–80
multisandwich complexes, 387
oxidative fusion, 388, 398, 422
properties, 376
reactions, 377
reactivity, 374–6
reagents for organic synthesis, 393
sandwich complexes, 375, 376
stabilizing organic groups, 416
structure, 374–6
synthesis, 376, 377
metal-promoted oxidative fusion, 377
tetracarbon, 421–4
tetradecker sandwich complexes, 381
tricarbon, 419–21
triple-decker sandwich complexes, 381
wedged, 398, 399
closo-Metallacarbaboranes
synthesis, 207, 208
nido-Metallacarbaboranes
synthesis, 207, 208
Metallation
crown ethers
organomagnesium compounds, 64
Hauser bases, 64
Metalloporphyrins
metal–indium bonds, 566, 567
metal–thallium bonds, 569
Methane, dilithio-, 21
Molybdenum
dicarbon metallacarbaboranes
12-vertex, 400, 418
monocarbon metallacarbaboranes
12-vertex, 378
Molybdenum complexes
indium bonds, 557
thallium bonds, 567
Multilayer sandwich compounds
synthesis, 199, 200

Naphthalene, 1-amino-
dilithiation, 17
Nickel
dicarbon metallacarbaboranes
7-vertex, 382
11-vertex, 396
12-vertex, 400
monocarbon metallacarbaboranes, 378
10-vertex, 378
12-vertex, 378
tricarbon metallacarbaboranes
7-vertex, 420
11-vertex, 420
Nickel complexes
bis(borole)
NMR, 204
catalysis
carbomagnesiation, 68
diarsaboranes, 266
diboranes, 311
indium bonds, 566
Nonaborates, pentadecahydro-
group 8 metals
metallaboranes, 338
group 9 metals
metallaboranes, 338, 339
group 11 metals
metallaboranes, 339
metallaboranes, 338, 339
Nonaborates, tridecahydro-
metallaboranes, 340–3
Nonaborates, undecahydro-
group 8 metals
metallaboranes, 344–6
group 9 metals
metallaboranes, 347, 348
group 10 metals
metallaboranes, 349
metallaboranes, 344–9

Octaborates
conjoined
metallaboranes, 337
Octaborates, dodecahydro-
group 7 metals
metallaboranes, 333
group 8 metals
metallaboranes, 334
group 9 metals
metallaboranes, 335
group 10 metals
metallaboranes, 336, 337
group 11 metals
metallaboranes, 337
metallaboranes, 333–7
Octaborates, octahydro-
metallaboranes, 337
Octaborates, tetradecahydro-
group 8 metals

metallaboranes, 332
group 9 metals
metallaboranes, 332
metallaboranes, 331–3
Oligodecker structures
electron count, 201
Oppolzer's magnesium
ene reactions, 79
Organoaluminum complexes
ab initio calculations, 441–3
bonding, 441–55
chemical trapping, 443–5
coordination, 454, 455
monomeric tricoordinate
bonding, 447–9
oligomers, 449, 450
preparation, 455–67
spectroscopy, 441–3
structure, 441–55
subvalent
structure, 441–7
Organoaluminum(I) complexes
structures, 445–7
Organoaluminum magnesium compounds
anthracene, 107
Organoaluminum reagents
in organic synthesis, 438, 439
reactions with transition metal compounds, 478–84
Organobarium compounds
dialkyl, 110, 111
diaryl, 110, 111
ion pairs, 111
Organobarium halides
preparation, 107–9
Organoberyllium complexes
alkenyl, 36–9
alkoxy, 42, 43
alkyl, 36–9
alkynyl, 36–9
amino, 42, 43
anionic, 42
anthracene, 36
hydrolysis, 36, 37
aryl, 36–9
bis(dienyl), 38
coordination, 39–41
cyclopentadienyl, 37
structure, 48–50
phosphino, 42, 43
radical, 40
spirobicyclic, 36
thio, 42, 43
x-ray crystal structure, 39
Organocalcium clusters, 58, 108
Organocalcium compounds
amides, 118
Schlenk-type equilibrium, 118
biradicals, 118
calcium–tellurium exchange, 109
cuprates, 120
dialkyl, 110, 111
diaryl, 110–12
Schlenk-type equilibrium, 109
Organocalcium(I) compounds
chemistry, 58
Organocalcium halides
polymeric, 108
preparation, 107–9
mechanism, 108
reactions
with organic compounds, 109
Organocalcium hydrides
aryl
formation, 118
properties, 118
Organocalcium oxides
preparation, 118
Organolithium complexes
amido
calculations, 30
inorganic salt complexes, 10
Lewis base donor complexes, 6
monomeric, structure, 9
structure, 5
arsenides, 10
imino, 2
calculations, 30
nitrogen compounds, 5
phosphides, 10
solvation, 19
structures
theory, 29
Organomagnesium alkoxides
properties, 101, 102
Organomagnesium alumoxanes, 106
Organomagnesium amides
crystal structures, 103
properties, 101–4
Organomagnesium clusters, 58
formation, 62
Organomagnesium compounds
heterobimetallic compounds, 104–7
nitrogen heterocycles
paramagnetic, 103
Schlenk equilibrium, 66, 75
selenocyanates, 104
thiocyanates, 104
Organomagnesium(I) compounds
chemistry, 58
Organomagnesium diazenides
properties, 104
Organomagnesium halides
chemistry, 58–84
preparation, 58–72
Organomagnesium hydrides
properties, 101
Organomagnesium reagents
clusters, 59
Organomagnesium tris(pyrazolyl)-hydroborates
preparation, 104
Organometallic alkali metals
magnesium alkoxides, 105
Organometallic halides
properties, 109, 110
reactivity, 109, 110
structures, 109, 110
Organopolyboranes
preparation, 245, 246
reactions, 246, 247
Organopolyboron hydrides
chemistry, 245–7
physical properties, 245
Organostrontium compounds
dialkyl, 110–12
diaryl, 110, 111
Organostrontium(I) compounds
chemistry, 58
Organostrontium halides
preparation, 107–9
Orthoboronation, 343
Orthomagnesiation
cubane, 64
Osmium
dicarbon metallacarbaboranes
7-vertex, 382
11-vertex, 396
12-vertex, 400
monocarbon metallacarbaboranes, 378
Osmium complexes
trinuclear clusters
metallaboranes, 296
Osmium(II) complexes
aluminum tetrahydride derivatives, 551
Oxaboranes
properties, 267
Oxaborolane
reactions, 173
Oxadiboririane
preparation, 171
Oxamagnesacycles
structures, 90
Oxametallaboranes
properties, 267

Palladium
dicarbon metallacarbaboranes
12-vertex, 400, 418
Palladium complexes
diarsaboranes, 266
diselenaboranes, 270
distibaundecaboranes, 267
ditelluraboranes, 270
heptahydrotriborates, 316, 317
nido-Pentaboranes, 327
structure, 319
Pentaborates, decahydro-
group 7 metals
metallaboranes, 323
group 9 metals
metallaboranes, 323
metallaboranes, 323
Pentaborates, heptahydro-
group 8 metals
metallaboranes, 327
group 9 metals
metallaboranes, 327
metallaboranes, 327
Pentaborates, nonahydro-
group 8 metals
metallaboranes, 323
group 9 metals
metallaboranes, 324
metallaboranes, 323, 324
Pentaborates, octahydro-
group 8 metals
metallaboranes, 326
group 9 metals
metallaboranes, 326
group 11 metals
metallaboranes, 326
group 12 metals
metallaboranes, 326
metallaboranes, 326
Pentaborates, pentahydro-
metallaboranes, 327
Pentadecker complexes
metal complexes, 199, 200

Phosphaboranes
properties, 263, 264
transition metal complexes, 265
Phosphacarboranes
properties, 265
Phosphadecaboranes
properties, 264
1*H*-1,2,5-Phosphadiborole, 2,5-dihydro-1-phenyl-
metal complexes, 214
1,2,5-Phosphadiborolenes
metal complexes, 214
Phosphadicarbadodecaboranes
properties, 265
Phosphadodecaboranes
properties, 264
1,2,5-Phosphasilaborolenes
metal complexes, 213
Phosphasilaboroles
formation, 169
Platinatetraboranes, 317
Platinum
dicarbon metallacarbaboranes
8-vertex, 396
10-vertex, 396
11-vertex, 396
12-vertex, 400
monocarbon metallacarbaboranes, 378
10-vertex, 378
11-vertex, 378
Platinum complexes
diarsaboranes, 266
gallium bonds, 557
indium bonds, 566
phosphaboranes, 260, 265
Plumbaboranes
properties, 259, 260
Plumbacarboranes
properties, 260
Polyboranes
rearrangements
tensor surface harmonic theoretical studies, 232
Polyhedral skeletal electron pair theory, 331
metallaboranes, 277
Porphyrin, octaethyl-
metal–indium bonds, 566
metal–thallium bonds, 569
Porphyrin, tetraphenyl-
metal–indium bonds, 566
metal–thallium bonds, 569
Potassium bisoctatetraene calcate
preparation, 117
Potassium, methyl-
preparation, 20
Potassium, pentamethylcyclopentadienyl-
pyridine solvate
structure, 22

Quadrupledecker complexes
transition metals, 199–201

Rhenium
dicarbon metallacarbaboranes
12-vertex, 400
Rhenium complexes
group 7 metals
direct bonds to gallium, 554
group 8 metals
direct bonds to gallium, 555
indium bonds, 560
thallium bonds, 567
Rhenium(III) complexes
aluminum tetrahydride derivatives, 550
Rhodium
dicarbon metallacarbaboranes
6-vertex, 382
7-vertex, 382
10-vertex, 396
11-vertex, 396
12-vertex, 400
monocarbon metallacarbaboranes
11-vertex, 378
12-vertex, 378
Rhodium, bis(triphenylphosphine)hydrido(azaundecaborane)-
preparation, 262
Rhodium complexes
metallaboranes, 340–2
thiaboranes, 268
Rieke's magnesium
Grignard reagent preparation, 61
preparation, 59
Ring-laddering
alkali metal organometallics, 2
Ring-stacking
alkali metal organometallics, 2
Rubidium, methyl-
preparation, 20
Ruthenium
dicarbon metallacarbaboranes
6-vertex, 382
7-vertex, 382
10-vertex, 396
11-vertex, 396
12-vertex, 400
monocarbon metallacarbaboranes, 378
12-vertex, 378
Ruthenium complexes
heptahydrodiborates
metallaboranes, 304
indium bonds, 562
tetranuclear clusters
metallaboranes, 300
thallium bonds, 568
thiaboranes, 268
Ruthenium(II) complexes
aluminum tetrahydride derivatives, 550
metallaboranes, 340–2
Ruthenium osmium
dicarbon metallacarbaboranes
7-vertex, 382
Ruthenium rhodium
dicarbon metallacarbaboranes
7-vertex, 382

Samarium
dicarbon metallacarbaboranes
12-vertex, 400, 418
Samarium(III) complexes
aluminum tetrahydride derivatives, 551
Scandium
dicarbon metallacarbaboranes
12-vertex, 400
Schlenk equilibrium
displacement
diorganylmagnesium compounds, 85
Grignard reagents
thermodynamic parameters, 76
Selenaboranes
properties, 270, 271
1,2,5-Selenasilaborolenes
metal complexes, 213
Silaboranes
properties, 259
Silacarboranes
properties, 259
Siladicarbadodecaboranes
preparation, 259
Silanes, vinyl-
hydroalumination, 471
Silver complexes
ruthenium clusters, 304
Sodium *t*-butoxide
structure, 11, 12
Sodium, indenyl-
structure, 24
Sodium magnesium compounds
alkynyl
preparation, 105
phenyl
preparation, 105
Sodium, methyl-
structure, 20
Sodium, pentamethylcyclopentadienyl-
pyridine solvate
structure, 22
Stannaboranes
properties, 259, 260
Stannacarboranes
properties, 260
Stibaboranes
properties, 267
Strontium, bis(pentamethylcyclopentadienyl)-, 112
Strontium, bis(triphenylmethyl)-
formation, 111
Strontium complexes
anthracene, 117
arene radical anion complexes, 117
chemistry, 57–120
Strontium, dialkynyl-
preparation, 111
Strontium, dicyclopentadienyl-, 112–16
Strontium, dimethyl-
structure, 110
Sulfones
dilithiation, 17
Superbases
alkali metal organometallics, 13

Tantalum
dicarbon metallacarbaboranes
12-vertex, 400
Tantalum complexes
hexahydrodiborates, 306
Telluraboranes
properties, 270, 271
Tetraazaborolines
preparation, 162

Tetraazadiborinanes
  properties, 185
Tetraborane(8) carbonyl
  electron diffraction, 245
Tetraboranes
  metallaboranes, 313
Tetraborates, dihydro-
  metallaboranes, 322
Tetraborates, heptahydro-
  metallaboranes, 321
Tetraborates, hexahydro-
  group 8 metals
    metallaboranes, 321
  group 9 metals
    metallaboranes, 321
  metallaboranes, 321
Tetraborates, nonahydro-
  group 5 metals
    metallaboranes, 319
  group 8 metals
    metallaboranes, 319
  metallaboranes, 319
Tetraborates, octahydro-
  group 8 metals
    metallaboranes, 319
  group 9 metals
    metallaboranes, 320, 321
  metallaboranes, 319–21
Tetraborates, tetrahydro-
  metallaboranes, 322
Tetraborocine
  preparation, 154
Tetragallium tetraiodide
  structure, 527
Thallacarboranes
  properties, 258, 259
Thallium complexes
  arenes, 534–6
  arsenic ligands, 518–24
  carboboranes, 532–4
  cyclopentadienyl, 529–31
  group 6 metals
    direct bonds to thallium, 567
  group 7 metals
    direct bonds to thallium, 567
  group 8 metals
    direct bonds to thallium, 567, 568
  group 9 metals
    direct bonds to thallium, 569
  group 10 metals
    direct bonds to thallium, 569
  group 11 metals
    direct bonds to thallium, 569
  low oxidation state, 528–36
  metal bonds, 567–9
  metal–metal bonds, 537–40
  nitrogen ligands, 518–24
  oxygen ligands, 524–8
  phosphorus ligands, 518–24
  selenium ligands, 524–8
  sulfur ligands, 524–8
  tellurium ligands, 524–8
  transition metal bonds, 545–69
Thallium compounds, 503–40
  anionic, 514
  cationic, 514
  halogen complexes, 511–13
  hydrides, 514–17
  pseudohalogen complexes, 511–13
  trialkyl, 504–11
  triaryl, 504–11
Thiaboranes
  properties, 268, 269
Thiacarboranes
  properties, 269
1,2,5-Thiadiborole, 2,5-dihydro-
  metal complexes, 214
1,2,5-Thiadiborolene ligands
  metal complexes, 199
1,2,5-Thiadiborolenes
  metal complexes, 214
Thiametallacarbaboranes
  preparation, 379
1,2,5-Thiasilaborolenes
  metal complexes, 213
Tin, bis[(trimethylsilyl)methyldicarbahexaborane]-
  preparation, 260
Titanium
  dicarbon metallacarbaboranes
    7-vertex, 382
Titanium compounds
  catalysts
    Grignard reagents, reaction with alkynes, 69
    Grignard reagents, reaction with dienes, 69
Titanium(III) aluminohydride, biscyclopentadienyl
  synthesis, 546
Titanocene
  alkylated
    aluminum tetrahydride derivatives, 548
Transition metal complexes
  boryl, 290–3
  metal bonds, 545–69
  metallacarbaboranes, 373–424
1,2,3-Triboracyclopentenyl ligands
  metal complexes, 198
Triboranes
  metallaboranes, 313–18
Triborates, heptahydro-
  group 8 metals
    metallaboranes, 316
  group 9 metals
    metallaboranes, 316
  group 10 metals
    metallaboranes, 316
  group 11 metals
    metallaboranes, 317
  metallaboranes, 316, 317
Triborates, hexahydro-
  group 8 metals
    metallaboranes, 317
  metallaboranes, 317
Triborates, octahydro-
  group 5 metals
    metallaboranes, 315
  group 6 metals
    metallaboranes, 315
  group 8 metals
    metallaboranes, 315
  group 9 metals
    metallaboranes, 316
  group 10 metals
    metallaboranes, 316
  metallaboranes, 315, 316
Triborates, trihydro-
  metallaboranes, 318
Triorganostannates
  structures, 14
1,3,5,2,4,6-Triphosphatriborinane
  metal complexes, 211, 212
Tripledecker complexes
  34 electron, 199
  boron rings
    metal complexes, 198, 199
  doubly opened, 199
  slipped, 199
Tungsten
  dicarbon metallacarbaboranes
    12-vertex, 400, 418
Tungsten complexes
  monohydroboranes, 292
Tungsten, hexaphenoxy-
  in organometallic synthesis, 84
Tungsten(IV) complexes
  aluminum tetrahydride derivatives, 548

Ultrasonic radiation
  magnesium
    Grignard reagent preparation, 60

Vanadium
  dicarbon metallacarbaboranes
    7-vertex, 382
Vanadium complexes
  tetrahydroborates, 283

Wade's rules
  tripledecker complexes, 198

Ytterbium
  dicarbon metallacarbaboranes
    12-vertex, 400, 418
Yttrium
  dicarbon metallacarbaboranes
    7-vertex, 382

Ziegler nickel effect
  mechanisms, 483, 484
  organoaluminum compounds, 478
Ziegler–Natta catalysis
  mechanisms, 482, 483
Ziegler–Rienäcker cyclization
  alkadienes, 470
Zinc compounds
  organomagnesium, 106
Zincates
  structures, 29
Zirconacyclopropane
  carbomagnesiation
    catalysis, 67
Zirconium
  dicarbon metallacarbaboranes
    7-vertex, 382
    12-vertex, 400
Zirconium complexes
  binuclear polyhydride, 280
  tetrahydroborates, 280
  trihydroborates, 288
Zirconium(IV) complexes
  triscyclopentadienyl
    aluminum tetrahydride derivatives, 548
Zirconocene
  carbomagnesiation
    catalysis, 67